FLORA OF CYPRUS

VOLUME ONE

ONOBRYCHIS VENOSA

FLORA OF CYPRUS

VOLUME ONE

R. D. MEIKLE
The Herbarium, Royal Botanic Gardens, Kew

1977
Published by The Bentham-Moxon Trust
Royal Botanic Gardens, Kew

Printed by Robert MacLehose and Company Limited
Printers to the University of Glasgow

PREFACE

ALMOST two hundred years have passed since John Sibthorp, John Hawkins and Ferdinand Bauer landed at Larnaca to begin the first scientific survey of Cyprian plants. Since then a succession of travellers and resident collectors have shown that the fields, forests and mountains of Cyprus shelter an unusually interesting and varied flora. In this respect, the absence from Cyprus of several species, widespread and locally common elsewhere in the eastern Mediterranean, is of no less scientific interest to phyto-geographers than the presence of numerous endemics, including the remarkably isolated *Bosea cypria*. But, until now, students of the Cyprus flora, and of Mediterranean floristics as a whole, have been handicapped by the lack of a comprehensive, up-to-date, descriptive account of the island's plants, and have had, perforce, to fall back for guidance on Boissier's *Flora Orientalis* or Post's *Flora of Syria, Palestine and Sinai*, works which, however excellent, cannot be regarded as very complete or convenient when put to this more restricted use. Holmboe's *Studies on the Vegetation of Cyprus* (Bergen, 1914), an outstandingly accurate and painstaking compilation, was never intended as a Flora, and in the absence of keys and descriptions, was useful only to those who were already well acquainted with the flora of the region, or who had access to the major European herbaria. The foresters alone have been better served: Mrs Esther F. Chapman's *Cyprus Trees and Shrubs* (Nicosia, 1949) deals very ably and exhaustively with the trees and woody plants of the island; indeed, the main object of the present work has been to supplement this, and to provide an equal measure of guidance for those who are as much concerned with the herbaceous as with the ligneous element in the Cyprus flora. Volume I accomplishes roughly half this task; the whole work, it is hoped, will be completed in a second volume.

The practical importance of a descriptive Flora, whether of Cyprus or of any other part of the World, cannot be overstressed. Everywhere (and very patently in parts of Cyprus), disregard for the vulnerability of the Earth's mantle of vegetation has led to impoverishment, erosion and desolation. Only now is positive action being taken to halt the progress of destruction, and to restore, as far as is possible, former verdure and fertility. Progress in this direction inevitably calls for a sound and detailed knowledge of the plants involved, for the science of Conservation, like every other science, can succeed only if it is based upon a foundation of established fact. It is hoped that this volume, and its successor, will be found to provide this essential basis.

J. P. M. BRENAN
Director, Royal Botanic Gardens, Kew

ACKNOWLEDGEMENTS

THE thanks of the author are due, in the first place, to the Government of Cyprus, and in particular to the Ministry of Agriculture and Natural Resources, Nicosia, for their patient assistance and generosity in the preparation of what has necessarily been a protracted undertaking. A special tribute must be paid to the Department of Forests for providing accommodation, guidance and transport, without which, in 1962, an extensive field survey of the island would have been largely impossible.

The author is no less grateful to the Ministry of Overseas Development for negotiating the printing of Volume One at a time when international events seemed certain to frustrate the outcome of many years' labour. In this connection, the keen interest and energetic assistance of Mr W. Hobman (of O.D.M.) is especially appreciated.

The kindness of the Government of the Republic of Iraq in permitting the re-use of twenty-seven of the excellent line illustrations, already published in the Flora of Iraq, Vols. 2 & 3 (Baghdad 1966 & 1974), has substantially enhanced the appearance and general usefulness of this volume, especially with regard to the families *Leguminosae* and *Rosaceae*.

Finally the sincere gratitude of the author is due to successive Directors and Keepers of the Herbarium of the Royal Botanic Gardens, Kew, and to botanists at Kew, the British Museum (Natural History), Edinburgh, and indeed throughout the world, for assistance given, either directly or indirectly, knowingly or unknowingly, in the preparation of this work. The same measure of appreciation is extended to all those institutions, at home or abroad, which have facilitated research on the plants of Cyprus by the loan of types and other important specimens.

R. D. M.

CONTENTS

"........................ in altos
Idaliae lucos, ubi mollis amaracus illum
floribus et dulci adspirans complectitur umbra."

VIRGIL, *Aeneid 1*, 692-694.

ILLUSTRATIONS

ILLUSTRATIONS

INTRODUCTION

Cyprus, the third largest island in the Mediterranean, is situated just outside the Gulf of Iskenderun, about 40 miles S. of Turkey and 65 miles W. of Syria, between 34°33′–35°41′N. and 32°17′–34°35′E. The island, which is extremely irregular in outline, with numerous bays and headlands, has an area of approximately 3,572 square miles (Sicily 9,831 sq. miles; Sardinia 9,196 sq. miles; Crete 3,207 sq. miles) with its greatest length from W.S.W. to E.N.E. approximately 140 miles, and its greatest width from N. to S. approximately 60 miles. About 50 miles of its length is occupied by the narrow Karpas peninsula, pointing N.E. towards Antakya and the Amanus Range.

TOPOGRAPHY

The outstanding physical features of the island can be summarized under 4 headings:

1. The Coastal Belt

The coasts of Cyprus are almost everywhere low and shelving. Sea-cliffs of any magnitude are extremely rare, and only at two places, at Sotira (1,193 ft.) on the N.E. side of the Akamas peninsula, and between Pomos Point and Kokkina Point, on the northern edge of the Troödos Range, do mountains descend abruptly into the sea. Elsewhere the island is bounded by a low-lying coastal belt, wide and flat around Morphou and Famagusta, but sometimes very narrow, rising abruptly to the steep Kyrenia Range in the N. or into more rounded, garigue-covered hillsides in the S. and W.

Many areas of the coastal belt are fertile and tilled almost to the edge of the sea, but extensive stretches of undisturbed, natural vegetation remain in the Akamas peninsula and in parts of the elongate Karpas peninsula in the E. For the most part the shores are rocky or stony, but small sandy bays are found all round the island, with extensive strands, backed by sand-dunes or salt-flats, about Famagusta and locally in the Karpas peninsula. Salt-lakes, fringed by salt-marshes and other saline habitats, occur at Larnaca and Limassol. Interesting fresh-water marshes (now extensively drained) are found near Phasouri S.W. of Limassol.

2. The Kyrenia (or Northern) Range

This spectacular, precipitous ridge, approximately 50 miles long, runs W. to E. like a high wall, roughly parallel with, and close to the N. coast, starting abruptly above Vasilia at its western end, and petering out gradually in the Karpas peninsula at Yioti near Ephtakomi. It is composed of hard Permo-Carboniferous–Cretaceous limestones and marbles, with scattered basaltic sills and dykes, rising in jagged pinnacles and crests to peaks at Kornos (3,104 ft.), Kyparissovouno (3,357 ft.), Trypa Vouno (3,068 ft.), Buffavento (3,131 ft.), Pentadaktylos (2,430 ft.), Yaïla (3,067 ft.), Olymbos (2,427 ft.) and Sina Oros (2,374 ft.).

The Kyrenia Range is for the greater part uncultivated, with extensive areas of *Pinus brutia*-*Cupressus sempervirens* forest on the upper slopes. Much of the north-facing side is clothed with dense, well-developed maquis, including an abundance of *Arbutus andrachne*, *Styrax* and *Pistacia*, while the numerous cliffs harbour a rich chasmophytic flora, including many endemics.

The south-facing slopes are much hotter and drier, with a sparser and rather less varied flora. The lower slopes of the Range, especially on the southern side, are flanked by extensive low, undulating hummocks of Miocene marls, sandstones and conglomerates. This Kythrean formation, despite its barren appearance, shelters a rich and characteristic flora, including such plants as *Iberis odorata*, *Pteranthus dichotomus*, *Daucus durieua*, *Cotula tridentata*, *Salvia crassifolia* and many others, restricted to the formation or rarely found off it. The Kythrean marls extend eastwards beyond the Kyrenia Range almost to the tip of the Karpas peninsula.

Though several important springs rise in the Kyrenia Range, there are few streams, and a notable deficiency of wet ground.

3. The Troödos (or Southern) Range

This is predominantly igneous, consisting of rounded masses of pillow lavas (characterized by an abundance of *Lavandula stoechas*) in the lower part, rising to steep, rocky, but rarely precipitous, peaks of gabbro, diabase and serpentine at the centre of the Range. Khionistra (Olympus) attains 6,401 ft. and is the highest point on the island; other high peaks are Tripylos (4,619 ft.), Madhari (5,289 ft.), Papoutsa (5,098 ft.) and Kionia (4,660 ft.), but much of the Range lies above 4,000 ft. and snow falls regularly in winter on the upper slopes, persisting until summer on shaded, north-facing parts of Khionistra.

Many endemics (or near-endemics) are confined to the Troödos Range, mostly at altitudes above 3,000 ft., with some, especially the perennial *Alyssum troodi* and *A. cypricum*, confined to the serpentine. It is of particular interest that the only allies of these two species on the island, namely *Alyssum chondrogynum* and *A. akamasicum*, are likewise restricted to small, isolated outcrops of serpentine in the Akamas and above Limassol.

The lower slopes of Troödos are well covered with forests of *Pinus brutia*, which reach their finest development in the valleys radiating from Stavros tis Psokas at the western (Paphos Forest) end of the Range. Here also, on the summit and slopes of Tripylos, is the famous Cedar forest, composed of the endemic *Cedrus libani* ssp. *brevifolia*. The topmost slopes of Khionistra, above 5,000 ft., are clothed with an open woodland of the distinctive *Pinus nigra* ssp. *pallasiana* (*P. nigra* var. *caramanica*) mixed with colonies of the Cypress-like *Juniperus foetidissima*. The closely related *Juniperus excelsa* is apparently restricted to Madhari and Papoutsa.

Extensive elevated areas of soft chalks and limestones (of the Lapithos and Dhali groups) abut the southern and eastern flanks of the Troödos Range, rising to over 3,000 ft. about Pano Panayia and Omodhos. Much of this ground has been planted with vineyards, but some areas (especially around Khrysorroyiatissa Monastery) are still clothed with remnants of *Quercus infectoria* woodland; in such places, as well as on the borders of vineyards, or on patches of waste ground, are to be found a range of plants rare or wanting in other parts of the island.

4. The Mesaoria or Central Plain

This fertile, tree-less plain runs right across the island, for a distance of about 55 miles from Morphou Bay in the W., to Famagusta Bay in the E., effectively separating the Kyrenia and Troödos Ranges. Much of the plain lies at levels below 500 ft., with a great part of the flatter, eastern half (the true Mesaoria) at less than 200 ft. The western half, towards Morphou, is higher and more diversified, with isolated ridges and hummocks rising to more than 1,000 ft. about Xeri and Ayia Marina. A large part of the Central Plain, especially towards its western and eastern ends, consists of Recent alluvium, the rest is composed of Pleistocene sands, marls, gravels and conglomerates interspersed with older formations of chalk and gypsum. The gypsum areas, which often harbour distinct plant communities in other parts of the Mediterranean region, have not, as yet, been studied in detail. Much of the Mesaoria consists of cereal fields, once notable for a rich and colourful weed flora, but now, through the general use of selective herbicides, almost weed-free, and, for the botanist, uninvitingly monotonous. Some of the segetals are now very local and rare, but many manage to survive, in greatly reduced quantity, by roadsides and field-margins, or on patches of waste or fallow ground.

The Mesaoria is traversed by several rivers, all seasonal and normally dry or almost dry during the summer, with a very limited riverine flora; there are, however, some interesting areas of marshy ground near Famagusta (at the mouth of the Pedieos River) and about Syrianokhori near Morphou.

CLIMATE

The climate of Cyprus may be described as arid Mediterranean, with a relatively short, cool, wet winter, followed by a long, dry, very hot summer. There is, however, sufficient diversity in topography and micro-climates to allow for an unexpectedly large range of plant communities.

The annual rainfall for the whole island is, on an average, about 500 mm. (20 inches), most of it falling between November and March, but local differences are very considerable, more than 1,000 mm. (40 inches) falling on the topmost slopes of Khionistra, and less than 254 mm. (10 inches) in parts of the western Mesaoria near Morphou. The wettest areas are mostly the highest, though much of the low-lying Karpas peninsula has an average annual rainfall exceeding 500 mm. (20 inches), considerably in excess of the Mesaoria, where it is generally less than 350 mm. (14 inches).

Although rainfall averages have some significance, it must be added that the variation in annual precipitation is very great, and prolonged droughts are not infrequent: one notorious instance, in the early part of the 4th century A.D., reputed by some authors to have lasted 36 years, is said to have depopulated the island; and the almost complete lack of rain from the spring of 1972 to the autumn of 1973 caused extensive damage, not only to crops but to mature *Pinus brutia* woodland and to *Arbutus* and *Myrtus* on the Kyrenia Range. In the Akamas, where *Cistus creticus* and *Cistus salviifolius* were apparently uninjured by the drought, it was noticed (March 1974) that large populations of *Cistus monspeliensis* were dead or almost dead.

Temperature varies, with season and altitude, the monthly mean maximum (averaged over 5 years) for Nicosia (536 ft. alt.) ranging from 60·2° F (15·65° C) in January to about 98·33° F (36·85° C) in August. The mean monthly maximum for Amiandos (4,450 ft. alt.) over the same period, was 43·21° F (7·35° C) in January, to 81·62° F (27·55° C) in August. More important than the averages, so far as plants are concerned, are the extremes: the lowest January temperature at Amiandos, averaged over 5 years, is 25° F (−3·85° C), exceptionally descending to 20° F (−6° C), that of Nicosia, for the same period, being 32·8° F (0·45° C), exceptionally descending to 31° F (−0·6° C). The highest August temperature at Amiandos, again averaged over 5 years, is 89·4° F (31·5° C) exceptionally rising to 95° F (35° C), that of Nicosia for the same period being 108° F (42·20° C) exceptionally rising to 112° F (44° C). Snow lies on the upper slopes of Khionistra, above 4,000 ft., for an average of 10 weeks each winter, whereas at Nicosia snow is rare, and seldom lies; frosts are occasional in the lowlands, and sometimes cause considerable damage. In favoured coastal areas, as around

Paphos, bananas can be grown as a commercial crop, though they are, on occasion, damaged by exceptional cold.

Local divergence in relative humidity and insolation leads to striking differences in the vegetation of south-facing and north-facing slopes; this is particularly marked in the Kyrenia Range, but is also noticeable, especially in late spring, on the low ridges and hummocks which are dotted over the lowlands of Cyprus, the hotter side being brown and desiccated, while the shaded side remains green.

The change from cool, moist conditions to the warmth and dryness of summer is generally rather abrupt; the Cyprus spring is short, and though a few plants begin to flower in January or February, March and April are the most rewarding months for the botanist. By May the lowlands are already becoming dry, and by July even the high mountain flora is past its best, only a small number of plants being sufficiently resistant to drought and high temperatures to flourish in August or September, though by the end of the latter month, and in October, a few autumn-flowering plants (mostly with tubers or bulbs) have appeared in advance of winter rains.

CHANGES IN THE VEGETATION

If the comments of Eratosthenes (275–195 B.C.) are to be believed, most of Cyprus, even the Mesaoria, was heavily forested in antiquity, and considerable remnants of these forests survive on the Troödos and Kyrenia Ranges, and locally at lower altitudes, about 17% of the whole island being classified as woodland. Where the forest has been destroyed, tall (4–6 m. high) shrub communities of *Arbutus andrachne*, *Pistacia terebinthus*, *Olea europaea*, *Styrax officinalis* and *Quercus coccifera* may survive, but such *maquis* is uncommon; over most of the island untilled ground bears a grazed covering of *garigue*, largely composed of low (less than 3 m. high) bushes of *Cistus*, *Genista sphacelata*, *Calycotome villosa*, *Lithospermum hispidulum*, *Phagnalon rupestre* and, locally, *Pistacia lentiscus* associated with a wide variety of annual and perennial herbs. Where grazing is excessive this covering is soon reduced, and an impoverished *batha* remains, consisting principally of *Sarcopoterium spinosum*, *Fumana* spp., *Micromeria* spp., *Thymus capitatus* and a few stunted herbs, so sparse that, in the distance, hillsides may appear completely denuded of vegetation. With protection against continued grazing, garigue will in time re-assert itself, but the process is a slow one, and the regeneration of forest, if not deliberately accelerated by sowing or planting, becomes, at most, a remote possibility.

THE BOTANICAL SUBDIVISION OF THE ISLAND

To illustrate the internal distribution of each species, the island has been divided (*Map 1.*) into 8 Botanical Divisions, corresponding very roughly to natural phytogeographic areas. Wherever possible the boundaries of the Divisions have been drawn to follow roads or rivers, which are recognizable in the field, but where this has proved impractical, recourse has had to be made to contour lines or some other expedient.

Since road systems are subject to change, it should be noted that in demarcating the Botanical Divisions, use was made of the second edition (1950, revised 1958) of the Survey of Cyprus Administration Map (scale: 4 miles to 1 inch).

Map 1. Cyprus: Botanical Divisions

Boundaries of Divisions

Division 1. Bounded by road from N. coast at Argaka junction to Ayios Merkourios, Abdoulinas junction, Jemali Bridge, Sarama, Simou, Lasa, Ayios Dhimitrianos, to about ¼ mile N.E. of Melamiou then S.S.W. down the Ezuza (Ezousas) River to the coast S.W. of Akhelia.

A very heterogeneous area topographically, geologically and floristically, with much natural (but often heavily grazed) vegetation. Mostly hilly, with deep narrow gorges cut through chalk, limestone or sandstone, and with interesting areas of serpentine in the west about Smyies. The coasts are generally low, and sandy or rocky, except for a limited range of steep cliffs just N.W. of the Baths of Aphrodite. Division 1 is still incompletely known botanically, the hinterland, from Stroumbi to Smyies, particularly calling for attention. Plants of special interest in the Division include: *Alyssum akamasicum*, *Arenaria rhodia* ssp. *cypria*, *Cistus monspeliensis*, *Cistus ladanifer* (? introduced), *Pistacia* × *saportae*, *Trifolium argutum*, *Arbutus unedo*, *Phlomis lunariifolia*, *Tulipa cypria*, *Scilla cilicica*, *Maillea crypsoides*.

Division 2. Bounded partly by the road from Argaka junction to Melamiou (see Div. 1) and by the N. coast from Argaka–Polis junction to Lefka–Karavostasi junction, thence by road to Lefka, Ayios Yeoryios, Katydhata, Evrykhou to Galata then eastwards from Evrykhou–Ayios Theodhoros junction in a line connecting the outposts of the 2,000′ contour to Panayia, thence by road via Ayios Epiphanios to Klirou, by a line connecting Klirou and Kambia, and on by road to Analiondas, Kataliondas, Lythrodonda, Pano Lefkara, Vavia, Layia, Ora, Melini, Arakapas, Athrakos, Kalokhoria, Ayios Mamas, Perapedhi, S.W. via Kilani and Vouni to Ayios Amvrosios, then via Mallia and Arsos to Ayios Nikolaos, on through Philousa and Arminou to Ayios Ioannis along a line connecting Ayios Ioannis to Vrecha, thence via Kilinia, Ayios Photios, Statos, Lapithiou, Kannaviou to ¼ mile N.E. of Melamiou and boundary of Division 1.

Mostly the Troödos Range, an area of high, but usually rounded or pyramidal, igneous peaks, with much of the ground above 4,000 feet, reaching 6,401 feet at the summit of Khionistra. The highest ground, which has a rich endemic flora, has been repeatedly explored, but much of the Paphos Forest (at the western end of the Range) would repay further attention, particularly the more inaccessible peaks and valleys. The chalk foothills about Pano Panayia and Omodhos, and the fertile, nut-growing valleys between Sarandi and Palekhori are also incompletely known. The spectacular, cliff-walled ravines of the Peristerona River and adjacent rivers, though difficult to penetrate, are inviting ground for the botanist, though for much of their length they fall within Division 6.

Plants of outstanding interest are too numerous to list, since the whole area abounds in endemics; the two most famous are the Golden Oak, *Quercus alnifolia*, and the Cedar, *Cedrus libani* ssp. *brevifolia*.

Division 3. Bounded in the W. by boundary of Division 1, in the N. chiefly by the boundary of Division 2, extended eastwards by the road from Pano Lefkara through Delikipo to Kornos and N. to the Nicosia District boundary thence S.E. along boundary to Tremithios R. and down the river to the coast S.S.W. of Kiti.

A region of rounded chalk hills, gradually rising inland towards the Troödos Range. Much of the ground consists of carefully tended vineyards, and the coastal belt is extensively cultivated except in the W., where, about Pissouri, there are heavily grazed hills of calcareous marls and limestones. Areas of special interest are the Akrotiri peninsula and Salt Lake. The marshes (now partly drained) about Phasouri are also floristically rich, and the serpentine about Louvaras and Kakomallis. Further work on the chalk slopes about Lefkara is also likely to be rewarding.

Stavrovouni Monastery, situated on the top of a steep diabase hill (2,258 ft. alt.) is of considerable historical interest, since it was visited by Sibthorp and almost all the early travellers to Cyprus, and is the classical locality for many first records of Cyprus plants.

Noteworthy plants of Division 3 are: *Alyssum chondrogynum*, *Linum maritimum*, *Fagonia cretica*, *Erodium crassifolium*, *Astragalus macrocarpus* ssp. *lefkarensis*, *Neurada procumbens*, *Artedia squamata*, *Centaurea veneris*, *Marsdenia erecta*, *Convolvulus* × *cyprius*, *Ipomoea sagittata*, *Euphorbia thompsonii* and *Cladium mariscus*.

Division 4. Bounded in the W. by the Tremithios R. (see Div. 3), in the N.W. by the Nicosia boundary line to the Nicosia–Larnaca road N.W. of Goshi, thence by road to Pyroi, Athienou, Lysi, Kondea, Prastio, Styllos, Engomi and due E. to the coast 2 miles S. of Salamis.

Mostly cultivated or heavily grazed, with typical Mesaoria cornfields in the N. and numerous barren, eroded chalk or limestone hills in the S. Larnaca Salt Lake provides a habitat for interesting *Limonium* species and other halophytes. Interesting plants from the Larnaca neighbourhood include *Hymenolobus procumbens*, *Matthiola fruticulosa*, *Erodium crassifolium* and *Urginea undulata*. *Crambe hispanica*, *Galium pisiferum* and *Scilla hyacinthoides* have been recorded from Cape Greco, and *Ipomoea stolonifera* from sandhills at Famagusta. There are old records for *Anemone hortensis* from Phurni and for *Helianthemum chamaecistus* from hills near Larnaca.

Division 5. Bounded in the S. by the boundaries of Divisions 3 and 4, in the W. by the boundaries of Division 2, and by the road from Klirou to Strovolos, E.S.E. to junction with Limassol road, N. to junction with Larnaca road at Eylenja, then N. of Eylenja by a line connecting points 1 mile E. of Palouriotissa to Beuyuk Kaimakli, Omorphita to Nicosia–Kyrenia road about ¼ mile S.W. of Trakhonas, then N.W. along Kyrenia road to Aghirda junction, then E. along road via Kato Dhikomo to Sykhari and Koutsovendis, S. to a point about 2 miles N., of Mia Milea, then eastwards along the 500′ contour line to the Yerani–Kantara–Dhavlos road, then S. by Yerani–Ayios Elias–Monarga road to the coast about ¼ mile S.W. of Boghaz.

Almost wholly occupied by the flat cornfields of the Mesaoria, with interesting weed communities on waste and fallow land. The Kythrean hummocks flanking the Kyrenia Range in the N. of the Division have a rich and characteristic flora (see p. 2) and would repay further investigation.

Division 6. Bounded in the S. and E. by the boundaries of Divisions 2 and 5, in the N. by the road through Aghirda, Krini, Ayios Ermolaos, Kondemenos, Asomatos, Myrtou, Panagra to the coast 1 mile W. of the Kyrenia–Vasilia junction.

Heavily cultivated, with cornfields in the centre and E. and extensive Citrus groves about Morphou. The most interesting areas botanically are about Kormakiti and Ayia Irini, in the driest region of the island. Here are *Argyrolobium uniflorum*, *Cyclamen graecum*, *Achillea santolina*, *Convolvulus oleifolius* var. *pumilus* and other rare plants. *Tulipa cypria* is locally abundant in fields about Dhiorios and Myrtou. Marshy ground about Syrianokhori has furnished some interesting records and calls for further investigation. The deep river gorges S. of Peristerona (see notes on Division 2) are also attractive and largely unexplored.

Division 7. Bounded in the W. and S. by the boundaries of Divisions 5 and 6, in the E. by road from Yerani and Kantara to Dhavlos and due N. to the sea.

Includes almost all the Kyrenia Range and the fertile coastal belt between Lapithos and Dhavlos with Kyrenia at its centre. The rugged limestone cliffs and pinnacles of the Kyrenia Range harbour the richest flora in the whole island, and have been, for many years, the favourite resort of botanists. The number of local endemics and rarities is too large to list, but it should be noted that while the area about Kyrenia and St. Hilarion has been thoroughly explored, there is scope for additional work at the western and eastern ends of the Range. The rediscovery of *Zelkova abelicea* (*Abelicea cretica*), recorded by Kotschy from " der Nordseite der Gebirge gegen Melandrina " [a short distance E. of Ayios Amvrosios] would be particularly welcome.

Division 8. Bounded in the W. by the boundaries of Divisions 5 and 7, elsewhere by the sea.

The Karpas peninsula, an area of fertile cultivated fields, low hills and extensive sandy or rocky shores, was formerly outside the range of travellers, and is still imperfectly known. The flora rather lacks distinction, but includes a miscellany of rare plants such as *Fumaria gaillardotii*, *Enarthrocarpus arcuatus*, *Helianthemum ledifolium*, *Aïzoön hispanicum* and *Trifolium globosum*. *Rosmarinus officinalis*, very rare as a wild plant in Cyprus, grows in some abundance on rocky shores between Yialousa and Rizokarpaso. *Juniperus phoenicea* forms extensive thickets about Apostolos Andreas Monastery. The maritime vegetation is of particular interest, and requires further investigation. Swampy ground about Galatia, and the moist Ronnas valley, are also likely to prove interesting.

HISTORY OF BOTANICAL EXPLORATION OF CYPRUS

Detailed historical information for the period up to 1905 is to be found in Holmboe, Studies on the Vegetation of Cyprus, 13–23 (1914); it is consequently unnecessary to go over the same ground again, save only to furnish some additional data relating to the more important of these earlier investigations. Since 1905 so much work has been done in the field that it is impossible to give more than a very general survey of collectors and itineraries. A list of collectors whose specimens have been examined in the preparation of this Flora will be found in Appendix II, p. 803.

Although references to the Cyprus flora, or rather to plants of economic importance in the Cyprus flora, go back as far as Homer, it may be said that scientific investigation, in the modern sense, really begins on 8th April, 1787, with the arrival at Larnaca of John Sibthorp (1758–96) and his companions John Hawkins (? 1758–1841) and the artist Ferdinand Bauer (1760–1826). Their (several or joint) itineraries were as follows:

April 8–12, 1787.	Larnaca
,, 12–13,	Stavrovouni
,, 13–17,	Larnaca
,, 17–18,	Ormidhia to Famagusta
,, 18–20,	" Upsera " (? Gypsos) and Antiphonitis Church
,, 20–22,	Kyrenia and Bellapais
,, 22–23,	Nicosia
,, 23–27,	Larnaca
,, 27,	Kiti, " Magado " (? Mazotos), Moni
,, 28,	Moni, Amathus, Limassol

April 29,	Limassol, Mesapotamos Monastery
,, 30,	Mesapotamos Monastery, Troödhitissa Monastery
May 1,	Summit of Khionistra, Ayios Nikolaos
,, 2,	Ayios Nikolaos, Peristerona, Arkhangelos Monastery
,, 3,	Arkhangelos Monastery, Larnaca
,, 4–8,	Larnaca
,, 8–11,	[visit to Rhodes]
,, 11,	Paphos, Ktima
,, 12,	Ktima, Polis
,, 13,	Polis, Fontana Amorosa, Polis, Paphos
,, 14,	Departure from Paphos.

H. Stuart Thompson (1906) estimated that Sibthorp's Cyprus collections did not exceed 313 species, and it has since become evident that some of the plants said to have been collected " in insulâ Cypro " were in fact found elsewhere. Sibthorp did not keep exact field notes, and only a few of the plants collected are mentioned in his diary (published in Walpole, Travels, 14–27; 1820). Most of the plants described and figured in the posthumous *Flora Graeca* are not precisely localized, and there is reason to suspect that many of the references to Cyprus are little more than guesswork on the part of the editors. John Hawkins, writing to one of these editors, Sir J. E. Smith (letter dated 13 February 1800) comments, " It is certainly a pity that Dr. Sibthorp did not mark all his specimens, or the drawings, but he trusted to memory, and dreamed not of dying ".

The collections made by the Austrian collector and explorer, Karl Georg Theodor Kotschy (1813–66) were much more extensive and more carefully annotated. Kotschy visited Cyprus on three separate occasions. His first visit, apparently from 18th September to (?) the end of November, 1840, was, for obvious reasons, not very productive, though, from the few records quoted in *Die Insel Cypern*, it is evident that he travelled extensively in the island, visiting Troödhitissa Monastery, Evrykhou, Khionistra, Prodhromos, Livadhia, Limassol and Paphos. His second stay in the island, in the spring of 1859, was brief but rewarding; his itinerary was then as follows:

March 31, 1859.	Larnaca
April 1,	Phurni, Arkhangelos, Athienou
,, 2,	Athienou, Nicosia
,, 3,	Mia Milea, Ayios Khrysostomos Monastery, Buffavento, Kythrea
,, 4,	Kokkini Trimithia, Peristerona
,, 5,	Galata, Prodhromos
,, 6,	Khionistra
,, 7,	Troödhitissa Monastery
,, 8,	Omodhos, Erimi, Limassol
,, 9,	Limassol, Moni, Maroni, Mazotos
,, 10,	Mazotos, Larnaca.

Kotschy's most important contribution to Cyprus botany was, however, the outcome of his third visit, in 1862, when, accompanied by Dr. Franz J. A. N. Unger (1800–70), he accumulated the data subsequently published in *Die Insel Cypern* (Vienna, 1865). Their itinerary was as follows:

March 26, 1862.	Larnaca
,, 28,	Larnaca to Famagusta via Avgorou
,, 29,	Paralimni, Cape Greco
,, 30,	Cape Greco, Ayia Napa

April	2,	Stavrovouni and Lefkara
,,	3,	Lefkara and Makheras Monastery
,,	4,	Makheras Monastery
,,	5,	Makheras to Philani
,,	6–8,	Larnaca
,,	10,	Livadhia (near Larnaca)
,,	12,	Kythrea
,,	13,	Pentadaktylos
,,	15,	Ayios Khrysostomos Monastery, Buffavento
,,	16,	Ayios Khrysostomos Monastery, Kythrea, Dhikomo
,,	17,	Kyrenia
,,	18,	Kyrenia, Lapithos
,,	19,	Kyrenia, Melandrina (E. of Ayios Amvrosios), Antiphonitis Church
,,	20,	Antiphonitis Church, Kyrenia
,,	21,	Lapithos
,,	22,	Famagusta
,,	23,	Boghaz, Syngrasis, Vatili
,,	24,	Vatili, Athienou, Troulli
,,	25–26,	? Larnaca
,,	27,	Kiti, Mazotos
,,	28–29,	Moni, Amathus
,,	30,	Limassol, Kolossi
May	1,	Cape Gata, Ayios Nikolaos Church, Lania
,,	2,	Kolossi, Episkopi
,,	3,	Limassol
,,	4,	Pissouri, Kouklia
,,	5,	Kouklia, Yeroskipos, Paphos
,,	6,	Paphos, Ktima to Khrysokhou
,,	7,	Khrysokhou to Khrysorroyiatissa Monastery
,,	8,	Khrysorroyiatissa Monastery
,,	9,	Vrecha to " Jophyri " (near Arsos)
,,	11,	Episkopi
,,	12,	Prodhromos
,,	13,	Prodhromos, Damaskinari
,,	14,	Lemithou, Tris Elies, Prodhromos
,,	15,	Prodhromos, Lemithou
,,	16,	Prodhromos, Troödhitissa Monastery
,,	17,	Prodhromos, Phini
,,	18,	Phini, Prodhromos
,,	19,	Prodhromos
,,	20,	Troödos, Khionistra
,,	22,	Prodhromos, Galata
,,	23,	Khionistra
,,	28,	Larnaca.

According to Holmboe, Kotschy collected no fewer than 1,050 flowering plants and ferns on the island, and these collections, supplemented by his observations, and Unger's notes on the geology, vegetation and non-vascular plants, provided a solid basis of information, far in advance of contemporary knowledge of adjacent areas, upon which later workers have been able to build with a high degree of confidence. Most of Kotschy's records have been confirmed since, and so reliable is his word in general, that one may still hope to see the rediscovery of *Anemone hortensis* and *Zelkova abelicea* (*Abelicea cretica*) and a few other plants which have not been seen since 1862, and which have to be regarded as doubtful records.

If the itineraries of Sibthorp and Kotschy are united, it will be seen that, by 1862, most of the island, except the Karpas peninsula, had at least been visited by botanists. The Karpas was to be examined eighteen years later by the German, Paul Ernst Emil Sintenis (1847–1907) and the Italian, Gregorio Rigo (1841–1922). The details of their journeyings in Cyprus were recorded by Sintenis in diary form in the *Oesterreichische botanische Zeitschrift* vols. 32 & 33 (1881–82), but incompletely so, for the entries, which begin on 17th Febr. 1880, with the arrival of Sintenis in Larnaca, end abruptly on 29th April of the same year, though it is clear that collecting continued until the end of July, or possibly even as late as September. Normally one might be able to piece together the missing data from herbarium labels, but, unfortunately, in common with other professional collectors of the period, Sintenis (or Sintenis and Rigo) appears to have re-arranged the collections after they had been identified, putting together, under a single number, locality and date, material which was thought to belong to one species, but which, in some instances, includes more than one species, and which includes specimens which must have been collected in different parts of the island, on different dates. This policy of amalgamating " duplicates " means that, after 29th April, it is impossible to provide more than a sketchy outline of their movements, as follows:

Febr.	17, 1880.	Sintenis arrives at Larnaca
,,	18–28,	Larnaca area
,,	29,	Klavdhia to Pyrga
March	1,	Stavrovouni, Pyrga
,,	2,	Pyrga, Psevdhas, Kalokhorio, Larnaca
,,	3–7,	Larnaca
,,	8,	Livadhia, Atherica, Xylotymbo
,,	9,	Ormidhia
,,	10,	Larnaca
,,	11–19,	Larnaca
,,	20,	To Athienou via Phurni
,,	21,	Athienou, Tymbou, Palekythro, Kythrea
,,	22–23,	Kythrea to Nicosia
,,	24–26,	Kephalovryso, Ayios Khrysostomos Monastery, Kythrea
,,	27,	Bellapais
,,	28–31,	About Kythrea
April	1,	joined by Rigo in Kythrea
,,	2–5,	about Kythrea and Pentadaktylos
,,	6,	Kythrea to Ayios Khrysostomos Monastery
,,	7,	Ayios Khrysostomos Monastery, Buffavento, Kythrea
,,	8–13,	about Kythrea
,,	14,	Kythrea to Lefkoniko
,,	15,	Lefkoniko to " Arthena " (? Ardhana)
,,	16–18,	Kantara Monastery
,,	19,	" Arthena " (? Ardhana)
,,	20,	Kantara Castle, Dhavlos
,,	21,	Dhavlos to Komi Kebir
,,	22,	Ephtakomi, Leonarisso, Ayios Andronikos, Yialousa
,,	23,	Eleoussa, Rizokarpaso
,,	24–25,	Paleokhori to Apostolos Andreas Monastery
,,	26,	Cape Andreas and Aphendrika (Epiotissa)

April	27,	Rizokarpaso
,,	28,	Galinoporni, Ayios Symeon, Leonarisso
,,	29,	Leonarisso, Kantara—end of published diary
May	1,	Kantara
,,	2,	Kantara
,,	4,	Akanthou, Phlamoudhi, Kantara
,,	5,	Akanthou, near Kalogrea, Pentadaktylos
,,	22,	between Kythrea and Nicosia
,,	23,	near Nicosia
,,	24,	Kythrea
,,	25,	Kythrea
,,	26,	Kyrenia, Ayios Epiktitos, Pentadaktylos
,,	27,	Ayios Epiktitos, above Bellapais
,,	28,	near Kyrenia
,,	29,	St. Hilarion
,,	31,	Lapithos
June	1,	Ayios Yeoryios near Kyrenia
,,	2,	Kyrenia
,,	5–8,	Kythrea
,,	14,	Peristerona
,,	15–16,	Galata
,,	17,	Galata, Troödos, Prodhromos
,,	18–21,	Troödos, Prodhromos
,,	23,	between Galata and Evrykhou
,,	26–27,	Spilia, Adelphi
,,	28–30,	Prodhromos
July	2,	Omodhos
,,	7–9,	Prodhromos
,,	12,	Dheftera
,,	14–18,	Larnaca.

Apart from one Sintenis gathering from Larnaca, dated 17th August, and another from the same area, dated 12th September, it would appear that collecting ceased about the middle of July, 1880.

The next major collection was that of Jens Holmboe (1880–1943) used in drawing up the detailed survey of the island's flora published as "Studies on the Vegetation of Cyprus" (1914). Holmboe outlines his itinerary on pp. 21–23 of this work, and it can be summarized as follows:

March	3, 1905.	arrived Larnaca
,,	3–9,	Ayia Napa and Ormidhia
,,	10–13,	Larnaca
,,	14–18,	Cape Kiti, Mazotos, Alethriko
,,	19–23,	Larnaca
,,	24–April 2,	Nicosia
April	3–4,	Vatili
,,	5–6,	Kouklia
,,	7–8,	Paralimni, Ayia Napa, Cape Greco
,,	9–10,	Varosha
,,	11–12,	Salamis
,,	13–15,	Ayios Theodhoros, Valia, Cape Elea
,,	16–17,	Ephtakomi
,,	18,	Komi Kebir
,,	19,	Dhavlos
,,	20–21,	Ardhana, Kantara Castle
,,	22,	Lefkoniko

April	23,	Pyla
,,	24–28,	Larnaca
,,	29,	Alethriko
,,	30,	Kophinou
May	2–4,	Kalavasos
,,	5–7,	Limassol
,,	8–9,	Akrotiri
,,	10–11,	Episkopi
,,	12–13,	Alekhtora
,,	14–16,	Ktima
,,	17–19,	Tsadha, Ayios Neophytos Monastery
,,	20,	Khoulou
,,	21,	Phiti
,,	22–23,	Polis, Fontana Amorosa
,,	24–25,	Yialia
,,	26,	Lefka
,,	27,	Peristerona
,,	28–31,	Nicosia
June	1,	Aghirda, St. Hilarion
,,	2,	Kyrenia
,,	3–4,	Lapithos, Kornos
,,	5,	Ayios Epiktitos
,,	6,	Ayios Amvrosios
,,	7,	Kythrea, Antiphonitis Church
,,	8,	Nicosia
,,	9–12,	Larnaca
,,	13,	Nicosia
,,	14,	Peristerona
,,	15,	Evrykhou
,,	16–17,	Troödos
,,	18–25,	Prodhromos and adjacent peaks
,,	26,	Kalopanayiotis
,,	27–July 1,	Kykko Monastery
July	2,	Kambos
,,	3,	Exo Mylos (camp in Paphos Forest)
,,	4,	Stavros tis Psokas
,,	5–7,	Pano Panayia
,,	8,	Kaminaria
,,	9,	Prodhromos
,,	10 – 14,	Troödos, Khionistra
,,	15,	Lania
,,	16–21,	Limassol
,,	22–25,	Larnaca
,,	26,	Alethriko
,,	27,	Lefkara
,,	28–30,	Vavatsinia, Prophitis Elias Monastery
,,	31–Aug. 2,	Makheras Monastery
Aug.	3–4,	Pharmakas
,,	5–7,	Palekhori
,,	8–9,	Agros, Papoutsa
,,	10,	Phasoula
,,	11–13,	Limassol
,,	14–16,	Kilani, Omodhos, Vasa, Potamiou, Vouni
,,	17–18,	Pano Platres, Moniatis, Perapedhi
,,	19–21,	Troödos, Khionistra

Aug. 22–23, Kyperounda, Madhari
,, 24, Alithinou
,, 25, Palekhori
,, 26–27, Klirou
,, 28–Sept. 7, Nicosia
Sept. 8–13, Larnaca
,, 13–14, Famagusta
,, 15–17, Cape Greco
,, 18, Famagusta
,, 19–20, Nicosia
,, 21–24, Aghirda, Bellapais
,, 25–26, Nicosia
,, 27–30, Larnaca
Oct. 1, leaves Larnaca for Europe.

By 1905 the work of A. G. & M. E. Lascelles and other residents on the island was already becoming an important factor in the elucidation of the Cyprus flora, and Holmboe's expedition may be said to close the period of initial research by visiting botanists. Apart, however, from the extensive investigations subsequently conducted by resident collectors and the staff of the Departments of Agriculture and Forests, two important expeditions must be mentioned: those of Manoog Haradjian (died c. 1914) and of Harald Lindberg (1871–1963).

Haradjian visited Cyprus during two successive years, and a brief outline of his itineraries (as given by K. H. Rechinger in Arkiv för Botanik, ser. 2, 1: 415; 1950) is as follows:

April 7–20, 1912. Larnaca
,, 22, Famagusta
,, 25–May 7, Rizokarpaso
May 10–22, Komi Kebir
,, 26–30, Armenian Monastery near Halevga
June 1–25, Troödos
April 25–May 5, 1913. Limassol
May 15–27, Paphos
,, 28–30, Stroumbi
June 1–7, Polis, Cape Arnauti
,, 7–16, Lyso
,, 18–22, Pano Panayia
,, 25–July 3, Kykko Monastery
July 4–10, Troödos
,, 11, Troödos, Galata, Nicosia
August Larnaca.

The collections made by Harald Lindberg in 1939 are fully analyzed in his "Iter Cyprium Contributio ad Cognitionem Florae Insulae Cypri", published in Act. Soc. Sci. Fenn., n.s.B., 2(7): 1–37 (1946). Lindberg includes a synopsis of his excursions in the introduction:

June 2, 1939. Famagusta
,, 4, Athalassa
,, 6, Athalassa
,, 8, Nicosia
,, 9, Boghazi, Kyrenia
,, 10, Trikomo
,, 11, Boghazi

June	12,	Kythrea
,,	13,	Lapithos
,,	15,	Evrykhou, Troödos
,,	16,	Troödos
,,	17,	Troödos, Khionistra
,,	18,	Platania
,,	19,	Prodhromos
,,	20,	Troödos
,,	21,	Kato Platres, Mesapotamos Monastery
,,	22,	Troödos
,,	23,	Platania, Kakopetria, Galata
,,	25,	Larnaca
,,	26,	Kouklia (Famagusta), Prastio
,,	27–29,	Larnaca, Perivolia
,,	30,	Stavrovouni
July	1,	Larnaca
,,	2–3,	Limassol, Akrotiri, Episkopi, Kolossi
,,	4,	Athalassa
,,	5,	Armenian Monastery, Kythrea
,,	6–7,	St. Hilarion
,,	8–9,	Famagusta, Salamis
,,	10,	Paralimni, Ayia Napa
,,	11,	Ayios Theodhoros
,,	13,	Bellapais
,,	15,	Ambelikou, Kambos, Kykko Monastery
,,	16,	Kykko Monastery, Milikouri
,,	18,	Argaki tou Kastarakia
,,	19,	Kannaviou, Ayia, Stavros tis Psokas
,,	20,	Akhelia
,,	21–22,	Ayios Neophytos Monastery, Stephani
,,	22,	Mesoyi, Stroumbi, Symvoulo.

Since 1939 the development of the road system within the island, and the widespread availability of private and public motor transport, has meant that itineraries in the old sense are no longer traceable, since a collector can, within a fortnight, have travelled over most of the area, sometimes paying short visits, in a single day, to many widely separated localities. The only areas which are likely to have escaped the attention of recent botanical collectors are those which are remote from motorable roads; in such areas new discoveries are still likely to be made; moreover, the very ease and speed of present-day transport means that plants are often overlooked *en route* which would scarcely have escaped the notice of a collector travelling on foot or by donkey.

Although the number of collections made since 1939 is too great for detailed enumeration, some mention must be made of those collectors whose contributions have been of particular importance in making possible the compilation of this Flora. Foremost amongst these is Mr. A. Syngrassides, formerly of the Cyprus Department of Agriculture, whose extensive collectings, made in all parts of the island between 1932 and 1939, not only added numerous new records to the flora but provided a sound basis of data for future floristic reasearch. Many of Mr. Syngrassides' specimens are cited by A. K. Jackson and W. B. Turrill in their " Additions to the Flora of Cyprus ", published in the *Kew Bulletin* between 1934 and 1939.

Another important contributor to our knowledge of the Cyprus flora was Mrs. E. W. Kennedy, whose meticulously prepared specimens, together with their invaluable field notes, were collected between 1936 and 1955 chiefly in the neighbourhood of Platres, Troödos and Kyrenia, with occasional additions from other parts of the island.

Dr. P. H. Davis, better known for his work on the flora of Turkey, traversed the island in 1940–41, and, with characteristic thoroughness, not only discovered several new plants, but considerably extended the distributions of many species previously known only from one or two localities.

The collections of Mr. G. A. Mavromoustakis, made between 1939 and 1948, are particularly rich in specimens from the Limassol area, a neighbourhood rather under-worked previously, and still worth further examination.

Mrs. E. F. Chapman, whose " Cyprus Trees and Shrubs " (1949) is such an accurate guide to the woody plants of the island, made repeated collections of herbs, trees and shrubs during her residence in Cyprus, between 1930 and 1950.

Amongst more recent workers, special mention must be made of the large amount of carefully selected material sent to Kew between 1951 and 1957 (and again in 1966) by Mr. L. F. H. Merton, an ecologist for some years on the staff of the Department of Agriculture, Nicosia, and in charge of the herbarium there. Mr. Merton was a useful and valued helper in the early stages of the preparation of this work.

Mr. Edwards C. Casey and his sister Mercy did valuable work, especially in the Kyrenia area, in the period immediately following the War, and their collections are supplemented by numerous specimens submitted between 1955 and 1960 by Miss G. E. Atherton and Mrs. N. Macdonald.

Visits to Cyprus by the author of this Flora in 1962 and 1974 have furnished some additional records, as well as information about areas which had not been much visited previously. It may be said, with a measure of confidence, that the content of the Cyprus flora is now fairly thoroughly known, so that the number of new species still undetected is likely to be small. It is, however, no less true to say that several widespread Mediterranean species, at present unrecorded, are still likely to be found, and that the distributions of many species within the island are as yet often very imperfectly plotted. The purpose of a Flora is to provide a summary of work done, to draw attention to gaps in our knowledge, and to stimulate further investigation. The author of this work will be satisfied if these ends are served, even if it involves substantial amendment of the data presented here.

FORMAT OF THE FLORA

The systematic arrangement of the Flora is based upon that followed in the *Genera Plantarum* (1862–83) of J. D. Hooker and G. Bentham. This is the system used (with some modifications) in the Herbarium at Kew, and has been adopted here solely for practical reasons, and for the convenience of the compiler.

Names and Synonyms

Names in heavy (bold) type are those of indigenous or naturalized genera and species; names in small capitals are those of introductions or unestablished escapes from cultivation; names in italics are synonyms; names enclosed in square brackets are errors or misidentifications. Abbreviated references to place and date of publication follow each name and author citation. A short list of literature references is given in Appendix I, p. 801.

References are generally restricted to original place of publication, to works of general interest for the area (e.g. Boissier, *Flora Orientalis*; Post, *Flora of Syria, Palestine and Sinai*, ed. 2, etc.) or to literature dealing particularly with Cyprus (e.g. Unger & Kotschy, *Die Insel Cypern*; Holmboe, *Studies on the Vegetation of Cyprus* etc.). Reference is also made to useful and up-to-date revisions and monographs, or to any publication likely to be of value or interest to students of the Cyprus flora.

Types

Wherever possible the type of the name used is cited, but without any indication of the status of the type. Originally it was intended to include after each type citation the code name (or names) of the institution (or institutions) holding the holotype, isotypes or syntypes (based on the herbarium abbreviations listed in Index Herbariorum, ed. 5, Regnum Vegetabile 31: 205–228; 1964). It was found that this, at least so far as concerned Linnean names, involved an unacceptable measure of speculation, so the plan had partly to be abandoned. Where the location of a type is known, or can be inferred with a reasonable degree of certainty, the code reference is given, with an added exclamation mark where the type specimen has been seen.

Habitat

This ecological data (including altitude and date of flowering) is almost wholly taken from herbarium labels. It should be accurate so far as it goes, but may have to be amended and augmented as additional data becomes available.

Distribution

As regards internal distribution, records are cited under their botanical subdivisions (see Introduction pp. 4–8). Generally a locality is given, together with year of collection, collector's name and number, and an exclamation mark where, as mentioned above, the cited specimen has been examined. Where there are more than four independent records for a single subdivision, the plant is sometimes cited as " frequent " or " common " in that subdivision and precise localities and dates of collection omitted, even then a selection of collectors' names and numbers is generally given, to allow for naming or re-naming in herbaria holding duplicates of these specimens. Place names are taken from the second (revised) edition of the Survey of Cyprus 1/253, 440 Administration Map.

As regards external distribution, the provision of exact information would impose an impossible burden on the compiler, since it would often involve careful examination and identification of material otherwise of little relevance to the Flora. The data provided, though in many instances inexact or incomplete, should give a general idea of the external range of each genus and species.

THE SCOPE OF THE FLORA

The Flora has been planned to include all native and naturalized vascular plants (flowering plants and ferns). Introduced plants pose a problem which cannot be solved to everyone's satisfaction: the policy adopted here is to include in detail those species which are now so widely established that they might almost be regarded as naturalized, and to make briefer mention of those species which are well known in gardens, though evidently dependent upon man for their continuing survival. It is obviously impossible to include all cultivated plants, since there are no adequate records of such plants; moreover in certain genera (e.g. *Eucalyptus*, *Acacia*) new introductions are constantly being added, while older introductions are lost to cultivation. A comprehensive account of the cultivated plants of Cyprus would be valuable, but it should rightly be regarded as supplementary to the present work.

The Flora includes keys to genera and species, but a key to families has been deferred to a second volume, when all the text will be complete. While keys to families are a traditional adjunct to Floras, their importance is questionable. Most readers of technical, scientific works of this kind are normally able to recognize a family without having to resort to a key. Where a difficulty arises with the identification of the family, reference can be made to Hutchinson, J., Families of Flowering Plants (ed. 2, 2 vols., Oxford, 1959), Lawrence, G. H. M., Taxonomy of Vascular Plants (New York, Macmillan, 1951), Benson, L. B., Plant Classification (Boston, D. C. Heath & Co., 1957) or Davis, P. H. & Cullen, J., The Identification of Flowering Plant Families (Oliver & Boyd, 1965). For the same reasons, a glossary of botanical terms is not included in this volume; comprehensive glossaries are included in Lawrence (above), and in Stearn, W. T., Botanical Latin (ed. 2, David & Charles, 1973). Jackson, B. D., Glossary of Botanic Terms (4th ed. Duckworth, 1928 repr. 1971) remains the standard English guide to botanical terminology.

FAMILIES IN VOL. I

1. Pinaceae
2. Cupressaceae
3. Ephedraceae
4. Ranunculaceae
5. Paeoniaceae
6. Berberidaceae
7. Papaveraceae
8. Fumariaceae
9. Cruciferae
10. Capparaceae
11. Resedaceae
12. Cistaceae
13. Violaceae
14. Polygalaceae
15. Frankeniaceae
16. Caryophyllaceae
17. Illecebraceae
18. Portulacaceae
19. Tamaricaceae
20. Elatinaceae
21. Guttiferae
22. Malvaceae
22a. Sterculiaceae
23. Tiliaceae
24. Linaceae
25. Zygophyllaceae
26. Geraniaceae
27. Oxalidaceae
28. Rutaceae
28a. Simaroubaceae
28b. Meliaceae
29. Rhamnaceae
30. Vitaceae
30a. Sapindaceae
31. Aceraceae
32. Anacardiaceae
33. Leguminosae
34. Rosaceae
35. Saxifragaceae
36. Crassulaceae
37. Hamamelidaceae
38. Callitrichaceae
39. Myrtaceae
40. Lythraceae
40a. Punicaceae
41. Onagraceae
42. Datiscaceae
43. Cucurbitaceae
43a. Cactaceae
44. Aizoaceae
45. Umbelliferae
46. Araliaceae
47. Caprifoliaceae
48. Sambucaceae
49. Rubiaceae
50. Theligonaceae

SPERMATOPHYTA

GYMNOSPERMAE

Pilger in Engler et Prantl, Pflanzenfam., ed. 2, 13: 121–403 (1926); Pilger et Melchior in Engler, Syllabus der Pflanzenfam., ed. 12, 1: 312–343 (1954).

1. PINACEAE

Evergreen or, less commonly, deciduous trees; buds scaly; leaves with resin canals, linear, linear-oblong or subulate, spirally arranged, or bifarious, or sometimes whorled or clustered; flowers unisexual in cones of spirally arranged scales; male cones with 2 pollen-sacs on the lower (abaxial) surface of each scale; female cones consisting of spirally arranged, more or less conspicuous bracts and fertile scales, the latter with 2 ovules on their upper (adaxial) surface. Fruit a woody cone, the scales and bracts persistent or caducous; seeds usually winged; cotyledons generally numerous.

The family includes 10 genera widely distributed in the northern hemisphere.

Leaves in fascicles of 2 (in our area) or more; cones usually patent, deflexed or pendulous; cone-scales persistent, not falling apart at maturity - - - - **2. Pinus**
Leaves not in fascicles, scattered on long shoots, whorled on short lateral spurs; cones erect, barrel-shaped; cone-scales caducous, falling apart at maturity - **1. Cedrus**

1. CEDRUS *Trew*

Cedr. Lib. Hist., 1: 6 (1757) nom. cons.

Evergreen trees; leaves linear-subulate, scattered on long shoots, whorled on short lateral spurs; cones terminal on short shoots, erect; male cones cylindrical or sausage-shaped; female cones large, blunt, barrel-shaped, the cone-scales closely and evenly overlapping, falling away from the persistent central axis at maturity; bracts small or apparently wanting; seeds winged; cotyledons 9–10.

Three species with disjunct distributions in N. Africa, western Asia, Afghanistan and the Himalayas.

1. C. libani *A. Rich.* in Bory, Dict. Class. Hist. Nat., 3: 299 (1823); G. Don in Loudon, Hort. Brit., 388 (1830); Loudon, Arb. et Frut. Brit., 4: 2402 (1838); Post, Fl. Pal., ed. 2, 2: 798 (1933); Davis, Fl. Turkey, 1: 71 (1965).
Pinus cedrus L., Sp. Plant., ed. 1, 1001 (1753).
Cedrus libanotica Link, Handbuch, 2: 480 (1831).
C. libanitica Trew ex Pilger in Engl. et Prantl, Pflanzenfam., ed. 2, 13: 329 (1926).

A massive tree 20–30 m. high with a thick trunk, pale greyish-brown, fissured bark, and spreading branches, often forming distinct horizontal (or tabular) layers in mature specimens; twigs rough with fissured or flaking greyish-brown or fuscescent bark; current year's long shoots glabrous or pubescent, angled, pale brown; buds small, blunt, with closely imbricate, brown scales; leaves subulate, 8–25 mm. long, 1–2 mm. wide, patent or erecto-patent, straight or somewhat incurved, green or glaucous, apex acute or rather blunt, sometimes shortly spinose; male cones bluntly cylindrical or sausage-shaped, 3–5 cm. long, 0·8–1·3 cm. wide, scales bluntly obovate, about 4 mm. long, pale brown, loosely imbricate with a suborbicular terminal appendage about 2 mm. diam., the margins of the appendage minutely, but distinctly, erose; seed-bearing cones maturing in 2 years, barrel-shaped, 8–11 cm. long, 5–6 cm. wide, pale or rather dark greyish-brown, consisting of numerous, closely overlapping, fan-shaped scales 4–4·5 cm. wide and about 2 cm. long, apical margin of scales smooth or slightly erose; seed obovate or pear-shaped, a little compressed, about 10–13 mm. long, 5–6 mm. wide, with a conspicuous brown, membranous, truncate wing, about 2 cm. long and almost as wide.

ssp. **brevifolia** (*Hook. f.*) *Meikle* **comb. nov.**

C. libani A. Rich. var. *brevifolia* Hook. f. in Journ. Linn. Soc. Bot., 17: 518 (1880); Boiss., Fl. Orient., 5: 699 (1884); Osorio-Tafall et Seraphim, List Vasc. Plants Cyprus, 4 (1973).

C. brevifolia (Hook. f.) Henry in Elwes et Henry, Trees of Great Britain and Ireland, 3: 467 (1908); Lindberg f., Iter Cypr., 5 (1946); Rehder, Bibl. Cult. Trees and Shrubs, 32 (1949); Elektra Megaw, Wild Flowers of Cyprus, 18, t. 40 (1973).

C. libanotica Link ssp. *brevifolia* (Hook. f.) Holmboe, Veg. Cypr., 29 (1914); Chapman, Cyprus Trees and Shrubs, 28 (1949).

C. libanitica Trew ex Pilger ssp. *brevifolia* (Hook. f.) O. Schwarz in Fedde Repert., 54: 29 (1951).

[*C. libani* (non A. Rich.) Post in Mém. Herb. Boiss., 18: 100 (1900).]

TYPE: Cyprus; a specimen communicated by Sir Samuel Baker and exhibited at the Linnean Society of London on Nov. 20, 1879. Part of this exhibit is preserved in the Herbarium at Kew (K !).

Leaves generally shorter and thicker than those of ssp. *libani*, mostly less than 15 mm. long (rarely up to 20 mm.), 1·5–2 mm. wide, often distinctly incurved, with shortly acute or bluntish apices. Cones usually less than 8 cm. long and 5 cm. wide, but not uniformly distinct from those of ssp. *libani*. *Plate 1*.

HAB.: Stony mountainsides on igneous formations; 3,500–4,600 ft. alt.; cones mature in late autumn and winter.

DISTR.: Division 2, locally abundant. Endemic.

2. Slopes and summit of Tripylos Peak, Paphos Forest. *Post* ! *Druce* ! *Davis* 1768 ! 3461 ! *Lindberg f.* ! *R. R. Waterer* ! *G. W. Chapman* ! *Meikle* 2662 ! *Economides* in ARI 960 ! *W. R. Price* 1065 ! etc.

NOTES: Holmboe and others have commented on the close affinities of all the Mediterranean Cedars. The Cyprus Cedar, though usually distinguishable by its short, thick, bluntish leaves, does, on occasion, approach other forms of *C. libani* so closely that separation at species level is difficult to justify. In overall habit many of the Cyprus trees approach Turkish forms of *C. libani* (*C. libanitica* Trew ex Pilger ssp. *stenocoma* O. Schwarz), having narrowly pyramidal, tapering crowns; this may, however, be no more than a passing characteristic of immaturity; many of the Cyprus trees are young, and of the relatively few aged specimens seen several have developed the horizontal, tabular system of branching associated with *C. libani* ssp. *libani*.

2. PINUS *L.*

Sp. Plant., ed. 1, 1000 (1753).
Gen. Plant., ed. 5, 434 (1754).

Evergreen trees; branches whorled; leaves spirally arranged, subulate, in clusters of 2, 3 or 5, generally semi-circular or crescent-shaped in transverse section in the 2-leaved species, base enclosed in a persistent, scaly sheath; male cones produced in clusters around the base of the young shoots, yellow or reddish, cylindrical, ovoid or suborbicular; female cones usually terminal or subterminal; bracts minute, absent in the mature cones; cone-scales persistent, not falling apart at maturity, often thick and woody with a conspicuous, sometimes mucronate, umbo or protuberance at the centre of the exposed part (apophysis) of each scale; cones usually opening and releasing the seeds at maturity, sometimes remaining closed and falling as a whole to the ground; seeds winged or unwinged, sometimes with a hard, woody testa.

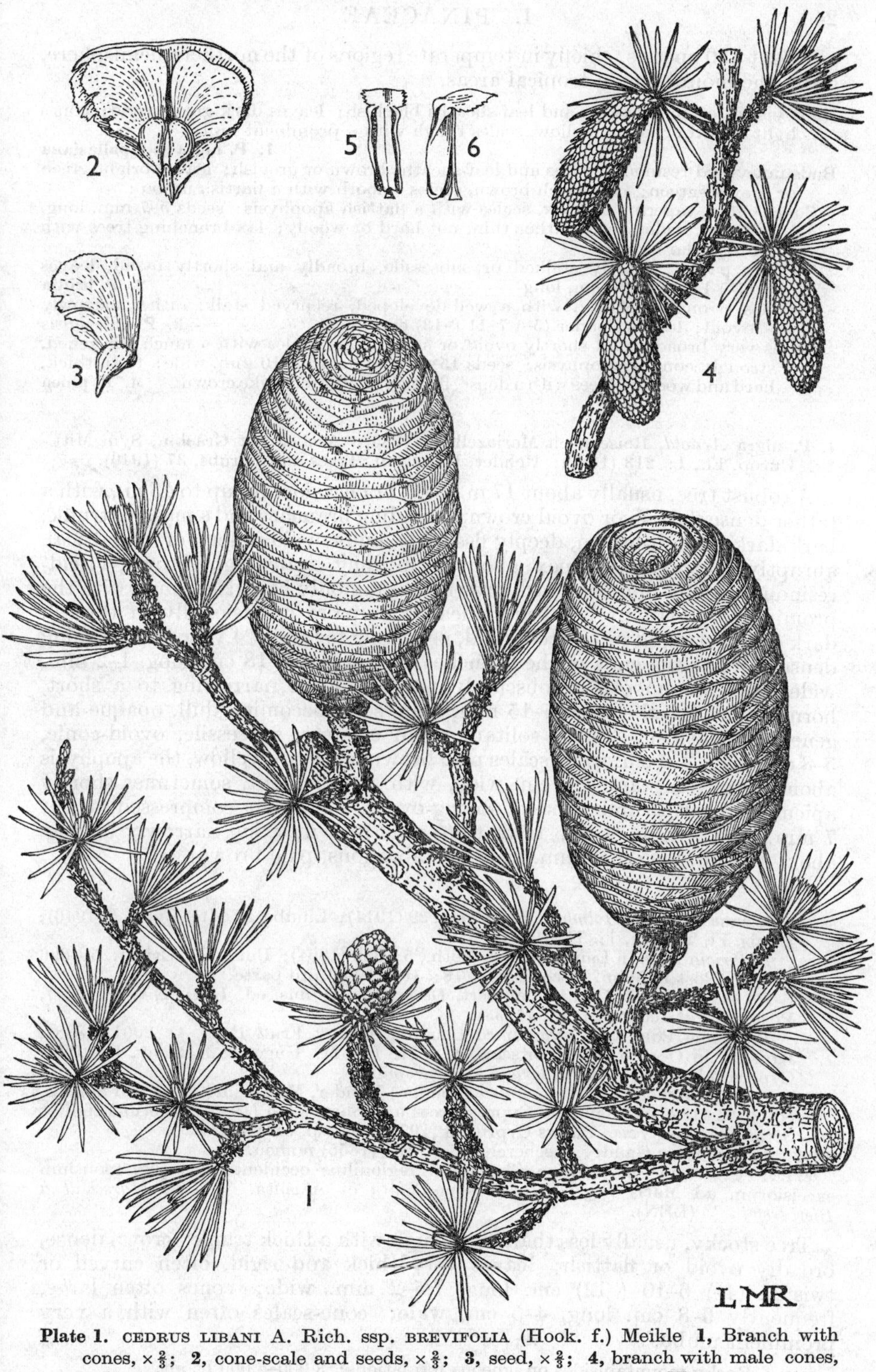

Plate 1. CEDRUS LIBANI A. Rich. ssp. BREVIFOLIA (Hook. f.) Meikle **1,** Branch with cones, $\times \frac{2}{3}$; **2,** cone-scale and seeds, $\times \frac{2}{3}$; **3,** seed, $\times \frac{2}{3}$; **4,** branch with male cones, $\times \frac{2}{3}$; **5,** scale from male cone and pollen-sacs. (*Chapman* 165; *Waterer*.

About 120 species, chiefly in temperate regions of the northern hemisphere, or on the mountains in tropical areas.

Buds viscid-resinous; twigs and leaf-sheaths blackish; leaves dark green, rigid; cones light brown or tawny-yellow, scales rough with a prominent umbo **1. P. nigra** ssp. **pallasiana**

Buds not viscid-resinous; twigs and leaf-sheaths brown or greyish; leaves bright green or greyish-green; cones rich brown, scales smooth with a flattish umbo :

Cones ovoid, tapering to apex, scales with a flattish apophysis; seeds 5–7 mm. long, 2–4 mm. wide; testa rather thin, not hard or woody; lax-branching trees with an irregular crown:

Cones patent, shortly stalked or subsessile, broadly and shortly ovoid, leaves (7–) 11–16 (–20) cm. long - - - - - - - - **2. P. brutia**

Cones strongly deflexed with a well-developed, recurved stalk, rather narrowly ovoid; leaves slender (5–) 7–11 (–13) cm. long - - - **3. P. halepensis**

Cones very broadly and shortly ovoid or subglobose, scales with a much thickened, strongly convex apophysis; seeds 15–20 mm. long, 8–10 mm. wide; testa thick, hard and woody; trees with a dense, flattish or umbrella-like crown **4. P. pinea**

1. P. nigra *Arnold*, Reise nach Mariazell, 8 (1785); Aschers. et Graebn., Syn. Mitt.-Europ. Fl., 1: 213 (1897); Rehder, Bibl. Cult. Trees and Shrubs, 37 (1949).

A robust tree, usually about 17 m. high, or occasionally up to 45 m., with a rather dense flattish or ovoid crown, spreading branches and a massive trunk; bark dark greyish-brown, deeply fissured; buds ovoid-oblong or cylindrical, abruptly contracted to apex, about 2 cm. long, scales light brown, viscid-resinous; young branches dark brown or commonly blackish, rough with the prominent, persistent leaf-bases; leaves in pairs (occasionally in threes), dark green, straight or twisted, rigid, persisting for about 4 years and forming dense tufts at the ends of the branches, about (4–) 6–18 cm. long, 1–2 mm. wide, margins smooth or obscurely toothed, apex narrowing to a short, horny point, basal sheath 5–15 mm. long, soon becoming dull, opaque and generally blackish. Cones solitary or in clusters, subsessile, ovoid-conic, 5–8 cm. long, 3–5 cm. wide, scales pale brown or tawny-yellow, the apophysis about 1 cm. long and 1·5 cm. wide, with a prominent, sometimes shortly apiculate, central umbo; seed oblong-ovoid, somewhat compressed, about 7 mm. long, 4 mm. wide; testa thin, pale brown; wing narrowly oblong, about 15 mm. long, 6–7 mm. wide, membranous, pale brown.

ssp. **pallasiana** (*D. Don*) *Holmboe*, Veg. Cypr., 29 (1914); Lindberg f., Iter Cypr., 6 (1946); Davis, Fl. Turkey, 1: 74 (1965).

P. laricio Poir. in Lam., Encycl. Méth., 5: 339 (1804); Boiss., Fl. Orient., 5: 696 (1882); Post in Mém. Herb. Boiss., 18: 100 (1900) pro parte.

P. pallasiana D. Don in Lambert, Descr. Gen. Pinus, ed. 1, 2: 1, t. 1 (1824); Post, Fl. Pal., ed. 2, 2: 798 (1933).

P. laricio Poir. var. *caramanica* Loudon, Arb. et Frut. Brit., 4: 2001 (1838); Kotschy in Oesterr. Bot. Zeitschr., 12: 278 (1862); Unger et Kotschy, Die Insel Cypern, 214 (1865)—as *P. laricio* var. *caramana*.

P. nigra Arnold var. *caramanica* (Loudon) Rehder, Man. Cult. Trees and Shrubs, ed. 1, 61 (1927); Chapman, Cyprus Trees and Shrubs, 26 (1949); Osorio-Tafall et Seraphim, List Vasc. Plants Cyprus, 3 (1973)—as ssp. *caramanica*.

P. caramana Gaudry, Recherches Sci., 197 (1855) nomen.

TYPE: Crimea; "in Tauriae Chersonesi regionibus occidentalibus jugi montium excelsiorum ad Maris Nigri littora usque Yalta et Alushta. *Pallas, Marschall à Bierberstein*." (LEN).

Tree stocky, usually less than 30 m. high with a thick trunk; crown dense, broadly ovoid or flattish; leaves very thick and rigid, often curved or twisted (4–) 6–10 (–12) cm. long, 1·5–2 mm. wide; cones often large, frequently 6–8 cm. long, 4–5 cm. wide; cone-scales often with a very prominent umbo.

HAB.: **Rocky mountainsides on igneous formations; 3,900–6,000 ft. alt.**

DISTR.: Division 2, locally abundant. Also Balkan peninsula, S. Carpathians, Crimea, Transcaucasia, Turkey, N. Syria.

2. Upper slopes of Khionistra, forming forests (with *Juniperus foetidissima* Willd.). *Kotschy* 759; *Post* s.n. ! *Wyatt* 27 ! *Syngrassides* 718 ! *Kennedy* 44–53 ! *Merton* 3183 ! *D. P. Young* 7323 !

NOTES: Cyprus specimens of *P. nigra* ssp. *pallasiana* are, on the whole, distinguishable by their short, thick, rigid leaves. In this respect, they differ slightly from Crimean ssp. *pallasiana*, and might be deserving of separate nomenclatural recognition were it not for the fact that Turkish populations provide a series of interlinking forms. The subdivision of *P. nigra* into numerous subspecies or varieties cannot be said to be very satisfactory, or very serviceable.

2. P. brutia *Tenore*, Fl. Nap. Prodr., LXXII (1811); Fl. Neap. Syll. in Fl. Nap., 4: 136 (1830); Boiss., Fl. Orient., 5: 696 (1882); Post in Mém. Herb. Boiss., 18: 100 (1900); H. Stuart Thompson in Journ. Bot., 44: 341 (1906); Druce in Proc. Linn. Soc., sess. 141: 52 (1930); Post, Fl. Pal., ed. 2, 2: 797 (1933); Czeczott in Fedde Repert. Beih., 107: 254 (1939); Feinbrun in Bull. Res. Council Israel, 7D: 146 (1959); Davis, Fl. Turkey, 1: 74 (1965); Osorio-Tafall et Seraphim, List Vasc. Plants Cyprus, 3 (1973).

P. halepensis Mill. ssp. *brutia* (Ten.) Holmboe, Veg. Cypr., 29 (1914); Chapman, Cyprus Trees and Shrubs, 27 (1949).

[*P. maritima* (non Mill.) Sibth. et Sm., Fl. Graec. Prodr., 2: 247 (1816); Poech, Enum. Plant. Ins. Cypr., 12 (1842); Kotschy in Oesterr. Bot. Zeitschr., 12: 276–278 (1862); Unger et Kotschy, Die Insel Cypern, 214 (1865) pro parte quoad pl. cypr.]

[*P. halepensis* (non Mill.) Holmboe, Veg. Cypr. 29 (1914); Druce in Proc. Linn. Soc., sess. 141: 51 (1930); R. M. Nattrass, First List Cypr. Fungi, 26–28, 42–43 (1937) pro parte quoad pl. cypr.]

TYPE: Italy; "in nemoribus Calabriae occidentalis: *Aspromonte*." (NAP—also figured in Fl. Nap. Atlas, 4: t. 200; 1811–38.)

Tree 12–25 (–40) m. high with an irregular, pyramidal or bluntly ovoid or flattened crown and rather lax branching; bark greyish-brown, deeply fissured with age; buds ovoid-conic, about 1 cm. long, scales pale brown with silvery margins, not viscid-resinous; young branches greyish or pale greyish-brown, at first rough with prominent leaf-bases, but becoming smooth in the third year; leaves in pairs, bright green or greyish-green, straight or twisted, persisting for 2 years and forming rather lax tufts at the ends of the branches, about (7–) 11–16 (–20) cm. long, 1–1·5 mm. wide, margins minutely and often obscurely serrulate especially in the upper half of the leaf, apex narrowing to a short, sharp, horny apiculus, basal sheath 5–10 (–14) mm. long, scales membranous, golden-brown with whitish fimbrillate margins. Cones solitary or paired (occasionally more in a cluster), patent with a short thick stalk (less than 1 cm. long) or subsessile, ovoid-conic, 5–6 (–8) cm. long, 4–4·5 (–5) cm. wide, scales rich lustrous brown, the apophysis about 1 cm. long, 1·5 cm. wide, with a flattened, greyish, central umbo; seed oblong-ovoid, somewhat compressed, about 7 mm. long, 3–3·5 mm. wide; testa blackish-brown, rather thin; wing oblong, 20–23 mm. long, 7–8 mm. wide, membranous, pale brown.

HAB.: Well-drained ground on calcareous, or acid, sedimentary and igneous formations throughout the island; sea-level to 4,000 ft. alt.; fruit matures Aug.–Sept.

DISTR.: Divisions 1–3, 5–8, probably throughout the island, often common and forming forests. Also S. Italy, Crete, Turkey, N. Iraq, W. Syria.

1. Ktima, 1932; *Nattrass* 685. Common in the Akamas, 1962 and 1974, *Meikle* !
2. Common; Stavros tis Psokas, Platres, Kakopetria, Madhari, Platania etc. *Kotschy* 760 ! *Syngrassides* 951 ! *Kennedy* 47 ! 54 ! *Meikle* 2839 ! etc.
3. Alethriko, 1905, *Holmboe* 205; Pissouri, 1967, *Merton* in ARI 681 !
5. Salamis, 1905, *Holmboe* 468.
6. Nicosia, 1932, *Nattrass* 347; Myrtou, 1933; *Nattrass* 393. Common about Dhiorios and Kormakiti, 1962 and 1974, *Meikle* !
7. Pentadaktylos, 1880, *Sintenis & Rigo* 715 ! Common on the Northern Range, 1962 and 1974, *Meikle* !
8. Common about Akradhes, 1962, *Meikle* !

3. **P. halepensis** *Mill.*, Gard. Dict., ed. 8, no. 8 (1768); Boiss., Fl. Orient., 5: 695 (1882); Post, Fl. Pal., ed. 2, 2: 797 (1933); Czeczott in Fedde Repert. Beih., 107: 255 (1939); Chapman, Cyprus Trees and Shrubs, 27 (1949); Davis, Fl. Turkey, 1: 75 (1965); Zohary, Fl. Palaest., 1: 18, t. 16 (1966); Osorio-Tafall et Seraphim, List Vasc. Plants Cyprus, 3 (1973).

TYPE: England; Sussex, cultivated at Goodwood from seed originally collected at Aleppo and sent by Consul Cox to Philip Miller in 1732. The type is illustrated in Miller, Figures of Plants, 2: t. 208 (1760)!

Tree 13–20 (–30) m. high, with an irregular pyramidal, ovoid or flattened crown and rather lax branching; bark greyish-brown, fissured with age; buds narrowly ovoid, 8–10 mm. long, scales brown with silvery, fimbriate margins, not viscid-resinous; young branches greyish, at first rough with prominent leaf-bases, becoming rather smooth in the third year; leaves in pairs, greyish-green, straight or less often twisted, slender, persisting for 2 years and forming rather lax tufts at the ends of the branches, (5–) 7–11 (–13) cm. long, less than 1 mm. wide, margins smooth or very minutely serrulate in the upper half of the leaf, apex narrowing to a short, not very sharp or horny apiculus; basal sheaths 5–9 mm. long, dull, whitish or brownish with fimbrillate margins. Cones solitary or in pairs, strongly deflexed with a well-developed recurved stalk 1–2 cm. long, rather narrowly ovoid, 5–11 cm. long, 2·5–4 cm. wide, scales rich brown, lustrous or rather dull, the apophysis about 8 mm. long, 10–13 mm. wide, with a flattened or slightly prominent, greyish, central umbo; seed oblong-ovoid, somewhat compressed, 5–7 mm. long, 3 mm. wide; testa blackish-brown, rather thin; wing oblong, about 20 mm. long, 6 mm. wide, membranous, pale brown.

HAB.: In plantations or by roadsides in lowland regions, almost certainly introduced.

DISTR.: Divisions 1, 3–6, 8, planted. Widespread in the Mediterranean region, but more common in the West than in the East.

1. Ktima, *Chapman*.
3. Phasouri, *Chapman*.
4. Near Famagusta, *Chapman*.
5. Salamis, *Chapman*.
6. Panayia, *Chapman* 216.
8. Valia Forest, *Chapman*.

NOTES: Most Cyprus records for *P. halepensis* Mill. refer to its close ally, here distinguished as *P. brutia* Ten. Mrs. Chapman is undoubtedly correct in regarding it as an introduction, and all wild lowland populations of Pines, so far examined, have been found to consist solely of *P. brutia*. Since, however, *P. halepensis* is evidently not uncommon as a planted tree, and since it is apparently native in areas adjacent to Cyprus, it has been given rather fuller treatment than it would otherwise merit.

4. **P. pinea** *L.*, Sp. Plant., ed. 1, 1000 (1753); Walpole, Mem. Europ. Asiat. Turkey, ed. 1, 82 (1817); Boiss., Fl. Orient., 5: 694 (1884); Holmboe, Veg. Cypr., 28 (1914); Post, Fl. Pal., ed. 2, 2: 797 (1933); Chapman, Cyprus Trees and Shrubs, 26 (1949); Davis, Fl. Turkey, 1: 75 (1965); Osorio-Tafall et Seraphim, List Vasc. Plants Cyprus, 4 (1973).

TYPE: "*in* Italia".

A robust tree, 20–25 m. high with ascending branches and characteristic, dense, flattish, umbrella-like crown; bark reddish-brown, coarsely and deeply fissured with age; buds narrowly ovoid, 6–12 mm. long, scales dark brown, whitish-margined, recurved, fimbriate, not viscid-resinous; young branches pale brown, at first rough with prominent leaf bases, becoming smooth in the third or fourth year; leaves in pairs, bright green, straight or slightly twisted, rather rigid, persisting for 2 or 3 years and forming dense tufts at the ends of the branches, 8–12 (–15) cm. long, about 2 mm. wide, margins minutely serrulate in the upper half of the leaf, apex narrowing to a short, very sharp, point; basal sheaths 8–10 mm. long, whitish or pale brown with fimbrillate margins. Cones solitary or clustered, erect on short, thick stalks, broadly ovoid or subglobose, 10–15 cm. long, 8–13 cm. wide, scales rich brown, apophysis thick, strongly convex, about 1·5 cm. long and up to 3·5 cm. wide with a prominent, greyish, central umbo; seeds large, oblong-ovoid, 15–20 mm. long, 8–10 mm. wide, with a thick, woody, blackish testa and a large, edible kernel; wing very short, rarely more than 8 mm. long, sometimes 15 mm. wide.

HAB.: In gardens and plantations in lowland areas, often on sandy ground; planted.

DISTR.: Divisions 2, 5 and probably elsewhere, but only as an introduction. Mediterranean region and Portugal.

2. Kato Platres, 1879, *Baker* sec. *Holmboe*; Moniatis, 1905, *Holmboe* 1157 !
5. Salamis Plantations, *Chapman*, and, same locality, *A. Genneou* in ARI 1266 !

NOTES: The Stone Pine, nowhere native in Cyprus, is seldom collected or recorded, and the above distribution is very incomplete. Mrs. Chapman (Cyprus Trees and Shrubs, 26) says it " has been planted extensively in gardens and avenues ". It is valued as a sand-binder in maritime areas, and the large, nut-like seeds have edible, nutritious kernels.

Two other Pines, both with leaves in threes, are also cultivated in Cyprus: they are P. RADIATA *D. Don* (in Trans. Linn. Soc., 17: 442; 1837), the Monterey Pine, a native of California, sometimes more than 30 m. high, with slender, bright green leaves 10–15 cm. long, and deflexed, markedly asymmetric cones (with the upper half much more convex than the lower), 8–11 cm. long, 5–7 cm. wide. The unopened cones often remain attached to the branches for many years. The second cultivated species, P. CANARIENSIS *C. Smith* (in Buch, Phys. Beschr. Canar. Ins., 159: 1825), from the Canary Islands, is at once distinguished by the very long leaves, commonly 25–30 cm. in length; the cones are broadly ovoid, 9–15 cm. long, 8–9 cm. wide, with thick scales and a prominent umbo. It is reported by Frangos (Cypr. Journ., 18: 87; 1923) from Nicosia and Government House gardens.

2. CUPRESSACEAE

Evergreen, resinous trees and shrubs; leaves decussate or whorled, often subulate-acicular in juvenile growths, but frequently becoming small, scale-like and imbricate in mature growths; inflorescences monoecious or dioecious, forming small axillary or terminal cones or clusters; cone-scales opposite or 3-whorled, often peltate, the males bearing 2–6 free pollen-sacs at the base of their lower (abaxial) side, the females with 1–many erect ovules in 1 or more rows at the base of their upper (adaxial) side. Fruit a small woody cone or fleshy galbulus, the cone-scales free or united at maturity; seeds free or rarely several together in a woody endocarp, winged or unwinged, cotyledons usually 2, sometimes 3–6.

The family comprises 10 genera with a cosmopolitan distribution.

Cones woody, the peltate scales separating at maturity - - - **1. Cupressus**
Cones fleshy, berry-like, the scales united at maturity - - - **2. Juniperus**

1. CUPRESSUS *L.*

Sp. Plant., ed. 1, 1002 (1753).
Gen. Plant., ed. 5, 431 (1754).

Evergreen, aromatic trees or rarely shrubs; branchlets of mature growths pinnately divided and covered with closely adpressed, scale-like leaves, imbricated in 4 ranks or subterete, rarely somewhat flattened; inflorescences small, solitary, terminal, the male cones ovoid or oblong, pallid, consisting of numerous scales, each bearing 2–6 pendulous, globose or obovoid pollen-sacs; female cones subglobose, with fewer, coriaceous scales, each bearing numerous ovules in several rows. Mature cones globose with 6–12 woody, umbonate scales, separating in the second year, when fully ripe; seeds angular, compressed, narrowly winged; cotyledons 3–4.

About 20 species widely distributed in the northern hemisphere.

1. C. sempervirens *L.*, Sp. Plant., ed. 1, 1002 (1753); Sibth. et Sm., Fl. Graec. Prodr., 2: 248 (1816); Poech, Enum. Plant. Ins. Cypr., 12 (1842); Boiss., Fl. Orient., 5: 705 (1884); Holmboe, Veg. Cypr., 28 (1914); Post, Fl. Pal., ed. 2, 2: 800 (1933); Chapman, Cyprus Trees and Shrubs, 24 (1949); Davis, Fl. Turkey, 1: 76 (1965); Zohary, Fl. Palaest., 1: 19, t. 17 (1966); Osorio-Tafall et Seraphim, List Vasc. Plants Cyprus, 4 (1973).

Tree 15–30 m. high; bark greyish-brown, lightly fissured, branches spreading or erect and fastigiate, reddish-brown or greyish and rather scurfy; leaves very small, scale-like, dull green, broadly rhomboid-ovate in outline, closely adpressed to the branchlets in decussately imbricating pairs; the ultimate divisions of the branchlets subterete and 1–1·5 mm. diam.; inflorescences monoecious; male flowers terminal on the branchlets, forming yellowish-brown, narrowly and bluntly ovoid cones, 3–5 mm. long, 2–3 mm. wide, each cone consisting of numerous peltate, membranous, papillose scales about 1–1·5 mm. wide; female cones inconspicuous, terminal on short lateral shoots towards the base of the branchlets, club-shaped or subglobose and consisting of about 12 overlapping, decussately imbricate coriaceous scales about 1·5 mm. diam. Mature cones subglobose, woody, about 2–3 cm. diam., scales becoming brownish and separating when fully ripe, the apophyses irregularly pentagonal or hexagonal, rugulose, each with a ± distinct, mucronate central umbo; seeds dark reddish-brown, broadly and bluntly elliptic-oblong, about 4 mm. long, 3 mm. wide, strongly compressed or flattened with a narrow marginal wing.

var. **sempervirens**

C. pyramidalis Targ.-Tozz. in Ann. Mus. Fis. Stor. Nat. Firenze, 2, 2: 73, 76, t. 10 (1810).

C. fastigiata DC. in Lam. et DC., Fl. Franç., ed. 3, 6: 336 (1815); Unger et Kotschy, Die Insel Cypern, 213 (1865).

C. sempervirens L. var. *pyramidalis* (Targ.-Tozz.) Nyman, Consp. Fl. Europ., 675 (1881).

C. sempervirens L. f. *pyramidalis* (Targ.-Tozz.) Holmboe, Veg. Cypr., 28 (1914).

TYPE: " *In* Creta ".

Branches erect, fastigiate; tree with a narrow columnar outline.

HAB.: Planted in gardens, parks, cemeteries or on the borders of fields etc., occasionally wild on calcareous hillsides; sea-level to 2,500 ft. alt.; fl. March–May.

DISTR.: All Divisions, but generally planted. Throughout the Mediterranean region and elsewhere as a planted tree. Recorded as wild in Division 7.

7. Kantara Monastery, 1905, *Holmboe*; Ayios Nikolaos and Trypimeni (? 1905) *G. Vassiliou* in *Holmboe*.

var. **horizontalis** (*Mill.*) *Aiton*, Hort. Kew., ed. 1, 3: 372 (1789).

C. horizontalis Mill., Gard. Dict., ed. 8, no. 2 (1768); Kotschy in Oesterr. Bot. Zeitschr., 12: 276 (1862); Unger et Kotschy, Die Insel Cypern, 213 (1865); Sintenis in Oesterr. Bot. Zeitschr., 31: 392, 394 (1881), 32: 192 (1882).

TYPE: Not indicated.

Branches spreading; tree with an irregular, pyramidal outline.

HAB.: On calcareous hillsides, 500–3,900 ft. alt.; fl. March–May.

DISTR.: Divisions 1, 2, 5, 7, 8, occasionally planted. Eastern Mediterranean region to Iran and Iraq.

1. Near Ayios Neophytos, 1905, *Holmboe*, and, 1974, *Meikle* !
2. Ayia Irini, by Kryos Potamos near Platres, 1939 and 1955, *Kennedy* 1532 ! 1870 ! Planted originally, but apparently now naturalized.
5. Athalassa, 1936, *Syngrassides* 953 !
7. Common on the Northern Range; 1787, *Sibthorp*; 1862, *Kotschy*; *Sintenis & Rigo* 506 ! 1905, *Holmboe* etc.
8. Kantara Castle, 1880, *Sintenis & Rigo*, and, 1905, *Holmboe*.

In addition to the native *Cupressus sempervirens*, three exotic Cypresses are occasionally planted in Cyprus. Two of these, *C. arizonica* Greene (in Bull. Torrey Bot. Club, 9: 64; 1882) and *C. guadalupensis* S. Wats. (in Proc. Amer. Acad., 14: 300; 1879) are distinguished from *C. sempervirens* by their glaucous or pruinose twigs, foliage and cones. In *C. arizonica* (from Arizona, Texas and N. Mexico) the leaves are often sharply acute, with a distinct resin-pit at the centre of their dorsal surface; the cones are smallish seldom more than 2·5 cm. diam., with sharply umbonate scales. *C. guadalupensis*, from Guadalupe Island, Lower California, has blunter leaves, usually without a dorsal resin-pit; its cones are larger (2·5–3·5 cm. diam.) and more bluntly umbonate. In *C. guadalupensis*, and in some forms of *C. arizonica*, the bark of the branches comes off in flakes exposing a smooth, grey or red, underlayer.

The third species, *C. macrocarpa* Hartw., (in Journ. Hort. Soc., 2: 187; 1847), from the Monterey Peninsula, California, resembles *C. sempervirens* more closely, and might pass as a subspecies of the latter were it not for its very different distribution. It is altogether a larger tree, with, at maturity, a tabular mode of branching not unlike the Cedar of Lebanon; the branchlets are often coarser, and the cones are larger (up to 3·5 cm. diam.), usually with fewer scales. The Monterey Cypress is popular as a quick-growing hedge or shelter plant in many maritime parts of western Europe, as well as in tropical and South Africa and in Australasia.

2. JUNIPERUS *L.*

Sp. Plant., ed. 1, 1038 (1753).
Gen. Plant., ed. 5, 461 (1754).

Evergreen, aromatic trees and shrubs; leaves opposite or 3-whorled, subulate-acicular in juvenile and mature growths, or often becoming scale-like and indistinguishable from those of *Cupressus* in mature growths; inflorescences monoecious or dioecious, axillary or terminal; male cones resembling those of *Cupressus*; female cones surrounded at the base with small, persistent, scale-like bracts; cone-scales 3–8, united, becoming fleshy in mature fruit and forming a berry-like galbulus; seeds 1–12 in each cone, without wings.

About 60 species, mostly in temperate regions of the northern hemisphere.

Leaves always subulate-acicular, 3-whorled - - - - - - **1. J. oxycedrus**
Leaves of mature growths scale-like, decussate:
  Fruits red-brown at maturity; leaves blunt, very closely adpressed-imbricate; a lowland plant - - - - - - - - - **2. J. phoenicea**
  Fruits purplish-black or violet, pruinose, at maturity; leaves acute or subacute, often rather loosely adpressed-imbricate; mountain plants:
    Ultimate divisions of branchlets often distinctly 4-sided; leaves with or without an obscure resin-pit on the dorsal surface; seeds 1–2 (–3) in each fruit **3. J. foetidissima**
    Ultimate divisions of branchlets subterete or irregularly angular; leaves generally with a distinct and active dorsal resin-pit; seeds 4–6 (–9) in each fruit **4. J. excelsa**

1. J. oxycedrus *L.*, Sp. Plant., ed. 1, 1038 (1753); Boiss., Fl. Orient., 5: 707 (1884); Post, Fl. Pal., ed. 2, 2: 801 (1933); Lindberg f., Iter Cypr., 6 (1946); Chapman, Cyprus Trees and Shrubs, 25 (1949); Davis, Fl. Turkey, 1: 80 (1965); Zohary, Fl. Palaest., 1: 20, t. 18 (1966); Osorio-Tafall et Seraphim, List Vasc. Plants Cyprus, 4 (1973).

J. rufescens Link ex Endl., Syn. Conif., 11 (1847); Unger et Kotschy, Die Insel Cypern, 212 (1865).

J. oxycedrus L. ssp. *rufescens* (Link ex Endl.) Holmboe, Veg. Cypr. 28 (1914).

TYPE: Spain, S. France; " *in* Hispania, G. Narbonensi ".

A rounded or prostrate shrub or small tree, up to 6 m. high; bark greyish or reddish, often rough and scurfy; stems of current year angular, glabrous, reddish; leaves 3-whorled, subulate-acicular in young and mature growths, pungent, 10–20 mm. long, 1–2 mm. wide at base, slightly keeled below, golden-green, glabrous, with two whitish stomatal bands; inflorescences dioecious, axillary; male cones with a very short, scaly-bracteate stipe, shortly oblong-ovoid, brownish, about 3 mm. long, 2 mm. wide, scales membranous, deltoid, about 2 mm. wide, 1·5 mm. long, apex acute, margins erose, covering 3–5 large, ovoid, yellow pollen-sacs; female cones with longer stipes, subglobose, about 2 mm. diam., composed of 6, shortly mucronate, united scales. Ripe fruit globose, about 1 cm. diam., lustrous reddish-brown; seeds usually 3 (or fewer by abortion), narrowly ellipsoid or spindle-shaped, about 7 mm. long, 2–3 mm. wide; testa thin, dark brown.

HAB.: On rocky mountainsides, often in Pine forest, 3,000–5,000 ft. alt.; fl. April–May.

DISTR.: Division 2, local. S. Europe eastwards to Caucasus and N. Iran.

2. Troödos, Prodhromos, between Livadhia and Khandria, Platania, Tripylos, etc. *Kotschy* 758; *Sintenis & Rigo* 717 ! *C. Norman* 417 ! *Syngrassides* 1680 ! *Lindberg f.* ! *Merton* 2389 ! 3179 ! *D. P. Young* 7284 ! *Meikle* 2838 ! etc.

NOTES: One sheet at Kew (*C. B. Ussher* 37) is labelled " Cape Cormakiti, Cyprus, Aug. 1929 ", perhaps by error, since *J. oxycedrus* is otherwise known only from altitudes above 3,000 ft. The specimen seems to be correctly identified, but the possibility remains that it may be the closely allied *J. oxycedrus* L. ssp. *macrocarpa* (Sibth. et Sm.) Ball, a maritime plant, differing chiefly in its larger fruits (10–20 mm. diam.) and broader leaves.

2. J. phoenicea *L.*, Sp. Plant., ed. 1, 1040 (1753); Gaudry, Recherches Sci., 191 (1855); Kotschy in Oesterr. Bot. Zeitschr., 12: 275–276 (1862); Unger et Kotschy, Die Insel Cypern, 213 (1865); Sintenis in Oesterr. Bot. Zeitschr., 31: 255–256 (1881), 32: 125, 293 (1882); Boiss., Fl. Orient., 5: 710 (1884); Holmboe, Veg. Cypr., 28 (1914); Post, Fl. Pal., ed. 2, 2: 802 (1933); Lindberg f., Iter Cypr., 6 (1946); Davis, Fl. Turkey, 1: 82 (1965); Zohary, Fl. Palaest., 1: 20, t. 19 (1966); Osorio-Tafall et Seraphim, List Vasc. Plants Cyprus, 4 (1973).

TYPE: " *in* Europa *australi*, Monspelii ".

Shrub or small tree up to 10 m. high; bark grey-brown, peeling off in narrow strips; twigs dull brown with flaking bark; leaves rhomboid-ovate, about 1 mm. long and almost as wide, blunt, very regularly and closely imbricate giving the ultimate divisions a terete, vermicular appearance; resin-pit on dorsal surface of leaf sometimes distinct and active, sometimes rather obscure; inflorescences monoecious or dioecious; male cones terminal on the ultimate divisions of the branchlets, shortly clavate or ovoid, about 4 mm. long, 3 mm. wide, pale brown, scales suborbicular, 1·5–2 mm. diam., apex rounded or very obtuse, margins whitish, erose; pollen-sacs 3–5, large, yellow; female cones terminal on short axillary spurs, subglobose, about 3 mm. diam., consisting of 6–8 united, blackish-violet, pruinose, shortly acute scales. Ripe fruit globose, about 1 cm. diam., dark reddish-brown, lustrous or slightly pruinose; seeds 3–6, irregularly and narrowly ovoid-ellipsoid, about 7 mm. long, 3 mm. wide, compressed or angled; testa pale brown, hard, minutely rugulose.

HAB.: In Pine forest on dry rocky ground; on dry slopes, occasionally on sandy ground by the sea; sea-level to 1,000 ft. alt.; fl. Febr.–April.

DISTR.: Divisions 1, 3–6, 8, locally common. Widely distributed in the Mediterranean region.

1. Near Toxeftera (Akamas), 1962, *Meikle* 2025 ! Maa Beach, 1967, *Economides* in ARI 977 !
3. Limassol, 1932, *Syngrassides* 415 ! Akrotiri, 1939, *Lindberg f.* s.n. ! Cape Gata, 1941, *Davis* 3545 !

4. Near Ormidhia, 1880, *Sintenis & Rigo* 716 ! Dhekelia, 1880, *Sintenis & Rigo* ! Cape Pyla, 1880, *Sintenis & Rigo*; Cape Greco, 1905, *Holmboe* 421; Ayia Napa, 1939, *Lindberg f.* s.n.
5. Near Salamis, 1962, *Meikle* !
6. Kormakiti, 1967, *Merton* in ARI 135 ! Ayia Irini, 1974, *Sir C. Barclay* 1076 !
8. Cape Andreas, 1880, *Sintenis & Rigo* and, 1950, *F. J. F. Barrington* ! Apostolos Andreas Forest, 1937, *Syngrassides* 1659 ! Valia Forest, 1938, *Chapman* 337 ! Ayios Theodhoros, 1939, *Lindberg f*; Akradhes, 1962, *Meikle* !

NOTES: The subulate, 3-whorled, juvenile foliage is sometimes (cf. *Syngrassides* 1659) found mixed with typical scaly leaves in mature plants.

3. J. foetidissima *Willd.*, Sp. Plant., 4: 853 (1806); Kotschy in Oesterr. Bot. Zeitschr., 12: 279 (1862); Unger et Kotschy, Die Insel Cypern, 213 (1865); Boiss., Fl. Orient., 5: 710 (1884); Holmboe, Veg. Cypr., 28 (1914); Post, Fl. Pal., ed. 2, 2: 802 (1933); Lindberg f., Iter Cypr., 6 (1946); Chapman, Cyprus Trees and Shrubs, 25 (1949); Davis, Fl. Turkey, 1: 82 (1965); Osorio-Tafall et Seraphim, List Vasc. Plants Cyprus, 4 (1973).

TYPE: " *in* Armenia " (B).

An erect tree, 10–15 (–20) m. high; bark grey, fibrous, peeling off in strips; twigs spreading or ascending, glabrous, reddish-brown; leaves dark green, scale-like, those of juvenile growths or immature trees usually subulate, rather spreading, decussately 4-ranked, 3–4 mm. long, 1 mm. wide at base, those of mature growths closely adpressed, often rather regularly 4-ranked, shortly acute, about 1·5 mm. long and almost as wide, generally with a very inconspicuous and inactive dorsal resin-pit; inflorescences usually dioecious; male cones terminal on the branchlets, ovoid, about 3 mm. long, 2–2·5 mm. wide, pale yellowish-brown with about 8–12 scales; scales suborbicular about 2·5 mm. diam., with membranous, erose margins, each covering about 4 relatively large, oblong-obovoid pollen-sacs; female cones axillary on scaly peduncles about 4 mm. long, dark violet-purple or blackish, subglobose, about 3 mm. diam., of 6 united scales. Ripe fruits dark purple or violet, usually pruinose, globose, about 8–10 mm. diam.; seeds (1–) 2 in each fruit, large, hemispherical, about 5–6 mm. diam.; testa shining, pale brown.

HAB.: Rocky mountain slopes at high altitudes on igneous formations, often in open *Pinus nigra* forest; 5,000–6,400 ft. alt.; fl. Dec.–June.

DISTR.: Division 2, locally common on Khionistra. Balkans, Turkey, Transcaucasia, Syria.

2. Locally common on Khionistra above 5,000 ft. alt. Prodhromos, Pasha Livadhi, Kryos Potamos, Kharchis, Asprokremnos, Kannoures etc. *Kotschy* 757 ! *Sintenis & Rigo* 714 ! *Sir R. Biddulph* ! *Kennedy* 37–43 ! *Lindberg f.* ! *Davis* 3521 ! *Merton* 3182 ! 3182A ! *D. P. Young* 7264 ! 7333 ! *Leontiades & Papadakis* T1–T4 ! *Genneou* in ARI 1279 !

NOTES: Recorded by E. Chapman (Cyprus Trees and Shrubs, 25) from the peak of Madhari, in Adelphi Forest, but, in the absence of specimens, presumably the following, *J. excelsa*, rather than *J. foetidissima*, which would consequently appear to be restricted to the topmost slopes of Khionistra.

4. J. excelsa *M. Bieb.*, Beschr. Länd. Flüss. Terek Kur Casp. Meere, 204 (1800), Fl. Taur.-Cauc., 2: 425 (1808); Post, Fl. Pal., ed. 2, 2: 802 (1933); Davis, Fl. Turkey, 1: 84 (1965); Osorio-Tafall et Seraphim, List Vasc. Plants Cyprus, 4 (1973).

TYPE: Crimea, *Pallas* (LE).

Erect, pyramidal tree up to 20 m. high, or exceptionally up to 30 m.; bark greyish, fibrous, peeling in strips; twigs fuscous or reddish-brown, spreading or ascending; leaves ovate, acute, 1–1·5 mm. long, 1 mm. wide at base, rather closely adpressed, but not regularly 4-ranked, distinctly greyish or glaucous, generally with a distinct and often active dorsal resin-pit; inflorescences usually dioecious; male cones terminal on the branchlets, ovoid, about 3 mm. long, 2–2·5 mm. wide, pale yellowish-brown with about 8–12 scales; scales suborbicular, about 2·5 mm. diam., with membranous, erose margins; pollen-sacs as in *J. foetidissima*; female cones axillary on short scaly peduncles about 2–3 mm. long, dark violet-purple or blackish, subglobose, about 3 mm. diam., of 6–9 united, acute scales. Ripe fruits dark purple or violet, usually conspicuously pruinose, globose, 1–1·2 cm. diam.; seeds 7–9 in each fruit, small, angular, ovoid (roughly the size and shape of grape stones), about 4 mm. long, 2–3 mm. wide; testa shining, brown.

HAB.: Rocky mountain slopes on igneous formations; 4,000–5,280 ft. alt.; fl. Dec.–June.

DISTR.: Division 2, locally common. Also Balkans, Crimea, Turkey, Syria and eastwards to Afghanistan and Central Asia.

2. Papoutsa, 1962, *Meikle* 2933 ! and, 1964, *Leontiades & Papadakis* P1–P4 ! Madhari, 1964, *Leontiades & Papadakis* M1–M4 !

NOTES: Apparently absent from Khionistra, but easily overlooked as a form of *J. foetidissima*. It is normally distinguishable (apart from its seeds) by its grey colour, and slender branchlets.

JUNIPERUS VIRGINIANA *L.* (Sp. Plant., ed. 1, 1039; 1753), the Eastern Red Cedar or Pencil Cedar, a very variable tree, up to 30 m., with glaucous branchlets, the juvenile, acicular leaves opposite, the mature foliage scale-like, with acute, often rather loosely adpressed scales, and with small, violet, pruinose, 1–2-seeded fruits about 6 mm. diam., is cultivated in Cyprus gardens (Frangos in Cypr. Journ., 18: 88; 1923) and has been planted at Salamis (Bovill, Report Plant. Work Cyprus, 14: 1915). It is a native of eastern North America, from Canada to Florida, and is represented in gardens throughout temperate regions of the world by a wide range of named cultivars.

JUNIPERUS OCCIDENTALIS *Hook.* (Fl. Bor.-Amer., 2: 166; 1838), the Western Juniper, is also reported as having been planted at Salamis (Bovill, ibid., 14; 1915); it is a smaller tree than *J. virginiana*, rarely more than 15 m. high, and sometimes no more than a shrub; the mature scaly leaves are grey-green in colour, with a conspicuous dorsal resin-gland, and are normally arranged in threes, not decussately opposite as in the other species described above. The fruits are small, about 1 cm. diam., and violet-blue with a glaucous bloom. *J. occidentalis* is a native of western N. America, from British Columbia south to California. It is rarely cultivated.

3. EPHEDRACEAE

Much-branched, erect or sprawling shrubs or woody climbers; branches green, equisetoid, not resinous; leaves opposite or whorled, connate towards base, reduced to membranous sheaths or scales; inflorescences dioecious, solitary or whorled, axillary, forming small cones consisting of superimposed, decussate bracts; male flowers with 2 (or more) connate scales (the perianth) surrounding the base of an exserted column bearing 2–8 microsporangia (or anthers); microsporangia 2–3-locular, dehiscing by an apical pore; female flowers solitary or in groups, the ovule surrounded by 2 pairs of bracts forming outer and inner integuments, the latter tubular and elongated into a style-like micropyle projecting at the top of the flower. Fruit coriaceous or fleshy, berry-like; seed with 2 cotyledons.

One genus in warm temperate areas of Europe, Asia, N. Africa, Atlantic Islands and S. America.

1. EPHEDRA *L.*

Sp. Plant., ed. 1, 1040 (1753).
Gen. Plant., ed. 5, 462 (1754).
O. Stapf in Denkschr. Math.-Nat. Cl. Akad. Wiss. Wien, 56: 1–112 (1889).

Description as for family; the genus includes about 40 species.

Stems scrambling or climbing, loosely and irregularly branched; female inflorescences 2-flowered, bracts connate almost to apex - - **1. E. fragilis ssp. campylopoda**
Stems erect, scoparioid, with numerous, crowded branches; female inflorescences 1-flowered, bracts connate for about one-third of their length - **2. E. major**

1. E. fragilis *Desf.*, Fl. Atlant., 2: 372 (1799); Unger et Kotschy, Die Insel Cypern, 214 (1865); Boiss., Fl. Orient., 5: 714 (1884); Chapman, Cyprus Trees and Shrubs, 23 (1949).

Erect, sprawling or climbing shrub up to 2 m. high; stems woody and covered with rough, fissured fuscous-brown bark in the lower part, loosely and irregularly branched above; branches opposite or whorled, dull green or greyish-green, longitudinally striate, smooth or minutely scabridulous; leaf-scales at first deltoid, about 2 mm. long and connate for the greater part of their length, soon reduced to a brownish or fuscous nodal zone; male inflorescences opposite or whorled, solitary or in clusters, ovoid, about 5 mm. long, 3 mm. wide, with 6–8 connate, decussate, concave, acute or obtuse, brownish, membranous bracts 1·5–2 mm. long and almost as wide, subtending 4–6 (–8) flowers; perianth membranous, inflated, brownish, about 2 mm. long, 1 m. wide; staminal column shortly exserted; anthers 4–6, subglobose or oblong-ovoid, about 0·5 mm. long and wide, pale yellow; female inflorescences solitary or whorled, simple or branched, with erect or reflexed, remotely scaly peduncles 0·2–1 (–1·5) cm. long; flowers solitary or paired, surrounded by 6–8 connate bracts, the innermost much longer than the others, 6–7 mm. long, 3 mm. wide, connate almost to apex; outer integument dark brown or fuscous, equalling or distinctly exceeding the innermost bracts; inner integument membranous, forming a slender, straight micropylar tube 1–2 mm. long. Fruit red, fleshy, consisting of accrescent, succulent bracts enveloping wholly or partly the persistent integuments and a spindle-shaped seed about 5–7 mm. long and 3 mm. wide; testa membranous, pale brown.

ssp. **campylopoda** (*C. A. Mey.*) *Aschers. et Graebn.*, Syn. Mitt.-Europ. Fl., 1: 258 (1897); Holmboe, Veg. Cypr., 28 (1914); Lindberg f., Iter Cypr., 5 (1946); Chapman, Cyprus Trees and Shrubs, 23 (1948).

E. campylopoda C. A. Mey., Vers. Mon. Gatt. Ephedra, 73 (1846); Boiss., Fl. Orient., 5: 715 (1884); Post in Mém. Herb. Boiss., 18: 100 (1900); H. Stuart Thompson in Journ. Bot., 44: 341 (1906); Post, Fl. Pal., ed. 2, 2: 804 (1933); Davis, Fl. Turkey, 1: 85 (1965); Zohary, Fl. Palaest., 1: 22, t. 22 (1966); Osorio-Tafall et Seraphim, List Vasc. Plants Cyprus, 5 (1973).

E. fragilis Desf. var. *campylopoda* (C. A. Mey.) Stapf in Denkschr. Math.-Nat. Cl. Akad. Wiss. Wien., 56: 56 (1889).

TYPE: Not indicated; probably best typified by Sieber material from Crete.

Stems scrambling or climbing, slender and flexuous; male inflorescences solitary or few at each node; female flowers generally in pairs, outer integument plano-convex, equalling or sometimes slightly shorter than uppermost pair of bracts; micropylar tube seldom more than 1 mm. long; integumented seed almost totally enveloped by the fleshy fruiting bracts.

HAB.: Rocky slopes, or commonly growing through shrubs in garigue; occasionally on walls; sea-level to 3,000 ft. alt.; fl. Jan.–July.

DISTR.: Divisions 1–5, 7, 8. Also Balkan Peninsula, Turkey, Syria, Palestine.

1. Near Ayios Yeoryios (Akamas), 1962, *Meikle* 2096 ! 2133 !

2. Kykko, 1939, *Lindberg f.* s.n.; Kalopanayiotis, 1963, *A. Genneou* in ARI 1277 !
3. Limassol district, 1901, *Gennadius* s.n. ! Kophinou, 1901, *A. G. & M. E. Lascelles* s.n. ! Kalavasos, 1905, *Holmboe* 608; Between Limassol and Amathus, 1939, *Lindberg f.* s.n. ! Lefkara, 1941, *Davis* 2790 !
4. Ayia Napa, 1905, *Holmboe* 60; Larnaca, 1905, *Holmboe* 115; Troulli, 1962, *Meikle* 2962 !
5. Athalassa, 1939, *Lindberg f.* s.n.
7. Mylous Forest Station, Kyrenia District, 1932, *Syngrassides* 289 ! Kantara, 1933, *Chapman* 101; Above Sisklipos, 1941, *Davis* 2109 ! Glykyotissa Island, 1955, *G. E. Atherton* 333 ! Kyrenia, 1955, *G. E. Atherton* 950 ! Kyrenia Pass, 1956, *Poore* 42 ! Akanthou, 1967, *Merton* in ARI 779 !
8. Boghaz, 1862, *Kotschy* s.n.; Ayios Theodhoros, 1939, *Lindberg f.* s.n. !

NOTES: Here treated as a subspecies, but with considerable claims to specific rank. *E. fragilis* Desf. ssp. *fragilis*, which is widely distributed in the western Mediterranean (Portugal to Sicily and North Africa), has thicker, more erect, more fragile stems, larger clusters of male inflorescences and 1-flowered female inflorescences with the outer integument distinctly longer than the uppermost bracts, and the micropylar tube often 2 mm. long. The distinction would be wholly satisfactory were it not that 2-flowered female inflorescences are found on some Spanish specimens otherwise agreeing with *E. fragilis* ssp. *fragilis*, while Palestine specimens of ssp. *campylopoda* may, on occasion, have 1-flowered female inflorescences.

2. E. major *Host*, Fl. Austr., 2: 671 (1832); Davis, Fl. Turkey, 1: 85 (1965).

E. nebrodensis Tineo in Guss., Fl. Sic. Syn., 2: 637 (1844); Boiss., Fl. Orient., 5: 714 (1884).

E. procera (? Fisch. et Mey.) Chapman, Cyprus Trees and Shrubs, 23 (1945).

E. major Host. ssp. *procera* (? Fisch. et Mey.) Markgraf in Feddes Repert., 70: 1 (1964); Fl. Europ., 1: 40 (1964); Osorio-Tafall et Seraphim, List Vasc. Plants Cyprus, 5 (1973).

TYPE: Yugoslavia; " in Dalmatiae montibus *Portenschlag* " (W).

Erect, scoparioid shrub less than 1 m. high; lower parts of stem covered with greyish roughly fissured bark, upper parts glaucous-green with numerous, crowded, smooth or minutely scabridulous branches 1–2 mm. diam.; leaf-scales at first deltoid, about 1·5 mm. long and connate for the greater part of their length, soon reduced to a conspicuous blackish or fuscous nodal zone; male inflorescences unilateral or occasionally opposite, solitary, subglobose, about 4 mm. diam., sessile, with 4–6, free or very slightly connate, decussate, concave, ovate-deltoid, acute, membranous, brownish bracts, about 1·5 mm. long and wide, subtending an equal number of flowers; perianth membranous, inflated, brownish, about 2 mm. long and about as wide; staminal column slightly shorter than perianth; anthers 5–8, subglobose, about 0·5 mm. diam., yellow; female inflorescences unilateral or opposite, solitary, subsessile or with a very short peduncle; flowers solitary, surrounded by 6 bracts, the innermost much larger than the others, about 4·5 mm. long, 2·5 mm. wide, connate for about one-third of its length with a free, acute apex, margins narrowly hyaline-membranous; outer integument almond-shaped, dark brown or fuscous, distinctly exceeding the innermost bracts, 5–7 mm. long, 2·5 mm. wide, longitudinally sulcate; inner integument membranous, forming a straight micropylar tube about 0·8 mm. long. Fruit red or yellow, fleshy, consisting of accrescent, succulent bracts partly enveloping the persistent integuments; ripe seed not seen.

HAB.: Rocky slopes on igneous and limestone formations; 2,500–3,000 ft. alt.; fl. April–May.

DISTR.: Division 2, rare. Mediterranean region and eastwards to Transcaucasia, Iran, Afghanistan, Baluchistan and Central Asia.

2. Kato Platres, 1930, *C. B. Ussher* 90 ! Stavros tis Psokas, 1934, *G. W. Chapman* in *Chapman* 141 ! 142 ! and, 1946, *Chapman* 1708 ! 1709 ! Perapedhi, 1941, *Davis* 2083 !

NOTES: It is doubtful if the Cyprus plant can be equated with the more robust, loosely branched *E. procera* Fisch. et Mey., originally described from specimens collected in

Transcaucasia and N. Iran. In habit and general appearance all the Cyprus material agrees better with *E. major* from the Balkan peninsula and Turkey. Though the fruits (outer integuments) are rather longer and proportionately narrower than is apparent in specimens from Spain and the western Mediterranean.

EPHEDRA ALATA *Decne.* and E. ALTE *C. A. Mey.* are both recorded from Cyprus, without further explanation, in Flora of Iraq, 2: 84, 88 (1966); it is presumably on the basis of these references that the two names are included in Osorio-Tafall et Seraphim, List Vasc. Plants Cyprus, 5 (1973). I have no satisfactory evidence that either species is found on the island, and suspect both are recorded by error.

ANGIOSPERMAE

DICOTYLEDONES

4. RANUNCULACEAE

Annual or perennial herbs, rarely shrubs or woody climbers (*Clematis*). Leaves often compound, arranged spirally, or occasionally opposite, normally exstipulate. Flowers actinomorphic or zygomorphic; sepals free, hypogynous, often petaloid; petals sometimes absent, if present free, hypogynous, all or some of them frequently nectariferous; stamens numerous, hypogynous. Carpels indefinite, usually free and arranged spirally; ovules 1–numerous. Fruits various, often achenes or follicles; seeds with endosperm, exarillate.

About 50 genera, distributed throughout the world, but chiefly in temperate regions.

Woody climbers; leaves opposite, with twisting petioles - - - **1. Clematis**
Erect or spreading herbs; leaves alternate or spirally arranged; petioles not twisting:
  Flowers zygomorphic, spurred; fruit consisting of 1 or more follicles **8. Delphinium**
  Flowers actinomorphic:
    Fruit an achene:
      Sepals petaloid; petals absent - - - - - - **2. Anemone**
      Sepals not petaloid; petals present:
        Petals without a basal nectary pouch - - - - - **3. Adonis**
        Petals with a basal nectary pouch:
          Sepals spurred; leaves entire, linear - - - - **4. Myosurus**
          Sepals not spurred; leaves lobed or serrate, not entire:
            Carpels with a single loculus - - - - - **6. Ranunculus**
            Carpels with a central, fertile loculus flanked by 2 empty, sterile loculi **5. Ceratocephala**
    Fruit consisting of 2–several, ± fused follicles - - - - - **7. Nigella**

1. CLEMATIS *L.*

Sp. Plant., ed. 1, 543 (1753).
Gen. Plant., ed. 5, 242 (1754).

Woody climbers, or rarely erect herbs or shrubs. Leaves opposite, usually compound, with twisting or clasping petioles and petiolules. Flowers actinomorphic, solitary or in panicles; sepals normally 4, valvate in bud; petals absent or reduced to staminodes transitional to stamens. Carpels numerous, normally uniovulate by abortion; achenes forming a rounded head, often crowned with accrescent, plumose styles.

More than 200 species, mostly in temperate regions of the N. and S. hemispheres.

Flowers large (4–8 cm. diam.) solitary; sepals yellow or violet-purple:
Sepals violet or violet-purple; involucre absent; achenes without a plume **1. C. viticella**
Sepals pale yellow; peduncle terminating in a cup-like involucre; achenes plumose **2. C. cirrhosa**
Flowers small (1·5–2 cm. diam.) in branched panicles; sepals creamy or greenish-white **3. C. vitalba**

1. C. viticella *L.*, Sp. Plant., ed. 1, 543 (1753); Boiss., Fl. Orient., 1: 2 (1867); Rechinger f. in Arkiv för Bot., ser. 2, 1: 421 (1949); Davis, Fl. Turkey, 1: 139 (1965); Osorio-Tafall et Seraphim, List Vasc. Plants Cyprus, 41 (1973).
TYPE: " *in* Italiae, Hispaniae *sepibus* ".

Scandent, evergreen shrub; stems pale brown, furrowed; leaves opposite with twisting petioles and petiolules, pinnate, the 3–5 pinnae further divided into 3-foliolate pinnules; leaflets very variable in shape and indumentum, normally ovate, obtuse, 1·5–3 cm. long, 1–2 cm. wide, entire, or deeply and irregularly lobed, glabrous or puberulous with conspicuously reticulate venation; flowers large, violet or purple, axillary; peduncles puberulous, 5–10 cm. long; calyx campanulate; sepals 4, ovate, acute, 2–4 cm. long, 1·5–2 cm. wide, glabrous or subglabrous internally, lanuginose externally with a narrow, triangular, subglabrous median stripe; stamens glabrous; styles glabrous, 5–10 mm. long, caducous. Achenes rather large, 7–10 mm. long, brown, pubescent or subglabrous with a sharp, slightly curved beak.

HAB.: Woodland margins; 4,300 ft. alt.; fl. May–July.

DISTR.: Division 2, rare. Central and S.E. Europe eastwards to Turkey, Syria and N.W. Iran.

2. Pedhoulas, 1930, *Wall* s.n. Prodhromos, 1955, *G. E. Atherton* 626 !

NOTES: Rechinger f. expresses some doubt about the accuracy of the first localization; Dr. Erik Wall collected in the neighbourhood of Pano Platres.

2. C. cirrhosa *L.*, Sp. Plant., ed. 1, 544 (1753); Unger et Kotschy, Die Insel Cypern, 314 (1865); Boiss., Fl. Orient., 1: 2 (1867); Holmboe, Veg. Cypr., 79 (1914); Post, Fl. Pal., ed. 2, 1: 3 (1932); Lindberg f., Iter Cypr., 16 (1946); Davis, Fl. Turkey, 1: 140 (1965); Zohary, Fl. Palaest., 1: 199, t. 288 (1966); Osorio-Tafall et Seraphim, List Vasc. Plants Cyprus, 41 (1973); Elektra Megaw, Wild Flowers of Cyprus, 7, t. 1 (1973).
TYPE: Spain; " *in* Baetica, *arbores operiens & deprimens* ".

Scandent evergreen shrub; stems reddish-brown, striate; leaves of primary shoots opposite, simple or trifoliolate, with ovate, toothed or irregularly lobed, glabrous leaflets, 3–5 cm. long and 2–4 cm. wide; petioles 5–7 cm. long, strongly twisting; petiolules 0·5–1 cm. long; leaves of secondary growths simple, clustered, similar in outline to leaflets of primary shoots, but sometimes rather larger; flowers large, solitary or rarely in pairs, pendulous, arising from the axils of the clustered leaves; peduncle slender, glabrous, 1–5 cm. long, terminating in a pubescent, cup-like involucre, formed by two fused bracteoles; pedicel rather stout, 5–10 mm. long, tomentose, accrescent in fruit; calyx campanulate, pale yellow; sepals ovate or elliptical, 2–3 cm. long, 1–2 cm. wide, pubescent externally, glabrous or glabrescent internally; stamens glabrous, 1–2 cm. long. Achenes brown, pubescent, plumose, plume 3–6 cm. long. *Plate 2.*

HAB.: In thickets and on hedges; on north-facing limestone or igneous slopes and precipices; sea-level to 3,000 ft. alt.; fl. Dec.–Febr.

DISTR.: Divisions 1–3, 6–8, common in Division 7. Mediterranean area eastwards to Turkey and Syria.

1. Ayios Yeoryios (Akamas), 1962, *Meikle* ! Near Polis, 1962, *Meikle* !
2. Phini, 1850, *Kotschy*. Evrykhou, 1850, *Kotschy* 37; Kykko, 1905, *Holmboe* 1032; Perapedhi, 1940, *Davis* 2081 ! Kionia, 1941, *Kennedy & Davis* 2696 !
3. Khirokitia Bridge, 1898, *A. Foggie* s.n. !

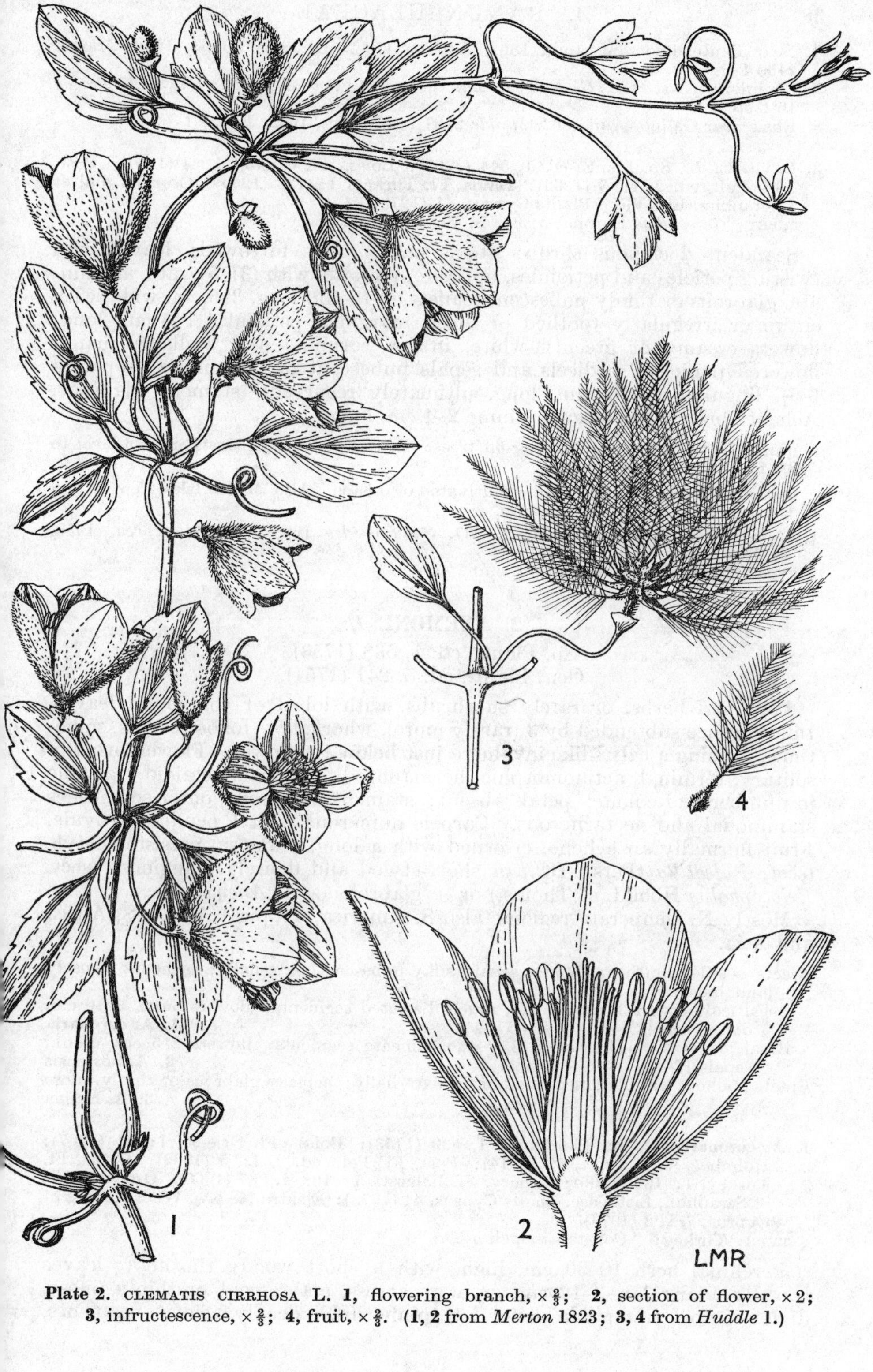

Plate 2. CLEMATIS CIRRHOSA L. **1**, flowering branch, ×⅔; **2**, section of flower, ×2; **3**, infructescence, ×⅔; **4**, fruit, ×⅔. (**1, 2** from *Merton* 1823; **3, 4** from *Huddle* 1.)

6. Near Panteleimon (Myrtou), 1862, *Kotschy* 479 ! Dhiorios Forest, 1956, *Merton* 2460 !
7. Common; *Kotschy. A. G. & M. E. Lascelles* s.n. ! *Rev. A. Huddle* 1 ! *J. A. Tracey* 16 ! etc.
8. Elisis near Galinoporni, 1962, *Meikle* 2487 ! Akrades, 1962, *Meikle* !

3. C. vitalba *L.*, Sp. Plant., ed. 1, 544 (1753); Boiss., Fl. Orient., 1: 4 (1867); Post, Fl. Pal., ed. 2, 1: 3 (1932); Davis, Fl. Turkey, 1: 138 (1965); Osorio-Tafall et Seraphim, List Vasc. Plants Cyprus, 41 (1973).
TYPE: " *in sepibus* Europae *australis*, Virginiae, Jamaicae ".

Scandent deciduous shrub; stems pale brown, furrowed; leaves with twisting petioles and petiolules, opposite, pinnate, with (3)–5 ovate, acuminate, glabrous or thinly pubescent leaflets, 6–10 cm. long, 3–6 cm. wide, with entire or irregularly toothed or lobed margins; petiolules 1–5 cm. long; flowers creamy or greenish-white, in lax terminal and axillary, many-flowered panicles; pedicels and sepals pubescent or lanuginose; sepals 4 (–5), lingulate, 7–10 mm. long, ultimately recurved; stamens glabrous. Achenes brown, pubescent; plume 2–4 cm. long.

HAB.: In thickets and climbing on trees; an escape from cultivation; sea-level to 5,000 ft. alt.; fl. June–August.

DISTR.: Divisions 2 and 7, rare; cultivated or escape. Also British Isles, Central and S. Europe eastwards to Afghanistan.

2. Government Cottage, Troödos, 1937, *Syngrassides* 1693 ! Troödos area, 1950, *F. M. Probyn* s.n. !
7. Near Kyrenia, 1939, *Lindberg f.* s.n. !

2. ANEMONE *L.*

Sp. Plant., ed. 1, 538 (1753).
Gen. Plant., ed. 5, 241 (1754).

Perennial herbs, or rarely subshrubs, with lobed or compound leaves. Inflorescence subtended by 3 (rarely more), whorled, ± foliose bracts, sometimes forming a calyx-like involucre just below the flower. Flowers usually solitary, terminal, actinomorphic, hermaphrodite; sepals petaloid, variable in number and colour; petals absent; stamens numerous, outer sometimes staminodal and nectariferous. Carpels numerous with 1 pendulous ovule. Fruit normally an achene, crowned with a long, plumose, persistent style (sect. *Pulsatilla* (Pers.) DC.) or short-styled and densely lanuginose (sect. *Eriocephalus* Hook.f. et Thoms.) or ± glabrous (sect. *Anemone*).

Mostly N. temperate regions, also S. America, Tropical Africa, S. Africa and Asia.

Bracts sessile, much reduced; sepals silky-pubescent externally; achenes densely lanuginose:
  Basal leaves with 3, petiolulate, much dissected segments; flowers large, 3–10 cm. diam., sepals 5–6 - - - - - - - - - - - **1. A. coronaria**
  Basal leaves 3-lobed or with 3 sessile, cuneate segments; flowers 2–5 cm. diam., sepals 12–18 - - - - - - - - - - - - **2. A. hortensis**
Bracts stalked, leaf-like; sepals glabrous externally; achenes glabrous or thinly pilose **3. A. blanda**

1. A. coronaria *L.*, Sp. Plant., ed. 1, 539 (1753); Boiss., Fl. Orient., 1: 11 (1867); Holmboe, Veg. Cypr., 79 (1914); Post, Fl. Pal., ed. 2, 1: 5 (1932); Davis, Fl. Turkey, 1: 136 (1965); Zohary, Fl. Palaest., 1: 198, t. 287 (1966); Osorio-Tafall et Seraphim, List Vasc. Plants Cyprus, 41 (1973); Elektra Megaw, Wild Flowers of Cyprus, 7, t. 2 (1973).
TYPE: Turkey; " Constantinopoli *allata* ".

Perennial herb 10–30 cm. high, with a short woody rhizome; leaves broadly triangular, 3–12 cm. or more across, glabrescent or thinly pilose, divided into 3, triangular, petiolulate, pinnatifid or deeply laciniate segments, ultimate divisions narrow, variously crenate or serrate; petioles 3–7 cm. long, thinly adpressed-pilose, particularly towards apex; petiolules variable in length, sometimes as much as 3–4 cm. long, usually somewhat channelled; flowers solitary, scape softly pubescent, particularly towards apex, where it is furnished with an involucre of 3 deeply laciniate bracts 1–2·5 cm. long; sepals usually 6, ovate, 1·5–5 cm. long, 1–3·5 cm. wide, acute or obtuse, silky-pubescent externally, ± glabrous internally, scarlet (var. *coronaria*), rose-pink (var. *rosea* (Hanry) Rouy et Fouc.), blue (var. *cyanea* (Risso) Ardoino), white (var. *alba* (Goaty et Pons) Burnat) or in various intermediate shades, sometimes parti-coloured, with a white or pale base; dwarf forms with small flowers (var. *parviflora* Boiss.) are frequent in many parts of the eastern Mediterranean; stamens numerous; filaments pink, violet or red; anthers purplish or black; styles filiform 1–2 mm. long, blackish, marcescent. Torus ovoid; achenes densely lanuginose.

HAB.: On grassy slopes and hillsides; in cultivated and fallow fields; by roadsides; sea-level to 2,900 ft. alt.; fl. Jan.–April.

DISTR.: Divisions 1–3, 5–8, common in Division 7 (Kyrenia). Mediterranean area eastwards to Iran and Iraq.

1. Near Ayios Yeoryios (Akamas), 1962, *Meikle* (var. *coronaria*) ! Near Peyia, 1962, *Meikle* 2128 (var. *parviflora* Boiss.) !
2. Saittas, 1932, *Syngrassides* 345 (var. *parviflora* Boiss.) ! Karavostasi, 1932, *Syngrassides* 256 (var. *parviflora* Boiss.) ! Perapedhi, 1937, *Kennedy* 446A (? var. *cyanea* (Risso) Ardoino) !
3. Trimiklini, 1932, *Syngrassides* 345 (var. *parviflora* Boiss.) ! Near Skarinou, 1956, *Merton* 2481 !
5. Athalassa, 1938, *Syngrassides* 43 (var. *parviflora* Boiss.) !
6. Myrtou, 1933, *Syngrassides* 62 (near var. *parviflora* Boiss.) ! Ayios Panteleimon Forest, 1956, *Merton* 2462 !
7. Common; *Sintenis & Rigo* 77 (var. *parviflora* Boiss.) ! *Mrs. Stagg* s.n. (var. *coronaria*) ! *C. H. Wyatt* s.n. (var. *alba* (Goaty et Pons) Burnat and var. *rosea* (Hanry) R. & F., both tending towards var. *parviflora* Boiss.) ! *Lady Loch* 23 (var. *coronaria*) ! 24 (var. *cyanea* (Risso) Ardoino) ! 34 (var. *rosea* (Hanry) R. & F. growing with var. *coronaria*) ! *Casey* 232 ! 363 (var. *coronaria*) ! *Probyn* 50 ! *G. E. Atherton* 781 ! 804 (var. *cyanea* (Risso) Ardoino) ! 807 (var. *cyanea* (Risso) Ardoino) ! 865 (var. *coronaria*) ! 979 (var. *coronaria*) ! 1023 (var. *rosea* (Hanry) R. & F.) ! 1024 (var. *cyanea* (Risso) Ardoino) !
8. Apostolos Andreas, 1941, *Davis* 2315 (var. *rosea* (Hanry) R. & F.) !

NOTES: Most varieties of *A. coronaria* have been described from southern European material and it is unlikely that the same variants occur in populations of *A. coronaria* from Cyprus. The value of these infraspecific categories is doubtful; var. *parviflora* Boiss. is almost certainly nothing more than a depauperate state of the typical plant.

2. A. hortensis *L.*, Sp. Plant., ed. 1, 540 (1753); Holmboe, Veg. Cypr., 80 (1914); Osorio-Tafall et Seraphim, List Vasc. Plants Cyprus, 41 (1973).

A. stellata Lam., Encycl. Méth., 1: 166 (1783); Unger et Kotschy, Die Insel Cypern, 314 (1865); Boiss., Fl. Orient, 1: 12 (1867).

TYPE: " *ad* Rhenum *& in* Italia ".

Perennial herb 10–30 cm. high with a short woody rhizome; leaves glabrous with scabridulous margins, lowermost reniform, 2·5–7 cm. wide, ternate or divided almost to the base into 3, broad, cuneate, bluntly or acutely lobed segments; upper leaves ternate, finely dissected, the ultimate divisions narrow-lanceolate, acuminate; petioles 3–7 cm. long, adpressed-pilose towards apex; flowers solitary, scapose, scape usually much longer than the leaves, densely adpressed-pilose towards apex; involucre usually situated half-way up the scape, or at a considerable distance from the flower, with 3, lanceolate, entire or irregularly toothed bracts, 1–2·5 cm. long; sepals usually 12–18, lanceolate, 1–2·5 cm. long, 5–7 mm. wide, obtuse or acute, silky-pubescent externally, ± glabrous internally, rosy-mauve, sometimes with a paler base; stamens numerous, filaments pinkish; anthers violet or dark greenish-black; styles blue-black or violet, marcescent. Torus ovoid; achenes densely lanuginose.

HAB.: Hillsides; c. 300–400 ft. alt.; fl. Febr.–April.

DISTR.: Division 4, rare. Northern Mediterranean region from France to Albania and Yugoslavia.

4. Hills above Phurni (between Larnaca and Athienou), 1859, *Kotschy* 13A.

NOTES: Not seen in Cyprus since recorded by Kotschy, and in view of its western Mediterranean distribution an unlikely plant in our area. Probably a form of *A. coronaria* L., which latter is not recorded by Kotschy.

3. A. blanda *Schott et Kotschy* in Oesterr. Bot. Wochenbl., 4: 129 (1854); Unger et Kotschy, Die Insel Cypern, 314 (1865); Boiss., Fl. Orient., 1: 13 (1867); Holmboe, Veg. Cypr., 80 (1914); Post, Fl. Pal., ed. 2, 1: 5 (1932); Davis, Fl. Turkey, 1: 135 (1965); Osorio-Tafall et Seraphim, List Vasc. Plants Cyprus, 41 (1973); Elektra Megaw, Wild Flowers of Cyprus, 7, t. 3 (1973).

TYPE: Turkey; " In monte Tauro, aestate 1836 " *Kotschy* 6 (isotype, K) !

Perennial herb 7–18 cm. high with a short woody rhizome; leaves broadly triangular or reniform, glabrous or sparsely adpressed-pilose, often purplish below, 2·5–5·5 cm. wide, ternately divided into subsessile or shortly petiolulate, cuneate, irregularly lobed segments; lobes acute or obtuse; petioles 5–15 cm. long, glabrous below, thinly pilose towards apex; flowers solitary, scape often tinged with purple, usually subglabrous below involucre, ± densely pubescent above involucre; involucre situated at a considerable distance (2–5 cm.) below the flower, consisting of 3, leaf-like, 3-fid or deeply lobed, distinctly stalked bracts, stalks 3–20 mm. long; sepals 9–15, lanceolate, 1–2 cm. long, 3–7 mm. wide, obtuse or acute, glabrous on both sides, blue, mauve, white or pink; stamens numerous, filaments whitish; anthers pale yellow; stigma minute, spathulate, sessile. Torus subglobose, achenes glabrous or thinly pilose.

HAB.: On rocky calcareous mountainsides, usually on North-facing slopes; very rarely on igneous formations; 2,400–4,000 ft. alt.; fl. Febr.–April.

DISTR.: Divisions 2 and 7, common in Division 7 on Northern Range 2,600–3,100 ft. alt., especially at Castello Regina (Buffavento Castle). Also Greece, Turkey, Caucasus and Syria.

2. North slopes of Madhari, by Yironas R., 1962, *Meikle* 2830 ! and, 1974, *Meikle* 4053 !
7. Common; *Kotschy* 394, 413; *Sintenis & Rigo* 76 ! *Syngrassides* 1425 ! *H. Roger-Smith* s.n. ! *Lady Loch* 38 ! *Davis* 2176 ! *H. Duke-Woolley* 50 ! *Casey* 270 ! *F. M. Probyn* s.n. ! 51 B ! *Meikle* !

3. ADONIS *L.*

Sp. Plant., ed. 1, 547 (1753).
Gen. Plant., ed. 5, 242 (1754).

Perennial or annual herbs; leaves alternate, 2- or 3-pinnate, with narrow, linear segments. Flowers usually solitary, terminal; sepals often coloured; petals 5–20, conspicuous, red or yellow, often blotched at the base, not nectariferous; stamens numerous; style persistent. Carpels numerous, containing 1 pendulous ovule. Fruit forming an ovoid or cylindrical head of wrinkled, beaked achenes.

Europe and Temperate Asia. All the Cyprian species belong to the section *Adonis*, consisting of annual, agrestal herbs with (5) 6–7 (–10) petals. The distinctions between the species of this section are obscure, and complete material, with ripe fruits, is necessary for accurate determination.

Achenes without a dorsal tooth or transverse ventral ridge - - **1. A. annua**
Achenes with a dorsal tooth and transverse ventral ridge:
Achenes c. 15–30, each 4–6 mm. long, 2–4 mm. wide, beak conspicuous, ascending, remote from dorsal tooth - - - - - - - **2. A. aestivalis**

Achenes c. 30–50, each about 2·5 mm. long, 2 mm. wide, beak rather short, pressed against dorsal tooth:
Transverse ventral ridge of achene obscurely sulcate or dentate; plants 15–30 cm. high (flowers red, yellow or orange) - - - - **3. A. microcarpa**
Transverse ventral ridge of achene conspicuously toothed; plants seldom more than 12 cm. high (? flowers always yellow) - - - - - **4. A. dentata**

1. A. annua *L.*, Sp. Plant., ed. 1, 547 (1753); Davis, Fl. Turkey, 1: 143 (1965); Zohary, Fl. Palaest., 1: 213, t. 314 (1966); Osorio-Tafall et Seraphim, List Vasc. Plants Cyprus, 41 (1973).
A. autumnalis L., Sp. Plant., ed. 2, 771 (1762); Boiss., Fl. Orient., 1: 16 (1867); Holmboe, Veg. Cypr., 80 (1914); Post, Fl. Pal., ed. 2, 1: 6 (1932).
TYPE: " *inter segetes* Europae *australis* ".

An erect, simple or branched, glabrous or thinly hirsute annual, 30–60 cm. high; leaves 2–5 cm. long, 3-pinnate, the ultimate segments narrowly linear, acuminate, about 5 mm. long; basal leaves with petioles 2–3 cm. long; upper leaves sessile and ± amplexicaul; flowers solitary at the ends of the branches; peduncle at first 1–3 cm. long, lengthening considerably in fruit and ultimately 7–12 cm. or more in length; flowers cup-shaped, c. 1–2 cm. diam.; sepals 5, patent or slightly reflexed, obovate, 5–12 mm. long, distinctly saccate at the base, glabrous or thinly pilose, greenish-bronze or reddish; petals obovate, 7–10 mm. long, 5–6 mm. wide, concave, rounded at the apex, crimson-scarlet (or rarely yellow), often blotched black at the base; anthers and filaments dark violet. Achenes about 15–30 forming an ovoid or shortly cylindrical head 1–3 cm. long; individual achenes rather large, 3–5 mm. long, 2–3 mm. wide, surface distinctly wrinkled, the persistent style forming a rigid, patent or ascending beak 1·5–2 mm. long; dorsal and ventral surface of achene without any distinct tooth or transverse ridge.

HAB.: In cultivated fields and damp meadows; sea-level to 3,000 ft. alt.; fl. Febr.–May.

DISTR.: Divisions 1–3, 6 and 7. Central and S. Europe to Iran and Iraq.

1. Dhrousha (Akamas), 1941, *Davis* 3287 ! Above Baths of Aphrodite, 1962, *Meikle* 2215 !
2. Perapedhi, 1938, *Kennedy* 1166 ! Mandria–Platres, Limassol District, 1939, *Kennedy* 1749 ! Aphamis, Limassol District, 1939, *Kennedy* 1748 !
3. Zakaki, 1955, *Merton* 2048 !
6. Plains around Nicosia, 1934, *Elizabeth Haig* 20 ! Myrtou, 1956, *Merton* 2512 !
7. Kyrenia, 1956, *G. E. Atherton* 982 !

2. A. aestivalis *L.*, Sp. Plant., ed. 2, 771 (1762); Unger et Kotschy, Die Insel Cypern, 315 (1865); Boiss., Fl. Orient., 1: 17 (1867); Holmboe, Veg. Cypr., 80 (1914); Post, Fl. Pal., ed. 2, 1: 6 (1932); Davis, Fl. Turkey, 1: 144 (1965); Zohary, Fl. Palaest., 1: 212, t. 311 (1966).
TYPE: " *inter segetes* Europae *australis* ".

Similar in habit and general appearance to *A. annua* L.; flowers rather large, opening wide (not cup-shaped as in *A. annua* L.), 2·5–4 cm. diam.; sepals spreading or adpressed to the petals, obovate, obtuse, saccate, greenish-bronze or reddish, glabrous or thinly pilose; petals flat, oblanceolate, rounded at the apex, scarlet (var. *aestivalis*) or yellow (var. *citrina* Hoffm.) often blotched with black at the base; anthers blackish or dark violet, about 1·2 mm. long. Achenes about 15–30, forming an ovoid or shortly cylindrical head 1–3 cm. long; individual achenes rather large, 4–6 mm. long, 2–4 mm. wide, surface distinctly wrinkled, style forming a rigid ascending beak, 1·5–2 mm. long, dorsal margin of achene distinctly toothed, ventral surface with a distinct transverse ridge a short distance above the base, the ridge often furnished with a prominent central tooth.

HAB.: Cultivated fields; 4,000–5,000 ft. alt.; fl. March–May.

DISTR.: Division 2, rare. S. Europe and Mediterranean region eastwards to W. Himalayas.

2. Fields near Prodhromos, 1862, *Kotschy* 794.

NOTES: Not seen since recorded by Kotschy and possibly a misidentification, but likely to occur in the area.

3. A. microcarpa *DC.*, Regn. Veg. Syst. Nat., 1: 223 (1817); Unger et Kotschy, Die Insel Cypern, 315 (1865); Boiss., Fl. Orient., 1: 18 (1867); Davis, Fl. Turkey, 1: 144 (1965); Osorio-Tafall et Seraphim, List Vasc. Plants Cyprus, 41 (1973).

A. cupaniana Guss., Fl. Sic. Syn., 2: 37 (1843); Post, Fl. Pal., ed. 2, 1: 6 (1932); Zohary, Fl. Palaest., 1: 212, t. 312 (1966).

A. aestivalis L. ssp. *microcarpa* (DC.) Holmboe, Veg. Cypr., 80 (1914).

TYPE: " inter segetes Hispaniae circa Tudelam, *Dufour* " (G, BM !).

Similar to *A. annua* L., but with elongate, cylindrical fruiting heads 2–3 cm. long, consisting of about 30–50 small wrinkled achenes, 2·5 mm. long and 2 mm. wide, each achene shortly beaked, with the dorsal tooth adnate to the ascending beak, and the ventral transverse ridge usually ± conspicuously developed. The flowers are intermediate in size between those of *A. aestivalis* and *A. annua*, normally rather cup-shaped, with spreading or irregularly reflexed sepals and concave, scarlet, orange or yellow petals. *A. microcarpa* tends to form low, spreading plants, 15–30 cm. high, with several suberect branches springing from near the base, but (like all annual *Adonis* species) it is variable in habit.

HAB.: In cultivated and fallow fields; occasionally in forest clearings; sea-level to 1,500 ft. alt.; fl. Febr.–April.

DISTR.: Divisions 1, 3, 4, 6–8. S. Europe, Mediterranean region to Iraq; Atlantic Islands.

1. Near Peyia, 1962, *Meikle* 2026 !
3. Zakaki, 1955, *Merton* 2048A ! Morokambos (Akhelia), 1956, *Merton* 2483 !
4. Near Larnaca, 1862, *Kotschy* 90 *teste Boiss.*; Golf Ground (Larnaca), 1900, *A. G. & M. E. Lascelles* s.n. ! Phaneromene near Larnaca, 1905, *Holmboe* 123; Ayia Napa, 1905, *Holmboe* 24, 25. Ayios Antonios, Sotira, 1952, *Merton* 1424 !
6. Monastery farm 1 m. S.W. of Nicosia, 1927, *Rev. A. Huddle* 4 ! 6 ! Fields near Nicosia, 1934, *E. Haig* 13 ! Keumurju, Kyrenia Distr., 1936, *Syngrassides* 1160 ! Morphou Central Experimental Farm, 1951, *Merton* 863 !
7. Kantara monastery, 1880, *Sintenis & Rigo* 84 ! Kyrenia, 1948, *H. Duke-Woolley* 36 ! and 1956, *G. E. Atherton* 1087 ! Ayios Epiktitos, 1951, *Casey* 1125 ! Kambyli, Ayios Panteleimon Monastery, 1956, *Merton* 2523 !
8. Cape Andreas, 1934, *Syngrassides* 1396 ! 1397 ! Karpas, 1937, *H. Roger-Smith* s.n. ! Apostolos Andreas to Khelones, 1941, *Davis* 2292 !

4. A. dentata *Del.*, Fl. Égypt., 143, t. 53, fig. 1 (1813); Boiss., Fl. Orient., 1: 18 (1867); Post, Fl. Pal., ed. 2, 1: 7 (1932); Davis, Fl. Turkey, 1: 144 (1965); Zohary, Fl. Palaest., 1: 213, t. 313 (1966); Osorio-Tafall et Seraphim, List Vasc. Plants, Cyprus, 41 (1973).

A. aestivalis L. ssp. *dentata* (Del.) Holmboe, Veg. Cypr., 80 (1914).

TYPE: Egypt; " auprès de la colonne de Pompée à Alexandrie " *Delile* (MPU).

A bushy, much-branched annual, seldom more than 15 cm. high, with suberect or almost horizontal branches (starved specimens sometimes unbranched and erect); leaves similar to those of *A. aestivalis* L. but usually rather smaller, with the ultimate divisions rather longer, and with the margins distinctly recurved, so as to give them a more delicate, Fennel-like appearance; flowers cup-shaped, rather small, usually 1–1·5 (–2) cm. diam.; sepals obovate, about half as long as petals, yellowish stained red or violet, glabrous or subglabrous, often irregularly laciniate at the apex; petals 5–7 (–10) mm. long, broadly obovate, frequently (? always) yellow, often blotched black at the base. Fruiting heads elongate, cylindrical, about 2–3 cm. long, consisting of 40–50 or more neatly imbricated achenes; individual achenes rather small, about 2·5 mm. long and 2 mm. wide, strongly sulcate, with a short up-turned beak pressed closely against the prominent dorsal tooth, ventral ridge conspicuously developed, bordered with well-marked teeth.

HAB.: In sandy or rocky cultivated or fallow fields; sea-level to 1,200 ft. alt.; fl. March–May.

DISTR.: Divisions 4–7. Also N. Africa, Egypt, Palestine, Turkey, Iran and Arabia.

4. Larnaca, 1859, *Kotschy teste Holmboe*, and 1880, *Sintenis teste Holmboe*; Akhyritou, 1962, *Meikle* 2967 !
5. Near Athalassa, 1933, *Syngrassides* 5 ! Petra tou Dhiyeni, 1951, *Merton* 929 !
6. Near Nicosia, 1930, *F. A. Rogers* 0662 ! Ayia Irini near Morphou, 1941, *Davis* 2536 !
7. Kythrea, 1927, *Rev. A. Huddle* 5 ! Lakovounara, 1950, *Chapman* 136 !

4. MYOSURUS *L.*

Sp. Plant., ed. 1, 284 (1753).
Gen. Plant., ed. 5, 137 (1754).

Small annual herbs with a basal tuft of narrow, entire leaves. Flowers actinomorphic, solitary; sepals spurred; petals tubular, inconspicuous. Fruits forming an elongate spike; achenes very numerous, small, flattened.

About 10 species in N. temperate regions, N. and S. America and New Zealand.

1. **M. minimus** *L.*, Sp. Plant., ed. 1, 284 (1753); Boiss., Fl. Orient., 1: 19 (1867); Holmboe, Veg. Cypr., 80 (1914); Post, Fl. Pal., ed. 2, 1: 8 (1932); Davis, Fl. Turkey, 1: 198 (1965); Zohary, Fl. Palaest., 1: addenda (1966); Osorio-Tafall et Seraphim, List Vasc. Plants Cyprus, 42 (1973).

TYPE: " *in Europae collibus apricis aridis* ".

A small glabrous annual, 2–10 (–13) cm. high; leaves forming a basal tuft, narrowly linear, 15–70 mm. long, 1–1·5 mm. wide, entire, rather fleshy, blunt at apex; flower-stalks numerous, erect, unbranched, leafless; flowers small, sepals 5 (–7), about 2–5 mm. long, concave, obtuse, pale greenish-yellow, with a relatively long basal spur, which is usually adpressed to the upper part of the flower-stalk; petals 5, tubular, inconspicuous, about as long as the sepals; stamens 5–10. Infructescence elongate, caudate (whence the generic name) 2·5–6 cm. or more in length; achenes numerous, pale brown, rectangular in outline, flattened, about 1·5 mm. long, with a short, upturned beak.

HAB.: Seasonally flooded ground in *Pinus nigra* forest; c. 5,000 ft. alt.; fl. May–June.

DISTR.: Division 2, rare. Europe, Orient and N. Africa, naturalized elsewhere.

2. Above Prodhromos, 1905, *Holmboe* 963; Damaskinari, 1957, *Merton*, and, 1962, *Meikle & Merton* in *Meikle* 2781 ! Prodhromos, towards Khionistra, 5,200 ft. alt., 1961, *D. P. Young* 7452 ! All records are from one locality.

5. CERATOCEPHALA *Moench*

Meth., 218 (1794).

Small tufted annuals with palmatisect leaves; petals and sepals 5, stamens few, usually about 6–10. Achene with a long beak, 3-locular, the two lateral loculi sterile and empty, forming distinct basal auricles.

Central and S. Europe, Asia Minor eastwards to N. India; 2 species.

1. **C. falcata** (*L.*) *Pers.*, Syn. Plant., 1: 341 (1805); Unger et Kotschy, Die Insel Cypern, 318 (1865); Boiss., Fl. Orient., 1: 58 (1867); Holmboe, Veg. Cypr., 80 (1914); Post, Fl. Pal., ed. 2, 1: 17 (1932); Davis, Fl. Turkey, 1: 198 (1965); Zohary, Fl. Palaest., 1: 210, t. 309 (1966); Osorio-Tafall et Seraphim, List Vasc. Plants Cyprus 42 (1973).

Ranunculus falcatus L., Sp. Plant., ed. 1, 556 (1753).

Ceratocephala falcata (L.) Pers. var. *exscapa* Boiss., Fl. Orient., 1: 58 (1867); Post, Fl. Pal., ed. 2, 1: 17 (1932).

TYPE: " *inter segetes* Europae *australis* ".

A small villose or subglabrous annual (1–) 3–5 (–12) cm. high; leaves all basal, 1–3 cm. long, deeply palmatisect, the ultimate divisions linear, obtuse or subacute; scapes usually rather longer than the leaves, sometimes very short or obsolete; flowers solitary, 8–15 mm. diam.; sepals 3–5 mm. long, 1·5–2 mm. wide with a narrow scarious margin; petals obovate, obtuse, 4–7 mm. long, distinctly clawed, claw with a small scale-covered nectary. Infructescence conspicuous, ovoid or cylindrical, achenes numerous, 5–10 mm. long, including the long sickle-shaped beak; seed pale brown, narrowly ovoid, flattened, about 2 mm. long.

HAB.: In cultivated and fallow fields on igneous and sedimentary formations; 100–3,700 ft. alt.; fl. March–April.

DISTR.: Divisions 2, 5 and 6, common in Division 6 (mainly about Nicosia). S. Europe, N. Africa and western Asia.

2. Phini, 1862, *Kotschy* 882; Platres, 1938, *Kennedy* 1168 !
5. Between Nicosia and Kythrea, 1880, *Sintenis & Rigo* 83 ! Near Orta Keuy, 1962, *Meikle* 2970 !
6. Common; *Kotschy* 82; *A. Foggie* ! *A. G. & M. E. Lascelles* s.n. ! *Holmboe* 274, 301; *Syngrassides* 774 !

6. RANUNCULUS *L.*

Sp. Plant., ed. 1, 548 (1753).
Gen. Plant., ed. 5, 243 (1754).

Annual or perennial, terrestrial or aquatic herbs. Leaves various, often compound and deeply lobed or dissected, alternate or spirally arranged, mostly exstipulate (stipulate in the subgenus *Batrachium*). Flowers regular, usually yellow, sometimes white or red; sepals 3–5, free, without a basal spur, spreading or reflexed in anthesis; petals 5–8 or more, with a basal nectariferous pit, usually covered by a scale; stamens spirally arranged, indefinite, often numerous. Carpels spirally arranged, usually numerous, uniovulate; achenes compressed or subglobose, often with a prominent persistent style, surface smooth, wrinkled, rugose or occasionally spinose.

About 300 species with a cosmopolitan distribution, though most abundant in N. Temperate regions. Many of the species are acrid or caustic, and poisonous to stock.

Aquatic herbs with stipulate, finely dissected submerged leaves; petals white or yellow only at the base; carpels transversely wrinkled:
- Floating leaves absent; flowers usually large; peduncles generally much longer than leaves, seldom recurved - - - - - **1. R. peltatus ssp. sphaerospermus**
- Floating leaves present, divided into 3–5 bluntly lobed segments; flowers rather small; peduncles often shorter than floating leaves, usually recurved **1. R. peltatus var. microcarpus**

Terrestrial herbs; leaves exstipulate; petals usually yellow or sometimes coloured; achenes not transversely wrinkled:
- Leaves all undivided, entire or shortly crenate or dentate:
  - Leaves and stems glabrous, rather fleshy; flower 3·5–6 cm. diam., sepals 3 **2. R. ficaria ssp. ficariiformis**
  - Leaves and stems hairy, not fleshy; flowers 1–1·5 cm. diam., sepals 5 **3. R. bullatus ssp. cytheraeus**
- All or some of the leaves compound or deeply lobed:
  - Flowers very large, 2·5–7 cm. diam., petals often red, pink or white, anthers violet-purple - - - - - - - - - - - - - - **6. R. asiaticus**
  - Flowers smaller, rarely exceeding 3·5 cm. diam.; petals and anthers yellow:
    - Flowers solitary, axillary or leaf-opposed; annual plants; flowers small and inconspicuous (except in 15 & 16):
      - Achenes echinate or strongly muricate:
        - Basal leaves spathulate-cuneate, dentate at apex; upper leaves with linear divisions; achenes with long slender spines; achene-margin at right angles to surface - - - - - - - - **18. R. arvensis**
        - Basal leaves reniform, deeply 3–7-lobed; upper leaves sometimes simple; achenes shortly echinate-muricate, with a broad flat marginal wing **17. R. muricatus**

Achenes muriculate, papillose or smooth, not echinate or strongly muricate:
Peduncle shorter than adjacent leaf, incrassate in fruit; achenes narrowly bordered, with a short beak - - - - - - **13. R. chius**
Peduncle as long as, or longer than adjacent leaf, slender, not incrassate:
Achene narrowly bordered, and with a short (usually upturned) beak; upper leaves often simple or rather coarsely laciniate **15. R. marginatus var. trachycarpus**
Achene distinctly winged, with a prominent (often deflexed) beak; upper leaves laciniate with numerous linear-lanceolate lobes **16. R. cornutus**
Flowers terminal; perennial plants; flowers usually conspicuous:
Inflorescence scapose, very sparsely branched, usually leafless or almost so; rootstock wholly or partly tuberous (grumose):
Rootstock crowned with a " collar " of interwoven fibres (old, decayed leaf-bases), lower leaves often simple, broadly ovate-cuneate **7. R. paludosus**
Rootstock without a collar of fibres:
Leaves broadly reniform, palmately divided into blunt lobes; plant glabrescent, often tinged purple; petals 5–12 or more **8. R. cadmicus var. cyprius**
Leaves deltoid, pinnately divided into small ovate or cuneate lobes; petals normally 5:
Flowers 2·5–3·5 cm. diam.; leaves pinnate with broad, dentate or lobed segments - - - - - - - - **19. R. sp. (or hybrid)**
Flowers less than 2·5 cm. diam.; leaves 2–3-pinnate:
Sepals patent or adpressed to petals; tubers clavate **10. R. millefoliatus ssp. leptaleus**
Sepals reflexed; tubers elongate-fusiform - - **4. R. isthmicus**
Inflorescence branched, often leafy:
Plant glabrous or glabrescent; radical leaves broadly reniform or almost orbicular, rather leathery - - - - - - **9. R. kykkoënsis**
Plant pubescent or villose; radical leaves broadly deltoid, deeply lobed or dissected, not leathery:
Radical leaves 2–3-pinnate, villose; peduncles divaricate, conspicuously incrassate in fruit; achenes forming a cylindrical head **11. R. millefolius**
Radical leaves broadly deltoid, 3-lobed; peduncles erect or ascending, not incrassate:
Fruiting-heads shortly cylindrical; achenes with a long, slender, straight or slightly recurved beak; root-tubers short, about 4–10 cm. long - - - - - - - - - - **5. R. rumelicus**
Fruiting-heads spherical; achenes with a short, triangular or hooked beak; root-tubers elongate, 4–12 cm. long, or wanting:
Achenes with a pale margin; roots intermixed with elongate cylindrical tubers - - - - - - **12. R. neapolitanus**
Achenes uniformly brown, without a margin; roots all fibrous, or at most slightly thickened - - **14. R. constantinopolitanus**

SUBGENUS 1. **Batrachium** (*DC.*) *A. Gray* Aquatic perennials (or rarely annuals) with the submerged foliage dissected into capillary lobes; flowers solitary, leaf-opposed; petals white; nectary uncovered or incompletely covered; achene transversely wrinkled.

1. R. peltatus *Schrank*, Baier. Fl., 2: 103 (1789); Meikle in Notes Roy. Bot. Gard. Edinb., 23: 14 (1959).

[*R. aquatilis* (non L.) Boiss., Fl. Orient., 1: 23 (1867) pro parte excl. var. β *submersus* Gren. et Godr.]

Perennial or annual aquatic, with slender or robust stems sometimes 1 m. or more in length; floating leaves present or absent; submerged leaves 1–5 cm. or more long, normally shortly petiolate or subsessile, with numerous slender, flaccid or rather rigid, capillary segments; stipules blunt, membranous, often thinly pilose, to about 8 mm. long; floating leaves (when present) long-stalked, generally reniform or semicircular in outline, ± hairy below, 1–3 cm. or more wide, divided almost half way into 3 (–5) bluntly lobed segments, basal margin of lamina truncate or forming a wide sinus;

peduncles usually longer than the submerged leaves, though not infrequently shorter than the floating leaves, tapering slightly from base to apex, often distinctly recurved after anthesis; flowers variable in size, with contiguous or over-lapping, obovate-cuneate, white, yellow-clawed petals 0·5–1 cm. long, 0·4–0·8 cm. wide; nectary-pit lunate, pyriform or almost orbicular; receptacle pilose, subspherical or ovoid. Achenes glabrous or pilose, 0·8–2 mm. long, with a distinctly convex dorsal margin and a rounded or obscurely beaked apex.

ssp. **peltatus**

Floating leaves present; achenes up to 2 mm. long.

var. **microcarpus** *Meikle* in Notes Roy. Bot. Gard. Edinb., 23: 15 (1959).

[*R. aquatilis* L. ssp. *confusus* (non Godr.) Holmboe, Veg. Cypr., 80 (1914).]

[*R. saniculifolius* (non Viv.) C. D. K. Cook in Mitt. Bot. Staats. München, 6: 111 (1966) pro parte quoad pl. cypr.; Osorio-Tafall et Seraphim, List Vasc. Plants Cyprus, 42 (1973).]

TYPE: Cyprus; Ayios Antonios, Sotira, 400 ft. alt.; in a pool on the limestone rocks, Febr. 29, 1952, *Merton* 624 (K !).

Achenes very small, up to 1 mm. long, glabrous or rarely pilose; plant annual, generally slender, with small flowers, the petals rarely exceeding 8 mm. in length; peduncles equalling or often shorter than the floating leaves, submerged leaves with very slender, collapsing segments.

HAB.: In pools on limestone rocks near the sea, or more rarely in ponds and drains inland; sea-level to 900 ft. alt.; fl. Febr.–April.

DISTR.: Divisions 1, 4, 5, 6, 8. Also in Malta and perhaps elsewhere in the Mediterranean region.

1. Roadside near Peyia, 1962, *Meikle* 2024 ! Erimidhes near Ayios Yeoryios (Akamas), 1962, *Meikle* 2144 !
4. Paralimni, 1905, *Holmboe* 420; Ayios Antonios, Sotira, 1952, *Merton* 624 !
5. Prastio, 1962, *Meikle* 2618 !
6. Near Prophitis Elias Monastery, beyond Skylloura, 1957, *Merton* 2888 !
8. Cape Elea, 1905, *Holmboe* 512; Galatia, 1934, *Syngrassides* 1398 ! and 1962, *Meikle* 2528 !

NOTES: Holmboe states that his specimen (*J.H.* 420) from Paralimni has hispid achenes, a characteristic found also in one specimen (*Meikle* 2528) from Galatia. Otherwise the Cyprus material has glabrous fruits, and resembles *R. confusus* or *R. baudotii* Godr. in this respect, differing markedly, however, in the relatively short, recurving peduncles, and in the total absence of any " intermediate ", laciniate leaves. *R. saniculifolius* Viv., with which Dr. C. D. K. Cook identifies the plant, is a dubious species, known only from a short description and an imperfect illustration; it might be either *R. peltatus* var. *peltatus* or *R. baudotii* (which I prefer to regard as a third subspecies of *R. peltatus*), but scarcely deserves recognition as an independent species.

ssp. **sphaerospermus** (*Boiss. et Blanche*) *Meikle* in Notes Roy. Bot. Gard. Edinb., 23: 16 (1959).

R. sphaerospermus Boiss. et Blanche in Boiss., Diagn. 2, 5: 6 (1856); C. D. K. Cook in Mitt. Bot. Staats. München, 6: 118 (1966); Zohary, Fl. Palaest., 1: 209, t. 307 (1966); Osorio-Tafall et Seraphim, List Vasc. Plants Cyprus, 42 (1973).

R. aquatilis L. var. *sphaerospermus* (Boiss. et Blanche) Boiss., Fl. Orient., 1: 23 (1867); Post in Mém. Herb. Boiss., 18: 89 (1900).

[*R. aquatilis* L. ssp. *circinatus* (non Sibth.) Holmboe, Veg. Cypr., 80 (1914).]

[*R. circinatus* (non Sibth.) Dinsmore in Post, Fl. Pal., ed. 2, 1: 9 (1932).]

? *R. pantothrix* (vix Brot.) Kotschy in Unger et Kotschy, Die Insel Cypern, 315 (1865).

? *R. paucistamineus* (vix Tausch) Holmboe, Veg. Cypr., 80 (1914).

TYPE: " in Syria in paludosis mari vicinis, circa Tripolin et Alexandrette ", *Blanche* (G).

Floating leaves absent, submerged leaves (0·5–) 1–2·6 (–6) cm. long, very shortly petiolate or subsessile, with numerous rather rigid, diverging segments, forming ± dense spherical tufts; peduncles (2·5–) 3–6 (–8) cm. long, often greatly exceeding the leaves; flowers usually large, 1·5–2·5 (–3) cm. diam., with broad, obovate petals. Achenes very small, about 0·7–1 mm. long, subglobose, glabrous or pilose at apex.

HAB.: In drains, pools and swamps; sea-level to 100 ft. alt.; fl. March–May.

DISTR.: Divisions 4, 5, 6. Aegean region eastwards through Turkey and Iraq to Kashmir and Nepal.

4. ? Near Kato Vlakhos [" Cataloco "] on the road to Dhali, 1862, *Kotschy*; Kouklia, 1905, *Holmboe* 382; Mouth of Pedieos R., 1955, *Merton* 2263 ! Akhyritou, 1948, *Mavromoustakis* s.n. ! Akhyritou, 1952, *Merton* 708 ! and, 1962, *Meikle* 2610 !
5. ? Vatili, 1905, *Holmboe* 348; near Engomi, 1949, *Mercy M. Casey* in *Casey* 553 !
6. Syrianokhori marshes, 1935, *Syngrassides* 776 ! and, 1952, *Merton* 792 ! 807 !

NOTES: Variable in general robustness, but otherwise exceptionally uniform (for this group) in its vegetative and floral characters. Small forms may have been misidentified with *R. trichophyllus* Chaix by Kotschy, Holmboe and others, but in the latter species, the peduncles are always short and recurved, and not elongate as in *R. peltatus* ssp. *sphaerospermus*. True *R. trichophyllus* occurs in the eastern Mediterranean region, and may yet be found in Cyprus.

SUBGENUS 2. **Ficaria** (*Huds.*) *L. Benson* Perennial terrestrial herbs; leaves simple, entire or subentire; sepals 3, greenish, caducous; petals 7–15 or more, yellow; nectary scale forming a shallow pocket. Achenes ± spherical, cotyledon apparently solitary.

2. R. ficaria *L.*, Sp. Plant., ed. 1, 550 (1753); Boiss., Fl. Orient., 1: 24 (1867); Davis, Fl. Turkey, 1: 193 (1965).

A glabrous perennial herb 5–30 cm. high; roots fibrous, scattered amongst clusters of fleshy, clavate or fusiform tubers; stems erect or decumbent; leaves deltoid or reniform, 1–6 cm. long and about as wide, entire, undulate or sometimes with shallow teeth; petiole 3–10 cm. or more in length, sheathing at the base; flowers 1·5–6 cm. diam., solitary, terminal; sepals 3, broadly and bluntly ovate, deeply concave, 5–10 mm. long, 4–8 mm. wide; petals 8–15, narrowly ovate, 10–18 mm. long, 4–6 mm. wide, bright golden yellow internally, stained bronze or green externally; stamens numerous; anthers elongate, 2–5 mm. long. Achenes numerous (in var. *fertilis* Clapham) or relatively sparse (var. *ficaria*), subspherical, 2–3 mm. diam., glabrous or thinly pubescent, with a very short beak; receptacle ± spherical, hairy.

Two distinct varieties occur (often intermixed) in western Europe: the diploid var. *fertilis* Clapham without axillary bulbils, and with numerous fertile achenes; and the tetraploid var. *ficaria* with axillary bulbils, and fewer, largely infertile, achenes. These two varieties are found throughout Europe, but are absent from Cyprus, and from most of Asia and N. Africa, where they are replaced by the following subspecies.

ssp. **ficariiformis** (*F. Schultz*) *Rouy et Fouc.*, Fl. France, 1: 73 (1893); Davis, Fl. Turkey, 1: 193 (1965); Zohary, Fl. Palaest., 1: 202, t. 290 (1966); Osorio-Tafall et Seraphim, List Vasc. Plants Cyprus, 42 (1973).

Ficaria grandiflora Rob., Plant. Phan. Toul., 57 and 112 (1838).

Ranunculus ficaria L. var. *calthifolius* Guss., Fl. Sic. Syn., 2: 41 (1843) pro parte.

R. calthifolius Jord., Obs. Plant. Nouv., 6: 2 (1847); Boiss., Fl. Orient., 1: 24 (1867); Post, Fl. Pal., ed. 2, 1: 9 (1932).

R. ficariiformis F. Schultz in Arch. de Flore, 260 (1858).

R. ficaria L. var. *grandiflorus* (Rob.) Strobl in Oesterr. Bot. Zeitschr., 28: 109 (1878).

R. ficaria L. ssp. *R. calthifolius* (Jord.) Holmboe, Veg. Cypr., 80 (1914).

TYPE: France, Toulon; " les champs humides, très commun, Mars, avril " (? TLON).

Similar to *R. ficaria* L. ssp. *ficaria* in general appearance, but altogether larger, without axillary bulbils, and generally more erect in habit; leaves 4–6 cm. long and 4–5 cm. wide, sometimes broadly oblong with a truncate base; flowers 3·5–6 cm. diam.; petals 9–15 or more; anthers 4–5 mm. long.

HAB.: In damp pastures, and on streamsides; sea-level–4,000 ft. alt.; fl. Febr.–March.

DISTR.: Divisions 1–3, 5, 7 and 8, common in Divisions 2 (around Troödos) and 7. S. Europe, western Asia, N. Africa.

1. Near Ayios Yeoryios (Akamas), 1962, *Meikle* 2094 ! Near Polis, 1962, *Meikle* !
2. Common; *Kennedy* 425 ! 1569 ! 1733 ! 1734 ! *G. E. Atherton* 1170 !
3. Near Klavdhia, 1880, *Sintenis & Rigo* 908 ! Alethriko, 1905, *Holmboe* 240.
5. Kythrea, 1932, *Syngrassides* 56 ! Mia Milea, 1956, *Merton* 2506 !
7. Common; *Sintenis & Rigo* 908 A ! *Mrs. Stagg* ! *Syngrassides* 1441 ! *Davis* 2502 ! *Casey* 1137 ! *G. E. Atherton* 872 ! 993 !
8. Ronnas Valley near Rizokarpaso, 1962, *Meikle* !

SUBGENUS 3. **Ranunculus** Annual or perennial terrestrial herbs; sepals 5, deciduous; petals yellow, rarely red, pink or white, nectary usually covered with a nectary-scale; achenes usually compressed, not transversely wrinkled.

3. R. bullatus *L.*, Sp. Plant., ed. 1, 550 (1753); Boiss., Fl. Orient., 1: 25 (1867); Holmboe, Veg. Cypr., 81 (1914).

Perennial; roots fibrous, scattered amongst clusters of elongate-fusiform tubers; leaves 3–6, forming a basal rosette, and springing from a fibrous " collar " of decayed petioles; petiole 1–3 cm. long, sometimes almost wanting, flattened, rather densely covered with long, silky hairs; lamina broadly obovate, or almost orbicular, somewhat bullate, (1·5–) 2·5–8 cm. long, (1–) 1·5–7 cm. wide, subglabrous or thinly villose above, rather densely villose below, margin bluntly serrate or crenate, especially in the upper part, apex obtuse or truncate, with prominent teeth, base rounded, cuneate or truncate; flowers violet-scented, solitary; scape erect, naked, unbranched or sometimes branched above (3–) 10–16 (–20) cm. long, with spreading, silky hairs; sepals 5, spreading, bluntly ovate, 5–8 mm. long, thinly villose, petals (5–) 8–12 (–15), yellow, obovate, (5–) 8–18 mm. long; receptacle elongate, glabrous. Achenes numerous, very small, 1–1·5 mm. long, inflated, glabrous, smooth, with a short beak.

ssp. **cytheraeus** (*Hal.*) *Vierhapper et Rechinger f.* in Oesterr. Bot. Zeitschr., 84: 129 (1935); Osorio-Tafall et Seraphim, List Vasc. Plants Cyprus, 41 (1973).

R. bullatus L. var. *cytheraeus* Hal. in Verh. Zool.-Bot. Ges. Wien, 54: 483 (1914).

R. bullatus L. var. *cuneifolius* Coust. et Gandoger in Bull Soc. Duffour, 4: 36 (1914); Gandoger, Fl. Cret., 6 (1916).

R. bullatus L. f. *cuneifolius* (Coust. et Gandoger) Coust. et Gandoger in Bull. Soc. Bot. Fr., 63: 13 (1916).

R. bullatus L. var. *cyrenaicus* Pamp. in Nuov. Giorn. Bot. Ital., 26: 212 (1919).

TYPE: " In insula Cythaera (Leonis) " (? W).

Altogether smaller than var. *bullatus*, with hairy or subglabrous leaves, 1·5–4 cm. long and 1–3 cm. wide, cuneate, truncate or rounded at the base, with coarsely dentate or subentire margins; scapes often spreading or decumbent, bracteate, usually branched above, 2–many-flowered; flowers small, about 1–1·5 cm. diam., with 5 (rarely 6) shortly obovate-cuneate petals.

HAB.: On rocky limestone hillsides; sea-level to 1,000 ft. alt.; fl. Nov.–Jan.

DISTR.: Divisions 1, 4–7, common in Division 7. Also Crete, Cythera, Rhodes, W. Turkey and Cyrenaica.

1. Paphos, rocks at Ktima, 1840, *Kotschy*.
4. Kalokhorio near Larnaca, 1862, *Kotschy*; Ormidhia near Famagusta, 1862, *Kotschy*; Ayia Napa, 1940, *Davis* 2052 !
5. Mile 5, Nicosia–Famagusta road, *Merton* 1094 ! Also Mesaoria between Nicosia and Lefkoniko, 1940, *Davis*.
6. Near Nicosia, 1940, *Davis* 2069 K !
7. Common; *Lady Loch* 7 ! *Kennedy* 1346 ! *Davis* 2001 ! 2105 ! *Casey* 197 ! *G. E. Atherton* 760 ! 766 ! 775 ! 782 !

4. **R. isthmicus** *Boiss.*, Diagn., 1, 6: 4 (1846 or 1847); Fl. Orient., 1: 28 (1867); Davis, Fl. Turkey, 1: 185 (1965); Osorio-Tafall et Seraphim, List Vasc. Plants Cyprus, 42 (1973).

[*R. cicutarius* (non Schlecht.) H. Stuart Thompson in Journ. Bot., 44: 275 (1906); Holmboe, Veg. Cypr., 81 (1914).]

TYPE: Greece; " in declivibus Isthmi Corinthiaci paulo supra *Calamathi* ad Corinthum eundo ", April 1842, *Boissier* (G, K !).

Dwarf, tufted perennial 2–4 cm. high; roots fibrous intermixed with clustered, elongate-fusiform, fleshy tubers 3–5 cm. long; leaves mostly radical, broadly triangular, 1–3 cm. long and about the same in width, pinnately 3-fid, the segments deeply and irregularly lobed and toothed, subglabrous or adpressed-pilose; petiole 1·5–3 cm. long, subglabrous or adpressed-pilose, with a wide, sheathing, membranous base; juvenile leaves cuneate-spathulate about 1 cm. long, dentate at apex; inflorescence scapose (or very sparsely branched); scape 1–2 cm. long, usually shorter than the leaves, densely adpressed-pilose; flowers solitary, yellow (or tinged bronze ?) about 1·5 cm. diam.; sepals ovate-acute, about 5 mm. long, distinctly reflexed, adpressed-pilose; petals broadly ovate, many-veined, about 7–8 mm. long; stamens slightly exceeding the carpels; anthers elongate-oblong, yellow; carpels (? beaked). " Fruiting head ovate-orbicular, 10–15 mm. Achenes tubercled . . . beak as long as achene, wide, falcate."

HAB.: Rocky or grassy places; sea-level to 500 ft. alt.; fl. Jan.–March.

DISTR.: Divisions 4 and 5, rare. Sicily, Greece and Turkey.

4. " Golf Ground [Larnaca] ", 1900, *A. G. & M. E. Lascelles* s.n. !
5. Laxia [" Latchia "], 1932, *Syngrassides* 79 !

5. **R. rumelicus** *Griseb.*, Spic. Fl. Rum. et Bith., 1: 305 (1843); Boiss., Fl. Orient., 1: 30 (1867); Davis, Fl. Turkey, 1: 178 (1965); Osorio-Tafall et Seraphim, List Vasc. Plants Cyprus, 41 (1973).

TYPE: Greece; near Thessaloniki, *Friedrichsthal* (GOET).

Perennial; roots fibrous, intermixed with clustered turbinate or broadly fusiform tubers 4–10 mm. long; stems erect, 13–30 (–40) cm. high, sparingly branched above, ± densely clothed with erecto-patent or subadpressed silky hairs; petioles of radical leaves 2–7 cm. long, dilated and membranous at the base, thinly or densely clothed with long, spreading or subadpressed silky hairs, lamina thinly adpressed-pilose, broadly ovate-deltoid in outline, 2–5 cm. long, 2–6 cm. wide, 3-lobed, sinuses narrow extending to base of leaf or at least half-way to base, lobes deeply and bluntly toothed, the median often conspicuously narrowed towards its base, or even shortly petiolulate; stem leaves few, much reduced and bract-like, sessile or shortly petiolate, irregularly lobed, often with narrowly oblong or oblanceolate lobes, adpressed-pilose; flowers medium to large, 2–3 (–4) cm. diam.; peduncles 2–8 cm. long, adpressed-pilose; sepals 5, rather weakly and irregularly reflexed, ovate, 5–8 mm. long, 3–4 mm. wide, obtuse or subacute, hairy externally, with a distinct membranous margin; petals golden-yellow within, distinctly veined externally, 8–17 mm. long, 7–11 mm. wide, ovate-cuneate with a rounded apex; stamens not very numerous, equalling or rather shorter than carpels. Achenes forming a short cylindrical head, 10–16 mm. long, 6–8 mm. wide, individually strongly flattened, about 3 mm. long, 1·5 mm. wide, with a slender, straight or recurved beak, lateral surfaces shortly and sparsely tuberculate, abaxial margin distinctly gibbous.

HAB.: In *Lolium perenne* turf in hazel groves; c. 3,000 ft. alt.; fl. April–May.

DISTR.: Division 2, rare. Also S.E. Europe, Aegean Islands and Turkey.

2. Askas, beyond Palekhori, 1957, *Merton* 3131 !

NOTES: A rather slender form, with smallish flowers, but leaf-shape and indumentum that of *R. rumelicus* and not that of the closely allied *R. reuterianus* Boiss.

6. R. asiaticus *L.*, Sp. Plant., ed. 1, 552 (1753); Unger et Kotschy, Die Insel Cypern, 316 (1865); Boiss., Fl. Orient., 1: 31 (1867); Holmboe, Veg. Cypr., 81 (1914); Post, Fl. Pal., ed. 2, 1: 11 (1932); Davis, Fl. Turkey, 1: 183 (1965); Zohary, Fl. Palaest., 1: 204, t. 295 (1966); Osorio-Tafall et Seraphim, List Vasc. Plants Cyprus, 41 (1973); Elektra Megaw, Wild Flowers of Cyprus, 7, t. 4 (1973).

TYPE: " *in* Asia *vel* Mauritania "—a " double "-flowered garden form, probably with red petals (LINN).

Perennial; roots fibrous, scattered among elongate-fusiform, clustered tubers; stem erect or suberect, 5–30 cm. high, sparsely branched, ± villose; radical leaves broadly ovate or deltoid, petiolate, normally ternate or deeply 3-lobed; petiole 1–9 cm. long, villose or subglabrous, flattened above, base expanded into a sheath; leaf-segments variable, broadly or narrowly ovate-cuneate with an irregularly toothed and lobed margin, or finely dissected, with narrow linear-lanceolate lobes (var. *tenuilobus* DC.), glabrous or ± villose; stem leaves sessile or shortly stalked, similar to the radical leaves, but usually more deeply and narrowly dissected with lanceolate or linear segments; flowers solitary, 2·5–7 cm. diam., red (var. *sanguineus* (Mill.) DC.); yellow (var. *flavus* Dörfl.), white (var. *albus* Hayek) or tinged with pink; sepals 5, often coloured, narrowly ovate-acuminate, 10–15 mm. long, thinly villose externally, glabrous internally, with a narrow scarious margin; petals 5–9 or more, broadly cuneate, 12–35 mm. long, 8–30 mm. wide, margins often irregularly undulate-dentate; stamens numerous, equalling or slightly shorter than the carpels at anthesis; anthers narrowly oblong, about 3 mm. long, dark violet-purple; receptacle elongate-conical, glabrous, lengthening in fruit. Achenes very numerous, 3–4 mm. long, strongly compressed, papery, with a minutely papillose surface, and a sharp, falcate beak.

HAB.: On rocky or grassy hillsides, on borders of fields, by roads, in ditches, and occasionally on mud-flats; sea-level—2,200 ft. alt.; fl. March–May.

DISTR.: All Divisions, except 2, common in Divisions 4–6, 8, very common in Division 7. Eastern Mediterranean region and eastwards to W. Iran and N. Iraq.

1. Polis, 1937, *Dr. H. Roger-Smith* s.n. (var. *albus* Hayek and var. *flavus* Dörfl.) ! Smyies–Neokhorio (Akamas), 1948, *Kennedy* 1606 (? var. *flavus* Dörfl.) ! Near Ayios Yeoryios (Akamas), 1962, *Meikle* 2085 (var. *flavus* Dörfl.) !
3. Pissouri, 1959, *Sparrow* 14 (var. *albus* Hayek) !
4. Common; *Kotschy* 263 ! 264 ! *Holmboe* 29 (var. *flavus* Dörfl.); *H. Duke Woolley* 42 (var. *tenuilobus* DC.) ! *Chapman* 4 (var. *albus* Hayek) ! *N. Macdonald* 9 (var. *flavus* Dörfl.) !
5. Common; *Sintenis & Rigo* 91 C (var. *tenuilobus* DC.) ! *Holmboe* 311 b (var. *tenuilobus* DC.); *Rev. A. Huddle* 2 ! *Syngrassides* 48 (var. *tenuilobus* DC.) ! 137 (var. *tenuilobus* DC.) ! *Merton* 2532 !
6. Common; *Holmboe* 311a (var. *sanguineus* (Mill.) DC.); *Norman* 365 ! *F. A. Rogers* 0656 (var. *sanguineus* (Mill.) DC. and var. *flavus* Dörfl.) ! *Syngrassides* 770 (var. *sanguineus* (Mill.) DC.) ! *Davis* 2559 (var. *sanguineus* (Mill.) DC.) ! *F. M. Probyn* 51a !
7. Very common; *Sintenis & Rigo* 91 ! 91b (var. *sanguineus* (Mill.) DC.) ! *Rev. A. Huddle* 3 (var. *flavus* Dörfl.) ! *C. H. Wyatt* 10 (? var. *flavus* Dörfl.) ! *H. Duke-Woolley* 86 (? var. *flavus* Dörfl.) ! *Casey* 154 (var. *albus* Hayek) ! 442 ! 480 ! 535 (var. *sanguineus* (Mill.) DC.) ! 649 ! *G. E. Atherton* 1063 (var. *flavus* Dörfl.) !
8. Common; *Holmboe* 493 (var. *albus* Hayek); *Dr. H. Roger-Smith* s.n. (? var. *tenuilobus* DC.) ! *Davis* 2316 (var. *flavus* Dörfl.) ! 2356 (var. *flavus* Dörfl.) ! 2424 (var. *tenuilobus* DC.) !

7. R. paludosus *Poir.*, Voy. en Barbarie, 2: 184 (1789): Davis, Fl. Turkey, 1: 171 (1965); Zohary, Fl. Palaest., 1: 203, t. 292 (1966); Osorio-Tafall et Seraphim, List Vasc. Plants Cyprus, 41 (1973).

R. flabellatus Desf., Fl. Atlant., 1: 438, t. 114 (1798); Holmboe, Veg. Cypr., 81 (1914); Post, Fl. Pal., ed. 2, 1: 11 (1932).

[*R. chaerophyllos* (non L.) Kotschy in Unger et Kotschy, Die Insel Cypern, 315 (1865); Boiss., Fl. Orient., 1: 31 (1867).]

TYPE: Algeria–Tunisia, not definitely localized, *Poiret* (P).

Perennial; roots fibrous intermixed with clusters of small, fleshy, clavate tubers; stem erect, 5–25 cm. high, unbranched or sparsely branched above; leaves mostly radical, pilose or subglabrous, arising from a collar of interwoven fibres; petiole (1–) 3–8 (–10) cm. long, flattened above, ± densely covered with erect or spreading hairs, rarely subglabrous; lowermost radical leaves often simple, broadly ovate-cuneate, about 2·5 cm. long and 2 cm. wide, with a deeply toothed apical margin; upper radical leaves deeply 3-lobed or 3-fid, the segments subdivided into ovate or lanceolate, acuminate lobes; stem leaves shortly stalked or subsessile, divided into elongate, narrow-lanceolate or linear lobes, or frequently simple, linear and entire; flowers solitary, medium size, 1·5–3·5 cm. diam., bright golden-yellow; sepals 5, spreading, narrowly ovate-acuminate, about 1 cm. long, ± densely adpressed-pilose externally; petals normally 5, rather broadly ovate-cuneate, with a rounded, entire apex; stamens numerous, longer than the carpels; anthers yellow, narrowly oblong, about 2–3 mm. long; receptacle elongate-conical, lengthening in fruit. Achenes very numerous, forming a dense cylindrical head 1–2 cm. long, strongly compressed, about 2 mm. long with a sharp, ascending beak.

HAB.: In damp places, open woodland, maquis and on grassy and stony hillsides; sea-level–5,000 ft. alt.; fl. Febr.–April.

DISTR.: All Divisions, common in Divisions 2 and 7. Western, Central and S. Europe and Mediterranean region.

1. Neokhorio, 1962, *Meikle* 2208 !
2. Common; *Kotschy* 702 ! *Kennedy* 429 ! 430 ! *Merton* 2196 ! *N. Macdonald* 62 ! *Meikle* 2728 !
3. Mazotos, 1905, *Holmboe* 173; Kandou, 1963, *Suart* 2 !
4. ? Athna ["Athona"], *G. C. Druce* 11 ! Larnaca–Famagusta road, Mile 3, 1950, *Chapman* 442 !
5. Athalassa (Nicosia), 1933, *Syngrassides* 1370 ! By the bridge before Pyroi, 1936, *Syngrassides* 1048 !
6. Kokkini Trimithia, 1933, *Syngrassides* 334 ! Aghirda, 1941, *Davis* 2488 ! Nicosia–Myrtou road, at Mile 3, 1950, *Chapman* 455 ! Ayia Irini, 1951, *Merton* 149 !
7. Common; *Sintenis & Rigo* 90 ! *A. G. & M. E. Lascelles* s.n. ! *Davis* 2861 ! *Casey* 289 ! *Chapman* 367 ! *G. E. Atherton* 943 ! 953 ! 954 ! 1156 ! *Merton* 2508 !
8. Ayios Andronikos, 1880, *Sintenis & Rigo* 90A ! Near Galinoporni, 1962, *Meikle* 2485 !

8. **R. cadmicus** *Boiss.*, Diagn., 1, 1: 65 (1843); Fl. Orient., 1: 32 (1867); Davis, Fl. Turkey, 1: 182 (1965).

Perennial; roots fibrous intermixed with clustered, elongate-fusiform, fleshy tubers; stems erect or spreading, often rather flexuous, simple or sparsely branched above, glabrous or very thinly hairy, frequently tinged purple; radical leaves often intermixed with 2–3 or more linear, obtuse or subacute membranous sheaths 20–30 mm. long and 4–5 mm. wide; petioles very variable in length, 2–10 cm. or more long, glabrous or thinly hairy, often purple-tinged towards the apex, distinctly flattened and sheathing towards the base; lamina rather fleshy, green on both sides or purplish below, broadly reniform 1·5–3·5 cm. diam., divided to the base into 3–7 bluntly lobed segments; stem leaves few, sessile or subsessile, 3–5-fid with linear or narrowly oblanceolate, blunt lobes; flowers solitary, 1·5–2·5 cm. diam., yellow; sepals 5, spreading, ovate, 4–7 mm. long, glabrous or adpressed-villose, frequently stained purple; petals 5–12 or more, narrowly ovate, 7–12 mm. long, with a rounded apex; receptacle elongate-conical, glabrous or thinly hairy. Achenes 2–3 mm. long, numerous, forming a small spherical head about 6 mm. diam., each achene strongly compressed, with a short, sharp, ascending beak.

var. **cyprius** *Boiss.*, Fl. Orient., 1: 32 (1867).

R. cadmicus Boiss. ssp. *cyprius* (Boiss.) Vierh. in Oesterr. Bot. Zeitschr., 89: 130 (1935); Osorio-Tafall et Seraphim, List Vasc. Plants Cyprus, 41 (1973).

[*R. cadmicus* (vix Boiss.) Kotschy in Unger et Kotschy, Die Insel Cypern, 316 (1865); Holmboe, Veg. Cypr., 81 (1914); Lindberg f., Iter Cypr., 16 (1946).]

TYPE: Cyprus; " in summo monte Troodos Cypri, *Kotschy* 719 " (W, K ! etc.).

Leaves normally stained violet-purple below (or sometimes on both sides); 5–7-lobed; basal sheaths broad and conspicuous. Achenes larger and with a longer beak than in typical var. *cadmicus*.

HAB.: Rocky slopes at high altitudes; 4,900–6,200 ft. alt.; fl. March–May.

DISTR.: Division 2, common locally on the high parts of Troödos. The variety is endemic to Cyprus; aggregate *R. cadmicus* is found also in Turkey.

2. Common locally about Troödos, *Kotschy* 719 ! *G. C. Druce* ! *Tracey* 17 ! *Kennedy* 431–433 ! 436–443 ! *Lindberg f.* s.n. ! *Davis* 3165 !

NOTES: Until more is known about the variability of *R. cadmicus* var. *cadmicus* it is impossible to decide what status should be given to the Cyprian plant. Turkish material of *R. cadmicus* differs (perhaps constantly) in having 3-lobed leaves, apparently green on both sides, and in lacking the conspicuous basal sheaths of var. *cyprius*.

9. R. kykkoënsis *Meikle* in Hooker's Icon. Plant., 37: t. 3649 (1969).

TYPE: Cyprus; by Kykko–Kambos road, just south of Selladhi tou Petrou, 975 m. alt., 25th March, 1960, *Philip Oswald* s.n. (K !).

Glabrous or glabrescent perennial (? 40–50 cm. high); stems erect, terete or slightly sulcate; leaves mostly radical, rather leathery, glabrous, broadly reniform or almost orbicular, 6–7 cm. diam., deeply 3-lobed, with broad, overlapping, bluntly toothed segments; petioles glabrous, 9–10 cm. long; inflorescence terminal, branched, cymose; peduncles with thin arachnoid indumentum; bracts foliose, palmately lobed with 3–7 lobes; flowers medium, 1·5–2 cm. diam., yellow; sepals broadly and bluntly ovate, glabrous or subglabrous about 5 mm. long; petals bluntly obovate-cuneate, 8–10 mm. long, many-veined; stamens numerous, slightly shorter than the carpels; anthers shortly oblong or almost square, about 1 mm. long, yellow; carpels forming a globose head, strongly compressed, glabrous, punctulate, about 4·5 mm. long, 3 mm. wide, apical beak about 1·5 mm. long. Ripe fruit not seen.

HAB.: On steep, moist, shaly slopes, 2,000–4,000 ft. alt.; fl. March–April.

DISTR.: Division 2, very rare. Endemic.

2. Kykko Monastery, 1939, *Lady Loch* 39 ! Kykko–Kambos road, just south of Selladhi tou Petrou, 1960, *Philip Oswald* s.n. !

10. R. millefoliatus *Vahl*, Symb. Bot., 2: 63 (1791); Boiss., Fl. Orient., 1: 35 (1867); ? Post, Fl. Pal., ed. 2, 1: 12 (1932); Davis, Fl. Turkey, 1: 185 (1965).

Perennial; roots fibrous, intermixed with clustered, clavate tubers; stem erect, unbranched or sparsely branched above, (2–) 12–20 (–30) cm. high, adpressed-villose, especially towards the apex; radical leaves petiolate, petioles 3–8 cm. long, slender, subglabrous or thinly hirsute, sheathing at the base; lamina ± glabrous, elongate-deltoid, 2–5 cm. long, 1·5–3 cm. wide at the base, (2–) 3-pinnate, the ultimate segments lanceolate or linear, acuminate or rather blunt, stem leaves few, shortly petiolate or subsessile, pinnately or irregularly divided into long, lanceolate or linear segments; flowers 1·5–3 cm. diam., golden yellow; sepals 5, spreading, ovate-acuminate, about 1 cm. long, usually quite glabrous, though occasionally with a few scattered hairs, bronze-tinged, with a narrow scarious margin; petals usually 5, broadly or narrowly ovate-cuneate, 8–18 mm. long, with a rounded apex; stamens numerous, longer than the carpels; anthers yellow, narrowly linear-oblong, about 4–5 mm. long; receptacle elongate-conical, lengthening in fruit, glabrous. Achenes forming a cylindrical head, numerous, strongly compressed, about 4–5 mm. long, with a strongly hooked beak and a gibbous base.

ssp. **leptaleus** (*DC.*) *Meikle* in Kew Bull., 1956: 545 (1957); Osorio-Tafall et Seraphim, List Vasc. Plants Cyprus, 41 (1973).

R. leptaleus DC., Regn. Veg. Syst. Nat., 1: 258 (1817); Unger et Kotschy, Die Insel Cypern, 316 (1865); Boiss., Fl. Orient, 1: 35 (1867); Holmboe, Veg. Cypr., 81 (1914).

[*R. millefoliatus* (non Vahl) Kotschy in Unger et Kotschy, Die Insel Cypern, 315 (1865).]

TYPE: " in insula Cypro ", 1787, *Labillardière* (FI).

A smaller, more slender plant than typical *R. millefoliatus*, (2–) 4–8 (–16) cm. high; leaves 2 or rarely 3-pinnate, ultimate segments rather blunt, obovate or broadly oblanceolate; flowers 1·5–2 cm. diam., petals rather narrowly ovate-cuneate.

HAB.: Grassy hillsides and pastures, sometimes in rock crevices; often very common on Kythrean marls; sea-level to 3,000 ft. alt.; fl. Febr.–May.

DISTR.: Divisions 3–8, locally common in Division 7, around Kyrenia and Pentadaktylos. Endemic.

3. Near Klavdhia, 1880, *Sintenis & Rigo* 87 ! Skarinou, 1941, *Davis* 2684 ! Lefkara, 1941, *Kennedy & Davis* 2749 !
4. Ormidhia, 1880, *Sintenis & Rigo*.
5. Pyroi near Larnaca, 1936, *Syngrassides* 1052 ! Mia Milea, 1951, *Merton* 124 ! Near Kato Dhikomo, 1962, *Meikle* 2971 !
6. Nicosia, 1927, *Rev. A. Huddle* 11 ! Near Nicosia, 1940, *Davis* 2133 K ! Between Ayia Irini and Kormakiti, 1941, *Davis* 2125 !
7. Locally common; *Kotschy*; *Sintenis & Rigo* 87 (a) ! *Rev. A. Huddle* s.n. ! *Miss E. Haig* 2 ! *C. H. Wyatt* 58 ! *Lady Loch* 68 ! *Davis* 2164 ! *Collingwood Ingram* 40 ! *H. Duke-Woolley* 24 ! etc.
8. Koma tou Yialou, 1935, *Syngrassides* 707 ! Apostolos Andreas, 1941, *Davis* 2284 !

NOTES: The Cyprus specimens differ constantly and uniformly from typical *R. millefoliatus* Vahl in the characters set out above. The achene characters given by Boissier are less satisfactory, and it is doubtful if ssp. *leptaleus* and ssp. *millefoliatus* differ in this respect. Vahl's type is from " cultis circa promontorium Carthaginis Tuneti ", and represents a finely dissected form of the species; many S. European specimens have broader leaf-segments than the type, and approach ssp. *leptaleus*, though differing in flower-size and general degree of robustness.

11. R. millefolius *Banks et Soland.* in Russell, Nat. Hist. Aleppo, ed. 2, 2: 254 (1794) excluding reference to Barrellier, Icones, t. 1153 (1714); Davis, Fl. Turkey, 185 (1965); Zohary, Fl. Palaest., 1: 203, t. 293 (1966); Osorio-Tafall et Seraphim, List Vasc. Plants Cyprus, 41 (1973).

R. orientalis L., Sp. Plant., ed. 1, 555 (1753) non auct. nomen confusum.

R. myriophyllus DC., Regn. Veg. Syst. Nat., 1: 257 (1817); Unger et Kotschy, Die Insel Cypern, 316 (1865); Boiss., Fl. Orient., 1: 36 (1867); Holmboe, Veg. Cypr., 81 (1914); Post, Fl. Pal., ed. 2, 1: 12 (1932).

TYPE: Syria; Aleppo, *Alexander Russell* (BM).

Perennial; roots fibrous, intermixed with elongate, fusiform, clustered tubers; stems erect, (10–) 15–20 (–30) cm. high, usually much-branched above the middle, ± densely clothed with long, white, ascending hairs; radical leaves petiolate, petioles 3–8 mm. long, ± densely villose, sheathing at the base; lamina narrowly deltoid, 2–3-pinnately dissected, ultimate segments ovate-cuneate or linear-lanceolate, acute or acuminate, ± villose above and below; stem leaves sessile or shortly stalked, the uppermost divided almost to the base into linear or narrowly lanceolate segments; flowers medium, 1·5–2·5 cm. diam., yellow; peduncles 4–10 cm. long, strongly divaricate and accrescent in fruit, often distinctly swollen towards apex; sepals 5, spreading, ovate-acuminate, about 8–12 mm. long, usually covered with long white hairs; petals 5 (rarely 6 or more), broadly ovate-cuneate with a rounded apex; stamens numerous, longer than the carpels;

anthers narrowly oblong, about 2–3 mm. long, yellow; receptacle glabrous, elongate-conical, lengthening in fruit. Achenes numerous, forming a cylindrical head, 1–2 cm. long, strongly compressed, minutely punctate, and often sparsely tuberculate, 3–4 mm. long, with a distinct margin, a sharp ascending or recurved beak, and a distinctly gibbous base.

HAB.: On rocky, calcareous ground; 250–3,000 ft. alt.; fl. March–April.

DISTR.: Divisions 5, 7, rare. Also Turkey, Syria, Palestine, Iraq and Iran.

5. Mile 6, Nicosia–Famagusta Road, 1951, *Merton* 986 ! Ayios Sozomenos, 1957, *Merton* 3097 !

7. Buffavento, 1859, *Kotschy* 266 ! Ayios Khrysostomos Monastery, *Kotschy* 626.

12. R. neapolitanus *Tenore*, Ind. Sem. Hort. Bot. Neap., 11 (1825); Boiss., Fl. Orient., 1: 38 (1867); Post, Fl. Pal., ed. 2, 13 (1932); Vierhapper in Oesterr. Bot. Zeitschr., 84: 132 (1935); Davis, Fl. Turkey, 1: 165 (1965); Osorio-Tafall et Seraphim, List Vasc. Plants Cyprus, 41 (1973).

Perennial; roots fibrous intermixed with clustered elongate-fusiform tubers 4–12 cm. long; stem erect, sparingly branched above, 20–40 cm. high, with spreading or adpressed hairs; radical leaves arising from a loose fibrous collar; petioles 5–10 cm. long, slender, pilose, sheathing at the base; lamina broadly deltoid, 3–5 cm. long, 3–6 cm. wide, deeply 3-lobed, sinus narrow extending at least ⅔ way to the base of the leaf, lobes crenate or rather bluntly dentate; stem leaves few, shortly petiolate or subsessile, palmatisect, with narrowly cuneate or linear, toothed or entire segments; flowers about 2 cm. diam., golden-yellow; sepals 5, strongly reflexed, ovate-acuminate, about 6–8 mm. long, pilose. Achenes forming a spherical head, numerous, round, strongly compressed, 3–4 mm. long, dark brown and minutely papillose with a paler margin; beak short, straight or hooked.

var. **adpresse-pilosus** *Freyn emend. Vierhapper* in Oesterr. Bot. Zeitschr., 84: 132 (1935).

R. neapolitanus Ten. ssp. *tommasinii* (Reichb.) Vierh. var. *adpresse-pilosus* Freyn emend. Vierhapper in Oesterr. Bot. Zeitschr., *loc. cit.*, 132.

R. neapolitanus Ten. β *adpresse-pilosus* αα *brevirostris* Freyn in Flora, 63: 216 (1880).

TYPE: " Orient " (G).

Achene with a short, straight beak; indumentum adpressed.

HAB.: Limestone hillsides; 500–1,000 ft. alt.; fl. April–May.

DISTR.: Division 7, rare. Also in Italy, Balkan peninsula, Malta, Crete, Aegean Islands, Crimea, Turkey and Syria.

7. Above Lapithos, 1880, *Sintenis & Rigo* 620 !

NOTES: Typical *R. neapolitanus* with a longer, hooked beak and spreading indumentum has not been found on the island.

13. R. chius *DC.*, Regn. Veg. Syst. Nat., 1: 299 (1817); Boiss., Fl. Orient., 1: 54 (1867); Holmboe, Veg. Cypr., 82 (1914); Post, Fl. Pal., ed. 2, 1: 15 (1932); Lindberg f., Iter Cypr., 16 (1946); Davis, Fl. Turkey, 1: 189 (1965); Zohary, Fl. Palaest., 1: 207, t. 301 (1966); Osorio-Tafall et Seraphim, List Vasc. Plants Cyprus, 42 (1973).

R. incrassatus Guss., Fl. Sic. Syn., 2: 50 (1844); Unger et Kotschy, Die Insel Cypern, 318 (1865).

[? *R. parviflorus* (vix L.) Osorio-Tafall et Seraphim, List Vasc. Plants Cyprus, 42 (1973).]

var. **chius**

TYPE: " in insula Scio Archipelagi Graeci, *Olivier* " (P).

Diffuse or ascending annual (6–) 10–20 (–30) cm. high; roots fibrous; stem ± pilose with patent hairs; radical leaves with hairy petioles 4–8 cm. long; lamina reniform about 2–3·5 cm. wide with 3–5 bluntly toothed lobes;

lower stem leaves similar but smaller, upper stem leaves often paired, ovate-acuminate and entire, 1·5–3 cm. long, 1–1·5 cm. wide, with a short petiole; inflorescence leaf-opposed, flowers small, 5–7 mm. diam., solitary on short stout peduncles 3–5 mm. long; sepals 5, irregularly reflexed, narrowly ovate, 2–3 mm. long, ± villose externally; petals yellow, narrowly ovate, 2–3 mm. long, 5-veined, usually ± erect; stamens inconspicuous, shorter than the carpels; anthers elongate-oblong; receptacles ovoid, glabrous or thinly hairy; peduncles lengthening and thickening in fruit, narrowly clavate, 1–2 (–4) cm. long. Achenes forming a spherical head, rather few (6–20), strongly compressed, 3–4 mm. long, surface dark brown, muricate, with a paler, smooth margin and a relatively long, falcate beak.

HAB.: Moist fields, streamsides and shady ground; sea-level to 5,000 ft. alt.; fl. Febr.–May.

DISTR.: Divisions 1–4, 6–8, common in Division 7. Also E. Mediterranean region, Palestine and Turkey.

1. Near Avgas R., Ayios Yeoryios (Akamas), 1962, *Meikle* 2008 ! Near Polis, 1962, *Meikle* !
2. Kryos Potamos (near Platres), 1937, *Kennedy* 426 ! Prodhromos, 1961, *D. P. Young* 7345 !
3. Garden of Stavrovouni Monastery, 1880, *Sintenis & Rigo* 88 !
4. Perivolia near Larnaca, 1905, *Holmboe* 160; Famagusta Castle, 1936, *Syngrassides* 1068 !
6. Myrtou–Panagra, 1934, *Syngrassides* 1359 ! Ayia Irini (Morphou), 1941, *Davis* 2572 ! Dhiorios, 1962, *Meikle* !
7. Common; *Kotschy* 354 ! *Sintenis & Rigo* 88a ! *Lady Loch* 70 ! *H. Duke-Woolley* 55 ! *G. E. Atherton* 994 B ! 1180 ! *Casey* 344 ! 1693 !
8. Ayios Andronikos (Karpas), 1938, *Syngrassides* 1791 ! Between Yialousa and Ayios Andronikos, 1941, *Davis* 2398 !

var. **leiocarpus** *P. H. Davis* in Kew Bull., 1949: 98 (1949).

TYPE: Cyprus; Tripylos (Paphos forest), damp roadside in Cedar forest, May 17, 1941, *Davis* 3487 (K !).

Achenes smooth, otherwise similar to var. *chius*.

HAB.: Damp roadsides and shaded ground in Cedar forest, 3,000–4,000 ft. alt.; fl. March–May.

DISTR.: Division 2, rare. Cyprus, Turkey.

2. Tripylos (Paphos Forest), 1941, *Davis* 3487 ! and 1962, *Meikle* 2661 !

14. R. constantinopolitanus (*DC.*) *Urv.* in Mém. Soc. Linn. Par., 1: 317 (1822); Boiss., Fl. Orient., 1: 49 (1867); Davis, Fl. Turkey, 1: 169 (1965); Zohary, Fl. Palaest., 1: 204, t. 296 (1966); Osorio-Tafall et Seraphim, List Vasc. Plants Cyprus, 41 (1973).

R. lanuginosus L. var. *constantinopolitanus* DC., Regn. Veg. Syst. Nat., 1: 281 (1817).

TYPE: Aubriet drawing (unpublished) in bibl. Jussieu (P).

An erect perennial 40–70 cm. high; rootstock erect, roots elongate, fibrous or rather fleshy; radical leaves broadly deltoid, 3–10 cm. long, 5–13 cm. wide, deeply 3-lobed or almost ternate; segments broadly obovate, irregularly toothed and lobed, with acute or rather blunt teeth, ± densely adpressed-pilose on both surfaces; petioles 10–15 cm. long or longer, densely clothed with adpressed or patent hairs; stem leaves sessile or shortly stalked, the lowermost divided into 3 linear, entire or irregularly dentate segments, 2–5 cm. long, 0·3–0·5 cm. wide, the uppermost linear, simple and entire; inflorescence lax, much-branched, with slender, adpressed-pilose peduncles 3–10 cm. long; flowers medium, 1·5–3 cm. diam., yellow; sepals ovate-acuminate, 5–7 mm. long, glabrous internally, adpressed-pilose externally, distinctly reflexed; petals usually 5, rather broadly cuneate, 7–15 mm. long, 5–12 mm. wide, margins entire; stamens numerous, much longer than the carpels at anthesis; anthers narrowly oblong, about 2 mm. long, yellow; receptacle clavate, furrowed, glabrous or thinly hairy, especially towards the base. Achenes about 12, forming a spherical head; each achene about 2–3 mm. long, strongly compressed, with the dorsal margin distinctly gibbous; surface minutely papillose; beak short, strongly hooked.

HAB.: Shady thickets on calcareous mountainside; c. 1,500 ft. alt.; fl. March–April.

DISTR.: Division 7, rare. Also Romania, Macedonia, Crimea, Caucasus, Turkey, Syria, Palestine eastwards to Iran.

7. Kantara Road, just before Kantara, 1951, *Merton* 244 ! 761 ! N. side Sina Oros near Kantara (probably same locality as previous), 1974, *Meikle* 4018 !

15. **R. marginatus** *Urv.* in Mém. Soc. Linn. Par., 1: 318 (1822); Boiss., Fl. Orient., 1: 55 (1867); Davis, Fl. Turkey, 1: 187 (1965); Zohary, Fl. Palaest., 1: 205, t. 298 (1966).

Erect or diffuse annual, 10–40 cm. high with glabrous or thinly pilose stems and fibrous roots; radical leaves with glabrous or pilose petioles, 3–6 cm. long; lamina reniform, 2–5 cm. wide with 3–5 crenate or bluntly toothed lobes; lower stem leaves often ternate, with broadly cuneate lobes; upper stem leaves deeply and irregularly lobed, or simple, with a narrowly lanceolate lamina; inflorescence leaf-opposed, flowers rather small, about 1–1·5 cm. diam., solitary on slender peduncles 1·5–6 cm. long; sepals 5, strongly reflexed, ovate-acuminate, 3–8 mm. long, glabrous or thinly pilose externally; petals 5, yellow, 5–7 cm. long, many-veined, spreading; stamens numerous, about as long as the carpels ; anthers elongate-oblong; receptacle elongate-conical, pilose. Achenes forming a spherical head, numerous, strongly compressed, about 4 mm. long, surface smooth (var. *marginatus*) or muriculate (var. *trachycarpus* (Fisch. et Mey.) Azn.), with a distinct 2-grooved margin, and a prominent upturned beak about 1 mm. long.

var. **trachycarpus** (*Fisch. et Mey.*) *Azn.* in Magyar Bot. Lapok, 1: 297 (1902); Rechinger f., Fl. Aegaea, 193 (1943); Davis, Fl. Turkey, 1: 187 (1965); Osorio-Tafall et Seraphim, List Vasc. Plants Cyprus, 42 (1973) [as ssp. *trachycarpus*].

R. trachycarpus Fisch. et Mey., Ind. Sem. Hort. Petrop., 3: 46 (1837); Unger et Kotschy, Die Insel Cypern, 316 (1865) pro parte; Boiss., Fl. Orient., 1: 55 (1867); Post, Fl. Pal., ed. 2, 1: 15 (1932); Rechinger f. in Arkiv för Bot., 1: 422 (1950).

[*R. trilobus* (non Desf.) Kotschy in Unger et Kotschy, Die Insel Cypern, 316 (1865); Holmboe, Veg. Cypr., 81 (1914).]

R. trilobus Desf. var. *tripetalus* Holmboe, Veg. Cypr., 82 (1914).

R. troodi Lindberg f., Iter Cypr., 16 (1946).

[*R. lomatocarpus* (non Fisch. et Mey.) Holmboe, Veg. Cypr., 81 (1914) pro parte quoad spec. *Holmboe* 399 fide Rechinger f. in Arkiv för Bot., 1: 422 (1950).]

TYPE: U.S.S.R.; " In Tauria, in regionibus transcaucasicis et in provincia Lenkoran " (LE).

Achenes distinctly muriculate, otherwise resembling var. *marginatus*. *Plate 3, figs. 1, 2.*

HAB.: On wet (sometimes brackish) ground by streams or springs; sea-level to 4,700 ft. alt.; fl. April–June.

DISTR.: Divisions 2, 4, 8, locally common in Division 2. Crimea, Caucasus, Greece, Aegean Islands, Turkey, Palestine, Syria, Iran.

2. Locally common; *Holmboe* 898 (type of *R. trilobus* var. *tripetalus*); *Rev. A. Huddle* 8 ! *Lindberg f.* s.n. (type of *R. troodi*); *Merton* 2396 (var. *tripetalus*) ! 2799 (var. *tripetalus*) ! *Meikle* 2757 ! *Suart* 5 !
4. Kouklia, 1905, *Holmboe* 399.
8. About Komi Kebir, 1912, *Haradjian* 318; between Yialousa and Ayios Andronikos, 1941, *Davis* 2397 ! Galatia, 1962, *Meikle* 2527 !

NOTES: *R. trilobus* Desf. var. *tripetalus* Holmboe is a starved form of *marginatus* var. *trachycarpus*; Holmboe evidently did not know the latter plant.

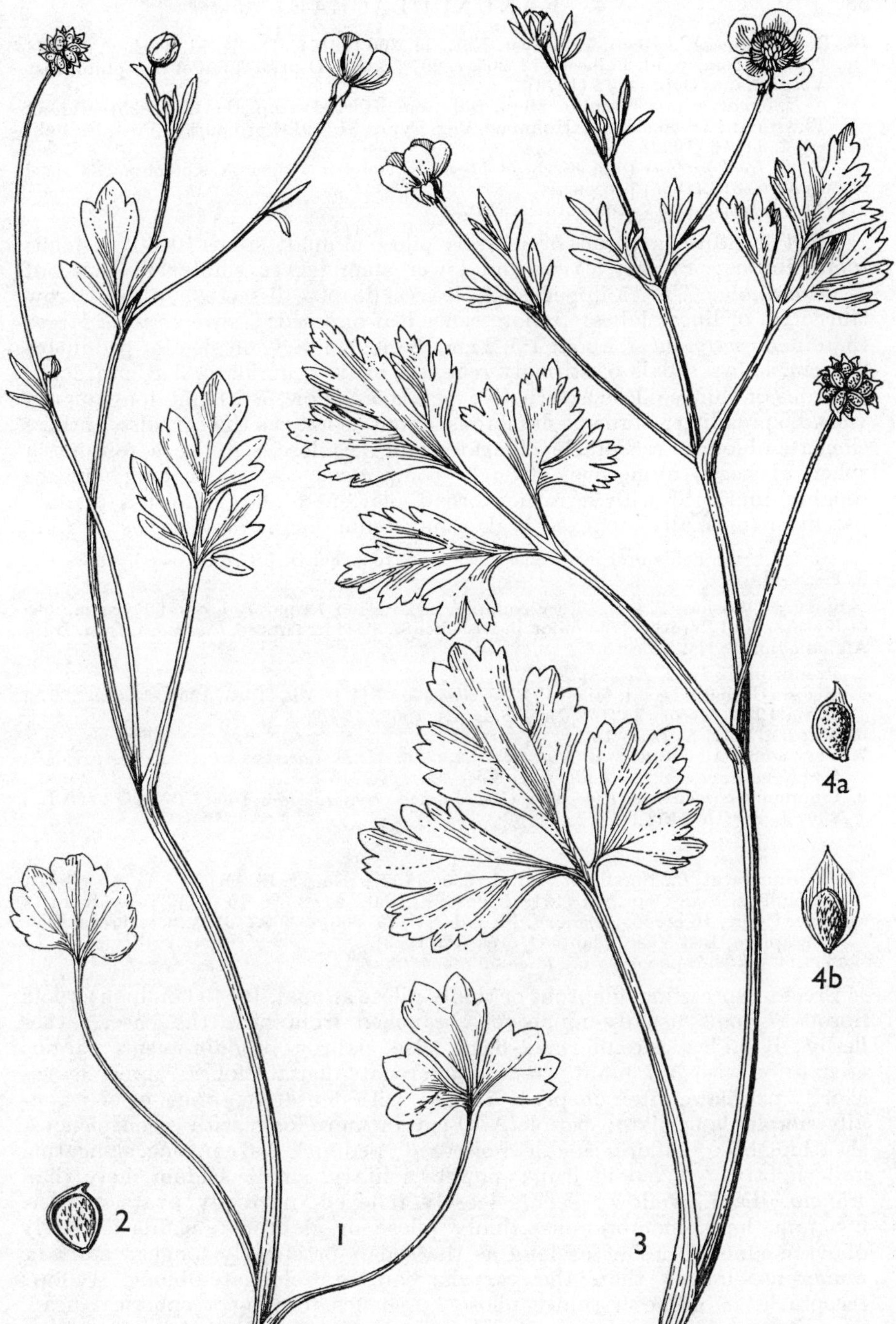

Plate 3. RANUNCULUS MARGINATUS Urv. var. TRACHYCARPUS (Fisch. et Mey.) Azn. (figs. **1, 2**); **1,** habit, $\times \frac{2}{3}$; **2,** achene, $\times 3$. R. CORNUTUS DC. (figs. **3, 4**); **3,** habit, $\times \frac{2}{3}$; **4a,** achene, $\times 3$; **4b,** achene, $\times 3$. (**1, 2** from *Lindberg* f. s.n. & *Barbey* 12; **3, 4a** from *Casey* 1690; **4b** from *Sintenis & Rigo* 186.)

16. R. cornutus *DC.*, Regn. Veg. Syst. Nat., 1: 300 (1817); Davis, Fl. Turkey, 1: 188 (1965); Zohary, Fl. Palaest., 1: 205, t. 297 (1966); Osorio-Tafall et Seraphim, List Vasc. Plants Cyprus, 42 (1973).

R. lomatocarpus Fisch. et Mey., Ind. Sem. Hort. Petrop., 1: 36 (1835); Boiss., Fl. Orient., 1: 56 (1867); Holmboe, Veg. Cypr., 81 (1914) pro parte; Post, Fl. Pal., ed. 2, 1: 16 (1932).

[*R. trachycarpus* (non Fisch. et Mey.) Kotschy in Unger et Kotschy, Die Insel Cypern, 316 (1865) pro parte.]

TYPE: " In Oriente " *Labillardière* (Fl).

Erect or diffuse, glabrous or sparsely pilose annual; stems 10–40 cm. high; roots fibrous; radical leaves and lower stem leaves similar to those of *R. marginatus* Urv.; upper stem leaves deeply dissected, with narrow lanceolate or linear lobes; inflorescence leaf-opposed; flowers rather larger than in *R. marginatus*, about 1·5–2 cm. diam., solitary, on slender peduncles 2–6 cm. long; sepals 5, strongly reflexed, ovate-acuminate, 3–8 mm. long, glabrous or thinly pilose externally; petals 5, yellow, 7–10 mm. long, many-veined, spreading; stamens numerous, about as long as the carpels; anthers elongate-oblong; receptacle elongate-conical, pilose. Achenes forming a spherical head, numerous, strongly compressed, 5–8 mm. long, surface roughly muricate with a rather broad, flattened margin and a distinct patent or (normally) deflexed beak about 2 mm. long. *Plate 3, figs. 3, 4.*

HAB.: In wet fields and in ditches; sea-level to 4,500 ft. alt., but usually lowland; fl. Febr.–May.

DISTR.: Divisions 2, 5–8. Very common in Division 7 (mainly around Kyrenia, also at Kythrea and Lapithos), common in Division 8. Also in Turkey, Caucasus, Iran, Iraq, Afghanistan, Syria, Palestine.

2. Prodhromos, 1862, *Kotschy* 886 !
5. Nicosia–Limassol road, Mile 14, 1950, *Chapman* 541 ! Mia Milea, 1956, *Merton* 2507 ! Mora, 1956, *Merton* 2531 ! Near Salamis, 1962, *Meikle* !
6. Nicosia, 1934, *E. Haig* 14 !
7. Very common; *Sintenis & Rigo* 86 ! *A. G. & M. E. Lascelles* s.n. ! *Rev. A. Huddle* 9 ! *Syngrassides* 62 ! *Lady Loch* 69 ! *Merton* 2246 ! etc.
8. Common; *Sintenis & Rigo* 86 ! *Druce* s.n. ! *Syngrassides* 1060 ! *Davis* 2289 K ! 2301 ! *Meikle* 2452 !

17. R. muricatus *L.*, Sp. Plant., ed. 1, 555 (1753); Boiss., Fl. Orient., 1: 56 (1867); Holmboe, Veg. Cypr., 81 (1914); Post, Fl. Pal., ed. 2, 1: 16 (1932); Lindberg f., Iter Cypr., 16 (1946); Zohary, Fl. Palaest., 1: 206, t. 300 (1936); Osorio-Tafall et Seraphim, List Vasc. Plants Cyprus, 42 (1973).

TYPE: " *in* Europae *australis fossis & humentibus* ".

Erect or sprawling, glabrous or thinly pilose annual, 10–40 cm. high; roots fibrous; stems usually numerous, branched from near the base, rather fleshy; basal leaves reniform, 2–6 cm. wide, glabrous or glabrescent, margins crenate or with 3–7 blunt, crenate or crenate-dentate lobes; upper leaves shortly petiolate, often deeply 3–5-lobed, with dentate margins, or occasionally simple, lanceolate; petiole 2–10 cm. or more long, with a conspicuous, sheathing base; inflorescence leaf-opposed; peduncle 2–6 cm. long, elongating and slightly deflexed in fruit; flowers solitary, small, seldom more than 1·5 cm. diam., yellow; sepals loosely reflexed, narrowly ovate, obtuse, 6–8 mm. long, membranous, thinly pilose or glabrous; petals narrowly obovate-cuneate, about as long as the sepals or slightly longer; stamens numerous, longer than the carpels; anthers elongate-oblong, yellow; receptacle ± globose, thinly pilose. Achenes forming a spherical head, numerous, 5–7 mm. long, surface densely muricate-echinate, margin with a distinct flattened wing; beak prominent, usually deflexed, 2–4 mm. long.

HAB.: Damp fields, roadsides and ditches; sea-level to 5,000 ft. alt.; fl. March–May.

DISTR.: Divisions 1–5, 7, 8, common in Divisions 2 and 7. S. Europe, Mediterranean region, Turkey, Palestine, eastwards to Turkestan and south to Arabia; casual elsewhere.

1. Near Neokhorio, 1962, *Meikle* 2193 ! Near Ayios Yeoryios (Akamas), 1962, *Meikle* !
2. Common; *Kennedy* 428 ! *Lindberg f.* s.n. *Mavromoustakis* s.n. ! *F. J. F. Barrington* s.n. ! *D. P. Young* 7344 !
3. Near Kolossi, 1862, *Kotschy*; Akrotiri Bay, 1948, *Mavromoustakis* 60 ! Cherkez, 1954, *Mavromoustakis* 12 !
4. Ayia Napa, 1862, *Kotschy*; Larnaca, 1905, *Holmboe* 117.
5. Ditches near Kythrea, 1880, *Sintenis & Rigo* 89 ! Near Salamis, 1962, *Meikle* !
7. Common; *Syngrassides* 25 ! *Casey* 491 ! 1691 ! *G. E. Atherton* 1148 ! 1261 ! 1393 ! *Merton* 2726 !
8. Platanisso, 1950, *Chapman* 244 !

18. R. arvensis *L.*, Sp. Plant., ed. 1, 555 (1753); Boiss., Fl. Orient., 1: 57 (1867); Holmboe, Veg. Cypr., 81 (1914); Post, Fl. Pal., ed. 2, 1: 16 (1932); Lindberg f., Iter Cypr., 16 (1946); Zohary, Fl. Palaest., 1: 207, t. 302 (1966); Osorio-Tafall et Seraphim, List Vasc. Plants Cyprus, 42 (1973).

TYPE: " *in* Europae *australioris agris* ".

Erect or diffuse, glabrous or thinly pilose annual, 8–30 cm. high; stems simple or branched; basal leaves simple or deeply 3-lobed, 1–3 cm. long, 0·5–2 cm. wide; lamina spathulate-cuneate, apex sharply toothed, base gradually narrowing to a petiole 1–3 cm. or more long; upper leaves shortly petiolate, irregularly laciniate with narrow cuneate or linear, entire or remotely dentate divisions 2–4 cm. long; uppermost leaves often simple, sessile or subsessile, linear, entire; inflorescence leaf-opposed; peduncle slender, 2–3 cm. long, accrescent in fruit; flowers solitary, small, 0·5–1·5 cm. diam., pale yellow; sepals 5, erect or patent, narrowly ovate, obtuse, about 5 mm. long, membranous, usually ± pilose; petals 5, narrowly ovate-cuneate, slightly longer than the sepals; stamens about as long as the carpels; anthers elongate-oblong, yellow; receptacle small, conical, pilose. Achenes 2–6, strongly compressed, about 7 mm. long; surface echinate, with rigid spines; margin with a narrow wing at right angles to the surface.

HAB.: Cultivated fields and waste ground; sea-level to 5,000 ft. alt.; fl. March–June.

DISTR.: All Divisions, common in Divisions 2 and 7. Temperate Europe and W. Asia, eastwards to India.

1. Yiolou above Baths of Aphrodite, 1962, *Meikle* 2226 !
2. Common; *Kotschy*; *A. G. & M. E. Lascelles* s.n. ! *Holmboe* 893; *Kennedy* 427 ! *Lindberg f.* s.n. *Davis* 3100 ! etc.
3. Morokambos (Akhelia), 1956, *Merton* 2502 ! Pyrgos, 1963, *Suart* 4 !
4. Phaneromene near Larnaca, 1905, *Holmboe* 122.
5. Kythrea, 1880, *Sintenis & Rigo* 85 ! Mora, 1956, *Merton* 2530 !
6. Dhiorios, 1962, *Meikle* !
7. Common; *Rev. A. Huddle* 7 ! *Syngrassides* 683 ! *Collingwood Ingram* 35 ! *G. E. Atherton* 918 ! 944 ! 1186 ! 1217 !
8. Between Rizokarpaso and Apostolos Andreas, 1938, *Syngrassides* 1776 ! Yialousa, 1941, *Davis* 2392 ! Ayios Philon near Rizokarpaso, 1941, *Davis* 2224 !

19. R. sp. indet.

Tufted perennial, 10 cm. high; roots fibrous, intermixed with clustered, shortly fusiform tubers; leaves all radical, broadly triangular, 1·5–2 cm. long, 2–3 cm. wide, pinnately 3-foliolate, leaflets ± ovate, irregularly toothed and lobed, subglabrous; petiole about 2 cm. long; flowers large, about 3 cm. diam., yellow (or reddish), solitary, scapose; scape about 10 cm. long, shortly tomentose, terete; sepals patent or irregularly reflexed, narrowly ovate-acuminate, about 8 mm. long, glabrous with a membranous margin; petals c. 1·5 cm. long, ovate-cuneate, many-veined; stamens numerous with elongate-oblong (? yellow) anthers slightly shorter than the purple-tinged carpels. Fruit not seen.

HAB.: Grassy hillsides and pastures; c. 300 ft. alt.; fl. Febr.–March.

DISTR.: Division 3, rare.

3. Near Klavdhia (Larnaca), 1880, *Sintenis & Rigo* 87 (mixed with specimens of *R. millefoliatus* Vahl ssp. *leptaleus* (DC.) Meikle) !

NOTES: Allied to *R. millefoliatus* ssp. *leptaleus*, but distinguished by its large flowers and leaflets. The single specimen collected looks almost as if it might be *R. asiaticus* × *R. millefoliatus* ssp. *leptaleus*, particularly as the petals and carpels seem to have been tinged red or purple. The two species sometimes grow together, and the hybrid should be looked for.

7. NIGELLA *L.*

Sp. Plant., ed. 1, 534 (1753).
Gen. Plant., ed. 5, 238 (1754).

Annual herbs with spirally arranged leaves usually divided pinnately into narrow linear or lanceolate segments. Flowers solitary, terminal, actinomorphic; sepals 5, ± petaloid; petals 5–8, nectariferous, usually smaller than the sepals, unguiculate, ± 2-lobed, with the lobes variously subdivided; stamens numerous, hypogynous. Fruit consisting of 2–10, more or less fused follicles; styles persistent, accrescent, crowning the ripe fruit.

About 20 species in Europe, the Mediterranean region and eastwards to Central Asia.

Petals (or nectaries) longer than the sepals; fruit dehiscent along external and internal sutures:
  Petals 6–8 mm. long without a distinct claw; outer lip divided into 2 narrow-lanceolate lobes; inner lip 2–3 mm. long - - - - - **1. N. nigellastrum**
  Petals 8–9 mm. long with a distinct claw; outer lip divided into 2 broadly lanceolate lobes; inner lip minute, about 1 mm. long - - - - **2. N. unguicularis**
Petals (or nectaries) shorter than the sepals; fruit dehiscent along the internal sutures only:
  Flowers blue or bluish; sepals glabrous, smooth or scabridulous; seeds trigonous, globose or ovoid:
    Flowers with a well-developed involucre of leafy bracts:
      Fruits star-shaped, beaks patent; anthers crimson-purple **4. N. fumariifolia**
      Fruits broadly ovoid or subglobose, bladder-like, beaks erect or spreading; anthers yellow or greenish - - - - - - - **6. N. damascena**
    Flowers without a well-developed involucre:
      Leaves and stems pubescent; fruit ovoid or oblong, carpels connate almost to apex - - - - - - - - - - - - - - **5. N. sativa**
      Leaves and stems glabrous or scabrid, not pubescent; fruit obconical, carpels connate for about ⅔ their length - - - - - - - **3. N. arvensis**
  Flowers yellow or yellowish; sepals pubescent; seeds flattened, discoid **7. N. ciliaris**

SUBGENUS 1. **Garidella** (*L.*) *Brand* Fruit 2 (–5)-locular, dehiscing along external and internal sutures; petals as large as or larger than sepals; styles small.

1. **N. nigellastrum** (*L.*) *Willk.* in Willk. et Lange, Prodr. Fl. Hisp., 3: 963 (1880); Holmboe, Veg. Cypr., 77 (1914); Lindberg f., Iter Cypr., 16 (1946); Davis, Fl. Turkey, 1: 104 (1965); Osorio-Tafall et Seraphim, List Vasc. Plants Cyprus, 40 (1973).
   Garidella nigellastrum L., Sp. Plant., ed. 1, 425 (1753); Boiss., Fl. Orient., 1: 64 (1867).
   TYPE: France; " *in* Galloprovincia ".

Slender, erect, annual herb, 15–40 cm. high, growing from a tough, elongate tap-root; stems glabrous, distinctly sulcate, sparsely branched; basal leaves 1–2-pinnate, with narrow linear segments, glabrous, 5–8 cm. long, including a petiole 2–4 cm. long; stem leaves irregularly pinnatisect, with long, pointed linear segments and short petioles; uppermost leaves small, often sessile or subsessile, irregularly lacinate, with narrow, linear segments; flowers solitary on long, slender, almost leafless stalks; sepals greenish tinged purple or blue, ovate, carinate, acuminate, 4–5 mm. long, glabrous;

petals 5–6, white, blotched with blue, 6–8 mm. long, 2-lipped; outer lip 4–6 mm. long, with 2 deeply cleft, narrow-lanceolate lobes, pilose towards the base, and distinctly gibbous; inner lip undivided, ovate, about 2 mm. long; claw slender, about 2–3 mm. long; stamens 3–4 mm. long, slightly longer than the carpels, with shortly oblong, purple-tinged anthers; carpels usually 2, greenish, verruculose. Fruit broadly oblong, 4–8 mm. long, 4–6 mm. wide, pale brown, smooth, emarginate, crowned with 2 short, persistent styles; seeds blackish, rugose, angled.

HAB.: Cultivated fields, waste ground, or amongst garigue on hillsides; sea-level to 2,000 ft. alt.; fl. March–June.

DISTR.: Divisions 1, 3–5, 7, common in Divisions 3 and 5, very common in Division 7. Spain, S. France, Italy, Greece, Turkey to Caucasus, Iran and Iraq.

1. Ayios Neophytos, 1939, *Lindberg f.* s.n.
3. Common; *Kotschy* 631 ! *Holmboe* 588 ! 609; *Davis* 3588 ! *Mavromoustakis* 3/57 !
4. Near Larnaca, 1862, *Kotschy* 283; Cape Greco, 1905, *Holmboe* 418; near Famagusta, 1939, *Lindberg f.* s.n.
5. Common; *Lindberg f.* s.n.; *Mavromoustakis* s.n. ! *Chapman* 568 ! 636 ! 673 ! *Merton* 2763 !
7. Very common; *Kotschy* 469; *Sintenis & Rigo* 81 ! *Pichler* s.n. ! *A. G. & M. E. Lascelles* s.n. ! *Tracey* 21 ! etc.

2. N. unguicularis (*Poir.*) *Spenner*, Mon. Gen. Nigellae, 12 (1829).
Garidella unguicularis Poir., Encycl. Méth. Suppl., 2: 709 (1812); Boiss., Fl. Orient., 1: 64 (1867); Post, Fl. Pal., ed. 2, 1: 18 (1932); Rechinger f. in Arkiv för Bot., 1: 421 (1950); Davis, Fl. Turkey, 1: 104 (1965); Zohary, Fl. Palaest., 1: 192, t. 279 (1966); Osorio-Tafall et Seraphim, List Vasc. Plants Cyprus, 40 (1973).
TYPE: " Dans l'Orient " coll. *André Michaux* (P).

Very similar to *N. nigellastrum* in habit and vegetative characters; flowers terminal, solitary; sepals greenish, ovate-acuminate, carinate, about 5 mm. long, slightly scabrid; petals 5, glabrous or very thinly pilose, white blotched crimson; claw slender, 8–9 mm. long; limb 2-lipped, outer lip broadly cordate, about 6 mm. long and 5 mm. wide, divided into 2 broadly lanceolate lobes; inner lip minute, about 1 mm. long; stamens numerous; anthers yellow or tinged crimson; ovary densely verruculose. Fruit 2–3-locular, similar to that of *N. nigellastrum*; seeds blackish or dark brown, strongly rugose, distinctly angular.

HAB.: Cultivated fields and waste ground; 3,500–4,300 ft. alt.; fl. May–June.

DISTR.: Division 2, rare. Also in Syria, Palestine, Turkey and Iraq.

2. Kykko, 1913, *Haradjian* 949 !

SUBGENUS 2. **Nigella** Fruit 2–14-locular, dehiscing along the internal sutures; petals inconspicuous, much smaller than the sepals; styles conspicuous, becoming hard and beak-like in the fruit.

3. N. arvensis *L.*, Sp. Plant., ed. 1, 534 (1753); Boiss., Fl. Orient., 1: 65 (1867); Holmboe, Veg. Cypr., 77 (1914); Post, Fl. Pal., ed. 2, 1: 19 (1932); Davis, Fl. Turkey, 1: 100 (1965); Zohary, Fl. Palaest., 1: 192, t. 280 (1966); Arne Strid in Opera Bot., no. 28, 1–169 (1970); Osorio-Tafall et Seraphim, List Vasc. Plants Cyprus, 40 (1973).

Erect or spreading annual (5–) 12–30 (–40) cm. high, with a tough taproot; stems glabrous, distinctly ridged, smooth or slightly scabrid, ascending or divaricate; leaves irregularly pinnatisect, with narrow, linear, acuminate segments, about 3 cm. long and 2 cm. wide, but often much less, shorter than the internodes or equalling them, green or sometimes glaucescent, smooth or scabridulous; basal leaves with a distinct petiole, stem leaves sessile or almost so; flowers solitary, sometimes distinctly pedunculate, or again with a short, leafy involucre; sepals greenish, whitish or pale blue, distinctly veined, glabrous, or with the veins scabridulous, broadly ovate or cordate, about

1 cm. long, with a cuspidate apex, and with the base narrowed abruptly into a claw about 5 mm. long; petals 5–8, whitish or yellowish with reddish or purple horizontal stripes, glabrous or thinly pilose, gibbous towards the base; 2-lipped; outer lip about 4 mm. long, with 2 deeply cleft, ovate, convex lobes, each lobe terminating in an elongate, filiform appendage; inner lip about 2 mm. long, ovate, with a long, filiform apical appendage; claw about 2 mm. long; stamens numerous, 7–10 mm. long, slightly shorter than the carpels, with narrow-oblong, yellow or purple-tinged anthers about 1·5–2 mm. long and an excurrent, cuspidate or aristate connective. Fruit narrowly obconical, consisting of 3–8 (usually 5) partly fused, smooth or verruculose carpels 1·5–2·5 cm. long, strongly 3-nerved dorsally; beaks erect or spreading, about 1 cm. long; seeds trigonous, black, finely papillose.

var. **glauca** *Boiss.*, Fl. Orient., 1: 66 (1867); Davis, Fl. Turkey, 1: 101 (1965).
N. tuberculata Griseb., Spic. Fl. Rum. et Bith., 1: 317 (1813).
N. arvensis (vix L.) Sibth. et Sm., Fl. Graec., 6: 9, t. 512 (1826).
N. arvensis L. var. *microcarpa* Boiss., Fl. Orient., 1: 66 (1867); Davis, Fl. Turkey, 1: 102 (1965).
[*N. arvensis* L. ssp. *cretica* (non *N. cretica* Mill.) Holmboe, Veg. Cypr., 77 (1914).]
TYPE: Turkey; " in cultis siccis et collibus . . . Armeniae ", *Calvert* (K !).

Stems spreading; leaves glaucescent, usually shorter than the internodes; segments rather short and broad; flowers rather small, fruits strongly verruculose.

HAB.: Fields, waste places etc. at low altitudes; fl. April–Aug.

DISTR.: Division ? Eastern Mediterranean region, to Iran and Iraq. Cyprus, 1787, *Sibthorp*.

NOTES: Not since recorded from Cyprus, but may yet be refound in cultivated ground or on pathsides.

4. **N. fumariifolia** *Kotschy* in Unger et Kotschy, Die Insel Cypern, 319 (1865); Boiss., Fl. Orient., 1: 66 (1867); Holmboe, Veg. Cypr., 77 (1914); Lindberg f., Iter Cypr., 16 (1946); Davis, Fl. Turkey, 1: 103 (1965); Osorio-Tafall et Seraphim, List Vasc. Plants Cyprus, 40 (1973).
N. fumariifolia Kotschy var. *normalis* Terracino in Bol. Ort. Bot. Palermo, 2: 40 (1898).
[*N. arvensis* L. var. *microcarpa* (non Boiss.) Sintenis in Oesterr. Bot. Zeitschr., 32: 293 (1882).]
[*N. stellaris* (non Boiss.) H. Stuart Thompson in Journ. Bot., 44: 275 (1906).]
TYPE: Cyprus; " in den Thalern bei Episcopi ", *Kotschy* 655 ! (also cited in original description " am Dorfe Fini ober Omodos ", May 17, 1862, *Kotschy* 741 !) (W, K ! etc.).

Spreading, ascending or prostrate annual 2–12 cm. high, with a tough tap-root; stems several, branched from the base, glaucescent, angular, glabrous; radical leaves distinctly petiolate, 2–7 cm. long, glabrous, glaucous, pinnately divided into 5–9 cuneate segments, each segment deeply lobed, with lanceolate or linear-lanceolate, obtuse, acute or acuminate lobes; stem leaves similar to radical leaves but sessile or subsessile, about 1–1·5 cm. long, much shorter than the internodes; flowers solitary, terminal, with a leafy involucre; sepals greenish, whitish or pale blue with greenish veins, glabrous, ovate or subcordate, 7–10 mm. long and about 5 mm. wide, with a blunt or mucronate apex, and with the base narrowed abruptly into a slender claw 2–4 mm. long; petals 5–8, yellowish tinged purple, glabrous or sparsely pilose, similar in structure and size to those of *N. arvensis* L.; stamens 5–8 mm. long, distinctly shorter than the carpels, with narrow-oblong, crimson-purple anthers 1·5–2 mm. long, and a muticous or slightly excurrent connective. Fruit star-shaped, carpels 5, fused for about 5–8 mm. from the base, verruculose; beaks patent, 1·5–2·5 cm. long, strongly verruculose on the basal part of the upper (inner) surface, smooth or sulcate above.

HAB.: Fallow fields and grassy hillsides on calcareous and igneous formations; sea-level to 2,000 ft. alt.; fl. April–May.

DISTR.: All Divisions, common in Divisions 3 and 7. Also in Rhodes, Karpathos, Khasos, Paros and Antiparos.

1. Dhrousha (Akamas), 2 miles S. of village, 1941, *Davis* 3229 ! Akamas, ? 1948, *Kennedy* 1599 !
2. Phini below Troödhitissa Monastery, 1862, *Kotschy* 655, 741 ! Anadhiou, 1901, *A. G. & M. E. Lascelles* s.n. ! Pano Panayia, 1905, *Holmboe* 1046 !
3. Common; *Kotschy*; *Holmboe* 599; *Syngrassides* 981 ! *Davis* 3446 ! *Chapman* 563 !
4. Cape Greco, 1905, *Holmboe* 417, and, 1958, *N. Macdonald* 60 ! 101 !
5. Between Lefkoniko and Ardhana, 1880, *Sintenis & Rigo* 80 ! Ayios Varnavas Monastery, 1948, *Mavromoustakis* s.n. ! Athalassa, 1950, *Chapman* 640 !
6. Vizakia, 1936, *Syngrassides* 1229 !
7. Common; *Davis* 3618 ! *Kennedy* 1648 ! *Chapman* 683 ! *Probyn* 91 ! etc.
8. Cape Andreas, 1880, *Sintenis & Rigo* 79 ! Rizokarpaso, 1912, *Haradjian* 135 ! Cape Andreas, 1948, *Mavromoustakis* s.n. !

NOTES: The interior lobe of the petal is sometimes aristate, and not always mucronate as figured by Holmboe (Veg. Cypr., 78); the other characters he gives will, however, serve to distinguish *N. fumariifolia* from the closely allied *N. stellaris* Boiss.

5. N. sativa *L.*, Sp. Plant., ed. 1, 534 (1753); Boiss., Fl. Orient., 1: 68 (1867); Post, Fl. Pal., ed. 2, 1: 20 (1932); Lindberg f., Iter Cypr., 16 (1946); Davis, Fl. Turkey, 1: 103 (1965); Osorio-Tafall et Seraphim, List Vasc. Plants Cyprus, 40 (1973).
TYPE: " *in* Aegypto, Creta ".

Erect pubescent annual, 20–40 cm. high, with a tough tap-root; branches ascending, distinctly furrowed; radical leaves 6–8 cm. long, with a petiole about 2–3 cm. long, pinnately divided, with 3–9 deeply dissected lobes; ultimate segments narrowly lanceolate, acute or acuminate; stem-leaves similar to radical leaves but sessile or subsessile, 2–6 cm. long, as long as or longer than the internodes; flowers solitary, terminal; sepals whitish or pale blue, tinged green, glabrous or sparsely pubescent, obovate, 10–15 mm. long, 5–8 mm. wide, with a blunt or mucronate apex and a short basal claw 2–3 mm. long; petals usually about 8, similar in size and structure to those of *N. arvensis* L., but with the lobes of the outer and inner lips acuminate and not filiform or aristate; stamens 8–10 mm. long, as long as or longer than the carpels, with narrow-oblong, yellow anthers 1·5–2 mm. long, connective scarcely excurrent. Fruit broadly ovoid, truncate above, 1–1·5 cm. long, about 1 cm. wide, carpels 4–8, connate almost to the apex, sparsely verruculose; beak slender, about 1 cm. long, erect; seeds trigonous, black, minutely papillose.

HAB.: Cultivated fields and waste ground; sea-level to c. 1,000 ft. alt.; fl. April–Aug.

DISTR.: Divisions 2 and 7, rare. Throughout W. Asia, from Turkey to India, usually cultivated or subspontaneous.

2. Kambos, 1939, *Lindberg f.* s.n. !
7. Lapithos, 1880, *Sintenis & Rigo* 619 ! Ayios Paulos near Lapithos, 1901, *A. G. & M. E. Lascelles* s.n. ! Kythrea, 1937, *Syngrassides* 1653 ! Ayios Epiktitos, 1951, *Casey* 1128 !

NOTES: Seeds used as a condiment.

6. N. damascena *L.*, Sp. Plant., ed. 1, 534 (1753); Boiss., Fl. Orient., 1: 68 (1867); Holmboe, Veg. Cypr., 77 (1914); Post, Fl. Pal., ed. 2, 1: 20 (1932); Davis, Fl. Turkey, 1: 103 (1965); Osorio-Tafall et Seraphim, List Vasc. Plants Cyprus, 40 (1973).
[*N. elata* (non Boiss.) Kotschy in Unger et Kotschy, Die Insel Cypern, 319 (1865).]
TYPE: " *inter segetes* Europae *australis* ".

An erect or spreading, glabrous annual, 20–40 cm. high, with a tough tap-root; stem distinctly sulcate; radical leaves 6–11 cm. long, with a petiole 3–5 cm. long, pinnately divided, with 5–13 deeply dissected lobes; ultimate segments linear-lanceolate, acuminate; stem leaves progressively smaller from the lower part of the stem upwards, usually shorter than the internodes, the uppermost sessile or subsessile, with narrow-linear or almost filiform, scabridulous segments; flowers solitary, terminal, with a large, spreading, leafy involucre of pinnately compound, filiform segments; sepals blue or white, minutely puberulous or papillose, obovate, 10–15 mm. long, 7–10 mm. wide, with a blunt or subacute apex, and a short basal claw 2–3 mm. long; petals about 8, purplish-bronze, pilose, similar in size and shape to those of *N. arvensis* L., but with the lobes of the outer and inner lips broadly ovate, and not filiform or aristate; stamens relatively few, about 1 cm. long; anthers shortly oblong, about 1–1·5 mm. long, greenish; connective not excurrent. Fruit broadly ovoid or subglobose, bladder-like, truncate above, 2–3 cm. long, about 2 cm. wide, carpels 5–6, connate almost to the apex, smooth; beak 1·5–2 cm. long, slender, erect or spreading; seeds obscurely angled, black, transversely wrinkled, papillose.

HAB.: Cultivated fields and waste ground, by pathsides etc.; sea-level to 2,000 ft. alt.; fl. April–June.

DISTR.: Divisions 1, 3, 7, common in Division 1. S. Europe, western Asia, N. Africa.

1. Common; *Holmboe* 729; *Post*; *Davis* 3219 ! *Kennedy* 1600 ! *Merton* 2166 !
3. Between Kouklia and Paphos, 1862, *Kotschy* 655 (as *N. elata* Boiss.); Episkopi near Limassol, 1954, *Merton* 1885 !
7. Between Kalogrea and Lefkoniko, *Sintenis & Rigo* 537 ! Kyrenia, 1922, *J. A. Tracey* 20 ! Karmi, 1948, *Kennedy* 1649 ! 1650 ! Ayios Epiktitos 1955, *G. E. Atherton* 231 !

7. **N. ciliaris** *DC.*, Regn. Veg. Syst. Nat., 1: 327 (1817); Boiss., Fl. Orient., 1: 70 (1867); Holmboe, Veg. Cypr., 79 (1914); Post, Fl. Pal., ed. 2, 1: 21 (1932); Zohary, Fl. Palaest., 1: 194, t. 281 (1966); Osorio-Tafall et Seraphim, List Vasc. Plants Cyprus, 40 (1973).

TYPE: Lebanon; " in Oriente prope Barut, *Labillardière* " (Fl).

Thinly pubescent annual, 12–40 (–50) cm. high with a tough tap-root; stems erect or spreading, sulcate; leaves very similar to those of *N. damascena* L., with the ultimate segments rather broader and distinctly scabrid-pubescent; flowers solitary, terminal, usually pedunculate, without an involucre; sepals yellowish, pilose, 10–15 mm. long, 6–8 mm. wide, ovate-acuminate, with a long, slender acumen, narrowed to the base, or very shortly clawed; petals 5–8, about 7 mm. long, pilose, 2-lipped; outer lip 2-lobed, each lobe with a short outer tooth and an elongate, linear or filiform inner tooth; inner lip ovate-acute, very much shorter than the outer; stamens numerous, about 7 mm. long, shorter than the carpels; anthers yellow, elongate-oblong, about 2 mm. long, with an excurrent, cuspidate, ± scabridulous connective. Carpels 5–13, connate for about $\frac{1}{3}$ their length, spreading or almost horizontal in fruit, about 2 cm. long and 1 cm. wide, thinly pilose and closely striate, with slender, straight or reflexed beaks about 1·5 cm. long; seeds flat, orbicular or almost so, about 6 mm. diam., with a shining, black, finely reticulate centre and a narrow, undulate, brownish-black wing.

HAB.: Cultivated fields and waste ground; sea-level to 500 ft. alt.; fl. April–May.

DISTR.: Divisions 3, 8, rare. Also in Syria and Palestine.

3. Amathus, 1862, *Kotschy* 592.
8. Between Rizokarpaso and Galinoporni, 1880, *Sintenis & Rigo* 78 ! Near Rizokarpaso, 1912, *Haradjian* 22.

8. DELPHINIUM *L.*

Sp. Plant., ed. 1, 530 (1753).
Gen. Plant., ed. 5, 236 (1754).
E. Huth in Engl. Bot. Jahrb., 20: 322 (1895).

Annual or perennial herbs with palmately, pinnately or irregularly lobed or dissected leaves. Flowers zygomorphic, borne in racemes or panicles;

sepals 5, blue, purple, pink or white, the 2 uppermost usually connate towards the base and forming a hollow spur. Petals usually 4, sometimes fewer, small, free or connate, the 2 uppermost with spur-like appendages enclosed within the calyx-spur; stamens numerous. Fruit consisting of 1–5 free or partly fused follicles.

About 200 species in north temperate regions.

Spur slender, tapering, as long as, or longer than sepals:
 Leaves divided into numerous, filiform segments; carpels solitary:
  Sepals generally 7–12 mm. long; flowers small, often pallid - **1. D. phrygium**
  Sepals generally 12–18 mm. long; flowers showy, often bright blue or pink **2. D. ambiguum**
 Leaves not divided into filiform segments; carpels several (often 3); plant annual or perennial:
  Lower leaves irregularly lobed, upper usually simple; petals glabrous; plant annual **3. D. peregrinum**
  Lower and upper leaves palmatilobed; petals hairy; plant perennial **4. D. caseyi**
Spur short, blunt, much shorter than the sepals; leaves large, palmatilobed; flowers large - - - - - - - - - - **5. D. staphisagria**

SUBGENUS 1. **Consolida** (*DC.*) *Dalla Torre et Harms* Carpels solitary; petals connate; seeds triquetrous, testa with transverse scales.

1. D. phrygium *Boiss.* in Ann. Sci. Nat., ser. 2, 16: 363 (1841), Fl. Orient., 1: 79 (1867); Holmboe, Veg. Cypr., 77 (1914).
Consolida phrygia (Boiss.) Soó in Oesterr. Bot. Zeitschr., 71: 245 (1922); Davis, Fl. Turkey, 1: 125 (1965); Osorio-Tafall et Seraphim, List Vasc. Plants Cyprus, 41 (1973).
TYPE: Turkey; Phrygia, *Aucher-Éloy* 71 (P).

Erect, sparsely branched annual to about 1 m. high; stems slightly sulcate, pubescent; leaves pubescent, the lowermost distinctly stalked, usually withered at time of flowering; the upper sessile or subsessile, repeatedly and irregularly dissected into narrowly linear or filiform segments, forming spreading tufts 3–6 cm. long, as long as or rather shorter than the internodes; inflorescence an unbranched or sparsely branched, congested or rather lax raceme, about 7–12 cm. long; rhachis and pedicels glandular-pubescent; flowers purple, mauve or whitish, rather small; pedicels 3–4 mm. long; bracteoles linear-subulate, those of the lower flowers about as long as, or slightly longer than the pedicel; sepals 7–12 mm. long, about as long as the spur, narrowly obovate-acute, pubescent; petals glabrous, rather longer than the sepals; stamens about 12, filaments 4–6 mm. long, minutely puberulous, distinctly flattened and dilated towards the base; anthers rounded-oblong, about 1 mm. long, yellow; ovary about 4 mm. long, narrowly ovoid, pubescent; style and stigma short, narrowly winged. Fruit 1–1·5 cm. long, densely pubescent; seeds dark brown or blackish with conspicuous transverse scales.

HAB.: Cultivated fields and waste places; c. 3,500 ft. alt.; fl. June–July.

DISTR.: Division 2, rare. Also Greece, Turkey, Thasos and Rhodes.

2. In fields near Spilia at Mt. Adelphi, 1880, *Sintenis & Rigo* 852 !

2. D. ambiguum *L.*, Sp. Plant., ed. 2, 749 (1762).
Consolida ambigua (L.) P. W. Ball in Feddes Repert., 66: 151 (1962); Davis, Fl. Turkey, 1: 126 (1965).
[*D. ajacis* (non L.) Boiss., Fl. Orient., 1: 79 (1867); Post, Fl. Pal., ed. 2, 1: 21 (1932).]
TYPE: "*in* Mauritania".

Erect annual, 20–60 (–100) cm. high; stems simple or branched above, thinly crispate-pubescent, often stained purple towards base, rather densely leafy; branches divaricate or suberect; leaves broadly ovate in outline, 2–6 (or more) cm. long and almost as wide, pinnately dissected into numerous narrowly linear or filiform segments, the basal with a slender petiole 2–3 cm. long, the upper sessile or subsessile, and equalling or exceeding the internodes; inflorescence a lax terminal raceme 8–20 (or more) cm. long, rhachis and pedicels subglabrous or ± densely glandular-pubescent; bracteoles linear-subulate, attached at or below the middle of the pedicel, 3–5 mm. long, generally falling short of the pedicel-apex; pedicels (0·5–) 1–1·5 (–3) cm. long, ascending or suberect; flowers blue, pink or white, showy; sepals ovate, 12–18 mm. long, the 2 lateral 10–14 mm. wide, the others seldom exceeding 8 mm. in width, glabrous or subglabrous internally, thinly crispate-pubescent externally, apex acute or obtuse, base narrowed rather abruptly to a very short claw; petals glabrous, usually rather pallid, equalling or a little longer than the sepals; spur straight or distinctly upcurved, 15–20 mm. long, tapering to apex; stamens 12–15, filaments about 6 mm. long, glabrous or sparsely ciliate, strongly flattened and dilated towards the base; anthers broadly oblong or suborbicular, about 1·2 mm. long, greenish-yellow; ovary about 4 mm. long, narrowly ovoid, densely pubescent; style short, narrowly winged. Fruit to 20 mm. long, 4 mm. wide, very shortly and sparsely pubescent, obscurely venulose; seeds angular, about 2 mm. long and almost as wide, testa blackish with conspicuous transverse scales.

HAB.: Cultivated fields; c. 500 ft. alt.; fl. May–June.

DISTR.: Division 3, rare, possibly introduced. Mediterranean region, but introduced and naturalized elsewhere.

3. Anglisidhes, by margin of barley-field, 1962, *Meikle* 2959 !

SUBGENUS 2. **Delphinium** Carpels several; petals 4, free.

3. **D. peregrinum** *L.* Sp. Plant., ed. 1, 531 (1753); Boiss., Fl. Orient., 1: 87 (1867); Holmboe, Veg. Cypr., 77 (1914); Post, Fl. Pal., ed. 2, 1: 24 (1932); Davis, Fl. Turkey, 1: 117 (1965); Zohary, Fl. Palaest., 1: 196, t. 283 (1966); Osorio-Tafall et Seraphim, List Vasc. Plants Cyprus, 40 (1973).

Erect or spreading, much-branched annual (12–) 20–50 cm. high; branches divaricate, glabrous or puberulous; leaves glaucous, subglabrous or pubescent, lower shortly petiolate, 3–5 cm. long and 4–6 cm. wide, fan-shaped, 3-fid, with each segment pinnately or irregularly divided into linear, acute lobes, 10–15 mm. long and 2–4 mm. wide; upper leaves sessile or subsessile, 3-lobed or simple, with linear or linear-lanceolate segments or leaf-blades, 2–3 cm. long about the middle of the stem, becoming progressively smaller, and finally subulate and bract-like towards the apex; flowers violet-blue, or with the petals white or yellowish towards the base, forming lax or rather dense, ± branched racemes; pedicel 4–5 mm. long; bracts subulate, about as long as the pedicel, or a little longer; spur about 1·5 cm. long, tapering to a blunt or acute apex; sepals 5–8 mm. long, narrowly ovate, pubescent externally; petals glabrous, lateral pair 7–9 mm. long, each with a narrowly obovate, blunt lamina, tapering gradually to a slender claw 4–5 mm. long; upper petals blunt or rounded, about 5 mm. long (excluding spur) with a sharp, subulate apical lobe 3–4 mm. long; stamens 12–15, about 5 mm. long; filaments dilated towards base; anthers about 1 mm. long, rounded-oblong; receptacle glabrous; carpels 3, shorter than the stamens, glabrous or pubescent. Follicles 5–8 mm. long, connivent, pale brown, glabrous or thinly hairy; seeds dull brown, with numerous transverse, membranous scales.

var. **eriocarpum** *Boiss.*, Fl. Orient., 1: 87 (1867); Lindberg f., Iter Cypr., 16 (1946).

D. eriocarpum (Boiss.) Hal., Consp. Fl. Graec., 1: 32 (1900); Post, Fl. Pal., ed. 2, 1: 25 (1932).

D. peregrinum L. ssp. *eriocarpum* (Boiss.) Holmboe, Veg. Cypr., 77 (1914).

TYPE: U.S.S.R.; "in provinciis transcaucasiis versus fines turcicas", *Nordmann* (LE).

Stems and leaves closely crispate-pubescent; carpels sparsely pilose with long white hairs.

HAB.: Cultivated fields, roadsides, grassy steppes and stony hillsides; sea-level to 3,200 ft. alt.; fl. April–July.

DISTR.: Divisions 1–5, 7, common in Divisions 2 and 7. Also Greece, Turkey, Syria, Palestine, Iraq.

1. Akamas, 1905, *Holmboe* 784, and 1948, *Kennedy* 1645 ! Lyso ["Lisso"], 1913, *Haradjian* 865 !
2. Common; *Sintenis & Rigo* 850 ! *Holmboe* 1045 ! *Kennedy* 444 ! 445 ! *Syngrassides* 1671 ! *Davis* 1841 ! *Merton* 2819 !
3. Trimiklini towards Limassol, 1935, *Syngrassides* 751 !
4. Larnaca aerodrome, 1950, *Chapman* 702 ! Between Cape Greco and Ayia Napa, 1958, *N. Macdonald* 75 !
5. Mia Milea, 1950, *Chapman* 678 !
7. Common; *A. G. & M. E. Lascelles* s.n. ! *Lindberg f.* s.n. ! *Casey* 880 ! *G. E. Atherton* 93 ! 248 ! 249 ! 404 !

NOTES: All the specimens seen have pilose follicles and fall under var. *eriocarpum* Boiss., the prevailing variety in the Middle East.

4. **D. caseyi** *B. L. Burtt* in Kew Bull., 1954: 67 (1954); Osorio-Tafall et Seraphim, List Vasc. Plants Cyprus, 40 (1973).

TYPE: Cyprus; Northern range, rocky summit S.W. of Hilarion Castle, June 18, 1950, *Casey* 1054 (K !).

Erect perennial with a thickened, rather woody rootstock; stem 50–85 cm. high, pubescent; radical leaves mostly withered at time of flowering, distinctly stalked, with a petiole up to 20 cm. long; lamina reniform, thinly pubescent or subglabrous above, ± crispate-pubescent below, about 5·5–7·5 cm. long and 10–14 cm. wide, divided almost to the base into 3 segments, with each segment deeply 2 or 3-times lobed, the ultimate lobes linear-oblong, apiculate; stem leaves smaller and more shortly stalked, the uppermost sometimes divided into 3–5 linear-lanceolate, entire or sparsely lobed segments; inflorescences terminal and axillary, forming simple or sparingly branched, rather dense-flowered racemes up to 30 cm. long; rhachis densely crispate-pubescent; bracts linear, the lower about 1 cm. long, the uppermost minute, scarcely more than 3 mm. long; pedicels 5–7 mm. long, densely pubescent, thickened and obconical near the apex; flowers violet or violet-purple, sepals shortly pubescent externally, about 8–10 mm. long, spur about 12 mm. long, rather blunt, pubescent; petals notched or shortly and unequally 2-lobed at apex, lateral pair about 9 mm. long and 4 mm. wide, with apical tufts of long white hairs, and with pubescent claws about 4 mm. long and 1·5 mm. wide; upper petals axe-shaped, about 6 mm. long, glabrous, with a slender spur about 12 mm. long; stamens 40–50, the outer 6 mm. long, with gradually tapering filaments; the inner rather longer, with the filaments dilated at the base; filaments purplish towards apex, pale below, thinly and shortly pilose; anthers yellowish tinged purple, oblong, about 1 mm. long, minutely pilose; carpels 3, glabrous or very thinly pilose, about 3 mm. long. Follicles glabrous; seeds dark brown with membranous scales.

HAB.: In fissures on the summits of high limestone mountains; c. 3,000 ft. alt.; fl. June–July.

DISTR.: Division 7, rare. Endemic.

7. Northern Range, rocky summit S.W. of Hilarion Castle, 1950, *Casey* 1054 ! Kyparissovouno, 1962, *Casey* in *Meikle* 2426 !

5. **D. staphisagria** *L.*, Sp. Plant., ed. 1, 531 (1753); Boiss., Fl. Orient., 1: 94 (1867); Holmboe, Veg. Cypr., 77 (1914); Post, Fl. Pal., ed. 2, 1: 26 (1932); Davis, Fl. Turkey, 1: 117 (1965); Osorio-Tafall et Seraphim, List Vasc. Plants Cyprus, 40 (1973).

TYPE: "*in* Istria, Dalmatia, Calabria, Apulia, Creta, Galloprovincia".

Stout, erect biennial, 30–100 cm. high; rootstock woody; stem unbranched or very sparingly branched, distinctly sulcate, covered with long, spreading, silky hairs; leaves distinctly stalked, those near the base of the stem with petioles up to 15 cm. or more in length, lamina broadly reniform or almost orbicular, 10–25 cm. diam., pubescent on both surfaces, deeply divided into 5–7 sharply lobed, palmate or pedate segments; upper leaves smaller, often consisting of 5 narrow, linear-lanceolate, entire segments. Inflorescences terminal and axillary, rather lax; rhachis densely villose; lower bracts 3-fid, with linear segments 5–10 mm. long, upper bracts frequently simple; pedicels 1–2 cm. long, densely villose; flowers dark or bright blue, rather large, sepals obovate, obtuse, pubescent externally, about 1·5 cm. long; spur very short, 2–4 mm. long, truncate at the apex; petals connate at the base, slightly notched at the apex, whitish or tinged blue, the lateral pair narrowly obovate, about 1·5 cm. long, minutely papillose and thinly pubescent, the upper pair smooth, glabrous or very thinly pubescent, with short, blunt spurs; stamens about 40, the longest 8–9 mm. long, with yellowish or bluish-tinged, oblong anthers about 1·5 mm. long; filaments bluish-tinged towards apex, pallid and distinctly dilated towards base; carpels 3, densely villose, about 5 mm. long. Follicles 15–20 mm. long, pale brown, somewhat inflated, thinly hairy, containing a few (1–5) large, brown, reticulate-foveolate seeds about 5–8 mm. long.

HAB.: Rocky slopes and screes on calcareous or igneous formations; 500–3,500 ft. alt.; fl. May–June.

DISTR.: Divisions 2, 7, 8, rare. Common in many parts of the Mediterranean region, also Spain, Portugal and Atlantic Islands.

2. Moutoullas [" Mudulla "] in the Marathasa valley, 1905, *Holmboe* 999.
7. Phterykha village, 6 m. W. of Kyrenia, 1955, *Casey* 1675 !
8. Yioti near Ephtakomi, 1962, *Meikle* 2517 !

5. PAEONIACEAE

Herbs or shrubs with spirally arranged, exstipulate, compound leaves. Flowers hermaphrodite, terminal, mostly solitary, actinomorphic, showy; sepals about 5, free, unequal, often passing into petals; petals 5–13, free; stamens numerous, hypogynous, free or shortly fused at the base, spirally arranged; carpels 2–5, free, fleshy, inserted on a fleshy disk; ovules several, with thick, well-developed integuments. Fruit a thick-walled follicle, often hairy.

The family comprises one genus only.

1. PAEONIA *L.*

Sp. Plant., ed. 1, 530 (1753).
Gen. Plant., ed. 5, 235 (1754).

Characters those of the family. The genus includes about 50 species, natives of central and southern Europe, temperate Asia and western N. America.

1. **P. mascula** (*L.*) *Mill.*, Gard. Dict., ed. 8, no. 1 (1768); Davis, Fl. Turkey, 1: 204 (1965); Zohary, Fl. Palaest., 1: 220, t. 323 (1966); Osorio-Tafall et Seraphim, List Vasc. Plants Cyprus, 42 (1973); Elektra Megaw, Wild Flowers of Cyprus, 8, t. 5 (1973).

P. officinalis L. var. *mascula* L., Sp. Plant., ed. 1, 530 (1753).

P. corallina Retz., Obs. Bot., 3: 34 (1783); Unger et Kotschy, Die Insel Cypern, 319 (1865); Boiss., Fl. Orient., 1: 97 (1867); Holmboe, Veg. Cypr., 76 (1914); Post, Fl. Pal., ed. 2, 1: 26 (1932); Lindberg f., Iter Cypr., 16 (1942).

P. corallina Retz. var. *orientalis* Thiébaut, Mém. Inst. d'Égypte, 31: 37 (1936).

P. arietina Anders. var. *orientalis* (Thiébaut) F. C. Stern, A Study of the Genus *Paeonia*, 81 (1946).

[*P. decora* (non Anders.) Boiss., Fl. Orient., 1: 98 (1867) pro parte quoad plantam cypriam.]

[*P. corallina* Retz. var. *triternata* (non (Pall.) Boiss.) Post in Mém. Herb. Boiss., 18: 90 (1900); H. Stuart Thompson in Journ. Bot., 44: 275 (1906).]

TYPE: Probably the illustration of *Paeonia mas* in Lobel's *Plantarum seu Stirpium Icones*, 684, t. 832 (1581).

Erect, glabrous or pilose perennial herb; roots thick, woody; stem distinctively sulcate, 30–60 cm. high, pale green stained rose or purple-red, arising from 4–7 lingulate, chartaceous, purplish basal sheaths, 1–7 cm. long and 1–1·5 cm. wide; leaves 4–7, irregularly ternate or biternate, about 25 cm. long, ultimate segments 12–21, elliptic, oblong or elliptic-lanceolate, often distinctly asymmetrical, 6–14 cm. long, 2·5–7 cm. wide, bright green tinged purple or pink, glabrous above, pilose or glabrous below, margins entire, apex acute or cuspidate; petioles 5–12 cm. long, petiolules variable, up to 11 cm. long; flower solitary, shallowly cup-shaped, sweetly scented, 6·5–10 cm. diam.; peduncle 3–12 cm. long; sepals 3–5, very unequal and irregular, 1–3·5 cm. long, obtuse or acuminate, flat or concave, frequently cucullate or foliose; petals 5–6, pale magenta-red or rosy-purple, concave, broadly obovate or orbicular, 2·5–5 cm. long, 2–5 cm. wide, margins frequently erose; stamens numerous, filaments 5–11 mm. long, magenta-purple, glabrous; anthers narrowly oblong, 3–5 mm. long; carpels 2–3, densely lanuginose, about 1·5 cm. long; stigmas almost sessile, flattened, magenta-purple, often twisted circinnately. Follicle lanuginose, about 3 cm. long, 1–1·5 cm. wide, recurved at maturity; seeds black, shining, about 5–6 mm. long.

HAB.: In Pine forest or on shaded screes at high altitudes; 4,500–5,700 ft. alt.; fl. April–June.

DISTR.: Division 2, locally common about Prodhromos, also on Madhari and Papoutsa, usually above 4,500 ft. alt. Europe eastwards to Iran.

2. Locally common; Prodhromos, *Kotschy* 701 ! *Sintenis & Rigo* 854 ! *C. B. Ussher* 61 ! *Syngrassides* 1570 ! *Kennedy* 446 ! 1624 ! etc. Papoutsa, *Post*, also *Davis* 3067 ! Madhari, *Davis* 1882 ! *Meikle* 2831 !

6. BERBERIDACEAE

Perennial herbs or shrubs; leaves alternate or radical, usually exstipulate, simple or compound. Flowers hermaphrodite, actinomorphic, hypogynous, solitary or in cymes or false racemes; perianth in 2 or more series; petals and sepals each 4–6, free, often undifferentiated, petals frequently nectariferous; stamens 4–18, usually in 2 whorls, the outer whorl opposite the petals; anthers 2-celled, dehiscing longitudinally or by flap-like valves. Ovary superior, 1- or rarely 2-locular, ovules usually several, ascending; style short or absent; fruit usually a berry; seed sometimes arillate, with copious endosperm; embryo small, straight.

About 12 genera chiefly in temperate regions of the northern hemisphere.

Shrubs with spinose branches; fruit a juicy berry - - - - - 1. Berberis
Unarmed herbs; fruit an inflated, irregularly dehiscing capsule:
Leaves pinnatisect; bracts brownish, scarious - - - - - 2. Bongardia
Leaves pinnately ternate, biternate or palmate; bracts herbaceous, foliose
3. Leontice

1. BERBERIS *L.*

Sp. Plant., ed. 1, 330 (1753).
Gen. Plant., ed. 5, 153 (1754).

Shrubs with smooth or spinose branches and yellow wood; leaves simple, often fasciculate and spinose-dentate. Flowers yellow, solitary, fasciculate or racemose, terminal or lateral on short, leafy shoots; sepals 6, petaloid; petals 6, each with 2 glands at the base; stamens 6, free, jerking inwards if touched at the base; anthers dehiscing by apical valves. Ovary 1-locular, with several ovules; fruit a berry.

About 160 species, found in Europe, Asia and S. America, with outlying species in Tropical Africa and N. America.

1. **B. cretica** *L.*, Sp. Plant., ed. 1, 331 (1753); Unger et Kotschy, Die Insel Cypern, 320 (1865); Boiss., Fl. Orient., 1: 103 (1867); Holmboe, Veg. Cypr., 82 (1914); Post, Fl. Pal., ed. 2, 1: 29 (1932); Lindberg f., Iter Cypr., 16 (1946); Davis, Fl. Turkey, 1: 210 (1965); Osorio-Tafall et Seraphim, List Vasc. Plants Cyprus 43 (1973).
TYPE: " *in* Creta ".

Spreading, glabrous shrub about 1 m. high; branches reddish-brown, slightly sulcate, with peeling bark and sharp, rigid, 3-partite spines; central spine 1·5–2 cm. long, lateral spines slightly shorter; leaves in fascicles of 5–9, sessile, oblanceolate, 12–18 mm. long, 4–8 mm. wide, somewhat coriaceous, margins entire, slightly recurved; inflorescence 5–12-flowered, fasciculate or shortly racemose, pedicels 3–9 mm. long; corolla shallowly cup-shaped, 7–10 mm. diam.; sepals 6, the longest about 6 mm. long; petals 6, obovate, 5–7 mm. long, 2–3 mm. wide, with 2 dark-coloured glands on either side of the midrib towards the base of the claw; stamens usually 6, about 3·5–5 mm. long; anthers about 1 mm. long; ovary about 3 mm. long. Fruit an oblong or ovoid berry about 7–9 mm. long and 3–5 mm. wide, red, later blue-black, pruinose, crowned with the persistent stigma; seed clavate, dark brown, about 5 mm. long; testa slightly wrinkled. *Plate 4.*

HAB.: Pine forests and screes at high altitudes on the Troödos Range; 3,000–6,400 ft. alt.; fl. May–July.

DISTR.: Division 2, locally common on Khionistra, also on Madhari and Papoutsa, 3,000–6,400 ft. alt. Crete, Greece, Aegean Islands, Turkey.

2. Locally common; *Sibthorp*; *Kotschy*; *Sintenis & Rigo* 699 ! *Holmboe*; *Syngrassides* 750 ! *Kennedy* 450 ! 451 ! 452 ! 453 ! *Lindberg f.* s.n. ! *Davis* 3509 ! etc.

2. BONGARDIA *C. A. Mey.*

Verz. Pflanz. Cauc., 174 (1831).

Rhizomatous perennial herb with radical, pinnatisect leaves. Flowers yellow, in loose leafless panicles; sepals 6, concave, obtuse; petals 6, obovate, imbricate, with a small nectary-pore at the base, but without a nectary-scale; stamens 6, opposite the petals; anthers dehiscing by valves; ovary 1-locular, ovules usually 5–6, ascending; stigma plicate or slightly lobed. Fruit an inflated capsule dehiscing irregularly when ripe.

One species ocurring from the Aegean Islands eastwards to the north-western borders of India.

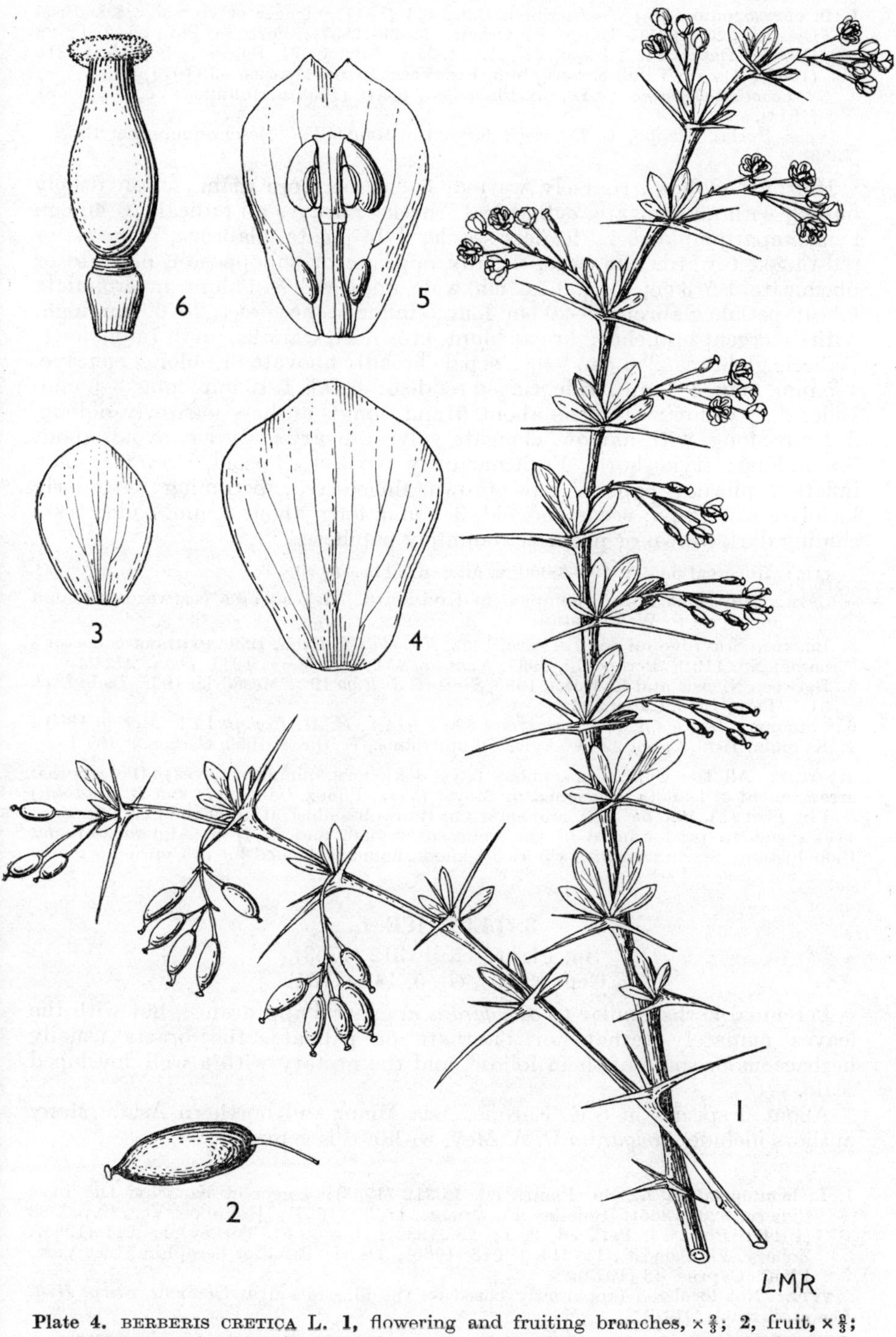

Plate 4. BERBERIS CRETICA L. **1**, flowering and fruiting branches, $\times \frac{2}{3}$; **2**, fruit, $\times \frac{2}{3}$; **3**, outer sepal, $\times 9$; **4**, inner sepal, $\times 9$; **5**, petal and stamen, $\times 9$; **6**, gynoecium, $\times 9$. (*Lindberg* f. s.n. and *Kennedy* 452.)

1. **B. chrysogonum** (*L.*) *Endl.*, Ench. Bot., 441 (1841); Unger et Kotschy, Die Insel Cypern, 320 (1865); Boiss., Fl. Orient., 1: 99 (1867); Post, Fl. Pal., ed. 2, 1: 28 (1932); Davis, Fl. Turkey, 1: 212 (1965); Zohary, Fl. Palaest., 1: 215, t. 316 (1966); Osorio-Tafall et Seraphim, List Vasc. Plants Cyprus, 43 (1973).

Leontice chrysogonum L., Sp. Plant., ed. 1, 312 (1753); Holmboe, Veg. Cypr., 82 (1914).

TYPE: Syria; Aleppo, *L. Rauwolf*; figured in Rauwolf's "Beschreibung der Raiss" (1583).

Rhizome globose, coarsely warted, 5 cm. or more diam., often deeply buried, with an elongate, cylindrical "neck"; leaves all radical, 20–40 cm. long, imparipinnatisect; leaflets sessile, 8–14-jugate, glabrous, glaucous or red-tinged towards the base, usually opposite or subopposite, obovate or obcuneate, 1·5–3 cm. long, 0·7–2 cm. wide, apex with 3–6 blunt, mucronulate lobes; petiole glabrous, 8–20 cm. long; inflorescence erect, 20–60 cm. high, with divergent branches; bracts blunt, brownish, scarious, up to 7 mm. long; pedicels glabrous, 2–6 cm. long; sepals broadly obovate or oblong, concave, 4–5 mm. long, 3–4 mm. wide, tinged reddish; petals 4–10 mm. long, 3–5 mm. wide, flat, yellow; stamens about 5 mm. long; anthers narrowly oblong, 2–4 mm. long, with narrow, elongate valves; ovary glabrous, ovoid, about 5 mm. long; style short. Fruit narrowly ovoid or ellipsoid, 1–1·5 cm. long, inflated, plicate, papery, pale brown tinged red, becoming irregularly laciniate when ripe; seeds obovoid, 3–4 mm. long, about 3 mm. wide; testa shining dark brown or pruinose, minutely papillose.

HAB.: In cereal fields; 100–3,000 ft. alt.; fl. Febr.–April.

DISTR.: Divisions 3, 5–7, common in Division 6. Aegean area eastwards through Turkey and Syria to W. Pakistan.

3. Between Stavrovouni and Lefkara, 1862, *Kotschy*; Silikou, 1930, (without collector's name) No. 119 ! Perapedhi, 1937, *Kennedy* 454 ! Lefkara, 1941, *Davis* 2724 !
5. Between Nicosia and Kythrea, 1880, *Sintenis & Rigo* 19 ! Mesaoria, 1938, *Lady Loch* 31 ! Pera, 1949, *Casey* 303 !
6. Common; *Druce* s.n. ! *Syngrassides* 890 ! 944 ! *F. M. Probyn* 14 ! *Merton* 1860 !
7. Kyrenia, 1949, *Casey* 298 ! Ayios Dhimitrianos, Kythrea, 1950, *Chapman* 167 !

NOTES: All the Cyprus specimens have distichous, pinnate leaves; the whorled arrangement of leaflets mentioned by Meyer (Verz. Pflanz. Cauc., 174 *sub B. rauwolfii*) and by Post (Fl. Pal. *loc. cit.*), and shown in Rauwolf's illustration of the plant (*loc. cit.*, 119) seems to predominate in the Caucasus region, and in Iran, Afghanistan and Baluchistan. Specimens with whorled leaflets should be looked for in Cyprus.

3. LEONTICE *L.*

Sp. Plant., ed. 1, 312 (1753).
Gen. Plant., ed. 5, 147 (1754).

Perennial herbs similar to *Bongardia* in general appearance, but with the leaves pinnately ternate or biternate or palmate, the bracts usually herbaceous or some of them foliose, and the nectary with a well-developed scale.

About 6 species in S.E. Europe, Asia Minor and northern Asia. Many authors include *Bongardia* C. A. Mey. within this genus.

1. **L. leontopetalum** *L.*, Sp. Plant., ed. 1, 312 (1753); Unger et Kotschy, Die Insel Cypern, 320 (1865); Boissier, Fl. Orient., 1: 99 (1867); Holmboe, Veg. Cypr., 82 (1914); Post, Fl. Pal., ed. 2, 1: 28 (1932); Davis, Fl. Turkey, 1: 211 (1965); Zohary, Fl. Palaest., 1: 214, t. 315 (1966); Osorio-Tafall et Seraphim, List Vasc. Plants Cyprus, 43 (1973).

TYPE: Not localized (apparently based on the illustration in Dodoens, *Stirp. Hist. Pempt. Sex*, 69; 1616).

Erect, glabrous perennial, 15–60 cm. high; rhizome globose or subglobose, coarsely warted or tuberculate, 5 cm. or more diam., often deeply buried;

leaves radical and cauline, 7–30 cm. long, glaucous, ternate, with the long-stalked divisions again ternate or biternate, or sometimes pinnate, ultimate divisions suborbicular or broadly obovate, entire, obtuse, emarginate or even 3-lobed, 2–8 cm. long and 1·5–6 cm. wide; petioles glabrous, 5–20 cm. long, petiolules 3–10 cm. long, leaflets sessile or shortly stalked; basal sheaths conspicuously ribbed, 2–5 cm. long; inflorescence broadly pyramidal or corymbose, 8–20 cm. long, 14–30 cm. wide; branches divaricate; basal bracts foliose, stalked, simple or ternate, up to about 15 cm. long; upper bracts simple, sessile, herbaceous, broadly obovate, 1–2 cm. long; flowers yellow; pedicels 7–20 mm. long or longer, lengthening considerably in fruit; sepals and petals together 6–8, obovate or oblong, 5–7 mm. long, 3–4 mm. wide; petals with a broadly obovate, suborbicular or retuse, convolute nectary-scale about 1–2 mm. long; stamens 6, opposite the innermost perianth segments, 3–4 mm. long; anthers 1–2 mm. long, with narrowly obovate valves; ovary 2–3 mm. long, style short, stigma truncate; ovules 2–4. Fruit obovoid or ellipsoid, 3–4 cm. long, 2–3 cm. wide, bladder-like, distinctly reticulate-veined, papery, rupturing irregularly when ripe; seeds globose 5–7 mm. diam.; testa dark brown, rather dull, or distinctly pruinose.

HAB.: In cereal fields; sea-level to 1,500 ft. alt.; fl. Febr.–April.

DISTR.: Divisions 1, 3–7, common in Divisions 6 and 7. S.E. Europe and western Asia, from Greece to Iran and Iraq.

1. Polis, 1933, *Nattrass* 303; Paphos, 1960, *C. E. H. Sparrow* 47 !
3. Near Stavrovouni, 1787, *Sibthorp*; Mazotos, 1905, *Holmboe*; Lefkara, 1935, *Nattrass* 565; Kophinou, 1937, *Nattrass* 845.
4. Near Larnaca, 1782, *Sestini*; Near Larnaca, near the Phaneromene, 1862, *Kotschy*; Phaneromene near Larnaca, 1905, *Holmboe* 139.
5. Between Nicosia and Kythrea, 1880, *Sintenis & Rigo* 20 ! Salamis, 1937, *Roger-Smith* s.n. !
6. Common; *Druce* s.n. ! *Syngrassides* 1406 ! *Lady Loch* 32 ! *Davis* 2131 ! *F. M. Probyn* 11 !
7. Common; *Chapman* 166 ! *G. E. Atherton* 1104 ! *Syngrassides* 1036 ! *Casey* 236 ! 758 ! *G. E. Atherton* 934 !

7. PAPAVERACEAE

Annual or perennial herbs (or rarely shrubs), often with milky or coloured juice. Leaves alternate or spirally arranged, exstipulate, usually deeply lobed or divided. Flowers solitary, or in racemes or cymes, actinomorphic, hypogynous or rarely perigynous, hermaphrodite, the parts arranged in whorls; sepals 2–3, usually free (sometimes calyptrate), caducous; petals 4–6 or 8–12, 2-seriate, free, imbricate and often crumpled in bud, rarely wanting; stamens usually numerous (4 only in *Hypecoum*), free; anthers 2-celled, dehiscing longitudinally; nectaries absent; ovary of 2 or more united carpels, 1-locular with parietal placentation; style often wanting (present in *Hypecoum*); stigmas as many as the carpels, opposite to or alternating with the placentas; ovules usually numerous. Fruit a capsule or lomentum (*Hypecoum*), opening by valves or pores, rarely indehiscent; seeds with a small embryo and oily or mealy endosperm.

About 30 genera, chiefly in temperate and subtropical regions of the northern hemisphere.

Fruit opening by small pores situated below the stigmatic disk, usually globose, ovoid or shortly cylindrical - - - - - - - - - - - **1. Papaver**

Fruit dehiscing longitudinally or transversely, elongate, linear or narrowly cylindrical:
Flowers solitary, terminal, axillary or leaf-opposed, usually rather large; petals all ± similar; stamens numerous:
Sepals 2, free, caducous:
Fruit opening by 2 valves; petals yellow, orange or scarlet - **2. Glaucium**
Fruit opening by 3–4 valves; petals violet or purple - - **3. Roemeria**
Sepals united into a pointed conical circumscissile cap or calyptra
4a. ESCHSCHOLZIA
Flowers in cymes, individually small; petals yellow or orange, the outer differing from the inner; stamens 4 - - - - - - - - - **4. Hypecoum**

1. PAPAVER *L.*

Sp. Plant., ed. 1, 506 (1753).
Gen. Plant., ed. 5: 224 (1754).
Fedde in Engl. Pflanzenr., 40 (IV, 104): 288 (1909).

Annual or perennial herbs with milky or yellowish latex; leaves generally pinnately dissected or lobed; flowers pedunculate, solitary or in raceme-like cymes, often showy; buds frequently nodding; sepals 2 (very rarely 3), concave, imbricate, caducous; petals 4 (rarely 5 or 6), obtuse, crumpled in bud; stamens numerous; ovary usually subglobose, ovoid or shortly cylindrical; style 0; stigmas sessile, radiating from the centre of a relatively broad, flat, convex or conical disk on the top of the ovary. Capsule subglobose, ovoid, clavate or cylindrical, crowned with the persistent stigmatic disk, dehiscing by short pores situated below the disk. Seeds small, reniform.

About 100 species, chiefly in temperate regions of the northern hemisphere.

Leaves deeply once or twice pinnatifid or pinnatisect, ± strigose; flowers red or purple-crimson:
Capsule glabrous:
Stems spreading or decumbent, much branched at the base; flowers small, petals 1–2 cm. long; capsule turbinate or broadly clavate - - **3. P. postii**
Stems erect or suberect ± branched above; flowers rather large, petals 2–4 cm. long, capsule subglobose to clavate:
Peduncles with spreading hairs, capsule obovoid-oblong, not twice as long as wide
1. P. rhoeas
Peduncles with adpressed hairs, capsule narrowly obovoid-oblong, more than twice as long as wide - - - - - - - - - **2. P. dubium**
Capsule ± bristly:
Capsule narrowly oblong or clavate, sparsely bristly with the bristles often confined to the upper part; peduncles conspicuously thickened in fruit **4. P. minus**
Capsule ovoid or ellipsoid, densely bristly; peduncles not conspicuously thickened in fruit - - - - - - - - - - - **5. P. hybridum**
Leaves shortly pinnatifid or irregularly toothed, subglabrous and ± glaucous; flowers pink or pale purple:
Capsule obovoid; peduncle with adpressed or ascending bristles - **6. P. setigerum**
Capsule clavate; peduncle glabrous or almost so - - - - - **7. P. gracile**

1. P. rhoeas *L.*, Sp. Plant., ed. 1, 507 (1753); Unger et Kotschy, Die Insel Cypern, 321 (1865); Boissier, Fl. Orient., 1: 113 (1867); Fedde in Engl., Pflanzenr., 40 (IV, 104): 293 (1909); Post, Fl. Pal., ed. 2, 1: 35 (1932); Osorio-Tafall et Seraphim, List Vasc. Plants Cyprus, 43 (1973).

Erect or spreading annual 10–30 cm. or more high; stems hispid, with patent (or rarely ± adpressed hairs); basal leaves narrowly obovate or oblanceolate, 7–20 cm. long, irregularly pinnatisect or pinnatipartite, hispid on both sides or subglabrous above; petiole 3–10 cm. long, narrowly winged; stem leaves similar, sometimes very shortly lobed, petiole short or almost absent; flowers solitary; peduncle slender, 10–20 cm. or more long, hispid, with patent (or rarely ± adpressed) hairs; buds nodding at first, later erect, ovoid, obtuse or rounded at apex, 10–25 mm. long, 5–20 mm. wide, usually hispid; petals broadly cuneate, 2–4 cm. long and 2–7 cm. or more wide, scarlet or crimson, with or without a black basal blotch; filaments 5–7 mm.

long, dark purple; anthers oblong or almost globose, 1–2 mm. long. Capsule glabrous, subglobose, turbinate or broadly clavate, 10–20 mm. long, 7–10 mm. wide; stigmatic rays (5–) 8–15 (–18); seeds reniform, about 0·75 mm. long, dark brown, often pruinose; testa reticulate-foveolate.

var. **oblongatum** *Boiss.*, Fl. Orient., 1: 113 (1867) pro parte excl. spec. Haussknecht.; Rechinger f. in Arkiv för Bot., ser. 2, 1: 422 (1950).

P. subpiriforme Fedde in Bull. Herb. Boiss., ser. 2, 5: 169 (1905); Feinbrun in Israel Journ. Bot., 12: 82 (1963); Zohary, Fl. Palaest., 1: 226, t. 330 (1966).

P. rhoeas L. sec. Holmboe, Veg. Cypr., 83 (1914); Lindberg f., Iter Cypr., 17 (1946).

P. rhoeas L. ssp. *humile* Holmboe, Veg. Cypr., 83 (1914).

? *P. humile* Fedde in Bull. Herb. Boiss., ser. 2, 5: 446 (1905); Osorio-Tafall et Seraphim, List Vasc. Plants Cyprus, 43 (1973).

[*P. rhoeas* L. var. *pryorii* (non Druce) Druce in Rep. B.E.C., 9: 469 (1930).]

TYPE: " in Syria circa Sidonem " *Blanche* (G).

Capsule obovate-oblong or turbinate, attenuate towards the base; otherwise resembling *P. rhoeas* L. var. *rhoeas*.

HAB.: Cultivated fields and waste ground, occasionally on sandy seashores; sea-level to 3,750 ft. alt.; fl. Febr.–June.

DISTR.: Divisions 1–7, common in most areas. Throughout temperate Europe and the Mediterranean; Atlantic Islands, and naturalized elsewhere, the var. *oblongatum* predominant in Syria and Palestine, but occurring elsewhere with typical *rhoeas*.

1. Ayios Yeoryios (Akamas), 1962, *Meikle* ! Polis area, 1962, *Meikle* !
2. Common; *Druce* s.n.; *Syngrassides* 1637 ! *Kennedy* 1169 ! 1170 ! 1799 ! 1819 ! etc.
3. Kato Lefkara, 1937, *Syngrassides* 1513 !
4. Common; *Kotschy* 19; *Sintenis & Rigo* 936 ! *Holmboe* 15, 142; *N. Macdonald* 28 !
5. Dhali, 1936, *Syngrassides* 1180 ! Mia Milea, 1950, *Chapman* 501 !
6. Nicosia, 1932, *Syngrassides* 234 ! *Merton* 3079 !
7. Common, *Kotschy* 473; *Druce* 20 ! *Chapman* 275 ! *G.E. Atherton* 973 ! 974 ! etc.

NOTES: Most of the material seen undoubtedly belongs to the var. *oblongatum* Boiss., though the distinctions between this and typical *rhoeas* are not always clear. *Chapman* 275 (from Kharcha) differs in having the rather soft hairs of the peduncle distinctly adpressed, and in some respects resembles *P. syriacum* Boiss. et Blanche (*P. rhoeas* L. var. *syriacum* (Boiss. et Blanche) Boiss.). Further material from this locality, with flowers and fruits, would be useful, also additional material of Holmboe's ssp. *humile* from seashores between Xylophagou and Ayia Napa.

2. P. dubium *L.*, Sp. Plant., ed. 1, 1196 (1753); Unger et Kotschy, Die Insel Cypern, 321 (1865); Boiss., Fl. Orient., 1: 115 (1867); Fedde *loc. cit.*, 313 (1909); Holmboe, Veg. Cypr., 83 (1914); Post, Fl. Pal., ed. 2, 1: 35 (1932); Osorio-Tafall et Seraphim, List Vasc. Plants Cyprus 43 (1973).

TYPE: " *Inter* Sueciae, Angliae *segetes* ".

Erect annual, 15–50 cm. or more high, superficially resembling *P. rhoeas* L. but with the leaves more regularly pinnatisect, and with the indumentum of the peduncle adpressed; the buds are acuminate towards the apex, often distinctly pointed, and the flowers are a paler salmon-red, with smaller petals, 2–3 cm. long and 2–5 cm. wide, usually without any basal markings. Capsule glabrous, obovate-oblong, about 20 mm. long and 7–8 mm. wide; stigma convex, usually 5–9-rayed; seeds rather larger than those of *P. rhoeas*, conspicuoously reticulate-foveolate.

HAB.: Cultivated fields and waste ground; c. 3,000 ft. alt.; fl. March–June.

DISTR.: Division 2, rare, perhaps erroneously recorded. Europe, Mediterranean region, Atlantic Islands and Ethiopia.

2. Prodhromos, Lemithou and Tris Elies, 1862, *Kotschy* 865.

NOTES: Not recorded since, and perhaps a misidentification; the following (*P. postii* Fedde), which resembles *P. dubium*, may have been the plant seen by Kotschy.

3. P. postii *Fedde* in Bull. Herb. Boiss., ser. 2, 5: 447 (1905), and in Engl. Pflanzenr., 40 (IV, 104): 323 (1909); Post, Fl. Pal., ed. 2, 1: 35 (1932); Burtt et Davis in Kew

Bull., 1949: 98 (1949); Meikle in Hooker's Icones, 36: t. 3573 (1962); Osorio-Tafall et Seraphim, List Vasc. Plants Cyprus, 44 (1973).

TYPE: Syria; Jebel Ansariyeh [Nusairy Mts.], top of southern spur, Aug. 4, 1890, *Post* ! (BE, K !).

Spreading or decumbent, much-branched annual, 6–20 cm. high; stems slender, with sparse, spreading or ascending hairs; basal leaves subglabrous or thinly strigose, forming a rosette, narrowly lanceolate, 2–8 cm. long, 0·5–1·5 cm. wide, deeply pinnatifid, the lobes narrowly ovate, acute, entire or with a few irregular serrations; stem leaves few, similar to the basal leaves but smaller, often more deeply and sharply lobed, and more distinctly strigose; flower solitary, peduncle slender, often rather flexuous, with patent or adpressed indumentum; buds ovoid, blunt, 7–9 mm. long, 4–5 mm. wide, subglabrous or very thinly pilose, sepals sometimes spotted purple; petals pale scarlet or light orange-red, orbicular-cuneate, 1–2 cm. long, 1–2·5 cm. wide; filaments violet-purple, about 5 mm. long; anthers oblong, yellow or tinged fuscous, less than 1 mm. long. Capsule glabrous, turbinate or broadly clavate, 6–8 mm. long, 3–5 mm. wide; stigma rays usually 5–6; seeds reniform, brown, about 0·7 mm. long; testa minutely reticulate-foveolate.

HAB.: On screes and slopes, or by roadsides, usually at high altitudes on igneous formations; 1,500–3,800 ft. alt.; fl. April–May.

DISTR.: Division 2, locally common. Also Syria and Turkey.

2. Locally common; Kambos village on the way to Karavostasi, 1937, *Syngrassides* 1599 ! Tripylos (Paphos forest), between Kykko and Stavros tis Psokas, 1941, *Davis* 3491 ! Vroisha, 1953, *Kennedy* 1798 ! At intervals from Stavros tis Psokas to Kykko Monastery, 1962, *Meikle* 2644 ! 2670 !

4. **P. minus** (*Boiv.*) *Meikle* in Kew Bull., 1956: 454 (1957).

Closterandra minor Boiv. in Bélanger, Voy. aux Indes-Orient., t. 3, fig. B (1834–1836).

P. belangeri Boiss., Fl. Orient., 1: 117 (1867); Fedde, *loc. cit.*, 330 (1909).

P. desertorum Grossh. in Grossh. et Schischk., Sched. Herb. Plant. Orient. Exsicc., 1: 23 (1924).

[*P. argemone* L. var. *glabrum* (non Koch) A. K. Jackson et Turrill in Kew Bull., 1939: 474 (1939).]

[*P. argemone* (non L.) Osorio-Tafall et Seraphim, List Vasc. Plants Cyprus, 44 (1973).]

TYPE: Iran; *Bélanger* (? P).

A decumbent or ascending annual, 7–25 cm. high, usually branched from the base (or unbranched and ± erect in starved specimens); stems with adpressed hairs; basal leaves forming a distinct rosette, narrowly ovate or lanceolate, pinnatisect or bipinnatisect, 2–8 cm. long, 2–4 cm. wide, thinly hispid on both sides or subglabrous above, ultimate divisions acuminate, often bristle-pointed; petioles 1–4 cm. long, narrowly winged; stem leaves smaller than the basal, deltoid, 0·5–6 cm. long, 0·5–5 cm. wide, sessile or subsessile, the basal divisions often elongate and strongly developed, giving the leaf a tripartite appearance; flowers small, solitary, funnel-shaped; peduncle adpressed-hispid, 2–10 cm. long, rather stout and thickening conspicuously in fruit; buds narrowly oblong or almost clavate, 1–1·5 cm. long, 0·5–0·8 cm. wide, thinly adpressed-hispid, usually emarginate at apex, often tinged reddish; petals obcuneate, 1·5–2·5 cm. long, 1–1·5 cm. wide, crimson or dull purplish-red with a dark basal blotch; filaments purple, 5–8 mm. long, distinctly flattened above; anthers yellowish, almost orbicular, about 1 mm. diam. Capsule cylindrical-clavate, 1·5–2·5 cm. long, about 0·5 cm. wide, thinly hispid or almost glabrous, with 4–6 well-marked ribs; stigmatic rays blue (3–) 4–5 (–6), extending beyond the margin of the convex, umbonate disk; seeds reniform, brown, about 0·7 mm. long, indistinctly reticulate-foveolate. *Plate 5.*

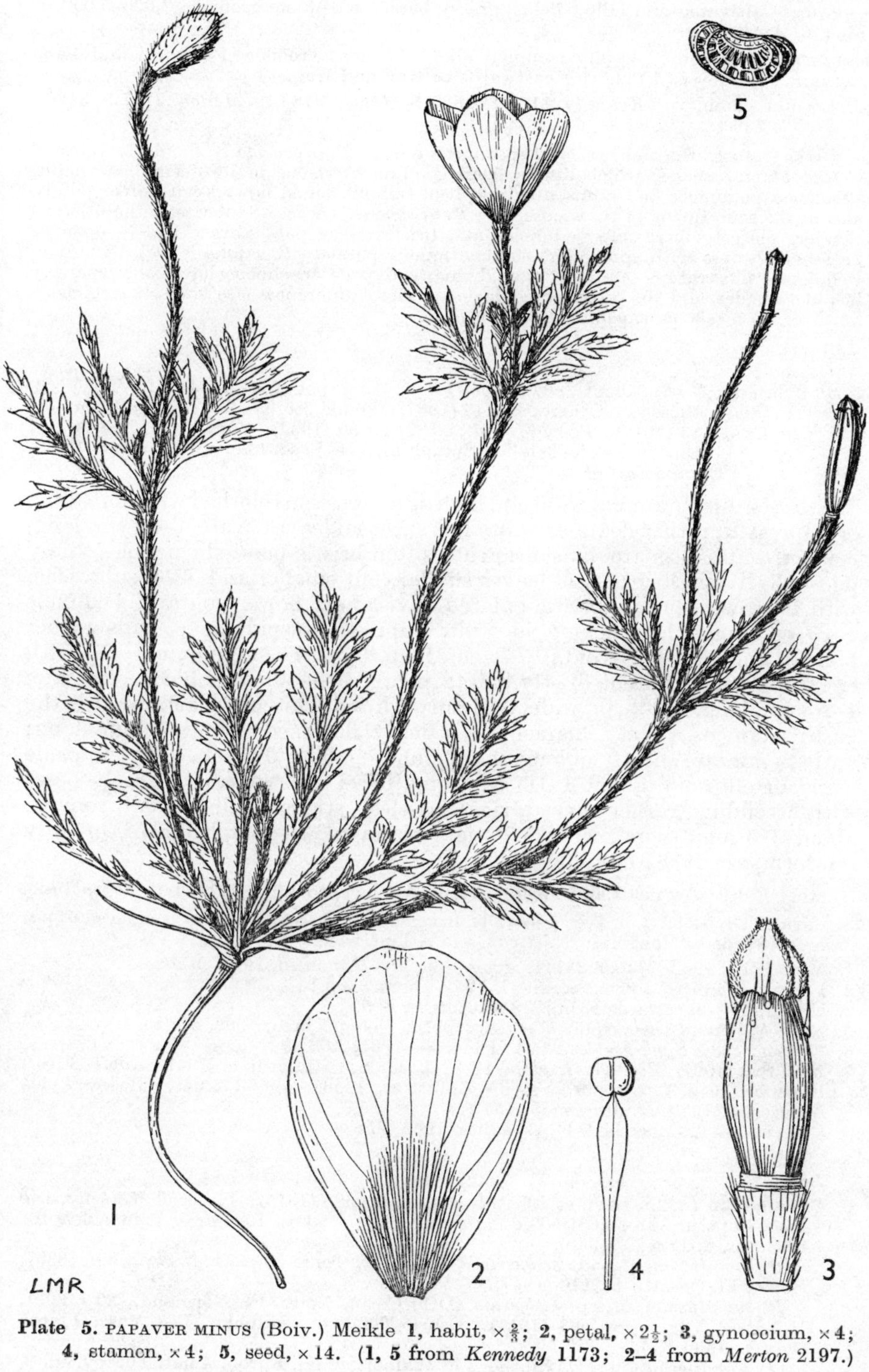

Plate 5. PAPAVER MINUS (Boiv.) Meikle **1**, habit, $\times\frac{2}{3}$; **2**, petal, $\times 2\frac{1}{2}$; **3**, gynoecium, $\times 4$; **4**, stamen, $\times 4$; **5**, seed, $\times 14$. (**1, 5** from *Kennedy* 1173; **2–4** from *Merton* 2197.)

HAB.: Cultivated and fallow fields, grassy banks and waste ground; 2,000–5,000 ft. alt.; fl. April–May.

DISTR.: Division 2, locally common about Platres, Troödos, Palekhori and Pano Lefkara. Turkey and Palestine eastwards to Iran and Iraq.

2. Locally common; *Kennedy* 1173 ! 1820 ! *Davis* 3180 ! *Merton* 2197 ! 3130 ! *Meikle* 836 !

NOTES: Most, if not all, oriental specimens named *P. argemone* L. should be referred to the above species, which differs from genuine *argemone* in its decumbent habit, thickened peduncles and crimson or purplish, funnel-shaped flowers with erect petals, also in the sculpturing of the seed-coat. *P. argemone* is erect or suberect in habit, with slender, scarcely incrassate peduncles and brick-red or pale scarlet, widely open or flattened flowers with spreading, non-contiguous petals. *P. minus* is well figured in Bélanger's illustration (*op. cit.*), though many Cyprus specimens have very sparsely hispid capsules and finely dissected leaves; these differences are scarcely sufficient, however, to merit nomenclatural recognition.

5. P. hybridum *L.*, Sp. Plant., ed. 1, 506 (1753); Unger et Kotschy, Die Insel Cypern, 321 (1865); Boiss., Fl. Orient., 1: 117 (1867); Fedde, *loc. cit.*, 332 (1909); Holmboe, Veg. Cypr., 83 (1914); Post, Fl. Pal., ed. 2, 1: 36 (1932); Zohary, Fl. Palaest., 1: 228, t. 336 (1966); Osorio-Tafall et Seraphim, List Vasc. Plants Cyprus, 44 (1973).

TYPE: " *in* Europa *australiore* ".

Erect or diffuse annual to 40 cm. high or more, stems clothed with spreading or adpressed, rather flexuous white hairs; basal leaves ovate, 6–20 cm. long, 2–6 cm. wide, ultimate divisions acute, often bristle-pointed; petioles thinly pilose, 3–10 cm. long; stem leaves similar, but smaller and often subsessile, with narrower, linear, bristle-pointed divisions; flower solitary, peduncle 4–15 cm. long, lengthening in fruit, adpressed-hispidulous; buds rather bluntly and broadly ovoid, 0·7–1 cm. long, 0·5–0·7 cm. wide, covered with erecto-patent or patent bristly hairs; petals rather small, obovate-cuneate, 1–2·5 cm. long, 0·7–2 cm. wide, dull purplish-red, blotched black towards the base; stamens violet; filaments distinctly flattened, about 7 mm. long; anthers minute, shortly oblong or subglobose, about 0·5 mm. long. Capsule ovoid or ellipsoid, about 1–1·5 cm. long, 0·7–1 cm. wide, ± densely beset with ascending, rather curved, rigid bristles; stigmatic disk small, convex, about 3–5 mm. diam., with 5–9 stigmatic rays; seeds dull brown, broadly reniform, about 0·7 mm. long.

HAB.: Cultivated fields and waste ground; sea-level to 4,000 ft. alt.; fl. Febr.–June.

DISTR.: Divisions 1–3, 5–8, generally frequent. Also western and central Europe; Asia eastwards to Mongolia; N. Africa and Atlantic Islands.

1. Near Polis, 1962, *Meikle* 2314! Ayios Yeoryios (Akamas), 1962, *Meikle* !
2. Papoutsa (Agros) above Palekhori, 1941, *Davis* 3102 !
3. Near Limassol towards Erimi, 1859, *Kotschy* 495.
5. Near Athienou, 1880, *Sintenis & Rigo* 147 ! Salamis area, 1962, *Meikle* !
6. Nicosia, 1932, *Syngrassides* 235 ! Dhiorios, 1962, *Meikle* !
7. Kythrea, 1927, *Rev. A. Huddle* 13 ! Larnaka-tis-Lapithou, 1941, *Davis* 3010 ! Kyrenia, 1949, *Casey* 440 ! Ayios Epiktitos, 1956, *Casey* 1703/4 ! Below Ayios Amvrosios, 1960, *N. Macdonald* 154 !
8. Gastria, 1941, *Davis* 2416 ! Akradhes, 1962, *Meikle* !

6. P. setigerum *DC.* in Lam. et DC., Fl. Franç., 6: 585 (1815); Unger et Kotschy, Die Insel Cypern, 321 (1865); Fedde, *loc. cit.*, 342 (1909); Rechinger f. in Arkiv för Bot., ser. 2, 1: 422 (1950).

P. somniferum L. var. *setigerum* (DC.) Elk., Tent. Mon. Gen. Pap., 30 (1839); Boiss., Fl. Orient., 1: 116 (1867).

P. somniferum L. ssp. *setigerum* (DC.) Corb., Nouv. Fl. Normand., 30 (1893); Holmboe, Veg. Cypr., 83 (1914); Osorio-Tafall et Seraphim, List Vasc. Plants Cyprus, 43 (1973).

[*P. somniferum* (non L.) Sibthorp in Walpole, Mem. Europ. Asiat. Turkey, ed. 2, 18 (1818).]

TYPE: France; " decouverte par M. Requien dans l'île du Levant (l'une des îles d'Hyères) . . . cultivé à Avignon " (P, G, K !).

Erect annual 30–60 cm. high, stems terete, usually glabrous; basal leaves glaucous, hispid on midrib and nerves below; narrowly obovate or oblanceolate, 5–12 cm. or more long, 1·5–3 cm. wide, coarsely toothed and lobed, each lobe terminating in a short, rigid bristle; petiole short, winged; stem leaves similar, sessile or ± amplexicaul, becoming progressively smaller above; flower solitary; peduncle stout, rather rigid, 10–20 cm. or more in length, sparsely clothed with adpressed or ascending bristles; bud bluntly ovoid, 1–2 cm. long, 0·6–1·5 cm. wide, thinly adpressed-hispid; petals pale purple, with a darker blotch towards the base, 5 cm. long, 3–6 cm. wide; stamens dark violet-purple; filaments flattened, a little dilated above, about 1 cm. long; anthers narrowly oblong, 2–3 mm. long. Capsule obovoid, 2–2·5 cm. long, 0·7–1·5 cm. wide; stigmatic rays (5–) 7–8 (–10), not extending quite to the margin of the disk; seeds broadly reniform, 0·6–0·7 mm. long, very dark brown, strongly reticulate-foveolate.

HAB.: In vineyards and on stony hillsides, usually at high altitudes; 1,000–4,000 ft. alt.; fl. April–June.

DISTR.: Divisions 2, 3, 7, common in Division 2 around the Troödos massif; also found between Ayios Theodhoros and Zoöpiyi. Mediterranean region and Atlantic Islands.

2. Common; *Kotschy* 856; *Kennedy* 1171 ! 1172 ! *Davis* 3469 ! *Merton* 2362 ! *Meikle* 2868 !
3. W. of Stavrovouni, 1957, *Merton* 3161 !
7. Between Antiphonitis Church and Bellapais, 1787, *Sibthorp*; Near the Armenian Monastery (E. of Pentadaktylos), 1912, *Haradjian* 356, 357; Lapithos, 1936, *R. M. Nattrass*.

NOTES: Very closely allied to the cultivated Opium Poppy (*P. somniferum* L.) but distinguished by the bristle-pointed lobes of the leaves; by the obovoid capsule, small stigmatic disk, and (usually) by the adpressed or ascending indumentum of the peduncle.

7. **P. gracile** *Auch. ex Boiss.* in Ann. Sci. Nat., ser. 2, 16: 372 (1841); Boiss., Fl. Orient., 1: 115 (1867); Fedde, *loc. cit.*, 343 (1909); Holmboe, Veg. Cypr., 83 (1914); Davis, Fl. Turkey, 1: 230 (1965); Osorio-Tafall et Seraphim, List Vasc. Plants Cyprus, 43 (1973).

TYPE: Turkey; " in montibus Cariae ", *Aucher-Éloy* 372 (G, K !).

Erect or sprawling annual 20–40 cm. high, stems rather sparingly branched, glabrous or very sparsely hispid towards the base, distinctly furrowed; leaves glaucous, glabrous or hispid on the midrib and veins, the basal leaves obovate or oblanceolate, 3–8 cm. long, 2–3 cm. wide, irregularly pinnatilobed with shortly toothed lobes; petiole short, usually ± winged; stem leaves similar to the basal, sessile or distinctly amplexicaul; flower solitary on a long, erect, glabrous or subglabrous peduncle; buds ovate, sharply acuminate, 1–2 cm. long, 0·7–1 cm. wide, glabrous or subglabrous; petals pink or rosy-mauve, cuneate, about 2 cm. long and 1·5 cm. wide; filaments violet about 0·8 mm. long; anthers violet, oblong, about 0·8 mm. long. Capsule glabrous, broadly clavate, about 1·5 cm. long and 0·7 cm. wide; stigmatic rays 5–9 not extending to margin of the disk; seeds dark brown, broadly reniform, about 0·7 mm. long; testa reticulate-foveolate.

HAB.: Dry, stony slopes, occasionally on walls; 1,000–3,000 ft. alt.; fl. April–June.

DISTR.: Divisions 2, 7. Also Turkey and eastern Aegean Islands.

2. Between Ayios Theodhoros and Zoöpiyi, 1941, *Davis* 3070 ! Makheras road, 1949, *Casey* 705 ! Phlevas near Pyrgos, on wall of old forestry hut, 1962, *Meikle* 2719 !
7. St. Hilarion; *Sintenis & Rigo* 522 ! *Holmboe* 825; *J. A. Tracey* 18 ! *Kennedy* 1735, 1736 ! *Casey* 645 ! Karmi, 1974, *G. Joscht* 7170 !

2. GLAUCIUM *Mill.*

Gard. Dict. Abridg., ed. 4 (1754).
Fedde in Engl., Pflanzenr. 40 (IV, 104): 221 (1909).

Erect or ascending annual, biennial or perennial herbs; stems and leaves usually glaucous, glabrous or ± pilose; radical leaves petiolate, generally pinnatifid; cauline leaves sessile, often amplexicaul, ± lobed or toothed. Flowers solitary, axillary and terminal, usually yellow or reddish, showy; sepals 2, caducous; petals 4 with convolute aestivation; stamens numerous; stigma ± sessile, distinctly 2-lobed. Fruit an elongate, linear, pod-like capsule, 2-valved, dehiscence from apex to base; seeds subglobose or reniform, numerous.

About 20 species chiefly in the Mediterranean region eastwards to Afghanistan.

Capsule adpressed-pilose, 11–15 cm. long; annual with red, yellow or variously blotched flowers - - - - - - - - - - - - **1. G. corniculatum**
Capsule glabrous, smooth or verruculose, up to 30 cm. or more in length; biennial or perennial herb with yellow flowers - - - - - - - **2. G. flavum**

1. G. corniculatum (*L.*) *J. H. Rudolph,* Diss. Fl. Jen. Plant., 13 (1781); Boiss., Fl. Orient., 1: 119 (1867); Fedde, *loc. cit.* 223 (1909); Holmboe, Veg. Cypr., 83 (1914); Post, Fl. Pal., ed. 2, 1: 38 (1932); Lindberg f., Iter Cypr., 16 (1946); Davis, Fl. Turkey, 215 (1965); Zohary, Fl. Palaest., 1: 231, t. 339 (1966).
Chelidonium corniculatum L., Sp. Plant., ed. 1, 506 (1753).
TYPE: " *in* Hungaria, Bohemia, Monspelii ".

Erect or spreading, glaucous annual, 7–30 cm. high, with ascending or almost horizontal, sparsely or ± densely and roughly pilose branches; radical leaves shortly pilose, 4–20 cm. long, 1·5–7 cm. wide, often forming a distinct rosette, petiolate, oblanceolate or narrowly obovate, pinnatisect, the oblong lobes sharply and irregularly dentate; petiole 2–6 cm. long, narrowly winged; cauline leaves rarely more than 10 cm. long, sessile, amplexicaul, the basal lobes extending beyond the stem; flowers solitary, terminal and axillary; peduncle 1–10 cm. long, subglabrous or sparsely crispate-pilose, usually accrescent in fruit; buds ovoid, 1·5–3·5 cm. long, 1–1·5 cm. wide, often acutely cuspidate or rostrate; sepals glabrous or sparsely pilose, with a narrow pale margin; petals red, orange or yellow, usually blotched at the base, broadly obovate-cuneate, 2–4 cm. long, 1·5–3·5 cm. wide; stamens numerous, filaments yellow, 1–1·5 cm. long; anthers greenish or fuscous, narrowly oblong, about 2 mm. long. Capsule usually curved, 11–15 cm. or more in length, about 3–4 mm. wide, adpressed-pilose; seeds dull brown, subglobose, 1–1·5 mm. diam., distinctly reticulate-foveolate.

var. **corniculatum**

G. phoeniceum Crantz, Stirp. Austr., ed. 1, fasc. 2: 133 (1763); Unger et Kotschy, Die Insel Cypern, 321 (1865).
G. corniculatum (L.) J. H. Rudolph var. *phoeniceum* (Crantz) DC., Regn. Veg. Syst. Nat., 1: 96 (1817); Boiss., Fl. Orient., 1: 119 (1867); Fedde, *loc. cit.*, 223 (1909); Post, Fl. Pal., ed. 2, 1: 38 (1932).
G. corniculatum (L.) J. H. Rudolph ssp. *phoeniceum* (Crantz) Holmboe, Veg. Cypr., 83 (1914); Osorio-Tafall et Seraphim, List Vasc. Plants Cyprus, 43 (1973).

Petals bright red with a violet basal blotch.

HAB.: Cultivated fields and waste ground, sometimes on grassy banks or dry stony hills; sea-level to 4,000 ft. alt.; fl. March–May.

DISTR.: Divisions 2, 3, 5–8. Also Mediterranean region, central and southern Europe eastwards to the Caucasus and Iran; Atlantic Islands.

2. Mavrovouni, 1932, *Syngrassides* 92 ! Amiandos, 1955, *Merton* 2322 ! Xeros-Yialia road, Mile 42, 1957, *Merton* 2951 !

3. Amathus, 1963, *J. B. Suart* 66 !
5. Near Athienou, 1880, *Sintenis & Rigo* 150b; Vatili, 1905, *Holmboe* 313.
6. Nicosia, 1905, *Holmboe* 285.
7. Kythrea, 1880, *Sintenis & Rigo* 150 !
8. Ovgoros, 1941, *Davis* 2477 !

var. **tricolor** (*Bernh. ex Spreng.*) *Ledeb.*, Fl. Ross., 1: 93 (1841); Boiss., Fl. Orient., 1: 120 (1867); Fedde, *loc. cit.*, 225 (1909).

G. tricolor Bernh. ex Spreng., Syst. Veg., 4 append.: 203 (1827).

G. corniculatum (L.) J. H. Rudolph ssp. *tricolor* (Bernh. ex Spreng.) Holmboe, Veg. Cypr., 83 (1914); Osorio-Tafall et Seraphim, List Vasc. Plants Cyprus, 43 (1973).

[*G. phoeniceum* (non Crantz) Sibth. et Sm., Fl. Graec., 5: 72, t. 489 (1825).]

TYPE: " Thuringia et Podolia ", *Bernhardi* (JE).

Flowers rather large; petals red or orange, basal blotch with a white margin.

HAB.: Cultivated fields and waste ground; sea-level to ? 500 ft. alt.; fl. March–May.

DISTR.: Division 1 or 2. General distribution similar to that of var. *corniculatum*.

1 or 2. Between Yialia and Polis tou Khrysokhou, 1905, *Holmboe* 792.

var. **flaviflorum** *DC.*, Regn. Veg. Syst. Nat., 2: 97 (1817); Boiss., Fl. Orient., 1: 120 (1867); H. Stuart Thompson in Journ. Bot., 44: 276 (1906); Fedde, *loc. cit.* 224 (1909).

TYPE: U.S.S.R.; " In Tauria collegit Cl. *Steven* " (G).

Flowers yellow.

HAB.: Cultivated fields and waste ground; 500–5,700 ft. alt.; fl. March–July.

DISTR.: Divisions 2, 3, 6. Chiefly in the eastern Mediterranean region and western Asia.

2. Olympus Camp Hotel, Troödos Pass, 1954, *Kennedy* 1821 !
3. Nata, 1956, *Poore* 20 !
6. Nicosia, near the Tannery, 1901, *A. G. & M. E. Lascelles* s.n. !

2. G. flavum *Crantz*, Stirp. Austr., ed. 1, fasc. 2: 133 (1763); Fedde, *loc. cit.*, 232 (1909); Post, Fl. Pal., ed. 2, 1: 39 (1932); Lindberg f., Iter Cypr., 17 (1946); Rechinger f. in Arkiv. för Bot., ser. 2, 1: 422 (1950); Zohary, Fl. Palaest., 1: 233, t. 343 (1966); Osorio-Tafall et Seraphim, List Vasc. Plants Cyprus, 43 (1973).

G. luteum Scop., Fl. Carniol., ed. 2, 1: 369 (1772); Boiss., Fl. Orient., 1: 122 (1867).

Erect or spreading, glaucous biennial or perennial, with a thick, fleshy taproot; branches glabrous or thinly hispid; radical leaves shortly pilose, 12–30 cm. or more long, 4·5–7 cm. wide, forming a rosette, obovate or oblanceolate, pinnatisect, with oblong, acute or rather blunt lobes; stem leaves broadly oblong or the uppermost deltoid, 4–12 cm. long, 3–6 cm. wide, sessile, amplexicaul, irregularly lobed, with the basal lobes extending beyond the stem; flowers solitary, axillary and terminal; peduncle very short, seldom more than 1–2 cm. long, lengthening in fruit, glabrous or shortly hispid-pilose; buds ovoid, 1·5–2·5 cm. long, about 1 cm. wide, acuminate or shortly rostrate; sepals glabrous or sparsely pilose with a very narrow, pale margin; petals yellow, broadly obovate-cuneate, 2–3·5 cm. long, 2–3 cm. wide; stamens numerous; filaments yellow, 1–1·5 cm. long; anthers yellow, narrowly oblong, about 2 mm. long. Capsule usually curved, up to 30 cm. or more in length, 5–8 mm. wide, glabrous, smooth or ± verruculose; seeds dark blackish-brown, reniform, about 2 mm. long and 1·2–1·5 mm. wide, distinctly reticulate-foveolate.

var. **leiocarpum** (*Boiss.*) *Stoj. et Stef.*, Fl. Bulg., ed. 1, 471 (1924); Turrill in Kew Bull., 1933: 174 (1933).

G. leiocarpum Boiss., Fl. Orient., 1: 122 (1867); Fedde, *loc. cit.* 234 (1909); Post, Fl. Pal., ed. 2, 1: 39 (1932).

TYPE: Turkey; " in collibus siccis et regione montana Anatoliae in monte Tmolo Lydiae ", *Boissier* (G).

Plant more slender than var. *flavum*, with rather smaller leaves; capsules slender, seldom more than 20 cm. long, usually smooth; sepals often subglabrous; petals sometimes reddish or tawny-yellow.

HAB.: Sandy ground near the sea; about sea-level; fl. May–Aug.

DISTR.: Division 4, common locally around Larnaca. Widespread in the eastern Mediterranean region and extending through Turkey to Iran. Typical var. *flavum* replaces var. *leiocarpum* in the western Mediterranean region and Atlantic Islands, and along the coasts of western Europe, also occurring inland in central Europe eastwards to Transcaucasia and the Black Sea.

4. Common locally; *Sintenis & Rigo* 897 ! *Haradjian* 1021 ! *Syngrassides* 1823 ! *Lindberg f.* s.n. ! *Merton* 2660 !

3. ROEMERIA *Medik.*

in Usteri, Ann., 3: 15 (1792).

Fedde in Engl. Pflanzenr., 40 (IV, 104): 238 (1909).

Annual herbs with a slender tap-root; leaves pinnatisect, the segments often divided into narrow linear lobes. Inflorescence leaf-opposed; flowers solitary, red or violet-purple; sepals 2, pilose, caducous; petals 4, crumpled in bud; stamens numerous; style short, stigma capitate. Capsule narrowly cylindrical, dehiscing by 3–4 valves from apex to base. Seeds reniform, numerous.

About 6 species in the Mediterranean region and western Asia, sometimes occurring as weeds elsewhere.

1. **R. hybrida** (*L.*) *DC.*, Regn. Veg. Syst. Nat., 2: 92 (1821); Unger et Kotschy, Die Insel Cypern, 322 (1865); Boiss., Fl. Orient., 1: 118 (1867); Fedde, *loc. cit.*, 239 (1909); Holmboe, Veg. Cypr., 83 (1914); Post, Fl. Pal., ed. 2, 1: 37 (1932); Zohary, Fl. Palaest., 1: 229, t. 337 (1966); Osorio-Tafall et Seraphim, List Vasc. Plants Cyprus, 43 (1973).

Chelidonium hybridum L., Sp. Plant., ed. 1, 506 (1753).

Glaucium violaceum (Lam.) Juss., Gen. Plant., 236 (1789); Sibth. et Sm., Fl. Graec. Prodr., 1: 358 (1809), Fl. Graec., 5: 73, t. 490 (1825).

Chelidonium violaceum Lam., Fl. Franç., ed. 1, 3: 169 (1778).

TYPE: " *in* Europa *australiore* ".

Erect or spreading, glabrous or ± hirsute annual, 10–40 cm. high; radical leaves glabrous or hispidulous, narrowly ovate, 4–20 cm. long, 1–4 cm. wide, bipinnatisect, with linear, cuspidate divisions; cauline leaves similar but smaller and shortly stalked or sessile, the uppermost often deltoid in outline with the basal pinnae much longer than the others; flowers solitary, terminal; buds narrowly oblong or obovoid, 1–2 cm. long, 0·5–1 cm. wide; sepals thinly pilose or subglabrous; petals violet or purple-red, broadly obovate-cuneate, 2–4 cm. long, 1·5–3 cm. wide, stamens numerous; filaments violet, about 1 cm. long; anthers shortly oblong, about 1 mm. long. Capsule narrowly cylindrical, straight or curved, 4–8 cm. or more in length, 2–3 mm. wide, bristly towards the apex (var. *hybrida*) or throughout (var. *eriocarpa* DC.); seeds reniform, about 1 mm. long, brown, pruinose, strongly reticulate-foveolate.

HAB.: Cultivated fields and waste ground; sea-level to 4,000 ft. alt.; fl. March–May.

DISTR.: Divisions 2, 3, 5–8. Also throughout the Mediterranean region and southern Europe, eastwards to Iran; casual elsewhere.

2. Platres, 1938, *Kennedy* 1174 !
3. Amathus, 1862, *Kotschy* 584; Alethriko, 1905, *Holmboe* 215; Silikou, 1930, *C. B. Ussher* 101 ! 120 ! Trimiklini, 1941, *Davis* 3125 !
5. Lakatamia, 1932, *Syngrassides* 366 ! Pera Khorio near Nisou, 1934, *Syngrassides* 1128 ! Mile 14, Nicosia–Limassol Rd., 1950, *Chapman* 544 !
6. Mesaoria, near Kykko, 1952, *Probyn* 27 ! 45 !
7. Near Ayios Khrysostomos, 1862, *Kotschy* 409; Lapithos, 1862, *Kotschy* 479.
8. Rizokarpaso, 1880, *Sintenis & Rigo* 149 !

4. HYPECOUM *L.*

Sp. Plant., ed. 1, 124 (1753).
Gen. Plant., ed. 5, 60 (1754).
Fedde in Engl. Pflanzenr., 40 (IV, 104): 85 (1909).

Low-growing, annual herbs with watery latex; leaves pinnatisect, with narrow divisions, usually ± glabrous and distinctly glaucous. Inflorescence cymose; flowers small, yellow; sepals 2, small, caducous; petals 4, in 2 series, the outer generally 3-lobed or entire, the inner usually deeply 3-fid with narrow lateral lobes and a spathulate middle lobe, the middle lobe often fringed with hairs; stamens 4, filaments ± winged; anthers yellow. Capsule narrowly cylindrical, either with marked transverse articulations, and dehiscing transversely into numerous segments, or without transverse articulations and dehiscing into 2 longitudinal valves. Seeds numerous, ovoid or D-shaped.

About 15 species in the Mediterranean region and Asia Minor, a few extending eastwards to Mongolia and N. China.

Fruiting pedicels erect or spreading; outer petals distinctly 3-lobed; capsule with prominent transverse articulations:
Inflorescence sparsely branched; bracts foliose; petals usually pale or clear yellow **1. H. procumbens**
Inflorescence repeatedly branched; bracts linear or filiform; petals usually orange-yellow - - - - - - - - - - - - **2. H. imberbe**
Fruiting pedicels sharply reflexed; outer petals narrowly rhombic, not distinctly 3-lobed; capsule without transverse articulations - - - - - - **3. H. pendulum**

1. **H. procumbens** *L.*, Sp. Plant., ed. 1, 124 (1753); Boiss., Fl. Orient., 1: 124 (1867); Fedde, *loc. cit.*, 87 (1909); Post, Fl. Pal., ed. 2, 1: 40 (1932); C. E. Gresham in New Flora and Silva, 5: 272 (1933); Burtt et Davis in Kew Bull., 1949: 99 (1949); Davis, Fl. Turkey, 1: 236 (1965); Zohary, Fl. Palaest., 1: 234, t. 344 (1966); Osorio-Tafall et Seraphim, List Vasc. Plants Cyprus, 44 (1973).

TYPE: " *Inter Archipelagi*, Narbonae & *Salmanticenses segetes* ".

Decumbent, spreading or erect, glabrous or subglabrous, ± glaucous annual, 5–20 cm. high; leaves forming a basal tuft, narrowly obovate or oblanceolate, 4–18 cm. long, 1·5–3 cm. wide, bipinnatisect, the pinnules with small, ovate-lanceolate, acute or acuminate lobes, petioles 1–8 cm. long; inflorescence a few (1–5)-flowered cyme; peduncle 5–16 cm. long, relatively stout, leafless; bracts foliose, about 1–2 cm. long, bipinnatisect, with narrow-ovate or lanceolate divisions; pedicels 7–12 mm. long, slender, accrescent in fruit; flowers pale or bright yellow; sepals ovate-acute, about 4–5 mm. long, 1·5–2 mm. wide; 2 exterior petals broadly rhombic, 3-lobed, 8–10 mm. long, 5–8 mm. wide, with the median lobe rather longer than the two laterals; 2 interior petals deeply 3-lobed, the 2 lateral lobes glabrous, oblong, obtuse, about 4 mm. long and 1–1·5 mm. wide; median lobe spathulate, ciliate, about 5–7 mm. long and 1–2 mm. wide; stamens 4–6 mm. long, the filaments narrowly winged; anthers narrowly oblong about 1 mm. long; ovary narrowly cylindrical, glabrous, about 6 mm. long, crowned with 2 short, filiform stigmas. Capsule linear, compressed, striate, often curved, about 5–8 cm. long, 3–4 mm. wide, tapering at both ends, distinctly narrowed between the numerous, distinct, transverse articulations; seeds about 1·3 mm. long, 1 mm. wide; testa blackish, rugulose.

HAB.: Sandy places near the sea; about sea-level; fl. Febr.–May.

DISTR.: Divisions 1–3, 5, 7, 8, uncommon. Maritime regions in the Mediterranean region and Portugal.

1. Skali and Ayios Yeoryios (Akamas), 1962, *Meikle* 2102 !
2. Yialia, 1962, *Meikle* ! Pomos Point, 1962, *Meikle* !

3. Curium, 1939, *Kennedy* 1341 !
5. Salamis, 1962, *Meikle* !
7. Kyrenia, 1932, *C. E. Gresham* s.n.
8. Ayios Philon near Rizokarpaso, 1941, *Davis* 2208 !

NOTES: Apparently confined to sand-dunes and maritime districts in Cyprus.

2. **H. imberbe** *Sibth. et Sm.*, Fl. Graec. Prodr., 1: 107 (1806), Fl. Graec., 2: 47, t. 156 (1813); Holmboe, Veg. Cypr., 84 (1914); Davis, Fl. Turkey, 1: 237 (1965).
H. grandiflorum Benth., Cat. Pyr., 91 (1826); Unger et Kotschy, Die Insel Cypern, 322 (1865); Boiss., Fl. Orient., 1: 125 (1867); Fedde, *loc. cit.*, 91 (1909); Holmboe, Veg. Cypr., 84 (1914); Post, Fl. Pal., ed. 2, 1: 40 (1932).
TYPE: " In insulâ Cypro ", 1787, *Sibthorp* (OXF !).

Very similar to *H. procumbens* in general appearance, but with the ultimate divisions of the leaves usually narrower, often linear-lanceolate; and with the bracts much less foliose and conspicuous, usually with linear or filiform divisions; inflorescence rather lax, 2 or 3 times dichotomously branched, the pedicels often exceeding 1 cm. in length; flowers larger, orange-yellow; sepals acuminate; exterior petals distinctly 3-lobed, 1 cm. or slightly more in length and about the same in width; inner petals 3-lobed and ciliate as in *H. procumbens*. Seeds scarcely distinguishable from those of *H. procumbens*.

HAB.: Cultivated fields and waste ground, usually inland; sea-level to 1,000 ft. alt.; fl. Jan.–May.

DISTR.: Divisions 2–7, common in Divisions 4 and 6. Throughout the Mediterranean region and S. Europe, eastwards through Turkey to Iraq and Iran.

2. Skouriotissa near Lefka, 1932, *Syngrassides* 108 ! Kalokhorio, Makheras, 1950, *Chapman* 295 (mixed with *H. pendulum* L.) !
3. Near Amathus, ? 1948, *Mavromoustakis* s.n. !
4. Common; *Kotschy* 46; *Sintenis & Rigo* 958 ! *Holmboe* 370; *H. Roger-Smith* s.n. ! *Collingwood Ingram* 17 !
5. Koutsovendis, 1927, *Rev. A. Huddle* 14 ! Pera, 1949, *Casey* 390 ! Salamis, 1962, *Meikle* !
6. Common; *Syngrassides* 371 ! *Miss Godman* 1 ! *Davis* 2138 ! 2560 ! *Chapman* 530 !
7. Kyrenia, 1956, *G. E. Atherton* 922 !

NOTES: The plant figured in Sibthorp & Smith, Flora Graeca 2: 47, t. 156 (1813) is certainly not identical with the yellow and violet-petalled *H. aegyptiacum* (Forssk.) Aschers. et Schweinf. as has been supposed by many authors. Examination of type material shows that it is conspecific with the plant formerly known as *H. grandiflorum* Benth.

3. **H. pendulum** *L.*, Sp. Plant., ed. 1, 124 (1753); Boiss., Fl. Orient., 1: 125 (1867); Fedde, *loc. cit.*, 95 (1909); Holmboe, Veg. Cypr., 84 (1914); Post, Fl. Pal., ed. 2, 1: 41 (1932); Zohary, Fl. Palaest., 1: 236, t. 348 (1966); Osorio-Tafall et Seraphim, List Vasc. Plants Cyprus 44 (1973).
TYPE: " *In* Galloprovincia ".

Frequently forming mixed populations with *H. imberbe* Sibth. et Sm. (*H. grandiflorum* Benth.) and very similar in habit and size, but with the ultimate divisions of the leaves even narrower, linear-acuminate or filiform and Fennel-like; bracts also linear or filiform; flowers smaller and paler than those of *H. imberbe*; sepals ovate-acute or deltoid about 1·5 mm. long, often shortly and irregularly serrulate; exterior petals oblanceolate or narrowly rhombic, 6–8 mm. long, 3–5 mm. wide; inner petals 3-lobed, lateral lobes lanceolate, obtuse or shortly acute, 4–5 mm. long, 1–2 mm. wide, median lobe ligulate, about 1 mm. wide, apex obtuse, ciliate; pedicels 1 cm. or more in length, lengthening and strongly reflexed in fruit. Capsule distinctly 4-angled, 4–6 cm. long, 2–5 mm. wide, gradually tapering towards apex and base, longitudinally ribbed, but without distinct transverse articulations; seeds similar to those of *H. procumbens*.

HAB.: Sandy fields and seashores, sometimes in dry fields inland; sea-level to 2,000 ft. alt.; fl. Febr.–May.

DISTR.: Divisions 2, 7, 8, rare. Widespread in southern Europe, N. Africa and western Asia.

2. Kalokhorio, Makheras, 1950, *Chapman* 295 (with *H. imberbe* Sibth. et Sm.) !
7. In fields between Akanthou and Kalogrea, 1880, *Sintenis & Rigo* 148 !
8. Near the seashore at Yialousa, 1880, *Sintenis*.

4a. ESCHSCHOLZIA CALIFORNICA *Chamisso* in Nees, Horae Phys. Berol., 73–74 (1820).

An erect or sprawling, glabrous, glaucous annual or perennial 20–60 cm. high; taproot thick, fleshy; leaves dissected into numerous narrow segments; flowers solitary, long-pedunculate, showy; sepals united into a narrow conical, acuminate, circumscissile cap or calyptra; petals 4, golden-yellow or orange, 2–6 cm. long, 2–3 cm. wide. Fruit an elongate, linear, 2-valved, prominently ribbed capsule 5–10 cm. long, 0·4–0·6 cm. wide; seeds 1·2–1·5 mm. long; testa grey-brown, reticulate.

The Californian Poppy, a native of western North America, introduced into horticulture by David Douglas in 1826 and now found as a garden escape in many European countries. It has recently spread from gardens in the Amiandos area, and is now locally abundant there by roadsides and on waste ground (Pano Amiandos, 1968, *Economides* in ARI 1233 ! 1970, *E. Chicken* 56 !).

8. FUMARIACEAE

Annual or perennial herbs with colourless juice; leaves alternate or spirally arranged, usually pinnatisect or deeply dissected. Flowers in racemes or spikes, zygomorphic, hypogynous, hermaphrodite, the parts arranged in whorls; sepals 2, small, caducous; petals 4, 2-seriate, free or ± cohering, imbricate, outer often saccate or spurred; stamens 6, 3 on either side of the pistil, filaments ± connate; central anther of each bundle 2-locular, lateral anthers 1-locular, dehiscing longitudinally; 1 or 2 nectary-glands often at base of androecium; ovary of 2 united carpels, 1-locular with parietal placentation; style 1, slender; stigmas 2–8; ovules 1–many. Fruit a capsule, opening by valves, or a nut; seeds with a small embryo and copious, soft, watery or fleshy endosperm.

About 17 genera, chiefly in temperate regions of the northern hemisphere.

Fruits (or some of them) dehiscent; perennial herbs:
- Fruits uniform, 2–more-seeded; mostly erect or spreading perennials with tuberous roots - - - - - - - - - - - - - **1. Corydalis**
- Fruits dimorphic, the upper, 1–2-seeded capsules, the lower, 1-seeded nutlets; slender tendril-bearing climbers - - - - - - - - - - **2. Ceratocapnos**

Fruits indehiscent, forming 1-seeded nutlets; annual herbs without tendrils **3. Fumaria**

1. CORYDALIS *Vent.*

Choix de Plantes, 19 (1803) nom. cons.

Glabrous perennial (or rarely annual) herbs, often with thick, tuberous roots; stems erect, spreading or scandent; leaves usually compound, variously divided. Inflorescence racemose; flowers zygomorphic, the upper petal spurred; upper group of stamens with the base prolonged into a spur. Ovules numerous; fruit a 2–many-seeded capsule; seeds arillate.

About 300 species, chiefly in the northern hemisphere.

1. C. rutifolia (*Sibth. et Sm.*) *DC.*, Reg. Veg. Syst. Nat., 2: 115 (1821) as *C. rutaefolia*; Boiss., Fl. Orient., 1: 126 (1867); Holmboe, Veg. Cypr. 84 (1914); Post, Fl. Pal., ed. 2, 1: 42 (1932); Lindberg f., Iter Cypr., 16 (1946); Davis, Fl. Turkey, 1: 241 (1965); Osorio-Tafall et Seraphim, List Vasc. Plants Cyprus, 44 (1973).

Fumaria rutifolia Sibth. et Sm., Fl. Graec. Prodr., 2: 49 (1813), Fl. Graec., 7: 62, t. 667 (1830).

Cryptoceras rutifolium Schott et Kotschy in Oesterr. Bot. Wochenbl., 4: 121 (1854); Unger et Kotschy, Die Insel Cypern, 322 (1865).

TYPE: " E Cypro ", 1787, *Sibthorp* s.n. (OXF).

Erect, unbranched perennial, 6–20 cm. high; root a deeply buried, spherical, roughly fissured, solid tuber 2·5–5 cm. diam.; stem without basal scales; leaves usually 2, or occasionally 3, opposite or whorled, pinnately ternate or biternate, the ultimate leaflets rather glaucous, glabrous, obovate or suborbicular, entire or sometimes irregularly lobed, obtuse, 1–2 cm. long, 0·5–1·5 cm. wide; petiole up to about 1 cm. long, often very short; petiolules much longer, those of the median leaflets often 1·5–2 cm. or more in length; inflorescence a lax, rather short raceme of (2–) 4–8 (–12) flowers; bracts broadly obovate, entire, about 1 cm. long and 0·6–0·9 cm. wide; pedicels ascending, 0·4–2 cm. long; sepals minute, membranous, caducous, about 1–2 mm. long and the same in width, distinctly 2-lobed, the lobes minutely and irregularly toothed; corolla pink, 1·5–3 cm. long, spur upturned except for the shortly revolute apex; basal petal with a sharply reflexed lip; stamens 0·7–1 cm. long; filaments flattened, winged; anthers yellow, about 1 mm. long. Capsule ovoid, about 1 cm. long; seeds reniform, flattened, about 2·5 mm. wide, shining black, minutely reticulate, with a conspicuous pale brown aril.

HAB.: Screes and mountain summits; 4,500–6,400 ft. alt.; fl. April–May.

DISTR.: Division 2, locally common on Khionistra, and on Tripylos Peak. Throughout western Asia eastwards to Iran; Syria.

2. Locally common; *Sibthorp*; *Kotschy* 718 ! *Holmboe* 969; *Syngrassides* 898 ! *Kennedy* 455–458 ! 462 ! 463 ! *Davis* 316 ! *Merton* 3173 ! *Meikle* 2723 ! etc.

2. CERATOCAPNOS *Durieu*

in Parlatore (ed.), Giorn. Bot. Ital., 1: 336 (1844).

Slender, perennial, tendril-bearing climbers, similar to *Corydalis*, with racemose inflorescences of spurred flowers. Fruits dimorphic, the lower 1-seeded, indehiscent, the upper 1–2-seeded, dehiscent, with an elongate, ensiform beak; seeds exarillate.

Two species in Palestine, Syria and North Africa.

1. C. palaestinus *Boiss.*, Diagn., 1, 8: 12 (1849); Fl. Orient., 1: 132 (1867); Post, Fl. Pal., ed. 2, 1: 43 (1932); B. L. Burtt in Kew Bull., 1954: 68 (1954); Zohary, Fl. Palaest., 1: 237, t. 349 (1966); Osorio-Tafall et Seraphim, List Vasc. Plants Cyprus, 44 (1973).

TYPE: " In rupestribus umbrosis totius regionis montanae Palaestinae, in via inter *Ramla* et *Jerusalem* ad rupes et muros, in monte *Garizim* supra *Naplouse*, in montibus *Carmel* et *Thabor*. Fl. Maio." *Boissier* (G).

Slender, twining perennial, the leaves usually terminating in threadlike, branched tendrils; stems glabrous, slightly sulcate; leaves biternate with petioles varying from 2–20 mm. or more in length; leaflets glabrous, rather glaucous below, entire, broadly ovate or suborbicular, 10–20 mm. long, 8–18 mm. wide, with a rounded or mucronate apex; petiolules varying greatly in length; inflorescence a rather short, crowded raceme, 1–2 cm. long, with a slender peduncle 10–30 mm. or more in length, raceme and peduncle lengthening greatly in fruit; bracts minute, deltoid-lanceolate, 1–1·5 mm. long, frequently streaked red; pedicels 1–2 mm. long; flowers very similar to those of *Corydalis* or *Fumaria* in general appearance, 5–7 mm.

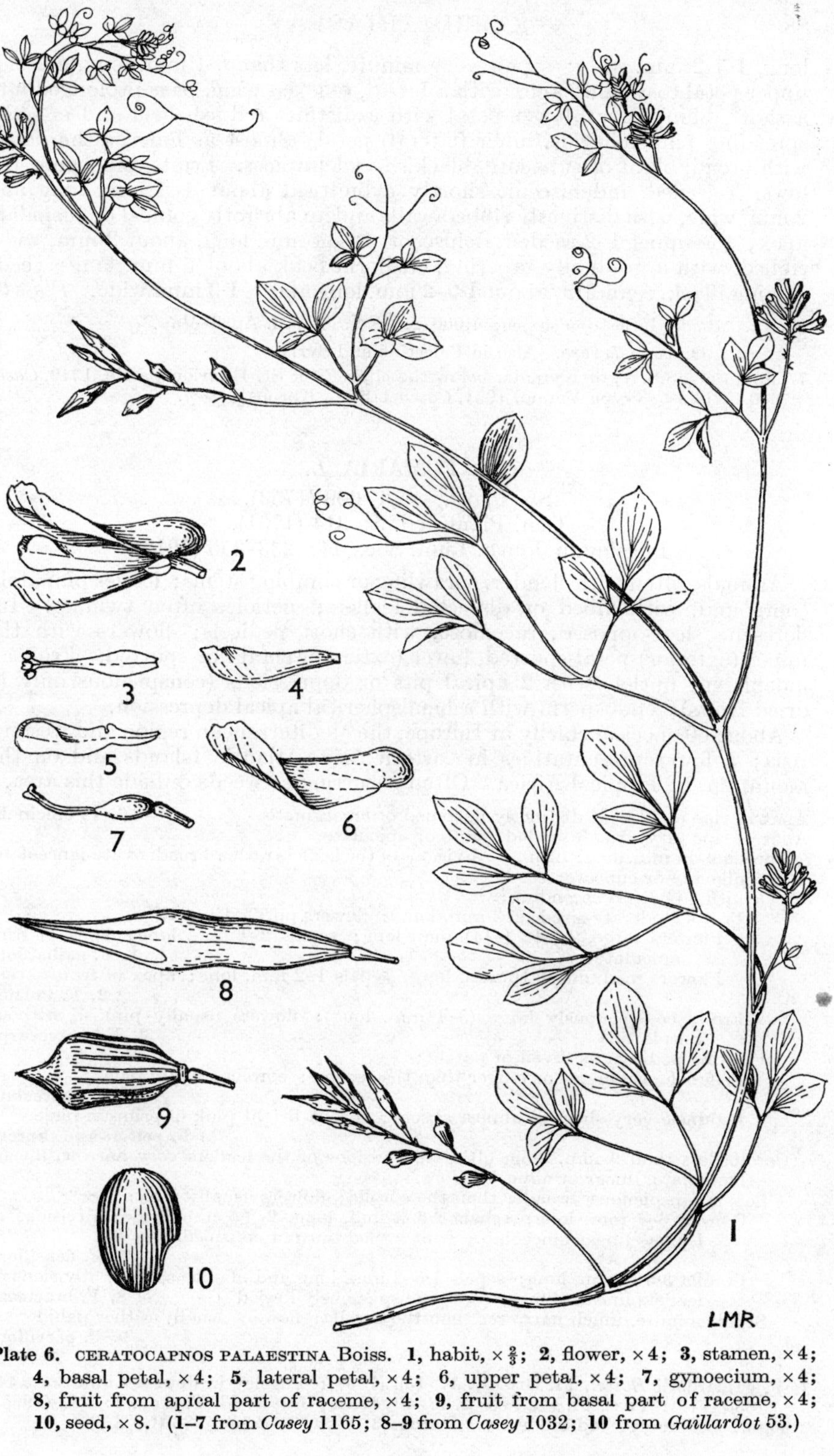

Plate 6. CERATOCAPNOS PALAESTINA Boiss. **1,** habit, ×⅔; **2,** flower, ×4; **3,** stamen, ×4; **4,** basal petal, ×4; **5,** lateral petal, ×4; **6,** upper petal, ×4; **7,** gynoecium, ×4; **8,** fruit from apical part of raceme, ×4; **9,** fruit from basal part of raceme, ×4; **10,** seed, ×8. (**1–7** from *Casey* 1165; **8–9** from *Casey* 1032; **10** from *Gaillardot* 53.)

long, 1·5–2 mm. wide; sepals very minute, less than 0·4 mm. long, deltoid; upper petal rosy-pink, apex with a broad, reflexed wing, base prolonged into a short, blunt spur; lower petal with a distinct reddish keel and erect or spreading pink margins, inner (lateral) petals almost as long as the outer, with broad, blunt or subacute, blackish-violet apices. Fruits dimorphic, the lower 1-seeded, indehiscent, shortly cylindrical about 4–5 mm. long and 2 mm. wide, with distinctly ribbed walls and an abruptly conical or cuspidate apex; the upper 1–2-seeded, dehiscent, 12–14 mm. long, about 2 mm. wide, ribbed, with a gradually tapering, ensiform beak about 5 mm. long; seeds shining black, reniform, about 1·5–2 mm. long and 1–1·3 mm. wide. *Plate 6.*

HAB.: Rocky limestone slopes; 1,000–2,000 ft. alt.; fl. April–May.

DISTR.: Division 7, rare. Also in Palestine and Syria.

7. Mountains S.S.W. of Kyrenia, below the cliffs E. of St. Hilarion Castle, 1949, *Casey* 640 ! 1032 ! Trypa Vouno, 1951, *Casey* 1165 ! Kornos, *Casey*.

3. FUMARIA *L.*

Sp. Plant., ed. 1, 699 (1753).
Gen. Plant., ed. 5: 314 (1754).
Pugsley in Journ. Linn. Soc., 44: 233–355 (1919).

Annuals, often with slender, sprawling or climbing stems; leaves pinnately compound, with lobed or dissected leaflets; petioles often twining. Inflorescence leaf-opposed, racemose, with short pedicels; flowers with the upper (exterior) petal spurred, lower (exterior) petal not spurred. Fruit an indehiscent nutlet, with 2 apical pits or depressions (conspicuous only in dried fruits); endosperm with a hemispherical apical depression.

About 60 species, chiefly in Europe, the Mediterranean region and western Asia; a few representatives in eastern Asia, Atlantic Islands and on the mountains of Tropical Africa. Often occurring as weeds outside this area.

Apex of ripe (dry) fruit distinctly flattened or emarginate - - - **6. F. officinalis**
Apex of ripe (dry) fruit rounded, acute or apiculate:
  Corolla 9–14 mm. long; ultimate divisions of the leaflets rather broad, ovate-lanceolate, obovate or cuneate:
    Fruiting pedicels ascending:
      Fruits moderate-sized (2–3 mm. long); flowers purple-tipped:
        Flowers large, 10–12 (–14) mm. long; sepals 2–4 mm. long; apex of fruit apiculate - - - - - - - - - - **1. F. gaillardotii**
        Flowers medium, 9–10 mm. long; sepals 1–2 mm. long; apex of fruit obtuse **2. F. judaica**
      Fruits conspicuously large (3–4 mm. long); flowers usually pinkish without purple tips - - - - - - - - - - **3. F. macrocarpa**
    Fruiting pedicels recurved or patent:
      Peduncle as long as or longer than the raceme; corolla usually pale **4. F. capreolata**
      Peduncle very short or almost absent; corolla bright pink or crimson-pink **5. F. petteri ssp. thuretii**
  Corolla less than 9 mm. long, ultimate divisions of the leaflets very narrow, linear-lanceolate, linear or almost filiform:
    Sepals conspicuous, broader than the corolla; flowers usually pink:
      Corolla 6–7 mm. long; sepals 2·5–3 mm. long, 2–2·5 mm. wide; divisions of leaflets linear-lanceolate; fruit keeled but not conspicuously so **7. F. densiflora**
      Corolla 3·5–5 mm. long; sepals 1·5–2 mm. long and about as wide; divisions of leaflets linear-filiform; fruit conspicuously keeled - - **8. F. bracteosa**
    Sepals minute, much narrower than the corolla; flowers usually rather pallid **9. F. parviflora**

1. **F. gaillardotii** *Boiss.*, Fl. Orient., 1: 139 (1867); Pugsley in Journ. Linn. Soc., 44: 265 (1919); Post, Fl. Pal., ed. 2, 1: 46 (1932); Davis, Fl. Turkey, 1: 243 (1965).
TYPE: Lebanon; " ad Sidonem ", March 13, 1860, *Gaillardot* 685 (G, K !).

Stems diffuse or scrambling, 20–50 cm. (or more) high, robust, sparingly branched; leaves glaucescent, bipinnatisect, 5–12 cm. long, petiole of basal leaves 2–5 cm. (or more) long, channelled above; upper leaves sessile or subsessile; petiolules well developed, slender, sometimes twining, 1–3 cm. (or more) long; leaflets broadly cuneate, about 1 cm. long and 1 cm. wide, with flat, oblanceolate or oblong, acute or obtuse lobes; racemes rather lax, 3–4 cm. long in flower, lengthening to 6 cm. in fruit; peduncles stout, 1–2 cm. long; bracts whitish with a greenish midrib, oblong, acute or cuspidate, 3–5 mm. long, equalling or exceeding the pedicels; flowers large, 10–12 (–14) mm. long, pale purplish-pink; sepals rather large, broadly ovate or suborbicular, acute, 2–4 mm. long, about 2 mm. wide, white, membranous, with a greenish or purplish midrib and sharply laciniate margins; upper (adaxial) petal tipped dark purple, with a wide, reflexed marginal wing, lower (abaxial) petal narrowly winged, with a greenish keel and purple-stained tip, inner petals tipped dark purple. Fruiting pedicels suberect; fruits compressed-globose, 2·5–3·5 mm. diam., distinctly rugulose-verruculose, apex rounded or almost truncate, minutely apiculate, margins strongly keeled; apical pits shallow, rather obscure.

HAB.: Amongst weeds under trees; sea-level to 400 ft. alt.; fl. March–April.

DISTR.: Division 8, rare. Also Italy, Sicily, Yugoslavia, Greece, Crete, Aegean Islands, Turkey, Syria, Palestine and Egypt.

8. Ronnas valley near Rizokarpaso, 1957, *Merton* 2913 !

2. F. judaica *Boiss.*, Diagn., 1, 8: 15 (1849) emend. Pugsley in Journ. Linn. Soc., 47: 434 (1927); Boiss., Fl. Orient., 1: 138 (1867); H. Stuart Thompson in Journ. Bot., 44: 276 (1906); Holmboe, Veg. Cypr., 84 (1914); Post, Fl. Pal., ed. 2, 1: 46 (1932); Davis, Fl. Turkey, 1: 244 (1965); Zohary, Fl. Palaest., 1: 238, t. 350 (1966); Osorio-Tafall et Seraphim, List Vasc. Plants Cyprus, 44 (1973).

TYPE: " In Palaestina " April 1846. *Boissier* (G, K !).

Stem diffuse or suberect, 20–50 cm. high, sparingly branched towards the base; leaves glaucescent, bipinnatisect, 5–10 cm. long; petioles trigonous, channelled above, 3–7 cm. long, petiolules up to 1 cm. long, slender, sometimes twining; leaflets broadly cuneate, about 1 cm. long and 1 cm. wide, with flat, broadly lanceolate, acute or rather blunt lobes; racemes about 1–2 cm. long in flower, with a peduncle about 1–1·5 cm. long; raceme and peduncle lengthening greatly in fruit; bracts pale, lanceolate-acuminate, about 2–2·5 mm. long, as long as or rather shorter than the pedicels; flowers 9–10 mm. long, white or pink-tinged; sepals small, ovate-lanceolate, about 1–2 mm. long, irregularly toothed; outer petals with a distinct marginal wing, apex tipped with green; inner petals with a small purple apex. Fruiting pedicels erect or ascending, fruits compressed-globose, slightly keeled, 2·5–3 mm. diam., distinctly rugulose-verruculose, apex obtuse, with well-marked apical pits.

HAB.: Cultivated and waste ground; often on steep, stony slopes on limestone formations; sea-level to 1,500 ft. alt.; fl. Febr.–May.

DISTR.: Divisions 1, 4, 5, 7, 8, locally common around Famagusta in Division 4; common in Division 7. Yugoslavia, Albania, Greece, Aegean Islands, Palestine and Egypt.

1. Ayios Yeoryios Island (Akamas), 1962, *Meikle* 2163 ! Neokhorio (Akamas), 1962, *Meikle* 2204 !
4. Locally common; *Haradjian* 122 ! *Syngrassides* 1075 ! *Mavromoustakis* 29 !
5. Salamis, 1928, *Druce* s.n. !
7. Common; *A. G. & M. E. Lascelles* s.n. ! *Druce* s.n. ! *Chapman* 126 ! *Casey* 1106 ! *G. E. Atherton* 910 !
8. Rizokarpaso, 1912, *Haradjian* 262 ! Ayios Andronikos, Karpas, 1938, *Syngrassides* 1793 !

3. F. macrocarpa *Parl.*, Plant. Nov., 5 (1842); Boiss., Fl. Orient., 1: 137 (1867); Pugsley *loc. cit.*, 268 (1919); Post, Fl. Pal., ed. 2, 1: 45 (1932); Lindberg f., Iter

Cypr., 16 (1946); Davis, Fl. Turkey, 1: 244 (1965); Zohary, Fl. Palaest., 1: 238, t. 351 (1966); Osorio-Tafall et Seraphim, List Vasc. Plants Cyprus, 44 (1973).

F. oxyloba Boiss., Diagn., 1, 8: 14 (1849); Unger et Kotschy, Die Insel Cypern, 322 (1865).

F. macrocarpa Parl. var. *oxyloba* (Boiss.) Hammar in Act. Reg. Soc. Sci. Ups., ser. 3, 2: 45 (1857); Holmboe, Veg. Cypr., 84 (1914).

F. macrocarpa Parl. var. *laxa* Sintenis in Oesterr. Bot. Zeitschr., 32: 54, 122 (1882).

TYPE: " In insula Chio Archipelagi invenit cl. *Aucher-Éloy* " (Fl).

Very similar to *F. judaica* Boiss. in general appearance, but with the leaflets more acutely and narrowly lobed; the sepals usually much narrower, often almost linear, and the fruit exceptionally large, 3–4 mm. diam., tuberculate-rugose with an indistinct keel. The lower (exterior) petal is often sharply reflexed, and rather broadly winged towards the apex, giving it a spathulate outline.

HAB.: Cultivated and waste ground, occasionally on stony slopes or screes on limestone formations; sea-level to 4,000 ft. alt.; fl. Febr.–May.

DISTR.: Divisions 1–4, 6–8, common in Division 7 along the Northern Range. Eastern Mediterranean from the Dalmatian coast to Libya. Yugoslavia, Greece, Crete, Aegean Islands, Turkey and Palestine.

1. Paphos, Ayios Neophytos, 1939, *Lindberg f.* s.n.; Baths of Aphrodite, Akamas, 1957, *Merton* 3057 ! Ayios Yeoryios (Akamas), 1962, *Meikle* 2106 !
2. Near Prodhromos, 1862, *Kotschy* 907. Lefkara, 1941, *Kennedy & Davis* 2716 ! Askas beyond Palekhori, 1957, *Merton* 3013 !
3. Khoulou, 1905, *Holmboe* 755; Amathus, 1963, *Suart* 41 !
4. Larnaca, 1894, *Post* 178 ! Famagusta, 1936, *Syngrassides* 1072 ! Cape Greco, 1960, *N. Macdonald* 98 !
6. W. of Orga, 1956, *Merton* 2606 !
7. Common; *Sintenis & Rigo* 146 ! *Syngrassides* 904 ! 1440 ! *Casey* 310 ! 389 ! 1167 ! *Merton* 772 ! *Meikle* 2371 !
8. Rizokarpaso, 1930, *Druce* s.n. ! Apostolos Andreas Forest, 1957, *Merton* 2936 !

4. F. capreolata *L.*, Sp. Plant., ed. 1, 701 (1753); Boiss. Fl. Orient., 1: 136 (1867); Pugsley, *loc. cit.*, 269 (1919); Post, Fl. Pal., ed. 2, 1: 45 (1932); Davis, Fl. Turkey, 1: 244 (1965); Zohary, Fl. Palaest., 1: 238, t. 352 (1966); Osorio-Tafall et Seraphim, List Vasc. Plants Cyprus, 44 (1973).

TYPE: " *In* G. Narbonensi, Anglia ".

Stem scrambling or diffuse, 30 cm. or more in length, much branched towards the base; leaves pale green or slightly glaucous, bipinnate, 5–8 cm. or more in length; petiole, petiolule and rhachis slender, often twining; leaflets broadly cuneate, 1–1·5 cm. long and about as wide, with broad, flat, rather blunt lobes; racemes 1–2·5 cm. long in flower with a stout peduncle 1–3 cm. long; raceme and peduncle lengthening considerably in fruit; bracts pallid, ovate-lanceolate, about 2–3 mm. long, as long as or rather longer than the pedicels; flowers white or pinkish, medium-sized, about 9–10 mm. long; sepals large, ovate-acuminate, about 5 mm. long and 3 mm. wide, much wider than the corolla, denticulate towards the base; petals whitish or pale pink tipped maroon, the upper narrowly winged. Fruiting pedicels strongly recurved; fruits compressed-globose, obscurely keeled, about 2·5 mm. diam., smooth or minutely rugulose, apex obtuse, with well-marked apical pits.

HAB.: On walls or stony slopes, sometimes as a weed of cultivated ground; near sea-level; fl. March–May.

DISTR.: Division 7, rare. Central and southern Europe; Mediterranean region.

7. Near Kyrenia, 1880, *Sintenis & Rigo* 521 ! Kyrenia, 1949, *Casey* 455 ! Karavas, 8 m. W. of Kyrenia, 1951, *Merton* 299 !

5. F. petteri *Reichb.*, Icones Fl. Germ., 3: 1 (1838); Pugsley, *loc. cit.*, 290 (1919).

Stems erect or diffuse, 20 cm. or more in length, repeatedly branched; leaves irregularly 2–3-pinnate, glaucescent, 5 cm. or more long, petiole

2–3 cm. long; leaflets broadly cuneate, 0·5–1 cm. long and about as wide, deeply divided into numerous, rather unequal, linear-oblong, blunt, subacute or mucronate lobes; racemes lax or dense, many-flowered, 2–4 cm. long in flower, almost sessile or with a very short, stout peduncle rarely more than 7 mm. long; bracts purple-tinged, narrowly lanceolate-acuminate, 1·5–2 mm. long, about as long as the pedicels; flowers medium, about 9–10 mm. long; sepals broadly ovate, 2·5–3 mm. long, 1·5–2 mm. wide, irregularly denticulate, whitish or pale pink with a greenish median nerve; petals bright crimson-pink, tipped dark purple, upper petal with conspicuous reflexed margins. Fruiting pedicels short, erecto-patent, patent or recurved; fruit medium or small, compressed-globose, obscurely keeled, about 2–3 mm. long and 1·5–2 mm. wide, minutely rugulose, apex subacute or obtuse with small apical pits.

ssp. **thuretii** (*Boiss.*) *Pugsley* in Journ. Linn. Soc., 50: 550 (1937); Davis, Fl. Turkey, 1: 245 (1965); Osorio-Tafall et Seraphim, List Vasc. Plants Cyprus, 44 (1973).

F. thuretii Boiss., Diagn., 2, 1: 15 (1853); Fl. Orient., 1: 137 (1867); Unger et Kotschy, Die Insel Cypern, 323 (1865); Holmboe, Veg. Cypr., 84 (1914); Post, Fl. Pal., ed. 2, 1: 45 (1932); Zohary, Fl. Palaest., 1: 239, t. 353 (1966).

TYPE: " In insulis Principum prope Byzantium (cl. *Thuret*) " (G !).

Differs from typical *F. petteri* Reichb. in the leaflets being divided into more numerous, narrower lobes; the inflorescence more dense, and the fruits smaller, about 2 mm. long and 1·5 mm. wide.

HAB.: Cultivated and waste ground, roadsides, screes etc.; usually at high altitudes on igneous formations; 3,000–5,000 ft. alt.; fl. April–June.

DISTR.: Division 2, locally common around Prodhromos; also found at Amiandos, Papoutsa Peak and Stavros tis Psokas. Yugoslavia, Greece, Aegean Islands, Turkey and Palestine.

2. Locally common; *Kotschy* 911; *Sintenis & Rigo* 787 ! *Kennedy* 1176 ! *Davis* 3092 ! 3472 ! *D. P. Young* 7395 ! *Meikle* !

NOTES: The number 911 is attached by Kotschy to two distinct collectings, one identified as *F. thuretii*, the other as *F. officinalis*.

6. **F. officinalis** *L.*, Sp. Plant., ed. 1, 700 (1753); Unger et Kotschy, Die Insel Cypern, 323 (1865); Boiss. Fl. Orient., 1: 134 (1867); Holmboe, Veg. Cypr., 84 (1914); Pugsley, *loc. cit.*, 302 (1919); Post, Fl. Pal. ed. 2, 1: 44 (1932); Davis, Fl. Turkey, 1: 246 (1965); Osorio-Tafall et Seraphim, List Vasc. Plants Cyprus, 44 (1973).

TYPE: " *In* Europae *agris, cultis* ".

Stems erect or diffuse, 20 cm. or more in length, repeatedly branched; leaves irregularly 2–3-pinnate, glaucescent, 5–8 cm. or more long; petiole 1–5 cm. or more long; leaflets rather narrowly cuneate, 4–10 mm. long or longer, 3–8 mm. wide, deeply divided into 3–7 unequal, narrow, linear, acute lobes; racemes dense, many-flowered, 1–5 cm. long in flower, peduncle short, stout, about 1–1·5 cm. long; bracts purple-tinged, narrowly oblong-acute, 1–2·5 mm. long, minutely denticulate or cuspidate at apex, about as long as, or a little longer than the pedicels; flowers medium, about 7–9 mm. long; sepals broadly ovate 2·5–3 mm. long, 1·5–2 mm. wide, denticulate, pink or purplish; petals pink or purple-pink, tipped dark purple; upper petal with conspicuously reflexed margins. Fruiting pedicels 2–4 mm. long, ascending; fruit medium, about 2 mm. long and 2·5 mm. wide, ± smooth or minutely rugulose, apex distinctly emarginate.

HAB.: Cultivated fields and waste ground; sea-level to c. 5,000 ft. alt.; fl. March–June.

DISTR.: Divisions 1–3. Europe; Mediterranean region; Asia east to Iran; Atlantic Islands.

1. Stroumbi near Paphos, 1913, *Haradjian* 784 !

2. Prodhromos, 1862, *Kotschy* 911; Kato Lefkara, 1937, *Syngrassides* 1512 ! **Askas** beyond Palekhori, 1957, *Merton* 3114 !
3. Yermasoyia R., 1948, *Mavromoustakis* 2 ! Kolossi, 1971, *Townsend* 71/120 !

NOTES: See comment on *Kotschy* 911 under 5. *F. petteri* ssp. *thuretii*.

7. **F. densiflora** *DC.*, Cat. Hort. Monsp., 113 (1813); Holmboe, Veg. Cypr., 84 (1914); Post, Fl. Pal., ed. 2, 1: 44 (1932); Lindberg f., Iter Cypr., 16 (1946); Davis, Fl. Turkey, 1: 246 (1965); Zohary, Fl. Palaest., 1: 239, t. 355 (1966); Osorio-Tafall et Seraphim, List Vasc. Plants Cyprus, 44 (1973).

F. micrantha Lag., Elench. Hort. Matrit., 21 (1816); Boiss., Fl. Orient., 1: 136 (1867); Pugsley, *loc. cit.*, 299 (1919).

TYPE: France; " in agris circa Telonem detexit cl. Ziz " (G).

Stems diffuse or suberect, 15–30 cm. or more in length, much-branched particularly at the base; leaves 2–3-pinnate, glaucescent or purplish, 2–6 cm. long, petiole 1–5 cm. long; leaflets rather narrowly cuneate, about 4–10 mm. long and 3–8 mm. wide, deeply divided into 5–9 unequal, very narrow, linear, acute lobes; racemes dense, many-flowered, 1–2·5 (–4) cm. long in flower; peduncle very short, stout, about 4–10 mm. long, bracts purplish, tongue-shaped, often mucronate or shortly cuspidate and minutely denticulate at the apex, 1·5–2·5 mm. long, as long as or rather longer than the pedicels; flowers small, 5–7 mm. long; sepals broadly ovate or almost orbicular, wider than the corolla, about 2·5–3 mm. long, and 2–2·5 mm. wide, irregularly denticulate, pale pink or purplish; petals bright purplish-pink or crimson, tipped dark purple; upper petal with a greenish keel and distinctly reflexed margins. Fruiting pedicels 2–3 mm. long, patent or ascending; fruit subglobose about 2–2·5 mm. diam., distinctly rugulose, apex rounded, with shallow pits.

HAB.: Cultivated fields and waste ground; sea-level to 2,000 ft. alt.; fl. Jan.–May.

DISTR.: Divisions 1–7. Central and southern Europe; Mediterranean region east to Iran.

1. Ayios Yeoryios (Akamas), 1962, *Meikle* 110 !
2. Galata, Mile 37, Nicosia–Troödos Road, 1951, *Merton* 90 !
3. Limassol, 1953, *Mavromoustakis* 2 ! Kolossi, 1971, *Townsend* 71/121 !
4. The Phaneromene near Larnaca, 1905, *Holmboe* 129a.
5. Kythrea, 1939, *Lindberg f.* s.n.
6. Near Nicosia, 1934, *E. Haig* 8 ! Near Nicosia, by Larnaca road, 1941, *Davis* 2140A !
7. Kyrenia, 1948, *Collingwood Ingram* 36 ! and, 1949, *Casey* 222 ! also, 1956, *G. E. Atherton* 843 ! 909 !

8. **F. bracteosa** *Pomel*, Nouv. Mat. Fl. Atlant., 239 (1874); Pugsley, *loc. cit.*, 300 (1919); Druce in Rep. B. E. C., 9: 469 (1931).

F. parlatoriana Kralik, Plant. Aegypt. Exs. (nomen); Post, Fl. Pal., ed. 2, 1: 44 (1932)—as *F. Parlatorians*.

F. micrantha Lag. var. *parlatoriana* Boiss., Fl. Orient., 1: 136 (1867).

TYPE: Algeria; " cultures aux environs d'Oran " *Pomel* (? P).

Stems diffuse, 10–30 cm. high, branched chiefly at the base; leaves glaucous, similar to those of *F. densiflora* DC. but with the ultimate lobes even narrower and finely linear-setaceous; racemes very dense in flower, 1–2 cm. long; peduncles short, stout, usually less than 5 mm. long; bracts large, purplish, oblong-acute, minutely denticulate, usually more than 2 mm. long, and longer than the pedicels; flowers very small, 3·5–5 mm. long; sepals broadly ovate or suborbicular, wider than the corolla, about 1·5–2 mm. long and almost as wide, irregularly denticulate, purplish; petals purplish-pink tipped dark purple; upper petal with a greenish keel and shortly reflexed, dark purple margins, and with a very short spur. Fruiting pedicels 1·5–2 mm. long, patent or ascending; fruit subglobose, about 2 mm. diam., rugulose and distinctly keeled; apex rounded or slightly flattened, with indistinct pits.

HAB.: Cultivated and waste ground; sea-level to c. 600 ft. alt.; fl. Jan.–May.

DISTR.: Divisions 1, 5–8. Also N. Africa, Egypt, Palestine and eastwards to Iraq.

1. Paphos, 1928 or 1930, *Druce* s.n.; Near Polis, 1962, *Meikle* 2264 !
5. Kythrea, 1880, *Sintenis & Rigo* 145 (mixed with *F. parviflora* Lam.) ! Mile 3, Nicosia–Famagusta Rd., 1950, *Chapman* 377 !
6. Near Nicosia on Larnaca Road, 1941, *Davis* 2140 (mixed with *F. parviflora* Lam.) !
7. Kyrenia, 1956, *G. E. Atherton* 851A !
8. Near Apostolos Andreas, 1962, *Meikle* 2460 !

9. F. parviflora *Lam.*, Encycl. Méth., 2: 567 (1788); Boiss., Fl. Orient., 1: 135 (1867); Holmboe, Veg. Cypr., 84 (1914); Pugsley, *loc. cit.*, 322 (1919); Post, Fl. Pal., ed. 2, 1: 44 (1932); Lindberg f., Iter Cypr., 16 (1946); Davis, Fl. Turkey, 1: 247 (1965); Zohary, Fl. Palaest., 1: 240, t. 356 (1966); Osorio-Tafall et Seraphim, List Vasc. Plants Cyprus, 44 (1973).

F. asepala (? Boiss.) A. K. Jackson et Turrill in Kew Bull., 1939: 474 (1939); Osorio-Tafall et Seraphim, List Vasc. Plants Cyprus, 44 (1973).

TYPE: France; "Provinces meridionales" and "aux environs de Paris" *Lamarck* (P).

Stems diffuse or suberect, 5–25 cm. long, several times branched; leaves glaucous, 1·5–4 cm. long, with a petiole 0·5–2 cm. long, 2–3-pinnate, the ultimate segments divided into very narrow linear, slightly channelled lobes; racemes rather dense in flower, 1–1·5 (–2) cm. long, peduncle very short or almost absent; bracts pallid or tinged purple, oblong-acute, minutely denticulate, 1·5–2 mm. long, usually slightly longer than the pedicels; flowers small, 4–5 mm. long; sepals minute, about 0·5 mm. long, ± deltoid-acute with an irregularly denticulate margin; petals whitish or tinged pinkish-purple, inner petals tipped dark purple; upper petal with a greenish keel, the winged margins strongly reflexed especially towards the apex; basal petal distinctly winged, spathulate. Fruiting pedicels short, usually less than 2 mm. long, patent or ascending; fruit subglobose, about 2 mm. long, strongly verruculose-rugulose, apex acute or apiculate with indistinct pits.

HAB.: Cultivated and waste ground; sea-level to 6,300 ft. alt.; fl. Jan.–July.

DISTR.: Divisions 1–8, locally common in Division 7. Central and southern Europe, Mediterranean region, Asia Minor and Orient E. to Afghanistan, Atlantic Islands; introduced elsewhere.

1. Ayios Minas near Smyies, 1962, *Meikle* 2173 ! Near Polis, 1962, *Meikle* 2313 !
2. Makheras Monastery, 1937, *Syngrassides* 1623 ! Platres, 1938, *Kennedy* 1175 ! Khionistra, 1938, *Kennedy* 1177 ! Near Kalopanayiotis, 1939, *Lady Loch* 63 !
3. Stavrovouni, near the monastery, 1939, *Lindberg f.* s.n.
4. Paralimni, 1940, *Davis* 2057 !
5. Kythrea, 1880, *Sintenis & Rigo* 145 (mixed with *F. bracteosa* Pomel) !
6. Near Nicosia on Larnaca road, 1941, *Davis* 2140 (mixed with *F. bracteosa* Pomel) ! Morphou, 1951, *Merton* 138 ! Above Dheftera, 1952, *Probyn* 71 !
7. Locally common; *Syngrassides* 1031 ! *Casey* 248 ! *Chapman* 91 ! *G. E. Atherton* 801 ! 844 ! 851B !
8. Apostolos Andreas Monastery, 1938, *Syngrassides* 1773 !

NOTES: *Kennedy* 1177, superficially resembling *F. asepala* Boiss., is without the relatively broad, blunt seaf-segments of that species; it also lacks the very broadly winged lower petal of *F. asepala*. Additional material should be looked for on Khionistra, since the amount available is too sparse for adequate identification.

[PLATYCAPNOS SPICATA (*L.*) *Bernh.* from S.W. Europe and the western Mediterranean has been recorded (as *Fumaria spicata* L.) from "Larnica and Limosol" in the list of plants collected in Cyprus by Dr. Hume, published in Walpole's Memoirs relating to European and Asiatic Turkey, ed. 1 (1817), p. 254. The occurrence of this distinctive species on the island is unlikely, and it is probable that some other *Fumaria* species was misidentified.]

9. CRUCIFERAE

O.E. Schulz in Engl. et Prantl, Pflanzenfam., ed. 2, 17b: 227–658 (1936).

Annual, biennial or perennial herbs or (rarely) subshrubs; leaves exstipulate, alternate or basal, occasionally opposite, simple, entire or variously lobed or dissected. Flowers nectariferous, usually in racemes, rarely solitary or scapose, hermaphrodite and generally actinomorphic; sepals 4, free, imbricate in 2 whorls; petals usually 4, cruciately arranged, imbricate or contorted, frequently produced at the base into a long claw, normally equal (unequal in *Iberis* and *Calepina*); stamens typically 6, in 2 whorls, tetradynamous, the filaments of each of the inner pairs sometimes connate; anthers 2- or rarely 1-locular, dehiscing longitudinally; torus often furnished with 1–4 sessile nectary glands commonly placed opposite the sepals; ovary superior generally sessile, apparently of 2 united carpels, divided into 2 loculi by a false, membranous septum; placentation parietal, ovules few or many; style 1, sometimes obsolete; stigmas 1 or 2; fruit a valvately dehiscent (elongate) silique or a (short) silicle, sometimes indehiscent or transversely jointed. Seeds without endosperm (or with a little endosperm); cotyledons accumbent (with their margins against the radicle), incumbent (cotyledons face to face with the back of one against the radicle) or variously folded.

A large family of about 350 genera and 2,500 species most abundantly represented in temperate regions of the northern hemisphere, but occurring as natives or introductions in most regions of the world.

Petals white, pink, purple or mauve, not yellow or yellowish:
  Leaves clothed with a ± dense tomentum of compound hairs:
    Basal leaves forming a distinct rosette; cauline leaves usually amplexicaul or auriculate - - - - - - - - - - - - - **33. Arabis**
    Basal leaves not forming a distinct rosette; cauline leaves usually narrowed to a sessile or shortly petiolate base - - - - - - - **36. Matthiola**
  Leaves glabrous or pilose, not tomentose; hairs usually simple or bifurcate:
    Petals small, less than 5 mm. long; individual flowers inconspicuous:
      Basal leaves pinnate, pinnatisect, deeply lobed or divided:
        Fruits reniform or cordate; inflorescences usually short, mostly leaf-opposed **19. Coronopus**
        Fruits not reniform or cordate; inflorescences usually elongate, rarely leaf-opposed:
          Fruit an elongate, linear silique:
            Leaves pinnate; leaflets broad, ovate, oblong or almost orbicular:
              Leaves mostly basal, forming a distinct rosette - - **32. Cardamine**
              Leaves scattered along the stems, not forming a distinct basal rosette **35. Nasturtium**
            Leaves pinnatifid or pinnatisect with narrow acute lobes **41. Torularia**
          Fruit a short silicle:
            Silicle distinctly flattened or compressed; inflorescence an unbranched or sparingly branched raceme:
              Seeds 1 or 2 in each loculus:
                Seeds 1 (rarely 2) in each loculus; filaments without appendages **16. Lepidium**
                Seeds 2 in each loculus; filaments with scale-like appendages **17. Teesdalia**
              Seeds several or numerous (3 or more) in each loculus:
                Silicle deltoid or deltoid-obcordate, distinctly notched at apex; cauline leaves auriculate - - - - - - **23. Capsella**
                Silicle elliptical or ovate, not notched at apex; cauline leaves not auriculate - - - - - - - - **24. Hymenolobus**
            Silicle not flattened or compressed, usually globose; inflorescence a much-branched panicle - - - - - - - - **10. Crambe**

Basal leaves simple, entire or variously toothed, not deeply lobed or divided:
Fruit an elongate, linear silique:
Petals white:
Cauline leaves distinctly auriculate, oblong - - - **33. Arabis**
Cauline leaves not auriculate:
Silique distinctly torulose; fruiting pedicels short, thick **41. Torularia**
Silique not torulose; fruiting pedicels elongate, slender **42. Arabidopsis**
Petals mauve, pink or purple - - - - - - - **37. Malcolmia**
Fruit a short silicle:
Cauline leaves absent; basal leaves forming a small rosette **31. Erophila**
Cauline leaves present:
Flowers in a much-branched terminal panicle:
Cauline leaves amplexicaul or auriculate; silicle 2–4 mm. long **18. Cardaria**
Cauline leaves tapering to base, not amplexicaul or auriculate; silicle 1·5–2 mm. long - - - - - - - - **16. Lepidium**
Flowers in unbranched or sparingly branched racemes:
Silicle ovoid or subglobose, not compressed or flattened **9. Calepina**
Silicle distinctly compressed or flattened:
Silicle without a distinct apical notch, ovate, elliptical or orbicular **29. Lobularia**
Silicle with a distinct apical notch, often winged:
Infructescence an elongate raceme:
Silicle deltoid or deltoid-obcordate, not winged **23. Capsella**
Silicle ovate, oblong, obcordate or suborbicular, usually with a distinct wing, at least in the upper part:
Cauline leaves auriculate; basal leaves long-stalked, often spathulate; seeds several in each loculus of the silicle **22. Thlaspi**
Cauline leaves not auriculate; basal leaves not long-stalked or spathulate; seeds 1 or 2 in each loculus:
Seeds 1 (rarely 2) in each loculus; filaments without appendages - - - - - **16. Lepidium**
Seeds 2 in each loculus; filaments with scale-like appendages - - - - - - **17. Teesdalia**
Infructescence sub-umbellate or shortly racemose; leaves narrow, toothed towards apex - - - - - - **20. Iberis**

Petals rather large, more than 5 mm. long, individual flowers conspicuous:
Lower leaves compound or deeply lobed:
Leaves thick, fleshy; flowers lilac or violet-tinged, seashore plant **14. Cakile**
Leaves not thick or fleshy:
Inflorescence wholly or partly bracteate - - - **8. Enarthrocarpus**
Inflorescence ebracteate:
Fruit usually containing 2 seeds (occasionally 1), 2-jointed, dehiscing transversely - - - - - - - - - - **12. Didesmus**
Fruit containing more than 2 seeds:
Lower leaves pinnate or bipinnate with numerous small leaflets, or else deeply divided into narrow, linear segments:
Fruit dehiscing longitudinally; petals white - - **32. Cardamine**
Fruit 2-jointed, upper joint indehiscent; petals mauve or pale purple (rarely white) - - - - - - **13. Erucaria**
Lower leaves coarsely pinnate or lyrate-pinnatisect or lobed, with few oblong or ovate lateral leaflets or segments:
Fruit dehiscing longitudinally - - - - - - **5. Eruca**
Fruit dehiscing transversely or indehiscent - - **7. Raphanus**

Lower leaves simple, entire or shortly lobed:
Lower leaves cordate, reniform or suborbicular; whole plant smelling of garlic when crushed - - - - - - - - - **39. Alliaria**
Lower leaves not as above; plant not smelling of garlic:
Leaves thick, fleshy, ± glaucous; silique 6–10 mm. wide - **1. Brassica**
Leaves not thick or fleshy; silique less than 6 mm. wide:
Cauline leaves sessile, amplexicaul or distinctly auriculate, not tapering to base:
Basal leaves and cauline leaves ± hairy - - - - **33. Arabis**
Basal leaves ± hairy; cauline leaves glabrous - - **34. Turritis**
Cauline leaves narrowed to base, not amplexicaul or auriculate **37. Malcolmia**

Petals yellow or yellowish:
Flowers in small axillary clusters; fruit a short, curved silique - **40. Sisymbrium**
Flowers in terminal or leaf-opposed racemes:
Fruit an elongate, usually linear, silique or loment:
Cauline leaves sessile, conspicuously auriculate:
Basal leaves ± hairy, toothed or lobed; cauline leaves glabrous; silique not conspicuously 4-angled - - - - - - - - - **34. Turritis**
All the leaves glabrous and entire; siliques conspicuously 4-angled **15. Conringia**
Cauline leaves not as above:
Inflorescence wholly or partly bracteate - - - - **8. Enarthrocarpus**
Inflorescence ebracteate:
Leaves and stems clothed with adpressed bifurcate hairs; silique ± 4-angled **38. Erysimum**
Leaves clothed with simple, unbranched hairs or glabrous:
Fruit (silique) erect, adpressed to rhachis:
Silique gradually tapering from base to apex - **40. Sisymbrium**
Silique ± parallel-sided, abruptly narrowed at apex:
Silique distinctly 4-angled; beak usually empty - **1. Brassica**
Silique not 4-angled; beak usually containing 1 or more seeds **3. Hirschfeldia**
Fruit spreading, not adpressed to rhachis:
Fruit indehiscent or breaking transversely into 1-seeded joints **7. Raphanus**
Fruit dehiscing longitudinally:
Seeds in 1 row in each loculus of the silique:
Silique terminating in a distinct beak; beak often with 1 or more seeds:
Valves of silique with a prominent midrib; lateral veins obscure; sepals ± erect - - - - - - - **1. Brassica**
Valves of silique 3–7-veined, lateral veins conspicuous; sepals ± spreading - - - - - - - - **2. Sinapis**
Silique without a distinct beak - - - - **40. Sisymbrium**
Seeds in 2 rows in each loculus of the silique - - **4. Diplotaxis**
Fruit a silicle:
Silicle distinctly flattened:
Leaves clothed with compound stellate hairs:
Silicle large, 1·5–2 cm. long - - - - - - **27. Fibigia**
Silicle less than 1 cm. long:
Silicle pendulous, distinctly winged - - - - - **30. Clypeola**
Silicle erect or spreading, not or indistinctly winged - **28. Alyssum**
Leaves clothed with simple or occasionally bifurcate hairs, or glabrous:
Silicles paired; lower leaves usually serrate or shortly lobed **21. Biscutella**
Silicles solitary; lower leaves deeply lobed or dissected - **16. Lepidium**
Silicle not flattened:
Silicle 2-jointed:
Upper joint flattened, oblong-elliptical or tongue-shaped; leaves pinnately or bipinnately divided into linear segments - - **6. Carrichtera**
Upper joint globose, rugulose-sulcate; leaves not divided into linear segments - - - - - - - - - - - **11. Rapistrum**
Silicle not jointed:
Silicle dehiscent, ovoid or ellipsoid; valves smooth or obscurely veined **43. Camelina**
Silicle indehiscent; valves rugose or reticulate:
Silicle reticulate; fruiting pedicel long, slender - - - **26. Neslia**
Silicle coarsely tuberculate-rugose, fruiting pedicel short, thick **25. Ochthodium**

TRIBE 1. **Brassiceae** *DC. emend. Hayek* Annual, biennial or perennial herbs, glabrous or with simple hairs. Fruit usually a silique, the lower part dehiscing longitudinally, the upper part forming an indehiscent beak often containing 1 or more seeds; fruit rarely with transverse joints or indehiscent. Lateral nectary-glands situated on the inner side of the shorter stamens; median glands situated between the longer stamens.

1. BRASSICA *L.*

Sp. Plant., ed. 1, 666 (1753).
Gen. Plant., ed. 5, 299 (1754).

Annual, biennial or perennial herbs, stems and leaves glabrous or with simple hairs; leaves alternate, simple or pinnatipartite, the basal sometimes crowded and forming a rosette. Inflorescence racemose, ebracteate; sepals usually erect or suberect, the outer oblong, apex obtuse and ± cucullate; inner sepals rather broader than the outer, ovate, subacute, somewhat saccate at the base; petals obovate, yellow (or rarely white) narrowed into a long basal claw. Stamens 6; anthers oblong, obtuse or subacute at apex; ovary cylindrical 5–45-ovulate, generally narrowed above into a distinct style; stigma capitate or ± bilobed. Siliques linear or oblong, straight or curved, biconvex in transverse section, usually compressed dorsally, and prolonged into a conical, cylindrical or filiform beak; valves with a prominent median nerve and more slender lateral nerves. Seeds globose or ovoid, uniseriate; cotyledons longitudinally conduplicate, distinctly emarginate at apex.

About 40 species chiefly in the Mediterranean area.

Leaves fleshy, thick, obovate-orbicular; petals white; silique 4–7·5 cm. long, 6–10 mm. wide - - - - - - - - - - - - - - 1. **B. hilarionis**
Leaves not fleshy; petals yellow; silique 1–4 cm. long, 1·5–2·5 mm. wide:
Siliques spreading, 3·5–4 cm. long; petals pale yellow - - 2. **B. tournefortii**
Siliques adpressed to rhachis, 1–1·8 cm. long; petals bright yellow - 3. **B. nigra**

1. **B. hilarionis** *Post* in Mém. Herb. Boiss., no. 18: 90 (1900); Holmboe. Veg. Cypr., 88 (1914); Osorio-Tafall et Seraphim, List Vasc. Plants Cyprus, 45 (1973).
B. cretica Lam. var. δ *hilarionis* (Post) O. E. Schulz in Engl., Pflanzenr., 70 (IV, 105): 37 (1919); Davis, Fl. Turkey, 1: 264 (1965).
B. cretica Lam. ssp. *hilarionis* Lindberg f., Iter Cypr., 17 (1946).
[*B. cretica* (non Lam., Encycl. Méth., 1: 747; 1785) Kotschy in Unger et Kotschy, Die Insel Cypern, 332 (1865); Boiss., Fl. Orient. Suppl., 66 (1888).]
TYPE: Cyprus; " in rupium fissuris ad castellum sancti Hilarionis Cypri ", 1894, *Post* s.n. (BE).

An erect or spreading, glabrous perennial 50–80 cm. high; stems stout, often distinctly ridged; leaves rather fleshy, thick, ± glaucous, the lower often forming a basal rosette, petiolate, obovate-orbicular, 5–20 cm. long, 3·5–14 cm. wide, margin undulate-crenate or bluntly dentate, occasionally deeply notched on either side towards base, often auriculate; petioles flattened and often narrowly winged, 1–8 cm. long; upper leaves smaller, becoming oblong-lanceolate, ± entire, auriculate or amplexicaul; inflorescences terminal and lateral, laxly racemose; pedicels 7–15 mm. long, elongating and patent or ascending in fruit; flowers white; sepals oblong-lanceolate, 8–10 mm. long, 2–3 mm. wide, obtuse, with membranous margins, often tinged pink or purple; petals obovate, blunt, entire, 1·5–2·5 cm. long, 4–6 mm. wide, narrowed into a basal claw; stamens 1·2–1·4 cm. long; anthers linear, 3–5 mm. long; stigma retuse. Silique erecto-patent, broadly linear, 4–7·5 cm. long (including beak), 6–10 mm. wide; beak 7–9 mm. long, tapering, generally empty; valves 1-nerved, ± smooth or often reticulately rugose; seeds subglobose, dull brown, 1·5–2 mm. diam.; testa minutely reticulate. *Plate 7.*

HAB.: In fissures of limestone cliffs; 1,000–2,800 ft. alt.; fl. March–May.

DISTR.: Division 7, locally common, Northern Range, Kornos to Yaïla Peaks. Endemic.

7. Locally common; *Kotschy* 397; *Sintenis & Rigo* 631 ! *Post* s.n. *Holmboe* 820; *Kennedy* 482 ! 1656 ! *Davis* 2499 ! 2971 ! 3029 ! 3597 ! etc.

Plate 7. BRASSICA HILARIONIS Post **1**, flowering stem, $\times\frac{2}{3}$; **2**, leaf, $\times\frac{2}{3}$; **3**, fruits, $\times\frac{2}{3}$; **4**, petal, $\times 1$; **5**, sepal, $\times 1$; **6**, gynoecium, $\times 1$; **7**, stamen, $\times 1$; **8**, seed, $\times 8$. (**1**, **2**, **4–7** from *Davis* 2499; **3** from *Sintenis* & *Rigo* 631; **8** from *Kennedy* 482.)

2. B. tournefortii *Gouan*, Illustr. Observat. Bot., 44, t. 20A (1773); Boiss., Fl. Orient., 1: 393 (1867); Holmboe, Veg. Cypr., 88 (1914); O. E. Schulz in Engl., Pflanzenr., 70 (IV, 105): 67 (1919); Davis, Fl. Turkey, 1: 265 (1965); Zohary, Fl. Palaest., 1: 309 (1966); Osorio-Tafall et Seraphim, List Vasc. Plants Cyprus, 45 (1973).

TYPE: Not localized, *in herb. Gouan* (K !).

An erect, simple or branched annual herb, 10–70 cm. high; stem straight or flexuous, striate, glabrous and often violet-tinged above, hispid below; radical leaves sessile or shortly petiolate, usually forming a rosette, oblanceolate or obovate (1·5–) 3–11 (–24) cm. long, 0·5–3 (–5·5) cm. wide, obtuse, lyrate-pinnatisect, lateral segments oblong, repand-dentate, hispid, especially along the nerves and veins; upper leaves few, smaller, sessile or subsessile, 10–30 mm. long, 1–10 mm. wide, acute, entire, dentate (or occasionally ± pinnatisect), glabrous or thinly hispid; inflorescences terminal and lateral, loosely racemose; flowers pale yellow or primrose-yellow; pedicels ascending, 1–6 mm. long, greatly elongating in fruit; sepals ovate or oblong-lanceolate, 3–4 mm. long, 1–1·5 mm. wide, acute or subacute, glabrous (or very sparsely hispid) with membranous margins; petals oblanceolate or obovate, clawed, 4–7 mm. long, 1·5–2 mm. wide, apex obtuse, rounded, entire; stamens 2–4 mm. long; anthers linear-oblong, about 1 mm. long; stigma obscurely retuse. Silique straight, ascending, linear, compressed, slightly constricted between the seeds, 3·5–4 cm. long (including beak), 1·5–2·5 mm. wide, median nerve prominent, lateral veins rather indistinct; beak 1·2–1·5 cm. long, slender, tapering, usually containing 1 or sometimes 2–3 seeds; seeds purplish-brown, ovoid-globose, about 1 mm. diam.; testa minutely foveolate.

HAB.: Sandy seashores; occasionally a weed in gardens; near sea-level; fl. Febr.–May.

DISTR.: Divisions 2–4, 5, 8. Mediterranean region eastwards to the Caspian and Baluchistan, adventive elsewhere.

2. Yialia, 1955, *Merton* 2110 ! and, 1962, *Meikle* 2287 !
3. Near Amathus, 1947, *Mavromoustakis* s.n. !
4. Near Dhekelia [" Redgelia "], Cape Pyla, 1880, *Sintenis & Rigo* 849 ! Ayia Napa, 1905, *Holmboe* 14; Famagusta, 1936, *Syngrassides* 1066 ! Near Ayios Memnon, 1948, *Mavromoustakis* s.n. !
5. Salamis, 1957, *Merton* 3205 !
8. Ayios Philon near Rizokarpaso, 1941, *Davis* 2218 !

3. B. nigra (*L.*) *Koch* in Röhling, Deutsch. Fl., ed. 3, 4: 713 (1833); Boiss., Fl. Orient., 1: 390 (1867); Holmboe, Veg. Cypr., 88 (1914); O. E. Schulz in Engl., Pflanzenr., 70 (IV, 105): 75 (1919); Post, Fl. Pal., ed. 2, 1: 120 (1932); Rechinger f. in Arkiv för Bot., ser. 2, 1: 422 (1950); Davis, Fl. Turkey, 1: 265 (1965); Zohary, Fl. Palaest., 1: 309, t. 458 (1966); Osorio-Tafall et Seraphim, List Vasc. Plants Cyprus, 45 (1973).

Sinapis nigra L., Sp. Plant., ed. 1, 668 (1753).

TYPE: " *per* Scaniam, Germaniam, Belgium, Angliam etc." (BM !).

An erect annual or biennial herb to 1 m. or more in height, glabrous or sometimes with the lower parts of the stems and lower leaves sparingly hispid; leaves scattered, petiolate, petioles 1·5–4 (–7) cm. long; lower leaves broadly ovate, 14–24 cm. or more long, 8–15 cm. or more wide, lyrate-pinnatisect, terminal lobe very large, margins entire or irregularly denticulate or lobed; upper leaves few, ovate oblong or lanceolate, 2–10 cm. long, 1–6 cm. wide, pinnatisect or irregularly lobed, entire or denticulate; inflorescences terminal and lateral, densely racemose; pedicels 1·5–5 mm. long; sepals oblong, obtuse, 3–5 mm. long, about 1–1·5 mm. wide, glabrous, with membranous margins; petals bright yellow, spathulate-obovate, 6–9 mm. long, 3–3·5 mm. wide, obtuse or retuse; stamens 6–8 mm. long; anthers linear-oblong, 1–1·5 mm. long; stigma rounded or obscurely retuse. Silique erect, adpressed to rhachis, 10–18 mm. long, 2–2·5 mm. wide, distinctly

4-angled; beak slender, seedless, about 2–3 mm. long; seeds dark brown, ± globose, about 1 mm. diam.; testa reticulate-foveolate.

HAB.: Cultivated and waste ground; sea-level to 2,000 ft. alt.; fl. March–June.

DISTR.: Divisions 1, 3, 7, rare. Europe, Mediterranean region and Atlantic Islands; naturalized elsewhere.

1. Stroumbi near Paphos, 1913, *Haradjian* 789 !
3. Kophinou, 1905, *Holmboe* 574.
7. Lambousa, 1955, *G. E. Atherton* 343 (in part) !

BRASSICA OLERACEA *L.*, Sp. Plant., ed. 1, 667 (1753), the Cabbage, has been recorded from Cyprus by Kotschy (Die Insel Cypern, 331) and by Stuart Thompson (Journ. Bot., 44: 277), both on the authority of a statement said to have been published by Gaudry (Recherches Scientifiques en Orient, 185; 1855). Gaudry, however, merely states that the Cabbage (amongst other western European vegetables) is cultivated in Cyprus, and makes no claim that the species is indigenous in the island. The quotation from Columella cited by Gaudry and by Kotschy—" Est et (Brassica) Cyprii generis ex albo rubicunda, levi et tenerrimo folio "—is also misleading, for Columella (De Re Rustica, lib. XI, cap. III) is referring not to a *Brassica*, but to a *Lactuca*.

2. SINAPIS *L.*

Sp. Plant., ed. 1, 668 (1753).
Gen. Plant., ed. 5, 299 (1754).

Annual herbs, very similar to *Brassica* in general appearance but with the valves of the silique distinctly 3–7-nerved, the lateral nerves often as distinct as the median. The sepals are said to be spreading in *Sinapis* and ascending or erect in *Brassica*, but this distinction is not always clear.

Ten species chiefly in Europe and the Mediterranean region.

Beak of fruit conical, tapering, shorter than the valves; ovules 8–17; upper leaves narrowly ovate or lanceolate - - - - - - - - - **1. S. arvensis**
Beak of fruit flattened, sword-shaped, equalling or exceeding the valves; ovules 4–8; all the leaves pinnatisect - - - - - - - - - - - - **2. S. alba**

1. **S. arvensis** *L.*, Sp. Plant., ed. 1, 668 (1753); Unger et Kotschy, Die Insel Cypern, 332 (1865); Boiss., Fl. Orient., 1 : 394 (1867); Holmboe, Veg. Cypr., 89 (1914); O. E. Schulz in Engl., Pflanzenr., 70 (IV, 105): 119 (1919); Post, Fl. Pal., ed. 2, 1: 123 (1932); Lindberg f., Iter Cypr., 17 (1946); Davis, Fl. Turkey, 1: 266 (1965); Zohary, Fl. Palaest., 1: 311, t. 460 (1966); Osorio-Tafall et Seraphim, List Vasc. Plants Cyprus, 45 (1973).

Erect, simple or branched annual up to 80 cm. or more in height; stems indistinctly furrowed, glabrescent, or hispid particularly towards the base, often stained purple; leaves scattered, obovate or lyrate-pinnatifid, glabrescent or hispid especially on midrib and nerves, the lower leaves distinctly petiolate, 8–20 cm. or more long, 2–8 cm. or more wide, with bluntly and irregularly dentate margins; the upper leaves sessile or subsessile, smaller and narrower, irregularly toothed or lobed; inflorescence rather densely racemose; pedicels 2–3 mm. long, lengthening slightly and thickening considerably in fruit; sepals oblong, about 5 mm. long, concave, glabrous or sparsely hispid; petals bright yellow or lemon-yellow, broadly obovate, 8–12 mm. long, about 5 mm. wide, narrowed to a long slender claw; stamens about 5 mm. long; anthers oblong-linear, about 1·5–2 mm. long; stigma rather large, flattened or retuse. Siliques ascending 2·5–4 (–5·5) cm. long (including beak), 2–3 mm. wide, cylindrical, obscurely or distinctly ribbed, glabrous, thinly hispid or occasionally densely hispid with retrorse bristles; beak conical, tapering, straight, 1–2 cm. long, usually containing a single seed; seeds globose, about 1·5 mm. diam., fuscous or dull brown; testa minutely foveolate.

var. **arvensis**
TYPE: " *in agris* Europae ".

Siliques erecto-patent, glabrous or subglabrous, 2·5–4 cm. long.

HAB.: Cultivated and waste ground; sea-level to 4,500 ft. alt.; fl. March–Sept.

DISTR.: Divisions 2, 4, 6–8. Throughout Europe and the Mediterranean area, eastwards to Siberia and Afghanistan; Atlantic Islands; adventive elsewhere.

2. Troödhitissa Monastery, 1853–4, *Gaudry*; Prodhromos, 1955, *G. E. Atherton* 662 !
4. Paralimni, 1940, *Davis* 2068 !
6. Krini, 1937, *Syngrassides* 1424 (in part) ! Nicosia, 1962, *Meikle* 2550 !
7. Myrtou, 1936, *Syngrassides* 1218 ! Kythrea, 1939, *Lindberg f.* s.n.; Bellapais, 1956, *G. E. Atherton* 858 ! Kantara, 1962, *Meikle* 2950 !
8. Apostolos Andreas, Karpas, 1938, *Syngrassides* 1774 !

var. **orientalis** (*L.*) *Koch et Ziz*, Cat. Plant. Palat., 12 (1814); O. E. Schulz in Engl., Pflanzenr., 70 (IV, 105): 123 (1919); Lindberg f., Iter Cypr., 17 (1946).
Sinapis orientalis L., Cent. I Plant., 20 (1755).
TYPE: " *in* Oriente ".

Ovary (excluding beak) ± densely covered with retrorse, white bristles; fruit usually ± bristly; otherwise similar to *S. arvensis* L. var. *arvensis*.

HAB.: Cultivated and waste ground; sea-level to 3,750 ft. alt.; fl. March–July.

DISTR.: Divisions 1, 2, 6, 7. Distribution that of var. *arvensis*.

1. Ayios Neophytos, 1939, *Lindberg f.* s.n.
2. Platres, 1938, *Kennedy* 1235 !
6. Krini, 1937, *Syngrassides* 1424 (in part; mixed with var. *arvensis* above) !
7. Kythrea, 1939, *Lindberg f.* s.n.; Kythrea, Ayios Dhimitrianos, 1950, *Chapman* 133 ! Kyrenia, Loukkos near Thermia, 1955, *G. E. Atherton* 755 ! and, 1956, 798 !

NOTES: var. *orientalis* and the following variety both intergrade in Cyprus with typical *S. arvensis*.

var. **schkuhriana** (*Reichb.*) *Hagenb.*, Fl. Basil. Suppl., 138 (1843); O. E. Schulz in Engl., Pflanzenr., 70 (IV, 105): 123 (1919).
Sinapis schkuhriana Reichb., Icon. Fl. Germ., 2: 20, t. 4425 b. (1814).
TYPE: Not localized, presumably the plant illustrated in Reichenbach's *Icones* (*loc. cit. supra*).

Siliques elongate, 4–5·5 cm. long, slender, often torulose and irregularly bent or twisted when mature. Otherwise resembling *S. arvensis* L. var. *arvensis*.

HAB.: Cultivated and waste ground; sea-level to 2,000 ft. alt.; fl. April–May.

DISTR.: Divisions 1, 6, rare. Distribution that of var. *arvensis*.

1. Stroumbi near Paphos, 1913, *Haradjian* 754 !
6. Peristerona (Morphou), 1936, *Syngrassides* 1213 !

2. S. alba *L.*, Sp. Plant., ed. 1, 668 (1753); Unger et Kotschy, Die Insel Cypern, 332 (1865); Boiss., Fl. Orient., 1: 394 (1867); Holmboe, Veg. Cypr., 89 (1914); O. E. Schulz in Engl., Pflanzenr., 70 (IV, 105): 129 (1919); Post, Fl. Pal., ed. 2, 1: 123 (1932); Lindberg f., Iter Cypr., 17 (1946); Davis, Fl. Turkey, 1: 266 (1965); Zohary, Fl. Palaest., 1: 311, t. 459 (1966); Osorio-Tafall et Seraphim, List Vasc. Plants Cyprus, 45 (1973).
TYPE: " *in agris* Belgii, Angliae, Galliae ".

Erect annual herb to 80 cm. or more in height; stems obscurely ribbed, usually clothed with scattered retrorse bristles; leaves all pinnatisect or pinnatifid with rather broad, obtuse or truncate, bluntly toothed or sinuate lobes, hispidulous, especially on midrib and nerves; the lower leaves usually forming a basal rosette, 5–10 cm. long and 2–3 cm. wide, with a petiole 1–4 cm. long; the upper often larger, to 15 cm. or more long and 8–10 cm. or more wide, distinctly petiolate or (rarely) subsessile; inflorescence rather densely racemose; pedicels 6–12 mm. or more long, patent, rather slender, becoming slightly thicker, but scarcely elongating, in fruit; sepals narrowly oblong, 6–7 mm. long, concave, glabrous or sparsely hispid; petals pale yellow, 10–15 mm. long, obovate, rounded at apex; stamens about as long as the sepals; anthers linear-oblong about 2 mm. long; stigma flattened or slightly retuse. Siliques patent or ascending, 4–5 cm. long (including beak), 5–6 mm. wide; valves convex, torulose, 1–2 cm. long, hispid with rigid bristles; beak flattened, sword-shaped, straight or curved, tapering to the persistent stigma, usually enclosing 1 seed; seeds globose, light brown when ripe; testa minutely foveolate.

HAB.: Cultivated and waste ground; sea-level to 3,950 ft. alt.; fl. Febr.–July.

DISTR.: Divisions 1–4, 6–8. Widespread in Europe and western Asia, adventive elsewhere.

1. Paphos, 1862, *Kotschy*. Stroumbi, near Paphos, 1913, *Haradjian* 772 !
2. Lefka, 1932, *Syngrassides* 257 ! Platres, 1938, *Kennedy* 1236 !
3. Near Limassol, 1862, *Kotschy*; Near Stavrovouni Monastery, 1862, *Kotschy* 202.
4. Near Larnaca, 1862, *Kotschy* 48 ! Near Larnaca, 1880, *Sintenis & Rigo*; Near Athienou, 1880, *Sintenis & Rigo*; Phaneromene near Larnaca, 1905, *Holmboe* 255.
6. Nicosia, 1937, *Syngrassides* 1404 ! and, 1950, *Chapman* 464 !
7. Near Kythrea, 1939, *Lindberg f.* s.n. ! Aghirda, 1941, *Davis* 2113 ! Kythrea, Ayios Dhimitrianos, 1950, *Chapman* 132 ! Thermia, 1956, *G. E. Atherton* 878 !
8. Near Kantara monastery, 1880, *Sintenis & Rigo* 266 !

3. HIRSCHFELDIA *Moench*
Meth., 264 (1794).

Similar to *Brassica* and *Sinapis* in general appearance; sepals erect or almost so, petals small, yellow or white; ovary with 8–13 ovules; stigma small, capitate. Silique small, linear, valves distinctly 3-veined when young, becoming ± smooth at maturity; beak short, conical or ovoid, slightly swollen, containing 0–2 seeds. Seeds small, ovoid or oblong, dark brown; testa minutely foveolate.

Two species, one in Europe, Mediterranean region and Atlantic Islands, the other in Socotra.

1. **H. incana** (*L.*) *Lagrèze-Fossat*, Fl. Tarn et Garonne, 19 (1847); O. E. Schulz in Engl., Pflanzenr., 70 (IV, 105): 137 (1919); Post, Fl. Pal., ed. 2, 1: 120 (1932); Rechinger f. in Arkiv för Bot. ser. 2, 1: 422 (1950); Davis, Fl. Turkey, 1: 267 (1965); Zohary, Fl. Palaest., 1: 312, t. 461 (1966); Osorio-Tafall et Seraphim, List Vasc. Plants Cyprus, 45 (1973).

Sinapis incana L., Cent. I. Plant., 19 (1755).

Hirschfeldia adpressa Moench, Meth., 264 (1794); Boiss., Fl Orient., 1: 390 (1867); H. Stuart Thompson in Journ. Bot., 44: 277 (1906).

Erucastrum incanum (L.) Koch, Syn. Fl. Germ. et Helv., ed. 1, 56 (1835); Holmboe, Veg. Cypr., 88 (1914); Lindberg f., Iter Cypr., 17 (1946).

TYPE: " *in* Gallia, Lusitania, Hispania ".

Erect or spreading annual or biennial (20–) 40–70 (–100) cm. high; stems usually much-branched, terete or slightly ribbed, often clothed, particularly towards the base, with short, white, retrorse hairs or bristles; radical leaves forming a rosette, rather irregularly lyrate-pinnatisect with obtuse or rounded, unevenly dentate lobes, 8–30 cm. long, 2–7 cm. wide; petiole 1–3 cm. or more long, stem leaves few, diminishing rapidly in size upwards and becoming ovate, lanceolate or linear-lanceolate, subentire or irregularly lobed; flowers in rather dense racemes, the rhachis lengthening greatly in fruit; pedicels 1–2 mm. long, thickening considerably, but lengthening only a little in fruit; sepals oblong-ovate, subacute, 3–4 mm. long, about 1 mm. wide, glabrous or thinly hairy; petals yellow, obovate, obtuse, about 7 mm.

long and 3 mm. wide, claw rather shorter than the lamina; stamens about 5 mm. long; anthers ovate, about 1 mm. long; stigma small, capitate. Siliques linear, erect and adpressed to rhachis, or sometimes irregularly twisted, about 1–1·5 cm. long (including beak) and 1–1·5 mm. wide; terete or indistinctly torulose, glabrous or sparsely pilose; beak 3–4 mm. long, conical or ovoid, acute, often distinctly constricted at the base, 0–1-seeded; seeds ovoid or oblong, about 1 mm. long and 0·7 mm. wide, dark brown; testa minutely foveolate.

HAB.: Roadsides, waste places and rocky hillsides; sea-level to 3,000 ft. alt.; fl. April–July.

DISTR.: All Divisions. S. Europe, Mediterranean region and Atlantic Islands; adventive elsewhere.

1. Dhrousha, 1941, *Davis* 3294 !
2. Near Omodhos, 1880, *Sintenis & Rigo* 817 ! 918 ! Platres, 1937, *Kennedy* 481!
3. Stavrovouni, 1937, *Syngrassides* 1490 ! Near Kornos on the main road to Limassol, 1938, *Syngrassides* 1803 !
4. Near Larnaca, 1939, *Lindberg f.* s.n.; Cape Greco, 1958, *N. Macdonald* 42 !
5. Salamis Forest, 1934, *Syngrassides* 482 !
6. Koutraphas station, 1936, *Syngrassides* 1018 !
7. Near Kythrea, 1880, *Sintenis & Rigo* 276 ! Kyrenia, 1955, *Mapple* 36 ! and, 1955, *G. E. Atherton* 94 !
8. About Komi Kebir, 1912, *Haradjian* 293.

4. DIPLOTAXIS *DC.*

Regn. Veg. Syst. Nat. 2: 628 (1821).

Annual or perennial herbs with irregularly lobed or pinnatifid leaves; inflorescence racemose, ebracteate; sepals usually ± spreading, the inner rather broader than the outer; petals clawed, yellow, less often white or tinged lilac or purple; stamens 6; ovary cylindrical, 16–250-ovulate, generally narrowed above into a distinct style; stigma usually bilobed. Siliques linear, straight or slightly curved and generally distinctly compressed dorsally; valves 1-veined, beak short or absent. Seeds ovoid, arranged in 2 rows in each loculus.

Twenty three species chiefly in central and southern Europe and the Mediterranean region, extending eastwards to India; adventive elsewhere.

1. **D. viminea** (*L.*) *DC.*, Regn. Veg. Syst. Nat., 2: 635 (1821); Boiss., Fl. Orient., 1: 387 (1867); O. E. Schulz in Engl., Pflanzenr., 70 (IV, 105): 177 (1919); Post, Fl. Pal., ed. 2, 1: 118 (1932); Davis, Fl. Turkey, 1: 268 (1965); Zohary. Fl. Palaest., 1: 308, t. 456 (1966).

Sisymbrium vimineum L., Sp. Plant., ed. 1, 658 (1753).

TYPE: " Parisiis & *ad* Tyberis *margines, inque* Moravia ".

A slender, spreading, glabrous or subglabrous annual 10–30 cm. high; stems obscurely sulcate or subterete; leaves mostly radical, often forming a basal rosette, oblanceolate, 4–10 cm. long, rather regularly and sinuately pinnatifid, lobes blunt, entire or remotely toothed; inflorescences springing from the basal rosette, loosely racemose, the rhachis lengthening greatly in fruit; peduncle 4–9 cm. long; flowers minute, at first sessile or subsessile, the pedicels elongating after fertilization and ultimately 5–15 mm. or more long in fruit; sepals oblong, obtuse or subacute, 2·5–3·5 mm. long, about 1 mm. wide; petals pale yellow, 3–4 mm. long, about 1 mm. wide, oblanceolate or obovate, obtuse, gradually narrowed to a short claw; stamens about as long as the sepals; anthers ovate about 0·5 mm. long; style about 0·5 mm. long; stigma obscurely bilobed; ovary 25–39-ovulate. Siliques ascending (or rarely patent), 1·5–4 cm. long, 1–2 mm. wide, linear, straight or very slightly curved, compressed, tapering to base and apex; style persisting as a short, seedless beak; seeds ovoid, about 1 mm. long, pale brown; testa smooth or very obscurely foveolate.

HAB.: Fallow fields and waste ground; c. 500 ft. alt.; fl. March–June.

DISTR.: Divisions 5, 7, rare. Widespread in the Mediterranean area, naturalized elsewhere in Europe.

5. Between Nicosia and Kythrea, 1880, *Sintenis & Rigo* 628 !
7. Kyrenia, in an olive grove, 1948 and 1950, *Casey* 605 ! 972 !

5. ERUCA *Mill.*

Gard. Dict. Abridg., ed. 4 (1754).

Annual, biennial or perennial herbs usually with pinnatifid or pinnatisect leaves; inflorescence ebracteate; sepals erect, subequal, outer ± cucullate at apex, inner distinctly saccate at base; petals long-clawed, white or yellow, or occasionally purplish, often with conspicuous dark venation; stamens 6; stigma small, bilobed. Siliques straight, narrowly oblong or ellipsoid; valves convex, 1-nerved; beak sword-shaped, flattened, seedless. Seeds biseriate (or rarely almost uniseriate), ovoid or subglobose.

Five species in Europe, Asia, N. Africa and Atlantic Islands.

1. **E. sativa** *Mill.*, Gard. Dict., ed. 8, no. 1 (1768); Boiss., Fl. Orient., 1: 396 (1867); O. E. Schulz in Engl., Pflanzenr., 70 (IV, 105): 181 (1919); Post, Fl. Pal., ed. 2, 1: 124 (1932); Rechinger f. in Arkiv för Bot., 1: 422 (1950); Davis, Fl. Turkey, 1: 269 (1965); Zohary, Fl. Palaest., 1: 313, t. 462 (1966); Osorio-Tafall et Seraphim, List Vasc. Plants Cyprus, 45 (1973).

Brassica eruca L., Sp. Plant., ed. 1, 667 (1753).

Erect or spreading annual or biennial, (7–) 15–40 (–60) cm. or more high; stems indistinctly striate, glabrous or rather thinly hispidulous especially towards the base; leaves 4–14 cm. or more long, 2–10 cm. or more wide, variable in outline, sometimes almost simple and broadly obovate, frequently lyrate-pinnatifid or irregularly pinnatisect, with broad or narrow, acute or obtuse, entire or toothed lobes; basal leaves distinctly stalked, sometimes forming a rosette; upper leaves sessile and sometimes auriculate; flowers forming a rather lax raceme, the rhachis elongating considerably in fruit; pedicels 2–5 mm. long, lengthening a little and thickening considerably in fruit; sepals narrowly oblong, 8–12 mm. long, about 2 mm. wide, acuminate, glabrous or hispidulous-pilose, often stained purple; petals pale yellow or white, with dark veins, obovate, about 15–22 mm. long, 4–8 mm. wide, apex obtuse or slightly emarginate; base narrowed to a long, slender claw; stamens 10–14 mm. long; anthers linear, about 2 mm. long; style 6–7 mm. long, usually longer than the ovary. Silique ascending or erect, 1·5–3 cm. long (including beak), 2–5 mm. wide; beak 6–10 mm. or more long, flattened, sword-shaped; valves distinctly or obscurely 1-nerved; seeds broadly ovoid 1·5–2·5 mm. long; testa pale brown, minutely foveolate.

var. sativa

E. cappadocica Reut., Cat. Grain. Genève 1857: 2 (1858); Boiss., Fl. Orient., 1: 396 (1867).

TYPE: " *in* Helvetia ".

Siliques 2–3 cm. long, 4–5 mm. wide, usually glabrous; beak normally shorter than valves. A robust, erect or suberect, leafy plant with large flowers; stems, leaves and calyx usually glabrous or rather sparsely hispid; leaves generally lyrate with a large terminal lobe and blunt lateral lobes.

HAB.: A weed in gardens, escaped from cultivation; sea-level to 3,750 ft. alt.; fl. March–June.

DISTR.: Divisions 2, 3, rare. Also in central and southern Europe, western Asia, N. Africa and Atlantic Islands.

2. Platres, 1939, *Kennedy* 1549 !
3. Limassol, 1913, *Haradjian* 586.

NOTES: The cultivated plant, used as a salad.

var. **eriocarpa** (*Boiss.*) *Post*, Fl. Pal., ed. 1, 79 (1896), ed. 2, 1: 124 (1932); O. E. Schulz in Engl., Pflanzenr., 70 (IV, 105): 185 (1919).

E. cappadocica Reut. var. *eriocarpa* Boiss., Fl. Orient., 1: 396 (1867).

Brassica eruca L. var. β Sibth. et Sm., Fl. Graec., 7: 43, t. 647 (1830).

E. sativa Mill. var. *longirostis* (Uechtritz) Rouy, Excurs. Bot. Exp., 2, 1881–1882: 52 (1883); O. E. Schulz in Engl., Pflanzenr., 70 (IV, 105): 185 (1919); Rechinger f. in Arkiv för Bot., ser. 2, 1: 422 (1950).

TYPE: " In insula Cypro ", *Sibthorp* (OXF).

Siliques 1·5–2 cm. long, 2–4 mm. wide, usually ± hispid, but sometimes glabrous or almost so; beak almost as long as the valves, or even a little longer. Stems usually spreading or decumbent, often hispid-pilose, especially towards the base; leaves frequently forming a loose basal rosette, often deeply and regularly pinnatisect, with a small or narrow terminal lobe, and acute, dentate lateral lobes. Leaves and calyx often hispid-pilose.

HAB.: Roadsides, waste ground and stony fields; sea-level to 500 ft. alt.; fl. Febr.–April.

DISTR.: Divisions 4–7. Throughout the Mediterranean region and western Asia.

4. Cape Pyla, 1880, *Sintenis & Rigo*; about Larnaca, 1912, *Haradjian* 56.
5. Mandres near Mia Milea, 1933, *Syngrassides* 323 ! Pyroi, 1936, *Syngrassides* 1053 !
6. Beuyuk Kaimakli, Nicosia, 1934, *Syngrassides* 1347 !
7. Above Kythrea, 1880, *Sintenis & Rigo* 997 teste *Schulz*; Kyrenia, 1950, *Casey* 972 !

NOTES: The " wild " form, apparently indigenous in the area, but perhaps also occurring as a reversion from cultivated stock.

6. CARRICHTERA DC.

in Mém. Mus. Hist. Nat. Paris, 7: 244 (1821) nom. cons.

Annual herb; leaves bipinnatisect, petiolate; racemes leaf-opposed, lax; pedicels very short; sepals erect, the outer linear-oblong, the inner slightly broader, obscurely saccate at base; petals small, narrowly obovate, tapering to a filiform claw; stamens 6, anthers linear-oblong, mucronate; stigma sessile; ovary 2-jointed, lower joint containing 6 pendulous ovules, upper joint sterile. Fruit pendulous with an ovoid-cylindrical, 2-locular, 2-valved, dehiscent lower joint, and a conspicuous oblong-elliptical, flattened, foliaceous upper joint. Seeds subglobose; cotyledons emarginate, longitudinally conduplicate.

One species widespread in the Mediterranean region and eastwards to Iran; also in the Canary Islands.

1. **C. annua** (*L.*) *DC.* in Mém. Mus. Hist. Nat. Paris, 7: 250 (1821); O. E. Schulz in Engl., Pflanzenr., 84 (IV, 105): 41 (1923); Post, Fl. Pal., ed. 2, 1: 125 (1932); O. E. Schulz in Engl. et Prantl., Pflanzenfam., ed. 2, 17b: 375 (1936); Zohary, Fl. Palaest., 1: 314, t. 463 (1966).

Vella annua L., Sp. Plant., ed. 1, 641 (1753).

C. vellae DC., Regn. Veg. Syst. Nat., 2: 641 (1821); Boiss., Fl. Orient., 1: 397 (1867).

TYPE: Spain; " *in Hispania ad pagum* Elda *septem sub Horivela leucis hispanicis* " (BM !).

An erect or spreading, branched annual, usually less than 30 cm. high; stems thinly hispid, with patent or deflexed bristles; leaves ovate-oblong, petiolate, 3–7 cm. long, 1·5–4 cm. wide, pinnately or bipinnately divided into narrow linear or linear-oblong, obtuse segments; petiole 1–2·5 cm. long;

racemes leaf-opposed, ebracteate, lax, lengthening greatly in fruit; pedicels slender, short, seldom more than 2 mm. long, at first erecto-patent, later becoming decurved; sepals thinly hispid, the outer linear-oblong, about 3·5 mm. long and 0·8 mm. wide, the inner about 3·5 mm long and 1 mm. wide; petals pale yellow, veined dark purple, obovate, about 8 mm. long, and 1·8 mm. wide near apex, gradually tapering to a long filiform claw; stamens 3·5–4·5 mm. long, filaments flattened; anthers linear-oblong, about 0·8 mm. long, with a distinct, papillose mucro; ovary about 3·5 mm. long. Fruit pendulous; lower joint distinctly hispid, upper joint glabrous or subglabrous; seeds about 3 in each loculus, c. 1 mm. diam.; testa brown, smooth.

HAB.: Waste ground, rocky hillsides and cliffs; 300–1,000 ft. alt.; fl. March.

DISTR.: Divisions 1, 3, rare. Mediterranean region eastwards to Iran; N. Africa; Atlantic Islands, adventive elsewhere.

1. Smyies (Akamas), 1962, *Meikle* 2199 !
3. Episkopi, 1960, *N. Macdonald* 43 !

7. RAPHANUS *L.*

Sp. Plant., ed. 1, 669 (1753).
Gen. Plant., ed. 5, 300 (1754).

Annual or biennial (or rarely short-lived perennial) herbs; rootstock sometimes thickened; stems erect, branched, often hispid-pilose, especially towards the base; leaves lyrate-pinnatisect; inflorescence ebracteate; sepals usually erect, the outer cucullate at the apex, the inner distinctly saccate at the base; petals long-clawed, obovate; stamens 6, anthers narrowly oblong or linear; ovary cylindrical, 2–20-ovulate; stigma minute, hemispherical or retuse. Fruit elongate, cylindrical, jointed, the lowest joint (corresponding to the valves of a silique) usually very short or obsolete, empty; upper part of fruit either indehiscent and cylindrical, or breaking transversely at constrictions into 1-seeded joints. Seeds ovoid or almost spherical, brown, testa reticulate.

About 8 species, chiefly in the Mediterranean area.

Fruits slender, 2–6 mm. diam., markedly constricted between the seeds, and breaking readily into 1-seeded joints; petals white, lilac (or yellow) with conspicuous dark venation - - - - - - - - - - - - - - **1. R. raphanistrum**
Fruits thick, 5–20 mm. diam., not distinctly constricted between the seeds, and not breaking into 1-seeded joints; petals usually purple or mauve, not conspicuously veined - - - - - - - - - - - - - - **2. R. sativus**

1. R. raphanistrum *L.*, Sp. Plant., ed. 1, 669 (1753); Unger et Kotschy, Die Insel Cypern, 334 (1865); Boiss., Fl. Orient., 1: 401 (1867); Holmboe, Veg. Cypr., 92 (1914); O. E. Schulz in Engl., Pflanzenr., 70 (IV, 105): 196 (1919); Post, Fl. Pal., ed. 2, 1: 127 (1932); Davis, Fl. Turkey, 1: 270 (1965); Zohary, Fl. Palaest., 1: 325, t. 475 (1966); Osorio-Tafall et Seraphim, List Vasc. Plants Cyprus, 45 (1973).
TYPE " *inter segetes* Europae ".

An erect or spreading annual (15–) 25–40 (–60) cm. high with a slender taproot; stems much branched above, terete or slightly furrowed, usually clothed with scattered, retrorse, conical bristle, sometimes subglabrous, ± pruinose above, often tinged with violet towards the base; lower leaves lyrate-pinnatisect, hispidulous, 5–24 cm. or more long, 3–10 cm. or more wide, with a distinct petiole 2–4 cm. long; lateral lobes oblong-ovate, blunt, obscurely toothed or crenate, terminal lobe broadly ovate or obovate; upper leaves smaller, often almost simple, shortly petiolate; inflorescence rather loosely racemose, the rhachis slender and somewhat flexuous, lengthening considerably in fruit; pedicels slender 1–2 cm. long, thickening slightly in fruit; sepals 8–10 mm. long, linear or narrowly oblong, the outer rather blunt at apex, the inner acuminate, glabrous or hispid-pilose, often stained purple; petals usually white (or occasionally tinged lilac) in Cyprus specimens, with conspicuous dark veins; stamens about 1 cm. long; anthers yellow, about 2 mm. long; stigma small, obscurely retuse. Fruit spreading, 2–6 cm. long, 2–6 mm. wide, longitudinally ridged, with a slender terminal beak 1–2 cm. long, markedly constricted between the 2–7 or more seeds, and breaking readily into 1-seeded joints; seeds broadly ovoid, reddish-brown; testa distinctly and uniformly reticulate.

HAB.: Cultivated and waste ground; sea-level to 800 ft. alt.; fl. Febr.–May.

DISTR.: Divisions 2, 4–7, locally common in Division 4. Throughout temperate Europe, W. Asia, N. Africa and Atlantic Islands; introduced elsewhere.

2. Yialia, 1955 and 1956, *Merton* 2109 ! 2498 !
4. Locally common around Larnaca; *Kotschy*; *Sintenis*; *Holmboe* 262; *Nattrass* 312. Paralimni, 1940, *Davis* 2055 !
5. Kythrea, 1880, *Sintenis* 268A ! and, 1932, *Syngrassides* 265 !
6. Nicosia, 1862, *Kotschy*.
7. Kyrenia, 1949, *Casey* 292 ! Kyrenia, Loukkos near Thermia, 1955 and 1956, *Atherton* 363 ! 797 ! Lapithos Reformatory School, 1955, *G. E. Atherton* 690 !

NOTES: Yellow-flowered variants of this species have not been recorded from Cyprus.

2. **R. sativus** *L.*, Sp. Plant., ed. 1, 669 (1753); Unger et Kotschy, Die Insel Cypern, 334 (1865); Boiss., Fl. Orient., 1: 400 (1867); Post, Fl. Pal., ed. 2, 1: 127 (1932); Davis, Fl. Turkey, 1: 270 (1965).
TYPE: Not localized.

Similar to *R. raphanistrum* in general appearance, but usually a stouter plant, up to 1 m. high with a fleshy fusiform taproot; stems glabrous or very sparsely hispid; leaves rather fleshy, pale green, often with reddish veins and margin, thinly hispidulous or almost glabrous; petals mauve, pink or purple (occasionally white). Fruit spreading or ascending, oblong-conical, somewhat inflated, 3–6 cm. or more long (including beak), 5–20 mm. wide, externally pale brown, smooth or obscurely ridged, spongy internally, not markedly constricted between the seeds nor breaking into 1-seeded joints; beak slender, subulate, 1–3 cm. long; seeds broadly ovoid, about 3 mm. long, brown, with a finely reticulate testa.

HAB.: Cultivated and waste ground; not indigenous; sea-level to 1,000 ft. alt.; fl. Febr.–May.

DISTR.: Divisions 2–5, 7. Cultivated almost throughout the world; natural distribution unknown.

2. Platres, 1938, *Kennedy* 1234 !
3. Limassol, Stavrovouni etc., 1928 or 1930, *Druce* s.n.
4. Larnaca, 1935, *Nattrass* 566; Perivolia, 1938, *Syngrassides* 1754 !
5. Near Kythrea, 1880, *Sintenis & Rigo* 268 ! Kythrea, 1936 and 1937, *Roger-Smith* s.n. !
7. Kyrenia, 1956, *G. E. Atherton* 1004 ! 1079 !

NOTES: Cultivated throughout the island (*teste* Syngrassides); escaping to wild ground occasionally.

Two of the specimens cited above (*Roger-Smith*, March 1936 and *G. E. Atherton* 1004) look somewhat intermediate between *R. raphanistrum* L. and *R. sativus* L., and may possibly be the hybrid *R.* × *micranthus* (Uechtr.) O. E. Schulz. Without knowing more about the population from which they were selected it is, however, impossible to establish hybridity. *R.* × *micranthus* should be looked for by collectors.

8. ENARTHROCARPUS *Labill.*

Nouv. Bull. Sci. Soc. Philomatique, 57 (1812).
Icon. Plant. Syr. Rar., 5: 4, t. 2 (1812).

Annual herbs; stems erect or spreading, mostly branched from the base; leaves pinnatisect, the uppermost often ± simple and irregularly lobed; inflorescence ± bracteate, elongating considerably in fruit; pedicels short;

calyx erect; petals yellow or white tinged or veined purple, long-clawed; stamens 6; ovary 3–15-ovulate; style short, indistinct, stigma capitate or somewhat retuse. Silique cylindrical or compressed, straight or curved, 2-articulate, without a septum; lower joint short, usually 1–3-seeded; upper joint caducous, usually elongate and constricted about the seeds, with swollen nodes between the seeds, occasionally constricted between the seeds; beak subulate or elongate-conical, rather short; seeds brown, broadly oblong or ovoid.

Five species, chiefly in the eastern Mediterranean region and North Africa.

Silique 4–9 cm. long, cylindrical, conspicuously torulose; petals 9–15 mm. long; upper half of inflorescence ebracteate - - - - - - - - **1. E. arcuatus**
Silique 2·5–4 cm. long, compressed, not or obscurely torulose; petals about 7 mm. long; inflorescence bracteate almost to apex - - - - - - **2. E. lyratus**

1. E. arcuatus *Labill.* in Nouv. Bull. Sci. Soc. Philomat., 57 (1812), Icon. Plant. Syr. Rar., 5: 4, t. 2 (1812); Unger et Kotschy, Die Insel Cypern, 333 (1865); Boiss., Fl. Orient., 1: 899 (1867); Holmboe, Veg. Cypr., 91 (1914) excl. syn.; O. E. Schulz in Engl., Pflanzenr., 70 (IV, 105): 211 (1919); Post, Fl. Pal., ed. 2, 1: 126 (1932); Davis, Fl. Turkey, 1: 271 (1965); Zohary, Fl. Palaest., 1: 327, t. 478 (1966); Osorio-Tafall et Seraphim, List Vasc. Plants Cyprus, 45 (1973).
TYPE: " In Libano ", *Labillardière* (FI).

Erect or spreading annual 10–40 cm. high; stems much-branched towards the base, angled, rather sparsely hispidulous-pilose; basal leaves thinly hispidulous, 7–10 cm. long, 1·5–3 cm. wide, lyrate-pinnatisect, with the large terminal lobe and small oblong lateral lobes irregularly dentate; petioles 2–4 cm. long; upper leaves similar but smaller, sessile or subsessile, the uppermost often obovate, dentate or irregularly lobed; inflorescence rather densely racemose, elongating greatly in fruit, the lower flowers subtended by oblanceolate or obovate, irregularly dentate, herbaceous bracts, the upper flowers ebracteate; pedicels short, about 5 mm. long, thickening considerably, but lengthening only slightly in fruit; sepals erect, narrowly oblong, obtuse at apex, about 7–8 mm. long and 2 mm. wide; petals narrowly obovate, about 9–15 mm. long (including claw) and 3–6 mm. wide, pale yellow or whitish with darker venation; claw slender, about 5–7 mm. long; stamens 7–10 mm. long; anthers narrowly oblong, 2–3 mm. long, deeply cleft at base; ovary subulate, stigma flattened or slightly retuse. Silique curved, cylindrical, 4–9 cm. long, 3–5 mm. diam., longitudinally sulcate, conspicuously torulose, pale brown and somewhate woody when ripe, usually clothed, at least in the upper part, with short ascending or adpressed bristles; articulation a short distance above base of silique, lower joint 1-seeded; beak about 4–9 mm. long, elongate-conical; seeds in the constricted portions of the silique; broadly oblong or almost square, about 1·5 mm. long, reddish-brown; testa minutely rugulose.

HAB.: On rocky pastures by the sea, or in fissures of sea cliffs; sea-level to c. 100 ft. alt.; fl. March–May.

DISTR.: Divisions 4, 8, rare. Also in Greece, Crete, Turkey, Aegean Islands, Syria, Palestine and Egypt.

4. Cape Greco, 1962, *Meikle* 2609 !
8. Cape Andreas, 1880, *Sintenis & Rigo* 265 ! and, 1962, *Meikle* 2470 !

NOTES: The earlier unlocalized record from Cyprus, usually attributed to Endlicher (Gen. Plant., 886), is in fact based on a misprint in Labillardière's paper (Nouv. Bull. Sci. Soc. Philomatique, 57; 1812) where the Tournefort record from Crete is erroneously quoted as " inque Cypri insulâ (C. B.[auhin] Tournef.)".

2. E. lyratus (*Forssk.*) *DC.*, Regn. Veg. Syst. Nat., 2: 661 (1821); Boiss., Fl. Orient., 1: 399 (1867), Fl. Orient. Suppl., 67 (1888); Holmboe, Veg. Cypr., 92 (1914);

Post, Fl. Pal., ed. 2, 1: 126 (1932); Zohary, Fl. Palaest., 1: 328 (1966); Osorio-Tafall et Seraphim, List Vasc. Plants Cyprus, 45 (1973).

Raphanus lyratus Forssk., Fl. Aegypt.-Arab., 119 (1775).

[*Enarthrocarpus strangulatus* (non Boiss.) Sintenis in Oesterr. Bot. Zeitschr., 32: 121 (1882).]

TYPE: Egypt; " Prope *Pyramides Gizenses* ad agrorum margines & in insula *Djesiret el dáhab* inter arundines copiose " (C).

Similar to *E. arcuatus* Labill. in general appearance, but often more robust, up to 1 m. high; stems subglabrous or thinly hispidulous, terete or obscurely angled; leaves similar to those of *E. arcuatus*; inflorescence bracteate almost to apex; flowers small in dense racemes; sepals narrowly oblong, obtuse, about 4 mm. long, often ± densely pilose; petals yellow, stained purple towards the base, narrowly obovate, about 7 mm. long (including claw), 1–1·5 mm. wide; stamens about 5 mm. long; anthers shortly oblong, about 1 mm. or less in length; ovary subulate; stigma capitate. Silique curved or almost straight, distinctly flattened, 2·5–4 cm. long, 2–4 mm. wide, longitudinally sulcate, but not regularly or conspicuously torulose, pale brown and somewhat woody, glabrous or very sparsely hispidulous; articulation usually 6–10 mm. above the base of the silique, the lower joint 1–3-seeded; beak 5–8 mm. long, conical, ± flattened; seeds oblong, about 2·5 mm. long and 1·5 mm. wide, reddish-brown; testa smooth or very obscurely rugulose.

HAB.: Roadsides and waste ground, probably an adventive; c. 500 ft. alt.; fl. March–May.

DISTR.: Division 5, rare. Also in Palestine, Egypt and Arabia; adventive elsewhere.

5. Near Kythrea, 1880, *Sintenis & Rigo* 256 !

NOTES: Not seen since and should be looked for.

9. CALEPINA *Adans.*

Fam. Plant., 2: 423 (1763).

Annual or biennial, glabrous herbs; basal leaves usually lyrate-pinnatisect, often forming a rosette; stem leaves usually simple, sessile and amplexicaul. Flowers small in rather dense racemes which elongate greatly in fruit; petals white (or rarely tinged pink), unequal; stamens 6. Fruit a caducous, indehiscent, ovoid or subglobose, apiculate, 1-locular silicle. Seed solitary, pendulous; cotyledons with margins inflexed.

One species, widespread throughout the Mediterranean region and eastwards to Iran and the Caspian; adventive elsewhere.

1. **C. irregularis** (*Asso*) *Thellung* in Schinz et Keller, Fl. Schweiz, ed. 2, 1: 218 (1905); O. E. Schulz in Engl. Pflanzenr., 70 (IV, 105): 225 (1919); Post, Fl. Pal., ed. 2, 1: 131 (1932); Davis, Fl. Turkey, 1: 271 (1965); Zohary, Fl. Palaest., 1: 324, t. 473 (1966).

Myagrum irregulare Asso, Syn. Stirp. Arag., 82 (1779).

C. corvini (All.) Desv. in Journ. de Bot., 3: 158 (1814); Boiss., Fl. Orient., 1: 409 (1867); Druce in Rep. B.E.C., 9: 469 (1931).

Crambe corvini All., Fl. Ped., 1: 256 (1785).

TYPE: Spain; Aragon: " locis gramineis, humidiusculis *en Hijar* ", *Asso y del Rio* (P).

A slender, erect or spreading, glabrous annual or biennial; stems subterete or obscurely angled, usually branched at least near the base; basal leaves often forming a rosette, obovate, generally lyrate-pinnatisect, 3–12 cm. long, 0·6–2 cm. wide, terminal lobe large, rounded, entire or irregularly and bluntly toothed, lateral lobes obtuse, deltoid or broadly oblong; petiole 1–3 cm. or more long; stem leaves smaller, usually oblong or obovate, entire or irregularly toothed or lobed, sessile, amplexicaul, with acute, sagittate auricles; flowers small, forming rather dense corymbs, the rhachis elongating in fruit to form a lax raceme; pedicels slender, about 2 mm. long, lengthening to 5–6 mm. in fruit; sepals erect, broadly oblong, obtuse, 1–1·5 mm. long, not saccate at base; petals obovate, white (or tinged pink), unequal, the outer 2·5–3 mm. long, the inner about 2 mm. long; stamens 1–1·5 mm. long, filaments distinctly flattened; anthers shortly oblong, about 0·3 mm. long; ovary ovoid about 1·5 mm. long, stigma minute, sessile. Silicle ascending, broadly ovoid, or sometimes almost globose, 2·5–4 mm. long, 2–3 mm. wide, apiculate at apex, obscurely 4-nerved, conspicuously reticulate when ripe; seed pale brown; testa minutely rugulose.

HAB.: Cultivated and waste ground, roadsides etc.; sea-level to 2,000 ft. alt.; fl. March–May.

DISTR.: Divisions 2, 6. Widespread in the Mediterranean region and western Asia.

2. Lefkara, 1941, *Kennedy & Davis* 2713 ! same locality, 1955, *Merton* 2040 !
6. Myrtou, 1928 or 1930, *Druce* s.n.

10. CRAMBE *L.*

Sp. Plant., ed. 1, 671 (1753).
Gen. Plant., ed. 5, 301 (1754).

Annual or perennial herbs, sometimes suffruticose, and often with a thick fleshy taproot; stems usually tall and branched; basal leaves large, lyrate-pinnatisect or occasionally simple; flowers small, white (or rarely yellowish), numerous, in widely branching panicles; sepals erect, not saccate at base; petals obovate, with a short claw; stamen 6, the inner filaments normally with a tooth-like appendage; anthers oblong. Silicle usually with a short, seedless basal joint forming a gynophore; upper joint spherical or ovoid, indehiscent, 1-seeded; seed erect.

About 17 species in Europe, Asia, N. and E. Africa, and Atlantic Islands.

1. **C. hispanica** *L.*, Sp. Plant., ed. 1, 671 (1753); Unger et Kotschy, Die Insel Cypern, 332 (1865); Boiss., Fl. Orient., 1: 408 (1867); Holmboe, Veg. Cypr., 91 (1914); Zohary, Fl. Palaest., 1: 323, t. 472 (1966); Osorio-Tafall et Seraphim, List Vasc. Plants Cyprus, 46 (1973).

TYPE: "*in* Hispania".

An erect annual, 30–100 cm. high; stem simple or sparsely branched, prominently ridged, and clothed with retrorse or spreading bristles; basal leaves sparsely adpressed-strigose, 4–12 cm. long, 2–8 cm. wide, with few (1–2 pairs), bluntly deltoid lateral lobes and a large rounded or reniform, dentate or crenate terminal lobe; upper stem leaves similar, but usually much smaller, and often lacking lateral lobes; inflorescence forming a large, much-branched panicle; flowers at first in rather dense corymbs, the rhachis elongating greatly in fruit; pedicels suberect, 3–5 mm. long, slender, lengthening to 8 mm. or more in fruit; sepals oblong, about 1·5–2 mm. long and 1 mm. wide, obtuse at apex, glabrous; petals white or tinged purplish at base, 3–3·5 mm. long, rounded or very slightly emarginate at apex, shortly clawed; stamens about 2 mm. long, inner (longer) filaments with a linear appendage near apex; anthers oblong, about 0·7–1 mm. long; stigma minute, sessile. Silicle globose, about 2–3 mm. diam., smooth, pale brown, blackish or greenish, the lower joint forming an inconspicuous, sulcate gynophore about 1 mm. long; seed subglobose, about 1·5 mm. diam., brown; testa minutely rugulose.

HAB.: Cornfields; rocky pastures, or in maritime garigue; sea-level to 1,000 ft. alt.; fl. March–May.

DISTR.: Divisions 4, 7, rare. Also occurs locally almost throughout the Mediterranean region.

4. E. side of Cape Greco, 1862, *Kotschy* 125.
7. Larnaka tis Lapithou, hills below the village, 1941, *Davis* 3012 !

11. RAPISTRUM *Crantz*

Class. Cruciform. Emend., 105 (1769) nom. cons.

Erect or spreading annual or perennial herbs; stems usually much-branched; basal leaves lyrate-pinnatipartite or sometimes irregularly lobed; upper leaves often simple, hastate or irregularly toothed or lobed. Inflorescence ebracteate, flowers at first in rather dense corymbs, the rhachis elongating greatly in fruit; sepals erect or spreading, the inner linear, the outer oblong, obtuse at apex, slightly saccate at base; petals yellow, obovate, blunt or truncate at apex, narrowed towards the base into a slender claw; stamens 6; anthers oblong; ovary shortly cylindrical; style distinct; stigma bilobed or retuse. Silicle 2-jointed, rather woody, the lower joint shortly cylindrical or sometimes obsolete, indehiscent or tardily dehiscent, 1 (–3)-seeded; the upper joint caducous, globose, usually distinctly wider than the lower, rugose or costate, indehiscent, 1-seeded, crowned with a short straight beak. Seeds ovoid, brown, pendulous in the lower joint, erect in the upper.

Three species in Europe, the Mediterranean region and western Asia; adventive elsewhere.

1. **R. rugosum** (*L.*) *All.*, Fl. Ped., 1: 257, t. 78 (1785); Boiss., Fl. Orient., 1: 404 (1867); Holmboe, Veg. Cypr., 91 (1914); O. E. Schulz in Engl., Pflanzenr., 70 (IV, 105): 254 (1919); Post, Fl. Pal., ed. 2, 1: 128 (1932); Lindberg f., Iter Cypr., 17 (1946); Davis, Fl. Turkey, 1: 273 (1965); Zohary, Fl. Palaest., 1: 321, t. 470 (1966); Osorio-Tafall et Seraphim, List Vasc. Plants Cyprus, 46 (1973).

Myagrum rugosum L., Sp. Plant., ed. 1, 640 (1753).

[*Rapistrum rugosum* (L.) All. var. *orientale* (non (L.) Arcang.) Rechinger f. in Arkiv för Bot., ser. 2, 1: 423 (1950).]

TYPE: " *in* Europa *australi* ".

An erect or spreading annual (or sometimes biennial) 20–100 cm. high, with a slender taproot; stems much branched, bluntly angled, glabrous or sparsely hispidulous, especially towards the base, often stained purple; basal leaves obovate, 8–15 cm. long, 3–6 cm. wide, lyrate-pinnatipartite, or with a few scattered sinuate lobes, glabrous or thinly pilose, apex usually blunt or rounded, base narrowed to a petiole 1–5 cm. long; upper leaves about 3–7 cm. long, 1–3 cm. wide, lanceolate, oblong or narrowly obovate, entire, or bluntly lobed or toothed, shortly petiolate; inflorescence much-branched; flowers forming rather dense corymbs, shortly pedicellate; sepals erecto-patent, 4–5 mm. long, glabrous or thinly pilose, the outer linear-oblong with a blunt or slightly cucullate apex, the inner oblong, narrowed to a blunt apex; petals yellow, 6–11 mm. long, 3–4 mm. wide; stamens 4–8 mm. long; anthers about 1–1·5 mm. long; ovary about 1–1·5 mm. long, glabrous or ± densely hirsute; style about 1·5 mm. long, stigma retuse. Fruit erect, often adpressed to rhachis, with a thickened pedicel 2–4 mm. long; silicle glabrous (var. *venosum* (Pers.) DC.) or ± hirsute, 8–10 mm. long; lower joint cylindrical, sulcate, 3–4 mm. long, 1–1·5 mm. wide, 1 (–3)-seeded; upper joint 2·5–4 mm. long, 3–4 mm. wide, globose or almost so, distinctly rugose-sulcate, 1-seeded, terminating in a sharp, straight, elongate-conical or subulate beak 2–4 mm. long; seed ovoid, light brown, 1·5–2 mm. long, about 1 mm. wide; testa smooth.

HAB.: Cultivated fields and waste ground; sea-level to 5,700 ft. alt.; fl. March–July.

DISTR.: Divisions 1–4, 7, common in Division 7. Also in central Europe, Mediterranean region, Atlantic Islands and western Asia to Iran and the Caspian; adventive elsewhere.

1. Stroumbi, 1913, *Haradjian* 712 ! Lyso [Lisso], 1913, *Haradjian* 832, 866; Polis, 1962, *Meikle* 2309 !

2. Troödos village, near the Post Office, 1955, *Merton* 2336 ! Troödos Pass, 1955, *Kennedy* 1827 !
3. Akhelia, 1939, *Lindberg f.* s.n.; Moni (Limassol), 1964, *Suart* 215 !
4. Kouklia, 1905, *Holmboe* 385.
7. Common; *Sintenis & Rigo* 629 ! *Syngrassides* 1219 ! 1531 ! *Chapman* 362 ! *Casey* 1163 !

[RAPISTRUM PERENNE (*L.*) *All.* is recorded from Cyprus by Sibthorp (Fl. Graec. Prodr., 2: 3; 1813), but as Homboe indicates (Veg. Cypr., 91; 1914), the occurrence of this European species is very unlikely, and the record is almost certainly a *lapsus calami* or an error in localization.]

12. DIDESMUS *Desv.*

Journ. de Bot., ser. 2, 3: 160, t. 25, fig. 11 (1814).

Annual herbs with a slender taproot; stems much branched, subterete or obscurely angular; leaves scattered, irregularly lyrate-pinnatisect or sometimes simple; inflorescence ebracteate, rather densely corymbose; rhachis elongating greatly in fruit; pedicels short; sepals ± erect, the outer linear-oblong, the inner oblong-ovate, blunt and hyaline-margined at apex, slightly saccate at base; petals white, with a slender purplish or mauve claw, obtuse, truncate or slightly retuse at apex; stamens 6; anthers oblong; ovary narrowly ovoid; style distinct; stigma capitate. Silicle 2-jointed, rather woody, the joints indehiscent; lower joint persistent, turbinate or distinctly 4–6-angled, about as broad as long, 1-seeded; upper joint caducous, ovoid or hemispherical, usually 1-seeded, terminating in a sharp conical or subulate beak. Seed ovoid, brown.

Two species in the Mediterranean region.

1. **D. aegyptius** (*L.*) *Desv.*, Journ. de Bot., ser. 2, 3: 160, t. 25, fig. 11 (1814); Unger et Kotschy, Die Insel Cypern, 333 (1865); Boiss., Fl. Orient., 1: 404 (1867); Holmboe, Veg. Cypr., 91 (1914); O. E. Schulz in Engl., Pflanzenr., 70 (IV, 105): 262 (1919); Post, Fl. Pal., ed. 2, 1: 129 (1932); Osorio-Tafall et Seraphim, List Vasc. Plants Cyprus, 46 (1973).

Myagrum aegyptium L., Sp. Plant., ed. 1, 641 (1753).

Rapistrum aegyptium (L.) Crantz, Class. Cruciform., 106 (1769); Hayek, Prodr. Fl. Pen Balc., 1: 459 (1925); Rechinger f., Fl. Aegaea, 236 (1943).

Spreading annual 10–40 cm. high; stems much-branched, obscurely angular, glabrous or sparsely and minutely scabrid-hispidulous; leaves 2–8 cm. or more long, 1·5–6 cm. wide, 1–2-pinnatisect or irregularly laciniate, with oblong or linear, entire or serrate, obtuse or acute lobes; upper leaves, or some of them, frequently simple, oblong, entire or toothed; petioles 1–3 cm. or more long; flowers in rather dense corymbs, the rhachis subglabrous, or ± densely covered with short, retrorse bristles; pedicels 2–5 mm. long, thickening considerably in fruit; sepals 4–5 mm. long, glabrous or thinly pilose, often tinged violet; petals white, narrowly obovate, 6–10 mm. long, apex rounded or slightly retuse; stamens 4–6 mm. long, filaments tinged purple; anthers yellow about 1·5 mm. long. Silicles erect, 6–10 mm. long, 3–4 mm. wide, divided transversely into 2 ± equal joints, sometimes distinctly constricted at articulation; lower joint prominently 4–6 ribbed or angled, the ribs often with conspicuous gibbosities towards the base; upper joint ovoid or conical, narrowing to a slender conical or subulate, seedless beak 2·5–4 mm. long; seed ovoid, about 1·5 mm. long; testa light brown, smooth or very obscurely rugulose.

var. aegyptius

TYPE: " *in* Aegypto ", *Hasselquist* (LINN).

Leaves either pinnatisect, with 3–5 narrowly oblong or oblanceolate, entire or serrate lobes, or simple and oblong; rhachis usually subglabrous or very sparsely scabrid-hispidulous.

HAB.: On rocky slopes or amongst garigue on limestone; sea-level to c. 500 ft. alt.; fl. March–June.

DISTR.: Divisions 1, 2, 4. Also in Greece, Crete, Aegean Islands, Turkey, Syria, Palestine, Egypt, Libya and Tunisia.

1. Akamas, 1941, *Davis* 3333 ! Akamas, Khrysokhou Bay, 1948, *Kennedy* 1702 !
2. Vouni, 1963, *Townsend* s.n. !
4. Cape Greco, 1862, *Kotschy* 117A, and, 1958, *N. Macdonald* 29 !

var. **tenuifolius** (*Sibth. et Sm.*) *Heldr.* in Bull. Herb. Boiss., 6: 233 (1898); Holmboe, Veg. Cypr., 91 (1914); O. E. Schulz in Engl., Pflanzenr., 70 (IV, 105): 264 (1919); Post, Fl. Pal., ed. 2, 1: 124 (1932); Rechinger f. in Arkiv för Bot., ser. 2, 1: 423 (1950).

Bunias tenuifolia Sibth. et Sm., Fl. Graec. Prodr., 2: 3 (1813), Fl. Graec., 7: t. 613 (1830).

Didesmus tenuifolius (Sibth. et Sm.) DC., Regn. Veg. Syst. Nat., 2: 659 (1821); Unger et Kotschy, Die Insel Cypern, 333 (1865); Boiss., Fl. Orient., 1: 405 (1867).

[*D. bipinnatus* (non (Desf.) DC.) Unger et Kotschy, Die Insel Cypern, 333 (1865).]

TYPE: " in insulâ Cypro ", 1787, *Sibthorp* (OXF).

Leaves pinnatisect or bipinnatisect, with numerous linear or narrowly oblong, entire or irregularly serrate lobes; rhachis often ± densely clothed with short retrorse bristles. Usually a more slender, flexuous plant than var. *aegyptius*.

HAB.: On rocky limestone or sandstone slopes, in garigue; sometimes a weed in cultivated fields; sea-level to c. 1,000 ft. alt.; fl. Jan.–June.

DISTR. Divisions 3–7, locally common in Division 5. Greece and N. Africa.

3. Near Limassol, 1840, *Kotschy* 48.
4. Cape Greco, 1862, *Kotschy* 117 ! Between Larnaca and Arpera Chiftlik, 1862, *Kotschy* 264; between Livadhia and Dhekelia, 1880, *Sintenis & Rigo*.
5. Locally common (around Athalassa); *Sintenis & Rigo* 255 ! *Syngrassides* 44 ! 1126 ! *Chapman* 639 ! *Merton* 2902 ! *Meikle* 2972 !
6. Near Nicosia, 1880, *Sintenis* 255a ! and, 1934, *E. Haig* 7 ! Ayia Irini (Morphou), 1941, *Davis* 2126a ! Near Nicosia by Larnaca Rd., 1941, *Davis* 2139 !
7. N. side of Pentadaktylos, 1862, *Kotschy* s.n.; Kyrenia, 1949, *Casey* 246 !

13. ERUCARIA *Gaertn.*

Fruct., 2: 298, t. 143, fig. 8 (1791).

Annual herbs with erect or spreading, glabrous or thinly hispidulous stems; leaves scattered, generally pinnatisect or irregularly laciniate. Inflorescence ebracteate, at first rather densely corymbose, elongating greatly in fruit; sepals erect, oblong, subacute at apex, often subsaccate at base; petals mauve, purplish or white, obovate, truncate or obtuse at apex, narrowed rather abruptly into a slender claw; stamens 6; anthers oblong; ovary narrowly cylindrical; style usually distinct; stigma capitate. Silique longitudinally nerved, transversely 2-jointed; lower joint cylindrical, dehiscent, 2-valved, 1–8-seeded; upper joint ovoid or ensiform, indehiscent, 1–4-seeded, terminating in a long or short seedless beak. Seeds oblong or ellipsoid, brown; testa smooth.

About 8 species in the Mediterranean region eastwards to Iran and Iraq.

1. **E. hispanica** (*L.*) *Druce* in Rep. B.E.C., 3: 418 (1914); Post, Fl. Pal., ed. 2, 1: 109 (1932); Davis, Fl. Turkey, 1: 274 (1965); Zohary, Fl. Palaest., 1: 317, t. 466 (1966); Osorio-Tafall et Seraphim, List Vasc. Plants Cyprus, 46 (1973).

Sinapis hispanica L., Sp. Plant., ed. 1, 669 (1753).

Erucaria myagroides (L.) Halácsy, Consp. Fl. Graec., 1: 123 (1900); O. E. Schulz in Engl., Pflanzenr., 84 (IV, 105): 19 (1923); Lindberg f., Iter Cypr., 17 (1946).

Bunias myagroides L., Mantissa 1, 96 (1767).

Erucaria aleppica Gaertn., Fruct. 2: 298, t. 143, fig. 9 (1791); Unger et Kotschy, Die Insel Cypern, 334 (1865); Boiss., Fl. Orient., 1: 365 (1867); Holmboe, Veg. Cypr., 92 (1914).

TYPE: " *in* Hispania " (BM !).

Erect or spreading, glabrous or very sparsely hispidulous annual 20–40 cm. high; stems subterete or obscurely angled; lower leaves 10–15 cm. long, 3–8 cm. wide, rather irregularly pinnatisect or laciniate, with 7–13 or more linear or narrowly oblong, entire or irregularly lobed divisions; petiole 2–4 cm. long; upper leaves smaller and shortly petiolate, irregularly pinnatisect, or sometimes entire, linear or oblong; pedicels very short, about 1–2 mm. long, thickening in fruit; sepals 5–8 mm. long, 1–2·5 mm. wide, glabrous or thinly hirsute, often tinged violet; petals mauve or pale purple (rarely white), obovate, 1–1·5 cm. long (including claw), 2·5–4 mm. wide, claw slender 5–8 mm. long; stamens 7–10 mm. long; anthers narrowly oblong about 1·5 mm. long. Silique erect or adpressed to rhachis, 1–1·5 (–2) cm. long, 1·5–2 mm. wide, glabrous, distinctly longitudinally ribbed; apex usually abruptly narrowed to a short beak (or persistent style) about 2 mm. long; lower joint 4–8-seeded; upper joint 1–4-seeded; seeds reddish-brown, about 1·3 mm. long and 1 mm. wide; testa smooth; cotyledons replicate (or occasionally recurved or straight).

HAB.: Cultivated fields and waste ground, occasionally on grassy or sandy ground by the sea; sea-level to 1,200 ft. alt.; fl. Jan.–June.

DISTR.: Divisions 1–7, common in Divisions 4–6. Chiefly in the eastern Mediterranean region, from Greece to Egypt; adventive elsewhere.

1. Ayios Yeoryios Island (Akamas), 1962, *Meikle* 2161 !
2. Prodhromos, 1961, *D. P. Young* 7393 !
3. Near Amathus, 1947, *Mavromoustakis* 16 ! and, 1960, *N. Macdonald* 15 ! Pyrgos, 1963, *Suart* 21 !
4. Common around Larnaca; *Syngrassides* 999 ! 1080 ! *Kennedy* 1550 ! *Lindberg f.* s.n. ! *Davis* 2060 ! *N. Macdonald* 29 !
5. Common; *Sintenis & Rigo* 262 ! *Syngrassides* 377 ! 413 ! *Davis* 2135 ! *Mavromoustakis* 26 ! *Chapman* 375 !
6. Locally common around Nicosia; *Sintenis & Rigo* 262A ! *Holmboe* 267; *Davis* 2135 ! *Chapman* 500 !
7. Kythrea, 1927, *Rev. A. Huddle* 15 ! Kyrenia, 1938, *Wyatt* 49 ! Kyrenia, 1949, *Casey* 246 !

NOTES: One specimen (*Meikle* 2161) from Ayios Yeoryios Island (Akamas) is exceptional in having very coarsely lobed fleshy leaves, and siliques 3–3·5 mm. wide; it may merit recognition as a distinct variety, but needs further study in the field.

14. CAKILE *Mill.*

Gard. Dict. Abridg., ed. 4 (1754).

Erect or spreading, annual or biennial herbs with fleshy leaves and stems; stems usually much-branched, glabrous; leaves scattered, obovate or oblanceolate, variously toothed, lobed or pinnatisect, glabrous; inflorescence ebracteate, at first rather densely corymbose, the rhachis elongating greatly in fruit; sepals erect, oblong or broadly linear, obtuse at apex, subsaccate at base; petals purplish or white, obovate, apex obtuse or ± retuse, base narrowed to a short claw; stamens 6; anthers oblong; ovary ± compressed, cylindrical, gradually narrowed above; style scarcely differentiated; stigma minute, capitate. Silique softly woody or spongy, 2-jointed, the lower joint turbinate, persistent, sometimes with 2 lateral horns, occasionally obsolescent, 0–1 (–2)-seeded; upper joint caducous, ovoid or elongate-pyramidal, usually 1–2-seeded. Seeds rather large, flattened, oblong.

About 6 species, mostly maritime, in Europe, Asia, America and Australia; 1 non-maritime species in Arabia.

1. **C. maritima** *Scop.*, Fl. Carniol., ed. 2, 2: 35 (1772); Boiss., Fl. Orient., 1: 365 (1867); Holmboe, Veg. Cypr., 91 (1914); O. E. Schulz in Engl., Pflanzenr., 84 (IV, 105): 19 (1923); Post, Fl. Pal., ed. 2, 1: 108 (1932); Lindberg f., Iter Cypr., 17 (1946); Davis, Fl. Turkey, 1: 275 (1965); Zohary, Fl. Palaest., 1: 320, t. 469 (1966); Osorio-Tafall et Seraphim, List Vasc. Plants Cyprus, 46 (1973).

Bunias cakile L., Sp. Plant., ed. 1, 670 (1753).

TYPE " *in* Europae, Africae, Americae *maritimis* ".

Spreading annual or biennial with a slender, branching rootstock; stems subterete, much-branched; leaves 4–8 cm. long, 2–5 cm. wide, irregularly pinnatisect with 5–9 oblong or linear, obtuse, entire lobes; petioles about 1–3 cm. long; pedicels 2–6 mm. long, thickening but scarcely lengthening in fruit; sepals glabrous, 4–5 mm. long, 1·5–2 mm. wide, purplish, with a narrow, hyaline margin; petals pale mauve (or white), obovate, 5–8 (–11) mm. long, 2–3 mm. wide, apex usually rounded or slightly emarginate, base gradually narrowed to a short claw; stamens 4–6 mm. long; anthers about 1 mm. long. Siliques patent or ascending, narrowly ovate or lanceolate, 1·5–2·5 cm. long, 4–6 mm. wide, pale brown, spongy and buoyant; articulation prominent, inversely V-shaped; lower joint indehiscent, ± turbinate, with 2 lateral, slightly reflexed horns near its apex, obscurely nerved, 1-seeded or occasionally abortive; upper joint indehiscent, caducous, narrowly ovoid-acuminate, subtetragonous, usually much longer than the lower joint, narrowed towards the apex into a slender, flattened beak, 1–2-seeded; seeds pendulous in lower joint, erect in upper, oblong, flattened, about 3–4 mm. long and 1·5 mm. wide, reddish-brown; testa minutely rugulose.

HAB.: On sandy seashores; sea-level or a little above sea-level; fl. Febr.–July.

DISTR.: Divisions 4, 5, 7, 8. Coasts of Europe, N. Africa and western Asia.

4. Paleon Phaleron, Famagusta, by the coast, 1934, *Syngrassides* 857 ! and, same locality, 1938, *Syngrassides* 1834 !
5. Salamis, 1905, *Holmboe* 447; Trikomo, on the shore, 1939, *Lindberg f.* s.n.
7. Kyrenia, 1955, *G. E. Atherton* 47 ! 164 ! and, 1956, *G. E. Atherton* 1305 !
8. Ayios Philon near Rizokarpaso, 1941, *Davis* 2231 ! Near Galinoporni, 1962, *Meikle* 2498 !

15. CONRINGIA *Adans.*

Fam. Plant., 2: 418 (1763).

Annual or rarely biennial, glabrous or, exceptionally, glandular-pilose herbs; leaves ovate or oblong-elliptic, entire, rather fleshy, sessile, cordate-amplexicaul, often glaucous; inflorescence at first rather dense, the rhachis elongating greatly in fruit; pedicels short; sepals erect or erecto-patent, the outer narrowly oblong, obtuse or cucullate at the apex, the inner somewhat broader, obtuse or acute at apex, distinctly saccate at base; petals white, yellowish or purplish, obovate, obtuse at apex, gradually narrowed to a slender claw; stamens 6; anthers oblong, obtuse; ovary narrowly cylindrical, 4-angled or flattened; style short, inconspicuous; stigma capitate or retuse. Siliques linear, elongate, dehiscent, 2-locular; beak very short, seedless. Seeds 1-seriate, oblong or ellipsoid, brown.

About 6 species in Europe, western and central Asia and in the Mediterranean area.

1. **C. orientalis** (*L.*) *Dum.*, Florula Belg., 123 (1827); Boiss., Fl. Orient., 1: 210 (1867); O. E. Schulz in Engl., Pflanzenr., 84 (IV, 105): 86 (1923); Post, Fl. Pal., ed. 2, 1: 67 (1932); Davis, Fl. Turkey, 1: 276 (1965); Zohary, Fl. Palaest., 1: 303, t. 450 (1966).

Brassica orientalis L., Sp. Plant., ed. 1, 666 (1753).

TYPE: " *in* Oriente ".

Erect, sparsely branched, glabrous annual, 15–50 cm. high; stems terete; leaves scattered, oblong-elliptic, obtuse, glaucous, 2–10 cm. long, 1–5 cm.

wide, margins entire or undulate; lower leaves narrowed to a sessile or subsessile base; upper leaves cordate-amplexicaul; inflorescence nodding (erect in fruit), rather dense, overtopped by the developing siliques; pedicels 3–9 cm. long, lengthening and thickening somewhat in fruit; sepals erect, 5–7 mm. long, the inner acute, with a short horn just below the apex; petals pale yellow, narrowly obovate, 7–13 mm. long, 1·5–3 mm. wide, rounded at apex, gradually narrowed to a long basal claw; stamens 5–9 mm. long; ovary 4-angled, 30–40-ovulate; style slightly narrowed, about 1 mm. long; stigma flattened, capitate. Siliques spreading or ascending, rarely pendulous, 8–13 cm. long, 2–2·5 mm. wide, conspicuously and sharply 4-angled, midrib prominent on each valve, veins obscure; beak short, narrowly conical, topped by the persistent stigma; seeds pendulous, ellipsoid or oblong, slightly furrowed longitudinally, about 2 mm. long and 1 mm. wide, dull brown; testa covered with small, rounded papillae which become more conspicuous when the seed is wetted.

HAB.: Cultivated and waste ground; sea-level to 1,000 ft. alt.; fl. March–June.

DISTR.: Division 7, rare. Central and southern Europe and Mediterranean region east to Iran; adventive elsewhere.

7. Near Kephalovryso [" Kephalorisso "], Kythrea, 1880, *Sintenis & Rigo* 625 ! Kyrenia, 1949, *Casey* 698 !

TRIBE 2. **Lepidieae** *DC.* Annual or perennial herbs or subshrubs with simple or, occasionally, branched hairs; filaments sometimes with an appendage; median nectary-glands often wanting or united with the lateral glands; ovary sessile or shortly stalked; stigma capitate or obscurely 2-lobed; fruit a silicle, usually compressed laterally and with a narrow septum, sometimes not or slightly compressed, rarely indehiscent.

16. LEPIDIUM *L.*

Sp. Plant., ed. 1, 643 (1753).
Gen. Plant., ed. 5, 291 (1754).
Thellung in Mitt. Bot. Mus. Univers. Zürich, 28: 1–340 (1906).

Annual, biennial or perennial herbs or subshrubs; stems erect or decumbent, terete or angled, glabrous or with simple hairs; leaves sometimes forming a basal rosette, or scattered along the stems, entire or variously lobed and toothed, occasionally subulate; inflorescence a branched or unbranched, terminal or lateral, ebracteate raceme, frequently corymbose at first and elongating in fruit; sepals usually spreading, concave, equal or subequal, not saccate at base; petals longer or shorter than the sepals, sometimes obsolete, often distinctly clawed, white, pink, mauve or yellow; stamens 2, 4 or 6; filaments without appendages; nectary-glands 4–6, tuberculate or elongate; style short or absent; stigma capitate or obscurely 2-lobed. Silicle compressed laterally, 2-locular, normally dehiscent, elliptic, oblong or obcordate, the valves often strongly keeled or winged. Seeds pendulous, solitary (or rarely 2) in each loculus, ellipsoid or ovoid.

About 130 species scattered throughout the world.

Fruits 4–7 mm. long with a distinct apical notch, strongly flattened or winged; lower leaves irregularly lobed or pinnatipartite; annual herbs:
- Leaves ± uniform, irregularly lobed or pinnatipartite; flowers white or tinged mauve:
  - Silicles imbricate forming dense caudate racemes terminating in a spine; flowers persistent, forming elongate racemes - - - - - - **1. L. spinosum**
  - Silicles rather distant, forming lax racemes without an apical spine; flowers not persistent in corymbs - - - - - - - - - - **2. L. sativum**
- Leaves conspicuously dimorphic, the lower pinnatipartite, the upper simple, entire, ovate-amplexicaul; flowers yellow - - - - - - **3. L. perfoliatum**

Fruits 1·5–2 mm. long, without a distinct apical notch, not strongly flattened or winged; lower leaves usually simple, ovate-elliptic, serrate; perennial herbs **4. L. latifolium**

1. L. spinosum *Ard.*, Animadv. Bot. Spec. Alt. 34, t. 16 (1764); Sibth. et Sm., Fl. Graec., 7: 15, t. 617 (1830); Thellung in Mitt. Bot. Mus. Univers. Zürich, 28: 117 (1906); Holmboe, Veg. Cypr., 91 (1914); Post, Fl. Pal., ed. 2, 1: 105 (1932); Davis, Fl. Turkey, 1: 280 (1965); Zohary, Fl. Palaest., 1: 300, t. 445 (1966); Osorio-Tafall et Seraphim, List Vasc. Plants Cyprus, 46 (1973).

L. cornutum Sibth. et Sm., Fl. Graec. Prodr., 2: 6 (1813); Boiss., Fl. Orient., 1: 354 (1867).

Notoceras cardaminefolium DC., Regn. Veg. Syst. Nat., 2: 205 (1821) pro parte quoad plantam cypriam; ? Unger et Kotschy, Die Insel Cypern, 324 (1865).

TYPE: Orient, *Donati* (TO).

Glabrous or subglabrous annual with a slightly thickened taproot; stem branched or unbranched, 10–40 cm. high, erect, rather rigid, subterete; basal leaves petiolate, 4–10 cm. long, 1·5–3 cm. wide, narrowly obovate, pinnatipartite with 13–19 or more distant, rather unequal segments, the lower segments ovate, shortly toothed on the antical margin, the upper segments elongate-acuminate, frequently somewhat falcate, the antical margin often deeply and sharply incised; upper leaves often simple, linear or linear-lanceolate, acuminate, sometimes irregularly and acutely lobed; individual flowers persistent, forming many-flowered, elongate, cylindrical, terminal and axillary racemes 3–5 cm. or more long; pedicels about 1 mm. long, thickening somewhat in fruit; sepals ovate-elliptic about 1–1·5 mm. long and 1 mm. wide, with a narrow hyaline margin; petals white or pinkish, narrowly obovate, obtuse, about 2–2·5 mm. long gradually narrowed below to a short claw; stamens 6, about 1·8 mm. long; anthers shortly oblong, about 0·5 mm. long, often tinged violet; style about 0·5 mm. long; stigma capitate. Silicles held erect, often imbricate, forming dense, caudate racemes up to 10 cm. or more long, terminating in a sharp, pale spine (formed by the indurated rhachis), 1–4 cm. or more long; silicles violet-tinged, oblong-ovate, about 5–7 mm. long and 3–4 mm. wide, valves distinctly reticulate-veined, margins sharply keeled; apex produced into 2 erect, obtuse or subacute lobes 1–2 mm. long; seeds narrowly ovoid, about 2 mm. long and 1 mm. wide, slightly compressed, reddish or blackish-brown; testa mucilaginous, bluntly papillose; cotyledons entire.

HAB.: Waste ground and fallow fields, often in moist soil; ? alt.; fl. March–June.

DISTR.: " In insulâ Cypro, 1787, *Sibthorp* s.n. Eastern Mediterranean from Greece to Palestine; adventive elsewhere.

NOTES: Not seen since recorded by Sibthorp. The identity of the plant recorded by Kotschy under the name *Notoceras cardaminefolium* (" Messaria häufig, bei Strullos [Troulli]") is uncertain.

2. L. sativum *L.*, Sp. Plant., ed. 1, 644 (1753); Unger et Kotschy, Die Insel Cypern, 331 (1865); Boiss., Fl. Orient., 1: 354 (1867); Thellung in Mitt. Bot. Mus. Univers. Zürich, 28: 121 (1906); Holmboe, Veg. Cypr., 90 (1914); Post, Fl. Pal., ed. 2, 1: 105 (1932); Davis, Fl. Turkey, 1: 281 (1965); Zohary, Fl. Palaest. 1: 301, t. 447 (1966); Osorio-Tafall et Seraphim, List Vasc. Plants Cyprus, 46 (1973).

TYPE: not localized.

An erect, simple or sparingly branched annual, 15–50 cm. high, usually glabrous towards the base, sometimes thinly pubescent above; taproot slender; stem terete, smooth or obscurely striate; leaves forming a loose basal rosette, scattered above; lower leaves 3–6 cm. long, 1·5–2·5 cm. wide, 2–3-pinnatipartite, ultimate segments very variable in outline, linear to obovate, entire or irregularly dentate, obtuse or acute; petiole 1·5–2·5 cm. long; upper leaves 2–4·5 cm. long, 1·5–2 cm. wide, oblanceolate and irregularly lobed or frequently pinnatipartite with narrowly elliptic or linear, subacute and usually entire segments; shortly petiolate or sessile; uppermost leaves often simple, linear; inflorescences terminal and axillary, at first densely corymbose, becoming lax and elongate in fruit; flowers whitish or tinged pink or lilac; pedicels 0·2 mm. long, elongating in fruit; sepals oblong-obovate, 1–1·5 mm. long, 1–1·5 mm. wide, rounded or truncate with membranous margins; petals with a slender claw, oblanceolate to obovate, 1·5–2·5 mm. long, about 1 mm. wide, white or pinkish-mauve; stamens 6, about 1–1·5 mm. long; anthers shortly oblong, violet, 0·5 mm. long; style about 0·5 mm. long; stigma capitate. Silicle held erect or suberect, broadly oblong-ovoid, flattened, 4–6 mm. long and 3–4 mm. wide, distinctly winged, apex shortly notched, sinus about the same length as the style or a little shorter; seeds light reddish-brown, narrowly ovoid, about 2 mm. long and 1 mm. wide, often distinctly curved; testa minutely papillose; cotyledons normally 3-lobed.

HAB.: Cultivated and waste ground; sea-level to c. 500 ft. alt.; fl. March–June.

DISTR.: Division 4, rare. Also Egypt and Palestine, eastwards to India, cultivated and naturalized in most temperate regions.

4. Near Athienou, 1880, *Sintenis & Rigo* 272 !

NOTES: Also recorded, without locality, by Sibthorp. *Lepidium sativum* (" Cress ") is commonly grown as a salad plant.

3. L. perfoliatum *L.*, Sp. Plant., ed. 1, 643 (1753); Boiss., Fl. Orient., 1: 362 (1867); Thellung in Mitt. Bot. Mus. Univers. Zürich, 28: 147 (1906); Holmboe, Veg. Cypr., 91 (1914); Post, Fl. Pal., ed. 2, 1: 107 (1932); Davis Fl. Turkey, 1: 282 (1965); Osorio-Tafall et Seraphim, List Vasc. Plants Cyprus, 46 (1973).

TYPE: " *in* Persia, Syria ".

An erect, simple or sometimes branched, glabrous or thinly pubescent annual; taproot slender; stems terete, often obscurely striate; leaves ± glaucous, sometimes forming a loose basal rosette, the lower oblong, 1–8 cm. long, 7–20 mm. wide, deeply 2–3-pinnatipartite, segments slender, linear, acute, entire or toothed; upper leaves simple, ovate-acute, entire, 5–30 mm. long, 6–15 mm. wide, with 2 broad, overlapping basal auricles, giving the leaf a perfoliate appearance; inflorescence much branched, with terminal and lateral racemes; flowers at first rather densely corymbose, the rhachis elongating greatly in fruit; pedicels 2–5 mm. long; sepals oblong-obovate, 0·5–1 mm. long, 0·5 mm. wide, apex rounded, margins narrowly membranous; petals minute, yellowish, oblanceolate, about 1 mm. long and 0·5 mm. wide, apex rounded, entire, base narrowed to a short claw; stamens 6, about 1 mm. long; anthers yellow, suborbicular, minute; style very short, about 0·3 mm. long; stigma capitate. Fruits patent or almost so with slender pedicels 3–6 mm. long; silicle flattened, ovate-orbicular, about 4 mm. long and 3–4 mm. wide; seeds dark dull brown, obovate, 2 mm. long, 1·5 mm. wide, with a narrow, pale brown, flattened margin; testa minutely papillose.

HAB.: Cultivated and waste ground; sea-level to 4,500 ft. alt.; fl. April–July.

DISTR.: Division 2, rare. Throughout S.E. Europe and western Asia.

2. Prodhromos, 1961, *D. P. Young* 7298 !

Also recorded from Cyprus, without locality, by Thellung (*op. cit.* 149).

4. L. latifolium *L.*, Sp. Plant., ed. 1, 644 (1753); Unger et Kotschy, Die Insel Cypern, 331 (1865); Boiss., Fl. Orient., 1: 359 (1867); Thellung in Mitt. Bot. Mus. Univers. Zürich, 28: 158 (1906); Holmboe, Veg. Cypr., 90 (1914); Post, Fl. Pal., ed. 2, 1: 107 (1932); Lindberg f., Iter Cypr., 17 (1946); Davis, Fl. Turkey, 1: 284 (1965); Zohary, Fl. Palaest., 1: 300, t. 443 (1966); Osorio-Tafall et Seraphim, List Vasc. Plants Cyprus, 46 (1973).

TYPE: " *juxta fluvios* Monspelii; *inque quibusdam* Angliae *locis* " (BM).

An erect, simple or branched perennial herb, 50–110 cm. high; roots thick, woody, branched; stems terete, obscurely striate, glabrous or glabrescent; leaves scattered, petiolate, the lower elliptic-ovate, 3·5–11 cm. long, 1·5–5 cm. wide, acute or acuminate, serrate (or occasionally lobed) glabrous above, subglabrous or thinly pubescent on lower surface; petioles 2·5–6 cm. long; upper leaves elliptic, sometimes very narrowly so, acute or acuminate, glabrous or pubescent on lower surface, margins serrate or entire; inflorescence a dense, much-branched, ovoid, pyramidal or corymbose panicle; flowers very numerous; pedicels very slender, 0·5–5 mm. long, not thickening or lengthening much in fruit; sepals suborbicular, concave, 0·5–1 mm. long, with a broad, scarious margin; petals white or whitish, distinctly clawed, obovate or suborbicular, 1·5–2·0 mm. long and 1–1·5 mm. wide, apex rounded, entire; stamens 6, about 1 mm. long; anthers yellow, suborbicular, minute; style very short (almost absent); stigma relatively large, capitate. Fruits forming dense panicles, individually held erect or erecto-patent; silicle small, 1·5–2 mm. long, 1–1·5 mm. wide, somewhat compressed, ovate, elliptical or almost orbicular; apex flattened or at most very slightly retuse; seed broadly ellipsoid or suborbicular, flattened, about 0·7 mm. long; testa dull reddish-brown, minutely papillose.

HAB.: Marshy ground by streams and ditches, sometimes on sandy seashores; sea-level to 4,500 ft. alt.; fl. May–Sept.

DISTR.: Divisions 1–4, common in Divisions 2 and 4. Temperate Europe, Asia and N. Africa.

1. Stroumbi, 1913, *Haradjian* 717 !
2. Common; *Haradjian* 952 ! *Kennedy* 1740 ! *Davis* 1850 ! *G. E. Atherton* 623 !
3. Limassol, 1801, *Hume* s.n.
4. Locally common around Larnaca; *Post* 204 ! *Sintenis & Rigo* 896 ! *Syngrassides* 638 ! *Lindberg f.* s.n. !

17. TEESDALIA *R. Br.*

in Aiton, Hort. Kew., ed. 2, 4: 83 (1812).

O. E. Schulz in Engl. et Prantl, Pflanzenfam., ed. 2, 17b: 448–450 (1936).

Glabrous or subglabrous annuals; indumentum of simple hairs; leaves forming a distinct basal rosette, usually pinnatifid; inflorescence ebracteate, flowers small; sepals ± spreading, not saccate; petals white, sometimes in unequal pairs; stamens 4 or 6, filaments with a basal appendage; ovary 4-ovulate; style short, stigma capitate. Fruit an orbicular-obcordate silicle, valves laterally compressed, narrowly winged, concave above, convex below; seeds 2 in each loculus; cotyledons accumbent.

Two species in Europe and the Mediterranean area.

1. **T. coronopifolia** (*Bergeret*) *Thell.* in Feddes Repert., 10: 289 (1912); O. E. Schulz in Engl. et Prantl, Pflanzenfam., ed. 2, 17b: 448–450 (1936); Davis, Fl. Turkey, 1: 341 (1965).

Thlaspi coronopifolium Bergeret, Phytonom. Univ., 3: 29 (1784).

Teesdalia lepidium DC., Regn. Veg. Syst. Nat., 2: 392 (1821); Boiss., Fl. Orient., 1: 339 (1867).

T. regularis Sm. in Trans. Linn. Soc., 11: 286 (1815); Post, Fl. Pal., ed. 2, 1: 101 (1932).

TYPE: France; " aux environs de Paris " (STR).

Slender, erect annual 5–15 cm. high; leaves mostly confined to a basal rosette, linear, pinnatifid, 20–40 mm. long, 2–5 mm. wide, bright green, glabrous, with remote, acute, tooth-like lobes; flowering stems unbranched or sparingly branched, leafless or with a few small, bract-like leaves at the base of the lateral branches; inflorescence at first crowded and corymbose, elongating to 3–6 cm. or more in fruit; flowers inconspicuous; pedicels 0·5–1 mm. long, lengthening to 4–5 mm. in fruit; sepals oblong, 0·8 mm. long, 0·4 mm. wide, obtuse; petals subequal, narrowly obovate, about 1 mm. long, 0·3 mm. wide, apex obtuse, base tapering to a short, inconspicuous claw; stamens 4, about 0·8 mm. long, filaments filiform, glabrous, basal scale ovate, about 0·2 mm. long and almost as wide, anthers suborbicular, about 0·2 mm. diam.; ovary suborbicular, about 1 mm. diam., style very short or almost wanting, stigma capitate. Fruit suborbicular, 3–4 mm. diam., with a spreading, almost horizontal, pedicel, narrowly winged with a retuse apex, valves glabrous, indistinctly reticulate-veined; seeds compressed, suborbicular, about 1·2 mm. diam.; testa pale creamy-brown, minutely foveolate.

HAB.: On gravelly slopes and summits; c. 4,000–4,500 ft. alt.; fl. March–May.

DISTR.: Division 2, rare. Widely distributed in the Mediterranean area.

2. Khorteri near Stavros tis Psokas, 1962, *Meikle* 2708 ! Selladhi tou Mavrou Dasous, above Spilia, 1962, *Meikle* !

18. CARDARIA *Desv.*

Journ. de Bot., ser. 2, 3: 163 (1814).

Thellung in Mitt. Bot. Mus. Univers. Zürich, 28: 1–340 (1906).

A small genus of perennial, rhizomatous herbs, closely allied to *Lepidium* and often included in that genus. Stems erect, leafy; inflorescences branched, ebracteate, forming corymbose panicles; petals white; stamens 6, without appendages; ovules 2 (or sometimes 4); style distinct, persistent; stigma relatively large, capitate. Silicle indehiscent, cordate or ovate, narrowed to an acute or obtuse (not emarginate) apex, often somewhat inflated; seeds solitary in each loculus, ovoid or oblong; testa brown.

One species (including several subspecies) widely distributed in central and southern Europe and western Asia to Transcaspia and Afghanistan; adventive elsewhere.

1. C. draba (*L.*) *Desv.* in Journ. de Bot., ser. 2, 3: 163 (1814); Rechinger f., Fl. Aegaea, 239 (1943); Davis, Fl. Turkey, 1: 285 (1965); Zohary, Fl. Palaest., 1: 302, t. 448 (1966); Osorio-Tafall et Seraphim, List Vasc. Plants Cyprus, 46 (1973).

Lepidium draba L., Sp. Plant., ed. 1, 645 (1753); Boiss., Fl. Orient., 1: 356 (1867); Holmboe, Veg. Cypr., 90 (1914); Post, Fl. Pal., ed. 2, 1: 106: (1932).

An erect, subglabrous or ± densely grey-pubescent herb with a far-creeping, woody rootstock; stems striate, 16–45 cm. high; basal leaves withering before the flowers open, narrowly obovate, 5–10 cm. long, 1·5–3 cm. wide, gradually narrowed to a petiole 2–5 cm. long, leaf-margin entire or variously toothed; upper leaves sessile, generally with conspicuous sagittate auricles, oblong, elliptic or frequently panduriform, 2–8 cm. long, 1–3 cm. wide, apex acute or obtuse, margins subentire or coarsely and irregularly dentate; inflorescence rather densely corymbose-paniculate; pedicels 0·2–1 cm. long, pubescent or glabrous; sepals ± erect, broadly oblong-obovate, concave, obtuse, about 1·5–2 mm. long, 1 mm. wide, with a broad membranous margin; petals oblanceolate or obovate, 2·5–3·5 mm. long, 1–1·5 mm. wide, apex rounded, entire, base narrowed to a slender claw; stamens 1·5 mm. long; anthers yellow, minute, shortly oblong. Infructescence rather densely paniculate, fruiting pedicels slender, patent, 1–1·5 cm. long; silicles pale brown, cordate or ovoid, ± inflated, 2–4 mm. long and about as wide; style persistent, about 1–1·5 mm. long; seeds oblong or ovoid, about 2 mm. long; testa dark brown, minutely rugulose.

ssp. draba

TYPE: " *in* Germania, *praesertim* Austria, Gallia, Italia ".

Silicle cordate, inflated, base distinctly emarginate; leaves usually oblong or broadly elliptical, often rather grey-pubescent.

HAB.: Cultivated and waste ground; sea-level to 5,600 ft. alt.; fl. March–June.

DISTR.: Divisions 2, 5–8, common in Division 5. Widespread in Europe and western Asia; adventive almost throughout the temperate regions of the world.

2. Troödos, 1914, *Feilden* s.n. ! Platres, 1938, *Kennedy* 1241 ! Perapedhi, 1941, *Davis* 3089 !
5. Common; *Sintenis & Rigo* 273 ! *Syngrassides* 223 ! 418 ! *Mavromoustakis* s.n. ! *Chapman* 567 !
6. Nicosia, 1928 or 1930, *Druce* s.n.; between Morphou and Syrianokhori, 1941, *Davis* 2624 !
7. Ayios Dhimitrianos, Kythrea, 1950, *Chapman* 170 ! Myrtou, 1956, *Merton* 2513 !
8. Ronnas Valley, near Rizokarpaso, 1957, *Merton* 2928 !

ssp. **chalepensis** (*L.*) *O. E. Schulz* in Engl. et Prantl, Pflanzenfam., ed. 2, 17b: 417 (1936); Rechinger f. in Arkiv för Bot., ser. 2, 1: 423 (1950).

Lepidium chalepense L., Cent. II Plant., 23 (1756); Post in Mém. Herb. Boiss., 18: 91 (1900).

L. draba L. ssp. *chalepense* (L.) Holmboe, Veg. Cypr., 90 (1914).

TYPE: " *in* Oriente " (LINN !).

Silicle ovate, elliptical or sub-deltoid, narrowed or truncate at the base, but not emarginate, usually much compressed; leaves often narrowly oblong or elliptical, glabrous, subglabrous or thinly pubescent.

HAB.: Cultivated and waste ground; sea-level to 6,400 ft. alt.; fl. March–July.

DISTR.: Divisions 2, 6, rare. Western Asia.

2. Troödos, 1912, *Haradjian* 498.
6. Nicosia, 1894, *Post* s.n.; Nicosia, outside the walls, 1905, *Holmboe* 293.

NOTES: Records for *Cardaria draba* (L.) Desv. ssp. *chalepensis* (L.) O. E. Schulz are all somewhat suspect. Druce (Rep. B.E.C., 9: 469; 1931) states that, in Cyprus, " the common plant is *L. chalepense* L.", but all the material at Kew is clearly referable to typical *C. draba*, and it is clear that ssp. *chalepensis*, if present at all, is not as abundant as typical *C. draba*. Holmboe admits (Veg. Cypr., 90) that his specimens were without mature silicles, and, in this condition, they can scarcely be assigned with certainty to the subspecies. The Post and Rechinger f. records are more likely to be correct, though a study of Palestine material shows that Post was prepared to apply the name *chalepensis* to plants which most authors would assign to typical *C. draba*.

19. CORONOPUS *J. G. Zinn*

Cat. Hort. Acad. Gotting., 325 (1757) nom. cons.

Annual, biennial or perennial herbs with decumbent or suberect stems, leaves usually 1–2-pinnatisect, sometimes simple and entire; inflorescence racemose, ebracteate, axillary or leaf-opposed; flowers small; sepals ± spreading; petals white or tinged violet; stamens 6, 4 or 2; filaments sometimes with an appendage; ovary flattened, ovoid, reniform or almost spherical, 2-locular, with 1 ovule in each loculus; style rather short, sometimes absent; stigma capitate or ± 2-lobed. Silicle broadly ovate, cordate or reniform, indehiscent or splitting longitudinally into two 1-seeded valves. Seeds ovoid or hemispherical; testa brown.

Eight species scattered throughout the world in temperate and tropical regions.

1. **C. squamatus** (*Forssk.*) *Aschers.*, Fl. Prov. Brandenb., 1: 62 (1860); Post, Fl. Pal., ed. 2, 1: 108 (1932); Rechinger f. in Arkiv för Bot., ser. 2, 1: 423 (1950); Davis, Fl. Turkey, 1: 286 (1965); Zohary, Fl. Palaest., 1: 303, t. 449 (1966); Osorio-Tafall et Seraphim, List Vasc. Plants Cyprus, 46 (1973).

Lepidium squamatum Forssk., Fl. Aegypt.-Arab., 117 (1775).

Cochlearia coronopus L., Sp. Plant., ed. 1, 648 (1753).

Coronopus procumbens Gilib., Fl. Lith., 2: 52 (1781); O. E. Schulz in Engl. et Prantl. Pflanzenfam., ed. 2, 17b: 411 (1936).

Coronopus ruellii All., Fl. Ped., 1: 256 (1785).

Senebiera coronopus (L.) Poir. in Lam., Encycl. Méth., 7: 76 (1806); Unger et Kotschy, Die Insel Cypern, 334 (1865); Boiss., Fl. Orient., 1: 363 (1867).

Coronopus verrucarius Muschler et Thell. in Mitt. Bot. Mus. Univers. Zürich, 28: 318 (1906); Holmboe, Veg. Cypr., 91 (1914)—based on *Nasturtium verrucarium* Garsault, Descr. Plant., 412 (1767), *nomen illegitimum*.

TYPE: Egypt; " Alexandriae in desertis vicinis " *Forsskål* (C).

A prostrate or decumbent, glabrous or minutely scabridulous annual or biennial, 4–10 (–25) cm. high; leaves at first forming a rosette around a central inflorescence which terminates the growth of the main axis; rosette-leaves petiolate, 4–10 cm. long, 2–3 cm. wide, deeply pinnatisect, with slightly runcinate linear-lanceolate or oblanceolate, acute segments which are entire or acutely lobed on the abaxial (and sometimes the adaxial) margins; petiole 2·5–5 cm. long; lateral or secondary stems elongate, compressed, smooth or striatulate, bearing scattered petiolate leaves similar to those of the basal rosette, but usually smaller, with shorter petioles, and often simply pinnatisect with entire, oblanceolate segments; primary and secondary inflorescences very shortly racemose, often subcorymbose, usually less than 1 cm. long, elongating somewhat in fruit; secondary inflorescences leaf-opposed; pedicels very short, patent or ascending, or flowers subsessile; sepals oblong, obtuse, about 1 mm. long, and 0·7 mm. wide, with a narrow membranous margin; petals obovate, obtuse, slightly longer than the sepals, narrowed to the base but not clawed; stamens 6, filaments less than 1 mm. long, flattened but not appendaged; anthers minute, oblong, fuscous-violet or yellow; ovary scaly, style very short; stigma capitate. Silicle reniform, indehiscent, pale dull brown, about 2–3 mm. long and 3–4 mm. wide, conspicuously reticulate-rugose with raised, transverse lateral ridges, crowned with the persistent style and stigma; seed ovoid-hemispherical, about 2 mm. long and 1 mm. wide; testa pale brown, minutely pitted.

HAB.: Moist, frequently brackish, ground; sea-level to 3,700 ft. alt.; fl. March–June.

DISTR.: Divisions 2, 4–6. A common weed throught temperate Europe and the Mediterranean region; adventive elsewhere.

2. Pano Platres, 1930, *Wall* s.n.
4. Along the R. Pedieos east of Famagusta, towards Vatili, 1862, *Kotschy*.
5. About Kythrea, 1880, *Sintenis & Rigo* 279 ! and, 1962, *Meikle* 2559 ! Salamis, 1962, *Meikle* 2568 !
6. Nicosia moat, Famagusta gate, 1952, *Merton* 1434 !

20. IBERIS *L.*

Sp. Plant., ed. 1, 648 (1753).
Gen. Plant., ed. 5, 292 (1754).

Annual or perennial herbs or subshrubs; stems and leaves glabrous or with short, simple hairs; leaves usually linear, oblanceolate or narrowly spathulate, entire or variously toothed or lobed; inflorescence racemose or subumbellate; sepals usually ± spreading; petals white, pink or purplish, unequal, the two on the outer side of the inflorescence usually much longer than the two inner; stamens 6, without appendages; ovary 2-locular with 1 ovule in each loculus; style usually distinct; stigma capitate or 2-lobed. Silicle ovate, obovate, broadly elliptical or suborbicular, markedly flattened, the valves keeled and often winged at the notched apex. Seed relatively large, ovate or oblong, flattened, sometimes winged; testa brown.

About 30 species in temperate Europe and the Mediterranean region.

1. **I. odorata** *L.*, Sp. Plant., ed. 1, 649 (1753); Boiss., Fl. Orient., 1: 335 (1867); Post, Fl. Pal., ed. 2, 1: 101 (1932); Zohary, Fl. Palaest., 1: 298, t. 441 (1966).

I. acutiloba Bertol., Misc. Bot., 2: 12, t. 2, fig. 1 (1843); Holmboe, Veg. Cypr., 90 (1914); O. E. Schulz in Engl. et Prantl, Pflanzenfam., ed. 2, 17b: 431 (1936); Davis, Fl. Turkey, 1: 311 (1965); Osorio-Tafall et Seraphim, List Vasc. Plants Cyprus, 46 (1973).

TYPE: origin uncertain, apparently cultivated by Clusius and illustrated—as *Thlaspi IIII parvum odorato flore*—in Clusius, Rariorum Plantarum Historia, lib. 5, CXXXII (1601) !

An erect or decumbent, thinly strigillose annual, 4–15 cm. high, sometimes unbranched, but usually much-branched from the base, each plant often forming a loose, hemispherical tuft; taproot slender; leaves rather numerous, scattered, ± uniformly oblanceolate, 1·5–4 (–6) cm. long, 3–9 mm. wide, entire below, bluntly toothed or pinnatifid towards apex, gradually narrowed to the base; inflorescence subumbellate or very shortly racemose; flowers small, inconspicuous; pedicels 1–3 mm. long; sepals usually tinged violet, bluntly oblong or obovate, about 2 mm. long and 1 mm. wide, glabrous or shortly hispid, apex often tipped with short, glandular papillae; petals white or pale mauve, oblanceolate or obovate, 2–3·5 mm. long, narrowed to the base but without a distinct claw; stamens 2·5–3·5 mm. long; filaments flattened and ± winged, often tinged violet; anthers minute, yellow, about 0·5 mm. long; ovary broadly elliptical, about 2 mm. long and 1·5 mm. wide, notched at the apex and with a relatively thick, cylindrical style about 0·5 mm. long; stigma capitate, inconspicuous. Silicle broadly ovate, 5–6 mm. long, 4–5 mm. wide, glabrous, pale brown, frequently tinged and veined violet, apex with a deep V-shaped notch formed between the acute triangular wings which extend from the apices of the valves; style persistent, shorter than the apical notch; seed oblong, flat, about 2 mm. long and 1 mm. wide, very narrowly winged; testa pale dull brown, minutely reticulate.

HAB.: On rocky limestone or sandstone hillsides; to 2,500 ft. alt.; fl. March–May.

DISTR.: Divisions 5, 7. Also in the eastern Mediterranean from Greece to Egypt and eastwards to Iraq.

5. Between Nicosia and Vouno, 1941, *Davis* 2820 !
7. Near Kythrea, 1880, *Sintenis & Rigo* 275 ! Buffavento, 1941, *Davis* 2816 ! and, 1948, *Duke-Woolley* 5 !

21. BISCUTELLA *L.*

Sp. Plant., ed. 1, 652 (1753).
Gen. Plant., ed. 5, 294 (1754).
Machatschki-Laurich in Botanisch. Archiv., 13: 1–110 (1926).

Erect or spreading, annual or perennial, herbs and subshrubs; leaves often forming a loose basal rosette, scattered and rather sparse above, simple, pinnatifid or pinnatisect, or sometimes toothed; upper leaves sessile and frequently amplexicaul; inflorescence ebracteate; flowers rather small, often distinctly pedicellate forming rather loose racemes or panicles; sepals ovate or oblong, spreading, glabrous or hairy, sometimes saccate at base; petals yellow, obovate, narrowed to base or sometimes clawed; stamens 6, filaments linear, sometimes winged, without appendages; ovary shortly stalked; style distinct; stigma capitate or obscurely 2-lobed. Silicle strongly flattened with 2 orbicular or suborbicular 1-seeded valves which break away from the septum from the base upwards. Seed central in each valve, relatively large, flattened, brown.

About 40 species chiefly in central and southern Europe and the Mediterranean region.

1. **B. didyma** *L.*, Sp. Plant., ed. 1, 653 (1753); Holmboe, Veg. Cypr., 90 (1914); Post, Fl. Pal., ed. 2, 1: 97 (1932); O. E. Schulz in Engl. et Prantl, Pflanzenfam., 17b: 436 (1936); Davis, Fl. Turkey, 1: 314 (1965); Zohary, Fl. Palaest., 1: 298, t. 442 (1966); Osorio-Tafall et Seraphim, List Vasc. Plants. Cyprus, 46 (1973).

An erect, simple or branched, hispid annual herb, 4–25 (–50) cm. high; basal leaves sessile, forming a loose rosette, elliptic or oblanceolate, 1–6 cm. long, 3–17 mm. wide, margins dentate or serrate, or occasionally entire or subentire, apex acute, base gradually narrowed; cauline leaves few, much reduced, oblong-lanceolate, sessile, semiamplexicaul, with blunt auricles, apex acute, margins entire or irregularly dentate; the uppermost cauline leaves linear-lanceolate, frequently entire; inflorescence rather sparsely branched, or sometimes an unbranched raceme; pedicels about 2 mm. long, elongating considerably in fruit; sepals elliptic or oblong 1·5–2 mm. long, 0·5–1 mm. wide, obtuse or acute, entire, sometimes with a narrow membranous margin, usually glabrous or subglabrous, often tinged purple; petals pale yellow, sometimes tinged lilac with age, obovate or oblanceolate, 2–3 mm. long, about 1 mm. wide, obtuse and entire, or sometimes a little retuse at apex, gradually narrowed to base; stamens 2–3 mm. long, filaments flattened; anthers ovoid, minute; style slender, 3–4 mm. long, persistent; stigma small, capitate. Silicle glabrous, scabridulous or with a shortly ciliate margin; valves 4–5 mm. diam; seed pale brown, about 1·5–2 mm. diam.; testa minutely punctate.

var. **didyma**

B. ciliata DC. in Ann. Mus. Hist. Nat. Par., 18: 297 (1811).

B. didyma L. var. *ciliata* (DC.) Vis., Fl. Dalm., 3: 113 (1850); Post, Fl. Pal., ed. 2, 1: 97 (1932).

TYPE: " *in* Germania, Gallia, Italia ".

Margin of valves shortly and densely ciliate, disk glabrous or subglabrous.

HAB.: Rocky slopes, stony pastures, sometimes on arable or waste land; sea-level to 500 ft. alt.; fl. Febr.–May.

DISTR.: Division 7. Throughout the Mediterranean region.

7. Kyrenia, 1933, *Syngrassides* 1377 ! Vasilia, 1941, *Davis* 2200 ! 6 miles W. of Kyrenia, 1956, *G. E. Atherton* 991 !

var. **columnae** (*Ten.*) *Hal.*, Consp. Fl. Graec., 1: 105 (1900); Post, Fl. Pal., ed. 2, 1: 97 (1932).

B. columnae Ten., Cat. Hort. Neap. Append., ed. 1, 58 (1815); Unger et Kotschy, Die Insel Cypern, 329 (1865); Boiss., Fl. Orient., 1: 321 (1867) pro parte.

B. didyma L. ssp. *columnae* (Ten.) Holmboe, Veg. Cypr., 90 (1914).

[*B. laevigata* (non L.) Sibth. et Sm., Fl. Graec. Prodr., 2: 16 (1813); Sintenis in Oesterr. Bot. Zeitschr., 32: 398 (1882).]

TYPE: Based on *Ion Draba Alyssoides Apula Spicata* in Columna, Ecphrasis, 283, t. 285 fig. 1 (1606) !

Margin of valves shortly and densely ciliate, disk scabridulous.

HAB.: Rocky slopes and stony pastures; sea-level to 3,100 ft. alt.; fl. Febr.–May.

DISTR.: Divisions 2–8. Italy, Greece and Asia Minor eastwards to Iraq and Iran.

2. Near Omodhos, 1859, *Kotschy* 426 (under var. *contracta* Boiss., *nom. ined.*); on the way from Lefkara to Makheras, 1862, *Kotschy* 212 (under var. *microcarpa* Boiss.); Platres, 1937, *Kennedy* 497 (in part) ! Gourri, 1950, *Chapman* 325 !
3. Stavrovouni, 1880, *Sintenis & Rigo*; Akrotiri Forest, 1939, *Mavromoustakis* 119 ! Yerasa, 1948, *Mavromoustakis* s.n. (in part) ! Akrotiri, 1960, *N. Macdonald* 31 !
4. Near the Salt Lake at Larnaca, 1862, *Kotschy* 6; a few miles south of Larnaca, 1880, *Sintenis & Rigo;* Larnaca, 1894, *Post*; Ayia Napa, 1905, *Holmboe* 17.
5. Near Lefkoniko, 1880, *Sintenis & Rigo*; Pera, 1908, *Reid* s.n. ! Mile 5, Nicosia–Famagusta Rd., 1950, *Chapman* 42 !
6. Nicosia, 1894, *Post*; Between Myrtou and Dhiorios, 1933, *Syngrassides* 41a ! Kokkini Trimithia–Kafkalou, 1933, *Syngrassides* 322 (in part) !
7. Near Kythrea, 1880, *Sintenis & Rigo* 271 (in part) ! Kantara, 1941, *Davis* 2465 ! Kythrea above Sirgania, 1941, *Davis* 2922 ! Kyrenia mts., 1948, *Collingwood Ingram* 65 !
8. Koma tou Yialou, 1950, *Chapman* 220 (in part) !

var. **lejocarpa** *Vis.*, Fl. Dalm., 3: 113 (1850); Post, Fl. Pal., ed. 2, 1: 97 (1932).

B. leiocarpa DC. in Ann. Mus. Hist. Nat. Par., 18: 299 (1811).

TYPE: Yugoslavia; near Cittavecchia, Lesina, *Nisiteo* (PAD).

Valves quite glabrous.

HAB.: Rocky slopes and stony pastures, occasionally on sand-dunes; sea-level to 3,100 ft. alt.; fl. Febr.–May.

DISTR.: Divisions 2, 3, 6–8, common in Division 7. Also in eastern Mediterranean region from Yugoslavia and Greece eastwards to Iraq and Iran.

2. Platres, 1937, *Kennedy* 497 (in part) ! Platres, 1938, *Kennedy* 1240 !
3. Near Paramytha, 1948, *Mavromoustakis* s.n. ! Yerasa, 1949, *Mavromoustakis* s.n. (in part) !
6. Between Myrtou and Dhiorios, 1933, *Syngrassides* 41 ! Kokkini Trimithia–Kafkalou, 1933, *Syngrassides* 322 (in part) !
7. Common locally between Pentadaktylos & Yaïla Peak, *Sintenis & Rigo* 271 (in part) ! *Davis* 2892 ! *Casey* 231 ! 975 ! *Chapman* 121 !
8. Ayios Philon near Rizokarpaso, 1941, *Davis* 2279 ! Koma tou Yialou, 1950, *Chapman* 220 (in part) !

NOTES: Frequently intermixed with var. *columnae* (Ten.) Hal. and scarcely distinct from this, or from typical var. *didyma*.

22. THLASPI *L.*

Sp. Plant., ed. 1, 645 (1753).
Gen. Plant., ed. 5, 292 (1754).

Annual, biennial or perennial, glabrous or subglabrous herbs; stems simple or branched, terete or ± sulcate; basal leaves often forming a rosette, simple, ovate, obovate, elliptical, oblong or spathulate, frequently narrowed to a distinct petiole; cauline leaves rather numerous, usually sessile and amplexicaul, with acute or obtuse auricles; inflorescence ebracteate, many-flowered, often densely corymbose at first, but lengthening greatly in fruit; flowers with slender, patent or spreading pedicels; sepals spreading, shortly oblong, often obtuse or rounded at apex, not saccate at base; petals white, purplish or occasionally yellowish, narrowed to the base or sometimes clawed; stamens 6, without appendages; ovary sessile, ellipsoid or oblong, 2–16-ovulate; style usually ± distinct; stigma capitate. Silicle elliptical, obcordate or suborbicular, strongly flattened and often distinctly winged, with a conspicuous apical notch; valves smooth or obscurely veined, dehiscing from base to apex. Seeds usually 2–4 in each loculus; testa brown.

About 60 species chiefly in temperate Europe and western Asia, but occurring in eastern Asia, N. America and (1 species) in temperate S. America.

Tufted perennial with a tough rootstock; petals tinged mauve or purple; silicles narrowly obcordate, almost twice as long as wide - - - **1. T. cyprium**
Slender annual; petals white; silicles broadly obcordate, often as wide as or wider than long - - - - - - - - - - - - - **2. T. perfoliatum**

1. T. cyprium *Bornm.* in Mitt. Thüring. Bot. Ver., 38: 54 (1929); Osorio-Tafall et Seraphim, List Vasc. Plants Cyprus, 47 (1973).

[*T. violascens* Schott et Kotschy in Unger et Kotschy, Die Insel Cypern, 328 (1865) pro parte quoad spec. cypr. *Kotschy* 717 excl. spec. typ. turc. *Kotschy* 70a; H. Stuart Thompson in Journ. Bot., 44: 276 (1906); Holmboe, Veg. Cypr., 90 (1914).]

[*T. densiflorum* Boiss. et Kotschy in Unger et Kotschy, Die Insel Cypern, 328 (1865) pro parte quoad spec. cypr. *Kotschy* 717a excl. spec. typ. turc. *Kotschy* 62; Boiss., Fl. Orient., 1: 328 (1867) pro parte; Holmboe, Veg. Cypr., 90 (1914); Lindberg f., Iter Cypr., 17 (1946).]

TYPE: " Insulae Cypri in monte ' Troödos ' alt. 5,000 et 6,000'," 1862, *Kotschy* 717 (K ! W, G).

A tufted perennial 4–12 cm. high; roots elongate, tough; stems branched at base, the lower parts often rather woody and clothed with the fibrous remnants of decayed petioles, the upper parts leafy; basal leaves rather fleshy, forming a loose rosette from which arise several flowering stems; lamina oblong or spathulate, 7–35 mm. long and 5–20 mm. wide, rounded or obtuse at apex, abruptly contracted at base into a petiole 5–25 mm. long, ± glaucous, often tinged crimson-purple especially on midrib and on lower surface; leaf-margin entire or obscurely undulate; nervation indistinct;

cauline leaves smaller, sessile, amplexicaul; lamina oblong-sagittate, 7–25 mm. long, 3–15 mm. wide, apex blunt or acute, base with 2 short, blunt or acute auricles; inflorescence shortly stalked, at first rather densely corymbose, elongating later to form a rather short, cylindrical, racemose infructescence; pedicels slender, 3–5 mm. long in flower, lengthening to 7–10 mm. in fruit; sepals oblong, obtuse, 1·5–2 mm. long, about 1 mm. wide, normally stained violet-purple; petals lilac, mauve or purplish, oblanceolate, about 2·5 mm. long and 1 mm. wide, gradually narrowed to base; stamens slightly shorter than petals; filaments flattened; anthers minute, oblong, yellow (or perhaps sometimes purplish); ovary narrowly flattened-ellipsoid about 2 mm. long and 1 mm. wide, emarginate at apex; style about 0·5 mm. long; stigma capitate. Fruiting pedicels patent, rigid; silicle narrowly obcordate about 8–10 mm. long and 4–5 mm. wide, lower surface distinctly convex, upper surface ± flattened or concave, margin distinctly winged towards apex, with 2 acute, deltoid terminal lobes; sinus rather wide, enclosing the short style; seeds 1–4 in each loculus, flattened, ovoid, about 1·3 mm. long and 1 mm. wide; testa bright fulvous-brown, obscurely rugulose. *Plate 8.*

HAB.: Moist, shady fissures in igneous rocks at high altitudes; 3,000–6,200 ft. alt.; fl. Febr.–May.

DISTR.: Division 2, very common locally on the central Troödos massif (Khionistra, Kryos Potamos, Troödos, Platres, Prodhromos), also found at Papoutsa and Kionia. Endemic.

2. Very common; *Kennedy* 498–509 ! *Davis* 1859K ! 3068 ! 3163 ! 3210 ! *Mavromoustakis* s.n. ! *Meikle* 2780 ! 2845 !

NOTES: Distinguished from *T. violascens* Schott & Kotschy by its short infructescences and fulvous-brown seeds; it is also very closely allied to *T. microstylum* Boiss. from Syria and the Lebanon, but differs in its narrower fruits with shorter styles, and in its distinctly purple-tinged leaves and flowers.

2. T. perfoliatum *L.*, Sp. Plant., ed. 1, 646 (1753); Boiss., Fl. Orient., 1: 325 (1867); Holmboe, Veg. Cypr., 90 (1914); Post, Fl. Pal. ed. 2, 1: 98 (1932); O. E. Schulz in Engl. et Prantl, Pflanzenfam., ed. 2, 17b: 445 (1936); Davis, Fl. Turkey, 1: 333 (1965); Zohary, Fl. Palaest., 1: 296, t. 438 (1966); Osorio-Tafall et Seraphim, List Vasc. Plants Cyprus, 47 (1973).

An erect or spreading, glabrous annual 3–15 cm. high, simple or branched from the base; radical leaves oblong, elliptic, suborbicular or occasionally almost reniform, 0·3–2·2 cm. long, 2–14 mm. wide, margin entire or obscurely dentate, apex obtuse or rounded, base abruptly or gradually narrowed to a petiole 0·2–3·5 cm. long; cauline leaves sessile, amplexicaul, oblong-lanceolate, 0·3–2·5 cm. long, 2–10 mm. wide, margin entire or obscurely dentate, apex acute or obtuse, base with rounded or acute, deltoid auricles; inflorescence at first corymbose, later elongating to form a rather lax raceme; pedicels slender, 1–3 mm. long in flower, elongating considerably in fruit; sepals oblong or oblong-ovate, 1·5–2·5 mm. long, 0·5–1 mm. wide, obtuse, with broad membranous margins; petals white, in unequal pairs, oblong-lanceolate or spathulate, 1·5–5 mm. long, 0·5–1·5 mm. wide, obtuse and entire at apex, narrowed gradually or ± abruptly to base; stamens 1–1·5 mm. long; filaments flattened; anthers yellow, minute, oblong-ovoid; ovary sessile, flattened, obcordate, 1–2 mm. long; style short or obsolete; stigma capitate. Silicle broadly obcordate, 4–6·5 mm. long, 3·5–6 mm. wide, the lower surface convex, the upper ± flattened, margin distinctly winged especially above, with 2 rather obtuse terminal lobes; style shorter than or equalling the sinus; seeds 1–3 (–4) in each loculus, rather flattened, ovate, 1–1·5 mm. long, 0·8–1 mm. wide; testa pale brown, minutely reticulate-foveolate.

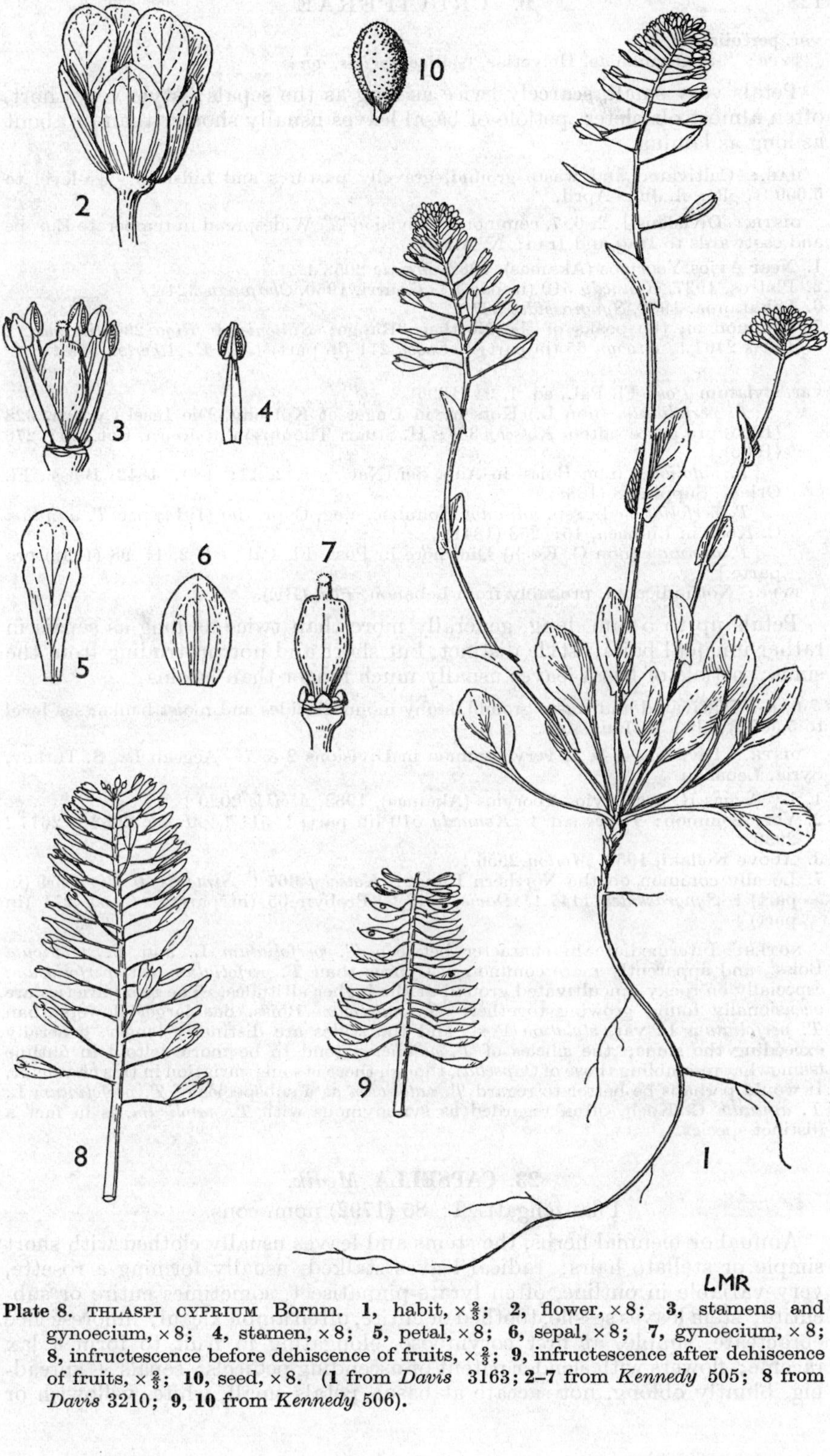

Plate 8. THLASPI CYPRIUM Bornm. **1**, habit, ×⅔; **2**, flower, ×8; **3**, stamens and gynoecium, ×8; **4**, stamen, ×8; **5**, petal, ×8; **6**, sepal, ×8; **7**, gynoecium, ×8; **8**, infructescence before dehiscence of fruits, ×⅔; **9**, infructescence after dehiscence of fruits, ×⅔; **10**, seed, ×8. (**1** from *Davis* 3163; **2–7** from *Kennedy* 505; **8** from *Davis* 3210; **9, 10** from *Kennedy* 506).

var. **perfoliatum**

TYPE: " *in* Germaniae. Helvetiae, Galliae *apricis, agris* ".

Petals very small, scarcely twice as long as the sepals; style very short, often almost obsolete; petiole of basal leaves usually shorter than or about as long as lamina.

HAB.: Cultivated and waste ground, gravelly pastures and hillsides; sea-level to 5,600 ft. alt.; fl. Jan.–April.

DISTR.: Divisions 1, 2, 6, 7, common in Division 7. Widespread in temperate Europe and eastwards to Iraq and Iran; N. Africa.

1. Near Ayios Yeoryios (Akamas), 1962, *Meikle* 2058 !
2. Platres, 1937, *Kennedy* 510 (in part) ! Gourri, 1950, *Chapman* 324 !
6. Lakatamia, 1932, *Syngrassides* 367 !
7. Common on the peaks of the Northern Range; *Sintenis & Rigo* 254 (in part) ! *Davis* 2467 ! *Probyn* 65 (in part) ! *Casey* 271 (in part) ! *G. E. Atherton* 873 !

var. **stylatum** *Post*, Fl. Pal., ed. 1, 94 (1896).

[*T. perfoliatum* (non L.) Kotschy in Unger et Kotschy, Die Insel Cypern, 328 (1865) pro parte saltem *Kotschy* 367; H. Stuart Thompson in Journ. Bot., 44: 276 (1906).]

[*T. natolicum* (non Boiss. in Ann. Sci. Nat., ser. 2, 17: 180; 1842) Boiss., Fl. Orient. Suppl., 58 (1888).]

[*T. perfoliatum* L. ssp. *annuum* Holmboe, Veg. Cypr., 90 (1914) nec *T. annuum* C. Koch in Linnaea, 15: 258 (1841).]

[*T. annuum* (non C. Koch) Dinsmore in Post, Fl. Pal., ed. 2, 1: 98 (1932) pro parte.]

TYPE: Not indicated, probably from Lebanon, *Post* (BE).

Petals up to 5 mm. long, generally more than twice as long as sepals in rather unequal pairs; style distinct, but short and not protruding from the sinus; petiole of basal leaves usually much longer than lamina.

HAB.: Cultivated and waste ground, stony mountainsides and moist banks; sea-level to 5,500 ft. alt.; fl. Jan.–May.

DISTR.: Divisions 1–3, 7, very common in Divisions 2 & 7. Aegean Is., S. Turkey, Syria, Lebanon.

1. By Avgas R. near Ayios Yeoryios (Akamas), 1962, *Meikle* 2020 !
2. Very common; *Druce* s.n. ! *Kennedy* 510 (in part) ! 511 ! *Meikle* 2663 ! 2677 ! 2939 !
3. Above Kellaki, 1956, *Merton* 2556 !
7. Locally common on the Northern Range; *Kotschy* 367 ! *Sintenis & Rigo* 254 (in part) ! *Syngrassides* 1147 ! *Davis* 2149 ! Probyn 65 (in part) ! *Casey* 271 (in part) !

NOTES: Intermediate in character between *T. perfoliatum* L. and *T. natolicum* Boiss., and apparently more common in Cyprus than *T. perfoliatum* var. *perfoliatum*, especially on rocky, uncultivated ground at the higher altitudes. The two varieties are occasionally found growing together. *T. natolicum* Boiss. has larger flowers than *T. perfoliatum* L. var. *stylatum* Post, and the styles are distinctly longer, generally exceeding the sinus; the silicles of *T. natolicum* tend to be more deltoid in outline (somewhat resembling those of *Capsella*), though there is some variation in this character. It would perhaps be better to regard *T. natolicum* as a subspecies of *T. perfoliatum* L. *T. annuum* C. Koch, often regarded as synonymous with *T. natolicum*, is in fact a distinct species.

23. CAPSELLA *Medik.*

Pflanzengatt., 1: 85 (1792) nom. cons.

Annual or biennial herbs, the stems and leaves usually clothed with short simple or stellate hairs; radical leaves stalked, usually forming a rosette, very variable in outline, often lyrate-pinnatisect, sometimes entire or subentire; stem leaves sessile, toothed or entire, often amplexicaul; inflorescence ebracteate, simple, at first corymbose, elongating in fruit to form a lax raceme; flowers with slender patent or ascending pedicels; sepals ± spreading, bluntly oblong, not saccate at base; petals small, white, yellowish or reddish, obovate, narrowed to base but scarcely clawed; stamens 6, without appendages; ovary sessile, obovate or obcordate, 12–24-ovulate; style short but usually distinct; stigma capitate. Silicle deltoid or deltoid-obcordate, strongly flattened and keeled, but not distinctly winged; valves ± distinctly veined, dehiscing from apex to base. Seeds 10 or more in each loculus, ovoid; testa brown.

About 5 closely allied species, mostly natives of temperate Europe and the Mediterranean region, but found as weeds in almost every part of the world.

1. C. bursa-pastoris (*L.*) *Medik.*, Pflanzengatt., 1: 85 (1792); Unger et Kotschy, Die Insel Cypern, 331 (1865); Boiss., Fl. Orient., 1: 340 (1867); Holmboe, Veg. Cypr., 91 (1914); Post, Fl. Pal., ed. 2, 1: 102 (1932); O. E. Schulz in Engl. et Prantl, Pflanzenfam., ed. 2, 17b: 454 (1936); Davis, Fl. Turkey, 1: 343 (1965); Zohary, Fl. Palaest., 1: 294, t. 435 (1966); Osorio-Tafall et Seraphim, List Vasc. Plants. Cyprus, 47 (1973).

Thlaspi bursa-pastoris L., Sp. Plant., ed. 1, 647 (1753).

Erect or spreading, green or subglaucous, annual herb, 30 cm. or more in height; taproot slender, elongate; stems branched particularly towards base, obscurely ridged, glabrous or clothed with simple and stellate hairs; basal leaves simple, subglabrous or with simple and stellate pubescence, usually forming a rosette; lamina oblanceolate, 3–10 cm. or more long, about 1–2 cm. wide, subentire, irregularly toothed or frequently lyrate-pinnatisect, with acute, entire or toothed lobes, base narrowed gradually into a petiole 1–4 cm. long; stem leaves smaller than basal leaves, sessile, amplexicaul, linear-lanceolate or narrowly oblong, acute or acuminate, margins entire or toothed. Flowers small, rather densely corymbose at first, the inflorescence lengthening greatly in fruit; pedicels slender, patent or ascending, elongating to 2 cm. or more in fruit; sepals oblong, blunt, 1–2 mm. long, about 1 mm. wide, glabrous or thinly pubescent, green or tinged reddish-purple; petals oblanceolate, obtuse, 1–3 mm. long, less than 1 mm. wide, white or tinged reddish-purple, occasionally wanting; stamens about 1·5 mm. long, filaments flattened; anthers yellow, ovoid-oblong, minute; ovary about 1–1·5 mm. long, 0·8 mm. wide. Silicle deltoid or deltoid-obcordate, 4–9 mm. long, 4–7 mm. wide, deeply or shallowly notched at apex; valves glabrous, obscurely or distinctly reticulate-veined; seeds narrowly ovoid-oblong, about 1 mm. long and 0·6 mm. wide; testa brown, minutely reticulate-foveolate.

var. bursa-pastoris
TYPE: " *in* Europae *cultis ruderatis* ".

Silicle deltoid-obcordate, usually longer than wide, lateral margins straight or convex.

HAB.: Cultivated and waste ground; sea-level to 5,000 ft. alt.; fl. Jan.–July.

DISTR.: Divisions 4, 5, rare. Cosmopolitan weed.

4. Akhyritou, 1948, *Mavromoustakis* s.n. !
5. Kythrea, 1930, *Druce* s.n. (as *Bursa mediterranea* (E. Almq.) Druce) !

var. brachycarpa *Heldr.*, Herb. Graec. Norm. no. 1212b. (? 1896).

[*Bursa rubella* Druce in Rep. B.E.C., 9: 469 (1931) nec *Capsella rubella* Reut. in Compt. Rend. Soc. Hallér., 18 (1853–54).]

Bursa concava Druce in Rep. B.E.C., 9: 469 (1931) ? *Capsella concava* E. Almq.

[*Capsella rubella* (non Reut.) Lindberg f., Iter Cypr., 17 (1946).]

TYPE: Greece; " in collibus saxosis (Lycabetto etc.) prope Athenas et maritimis (pr. Piraeum et Eleusin) " March 27–May 5, 1894, *Heldreich* herb. norm. 1212b (K !).

Silicle deltoid, about as wide as long, lateral margins straight or concave; petals longer than sepals.

HAB.: Cultivated and waste ground; sea-level to 5,000 ft. alt.; fl. Jan.–July.

DISTR.: Divisions 1–8, common in Divisions 2 and 7. Eastern Mediterranean from Greece to Palestine.

1. Peyia, 1962, *Meikle* 2027 ! Near Polis, *Meikle* !
2. Common (Amiandos, Platres, Prodhromos); *Druce* s.n. ! *Lindberg f.* s.n. ! *Mavromoustakis* s.n. ! *Merton* 2202 ! *Kennedy* 1824 !
3. Limassol, 1928, *Druce* s.n. !
4. Larnaca, 1928, *Druce* s.n. ! Paralimni, 1940, *Davis* 2059 !
5. Malounda, 1950, *Chapman* 309 ! Near Kythrea, 1962, *Meikle* ! Salamis, 1962, *Meikle* !
6. Nicosia, Department of Agriculture grounds, 1957, *Merton* 3075 ! Dhiorios, 1962, *Meikle* !
7. Common; *Druce* s.n. ! *Syngrassides* 1321 ! *Mavromoustakis* s.n. ! *G. E. Atherton* H. 786 ! 822 !
8. Near Rizokarpaso, 1941, *Davis* 2306 ! Akradhes, 1962, *Meikle* !

NOTES: Aggregate *C. bursa-pastoris* has also been recorded from Prodhromos (April 6, 1859, *Kotschy* s.n.) and from Nicosia (March 1934, *Nattrass* 422), and Holmboe (Veg. Cypr., 91) remarks that the plant is common on the island, though it would seem that the majority of specimens fall under var. *brachycarpa* rather than under typical *C. bursa-pastoris* (L.) Medik. *C. rubella* Reut., with fruits similar to those of var. *brachycarpa*, but with very small, red-tinged flowers, the petals scarcely exceeding the sepals, is said to have been found in Cyprus (Druce, Rep. B.E.C., 9: 469 and Lindberg f., Iter Cypr. 17) but these records are errors for var. *brachycarpa*. True *C. rubella* is a plant of Central Europe and the Mediterranean coasts of France, Italy and Yugoslavia; I have not seen genuine material from Cyprus, though it may occur as an introduction. No attempt has been made to determine *C.* (*Bursa*) *concava* E. Almq. (Larnaca, Paphos); *C.* (*Bursa*) *patagonica* E. Almq. (" Gambo " [? Kambos]; " Ag. Neophilus " [? Ayios Neophytos]); *C.* (*Bursa*) *batavorum* E. Almq. (Nicosia); *C.* (*Bursa*) *mediterranea* E. Almq. (" Pedaila " [? Pedhoulas]; Kythrea); *C.* (*Bursa*) *treviorum* E. Almq. (" Gambo " [? Kambos] valley) and *C.* (*Bursa*) *turoniensis* E. Almq. (Makheras), all recorded from Cyprus by Druce (Rep. B.E.C., 9: 469). The taxonomic value of these microspecies is extremely dubious, for it must be admitted that even the three larger aggregates—*C. bursa-pastoris* var. *bursa-pastoris*, *C. bursa-pastoris* var. *brachycarpa* and *C. rubella*—are not always clearly demarcated.

24. HYMENOLOBUS *Nutt.*

ex Torrey et Gray, Fl. N. America, 1: 117 (1838).

Slender, glabrous or subglabrous annual herbs with erect, spreading or decumbent, branched or unbranched stems; basal leaves often forming a loose rosette, obovate or oblanceolate, entire or variously lobed or pinnatisect; cauline leaves rather few, frequently entire or subentire, lanceolate or almost linear; inflorescence ebracteate, simple, at first corymbose, elongating in fruit to form a very lax raceme; flowers with slender, fiilform, patent or ascending pedicels; sepals spreading, bluntly oblong, not saccate at base; petals minute, white, obovate, narrowed to base; stamens 6, filaments flattened, without appendages; stigma sessile, capitate; ovary sessile, ovate, 12–20-ovulate. Silicle ovoid or ellipsoid, distinctly compressed; valves convex, reticulate-veined, dehiscing from apex to base. Seeds 6–10 in each loculus, ovoid; testa brown.

Two or possibly three closely related species scattered over temperate Europe, Asia and America.

1. H. procumbens (*L.*) *Schinz et Thell.* in Viertelj. Naturf. Ges. Zürich, 66: 285 (1921); O. E. Schulz in Engl. et Prantl, Pflanzenfam., ed. 2, 17b: 457 (1936); Davis, Fl. Turkey, 1: 342 (1965); Zohary, Fl. Palaest., 1: 295, t. 436 (1966); Osorio-Tafall et Seraphim, List Vasc. Plants Cyprus, 47 (1973).

Lepidium procumbens L., Sp. Plant., ed. 1, 643 (1753).

Capsella procumbens (L.) Fries, Novit. Fl. Suec. Mant., 1: 14 (1832); Boiss., Fl. Orient., 1: 340 (1867); Holmboe, Veg. Cypr., 91 (1914); Post, Fl. Pal., ed. 2, 1: 102 (1932).

Hutchinsia procumbens (L.) Desv., Journ. de Bot., 3: 168 (1814).

Hornungia procumbens (L.) Hayek, Prodr. Fl. Pen. Balc., 1: 480 (1925) pro parte.

TYPE: France; " Monspelii ".

Slender, spreading, erect or decumbent, glabrous or subglabrous annual, 4–15 (–25) cm. high; stems usually branched, especially at the base, sometimes unbranched; basal leaves often forming a loose rosette, narrowly obovate, entire, lobed or pinnatisect, 1·5–3 cm. long, 5–10 mm. wide, petiole slender about 1 cm. long; cauline leaves smaller, sessile or shortly stalked, frequently linear-lanceolate and entire; flowers minute; sepals bluntly oblong, less than 1 mm. long and about 0·5 mm. wide; petals white, about 1 mm. long and 0·5 mm. wide, obovate, obtuse, narrowed to base; stamens a little longer than petals; anthers minute, shortly oblong or almost globose, yellow. Silicles ovoid or bluntly ellipsoid, 2·5–4 mm. long, about 2 mm. wide, distinctly flattened; fruiting pedicels slender, patent or ascending, about 0·5–1 cm. long; seeds numerous, minute, ovoid, about 0·4 mm. long and 0·2 mm. wide; testa tawny-brown, distinctly, but minutely, reticulate-foveolate.

HAB.: In saline or brackish marshes; near sea-level; fl. Febr.–April.

DISTR.: Divisions 4, 5. Europe, Asia, N. Africa, temperate N. & S. America, Australia, New Zealand.

4. On the shore of the Salt Lake at Larnaca, 1905, *Holmboe* 109b.
5. By the bridge before Pyroi, 1936, *Syngrassides* 1045 !

TRIBE 3. **Euclidieae** *DC.* Mostly annual or biennial herbs with simple or branched hairs; filaments linear; median nectary-glands usually wanting; ovary sessile or very shortly stalked; stigma capitate or 2-lobed; fruit a hard, nut-like silicle, indehiscent or tardily dehiscent, not or only a little compressed; septum broad.

25. OCTHODIUM *DC.*

Regn. Veg. Syst. Nat., 2: 423 (1821).

Robust biennial herb; stems branched, leafy; lower leaves lyrate-pinnatisect; upper leaves frequently linear-lanceolate, entire; inflorescence at first rather loosely corymbose, elongating greatly after fertilization and finally becoming a lax raceme; sepals spreading, oblong or ovoid; petals yellow, longer than sepals, narrowed to base or with a short claw, obtuse or slightly retuse at apex; stamens 6, anthers ovoid; ovary sessile, 2-locular, with a solitary ovule in each loculus; style obscure, conical; stigma slightly 2-lobed. Silicle indehiscent, subglobose, indistinctly quadrangular, with thick tuberculate-rugose walls, usually crowned with the persistent style and stigma. Seeds ovoid-oblong, distinctly flattened; testa brown; cotyledons obliquely accumbent.

One species in the eastern regions of the Mediterranean.

1. O. aegyptiacum (*L.*) *DC.*, Regn. Veg. Syst. Nat., 2: 423 (1821); Boiss., Fl. Orient., 1: 369 (1867); Post, Fl. Pal., ed. 2, 1: 112 (1932); Davis, Fl. Turkey, 1: 347 (1965); Zohary, Fl. Palaest., 1: 257, t. 374 (1966).
Bunias aegyptiaca L., Syst. Nat., ed. 12, 3: 231 (1768).
TYPE: " *in* Aegypto. *Zoega* ".

Much-branched, spreading biennial 20–60 cm. or more in height; stems distinctly sulcate, glabrous or sparsely hispid, often tinged violet towards base; basal leaves 5–12 (–20) cm. long, 2·5–5 (–8) cm. wide, variable in outline, usually lyrate-pinnatisect, with a large, dentate or undulate, rounded or acute terminal lobe and 4–6 smaller, acute or rather blunt lateral lobes, glabrous or sparsely hispid; petiole 1–3 cm. or more long; lower cauline leaves similar to basal leaves, but usually much smaller and more shortly stalked; upper cauline leaves often simple, entire, lanceolate or linear-lanceolate, sometimes hastate-lobed towards the base, generally with distinct petioles; flowers in small corymbs; pedicels slender, about 3–4 mm. long, lengthening considerably in fruit; sepals oblong or ovoid, obtuse or subacute, 1·5–2 mm. long, about 1 mm. wide, pale greenish yellow with a narrow hyaline margin; petals yellow, obovate, about 3 mm. long and 1·5 mm. wide; stamens about 1·5 mm. long, filaments linear, flattened. Silicle about 5 mm. long, slightly less in width, pale brown, coarsely tuberculate-rugose, obscurely 4-angled; seed about 2·5 mm. long, 2 mm. wide, plano-convex; testa minutely longitudinally rugulose.

Hab.: Cultivated and waste ground; c. 3,000 ft. alt.; fl. April–June.

DISTR.: Division 2, rare. Eastern Mediterranean region.

2. Khrysorroyiatissa, 1941, *Davis* 3434 !

26. NESLIA *Desv.*

Journ. de Bot., ser. 2, 3: 162 (1814) nom. cons.
Vierhapper in Oesterr. Bot. Zeitschr., 70: 167 (1921).

Annual or biennial herbs with branched, leafy stems; basal leaves oblanceolate, entire or remotely toothed, gradually tapering to a short flattened petiole; stem leaves lanceolate, oblanceolate-sagittate, entire or subentire, sessile, with projecting, acuminate auricles; inflorescence paniculate, often much-branched, ebracteate; flowers rather small, with slender pedicels, at first forming a loose corymb, the rhachis elongating greatly after fertilization and finally forming a lax raceme; sepals erect or suberect, oblong, obtuse, slightly gibbous at base but scarcely saccate; petals narrowly obovate or oblanceolate, gradually tapering to base; stamens 6, filaments linear, flattened; anthers yellow, shortly oblong; ovary ovoid, subsessile or shortly stipitate, 1-locular, 2–4-ovulate; ovules pendulous; style distinct, slender; stigma capitate. Silicle globose or subglobose, indehiscent, 1-seeded. Seed ovoid; testa brown, cotyledons incumbent.

Two species in Europe, Mediterranean region and western Asia, naturalized elsewhere.

Style articulated above base; persistent style-base prominent, apiculate; silicle ± tapering at base - - - - - - - - - - - **1. N. apiculata**
Style articulated almost at base; style-base indistinct or wanting; silicle abruptly constricted at base - - - - - - - - - - - **2. N. paniculata**

1. N. apiculata *C. A. Mey.* in Fisch., Mey. et Avé-Lall., Ind. Sem. Hort. Petrop., 8: 68 (1842); Rechinger f. in Arkiv för Bot., ser. 2, 1: 423 (1950); Davis, Fl. Turkey, 1: 350 (1965); Zohary, Fl. Palaest., 1: 293, t. 434 (1966); Osorio-Tafall et Seraphim, List Vasc. Plants Cyprus, 47 (1973).

[*N. paniculata* (non (L.) Desv.) Kotschy in Unger et Kotschy, Die Insel Cypern, 331 (1865); Boiss., Fl. Orient., 1: 371 (1867) pro parte; Holmboe, Veg. Cypr., 91 (1914); Osorio-Tafall et Seraphim, List Vasc. Plants Cyprus, 47 (1973).]

[*Vogelia paniculata* (non (L.) Hornem.) Dinsmore in Post, Fl. Pal., ed. 2, 1: 113 (1932) pro parte.]

TYPE: Arabia; "ad radices montis Sinai, locis graminosis" May 6, 1835, *Schimper* 158 (LE, K !).

Erect, branching annual or biennial, 20–50 cm. high; stems obscurely angled, subglabrous or often ± densely clothed with mixed simple and branched, bristly hairs; leaves variable in size, the lowermost 2–7 cm. or more long, 4–15 mm. wide, oblanceolate, acute, entire or remotely toothed, narrowed towards base into a short flattened petiole, lamina clothed with sparse, short, branched hairs; upper leaves progressively smaller, lanceolate or oblanceolate-sagittate, sessile, with projecting auricles; inflorescence a lax, spreading panicle with rather flexuous branches; flowers small; pedicels 3–5 mm. long; sepals about 2 mm. long and 1 mm. wide, oblong or ovate, obtuse, usually clothed at least with a few long, white hairs; petals yellow, about 3 mm. long and 0·6–1 mm. wide, gradually narrowed to base; stamens about 2 mm. long; anthers minute; ovary about 1 mm. long, narrowed at base into a short gynophore; style about 1 mm. long, persistent, articulate about 0·4 mm. above base. Silicle subglobose, slightly compressed, about 3 mm. long and 3·5 mm. wide, conspicuously reticulate-veined; 4-costate, with 2 prominent lateral ribs and 2 indistinct median ribs, usually crowned with the indurated style or the apiculate style-base and narrowed at the base into a short stipe; seed ovoid; testa rather pale brown, irregularly reticulate.

HAB.: Cultivated fields, waste ground, roadsides etc.; sea-level to 3,700 ft. alt.; fl. Febr.–June.

DISTR.: Divisions 1–8. Throughout the eastern Mediterranean region and the Middle East to the borders of India; colonist elsewhere.

1. Dhrousha, 1941, *Davis* 3231 !
2. Panayia, 1913, *Haradjian* 905. Platres, 1938, *Kennedy* 1242 ! 1243 ! Gourri, 1950, *Chapman* 311 ! Pomos, 1962, *Meikle* 2257 !
3. Polemidhia, 1948, *Mavromoustakis* s.n. ! Akrotiri Bay, 1948, *Mavromoustakis* s.n. ! Yerasa, 1948, *Mavromoustakis* s.n. !
4. Near Larnaca, 1862, *Kotschy* 80; near Larnaca, 1880, *Sintenis & Rigo* 848 ! Larnaca, 1905, *Holmboe* 114; Famagusta Castle, 1936, *Syngrassides* 1157 ! Kondea–Pergamos Rd., 1954, *Merton* 1845 !
5. Ayia Katerina, near Salamis, 1948, *Mavromoustakis* s.n. !
6. Nicosia, 1935, *Nattrass* 556; Kondemonos, 1935, *Syngrassides* 636 ! Peristerona, 1935, *Syngrassides* 773 ! Ayia Irini, 1941, *Davis* 2556 ! Nicosia, 1959, *Oswald* 28 !
7. Kazaphani, 1949, *Casey* 235 !
8. Ayios Theodhoros, 1947, *Mavromoustakis* s.n. !

2. N. paniculata (*L.*) *Desv.* in Journ. de Bot., ser. 2, 3: 162 (1814); Boiss., Fl. Orient., 1: 371 (1867) pro parte; Davis, Fl. Turkey, 1: 350 (1965); Zohary, Fl. Palaest., 1: 293, t. 434 (1966).

Myagrum paniculatum L., Sp. Plant., ed. 1, 641 (1753).

Vogelia paniculata (L.) Hornem., Hort. Hafn., 2: 594 (1815).

TYPE: " *in Europa, juxta agros* ".

Very similar to *N. apiculata* in general appearance; stems and leaves often rather sparsely pilose; inflorescence less branched than in *N. apiculata*, with straight, ascending branches. Silicle smaller, about 2 mm. long and slightly more in width, abruptly constricted at apex and base, less distinctly compressed than in *N. apiculata*, venation more minutely and obscurely reticulate, and median ribs wanting or indistinct; style articulated almost at its base, with the persistent style-base forming, at most, a tiny apiculus.

HAB.: Cultivated and waste ground; c. sea-level; fl. Febr.–June.

DISTR.: Division 4. Europe and N. Asia, generally replaced by *N. apiculata* in the eastern Mediterranean region and Orient.

4. Near Larnaca, 1880, *Sintenis & Rigo* 848 (in part) ! Famagusta, 1957, *Merton* 2890 !

TRIBE 4. **Alysseae** *Gren. et Godr.* Annual or perennial herbs with simple, branched or stellate indumentum; nectary-glands 1 (or rarely 2) on either side of the shorter stamens; median nectary-glands absent. Filaments with or without appendages; ovary sessile or shortly stipitate; style usually distinct; stigma truncate, capitate or 2-lobed. Fruit a broad, flattened or sometimes inflated, usually dehiscent silicle, rarely a linear silique; septum broad. Cotyledons accumbent.

27. FIBIGIA *Medik.*

Pflanzengatt., 1: 90, t. 2, fig. 23 (1792).

Erect or spreading perennials, ± densely clothed with stellate, simple or mixed simple-stellate indumentum; stems obscurely angled, leafy; leaves oblanceolate or obovate, tapering to the base, the lowermost usually shortly petiolate; leaf-margins entire, sinuate or irregularly dentate or lobed. Inflorescence sparsely branched, ebracteate; flowers subsessile or shortly pedicellate, at first forming a dense corymb, often overtopped by uppermost stem leaves, rhachis elongating rapidly after fertilization and finally forming a long rather rigid raceme; sepals erect, narrowly oblong or ovate, indistinctly saccate or gibbous at base; petals yellow or purple, narrowly obovate, tapering gradually to base, but not abruptly clawed; stamens 6, the outer filaments often with a small tooth, the inner distinctly broadened and flattened towards apex; ovary narrowly ovate, 4–18-ovulate; stigma capitate. Silicle large, elliptical or almost orbicular, flattened; valves often densely pilose; septum shining, papery; seeds suborbicular, brown, flattened, often with a conspicuous marginal wing.

About 12 species chiefly in the eastern Mediterranean region.

1. F. eriocarpa (*DC.*) *Boiss.*, Fl. Orient., 1: 258 (1867); Post, Fl. Pal., ed. 2, 1: 82 (1932); Davis, Fl. Turkey, 1: 357 (1965).
Farsetia eriocarpa DC., Regn. Veg. Syst. Nat., 2: 288 (1821).
TYPE: " in insula Cypro " *Labillardière* (FI).

A robust, erect perennial, 30–60 cm. or more in height, with grey, densely stellate-tomentose leaves and stems; sterile shoots forming a loose basal tuft bearing narrow, oblanceolate, acute leaves 3–7 cm. or more long, 5–15 mm. wide, gradually tapering at the base into a petiole 1–2 cm. long; leaf margin entire or irregularly and remotely toothed; leaves of the flowering shoots similar, but often rather larger (up to 11 cm. long and 2 cm. wide) and tapering to a sessile base, becoming progressively smaller towards apex of stem, and commonly forming an involucre-like cluster around the base of the young inflorescence; flowers very shortly pedicellate; sepals narrowly oblong, about 7 mm. long and 2 mm. wide, densely clothed externally with intermixed stellate and simple hairs, and with a broad hyaline margin; petals yellow, narrowly oblanceolate, about 10 mm. long and 2·5 mm. wide, apex rounded or truncate, gradually narrowed to base; stamens 4–5 mm. long, the filaments of the shorter pair often with a blunt lateral tooth, those of the longer ones broadly and asymmetrically winged; anthers yellow, oblong, about 1 mm. long; ovary about 2 mm. long, densely pubescent; style about 1·5 mm. long; stigma capitate. Silicle broadly elliptical, 1·5–2 cm. long, 1–1·3 cm. wide, densely grey pilose-tomentose, with a layer of short, branched hairs overtopped by a layer of spreading, silky, simple hairs; valves rather rigid, dehiscing from the base upwards; seeds dull, rather dark brown, suborbicular, flat, about 5 mm. diam., with a membranous wing about 1·3 mm. diam.; testa minutely reticulate.

HAB.: Not known.

DISTR.: Division 5 or 6. Greece, Turkey, Caucasus and Syria.
5. or 6. Near Nicosia, 1930, *Rogers* 0672 ! See note.
Also (? Division), Cyprus, 1787, *Labillardière*.

NOTES: Not collected recently, and perhaps an error. Several Rogers specimens, purporting to have come from Cyprus, and so labelled, were in fact collected in the neighbourhood of Iskenderun (Alexandretta) in Turkey.

28. ALYSSUM *L.*

Sp. Plant., ed. 1, 650 (1753).
Gen. Plant., ed. 5, 293 (1754).

Annual or perennial herbs or subshrubs clothed with simple, branched or stellate indumentum; stems leafy; leaves simple; inflorescence ebracteate, flowers usually small; sepals erect or spreading, not saccate at base; petals yellow, ± clawed; stamens 6, usually free, with all or some of the filaments winged or with toothed appendages; ovary sessile, flask-shaped, (1–) 2–16-ovulate; style generally short; stigma small, truncate, capitate or ± 2-lobed. Silicle suborbicular, ovate or elliptical, flattened or rarely inflated, valves ± distinctly reticulate-veined. Seeds orbicular or ovate, usually winged.

About 100 species in Europe, the Mediterranean area and temperate Asia.

Subshrubs or perennials with woody rootstocks:
 Silicle ovate or obovate-elliptical, narrowed at both ends:
  Silicle glabrous, 7–8 mm. long, 3–4 mm. wide; leaves rather broadly obovate-spathulate - - - - - - - - **1. A. troodi**
  Silicle sparsely stellate-pilose, about 4 mm. long and 2·5 mm. wide; leaves narrowly obovate-spathulate - - - - - - - **2. A. cypricum**
 Silicle obovate-suborbicular, rounded, truncate or retuse at apex:
  Silicle glabrous or almost so, tuberculate-papillose; a robust, erect subshrub **3. A. chondrogynum**
  Silicle stellate-pilose; a weak, straggling perennial - - **4. A. akamasicum**
Annual herbs with suborbicular fruits:
 Silicle stellate-pilose:
  Infructescence elongate, racemose:
   Style stellate-pilose; fruits with a uniform indumentum of stellate hairs **5. A. minus**
   Style glabrous; fruits with a dimorphic indumentum of long, simple or forked hairs, and shorter stellate hairs - - - - - - **6. A. strigosum**
  Infructescence umbellate, or at most very shortly racemose **7. A. umbellatum**
 Silicle glabrous:
  Leaves rather loosely and unevenly stellate-pilose; silicle about 3·5 mm. diam.; seed fulvous-brown with a narrow inconspicuous wing - - **8. A. minutum**
  Leaves densely and closely stellate-pilose; silicle about 4 mm. diam.; seed fuscous-brown with a conspicuous pale wing - - - - - - **9. A. foliosum**

1. A. troodi *Boiss.*, Fl. Orient. Suppl., 49 (1888); Holmboe, Veg. Cypr., 89 (1914); Lindberg f., Iter Cypr., 17 (1946); Burtt et Davis in Kew Bull., 1949: 102 (1949); Osorio-Tafall et Seraphim, List Vasc. Plants Cyprus, 47 (1973).

A. coriaceum Nyárády in Bull. Grăd. Bot. Univ. Cluj, 9: 46 (1929).

[*A. alpestre* L. β *obtusifolium* Fenzl sec. Kotschy in Unger et Kotschy, Die Insel Cypern, 326 (1865); H. Stuart Thompson in Journ. Bot., 44: 276 (1906).]

TYPE: " in monte Troodos Cypri " *Sintenis & Rigo* 844 (K ! G etc.).

A much-branched, spreading subshrub, 10–25 cm. high, with a tough woody rootstock; leaves of the sterile shoots persistent, forming rather dense terminal clusters, covered with a close indumentum of whitish or silvery stellate hairs, obovate-spathulate, 5–10 mm. long, 2–7 mm. wide, margin entire, apex obtuse or subacute, base gradually tapering to a short petiole; leaves of flowering shoots similar, but rather distant and less densely hairy, often falling before the fruits ripen; inflorescence rather densely corymbose, terminal or with a few axillary branches; pedicels spreading 2–4 mm. long; sepals erect, oblong, concave, about 2 mm. long and 1 mm. wide, ± densely stellate-pilose externally; petals golden-yellow, narrowly oblanceolate, about 3 mm. long and 1 mm. wide, apex obtuse, gradually tapering to base; stamens with distinct winged appendages, about 2·5 mm. long; anthers oblong, minute, yellow; ovary flask-shaped; about 1 mm. long and 0·7 mm. wide, glabrous; style about 0·7 mm. long, stigma capitate. Silicle elliptic, flattened, glabrous, about 7–8 mm. long and 3–4 mm. wide, tapering at both ends; seeds ovate, flattened, about 4 mm. long and 2·5 mm. wide; testa dull purplish-brown, minutely foveolate.

HAB.: Rocky mountainsides on ultrabasic (serpentine) formations; 4,000–6,400 ft. alt.; fl. May–July.

DISTR.: Division 2, very common locally on Khionistra. Endemic.

2. Locally common around Troödos; also found at Prodhromos; *Sintenis & Rigo* 844 ! *Haradjian* 468 ! 980 ! *Kennedy* 493 ! 494 ! 495 ! *Merton* 2338 ! 2436 !

2. A. cypricum *Nyárády* in Bull. Grăd. Bot. Univ. Cluj, 7: 156 (1928); Davis, Fl. Turkey, 1: 397 (1965); Osorio-Tafall et Seraphim, List Vasc. Plants Cyprus, 47 (1973).

A. virgatum Nyárády var. *mutabile* Nyárády in Bull. Grăd. Bot. Univ. Cluj, 7: 116 (1928).

[*A. argenteum* (non Wittm.) Kotschy in Unger et Kotschy, Die Insel Cypern, 327 (1865).]

[*A. condensatum* (non Boiss. et Hausskn.) Boiss., Fl. Orient. Suppl., 49 (1888); Holmboe, Veg. Cypr., 89 (1914); Lindberg f., Iter Cypr., 17 (1946).]

A. murale (non W. et K.) Boiss., Fl. Orient. Suppl., 50 (1888).]

[? *A. alpestre* L. var. *suffrutescens* (non Boiss.) Post in Mém. Herb. Boiss., 18: 90 (1900).]

TYPE: Cyprus; Troödos, at Troödhitissa Monastery, 1,300 m. alt., Oct. 11, 1893, *E. Deschamps* s.n. (BRNM).

A much-branched, spreading or tufted subshrub, 10–25 cm. high, with a tough, woody rootstock; leaves of the sterile shoots persistent, seldom forming marked terminal clusters as in *A. troodi* Boiss., covered with a close indumentum of whitish or silvery stellate hairs, narrowly obovate-spathulate, 2–10 mm. long, 1·5–4 mm. wide, margins entire, apex obtuse or subacute, base gradually tapering to a short petiole; leaf-margins often suberect, giving the leaf a keeled appearance; leaves of flowering shoots somewhat caducous, similar to those of the sterile shoots but usually rather larger, flatter and less densely stellate-hairy; inflorescence rather lax, usually much-branched, forming a loose open corymb; pedicels spreading, 2–4 mm. long; sepals erect, rather concave, about 1·5–2 mm. long and 1 mm. wide, ± densely stellate-pubescent externally, apex subacute; petals golden-yellow, oblanceolate or narrowly obovate, obtuse, about 3 mm. long and 1–1·5 mm. wide, gradually tapering to base; stamens with distinct winged appendages, about 2·5 mm. long; anthers oblong, minute, yellow; ovary flask-shaped, about 0·8 mm. long and 0·5 mm. wide, densely stellate-pubescent; style about 0·7 mm. long, glabrous, stigma capitate. Silicle ovate or obovate-elliptical, flattened, sparsely stellate-pubescent, about 4 mm. long and 2·5 mm. wide, narrowed but obtuse at both ends; seeds ovate, flattened, about 1·5 mm. long and 1·2 mm. wide, narrowly winged especially towards hilum; testa dull brown, minutely foveolate.

HAB.: Rocky mountainsides, on ultrabasic (serpentine) formations; 3,500–6,000 ft. alt.; fl. April–June.

DISTR.: Division 2, common. Also Turkey.

2. Common on the higher slopes of central Troödos, around Prodhromos, Khionistra, Troödos & Amiandos; *Kotschy* 703 ! *Sintenis & Rigo* 843 ! *Kennedy* 487–492 ! *Davis* 1784 ! 1867 ! 3156 ! 3207 !

3. A. chondrogynum *B. L. Burtt* in Kew Bull., 1949: 101 (1949); Osorio-Tafall et Seraphim, List Vasc. Plants Cyprus, 47 (1973).

TYPE: Cyprus; above Yerasa (distr. Limassol) at 2,000 ft. alt., April 10, 1941, *Davis & Kennedy* in *Davis* 3083 (K !).

A much-branched, erect or spreading subshrub, 20–30 cm. high, with a stout, woody rootstock; stems and leaves densely clothed with adpressed, silvery, stellate hairs; leaves flat or slightly keeled, obovate-spathulate or suborbicular, about 10 mm. long and 8 mm. wide, apex obtuse, sometimes with a short recurved mucro, base contracted abruptly to a short petiole; leaf-margins entire; inflorescence a rather loosely branched, terminal corymb; flowers at first rather crowded, becoming more distant after fertilization; pedicels at first ascending, about 2·5 mm. long, later patent and lengthening to about 5 mm.; sepals erect, oblong, obtuse, boat-shaped, 2·5 mm. long, 1·3 mm. wide, densely stellate-pilose externally; petals golden-yellow, oblanceolate, obtuse, 3·25 mm. long, 0·75 mm. wide, glabrous,

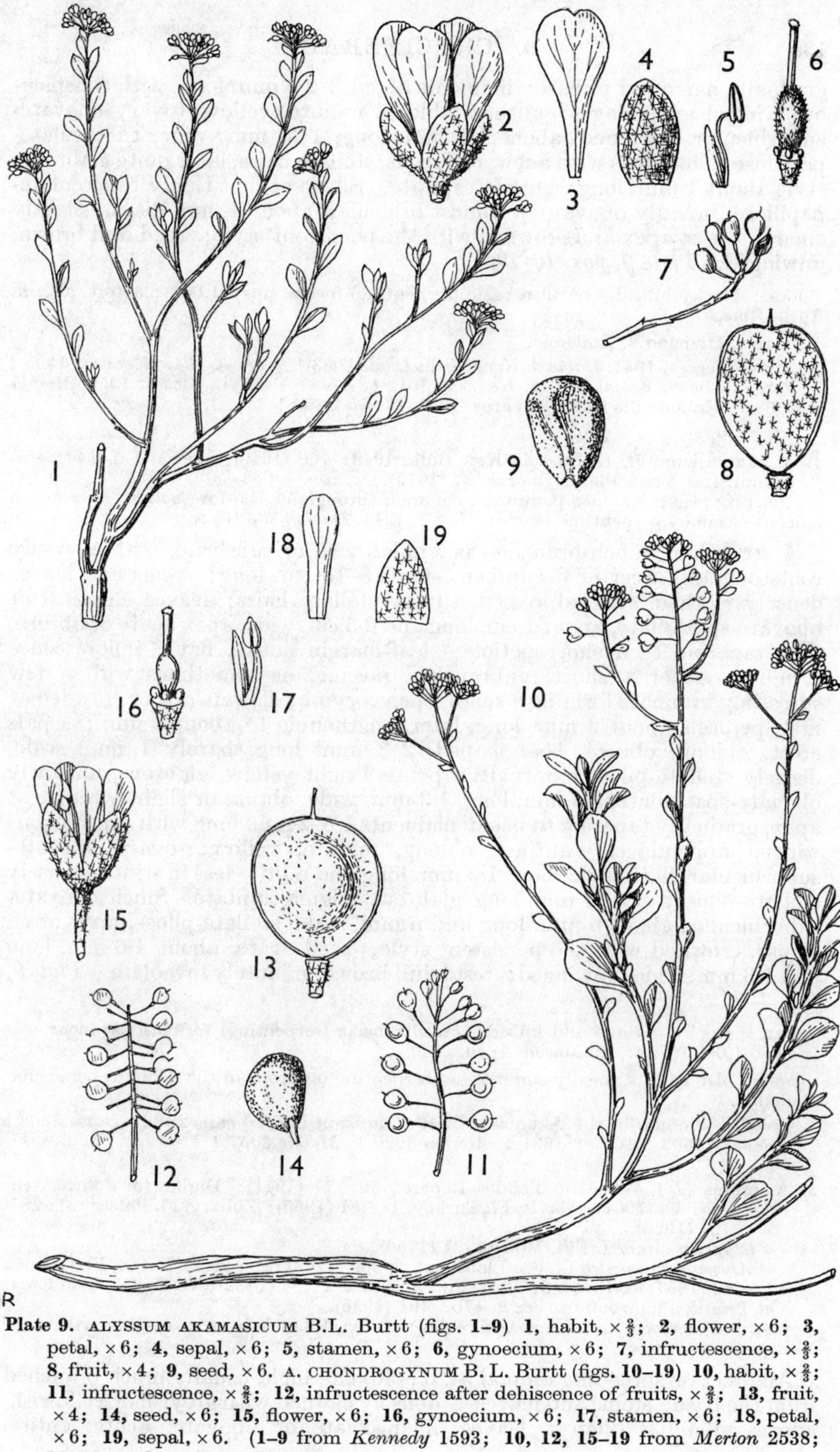

Plate 9. ALYSSUM AKAMASICUM B.L. Burtt (figs. **1–9**) **1**, habit, $\times\frac{2}{3}$; **2**, flower, $\times 6$; **3**, petal, $\times 6$; **4**, sepal, $\times 6$; **5**, stamen, $\times 6$; **6**, gynoecium, $\times 6$; **7**, infructescence, $\times\frac{2}{3}$; **8**, fruit, $\times 4$; **9**, seed, $\times 6$. A. CHONDROGYNUM B. L. Burtt (figs. **10–19**) **10**, habit, $\times\frac{2}{3}$; **11**, infructescence, $\times\frac{2}{3}$; **12**, infructescence after dehiscence of fruits, $\times\frac{2}{3}$; **13**, fruit, $\times 4$; **14**, seed, $\times 6$; **15**, flower, $\times 6$; **16**, gynoecium, $\times 6$; **17**, stamen, $\times 6$; **18**, petal, $\times 6$; **19**, sepal, $\times 6$. (**1–9** from *Kennedy* 1593; **10, 12, 15–19** from *Merton* 2538; **11, 13, 14** from *Kennedy* 1595.)

gradually narrowed to base; filaments about 2–2·5 mm. long, with conspicuous winged appendages; anthers oblong, minute, yellow; ovary obovate-suborbicular, flattened, about 1·5 mm. long, 1·25 mm. wide, tuberculate-papillose, sometimes with a few, caducous, stellate hairs, later quite glabrous; style about 1 mm. long; stigma capitate. Silicle at first thinly tuberculate-papillose, broadly obovate or almost orbicular, about 4 mm. diam., slightly emarginate at apex and crowned with the persistent style; seed dull brown, unwinged. *Plate 9, figs. 10–19.*

HAB.: Rocky hillsides on ultrabasic (serpentine) formations; 1,000–2,300 ft. alt.; fl. April–June.

DISTR.: Division 3. Endemic.

3. Above Yerasa, 1941, *Davis & Kennedy* in *Davis* 3083 ! Yerasa, 1948, *Kennedy* 1595 ! Parekklisha to Kellaki, 1948, *Kennedy* 1612 ! Above Panayia Glossa, 1956, *Merton* 2538 ! Kakomallis near Louvaras, 1962, *Meikle* 2886 !

4. A. akamasicum *B. L. Burtt* in Kew Bull., 1949: 100 (1949); Osorio-Tafall et Seraphim, List Vasc. Plants Cyprus, 47 (1973).

TYPE: Cyprus; Akamas peninsula, common throughout the forest area, growing on igneous—usually serpentine—scree; May 3, 1941, *Davis* 3308 (K !).

A straggling, much-branched perennial herb or subshrub, with a woody rootstock and erect or decumbent stems 8–15 cm. long; stems and leaves densely clothed with adpressed, silvery stellate hairs; leaves elliptical or obovate-spathulate, about 1 cm. long and 0·5 cm. wide, apex acute or obtuse, base tapering to a short petiole; leaf-margin entire, flat; inflorescence terminal, either a short, unbranched raceme, or sometimes with a few spreading branches forming a small, open corymb; flowers rather crowded at first, pedicels about 2 mm. long, later lengthening to about 5 mm.; sepals erect, oblong, obtuse, boat-shaped, 2–3 mm. long, barely 1 mm. wide, densely stellate-pilose externally; petals bright yellow, glabrous, narrowly obovate-spathulate, 3·3 mm. long, 1·3 mm. wide, obtuse or slightly retuse at apex, gradually tapering to base; filaments 1·5–2 mm. long with conspicuous winged appendages; anthers oblong, minute, yellow; ovary obovate-suborbicular, flattened, about 1·5 mm. long and a little less in width, densely stellate-pilose; style 1 mm. long, glabrous, stigma capitate. Silicle obovate-suborbicular, about 6 mm. long and 5 mm. wide, stellate-pilose, apex often retuse, crowned with the persistent style; seed ovate, about 1·6 mm. long and 1·3 mm. wide, unwinged; testa dull brown, minutely foveolate. *Plate 9, figs. 1–9.*

HAB.: Rocky garigue and hillsides on ultrabasic (serpentine) formations; near sea-level to 1,000 ft. alt.; fl. March–April.

DISTR. Division 1, locally common on serpentine outcrops in the Akamas peninsula. Endemic.

1. Locally common in the Akamas; *G. W. Chapman* in *Chapman* 284 ! *Davis* 3308 ! *Kennedy* 1593 ! 1607 ! 1631 ! *Merton* 3020 ! *Meikle* 2057 !

5. A. minus (*L.*) *Rothm.* in Feddes Repert., 50: 77 (1941); Dudley in Journ. Arn. Arb., 45: 63 (1964); Davis, Fl. Turkey, 1: 381 (1965); Zohary, Fl. Palaest., 1: 287, t. 424 (1966).

Clypeola minor L., Fl. Monsp., 21 (1756).

Alyssum campestre L., Sp. Plant., ed. 2, 909 (1763) pro parte; Boiss., Fl. Orient., 1: 283 (1867) excl. var. β; Post, Fl. Pal., ed. 2, 1: 87 (1932); O. E. Schulz in Engl. et Prantl, Pflanzenfam., ed. 2, 17b: 491 (1936).

TYPE: Described from material collected by Magnol in the neighbourhood of Montpellier.

An erect or spreading annual 3–20 (–30) cm. high, usually much branched from the base; stems and leaves ± densely clothed with greyish or yellowish, adpressed-stellate hairs; leaves oblanceolate or obovate, acute, entire,

0·8–2·5 mm. long, 3–10 mm. wide, the larger leaves usually near the apex of the stem; inflorescence at first rather densely corymbose, elongating after fertilization and ultimately forming a raceme 3–10 cm. or more long; pedicels 2–4 mm. long, densely pilose; flowers inconspicuous; sepals caducous, ovate, about 1·5 mm. long and 1 mm. wide, concave, subacute, pilose externally; petals yellowish, narrowly oblanceolate or almost ligulate, about 1·8–2 mm. long and 0·5 mm. wide, apex obtuse, truncate or ± emarginate, sometimes sparsely pilose; stamens about 1·8 mm long, the filaments of the shorter stamens with a conspicuous, often toothed appendage; filaments of the longer stamens flattened, toothed or entire; anthers oblong, minute, yellow; ovary ± ovate, about 1·8 mm. long, densely pilose; style persistent, usually pilose, about 1 mm. long; stigma truncate. Silicle flattened, suborbicular, 3–4 mm. diam., with a distinct marginal rim, densely and regularly (or somewhat irregularly) stellate-pilose; seeds suborbicular, about 1·5 mm. diam.; testa fulvous-brown, minutely papillose, with a narrow, pale marginal wing.

HAB.: Stony hillsides in open garigue, vineyards and waste ground; 2,000–3,500 ft. alt.; fl. Febr.–May.

DISTR.: Division 2, often mixed with *A. strigosum* Banks et Sol. Widespread in the Mediterranean region and western Asia.

2. Platres (with *A. strigosum*), 1937, *Kennedy* 513A ! Phini near Platres (with *A. strigosum*), 1939, *Kennedy* s.n. ! Above Palekhori, 1957, *Merton* 3011 ! Chakistra, 1962, *Meikle* 2725 !

NOTES: Most of the specimens resemble *A. minus* (L.) Rothm. var. *micranthum* (C. A. Mey.) Dudley in having stellate hairs with unequal rays on the fruits and pedicels; in other respects, however, they agree with typical var. *minus*, and it is doubtful if the two varieties can be distinguished in our area. Indeed so close are the resemblances between *A. minus* and the following species, *A. strigosum*, that it might be better to regard both as subspecies of *A. minus*. It would appear that *A. minus* (in the strict sense) is confined to higher altitudes on the Troödos range, whereas *A. strigosum* descends almost to sea-level, and is found on a variety of geological formations.

6. A. strigosum *Banks et Sol.* in Russell, Nat. Hist. Aleppo, ed., 2, 257 (1794); Dudley in Journ. Arn. Arb., 45: 67 (1964); Davis, Fl. Turkey, 1: 382 (1965).

A. micropetalum Fisch. ex DC., Regn. Veg. Syst. Nat., 2: 313 (1821).

A. campestre L. var. *micropetalum* (Fisch. ex DC.) Koch in Flora, 24: 463 (1841); Boiss., Fl. Orient., 1: 284 (1867).

A. cedrorum Schott et Kotschy in Oesterr. Bot. Wochenbl., 11: 269 (1854).

A. strigosum Banks et Sol. ssp. *cedrorum* (Schott et Kotschy) Dudley in Journ. Arn. Arb., 45: 67 (1964); Davis, Fl. Turkey, 1: 383 (1965); Osorio-Tafall et Seraphim, List Vasc. Plants Cyprus, 47 (1973).

[*A. hirsutum* (non M. Bieb.) Kotschy in Unger et Kotschy, Die Insel Cypern, 326 (1865); Boiss., Fl. Orient., 1: 284 (1867) pro parte quoad spec. cypr.]

[*A. campestre* L. ssp. *hirsutum* (non M. Bieb.) Holmboe, Veg. Cypr., 89 (1914).]

Type: Syria; near Aleppo, *Russell* (BM).

Differs from *A. minus* (L.) Rothm. in having dimorphic indumentum on the fruits and pedicels, with long, simple or forked, often somewhat tuberculate-based hairs intermingled with shorter stellate hairs; this indumentum gives the fruits a rather shaggy appearance. The styles of *A. strigosum* are normally glabrous (and not stellate-pilose as in *A. minus*) and generally a little longer. The fruits are commonly rather larger than those of Cyprus forms of *A. minus*.

HAB.: Open stony hillsides, roadsides, waste land, vineyards, cultivated fields, occasionally on sand-dunes; sea-level to 4,000 ft. alt.; fl. Febr.–May.

DISTR.: Divisions 2–8, common in Divisions 2, 3 and 7. Also widespread in S.E. Europe and Asia Minor eastwards to Iran.

2. Common; *Reid* s.n. ! *Kennedy* 513 ! 1237 ! 1825 ! 1826 ! *N. Macdonald* 8 ! *Meikle* 2931 !

3. Common; *Syngrassides* 14 ! *Kennedy* 1596 ! *N. Macdonald* 14 ! *Barrington* 16 ! *Meikle* 2979 !

4. Near Larnaca, 1862, *Kotschy* 14; near Larnaca, 1880, *Sintenis & Rigo*; S.W. of Larnaca, 1967, *Merton* in ARI 385 !
5. Near Tymbou, by the bed of Yialias [Idalion] R., 1880, *Sintenis & Rigo*; Denarga, near Lefkoniko, 1962, *Meikle* 2964 !
6. Orga, 1941, *Davis* 2101 ! Ayia Irini (Morphou), 1941, *Davis* 2122 ! Nicosia, 1955, *Baker* 156 !
7. Common; *Sintenis & Rigo* 269 ! *Davis* 2098 ! *Chapman* 102 ! *G. E. Atherton* 854 ! *Casey* 230 !
8. N. end of the Karpas, between Apostolos Andreas Monastery and Aphendrika, 1880, *Sintenis & Rigo*; Koma tou Yialou, 1935, *Syngrassides* 711 ! Ayios Philon, near Rizokarpaso, 1941, *Davis* 2215 !

7. **A. umbellatum** *Desv.*, Journ. de Bot., ser. 2, 3: 184 (1814); Boiss., Fl. Orient, 1: 282 (1867); Post, Fl. Pal., ed. 2, 1: 86 (1932); Davis, Fl. Turkey, 1: 379 (1965); Osorio-Tafall et Seraphim, List Vasc. Plants Cyprus, 47 (1973).

TYPE: Cultivated in the Jardin des Plantes, Paris (P).

A prostrate or decumbent annual, branches densely stellate-pubescent, 1·5–5 cm. long, usually radiating from rootstock, or plant occasionally unbranched; leaves oblanceolate, 8–20 mm. long, 1·5–3 mm. wide, greyish with a dense indumentum of adpressed-stellate hairs, apex acute or subacute, base tapering to an indistinct petiole; inflorescence umbellate, of 8–12 (or more) inconspicuous flowers; pedicels at first 1·5–2 mm. long, lengthening to 6 mm. or more in fruit, densely and rather unevenly stellate-pilose; sepals oblong, about 2 mm. long and 1 mm. wide, concave, subacute, stellate-pilose externally, with a narrow membranous margin; petals pale dull yellow, narrowly obovate, 2·5 mm. long, 0·8 mm. wide, apex rounded or retuse, base tapering to a slender claw; stamens about 2·5 mm. long, filaments with a linear, partly free or adnate appendage, 1–1·5 mm. long; anthers suborbicular, about 0·3 mm. diam.; ovary about 1·4 mm. long and 1 mm. wide, densely stellate-pubescent, style about 1 mm. long; stigma truncate. Infructescence umbellate or very shortly racemose; fruits flattened, suborbicular, about 3–4 mm. diam., rather densely stellate-pubescent, with a conspicuous persistent style; seeds suborbicular, 1·3–1·5 mm. diam.; testa pale brown, minutely papillose, with a very narrow paler wing.

HAB.: Dry, stony summits and mountainsides; c. 4,000 ft. alt.; fl. March-May.

DISTR.: Division 2, rare. Crimea, Greece, Aegean Is., Turkey and Syria.

2. Khorteri, above Stavros tis Psokas, 1962, *Meikle* 2709 !

8. **A. minutum** *Schlechtend. ex DC.*, Regn. Veg. Syst. Nat., 2: 316 (1821); Boiss., Fl. Orient., 1: 281 (1867); Post, Fl. Pal., ed. 2, 1: 86 (1932); Davis, Fl. Turkey, 1: 377 (1965); Osorio-Tafall et Seraphim, List Vasc. Plants Cyprus, 47 (1973).

[*A. fulvescens* Sibth. et Sm., Fl. Graec. Prodr., 14 (1813) pro parte quoad spec. cypr.; Unger et Kotschy, Die Insel Cypern, 327 (1865); Boiss., Fl. Orient., 1: 280 (1867); Holmboe, Veg. Cypr., 89 (1914); A. K. Jackson et Turrill in Kew Bull., 1938: 464 (1938).]

TYPE: "ad rupes Sibiriae?" *Pallas* sec. DC. (LE). Stated by C. A. Meyer to be a *Steven* specimen, originally collected "in Taurien, beim Dorfe Tschorguna".

A slender annual, 2–6 (–10) cm. high, erect and unbranched, or with several spreading branches from the base; stems and leaves ± densely clothed with rather loose, unevenly branched, spreading, stellate hairs; basal leaves obovate-spathulate, 6–15 mm. long, entire, obtuse or subacute at apex, narrowed to a short petiole at base; upper leaves gradually shorter and narrower, the uppermost often narrowly oblanceolate; inflorescence a small dense terminal corymb, elongating after fertilization to form a short raceme; pedicels very short in flower, lengthening to 3–5 mm. and becoming patent or almost patent in fruit; flowers very small; sepals persistent, ovate, concave, 2·5 mm. long, about 1·5 mm. wide, stellate-pilose externally; petals yellow, ligulate, thinly stellate-pilose, about 3·5 mm. long, 1–1·5 mm. wide, apex truncate or often slightly retuse; longer stamens about 2·5 mm.

long, filaments with an inconspicuous adnate appendage; shorter stamens about 2 mm. long, appendage adnate, forming a conspicuous tooth; anthers shortly oblong, minute, yellow; ovary glabrous, flattened, obovate, about 1 mm. long and 0·8 mm. wide; style rather less than 1 mm. long; stigma truncate. Silicle flattened, suborbicular about 3·5 mm. diam., apex slightly retuse, crowned with the persistent style; valves glabrous and rather lustrous; seeds ovate, about 1·2 mm. long and 1 mm. wide; testa minutely rugulose, bright fulvous-brown, with a very narrow marginal wing.

HAB.: Rocky mountainsides, often in open Pine forest; 3,000–4,500 ft. alt.; fl. Febr.–May.

DISTR.: Division 2, rare. Also in Yugoslavia, Greece, Bulgaria, S. Russia, Turkey and Syria.

2. Around Prodhromos, 1862, *Kotschy* 808; Platres, 1937, *Kennedy* 512 (growing with *A. strigosum* and *A. foliosum*) !

NOTES: The type of *A. fulvescens* Sibth. et Sm. is presumably the Greek and Aegean plant with relatively large flowers and a long style. This species is not found in Cyprus, and records for it should probably be referred either to *A. minutum* or to *A. foliosum* (below). For a note on *A. fulvescens* in Cyprus see B. L. Burtt in Kew Bull., 1954: 68 (1954).

9. A. foliosum *Bory et Chaub.*, Expéd. Sci. Morée, 185 (1832), atlas, t. 23, fig. 1 (1835); Boiss., Fl. Orient., 1: 282 (1867); Davis, Fl. Turkey, 1: 377 (1965); Osorio-Tafall et Seraphim, List Vasc. Plants Cyprus, 47 (1973).
TYPE: Greece; Figalia, *Bory de St. Vincent & Chaubard* (P).

A slender annual, similar in habit and appearance to *A. minutum* Schlechtend., but with distinctive indumentum and larger silicles. Stems and leaves covered with close, adpressed, silvery stellate hairs; the basal leaves narrowly spathulate obovate, about 10 mm. long and 3–5 mm. wide, acute, narrowed at base into a short petiole; upper leaves lanceolate or oblanceolate, acute, about 5–8 mm. long and 1·5–3 mm. wide; inflorescence a dense terminal corymb, lengthening in fruit to form a short raceme; flowers small; sepals ovate, concave, about 1·5–2 mm. long and 1 mm. wide, stellate-pilose externally, not so persistent as in *A. minutum*; petals yellow, narrowly oblong, about 2 mm. long and 1 mm. wide, slightly retuse at apex; stamens about 1·5 mm. long, filaments flattened, appendage absent or inconspicuous; anthers shortly oblong, minute, yellow; ovary flask-shaped, glabrous, about 0·5 mm. long and 0·3 mm. wide; style very short and relatively thick; stigma truncate. Silicle slightly inflated, suborbicular, about 4 mm. diam., apex truncate or very slightly retuse, crowned with the persistent style; valves often tinged purple, scarcely lustrous; seeds ovate, about 1·5 mm. long and 1 mm. wide; testa minutely rugulose, fuscous or very dark brown with a direct pale brown marginal wing.

HAB.: Dry stony summits and mountainsides; 3,000–4,000 ft. alt.; fl. Febr.–April.

DISTR.: Division 2, rare. Greece, Aegean Is. and Turkey.

2. Platres, 1937, *Kennedy* 512a (growing with *A. minutum* & *A. strigosum*) ! Khorteri, above Stavros tis Psokas, 1962, *Meikle* (with *A. umbellatum*) !

29. LOBULARIA *Desv.*

Journ. de Bot., ser. 2, 3: 162 (1814) nom. cons.

Annual or (more rarely) perennial herbs or subshrubs; stems and leaves clothed with medifixed, bipartite hairs; leaves narrow, entire; inflorescence ebracteate or with the lower flowers subtended by leaf-like bracts; sepals ± spreading, not saccate at base; petals small, white or purplish, oblong, obovate or suborbicular, short-clawed; stamens 6, filaments flattened but without appendages; ovary 2–10-ovulate; style short, persistent; stigma capitate. Silicle ovate or suborbicular, flattened, the valves indistinctly veined, with a slender (often obscure) midrib. Seeds compressed, broadly ovate or suborbicular with a distinct marginal wing.

Five species in the Mediterranean region and Atlantic Islands.

Silicle about 2·3 mm. long and 2 mm. wide, 2-seeded; petals 2·5–3 mm. long; leaves tapering towards apex and base - - - - - - - **1. L. maritima**

Silicle 4–5 mm. long, 3–4 mm. wide, c. 6–10-seeded; petals about 2 mm. long; leaves rather shortly acute at apex - - - - - - - **2. L. libyca**

1. L. maritima (*L.*) *Desv.* in Journ. de Bot., ser. 2, 3: 162 (1814); Post, Fl. Pal., ed. 2, 1: 89 (1932); O. E. Schulz in Engl. et Prantl, Pflanzenfam., ed. 2, 17b: 495 (1936); A. K. Jackson et Turrill in Kew Bull., 1938: 461 (1938); Davis, Fl. Turkey, 1: 360 (1965); Osorio-Tafall et Seraphim, List Vasc. Plants, Cyprus, 47 (1973).

Clypeola maritima L., Sp. Plant., ed. 1, 652 (1753).

Alyssum maritimum (L.) Lam., Encycl. Méth., 1: 98 (1783); Druce in Rep. B.E.C., 9: 469 (1931).

Koniga maritima (L.) R. Br. in Denh. et Clapp., Narr. Exp. Afr., 2: 214–217 (1826); Boiss., Fl. Orient., 1: 289 (1867).

TYPE: " *in* Gallia Narbonensi " [cult. Uppsala] (LINN !).

Much-branched annual or short-lived perennial herb 4–20 (–30) cm. high with ascending or spreading stems; leaves sessile, oblanceolate, lanceolate or linear-lanceolate, 13–50 mm. long, 1·5–7 mm. wide, tapering to apex and base, ± densely clothed with adpressed, greyish hairs; inflorescence at first densely corymbose, overtopped by the uppermost cauline leaves, lengthening greatly after fertilization, and finally forming a rather lax cylindrical raceme; lowermost flowers sometimes subtended by leafy bracts; sepals caducous, oblong, obtuse, about 1·7 mm. long and 1 mm. wide, thinly hairy externally; petals suborbicular, white or purplish, entire, about 2·5–3 mm. long and 2–2·5 mm. wide, abruptly narrowed into a claw about 1 mm. long; stamens about 1·8 mm. long; filaments flattened; anthers oblong, minute, yellow; ovary flask-shaped, compressed, about 1·3 mm. long; style short; stigma relatively large, capitate. Silicle suborbicular, flattened, about 2·3 mm. long and 2 mm. wide, crowned with the persistent style; valves glabrous or thinly pilose; seeds broadly ovate, about 1·2 mm. long and 1 mm. wide; testa minutely verruculose, fulvous-brown with a pale marginal wing.

HAB.: Dry, stony hillsides, probably always an escape from cultivation; 1,000–3,750 ft. alt.; fl. Febr.–June.

DISTR.: Divisions 2, 7. Mediterranean region, Atlantic Is., and extensively cultivated and naturalized in other parts of the world.

2. Platres, 1937, *Kennedy* 486 !

7. Bellapais, 1928 or 1930, *Druce* s.n. (probably introduced).

2. L. libyca (*Viv.*) *Webb et Berth.*, Hist. Nat. Îles Canar., 3, pt. 2: 90 (1837); Post, Fl. Pal., ed. 2, 1: 89 (1932); O. E. Schulz in Engl. et Prantl, Pflanzenfam., ed. 2, 17b: 495 (1936)—as *L. libyca* (Viv.) Meisn.; Zohary, Fl. Palaest., 1: 289, t. 426 (1966).

Lunaria libyca Viv., Fl. Lib. Spec., 34, t. XVI, fig. 1 (1824).

Koniga lybica (Viv.) R. Br. in Denh. et Clapp., Narr. Exp. Afr., 2: 214–217 (1826); Boiss., Fl. Orient. Suppl., 52 (1888).

TYPE: Libya; " in Magna Syrte " *P. Della Cella* (GE).

A straggling annual, much-branched towards the base, 4–20 cm. or more in height; stems rather sparsely adpressed-pilose, often tinged purple; leaves scattered and rather few, sessile, oblanceolate or narrowly oblong, 10–25 mm. long, 1·5–2 mm. wide, rather shortly acute at apex, tapering to base, adpressed-pilose; inflorescence at first densely corymbose, later elongate-racemose, the lowermost flowers often bracteate; flowers small, inconspicuous; sepals rather persistent, obovate, acute, about 1 mm. long and 0·8 mm. wide, pilose externally and often tinged purple; petals white, suborbicular, entire, about 2 mm. long and 1·5 mm. wide, abruptly narrowed to a claw about 0·7 mm. long; stamens about 1·2 mm. long; filaments flattened; anthers oblong, minute, yellow; ovary flask-shaped, compressed, about 0·7 mm. long and 0·5 mm. wide, densely adpressed-pilose; style very short, stigma capitate. Silicle rather large, ovate, flattened, about 4–5 mm. long and 3–4 mm. wide, crowned with the persistent style; valves subglabrous or adpressed-pilose; seeds ovate, flattened, about 1·2 mm. long and 1 mm. wide; testa fulvous-brown, minutely rugulose, with a paler marginal wing.

HAB.: Dry, sandy places; ? altitude; fl. Febr.–May.

DISTR.: ? Division. Also in S. Aegean, Palestine, Egypt, N. Africa, S. Spain and Atlantic Is.

? Division. "Insulae Cypri," 1880, *Sintenis & Rigo* s.n. sec. *Boissier*.

NOTES: Not collected since and should be looked for.

30. CLYPEOLA *L.*

Sp. Plant., ed. 1, 652 (1753).
Gen. Plant., ed. 5, 293 (1754).
Chaytor et Turrill in Kew Bull., 1935: 1–24 (1935).

Slender annual herbs; leaves and stems ± densely clothed with stellate hairs; leaves oblanceolate- or obovate-spathulate, entire; inflorescence ebracteate; flowers small; sepals ± spreading, not saccate at base; petals inconspicuous, yellow or whitish, oblanceolate or almost linear or spathulate; stamens 6, free, filaments flattened towards base, all or some of them with a conspicuous toothed appendage; ovary 1 (–2)-ovulate; style often very short; stigma capitate or truncate. Silicle indehiscent, 1-seeded, pendulous on a short pedicel, broadly ovate or suborbicular, often shortly emarginate at apex, and frequently with a distinct marginal wing. Seeds ovate, unwinged.

About 8 species in the Mediterranean region and western Asia.

1. **C. jonthlaspi** *L.*, Sp. Plant., ed. 1, 652 (1753); Boiss., Fl. Orient., 1: 308 (1867); Holmboe, Veg. Cypr., 91 (1914); Post, Fl. Pal., ed. 2, 1: 94 (1932); O. E. Schulz in Engl. et Prantl, Pflanzenfam., ed. 2, 17b: 497 (1936); Davis, Fl. Turkey, 1: 410 (1965); Zohary, Fl. Palaest., 1: 290, t. 428 (1966); Osorio-Tafall et Seraphim, List Vasc. Plants Cyprus, 48 (1973).

A slender, erect herb, 8–20 cm. or more high, often much branched at the base, sometimes (especially in starved specimens) unbranched; stems terete, covered with closely adpressed stellate hairs; leaves sessile, stellate-pilose, narrowly oblanceolate or oblanceolate-spathulate, or occasionally almost obovate, 2–20 mm. long, 1–4 mm. wide, acute or obtuse at apex, gradually tapering to base; inflorescence at first short and congested, later lengthening and forming a long cylindrical raceme; pedicels 1–2 mm. long, sharply recurved in fruit; sepals concave, ovate or oblong, acute or obtuse, 0·9–1·75 mm. long, 0·7–0·9 mm. wide, stellate-pilose externally; petals oblanceolate or almost linear, entire and obtuse or shortly emarginate at apex, tapering to base, 0·8–1·8 mm. long, about 0·7–1 mm. wide, glabrous; stamens 1·3–1·5 mm. long, filaments each with a conspicuous toothed appendage; anthers almost globose, minute, orange-yellow; ovary ovate, flattened, about 1 mm. long; style very short; stigma small, capitate. Silicle ovate, obovate or almost orbicular, 1·7–5·3 mm. long, 1·5–5 mm. wide, apex emarginate, margin distinctly winged, entire, the whole silicle pilose or glabrous, or the disk or wing pilose and the rest of the silicle glabrous; seed ovate, compressed, 1–1·8 mm. long, 0·8–1·3 wide; testa fulvous-brown, minutely rugulose.

var. **jonthlaspi**

var. *intermedia* Hal., Consp. Fl. Graec., 1: 117 (1900); Chaytor et Turrill in Kew Bull., 1935: 10 (1935) as var. 4.

TYPE: France; "*in collibus sterilibus soli orienti expositis* Anversae *Valuensium*, Monspeliis, *in agro* Narbonensi *&* Nemausensi" (BM).

Silicle 3·5 mm. or more diam.; disk glabrous or almost glabrous, wing hairy.

HAB.: Stony hillsides, roadsides etc.; 2,000–2,500 ft. alt.; fl. Febr.–May.

DISTR.: Division 7, rare. Widespread in the Mediterranean region and western Asia.

7. Pentadaktylos, 1880, *Sintenis & Rigo* 270 partly ! Yaïla (Kyrenia Range) at Halevga Forest Station, 1941, *Davis* 2893 !

var. **lasiocarpa** *Guss.*, Fl. Sic. Prodr., 2: 197 (1828); Chaytor et Turrill in Kew Bull., 1935: 6 (1935) as var. 1.

TYPE: Sicily; Polizzi, Nicosia, *Gussone* (NAP).

Silicle 3·5 mm. or more diam.; disk and wing hairy.

HAB.: Stony hillsides; 2,000–2,500 ft. alt.; fl. Febr.–May.

DISTR.: Division 7, rare. Widespread in the Mediterranean region.

7. Pentadaktylos, 1880, *Sintenis & Rigo* 270 partly ! Above Aghirda, 1941, *Davis* 2492 ! Kyrenia, 1949, *Casey* 432 !

var. **hispida** (*Presl*) *Reynier* in Bull. Géogr. Bot., 21: 290 (1911); Chaytor et Turrill in Kew Bull., 1935: 9 (1935) as var. 3.

C. hispida Presl, Bot. Bemerk., 9 (1844).

TYPE: Sinai Peninsula; "ad radices montis Sinai", April 19, 1835, *Schimper* 415.

Silicle 1·7–2·5 mm. diam.; disk and wing hairy.

HAB.: Stony mountain slopes and summits on igneous formations; 2,500–4,600 ft. alt.; fl. April–June.

DISTR.: Division 2. Widespread in the Mediterranean region, also in Switzerland, Iraq and Arabia.

2. Just above Palekhori, 1957, *Merton* 3007 ! Mandria, 1960, *N. Macdonald* 46 ! Tripylos, 1962, *Meikle* 2721 !

var. **glabra** (*Boiss.*) *Fiori* in Fl. Ital. Exsicc. no. 1278 (1910) et in Nuov. Giorn. Bot. Ital. N.S. 17: 611 (1910); Chaytor et Turrill in Kew Bull., 1935: 13 (1935) as var. 12.

C. glabra Boiss. in Ann. Sci. Nat., ser. 2, 17: 173 (1842).

TYPE: Greece; "in parte superiori montium Atticae Pentelicus, Parnes" *Boissier* (G).

Silicle 2·5 mm. or less in diam.; disk and wing glabrous.

HAB.: Rocky mountain slopes under Pine trees; 4,000 ft. alt.; fl. April–June.

DISTR.: Division 2, rare. Also Spain and Greece.

2. Platres, 1948, *Kennedy* 1238 !

NOTES: Aggregate *C. jonthlaspi* is also recorded from Prodhromos (*Kotschy*), Kythrea, March 21, 1880 (*Sintenis & Rigo*) and near Rizokarpaso, Apr. 26, 1880 (*Sintenis & Rigo*). In the absence of specimens, it is impossible to refer these records to any particular variety.

TRIBE 5. **Drabeae** *O. E. Schulz* Annual or perennial herbs with simple, branched or stellate indumentum; lateral nectary glands 1 (or rarely 2), on either side of the shorter stamens, often united; median nectary glands absent or present, free or united with the lateral glands; filaments sometimes dilated towards base, the inner occasionally with winged or toothed appendages; ovary sessile or sometimes shortly stipitate; style ± distinct; stigma truncate, capitate or occasionally ± 2-lobed. Fruit a flattened, or sometimes inflated, spherical, ovoid or rarely elongate dehiscent silicle; septum broad. Cotyledons accumbent.

31. EROPHILA *DC.*

Regn. Veg. Syst. Nat., 2: 856 (1821) nom. cons.
Winge in Compt. Rend. Trav. Lab. Carlsb., 23: 41–73 (1940).

Slender annual herbs with simple or variously branched indumentum, and with the leaves all basal, forming a small rosette; inflorescence ebracteate; flowers small; pedicels lengthening greatly after fertilization; sepals erecto-patent, not saccate at base; petals white, deeply bifid; stamens 6, without appendages; ovary sessile, 10–60-ovulate, ovoid or elliptical, compressed or somewhat inflated; style very short; stigma truncate or capitate. Silicle ovate, elliptical or occasionally elongate-oblong, compressed or inflated, valves with a rather obscure midrib and lateral nerves; seeds numerous, minute, in 2 ranks in each loculus.

About 3 species in Europe, W. Asia and N. Africa. *E. verna* (L.) Chevall. is frequently subdivided into a number of ill-marked microspecies.

Leaves broadly oblanceolate, oblong or obovate-elliptical, entire or occasionally with a few small teeth; silicle 3·5–7 mm. long, oblanceolate or elliptical **1. E. verna**
Leaves narrowly linear or linear-oblanceolate, often deeply toothed; silicle 2·5–3·5 (–4) mm. long, broadly elliptical - - - - - - - **2. E. minima**

1. E. verna (*L.*) *Chevall.*, Fl. Paris, 2: 898 (1827); Post, Fl. Pal., ed. 2, 1: 92 (1932); O. E. Schulz in Engl., Pflanzenr., 89 (IV, 105): 345 (1927); Davis, Fl. Turkey, 1: 421 (1965); Zohary, Fl. Palaest., 1: 291, t. 431 (1966); Osorio-Tafall et Seraphim, 48 (1973).

Draba verna L., Sp. Plant., ed. 1, 642 (1753); Sibth. et Sm., Fl. Graec. Prodr., 2: 4 (1813).

Erophila vulgaris DC., Regn. Veg. Syst. Nat., 2: 356 (1821); Unger et Kotschy, Die Insel Cypern, 327 (1865); Boiss., Fl. Orient., 1: 304 (1867).

Draba verna L. ssp. *vulgaris* (DC.) Holmboe, Veg. Cypr., 90 (1914).

Erophila praecox (Steven) DC. sec. O. E. Schulz in Engl., Pflanzenr., 89 (IV, 105): 367 (1927).

Erophila praecox (Steven) DC. var. *virescens* (Jord.) O. E. Schulz sec. Druce in Rep. B.E.C., 9: 469 (1931).

TYPE: " *vulgaris per* Europam " (BM).

Leaves varying greatly in size and shape, from 5–30 mm. long and 2–10 mm. wide, oblanceolate, oblong, obovate or elliptical, entire or irregularly toothed, sessile or with a short flattened petiole, clothed with an indumentum of simple, forked or mixed simple and forked hairs; inflorescences 1–several from each rosette, 3–18-flowered, at first corymbose, becoming lax and racemose after fertilization, the pedicels lengthening to 1–2 cm.; sepals oblong, subacute, deeply concave, about 1·5 mm. long and 1 mm. wide, sparsely pilose externally; petals white, obovate, deeply 2-lobed, about 2 mm. long and 1·5 mm. wide, tapering to the base but scarcely clawed; stamens about 1 mm. long, filaments flattened; anthers shortly oblong, minute, yellow; ovary ovate-elliptical, compressed, about 1·5 mm. long and 1·2 mm. wide; style almost obsolete; stigma capitate. Silicle compressed, 3·5–7 mm. long, 2·5–3·5 mm. wide, elliptical, ovate, obovate or elliptic-oblong, containing 15–20 or more seeds, valves obscurely reticulate-veined, glabrous; seeds compressed, about 0·7 mm. long and 0·4 mm. wide; testa brown, minutely rugulose.

HAB.: Dry open hillsides, roadsides, waste ground, sand-dunes; sea-level to 6,400 ft. alt.; fl. Jan.–May.

DISTR.: Divisions 1–8, common in Divisions 2 and 7. Widespread in Europe and western Asia; introduced elsewhere.

1. Yiolou, above Baths of Aphrodite, 1962, *Meikle* 2225 !
2. Common; *Druce* s.n. ! *Syngrassides* 349 ! 947 ! *Kennedy* 483 ! 484 ! *Chapman* 320 ! *Merton* 3127 !
3. Stavrovouni, 1880, *Sintenis & Rigo* 1011 ! Limassol, 1928 or 1930, *Druce* s.n.

4. Athna Forest, 1952, *Merton* 1000b ! Pergamos, 1954, *Merton* 1840b !
5. Mile 12, Nicosia–Famagusta Rd., north side, 1951, *Merton* 6 ! Athalassa, 1962, *Meikle* 2982 !
6. Nicosia–Troödos Rd., 1951, *Merton* 99 !
7. Common; *Davis* 2884 ! *Duke-Woolley* 91 ! *G. E. Atherton* 885 ! 1291 ! *Casey* 255 !
8. Apostolos Andreas Forest, 1956, *Merton* 2473 !

2. E. minima *C. A. Mey.*, Verz. Pflanz. Cauc., 184 (1831); Boiss., Fl. Orient., 1: 303 (1867); O. E. Schulz in Engl., Pflanzenr., 89 (IV, 105): 371 (1927); Post, Fl. Pal., ed. 2, 1: 92 (1932); Davis, Fl. Turkey, 1: 422 (1965); Zohary, Fl. Palaest., 1: 292, t. 432 (1966).

TYPE: Caucasus; " in collibus et campis provinciae Baku ", *C. A. Meyer* (LE).

Very similar to *E. verna* in general aspect, but usually a smaller and more slender plant; leaves sparsely pilose with simple hairs, or almost glabrous, linear or linear-lanceolate, entire or (more usually) deeply toothed or lobed, about 10–20 mm long and 2–5 mm. wide; inflorescences and flowers barely distinguishable from those of *E. verna*; sepals often stained purple or violet. Silicle small, ovate-elliptic or broadly elliptical, 2·5–3·5 (–4) mm. long, 1·5–2·5 mm. wide, containing 10–15 seeds; seeds ovate, 0·6 mm. long, 0·4 mm. wide; testa dull fuscous-brown, minutely rugulose.

HAB.: Moist turfy ground between rocky outcrops; 300–600 ft. alt.; fl. Jan.–March.

DISTR.: Divisions 4, 5. Widespread in the eastern Mediterranean and western Asia.

4. Athna Forest, 1952, *Merton* 1000a !
5. Mile 12, Nicosia–Famagusta Rd., 1951, *Merton* 5 ! Athalassa, 1962, *Meikle* 2981 !

TRIBE 6. **Arabideae** *DC.* Annual or perennial, glabrous or pilose herbs; nectary glands usually conspicuous; free or variously united. Filaments linear, the inner sometimes dilated towards the base, rarely with a toothed appendage; ovary sessile or occasionally shortly stipitate; style usually short; stigma truncate, capitate or ± 2-lobed. Fruit a linear or narrowly elongate dehiscent silique, rarely a silicle; septum narrow. Cotyledons accumbent.

32. CARDAMINE *L.*

Sp. Plant., ed. 1, 655 (1753).
Gen. Plant., ed. 5, 295 (1754).

Annual, biennial or perennial herbs, glabrous or with an indumentum of simple hairs; stems branched or unbranched; leaves usually petiolate, trifoliolate or pinnate, alternate or rarely opposite or whorled; inflorescence an ebracteate raceme; sepals erecto-patent, sometimes cucullate at apex, not, or very slightly, saccate at base; petals white, purple or rarely yellowish, occasionally wanting; lamina rounded or emarginate at apex, tapering to a narrow basal claw; stamens 4–6, rarely fewer, filaments without appendages; ovary sessile, 4–40-ovulate; style usually short; stigma truncate or ± 2-lobed. Silique compressed, valves inconspicuously veined, usually coiling suddenly from the base at dehiscence and ejecting the seeds. Seeds in a single row in each loculus, elliptic, ovate or oblong, occasionally with a narrow wing.

About 130 species occurring in most regions of the world, but usually at high altitudes within the tropics.

Flowering stems with few leaves; leaflets entire or obscurely lobed; petals minute, rarely more than 2·7 mm. long; silique 1–1·5 mm. wide - - **1. C. hirsuta**
Flowering stems leafy; leaflets distinctly 3- or more lobed; petals 4–6 mm. long; silique 2·5–4 mm. wide - - - - - - - **2. C. graeca**

1. C. hirsuta *L.*, Sp. Plant., ed. 1, 655 (1753); Unger et Kotschy, Die Insel Cypern, 326 (1865); Boiss., Fl. Orient., 1: 160 (1867); Holmboe, Veg. Cypr., 86 (1914); Post, Fl. Pal., ed. 2, 1: 59 (1932); O. E. Schulz in Engl. et Prantl, Pflanzenfam., ed. 2, 17b: 530 (1936); Lindberg f., Iter Cypr., 17 (1946); Davis, Fl. Turkey, 1: 443 (1965); Zohary, Fl. Palaest., 1: 278, t. 409 (1966); Osorio-Tafall et Seraphim, List Vasc. Plants Cyprus, 48 (1973).

C. hirsuta L. var. *glabra* Sintenis in Oesterr. Bot. Zeitschr., 31: 229 (1881).

TYPE: Not precisely localized (BM).

A slender annual, 5–20 cm. or more high, glabrous or ± pilose with scattered, simple hairs; leaves mostly basal, forming a well-marked rosette, petiolate, 2–7 cm. or more long, 0·5–2 cm. wide, pinnate, with 5–10 opposite or alternate, rounded, entire or obscurely undulate or lobed leaflets 2–10 mm. wide; terminal leaflet rather larger than the laterals and sometimes obscurely 3-lobed; flowering stem erect, obscurely angled, bearing 1–3 sessile or shortly stalked pinnate leaves similar to the basal leaves, but with rather narrower leaflets; inflorescence at first corymbose, later developing into a lax raceme, the flowers overtopped by the young fruits; flowers small, inconspicuous; sepals about 1·7 mm. long and 1 mm. wide, oblong, obtuse or subacute, concave, glabrous or thinly pilose externally; petals white, glabrous, narrowly obovate, 1·5–2·5 (–3) mm. long, about 1·2 mm. wide, rounded at apex, gradually narrowed to a short basal claw; stamens 4 (or occasionally more) about 2 mm. long; filaments flattened; anthers minute, yellow; ovary cylindrical, about 2–2·5 mm. long; style obsolete; stigma truncate. Silique held erect on a pedicel about 5–10 mm. long, glabrous, compressed or slightly torulose, 15–25 mm. long, 1–1·5 mm. wide, abruptly narrowed at apex; seeds oblong, flattened, about 1–1·3 mm. long and 0·8 mm. wide; testa bright brown, minutely reticulate.

HAB.: Moist shady ground; sometimes a weed of cultivation; sea-level to 5,000 ft. alt.; fl. Jan.–May.

DISTR.: Divisions 1–3, 8. A widespread weed in Europe and Asia, naturalized in most temperate regions of the world.

1. Near Smyies (Akamas), 1962, *Meikle* 2175 ! Neokhorio, 1962, *Meikle* 2205 !
2. Above Prodhromos, 1859, *Kotschy* 1007; Platres, 1939, *Kennedy* 1340 ! Kambos, 1939, *Lindberg f.* s.n.; above Kalokhorio, Makheras, 1950, *Chapman* 299 ! Polystipos, 1957, *Merton* 3118 !
3. Stavrovouni, 1880, *Sintenis & Rigo* 847 !
8. Ayios Philon near Rizokarpaso, 1941, *Davis* 2259 ! Apostolos Andreas Monastery, 1957, *Merton* 2876 !

2. C. graeca *L.*, Sp. Plant., ed. 1, 655 (1753); Boiss., Fl. Orient., 1: 164 (1867); Holmboe, Veg. Cypr., 86 (1914); Post, Fl. Pal., ed. 2, 1: 59 (1932); O. E. Schulz in Engl. et Prantl., Pflanzenfam., ed. 2, 17b: 531 (1936); Davis, Fl. Turkey, 1: 442 (1965); Osorio-Tafall et Seraphim, List Vasc. Plants Cyprus, 48 (1973).

Pteroneuron graecum (L.) DC., Regn. Veg. Syst. Nat., 2: 259 (1831); Unger et Kotschy, Die Insel Cypern, 326 (1865).

TYPE: " *in* Sicilia, Corsica, *insulis* Graeciae " (LINN !).

An erect or spreading, subglabrous or thinly hairy biennial or annual, 5–20 cm. high; stems usually branched from base, branches sulcate, rather flexuous; leaves 2–8 cm. long, 0·8–2·8 cm. wide, petiolate, pinnate, scattered along the stems, and not (at least in flowering specimens) forming a distinct basal rosette; leaflets 5–10, ovate-cuneate, 4–15 mm. long, 3–12 mm. wide, deeply cleft into 3–5 (or occasionally more) blunt lobes; inflorescence at first forming a lax corymb, elongating later and finally racemose, young fruits usually shorter than or equalling the inflorescence; flowers rather small; sepals narrowly oblong, 3–4 mm. long, 1–1·5 mm. wide, concave and ± cucullate at apex, glabrous or sparsely pilose; petals white, narrowly obovate, 4–6 mm. long, 2–2·5 mm. wide, rounded at apex, gradually narrowed to a short basal claw; stamens about 5 mm. long; filaments flattened; anthers oblong, about 1 mm. long, yellow; ovary glabrous or pilose, compressed-cylindrical, about 5 mm. long; style short, thick; stigma capitate. Silique erect or spreading, on a pedicel about 8 mm. long, linear, strongly compressed, not torulose, 30–40 mm. long, 2·5–4 mm. wide, gradually narrowing to a long apical beak, valves not coiling upwards at dehiscence; seeds oblong, flattened, about 4 mm. long and 2–2·5 mm. wide; testa dark reddish-brown, minutely reticulate and rugulose, often narrowly winged on margin opposite hilum.

HAB.: Moist, shady screes, banks and forest clearings; 2,000–4,500 ft. alt.; fl. March–May.

DISTR.: Division 2. Widespread in the Mediterranean region and S.E. Europe, from Corsica eastwards to Turkey and Palestine.

2. Papoutsa, above Palekhori, 1941, *Davis* 3098 ! Near Stavros tis Psokas, 1962, *Meikle* 2697 ! Between Stavros tis Psokas and Kykko Monastery, 1962, *Meikle* !

2. or 3. " On the way down from Maschera [Makheras] towards the monastery at Herakli [Kellaki] in pine forests in the region of the sea ", 1862, *Kotschy*.

33. ARABIS *L.*

Sp. Plant., ed. 1, 664 (1753).
Gen. Plant., ed. 5, 298 (1754).

Annual, biennial or perennial herbs, with simple, bifurcate or stellate indumentum, basal leaves usually forming a rosette; stem leaves sessile, amplexicaul and often auriculate, rarely narrowed to base. Inflorescence usually ebracteate, at first corymbose or clustered, later racemose; sepals erect or spreading, the inner often distinctly saccate at base; petals white, pink, purplish, violet or rarely yellowish, rounded or sometimes emarginate at apex, narrowed to a distinct basal claw; stamens 6, filaments usually linear, without appendages; anthers oblong; ovary narrow, linear or cylindrical, sessile or very shortly stalked, 20–60-ovulate; style usually short or wanting; stigma capitate or truncate, sometimes obscurely 2-lobed. Silique linear, flattened; valves reticulate-veined, midrib often obscure. Seeds numerous, in 1 or 2 rows in each loculus, ovate or oblong, compressed, partly or wholly winged, or unwinged.

About 100 species in temperate Europe, Mediterranean area and N. America, also on mountains of tropical Africa.

Leaves ± densely stellate-tomentose; flowers large; petals 10–18 mm. long:
  Leaves 2–8 cm. long, softly tomentose with long hairs; petals white or pale pink or pink-tinged - - - - - - - - - - **1. A. cypria**
  Leaves 0·8–2 (–4) cm. long, densely tomentose with short hairs; petals pink, lilac or purplish, rarely white - - - - - - **2. A. purpurea**
Leaves subglabrous or sparsely stellate-pubescent; flowers small:
  Petals purple or violet, 7–9 mm. long; stem unbranched or branched only at base **3. A. verna**
  Petals white, about 3·5 mm. long; stem usually much-branched, especially towards base - - - - - - - **4. A. kennedyae**

1. **A. cypria** *Holmboe*, Veg. Cypr., 85 (1914); Lindberg f., Iter Cypr., 17 (1946); Rechinger f. in Arkiv för Bot., ser. 2, 1: 422 (1950); Osorio-Tafall et Seraphim, List Vasc. Plants Cyprus, 48 (1973); Elektra Megaw, Wild Flowers of Cyprus, 8, title page (1973).

[*A. albida* (non Stev.) Sintenis in Oesterr. Bot. Zeitschr., 32: 53 (1882); H. Stuart Thompson in Journ. Bot., 44: 276 (1906).]

[*A. billardieri* (non DC.) Kotschy in Unger et Kotschy, Die Insel Cypern, 325 (1865).]

[*A. albida* Stev. var. *billardieri* (non (DC.) Boiss.) Sintenis in Oesterr. Bot. Zeitschr., 32: 22, 194 (1882); Post et Autran in Bull. Herb. Boiss., 7: 149 (1899).]

[*A. deflexa* (non Boiss.) Boiss., Fl. Orient. Suppl., 34 (1888).]

TYPE: Cyprus; St. Hilarion Castle, 1905, *Holmboe* 814 (O).

An erect, tufted perennial, 15–25 cm. high, rootstock rather woody, slender, shortly creeping; basal leaves crowded into rosettes, densely and softly tomentose with long whitish or greyish stellate hairs, obovate-spathulate, 2–8 cm. long, 1–2 cm. wide, apex rounded or obtuse, base gradually narrowed to a flattened petiole 1–2·5 cm. long, margins sinuate or with several deep, blunt lobes; flowering stems bearing a few, scattered, oblong, acute or obtuse, semi-amplexicaul leaves 1–1·5 cm. long and about 0·5 cm. wide, with entire or sinuate margins; inflorescence at first corymbose, later forming a lax raceme or sparsely branched panicle; rhachis and pedicels glabrous or thinly tomentose (especially towards base of inflorescence); pedicels slender, patent or ascending, 5–12 mm. long; inner sepals boat-shaped, acute, concave, about 6 mm. long and 2–3 mm. wide, conspicuously saccate at base, glabrous or thinly hairy; outer sepals linear, obtuse, concave, about 5 mm. long and 1–1·5 mm. wide; petals white, pale pink or with the lower part of the lamina tinged pink, obovate, 10–18 mm. long, 5–8 mm. wide, rounded at apex, gradually tapering to base; stamens 5–6 mm. long, filaments flattened; anthers oblong, yellow, about 1·5 mm. long and 0·7 mm. wide; ovary cylindrical, glabrous, about 6 mm. long and 0·7 mm. wide; style short but distinct, about 0·8–1 mm. long; stigma rather large, capitate. Siliques spreading or slightly deflexed, often secund, straight or slightly curved, 20–40 mm. long, 2·5–3 mm. wide, apex obtuse or shortly acute, crowned by persistent style and stigma, base rather abruptly narrowed to pedicel; valves pale brown, obscurely reticulate-veined; seeds in 2 ranks in each loculus, broadly ovate, about 0·7 mm. long and 0·6 mm. wide, without a marginal wing; testa dark brown, minutely foveolate.

HAB.: Shaded limestone cliffs and rocks; 1,000–3,000 ft. alt.; fl. March–April.

DISTR.: Divisions 7, 8. Throughout the Northern Range, locally common at St. Hilarion, Pentadaktylos, Buffavento and Halevga. Endemic.

7. Common; *Kotschy* 438 ! *A. G. & M. E. Lascelles* s.n. ! *Kennedy* 474 ! 1075 ! 1568 ! *Davis* 1749 ! 2504 ! 2793 ! 2805 ! 2967 ! 3623 !

8. Kantara, 1880, *Sintenis & Rigo* 259 ! Near Komi Kebir, 1912, *Haradjian* 334; Kantara Castle, 1933, *Chapman* 34 !

NOTES: A specimen from Nikos (Troödos range) collected in May, 1937, by Dr. H. Roger-Smith seems to be *A. cypria* rather than *A. purpurea* Sibth. & Sm., though *A. purpurea* is alone recorded for the region. The record is omitted, however, pending further investigation.

2. A. purpurea *Sibth. et Sm.*, Fl. Graec. Prodr., 2: 28 (1813), Fl. Graec., 7: 40, t. 643 (1830); Unger et Kotschy, Die Insel Cypern, 325 (1865): Boiss., Fl. Orient., 1: 176 (1867); Holmboe, Veg. Cypr., 85 (1914); Lindberg f., Iter Cypr., 17 (1946); Rechinger f. in Arkiv för Bot., ser. 2, 1: 422 (1950); Osorio-Tafall et Seraphim, List Vasc. Plants Cyprus, 48 (1973).

[*A. albida* (non Stev.) Poech, Plant. Ins. Cypr., 31 (1842).]

TYPE: Turkey; " In Olympi Bithyni cacumine ", 1787, *Sibthorp*. See Notes.

An erect tufted perennial 6–25 cm. high, with a tough, woody, creeping rootstock; stems and leaves ± densely stellate-tomentose, with short, soft greyish hairs; basal leaves forming rather loose rosettes, obovate-spathulate 0·8–2 (–4) cm. or more long, 0·4–0·7 (–1·4) cm. wide, apex acute or obtuse, margin entire or irregularly and bluntly sinuate-serrate, base tapering gradually to a slender, flattened petiole 0·4–1·4 (–3) cm. long; stem leaves few, usually rather narrow, oblong or oblanceolate, sessile, amplexicaul, with short, acute or obtuse auricles; inflorescence at first corymbose, later developing into a lax unbranched or very sparingly branched raceme; rhachis and pedicels glabrous or almost so, pedicels 0·5–1 cm. long, ascending or patent; inner sepals boat-shaped, acute, 6–7 mm. long, 2–3 mm. wide, thinly pubescent externally; outer sepals much narrower, linear, concave, about 5 mm. long and 1 mm. wide; petals pink, lilac or purplish, rarely white, narrowly obovate, 10–15 mm. long and about 5 mm. wide, apex rounded, base tapering gradually to a slender claw; stamens 4–7 mm. long, filaments flattened, conspicuously dilated towards base; anthers yellow, oblong, about 1 mm. long; ovary cylindrical about 6 mm. long; style absent or very short; stigma capitate. Siliques patent or slightly deflexed, usually secund, straight or curved, 25–60 mm. long, 1·5–2·5 mm. wide, rather abruptly narrowed at apex and base, frequently tinged purple; seeds small, in 2 ranks in each loculus, ovate, flattened, about 0·8 mm. long and 0·5 mm. wide, not winged; testa dark brown, minutely foveolate.

HAB.: Shaded cliffs and rocks usually on igneous formations; 2,000–4,500 ft. alt.; fl. March–May.

DISTR.: Divisions 1, 2. Widespread and locally abundant on the Troödos range. Endemic.

1. Lyso, 1913, *Haradjian* 850 !
2. Common; *Kotschy* 256 ! 745 ! *Ussher* 41 ! *Kennedy* 471 ! 475–480 ! *Merton* 2488 ! etc. On chalk near Pano Panayia, 1974, *Meikle* !

NOTES: This, in common with other localizations in the *Flora Graeca*, is an obvious error; *Arabis purpurea* has not been recorded outside Cyprus.

3. A. verna (*L.*) *R. Br.* in Ait. f., Hort. Kew., ed. 2, 4: 105 (1812); Unger et Kotschy, Die Insel Cypern, 325 (1865); Boiss., Fl. Orient., 1: 168 (1867); Holmboe, Veg. Cypr., 84 (1914); Post, Fl. Pal., ed. 2, 1: 60 (1932); O. E. Schulz in Engl. et Prantl, Pflanzenfam., ed. 2, 17b: 545 (1936); Lindberg f., Iter Cypr., 17 (1946); Davis, Fl. Turkey, 1: 428 (1965); Zohary, Fl. Palaest., 1: 279, t. 410 (1966); Osorio-Tafall et Seraphim, List Vasc. Plants Cyprus, 48 (1973).

Hesperis verna L., Sp. Plant., ed. 1, 664 (1753).

TYPE: " *in* Galloprovinciae *maritimis* " (LINN !).

An erect annual, (2–) 10–15 (–30) cm. high with slender branched roots; stems and leaves rather thinly stellate-pilose; basal leaves forming a compact rosette, obovate-oblong, 1–5 (–8) cm. long, 0·6–1·5 (–3) cm. wide, apex rounded or obtuse, margins bluntly and often obscurely serrate, base tapering gradually to a short flattened petiole; stem leaves few or often wanting, much smaller than basal leaves, up to about 3 cm. long and 2 cm. wide, ovate, acute, sessile and sometimes amplexicaul, but rarely auriculate; inflorescence unbranched or branched only at the base, at first clustered, later forming a rather lax 2–9-flowered raceme; flowers sessile or with short pedicels (up to about 4 mm. long); inner sepals thinly pilose, narrowly oblong, concave, about 5 mm. long and 1·5 mm. wide, apex obtuse, base distinctly saccate; petals purple or violet, narrowly oblanceolate, 7–9 mm. long, 1·5–2 mm. wide, apex rounded, base gradually tapering to a long claw; stamens 4–6 mm. long, filaments flattened, slightly dilated towards base; anthers yellow, oblong, about 1 mm. long; ovary cylindrical, glabrous or pubescent, about 5 mm. long, 0·5 mm. wide, stigma sessile, capitate. Siliques erect or ascending, straight or curved, 30–40 mm. long, about 1·5 mm. wide, abruptly narrowed at apex and base; valves pale brown, glabrous (var. *verna*; var. *liocarpa* Lindb. f.) or pubescent (var. *dasycarpa* Godr. ex Rouy et Fouc.), with ± distinct midrib and veins; seeds in a single row in each loculus, ovate, compressed, about 1 mm. long, not winged; testa dark brown, minutely rugulose.

HAB.: Shaded cliffs and rocks; 500–4,000 ft. alt.; fl. Febr.–May.

DISTR.: Divisions 1–3, 7, very common in Division 7, along the Northern Range. Widespread in the Mediterranean region.

1. Near Ayios Yeoryios, 1960, *N. Macdonald* 55 ! Yiolou above Baths of Aphrodite, 1962, *Meikle* 2227 !
2. Below Makheras monastery, 1859, *Kotschy* 422; valley of Kryos Potamos, 1937, *Kennedy* 473 ! Platres, 1938, *Kennedy* 1727 ! 1728 ! Kalokhorio, Makheras, 1950, *Chapman* 305 !

3. Stavrovouni, above Ayia Varvara monastery, 1862, *Kotschy* 199 !
7. Common; *Sintenis & Rigo* 258 ! *Syngrassides* 1146 ! 1438 ! *Lindberg f.* s.n. ! *Mavromoustakis* 31 !

NOTES: Varieties with glabrous and pubescent fruits are both found on the island, sometimes growing intermixed.

4. **A. kennedyae** *Meikle* in Hooker's Icones Plant., 36: t. 3574 (1962).
[? *A. montbretiana* (vix Boiss.) Kotschy in Oesterr. Bot. Zeitschr., 12: 276 (1862).]
[*A. auriculata* (non Lam.) A. K. Jackson et Turrill in Kew Bull., 1939: 474 (1939).]
TYPE: Cyprus; Xerokolymbos, 4,350 ft. alt., April 6, 1938, *Kennedy* 1228 (K !).

Erect annual or biennial up to 30 cm. high with a slender taproot; stems terete or obscurely sulcate, shortly and sparsely stellate-pubescent, frequently tinged purple; basal leaves forming a loose rosette, thin, papery, pubescent, obovate-oblong, up to 6 cm. long and 2 cm. wide, margins subentire or bluntly and irregularly sinuate-dentate, apex obtuse, base gradually tapering to a short flattened petiole or sessile; stem leaves few, shorter than basal leaves, and often proportionally wider, sessile, amplexicaul, with short, spreading auricles; inflorescence consisting of a long terminal and shorter axillary racemes, the latter arising from the basal part of the central axis and giving the whole plant, especially when in fruit, a broadly conical or pyramidal outline; pedicels short, 1–3 mm. long, lengthening and thickening considerably in fruit; flowers small, inconspicuous, sepals subglabrous or thinly pubescent, narrowly oblong-lanceolate, 2–2·5 mm. long, about 1 mm. wide, concave, the inner pair distinctly saccate at base; petals white, glabrous, narrowly obovate or oblanceolate, about 3·5 mm. long and 1·5 mm. wide, apex rounded, base tapering to a long, slender claw; stamens 3–3·5 mm. long, filaments slender, flattened; anthers oblong, about 0·5 mm. long, yellow; ovary cylindrical, about 2·5 mm. long, pubescent; stigma sessile or almost so, capitate. Silique ascending, straight or slightly curved, 25–40 mm. long, about 1·5 mm. wide, obtuse at apex and base, valves pubescent, with obscure or ± distinct longitudinal veins, often tinged purple; seeds in a single row in each loculus, broadly oblong, about 1·3 mm. long and 1 mm. wide, not winged; testa rather dark brown, almost black around periphery, minutely foveolate.

HAB.: Shaded rocky ground by streams; 3,900–4,300 ft. alt.; fl. April–May.
DISTR.: Divisions 2, [7]. Endemic.
2. Xerokolymbos, 1938, *Kennedy* 1228 ! Kryos Potamos, 1938, *Kennedy* 1229 ! and, 1939, *Kennedy* 1548 ! Psilodendron, by Kryos Potamos, 1962, *Meikle* 2785 !
[7.] [" Castello della Regina " (Buffavento Castle), 1862, *Kotschy* s.n.]

NOTES: Closely allied to *A. montbretiana* Boiss., but differing in habit and shape of silique. The identity of Kotschy's *A. montbretiana* is doubtful, for it is difficult to believe that his record was a misidentification of the very distinct *A. cypria*, also recorded from Buffavento; the record for *A. montbretiana* is not repeated in " Die Insel Cypern "; it has been cited here simply to indicate the possible, but improbable, occurrence of *A. kennedyae* in this locality.

34. TURRITIS *L.*

Sp. Plant., ed. 1, 666 (1753).
Gen. Plant., ed. 5, 298 (1754).

Erect, glaucous, biennial herbs; basal leaves and lower part of stem usually clothed with furcate or stellate hairs, upper part of stem, and upper stem leaves glabrous, the stem leaves usually numerous, sessile and distinctly auriculate; inflorescence lengthening greatly after fertilization; sepals ± erect, the inner scarcely saccate; petals white, yellowish or rarely pink-tinged, narrowly ovate-spathulate, rounded at apex, gradually narrowed to a basal claw; stamens 6; filaments linear; anthers narrow, acute at apex; ovary sessile, narrowly cylindrical; style short; stigma capitate or slightly 2-lobed; ovules very numerous (130–200). Silique very long and narrow, valves convex or slightly angled, midrib usually prominent. Seeds 1–2-ranked in each loculus, small, ovate or oblong, not winged, or occasionally with a narrow wing.

Three species in temperate Europe, Asia and mountains of Tropical Africa; N. America (? introduced) and as an introduction elsewhere.

1. T. laxa (*Sibth. et Sm.*) *Hayek*, Prodr. Fl. Pen. Balc., 1: 402 (1925); O. E. Schulz in Engl. et Prantl, Pflanzenfam., ed. 2, 17b: 549 (1936); Davis, Fl. Turkey, 1: 429 (1965); Zohary, Fl. Palaest., 1: 281, t. 414 (1966); Osorio-Tafall et Seraphim, List Vasc. Plants Cyprus, 48 (1973).

Arabis laxa Sibth. et Sm., Fl. Graec. Prodr., 2: 28 (1818); Boiss., Fl. Orient. Suppl., 32 (1888); Holmboe, Veg. Cypr., 84 (1914); Post, Fl. Pal., ed. 2, 1: 60 (1932); Lindberg f., Iter Cypr., 17 (1946); Rechinger f. in Arkiv för Bot., ser. 2, 1: 422 (1950).

A. cremocarpa (Boiss. et Bal.) Kotschy in Oesterr. Bot. Zeitschr., 12: 278 (1862); Unger et Kotschy, Die Insel Cypern, 325 (1865).

A. laxa Sibth. et Sm. var. *cremocarpa* (Boiss. et Bal.) Boiss., Fl. Orient., 1: 168 (1867); Holmboe, Veg. Cypr., 84 (1914); Lindberg f., Iter Cypr., 17 (1946).

[*Turritis glabra* (non L.) Sibth. in Walpole, Travels, 22 (1820); Unger et Kotschy, Die Insel Cypern, 325 (1865); H. Stuart Thompson in Journ. Bot., 44: 276 (1906).]

[*Arabis glabra* (non (L.) Bernh.) Holmboe, Veg. Cypr., 84 (1914).]

TYPE: Greece; " in agro Laconico " *Sibthorp* (OXF).

Erect, somewhat glaucous, unbranched or sparsely branched biennial, 25–40 cm. high; basal leaves ± densely stellate-pubescent, forming a loose rosette, oblanceolate or narrowly obovate, 3–8 cm. long, 1–2·5 cm. wide, apex obtuse or rounded, margins sinuate-lobed or occasionally subentire, base tapering gradually to a short, flattened petiole; flowering stem and stem leaves glabrous, often tinged purple; stem leaves numerous, sessile, amplexicaul, ovate-acute, 2–5 cm. long, 0·6–1·8 cm. wide, with 2 conspicuous, obtuse, basal auricles, margins entire or sinuate-dentate; inflorescence terminal and axillary, at first corymbose, later forming a lax raceme; pedicels slender, 0·8–1 cm. long, thickening considerably in fruit; sepals glabrous, narrowly oblong, obtuse, about 5 mm. long and 1·5 mm. wide, tinged purple, the inner sepals slightly saccate at base; petals creamy-white, oblanceolate, about 6 mm. long and 2 mm. wide, apex obtuse, base narrowed gradually to a rather short claw; stamens 5–6 mm. long; filaments linear, flattened; anthers purplish, narrowly oblong, usually curved, about 1·8 mm. long, tapering to a pointed apex; ovary narrowly compressed-cylindrical, about 4 mm. long and 0·5 mm. wide; style short but distinct; stigma capitate. Siliques straight or slightly curved, spreading or pendulous, widely spaced, narrowly linear, 60–100 mm. long, 1·5–2 mm. wide; valves with a prominent midrib, dark purplish; seeds in a single row in each loculus, oblong, 1·8 mm. long, 1 mm. wide, without a wing or sometimes narrowly winged on the part of the margin opposite the hilum; testa bright brown, minutely foveolate.

HAB.: Shady rocks and cliffs at high altitudes; 2,000–6,400 ft. alt.; fl. Febr.–June.

DISTR.: Division 2, common, mainly in the central Troödos region; locally abundant around Platres, Troödos, Kykko and Stavros tis Psokas. Also in Greece, Turkey, Syria and Palestine.

2. Common; *Kotschy* 726 ! *Sintenis & Rigo* 846 ! *Syngrassides* 945 ! 1608 ! *Meikle* 2667 ! *J. B. Suart* 115 !

35. NASTURTIUM *R. Br.*

in Ait. f., Hort. Kew., ed. 2, 4: 109 (1812) nom. cons.

Perennial, glabrous or glabrescent herbs; stems sprawling or ascending; leaves normally pinnate; inflorescence ebracteate; flowers white or tinged lilac; sepals ± erect, inner slightly saccate; petals rounded at apex, gradually tapering to base; stamens 6, filaments linear without appendages; ovary elongate, c. 20–40-ovulate; style short but distinct, stigma shallowly 2-lobed. Silique linear, distinctly compressed, valves rather thin, indistinctly veined. Seeds in 1 or 2 ranks in each loculus; testa distinctly reticulate.

Two species in Europe, W. Asia, Africa and N. America; introduced elsewhere.

1. **N. officinale** *R. Br.* in Ait. f., Hort. Kew., ed. 2, 4: 110 (1812); Unger et Kotschy, Die Insel Cypern, 324 (1865); Boiss., Fl. Orient., 1: 178 (1867); O. E. Schulz in Engl. et Prantl, Pflanzenfam., ed. 2, 17b: 353 (1936); Davis, Fl. Turkey, 1: 430 (1965); Zohary, Fl. Palaest., 1: 277, t. 408 (1966); Osorio-Tafall et Seraphim, List Vasc. Plants Cyprus, 48 (1973).

Sisymbrium nasturtium-aquaticum L., Sp. Plant., ed. 1, 657 (1753).

Cardamine fontanum Lam., Encycl. Méth., 2: 185 (1786).

Nasturtium fontanum (Lam.) Aschers., Fl. Brandenb., 32 (1864); Holmboe, Veg. Cypr., 86 (1914); Post, Fl. Pal., ed. 2, 1: 62 (1932); Lindberg f., Iter Cypr., 17 (1946).

N. officinale R. Br. var. *asarifolium* Kralik ex Rouy et Fouc., Fl. France, 1: 204 (1893).

N. officinale R. Br. var. *simplicifolium* Neum., Sver. Fl., 455 (1901).

N. fontanum (Lam.) Aschers. var. *simplicifolium* (Neum.) Holmboe, Veg. Cypr., 86 (1914).

TYPE: Not precisely localized (BM !).

A decumbent or ascending, glabrous, perennial herb, usually 10–60 cm. high, but variable in height and habit; stems angular, ± hollow, rooting at the lower nodes; leaves lyrate-pinnate or occasionally lacking the lateral leaflets (var. *asarifolium* Kralik ex Rouy et Fouc.; var. *simplicifolium* Neum.), 2–12 cm. or more long, 1–12 cm. or more wide, distinctly petiolate or the upper subsessile, the petiole-base flanked by 2 inconspicuous, narrow, acute auricles; lateral leaflets 2–8, irregularly placed, ovate, oblong or broadly elliptical, 0·5–6 cm. or more long and 0·5–2 cm. or more wide, entire or obscurely sinuate-serrate; terminal leaflet 1–6 cm. or more long and about as wide or sometimes wider, orbicular or broadly ovate, entire or sinuate-dentate; inflorescence ebracteate, at first rather densely corymbose, later elongating into a lax raceme; pedicels slender, ascending, 2–5 (–8) mm. long, lengthening considerably in fruit and becoming patent or slightly deflexed; sepals 2–3 mm. long, 1–1·5 mm. wide, glabrous, concave, the outer rather obtuse at apex, the inner frequently acute and cucullate with a slightly saccate base; petals white, 5–6 mm. long, 2–2·5 mm. wide, tapering to a distinct claw; stamens 2·5–3 mm. long, filaments flattened, linear, tinged purple or lilac; anthers shortly oblong, about 0·4 mm. long, yellow; ovary linear, compressed, 2·5–3 mm. long, style short but distinct, stigma rather large, capitate. Silique patent or often slightly deflexed, on a slender stalk 4–8 mm. long, linear, compressed, 10–20 mm. long, about 2–2·5 mm. wide, rather abruptly contracted at base and apex; valves thin, pale brown or slightly purplish, indistinctly veined, with the seeds forming a distinct raised pattern; seeds in 2 ranks in each loculus, broadly ovoid or almost spherical, about 1 mm. diam.; testa rich brown, rather coarsely and conspicuously reticulate.

HAB.: Marshy ground by springs and streams; sea-level to 3,900 ft. alt.; fl. March–May.

DISTR.: Divisions 2, 4, 7, 8, common in Division 7. Widely distributed in Europe, W. Asia and N. Africa and introduced into many areas of the world.

2. Near Prodhromos, 1862, *Kotschy* 792; Kilani, 1905, *Holmboe* 1149 (var. *simplicifolium* Neum.); Kryos Potamos, 1938, *Kennedy* 1230 ! Kambos, below the village, 1939, *Lindberg f.* s.n. !
4. Kouklia in the Mesaoria, 1905, *Holmboe* 366.
7. Common; *Sintenis & Rigo* 627 ! *Syngrassides* 95 ! 266 ! 1203 ! *Merton* 2030 ! 2244 ! 2976 !
8. Ronnas Valley near Rizokarpaso, 1957, *Merton* 2924 !

NOTES: Only one species seems to occur in Cyprus, though the closely related *N. microphyllum* (Boenn.) Reichb., with narrower fruits, and with the more minutely reticulate seeds in a single rank in each loculus, should be looked for. The var. *asarifolium* Kralik (var. *simplicifolium* Neum.) is scarcely more than a seasonal condition, and does not merit taxonomic recognition.

TRIBE 7. **Matthioleae** *O. E. Schulz* Annual, biennial or perennial herbs or subshrubs clothed with simple, branched or stellate hairs, sometimes glandular, rarely glabrous. Sepals usually erect; nectary glands variously arranged; filaments linear, the inner often dilated and sometimes connate in pairs, the outer occasionally toothed; anthers truncate or pointed; ovary sessile; style indistinct or wanting; stigma often 2-lobed, sometimes with conspicuous dorsal appendages, rarely capitate. Silique linear or elongate, rarely short, dehiscent, indehiscent or sometimes breaking transversely into segments. Cotyledons accumbent.

36. MATTHIOLA *R. Br.*

in Ait. f., Hort. Kew., ed. 2, 4: 119 (1812) nom. cons.

Annual, biennial or perennial herbs and subshrubs with stellate or branched indumentum sometimes intermixed with stalked glands; leaves entire, sinuate or pinnatifid, usually grey-hairy, sometimes glabrous; inflorescence usually a lax raceme, flowers sessile or pedicellate, sepals erect, the inner saccate at base; petals purple, violet, pink, white or brownish, rounded, emarginate or acute at apex, tapering to a slender, clawed base; stamens 6, filaments free, anthers elongate, linear, with a pointed apex; ovary cylindrical, ovules numerous; style usually obsolete, stigma 2-lobed, often with 2 horn-like dorsal appendages. Silique dehiscent, linear, usually elongate and ± compressed, distinctly or obscurely 1-veined. Seeds in a single rank in each loculus, oblong, ovate or almost orbicular, usually much compressed, sometimes with a distinct marginal wing.

About 50 species, chiefly in the Mediterranean area, with a few outlying representatives in Asia and Africa.

Flowers distinctly pedicellate; siliques strongly compressed:
  Leaves entire or almost entire; siliques eglandular - - - - - **1. M. incana**
  Leaves sinuate or pinnatifid; siliques glandular - - - - - **2. M. sinuata**
Flowers sessile or very shortly pedicellate; siliques terete or subterete:
  Perennial herb, suffruticose at base; leaves narrow, linear, entire or sparsely lobed; stigma-lobes with small, inconspicuous dorsal appendages **3. M. fruticulosa**
  Annual herbs with lanceolate, oblanceolate or oblong, sinuate, pinnatifid or irregularly lobed leaves; dorsal appendages of stigma conspicuous, forming long horns:
    Stigma much shorter than its dorsal appendages; siliques apparently 2-horned at apex; indumentum rather short and close - - - **4. M. longipetala**
    Stigma as long or longer than its dorsal appendages; siliques apparently 3-horned at apex; indumentum rather loose and woolly - - **5. M. tricuspidata**

1. M. incana (*L.*) *R. Br.* in Ait. f., Hort. Kew., ed. 2, 4: 119 (1812); Unger et Kotschy, Die Insel Cypern, 323 (1865); Boiss., Fl. Orient., 1: 148 (1867); Holmboe, Veg. Cypr., 86 (1914); Post, Fl. Pal., ed. 2, 1: 54 (1932); O. E. Schulz in Engl. et Prantl,

Pflanzenfam., ed. 2, 17b: 563 (1936); Davis, Fl. Turkey, 1: 448 (1965); Osorio-Tafall et Seraphim, List Vasc. Plants Cyprus, 48 (1973).

Cheiranthus incanus L., Sp. Plant., ed. 1, 662 (1753).

TYPE: " *in* Hispaniae *maritimis* ".

A robust, erect, branched perennial, 25–50 (–80) cm. high, with a thick, tough taproot; lower part of stem becoming woody and leafless, with prominent leaf-scars; upper part of stem and the leaves densely clothed with short, stellate, greyish hairs; leaves rather crowded towards the ends of the branches, sessile or shortly and indistinctly stalked, narrowly oblanceolate, 2–8 cm. long, 0·5–2 cm. wide, obtuse or rounded at apex, margins entire or almost so, tapering gradually to base; inflorescence a lax raceme; flowers showy, distinctly pedicellate, pedicels 5–15 mm. or more long; sepals pubescent externally, narrowly oblong or linear, 10–15 mm. long, 2–3 mm. wide, apex obtuse, bases of the inner sepals distinctly saccate; petals cerise-red, purple, pink, mauve or white, obovate, about 2·5 cm. long and 1 cm. wide, apex rounded, base forming a slender claw, about 1 cm. long; stamens very unequal, the 4 longer about 12 mm. long, the 2 shorter about 7 mm. long; filaments of the longer stamens distinctly flattened, usually with 1 small lateral tooth; anthers linear, tapering to a point, about 5 mm. long and 1 mm. wide, yellow; ovary narrowly cylindrical, 10–15 mm. long, about 1·5 mm. diam., densely tomentose; stigma obscurely 2-lobed, the dorsal appendages at first inconspicuous, developing as the fruit ripens, though never very large. Silique ascending or almost erect, straight or ± curved, linear, 8–12 cm. long, 0·3–0·4 cm. wide, distinctly compressed, valves ± densely grey-pubescent, eglandular, midrib usually distinct, lateral veins obscure; seeds orbicular, flattened, about 2 mm. diam.; testa minutely papillose, dark fuscous-brown with a narrow, pale, marginal wing.

HAB.: Rocky seashores and maritime cliffs; about sea-level; fl. Febr.–May.

DISTR.: Divisions 4, [7], rare. Widespread in the Mediterranean region, from Spain eastwards to the Aegean Islands and Cyprus, also in the Atlantic Islands. Much cultivated, and often an escape.

4. Seashore at Cape Greco, rare, 1862, *Kotschy* 154a.

[7. Temblos, 1956, *G. E. Atherton* 867 ! Probably a garden escape.]

2. M. sinuata (*L.*) *R. Br.* in Ait. f., Hort. Kew., ed. 2, 4: 120 (1812); Sibth. et Sm., Fl. Graec. Prodr., 2: 27 (1813), Fl. Graec., 7: 37, t. 640 (1830); Boiss., Fl. Orient., 1: 148 (1867); Holmboe, Veg. Cypr., 86 (1914); Post, Fl. Pal., ed. 2, 1: 54 (1932); O. E. Schulz in Engl. et Prantl, Pflanzenfam., ed. 2, 17b: 563 (1936); Davis, Fl. Turkey, 1: 448 (1965); Osorio-Tafall et Seraphim, List Vasc. Plants Cyprus, 48 (1973).

Cheiranthus sinuatus L., Sp. Plant., ed. 2, 926 (1763).

TYPE: " *in* Hispaniae, Monspelii *maritimis* " (LINN !).

Superficially similar to *M. incana* (L.) R. Br. in general habitat and appearance; a biennial or perennial herb; base of stem somewhat woody, but rarely forming an elongate stalk as in *M. incana*; upper parts of stem and leaves densely grey-tomentose; leaves lanceolate or oblanceolate, 5–11 (–20) cm. long, 1–3 (–4) cm. wide, apex obtuse, margin distinctly sinuate-lobed or occasionally subentire, base tapering to a distinct, flattened petiole in the lower leaves; upper leaves sessile or subsessile; inflorescence a lax raceme; rhachis and pedicels grey-tomentose, with scattered, shortly stalked glands; flowers showy, distinctly pedicellate, pedicels 5–10 mm. long; sepals linear or narrowly oblong, obtuse, 10–15 mm. long, 1·5–3 mm. wide, densely grey-tomentose externally; petals pink, purple, violet or white, about 16–20 mm. long, 6–10 mm. wide, apex rounded or emarginate, base tapering to a slender claw about 8 mm. long; stamens very unequal, the 4 longer about 8–9 mm. long, the 2 shorter 5–6 mm long; filaments of the longer stamens distinctly flattened, often with 1 small lateral tooth; anthers linear, tapering to a point, about 3·5 mm. long and 0·8 mm. wide, yellow; ovary narrowly cylindrical about 5–8 mm. long, 1 mm. diam., villose-tomentose; stigma obscurely 2-lobed, dorsal appendages inconspicuous even in ripe fruit. Silique spreading or ascending, straight or ± curved, linear, 8–12 (–15) cm. long, 0·3–0·6 cm. wide, distinctly compressed, valves densely grey-tomentose, generally dotted over with conspicuous stalked glands, midrib distinct, lateral veins obscure; seeds orbicular or broadly oblong, 2·5–3 mm. diam.; testa minutely papillose, dark brown, with a broad, conspicuous, pale brown marginal wing.

HAB.: Sandy seashores; about sea-level; fl. Febr.–May.

DISTR.: " In insulâ Cypro ", 1787, *Sibthorp* s.n. Widespread in the Mediterranean from Spain to Greece, rarer eastwards, but recorded from the Aegean Islands and Palestine; also in N. Africa (Morocco to Cyrenaica) and northwards along the western coast of Europe to the British Isles.

NOTES: Not collected since 1787, and perhaps an error in localization.

3. M. fruticulosa (*L.*) *Maire* in Jahandiez et Maire, Cat. Plant. Maroc., 2: 311 (1932); Davis, Fl. Turkey, 1: 449 (1965); Osorio-Tafall et Seraphim, List Vasc. Plants Cyprus, 48 (1973).

Cheiranthus fruticulosus L., Sp. Plant., ed. 1, 662 (1753).

Cheiranthus tristis L., Syst. Nat., ed. 10, 1134 (1759); Sp. Plant., ed. 2, 925 (1763).

Matthiola tristis (L.) R. Br. in Ait. f., Hort. Kew., ed. 2, 4: 120 (1812); Boiss., Fl. Orient., 1: 153 (1867); Holmboe, Veg. Cypr., 86 (1914); Post, Fl. Pal., ed. 2, 1: 55 (1932); O. E. Schulz in Engl. et Prantl, Pflanzenfam., ed. 2, 17b: 564 (1936).

Cheiranthus coronopifolius Sibth. et Sm., Fl. Graec. Prodr., 2: 25 (1813), Fl. Graec., 7: 35, t. 637 (1830).

Matthiola coronopifolia (Sibth. et Sm.) DC., Regn. Veg. Syst. Nat., 2: 173 (1821); Unger et Kotschy, Die Insel Cypern, 323 (1865).

TYPE: Spain; not precisely localized, *Loefling* (LINN !).

An erect, tufted perennial, 10–30 cm. or more high, with a tough woody taproot, and a rootstock ± densely clothed with old, partially decayed leaf-bases; stems simple or with a varying number of ascending branches, terete or obscurely angled, rather sparsely clothed with short, greyish stellate pubescence; basal leaves numerous, rather crowded, grey-pubescent, linear or narrowly oblanceolate, 20–50 mm. long, 2–6 mm. wide, apex obtuse or acute, margins entire or ± irregularly sinuate-lobed; base tapering to a very short, flattened petiole; stem leaves similar, but smaller and sometimes narrowly linear; inflorescence a rather lax, elongate raceme; flowers sessile or very shortly stalked; sepals grey-pubescent externally, linear-oblong, acute or obtuse, about 11 mm. long and 1·5–2 mm. wide, the inner pair markedly concave and distinctly saccate at base; petals brownish or sometimes tinged pink, strap-shaped, about 18–25 mm. long and 2·5–3 mm. wide, apex obtuse, margins distinctly undulate, base tapering gradually to a slender claw; stamens 8–11 mm. long, filaments linear, distinctly flattened, those of the longer stamens often with one margin gibbous; anthers linear, about 4 mm. long, gradually tapering to apex; ovary cylindrical, about 6 mm. long and 1 mm. diam., densely grey-tomentose; stigma sessile, short, appendages obscure. Silique shortly stalked, spreading or ascending, narrowly linear, slightly compressed and sometimes obscurely torulose, 60–100 mm. long, about 1·5 mm. wide, apex with 2, small, rather inconspicuous, erect or incurved horned appendages; valves pale brown, sparingly clothed with grey, stellate pubescence, and with obscure longitudinal veins; seeds oblong, flattened, about 2 mm. long and 1 mm. wide; testa minutely rugulose-papillose, olive-brown, with a narrow hyaline marginal wing.

HAB.: Dry rocky ground; sea-level to 500 ft. alt.; fl. Febr.–May.

DISTR.: Division 4, rare. Throughout the western Mediterranean from Spain to Greece, also in N. Africa; very rare in Turkey and Palestine.

4. Larnaca, 1862, *Kotschy* 130a. Waste ground near umber diggings, close to Troulli Mine, 1962, *Meikle* !

NOTES: The Sibthorp record mentioned by Kotschy (Die Insel Cypern, 323) should probably be referred to *M. longipetala* (Vent.) DC., so that admission of the species to the Cyprus flora rests solely on Kotschy's record and upon specimens seen (but unfortunately not collected) by the author in 1962. The species has been recorded from Turkey and Palestine, and should be looked for in Cyprus.

4. M. longipetala (*Vent.*) *DC.*, Regn. Veg. Syst. Nat., 2: 174 (1821); Holmboe, Veg. Cypr., 86 (1914); Zohary in Pal. Journ. Bot., J. series, 2: 157 (1941); Davis, Fl. Turkey, 1: 449 (1965); Zohary, Fl. Palaest., 1: 272, t. 399 (1966); Osorio-Tafall et Seraphim, List Vasc. Plants Cyprus, 48 (1973).

Cheiranthus longipetalus Vent., Descr. Plant. Nouv. Jard. Cels., t. 93 (1802).

Cheiranthus bicornis Sibth. et Sm., Fl. Graec. Prodr., 2: 26 (1813).

Matthiola bicornis (Sibth. et Sm.) DC., Regn. Veg. Syst. Nat., 2: 177 (1821); Boiss., Fl. Orient., 1: 155 (1867).

M. oxyceras DC., Regn. Veg. Syst. Nat., 2: 173 (1821); Unger et Kotschy, Die Insel Cypern, 324 (1865); Boiss., Fl. Orient., 1: 156 (1867).

M. tenella DC., Regn. Veg. Syst. Nat., 2: 169 (1821); Poech, Plant. Ins. Cypr., 31 (1842); Unger et Kotschy, Die Insel Cypern, 324 (1865).

M. bicornis (Sibth. et Sm.) DC. var. *oxyceras* (DC.) Bornm. in Beih. Bot. Centralbl., 27: 297 (1910).

M. longipetala (Vent.) DC. var. *bicornis* (Sibth. et Sm.) Zohary in Pal. Journ. Bot., J. series, 2: 157 (1941).

M. longipetala (Vent.) DC. var. *oxyceras* (DC.) Zohary, *loc. cit.* 157 (1941).

TYPE: Iraq; " in Oriente circa Bagdad " *Olivier* (G).

An erect annual herb, 13–30 cm. high, with a slender taproot; stems simple or branched, obscurely angled, sparingly or rather densely pubescent, with scattered, shortly stalked glands; leaves rather narrowly lanceolate, tapering to apex and to the shortly petiolate base, (2–) 3–10 (–14) cm. long, 0·4–1·5 cm. wide, subentire or irregularly lobed, or sometimes deeply pinnatifid with acute, toothed lobes; inflorescence a lax raceme; rhachis usually glandular, individual flowers sessile or shortly pedicellate; sepals narrowly oblong, acute, 8–11 mm. long, 1–2 mm. wide, densely pubescent externally, the inner pair distinctly saccate at base; petals very variable in colour, purple, purplish, livid, brownish-pink, pink or whitish, sometimes with a yellowish claw, equally variable in shape, ranging from oblong-obovate (var. *bicornis* (Sibth. et Sm.) Zohary) to almost linear (var. *longipetala*), about 18–20 mm. long, 2–7 mm. wide, margins flat or distinctly undulate, base tapering gradually or rather abruptly to a long claw; stamens very unequal, 3–8 mm. long, filaments linear, flattened, those of the longer stamens with one margin markedly gibbous; anthers linear, 1·5–2 mm. long, tapering to apex, yellow; ovary narrowly cylindrical, about 5 mm. long and 1 mm. diam., densely grey-pubescent; stigma short, 2-lobed, appendages obscure. Silique shortly stalked, ascending, narrowly linear, subterete, 40–90 mm. long, 2–2·5 mm. wide, pubescent and ± glandular, apex with 2 large, deflexed or spreading (var. *longipetala*; var. *oxyceras* (DC.) Zohary) or incurved (var. *bicornis* (Sibth. et Sm.) Zohary) horned appendages; valves rather thick, with obscure longitudinal venation; seeds oblong, flattened, about 1·8 mm. long and 1·5 mm. wide; testa minutely rugulose, fuscous-brown, with a very narrow hyaline marginal wing.

HAB.: Dry, eroded slopes, sandy or stony pastures, roadsides or waste land; sea-level to 1,000 ft. alt. (occasionally higher); fl. Febr.–May.

DISTR.: Divisions 1, 3–5, 7, common in Division 5. Also in Greece, Turkey, Palestine and Egypt eastwards to Iran and Arabia.

1. or 3. Between Stroumbi and Pissouri, 1934, *Syngrassides* 849 !
3. Near Limassol, 1859, *Kotschy* 457; Episkopi, 1963, *J. B. Suart* 70 !
4. Near Larnaca, 1859, *Kotschy* s.n.

5. Common (especially around Kythrea); *Sintenis & Rigo* 261 ! *A. G. & M. E. Lascelles* s.n. ! *Syngrassides* 304 ! 1794 ! *Sparrow* 5 !
7. Lakovounara Forest, 1950, *Chapman* 76 ! 159 ! 503 !

NOTES: Most of the Cyprus material falls within the range of var. *oxyceras* (DC.) Zohary, but this is scarcely more than one of the intermediates connecting the two extremes, var. *longipetala* and var. *bicornis*. It is doubtful if these varieties represent recognizable taxonomic groups.

5. **M. tricuspidata** (*L.*) *R. Br.* in Ait. f., Hort. Kew., ed. 2, 4: 120 (1812); Unger et Kotschy, Die Insel Cypern, 323 (1865); Boiss., Fl. Orient., 1: 154 (1867); Holmboe, Veg. Cypr., 87 (1914); Post, Fl. Pal., ed. 2, 1: 55 (1932); O. E. Schulz in Engl. et Prantl, Pflanzenfam., ed. 2, 17b: 564 (1936); Lindberg f., Iter Cypr., 17 (1946); Davis, Fl. Turkey, 1: 449 (1965); Zohary, Fl. Palaest., 1: 271, t. 396 (1966); Osorio-Tafall et Seraphim, List Vasc. Plants Cyprus, 48 (1973).

Cheiranthus tricuspidatus L., Sp. Plant., ed. 1, 663 (1753).

Matthiola tricuspidata (L.) R. Br. var. *integrifolia* G. Strobl in Verh. Zool.-Bot. Ges. Wien, 53: 448 (1903); Holmboe, Veg. Cypr., 87 (1914).

TYPE: " *in* Europae *australis maritimis* " (BM !).

An erect, diffuse or procumbent annual, (2–) 6–17 (–20) cm. high; stems often much branched near the base, ± densely and softly woolly, with scattered, shortly stalked glands; leaves tomentose, oblanceolate or narrowly obovate, 2–5 (–7) cm. long, 0·5–1·5 (–2·5) cm. wide, apex obtuse, margins rather regularly sinuate-pinnatifid, base narrowing to a short, flattened petiole; inflorescence at first rather densely corymbose, later lengthening to form a loose raceme; flowers sessile or shortly stalked; sepals narrowly oblong, tomentose externally, about 10 mm. long, 1·2–2 mm. wide, apex acute, bases of the inner pair of sepals slightly saccate; petals mauve or pink, obovate, 13–20 mm. long, 4–8 mm. wide, apex rounded or sometimes emarginate, base tapering to a slender claw about 1 cm. long; stamens very unequal, 4–9 mm. long, filaments linear, distinctly flattened, those of the longer stamens somewhat gibbous; anthers linear, about 3–4 mm. long, tapering to apex, yellow; ovary cylindrical, about 5 mm. long and 0·7 mm. diam., densely tomentose; stigma 2-lobed, lateral appendages at first obscure, developing in the fruit. Silique hard, woody, shortly stalked, ascending or sometimes almost patent, subterete, straight or slightly curved, 20–60 mm. long, about 2·5 mm. diam., terminating in a sharp, narrowly conical stigma, flanked by 2 prominent, horned lateral appendages; valves pale brown, rather loosely tomentose and sparsely glandular, with obscure longitudinal veins; seeds oblong, flattened, about 1·7 mm. long and 1–1·2 mm. wide; testa minutely papillose, fuscous, with a very narrow or almost obsolete marginal wing.

HAB.: Sandy seashores; about sea-level; fl. Febr.–May.

DISTR.: Divisions 1, 3–5, 8, common in Division 4. Widespread in Mediterranean region.

1. Near Ktima, 1862, *Kotschy* 663 ! Near Paphos, 1902, *A. G. & M. E. Lascelles* s.n. ! Ayios Yeoryios (Akamas), 1962, *Meikle* 2073 !
3. Akrotiri, 1950, *Barrington* 25 !
4. Common (especially around Famagusta); *Syngrassides* 856 ! 1152 ! *Mavromoustakis* 64 ! *N. Macdonald* 36 ! 111 !
5. Common at Salamis; *Holmboe* 445a, 445b; *Lindberg f.* s.n.; *Collingwood Ingram* 38 !
8. Cape Andreas, 1880, *Sintenis & Rigo* 260 ! Cape Andreas, 1962, *Meikle* 2915 !

TRIBE 8. **Hesperideae** *Prantl emend. O. E. Schulz* Annual or perennial herbs clothed with simple or variously branched hairs, sometimes glandular, rarely glabrous. Sepals erect; nectary glands variously arranged, filaments linear, the inner often dilated and occasionally connate; anthers blunt or acute; ovary sessile or stipitate; style present or absent; stigma truncate, capitate or 2-lobed. Silique ± elongate, dehiscent or indehiscent, sometimes breaking transversely into joints. Cotyledons incumbent or occasionally accumbent.

37. MALCOLMIA *R. Br.*

in Ait. f., Hort. Kew., ed. 2, 4: 121 (1812) nom. cons.

Mostly annual herbs with an indumentum of simple, medifixed or stellate hairs; leaves usually entire or occasionally pinnatifid; sepals erect, the inner often saccate at base; petals pink, violet, purple or white, rarely yellow, distinctly clawed; stamens free, the filaments of the inner pairs rather dilated; anthers oblong or linear, acute or obtuse; ovary sessile, 30–80-ovulate; style short or distinct; stigma truncate, pointed or capitate, with spreading or erect lobes. Silique usually elongate-cylindrical, straight or variously curved, the valves often thick and tardily dehiscent; seeds in one or rarely two ranks in each loculus, oblong or ovate; cotyledons incumbent.

About 25 species, occurring chiefly about the Mediterranean, but with outlying representatives in Iraq and Iran, and as far east as Central Asia. *M. nana* (DC.) Boiss. is often placed in the separate genus *Maresia* Pomel, but it is joined to *Malcolmia* through *M. ramosissima* (Desf.) Thell.

Style distinct; stigma obtuse or truncate; petals small (c. 4–5 mm. long) **1. M. nana**
Style indistinct or wanting; stigma pointed; petals usually more than 6 mm. long:
Petals 12–17 mm. long; fruits spreading with the fruiting pedicels usually thicker than the base of the silique - - - - - **2. M. flexuosa**
Petals 6–10 mm. long; fruits ascending or almost erect with the fruiting pedicels narrower than the base of the silique - - - - - **3. M. chia**

SECT. 1. **Sisymbrioideae** *Boiss.* Silique soon dehiscent; style distinct, stigma obtuse, truncate or capitate.

1. M. nana (*DC.*) *Boiss.*, Fl. Orient., 1: 222 (1867).
Sisymbrium ? nanum DC., Regn. Veg. Syst. Nat., 2: 486 (1821).
Malcolmia confusa Boiss., Fl. Orient., 1: 221 (1867); Holmboe, Veg. Cypr., 88 (1914); Post, Fl. Pal., ed. 2, 1: 72 (1932).
Maresia nana (DC.) Battandier, in Battandier et Trabut, Fl. Alg., 68 (1888); O. E. Schulz in Engl. et Prantl, Pflanzenfam., ed. 2, 17b: 580 (1936); Davis, Fl. Turkey, 1: 462 (1965); Zohary, Fl. Palaest., 1: 268, t. 391 (1966); Osorio-Tafall et Seraphim, List Vasc. Plants Cyprus, 49 (1973).

A slender, erect or spreading annual, 3–17 cm. or more high; stems usually much-branched towards the base, sometimes simple, ± densely clothed with short, grey, stellate pubescence (except in var. *glabra* Meikle); leaves stellate-pubescent (except in var. *glabra* Meikle), variable in size and shape, oblanceolate, obovate, lanceolate or elliptical, 5–20 (–40) mm. long, 2–6 (–8) mm. wide, margins entire, irregularly and bluntly serrate, or bluntly pinnatifid, the radical leaves tapering to a distinct petiole, the cauline leaves sessile or subsessile; inflorescence rather densely corymbose, at first often nodding, later erect and overtopped by the developing fruits; flowers small, shortly pedicellate, sepals narrowly oblong, about 1·5–2 mm. long and 0·5 mm. wide, densely stellate-pubescent externally (except in var. *glabra* Meikle), blunt at apex with a narrow membranous margin, the inner sepals distinctly saccate at base; petals mauve or pink, obovate, about 4–5 mm. long, 1–1·5 mm. wide, rounded at apex, base tapering gradually to a slender claw; stamens not markedly unequal, about 2·5 mm. long, outer pair with filiform or very narrowly linear filaments, inner pair with distinctly flattened and dilated filaments; anthers shortly oblong, minute, 0·1–0·2 mm. long, yellow; ovary cylindrical or slightly tapering towards apex, about 2·5 mm. long, 0·8 mm. wide, densely pubescent or glabrous (var. *glabra* Meikle); style short but distinct, glabrous; stigma obtuse or capitate.

Siliques distinctly stalked, ascending, spreading or rarely deflexed, forming a short, lax raceme, narrowly linear, compressed, about 15–20 (–25) mm. long and 1 mm. wide; valves thin, soon dehiscent, pale brown, pubescent or glabrous; seeds oblong, about 0·6 mm. long and 0·3 mm. wide, not much compressed; testa minutely rugulose, pale brown.

var. nana

TYPE: " in Sibiria orientali " *Merk* sec. *De Candolle* (G), but see note.

Stems, leaves and fruits ± densely and closely stellate-pubescent.

HAB.: On sand-dunes and sandy seashores; about sea-level; fl. Febr.–June.

DISTR.: Divisions 1–4, 7, 8. Widely distributed in the Mediterranean region and eastwards to the Caspian.

1. Amoudhi near Ayios Yeoryios (Akamas), 1962, *Meikle* 2078 !
2. Xeros–Polis road, 48½ miles from Nicosia, 1957, *Merton* 3106 ! Yialia, 1962, *Meikle* !
3. Amathus, 1960, *N. Macdonald* 19 !
4. Ayia Napa, 1905, *Holmboe* 18.
7. Near Vasilia, on seashore, 2 miles W. of village, 1941, *Davis* 2117 !
8. Dhavlos, 1880, *Sintenis & Rigo*. Ayios Philon near Rizokarpaso, 1941, *Davis* 2278 ! Ronnas Bay near Rizokarpaso, 1941, *Davis* 2346 !

NOTES: Boissier (Fl. Orient., 1: 222) notes that the Siberian localization of the type material is an error; *M. nana* is unknown in eastern Asia.

var. **glabra** *Meikle* in App. III, p. 806

TYPE: Cyprus; Yialia, on the seashore; March 15, 1962, *Meikle* 2195 (K !).

Stems, leaves and fruits glabrous or very nearly so.

HAB.: Sandy seashore; just above sea-level; fl. March.

DISTR.: Division 2, rare. ? Endemic.

2. Yialia, 1962, *Meikle* 2195 !

NOTES: Growing with ordinary var. *nana* on sandy seashore just above tide-level; there was no evidence of intergradation between glabrous and pubescent plants, and both were forming fairly large colonies consisting of numerous individuals.

SECT. 2. **Malcolmia** Silique tardily dehiscent; style obscure or absent; stigma conical, usually acute.

2. M. flexuosa (*Sibth. et Sm.*) *Sibth. et Sm.*, Fl. Graec., 7: 33, t. 634 (1830); Boiss., Fl. Orient., 1: 228 (1867); Holmboe, Veg. Cypr., 88 (1914); O. E. Schulz in Engl. et Prantl, Pflanzenfam., ed. 2, 17b: 569 (1936); Davis, Fl. Turkey, 1: 461 (1965); Osorio-Tafall et Seraphim, List Vasc. Plants Cyprus, 49 (1973).

Cheiranthus flexuosus Sibth. et Sm., Fl. Graec. Prodr., 2: 24 (1813).

TYPE: " in insula Cypro " *Sibthorp* (OXF).

An erect or spreading annual 4–15 (–20) cm. high, usually much-branched towards the base, but sometimes simple, especially in starved specimens; leaves and stems rather sparsely clothed with adpressed, bipartite, medifixed hairs; stems obscurely angled, often tinged purple or violet; leaves somewhat fleshy, lustrous above, obovate or spathulate, 20–50 mm. long, 3–15 mm. wide, apex obtuse or rounded, margins entire or remotely serrulate, base tapering to a slender flattened petiole; inflorescence a rather short, lax raceme; flowers conspicuous; pedicels 5 mm. or more long; sepals narrowly oblong, 8–9 mm. long, 1–1·5 mm. wide, with a distinct membranous margin, apex obtuse, bases of the inner pair of sepals distinctly saccate; petals pink, mauve, lilac or white, 12–17 mm. long, 5–7 mm. wide, apex rounded or slightly emarginate, base tapering to a slender claw; stamens not very unequal, about 8–10 mm. long; filaments flattened, those of the longer stamens distinctly dilated; anthers narrowly oblong, about 2·5–3 mm. long, slightly tapering towards an acute apex, yellow; ovary cylindrical, about

6 mm. long and 1–1·5 mm. diam., slightly tapering towards apex, adpressed-pubescent; style obscure or wanting; stigma conical, about 1·3 mm. long, with 2 erect, connivent stigmas. Fruiting raceme rather lax, about 4–9 cm. or more long with a flexuous rhachis, very sparsely adpressed-pubescent; siliques spreading almost horizontally, straight or slightly curved, somewhat compressed, 35–70 mm. long, about 2 mm. wide, tapering rather abruptly to a conical apex; fruiting pedicels about 5 mm. long, as thick as or thicker than the bases of the siliques; valves thick, tardily dehiscent, pale brown, rugulose, with a distinct midrib, glabrous or very sparsely pubescent; seeds ovate, much compressed, about 1·3 mm. long and 1 mm. wide; testa dark brown, minutely papillose.

HAB.: Seashores, coastal maquis, sometimes on walls or on limestone or sandstone rocks; usually near sea-level, but exceptionally inland up to 2,380 ft. alt.; fl. Febr.–May.

DISTR.: Divisions 6, 7, common along the northern shore or a few miles inland (especially around Kyrenia). Also in Greece, Crete, Aegean Islands and Turkey.

6. Near Orga, 1956, *Merton* 2580 !

7. Common; *Sintenis & Rigo* 630 ! *Kennedy* 1665 ! 1666 ! 1668 ! 1674 ! *Davis* 3049 ! *Chapman* 281 ! *G. E. Atherton* 916 ! 997 !

3. M. chia (*L.*) *DC.*, Regn. Veg. Syst. Nat., 2: 440 (1821); Unger et Kotschy, Die Insel Cypern, 329 (1865); Boiss., Fl. Orient., 1: 229 (1867); Holmboe, Veg. Cypr., 88 (1914); Post, Fl. Pal., ed. 2, 1: 75 (1932); Davis, Fl. Turkey, 1: 461 (1965); Zohary, Fl. Palaest., 1: 265, t. 387 (1966); Osorio-Tafall et Seraphim, List Vasc. Plants Cyprus, 49 (1973).

Cheiranthus chius L., Sp. Plant., ed. 1, 661 (1753).

Malcolmia micrantha Boiss. et Reut. in Boiss., Fl. Orient. Suppl., 45 (1888); Holmboe, Veg. Cypr., 88 (1914); Lindberg f., Iter Cypr., 17 (1946) pro parte quoad spec. cypr.

An erect, usually much-branched annual, (3–) 8–16 (–23) cm. high; stems terete or slightly ridged, often stained purple, rather sparsely clothed with adpressed bipartite, medifixed hairs; leaves rather thin, variable in size, from 5–40 mm. or more long, 3–15 mm. or more wide, thinly clothed with adpressed bipartite hairs occasionally intermixed with a few stellate hairs; lower leaves ovate, broadly elliptical or sometimes spathulate, apex blunt or rounded, margins entire or remotely serrate (rarely lyrate-pinnatifid), base tapering to a slender flattened petiole; upper leaves smaller and narrower, acute or acuminate, entire or very obscurely serrulate, shortly petiolate or subsessile. Flowers smaller than those of *M. flexuosa*, at first clustered, later crowning the elongate fruiting raceme, rarely overtopped by the young siliques; sepals narrowly oblong, about 4·5 mm. long and 1 mm. wide, rather densely adpressed-pubescent externally, with a narrow membranous margin; apex acute, bases of the inner pair of sepals distinctly saccate; petals purple, pink, lilac or white, 6–10 mm. long, 2·5–4 mm. wide, apex distinctly emarginate; stamens subequal, 3·5–4 mm. long, filaments flattened, those of the longer stamens distinctly dilated; anthers narrowly oblong, about 1–1·5 mm. long, slightly tapering towards an acute apex, yellow; ovary cylindrical, about 3·5–4 mm. long and 0·8 mm. diam., slightly tapering towards apex, adpressed-pubescent; style obscure or wanting, stigma conical, less than 1 mm. long, with 2 erect, connivent stigmas. Fruiting raceme elongate, lax, rhachis straight or slightly flexuous, thinly adpressed-pubescent; siliques erect or ascending, usually somewhat curved, slightly compressed, 20–30 mm. long, 1·5–2 mm. wide, tapering rather abruptly to a sharp conical apex; fruiting pedicels 4–5 mm. long, thickened, but usually not so thick as the bases of the siliques; valves thick, tardily dehiscent, rather dark purplish-brown, rugulose, with a distinct midrib, thinly adpressed-pubescent; seeds oblong, compressed, about 1·3 mm. long and 1 mm. wide; testa dark brown, minutely papillose.

var. **chia**

TYPE: The plate in Dillenius, Hortus Elthamensis, 180, t. 148, fig. 178 (1732). Lectotype selected by Dr. A. L. Stork (Opera Botanica, no. 33: 18; 1972).

Lower leaves shortly and remotely serrate or entire.

HAB.: Limestone cliffs and ravines, usually inland, occasionally near the sea; 50–3,000 ft. alt.; fl. Febr.–May.

DISTR.: Divisions 1, 7, locally common in Division 7 along the Northern Range. Also in Yugoslavia, Greece, Crete, Aegean Is., western Turkey and Palestine.

1. Aprilokambos near Ayios Yeoryios (Akamas), 1962, *Meikle* 2012 !
7. Common; *Sintenis & Rigo* 278 ! 278a ! *Kennedy* 1667 ! 1673 ! 1729 ! *Davis* 2115 ! 2519 ! 2996 ! *Casey* 132 !

var. **lyrata** (*Sibth. et Sm.*) *Boiss.*, Fl. Orient., 1: 229 (1867); Holmboe, Veg. Cypr., 88 (1914).

Cheiranthus lyratus Sibth. et Sm., Fl. Graec. Prodr. 2: 24 (1813).

M. lyrata (Sibth. et Sm.) Sibth. et Sm., Fl. Graec., 7: 34 (1830); Unger et Kotschy, Die Insel Cypern, 329 (1865).

TYPE: Cyprus, without locality, 1787, *Sibthorp* s.n. (OXF).

Lower leaves lyrate-pinnatifid, or with deeply lobed margins.

HAB.: Unknown.

DISTR.: Known only from the type material which is not precisely localized.

NOTES: Dr. A. L. Stork (Opera Botanica, no. 33: 23; 1972) says that " in the course of work on *M. chia* and its allies specimens were rarely found with lyrate basal leaves. Such specimens occur here and there throughout the distribution area."

38. ERYSIMUM *L.*

Sp. Plant., ed. 1, 660 (1753).
Gen. Plant., ed. 5, 296 (1754).

Annual or perennial herbs or subshrubs with leafy stems and branches clothed with bipartite, or variously compound, medifixed hairs; sepals erect, sometimes connate, the outer slightly gibbous or horned just below the apex, the inner more or less distinctly saccate at the base; petals usually yellow, occasionally purple, bronze or white, distinctly clawed; filaments free; anthers narrowly oblong or linear, usually rather blunt at apex; median nectary glands present, usually free; ovary sessile, cylindrical, 30–70-ovulate; style rather short; stigma capitate, retuse or sometimes 2-lobed. Silique linear, elongate, more or less compressed and often 4-angled; valves hairy, with prominent midribs. Seeds in a single row in each loculus, sometimes winged; cotyledons incumbent.

About 80 species, scattered over most temperate regions of the northern hemisphere, but chiefly in Europe, the Mediterranean region, N. and Central America.

1. **E. repandum** *L.*, Demonstr. Plant., 17 (1753), Amoen. Acad., 3: 415 (1756); Unger et Kotschy, Die Insel Cypern, 330 (1865); Boiss., Fl. Orient., 1: 189 (1867); Holmboe, Veg. Cypr., 87 (1914); Post, Fl. Pal., ed. 2, 1: 64 (1932); O. E. Schulz in Engl. et Prantl, Pflanzenfam., ed. 2, 17b: 578 (1936); Lindberg f., Iter Cypr., 17 (1946); Davis, Fl. Turkey, 1: 477 (1965); Zohary, Fl. Palaest., 1: 262, t. 380 (1966); Osorio-Tafall et Seraphim, List Vasc. Plants Cyprus, 49 (1973).

TYPE: Spain; cultivated at Uppsala (LINN !).

An erect, rather robust annual or biennial, 10–30 cm. or more high, usually simple or sparingly branched in the lower half, much branched above, or occasionally unbranched or branched from the base; stems rather woody, distinctly angled, ± densely clothed with short, adpressed, bifurcate hairs; basal leaves linear-lanceolate, 30–70 mm. or more long, 2·5–15 mm. wide, apex acute, margins coarsely repand- or runcinate-dentate, occasionally subentire or entire, base tapering to a short, flattened petiole; lamina rather sparsely clothed on both sides with short, adpressed, bifurcate and compound hairs; upper leaves similar to the lower in shape and indumentum but smaller; inflorescence at first densely corymbose, later lengthening to form a rather lax raceme; flowers inconspicuous, shortly pedicillate; sepals narrowly oblong, concave, 4–6 mm. long, 0·7–1·2 mm. wide, adpressed-pilose externally, acute but scarcely horned at apex, the inner pair slightly saccate at base; petals small, yellow, 6–9 mm. long, 1–1·5 (–2) mm. wide, apex rounded, base tapering to a long, slender claw; stamens subequal, 5–7 mm. long; filaments linear, flattened; anthers yellow, narrowly oblong, with a shortly apiculate apex; ovary cylindrical about 5 mm. long, 0·7 mm. diam., adpressed-pilose; style about 0·6 mm. long, almost as broad as the ovary; stigma truncate or distinctly retuse. Siliques spreading, sometimes almost horizontal, 4-angled, often conspicuously torulose, 35–70 mm. long, 1·2–1·5 mm. wide, with a short, thickened pedicel; apex of silique narrowing rather abruptly to the persistent style and stigma; valves with a prominent midrib, sparingly adpressed-pilose, tardily dehiscent; seeds oblong, 1·3 mm. long, about 0·6 mm. wide; testa bright fulvous-brown, minutely granulate, unwinged.

HAB.: Vineyards, fallow fields and stony mountainsides on igneous formations; 4,000–6,400 ft. alt.; fl. May–July.

DISTR.: Division 2, locally common (Prodhromos, Khionistra, Troödos). Widespread in central and southern Europe and the Mediterranean region, extending eastwards to Baluchistan and N.W. India.

2. Locally common; *Kotschy* 711; *Sintenis & Rigo* 845 ! *Kennedy* 1233 ! *Lindberg f.* s.n. ! *Davis* 1812 ! *Casey* 679 ! etc.

TRIBE 9. **Sisymbrieae** *DC. emend. O. E. Schulz* Annual, biennial or perennial herbs with simple, bifurcate or variously branched hairs, sometimes glandular, occasionally glabrous. Sepals spreading or ascending, rarely erect; nectary glands more or less symmetrical, the lateral and median glands often contiguous; filaments linear, sometimes dilated towards the base, not toothed; anthers blunt; ovary sessile; style short or occasionally wanting; stigma truncate, capitate or ± 2-lobed. Fruit a silique or a silicle. Cotyledons incumbent, sometimes with an obliquely bent radicle.

39. ALLIARIA *Scop.*

Fl. Carniol., ed. 1, 515 (1760).

Annual, biennial or perennial herbs with an indumentum of simple hairs; leaves rather large, simple, petiolate; flowers relatively small; sepals spreading, blunt, the inner not saccate at base; petals white, shortly clawed; anthers obtuse, with an exserted connective; ovary sessile, cylindrical, 4–18-ovulate; style short; stigma truncate or capitate, scarcely 2-lobed. Fruit a ± 4-angled, dehiscent silique; valves with 3 longitudinal nerves. Seeds large, in a single row in each loculus; cotyledons incumbent, sometimes obliquely so.

Two species, one in Europe, N. Africa and western Asia, the other in western Asia.

1. A. petiolata (*M. Bieb.*) *Cavara et Grande* in Bol. Ort. Bot. Nap., 3: 418 (1913); Davis, Fl. Turkey, 1: 480 (1965); Osorio-Tafall et Seraphim, List Vasc. Plants Cyprus, 49 (1973).

Arabis petiolata M. Bieb., Fl. Taur.-Cauc., 2: 126 (1808).

Erysimum alliaria L., Sp. Plant., ed. 1, 660 (1753).

Sisymbrium alliaria (L.) Scop., Fl. Carniol., ed. 2, 2: 26 (1772).

Alliaria officinalis Andrz. ex M. Bieb., Fl. Taur.-Cauc., 3: 445 (1819); Unger et Kotschy, Die Insel Cypern, 330 (1865); Boiss., Fl. Orient., 1: 212 (1867); Holmboe, Veg. Cypr., 87 (1914); O. E. Schulz in Engl., Pflanzenr., 86 (IV, 105): 21 (1924); Post, Fl. Pal., ed. 2, 1: 67 (1932); Lindberg f., Iter Cypr., 17 (1946).

TYPE: Caucasus; " habitat in Iberia ", *Steven* (LE).

An erect, unbranched or sparingly branched, biennial (or occasionally short-lived perennial) herb; stem erect, 20–100 cm. high, terete or obscurely ridged, the basal part fistulose, thinly pilose, pruinose and often tinged violet; leaves rather large, thin, smelling of garlic when crushed; basal leaves with a channelled petiole up to 16 cm. or more long, reniform or suborbicular, 5–12 cm. or more wide, with a sinuate or bluntly toothed margin, often thinly pilose; lower stem-leaves glabrous or subglabrous, often larger than the basal leaves, more shortly petiolate, reniform or rather bluntly cordate, with a wide basal sinus and a dentate or crenate margin; upper stem-leaves smaller, shortly petiolate, acutely cordate, often terminating in an elongate apical tooth; inflorescence at first densely corymbose, later forming a lax raceme, the lower flowers often subtended by small bract-like leaves; pedicels 1–2·5 mm. long; sepals caducous, oblong or narrowly ovate, 2–3·5 mm. long, 1–1·5 mm. wide, glabrous, sometimes tinged violet; petals obovate, 4–6 mm. long, about 2 mm. wide, apex rounded, base tapering to a short claw; stamens 2·5–4·5 mm. long; filaments linear, flattened; anthers shortly oblong, about 0·7–1 mm. long, yellow; ovary cylindrical, about 4 mm. long. Siliques patent or ascending, 40–60 mm. long, about 2 mm. wide, with a short, stout pedicel about 3–5 mm. long; valves greenish-brown or violet-tinged, distinctly 3-nerved, glabrous; seeds oblong-rhomboid, narrowed at both ends, 2·5–3·5 mm. long, about 1–1·5 mm. wide; testa dark fuscous-brown, minutely papillose, with numerous, conspicuous longitudinal ridges.

HAB.: Shaded ground under trees at high altitudes on igneous formations; 3,000–6,000 ft. alt.; fl. April–July.

DISTR.: Division 2, local. Also throughout Europe and the Mediterranean region eastwards to the Himalayas and Baluchistan.

2. Local; Prodhromos, Milikouri, Polystipos etc.; *Kotschy* 789; *Holmboe* 909; *Lindberg f.* s.n. ! *Merton* 2425 ! 3120 ! *Pierides* in *Merton* 3165 ! *Meikle* 2876 !

40. SISYMBRIUM *L.*

Sp. Plant., ed. 1, 657 (1753).
Gen. Plant., ed. 5, 296 (1754).
O. E. Schulz in Engl., Pflanzenr., 86 (IV, 105): 46 (1924).

Annual, biennial or perennial herbs, sometimes with a woody base; stems and leaves glabrous or clothed with simple (or rarely compound) hairs; leaves usually lyrate-pinnatisect or pinnatipartite; inflorescence generally ebracteate; flowers rather small; sepals erecto-patent, not or only slightly saccate at base; petals yellow, rarely white or purplish, rounded at apex, ± clawed at base; stamens 6, filaments usually dilated towards base; anthers oblong, obtuse; ovary cylindrical, sessile; ovules numerous; style short; stigma capitate or obscurely 2-lobed. Siliques linear, elongate, dehiscent; valves generally 3-nerved. Seeds normally in one rank in each loculus, small, oblong or ellipsoid; cotyledons incumbent.

About 80 species, chiefly in temperate Europe, Asia and S. America; many species widely naturalized.

Flowers in elongate, leafless terminal racemes:
Siliques 3–10 cm. long, erecto-patent or spreading:
Siliques 3–5 (–6) cm. long; leaves and stems glabrous or subglabrous - **1. S. irio**
Siliques 6–10 cm. long; leaves and stems hairy - - - - - **2. S. orientale**
Siliques 1·1–1·8 cm. long, erect and adpressed to rhachis - - **4. S. officinale**
Flowers in small axillary clusters - - - - - - - **3. S. polyceratium**

1. S. irio *L.*, Sp. Plant., ed. 1, 659 (1753); Boiss., Fl. Orient., 1: 217 (1867); Holmboe, Veg. Cypr., 87 (1914); O. E. Schulz in Engl., Pflanzenr., 86 (IV, 105): 89 (1924); Post, Fl. Pal., ed. 2, 1: 71 (1932); Davis, Fl. Turkey, 1: 485 (1965); Zohary, Fl. Palaest., 1: 253, t. 366 (1966); Osorio-Tafall et Seraphim, List Vasc. Plants Cyprus, 49 (1973).

TYPE: " *in* Europae *cultis* " (LINN !).

An erect, much-branched annual, 10–60 cm. or more high, stems subterete, glabrous or very sparsely clothed with short, adpressed pubescence, often tinged violet; basal leaves glabrous or subglabrous, distinctly petiolate, oblong, 10–16 cm. or more long, 2–6 cm. wide, pinnatisect, with 2–6 pairs of oblong, acute or acuminate, coarsely toothed lobes, terminal lobe often conspicuously elongate; petiole flattened, 2–3 cm. or more long; upper leaves shortly petiolate, rather irregularly pinnatisect, often with elongate, toothed or entire lobes; inflorescence ebracteate, at first densely corymbose, later forming a rather lax raceme, the inconspicuous flowers soon overtopped, by the developing fruits; pedicels filiform, erecto-patent, 3–6 mm. long, lengthening considerably in fruit; sepals oblong, blunt, 2–2·5 mm. long, about 1 mm. wide, glabrous or thinly pubescent externally; petals yellow, 2·5–3·5 mm. long, about 1 mm. wide, base tapering to a short claw; stamens subequal, 2·5–3·5 mm. long; anthers yellow, narrowly oblong, about 0·7 mm. long; ovary narrowly cylindrical, about 2·5 mm. long, 40–80-ovulate; style very short, stigma capitate. Siliques erecto-patent or spreading, straight or slightly curved, narrowly linear, 30–50 (–60) mm. long, 0·7–1·3 mm. wide; valves pale greenish-brown, distinctly torulose, 3-nerved; seeds oblong, about 0·8 mm. long, 0·3 mm. wide; testa bright reddish-brown, minutely granulate.

HAB.: Cultivated and waste ground; sea-level to 600 ft. alt.; fl. Febr.–May.

DISTR.: Divisions 4, 6, 7. Throughout the Mediterranean region and eastwards to Central Asia and the borders of India; adventive elsewhere.

4. East of the town of Larnaca, 1880, *Sintenis & Rigo*; Ormidhia, 1880, *Sintenis & Rigo*; near Athienou, 1880, *Sintenis & Rigo* 274 ! Near Akhyritou, 1948, *Mavromoustakis* s.n. !

6. Nicosia hospital, 1934, *Syngrassides* 1387 ! Nicosia, 1938, *Nattrass & Pappaioannou* in *Nattrass* 913, and, 1950, *Chapman* 494 !

7. Kythrea, 1880, *Sintenis & Rigo*; Kyrenia, *Casey* 272 ! *G. E. Atherton* 952 ! 1189 !

NOTES: All the Cyprus material probably belongs to the hexaploid race, with long, slender siliques; the type is also probably hexaploid. (See T. N. Khoshoo in Nature, 176: 608; 1955.)

2. S. orientale *L.*, Cent. II. Plant., 24 (1756); O. E. Schulz in Engl., Pflanzenr., 86 (IV, 105): 122 (1924); Post, Fl. Pal., ed. 2, 1: 69 (1932); A. K. Jackson et Turrill in Kew Bull., 1939: 474 (1939); Rechinger f. in Arkiv för Bot., ser. 2, 1: 422 (1950); Davis, Fl. Turkey, 1: 484 (1965); Zohary, Fl. Palaest., 1: 254, t. 369 (1966); Osorio-Tafall et Seraphim, List Vasc. Plants Cyprus, 49 (1973).

S. columnae Jacq., Fl. Austr., 4: 12, t. 323 (1776); Boiss., Fl. Orient., 1: 216 (1867).

TYPE: " *in* Oriente *Miller*."

An erect, rather sparsely branched annual 30–100 cm. or more high; stems subterete, rather flexuous, often tinged purple, clothed (like the leaves) with soft, spreading hairs; lower leaves forming a loose rosette, distinctly petiolate, 6–10 (–13) cm. or more long, 1·5–5 cm. or more wide, lyrate-pinnatisect or pinnatipartite with 4–5 pairs of acute or acuminate, oblong, toothed lateral lobes, and a broad oblong or deltoid, toothed terminal lobe; petiole channelled, 2–4 cm. or more long; upper leaves smaller, more shortly petiolate, with 2–3 pairs of lateral lobes and an elongate terminal lobe, or sometimes hastate or sagittate, with 2 spreading or deflexed lateral lobes; inflorescence at first corymbose, later forming a lax raceme, the flowers soon overtopped by the developing fruits; pedicels slender, 2–5 mm. long, lengthening and thickening considerably in fruit; sepals erecto-patent, narrowly oblong, obtuse, 4–5 mm. long, about 1–1·5 mm. wide, glabrous or pilose externally, with a wide membranous margin; petals yellow becoming white with age, 6–9 mm. long, 1·5–3 mm. wide, tapering gradually to a slender basal claw; stamens distinctly unequal, 3·5–6·5 mm. long; anthers yellow, narrowly oblong, about 2 mm. long; ovary sessile, narrowly cylindrical, pilose, about 4 mm. long, 80–100-ovulate; style short; stigma rather large, ± distinctly 2-lobed. Siliques erecto-patent, straight or slightly curved, subterete or obscurely 4-angled, 60–100 mm. long, 1·5–2 mm. wide; valves rather thick, longitudinally veined, pale or dark brown, glabrous or ± pilose; seeds rather shortly oblong, about 1·2 mm. long and 1 mm. wide; testa reddish-brown, minutely granulate.

HAB.: Cultivated and waste ground, roadsides etc.; sea-level to 4,000 ft. alt.; fl. April–Aug.

DISTR.: Divisions 2, 3, 7, common in Division 2. Throughout the Mediterranean region and eastwards to Afghanistan; widely naturalized elsewhere.

2. Common (especially around Platres); *Kennedy* 1232 ! 1822 ! *Davis* 3174 ! *Casey* 930 ! *Merton* 2105 !
3. Limassol, 1930, *Wall* s.n.
7. Kyrenia, 1950, *Casey* 1024 ! 1065 ! Kyrenia, Tchingen, 1955, *G. E. Atherton* 304 !

3. S. polyceratium *L.*, Sp. Plant., ed. 1, 658 (1753); Boiss., Fl. Orient., 1: 220 (1867); O. E. Schulz in Engl., Pflanzenr., 86 (IV, 105): 128 (1924); A. K. Jackson in Kew Bull., 1937: 341 (1937); Davis, Fl. Turkey, 1: 482 (1965); Osorio-Tafall et Seraphim, List Vasc. Plants Cyprus, 49 (1973).
TYPE: " *in* Helvetiae, Italiae *ruderatis* (LINN !).

An erect, spreading or almost prostrate, malodorous annual herb; stems 10–50 cm. or more long, commonly much-branched at the base, obscurely angled, usually glabrous and shining, sometimes thinly pilose; basal leaves forming a distinct rosette, petiolate, narrowly oblong, 5–12 cm. or more long, 1·5–3·5 cm. wide, coarsely runcinate-pinnatifid, with an acute deltoid apex and acute, entire or remotely toothed lateral lobes; petiole flattened, 2–5 cm. or more long; stem leaves smaller, shortly petiolate, hastate or with 1–2 pairs of lateral lobes; uppermost leaves small, usually simple and often subentire, bract-like; inflorescence axillary, consisting of a small cluster of minute, subsessile flowers, or sometimes of a solitary flower; sepals erecto-patent, ± hispid at apex, the outer narrowly oblong, 1·5–1·8 mm. long, 0·5 mm. wide, the inner broader, about 1 mm. wide; petals yellow, obovate, about as long as the sepals, tapering to a short basal claw; stamens subequal, 1·5–1·8 mm. long; anthers narrowly oblong, yellow, about 0·4 mm. long; ovary cylindrical, about 1·8 mm. long and 0·5 mm. diam., 40–55-ovulate; style very short; stigma truncate-capitate. Siliques in small clusters of 1–3 (–5), with short, stout pedicels, erect or spreading, usually curved or even falcate, about 12–18 mm. long, 0·7 mm. diam., gradually tapering from the base to the short, style-crowned apex; valves pale brown, convex, obscurely 3-nerved, frequently remaining attached along the base after dehiscence; seeds shortly oblong, 0·8 mm. long, 0·5 mm. wide; testa bright brown, minutely granulate.

HAB.: Cultivated and waste ground, roadsides etc.; sea-level to 200 ft. alt.; fl. Febr.–May.

DISTR.: Divisions 2, 3, rare. Common in the Mediterranean region, but becoming rarer in the east, and largely replaced by the closely related *S. runcinatum* Lag. in Asia.

2. Kato Yialia, 1934, *Syngrassides* 853 !
3. Limassol, 1930, *Wall* s.n.

4. **S. officinale** (*L.*) *Scop.*, Fl. Carniol., ed. 2, 2: 26 (1772); Boiss., Fl. Orient., 1: 220 (1867); O. E. Schulz in Engl., Pflanzenr., 86 (IV, 105): 137 (1924); Post, Fl. Pal., ed. 2, 1: 71 (1932); Lindberg f., Iter Cypr., 17 (1946); Davis, Fl. Turkey, 1: 483 (1965); Zohary, Fl. Palaest., 1: 255, t. 371 (1966); Osorio-Tafall et Seraphim, List Vasc. Plants Cyprus, 49 (1973).
Erysimum officinale L., Sp. Plant., ed. 1, 660 (1753).
TYPE: " *in* Europae *ruderatis* ".

An erect, thinly hispid-pilose annual or biennial herb, 20–90 cm. or more high, with a stout taproot; stems subterete, often stained purple, with numerous spreading branches, particularly in the upper half of the plant; basal leaves forming an overwintering rosette, oblong or obovate, petiolate, 8–20 (–30) cm. long, pinnatisect, with a bluntly deltoid or oblong, dentate terminal lobe, and 2–4 pairs of oblong, acute or obtuse, irregularly toothed lateral lobes; petiole flattened, 2–10 cm. or more long; stem leaves rather similar to the basal leaves but smaller and more shortly petiolate, the uppermost often subsessile, narrowly oblong, and irregularly hastate-lobed; inflorescence ebracteate, terminal, at first densely corymbose, later forming a narrow, elongate raceme; flowers inconspicuous, pedicels about 1–1·5 mm. long; sepals erecto-patent, shortly oblong, about 1·5–2 mm. long and 0·8 mm. wide, pilose externally, apex subacute; petals yellow, rather unequal, obovate, with a rounded or retuse apex and a shortly clawed base, the longer pair about 2·5–2·8 mm. long, 1–1·2 mm. wide, the shorter pair about 2 mm. long and 1 mm. wide; stamens unequal, 1·5–2 mm. long, filaments rather dilated towards base; anthers yellow, minute, shortly oblong or almost orbicular; ovary narrowly flask-shaped or almost cylindrical, about 1·5 mm. long and 0·3 mm. diam., thinly covered with long hairs; style obscure; stigma capitate. Siliques with short stout pedicels, erect, adpressed to rhachis, straight, rigid, 11–18 mm. long, about 1·5 mm diam. at base, tapering to a sharply pointed apex, thinly pilose or occasionally (var. *leiocarpum* DC.) glabrous, obscurely nerved, rather tardily dehiscent; seeds rather irregularly shaped, oblong and pointed, rounded or obliquely truncate at one or both ends, about 1–1·6 mm. long, 0·7–1 mm. wide; testa rich brown, minutely granulate.

HAB.: Waste ground, roadsides, stone terraces etc.; sea-level to 4,000 ft. alt.; fl. April–July.

DISTR.: Divisions 2, 7. Europe, N.W. Africa, Caucasus, N.W. Asia. Naturalized in many temperate parts of both hemispheres.

2. Moutoullas, 1937, *Syngrassides* 1556 ! Stavros tis Psokas, 1937, *Syngrassides* 1586 ! Kambos, 1937, *Lindberg f.* s.n.
7. Ayios Epiktitos, 1880, *Sintenis & Rigo* 257 !

NOTES: All the localized specimens belong to the hairy-fruited, typical *S. officinale* (L.) Scop. var. *officinale*, but the glabrous var. *leiocarpum* DC. has once been seen (comm. *B. Trenbath*) in an unlocalized Cyprus collecting.

41. TORULARIA (*Cosson*) *O. E. Schulz*

in Engl., Pflanzenr., 86 (IV, 105): 213 (1924).
Sisymbrium L. sect. *Torularia* Cosson, Compend. Fl. Atlant., 136, 139 (1882).

Erect or spreading annual, biennial and perennial herbs; stems and leaves clothed with simple or branched hairs; leaves usually runcinate-dentate or pinnatifid; inflorescence sometimes bracteate, flowers small; sepals caducous, suberect or erecto-patent, oblong, rather obtuse at apex, not saccate at base; petals sometimes wanting, when present white or purplish, spathulate or obovate, with a truncate or retuse apex; stamens 6; filaments slender; anthers oblong or ovoid, obtuse; lateral nectary glands in pairs, ± ovoid or globose, median glands wanting; ovary cylindrical, sessile, 5–35-ovulate; style short or obsolete; stigma flattened or slightly 2-lobed. Siliques linear, usually subterete, straight or curved, rather tardily dehiscent; valves with a distinct midrib and obscure lateral nerves. Seeds in one rank in each loculus; cotyledons incumbent, sometimes obliquely so.

About 12 species, chiefly in Central Asia.

1. **T. torulosa** (*Desf.*) *O. E. Schulz* in Engl., Pflanzenr., 86 (IV, 105): 214 (1924); Engl. et Prantl, Pflanzenfam., ed. 2, 17b: 628 (1936); Davis, Fl. Turkey, 1: 488 (1965); Zohary, Fl. Palaest., 1: 267, t. 390 (1966); Osorio-Tafall et Seraphim, List Vasc. Plants Cyprus, 49 (1973).

Sisymbrium torulosum Desf., Fl. Atlant., 2: 84, t. 159 (1798); Unger et Kotschy, Die Insel Cypern, 330 (1865); Holmboe, Veg. Cypr., 87 (1914).

Malcolmia torulosa (Desf.) Boiss., Fl. Orient., 1: 225 (1867); Post, Fl. Pal., ed. 2, 1: 74 (1932).

TYPE: Tunisia; "in arvis incultis prope Sbibam [Sbiba] in regno Tunetano" *Desfontaines* (P).

A spreading annual, 3–15 (–30) cm. high, much branched at the base; stems subterete usually clothed with simple, patent bristles intermixed with shorter, spreading, bifurcate bristles; basal leaves sparingly hispid-pilose, forming a loose rosette, linear-lanceolate, 20–50 (–70) mm. long, 2–10 mm. wide, often rather regularly pinnatifid, with sharp forward-pointing lobes, base of lamina merging with a short flattened petiole; cauline leaves similar, but usually rather smaller and sessile or subsessile; inflorescence ebracteate, at first densely corymbose, later forming a lax, elongate raceme; flowers small; pedicel short, stout, less than 1 mm. long; sepals oblong, obtuse, hispid externally and slightly cucullate at apex, about 2 mm. long and 0·8 mm. wide; petals white, about 2·3 mm. long and 0·7 mm. wide, tapering from a truncate apex to the base, but scarcely clawed; stamens unequal, 1·5–2 mm. long; filaments flattened; anthers yellow, ovoid, about 0·6 mm. long, the connective exserted and forming a short apiculus; ovary cylindrical, about 2 mm. long, 0·8 mm. diam.; style obsolete; stigma capitate. Siliques erecto-patent, rigid, narrowly cylindrical, about 15–20 mm. long and 1·3 mm. diam., straight or circinnately curved, tapering abruptly at apex; valves pale or dark purplish-brown, distinctly torulose, hispid-pilose or sometimes glabrous (var. *scorpiuroides* (Boiss.) O. E. Schulz); seeds shortly oblong, compressed, about 1 mm. long and 0·8 mm. wide; testa mid-brown, minutely granulate-papillose, with a blackish stripe marking the contiguous margins of the cotyledons.

HAB.: Cultivated and waste ground, or bare hillsides, frequent on Kythrean marls; 200–1,000 ft. alt.; fl. Febr.–April.

DISTR.: Divisions 3, 5–8, common in Division 5 on Kythrean marls. Widely distributed in N. Africa and the eastern Mediterranean region eastwards to N.W. India; adventive elsewhere.

3. Alethriko, 1905, *Holmboe* 248.

5. Common (especially in the Mia Milea area); *Sintenis & Rigo* 264 ! *Syngrassides* 315 ! 1132 ! *Chapman* 477 ! *Casey* 1327 ! *Meikle* 2969 !
6. On the way to Kaimakli from Nicosia, 1934, *Syngrassides* 1348 ! Near Philia, 1956, *Merton* 2756 !
7. Lakovounara Forest, 1950, *Chapman* 143 !
8. Near Yialousa, 1880, *Sintenis & Rigo*; Ovgoros, 1941, *Davis* 2478 !

NOTES: All the Cyprus material belongs to the hispid-fruited var. *torulosa*.

42. ARABIDOPSIS (*DC.*) *Heynh.*

in Holl et Heynh., Fl. Sachsen, 1: 538 (1842).

Sisymbrium L. sect. *Arabidopsis* DC., Regn. Veg. Syst. Nat. 2: 480 (1821).

Slender, erect annual, biennial and perennial herbs, clothed with simple and branched hairs; leaves oblong, the basal petiolate, the cauline sessile or amplexicaul, entire, toothed or lyrate-pinnatifid. Inflorescence sometimes bracteate; flowers small; sepals erect or spreading, oblong, sometimes slightly saccate at base; petals white or purplish (occasionally yellow), obovate-cuneate or spathulate, apex rounded or truncate; stamens 6, rarely 5 or 4; filaments slender; anthers oblong or ovoid, obtuse; median nectary glands present, sometimes joined with the lateral; ovary sessile or shortly stipitate, narrowly cylindrical, 20–72-ovulate; style short, rather thick; stigma truncate or capitate, rarely ± 2-lobed. Siliques with slender pedicels, subterete, midrib distinct, lateral nerves obscure and ± reticulate. Seeds in 1 or 2 rows in each loculus, ovoid; cotyledons incumbent, sometimes obliquely so.

About 13 species in Europe, Asia and Africa; naturalized elsewhere.

1. **A. thaliana** (*L.*) *Heynh.* in Holl et Heynh., Fl. Sachsen, 1: 538 (1842); Post, Fl. Pal., ed. 2, 1: 68 (1932); Davis, Fl. Turkey, 1: 489 (1965); Osorio-Tafall et Seraphim, List Vasc. Plants Cyprus, 49 (1973).

Arabis thaliana L., Sp. Plant., ed. 1, 665 (1753); Sibth. et Sm., Fl. Graec. Prodr., 2: 28 (1813); Unger et Kotschy, Die Insel Cypern, 326 (1865).

Sisymbrium thalianum (L.) J. Gay in Ann. Sci. Nat., ser. 1, 7: 399 (1826) in sched.; Boiss., Fl. Orient., 1: 214 (1867).

Stenophragma thalianum (L.) Čelak., Květ. Ok. Pražs., 75 (1870); Holmboe, Veg. Cypr., 87 (1914).

TYPE: "*in* Europae *septentrionalioris sabulosis*".

Slender, erect, branched or unbranched annual, 4–40 cm. high; stems slightly sulcate, glabrous or clothed with spreading, simple hairs, and often tinged violet below, glaucous and ± pruinose above; leaves rather densely clothed with stipitate, bifurcate hairs, occasionally subglabrous; basal leaves forming a distinct rosette, obovate or spathulate, 4–30 mm. or more long, 2–7 mm. or more wide, apex obtuse or rounded, margins entire or remotely and irregularly dentate, base tapering to a short, flattened petiole; cauline leaves few and remote, oblong, sessile, seldom more than 10 mm. long and 5 mm. wide, apex acute or obtuse, margins usually entire; inflorescence ebracteate, branched or unbranched, at first densely corymbose, later lengthening to form an elongate, lax raceme; inflorescence-branches erect or ascending; flowers very small and inconspicuous; sepals oblong, about 1·7 mm. long and 0·8 mm. wide, apex hispid-pilose, obtuse or subacute; inner pair of sepals slightly saccate at base; petals narrowly obovate, about 2·3 mm. long, 0·8 mm. wide, apex rounded, tapering towards the base, but not clawed; stamens usually 4 (occasionally 5–6), subequal, 1·3–1·5 mm. long; filaments linear, flattened; anthers ovate, obtuse, yellow, about 0·3 mm. long; ovary cylindrical, glabrous, about 1·8 mm. long, 0·4 mm. diam.; style short, stout; stigma truncate or capitate. Siliques linear, compressed, 8–15 (–20) mm. long, about 1 mm. wide, erecto-patent or spreading, on filiform pedicels 4–10 mm. or more long, apex of silique rather abruptly narrowed to the persistent style and stigma; valves glabrous, with a rather prominent midrib, lateral nerves very obscure; seeds in a single row in each loculus, minute, ovoid, about 0·4 mm. long and 0·3 mm. wide; testa light brown, minutely and densely granulate.

HAB.: Vineyards, roadsides, stony slopes etc., usually on igneous formations; 500–4,000 ft. alt.; fl. Febr.–May.

DISTR.: Divisions 2, 3, common at middle altitudes in Division 2. Widespread in Europe and temperate Asia; N. Africa and Atlantic Is.; introduced elsewhere.

2. Common; *Kennedy* 466 ! 1231 ! *Davis* 2739 ! 3093 ! *Merton* 3133 ! *Meikle* 2823c ! 2857 !
3. Stavrovouni, 1862, *Kotschy* 209; Limassol–Nicosia Rd., Mile 24, 1956, *Merton* 2479 !

43. CAMELINA *Crantz*

Stirp. Austr., ed. 1, fasc. 1: 17 (1762).

Annual or biennial herbs, stems and leaves glabrous or clothed with simple or branched hairs; leaves lanceolate, oblanceolate or spathulate, entire or variously toothed or lobed; cauline leaves sessile often auriculate; inflorescence ebracteate, flowers small; sepals erect, obtuse, the inner sometimes slightly saccate at base; petals yellow or white, ligulate or narrowly spathulate; stamens 6; filaments linear; anthers ovate, obtuse; ovary elliptic or obovate, 8–25-ovulate; style distinct; stigma capitate. Silicle inversely pyriform, obovoid or subglobose, ± inflated, dehiscent, crowned with the persistent style and stigma, midrib ± prominent, lateral nerves usually obscure. Seeds elongate-ellipsoid or oblong, occasionally flattened; cotyledons incumbent, or occasionally ± accumbent.

About 10 species in Europe, temperate Asia and N. Africa; often occurring as adventives elsewhere.

1. C. rumelica *Velenovský* in Sitz. Kgl. Boehm. Ges., 1887: 448 (1887); Fl. Bulg., 43 (1891); O. E. Schulz in Engl. et Prantl, Pflanzenfam., ed. 2, 17b: 648 (1936); Davis, Fl. Turkey, 1: 491 (1965).

C. sylvestris Wallr. var. *albiflora* Boiss., Fl. Orient., 1: 312 (1867).

C. microcarpa Andrzj. var. *albiflora* (Boiss.) Bornm. in Beih. Bot. Centralbl., 31, pt. 2: 187 (1914); Post, Fl. Pal., ed. 2, 1: 95 (1932).

[*Alyssum sativum* (non (L.) Scop.) Sibth. et Sm., Fl. Graec. Prodr., 2: 15 (1813).]

[*Camelina sativa* (non (L.) Crantz) Kotschy in Unger et Kotschy, Die Insel Cypern, 330 (1865); Holmboe, Veg. Cypr., 90 (1914); Druce in Proc. Linn. Soc. Session 141: 51 (1930) and in Rep. B.E.C., 9: 469 (1931).]

[*C. sylvestris* (non Wallr.) A. K. Jackson et Turrill in Kew Bull., 1939: 474 (1939).]

TYPE: Bulgaria; "In campestribus prope Sliven", *Velenovský*.

An erect annual, 14–30 (–60) cm. high, usually with several spreading or ascending branches; lower half of plant ± densely hispidulous, with numerous long, simple, bristly hairs intermixed with a few shorter, branched hairs; upper half of plant glabrous or subglabrous; basal leaves forming a distinct rosette, usually persisting until anthesis or even later, oblanceolate or spathulate, 1–10 cm. or more long, 0·5–2 cm. wide, apex obtuse or rounded, margin entire or denticulate, base narrowed gradually or abruptly to a short, flattened petiole; cauline leaves usually much smaller, generally 4 cm. or less in length, sessile, lanceolate, tapering to an acute apex, base with 2 small, pointed auricles, margins entire or denticulate; inflorescence at first densely corymbose, lengthening in fruit to form a lax raceme; flowers pedicellate; pedicels slender, about 3–5 mm. long; sepals sparingly clothed with long, slender, arachnoid hairs, the outer pair narrowly ovate-oblong, about 4 mm. long and 1·2 mm. wide, the inner pair narrowly oblong, obtuse, about 4 mm. long and 1 mm. wide, shortly saccate at base; petals whitish or pale yellow, rather distinctly veined, narrowly spathulate or almost ligulate, 5–7 (–9) mm.

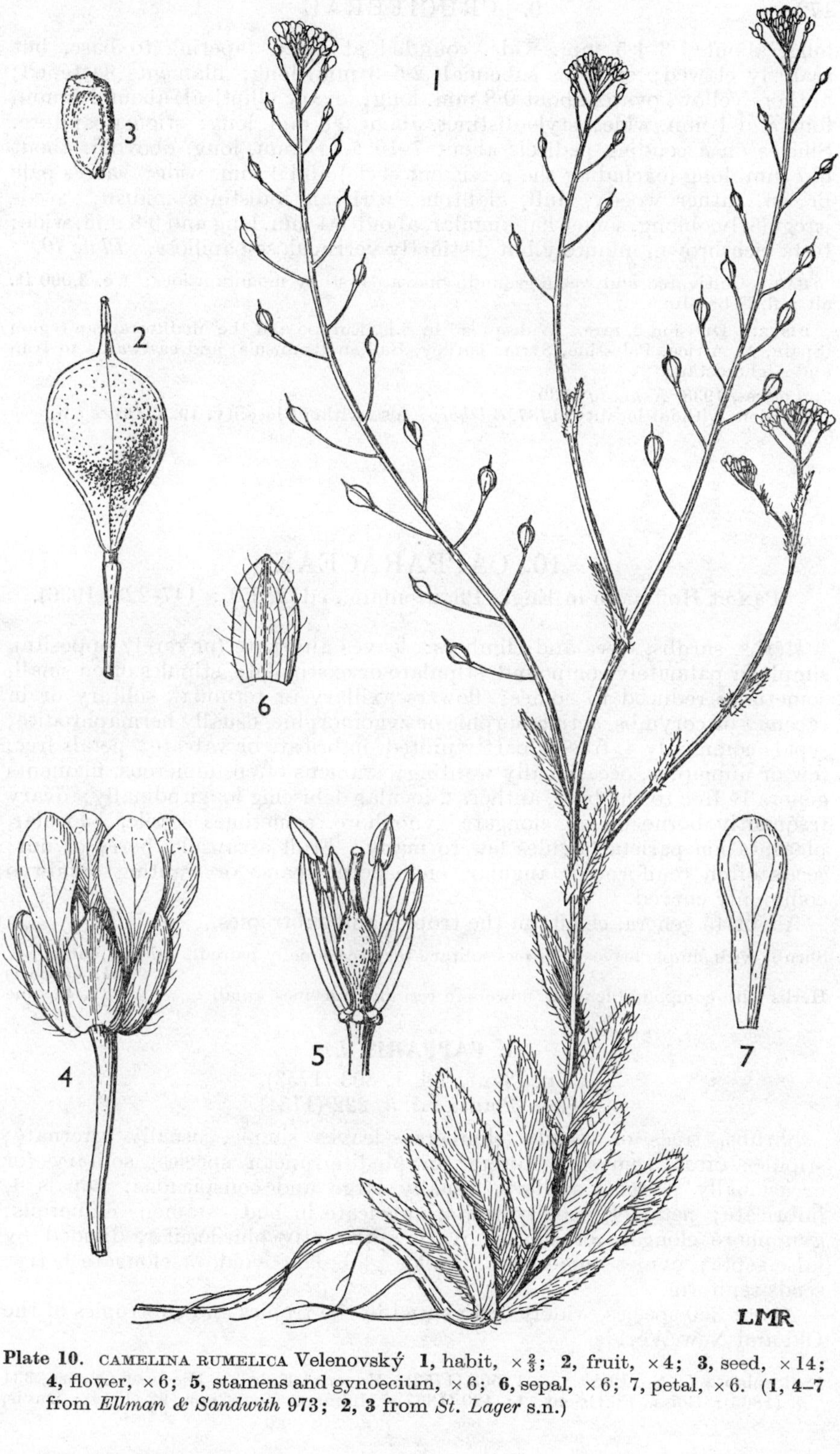

Plate 10. CAMELINA RUMELICA Velenovský **1**, habit, ×⅔; **2**, fruit, ×4; **3**, seed, ×14; **4**, flower, ×6; **5**, stamens and gynoecium, ×6; **6**, sepal, ×6; **7**, petal, ×6. (**1, 4–7** from *Ellman & Sandwith* 973; **2, 3** from *St. Lager* s.n.)

long, about 1·3–1·5 mm. wide, rounded at apex, tapering to base, but scarcely clawed; stamens subequal, 2·5–3 mm. long; filaments flattened; anthers yellow, ovate, about 0·8 mm. long; ovary elliptical, about 1·5 mm. long and 1 mm. wide; style distinct, about 0·8 mm. long; stigma capitate. Silicles on ascending pedicels about 7–10 (–14) mm. long, obovoid, about 6–7 mm. long (excluding the persistent style) and 4 mm. wide; valves pale brown, rather woody, dull, glabrous, with an indistinct midrib; seeds irregularly oblong, somewhat angular, about 1·4 mm. long and 0·8 mm. wide; testa rich brown, minutely but distinctly verruculose-papillose. *Plate 10.*

HAB.: Cultivated and waste ground, vineyards, stony mountainsides; ? c. 3,000 ft. alt.; fl. Febr.–June.

DISTR.: Division 2, rare. Widespread in S.E. Europe and the Mediterranean region (Spain, N. Africa, Palestine, Syria, Turkey, Balkan Peninsula) and eastwards to Iran and Afghanistan.

2. Platres, 1938, *Kennedy* 1239 !

? Division, without locality, 1787, *Sibthorp*; also without locality, 1928, *Druce* s.n.

10. CAPPARACEAE

Pax et Hoffmann in Engl., Pflanzenfam., ed. 2, 17b: 147–223 (1936).

Herbs, shrubs, trees and climbers; leaves alternate (or rarely opposite), simple or palmately compound, stipulate or exstipulate, stipules often small, sometimes reduced to spines; flowers axillary or terminal, solitary or in racemes or corymbs, actinomorphic or zygomorphic, usually hermaphrodite; sepals commonly 4, free or partly united, imbricate or valvate; petals free, few or numerous, occasionally wanting; stamens often numerous, filaments generally free to the base, anthers 2-locular dehiscing longitudinally; ovary frequently borne on an elongate gynophore, sometimes sessile, 1-locular, placentation parietal, ovules few to many. Fruit a capsule, berry or nut; seeds often reniform or angular, endosperm sparse or wanting; embryo commonly curved.

About 45 genera, chiefly in the tropics and subtropics.

Shrubs with simple leaves; flowers solitary (or occasionally paired), axillary, large **1. Capparis**

Herbs with compound leaves; flowers in terminal racemes, small - - **2. Cleome**

1. CAPPARIS *L.*

Sp. Plant., ed. 1, 503 (1753).
Gen. Plant., ed. 5, 222 (1754).

Shrubs, trees or woody climbers; leaves simple, usually alternate; stipules often spinose; flowers (in Mediterranean species) solitary (or occasionally paired), axillary, usually large and conspicuous; sepals 4, imbricate; petals 4, usually white, imbricate in bud; stamens numerous; gynophore elongate or short; ovary apparently plurilocular, divided by false septa; ovules numerous. Fruit a globular, ovoid, or elongate berry; seeds reniform.

About 350 species, widely distributed in the tropics and subtropics of the Old and New Worlds.

1. **C. spinosa** *L.*, Sp. Plant., ed. 1, 503 (1753); Unger et Kotschy, Die Insel Cypern, 334 (1865); Boiss., Fl. Orient., 1: 420 (1867); Holmboe, Veg. Cypr., 92 (1914); Davis,

Fl. Turkey, 1: 496 (1965); Zohary, Fl. Palaest., 1: 242, t. 358 (1966); Osorio-Tafall et Seraphim, List Vasc. Plants Cyprus, 45 (1973); Elektra Megaw, Wild Flowers of Cyprus, 8, t. 6 (1973).

C. ovata Desf., Fl. Atlant., 1: 404 (1798); Davis, Fl. Turkey, 1: 497 (1965); Osorio-Tafall et Seraphim, List Vasc. Plants Cyprus, 45 (1973).

A sprawling or ascending, much-branched shrub to about 5 feet high; stems terete, straight or flexuous, glaucous, often tinged purple, glabrous or tomentose, especially towards apex; leaves ovate, elliptical, obovate or almost orbicular, 1–5 cm. or more long, 0·8–5 cm. or more wide, acute, obtuse or slightly emarginate at apex, narrowed, truncate or occasionally subcordate at base; margin entire; midrib usually exserted as a short cusp or spine, lamina glabrous or glabrescent, sometimes thinly tomentellous, green or glaucous; petiole 0·5–1·5 cm. or more long, normally flanked at the base by 2 short, straight or falcate, yellow-brown stipular prickles; flowers produced at intervals along the stems; peduncles stout, usually rather longer than the subtending leaves; sepals glabrous, glabrescent or thinly tomentellous, the two outer sepals deeply concave, about 1·5–2 cm. long and 1–1·5 cm. wide, apex obtuse or subacute, inner pair rather smaller, oblong, obtuse, not markedly concave, about 1–1·8 cm. long and 0·8–1 cm. wide; petals white or tinged mauve or pink, broadly obovate or oblong, 1·5–3 cm. or more long, 1–2 cm. wide, obtuse at apex, narrowed to base; stamens very numerous, filaments usually longer than the petals, up to 5 cm. long, often tinged mauve or pink especially towards apex; anthers violet, narrowly oblong, about 2 mm. long and 1 mm. wide; gynophore 1·5–2·5 cm. or more long, lengthening in fruit; ovary cylindrical, about 5 mm. long and 1·5 mm. wide; stigma sessile, capitate. Fruit ovoid or ellipsoid, about 2·5–4 cm. long and 1·5–2·5 cm. wide; epicarp rather leathery, distinctly 5–6-ribbed, greenish, turning reddish when fully ripe; seeds numerous, reniform, about 3 mm. long and 2·5 mm. wide; testa smooth, rather dull, chocolate-brown.

var. **canescens** *Cosson*, Notes Plant. Nouv., fasc. 2: 28 (1849); Boiss., Fl. Orient., 1: 420 (1867).

C. sicula Duham., Traité des Arbres, ed. 2, 1: 159 (1800–1801); Post, Fl. Pal., ed. 2, 1: 135 (1932); Lindberg f., Iter Cypr., 17 (1946).

C. spinosa L. ssp. *sicula* (Duham.) Holmboe, Veg. Cypr., 92 (1914).

C. ovata Desf. var. *canescens* (Coss.) Heywood in Feddes Repert., 69: 56 (1964); Davis, Fl. Turkey, 1: 497 (1965).

TYPE: Sicily; " . . . frequemment aux environs d'Agrigente ", in herb. Duhamel (? P).

Branches (especially the young ones) often tomentose; leaves rather small, about 1·5–4 cm. long, ovate or elliptical, spine-tipped; stipular spines usually stout, falcate.

HAB.: Dry fields, waste land and rocky slopes; sea-level to 2,800 ft. alt.; fl. May–Aug.

DISTR.: Divisions 1–5, 7, common in Divisions 2 and 7. Widespread in the Mediterranean region.

1. Paphos, 1905, *Holmboe* 720, and 1913, *Haradjian* 623 !
2. Perapedhi, Platres; *Kennedy* 1039–1043 !
3. Limassol, 1801, *Hume* s.n. and, 1862, *Kotschy*; above Skarinou, 1941, *Davis* 3592a !
4. Larnaca, 1801, *Hume* s.n.; near Larnaca, 1880, *Sintenis* 785 !
5. Alambra, 1932, *Syngrassides* 384 ! Athalassa, 1939, *Lindberg f.* s.n. !
7. Common (especially around Kyrenia); *Mapple* 41 ! *G. E. Atherton* 59 ! 214 ! 395 !

2. CLEOME *L.*

Sp. Plant., ed. 1, 671 (1753).
Gen. Plant., ed. 5, 302 (1754).

Herbs or subshrubs; leaves usually compound, palmately 3–7-foliolate, occasionally simple, entire or variously toothed, stipulate or exstipulate; flowers usually in terminal, bracteate racemes; sepals 4, partly united;

petals 4, often coloured, convolute, imbricate or open in bud; stamens 4–6; gynophore often short; ovary 1-locular, ovules numerous. Fruit a sessile or stipitate, usually elongate, 2-valved capsule; seeds reniform or subglobose.

About 200 species, mostly in the tropical or subtropical regions of the Old and New Worlds; 2 species in Cyprus.

Sepals lanceolate, acuminate, ± densely clothed with stalked glands; pedicels glandular; plant erect with rather rigid, erecto-patent branches - **1. C. ornithopodioides**
Sepals oblong, obtuse or cuspidate, glabrous or with a very few sessile glands; pedicels ± glabrous; plant with slender, spreading or ascending branches **2. C. iberica**

1. C. ornithopodioides *L.*, Sp. Plant., ed. 1, 672 (1753); Boiss., Fl. Orient., 1: 411 (1867) pro parte, Fl. Orient. Suppl., 68 (1888) pro parte quoad pl. cypr.; Post, Fl. Pal., ed. 2, 1: 133 (1932); Davis, Fl. Turkey, 1: 498 (1965) pro parte.
C. cypria Čelak. in Oesterr. Bot. Zeitschr., 34: 114 (1884).
C. ornithopodioides L. ssp. *cypria* (Čelak.) Holmboe, Veg. Cypr., 92 (1914); Lindberg f., Iter Cypr., 17 (1946); Osorio-Tafall et Seraphim, List Vasc. Plants Cyprus, 45 (1973).
TYPE: Turkey; Constantinople, " circa Peram in agris " (LINN !).

An erect, much-branched annual, 7–90 cm. high, with a stout taproot; branches erecto-patent, rather rigid, striate, densely clothed with short stalked glands; leaves also glandular, exstipulate, the basal ones trifoliolate; leaflets narrowly oblong, oblanceolate or linear-oblong, 10–35 mm. or more long, 3–7 mm. or more wide, margins entire, obtuse or subacute at apex, tapering rather abruptly to the base; petiole 1–3 cm. long, slender, flattened or slightly channelled above; cauline leaves similar to basal leaves, often increasing in size upwards to the middle of the stems, and then diminishing rather suddenly towards the base of the inflorescence; inflorescence a lax raceme, lengthening greatly in fruit; flowers subtended by small, simple or trifoliolate, leaf-like bracts; pedicels filiform, glandular, about 8–10 (–15) mm. long; disk inconspicuous; sepals lanceolate, or acuminate, clothed with stalked glands, unequal, one pair 1·5–2·5 mm. long and about 0·8–1 mm. wide, the other pair 1–2 mm. long and about 0·8 mm. wide, free almost to base; petals white or yellowish, often with a central crimson stripe, becoming reddish all over with age, broadly obovate, clawed, about 2·5–4 mm. long, 1·5–2·5 mm. wide; claw about 1 mm. long; petal margin entire; stamens 6; filaments often pinkish or reddish, to about 5 mm. long; anthers yellow, oblong, about 0·8 mm. long, 0·4 mm. wide; ovary shortly cylindrical, about 0·8 mm. long, narrowing abruptly at apex; style obsolete, stigma truncate. Fruit pendulous, cylindrical, distinctly torulose, 20–30 mm. long, 2·5–3 mm. wide, base tapering to the apex of the pedicel or sometimes borne on a short gynophore, apex with a short, slender beak; valves ± densely glandular, longitudinally striate with anastomosing veins; seeds globose or slightly reniform, about 1·5 mm. diam.; testa dark brown (sometimes mottled), densely and conspicuously papillose. *Plate 11, figs. 1–5.*

HAB.: Vineyards, river-beds and stony gullies; (500–) 1,300–4,000 ft. alt.; fl. June–Sept.

DISTR.: Divisions 2, 6, common in Division 2. Also in Thrace, Turkey and the Caucasus; perhaps also in Syria and Lebanon.

2. Galata, Troödos, Platres, Mandria; *Sintenis & Rigo* 812 ! *A. G. & M. E. Lascelles* s.n. ! *Kennedy* 516–519 ! *Merton* 2818 !

6. Nicosia, Pedieos River, 1935, *Syngrassides* 652 !

NOTES: In view of the confused history of this species, only those specimens which have been examined are cited above.

2. C. iberica *DC.*, Prodr., 1: 240 (1824).
[*C. ornithopodioides* (non L.) Boiss., Fl. Orient., 1: 411 (1867) pro parte; Davis, Fl. Turkey, 1: 498 (1965) pro parte.]
TYPE: Caucasus; " In Iberia circa Tiflin [Tiflis] " (G).

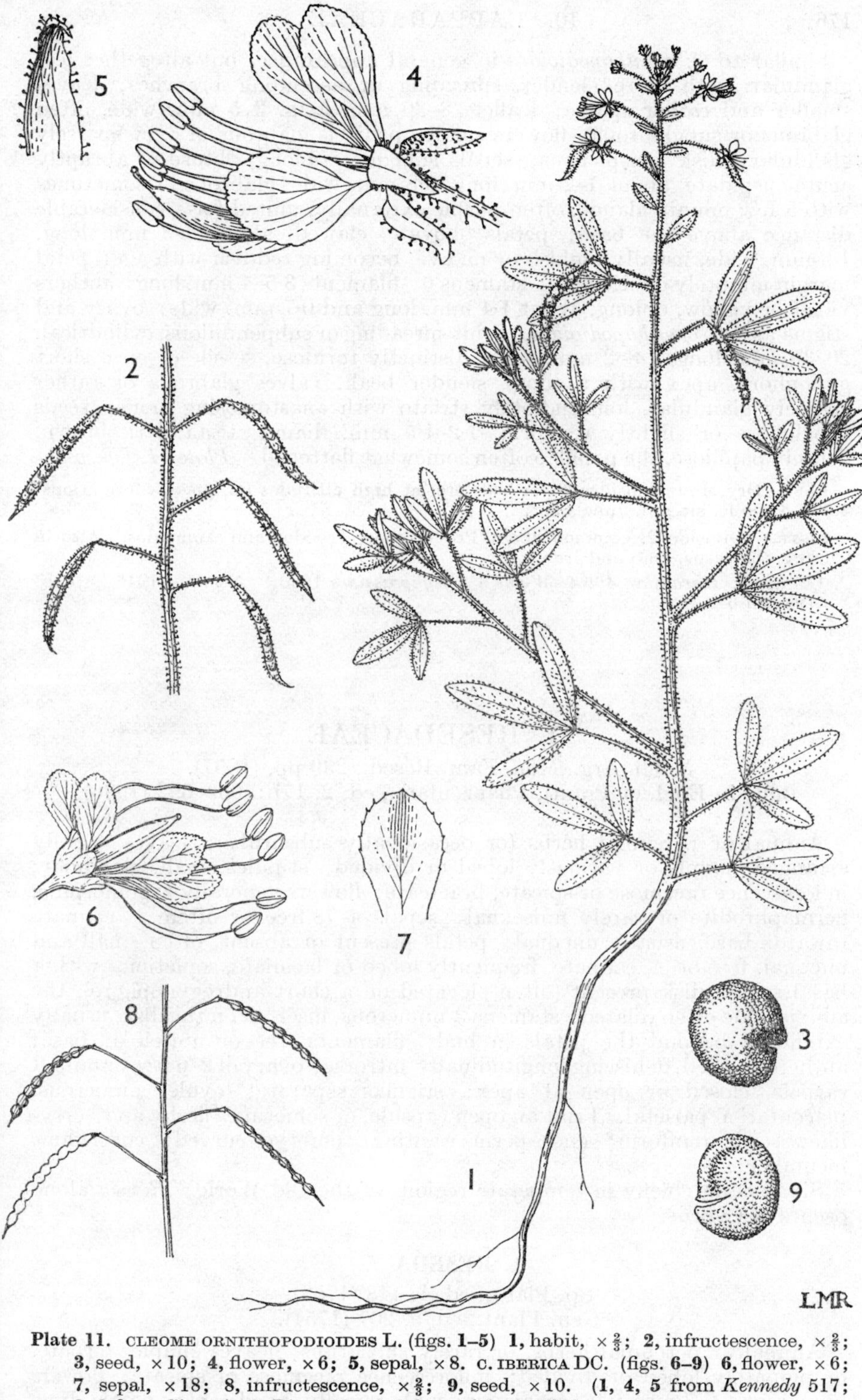

Plate 11. CLEOME ORNITHOPODIOIDES L. (figs. **1–5**) **1**, habit, $\times\frac{2}{3}$; **2**, infructescence, $\times\frac{2}{3}$; **3**, seed, $\times 10$; **4**, flower, $\times 6$; **5**, sepal, $\times 8$. C. IBERICA DC. (figs. **6–9**) **6**, flower, $\times 6$; **7**, sepal, $\times 18$; **8**, infructescence, $\times\frac{2}{3}$; **9**, seed, $\times 10$. (**1, 4, 5** from *Kennedy* 517; **2, 3** from *Sintenis & Rigo* 812; **6–9** from *Atherton* 640.)

Similar to *C. ornithopodioides* in general appearance, but altogether less glandular, with more slender, spreading or ascending branches; leaves smaller and rather sparse; leaflets 8–30 mm. long, 3–5 mm. wide, often glabrous or subglabrous; flowers small; pedicels glabrous or very sparsely glandular; disk conspicuous; sepals subequal, oblong, obtuse or abruptly acute-cuspidate, about 1–2 mm. long, 0·8 mm. wide, glabrous or sometimes with a few minute glands, often purple externally, united for a considerable distance above the base; petals obovate, clawed, about 2–3 mm. long, 1·8 mm. wide, usually pinkish or mauve, becoming reddish with age; petal margin minutely denticulate; stamens 6; filaments 3·5–4 mm. long; anthers violet or yellow, oblong, about 1·4 mm. long and 0·7 mm. wide; ovary and stigma as in *C. ornithopodioides*. Fruit spreading or subpendulous, cylindrical, 20–30 mm. long, 1·8–2 mm. wide, distinctly torulose, sessile or on a short gynophore, apex with a short, slender beak, valves glabrous or rather sparsely glandular, longitudinally striate with anastomosing veins; seeds subglobose or slightly reniform, 1·2–1·5 mm. diam.; testa dark brown, densely papillose, the papillae often somewhat flattened. *Plate 11, figs. 6–9.*

HAB.: Dry stony hillsides, river-beds etc. at high altitudes on igneous formations; 4,000–6,000 ft. alt.; fl. June–Sept.

DISTR.: Division 2, common about Prodhromos, Troödos and Amiandos. Also in Turkey, Caucasus, Iraq and Iran.

2. Common; *Haradjian* 466 ! *Wyatt* 6 ! *Syngrassides* 1690 ! *Kennedy* 1246 ! *G. E. Atherton* 640 !

11. RESEDACEAE

Muell. Arg., Mon. Fam. Réséd., 239 pp. (1857).
Bolle in Engl. et Prantl, Pflanzenfam., ed. 2, 17b: 659–692 (1936).

Annual or perennial herbs (or occasionally subshrubs); leaves spirally arranged, simple or variously lobed or divided; stipules small, glandular; inflorescence racemose or spicate, bracteate; flowers generally zygomorphic, hermaphrodite or rarely unisexual; sepals 4–7, free or often ± connate towards base, usually unequal; petals present or absent, often small and unequal, free or ± connate, frequently lobed or laciniate, sometimes with a basal scale; disk present (often elevated on a short androgynophore), the adaxial side often dilated; stamens 3-numerous, inserted on the disk, usually extending beyond the petals in bud; filaments free or united at base; anthers 2-celled, dehiscing longitudinally, introrse; ovary of 2–6 free or united carpels, closed or open at apex; stigmas separate; ovules numerous, placentation parietal. Fruit an open capsule, or sometimes fleshy and berry-like; seeds reniform; endosperm wanting; embryo curved; cotyledons incumbent.

Six genera, chiefly in temperate regions of the Old World; *Reseda* alone occurs in Cyprus.

1. RESEDA *L.*

Sp. Plant., ed. 1, 448 (1753).
Gen. Plant., ed. 5, 207 (1754).

Annual or perennial herbs, or rarely subshrubs; leaves simple, ternate, or pinnately lobed or divided; inflorescence racemose or spicate; flowers numerous, 4–7-merous; sepals unequal, usually persistent; petals free, white or yellow, unequal, entire, lobed or with fimbriate appendages; stamens 3–40, inserted on a zygomorphic disk, those on the abaxial side of the disk being longer and more numerous; ovary sessile or stipulate, unilocular, ± open at apex, with 3–4 (–6) apical stigmatiferous stylar lobes. Fruit a capsule, usually gaping at apex; seeds reniform.

About 50 species, chiefly in temperate Europe, Asia and North Africa.

Capsules erect or spreading; flowers scentless:
  Petals white; seeds distinctly papillose; capsule with 4 (rarely 5) apical lobes **1. R. alba**
  Petals yellow or greenish; seeds smooth, shining; capsule with 3 apical lobes:
    Leaves simple, entire; seeds dark brown - - - - - - **2. R. luteola**
    Leaves pinnatisect; seeds black - - - - - - - - **3. R. lutea**
Capsules pendulous; flowers scented:
  Calyx accrescent in fruit; cauline leaves mostly lobed - - - **4. R. orientalis**
  Calyx not (or very little) accrescent in fruit; cauline leaves mostly simple, entire **5. R. odorata**

1. R. alba *L.*, Sp. Plant., ed. 1, 449 (1753); Unger et Kotschy, Die Insel Cypern, 335 (1865); Boiss., Fl. Orient., 1: 425 (1867); Holmboe, Veg. Cypr., 92 (1914); Post, Fl. Pal., ed. 2, 1: 137 (1932); Davis, Fl. Turkey, 1: 500 (1965); Zohary, Fl. Palaest., 1: 332, t. 483 (1966); Osorio-Tafall et Seraphim, List Vasc. Plants Cyprus, 50 (1973).

[*R. undata* (non L.) Sibth. et Sm., Fl. Graec. Prodr., 1: 322 (1806); Poech, Enum. Plant. Ins. Cypr., 33 (1842).]

TYPE: France; " Monspelii ".

An erect or spreading, branched, glabrous perennial, 60 cm. or more in height, often woody towards the base; stems distinctly ribbed; leaves numerous, sessile or subsessile, often rather crowded towards the base of the plant, 4–10 cm. or more long, 1·5–5 cm. wide, deeply imparipinnatisect with a winged rhachis and 5–19 linear, entire, obtuse or acute lobes 1–7 cm. long, 0·3–0·6 cm. wide; inflorescence a dense terminal raceme, lengthening considerably in fruit; flowers very numerous; pedicels 2–6 mm. long; bracts small, subulate; sepals 5, subequal, connate at base, persistent, oblong, obtuse or subacute, concave with a slightly cucullate apex and a very narrow membranous margin, about 3 mm. long and 1 mm. wide; petals 5, white, subequal, obovate, 5–8 mm. long, 1·5–3 mm. wide, deeply divided at the apex into 3 linear, obtuse, entire or toothed, finger-like lobes; stamens about 12, filaments linear, flattened, about 3 mm. long, connate at the base and forming a cup around the base of the ovary; anthers oblong, about 1·5–2 mm. long and 1–1·5 mm. wide, almost closed at apex by 4 (or rarely 5) blunt stigmatiferous lobes. Capsule erect, oblong, about 7 mm. long and 3 mm. wide, conspicuously 4–5-angled; pericarp rugose with bulging seeds; seeds reniform, about 1·2 mm. long; testa dark brown, conspicuously papillose.

HAB.: Waste ground; sea-level to 1,000 ft. alt.; fl. Jan.–May.

DISTR.: Divisions 3, 4, common in Division 4 (especially around Larnaca). Widespread in the western Mediterranean region, becoming rarer eastwards and probably introduced in Cyprus; naturalized in many parts of temperate Europe and elsewhere.

3. Stavrovouni, 1928 or 1930, *Druce* s.n. (" forma *parvula* ").

4. Larnaca and Famagusta, *Kotschy* 56, 286; *Sintenis & Rigo*; *A. G. & M. E. Lascelles* s.n. ! *Haradjian* 81; *Syngrassides* 478 !

2. R. luteola *L.*, Sp. Plant., ed. 1, 448 (1753); Boiss., Fl. Orient., 1: 434 (1867); Holmboe, Veg. Cypr., 92 (1914); Post, Fl. Pal., ed. 2, 1: 140 (1932); Lindberg f., Iter Cypr., 17 (1946); Davis, Fl. Turkey, 1: 505 (1965); Zohary, Fl. Palaest., 1: 338, t. 493 (1966); Osorio-Tafall et Seraphim, List Vasc. Plants Cyprus, 50 (1973).

TYPE: " *in* Europa *ad vias & pagos* ".

A robust, erect perennial or biennial, 30–100 cm. or more high, unbranched or with several erect or erecto-patent branches especially towards the apex of the main axis just below the inflorescence; stems hollow, distinctly ribbed, leafy, basal leaves forming a distinct tuft or rosette, linear-oblong, sessile, (2–) 8–15 (–30) cm. long, 0·5–1·5 (–2) cm. wide, margin entire, flat or sometimes undulate (var. *crispata* (Link) Muell. Arg.), apex obtuse and mucronate or subacute; cauline leaves similar but usually rather shorter and narrower, often tapering to a rather acute point; inflorescence terminal, unbranched, densely racemose, to 40 cm. or more long, lengthening considerably in fruit; flowers very numerous, subsessile or with pedicels 1–5 mm. long; bracts small, subulate, 3–5 mm. long, 1–2·5 mm. wide at base; sepals 4, connate at base, persistent, bluntly oblong, concave, slightly cucullate at apex, about 1·5–2·5 mm. long, 0·8–1 mm. wide, with a very narrow membranous margin; petals (3–) 4 (–5), yellowish-green, unequal, the upper 3–4·5 mm. long, 1–1·5 mm. wide with a ± orbicular basal part and an obovate appendage deeply cleft at the apex into 6–8 or more linear, obtuse, finger-like lobes; lateral petals about 3 mm. long and less than 1·5 mm. wide, narrowly obovate, with the apex cleft into 2–3 finger-like lobes; basal petal oblong or lanceolate, undivided; stamens about 20, filaments about 1 mm. long, linear, tapering to apex; anther shortly oblong about 0·5 mm. long and 0·3 mm. wide; ovary ovoid, about 1–1·5 mm. long, 0·8 mm. wide, ± smooth or distinctly papillose externally with 3 conspicuous, pointed, stigmatiferous lobes. Capsule small, erect, ± spherical, 4–5 mm. long, about 4 mm. wide, deeply 3-lobed, gaping at apex, pericarp rugose with bulging seeds; seeds reniform or almost spherical, about 0·7 mm. diam.; testa smooth, shining, dark brown.

HAB.: Waste ground, roadsides etc.; sea-level to 3,000 ft. alt.; fl. April–July.

DISTR.: Divisions 1–4, 7. Europe, western and central Asia, N. Africa.

1. Lyso [" Lisso "], 1913, *Haradjian* 892.
2. Anadhiou, 1901, *A. G. & M. E. Lascelles* s.n. ! Kannaviou, in R. Ezuza, 1939, *Lindberg f.* s.n.; Stavros tis Psokas, 1962, *Meikle* 2744 !
3. Limassol, 1801, *Hume*; Anglisidhes, by roadside between Larnaca and Limassol, 1939, *Lindberg f.* s.n.
4. Larnaca, 1801, *Hume*.
7. Near Bellapais, 1880, *Sintenis & Rigo* 583 (var. *crispata* (Link) Muell. Arg.) !

3. R. lutea *L.*, Sp. Plant., ed. 1, 449 (1753); Unger et Kotschy, Die Insel Cypern, 335 (1865); Boiss., Fl. Orient., 1: 429 (1867); Holmboe, Veg. Cypr., 92 (1914); Post, Fl. Pal., ed. 2, 1: 138 (1932); Lindberg f., Iter Cypr., 17 (1946); Davis, Fl. Turkey, 1: 502 (1965); Zohary, Fl. Palaest., 1: 335, t. 488 (1966); Osorio-Tafall et Seraphim, List Vasc. Plants Cyprus, 50 (1973).

[*R. truncata* (non Fisch. et Mey.) H. Stuart Thompson in Journ. Bot., 44: 277 (1906).]

TYPE: " *in* Europae *australioris montibus cretaceis* ".

An erect or spreading, branched perennial herb 20–120 cm. high; stems leafy, sparsely or densely scabridulous, distinctly ribbed, often slightly woody towards the base; leaves 2–14 cm. long, 1·5–8 cm. wide, midrib and margin often scabridulous; lower half of lamina entire, tapering gradually to base; upper half deeply imparipinnatisect with 3–7 entire, flat or undulate (var. *crispa* West.), acute, linear-lanceolate segments, the lowermost segments sometimes lobed along the adaxial margin. Inflorescence terminal, ± densely racemose, 5–20 cm. or more long, lengthening considerably in fruit, rhachis scabridulous, flowers numerous, pedicels often scabridulous, 2–5 mm. long, 0·5 mm. wide; bracts linear-oblong, membranous, caducous, 1·5–2 mm. long, 0·4–0·5 mm. wide; sepals linear-oblong, 1–1·5 mm. long, 0·4 mm. wide, obtuse, glabrous, with a narrow membranous margin, scarcely accrescent in fruit; petals 6, yellow, upper pair 2·5–3 mm. long, 2–2·5 mm. wide, with a broadly oblong or suborbicular base and a deeply 3-lobed appendage, consisting of 2 lunulate lateral lobes and a shorter, linear, median lobe; lateral and basal petals usually similar, but smaller and narrower; disk distinctly gibbous on adaxial margin, pubescent or villous; stamens 14–17; filaments 1·5–2 mm. long, flattened, somewhat dilated towards apex, distinctly papillose; anthers about 1 mm. long, 0·6 mm. wide; ovary oblong, about 2 mm. long, 1 mm. wide; stigmatiferous lobes 3, broad, blunt, almost closing apex of ovary. Capsule suberect or spreading, oblong or obovoid, about 10–15 mm. long and 5–7 mm. wide, apical lobes indistinct, pericarp often strongly tuberculate-papillose; seeds reniform, about 1·5–2 mm. long; testa smooth, black, shining.

HAB.: Fields, waste ground, roadsides etc.; sea-level to 5,100 ft. alt.; fl. Febr.–Sept.

DISTR.: All Divisions, except 6; common in Divisions 2 and 7. Widespread throughout temperate Europe and the Mediterranean region eastwards to Iraq and Iran.

1. Ayios Neophytos Monastery, 1939, *Lindberg f.* s.n.; near Neokhorio (Akamas), 1962, *Meikle* 2191 !
2. Common (Troödos, Kakopetria etc.); *A. G. & M. E. Lascelles* s.n. ! *Haradjian* 521 ! *Kennedy* 515 ! *Merton* 2320 ! 2359 !
3. Lefkara, 1940, *Davis* 1910 !
4. Larnaca, 1880, *Sintenis & Rigo*. Phaneromene near Larnaca, 1905, *Holmboe* 132.
5. Lakatamia, 1932, *Syngrassides* 362 ! Near Apostolos Varnavas Monastery, 1948, *Mavromoustakis* s.n. !
7. Common locally around Kythrea and Kyrenia Pass; *Sintenis & Rigo* 40 ! *Syngrassides* 196 ! 359 ! *Lindberg f.* s.n. (as var. *crispa* West.); *Chapman* 90 ! *G. E. Atherton* 1036 ! 1141 ! 1374 ! *Merton* 2971 !
8. About Komi Kebir, 1912, *Haradjian* 322 !

4. R. orientalis (*Muell. Arg.*) *Boiss. ex Kotschy* in Unger et Kotschy, Die Insel Cypern, 335 (1865); Boiss., Fl. Orient., 1: 427 (1867); Holmboe, Veg. Cypr., 92 (1914); Post, Fl. Pal., ed. 2, 1: 138 (1932); Lindberg f., Iter Cypr., 17 (1946); Davis, Fl. Turkey, 1: 501 (1965); Zohary, Fl. Palaest., 1: 334, t. 487 (1966); Osorio-Tafall et Seraphim, List Vasc. Plants Cyprus, 50 (1973).

R. macrosperma Reichb. var. *orientalis* Muell. Arg., Mon. Fam. Réséd., 135 (1857).

[*R. mediterranea* (? L.) Poech, Enum. Plant. Ins. Cypr., 32 (1842).]

[*R. phyteuma* (non L.) H. Stuart Thompson in Journ. Bot., 44: 277 (1906).]

TYPE: Palestine; " circa Gaza " *Boissier* (G).

A decumbent, spreading or suberect, branched annual (5–) 10–30 (–80) cm. high, with a stout taproot; stems distinctly ribbed, ± scabrid; leaves numerous, sessile (1–) 3–6 (–8) cm. long, 0·5–3 cm. wide, with the midribs and margins ± scabrid; basal leaves often simple and entire, oblanceolate or linear-oblanceolate, obtuse or acute; cauline leaves deeply and rather irregularly 2–5-lobed in the upper half; inflorescence densely racemose, elongating greatly and becoming lax in fruit; flowers fragrant; rhachis and pedicels ± densely scabrid-hispidulous; pedicels 2–5 mm. long in flower, lengthening to 2 cm. or more in fruit; bracts subulate, herbaceous, scabrid-hispidulous; sepals connate at base, usually 6, linear, obtuse or subacute, about 3 mm. long, 0·5 mm. wide, persistent and considerably accrescent in fruit, the upper sepals normally rather longer then the lower, all papillose-tuberculate with a narrow membranous margin; disk conspicuous, densely clothed with long papillae; petals 6, white, upper pair about 2·5–3 mm. long and almost as wide, with a broadly oblong basal portion clothed with dense papillae and an appendage divided almost to the base into about 12 linear-spathulate, obtuse, minutely papillose lobes; lateral petals similar, but smaller and about 6–7-lobed; basal petals inconspicuous with an entire, linear-spathulate appendage; stamens about 14–21; filaments 2 mm. long, flattened, tapering to apex; anthers oblong, about 1 mm. long and 0·7 mm. wide; ovary oblong, about 2 mm. long and 1 mm. wide, distinctly papillose, with 3 acute stigmatiferous lobes. Capsule pendulous on a slender pedicel, oblong, about 10 mm. long and 7 mm. wide, apex shortly lobed; pericarp scabridulous, ± rugose with bulging seeds, slightly angled; seeds reniform, about 2 mm. long; testa dull purplish-brown, distinctly rugose. *Plate 12.*

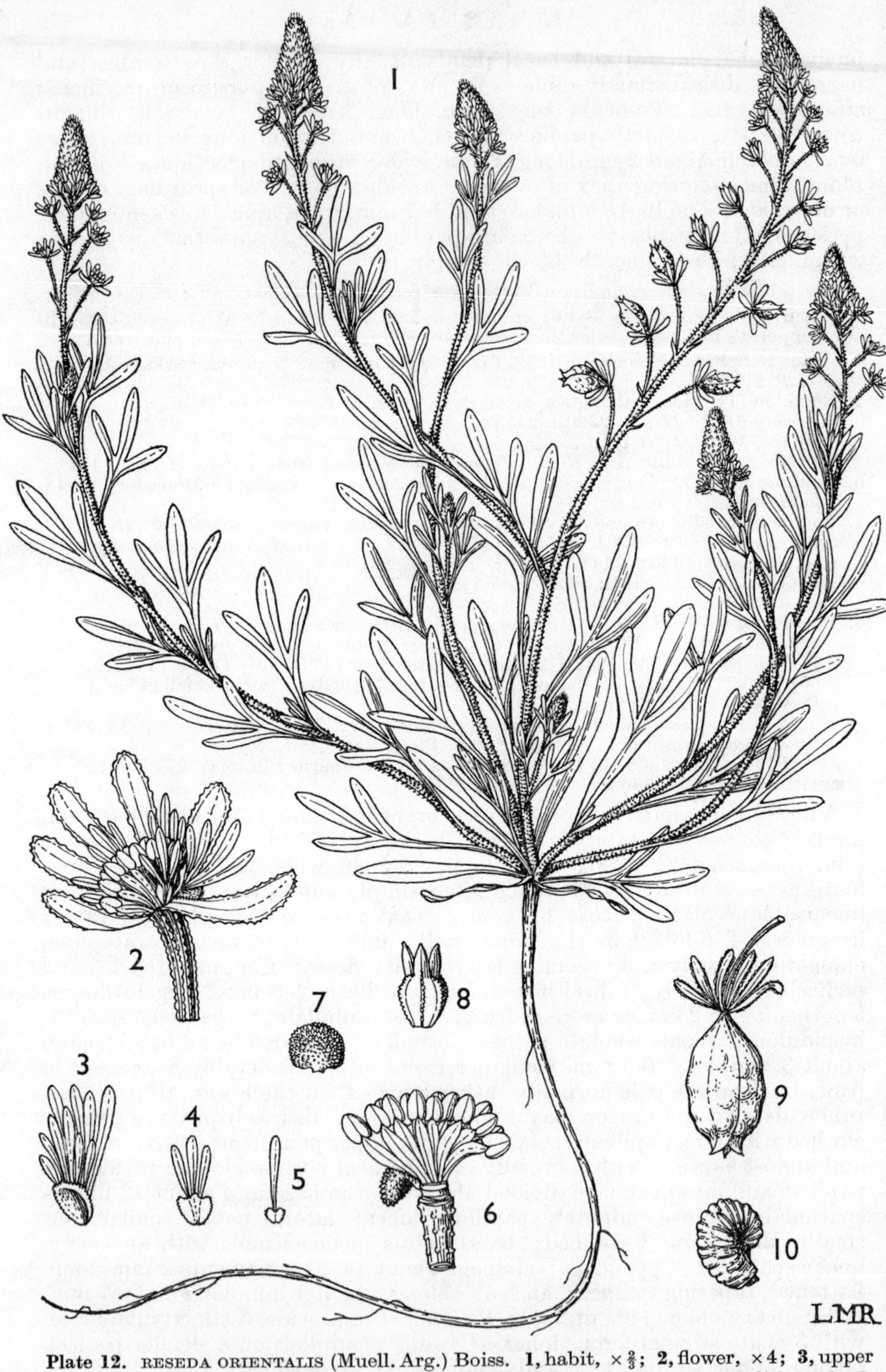

Plate 12. RESEDA ORIENTALIS (Muell. Arg.) Boiss. **1,** habit, $\times \frac{2}{3}$; **2,** flower, ×4; **3,** upper petal, ×4; **4,** median petal, ×4; **5,** basal petal, ×4; **6,** stamens and disk, ×4; **7,** disk, ×4; **8,** gynoecium, ×4; **9,** fruit, ×2; **10,** seed, ×8. (**1–8** from *Davis* 2558 and *Merton* 2867; **9–10** from *Haradjian* 243.)

HAB.: Cultivated fields and waste ground, roadsides, dry stony hillsides; sea-level to 2,000 ft. alt.; fl. Febr.–May.

DISTR.: Divisions 1–3, 5–8, common in Divisions 7 and 8. Also S. Greece, Crete, Turkey, Syria and Palestine.

1. Near Khrysokhou, 1862, *Kotschy* 670; Paphos, 1913, *Haradjian* 639 ! Ktima, 1913, *Haradjian* 658 !
2. Evrykhou, 1935, *Syngrassides* 824 ! Pyrgos, 1957, *Merton* 2867 !
3. Around Limassol, 1840, *Kotschy*; Petra tou Romiou, 1950, *Barrington* 67 !
5. Mia Milea–Mandres, 1933, *Syngrassides* 307 ! Athalassa, 1933, *Syngrassides* 4 ! Near Apostolos Varnavas Monastery, 1948, *Mavromoustakis* s.n. (intermixed with *R. lutea* L.) !
6. Near the river at Strovolos [" Strovilo "], 1901, *A. G. & M. E. Lascelles* s.n. ! Ayia Irini (Morphou), 1941, *Davis* 2558 !
7. Common; *Sintenis & Rigo* 39 ! 39a (intermixed with *R. lutea* L.) ! *Lindberg f.* s.n. ! *Chapman* 89 ! *Casey* 643 ! 1169 !
8. Common; *Haradjian* 243 ! *Syngrassides* 1779 ! *Davis* 2320 ! 2369 !

5. R. odorata *L.*, Syst. Nat., ed. 10, 1046 (1759); Unger et Kotschy, Die Insel Cypern, 335 (1865); Boiss., Fl. Orient., 1: 428 (1867); Holmboe, Veg. Cypr., 92 (1914); Post, Fl. Pal., ed. 2, 1: 138 (1932); Davis, Fl. Turkey, 1: 501 (1965); Zohary, Fl. Palaest., 1: 334 (1966); Osorio-Tafall et Seraphim, List Vasc. Plants Cyprus, 50 (1973).

TYPE: " *in* Aegypto ".

A decumbent or spreading herb, superficially similar to *R. orientalis* (Muell. Arg.) Boiss., but with the stems and leaves usually subglabrous or very sparsely scabridulous; all or most of the leaves simple, entire, lanceolate, oblanceolate or narrowly oblong, the uppermost cauline leaves sometimes 3-lobed towards the apex; inflorescence as in *R. orientalis* and flowers fragrant, but sepals smaller, linear, obtuse, to about 2·5 mm. long, not or only very slightly accrescent in fruit; petals smaller, 1·5–2·5 mm. long, about 1·5 mm. wide, with numerous, small, narrowly linear-spathulate lobes; filaments slightly dilated towards apex, about 2 mm. long; anthers shortly oblong, about 0·8 mm. long, 0·6 mm. wide, orange or pinkish-brown before dehiscence; ovary narrowed at base and shortly stipitate, apex with 3 conspicuous stigmatiferous lobes. Fruit pendulous as in *R. orientalis*, capsule often slightly stipitate; seeds reniform, about 1·5–2 mm. long; testa dull brown, distinctly rugose.

HAB.: In cornfields; near sea-level; fl. Febr.–May.

DISTR.: Division 4, rare. Native in Egypt and Cyrenaica, naturalized elsewhere; commonly cultivated.

4. About Larnaca, 1862, *Kotschy* (an escape from cultivation).

NOTES: Not recorded since from Cyprus but easily overlooked as a form of *R. orientalis*.

12. CISTACEAE

Grosser in Engl., Pflanzenreich, 14 (IV, 193): 161 pp. (1903).
Janchen in Engl. et Prantl, Pflanzenfam., ed. 2, 21: 289–313 (1925).

Herbs, subshrubs and shrubs, often with an indumentum of stellate hairs; leaves evergreen, usually opposite, simple, stipulate or exstipulate. Flowers solitary or in cymes or cymose racemes, actinomorphic, hermaphrodite; sepals 3–5, imbricate, contorted or convolute, frequently unequal, the outer shorter than the inner; petals 5 or 3 (sometimes wanting in cleistogamous flowers), free, imbricate or contorted, often caducous; stamens numerous, free, hypogynous, inserted on a disk-like projection of the receptacle; anthers 2-celled, laterally or introrsely dehiscent; ovary 1-locular, or incompletely 3–10-locular by the intrusion of placentae; placentation parietal; ovules few to many, orthotropous or rarely anatropous; funicle often strongly developed; style simple, with 1–5 stigmas. Fruit a leathery or woody loculicidal capsule; seeds usually small, angular, endosperm present; embryo variously curved or conduplicate, rarely ± straight.

About 8 genera in Europe and N. America; chiefly in the Mediterranean region.

Capsule 5–10-valved; flowers pink or white, never yellow - - - - **1. Cistus**
Capsule 3-valved; flowers (in the Cyprus species) yellow:
All the stamens fertile, with anthers:
Stigma sessile or subsessile - - - - - - - - - **2. Tuberaria**
Stigma not sessile; style distinct, elongate, often twisted at base **3. Helianthemum**
Outer stamens sterile, without anthers; filaments of sterile stamens nodulose or moniliform - - - - - - - - - - **4. Fumana**

1. CISTUS *L.*

Sp. Plant., ed. 1, 523 (1753).
Gen. Plant., ed. 5: 234 (1754).

Shrubs and subshrubs; leaves opposite, exstipulate; flowers usually in cymes, rarely solitary; sepals 3–5; petals 5; stamens numerous, all fertile; ovary 5- or 10-locular; style elongate or almost wanting; stigma capitate, with 5–10 lobes. Fruit a woody, loculicidal capsule; seeds numerous, small, angled; embryo circinnate, cotyledons linear.

About 16 species, almost wholly confined to the Mediterranean region and Atlantic Islands. Most of the species have showy, though ephemeral, flowers, and several exude an aromatic, viscid resin.

Flowers pink:
Leaves penninerved, not distinctly grey-tomentose; style elongate - **1. C. creticus**
Leaves longitudinally 3-nerved, grey-tomentose; style very short - **2. C. parviflorus**
Flowers white:
Leaves ovate or oblong; outer sepals distinctly cordate - - - **3. C. salviifolius**
Leaves linear-lanceolate or narrowly lanceolate; outer sepals not distinctly cordate:
Flowers small (seldom more than 2·5 cm. diam.) in scorpioid or umbel-like cymes; leaves with conspicuous venation - - - - - - **4. C. monspeliensis**
Flowers large (5–9 cm. or more diam.), solitary at the ends of lateral branchlets; leaves with inconspicuous venation - - - - - - **5. C. ladanifer**

1. C. creticus *L.*, Sp. Plant., ed. 2, 738 (1762); Unger et Kotschy, Die Insel Cypern, 336 (1865); Davis, Fl. Turkey, 1: 507 (1965).

C. pilosus L., Sp. Plant., ed. 2, 736 (1762) non *C. pilosus* L., Sp. Plant., ed. 1, 528 (1753).

C. incanus L., Sp. Plant., ed. 2, 737 (1762); ? *C. incanus* L., Sp. Plant., ed. 1, 524 (1753); Osorio-Tafall et Seraphim, List Vasc. Plants Cyprus, 72 (1973).

C. villosus L., Gen. Plant., ed. 6, last page (1764); Syst. Nat., ed. 12, 2: 366 (1767); Boiss., Fl. Orient., 1: 437 (1867); Grosser in Engl., Pflanzenr., 14 (IV, 193): 14 (1903); Holmboe, Veg. Cypr., 130 (1914); Janchen in Engl. et Prantl, Pflanzenfam., ed. 2, 21: 302 (1925); Post, Fl. Pal., ed. 2, 1: 142 (1932).

A much-branched shrub 20–150 cm. high, bark reddish-brown with vertical fissures exposing a lighter coloured underlayer; young shoots reddish, clothed with white glandular or eglandular hairs; leaves obovate or almost orbicular, acute or obtuse, 1–4 cm. or more long, 1–2 cm. wide, the uppermost usually rather narrower, sometimes oblanceolate; petiole 0·5–1 cm. long, winged and dilated below, the bases of the petioles are sometimes connate; lamina hoary with stellate hairs, frequently rugose with an undulate margin and conspicuously impressed venation; inflorescence terminal, forming a loose 1–3 (–6)-flowered cyme; peduncle usually rather short, 1–6 cm. long, pilose, glandular or eglandular; pedicels rarely more than 2 cm. long, usually shorter than the lanceolate or oblanceolate bracts; 2 outer sepals broadly ovate, 10–15 mm. long, 8–11 mm. wide, densely or sparsely villose, herbaceous, with an elongate, acuminate apex; 3 inner sepals similar but a little narrower and often rather hard and scarious in texture; petals purplish-pink, obovate, obtuse, 2·5–3 cm. long, 1·5–2 cm. wide; stamens yellow, 5–8 mm. long, filaments linear; anthers oblong about 0·8 mm. long, 0·3 mm. wide; ovary subglobose, 2·5–3 mm. diam., densely white-villose; style 2·5–4 mm. long, glabrous; stigma capitate. Capsule broadly ovoid or subglobose, 6–10 mm. long, about 6–8 mm. wide, dark brown, woody, thinly villose; seeds irregularly angular, about 1 mm. long; testa mid-brown, obscurely granulose.

var. **creticus**

C. villosus L. var. *creticus* (L.) Boiss., Fl. Orient., 1: 437 (1867); Holmboe, Veg. Cypr., 130 (1914); Post, Fl. Pal., ed. 2, 1: 142 (1932); Chapman, Cyprus Trees and Shrubs, 55 (1949).

C. polymorphus ssp. *villosus* b. *viscidus* β *rumelicus* Willk., Icon. et Descr. Plant., 2: 24, t. 82 (1857).

TYPE: " *in* Creta, Syria ".

Whole plant glandular, viscid and aromatic; leaves thinly hairy, rather small, broadly obovate, strongly rugose with crisped or undulate margins.

HAB.: On dry, stony hillsides and roadsides, in garigue and open woodland; sea-level to 5,500 ft. alt.; fl. Febr.–June.

DISTR.: All Divisions, very common in Divisions 2 and 7. Widespread in the central and eastern Mediterranean region from Corsica to Palestine.

1. On the way to Paphos, 1862, *Kotschy*; Ayios Yeoryios (Akamas), 1962, *Meikle* ! Polis, 1962, *Meikle* !
2. Common in central Troödos; *Kotschy* 713 ! *Syngrassides* 1103 ! *Wyatt* 25 ! *Merton* 2350 ! 2385 ! *G. E. Atherton* 532 !
3. Limassol, 1801, *Hume*; Stavrovouni, near Ayia Varvara, 1862, *Kotschy*; Cape Gata, 1862, *Kotschy*; Mazotos, 1905, *Holmboe* 177; Between Nisou and Stavrovouni, 1933, *Syngrassides* 39 !
4. Larnaca, 1801, *Hume*; Famagusta, 1862, *Kotschy*.
5. Athalassa, 1939, *Lindberg f.* s.n.
6. Dhiorios, 1962, *Meikle* !
7. Common; *Sintenis & Rigo* 221 (pro parte) ! *A. G. & M. E. Lascelles* s.n. ! *Tracey* 35 ! *G. E. Atherton* 158 ! 264 ! 266 !
8. Valia, 1905, *Holmboe* 502; Akradhes, 1962, *Meikle* !

var. **tauricus** (*Presl*) *Dunal* in DC., Prodr., 1: 264 (1824).

C. tauricus Presl, Delic. Prag., 24 (1822).

C. villosus L. var. *tauricus* (Presl) Grosser in Engl., Pflanzenr., 14 (IV, 193): 14 (1903); Holmboe, Veg. Cypr., 130 (1914); Chapman, Cyprus Trees and Shrubs, 55 (1949).

C. polymorphus ssp. *incanus* var. b. *orientalis* β *spathulifolius* Willk., Icon. et Descr. Plant., 2: 22, t. 80 A & B (1857); Unger et Kotschy, Die Insel Cypern, 335 (1865).

TYPE: " In Tauria " *Steven* (PRC).

Plant much less glandular than var. *creticus* and scarcely viscid; upper parts of stems and sepals often densely clothed with long white hairs; leaves frequently rather narrowly obovate, obscurely rugose and with flat or slightly crisped margins.

HAB.: On dry, stony hillsides and roadsides, in garigue and open woodland; 200–2,000 ft. alt.; fl. Febr.–June.

DISTR.: Divisions 1, 4, 7, 8. Mediterranean region, Crimea and Caucasus.

1. Dhrousha, 1941, *Davis* 3261 !
4. Near Ormidhia, 1787, *Sibthorp* sec. *Kotschy*.
7. Kazaphani, 1956, *G. E. Atherton* 931

8. Koronia Forest Station (Kantara), 1933, *Chapman* 43 ! and same locality, 1937, *Chapman* 236 !

NOTES: No clear distinction can be drawn in Cyprus between var. *creticus* and var *tauricus*. It is evident, however, that non-viscid plants do occur here and there, sometimes amongst populations of glandular, viscid var. *creticus*, and that these non-viscid plants are perhaps more frequent about the Northern Range than elsewhere. The distinction between the two varieties may depend, at least partly, upon the season when specimens are collected, for plants of *C. creticus* collected in the spring are generally much less viscid than those collected later in the year.

Recent authorities have named this species *C. incanus* L. (Sp. Plant., ed. 1, 524; 1753) but examination of type material in herb. Hort. Cliff. (BM) and in the Linnean herbarium raises doubts as to the correct application of this name. The type material appears to have been grown under glass and is somewhat etiolate and unsatisfactory. It may be abnormal *C. creticus*, but could also be a hybrid, most probably *C. albidus* L. × *C. crispus* L. On phytogeographic grounds this identification is not unlikely: Linnaeus says his *C. incanus* comes from " Hispania, G. Narbonensi "; it is perhaps of some significance that *C. creticus* is not reliably recorded from either Spain or France (see Grosser's comments in Engl., Pflanzenr. *loc. cit.*, 16).

2. C. parviflorus *Lam.*, Encycl. Méth., 2: 14 (1790); Unger et Kotschy, Die Insel Cypern, 336 (1865); Boiss., Fl. Orient., 1: 437 (1867); Grosser in Engl., Pflanzenr., 14 (IV, 193): 17 (1903); Holmboe, Veg. Cypr., 130 (1914); Janchen in Engl. et Prantl, Pflanzenfam., ed. 2, 21: 302 (1925); Lindberg f., Iter Cypr., 23 (1946); Chapman, Cyprus Trees and Shrubs, 55 (1949); Davis, Fl. Turkey, 1: 507 (1965); Osorio-Tafall et Seraphim, List Vasc. Plants Cyprus, 72 (1973).

[*C. incanus* (non L.) Sibth. et Sm., Fl. Graec. Prodr., 363 (1809); Poech, Enum. Plant. Ins. Cypr., 33 (1842).]

TYPE: Crete; " Tournefort a trouvé cette plante dans l'Isle de Candie " (P).

An erect or spreading bush 30–70 cm. high, bark fuscous-brown, rugose, distinctly grey-tomentose on the younger shoots; leaves ovate-oblong or elliptic, 1·5–4 cm. long, 0·7–2 cm. wide, acute or obtuse, very shortly and densely tomentose above, conspicuously grey-tomentose below, venation distinctly impressed above, prominently 3-nerved and finely reticulate below; petiole distinct, short, about 0·5–0·7 cm. long; flowers in short-stalked, few (c. 3–6)-flowered terminal cymes; peduncle and pedicels densely tomentose; bracts lanceolate or almost linear, distinctly longitudinally nerved, caducous; pedicels about 1 cm. long, articulated at the base; outer sepals ovate, 8–10 mm long, 4–5 mm. wide, villose on both surfaces; inner sepals distinctly broader, 6–7 mm. wide, with a cuspidate apex, villose on the outer surface only; petals pink, wedge-shaped, 1–1·5 cm. long, 0·8–1·2 cm. wide, with a truncate or slightly emarginate apex; stamens 4 mm. long; filaments linear, glabrous; anthers oblong. about 1·5 mm. long; ovary globose 2 mm. diam., densely villose; style very short, less than 0·5 mm. long; stigma capitate, rugose. Capsule broadly ovoid, about 8 mm. long and 5 mm. wide, thinly and shortly pilose, fuscous-brown, enveloped until fully ripe by the persistent withered sepals; seeds irregularly angular, about 1·2 mm. long and 1 mm. wide; testa closely rugose, dark brown.

HAB.: On dry stony slopes, by roadsides or on waste ground, in margins of open woodland on calcareous formations; sea-level to 3,000 ft. alt.; fl. Jan.–May.

DISTR.: Divisions 1, 3, 4, 6–8, very common in Division 7, common in Division 8. Also in Sicily, Greece, Aegean Islands and Turkey.

1. Ayios Yeoryios, Akamas, 1962, *Meikle* ! Polis, 1962, *Meikle* !
3. On the way from Episkopi to Paphos, 1862, *Kotschy*; Cape Gata, 1941, *Davis* 3585 !
4. Ormidhia, 1862, *Kotschy*.
6. Between Morphou and Panteleimon Monastery, 1862, *Kotschy*; Dhiorios, 1962, *Meikle* !
7. Very common; *Sintenis & Rigo* 221b ! *Tracey* 36 ! *Syngrassides* 22 ! 1452 ! *Casey* 179 ! *G. E. Atherton* 175 ! 267 ! 829 ! 1071 !
8. Common; *Holmboe* 536; *Druce* 51 ! *Davis* 2342 ! *Merton* 2947 ! *N. Macdonald* 15 !

C. parviflorus *Lam.* × **salviifolius** *L.*

Leaves less densely tomentose; flowers pale pink; outer sepals with a very broad base, almost cordate, clothed with long pale hairs.

HAB.: Growing with both parents, in garigue.

DISTR.: Division 8, frequent.

8. Frequent; Kantara, 1941, *Davis* 2475 ! and, 1951, *Probyn* s.n. ! Akradhes, 1957, *Merton* 2878 ! Eleousa Forest, 1957, *Merton* 2948 ! Peristeria Forest, Yialousa, 1957, *Poore* 50 ! Peristerges, Kilanemos, 1962, *Meikle* 2434 !

NOTES: Probably sterile, but little is known of the hybrid or its biology.

3. C. salviifolius *L.*, Sp. Plant., ed. 1, 524 (1753—as *C. Salvifolius*); Unger et Kotschy Die Insel Cypern, 337 (1865); Boiss., Fl. Orient., 1: 438 (1867); Grosser in Engl., Pflanzenr., 14 (IV, 193): 20 (1903); Holmboe, Veg. Cypr., 130 (1914); Janchen in Engl. et Prantl, Pflanzenfam., ed. 2, 21: 302 (1925); Post, Fl. Pal., ed. 2, 1: 142 (1932); Lindberg f., Iter Cypr., 23 (1946); Chapman, Cypus Trees and Shrubs, 55 (1949); Davis, Fl. Turkey, 1: 508 (1965); Zohary, Fl. Palaest., 2: 337, t. 496 (1972); Osorio-Tafall et Seraphim, List Vasc. Plants Cyprus, 72 (1973); Elektra Megaw, Wild Flowers of Cyprus, 8, t. 7 (1973).

C. salviifolius L. f. *brevipedunculatus* Willk., Icon. et Descr. Plant., 2: 38, t. 91, fig. 1 (1857); Grosser in Engl., Pflanzenr., 14 (IV, 193): 20 (1903); Holmboe, Veg. Cypr., 130 (1914).

C. salviifolius L. f. *longipedunculatus* Willk., Icon. et Descr. Plant., 2: 38, t. 91, fig. 1 (1857); Grosser in Engl., Pflanzenr., 14 (IV, 193): 20 (1903); Holmboe, Veg. Cypr., 130 (1914).

C. salviifolius L. var. *biflorus* Willk., Icon. et Descr. Plant., 2: 39, t. 92, fig. 2 (1857); Grosser in Engl., Pflanzenr., 14 (IV, 193): 21 (1903); Holmboe, Veg. Cypr., 130 (1914).

C. salviifolius L. var. *undulatifolius* Kotschy in Unger et Kotschy, Die Insel Cypern, 337 (1865).

TYPE: " *in* Italia, Sicilia, Narbona ".

An erect, much-branched shrub to 1 m. high, with fuscous or dark grey-brown minutely rugose bark; young shoots densely clothed with greyish-stellate hairs; leaves ovate or oblong, subacute or obtuse, 2–4 cm. long, 0·8–2 cm. wide, rugose-bullate and scabrid above, thinly pilose and distinctly veined below, margins often undulate or crimped; petiole distinct, usually about 5 mm. long; bracts linear-oblong, about 5–7 mm. long and 1–2 mm. wide, distinctly longitudinally 3-nerved, often rather densely pilose; flowers solitary or paired (rarely in threes), axillary; peduncle variable in length, from 2–8 cm. long, densely grey-pilose; pedicel 1–2 cm. long, frequently articulated near the middle; sepals shortly pubescent, the outer 2 conspicuously cordate, 0·8–1·5 cm. long, 0·6–1 cm. wide, the inner 3 semi-cordate or broadly ovate, concave, slightly smaller than the outer; petals white, wedge-shaped, emarginate, 1·5–2·5 cm. long, 0·8–1·4 cm. wide; stamens about 2·5 mm. long, filaments glabrous; anthers shortly oblong, about 0·8 mm. long and 0·6 mm. wide; ovary subglobose, about 1·5 mm. diam., densely hairy; stigma sessile or subsessile, capitate, rugose. Capsule globose, about 7 mm. diam., thinly hairy, dark brown, concealed by the persistent, accrescent, papery sepals; seeds subglobose, about 1·2 mm. diam.; testa dark brown, slightly rugose or almost smooth.

HAB.: Stony hillsides and waste places, in garigue and open Pine forest on calcareous and non-calcareous formations; sea-level to 4,500 ft. alt.; fl. Febr.–May.

DISTR.: All Divisions, except 5, very common in Division 7. Widespread in the Mediterranean region.

1. On the way to Paphos, 1862, *Kotschy;* Ayios Yeoryios (Akamas), 1962, *Meikle* ! Polis, 1962, *Meikle* !
2. Near Prodhromos, 1862, *Kotschy* 721 ! Platres, 1941, *Davis* 3529 ! Below Platania, 1955, *Merton* 2349 ! Prodhromos, 1955, *G. E. Atherton* 533 !
3. Near Ayia Varvara Monastery, just below Stavrovouni, 1862, *Kotschy*; around Episkopi towards Pissouri, 1862, *Kotschy*; Mazotos, 1905, *Holmboe* 192.

4. Cape Greco, 1862, *Kotschy* 172, also, 1962, *Meikle* !
6. Dheftera, 1928 or 1930, *Druce* s.n. ! Nicosia, 1931, *Rogers* 0638 ! 0645 ! 5242 ! Dhiorios, 1962, *Meikle* !
7. Very common; *A. G. & M. E. Lascelles* s.n. ! *Tracey* 37 ! *Syngrassides* 272 ! 1453 ! *G. E. Atherton* 144 ! 270 ! 897 ! 932 ! 1072 !
8. Yioti near Ephtakomi, 1905, *Holmboe* 537; Eleousa Forest, 1957, *Merton* 2946 ! Akradhes, 1962, *Meikle* !

4. C. monspeliensis *L.*, Sp. Plant., ed. 1, 524 (1753); Unger et Kotschy, Die Insel Cypern, 336 (1865); Boiss., Fl. Orient., 1: 438 (1867); Grosser in Engl., Pflanzenr., 14 (IV, 193): 17 (1903); Holmboe, Veg. Cypr., 130 (1914); Janchen in Engl. et Prantl, Pflanzenfam., ed. 2, 21: 303 (1925); Chapman, Cyprus Trees and Shrubs, 55 (1949); Osorio-Tafall et Seraphim, List Vasc. Plants Cyprus, 72 (1973).

Cistus fl. albo foliis linearibus Kotschy in Oesterr. Bot. Zeitschr., 12: 277–8 (1862).

TYPE: " *in* Narbonensi & Regno Valentino ".

An erect much-branched shrub, 30–100 cm. high; old branches glabrous, dark reddish brown, bark very lightly rugose; young shoots shortly and rather thinly pilose; leaves linear-lanceolate, 2–6 cm. long, 0·3–1 cm. wide, sessile, acute, margins strongly revolute; upper surface of lamina glabrous or subglabrous, bullate with deeply impressed venation, lower surface densely pilose-tomentose with prominent reticulate venation; bracts lanceolate-ovate 1–2·5 cm. long, 0·5–0·8 cm. wide, with 3–5 prominent longitudinal nerves, but not rugose or bullate like the leaves; flowers in a dense 3–10-flowered scorpioid or umbel-like cyme; peduncle slender, 2–4 cm. long, thinly pilose; pedicels short, seldom more than 0·8 cm. long, ± densely clothed with glands and long greyish hairs; sepals broadly ovate, often subcordate, 7–9 mm. long, 5–6 mm. wide, acuminate-cuspidate, the outer surface clothed with long whitish hairs; petals white, wedge-shaped, slightly emarginate, about 1 cm. long and 0·7–0·8 cm. wide; stamens to about 3 mm. long, filaments glabrous; anthers shortly oblong, about 1 mm. long; ovary subglobose about 1·5 mm. diam., shortly pilose; style short but distinct; stigma capitate, rugose. Capsule subglobose, about 4–5 mm. diam., dark brown, subglabrous, concealed within the accrescent, papery, persistent sepals; seeds obscurely angled, about 1·5 mm. long; testa fuscous, slightly rugulose.

HAB.: On rocky slopes and waste ground, in garigue; sea-level to 2,000 ft. alt.; fl. April–June.

DISTR.: Divisions 1–3, locally common in Division 1. Also in the Atlantic Islands, and widespread in the western Mediterranean region, becoming rarer eastwards and reaching its easternmost limit in Cyprus.

1. Common; *Holmboe* 779; *Haradjian* 823 ! *Syngrassides* 1722 ! *Davis* 3218 ! *Meikle* 2149 !
2. Lower Stavros valley, 1948 or 1949, *Chapman* 147 ! Jemali Bridge, 1962, *Meikle* 2768 !
3. Cape Gata, between Ayios Nikolaos and Lania [" Lamnias "], 1862, *Kotschy* 599 ! Cape Gata, 1941, *Davis* 3584 !

NOTES: All the Cyprus specimens examined belong to the small, narrow-leaved var. *minor* Willk. (Icon. et Descr. Plant., 2: 30; 1857). It is doubtful, however, if this variety can be satisfactorily distinguished from typical *monspeliensis*, nor does it appear to have a distinct geographic distribution.

C. monspeliensis *L.* × **parviflorus** *Lam.*

C. × *skanbergii* Lojac. in Naturalista Sic., 4: 95 (1885); Grosser in Engl., Pflanzenr., 14 (IV, 193): 29 (1903); Janchen in Engl. et Prantl, Pflanzenfam., ed. 2, 21: 304 (1925); Chapman, Cyprus Trees and Shrubs, 55 (1949).

Leaves lanceolate or oblanceolate 2–5 cm. long, 0·3–1 cm. wide, grey-green, shortly pilose above with distinctly impressed veins; tomentose and prominently reticulate-veined below; flowers pale pink, rather larger than those of *C. monspeliensis*; sepals broadly ovate, acuminate, clothed with long silky whitish hairs. Probably sterile.

HAB.: On rocky and sandy ground, in garigue and open woodland; sea-level to 1,000 ft. alt.; fl. March–May.

DISTR.: Divisions 1, 3, 4.

1. Smyies Forest Station, Akamas, 1937, *Chapman* 289 ! and 1941, *Davis*, 3300 ! Beyond Smyies, 1957, *Merton* 3016 ! Karavopetres (Akamas), 1962, *Meikle* 2056 !
3. Akrotiri peninsula, 1957, *Poore* 49 !
4. Larnaca Salt Lake plantation, 1955, *Merton* 2013 !

5. C. ladanifer *L.*, Sp. Plant., ed. 1, 523 (1753)—as *C. ladanifera*; Grosser in Engl., Pflanzenr., 14 (IV, 193): 23 (1903); Janchen in Engl. et Prantl, Pflanzenfam., ed. 2, 21: 303 (1925); Chapman, Cyprus Trees and Shrubs, 56 (1949)—all as *C. ladaniferus*; Osorio-Tafall et Seraphim, List Vasc. Plants Cyprus, 72 (1973).

TYPE: " *in* Hispaniae, Lusitaniae *collibus* ".

An erect or spreading shrub, up to 2m. or more in height; bark reddish-brown, lightly sulcate, young shoots glabrous, viscid with aromatic resin; leaves narrowly lanceolate, 4–10 cm. long, 0·8–2 cm. wide, subsessile, coriaceous, glabrous and shining above, greyish-tomentose below, venation indistinct; leaf-bases connate; flowers solitary at the ends of lateral, axillary branchlets; bracts numerous along the whole length of the flowering branchlets, the lowermost often foliose, lanceolate, yellowish and distinctly dilated at the base, 2–2·5 cm. long, 0·4–0·6 cm. wide, the upper broadly elliptical or sometimes broadly obovate, scarious, aromatic-viscid, margined with white silky hairs; bud globose, 1–1·5 cm. diam.; sepals orbicular, about 1 cm. diam., densely lepidote, fringed with white silky hairs; petals white, with or without a crimson blotch near the base, broadly wedge-shaped, 4–5 cm. long, 3–5 cm. wide; stamens about 5 mm. long, filaments glabrous; anthers narrowly oblong, about 2 mm. long and 0·8–1 mm. wide; ovary depressed globose, about 0·8 mm. diam., longitudinally sulcate, densely grey-tomentose; stigma almost sessile, capitate. Capsule globose, sulcate, about 1·3 cm. diam., dull brown, thinly scurfy-tomentose; seeds bluntly angular, about 1 mm. long; testa dull brown, minutely foveolate.

HAB.: Dry slopes and open forest; 800–2,600 ft. alt.; fl. April–May.

DISTR.: Divisions 1, 7, rare; probably an introduction. A native of S. France, Spain, Portugal, Morocco and Algeria.

1. Kephalovrysi, Akamas Forest, 1937, *Chapman* 292 !
7. Beyond Halevga, 1952, *Probyn* 92 !

NOTES: Presumably introduced and naturalized; other exotics have been planted in the vicinity of both localities.

C. ladanifer *L.* × **laurifolius** *L.*

C. × *cyprius* Lam., Encycl. Méth., 2: 16 (1790); Unger et Kotschy, Die Insel Cypern, 337 (1865); Boiss., Fl. Orient., 1: 439 (1867); Grosser in Engl., Pflanzenr., 14 (IV, 193): 25 (1903).

This hybrid, erroneously supposed to have originated in Cyprus, is known only from cultivated material, and is presumed to be of garden origin. It resembles *C. ladanifer* but the flowers are borne in small cymes, not solitarily, and the leaves are rather broader, duller, less resinous and ± distinctly longitudinally 3-nerved.

C. CRISPUS *L.*, a western Mediterranean species, has been recorded for Cyprus (Hume in Walpole, Mem. Europ. Asiat. Turkey, ed. 1, 254; 1817). The record is undoubtedly an error, and the plant observed was probably *C. creticus* L.

2. TUBERARIA (*Dunal*) *Spach*

Ann. Sci. Nat., ser. 2, 6: 364 (1838).

Helianthemum Mill. sect. *Tuberaria* Dunal in DC., Prodr., 1: 270 (1824).

Annual or perennial herbs; leaves longitudinally 3-nerved, the basal exstipulate, forming a rosette, the upper opposite, stipulate or exstipulate; flowers forming a cincinnus; sepals 5, the outer much smaller or much larger than the inner; petals 5; stamens numerous, all fertile; ovary incompletely 3-locular; ovules numerous; style short or wanting; stigma capitate, 3-lobed. Capsule 3-valved; seeds with a resupinate, clavate funicle; embryo twisted or curved, but not circinnate.

About 12 species, chiefly in the Mediterranean region.

1. T. guttata (*L.*) *Fourr.* in Ann. Soc. Linn. Lyon., n.s., 16: 340 (1868); Grosser in Engl. Pflanzenr., 14 (IV, 193): 56 (1903); Janchen in Engl. et Prantl, Pflanzenfam., ed. 2, 21: 307 (1925); Post, Fl. Pal., ed. 2, 1: 143 (1932); Davis, Fl. Turkey, 1: 523 (1965); Zohary, Fl. Palaest., 2: 337, t. 497 (1972); Osorio-Tafall et Seraphim, List Vasc. Plants Cyprus, 73 (1973).

Cistus guttatus L., Sp. Plant., ed. 1, 526 (1753); Sibth. et Sm., Fl. Graec. Prodr., 366 (1809), Fl. Graec., 5: 79 (1825).

Helianthemum guttatum (L.) Mill., Gard. Dict., ed. 8, no. 18 (1768); Boiss., Fl. Orient., 1: 440 (1867); Holmboe, Veg. Cypr., 131 (1914).

Tuberaria variabilis Willk., Icon. et Descr. Plant., 2: 73, t. 112 (1857); Unger et Kotschy, Die Insel Cypern, 337 (1865)—*sphalm.* Tubularia.

TYPE: " *in* G. Narbonensi, Italia ".

An erect, slender, branched or unbranched annual (6–) 15–20 (–30) cm. high; lower part of stem loosely villose with long white hairs, upper part of stem sometimes glabrous, or thinly villose, the hairs intermixed with short reddish glands; basal leaves sessile, forming a loose rosette, elliptic, obovate, or occasionally lanceolate, about 2–8 (–10) cm. long and 0·5–3 (–4) cm. wide, hairy on both sides, with simple hairs above and stellate hairs below; stem leaves smaller and narrower, the uppermost pair often stipulate; inflorescence a loose cincinnus; peduncles glabrous or ± glandular-hairy; pedicels slender, ebracteate, about 1 cm. long, glabrous or glandular-hairy, ascending or patent in flower, recurved, deflexed or arcuate in fruit; sepals hairy, persistent, the two outer narrowly ovate-lanceolate, about 2–2·5 mm. long and 1–1·5 mm. wide, the three inner boat-shaped 4–5 (–8) mm. long and 2–3 (–4) mm. wide, often heavily black-spotted; petals yellow, commonly purple- or brown-blotched near the base, fugacious, cuneate, about 6–8 (–13) mm. long and 4–5 (–10) mm. wide, apex often undulate or denticulate; stamens about 2·5 mm. long; anthers ovate, about 0·8 mm. long; ovary ovoid, hairy or subglabrous, about 1·2 mm. long and 0·8 mm. wide; stigma sessile, capitate. Capsule almost concealed by the persistent sepals, ovoid, ± glabrous, about 4 mm. long and 3 mm. wide; seeds ovoid, obscurely 3-angled, about 0·5 mm. long; testa mid-brown, papillose with short conical papillae.

HAB.: On dry hillsides, in open woodland, in fields, amongst maquis, and on sand-dunes; sea-level to 3,500 ft. alt.; fl. Febr.–May.

DISTR.: Divisions 1–3, 6, 8, common in Division 2. Widespread throughout the Mediterranean area, and north to England and Ireland.

1. Paphos, 1928 or 1930, *Druce* 45 ! Smyies, 1957, *Merton* 3017 ! Inia (Akamas), 1957, *Merton* 2023a (cleistogamous form) !
2. Common; *Syngrassides* 822 ! *Kennedy* 643 ! 644 ! 1334 ! *Davis* 3101 ! *Meikle* 2691 (large flowers) !
3. Stavrovouni, 1862, *Kotschy* 188 ! Akrotiri Forest, 1939, *Mavromoustakis* 108 !
6. Ayia Irini (Morphou), 1941, *Davis* 2581 ! and, 1956, *Merton* 670 ! Orga, 1956, *Poore* 2656 !

8. Cape Andreas, 1880, *Sintenis & Rigo* 215 ! Valia, near Ayios Theodoros, 1905, *Holmboe* 507; Cape Andreas, 1938, *Syngrassides* 1761 ! Ronnas Bay, near Rizokarpaso, 1941, *Davis* 2367 !

NOTES: Attempts to match Cyprus material of this polymorphic species with various western Mediterranean segregates are wholly unsatisfactory. Grosser (Engl., Pflanzenr., 14 (IV, 193): 56–57) lists *Kotschy* 188 under var. *eriocaulon* (Dunal) Grosser *and* (without explanation) also under var. *plantaginea* (Willd.) Grosser. *Sintenis & Rigo* 215 is cited as var. *plantaginea*. Holmboe (Veg. Cypr., 131) refers *Kotschy* 188 and his own specimen from Valia to *Helianthemum guttatum* (L.) Mill. ssp. *eriocaulon* (Dunal) Holmboe. Cyprus specimens do somewhat resemble *H. guttatum* var. *cavanillesii* Dunal (which Grosser unites with var. *eriocaulon*) in having rather narrow leaves and small flowers, but the resemblance is not exact, nor can it be agreed that var. *cavanillesii* (based on *Cistus serratus* Cav.) and var. *eriocaulon* are identical. There is considerable variation in Cyprus specimens of *T. guttata*, so much that it would be unwise to regard them as forming a homogeneous population; two features, however, do distinguish the majority of the specimens seen from typical western European *guttata*, namely the small flowers, and the general sparsity of hairs on the upper parts of the stems and on the peduncles. Whether such characters do justify varietal distinction is a matter which would require re-examination of the species throughout its range of distribution, and, until such detailed revision has been undertaken, it is probably best to leave the Cyprus material under *T. guttata* without attempting any more minute discrimination.

3. HELIANTHEMUM *Mill.*

Gard. Dict. Abridg., ed. 4 (1754).

Perennial or annual herbs or subshrubs; leaves all opposite or the upper sometimes alternate, stipulate or exstipulate; inflorescence a simple or compound cincinnus, bracteate or sometimes furnished with unmodified leaves; sepals 5, the 2 outer smaller than the inner, often strongly nerved, and accrescent in fruit; petals 5, yellow, white or pink, sometimes wanting in cleistogamous flowers; stamens usually numerous, all fertile and with anthers; ovary 1- or 3-locular, the placentae 2–12-ovulate; ovules orthotropous; style elongate, filiform; stigma capitate. Capsule 3-valved, ovoid or subglobose, often distinctly trigonous or triquetrous; seeds few or numerous; funicle resupinate; embryo curved or twisted, but not circinnate.

About 60 species, confined to the Old World, and chiefly in the Mediterranean region and the Middle East.

Perennial shrubs or subshrubs:
  Inflorescence compound, corymbose; a stout woody subshrub - **1. H. syriacum**
  Inflorescence a simple or very sparsely branched cincinnus:
    Petals distinctly longer than sepals; flowers pedicellate:
      Inner sepals about 6 mm. long, thinly pilose; a slender trailing or decumbent subshrub - - - - - - - - - - **2. H. chamaecistus**
      Inner sepals about 6–9 mm. long, densely white strigose along the nerves; a rather robust, sprawling or suberect subshrub - - - - **3. H. obtusifolium**
    Petals shorter than sepals or about as long; flowers sessile or subsessile **4. H. stipulatum**
Annual herbs:
  Inner sepals herbaceous, not conspicuously accrescent in fruit:
    Pedicels patent or arcuate in fruit; inner sepals 4–5 mm. long **5. H. salicifolium**
    Pedicels erect, straight and rather thick; inner sepals 8–15 mm. long **6. H. ledifolium**
  Inner sepals membranous, stramineous and conspicuously accrescent (and slightly inflated) in fruit - - - - - - - - - - - **7. H. aegyptiacum**

SECT. 1. **Argyrolepis** *Spach* Inflorescence cymose-corymbose, consisting of 2–4 cincinni; petals longer than sepals; style filiform, bent at the base; capsule trigonous containing few seeds, concealed by the persistent sepals.

1. **H. syriacum** (*Jacq.*) *Dum.-Cours.*, Le Botaniste Cultivateur, ed. 1, 3: 129 (1802); Meikle in Israel Journ. Bot., 19: 249–253 (1970); Zohary, Fl. Palaest., 2: 339 (1972).

Cistus syriacus Jacq. in L., Syst. Nat., ed. 14, 498 (1784); Jacq., Icones Plant. Rar., 1: 10, t. 96 (1784), Collectanea, 1: 33 (1787).

Helianthemum thibaudii Pers., Syn. Plant., 2: 79 (1806).

H. fasciculi Greuter in Boissiera, 13: 55 (1967).

[*Cistus racemosus* (non L.) Cav., Icones, 2: 33, t. 140 (1793).]

[*Helianthemum racemosum* (non (L.) Pers.) Pau in Inst. Catal. Hist. Nat., 202 (1916); Guinea, Cist. Españ., 68 (1954); Davis, Fl. Turkey, 1: 510 (1965); Osorio-Tafall et Seraphim, List Vasc. Plants Cyprus, 72 (1973).]

[*H. lavandulifolium* (non Mill.) Boiss., Fl. Orient., 1: 445 (1867); H. Stuart Thompson in Journ. Bot., 44: 277 (1906); Holmboe, Veg. Cypr., 131 (1914); Chapman, Cyprus Trees and Shrubs, 56 (1949); Rechinger f. in Arkiv för Bot., ser. 2, 1: 427 (1951).]

TYPE: Illustration in Jacquin, Icones Plant. Rar., 1: t. 96 !

A spreading or erect subshrub, to about 40 cm. high; branches and foliage ± densely grey-tomentose; leaves lanceolate or narrowly ovate, 1–3·5 cm. long, 0·2–0·8 cm. wide, margins strongly revolute, nervation sometimes prominent below; petiole short, seldom more than 4 mm. long; stipules filiform, caducous, hairy, about 5 mm. long; inflorescences lateral and terminal, each consisting of 2–3 rather condensed, many-flowered cincinni; pedicels short (seldom more than 5 mm. long), tomentose; outer sepals much shorter than inner, narrowly ovate, 3–4 mm. long, 1–1·5 mm. wide, conspicuously ciliate; inner sepals boat-shaped, about 6 mm. long and 3 mm. wide, canescent externally, thinly hairy internally; petals yellow, narrowly obovate, about 7–8 mm. long and 4–5 mm. wide; stamens up to 4 mm. long, filaments glabrous; anthers broadly oval, about 0·7 mm. long; ovary subglobose, about 1 mm. diam., densely white-hairy; style 4–5 mm. long, stigma capitate. Capsule trigonous, about 4·5 mm. long and 2 mm. wide, subglabrous towards base, pilose at apex; seeds angular, about 1 mm. long; testa bright brown, minutely and regularly verruculose.

HAB.: On rocky hillsides, pathsides, open woodland and sand-dunes; sea-level to 2,500 ft. alt.; fl. April–June.

DISTR.: All Divisions, common in Division 3. Widespread in the Mediterranean area.

1. Akamas, between Smyies and Ayia Nikola, 1941, *Davis* 3340 ! Beyond Smyies ridge, 1957, *Merton* 3019 !
2. Troödos area, 1950, *Probyn* s.n. (localisation doubtful).
3. Common; *Mavromoustakis* 24 ! 96 ! *Kennedy* 1565 ! *Davis* 3131 ! 3567 ! *Merton* 2771 ! *Meikle* 2954 ! *Merton* in ARI 709 !
4. Dhekelia, 1935, *Syngrassides* 639 ! Cape Greco, 1962, *Meikle* !
5. Near Sykhari, 1901, *A. G. & M. E. Lascelles* s.n. ! Athalassa, 1936, *Syngrassides* 1230 !
6. Between Ayia Irini and Kormakiti, 1941, *Davis* 2127 ! Dhiorios, 1962, *Meikle* 2417 !
7. Kyrenia, Upper Blackall's Valley, 1955, *G. E. Atherton* 279 !
8. Near Rizokarpaso, 1912, *Haradjian* 235 ! Ronnas Bay near Rizokarpaso, 1941, *Davis* 2352 ! Kantara monastery, 1941, *Davis* 2471 ! Near Kantara Rd., 1950 *Casey* 1049 !

SECT. 2. **Helianthemum** Flowers in simple, bracteate cincinni; petals exceeding sepals; capsule ovoid or subglobose, concealed within or exceeding the persistent sepals.

2. H. chamaecistus *Mill.*, Gard. Dict., ed. 8, no. 1 (1768); Grosser in Engl., Pflanzenr., 14 (IV, 193): 81 (1903); Holmboe, Veg. Cypr., 131 (1914); Post, Fl. Pal., ed. 2, 1: 147 (1932); Chapman, Cyprus Trees and Shrubs, 56 (1949).

H. vulgare Gaertn., Fruct., 1: 371 (1788); Unger et Kotschy, Die Insel Cypern, 338 (1865); Boiss., Fl. Orient., 1: 446 (1867); H. Stuart Thompson in Journ. Bot., 44: 277 (1906).

[*H. nummularium* (non Mill.) Coode et Cullen in Notes Roy. Bot. Gard. Edinb., 26: 198 (1965); Davis, Fl. Turkey, 1: 512 (1965) et auct. mult.]

TYPE: " on the chalky hills and banks in many parts of England " *Miller* (BM).

A decumbent or trailing subshrub with terete, thinly stellate-hairy branches up to 60 cm. long; leaves lanceolate, elliptic or ovate, 0·8–3 cm. long, 0·2–1·1 cm. wide, acute or obtuse, margins flat or slightly recurved, thinly clothed with forked hairs above, ± densely grey-tomentose below; petiole very short, usually less than 3 mm. long; stipules linear-lanceolate, ciliate, 3–8 mm. long; inflorescence terminal, a slender unbranched cincinnus; rhachis and pedicels ± densely stellate-tomentose; bracts linear, to about 5 mm. long; pedicels 0·5–1·5 cm. long, deflexed after flowering; 2 outer sepals very small, ovate-lanceolate, ciliate, 2–3 mm. long, about 1 mm. wide; inner sepals ovate, concave, thinly pilose, conspicuously nerved, about 6 mm. long and 3–5 mm. wide; petals yellow (occasionally white or pink) almost orbicular or broadly cuneate, overlapping, 1–1·3 cm. long, 0·8–1 cm. wide; stamens very numerous, about 5 mm. long; filaments glabrous; anthers broadly oblong, about 0·6 mm. long; ovary globose, about 1·5 mm. diam., densely tomentose; style about 4 mm. long, twisted at base, glabrous; stigma capitate. Capsule broadly ovoid, about 5 mm. long and 4 mm. wide, thinly tomentose, ± concealed by the persistent sepals; seeds bluntly angular, about 1·5–2 mm. long; testa dark brown, minutely verruculose.

HAB.: Dry hillsides and open grassy places.

DISTR.: Division 4, rare. Widespread in central and southern Europe, and throughout the Mediterranean region eastwards to Iran.

4. Near Larnaca, 1862, " not infrequent on conglomerate hills ", *Kotschy* 72.

NOTES: Not seen recently on the island, and perhaps a misidentification, though the species is recorded from adjacent countries and is not unlikely to occur in Cyprus. Kotschy cites his material under *H. vulgare* Gaertn. var. *microphyllum* Willk., a variant (not recognized by Grosser) with small leaves and flowers.

3. H. obtusifolium *Dunal* in DC., Prodr., 1: 281 (1824); Boiss., Fl. Orient., 1: 446 (1867); Grosser in Engl., Pflanzenr., 14 (IV, 193): 90 (1903); Holmboe, Veg. Cypr., 131 (1914); Janchen in Engl. et Prantl, Pflanzenfam., 21: 309 (1925); Chapman, Cyprus Trees and Shrubs, 56 (1949); Rechinger f. in Arkiv för Bot., ser. 2, 1: 427 (1950); Osorio-Tafall et Seraphim, List Vasc. Plants Cyprus, 72 (1973).

[*Cistus hirtus* (non L.) Sibth. et Sm., Fl. Graec. Prodr., 368 (1809), Fl. Graec., 6: 1 (1826).]

[*Helianthemum hirtum* (non (L.) Pers.) Poech, Enum. Plant. Ins. Cypr., 33 (1842).]

[*H. obovatum* (non Dunal) Kotschy in Unger et Kotschy, Die Insel Cypern, 339 (1865).]

[*H. appeninum* (non (L.) Lam. et DC.) Sintenis in Oesterr. Bot. Zeitschr., 32: 192 (1882).]

TYPE: " in ins. Cypro " *Rostung* [? *Casimir Rostan*] (G).

A sprawling or suberect subshrub; branches terete, to about 25 cm. long, ± densely white-tomentose, at least towards the extremities; leaves lanceolate, oblanceolate, or narrowly oblong-elliptic, 1·5–2·5 cm. long, 0·2–0·4 cm. wide, apex usually obtuse, margins distinctly recurved, base narrowed to a short but distinct petiole; upper leaf-surface subglabrous or thinly pilose, lower leaf-surface usually densely stellate-tomentose; stipules linear-lanceolate, about 4–5 mm. long and 1–1·5 mm. wide; inflorescence a simple, bracteate cincinnus; rhachis densely stellate-tomentose; bracts linear or linear-lanceolate, about 5 mm. long; pedicels flexuous, thinly stellate-tomentose, 5–10 mm. long, lengthening in fruit; 2 outer sepals lanceolate, 4–5 mm. long, about 1·5 mm. wide, margin with long white hairs; 3 inner sepals broadly boatshaped, 6–9 mm. long, 4–6 mm. wide, the prominent purple-black nerves clothed with long white hairs; petals pale, creamy yellow, broadly cuneate or almost rounded, to 1·5 cm. long and about as wide; stamens about 4 mm. long, filaments glabrous; anthers oblong, about 0·3 mm. long; ovary globose, about 1·5 mm. diam., densely tomentose, style 3–5 mm. long, twisted at base; stigma capitate. Capsule subtrigonous, about 5 mm. long and 4 mm. wide, thinly and shortly hairy; seeds bluntly angular, about 1·5 mm. long; testa brown, minutely verruculose. *Plate 13.*

HAB.: On dry rocky ground and hillsides, in garigue; sea-level to 3,000 ft. alt.; fl. Febr.–May.

DISTR.: Divisions 1, 3–8, common in Division 7. Endemic.

1. Ayios Yeoryios (Akamas), 1962, *Meikle* ! Polis, 1962, *Meikle* !
3. Near Limassol, on the way towards Omodhos, 1859, *Kotschy* 488; Limassol–Nicosia Rd., Mile 35, 1938, *Chapman* 344 ! Akrotiri, 1960, *N. Macdonald* 30 !
4. About Larnaca towards Livadhia, 1862, *Kotschy* 311 ! Ayia Napa, 1905, *Holmboe* 58; Mile 17, Nicosia–Larnaca Rd., 1950, *Chapman* 402 !
5. 3 miles N. of Lefkoniko, 1938, *Syngrassides* 1800 ! Pera, 1949, *Casey* 473 ! Salamis, 1962, *Meikle* !
6. Nicosia, 1931, *Rogers* 1595 ! Govt. House gardens, Nicosia, 1932, *Syngrassides* 499 ! Kokkini Trimithia–Kafkalou road, 1933, *Syngrassides* 332 ! Ayia Irini (Morphou), 1941, *Davis* 2577 ! Dhiorios, 1962, *Meikle* !
7. Common along the Northern Range; *Syngrassides* 921 ! *Davis* 2508 ! *Chapman* 181 ! *Casey* 1311 ! *G. E. Atherton* 1059 ! 1099 ! 1286 !
8. Rizokarpaso, 1880, *Sintenis & Rigo* 219 ! Near Komi Kebir, 1912, *Haradjian* 264; Valia, 1938, *Chapman* 335 ! Akradhes, 1962, *Meikle* !

SECT. 3. **Eriocarpum** *Dunal* Perennial herbs or subshrubs. Flowers in simple or paired, bracteate cincinni; petals usually shorter than sepals, sometimes wanting in cleistogamous flowers; capsule globose-trigonous, pubescent or hirsute.

4. H. stipulatum (*Forssk.*) *C. Christens.* in Dansk Bot. Arkiv, 4, 3: 20 (1922); Davis, Fl. Turkey, 1: 511 (1965); Osorio-Tafall et Seraphim, List Vasc. Plants Cyprus, 72 (1973).

Cistus stipulatus Forssk., Fl. Aegypt.-Arab., 100 (1775).

Helianthemum lippii (L.) Dum.-Cours. var. *ehrenbergii* (Willk.) Boiss., Fl. Orient., 1: 443 (1867), Fl. Orient. Suppl., 71 (1888).

[*H. pulverulentum* (non Lam. et DC.) Kotschy in Unger et Kotschy, Die Insel Cypern, 338 (1865); H. Stuart Thompson in Journ. Bot., 44: 277 (1906).]

[*H. lippii* (L.) Dum.-Cours. var. *ellipticum* (non (Desf.) Boiss.) Sintenis in Oesterr. Bot. Zeitschr., 32: 364 (1882).]

[*H. ellipticum* (non (Desf.) Pers.) Grosser in Engl., Pflanzenr., 14 (IV, 193): 96 (1903); Holmboe, Veg. Cypr., 131 (1914); Janchen in Engl. et Prantl, Pflanzenfam., ed. 2, 21: 310 (1925); Post, Fl. Pal., ed. 2, 1: 146 (1932); Lindberg f., Iter Cypr., 23 (1946); Chapman, Cyprus Trees and Shrubs, 56 (1949); Rechinger f. in Arkiv för Bot., ser. 2, 1: 427 (1950).]

TYPE: Egypt; " Alexandriae circa Salinas in collibus siccissimis " *Forsskål* (C).

A spreading, intricately branched subshrub to about 25 cm. high, mature twigs with glabrous, grey-brown bark, young shoots and foliage densely grey-tomentose; leaves seldom more than 10 mm. long and 5 mm. wide, variable in shape, ovate, elliptical, oblong, lanceolate, oblanceolate or almost linear, with the margins shortly recurved or very strongly revolute; apex acute or obtuse, base narrowed to a very short petiole; lamina very closely and evenly grey-tomentose on both sides; stipules lanceolate or narrowly triangular, 1–4 mm. long, usually tipped with a few long hairs; inflorescence a simple cincinnus; rhachis densely tomentose; flowers sessile or almost sessile; bracts lanceolate, often sharply 2–3 lobed, 1–4 mm. long; 2 outer sepals linear, about 4 mm. long and 1 mm. wide, 3 inner sepals broadly ovate, acute, 4–5 mm. long (increasing to 6–7 mm. in fruit), 3–4 mm. wide (increasing to 5 mm. in fruit), strongly 5-nerved, densely tomentose-villose with short and long hairs intermixed; petals yellow, usually shorter than sepals, often very short (less than 3 mm. long) or occasionally absent in cleistogamous flowers; stamens numerous, about 3 mm. long; filaments glabrous; anthers shortly oblong, 0·5 mm. long; ovary broadly ovoid, 1·8 mm. long and 1·2 mm. wide, densely villose; style straight, about 2 mm. long, glabrous; stigma rather small, truncate-capitate. Capsule subglobose, 4–5 mm. diam., densely

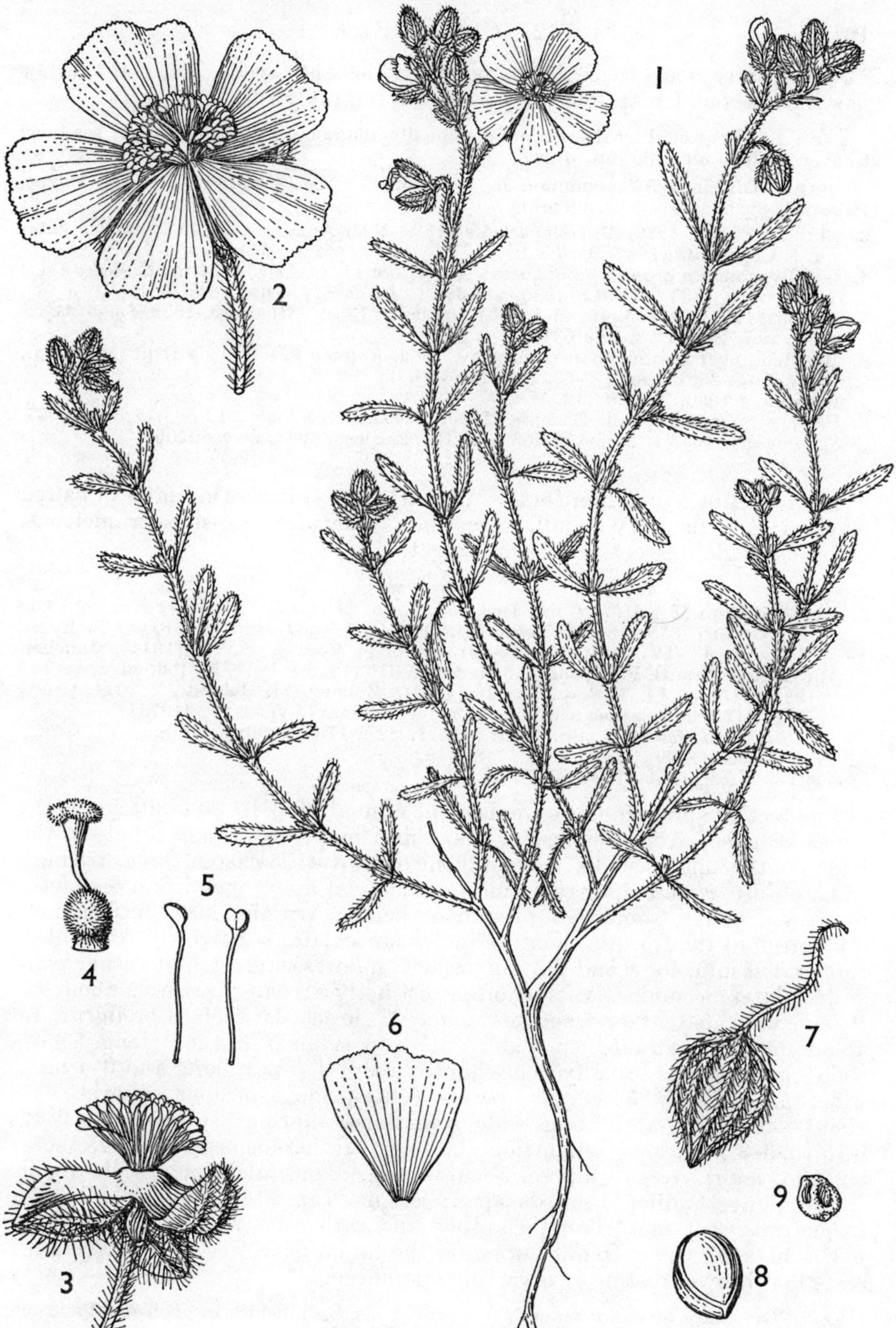

Plate 13. HELIANTHEMUM OBTUSIFOLIUM Dunal **1,** habit, ×⅔; **2,** flower, ×2; **3,** sepals and stamens, ×2; **4,** gynoecium, ×4; **5,** stamens, front and side view, ×4; **6,** petal, ×2; **7,** fruit enclosed by sepals, ×2; **8,** fruit, ×2; **9,** seed, ×9. (**1–6** from *Chapman* 402; **7–9** from *Syngrassides* 921.)

villose, partly concealed by the persistent sepals; seeds ± ovate, rather flattened, about 1 mm. long; testa brown, minutely rugulose.

HAB.: Sandy ground by the sea, or occasionally inland on dry, rocky slopes; sea-level to about 550 ft. alt.; fl. Jan.–July.

DISTR.: Divisions 3–8, common in Division 4. Also eastern N. Africa, Egypt, Palestine, Syria and southern Turkey.

3. Akrotiri Forest, 1939, *Mavromoustakis* 115 ! Near Amathus, 1947, *Mavromoustakis* s.n. ! Cape Gata, 1962, *Meikle* !
4. Locally common around Famagusta; *Haradjian* 112; *A. G. & M. E. Lascelles* s.n. ! *Syngrassides* 1335 ! 1336 ! *Chapman* 330 ! *Lindberg f.* s.n. ! *Meikle* !
5. Salamis, 1905, *Holmboe* 451 and, 1934, *Chapman* 135 ! Athalassa, 1932, *Syngrassides* 410 ! and, 1950, *Chapman* 631 !
6. Morphou, near Panteleimon monastery, 1862, *Kotschy* 924 ! Ayia Irini (Morphou), 1941, *Davis* 2561 ! and, 1953, *Kennedy* 1801 !
7. Kyrenia, ? 1930, *Tracey* 39 !
8. Between Yialousa and Eleousa, 1880, *Sintenis & Rigo* 212 ! Boghaz, 1937, *Syngrassides* 1645 ! Ayios Philon near Rizokarpaso, 1941, *Davis* 2206 !

SECT. 4. **Brachypetalum** *Dunal* Annual herbs. Flowers in simple or paired cincinni; petals very small, sometimes wanting; capsule triquetrous, glabrous or glabrescent, rarely pubescent.

5. **H. salicifolium** (*L.*) *Mill.*, Gard. Dict., ed. 8, no. 21 (1768); Unger et Kotschy, Die Insel Cypern, 338 (1865); Boiss., Fl. Orient., 1: 441 (1867); Grosser in Engl., Pflanzenr., 14 (IV, 193): 104 (1903); Holmboe, Veg. Cypr., 131 (1914); Janchen in Engl. et Prantl, Pflanzenfam., ed. 2, 21: 310 (1925); Post, Fl. Pal. ed. 2, 1: 144 (1932); Davis, Fl. Turkey, 1: 516 (1965); Zohary, Fl. Palaest., 2: 344, t. 508 (1972); Osorio-Tafall et Seraphim, List Vasc. Plants Cyprus, 72 (1973).

Cistus salicifolius L., Sp. Plant., ed. 1, 527 (1753); Sibth. et Sm., Fl. Graec. Prodr., 367 (1809), Fl. Graec., 5: 80 (1825).

TYPE: "*in* Lusitania, Hispania".

An erect or spreading, grey-pubescent annual, (6–) 10–20 (–30) cm. high; stems branched from the base or sometimes simple or branched only in the region of the inflorescence; leaves oblong or obovate, 5–20 mm. long, 3–8 mm. wide, obtuse, rounded, or sometimes subacute at apex; nervation distinctly impressed above, somewhat prominent below; petiole short but distinct, seldom more than 5 mm. long; stipules lanceolate or narrowly triangular, to about 8 mm. long and 2·5 mm. wide; inflorescence at first rather congested, later becoming lax and often distinctly arcuate; pedicels about 5–10 cm. long, patent or frequently curved; bracts lanceolate or linear, to about 1 cm. long and 0·4 cm. wide; outer sepals linear, 3–4 mm. long, 1 mm. wide; inner sepals broadly boat-shaped, about 4–5 mm. long and 3–4 mm. wide, prominently 3-nerved; petals yellow, inconspicuous, obovate, to about 4 mm. long and 3 mm. wide, sometimes wanting; stamens (7–) 9–12 (–15), 2·5–3 mm. long; filaments glabrous; anthers shortly oblong, about 0·2 mm. long; ovary globose, about 1 mm. diam., glabrous; style about 1 mm. long, swollen towards apex; stigma capitate. Capsule globose-triquetrous, 3–4 mm. diam., glabrous and rather lustrous; seeds angled, pyriform, with the extremity opposite the hilum narrowed to a projection; testa mid-brown, distinctly crystalline-papillose.

HAB.: On sandy or rocky ground, in sterile fields, on hillsides in open woodland or amongst garigue; sea-level to 3,000 ft. alt.; fl. Febr.–May.

DISTR.: All Divisions, common in Divisions 4 and 7. Widespread in the Mediterranean region eastwards to Iran.

1. Ayios Yeoryios (Akamas), 1962, *Meikle* ! Polis, 1962, *Meikle* !
2. Mandria–Perapedhi, 1937, *Kennedy* 647 ! East of Mandria, 1938, *Kennedy* 1333 !
3. Yermasoyia Potamos, near Limassol, 1948, *Mavromoustakis* s.n. ! Near Kouklia (Paphos), 1960, *N. Macdonald* 24 ! Akrotiri, 1960, *N. Macdonald* 28 !

4. Common along or near the coast; *Kotschy* 54, 171; *Sintenis & Rigo* 214 ! *Holmboe* 49. *Druce* 43 ! *Chapman* 10 ! *Merton* 2013 !
5. Salamis, 1905, *Holmboe* 451 and, 1962, *Meikle* ! Mia Milea, 1933, *Syngrassides* 310 !
6. Ayia Irini (Morphou), 1941, *Davis* 2585 ! Dhiorios, 1962, *Meikle* !
7. Common; *Sintenis & Rigo* 214a ! *Wyatt* 14 ! *Chapman* 61 ! 280 ! *G. E. Atherton* 1016 ! 1075 ! 1295 !
8. North part of Karpas, between Apostolos Andreas Monastery and Aphendrika, 1880, *Sintenis & Rigo*; Apostolos Andreas Monastery, 1941, *Davis* 2293 (form with very large capsules) ! Ronnas Bay near Rizokarpaso, 1941, *Davis* 2349 ! Akradhes, 1962, *Meikle* !

var. glabrum *Meikle* in App. III, p. 806

TYPE: Cyprus; Kokkini Trimithia, April 7, 1933, *Syngrassides* 331 (K !)

Whole plant glabrous; seeds perhaps a little smaller than in typical *H. salicifolium.*

HAB.: On rocky ground, c. 500 ft. alt.; fr. April.

DISTR.: Division 6, rare. Known only from the single type specimen.

6. Kokkini Trimithia, 1933, *Syngrassides* 331 !

NOTES: So distinct on account of its glabrous leaves and stems that it requires taxonomic recognition.

6. H. ledifolium (*L.*) *Mill.*, Gard. Dict., ed. 8, no. 20 (1768); Unger et Kotschy, Die Insel Cypern, 338 (1865); Grosser in Engl., Pflanzenr., 14 (IV, 193): 101 (1903); Holmboe, Veg. Cypr., 131 (1914); Janchen in Engl. et Prantl, Pflanzenfam., ed. 2, 21: 310 (1925); Post, Fl. Pal., ed. 2, 1: 143 (1932); Davis, Fl. Turkey, 1: 515 (1965); Zohary, Fl. Palaest., 2: 343, t. 506, 506a (1972); Osorio-Tafall et Seraphim, List Vasc. Plants Cyprus, 72 (1973).

Cistus ledifolius L., Sp. Plant., ed. 1, 527 (1753).

Helianthemum niloticum (L.) Pers., Syn. Plant., 2: 78 (1807); Boiss., Fl. Orient., 1: 441 (1867); Sintenis in Oesterr. Bot. Zeitschr., 32: 398 (1882).

Cistus niloticus L., Mantissa altera, 246 (1771).

An erect or spreading annual (10–) 20–30 (–40) cm. high; stems terete or obscurely angled, stellate-pubescent or sometimes rather roughly stellate-pilose; leaves linear, oblong or obovate, 1–4 cm. long, 0·2–1·3 cm. wide, acute or obtuse, usually densely pubescent, nervation often distinctly impressed; petiole short, seldom more than 5 mm. long; stipules conspicuous, lanceolate, to 2 cm. long and 0·5 cm. wide, margins often distinctly reflexed or revolute; inflorescence consisting of simple or (usually sparingly) branched cincinni; flowers close together or remote; rhachis straight and rigid or flexuous; bracts leaf-like, to 3 cm. or more in length, often longer than the opposed flowers; pedicels short, rigid, densely hairy, usually about 5 mm. long; outer sepals lanceolate-acuminate, 8–12 mm. long, clothed with short stellate hairs and with longer simple hairs on the 3 prominent nerves; inner sepals similar, but much broader (to about 1 cm. wide) and a little longer (to about 1·5 cm. long) with 5 prominent nerves; petals yellow, small, narrowly cuneate, about 7 mm. long and 4 mm. wide; stamens about 4 mm. long; filaments glabrous; anthers rounded or shortly oblong about 0·2 mm. long; ovary globose-triquetrous, 2·5–3 mm. diam., glabrous with ciliate angles, or pubescent at apex or sometimes densely hairy all over; style 1 mm. long, stout, expanded above; stigma capitate. Capsule usually markedly triquetrous, longer or shorter than the persistent sepals, to about 1·5 cm. long and 1·2 cm. wide; seeds inversely pyriform, rather flattened, 0·8–1·2 mm. long; testa mid-brown, smooth, sparsely foveloate, or ± densely covered with short crystalline papillae.

ssp. ledifolium

Plant usually erect and rather robust; inflorescence usually straight and rigid; flowers approximate or remote; capsule glabrous with ciliate angles,

0·6–1·5 cm. long, 0·3–1·2 cm. wide, shorter or longer than the calyx; seeds usually smooth or foveolate, or very sparsely papillose.

var. **microcarpum** *Coss. ex Willk.*, Icon. et Descr. Plant., 2: 87 (1856); Grosser in Engl., Pflanzenr., 14 (IV, 193): 103 (1903).

Capsule shorter than calyx, 6–7 mm. long, 3–5 mm. wide; seeds often rather small.

f. **laxiflorum** *Grosser* in Engl., Pflanzenr., 14 (IV, 193): 103 (1903).
f. *dissitiflorum* Willk., Icon. et Descr. Plant., 2: 87 (1856); Unger et Kotschy, Die Insel Cypern, 338 (1865).
TYPE: Spain; " in agro Madritense prope *Chamartin* ", *Bourgeau* 2111 (K !).

Flowers remote; flowering branches and rhachis usually slender.

HAB.: Dry stony or sandy ground at low altitudes; fl. March–May.

DISTR.: Divisions 2, 4. The type variety is widespread in the Mediterranean area eastwards to Iran and Arabia; var. *microcarpum* seems particularly prevalent in North Africa and the Orient.

2. Near Prodhromos, 1862, *Kotschy* 801a.
4. Around Larnaca, 1862, *Kotschy* 138.

ssp. **lasiocarpum** (*Desf. ex Jacques et Herincq*) *Nyman*, Consp. Fl. Europ., 73 (1878).
H. lasiocarpum Desf. ex Jacques et Herincq, Man. Gén. Plant., 1: 120 (1847); Grosser in Engl., Pflanzenr., 14 (IV, 193): 103 (1903); Janchen in Engl. et Prantl, Pflanzenfam., ed. 2, 21: 310 (1925); Post, Fl. Pal. ed. 2, 1: 144 (1932).
H. ledifolium (L.) Mill. var. *lasiocarpum* (Desf. ex Jacques et Herincq) Bornm. in Engl. Bot. Jahrb., 59: 373 (1925); Davis, Fl. Turkey, 1: 516 (1965).
TYPE: Transcaucasia; " ad radices montium Garial et Schuschka in Georg. Cauc.", *Hohenacker* (LE, K !).

A more slender plant than ssp. *ledifolium*, the leaves usually narrower, the flowers remote, and the rhachis usually flexuous. The capsule is generally pubescent all over (but *occasionally* glabrous except at the angles) and the seeds are ± densely crystalline-papillose.

HAB.: Dry stony or sandy ground at low altitudes; fl. March–May.

DISTR.: Division 8, rare. Also in Yugoslavia, Bulgaria, Greece, Caucasus, Turkey, Palestine, Iraq and Iran.

8. Cape Andreas, 1880, *Sintenis & Rigo* 217 !

NOTES: Recognizably distinct from ssp. *ledifolium* and with a very definite distribution, but none of the characters cited above will, in isolation, be found reliable and constant, and the plant's affinities with *ledifolium* are so obvious that it seems best to regard it as a subspecies. *H. ledifolium* ssp. *lasiocarpum* has not been collected in Cyprus since 1880, and should be looked for. A record for *H. ledifolium* from Kyrenia (C. E. Gresham in New Flora & Silva, 5: 272; 1933) also requires investigation; the species has not been recorded from this well-worked area by any other collector.

7. **H. aegyptiacum** (*L.*) *Mill.*, Gard. Dict., ed. 8, no. 23 (1768); Unger et Kotschy, Die Insel Cypern, 338 (1865); Boiss., Fl. Orient., 1: 442 (1867); Grosser in Engl., Pflanzenr., 14 (IV, 193): 106 (1903); Holmboe, Veg. Cypr., 131 (1914); Janchen in Engl. et Prantl, Pflanzenfam., ed. 2, 21: 310 (1925); Post, Fl. Pal., ed. 2, 1: 144 (1932); Davis, Fl. Turkey, 1: 517 (1965); Zohary, Fl. Palaest., 2: 345, t. 509 (1972); Osorio-Tafall et Seraphim, List Vasc. Plants Cyprus, 72 (1973).
Cistus aegyptiacus L., Sp. Plant., ed. 1, 527 (1753).
TYPE: " *in* Aegypto. *B. Jussieu* ".

A slender, erect, simple or sparingly branched annual (5–) 10–12 (–15) cm. high; stems terete, sparsely crispate-pubescent; leaves linear, 10–15 (–20) mm. long, 1·5–4 mm. wide, sparingly pilose, with a deeply impressed midrib and strongly revolute margins; stipules linear, to about 5 mm. long; petiole very short, rarely more than 2–3 mm. long; inflorescence a rather lax cincinnus; rhachis very slender, sometimes flexuous, densely pubescent; bracts very similar to leaves, flanked by small stipules; pedicels 5–8 mm.

long, arcuate or sometimes pendulous, densely pubescent; calyx somewhat inflated in fruit, the outer 2 sepals herbaceous, linear-lanceolate, about 3 mm. long, thinly pilose, the inner 3 membranous, boat-shaped, ciliate along the nerves, about 7 mm. long and 4 mm. wide, enlarging in fruit to 10 mm. or more long and 5–8 mm. wide and becoming stramineous with conspicuous purple nervation; petals yellow, narrowly cuneate, shorter than inner sepals (about 5–6 mm. long and 2–4 mm. wide); stamens about 2·5 mm. long; filaments glabrous; anthers minute, shortly oblong; ovary globose, about 1·3 mm. diam., densely hairy; style about 0·8 mm. long, glabrous, expanding above; stigma capitate. Capsule ovoid, about 7 mm. long and 5 mm. wide, thinly pubescent, enclosed within the persistent, accrescent calyx; seeds shortly obpyriform, about 1·5 mm. long, angular; testa fuscous-brown, densely crystalline-papillose.

HAB.: On dry hillsides and mountain slopes, 500–3,500 ft. alt.; fl. March–June.

DISTR.: Divisions 2, 3, 5, common in Division 2. Widespread in the Mediterranean region and eastwards to Iran.

2. Common; Evrykhou, 1935, *Syngrassides* 828 ! Platres, 1937, *Kennedy* 645 ! 646 ! Mandria, 1939, *Kennedy* 1333a ! Gourri, 1950, *Chapman* 330 !
3. Stavrovouni, near Ayia Varvara, 1862, *Kotschy* 207.
5. Athalassa, 1962, *Meikle* 2983 !

4. FUMANA (*Dunal*) *Spach*

Ann. Sci. Nat., ser. 2, 6: 359 (1836).

Helianthemum sect. *Fumana* Dunal in DC., Prodr., 1: 274 (1824).

Low subshrubs; leaves alternate or opposite, usually small and narrow, stipulate or exstipulate; inflorescence a cincinnus, or flowers sometimes solitary and leaf-opposed; sepals 5, the 2 outer smaller than the inner; petals 5, yellow or orange, blotched or unblotched; stamens numerous, the outermost sterile, without anthers and often with the filaments distinctly nodulose or moniliform; ovary usually 3-locular, the placentae 2–4-ovulate; ovules anatropous; style elongate, filiform; stigma usually capitate, indistinctly 3-lobed. Capsule 3-valved, ovoid or subglobose, often distinctly trigonous; seeds few; embryo circumflexed, cotyledons linear.

About 10 species chiefly in the Mediterranean region and western Asia.

Flowers solitary, leaf-opposed; petals 8–10 (–15) mm. long - - **1. F. arabica**
Flowers in an elongate, bracteate cincinnus; petals 5–8 mm. long - **2. F. thymifolia**

1. F. arabica (*L.*) *Spach* in Ann. Sci. Nat., ser. 2, 6: 359 (1836); Boiss., Fl. Orient., 1: 449 (1867); Grosser in Engl., Pflanzenr., 14 (IV, 193): 124 (1903); Janchen in Engl. et Prantl, Pflanzenfam., ed. 2, 21: 311 (1925); Post, Fl. Pal., ed. 2, 1: 148 (1932); Davis, Fl. Turkey, 1: 518 (1965); Zohary, Fl. Palaest., 2: 345, t. 510 (1972); Osorio-Tafall et Seraphim, List Vasc. Plants Cyprus, 73 (1973).

Cistus arabicus L., Sp. Plant., ed. 2, 745 (1762).

Helianthemum arabicum (L.) Pers., Syn. Plant., 2: 80 (1807); Holmboe, Veg. Cypr., 131 (1914); Lindberg f., Iter Cypr., 23 (1946); Chapman, Cyprus Trees and Shrubs, 57 (1949).

Wiry subshrub (6–) 30–60 cm. high; branches decumbent, spreading or suberect, ± densely clothed with short glandular hairs sometimes intermixed with long eglandular hairs; leaves subsessile or very shortly stalked, alternate, lanceolate, linear or narrowly elliptical, 2–14 mm. long, 1–3 mm. wide, subglabrous, or shortly glandular-pilose or occasionally ± densely canescent with long pale hairs; margins sharply recurved; stipules subulate or narrowly triangular to about 3 mm. long and 1 mm. wide; flowers solitary, leaf-opposed; pedicel slender, usually patent and arcuate towards apex, 1–1·5 mm. long, shortly glandular-pilose, usually reddish;

outer sepals subulate, 4–5 mm. long, 1–1·5 mm. wide, reddish, glandular-pilose; inner sepals boat-shaped, 8–10 mm. long, 4–5 mm. wide, glandular-pilose externally, with 5 red nerves; petals yellow, often orange-blotched at base, as long as, or rather longer than the sepals, cuneate, about 6–8 mm. wide; stamens numerous, the outermost sterile and without anthers, about 2 mm. long, filaments glabrous, distinctly nodulose or almost moniliform, twisted; inner stamens to about 3 mm. long; anthers minute, shortly oblong; ovary globose, about 1·2 mm. diam., style 3–4 mm. long, slender, glabrous, usually bent at base; stigma capitate. Capsule triquetrous, brown, shining, ± enveloped by the persistent, reddish, papery sepals, about 7 mm. long and 6 mm. wide; seeds 6–9, ovoid, angular, about 2·5 mm. long and 2 mm. wide; testa dull brown (or sometimes with a cream-coloured coating) rather regularly foveolate.

var. **arabica**
TYPE: " in Arabia ".

Plant usually 30 cm. or more high; leaves and stems shortly glandular-pilose; leaves narrow, linear or linear-lanceolate.

HAB.: On dry hillsides, in garigue; just above sea-level to 2,500 ft. alt.; fl. Febr.–May.

DISTR.: Divisions 1–3, 5, 7, 8, common in Division 7. Mediterranean region from Italy and E. Algeria east to Iran.

1. Lyso, [" Lisso "], 1913, *Haradjian* 864 !
2. Near Philani, 1908, *Reid* s.n. ! Perapedhi, 1937, *Kennedy* 652 ! Troödos area, 1951, *Probyn* s.n. !
3. Akrotiri Forest, 1939, *Mavromoustakis* 97 (in part) ! Yerasa, 1948, *Mavromoustakis* s.n. !
5. Dheftera, 1928 or 1930, *Druce* 34 ! Syngrasis, 1936, *Syngrassides* 324 !
7. Common; *Sintenis & Rigo* 213 ! *Casey* 985 ! 1310 ! *Chapman* 186 ! 511 ! *G. E. Atherton* 1090 ! 1140 ! 1250 !
8. Karpas, 1930, *Ussher* 117 ! Between Ayios Philon & Rizokarpaso, 1941, *Davis* 2247 ! Ronnas Bay near Rizokarpaso, 1941, *Davis* 2340 !

var. **incanescens** *Hausskn.* in Mitt. Thür. Bot. Ver. N. Folge, 5: 43 (1893); Grosser in Engl., Pflanzenr., 14 (IV, 193): 125 (1903); Lindberg f., Iter. Cypr., 23 (1946); Davis, Fl. Turkey, 1: 518 (1965).
TYPE: Greece; " in saxosis distr. Laurion ad Kamariza ", *Haussknecht* (JE).

A smaller plant than the type (less than 30 cm. high, and often quite dwarf); leaves and stems with long eglandular hairs mixed with the short glandular indumentum; leaves often rather short and elliptical.

HAB.: On dry rocky slopes, in garigue; sea-level to 3,800 ft. alt.; fl. March–July.

DISTR.: Divisions 1, 2, 7, 8, common in Division 2. Also in Greece, Crete, Turkey, Syria and Iraq, probably widespread in the Middle East.

1. Ayios Neophytos, 1939, *Lindberg f.* s.n. ! Psindron, near Ayios Yeoryios (Akamas), 1962, *Meikle* 2126 !
2. Locally common around Platres and Galata-Kakopetria; *Sintenis & Rigo* 212b ! *Chapman* 199 ! *Kennedy* 648 ! 649 ! 651 !
7. Near Kythrea, 1880, *Sintenis & Rigo* 216 ! Kyrenia, Mylos Plantation, 1949, *Casey* 431 ! Kyrenia Range, near Tchiklos, 1955, *G. E. Atherton* 257 ! Boghaz, 1955, *G. E. Atherton* 790 !
8. Koronia, 1937, *Chapman* 226 ! 227 ! 245 ! 250 ! 254 !

NOTES: In the extreme forms these two varieties are very easily distinguished, var. *arabica* having narrow almost glabrous leaves, while those of var. *incanescens* are short and elliptical, and conspicuously grey-hairy. The two varieties also appear to be separated ecologically, var. *arabica* preferring the lower altitudes, whilst var. *incanescens* is rarely seen below 1,000 ft. alt. The two are connected, however, by intermediates: (a) with sparse eglandular indumentum or (b) with dense, and rather conspicuous glandular indumentum, and it is impossible to consider them more than variations of a single species. Most of the specimens cited above have been examined, *F. arabica* sens. lat. is also recorded from Valia near Ayios Theodhoros (1905, *Holmboe* 503); Rizokarpaso

(1912, *Haradjian* 533); Pano Platres (1930, *Wall* s.n.); between Ayios Neophytos Monastery and Stephani (1939, *Lindberg f.*); Pentadaktylos (1939, *Lindberg f.*); Galata (1939, *Lindberg f.*); Mesopotamos, near cataract (1939, *Lindberg f.*); Sitjes, Kato Platres (1939, *Lindberg f.*) and near Larnaca (1939, *Lindberg f.*).

2. F. thymifolia (*L.*) *Verlot*, Cat. Plant. Dauph., 43 (1872); Grosser in Engl., Pflanzenr., 14 (IV, 193): 129 (1903); Janchen in Engl. et Prantl, Pflanzenfam., ed. 2, 21: 312 (1925); Post, Fl. Pal., ed. 2, 1: 148 (1932); Rechinger f. in Arkiv för Bot., ser. 2, 1: 427 (1950); Davis, Fl. Turkey, 1: 520 (1965); Zohary, Fl. Palaest., 2: 346, t. 511 (1972); Osorio-Tafall et Seraphim, List Vasc. Plants Cyprus, 73 (1973).

Cistus thymifolius L., Sp. Plant., ed. 1, 528 (1753).

Helianthemum thymifolium (L.) Pers., Syn. Plant., 2: 78 (1807); Holmboe, Veg. Cypr., 132 (1914); Lindberg f., Iter Cypr., 23 (1946); Chapman, Cyprus Trees and Shrubs, 57 (1949).

Fumana glutinosa (L.) Boiss., Fl. Orient., 1: 449 (1867); H. Stuart Thompson in Journ. Bot., 44: 277 (1906).

Helianthemum glutinosum (L.) Benth., Cat. Pyr., 85 (1826); Sintenis in Oesterr. Bot. Zeitschr., 32: 192, 259 (1882).

[*Fumana spachii* (non Gren. et Godr.) Kotschy in Unger et Kotschy, Die Insel Cypern, 339 (1865); H. Stuart Thompson in Journ. Bot., 44: 277 (1906).]

A low, spreading, much-branched subshrub, seldom more than 25 cm. high; old branches with rough, fuscous-brown bark, young shoots subglabrous or ± densely glandular, pubescent; leaves rather dense, linear, lanceolate, oblanceolate or narrowly obovate in outline, 3–15 (–20) mm. long, 1–2 mm. wide, acute or subobtuse, often mucronate, glabrous or ± densely glandular-pubescent, margins strongly revolute, often almost meeting along the midrib on the lower surface of the lamina; stipules linear or narrowly triangular, sometimes almost as long as the leaf, or again very short and inconspicuous; petiole to about 1·5 mm. long, frequently very short or almost wanting; uppermost leaves sometimes alternate, the lower normally opposite; inflorescence a lax terminal cincinnus; rhachis and pedicels generally densely glandular-pubescent; bracts much shorter than pedicels, seldom more than 5 mm. long, flanked by small stipules; pedicels patent or arcuate, 5–15 mm. long; outer sepals lanceolate or lingulate, 1–3 mm. long, 1–2 mm. wide, densely glandular pubescent; inner sepals broadly boat-shaped 4–7 mm. long, 3–6 mm. wide, distinctly 5–6-nerved, acute or obtuse; petals yellow, cuneate, 5–8 mm. long, 4–6 mm. wide, sometimes red-blotched at base; outer stamens sterile, filaments glabrous, slightly nodulose, about 1·5 mm. long; inner stamens about 2·5 mm. long; anthers minute, shortly oblong; ovary globose, about 1·2 mm. diam., subglabrous at base, often shortly pubescent at apex; style glabrous, about 2 mm. long, expanding towards apex; stigma capitate. Capsule generally rather shorter than the persistent calyx, ovoid, triquetrous, about 3–5 mm. long and 3–4 mm. wide, pale shining brown, containing 6 seeds; seeds about 2·5 mm. long and 1·5 mm. wide, shaped like the segments of an orange; testa dull brown, rather regularly foveolate.

var. **thymifolia**

Helianthemum thymifolium (L.) Pers. var. *glutinosum* (L.) Lindberg f., Iter Cypr., 23 (1946).

TYPE: " *in* G. Narbonensi, Hispania ".

Leaves ± densely glandular-pubescent, oblanceolate or narrowly obovate, generally under 1 cm. long; stipules often almost as long as the leaves, so that the foliage appears to be in whorls or tufts; flowers rather small; inner sepals obtuse or subacute.

HAB.: On dry rocky ground, or on fixed sand-dunes; sea-level to over 3,000 ft. alt.; fl. Febr.–June.

DISTR.: Divisions 1–8, common in Division 7. Widespread in the Mediterranean region and eastwards to the borders of Iran.

1. Akamas, 1937, *Chapman* 286 !
2. Platres, 1937, *Kennedy* 650 !
3. Limassol, 1913, *Haradjian* 563 ! Stavrovouni, 1933, *Syngrassides* 1376 ! Akrotiri Forest, 1939, *Mavromoustakis* 97 (in part) !
4. Near Larnaca, 1912, *Haradjian* 58; near Larnaca, 1939, *Lindberg f.* s.n.; Cape Greco, 1958, *N. Macdonald* 14 !
5. Salamis, 1931, *Tracey* 38 ! Athalassa, 1939, *Lindberg f.* s.n.
6. Kokkini Trimithia, 1933, *Syngrassides* 318 ! Ayia Irini (Morphou), 1941, *Davis* 2590 !
7. Common; *Kotschy* 516 ! *Sintenis & Rigo* 218 ! *Chapman* 60 ! 180 ! 510 ! *G. E. Atherton* 1089 ! 1364 !
8. Near Rizokarpaso, 1912, *Haradjian* 251; Koronia (near Kantara), 1937, *Chapman* 253 ! Ronnas Bay near Rizokarpaso, 1941, *Davis* 2343 !

var. **laevis** (*Cav.*) *Grosser* in Engl., Pflanzenr., 14 (IV, 193): 130 (1903); Rechinger f. in Arkiv för Bot., ser. 2, 1: 427 (1950); Post, Fl. Pal., ed. 2, 1: 149 (1932).

Cistus laevis Cav., Icones, 2: 35, t. 145, f. 1 (1793).

Helianthemum glutinosum (L.) Benth. var. *laeve* (Cav.) Benth., Cat. Pyr., 85 (1826).

Fumana glutinosa (L.) Boiss. var. *viridis* (Ten.) Boiss., Fl. Orient., 1: 449 (1867).

Helianthemum viride Ten., Fl. Nap., 1: 299, t. 47 (1811–1815).

Fumana laevis (Cav.) Sennen in Ann. Soc. Linn. Lyon, n.s., 71: 9 (1925).

TYPE: Spain; " in collibus Enguerae et in *Collado* de Bocayrente " *Cavanilles* (MA).

Leaves (except the uppermost) subglabrous and eglandular, linear or very narrowly oblanceolate, often more than 1 cm. long; stipules (of the upper leaves) usually very small, so that the leaves are clearly opposite, and not tufted in appearance; flowers often larger than in var. *thymifolia*; inner sepals acute or subacute.

HAB.: Similar to that of var. *thymifolia*; fl. Febr.–June.

DISTR.: Divisions 2, 8. External distribution similar to that of var. *thymifolia*.

2. Kykko Monastery, [" Ticco "], 1913, *Haradjian* 948; Statos, 1941, *Davis* 3396 !
8. Near Rizokarpaso, 1912, *Haradjian* 233.

NOTES: The two varieties are almost always clearly distinguishable, and may well merit subspecific (or possibly specific) differentiation. On the evidence available, however, it would be unwise to raise them beyond the rank of variety, but they should be given attention in the field. Var. *laevis* is apparently rare in Cyprus.

13. VIOLACEAE

H. Melchior (and W. Becker) in Engl. et Prantl, Pflanzenfam., ed. 2, 21: 329–377 (1925).

Perennial (or rarely annual) herbs, shrubs and subshrubs; leaves usually alternate, simple, stipulate; flowers solitary, axillary or in racemose cymes or panicles, actinomorphic or zygomorphic, sometimes cleistogamous; sepals 5, imbricate, persistent; petals 5, hypogynous, imbricate or contorted, often unequal, the lowermost larger and frequently spurred; stamens 5, hypogynous or slightly perigynous; anthers erect, introrse, generally dehiscing longitudinally, connivent or connate and forming a ring around the ovary; filaments short or wanting, connective often produced and dilated; staminodes sometimes present; ovary free, sessile, 1-locular with 3–5 parietal placentas; style usually simple sometimes wanting, stigma terminal often conspicuously thickened. Fruit a loculicidal capsule (commonly dehiscing explosively) or a berry; seeds occasionally winged, sometimes carunculate; endosperm fleshy, embryo straight.

Widely distributed through tropical and temperate regions, and not infrequently woody or arborescent, especially in the tropics. The family, which consists of about 16 genera, is represented in Cyprus only by *Viola*.

1. VIOLA *L.*

Sp. Plant., ed. 1, 933 (1753).
Gen. Plant., ed. 5, 402 (1754).
W. Becker in Beih. Bot. Centralbl., 26: 1–44 (1909), 289–390 (1910).

Perennial or annual herbs, rarely subshrubs; leaves alternate; stipules persistent, sometimes foliaceous; flowers solitary, axillary, often cleistogamous; sepals 5, commonly produced backwards from the point of insertion to form appendages; petals 5, spreading, unequal, the lowermost larger than the others and spurred or saccate at the base; stamens subsessile, the 2 lowermost frequently with basal spurs running back into the petal spur; anthers with conspicuous membranous connectives. Fruit a 3-valved capsule, frequently dehiscing explosively; seeds rather large, carunculate.

About 400 species, the majority in temperate regions of the northern hemisphere, but with a cosmopolitan distribution.

Perennial species; leaves cordate:
Sepals obtuse; plants ± pubescent; flowers scented; stipules membranous:
Stipules ovate or ovate-lanceolate, shortly glandular-fimbriate; stolons rooting at tip - - - - - - - - - - - - - **1. V. odorata**
Stipules linear-lanceolate with long-ciliate margins; stolons not (or tardily) rooting at tip - - - - - - - - - - - - - - **2. V. alba**
Sepals acute; plant glabrous; flowers scentless; stipules not membranous **3. V. sieheana**
Annuals; leaves not cordate:
Stipules with foliaceous mid-lobe and several palmately arranged lateral lobes; leaves usually crenate-serrate - - - - - - - **4. V. arvensis ssp. kitaibeliana**
Stipules not foliaceous, with a few irregular lateral lobes or none:
Sepals narrowly oblong; mid-lobe of stipule usually with a slender stipe; plant subglabrous or sparsely puberulous - - - - - **5. V. heldreichiana**
Sepals ovate; mid-lobe of stipule not stipitate; plant usually distinctly villose **6. V. parvula**

SECT. 1. **Viola.** Fertile flowers of two kinds: normal, with petals (chasmogamous), and modified, apetalous (cleistogamous); style hooked or obliquely truncate, with a subglobose head; stipules usually small, not leaf-like.

1. **V. odorata** *L.*, Sp. Plant., ed. 1, 934 (1753); Unger et Kotschy, Die Insel Cypern, 339 (1865); Boiss., Fl. Orient., 1: 458 (1867); Holmboe, Veg. Cypr., 132 (1914); Post, Fl. Pal., ed. 2, 1: 150 (1932); Lindberg f., Iter Cypr., 24 (1946); Davis, Fl. Turkey, 1: 525 (1965); Zohary, Fl. Palaest., 2: 334 (1972); Osorio-Tafall et Seraphim, List Vasc. Plants Cyprus, 73 (1973).
[*V. suavis* (non M. Bieb.) Kotschy in Oesterr. Bot. Zeitschr., 12: 278 (1862).]
TYPE: " *in* Europae *nemoribus* ".

A perennial with a short thick rhizome and long creeping flowerless stolons rooting at the tip; leaves bluntly cordate or almost reniform, 2·5–8 cm. long, 2–6 cm. wide, thinly puberulous or subglabrous, basal sinus usually wide; leaf-margins rather bluntly crenate-serrate; petiole to 16 cm. long, or sometimes longer, thinly deflexed-puberulous and distinctly channelled on the adaxial surface; stipules membranous, ovate or ovate-lanceolate, 0·8–1·5 cm. long, 0·4–0·8 cm. wide, margins shortly glandular-fimbriate; pedicels equalling or a little longer than the leaves, with two small linear bracteoles about the middle; flowers held horizontally or nodding, sweet-scented; sepals oblong, blunt, 5–7 mm. long, 2–4 mm. wide, glabrous or subglabrous with distinct spreading basal appendages; petals (in Cyprus specimens) usually rich violet-blue, occasionally paler or white, the upper four petals narrowly obovate-oblong, about 1·5 cm. long and 0·4 cm. wide, lateral petals shortly bearded towards base, the basal petal about 0·6–0·8 cm. wide, produced into a straight, blunt spur 0·4–0·5 cm. long; anthers about 2 mm. long, a little longer than the brownish, expanded connective; ovary hemispherical, about 1·8 mm. diam., pubescent; style about 3 mm. long, with an acute deflexed apex. Capsule (developed from cleistogamous flower) subglobose, 5–7 mm. diam., obscurely trigonous, shortly pubescent, and commonly covered with small violet blotches; seeds ovoid, about 3 mm. long and 1·5 mm. wide, conspicuously caruncülate; testa pale creamy-brown, smooth.

HAB.: In shaded places under rocks and bushes, 1,500–3,500 ft. alt.; fl. Febr.–April.

DISTR.: Division 2, common. Widespread in Europe and western Asia.

2. Common on the Troödos massif; *Kotschy* 753 ! *Syngrassides* 44 ! *Kennedy* 642 ! 1800 ! *Merton* 1962 ! 2491 ! 3191 ! *Meikle* 2655 !

2. V. alba *Bess.*, Primit. Fl. Galic., 1: 171 (1809); Boiss., Fl. Orient., 1: 458 (1867); Post, Fl. Pal., ed. 2, 1: 150 (1932).

V. dehnhardtii Ten., Cat. Sem. Hort. Neap., 12 (1830).

V. thessala Boiss. et Sprun. in Boiss., Diagn. 1, 8: 51 (1849); Lindberg f., Iter Cypr., 24 (1946).

V. alba Bess. ssp. *dehnhardtii* (Ten.) W. Becker in Ber. Bayr. Bot. Ges., 8: 257 (1902); Davis, Fl. Turkey, 1: 526 (1965); Osorio-Tafall et Seraphim, List Vasc. Plants Cyprus, 73 (1973).

TYPE: Czechoslovakia; " In montosis et sylvaticis versus Duklam " (? KW).

A sprawling perennial, with a short thick rhizome and long slender stolons, not rooting or rooting tardily, and commonly bearing flowers during their first season of growth; leaves tougher in texture than those of *V. odorata*, often persisting through the winter, and generally narrower, more acute and often (together with the petioles) distinctly pilose with short stiff hairs, but sometimes glabrous or subglabrous; stipules of the spring growths linear-lanceolate, much narrower than those of *V. odorata*, frequently pilose and with long-ciliate margins; flowers sweet-scented; sepals obtuse, often ciliate, frequently shorter and broader than in *V. odorata*; petals violet, mauve or white; stamens, style and ovary as in *V. odorata*. Capsule subglobose, often distinctly pilose; seeds as in *V. odorata*.

HAB.: In shaded places under trees or rocks, or by streamsides; 2,000–3,500 ft. alt.; fl. Febr.–March.

DISTR.: Division 2. Widespread in Europe and eastwards to Iraq and Iran.

2. Saïttas, 1932, *Syngrassides* 339 ! and 1962, *Meikle* 2791 ! Perapedhi, 1932, *Syngrassides* 392 ! Platres, 1936, *Kennedy* 641a ! Platania, 1939, *Lindberg f.* s.n. ! Stavros tis Psokas, 1962, *Meikle* 2701 !

NOTES: Material is insufficient for infraspecific determinations; it must moreover be admitted that the differences between ssp. *alba*, ssp. *dehnhardtii* (Ten.) W. Becker and ssp. *thessala* (Boiss. et Sprun.) Hayek are somewhat tenuous.

3. V. sieheana *W. Becker* in Bull. Herb. Boiss., ser. 2, 2: 751 (1902); Rechinger f. in Arkiv för Bot., ser. 2, 1: 427 (1950); Davis, Fl. Turkey, 1: 527 (1965); Osorio-Tafall et Seraphim, List Vasc. Plants Cyprus, 73 (1973).

V. sieheana W. Becker var. *oblongifolia* W. Becker, Viol. exsicc., 135 (1905); Beih. Bot. Centralbl., 26, Abt. 2: 298 (1910).

V. sieheana W. Becker var. *oblongifolia* W. Becker f. *grandistipulata* W. Becker in Beih. Bot. Centralbl., 26, Abt. 2: 298 (1910).

V. cypria Lindberg f., Iter Cypr., 23 (1946).

[*V. canina* (non L.) Sibth. et Sm., Fl. Graec. Prodr., 146 (1806); Poech, Enum.

Plant. Ins. Cypr., 33 (1842); Unger et Kotschy, Die Insel Cypern, 339 (1865) including δ *macrantha* and γ *sylvestris*.]

[*V. sylvatica* (non Fries) Boiss., Fl. Orient., 1: 459 (1867) pro parte.]

[*V. sylvatica* Fries var. *riviniana* Boiss., Fl. Orient., 1: 459 (1867) pro parte—non *V. riviniana* Reichb.]

[*V. silvestris* Lam. ssp. *riviniana* Holmboe, Veg. Cypr., 132 (1914)—non *V. riviniana* Reichb.]

[*V. silvestris* (non Lam.) Nattrass, First List Cypr. Fungi, 21, 33 (1937).]

[*V. riviniana* (non Reichb.) Druce in Rep. B.E.C., 9: 470 (1931).]

[*V. mirabilis* (non L.) Druce in Rep. B.E.C., 9: 470 (1931).]

[*V. riviniana* Reichb. ssp. *neglecta* Post, Fl. Pal., ed. 2, 1: 150 (1932); Rechinger f. in Arkiv för Bot., ser. 2, 1: 427 (1950)—non *V. neglecta* Schmidt.]

TYPE: Turkey, Cilicia; " Kagiraki, ad ripas, 600 m." 1895, *Siehe* 131 (K !).

A loosely tufted perennial 3–20 cm. high, without stolons; rootstock elongate, woody; stems spreading or suberect, distinctly angled, glabrous, arising from an ill-defined central rosette of small long-stalked basal leaves; leaves broadly cordate to oblong-cordate, acute or obtuse, 1–5 cm. long, 1–4 cm. wide, glabrous, with rather distinct radiating nervation, a shallow basal sinus, and a shallowly crenate-serrate margin; petioles glabrous, channelled above, up to about 8 cm. long becoming shorter towards apex of stem; stipules of the basal rosette usually brownish, papery, ovate-lanceolate, distinctly glandular-ciliate; stipules of the cauline leaves progressively larger towards apex of stem, becoming green and foliaceous, linear-lanceolate, 1–2 cm. long, 0·3–0·5 cm. wide, subentire or irregularly ciliate-serrate; flowers held horizontally or nodding, scentless; pedicels slender, 2–10 cm. long, with 2 linear bracteoles, 5–7 mm. long, usually well above the middle; sepals linear-lanceolate, acuminate, 6–10 mm. long, 1·5–2 mm. wide, with conspicuous, truncate basal appendages; petals lavender-blue or lilac (occasionally white), distinct or overlapping, 0·8–2 cm. long, 0·3–1·2 cm. wide, the lateral petals shortly bearded near the base, the basal petal often not very much wider than the others, finely pencilled with black lines towards the centre, and continued into a blunt, straight or upturned spur about 1 cm. long; anthers oblong, about 2 mm. long and 1·5 mm. wide, with papillose-ciliate margins, and an ovate, acute connective about 1·5–2 mm. long; ovary conical, glabrous, about 2 mm. long; style about 2 mm. long, with a deflexed apical beak. Capsule narrowly ellipsoid, acute, trigonous, 1–1·5 cm. long, about 0·5 cm. wide, glabrous; seeds ovoid, about 2·3 mm. long and 1·2 mm. wide with a small caruncle; testa pale brown, glabrous, smooth. *Plate 14.*

HAB.: In shaded places under trees and by streams, 2,500–6,400 ft. alt.; fl. March–May and occasionally also during autumn and winter.

DISTR.: Division 2, common. Also in Turkey, Iran and the Caucasus area.

2. Very common locally on the central Troödos massif; *Kotschy* 708 ! *Sintenis & Rigo* 780 ! *Holmboe* 855, 868; *Syngrassides* 45 ! 740 ! *Kennedy* 638–641 ! 1507 ! *Davis* 1089 ! 3099 ! 3154 ! *Merton* 3115 ! 3203 ! *Meikle* 2654 ! 2674 !

NOTES: The flowering stems of this plant are very similar to those of the European *V. montana* L., from which, however, *V. sieheana* differs in having a rosette of basal leaves flanked with conspicuous, glandular-ciliate stipules.

SECT. 2. **Melanium** *Ging.* Flowers all normal, with petals; style usually with a subglobose head bearing a stigmatic cavity on its undersurface; stipules generally conspicuous, often foliaceous and pinnatisect.

4. **V. arvensis** *Murr.*, Prodr. Stirp. Gotting., 73 (1770); W. Becker in Mitt. Thür. Bot. Ver., 19: 42 (1904).

An erect or spreading, branched or unbranched herb, 5–50 cm. high; stems angular, thinly pubescent or subglabrous; basal leaves pubescent or subglabrous, broadly oblong or suborbicular, 0·8–2 cm. long, 0·5–1·5 cm.

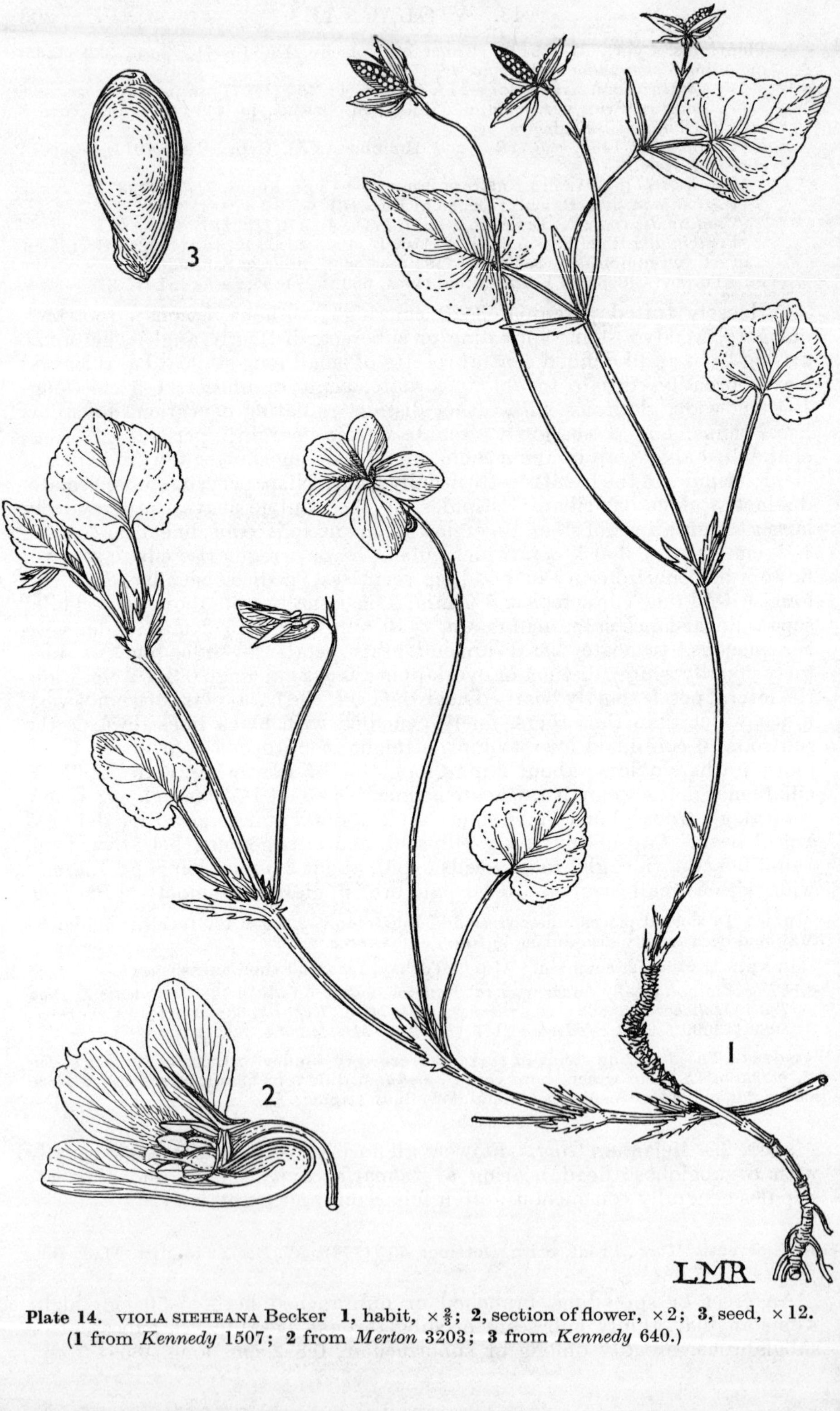

Plate 14. VIOLA SIEHEANA Becker **1**, habit, $\times\frac{2}{3}$; **2**, section of flower, $\times 2$; **3**, seed, $\times 12$.
(**1** from *Kennedy* 1507; **2** from *Merton* 3203; **3** from *Kennedy* 640.)

wide, margins distinctly crenate; petiole 1–4 cm. long, or sometimes longer; stipules conspicuous, with a stipitate, foliaceous mid-lobe, often approaching the leaf in size and shape, and with 2–5 linear lateral lobes arranged palmately or pinnately at the base of or along the lower part of the stipe of the mid-lobe, upper leaves progressively longer, narrower and more acute, sometimes up to 7 cm. long, tapering gradually at the base to a short, flattened petiole, margins rather remotely serrate-crenate; uppermost leaves frequently almost linear, with very remote and obscure marginal serrations; flowers all chasmogamous; pedicels to about 7 cm. long, generally rather longer than the subtending leaves, with two small bracteoles near the apex; sepals narrowly ovate, acute, 0·5–1·2 cm. long, subglabrous or ciliate, with well-developed, truncate basal appendages; petals milky-white, yellowish or occasionally tinged blue, sometimes shorter than sepals, or sometimes a little longer, the two lateral petals, and the much broader basal petal, shortly bearded towards the base; spur often a little longer than the basal appendages of the calyx, frequently up-turned at the base; anthers broadly oblong, about 1 mm. long, glabrous or ciliolate, with an expanded brownish connective; ovary about 1·5 mm. long, glabrous; style about 1 mm. long, with a swollen, globose, papillose apex, on the lower side of which is the small stigmatic cavity. Capsule 3–8 mm. long, markedly trigonous, acute, glabrous or subglabrous, dehiscing explosively; seeds ellipsoid-ovoid, 1–1·6 mm. long, 0·5–1 mm. wide, with a small caruncle; testa glossy brown, smooth or sometimes slightly roughened.

ssp. **kitaibeliana** (*Roem. et Schult.*) *W. Becker* in Mitt. Thür. Bot. Ver., 19: 42 (1904).
V. kitaibeliana Roem. et Schult., Syst. Veg., 5: 383 (1819); Post., Fl. Pal., ed. 2, 1: 151 (1932); A. K. Jackson et Turrill in Kew Bull., 1938: 461 (1938); Davis, Fl. Turkey, 1: 531 (1965); Zohary, Fl. Palaest., 2: 334, t. 491 (1972); Osorio-Tafall et Seraphim, List Vasc. Plants Cyprus, 73 (1973).
[*V. tricolor* L. var. *arvensis* Kotschy in Unger et Kotschy, Die Insel Cypern, 340—non *V. arvensis* Murr.]
[*V. occulta* (non Lehm.) Boiss., Fl. Orient., 1: 467 (1867) pro parte quoad spec. cypr.; Holmboe, Veg. Cypr., 132 (1914).]
TYPE: Hungary; " in Pannonia " *Kitaibel* (M).

Smaller and more slender than ssp. *arvensis*; lateral lobes of stipules usually numerous (even in small specimens) and palmately arranged at the base of the stipitate mid-lobe; petals usually (but not always) shorter than sepals. Capsule small.

HAB.: In vineyards, gardens and fields; sometimes on grassy hillsides or on bare sandy places; c. 2,000–4,000 ft. alt.; fl. March–May.

DISTR.: Division 2, common. Widespread in the Mediterranean area.

2. Common in the central and eastern parts of the Division, especially around Platres; *Kotschy* 710 ! *Kennedy* 636 ! 637 ! 637A ! 1564 ! 1980 ! *Davis* 3094 ! *Merton* 3146 ! *N. Macdonald* 181 ! *Meikle* 2934 !

NOTES: Intergrades with typical *V. arvensis* in some parts of Europe, but tolerably uniform and distinct in Cyprus. True *V. occulta* Lehm., which occurs in Turkey, has not yet been discovered in Cyprus. It can be recognized by the ebracteolate pedicels and by the very much enlarged basal appendages of the lowermost sepals. In *V. occulta* the leaves are frequently entire or subentire.

5. **V. heldreichiana** *Boiss.*, Diagn., 1, 8: 53 (1849), Fl. Orient., 1: 466 (1867); Holmboe, Veg. Cypr., 132 (1914); Lindberg f., Iter Cypr., 24 (1946); Davis, Fl. Turkey, 1: 531 (1965); Osorio-Tafall et Seraphim, List Vasc. Plants Cyprus, 73 (1973).
TYPE: Turkey; " in lapidosis inter frutices prope pagum *Karabahir* ad radices montis *Climacis Lyciae* " *Heldreich* (G).

A very small, subglabrous or sparsely puberulous annual, 1–8 cm. high (to 12 cm. in cultivation); stems unbranched or with a few decumbent or spreading basal branches, often tinged purple; basal leaves orbicular or spathulate, usually less than 8 mm. wide, entire or obscurely crenate, often purplish, with a slender petiole 0·5–1·5 cm. long; upper leaves narrower, oblong, lanceolate or occasionally almost linear, obtuse or acute, generally subentire, tapering to a distinct, slender petiole; stipules much reduced, consisting of a stipitate, entire, narrowly spathulate or oblanceolate mid-lobe with or without 1 or 2 minute basal teeth or filiform lobes; flowers all chasmogamous, minute; pedicels longer than leaves, divaricate with 2 minute bracteoles; sepals narrowly oblong, acute, about 3–5 mm. long, basal appendages of the lowermost sepals well-developed, the remainder very short; petals white or tinged mauve, shorter than sepals, distinctly bearded towards base; spur about as long as basal appendages; anthers oblong, about 0·7 mm. long with a dilated brown connective; ovary less than 1 mm. long, glabrous; style about 1 mm. long, with a very swollen, globose, papillose apex. Capsule trigonous, about 4 mm. long; seeds ellipsoid about 1·2 mm. long and 0·7 mm. wide, with a very small caruncle; testa smooth, shining brown.

HAB.: On mountainsides under rocks and bushes, or on gravel in dried-up stream-beds; 3,500–6,400 ft. alt.; fl. March–Aug.

DISTR.: Division 2, common. Also in Crete and Turkey.

2. Tripylos, Kionia and locally very common on the upper slopes of central Troödos; *Sintenis & Rigo* 781 (as *V. parvula* Tineo) ! *Holmboe* 981, 1080; *Kennedy* 628–634 ! (see notes) *Lindberg f.* s.n. ! *Davis* 1803 ! 1817 ! 2705 ! 3208 ! 3520 ! *Merton* 1586 ! *Meikle* 2680 ! 2774 !

NOTES: Amongst the specimens of *Kennedy* 631 are several with exceptionally large flowers, the petals exceeding the sepals, and approaching *V. alajensis* W. Becker and *V. mercurii* Orph. (both regarded as forms of *V. heldreichiana* by Becker in Beih. Bot. Centralbl., 26: 341; 1910). The Cyprus specimens are too depauperate for accurate determination, but collectors should look out for such large-flowered plants in the neighbourhood of the upper reaches of the Kryos Potamos. It is noticeable that these forms tend to be unbranched, and paler green than the common, small-flowered form, and that the spur frequently exceeds the basal appendages of the calyx. Additional material (*Meikle* 2680 !) from the summit of Tripylos Peak also tends to have rather larger flowers than typical *V. heldreichiana*, but is, in other respects, indistinguishable from the latter species.

6. V. parvula *Tineo*, Plant. Rar. Sic. Pug., 5 (1817); Boiss., Fl. Orient., 1: 466 (1867); Holmboe, Veg. Cypr., 132 (1914); Post, Fl. Pal., ed. 2, 1: 151 (1932); Lindberg f., Iter Cypr., 24 (1946); Davis, Fl. Turkey 1: 530 (1965); Osorio-Tafall et Seraphim, List Vasc. Plants Cyprus, 73 (1973).

V. parvula Tineo var. *subarachnoidea* Lindberg f., Iter Cypr., 24 (1946).

TYPE: Sicily; " *in Nebrodibus*. Sutta la Curma grande " *Tineo* (G).

A very small annual, 1–6 (–13) cm. high; whole plant sparsely clothed with long white hairs, or sometimes subglabrous; stems unbranched, erect, or frequently with numerous spreading branches, forming low rounded tufts; basal leaves spathulate-obovate or suborbicular, 5–15 mm. long 4–10 mm. wide, margin entire or very obscurely dentate; petiole 3–10 mm. long; upper leaves longer and narrower, elliptic, narrowly obovate or oblanceolate, usually entire, gradually tapering to petiole; stipules with a narrowly oblong or ligulate mid-lobe and 1–3 acute lateral lobes, or sometimes without lateral lobes; flowers all chasmogamous; pedicels short, slender, 2-bracteolate, equalling or slightly exceeding the leaves; sepals ovate, acute, 3–5 mm. long, 1·5–3 mm. wide, often densely villose; basal appendages rather strongly developed, usually longer than the short blunt spur; petals equalling or shorter than sepals, creamy-white flushed blue or purple; anthers oblong, about 0·5 mm. long with a short, dilated, brown connective; ovary glabrous about 0·5 mm. long; style very short, about 0·3 mm. long, bent upwards and with a small globose, papillose apex. Capsule trigonous, 3–4 mm. long;

seeds narrowly ellipsoid, about 1·2 mm. long and 0·8 mm. wide, caruncle minute; testa smooth, shining brown.

HAB.: Gravelly ground on mountain summits, often under bushes of *Berberis cretica*; 6,200–6,400 ft. alt.; fl. April–June.

DISTR.: Division 2, locally common. Also Spain, Corsica, Sicily, S. Italy, Greece, Turkey and Syria.

2. Locally common on the summit of Khionistra; *Holmboe* 980; *Kennedy* 635 ! *Lindberg f.* s.n. ! *Davis* 3162 ! *Merton* 3168 ! *Meikle* 2858 !

14. POLYGALACEAE

Herbs, shrubs, climbers or small trees; leaves usually alternate, simple, exstipulate; flowers hypogynous, bisexual, zygomorphic, solitary or in spikes or racemes; pedicels often articulated, each frequently subtended by a bract and sometimes furnished with 2 bracteoles; sepals generally 5, imbricate, free or partly united, commonly unequal, the 2 inner often large and petaloid; petals 3–5, frequently adnate basally to stamens, lowermost (or median) petal often concave with a fringed crest or keel; stamens usually 8 (rarely 3–7), monadelphous or rarely free; anthers basifixed, usually 1-locular, rarely 2-locular, dehiscing by a terminal pore, or rarely splitting longitudinally; ovary generally 2-locular (rarely 1-locular or 3–5-locular); style simple; stigmas or stigma-lobes as many as the loculi; placentation axile, ovules usually solitary in each loculus, pendulous, anatropous. Fruit a capsule, samara or drupe; seed often hairy and with a conspicuous caruncle.

About 20 genera, widely distributed in temperate and tropical regions of both hemispheres.

1. POLYGALA *L.*

Sp. Plant., ed. 1, 701 (1753).
Gen. Plant., ed. 5, 315 (1754).

Annual or perennial herbs or subshrubs; leaves generally alternate; flowers in terminal or lateral racemes (rarely solitary); sepals conspicuously unequal, the 2 inner larger than the 3 outer, and petaloid, all free or the 2 anterior sometimes partly united; petals basically 5, but usually only 3 developed, the lowermost concave with a conspicuous fimbriate crest, the 2 lateral absent or vestigial, the 2 uppermost joined at the base to the lowermost; stamens usually 8 (or 6 stamens and 2 staminodes) monadelphous; ovary 2-locular. Fruit a flattened 2-locular capsule; seed solitary in each loculus of the capsule, usually covered with silky hairs, conspicuously carunculate.

About 500 species widely distributed in tropical and temperate regions.

Corolla distinctly shorter than the 2 enlarged inner sepals; plant annual **1. P. monspeliaca**
Corolla distinctly longer than the 2 enlarged inner sepals; plant perennial **2. P. venulosa**

1. **P. monspeliaca** *L.*, Sp. Plant., ed. 1, 702 (1753); Boiss., Fl. Orient., 1: 469 (1867); Chodat, Mon. Polyg., 480 (1891); Holmboe, Veg. Cypr., 118 (1914); Post, Fl. Pal., ed. 2, 1: 152 (1932); Davis, Fl. Turkey, 1: 538 (1965); Zohary, Fl. Palaest., 2: 294, t. 433 (1972); Osorio-Tafall et Seraphim, List Vasc. Plants Cyprus, 66 (1973).

P. glumacea Sibth. et Sm., Fl. Graec. Prodr., 2: 52 (1813), Fl. Graec., 7: 66, t. 670 (1830); Unger et Kotschy, Die Insel Cypern, 356 (1865).

TYPE: France; " Monspelii, *in collibus sterilibus* ".

An erect or spreading, branched or unbranched annual 7–20 (–30) cm. high; stem subterete or obscurely angular, ± distinctly crispate-pubescent; lower leaves frequently opposite, rather crowded, sessile, oblong or elliptic, 0·4–1 cm. long, 0·2–0·5 cm. wide, glabrous or puberulous along margins and midrib; upper leaves alternate, sessile, lanceolate-acuminate, 1–3·5 cm. long, 0·1–0·5 cm. wide; inflorescence terminal, forming a lax raceme 5–15 cm. long, and lengthening in fruit; pedicels 1–2 mm. long, subtended by 1 narrowly ovate, acuminate bract about 1·6 mm. long and 0·5 mm. wide and 2 smaller bracteoles about 1·2 mm. long and 0·4 mm. wide; outer sepals narrowly lanceolate, acuminate, about 2·5 mm. long and 0·5 mm. wide, glabrous; inner sepals ovate-elliptic, 7–10 mm. long, 3–5 mm. wide, distinctly veined; petals whitish, the 2 uppermost about 4 mm. long and 0·5–1 mm. wide, not overlapping, glabrous, minutely papillose at apex, connate along their lower margins to the basal petal which is about 2·5 mm. long and conspicuously fimbriate-papillose at apex; filament-sheath about 2·5 mm. long, adnate to basal petal for most of its length, ciliate along the margin; anthers 8, yellow, oblong, about 0·2 mm. long and 0·1 mm. wide; style glabrous about 0·6 mm. long, produced beyond the stigma to form a small, ovate, concave appendage. Capsule narrowly obcordate, 5–6 mm. long, 2·5–3 mm. wide, with a distinct marginal wing; seed oblong, about 2·5 mm. long and 1 mm. wide, densely white-hairy; testa fuscous; caruncle about 0·5 mm. long, yellowish-brown, base bluntly 3-lobed, apex crowned with a tuft of hairs.

HAB.: On grassy slopes; in garigue; by roadsides or in shallow depressions on sand or clay; sea-level to c. 3,000 ft. alt.; fl. Febr.–May.

DISTR.: Divisions 1, 2, 4–8, common in Division 7. Widespread in the Mediterranean region and eastwards to Iraq.

1. Dhrousha (Akamas), 1941, *Davis* 3256 ! Fontana Amorosa, 1962, *Meikle* 2290 !
2. Ayios Nikolaos (Paphos Distr.), 1960, *N. Macdonald* 165 !
4. Around Larnaca, 1862, *Kotschy* 316; Ayia Napa, 1905, *Holmboe* 68; Cape Greco, 1905, *Holmboe* 419.
5. Kythrean hummocks above Mia Milea, 1957, *Merton* 2962 !
6. Syrianokhori, 1935, *Syngrassides* 802 ! Makhedonitissa Monastery [" Machinoditissa "], 1936, *Syngrassides* 1107 !
7. Common; *Sintenis & Rigo* 3 ! *Davis* 2458 ! 2775 ! 2843 ! 3008 ! *Casey* 646 ! *Chapman* 63 ! *Meikle* 2360 !
8. Between Ayios Philon and Rizokarpaso, 1941, *Davis* 2227 !

2. P. venulosa *Sibth. et Sm.*, Fl. Graec. Prodr., 2: 52 (1813), Fl. Graec., 7: 65, t. 669 (1830); Unger et Kotschy, Die Insel Cypern, 356 (1865); Boiss., Fl. Orient., 1: 473 (1867); Chodat, Mon. Polyg., 433 (1891); Holmboe, Veg. Cypr., 118 (1914); Davis, Fl. Turkey, 1: 536 (1965); Osorio-Tafall et Seraphim, List Vasc. Plants Cyprus, 66 (1973).

TYPE: Cyprus, Greece; " In insulae Cypri, nec non in Argolidis et Lacaoniae, montibus " *Sibthorp* (OXF).

An erect, spreading or decumbent perennial, (3–) 10–25 (–30) cm. high; rootstock woody; stems angular, shortly and often rather sparsely crispate-pubescent; leaves alternate, sessile, the lower narrowly obovate, subacute or obtuse, 0·5–1 cm. long, 0·3–0·5 cm. wide, glabrous or pubescent on margin and midrib; upper leaves similar but oblanceolate or narrowly elliptical, acute, 1–2 (–3) cm. long, 0·5–0·8 mm. wide; uppermost leaves often very narrow, sometimes almost linear; inflorescence terminal, at first forming a rather dense raceme about 2–5 cm. long, lengthening to 8–10 cm. or more in fruit and becoming lax; pedicels 1–2 mm. long, subtended by 1 oblong-lanceolate, concave, ciliate bract about 1–1·5 mm. long and 0·8 mm. wide and 2 smaller, subglabrous or sparsely ciliate bracteoles about 0·8–1 mm. long and 0·5 mm.

wide; outer sepals narrowly oblong, 2–3 mm. long, 0·8–1 mm. wide, glabrous or sparsely ciliate; inner sepals narrowly obovate or elliptical, 5–8 mm. long, 2–4 mm. wide, enlarging in fruit, glabrous or very sparsely ciliate, distinctly green-veined; petals pale mauve, pale pink or whitish, the 2 uppermost 1·2–1·7 cm. long, 0·3–0·4 cm. wide towards apex and overlapping, connate along their lower margins with the basal petal; basal petal about 1–1·2 cm. long, conspicuously papillose-fimbriate at apex and with the lateral margins distinctly ciliate; filament-sheath about 8 mm. long, adnate to basal petal for most of its length, margin ciliate; anthers 8, long, yellow, about 0·5 mm. long, opening by an apical pore; ovary obovate, about 2 mm. long and 1 mm. wide; style about 0·5 cm. long, produced beyond the stigma as an ovate appendage; lower surface of style, immediately below the stigma, densely bearded. Capsule obcordate, 5–6 mm. long, 4–5 mm. wide, with a distinct marginal wing; seeds oblong-ovoid, about 2·5 mm. long and 1·2 mm. wide, densely white hairy; testa brown or blackish; caruncle about 0·8 mm. long, yellowish-brown, base sharply 3-lobed, apex crowned with a tuft of hairs.

HAX.: On dry grassy hillsides; in garigue; in open Pine forest; on sand-dunes; sea-level to 2,600 ft. alt.; fl. Febr.–May.

DISTR.: Divisions 1, 3, 6–8, common in Division 7. Also in Greece, Crete, Rhodes and the Aegean Islands.

1. Dhrousha (Akamas), 1941, *Davis* 3236 ! Smyies (Akamas), 1948, *Kennedy* 1592 ! Terra, near Khrysokhou, 1955, *Merton* 2117 ! Karavopetres near Ayios Yeoryios (Akamas), 1962, *Meikle* 2063 !
3. Above Trimiklini, 1941, *Davis* 3145 ! Pano Polemidhia, 1948, *Mavromoustakis* s.n. ! Near Souni, 1960, *N. Macdonald* 122 ! Kandou, 1963, *J. B. Suart* 50 !
6. Dhiorios, 1952, *Probyn* 83 !
7. Common (locally around Kyrenia, Kantara and Halevga); *Sintenis & Rigo* 2 ! *Syngrassides* 1539 ! *Davis* 2438 ! 2827 ! *G. E. Atherton* 1032 ! 1242 !
8. Ronnas Bay near Rizokarpaso, 1941, *Davis* 2348 !

POLYGALA APOPETALA *T. S. Brandeg.*, a tall shrubby species from southern California, was planted at Salamis sometime prior to 1915, but " without any marked success " (Bovill, A Report on Plantation Work in Cyprus, 14; 1915).

15. FRANKENIACEAE

Niedenzu in Engl. et Prantl, Pflanzenfam., ed. 2, 21: 276–281 (1925).

Annual or perennial herbs or subshrubs. Stems much-branched. Leaves opposite, entire, often small and ericoid, exstipulate; flowers solitary or in terminal or axillary cymes, hermaphrodite or rarely unisexual, actinomorphic; sepals 4–6, persistent, connate, forming a tubular, shortly lobed calyx; petals equal in number to sepals, free or united, hypogynous, clawed, with a scale-like appendage on the claw; stamens usually 6, hypogynous, in 2 whorls, free or connate towards the base; anthers 2-locular, extrorse, versatile, dehiscing longitudinally; ovary superior, 1-locular, sessile, with 2–4 parietal placentae; ovules usually numerous, long-funiculate; style I, slender, divided above into 2–4 stigmatic arms. Fruit a loculicidal capsule enclosed in the persistent calyx. Seeds ovoid or oblong, endospermous; embryo straight.

A family of 4 genera, represented in every continent, and particularly characteristic of areas having a Mediterranean type of climate. Most of the species show a distinct preference for saline habitats.

1. FRANKENIA *L.*

Sp. Plant., ed. 1, 331 (1753).
Gen. Plant., ed. 5, 154 (1754).

Herbaceous annuals or subshrubby perennials; stems terete; leaves opposite, decussate, usually patent or recurved and sessile or shortly petiolate; flowers solitary or in cymes; sepals and petals usually 5 (sometimes 4); petals free, imbricate; stamens 6 (sometimes 4–5); ovary of 3 or 2 carpels, each with a prominent midrib; all the placentae bearing ovules. Capsules angular.

About 44 species with a cosmopolitan distribution; particularly abundant in the Mediterranean area, South Africa, Australia, California and temperate South America.

Leaves obovate, often retuse; flowers solitary; annual herb - **1. F. pulverulenta**
Leaves linear or oblong-linear; flowers in cymes; perennial herb or subshrub **2. F. hirsuta**

1. F. pulverulenta *L.*, Sp. Plant., ed. 1, 332 (1753); Boiss., Fl. Orient., 1: 779 (1867); Holmboe, Veg. Cypr., 129 (1914); Post, Fl. Pal., ed. 2, 1: 225 (1932); Lindberg f., Iter Cypr., 23 (1946); Davis, Fl. Turkey, 2: 352 (1967); Zohary, Fl. Palaest., 2: 364, t. 530 (1972); Osorio-Tafall et Seraphim, List Vasc. Plants Cyprus, 71 (1973).
TYPE: " *in* Susexiae, Narbonae, Italiae, Apuliae *littoribus* ".

Procumbent, much-branched annual; stems very shortly and thinly white-pilose or pulverulent; leaves obovate, 2–7 mm. long, 1–3 mm. wide, subsessile or very shortly petiolate, apex usually retuse, margins flat or a little recurved, upper surface of lamina sparsely pulverulent, lower surface more densely so; petiole 0·5–1·5 mm. long, flattened, often ciliate. Flowers solitary, sessile in the forks of the branches or axillary; calyx about 4 mm. long, narrowly cylindrical, strongly ribbed, pulverulent, with 5 acute, triangular lobes 0·5–1 mm. long; petals 5, pink, narrowly spathulate, about 5 mm. long and 1–1·5 mm. wide at the obtuse, erose apex, appendage narrowly elliptical; stamens 3·5–4 mm. long, with membranous, distinctly dilated filaments and shortly oblong anthers about 0·4 mm. long; ovary fusiform, distinctly triquetrous, about 1·5 mm. long and 0·4 mm. wide; style 1–1·5 mm. long; seeds ellipsoid, slightly compressed, about 0·5 mm. long and 0·3 mm. wide; testa pale brown, minutely papillose all over or at least towards distal end of seed.

HAB.: Sandy or rocky ground by the sea; occasionally on sandy ground inland; sea-level to about 500 ft. alt.; fl. March–Sept.

DISTR.: Divisions 1, 4, 6, 8, common in Division 4. Throughout the Mediterranean region and eastwards to Central Asia and the Punjab; south to Senegal; introduced elsewhere.

1. Ktima, 1913, *Haradjian* 674 !
4. Common; *Kotschy* s.n.; *Holmboe* 427; *Syngrassides* 1768 ! *Kennedy* 1743 ! 1746 ! 1747 ! *Chapman* 28 ! 436 ! *Merton* 2253 ! etc.
6. Near Nicosia, 1889, *Pichler* s.n. ! Dhiorios, 1962, *Meikle* !
8. Near Yialousa, 1880, *Sintenis & Rigo* 35 ! Cape Andreas, 1938, *Syngrassides* 1768 !

2. F. hirsuta *L.*, Sp. Plant., ed. 1, 331 (1753); Boiss., Fl. Orient, 1: 780 (1867); Holmboe, Veg. Cypr., 129 (1914); Post, Fl. Pal., ed. 2, 1: 225 (1932); Lindberg f., Iter Cypr., 23 (1946); Davis, Fl. Turkey, 2: 352 (1967); Zohary, Fl. Palaest., 2: 363, t. 529 (1972); Osorio-Tafall et Seraphim, List Vasc. Plants Cyprus, 71 (1973).

Procumbent, trailing, perennial herb or subshrub; stems ± densely clothed with long or short whitish hairs; leaves greyish, linear-oblong, 3–6 mm. long, 0·5–1 mm. wide, straight or more often arcuate, apex acute or obtuse, upper surface glabrous or pubescent, lower surface ± densely pulverulent, margins strongly revolute; petiole very short, often reddish, flattened and distinctly ciliate; lower leaves usually subtending condensed leafy shoots; flowers in rather dense terminal or lateral cymes furnished with numerous leaf-like bracts; calyx about 5 mm. long, narrowly cylindrical, strongly ribbed, subglabrous or ± densely hirsute or hispid, with 5 acute triangular lobes about 1 mm. long; petals 5, white or pink, narrowly spathulate, about 6–7 mm. long and 1·5 mm. wide at the obtuse apex, appendage narrowly elliptical; stamens c. 5 mm. long, with membranous, distinctly dilated filaments and shortly oblong anthers about 0·5 mm. long; ovary fusiform, distinctly triquetrous, about 1·5–2 mm. long and 1 mm. wide; style about 5 mm. long; style-arms 3, about 1 mm. long, stigmas oblique. Capsule 2·5–3 mm. long; seeds narrowly ellipsoid, compressed, about 0·7 mm. long and 0·2 mm. wide; testa brown, smooth.

var. **brevipes** *Hausskn.* in Mitt. Thür. Bot. Ver. n.f., 5: 46 (1893)—[as var. *previpes*].
var. *intermedia* Boiss., Fl. Orient., 1: 780 (1867) pro parte; Lindberg f., Iter Cypr., 23 (1946).
[? *F. laevis* (non L.) Sibth. et Sm., Fl. Graec. Prodr., 1: 243 (1806); Unger et Kotschy, Die Insel Cypern, 340 (1865).]
[? *F. hirsuta* L. ssp. *laevis* (non *F. laevis* L.) Holmboe, Veg. Cypr., 129 (1914).]
TYPE: Greece; " Laurion in saxas maritimas ", Oct. 5, 1885, *Haussknecht* s.n. (K !).

Stems rather sparsely clothed with shortish hairs; leaves usually crowded, straight or slightly arcuate; inflorescence congested, subcapitate; calyx often subglabrous or very sparingly hispid.

HAB.: On sandy banks and seashores; sea-level; fl. March–Sept.

DISTR.: Division 4, rare. Also in Greece, Aegean Islands and Crete.
4. Larnaca, 1860, *Hooker f. & Hanbury* s.n.! Larnaca, 1939, *Lindberg f.* s.n.; near Famagusta, 1939, *Lindberg f.* s.n.

NOTES: Perhaps only an habitat form of the following (var. *hispida* Boiss.) from which it differs most obviously in its slender habit, small, straight, crowded leaves and condensed, few-flowered inflorescences. *F. intermedia* DC. is synonymous with *F. hirsuta* L. (var. *hirsuta*).

var. **hispida** (*DC.*) *Boiss.*, Fl. Orient., 1: 780 (1867); Lindberg f., Iter Cypr., 23 (1946).
F. hispida DC., Prodr., 1: 349 (1824).
F. hirsuta L. ssp. *hispida* (DC.) Holmboe, Veg. Cypr., 129 (1914).
F. hirsuta (non L.) Sibth. et Sm., Fl. Graec. Prodr., 1: 243 (1806), Fl. Graec., 4: 36, t. 343 (1823); Unger et Kotschy, Die Insel Cypern, 340 (1865).
TYPE: Based on the Cretan specimen sent to C. Bauhin by A. Gillenius.

Stems densely clothed with white bristly hairs, especially in the upper parts; leaves frequently 5–6 mm. long; calyx generally rather conspicuously hairy; petals (in Cyprus specimens) often white, occasionally pale pink.

HAB.: On sandy banks and seashores, or by margins of salt lakes; sea-level; fl. March–Sept.

DISTR.: Divisions 1, 4, 5, 7, 8, common in Divisions 4, 7. Widespread in the eastern Mediterranean, from Greece to Egypt.
1. Ktima, 1950, *F. J. F. Barrington* 52 !
4. Common; *Kotschy* 244 ! *Post* 249 ! *Sintenis & Rigo* 34 ! *Chapman* 333 ! 408 ! 687 ! *Syngrassides* 1833 ! *Kennedy* 1745 ! etc.
5. Kharanka near Trikomo, 1905, *Holmboe* 471.
7. Common; *Syngrassides* 10 ! 1298 ! *Casey* 806 ! *G. E. Atherton* 169 ! 1227 ! 1339 !
8. Near Yialousa, 1880, *Sintenis & Rigo* 34 ! Cape Andreas, 1880, *Sintenis & Rigo* s.n.; Cape Andreas, 1948, *Mavromoustakis* s.n. !

NOTES: Typical *Frankenia hirsuta* L. (var. *hirsuta*), with short, close indumentum, has not been found in Cyprus, and is primarily a plant of the western Mediterranean, where it is locally abundant from France eastwards to Italy. It is the Apulian plant referred to by Linnaeus (Sp. Plant., ed. 1, 332) and figured (badly) in Micheli, Gen.

Plant. Nov., 23 t. 22, f. 2; this same plant was subsequently named *F. intermedia* by de Candolle, while the Cretan plant mentioned by Linnaeus is identical with de Candolle's *F. hispida* and our *F. hirsuta* var. *hispida*.

16. CARYOPHYLLACEAE

Pax et Hoffman in Engl. et Prantl, Pflanzenfam., ed. 2, 16c: 275–367 (1934) —excl. tribes *Paronychieae* and *Pterantheae*.

Annual or perennial herbs, or rarely subshrubs; stems often swollen at the nodes; leaves opposite, decussate, simple, entire, commonly connected at the base by a transverse line; stipules scarious or absent. Flowers actinomorphic, normally hermaphrodite, in simple or compound cymose inflorescences or occasionally solitary and terminal; bracts sometimes present; calyx of 5 (rarely 4) free or connate sepals; petals as many as sepals, free (or very rarely united), sometimes minute or absent; stamens generally twice as many as petals, or equal in number to them, in 1–2 series; filaments free or partly connate, anthers 2-locular, dehiscing longitudinally. Ovary superior, sessile or stipitate on a short anthophore, 1-locular or partly divided at the base; placentation free-central (occasionally basal), ovules usually numerous. Styles and stigmas 2–5, free or partly connate. Fruit a capsule, dehiscing by valves or apical teeth. Seeds often with an excentric or curved peripheral embryo surrounding a central mass of endosperm.

About 60 genera widespread in temperate regions, particularly of the northern hemisphere; rare in the tropics. The family is commonly united with *Illecebraceae* (p. 279), the latter consisting of plants with apetalous flowers and indehiscent achaenial fruits.

In its restricted sense the *Caryophyllaceae* includes many decorative plants, of which the best known are the Carnation (*Dianthus caryophyllus* L.) and the Pink (*Dianthus plumarius* L.), though several species of *Gypsophila*, *Lychnis*, *Minuartia*, *Saponaria* and *Silene* are also popular in cultivation.

Sepals united for more than half their length:
- Styles 2:
  - Petals without coronal scales:
    - Capsule splitting into 2 valves; calyx-tube very narrow, 1–1·5 mm. diam.; petal limb very small - - - - - - - - - - - **1. Velezia**
    - Capsule opening by 4 apical teeth or by 4 valves; calyx-tube usually exceeding 1–1·5 mm. diam.; petal limb conspicuous:
      - Fruiting calyx cylindrical or narrowly ovoid:
        - Seeds peltately attached, flattish; embryo straight:
          - Calyx not membranous, nor with membranous or scarious commissures, many-nerved - - - - - - - - - - **2. Dianthus**
          - Calyx membranous, or with membranous or scarious commissures, 5–15-veined:
            - Inflorescence enveloped by broad membranous or scarious, ecostate bracts - - - - - - - - - - **3. Kohlrauschia**
            - Inflorescence not enveloped by membranous or scarious bracts, or bracts costate - - - - - - - - - - **4. Petrorhagia**
        - Seeds not peltately attached; funicle marginal, embryo curved **6. Saponaria**
      - Fruiting calyx campanulate, urceolate or turbinate, not cylindrical or narrowly ovoid:
        - Calyx with 5 distinctly winged ribs - - - - - - - **7. Vaccaria**
        - Calyx not winged:
          - Capsule with 4 apical teeth; seeds peltately attached, embryo straight **4. Petrorhagia**

Capsule deeply divided into 4 valves; seeds not peltately attached, embryo curved - - - - - - - - - - - **5. Gypsophila**
Petals with coronal scales - - - - - - - - - - **6. Saponaria**
Styles 3 (–5); petals usually with conspicuous coronal scales; capsule opening by (3) 6 (10) apical teeth - - - - - - - - - - **8. Silene**
Sepals free, or united only at base:
Stipules absent:
Capsule opening by twice as many teeth or valves as there are styles:
Capsule opening by apical teeth, not dividing into valves:
Petals bifid or deeply emarginate; seeds attached marginally; capsule elongate often curved - - - - - - - - - - **10. Cerastium**
Petals entire or denticulate, not 2-lobed; capsule generally straight:
Petals denticulate; flowers in umbel-like cymes; seeds attached peltately **9. Holosteum**
Petals entire; flowers not in umbel-like cymes; seeds attached marginally **12. Arenaria**
Capsule dividing to beyond the middle into 6 valves; petals (when present) deeply bifid - - - - - - - - - - **11. Stellaria**
Capsule opening by the same number of teeth or valves as there are styles:
Capsule dividing into 3 valves - - - - - - - - **13. Minuartia**
Capsule dividing into 4–5 valves - - - - - - - - **14. Sagina**
Stipules present, usually scarious and conspicuous:
Leaves linear, often fleshy; flowers usually fairly conspicuous, often with pink or mauve petals - - - - - - - - - - - **15. Spergularia**
Leaves obovate or suborbicular, not fleshy; flowers small, with minute, hyaline petals - - - - - - - - - - - - **16. Polycarpon**

TRIBE 1. **Caryophylleae** Stipules wanting. Sepals connate, forming a tubular 4–5-toothed calyx; petals with or without coronal scales; stamens and ovary commonly raised on a short anthophore; styles free to the base.

1. VELEZIA *L.*

Sp. Plant., ed. 1, 332 (1753).
Gen. Plant., ed. 5, 155 (1754).

Annual herbs; stems rather rigidly dichotomous; leaves narrow, subulate. Flowers sessile or subsessile; calyx narrowly cylindrical with 5 apical teeth; petals 5, sometimes hairy at the apex of the claw, but without coronal scales; stamens 5 (or 10), styles 2. Capsule linear, 4-toothed, splitting into 2 valves; seeds rather few.

Four species in the Mediterranean area and western Asia.

1. **V. rigida** *L.*, Sp. Plant., ed. 1, 332 (1753); Boiss., Fl. Orient., 1: 478 (1867); Holmboe, Veg. Cypr., 75 (1914); Post, Fl. Pal., ed. 2, 1: 154 (1932); Lindberg f., Iter. Cypr., 34 (1946); Zohary, Fl. Palaest., 1: 111 (1966); Davis, Fl. Turkey, 2: 137 (1967); Osorio-Tafall et Seraphim, List Vasc. Plants Cyprus, 37 (1973).

TYPE: " *in* Europa *australi* ".

Erect or spreading, glandular or glabrescent annual, 7–30 cm. high; stems terete, rigid; leaves linear-subulate, 1–1·5 cm. long, about 1 mm. wide, connate at base to form a short sheath, margins entire, apex subacute, sometimes a little cucullate; flowers axillary and terminal, solitary or in groups of 2–3, erecto-patent; bracts 1–2, lanceolate-acuminate, 2·5–5 mm. long; calyx 10–12 mm. long, 1–1·5 mm. wide, 15-nerved, calyx-teeth acuminate, keeled, about 1 mm. long; petals pink or reddish, with a spathulate-cuneate, emarginate limb, about 3 mm. long, and a very slender claw 10–13 mm. long; stamens 5, very unequal, sometimes as long as the claw, sometimes less than half as long; anthers shortly oblong, about 0·3 mm. long and 0·2 mm. wide, yellow; ovary narrowly oblong-cylindrical, about 2·5 mm. long and 0·5 mm. wide, glabrous, borne (with the stamens) on a glabrous anthophore about 0·5 mm. long; styles 2, filiform, about 8 mm. long.

Capsule translucent, about 10 mm. long and 1 mm. wide; seeds narrow about 2 mm. long and 0·5 mm. wide, with an apiculate apex and narrow recurved wings; testa dark brown, minutely papillose.

HAB.: On rocky ground; in garigue and on grassy slopes; 200–6,200 ft. alt.; fl. April–July.

DISTR.: Divisions 1–3, 5, 7, common in Division 2. Widespread in S. Europe and the Mediterranean area, and eastwards to the Tian-Schan range (Central Asia) and Chitral.

1. Akamas, 1905, *Holmboe* 783; Akamas, 1941, *Davis* 3328 ! Akamas, Kakoscaliou, 1949, *Casey* 757 !
2. Common; *Kotschy* 861; *Syngrassides* 1603 ! *Kennedy* 411 ! 412 ! 1186A ! 1187 ! *Merton* 2375 ! *Meikle* 2639 ! 2814 !
3. Petra tou Romiou, 1950, *F. J. F. Barrington* 14.
5. About Kythrea, 1880, *Sintenis & Rigo* 238 !
7. Near the Armenian Mon., 1912, *Haradjian* 392 !

2. DIANTHUS *L.**

Sp. Plant., ed. 1, 409 (1753).
Gen. Plant., ed. 5, 191 (1754).
F. N. Williams in Journ. Linn. Soc. Bot., 29: 346–478 (1893).

Annual or short-lived perennial herbs or subshrubs; leaves opposite and decussate, frequently linear, exstipulate, connate basally and forming a sheath, frequently 5-nerved; flowers 5-merous, solitary, clustered or capitate, surrounded at the base by one to several pairs of imbricate calyx-bracts and, in the species with capitate inflorescences, with the capitulum subtended by inflorescence bracts; calyx gamosepalous, 5-toothed, separated from the other floral parts by an anthophore; petals and stamens united at the extreme base into an annular ring surrounding the short gynophore; petal clawed, the claw bearing two more or less conspicuous longitudinal ridges; ovary unilocular. Seeds black, orbicular, flattened, with a central hilum and straight embryo; testa with raised, polygonal (" armadillo ") sculpturing.

About 300 species, chiefly in Europe, the Mediterranean region and Asia; also Tropical and South Africa and extreme N.W. of N. America.

Calyx-bracts 4, aristate, with a subulate arista 5–12 mm. long; plants annual **1. D. tripunctatus**
Calyx-bracts (4) 6–18, obtuse, or shortly cuspidate or mucronate; plants perennial:
Herbaceous; calyx 1·2–1·6 cm. long, ovoid - - - **2. D. strictus** var. **troodi**
Suffruticose; calyx 2·2–2·5 cm. long, cylindrical - - - - **3. D. cyprius**

1. D. tripunctatus *Sibth. et Sm.*, Fl. Graec. Prodr., 1: 286 (1809), Fl. Graec., 4, 86: t. 398 (1823); Unger et Kotschy, Die Insel Cypern, 348 (1865)—as *D. tricuspidatus*; Boiss., Fl. Orient. 1: 482 (1867); Post, Fl. Pal., ed. 2, 1: 155 (1932); Holmboe, Veg. Cypr., 76 (1914); Pinto da Silva, A. N. Teles et M. da Silva in Agron. Lusit., 14: 11 (1952); Zohary, Fl. Palaest., 1: 109, t. 146 (1966); Davis, Fl. Turkey, 2: 106 (1967); Osorio-Tafall et Seraphim, List Vasc. Plants Cyprus, 37 (1973).

TYPE: " In insulâ Cypro " *Sibthorp* (OXF, K !).

A glabrous annual up to 60 cm. high with a slender taproot; basal leaves forming a rosette in young plants; flowering stems 1–several, rigid, leafy, terete, with conspicuously swollen nodes, freely and divaricately branched, branches generally incurved; leaves of the basal rosette thin, narrowly spathulate, 30–50 mm. long, 3–8 mm. wide, obtuse, base narrowed and long-ciliolate on the margin, sheath very short, membranous; lower cauline leaves linear-lanceolate, upper narrowly linear, up to 100 mm. long and 2·5 mm. wide, often exceeding the corresponding internode in length, long-acuminate, conspicuously 3-nerved; sheaths membranous, short; flowers solitary, long-pedicelled; calyx-bracts 4, elliptical, acuminate with a long subulate arista

* by Sheila S. Hooper

5–12 mm. long, stiff, scarious except for a broad membranous margin in the upper half, surface verruculose, nerved only at the apex and on the arista; calyx narrowly conical with a cuneate base, 16–20 mm. long, 3–4 mm. wide, surface regularly verruculose, prominently nerved, nerves about 40; calyx-teeth lanceolate 3·5–5 mm. long, with a membranous margin; petal-lamina pink, marked with three dark lines in the centre above, pale beneath, obovate-cuneate, 6–8 mm. long, 3·5–5 mm. wide, sharply dentate on the antical margin, sparsely long pilose or smooth; anthers c. 1·5 mm. long. Capsule ovoid, not exceeding the calyx; seeds 2 mm. long, 1·25 mm. wide; testa minutely verruculose, shining.

HAB.: Dry exposed situations, generally in coastal regions, rare; fl. ? May–August.

DISTR.: Division 1, rare. Also Portugal, Elba, Algeria, Calabria and several localities from S. Greece to Palestine.

1. Timi near Paphos, 1905, *Holmboe* 713. Akamas, 1905, *Holmboe* 782 !

NOTES: Little collected, and should be looked for in western coastal areas.

2. D. strictus *Banks et Sol.* in Russell, Nat. Hist. Aleppo, 2: 252 (1794); Davis, Fl. Turkey, 2: 106 (1967); Osorio-Tafall et Seraphim, List Vasc. Plants Cyprus, 37 (1973).

D. multipunctatus Ser. in DC., Prodr., 1: 362 (1824); Boiss., Fl. Orient., 1: 482 (1867); Holmboe, Veg. Cypr., 76 (1914); Post, Fl. Pal., ed. 2, 1: 156 (1932); Lindberg f., Iter Cypr., 14 (1946); Rechinger f. in Arkiv för Bot., ser. 2, 1: 421 (1950).

[*D. sulcatus* (non Boiss.) H. Stuart Thompson in Journ. Bot., 44: 277 (1906).]
[*D. quadrilobus* (non Boiss.) Holmboe, Veg. Cypr., 76 (1914).]

Glaucous, erect or semi-decumbent perennial herb, 15–40 (–80) cm. high, becoming shortly woody at the base; sterile shoots short, evanescent or marcescent; flowering stems slender, branched in the upper half, at first leafy below, with swollen nodes and glabrous or occasionally hairy internodes; leaves on the sterile shoots thin, narrowly obovate or elliptic, c. 30 mm. long and 5 mm. wide, obtuse or rarely subacute, glabrous or sparsely hairy, with a distinct midrib and two pairs of indistinct lateral nerves, base narrowing gradually into a flattened, 3-nerved, ciliolate petiole; sheaths short, membranous; cauline leaves erect or appressed, the lower narrowly elliptic, the upper linear-lanceolate, up to 50 mm. long and 1–2·5 mm. wide, gradually acuminate, 3-nerved, generally pilose on the upper surface, with a narrowly membranous border distinctly ciliolate towards the base and with a short membranous sheath; flowers solitary or paired; pedicels angular; calyx-bracts 4–8, lowermost pair narrowly elliptic, shortly cuspidate with a broad membranous margin, upper pairs obovate, obtuse, apiculate or mucronate, verruculose, scarious with a broad, regular, membranous margin and prominent purple nerves in the upper half; calyx ovoid, 12–16 mm. long and 3·5–5 mm. wide, hard, verruculose, with prominent green or purple nerves on the teeth and running down the sutures; calyx-teeth lanceolate-apiculate, with a wide, membranous, ciliolate border; petals pink with an irregular band of dark red outlining the buff or greenish base of the lamina, tinged greenish-yellow beneath, cuneate; lamina 5–10 mm. long, 4–9 mm. wide, coarsely triangular-dentate on the antical margin, long hairy over the distal half; anthers 2·5–3·0 mm. long, pink. Capsule ovoid; seeds broadly elliptic, flattened, c. 3 mm. long, 1·5 mm. wide; testa black.

var. troodi (*Post*) *S. S. Hooper* **comb. nov.**

D. multipunctatus Ser. var. *troodi* Post in Mém. Herb. Boiss., 18: 91 (1900); Holmboe, Veg. Cypr. 76 (1914); Lindberg f., Iter Cypr., 14 (1946).

D. strictus Banks et Sol. ssp. *troodi* (Post) Osorio-Tafall et Seraphim, List Vasc. Plants Cyprus, 37 (1973) nom. invalid.

TYPE: Cyprus; " in rupestribus cacuminis Troodi ", 1894, *Post* 910 (BEI).

Differs from the typical variety in the short cusp to the calyx-bracts and the large size of the petals. The calyx-bracts are generally 6 or 8 in number rather than 4 or 6 as in the typical variety. *Plate 15, figs. 6, 7.*

HAB.: Fallow fields, roadsides, waste places and rocky ground, often on calcareous soil; sea-level to near the summit of Khionistra (6,400 ft.); fl. May–Sept. (Oct.).

DISTR.: Divisions 1, 2, 4–8, common in Division 4, very common in Divisions 2, 7. The variety endemic; typical *D. strictus* in Aegean Islands, Lebanon, Syria and Turkey.

1. Paphos, Ayios Neophytos, 1939, *Lindberg f.* s.n. !
2. Very common, *Post* s.n.; *Haradjian* 254 ! 515 ! *Kennedy* 391 ! 400–407 ! *Davis* 1792 ! 1857 ! 2082 ! 3587 ! etc.
4. Common; *Haradjian* 1019 ! *Syngrassides* 431 ! *Lindberg f.* s.n. ! *N. Macdonald* 77 ! *Meikle* 2596 !
5. Athalassa, 1932, *Syngrassides* 404 ! 411 ! Mile 3, Nicosia–Famagusta Road, 1950, *Chapman* 712 !
6. Nicosia, 1958, *P. H. Oswald* 14 !
7. Very common; *Sintenis & Rigo* 241 ! *Syngrassides* 988 ! *Kennedy* 345 ! 1188 ! 1623 ! *Casey* 57 ! *G. E. Atherton* 101 ! 137 ! 178 ! 763 ! *Meikle* 2949 ! etc.
8. Rizokarpaso, 1912, *Haradjian* 254 ! Ayios Theodhoros, 1934, *Syngrassides* 988 ! Ayios Bakos, near Yialousa, 1962, *Meikle* 2439 !

NOTES: At altitudes of 3,000 ft. and above in the Troödos Mountains the plant exhibits a somewhat dwarf habit with a tendency to divaricate branching of the inflorescence and hairy stems. From the description, the type of var. *troodi* would appear to belong to this high altitude form, but it cannot be clearly separated from the lowland form in Cyprus and all the Cyprus material has therefore been assigned to this variety.

Individual gatherings from Cyprus exhibit the sessile axillary flowers characteristic of var. *axilliflorus* (Fenzl) Eig, but this appears to be merely an accidental form, produced either after the destruction of the upper part of the inflorescence by grazing, or by other means, or on shoots produced late in the flowering season.

3. D. cyprius *A. K. Jackson et Turrill* in Kew Bull., 1938: 462 (1938); Chapman, Cyprus Trees and Shrubs, 38 (1949); Seligman in Bull. Alp Gard. Soc., 19: 55 (1951).

D. pendulus Boiss. et Blanche ssp. *cyprius* (A. K. Jackson et Turrill) Osorio-Tafall et Seraphim, List Vasc. Plants Cyprus, 37 (1973) nom. invalid.

TYPE: Cyprus; Kyrenia, 1,100 ft. alt., November 14, 1936, *Kennedy* 408 (K !).

Glabrous, suffruticose perennial, branches at maturity often pendulous and curved, up to 1 m. long, bearing an apical tuft of erect leaves and axillary flowering stems, woody below with a greyish bark marked by annular scars; flowering stems erect, leafy, branched only from the uppermost nodes, with at least the upper internodes conspicuously angled and the nodes swollen; leaves a pale, sometimes glaucous, green, linear-lanceolate or very narrowly elliptic, 60–90 mm. long, 2·5–3·5 mm. wide, gradually narrowed in the upper third to an acuminate apex, midrib and two pairs of marginal nerves prominent beneath, midrib impressed above, margin membranous and minutely denticulate at the base, sheath membranous, 1–1½ times as long as broad, leaves on the peduncles similar but smaller and progressively reduced in size upwards, generally equalling or slightly exceeding the corresponding internode; inflorescence branches strict, short, flowers clustered, sometimes subcorymbose; calyx-bracts 6–9 pairs, investing the short pedicel, uppermost pair the largest, elliptic, c. 8 mm. long, 5 mm. wide, herbaceous, margin very narrowly membranous, densely ciliolate, apex minutely apiculate, lower pairs successively smaller, lowermost deltoid with a broad membranous margin; calyx narrowly elliptical, closely invested by the calyx-bracts below, c. 22–25 mm. long, 4 mm. wide, finely striate, teeth narrowly lanceolate, 5 mm. long, 2 mm. wide, acute; petals pink with red flecking, lamina obovate, 8–12 mm. long, 6–10 mm. wide, tinged green at the base, fimbriate on the antical margin, barbellate; claw slender c. 2·5 cm. long, shortly exserted; anthophore 3 mm. long; anthers 2·5 mm. long; gynophore 1·5 mm. long. Capsule cylindrical, 20–23 mm. long, 5 mm. wide, shortly exserted; seeds 3 mm. long, 2 mm. wide; testa black, minutely verruculose. *Plate 15, figs. 1–5.*

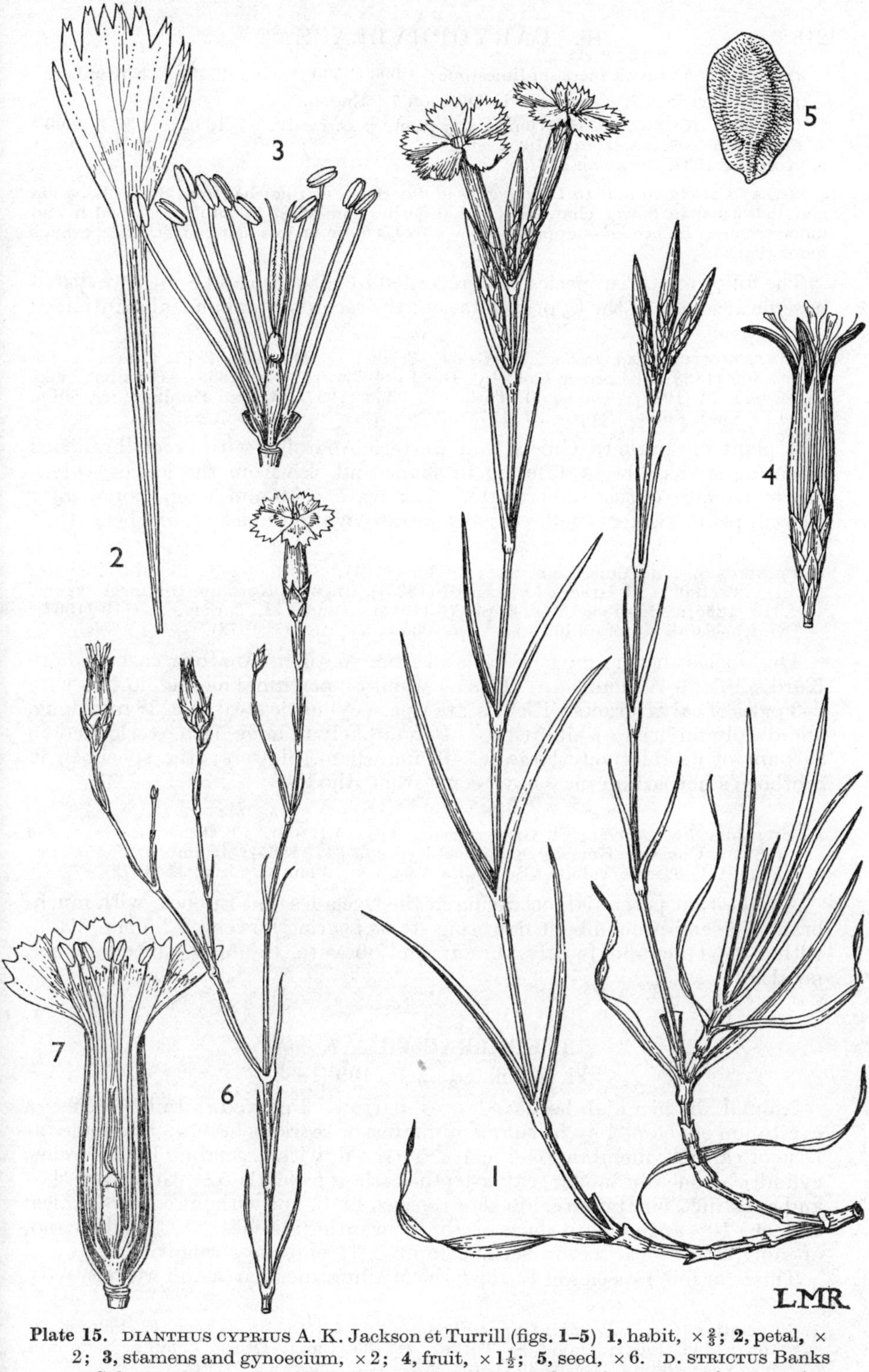

Plate 15. DIANTHUS CYPRIUS A. K. Jackson et Turrill (figs. **1–5**) **1,** habit, $\times \frac{2}{3}$; **2,** petal, $\times 2$; **3,** stamens and gynoecium, $\times 2$; **4,** fruit, $\times 1\frac{1}{2}$; **5,** seed, $\times 6$. D. STRICTUS Banks et Sol. var. TROODI (Post) S. S. Hooper (figs. **6–7**) **6,** flowering stem, $\times \frac{2}{3}$; **7,** section of flower, $\times 2$. (**1** from *Atherton* 762; **2, 3** from specimens grown at R.B.G. Kew; **4, 5** from *Kennedy* 408; **6** from *Kennedy* 400; **7** from *Kennedy* 391.)

HAB.: Cliffs and rock faces on limestone; 1,000–2,500 ft. alt.; fl. June–November.

DISTR.: Divisions 7, 8, common in Division 7. Endemic.

7. Common; Kornos, St. Hilarion, Halevga etc.; *Davis* 1923 ! 1992 ! 2032 ! 2800 ! *Casey* 136 ! *G. E. Atherton* 762 ! etc.
8. Koronia, 1937, *Chapman* 233 !

NOTES: Closely similar to *Dianthus pendulus* Boiss. et Blanche (from Syria, Lebanon and Palestine) in flower characters but differing somewhat in habit, leaf width and inflorescence; *D. pendulus* appears always to be more or less asperulous, and perhaps more glaucous.

The following three species, first recorded by Sibthorp and Smith, are listed in some accounts of the Cyprus flora, but the records remain unsubstantiated:

D. CINNAMOMEUS *Sibth. et Sm.*, Fl. Graec. Prodr., 1: 287 (1809), Fl. Graec., 4: 88, t. 400 (1823); Unger et Kotschy, Die Insel Cypern, 348 (1865); Holmboe, Veg. Cypr., 76 (1914); Davis, Fl. Turkey, 2: 114 (1967); Osorio-Tafall et Seraphim, List Vasc. Plants Cyprus, 37 (1973).

A plant of southern Greece and western Anatolia, with erect, branched flowering stems, bracts differing in shape and size from the leaves, calyx-bracts obovate, obtuse, mucronate, calyx ovate, c. 19 mm. long, prominently nerved, petals rather small, coarsely serrate, white tinged bronze beneath.

D. CRINITUS *Sm.* in Trans. Linn. Soc., 2: 300 (1794); Sibth. et Sm., Fl. Graec. Prodr., 1: 287 (1809), Fl. Graec., 5: 1, t. 401 (1825); Unger et Kotschy, Die Insel Cypern, 348 (1865); Holmboe, Veg. Cypr. 76 (1914); Davis, Fl. Turkey, 2: 119 (1967); Osorio-Tafall et Seraphim, List Vasc. Plants Cyprus, 37 (1973).

This species occurs on Rhodes, and from southern Anatolia eastwards to Kurdistan and W. Pakistan. It is a caespitose perennial having flowers with 2–3 pairs of calyx-bracts with aristate tips, a cylindrical calyx c. 28 mm. long, and deeply fimbriate white petals. It is unlikely to have been overlooked on account of its relatively large (c. 40 mm. diam.) flowers; the specimen in Sibthorp's herbarium may have come from Rhodes.

D. DIFFUSUS *Sibth. et Sm.*, Fl. Graec. Prodr., 1: 285 (1809), Fl. Graec. 4: 85, t. 396 (1823); Unger et Kotschy, Die Insel Cypern., 347 (1865); Holmboe, Veg. Cypr., 76 (1914); Osorio-Tafall et Seraphim, List Vasc. Plants Cyprus, 37 (1973).

A pubescent perennial, occurring in the Cyclades and Euboea, with much-branched semi-procumbent flowering stems bearing flowers 2–2·5 cm. diam., with 2 or 4 acuminate calyx-bracts and obovate, dentate, barbellate, red petals.

3. KOHLRAUSCHIA *Kunth*

Fl. Berol., ed. 2, 1: 109 (1838).

Annual or biennial herbs; leaves narrow, 3-nerved. Inflorescence a capitulum enveloped by broad membranous or scarious bracts; bracteoles at base of calyx 2, membranous; calyx 5-nerved, with ascending lateral veins, cylindrical, membranous; calyx-teeth 5, short; petals 5, with a long claw and a distinct, emarginate, lobed or toothed limb, but without coronal scales; stamens 10; anthophore short or almost wanting; styles 2. Capsule ovoid, opening by 4 apical teeth; seeds compressed; embryo straight.

Three (or four) species in Europe, the Mediterranean area and western Asia.

1. **K. velutina** (*Guss.*) *Reichb.*, Icones Fl. Germ., 6: 43 (1844); Zohary, Fl. Palaest., 1: 106, t. 142 (1966); Davis, Fl. Turkey, 2: 135 (1967); Osorio-Tafall et Seraphim, List Vasc. Plants Cyprus, 36 (1973).
   Dianthus velutinus Guss., Fl. Sic. Prodr., 1: 493 (1827).

Tunica velutina (Guss.) Fisch. et Mey., Ind. Sem. Hort. Petrop., 6: 66 (1840); Boiss., Fl. Orient., 1: 516 (1867); Holmboe, Veg. Cypr., 76 (1914); Post, Fl. Pal., ed. 2, 1: 160 (1932); Lindberg f., Iter Cypr., 16 (1946).

[*T. prolifera* (non (L.) Kunth) Sintenis in Oesterr. Bot. Zeitschr., 31: 393 (1881).]

TYPE: Sicily; " Val di Mazzara, e Val di Noto, Madonie ", *Gussone* (FI, ? NAP).

Erect annual or biennial, 10–30 (–60) cm. high; stems distinctly sulcate, glandular-pubescent especially in the lower half or occasionally glabrous or subglabrous; basal leaves forming a loose tuft, narrowly oblong-linear, obtuse or shortly acute, 20–60 mm. long, 2–7 mm. wide, scabridulous-pubescent or glabrous, amplexicaul at base; cauline leaves narrower and tapering to a slender acumen, connate at the base and forming a sheath usually more than twice as long as wide. Inflorescence a solitary, terminal capitulum, comprising 3–6 or more shortly pedicellate or subsessile flowers, surrounded by conspicuous, scarious bracts; bracts broadly ovate, 1–1·7 cm. long, 0·5–1 cm. wide, apex sharply cuspidate; calyx narrowly cylindrical, 10–12 mm. long, 1–2 mm. wide, membranous, glabrous, with 5 shallow, blunt teeth; petals pink, pencilled crimson at base of limb, narrowly spathulate, about 15 mm. long and 3–4 mm. wide across limb, with a shortly 2-lobed apex and the base tapering to a slender claw; stamens 10, unequal, filaments sometimes as long as the petals, sometimes about half as long; anthers yellow, oblong, about 0·7 mm. long and 0·5 mm. wide; anthophore almost wanting, but filaments united at base to form a cup about 1 mm. long around base of ovary; ovary narrowly oblong-cylindrical, 4–5 mm. long and about 1 mm. wide; styles 7–9 mm. long, filiform. Capsule narrowly ovoid, 5–6 mm. long, 3 mm. wide; seeds cymbiform, about 1 mm. long and 0·7 mm. wide; testa blackish, dull, covered with short, sharp conical papillae interconnected by radiating basal ridges.

HAB.: Grassy and rocky hillsides; in garigue; in fallow fields, vineyards and on bare sandy ground; sea-level to 4,500 ft. alt.; fl. Febr.–May.

DISTR.: All Divisions, common in Divisions 2, 6, 7. Mediterranean region eastwards to Turkey and Palestine.

1. Avgas, near Ayios Yeoryios (Akamas), 1962, *Meikle* 2010 !
2. Common; *Kotschy* 218, 810; *Syngrassides* 819 ! *Kennedy* 410 ! 1189 ! 1190 ! 1191 !
3. North side of Stavrovouni Mon. (" Sta. Croce "), 1862, *Kotschy* 203.
4. Ayia Napa, 1905, *Holmboe* 48; Larnaca, 1912, *Haradjian* 28 !
5. Salamis, 1971, *Townsend* 71/108 !
6. Common; *Syngrassides* 1140 ! *Davis* 2538 ! *Chapman* 301 ! *Merton* 2588 !
7. Common; *Sintenis & Rigo* s.n.; *Casey* 400 ! *G. E. Atherton* 1194 !
8. Near Rizokarpaso, 1912, *Haradjian* 178; Ayios Philon near Rizokarpaso, 1941, *Davis* 2257 !

4. PETRORHAGIA (*Ser.*) *Link*

Handbuch, 2: 235 (1831).

Gypsophila L. sect. *Petrorhagia* Ser. in DC., Prodr., 1: 354 (1824) pro parte.

[*Tunica* auct. non Ludwig, Def. Gen. Plant., ed. 3, 298 (1760).]

Annual and perennial herbs; leaves 1–5-nerved, narrow, sometimes subulate. Inflorescence cymose or capitate, ebracteate or flowers subtended, but not enveloped, by scarious, costate bracts; calyx bracteolate or ebracteolate, distinctly 5–15-nerved, cylindrical or campanulate, with membranous commissures; calyx-teeth 5; petals 5, entire or emarginate, without a distinct limb and claw, and without coronal scales; stamens 10; anthophore developed or almost wanting; styles 2. Capsule ovoid or ellipsoid, opening by 4 apical teeth; seeds compressed; embryo straight.

About 20 species in Europe, the Mediterranean area and western Asia.

Calyx 15-nerved; calyx-teeth cuspidate; petals entire - - - **1. P. cretica**
Calyx 5-nerved; calyx-teeth subacute; petals bifid - - - - **2. P. kennedyae**

1. P. cretica (*L.*) *P. W. Ball et Heywood* in Bull. Brit. Mus. Bot., 3: 142 (1964).

Saponaria cretica L., Sp. Plant., ed. 2, 584 (1762) excl. syn.

Tunica pachygona Fisch. et Mey., Ind. Sem. Hort. Petrop., 4: 50 (1837); Boiss., Fl. Orient., 1: 522 (1867); Holmboe, Veg. Cypr., 76 (1914); Post, Fl. Pal., ed. 2, 1: 161 (1932); Lindberg f., Iter Cypr., 16 (1946); Zohary, Fl. Palaest., 1: 105, t. 141 (1966); Davis, Fl. Turkey, 2: 133 (1967); Osorio-Tafall et Seraphim, List Vasc. Plants Cyprus, 37 (1973).

TYPE: Crete; " *in* Cretae *aridis* " (LINN), but see note.

Erect, much-branched annual (or biennial) 10–25 cm. high; stems glabrous towards base, densely glandular in the middle region, and again glabrous towards apex; basal leaves withered at flowering time, forming a loose tuft, strap-shaped or obovate, obtuse or subacute, 3-nerved, 1–4 cm. long, 0·2–0·5 cm. wide, margins shortly scabridulous; cauline leaves narrower, linear-subulate, 1–2·5 cm. long, 0·15–0·2 mm. wide, remotely scabridulous, slightly keeled, often tinged purplish; inflorescence repeatedly dichotomous, forming a lax, open cyme; flowers erect, solitary at the ends of slender pedicels 1–2·5 cm. long; calyx ebracteolate, narrowly campanulate, glabrous, often purple-tinged, 5–9 mm. long, 3–4 mm. wide, 15-nerved, with the nerves running into each shallow, triangular calyx-tooth in groups of 3, the median projecting as a distinct cusp beyond the apex of each calyx-tooth; petals strap-shaped or narrowly oblanceolate, entire, 7–9 mm. long, 1·5 mm. wide, white with 3 buff- or reddish-tinged nerves above, often stained reddish below; stamens unequal, 5–7 mm. long; anthers oblong, about 0·5 mm. long, yellow; ovary turbinate, about 5 mm. long, 2–2·5 mm. wide; styles filiform, about 4 mm. long. Capsule 7–8 mm. long, 3–4 mm. wide, apical teeth acute, slightly recurved; seeds strongly flattened, about 2·5 mm. long, 1·5 mm. wide; testa dark brown, minutely and bluntly papillose.

HAB.: Dry, rocky slopes, usually on igneous formations; 500–3,000 ft. alt.; fl. April–June.

DISTR.: Divisions 2–4, common in Division 2. Also in Greece, Turkey, Caucasus area, N. Iran, Syria and Palestine.

2. Common; *Sintenis & Rigo* 960 ! *Kennedy* 1035 ! *Davis* 3445 !

3. Between Parekklisha and Kellaki, 1948, *Kennedy* 1615 ! Between Vasa and Asgata, 1950, *F. J. F. Barrington* 57.

4. Troulli Mine, 1962, *Meikle* 2961 !

NOTES: The description of *Saponaria cretica* L. (Sp. Plant., ed. 2, 584) is evidently based on the specimen still preserved in the Linnean herbarium, and not on " *Saxifraga altera* " of Alpinus, cited in synonymy by Linnaeus. In the circumstances one is reluctantly obliged to reject the well-known name *Tunica pachygona* Fisch. et Mey. as a later synonym, a course of action made all the more regrettable by the fact that *Petrorhagia cretica* has not been recorded from Crete, nor is it at all likely that the holotype is correctly localized. Were it at all possible, one would much prefer to follow tradition, and use the Alpinus synonym as a basis for the species. This could only be done by rejecting *Saponaria cretica* L. as a *nomen confusum*, consisting (as it does) of two conflicting elements.

2. P. kennedyae (*A. K. Jackson et Turrill*) *Meikle* **comb. nov.**; Osorio-Tafall et Seraphim, List Vasc. Plants Cyprus, 37 (1973) nom. invalid.

Tunica kennedyae A. K. Jackson et Turrill in Kew Bull., 1938: 462 (1938); Lindberg f., Iter Cypr., 15 (1946); Rechinger f. in Arkiv för Bot., ser. 2, 1: 421 (1950).

TYPE: Cyprus; Platres, 900–1,000 m. alt., *Kennedy* 1037 (K !).

Erect annual (or biennial) 5–25 cm. high, occasionally unbranched but usually with numerous ascending branches; stems glabrous below, distinctly glandular towards apex, dull green, tinged purplish; basal leaves usually withered at flowering-time, forming a loose tuft, oblong or narrowly obovate 0·5–1·8 cm. long, dull green above, purplish below, apex obtuse, margins entire, ciliate towards base; cauline leaves erect, narrowly linear, 3-nerved, 1–1·8 cm. long, 0·5–1 mm. wide, apex subacute, margins ciliate at base;

inflorescence repeatedly dichotomous, lax, pedicels 1–2 cm. long, viscid-glandular; flowers solitary, ebracteolate; calyx narrowly cylindrical-turbinate, 5–6 mm. long, 1·5–2 mm. wide, 5-nerved with membranous commissures, glandular, often tinged purplish; calyx-teeth triangular, acute but not cuspidate; petals narrowly oblanceolate, 6–9 mm. long, 1·5–2 mm. wide, 3-nerved, apex deeply bifid, white or pale pink above, stained or striped reddish below; stamens unequal, 2–5 mm. long; anthers oblong, 0·3 mm. long, yellow; styles 1–1·5 mm. long; ovary oblong-ellipsoid, 2–4 mm. long, 1–1·5 mm. wide. Capsule narrowly oblong, 7–8 mm. long, 2–2·5 mm. wide, projecting beyond persistent calyx; seeds 1–1·3 mm. long, 0·5–0·8 mm. wide, flattened, often concave; testa fuscous, distinctly cerebriform-rugulose.

HAB.: On dry stony or sandy hillsides, on igneous formations, under *Pinus*, or in garigue; 600–6,400 ft. alt.; fl. April–July.

DISTR.: Divisions 2, 4, locally common in Division 2. Endemic.

2. Common; *Sintenis & Rigo* 764 ! *Kennedy* 360 ! 361 ! 409 ! 1036 ! 1037 ! 1038 ! *Davis* 1759 ! 3450 ! *Merton* 2408 ! *Meikle* 2808 !

4. Nicosia–Limassol road, at Kakoradjia, 1962, *Meikle* 2955 !

5. GYPSOPHILA *L.*

Sp. Plant., ed. 1, 406 (1753).
Gen. Plant., ed. 5, 191 (1754).
Y. I. Barkoudah, A Revision of *Gypsophila*, *Bolanthus*, *Ankyropetalum* and *Phryna* in Wentia, 9: 1–203 (1962).

Annual or perennial herbs, rarely subshrubs; stems glabrous, or glandular-pilose, often glaucous; leaves usually narrow, lanceolate or linear, occasionally broader or even spathulate; flowers usually small, often numerous in dichasial cymes, or inflorescences rarely capitate; bracts mostly scarious, bracteoles absent; calyx campanulate or turbinate, seldom cylindrical, with 5 teeth or lobes, hyaline between the nerves; petals 5, white or pink, inserted on a cup-shaped disk, cuneate, without coronal scales; stamens 10 (rarely 5), introrse; anthophore very short or wanting; ovary globose, ovoid or oblong, 1-locular; ovules 4–36; styles 2 (very occasionally 3). Capsule splitting deeply into 4 valves; seeds auriculate, compressed; testa usually tuberculate, rarely smooth; hilum marginal; embryo circular with a prominent radicle.

About 126 species widely distributed in Europe and Asia, but most abundantly represented in the area comprising Turkey, the Caucasus, North Iraq and North Iran.

1. **G. pilosa** *Huds.* in Phil. Trans. Roy. Soc., 56: 252 (1767); Barkoudah, Rev. *Gypsophila* etc., 151 (1962); Zohary, Fl. Palaest., 1: 101, t. 136 (1966); Davis, Fl. Turkey, 2: 170 (1967).

Silene porrigens Gouan ex L., Syst. Nat., ed. 12, 3: 230 (1768).

Gypsophila porrigens (L.) Boiss., Fl. Orient., 1: 557 (1867); Druce in Rep. B.E.C., 9: 469 (1931); Post, Fl. Pal., ed. 2, 1: 168 (1932); Osorio-Tafall et Seraphim, List Vasc. Plants Cyprus, 36 (1973).

TYPE: Cultivated in Chelsea Physic Garden (BM).

An erect annual, 15–80 cm. high; stem unbranched or sparingly branched, glabrous near base, hispid-pilose above with long spreading hairs; basal leaves oblong or lanceolate, 3–9 cm. long, 1–2 cm. wide, acuminate or occasionally subobtuse, generally glabrous, 3–5-nerved, sessile and slightly connate at base; upper leaves similar, but generally narrower, often almost linear, and frequently pilose; inflorescence a lax, spreading, many-branched dichasium; bracts linear, herbaceous, pilose, about 0·3–0·8 cm. long; pedicels capillary, 1–4 cm. long, glabrous; calyx narrowly campanulate or almost cylindrical, 3–6 mm. long and 2–3 mm. wide, glandular-pilose, with 5 broad green nerves; calyx-teeth narrowly ovate, apiculate, about 1·5–2·5 mm. long and 0·8 mm. wide at base, with narrow membranous margins; petals white or pale pink, linear-oblong, 7–8 mm. long and about 1 mm. wide, with a distinct constriction between the limb and the claw, and with the apex emarginate or shallowly 2-lobed, the middle parts of the claw sometimes connate; stamens 2–5 mm. long; filaments glabrous; anthers almost orbicular, yellow, about 0·5 mm. diam.; ovary broadly ellipsoid or ovoid, about 1·5 mm. long and 1 mm. wide; styles 1·5 mm. long, divergent; ovules 16–20. Capsule broadly ovoid, 5–7 mm. long, about 5 mm. wide, deeply 4-lobed; seeds about 1·5 mm. long and a little less in width; testa dark blackish-brown with rows of acute tubercles.

HAB.: Waste ground, cultivated ground and earthy banks; sea-level to 4,600 ft. alt.; fl. April–August.

DISTR.: Divisions 2, 4, rare. Widespread in western Asia from Turkey and Palestine to Afghanistan.

2. Platres, 1938, *Kennedy* 1186 !
4. Larnaca, 1928 or 1930, *Druce* s.n.

NOTES: Possibly an unestablished alien in Cyprus, as in other parts of Europe and Asia, but the two records cited above suggest that it may yet be found elsewhere on the island.

6. SAPONARIA *L.*

Sp. Plant., ed. 1, 408 (1753).
Gen. Plant., ed. 5, 191 (1754).

Erect or decumbent perennial, or rarely annual, herbs; leaves obovate, oblong or spathulate, not subulate or linear; flowers solitary or in cymes; calyx ovoid or cylindrical, obscurely nerved, 5-toothed; petals 5, claw and limb distinct or indistinct, the latter entire or emarginate, with or without coronal scales; stamens 10; styles usually 2; anthophore usually short; ovary 1-locular. Capsule ovoid or oblong, opening by 4 short teeth; seeds with marginal funicle, embryo curved.

About 24 species in Europe and Asia. The genus resembles *Silene*, differing primarily in the number of styles and capsule-teeth.

Perennial with woody rootstock; flowers 2–3 cm. diam.; calyx about 2·5 cm. long - **1. S. cypria**
Annual; flowers less than 1 cm. diam.; calyx 6–7 mm. long - - **2. S. orientalis**

1. S. cypria *Boiss.*, Fl. Orient. Suppl., 83 (1888); Lindberg f., Iter Cypr., 15 (1946); Osorio-Tafall et Seraphim, List Vasc. Plants Cyprus, 36 (1973).

S. cyprica Post et Autran in Bull. Herb. Boiss., 5: 755 (1897) and in Bull. Herb. Boiss., 7: 161 (1899).

S. depressa Biv. ssp. *cypria* (Boiss.) Holmboe, Veg. Cypr., 74 (1914).

TYPE: Cyprus; " in monte Troödos Cypri circa Prodromo ", *Sintenis & Rigo* 767 (G, K !).

An erect or spreading perennial, 8–20 (–35) cm. high, with a thick, branched woody rootstock; basal leaves loosely clustered, glabrous or thinly ciliate at base of petiole, spathulate, obovate or oblanceolate, 2·5–5 cm. long, 0·5–1·5 cm. wide, apex rounded or subacute, base gradually narrowed to a petiole; lamina rather fleshy, glaucous; cauline leaves distinctly smaller and narrow, the upper (and the upper parts of the stem) densely glandular-villose; flowers solitary, or in lax cymes, occasionally congested, opening in the evening, closing about noon of the following day; calyx cylindrical, about 2·5 cm. long and 0·5 cm. wide, densely glandular-villose, obscurely 15-nerved; calyx-teeth triangular-cuspidate, about 2 mm. long and 1·5 mm. wide; petals pink, 3–4 cm. long, limb 0·7–1 cm. wide, deeply and acutely emarginate; claw slender 2–3 cm. long, slightly exserted from calyx; coronal scales 1–2 mm. long, subulate-bifurcate; stamens 2–2·5 cm. long; anthers oblong, yellow, about 1·5 mm. long and 1 mm. wide; ovary cylindrical, 5–7 mm. long, 1·5–2 mm. wide, glabrous; styles 1–1·5 cm. long; anthophore very short. Capsule cylindrical, about 2 cm. long, 0·4–0·5 cm. wide, mostly concealed by persistent calyx, opening by 4 deeply cleft, subacute teeth; seeds compressed, reniform, about 2 mm. long, and a little less in width; testa fuscous, closely and regularly verruculose.

HAB.: On screes and rocky mountainsides, in open or under Pines, or by streamlets; confined to igneous formations; (3,400–) 4,500–6,400 ft. alt.; fl. July–September.

DISTR.: Division 2, locally common in the Troödos area. Endemic.

2. Common; *Sintenis & Rigo* 769 ! *Holmboe* 1160; *Syngrassides* 722 ! 742 ! *Kennedy* 324–326 ! 328–343 ! *Davis* 1857 ! *Merton* 1954 !

2. S. orientalis *L.*, Sp. Plant., ed. 1, 409 (1753); Boiss., Fl. Orient., 1: 531 (1867); Holmboe, Veg. Cypr., 75 (1914); Post, Fl. Pal., ed. 2, 1: 164 (1932); Lindberg f., Iter Cypr., 15 (1946); Davis, Fl. Turkey, 2: 142 (1967); Osorio-Tafall et Seraphim, List Vasc. Plants Cyprus, 36 (1973).

TYPE: " in Oriente ".

An erect or spreading annual, forming small bushy plants 2–20 (–30) cm. high; stems repeatedly dichotomous, rather sparsely glandular, often tinged crimson-purple; leaves elliptical, oblong or obovate-spathulate, 0·8–2 (–3) cm. long, 0·3–1 cm. wide, apex acute, base narrowed to a short petiole, lamina glabrous, rather glaucous; flowers in lax dichasia; calyx narrowly ovoid-cylindrical, ± densely covered with conspicuous, long, glandular hairs, 6–7 mm. long, 2·5–3 mm. wide; calyx-teeth narrowly triangular, acute, 1·5–2 mm. long, about 1–1·5 mm. wide, with narrow hyaline, glandular-papillose margins; petals pink with 3 darker nerves, narrowly oblanceolate, about 8 mm. long and 1–1·5 mm. wide near the emarginate apex, limb and claw not clearly distinguished, coronal scales inconspicuous or wanting; stamens unequal, 3·5–4·5 mm. long; anthers yellow, shortly oblong, about 0·4 mm. long; ovary ovoid 2·5–3 mm. long, 1·5–2 mm. wide, styles about 1·5 mm. long. Capsule narrowly ovoid-cylindrical, 6–7 mm. long, 2–2·5 mm. wide, slightly exceeding the persistent calyx, teeth subacute, about 1 mm. long; seeds reniform, slightly compressed, about 1 mm. long, and a little less in width; testa fuscous, regularly reticulate-verruculose.

HAB.: On rocky and gravelly slopes; in dried-up stream-beds; by pathsides; 5,500–6,400 ft. alt.; fl. June–August.

DISTR.: Division 2, locally common in the Troödos area. Also in Balkans, Caucasus area, Turkey, Syria and Iran, occasionally adventive elsewhere.

2. Common; *Holmboe* 1082; *Kennedy* 383 ! 383a ! 1687 ! 1688 ! *Davis* 1791 ! 1861 ! *Merton* 1938 !

7. VACCARIA *Wolf*

Gen. Plant. Vocab. Char. Def., 111 (1776).

Saponaria L., Sp. Plant., ed. 1, 409 (1753) pro parte.

Erect annual herbs; leaves ovate or lanceolate, sessile; flowers in lax terminal cymes; epicalyx wanting; calyx ± inflated below, whitish, with 5 conspicuous green wings; petals 5; coronal scales wanting; stamens 10; styles 2. Capsule opening by 4 valves, exocarp hard and thick, endocarp membranous, dehiscing irregularly. Seeds with a marginal funicle.

Probably one species only, though as many as 4 species have been recognized by Pax and Hoffmann (Pflanzenfam., ed. 2, 16c: 361) and 3 by Boissier (Fl. Orient., 1: 525).

1. V. pyramidata *Medik.*, Phil. Bot., 1: 96 (1789); Zohary, Fl. Palaest., 1: 104, t. 140 (1966); Davis, Fl. Turkey, 2: 177 (1967); Osorio-Tafall et Seraphim, List Vasc. Plants Cyprus, 36 (1973).

Saponaria vaccaria L., Sp. Plant., ed. 1, 409 (1753); Boiss., Fl. Orient., 1: 525 (1867); Holmboe, Veg. Cypr., 75 (1914); Lindberg f., Iter Cypr., 15 (1946).

Saponaria segetalis Neck., Delic. Gallo-Belg., 1: 194 (1768) nom. illeg.

Vaccaria segetalis (Neck.) Garcke in Aschers., Fl. Brandenb., 1: 84 (1860); Post, Fl. Pal., ed. 2, 1: 162 (1932); Rechinger f. in Arkiv för Bot., ser. 2, 1: 421 (1951).

TYPE: " *inter segetes* Galliae, Germaniae ".

Erect annual, 20–100 cm. high; stems glabrous, terete or ± sulcate above, often distinctly swollen at nodes, generally unbranched except at the initiation of the inflorescence; leaves glabrous, glaucous, ovate, oblong or lanceolate, 3–15 (–20) cm. long, 0·8–4 (–5) cm. wide, apex acute, base sessile, amplexicaul, the basal margins of each pair of leaves frequently connate; inflorescence lax, much-branched, pedicels 2–5 cm. long, often purple-tinged, subtended by lanceolate-subulate, scarious, purple-tinged bracts 2–10 mm. long, 0·2–1 mm. wide; calyx at first ± cylindrical, 10–15 mm. long, 3–10 mm. wide, glabrous, pale whitish-green with 5 conspicuous darker green wings, later accrescent, and becoming urceolate with a constricted mouth; calyx-teeth ovate-deltoid, acute, 1–2 mm. long, 1–1·5 mm. wide, with broadly membranous margins and a purple-stained base; petals pink (or sometimes white), 1·5–2 (–2·5) cm. long, 0·5–0·8 cm. wide, limb usually distinct, obcordate or obovate with a shallow emarginate apex, claw slender, with 2 distinct longitudinal wings, joined at base with the bases of filaments and other petals to form a short tube (c. 6 mm. high) around anthophore; stamens subequal, about 1·5 cm. long; anthers yellow, oblong, about 1 mm. long and 0·8 mm. wide; ovary oblong, 3–4 mm. long, 1·5–2 mm. wide; styles 10–12 mm. long. Capsule broadly ovoid or almost globose, 7–10 mm. long, 6–7 mm. wide, exocarp hard, bright yellow, dehiscing by 4 ovate-triangular teeth 1–2 mm. long, endocarp white, membranous, dehiscing irregularly; seeds almost globose, 1·5–2 mm. diam.; testa dark brown, densely papillose-verruculose.

HAB.: Cultivated and fallow fields; sea-level to 5,500 ft. alt.; fl. Febr.–July.

DISTR.: Divisions 1–3, 5, 7, 8, common in Divisions 2 and 7. Widespread in warm-temperate areas of Europe, Asia, and North Africa; adventive elsewhere.

1. Stroumbi (Paphos), 1913, *Haradjian* 713.
2. Common; *Holmboe* 860; *Syngrassides* 255 ! *Lindberg f.* s.n.; *F. M. Probyn* 95 ! *Meikle* 2272 !
3. Alethriko, 1905, *Holmboe* 219; Limassol, 1913, *Haradjian* 564 !
5. Near Kythrea, 1880, *Sintenis & Rigo* 240 ! Near Pyroi, 1934, *Syngrassides* 479 ! Near Varvaras, 1948, *Mavromoustakis* s.n. !
7. Common; *Syngrassides* 457 ! *Davis* 2183 ! *Casey* 545 ! *G. E. Atherton* 1181 !
8. Between Apostolos Andreas and Khelones, 1941, *Davis* 2288 !

8. SILENE *L.*

Sp. Plant., ed. 1, 416 (1753).
Gen. Plant., ed. 5, 193 (1754).
Rohrbach, Mon. Gatt. Silene (1868).
Chowdhuri in Notes Roy. Bot. Gard. Edinb., 22: 221–278 (1957).

Annual, biennial or perennial herbs or small shrubs. Leaves entire, sessile or petiolate; inflorescence usually compound, very variable in structure, or flowers occasionally solitary; calyx gamosepalous with 5 short teeth; calyx tube 10–20 (–60)-nerved, sometimes strongly inflated; epicalyx wanting; petals 5, normally free, limb and claw distinct; stamens 10; styles 3 (occasionally 4 or 5). Fruit an incompletely or impermanently 3- (less often

1- or 4–5-) locular capsule, the dissepiments generally disappearing before maturity; capsule teeth splitting on dehiscence, and normally 6 (occasionally 8 or 10), double the number of styles, or capsules occasionally dehiscing irregularly. Seeds with a marginal funicle.

About 440 species, widely distributed in the temperate regions of the northern hemisphere, but also occurring in Tropical Africa and South Africa, and common as introduced weeds in other parts of the world. The genus is particularly well represented in the Mediterranean area and in western Asia. In many of the species the petals are shrivelled and inconspicuous during the day, expanding only towards nightfall, when the flowers often become noticeably fragrant.

Flowers in dense globose clusters or verticillasters spaced along the flowering stem - **2. S. gigantea**
Flowers not in dense globose clusters or verticillasters:
  Calyx distinctly inflated at anthesis:
    Calyx 20-nerved; petals very deeply bilobed; plant perennial - **6. S. vulgaris**
    Calyx 10-nerved; petals shortly lobed; plant annual - - - **15. S. behen**
  Calyx not distinctly inflated at anthesis:
    Plants perennial, with hard, woody rootstocks:
      Leaves bordered with several rows of short, stiff, recurved hairs, otherwise glabrous; plant suffruticose; leaves in tufts at ends of woody branches **1. S. fruticosa**
      Leaves pubescent all over; plant not suffruticose, nor with leaves in tufts at the end of woody branches:
        Flowers numerous, in large spreading panicles; petals white or greenish:
          Petal-lobes rounded, less than 1 cm. long; calyx thinly pubescent, often purplish; flowers erect - - - - - - - - **3. S. italica**
          Petal-lobes narrowly oblong or strap-shaped, 1–1·3 cm. long; calyx glabrous (calyx-teeth ciliate), pallid; flowers nodding - **5. S. longipetala**
        Flowers rather few and distant in very lax, narrow panicles; petals pink or pink-tinged - - - - - - - - - - - **4. S. galataea**
    Plants annual; base seldom hard or woody:
      Calyx 30- or 60-nerved; calyx-teeth long, narrow, subulate or almost so; anthophore very short:
        Calyx 30-nerved; capsule 1·6–2 cm. long with a slender neck; flowers usually deep reddish-pink - - - - - - - - - - **25. S. conoidea**
        Calyx 60-nerved; capsule about 1 cm. long, without a slender neck; flowers usually pale pink - - - - - - - - **26. S. macrodonta**
      Calyx 10-nerved, teeth not normally subulate (except in *S. tridentata*):
        Cauline leaves linear or subulate, with a long slender acumen:
          Capsule broadly ovoid or subglobose; seeds papillose; coronal scales triangular, acuminate - - - - - - - - **13. S. cretica**
          Capsule narrowly ovoid or ellipsoid; seeds bluntly ridged; coronal scales oblong, blunt - - - - - - - - - - **14. S. gemmata**
        Cauline leaves broader, not linear or subulate, nor with a long, slender acumen:
          Anthophore very short (less than 2 mm. long) or wanting; calyx rather abruptly constricted at base, or rounded:
            Plant quite glabrous, glaucous; flowers small, pink - **16. S. laevigata**
            Plant pubescent or glandular-pilose:
              Cauline leaves usually less than 1 cm. long, fleshy; flowers small, pink; stems spreading; seaside plant - - - - - **12. S. sedoides**
              Cauline leaves usually exceeding 1 cm. in length, not fleshy:
                Calyx distinctly constricted at mouth in fruiting specimens:
                  Petals bright carmine; inflorescence regularly dichotomous with spreading branches; calyx distinctly ribbed - **17. S. papillosa**
                  Petals paler pink or whitish, sometimes blotched crimson, or wanting; inflorescence not regularly dichotomous:
                    Calyx-teeth subulate, about 5 mm. long, conspicuous; petals small, inconspicuous - - - - - **22. S. tridentata**
                    Calyx-teeth not subulate; petals usually conspicuous:
                      Calyx distinctly ribbed; petals whitish; flowers usually distant - - - - - - - **18. S. dichotoma**
                      Calyx not ribbed; petals pink or crimson-blotched; flowers usually numerous in a raceme-like monochasium **21. S. gallica**

Calyx not much constricted at mouth in fruiting specimens:
Capsule narrowly ovoid-cylindrical; calyx with distinct anastomosing veins; seeds not winged; flowers in long raceme-like monochasia - - - - - - - - **20. S. nocturna**
Capsule broadly ovoid; calyx without distinct anastomosing veins; seeds winged; inflorescence often irregular - **24. S. apetala**
Anthopore more than 2 mm. long; calyx often tapering gradually to base:
Inflorescence regularly dichotomous with spreading branches; calyx distinctly ribbed - - - - - - - - **17. S. papillosa**
Inflorescence not regularly dichotomous with spreading branches; calyx not distinctly ribbed:
Inflorescence a rather crowded, much-branched terminal cluster of dichasia:
Calyx whitish (sometimes tinged purple) papery; flowers small; petals seldom more than 1 cm. long; plant subglabrous, glaucous **8. S. rubella**
Calyx purplish, glandular-hairy; flowers rather large; petals 1·3–1·5 cm. long; plant pubescent or glandular-hairy:
Bracts herbaceous, green or purplish; calyx 1–1·3 cm. long; seeds bluntly ridged or striate - - - - - **9. S. fuscata**
Bracts membranous, whitish; calyx 1·4–1·8 cm. long; seeds bluntly papillose - - - - - - - - **10. S. aegyptiaca**
Inflorescence not crowded, either monochasial and raceme-like, or a lax dichasium:
Calyx at anthesis rather narrowly clavate, tapering to base:
Flowers shortly pedicellate or subsessile in narrow raceme-like monochasia:
Basal leaves forming persistent tufts or rosettes at anthesis; capsule narrowly ovoid-oblong, 7 mm. long and 4 mm. wide; seeds not winged - - - - - - **19. S. discolor**
Basal leaves not forming tufts or rosettes at anthesis; capsule broadly ovoid or subglobose, 5–7 mm. long, 5–6 mm. wide; seeds winged - - - - - - - **23. S. colorata**
Flowers distinctly pedicellate, in lax dichasia or monochasia:
Inflorescence repeatedly branched; petals deeply bifid, pale pink or whitish; plant erect or suberect - - **7. S. kotschyi**
Inflorescence monochasial, or sparsely and irregularly branched; petals bright pink, truncate or shallowly emarginate; plant usually sprawling - - - - **11. S. fraudatrix**
Calyx at anthesis broadly oblong or turbinate, abruptly narrowed at base; petals often minute or wanting - - **24. S. apetala**

SECT. 1. **Paniculatae** *Boiss.* Inflorescence paniculate with opposite branches; flowers hermaphrodite, erect; calyx 10-nerved, clavate, ± constricted below capsule; petal-claw glabrous or ciliate. Perennial plants with woody rootstocks and persistent basal leaves.

1. S. fruticosa *L.*, Sp. Plant., ed. 1, 417 (1753); Sibth. et Sm., Fl. Graec. Prodr., 1: 299 (1809), Fl. Graec., 5: 20, t. 428 (1825); Boiss., Fl. Orient. 1: 633 (1867); Holmboe, Veg. Cypr., 74 (1914); Rechinger f. in Arkiv för Bot., ser. 2, 1: 421 (1950); Davis, Fl. Turkey, 2: 193 (1967); Osorio-Tafall et Seraphim, List Vasc. Plants Cyprus, 35 (1973).

TYPE: " *in* Sicilia ".

A loosely tufted suffruticose perennial, with a stout, woody, branched base; stems erect, 15–50 cm. high, thinly crisped-pubescent below, glabrescent above; basal leaves deep green or glaucous-green, glabrous except at margin, forming loose persistent tufts, obovate-spathulate, (1·5–) 4–8 cm. long, (0·3–) 1–2·5 cm. wide, apex acute, sometimes apiculate, base gradually narrowed to a petiole 0·5–3 cm. long, leaf-margins with several rows of short, stiff, recurved hairs; cauline leaves similar to basal, usually smaller and narrower, and shortly petiolate or subsessile; inflorescence a lax or rather congested, few- to many-flowered panicle 5–30 cm. long, with 1–4 pairs of opposite, ascending branches, subtended by oblanceolate, leafy bracts; flowers erect, rather large and conspicuous; pedicels usually less than 6 mm.

long, ± densely glandular-pubescent; calyx narrowly clavate, 1·8–2·5 cm. long, 0·3–0·4 cm. wide at apex, 10-nerved, ± densely glandular-pilose, often stained purplish; calyx-teeth ovate-triangular, subacute, about 3 mm. long and 2 mm. wide, with broadly membranous, ciliate margins; petals pale or deep pink internally, often stained brownish externally, 2·5–3 cm. long, 0·7–0·8 cm. wide towards apex, limb deeply bifid, with oblong, rounded lobes about 3 mm. long, gradually tapering (without distinct auricles) into a glabrous claw about 1·5 cm. long; coronal scales reduced to 2 narrow, slightly toothed wings, or sometimes wanting; anthophore about 8 mm. long; stamens 1·5–1·8 cm. long, filaments glabrous; anthers oblong, about 2·5 mm. long and 1·5 mm. wide; ovary narrowly ovoid-cylindrical, 5–7 mm. long, 2–4 mm. wide; styles 3, 10–12 mm. long. Capsule ovoid-urceolate, 12–14 mm. long, 6–8 mm. wide, capsule-teeth only slightly reflexed; seeds reniform 1·5–1·8 mm. long, about 1·5 mm. wide, dorsal surface canaliculate; testa blackish, sharply papillose.

HAB.: On cliffs; in rock-crevices; amongst boulders, or on ancient walls; usually, but not always, on limestone rocks; sea-level to 2,000 ft. alt.; fl. April–June.

DISTR.: Divisions 1, 8, locally common in the Akamas, and in the Kantara–Yioti area. Also in Sicily, Crete, Malta, Greece, Aegean Islands and Cyrenaica.

1. Locally common; *Sibthorp* s.n. ! *Kotschy* 662 ! *Holmboe* 718; *Davis* 1777 ! 3211 ! 3343 ! *Merton* 3048 ! *Meikle* 2035 !

8. Locally common; *Haradjian* 279 ! *Chapman* 238 ! *Kennedy* 1570 ! *Davis* 2453 ! *Meikle* 2519 !

NOTES: Typical material from Sicily tends to have rather narrower, more acuminate leaves than Cyprus specimens; there is, however, some overlapping in this character. The degree of development of the coronal scales varies greatly from specimen to specimen.

2. S. gigantea *L.*, Sp. Plant., ed. 1, 418 (1753) emend. Sp. Plant., ed. 2, 598 (1762); Sibth. et Sm., Fl. Graec. Prodr., 1: 299 (1809), Fl. Graec., 5: 22, t. 432 (1825); Boiss., Fl. Orient., 1: 645 (1867); Holmboe, Veg. Cypr., 84 (1914); Post, Fl. Pal., ed. 2, 1: 187 (1932); Lindberg f., Iter Cypr., 15 (1946); Davis, Fl. Turkey, 2: 192 (1967); Osorio-Tafall et Seraphim, List Vasc. Plants Cyprus, 35 (1973).

TYPE: " *in* Africa ", according to Linnaeus (Sp. Plant. ed. 2, 598), but really based on Tournefort's *Lychnis graeca, sedi arborescentis folio et facie, flore albo*, and probably from the Greek Islands, or possibly from Crete (P).

Monocarpic, with a thick woody rootstock; stems 30–100 cm. high, densely pubescent below, glabrous and very viscid above, branched chiefly at the base; basal leaves glaucous, pubescent, forming a neat *Sempervivum*-like rosette, spathulate, 4–16 cm. long, 1·5–6 cm. wide, apex rounded, mucronate, margins often distinctly ciliate, base gradually narrowing to a petiole 1·5–7·5 cm. long; cauline leaves usually narrower, obovate-spathulate or obovate, often acute, subsessile or with short petioles, the uppermost frequently oblanceolate; inflorescence 20–50 cm. long, unbranched or with a few pairs of ascending branches; flowers numerous in dense widely separated globose clusters or verticillasters, subtended by small, viscid-pubescent bracts and bracteoles; pedicels 0·5–1·5 cm. long, densely glandular-pubescent; calyx narrowly clavate, about 10 mm. long and 3 mm. wide at apex, membranous, obscurely 10-nerved, thinly glandular-pubescent; calyx-teeth bluntly ovate-triangular, about 2 mm. long and 1·5 mm. wide with hyaline, ciliate margins; petals greenish-yellow or whitish, 1·2–1·5 cm. long, 0·5–0·7 cm. wide towards apex, limb bifid almost to base, with narrowly oblong, rounded lobes about 4 mm. long, auricles absent; coronal scales reduced to two minute callosities on each petal; claw gradually tapering to base, about 6 mm. long with glabrous margins; anthophore 4–5 mm. long; stamens about 8 mm. long, filaments glabrous, filiform; anthers oblong, about 2 mm. long and 1 mm. wide, yellowish or olivaceous; ovary ovoid-cylindrical, about

3 mm. long and 1·5 mm. wide; styles 3, 1–2·5 cm. long, far exceeding the petals. Capsule broadly ovoid, about 8–10 mm. long and 6–7 mm. wide, capsule teeth only slightly reflexed; seeds reniform, about 1·5 mm. long and 1·3 mm. wide, dorsal surface canaliculate; testa dark brown or ashy-brown, with numerous very slender ridges arranged in regular, radiating rows.

HAB.: In fissures of limestone rocks; 2,000–3,000 ft. alt.; fl. May–July.

DISTR.: Division 7, locally common about St. Hilarion, Buffavento, Pentadaktylos, Yaïla. Also in Yugoslavia, Albania, Bulgaria, Greece, Crete, Aegean Islands, W. and S. Turkey and Lebanon.

7. Common; *Kotschy* 458; *Sintenis & Rigo* 664 ! *Kennedy* 1178 ! *Holmboe* 818. *Davis* 3626 ! 3603 !

S. PARADOXA *L.*, Sp. Plant., ed. 2, 1673 (1763); Sibth. et Sm., Fl. Graec. Prodr., 1: 298 (1806); Unger et Kotschy, Die Insel Cypern, 350 (1865); Boiss., Fl. Orient., 1: 633 (1867); Holmboe, Veg. Cypr., 74 (1914); Osorio-Tafall et Seraphim, List Vasc. Plants Cyprus, 35 (1973).

Recorded from Cyprus by Sibthorp, but probably an error. It much resembles the following (*S. italica* L.) but has narrower leaves and larger flowers, with the calyx sometimes 3 cm. long; the coronal scales are better developed than in *S. italica*, being lanceolate and acute. *S. paradoxa* is widely distributed in southern Europe eastwards to Greece and N.W. Turkey. It has not been found in areas adjacent to Cyprus.

3. S. italica (*L.*) *Pers.*, Syn. Plant., 1: 498 (1805); Boiss., Fl. Orient., 1: 631 (1867); Post in Mém. Herb. Boiss., 18: 91 (1900); H. Stuart Thompson in Journ. Bot., 44: 278 (1906); Holmboe, Veg. Cypr., 74 (1914); Post, Fl. Pal., ed. 2, 1: 185 (1932); Zohary, Fl. Palaest., 2: 84, t. 98 (1966); Davis, Fl. Turkey, 2: 192 (1967); Osorio-Tafall et Seraphim, List Vasc. Plants Cyprus, 35 (1973).

Cucubalus italicus L., Syst. Nat., ed. 10, 1030 (1759).

TYPE: " *in* Italia ".

Loosely tufted perennial with a branched woody rootstock; stems 30–50 cm. high, erect or ascending, thinly grey-pubescent towards base, glabrescent and viscid-glandular above; basal leaves in tufts, persistent at flowering time; lamina obovate-elliptical, 2–4 cm. long, 1–2 cm. wide, acute, pubescent, gradually narrowing at base, and decurrent along a petiole 1–4 cm. long; cauline leaves widely spaced, narrowly obovate or oblanceolate, becoming narrower and less distinctly petiolate upwards; inflorescence a lax pyramidal panicle 10–20 cm. long, with 2–4 (–5) pairs of opposite spreading branches subtended by narrow, lanceolate, leafy bracts; flowers erect, relatively large and conspicuous; pedicels 1–5 mm. long; calyx narrowly clavate, 1·5–2·5 cm. long, 0·3–0·4 cm. wide at apex, 10-nerved, thinly glandular-pubescent, often stained purplish; calyx-teeth ovate-triangular, subacute, about 2 mm. long and 1·5 mm. wide, with membranous, ciliate margins; petals white (sometimes with greenish veins), 2–2·5 cm. long, 0·6–0·8 cm. wide near apex, limb deeply bifid, with rounded lobes less than 1 cm. long, shortly and bluntly auriculate at base, with the coronal scales reduced to 2 inconspicuous callosities, claw 1·3–1·5 cm. long, gradually tapering to base, margins glabrous; anthophore 1·3–1·5 cm. long; stamens 1·5–2 cm. long, filaments glabrous; anthers narrowly oblong, 2·5–3 mm. long, 1–1·5 mm. wide, fuscescent; ovary cylindrical, 5–6 mm. long, 1·5 mm. wide; styles 3, 7–9 mm. long. Capsule ovoid-urceolate, 1–1·2 cm. long, 0·5–0·6 cm. wide, capsule teeth slightly reflexed; seeds reniform, 1 mm. long, 0·8 mm. wide, dorsal surface flattened or canaliculate; testa blackish, sharply papillose.

HAB.: In cultivated ground; on rocky hillsides, or in rock-crevices; ? alt.; fl. April–July.

DISTR.: Division 1, rare. Widespread in south and central Europe. Also in Caucasus area, Turkey, Syria, Lebanon, Iran and North Africa.

1. Paphos, 1894, *Post*.

NOTES: Not seen since, and record possibly erroneous, though it is hard to imagine any other species with which it could have been confused.

4. S. galataea *Boiss.*, Fl. Orient. Suppl., 102 (1888); Holmboe, Veg. Cypr., 74 (1914); Lindberg f., Iter Cypr., 15 (1946); Osorio-Tafall et Seraphim, List Vasc. Plants Cyprus, 35 (1973).

[*S. pseudo-atocion* (non Desf.) H. Stuart Thompson in Journ. Bot., 44: 278 (1906).]

TYPE: Cyprus; " in vineis Cypri circa Galata ", *Sintenis & Rigo* 768 (G, K !).

A loosely tufted perennial with a branched woody rootstock; stems 20–50 cm. high, sprawling or ascending, purplish, grey-pubescent below, glabrescent and viscid-glandular above; basal leaves in very loose tufts, often persistent at flowering time; lamina obovate-elliptical, 2–5 cm. long, 1–2 cm. wide, acute, thinly pubescent, dull green or purplish, narrowing gradually at base and decurrent along a petiole 2–5 cm. long; cauline leaves widely spaced, often rather broadly ovate or elliptical, 1·5–4 cm. long, 0·8–1·5 cm. wide, becoming more shortly petiolate or subsessile upwards; inflorescence a very lax panicle up to 40 cm. long, with 2–5 pairs of elongate, widely spreading branches subtended by small ovate or lanceolate leafy bracts; flowers erect (? or occasionally slightly nodding); pedicels 1–10 mm. long; calyx narrowly clavate, about 1·5 cm. long, 0·4–0·5 cm. wide at apex, 10-nerved, thinly pubescent, purplish; calyx-teeth ovate-triangular, subacute, about 2 mm. long and 1·5 mm. wide, with membranous, ciliate margins; petals rose-pink internally, stained brownish externally, 1·5 cm. long, 0·5–0·8 cm. wide near apex, limb deeply bifid with oblong, rounded lobes, scarcely auriculate at base, coronal scales reduced to 2 small crests, claw about 8 mm. long, gradually tapering from apex to base, margins glabrous; anthophore about 5 mm. long; stamens about 10 mm. long, alternate ones with filaments adnate to the petal-claw for a short distance above base; margins of filaments glabrous; anthers yellowish-fuscescent, oblong, 3 mm. long, 1·5 mm. wide; ovary cylindrical, 5–6 mm. long, 1·5 mm. wide; styles 3, 5–6 mm. long. Capsule narrowly ovoid-urceolate, 8–9 mm. long, about 4 mm. wide, capsule-teeth slightly reflexed; seeds reniform, 1 mm. long, 0·8 mm. wide; dorsal surface canaliculate; testa dark reddish-brown, sharply papillose.

HAB.: On dry rocky slopes under *Pinus*; roadside banks; in vineyards; usually on igneous formations, but also on chalk; 2,500–4,000 ft. alt.; fl. April–June.

DISTR.: Division 2, locally common in Stavros tis Psokas, Vroisha, Tripylos area, also at Prodhromos & Galata. Endemic.

2. Common; *Sintenis & Rigo* 768 ! *Syngrassides* 1596 ! *Kennedy* 1506 ! *Davis* 3490 ! *Merton* 2873 ! *Meikle* 715 !

NOTES: Closely allied to *S. italica* (L.) Pers. and certainly belonging to sect. *Paniculatae*, but placed in sect. *Siphonomorpha* Otth by Chowdhuri (*op. cit.*, 234 250).

SECT. 2. **Lasiostemones** *Boiss.* Similar in habit and inflorescence to sect. *Paniculatae*, but calyx broader, stamens hairy at base; petals with ciliate claw; flowers often nodding.

5. S. longipetala *Vent.*, Jard. Cels., 83, t. 83 (1802); Sibth. et Sm., Fl. Graec., 5: 13, t. 419 (1825); Boiss., Fl. Orient., 1: 636 (1867); Holmboe, Veg. Cypr., 74 (1914); Post, Fl. Pal., ed. 2, 1: 185 (1932); Lindberg f., Iter Cypr., 15 (1946); Zohary, Fl. Palaest., 1: 84, t. 99 (1966); Davis, Fl. Turkey, 2: 194 (1967); Osorio-Tafall et Seraphim, List Vasc. Plants Cyprus, 35 (1973).

TYPE: Syria; " aux environs d'Alep " *Bruguière et Olivier* (P).

Tufted perennial with a branched woody rootstock; stems 30–80 cm. high, erect, shortly puberulous below, glabrescent and very viscid above; basal leaves in loose tufts, persistent at flowering time; lamina oblong-obovate, 3–7 cm. long, 1–3 cm. wide, acute, thinly puberulous, glaucous, narrowing gradually at base and decurrent along a petiole 1–3 cm. long; cauline leaves rather closely spaced, similar to basal, but shortly petiolate or subsessile; inflorescence a very large, lax, spreading, dome-shaped or broadly pyramidal panicle, with 2–5 (or more) pairs of very elongate lateral branches much overtopping the central axis; bracts linear-lanceolate, the upper with distinct membranous, ciliate margins; buds erect, open flowers nodding; pedicels very short, seldom more than 4 mm. long (except for the terminal flower of the primary axis); calyx broadly clavate or narrowly turbinate, 1–1·3 cm. long, 0·3–0·4 cm. wide near apex, 10-nerved, glabrous, whitish with pale green nerves; calyx-teeth broadly ovate, about 2 mm. long and almost as wide at base, with membranous, ciliate margins; petals pale greenish-white, 1·5 cm. long, 0·5 cm. wide near apex, limb divided almost to base into 2 narrowly oblong or strap-shaped lobes 1–1·3 cm. long, basal auricles inconspicuous, coronal scales lingulate, about 1 mm. long, claw oblong, about 5 mm. long and 3 mm. wide, narrowing rather abruptly at base, hairy towards base, with long-ciliate margins; anthophore about 4 mm. long; stamens about 8 mm. long, alternate ones with filaments adnate to petal-claw for a short distance above base; filaments hairy towards base; anthers narrowly oblong, about 2·5 mm. long and 1 mm. wide; ovary oblong, about 3·5 mm. long and 2·5 mm. wide; styles 3, about 6 mm. long, hairy in lower half. Capsule broadly ovoid, 6–7 mm. long, 5–6 mm. wide, capsule-teeth slightly reflexed; seeds reniform, large, about 2·5 mm. long and 2 mm. wide; dorsal surface flattish or canaliculate; testa dull brown, covered with regular rows of short ridges.

HAB.: On open stony slopes under *Pinus*; in vineyards and cornfields, on igneous and limestone formations; 300–5,700 ft. alt.; fl. March–July.

DISTR.: Divisions 1–4, 6, common in Division 2. Also Greece, Turkey, Syria, Palestine and Iraq.

1. Ktima, 1913, *Haradjian* 670 !
2. Common; *Sintenis & Rigo* 767 ! 767a ! *Holmboe* 767; *Syngrassides* 1520 ! *Kennedy* 353–356 ! *Davis* 3433 ! *Merton* 2368 !
3. Between Yerasa and Louvaras, 1962, *Meikle* 2894 !
4. Larnaca, 1862, *Kotschy* 55.
6. Between Orta Keuy and Ayios Pavlos (Nicosia), 1902, *A. G. & M. E. Lascelles* !

S. OTITES (*L.*) *Wibel*, Primit. Fl. Werthem., 241 (1799); Sibth. et Sm., Fl. Graec. Prodr., 1: 301 (1809); Boiss., Fl. Orient., 1: 606 (1867); Holmboe, Veg. Cypr., 73 (1914); Osorio-Tafall et Seraphim, List Vasc. Plants Cyprus, 35 (1973).

Recorded from Cyprus by Sibthorp, but probably an error, as the plant has not been found since, nor is it known from adjacent areas. It is an erect, glaucous biennial or short-lived perennial, 10–60 (–100) cm. high, with small, whitish, usually unisexual flowers, 5–7 mm. long, in dense interrupted clusters, the whole inflorescence often forming an elongate verticillaster. *Silene otites* is widespread in Europe, also occurring in the Caucasus area, central Turkey and northern Iran.

SECT. 3. **Inflatae** *Boiss.* Perennials; stems branched from the middle upwards; inflorescence a dichasial cyme; calyx 10–20-nerved, inflated, usually glabrous and glaucous with reticulately anastomosing nervation; petal-limb cleft or emarginate, coronal scales usually small.

6. S. vulgaris (*Moench*) *Garcke*, Fl. Nord-Mitt.-Deutschl., ed. 9, 64 (1869); Marsden-Jones et Turrill, Bladder Campions (1957); Zohary, Fl. Palaest., 1: 85, t. 101

(1966); Davis, Fl. Turkey, 2: 213 (1967); Osorio-Tafall et Seraphim, List Vasc. Plants Cyprus, 35 (1973).

Behen vulgaris Moench, Meth., 709 (1794).

Cucubalus behen L., Sp. Plant., ed. 1, 414 (1753) nec *Silene behen* L.

Cucubalus venosus Gilib., Fl. Lith., 2: 165 (1781) nom. illeg.

Cucubalus inflatus Salisb., Prodr., 302 (1796) nom. illeg.

Silene cucubalus Wibel, Primit. Fl. Werthem., 241 (1799).

Silene inflata Sm., Fl. Brit., 467 (1800); Unger et Kotschy, Die Insel Cypern, 348 (1865); Boiss., Fl. Orient., 1: 628 (1867).

Silene venosa Aschers., Fl. Brandenb., 2: 23 (1859); Holmboe, Veg. Cypr., 72 (1914); Post, Fl. Pal., ed. 2, 1: 184 (1932); Lindberg f., Iter Cypr., 15 (1946).

TYPE: " *in* Europae *septentrionalioris pratis siccis* ".

An erect or spreading perennial up to 70 cm. high; rootstock branched, often woody, sometimes stoloniferous; stems glabrous or occasionally pubescent, not viscid, often glaucous; leaves glabrous or pubescent, often glaucous, rather crowded towards the base of the plant, but not forming distinct tufts, the lower oblanceolate or narrowly oblong, 2·5–8 cm. long, 0·8–2 (–3·5) cm. wide, apex acute, margin often minutely toothed or papillose, base tapering to a short flattened petiole, or leaves subsessile; upper cauline leaves similar but often lanceolate and more acuminate, and normally sessile; inflorescence a lax dichasial cyme, 7–20 (–40) cm. long, consisting of 9–30 or more slightly nodding hermaphrodite or unisexual flowers, or, in depauperate specimens, flowers occasionally fewer or solitary; bracts small, ovate or lanceolate, herbaceous or the uppermost scarious; pedicels glabrous or sometimes puberulous, 1–3 (–6) cm. long; calyx broadly ovoid or subglobose, inflated, persistent, often purplish, 1·5–2 cm. long, 1–1·5 cm. wide, distinctly constricted towards apex, 20-nerved with anastomosing nerves; calyx-teeth broadly triangular, 2–3 mm. long and about as wide, with a narrow puberulous margin; petals white or sometimes pink or reddish, 1·5–2 cm. long, 0·7–0·9 cm. wide towards apex, limb bifid almost to base, with narrowly obcuneate, rounded lobes 5–7 mm. long; auricles conspicuous, rounded, membranous; coronal scales reduced to two small callosities on each petal; claw 1–1·5 cm. long, gradually tapering from apex to base, glabrous; anthophore 3–3·5 mm. long; stamens 1·5–2·3 mm. long, filaments glabrous, often purplish; anthers narrowly oblong, about 3 mm. long and 1 mm. wide, purplish; ovary narrowly ovoid-cylindrical, 4–5 mm. long, about 3 mm. wide, capped with a distinct dome-like stylophore; styles 3, about 1 cm. long. Capsule broadly ovoid, about 1–1·3 cm. long and 0·7–1 cm. wide, usually with a narrow mouth. Capsule teeth normally only a little reflexed; seeds reniform, about 1·5 mm. long and 1–1·2 mm. wide, dorsal surface flattened or slightly canaliculate; testa grey-brown, strongly papillose or minutely tuberculate, tubercles rarely flattened.

The above description relates to *Silene vulgaris* in the aggregate sense; from material available further analysis, except in a very few instances, is impracticable (see Marsden-Jones and Turrill, The Bladder Campions, 133).

HAB.: In cultivated and fallow fields; by roadsides; on open stony mountain-sides; occasionally on walls; sea-level to 6,400 ft. alt.; fl. March–July (and occasionally at other seasons).

DISTR.: All Divisions, common in Division 2. Throughout temperate Europe, the Mediterranean area, Middle East, Central Asia eastward to Kamtschatka.

1. Lyso, 1913, *Haradjian* 835 ! and, 1938, *Nattrass* 931.
2. Common; *Holmboe* 766; *Kennedy* 352 ! *Lindberg f.* s.n.; *Davis* 1797 ! *Merton* 3196 !
3. Limassol, 1913, *Haradjian* 611 ! Akrotiri, 1927, *Rev. A. Huddle* 20 !
4. Near Larnaca, 1862, *Kotschy* 288 ! Paralimni, 1940, *Davis* 2054 !
5. Kythrea, 1880, *Sintenis & Rigo* 997 ! Athalassa, 1935, *Nattrass* 629; Kythrea, 1950, *Chapman* 172 !
6. Syrianokhori, 1935, *Syngrassides* 640 ! Nicosia, 1931, *Nattrass* 41; near Dhiorios, 1962, *Meikle* 2334 !

7. Kyrenia, *J. A. Tracey* 11 ! Kyrenia, 1955 and 1956, *G. E. Atherton* 204 ! 870 ! Bellapais, 1956, *G. E. Atherton* 860 !
8. Between Apostolos Andreas and Khelones, 1941, *Davis* 2290 !

ssp. **macrocarpa** *Turrill* in Hooker's Icon. Plant., 36: t. 3551 (1956).
TYPE: Cult. Kew " from seeds collected at Khionistra, 1620 m., 5 July 1938, *Mrs. E. W. Kennedy*, K.2312, plant 10 " (K !).

Said to differ from typical *S. vulgaris* in having subterranean stolons and a larger, flask-shaped capsule, 1·3 cm. long and 0·8 cm. wide. The petals are usually white, but sometimes pink or purplish.

HAB.: Probably similar to that of typical *S. vulgaris*.

DISTR.: Division 2, possibly common on the island. (See note.)

2. Khionistra, 1938, *Kennedy* 2312 !

NOTES: Possibly also Paralimni, Kyrenia, Larnaca, Kythrea, Limassol, St. Hilarion, Kryos Potamos, Platres, where stoloniferous plants are thought, or known, to occur, but definite identifications are not given either in Hooker's Icones, *loc. cit.* (1956) nor in Marsden-Jones & Turrill, The Bladder Campions, 131–133 (1957). *S. vulgaris* ssp. *macrocarpa* is said (The Bladder Campions, p. 133) to occur in Britain, at Plymouth Hoe, where it apparently interbreeds with *S. vulgaris* ssp. *vulgaris* or with *S. maritima* With.

ssp. **commutata** (*Guss.*) *Hayek*, Prodr. Fl. Pen. Balc., 1: 258 (1924).
S. commutata Guss., Fl. Sic. Prodr., 1: 499 (1827); Boiss., Fl. Orient., 1: 269 (1867); Marsden-Jones et Turrill, Bladder Campions, 106 (1957).
? *S. inflata* Sm. var. *major* Kotschy in Unger et Kotschy, Die Insel Cypern, 348 (1865) nomen.
S. venosa Aschers. var. *commutata* (Guss.) Holmboe, Veg. Cypr., 72 (1914).
TYPE: Sicily; " in glareosis, et rupibus montosis; Busambra, Pizzuta, Favignana " *Gussone* (Fl, ? NAP).

Differs from *S. vulgaris* ssp. *vulgaris* in having broadly ovate, fleshy leaves, 4·5–7 cm. long and 2·4–3·5 cm. wide, with acute, apiculate apices and rounded or cordate bases. The seeds of typical ssp. *commutata* have strongly flattened (" armadillo ") tubercles. The flowers are said to be somewhat larger than those of ssp. *vulgaris* and frequently very numerous.

HAB.: Not known.

DISTR.: Cyprus, without locality, *Kotschy* Suppl. 480 sec. Boissier. Also in Sicily, Caucasus area, Turkey, Palestine, ? Crete, ? Greece.

NOTES: Boissier's record cannot be traced to its source, nor do any of the Cyprus specimens examined resemble *S. vulgaris* ssp. *commutata*. *Kotschy* 288, from the vicinity of Larnaca, is labelled var. *major*, and may have been considered identical with ssp. *commutata* by Boissier, though the Kew specimen in no way resembles the Sicilian plant.

f. **colorata** (*Hampe*) *Hayek*, Prodr. Fl. Pen. Balc., 1: 257 (1924).
S. inflata Sm. var. *colorata* Hampe in Linnaea, 11: 46 (1837).
S. inflata Sm. var. *rubriflora* Boiss., Fl. Orient., 1: 629 (1867).
S. venosa Aschers. var. *rubriflora* (Boiss.) Holmboe, Veg. Cypr., 72 (1914).
TYPE: Germany; Harz Mts., *Hampe* (B, L).

Flowers tinged pink or purplish-red.

HAB.: Not known.

DISTR.: Division 2.

2. Phiti, 1905, *Holmboe* 766. " Not rare " according to Holmboe.

NOTES: Of little taxonomic significance; pink- or purple-petalled forms are found within a number of different infraspecific variants of *S. vulgaris*, and have been reported from Cyprus.

S. THYMIFOLIA *Sibth. et Sm.*, Fl. Graec. Prodr., 1: 292 (1809), Fl. Graec., 5: 8, t. 411 (1825); Boiss., Fl. Orient., 1: 648 (1867); Osorio-Tafall et Seraphim, List Vasc. Plants Cyprus, 35 (1973).

Recorded from sandy seashores in Cyprus by Sibthorp and Smith, but almost certainly an erroneous localization; *S. thymifolia* is confined to N.W. Turkey and the Black Sea coast, and is conspecific with *S. pontica* Brandza.

SECT. 4. **Rigidulae** *Boiss.* Plants rather rigid, annual; inflorescence irregularly dichotomous, ultimate branches and pedicels often filiform; calyx-apex not constricted in fruit; seeds reniform with flat faces.

7. **S. kotschyi** *Boiss.*, Diagn., 1, 1: 40 (1843), Fl. Orient., 1: 602 (1867); Post, Fl. Pal., ed. 2, 1: 180 (1932); Davis, Fl. Turkey, 2: 228 (1967); Osorio-Tafall et Seraphim, List Vasc. Plants Cyprus, 35 (1973).

Erect or spreading annual, 10–40 cm. high; stems usually much branched especially in the upper half, scabridulous-pubescent towards base, glabrescent and very viscid above; basal leaves mostly withered at flowering time, narrowly oblanceolate, 1·5–4 cm. long, 0·2–0·4 cm. wide, apex acute, base tapering to an indistinct petiole; surfaces of lamina greyish, ± densely scabridulous-pubescent; cauline leaves narrower, sessile, sometimes fasciculate, the upper linear-subulate, 1–2·5 cm. long, 0·1–0·2 cm. wide, with a distinct hardened apiculus, and often strongly recurved margins; inflorescence repeatedly, but somewhat unequally, dichotomously branched, branches and pedicels usually very slender; bracts subulate, about 5 mm. long and less than 1 mm. wide, herbaceous; pedicels 0·8–4 cm. or more in length; calyx clavate, about 10 mm. long and 2–3 mm. wide at anthesis, often conspicuously striped, with 10 purplish or reddish nerves, calyx-teeth narrowly triangular, acuminate, 1·5–2 mm. long, 1–1·5 mm. wide; petals pink, yellowish-livid or white, 10–12 mm. long, 3–4 mm. wide near apex, deeply bifid with narrowly oblong lobes about 3 mm. long; coronal scales conspicuous, forming 2 roughly triangular flaps, rather less than 1 mm. long; auricles inconspicuous; claw gradually tapering to base, about 6–8 mm. long, with ciliate margins; stamens 10–12 mm. long, filaments distinctly hairy towards base; anthers oblong, about 1·3 mm. long and 0·8 mm. wide; anthophore slender, 4–5 mm. long; ovary cylindrical 2·5–3 mm. long, 1·5–2 mm. wide; styles 3, about 8 mm. long. Capsule oblong-ellipsoid, 5–7 mm. long, 2·5–3·5 mm. wide, with short, triangular, slightly reflexed teeth; seeds reniform about 1 mm. long and 0·6 mm. wide, dorsal surface canaliculate, faces flat; testa dark brown with numerous minute radiating ridges.

var. **maritima** *Boiss.*, Fl. Orient., 1: 602 (1867); Post, Fl. Pal., ed. 2, 1: 180 (1932); Davis, Fl. Turkey, 2: 228 (1967).

Silene stenocalyx Lindberg f., Iter Cypr., 15 (1946).

S. kotschyi Boiss. var. *stenocalyx* (Lindberg f.) Chowdhuri in Notes Roy. Bot. Gard. Edinb., 22: 276 (1957).

TYPE: Turkey; " in arenosis maritimis Ciliciae ad Mersina ", June 2, 1855, *Balansa* 801 (G, K !).

Usually shorter than *S. kotschyi* var. *kotschyi*, often much branched from near the base; internodes and pedicels also rather short.

HAB.: On sand-dunes and sandy seashores; about sea-level; fl. June–July.

DISTR.: Divisions 4, 5, rare. Also Turkey.

4. Near Famagusta, 1939, *Lindberg f.* s.n. !
5. Salamis near Famagusta, 1939, *Lindberg f.* s.n. !

NOTES: None of the characters purporting to distinguish *S. stenocalyx* will bear examination; the Lindberg Cyprus material closely resembles a recent sheet (*E. R. B. Little* 7) of *S. kotschyi* var. *maritima* from Antalya, Turkey.

SECT. 5. **Atocion** *Otth* Annuals; inflorescence a lax or crowded dichasium, with equal or unequal branches; calyx 10-nerved not constricted at apex in fruit. Seeds often subglobose, or with convex faces, frequently with a deeply excavate hilum.

8. **S. rubella** *L.*, Sp. Plant., ed. 1, 419 (1753); Sibth. et Sm., Fl. Graec. Prodr., 297 (1809), Fl. Graec., 5: 19, t. 426 (1825); Boiss., Fl. Orient., 1: 598 (1867); Holmboe, Veg. Cypr., 73 (1914); Post, Fl. Pal., ed. 2, 1: 179 (1932); Zohary, Fl. Palaest., 1: 90, t. 111 (1966); Osorio-Tafall et Seraphim, List Vasc. Plants Cyprus, 35 (1973).

TYPE: " in Lusitania ".

Erect, branched or unbranched, glaucescent annual, 15–50 cm. high; stems subglabrous or sparsely deflexed-pubescent below, more densely crispate-pubescent above; basal leaves obovate-spathulate, 2–3 cm. long, 0·6–1 cm. wide, glabrous with sparingly ciliate margins, apex obtuse, base narrowing gradually to a short, flattened, ciliate petiole; cauline leaves often larger, narrowly oblong, 2–6 cm. long, 0·5–2 cm. wide, sessile and shortly connate at the base; inflorescence a ± condensed, unequally branched dichasium; bracts herbaceous, foliaceous; pedicels short, 0·5–1·5 cm. long, crispate-pubescent; flowers erect; calyx rather broadly clavate, whitish or purplish-tinged, papery, 1–1·4 cm. long, 0·3–0·4 cm. wide, minutely adpressed-pubescent, obscurely or conspicuously 10-nerved, not much constricted at apex; calyx-teeth broadly and bluntly oblong, about 1·5 mm. long and 1 mm. wide, rather cucullate at apex, with strongly ciliate margins; petals rich rose-pink, about 10 mm. long, joined together by the lateral fusion (? permanent) of the coronal scales; limb 3–4 mm. long, 1·5–2 mm. wide at emarginate apex; auricles absent or insignificant; coronal scales very prominent, ovate, acute, about 1 mm. long, the lateral (outer) margin fused to that of the adjacent petal; claw tapering from apex to base; stamens about 6 mm. long; filaments glabrous; anthers oblong, small, about 0·7 mm. long and 0·3 mm. wide, fuscous-brown; anthophore 4–5 mm. long, minutely pubescent, ovary cylindrical, 3·5–4 mm. long, 1–1·5 mm. wide; stylophore prominent, dome-shaped; styles 3, rather short, 2·5–3 mm. long. Capsule shortly oblong, about 7 mm. long and 5 mm. wide, with a wide mouth and short, sharply recurved capsule-teeth; seeds reniform about 1 mm. long, 0·8 mm. wide, dorsal surface flattish or canaliculate, lateral surfaces and ventral surface deeply excavate; testa dark brown, with numerous, fine, sharp, regular ridges.

HAB.: In cornfields, often on moist ground or by edges of irrigation channels; sea-level to 500 ft. alt.; fl. Febr.–May.

DISTR.: Divisions 4, 5, 7, 8. Also [Rhodes], Syria, Palestine, Iraq, Egypt and N. Africa and from Portugal eastwards to Sicily.

4. Larnaca, 1880, *Sintenis & Rigo*; Phaneromene, Larnaca, 1905, *Holmboe* 126b; Paralimni, 1928 or 1930, *Druce* 63 ! Near Goshi, 1974, *Meikle* 4089 !
5. Near Kythrea, 1880, *Sintenis & Rigo* 251 ! Near Lefkoniko, 1880, *Sintenis & Rigo*; Mora, 1956, *Merton* 2562 !
7. W. of Kyrenia, 1956, *G. E. Atherton* 988A !
8. Near Ayios Andronikos, 1880, *Sintenis & Rigo*.

9. **S. fuscata** *Link* in Brot., Fl. Lusit., 2: 187 (1804); Boiss., Fl. Orient., 1: 600 (1865); Post, Fl. Pal., ed. 2, 1: 179 (1932); Rechinger f. in Arkiv för Bot., ser. 2, 1: 421 (1950); Zohary, Fl. Palaest., 1: 89, t. 110 (1966); Davis, Fl. Turkey, 2: 230 (1967); Osorio-Tafall et Seraphim, List Vasc. Plants Cyprus, 35 (1973).

TYPE: Portugal; " in collibus circa Cabeça de Montachique necnon circa Olidos, et alibi in Extremadura ", *Brotero* (LISU).

Erect, branched or unbranched annual, 15–50 cm. high; stems pubescent towards base of plant, with long, scattered, white, multicellular hairs, glandular-viscid above, with numerous shorter, brownish, stalked glands; basal leaves oblong-spathulate, 4–5 cm. long, 1–1·5 cm. wide, glabrous or thinly pubescent, apex rounded, base narrowing gradually to a distinct flattened petiole; cauline leaves narrowly oblong, 4–7 cm. long, 1–2 cm. wide, apex subacute, margins shortly papillose, often undulate, bases shortly connate, ciliate; inflorescence a rather condensed, unequally branched dichasium; bracts herbaceous, foliaceous, glandular-pubescent; pedicels 0·5–2·5 cm. long, densely viscid-glandular; flowers erect; calyx narrowly clavate or subcylindrical, 1–1·3 cm. long, 0·3–0·4 cm. wide, purplish, with a sparse indumentum of brownish glandular hairs, inconspicuously 10-nerved, not much constricted at apex; calyx-teeth broadly and bluntly oblong, about 1·5 mm. long and 1 mm. wide, rather cucullate at apex and with strongly ciliate margins; petals rose-pink, 1·3–1·4 cm. long, joined together by the fusion of coronal scales and claws; limb oblong about 5 mm. long and 1·5–2 mm. wide near the truncate or slightly emarginate apex; auricles absent or inconspicuous; coronal scales conspicuous, each pair connate, about 2 mm. long and 1·5 mm. wide with a shortly bifid apex, and with their lateral margins connate to those of the adjacent petals as in *S. rubella*; claws wide, membranous, connate to base; stamens 11–12 mm. long; filaments glabrous; anthers oblong, small, about 0·7 mm. long and 0·3 mm. wide, fuscous brown; anthophore rather stout, about 6 mm. long, with 2 narrow translucent wings; ovary narrowly ovate or cylindrical, 3·5 mm. long, 1–1·5 mm. wide; stylophore inconspicuous; styles 3, about 5–7 mm. long. Capsule ovoid, about 8 mm. long and 4–5 mm. wide, with a wide mouth and short, strongly recurved capsule-teeth; seeds reniform about 1 mm. long and 0·8 mm. wide, dorsal surface flattish or canaliculate; lateral and ventral surfaces deeply excavate; testa blackish-brown, with numerous regular, radiating rows of short, low ridges.

HAB.: Fields and cultivated ground; sea-level–1,500 ft. alt.; fl. Febr.–May.

DISTR.: Divisions 3, 7, rare. Also Crete, Turkey, Syria, Palestine, Algeria, Portugal, Spain, Italy, Sardinia and Sicily.

3. Limassol, 1913, *Haradjian* 621; 1 m. S. of Kato Lefkara, 1974, *Meikle* 5008 !
7. Six miles W. of Kyrenia, 1956, *G. E. Atherton* 987 ! 988 !

NOTES: The H. Stuart Thompson record for *S. damascena* mentioned by Rechinger f. (Arkiv för Bot. *loc. cit.*) is referable to *S. colorata* Poir. and not to *S. fuscata* Link.

10. S. aegyptiaca (*L.*) *L.f.*, Suppl. Plant., 241 (1781); Holmboe, Veg. Cypr., 73 (1914); Post, Fl. Pal., ed. 2, 1: 179 (1932); Rechinger f. in Arkiv för Bot., ser. 2, 1: 421 (1950); Zohary, Fl. Palaest., 1: 89, t. 109 (1966); Davis, Fl. Turkey, 2: 229 (1967); Osorio-Tafall et Seraphim, List Vasc. Plants Cyprus, 35 (1973).

Cucubalus aegyptiacus L., Sp. Plant., ed. 1, 415 (1753).

Silene atocion Juss. ex Jacq., Hort. Vindob., 3: 19, t. 32 (1776); Unger et Kotschy, Die Insel Cypern, 351 (1865); Boiss., Fl. Orient., 1: 600 (1867).

S. orchidea L.f., Suppl. Plant., 241 (1781); Sibth. et Sm., Fl. Graec. Prodr., 1: 297 (1806), Fl. Graec., 5: 19, t. 427 (1825).

TYPE: Egypt; " in Aegypto ", *Hasselquist* (specimen not known to be in existence).

Erect annual, 10–30 cm. high, usually much branched from base; stems pubescent below, glandular-viscid above; leaves variable in shape, glabrous or minutely scabridulous, the basal generally spathulate, 1·5–6 cm. long, 0·8–2 cm. wide, apex obtuse, mucronulate, base narrowing rather abruptly to a slender ciliate petiole 2·5–3 cm. long; cauline leaves often larger, oblong or obovate, acute, often purplish with undulate margins, the uppermost sessile or subsessile, oblong-elliptic; inflorescence a rather congested dichasium of erect flowers; bracts ciliate, generally membranous, broadly ovate, 2–4 mm. long, 2–3 mm. wide; pedicels usually short, less than 8 mm. long, glandular-viscid; calyx cylindrical-clavate, 1·4–1·8 cm. long, 0·3–0·4 cm. wide at apex, commonly purplish, ± densely glandular-pubescent; calyx-teeth ovate-deltoid, subacute, about 2·5 mm. long and 1·5 mm. wide, often tipped dorsally with a small tuft of long, translucent, several-celled hairs; petals bright pink, about 1·5 cm. long and 0·2–0·3 cm. wide near apex; limb deeply bifid with broadly oblong lobes about 1·5 mm. long; auricles strongly developed, narrowly oblong, obtuse, about 1·5–2 mm. long, forming a finger-like lobe on either side of the limb; coronal scales about 1 mm. long with a dilated base, and an elongate, subulate apex; claw slender, 8–9 mm. long, tapering gradually from apex to base; stamens 9–10 mm. long, filaments glabrous; anthers oblong, violet, about 1·5 mm. long and 0·8 mm. wide; anthophore slender, 8–9 mm. long; ovary cylindrical, about 3·5 mm. long and 1 mm. wide, slightly constricted at apex with a distinct pulviniform stylophore; styles 3, about 5 mm. long. Capsule broadly ovoid, 5–7 mm. long, about 5 mm. wide, with thin, papery walls; capsule teeth about 1 mm. long, strongly recurved; seeds reniform or almost globose, a little more than 1 mm. long, a little less than 1 mm. wide, dorsal and lateral surfaces convex, ventral surface deeply excavate about hilum; testa dark brown, with radiating rows of well-marked, bluntish papillae.

HAB.: Moist cultivated ground, vineyards, olive-groves, cereal fields, roadsides etc.; sea-level to about 2,000 ft. alt.; fl. Febr.–July.

DISTR.: Divisions 3–8, common in Divisions 3, 5. Widespread in Turkey, Syria, Palestine, southwards to Egypt.

3. Common; *Kotschy* 467; *Davis* 2732 ! *Mavromoustakis* s.n. ! *Poore* 45 ! *N. Macdonald* 12 !
4. Skarinou (Larnaca Distr.), 1941, *Davis* 2681 !
5. Common; *Sintenis & Rigo* 250 ! *Druce* 65 ! *Casey* 316 ! *Chapman* 539 ! *Meikle* 2561 !
6. Nicosia, 1932, *Syngrassides* 68 !
7. Halevga, 1912, *Haradjian* 380 ! Bellapais, 1948, *Collingwood Ingram* 24 ! Temblos, 1956, *G. E. Atherton* 1111 ! Thermia, 1956, *G. E. Atherton* 874 !
8. Ayios Andronikos, 1880, *Sintenis & Rigo*.

NOTES: Hitherto so confused with the succeeding species that, in the absence of specimens, records from Division 7 are unreliable. *Sintenis & Rigo* 250 includes both *S. aegyptiaca* and *S. fraudatrix*, and it is evident that these collectors followed the regrettable practice of putting together under one number specimens which were collected in various localities, at different dates, but which they considered taxonomically conspecific. Material of *S. fraudatrix* from " rocks near Akanthou ", has been re-numbered 250A. It is unlikely that true *S. aegyptiaca* was collected on rocks in this same locality, though a specimen of *S. aegyptiaca* at Kew is so labelled as to make this appear to be the case.

11. S. fraudatrix *Meikle* in Hooker's Icon. Plant., 37: t. 3650 (1969); Osorio-Tafall et Seraphim, List Vasc. Plants Cyprus, 35 (1973).

[*S. atocion* (non Juss. ex Jacq.) Sintenis in Oesterr. Bot. Zeitschr., 31: 326, 392 (1881), 32: 293 (1882) pro parte.]

TYPE: Cyprus; Yaïla above Halevga, 2,000–3,000 ft. alt., March 27, 1941, *Davis* 2846 (K !).

Slender annual; stems 5–15 cm. high, suberect or sprawling, usually much branched at base, thinly pubescent in the lower half, glandular-pubescent above; basal leaves generally present at flowering-time, oblong-spathulate, 2–4 cm. long, 0·5–0·8 cm. wide, glabrous, often minutely scabridulous, apex acute, base tapering to a flattened ciliate petiole 1–2 cm. long; cauline leaves very similar to the basal, but more shortly petiolate, the uppermost sessile or subsessile, rather narrowly oblong-ovate; inflorescence a very lax monochasium or irregular dichasium, or flowers occasionally solitary in weak plants; bracts herbaceous, glandular-pubescent, narrowly ovate, the lower foliaceous, the upper minute, sometimes less than 1 mm. long; pedicels elongate, 0·8–3 cm. long, glandular-pubescent; flowers erect; calyx narrowly clavate or subcylindrical, 1·2–1·5 cm. long, 0·2–0·3 cm. wide, thinly glandular-pubescent, green or purplish, calyx-teeth bluntly oblong, about 2·5 mm. long, 1·5–2 mm. wide, with a broad, membranous, ciliate margin; petals bright pink, 1·5–1·8 cm. long, limb about 5 mm. wide near the truncate or shallowly emarginate apex, auricles reduced to 2 small, acute teeth; claw about 8 mm. long, glabrous, tapering gradually from apex to base; coronal-scales very prominent, oblong, subobtuse, about 1·5 mm. long and 0·8 mm. wide; stamens about 9 mm. long, filaments glabrous; anthers oblong, about 1 mm. long and 0·8 mm. wide, yellow; anthophore 8–9 mm. long, slender, glabrous; ovary shortly cylindrical, about 2·5 mm. long and 1·4 mm. wide; styles 3, about 5 mm. long. Capsule oblong, 6–7 mm. long, 4–5 mm. wide, walls thin and papery, teeth about 1 mm. long, strongly recurved, forming a wide mouth; seeds reniform, about 1 mm. long and 0·8 mm. wide, dorsal surface flattened or even slightly canaliculate, sides flattened, ventral surface strongly excavate about hilum; testa dull grey-brown, central part of lateral surface flattish, reticulate, marginal and dorsal surfaces with regular rows of acute papillae, arising from expanded, radially ridged bases.

HAB.: On rocky limestone hillsides, or by roadsides in Cypress forest; 1,300–3,000 ft. alt.; fl. March–May.

DISTR.: Division 7, locally common in Halevga, Kharcha, Antiphonitis Mon., Akanthou area. Endemic.

7. Locally common; *Sintenis & Rigo* 242 ! 250A ! *Davis* 2846 ! *F. M. Probyn* 21 ! 61 ! *Merton* 2270 !

NOTES: Obviously related to *S. aegyptiaca* (L.) L.f., *S. atocioides* Boiss., *S. pseudo-atocion* Desf., *S. insularis* Barbey and *S. delicatula* Boiss., but distinguished from these either by habit and shape of petals, or by length of calyx and shape of seeds. It is at present known only from a limited area in the eastern half of the Northern Range and should be looked for elsewhere.

12. S. sedoides *Poir.*, Voy. en Barbarie, 2: 164 (1789); Boiss., Fl. Orient., 1: 598 (1867;) Holmboe, Veg. Cypr., 73 (1914); Post, Fl. Pal., ed. 2, 1: 179 (1932); Zohary, Fl. Palaest., 1: 90, t. 112 (1966); Davis, Fl. Turkey, 2: 230 (1967); Osorio-Tafall et Seraphim, List Vasc. Plants Cyprus, 35 (1973).

TYPE: Tunisia; probably from the vicinity of Cap Bon, *Poiret* (P).

A small tufted annual, 3–15 cm. high; stems usually much branched from the base, spreading, subglabrous or thinly strigillose in the lower part, viscid-glandular above; basal leaves fleshy, subglabrous, sometimes purplish, obovate-spathulate, 1–3 cm. long, 0·4–0·8 cm. wide, apex obtuse or rounded, base narrowing gradually to a flattened petiole; cauline leaves smaller, often densely strigillose or glandular; inflorescence irregularly dichotomous or racemose; bracts herbaceous, foliaceous, glandular; pedicels slender, 0·4–1·5 cm. long, glandular, at first erect, becoming patent or erecto-patent in fruit; calyx subcylindrical, about 7 mm. long and 2 mm. wide, viscid-glandular, often purplish, scarcely constricted at apex, nerves 10, scarcely visible, calyx-teeth oblong, obtuse, 1·5 mm. long, 0·7 mm. wide, slightly cucullate, with ciliate margins; petals bright pink, about 7 mm. long and 0·7 mm. wide at the emarginate apex, limb about 1·5 mm. long, auricles wanting, coronal scales conspicuous, narrowly oblong, subacute, about 0·7 mm. long; stamens 3–4 mm. long, filaments glabrous; anthers oblong, about 0·4 mm. long, yellow; anthophore glabrous, rather stout, about 1·5 mm. long; ovary oblong-cylindrical, about 2 mm. long and 0·8 mm. wide; stylophore indistinct; styles 3, 2–2·5 mm. long. Capsule narrowly oblong, 4–6 mm. long, 2–3 mm. wide, capsule-teeth about 0·5 mm. long, slightly recurved; seeds reniform about 0·6 mm. long, 0·5 mm. wide, dorsal surface strongly canaliculate, lateral surfaces flattish, ventral surface slightly excavate; testa blackish with regular, radiating rows of short, low, blunt ridges.

HAB.: On rocky ground by the sea; sea-level–50 ft. alt.; fl. April–June.

DISTR.: Divisions 1, 3, 4, 7, 8. Widely distributed in the Mediterranean area.

1. Baths of Aphrodite, Akamas, 1949, *Casey* 760 !
3. Cape Gata, 1905, *Holmboe* 678.
4. Cape Greco, 1958, *N. Macdonald* 57 !
7. Kyrenia, 1880, *Sintenis & Rigo* 243A ! Kyrenia Castle, 1941, *Davis* 3047 ! Kyrenia, 1949, *Casey* 820 ! and, 1956, *G. E. Atherton* 1335 !
8. Cape Andreas, 1880, *Sintenis & Rigo* 243 ! Cape Elea, 1905, *Holmboe* 510; near Rizokarpaso, 1912, *Haradjian* 224 ! Cape Andreas, 1962, *Meikle* 2463 !

SECT. 6. **Behenantha** *Otth* Annuals or biennials, rarely perennials; plants glabrous or sparsely pubescent below, always glabrous and often viscid above; inflorescence dichasial or monochasial; calyx glabrous, often glaucous, constricted at apex in fruit.

13. S. cretica *L.*, Sp. Plant., ed. 1, 420 (1753); Sibth. et Sm., Fl. Graec. Prodr., 1: 296 (1809), Fl. Graec., 5: 16 (1825); Unger et Kotschy, Die Insel Cypern, 349 (1865); Boiss., Fl. Orient., 1: 584 (1867); Holmboe, Veg. Cypr., 73 (1914); Post, Fl. Pal., ed. 2, 1: 173 (1932); Davis, Fl. Turkey, 2: 232 (1967); Osorio-Tafall et Seraphim, List Vasc. Plants Cyprus, 35 (1973).

TYPE: "*in* Creta".

An erect, branched or unbranched annual (15–) 30–50 cm. high, branches usually erect or sharply ascending, stems thinly pubescent towards base, glabrescent and viscid above; basal leaves obovate-spathulate, 1–4 cm. long, 0·5–1·5 cm. wide, glabrescent or thinly pubescent, apex subacute or obtuse, base tapering gradually to a short flattened ciliate petiole; cauline leaves sessile or subsessile, progressively narrower and more acute, the middle usually lanceolate or linear-lanceolate, 3–6 cm. long, 0·3–0·5 cm. wide, the uppermost linear-subulate with a long slender acumen; inflorescence a lax, erect, unequally branched dichasium; bracts herbaceous, foliaceous; pedicels slender, 2–4 cm. long (or sometimes longer); calyx at first narrowly ovoid, 1–1·3 cm. long, 0·3–0·4 cm. wide, glabrous, with 10 prominent, obscurely branched nerves; calyx-teeth ovate-acuminate, about 3 mm. long and 2 mm. wide, with wide membranous, overlapping, glabrous margins, and a sharp nervose tip; calyx becoming swollen, ovoid or almost subglobose in fruit; petals pink, 1–1·3 cm. long, 0·5–1 cm. wide at apex, bifid with bluntly oblong lobes about 1·5 mm. long; auricles not developed; coronal scales conspicuous, narrowly triangular, acuminate, about 1 mm. long, with irregularly denticulate margins or subentire; claw glabrous, tapering from apex to base; stamens 5–7 mm. long; filaments glabrous; anthers oblong, about 1 mm. long and 0·5 mm. wide; anthophore stout, short, about 1·5–2 mm. long; ovary ovoid-ellipsoid, 2·5–3 mm. long, 2 mm. wide; stylophore conspicuous, dome-shaped; styles 3, rather short and flattened, 2–3 mm. long. Capsule broadly ovoid or subglobose, 8–14 mm. long, 5–8 mm. wide, much constricted at apex, with 6 shortly reflexed teeth; seeds reniform, 1·5 mm. long, 1·2 mm. wide, dorsal surface shallowly canaliculate, lateral surfaces flattish, ventral surface deeply excavate about hilum; testa brown, papillose, the papillae with broad, often radially ridged bases.

HAB.: In cultivated fields; sea-level to 800 ft. alt.; fl. April–June.

DISTR.: Division 6, rare. S. Europe, N. Africa, Turkey, Palestine.

6. Near Panteleimon Mon., 1862, *Kotschy* 951; Nicosia, 1905, *Holmboe* 269.

14. S. gemmata *Meikle* in Hooker's Icon. Plant., 37: t. 3651 (1969); Osorio-Tafall et Seraphim, List Vasc. Plants Cyprus, 35 (1973).

TYPE: Cyprus; between Platres and Aphamis, 3,000–3,400 ft. alt., in garigue on igneous mountainside; flowers vivid carmine; April 22, 1941, *Davis* 3183 (K !).

An erect annual (6–) 10–25 (–30) cm. high; stems unbranched, or occasionally much-branched near base, often purplish, rather densely white pubescent in the lower part, glabrous and very viscid above; basal leaves forming a loose tuft, ± persistent until anthesis, oblanceolate or narrowly spathulate, 1–4 cm. long, 0·15–0·4 cm. wide, apex subacute, obtuse or mucronulate, base tapering gradually to a short ciliate petiole, lamina often purplish, ± distinctly white-pubescent; cauline leaves linear-subulate, sessile, 2·5–4 cm. long, glabrous or thinly puberulous, apex acute, bases connate to form a shallow cup; inflorescence a lax, irregular dichasium; bracts subulate, herbaceous, foliaceous; pedicels slender 2·5–6 cm. long, very viscid; flowers erect; calyx broadly clavate or narrowly ovoid, 1·2–1·4 cm. long, 0·2–0·3 cm. wide, glabrous, often purplish, with 10 prominent, obscurely branched nerves; calyx-teeth ovate, acute, about 2 mm. long and 1·5 mm. wide at base, margins membranous, minutely papillose-denticulate, glabrous; petals vivid carmine-pink, 10–12 mm. long, 2–3 mm. wide at the deeply emarginate apex; auricles wanting; coronal scales conspicuous, oblong, obtuse or subacute, about 0·7 mm. long, 0·5 mm. wide; claw glabrous, tapering gradually from apex to base; stamens 3·5–8 mm. long, filaments glabrous; anthers oblong, 0·8 mm. long, 0·5 mm. wide; anthophore rather stout, glabrous, about 4 mm. long; ovary narrowly oblong about 4 mm. long and 1·5 mm. wide; styles 3, flattened, 1·5–2 mm. long. Capsule narrowly ovoid or ellipsoid 8–10 mm. long, 3–4 mm. wide, distinctly constricted at apex, with 6 slightly recurved teeth; seeds reniform, 0·8–1 mm. long, 0·5–0·7 mm. wide; dorsal surface shallowly canaliculate, lateral surfaces flattish, ventral surface distinctly excavate about hilum; testa brown, regularly striate, with radiating rows of short, low, blunt ridges.

HAB.: On rocky mountainsides in garigue; on earthy banks, or in fields, usually on igneous formations, occasionally on chalk; 2,500–3,600 ft. alt.; fl. April–May.

DISTR.: Divisions 2 [7], locally common in Division 2, Platres–Mandria–Perapedhi. Endemic.

2. Locally common; *Kennedy* 359 ! 1184 ! 1185 ! 1679–1682 ! *Davis* 3123 ! 3183 ! 3361 ! *Meikle* 2803 ! 2864 !
[7. Kyrenia, 1931, *J. A. Tracey* 12 !]

NOTES: Apart from one very suspect record from Kyrenia, this attractive, jewel-like (hence, *gemmata*) species is confined to a limited area of Cyprus, south and south-west of the main Troödos range. It should be looked for elsewhere in the vicinity of the Troödos range, especially near the junction of igneous and chalk formations.

15. S. behen *L.*, Sp. Plant., ed. 1, 418 (1753); Boiss., Fl. Orient., 1: 583 (1867); Holmboe, Veg. Cypr., 73 (1914); Post, Fl. Pal., ed. 2, 1: 173 (1932); Lindberg f., Iter Cypr., 15 (1946); Zohary, Fl. Palaest., 1: 90, t. 113 (1966); Davis, Fl. Turkey, 2: 231 (1967); Osorio-Tafall et Seraphim, List Vasc. Plants Cyprus, 35 (1973).
TYPE: " *in* Creta ".

An erect, glabrous, glaucous annual, 10–30 (–40) cm. high; stems unbranched or branched from near the base; basal leaves mostly withered at flowering time, usually much smaller than the lower cauline leaves; cauline leaves rather crowded, obovate, 4–10 cm. long, 0·6–2·5 cm. wide, apex acute, base tapering to a flattened petiole, or (in the upper parts of the stem) leaves sessile or subsessile; inflorescence a lax, spreading dichasium; bracts lanceolate, foliaceous; pedicels short, 0·5–1 cm. long, sometimes almost wanting, often elongating in fruit; flowers slightly nodding; calyx narrowly ovoid, 1·2–1·3 cm. long, 0·6–0·8 cm. wide, glabrous, whitish, somewhat inflated, constricted at apex, 10-nerved with bronze-purplish anastomosing nerves and veins; calyx-teeth broadly triangular, about 1 mm. long and 1·5 mm. wide at base, slightly cucullate, with narrow, membranous, minutely pubescent margins, and an acute, pubescent apex; petals pale pink, 1–1·3 cm. long, 0·2–0·3 cm. wide at the bifid apex, lobes broadly oblong, rounded, about

1·5 mm. long; auricles and coronal scales obsolete; claw relatively broad, about 1 cm. long, tapering from apex to base; stamens 6–8 mm. long, filaments glabrous; anthers oblong, about 1·5 mm. long and 1 mm. wide; anthophore short, stout, about 1 mm. long; ovary oblong-ellipsoid, about 4 mm. long and 2·5 mm. wide; styles 3, about 4 mm. long. Capsule broadly ovoid or subglobose, 8–9 mm. long, 7–8 mm. wide, with a smallish mouth, and slightly recurved teeth about 2 mm. long; seeds reniform, about 1·5 mm. long and 1·2 mm. wide; dorsal surface flattish, lateral surfaces flattish, ventral surface deeply excavate about hilum; testa greyish-fuscous, with regular, radiating rows of short ridges, each ridge elevated centrally into a short conical papilla.

HAB.: In cultivated and fallow fields, occasionally on dry mountain slopes; sea-level to 3,750 ft. alt.; fl. Febr.–May.

DISTR.: Divisions 1–3, 5–8, common in Division 7. Also Italy, Sicily, Sardinia, Malta, Greece, Crete, Aegean Islands, Turkey, Syria and Palestine; not infrequent as an introduction elsewhere.

1. Ktima, 1913, *Haradjian* 687; Ayios Neophytos, 1939, *Lindberg f.* s.n.; Peyia, 1962, *Meikle* 2051 !
2. Platres, 1938, *Kennedy* 1183 ! Troödos, 1939, *Lindberg f.* s.n.
3. Limassol, 1913, *Haradjian* 574; near Phasouri, 1938, *Syngrassides* 1802 ! Perapedhi, 1941, *Davis* 3084 ! Below Kouklia (Paphos), 1960, *N. Macdonald* 26 !
5. Near Kythrea, 1880, *Sintenis & Rigo* 249 !
6. Nicosia, 1927, *Rev. A. Huddle* 21 ! Kormakiti, 1936, *Syngrassides* 1221 !
7. Common; *A. G. & M. E. Lascelles* s.n. ! *H. Duke-Woolley* 58 ! *G. E. Atherton* 1117 ! *Casey* 1701 !
8. Between Ayios Andronikos and Yialousa, 1880, *Sintenis & Rigo* 249A ! Near Rizokarpaso, 1912, *Haradjian* 175, 224; between Apostolos Andreas and Khelones, 1941, *Davis* 2302 !

16. S. laevigata *Sibth. et Sm.*, Fl. Graec. Prodr., 1: 295 (1809), Fl. Graec., 5: 13, t. 418 (1825); Unger et Kotschy, Die Insel Cypern, 350 (1865); Boiss., Fl. Orient., 1: 585 (1867); Holmboe, Veg. Cypr., 73 (1914); Lindberg f., Iter Cypr., 15 (1946); Osorio-Tafall et Seraphim, List Vasc. Plants Cyprus, 35 (1973).

TYPE: Cyprus; " in insulae Cypri montosis " *Sibthorp* (OXF ! K !).

Erect or decumbent, glaucous annual, 6–20 (–27) cm. high; stems smooth, glabrous, simple or often much-branched; leaves glabrous, fleshy, tinged purple, the basal tufted, persisting until anthesis, broadly spathulate or almost orbicular, 1·5–3·5 cm. long, 0·5–3 cm. wide, apex rounded or subacute, base tapering rather abruptly to a short flattened petiole; cauline leaves similar to basal, usually more acute, and sessile; inflorescence a lax, spreading dichasium; bracts ovate, acute, foliaceous; pedicels slender, glabrous, 0·5–1 cm. long, lengthening to 1·5 cm. or more in fruit and thickening; flowers erect; calyx broadly ovate, very slightly inflated, constricted at apex, 7–8 mm. long, 4–5 mm. wide, glabrous, purplish, with indistinct anastomosing venation; calyx-teeth ovate-triangular, about 1 mm. long and 1 mm. wide, distinctly cucullate, with narrow, membranous, pubescent margins; petals pink, 9–10 mm. long, about 1·5–2 mm. wide near the deeply bifid apex, lobes narrowly oblong, about 2 mm. long and 0·7 mm. wide, rounded at apex; auricles and coronal scales obsolete; claw about 7 mm. long, widening slightly from apex to about the middle, where it is as wide as, or wider than, the limb, then tapering to base; stamens 4–5 mm. long, filaments glabrous; anthers yellow, shortly oblong, about 0·5 mm. long and 0·4 mm. wide; anthophore almost wanting, about 0·5 mm. long; ovary oblong-elliptical 2·5 mm. long, 1·5 mm. wide; styles 3, 2·5–3 mm. long. Capsule ovoid, about 8 mm. long and 6 mm. wide, constricted at mouth with spreading teeth about 1 mm. long; seeds reniform, about 1 mm. long and 0·8 mm. wide, dorsal surface flattish, lateral surfaces concave, ventral surface deeply indented at hilum; testa dark brown, with regular rows of short, low, radiating ridges, each ridge elevated centrally into a very short conical papilla, and often with a small blackish point (or gland) at either extremity.

HAB.: In *Pinus* woodland, or in garigue on dry igneous mountainsides; 3,000–6,400 ft. alt.; fl. March–June.

DISTR.: Divisions 2, [4], locally common in Division 2 in Troödos–Platres area, and Zoopiyi–Palekhori area, also at Kambos. Endemic.

2. Common; *Sintenis & Rigo* 766 ! *Kennedy* 346–349 ! 1179–1182 ! *Davis* 1847 ! 2742 ! 3077 ! 3104 ! 3185 ! 3382 ! *Meikle* 2664 !

[4. Seashore near Larnaca, 1862, *Kotschy* 280.]

NOTES: Apart from Kotschy's Larnaca record, all the specimens of *S. laevigata* come from high altitudes on the Troödos range. The Kotschy specimen has not been seen, and the identification must be considered very dubious, though it is difficult to imagine what could have been confused with this very distinct endemic.

SECT. 7. **Lasiocalycinae** *Boiss.* Annual plants; calyx papillose or scabrid with bulbous-based hairs, distinctly constricted at apex in fruit.

17. S. papillosa *Boiss.*, Diagn., 1, 1: 39 (1843), Fl. Orient., 1: 587 (1867); Post, Fl. Pal., ed. 2, 1: 174 (1932); A. K. Jackson et Turrill in Kew Bull., 1939: 475 (1939); Zohary, Fl. Palaest., 1: 92, t. 116 (1966); Davis, Fl. Turkey, 2: 233 (1967); Osorio-Tafall et Seraphim, List Vasc. Plants Cyprus, 35 (1973).

TYPE: Turkey; " in arenosis *Cariae*. Aucher pl. exs. No. 483 " (G).

Erect or spreading annual 15–30 cm. high; stems puberulous with recurved white hairs towards base, glabrescent above, branches divaricate, with very long internodes; leaves bright green, the basal narrowly obovate, 1–5 cm. long, 0·3–0·8 cm. wide, thinly strigillose, apex acute or subacute, base tapering to a very short flattened petiole, cauline leaves narrowly oblong or linear-lanceolate, usually less than 0·5 cm. in width; inflorescence a very open unequally branched dichasium, with very long glabrous or subglabrous, rather rigid, divaricate branches; bracts narrowly linear-lanceolate or almost subulate, foliaceous; pedicels very short, rarely exceeding 2 mm. at anthesis, elongating a little in fruit; calyx narrowly ellipsoid-cylindrical, 1·2–1·4 cm. long, 0·2–0·3 cm. wide, conspicuously 10-nerved, the nerves clothed with long, multicellular hairs; calyx-teeth linear-subulate, about 3 mm. long and 0·5 mm. wide, margins narrowly membranous, glandular-ciliate; petals bright carmine-pink about 13 mm. long, 2–3 mm. wide near the deeply emarginate apex; auricles distinct, ovate-acute, about 0·3 mm. long; coronal scales oblong-ovate, subacute, about 0·5 mm. long; claw slender, about 8 mm. long, tapering gradually from apex to base; stamens 6–9 mm. long, filaments glabrous; anthers narrowly oblong, about 1 mm. long and 0·5 mm. wide; anthophore glabrous, 3–4 mm. long; ovary ovoid, about 3 mm. long and 1·5 mm. wide; styles 3, about 1 cm. long. Capsule ovoid, about 8 mm. long and 5–6 mm. wide, dehiscing irregularly, often splitting to half-way into two (or three) longitudinal divisions, but without the usual capsule-teeth; capsule-wall rugulose externally; seeds reniform, about 1 mm. long and 0·8 mm. wide, dorsal surface flattish or shallowly canaliculate, lateral surfaces flattish, ventral surface deeply indented about hilum; testa fuscous, with rather irregular rows of short, low, blunt ridges, sometimes elevated into small blunt papillae.

HAB.: On grassy ground; in vineyards; in garigue; on calcareous or (more commonly) igneous formations; 50–3,250 ft. alt.; fl. March–June.

DISTR.: Divisions 1–3, common in Division 2. Also Turkey, Syria, Lebanon.

1. Dhrousha, 1941, *Davis* 3256A ! Smyies (Akamas) 1948, *Kennedy* 1609 ! Karavopetres, near Ayios Yeoryios (Akamas), 1974, *Meikle* 4005 !

2. Common; *Kennedy* 357 ! 1683 ! *Davis* 3201 ! *Merton* 393 ! 2199 ! 2298 ! *Meikle* 2693 !

3. Kellaki [Colloci], 1894, *Post* s.n. !

SECT. 8. **Erecto-refractae** *Chowdhuri* Pubescent or glandular-pubescent annuals; stems erect or ascending, usually much-branched, leafy; inflorescence a lax dichasium, or a raceme-like monochasium; pedicels generally long; flowers erect; fruits horizontal or pendulous, rarely erect; calyx constricted at mouth in fruit; petals bipartite.

S. PENDULA *L.*, Sp. Plant., ed. 1, 418 (1753); Sibth. et Sm., Fl. Graec. Prodr., 291 (1809); Boiss., Fl. Orient., 1: 590 (1867); Holmboe, Veg. Cypr., 72 (1914); Osorio-Tafall et Seraphim, List Vasc. Plants Cyprus, 35 (1973).

A pubescent annual, with spreading or ascending stems 15–40 cm. long, and numerous oblong or obovate, acute or subacute leaves 2–7 cm. long and 1–2·5 cm. wide; inflorescence a lax monochasium; flowers remote, at first erect, becoming pendulous after fertilization; calyx 1·4–1·8 cm. long, 0·7–0·8 cm. wide, inflated, pallid, pleated, with 10 conspicuous greenish or purplish nerves; petals pink, 1·7–2 cm. long, with a broad deeply emarginate limb. Capsule 1–1·2 cm. long, 0·6–0·8 cm. wide; seeds subglobose, about 1 mm. long and almost as wide, with a broad flattish or convex dorsal surface and flattish or convex lateral surfaces; testa blackish with radiating rows of conspicuous papillae.

Recorded for Cyprus by Sibthorp, but not seen since his visit to the island, and, if not an error, then no more than a fleeting, casual occurrence, and certainly not a permanent feature of the flora. *S. pendula* is a native of S. Italy, Sicily and Corsica, but makes sporadic appearances as a casual in many parts of Europe.

SECT. 9. **Dichotomae** (*Rohrb.*) *Chowdhuri* Pubescent annuals (or biennials); inflorescence a dichasium or racemose monochasium; pedicels short, spreading or horizontal at anthesis, becoming stouter and erect in fruit; calyx not inflated, constricted at mouth in fruit. Seeds reniform or subglobose, not winged.

18. S. dichotoma *Ehrh.*, Beitr. Naturk., 7: 144 (1792); Boiss., Fl. Orient., 1: 588 (1867); Post, Fl. Pal., ed. 2, 1: 175 (1932); Davis, Fl. Turkey, 2: 234 (1967); Osorio-Tafall et Seraphim, List Vasc. Plants Cyprus, 36 (1973).

Erect or spreading annual (or ? biennial); stems often purplish, unbranched or with several branches from the base, thinly or rather densely crisped-pubescent; basal leaves oblanceolate, spathulate or broadly obovate, often forming a distinct crowded tuft, 5–10 cm. long, 1–3 cm. wide, apex acute or subacute, base tapering to a rather long flattened petiole, lamina and petiole ± densely pubescent, occasionally glabrous or almost so; cauline leaves generally narrower, oblanceolate, elliptical or narrowly ovate or obovate, often distinctly longitudinally 3-nerved, the uppermost sometimes linear-lanceolate; inflorescence once or repeatedly dichotomous, with elongate, racemose monochasial branches; lower bracts narrowly lanceolate, upper membranous or scarious, glabrous or ciliate; flowers subsessile or very shortly pedicellate, spreading or nodding, becoming erect in fruit; calyx ± cylindrical, 1·2–1·5 cm. long, 0·2–0·3 cm. wide, pallid, with 10 strongly ribbed, greenish or purplish nerves, subglabrous or ± densely pilose with long, spreading, multicellular hairs; calyx-teeth narrowly triangular, about 2·5 mm. long and 1 mm. wide at base, margins narrowly membranous, shortly papillose-ciliate; petals whitish, 1·6–1·8 cm. long, 0·4–1 cm. wide near apex of deeply bifid limb, lobes 1·5–5 mm. wide, narrowly oblong to suborbicular; auricles obsolete; coronal scales shortly oblong, about 0·5 mm. long, or sometimes reduced to two small callosities; claw 0·8–1·2 cm. long, glabrous, tapering from apex to base; stamens very long, often to 1·5 cm. or even longer, filaments slender, glabrous; anthers long exserted, narrowly oblong, about 2–2·5 mm. long and 1 mm. wide; anthophore very short and stout, 1–1·5 mm. long; ovary narrowly ovoid, 3–4 mm. long, about 1·5 mm. wide; stylophore distinct, subconical; styles 3, 10–13 mm. long. Capsule ovoid, 8–10 mm. long, 6 mm. wide, shorter or slightly longer than the persistent calyx; capsule-walls slightly rugulose; seeds reniform 1–1·5 mm. long, 0·8–1 mm. wide, dorsal surface slightly convex, lateral surfaces flattish or concave, ventral surface abruptly excavate about hilum; testa greyish-fuscous with radiating rows of conical papillae arising from broad rounded or elongate, radially ridged bases.

ssp. **racemosa** (*Otth*) *Hayek*, Prodr. Fl. Pen. Balc., 1: 261 (1924).

S. racemosa Otth in DC., Prodr., 1: 384 (1824); Boiss., Fl. Orient., 1: 589 (1867).

S. sibthorpiana Reichb., Fl. Germ. Excurs., 815 (1832); Unger et Kotschy, Die Insel Cypern, 349 (1865); Post, Fl. Pal., ed. 2, 1: 175 (1932).

S. racemosa Otth var. *sibthorpiana* (Reichb.) Boiss., Fl. Orient., 1: 589 (1867); Holmboe, Veg. Cypr., 72 (1914).

TYPE: Turkey; "in agro Cariensi", 1787, *Sibthorp* (OXF !)—illustrated in Fl. Graeca, t. 414, as *S. divaricata* !

Basal leaves often forming a distinct tuft at anthesis; inflorescence often several times dichotomously divided; flowers usually remote; petals with coronal scales (var. *racemosa*) or coronal scales reduced to two small callosities (var. *sibthorpiana* (Reichb.) Boiss.). Capsule-walls thick and often dehiscing irregularly; seeds often strongly papillose.

HAB.: In cultivated ground, and on rocky slopes; sea-level–1,000 ft. alt.; fl. March–May.

DISTR.: Division 3, rare. Also eastern and south-eastern Europe, Turkey, Syria, Palestine, eastwards to Iraq. Commonly introduced elsewhere.

3. Near Limassol, 1859, *Kotschy* 466; Kholetria near Kelokedhara, at Kryanera, 1941, *Davis* 3427 !

NOTES: The Davis specimen cited above is typical ssp. *racemosa* (Otth) Hayek var. *racemosa*, with well-formed coronal scales; these scales show great variation in degree of development, and it is doubtful if var. *sibthorpiana* (Reichb.) Boiss. can be satisfactorily distinguished, even at varietal level.

SECT. 10. **Scorpioideae** (*Rohrb.*) *Chowdhuri* Annuals; inflorescence a monochasial cyme, rarely a dichasium below; flowers ± erect; calyx not constricted at apex in fruit; capsule oblong or ovoid-oblong; seeds reniform or subglobose, not winged.

19. S. discolor *Sibth et Sm.*, Fl. Graec. Prodr., 1: 292 (1809), Fl. Graec., 5: 8, t. 410 (1825); Boiss., Fl. Orient., 1: 592 (1867); Holmboe, Veg. Cypr., 72 (1914); Davis, Fl. Turkey, 2: 237 (1967); Osorio-Tafall et Seraphim, List Vasc. Plants Cyprus, 36 (1973).

S. nicaeënsis All. var. *latifolia* Boiss., Fl. Orient., 1: 592 (1867).

[*S. nicaeënsis* (non All.) Sintenis in Oesterr. Bot. Zeitschr., 32: 263, 264, 398 (1882); Holmboe, Veg. Cypr., 73 (1914); ? Post, Fl. Pal., ed. 2, 1: 176 (1932).]

TYPE: "In insulâ Cypro" *Sibthorp* (OXF !).

Erect or decumbent, glandular-pubescent annual, 3–15 cm. high; basal leaves forming a loose rosette, obovate or oblanceolate, 1–3 cm. long, 0·4–0·8 cm. wide, apex obtuse or rounded, base tapering gradually to a broad flattened petiole; cauline leaves few, small, ovate-lanceolate or narrowly oblong, sessile or subsessile; inflorescence a raceme-like, 2–6-flowered monochasium; flowers ± erect; bracts narrowly lanceolate, foliaceous; pedicels 2–12 mm. long, lengthening a little in fruit; calyx narrowly clavate, 1–1·2 cm. long, rather densely pubescent with 10 distinct purplish nerves; calyx-teeth oblong, about 2·5 mm. long, 1 mm. wide, apex obtuse, margins narrowly membranous, ciliate; petals pale pink stained greenish or brownish externally,

1·2–1·3 cm. long, 0·2–0·3 cm. wide near apex, limb very deeply bifid, with narrowly oblong lobes 4–5 mm. long, about 1 mm. wide; auricles very small, truncate; coronal scales conspicuous, each pair connate along the interior margins, forming a single, 4-lobed, oblong scale, about 1 mm. long and 1·3 mm. wide; claw oblong, about 5 mm. long, narrowing rather abruptly to base; stamens about 6 mm. long; anthers oblong, about 1 mm. long and 0·7 mm. wide; anthophore 3–3·5 mm. long, pubescent; ovary narrowly oblong, 2–2·5 mm. long; stylophore distinct, depressed-conical; styles 3, about 4·5 mm. long. Capsule narrowly ovoid-oblong, about 7 mm. long and 4 mm. wide, teeth triangular, distinctly recurved; seeds small, reniform, about 0·8 mm. long, 0·6 mm. wide, dorsal surface shallowly canaliculate, lateral surfaces flattened or slightly convex, ventral surface abruptly excavate about hilum; testa fuscous, with numerous very low, radiating ridges.

HAB.: On sandhills and sandy shores, or on sandy ground near the sea; at sea-level and a little above it; fl. March–May.

DISTR.: Divisions 1, 2, 6, 7. Also Crete, Melos, S. Turkey and Rhodes.

1. Paphos, 1930, *Druce* 61 ! Kapsala near Ayios Yeoryios (Akamas), 1962, *Meikle* 2074 !
2. Yialia, 1955, *Merton* 2100 !
6. Syrianokhori, 1935, *Syngrassides* 769 ! Ayia Irini (Morphou), 1941, *Davis* 2605 ! Near Syrianokhori, 1956, *Merton* 2624 !
7. Dhavlos, 1880, *Sintenis & Rigo* 247 ! Ayios Epiktitos, 1880, *Sintenis & Rigo* 247A ! Between Ayios Epiktitos and Ayios Amvrosios, 1941, *Davis* 3622 ! Six miles E. of Kyrenia, 1956, *Casey* 1700 !

NOTES: Much confused with *S. nicaeënsis* All. which is, however, a more robust plant, with narrower leaves, and with the lower flowers of each inflorescence in clusters, and not singly as in *S. discolor*. It is doubtful if true *S. nicaeënsis* occurs further east than Greece; the d'Urville record from Melos, attributed to *S. discolor* by Rohrbach (Mon. Silene, 110), has not been confirmed. *S. discolor* is much more closely related to *S. colorata* Poir. from which, however, it differs in habit, capsule and seed.

20. S. nocturna *L.*, Sp. Plant., ed. 1, 416 (1753); Unger et Kotschy, Die Insel Cypern, 349 (1865); Boiss., Fl. Orient., 1: 595 (1867); Holmboe, Veg. Cypr., 72 (1914); Post, Fl. Pal., ed. 2, 1: 177 (1932); Zohary, Fl. Palaest., 1: 96, t. 125 (1966); Davis, Fl. Turkey, 2: 237 (1967); Osorio-Tafall et Seraphim, List Vasc. Plants Cyprus, 36 (1973).

An erect or spreading, pubescent annual, (7–) 10–30 (–50) cm. high; stems somewhat viscid, usually greenish, much-branched in well-developed specimens; basal leaves obovate-spathulate, 1–4 cm. long, 0·3–1·5 cm. wide, apex rounded, base tapering to a flattened petiole; cauline leaves narrower, often oblanceolate, or almost linear-oblong, sessile or subsessile; inflorescence a raceme-like monochasium, occasionally bifurcate below; flowers erect, crowded, and mostly unilateral above, widely spaced towards base of inflorescence; bracts linear, foliaceous, the lower large and conspicuous; pedicels up to 2 cm. long, or the upper flowers sessile or subsessile; calyx cylindrical or narrowly ovoid, 0·6–1·5 cm. long, 0·3–0·4 cm. wide, greenish, 10-nerved, with the nerves distinctly anastomosing above; calyx-teeth narrowly ovate-triangular, 2–2·5 mm. long, about 1 mm. wide at base, margins narrowly membranous, ciliate; petals whitish or mauve-pink within; bluish or bluish-green externally, often much reduced, from 2–2·5 mm. long in short-petalled specimens, to 1·3–1·5 cm. long in long-petalled specimens, 0·5–5 mm. wide at the deeply bifid or emarginate apex, lobes minute, or bluntly oblong, to 3 mm. long and 1 mm. wide; auricles obsolete; coronal scales absent (in short-petalled flowers) or well-developed and connate into a toothed ring, about 8 mm. long, at base of petal-limb; petal-claws often connate for all or part of their length, varying from 1–8 mm. long, and to about 1 mm. wide; stamens 3–8 mm. long, filaments glabrous; anthers 0·8 mm. long, 0·5 mm. wide; anthophore 1–2 mm. long, puberulous; ovary narrowly ovoid-ellipsoid, 2–3 mm. long, 0·8–1 mm. wide; stylophore forming a distinct cap; styles 3, 1·5–2·5 mm. long. Capsule 6–10 mm. long, 3–5 mm. wide, narrowly ovoid-cylindrical, teeth shortly triangular, a little recurved; seeds very small, reniform, about 0·8 mm. long and 0·5 mm. wide, dorsal surface slightly canaliculate, lateral surfaces concave or excavate, ventral surface abruptly excavate about hilum; testa fuscous, with low radiating ridges, which are elongate on the lateral surfaces, but short and sub-papillose on the dorsal surface.

var. **nocturna**
TYPE: " *in* Italia, Pensylvania ".

Petals exceeding calyx, with well-developed coronal scales; inflorescence rather dense; capsule rather large, about 10 mm. long.

HAB.: By roadsides; in fallow and cultivated fields or on waste ground; occasionally on mountain slopes; to 800 ft. alt. or higher; fl. March–June.

DISTR.: Divisions 1, 4, 5, 7. Widespread in the Mediterranean area; often introduced elsewhere.

1. Ayios Neophytos [? Neophilos], 1928 or 1930, *Druce* 57 !
4. Near Larnaca, 1862, *Kotschy* 131.
5. Perakhorio, 1955, *Merton* 2045 !
7. Pentadaktylos, 1880, *Sintenis & Rigo* 245b ! Near Bellapais, 1952, *Casey* 1239 !

var. **brachypetala** (*Rob. et Cast.*) *Benth.*, Cat. Pyr., 122 (1826); Boiss., Fl. Orient., 1: 596 (1867); Lindberg f., Iter Cypr., 15 (1946).
S. brachypetala Rob. et Cast. in Lam. et DC., Fl. Franç., ed. 3, 6: 607 (1815); Sintenis in Oesterr. Bot. Zeitschr., 32: 54 (1882).
TYPE: France; " auprès de Marseille, notamment au château Borelli et le long du chemin qui conduit de la porte Saint-Victor à la batterie d'Eaudroume " *Robillard et Castagne* (G).

Petals shorter than calyx, without coronal scales; capsule small, usually less than 1 cm. long; plant less robust than var. *nocturna*, with lax, few-flowered monochasia.

HAB.: By roadsides, in fallow or cultivated fields or on waste ground, occasionally on rocky hillsides amongst garigue; sea-level to 4,000 ft. alt.; fl. March–June.

DISTR.: Divisions 1, 2, 6, 7, common in Division 7. Distribution similar to that of var. *nocturna*, but perhaps commoner in the eastern Mediterranean than the latter.

1. Near Smyies (Akamas), 1941, *Davis* 3306 !
2. Mesopotamos Mon., 1939, *Lindberg f.* s.n. !
6. Nicosia, 1950, *Chapman* 682 !
7. Common; *Sintenis & Rigo* 245 ! *Casey* 444A ! 1261 ! *G. E. Atherton* 1399 !

SECT. 11. **Silene** Annuals; inflorescence a monochasial cyme, sometimes forked below; flowers ± erect; calyx constricted at apex in fruit; anthophore pubescent. Capsule ovoid or subglobose; seeds reniform or subglobose, not winged.

21. S. gallica *L.*, Sp. Plant., ed. 1, 417 (1753); Boiss., Fl. Orient., 1: 590 (1867); Holmboe, Veg. Cypr., 72 (1914); Post, Fl. Pal., ed. 2, 1: 175 (1932); Lindberg f., Iter Cypr., 15 (1946); Zohary, Fl. Palaest., 1: 96, t. 126 (1966); Davis, Fl. Turkey, 2: 238 (1967); Osorio-Tafall et Seraphim, List Vasc. Plants Cyprus, 36 (1973).
TYPE: France; " *in* Gallia ".

Erect or sprawling annual, 15–40 cm. high; stems unbranched or with several branches from base, pubescent, or thinly villose, viscid especially above; basal leaves obovate, 2–10 cm. long, 0·8–2 cm. wide, obtuse or subacute, thinly pilose or glabrescent, base narrowed to an indistinct, flattened petiole; cauline leaves similar; inflorescence an unbranched, or basally forked raceme-like monochasium; flowers erect; bracts linear-lanceolate, foliaceous, viscid-pilose; pedicels at first very short, barely 2 mm.

long, the lowermost lengthening considerably (sometimes to 2 cm.) in fruit; calyx subcylindrical, 6–8 mm. long, 2–3 mm. wide, glandular-pilose or pubescent, with 10 conspicuous, usually purplish, nerves; calyx-teeth linear-subulate, 2·5 mm. long, 0·5 mm. wide, with very narrowly membranous, long-ciliate margins; petals pink, or pink with the margin or the base of the limb white, or whitish with a crimson spot near the base of the limb (var. *quinquevulnera* (L.) Mert. et Koch), 8–13 mm. long, 2–3·5 mm. wide near the deeply bifid apex, lobes rounded, about 1·5 mm. long and 1·5 mm. wide; auricles obsolete; coronal scales narrowly oblong, acute, 1–1·3 mm. long, about 0·5 mm. wide; claws 5–6 mm. long, gradually tapering from apex to base, sometimes partially connate, especially towards apex; stamens 4–5 mm. long, filaments pubescent towards base; anthers purplish, oblong, about 1 mm. long and 0·5 mm. wide; anthophore very short, 0·5–1 mm. long, pubescent; ovary narrowly ovoid or lageniform, 2–2·5 mm. long, 0·8–1 mm. wide; stylophore distinct, dome-shaped; styles 3, 1·5–3 mm. long. Capsule ovoid, 6–8 mm. long, 4–5 mm. wide; capsule-teeth narrowly triangular, distinctly recurved; seeds reniform, 1 mm. long, 0·8 mm. wide, dorsal surface broad, flattish, lateral surfaces distinctly excavate, ventral surface abruptly excavate about hilum; testa fuscous, with a regular series of bluntish ridges on the lateral surfaces, and low, broad-based, furrowed papillae on the dorsal surface.

HAB.: In cultivated fields or on waste ground, often on moist ground by streams or springs, occasionally on sandy seashores; sea-level to 3,750 ft. alt.; fl. March–May.

DISTR.: Divisions 2, 4, 6, common in Division 2. Widespread in central and southern Europe, and the Mediterranean region, eastwards to Iran; occurring as a weed almost throughout the world.

2. Common; *Kennedy* 1503 ! 1504 ! *Davis* 3189 ! *Merton* 2099 ! 2297 ! *Meikle* 2239 !

4. Paralimni, 1928 or 1930, *Druce*.

6. Syrianokhori, 1935, *Syngrassides* 790 !

NOTES: The species has been divided into numerous varieties and microspecies by Jordan and other authors, but the value of these segregates is dubious. Var. *quinquevulnera* (L.) Mert. et Koch, recorded from Syrianokhori by Syngrassides, is perhaps the most noteworthy of the variants, and was at one time cultivated in gardens in western Europe. It occurs, not infrequently, as an odd plant or two in populations of typical *S. gallica*, and, to judge from collectors' notes, is probably fairly widely distributed in Cyprus.

22. S. tridentata *Desf.*, Fl. Atlant., 1: 349 (1798); Zohary, Fl. Palaest., 1: 97, t. 128 (1966); Osorio-Tafall et Seraphim, List Vasc. Plants Cyprus, 36 (1973).

S. cerastoides (non L.) Lindberg f., Iter Cypr., 15 (1946).

TYPE: " in arvis Algeriae " *Desfontaines* (P).

Erect annual, 8–15 cm. high; stems unbranched, or with numerous spreading or erect branches, ± densely hispid-pubescent, often purplish; basal leaves oblanceolate or narrowly spathulate, 2–4 cm. long, 0·5–0·8 cm. wide, obtuse or subacute, pubescent, base narrowed to a short flattened petiole; cauline leaves usually narrower, often linear, and more acute, sessile and amplexicaul; inflorescence a monochasium, sometimes forked at base; flowers erect, rather distant; bracts linear-lanceolate, foliaceous, often hispid-pubescent; pedicels at first very short, but lengthening a little (to about 2–3 mm.) in fruit; calyx ovoid, 10–13 mm. long, 3–5 mm. wide, hispid-pubescent or sometimes subglabrous, with 10 very prominent raised nerves and obscurely anastomosing veins; calyx-teeth subulate, about 5 mm. long and 1 mm. wide at base; petals often connate for part of their length, small, inconspicuous, pinkish, about 10 mm. long and 1–1·5 mm. wide near the deeply bifid apex, lobes narrowly oblong, about 1–1·5 mm. long and 0·3 mm. wide; auricles obsolete; coronal scales narrowly oblong, about 1 mm. long and 0·5 mm. wide; claw 6–7 mm. long, gradually tapering from apex to base; stamens 4–6 mm. long, filaments glabrous; anthers purplish, oblong, about 1 mm. long and 0·5 mm. wide; anthophore very short or almost obsolete; ovary narrowly ovoid or pyriform, about 3·5 mm. long and 2 mm. wide; stylophore distinct, dome-shaped; styles 3, 2–2·5 mm. long. Capsule ovoid, about 6–7 mm. long and 5 mm. wide; capsule teeth narrowly triangular, about 0·8 mm. long, distinctly recurved; seeds reniform, about 0·8 mm. long and 0·6 mm. wide, dorsal surface broad, shallowly furrowed, lateral surfaces distinctly excavate, ventral surface abruptly excavate about hilum; testa greyish, with short, regular, obtuse, radiating ridges on the lateral surfaces, and a bluntly ridged dorsal surface.

HAB.: In dry fields or on hillsides, often on sandy ground; sea-level to ? 1,000 ft. alt.; fl. March–May.

DISTR.: Divisions 5–7, rare. Spain; North Africa, from Morocco to Egypt; Palestine.

5. Athalassa, 1939, *Lindberg f.* s.n.; sands E. of Trikomo, 1970, *A. Hansen* 134 !
6. Ayia Irini (Morphou), 1941, *Davis* 2555 !
7. Pentadaktylos, 1880, *Sintenis & Rigo* 244A.

NOTES: Easily confused with *S. gallica*, and perhaps more widespread in Cyprus than the records would suggest. True *Silene cerastoides* L. is spreading or decumbent in habit, with shorter calyx-teeth; more conspicuous, anastomosing calyx-veins; longer, more slender pedicels, and distinctly smaller capsules.

SECT. 12. **Dipterospermae** (*Rohrb.*) *Chowdhuri* Annuals; inflorescence usually monochasial; calyx not much constricted at mouth in fruit. Seeds reniform, dorsal surface deeply channelled with 2 marginal wings.

23. S. colorata *Poir.*, Voy. en Barbarie, 2: 163 (1789); Holmboe, Veg. Cypr., 72 (1914); Post, Fl. Pal., ed. 2, 1: 178 (1932); Zohary, Fl. Palaest., 1: 97, t. 129 (1966); Davis, Fl. Turkey, 2: 239 (1967); Osorio-Tafall et Seraphim, List Vasc. Plants Cyprus, 36 (1973).

S. bipartita Desf., Fl. Atlant., 1: 352 (1798); Boiss., Fl. Orient., 1: 597 (1867); Sintenis in Oesterr. Bot. Zeitschr., 32: 398 (1882).

S. leucophaea Sibth. et Sm., Fl. Graec. Prodr., 1: 297 (1809), Fl. Graec., 5: 17, t. 424 (1825); Unger et Kotschy, Die Insel Cypern, 350 (1865).

[*S. vespertina* (non Retz.) Kotschy in Unger et Kotschy, Die Insel Cypern, 349 (1865); H. Stuart Thompson in Journ. Bot., 44: 277 (1906).]

[*S. palestina* Boiss. var. *damascena* (non Boiss. et Gaill.) Rohrb., Mon. Silene, 103 (1868); H. Stuart Thompson in Journ. Bot., 44: 277 (1906).]

[*S. damascena* (non Boiss. et Gaill.) Holmboe, Veg. Cypr., 72 (1914).]

Erect or spreading annual 10–30 (–50) cm. high; stems branched or unbranched, usually purplish, pubescent or sometimes almost glabrous; basal leaves narrowly obovate or spathulate, or occasionally linear, 2–5 cm. long, 0·5–2 cm. wide, thinly pubescent, apex obtuse or subacute, occasionally acute, base narrowed to a flattened petiole; cauline leaves similar, but narrower, often oblanceolate or linear, and obscurely petiolate or subsessile; inflorescence a raceme-like monochasium, occasionally bifurcate especially towards base; flowers rather distant, erect; bracts small, lanceolate or linear, foliaceous; pedicels slender, 0·5–1·5 cm. long, spreading or suberect in fruit; calyx narrowly clavate, 1–1·5 cm. long, 0·2–0·4 cm. wide, subglabrous or pilose, with 10 prominent greenish or reddish nerves, becoming rather broadly clavate in fruit; calyx-teeth ovate, subacute, about 2 mm. long and 1·2 mm. wide with membranous, ciliate margins; petals pink, 1·3–1·5 cm. long, 0·5–0·8 cm. wide near the deeply bifid apex, lobes oblong, blunt, 3–4 mm. long, 2–3 mm. wide; auricles indistinct; coronal scales conspicuous, oblong, obtuse or subacute, about 1·5 mm. long and nearly 1 mm. wide; claw about 1 cm. long, tapering gradually to base; stamens 5–7 mm. long, filaments glabrous; anthers oblong, fuscous, about 1·5–2 mm. long and 1 mm. wide; anthophore minutely pubescent, about 5–7 mm. long; ovary narrow oblong, about 4 mm. long; stylophore indistinct; styles 3, about 4–5 mm. long. Capsule broadly ovoid, almost subglobose, 5–7 mm. long and 5–6 mm. wide, capsule teeth triangular about 2 mm. long and 1·5 mm. wide at base, distinctly recurved; seeds broadly reniform, 1–1·5 mm. long, 1–1·2 mm. wide, dorsal surface deeply channelled with conspicuous straight or undulate marginal wings; lateral surfaces flattened or a little excavate; ventral surface excavate about hilum; testa dark brown, with numerous fine radiating ridges.

var. **colorata**

S. bipartita Desf. var. *stenophylla* Boiss., Fl. Orient., 1: 597 (1867).

S. colorata Poir. var. *stenophylla* (Boiss.) Dinsm. in Post, Fl. Pal., ed. 2, 1: 178 (1932).

[*S. oliveriana* (non Otth) Kotschy in Unger et Kotschy, Die Insel Cypern, 349 (1865).]

TYPE: North Africa; probably near Tunis, 1785–86, *Poiret* (P).

Leaves spathulate or oblanceolate, at least towards base; calyx usually reddish-nerved; petals usually bright pink.

HAB.: Dry sandy or stony ground; arable fields; waste places; sea-level to 1,800 ft. alt.; fl. Febr.–May.

DISTR.: Divisions 3–5, 7, 8, common in Divisions 4, 7. Widespread in Mediterranean area.

3. Near the base of Stavrovouni, 1787, *Sibthorp*.
4. Common; *Kotschy* 64 ! 91 ! 95 ! 132 ! *Sintenis & Rigo* 248 ! *Holmboe* 126a; *Syngrassides* 1042 ! *Mavromoustakis* s.n. !
5. Salamis, 1949, *Mercy Casey* in *Casey* 550 ! and, 1957, *Merton* 3206 !
7. Common; *J. A. Tracey* 14 ! *Collingwood Ingram* 41 ! *Casey* 444 ! *G. E. Atherton* 921 ! 960 !
8. Ayios Philon near Rizokarpaso, 1941, *Davis* 2257 ! Apostolos Andreas, 1948, *Mavromoustakis* s.n. !

var. **decumbens** (*Biv.*) *Rohrb.*, Mon. Silene, 115 (1868); Post, Fl. Pal., ed. 2, 1: 178 (1932).

S. decumbens Biv., Sic. Plant. Cent. 1, 75, t. 6 (1806).

S. colorata Poir. ssp. *decumbens* (Biv.) Holmboe, Veg. Cypr., 73 (1914); Osorio-Tafall et Seraphim, List Vasc. Plants Cyprus, 36 (1974).

S. bipartita Desf. var. *eriocaulon* Boiss., Fl. Orient., 1: 597 (1867).

TYPE: Sicily; " in pascuis arenosis maritimis " *Bivona* (? Fl).

Stems spreading or decumbent, often pubescent; leaves frequently rather fleshy, ± pubescent; flowers relatively few; calyx sometimes villose; petals pink.

HAB.: Sandy seashores and sandy fields; about sea-level; fl. Febr.–May.

DISTR.: Divisions 3–5, 8. Widespread in the Mediterranean area.

3. Amathus, 1947, *Mavromoustakis* s.n. ! Akrotiri, 1950, *F. J. F. Barrington* 26 ! Episkopi bay, 1950, *F. J. F. Barrington* 70 !
4. Ayia Napa, 1905, *Holmboe* 5; Famagusta, 1933, *Syngrassides* 47 ! Perivolia village (Larnaca Distr.), 1938, *Syngrassides* 1753 !
5. Salamis, 1905, *Holmboe* 436.
8. Cape Andreas, 1880, *Sintenis & Rigo* 248b, 252c ! Khelones near Rizokarpaso, 1941, *Davis* 2286 !

NOTES: A maritime form of *S. colorata*, usually easy to recognize, but intergrading with the typical var. *colorata* in some areas. Plants with villose calyces are sometimes called var. *distachya* (Brot.) Rohrb., but it is doubtful if this epithet can be accurately applied to any variety occurring in Cyprus.

var. **oliveriana** (*Otth*) *Muschler*, Man. Fl. Egypt, 1: 338 (1912).

S. oliveriana Otth in DC., Prodr., 1: 373 (1824); Boiss., Fl. Orient., 1: 597 (1867); Post, Fl. Pal., ed. 2, 1: 179 (1932).

S. colorata Poir. ssp. *oliveriana* (Otth) Rohrb., Mon. Silene, 116 (1868); Holmboe, Veg. Cypr., 73 (1914); Osorio-Tafall et Seraphim, List Vasc. Plants Cyprus, 36 (1973).

TYPE: " circa Alep [Aleppo] et Mossul " *Olivier* (P).

Leaves linear, becoming almost subulate in upper parts of stem; calyx often green-nerved, but sometimes with reddish nerves; petals often whitish or very pale pink; stems and calyx often glabrescent.

HAB.: In sandy or stony fields; sea-level to 500 ft. alt.; fl. Febr.–May.

DISTR.: Division 5 or 6. Also Turkey, Syria, Palestine, Iraq, Iran, Arabia and Egypt.

5 or 6. Peristerona, 1935, *Syngrassides* 777 ! Mesaoria, without detailed locality, 1952, *F. M. Probyn* 5 !

24. S. apetala *Willd.*, Sp. Plant., 2, 1: 703 (1799); Boiss., Fl. Orient., 1: 596 (1867); Holmboe, Veg. Cypr., 73 (1914); Post, Fl. Pal., ed. 2, 1: 178 (1932); Zohary, Fl. Palaest., 1: 98, t. 130 (1966); Davis, Fl. Turkey, 2: 239 (1967).
S. rigoi Sintenis in Oesterr. Bot. Zeitschr., 31: 393 (1881), 32: 123 (1882) nomen.

Erect or spreading annual (3–) 7–30 cm. high; stems unbranched or with numerous ascending branches, softly adpressed-pubescent; basal leaves narrowly oblanceolate or linear, acute, 2–5 cm. long, 0·3–0·5 cm. wide, dull green, adpressed-pubescent, base tapering to a short flattened petiole; cauline leaves often rather larger than basal, frequently very narrowly oblanceolate or linear, sessile; inflorescence a raceme-like monochasium, commonly forked at the base or otherwise irregular; flowers erect; pedicels slender, to about 1·7 cm. long; bracts linear-lanceolate, foliaceous; calyx oblong or turbinate, 6–8 mm. long, 4–5 mm. wide, shortly pubescent, with 10 greenish or reddish, conspicuously or obscurely hairy nerves; calyx-teeth acuminate, 3 mm. long, 1 mm. wide at base, distinctly keeled; petals free or connate, white or pinkish, or occasionally absent, to about 8 mm. long, but often shorter than calyx, deeply bifid, with blunt oblong lobes about 1·5 mm. long and 0·5 mm. wide; auricles obsolete; coronal scales prominent, blunt, about 0·5 mm. long and the same in width; stamens 3–5 mm. long; filaments glabrous; anthers oblong, about 0·5 mm. long and 0·3 mm. wide; anthophore pubescent, about 3 mm. long; ovary narrowly oblong, about 3 mm. long, 1·5 mm. wide; stylophore distinct, dome-shaped; styles 3, about 2 mm. long. Capsule broadly ovoid, about 5 mm. long and 4 mm. wide; capsule-teeth shortly triangular, about 0·5 mm. long, recurved; seeds reniform, about 1 mm. long and almost as wide; dorsal surface deeply channelled, with conspicuous straight or undulate marginal wings; lateral surfaces flattish or slightly excavate; ventral surface abruptly excavate about hilum; testa dark brown, with numerous, fine, radiating ridges.

var. **apetala**
TYPE: Not localized (B).

Petals shorter than calyx or wanting; calyx-nerves greenish, not conspicuously pilose.

HAB.: Sandy or rocky ground, or in cultivated land, often near the sea; sea-level to ? 500 ft. alt.; fl. Febr.–May.

DISTR.: Divisions 1, 4–8. Widespread in the Mediterranean area.

1. Ayios Yeoryios Island (Akamas), 1962, *Meikle* 157 !
4. Ayia Napa, 1905, *Holmboe* 66; Ayios Antonios, Sotira, 1952, *Merton* 1343 !
5. Between Mia Milea and Koutsovendis, 1974, *Meikle & Osorio-Tafall* in *Meikle* 4076 !
6. Nicosia, 1950, *Chapman* 498 !
7. Pentadaktylos, 1880, *Sintenis & Rigo* 244 !
8. Khelones near Rizokarpaso, 1941, *Davis* 2304 !

var. **grandiflora** *Boiss.*, Fl. Orient., 1: 597 (1867); Holmboe, Veg. Cypr., 73 (1914); Post, Fl. Pal., ed. 2, 1: 178 (1932).
TYPE: Greece; Attica, *Spruner* (K !).

Petals longer than calyx, whitish or pinkish.

HAB.: Often growing with var. *apetala*, and in the same habitats.

DISTR.: Divisions 4, 6, 8, rare. General distribution that of var. *apetala*.
4. At the Phaneromene near Larnaca, 1905, *Holmboe* 146.
6. Syrianokhori, 1936, *Syngrassides* 966 !
8. Rizokarpaso, 1941, *Davis* 2298 !

NOTES: Intergrades with typical *apetala* and of minimal taxonomic significance.

var. **alexandrina** *Aschers.* in Aschers. et Schweinf., Illustr. Fl. d'Égypte, 46 (1887).
TYPE: Egypt; Alexandria, *Ehrenberg* (B).

Calyx-nerves conspicuously pilose, often reddish; petals pink, longer than calyx; plants usually small and slender.

HAB.: On rocky slopes; in fallow or waste land; 2,000–3,000 ft. alt.; fl. Febr.–May.

DISTR.: Divisions 5, 7, common in Division 7. Also Egypt and North Africa.
5. Mile 3, Nicosia–Famagusta Road, 1950, *Chapman* 395 !
7. Common about Pentadaktylos, Buffavento and Halevga; *Sintenis & Rigo* 244b. ! *Davis* 2812 ! *H. Duke-Woolley* 26 ! *F. M. Probyn* 63 ! *Casey* 1284 !

NOTES: Apparently most common, in Cyprus, on the Northern (Kyrenia) Range, and easily recognized by its slender habit and conspicuous calyx-nerves; it may be deserving of more than varietal status.

SECT. 13. **Conoïmorphae** *Otth* Annuals; inflorescence usually dichasial; flowers erect; calyx 20–30-nerved or 60-nerved, nerves parallel, greenish.

25. S. conoidea *L.*, Sp. Plant., ed. 1, 418 (1753); Boiss., Fl. Orient., 1: 580 (1867); Holmboe, Veg. Cypr., 72 (1914); Post, Fl. Pal., ed. 2, 1: 172 (1932); Rechinger f. in Arkiv för Bot., ser. 2, 1: 421 (1950); Zohary, Fl. Palaest., 1: 99, t. 132 (1966); Davis, Fl. Turkey, 2: 241 (1967); Osorio-Tafall et Seraphim, List Vasc. Plants Cyprus, 36 (1973).
TYPE: Spain; " *inter segetes* Hispaniae ".

Erect, unbranched or sparingly branched annual, (10–) 15–30 (–50) cm. high; stems purplish and glabrescent or sparsely puberulous below, greenish and densely viscid-pubescent above; basal leaves mostly withered at anthesis, narrowly oblong or oblanceolate, 1–4 cm. long, 0·4–0·6 cm. wide, subglabrous or thinly pubescent, apex obtuse or subacute, base narrowed to an indistinct petiole; cauline leaves larger, 3–8 cm. long, 0·4–1·3 (–2·5) cm. wide, oblong or linear, usually tapering to an acute apex, sessile and amplexicaul at base, subglabrous or thinly pubescent; inflorescence a somewhat irregular dichasium, or flowers occasionally solitary; bracts linear-subulate, foliaceous, up to 3 cm. long, often densely viscid-glandular; pedicels at first short, usually less than 1 cm. long, lengthening to 1–1·5 cm. in fruit, densely viscid-glandular; calyx at first narrowly oblong, or cylindrical, 2–2·5 cm. long, 0·5–0·7 cm. wide, becoming widely ovoid in fruit, viscid-glandular, 30-nerved, the nerves greenish, the areas between the nerves whitish and membranous in fruiting specimens; calyx-teeth narrowly lanceolate-subulate, 5–7 mm. long, about 1·5 mm. wide at base, margins membranous, long-ciliate; petals magenta-carmine, limb 4–7 mm. long, 2·5–5 mm. wide, obtuse, acute or deeply emarginate; auricles oblong, blunt, about 0·5–0·8 mm. long and almost as wide; coronal scales oblong, about 2 mm. long and 0·8–1 mm. wide, apex blunt, often irregularly toothed; petal-claws usually connate, at least towards apex, 1·8–2 cm. long; stamens 1·3–2 cm. long, filaments glabrous above, thinly hairy near base; anthers yellowish, oblong, about 2 mm. long and 1 mm. wide; anthophore short, 1–2 mm. long, glabrous; ovary narrowly oblong, about 5 mm. long and 1·5–2 mm. wide; styles 3, about 8 mm. long. Capsule flask-shaped, 1·6–2 cm. long, 0·8–1 cm. wide near base, with a long, slender, tapering neck; capsule teeth 6, about 1·5 mm. long and 1 mm. wide at base, slightly recurved; seeds reniform, about 1·3 mm. long and 1 mm. wide, dorsal surface flattish, lateral surfaces flattish, ventral surface excavate about hilum; testa greyish, tuberculate-papillose, with the bases of the papillae minutely ridged.

HAB.: In vineyards, cornfields and fallows; sea-level to 3,700 ft. alt.; fl. March–May.

DISTR.: Division 2–5, 7, common in Division 2. Southern Europe from Spain eastwards through Asia Minor to the Western Himalayas; often a casual weed of cultivated land.

2. Platres, Trimiklini etc., *Kennedy* 344 ! 358 ! *Davis* 3128 ! 3476 ! *N. Macdonald* 136 !
3. Silikou, 1930, *C. B. Ussher* 96 !
4. Near Larnaca, 1859, *Kotschy* s.n.
5. Near Sha, 1956, *Merton* 2496 ! Between Kornos and Delikipo, 1974, *Osorio-Tafall & Seraphim* s.n. !
7. Melounda, 1950, *Chapman* 316 !

26. S. macrodonta *Boiss.*, Diagn., 1, 1: 37 (1843); Fl. Orient., 1: 580 (1867); Holmboe, Veg. Cypr., 72 (1914); Post, Fl. Pal., ed. 2, 1: 172 (1932); Rechinger f. in Arkiv för Bot., ser. 2, 1: 421 (1950); Zohary, Fl. Palaest., 1: 99, t. 133 (1966); Davis, Fl. Turkey, 2: 239 (1967); Osorio-Tafall et Seraphim, List Vasc. Plants Cyprus, 36 (1973).

? *S. conica* (non L.) Sibth. et Sm., Fl. Graec. Prodr., 1: 296 (1809) pro parte quoad pl. cypr.

TYPE: Turkey; " in cultis planitiei Cariensis ad meridiem montis Cadmi sitae ubi Junio 1842 fructiferam legi " *Boissier* (G).

Erect annual, 5–30 cm. high, usually much branched from base: basal leaves usually present, but often withered, at anthesis, narrowly oblong or oblanceolate, 2–4 cm. long, 0·4–0·7 cm. wide, thinly glandular-pubescent or subglabrous, apex obtuse or subacute, base tapering to an indistinct flattened petiole; cauline leaves 1–6 cm. long, 0·2–0·7 cm. wide, linear, acute or acuminate, sessile, often tinged purplish, subglabrous or ± densely glandular-pubescent; inflorescence rather irregularly dichasial, usually viscid-glandular; bracts linear-subulate, 0·5–1·5 cm. long, foliaceous; pedicels usually less than 0·5 cm. long at anthesis, lengthening to 1–1·5 cm. in fruit; calyx pale green, cylindrical, 1–1·3 cm. long, 0·3–0·4 cm. wide, becoming narrowly ovoid in fruit, calyx-nerves 60, intervening spaces scarcely visible; calyx-teeth triangular-subulate, about 6 mm. long and 1·5 mm. wide at base, margins membranous, long-ciliate; petals pale or clear pink, free, 1–1·3 cm. long, 0·2–0·3 cm. wide near the distinctly emarginate apex; auricles conspicuous, oblong, about 0·5 mm. long and rather less in width; coronal scales usually connate in pairs along their inner margins, forming an emarginate scale 1 mm. long, 1·3 mm. wide; claw 8–9 mm. long, tapering gradually to base; stamens 6–10 mm. long, filaments thinly hairy towards base; anthers yellowish, oblong, about 1·3 mm. long and 0·6 mm. wide; anthophore very short, seldom exceeding 1 mm., glabrous; ovary narrowly oblong, about 3 mm. long and 1·3 mm. wide; styles 3, 7–9 mm. long. Capsule narrowly ovoid, about 1 cm. long and 0·4–0·5 cm. wide near base, neck not very slender or elongate; capsule-teeth triangular, less than 1 mm. long and about 0·6 mm. wide at base, distinctly recurved; seeds reniform, less than 1 mm. long and 0·8 mm. wide, dorsal surface slightly channelled, lateral surfaces flattened, ventral surface excavate about hilum; testa pale grey, with much flattened, obscure radiating ridges, minutely sinuous along their bases.

HAB.: In cultivated and fallow fields, usually on sandy soils; on fixed sand-dunes; occasionally amongst garigue on dry, rocky hillsides; sea-level to 3,000 ft. alt.; fl. Febr.–April.

DISTR.: Divisions 2–8. Also Crete, Turkey, Syria and Palestine.

2. Evrykhou, 1935, *Syngrassides* 825 ! Platres, 1939, *Kennedy* 1505 ! 1505b ! 48½ Mile Post on Xeros–Polis road, 1957, *D. R. Harris* 3107 !
3. Limassol, 1859, *Kotschy* 435; Alethriko, 1905, *Holmboe* 245.

4. Larnaca, 1862, *Kotschy* 62; near Famagusta, 1912, *Haradjian* 96; Paralimni, 1928 or 1930, *Druce* 70 !
5. Before Pyroi from Nicosia, 1932, *Syngrassides* 471 ! 1111 !
6. Ayia Irini (Morphou), 1941, *Davis* 2554 ! and, 1962, *Meikle* 2378 !
7. Pentadaktylos, 1880, *Sintenis & Rigo* 252b ! Melounda, 1950, *Chapman* 315 !
8. Between Yialousa and Eleousa, 1880, *Sintenis & Rigo* 252 ! Ayios Philon near Rizokarpaso, *Davis* 2223 !

NOTES: Often misidentified as *S. conica* L. by the older authors, an error which probably accounts for Sibthorp's record for the latter from Cyprus. True *S. conica* is at once distinguished by having 30 (not 60) calyx-nerves, the individual nerves separated by pale, membranous tissue in the fruiting-calyx; otherwise it much resembles *S. macrodonta*, though the flowers are normally rather larger and with longer pedicels.

TRIBE 2. **Alsineae** *DC.* Stipules absent (or small and scarious in *Spergularia*); sepals free to the base, or connate at the base only; petals without coronal scales, and not clearly divided into limb and claw; anthophore absent or very indistinct; styles free to the base or almost to the base.

9. HOLOSTEUM *L.*

Sp. Plant., ed. 1, 88 (1753).
Gen. Plant., ed. 5, 98 (1754).

Annual herbs; stems and leaves often viscid-pubescent especially towards apex of plant; inflorescence an umbel-like terminal cyme; sepals 5, free; petals 5, entire or irregularly toothed, but not distinctly 2-lobed; stamens hypogynous, variable in number, either 3–5, or up to 10; ovary 1-locular, ovules numerous; styles 3, occasionally 4–5. Fruit an ovoid or cylindrical capsule, normally opening by 6 short recurved teeth; seeds attached centrally (peltate), papillose.

About 6 species in Europe, North Africa and Asia, extending eastwards to Turkestan and Kashmir.

1. **H. umbellatum** *L.*, Sp. Plant., ed. 1, 88 (1753); Unger et Kotschy, Die Insel Cypern, 346 (1865); Boiss., Fl. Orient., 1: 709 (1867); Holmboe, Veg. Cypr., 72 (1914); Post, Fl. Pal., ed. 2, 1: 202 (1932); Zohary, Fl. Palaest., 1: 119, t. 164 (1966); Davis, Fl. Turkey, 2: 86 (1967); Osorio-Tafall et Seraphim, List Vasc. Plants Cyprus, 38 (1973).

TYPE: " *in* Germaniae, Galliae *arvis* ".

Erect or spreading annual, 3–20 (–30) cm. high; stems unbranched or branched at the base, glabrous or sometimes glandular-pubescent; leaves oblong, oblanceolate, obovate or ovate, 0·8–3 cm. long, 0·3–1 cm. wide, glaucous, glabrous or glandular-pubescent, acute or obtuse, the basal with a short flattened petiole, the cauline sessile and shortly connate; inflorescence an umbellate cyme, at the apex of an erect, naked peduncle 2–20 cm. long; flowers at first erect or spreading, later often deflexed; bracts ovate, acute, 1–3 mm. long, 0·8–2 mm. wide, scarious or thinly herbaceous; pedicels slender, 0·8–4 cm. (occasionally more) in length, glabrous (var. *umbellatum*) or glandular-pubescent (var. *glandulosum* Vis.); sepals narrowly ovate, concave, 2–5 mm. long, 1·5–3 mm. wide, glabrous or thinly glandular, apex acute or obtuse, margins narrowly membranous; petals white or tinged pink, oblanceolate or narrowly obovate, 3–7 mm. long (or longer), 1·5–3 mm. wide, apex rounded, subentire or irregularly toothed, base glabrous or bearded; stamens usually 3–4, sometimes more, filaments glabrous, 2–3 mm. long; anthers square or rounded, about 0·5 mm. long and almost as wide, yellow; ovary subglobose, about 2 mm. diam.; styles 3–4, about 0·5–0·8 mm. long. Capsule 3–8 mm. long, 3–5 mm. wide, with 6 ovate-triangular, strongly recurved teeth; seeds shortly oblong, about 0·8 mm. long and 0·5 mm. wide, furrowed on the dorsal side and keeled on the ventral side; testa brown, with rows of blunt, ridged tubercles.

HAB.: On screes and dry slopes; 3,000–5,000 ft. alt.; fl. Febr.–April.

DISTR.: Division 2. Widespread in Europe, the Mediterranean area and Asia, east to Afghanistan.

2. Vrysi tou Damaskinari, Troödos ("Quelle Ta Maschinari"), 1862, *Kotschy*; Mandria, 1938, *Kennedy* 1218 ! Between Agros and Palekhori, near Papoutsa, 1941, *Davis* 3118 ! Tripylos, Cedar valley, 1959, *P. H. Oswald* 87 ! Madhari, above Lagoudhera, 1974, *Meikle* 4051 !

NOTES: The Cyprus specimens are all, to some degree, glandular and should probably come under *H. umbellatum* L. var. *glandulosum* Vis. No clear line can, however, be drawn between the glabrous and glandular variants, and it is considered better not to give them formal recognition here. *H. umbellatum* is an extremely variable species.

10. CERASTIUM *L.*

Sp. Plant., ed. 1, 437 (1753).
Gen. Plant., ed. 5, 199 (1754).

Annual or perennial herbs, or rarely subshrubs; leaves usually sessile and often pilose or glandular; inflorescence usually branched, cymose or flowers occasionally solitary; sepals 4 or 5, free, with membranous margins; petals white, 4–5, free, deeply 2-lobed, or emarginate, sometimes wanting; stamens 10 (or 8) or occasionally 5 or fewer; ovary 1-locular, ovules numerous; styles generally 5 (sometimes 6, 4 or 3). Capsule oblong or cylindrical, often curved and usually exceeding the calyx, opening by twice as many short teeth as there are styles; seeds reniform or subglobose; hilum marginal.

About 100 species, chiefly in temperate regions of the Old World, but widely distributed as weeds of cultivation elsewhere.

Bracts and bracteoles entirely herbaceous, or with a very narrow membranous margin:
  Leaves glabrous or glabrescent, linear, usually acute or acuminate - **1. C. dubium**
  Leaves hairy or glandular, oblong, elliptical, obtuse or subacute:
    Calyx eglandular, clothed with long, white silky hairs; capsule-teeth very short or obsolete - - - - - - - - - - **2. C. illyricum ssp. pilosum**
    Calyx normally glandular, with shorter, less conspicuous eglandular hairs; capsule-teeth well developed:
      Sepals 8–12 mm. long; capsule 15–25 mm. long - - **3. C. dichotomum**
      Sepals less than 8 mm. long; capsule 4–10 mm. long:
        Pedicels usually shorter than calyx; flowers often in dense clusters or condensed cymes - - - - - - - - - - - **4. C. glomeratum**
        Pedicels usually longer than calyx; flowers in open repeatedly branched cymes:
          Calyx with long eglandular hairs and shorter glandular hairs intermixed; fruiting pedicels curved just below apex **5. C. brachypetalum ssp. roeseri**
          Calyx indumentum wholly glandular, without long eglandular hairs; fruiting pedicels straight, often deflexed - - - - - **7. C. fragillimum**
Bracts and bracteoles with a conspicuous, membranous-hyaline tip **6. C. semidecandrum**

1. C. dubium (*Bast.*) *Schwarz* in Mitt. Thür. Bot. Ges., 1: 98 (1949); Zohary, Fl. Palaest., 1: 120, t. 165 (1966).

Stellaria dubia Bast., Fl. Maine-et-Loire Suppl., 24 (1812).

Cerastium anomalum (non Schrank) Waldst. et Kit. in Willd., Sp. Plant., 2: 812 (1799); Boiss., Fl. Orient., 1: 714 (1867); Post in Mém. Herb. Boiss., 18: 92 (1900); Holmboe, Veg. Cypr., 71 (1914); Post, Fl. Pal., ed. 2, 1: 203 (1932); Davis, Fl. Turkey, 2: 75 (1967); Osorio-Tafall et Seraphim, List Vasc. Plants Cyprus, 39 (1973).

TYPE: France; "A la Challoire, aux Pont-aux-Filles, en Reculée, à Ste.-Gemmes, etc.", *Bastard* (? ANG).

An erect annual, 4–20 (–30) cm. high; stems unbranched or branched at the base, glabrous or puberulous; basal leaves narrowly linear-oblanceolate, 10–12 mm. long, 1·5–3 mm. wide, subacute, glabrescent, mostly withered at anthesis; cauline leaves linear, 15–30 mm. long, 2–3 mm. wide, acute; inflorescence a lax rather irregular dichasium; flowers erect; bracts linear, about 5 mm. long, herbaceous, foliaceous, usually ± glandular-pubescent; pedicels slender, 0·3–1·5 cm. long, glandular-pubescent; sepals narrowly ovate, about 5 mm. long and 1·5–2 mm. wide, thinly glandular-pubescent with an acute apex, the inner with narrow hyaline margins; petals narrowly cuneate, about 5–6 mm. long and 2–3 mm. wide at the deeply bilobed apex, lobes oblong, obtuse, about 1·5 mm. long; stamens 2·5–3 mm. long, filaments glabrous; anthers almost orbicular, about 0·3 mm. diam., yellow; ovary shortly oblong, about 1·5 mm. long and 1 mm. wide; styles 3, 1–1·5 mm. long. Capsule cylindrical, straight, 0·8–1 cm. long, 0·2–0·3 cm. wide, opening by 6 short erect or slightly recurved teeth; seeds rounded or reniform, compressed, about 0·7 mm. long and 0·5 mm. wide; testa pale brown, covered with sharp conical, ridged papillae.

HAB.: In moist fields and on damp waste ground; ? alt.; fl. April–June.

DISTR.: Division 2, rare. Also widely distributed in Europe and in western Asia eastwards to Iran.

2. Troödos, 1894, *Post* s.n.

NOTES: Not collected since. Possibly an error due to confusion with *Stellaria cilicica* Boiss. et Bal., which is frequent in the area and superficially similar to *Cerastium dubium*, and which is not recorded by Post. There is, however, no fundamental objection to the possibility of the species occurring in Cyprus, and the description has been included as a guide to searchers.

2. C. illyricum *Ard.*, Animadv. Bot. Spec., 2: 26, t. 11 (1764); Boiss., Fl. Orient., 1: 720 (1867); Post, Fl. Pal., ed. 2, 1: 204 (1932).

An erect or spreading annual 2–10 (–15) cm. high; stems usually much branched, purplish, ± densely clothed with deflexed silky hairs; lower cauline leaves oblong-spathulate, 1–2 cm. long, 0·4–0·7 cm. wide, obtuse or subacute, ± hairy, base tapering to a short flattened petiole; upper cauline leaves usually smaller and more acute, sessile or subsessile; inflorescence cymose, lax or congested, regularly or irregularly dichotomous; bracts herbaceous, foliaceous; pedicels sometimes very short, or occasionally almost 5 cm. long, clothed with silky deflexed hairs; sepals lanceolate, acuminate, deeply concave, 5–10 mm. long, 1·3–1·5 mm. wide, with broad or narrow hyaline margins, and with the dorsal surface ± densely covered with long, spreading hairs; petals shorter than, or equalling, or exceeding sepals, sometimes wanting, oblong-obovate, suddenly dilated just above point of attachment, apex shortly bilobed, lobes obtuse; stamens 5–10, almost as long as petals; anthers yellow, oblong, about 0·3 mm. long; ovary ovoid 2–3 mm. long and 1–2 mm. wide; styles 5, 1–4 mm. long. Capsule narrowly cylindrical, 5–8 (–12) mm. long, 1·5–3 mm. wide at base; capsule-teeth minute or obsolete; seeds about 0·5 mm. long and 0·3 mm. wide, rather irregularly reniform, not much compressed; testa rich brown, minutely tuberculate.

ssp. **pilosum** *Rouy et Fouc.*, Fl. France, 3: 222 (1896).

C. pilosum Sibth. et Sm., Fl. Graec. Prodr., 1: 316 (1809), Fl. Graec., 5: 41, t. 454 (1825) excl. syn.

C. comatum Desv. in Journ. de Bot., 3: 228 (? 1816); Lonsing in Feddes Repert., 46: 147 (1939).

C. illyricum ssp. *comatum* (Desv.) Sell et Whitehead in Feddes Repert., 69: 15 (1964); Davis, Fl. Turkey, 2: 85 (1967); Osorio-Tafall et Seraphim, List Vasc. Plants Cyprus, 39 (1973).

[*C. illyricum* (non Ard.) Unger et Kotschy, Die Insel Cypern, 347 (1865); Sintenis in Oesterr. Bot. Zeitschr., 32: 51 (1882); Holmboe, Veg. Cypr., 72 (1914); Lindberg f., Iter Cypr., 14 (1946); Rechinger f. in Arkiv för Bot., ser. 2, 1: 421 (1950).]

TYPE: "In Peloponnesi et insulae Cypri montibus" *Sibthorp* (OXF).

Fruiting pedicels 2–20 mm. long, usually erect; sepals with a narrow hyaline margin, or sometimes wholly herbaceous, conspicuously hairy, with the hairs projecting beyond the calyx; petals usually much shorter than sepals; styles about 1 mm. long.

HAB.: On rocky slopes, and on roadside banks, occasionally on gravel banks by river-margins; 500–3,400 ft. alt.; fl. March–May.

DISTR.: Divisions 2, 3, 7, common in Division 7. Also in Corsica, E. Greece, Crete, Aegean Islands, Turkey and Palestine.

2. Stavros tis Psokas, 1928 or 1930, *Druce* s.n.; valley of Kryos Potamos, 1937, *Kennedy* 365 ! Between Ambelikou and Kambos, 1939, *Lindberg f.* s.n.; Paliambela, 1962, *Meikle* 2250 !
3. Alethriko, 1905, *Holmboe* 216.
7. Common; *Kotschy* 361. *Sintenis & Rigo* 225 ! *Davis* 2810 ! 2835 ! 2886 !

NOTES: Typical *C. illyricum* with larger flowers, less hairy sepals with broader hyaline margins, and longer petals, is apparently confined to islands off the west (Ionian) coast of Greece.

3. C. dichotomum *L.*, Sp. Plant., ed. 1, 438 (1753); Boiss., Fl. Orient., 1: 721 (1867); Holmboe, Veg. Cypr., 71 (1914); Post, Fl. Pal., ed. 2, 1: 204 (1932); Lindberg f., Iter Cypr., 14 (1946); Zohary, Fl. Palaest., 1: 120, t. 166 (1966); Davis, Fl. Turkey, 2: 81 (1967); Osorio-Tafall et Seraphim, List Vasc. Plants Cyprus, 39 (1973).
TYPE: Spain; " *inter segetes* Hispaniae ".

An erect viscid-pubescent annual, 7–20 (–25) cm. high; stems unbranched or with divaricate branches; leaves oblong-linear, 1–3 cm. long, 0·3–0·8 cm. wide, grey-green, glandular-pubescent, apex acute or subacute, base of lowest leaves tapering to a flattened petiole, otherwise leaves sessile but scarcely connate; inflorescence dichotomously branched, at first rather congested, later lax; flowers ± erect; bracts herbaceous, foliaceous, densely glandular; pedicels 3–15 mm. long, densely glandular; sepals lanceolate, shallowly concave, 8–12 mm. long, 2–3 mm. wide, acuminate, the outer entirely herbaceous, the inner with a narrow membranous margin, all ± densely glandular-pubescent; petals obovate, 6–7 mm. long, 3–4 mm. wide, shortly bifid, with a narrow sinus and two broad, blunt lobes; stamens 5–10, filaments about 3 mm. long; anthers yellow, oblong, about 0·5 mm. long and 0·3 mm. wide; ovary ovoid, about 2·5 mm. long and 1·8 mm. wide at base; styles 5, less than 1 mm. long, stigmas recurved. Capsule narrowly flask-shaped, 1·5–2·5 cm. long, 0·3–0·4 cm. wide at base, neck distinctly curved often splitting longitudinally from the apex downwards; capsule-teeth narrowly triangular, about 1 mm. long and 0·3 mm. wide at base, remaining erect even after dehiscence; seeds subglobose about 1 mm. diam.; testa rich brown, acutely tuberculate.

HAB.: On cultivated ground; in fallow fields; occasionally on grassy banks, or on dry mountain slopes; sea-level to 5,500 ft. alt.; fl. Dec.–May.

DISTR.: Divisions 2, 4, 6, 7, common in Division 2. Also Spain, S. France, Greece, Bulgaria, and widespread in western Asia from Turkey to Afghanistan.

2. Common; *Holmboe* 1081; *Syngrassides* 107 ! *Kennedy* 363 ! 364 ! 1219 ! *Davis* 3055 ! *Chapman* 314 ! *Merton* 2486 ! 3145 !
4. Paralimni, 1940, *Davis* 2063 !
6. Near Dhiorios, 1962, *Meikle* 2336 !
7. Buffavento, 1941, *Davis* 2182 !

4. C. glomeratum *Thuill.*, Fl. Par., ed. 2, 226 (1798); Holmboe, Veg. Cypr., 71 (1914); Lindberg f., Iter Cypr., 14 (1946); Zohary, Fl. Palaest., 1: 121, t. 167 (1966); Davis, Fl. Turkey, 2: 82 (1967); Osorio-Tafall et Seraphim, List Vasc. Plants Cyprus, 39 (1973).
C. viscosum (? L.) Sibth. et Sm., Fl. Graec. Prodr. 1: 315 (1809); Unger et Kotschy, Die Insel Cypern, 347 (1865); Boiss., Fl. Orient., 1: 722 (1867); Post, Fl. Pal., ed. 2, 1: 205 (1932) [err. " *viscosm* "].

TYPE: France; "se trouve dans le bois de Boulogne; à Vincennes et ailleurs" *Thuillier* (G).

Erect or decumbent annual, 7–30 (–45) cm. high; stems with spreading glandular and eglandular hairs, usually much branched at the base, often tinged purplish; leaves bright pale green, 5–25 (–30) mm. long, 3–12 (–18) mm. wide, the lowermost broadly obovate or spathulate, with a short flattened petiole, the upper oblong or elliptic, all obtuse and clothed with long eglandular, and shorter glandular hairs; inflorescence usually a dense cymose cluster, occasionally a more open cyme with rather stiff, erecto-patent branches; pedicels usually less than 1 cm. long, glandular-hairy; bracts herbaceous, foliaceous, glandular-hairy; flowers erect or horizontal, not distinctly nodding; sepals lanceolate, acuminate, shallowly concave, about 5 mm. long and 1–1·5 mm. wide, glandular-hairy, often purplish-tipped, the outer almost entirely herbaceous, the inner with a distinct, narrow, hyaline margin; petals broadly oblanceolate, as long as, or sometimes a little longer than sepals, or shorter than sepals, or rarely wanting (f. *apetalum* (Dum.) Briq.), about 2–2·5 mm. wide, apex bilobed, with a very narrow sinus about 1·5 mm. long and blunt or subacute lobes, petal-base usually with a few marginal hairs; stamens about 2 mm. long, often sparsely hairy at base of filament; anthers yellow, oblong, about 0·5 mm. long, and a little less in width; ovary broadly oblong or almost globose, 1·5–2 mm. long; styles 5, about 0·5 mm. long. Capsule narrowly cylindrical, 7–10 mm. long; 1·5–2 mm. wide, generally distinctly curved, with 10 slender, erect teeth about 0·7 mm. long; seeds subglobose, obscurely angled, about 0·4–0·5 mm. long; testa light brown, shortly and rather bluntly tuberculate.

HAB.: Usually in moist ground by springs and streams, occasionally on roadside banks or in fields; sea-level to 4,000 ft. alt.; fl. Febr.–June.

DISTR.: Divisions 1, 2, 6–8, common in Divisions 2, 7. Also throughout Europe and western Asia, and as an introduction in almost every part of the world.

1. Toxeftera near Ayios Yeoryios (Akamas), 1962, *Meikle* 2002 ! Smyies, 1962, *Meikle* 2181 !
2. Common; *Kotschy* 838a; *Lindberg f.* s.n.; *Kennedy* 3721 !
6. Ayia Irini, 1941, *Davis* 2594 !
7. Common; *Sintenis & Rigo* 224 ! 226 ! *Davis* 2884 ! *Casey* 1289 ! *G. E. Atherton* 1007 !
8. Near Rizokarpaso, 1941, *Davis* 2305 !

5. C. brachypetalum *Pers.*, Syn. Plant., 1: 520 (1805); Unger et Kotschy, Die Insel Cypern, 347 (1865); Boiss., Fl. Orient., 1: 723 (1867); Holmboe, Veg. Cypr., 71 (1914).

An erect, branched or unbranched annual, 5–25 cm. high; stems and leaves ± densely clothed with long, spreading silvery hairs, sometimes intermixed with shorter glandular hairs; basal part of stems often purplish; leaves pale greyish-green, 0·5–2·5 cm. long, 0·2–0·8 cm. wide, the lowermost spathulate, with a distinct, but short, flattened petiole, the upper oblong or elliptic, sessile, all blunt or subacute; inflorescence a repeatedly dichotomous cyme; bracts herbaceous, villose or glandular-villose, foliaceous; pedicels slender, 0·5–2·5 (–3) cm. long, bent near apex so that the flowers are nodding; sepals lanceolate, acuminate or subacute, 3·8–6 mm. long, 1–1·5 mm. wide, shallowly concave, ± densely clothed with long whitish hairs, glandular or eglandular, margins narrowly hyaline; petals 3·5–5 mm. long, 2–3 mm. wide, obovate, deeply bilobed, with a narrow sinus 1·5–2 mm. long, and rounded or subacute lobes; petal-base often ciliate; stamens about 2 mm. long, filaments glabrous or slightly hairy at base; anthers yellow, oblong, about 0·5 mm. long, and slightly less in width; ovary subglobose, about 1 mm. diam.; styles 5, about 1 mm. long. Capsule cylindrical, straight or curved, 5–9 mm. long, 2–2·5 mm.

wide; capsule-teeth slender, erect, about 0·8 mm. long; seeds angled, 0·4–0·6 mm. long, about 0·4 mm. wide; testa dull brown, minutely, rather sharply tuberculate.

ssp. **roeseri** (*Boiss. et Heldr.*) *Nyman*, Consp. Fl. Europ., 1: 109 (1878); Davis, Fl. Turkey, 2: 82 (1967); Osorio-Tafall et Seraphim, List Vasc. Plants Cyprus, 39 (1973).

C. roeseri Boiss. et Heldr. in Boiss., Diagn., 2, 1: 93 (1854); Lindberg f., Iter Cypr., 14 (1946).

C. brachypetalum Pers. var. *roeseri* (Boiss. et Heldr.) Boiss., Fl. Orient., 1: 723 (1867).

TYPE: Greece; " in lapidosis regionis alpinae *Parnassi* " *Heldreich* (G, K !).

Whole plant viscid-glandular; inflorescence with divaricate branches in fully developed specimens; petals equalling or shorter than sepals.

HAB.: On moist roadside banks; on screes; in garigue on open mountain slopes; by mountain streams; usually in rather damp places; 2,000–5,500 ft. alt.; fl. Febr.–June.

DISTR.: Divisions 2, 3, 7, very common in Division 2. Also Italy, Sicily, Balkan peninsula, Crete, Aegean Islands, Turkey, Lebanon and Iraq.

2. Common; *Kotschy* 838; *Kennedy* 366 ! 367 ! 371a ! 1222 ! *Davis* 2741 ! 3116 ! 3160 ! 3492 ! *Meikle* 2668 ! 2684 ! 2707 ! 2754 ! 2823 ! 2940 !
3. Stavrovouni, 1880, *Sintenis & Rigo* 1012 !
7. Yaïla, 1941, *Davis* 2836 !

NOTES: Differs from typical *brachypetalum* chiefly in its viscid-glandular indumentum.

6. C. semidecandrum *L.*, Sp. Plant., ed. 1, 438 (1753); Boiss., Fl. Orient., 1: 723 (1867); Holmboe, Veg. Cypr., 71 (1914); Post, Fl. Pal., ed. 2, 1: 205 (1932); Davis, Fl. Turkey, 2: 83 (1967); Osorio-Tafall et Seraphim, List Vasc. Plants Cyprus, 39 (1973).

TYPE: " *in campis apricis sterilissimis* Europae *borealis* ".

Erect or decumbent annual, 1·5–15 (–20) cm. high; stems usually much branched from base, often purplish, and generally viscid-glandular; leaves usually hairy at least above, the basal narrowly spathulate or oblanceolate 1–2·5 cm. long, 0·2–0·4 cm. wide, apex obtuse, base tapering to a short flattened petiole; cauline leaves sessile, oblong or elliptical, 0·5–1·5 cm. long, 0·2–0·5 cm. wide, obtuse or subacute; inflorescence a rather condensed, irregularly branched cyme; bracts ovate, 2–4 mm. long, 1–2 mm. wide, upper third membranous, hyaline; pedicels slender, 3–12 mm. long, patent or deflexed in fruit, generally viscid-glandular; sepals ovate, acute or subacute, 2·5–3 mm. long, 1–1·5 mm. wide, with a rather wide hyaline margin and apex, the dorsal surface often glandular-pilose; petals obovate or oblong-obovate, about 3 mm. long and 1·5 mm. wide, apex shortly bilobed, with a narrow sinus less than 0·5 mm. long and obtuse or subacute lobes, base of petal glabrous; stamens 4–5, usually about 2 mm. long, filaments glabrous; anthers rounded, about 0·3 mm. diam., yellow; ovary globose about 1 mm. diam.; styles 5, about 0·5 mm. long; capsule cylindrical, 4–7 mm. long, 1·5–2·5 mm. wide; capsule-teeth slender, erect, about 0·8 mm. long; seeds small, angled, about 0·5 mm. long and 0·4 mm. wide, narrowed towards hilum; testa bright brown, minutely and very bluntly tuberculate.

HAB.: Sandy fields, sand-dunes and dry slopes; generally lowland; fl. ? Dec.–May.

DISTR.: Division 6, rare. Throughout Europe and eastwards through Turkey and Lebanon to Iraq.

6. Nicosia, 1905, *Holmboe* 290.

NOTES: Perhaps a fleeting introduction, but more probably overlooked by other collectors, as the plants are often very small and inconspicuous. It should be searched for in sandy fields during winter and early spring, and is easily identified by its hyaline, membranous-margined bracts.

7. **C. fragillimum** *Boiss.*, Diagn., 1, 1: 54 (1843), Fl. Orient., 1: 724 (1867); Holmboe, Veg. Cypr., 71 (1914); Post, Fl. Pal., ed. 2, 1: 205 (1932); Lindberg f., Iter Cypr., 14 (1946); Davis, Fl. Turkey, 2: 84 (1967); Osorio-Tafall et Seraphim, List Vasc. Plants Cyprus, 39 (1973).

TYPE: Turkey; " ad umbram *Juniperorum* in regione alpina, Cadmus, Mesogis, Tmolus; legi Jun. 1842 ", *Boissier* (G).

Erect, viscid-glandular annual, 4–10 (–15) cm. high; stems unbranched or branched at the base, brittle, often purplish; leaves densely glandular, often purplish, the basal spathulate, 1–2 (–2·5) cm. long, about 0·5–0·6 cm. wide, obtuse or subacute, tapering to a distinct flattened petiole; upper leaves sessile, oblong or elliptical, 0·8–1 (–1·5) cm. long, 0·2–0·6 cm. wide, obtuse or subacute, usually in rather distant pairs; inflorescence a very lax, divaricate-branched cyme; bracts minute, oblong, usually under 2 mm. long and 1 mm. wide, herbaceous, glandular; pedicels very slender, 0·5–1·5 cm. long, at first spreading, later sharply deflexed, so that the fruits are pendulous; sepals ovate, 4–5 mm. long, about 2 mm. wide, obtuse or subacute, densely glandular, the outer entirely herbaceous or with a very narrow membranous margin, the inner broadly membranous-margined; petals narrowly obovate, about 5 mm. long and 2–2·5 mm. wide, apex shortly bilobed, with a rather wide sinus and subacute or obtuse lobes, base glabrous; stamens 10, to about 3 mm. long, filaments glabrous; anthers suborbicular, less than 0·5 mm. diam., yellow; ovary globose, about 1·5 mm. diam.; styles 5, about 1 mm. long. Capsule cylindrical, straight or slightly curved, 5–7 mm. long, 2–3 mm. wide; capsule-teeth slender, about 0·8 mm. long; seeds suborbicular, compressed, about 1 mm. diam.; testa mid-brown, with relatively coarse, flattened tubercles.

HAB.: On gravelly or earthy banks; on open mountainside; on screes; or by mountain streams and springs; 3,600–6,400 ft. alt.; fl. April–July.

DISTR.: Division 2, locally very common, Stavros tis Psokas–Tripylos–Troödos area, Madhari and Papoutsa. Also Turkey, Syria and Lebanon.

2. Common; *Kotschy* 715 ! *Holmboe* 975; *Kennedy* 368 ! 369 ! 370 ! 371 ! 374 ! 1220 ! 1221 ! *Davis* 1917 ! 3113 ! 3148 ! *Merton* 375 ! 3172 ! *Meikle* 2706 ! 2941 !

[C. FONTANUM *Baumg.* (*C. vulgatum* auct., *C. holosteoides* Fr.) is recorded, under the name *C. vulgatum*, from Kythrea, by Sintenis (Oesterr. Bot. Zeitschr., 31: 326; 1881) and from " Kuklia " [Kouklia, probably the village near Paphos] by Druce (Rep. B.E.C., 9: 469; 1931). The first record is almost certainly a slip of the pen for *C. viscosum* (=*C. glomeratum* Thuill.) which Sintenis and Rigo collected at Kythrea; the second record is more obscure, but in view of the fact that *C. fontanum* has not since been recorded from Cyprus, and is generally absent from this area of the Mediterranean, it must be regarded with the gravest suspicion. *C. fontanum*, a sprawling or creeping perennial, with the upper bracts narrowly hyaline-margined, and with a long curved capsule, might possibly be found high up on the Troödos range, but it is not likely to occur in lowland Cyprus].

11. STELLARIA *L.*

Sp. Plant., ed. 1, 421 (1753).
Gen. Plant., ed. 5, 193 (1754).

Annual or perennial herbs, often with fragile, sprawling stems; leaves sometimes petiolate; flowers in dichasial cymes, rarely solitary; bracts membranous or herbaceous; sepals 5, free; petals usually 5, white, deeply bifid, occasionally wanting; stamens 10 or fewer; nectaries generally present; ovary 1-locular; styles 3. Fruit a globose or subcylindrical capsule, opening by 6 valves; seeds reniform or subglobose.

About 100 species, distributed throughout the world.

Lower leaves distinctly petiolate, ovate:
Upper parts of stem glandular-pubescent; petals as long as sepals; stamens 9–10
1. S. media ssp. cupaniana

Upper parts of stem glabrous or with a single line of hairs; stamens (1–) 2–7 (–8):
Petals equalling or slightly shorter than sepals; stamens (3–) 5–7 (–8) **1. S. media ssp. media**
Petals absent or minute; stamens (1–) 2–3 - - - **1. S. media ssp. apetala**
Lower leaves sessile, linear-acuminate - - - - - - **2. S. cili ica**

1. S. media (*L.*) *Vill.*, Hist. Plant. Dauph., 3: 615 (1789); Boiss., Fl. Orient., 1: 707 (1867); Holmboe, Veg. Cypr., 70 (1914); Post, Fl. Pal., ed. 2, 1: 201 (1932); Zohary, Fl. Palaest., 1: 117, t. 162 (1966); Davis, Fl. Turkey, 2: 69 (1967); Osorio-Tafall et Seraphim, List Vasc. Plants Cyprus, 38 (1973).
Alsine media L., Sp. Plant., ed. 1, 272 (1753).

Sprawling annual; stems terete, (5–) 15–25 (–40) cm. high, usually with a single line of hairs, but occasionally glabrous or pubescent-glandular all over; leaves ovate, acute, bright or pale green, 0·5–4 cm. long, 0·5–3 cm. wide, glabrous or thinly pubescent, the lower usually with a distinct flattened petiole often almost as long as or longer than the lamina, the upper shortly petiolate to sessile; inflorescence few- or many-flowered, sometimes leafy, or cymose with the leaves reduced to small herbaceous bracts; pedicels variable in length, erect, spreading or sharply deflexed, glabrous or glandular-hairy; sepals ovate or lanceolate, 2–5 mm. long, 1–2 mm. wide, apex acute or subobtuse, outer sepals usually herbaceous, glabrous or glandular-hairy, inner often whitish, membranous; petals as long as the sepals, or shorter, or wanting, deeply 2-lobed; stamens 1–10, filaments shorter than petals; anthers purplish or violet; ovary subglobose. Capsule ovoid, rather longer than the calyx; seeds reniform; testa light or dark brown, acutely or bluntly papillose, or tuberculate.

ssp. media
TYPE: " in Europae cultis ".

Stems glabrous except for a longitudinal line of hairs; inflorescence usually rather lax; outer sepals ovate-lanceolate, glandular-hairy externally, 4–5 mm. long, 1·8–2 mm. wide, apex subacute or apparently so; inner sepals a little shorter and narrower; petals equalling or a little shorter than sepals; stamens (3–) 5–7 (–8), filaments 1·8–2 mm. long; anthers purplish, suborbicular, about 0·4 mm. diam. Capsule ovoid, about 5 mm. long, 2·5 mm. wide; seeds reniform, about 1 mm. long and wide; testa dark brown, bluntly tuberculate.

HAB.: In fields, gardens and waste ground; 3,000–5,400 ft. alt.; fl. Dec. to July.

DISTR.: Division 2, apparently uncommon. A widespread weed in temperate regions of the world.

2. Stavros tis Psokas, 1974, *Meikle* ! Prodhromos, 1974, *Meikle* !

NOTES: Generally replaced by ssp. *cupaniana* and ssp. *apetala* in Cyprus; it could not be found anywhere on the island during the author's 1962 visit, but had evidently established itself since, and was seen on several occasions in 1974, though only at fairly high altitudes in the Troödos area. It is evidently on the increase, and may be expected to become frequent in other Divisions.

ssp. **cupaniana** (*Jord. et Fourr.*) *Nyman*, Consp. Fl. Europ., 1: 111 (1878).
Alsine cupaniana Jord. et Fourr., Brev. Plant. Nov., 2: 19 (1868).
Stellaria media (L.) Vill. var. *pubescens* Post, Fl. Pal., ed. 1, 154 (1896).
S. media (L.) Vill. ssp. *postii* Holmboe, Veg. Cypr., 70 (1914); Davis, Fl. Turkey, 2: 70 (1967).
S. neglecta Weihe ssp. *postii* (Holmboe) Rechinger f. in Arkiv för Bot., ser. 2, 1: 421 (1949).
S. media (L.) Vill. var. *major* (non Koch) Sintenis in Oesterr. Bot. Zeitschr., 31: 392 (1881); H. Stuart Thompson in Journ. Bot., 44: 278 (1906).
TYPE: Sicily; " in ruderalibus Siciliae: circa Panormum " ? coll. (? AWH).

Stems ± densely glandular-pubescent all round in the upper part;

inflorescence becoming cymose with age as in *S. media* (L.) Vill. ssp. *neglecta* (Weihe) Murb.; outer sepals lanceolate, glandular-pubescent externally, 4–5 mm. long, about 2 mm. wide, apex obtuse or subacute; inner sepals membranous, thinly glandular-pubescent externally, a little shorter than outer sepals; petals 4–5 mm. long, about 2·5 mm. wide, divided almost to the base into 2 narrowly oblong, obtuse lobes; stamens 9 or 10, filaments 2–2·5 mm. long; anthers livid or purplish, suborbicular, about 0·3–0·5 mm. diam.; ovary subglobose, about 1·5 mm. diam. Capsule ovoid, 5–6 mm. long, 3–4 mm. wide; seeds reniform, about 1·4 mm. long and 1·3 mm. wide; testa rich brown with darker acute tubercles.

HAB.: On stony hillsides; by roadsides; in cultivated and waste ground; 200–4,500 ft. alt.; fl. Jan.–April.

DISTR.: Divisions 1–4, 6–8, common in Division 7. Widespread in the Mediterranean area and particularly frequent in the eastern Mediterranean.

1. Above Peyia, 1962, *Meikle* 2052 !
2. Platres, 1938, *Kennedy* 1217 ! Kionia, 1941, *Davis* 2708 ! Lefkara, 1941, *Davis* 2748 !
3. Alethriko, 1905, *Holmboe* 220.
4. Kouklia (Mesaoria), 1905, *Holmboe* 365.
6. Ayia Irini, 1941, *Davis* 2595 ! Orga, 1941, *Davis* 2100 !
7. Common; *Sintenis & Rigo* 227 ! *Syngrassides* 712 ! *Davis* 2100 ! 2826 ! *Casey* 224 ! *G. E. Atherton* 832 !
8. Between Rizokarpaso and Ayios Philon, 1941, *Davis* 2708 !

NOTES: *S. media* ssp. *cupaniana* and ssp. *postii* cannot be maintained as distinct even at varietal level, a conclusion reached by N. Y. Sandwith and N. D. Simpson in 1941 (Journ. Bot., 89: 38) and confirmed by subsequent examination of herbarium material. The relationship between ssp. *cupaniana* and ssp. *neglecta* (Weihe) Murb. is undoubtedly close, but the two are readily separable by indumentum characters.

ssp. **apetala** *Čelak.*, Fl. v. Böehm., 870 (1881).

S. glabella Jord. et Fourr., Brev. Plant. Nov., 2: 20 (1866) sens. ampl.

S. media (L.) Vill. var. *glabella* (Jord. et Fourr.) Briq., Prodr. Fl. Corse, 1: 499 (1910) sens. ampl.

? *S. apetala* Ucria in Roemer, Arch., 1, 1: 68 (1796).

S. apetala Murb. in Act. Univ. Lund, 8: 37 (1897) pro parte quoad spec. cypr.

S. media (L.) Vill. ssp. *apetala* (Ucria) Holmboe, Veg. Cypr., 71 (1914) pro parte quoad spec. cypr.

[*S. media* (L.) Vill. ssp. *neglecta* (non (Weihe) Murb.) Lindberg f., Iter Cypr., 15 (1946).]

TYPE: Czechoslovakia; near Prague, *Tausch.*

Stems glabrous except for a longitudinal line of hairs; inflorescence usually crowded; pedicels often short; flowers rather small; outer sepals lanceolate, glandular hairy (or occasionally glabrous) externally, 2–3 mm. long and about 1 mm. wide, apex often subacute or apparently so; inner sepals a little shorter than outer, almost membranous, thinly pubescent or glabrous; petals wanting or very minute; stamens (1–) 2–3, filaments 0·8–1 mm. long; anthers purplish, suborbicular, about 0·3 mm. diam. Capsule ovoid, about 3 mm. long, 1·5–2 mm. wide; seeds reniform, 0·8–1 mm. long, about 0·8 mm. wide; testa light brown, rather bluntly tuberculate.

HAB.: In fields, gardens and waste ground, on sandy or gravelly river-margins; a common weed; 200–5,600 ft. alt.; fl. Dec. to July.

DISTR.: Divisions 1, 2, 4, 6, 7. Widely distributed in Europe and the Mediterranean area.

1. Peyia, 1962, *Meikle* 2052 !
2. Troödos, near " Olympus Camp Hotel ", 1939, Lindberg f. s.n. ! Paliambela, 1962, *Meikle* 2252 ! Prodhromos, 1961, *D. P. Young* 7301 !
4. Paralimni, 1940, *Davis* 2061 !
6. Ayia Irini, 1941, *Davis* 2595 ! Nicosia, 1950, *Chapman* 495 !
7. Near Kythrea, 1880, *Sintenis & Rigo* 227; Kythrea, 1933, *Syngrassides* 54 !

NOTES: The Kew material of *Sintenis & Rigo* 227 is referable to ssp. *cupaniana*, but as these two collectors frequently bulked specimens from distinct localities under the same number it is not improbable that the material seen by Murbeck was correctly identified as *S. apetala*. *S. media* ssp. *apetala* Čelak. and *S. apetala* Ucria are probably the same, though the original description of the latter is altogether too imperfect to allow for satisfactory identification.

2. **S. cilicica** *Boiss. et Bal.* in Boiss., Diagn., 2, 5: 59 (1856); Unger et Kotschy, Die Insel Cypern, 346 (1865); Boiss., Fl. Orient., 1: 706 (1867); Holmboe, Veg. Cypr., 71 (1914); Post, Fl. Pal., ed. 2, 1: 201 (1932); Rechinger f. in Arkiv för Bot., ser. 2, 1: 421 (1950); Davis, Fl. Turkey, 2: 72 (1967); Osorio-Tafall et Seraphim, List Vasc. Plants Cyprus, 38 (1973).

TYPE: Turkey; "in herbidis faucis *Guzel Déré* supra *Sedichig* in regione littorali Ciliciae. Floret Aprili". *Balansa* (G, K !).

A lax, sprawling annual, (8–) 15–20 (–30) cm. high, usually much branched at the base; stems glabrous or sparsely scabridulous, strongly ribbed-angular; leaves linear, acuminate, sessile, bright green, 1·5–5 cm. long, 0·1–0·5 cm. wide, glabrous or minutely scabridulous, with a prominent midrib and a very narrow membranous-papillose margin; inflorescence repeatedly dichotomous, with spreading branches; bracts foliose; pedicels 1–3 cm. long, curved at apex especially at fruiting stage; outer sepals narrowly ovate, acuminate, 4–5 mm. long, 1–2 mm. wide, glabrous, concave, with a narrow membranous margin; inner sepals a little shorter and wider than outer, with a broader membranous margin; petals white, obovate, 4–5 mm. long, about 2 mm. wide, shortly bilobed, with rounded lobes and a narrow sinus 1–1·5 mm. long; stamens 9–10; filaments slender, glabrous, about 1–1·5 mm. long; anthers yellow, suborbicular, about 0·3 mm. diam.; ovary globose, glabrous, about 1·5 mm. diam.; styles 3, recurved, about 1 mm. long. Capsule subglobose, splitting to base long before seeds are ripe, about 5 mm. across, surrounded by the persistent sepals and petals; seeds few, large, flattened-reniform, about 2 mm. long and 1·5 mm. wide; testa bright brown, acutely tuberculate, with conspicuous radially ridged tubercles.

HAB.: On shaded rocky slopes, often under *Quercus alnifolia*; 2,500–5,000 ft. alt.; fl. March–May (–July).

DISTR.: Division 2, locally common about Tripylos–Khionistra–Papoutsa–Kionia. Also in Turkey and Lebanon.

2. Locally common; *Kotschy* 912; *Druce* 98 ! 99 ! *Kennedy* 380 ! 381 ! 1689 ! 1690 ! *Davis* 1760 ! 2738 ! 3114 ! 3481 ! *Meikle* 2752 !

12. ARENARIA *L.*

Sp. Plant., ed. 1, 423 (1753).
Gen. Plant., ed. 5, 193 (1754).
J. McNeill in Notes Roy. Bot. Gard. Edinb., 24: 241–426 (1963).

Annual, biennial or perennial herbs, or rarely subshrubs, usually less than 30 cm. high; leaves ovate, lanceolate, linear or subulate, petiolate or sessile; flowers solitary or in cymes, generally 5-merous (occasionally 4-merous), bracts herbaceous; sepals free; petals usually white, entire or retuse but not bilobed; stamens 10 (occasionally fewer) inserted on a glandular perigynous disk; ovary 1-locular, many-ovulate; styles 3 (rarely 2, 4 or 5). Capsule subglobose, oblong or pyriform, opening by twice as many teeth as there are styles; seeds usually reniform (sometimes globose) smooth or tuberculate.

About 160 species with a cosmopolitan distribution.

Petals longer than sepals; pedicels elongate 0·8–2·5 cm. long:
Bracts minute, subulate; sepals acute, with a narrow membranous margin
1. A. rhodia ssp. cypria
Bracts conspicuous, oblanceolate or linear-subulate; sepals acuminate with a distinct membranous margin - - - - - - - **2. A. pamphylica** ssp. **kyrenica**
Petals shorter than sepals; pedicels usually less than 0·8 cm. long:

Flowers in lax cymes; sepals 2–2·5 mm. long and about 0·8 mm. wide; leaves ovate **3. A. leptoclados**
Flowers in crowded cymes; sepals 5–6 mm. long and 2–3 mm. wide; leaves oblanceolate - - - - - - - - - - **4. A. saponarioides**

1. A. rhodia *Boiss.*, Diagn., 1, 1: 52 (1843).
[*A. oxypetala* (non Sibth. et Sm.) Boiss., Fl. Orient., 1: 704 (1867) pro parte.]

An erect, or sprawling, glandular-pubescent, viscid annual 4–15 (–30) cm. high; stems terete, slender, fragile; leaves rather remote, broadly obovate-orbicular, 2–8 (–25) mm. long, 1·5–8 (–18) mm. wide, apex acute or obtuse, base narrowed gradually or abruptly to a distinct flattened petiole usually less than 10 mm. long; inflorescence a lax, repeatedly dichotomous or irregularly branched cyme with erect or spreading branches; bracts minute, subulate, generally less than 2 mm. long and 1 mm. wide, often barely visible to the naked eye; pedicels slender, filiform, 1–2·5 cm. long; sepals narrowly ovate or lanceolate, 3–4 mm. long, 1–1·5 mm. wide, acute, greenish or purple-tinged, glandular, with a very narrow, indistinct membranous margin; petals white, narrowly obovate or oblanceolate, 4–6 mm. long, 1–1·5 mm. wide, apex obtuse, tapering very gradually to a rather wide base; filaments slender, 1–2·5 mm. long; anthers shortly oblong, yellow, about 0·5 mm. long; ovary broadly pyriform, 1·5–2 mm. long, about 1·3 mm. wide; styles 3, about 1 mm. long. Capsule narrowly pyriform 4–5 mm. long, 1·5–2 mm. wide; capsule-teeth short, triangular, acute; seeds reniform, 0·5–0·6 mm. long, 0·4–0·5 mm. wide; testa dark blackish-brown, bluntly or rather sharply tuberculate.

ssp. **cypria** (*Holmboe*) *McNeill* in Notes Roy. Bot. Gard. Edinb., 24: 279 (1963); Osorio-Tafall et Seraphim, List Vasc. Plants Cyprus, 37 (1973).
A. cypria Holmboe, Veg. Cypr., 68 (1914); Lindberg f., Iter Cypr., 14 (1946).
A. sp. Rechinger f. in Arkiv för Bot., ser. 2, 1: 420 (1950).
[*A. pamphylica* (non Boiss. et Heldr.) Kotschy in Unger et Kotschy, Die Insel Cypern, 346 (1865); Boiss., Fl. Orient. Suppl., 116 (1888).]

TYPE: Cyprus; rocky places on the Troödos mountains, near Prodhromos, 1905, *Holmboe* 888 (O).

A more robust plant than ssp. *rhodia*, with spreading stems and broadly ovate or suborbicular, greyish-green, ciliate leaves frequently more than 6 mm. wide; flowers with long pedicels, and with a green (not purplish) calyx; seeds usually acutely tuberculate; plants often forming intricate mats.

HAB.: On screes and in clefts of serpentine rocks; occasionally on walls; 50–6,000 ft. alt.; fl. April–July.

DISTR.: Divisions 1, 2, locally common in Division 2. Endemic.

1. Between Smyies and Dhrousha, 1941, *Davis* 3311 ! Ayia Nicola above Neokhorio, 1941, *Davis* 3325 ! Karavopetres, near Lara (Akamas), 1974, *Meikle* 4006 !
2. Locally very common; *Kotschy* 868; *Sintenis & Rigo* 763 ! *Kennedy* 373 ! 375–379 ! 1214–1216 ! *Davis* 1782 ! 2048 ! 3170 ! *Meikle* 2840 !

2. A. pamphylica *Boiss. et Heldr.* in Boiss., Diagn., 1, 8: 102 (1849); Davis, Fl. Turkey, 2: 27 (1967).

Resembling *A. rhodia* Boiss. ssp. *cypria* (Holmboe) McNeill, but a more erect plant, with smaller leaves, larger, almost foliaceous bracts, and acuminate sepals.

Plant annual, 4–15 (–25) cm. high; stems thinly glandular-pubescent; leaves rather broadly ovate, usually less than 1 cm. long and 0·5 cm. wide (occasionally up to 1·5 cm. long and 0·6 cm. wide), thinly glandular-pubescent, apex acute, base gradually or abruptly narrowed to a flattened petiole generally less than 0·3 cm. long; inflorescence a lax, repeatedly dichotomous cyme; bracts oblanceolate or linear-subulate, almost foliaceous, 2–8 mm. long, 0·5–2 mm. wide; pedicels slender, glandular-pubescent, 0·8–1·5 cm.

long; sepals lanceolate-acuminate, 2–3 mm. long, 0·5–0·8 mm. wide, thinly glandular-pubescent, with a slender acuminate apex and wide or narrow membranous margins; petals white, oblong, 3–4·5 mm. long, 0·8–1·5 mm. wide, apex rounded; stamens 1·5–2 mm. long, filaments glabrous; anthers suborbicular, whitish-yellow, 0·2–0·4 mm. diam.; ovary subglobose, glabrous, 0·6–1 mm. diam.; styles 0·5–0·8 mm. long. Capsule oblong, 2–2·5 mm. long, 1–1·5 mm. wide, with bluntly triangular, slightly recurved teeth about 0·6 mm. long; seeds reniform 0·3–0·5 mm. long, 0·2–0·3 mm. wide; testa black, rather glossy, with acute, finely verruculose tubercles.

ssp. **kyrenica** *McNeill* in Notes Roy. Bot. Gard. Edinb., 24: 281 (1963); Osorio-Tafall et Seraphim, List Vasc. Plants Cyprus, 37 (1973).

[*A. oxypetala* (non Sibth. et Sm.) Kotschy in Unger et Kotschy, Die Insel Cypern, 346 (1865); Sintenis in Oesterr. Bot. Zeitschr., 31: 393 (1881), 32: 54, 193 (1882); Boiss., Fl. Orient. Suppl., 116 (1888); Holmboe, Veg. Cypr. 69 (1914); Rechinger f. in Arkiv för Bot. ser. 2, 1: 420 (1950).]

[*A. muralis* (non (Link) Sieb.) Lindberg f., Iter Cypr., 14 (1946); Osorio-Tafall et Seraphim, List Vasc. Plants Cyprus, 37 (1973).]

TYPE: Cyprus; Kornos Peak, 1,000–2,000 ft. alt. above Larnaka-tis-Lapithou, April 3, 1941, *Davis* 2990 (E ! K !).

Sepals with distinct, rather wide, membranous margins; plant generally more robust than ssp. *pamphylica*, with larger leaves and flowers, and rather larger seeds; plants often forming lax spreading tufts.

HAB.: On rocks and walls, less often on banks or in pastures, generally on limestone, but occasionally on sandstone; sea-level to 3,000 ft. alt.; fl. March–July.

DISTR.: Divisions 6, 7, locally very common in Division 7. Also in Syria.

6. Near Orga, 1956, *Poore* 2647 !
7. Locally very common; *Kotschy* 444; *Sintenis & Rigo* 233 ! 233A ! *Kennedy* 1571 ! 1572 ! *Davis* 1753 ! 2511 ! 2771 ! 2806 ! 2849 ! 2990 ! 3011 ! *Merton* 2095 ! *G. E. Atherton* 1349A ! *Meikle* 2364 !

3. A. leptoclados (*Reichb.*) *Guss.*, Fl. Sic. Syn., 2: 824 (1845); Boiss., Fl. Orient., 1: 701 (1867); Post, Fl. Pal., ed. 2, 1: 200 (1932); Zohary, Fl. Palaest., 1: 112, t. 152 (1966); Davis, Fl. Turkey, 2: 28 (1967); Osorio-Tafall et Seraphim, List Vasc. Plants Cyprus, 37 (1973).

A. serpyllifolia L. var. *leptoclados* Reichb., Icones Fl. Germ. et Helv., 5: 32, t. 4941β (1841).

Erect or spreading annual, 2–15 (–20) cm. high; stems slender, subglabrous or shortly scabrid-pubescent or ± glandular-viscid; leaves ovate, 2–5 mm. long, 1·5–4 mm. wide, the lowermost with a flattened petiole 1–3 mm. long, the upper sessile or subsessile with a thinly puberulous or scabridulous lamina, an acute apex and an abruptly narrowed base; inflorescence a lax, repeatedly and somewhat irregularly dichotomous cyme; bracts ovate, herbaceous, foliaceous; pedicels slender, often arcuate-erect, about 5 mm. long, subglabrous, scabrid-pubescent or distinctly glandular; outer sepals lanceolate, acuminate, 3-nerved, about 2–2·5 mm. long and 0·8 mm. wide, pubescent or glandular-pubescent externally, with a very narrow membranous margin; inner sepals a little wider, narrowly ovate-lanceolate, with one or both margins distinctly membranous; petals white, oblong, about 1·5 mm. long and 0·6 mm. wide, obtuse or rounded at apex; stamens 0·5–0·8 mm. long, filaments very slender, glabrous; anthers minute, yellow, about 0·2 mm. diam.; ovary oblong-globose, 1–1·2 mm. long, 0·8 mm. wide; styles 3, about 0·6 mm. long. Capsule narrowly pyriform, about 2·5 mm. long and 1·5 mm. wide at base, with straight or slightly reflexing triangular capsule-teeth about 0·5 mm. long; seeds orbicular-reniform, about 0·4 mm. long and 0·3 mm. wide; testa dark, lustrous blackish-brown, lateral surfaces with flattish, radiating, elongate tubercles, dorsal surface acutely or bluntly tuberculate.

var. **viscidula** (*Rouy et Fouc.*) *F. N. Williams* in Journ. Linn. Soc., 33: 368 (1898).
A. serpyllifolia L. var. *viscidula* Rouy et Fouc., Fl. France, 3: 242 (1896).
[*A. leptophylla* (non Cham. et Schlecht.) Kotschy in Unger et Kotschy, Die Insel Cypern, 346 (1865).]
[*A. serpyllifolia* (non L.) Sintenis in Oesterr. Bot. Zeitschr., 31: 326 (1881), 32: 123 (1882).]
[*A. serpyllifolia* L. ssp. *leptoclados* (non (Reichb.) Rouy et Fouc.) Holmboe, Veg. Cypr., 68 (1914).]
[*A. serpyllifolia* L. var. *viscida* (non (Lois.) DC.) Lindberg f., Iter Cypr., 14 (1946).]
TYPE: Not designated (? LY).

Upper parts of stems, also pedicels and calyx ± densely clothed with viscid stalked-glands; otherwise as in ssp. *leptoclados*.

HAB.: On dry rocky slopes; on roadside banks; on sand-dunes or in sandy fields; in gravelly river-beds; sea-level to 6,100 ft. alt.; fl. Febr.–June.

DISTR.: Divisions 1–8, common in Divisions 2, 7. Widespread in the Mediterranean area and eastwards to Iran and Afghanistan.

1. Peyia, Kalifes, 1962, *Meikle* 2045 !
2. Common; *Sintenis & Rigo* 223 ! *Kennedy* 1204 ! 1205 ! *Lindberg f.* s.n. ! *Meikle* 2824 !
3. Alethriko, 1905, *Holmboe* 244; Skarinou, 1941, *Davis* 2688 !
4. Larnaca, 1862, *Kotschy*.
5. Lefkoniko, 1880, *Sintenis & Rigo*.
6. Ayia Irini, 1941, *Davis* 2533 !
7. Common; *Sintenis & Rigo*; *Davis* 2463 ! 2487 ! 2840 ! 2887 ! *Casey* 1226 !
8. Near Apostolos Andreas, 1941, *Davis* 2324 !

NOTES: The glandular variety replaces typical, scabridulous-pubescent *A. leptoclados* in Cyprus, as in many other Mediterranean countries; the density of the glands varies considerably from individual to individual.

4. **A. saponarioides** *Boiss. et Bal.* in Boiss., Diagn., 2, 6: 35 (1859); Davis, Fl. Turkey, 2: 30 (1967).
A. nana (non Willd.) Boiss., Diagn., 1, 8: 103 (1849), Fl. Orient. Suppl., 116 (1888).
A. macrosepala Boiss. var. *minor* Boiss., Fl. Orient., 1: 703 (1867).
A. boissieri Pax in Engl., Bot. Jahrb., 18: 30 (1893).
A. macrosepala Boiss. ssp. *saponarioides* (Boiss. et Bal.) Holmboe, Veg. Cypr., 69 (1914).
A. saponarioides Boiss. et Bal. ssp. *boissieri* (Pax) McNeill in Notes Roy. Bot. Gard. Edinb., 24: 289 (1963); Osorio-Tafall et Seraphim, List Vasc. Plants Cyprus, 37 (1973).
[*A. macrosepala* (non Boiss.) Rechinger f. in Arkiv för Bot., ser. 2, 1: 420 (1950).]
TYPE: Turkey; "in Phrygia inter Ouchak et Yachamichlarkeui", June 6, 1857, *Balansa* 1297 (G, K !).

A dwarf viscid-glandular annual, usually less than 4 cm. high; stems unbranched or with several ascending branches; basal leaves narrowly oblanceolate, 0·3–1 (–1·5) cm. long, 0·1–0·3 cm. wide, prominently 3 (–5)-nerved, apex obtuse or subacute, base tapering to a short, flattened, often indistinct petiole; upper leaves similar, frequently a little larger than the basal and usually sessile or subsessile; inflorescence a crowded, repeatedly dichotomous cyme; bracts herbaceous, linear-subulate or narrowly oblong, 3–5 mm. long, 0·5–1 mm. wide, foliaceous; pedicels very short, generally less than 3 mm. long, sometimes almost wanting; sepals ovate or oblong, 5–6 mm. long, 2–3 mm. wide, 3-nerved, acute or acuminate; petals white, broadly oblong-elliptic, about 3·5 mm. long, 1·8 mm. wide, apex obtuse or subacute; stamens about 2 mm. long, filaments glabrous; anthers yellow, almost orbicular, about 0·5 mm. diam.; ovary subglobose, around 2 mm. diam.; styles 0·8 mm. long. Capsule shortly oblong, about 4 mm. long and 3 mm. wide, capsule-teeth triangular, about 1 mm. long, not reflexing; seeds broadly reniform, about 1 mm. long and 0·8 mm. wide; testa rich brown, with short, rather sinuous, thinly verruculose, low ridges.

HAB.: On screes and rocky mountain slopes; restricted to serpentine formations; 4,900–6,200 ft. alt.; fl. April–July.

DISTR.: Division 2, locally common Khionistra–Troödos–Amiandos. Also in Turkey and Crete.

2. Locally common; *Sintenis & Rigo* 762 ! *Kennedy* 389 ! 1192–1196 ! *Davis* 1858 ! 3512 ! *Merton* 2201 ! *Meikle* 2879 !

NOTES: The distinctions between ssp. *saponarioides* and ssp. *boissieri* (Pax) McNeill are not wholly satisfactory; the Cyprus material agrees rather better with ssp. *saponarioides* than with the one (very immature) Kew sheet of ssp. *boissieri* (Crete; " in cacumine Authenti montium Lassiti ", 1846, *Heldreich* s.n.).

A. CRETICA *Spreng.*, Syst. Veg., 2: 396 (1825); Boiss., Fl. Orient., 1: 699 (1867); Holmboe, Veg. Cypr., 69 (1914); Osorio-Tafall et Seraphim, List Vasc. Plants Cyprus, 37 (1973).
[*A. ciliata* (non L.) Sibth. et Sm., Fl. Graec. Prodr., 1: 303 (1809), Fl. Graec., 5: 27, t. 438 (1825); Unger et Kotschy, Die Insel Cypern, 346 (1865).]

Recorded, " in rupibus insulae Cypri " by Sibthorp and Smith, but not seen since, and almost certainly an error. *A. cretica* (a native of Greece, Crete and Albania) is a creeping, mat-forming perennial with small obovate or oblanceolate leaves and relatively large white flowers (petals up to 1 cm. long) either solitary or in small, few-flowered cymes. It is an attractive, conspicuous plant, and not one that is likely to be overlooked.

13. MINUARTIA *L.*

Sp. Plant., ed. 1, 89 (1753).
Gen. Plant., ed. 5, 39 (1754).
Alsine (non L.) Gaertn., Fruct., 2: 223 (1791) et auct. mult.
J. McNeill in Notes Roy. Bot. Gard. Edinb., 24: 241–426 (1963).

Very similar to *Arenaria* in general appearance; leaves commonly linear or subulate; flowers generally 5-merous, solitary or in cymes; stamens 10 (occasionally fewer), usually inserted on a glandular hypogynous disk; ovary 1-locular, many-ovulate; styles 3. Capsule ovoid or oblong generally splitting almost or quite to the base into 3 valves; seeds reniform, smooth or tuberculate.

About 100 species with a wide distribution in the northern hemisphere, but most abundantly represented in Europe and western Asia.

Perennial with a tough woody rootstock and prostrate stems; petals pink; seashore plant - - - - - - - - - - - - - - **1. M. geniculata**
Annuals; petals white, occasionally stained or streaked reddish externally:
  Leaves setaceous in dense fascicles; petals conspicuous, often stained red externally **2. M. picta**
  Leaves not setaceous nor in dense fascicles; petals always white, often minute:
    Sepals indurated, generally erect and connivent at anthesis; pedicels very short, much shorter than calyx:
      Sepals with a distinct white membranous margin; base of calyx rounded **3. M. sintenisii**
      Sepals with an inconspicuous margin; base of calyx truncate:
        Bracts generally shorter than flower-clusters; flowers shortly but distinctly pedicellate - - - - - - - - - - - - **4. M. globulosa**
        Bracts generally longer than flower-clusters; flowers indistinctly pedicellate or subsessile - - - - - - - - - - **5. M. intermedia**
    Sepals not indurated, generally spreading at anthesis; pedicels longer than calyx (except in 9. *M. mediterranea*):
      Petals as long as sepals or a little longer; leaves often rather fleshy; normally a seashore plant - - - - - - - - - - **6. M. thymifolia**
      Petals shorter than sepals; leaves not fleshy; plants not normally restricted to seashores:
        Sepals broadly ovate or oblong-ovate, obscurely nerved; pedicels conspicuously elongate and slender - - - - - - **7. M. subtilis ssp. filicaulis**

Sepals linear-lanceolate, distinctly 3-nerved; pedicels not conspicuously elongate and slender:
Inflorescence lax, cymose; pedicels generally longer than calyx **8. M. hybrida**
Inflorescence congested, fasciculate; pedicels of the upper flowers usually shorter than calyx - - - - - - - **9. M. mediterranea**

SUBGENUS 1. **Rhodalsine** (*J. Gay*) *McNeill* Cotyledons accumbent; stamens inserted at two levels, the inner whorl adnate to the petals; the outer arising below them on a very short calyx-tube; petals often pink; leaves linear to ovate.

1. M. geniculata (*Poir.*) *Thell.* in Mém. Soc. Nat. Cherbourg, 4, 38: 231 (1912); A. K. Jackson in Kew Bull., 1933: 408 (1933); Osorio-Tafall et Seraphim, List Vasc. Plants Cyprus, 38 (1973).
Arenaria geniculata Poir., Voy. en Barbarie, 2: 166 (1789).
A. procumbens Vahl, Symb. Bot., 2: 50, t. 33 (1791).
Alsine procumbens (Vahl) Fenzl, Geogr. Verbreit. Alsin., 87 (1833); Boiss., Fl. Orient., 1: 671 (1867).
Rhodalsine geniculata (Poir.) F. N. Williams in Bull. Herb. Boiss., 6: 3 (1898).
TYPE: Algeria; " environs de la Calle " *Poiret* (P).

Prostrate or decumbent perennial with a tough, woody rootstock; stems 15–30 (–50) cm. long, slightly swollen at the nodes; leaves linear, oblanceolate, obovate or almost orbicular, obtuse or acute, 1-nerved, 0·5–2 (–3·5) cm. long, 0·1–0·7 cm. wide, subglabrous or ± densely glandular-pubescent, sessile or with an indistinct flattened petiole; inflorescence a lax, rather irregularly branched cyme; bracts foliaceous, often densely glandular-pubescent; pedicels slender, erect or spreading, 0·5–1·5 (–2) cm. long, often glandular-pubescent; sepals ovate, sometimes narrowly so, about 3 mm. long and 1·5–2 mm. wide, obscurely nerved, glandular-pubescent externally, apex obtuse or subacute, margins of outer sepals narrowly membranous, those of inner sepals rather widely so; petals pink (in Cyprus specimens) or white, broadly ovate, obtuse, about 2·5 mm. long and 2 mm. wide, 7–9-nerved; filaments glabrous, those of sepal-opposed stamens about 1·5 mm. long, those of petal-opposed stamens about 1 mm. long; anthers yellow, oblong, about 0·7 mm. long and 0·5 mm. wide; ovary shortly stipitate, subglobose, about 1 mm. diam.; styles about 1 mm. long. Capsule oblong, about 3·5 mm. long and 2·5 mm. wide, splitting to the base into 3 boat-shaped valves; seeds broadly reniform, about 0·7–0·8 mm. long and 0·6–0·7 mm. wide; testa dark blackish-brown, lustrous, with low, sinuous, radiating ridges.

HAB.: On rocky and sandy seashores; near sea-level; fl. March–July.

DISTR.: Division 7, rare, Kyrenia area only. Widespread in the Mediterranean area.

7. Kyrenia; *Syngrassides* 12 ! *Casey* 1161 ! *G. E. Atherton* 13 ! 110 ! 1341 !

NOTES: Typical *M. geniculata* is said to have white flowers, and some authors have distinguished the pink-flowered plant (which alone occurs in Cyprus) as *M. procumbens* (Vahl) Graebn. A. K. Jackson concludes that the two supposed species cannot be distinguished specifically (Kew Bull., 1933: 408) and his conclusion has been accepted here.

SUBGENUS 2. **Spergella** (*Fenzl*) *McNeill* Annuals; seeds pyriform, cotyledons accumbent; flowers often tinged or stained red; petals exceeding sepals; leaves in whorl-like fascicles, filiform or setaceous.

2. M. picta (*Sibth. et Sm.*) *Bornm.* in Beih. Bot. Centralbl., 28, 2: 148 (1911); Zohary, Fl. Palaest., 1: 116, t. 159 (1966); Davis, Fl. Turkey, 2: 43 (1967); Osorio-Tafall et Seraphim, List Vasc. Plants Cyprus, 38 (1973).
Arenaria picta Sibth. et Sm., Fl. Graec. Prodr., 1: 304 (1809), Fl. Graec., 5: 28, t. 440 (1825).
Arenaria filiformis Labill., Icon. Plant. Syr. Rar., 4: 8, t. 3, fig. 2 (1812).

Alsine picta (Sibth. et Sm.) Fenzl, Geogr. Verbreit. Alsin., 57 (1833); Unger et Kotschy, Die Insel Cypern, 345 (1865); Boiss., Fl. Orient., 1: 687 (1867); Holmboe, Veg. Cypr., 70 (1914); Post, Fl. Pal., ed. 2, 1: 197 (1932).

Arenaria cerignensis Sibth. in Walpole, Mem. Europ. Asiat. Turkey, ed. 2, 1: 18 (1818) nom. nud.

TYPE: Cyprus; " in insulae Cypri campestribus ", 1787, *Sibthorp* (OXF).

An erect, viscid-glandular annual, 3–12 cm. high; stem short, seldom exceeding 2 cm., normally unbranched, often tinged reddish, densely covered with erect, setaceous, glaucous or purplish leaves 0·7–2·5 cm. long and less than 0·5 mm. wide, sparingly glandular or eglandular with sharp cuspidate tips; inflorescences arising laterally from the axils of the upper leaves, forming lax cymes at the apices of relatively long (2–6 cm.) naked, slender, ± glandular peduncles; bracts minute, ovate-triangular, 0·5–0·6 mm. long and about as wide at the base, with broad membranous margins and a stout greenish or purplish midrib; pedicels very slender, 0·5–2 cm. long, usually viscid-glandular; sepals oblong, rounded at apex, 3–4·5 mm. long, 1·5–2 mm. wide, with a broad greenish or purplish nerve and wide membranous margins; petals white internally, often stained or striped red externally, oblong or oblanceolate, 6–8 mm. long, 2·5–3 mm. wide, 5–9-nerved with a shallow emarginate or truncate apex; stamens 2-seriate, sepal-opposed filaments about 3 mm. long with a conspicuous glandular base, petal-opposed filaments about 2 mm. long without a conspicuous basal gland; anthers oblong, about 0·8 mm. long and 0·6 mm. wide, yellow or livid; ovary oblong, about 1·4 mm. long and 1·2 mm. wide; styles erect, about 1·2 mm. long. Capsule oblong, about 4 mm. long and 3 mm. wide, splitting to the base into 3 shallowly boat-shaped valves; seeds irregularly pyriform with a conspicuous radicular prominence, about 0·8 mm. long and 0·6 mm. wide; testa dark brown, with distinct, blunt, radiating ridges.

HAB.: Dry, rocky hillsides on limestone or igneous formations; on sandy ground or amongst shingle in river-beds; a local weed of cultivated ground; sea-level to 4,000 ft. alt.; fl. Jan.–May.

DISTR.: Divisions 1–7, common in Divisions 2, 7. Also Turkey, Syria, Palestine, Sinai, Egypt, Iraq, Iran, Afghanistan, ? Arabia.

1. Smyies, 1962, *Meikle* 2198 !
2. Common; *Rev. A. Huddle* 23 ! *Kennedy* 382A ! *F. M. Probyn* 10 ! *Meikle* 2641 ! 2813 !
3. Alethriko, 1905, *Holmboe* 224; Malounda, 1950, *Chapman* 297 !
4. Larnaca, 1862, *Kotschy*; Paralimni, 1928 or 1930, *Druce* s.n. ! Famagusta, 1936, *Syngrassides* 878 ! Skarinou, 1941, *Davis* 2680 !
5. Vatili, 1905, *Holmboe* 320.
6. Kokkini-Trimithia, 1933, *Syngrassides* 317 ! Near Nicosia, 1941, *Davis* 2168 ! Syrianokhori-Ayia Irini, 1941, *Davis* 2604 !
7. Common; *Sibthorp* s.n.; *Kotschy* 408. *Sintenis & Rigo* 232 ! *Mavromoustakis* 32 ! *G. E. Atherton* 1130 !

NOTES: Very variable in petal-colour; in some populations all the flowers are pure white, in others the petals are deep red externally, or flushed with red, but in most localities mixed populations are the rule. *M. picta* shows an equal diversity in its choice of habitats; like some other *Minuartia* species it seems to thrive particularly well on mineral-rich " spoil-banks " in mining areas.

SUBGENUS 3. **Minuartia** Annuals or perennials; seeds reniform or orbicular, cotyledons incumbent; petals usually white, shorter than or exceeding sepals; sepals 1–3-nerved, acute or acuminate, often indurated; stamens in 1–2 series; leaves generally subulate or linear, not normally in fascicles or apparent whorls.

3. M. sintenisii (*Lindberg f.*) *Rechinger f.* in Arkiv för Bot., ser. 2, 1: 421 (1951); Osorio-Tafall et Seraphim, List Vasc. Plants Cyprus, 38 (1973).

Alsine sintenisii Lindberg f., Iter Cypr., 14 (1946).

Minuartia meyeri (Boiss.) Bornm. var. *cypricola* Mattf. in Beih. Feddes Rep., 15: 60 (1922).

[*Alsine meyeri* (non Boiss.) Boiss., Fl. Orient. Suppl., 114 (1888).]

[*A. brevis* (non Boiss.) Holmboe, Veg. Cypr., 69 (1914).]

TYPE: Cyprus; Troödos, " in campo sterili in pineto juxta ' Olympus Camp Hotel ' ", June 15, 1939, *Lindberg f.* s.n. (H, K !).

Erect, decumbent or prostrate annual 1–6 cm. high, usually branched from the base; branches shortly glandular-pubescent, greenish or purplish; leaves linear, 3-nerved, 0·2–1 cm. long, 0·1–0·2 cm. wide, acute, glabrescent or sparsely glandular-pubescent; inflorescences terminal and lateral, each branch bearing 2–6 fascicles or congested cymes of greenish or purplish flowers; bracts herbaceous, foliaceous, up to about 0·5 cm. long; pedicels 1–3 mm. long, glandular; sepals lanceolate, acuminate, 4–5 mm. long, 0·8 mm. wide, conspicuously 3-nerved, shortly glandular-pubescent with a very narrow membranous margin; petals white, minute, bluntly ovate, about 1·2 mm. long and 0·8 mm. wide; stamens subequal, about 1·3 mm. long, filaments slightly expanded at base; anthers suborbicular, about 0·2 mm. diam., yellow; ovary ovoid, about 1·2 mm. long and 1 mm. wide; styles recurved, about 0·4 mm. long. Capsule oblong, about 2·5 mm. long and 1·4 mm. wide, divided to the base into 3 oblong, shallowly concave valves; seeds relatively large, orbicular-reniform, about 1 mm. diam.; testa pale brown with darker flattened (" armadillo ") tubercles.

HAB.: Dry mountain slopes on igneous formations; 4,000–6,400 ft. alt.; fl. May–July.

DISTR.: Division 2, locally common Prodhromos–Khionistra–Troödos area, also at Livadhia and Papoutsa. Endemic.

2. Locally common; *Sintenis & Rigo* 760 ! *Holmboe* 879, 983; *Kennedy* 387 ! 388 ! 390 ! 1201A ! 1202 ! 1203 ! 1206 ! *Meikle* 2938 !

4. M. globulosa (*Labill.*) *Schinz et Thell.* in Bull. Herb. Boiss., ser. 2, 7: 403 (1907); Rechinger f. in Arkiv för Bot., ser. 2, 1: 421 (1950); Zohary, Fl. Palaest., 1: 115, t. 157 (1966); Davis, Fl. Turkey, 2: 54 (1967); Osorio-Tafall et Seraphim, List Vasc. Plants Cyprus, 38 (1973).

Arenaria globulosa Labill., Icones Plant. Syr. Rar., 4: 6, t. 3, fig. 1 (1812).

Alsine smithii Fenzl, Geogr. Verbreit. Alsin., 57 (1833); Boiss., Fl. Orient., 1: 683 (1867), Fl. Orient. Suppl., 114 (1888).

Alsine globulosa (Labill.) Hal. in Denkschr. Akad. Wiss. Wien, 41: 491 (1894); Holmboe, Veg. Cypr., 69 (1914); Post, Fl. Pal., ed. 2, 1: 196 (1932).

TYPE: Lebanon; " in Libano juxta Tripolim " *Labillardière* (P, K !).

Erect, decumbent or sprawling annual, 2–15 cm. high; stems usually much branched at base, greenish or purplish, ± densely glandular-pubescent; leaves linear, 10–25 mm. long, 0·8–2 mm. wide, acuminate, prominently 5 (–7)-nerved, thinly glandular-pubescent; inflorescences terminal and lateral, each branch bearing 3–7 or more small, few-flowered fascicles of greenish or purplish flowers; bracts herbaceous, foliaceous, 0·3–1·5 cm. long; pedicels short, seldom exceeding 3 mm. in length, glandular-pubescent; flowers at first erect, later often patent or nodding; sepals lanceolate, acute, 4–5 mm. long, 1–1·3 mm. wide, 3-nerved, the outer glandular-pubescent with an indistinct margin, the inner thinly glandular or glabrescent with a narrow membranous margin; petals white, narrowly ovate-elliptical, about 2 mm. long and 0·8 mm. wide; stamens subequal, about 1·5 mm. long, filaments slightly expanded at base; anthers oblong, about 0·4 mm. long and 0·3 mm. wide, yellow; ovary ovoid, about 2 mm. long and 0·8 mm. wide; styles recurved about 0·8 mm. long. Capsule oblong, 3–4 mm. long, 1–1·5 mm. wide; seeds orbicular-reniform, about 0·8 mm. diam.; testa bright brown with numerous, prominent, flat-topped tubercles.

HAB.: Dry stony mountain slopes; banks and edges of cultivated ground; 1,800–4,000 ft. alt.; fl. May–July.

DISTR.: Divisions 1, 2, common in Division 2. Also Yugoslavia, Greece, Turkey and Palestine.

1. Between Dhrousha and Kato Arodhes, 1941, *Davis* 3281 !
2. Common; *Sintenis & Rigo* 761 ! *Kennedy* 1197–1200 ! *Davis* 1880 ! 3379 !

5. M. intermedia (*Boiss.*) *Hand.-Mazz.* in Ann. Nat. Mus. Wien, 26: 148 (1912); Davis, Fl. Turkey, 2: 55 (1967); Osorio-Tafall et Seraphim, List Vasc. Plants Cyprus, 38 (1973).

Alsine intermedia Boiss., Fl. Orient., 1: 685 (1867); Sintenis in Oesterr. Bot. Zeitschr., 31: 392 (1881), 32: 54 (1882); Holmboe, Veg. Cypr., 70 (1914); Post, Fl. Pal., ed. 2, 1: 196 (1932).

Minuartia decipiens (Fenzl) Bornm. ssp. *damascena* McNeill var. *damascena*, McNeill in Notes Roy. Bot. Gard. Edinb., 24: 362 (1963); Osorio-Tafall et Seraphim, List Vasc. Plants Cyprus, 38 (1973).

[*Alsine montana* (non (L.) Fenzl) Lindberg f., Iter Cypr., 14 (1946).]

TYPE: Turkey; " défilé des Portes Ciliciennes ", June 23, 1855, *Balansa* 598 (G, K !).

An erect or sprawling annual, 3–12 cm. high; stems unbranched or branched from near the base, crispate-pubescent, often tinged purplish; leaves linear, acuminate, 1–1·5 cm. long, 0·1–0·15 cm. wide, prominently 3-nerved, thinly glandular-pubescent or subglabrous; inflorescences lateral and terminal, often arranged spicately, or occasionally in terminal corymbose heads, each inflorescence consisting of 3–7 or more flowers arranged in dense glomerules; bracts herbaceous, foliaceous, usually exceeding the inflorescence-clusters, 3–7-nerved, subglabrous or thinly glandular-pubescent; pedicels very short or almost wanting; flowers mostly erect; calyx abruptly narrowed or truncate at base; sepals linear-lanceolate, acuminate, slightly unequal, 4–5 mm. long, 0·8 mm. wide, prominently 3-nerved, thinly glandular-pubescent or subglabrous, with narrow membranous margins; petals minute, ovate, acute, white, about 0·6 mm. long and 0·3–0·4 mm. wide; stamens 0·8–1 mm. long, subequal, alternate filaments expanded at base, sometimes with distinct basal lobes; anthers suborbicular, yellow, about 0·3 mm. diam.; ovary oblong-ovoid, 1·3 mm. long, 0·8 mm. wide; styles 0·5 mm. long. Capsule oblong, 3-sided, about 3 mm. long and 1·3 mm. wide, splitting to base into 3 flattish oblong valves; seeds G-shaped, 0·6–0·8 mm. long, 0·5–0·6 mm. wide; testa light brown, the dorsal surface covered with distinct or obscure, truncate, darker brown tubercles.

HAB.: Rocky slopes on limestone formations; 1,000–3,000 ft. alt.; fl. April–June.

DISTR.: Division 7. Also Turkey, Caucasus area, Syria, Iraq and Iran.

7. Pentadaktylos Peak, 1880, *Sintenis & Rigo*; Ayios Khrysostomos Mon., 1880, *Sintenis & Rigo* (both combined as " in montibus circa Kythream, 1880 ", *Sintenis & Rigo* 236 !) St. Hilarion, 1939, *Lindberg f.* s.n. Above Kyrenia Pass, 1953, *Casey* 1307 ! Buffavento, 1974, *Meikle* 4097 !

NOTES: The calyx in the few Cyprus specimens examined is more densely glandular than is usual in typical *M. intermedia*, and in this respect the Cyprus plants resemble the closely allied *M. decipiens* (Fenzl) Bornm. In all other respects they agree better with *M. intermedia*, and I am inclined to follow Mattfeld (Feddes Repert. Beih., 15: 67; 1922) in allowing them to remain under this name.

6. M. thymifolia (*Sibth. et Sm.*) *Bornm.* in Beih. Bot. Centralbl., 31, 2: 193 (1914); Davis, Fl. Turkey, 2: 65 (1967); Osorio-Tafall et Seraphim, List Vasc. Plants Cyprus, 38 (1973).

Arenaria thymifolia Sibth. et Sm., Fl. Graec. Prodr., 1: 305 (1809), Fl. Graec., 5: 29, t. 441 (1825).

Alsine thymifolia (Sibth. et Sm.) Fenzl, Geogr. Verbreit. Alsin., 57 (1833); Boiss., Fl. Orient., 1: 687 (1867); Holmboe, Veg. Cypr., 70 (1914); Post, Fl. Pal., ed. 2, 1: 197 (1932).

TYPE: Crete; " in insulâ Cretâ ", *Sibthorp* (OXF). See Notes.

An erect or spreading annual, 2–10 cm. high, stems usually much branched at or near the base, glabrous below, minutely viscid-glandular above; leaves narrowly oblong, subspathulate or linear, 0·5–1 (–1·5) cm. long, less than 0·2 cm. wide, obtuse, glabrous or the upper thinly glandular, sessile and connate at the base, often rather fleshy; inflorescence a lax repeatedly dichotomous cyme; bracts oblong or narrowly ovate, subacute, 2–5 mm. long, 0·5–1 mm. wide, obscurely 3-nerved, glandular, with a narrow membranous margin; pedicels slender, 0·5–1·5 cm. long, thinly viscid-glandular, erect or spreading; sepals narrowly ovate, acute, 2·5–3 mm. long, 0·8–1 mm. wide, distinctly 3-nerved, minutely glandular, with a distinct membranous margin; petals white, broadly oblong or suborbicular, about as long as sepals or a little longer, 1·5–2 mm. wide, apex rounded; stamens 1·5–2 mm. long, filaments not much expanded at base; anthers oblong, reddish, about 0·5 mm. long and 0·3 mm. wide; ovary pyriform about 1·3 mm. long and 1 mm. wide; styles 0·8–1 mm. long. Capsule oblong, 2–3 mm. long, 1–1·5 mm. wide, split to the base into 3 oblong valves with acute, recurved apices; seeds orbicular-reniform, about 0·4 mm. diam.; dorsal surface flattened or slightly channelled; testa dull brown with numerous low, short, radiating ridges.

HAB.: On sandy and rocky ground near the sea; sea-level to 150 ft. alt.; fl. Jan.–April.

DISTR.: Divisions 1, 4, 6–8, common in Division 8. Also in Rhodes and Lebanon, ? Turkey, ? Crete.

1. Cape Arnauti, 1962, *Meikle* 2298 !
4. Near Ayia Napa, 1905, *Holmboe* 16.
6. Between Orga and Vasilia, 1941, *Davis* 2116 ! Liveras, 1956, *Merton* 2510 ! Orga–Liveras road, 1956, *Merton* 2855 !
7. Hartcha [? Kharcha], 1950, *Chapman* 262 ! Kyrenia, 1951, *Casey* 1174 ! and, 1952, *Casey* 1217 ! and, 1956, *G. E. Atherton* 882 !
8. Common; *Sintenis & Rigo* 228 ! *Druce* 88 ! 89 ! *Davis* 2262 ! 2303 ! *Merton* 2472 !

NOTES: It is very probable that Sibthorp's type material came from Cyprus and not from Crete as recorded; no other collector has succeeded in re-finding it in Crete, and experience proves that the *Flora Graeca* is not at all reliable as regards localizations. The Turkish record is also suspect, though there is good reason for supposing that *M. thymifolia* may occur on the Cilician coast, especially as it is so frequent along the north coast of Cyprus.

7. **M. subtilis** (*Fenzl*) *Hand.-Mazz.* in Ann. Nat. Mus. Wien, 26: 148 (1912); Davis, Fl. Turkey, 2: 66 (1967).

Alsine subtilis Fenzl in Kotschy Plant. Pers. Bot. Ed. R. F. Hohenacker No. 272 (1846) nomen.

Alsine tenuifolia (L.) Crantz var. *γ subtilis* Fenzl, Diagn. Plant. Orient., 9 (1860); Boiss., Fl. Orient., 1: 687 (1867).

Alsine lydia Boiss. var. *kotschyana* Boiss., Diagn., 2, 1: 86 (1853).

Erect or spreading, slender annual, 1·5–13 cm. high; stems repeatedly branched, often tinged purplish, glabrous or sparingly glandular; leaves linear 1·5–12 mm. long, 0·3–1 mm. wide, obtuse or subacute, sessile and connate at the base, glabrous or sparingly glandular; inflorescence a lax repeatedly dichotomous cyme; bracts herbaceous, foliaceous, glandular or eglandular, usually less than 5 mm. long, with a very narrow membranous margin; pedicels filiform, 0·5–2 cm. long, glandular or eglandular; sepals narrowly or rather broadly ovate or oblong-ovate, acute or obtuse, 2–2·5 mm. long and 0·8 mm. wide, obscurely 3-nerved, or without distinct nerves, and with a rather distinct membranous margin, dorsal surface glabrous or glandular; petals white, oblong, 0·8–1·5 mm. long, 0·5–0·8 mm. wide, apex rounded; stamens unequal 0·6–1·2 mm. long, filaments not much expanded at base; anthers yellow, suborbicular, about 0·3 mm. diam.; ovary broadly ovoid, about 0·8–1 mm. long and 0·6–1 mm. wide; styles about 0·5 mm. long. Capsule oblong, 3 mm. long, about 1·8 mm. wide; seeds reniform or almost orbicular, about 0·5 mm. diam.; testa dull brown, minutely verruculose-tuberculate with blunt or acute tubercles.

ssp. **filicaulis** (*Lindberg f.*) *McNeill* in Notes Roy. Bot. Gard. Edinb., 24: 391 (1963); Osorio-Tafall et Seraphim, List Vasc. Plants Cyprus, 38 (1973).

Alsine filicaulis Lindberg f., Iter Cypr., 14 (1946).

Minuartia filicaulis (Lindberg f.) Rechinger f. in Arkiv för Bot., ser. 2, 1: 420 (1950).

[*Alsine tenuifolia* (L.) Crantz ssp. *kotschyana* (non *A. lydia* Boiss. var. *kotschyana* Boiss.) Holmboe, Veg. Cypr., 70 (1914).]

TYPE: Cyprus; Troödos, " in terra nuda humidiuscula in declivi in cacumine montis Chionistra ", June 17, 1939, *Lindberg f.* s.n. (H).

Sepals oblong-ovate, obtuse or subacute, about 2 mm. long and 0·8 mm. wide; petals 1–1·5 mm. long, 0·8 mm. wide; stamens almost as long as petals. Seeds very obscurely verruculose, scarcely tuberculate. Plants usually neater and more compact than in typical ssp. *subtilis*, and often with very long slender spreading pedicels, but variable in both of these characters.

HAB.: Moist ground by mountain streams and rivulets, and in dried-up torrent beds; 4,000–6,400 ft. alt.; fl. April–July.

DISTR.: Division 2, locally common Khionistra–Troödos–Platania. Endemic.

2. Locally common; *Sintenis & Rigo* 765 ! *Holmboe* 976; *Kennedy* 384–386 ! 1211–1213 ! *Davis* 1863 ! *Merton* 3192 !

NOTES: Lindberg f. distinguishes three forms of his *Alsine filicaulis*: f. *typica* (nowadays, correctly f. *filicaulis*), f. *subglabra* and f. *glaberrima*—these simply represent different degrees of glandulosity, and as there is considerable variation in this respect, even within a single population of the plant, as well as a complete range of specimens from very glabrous to very glandular, it is neither necessary nor desirable that formal taxonomic recognition should be given to these forms, nor that they should be republished as new combinations under *M. subtilis* (Fenzl) Hand.-Mazz.

8. **M. hybrida** (*Vill.*) *Schischk.* in Komarov, Fl. U.R.S.S., 6: 488 (1936); Zohary, Fl. Palaest., 1: 114, t. 155 (1966); Davis, Fl. Turkey, 2: 66 (1967); Osorio-Tafall et Seraphim, List Vasc. Plants Cyprus, 38 (1973).

Arenaria hybrida Vill., Prosp. Plant. Dauph., 48 (1779).

Arenaria tenuifolia L., Sp. Plant., ed. 1, 424 (1753); Sibth. et Sm., Fl. Graec. Prodr., 305 (1809).

Alsine tenuifolia (L.) Crantz, Instit., 2: 407 (1766); Unger et Kotschy, Die Insel Cypern, 345 (1865); Boiss., Fl. Orient., 1: 685 (1867); Holmboe, Veg. Cypr., 70 (1914); Post, Fl. Pal., ed. 2, 1: 196 (1932).

A. tenuifolia (L.) Crantz var. *viscosa* Boiss., Fl. Orient., 1: 686 (1867); Lindberg f., Iter Cypr., 14 (1946)—non *A. viscosa* Schreb.

Minuartia tenuifolia (L.) Hiern in Journ. Bot., 37: 321 (1899)—non *M. tenuifolia* Nees ex Mart.

Alsine tenuifolia (L.) Crantz var. *hispida* Boiss. ex Kotschy in Unger et Kotschy, Die Insel Cypern, 345 (1865)—nomen.

TYPE: France; Dauphiné, " sortant par la porte de la Graille, suivant l'Ilfere jusqu'à sa jonction avec le Drac, remontant ce torrent pour revenir par la Rondeau. Près le chemin de la Butte " *Villars* (? P, GR).

An erect or spreading annual, 2–15 (–25) cm. high; stems glabrous or almost so, often purplish, usually much branched from the base; leaves linear-subulate, 0·5–1 (–1·5) cm. long, generally less than 0·1 cm. wide at base, sharply acuminate, sessile and distinctly connate at base, glabrous or at most very sparsely glandular; inflorescence repeatedly branched, forming a lax, spreading cyme; bracts herbaceous, foliaceous, subulate, glabrous or very thinly glandular, with a distinct membranous margin; pedicels slender, filiform, erect or erecto-patent, 0·5–2 (–2·5) cm. long, generally much longer than the calyx; sepals linear-lanceolate, 4–5 mm. long, 0·6–0·8 mm. wide, distinctly 3-nerved, glandular or eglandular externally, with a slender acumen and narrowly membranous margins; petals white, oblong-ovate, 1·5–2 mm. long, 1 mm. wide; stamens almost as long as petals, alternate filaments shortly but distinctly expanded at the base; anthers yellow or brownish, suborbicular, about 0·5 mm. diam.; ovary ovoid, about 0·8 mm. long and 0·6 mm. wide; styles 0·5 mm. long. Capsule oblong about 5 mm. long and 1·4 mm. wide at base, distinctly overtopping the sepals at maturity; seeds orbicular-reniform, about 0·4 mm. diam.; testa light brown, covered with short, blunt tubercles.

HAB.: In cultivated and fallow fields; in vineyards; on rocky slopes; in dried-up river beds; sea-level to 4,600 ft. alt.; fl. Febr.–June.

DISTR.: Divisions 1–7, common in Divisions 2, 7. Widespread in the Mediterranean area and eastwards to Afghanistan.

1. Peyia Forest, Kalifes, 1962, *Meikle* 2046 !
2. Common; *Kotschy* 898; *Kennedy* 1207–1210 ! 1830 ! *Meikle* 2722 ! 2798 !
3. Alethriko, 1905, *Holmboe* 243; Skarinou, 1941, *Davis* 2687 !
4. Between Cape Pyla and Ormidhia, 1880, *Sintenis & Rigo*.
5. Lefkoniko, 1880, *Sintenis & Rigo*.
6. Astromeritis to Katokopia, 1936, *Syngrassides* 1314 !
7. Common; *Sintenis & Rigo* 231 ! *Davis* 2150 ! 2464 ! *Casey* 259 ! 340 ! 401 ! 1105 !

9. **M. mediterranea** (*Ledeb.*) *K. Malý* in Glasnik Muz. Bosn. Herceg., 20: 363 (1908); McNeill in Notes Roy. Bot. Gard. Edinb., 24: 396 (1963); Zohary, Fl. Palaest., 1: 114, t. 154 (1966); Davis, Fl. Turkey, 2: 67 (1967); Osorio-Tafall et Seraphim, List Vasc. Plants Cyprus, 38 (1973).

Arenaria mediterranea Ledeb. in Link, Enum. Hort. Berol., 1: 431 (1821).

Alsine mediterranea (Ledeb.) J. Malý, Enum. Plant. Austr., 296 (1848); Post, Fl. Pal., ed. 2, 1: 197 (1932).

Alsine tenuifolia (L.) Crantz var. *mucronata* Boiss., Fl. Orient., 1: 686 (1867) nom. illeg.

TYPE: ? cultivated in Bot. Gard. Berlin (? B).

An erect or spreading annual, much resembling *M. hybrida*; stems 2·5–8 (–12) cm. long, usually branched at or near the base, commonly purplish; leaves linear-subulate, 0·5–1 (–1·5) cm. long, generally less than 0·1 cm. wide at base, sharply acuminate, sessile and connate at base, glabrous or rarely very sparsely glandular; inflorescence generally congested, with the flowers in ± dense lateral and terminal fascicles; bracts herbaceous, foliaceous, lanceolate-subulate, 2–5 mm. long, glabrous or sparsely glandular with a distinct membranous margin; pedicels 1–6 mm. long, erect, the upper pedicels in each inflorescence normally shorter than the calyx; sepals narrowly lanceolate, about 3·5 mm. long and 1 mm. wide, distinctly 3-nerved, glandular or eglandular externally with a very slender (often purple-tinged) acumen and narrow membranous margins; petals white, oblong, subacute, about 1·5 mm. long and 0·5 mm. wide; stamens about as long as petals, alternate filaments distinctly expanded at base; anthers yellow, suborbicular, about 0·3 mm. diam.; ovary ovoid, about 0·8 mm. long and 0·6 mm. wide; styles 0·5 mm. long. Capsule oblong, 2·5–3 mm. long, 1–1·3 mm. wide, generally rather shorter than sepals, but not infrequently equalling or even exceeding them; seeds orbicular-reniform, about 0·4 mm. diam.; testa light brown, with low, rounded ridges, or minutely tuberculate especially on the dorsal surface.

HAB.: In fields, pastures and on dry, open hillsides; sea-level to ? 2,400 ft. alt.; fl. ? Febr.–June.

DISTR.: Divisions 1, 3, 5, 7. Throughout the Mediterranean area.

1. Paphos, 1928 or 1930, *Druce* s.n. !
3. Stavrovouni, 1928 or 1930, *Druce* s.n. !
5. Mile 3, Nicosia–Famagusta Road, 1950, *Chapman* 397 !
7. Pentadaktylos, 1880, *Sintenis & Rigo* 231 !

NOTES: Probably much more common in Cyprus than would appear from the above records, for *M. mediterranea* is an inconspicuous plant, easily confused with *M. hybrida*, from which it is most readily distinguished by its congested inflorescences and relatively short capsules. The localization of the Sintenis and Rigo specimen is dubious in view of the fact that these collectors often amalgamated under one number material collected from several distinct areas.

14. SAGINA *L.*

Sp. Plant., ed. 1, 128 (1753).
Gen. Plant., ed. 5, 62 (1754).

Annuals or perennials; mostly small with slender, erect, decumbent or prostrate stems; leaves linear, subulate or setaceous, connate at base, exstipulate; inflorescence cymose or flowers solitary, axillary or terminal; sepals 4 or 5, free; petals 4 or 5, or sometimes wanting, white, entire, often very small; stamens as many as or twice as many as styles; ovary 1-locular; styles 4 or 5, alternating with sepals; ovules numerous. Fruit a capsule, splitting to the base into 4 or 5 valves; seeds orbicular-reniform, minute.

About 30 species, chiefly in the northern hemisphere, but several now widespread as weeds.

Leaves with distinct, short awns; pedicels and sepals frequently glandular **1. S. apetala**
Leaves blunt or mucronulate, not awned; pedicels and sepals normally glabrous **2. S. maritima**

1. S. apetala *Ard.*, Animadv. Bot. Spec. Alt., 2: 22 (1764); Boiss., Fl. Orient., 1: 663 (1867); Holmboe, Veg. Cypr., 67 (1914); Post, Fl. Pal., ed. 2, 1: 192 (1932); Zohary, Fl. Palaest., 1: 121, t. 168 (1966); Davis, Fl. Turkey, 2: 91 (1967); Osorio-Tafall et Seraphim, List Vasc. Plants Cyprus, 39 (1973).
[*S. procumbens* (non L.) Druce in Rep. B.E.C., 9: 470 (1931) pro parte.]
TYPE: Italy; possibly in or near Padua, but not localized, *Arduino* (? FI, PAD).

A slender annual 1–8 cm. high; stems simple, erect or more usually much branched from the base, the central stem erect, the lateral stems decumbent or ascending; leaves linear-subulate, 5–15 mm. long, generally less than 0·8 mm. wide, apex shortly awned, bases connate forming a short membranous cup frequently with ciliate margins; flowers solitary; pedicels slender, ± erect, 0·5–1·8 cm. long, glabrous or thinly glandular; sepals 4, concave, ovate, slightly cucullate, spreading or adpressed in fruit, glabrous or thinly glandular, the outer 2–2·5 mm. long and 1 mm. wide, the inner slightly smaller, apex obtuse or subobtuse, margins narrowly membranous; petals wanting; stamens 4, about 1 mm. long, filaments expanded slightly at base; anthers yellow, ± orbicular, minute, about 0·2 mm. diam.; ovary broadly ovoid or subglobose, about 1 mm. diam.; styles 4, each about 0·3 mm. long. Capsule ovoid, about 2 mm. long and 1·5 mm. wide, with 4 narrowly ovate, boat-shaped valves, attenuate and truncate at the apex; seeds pyriform, obscurely angular, about 0·3 mm. long and 0·2 mm. wide; testa dark brown, minutely and irregularly rugulose.

HAB.: In cultivated fields or gardens, or on bare open ground, usually on sandy soils; sea-level to ? 2,000 ft. alt.; fl. March–May.

DISTR.: Divisions 1, 2, 5–7. Widespread in Europe and the Mediterranean area.

1. Ktima, 1928 or 1930, *Druce* 75 !
2. Phiti [Phyti], 1905, *Holmboe* 772; Pomos Point, 1962, *Meikle* 2247 !
5. Athalassa, 1957, *Merton* 2906 !
6. Dhiorios, 1962, *Meikle* 2424 !
7. About Kantara Mon., 1880, *Sintenis & Rigo* 229 ! Kyrenia, 1949, *Casey* 260 !

NOTES: N. Jardine has annotated the Pomos and Dhiorios gatherings: " This is the Mediterranean race [of *S. apetala* Ard.]—*S. melitensis* Gulia ex Duthie ".

2. S. maritima *G. Don*, Herb. Brit., fasc. 7, no. 155 (1806); Sm. et Sowerby, Engl. Bot., 31: t. 2195 (1810); Unger et Kotschy, Die Insel Cypern, 345 (1865); Boiss., Fl. Orient., 1: 663 (1867); Holmboe, Veg. Cypr., 68 (1914); Post, Fl. Pal., ed. 2, 1:

192 (1932); Davis, Fl. Turkey, 2: 91 (1967); Osorio-Tafall et Seraphim, List Vasc. Plants Cyprus, 39 (1973).

[*S. procumbens* (non L.) Druce in Rep. B.E.C., 9: 470 (1931) pro parte.]

TYPE: Scotland; " on the sea coast not infrequent, in Angus-shire, Isle of Sky, near Aberdeen, Queensferry, and Edinburgh ", *G. Don* (K !).

Erect or decumbent annual, usually much branched at the base, with numerous glabrous, radiating branches; leaves linear-lanceolate or narrowly oblong, rather fleshy, 3–10 mm. long, 0·5–1·5 mm. wide, apex obtuse, blunt or very shortly mucronulate, but not awned, bases connate forming a distinct, usually glabrous, membranous cup; flowers solitary; pedicels slender, erect, 0·5–2 cm. long, usually glabrous; sepals 4, broadly oblong, concave, slightly cucullate, usually spreading in fruit, glabrous, the outer 1·5–2 mm. long, 1–1·3 mm. wide, the inner slightly smaller, apex very obtuse or rounded, margins narrowly membranous often purplish; petals wanting; stamens 4, about 0·8 mm. long, filaments distinctly expanded at base; anthers yellow, minute, orbicular, about 0·2 mm. diam.; ovary subglobose, about 1 mm. diam.; styles 4, about 0·3 mm. long. Capsule oblong-ovoid, about as long as the persistent sepals or a little longer, and about 2 mm. wide, with 4 ovate, boat-shaped valves, attenuate and truncate at the apex; seeds pyriform, obscurely angular, about 0·3 mm. long and 0·2 mm. wide; testa dark brown, minutely and bluntly rugulose.

HAB.: Sandy or rocky places, or on walls, near the sea; about sea-level; fl. ? Febr.–May.

DISTR.: Divisions 1, 4, 7. Widespread along the Mediterranean coasts and north along the Atlantic coasts of Europe to Norway; Atlantic Islands.

1. Ktima, 1928 or 1930, *Druce* 75 !
4. Larnaca, 1862, *Kotschy* 318; Larnaca, 1905, *Holmboe* 108.
7. Kyrenia, 1956, *Casey* 1736A !

15. SPERGULARIA (*Pers.*) *J. & C. Presl*

Fl. Čech., 94 (1819) nom. cons.

Arenaria L. sect. *Spergularia* Pers., Syn. Plant., 1: 504 (1805).

P. Monnier in Bull. Soc. Sci. Nat. Maroc, 35: 145–163 (1956).

Annual, biennial or perennial herbs, occasionally with woody rootstocks; stems erect, ascending, decumbent or prostrate, ± oval in section, swollen at the nodes; leaves linear, often rather fleshy, opposite-decussate, or apparently fascicled, with 4 conspicuous, scarious stipules united in pairs, and sometimes connate at the base on either side of the stem between the leaves; inflorescence cymose or pseudo-racemose; flowers 5-merous, hermaphrodite or unisexual; sepals divided almost to base, obscurely nerved; petals entire, often pink or mauve, sometimes very much reduced or nearly wanting; stamens 2–5 or 10, unequal; anthers medifixed, introrse; ovary 1-locular; styles 3, free or partly united. Capsule subglobose or ovoid-trigonous, dividing almost to base into 3 valves; seeds numerous, ovoid, pyriform or subtrigonous, sometimes winged; testa smooth, rugose or variously tuberculate; embryo curved.

About 25 species with a cosmopolitan distribution; many of the species are markedly halophytic.

All the seeds without wings; leaves mucronate, cuspidate or shortly aristate, not very fleshy:
- Bracts wanting or very short; pedicels very slender; inflorescence a lax repeatedly dichotomous cyme - - - - - - - - - - **1. S. diandra**
- Bracts linear, foliaceous; pedicels not very slender; inflorescence with the ultimate branches usually elongate and racemose - - - - - - **2. S. bocconii**

Some of the seeds, or all the seeds winged (rarely all wingless); leaves blunt, acute or almost mucronulate, fleshy - - - - - - - - - **3. S. marina**

1. S. diandra (*Guss.*) *Heldr. et Sart.*, Herb. Graec. Norm. no. 492 (1855); Boiss., Fl. Orient., 1: 733 (1867); Holmboe, Veg. Cypr., 68 (1914); Post, Fl. Pal., ed. 2, 1: 207 (1932); Zohary, Fl. Palaest., 1: 124, t. 173 (1966); Davis, Fl. Turkey, 2: 95 (1967); Osorio-Tafall et Seraphim, List Vasc. Plants Cyprus, 39 (1973).

Arenaria diandra Guss., Prodr. Fl. Sic., 1: 515 (1827).

TYPE: Sicily; " in argillosis inundatis maritimis; *Girgenti, Sciacca, Marina di Menfrici, Comiso a S. Lorenzo* ", *Gussone* (FI, NAP).

Slender, ascending or spreading annual, (3–) 6–15 (–30) cm. high; stems usually much branched at or near the base, occasionally unbranched, glabrous or subglabrous below, often ± densely glandular-viscid near apex, frequently tinged reddish or purplish; leaves slender, 8–20 (–30) mm. long, 0·3–1 mm. wide, sometimes a little fascicled, mucronate or very shortly cuspidate; stipules small, ovate-deltoid, 1–3 mm. long, 1–2 mm. wide, acute, very shortly connate; inflorescence a lax, repeatedly dichotomous terminal or subterminal cyme the ultimate branches occasionally racemose; bracts wanting, or very short and subulate; pedicels slender, viscid-glandular, 3–10 mm. long; sepals oblong, concave, 1·5–3 mm. long, 0·7–1 mm. wide, obtuse, with a broad greenish or purplish median band, glandular externally, and with distinct membranous margins; petals pink, oblong, obtuse, 2–2·5 mm. long, usually a little shorter than the sepals, obscurely 2–4 (–5)-veined; stamens 2–3, rarely more, 1–1·5 mm. long, filaments glabrous; anthers oblong, yellow, about 0·4 mm. long and 0·3 mm. wide; ovary subglobose, about 1 mm. diam., shortly stipitate or subsessile; styles about 0·3 mm. long, free, recurved. Capsule broadly ovoid, 2–2·5 mm. long, 1·5–2 mm. wide, often blackish, valves 3 (occasionally 2) broadly boat-shaped; seeds all without wings, elongate-pyriform, about 0·5–0·6 mm. long and 0·4 mm. wide; testa blackish-brown, minutely rugulose, sometimes acutely papillose-tuberculate especially on the dorsal surface.

HAB.: Sandy or rocky seashores, or inland as a weed in gardens and fields; occasionally on walls, rooftops or by roadsides; sea-level to 3,800 ft. alt.; fl. Febr.–June.

DISTR.: Divisions 1, 2, 4, 6–8. Widespread in the Mediterranean area and eastwards to Afghanistan; Atlantic Islands.

1. Ayios Yeoryios (Akamas), 1962, *Meikle* 2113 !
2. Platres, 1940, *Kennedy* 1691 !
4. Paralimni, 1905, *Holmboe* 426; Larnaca, 1928 or 1930, *Druce* 82 ! Cape Greco, 1962, *Meikle* 2601 !
6. Nicosia, 1950, *Chapman* 497 !
7. Kythrea, at Sirgania, 1941, *Davis* 2938 ! Kyrenia, 1949, *Casey* 258 !
8. Ayios Philon near Rizokarpaso, 1941, *Davis* 2266 ! Platanisso, 1950, *Chapman* 245 !

2. S. bocconii (*Scheele*) *Aschers. et Graebn.*, Syn. Mitt.-Europ. Fl., 5, 1: 849 (1919); Zohary, Fl. Palaest., 1: 125, t. 175 (1966); Davis, Fl. Turkey, 2: 95 (1967); Osorio-Tafall et Seraphim, List Vasc. Plants Cyprus, 39 (1973).

Alsine bocconii Scheele in Flora, 26: 431 (1843).

Spergularia rubra (L.) J. & C. Presl var. *atheniensis* Heldr. et Sart. in Herb. Graec. Norm. no. 590 (1856) nomen.

S. atheniensis Aschers. in Schweinf., Beitr. Fl. Aeth., 267, 305 (1867); Lindberg f., Iter Cypr. 15 (1946).

[*Arenaria rubra* (non L.) Sibth. et Sm., Fl. Graec. Prodr., 1: 304 (1809).]

[*S. rubra* (non (L.) J. & C. Presl) Boiss., Fl. Orient., 1: 732 (1867) pro parte; Post in Mém. Herb. Boiss., 18: 92 (1900); H. Stuart Thompson in Journ. Bot., 44: 278 (1906); Post, Fl. Pal., ed. 2, 1: 206 (1932) pro parte; Rechinger f. in Arkiv för Bot., ser. 2, 1: 420 (1950).]

[*S. campestris* (non (L.) Aschers.) Holmboe, Veg. Cypr., 68 (1914).]

TYPE: Corsica; Calvi, 1825, *Soleirol* (? P).

Annual or biennial, 5–25 cm. high; stems slender, but more robust than those of *S. diandra*, erect, spreading, decumbent or prostrate, usually much branched at or near the base, glabrous or subglabrous below, somewhat glandular towards apex, bright or dull green, often tinged purplish; leaves slender, rather fleshy, 1–2·5 (–4·5) cm. long, usually less than 1 mm. wide, apex distinctly cuspidate or shortly aristate; lower leaves generally glabrous, the uppermost sometimes thinly glandular; stipules conspicuous, broadly ovate-deltoid, 2–3·5 mm. long, 1·5–2·5 mm. wide, acute, shortly connate; inflorescence terminal or subterminal, forming a rather congested, or rarely lax cyme, the ultimate branches generally somewhat elongate and racemose; bracts linear, foliaceous, 1–3 mm. long; pedicels short, slender, 2–8 (–15) mm. long, ± densely glandular; sepals narrowly oblong, concave, 2·5–4·5 mm. long, 1–1·5 mm. wide, obtuse or subacute, with a narrow greenish or purplish median band, glandular externally, and with distinct, rather broad membranous margins; petals pink (or occasionally white or whitish), ovate or oblong-ovate, 2–2·5 mm. long, 1–1·5 mm. wide, obtuse or subacute, obscurely 4–6-veined; stamens 7–8 in specimens examined (often fewer) 1·5–2 mm. long, filaments glabrous, distinctly expanded towards base; anthers oblong, 0·4–0·6 mm. long, 0·3–0·5 mm. wide, yellow; ovary ovoid 1·4–2·5 mm. long, 0·8–1·2 mm. wide, shortly stipitate; styles about 0·4 mm. long, free, recurved. Capsule ovoid, 4–4·5 mm. long, 2·5–3 mm. wide, valves 3, broadly boat-shaped; seeds all without wings, broadly pyriform, somewhat compressed, 0·4–0·6 mm. long, 0·4–0·5 mm. wide; testa brown, smooth or minutely rugulose, with numerous distinct, truncate tubercles especially on the dorsal surface.

HAB.: Saltmarshes and seashores, or as a roadside, garden or field weed; occasionally on old walls especially near the sea; sea-level to 2,500 ft. alt.; fl. Febr.–June (and sporadically at other times).

DISTR.: Divisions 1, 2, 4–7. Widespread in the Mediterranean area; S.W. Europe and S. Britain; eastwards to Iran; Atlantic Islands.

1. Ktima, 1913, *Haradjian* 673 !
2. Kambos, 1939, *Lindberg f.* s.n.; Pomos Point, 1962, *Meikle* 2248 !
4. Larnaca, 1889, *Pichler* s.n. !
5. Salamis, 1962, *Meikle* 2586 !
6. Nicosia, 1905, *Holmboe* 270; Nicosia, 1950, *Chapman* 497B !
7. Kyrenia, 1949, *Casey* 392 ! and, 1955, *G. E. Atherton* 208 ! and, 1956, *G. E. Atherton* 1296 ! 1342 !

3. **S. marina** (*L.*) *Griseb.*, Spic. Fl. Rum. et Bith., 1: 213 (1843); Unger et Kotschy, Die Insel Cypern, 344 (1865); H. Stuart Thompson in Journ. Bot., 44: 278 (1906); Post, Fl. Pal., ed. 2, 1: 207 (1932); Davis, Fl. Turkey, 2: 94 (1967); Osorio-Tafall et Seraphim, List Vasc. Plants Cyprus, 39 (1973).

Arenaria rubra L. var. *marina* L., Sp. Plant., ed. 1, 423 (1753).

Arenaria marina (L.) All., Fl. Ped., 2: 114 (1785).

Spergularia salina J. & C. Presl., Fl. Čech., 95 (1819); Holmboe, Veg. Cypr., 68 (1914); Lindberg f., Iter Cypr., 15 (1946); Rechinger f. in Arkiv för Bot., ser. 2, 1: 420 (1950); Zohary, Fl. Palaest., 1: 124, t. 172 (1966).

[*Spergularia media* (non (L.) Presl) Boiss., Fl. Orient., 1: 733 (1867).]

[*S. diandra* (non (Guss.) Heldr. et Sart.) Sintenis in Oesterr. Bot. Zeitschr., 32: 364 (1882).]

TYPE: Not localized; "*in litoribus marinis*".

An erect or spreading annual or biennial, 5–20 (–30) cm. high; stems usually branched from near the base, occasionally unbranched, glabrous or subglabrous, or glandular above in mature plants, often tinged purplish; leaves slender, 1–2·5 (–5) cm. long, 0·1–0·15 cm. wide, distinctly fleshy, acute or mucronulate, but not cuspidate or aristate, glabrous or glandular, sometimes a little fascicled; stipules broadly ovate-deltoid, 2–3 mm. long and almost as wide, acute or subacute, distinctly connate for about half their length; inflorescence variable, irregularly cymose, usually somewhat congested, but not infrequently lax; bracts subulate, foliaceous, 1–3 mm. long; pedicels 1–10 (–15) mm. long, ± glandular; sepals narrowly ovate 3–4 (–5) mm. long, 1·5–2·5 mm. wide, obtuse or subacute, with a rather narrow greenish or purplish median band, glandular externally, and with broad membranous margins; petals pink or lilac, ovate or ovate-oblong, about 3 mm. long and 2 mm. wide, obtuse, 5–9-veined, occasionally much reduced or wanting; stamens (2–) 4–5 (–10), about 2–3 mm. long, filaments glabrous, distinctly expanded towards base; anthers yellow, oblong, about 1 mm. long and 0·7 mm. wide; ovary ovoid 1·5–2 mm. long, about 1 mm. wide, very shortly stipitate; styles about 0·6 mm. long, free, recurved. Capsule ovoid, 3–6 mm. long, 3–4 mm. wide, shorter or longer than the calyx; valves 3, broadly boat-shaped; seeds broadly pyriform, distinctly compressed, winged or unwinged, the unwinged about 7 mm. long and 5 mm. wide, the winged about 1·2 mm. diam.; testa brown, smooth or with a varying number of hyaline, truncate tubercles, particularly numerous on the dorsal surface of wingless seeds, often missing in winged seeds; wings striate, with a laciniate-fimbriate margin.

HAB.: Seashores and brackish marshy ground; sea-level to ? 500 ft. alt.; fl. Febr.–June, or sporadically to September.

DISTR.: Divisions 1, 3–8, common in Division 4. Widespread in temperate coastal areas of the northern hemisphere and (? native) in Australasia.

1. Ktima, 1913, *Haradjian* 673A (see Notes).
3. Near Limassol, 1939, *Lindberg f.* s.n.
4. Common; *Sintenis & Rigo* 977 ! *Holmboe* 354, 425; *Kennedy* 1542 ! *Lindberg f.* s.n. ! *Chapman* 432 !
5. Vatili, 1905, *Holmboe* 397; by bridge before Pyroi, 1934, *Syngrassides* 461 ! and, 1936, *Syngrassides* 1051 !
6. Yerolakkos, 1932, *Syngrassides* 66 ! Syrianokhori, 1941, *Davis* 2621 !
7. Near Ayios Khrysostomos Mon., 1862, *Kotschy* 388; Karakoumi, 1956, *G. E. Atherton* 1226 ! East of Kyrenia, 1956, *G. E. Atherton* 1355 !
8. Between Yialousa and Eleousa, 1880, *Sintenis & Rigo* 977 !

NOTES: More decidedly maritime than either of the preceding species.

The significance of f. *leiosperma* (Kindb.) Holmboe, (Veg. Cypr., 68) and f. *heterosperma* (Fenzl) Holmboe (*op. cit.*, 68) is very dubious, and no attempt has been made to maintain the distinction here. The Kew specimen of *Haradjian* 673 is wholly *S. bocconii*, but Rechinger f. (Arkiv för Bot., ser. 2, 1: 420) states that the material he has seen is partly *S. salina*; the statement that this Haradjian number also includes *S. rubra* is clearly an error, and this record must be referred to *S. bocconii*.

TRIBE 3. **Polycarpeae** (*DC.*) *Benth.* Stipules present; sepals free to the base or connate at the base only; petals like those of tribe *Alsineae*, often very small, stamens 5 or fewer; styles 2–3, united in their lower part.

16. POLYCARPON *L.*

Syst. Nat., ed. 10, 881 (1759).

Annual, biennial or perennial herbs; stems usually scabridulous; leaves obovate to suborbicular, ± petiolate, often apparently verticillate; stipules scarious; flowers small in much-branched cymes with scarious bracts; sepals 5, generally carinate, ± cucullate; petals 5, small, entire or emarginate; stamens 5 or fewer, filaments shortly connate at base; styles united below, 3-fid above; ovary 1-locular, multiovulate. Fruit a capsule, dividing almost to the base into 3 spirally twisting valves; seeds convex dorsally, straight ventrally.

About 16 species, some highly critical; widely distributed in the warmer regions of the world.

1. **P. tetraphyllum** (*L.*) *L.*, Syst. Nat., ed. 10, 881 (1759); Hume in Walpole, Mem. Europ. Asiat. Turkey, ed. 1, 254 (1817); Boiss., Fl. Orient., 1: 735 (1867); Holmboe, Veg. Cypr., 67 (1914); Post, Fl. Pal., ed. 2, 1: 209 (1932); Lindberg f., Iter Cypr., 15 (1946); Zohary, Fl. Palaest., 1: 126, t. 177 (1966); Davis, Fl. Turkey, 2: 96 (1967); Osorio-Tafall et Seraphim, List Vasc. Plants Cyprus, 39 (1973).

Mollugo tetraphylla L., Sp. Plant., ed. 1, 89 (1753).

[*Polycarpon alsinefolium* (vix DC.) Sintenis in Oesterr. Bot. Zeitschr., 32: 397 (1882).]

TYPE: " *in* Italiae, Narbonae *vineis* ".

Erect, spreading or decumbent annual, 5–10 (–15) cm. high; stems much branched, glabrous, with longitudinal, scabridulous ridges; leaves obovate to suborbicular, obscurely nerved, 4–15 mm. long, 1–7 mm. wide, the lower usually opposite, the upper often in apparent whorls of 4, lamina green or purplish, minutely scabridulous, with a very narrow hyaline, serrulate margin, apex generally rounded or at most subacute, base narrowing to short petiole, usually less than 3 mm. long, or upper leaves subsessile; stipules white, ovate-deltoid, 1–2 mm. long, 0·8–1·5 mm. wide, the apex generally divided into 2 slender tails; inflorescence a much-branched, lax or rather congested cyme; inflorescence-branches sometimes distinct, up to 1·5 cm. long; pedicels usually very short, rarely exceeding 3 mm.; bracts white, conspicuous, resembling the stipules; sepals strongly keeled, with hyaline-serrulate keels, the outer narrowly ovate-acuminate, 1·5–2 mm. long, 0·5–0·8 mm. wide, with a wide membranous margin, the inner similar, but larger, 2–2·5 mm. long and 0·8–1 mm. wide; petals membranous, hyaline, narrowly oblong, about 1 mm. long and 0·3–0·4 mm. wide, apex truncate or slightly emarginate; stamens 3 (in Cyprus specimens) or perhaps occasionally 1–2 or 4–5, as elsewhere, concealed within the inner sepals, about 0·5 mm. long; anthers yellow, suborbicular, about 0·2 mm. diam.; ovary broadly ovoid, 0·6–1 mm. long, 0·5–0·8 mm. wide near base; style about 0·2 mm. long, 3-fid above. Capsule 1·5–2 mm. long, about 1 mm. wide, walls of valves distinctly incurved after dehiscence; seed 0·4–0·5 mm. long, 0·3–0·4 mm. wide; testa pale brown, strongly papillose-tuberculate, with whitish tubercles.

HAB.: On coastal sand-dunes; as a weed in gardens and sandy fields; on waste ground; on threshing floors; on walls and by paths and roadsides; occasionally on stony mountain-sides; sea-level to 4,300 ft. alt.; fl. Febr.–Aug.

DISTR.: Divisions 1–8, common in Divisions 2, 6, 7. Central Europe and Mediterranean area, now cosmopolitan as a weed.

1. Skali near Ayios Yeoryios (Akamas), 1962, *Meikle* 2104 !
2. Common; *Sintenis & Rigo* 783 ! *Haradjian* 938 ! *Kennedy* 415 ! 1543 ! *Davis* 1773 ! *N. Macdonald* 169 !
3. Limassol, 1801, *Hume* s.n.
4. Larnaca, 1801, *Hume* s.n.
5. Salamis, 1962, *Meikle* 2591 !
6. Common; *Holmboe* 276; *Rev. A. Huddle* 24 ! *Syngrassides* 1616 ! *Lindberg f.* s.n. ! *Davis* 2335 !
7. Common; *Sintenis & Rigo* 237 ! *Syngrassides* 18 ! *Davis* 3023 ! *Casey* 576 ! *G. E. Atherton* 195 !
8. Cape Andreas, 1948, *Mavromoustakis* s.n. ! Platanisso, 1950, *Chapman* 246 !

NOTES: The Cyprus material is tolerably uniform, and can be referred to *P. tetraphyllum* (L.) L. with a reasonable degree of certainty. It is questionable if the closely allied *P. diphyllum* Cav. and *P. alsinefolium* DC. merit even varietal status; the latter is almost certainly just a maritime form of *P. tetraphyllum*.

17. ILLECEBRACEAE

CARYOPHYLLACEAE subfam. PARONYCHIOIDEAE Pax et Hoffm. in Engl. et Prantl, Pflanzenfam. ed. 2, 16C: 297–314 pro parte excl. trib. POLYCARPEAE et SPERGULEAE.

Annual or perennial herbs, occasionally suffruticose or woody at the base; leaves generally opposite, small, simple, mostly entire; stipules often conspicuous, scarious, frequently connate, rarely absent; flowers small, usually greenish and inconspicuous, actinomorphic, generally hermaphrodite, in cymes or clusters, often with conspicuous bracts; sepals (4–) 5 (–6) united below, persistent, sometimes hard or coriaceous, imbricate; petals absent; stamens 10 (rarely 2–3), perigynous, sometimes all fertile, or alternate ones transformed into sterile staminodes, the 5 fertile opposite the sepals; filaments free or united at the base, anthers 2-thecous, dehiscing longitudinally; ovary free, 1-locular; ovule 1 (rarely 2), basal, erect or pendulous; styles 1–3. Fruit indehiscent, 1-seeded, generally enclosed in the persistent calyx; seed with copious or scanty endosperm; embryo commonly curved and peripheral, occasionally straight and central.

About 23 genera widely distributed in temperate and tropical regions.

The family is often united with *Caryophyllaceae*, but the apetalous flowers, perigynous stamens and indehiscent fruits provide obvious, if somewhat artificial distinctions.

Flowers sessile in groups of 3 at the apex of a conspicuous flattened peduncle; leaves in apparent whorls - - - - - - - - - - **3. Pteranthus**
Flowers not as above; leaves opposite, not whorled:
Leaves stipulate, usually oblong or elliptic:
Inflorescences with conspicuous silvery-membranous bracts - - **1. Paronychia**
Inflorescences without conspicuous bracts - - - - - - **2. Herniaria**
Leaves without stipules, subulate - - - - - - - - **4. Scleranthus**

1. PARONYCHIA *Mill.*

Gard. Dict. Abridg., ed. 4 (1754).

M. N. Chaudhri, A Revision of the Paronychiinae, 80–297 (1968).

Annual or perennial herbs, sometimes woody at the base or suffruticose; stems prostrate or spreading, rarely erect, usually much branched from the base; leaves opposite, sometimes unequal, elliptical, oblong or (rarely) linear; stipules conspicuous often silvery and shining; flowers small, in axillary and terminal clusters with large, conspicuous, silvery-membranous bracts; sepals 5, often cucullate or shortly awned and with membranous margins; stamens 5; staminodes 5, filiform, sometimes absent; styles short, free or connate; stigmas 2, elongate or clavate. Fruit a utricle enclosed in the persistent calyx; seed oblong, ovoid or globose; embryo curved.

About 109 species, widely distributed in most regions of the world except central and south-east Asia and southern Africa.

Leaves shortly mucronate, subglabrous or minutely and sparsely strigose; sepals with a conspicuous hyaline-membranous margin and a distinct dorsal awn **1. P. argentea**
Leaves acute or obtuse, but not mucronate, rather densely adpressed-pubescent; sepals uniformly green, without a membranous margin or awn - **2. P. macrosepala**

1. P. argentea *Lam.*, Fl. Franç., 3: 230 (1778); Unger et Kotschy, Die Insel Cypern, 344 (1865); Boiss., Fl. Orient., 1: 745 (1867); Holmboe, Veg. Cypr., 67 (1914); Post, Fl. Pal., ed. 2, 1: 214 (1932); Lindberg f., Iter Cypr., 15 (1946); Zohary, Fl. Palaest., 1: 131, t. 186 (1966); Davis, Fl. Turkey, 2: 252 (1967); M. N.

Chaudhri, A Revision of the Paronychiinae, 211 (1968); Osorio-Tafall et Seraphim, List Vasc. Plants Cyprus, 40 (1973).

Illecebrum paronychia L., Sp. Plant., ed. 1, 206 (1753).

Paronychia hispanica DC. var. *rotundata* DC. in Lam., Encyl. Méth., 5: 24 (1804).

P. argentea Lam. var. *rotundata* (DC.) Chaudhri, A Revision of the Paronychiinae, 215 (1968).

TYPE: " *in* Hispania & Narbona ".

Prostrate perennial with a thick woody rootstock, stems 10–30 (–40) cm. long, much branched, subterete, often reddish, with a narrow unilateral band of short hairs, otherwise glabrous; leaves bluish-green, oblong-elliptic, linear-lanceolate or obovate, 3–12 mm. long, 1·5–5 mm. wide, subglabrous or sparsely and shortly strigose, apex shortly mucronate; stipules silvery-membranous, broadly or narrowly ovate-acuminate, about 5 mm. long, 2–5 mm. wide, shorter than, or rarely longer than the leaves; flowers very numerous, in dense lateral and terminal clusters 12–15 mm. diam.; bracts very conspicuous, silvery-white, ovate, acute, about 5 mm. long, 3–4 mm. wide, much exceeding and concealing the small flowers; sepals erecto-patent, oblong, cucullate, pilose externally with straight (not hooked) hairs, and with a conspicuous smooth or very shortly scabrid subapical awn, the 3 outer sepals about 1·5 mm. long, 0·8 mm. wide with broad hyaline-membranous margins and a greenish or reddish median stripe, the 2 inner often rather narrower, with narrower hyaline margins; stamens 5, alternating with 5 filiform staminodes about 0·4 mm. long; anthers oblong, about 0·5 mm. long, 0·3 mm. wide; ovary globose, compressed laterally, about 0·6 mm. diam., thinly hispid-papillose, ovule pendulous from a basal funicle; stigmas 2, narrowly clavate, about 0·3 mm. long. Fruit a membranous, shortly pilose utricle, about 1·2 mm. diam.; seed compressed globose, about 1 mm. diam.; testa dark shining brown, very slightly rugulose.

HAB.: Sand-dunes, roadsides, waste ground and dry, rocky or gravelly slopes on calcareous or non-calcareous formations; sea-level to 4,600 ft. alt.; fl. Dec.–July.

DISTR.: Divisions 1–8, common. Throughout the Mediterranean region.

1. Skali near Ayios Yeoryios (Akamas), 1962, *Meikle* 2082 ! Polis, 1962, *Meikle* !
2. Common; *Kennedy* 416 ! 417 ! 418 ! 1158 ! 1159 ! *Lindberg f.* s.n. !
3. Common; *Kotschy* 610, 625; *Roger-Smith* ! *Mavromoustakis* 17 ! *Barrington* 42 !
4. Common; *Kotschy* 17; *Sintenis & Rigo*; *Syngrassides* 473 ! *N. Macdonald* 34 ! *Genneou* in ARI 1364 !
5. Common; *A. G. & M. E. Lascelles* ! *Syngrassides* 71 ! *Chapman* 718 ! *Merton* in ARI 173 !
6. Nicosia, 1929, *C. B. Ussher* 33 ! Ayia Irini, 1941, *Davis* 2124 !
7. Common; *Sintenis & Rigo* 235 ! *Syngrassides* 4 ! *Kennedy* 1546 ! *Davis* 2177 ! *G. E. Atherton* 1096 ! 1143 ! *Economides* in ARI 1124 !
8. Cape Andreas, 1880, *Sintenis & Rigo* 235A ! and, 1941, *Davis* 2326 ! Ayios Philon near Rizokarpaso, 1941, *Davis* 2275 !

NOTES: *Kennedy* 1158 (collected Platres, June 6, 1938) is exceptional in having crowded, almost imbricate leaves considerably exceeded by long, narrow, acuminate stipules. It is presumably *P. argentea* Lam. var. *scariossima* Post (Journ. Linn. Soc. Bot., 24: 423; 1888), but it is questionable if this is more than a seasonal condition of the typical plant. *Meikle* 2911 (from Cape Gata, May 9, 1962) has primary, flowering stems like those of typical *argentea* and secondary, basal growths resembling var. *scariosissima*.

2. P. macrosepala *Boiss.*, Diagn. 1, 3: 11 (1843), Fl. Orient., 1: 745 (1867) et Suppl., 123 (1888); Holmboe, Veg. Cypr., 67 (1914); Post, Fl. Pal., ed. 2, 1: 213 (1932); Davis, Fl. Turkey, 2: 254 (1967); M. N. Chaudhri, A Revision of the Paronychiinae, 272 (1968); Osorio-Tafall et Seraphim, List Vasc. Plants Cyprus, 40 (1973).

P. capitata L. ssp. *macrosepala* (Boiss.) Maire et Weiller in Maire, Fl. Afr. Nord, 9: 29 (1963); Greuter in Candollea, 20: 175 (1965).

P. capitata (non L.) Sibth. et Sm., Fl. Graec. Prodr., 1: 163 (1806), Fl. Graec., 3: 41, t. 247 (1819); Unger et Kotschy, Die Insel Cypern, 344 (1865); H. Stuart Thompson in Journ. Bot., 44: 278 (1906).

TYPE: Greece; " loco in schedulis non commemorato ", *Boissier* (G).

Densely tufted perennial with a woody rootstock; stems sprawling or prostrate, rarely exceeding 12 cm. in length, much branched, pubescent all over; leaves often crowded, oblanceolate or narrowly elliptic, acute or obtuse but not mucronate, 3–7 mm. long, 1·5–3 mm. wide, rather densely adpressed-pubescent on both surfaces; stipules narrowly ovate-acuminate, 2·5–3 mm. long, 1·5–2 mm. wide, usually adpressed to stem and shorter than the leaves; inflorescences lateral and terminal, capitate, about 10 mm. diam., often rather irregular in outline; bracts conspicuous, silvery white, broadly ovate or suborbicular, about 4–5 mm. long and almost as wide, acute or rounded, not quite concealing the flowers; sepals erect, usually somewhat recurved at the apex, narrowly oblong-lanceolate, very unequal, the longest about 3·5 mm. long, 0·7 mm. wide, the shortest about 2 mm. long and 0·5 mm. wide, all pubescent, herbaceous, acute or subacute or obtuse, without membranous margins or an apical awn; stamens 5; filaments very short or almost wanting; anthers yellow, oblong, about 0·6 mm. long, 0·3–0·4 mm. wide; staminodes absent; ovary compressed-globose, about 0·4 mm. diam., glabrous; stigmas 2, filiform, about 0·2–0·3 mm. long. Fruit a membranous, glabrous utricle about 1·3 mm. diam.; seed oblong-ovoid, compressed, about 1·2 mm. long, 0·8 mm. wide; testa smooth, shining, dark brown.

HAB.: Sand-dunes, dry rocky and stony slopes on calcareous and non-calcareous formations; sea-level to 4,800 ft. alt.; fl. Febr.–June.

DISTR.: Divisions 1, 2, 3, 6, 7, 8. Widespread in the eastern Mediterranean region, from Greece to Palestine.

1. Akamas, 1941, *Davis* 3318 ! and, 1948, *Kennedy* 1647 ! Peyia, Kalifes, 1962, *Meikle* 2050 !
2. Khionistra, 1939, *Kennedy* 1541 !
3. Between Nisou and Stavrovouni, 1933, *Syngrassides* 35 ! Skarinou, 1941, *Davis* 3567 !
6. Ayia Irini, 1941, *Davis* 2524 !
7. Ayios Panteleimon Mon., 1862, *Kotschy* 941; Ayios Khrysostomos Mon., 1880, *Sintenis & Rigo* 235B ! Buffavento, 1939, *Kennedy* 1545 ! Yaïla, 1941, *Davis* 2901 !
8. Near Apostolos Andreas Mon., 1941, *Davis* 2314 ! Ronnas Bay near Rizokarpaso, 1941, *Davis* 2350 ! Cape Andreas, 1948, *Mavromoustakis* s.n. !

2. HERNIARIA *L.*

Sp. Plant., ed. 1, 218 (1753).
Gen. Plant., ed. 5, 103 (1754).
M. N. Chaudhri, A Revision of the Paronychiinae, 306–398 (1968).

Prostrate or ascending annual or perennial herbs; stems usually much branched; leaves opposite (sometimes apparently alternate), entire; stipules small; flowers small, inconspicuous, in lateral, leaf-opposed or terminal clusters, without conspicuous bracts; sepals (4–) 5 (–6), erect, subequal or unequal; stamens 2–5, staminodes usually 4–5, filiform, inconspicuous or sometimes wanting; style very short or absent; stigmas 2. Fruit an ovoid, membranous utricle enclosed within, or sometimes protruding from the persistent calyx; seed compressed, globose or ovoid; embryo curved.

About 47 species, chiefly in Europe and S.W. Asia.

Calyx densely hispidulous with long, spreading, frequently hooked hairs; plants grey-green in appearance; sepals unequal; stamens usually 2 - - **2. H. cinerea**
Calyx without hooked hairs; plants not grey-green; sepals subequal; stamens 2–5:
Plants bright pale green; leaves often turning reddish with age; flowers about 1 mm. long; calyx thinly and rather unevenly pilose; stamens usually 2–3; seeds smooth - - - - - - - - - - - - - - **1. H. hirsuta**
Plants dark bluish-green; leaves not often turning reddish with age; flowers about 0·7 mm. long; calyx rather densely and uniformly pilose; stamens 5; seeds shagreened - - - - - - - - - - - - **3. H. micrantha**

1. **H. hirsuta** *L.*, Sp. Plant., ed. 1, 218 (1753); Clarke, Travels, ed. 2, 2: 270 (1816); Boiss., Fl. Orient., 1: 740 (1867); Sintenis in Oesterr. Bot. Zeitschr., 32: 398 (1882); Holmboe, Veg. Cypr., 67 (1914); Post, Fl. Pal., ed. 2, 1: 211 (1932); Lindberg f., Iter Cypr., 15 (1946); Zohary, Fl. Palaest., 1: 133, t. 191 (1966); Davis, Fl. Turkey, 2: 248 (1967); M. N. Chaudhri, A Revision of the Paronychiinae, 337 (1968); Osorio-Tafall et Seraphim, List Vasc. Plants Cyprus, 39 (1973).
TYPE: " *in* Angliae, Hispaniae, Italiae *agris* ".

Prostrate annual; stems usually 10–15 cm. long, occasionally up to 30 cm., subterete or somewhat angled, thinly pubescent, often stained reddish-purple; leaves opposite towards base of plant, apparently alternate above, oblanceolate or obovate, 4–7 mm. long, 1·5–3·5 mm. wide, acute or obtuse, subsessile or very shortly petiolate, subglabrous or thinly pilose especially along the margins, commonly turning reddish with age; flowers in dense lateral and terminal clusters, sometimes contiguous along the whole length of the branches, but more often separated by short lengths of internode; bracts resembling the stipules, but smaller and much shorter than the flowers; calyx bright pale green, about 1 mm. long; sepals 5, narrowly ovate, acute or obtuse, subequal or rather inconspicuously unequal, about 0·7 mm. long, 0·3 mm. wide, rather thinly pilose externally, with slender, unequal, straight or curved (but not hooked) hairs, the basal (connate) part of the calyx frequently subglabrous, the sepal-tips often with a few relatively long, coarse bristles; stamens 2–3 (–5); anthers suborbicular, yellow, about 0·2 mm. diam.; filaments very short or almost wanting; ovary globose, about 0·3 mm. diam., verruculose-papillose in the upper half; style very short but usually distinct; stigmas 2, minute, narrow, erect or suberect. Utricle ovoid, enclosed within the persistent calyx; seed compressed-subglobose, about 0·7 mm. diam.; testa dark blackish-brown, shining, smooth.

HAB.: Bare stony ground, roadsides, screes; 250–5,650 ft. alt., but apparently commoner above 2,500 ft. alt.; generally on igneous formations; fl. (April) May–Oct.

DISTR.: Divisions 2, 4, 7, [8]; frequent in Division 2. Central and southern Europe; widespread in the Mediterranean region and extending to Afghanistan, western Pakistan and Ethiopia.

2. Frequent: Troödos, 1894, *Post*; Philani, 1908, *Clement Reid* s.n.! Platres, 1936–1960, *Kennedy* 421! 1160! 1547! 1833! 1834! *Davis* 3477! *N. Macdonald* 171! Khionistra, 1938, *Kennedy* 1223! Kryos Potamos, 1938, *Kennedy* 1225! Kambos, 1939, *Lindberg f.* s.n.; Agros, 1940, *Davis* 1887! Mesopotamos, 1941, *Davis* 3451! Papoutsa Peak, 1962, *Meikle* 2942!
4. Pergamos, 1955, *Merton* 2211!
7. Near Kythrea, 1880, *Sintenis & Rigo* 234! and, 1974, *Meikle* 4068!
[8. Near Cape Andreas, April 26, 1880, *Sintenis & Rigo*.]

NOTES: The Sintenis & Rigo record from near Cape Andreas may be correct, but it is more likely that their plant was *H. cinerea* DC., which generally replaces *H. hirsuta* in lowland areas of Cyprus. It is clear, from herbarium material, that Sintenis & Rigo were not well acquainted with the intricacies of *Herniaria* taxonomy.

Chaudhri (Revision, 338) says: " stamens 2–5, mostly in the order of frequency 5, 4, 3, and 2 ". All the Cyprus material examined has either 2, or, rather more frequently, 3 stamens.

2. **H. cinerea** *DC.* in Lam. et DC., Fl. Franç., 5: 375 (1815); Boiss., Fl. Orient., 1: 738 (1867); Holmboe, Veg. Cypr., 67 (1914); Post, Fl. Pal., ed. 2, 1: 211 (1932); Lindberg f., Iter Cypr., 15 (1946); Rechinger f. in Arkiv för Bot., 2, 1: 420 (1950); M. N. Chaudhri, A Revision of the Paronychiinae, 347 (1968).
[*H. incana* (non Lam.) Sintenis in Oesterr. Bot. Zeitschr., 31: 326 (1881), 32: 54 (1882); Osorio-Tafall et Seraphim, List Vasc. Plants Cyprus, 39 (1973).]
TYPE: France; Montpellier, " entre le Crès et Castelnau ", *de Candolle* (G).

Very similar in habit and general appearance to *H. hirsuta* L., but usually more densely hispid-pilose, and consequently grey-green rather than bright pale green in overall colour, the leaves and stems generally remaining green and not turning reddish or purplish with age; leaves opposite towards base of plant, apparently alternate above, oblanceolate or narrowly obovate, 3–6 mm. long, 1–3 mm. wide, acute or obtuse, subsessile or shortly petiolate, rather densely clothed with subadpressed strigose hairs; flowers in dense lateral and terminal clusters as in *H. hirsuta*, with similar bracts; calyx grey-green, often densely hispid-pilose, with short, slender, spreading, often hooked hairs much overtopped by numerous, spreading bristles, which are commonly hooked at the apex, especially in the middle and lower zones of the calyx; sepals 5, ovate, acute or subacute, distinctly unequal, the 2 outer up to 1 mm. long and 0·5 mm. wide, the 3 inner about 0·7 mm. long and 0·3 mm. wide; stamens 2 (? sometimes, but rarely, 3); anthers suborbicular, yellow, about 0·2 mm. diam.; filaments very short; ovary as in *H. hirsuta*, but style rarely at all developed; stigmas 2, rather broadly clavate, spreading. Utricle ovoid, enclosed within the persistent calyx; seed perhaps a little smaller than that of *H. hirsuta*, but otherwise very similar in shape and colour, and scarcely distinguishable.

HAB.: Sandy and stony fields and fallows, sometimes a weed in gardens, or amongst the stones on threshing floors; sea-level to about 900 ft. alt.; fl. February–May.

DISTR.: Divisions 1, 4, 5, 6, locally common. Also throughout the Mediterranean region and eastwards to Afghanistan and Pakistan; Atlantic Islands; occasionally adventive elsewhere.

1. Near Skali, Ayios Yeoryios (Akamas), 1962, *Meikle* 2103 ! Paphos, 1974, *Meikle* 4012 !
4. Famagusta, 1936, *Syngrassides* 1076 ! Mile 3, Larnaca–Famagusta Road, 1950, *Chapman* 448 ! 714 !
5. Near Lefkoniko, 1880, *Sintenis & Rigo* 234 ! Athalassa, 1933, *Syngrassides* 1373 ! Mia Milea, 1950, *Chapman* 488 ! Xeri, 1959, *P. H. Oswald* 117 !
6. Nicosia, 1905, *Holmboe* 276, and 1937, *Syngrassides* 1615 ! Ayia Irini, 1941, *Davis* 2534 !

NOTES: Best distinguished from *H. hirsuta* by its densely hispid calyces, with generally a fairly large number of the hairs sharply hooked at the tip. It seems to have a distinct habitat and distribution in Cyprus, being exclusively lowland, and commonly associated with cultivation.

3. **H. micrantha** *A. K. Jackson et Turrill* in Kew Bull., 1939: 478 (1939); Davis, Fl. Turkey, 2: 247 (1966); M. N. Chaudhri, A Revision of the Paronychiinae, 333 (1968); Osorio-Tafall et Seraphim, List Vasc. Plants Cyprus, 39 (1973).

[*H. parnassica* (non Heldr. et Sart. ex Boiss.) Lindberg f., Iter Cypr., 15 (1946); Osorio-Tafall et Seraphim, List Vasc. Plants Cyprus, 39 (1973).

TYPE: Cyprus; Kryos Potamos, 5,600 ft. alt., June 3, 1937, *Kennedy* 419 (K !).

Prostrate annual or biennial (occasionally perennating and developing a tough woody rootstock), superficially resembling *H. hirsuta* and *H. cinerea*, but generally more compact and more glabrous, with smaller leaves and flowers; stems much branched, slender, 4-angled, 3–15 cm. long, subglabrous or thinly and shortly pubescent, generally stained purple; leaves broadly oblong-obovate, sometimes almost orbicular, 2·5–5 mm. long, 2–2·5 mm. wide, obtuse, subsessile or shortly petiolate, rather thick, bluish-green, the lower often glabrous or subglabrous, the upper thinly hispidulous especially along the margins and towards the apex; flowers minute, in dense, contiguous, lateral and terminal clusters; sepals 5, subequal, ovate, acute or obtuse, about 0·7 mm. long, 0·4 mm. wide, clothed externally with numerous, rather uniform, short, white, spreading hairs; stamens 5; anthers suborbicular, about 0·2 mm. diam., yellow; filaments very short or almost absent; ovary globose, about 0·3 mm. diam., verruculose-papillose in the upper half; style not developed; stigmas 2, shortly and broadly clavate, spreading or suberect. Utricle ovoid, 0·7–0·8 mm. long, 0·5 mm. wide, enclosed within, or very slightly protruding from the glabrescent, persistent calyx; seed compressed-subglobose, about 0·6–0·7 mm. diam.; testa rich dark brown, shining, distinctly shagreened with anastomosing furrows.

HAB.: Rocky slopes, banks, dry stream beds or bare earth, frequently near water, or where water has stood in the winter; 5,400–6,400 ft. alt.; fl. May–Nov.

DISTR.: Division 2, locally frequent. Also Turkey and eastern Aegean islands.

2. At high altitudes on Khionistra (Kannoures Springs, Kryos Potamos, Prodhromos, Damaskinari), locally frequent. *Haradjian* 512 ! *Kennedy* 419 ! 420 ! 1161 ! 1162 ! 1163 ! 1164 ! 1165 ! 1224 ! 1244 ! 1245 ! 1255 ! *Lindberg f.* s.n. ! *Davis* 3519 ! 3524 ! *D. P. Young* 7261 ! 7343 ! *Meikle* 2782 !

NOTES: All the specimens examined have 5 well-developed stamens, and distinctly shagreened seeds; in these respects *H. micrantha* stands apart from the two other *Herniaria* species found in Cyprus.

HERNIARIA GLABRA *L.*, Sp. Plant., ed. 1, 218 (1753), with glabrous leaves and sepals, is recorded in Osorio-Tafall et Seraphim, List Vasc. Plants Cyprus, 39 (1973); apart from this record, I have no evidence of the occurrence of the species in Cyprus.

3. PTERANTHUS *Forssk.*

Fl. Aegypt-Arab., 36 (1775).

Murbeck in Lunds Univ. Årsskr. N.F., Avd. 2, 2 (6): 1–20 (1906).

B. L. Burtt et P. Lewis in Kew Bull., 7: 339–341 (1952).

Low annual herb; stems much branched; leaves opposite (or in apparent whorls through the suppression of axillary branches), linear, stipulate; flowers sessile in groups of 3 at the apex of a deciduous, conspicuously enlarged and flattened peduncle; lateral flowers subtended by two (or sometimes more) sterile, modified shoots clothed with numerous, rigid, falcate leaves; sepals 4, united near base, boat-shaped, rigid, erect, awned, the 2 exterior sepals of the central flower conspicuously winged at the apex, the outermost sepal of the lateral flowers similarly winged; petals absent; stamens 4, united towards base, opposite the sepals; staminodes absent; anthers 2-thecous, versatile, dorsifixed; ovary globose; style slender; stigmas 2, filiform; ovule 1. Fruit a membranous utricle containing 1 rather large seed; embryo erect, straight, central.

One species widely distributed in North Africa and south-western Asia, from Cyprus eastwards to Arabia and Pakistan.

1. **P. dichotomus** *Forssk.*, Fl. Aegypt.-Arab., 36 et LXII (1775); Holmboe, Veg. Cypr., 67 (1914); Post, Fl. Pal., ed. 2, 1: 217 (1932); Rechinger f. in Arkiv för Bot., ser. 2, 1: 420 (1950); Zohary, Fl. Palaest., 1: 136, t. 195 (1966); Osorio-Tafall et Seraphim, List Vasc. Plants Cyprus, 40 (1973).

Camphorosma pteranthus L., Mantissa 1, 41 (1767); Sibth. et Sm., Fl. Graec. Prodr., 1: 105 (1806), Fl. Graec., 2: 43, t. 153 (1813).

Pteranthus echinatus Desf., Fl. Atlant., 1: 144 (1798); Unger et Kotschy, Die Insel Cypern, 344 (1865); Boiss., Fl. Orient., 1: 752 (1867).

TYPE: Egypt; " in desertis Káhirinis ", *Forsskål* (C).

Sprawling, much-branched annual; stems 4–15 (–20) cm. long, angled or longitudinally costate, glabrous or sparsely scabridulous; leaves linear, spreading, 5–20 mm. long, 1–1·5 mm. wide, acute or obtuse, mucronate, glabrous, rather fleshy, strictly opposite, but usually in apparent whorls of 4–5; stipules narrowly ovate, about 1 mm. long, 0·5 mm. wide, irregularly lacerate and ciliate, white, membranous, persistent; inflorescence a branched dichasium; flowers sessile in groups of 3 at the apex of a very conspicuous, deciduous, flattened, obovate, papillose peduncle about 10 mm. long and 5–7 mm. wide; lateral flowers at right-angles to the central, each subtended by 3 (sometimes more) broadly ovate, membranous bracts and by 2 (or more) modified shoots 5–8 mm. long covered with falcate-awned, imbricate, scale-like, papillose leaves with wide membranous margins; central flower without these accessories; sepals 4, erect, rigid, persistent, boat-shaped, imbricate, papillose, the interior about 6 mm. long and 1 mm. wide, gibbous towards the base, narrowly membranous-aristate towards apex, with a conspicuous, falcate apical awn; the exterior sepals heterogeneous: those of the central flower matching, similar to the interior but about 7 mm. long and 2 mm. wide, with a very conspicuous white-membranous apical crest; those of the lateral flowers unlike, the outermost with a wide membranous crest, the other falcate-awned and closely resembling the interior sepals; stamens 4, opposite the sepals, filaments linear-filiform, about 2–3 mm. long, joined for a short distance above the base; anthers oblong, yellow, about 0·5 mm. long, 0·4 mm. wide; ovary globose or subglobose, about 2 mm. diam., obscurely rugulose-papillose in the upper half; style slender, about 0·5 mm. long; stigmas 2, filiform, about 0·3 mm. long, spreading or recurved. Fruit a membranous, subglobose utricle, about 2·5 mm. diam., enclosed within the persistent calyx; seed obovate or broadly clavate, about 2·3 mm. long, 1·3 mm. wide, hilum deeply excavate; testa pale ochraceous-brown with a very prominent red blotch on one side of the seed; embryo large, straight, central, erect; endosperm scanty.

HAB.: Banks, bare slopes and fallow fields, usually on clays and sandstones of the Kythrean formation; sea-level to 1,000 ft. alt.; fl. March–May.

DISTR.: Divisions 3, 5, 7, 8, rather rare. Distribution outside Cyprus that of the genus.

3. Parekklisha–Kellaki road, on serpentine, 1948, *Kennedy* 1614 !
5. Near Lefkoniko, 1938, *Syngrassides* 1795 ! Near Kythrea, 1950, *Chapman* 110 ! 516 !
7. Mountains above Kythrea, 1880, *Sintenis & Rigo* 21 ! Ayios Khrysostomos Mon., 1889, *Pichler* s.n. !
8. Near Rizokarpaso, 1912, *Haradjian* 137 ! Between Ayios Philon and Rizokarpaso, 1941, *Davis* 2238 ! Apostolos Andreas Mon., 1941, *Davis* 2331 ! Matakos near Galinoporni, 1962, *Meikle* 2444 !

NOTES: The complexities and irregularities of the inflorescences of *Pteranthus* are well described by Murbeck and by Burtt & Lewis in the papers cited above; that the falcate-leaved structures subtending the lateral flowers are modified branches is evident from the fact that, in one of the Cyprus specimens examined, the apical " leaves " of a branch have become partly sepaloid, and surround 4 sterile stamens and an atrophied ovary.

4. SCLERANTHUS *L.*

Sp. Plant, ed. 1, 406 (1753).
Gen. Plant., ed. 5, 190 (1754).
W. Rössler in Oesterr. Bot. Zeitschr., 102: 30–72 (1955).
P. D. Sell in Feddes Repert., 68: 167–169 (1963).

Erect or decumbent annual or perennial herbs; stems angled, bifariously pubescent; leaves opposite, subulate, expanding into papery or membranous, connate bases; stipules absent; flowers greenish, inconspicuous, in lax or dense terminal and lateral cymes; bracts resembling the leaves; sepals 5, united below into a turbinate tube which enlarges and hardens around the fruit; petals absent; stamens 1–10, frequently sterile; styles 2, free, filiform; ovule solitary, pendulous from a basal funicle. Fruit a membranous utricle surrounded by the persistent, indurated calyx; embryo curved, peripheral; endosperm abundant.

About 10 species widely distributed in Europe, Asia, Africa and Australasia.

1. **S. annuus** *L.*, Sp. Plant., ed. 1, 406 (1753); Unger et Kotschy, Die Insel Cypern, 345 (1865); Boiss., Fl. Orient., 1: 750 (1867); Holmboe, Veg. Cypr., 67 (1914); Post, Fl. Pal., ed. 2, 1: 216 (1932); Davis, Fl. Turkey, 2: 264 (1967).

Erect, sprawling or prostrate annual; stems much branched, slender, angled, bifariously pubescent, often purplish, 1–15 (–25) cm. long; leaves subulate-filiform, 4–11 (–15) mm. long, about 1–2 mm. wide, glabrous, acute or obtuse, bases dilated, membranous, ciliate, connate; flowers very numerous, in rather dense terminal and lateral cymes, subsessile or shortly pedicellate; bracts filiform-subulate, closely resembling the leaves, but usually smaller and more patent; calyx 2–4·5 mm. long, divided from slightly below the middle into 5 subequal, narrow, subulate-deltoid, erect or somewhat recurved, subacute, keeled sepals, 1·5–2·5 mm. long, 0·7–1 mm. wide at base, with narrow membranous margins; basal part turbinate, shallowly sulcate, 1·2–2 mm. long, 1·2–1·5 mm. wide; stamens varying in number, commonly 3–4, sometimes fewer, flanked by sterile, anther-less filaments; anthers very small, suborbicular, about 0·25 mm. diam., reddish, often malformed and apparently non-functional; ovary pyriform, about 1–2 mm. long, 0·8–1·2 mm. wide; styles filiform, 0·5–0·8 mm. long; stigmas slightly swollen or sub-clavate at apex. Fruit crowned with the erect or slightly spreading sepals; seed subglobose, 0·5–0·7 mm. diam.; testa smooth, pale brown.

ssp. **delortii** (*Gren.*) *Meikle* **comb. nov.**

S. collinus Hornung et Opiz, Naturalientausch, 10: 232 (1825); Post, Fl. Pal., ed. 2, 1: 216 (1932).

S. verticillatus Tausch in Flora, 12 (1) Erganz.: 50 (1829); Boiss., Fl. Orient., 1: 750 (1867).

S. delortii [" *delorti* "] Gren. in F. W. Schultz (ed.), Arch. de Flore, 206 (1852).

S. verticillatus Tausch. ssp. *delortii* (Gren.) Nyman, Consp. Fl. Europ., 257 (1879).

S. annuus L. ssp. *verticillatus* (Tausch) Arcang., Compend. Fl. Ital., ed. 1, 110 (1882); P. D. Sell in Fl. Europ., 1: 149 (1964); Davis, Fl. Turkey, 2: 264 (1967); Osorio-Tafall et Seraphim, List Vasc. Plants Cyprus, 40 (1973).

S. ruscinonensis Gillot et Coste in Bull. Soc. Bot. Fr., 38: CXXIII (1892).

S. annuus L. ssp. *ruscinonensis* (Gillot et Coste) P. D. Sell in Feddes Repert., 68: 169 (1963).

S. polycarpos (vix L.) Sibth. et Sm., Fl. Graec. Prodr., 1: 278 (1809); DC., Prodr., 3: 378 (1828); W. Rössler in Oesterr. Bot. Zeitschr., 102: 51 (1955).

S. annuus L. ssp. *polycarpos* (vix L.) Thell. in Schinz et Keller, Fl. Schweiz, ed. 3, 2: 109 (1914); P. D. Sell in Fl. Europ., 1: 149 (1964).

TYPE: France; Narbonne, 1846, *Delort* (? P).

Smaller in all its parts than ssp. *annuus*; leaves 4–6 (–10) mm. long; fruit (including persistent calyx) 1·5–3 mm. long, with erect, very narrowly membranous-edged sepals.

HAB.: Roadsides and bare gravelly places; 2,000–5,000 ft. alt.; fl. April–June.

DISTR.: Division 2 [4], rather rare. Widespread in southern Europe and the Mediterranean region eastwards to the Caucasus.

2. Troödos, 1862, *Kotschy* 812; Prodhromos, 1880, *Sintenis & Rigo* 770 ! Platres, 1938, *Kennedy* 1349 ! 1350 ! 1351 ! Papoutsa, 1941, *Davis* 3091 ! and, 1940, *Meikle* 2936 ! Kalokhorio, Makheras, 1950, *Chapman* 449 ! Tripylos, 1962, *Meikle* 2683 ! Ayios Nikolaos Mon. near Kakopetria, 1962, *Meikle* 2854 !

[4. Larnaca, 1862, *Kotschy* 264a—see Notes.]

NOTES: The Cyprus material is uniform, and uniformly distinct from typical *S. annuus* (ssp. *annuus*) in its small size and small flowers. In other areas the two subspecies converge, nor can any satisfactory distinctions be drawn between *S. collinus*, *S. ruscinonensis*, *S. polycarpos* and the other segregates into which the rather variable *S. annuus* has been subdivided. Typical *S. annuus* ssp. *delortii* has rather smaller flowers than those of the Cyprus plant, but is only one of the many intermediates linking the very small-flowered *S. ruscinonensis* with the relatively large-flowered *S. annuus*.

Kotschy's record for *S. annuus* " Bei Larnaca an steinigen Platzen ostlich von der Marina n. 264a " calls for re-investigation. *S. annuus* ssp. *delortii* would appear to be a plant of high altitudes on the island.

18. PORTULACACEAE

Pax et Hoffmann in Engl. et Prantl, Pflanzenfam., ed. 2, 16C: 234–262 (1934).

Annual, biennial or perennial herbs, or occasionally subshrubs; stems and leaves often succulent; leaves opposite or alternate, simple, usually entire, exstipulate or with scarious or bristly stipules; flowers solitary or in cymes or racemes, actinomorphic, bisexual; sepals usually 2, free or united, persistent or deciduous; petals (2–) 4–6, occasionally more, free or connate at the base, often caducous; stamens as many as petals, opposite to them, occasionally fewer or more numerous, free; anthers 2-locular; ovary superior or half-inferior, 1-locular; placentation basal; ovules few to many; styles generally united below; stigmas 2–5 (sometimes more). Fruit a capsule dividing into valves, or circumscissile, rarely indehiscent; seeds globose or reniform with the embryo often curved and peripheral; endosperm generally copious.

About 16 genera, widely distributed, but most abundantly represented in North and South America.

1. PORTULACA *L.*

Sp. Plant., ed. 1, 445 (1753).
Gen. Plant., ed. 5, 204 (1754).

Succulent herbs or subshrubs; leaves opposite or alternate, often forming an involucre-like cluster below the flowers; stipules scarious or bristly, frequently very small; flowers solitary or in cymes; sepals 2, imbricate, the anterior larger and overlapping the posterior, deciduous; petals 4–6, free or united below, deliquescent; stamens 4 or more, sometimes numerous, filaments often hairy below; ovary half-inferior; styles divided above into 3 or more branches. Fruit a circumscissile capsule; seeds numerous, with a peripheral embryo.

About 100 species widely distributed in tropical and subtropical regions.

1. P. oleracea *L.*, Sp. Plant., ed. 1, 445 (1753); Unger et Kotschy, Die Insel Cypern, 343 (1865); Boiss., Fl. Orient., 1: 757 (1867); Post, Fl. Pal., ed. 2, 1: 220 (1932); Zohary, Fl. Palaest., 1: 78, t. 95 (1966); Davis, Fl. Turkey, 2: 13 (1967); Osorio-Tafall et Seraphim, List Vasc. Plants Cyprus, 34 (1973).

P. oleracea L. ssp. *silvestris* (DC.) Thell. in Mém. Soc. Nat. Cherbourg, 38: 222 (1911); Lindberg f., Iter Cypr., 14 (1940).

P. oleracea L. var. *sylvestris* DC., Prodr., 3: 353 (1828).

TYPE: " *in* Europa australi, India, *Ins.* Ascensionis, America ".

Erect, prostrate or decumbent, succulent annual 10–30 (–40) cm. high; stems usually much branched, glabrous; leaves alternate, sessile or very shortly petiolate, glabrous, obovate or oblong-cuneate, 5–50 mm. long, 3–20 mm. wide, apex rounded (or occasionally emarginate), margins entire; stipules reduced to a very small cluster of bristles; flowers 1–6 (occasionally more) terminally or in the axils of the upper branches, sessile, generally enveloped by an involucre-like cluster of leaves; sepals 2, imbricate, about 4–5 mm. long and 3–4 mm. wide, distinctly keeled, the outer slightly cucullate with a narrow membranous margin, the inner more strongly cucullate with a wider membranous margin; petals 5–6, yellow, shortly connate at base, ovate, cucullate, about 3–4 mm. long and 2–3 mm. wide; stamens 8–12 (–15), filaments slender, glabrous, about 1 mm. long; anthers oblong, about 0·5 mm. long and 0·3 mm. wide, yellow; ovary half-inferior, about 2 mm. long and 1·5 mm. wide, upper half conical; style about 1 mm. long, with 3–5 recurved stigma-branches about 0·8 mm. long. Capsule ovoid-conical, about 7–8 mm. long and 5–6 mm. wide; seeds compressed, ovoid-reniform, about 1 mm. long and 0·8 mm. wide; testa dark blackish-brown, distinctly scabrid-verruculose.

HAB.: Roadsides, cultivated and waste land; sea-level to 3,750 ft. alt.; fl. May–September.

DISTR.: Divisions 2, 6, 7, common in Division 7. A cosmopolitan weed.

2. Platres, 1937, *Kennedy* 1034 !
6. Nicosia, 1862, *Kotschy*.
7. Common; *G. E. Atherton* 54 ! 322 ! 384 ! *Casey* 1736 ! 1737 !

NOTES: Commonly cultivated as a salad herb and as a substitute for Spinach, and probably always a relic of cultivation in Cyprus. The form grown in gardens, with large, often emarginate leaves, and the sepals frequently winged dorsally, is sometimes distinguished as ssp. *sativa* (Haw.) Thell.

19. TAMARICACEAE

Niedenzu in Engl. et Prantl, Pflanzenfam., ed. 2, 21: 282–289 (1925).

Trees and shrubs often with small, scale-like, alternate exstipulate leaves; flowers actinomorphic, usually bisexual, often small, solitary or in spikes or racemes; sepals 4–5 (–6), imbricate, free or partly connate; petals as many as sepals, free or at most very shortly connate at base, imbricate, inserted below a nectariferous disk; stamens as many as or twice as many as petals, rarely more numerous, alternate with petals or diplostemonous, free or connate at base, anthers 2-locular, dehiscing longitudinally; ovary superior, 1-locular, with parietal or basal placentation; ovules ascending, few to many; styles (2–) 3–4 (–5) as many as placentas, free or united for part of their length, occasionally wanting. Fruit a capsule, dehiscing into valves; seeds with a tuft or covering of hairs, with or without endosperm; embryo straight.

Four genera in the Mediterranean region, Central Asia, Africa and Europe.

1. TAMARIX *L.*

Sp. Plant., ed. 1, 270 (1753).
Gen. Plant., ed. 5, 131 (1754).

B. Baum, Monographic Revision of the Genus Tamarix, III + 193 pp. (1966).

Leaves small, scale-like; flowers in catkin-like racemes or spikes; sepals 4–5 (–6), free or almost so; petals free or slightly connate at base; stamens (4–) 5–10 (–14) haplostemonous or diplostemonous, free or almost free; anthers generally extrorse; ovary tapering to a conical apex; styles (2–) 3–4 (–5) short and rather thick; ovules numerous, basal or sub-basal. Seeds with an apical tuft of hairs, without endosperm.

About 75 species in the Mediterranean region, central and southern Asia, South Africa and western Europe.

Inflorescence arising from year-old twigs:
  Sepals, petals, stamens and styles normally 4:
    Inflorescences (3·5–) 5–12 (–16) cm. long, 0·5–0·8 cm. wide; bark of twigs reddish-brown; large shrub or small tree to 15 m. high - - - **1. T. tetragyna**
    Inflorescence seldom exceeding 3 cm. long, 0·5 cm. wide; bark of twigs blackish-brown; shrub to 3 m. high - - - - - - - **2. T. tetrandra**
  Sepals, petals and stamens normally 5; styles normally 3 - - **3. T. hampeana**
Inflorescences arising from current year's shoots; sepals, petals and stamens normally 5, styles normally 3 - - - - - - - - **4. T. smyrnensis**

1. T. tetragyna *Ehrenb.* in Linnaea, 2: 247–258 (1827); Unger et Kotschy, Die Insel Cypern, 354 (1865) [—as *T. tetragona*]; Boiss., Fl. Orient., 1: 768 (1867); Holmboe, Veg. Cypr., 130 (1914); Post, Fl. Pal., ed. 2, 1: 222 (1932); Chapman, Cyprus Trees and Shrubs, 53 (1949); Baum, Mon. Rev. Gen. Tamarix, 113, fig. 46 (1966); Zohary, Fl. Palaest., 2: 359, t. 524, 524a (1972); Osorio-Tafall et Seraphim, List Vasc. Plants Cyprus, 71 (1973).

T. meyeri Boiss., Diagn., 1, 10: 9 (1849); Unger et Kotschy, Die Insel Cypern, 354 (1865); Lindberg f., Iter Cypr., 23 (1946); Chapman, Cyprus Trees and Shrubs, 53 (1949); Osorio-Tafall et Seraphim, List Vasc. Plants Cyprus, 71 (1973).

T. tetragyna [*tetragona*] Ehrenb. var. *meyeri* (Boiss.) Boiss., Fl. Orient., 1: 768 (1867); Sintenis in Oesterr. Bot. Zeitschr., 32: 399 (1882).

T. tetragyna Ehrenb. ssp. *meyeri* (Boiss.) Holmboe, Veg. Cypr., 130 (1914).

[*T. mannifera* (non Ehrenb.) H. Stuart Thompson in Journ. Bot., 44: 278 (1906).]

TYPE: Egypt; " ad lacum Menzaleh " *Ehrenberg* (B, K !).

A spreading shrub or small tree 5–15 m. high; twigs glabrous, dark reddish-brown with conspicuous leaf-scars; leaves distinctly glaucous, those of the primary shoots narrowly lanceolate-deltoid, 5–6 mm. long, 2–3 mm. wide, those of secondary and subordinate shoots much smaller, often less than 1 mm. long and 0·5 mm. wide, acute, ± adpressed; inflorescences rather densely spicate, (3·5–) 5–12 (–16) cm. long, 0·5–0·8 cm. wide, arising copiously from the year-old branches with or slightly in advance of the leaves; peduncles 0·5–2 (–3) cm. long, ± densely covered with pallid or greenish bracts; floral bracts linear-oblong 2–5 (–8) mm. long, 0·5–1 mm. wide, obtuse or subacute, often distinctly reflexed, whitish; pedicels 1–2 mm. long; sepals normally 4 (sometimes 5–6), broadly ovate or almost suborbicular, 2–3 mm. long and about 2 mm. wide, obtuse or subacute, concave, free almost to the base, greenish with a pale margin; petals normally 4 (5–6) pale pink or whitish, spreading or slightly reflexed, oblong-obovate, 3–4 mm. long, about 2 mm. wide, subacute or rounded, obscurely penninerved; disk about 1·3 mm. diam., normally with 8 blunt lobes, glabrous; stamens normally 4 inserted between alternate lobes of the disk; filaments glabrous, about 2·5 mm. long; anthers suborbicular, exapiculate, 0·8–1 mm. diam.; ovary ± flask-shaped, with a rather short neck, about 2 mm. long and 1·2 mm. wide at base; styles normally 4, very short, stigmas broadly spathulate, about 0·8 mm. long and 0·6 mm. wide. Capsule narrowly ovoid, about 6 mm. long and 2·5 mm. wide, with a slender beak; seeds oblong, about 0·6 mm. long and 0·3 mm. wide; testa smooth, bright brown; apical coma 3–4 mm. long, of whitish hairs with thickened, slightly roughened bases.

HAB.: By streamsides; in fresh or brackish marshes or moist places; in moist gullies on calcareous marls; occasionally on sandy seashores; sea-level to 1,000 ft. alt.; fl. Febr.–April.

DISTR.: Divisions 3–8, common in Divisions 4, 6. Caucasus, Turkey, Palestine, Egypt and eastwards through Iraq and Iran to Afghanistan.

3. Limassol, 1932, *Syngrassides* 391 ! Akrotiri, 1938, *Syngrassides* 1807 ! Kannaviou, 1939, *Lindberg f.*
4. Common; *Kotschy* 243 ! *Holmboe* 383, 429; *Syngrassides* 472 ! *Lindberg f.* s.n.; *Merton* 1872 ! etc.
5. Xeri, 1932, *Syngrassides* 249 ! Near Leonarisso by Trikomo road, 1933, *Chapman* 23 ! Salamis, 1934, *Chapman* 136 ! and, 1939, *Lindberg f.* s.n.
6. Common; *Kotschy* 961; *Syngrassides* 1542 ! *Davis* 2575 ! *Merton* 2593 !
7. Larnaca-tis-Lapithou, 1941, *Davis* 3017 ! Near Panagra, 1962, *Meikle* 2357 !
8. Between Galinoporni and Leonarisso (at ? Elisis), 1880, *Sintenis & Rigo* 14 ! Patriki, 1938, *Syngrassides* 1801 ! Gastria, 1941, *Davis* 2411 !

2. T. tetrandra *Pall. ex M. Bieb.*, Fl. Taur.-Cauc., 1: 247 (1808) pro parte; Unger et Kotschy, Die Insel Cypern, 354 (1865); Boiss., Fl. Orient., 1: 769 (1867); Holmboe, Veg. Cypr., 130 (1914); Post, Fl. Pal., ed. 2, 1: 223 (1932); Chapman, Cyprus Trees and Shrubs, 54 (1949); Baum, Mon. Rev. Gen. Tamarix, 115, fig. 47 (1966); Davis, Fl. Turkey, 2: 349 (1967); Osorio-Tafall et Seraphim, List Vasc. Plants Cyprus, 71 (1973).

T. parviflora DC. var. *cypria* Kotschy in Unger et Kotschy, Die Insel Cypern, 354 (1865).

[*T. parviflora* (non DC.) Boiss., Fl. Orient., 1: 769 (1867) pro parte quoad pl. Cypr.; Holmboe, Veg. Cypr., 130 (1914); Chapman, Cyprus Trees and Shrubs, 54 (1949).]

TYPE: U.S.S.R., Crimea; " e Tauria ", *Pallas*.

Erect or spreading shrub 1·5–3 m. high; twigs dark blackish brown, glabrous, with rather small leaf-scars; leaves of the primary shoots ovate-oblong, generally less than 2 mm. long and 1 mm. wide, ± adpressed, pale green, with distinct glandular pits, and often with a sharply acute, brown, claw-like apex; leaves of secondary and subordinate shoots much smaller, usually less than 1 mm. long, distinctly adpressed, normally without the brown apical appendage; flowers in rather dense spikes arising from year-old twigs, with or somewhat in advance of the leaves; spikes seldom more than 3 cm. long, about 0·5 cm. wide, sessile or very shortly pedunculate, the peduncle ± densely covered with ovate, acute, brownish bracts; floral bracts oblong, acute, about 1·2 mm. long and 0·6 mm. wide, concave, greenish with narrow pale margins; pedicels 1–1·5 mm. long, glabrous; sepals 4, broadly ovate or suborbicular, about 1·5 mm. long and almost as wide, free almost to base, obtuse or acute, greenish with a distinct pale margin; petals 4, whitish or pale pink, oblong, about 2·5 mm. long and 1·5 mm. wide, concave, erect or spreading, apex acute or rounded, obscurely penninerved; disk about 1·5 mm. diam., bluntly 8-lobed; stamens 4 inserted between alternate lobes; filaments about 2 mm. long, distinctly thickened towards base; anthers cordate about 0·8 mm. long and almost as wide, pinkish, exapiculate; ovary flask-shaped, about 1·2 mm. long and almost as wide, with a short neck; styles (3) 4, about 0·7 mm. long, with slightly thickened stigmas. Capsule elongate flask-shaped, about 3 mm. long and 1 mm. wide; seed oblong, about 0·8 mm. long and 0·4 mm. wide; testa bright brown, smooth; apical coma about 2·5 mm. long, spreading, hairs whitish, distinctly scabridulous towards base.

HAB.: In rocky or gravelly river beds, or on rocky river banks; sea-level to c. 1,500 ft. alt.; fl. March–April.

DISTR.: Divisions 1–4, 7, common in Divisions 2, 3. Also S. Russia, Bulgaria, Yugoslavia, Greece, Crete, Turkey, Palestine, Egypt.

1. Limni marshes, 1938, *Chapman* 340 ! Polis marshes, 1938, *Chapman* 354 ! Toxeftera, near Ayios Yeoryios (Akamas) 1962, *Meikle* 2090 !
2. Common; *Kotschy* 698 ! *Chapman* 356 ! 358 ! 359 ! 361 !
3. Common; *Kotschy* 258 ! *Holmboe* 623; *Syngrassides* 1484 ! 1505 ! 1804 ! 1811 ! *Davis* 2789 ! *Merton* 2539 !
4. Cyprus (probably Larnaca), 1860, *Hooker & Hanbury* s.n. ! Between Larnaca and Stavrovouni above Kalokhorio, 1862, *Kotschy* 226.
7. Thermia, 1956, *G. E. Atherton* 1208 !

3. **T. hampeana** *Boiss. et Heldr.* in Boiss., Diagn., 1, 10: 8 (1849); Fl. Orient., 1: 767 (1867); Baum, Mon. Rev. Gen. Tamarix, 105, fig. 42 (1966); Davis, Fl. Turkey, 2: 349 (1967).

TYPE: Greece; " in maritimis Atticae ad *Phalerum* (Spruner, Heldreich) " (G, K !).

A sprawling shrub about 7 m. high; twigs dark brown, similar to those of *T. tetragyna* Ehrenb.; leaves bright green (glaucous on drying), those of the primary shoots linear-deltoid, about 4 mm. long and 1·5 mm. wide, those of secondary and subordinate shoots much smaller, narrowly linear-deltoid, acuminate, minutely and bluntly papillose at least when young, ± adpressed; inflorescences rather densely spicate, about 2·5–5 cm. long and 0·6–0·8 cm. wide, arising rather abundantly from the year-old branches, or very occasionally from current year's growths, about the same time as the leaves; peduncles generally less than 1 cm. long, ± densely covered with brownish, ovate bracts; floral bracts linear, about 3 mm. long and 0·6 mm. wide, with an acute, ± cuspidate apex, flattish, pallid, usually erect; pedicels about 1 mm. long; sepals 5, ovate or broadly oblong, about 1·5–2 mm. long and 1 mm. wide, connate for about $\frac{1}{3}$ of their length from the base, apex obtuse or rounded, greenish with a narrow pale margin; petals 5, pink, oblong, about 3 mm. long and 1·5 mm. wide, erect, concave and with a slightly cucullate, obtuse or rounded apex, obscurely penninerved with sharply ascending nerves; disk concave, about 0·8 mm. diam., with 5 rounded, shallow lobes, glabrous; stamens 5 inserted at the apex of each lobe of the disk; filaments glabrous about 2·5 mm. long; anthers suborbicular, about 0·7–0·8 mm. diam., yellowish or reddish, rounded or very obscurely apiculate; ? pollen defective; ovary flask-shaped, with a rather long neck, about 2 mm. long and 1 mm. wide at base; styles 3, distinct, about 0·7 mm. long, free except at base; stigmas rounded-spathulate about 0·5 mm. diam. Capsule and seeds not seen.

HAB.: By roadside near river-bank; 400–500 ft. alt.; fl. March.

DISTR.: Division 7, rare. Also S. Greece, Aegean area, Palestine.

7. Near Panagra, by stream E. of Myrtou–Panagra road, about $\frac{1}{4}$ mile S. of the latter village, 1962, *Meikle* 2377 !

T. GALLICA L. var. MANNIFERA *Ehrenb.* (in Linnaea, 2: 270; 1827) and T. APHYLLA (*L.*) *Karst.* (Deutsch Fl., 641; 1882) are recorded in Osorio-Tafall et Seraphim, List Vasc. Plants Cyprus, 72 (1973); I have no satisfactory evidence of the occurrence of either species in Cyprus.

4. T. smyrnensis *Bunge*, Tent. Gen. Tamaric., 53 (1852); Unger et Kotschy, Die Insel Cypern, 354 (1865); Baum, Mon. Rev. Gen. Tamarix, 45, fig. 16 (1966); Davis, Fl. Turkey, 2: 350 (1967).

[*T. pallasii* (non Desv.) Boiss., Fl. Orient., 1: 773 (1867) pro parte; Holmboe, Veg. Cypr., 130 (1914); Lindberg f., Iter Cypr., 23 (1946); Chapman, Cyprus Trees and Shrubs, 53 (1949).]

TYPE: Turkey; " in arenosis humidis Smyrnae (Fleischer !) " (P, K !).

A lax shrub or small tree, to about 4 m. high; twigs dark reddish-brown, glabrous, often slender, leaf-scars rather small; leaves bright or glaucous green, those of the primary shoots widely spaced, deltoid, acuminate, 1–2 mm. long, about 1 mm. wide, those of secondary and subordinate shoots rather smaller, overlapping, often glandular-pitted and minutely papillose, ± adpressed; inflorescences lateral arising from the current year's shoots, 1–2·5 cm. long, 0·2–0·5 cm. wide, forming lax or rather dense sub-terminal panicles; peduncles very short, generally less than 5 mm. long, with a few scattered acuminate bracts; floral bracts ± erect, narrowly ovate, about 1 mm. long, 0·5 mm. wide, apex acuminate-cuspidate, margins often erose; pedicels slender, about 1 mm. long; sepals 5, broadly ovate, about 0·8 mm. long and 0·6 mm. wide, subacute, connate for about $\frac{1}{3}$ of their length, with a narrow greenish central stripe and broad membranous, often erose margins; petals 5, erect, obovate, pink or whitish, 1–1·5 mm. long, about 0·8 mm. wide, 1-nerved or very obscurely penninerved, concave, apex rounded; disk about 0·6 mm. diam., with 10 rounded lobes; stamens 5, inserted between alternate, more deeply cut lobes; filaments 2–2·5 mm. long, scarcely widened at base; anthers orbicular-cordate, about 0·5 mm. diam., exapiculate or minutely apiculate, yellow or pink; ovary ovoid-conical, slightly more than 1 mm. long and 0·7 mm. wide at base; styles 3, about 0·3 mm. long with spathulate stigmas about 0·2 mm. long. Capsule small, about 2·5–3 mm. long and 0·7 mm. wide, the base partly concealed by the persistent petals; seeds oblong, about 0·5 mm. long and 0·2 mm. wide; testa pale brown, smooth; apical coma 1–1·5 mm. long, with spreading whitish hairs slightly swollen and scabridulous at the base.

HAB.: In river beds, and by rivers and streamsides; on marshy ground, or moist places in sand-dunes or on rocky shores; sea-level to 3,100 ft. alt.; fl. April–June.

DISTR.: Divisions 1–3, 5–7, common in Divisions 1–3. South-east Europe and eastern Mediterranean region eastwards to Afghanistan.

1. Common; *Haradjian* 798 ! *Chapman* 362 ! *Lindberg f.* s.n.; *Merton* 2164 !
2. Common; *Sintenis & Rigo* 740 ! *Syngrassides* 965 ! *Kennedy* 627 ! *Chapman* 361 ! *Lindberg f.* s.n. !
3. Common; *Kotschy* 572 ! *Holmboe* 236, 600, 622; *Syngrassides* 1499 ! *Davis* 3541 !
5. " Road between Geunely [Geunyeli] and Photta ", 1902, *A. G. & M. E. Lascelles* s.n. ! Athalassa, 1932, *Syngrassides* 403 ! Salamis, 1933, *Chapman* 70 !
6. Nicosia, by Pedieos R., 1939, *Lindberg f.* s.n. !
7. Near Kythrea, 1880, *Sintenis & Rigo* 505 ! Path through hummocks to Pentadaktylos, 1902, *A. G. & M. E. Lascelles* s.n. ! Kornos [Agricultural] Station, 1934, *Syngrassides* 434 ! Kyrenia, 1955, *G. E. Atherton* 48 !

20. ELATINACEAE

Niedenzu in Engl. et Prantl, Pflanzenfam., ed. 2, 21: 270–276 (1925).

Annual or perennial herbs or subshrubs, often very small; leaves simple, opposite or whorled, generally entire, stipulate; flowers solitary or in cymes, axillary, bisexual, actinomorphic, usually small; sepals 3–5, free or connate at the base, imbricate; petals as many as sepals, free, imbricate; stamens hypogynous, as many as or twice as many as petals; anthers introrse, versatile; ovary superior, 3–5-locular with axile placentation; styles 3–5, free; ovules numerous. Fruit a septicidal capsule; seeds often curved; endosperm wanting or scanty; cotyledons short.

Two genera and about 40 species with a cosmopolitan distribution.

1. ELATINE *L.*

Sp. Plant., ed. 1, 367 (1753).
Gen. Plant., ed. 5, 172 (1754).

Small aquatic or amphibious herbs; leaves and stems glabrous or subglabrous; leaves opposite or verticillate; flowers very small, generally solitary; sepals 3–4 (occasionally 2), obtuse, ecostate, often membranous towards base; ovary globose. Capsule membranous; seeds with a conspicuously sculptured testa.

About 15 species distributed throughout the world.

1. E. macropoda *Guss.*, Fl. Sic. Prodr., 1: 475 (1827); Zohary, Fl. Palaest., 2: 366, t. 533 (1972).

E. campylosperma Seub. in Walpers Repert. Bot. Syst., 1: 284 (1842); Boiss., Fl. Orient., 1: 781 (1867); Post, Fl. Pal., ed. 2, 1: 226 (1932).

[*E. hydropiper* (non L.) Osorio-Tafall et Seraphim, List Vasc. Plants Cyprus, 71 (1973).]

TYPE: Sicily; " in calcareis hyeme inundatis; *Modica, Scili, Spaccaforno, Ragusa, Siracusa, Noto* ", *Gussone* (FI, ? NAP).

Prostrate creeping annual, forming flat mossy patches on mud at the bottom of shallow pools; stems often reddish, rooting at the nodes; leaves deep green, narrowly oblong or oblanceolate-spathulate, 3–10 mm. long, 1–1·8 mm. wide, entire, obscurely penninerved, apex obtuse or rounded, base tapering to a flattened petiole up to 7 mm. long; stipules interpetiolar, ovate-deltoid, about 0·5 mm. long and almost as wide; flowers solitary, axillary; peduncle at first about 1·5 mm. long, lengthening to 5–10 mm. in fruit; sepals 4, persistent, oblong, obtuse, connate at base, alternate pairs slightly unequal, one sepal about 2 mm. long and 1 mm. wide, the adjacent one about 1·5 mm. long and 0·8 mm. wide; petals 4, white or pinkish, oblong, obtuse, about 1·5 mm. long and 0·8 mm. wide, membranous with a distinct midrib vanishing below the apex, margins minutely denticulate; stamens 8, alternately about 1 mm. long and 0·8 mm. long; filaments flattened; anthers suborbicular, about 0·2–0·3 mm. diam.; ovary depressed-globose, about 1·3 mm. diam.; styles 4, about 0·25 mm. long, slightly recurved; stigma subcapitate. Capsule about 2 mm. diam., depressed-globose, walls thin, dehiscing cruciately into 4 parts; seeds cylindrical, comma-shaped (in Cyprus specimens) or occasionally almost straight, about 0·7 mm. long; testa brown, with a conspicuous network of squarish ribs.

HAB.: In shallow pools on limestone rocks near the sea; sea-level to about 200 ft. alt.; fl. Febr.–April. Often associated with *Ranunculus aquatilis* aggr., *Limosella* and *Crassula* (*Bulliarda*) *vaillantii*.

DISTR.: Division 1, rare. Widespread in the Mediterranean area.

1. Erimidhes, near Ayios Yeoryios (Akamas), 1962, *Meikle* 2141 ! By cross-roads between Neokhorio and Baths of Aphrodite, 1962, *Meikle* 2236 !

NOTES: Sometimes distinguished from typical *E. macropoda* by its distinctly curved seeds, and named *E. campylosperma* Seub. The distinction is not valid, for forms with straightish and distinctly curved seeds occur in the same locality. Some authors unite *E. macropoda* with *E. hydropiper* L., but the latter is readily distinguished by its sessile or subsessile flowers, and by the narrowly rectangular (not squarish) sculpturing of the normally strongly curved seeds.

21. GUTTIFERAE

by N. K. B. Robson.

Engler in Engl. et Prantl, Pflanzenfam., ed. 2, 21: 154 (1925).

Trees, shrubs, herbs and (rarely) climbers, glabrous or with an indumentum of simple or stellate hairs, with various glandular secretions; leaves opposite or rarely whorled or alternate, simple, almost always exstipulate, entire or rarely gland-fringed; flowers in terminal or sometimes axillary cymes or panicles, or sometimes solitary, actinomorphic, hermaphrodite or unisexual; sepals (2) 4–5 (6), imbricate, free or united at the base or rarely completely united in bud, persistent; petals 4–5 (–10), free, contorted in bud, deciduous or persistent, sometimes with nectariferous scales; stamens in 2 diplostemonous whorls of fascicles, often modified by various fusions and by sterilisation or suppression of the antisepalous whorl, sometimes deciduous; filaments apparently free or variously united; anthers 2-locular, dorsifixed, introrse, usually dehiscing longitudinally; ovary superior, 1–12-locular; placentation axile, parietal or rarely basal; ovules 1–many on each placenta, erect or ascending; styles 1–6, free or ± united, or absent, stigmas 1–12, variously shaped. Fruit a septicidal or septicidal and loculicidal capsule, or baccate or drupaceous; seeds cylindric or ellipsoid, sometimes with sculptured testa, sometimes winged or arillate or rarely carunculate; endosperm absent; embryo straight or slightly curved; cotyledons flat or not, or absent.

About 44 genera and 1050 species widely distributed in tropical regions, with 2 genera also distributed in the temperate zone.

1. HYPERICUM *L.*

Sp. Pl., ed. 1, 783 (1753).
Gen. Pl., ed. 5, 341 (1754).
R. Keller in Engl. et Prantl, Pflanzenfam., ed. 2, 21: 175–183 (1925).

Annual or perennial herbs or shrubs; leaves opposite or whorled, with translucent glands containing essential oils and sometimes black or red glands containing hypericin; inflorescences terminal, usually dichasially cymose; flowers bisexual; sepals (4) 5; petals (4) 5, yellow, often tinged red; antisepalous fascicles absent or rarely staminodial, antipetalous fascicles free or variously united, filaments apparently free or variously united; ovary 3–5-locular or 1-locular with (2) 3–5 parietal placentas; styles present, free or ± united. Fruit a septicidal capsule or rarely tardily dehiscent or baccate; seeds with sculptured testa, sometimes winged.

About 400 species, chiefly in temperate and subtropical regions of both hemispheres.

Stamen-fascicles 5, deciduous; broad-leaved shrubs; black glands absent — **1. H. hircinum**
Stamen-fascicles 3 (4), usually persistent; perennial herbs or narrow-leaved shrublets; black glands present:
- Leaves in whorls of 3; shrublet; petals and stamens deciduous — **2. H. empetrifolium**
- Leaves opposite; perennial herbs; petals and stamens persistent:
  - Black glands absent from leaves, petals and anthers; seed papillose; stem glabrous, leaves ± pubescent - - - - **3. H. confertum ssp. stenobotrys**
  - Black glands present at least on anthers and leaves; seeds reticulate or striate; stem and leaves both glabrous or pubescent:
    - Stem and leaves pubescent - - - - - - - **4. H. lanuginosum**
    - Stem glabrous; leaves glabrous or with undulate epidermis:
      - Stem eglandular; sterile axillary shoots not usually developed:
        - Stems erect or decumbent; leaves 13–60 mm. long, ovate or triangular-lanceolate; styles 3 - - - - - - - **5. H. perfoliatum**
        - Stems prostrate or ascending, rooting; leaves 5–7 mm. long, oblanceolate-spathulate; styles 5 (or rarely 3) - - - - - **6. H. repens**
      - Stem with black glands (sometimes confined to the raised lines); sterile axillary shoots usually developed:
        - Leaves sessile or subsessile, with margin plane; inflorescence branches ascending - - - - - - - - - - **7. H. perforatum**
        - Leaves amplexicaul, with margin undulate; inflorescence branches widely spreading - - - - - - - - - - **8. H. triquetrifolium**

In addition to the above species, the following have been recorded from Cyprus, probably in error:

H. NUMMULARIUM *L.*, Sp. Plant., ed. 1, 787 (1753); Hume in Walpole, Mem. Europ. Asiat. Turkey, ed. 1, 254 (1817) [" nummularia "].

H. HYSSOPIFOLIUM *Chaix* in Vill., Hist. Plant. Dauph., 1: 329, t. 44 (1786); Post in Mém. Herb. Boiss., 18: 92 (1900); H. Stuart Thompson in Journ. Bot., 44: 278 (1906); Holmboe, Veg. Cypr., 126 (1914); Osorio-Tafall et Seraphim, List Vasc. Plants Cyprus, 70 (1973).

SECT. 1. **Androsaemum** (*Duhamel*) *Godron* Glabrous shrubs, without black glands; stems 2-lined, with lines decurrent from between leaves, or 4-angled; leaves opposite; flowers few, in terminal cymes; petals and stamens deciduous; stamen-fascicles 5; styles 3. Fruit ± fleshy at first, tardily dehiscent, not vittate. Seeds reticulate, winged.

1. **H. hircinum** *L.*, Sp. Plant., 784 (1753); Sibth. et Sm., Fl. Graec., 8: 53, t. 773 (1833); Boiss., Fl. Orient., 1: 788 (1867); Post, Fl. Pal., ed. 2, 1: 228 (1932); Lindberg f., Iter Cypr., 23 (1946); Sauvage in Bull. Soc. Sci. Nat. Maroc, 38: 122 (1958);

Rechinger f. in Arkiv för Bot., ser. 2, 5: 291 (1960); Zohary, Fl. Palaest., 1: 221, t. 324 (1966); Davis, Fl. Turkey, 2: 366 (1967); Osorio-Tafall et Seraphim, List Vasc. Plants Cyprus, 70 (1973).

Androsaemum hircinum (L.) Spach, Hist. Nat. Vég., 5: 419 (1836).

Androsaemum foetidum Willk. et Lange, Prodr. Fl. Hisp., 3: 589 (1878) in synon.

TYPE: " *in* Sicilia, Calabria, Creta ".

A shrub 20–200 cm. high; stems erect, 2-lined or 4-angled, much branched; leaves 2–7·5 cm. long, (0·7–) 1–3 cm. wide, sessile or subsessile, narrowly lanceolate to broadly ovate, subacute or obtuse or rounded, truncate or subcordate, with densely reticulate venation below, when crushed often smelling of goats; sepals 3–7 mm. long, 1–3 mm. wide, somewhat unequal, ovate-lanceolate to lanceolate or narrowly oblong, acute, deciduous after flowering; petals golden yellow, 15–25 mm. long, 7–10 mm. wide, oblanceolate to narrowly obovate, eglandular; stamens exceeding petals, c. 18–28 mm. long; anthers eglandular; ovary ovoid-ellipsoid, 3–3·5 mm. long, 2·5–3 mm. wide; styles c. 3–5 times longer than the ovary, slender, suberect. Capsule ellipsoid to subcylindric, 8–13 mm. long, 4–6 mm. wide, subcoriaceous; seeds oblong or subcylindric, c. 1·3 mm. long, 0·4 mm. wide, apex shortly winged; testa mid-brown, reticulate.

HAB.: In damp or shady places, often beside streams; c. 800 ft. alt.; fl. June–August.

DISTR.: Division 7, rare. Scattered through the Mediterranean region; Arabia; Morocco.

7. Lapithos, Ayia Paraskevi, 1929, *Frangos* 87 ! 1939, *Lindberg f.* s.n. ! 1954 and 1955, *Merton* 1710 ! 2235 ! Ayios Amvrosios, 1937, *Syngrassides* 1461 ! 1728 !

NOTES: *Hypericum hircinum* appears to be confined in Cyprus to the Northern Range. The Cyprus plants resemble those from Greece and Crete in having relatively short and broad leaves with conspicuous venation.

SECT. 2. **Coridium** *Spach* Shrubs or shrublets, glabrous or with leaves papillose, with black glands confined to the sepal margin or absent; stems 3–4-lined, with lines decurrent from the leaf bases; leaves whorled; flowers in panicles or cymes or rarely solitary; petals and stamens persistent or deciduous; stamen-fascicles 3; styles 3. Fruit capsular; valves with longitudinal vittae and/or oblique vesicles; seeds papillose, not winged.

2. H. empetrifolium *Willd.*, Sp. Plant., 3: 1452 (1802); Sibth. et Sm., Fl. Graec., 8: 54, t. 774 (1833); Poech, Enum. Plant. Ins. Cypr., 35 (1842); Unger et Kotschy, Die Insel Cypern, 353 (1865); Boiss., Fl. Orient., 1: 792 (1867); Gennadius in Cypr. Journ., 1, 5: 7 (1904); Holmboe, Veg. Cypr., 126 (1914); Gresham in New Flora and Silva, 5: 271 (1933); Davis, Fl. Turkey, 2: 385 (1967); Osorio-Tafall et Seraphim, List Vasc. Plants Cyprus, 71 (1973).

[*H. coris* (non L.) Sibth et Sm., Fl. Graec., 8: 56, t. 777 (1833).]

TYPE: " *in* Oriente " (B).

A shrub or shrublet c. 10–60 cm. high; stems erect, 3-lined, with strict branches; leaves in whorls of 3, 4–12 mm. long, 0·7–2 mm. wide, linear, with revolute margins, ericoid, discolorous, glabrous; sepals 1–2·5 mm. long, 0·5–1 mm. wide, oblong to elliptic, rounded, with sessile black marginal glands; petals sometimes tinged red in bud, 5–10 mm. long, 2–3 mm. wide, oblong-elliptic, with pellucid glandular lines and sometimes dots, deciduous; stamens c. 4–8 mm. long, deciduous; anthers with an orange gland; ovary c. 1·5 mm. long, c. 1 mm. wide, ovoid; styles c. 2·5 mm. long, spreading. Capsule narrowly ovoid, 4–6 mm. long, 3–4·5 mm. wide; valves with oblique vesicles; seeds oblong, subcylindric, c. 1·2 mm. long, 0·4 mm. wide; testa bright brown, densely and conspicuously papillose.

HAB.: In coniferous woods, garigue or among rocks, in calcareous soil; ? alt.; fl. May–June.

DISTR.: Division 7 ?. N. Albania, Greece, Aegean, W. Turkey, Cyrenaica.

7. Lapithos, 1904, *Gennadius*; Kyrenia area, 1933, *Gresham*.

NOTES: Until *H. empetrifolium* was recorded from Lapithos by Gennadius all the references to its occurrence in Cyprus were based on Sibthorp's " nec non in insulâ Cypro ". More recently, Gresham included it in a list of plants from the Kyrenia area. I have seen no specimens from Cyprus, but expect that they would be the typical erect variety (Fl. Graec., t. 777), not the procumbent montane one (Fl. Graec., t. 774).

SECT. 3. **Taeniocarpium** *Jaub. et Spach* Perennial herbs, sometimes woody at the base, with black glands confined to the margin of sepals and petals and sometimes the leaf apex; stems terete or 2-lined, with lines decurrent from the leaf bases; leaves opposite; flowers in panicles or cymes or solitary; petals and stamens persistent; stamen-fascicles 3; styles 3. Fruit capsular; valves longitudinally vittate; seeds rugulose to papillose, not winged.

3. H. confertum *Choisy*, Prodr. Monogr. Hypér., 55, t. 8 (1821); Boiss., Fl. Orient., 1: 797 (1867); Post in Mém. Herb. Boiss., 18: 92 (1900); H. Stuart Thompson in Journ. Bot., 44: 278 (1906); Post, Fl. Pal., ed. 2, 1: 231 (1932); Gresham in New Flora and Silva, 5: 271 (1933); Rechinger f. in Arkiv för Bot., ser. 2, 5: 290 (1960).

A herb 10–35 cm. high; stems erect or ascending from a rooting and branching base, glabrous to pubescent, with axillary short shoots; leaves on main stem 7–15 (–20) mm. long, 1–3 (–4) mm. wide, lanceolate to oblong-linear, with ± revolute margins, pruinose to pubescent; leaves on axillary shoots smaller and crowded; inflorescence narrowly pyramidal to cylindric or interrupted-spicate, 3–20 (–33)-flowered; sepals 4–7 mm. long, 1–3 mm. wide, lanceolate to oblong or ovate, acute to rounded, 3–5-ribbed, with margin black-glandular-ciliate to -fimbriate; petals tinged red or red-veined, 7–16 mm. long, 4–6 mm. wide, obovate to elliptic-obovate, with pellucid glandular lines and dots, and sometimes marginal black glandular cilia towards the apex; stamens c. 6–8 mm. long; anthers with an amber gland; ovary 2–3 mm. long, 1·5–1·8 mm. wide, ovoid; styles c. 6 mm. long, spreading. Capsule narrowly ovoid, 6–9 mm. long, 2·5–5 mm. wide.

ssp. **stenobotrys** (*Boiss.*) *Holmboe*, Veg. Cypr., 126, t. 40 (1914); Post, Fl. Pal., ed. 2, 1: 231 (1932); Warr et Gresham in Journ. R.H.S., 59: 71 (1934); Lindberg f., Iter Cypr., 23 (1946); Rechinger f. in Arkiv för Bot., ser. 2, 5: 291 (1960); Davis, Fl. Turkey, 2: 380 (1967); Osorio-Tafall et Seraphim, List Vasc. Plants Cyprus, 71 (1973).
H. stenobotrys Boiss., Diagn., 1, 8: 115 (1849).
H. confertum Choisy var. *stenobotrys* (Boiss.) Boiss., Fl. Orient., 1: 797 (1867); Fl. Orient. Suppl., 128 (1888); H. Stuart Thompson in Journ. Bot., 44: 278 (1906).
TYPE: Syria or Turkey; " in sylvis jugi Cassii ", *Boissier* (G).

Stems usually glabrous (always so in Cyprus); inflorescence cylindric or interrupted spicate; sepals oblong to ovate, obtuse to rounded, with margin ciliate (not fimbriate). Seeds curved-cylindric, 2·2–2·5 mm. long, 0·6–0·8 mm. wide; testa mid-brown, densely rugulose.

HAB.: On rocky ground in Pine woods; 4,000–6,604 ft. alt.; fl. June–July.

DISTR.: Division 2, common in the area above 4,000 ft. alt. South and west Turkey, Syria, Lebanon.

2. Common; *Sintenis & Rigo* 691 ! *Post* s.n. ! *Holmboe* 858 ! *Haradjian* 395 ! 981 ! *Syngrassides* 721 ! *Kennedy* 624 ! 625 ! *Lindberg f.* s.n. ! *Davis* 1808 ! *Mavromoustakis* 28 ! *Casey* 839 ! *Merton* 2441 !

NOTES: Typical ssp. *confertum* occurs in western Turkey, whereas ssp. *stenobotrys* is found in south Turkey (Taurus, Amanus), Syria, the Lebanon and Cyprus. Some plants ± intermediate between the subspecies have been collected in Central Anatolia.

SECT. 4. **Adenosepalum** *Spach* Perennial herbs, usually ± pubescent, with black or reddish (intramarginal and sometimes superficial) glands present on leaves, sepals, anthers and sometimes petals; stems usually terete, usually without axillary shoots; leaves opposite; flowers in pyramidal panicles, corymbs or rarely solitary; petals and stamens persistent; stamen-fascicles 3; styles 3. Fruit capsular, valves longitudinally vittate; seeds linear-foveolate to subscalariform, not winged.

4. **H. lanuginosum** *Lam.*, Encycl. Méth., 4: 171 (1797); Boiss., Fl. Orient., 1: 807 (1867); Holmboe, Veg. Cypr., 127 (1914); Post, Fl. Pal., ed. 2, 1: 233 (1932); Rechinger f. in Arkiv för Bot., ser. 2, 5: 292 (1960); Zohary, Fl. Palaest., 1: 223, t. 329 (1966); Robson in Notes Roy. Bot. Gard. Edinb., 27: 196 (1967); Davis, Fl. Turkey, 2: 386 (1967); Osorio-Tafall et Seraphim, List Vasc. Plants Cyprus, 71 (1973).

Herb 12–80 cm. high; stems erect or rarely decumbent but not rooting, whitish-crisped-pubescent to -puberulous or rarely glabrous, terete; leaves 15–60 mm. long, 10–25 mm. wide, ovate to lanceolate or oblong, densely whitish-pubescent to scabridulous; inflorescence pyramidal or subcylindric, many-flowered, glabrous; sepals 2–4·5 mm. long, 1–2·5 mm. wide, broadly ovate to oblong or rarely lanceolate, obtuse to rounded, with margin black-glandular-denticulate to -ciliate or with sessile glands, without superficial black glands; petals rarely red-veined, 7–12 mm. long, 3–5 mm. wide, elliptic to oblong, completely eglandular or with pellucid dots and rarely a few marginal or subapical black dots; stamens 4–8 mm. long; anthers with a black gland; ovary c. 2 mm. long, 1–1·5 mm. wide, ovoid; styles 3·5–5 mm. long, spreading. Capsule broadly ovoid, 4–6 mm. long, 2·5–4 mm. wide; seeds oblong, subcylindric, about 0·5 mm. long, 0·3 mm. wide; testa subfuscous, closely rugulose-reticulate or subscalariform.

var. lanuginosum

H. gracile Boiss., Diagn., 2, 5: 70 (1856).

H. lanuginosum Lam. var. *gracile* (Boiss.) Boiss., Fl. Orient., 1: 808 (1867); Fl. Orient Suppl., 129 (1888); Post, Fl. Pal., ed. 2, 1: 232 (1932).

H. lanuginosum Lam. subsp. *gracile* (Boiss.) Holmboe, Veg. Cypr., 129 (1914) in adnot.

H. lanuginosum Lam. subsp. *millepunctatum* Holmboe, Veg. Cypr., 129 (1914) pro parte excl. syn. *H. supinum* Vis. et spec. *Sintenis* 602; Lindberg f., Iter Cypr., 23 (1946).

[*H. atomarium* (non Boiss.) Osorio-Tafall et Seraphim, List Vasc. Plants Cyprus 71 (1973).]

TYPE: Lebanon ?; " Levant " *in Hb. Jussieu* (P).

Plant usually robust; stem and leaves pubescent or puberulous; sepals with margin regularly glandular.

HAB.: Dry calcareous soil, in shade or in the open in damper habitats; alt. c. 1,000 ft.; fl. May–June.

DISTR.: Division 7, rare. South Turkey (Taurus, Amanus), Syria, Lebanon, Palestine, Sinai.

7. Rare; Lapithos, 1880, *Sintenis & Rigo* 616 ! Same locality, 1939, 1959 *Lindberg f.* s.n. ! *Mapple* 92 !

NOTES: Continuous variation in sepal width, sepal dentation and stoutness of the inflorescence make it impossible to recognise *H. gracile*, even as a variety. Likewise, the presence or absence of numerous pellucid glands in the leaves is apparently uncorrelated with other characters in such a way as to allow the recognition of Holmboe's ssp. *millepunctatum*. In Cyprus, *H. lanuginosum* has been found only in the mountains above Lapithos, where it is represented by the typical variety. Two other varieties occur in south Turkey.

SECT. 5. **Drosocarpium** *Spach* Perennial herbs, almost always glabrous, with black (intramarginal and sometimes superficial) glands present on leaves, sepals, petals and anthers; stems terete or slightly 2-lined, usually without axillary shoots; leaves opposite; flowers in broadly pyramidal panicles or corymbs; petals and stamens persistent; stamen-fascicles 3 (4–5); styles 3 (4–5). Fruit capsular, valves with vesicles or interrupted longitudinal vittae. Seeds longitudinally ribbed, transversely striate, not winged.

5. **H. perfoliatum** *L.*, Syst. Nat., ed. 12, 510 (1767); Burtt et Davis in Kew Bull. 1: 102 (1949); Rechinger f. in Arkiv för Bot., ser. 2, 1: 427 (1950); Davis, Fl. Turkey, 2: 389 (1967); Osorio-Tafall et Seraphim, List Vasc. Plants Cyprus, 71 (1973).
H. ciliatum Lam., Encycl. Méth., 4: 170 (1797); Boiss., Fl. Orient., 1: 812 (1867).
H. dentatum Loisel., Fl. Gall., ed. 1, 2: 499, t. 17 (1807).
H. myrtifolium Spach, Hist. Nat. Vég., 5: 399 (1836) non Lam. (1797).
TYPE: Without locality.

A herb 15–80 cm. high, glabrous; stems erect or decumbent but not rooting, terete or slightly 2-lined; leaves 13–60 mm. long, 7–24 mm. wide, ovate to triangular-lanceolate or rarely linear, usually amplexicaul, with obscure reticulate venation, at least the upper ones densely pellucid-gland-dotted, and often with some marginal black glandular cilia; inflorescence subcorymbose at first, elongating by monochasial branching, many-flowered; sepals 3–4 mm. long, 1–2 mm. wide, oblong, subacute to rounded, with margin densely and irregularly black-glandular-ciliate, with ± numerous superficial black streaks and dots mostly in 2 vertical rows; petals sometimes red-tinged, 9–14 mm. long, 4–7 mm. wide, obovate, with pellucid dots and streaks, with intramarginal black dots, small marginal reddish-glandular cilia, and sometimes with superficial black dots or streaks towards the apex; stamens 6–9 mm. long; anthers with a black gland; ovary 3 mm. long, c. 2 mm. wide, ovoid; styles 6–8 mm. long, ascending. Capsule broadly ovoid, 5–6 mm. long, 4–4·5 mm. wide, valves with dorsal vittae and lateral vesicles; seeds oblong, subcylindric, 1·2 mm. long, 0·5 mm. wide; testa pale brown, with regular longitudinal ridges.

HAB.: In shady places, among rocks; 2,000–2,500 ft. alt.; fl. April–June.

DISTR.: Division 2, rare. South Europe, Aegean Islands, extreme W. and S. Turkey, N.W. Africa, Madeira, Canary Is.

2. Rare, Ayios Theodhoros R., near Stavros tis Psokas c. 2,500 ft. alt. 1940, *Davis* 1765 ! and, 1962, *Meikle* 2696 !

NOTES: *H. perfoliatum* reaches its easternmost limit in Cyprus, where it has been found twice in the south-west of the island. A narrow-leaved form occurs in two places in the Manavgat area of southern Anatolia, but in Turkey the species is otherwise confined to the extreme west. It gives place, eastwards, to the closely related *H. montbretii* Spach, which occurs as far south as the Lebanon.

SECT. 6. **Oligostema** (*Boiss.*) *Stefanoff* Perennial herbs, almost always glabrous, with black (intramarginal and superficial) glands present on leaves, sepals, petals and anthers; stems 2-lined, sometimes with axillary shoots; leaves opposite; flowers in corymbs or solitary; petals and stamens persistent; stamen fascicles 3–5; styles 3–5. Fruit capsular, valves with longitudinal vittae; seeds linear-foveolate.

6. **H. repens** *L.*, Cent. I Plant., 287 (1755), Fl. Palaest., 27 (1756), Sp. Plant., ed. 2, 1103 (1763); Hume in Walpole, Mem. Europ. Asiat. Turkey, ed. 1, 253 (1817).
H. tenellum Clarke, Travels, ed. 1, 2 (1): 644 (1812), 2 (3): 720 (1816).
H. modestum Boiss., Fl. Orient. Suppl., 127 (1888); Holmboe, Veg. Cypr., 126, t. 39 (1914); Gresham in New Flora and Silva, 5: 271 (1933); Warr et Gresham in Journ. R.H.S., 59: 71 (1934); Lindberg f., Iter Cypr., 23 (1946); Osorio-Tafall et Seraphim, List Vasc. Plants Cyprus, 71 (1973).
[*H. leprosum* (non Boiss.) R. Keller in Engl. et Prantl, Pflanzenfam., ed. 1, 3, 6: 213 (1895).]
TYPE: " *in* Palaestina & *adjacentibus*. *D. Hasselquist* " (LINN).

A herb 1–20 cm. long; stems prostrate or ascending, rooting, 2-lined; leaves 2–9 mm. long, 1–4 mm. wide, mostly oblanceolate-spathulate, lower ones ± elliptic, rounded to apiculate at the apex, narrowly cuneate at the base, with a row of intramarginal black dots, or often with 1–2 subapical ones, or sometimes with only a few scattered, black dots; epidermis undulate or finely papillose, often glaucous; inflorescence 1–8-flowered, usually condensed; sepals 3–5 mm. long, 1–2 mm. wide, oblong to linear or linear-spathulate, acute to obtuse or apiculate, entire or subentire, with intramarginal black dots and often subapical superficial ones; petals red-tinged, 6–10 mm. long, 3–5 mm. wide, obovate-elliptic to oblanceolate, with pellucid dots and streaks and marginal black dots; stamens 4–7 mm. long; anthers with a black gland; ovary 2–2·5 mm. long, 1–1·7 mm. wide, ovoid or ellipsoid; styles 2·5–3·5 mm., ascending. Capsule 3–4 mm. long, 2–3 mm. wide, broadly ovoid; seeds shortly curved-cylindric, about 1·2 mm. long, 0·5 mm. wide; testa subfuscous, regularly foveolate.

HAB.: In dry stony ground and grazed pastures; alt. 0–4,800 ft.; fl. May–July.

DISTR.: Divisions 2, 3, 5, 7. Endemic.

2. Kryos Potamos, 1937, *Kennedy* 621 ! Platres 1937, *Kennedy* 622 ! 623 !
3. Trimiklini, 1959, *Oswald* 164 ! Kakomallis, 1962, *Meikle* 2899 !
5. Mia Milea, 1950, *Chapman* 487 ! 672 !
7. Kythrea, 1880, *Sintenis & Rigo* 582 ! Kyrenia, 1938, *Wyatt* 33 ! 1939, *Lindberg f.* s.n. !

NOTES: It may be assumed that the type collecting of *H. repens* L. was made in Cyprus by Hasselquist during his short stay in the island, from 28th May to early July, 1751. If so, *H. repens*, and not *Quercus alnifolia*, would be the first endemic to have been collected on the island.

SECT. 7. **Hypericum** Glabrous perennial herbs, with intramarginal and sometimes superficial black glands present on stems, leaves, anthers, and sometimes sepals and petals; stems 2–4-lined, with axillary shoots; leaves opposite; flowers in corymbose or pyramidal cymes or panicles; petals and stamens persistent; stamen-fascicles 3; styles 3. Fruit capsular, valves with longitudinal vittae or with dorsal vittae and lateral vesicles; seeds reticulate-pitted, not winged.

7. **H. perforatum** *L.*, Sp. Plant., ed. 1, 785 (1753); Boiss., Fl. Orient., 1: 809 (1867); Post in Mém. Herb. Boiss., 18: 92 (1900); Gennadius in Cypr. Journ., 1, 5: 7 (1904); H. Stuart Thompson in Journ. Bot., 44: 278 (1906); Holmboe, Veg. Cypr., 129 (1914); Post, Fl. Pal., ed. 2, 1: 234 (1932); Lindberg f., Iter Cypr., 23 (1946); Rechinger f. in Arkiv för Bot., ser. 2, 1: 427 (1950); Davis, Fl. Turkey, 2: 400 (1967); Osorio-Tafall et Seraphim, List Vasc. Plants Cyprus, 71 (1973).

H. perforatum L. var. *angustifolium* DC. in Lam. et DC., Fl. Franç., 5: 630 (1815); Druce in Rep. B.E.C., 9: 470 (1931).

H. (?) sp. aff. *H. perforato* sensu Kotschy in Unger et Kotschy, Die Insel Cypern, 353 (1865).

TYPE: " *In* Europae *pratis* ".

A herb 10–110 cm. high; stems erect or decumbent and rooting, 2-lined, with branches ± ascending; leaves 5–35 mm. long, 1–13 mm. wide, narrowly ovate or lanceolate to elliptic-oblong or linear, sessile or subsessile, with plane margins and lax reticulate venation, often with numerous large pellucid dots and sometimes a few superficial black ones; inflorescence corymbose or broadly pyramidal, many-flowered; sepals 3–6 mm. long, 0·5–2 mm. wide, lanceolate or elliptic to oblong or linear, acute to acuminate or shortly aristate, entire, sometimes with a few superficial black dots; petals not red-tinged, (5–) 8–15 mm. long, 3–6 mm. wide, oblanceolate, with pellucid dots and streaks, a few marginal or intramarginal black dots, and sometimes with superficial black streaks; stamens 5–10 mm. long; anthers with a black gland; ovary 2–2·5 mm. long, 1–1·5 mm. wide, ovoid; styles 5–8 mm. long, spreading. Capsule ovoid to pyramidal, (4–) 5–9 mm. long, 3–5 mm. wide, with dorsal vittae and lateral vesicles; seeds oblong, subcylindric, 0·8–0·9 mm. long, 0·4 mm. wide; testa subfuscous, regularly foveolate.

HAB.: In cultivated ground and on dry hillsides in mesophytic regions, near water elsewhere; 2,000–5,000 ft. alt.; fl. June–July.

DISTR.: Divisions 1, 2, 7, common in Division 2. Europe, N. Africa and the Atlantic Islands westward to W. Pakistan, Central Asia, China and Manchuria; introduced elsewhere.

1. Lyso, 1913, *Haradjian* 851 !
2. Common in Troödos area; *Druce* s.n. ! *Syngrassides* 738 ! *Lindberg f.* s.n. ! *Davis* 1787A ! 3538 ! *Casey* 842 ! *Merton* 2304 ! *G. E. Atherton* 589 ! 667 !
7. Armenian Monastery, c. 2,500 ft. alt., 1912, *Haradjian* 348 !

NOTES: The Cyprus plants all seem to belong to the narrow-leaved form that has been named var. *angustifolium* DC. In other regions this form tends to intergrade with the typical broad-leaved form, although there is evidence that they differ genetically to some extent. It seems undesirable, however, to give these forms taxonomic status at present.

8. **H. triquetrifolium** *Turra*, Farsetia Nov. Plant. Gen., 12 (1765); Post, Fl. Pal., ed. 2, 1: 233 (1932); Rechinger f. in Arkiv för Bot., ser. 2, 5: 294 (1960); Zohary, Fl. Palaest., 1: 223, t. 328 (1966); Davis, Fl. Turkey, 2: 400 (1967); Osorio-Tafall et Seraphim, List Vasc. Plants Cyprus, 71 (1973).

H. crispum L., Mantissa I, 106 (1767); Sibth. et Sm., Fl. Graec., 8: 55, t. 776 (1833); Unger et Kotschy, Die Insel Cypern, 353 (1865); Boiss., Fl. Orient., 1: 806 (1867); Post in Mém. Herb. Boiss., 18: 92 (1900); H. Stuart Thompson in Journ. Bot., 44: 278 (1906); Holmboe, Veg. Cypr., 127 (1914); Lindberg f., Iter Cypr., 23 (1946).

H. sp., Clarke, Travels, ed. 1, 2, 3: 720 (1816).

TYPE: " Habitat in Grecia, Sicilia, Calabria ".

A herb 15–55 cm. high; stems erect or decumbent and sometimes rooting, 2-lined, with branches widely spreading usually forming a pyramid; leaves 3–20 mm. long, 1–9 mm. wide, triangular-lanceolate or rarely narrowly ovate to linear-oblong, amplexicaul, with undulate margins and usually obsolete reticulate venation, sometimes with medium to small pellucid dots, but without superficial black ones; inflorescence pyramidal, many-flowered; sepals 1–2·2 mm. long, 0·5–1 mm. wide, oblong to ovate-oblong, acute to apiculate or rounded, entire or denticulate, without black dots; petals not red-tinged, 5–7 mm. long, 1–1·5 mm. wide, linear-oblanceolate, with pellucid streaks and dots, without black dots or occasionally with a superficial one; stamens 2·5–5 mm. long; anthers with a black gland; ovary 1·5–2 mm. long, 0·5–1 mm. wide; styles 2·5–3 mm. long. Capsule ovoid, 3–5 mm. long, 2–3 mm. wide, with longitudinal vittae and occasionally a few lateral vesicles; seeds rather irregularly oblong, subcylindric, straight or slightly curved, about 1·5 mm. long, 0·5–0·6 mm. wide; testa subfuscous, distinctly foveolate or transversely rugulose. *Plate 16.*

HAB.: In dry, open stony or sandy ground and cultivated fields; sea-level to 4,500 ft. alt.; fl. May–September.

DISTR.: Divisions 1, 2, 4, 6, 7. Also Sicily, S. Italy and Tunisia eastward to Iraq and S. Iran; naturalised further west in the Mediterranean region.

1. Cape Arnauti, 1913, *Haradjian* 811 ! Yeroskipos, 1939, *Lindberg f.* s.n. !
2. Platres, *Ussher* 84B ! *Kennedy* 626 ! Kryos Potamos, *Kennedy* 1332 ! Prodhromos, *G. E. Atherton* 588 !
4. Athienou, 1860, *Hooker & Hanbury* s.n. ! 1862, *Kotschy* 972 !
6. Nicosia district, 1932, *Syngrassides* 195 ! 1958, *Oswald* 13 !
7. Kyrenia, *Sintenis & Rigo* 581 ! *Casey* 15 ! *Mapple* 81 ! *G. E. Atherton* 85 ! 139 ! 217 ! 362 !

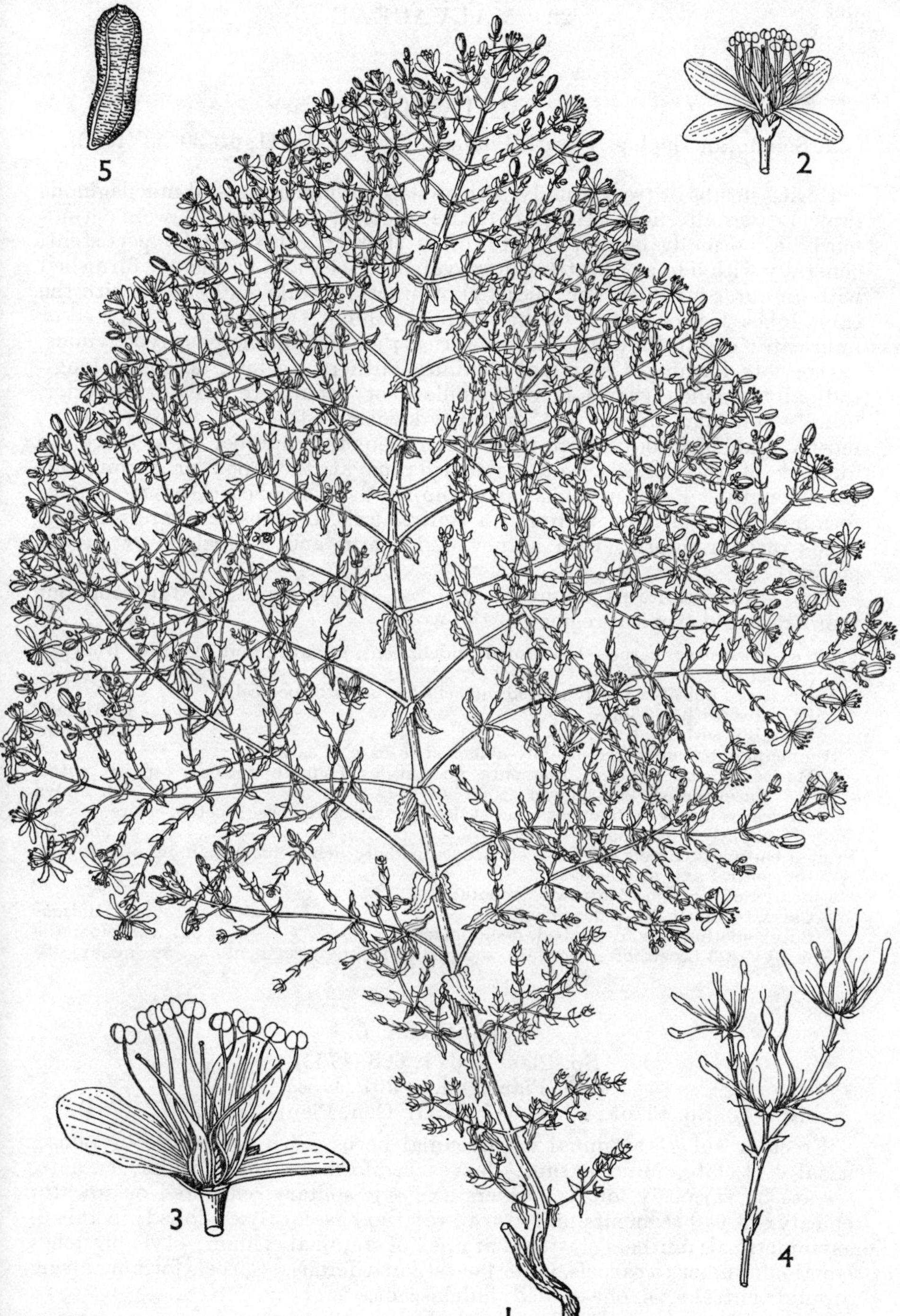

Plate 16. HYPERICUM TRIQUETRIFOLIUM Turra **1**, habit, $\times\frac{2}{3}$; **2**, flower, $\times 2$; **3**, section of flower, $\times 4$; **4**, fruits, $\times 2$; **5**, seed, $\times 10$. (**1, 3** from *Mapple* 81; **2** from *Atherton* 85; **4, 5** from *Atherton* 362.)

22. MALVACEAE

K. Schum. in Engl. et Prantl, Pflanzenfam., ed. 1, III, 6: 30–53 (1890).

Herbs, shrubs or trees usually with stellate indumentum and mucilaginous sap. Leaves alternate, stipulate, entire, toothed or lobed. Flowers actinomorphic, normally hermaphrodite, often showy; calyx usually persistent, generally with 5 free or partly united, valvate lobes, and frequently furnished with an outer involucre (or epicalyx) of bracts; petals 5, alternate with the calyx lobes, free or adnate to the staminal column; aestivation contorted or imbricate; corolla limb often asymmetrical; stamens numerous, hypogynous, ± connate forming a staminal column; anthers 1-celled, dehiscing longitudinally; pollen grains large, papillose or echinulate; ovary superior, usually 5-locular, rarely 1–4 or many-locular; styles ± connate below, mostly branched above, with the branches equal in number to, or twice the number of the carpels, and stigmatiferous at the apex or throughout their length. Fruit a schizocarp composed of free cocci or mericarps, forming a ring or cluster around a central axis, or capsular, rarely fleshy. Seeds often reniform, with scanty endosperm and a straight or curved embryo.

The family contains about 80 genera, and is represented throughout temperate and tropical regions of the world.

Fruit a schizocarp; carpels 1–2-seeded, indehiscent, separating at maturity from the central axis:
- Segments of epicalyx partly united, attached to apex of pedicel:
  - Epicalyx with 6–9 lobes - - - - - - - - - - - **1. Althaea**
  - Epicalyx with 3 lobes - - - - - - - - - - - **2. Lavatera**
- Segments of epicalyx free, usually attached to base of calyx:
  - Mericarps not inflated, filaments and styles connate to near apex; petals conspicuously emarginate - - - - - - - - - **3. Malva**
  - Mericarps inflated; filaments and styles free almost to base; petals entire or very slightly emarginate - - - - - - - - - **4. Malvella**

Fruit a capsule; carpels several-seeded, loculicidally dehiscent, remaining attached to the central axis at maturity:
- Epicalyx of 5 or more narrow segments:
  - Calyx regularly 5-lobed, persistent - - - - - - - **5. Hibiscus**
  - Calyx spathaceous, 0–3-lobed, deciduous - - - - - 5A. ABELMOSCHUS
- Epicalyx of 3 broad, often cordate, toothed or laciniate segments 5B. GOSSYPIUM

1. ALTHAEA *L.*

Sp. Plant., ed. 1, 686 (1753).
Gen. Plant., ed. 5, 307 (1754).

Alcea L., Sp. Plant., ed. 1, 687 (1753), Gen. Plant., ed. 5, 307 (1754).

Erect or suberect annual or perennial herbs, with tomentose or pilose, usually stellate, indumentum; leaves reniform or suborbicular, crenate, serrate or variously lobed; flowers axillary, solitary, clustered or spicate; epicalyx of 6–9 segments united towards the base; calyx 5-lobed; petals 5, asymmetrical; anthers clustered at apex of staminal column; style branches equal in number to carpels, inner face stigmatiferous. Carpels forming a ring around central axis, one-seeded, indehiscent.

About 20 species in Europe, North Africa and western Asia.

Flowers long-pedicellate, petals entire, 1–2 cm. long; annual herb 10–50 cm. high **1. A. hirsuta**

Flowers subsessile, inflorescence spicate, petals emarginate, 4–5 cm. long; robust virgate perennial, about 100 cm. high - - - - - - - - **2. A. setosa**

1. A. hirsuta *L.*, Sp. Plant., ed. 1, 687 (1753); Boiss., Fl. Orient., 1: 824 (1867); Holmboe, Veg. Cypr., 125 (1914); Post, Fl. Pal., ed. 2, 1: 239 (1932); Rechinger f. in Arkiv för Bot., ser. 2, 1: 427 (1950); Davis, Fl. Turkey, 2: 420 (1967); Zohary, Fl. Palaest., 2: 329, t. 485 (1972); Osorio-Tafall et Seraphim, List Vasc. Plants Cyprus, 70 (1973).

[*Malva althaeoides* (non Cav.) Kotschy in Unger et Kotschy, Die Insel Cypern, 352 (1865).]

TYPE: " *in* Gallia, Italia, Hispania ".

An erect, sparingly branched, strigose annual 10–50 cm. high; leaves variable, 1–4 cm. wide, the lowermost suborbicular or reniform with crenate margins, the upper sometimes divided almost to the base into 3–5 narrow-lanceolate, irregularly toothed lobes; lamina glabrescent above, strigose below, especially on nerves and veins; petiole 2–6 cm. long; stipules herbaceous, narrowly ovate or lanceolate, acuminate; flowers solitary, axillary, with a divaricate pedicel 1–6 cm. long; calyx and epicalyx accrescent in fruit; epicalyx normally 9-lobed, the lobes narrowly lanceolate, acuminate; calyx lobes similar to those of the epicalyx, but the lobes longer and narrower, 1–2 cm. in length; petals bluish or rosy-mauve, cuneate, entire, about as long as calyx lobes. Carpels pale brown, reniform, 2–3 mm. long, glabrous, laterally sulcate, with regular radiating furrows; seed dark brown; testa minutely rugulose.

HAB.: In cultivated or fallow fields; occasionally in garigue or on rocky shores; sea-level to 2,000 ft. alt.; fl. March–May.

DISTR.: Divisions 1, 3, 5, 7, 8, common in Divisions 1 and 7. Also Central and S. Europe, Mediterranean area and eastwards to Iran.

1. Common; *Haradjian* 747 ! *Syngrassides* 1839 ! *Davis* 3331 ! 3243 ! *Casey* 1012 ! *Meikle* 2270 !
3. Petra tou Romiou, 1955, *Merton* 2133 !
5. Near Lefkoniko, 1880, *Sintenis & Rigo* (sec. Holmboe).
7. Common; *Kotschy* 468 ! *Syngrassides* 884 ! *Wyatt* 40 ! *G. E. Atherton* 1313 ! 1329 ! *Meikle* 2399 !
8. Ayios Andronikos, 1880, *Sintenis & Rigo* 203 ! Eastern tip of the Karpas, 1880, *Sintenis & Rigo*; Rizokarpaso, 1912, *Haradjian* 232.

2. A. setosa *Boiss.*, Diagn., 1, 8: 107 (1849); Post, Fl. Pal., ed. 2, 1: 241 (1932).

Alcea setosa (Boiss.) Alef. in Oesterr. Bot. Zeitschr., 12: 255 (1862); Zohary in Bull. Res. Council Israel, 11 D: 228 (1963); Davis, Fl. Turkey, 2: 414 (1967); Zohary, Fl. Palaest., 2: 325, t. 479 (1972); Osorio-Tafall et Seraphim, List Vasc. Plants Cyprus, 70 (1973).

Alcea pontica Janka in Oesterr. Bot. Zeitschr., 23: 201 (1873).

Althaea pontica (Janka) Baker f. in Journ. Bot., 28: 144 (1890).

[*Althaea rosea* (non (L.) Cav.) Druce in Rep. B.E.C., 9: 470 (1931).]

[*Althaea lavateriflora* (non DC.) Lindberg f., Iter Cypr., 22 (1946).]

TYPE: Palestine; " in collibus Palestiniae in prov. Samaria ", 1846, *Boissier* (G).

Erect, unbranched perennial, 1 m. or more high, with subglabrous or strigose, purple-blotched stems; basal leaves suborbicular, about 20 cm. diam., with 7 broad, obtuse or subacute lobes, thinly strigose above, more densely stellate-pilose below; cauline leaves variable, suborbicular or reniform, 5–8 cm. or more wide, densely stellate-pilose or almost tomentose on both sides, with 5–7 shallow or deeply cut, acute or obtuse, crenate or dentate lobes; petioles of basal leaves 20–30 cm. long, thinly strigose, those of the cauline leaves much shorter, 6–10 cm. long, often densely strigose; stipules pale brown, chartaceous, caducous, laciniate, with 3–5 narrowly triangular or digitate lobes; flowers showy, solitary or rarely geminate in the axils of the upper leaves, forming a long spicate inflorescence, lower flowers shortly pedicellate, upper almost sessile; epicalyx densely stellate-tomentose, with 6 narrow triangular lobes; calyx similarly clothed, with 5 ovate-acute or triangular lobes 1–1·5 cm. long, slightly longer than the epicalyx; petals obcordate, 4–5 cm. long, rosy-purple with a paler base. Carpels pale brown, broadly lunate or almost orbicular, 4–5 mm. diam., deeply grooved dorsally and ± pilose, laterally ± glabrous and sulcate; seed dark brown, smooth or thinly pubescent on dorsal margin. *Plate 17.*

HAB.: On waste ground; roadsides and stony hillsides; 500–5,600 ft. alt.; fl. May–June.

DISTR.: Divisions 2 and 7, rare. Crete, western Turkey, Syria and Palestine.

2. Makheras, 1928 or 1930, *Druce* s.n.; Troödos, near Olympus Camp Hotel, 1939, *Lindberg f.* s.n.
7. Locally common around Lapithos, 1880, *Sintenis & Rigo* 670 ! *Lindberg f.* s.n. ! *Merton* 2219 !

2. LAVATERA *L.*

Sp. Plant., ed. 1, 690 (1753).
Gen. Plant., ed. 5, 308 (1754).

Herbs or shrubs, similar in general appearance to *Althaea* and *Malva*, but with a 3-lobed epicalyx; the central axis of the receptacle is frequently conspicuous, conical or dome-shaped, rising above the level of the carpels.

Chiefly Europe, Middle East and N. Africa, with outlying representation in Atlantic Islands, Central Asia, India, N.E. tropical Africa, Australia and California. About 40 species.

Flowers in axillary clusters, rarely solitary, subsessile or shortly pedicellate; petals about 1·5 cm. long - - - - - - - - - - - **1. L. cretica**
Flowers solitary, axillary, petals about 2–3 cm. long:
 Flowers subsessile or shortly pedicellate; shrub 1–3 m. high - - **2. L. bryoniifolia**
 Flowers with long, divaricate pedicels, herb 20–100 cm. high - - **3. L. punctata**

1. L. cretica *L.*, Sp. Plant., ed. 1, 691 (1753); Unger et Kotschy, Die Insel Cypern, 351 (1865); Boiss., Fl. Orient., 1: 822 (1867); Holmboe, Veg. Cypr., 125 (1914); Post, Fl. Pal., ed. 2, 1: 237 (1932); Lindberg f., Iter Cypr., 22 (1946); Davis, Fl. Turkey, 2: 409 (1967); Zohary, Fl. Palaest., 2: 321, t. 473 (1972); Osorio-Tafall et Seraphim, List Vasc. Plants Cyprus, 70 (1973).

TYPE: " *in* Creta " (LINN !).

Erect or diffuse perennial herb, 15–100 cm. high, sometimes woody at the base; stems thinly stellate-pubescent, sometimes rather scabrid; lower leaves suborbicular 5–15 cm. diam., with 5–7 shallow, obtuse, crenate or dentate lobes, glabrescent or thinly stellate-puberulous above, stellate-pubescent below; upper leaves smaller and usually more deeply cut with 5–7 acute, crenate or dentate, triangular lobes; petioles long, pubescent; stipules deltoid, 2–8 mm. long, scarious or membranous, caducous; flowers subsessile or shortly pedicellate, geminate or in clusters of 3–7 or more in the axils of the upper leaves, rarely solitary; calyx and epicalyx hirsute, accrescent in fruit; epicalyx 4–7 mm. long with 3 broadly ovate, obtuse or subacute lobes; calyx 6–9 mm. long with 5 acuminate or cuspidate lobes; petals 5, violet or mauve, narrowly obcordate, about 1·5 cm. long, deeply lobed at the apex, with a bearded claw. Carpels reniform, about 3 mm. long, with a close-fitting, glabrous or pubescent, mid-brown pericarp, the dorsal surface obscurely rugulose; lateral surfaces regularly sulcate with radiating furrows; seed mid-brown; testa distinctly papillose.

HAB.: Roadsides, waste ground or edges of cultivated fields, occasionally on sand-dunes; sea-level to 3,500 ft. alt.; fl. Febr.–May (or throughout the summer on irrigated land).

DISTR.: Divisions 1, 2, 4–8, locally very common. Throughout the Mediterranean area; Atlantic Islands.

1. Near Polis, 1962, *Meikle* 2310 !
2. Kato Lefkara, 1937, *Syngrassides* 1523 ! Kryos Potamos, 1938, *Kennedy* 1331 ! Platres, 1955, *Kennedy* 1837 !
4. Near Larnaca, 1862, *Kotschy* 300; Famagusta–Phaleron, 1936, *Syngrassides* 897 !

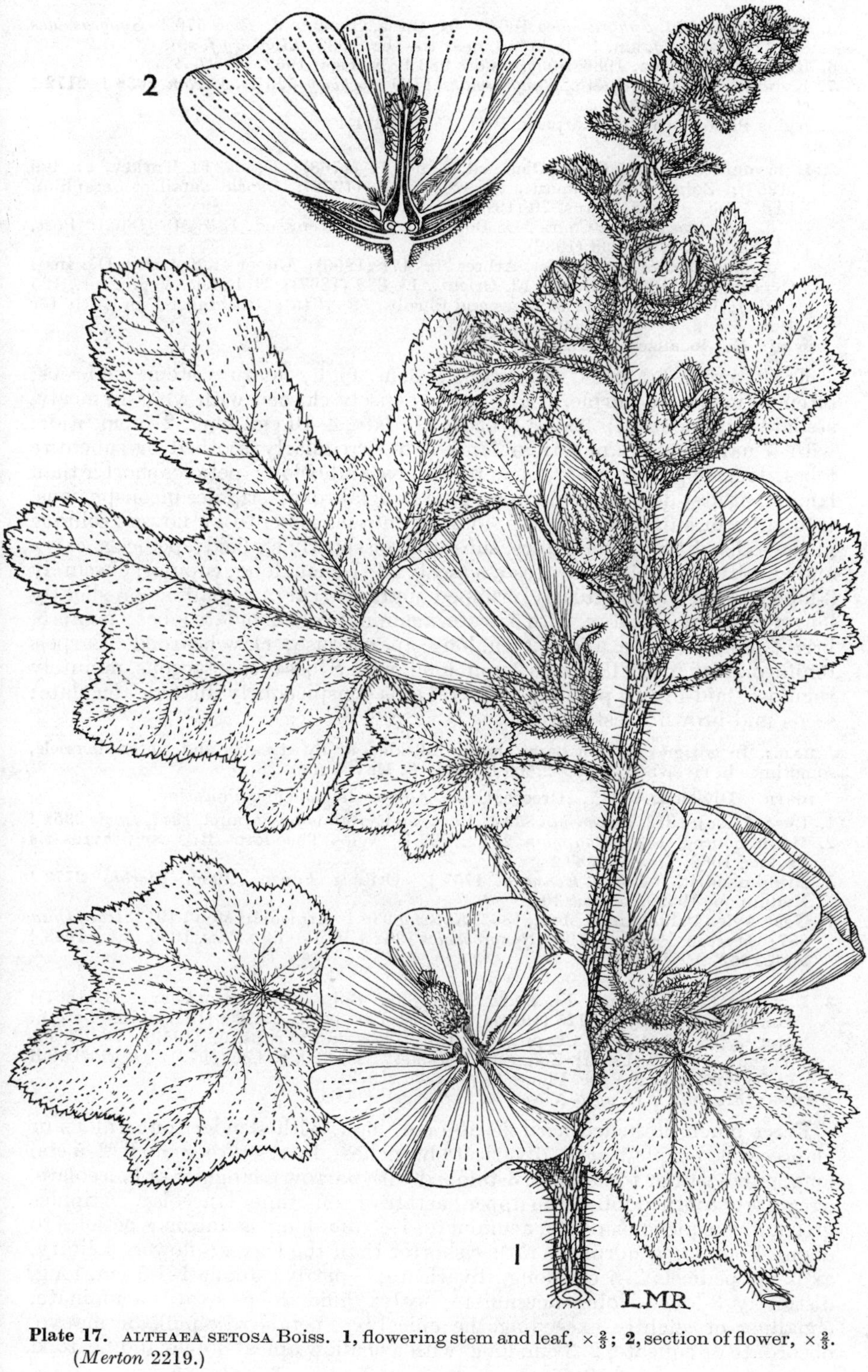

Plate 17. ALTHAEA SETOSA Boiss. **1,** flowering stem and leaf, $\times \frac{2}{3}$; **2,** section of flower, $\times \frac{2}{3}$. (*Merton* 2219.)

5. Syngrasis, 1936, *Syngrassides* 1004 ! Kythrea, *Sintenis & Rigo* 576 ! *Syngrassides* 883 ! *Lindberg f.* s.n. ! Trikomo, near the sea, 1939, *Lindberg f.* s.n.
6. Kokkini Trimithia, 1936, *Syngrassides* 982 ! Nicosia, 1962, *Meikle* !
7. Kyrenia, Kazaphani, etc., *Syngrassides* 1459 ! *Casey* 526 ! *Merton* 2036 ! **2172** ! *G. E. Atherton* 1005 !
8. Ayios Philon near Rizokarpaso, 1941, *Davis* 2221 !

2. L. bryoniifolia *Mill.*, Gard. Dict., ed. 8, no. 11 (1768); Davis, Fl. Turkey, 2: 409 (1967); Zohary, Fl. Palaest., 2: 320, t. 470 (1972); Osorio-Tafall et Seraphim, List Vasc. Plants Cyprus, 70 (1973).

L. tomentosa Dum.-Cours., Le Botaniste Cultivateur, ed. 1, 3: 48 (1802); Post, Fl. Pal., ed. 2, 1: 238 (1932).

L. unguiculata Desf., Hist. Arbres, 1: 471 (1809); Unger et Kotschy, Die Insel Cypern, 351 (1865); Boiss., Fl. Orient., 1: 822 (1867); Holmboe, Veg. Cypr., 125 (1914); Chapman, Cyprus Trees and Shrubs, 52 (1949); Rechinger f. in Arkiv för Bot., ser. 2, 1: 427 (1950).

TYPE: Not localized (BM).

Erect, much-branched subshrub, 1–3 m. high; mature stems glabrous, often stained rosy-purple, young stems densely clothed with whitish, mealy, stellate indumentum; leaves broadly cordate, 4–10 cm. long, 2–6 cm. wide, with 5 narrowly triangular or lanceolate, irregularly toothed or subentire lobes, densely pubescent or tomentose above and below; petioles shorter than lamina, at first densely tomentose, later subglabrous; stipules inconspicuous, triangular or subulate, 2–3 mm. long, caducous, tomentose; flowers solitary in the axils of the upper leaves, subsessile or shortly stalked; pedicels, calyx and epicalyx densely tomentose; epicalyx 1–1·2 cm. long, cupular, obscurely 3-lobed, with an irregularly crimped margin; calyx equalling or slightly longer than the epicalyx, with 5 ovate, acuminate or cuspidate lobes; petals 5, bright pink, cuneate, about 2 cm. long, apex retuse; claw bearded. Carpels reniform, rather small, about 2 mm. long, with a loose, chartaceous, minutely rugulose, mid-brown pericarp; central axis conspicuously dilated, apiculate; seeds mid-brown; testa papillose.

HAB.: In garigue and on bare, chalky hillsides, or by streams and water-channels, sometimes in river-beds; 500–2,500 ft. alt.; fl. May–June.

DISTR.: Divisions 1–3, 7. Greece, Crete, Aegean Islands and Palestine.

1. Lyso [Lisso], 1913, *Haradjian* 869 ! Between Evretou and Sarama, 1941, *Davis* 3351 !
2. Moummouros, 1937, *Chapman* 224 ! 293 ! Ayios Theodoros R. near Stavros tis Psokas, 1962, *Meikle* 2689 !
3. Arkhimandrita, 1952, *Kennedy* 1757 ! Oritaes Forest, 1956, *Merton* 2773 ! Evdhimou, 1959, *Sparrow* 40 !
7. Near Ayios Panteleimon Mon., 1862, *Kotschy* 936 ! Armenian Mon., 1912, *Haradjian* 387; Myrtou–Panagra, 1934, *Syngrassides* 1362 ! Lefkoniko Pass, 1950, *Casey* 1043 ! and, 1962, *Meikle* 2948 !

3. L. punctata *All.*, Auct. Fl. Ped., 26 (1789); Boiss., Fl. Orient., 1: 823 (1867); Holmboe, Veg. Cypr., 125 (1914); Post, Fl. Pal., ed. 2, 1: 238 (1932); Lindberg f., Iter Cypr., 22 (1946); Chapman, Cyprus Trees and Shrubs, 52 (1949); Davis, Fl. Turkey, 2: 410 (1967); Zohary, Fl. Palaest., 2: 320, t. 471 (1972); Osorio-Tafall et Seraphim, List Vasc. Plants Cyprus, 70 (1973).

TYPE: France; " In agro *Nicaeensi* frequens " (TO).

Erect much-branched herb 20–100 cm. high, with slender, puberulous or glabrescent, purple-tinged, divaricate branches; leaves rather small, 2–5 cm. long, pubescent or tomentose, 3–5-lobed with narrow triangular or lanceolate, irregularly toothed lobes, the upper hastate or sometimes lanceolate; stipules green, inconspicuous, ovate, acuminate, 1–2 mm. long, caducous; petioles to about 4 cm. long, normally rather shorter than the leaves; flowers solitary, axillary, pedicels 3–8 cm. long, divaricate; epicalyx about 1–1·5 cm. long, distinctly 3-lobed, lobes acuminate; calyx 5-fid, lobes ovate, acuminate, equalling or slightly exceeding the epicalyx; petals rose-pink or mauve, obcordate or cuneate, 2–3 cm. long, with a shallow apical notch, and bearded claw. Carpels reniform 2–2·5 mm. long, with a loose, membranous, ribbed, pale brown pericarp; seed dark brown; testa smooth.

HAB.: Olive groves, orchards and field borders; sea-level to 200 ft. alt.; fl. May–August.

DISTR.: Divisions 3 and 7, common in Division 7. Mediterranean area, also (? native) Caucasia and Iran.

3. Akhelia, 1939, *Lindberg f.* s.n. !
7. Common on the coastal lowlands; *Sintenis & Rigo* 618 ! *Holmboe* 830; *Syngrassides* 973 ! *G. E. Atherton* 374 ! *Mapple* 20 ! *Casey* 745 ! 1676 !

3. MALVA *L.*

Sp. Plant., ed. 1, 687 (1753).
Gen. Plant., ed. 5, 308 (1754).

Herbs and subshrubs, superficially resembling *Althaea* and *Lavatera* but with an epicalyx of 1–3 free segments, usually adnate to the base of the calyx; central axis of the receptacle rarely conspicuous, not raised above the level of the carpels.

Europe and Asia, south to tropical Africa, India and Malaysia; introduced elsewhere. About 30 species.

Flowers solitary, axillary; annual herbs:
  Stems patently strigose; erect herbs, 10–60 cm. high; basal leaves broadly cordate or almost orbicular:
    Petals 8–15 mm. long, about as long as calyx lobes - **1. M. cretica var. cretica**
    Petals 18–20 mm. long, much longer than calyx lobes **1. M. cretica var. althaeoides**
  Stems adpressed-strigose; decumbent or suberect herbs, 5–15 cm. high, all the leaves deeply laciniate - - - - - - - - - - **2. M. aegyptia**
Flowers clustered, axillary:
  Epicalyx with ovate or broadly lanceolate or oblanceolate segments:
    Petals 15–20 mm. long; plant 30–150 cm. high - - - - **3. M. sylvestris**
    Petals 7–10 mm. long; plant 7–40 cm. high - - - - - **4. M. nicaeënsis**
  Epicalyx with linear or filiform segments:
    Leaves with shallow, obtuse, crenate or shortly dentate lobes:
      Dorsal surface of carpels obscurely reticulate; calyx not noticeably accrescent in fruit - - - - - - - - - - **5. M. neglecta**
      Dorsal surface of carpels strongly reticulate; calyx conspicuously accrescent in fruit - - - - - - - **6. M. parviflora var. parviflora**
    Leaves deeply incised, with sharply dentate lobes **6. M. parviflora var. oxyloba**

1. M. cretica *Cav.*, Diss., 2: 67 (1786); Unger et Kotschy, Die Insel Cypern, 352 (1865); Boiss., Fl. Orient., 1: 818 (1867); Holmboe, Veg. Cypr., 125 (1914); Davis, Fl. Turkey, 2: 405 (1967); Osorio-Tafall et Seraphim, List Vasc. Plants Cyprus, 69 (1973).

An erect, patently strigose, or glabrescent annual, 10–60 cm. high; stems simple, or branched at the base in lush specimens, often purple-tinged; basal leaves broadly cordate or suborbicular, 1·5–4 cm. diam., glabrescent above, thinly strigose below, margin crenate; cauline leaves deeply cut into 5–7 lanceolate or ovate, acute or obtuse, crenate-serrate lobes; uppermost leaves divided almost to the base into 3 lanceolate or linear-lanceolate, remotely serrate or subentire lobes; petioles of basal leaves longer than lamina, upper leaves shortly stalked or subsessile; stipules 3–5 mm. long, ovate or lanceolate, acute or obtuse, sometimes minutely dentate, green, persistent; flowers solitary, axillary; pedicels 1·5–8 cm. long, longer than the subtending leaf, divaricate or almost patent in fruit; segments of the epicalyx hispid, linear-lanceolate, 3–8 mm. long, attached to the base of the calyx; calyx with 5 subulate lobes 8–15 mm. long; petals mauve-pink, about as long as the calyx lobes, or much longer in var. *althaeoides*. Carpels reniform, about

2 mm. long, mid-brown, with distinct dorsal and lateral furrows; seeds dark brown; testa papillose.

var. **cretica**

TYPE: Crete; " apud de Jussieu in herbario D. Isnardi. Unicum exemplar " (P).

Petals equalling or a little longer than calyx-lobes, generally less than 15 mm. long; plant more slender than var. *althaeoides*.

HAB.: Rocky pastures and open garigue; sea-level to 1,600 ft. alt.; fl. April–June.

DISTR.: Divisions 3 and 7, frequent in Division 7. Widespread in Mediterranean area.

3. Around Limassol, towards Kolossi, 1862, *Kotschy*; Cape Gata, 1962, *Meikle* !
7. Frequent about Kyrenia; *Sintenis & Rigo* 578 ! *Tracey* 33 ! *Wyatt* 12 ! *Merton* 2682 ! 2689 !

var. **althaeoides** (*Cav.*) *Gavioli* in Cavanillesia, 2: 84 (1929).

M. althaeoides Cav., Icones, 2: 30, t. 135 (1793).

TYPE: Spain; " prope Enguerae oppidum in tractu vulgo Lloruz " (MA).

More robust than *M. cretica* Cav. var. *cretica*, with larger flowers; petals 18–20 mm. long, almost twice the length of the calyx lobes. The stem is often conspicuously speckled with small, dark purple, scabrid points.

HAB.: Not known; near sea-level; fl. March.

DISTR.: Division 4, rare. Also S.E. Spain, S. Italy and Malta.

4. Around Larnaca, 1880, *Sintenis & Rigo* 200 !

NOTES: An interesting record for a plant long supposed to occur only in S.E. Spain, but more recently reported from Italy and Malta. *Sintenis & Rigo* 200 is labelled *Malva sylvestris*, but of the three sheets so numbered and labelled in herb. Kew. one only consists of *M. sylvestris* (var. *oxyloba* Post); a second consists of a mixture of *M. sylvestris* and *M. cretica* var. *althaeoides*, and the third is entirely *M. cretica* var. *althaeoides*. There are no subsequent records for the variety from Cyprus, and it may have been a mere casual, which did not persist, though it is also possible that *M. cretica* var. *althaeoides* occurs sporadically as a mutant in populations of typical *M. cretica*.

2. **M. aegyptia** *L.*, Sp. Plant., ed. 1, 690 (1753); Boiss., Fl. Orient., 1: 818 (1867), Fl. Orient. Suppl., 131 (1888); Holmboe, Veg. Cypr., 125 (1914) as *M. aegyptica*; Post, Fl. Pal., ed. 2, 1: 235 (1932); Rechinger f. in Arkiv för Bot., ser. 2, 1: 427 (1950); Davis, Fl. Turkey, 2: 406 (1967); Zohary, Fl. Palaest., 2: 316, t. 463 (1972); Osorio-Tafall et Seraphim, List Vasc. Plants Cyprus, 69 (1973).

TYPE: " *In* Aegypto " (LINN !).

Much branched decumbent or suberect annual, 5–15 cm. high; stems adpressed-strigose; leaves broadly cordate, 1·5–3 cm. wide, glabrous above, thinly strigose below, divided to the base into 5 (rarely 3) cuneate, deeply laciniate, acutely lobed segments; petioles to about 3·5 cm. long, as long as or longer than the lamina; stipules ovate, 2–5 mm. long, acute, green, persistent, sometimes with one or two distinct teeth; flowers solitary, axillary, usually crowded and ± corymbose towards the apex of the stems; pedicels 2–20 mm. long; epicalyx with 1–2, or rarely 3, linear, hispid segments 2–5 mm. long, attached to the apex of the pedicel; calyx 5–8 mm. long with 5 broad, ovate or triangular, acuminate, somewhat accrescent lobes; petals pink, narrowly cuneate, about as long as the calyx. Carpels reniform, about 1·5–2 mm. long, pale brown, with distinct lateral and dorsal furrows, minutely papillose; seeds mid-brown; testa smooth.

HAB.: In rocky pastures; by roadsides, or occasionally on shaded banks in *Pinus* woodland; sea-level to 1,700 ft. alt.; fl. March–April.

DISTR.: Divisions 1, 3–8. Mediterranean area eastwards to Central Asia.

1. Akamas, 1948, *Kennedy* 1646 !
3. Akrotiri Forest, 1939, *Mavromoustakis* 120 !
4. Neck of Cape Greco, 1958, *N. Macdonald* 6 !

5. Near Lefkoniko, 1880, *Sintenis & Rigo* 202 ! Vatili, 1905, *Holmboe* 342; Athalassa, 1936, *Syngrassides* 1125 ! Lapathos, 1962, *Meikle* 2965 !
6. Cape Kormakiti, 1960, *N. Macdonald* 74 ! Dhiorios, 1962, *Meikle* 2324 !
7. Above Kyrenia Pass, 1953, *Casey* 1298 !
8. Cape Andreas, 1938, *Syngrassides* 1766 ! Kantara, 1941, *Davis* 2466 ! Koma tou Yialou, 1950, *Chapman* 224 !

3. M. sylvestris *L.*, Sp. Plant., ed. 1, 689 (1753); Hume in Walpole, Mem. Europ. Asiat. Turkey, ed. 1, 254 (1817); Unger et Kotschy, Die Insel Cypern, 352 (1865); Boiss., Fl. Orient., 1: 819 (1867); Holmboe, Veg. Cypr., 125 (1914); Post, Fl. Pal., ed. 2, 1: 236 (1932); Davis, Fl. Turkey, 2: 406 (1967); Zohary, Fl. Palaest., 2: 316, t. 464 (1973); Osorio-Tafall et Seraphim, List Vasc. Plants Cyprus, 70 (1973).

M. sylvestris L. var. *eriocarpa* Boiss., Fl. Orient., 1: 819 (1867); Post, Fl. Pal., ed. 2, 1: 236 (1932); Lindberg f., Iter Cypr., 22 (1946).

Diffuse or erect, subglabrous or thinly pubescent perennial, 30–150 cm. high; leaves broadly cordate or almost orbicular, to 10 cm. diam., with 3–7 lobes; petiole to 15 cm. long, often much longer than the lamina; stipules triangular, 4–8 mm. long, membranous, greenish, persistent, with ciliate, entire or irregularly toothed margins; flowers axillary, in clusters of 2–5; pedicels rather unequal, to 2 cm. or more long; epicalyx and calyx accrescent in fruit; epicalyx with 3 ovate-lanceolate, pubescent segments, 3–4 mm. long; calyx 4–7 mm. long, pubescent, with 5 triangular lobes; petals mauve, paler towards base, obcordate, 1·5–2 cm. or more long, with a distinct apical notch and a bearded claw. Carpels reniform, about 3 mm. long, pale brown, glabrous or pubescent, lateral surfaces with radiating furrows, dorsal surface conspicuously reticulate; seeds dark brown; testa minutely rugulose.

var. **sylvestris**
TYPE: "*in* Europae *campestribus*" (LINN !).

Leaves with 5–7, rather shallow, blunt, crenate or crenate-dentate lobes; calyx-lobes broadly triangular.

HAB.: Waste ground, cultivated and fallow fields; roadsides; 2,000–4,500 ft. alt.; fl. April–Sept.

DISTR.: Divisions 1, 2, 6 and 7, common in Division 2. Throughout Europe, N. Africa and temperate Asia; introduced elsewhere.

1. Ayios Yeoryios (Akamas), 1962, *Meikle* ! Polis, 1962, *Meikle* !
2. Common about Platres and Prodhromos; *Syngrassides* 1598 ! 1627 ! *Davis* 3418 ! *G. E. Atherton* 553 ! *Kennedy* 1329 ! 1330 ! 1835 ! 1836 ! 1840 ! 1841 ! *Merton* 2358 ! 2367 ! 2391 ! 2392 !
6. Dhiorios, 1962, *Meikle* !
7. Ayios Khrysostomos Mon., 1955, *Merton* 2035 !

var. **oxyloba** *Post*, Fl. Pal., ed. 1, 178 (1896).
TYPE: Palestine; Tel-er-Ramé (Jordan valley) *Post* (BEI).

Leaves with 5–7 acute, sharply serrate-dentate lobes; calyx-lobes acuminate.

HAB.: Waste ground, cultivated and fallow fields; roadsides; often near habitations; sea-level to 4,500 ft. alt.; fl. March–Sept.

DISTR.: Divisions 2, 4–7. Also Palestine and Syria.

2. Platres, 1938, *Kennedy* 1326 !
4. Mile 20, Nicosia–Larnaca Rd., 1950, *Chapman* 403 !
5. Near Kythrea, 1880, *Sintenis & Rigo* 200 pro parte ! Kythrea, 1936, *Syngrassides* 885 ! 887 ! Dhali, 1936, *Syngrassides* 900 !
6. Makhedonitissa Mon., ["Machinoditissa"], 1936, *Syngrassides* 891 ! Ayia Irini, Morphou, 1941, *Davis* 2552 ! Nicosia airport, 1955, *Merton* 2038 !
7. Ayios Panteleimon Mon., 1962, *Meikle* 2418 !

NOTES: *Malva sylvestris* var. *oxyloba* appears to be commoner in Cyprus (and in Palestine) than typical W. European var. *sylvestris*; both type and variety have glabrous and hairy-fruited forms, but carpel indumentum varies greatly, and is a character of doubtful taxonomic value. *M. sylvestris* var. *eriocarpa* Boiss. is an extreme variant, with tomentose carpels; it is recorded from Cyprus by Lindberg f.: " M. Troödos in declivi lapidoso siccissimo juxta monasterium Kykko. Kambos, in margine viae." It should be noted that most hairy-fruited specimens of *M. sylvestris* come from arid habitats, or poor, dry soils.

4. M. nicaeënsis *All.*, Fl. Ped., 2: 40 (1785); Boiss., Fl. Orient., 1: 819 (1867); Holmboe, Veg. Cypr., 126 (1914); Post, Fl. Pal., ed. 2, 1: 236 (1932); Davis, Fl. Turkey, 2: 406 (1967); Zohary, Fl. Palaest., 2: 317, t. 465 (1972); Osorio-Tafall et Seraphim, List Vasc. Plants Cyprus, 70 (1973).

[*M. montana* (non Forssk.) Hayek, Prodr. Fl. Pen. Balc., 1: 547 (1925); Rechinger f., Fl. Aegaea, 267 (1943).]

TYPE: France; " — primus reperit *Cl. Bellardi* in agro *Nicaeensi* prope portum *di Limpia* " (TO).

Straggling suberect perennial, 7–40 cm. high, with glabrous or sparsely pilose stems and petioles; leaves almost orbicular, 3–8 cm. diam., with 5–7 shallow, obtuse, shortly crenate-dentate lobes, subglabrous or thinly pubescent; petioles to 15 cm. or more long, usually much longer than the lamina; stipules membranous, broadly ovate, obtuse or subacute, 4–7 mm. long; flowers axillary, solitary, geminate or in clusters of 3–5, pedicels unequal, usually less than 2 cm. long, glabrous or thinly pubescent; epicalyx and calyx accrescent in fruit; epicalyx of 3 elliptic or narrowly ovate, subobtuse, entire or irregularly dentate segments, 5–8 mm. long; calyx with 5 broadly ovate, acute or subacute lobes, a little longer than segments of epicalyx; petals mauve, narrowly cuneate, emarginate, 7–10 mm. long, with a bearded claw. Carpels reniform, 2–3 mm. long, pale brown, glabrous, similar to those of *M. sylvestris*, with a prominently reticulate dorsal surface; seeds dark brown; testa minutely rugulose.

HAB.: Roadsides, waste ground, cultivated and fallow fields; sea-level to 3,700 ft. alt.; fl. Febr.–May (occasionally throughout the summer).

DISTR.: Divisions 2, 3, 5–7. Mediterranean area eastwards to Iraq and Iran; introduced elsewhere.

2. 3 miles from Lefka on the road to Kykko, 1927, *Rev. A. Huddle* 25 ! Platres, 1938, *Kennedy* 1325 !
3. Alethriko, 1905, *Holmboe* 223; Akrotiri peninsula, 1955, *Merton* 2069 !
5. Dhali, 1937, *Syngrassides* 1519 !
6. Makhedonitissa Mon., [" Machinoditissa "], 1936, *Syngrassides* 892 !
7. Common; *Syngrassides* 1436 ! *G. E. Atherton* 58 ! 285 ! 1002 ! 1128 ! 1224 ! *Merton* 2037 ! 2247 !

NOTES: The type of *M. montana* Forssk. has been examined by R. A. Blakelock; it is conspecific with *M. verticillata* L.

5. M. neglecta *Wallr.*, Syll. Plant. Nov., 1: 140 (1824); Davis, Fl. Turkey, 2: 407 (1967); Zohary, Fl. Palaest., 2: 317, t. 466 (1972); Osorio-Tafall et Seraphim, List Vasc. Plants Cyprus, 70 (1973).

[*M. pusilla* (non Sm.) Lindberg f., Iter Cypr., 22 (1946); Osorio-Tafall et Seraphim, List Vasc. Plants Cyprus, 70 (1973).]

TYPE: Germany; Regensburg, " In suburbiorum areis desertis ".

Similar in general habit to *M. nicaeënsis* All. Stems glabrous or shortly and ± stellately pubescent, leaves smaller, rarely more than 4 cm. diam.; calyx and epicalyx not noticeably accrescent; segments of epicalyx narrow, linear or filiform; calyx-lobes triangular, acuminate; petals mauve, narrowly cuneate, emarginate, with bearded claw. Carpels pubescent, with dorsal surface obscurely reticulate or almost smooth; seeds dark brown; testa minutely rugulose.

HAB.: Dry banks and roadsides; 3,800–6,300 ft. alt.; fl. May–Sept.

DISTR.: Divisions 1 and 2, common in Division 2. Widely distributed in Europe, N. Africa and temperate Asia; often adventive.

1. Ayios Neophytos, below the Monastery, 1939, *Lindberg f.* s.n.
2. Common (especially at high altitudes in the central Troödos area), *Syngrassides* 1580 ! *Kennedy* 1327 ! 1328 ! *Lindberg f.* s.n. ! *Casey* 1200 ! *G. E. Atherton* 617 ! *Merton* 2348 ! *D. P. Young* 7309 !

6. M. parviflora *L.*, Sp. Plant., ed. 2, 969 (1763); Boiss., Fl. Orient., 1: 820 (1867); Holmboe, Veg. Cypr., 126 (1914); Post, Fl. Pal., ed. 2, 1: 237 (1932); Davis, Fl. Turkey, 2: 407 (1967); Zohary, Fl. Palaest., 2: 318, t. 467 (1972); Osorio-Tafall et Seraphim, List Vasc. Plants Cyprus, 70 (1973).

M. flexuosa Hornemann, Hort. Hafn., 2: 655 (1815); Unger et Kotschy, Die Insel Cypern, 352 (1865).

Similar in habit to *M. nicaeënsis* and *M. neglecta*; often glabrous or subglabrous; flowers in short-stalked, dense, axillary clusters; pedicels seldom more than 1 cm. long; calyx conspicuously accrescent in fruit, with broad ovate or triangular lobes; segments of epicalyx linear or filiform; petals inconspicuous, 3–7 mm. long, scarcely exceeding calyx, with a glabrous claw. Carpels pale brown, glabrous or, less often, pubescent; dorsal surface strongly reticulate, dorsal margins distinctly ridged; seeds dark brown; testa minutely rugulose.

var. **parviflora**

TYPE: " In Barbaria " (LINN !).

Leaves with broad, blunt, crenate or shallowly toothed lobes; calyx strongly accrescent.

HAB.: Cultivated and fallow fields, roadsides and waste ground; sea-level to 3,500 ft. alt.; fl. Febr.–June.

DISTR.: Divisions 1–8. Europe, N. Africa and temperate Asia, often adventive.

1. Near Ayios Yeoryios (Akamas), 1962, *Meikle* 2105 ! 2112 !
2. Platres, 1955, *Kennedy* 1838 ! 1839 !
3. Akrotiri near Limassol, 1905, *Holmboe* 683.
4. Around Larnaca, 1862, *Kotschy* s.n.; near Old Larnaca, 1880, *Sintenis & Rigo*; Famagusta–Phaleron, 1936, *Syngrassides* 897 !
5. Dhali, 1937, *Syngrassides* 1517 !
6. Near Nicosia, 1880, *Sintenis & Rigo* 201 ! Syrianokhori, 1935, *Syngrassides* 759 ! Makhedonitissa Mon. [" Machinoditissa "], 1936, *Syngrassides* 889 ! 894 ! Nicosia, 1936, *Nattrass* 772; Nicosia, 1939, *Lindberg f.* s.n.
7. Near Kythrea, 1880, *Sintenis & Rigo*; above Kyrenia Pass, 1953, *Casey* 1309 !
8. Panayia tou Kantara, 1880, *Sintenis & Rigo.*

var. **oxyloba** (*Boiss.*) *Kristofferson* in Hereditas, 7: 346 (1926).

M. oxyloba Boiss., Diagn., 1, 8: 109 (1849), Fl. Orient., 1: 821 (1867), Fl. Orient. Suppl., 131 (1888); Holmboe, Veg. Cypr., 126 (1914); Post, Fl. Pal., ed. 2, 1: 237 (1932); Osorio-Tafall et Seraphim, List Vasc. Plants Cyprus, 70 (1973).

TYPE: Palestine; " in pinguibus planitiei Esdraelon Galilaei, et colui e seminibus palaestinis " (G).

Similar in habit to *M. parviflora* var. *parviflora*, with stems, petioles and leaves glabrous or subglabrous; leaves 3–6 cm. diam. with 5–7 deeply incised, sharply dentate lobes. Calyx accrescent, but not quite as conspicuously as in *M. parviflora* var. *parviflora*.

HAB.: As for *M. parviflora* var. *parviflora*; c. 600 ft. alt.; fl. May.

DISTR.: Division 7, rare. Also Sicily, Palestine, N. Africa and Atlantic Islands, rare and sporadic.

7. Kythrea, 1880, *Sintenis & Rigo* 577 !

M. ALCEA *L.*, recorded by Druce (Rep. B.E.C., 9: 470, 1931) from Platres, is an erect perennial with deeply lobed leaves and clusters of large pink flowers. It is a native of W. Europe, sometimes cultivated, and perhaps an escape in Cyprus. It has not been recorded for the island by any other author.

No description of M. CYPRIANA *D. Don* (ex Walpole, Mem. Europ. Asiat. Turkey, ed. 1, 254: 1817) has been traced, and the plant so named cannot be identified.

4. MALVELLA *Jaub. et Spach*

Illustr. Plant. Orient., 5: t. 444 (1855).

Perennial herb; epicalyx of 2–3 filiform segments attached to base of calyx or apex of pedicel; calyx with 5 acuminate lobes; petals 5, cuneate, asymmetrical, entire or slightly emarginate, adnate to base of staminal column; stamens numerous, free almost to the base; stigmas terminal, oblong. Carpels inflated, attached at one point only to the small central axis; seed solitary, pendulous.

One species, formerly placed under *Malva* or *Sida*, from both of which it differs in flower and fruit structure.

1. **M. sherardiana** (*L.*) *Jaub. et Spach*, Illustr. Plant. Orient., 5: t. 444 (1855); Boiss., Fl. Orient., 1: 834 (1867); Holmboe, Veg. Cypr., 126 (1914); Post, Fl. Pal., ed. 2, 1: 243 (1932); Lindberg f., Iter Cypr., 22 (1946); Davis, Fl. Turkey, 2: 408 (1967); Zohary, Fl. Palaest., 2: 319, t. 469 (1972); Osorio-Tafall et Seraphim, List Vasc. Plants Cyprus, 70 (1973).

Malva sherardiana L., Sp. Plant., ed. 2, 1675 (1763); Unger et Kotschy, Die Insel Cypern, 352 (1865).

TYPE: Turkey; " in Bithynia *Arduini* " (LINN !).

Prostrate or suberect, ± tomentose perennial; leaves reniform or suborbicular, 1·5–3 cm. wide, often somewhat plicate, with deeply impressed nerves; leaf-margin variable, subentire, crenate or occasionally dentate-lobed; petiole 1–3 cm. long; stipules green, narrowly lanceolate-acuminate, about 2 mm. long; flower solitary, axillary, on a slender flexuous pedicel 1·5–4 cm. or more long; epicalyx segments filiform, inconspicuous, 2–3 mm. long; calyx 5–8 mm. long, with acuminate, triangular lobes, accrescent in fruit; petals pink, ovate-cuneate, asymmetrical, entire or rarely a little emarginate at apex, claw glabrous. Carpels pale brown, inflated, densely pubescent; seeds subglobose, dark brown; testa densely papillose.

HAB.: Cultivated and waste ground; below 500 ft. alt.; fl. May–July.

DISTR.: Divisions 3 and 4, rare. Also in central Spain, Balkans, Crete, Aegean Islands and Asia eastwards to Iran.

3. Akhelia, 1939, *Lindberg f.* s.n. !
4. Between Larnaca and Aradhippou, 1862, *Kotschy*.

5. HIBISCUS *L.*

Sp. Plant., ed. 1, 693 (1753).
Gen. Plant., ed. 5, 310 (1754).

Herbs, shrubs, trees and climbers with simple or variously lobed leaves; flowers axillary or terminal, generally showy; epicalyx with 3–5, or numerous, free, or partly united segments, rarely absent; calyx 5-lobed, often accrescent and persistent; petals frequently blotched towards the base; staminal column truncate or toothed at the apex, antheriferous for a varying portion of its length; styles 5, ± connate; stigmas terminal, capitate or spathulate. Fruit a 5-celled, loculicidally dehiscent capsule, remaining attached to central axis at maturity; seeds reniform or subglobose, glabrous, pilose or lanuginose.

A very large genus, common in many tropical regions, less frequent, and often adventive, in temperate regions.

1. **H. trionum** *L.*, Sp. Plant., ed. 1, 697 (1753); Sibth. et Sm., Fl. Graec. Prodr., 2: 48 (1813), Fl. Graec., 7: 61, t. 666 (1830); Unger et Kotschy, Die Insel Cypern, 352 (1865); Boiss., Fl. Orient., 1: 840 (1867); Holmboe, Veg. Cypr., 126 (1914); Post, Fl. Pal., ed. 2, 1: 245 (1932); Lindberg f., Iter Cypr., 22 (1946); Davis, Fl. Turkey, 2: 402 (1967); Zohary, Fl. Palaest., 2: 312, t. 457 (1972); Osorio-Tafall et Seraphim, List Vasc. Plants Cyprus, 69 (1973).

TYPE: " *in* Italia, Africa " (LINN !).

A diffuse ± strigose annual, 30–40 cm. high; leaves variable, the basal broadly ovate or suborbicular, entire or remotely toothed, the upper broadly cordate, 3–8 cm. diam., palmately divided into 3–5 lanceolate or obovate lobes with crenate, dentate or laciniate margins, rarely entire or subentire; petioles 2–6 cm. long; stipules greenish, linear or filiform, 3–5 mm. long; flowers solitary, axillary, pedicels to 7 cm. or more in length; epicalyx with 8–12 slender, filiform, hispid segments, 5–7 mm. long; calyx 10–15 mm. long, pale green, veined purple, with 5 broad triangular lobes; petals pale yellow, stained purple externally, and blotched dark purple towards the base, 20–30 mm. long, asymmetrical, cuneate, entire; staminal column, filaments and styles reddish or dark purple; stigmas red or purple, capitate, conspicuously papillose. Capsule concealed within the accrescent, papery calyx, about 1·5 cm. long, densely villose, each of the five cells containing 6–8 dark brown, reniform, bluntly angular seeds, 2–3 mm. long.

HAB.: Cultivated and waste ground; sea-level to 5,000 ft. alt.; fl. June–Sept.

DISTR.: Divisions 2, 3, 6, 7, common in Division 7. A widespread weed in the tropics and subtropics.

2. Near Evrykhou, 1840, *Kotschy*; Trikoukkia Nursery Gardens, 1935, *Syngrassides* 746 ! Kannaviou, near the Ezuza R., 1939, *Lindberg f.* s.n.
3. Akhelia, 1939, *Lindberg f.* s.n.
6. Makhedonitissa Mon. [" Machinoditissa "], 1935, *Syngrassides* 795 !
7. Common about Kythrea; *Syngrassides* 524 ! 694 ! *Lindberg f.* s.n. ! Kyrenia, *G. E. Atherton* 367 ! *Casey* 1213 !

5a. ABELMOSCHUS ESCULENTUS (*L.*) *Moench* (*Hibiscus esculentus* L.), Gombo, Okra or Ladies' Fingers, an annual herb up to 2 m. high with large suborbicular, shallowly or deeply palmatilobed leaves and spicate inflorescences of axillary flowers with large yellow, purple-based petals, is commonly grown in Cyprus as a summer vegetable, the young capsules being succulent and mucilaginous. It is generally believed to be a native of tropical Asia, but is now widely cultivated (and sometimes naturalized) in tropical and subtropical countries.

5b. GOSSYPIUM *L.* Cotton. Three Cotton species, the short-stapled *G. herbaceum* L. (Levant Cotton) and *G. hirsutum* L. (American Upland Cotton) and the long-stapled *G. barbadense* L. (Sea Island Cotton) are recorded from Cyprus by Hume, Kotschy, Gennadius and others. In addition to the species, there are numerous hybrids and cultivars known by local names.

Cotton was an important crop in Cyprus during the 15th and 16th centuries, and perhaps earlier, if we can allow Gennadius' suggestion (Rep. Agric. Cyprus, pt. 2: 4 footnote; 1896) that the Karpas peninsula derives its name from the Sanskrit *kurpasa*, a cotton cloth known to the ancients.

Cotton is still grown, though of declining importance in the island's economy.

22a. STERCULIACEAE

K. Schum. in Engl. et Prantl., Pflanzenfam., ed. 1, III, 6: 69–99 (1890–93).

Trees, shrubs or rarely herbs, usually with stellate indumentum; leaves alternate or rarely subopposite, simple or palmately compound, with caducous stipules; inflorescence often axillary, or plants sometimes cauliflorous; flowers mostly actinomorphic, hermaphrodite or unisexual; sepals 3–5, generally connate below with valvate lobes; petals 5, often much reduced or wanting, contorted-imbricate, hypogynous, sometimes adnate to the base of the staminal tube; stamens in 2 whorls, or those of the outer whorl reduced to staminodes or wanting, connate into a tube, or occasionally free to the base; anthers 2-locular, the loculi sometimes confluent towards the apex; ovary superior, of (2–) 4–5 (or 10–12) more or less united carpels; placentation axile; ovules 2 or more in each loculus, rarely solitary, ascending or horizontal; styles generally connate, sometimes free, simple or divided apically into lobes. Fruit leathery, fleshy or papery (rarely woody), dehiscent or indehiscent, or occasionally dividing into mericarps; seeds with abundant or scanty endosperm; embryo curved or straight.

About 50 genera, chiefly in the tropics and subtropics, and comprising many ornamental plants as well as the economically valuable Cocoa (*Theobroma cacao* L.) and Cola (*Cola nitida* (Vent.) Schott et Endl.; *C. acuminata* (P. Beauv.) Schott et Endl.).

The family is not represented in Cyprus by any indigenous species, but the Australian *Brachychiton populneus* R. Br., Kurrajong or Bottle-tree, is extensively planted in towns and villages. It forms a moderately stout tree, 6–20 m. high, with the trunk conspicuously swollen at the base in young specimens. The leaves are long-petiolate, glabrous and shining, generally entire, ovate or ovate-lanceolate, 5–10 cm. long, with long acuminate apices, but not infrequently palmately 3- or 5-lobed, with short or relatively long lateral lobes; the flowers are unisexual and apetalous, in lax axillary panicles, with a broadly campanulate calyx, 1·5–2 cm. diam., creamy-white or yellowish externally, blotched reddish within, divided above into 5 acute, triangular lobes; the stamens are united into a tube, and the fruit consists of 1–5 stalked follicles, 4–10 cm. long, glabrous and leathery externally, hairy within, containing numerous rather large seeds.

Brachychiton populneus is a valuable shade tree, easily grown from seed, and with deep roots which do not interfere with those of other plants grown around its base. The seeds are reputedly injurious to stock, but the foliage is innocuous (see R. H. Anderson, The Trees of New South Wales, 7; 1947) and the species is said to be " extremely useful as a fodder tree ".

23. TILIACEAE

K. Schum. in Engl. et Prantl, Pflanzenfam., ed. 1, III, 6: 8–30 (1890).

Trees or shrubs, or rarely herbs; indumentum usually of branched hairs; leaves alternate, rarely opposite or subopposite, generally simple; stipules present; inflorescence cymose; flowers usually hermaphrodite, actinomorphic; sepals 5 (rarely 3–4) free or partly connate, normally valvate;

petals as many as sepals or sometimes wanting, hypogynous, free, contorted, imbricate or valvate; stamens often numerous, free or connate at base, sometimes fascicled; anthers 2-thecous, dehiscing longitudinally or by apical pores; ovary superior, of 2–10 united carpels, placentation axile or rarely parietal; ovules ascending or pendulous, 1–many in each loculus; style simple or stigmas rarely sessile. Fruit fleshy or dry, dehiscent or indehiscent; seeds with scanty or copious endosperm; embryo usually straight.

About 40 genera, widely distributed but with a preponderance of species in tropical and subtropical regions.

1. CORCHORUS *L.*

Sp. Plant., ed. 1, 529 (1753).
Gen. Plant., ed. 5, 234 (1754).

Herbs and subshrubs; indumentum of simple or branched hairs; leaves alternate, variously toothed or lobed, stipulate; inflorescences pedunculate, bracteate, often leaf-opposed; flowers small, hermaphrodite; sepals 5 (4); petals as many as sepals, yellow, usually with a short basal claw; stamens usually numerous, free, arising from the apex of a short anthophore; ovary 2–5-locular; placentation axile; ovules numerous; style glabrous, with a concave, capitate or very shortly lobed stigma. Capsule elongate and pod-like or subglobose, smooth, scabrid or echinate, 2–5-locular, dehiscing loculicidally, sometimes transversely septate, many-seeded; seeds pendulous or horizontal; embryo generally curved; cotyledons flat, foliaceous.

About 30 species widely distributed in the tropics and subtropics; the Jute of commerce comes chiefly from *C. capsularis* L., though other species are also occasionally cultivated for their fibres.

Capsule 5-valved, 10-ribbed; sepals glabrous except at base - - **1. C. olitorius**
Capsule 3-valved, obscurely 6-ribbed; sepals pubescent - - - **2. C. trilocularis**

1. C. olitorius *L.*, Sp. Plant., ed. 1, 529 (1753); Unger et Kotschy, Die Insel Cypern, 353 (1865); Boiss., Fl. Orient., 1: 845 (1867); Post, Fl. Pal., ed. 2, 1: 247 (1932); Davis, Fl. Turkey, 2: 424 (1967); Zohary, Fl. Palaest., 2: 310, t. 455 (1972).

[*C. trilocularis* (non L.) Lindberg f., Iter Cypr., 23 (1946).]

TYPE: " in Asia, Africa, America ".

Erect annual 1 (–2) m. high; stems glabrous, becoming woody with age, often rather sparingly branched; leaves ovate or oblong-ovate, 3–10 cm. long, 1–5 cm. wide, glabrous or thinly pubescent on nerves below, apex acute, base truncate, rounded or shortly cuneate, margins sharply serrate, the two lowest serrations produced into conspicuous filiform appendages; petiole 1–3 cm. long, channelled and strigose above, glabrous below; stipules filiform 3–15 mm. long; flowers solitary or in minute 2–3-flowered cymes opposite the upper leaves; peduncles very short, about 1 mm. long, glabrous; bracts linear-filiform, 2–4 mm. long; sepals boat-shaped, about 8 mm. long and 1·5 mm. wide, with a slender acumen, glabrous, or with the margins ciliate, especially towards the base; petals yellow, oblanceolate, 5–7 mm. long, about 2 mm. wide, tapering to a short ciliate claw; stamens 40 or more, about 5 mm. long, filaments glabrous, attached at the apex of a cupular, crenate torus about 0·5 mm. high; anthers oblong, brownish, about 0·5 mm. long and 0·3 mm. wide; ovary cylindrical, about 3·5 mm. long, 1·2 mm. wide, with 10 undulate ribs, minutely glandular and strigose, the base concealed in the cupular torus, apex truncate or rounded; style 0·8–1 mm. long; stigma capitate. Capsule dehiscing into 5 valves, pentagonal-cylindrical, about 60 mm. long and 5 mm. wide, straight or curved, 10-ribbed, subglabrous, with a distinct apical beak about 5–6 mm. long, the valves transversely and longitudinally septate within; seeds broadly spindle-shaped, about 3 mm. long and 1·5 mm. wide with 6 concave faces; testa fuscous, dull, minutely rugulose.

HAB.: In gardens and waste ground; sea-level to ? 1,000 ft. alt.; fl. May-Oct.

DISTR.: Divisions 5, 6, 7. A pantropical weed, frequently cultivated as a vegetable in the eastern Mediterranean area and western Asia; also producing a useful fibre.

5. Kythrea, 1934, *Syngrassides* 855 ! and, 1939, *Lindberg f.* s.n. !
6. Nicosia, 1862, *Kotschy*.
7. Kyrenia, 1948, *Casey* 123 ! and, 1955, *G. E. Atherton* 315 !

2. C. trilocularis *L.*, Syst. Nat., ed. 12, 2: 369 (1767), Mantissa Plant. 1, 77 (1767); Boiss., Fl. Orient., 1: 845 (1867); Post, Fl. Pal., ed. 2, 1: 247 (1932); Zohary, Fl. Palaest., 2: 310, t. 454 (1972); Osorio-Tafall et Seraphim, List Vasc. Plants Cyprus, 69 (1973).

TYPE: " *in* Arabia. *Forskahl* ".

Annual herb resembling *C. olitorius* L., but generally less robust, 30–100 cm. high, with erect, spreading or decumbent, pubescent branches; leaves oblong-lanceolate, 2·5–10 cm. long, 0·5–3 cm. wide, ± pubescent on both surfaces, apex acute or obtuse, base truncate or rounded, margins rather bluntly serrate, the two lowest serrations sometimes, but not always, produced into filiform appendages; petiole about 1 cm. long, channelled above, ± strigose-pubescent all over; stipules filiform, to about 5 mm. long; flowers solitary or in small 2–3-flowered cymes opposite the upper leaves; peduncles very short, about 1 mm. long, pubescent; bracts narrowly lanceolate, 1–2 mm. long, thinly pubescent; sepals boat-shaped, about 5 mm. long and 1·5 mm. wide, pubescent externally, apex cuspidate; petals yellow, narrowly obovate, about 5 mm. long, 2 mm. wide, tapering to a short ciliate claw; stamens about 30, 4–5 mm. long, filaments glabrous, attached at the apex of a very short cupular, crenate torus about 0·3 mm. high; anthers shortly oblong, pale, about 0·3 mm. long and a little less in width; ovary cylindrical, about 3 mm. long, 0·8 mm. wide, obscurely ribbed, pubescent, the base concealed in the cupular torus; apex truncate or rounded; style about 0·3 mm. long, stigma capitate. Capsule dehiscing into 3 valves, prismatic-cylindrical, 45–70 mm. long, 2–3 mm. wide, straight or slightly curved, obscurely 6-ribbed, thinly scabrid-pubescent with an apical beak about 5 mm. long, the valves transversely and longitudinally septate within; seeds roughly oblong, about 1–1·3 mm. long and 1 mm. wide with 6 ± rectangular faces; testa blackish, pruinose, dull, smooth.

HAB.: In gardens and waste ground; sea-level to ? 1,000 ft. alt.; fl.? May-Oct.

DISTR.: Division 5. A pantropical weed; cultivated as a vegetable in the eastern Mediterranean and western Asia.

5. Kythrea, 1934, *Syngrassides* 854 !

A few members of the family are also planted for ornament: the Silver Lime, *Tilia tomentosa* Moench is recorded from Nicosia (as *T. argentea* Desf.) by G. Frangos (in Cypr. Agric. Journ., 18, part 3: 88; 1923). It is a handsome tree with orbicular-cordate leaves, white-tomentose below, and with pendulous, conspicuously bracteate clusters of small, fragrant greenish-yellow flowers. *Tilia tomentosa* is a native of S.E. Europe and western Asia. One other member of the *Tiliaceae*, *Sparmannia africana* L. f., from South Africa, is also commonly cultivated in warm-temperate and subtropical gardens, or in greenhouses in colder regions. It is a shrub 10–15 feet in height, with large cordate, often palmately lobed, hairy leaves, and umbels of conspicuous flowers, 3–4 cm. across, the pure white petals contrasting with a central mass of purplish or crimson stamens; the fruits of *Sparmannia africana* are strongly echinate.

24. LINACEAE

H. Winkler in Engl. et Prantl, Pflanzenfam., ed. 2, 19a: 82–130 (1931) pro parte.

Trees, shrubs, herbs and climbers, glabrous or with an indumentum of simple hairs; leaves alternate or opposite, simple, stipulate (or rarely exstipulate), stipules deciduous, entire or divided or gland-like; flowers in terminal or axillary cymes, or sometimes solitary, actinomorphic, hermaphrodite; sepals 4–5, imbricate, free or partly united; petals often fugacious, contorted in bud, free or occasionally united at the base, frequently clawed; stamens twice (rarely thrice) as many as petals, or with alternate antipetalous stamens reduced to staminodes or sometimes wanting; filaments united in the lower part; anthers 2-thecous, dorsifixed, introrse, dehiscing longitudinally; ovary superior, 2–5-locular, the loculi sometimes almost completely subdivided by false septa; placentation axile; ovules usually 2 in each loculus, pendulous; styles 2–5, free or united at the base, stigmas capitate or clavate. Fruit a septicidal, or a septicidal and loculicidal capsule, sometimes indehiscent or occasionally drupaceous; seeds ± compressed, shining, exarillate; endosperm present or absent; embryo straight or slightly curved; cotyledons flat.

About 12 genera and 275 species widely distributed in temperate and tropical regions.

1. LINUM *L.*

Sp. Plant., ed. 1, 277 (1753).
Gen. Plant., ed. 5, 135 (1754).
H. Winkler in Engl. et Prantl, Pflanzenfam., ed. 2, 19a: 112–119 (1931).
P. H. Davis in Notes Roy. Bot. Gard. Edinb., 22: 135–161 (1957).

Annual, biennial or perennial herbs or subshrubs; leaves usually alternate, sessile, entire, frequently narrow and with parallel venation or with only the midrib prominent; stipules wanting or gland-like; inflorescence cymose, sometimes spicate or glomerulate; flowers 5-merous; stamens 5, alternating with tooth-like or filiform staminodes, staminal tube often glanduligerous externally; ovary 5-locular. Fruit dehiscing loculicidally and septicidally into 10, 1-seeded valves, rarely ± indehiscent; seeds flat, with copious or sparse endosperm.

About 200 species chiefly in temperate and subtropical regions of the northern hemisphere.

Flowers blue, pink or white:
  Pedicels slender, elongating in fruit; petals blue or white; leaves usually glabrous:
    Petals 7–8 mm. long, pale blue; stems usually branched at base; capsule dehiscing - - - - - - - - - - - - - **1. L. bienne**
    Petals 13–18 mm. long, blue or white; stems usually unbranched or sparingly branched at base; capsules indehiscent or tardily dehiscent **2. L. usitatissimum**
  Pedicels short, stout, not much elongating in fruit; petals pink; leaves usually ± hairy - - - - - - - - - - - - - **3. L. pubescens**
Flowers yellow:
  Flowers small, petals less than 8 mm. long; stigmas capitate or very shortly clavate:
    Leaves with smooth or minutely papillose margins; sepals distinctly shorter than petals; petals connate along part of their margins - - **4. L. trigynum**
    Leaves with scabridulous margins; sepals equalling or almost equalling, or exceeding petals; petals free:
      Inflorescence a lax, open cyme with slender branches - **5. L. corymbulosum**
      Inflorescence crowded, rigid, subspicate, branches very short **6. L. strictum** ssp. **spicatum**

Flowers medium to large, petals 1 cm. or more in length; stigmas clavate or almost linear:
Perennial with woody rootstock; inflorescence lax; sepals ovate, rather blunt; pedicels rather slender; gland-like stipules wanting - - - 7. **L. maritimum**
Annual; inflorescence rigid, dichotomous; sepals linear-subulate; pedicels very short; gland-like stipules present - - - - - - 8. **L. nodiflorum**

SECT. 1. **Linum** Annual, biennial or perennial herbs; pedicels elongating in fruit; flowers medium to large; petals blue, white, pink or red, free; stigmas elongate-clavate; leaves alternate, usually glabrous, without gland-like stipules.

1. **L. bienne** *Mill.*, Gard. Dict., ed. 8, no. 8 (1768).
L. angustifolium Huds., Fl. Angl., ed. 2, 1: 134 (1778); Unger et Kotschy, Die Insel Cypern, 366 (1865); Boiss., Fl. Orient., 1: 861 (1867); Holmboe, Veg. Cypr., 118 (1914); Post, Fl. Pal., ed. 2, 1: 252 (1932); Rechinger f. in Arkiv för Bot., ser. 2, 1: 426 (1950); Davis, Fl. Turkey, 2: 447 (1967); Zohary, Fl. Palaest., 2: 260, t. 374 (1972); Osorio-Tafall et Seraphim, List Vasc. Plants Cyprus, 65 (1973).
L. cribrosum Reichb., Icones Fl. Germ., 6: 63, t. 330, fig. 5158b (1844); Sintenis in Oesterr. Bot. Zeitschr., 31: 259 (1881), 32: 54 (1882).
TYPE: Yugoslavia; " received from Istria ", *Miller* (BM).

Erect or ascending perennial (or sometimes annual or biennial) (5–) 15–30 (–60) cm. high with a woody, tough rootstock; stems branched at base and rather sparingly above, or sometimes unbranched, glabrous, obscurely ridged longitudinally; basal leaves oblong, acute, 2–8 mm. long, 1–2 mm. wide, glabrous, crowded; upper leaves linear-acuminate, 8–25 mm. long, 2–4 mm. wide, more remote, normally with 3 parallel nerves; inflorescence an irregular, lax, branched, leafy cyme; flowers leaf- (or bract-) opposed; pedicels slender, ± erect, 1–1·5 cm. long at anthesis, lengthening to 2–3 cm. in fruit; sepals concave, glabrous, ovate-acuminate, 5–6 mm. long, 2–2·5 mm. wide, with membraneous margins, distinctly erose in the upper half, and with a prominent keeled midrib running into the apex, externally minutely serrulate in the upper part; petals free, pale blue, obovate, about 7–8 mm. long and 5 mm. wide with a cuneate base; stamens shortly united at base, filaments slender, about 4 mm. long, the basal, connate part bearing 5 prominent glands externally; anthers yellow, narrowly oblong, about 1·5 mm. long and 0·3 mm. wide; ovary spherical, about 2 mm. diam.; styles 2 mm. long, with elongate, narrowly linear-clavate stigmas about 0·6 mm. long. Capsule dehiscent, light brown, spherical, 5–6 mm. diam., surrounded by persistent, accrescent sepals; septa usually hairy; seeds oblong-elliptical, about 2·5 mm. long and 1·5 mm. wide; testa mid-brown, minutely faveolate.

HAB.: In marshy or rocky fields; by streams and water-channels, sometimes on dry mountainsides in garigue or in forest, occasionally on sand dunes; sea-level to 4,000 ft. alt.; fl. Jan.–June.

DISTR.: Divisions 1–8, common in Divisions 2, 7. Widespread in the Mediterranean area and northwards along the Atlantic coast of Europe to the British Isles; Atlantic Islands.

1. Stroumbi, 1941, *Davis* 3371 ! Akamas, Rigopoulos, N.E. of Neokhorio, 1957, *Merton* 3016 ! Ayios Minas near Smyies, 1962, *Meikle* 2169 !
2. Common; *Kotschy* 725 ! *Haradjian* 383; *Kennedy* 1321 ! *Merton* 2862 !
3. Amathus, 1941, *Mavromoustakis* s.n. !
4. Near Ormidhia, 1880, *Sintenis & Rigo* 938 ! Near Larnaca, 1912, *Haradjian* 13; Ormidhia, 1954, *Merton* 1839 !
5. Syngrasis, 1936, *Syngrassides* 1277 ! Vatili, 1905, *Holmboe* 347.
6. Skouriotissa, 1932, *Syngrassides* 93 ! Syrianokhori, 1935, *Syngrassides* 788 ! and, 1941, *Davis* 2607 !
7. Common; *Sintenis & Rigo* 938 ! *Syngrassides* 1196 ! *Davis* 2109 ! *Casey* 657 ! *Chapman* 366 ! *Merton* 1858 !

8. Cape Andreas, 1880, *Sintenis & Rigo* 205b; Apostolos Andreas, 1948, *Mavromoustakis* s.n. ! Koma tou Yialou, 1950, *Chapman* 229 !

2. L. usitatissimum *L.*, Sp. Plant., ed. 1, 277 (1753); Unger et Kotschy, Die Insel Cypern, 366 (1865); Boiss., Fl. Orient., 1: 860 (1867); Holmboe, Veg. Cypr., 117 (1914); Post, Fl. Pal., ed. 2, 1: 252 (1932); Lindberg f., Iter Cypr., 21 (1946); Davis, Fl. Turkey, 2: 448 (1967); Osorio-Tafall et Seraphim, List Vasc. Plants Cyprus, 65 (1973).

TYPE: " *Inter* segetes Europae *australis* ".

Larger in all its parts than *L. bienne* Mill., and not perennating; stems erect, to 1 m. high, generally unbranched at the base, or very sparingly branched; basal leaves oblong, 1–2 cm. long, 0·3–0·5 cm. wide, withering before the plants are mature; upper leaves linear, acute or acuminate, 1·5 5 cm. long, 0·3–0·5 cm. wide; inflorescence as in *L. bienne*; sepals ovate, acuminate or subacute, 0·5–0·8 cm. long, 0·3–0·5 cm. wide, glabrous or sometimes minutely ciliate; petals bright blue (sometimes white), obovate, 1·3–1·8 cm. long, 1–1·5 cm. wide with a cuneate base; stamens 7–8 mm. long; anthers about 2·5 mm. long and 0·8 mm. wide; ovary ovoid 4–5 mm. long, 2·5–3 mm. wide; styles about 4 mm. long; stigmas narrowly linear-clavate. Capsule subglobose, about 1 cm. diam., indehiscent with glabrous septa (var. *usitatissimum*; var. *vulgare* Boenningh.), indehiscent with ciliate or hairy septa (var. *bienne* Dierb.; var. *hibernum* Dierb.; var. *hyemale* A.DC.; var. *romanum* Heer) or dehiscent with ciliate or hairy septa (var. *humile* (Mill.) Pers.; var. *crepitans* Boenningh.; *L. humile* Mill.; *L. crepitans* Dum.); seeds oblong, 4–4·5 mm. long, about 2 mm. wide; testa rather dark brown, minutely faveolate.

HAB.: In fields, on waste ground or by roadsides, sea-level to ? 3,900 ft. alt.; fl. ? Febr.–June.

DISTR.: Divisions 2, 6, 8, common in Division 6. Not indigenous, but occurring as an escape from cultivation or as a casual. Widely cultivated in temperate regions of the world.

2. Platania, 1939, *Lindberg f.* s.n.
6. Common; *Post*, s.n.; *Holmboe*; *Nattrass* 99, 342.
8. Ephtakomi, 1880, *Sintenis & Rigo* 206 !

NOTES: *Linum usitatissimum* is nowhere known as a truly wild plant, but is generally considered a cultivated derivative of *L. bienne* Mill. The herbarium material of the cultivated plant is insufficient to allow for detailed taxonomic analysis, but it is clear that several varieties occur in Cyprus, having been introduced for agricultural purposes at different periods. The *Sintenis & Rigo* specimen cited above has been considered var. *humile* (Mill.) Pers. (*L. humile* Mill.) by the finders and subsequent authors, but the capsules of the specimen examined appear to be indehiscent or, at most, tardily dehiscent, and the plant is more likely to be *L. usitatissimum* var. *bienne* Dierb. The same is possibly true of the Post record for *L. humile*, but the specimen has not been seen. The statement in " The Handbook of Cyprus " 9th issue (1930), p. 177, that " practically all the flax grown in Cyprus is of the native variety, indigenous to the Mediterranean seaboard and botanically classed as ' macrosporea grandiflora ', blue blossom " is puzzling, for the only flax of the *usitatissimum* complex which can be termed " native " is *L. bienne* Mill., and it is unlikely that this would be named " macrosporea grandiflora " (a non-botanical usage); subsequent comments in " The Handbook " (p. 177) suggest that the two sub-varieties described are *L. usitatissimum* var. *bienne* (grown for fibre and seed) and var. *humile* (grown for seed only). The " Belgian, Riga and English pedigree seed " referred to on the same page would all be referable to *L. usitatissimum* var. *usitatissimum*.

SECT. 2. **Dasylinum** (*Planch.*) *Juz.* Annual, biennial or perennial herbs; pedicels short, not much elongating in fruit; flowers usually large; petals blue, pink or white, free or sometimes coherent; stigmas elongate-clavate or clavate; leaves alternate, usually hairy, without gland-like stipules.

3. L. pubescens *Banks et Soland.* in Russell, Nat. Hist. Aleppo, ed. 2, 2: 268 (1794); Boiss., Fl. Orient., 1: 860 (1867); Holmboe, Veg. Cypr., 117 (1914); Post, Fl.

Pal., ed. 2, 1: 252 (1932); Davis, Fl. Turkey, 2: 442 (1967); Zohary, Fl. Palaest., 2: 260, t. 373 (1972); Osorio-Tafall et Seraphim, List Vasc. Plants Cyprus, 65 (1973).

L. sibthorpianum Marg. et Reut. in Mém. Soc. Phys. Genèv., 8: 283 (1839).

[*L. hirsutum* (non L.) Sibth. et Sm., Fl. Graec. Prodr., 1: 215 (1806), Fl. Graec., 4: 2, t. 302 (1823); Unger et Kotschy, Die Insel Cypern, 366 (1865).]

[*L. viscosum* (non L.) Sibth. et Sm., Fl. Graec. Prodr., 1: 214 (1806).]

TYPE: Syria or Turkey; between El Ladhiqiya (Latakia) and Iskenderun, June 1770, *Russell* (BM).

Erect or ascending annual, 7–35 (–40) cm. high with a tough woody rootstock; stems rather stout, thinly hairy, much branched at the base, or occasionally simple, sparingly branched or commonly unbranched between base and inflorescence; leaves often rather crowded, oblong or oblong-lanceolate, 1–2·5 (–4) cm. long, 0·3–0·8 (–1·5) cm. wide, thinly hairy or subglabrous, glandular or eglandular, with 3–5 slender parallel nerves; inflorescence a divaricate, somewhat irregular, terminal cyme; bracts foliose; pedicels short, stout, generally less than 2 mm. long, not lengthening much in fruit; sepals narrowly linear-oblong, 8–10 mm. long, 2–3 mm. wide at base, rather abruptly narrowed to a subacute apex, thinly hairy and ± glandular, especially towards apex; petals pink, often tinged blue towards base, obovate-cuneate, 2–3 cm. long, 1–2 cm.wide with a rounded apex and distinctly clawed base; stamens unequal, 8–12 mm. long, distinctly connate at base into a short tube; filaments slender, glabrous; anthers oblong, about 2 mm. long and 0·8 mm. wide; ovary subglobose, about 1 mm. diam., glabrous; styles connate into a tube about 3–4 mm. long, free for 1·5–2 mm. in the upper part; stigmas narrowly linear-clavate, about 0·8 mm. long. Capsule woody, indehiscent or tardily dehiscent, globose, 3–4 mm. diam., pale brown and distinctly 10-ribbed, with a short apical beak formed by the persistent style-bases; seeds oblong-elliptical, about 2·5 mm. long and 1·5 mm. wide; testa dark brown, minutely shagreened.

HAB.: In fields, vineyards and amongst scrub on open hillsides; generally lowland; fl. March–June.

DISTR.: Divisions 3, 4. Widely distributed in the eastern Mediterranean region eastwards to N. Iraq.

3. Near Limassol, Kolossi, 1862, *Kotschy*.
4. Between Ormidhia and Famagusta, 1787, *Sibthorp*. [Between Famagusta and Nicosia, 1930, *F. A. Rogers* 0650 !—but see note.]

NOTES: The absence of recent records is very surprising, especially in view of the fact that *L. pubescens* is such a conspicuous plant. The Rogers record cited above must be regarded with suspicion, for close examination of the specimen shows that it is clearly a broken-off portion of another specimen (*Rogers* K. 34) said to have been collected at Aleppo in March 1930. It is now too late to find out if these specimens came from Cyprus or from Aleppo, but there can be no doubt that both had the same origin.

SECT. 3. **Linastrum** (*Planch.*) *Benth.* Annual herbs; flowers small; petals yellow, free or cohering; stigmas capitate or very shortly clavate; leaves alternate or the lowest sub-opposite, without gland-like stipules.

4. **L. trigynum** *L.*, Sp. Plant., ed. 1, 279 (1753); Davis, Fl. Turkey, 2: 436 (1967); Osorio-Tafall et Seraphim, List Vasc. Plants Cyprus, 64 (1973).

L. gallicum L., Sp. Plant., ed. 2, 401 (1762); Unger et Kotschy, Die Insel Cypern, 365 (1865); Boiss., Fl. Orient., 1: 851 (1867); Holmboe, Veg. Cypr., 117 (1914).

TYPE: France; " Monspelii " (Montpellier).

A slender, erect, glabrous annual 7–30 cm. high; stems usually much branched at the base, occasionally unbranched; basal leaves linear-oblong or oblanceolate, 2–10 mm. long, 1–2 mm. wide, acute or obtuse; upper stem leaves usually larger, linear, to about 25 mm. long and 2·5 mm. wide with a slender acuminate apex; inflorescence a lax, spreading cyme, with numerous filiform branches; bracts remote, foliaceous; pedicels usually less than 4 mm. long, not lengthening much in fruit; flowers small; sepals lanceolate, acuminate, 3–4 mm. long, about 1 mm. wide, obscurely 3-nerved, margins often conspicuously glandular; petals yellow, narrowly obovate, about 5 mm. long and 2 mm. wide, with a rounded or obtuse apex and a tapering base, the margins of adjacent petals cohering for about 1·5 mm. from a little way above the base; stamens about 3 mm. long; filaments distinctly expanded at base and connate to form a short tube, filiform above; staminodes short, filiform, sometimes wanting; anthers oblong, about 0·5 mm. long and 0·3 mm. wide; ovary globose, about 1 mm. diam.; styles 5, about 1 mm. long, free almost to base; stigmas shortly clavate or subcapitate. Capsule globose, about 2 mm. diam., dehiscent; seeds ovate, 1–1·2 mm. long, 0·8 mm. wide; testa light brown, minutely and irregularly reticulate-rugose.

HAB.: On dry, stony slopes, often, but not always, on serpentine formations; 300–2,000 ft. alt.; fl. March–May.

DISTR.: Divisions 1, 3, 7, 8. Widespread in Mediterranean area; ? N. Iran (Caspian coast); Atlantic Islands; Ethiopia.

1. Melanos, near Ayios Yeoryios (Akamas), 1962, *Meikle* 2150 !
3. Between Yerasa and Kalokhorio, 1962, *Meikle* 2893 !
7. Akanthou [Agathu], 1862, *Kotschy* s.n.
8. Cape Andreas, 1880, *Sintenis & Rigo* 210 ! Apostolos Andreas Forest, 1957, *Merton* 2930 !

NOTES: Rare in the eastern Mediterranean, and commonly confused with the following, *L. corymbulosum* Reichb.

5. **L. corymbulosum** *Reichb.*, Fl. Germ. Excurs., 834 (1832); Boiss., Fl. Orient., 1: 852 (1867); Holmboe, Veg. Cypr., 117 (1914); Post, Fl. Pal., ed. 2, 1: 250 (1932); Davis, Fl. Turkey, 2: 435 (1967); Zohary, Fl. Palaest., 2: 261, t. 377 (1972); Osorio-Tafall et Seraphim, List Vasc. Plants Cyprus, 65 (1973).
[*L. gallicum* (non L.) Rechinger f. in Arkiv för Bot., ser. 2, 1: 426 (1950).]
TYPE: Italy; " Bei Triest " (*Reichenbach*).

Similar to the preceding, but generally more robust, 10–30 (–40) cm. high; stems glabrous, branched or unbranched at the base, ± distinctly sulcate above; leaves linear-subulate, scabrid with short bristles, the lower about 5 mm. long and 1 mm. wide, the upper to about 20 mm. long and 2 mm. wide, erecto-patent and rather rigid; inflorescence a lax, open cyme with slender branches; bracts remote, foliaceous; pedicels 2–8 mm. long, not lengthening much in fruit; flowers small, but a little larger than in *L. trigynum*; sepals lanceolate, acuminate, concave, 4–5 mm. long, about 1·5 mm. wide, obscurely 3-nerved, margins usually conspicuously glandular; petals yellow, narrowly obovate, about 5 mm. long and 1·5 mm. wide, with a rounded apex and tapering base, free; stamens about 2·5 mm. long, filaments very shortly expanded at base and connate to form a short tube, filiform above; staminodes very short, filiform; anthers oblong, about 0·5 mm. long and 0·3 mm. wide; ovary globose, about 1 mm. diam.; styles 5, about 1 mm. long, free to base; stigmas capitate. Capsule globose, about 2·5 mm. diam., dehiscent; seeds oblong-ovate, about 1·5 mm. long and 0·8 mm. wide; testa light brown, minutely and rather regularly reticulate.

HAB.: On dry stony or rocky slopes in garigue; sea-level to 4,000 ft. alt.; fl. April–May.

DISTR.: Divisions 1–3, 7, 8, common in Divisions 7, 8. Widespread in the Mediterranean area and eastward to Iran and Afghanistan; Central Asia, Ethiopia.

1. Hovlon (Polis-Stroumbi), 1938, *Syngrassides* 1848 ! Dhrousha, 1941, *Davis* 3253 ! Akamas, 1941, *Davis* 3327 ! Below Stroumbi, 1941, *Davis* 3375 !
2. Platres, 1938, *Kennedy* 1322 !
3. Louvaras, 1962, *Meikle* 2904 !

7. Common; *Sintenis & Rigo* 207 ! *Casey* 184 ! 185 ! 620 ! *G. E. Atherton* 1379 !
8. Common; *Sintenis & Rigo* s.n.; *Haradjian* 197 ! *Mavromoustakis* s.n. ! *Meikle* 2486 !

6. L. strictum *L.*, Sp. Plant., ed. 1, 279 (1753); Boiss., Fl. Orient., 1: 852 (1867); Holmboe, Veg. Cypr., 117 (1914); Post, Fl. Pal., ed. 2, 1: 250 (1932); Davis, Fl. Turkey, 2: 436 (1967); Zohary, Fl. Palaest., 2: 262, t. 378 (1972).

An erect herb 5–40 cm. high; stems branched at base, or simple, generally unbranched above, obscurely ridged, glabrous; basal leaves usually very small, less than 5 mm. long and 1·5 mm. wide, oblong or oblanceolate, scabridulous on margins and surfaces, upper stem leaves ± erect, linear, acuminate, 10–25 mm. long, 1·5–2·5 mm. wide, scabridulous; inflorescence spreading and corymbose, or narrow and subspicate; pedicels very short, generally less than 1 mm. long, not lengthening much in fruit; bracts rigid, foliaceous; flowers small, rather crowded; sepals unequal or subequal, linear-subulate or linear-lanceolate, 4–6 mm. long, 1–1·5 mm. wide, 3-nerved, scabridulous, with glandular margins; petals yellow, free, oblanceolate, 3–5 nerved at base, about 5 mm. long and 1·5 mm. wide, apex rounded, tapering gradually to base; stamens 3·5 mm. long, filaments expanded at base and connate to form a very short tube, filiform above; anthers oblong, 0·5–0·7 mm. long and 0·3–0·4 mm. wide; staminodes very short, filiform; ovary globose, 0·7 mm. diam.; styles 5, 1·3 mm. long; stigmas capitate. Capsule globose, dehiscent, 2–3 mm. diam.; seeds oblong-elliptic, 1·3–1·6 mm. long, 0·8–1 mm. wide; testa shining brown, with minute, narrowly rectangular, regular reticulations.

ssp. **spicatum** (*Pers.*) *Lindberg f.* in Öfvers. Finsk. Vet. Soc. Förh., 48: 63 (1906); Holmboe, Veg. Cypr., 117 (1914); Osorio-Tafall et Seraphim, List Vasc. Plants Cyprus, 65 (1973).

L. strictum L. var. *spicatum* Pers., Syn. Plant., 1: 336 (1805); Boiss., Fl. Orient., 1: 852 (1867); Post, Fl. Pal., ed. 2, 1: 250 (1932); Zohary, Fl. Palaest., 2: 262 (1972).

[*L. strictum* (non L.) Kotschy in Unger et Kotschy, Die Insel Cypern, 365 (1865).]

TYPE: Portugal; " in Lusitania " (P).

Plant often rigidly erect, frequently unbranched; inflorescence narrow, subspicate; sepals markedly unequal, linear-lanceolate; petals usually 3-nerved at base; anthers about 0·5 mm. long. Capsule about 2 mm. diam., generally smaller than in typical *L. strictum*.

HAB.: On dry rocky ground in garigue or on sandy ground close to the sea; sea-level to about 1,000 ft. alt.; fl. March–May.

DISTR.: Divisions 1, 3–8, common in Division 7. Mediterranean region and eastwards to Iran and Afghanistan.

1. Neokhorio, 1941, *Davis* 3326 !
3. Akrotiri, 1963, *J. B. Suart* 31!
4. Ayios Memnon, 1948, *Mavromoustakis* s.n. ! Akhna Forest, 1952, *Merton* 731 ! Cape Greco, 1960, *N. Macdonald* 97A !
5. Between Nisou and Stavrovouni, 1933, *Syngrassides* 24 ! Athalassa, 1936, *Syngrassides* 1179 !
6. Ayia Irini (Morphou), 1941, *Davis* 2531 ! Orga, 1956, *Poore* 2651A !
7. Common; *Kotschy* 423; *Sintenis & Rigo* 208 ! *Chapman* 282 ! *G. E. Atherton* 339 ! 1346 ! *Meikle* 2359 !
8. Cape Andreas, 1880, *Sintenis & Rigo* s.n.; between Ayios Thyrsos and Rizokarpaso, 1938, *Syngrassides* 1770 ! Apostolos Andreas Forest, 1957, *Merton* 2927 !

NOTES: Typical ssp. *strictum* has not been found in Cyprus; it is generally rare in the eastern Mediterranean, though not infrequent in Turkey, Iraq and eastwards to Afghanistan, where forms intergrading with ssp. *spicatum* also occur. In the Mediterranean region proper the two subspecies are normally well differentiated.

SECT. 4. **Halolinum** (*Planch.*) *Meikle* **stat. nov.** in App. III, p. 806. Perennial herbs; flowers medium-sized; petals yellow, free; stigmas clavate; lower leaves often opposite; gland-like stipules wanting.

7. **L. maritimum** *L.*, Sp. Plant., ed. 1, 280 (1753); Boiss., Fl. Orient., 1: 851 (1867); Post, Fl. Pal., ed. 2, 1: 250 (1932); Lindberg f., Iter Cypr., 21 (1946); Zohary, Fl. Palaest., 2: 261, t. 376 (1972); Osorio-Tafall et Seraphim, List Vasc. Plants Cyprus, 65 (1973).

TYPE: " *in* Austria *ad Thermas badenses*, Monspelii ".

Erect, spreading or decumbent, glabrous perennial 20–60 (–100) cm. high with a tough woody rootstock; stems much-branched at base, unbranched or sparingly branched above; lower leaves opposite or subopposite, obovate or oblong, acute or obtuse, 4–10 mm. long, 2–4 mm. wide, upper leaves narrowly oblong or linear, acute, 10–30 mm. long, 1·5–3 mm. wide, distinctly 3-nerved, rather glaucous; inflorescence a lax, narrow or spreading, irregular cyme; pedicels rather slender, 2–15 mm. long, elongating slightly in fruit; bracts foliaceous; sepals unequal or subequal, ovate, rather blunt, 3–4 mm. long, 1·5–2 mm. wide, glabrous, with glandular margins; petals copper-orange in bud, later yellow, free, broadly obovate, 10–12 mm. long, 5–7 mm. wide, 5–7-nerved, apex rounded, tapering to base; stamens about 6 mm. long, filaments gradually expanded below and connate to form a distinct tube, filiform above; anthers oblong, about 1 mm. long and 0·6 mm. wide; staminodes short, filiform; ovary globose, 1·5–2 mm. diam.; styles free almost to base, about 1·5 mm. long; stigmas clavate, about 0·5 mm. long. Capsule globose, 3–4 mm. diam., dehiscent; seeds oblong-elliptic, about 1·8 mm. long and 1·2 mm. wide; testa rich brown with rather regular, narrowly rectangular reticulations.

HAB.: In fresh and brackish marshes amongst *Juncus* and *Schoenus* just above sea-level; fl. May–Sept.

DISTR.: Division 3, rare, only around Limassol Salt Lake. S.E. Europe and Mediterranean region.

3. Limassol Salt Lake; *Mavromoustakis* 6 ! 56 ! 83 ! 112 ! *Lindberg f.* s.n. ! *Davis* 3560 !

SECT. 5. **Limoniopsis** (*Planch.*) *Juz.* Annual or perennial herbs or sub-shrubs; pedicels short, not elongating much in fruit; flowers rather large, generally yellow; petals coherent; stigmas linear or elongate-clavate; leaves alternate or opposite towards base of stems, usually glabrous, with distinct gland-like stipules.

8. **L. nodiflorum** *L.*, Sp. Plant., ed. 1, 280 (1753); Sibth. et Sm., Fl. Graec. Prodr., 217 (1806), Fl. Graec., 4: 5, t. 307 (1823); Unger et Kotschy, Die Insel Cypern, 365 (1865); Boiss., Fl. Orient., 1: 853 (1867); Holmboe, Veg. Cypr., 117 (1914); Post, Fl. Pal., ed. 2, 1: 250 (1932); Davis, Fl. Turkey, 2: 434 (1967); Zohary, Fl. Palaest., 2: 263, t. 381 (1972); Osorio-Tafall et Seraphim, List Vasc. Plants Cyprus, 64 (1973).

TYPE: " *in* Italiae *pratis argillosis* ".

An erect annual, 4–15 (–50) cm. high; stem unbranched or sparingly branched below, distinctly ridged, scabridulous along the ridges; basal leaves opposite or subopposite, obovate-spathulate, 5–10 mm. long, 3–5 mm. wide, with a rounded apex and scabridulous margins; upper leaves alternate, becoming progressively narrower and more acute upwards, 10–30 (–40) mm. long, 2–8 mm. wide, 3-nerved, with scabridulous nerves and margins; stipules gland-like, rather conspicuous; inflorescence a lax, spreading cyme; flowers remote, rather large; pedicels very short and stout, generally less than 2 mm. long, or flowers subsessile; bracts foliaceous; sepals linear-subulate, 10–12 mm. long, 1·5–2 mm. wide at base, scabridulous, eglandular;

petals pale or bright yellow, obovate, 15–20 mm. long, 5–8 mm. wide, connate into a slender tube from the base to almost half way up; stamens subequal, 7–8 mm. long, connate into a basal tube 5–6 mm. long; anthers oblong, 1·3 mm. long, 0·3 mm. wide; ovary globose or subglobose, about 1·3 mm. diam.; styles 4–5 mm. long, connate at base, free above; stigmas linear-clavate, about 0·5 mm. long. Capsule globose, 4–5 mm. diam., dehiscent, with a short apical beak derived from the persistent style bases; seeds oblong-elliptic, 2·5 mm. long, 1·5 mm. wide; testa light brown, with minute, narrow rectangular reticulations.

HAB.: Dry stony, rocky or sandy ground, amongst garigue or in thin Pine forest on igneous or limestone formations; sea-level to ? 500 ft. alt.; fl. March–May.

DISTR.: Divisions 1–3, 8, common in Division 1. E. Mediterranean region and eastwards through Crimea to Iran.

1. Common; *Davis* 3310 ! *Merton* 3043 ! *Meikle* 2060 ! 2135 ! 2296 !
2. Prodhromos, 1862, *Kotschy* 833a. See note.
3. Near " Magado " (Mazotos), 1787, *Sibthorp* s.n.
8. Near Platanisso, 1880, *Sintenis & Rigo* 209 ! Yioti (" Juti ") near Ephtakomi, 1905, *Holmboe* 531; Shiramilia near Platanisso, 1962, *Meikle* 2509 !

NOTES: The Kotschy localization is dubious, for *Linum nodiflorum* is not generally found at any considerable altitude, and would not be likely to occur at Prodhromos.

25. ZYGOPHYLLACEAE

Engl. in Engl. et Prantl, Pflanzenfam., ed. 2, 19a: 144–184 (1931).

Herbs and shrubs, rarely trees; branches often jointed at the nodes; leaves opposite, or rarely alternate, pinnately compound or 2-foliolate (occasionally simple or 3-foliolate), leaflets entire, without gland dots; stipules paired, persistent, often spinescent; flowers solitary, paired or in cymes, generally hermaphrodite and actinomorphic; sepals 5 (rarely 4), free or connate at base, imbricate or rarely valvate; petals 5 (or 4) or occasionally wanting, hypogynous, free, imbricate or contorted, rarely valvate; disk usually present; stamens hypogynous, as many as the petals or up to 3 times as many, often in unequal whorls; filaments free; anthers introrse, 2-thecous, dehiscing longitudinally; ovary superior, sessile or shortly stipitate, 4–5-locular, usually furrowed, ridged or winged; style simple (rarely divided) or stigmas sessile; placentation axile; ovules 2 (or more, rarely 1) in each loculus. Fruit a loculicidal or septicidal capsule, or a schizocarp, rarely drupaceous; seed with a straight or slightly curved embryo; endosperm copious or scanty, rarely absent.

About 27 genera, chiefly in tropical or subtropical regions, often in arid climates.

Leaves alternate, laciniate; flowers whitish - - - - - - **1. Peganum**
Leaves opposite:
  Stipules spinose; leaves 3-foliolate (in our species); flowers magenta **2. Fagonia**
  Stipules not spinose; leaves not 3-foliolate:
    Leaflets paired, thick, succulent; flowers whitish; fruit not spinose; maritime shrub - - - - - - - - - - - - - - **3. Zygophyllum**
    Leaflets numerous, paripinnate, not succulent; flowers yellow; fruit spinose; annual weed - - - - - - - - - - - - **4. Tribulus**

TRIBE 1. **Peganeae** *Engl.* Leaves alternate, laciniate; stamens without basal scales. Fruit normally a 3-locular capsule, with numerous seeds in each loculus, rarely baccate.

1. PEGANUM *L.*

Sp. Plant., ed. 1, 444 (1753).
Gen. Plant., ed. 5, 204 (1754).

Perennial, branched, glabrous or puberulous herbs; leaves alternate, irregularly laciniate with narrow divisions, stipules setaceous; flowers solitary, leaf-opposed or lateral; sepals 4–5, open in bud, persistent, linear or sometimes divided pinnately into narrow divisions; petals 4–5, imbricate, spreading, subequal; stamens 12–15, inserted below the margin of the disk; anthers linear; ovary subglobose, 2–3-locular, ovules numerous in each loculus; style simple, 2–3-keeled above, the keels stigmatiferous. Fruit a 2–3-locular capsule, dividing into 2–3 valves, or rarely berry-like and indehiscent (sect. *Malacocarpus* (Fisch. et Mey.) Engl.). Seeds angular, with fleshy endosperm; embryo straight or curved.

Four species in southern Europe, North Africa, temperate Asia, and Mexico; *P. harmala* L. is occasionally found as a weed elsewhere. *P. crithmifolium* from Iran and the Caspian area, with indehiscent fruits, is sometimes placed in the separate genus *Malacocarpus* Fisch. et Mey.

1. **P. harmala** *L.*, Sp. Plant., ed. 1, 444 (1753); Unger et Kotschy, Die Insel Cypern, 362 (1865); Boiss., Fl. Orient., 1: 917 (1867); Holmboe, Veg. Cypr., 118 (1914); Post, Fl. Pal., ed. 2, 1: 272 (1932); Lindberg f., Iter Cypr., 21 (1946); Davis, Fl. Turkey, 2: 494 (1967); Zohary, Fl. Palaest., 2: 245, t. 352 (1972); Osorio-Tafall et Seraphim, List Vasc. Plants Cyprus, 65 (1973).

TYPE: " *in arena* Madritii, Alexandriae, Cappadociae, Galatia ".

Erect perennial 25–100 cm. high with a tough woody rootstock; stems glabrous, much branched; leaves glabrous, sessile, rather fleshy, 2–8 cm. long, irregularly divided into acute, linear lobes 0·5–4 cm. long, 0·1–0·5 cm. wide; stipules setaceous, 2–4 mm. long; flowers solitary, leaf-opposed, terminal and subterminal on the branches; pedicels stout, ebracteolate, 1–2 cm. long; sepals 5, linear (occasionally pinnatisect), acute, 10–15 mm. long and about 1 mm. wide, glabrous externally, minutely glandular internally; petals 5, whitish or greenish, oblong, 8–10 mm. long, 3–4 mm. wide, apex acute or subacute, base truncate or rounded; stamens 15, subequal, 8–9 mm. long, inserted with the petals below the margin of a crenate disk, filaments 3–4 mm. long, distinctly expanded at base, glabrous; anthers about 5 mm. long and 0·6 mm. wide; ovary subglobose, about 3 mm. diam., distinctly 3-lobed, glabrous; style 4 mm. long; stigmas about 3 mm. long. Fruit a globose, loculicidal capsule about 1–1·5 cm. diam.; seeds segment-shaped, about 3·5 mm. long and 2 mm. wide; testa brown, distinctly rugose.

HAB.: Roadsides and waste ground; sea-level to c. 1,000 ft. alt.; fl. May–July.

DISTR.: Divisions 1, 4, 6. S. Europe and Mediterranean region east to Central Asia and N.W. India; often a relic of cultivation.

1. Paphos on the way to Arodhes [Arora], 1862, *Kotschy*; Stephani near Ayios Neophytos, 1939, *Lindberg f.* s.n. !
4. Between Famagusta and Syngrasis, 1862, *Kotschy*; Famagusta, Othello's Tower, 1931, *J. A. Tracey* 58 !
6. Outside Nicosia, 1932, *Syngrassides* 417 ! Near Famagusta Gate, 1957, *Merton* 3163 !

TRIBE 2. **Zygophylleae** *Engl.* Leaves opposite; leaflets paired, pinnate or trifoliolate. Stamens with or without a basal scale. Fruit normally a capsule, rarely indehiscent; seeds with endosperm.

2. FAGONIA *L.*

Sp. Plant., ed. 1, 386 (1753).
Gen. Plant., ed. 5, 182 (1754).

Perennial, sprawling or prostrate, much-branched herbs; leaves opposite, with 1–3 leaflets; stipules usually spinescent; flowers solitary, arising between the stipules, generally conspicuous, with 5 pink or reddish, clawed, imbricate petals; sepals 5, imbricate, mucronate, usually deciduous; stamens 10; filaments slender, without basal scales; anthers short or elongate; ovary sessile, 5-locular, 5-angled; ovules 2, side by side near the base of each loculus, pendulous from ascending funicles; style subulate, 5-angled; stigma simple. Fruit a pyramidal, 5-angled or 5-winged capsule, each 1-seeded carpel separating from the central axis at maturity; papery exocarp and horny endocarp generally distinct; seeds with a mucilaginous testa and horny endosperm.

About 30 species in the Mediterranean area, Atlantic Islands, western Asia; N., N.E. and S.W. Africa; California; Mexico and Chile.

1. **F. cretica** *L.*, Sp. Plant., ed. 1, 386 (1753); Kotschy, Die Insel Cypern, 362 (1865); Boiss., Fl. Orient., 1: 905 (1867); Holmboe, Veg. Cypr., 118 (1914); Post, Fl. Pal., ed. 2, 1: 267 (1932); Chapman, Cyprus Trees and Shrubs, 46 (1949); Osorio-Tafall et Seraphim, List Vasc. Plants Cyprus, 65 (1973).

TYPE: " *in* Creta ".

Sprawling, scrambling or prostrate perennial 10–30 (sometimes more) cm. high with a woody base; stems jointed at the nodes, much-branched, angular, glabrous or with a few scattered hairs; leaves trifoliolate, leaflets narrowly lanceolate-elliptic, 5–12 (–20) mm. long, 1–4 (–6) mm. wide, glabrous, coriaceous, acute, with a short, pallid, spinose tip and a cuneate or tapering base; petiole 4–10 mm. long, distinctly channelled above; stipules spiny 2–5 (–10) mm. long; flowers solitary, arising between the stipules of the upper leaves; pedicels 3–5 mm. long, distinctly ribbed, glandular-hairy; sepals ovate, 2–2·5 mm. long, 1·5–2 mm. wide, acute or mucronate, glandular-hairy externally; petals bright magenta, obovate or suborbicular 6–8 mm. long, 4–5 mm. wide, apex rounded, base narrowed rather abruptly to a slender claw about 1·5 mm. long; stamens 5–6 mm. long, filaments slender, glabrous, attached with the petals below the margin of a small disk; anthers oblong, about 1 mm. long and 0·6 mm. wide; ovary ovoid, 2·5 mm. long, 1·8 mm. wide, strongly 5-angled, ciliate along the angles; style about 3 mm. long, tapering from base to apex; stigma very small, capitate. Capsule broadly ovoid-pyramidal, 6–8 mm. long, 5–7 mm. wide, deeply and sharply 5–angled, with a conspicuous apical cusp formed by the persistent style; seed D-shaped, flattened, about 4 mm. long and 2·5 mm. wide; testa dark brown, thickly mucilaginous; embryo straight with broad flat cotyledons.

HAB.: Dry chalky slopes; sea-level to c. 800 ft. alt.; fl. Febr.–June.

DISTR.: Divisions 1, 3, locally common on south coast. Widely distributed in the Mediterranean area; Atlantic Islands.

1. Psindron near Ayios Konon (Akamas), 1962, *Meikle* 2123 !
3. Locally common about Limassol and Episkopi; *Kotschy* 641 ! *A. G. & M. E. Lascelles* s.n. ! *Holmboe* 164, 710; *H. Roger-Smith* s.n.; *Syngrassides* 1759 ! *Miss M. Mackensie* in *Chapman* 352 !

3. ZYGOPHYLLUM *L.*

Sp. Plant., ed. 1, 385 (1753).
Gen. Plant., ed. 5, 182 (1754).

Woody herbs or subshrubs, often with spreading or prostrate branches, rarely annual herbs; stems jointed at the nodes; leaves opposite usually composed of 2 leaflets, rarely 1-foliolate or multifoliolate, generally fleshy; stipules paired, often spinescent; flowers solitary (rarely in pairs) arising from between the stipules; sepals 4–5, deciduous or persistent, imbricate; petals 4–5, clawed, imbricate or contorted, white or yellow, frequently with a red or purple basal blotch; stamens 8–10, inserted below margin of disk, filaments with a basal scale on the inner side; anthers oblong; ovary sessile, 4–5-angled, 2–5-locular, ovules 2 or more in each loculus; style subulate, stigma minute. Capsule 4–5-angled or 4–5-winged, indehiscent, or splitting septicidally into 4–5 divisions at maturity, or loculicidally down the middle of each carpel; endocarp and exocarp often distinct; seeds pendulous, 1–several in each loculus; endosperm sparse.

About 80 species in the Mediterranean area, temperate Asia, temperate and tropical Africa, and Australasia.

1. Z. album *L. f.*, Dec. Prim. Plant. Rar. Hort. Upsal., 11, t. 6 (1762); L., Sp. Plant., ed. 2, 551 (1762); Sibth. et Sm., Fl. Graec. Prodr., 273 (1809), Fl. Graec., 4: 64, t. 371 (1823); Unger et Kotschy, Die Insel Cypern, 363 (1865); Boiss., Fl. Orient., 1: 915 (1867); Holmboe, Veg. Cypr., 118 (1914); Post, Fl. Pal., ed. 2, 1: 270 (1932); Lindberg f., Iter Cypr., 21 (1946); Chapman, Cyprus Trees and Shrubs, 46 (1949); Davis, Fl. Turkey, 2: 492 (1967); Zohary, Fl. Palaest., 2: 254, t. 366 (1972); Osorio-Tafall et Seraphim, List Vasc. Plants Cyprus, 65 (1973).

TYPE: Egypt; " in Aegypto, unde missa semina anno 1760. a *D:no. Doctore Roquè* ".

Spreading much-branched shrub, up to 1·6 m. high; older stems and branches covered with rough, greyish bark, younger growths fleshy, ribbed, densely white-pubescent; leaves fleshy, glaucous, pubescent, composed of 2 blunt, oblong or clavate leaflets, each 5–10 mm. long; petiole resembling the leaflets, but often a little longer, especially on sterile growths; stipules triangular, membranous, 1–1·5 mm. long and almost as wide, densely white-pubescent externally; flowers solitary or sometimes in pairs, scattered throughout the length of the branches; pedicels short, stout, about 2 mm. long, distinctly ribbed, lengthening a little in fruit; sepals 5, oblong or broadly ovate, concave, blunt, about 3 mm. long and 2·5 mm. wide, pubescent and often tinged red or pink externally, deciduous; petals 5, white, ovate, about 5 mm. long and 2·5 mm. wide, apex acute, base tapering to a claw about 1·5 mm. long; stamens 10, about 3·5–4 mm. long; filaments glabrous, with an ovate basal scale about 2 mm. long and 0·8 mm. wide on the inner side; anthers oblong, about 1 mm. long and 0·8 mm. wide; disk conspicuous, prominently crenate; ovary subglobose, about 1·5 mm. diam., 5-ribbed, pubescent; style glabrous, about 1·2 mm. long, stigma small, capitate; ovules usually 2 in each loculus. Capsule obovoid, deeply and sharply 5-ribbed, about 8–10 mm. long and 6–7 mm. wide, dehiscing septicidally and loculicidally when mature, the separate valves falling away from the axis; seed ovoid-pyriform, about 2·5 mm. long and 1·5 mm. wide; testa dull brown, almost smooth.

HAB.: Rocky and sandy seashores and salt-marshes; about sea-level; fl. March–May.

DISTR.: Divisions 3–6, common in Division 4. Mediterranean area; Arabia; N.E. Tropical Africa.

3. Akrotiri, 1905, *Holmboe* 691; Limassol, 1939, *Lindberg f.* s.n.; Limassol Salt Lake, 1941, *Davis* 3549 !
4. Common; *Kotschy* 285 ! *Sintenis & Rigo* 514 ! *Holmboe* 103; *Syngrassides* 494 !
5. Salamis, 1934, *Chapman* 134 ! 1939, *Lindberg f.* s.n.; 1957, *Merton* 3219 ! 1962, *Meikle* !
6. Syrianokhori, 1935, *Syngrassides* 1333 ! Morphou Bay, 1948, *H. Duke-Woolley* 37 !

TRIBE 3. **Tribuleae** *Engl.* Leaves opposite, pinnate; alternate filaments glandular at base. Fruit a capsule, or splitting at maturity into 5 or more indehiscent, 1- or several-seeded mericarps; seeds without endosperm.

4. TRIBULUS *L.*

Sp. Plant., ed. 1, 386 (1753).
Gen. Plant., ed. 5, 183 (1754).

Sprawling or prostrate, usually annual, herbs; stems generally much branched; leaves opposite in unequal pairs, sometimes alternate by abortion, stipulate, pinnate, the leaflets opposite; flowers solitary, pedicellate, axillary or lateral, often yellow; sepals 5, imbricate, deciduous or persistent; petals 5, caducous, imbricate; disk 5–10-lobed; stamens 10, (or 5 by abortion) inserted below the margin of the disk, those opposite the petals often longer than those opposite the sepals, the latter sometimes sterile; alternate filaments externally glandular at base but without inner scales; ovary sessile, 5-lobed, with 3–5 ovules in each carpel, the ovules separated by transverse walls within each loculus; style simple, 5-angled; stigma 5-angled or 5-rayed. Fruit splitting into 5 indehiscent mericarps at maturity, the mericarps dorsally spinose, tuberculate or winged, rarely smooth; seeds without endosperm.

About 25 species widely distributed in tropical and warm-temperate regions, often in arid climates.

1. **T. terrestris** *L.*, Sp. Plant., ed. 1, 387 (1753); Clarke, Travels, ed. 2, 2: 724 (1816); Unger et Kotschy, Die Insel Cypern, 362 (1865); Boiss., Fl. Orient., 902 (1867); Holmboe, Veg. Cypr., 118 (1914); Post, Fl. Pal., ed. 2, 1: 266 (1932); Lindberg f., Iter Cypr., 21 (1946); Davis, Fl. Turkey, 2: 493 (1967); Zohary, Fl. Palaest., 2: 256, t. 370 (1972); Osorio-Tafall et Seraphim, List Vasc. Plants Cyprus, 65 (1973).

TYPE: " *in* Europa *australi ad semitas* ".

Sprawling, decumbent or prostrate annual; stems up to 50 cm. long, sulcate, much branched, especially at base, thinly or rather densely strigose-pubescent; leaves oblong in outline, about 1·5–5 (–9) cm. long, 1–1·5 cm. wide, paripinnate, with (3–) 5–6 (–7) pairs of subsessile or shortly petiolulate, oblong, obtuse, or acute, entire, sericeous-pilose or glabrescent leaflets, 3–15 mm. long, 2–7 mm. wide; petiole short, usually less than 1 cm. long, channelled above, hairy or glabrescent; stipules lanceolate-acuminate, often obliquely curved, 1·5–3 mm. long, 0·5–1·5 mm. wide, pubescent externally; flowers solitary, apparently axillary, but really lateral, arising from between the leaves in the upper parts of the stems; pedicels 0·5–1·5 cm. long, strigose-pubescent; sepals lanceolate, 4–5 mm. long, 1–1·5 mm. wide, strigose-pubescent externally, deciduous; petals yellow, narrowly obovate, 4·5–6 mm. long, 2·5–3 mm. wide, apex rounded or truncate, base tapering to a very short claw; stamens 2·5–3 mm. long, inserted with the petals below the margin of the disk; filaments glabrous, those of alternate (episepalous) stamens with a small, oblong, slightly emarginate, basal gland on the outer side; anthers oblong, about 1 mm. long and 0·6 mm. wide; disk conspicuously 5-lobed; ovary ovoid-conical, 2·5–3 mm. long, 2–2·5 mm. wide, tuberculate, ± densely adpressed-strigose; style about 0·5 mm. long, thick, glabrous; stigma 5-rayed; ovules usually 3 in each loculus. Fruit depressed globose, about 1 cm. diam.; mericarps pubescent or glabrescent, usually with 2 (or 4) stout, spreading dorsal spines and an intervening crest of smaller spines, inner faces wrinkled; seeds usually 3 in each mericarp, oblong, about 2·5 mm. long and 1·5 mm. wide; testa membranous, pale, almost smooth.

HAB.: In gardens and cereal fields; occasionally in garigue; sea-level to 3,300 ft. alt.; fl. May–Oct.

DISTR.: Divisions 1, 2, 4–7. Tropical and warm-temperate areas throughout the world.

1. Ayios Neophytos, 1939, *Lindberg f.* s.n.; Skoulli near Polis, 1939, *Lindberg f.* s.n.
2. Near Galata, 1880, *Sintenis & Rigo* 784 ! Evrykhou [Erucon], 1894, *Post*; Platres, 1937, *Kennedy* 514 !
4. Larnaca, 1905, *Holmboe*; between Prastio and Kouklia, 1939, *Lindberg f.* s.n.; Perivolia, 1932, *Lindberg f.* s.n.; Larnaca, 1939, *Lindberg f.* s.n.
5. Near Syngrasis, 1862, *Kotschy*.
6. Nicosia, 1939, *Lindberg f.* s.n. ! English school, Nicosia, 1958, *P. H. Oswald* 12 !
7. Lapithos, 1932, *Syngrassides* 238 ! Kyrenia, 1948, *Casey* 29 ! Kyrenia, 1948, *Casey* 83 ! Kyrenia near Thermia, 1955, *G. E. Atherton* 366 !

26. GERANIACEAE

Knuth in Engl., Pflanzenr., 53 (IV, 129) (1912).
Knuth in Engl. et Prantl, Pflanzenfam., ed. 2, 19a: 43–66 (1931).

Herbs or subshrubs, rarely arborescent; leaves opposite or alternate, simple or compound, stipulate, stipules often paired; flowers generally in bracteate cymes or umbels, sometimes solitary, actinomorphic or zygomorphic, hermaphrodite or rarely unisexual; sepals 5 (rarely 4), free or connate at base, imbricate or rarely valvate, the dorsal sepal sometimes spurred; petals 5 (or 4, or occasionally wanting), free, hypogynous, imbricate or less often contorted, commonly alternating with nectariferous glands; stamens usually 2–3 times as many as sepals, sometimes fewer by abortion, filaments often connate at the base, rarely free or united into bundles; anthers 2-locular, versatile, dehiscing longitudinally; ovary superior, 3–5-lobed; ovules pendulous, 1–2 (rarely more) in each loculus, placentation axile; styles divided wholly or partly into stigma-bearing arms. Fruit often beaked, splitting (often elastically) into 1 (rarely more)-seeded dehiscent portions at maturity; seeds usually without endosperm.

Five genera, widely distributed, but chiefly in temperate regions of the northern hemisphere and in South Africa.

Leaves palmate, or palmately dissected or lobed; beak of carpel not twisting spirally on dehiscence - - - - - - - - - - - - - - **1. Geranium**
Leaves pinnate, or pinnately dissected or lobed; beak of carpel twisting spirally on dehiscence - - - - - - - - - - - - - - **2. Erodium**

1. GERANIUM *L.*

Sp. Plant., ed. 1, 676 (1753).
Gen. Plant., ed. 5, 306 (1754).

Herbs or subshrubs; leaves opposite or alternate, palmately lobed or dissected, rarely dentate only; peduncles axillary, 1–2-flowered; flowers axillary, actinomorphic, often ornamental; sepals 5, imbricate; petals 5, imbricate, alternating with nectariferous glands; stamens 10, usually all fertile, free or connate at base; ovary 5-lobed, 5-locular, beaked; ovules 2 in each loculus; stigma-arms 5. Fruit 1-seeded, each carpel breaking away elastically and curling upwards at maturity, the apex of the beak often remaining attached to the central axis; seed generally becoming free from carpel, with little or no endosperm; embryo with induplicate-plicate, or convolute cotyledons.

About 300 species chiefly in temperate regions, or in mountains in the tropics.

Plant perennial with a tuberous rhizome; flowers large, petals more than 1 cm. long; leaves all basal - - - - - - - - - - - **1. G. tuberosum**

Plants annual; flowers small, petals less than 1 cm. long; stems leafy:
Sepals spreading; petals with a short claw; carpels breaking away at maturity with a portion of the beak attached:
Lower leaves divided to about half way into oblong-cuneate, shortly lobed divisions:
Carpels smooth, pubescent:
Petals distinctly emarginate, pale purple, to about 4 mm. long **2. G. pusillum**
Petals entire or indistinctly emarginate, bright pink or rosy-purple, to about 7 mm. long - - - - - - - - - - **3. G. rotundifolium**
Carpels glabrous, wrinkled; petals deeply emarginate or bifid, to about 5 mm. long - - - - - - - - - - - - - - **4. G. molle**
Lower leaves divided almost to base into deeply lobed or pinnatifid divisions:
Peduncles 5–10 mm. long; pedicels 10–15 mm. long; carpels glandular-pubescent **5. G. dissectum**
Peduncles 20–90 mm. long; pedicels to 50 mm. long; carpels subglabrous or sparsely strigose - - - - - - - - - **6. G. columbinum**
Sepals erect ± connivent; petals with a long slender claw; carpels breaking away at maturity without a portion of beak attached:
Leaves palmately divided to base; sepals strigose-glandular externally, without transverse wrinkles - - - - - - - - - - **7. G. purpureum**
Leaves not divided to base; sepals glabrous, transversely wrinkled **8. G. lucidum**

1. G. tuberosum *L.*, Sp. Plant., ed. 1, 680 (1753); Sibth. et Sm., Fl. Graec. Prodr., 2: 38 (1813), Fl. Graec., 7: 54, t. 659 (1830); Unger et Kotschy, Die Insel Cypern, 364 (1865); Boiss., Fl. Orient., 1: 872 (1867); Knuth in Engl., Pflanzenr., 53 (IV, 129): 96 (1912); Holmboe, Veg. Cypr., 115 (1914); Post, Fl. Pal., ed. 2, 1: 256 (1932); Davis, Fl. Turkey, 2: 462 (1967); Zohary, Fl. Palaest., 2: 228, t. 323 (1972); Osorio-Tafall et Seraphim, List Vasc. Plants Cyprus, 64 (1973).
TYPE: Not localized [" *in* Anglia ? " *Linnaeus*].

Erect perennial, 7–20 (–40) cm. high, arising from a brown subglobose or ovoid, tuberous rhizome 1–4 cm. long; leaves all basal, with long slender petioles, ± orbicular in outline, slightly glaucous, subglabrous with sparsely pilose margins, divided to the base into 5–9, ovate or lanceolate, bluntly or acutely, bipinnately dissected lobes 2·5–5 cm. long, 1–3 cm. wide; petioles channelled, 10–20 cm. long, sparsely or rather densely pubescent; flowering stems normally unbranched, pubescent or glabrescent, often tinged purplish, sulcate; inflorescence terminal, at first rather densely cymose, consisting of repeated dichotomies, with a 2 (or rarely 3)-flowered partial inflorescence at each dichotomy; primary bracts opposite, foliaceous, bipinnately dissected, flanked by paired, oblong, acute or obtuse, glabrous or ciliate, purplish stipules, 2–3 mm. long, 1·5–2 mm. wide, secondary bracts smaller, with narrower, linear or lanceolate, pinnate, or deeply serrate (or sometimes entire) lobes; peduncles slender, 1–7 cm. long, ± densely white-pubescent; pedicels usually less than 1·5 cm. long, very slender, pubescent, flanked at the base by small purplish, ciliate bracteoles; sepals ovate or oblong, blunt, about 6 mm. long and 2·5–3 mm. wide, pubescent or pilose externally, ± distinctly 5-nerved, with a caudate-cuspidate apex; petals deep pink veined purple or red, obovate, about 13 mm. long and 7–8 mm. wide, deeply emarginate, margins long-ciliate at base; stamens about 3 mm. long; filaments purple, free, tapering to apex from a wide base, the outer whorl ± glabrous, the inner distinctly pilose; anthers oblong about 1 mm. long, 0·5 mm. wide; ovary deeply 5-lobed, hairy, about 2·5 mm. wide and 1·5 mm. high, beak about 2·5 mm. long, stigma-arms about 2 mm. long, recurved, purple. Fruit 5–6 mm. wide, about 4 mm. high, pubescent, with a beak 1·5–2 cm. long, carpels almost smooth; seeds oblong, about 3 mm. long and 2 mm. wide; testa dark brown, minutely punctulate-reticulate.

HAB.: Cultivated and fallow fields, occasionally by riversides or in garigue; sea-level to 2,500 ft. alt.; fl. Febr.–April.

DISTR.: Divisions 1–4, 6–8, common in Division 7. Mediterranean area and eastwards to Central Asia.

1. Toxeftera (Akamas), 1962, *Meikle* 2098 ! Ayios Minas near Smyies, 1962, *Meikle* 2167 !

2. North side of Troödos, 1862, *Kotschy*; Kryos Potamos, 1939, *Kennedy* 597 !
3. Limassol, 1951, *Mavramoustakis* 31 !
4. Near Larnaca, 1880, *Sintenis & Rigo* 832 ! Phaneromene near Larnaca, 1905, *Holmboe*, 127; Larnaca, 1937, *Mrs. Stagg* s.n. !
6. Kykko Monastery farm, 1 m. S.W. of Nicosia, 1927, *Rev. A. Huddle* 29 !
7. Common; *Syngrassides* 1192 ! *Casey* 408 ! *G. E. Atherton* 969 ! 1064 ! 1097 !
8. Between Apostolos Andreas and Khelones, 1941, *Davis* 2291 ! Gastria, 1941, *Davis* 2430 !

2. G. pusillum *Burm. f.*, Sp. Geran., 27 (1759); L., Sp. Plant., ed. 2, 957 (1763); Unger et Kotschy, Die Insel Cypern, 364 (1865); Boiss., Fl. Orient., 1: 880 (1867); Knuth in Engl. Pflanzenr., 53 (IV, 129): 48 (1912); Holmboe, Veg. Cypr., 116 (1914); Post, Fl. Pal., ed. 2, 1: 257 (1932); Davis, Fl. Turkey, 2: 460 (1967); Osorio-Tafall et Seraphim, List Vasc. Plants Cyprus, 64 (1973).

TYPE: " in ANGLIA & GALLIA " (? G).

Erect or decumbent annual, 10–40 cm. high, stems usually much branched from base, pubescent or puberulous; basal leaves orbicular-reniform in outline, 1–4 cm. diam., pubescent, divided to about half way into 7–9 oblong, bluntly trifid lobes 3–6 mm. wide at base, 5–10 mm. wide at apex; petioles pubescent, 7–20 cm. long; upper leaves generally alternate, becoming progressively smaller and more shortly petiolate, with narrower more deeply cut lobes towards apex of stem; stipules brown or reddish, ovate-acuminate, often bifid, about 2 mm. long and 1·5 mm. wide, pubescent externally; inflorescences usually 2-flowered, arising singly from the axils of the upper leaves; peduncles ± patent, about 1 cm. long, shortly pubescent; pedicels slender, 8–12 mm. long, glandular-pilose, with an acute or indistinctly mucronate apex; petals a little longer than sepals, to about 4 mm. long and 2 mm. wide, pale purple, distinctly emarginate with a ciliate base; 5 fertile stamens about 2 mm. long, with hairy filaments wide at the base and narrowing rather abruptly just below apex; anthers suborbicular, about 0·5 mm. diam.; sterile stamens about half as long as the fertile stamens, usually glabrous with minute, atrophied anthers; ovary about 1·5 mm. wide, 1 mm. high, pubescent, with a short pubescent beak; stigma-arms about 0·5 mm. long. Fruit about 3·5 mm. wide, 2·5 mm. high, with a slender beak about 7–8 mm. long, carpels pubescent, not wrinkled; seeds oblong, about 1·6 mm. long, 1 mm. wide; testa dark brown, smooth.

HAB.: On hillsides in garigue, c. 2,000–5,000 ft. alt.; fl. ? March–May.

DISTR. Divisions 2, 3, 7, rare. Europe and Mediterranean area eastwards to the Himalayas.

2. Above Prodhromos in several places, 1862, *Kotschy* 704 !
3. North side of Stavrovouni [Sta. Croce], 1862, *Kotschy* 198.
7. Pentadaktylos Peak, 1862, *Kotschy* 365.

NOTES: Not collected in recent years, but easily overlooked. Holmboe (p. 116) comments on the confusion in the numbering of Kotschy's specimens. The specimen cited by Boissier (*Kotschy* 930) as *G. pusillum* is listed under *G. robertianum* L. var. *purpureum* Vill. by Kotschy (Die Insel Cypern, 365). The specimen cited as *G. rotundifolium* by Kotschy is correctly *G. pusillum*, but the Kew specimen is numbered *Kotschy* 704, and not *Kotschy* 933 as stated in *Die Insel Cypern*, p. 364. In view of the confusion, some doubt must attach to statements concerning the habitat and distribution of *G. pusillum*; there can be little doubt that the plant does occur in Cyprus, but it is very desirable that its distribution should be checked. *G. pusillum* is generally rare in the eastern Mediterranean, and is mostly found at considerable altitudes there.

3. G. rotundifolium *L.*, Sp. Plant., ed. 1, 683 (1753); Unger et Kotschy, Die Insel Cypern, 364 (1865); Boiss., Fl. Orient., 1: 880 (1867); Knuth in Engl., Pflanzenr., 53 (IV, 129): 55 (1912); Holmboe, Veg. Cypr., 116 (1914); Post, Fl. Pal., ed. 2, 1: 257 (1932); Lindberg f., Iter Cypr., 21 (1946); Davis, Fl. Turkey, 2: 459 (1967); Zohary, Fl. Palaest., 2: 229, t. 327 (1972); Osorio-Tafall et Seraphim, List Vasc. Plants Cyprus, 64 (1973).

TYPE: " *in* Europae *cultis* ".

An erect or decumbent annual, similar to *G. pusillum* in habit and general appearance, but generally more robust and more densely glandular-pubescent; lower leaves to 7 cm. diam., orbicular-reniform, divided to about half way into 7–9 bluntly toothed lobes, the base of the sinuses between the primary lobes often stained reddish; petioles long and slender; stipules purple, narrowly ovate-acuminate, 2–5 mm. long, 1–2 mm. wide; upper leaves often opposite, smaller, more deeply and more acutely lobed, with shorter petioles; inflorescences of paired flowers as in *G. pusillum*, but peduncles and pedicels often longer; flowers distinctly larger and more conspicuous; sepals oblong-ovate, 5–6 mm. long, about 2·5 mm. wide, 3 (or obscurely 5)-nerved, hairy externally with a conspicuous caudate-cuspidate tip; petals usually bright pink or rosy purple, narrowly obovate, 5–7 mm. long, 2·5–3 mm. wide, with an entire or very slightly emarginate apex and a slender tapering claw, sparsely pilose at the base; stamens all fertile, about 5 mm. long; filaments ciliate, gradually tapering from a wide base to a filiform apex; anthers broadly oblong, about 0·5 mm. long and almost as wide; ovary depressed globose, about 1·5 mm. diam., densely pubescent, beak about 3 mm. long, with a distinctly swollen base; stigma-arms purple, about 6 mm. long. Fruit 5–6 mm. diam. with a beak 12–17 (–20) mm. long, carpels smooth, thinly pubescent, often dark olive-brown when ripe; seeds broadly oblong or almost globose, 1·5–2 mm. long and nearly as wide; testa dark brown, distinctly reticulate.

HAB.: Gardens, fields and waste ground; roadsides; rocky hillsides and screes below cliffs, also on fixed sand-dunes; sea-level to 3,800 ft. alt.; fl. Febr.–July.

DISTR.: Divisions 1–8, common in Division 2. Europe and Mediterranean area eastwards to Central Asia and the Himalayas; Atlantic Islands; introduced elsewhere.

1. Avgas R., near Ayios Yeoryios (Akamas), 1962, *Meikle* 2009 !
2. Common; *Kotschy* 933 (see note); *Lindberg f.* s.n. ! *Merton* 2159 !
3. Limassol, 1954, *Mavromoustakis* 11 ! and, 1956, *Merton* 2477 ! Pyrgos, 1963, *J. B. Suart* 46 !
4. Famagusta, 1948, *Mavromoustakis* s.n. !
5. Athalassa, 1936, *Syngrassides* 1177 !
6. Limestone cliffs W. of Orga, 1956, *Merton* 2599 !
7. Lapithos, 1939, *Lindberg f.* s.n.; Kantara, 1941, *Davis* 2457 ! Kyrenia, 1949, *Casey* 240 ! Thermia, 1956, *G. E. Atherton* 881 !
8. Ayios Philon near Rizokarpaso, 1941, *Davis* 2220 ! Ronnas valley, Rizokarpaso, 1957, *Merton* 2922 !

NOTES: As mentioned under *G. pusillum*, Kotschy's record for *G. rotundifolium* from Prodhromos is dubious, and the occurrence of the plant in this area required confirmation.

4. **G. molle** *L.*, Sp. Plant., ed. 1, 682 (1753); Unger et Kotschy, Die Insel Cypern, 364 (1865); Boiss., Fl. Orient., 1: 882 (1867); Knuth in Engl., Pflanzenr., 53 (IV, 129): 57 (1912); Holmboe, Veg Cypr., 116 (1914); Post, Fl. Pal., ed. 2, 1: 257 (1932); Lindberg f., Iter Cypr., 20 (1946). Davis, Fl. Turkey, 2: 459 (1967); Zohary, Fl. Palaest., 2: 230, t. 328 (1972); Osorio-Tafall et Seraphim, List Vasc. Plants Cyprus, 64 (1973).
TYPE: " *in* Europa *ad plateas* " (LINN)

A decumbent or sprawling annual, very similar to *G. pusillum* in habit and general appearance, with thinly pubescent stems 10–40 cm. long, and pubescent, orbicular-reniform leaves, the lower 2·5–5 cm. wide, divided to about half way into 5–7 (–9) bluntly toothed, oblong-cuneate lobes; stipules larger and more conspicuous than in either *G. pusillum* or *G. rotundifolium*, pale brown, membranous, pubescent externally, ovate, 5–6 mm. long, 2mm. wide at base, the apex acute or shortly acuminate, often laciniate or frayed; upper leaves usually alternate, smaller and more deeply lobed than the lower, with shorter petioles; inflorescences of paired flowers as in *G. pusillum*, peduncles 1–3 cm. long, slender, glandular-pubescent, pedicels 0·8–1·5 cm. long, up-turned towards apex in fruit; flowers larger than those of *G.*

pusillum, but usually smaller and less conspicuous than those of *G. rotundifolium*; sepals oblong-ovate, about 4 mm. long and 2 mm. wide, glandular-pilose externally, 3-nerved, with an obtuse or subacute, shortly caudate-mucronate apex; petals pink, obovate, deeply emarginate or bifid, about 5 mm. long and 3 mm. wide, narrowing to a short, thinly pilose claw; stamens all fertile, about 3·5 mm. long; filaments glabrous or subglabrous, tapering from a wide base to a filiform apex; anthers shortly oblong, about 0·5 mm. long, and a little less in width; ovary depressed-globose, about 2 mm. wide and 1·5 mm. high, glabrous, beak short, thick, pubescent, about 1–1·5 mm. long; stigma-arms purple, relatively long, to about 1·5 mm. in length. Fruit about 3·5 mm. wide, 2 mm. high, with a beak 8–9 mm. long, carpels glabrous, distinctly wrinkled; seeds oblong or subglobose, about 1·5 mm. long and 1 mm. wide; testa dull brown, minutely and obscurely punctulate or almost smooth.

HAB.: Fields, gardens and roadsides, occasionally on rocky hills or on sand-dunes; sea-level to 4,500 ft. alt.; fl. Febr.–May.

DISTR.: Divisions 2–4, 6–8. Europe, North Africa, Atlantic Islands, Asia eastwards to the Himalayas; introduced elsewhere.

2. Pedhoulas, 1932, *Syngrassides* 39 ! Prodhromos, 1939, *Lindberg f.* s.n.
3. Alethriko, 1905, *Holmboe* 222.
4. Cape Pyla, 1880, *Sintenis & Rigo.*
6. Government House, Nicosia, 1927, *Rev. A. Huddle* 26 ! Nicosia, 1950, *Chapman* 467 !
7. Pentadaktylos 1880, *Sintenis & Rigo*; Kyrenia, 1949, *Casey* 332 ! and, 1956, *G. E. Atherton* 961 ! Temblos, 1956, *G. E. Atherton* 1205 !
8. Cape Andreas 1938, *Syngrassides* 1765 ! Ayios Philon near Rizokarpaso, 1941, *Davis* 2209 ! 2260 ! Platanisso, 1950, *Chapman* 250 !

5. G. dissectum *L.*, Cent. I. Plant., 21 (1755); Unger et Kotschy, Die Insel Cypern, 364 (1865); Boiss., Fl. Orient., 1: 881 (1867); Knuth in Engl., Pflanzenr., 53 (IV, 129): 51 (1912); Holmboe, Veg. Cypr., 116 (1914); Post, Fl. Pal., ed. 2, 1: 257 (1932); Lindberg f., Iter Cypr., 20 (1946); Davis, Fl. Turkey, 2: 461 (1967); Zohary, Fl. Palaest., 2: 229, t. 326 (1972); Osorio-Tafall et Seraphim, List Vasc. Plants Cyprus, 64 (1973).

TYPE: " *in* Europa *australiori* ".

A decumbent or sprawling annual; stems sparsely or densely glandular-strigose, often much branched from base; lower leaves orbicular-reniform in outline 4–6 cm. diam., divided almost to base into 5–7 pinnately dissected, cuneate lobes, secondary lobes narrowly oblong or linear, obtuse or subacute; petioles long and slender, sparsely strigose; stipules ovate-acuminate, 6–9 mm. long, 3–4 mm. wide, brownish, membranous, thinly strigose externally; upper leaves opposite or alternate, smaller than the lower and more shortly petiolate, deeply divided into linear, entire or irregularly pinnatisect lobes; inflorescences of paired flowers arising from the upper nodes; peduncles 5–10 mm. long, glandular-strigose, pedicels 10–15 mm. long, densely glandular towards apex; sepals narrowly ovate, 5–6 mm. long, about 2 mm. wide, accrescent in fruit, 3-nerved, glandular-pubescent externally, with a caudate-acuminate apex; petals bright rosy-purple, obovate, about 4 mm. long, 2 mm. wide, shallowly emarginate, distinctly ciliate-pilose at base; stamens all fertile, about 2·5 mm. long; filaments ciliate, wide at base, narrowing rather abruptly towards apex; anthers suborbicular, about 0·5 mm. diam.; ovary pubescent, about 1 mm. wide and almost as high, beak about 1 mm. long, pubescent, stigma-arms about 0·5 mm. long, purple. Fruit about 5 mm. wide and 4 mm. high, beak 1·3–1·5 mm. long, carpels smooth, glandular-pubescent; seeds subglobose, about 2 mm. long and 1·8 mm. wide; testa dark brown, strongly and conspicuously reticulate.

HAB.: Fields and waste ground, often in moist situations; sea-level to 4,500 ft. alt.; fl. Febr.–May.

DISTR.: Divisions 1, 2, 4, 7, 8, common in Division 7. Europe and Mediterranean area eastwards to Iran; Atlantic Islands; introduced elsewhere.

1. Near Pelathousa (Polis), 1962, *Meikle* 2271 !
2. Prodhromos, 1939, *Lindberg f.* s.n.; Prodhromos, 1955, *Merton* 2394 !
4. Cape Greco, 1862, *Kotschy* 150; Cape Pyla, 1880, *Sintenis & Rigo*; Kouklia (Mesaoria), 1905, *Holmboe* 393; Mile 3, Larnaca–Famagusta Road, 1950, *Chapman* 437 !
7. Common; *Kotschy* 420; *Sintenis & Rigo*; *Casey* 253 ! 1182 ! *G. E. Atherton* 369 ! *Merton* 2725 !
8. Ayios Andronikos (Karpas), 1938, *Syngrassides* 1792 ! Between Yialousa and Ayios Andronikos, 1941, *Davis* 2396 ! Rizokarpaso, Ronnas Valley, 1957, *Merton* 2921 !

6. G. columbinum *L.*, Sp. Plant., ed. 1, 682 (1753); Boiss., Fl. Orient., 1: 881 (1867); Knuth in Engl., Pflanzenr., 53 (IV, 129): 50 (1912); Post, Fl. Pal., ed. 2, 1: 257 (1932); A. K. Jackson et Turrill in Kew Bull., 1938: 464 (1938); Davis, Fl. Turkey, 2: 461 (1967); Zohary, Fl. Palaest., 2: 228, t. 325 (1972); Osorio-Tafall et Seraphim, List Vasc. Plants Cyprus, 64 (1973).

TYPE: " *in* Gallia, Helvetia, Germania ".

An erect or sprawling annual, 7–30 cm. high; stems usually much branched from base, adpressed-strigose; lower leaves orbicular-reniform 2–4 cm. diam., adpressed-strigose, divided two-thirds way or more in 5–7 oblong-cuneate lobes, the secondary lobes (or teeth) oblong, obtuse or acute; petioles long, slender (to 10 cm. or more), adpressed-strigose; stipules linear or almost filiform, 5–7 mm. long, to about 1 mm. wide at base, reddish, strigose; upper leaves similar, but more deeply divided into narrower lobes, and with shorter petioles, usually opposite; inflorescences of paired flowers arising from the upper nodes; peduncles 2–9 cm. long, elongating in fruit, thinly strigose or subglabrous; pedicels long and slender, to 5 cm. or more; sepals ovate, about 10 mm. long and 4 mm. wide, 3-nerved, adpressed-strigose externally, with a caudate-acuminate apex and membranous, ciliate margins; petals bright pink, conspicuous, obovate, 7–9 mm. long, 4 mm. wide, apex rounded, truncate, or irregularly toothed, not emarginate, base sparsely ciliate; stamens all fertile, about 4 mm. long; filaments glabrous or almost so, tapering from base to apex; anthers oblong, about 1 mm. long and 0·7 mm. wide; ovary about 1·5 mm. wide, 1 mm. high, minutely puberulous with a few scattered longer hairs, beak about 1 mm. long, puberulous; stigma-arms purplish, about 1 mm. long. Fruit hidden by the conspicuously accrescent calyx, about 6 mm. wide and 4 mm. high, beak 17–20 mm. long, carpels smooth, subglabrous or very sparsely strigose; seeds rounded or shortly oblong, about 2·5 mm. long and 2 mm. wide; testa dark blackish-brown, distinctly reticulate.

HAB.: Damp rocky banks and pastures; sea-level to c. 3,000 ft. alt.; fl. March–May.

DISTR.: Divisions 1, 2, 7, 8. Europe and Mediterranean area eastwards to Iran and Central Asia; introduced elsewhere.

1. Ayios Nikolaos (Akamas), 1950, *Casey* 1011 ! Between Androlikou and Prodhromi, 1957, *Merton* 3035 ! Mirillis R. near Ayios Yeoryios (Akamas) 1962, *Meikle* 2070 !
2. Yialia, 1955, *Merton* 2139 ! Ayios Nikolaos Monastery near Kakopetria, 1962, *Meikle* 2855 !
7. Above Antiphonitis Monastery, 1955, *Merton* 2283 ! Kyrenia, 1956, *G. E. Atherton* 1199 !
8. Cape Andreas, 1938, *Syngrassides* 1767 !

7. G. purpureum *Vill.*, Fl. Delph., 72 (1785), Hist. Plant. Dauph., 1: 272 (1786); Boiss., Fl. Orient., 1: 883 (1867); Sintenis in Oesterr. Bot. Zeitschr., 32: 291 (1882); Post, Fl. Pal., ed. 2, 1: 258 (1932); Rechinger f. in Arkiv för Bot., ser. 2, 1: 426 (1950); Davis, Fl. Turkey, 2: 456 (1967).

G. robertianum L. var. *purpureum* (Vill.) DC. in Lam. et DC., Fl. Franc., 4: 853 (1805); Knuth in Engl., Pflanzenr., 53 (IV, 129): 66 (1912).

G. modestum Jord., Ind. Sem. Jard. Bot. Grenoble, 16 (1849); Unger et Kotschy, Die Insel Cypern, 365 (1865).

G. robertianum L. ssp. *purpureum* (Vill.) Murb., Contr. Fl. Nord-Ouest Afr., 1: 52 (1897); Holmboe, Veg. Cypr., 116 (1914); Lindberg f., Iter Cypr., 21 (1946).

[*G. robertianum* (non L.) Kotschy in Unger et Kotschy, Die Insel Cypern, 365 (1865); Sintenis in Oesterr. Bot. Zeitschr., 32: 143 (1882); Osorio-Tafall et Seraphim, List Vasc. Plants Cyprus, 64 (1973).]

TYPE: France, Dauphiné: " *le Pont-de-Clais, au Buis, etc.*" (? GRM).

Sprawling or ascending, aromatic annual, 10–40 (–50) cm. high; stems fragile, much branched, thinly glandular-strigose or almost glabrous, often stained red; lower leaves palmate, usually of 5 divisions, the 2 lateral divisions joined basally by a short common petiolule; divisions ovate-rhomboid in outline (1–) 2–4 (–6) cm. long (0·6–) 1–2·5 (–3) cm. wide, pinnately divided into 3–7 narrowly oblong bluntly toothed lobes, lamina glabrous to thinly strigose, often reddish, base tapering to a ± distinct, slender petiolule; petiole long, slender, often exceeding 10 cm. in length, glabrous or thinly strigose; stipules oblong, blunt, about 2–3 mm. long, 2 mm. wide, membranous with strigose margins; upper leaves similar to lower but smaller and more shortly petiolate; flowers arising in pairs from the upper nodes; peduncles glandular-pubescent, 1–6 cm. long, pedicels often densely glandular, 3–10 mm. long; sepals ovate, about 6 mm. long, 2 mm. wide, concave, strongly 3-nerved, glandular-pubescent externally, with an acuminate-caudate apex; petals bright pink, obovate, about 7 mm. long, 2 mm. wide, apex rounded, base tapering to a long, slender, glabrous claw; stamens all fertile, about 7 mm. long; filaments shortly connate at base, tapering gradually to apex, glabrous; anthers suborbicular, about 0·5 mm. diam., pollen yellow; ovary about 1·7 mm. wide, 1·3 mm. high, glabrous; beak about 5 mm. long, thinly puberulous; stigma-arms purple, about 0·7 mm. long. Fruit about 4 mm. wide, 3 mm. high, enclosed within the persistent accrescent calyx, beak about 1·8 cm. long, carpels oblong, about 2·5 mm. long, 1·5 mm. wide, bright brown, glabrous, very conspicuously marked dorsally with transverse and reticulate wrinkles; carpels breaking away from the intact beak at maturity, but remaining temporarily attached to apex of beak by paired strands of silken threads joined to the ventral apex of each carpel; seeds oblong, about 2 mm. long and 1 mm. wide; testa bright brown, apparently smooth, but minutely pitted or striatulate under the microscope, with small scattered dark brown points.

HAB.: Damp, shaded rocky places in maquis or forest; sea-level to 4,500 ft. alt.; fl. Febr.—May.

DISTR.: Divisions 1, 2, 7, 8, common in Divisions 2, 8. S. Europe and Mediterranean area, along the Atlantic coast to the British Isles; eastwards to Iran. Also Atlantic Islands and tropical East Africa.

1. By Avgas R. near Ayios Yeoryios (Akamas), 1962, *Meikle* 2021 ! Near Neokhorio (Polis), 1962, *Meikle* !
2. Common; *Kotschy* 930; 931; *Lindberg f.* s.n.; *Davis* 2825 !
7. Pentadaktylos, 1880, *Sintenis & Rigo*; St. Hilarion, 1902, *A. G. & M. E. Lascelles* s.n.; W. of St. Hilarion, 1949, *Casey* 399 ! Above Antiphonitis Monastery, 1955, *Merton* 2284 !
8. Common; *Sintenis & Rigo* 198 ! *Holmboe* 504; *Syngrassides* 1763 ! *Chapman* 230 ! *Merton* 2925 !

NOTES: Closely allied to *G. robertianum* L. from which it differs chiefly in the smallness of the flowers, its yellow (not orange) pollen, and its strongly wrinkled carpels; *G. purpureum* is often a greener plant than *G. robertianum* and its leaflets are normally less dissected, with narrower, less markedly lobed divisions.

8. **G. lucidum** *L.*, Sp. Plant., ed. 1, 682 (1753); Boiss., Fl. Orient., 1: 884 (1867); Knuth in Engl., Pflanzenr., 53 (IV, 129): 63 (1912); Holmboe, Veg. Cypr., 116 (1914); Post, Fl. Pal., ed. 2, 1: 258 (1932); Lindberg f., Iter Cypr., 20 (1946); Davis, Fl. Turkey, 2: 456 (1967); Zohary, Fl. Palaest., 2: 231, t. 331 (1972); Osorio-Tafall et Seraphim, List Vasc. Plants Cyprus, 64 (1973).

TYPE: " *in Europae rupibus umbrosis.*"

Erect or spreading annual, 3–20 cm. high, usually much branched from base; stems and leaves glabrous or sparsely strigose, shining, bright green or often stained red; basal leaves orbicular-reniform, 0·5–3·5 (–5) cm. diam., divided more than half way into 5–7 oblong-cuneate, bluntly lobed segments; petioles 1–8 cm. long; stipules narrowly ovate-acuminate, 3–4 mm. long, 1 mm. wide at base, ± glabrous, membranous; upper leaves usually opposite, smaller than lower, with narrower more widely separated segments and with shorter petioles; flowers in pairs from most of the nodes; peduncles 1–3·5 cm. long, usually glabrous, pedicels slender 0·3–1·3 cm. long; sepals glabrous, unequal, connivent, the 3 outer ovate, about 5 mm. long, 2·5–3 mm. wide, 3-nerved, longitudinally plicate, with conspicuous transverse wrinkles, apex cuspidate, base gibbous, 2 inner lanceolate, about 5 mm. long and 1 mm. wide, membranous; petals bright pink, narrowly obovate, 5–10 mm. long, 0·8–2 mm. wide, with a rounded apex and a long slender claw; stamens all fertile, 5–8 mm. long; filaments free or ± connate, linear-lanceolate, glabrous, about 0·4 mm. wide at base; anthers small, about 0·4 mm. long and 0·3 mm. wide; ovary subglobose, about 0·8–1 mm. diam., beak 3–5 mm. long; stigma-arms about 0·5 mm. long. Fruit about 5 mm. wide, 3·5 mm. high, enclosed within the accrescent calyx, carpels oblong, about 2–2·5 mm. long, 1·5 mm. wide, breaking away from intact beak at maturity, but without attachment threads, light brown, conspicuously reticulate-rugose, base and ventral margin densely glandular-ciliate; seeds oblong, about 1·8 mm. long, 1·4 mm. wide; testa bright brown, smooth or minutely and obscurely striatulate.

HAB.: In rock crevices, or sometimes on stony slopes or summits, usually in shade; 1,000–4,700 ft. alt.; fl. March–May.

DISTR.: Divisions 1, 2, 7, common in Division 7. Europe and Mediterranean area, eastwards to western Himalayas; Atlantic Islands.

1. Yioulou Chiftlik above Baths of Aphrodite (Akamas), 1962, *Meikle* 2228 !
2. Between Lefka and Kykko, 1927, *Rev. A. Huddle* 30 ! Milikouri near Kykko Monastery, 1939, *Lindberg f.* s.n.; Papoutsa, 1941, *Davis* 3115 ! Askas, beyond Palekhori, 1957, *Merton* 3128 !
7. Common; *Sintenis & Rigo* 197 ! *Mavromoustakis* 29 ! *Davis* 2468 ! 2518 ! *Casey* 355 ! *G. E. Atherton* 1293 !

2. ERODIUM *L'Hérit.*

Geranologia, t. 1–6 (1787–88).
Aiton, Hort. Kew., ed. 1, 2: 414 (1789).

Annual or perennial herbs or rarely subshrubs; leaves opposite or alternate, pinnately lobed or dissected, sometimes almost entire; inflorescences usually umbellate, sometimes 1-flowered by abortion; flowers hermaphrodite or rarely unisexual; sepals 5, imbricate; petals 5, alternating with glands, hypogynous, imbricate, the 2 upper often differing from the 3 lower; fertile stamens 5, antisepalous, free or connate at the base, staminodes 5, antipetalous, sometimes reduced to scales; ovary 5-lobed, 5-locular, beaked, ovules 2 in each loculus; ripe carpels 1 (rarely 2)-seeded, breaking away with beak from central axis at maturity, the beak curling spirally and often pilose or plumose within; seed(s) remaining enclosed within carpel walls.

About 60 species chiefly in the Mediterranean area, though extending eastwards to China and Japan, and also represented in North and South America, Australasia, tropical East Africa and South Africa. Some species are cosmopolitan weeds.

Leaves pinnate, with distinct leaflets:
Leaflets deeply pinnatisect; apical pits of carpel not glandular-papillose within **1. E. cicutarium**

Leaflets often lobed but not deeply pinnatisect; carpel-pits glandular-papillose within - - - - - - - - - - - - - - **2. E. moschatum**
Leaves pinnatisect, pinnatifid, lobed or entire, not divided into distinct leaflets:
Beak of carpels shortly hairy or with long bristles towards the base only, not long-plumose throughout its length; rootstock not woody, roots not tuber-bearing; plant annual or biennial:
Flowers large, showy, more than 2 cm. diam.:
Leaves deeply pinnatisect; fruiting beak 10–11 cm. long; seeds solitary in each carpel - - - - - - - - - - - - - **3. E. ciconium**
Leaves lobed, but not pinnatisect; fruiting beak 6–7 (–9) cm. long; seeds 2 in each carpel - - - - - - - - - - - - - **4. E. gruinum**
Flowers small, less than 2 cm. diam.:
Fruiting beak 8–10 cm. long; basal leaves normally different from upper leaves **5. E. botrys**
Fruiting beak less than 6 cm. long; basal leaves and upper leaves not markedly different:
Leaves not distinctly glandular below; apical pits of carpel not bordered by a concentric ridge - - - - - - - - - - **6. E. laciniatum**
Leaves distinctly glandular below, with sessile, shining glands; apical pits of carpel bordered by a distinct concentric ridge - **7. E. malacoides**
Beak of carpel long-plumose throughout its length; rootstock woody, roots with scattered tubers; plant perennial - - - - - - - **8. E. crassifolium**

1. E. cicutarium (*L.*) *L'Hérit.* in Aiton, Hort. Kew., ed. 1, 2: 414 (1789); Unger et Kotschy, Die Insel Cypern, 363 (1865); Boiss., Fl. Orient., 1: 890 (1867); Knuth in Engl., Pflanzenr., 53 (IV, 129): 274 (1912); Holmboe, Veg. Cypr., 116 (1914); Post, Fl. Pal., ed. 2, 1: 259 (1932); Lindberg f., Iter Cypr., 20 (1946); Davis, Fl. Turkey, 2: 485 (1967); Zohary, Fl. Palaest., 2: 235, t. 337 (1972); Osorio-Tafall et Seraphim, List Vasc. Plants Cyprus, 64 (1973).
Geranium cicutarium L., Sp. Plant., ed. 1, 680 (1753).
TYPE: " *in* Europae *sterilibus cultis* ".

Erect, decumbent or sprawling annual or biennial, 5–30 cm. high; stems strigose, glandular or eglandular; basal leaves forming a loose rosette, oblong, pinnate, 2–12 (–15) cm. long, 0·8–4 cm. wide, leaflets deeply toothed or dissected into linear lobes, strigose, ± glandular; petioles 1–6 cm. long; stipules 2–5 mm. long, 1–2 mm. wide, acute, glabrous or with strigose margins; cauline leaves similar to basal; inflorescence umbellate, 2–7-flowered; peduncle 1–12 cm. long; bracts deltoid-ovate, 2–5 mm. long, 1·5–4 mm. wide, membranous, glabrous or thinly strigose, often reddish, frequently with a bifid apex; pedicels 2–15 mm. long at anthesis, erect, slender, lengthening and becoming sharply curved at apex in fruit, thinly strigose, ± glandular; sepals oblong-ovate, 3–5 mm. long, 1–2 mm. wide, 5-nerved, often stained reddish, with a cuspidate, bristle-pointed apex and narrow membranous margins, usually strigose and sometimes glandular; petals pink or purplish, the 2 upper oblong-elliptical, 6–7 mm. long, about 3 mm. wide, blotched purple towards base or unblotched, apex obtuse, base narrowed to a very short hairy claw, 3 lower petals about 5 mm. long, unblotched; stamens 4–5 mm. long, filaments oblong or narrowly ovate and connate at base, narrowing rather abruptly to a filiform apex; anthers oblong, about 1·2 mm. long and 0·8 mm. wide; staminodes subulate; ovary flask-shaped, villose, about 1·5 mm. long, 1·4 mm. high, beak about 1 mm. long; stigma-arms about 0·5 mm. long. Fruit oblong, about 5 mm. long and 3 mm. high, enclosed within persistent calyx before dehiscence, beak about 5 cm. long, tapering to a slender apex; carpels spindle-shaped, about 5 mm. long and 1·2 mm. wide, brown, with a slender pointed base, covered with suberect bristles, and with a blunt, thinly bristly or subglabrous apex with 2 conspicuous pits, beak strigose internally, with sparse scattered long hairs, the basal half closely spiralling; seed clavate, about 2·5 mm. long, 0·8–1 mm. wide; testa brown, apparently smooth or, under the microscope, minutely pitted and striatulate.

HAB.: On stony mountainsides, closely grazed pastures, in garigue, in fields and gardens, occasionally on walls; sea-level to over 5,000 ft. alt.; fl. Febr.–June.

DISTR.: Divisions 2–5, 7, 8, common in Divisions 2, 7. Widespread in Europe, temperate Asia, Mediterranean region and Atlantic Islands; introduced elsewhere.

2. Common; *Kotschy* 932; *Kennedy* 594 ! 596 ! *Lindberg f.* s.n.; *Casey* 307 !
3. Akrotiri Forest, 1939, *Mavromoustakis* 118 !
4. Larnaca aerodrome, 1950, *Chapman* 22 !
5. Nicosia–Famagusta Road, 1950, *Chapman* 372 ! Mia Milea, 1950, *Chapman* 674 ! Athalassa, 1962, *Meikle* 2980 !
7. Common; *Casey* 378 ! *Chapman* 142 ! *F. M. Probyn* 62 ! *G. E. Atherton* 1292 !
8. Apostolos Andreas, 1938, *Syngrassides* 1772 !

NOTES: The Cyprus population of this polymorphic species is unusually uniform, consisting of eglandular or very sparsely glandular forms referable to *E. cicutarium* ssp. *cicutarium*. Many of the plants are small, but this is probably a direct phenotypic response to inhospitable soil and climate, for exceptional specimens, growing under better conditions, are just as luxuriant as European material of *E. cicutarium*.

2. E. moschatum (*L.*) *L'Hérit.* in Aiton, Hort. Kew., ed. 1, 2: 414 (1789); Boiss., Fl. Orient., 1: 891 (1867); Knuth in Engl., Pflanzenr., 53 (IV, 129): 281 (1912); Druce in Rep. B.E.C., 9: 470 (1931); Post, Fl. Pal., ed. 2, 1: 260 (1932); Davis, Fl. Turkey, 2: 486 (1967); Zohary, Fl. Palaest., 2: 236, t. 338 (1972); Osorio-Tafall et Seraphim, List Vasc. Plants Cyprus, 64 (1973).

Geranium cicutarium L. β *moschatum* L., Sp. Plant., ed. 1, 680 (1753).

TYPE: " *in* Europae *sterilibus cultis* ".

A robust annual or biennial, up to 50 cm. high; stems decumbent or sprawling, thick and rather fleshy, ± strigose, with numerous small, scattered, sessile glands; leaves oblong, 10–30 cm. long, 3–8 cm. wide, pinnate, with 5–11 alternate or opposite, oblong, ± deeply lobed, serrate, shortly petiolulate or sessile leaflets, 1–4 cm. long, 0·4–3 cm. wide, lamina thinly strigose-glandular on both surfaces, apex rounded or less often acute, base broadly cuneate or rounded; petioles up to 15 cm. long; stipules ovate-oblong, 5–13 mm. long, 4–10 mm. wide, obtuse, papery or membranous, glabrous or with ciliate-strigose margins; inflorescence umbellate, 6–12 (or more)-flowered; peduncles 8–10 cm. long, elongating greatly in fruit, ± densely glandular-strigose; bracts broadly ovate, about 3 mm. long, 2·5 mm. wide, membranous, subglabrous or thinly glandular-strigose; pedicels slender, at first less than 1 cm. long, lengthening to 2 cm. or more in fruit, and becoming patent, with sharply up-curved tips, densely glandular; sepals oblong-ovate, concave, cucullate, 5–8 mm. long, 2–3 mm. wide, 5-nerved, glandular-pilose externally, with a cuspidate-caudate apex and membranous margins; petals mauve-pink, oblong, 5–7 mm. long, 2–3 mm. wide, apex rounded, base narrowed rather abruptly to a short pilose claw; stamens 3–4 mm. long, filaments free, with a broad flattened base and a subulate apex; anthers oblong, about 0·8 mm. long and 0·6 mm. wide; staminodes oblong, about 2 mm. long, 0·6 mm. wide, with an acute or emarginate apex; ovary about 2 mm. wide and 1·5 mm. high, densely villose, beak about 1·5 mm. long, thick, densely villose; stigma-arms about 0·5 mm. long. Fruit oblong, about 3 mm. wide, 5 mm. high, strigose, beak 4–5 cm. long, thinly adpressed-strigose; carpels narrowly spindle-shaped, about 6 mm. long, 1·5 mm. wide, with a slender pointed base, covered with suberect bristles, and with a blunt, thinly bristly apex with 2 conspicuous pits, glandular-papillose within, beak adpressed strigose, with a few longer hairs, the basal half closely spiralling; seeds clavate, about 3·5 mm. long and 1 mm. wide; testa bright brown, apparently smooth, but minutely pitted under the microscope.

HAB.: Waste ground, roadsides, gardens, fields; sea-level to about 1,500 ft. alt., but usually lowland; fl. Jan.–June.

DISTR.: Divisions 1, 4, 6, 7, 8, common in Division 7. Mediterranean region, north along the Atlantic coastal area to Britain; Atlantic Islands, Ethiopia; now widely distributed as an introduced weed elsewhere.

1. Aspro Potamos near Ayios Yeoryios (Akamas), 1962, *Meikle* 2088 ! Common in Polis area, 1962, *Meikle* !
4. Famagusta, 1928 or 1930, *Druce* s.n.
6. Court Garden (? Nicosia), 1901, *A. G. & M. E. Lascelles* s.n.; Nicosia, 1927, *Rev. A. Huddle* 32 ! and, 1956, *Merton* 2612 ! Common in Dhiorios area, 1962, *Meikle* !
7. Common; *Sintenis & Rigo* 192 ! *Casey* 245 ! *G. E. Atherton* 838 ! 842 ! 956 !
8. Ronnas Bay near Rizokarpaso, 1941, *Davis* 2339 !

NOTES: In view of the local abundance of this species, it is strange that it should have been overlooked by Sibthorp, Kotschy and Holmboe. Has *Erodium moschatum* become a common plant only in relatively recent times, or did these collectors confound it with *E. cicutarium*?

3. E. ciconium (*L.*) *L'Hérit.* in Aiton, Hort. Kew., ed. 1, 2: 415 (1789); Unger et Kotschy, Die Insel Cypern, 363 (1865); Boiss., Fl. Orient., 1: 891 (1867); Knuth in Engl., Pflanzenr., 53 (IV, 129): 260 (1912); Holmboe, Veg. Cypr., 116 (1914); Post, Fl. Pal., ed. 2, 1: 260 (1932); Davis, Fl. Turkey, 2: 479 (1967); Zohary, Fl. Palaest., 2: 237, t. 340, 340a (1972); Osorio-Tafall et Seraphim, List Vasc. Plants Cyprus, 64 (1973).

Geranium ciconium L., Cent. I Plant., 21 (1755), Sp. Plant., ed. 2, 952 (1763).

TYPE: " *In* Europa *australi*, Valesia Italiae, Monspelii ".

An erect annual or biennial, 20–60 (–100) cm. high; stems sparingly branched from base or often solitary, thinly strigose-pubescent, usually glandular; leaves oblong, 3–10 (–15) cm. long (excluding petiole) 4–10 cm. wide, bipinnatisect, the pinnules bluntly or acutely lobed, lamina thinly glandular-pubescent, greyish-green; petioles up to 30 cm. long, glandular-pubescent; stipules deltoid-ovate, 3–4 mm. long and almost as wide at base, acute, often cuspidate, shortly pubescent, brownish; inflorescence umbellate, 2–6-flowered; peduncles stout, 6–20 cm. long or longer, ± erect, densely glandular-pubescent; bracts oblong, acute, about 4–5 mm. long and 2 mm. wide, brownish, pubescent externally; pedicels 1–5 cm. long, densely glandular-pubescent, erect in flower, elongating, thickening and reflexing or spreading in fruit, with abruptly up-curved tips; flowers showy; sepals oblong-ovate, about 10 mm. long and 4–5 mm. wide at anthesis, accrescent in fruit, strongly 5-nerved, densely glandular externally, apex caudate-cuspidate; petals violet-blue, caducous, obovate, 12–15 mm. long, 8–10 mm. wide, rounded above, tapering to a very short, pilose basal claw; stamens free, about 9 mm. long, with the lower half of the filament oblong, about 1·7 mm. wide, ciliate, the upper half subulate, glabrous; anthers oblong, about 2 mm. long and 1 mm. wide; staminodes oblong, about 3 mm. long, 1·5 mm. wide at base, with an acute, ciliate apex; ovary about 3 mm. wide, 2 mm. high, densely villose, beak thick, conical, about 3·5 mm. long, densely pubescent; stigma-arms about 1 mm. long. Fruit about 6 mm. wide, 12–15 mm. high, with a beak 10–11 cm. long, carpels narrowly spindle-shaped, 12–15 mm. long, 2–3 mm. wide, tapering to a slender pointed base covered with suberect bristles, apex blunt with 2 deep, conspicuous pits, beak thinly adpressed-strigose externally, densely covered with ascending bristles within, those towards the middle and base of the beak often 2–3 cm. long, those in the upper half of the beak very short; seed clavate, 6–7 mm. long, 1·5–2 mm. wide; testa rich brown, apparently smooth, but, under the microscope, minutely pitted and regularly striatulate.

HAB.: Roadsides, fields and waste ground; sea-level to ? 2,000 ft. alt.; fl. Febr.–April.

DISTR.: Divisions 4–7, common in Division 6. Mediterranean area, S.E. Europe and eastwards to Iran and Turkestan.

4. Larnaca, 1859, *Kotschy* 493; and, 1880, *Sintenis & Rigo*; in fields near Ormidhia, 1880, *Sintenis & Rigo* 195 ! Phaneromene near Larnaca, 1905, *Holmboe* 261.

5. Mia Milea, 1950, *Chapman* 492 ! Athalassa Farm, 1962, *Meikle* 2975 !
6. Common; *Rev. A. Huddle* 31 ! 33 ! *Chapman* 466 ! *Casey* 1222 ! *F. M. Probyn* 44 ! *Merton* 2611 !
7. Pentadaktylos, 1880, *Sintenis & Rigo* 194 !

4. E. gruinum (*L.*) *L'Hérit.* in Aiton, Hort. Kew., ed. 1, 2: 415 (1789); Unger et Kotschy, Die Insel Cypern, 363 (1865); Boiss., Fl. Orient., 1: 892 (1867); Knuth in Engl., Pflanzenr., 53 (IV, 129): 258, fig. 34 (1912); Holmboe, Veg. Cypr., 116 (1914); Post, Fl. Pal., ed. 2, 1: 260 (1932).

Geranium gruinum L., Sp. Plant., ed. 1, 680 (1753); Davis, Fl. Turkey, 2: 478 (1967); Zohary, Fl. Palaest., 2: 238, t. 342 (1972); Osorio-Tafall et Seraphim, List Vasc. Plants Cyprus, 64 (1973).

TYPE: " *in* Creta ".

Erect or spreading annual or biennial, 10–30 (–50) cm. high; stems unbranched or branched especially at base, strigose with coarse, stiff, white hairs, eglandular or with scattered sessile glands; leaves rather variable in shape, the basal oblong-cordate 2–5 cm. long, 1–3 cm. wide, crenate or obscurely lobed, thinly strigose, with petioles 2–8 cm. long, the upper larger, up to 10 (–12) cm. long, 5 cm. wide, distinctly 3–lobed, the middle lobe much larger than the laterals and often distinctly pinnatifid, petioles to 15 cm. long, uppermost leaves in luxuriant plants sometimes pinnatisect, with broad oblong, sharply dentate lobes; stipules ovate, acute or acuminate, 5–10 mm. long, 3–8 mm. wide, conspicuous, membranous, glabrous or with minutely ciliolate margins; inflorescences umbellate, generally 2-flowered, but occasionally 1- or 3–5-flowered; peduncles 3–12 cm. long, erect, strigose with spreading hairs; bracts 4, connate in pairs, occasionally more, about 5 mm. long, 4 mm. wide, membranous, conspicuously costate, ciliolate; pedicels 2–4·5 cm. long, strigose, at first ± erect, later spreading with sharply up-turned tips; flowers showy; sepals oblong, about 10 mm. long and 3–4 mm. wide, strongly 5-nerved, strigose externally, apex conspicuously caudate, margins of inner sepals broadly membranous, ciliate, those of outer sepals more narrowly so and generally glabrous; petals lavender-blue or mauve, caducous, broadly obovate, 1·3–1·5 cm. long, 0·9–1 cm. wide, apex rounded, base narrowed to an indistinct, sparsely bearded or glabrous claw; stamens about 9 mm. long, filaments free, basal half oblong, about 2 mm. wide, narrowing suddenly (and sometimes denticulate) above to a slender subulate apex; anthers oblong, about 1·7 mm. long and 1·2 mm. wide; ovary about 3 mm. wide, 2·5 mm. high, densely pilose with stiff ascending hairs, beak thick, conical, about 4 mm. long, very shortly and densely pilose; stigma-arms about 1·5 mm. long. Fruit about 8 mm. wide, 1·5 mm. high, beak 6–7 (–9) cm. long, carpels narrowly spindle-shaped, about 1·5 mm. long, 2·5 mm. wide, distinctly pitted all over and with copious spreading bristles, base tapering to a slender point, apex with 2 conspicuous pits, each with a 2-ridged margin, beak adpressed-strigose within, the basal half with conspicuous stiff hairs 1–2 cm. long; seeds 2 in each carpel, irregularly oblong, about 5 mm. long and 2 mm. wide; testa rich brown, apparently smooth, but, under the microscope, minutely pitted and faintly striatulate.

HAB.: Dry hillsides, stony fields, roadsides and waste places; sea-level to 2,000 ft. alt.; fl. Febr.–April.

DISTR.: Divisions 1, 3, 4, 5, 7, 8, common in Division 7. Eastern Mediterranean from Sicily to Egypt and Cyrenaica; eastwards to Iran.

1. Cliffs above Peyia, 1962, *Meikle* 2040 !
3. Yerasa, 1947, *Mavromoustakis* s.n. !
4. Larnaca, 1859, *Kotschy* 423, and, 1880, *Sintenis & Rigo*, and, 1912, *Haradjian* 16 ! East of Paralimni, 1958, *N. Macdonald* 52b !
5. Athalassa, 1931, *J. A. Tracey* 30 !
7. Common; *Sintenis & Rigo* 194a ! *Syngrassides* 907 ! *Chapman* 131 ! 260 ! 317 ! *G. E. Atherton* 968 ! 1010 ! 1074 ! 1222 ! 1263 !
8. Rizokarpaso, 1941, *Davis* 2371 !

5. **E. botrys** (*Cav.*) *Bert.*, Amoen. Ital., 35 (1819); Boiss., Fl. Orient., 1: 892 (1867); Knuth in Engl., Pflanzenr., 53 (IV, 129): 256 (1912); Post, Fl. Pal., ed. 2, 1: 261 (1932); Davis, Fl. Turkey, 2: 478 (1967); Zohary, Fl. Palaest. 2: 239, t. 344 (1972).
Geranium botrys Cav., Diss., 4: 218, t. 90, fig. 2 (1787).

TYPE: " *in arenosis prope mare Hispanicum et Lusitanicum, atque in pratis Neapolitanis. V.S. communic. a D. de Jussieu* " (MA).

An erect or spreading annual or biennial, (7–) 15–30 (–50) cm. high; stems usually much branched, especially towards base, ± densely covered with retrorse bristles; basal leaves forming a loose rosette, oblong, 1·5–4 cm. long, 0·8–3 cm. wide, shortly pinnatifid with bluntly toothed lobes, or sometimes with the lobes barely distinct, apex obtuse, base truncate or shallowly cordate, lamina thinly strigose above and below; petiole 1–10 cm. long, with conspicuous retrorse bristles; cauline leaves oblong or narrowly deltoid, deeply pinnatisect, with 7–9 narrowly oblong, sharply toothed or laciniate lobes, lamina normally hispid, often glandular, petioles shorter than in basal leaves, the upper cauline leaves often subsessile; stipules ovate-oblong, acute, 3–5 mm. long, about 3 mm. wide, membranous with ciliate margins; inflorescences umbellate, 2 (–4)-flowered; peduncles 2–8 cm. long, often densely glandular, especially towards apex; bracts normally free, ovate, acute, 2–3 mm. long, 1·5–2 mm. wide, ciliate; pedicels 1–2 cm. long, often densely glandular, at first erect, later spreading with up-curved tips; sepals oblong, 7–8 mm. long, 2–3 mm. wide, obtuse, distinctly caudate-cuspidate, ± densely glandular externally, the 3 outer 5-nerved, the 2 inner 3-nerved with wide membranous, ciliate margins; petals violet or purplish, obovate, 7–8 mm. long, about 4 mm. wide, apex rounded, base tapering to a very short sparsely pilose claw; stamens about 4·5 mm. long, filaments free, ovate-oblong, and about 1·5 mm. wide for half their length, subulate above; anthers 1 mm. long, 0·8 mm. wide; staminodes lanceolate or narrowly deltoid, about 2 mm. long, 0·5 mm. wide at base; ovary subglobose, about 2 mm. diam., densely villose, beak about 1·5 mm. long, thick, densely pubescent; stigma-arms about 0·5 mm. long. Fruit about 7 mm. wide, 1 cm. high, concealed by persistent, accrescent sepals, beak up to 10 cm. long, carpels narrowly spindle-shaped, about 10 mm. long and 1–1·5 mm. wide, base tapering to a slender point, densely covered with suberect bristles, apex blunt with 2 deep pits, each pit surrounded by 3 distinct concentric ridges, beak covered with short ± adpressed bristles internally, with scattered longer bristles, up to 2 cm. long, in the lower half; seeds solitary in each carpel, clavate, 4·5 mm. long, 1–1·5 mm. wide; testa light brown, apparently smooth, but, under the microscope, minutely pitted and striatulate.

HAB.: By roadsides; 2,000 ft. alt.; fl. ? Febr.–April.

DISTR.: Division 2, rare. Widespread in Mediterranean area, particularly frequent in western half; Atlantic Islands.

2. Makheras Road, 1949, *Casey* 275 !

NOTES: The record is based on one depauperate, and not very characteristic, specimen, but there can be little doubt about the identification and the plant should be looked for, especially on the lower slopes of the Troödos Range.

6. **E. laciniatum** (*Cav.*) *Willd.*, Sp. Plant., 3: 633 (1800); Sibth. et Sm., Fl. Graec. Prodr., 2: 36 (1813), Fl. Graec., 7: 50, t. 655 (1830); Unger et Kotschy, Die Insel Cypern, 363 (1865); Boiss., Fl Orient., 1: 898 (1867); Knuth in Engl., Pflanzenr., 53 (IV, 129): 241 (1912); Holmboe, Veg. Cypr., 116 (1914); Post, Fl. Pal., ed. 2, 1: 161 (1932); Burtt et Lewis in Kew Bull., 9: 404 (1954); Davis, Fl. Turkey, 2: 480 (1967); Zohary, Fl. Palaest. 2: 240, t. 345 (1972); Osorio-Tafall et Seraphim, List Vasc. Plants Cyprus, 64 (1973).
Geranium laciniatum Cav., Diss., 4: 228, t. 113, fig. 3 (1787).
? *Geranium triangulare* Forssk., Fl. Aegypt.-Arab., 123 (1775).
? *Erodium triangulare* (Forssk.) Muschl., Man. Fl. Egypt., 1: 558 (1912).

[*E. hirtum* (non (Forssk.) Willd.) Kotschy in Unger et Kotschy, Die Insel Cypern, 363 (1865) *teste* Holmboe; Thompson in Journ. Bot., 44: 305 (1906).]

[*E. chium* (non (Burm. f.) Willd.) Sintenis in Oesterr. Bot. Zeitschr., 32: 364 (1882).]

[*E. laciniatum* (Cav.) Willd. var. *cavanillesii* (? (Willk.) Hal.) Lindberg f., Iter Cypr., 20 (1946).]

Sprawling or decumbent annual or biennial, 5–30 (–40) cm. high; stems usually much branched, often reddish towards base, ± densely clothed with short retrorse bristles; basal leaves ovate- or oblong-cordate, 1·5–7 cm. long, 1–4·5 cm. wide, deeply or shallowly pinnatifid, with 3–7 bluntly toothed lobes; lamina greyish-green, thinly or densely pilose with short, often curved, hairs; petioles to about 10 cm. long, retrorsely hispidulous; stipules rather conspicuous, ovate, acute or subacute, 2–5 mm. long, 1·5–4 mm. wide, membranous, glabrous or ciliate, frequently stained brownish; cauline leaves usually more deeply pinnatisect, often with narrow, oblong, distinctly toothed lobes, ± densely hispidulous, the petioles shorter than those of the basal leaves, uppermost cauline leaves almost sessile; inflorescences umbellate, 2–6 (–8)-flowered; peduncles 2–5 (–7) cm. long; bracts 2–4, shortly or sometimes distinctly connate at the base, or almost free, ovate, subacute, 2–5 mm. long, 2–4 mm. wide, membranous, ciliate or subglabrous, commonly stained brownish; pedicels slender, 0·3–20 mm. long, at first ± erect, later spreading with an up-curved apex, hispidulous, eglandular or glandular (var. *glanduloso-pilosum* Vierh.); sepals oblong, 4–5 mm. long, 1·5–2 mm. wide, hispidulous, eglandular or glandular (var. *glanduloso-pilosum* Vierh.) externally, the 3 outer prominently 5-nerved, with a blunt, cuspidate or mucronate apex and narrow membranous margins, the 2 inner less conspicuously nerved with a concave, sub-cucullate, distinctly mucronate or caudate-cuspidate apex, and rather wide membranous margins; petals pink or purple, obovate, about as long as sepals, 2–3 mm. wide, with the apex rounded and the base tapering to a very short pilose claw; stamens free, 4–5 mm. long, filaments oblong in lower half, about 0·5 mm. wide, tapering above, or abruptly contracted to a slender subulate apex; anthers oblong, about 0·8 mm. long and 0·5 mm. wide; staminodes lanceolate-oblong, 2–2·5 mm. long, 0·5 mm. wide, apex acute or acuminate; ovary subglobose, 1·8–2 mm. diam., adpressed-strigose, beak about 2·5 mm. long, rather slender, shortly pilose; stigma-arms 0·5 mm. long. Fruit 5 mm. wide, and about as high, beak (2–) 4–5 (–6) cm. long, slender, subglabrous or thinly strigose; carpels narrowly spindle-shaped, about 5 mm. long, 1 mm. wide, surface minutely scabrid-verruculose, ± densely clothed with suberect bristles, base slender, pointed, apex with 2 pits, without concentric ridges, beak ± adpressed-pilose within, with a few longer basal hairs; seeds clavate, about 2·5 mm. long, 0·8 mm. wide; testa light brown, apparently smooth, but, under the microscope, minutely pitted.

HAB.: On seashores and sand-dunes, occasionally on sandy ground inland; sea-level to 550 ft. alt.; fl. March–May.

f. laciniatum

TYPE: Not localized: " Communicatum a D. de Jussieu " (MA).

DISTR.: Divisions 4–8. Throughout the Mediterranean area and eastwards to Afghanistan.

4. Larnaca, 1862, *Kotschy* 10, and, 1939, *Lindberg f.* s.n.
5. Athalassa, 1950, *Chapman* 633 ! Mile 3, Nicosia–Famagusta road, 1950, *Chapman* 719 !
6. Ayia Irini (Morphou), 1941, *Davis* 2543 !
7. Mile 6, Kyrenia–Akanthou Road, 1950, *Chapman* 611 !
8. Seashore at Yialousa, 1880, *Sintenis & Rigo* 191 ! South of Galinoporni, 1962, *Meikle* 2502 !

f. glanduloso-pilosum *Vierh.* in Oesterr. Bot. Zeitschr., 64: 476 (1914).
TYPE: Crete; Tybaki–Klima, F.v. *Wettstein*. Knossos, *Hayek* (W).

DISTR.: Divisions 3, 7. General distribution similar to f. *laciniatum*.

3. Seashore, Limassol, 1939, *Lindberg f.* s.n.
7. Dhavlos [Davlu], 1880, *Sintenis & Rigo* 193 ! Ammos (Kyrenia District), on seashore, 1939, *Lindberg f.* s.n.; 6 m. E. of Kyrenia, on sandhills 1951, *Casey* 1170 ! Pachyammos, 7 m. E. of Kyrenia, 1955, *Merton* 2292 !

NOTES: The reasons for rejecting the name *Erodium triangulare* (Forssk.) Muschl., despite its antedating *E. laciniatum*, have been set out by Burtt & Lewis in Kew Bull., 9: 404 (1954); this same article also deals in some detail with the problem of distinguishing *E. laciniatum* from *E. pulverulentum* (Cav.) Willd. (*E. cavanillesii* Willk.). It is possible that some of the Cyprus material may approach *E. pulverulentum*, but none of the characters used in the past to distinguish this from *E. laciniatum* can be considered wholly satisfactory, nor does examination of the geographical distribution of the two segregates suggest that they can be considered more than minor variants of one polymorphic species. The distinctions between *E. laciniatum* f. *laciniatum* and *E. laciniatum* f. *glanduloso-pilosum* Vierh. are easily demonstrable, and all the Cyprus material so far examined fits neatly into eglandular and glandular categories, with glandular forms apparently common along the north coast of the island and eglandular forms more frequent elsewhere. But it is clear that a great deal of attention must be given to the ecology and genetics of *E. laciniatum* sens. lat. before the value of these subordinate taxa can be properly assessed.

7. E. malacoides (*L.*) *Willd.*, Phytogr., 1: 10 (1794); Sibth. et Sm., Fl. Graec. Prodr., 2: 37 (1813), Fl. Graec., 7: 53 (1830); Unger et Kotschy, Die Insel Cypern, 364 (1865); Boiss., Fl. Orient., 1: 893 (1867); Knuth in Engl., Pflanzenr., 53 (IV, 129): 245 (1912); Holmboe, Veg. Cypr., 116 (1914); Post, Fl. Pal., ed. 2, 1: 261 (1932); Davis, Fl. Turkey, 2: 480 (1967); Zohary, Fl. Palaest., 2: 242, t. 348 (1972); Osorio-Tafall et Seraphim, List Vasc. Plants Cyprus, 64 (1973).
Geranium malacoides L., Sp. Plant., ed. 1, 680 (1753).
TYPE: " *in* Italia G. narbonensis, Angliae *maritimis* ".

An erect or sprawling annual or biennial 6–40 (–50) cm. high; stems usually much branched, densely or sparsely clothed with spreading or retrorse bristles interspersed with shorter stalked glands; leaves broadly ovate-cordate, 1·5–6 cm. long, and almost as wide, the lower leaves obscurely pinnatifid, with 5–7 shallow, rounded, dentate lobes, apex obtuse or rounded, base distinctly cordate, lamina thinly adpressed-hispidulous above and below, with conspicuous, shining, sessile glands; upper leaves similar, but often small and rather acute, with sharply toothed lobes; petioles 1–5 cm. long, glandular-pilose, the uppermost leaves shortly petiolate or subsessile; stipules ovate, obtuse, 4–10 mm. long, 3–6 mm. wide, membranous, glabrous or hispid-pilose, often brownish; inflorescence umbellate (2–) 4–8-flowered; peduncles slender 2–4 (–7) cm. long, ± densely glandular-hispidulous; bracts 3–6 (occasionally more) free, ovate, about 2 mm. long and 1·5 mm. wide, membranous, ciliate; pedicels filiform, about 1 cm. long, ± densely glandular-pilose, erect at anthesis, later spreading with up-curved tips; sepals oblong, the outer about 5 mm. long (including tip) and 1·8 mm. wide, glandular-pilose externally, 5-nerved, apex obtuse, ± cucullate, conspicuously caudate-cuspidate, with a very narrow membranous margin, inner sepals narrower, 3-nerved, with a wide membranous margin; petals magenta-pink or purplish, about 6 mm. long and 3 mm. wide, apex rounded, base with a very short, sparsely pilose or subglabrous claw; stamens about 4 mm. long, filaments lanceolate-subulate, tapering from a base about 0·5 mm. wide to a slender apex; anthers oblong, about 1 mm. long, 0·8 mm. wide; staminodes oblong-ovate, about 2 mm. long, 0·8 mm. wide; ovary subglobose, 1–1·5 mm. diam., villose; beak narrowly conical, about 1·5 mm. long, shortly pubescent; stigma-arms 0·8–1 mm. long. Fruit about 4 mm. long, 3 mm. wide with a slender, subglabrous or glabrous beak 2–3 cm. long; carpels narrowly spindle-shaped, about 4 mm. long and 1 mm. wide, with a very slender pointed base and a blunt apex, ± densely covered with suberect hairs, apical pits glandular-papillose within, surrounded by a distinct concentric ridge, beak ± adpressed-pilose within, with a tuft of longer basal hairs; seed clavate 2·5–3 mm. long, about 0·8 mm. wide; testa bright brown, apparently smooth, but, under the microscope, minutely pitted.

HAB.: Roadsides, waste ground and cultivated fields; usually lowland, but recorded up to 4,500 ft. alt.; fl. Jan.–May.

DISTR.: Divisions 2–4, 6–8, common in Division 4, very common in Division 7. Throughout the Mediterranean area and eastwards to Iran; Atlantic Islands.

2. Near Prodhromos, 1862, *Kotschy* 933.
3. Stavrovouni, 1880, *Sintenis & Rigo*; Yerasa, 1962, *Meikle* 2916 !
4. Common; *Sibthorp*; *Kotschy* 140; *Sintenis & Rigo*; *Haradjian* 26 !
6. Government House, Nicosia, 1927, *Rev. A. Huddle* 34 ! Nicosia, 1930, *F. A. Rogers* 0724 ! and, 1950, *Chapman* 499 !
7. Very common; *Kotschy* 435; 527; *Sintenis & Rigo* 190 ! *Casey* 244 ! *G. E. Atherton* 830 ! 847 ! 849 !
8. Cape Andreas, 1938, *Syngrassides* 1762 ! Between Rizokarpaso and Apostolos Andreas Monastery, 1938, *Syngrassides* 1777 ! Platanisso, 1950, *Chapman* 249 !

8. E. crassifolium *L'Hérit.* in Aiton, Hort. Kew., ed. 1, 2: 414 (1789); Sweet, Geran., 2: t. 111 (1822–24); DC., Prodr., 1: 646 (1824).

Geranium hirtum (non Burm. f.) Forssk., Fl. Aegypt.–Arab., 123 (1775) nom. illeg.

Erodium hirtum Willd., Sp. Plant., 3: 632 (1800); Unger et Kotschy, Die Insel Cypern, 363 (1865); Boiss., Fl. Orient., 1: 894 (1867); H. Stuart Thompson in Journ. Bot., 44: 305 (1906); Knuth in Engl., Pflanzenr., 53 (IV, 129): 227 (1912); Post, Fl. Pal., ed. 2, 1: 262 (1932).

E. crassifolium L'Hérit. var. *salinarium* Sibth. ex DC., Prodr., 1: 646 (1824).

TYPE: " *Nat.* of the Island of Cyprus. *John Sibthorp*, MD." (OXF).

An erect or spreading perennial 8–20 (–30) cm. high, with a hard woody base and long woody roots swollen here and there to form conspicuous spherical tubers; stems much branched, sparsely or rather densely hispid-pubescent; basal leaves deltoid in outline, to about 5 cm. long and 3 cm. wide, deeply bipinnatisect, with 5–7 primary lobes subdivided into narrowly oblong or almost linear, acute, toothed, secondary lobes, lamina rather thick and fleshy, subglabrous or ± densely hispid; petiole 2–6 cm. long, subglabrous or hispid; upper cauline leaves similar to basal, but smaller and more shortly petiolate; stipules ovate, to about 5 mm. long and 4 mm. wide, membranous, brownish, ± densely pubescent externally; inflorescence umbellate, 2–6-flowered; peduncle 1–7 cm. long, subglabrous or ± densely pubescent; bracts 4–6 or more, free, ovate, rather blunt, 2–3 mm. long, 1·5–2 mm. wide, membranous, often pubescent externally; pedicels 0·8–1·5 cm. long, glabrous or pubescent; sepals oblong, about 6 mm. long, the outer about 2 mm. wide, 5-nerved, subglabrous or thinly pubescent externally with a mucronate apex, the inner about 1·5 mm. wide, 3-nerved, with a cucullate, mucronate apex and wide membranous margins; petals bright pink with a purple blotch at the base, obovate, about 8–10 mm. long, 6–8 mm. wide, apex rounded, base tapering to a very short pilose or glabrescent claw; stamens 4–5 mm. long, base of filament oblong, about 2 mm. long and 1 mm. wide, ciliate, narrowed abruptly to a slender subulate apex; anthers oblong, about 1·5 mm. long and 1 mm. wide; staminodes oblong-ovate, about 2 mm. long, 1 mm. wide, acute, long-ciliate; ovary about 1·5 mm. wide, 2 mm. high, ± densely pilose, beak thick, not tapering much from base to apex, shortly pubescent; stigma-arms short, clavate, up to about 5 mm. long. Fruit narrowly oblong, about 2·5 mm. wide, 5 mm. high, beak very slender, 5–10 cm. long, thinly pubescent; carpels narrowly spindle-shaped 5–6 mm. long, 1 mm. wide, thinly clothed with suberect hairs, tapering to a slender pointed base, apex blunt with 2 inconspicuous pits surrounded by double concentric ridges, beak internally densely plumose with long spreading hairs; seed narrowly clavate, 3–3·5 mm. long, 0·7 mm. wide; testa light brown, apparently smooth, but, under the microscope, minutely pitted.

HAB.: By margin of salt lake and on low chalk cliffs by the sea; near sea-level; fl. March–May.

DISTR.: Divisions 3, 4, rare. Crete, Morocco, Algeria, Tunisia, Libya, Egypt, Sinai, Palestine, Arabia, Iraq.

3. Episkopi cantonment on low chalk cliffs, 1967, *Merton* 622 !
4. Near Larnaca Salt Lake ["salinas Arnacae"], 1787, *Sibthorp*; and, 1952, *Merton* 840 !

27. OXALIDACEAE

Knuth in Engl., Pflanzenr., 95 (IV, 130) (1930).
Knuth in Engl. et Prantl, Pflanzenfam., ed. 2, 19a: 11–42 (1931).

Herbs, or rarely subshrubs or trees; stem sometimes fleshy and swollen or bulbiferous; leaves alternate, opposite, or radical, or in an apical tuft, digitately or pinnately compound, or occasionally 1-foliolate by reduction; leaflets entire or emarginate, stipules usually wanting; flowers solitary or several in cymose, umbellate or racemose inflorescences, actinomorphic or almost so, hermaphrodite, 5-merous, sometimes minute and cleistogamous, frequently heterostylous; sepals 5, free, imbricate or rarely valvate; petals 5, contorted or imbricate, free or shortly connate at base; stamens 10 (or 15) in 2 (or 3) series, hypogynous, filaments often connate at base; anthers 2-thecous, dehiscing longitudinally; ovary superior, 5-locular, ovules 1–many in each loculus, placentation axile, styles usually 5, free, stigmas capitate. Fruit a loculicidally dehiscing capsule, rarely fleshy and indehiscent, often distinctly 5-furrowed or 5-lobed; seeds with or without endosperm.

Eight genera, with a wide distribution, but most abundantly represented in tropical and subtropical regions. Hutchinson (Families of Flowering Plants, ed. 2, p. 497) reduces the extent of the family by transferring *Averrhoa* to a separate family *Averrhoaceae* and by placing *Dapania, Sarcotheca* and *Lepidobotrys* in the *Lepidobotryaceae*. Only one genus in Cyprus.

1. OXALIS *L.*

Sp. Plant., ed. 1, 433 (1753).
Gen. Plant., ed. 5, 198 (1754).

Annual or perennial herbs, or rarely subshrubs, sometimes with thick fleshy rhizomes or bulbils; leaves alternate, opposite, or all radical, or in an apical tuft, stipulate or exstipulate, digitately 3- (or many-)foliolate, or pinnate, or 1-foliolate by reduction; phyllodes occasionally present; inflorescence axillary or radical, cymose or umbellate, bracteate, or flowers solitary; petals equal or subequal, clawed, with the claws free or coherent; stamens 10, 2-seriate, filaments connate at base or free, all bearing introrse anthers; ovary 5-lobed, 5-locular, without a beak, ovules 1–many in each loculus; styles 5. Fruit a capsule; seeds often extruded explosively by means of an elastic integument or aril, with or without endosperm.

About 900 species widely distributed, but predominantly represented in South Africa, South America and Mexico.

Flowers small, petals 5–6 mm. long; plant with spreading or creeping stems, without bulbils - - - - - - - - - - - - **1. O. corniculata**
Flowers large, petals 1·8–2·5 cm. long; plant without aerial stem, bulbils present at base of leaves and along underground stem - - - - **2. O. pes-caprae**

1. O. corniculata *L.*, Sp. Plant., ed. 1, 435 (1753); Unger et Kotschy, Die Insel Cypern, 366 (1865); Boiss., Fl. Orient., 1: 867 (1867); Holmboe, Veg. Cypr., 117 (1914); Wiegand in Rhodora, 27: 120 (1925); Knuth in Engl., Pflanzenr., 95 (IV, 130): 146 (1930); Post, Fl. Pal., ed. 2, 1: 253 (1932); Lindberg f., Iter Cypr., 21 (1946); Eiten in Taxon, 4: 99–105 (1951); D. P. Young in Watsonia, 4: 51 (1958); Davis, Fl. Turkey, 2: 490 (1967); Zohary, Fl. Palaest., 2: 225, t. 321 (1972); Osorio-Tafall et Seraphim, List Vasc. Plants Cyprus, 64 (1973).

TYPE: " *in* Italia, Sicilia ".

Procumbent or sprawling annual with a slender taproot; stems 5–20 cm. or more long, usually much branched and often rooting at the nodes, sparsely or densely (f. *villosa* (M. Bieb.) Goiran) clothed with ± adpressed whitish hairs; leaves alternate, digitately 3-foliolate, leaflets obcordate, the median 0·5–1·8 cm. long, 0·5–2 cm. wide, glabrous or thinly pilose above, more densely pilose below, apex with a narrowly acute sinus and rounded lobes, lateral leaflets similar, subequal to or a little smaller than the median; petiole slender 2·5–8 cm. or more long; stipules adnate to base of petiole, oblong-auriculate, 2–3 mm. long, 1·5–2 mm. wide near apex, scarious, brownish, ± densely pilose; inflorescence axillary, umbellate, generally 2–3-flowered, peduncle 3–8 cm. or more long, thinly adpressed-pilose; bracts linear-subulate, 2–3 mm. long, 0·5 mm. wide at base, thinly pilose externally; pedicels 0·8–1·5 (–3) cm. long, distinctly reflexed in fruit; sepals narrowly oblong, about 4–4·5 mm. long, 1·5 mm. wide, thinly adpressed pilose, apex subacute or obtuse; petals yellow (sometimes purplish towards base), 5–6 mm. long, about 3 mm. wide, apex rounded, claws usually somewhat coherent; stamens in 2 series, about 4 mm. and 3 mm. long, filaments glabrous, subulate above, connate into a basal tube about 1 mm. long; anthers yellow, suborbicular, about 0·3 mm. diam.; ovary narrowly conical, 5-furrowed, about 1·5 mm. long, 0·5 mm. wide, shortly pubescent, styles about 1 mm. long, pubescent; stigmas minute, oblique, concave. Capsule cylindrical, 1–2 cm. long, 0·3–0·4 cm. wide, 5-furrowed, shortly pubescent; seeds ovate-elliptical, strongly compressed, about 1·3 mm. long and 0·8 mm. wide; testa dark brown, with conspicuous transverse ridges, elastic integument pouch-shaped, completely covering the seed and splitting at maturity along one lateral margin; endosperm copious.

HAB.: Gardens, waste ground and roadside banks, usually in moist situations; sea-level to 4,400 ft. alt.; fl. April–December.

DISTR.: Divisions 2–4, 6, 7, common in Divisions 2, 6. A cosmopolitan weed.

2. Common; *Kotschy* 53; *Lindberg f.* s.n.; *Davis* 3057 ! *Kennedy* 1842 !
3. Episkopi, 1905, *Holmboe* 707; between Silikou and Ayios Yeoryios, 1940, *Davis* 2072 !
4. Larnaca, 1905, *Holmboe* 249.
6. Common; *Syngrassides* 1816 ! 1825 ! 1828 ! *G. E. Atherton* 420 ! *P. H. Oswald* 135 !
7. Kyrenia, 1955, *G. E. Atherton* 211 ! Ayios Epiktitos, 1955, *G. E. Atherton* 236 !

2. O. pes-caprae *L.*, Sp. Plant., ed. 1, 434 (1753); Salter in Journ. S. Afr. Bot., 5: 47–52 (1939).

O. cernua Thunb., Diss. Oxalis, 14 (1781); H. Stuart Thompson in Journ. Bot., 44: 304 (1906); Holmboe, Veg. Cypr., 117 (1914); Knuth in Engl., Pflanzenr., 95 (IV, 130): 297 (1930); Post, Fl. Pal., ed. 2, 1: 254 (1932); Davis, Fl. Turkey, 2: 488 (1967); Zohary, Fl. Palaest., 2: 225, t. 320 (1972); Osorio-Tafall et Seraphim, List Vasc. Plants Cyprus, 64 (1973).

TYPE: The illustration in Burmann, Rar. Afr. Plant. Dec., 80, t. 29 (1738) !

An erect perennial with a slender, bulbilliferous underground stem, sometimes 25 cm. long, arising from a dark, glossy brown, pyriform tuber 2–3 cm. long, 1–1·5 cm. wide; leaves densely clustered at apex of stem,

3-foliolate, with widely obcordate leaflets 1–2 cm. long and 1·2–3·5 cm. wide, lamina glabrous or subglabrous above, ± pilose or glabrescent below, apex with a wide sinus and rounded lobes, the lobes of the lateral leaflets frequently somewhat unequal; petiole slender 5–15 cm. or more in length, subglabrous or sparsely pilose; stipules narrow, membranous, ciliate, adnate to base of petiole throughout their length; inflorescence umbellate, 3–many-flowered; peduncle stout, often 30 cm. long, usually rising well above the leaves; bracts linear-subulate, 4–5 mm. long, about 1 mm. wide at base, shortly pilose externally, the margins thickened and glandular towards apex; pedicels slender 1–3 cm. long, thinly pilose; sepals narrowly lanceolate, subacute, 7–8 mm. long, 1·5–2 mm. wide, thinly pilose externally with 2 small, linear, orange-brown apical glands; petals bright yellow, large and showy, 1·8–2·5 cm. long, 1–1·3 cm. wide, apex rounded, claws long, tapering, coherent; stamens in 2 series, about 0·6 and 0·8 mm. long, filaments flattened, glabrous, those of the longer stamens with a tongue-shaped basal appendage adnate to the outer surface; anthers oblong, about 1 mm. long and 0·8 mm. wide; ovary oblong-cylindrical, 2·5–3 mm. long, about 1·5 mm. wide, thinly pilose; styles short (in European and Mediterranean specimens), about 1 mm. long; stigmas fan-shaped, concave, long-papillose. Fruits rarely produced, but in one Cyprus specimen (*Chapman* 462) oblong-cylindrical, about 6 mm. long and 2 mm. wide, 5-furrowed, adpressed-pilose; seed (? immature) oblong-ovoid, about 1 mm. long, 0·8 mm. wide, not much compressed; testa bright brown, conspicuously rugose with transverse wrinkles, integument pouch-shaped, almost enclosing the seed; endosperm copious.

HAB.: Gardens, orchards, fields and waste ground, sea-level to 800 ft. alt. (? to higher altitudes); fl. Nov.–May.

DISTR.: Divisions 1, 3–7. A native of the Cape of Good Hope, now widespread in the Mediterranean area, Atlantic Islands and in many other warm-temperate and subtropical areas.

1. Ayios Minas near Smyies, 1962, *Meikle* 2172 ! Polis, 1962, *Meikle* !
3. Limassol, 1913, *Haradjian* 615 !
4. About Larnaca, 1880, *Sintenis & Rigo*, and, 1905, *Holmboe* 111; Larnaca, 1912, *Haradjian* 27 !
5. Third mile from Lefkoniko to Kyrenia, 1938, *Syngrassides* 1796 ! Salamis, 1962, *Meikle* ! Kythrea, 1962, *Meikle* !
6. Nicosia, 1905, *Holmboe* 111, and, 1950, *Chapman* 462 ! Dhiorios, 1962, *Meikle* !
7. Kyrenia, 1948, *Casey* 152 ! and, 1955, *G. E. Atherton* 787 !

NOTES: Generally believed to have been first introduced from the Cape area into Malta at some time about 1806; within 50 years of its first introduction it had become a common and persistent weed in many Mediterranean countries. The plant was first recorded from Cyprus by Sintenis and Rigo in 1880, and was, at that time, evidently common only in the neighbourhood of Larnaca, though by the end of the nineteenth century it had become common probably throughout the lowlands of the island. It is not mentioned by Kotschy nor by any earlier visitors to Cyprus. *Oxalis pes-caprae* is often said to be sterile outside the area of its natural distribution, but one Cyprus specimen does bear fruits, even if rather malformed ones, and from these it has been possible to extract well-developed seeds; it should be added that the South African material appears to be just as sterile as that from the Mediterranean countries.

"Double-flowered" (*flore-pleno*) *Oxalis pes-caprae* is frequently reported; it is commonly stated to be less robust than the normal plant, with darker coppery-yellow flowers; no Cyprus specimens of this variant have been seen.

OXALIS ACETOSELLA *L.*, Sp. Plant., ed. 1, 433 (1753), Wood Sorrel, a woodland plant with a wide distribution in Europe and Asia, is listed by Osorio-Tafall and Seraphim (List Vasc. Plants Cyprus, 64), but no specimen has been seen. It is readily distinguished from other species occurring in Cyprus by its white (or pinkish), purple-veined petals. Normally *O. acetosella* is a plant of cool, moist, shaded places.

28. RUTACEAE

Engl. in Engl. et Prantl, Pflanzenfam., ed. 2, 19a: 187–359 (1931).

Herbs, shrubs and trees; leaves generally opposite, simple or compound, dotted with translucent glands and pungent-aromatic; stipules absent; infloresences racemose, paniculate or cymose, seldom spicate or umbellate; flowers hypogynous or perigynous, mostly actinomorphic and hermaphrodite; sepals usually 4–5, free or united at base, generally imbricate; petals usually present, free, seldom connate at base, imbricate or rarely valvate; stamens as many as petals or twice as many, occasionally more numerous, usually free, sometimes united or in fascicles, inserted at the base, or along the rim, of a disk; anthers 2-thecous, introrse, longitudinally dehiscent, sometimes with basal appendages and a gland-tipped connective; ovary superior, syncarpous or with the carpels only partly united, rarely apocarpous, generally 4–5-locular, placentation axile, ovules generally 1–2 in each loculus, collateral or superimposed; styles free or united. Fruits various, often baccate, drupaceous and indehiscent, sometimes capsular or separating into mericarps; seeds with or without endosperm, embryo straight or curved.

About 140 genera widely distributed in temperate and tropical regions and particularly well represented in South Africa and Australia. Many of the species are ornamental or of economic importance.

Leaves pinnatipartite; flowers mostly 4-merous - - - - - - **1. Ruta**
Leaves simple or occasionally lobed; flowers 5-merous - - **2. Haplophyllum**

1. RUTA *L.*

Sp. Plant., ed. 1, 383 (1753) pro parte.
Gen. Plant., ed. 5, 180 (1754) pro parte.

Perennial herbs or subshrubs; leaves alternate, pinnatisect or pinnatipartite, gland-dotted, aromatic; flowers in bracteate cymes; calyx 4-partite (occacionally 5-partite); petals 4 (or sometimes 5) yellow, toothed or fimbriate, rarely entire, often hooded; stamens twice as many as petals, inserted at base of disk, the antipetalous stamens sometimes shorter than the antisepalous ones; ovary 4 (or 5)-lobed, ovules 6–many in each loculus, style central, stigma terminal. Fruit a 4 (or 5)-lobed capsule, splitting inwardly at apex; seeds angular, with endosperm; embryo somewhat curved.

Four species, widely distributed in southern Europe, the Mediterranean region and western Asia.

1. R. chalepensis *L.*, Mantissa I, 69 (1767); Hume in Walpole, Mem. Europ. Asiat. Turkey, ed. 1, 254 (1817); Boiss., Fl. Orient., 1: 922 (1867); Post, Fl. Pal., ed. 2, 1: 274 (1932); Chapman, Cyprus Trees and Shrubs, 47 (1949); Davis, Fl. Turkey, 2: 496 (1967); Zohary, Fl. Palaest., 2: 289, t. 427 (1972); Osorio-Tafall et Seraphim, List Vasc. Plants Cyprus, 65 (1973).

R. bracteosa DC., Prodr., 1: 710 (1824); Sintenis in Oesterr. Bot. Zeitschr., 31: 392 (1881); Rechinger f. in Arkiv för Bot., ser. 2, 1: 426 (1949).

R. chalepensis L. var. *bracteosa* (DC.) Boiss., Fl. Orient., 1: 922 (1867).

R. chalepensis L. ssp. *bracteosa* (DC.) Batt. in Batt. et Trabut, Fl. Alg., 180 (1888); Holmboe, Veg. Cypr., 118 (1914); Lindberg f., Iter Cypr., 21 (1946).

R. chalepensis L. var. *latifolia* (Salisb.) Fiori, Nuov. Fl. Anal. It., 2: 150 (1925); Post, Fl. Pal., ed. 2, 1: 275 (1932).

[*R. graveolens* (non L.) Sibth. in Walpole, Mem. Europ. Asiat. Turkey ed. 2, 25 (1818).]

[*R. angustifolia* (non Pers.) Kotschy in Unger et Kotschy, Die Insel Cypern, 362 (1865).]

TYPE: Cultivated at Uppsala (LINN !).

Erect, glabrous, glaucescent, foetid-smelling subshrub; stems leafy almost to apex, usually with numerous erect or suberect branches, thickly dotted with sessile glands; leaves impari-bipinnate, oblong in outline, 3–9 cm. long, 2–5 cm. wide, subsessile or with a petiole up to 4 cm. long, pinnae 5–9 or more, opposite or alternate, divided into 3–7 or more, narrowly obovate or oblong, obtuse, entire, pellucid-glandular leaflets 7–20 mm. long, 2–7 mm. wide; inflorescence much-branched, corymbose, 3–15 cm. diam.; bracts conspicuous, broadly ovate or cordate, 6–15 mm. long, 5–10 mm. wide; pedicels usually less than 1 cm. long, glabrous; sepals ovate or deltoid, 4–5 mm. long, 2–3 mm. wide, glabrous, with entire or undulate-dentate margins; petals conspicuous, bright greenish-yellow, obovate-oblong, strongly cucullate, 6–7 mm. long, 5 mm. wide, with deeply fimbriate margins and a basal claw about 1 mm. long; stamens subequal, inserted near the base of a well-developed disk, about 7 mm. long, filaments glabrous, subulate; anthers oblong, about 1·5 mm. long and 0·8 mm. wide; ovary subglobose, 2–2·5 mm. diam., rugose, sulcate, 4-lobed at apex, ovules about 12 in each loculus; style about 1–1·5 mm. long, not exceeding the connivent apical lobes of the ovary; stigma minute. Fruit a globose capsule 4–8 mm. diam. with 4 (or 5) erect, acute apical lobes 1·5–3 mm. long; seeds lithoid, 2–3 mm. long, about 1·3 mm. wide; testa fuscous, distinctly rugose-verruculose.

HAB.: Rocky ground and cliffs; sea-level to 2,300 ft. alt.; fl. Febr.–May.

DISTR.: Divisions 1, [3, 4] 7, 8, common in Divisions 1 and 7. Widespread in the Mediterranean area, also Atlantic Islands.

1. Common; *Sibthorp*; *Kotschy* 661 ! *Holmboe* 716; *Chapman* 280 ! *Lindberg f.* s.n.; *Meikle* 2033 !

[3. Limassol, 1801, *Hume*]

[4. Larnaca, 1801, *Hume*]

7. Common; *Sintenis & Rigo* 16 ! *Chapman* 131 ! 240 ! *Syngrassides* 1532 ! *Davis* 2008 ! 2985 ! *G. E. Atherton* 167 ! 340 ! 737 ! *Merton* 2227 !

8. Yioti [Juti] 1905, *Holmboe* 535; about Komi Kebir, 1912, *Haradjian* 328 !

2. HAPLOPHYLLUM *Adr. Juss.*

in Mém. Mus. Hist. Nat. Par., 12: 464 (1825) [as *Aplophyllum*] nom. cons.
Ruta L., Sp. Plant., ed. 1, 383 (1753) pro parte.

Perennial herbs, sometimes woody at the base; leaves simple or trisect; flowers in cymes; sepals 5; petals 5, entire; stamens 10, free or connate at base, the lower half of the filament often dilated and hairy; ovary 5-locular, ovules 2–6 in each loculus. Fruit a capsule, dehiscing inwardly at apex, rarely indehiscent.

A difficult group, comprising 100 or more critical species widely distributed in the eastern Mediterranean region and in western and central Asia. Engler (in Engl. et Prantl, Pflanzenfam., ed. 2, 19a: 244) regards it as a subgenus within *Ruta*, but Boissier and Post, and many others, are prepared to allow it generic status, mainly on account of its simple or trisect leaves and its 5-merous flowers with entire petals. As in *Ruta*, many of the species (but not all) are strongly and often unpleasantly aromatic.

1. **H. buxbaumii** (*Poir.*) *G. Don*, Gen. Syst., 1: 780 (1831); Boiss., Fl. Orient., 1: 937 (1867); Post, Fl. Pal., ed. 2, 1: 276 (1932); Davis, Fl. Turkey, 2: 503 (1967); Zohary, Fl. Palaest., 2: 291, t. 430 (1972); Osorio-Tafall et Seraphim, List Vasc. Plants Cyprus, 65 (1973).

Ruta buxbaumii Poir. in Lam., Encycl. Méth., 6: 336 (1804); Holmboe Veg. Cypr., 118 (1914); Engl. in Engl. et Prantl., Pflanzenfam., ed. 2, 19a: 244 (1931).

Ruta spathulata Sibth. et Sm., Fl. Graec., 4: 63, t. 370 (1823).

[*R. linifolia* (non L.) Sibth. et Sm., Fl. Graec. Prodr., 1: 273 (1806); Unger et Kotschy, Die Insel Cypern, 361 (1865).]

TYPE: North Africa; sent by Vahl to Lamarck (P).

Erect or sprawling perennial 20–50 cm. high; stems usually branched towards base, unbranched above, thinly pubescent or glabrescent, gland-dotted, often stained purplish below; leaves numerous, narrowly spathulate or oblanceolate, simple or occasionally 2- or 3-lobed towards apex of branches, 1·5–4 cm. long, 0·3–1 cm. wide, gland-dotted, subglabrous with ciliate margins and midrib, or very thinly pubescent, margins entire, apex acute or obtuse, base tapering to a short and indistinct petiole; inflorescence a very lax, spreading, many-flowered cyme, the primary branches often 12 cm. long, clothed with scattered linear-lanceolate leaves or leaf-like bracts, secondary and tertiary branches shorter, rather crowded towards apex of primary branches; pedicels very short, seldom more than 3 mm. long, bracteoles lingulate, 3–5 mm. long, glabrous or with ciliate margins; sepals oblong, blunt, glabrous, about 1 mm. long, 0·5 mm. wide; petals golden-yellow (often drying blackish), oblong, obtuse, concave, about 5–6 mm. long, 2 mm. wide; stamens free, about 4 mm. long, filaments conspicuously dilated and hairy at base, subulate above; anthers oblong, about 1·5 mm. long, 0·8 mm. wide; disk conspicuous, basin-shaped, with a spreading rim; ovary depressed-globose, 5-furrowed, about 1·2 mm. diam., verruculose, glabrous, ovules 2 in each loculus; style about 2·5 mm. long, bent towards base, distinctly thickened towards apex; stigma capitate. Capsule depressed globose, about 4 mm. wide, 2·5 mm. high, deeply 5-furrowed, strongly verruculose; seeds lithoid, about 1·6 mm. long, 1·3 mm. wide; testa fuscous, strongly rugose.

HAB.: In fields and grassy waste places; sea-level to 500 ft. alt.; fl. May–June.

DISTR.: Divisions 1, 4, 6, 7. Also Thrace, Rhodes, Turkey, Syria, Palestine, Iraq, N.W. Iran, Tunisia.

1. Near Ktima, 1787, *Sibthorp* s.n. ! Paphos, 1901, *A. G. & M. E. Lascelles* s.n. ! Near Ktima and eastwards to Timi, 1905, *Holmboe* 714.
4. Dhekelia, near Larnaca, 1935, *Syngrassides* 833 !
6. Minzeli near Nicosia, 1936, *Syngrassides* 989 ! Nicosia, western outskirts, 1962, *Meikle* 2947 !
7. South-west of Kyrenia, 1948, *M. Casey* in *Casey* 4 !

The two indigenous representatives of the family *Rutaceae*, both belonging to the tribe *Ruteae*, have dehiscent, 4- or 5-locular, capsular fruits, and seeds with endosperm. The cultivated *Citrus* species, generally believed to have originated in tropical and subtropical eastern Asia, belong to the distinct subfamily *Aurantioideae*, tribe *Citreae*, subtribe *Citrinae*, and have indehiscent 8–18-locular fruits (technically *hesperidia*) with leathery rinds and juicy pulp consisting of crowded masses of elongate pulp-vesicles arising from the dorsal walls of the loculi; the seeds are without endosperm, and are frequently polyembryonic. The Citron (*Citrus medica* L.) is known to have been in cultivation in the Mediterranean area since Classical times, and the Orange (almost certainly the Sour Orange) was recorded as growing at Famagusta in the year 1394, though both of these, and the Lemon, had probably been introduced into the island long before this date. But none of the Citrus fruits can be considered part of the indigenous or naturalized flora of Cyprus, and detailed consideration of their taxonomy would be out of place in these pages. From the economic aspect, the Sweet Orange (*Citrus sinensis* (L.) Osbeck) and the Lemon (*C. limon* (L.) Burm. f.) are the two species of outstanding importance; the Sour or Seville Orange (*C. aurantium* L.), the Grapefruit (*C. paradisi* Macf.) and the Tangerine or Mandarin (*C. deliciosa* Ten.) are of secondary importance, while the Sweet Lime (*C. limetta* Risso), Citron (*C.*

medica L.) and Bergamot Orange (*C. bergamia* Risso et Poit.—often considered a variety of the Sour Orange) are grown only to a very limited extent. The true Lime (*C. aurantifolia* (Christm.) Swingle) and the Shaddock or Pummelo (*C. grandis* (L.) Osbeck) are probably too tender to thrive except in very favoured situations.

The taxonomy of the genus *Citrus* is very complex and often controversial, but useful summaries of current views will be found in Swingle, W. T. " The Botany of Citrus and Its Wild Relatives " (Univ. of California, 1943) and in Tanaka, T. " Species Problem in Citrus " (Revisio Aurantiacearum IX, Tokyo, 1954). The history of cultivated *Citrus* is fully documented in Tolkowsky, F. " Hesperides " (London, John Bale, Sons & Curnow, Ltd., 1938).

28a. SIMAROUBACEAE

Engl. in Engl. et Prantl, Pflanzenfam., ed. 2, 19a: 359–405 (1931).

Trees and shrubs with bitter bark; leaves usually alternate, generally imparipinnate; stipules absent; flowers mostly unisexual, actinomorphic, in panicles, racemes or spikes; calyx gamosepalous, (3–) 5 (–8)-lobed or -partite, lobes imbricate or valvate; petals (3–) 5 (or more) rarely absent, free or united, imbricate or valvate; disk present; stamens inserted at base of disk, free, as many as, or twice as many as petals, or occasionally numerous; filaments sometimes with a basal scale or appendage; anthers 2-thecous, dehiscing longitudinally; ovary 2–5-locular, carpels united or free; styles 2–5; ovules mostly solitary or in pairs in each loculus; placentation axile. Fruit a capsule, samara or schizocarp, sometimes a berry or drupe; seed with a straight or curved embryo; endosperm absent or scanty.

About 32 genera, chiefly in the tropics, with a few temperate representatives, one of them cultivated in Cyprus.

Ailanthus altissima (*Mill.*) *Swingle* in Journ. Washington Acad. Sci., 6: 495 (1916) (*Toxicodendron altissimum* Mill., Gard. Dict., ed. 8, no. 10; 1768; *Ailanthus glandulosa* Desf. in Mém. Acad. Sci. Par., 1786: 265, pl. 8; 1788).

A deciduous tree 15–30 m. high; bark grey, fissured; branches with conspicuous leaf-scars; leaves imparipinnate, 40–50 (–100) cm. long, with 13–30 ovate, acute or acuminate, shortly stalked leaflets; leaflet-margins with 2–4 coarse teeth near the base, each with a large gland beneath; flowers unisexual (males and females usually on separate trees) in terminal panicles, small, greenish, the staminate malodorous. Fruit a narrow oblong, reddish samara about 3–4 cm. long tapering and twisted at both ends with 1 compressed seed at the centre.

A native of China, widely cultivated and naturalized in Europe and North America, and valued as a rapid-growing shade tree; the large bunches of red fruits are ornamental, but nectar from the flowers, especially from the male, can give a disagreeable flavour to honey.

The tree is frequently planted in Cyprus, and is recorded by Bovill (Rep. Plant. Work Cyprus; 1915) from Nicosia, Athalassa, Athna and Xylotymbou.

28b. MELIACEAE

Harms in Engl. et Prantl, Pflanzenfam., ed. 2, 19b, 1: 1–172 (1940).

Trees or shrubs, rarely subshrubs, often with hard, scented wood; leaves alternate, exstipulate, simple, 3-foliolate or 1–3-pinnate, leaflets entire or serrate; flowers actinomorphic, hermaphrodite, unisexual or sometimes polygamous, in axillary cymes or cymose panicles or fascicles; sepals (4–) 5 (–6), usually connate at base, imbricate or with open aestivation, not completely covering the corolla in bud; petals 4–5 (rarely fewer or more), free or partly connate, valvate, imbricate or contorted; stamens hypogynous, usually 8–10, the filaments generally connate to form a staminal tube; anthers 2-thecous, dehiscing longitudinally, often sessile within the staminal tube; disk present, intrastaminal, variable in structure; ovary superior, (2–) 4–5 (–20)-locular, ovules 1–many in each loculus, placentation axile, style 1, stigma often capitate or discoid, sometimes lobed. Fruit a loculicidal or septifragal capsule, a berry or a drupe; seeds usually arillate, sometimes winged, endosperm present or absent.

A family or about 50 genera and 800 or more species, almost wholly restricted to tropical and subtropical regions, and including a number of valuable timber trees, such as Mahogany (*Swietenia*) and African Mahogany (*Khaya*). Few members of the family are ornamental, but the so-called "Persian Lilac", *Melia azedarach* L. Sp. Plant., ed. 1, 384 (1753), from India, with large panicles of scented, lilac-coloured flowers, is extensively cultivated as a street tree in Cyprus and in many warm-temperate and tropical countries. *Melia azedarach* is a deciduous tree, up to 15 m. tall, with bipinnate (or rarely tripinnate) leaves 25–40 cm. long, 11–20 cm. wide, and ovate, acuminate, entire or serrate-crenate, glabrous or subglabrous leaflets, 3–5 cm. long, 1–3 cm. wide. The inflorescences are borne during April and May in the axils of the young shoots, and consist of numerous flowers in a lax cymose panicle; each flower has 5 narrowly oblong, spreading petals 6–10 mm. long and about 2 mm. wide, and a darker violet-purple staminal column, 5–7 mm. long. The fruit is a yellowish, plum-like drupe, about 2 cm. long and 1·5 cm. wide.

29. RHAMNACEAE

Suessenguth in Engl. et Prantl, Pflanzenfam., ed. 2, 20d: 7–173 (1953).

Trees, shrubs and woody climbers, rarely herbs; leaves alternate or opposite, simple, generally stipulate, frequently with conspicuous longitudinal nervation; flowers hermaphrodite or unisexual, or occasionally polygamous, actinomorphic, mostly in axillary cymes or corymbs, seldom large or brightly coloured; sepals 4–5, valvate, attached to an obconical, turbinate or cylindrical hypanthium; petals as many as sepals, or occasionally wanting, often cucullate or involute, clawed or sessile; stamens as many as petals and opposite to them, and sometimes enclosed by them, perigynous, usually arising from an intrastaminal disk which lines or rims the hypanthium; anthers 2-thecous, dehiscing longitudinally; ovary sessile, free or sunk in the disk, 2–4-locular, ovules 1 (rarely 2) in each loculus, erect, placentation basal, styles 1 (or 2), frequently lobed above, stigmas terminal, capitate or sometimes lateral along the style-branches. Fruit often drupaceous, or a capsule, or occasionally a samara; seeds not infrequently arillate, endosperm abundant or scanty, embryo large, straight.

About 58 genera and more than 900 species with a cosmopolitan distribution.

Leaves penninerved; stipules small, deciduous; branches ± straight; flowers polygamo-dioecious - - - - - - - - **1. Rhamnus**
Leaves longitudinally 3 (or more)-nerved; stipules usually persistent and spinescent; branches ± zig-zag; flowers hermaphrodite:
Fruit discoid with a broad leathery wing - - - - - - **2. Paliurus**
Fruit a fleshy, globose, ovoid or oblong drupe - - - - **3. Ziziphus**

1. RHAMNUS *L.*

Sp. Plant., ed. 1, 193 (1753).
Gen. Plant., ed. 5, 89 (1754).
excl. *Frangula* Mill., Gard. Dict. Abridg., ed. 4 (1754).

Trees and shrubs; winter buds scaly, terminal bud wanting; leaves deciduous or evergreen, frequently opposite, penninerved; stipules small, deciduous; flowers in cymes, racemes or clusters, polygamo-dioecious, usually 4-merous; sepals deltoid-ovate; petals usually 4, sometimes wanting, often cucullate, small, greenish, yellowish or whitish; stamens of female flowers reduced to staminodes; ovary free or basally adnate to hypanthium; style 3- or 4-lobed. Fruit drupaceous, with a fleshy or leathery exocarp enclosing 2–4 hard or cartilaginous, dehiscent or indehiscent pyrenes; seeds sulcate dorsally, germination epigeal, cotyledons thin, green, curved.

About 100 species, widely distributed in the northern hemisphere, with outlying representatives in south east Asia, tropical and South Africa and in South America.

Flowers in racemes; unarmed shrubs or small trees; leaves leathery, serrulate **1. R. alaternus**
Flowers in axillary clusters; thorny shrubs; leaves entire or subentire **2. R. oleoides ssp. graecus**

SECT. 1. **Alaternus** *DC.* Unarmed evergreen shrubs with leathery leaves; flowers usually 5-merous, inflorescences racemose.

1. R. alaternus *L.*, Sp. Plant, ed. 1, 193 (1753); Unger et Kotschy, Die Insel Cypern, 357 (1865); Boiss., Fl. Orient., 2: 14 (1872); Holmboe, Veg. Cypr., 125 (1914); Post, Fl. Pal., ed. 2, 1: 290 (1932); Chapman, Cyprus Trees and Shrubs, 50 (1949); Rechinger f. in Arkiv för Bot., ser. 2, 1: 427 (1950); Davis, Fl. Turkey, 2: 531 (1967); Zohary, Fl. Palaest., 2: 306, t. 448 (1972); Osorio-Tafall et Seraphim, List Vasc. Plants Cyprus, 69 (1973).
TYPE: " *in* Europa *australi* ".

Erect, unarmed, evergreen bush or small tree, 1–4 m. high, twigs greyish-brown; leaves ovate or elliptical, leathery, 2–4 cm. long, 1–2·5 cm. wide, dark lustrous green above, paler, often golden-green below, apex acute, margins thickened, narrowly revolute, with small, sharp, gland-tipped serrations; petiole short, generally less than 8 mm. long; stipules deltoid-acuminate, about 2 mm. long, 1 mm. wide at base, thinly puberulous, caducous; flowers dioecious; male inflorescences forming dense cylindrical racemes about 1–1·5 cm. long, peduncle very short or almost wanting, rhachis angular, shortly pubescent, bracts ovate-deltoid, about 8 mm. long and almost as wide, acute, shortly pubescent, often with dark sessile glands on the dorsal surface and margins; pedicels short, thick, about 1 mm. long, glabrous; hypanthium obconical, about 1·5 mm. long, 1 mm. wide at apex, glabrous; sepals 4 or 5, ovate, acute, 3-nerved, about 1·5 mm. long, 1 mm. wide at base, reflexed at anthesis; petals 4 or 5, greenish-yellow, small, narrowly oblong, about 1 mm. long, 0·4 mm. wide, apex subacute, base very shortly clawed; stamens 4 or 5, filaments glabrous, about 0·5 mm. long; anthers yellow, oblong, about 1 mm. long, 0·6 mm. wide; female inflorescences similar, rather shorter; sepals 4 or 5, similar to those of male flowers but erect or spreading at anthesis; petals minute or commonly wanting; stamens reduced to minute staminodes; ovary free, globose, glabrous or minutely pubescent, about 1·5 mm. diam., 2–3-locular, style about 0·8 mm. long, divided above into 2–3 filiform lobes, each 1–1·5 mm. long, stigma ovate, terminal and lateral. Fruit an obovoid 2- or 3-locular drupe 5–6 mm. long, 4–5 mm. wide, attached basally to the persistent hypanthium, exocarp thin, black in ripe fruit, endocarp bony, splitting from the base upwards along an internal suture at maturity: seeds obovoid, about 5 mm. long, 3·5 mm. wide, with a distinct median ridge on the ventral surface, and a broad median furrow, surrounded by a narrow horseshoe-shaped marginal furrow on the dorsal surface; testa pale brown, cerebriform-verruculose.

HAB.: In *Pinus* and *Cupressus* woodland and by streams in *Pistacia-Juniperus* scrub; sea-level to 3,500 ft. alt.; fl. March–April.

DISTR.: Divisions 1, 2, 7, 8. Throughout the Mediterranean area, but more frequent in the western half.

1. Lyso, 1913, *Haradjian* 874 ! Kavakas near Ayios Yeoryios (Akamas), 1962, *Meikle* 2117 !
2. Between Lefkara and Makheras Monastery, 1862, *Kotschy* 224; Pano Platres, 1930, *Wall* s.n.; Saïttas, 1932, *Syngrassides* 343 ! Ayios Mamas, 1956, *Merton* 2816 !
7. Ayios Amvrosios, 1937, *Syngrassides* 1727 ! Kyrenia Pass, 1957, *Merton* 2880 !
8. Rizokarpaso, *Chapman* 28.

NOTES: All the fruiting and flowering material seen is unusual in having 2-locular drupes and 4-merous flowers; outside Cyprus 5-merous flowers and 3-locular fruits appear to be the rule, but it is difficult to see any other distinctions between the Cyprus plant and specimens of *R. alaternus* from outside the island. Further investigation is required, for if it should prove that Cyprian *R. alaternus* is uniformly 4-merous and 2-locular, there would be good grounds for considering the Cyprus plant distinct, at least at subspecific level.

SECT. 2. **Cervispina** (*Moench*) *DC.* Thorny shrubs and subshrubs; flowers in axillary clusters, generally 4-merous; leaves usually small, deciduous or persistent.

2. **R. oleoides** *L.*, Sp. Plant., ed. 2, 279 (1762); Boiss., Fl. Orient., 2: 15 (1872); Holmboe, Veg. Cypr., 125 (1914).

Erect, spreading or prostrate, intricately branched shrub 0·5–2·5 m. high; branches often divaricate, terminating in a long, rigid thorn, bark grey, greenish or brown, glabrous or minutely puberulous; leaves alternate or clustered on short lateral spurs, variable, narrowly oblanceolate, oblong, obovate, obcordate or almost orbicular, 7–30 mm. long, 2–15 mm. wide, glabrous or ± pubescent, distinctly or indistinctly reticulate-veined below, apex acute, mucronate, obtuse, rounded or emarginate, margins usually entire or subentire, flat or very narrowly recurved, occasionally distinctly serrulate; petioles slender, 1–5 (–10) mm. long, sulcate above, glabrous or ± pubescent; stipules subulate, 0·8–1·5 mm. long, 0·3 mm. wide at base, glabrous or pubescent, brownish, caducous; flowers axillary, forming clusters on the lateral spurs, polygamo-dioecious; pedicels slender, 2–7 mm. long, glabrous or puberulous; hypanthium obconical 1–2 mm. long, 1–1·5 mm. wide at apex, glabrous or ± puberulous; sepals usually 4, spreading in male flower, ovate-deltoid, 3-nerved, 1·5–2 mm. long, 1–1·5 mm. wide at base, glabrous or ± puberulous; disk glabrous or pubescent; petals 4, or often wanting, greenish, minute, oblong, about 1 mm. long and 0·5 mm. wide, apex acute, base very shortly clawed; stamens usually 4, filaments glabrous, about 0·5 mm. long; anthers yellow, oblong, about 0·5–0·7 mm. long and 0·4–0·6 mm. wide; female flowers similar to male, but usually fewer in each cluster, slightly smaller and more shortly pedicellate; stamens rudimentary; ovary subglobose, 2-lobed, about 0·8 mm. high and 1 mm. wide, glabrous, style rather thick, about 0·6 mm. long, divided above into 2 filiform lobes about 1 mm. long; stigmas conspicuous, spathulate. Fruit a subglobose or broadly obovoid 2-locular drupe, 4–6 mm. high and almost as wide, often distinctly 2-lobed, exocarp thin, leathery, reddish in ripe fruit, endocarp bony, dehiscing upwards along the internal wall; seed oblong 4–5 mm. long, about 3 mm. wide, with an obscure ventral ridge and a deep, narrow, almost closed dorsal furrow; testa pale brown with minute, paler cerebriform blotches.

ssp. **graecus** (*Boiss. et Reut.*) *Holmboe*, Veg. Cypr., 125 (1914); Lindberg f., Iter Cypr., 22 (1946); Davis in Notes Roy. Bot. Gard. Edinb., 21: 115 (1952), Fl. Turkey, 2: 538 (1967); Osorio-Tafall et Seraphim, List Vasc. Plants Cyprus, 69 (1973).

R. graecus Boiss. et Reut. in Boiss., Diagn., 2, 5: 74 (1856); Kotschy in Oesterr. Bot. Zeitschr., 12: 277 (1862); Unger et Kotschy, Die Insel Cypern, 358 (1865).

R. oleoides L. var. *parvifolia* Kotschy in Unger et Kotschy, Die Insel Cypern, 358 (1865).

[*R. heldreichii* (non Boiss.) Sintenis in Oesterr. Bot. Zeitschr., 32: 20, 192 (1882); Chapman, Cyprus Trees and Shrubs, 51 (1949); Rechinger f. in Arkiv för Bot., ser. 2, 1: 427 (1950).]

[*R. punctatus* (non Boiss.) Boiss., Fl. Orient. Suppl., 156 (1888) quoad spec. cypr.; Post in Mém. Herb. Boiss., no. 18: 92 (1900).]

TYPE: Greece; "in *Palamede* Argolidis Spruner ! montibus Parnes et *Hymetto* Heldr. ! *Laconiâ* Heldr. ! *Cretâ* Raulin ! " (G).

Bark greyish-brown, often minutely fissured, branches usually very numerous, at first ± puberulous; leaves often rather small, normally entire, obscurely veined below at maturity; petioles ± pubescent; floral disk pubescent or tomentose; dorsal furrow of seed often shortly bifid at apex.

HAB.: On rocky hillsides and amongst coastal garigue; sea-level to 3,000 ft. alt.; fl. March–April.

DISTR.: Divisions 1–3, 5–8, common in Divisions 2, 7, 8. Greece, Aegean Islands, Turkey, ? Palestine; *R. oleoides* L. sens. lat. is widespread in the Mediterranean area.

1. Lyso, 1962, *Meikle* 2283 ! Ayios Yeoryios (Akamas), 1962, *Meikle* !
2. Common; *Sintenis & Rigo* 11a ! *Chapman* 86 ! 87 ! 274 ! 275 ! *Kennedy* 620 !
3. On the seashore near Kiti on the way to Mazotos, 1862, *Kotschy*; Mazotos, 1905, *Holmboe* 183; Kophinou, 1905, *Holmboe* 596.
5. Syngrasi, borders of reservoir, 1937, *Syngrassides* 1654 ! Salamis, 1962, *Meikle* !
6. Liveras, in sand dunes, 1962, *Meikle* 2416 ! Dhiorios, 1962, *Meikle* !
7. Common; *Syngrassides* 193 ! 1463 ! *Chapman* 229 ! *Merton* 2208 ! 2237 ! 2291 ! *G. E. Atherton* 696 !
8. Common; *Sintenis & Rigo* 11 ! *Chapman* 30 ! 49 ! 264 !

NOTES: Many of the Cyprus specimens have small leaves, and might be referred to var. *microphyllus* Hal. (*R. oleoides* L. ssp. *microphyllus* (Hal.) P. H. Davis), which is common in Crete; the differences between this variety and typical ssp. *graecus* are, however, so slight, and the range of variation in leaf-size so considerable, that one questions the taxonomic value of such distinctions, even at varietal or subspecific levels. On the whole, the Cyprus population of *R. oleoides* ssp. *graecus* is tolerably uniform, and most, if not all, the anomalous specimens are habitat or seasonal forms. Plants growing in shaded situations under rocks and trees often have exceptionally large, thin leaves, prominently veined below and not infrequently with crenate-serrate margins; similar aberrations are to be seen in the regenerating growths from cut-down or goat-grazed plants, and young, immature shoots and leaves are often strikingly different (especially when dried) from mature specimens.

R. PUNCTATUS *Boiss.* (Diagn., 1, 2: 4; 1843) has been admitted to the flora of Cyprus on the strength of three records:— a specimen said to have been collected on the island

(" probably near Larnaka "—*teste* Holmboe) by Labillardière; a Sintenis and Rigo specimen from Kantara (Boiss., Fl. Orient. Suppl., 156; 1888) and a statement by Post in Mém. Herb. Boiss., no. 18: 92 (1900). The Sintenis and Rigo specimen has been examined, and is *R. oleoides* L. ssp. *graecus* (Boiss. et Reut.) Holmboe. The statement by Post (" Montagnes de Chypre, partout ") is clearly erroneous, and should refer to *R. oleoides* ssp. *graecus* ; so that the admission of *R. punctatus* finally rests on the Labillardière record alone. Here too there is some confusion, for, in the original description of *R. punctatus*, Boissier says " Syriaca a Labillardière collecta ", and it was not until almost thirty years later (Fl. Orient., 2: 14; 1872) that he claimed the Labillardière record for Cyprus; in this he was followed by Holmboe, who claims to have seen the specimen in herb. Boissier, though, curiously enough, Boissier first refers the specimen to herb. De Candolle (Diagn., *loc. cit.*) and later (Fl. Orient. *loc. cit.*) to herb. Labillardière. In view of the fact that *R. punctatus* has been many times correctly recorded from Syria (and occurs also in S.E. Turkey and Palestine), and has not since been discovered in Cyprus, it is reasonably safe to assume that the original localization of the Labillardière specimen was correct, and that it came from Syria and not from Cyprus.

R. punctatus is easily distinguished from all forms of *R. oleoides* by its very thick, pubescent leaves with strongly revolute margins; it resembles *R. oleoides* in general habit.

2. PALIURUS *Mill.*

Gard. Dict. Abridg., ed. 4 (1754).

Shrubs and small trees; leaves ± distichous, alternate, simple, entire or crenate-serrate, with 3 conspicuous longitudinal nerves; stipules thorny, both straight, or one straight and the other arcuate; flowers in axillary (or rarely terminal) cymes or clusters, 5-merous, hermaphrodite; hypanthium obconical; sepals deltoid-ovate; petals small, greenish-yellow, cucullate, inserted with the stamens on the disk which lines the hypanthium; stamens often enclosed by petals; ovary superior or half-inferior, partly adnate to disk at its base, 2–3-locular; styles short; stigmas 2–3, oblong. Fruit drupaceous, 2–3-locular, broadly winged above the middle, exocarp leathery, reddish, endocarp woody; seeds 1 in each loculus, compressed; endosperm scanty.

About 8 species, one in the Mediterranean area and western Asia, the remainder in eastern Asia.

1. **P. spina-Christi** *Mill.*, Gard. Dict., ed. 8 (1768); Post, Fl. Pal., ed. 2, 1: 288 (1932); Chapman, Cyprus Trees and Shrubs, 50 (1949); Davis, Fl. Turkey, 2: 522 (1967); Zohary, Fl. Palaest., 2: 306, t. 449 (1972); Osorio-Tafall et Seraphim, List Vasc. Plants Cyprus, 68 (1973).

Rhamnus paliurus L., Sp. Plant., ed. 1, 194 (1753); Hume, in Walpole, Mem. Europ. Asiat. Turkey, ed. 1, 253 (1817).

Paliurus australis Gaertn., Fruct., 1: 203, t. 43 (1788); Holmboe, Veg. Cypr., 124 (1914).

P. aculeatus Lam., Illustr., 2: 347, t. 210 (1792); Unger et Kotschy, Die Insel Cypern, 357 (1865); Boiss., Fl. Orient., 2: 12 (1872).

Zizyphus paliurus (L.) Willd., Berl. Baumz., 415 (1796); Clarke, Travels, 2, 3: 724 (1816).

TYPE: " *in* Europa *australi* ".

Lax shrub or small tree to about 4 m. high, branches slender, ± zigzag, spreading or arching, at first thinly pilose, later glabrous, with dark grey-brown bark; leaves broadly ovate, asymmetrical, 2–5 cm. long, 1–4 cm. wide, with 3 conspicuous longitudinal nerves, glabrous, or thinly pilose along the nerves, apex acute or obtuse, base broadly cuneate, truncate or slightly cordate, margins minutely glandular-serrulate or sub-entire; petioles 0·5–1·5 cm. long, channelled above, thinly pilose; stipules thorny, with elongate, decurrent bases, one stipule straight and about 5–10 mm. long, the other arcuate and usually much shorter; flowers forming dense cymose clusters in the leaf-axils, peduncle very short or wanting, pedicels 5–8 mm. long, bracts oblong, about 1 mm. long, 0·5 mm. wide, scarious, thinly pilose;

hypanthium broadly obconical or almost flat; sepals ovate-deltoid, about 2 mm. long and almost as wide; petals golden-yellow, strongly cucullate, about 2 mm. long and 1 mm. wide, apex obtuse, base narrowed to a short claw; stamens about 2 mm. long, filaments glabrous; anthers oblong, about 0·5 mm. long and 0·3 mm. wide; disk conspicuous, flat, with a crenate margin; ovary partly immersed in disk, flask-shaped, about 0·8 mm. long, 0·5 mm. wide, styles free almost to base, stout, about 0·7 mm. long; stigmas small, oblique. Fruit discoid, about 2–3·5 cm. diam., with a broad marginal wing, and conspicuous radiating venation, central portion broadly obconical, about 1·5–1·8 cm. diam., corky; seed suborbicular or broadly obovate, strongly compressed, about 4·5 mm. long and 4 mm. wide, ventral surface with an obscure median ridge; testa rich brown, lustrous, minutely verruculose.

HAB.: Rocky hillsides and streamsides; ? alt.; fl. April–June.

DISTR.: Divisions 2, 3, 5–8. Mediterranean area and east to Iran.

2. Kambos, *Chapman* 299.
3. Limassol, 1801, *Hume* s.n.; On the south coast, *Kotschy*.
5. Syngrasi in the eastern Mesaoria, *Kotschy*; Salamis, 1905, *Holmboe* 470 ("Dry and stony fields, common everywhere in lower regions, *Holmboe* ").
6. Nicosia, *Michaelides* s.n.
7. Buffavento above Ayios Khrysostomos Monastery, *Kotschy*.
8. Near Comares [locality between Boghaz and Gastria], *Kotschy*.

NOTES: Not a single correctly identified specimen of this plant has been seen in any Cyprus collections examined, though, on the strength of statements by Holmboe and Chapman, one cannot expunge *Paliurus* from the Cyprus flora. It is odd that the plant is not listed by Sintenis and Rigo, nor collected by Syngrassides, nor included amongst pasture weeds in the " Report on Pasture Research, Survey and Development " (1958), though, significantly, *Ziziphus lotus*, evidently not well known to either Holmboe or Chapman, is listed in the Report as a widespread weed. Investigation may show that all records for *Paliurus spina-Christi* should be referred to *Ziziphus lotus*. *Paliurus* is also missing from Crete and almost all Judaea.

3. ZIZIPHUS *Mill.*

Gard. Dict. Abridg., ed. 4 (1754).

Shrubs, trees, or sometimes climbers; branches often zig-zag, frequently with whitish bark; leaves alternate, ± distichous, asymmetrical, with 3 or more prominent longitudinal nerves; stipules often thorny, one straight, the other often arcuate; flowers hermaphrodite in axillary fascicles or cymes, inconspicuous; hypanthium broadly obconical; sepals 5; petals 5 (rarely wanting), cucullate, often deflexed; stamens 5, inserted with the petals on the outer edge of the disk, and often concealed within the petals; disk conspicuous, 5-lobed, flattish; ovary sunk into the disk, adnate to it at least in the lower half, 2–3-locular, styles 2–3 (–4); stigmas small. Fruit a fleshy globose or oblong drupe, endocarp woody or bony, 1–3 (–4)-locular, each loculus 1-seeded; seeds plano-convex, endosperm wanting or very sparse.

About 50 species, chiefly in tropical Asia and America, but widely distributed throughout the warmer regions of the world.

Pedicels and calyx densely pubescent; fruit globose - - - - **1. Z. spina-Christi**
Pedicels and calyx glabrous or sparsely pilose:
  Fruit oblong or ovoid; leaves narrowly oblong; fertile twigs usually unarmed; a tree - - - - - - - - - - - - - **2. Z. zizyphus**
  Fruit globose; leaves rather broadly oblong; fertile twigs usually armed; a low bush - - - - - - - - - - - - - **3. Z. lotus**

1. Z. spina-Christi (*L.*) *Willd.*, Sp. Plant., 1: 1105 (1798); Boiss., Fl. Orient., 2: 13 (1872); Holmboe, Veg. Cypr., 124 (1914); Post, Fl. Pal., ed. 2, 1: 289 (1932);

Chapman, Cyprus Trees and Shrubs, 50 (1949); Zohary, Fl. Palaest., 2: 307, t. 450 (1972); Osorio-Tafall et Seraphim, List Vasc. Plants Cyprus, 69 (1973).

Rhamnus spina-Christi L., Sp. Plant., ed. 1, 195 (1753).

TYPE: " *in* Aethiopia, Palaestina ".

A round-headed evergreen (or facultatively deciduous) tree, 3–14 m. high, twigs with whitish or pale brown distinctively lenticellate bark, armed (var. *spina-Christi*) or unarmed (var. *inermis* Boiss.); leaves oblong-elliptic, 1·5–4·5 cm. long, 1–3 cm. wide, with 3 conspicuous longitudinal nerves, glabrous above, subglabrous or thinly crisped-pubescent below, apex obtuse or rounded, base truncate or obtuse, margins minutely crenate-serrate; petioles to about 1·5 cm. long, thinly pubescent; stipules (when present) short, broad-based, one straight, to about 1·8 cm. long, the other arcuate, to about 0·8 cm. long; flowers in dense cymose clusters, peduncles wanting or very short, pedicels slender, densely pubescent, about 3–6 mm. long, bracts oblong, acute, densely pubescent, about 1 mm. long and 0·5 mm. wide; sepals broadly ovate-deltoid, acute, about 1·8 mm. long and almost as wide, densely pubescent; petals yellowish, obovate, cucullate, about 1·2 mm. long, 1 mm. wide, glabrous, apex rounded or slightly retuse, base cuneate; stamens about 1·5 mm. long, filaments subulate, glabrous; anthers oblong, about 0·5 mm. long, and almost as wide; disk conspicuous, flattish, about 2·5 mm. diam., rugose, 5-lobed, the apex of each lobe shortly emarginate; ovary globose, almost wholly sunk in disk, style very short, divided above into 2 lobes, each less than 0·5 mm. long; stigmas minute, terminal, oblique. Fruit globose, 1·3–1·8 cm. diam., yellowish or reddish-brown when ripe, flesh sweetish or slightly astringent, endocarp very hard and woody, 2-locular; seeds obovate 6–7 mm. long, 5–6 mm. wide; testa smooth, bright brown, lustrous; endosperm scanty; cotyledons large, roundish.

HAB.: Arid plains and hillsides; planted in our area.

DISTR.: Division 5, rare. Syria, Palestine, Egypt, Ethiopia and Somaliland, eastwards to northern India; often planted.

5. A small wood of old trees between Salamis and Ayios Varnavas Monastery, 1862, *Kotschy*.

2. Z. zizyphus (*L.*) *Meikle* **comb. nov.**

Rhamnus zizyphus L., Sp. Plant., ed. 1, 194 (1753).

Z. jujuba Mill., Gard. Dict., ed. 8, no. 1 (1768); Davis, Fl. Turkey, 2: 524 (1967); Osorio-Tafall et Seraphim, List Vasc. Plants Cyprus, 69 (1973).

Zizyphus officinarum Medik., Bot. Beobacht., 333 (1783); Post, Fl. Pal., ed. 2, 1: 289 (1932).

Zizyphus sativa Gaertn., Fruct., 1: 202, t. 42 (1788); Holmboe, Veg. Cypr., 124 (1914); Chapman, Cyprus Trees and Shrubs, 50 (1949).

Ziziphus vulgaris Lam., Encycl. Méth., 3: 316 (1789); Unger et Kotschy, Die Insel Cypern, 357 (1865); Boiss., Fl. Orient., 2: 12 (1872).

TYPE: Probably the illustration in Dodoens, Stirp. Hist. Pempt. Sex, 795 (1583).

A large deciduous bush or round-headed tree, 3–7 m. high; twigs often distinctly zig-zag, the sterile ones often armed, the flowering shoots generally unarmed, bark dark greyish-brown; leaves ± distichous, narrowly oblong, asymmetric, 2–7 cm. long, 1–3 cm. wide, with 3 conspicuous longitudinal nerves, upper surface bright shining green, glabrous, lower surface subglabrous, the nerves often thinly pilose, apex obtuse or subacute, base obliquely cuneate or rounded, margins distinctly crenate-serrulate; petioles very short, usually less than 5 mm. long, subglabrous or ± pilose; stipules spinescent, slender or stout with an enlarged base, bright or dark brown, unequal, one straight and up to 2·5 cm. long, the other arcuate and usually less than 0·8 cm. long; flowers in dense axillary clusters, or occasionally solitary; peduncle wanting or very short, pedicels 1–3 mm. long, glabrous or very sparsely pilose; sepals ovate, acute, about 2 mm. long and 1·3 mm. wide at base, glabrous or very sparsely pilose externally; petals greenish-yellow, obovate, cucullate, about 1·5 mm. long, 1 mm. wide, apex rounded, base tapering to a short claw; stamens 2 mm. long, filaments glabrous, subulate; anthers oblong, about 0·6 mm. long, 0·4 mm. wide; disk flattish, conspicuous, about 2 mm. diam., obscurely pentagonal, with the lobes distinctly emarginate; ovary globose, about 0·5 mm. diam., almost wholly sunk in disk, styles 2, free almost to base, about 0·6 mm. long; stigmas minute, terminal, oblique. Fruit oblong or ovoid (occasionally subglobose) 1·5–3 cm. long, 1–2·5 cm. wide, reddish-brown when ripe, flesh sweet; endocarp very hard and woody, 1 (or 2)-locular; seed oblong, 6–7 mm. long, 4–5 mm. wide; testa dark brown, minutely punctulate; endosperm scanty; cotyledons large, roundish.

HAB.: In hedgerows and orchards etc. close to habitations; cultivated in our area.

DISTR.: Divisions 3, 7, 8. Cultivated in the Mediterranean region, probably native in central and eastern Asia.

3. Limassol, *Michaelides* sec. *Holmboe*.
7. Larnaka tis Lapithou, 1936, *Syngrassides* 1288 ! Trypimeni, *Chapman* 402.
8. Near Comares (between Boghaz and Gastria), 1862, *Kotschy*; Boghaz, 1938, *Chapman* 338 !

NOTES: The Trypimeni specimen (*Chapman* 402) is too imperfect for certain determination, but appears to belong here, rather than to *Z. lotus* (L.) Lam. as labelled.

3. Z. lotus (*L.*) *Lam.*, Encycl. Méth., 3: 317 (1789); Gaudry, Recherches Sci., 188 (1855); Unger et Kotschy, Die Insel Cypern, 357 (1865); Boiss., Fl. Orient., 2: 12 (1872); Holmboe, Veg. Cypr., 124 (1914); Post, Fl. Pal., ed. 2, 1: 289 (1932); Lindberg f., Iter Cypr., 22 (1946); Chapman, Cyprus Trees and Shrubs, 50 (1949); Zohary, Fl. Palaest., 2: 308, t. 452 (1972); Osorio-Tafall et Seraphim, List Vasc. Plants Cyprus, 69 (1973).

Rhamnus lotus L., Sp. Plant., ed. 1, 194 (1753).

TYPE: " *in* Regno Tunetano ".

Deciduous sprawling or ascending shrub 1–2 m. high, branches numerous, intricate, zig-zag, usually armed with slender spines, bark whitish or pale ashy grey; leaves oblong, 1–2 cm. long, 1–1·3 cm. wide, glabrous above and below, with 3 distinct longitudinal nerves, apex obtuse or subacuate, margin subentire or minutely and obscurely serrulate, base rounded or unequally and obliquely cuneate; petiole very short, usually less than 2 mm. long; stipules unequal, slender, from a slightly thickened base, one straight, 0·6–1·5 cm. long, the other arcuate, to about 0·5 cm. long; flowers in axillary clusters, peduncles very short or wanting, glabrous, pedicels 2–4 mm. long, glabrous; bracts minute, subulate, or wanting; hypanthium broadly obconical or saucer-shaped, glabrous; sepals ovate, acute, about 2 mm. long, 1·5 mm. wide at base, glabrous; petals yellowish, about 2 mm. long, 1 mm. wide at apex, strongly cucullate, apex rounded, base narrowing rather abruptly to a distinct claw; stamens 1·5 mm. long, filaments glabrous, subulate, anthers oblong, about 0·5 mm. long, 0·4 mm. wide; disk flattish or slightly concave, with a 10-lobed crenate margin; ovary completely sunk in disk, styles 2, stout, about 0·6 mm. long, stigmas minute, terminal. Fruit globose, 1–1·5 cm. diam., yellow or reddish-brown when ripe, flesh sweetish, endocarp hard, woody, subglobose or broadly ellipsoid, 1 (2)-locular; seed strongly compressed, broadly obovate or suborbicular, about 5 mm. long and 4·5 mm. wide; testa dark brown, smooth, lustrous; endosperm rather sparse, cotyledons broad, rounded.

HAB.: In dry fields and waste ground; by roadsides and tracks; sea-level to c. 1,500 ft. alt.; fl. end of May–June.

DISTR.: Divisions 2–7, common in Division 5. Portugal, Spain, Sicily, S. Greece, North Africa, ? Turkey, Palestine, Arabia.

2. Near Evrykhou, 1880, *Sintenis & Rigo* 507a ! Between Ambelikou and Kambos, 1939, *Lindberg f.* s.n.
3. Alethriko, 1905, *Holmboe* 1099.
4. About Larnaca Salt Lake, 1853, *Gaudry*; near Larnaca, and by the coast near Kiti, 1862, *Kotschy*; Meneou, 1905, *Holmboe* 1096; Famagusta 1939, *Antoniades* in *Chapman* 412 !
5. Common; *Syngrassides* 383 ! *Lindberg f.* s.n. ! *Merton* 2444 !
6. Frequent about Nicosia, 1962, *Meikle* !
7. Near Kythrea, 1880, *Sintenis & Rigo* 507 ! Kyrenia, 1955, *G. E. Atherton* 7 !

NOTES: Often troublesome as a weed of pastures and cereal fields; it seems probable that Holmboe and Chapman both misidentified this as *Paliurus spina-Christi*, which it much resembles in general habit. Holmboe's reference to " old trees " of *Zizyphus lotus* is obscure, for the species is normally a low sprawling shrub in Cyprus, and certainly not at all tree-like.

30. VITACEAE

Suessenguth in Engl. et Prantl, Pflanzenfam., ed. 2, 20d: 174–333 (1953).

Woody, often tendril-bearing, climbers, or less frequently shrubs or small trees; stems commonly jointed or swollen at the nodes, sometimes succulent; leaves alternate or the lower sometimes opposite, simple or pinnately or palmately compound, sometimes pellucid-punctate, exstipulate or with petiolar stipules; flowers hermaphrodite or unisexual, actinomorphic, usually small and numerous in leaf-opposed cymes, panicles, spikes or racemes; sepals 4–5 (–7), often connate at base; petals as many as sepals, valvate, free or variously connate, frequently caducuous; disk usually conspicuous, entire or lobed; stamens 4–5 (–7) free or connate at base, inserted at base of disk and opposite petals, anthers 2-thecous, introrse, dehiscing longitudinally; ovary superior, generally immersed in disk, 2 (–6)-locular, placentation axile, ovules 1–2 in each loculus; style simple, often short, sometimes wanting, stigma capitate or discoid. Fruit a berry; seeds with a small straight embryo and with copious (sometimes ruminate) endosperm.

About 12 genera and 700 species widely distributed in tropical and subtropical, or warm-temperate regions.

1. VITIS *L.*

Sp. Plant., ed. 1, 202 (1753).
Gen. Plant., ed. 5, 95 (1754).

Deciduous or rarely evergreen, tendril-bearing climbers; leaves simple or rarely palmately compound, usually lobed; flowers polygamo-dioecious or functionally dioecious, 5-merous, in leaf-opposed panicles; sepals minute or obsolete; petals connate at apex and falling off united at anthesis; ovary 2-locular, ovules 1–2 in each loculus; style short, conical, stigma minute. Fruit a fleshy 2–4-seeded berry; seeds usually pyriform with a narrowed basal beak, and with 2 furrows on the ventral surface.

About 60 species chiefly in temperate regions of the northern hemisphere.

1. V. vinifera *L.*, Sp. Plant., ed. 1, 202 (1753); Unger et Kotschy, Die Insel Cypern, 310 (1865); Boiss., Fl. Orient., 1: 955 (1867); Holmboe, Veg. Cypr., 125 (1914); Post, Fl. Pal., ed. 2, 1: 283 (1932); Chapman, Cyprus Trees and Shrubs, 51 (1949);

Davis, Fl. Turkey, 2: 522 (1967); Osorio-Tafall et Seraphim, List Vasc. Plants Cyprus, 69 (1973); D. Zohary et P. Spiegel-Roy in Science, 187: 321–323 (1975).
TYPE: " *in* Orbis *quatuor partibus temperatis* ".

Deciduous climber or trailer; bark light- or mid-brown, peeling off in long narrow strips; stems distinctly swollen at the nodes; tendrils branched above, arising opposite leaves; leaves broadly cordate, 4–15 cm. (or more) long, 5–15 cm. (or more) wide, distinctly or obscurely palmatilobed, or sometimes without obvious lobes, margins normally sharply and deeply dentate-serrate, upper surface of lamina glabrescent or thinly floccose-tomentellous, lower surface subglabrous or ± densely tomentose; petiole 3–8 cm. (or more) long, glabrous or tomentellous, canaliculate above; stipules scarious, caducous, oblong, 2–5 mm. long, 1·5–2·5 mm. wide, glabrous or tomentellous, with a blunt or rounded apex; flowers unisexual or hermaphrodite, very numerous in dense, narrowly pyramidal panicles 5–20 cm. (or more) long; rhachis subglabrous or thinly tomentellous; pedicels 3–5 mm. long, subglabrous, at first slender, thickening in fruit; bracteoles membranous, brownish, oblong-ovate, 1–2 mm. long, 0·5–1 mm. wide, subglabrous or thinly tomentellous; calyx minute, saucer-shaped, about 0·8–1 mm. diam., scarcely lobed, subglabrous; petals 5, greenish-white, oblong, 1–2 mm. long, 0·5–0·8 mm. wide, subglabrous, connate at apex, free at base, the free portions becoming revolute at anthesis, so that the calyptrate structure of fused petals is cast off as a whole; stamens of male flowers about 3 mm. long, filaments whitish, glabrous; anthers oblong, brownish-yellow, about 0·6 mm. long and 0·4 mm. wide; stamens of hermaphrodite flowers similar, but generally less than 2·5 mm. long, those of female flowers much reduced, usually less than 1·5 mm. long, with very slender filaments and aborted anthers; disk brownish-yellow with 5 conspicuous rounded lobes; ovary ovoid-conical, 1–1·5 mm. long, 0·5–0·8 mm. wide, style short, thick, usually less than 0·5 mm. long, stigma terminal. Fruit globose or oblong-ovoid, 0·8–4 cm. long (or sometimes even longer), to about 2·5 cm. wide, violet-purple, reddish, yellowish or greenish, sour or sweet when ripe; seed pyriform 5–8 mm. long, 4–5 mm. wide with a short or elongate basal beak, ventral furrows ± distinct, separated by a conspicuous raphe which runs from near the base of the seed over the slightly emarginate apex to a distinct, orbicular or elliptical sunken chalaza on the dorsal surface; testa pale or dark brown, hard, minutely rugulose-verruculose.

HAB.: In thickets and wooded ravines; subspontaneous on waste ground, or cultivated in gardens and vineyards; sea-level to 5,600 ft. alt.; fl. May–June.

DISTR.: Division 2, rare. Widely cultivated in the island, and commonly occurring as an escape from cultivation; probably indigenous in some localities, but its natural distribution is difficult to ascertain (see notes). Also considered indigenous in central and south-eastern Europe, the Caucasus, Turkey, and eastwards to Turkestan and Kashmir. Cultivated almost throughout the world.

2. Away from habitations, and probably indigenous in Tylliria, 1905, *Holmboe*; Stavros tis Psokas, *Chapman* 376; near Ayia, 1962, *Meikle* !

NOTES: The wild Grape (*Vitis vinifera* L. ssp. *sylvestris* (Gmel.) Berger; *V. sylvestris* Gmel.) has been distinguished from the cultivated Grape (*V. vinifera* L. ssp. *vinifera*) by its relatively slender branches, smaller leaves, dioecious inflorescences, smaller, violet-purple, sour berries, and smaller, thick seeds with a short, blunt beak. The leaves of the male plant of ssp. *sylvestris* are said to be normally distinctly lobed, those of the female plant to be without lobes or very obscurely lobed. The amount of material available for examination is too limited to allow for careful scrutiny of these distinctions, and the extreme variability of the cultivated Grape adds to the difficulties of making a satisfactory distinction between the two subspecies. Until a wide range of flowering and fruiting specimens of both wild and cultivated Grapes has been collected in Cyprus, it will be impossible to decide whether indigenous and subspontaneous plants can be satisfactorily distinguished, or to plot the natural distribution of the wild Grape. Cyprus would make an excellent centre for such studies, which should prove of the highest interest.

30a. SAPINDACEAE

Radlkofer in Engl., Pflanzenr., 98 (IV. 165): 1–1539 (1931–34).

Trees, shrubs or woody climbers, rarely herbs; leaves alternate, seldom opposite, commonly pinnately compound, mostly exstipulate. Flowers in cymes or panicles, actinomorphic or zygomorphic, polygamo-dioecious or functionally unisexual, rarely hermaphrodite, often very small and numerous; sepals usually 4–5, free or variously connate, valvate or imbricate; petals 1–5, sometimes wanting, free, equal or unequal, imbricate; disk usually present, sometimes unilateral; stamens 4–10, often 8, hypogynous, inserted within disk or unilaterally, filaments free, often hairy; anthers 2-thecous; ovary superior, 1–8-locular, commonly 3-locular, placentation axile, ovules 1–2 (occasionally many) in each loculus, style 1, simple or lobed above, terminal or gynobasic. Fruit capsular or indehiscent; seeds often arillate, without endosperm, embryo often with plicate or twisted cotyledons.

The family comprises about 130 genera and 1,000 species, most of them restricted to tropical or subtropical regions, and none indigenous in our area. *Dodonaea viscosa* (L.) Jacq. is, however, commonly grown as an evergreen hedge, and is not infrequent as an escape or relic of cultivation; it is an erect shrub 4–10 feet high, with simple, oblanceolate, glabrous, viscid leaves; small apetalous flowers and conspicuous, membranous, 2–3-winged fruits, which are often tinged rosy-purple. *D. viscosa* is found almost throughout the tropics and subtropics, and is generally reckoned to be one of the best hedging plants for warm, dry countries. Another widespread tropical plant, *Cardiospermum halicacabum* L., the Balloon-vine, an herbaceous, tendril-bearing climber, with biternate, deeply toothed leaves, small white flowers, and inflated bladder-like fruits, is occasionally grown as an ornamental, and survives as a garden reject or escape. Two other members of the *Sapindaceae*, *Koelreuteria paniculata* Laxm. and *Sapindus mukorossi* Gaertn., are listed by G. Frangos (Cypr. Agric. Journ., 18: 88; 1923) as cultivated in Nicosia: the former is a handsome small tree, with imparipinnate leaves, strongly toothed leaflets, and large panicles of showy golden-yellow flowers followed by conspicuous inflated fruits, not unlike those of *Cardiospermum*; it is a native of China. *Sapindus mukorossi*, also from China, has coriaceous, paripinnate leaves, entire leaflets, panicles of small brownish or greenish flowers, and large, fleshy, yellow or orange-brown, indehiscent fruits. The fruits are rich in a saponin, and were at one time used as a substitute for soap, a characteristic which *S. mukorossi* shares with other Soap-berries or *Sapindus* species.

31. ACERACEAE

Pax in Engl., Pflanzenr., 8 (IV, 163): 6–89 (1902).

Trees and shrubs; leaves exstipulate, opposite, simple, often palmately lobed, sometimes pinnately compound; flowers in racemes, panicles, corymbs or fascicles, actinomorphic, commonly polygamous, with rudimentary ovary present in male flower, occasionally hermaphrodite or unisexual; sepals 4–5, free or connate at base, imbricate; petals 4–5, sometimes wanting, free, imbricate; disk usually present, annular or lobed, occasionally reduced to small teeth, intrastaminal or extrastaminal; stamens 4–12, often 8, hypogynous or perigynous, filaments free; anthers 2-thecous; ovary 2-locular, usually compressed at right angles to septum, ovules 2 in each loculus, placentation axile, styles 2, free or partly connate, usually divergent, stigmas terminal. Fruit a samaroid schizocarp, splitting into 2 (–3) indehiscent parts at maturity; seeds often solitary by abortion, without endosperm; embryo with flattened or plicate cotyledons.

The family comprises 2 genera (*Acer* L. and *Dipteronia* Oliv.) and about 150 species, widely distributed in temperate regions of the northern hemisphere, with a few outlying representatives in south-eastern Asia.

1. ACER *L.*

Sp. Plant., ed. 1, 1054 (1753).
Gen. Plant., ed. 5, 474 (1754).

Characters those of the family; mericarps connate, winged unilaterally, not completely surrounded by a wing as in *Dipteronia*.

1. **A. obtusifolium** *Sibth. et Sm.*, Fl. Graec. Prodr., 1: 263 (1809), Fl. Graec., 4, t. 361 (1823); Unger et Kotschy, Die Insel Cypern, 355 (1865); Holmboe, Veg. Cypr., 123 (1914); Chapman, Cyprus Trees and Shrubs, 49 (1949); Zohary, Fl. Palaest., 2: 302, t. 443 (1972); Osorio-Tafall et Seraphim, List Vasc. Plants Cyprus, 68 (1973).

A. orientale L. var. *obtusifolium* (Sibth. et Sm.) Spach in Ann. Sci. Nat. Par., ser. 2, 2: 174 (1834); Lindberg f., Iter Cypr., 22 (1946).

A. syriacum Boiss. et Gaillardot in Boiss., Diagn. 2, 5: 72 (1856); Post, Fl. Pal., ed. 2, 1: 281 (1932).

A. syriacum Boiss. et Gaillardot var. *cyprium* [" *Cypria* "] Kotschy in Unger et Kotschy, Die Insel Cypern, 356 (1865); Boiss., Fl. Orient., 1: 952 (1867); Pax in Engl., Pflanzenr., 8 (IV, 163): 63 (1902).

[*A. creticum* (non L.) Unger et Kotschy, Die Insel Cypern, 356 (1865); Post in Mém. Herb. Boiss., 18: 92 (1900); H. Stuart Thompson in Journ. Bot., 44: 305 (1906).]

TYPE: Crete; " In montibus Sphacioticis Cretae " *Sibthorp* (OXF, K !). But see note.

Small evergreen tree or large bush, to about 10 m. high; bark smooth or finely striate, pale greyish-brown; twigs glabrous often distinctly lenticillate; buds ovoid, acute, 2–5 mm. long, closely covered with numerous ovate-deltoid, acute or subacute, minutely ciliate scales, the innermost developing into conspicuous, pale green, puberulous, oblanceolate cataphylls 1·5–3 cm. long, 0·4–0·8 cm. wide; leaves coriaceous, dark green, obscurely or distinctly 3-lobed, (2–) 4–5 (–6·5) cm. long, (2–) 4·5–6 (–7) cm. wide, glabrous on both surfaces, the primary and secondary nervation prominent below, impressed above, the tertiary nervation conspicuously and minutely reticulate, lobes ± deltoid, rounded or subacute, rarely acute, frequently very shallow, margins entire or ± sharply serrulate, base of leaf broadly cuneate, or rounded, or distinctly cordate; petiole (1–) 2–4 (–6) cm. long, glabrous, distinctly channelled above; flowers in a lax terminal corymbose cyme, bracts membranous, long-ciliate, narrowly ovate-lanceolate, 1–4 mm. long, 0·5–2 mm. wide, pedicels slender, glabrous, 0·8–1·5 (–2) cm. long; male and perfect flowers intermixed in same inflorescence; male flowers with 4–5 (sometimes more) oblong, obtuse or emarginate, yellow-green sepals, about 5 mm. long and 2·5 mm. wide; petals 4–10 (or more) similar to sepals, but usually broader, oblong-obovate, obtuse or emarginate, about 5 mm. long and 3·5 mm. wide; stamens 4–8 (–10) inserted towards the centre of a conspicuous, lobed disk, filaments glabrous, about 4 mm. long; anthers oblong, about 2·5 mm. long and 1–1·2 mm. wide; ovary rudimentary, pilose; perfect flowers resembling male flowers, but petals and sepals often rather smaller, filaments 1·5–2 mm. long, ovary thinly pilose, about 4 mm. wide, 2·5 mm. high, with 2 conspicuous, up-turned marginal wings about 3 mm. long, style about 1·5 mm. long, stigmas about 2–2·5 mm. long. Fruit with 2 spreading or suberect, strongly veined, papery wings 2–3 cm. long, 0·8–1 cm. wide, apex of wings rounded or very blunt; seed-bearing basal portion of fruit about 7 mm. long and almost as wide, glabrous, veiny, the dorsal area strongly compressed and keeled, the ventral area with two conspicuous gibbosities; seed angled, about 4 mm. long, 3 mm. wide; testa dark brown, rugulose; cotyledons circinnate-replicate.

HAB.: On rocky hillsides; in Pine forest; by field borders; often along the margins of rivulets and streams; sea-level to 4,400 ft. alt.; fl. Febr.–March.

DISTR.: Divisions 1, 2, 7, 8, common in Divisions 2, 7. Also Lebanon, ? Syria.

1. Lyso, 1913, *Haradjian* 875 ! Toxeftera, near Ayios Yeoryios (Akamas) 1962, *Meikle* 2004 !
2. Very common; *Sibthorp* s.n.; *Kotschy* 221; 829 ! *Holmboe* 1004; 1114; *Syngrassides* 1084 ! *Kennedy* 619 ! 1843 ! *Chapman* 220 ! *Davis* 3364 ! *Merton* 2993 !
7. Very common; *Kotschy* 477 ; *Sintenis & Rigo* ; *Holmboe* 816 ; *Syngrassides* 673 ! 1450 ! *Lindberg f.* s.n.; *G. E. Atherton* 1264 !
8. Near Eleousa Monastery, Karpas peninsula, 1880, *Sintenis & Rigo* 12 !

NOTES: The reasons for believing that the type of *A. obtusifolium* is a Cyprian (and not a Cretan) plant are fully set out by Holmboe (Veg. Cypr., 123), and are accepted here. The type specimens of *A. obtusifolium* match the Cyprus Maple very closely, and are quite distinct from any of the forms of *A. creticum* L. from Crete. *Acer syriacum* Boiss. et Gaillardot so closely resembles *A. obtusifolium* that it is questionable if it can be distinguished, even as a variety, from the latter.

The North American *Acer negundo* L., commonly cultivated in Europe as an ornamental, is reported from Alona village by G. Frangos (Cypr. Agric. Journ., 18: 87; 1923) and from Salamis plantation by Hutchins (Rep. Cypr. Forestry, 1909). It is at once distinguished from the indigenous species by its imparipinnate 3–5-foliolate leaves, unisexual, apetalous flowers, and fruits with distinctly incurved wings.

32. ANACARDIACEAE

Engl. in Engl. et Prantl, Pflanzenfam., ed. 1, III, 5: 138–178 (1892).

Trees and shrubs, often with resinous bark; leaves alternate, rarely opposite, simple or compound; stipules absent or very obscure; flowers hermaphrodite or unisexual, generally actinomorphic and small, in terminal or axillary panicles or spikes; sepals (3–) 5 (–7), rarely wanting, imbricate, connate at base and sometimes (together with other floral parts) adnate to gynoecium; petals equal in number to sepals or wanting, imbricate, free or rarely connate basally, usually hypogynous; stamens commonly twice as many as petals, in 2 whorls, sometimes equal in number to petals or more numerous, filaments free or connate basally, arising from the edge, or from under the edge of an annular intrastaminal disk, anthers 2-thecous, dehiscing longitudinally; gynoecium generally syncarpous, rarely pseudo-apocarpous; ovary 1-locular by abortion, often zygomorphic, rarely 2–5-locular; styles distinct or ± connate, stigmas terminal or lateral; ovules solitary, strictly axile, but often apparently apical or basal. Fruit usually a drupe, 1-seeded, with a resinous mesocarp; seed without endosperm, or endosperm scanty; embryo curved, cotyledons usually fleshy.

About 60 genera, chiefly in the tropics of the Old and New Worlds, but with a few representatives in temperate or subtropical regions.

Flowers without petals; inflorescences lateral or axillary - - - - 1. **Pistacia**
Flowers with petals; inflorescences terminal or subterminal:
Stamens 8–10; leaves and stems glabrous or subglabrous; fruit globose, glabrous 2. **Schinus**
Stamens 5; leaves and stems hairy or pubescent; fruit compressed, hairy 3. **Rhus**

1. PISTACIA *L.*

Sp. Plant., ed. 1, 1025 (1753).
Gen. Plant., ed. 5, 450 (1754).
Zohary in Pal. Journ. Bot., J. series, 5: 187–236 (1952).

Resinous trees and shrubs; leaves alternate, paripinnate or imparipinnate, rarely simple or trifoliolate; flowers small, unisexual, anemophilous, in axillary panicles or racemes; perigonium consisting of 1–3 small bracts and (1–) 2–7 unequal or subequal bracteoles (commonly termed sepals or tepals); male flowers with (3–) 4–5 (–8) stamens, filaments very short, inserted on the disk, anthers proportionally large, ovate or oblong, rudimentary pistil minute or wanting; female flowers with a subglobose or ovoid 1-locular ovary, style short, divided above, with (2–) 3 oblong or obovate, often slightly lobed or emarginate stigmas; rudimentary stamens wanting; ovule pendulous. Fruit a 1-seeded, globose or obovoid, laterally compressed drupe, mesocarp thin, endocarp hard, bony, rarely leathery; seeds often compressed laterally; testa membranous, endosperm wanting, cotyledons fleshy, plano-convex.

Eleven species, chiefly in the Mediterranean area and western Asia, but also in Mexico and southern U.S.A., Atlantic Islands, East Tropical Africa, Central Asia, China, and south-eastern Asia.

Leaves evergreen; rhachis winged:
Female inflorescences clustered, racemose - - - - - - 1. **P. lentiscus**
Female inflorescences paniculate - - - - - - - 1a. **P. × saportae**
Leaves deciduous; female inflorescences paniculate:
Rhachis winged; leaflets not mucronate; tree, 7–10 (–20) m. high 2. **P. atlantica**
Rhachis unwinged; leaflets mucronate; shrub, generally less than 6 m. high 3. **P. terebinthus**

SECT. 1. **Eu-lentiscus** *Zohary* Leaves paripinnate (or rarely with a reduced or rudimentary terminal leaflet), coriaceous, persistent; rhachis winged; leaflets obtuse or retuse, mucronate or muticous; female inflorescence fasciculate-racemose (rarely paniculate); germination epigeal.

1. **P. lentiscus** *L.*, Sp. Plant., ed. 1, 1026 (1753); Unger et Kotschy, Die Insel Cypern, 361 (1865); Boiss., Fl. Orient., 2: 8 (1872); Holmboe, Veg. Cypr., 122 (1914); Post, Fl. Pal., ed. 2, 1: 286 (1932); Lindberg f., Iter Cypr., 22 (1946); Chapman, Cyprus Trees and Shrubs, 48 (1949); Zohary in Pal. Journ. Bot., J. series, 5: 199 (1952); Davis, Fl. Turkey, 2: 544 (1967); Zohary, Fl. Palaest., 2: 299, t. 439 (1972); Osorio-Tafall et Seraphim, List Vasc. Plants Cyprus, 68 (1973).

TYPE: " *in* Italia, Sicilia, G. Narbonensi, Lusitania, Baetica, Regno Valentino. *In* Chio *insula ubi Mastichen fundit* ".

Evergreen shrub or small tree, 1–2 (–4) m. high, bark pale greyish-brown, distinctly lenticellate; leaves paripinnate, (3–) 5–10 cm. long, 3–7 cm. wide, with (1–) 2–4 (–7) pairs of thick, coriaceous, dark green, glabrous, ovate, oblong or elliptic leaflets 12–45 mm. long, (4–) 6–18 mm. wide, apex of leaflets generally obtuse or rounded, mucronate, margins entire, distinctly recurved; petiole usually less than 2 cm. long, channelled above and often narrowly winged; rhachis often rather conspicuously winged; male flowers forming narrow spicate racemes or panicles 1–2·5 cm. long, 0·5–0·8 cm. wide; peduncle 2–5 mm. long; bracts ovate, concave, about 1·5 mm. long and

0·8 mm. wide, acute, ciliate; bracteoles 5–6, ovate, acute, unequal, the smaller about 0·5 mm. long, the larger about 1 mm. long, glabrous or more often thinly ciliate; stamens 4–6, filaments very short; anthers oblong, apiculate, usually red or reddish, about 1·8 mm. long and 0·8 mm. wide; rudimentary pistil minute; female flowers forming small unbranched or sometimes slightly branched racemes 1–3 cm. long; peduncle often very short, generally less than 1 cm. long; bracts similar to those of the male flowers; pedicels very short and thick; bracteoles 4–5, connate basally; broadly ovate, obtuse or acute, about 0·7 mm. long and 0·5 mm. wide, subglabrous or thinly pubescent, with a membranous margin; ovary broadly ovoid, about 0·6 mm. long and almost as wide, glabrous, styles 3, divided almost to base, about 0·5 mm. long, stigmas obovate, retuse, about 0·6 mm. long, reflexed. Fruit red, turning lustrous black at maturity, lenticular or depressed-globose, about 5 mm. diam., crowned with the persistent styles and stigmas, mesocarp thinly fleshy, endocarp bony; seed strongly compressed, about 3·5 mm. long, 3 mm. wide; testa membranous, shining brown, minutely rugulose.

HAB.: On dry, rocky slopes and hillsides; on sand-dunes; in lowland Pine forest; sea-level to 2,500 ft. alt.; fl. Febr.–May.

DISTR.: Divisions 1–8, common in Divisions 3, 7, 8. Throughout the Mediterranean region, except Egypt; also Portugal and Atlantic Islands; the var. *emarginata* Engl. occurs in Tropical East Africa, from Somaliland south to Tanzania.

1. Cliffs above Peyia, 1962, *Meikle* 2036 ! Near Baths of Aphrodite (Polis area), 1962, *Meikle* ! Common, *Meikle* !
2. Makheras Monastery, 1862, *Kotschy*.
3. Common; *Sintenis & Rigo* 18 ! *Holmboe*; *Lindberg f.* s.n. ! *Meikle* !
4. Ayios Theodhoros (Famagusta), 1939, *Lindberg f.* s.n.
5. Between Boghaz and Trikomo, 1962, *Meikle* !
6. Liveras–Kormakiti area, 1937, *Syngrassides* 1423 !
7. Common; *Kotschy*; *Sintenis & Rigo*; *G. E. Atherton* 227 ! 253 ! 336 ! 1093 ! 1198 !
8. Common; *Sintenis & Rigo* 18 ! *Haradjian* 268 ! 295 ! *Chapman* 17; 38; *Meikle* !

NOTES: Variable in the number, size and shape of the leaflets. Gennadius (Report on the Agriculture of Cyprus, part II, 25; 1896) states that a natural variant occurring on the coast near Koma tou Yialou produced mastic " equally choice as that of Chios "; he also remarks that the male bush of *P. lentiscus* yields more gum than the female. Gaudry (Recherches Sci., 192; 1855) says that the mastic obtained from Cyprian *P. lentiscus* is yellow and less translucent than that of Chios, and is, on this account, less valued.

1a. P. × saportae *Burnat*, Fl. Alp.-Marit., 2: 54 (1896); Zohary in Pal. Journ. Bot., J. series, 5: 203 (1952); Zohary, Plant Life of Palestine, 58 (1962); Davis, Fl. Turkey, 2: 545 (1967); Zohary, Fl. Palaest., 2: 297, t. 436 (1972); Osorio-Tafall et Seraphim, List Vasc. Plants Cyprus, 68 (1973).—*P. lentiscus* L. × *P. terebinthus* L.

TYPE: France; Provence, valley of St.-Zacharie near Marseille, *Saporta et Marion* (type material figured in Ann. Sci. Nat., ser. 5, 14: pl. 1, 2; 1872).

A tall spreading shrub to 7 m. high; bark pale grey-brown; leaves persistent, coriaceous, paripinnate or imparipinnate, terminal leaflet often much reduced, lateral leaflets oblong or ovate, 1·5–4 cm. long, 0·8–2·5 cm. wide, glabrous, distinctly penninerved below, apex obtuse or subacute, mucronate, base cuneate, margins narrowly recurved; rhachis channelled above, very narrowly winged; petiole 2–4 cm. long; [male inflorescence consisting of a few clustered racemes or of a panicle, 3–8 cm. long, branches sparsely hairy; staminate flowers in small clusters, short-pedicelled, subtended by an ovate, herbaceous bract; perigonial bracteoles 4–6, unequal in length and breadth, ovate-lanceolate, 1 mm. long, stamens 4–6, 2 mm. long; anthers ovate, apiculate, filaments almost wanting; female inflorescences often in fascicles of simple racemes or in strictly branching panicles, up to 8 cm. long; flower subtended by an ovate-oblong, ciliolate bract and surrounded by 5–6 (or more) lanceolate, acuminate, ciliate perigonial bracteoles, the outer 1·5–2 mm. long, the inner shorter; pistil 1–2 mm. long, ovary globose, style tripartite, stigmas 3, obovate, one of them often smaller]. Fruit whitish (? turning black when ripe), depressed-globose, distinctly compressed, about 4 mm. long and 5 mm. wide, crowned with the persistent style-base, mesocarp thinly fleshy, endocarp bony; seed completely atrophied, or very imperfectly developed in all the specimens examined.

HAB.: On steep limestone hillside; 500 ft. alt.; fl. ? March–May.

DISTR.: Division 1, rare. Also France, Sardinia, Italy, Palestine, Morocco and Algeria.

1. Akamas Forest, 1955, *Chapman* s.n.; on the lower N. slopes of Sotira, west of the Baths of Aphrodite, 1962, *Meikle* 2288 !

NOTES: A solitary bush in an area where *P. lentiscus* and *P. terebinthus* are frequent, and almost certainly of hybrid origin. Unlike most specimens of *P.* × *saportae* hitherto described, the Cyprus specimen is nearer to *P. lentiscus* in general aspect than to *P. terebinthus*, though at once distinguished by its larger, longer-petiolate leaves and paniculate infructescences. Flowering specimens have not been collected, and the description of the floral parts is taken, with a very few modifications, from Zohary (Pal. Journ. Bot., J. series, 5: 203). Although *P.* × *saportae* is generally rare, there is every reason to believe that it might be found in other parts of Cyprus if looked for. The sterile fruits persist at a season when most *Pistacia* species are without fruits, and draw attention to the plant.

SECT. 2. **Butmela** *Zohary* Leaves imparipinnate, deciduous; rhachis winged; leaflets obtuse or muticous; female inflorescence paniculate; germination hypogeal.

2. **P. atlantica** *Desf.*, Fl. Atlant., 2: 364 (1800); Holmboe Veg. Cypr., 122 (1914); Chapman, Cyprus Trees and Shrubs, 48 (1949); Zohary in Pal. Journ. Bot., J. series, 5: 204 (1952); Davis, Fl. Turkey, 2: 545 (1967); Zohary, Fl. Palaest., 2: 297, t. 436 (1972).

P. atlantica Desf. var. *latifolia* DC., Prodr., 2: 61 (1825); Zohary in Pal. Journ. Bot., J. series, 5: 205 (1952).

P. mutica Fisch. et Mey. in Hoh., Enum. Plant. Talysch, 102 (1837); Boiss., Fl. Orient. Suppl., 154 (1888); Post, Fl. Pal., ed. 2, 1: 286 (1932).

P. atlantica Desf. ssp. *cypricola* Lindberg f. in Soc. Sci. Fenn. Årsbok, 20B: No. 7 (1942); Osorio-Tafall et Seraphim, List Vasc. Plants Cyprus, 68 (1973).

P. mutica Fisch. et Mey. ssp. *cypricola* (Lindberg f.) Lindberg f., Iter Cypr., 22 (1946).

[*P. terebinthus* (non L.) Gaudry, Recherches Sci., 193 (1855); Kotschy in Unger et Kotschy, Die Insel Cypern, 361 (1865).]

TYPE: Tunisia; Gofsa, " non longe ab urbe Cafsa distantibus " *Desfontaines* (P).

A robust, broad-crowned deciduous tree 7–10 (–20) m. high; bark ashy grey-brown, fissured; leaves imparipinnate, 10–15 (–20) cm. long, 5–12 cm. wide, divided into 3–5 pairs of leaflets; petiole 3–5 cm. long, distinctly flattened above, not winged, glabrous or minutely crisped-puberulous; rhachis narrowly but distinctly winged, minutely puberulous or subglabrous; leaflets lanceolate or narrowly oblong or ovate-oblong, 3–7 cm. long, 0·6–3 cm. wide, sessile or subsessile, glabrous or very thinly puberulous, commonly drying glaucous or grey-green, apex obtuse, not mucronate, base rounded or broadly cuneate; margins flat; male panicles at first very compact and dense, becoming a little looser at maturity, 3–10 cm. long, 2–5 cm. wide, peduncles and pedicels ± densely floccose-pubescent; bracts ovate, acute, concave, 3–4 mm. long, 1·5–2 mm. wide, lanuginose externally; flowers solitary or more usually in small clusters, pedicels very short or obsolete; perigonial bracteoles usually 5, unequal, narrowly oblong, acuminate, ranging from 1–2 mm. long and from 0·3–0·6 mm. wide, glabrous or lanuginose externally; stamens usually 5, filaments very short or obsolete; anthers oblong, incurved, about 1·5–2 mm. long, 1 mm. wide, often reddish, apex rounded or very obscurely apiculate, rudimentary pistil obsolete; female panicles lax, diffusely branched, peduncle short, branches glabrous or ± pubescent; bracts similar to those of the male panicles; flowers solitary or in small groups; pedicels short, about 1·5–2 mm long; perigonial bracteoles 6–8, the outer narrowly linear-ovate about 2 mm. long and 0·5 mm. wide, the inner ovate or oblong, about 1–1·5 mm. long, 0·5–0·8 mm. wide, acute, obtuse or rounded at the apex; ovary globose, glabrous, about 0·7 mm. diam., style very short, stigmas unequal, broadly spathulate, retuse, recurved, about 0·4–0·8 mm. long. Fruit obovoid or subglobose, a little compressed, 6–7 mm. long and almost as wide, slightly oblique, crowned with the persistent style-base; mesocarp fleshy, blue-black when ripe, dried specimens longitudinally wrinkled, especially when unripe; endocarp rather thick, bony; seed subglobose, 3–3·5 mm. diam.; testa brown, minutely rugulose.

HAB.: Cultivated in and near villages, and perhaps an ancient introduction; sea-level to 5,000 ft. alt., usually lowland; fl. Febr.–April.

DISTR.: Divisions 1–4, 7, common in Division 1. Also in Greece, Aegean Islands, Turkey, Syria, Palestine, Arabia, Crimea, Caucasus area, Iran, Baluchistan, western Pakistan, North Africa and Atlantic Islands.

1. Common; *Gaudry*; *Kotschy*; *Chapman* 143 ! 347 ! *Lindberg f.* s.n.; *Meikle* !
2. Phini, 1862, *Kotschy*; Troödos Forest, Platres Road, 1933, *Chapman* 98 !
3. Arkhimandrita, 1937, *Chapman* 307 !
4. Kiti, 1905, *Holmboe* 1097, and, 1974, *Meikle* !
7. Kambyli, 1955, *Merton* 1988 !

NOTES: It is probable that all the Cyprus material should be referred to *P. atlantica* Desf. var. *latifolia* DC. (Prodr., 2: 61; 1825) since most of it differs from typical *atlantica* in having broader, blunter leaflets. It is evident, however, that there is considerable intergradation between the typical species and the variety (cf. Zohary, Pal. Journ. Bot., J. series, 5: 205, 206), and the taxonomic significance of the distinction is questionable. I find that unripe fruits often become longitudinally wrinkled when dried, whereas ripe fruits, from the same localities, become reticulately wrinkled, so that it is inadvisable to place too much stress upon this character in separating var. *atlantica* from var. *latifolia*.

This species was frequently misnamed *P. terebinthus* by the older authors, and it will often be found that references to mastic-producing trees, especially to trees growing in the Ktima district, relate, not to *P. terebinthus* as stated, but to *P. atlantica*.

SECT. 3. **Pistacia** Leaves imparipinnate or paripinnate, deciduous; rhachis not winged; leaflets obtuse or acute, sometimes mucronate or caudate; female inflorescence paniculate; germination hypogeal (in *P. vera* and *P. palaestina*).

3. P. terebinthus *L.*, Sp. Plant., ed. 1, 1025 (1753); Boiss., Fl. Orient., 2: 6 (1872); Holmboe, Veg. Cypr., 123 (1914); Post, Fl. Pal., ed. 2, 1: 286 (1932); Lindberg f., Iter Cypr., 22 (1946); Chapman, Cyprus Trees and Shrubs, 48 (1949); Zohary in Pal. Journ. Bot., J. series, 5: 207 (1952).

P. palaestina (Boiss.) Kotschy in Unger et Kotschy, Die Insel Cypern, 361 (1865); Chapman, Cyprus Trees and Shrubs, 48 (1949); Zohary in Pal. Journ. Bot., J. series, 5: 207 (1952) pro parte quoad plant. cypr.

P. terebinthus L. ssp. *palaestina* (Boiss.) Engl. in DC., Mon. Phan., 4: 290 (1883); Lindberg f., Iter Cypr., 22 (1946); Davis, Fl. Turkey, 2: 546 (1967); Osorio-Tafall et Seraphim, List Vasc. Plants Cyprus, 68 (1973) pro parte quoad plant cypr.

[*P. vera* (non L.) Sintenis in Oesterr. Bot. Zeitschr., 32: 193 (1882).]

TYPE: "*in siccis lapidosis soli expositis* Galliae Narbonensis, Hispaniae, Lusitaniae, Italiae, Siciliae, Africae, Indiae, Syriae, Cypri, Cycladum, Mesopotamiae, Assyriae".

A deciduous shrub or small, spreading tree to about 6 m. high; bark dark ashy grey, often rough and lenticellate; leaves generally imparipinnate, sometimes with the terminal leaflet much reduced or wanting, (5–) 10–20 (–25) cm. long, 5–10 (–15) cm. wide, divided into 2–4 (–6) pairs of leaflets; petiole 2–5 cm. long, distinctly flattened above, but not winged, glabrous or sometimes thinly crisped-pubescent; rhachis slender, not winged, glabrous or thinly pubescent especially towards the axils of the leaflets; leaflets with distinct short petiolules, variable in outline, oblong or ovate, 2–6 cm. long, 0·8–3 cm. wide, glabrous or very thinly pubescent along midrib and margins, shining green and rather prominently penninerved above, a little paler below, apex obtuse, acute or sometimes acuminate, distinctly mucronate, base rounded or broadly cuneate, margins not recurved; inflorescences paniculate, appearing in spring just below the terminal tufts of young (red and translucent) foliage; male panicles dense, 5–10 cm. long, 2–5 cm. wide towards base, shortly pedunculate, peduncle and rhachis glabrous or pubescent; bracts boat-shaped, concave, 4–8 mm. long, 1–2·5 mm. wide, lanuginose externally, with an acute, brown-stained apex, bracteoles similar but smaller and narrower, about 3 mm. long and 0·5 mm. wide; flowers solitary or sometimes in groups of 2 or 3, very shortly pedicellate; perigonial bracteoles slightly unequal, linear, acute, 1–1·5 mm. long, about 0·3 mm. wide, glabrous or thinly lanuginose; stamens (4–) 5 (–6), filaments almost obsolete; anthers oblong, incurved, about 1·5 mm. long, 1 mm. wide, crimson, apex rounded or very obscurely apiculate; rudimentary pistil obsolete; female panicles lax, diffusely branched, peduncle short, branches glabrous or pubescent; bracts similar to those of male panicles; flowers solitary, distinctly pedicellate, perigonial bracteoles usually 8, ranging from 0·5–2·5 mm. long and 0·2–0·5 mm. wide, ovate to linear, with an acute or acuminate apex; ovary globose, glabrous, about 0·6 mm. diam., style about 0·5 mm. long, deeply 3-fid, stigmas somewhat unequal, obovate or spathulate, about 1 mm. long and 0·8 mm. wide, apex rounded or retuse. Fruit obovoid, 5–7 mm. long, about 5 mm. wide, slightly compressed and oblique, crowned with the persistent style-base; mesocarp thinly fleshy, reddish at first, blackish or blue-black at maturity, endocarp bony; seed subglobose about 4·5 mm. diam.; testa brown, very slightly roughened.

HAB.: On dry, rocky slopes and hillsides; in Pine forest; from just above sea-level to 4,500 ft. alt.; fl. March–April.

DISTR.: Divisions 1, 2, 3, 7, 8, common in Divisions 2, 7. Throughout the Mediterranean area.

1. Lyso, 1913, *Haradjian* 880 ! Near Ayios Neophytos, 1934, *Lindberg f.* s.n.; Akamas near Ayios Konon, 1957, *Poore* 57 ! Baths of Aphrodite, 1962, *Meikle* 2218 !
2. Common; *Kotschy*; *Holmboe* 1113; *Syngrassides* 1334 ! 1577 ! *Kennedy* 618 ! *Davis* 3391 ! *G. E. Atherton* 582 ! 1272 ! *Merton* 2815 ! 2833 !
3. Kalavasos, 1905, *Holmboe* 606.
7. Common; *Sintenis & Rigo* 17a; *Syngrassides* 669 ! 677 ! 1460 ! *Lindberg f.* s.n.; *Chapman* 44.
8. Between Rizokarpaso and Apostolos Andreas, 1933, *Syngrassides* 302 ! and, 1934, *Syngrassides* 517 ! Yialousa, 1950, *F. J. F. Barrington* s.n.; Ronnas Valley, Rizokarpaso, 1957, *Merton* 2931 !

NOTES: Although more than one author has reported both *P. terebinthus* L. and *P. palaestina* Boiss. from Cyprus, it can be stated very definitely that only one species of the *terebinthus* complex occurs on the island and that this, while differing from western Mediterranean *P. terebinthus* in its slightly smaller fruits and thinner, less noticeably discolorous leaflets, is equally distinct from typical *P. palaestina* in having, on the whole, imparipinnate leaves with a well developed terminal leaflet. While extreme forms of *P. terebinthus* and *P. palaestina* seem distinct, one suspects that careful sampling of the complete Mediterranean populations of the aggregate *terebinthus-palaestina* would show a cline, with extreme *terebinthus* at the western extremity of the aggregate's distribution, extreme *palaestina* at the eastern extremity of the area, and a variety of intermediates in Cyprus, Turkey, Crete, Greece and other " in-between " areas. Leaflet-shape, presence or absence of terminal leaflet, indumentum, and shape of inflorescences, the characters normally used to distinguish *P. terebinthus* and *P. palaestina*, simply do not work in Cyprus. It is just possible that valid characters might be found in texture and colour of leaflets and in fruit size, shape and sculpturing of pericarp (in dried specimens), but it is very doubtful if any of these characters, however valid, sustain more than subspecific discrimination.

A very curious form of *P. terebinthus*, with the leaves reduced to a single, enlarged terminal leaflet, has been collected in the Ayia valley by Merton (2833) and near Ayios Konon (Akamas) by Poore (57); apart from two leaves in the Merton specimen, which are 2-foliolate and 3-foliolate, all the material is uniformly unifoliolate. Such reductions in the number of leaflets have been reported in *P. khinjuk* Stocks and *P. vera* L., but not, it would appear, in *P. terebinthus*.

PISTACIA VERA *L.*, the Pistachio, a native of Central Asia, is recorded as a cultivated plant from Nicosia (*Bovill*), Limassol District (*D. Michaelides*), Ayios Neophytos Monastery (*Lindberg f.*), the Armenian Monastery (*Lindberg f.*), Larnaca, Pergamos, and is doubtless grown in other parts of the island. It is a tree about 3–8 m. high, with deciduous, long-petiolate leaves, imparipinnately divided into 3–5 (–7) large, often broadly ovate or suborbicular, conspicuously reticulate-veined, leathery leaflets; the leaf rhachis is unwinged or at most very obscurely winged; male and female inflorescences resemble those of *P. atlantica*, but the fruit is very much larger, 10–20 mm. long and 6–12 mm. wide, with the fleshy mesocarp separating readily at maturity from the bony endocarp.

Pistacia vera is commonly propagated by grafting on *P. terebinthus*. To obtain fertile specimens male and female trees must be grown in proximity, or scions of both sexes grafted on the stock, otherwise the fruits are empty and valueless. The tree is slow-growing and requires seven years or more to reach fruiting condition, though, once mature, it remains productive for many years.

2. SCHINUS *L.*

Sp. Plant., ed. 1, 388 (1753).
Gen. Plant., ed. 5, 184 (1754).
F. A. Barkley in Lilloa, 28: 5–110 (1957).

Trees or shrubs with armed or unarmed branches; leaves simple or pinnately compound, often with a distinctly winged rhachis; flowers polygamo-dioecious in spicate, racemose or paniculate inflorescences; bracts small, deltoid-ovate or deltoid-lanceolate; pedicels distinctly articulated; calyx of 4–5, blunt, imbricate lobes; petals 4–5, oblong, blunt or truncate, imbricate, inserted at the base of the disk; stamens 8–10, filaments subulate, anthers ovoid or oblong; disk 8–10-lobed, thick, saucer-shaped; ovary sessile, ovule 1, pendulous from apex of ovary, styles 3, free above, united below, stigmas subglobose or inverted V-shaped. Fruit a small globose drupe, exocarp papery, lustrous, separating from the resinous mesocarp at maturity; endocarp bony; seed lenticular, with a very thin testa; cotyledons flat, radicle ascending.

About 28 species, all from South (and Central) America; a few are cultivated as ornamentals, and occasionally naturalized, outside this area.

1. S. MOLLE *L.*, Sp. Plant., ed. 1, 388 (1753); Post, Fl. Pal., ed. 2, 1: 287 (1932); Barkley in Lilloa, 28: 184 (1957); Osorio-Tafall et Seraphim, List Vasc. Plants Cyprus, 68 (1973).

A graceful, round-headed evergreen tree, 5–15 m. high; branches slender, pendulous; leaves narrowly oblong in outline, 15–30 cm. long, impari- or paripinnately divided into 20–40 (or more) linear-lanceolate, acuminate, entire or serrulate, glabrous or subglabrous leaflets 20–50 mm. long, 5–6 mm. wide; rhachis flattened or very narrowly winged; inflorescence a lax, narrowly conical, terminal and subterminal panicle 20–30 cm. (or more) long; flowers small, numerous; sepals connate, 4–5-lobed, the lobes oblate, about 0·4–0·5 mm. long, 0·6–0·7 mm. wide, glabrous; petals whitish, oblong with a rounded apex, about 2 mm. long and 1–1·2 mm. wide; stamens of male flowers about 1·6 mm. long, filaments subulate; anthers oblong, about 0·6 mm. long, 0·4 mm. wide; stamens of hermaphrodite flowers much smaller; ovary globose, glabrous, about 0·8–1 mm. diam., styles about 0·2 mm. long, united towards base, stigmas subglobose. Fruit globose, rosy-red, 5–6 mm. diam.; mesocarp strongly aromatic; seed about 3·5 mm. diam.; testa whitish or pale brown, very thinly membranous.

HAB.: Cultivated for ornament in lowland areas; fl. April–Aug.

NOTES: A native of S. Brazil, Uruguay, Paraguay and northern Argentina, commonly grown throughout the tropics and subtropics and occasionally naturalized. The popular name, Pepper-tree, refers to the fruits, which are about the size and shape of a peppercorn. The specific epithet is not from the Latin *mollis* " soft ", but from " Molle ", the vernacular American-Indian name of the tree. *Schinus molle* is found in many of the lowland towns and villages of Cyprus.

Schinus terebinthifolius Raddi, from Venezuela, Brazil and Argentina, is also cultivated in the Mediterranean area, but less often than *S. molle*; it resembles a coarse form of *Pistacia lentiscus*, with obovate leaflets, flowers in axillary panicles, and clusters of bright red fruits. Specimens have been seen in the Municipal Gardens, Nicosia (*Meikle* 4085 !).

3. RHUS *L.*

Sp. Plant., ed. 1, 265 (1753).
Gen. Plant., ed. 5, 129 (1754).

Trees, shrubs and subshrubs with alternate, simple, trifoliolate or imparipinnate leaves; flowers polygamo-dioecious, small, usually numerous in branched paniculate inflorescences; calyx with 5 imbricate lobes; petals imbricate, generally exceeding the calyx; stamens 5, inserted on the underside of a ± conspicuous disk; filaments subulate; anthers oblong or ovoid, those of the female flowers very small; ovary ovoid or subglobose, ovule pendulous from a basal funicle; styles 3, free or partly united, stigmas globose. Fruit a globose or compressed drupe; exocarp glabrous or pilose; mesocarp resinous; endocarp crustaceous or bony; seed reniform, cotyledons flattish, radicle curved.

About 200 species, widely distributed through the warmer regions of the world, with a marked concentration of species in southern Africa.

1. **R. coriaria** *L.*, Sp. Plant., ed. 1, 265 (1753); Unger et Kotschy, Die Insel Cypern, 361 (1865); Boiss., Fl. Orient., 2: 4 (1872); Holmboe, Veg. Cypr., 123 (1914); Post, Fl. Pal., ed. 2, 1: 285 (1932); Lindberg f., Iter Cypr., 22 (1946); Chapman, Cyprus Trees and Shrubs, 49 (1949); Davis, Fl. Turkey, 2: 543 (1967); Zohary, Fl. Palaest., 2: 300, t. 440 (1972); Osorio-Tafall et Seraphim, List Vasc. Plants Cyprus, 68 (1973).

R. coriaria L. var. *humilior* (? Nicotra) Pojero, Fl. Sic., 1, 2: 13 (1891) pro parte quoad spec. cypr.

TYPE: " *in* Europa *australi* ".

An erect or spreading, deciduous shrub to about 2 m. high; twigs brown, pubescent or shortly pilose, exuding milky latex when broken; buds globose, densely woolly; leaves 6–20 (–30) cm. long, 4–8 (–12) cm. wide, narrowly oblong in outline, imparipinnately divided into 4–8 pairs of oblong or elliptical-ovate leaflets, 1–4 (–6) cm. long, 0·6–2 (–2·5) cm. wide, leaflets densely pubescent above and below, usually with the nerves distinctly impressed on the upper surface and prominent on the lower, apex of leaflets acute or more rarely obtuse, base sessile, rounded or broadly cuneate, occasionally tapering, margins serrate or crenate-serrate; petiole short, usually less than 3 cm. long, densely pilose, slightly flattened above; rhachis pilose, the uppermost part narrowly winged; inflorescence a narrowly conical or thyrsoid, terminal and subterminal panicle, 3–15 (–20) cm. long, 2·5–10 cm. wide, branches erecto-patent, peduncles and pedicels ± densely pilose-pubescent; bracts narrowly oblanceolate or lingulate, 3–4 mm. long, about 1 mm. wide, pubescent, apex blunt or acute; pedicels short, stout, seldom exceeding 2·5 mm. in length, bracteolate; sepals free almost to base, oblong, blunt, about 2–2·5 mm. long and 1–1·2 mm. wide, pubescent externally; petals glabrous or thinly pubescent, whitish, oblong, recurved, about 3·5 mm. long, 1·5 mm. wide, apex rounded, base narrowly cuneate; stamens of male flowers about 3 mm. long, filaments subulate, glabrous; anthers oblong, about 1·5 mm. long, 0·8 mm. wide; disk conspicuous, annular, 5-lobed, surrounding the rudimentary pistil; stamens of female flowers minute or wanting; ovary compressed, orbicular, pilose, about 1–1·5 mm. diam., styles 3, ± free, about 0·5 mm. long, stigmas terminal, globose. Fruit compressed, suborbicular, about 5 mm. diam., crimson, pubescent; mesocarp separating readily from hard, bony, dark brown endocarp at maturity; seed about 2·5 mm. wide, testa whitish or pale brown, membranous.

HAB.: On stony mountainsides and in vineyards; 2,000–6,000 ft. alt., fl. June–July (to Oct.).

DISTR.: Divisions 2, [7], locally very common on the Troödos Range. Widespread in the Mediterranean area and eastwards to Iran.

2. Common; *Gaudry*; *Kotschy*; *Sintenis & Rigo* 17 ! *Holmboe* 852; 1029; *Syngrassides* 734 ! *Kennedy* 614 ! 615 ! 616 ! 617 ! *Chapman* 104; *G. E. Atherton* 511 !

[7. Near Kyrenia in a moist situation, 1939, *Lindberg f.* s.n.; Boghazi, above Kyrenia, 1939, *Lindberg f.* s.n.]

COTINUS COGGYGRIA *Scop.* (*Rhus cotinus* L.), the Smoke-tree or Venetian Sumach, a common shrub in many parts of the Mediterranean area, is not indigenous in Cyprus, though reported (Bovill, 1915) as cultivated in Salamis plantation, and perhaps grown elsewhere. It is a rounded, much-branched deciduous bush 3–5 m. high, with simple, undivided, obovate or suborbicular leaves 3–8 cm. long, and conspicuous terminal panicles, the final branches of which are mostly sterile and thread-like, clothed with numerous spreading, silky hairs; these hairs become greyish or purplish in the infructescences and give the plant its nebulous, smoky appearance. *Cotinus coggygria* is popular as a garden shrub in Great Britain, and in many other countries of Europe and North America; several selected varieties with brightly coloured leaves or inflorescences are listed in horticultural literature.

33. LEGUMINOSAE

A. Burkart, Las Leguminosas Argentinas, ed. 2 (1952).
J. Hutchinson, Genera of Flowering Plants, 1: 221–489 (1964).

Trees, shrubs, herbs and climbers; leaves usually alternate, stipulate, pinnately or palmately compound; flowers terminal or axillary, solitary or in spikes, racemes or panicles, zygomorphic or actinomorphic, generally hermaphrodite; calyx usually gamosepalous, with (4–) 5 (–6) imbricate or valvate teeth; petals imbricate or valvate, typically 5, the 2 lower (abaxial) sometimes connate, occasionally all the petals connate basally and forming a tube; stamens as many as, or twice as many as petals, or numerous; filaments free or ± united or diadelphous; anthers 2-thecous, dehiscing longitudinally or by pores; ovary superior, 1-locular, style and stigma 1; ovules 2–many. Fruit generally a pod, dehiscent into 2 valves, but sometimes a loment, or dehiscing along one suture only, or tearing irregularly, or indehiscent; seeds sometimes arillate, often large, endosperm gelatinous or mucilaginous, often wanting.

About 550 genera and more than 13,000 species widely distributed in temperate and tropical areas, and often subdivided into 3 families or sub-families, *Mimosoideae* (*Mimosaceae*), *Caesalpinioideae* (*Caesalpiniaceae*) and *Papilionoideae* (*Papilionaceae*); these subfamilies are generally easy to recognize, though not demarcated by any single character or feature which can be consistently relied upon to distinguish them. A synopsis of the most useful distinctions will be found at the beginning of each subdivision after the name of the subfamily. The *Mimosoideae* and *Caesalpinioideae* are predominantly tropical in distribution, and sparingly represented in Cyprus, but the *Papilionoideae* form one of the largest and most conspicuous components of the island's flora.

Trees, shrubs or woody subshrubs:
  Twigs or branches spinose, or leaves terminating in a spine:
    Leaves simple or 1–3-foliolate:
      Flowers yellow:
        Leaves 3-foliolate; calyx circumscissile; pod densely hairy - **4. Calycotome**
        Leaves simple (or 1-foliolate); calyx not circumscissile, 5-toothed; pod adpressed-pubescent - - - - - - **5. Genista**
      Flowers pink, crimson or purple (rarely white):
        Leaves (at least the lower ones) 3-foliolate, glandular-pubescent; flowers usually pink; pod ovoid, usually shorter than the persistent calyx **6. Ononis**
        Leaves all simple (or 1-foliolate), thinly pubescent; flowers crimson; pod linear, elongate, terete or moniliform, much longer than calyx **26. Alhagi**
    Leaves pinnate or bipinnate with 5 or more leaflets:
      Flowers white or pink with conspicuous petals:
        A tree; stipules spinose; leaves imparipinnate - - - ROBINIA (p. 490)
        A dwarf subshrub; leaves paripinnate, rhachis terminating in a spine **18. Astragalus**
      Flowers yellow or greenish yellow:
        Petals large, conspicuous; pod linear often torulose or moniliform PARKINSONIA (p. 587)
        Petals small, inconspicuous (stamens sometimes numerous and conspicuous); flowers in dense spicate or globose inflorescences:
          Stamens 5–10, equalling or twice as many as calyx-teeth:
            Leaflets lanceolate or oblong, 2·5–12 mm. wide; pod flattened, 3–5 cm. wide - - - - - - - - GLEDITSIA (p. 587)
            Leaflets narrowly oblong or linear, 0·7–3 mm. wide; pod compressed or subterete, 1–2 cm. wide - - - - - **33. Prosopis**
          Stamens numerous, more than twice as many as calyx-teeth - **34. Acacia**
  Twigs, branches and leaves without spines:
    Leaves regularly 3-foliolate:
      Leaves and stems clothed with adpressed silvery hairs; flowers small, solitary, yellow; pod linear, compressed, 15–25 mm. long; a dwarf suberect or decumbent subshrub - - - - - **3. Argyrolobium**
      Leaves and stems not clothed with adpressed silvery hairs; plant not as above:
        Leaflets 20–60 mm. long; flowers dull yellowish; pod glabrous, narrowly oblong, 10–15 cm. long, pendulous - - - - - **1. Anagyris**
        Leaflets 2–5 (–8) mm. long; flowers bright yellow; pod glandular-pilose, 0·6–1·1 (–1·7) cm. long, not pendulous - - - - - **6. Ononis**
    Leaves pinnately or bipinnately 2–7-foliolate or many-foliolate, or simple (1-foliolate) or reduced to a leaf-like phyllode; not regularly 3-foliolate:
      Leaves simple, suborbicular or reniform:
        Leaves 2-lobed at apex; flowers not papilionaceous, individually large and showy - - - - - - - - - - BAUHINIA (p. 591)
        Leaves not 2-lobed at apex; flowers papilionaceous (with apparent standard and wings) individually small, in clusters on the leafless branches CERCIS (p. 591)
      Leaves pinnate or bipinnate, or reduced to a leaf-like phyllode, but not orbicular or reniform:

Flowers numerous in cylindrical spikes or globose heads; petals inconspicuous or wanting:
Inflorescences conspicuous; stamens very numerous, more than twice as many as calyx-teeth:
Filaments free or connate only at extreme base - - - **34. Acacia**
Filaments connate for some distance above base - ALBIZIA (p. 603)
Inflorescences inconspicuous; stamens equalling or twice as many as calyx-teeth:
Evergreen tree; leaves simply pinnate; leaflets oblong or obovate, 10–45 mm. or more in width - - - - - **32. Ceratonia**
Deciduous tree; leaves mostly bipinnate; leaflets lanceolate or oblong, 2·5–12 mm. wide - - - - - GLEDITSIA (p. 587)
Flowers not in cylindrical spikes or globose heads; petals usually conspicuous:
Flowers papilionaceous (with standard, wings and keel):
Pods somewhat compressed but not flattened or winged, narrowly linear:
Leaves paripinnate, leaflets 20–30 or more; claws of petals not exserted from calyx - - - - - - - - SESBANIA (p. 490)
Leaves imparipinnate; leaflets (3–) 5–7 (–9); claws of petals exserted from calyx - - - - - - - - **22. Coronilla**
Pods flattened or winged, not narrowly linear:
Leaflets (1–) 3–5, suborbicular or with an acuminate-cuspidate apex; flowers small, creamy-white or pale yellow - DALBERGIA (p. 583)
Leaflets 9–30 (or more), oblong, apex rounded or emarginate; flowers large, conspicuous:
Flowers white or pink, in compact, drooping racemes; pods flattened but not conspicuously winged - - ROBINIA (p. 490)
Flowers yellow, in open erect panicles; pod winged, like an *Acer* samara - - - - - - - TIPUANA (p. 583)
Flowers not papilionaceous (without distinct standard, wings and keel):
Leaves bipinnate; leaflets numerous, small; filaments red, very long and conspicuous - - - - - - - CAESALPINIA (p. 585)
Leaves simply pinnate; leaflets neither very numerous nor very small; filaments not very long or conspicuous - - - CASSIA (p. 589)
Annual or perennial herbs; not tree-like or suffruticose:
Plant without true leaves or leaf-like phyllodes; stipules enlarged and foliaceous; tendrils present - - - - - - - **30. Lathyrus**
Plant with leaves or leaf-like phyllodes:
Leaves palmately 5- (or more) foliolate; flowers in conspicuous terminal racemes **2. Lupinus**
Leaves not as above:
Leaves simple, or 1- or 3-foliolate:
Leaves (or phyllodes) simple, without leaflets:
Leaves all without tendrils; flowers in umbels - - - **20. Scorpiurus**
Leaves (phyllodes) or at least the upper ones with branched tendrils; flowers solitary or occasionally paired, axillary - - - **30. Lathyrus**
Leaves with 1 or 3 leaflets:
Climbing or twining plants; leaflets large, entire:
Keel strongly twisted; seeds without a conspicuous strophiole PHASEOLUS (p. 581)
Keel not twisted; seeds with a conspicuous white strophiole DOLICHOS (p. 582)
Erect, diffuse or prostrate plants, not climbing or twining:
Inflorescence an elongate spike or raceme:
Pod small but conspicuous, not concealed by calyx:
Pod falcately or circinnately curved - - - **8. Medicago**
Pod not conspicuously curved:
Flowers numerous in each inflorescence, small, rather crowded; pod globose or ovoid; plant glabrous or sparsely pubescent **10. Melilotus**
Flowers few in each inflorescence, not crowded, often rather large; pod linear, oblong or rhomboidal; plant generally glandular-pubescent - - - - - - - - - **6. Ononis**
Pod inconspicuous, generally quite concealed by the persistent, marcescent calyx and corolla - - - - - **11. Trifolium**
Inflorescence a compact head or umbel, or flowers solitary or few, not in elongate spikes or racemes:
Pod very short, inconspicuous, generally quite concealed by the persistent, marcescent calyx and corolla - - - **11. Trifolium**
Pod conspicuous, exserted from calyx; corolla generally not persistent:

Pod twisted into tight spiral coils, often spinose - **8. Medicago**
Pod straight or curved, but not twisted into spiral coils:
Plant geocarpic (pods maturing below the surface of the soil); gynophore elongating and becoming thread-like in fruit; flowers solitary, minute - - - **9. Factorovskya**
Plant not geocarpic (pods maturing above soil-level); gynophore not elongating or thread-like:
Pods torulose, dehiscing transversely into 1-seeded joints **22. Coronilla**
Pods not as above, dehiscing longitudinally or indehiscent:
Pods with a long falciform or subulate beak:
Flowers numerous in dense heads, leaflets large, entire **17. Aspalthium**
Flowers 1–2, axillary; leaflets not very large, distinctly toothed - - - - - - **7. Trigonella**
Pods without a terminal beak:
Pods subglobose or reniform, scarcely longer than wide, 1-seeded, blackish and reticulate-veined when ripe **8. Medicago**
Pods linear or oblong, longer than wide, straight or curved, not reniform:
Pods terete or subterete with 4 conspicuous longitudinal wings; flowers solitary or in pairs **16. Tetragonolobus**
Pods without longitudinal wings, sometimes compressed or flattened:
Plants glandular-pubescent, often viscid; pods short, oblong or ovoid, not much exceeding calyx **6. Ononis**
Plants glabrous or hairy, but not glandular or viscid; pods linear or narrowly oblong, much longer than calyx:
Petals whitish or tinged pink or purple; keel blunt or subacute - - - - - **14. Dorycnium**
Petals yellow:
Leaflets entire, the basal stipuloid; true stipules obsolete or wanting - - - **15. Lotus**
Leaflets toothed or serrate; true stipules present **7. Trigonella**

Leaves regularly or irregularly pinnate, with 2 or 4 or more leaflets, not simple nor regularly 1- or 3-foliolate:
Leaves paripinnate, the terminal leaflet replaced by a tendril, bristle or cusp:
Pods strongly compressed, containing 1–2 orbicular, flattened seeds. Small slender annuals; flowers solitary or in few-flowered racemes, usually whitish or pale blue; calyx-teeth subequal, long and slender **29. Lens**
Pods not strongly compressed, containing several (rarely 1–2) globose or angular seeds; calyx with the upper (adaxial) teeth usually distinctly shorter than the lower (abaxial):
Style tapering to apex, glabrous or bearded only about apex or on the abaxial (lower) surface; staminal sheath oblique at apex; leaflets often numerous - - - - - - - - **28. Vicia**
Style flattened or dorsiventrally compressed (sometimes subspathulate) towards apex, glabrous or bearded along the adaxial (upper) surface; staminal sheath truncate at apex; leaflets often 2 or 4, rarely numerous:
Stems strongly angular or winged; leaflets often parallel-nerved; style-margins without reflexed or connate wings **30. Lathyrus**
Stems terete or obscurely angled, not winged; leaflets pinnately nerved; style margins winged, the wings connate towards apex **31. Pisum**
Leaves imparipinnate, with a terminal leaflet:
Pod concealed within the persistent inflated calyx **12. Physanthyllis**
Pod conspicuous, protruding from calyx:
Pods dehiscing transversely into 1-seeded joints:
Inflorescences umbellate, or flowers solitary or in compact clusters:
Pod terete or subterete or compressed, but not flattened, without excavated sinuses:

Keel blunt at apex, inflorescence subtended by a bract-like floral leaf - - - - - - - - - 21. **Ornithopus**
Keel acute or rostrate; inflorescence without a subtending floral leaf - - - - - - - - 22. **Coronilla**
Pod flattened, deeply and regularly excavate with rounded sinuses 23. **Hippocrepis**
Inflorescence racemose - - - - - - - - 24. **Hedysarum**
Pods dehiscing longitudinally or indehiscent:
Pods suborbicular or circinnately curved, strongly compressed laterally:
Flowers in small clusters subtended by a bract-like floral leaf; leaves without true stipules; leaflets irregularly pinnate; valves of pod almost smooth - - 13. **Hymenocarpos**
Flowers in racemes; leaves stipulate; leaflets regularly pinnate; pod with foveolate or spinose valves - 25. **Onobrychis**
Pods linear, oblong or ovoid, not suborbicular or circinnately curved:
Leaflets toothed or laciniate - - - - - - - 27. **Cicer**
Leaflets entire (often emarginate at apex):
Flowers bright yellow, in umbels; pod linear, laterally compressed, with a curved rostrate apex 22. **Coronilla**
Flowers not bright yellow; not in umbels; pod not as above:
Anther-thecae confluent at apex; leaflets glutinous and gland-dotted - - - - - - - - 19. **Glycyrrhiza**
Anther-thecae distinct at apex; leaflets not gland-dotted or glutinous - - - - - - - 18. **Astragalus**

SUBFAMILY 1. **Papilionoideae** Mostly herbs and herbaceous climbers; leaves generally pinnate, sometimes palmate; flowers strongly zygomorphic, petals imbricate, the upper (adaxial) petal outermost, often distinctly broader than the others and called the *standard*; lateral petals (*wings*) usually narrow and directed forwards, the two lower (abaxial) petals generally connate along their basal margins forming the *keel*; stamens usually 10 (rarely 5), monadelphous or diadelphous, seldom free, anthers without an apical gland. Seeds without an areole, radicle commonly curved.

TRIBE 1. **Podalyrieae** *Benth.* Shrubs, rarely herbs; leaves palmately compound, rarely pinnate; stamens 10, free, or very shortly connate at base.

1. ANAGYRIS *L.*

Sp. Plant., ed. 1, 374 (1753).
Gen. Plant., ed. 5, 176 (1754).

Deciduous, malodorous shrubs; leaves palmately 3-foliolate, stipules connate; petals 5, standard shorter than wings or keel; stamens 10, free. Pod stipitate, flattened, septate, many seeded, coriaceous.

Two species (or possibly 2 subspecies of a single species) limited to the Mediterranean area, W. Asia and the Canary Islands.

1. **A. foetida** *L.*, Sp. Plant., ed. 1, 374 (1753); Sibthorp in Walpole, Travels, 22 (1820); Unger et Kotschy, Die Insel Cypern, 372 (1865); Boiss., Fl. Orient., 2: 24 (1872); Holmboe, Veg. Cypr., 101 (1914); Post, Fl. Pal., ed. 2, 1: 296 (1932); Lindberg f., Iter Cypr., 19 (1946); Chapman, Cyprus Trees and Shrubs, 44 (1949); Davis, Fl. Turkey, 3: 12 (1970); Zohary, Fl. Palaest. 2: 37, t. 48 (1972); Osorio-Tafall et Seraphim, List Vasc. Plants Cyprus, 55 (1973).
TYPE: " *in* Italiae, Siciliae, Hispaniae *montibus* ".

Spreading shrub, to about 3 m. high; bark of older branches dark ashy-grey; stipules 4–10 mm. long, 2–4 mm. wide, connate along one margin, forming a sheath around the stem opposite each leaf, caducous, leaving a conspicuous circular scar, acutely 2-lobed at apex, adpressed-sericeous along nerves externally; petiole 1–3 cm. long, adpressed-pubescent, channelled above, distinctly swollen at base; leaflets greyish-green, elliptic, oblong or

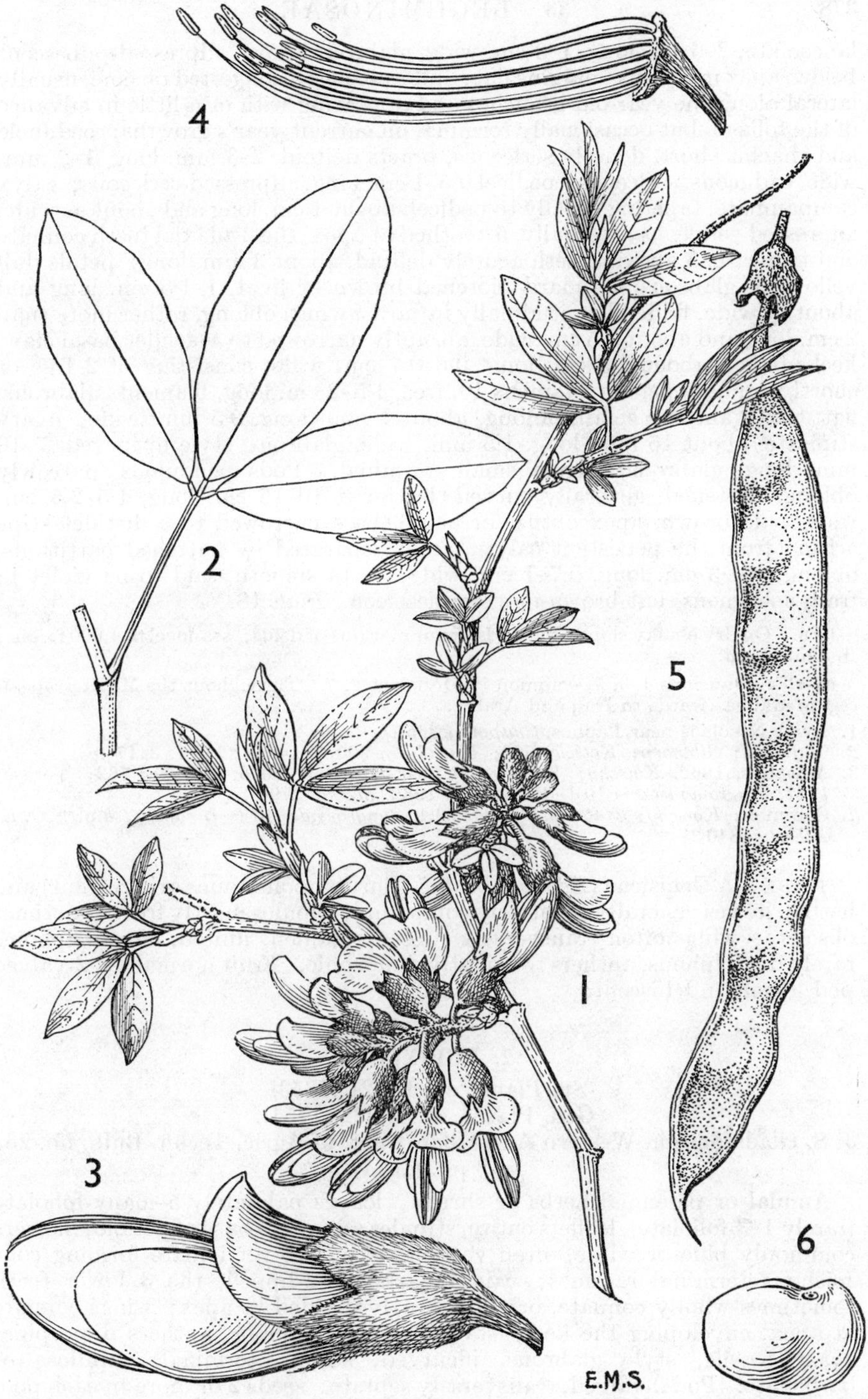

Plate 18. ANAGYRIS FOETIDA L. **1,** flowering shoot, ×1; **2,** leaf, ×1; **3,** flower, ×3; **4,** androecium, ×3; **5,** pod, ×1; **6,** seed, ×2. (**1, 3 & 4,** *Guest* 1328; **2,** *Haines* W. 1023; **6,** *Field & Lazar* 711.)

lanceolate, 2–6 cm. long, 1–3 cm. wide, glabrous above, adpressed-pubescent below, apex muticous, margins flat; inflorescence a congested raceme, usually lateral along the year-old branches and appearing with or a little in advance of the foliage, but occasionally terminal on current year's growths; peduncle and rhachis short, densely sericeous, bracts deltoid, 2–3 mm. long, 1–2 mm. wide, caducous, sericeous; pedicel 0·5–1 cm. long, adpressed-sericeous; calyx campanulate, tapering basally to pedicel, about 1 cm. long and about as wide, adpressed-sericeous externally, 5-toothed at apex, the 2 adaxial teeth connate and truncate, 3 abaxial teeth acutely deltoid, about 3 mm. long; petals dull yellowish, glabrous, standard blotched brown or livid, 1–1·5 cm. long and about as wide, folded longitudinally inwards, wings oblong, rather more than 2 cm. long and about 0·5 cm. wide, abruptly narrowed to a slender basal claw, keel oblong, about 2·5 cm. long, 0·6–0·7 cm. wide, consisting of 2 free or shortly connate petals; stamens 10, free, 1·5–2 cm. long, filaments glabrous, flattened; anthers small, oblong, about 1 mm. long, 0·5 mm. wide; ovary stipitate, about 15 mm. long, 1·5 mm. wide, glabrous, style upcurved, 7–10 mm. long, glabrous, stigma small, terminal. Pods pendulous, narrowly oblong, flattened, generally curved, leathery, 10–15 cm. long, 1·5–2·5 cm. wide, pale brown, apex obtuse or acute, base narrowed to a distinct stipe arising from the persistent calyx; seeds separated by flattened partitions, oblong, 1–1·3 cm. long, 0·7–1 cm. wide; testa smooth, said to be violet in fresh specimens, but brown in examples seen. *Plate* 18.

HAB.: On dry chalky slopes, by roadsides or margins of fields; sea-level to 4,000 ft. alt.; fl. Jan.–April.

DISTR.: Divisions 1–3, 7, common in Divisions 2, 7. Throughout the Mediterranean region and eastwards to Iraq and Arabia.

1. Ayios Nikolaos near Paphos, *Chapman* 365.
2. Common; *Sibthorp*; *Kotschy* 696; *Haradjian* 900 ! *Davis* 2733 ! 3417 !
3. Amathus, 1862, *Kotschy*; Kannaviou near Khoulou, 1905, *Holmboe* 762; Yerasa, 1947, *Mavromoustakis* 10 ! Perapedhi, 1962, *Meikle* 2799 !
7. Common; *Kotschy*; *Sintenis & Rigo* 669 ! *Syngrassides* 40 ! *Lindberg f.* s.n. ! *G. E. Atherton* 840 !

TRIBE 2. **Genisteae** (*Adans.*) *Benth.* Shrubs (sometimes arborescent) and herbs; leaves generally palmately compound, stipules mostly free, sometimes obsolete; wings often transversely rugose; stamens united, monadelphous, rarely diadelphous, anthers frequently dimorphic. Fruit normally a 2-valved pod, rarely indehiscent.

2. LUPINUS *L.*

Sp. Plant., ed. 1, 721 (1753).
Gen. Plant., ed. 5, 322 (1754).
J. S. Gladstones in Western Australian Dept. of Agric. Techn. Bull., no. 26: 48 pp. (1974).

Annual or perennial herbs or shrubs; leaves palmately 5–many-foliolate (rarely 1–3-foliolate), leaflets entire, stipules adnate to base of petiole; flowers commonly blue or white, often verticillate or sub-verticillate forming conspicuous terminal racemes; calyx markedly 2-lipped, the 3 lower teeth sometimes wholly connate, or very shortly divided at apex; wings connate at apex, enveloping the keel; stamens monadelphous, anthers dimorphic; ovary sessile, style glabrous, incurved, stigma terminal, papillose or barbellate. Pod 2-valved, transversely septate; seeds 2 or more in each pod, hilum oblong or linear, endosperm wanting.

About 150 species, chiefly in North America, with relatively few indigenous representatives in southern Europe, the Mediterranean area, and western

Asia. Some species are cultivated as decorative plants, others are useful as fodder or green manure, or for their edible seeds, but many are distinctly poisonous.

Leaves and stems clothed with long, spreading hairs; leaflets oblanceolate or narrowly obovate; lower lip of calyx deeply 3-lobed - - - - - **1. L. micranthus**
Leaves and stems shortly adpressed-pubescent; leaflets narrowly oblong or ligulate; lower lip of calyx shortly 3-toothed - - - - - **2. L. angustifolius**

1. L. micranthus *Guss.*, Fl. Sic. Prodr., 2: 400 (1828); Unger et Kotschy, Die Insel Cypern, 372 (1865); Davis, Fl. Turkey, 3: 39 (1970); Zohary, Fl. Palaest. 2: 43, t. 56 (1972); Osorio-Tafall et Seraphim, List Vasc. Plants Cyprus, 55 (1973).
L. hirsutus L. var. *micranthus* (Guss.) Boiss., Fl. Orient., 2: 28 (1872).
L. hirsutus L. ssp. *micranthus* (Guss.) Holmboe, Veg. Cypr., 101 (1914).
[*L. hirsutus* L., Sp. Plant., ed. 2, 1015 (1763); Boiss., Fl. Orient., 2: 28 (1872); Post, Fl. Pal., ed. 2, 1: 298 (1932) non L. Sp. Plant., ed. 1, 721 (1753).]

TYPE: Sicily; " in collibus, et planis aridis subarenosis Siciliae meridionalis; *Castel vetrano, Marsala, Mazzara, Selinunte, da Vittoria a Siracusa* " (Fl, ? NAP).

Erect annual 5–25 (–30) cm. high, ± densely clothed with long spreading hairs (at first silvery, becoming rusty-brown in dried specimens); stems unbranched or commonly branched from the base; leaves numerous, (5–) 7-foliolate, 3–7 cm. diam., leaflets oblanceolate or narrowly obovate 1–8 cm. long, 0·5–1·3 cm. wide, obtuse or subacute with a tapering base; petiole slender 2–9 cm. long; stipules narrow, caudate above, 1–2 cm. long; inflorescence 3–5 (–7) cm. long, often overtopped by foliage, flowers alternate or irregularly verticillate, often rather crowded; peduncle very short, pedicels usually less than 2 mm. long or almost obsolete; bracts linear, about 8 mm. long, 1 mm. wide, bracteoles linear, about 4 mm. long, 1 mm. wide, adnate to base of calyx; flowers bright blue; upper (adaxial) lip of calyx about 8 mm. long, divided into 2 acuminate teeth about 6 mm. long, and 2 mm. wide at base, lower (abaxial) lip about 10 mm. long, divided into 3 linear teeth 4–5 mm. long, 1 mm. wide at base; standard obovate, 10–13 mm. long, 5–6 mm. wide, glabrous, with 2 prominent longitudinal pleats, wings glabrous, about 12–14 mm. long, 5 mm. wide, glabrous, transversely rugose towards base, keel about 12 mm. long, 3 mm. wide, with an acuminate, up-curved apex; stamens alternately about 9 mm. and 11 mm. long, filaments connate for about 6 mm., slightly swollen just below anther; anthers of longer stamens minute, about 0·2 mm. long, 0·1 mm. wide, anthers of shorter stamens narrowly oblong, about 1·2 mm. long, 0·3 mm. wide; ovary densely hairy, about 5–6 mm. long and 1–1·5 mm. wide, style glabrous, up-curved, about 5 mm. long, stigma barbellate. Pod oblong, acute, 2·5–5 cm. long, 1–1·3 cm. wide, strongly compressed laterally, densely clothed with spreading hairs; seeds 2–4 in each pod, oblong-orbicular, flattened, about 6 mm. long and 5 mm. wide; testa pale creamy-brown, blotched darker brown, smooth.

HAB.: Amongst *Cistus* on wooded mountainsides; in cultivated fields; often on igneous formations; 100–3,000 ft. alt.; fl. March–May.

DISTR.: Divisions 2, 4, 6, common in Division 2. Widespread in the Mediterranean area; Portugal.

2. Common; *Kotschy* 223, 740 ! *Syngrassides* 1592 ! *F. J. F. Barrington* 51 ! *Meikle* 2242 !
4. Near Avgorou, 1862, *Kotschy* 88; Sotira, 1905, *Holmboe* 409.
6. Kormakiti–Liveras Road, 1954, *Merton* 1879 !

2. L. angustifolius *L.*, Sp. Plant., ed. 1, 721 (1753); Unger et Kotschy, Die Insel Cypern, 372 (1865); Boiss., Fl. Orient., 2: 28 (1872); Holmboe, Veg. Cypr., 101 (1914); Post, Fl. Pal., ed. 2, 1: 298 (1932): Davis, Fl. Turkey, 3: 39 (1970); Zohary, Fl. Palaest., 2: 43, t. 57 (1972); Osorio-Tafall et Seraphim, List Vasc. Plants Cyprus, 55 (1973).

TYPE: Locality not designated.

Erect annual or biennial 10–50 cm. high, stems unbranched or with numerous suberect or ascending, adpressed-pubescent branches; leaves numerous, (5–) 7-foliolate, 2–7 cm. diam., leaflets narrowly oblong or ligulate, 10–50 mm. long, 1–6 mm. wide, obtuse or rounded at apex, with a tapering base, upper surface glabrous or subglabrous, dark bluish-green, lower surface thinly adpressed-pubescent; petiole slender, 0·8–5 cm. long, adpressed-pubescent, channelled above; stipules linear-caudate, 0·5–1 cm. long; inflorescence usually a rather lax raceme, 5–10 cm. long, overtopping the foliage, or flowers occasionally solitary, axillary, or in short terminal heads; peduncle very short, flowers irregularly disposed in racemes, pedicels pubescent, usually less than 2·5 mm. long; bracts elliptical, 5–8 mm. long, 3–4 mm. wide, thinly adpressed-pilose externally, caducous, bracteoles narrowly obovate or oblanceolate, 1–2 mm. long, 0·5–0·8 mm. wide, adnate to upper part of calyx; flowers bright purplish-blue with a paler keel; upper (adaxial) lip of calyx 3–4 mm. long, divided almost to base into 2 acute teeth about 2 mm. long and 1·5 mm. wide at base, lower (abaxial) lip about 7 mm. long, shortly 3-toothed at apex; standard ovate, subacute, about 14 mm. long and 8 mm. wide, glabrous with 2 marked longitudinal pleats and reflexed margins, wings glabrous, about 13 mm. long, 6 mm. wide, distinctly rugose towards base, keel about 13 mm. long, 3 mm. wide, with an acuminate, dark-tipped, upcurved apex; stamens alternately about 10 mm. and 12 mm. long, filaments connate for about 6 mm., not distinctly swollen below anther; anthers of longer stamens minute, about 0·3 mm. long, 0·2 mm. wide, anthers of shorter stamens narrowly oblong about 1·5 mm. long, 0·3 mm. wide; ovary densely hairy, about 5 mm. long, 1·2 mm. wide, style glabrous, up-curved, about 7 mm. long, stigma barbellate. Pod oblong, acute, often beaked, 4–5 cm. long, 1–1·3 cm. wide, strongly compressed laterally and ± densely adpressed-pubescent; seeds 3–5 in each pod, oblong-orbicular, not much flattened, about 6 mm. long and 5 mm. wide; testa marbled cream, brown and grey, smooth.

HAB.: In arable fields and vineyards, or occasionally in garigue; chiefly on igneous formations; 2,000–4,000 ft. alt.; fl. April–May.

DISTR.: Divisions 2, 3, common in Division 2. Throughout the Mediterranean area.

2. Common; *Kotschy* 733 ! *Kennedy* 545 ! 1844 ! *Davis* 3188 ! *Casey* 718 ! *Meikle* 2724 !

3. Kellaki, 1956, *Merton* 2558 !

3. ARGYROLOBIUM *Eckl. et Zeyh.*

Enum. Plant. Afr. Austr. Extratrop., 184 (1836) nom. cons.

Shrubs and subshrubs, usually clothed with silky hairs; leaves palmately 3-foliolate; stipules free; flowers solitary or in racemes, terminal or leaf-opposed; bracts and bracteoles usually small; calyx 2-labiate, the upper (adaxial) lip 2-fid, the lower (abaxial) 3-fid, petals generally yellow, standard suborbicular, wings oblong, free or shortly connate at apex, keel obtuse, shorter than standard; stamens monadelphous forming a closed staminal tube, or sometimes with the tube split dorsally, or very rarely diadelphous; filaments alternately short and long, anthers of shorter stamens basifixed, those of the longer stamens versatile; style incurved, glabrous, stigma terminal, often oblique. Pod linear, compressed, eglandular, 2-valved, ± septate between the seeds.

About 70 species, chiefly in Africa (especially southern Africa), but with about 12 representatives in the Mediterranean area and eastwards to north-west India. Only one species native in Cyprus.

1. **A. uniflorum** (*Decne.*) *Jaub. et Spach*, Ill. Plant. Orient., 1: 115 (1843); Boiss., Fl. Orient., 2: 32 (1872); Post, Fl. Pal., ed. 2, 1: 299 (1932); Davis, Fl. Turkey, 3:

38 (1970); Zohary, Fl. Palaest., 2: 40, t. 51 (1972); Osorio-Tafall et Seraphim, List Vasc. Plants Cyprus, 55 (1973).

Cytisus uniflorus Decne. in Ann. Sci. Nat., ser. 2, 3: 265 (1835).

Genista uniflora (Decne.) Briquet, Études Cytises Alp. Mar., 188 (1894).

TYPE: Sinai, June 1832, *Bové* 197 (P, K !).

Small suberect or decumbent shrub, 5–20 cm. high, with a long, tough taproot; stems and leaves thinly or densely clothed with adpressed silvery hairs; leaflets oblong, oblanceolate or obovate, 4–8 (–10) mm. long, 1·5–3 mm. wide, tapering to base from an acute or subacute apex; petiole generally less than 5 mm. long; stipules linear, 0·5–1·3 mm. long; flowers solitary, yellow, leaf-opposed; pedicel very short, about 1 mm. long, densely hairy; bracts minute, narrowly ovate, about 0·5 mm. long, hairy externally, bracteoles barely visible, adnate to lower part of calyx; calyx 4–5 mm. long, 2·5–3 mm. wide, adpressed-hairy externally, deeply 2-lobed, upper (adaxial) lip with 2 narrowly lanceolate-deltoid teeth, lower (abaxial) lip unequally 3-fid, the marginal teeth narrowly deltoid, about 1 mm. long, the median tooth about 0·5 mm. long; standard oblong, obtuse, about 5 mm. long, 4 mm. wide, margins infolded over wings and keel, wings oblong, obtuse, about 4·5 mm. long, 1·3 mm. wide, auriculate and inconspicuously rugulose towards base, keel obtuse, glabrous, about 5 mm. long, 1·5–2 mm. wide, stamens about 5 mm. long, subequal, filaments connate for about half their length, generally split dorsally almost to base, or sometimes diadelphous; anthers of shorter stamens about 0·5 mm. long, 0·2 mm. wide, those of longer stamens about 0·2 mm. long and wide; ovary about 4 mm. long, 0·8 mm. wide, densely hairy; style about 2 mm. long, glabrous, stigma terminal, capitellate, papillose. Pod grey-brown, adpressed-hairy, linear, straight or slighty curved, 15–25 mm. long, 2·5 mm. wide, acute, distinctly torulose; seeds suborbicular, compressed, about 2 mm. diam.; testa yellow or brown, sometimes blotched black, smooth.

HAB.: On exposed limestone rocks and amongst coastal scrub; about sea-level; fl. March–April.

DISTR.: Division 6, rare. Also Turkey, Syria, Lebanon, Palestine, Sinai, Egypt and westwards to Algeria.

6. Near Syrianokhori, 1956, *Poore* in *Merton* 2629 ! Ayia Irini, 1962, *Meikle* 2386 !

NOTES: The Cyprus specimens are suberect and compact, with rather short internodes; they belong, very probably, to f. *compactum* Cavara (Bull. Ort. Bot. Nap., 9: 45; 1928), the type of which was found in Cyrenaica. The taxonomic significance of this form is, however, questionable. Briquet (Études Cytises Alp. Mar., 118–121) admits to the genus *Argyrolobium* only those species with the staminal tube deeply split dorsally, referring all others to *Genista* sections *Lotoides* (DC.) Briq. and *Eremolobium* (Spach) Briq. In doing so he appears to have overlooked the fact that the staminal tube of *A. uniflorum* is normally split dorsally, or that the androecium may even be diadelphous.

4. CALYCOTOME *Link*

in Schrad., Neues Journ., 2, 2: 50 (1808)—as *Calicotome*.

Spiny shrubs with divaricate, sulcate branches; leaves palmately 3-foliolate, stipules inconspicuous; flowers yellow, arising in clusters from short lateral spurs; bracts and bracteoles connate at apex of pedicel; calyx circumscissile, the upper half being pushed off by the expanding corolla and leaving a truncate, membranous basal cup; standard obovate, wings oblong-obovate, shorter than the up-curved keel; stamens monadelphous, forming a closed tube, filaments unequal, the shorter ones with basifixed anthers, the longer ones with smaller, versatile anthers; style glabrous, up-curved, stigma terminal or oblique. Pod linear-oblong, compressed, 2-valved, sutures, especially the dorsal suture, distinctly thickened or winged.

Seven species in the Mediterranean region.

1. C. villosa (*Poir.*) *Link* in Schrad., Neues Journ., 2, 2: 50 (1808); Boiss., Fl. Orient., 2: 36 (1872); Holmboe, Veg. Cypr., 103 (1914); Post, Fl. Pal., ed. 2, 1: 300 (1932); Lindberg f., Iter Cypr., 19 (1946); Chapman, Cyprus Trees and Shrubs, 45 (1949); Davis, Fl. Turkey, 3: 33 (1970); Zohary, Fl. Palaest., 2: 45, t. 59 (1972); Osorio-Tafall et Seraphim, List Vasc. Plant Cyprus, 55 (1973).

Spartium villosum Poir., Voy. en Barbarie, 2: 207 (1789).

Cytisus laniger DC. in Lam. et DC., Fl. Franç., 4: 504 (1805); Unger et Kotschy, Die Insel Cypern, 374 (1865); Sintenis in Oesterr. Bot. Zeitschr., 31: 226, 392, 394 (1881)—as *C. lanigerus*.

TYPE: Algeria–Tunisia; " sur les montagnes de la Numidie " *Poiret* (P).

Erect, rigid shrub to about 3 m. high; twigs spinose, at first densely white-pubescent, later grey-brown and distinctly sulcate or striate; leaves dark green or grey-green, petiole slender, usually less than 15 mm. long, channelled above, pubescent or subglabrous; stipules minute, pubescent; leaflets narrowly obovate or oblanceolate, 4–12 mm. long, 3–5 mm. wide, glabrous or thinly pubescent above, shortly adpressed-pubescent below, apex obtuse or subacute; flowers in loose fascicles along the branches; pedicels densely pubescent, about 3–5 mm. long; bracts and bracteoles united to form an oblong, obscurely 3-lobed, pubescent sheath, 3 mm. long, 2 mm. wide, at the base of the calyx; calyx at first subcylindrical, adpressed-pubescent, about 6 mm. long, 2·5 mm. wide, with 5 short, acute apical teeth, the upper two-thirds falling off as the bud expands, leaving a cup-shaped, truncate base about 2 mm. long; petals yellow, with free claws; standard oblong-obovate, glabrous, about 1·5 cm. long and 0·9 cm. wide, wings narrowly oblong, obtuse, glabrous, about 1·5 cm. long, 0·5 cm. wide, rugulose towards base, keel connate towards apex, basal margins free and ciliate towards the claws, about 1·2 cm. long and 0·3 cm. wide; stamens about 1·2 cm. long, filaments united to form a closed tube about 0·9 cm. long, 1·5 mm. wide; basifixed anthers narrowly oblong, about 1·2 mm. long, 0·5 mm. wide, versatile anthers about 0·5 mm. long and almost as wide; ovary linear, about 10 mm. long, 0·8 mm. wide, densely pilose, shortly stipitate, style upcurved, about 7 mm. long, glabrous, stigma terminal, capitate. Pod narrowly oblong, about 2·5–3·5 cm. long, 0·5–0·7 cm. wide, densely hairy with prominently thickened or slightly winged ventral and dorsal sutures; seeds oblong-orbicular, 2·5–4 mm. long, 2–3·5 mm. wide, distinctly compressed; testa yellow or brownish, smooth.

HAB.: Dry rocky fields and hillsides amongst maquis and garigue; sea-level to 3,600 ft. alt.; fl. Dec.–April.

DISTR.: Divisions 1–8, common in Division 7. Widespread and common in the Mediterranean region.

1. Lyso, 1913, *Haradjian* 881 ! Avgas near Ayios Yeoryios (Akamas), 1962, *Meikle* 2017 ! Common in Ayios Yeoryios and Polis areas, 1962, *Meikle* !
2. Platres, 1936, *Kennedy* 546 ! Kakopetria, 1939, *Lindberg f.* s.n.
3. Near Ayios Varvara, north of Stavrovouni, 1787, *Sibthorp*; Stavrovouni, 1880, *Sintenis & Rigo*.
4. Ayia Napa, 1905, *Holmboe* 61.
5. Near Lefkoniko, 1880, *Sintenis & Rigo*.
6. Dhiorios, 1962, *Meikle* !
7. Common; *Kotschy*; *Sintenis & Rigo* 424 ! *Haradjian* 881 ! *Syngrassides* 19 ! *Kennedy* 547 ! *G. E. Atherton* 901 !
8. Koronia, *Chapman* 12; Peristerges near Kilanemos, 1962, *Meikle* 2428 ! Akradhes, 1962, *Meikle* !

5. GENISTA *L.*

Sp. Plant., ed. 1, 709 (1753).
Gen. Plant., ed. 5, 318 (1754).

Spiny, or occasionally unarmed, shrubs and subshrubs; leaves usually simple, sometimes very much reduced or wanting, rarely 3-foliolate; stipules minute; flowers yellow, in racemes, capitula, fascicles or solitary; bracts and bracteoles generally small, often caducous; calyx campanulate or shortly cylindrical, 2-fid above, shortly 3-fid below; standard obovate-oblong, wings oblong, keel commonly deflexed after pollination, the claws frequently adnate to the base of the staminal tube; filaments forming a closed tube, anthers alternately long and basifixed and short and versatile; ovary sessile or subsessile, 2–many-ovulate; style upcurved, stigma capitate, terminal or rarely oblique. Pod dehiscing (often explosively) into 2 valves.

1. **G. sphacelata** *Decne.* in Ann. Sci. Nat., ser. 2, 4: 360 (1835) [as *G.* "*fasselata*"]; Boiss., Fl. Orient., 2: 39 (1872), Fl. Orient. Suppl., 159 (1888); Holmboe, Veg. Cypr., 102 (1914); Post, Fl. Pal., ed. 2, 1: 302 (1932) [as *G.* "*fasselata*"]; Lindberg f., Iter Cypr., 19 (1946); Chapman, Cyprus Trees and Shrubs, 44 (1949); Zohary, Fl. Palaest., 2: 46, t. 61 (1972) [as *G.* "*fasselata*"]; Osorio-Tafall et Seraphim, List Vasc. Plants Cyprus, 55 (1973) [as *G.* "*fasselata*"].

[*G. acanthoclada* (non DC.) Sintenis in Oesterr. Bot. Zeitschr., 32: 20, 194 (1882).]

[*Ulex europaeus* (non L.) Kotschy in Unger et Kotschy, Die Insel Cypern, 374 (1865).]

Erect or prostrate shrub, typically 1–2 m. high, but much smaller in the montane var. *crudelis*; branches numerous, divaricate, rigid, conspicuously striate, glabrous or glabrescent, dark green with black spinose tips; leaves sparse, simple, oblanceolate or linear, 5–12 mm. long, 1–2·5 mm. wide, acute or subacute, sessile or subsessile, adpressed-strigose, often ± carinate, and commonly reduced to black deltoid scales 1–2 mm. long and about as wide; stipules scarcely developed; flowers solitary, axillary, scattered along the branches; bracts minute, rarely foliose, generally blackish, deltoid, less than 1 mm. long; pedicels 1–1·5 mm. long, pubescent; bracteoles subulate, 0·5–1·5 mm. long, 0·5 mm. wide at base, brownish, pubescent; calyx campanulate, adpressed-pubescent, 2·5–3·5 mm. long, 2–2·5 mm. wide, divided above into 2 deltoid teeth about 1·5 mm. long, and below into 3 subulate teeth 1–1·2 mm. long; petals yellow, standard broadly oblong, 7–9 mm. long, 5–7 mm. wide, narrowed very abruptly to a slender claw, adpressed-pubescent externally, wings oblong, glabrous, 5–8 mm. long, 3–4 mm. wide, claw adnate to basal part of staminal tube, keel straight, blunt, 9–11 mm. long, 2–3 mm. wide, adpressed-pubescent towards apex, claws adnate to base of staminal tube, basal margins minutely pubescent; stamens 6–10 mm. long, staminal tube 3–6 mm. long, ± papillose externally; anthers subequal, mucronate, the longer about 1 mm. long and 0·5 mm. wide, the shorter about 0·8 mm. long, 0·5 mm. wide; ovary 2–2·5 mm. long, 0·8–1 mm. wide, adpressed-pubescent, style glabrous 4–6 mm. long, stigma minute, oblique. Pod broadly oblong, flattened, dark brown, beaked, about 7–9 mm. long, 4–6 mm. wide, adpressed-pubescent; seed usually solitary, suborbicular, flattened, about 2·5–3 mm. diam.; testa brown or livid, smooth, funicle well developed, rather thick.

var. **sphacelata**

G. sphacelata Decne. var. *bovilliana* Holmboe, Veg. Cypr., 102 (1914); Lindberg f., Iter Cypr., 19 (1946); Chapman, Cyprus Trees and Shrubs, 45 (1949).

TYPE: Palestine; Mt. Carmel, Sept. 1832, *Bové* 488 (P, K !).

Erect shrub 1–2 m. high; flowers usually about 1 cm. long.

HAB.: Dry rocky fields and hillsides amongst maquis and garigue; sea-level to 3,000 ft. alt.; fl. March–July.

DISTR.: Divisions 1–3, 6–8, common in Division 7. Also Aegean Islands, Lebanon and Palestine.

1. Stroumbi, 1933, *A. Foggie* 283 ! Near Ayios Yeoryios (Akamas), 1962, *Meikle* ! Near Polis, 1962, *Meikle* !
2. Near Lefkara, 1862, *Kotschy*; Mesopotamos Monastery, 1939, *Lindberg f.* s.n. ! Jemali Bridge, near Sarama, 1962, *Meikle* 2769 !

3. Between Episkopi and Pissouri, 1862, *Kotschy*.
6. Dhiorios, 1962, *Meikle* !
7. Common; *Sintenis & Rigo* 539 ! *Kotschy*; *Post* s.n. ! *Syngrassides* 278 ! *Kennedy* 549 ! *G. E. Atherton* 83 ! 133 !
8. Peristerges near Kilanemos, 1962, *Meikle* 2432 !

NOTES: Type material of *G. sphacelata* from Mt. Carmel is in no way distinct from the plant described as var. *bovilliana* by Holmboe (Veg. Cypr., 102), and his statement that the dwarf variety from Troödos represents true *sphacelata* is not correct; it is in fact this montane form which requires to be distinguished.

var. **crudelis** *Meikle* in App. III, p. 806.

TYPE: Cyprus; above Amiandos, 5,000 ft. alt., June 21, 1935, *A. Foggie* 361 (K !).

A dense, dwarf shrub, 5–17 cm. high, forming large mats or hemispherical cushions, with intricately branched, rigidly spinose branches; flowers often (but not always) rather smaller than those of var. *sphacelata*, commonly less than 1 cm. long.

HAB.: On bare rocky mountainsides, sometimes under *Pinus*; 3,900–5,500 ft. alt.; fl. June–July.

DISTR.: Division 2, common. Possibly endemic.

2. Common; *Holmboe*; *Haradjian* 494 ! *A. Foggie* 361 ! *Syngrassides* 731 ! *C. Wyatt* 18 ! *Kennedy* 548 !

NOTES: Although normally very distinct, var. *crudelis* tends to approach typical *G. sphacelata*, especially at the lower limits of its altitudinal range; it would be interesting to know if it retains its dwarf habit when cultivated at low altitudes.

GENISTA TINCTORIA *L.*, a rhizomatous, unarmed shrub, 30–70 cm. high, with oblong-lanceolate leaves and flowers in terminal racemes, is said to have been planted at Salamis (Bovill, Rep. Plant. Work Cyprus, 14; 1915) but probably has not survived. It is a common plant in many parts of Europe and western Asia.

TRIBE 3. **Trifolieae** (*Bronn*) *Benth.* Herbs or rarely subshrubs, leaves pinnately, or sometimes digitately 3-foliolate, rarely 1-foliolate or 5–7-foliolate, leaflets usually serrulate or denticulate with excurrent lateral veins; inflorescences axillary or terminal (pseudoterminal), capitate, umbellate or racemose, or flowers rarely solitary or few; stamens diadelphous or rarely monadelphous, free parts of filaments often dilated, anthers generally uniform. Pod inarticulate, 2-valved or indehiscent.

6. ONONIS *L.*

Sp. Plant., ed. 1, 716 (1753).
Gen. Plant., ed. 5, 321 (1754).
Širjaev in Beih. Bot. Centralbl., 49, 2: 381–665 (1932).

Annual or perennial herbs, or rarely shrubs, generally unarmed; stems and leaves usually glandular-pilose, rarely subglabrous; leaves generally petiolate, pinnately 3-foliolate, rarely unifoliolate by reduction or 2–5-foliolate, leaflets ± dentate, usually petiolulate; stipules herbaceous, adnate to the petiole or sometimes connate at base and bifid at apex; inflorescences axillary, mostly racemose with a muticous or aristate peduncle, occasionally subcapitate and sessile; pedicels frequently elongating after anthesis; floral leaves similar to, or markedly distinct from cauline leaves; calyx persistent, campanulate or rarely cylindrical, sometimes accrescent, calyx-teeth subequal or occasionally unequal; petals pink, yellow, white or parti-coloured, often veined or pencilled with a contrasting colour, standard obovate or suborbicular, glabrous or pilose with a short claw, wings obovate-oblong, keel incurved, generally rostrate, shorter than standard; stamens monadelphous, filaments glabrous, often dilated above, anthers alternately short and versatile, longer and basifixed; ovary shortly stipitate, 2–many-ovulate, style upcurved, glabrous, stigma small, capitate, terminal or oblique. Pod linear, oblong or rhomboidal, dehiscing into 2 valves, usually continuous within, rarely septate between the seeds and ± torulose; seeds reniform or suborbicular, smooth, punctate or tuberculate, rarely rugose.

About 70 species in Europe, western Asia, North Africa and the Atlantic Islands.

Inflorescence or flower distinctly pedunculate:
  Corolla yellow, yellowish or creamy-white:
    Shrubby perennials - - - - - - - - - [O. NATRIX p. 385]
    Herbaceous annuals:
      Pod torulose - - - - - - - - - **1. O. ornithopodioides**
      Pod not torulose:
        Peduncle distinctly aristate; pods oblong:
          Pod 20–25 mm. long, 5–7 mm. wide; corolla distinctly longer than calyx; flowers frequently in pairs - - - - - - **2. O. biflora**
          Pod 8–17 mm. long, 3–4 mm. wide; corolla equalling or shorter than calyx; flowers usually solitary:
            Keel sharply upcurved to a beak as long as or longer than the basal part; peduncle clothed with long spreading hairs; plant 15–40 cm. high **3. O. viscosa ssp. breviflora**
            Keel upcurved to a beak shorter than the basal part; peduncle shortly glandular-pubescent; plant 7–15 (–30) cm. high - **4. O. sicula**
        Peduncle exaristate, or arista very short; pods ovoid or ellipsoid **6. O. pubescens**
  Corolla pink or pink and white, peduncle exaristate; flower and fruit strongly reflexed **5. O. reclinata**
Inflorescence or flower sessile or subsessile:
  Corolla yellow or yellowish:
    Leaves 3-foliolate; subshrubby perennial - - - - **7. O. pusilla**
    Leaves 1-foliolate; maritime annual - - - - - - **9. O. variegata**
  Corolla pink or pink and white:
    Subshrubby spinescent perennials - - - **8. O. spinosa ssp. leiosperma**
    Herbaceous annuals:
      Cauline leaves 3-foliolate; calyx shortly glandular-pubescent or subglabrous:
        Standard glandular or hairy externally; stipules of floral leaves not whitish or conspicuous:
          Inflorescence a dense oblong spike; standard obovate-orbicular, 6–7 mm. wide; leaflets with 12–24 (–30) small teeth - - - **10. O. diffusa**
          Inflorescence a slender, rather lax spike; standard oblong-obovate, 4–5 mm. wide; leaflets with 10–20 conspicuous spreading teeth **11. O. serrata**
        Standard glabrous externally; stipules of floral leaves whitish and conspicuous **12. O. mitissima**
      Cauline leaves 1-foliolate; calyx with long spreading hairs [O. ALOPECUROIDES p. 400]

SECT. 1. **Natrix** *DC. emend. Griseb.* Inflorescence pedunculate, 1–2 (–3)-flowered. Pod linear, oblong or ovoid-oblong, reflexed, often many-seeded.

SUBSECT. 1. **Natrix** *Willk. emend. Širjaev* (as eu-Natrix) Subshrubs or perennial herbs; leaves (1–) 3 (–5–7)-foliolate; floral leaves 1–3-foliolate, rarely bract-like; corolla yellow, standard striped red or violet, glabrous, wings with a conjunctival tooth. Seed minutely punctate or tuberculate.

[O. NATRIX *L.* ssp. HISPANICA (*L. f.*) *P. Cout.* A bushy, much-branched subshrub, to about 60 cm. high; branches ascending or suberect, shortly glandular; leaves rather rigid, generally 3-foliolate, usually small, about 2–5 (–8) mm. long, ovate or ovate-oblong, or narrowly obovate, with a blunt or rounded apex; peduncles 1-flowered; calyx shortly glandular-pilose, with linear, 3-nerved teeth, usually 3–4 times longer than the tube; corolla yellow,

6–11 (–15) mm. long, glabrous. Pod linear, ± compressed, glandular-pilose, about 6–11 (–17) mm. long; seeds globose, fuscous, minutely tuberculate.

Recorded—as *O. crispa* L.—from Cyprus by Sibthorp (Fl. Graec. Prodr., 2: 58; 1813, Fl. Graec., 7: 74, t. 680; 1830), but not re-found by any other collector, and most probably an error of localization. Stuart Thompson's record for *O. natrix*, " above Dhikomo [" Dikoma "] *Lascelles* " (Journ. Bot., 44: 305; 1906) is an error, the plant collected being *O. pubescens* L. *O. natrix*, a widespread Mediterranean species, occurs in several countries adjacent to Cyprus, and may yet be re-discovered on the island, though, in view of its conspicuousness, it is not likely to have been overlooked.]

SUBSECT. 2. **Torulosae** *Širjaev* Annuals; leaves all 3-foliolate; peduncles 1–2-flowered, aristate; corolla yellow, wings with a conjunctival tooth. Pods torulose; seeds tuberculate.

1. O. ornithopodioides *L.*, Sp. Plant., ed. 1, 718 (1753); Sibth et Sm., Fl. Graec. Prodr., 2: 58 (1813), Fl. Graec., 7: 73, t. 679 (1830); Unger et Kotschy, Die Insel Cypern, 373 (1865); Boiss., Fl. Orient., 2: 59 (1872); Holmboe, Veg. Cypr., 103 (1914); Post, Fl. Pal., ed. 2, 1: 307 (1932); Širjaev in Beih. Bot. Centralbl., 49, 2: 486 (1932); Davis, Fl. Turkey, 3: 377 (1970); Zohary, Fl. Palaest., 2: 116, t. 169 (1972); Osorio-Tafall et Seraphim, List Vasc. Plants Cyprus, 58 (1973).

TYPE: Sicily; " *ad Capo passaro* ".

Erect annual 8–15 (–30) cm. high, usually much branched, with erect or ascending branches; stems slender, glandular-pilose; leaves dull green, thinly viscid-glandular; petioles slender, 0·5–2 cm. long; stipules herbaceous, auriculate, acute, 2–5 mm. long, 1·5–3 mm. wide; leaflets ovate or ovate-oblong, acute or obtuse, 0·5–2 cm. long, 0·3–1·3 cm. wide with a rounded or broadly cuneate base, venation distinct, extending into margin of leaflets; terminal leaflet usually larger than the laterals, with a petiolule 4–10 mm. long; inflorescences spaced along the greater length of the branches; peduncle 0·5–1·5 cm. long, elongating in fruit; pedicels curved, 1–2 mm. long; calyx 7–9 mm. long, glandular-pilose, the teeth linear-filiform, 4–6 times as long as the very short, campanulate tube; petals yellow, standard broadly obovate, 6–7 mm. long, 5–6 mm. wide, with a short claw, glabrous, usually embracing wings and keel, apex rounded, apiculate or muticous, wings oblong, 5–6 mm. long, 1·5–2 mm. wide, apex rounded, base distinctly clawed with a conspicuous auricle and conjunctival tooth, keel 5–6 mm. long, 2–3 mm. wide, apex almost vertical, shortly and bluntly beaked; stamens 5–6·5 mm. long, staminal tube glabrous 4–5 mm. long; anthers oblong, subequal 0·2–0·3 mm. long, 0·2 mm. wide; ovary shortly stipitate, about 4 mm. long, 0·5 mm. wide, densely glandular-pubescent, style glabrous, upcurved, about 2·5 mm. long, stigma terminal, minute, capitate. Pod linear, 1·5–2·8 cm. long, 0·2–0·3 cm. wide, strongly compressed and torulose, generally subfalcate, brownish, thinly glandular-pubescent, with an abruptly acute apex; seeds subglobose, c. 1·3 mm. diam.; testa brown, acutely scabridulous-tuberculate.

HAB.: Rocky hillsides, pathsides, waste ground and open places in Pine forest, on limestone and sandstone; 300–2,500 ft. alt.; fl. March–May.

DISTR.: Divisions 1–5, 7, 8, common in Division 7. Widespread in the Mediterranean region.

1. Baths of Aphrodite, 1962, *Meikle* 2223 !
2. N. of Kykko Monastery, 1949, *Oswald* 107 !
3. Episkopi, 1960, *N. Macdonald* 42 !
4. Akhna Forest, 1952, *Merton* 819 !
5. Between Ovgoros and Trikomo, 1941, *Davis* 2482 !
7. Common; *Sibthorp*; *Sintenis & Rigo* 407 ! *Davis* 2860 ! *Casey* 1303 ! *Merton* 3151 !
8. Valia, 1905, *Holmboe* 485.

SUBSECT. 3. **Biflorae** *Širjaev* Annuals; leaves 3-foliolate; peduncles well developed, (1–) 2-flowered; corolla yellow, glabrous; wings with a conjunctival tooth. Pod not torulose.

2. O. biflora *Desf.*, Fl. Atlant., 2: 143 (1800); Unger et Kotschy, Die Insel Cypern, 373 (1865); Boiss., Fl. Orient., 2: 60 (1872); Holmboe, Veg. Cypr., 103 (1914); Post, Fl. Pal., ed. 2, 1: 308 (1932); Širjaev in Beih. Bot. Centralbl., 49, 2: 489 (1932); Davis, Fl. Turkey, 3: 377 (1970); Zohary, Fl. Palaest., 2: 118, t. 172 (1972); Osorio-Tafall et Seraphim, List Vasc. Plants Cyprus, 58 (1973).
TYPE: North Africa; " in Barbaria " (P).

Erect annual, 10–40 (–50) cm. high, unbranched or with numerous erect or ascending branches; stems thinly glandular-pubescent; leaves dull green, thinly glandular; petioles flattened, ± winged, 1–1·5 cm. long; stipules large, herbaceous, acute or acuminate, 8–15 mm. long, 3–5 mm. wide, adnate to the petiole for the greater part of their length; leaflets thinly glandular, oblong or narrowly obovate, 1–2·5 cm. long, 0·5–1·5 cm. wide, irregularly serrate, with a rounded or subtruncate apex and a rounded or broadly cuneate base, venation rather indistinct; terminal leaflet a little larger than the laterals, with a distinct petiolule 3–10 mm. long; inflorescences spaced along the upper parts of the stems; peduncle 1–3 (–4) cm. long, elongating a little in fruit, 2-flowered, or often 1-flowered in Cyprus specimens, with a filiform arista about 5–10 mm. long; pedicels at first ± straight, later curved, about 3–5 mm. long; calyx 9–12 mm. long, glandular-pubescent, the teeth linear-filiform, 6–9 mm. long, much longer than the short narrowly campanulate tube; petals creamy-white or pale yellow, the keel tipped brown or reddish, standard short-clawed, suborbicular, 1·3–1·7 cm. long, 1–1·3 cm. wide, apex rounded or emarginate, often shortly apiculate, wings oblong, 1·3 cm. long, about 0·5 cm. wide towards the rounded apex, basal auricle and conjunctival tooth conspicuous, keel about 1·4 cm. long, 0·6 cm. wide, with the sharply rostrate apex curved almost at a right angle to the base; stamens about 1·3 cm. long, staminal tube glabrous, about 1 cm. long; anthers narrowly oblong, subequal, about 0·8 mm. long, 0·3 mm. wide; ovary linear, 3–4 mm. long, about 1 mm. wide, with a distinct stipe about 1·5–2 mm. long, ± densely glandular; style glabrous, about 6 mm. long, stigma small, terminal. Pod narrowly oblong, blunt or shortly acute, somewhat inflated, 2–2·5 cm. long, 0·5–0·7 cm. wide, valves pale brown, thinly glandular; seeds reniform, compressed, about 2·5 mm. wide and almost as long; testa fuscous or brownish, bluntly tuberculate.

HAB.: In cultivated fields and on waste ground; sea-level to 2,000 ft. alt.; fl. March–April.

DISTR.: Divisions 1–4, 6, 7, common in Divisions 3, 7. Spain, Sardinia, Italy, S. Turkey, Syria, Palestine, Iraq, North Africa.

1. Neokhorio (Akamas), 1962, *Meikle* 2190 !
2. Asinou valley, Troödos, 1953, *Kennedy* 1779 ! Near Vouni Palace, 1974, *Meikle* 4020 !
3. Common; *Kotschy* S.1006; *Holmboe* 221; *Davis & Kennedy* 2721 ! *Merton* 2039 !
4. Larnaca, 1859, *Kotschy* S.1006; near Old Larnaca, 1880, *Sintenis & Rigo* 403 ! Larnaca 1892–94, *Deschamps* 105 (see note); Phaneromene near Larnaca, 1905, *Holmboe* 137.
6. Peristerona (Morphou), 1932, *Syngrassides* 97 ! Kykko (Nicosia) 1936, *Syngrassides* 1163 ! Near Nicosia, 1937, *Syngrassides* 1411.
7. Common; *Sintenis & Rigo* 403A ! *Syngrassides* 1023 ! *Casey* 412 ! 1000 ! *G. E. Atherton* 1265 !

NOTES: Širjaev (l.c., 490) records *O. biflora* Desf. var. *maroccana* (Batt. et Pitard) Joh. et Maire on the strength of a specimen (*Deschamps* 105) collected at Larnaca. This specimen has not been examined, but, to judge from other material, the variety is of doubtful taxonomic value, being little more than a robust, large-flowered and large-fruited race of *O. biflora*; some of the Cyprus material is very luxuriant, but none examined equals North African var. *maroccana* in size of flower and fruit.

SUBSECT. 4. **Viscosae** *Širjaev emend. Meikle* Annuals; lower leaves 3-foliolate, upper sometimes 1-foliolate; peduncles aristate, 1 (–2)-flowered; corolla yellow, standard glabrous, wings with or without a conjunctival tooth. Pod oblong, not torulose; seeds tuberculate.

3. O. viscosa *L.*, Sp. Plant., ed. 1, 718 (1753); Širjaev in Beih. Bot. Centralbl., 49, 2: 521 (1932).

Viscid-pubescent annual, 15–45 cm. high; branches erect or ascending; leaves bright green, the lower usually 3-foliolate, the upper often 1-foliolate, sometimes all 1-foliolate; petiole about 0·5–2 cm. long; stipules large, 1–2 cm. long, 0·2–0·5 cm. wide, adnate to petiole for more than half their length, apex acuminate, entire or finely denticulate; leaflets large, oblong or obovate-oblong, 1·5–4 cm. long, 0·5–3 cm. wide, apex rounded or acute, denticulate except towards base; inflorescences spaced along the greater length of the branches; peduncles slender, longer or shorter than the subtending leaves, usually clothed with long spreading hairs, 1-flowered, with a filiform arista often more than 1 cm. long, pedicels well developed, 3–8 mm. long, at first erect or horizontal, later curved downwards; calyx glandular-pilose, 7–12 mm. long, teeth linear or narrowly lanceolate, much longer than the calyx-tube; corolla shorter or longer than calyx, standard often pencilled reddish, suborbicular, emarginate or apiculate, wings oblong or ovate-oblong, obtuse, with a prominent basal auricle and conjunctival tooth; keel equalling or shorter than wings, sharply upcurved to a prominent acute beak, upcurved part often as long as or longer than basal part. Pod oblong, pale brown, glandular-pubescent, slightly inflated, similar to that of *O. sicula* Guss.; seeds suborbicular; testa tuberculate.

ssp. **breviflora** (*DC.*) *Nyman*, Consp. Fl. Europ., 161 (1878); Širjaev in Beih. Bot. Centralbl., 49, 2: 524 (1932); Davis, Fl. Turkey, 3: 379 (1970); Zohary, Fl. Palaest., 2: 116, t. 169 (1972); Osorio-Tafall et Seraphim, List Vasc. Plants Cyprus, 58 (1973).

O. breviflora DC., Prodr., 2: 160 (1825); Boiss., Fl. Orient., 2: 60 (1872); Holmboe, Veg. Cypr., 103 (1914); Post, Fl. Pal., ed. 2, 1: 307 (1932); Lindberg f., Iter Cypr., 20 (1946); Rechinger f. in Arkiv för Bot., ser. 2, 1: 424 (1950).

O. viscosa L. var. β L., Sp. Plant., ed. 2, 1009 (1763).

TYPE: " Monspelii *&* *in* Hispania ".

Leaflets obtuse, rounded or retuse; peduncles equalling or often shorter than the leaves, arista very long, often 3–7 times as long as pedicel; calyx-teeth linear-subulate, tapering to a very slender apex; corolla distinctly shorter than calyx; standard oblong-orbicular, about 8–9 mm. long, 6 mm. wide, wings broadly oblong, 6–7 mm. long, 3–4 mm. wide, apex rounded, base with a conspicuous auricle and conjunctival tooth; keel 5–6 mm. long, up-curved at right angles to a sharp, prominent beak as long as, or often longer than the basal part; stamens 6 mm. long, staminal tube glabrous, 3–3·5 mm. long; anthers unequal, about 0·5 or 0·8 mm. long and 0·4 mm. wide; ovary oblong, 3–4 mm. long, 1–1·5 mm. wide, glandular-pubescent, style glabrous 3–4 mm. long, stigma small, terminal. Pod oblong (8–) 12–16 mm. long, 4 mm. wide, glandular-pubescent; seeds orbicular-reniform, 2 mm. diam.; teata dull brown, densely covered with small tubercles. *Plate 19, figs. 8–10.*

HAB.: In cultivated and fallow fields, by roadsides, or on rocky hillsides; sea-level to 2,000 ft. alt.; fl. March–June.

DISTR.: Divisions 1–7, common in Divisions 2, 7. Throughout the Mediterranean region and eastwards to Iraq.

1. Ayios Neophytos Monastery, 1939, *Lindberg f.* s.n.; Dhrousha, 1941, *Davis* 3228 !
2. Common; *Davis* 3410 ! 3432 ! *Merton* 2900 ! *N. Macdonald* 166 !
3. Skarinou, 1941, *Davis* 2686 ! Near Souni, 1960, *N. Macdonald* 44 !

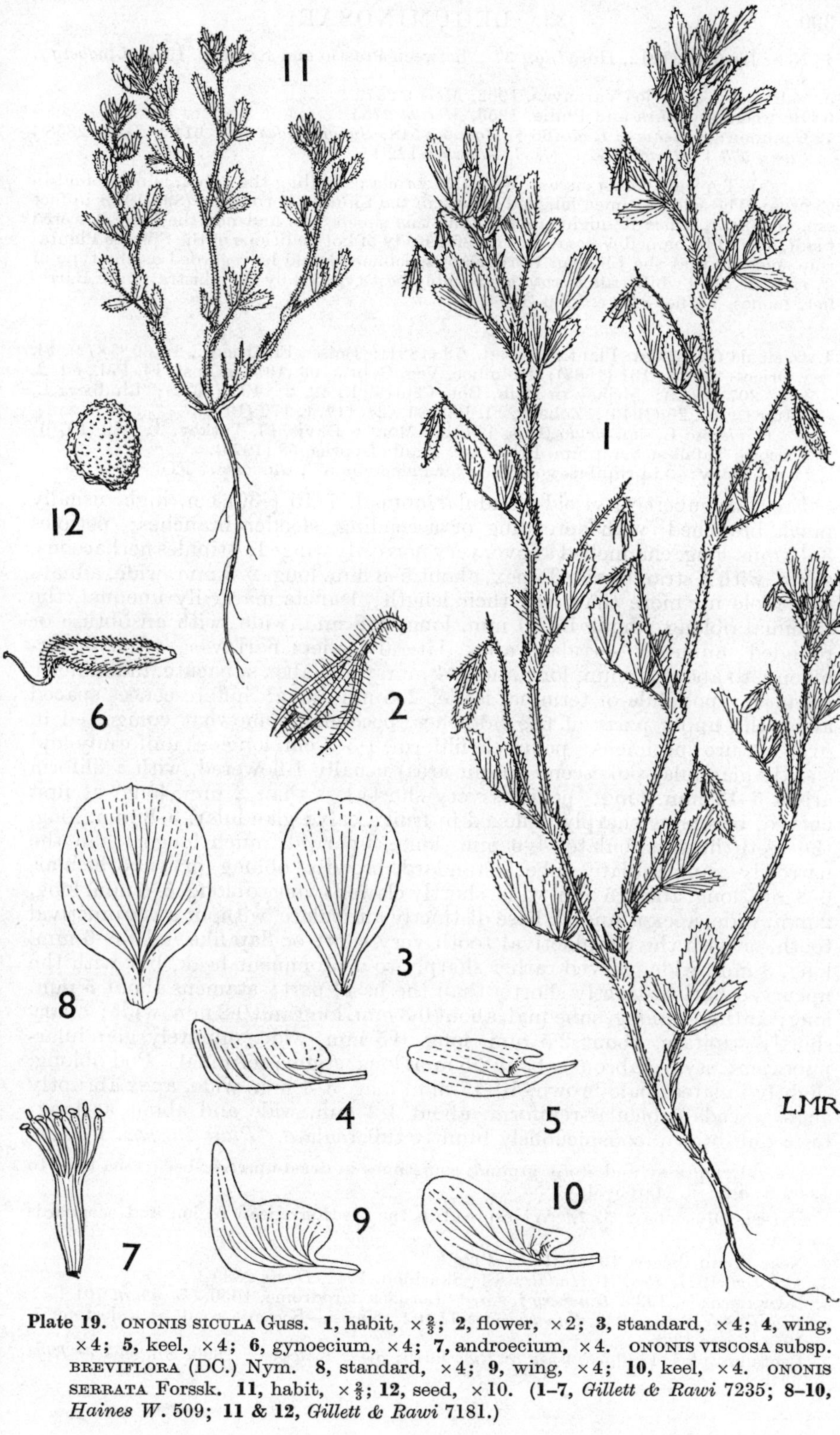

Plate 19. ONONIS SICULA Guss. **1,** habit, $\times\frac{2}{3}$; **2,** flower, $\times 2$; **3,** standard, $\times 4$; **4,** wing, $\times 4$; **5,** keel, $\times 4$; **6,** gynoecium, $\times 4$; **7,** androecium, $\times 4$. ONONIS VISCOSA subsp. BREVIFLORA (DC.) Nym. **8,** standard, $\times 4$; **9,** wing, $\times 4$; **10,** keel, $\times 4$. ONONIS SERRATA Forssk. **11,** habit, $\times\frac{2}{3}$; **12,** seed, $\times 10$. (**1–7,** *Gillett & Rawi* 7235; **8–10,** *Haines W.* 509; **11 & 12,** *Gillett & Rawi* 7181.)

4. Near Larnaca, 1912, *Haradjian* 37 ; between Prastio and Kouklia, 1939, *Lindberg f.* s.n.
5. Salamis, near Ayios Varnavas, 1962, *Meikle* 2579 !
6. Between Skylloura and Philia, 1956, *Merton* 2755 !
7. Common; *Sintenis & Rigo* 406 ! *Holmboe* 548; *Syngrassides* 58 ! 614 ! *Davis* 2858 ! *Casey* 577 ! *Merton* 557 ! *G. E. Atherton* 122 !

NOTES: Typical *Ononis viscosa*, with the corolla exceeding the calyx, is not found in Cyprus. The sole specimen labelled *viscosa* in the Linnean herbarium (896/19) is in fact ssp. *breviflora*, hence it might appear that this subspecies, and not the large-flowered plant, should be named *viscosa*; but careful study of both editions of the Species Plantarum suggests that the Linnean Herbarium specimen should be regarded as the type of *O. viscosa* var. β while true *viscosa* is probably best typified by the illustration in Barrelier, Icones, 75, no. 840, t. 1239 (1714).

4. O. sicula *Guss.*, Cat. Plant., Boccad., 78 (1821); Boiss., Fl. Orient., 2: 60 (1872), Fl. Orient. Suppl., 161 (1888); Holmboe, Veg. Cypr., 103 (1914); Post, Fl. Pal., ed. 2, 1: 307 (1932); Širjaev in Beih. Bot. Centralbl., 49, 2: 496 (1932); Lindberg f., Iter Cypr., 20 (1946); Zohary, Fl. Palaest., 2: 117, t. 170 (1972).

O. viscosa L. ssp. *sicula* (Guss.) Huber-Mor. in Davis, Fl. Turkey, 3: 380 (1970); Osorio-Tafall et Seraphim, List Vasc. Plants Cyprus, 58 (1973).

TYPE: Sicily; " in rupibus volcanicis ad Palagoniam ", *Gussone* (? FI).

Erect or suberect, viscid-glandular annual, 7–15 (–30) cm. high, usually much branched, with spreading or ascending, slender branches; petioles 3–10 mm. long, channelled above, very narrowly winged; stipules herbaceous, acute, with a straight-sided apex, about 5–8 mm. long, 2–3 mm. wide, adnate to petiole for more than half their length; leaflets markedly unequal, the terminal oblong, about 15–20 mm. long, 5–8 mm. wide, with an obtuse or rounded, minutely serrulate apex, lateral leaflets narrower, often linear-oblong, to about 15 mm. long and 3–4 mm. wide, apex subacute, indistinctly serrulate; petiolule of terminal leaflet 2–5 mm. long; inflorescences spaced along the upper parts of the branches, becoming somewhat comgested in over-mature specimens; peduncle filiform, 1·5–3 cm. long, ± uniformly and shortly glandular-pubescent (in our area) usually 1-flowered, with a filiform arista 5–13 mm. long; pedicels very short, less than 2 mm. long, at first curved, becoming sharply deflexed in fruit; calyx glandular, 5–8 mm. long, the teeth linear-subulate, 4–6 mm. long, 3-nerved, much longer than the narrowly campanulate tube; standard broadly oblong or suborbicular, 6–8 mm. long, about 5 mm. wide, shortly clawed, wings oblong, 5–6 mm. long, 2 mm. wide, apex rounded, base distinctly auriculate, without a conjunctival tooth, or with the conjunctival tooth very short or flap-like, keel 5–6 mm. long, 3 mm. wide, curved rather sharply to a prominent beak, but with the upcurved part distinctly shorter than the basal part; stamens about 5 mm. long; anthers oblong, subequal, about 0·3 mm. long and 0·2 mm. wide; ovary shortly stipitate, about 2·5 mm. long, 0·5 mm. wide, minutely glandular-pubescent, style glabrous, about 2·5 mm. long, stigma terminal. Pod oblong, slightly inflated, pale brown, 10–17 mm. long, 3–4 mm. wide, apex abruptly acute; seeds orbicular-reniform, about 1·2 mm. wide and about as long; testa dull brown, conspicuously bluntly tuberculate. *Plate 19, figs. 1–7.*

HAB.: Dry grassy and stony ground, sometimes in dried-up river-beds; sea-level to 1,000 ft. alt.; fl. March–May.

DISTR.: Divisions 2–5, 7. Widespread in the Mediterranean region and eastwards to Iran.

2. Near Vouni Palace, 1974, *Meikle* 4022 !
3. Akrotiri, 1927, *Rev. A. Huddle* 48 ! Skarinou, 1941, *Davis* 2689 !
4. Near Larnaca, 1939, *Lindberg f.* s.n. ! Larnaca aerodrome, 1950, *Chapman* 701 !
5. Nisou near Pera, 1936, *Syngrassides* 1131 ! Nicosia–Kyrenia road, at milestone 7, 1954, *Casey* 1326 !
7. Pentadaktylos, 1880, *Sintenis & Rigo* 994b; near Kalogrea, 1880, *Sintenis & Rigo* 994h !

SUBSECT. 5. **Reclinatae** *Širjaev emend. Meikle* Annuals; leaves mostly 3-foliolate; peduncles exaristate, 1-flowered; corolla pink or pinkish, standard glabrous, wings without a conjunctival tooth. Pods not torulose; seeds tuberculate.

5. O. reclinata *L.*, Sp. Plant., ed. 2, 1011 (1763); Boiss., Fl. Orient., 2: 61 (1872); Holmboe, Veg. Cypr., 103 (1914); Post, Fl. Pal., ed. 2, 1: 308 (1932); Širjaev in Beih. Bot. Centralbl., 49, 2: 501 (1932); Davis, Fl. Turkey, 3: 378 (1970).

[*O. cherleri* (non L.) Sibth. et Sm., Fl. Graec. Prodr., 2: 57 (1813), Fl. Graec., 7: 72, t. 677 (1830); Unger et Kotschy, Die Insel Cypern, 373 (1865).]

Erect viscid-glandular annual 3–15 cm. high, often bushy and much-branched from near the base; leaves dull green, mostly 3-foliolate, the upper, or rarely all, 1-foliolate; petioles short rarely exceeding 8 mm. long, slightly channelled above; stipules ovate or oblong, 2–2·5 mm. long, 1–1·5 mm. wide, acute or obtuse, sometimes obscurely denticulate towards apex, conspicuously veined; leaflets oblong or narrowly obovate, 3–10 mm. long, 2–6 mm. wide, the terminal usually a little larger than the laterals, apex of leaflets truncate or rounded, generally sharply denticulate, base narrowly cuneate, petiolule of terminal leaflet usually very short, seldom exceeding 3 mm. in length; inflorescences usually rather crowded towards the tips of the branches; peduncle usually less than 8 mm. long, shorter than subtending leaf, slender, glandular, 1-flowered; flower terminal, at first nodding, later distinctly reflexed; calyx 5–8 mm. long, ± densely glandular-pilose, the teeth linear or linear-subulate, entire or bluntly and obscurely toothed near apex, 4–6 mm. long, 3-nerved; corolla shorter than, or equalling, or slightly exceeding calyx; standard pink, broadly obovate, 5–8 mm. long, 3–4 mm. wide, apex rounded, ± apiculate, base narrowed to a very short broad claw, wings oblong, 4–6 mm. long, 1–2 mm. wide, apex rounded, base bluntly auriculate, conjunctival tooth absent, keel white, about 5 mm. long, 2–3 mm. wide, curved almost at a right angle to the short, rather blunt beak; stamens about 5 mm. long, staminal tube 3 mm. long, glabrous; anthers oblong, subequal, 0·3 mm. long, 0·2 mm. wide; ovary oblong, subsessile, 2–3 mm. long, 1 mm. wide, densely glandular-villose, style glabrous, about 3 mm. long, stigma terminal, capitate. Pod oblong, 5–8 mm. long, 3–4 mm. wide, pale or dark brown, apex blunt; seeds subglobose, 0·8–1·3 mm. diam.; testa pale brown, covered with blunt, glistening tubercles.

var. **minor** *Moris*, Fl. Sardoa, 1: 422 (1837); Boiss., Fl. Orient., 2: 61 (1872); Zohary, Fl. Palaest., 2: 119, t. 173 (1972).

O. mollis Savi in Mém. Soc. Ital. Mod., 9: 351, t. 8 (1802); Rechinger f. in Arkiv för Bot., ser. 2, 1: 424 (1950).

O. reclinata L. var. *mollis* (Savi) Heldr. in Bull. Herb. Boiss., 6: 292 (1898); Post, Fl. Pal., ed. 2, 1: 308 (1932); Širjaev in Beih. Bot. Centralbl., 49, 2: 502 (1932); Lindberg f., Iter Cypr., 20 (1946).

O. reclinata L. ssp. *mollis* (Savi) Bég. in Bull. Soc. Bot. Ital., 1912: 134 (1912); Holmboe, Veg. Cypr., 103 (1914); Osorio-Tafall et Seraphim, List Vasc. Plants Cyprus, 58 (1973).

TYPE: Sardinia; without precise locality, *Moris* (TO).

Leaves mostly 3-foliolate; leaflets often rather small and narrow, calyx lobes narrowly linear-subulate; corolla equalling or shorter than calyx. Pod about as long as calyx; seeds 0·8 mm. diam. (sometimes smaller).

HAB.: In stony or rocky fields, in dry pastures, on sandy or rocky seashores, amongst maquis on dry calcareous or igneous hillsides or in thin *Pinus* woodland; sea-level to 4,000 ft. alt.; fl. March–June.

DISTR.: Divisions 2–8, common in Divisions 7, 8. Throughout the Mediterranean area and eastwards to Iran, Ethiopia and Atlantic Islands.

2. Platres, 1938, *Kennedy* 1261 !

3. Skarinou near Lefkara, 1941, *Davis* 2677 ! Limassol aerodrome, 1950, *Chapman* 536 ! 4 miles N. of Kandou (Limassol), 1963, *J. B. Suart* 3 !
4. Larnaca aerodrome, 1950, *Chapman* 698 ! Cape Greco, 1960, *N. Macdonald* 104 !
5. Athalassa, 1939, *Lindberg f.* s.n. ! Near Salamis, 1948, *Mavromoustakis* s.n. ! Mile 3, Nicosia–Famagusta road, 1950, *Chapman* 396 ! Mia Milea, 1952, *Merton* 686 !
6. Syrianokhori, 1935, *Syngrassides* 615 ! 765 ! Between Skylloura and Pyrga, 1956, *Poore* in *Merton* 2761 ! Ayia Irini, 1962, *Meikle* 2385 !
7. Common; *Kotschy* 529; *Sintenis & Rigo* 405 ! *Syngrassides* 919 ! *Davis* 3025 ! *Casey* 664 ! *G. E. Atherton* 1370 !
8. Common; *Sintenis & Rigo*; *Holmboe* 486; *Haradjian* 167 ! 240 ! *Merton* 3070 !

NOTES: Var. *minor*, which predominates in the Mediterranean area, differs from typical *reclinata* only in the smaller size of its corolla, leaflets and pod; it is, however, variable in all these characters, and is connected to var. *reclinata* by a series of intermediate forms.

var. **monophylla** (*Bég.*) *Pamp.*, Prodr. Fl. Ciren., 248 (1930); Širjaev in Beih. Bot. Centralbl., 49, 2: 505 (1932).

O. reclinata L. ssp. *monophylla* Bég. in Bull. Soc. Bot. Ital., 1912: 134 (1912).

TYPE: Cyrenaica; Derna, Febr. 21, 1912, *A. Vaccari* (Fl).

Leaves uniformly 1-foliolate, larger than in var. *minor*; corolla equalling or slightly exceeding the linear calyx-lobes; pod slightly longer than calyx; seeds 1–1·3 mm. diam.

HAB.: Amongst garigue on steep rocky slope; 100 ft. alt.; fl. April.

DISTR.: Division 1, rare. Also Cyrenaica and (?) Egypt.

1. Kako Skala (Akamas), 1950, *Casey* 1010 ! and, 1954, *Casey* 1343 !

NOTES: A striking variant, with slender, sparingly branched stems, and unifoliolate leaves widely separated by long internodes; the calyx-teeth are distinctly wider than in var. *minor* and often (but not always) shortly lobed or bluntly toothed at the apex as in *O. dentata* Sol. ex Lowe (*O. reclinata* var. *tridentata* (Lowe) Lowe) from Madeira. In all essential characters, however, the plant resembles *O. reclinata*, and cannot be considered more than a variety of this species.

SUBSECT. 6. **Pubescentes** (*Širjaev*) *Meikle* **stat. nov.** (Series *Pubescentes* Širjaev in Beih. Bot. Centralbl., 49, 2: 518; 1932) Annuals; lower leaves 3-foliolate, upper often 1-foliolate, the uppermost (floral) leaves sometimes without leaflets; peduncle exaristate, 1-flowered; corolla yellow, standard glabrous, often tinged red, wings with a conjunctival tooth. Pods ovoid or ellipsoid, strongly compressed; seeds smooth.

6. O. pubescens *L.*, Mantissa Altera, 267 (1771); Boiss., Fl. Orient., 2: 62 (1872); Post, in Mém. Herb. Boiss., 18: 92 (1900); Holmboe, Veg. Cypr., 103 (1914); Post, Fl. Pal., ed. 2, 1: 308 (1932); Širjaev in Beih. Bot. Centralbl., 49, 2: 518 (1932); Lindberg f., Iter Cypr., 20 (1946); Davis, Fl. Turkey, 3: 379 (1970); Zohary, Fl. Palaest., 2: 119, t. 174 (1972); Osorio-Tafall et Seraphim, List Vasc. Plants Cyprus, 58 (1973).

[*O. natrix* (non L.) H. Stuart Thompson in Journ. Bot., 44: 305 (1906).]

TYPE: " in Europa *australi* ".

Erect, viscid-glandular annual, (7–) 15–30 cm. high, usually much branched, with erect, or spreading, villose-glandular branches; leaves mostly 3-foliolate; petiole 1–3 cm. long, strongly channelled above; stipules large, herbaceous, acuminate, entire 7–25 mm. long, 3–7 mm. wide, adnate to petiole for about half their length; leaflets oblong, serrulate, 8–30 mm. long, 2–15 mm. wide, apex and base rounded or subacute, terminal leaflet frequently much larger than the laterals, with a petiolule 2–5 mm. long; flowers crowded towards the apices of the branches, forming cylindrical, racemose inflorescences; peduncles slender, 0·5–1·5 (–2) cm. long, often shorter than the subtending floral leaf, exaristate; pedicels very short, curved downwards; calyx 1–1·5 cm. long, viscid-pilose, teeth linear-lanceolate 8–12 mm. long, 1·5–3·5 mm. wide, narrowing gradually to an acuminate, entire apex, calyx-tube short, campanulate, strongly veined; standard whitish or pale yellow tinged red, or rarely uniformly golden-yellow, suborbicular, 1–1·5 cm. long, 1–1·3 cm. wide, with a very short, broad basal claw, wings golden yellow, broadly oblong, 0·8–1 cm. long, 0·3–0·5 mm. wide, with a rounded apex and a narrow, plicate basal auricle and a prominent, rather blunt conjunctival tooth, keel 1–1·2 cm. long, 0·3–0·5 cm. wide, upper margins strongly gibbous, apex curving sharply to a long pointed beak; stamens 9–12 mm. long, staminal tube glabrous, 4–6 mm. long; anthers oblong, unequal, those of the shorter stamens 0·8–1 mm. long, 0·3 mm. wide, those of the longer stamens about 0·5 mm. long, 0·3 mm. wide; ovary oblong-ellipsoid, 2–3 mm. long, 1–2 mm. wide, densely glandular-hairy, style 8–10 mm. long, glabrous above, glandular towards base, stigma capitate, terminal or slightly oblique. Pod ovoid or ellipsoid, strongly compressed 1–1·3 cm. long, 0·5–0·7 cm. wide, with a slender acuminate apex, valves pale brown, glandular-pubescent; seeds subglobose, about 2·5 mm. diam., slightly compressed; testa dark brown, smooth.

HAB.: Fields, roadsides and waste ground, also amongst garigue on eroded or rocky chalk or sandstone hillsides; sea-level to 3,000 ft. alt.; fl. March–June.

DISTR.: Divisions 2, 3, 5–7, common in Division 7. Widespread in the Mediterranean region (absent from Italy, Sardinia, Sicily and Egypt) eastwards to Turkey and Iraq.

2. Pano Panayia, 1905, *Holmboe* 1055; above Arminou, 1941, *Davis* 3401 !
3. Limassol, 1894, *Post*; Stavrovouni, 1939, *Lindberg f.* s.n. !
5. Kambia above Dheftera, 1936, *Syngrassides* 1291 ! Salamis, 1939, *Lindberg f.* s.n. and, 1971, *Townsend* 71/116 !
6. Dhiorios, 1962, *Meikle* !
7. Common; *Sintenis & Rigo* 635 ! *Haradjian* 362 ! *Davis* 1748 ! 2956 ! *Casey* 194 ! 768 ! *Merton* 822 ! *G. E. Atherton* 246 ! 1358 ! 1378 !

SECT. 2. **Ononis** Peduncles quite wanting or very much reduced. Pods ovoid or ellipsoid, 2–4-seeded, not reflexed.

SUBSECT. 1. **Bugranoides** *Willk. emend. Širjaev* Shrubs or perennials; leaves 3-foliolate, or the floral leaves sometimes 1-foliolate; peduncles quite wanting; corolla glabrous, yellow. Pods straight; seeds tuberculate or smooth.

7. O. pusilla *L.*, Syst. Nat., ed. 10, 1159 (1759); Post, Fl. Pal., ed. 2, 1: 305 (1932); Rechinger f. in Arkiv för Bot., ser. 2, 1: 423 (1950); Davis, Fl. Turkey, 3: 380 (1970); Osorio-Tafall et Seraphim, List Vasc. Plants Cyprus, 58 (1973).

O. columnae All., Auct. Syn. Meth. Stirp. Hort. Reg. Taur., 77 (1773), Fl. Ped., 318, t. 20, f. 3 (1785); Boiss., Fl. Orient., 2: 57 (1872); Širjaev in Beih. Bot. Centralbl., 49, 2: 543 (1932); Lindberg f., Iter Cypr., 20 (1946).

O. columnae All. var. *orientalis* Širjaev in Beih. Bot. Centralbl., 49, 2: 545 (1932); A. K. Jackson in Kew Bull., 1937: 341 (1937).

[*O. cherleri* (non L. Syst. Nat. ed. 10, 1160; 1759) L., Sp. Plant., ed. 2, 1007 (1763).]

TYPE: Not localized.

Erect, spreading or decumbent, subshrubby perennial, 10–20 (–30) cm. high, usually much branched from near the base; branches ± densely glandular, with intermixed, longer, eglandular hairs; leaves mostly 3-foliolate, stipules persistent, often purplish, lanceolate, 4–10 mm. long, about 2 mm. wide, distinctly veined, apex acuminate, margins sharply serrulate; petioles 8–20 mm. long, glandular-pubescent, channelled above; leaflets subequal, oblong or obovate, 4–14 mm. long, 2–10 mm. wide, subglabrous or thinly glandular-pubescent, veins numerous, ± parallel, distinctly raised above and below and running into the sharply serrulate margins, apex rounded or rarely subacute, base cuneate, entire or subentire; terminal leaflet with a distinct petiolule 2–5 mm. long; flowers sessile or subsessile in congested or lax cylindrical inflorescences at the tips of the branches, generally much shorter than the subtending floral leaves; calyx 8–12 mm. long, teeth lanceolate-subulate, attenuate, 5–7 mm. long, pubescent or glandular-pubescent, calyx-tube narrowly campanulate, many-nerved; corolla yellow, very variable in size, sometimes equalling the calyx or even longer, often small and ± abortive, standard broadly obovate, 3–9 mm. long, 3–6 mm. wide, apex rounded, often apiculate, wings narrowly oblong 2·5–6 mm. long, 1–1·5 mm. wide, apex rounded, base with a short blunt auricle, conjunctival tooth short or sometimes wanting, keel 4–8 mm. long, 1·5–4 mm. wide, sharply upcurved to a short rather blunt beak; stamens 4–8 mm. long, staminal tube glabrous, 2–4 mm. long; anthers subequal or markedly unequal (especially in plants with small corollas), cordate-acute, 0·3–0·5 mm. long, 0·2–0·3 mm. wide; ovary oblong, 1·5–2 mm. long, 1–1·5 mm. wide, glandular-pubescent, style 2–5 mm. long, glabrous, stigma terminal, capitate. Pod ovoid-acute, compressed, pale brown, 7–9 mm. long, 5–6 mm. wide, glabrous or very sparsely pubescent; seeds subglobose-reniform, about 2 mm. diam.; testa dark brown, minutely and densely verruculose.

HAB.: Dry chalk and limestone hillsides, often under Pines, 1,100–3,000 ft. alt.; fl. May–July.

DISTR.: Divisions 2, 7, common in both Divisions. Throughout the Mediterranean region and eastwards to Iran, also in France, Switzerland, Austria and Hungary.

2. Common; *Haradjian* 901 ! 998 ! *Davis* 1837 ! 1907 ! 3363 !

7. Common; *Sintenis & Rigo* 404 ! *Haradjian* 351; *Lindberg f.* s.n.; *Davis* 3621 ! *Casey* 138 ! 1347 ! *Chapman* 652 !

NOTES: Forms with large and small corollas occur in our area, the former seemingly predominating on the Northern Range, the latter on the calcareous foothills of the Troödos Range, though neither is exclusively restricted to one or the other range. The distinctions between var. *pusilla* and var. *orientalis* (*O. columnae* var. *orientalis* Širjaev) are, as Širjaev admits, often obscured by intermediate forms; most of the Cyprus material is more robust than typical *pusilla*, with larger leaflets, but close examination of the species throughout its range shows a distinct cline, with small-leaved, dwarf forms at the western end and more robust forms in the east, without any break in the gradual transition from one extreme to the other.

SUBSECT. 2. **Ononis** Subshrubs or perennials; branches often spinescent; leaves 3-foliolate or 1-foliolate, floral leaves sometimes bract-like; flowers solitary or in pairs forming raceme-like inflorescences; corolla pink, standard hairy. Seeds verruculose or sometimes smooth.

8. **O. spinosa** *L.*, Sp. Plant., ed. 1, 716 (1753) excl. var. β; Širjaev in Beih. Bot. Centralbl., 49, 2: 575 (1932).

Erect or spreading, spinescent subshrub, to about 80 cm. high, branches with scattered or bifarious hairs or sometimes subglabrous, often purplish, branchlets normally terminating in a long, sharp, slender spine; leaves 3-foliolate or 1-foliolate, the basal often rather large, with leaflets frequently exceeding 2 cm. in length, the upper small, often 1-foliolate and bract-like, generally narrowly oblong, obtuse or rounded at apex, ± glandular-pubescent, with sharply denticulate margins; flowers scattered along the upper parts of branches and branchlets, solitary or very rarely in pairs, shortly pedicellate or subsessile; calyx 5–10 mm. long, ± densely glandular-pubescent, with narrow linear-subulate, subfalcate teeth 3–6 mm. long; corolla pink (sometimes white), longer or shorter than calyx, standard broadly obovate, subglabrous or glandular-pubescent externally, wings oblong, obtuse or rounded at apex, with a distinct basal auricle and conjunctival tooth; keel usually longer than wings, upcurved gradually or abruptly to a sharp, pointed beak; anthers narrowly oblong, 0·6–0·9 mm. long, 0·2–0·3 mm. wide; ovary oblong, 1·5–3 mm. long, 1–2 mm. wide, glandular-pubescent, style glabrous often more than 1 cm. long, stigma terminal, capitate. Pod ovoid, strongly compressed, 6–9 mm. long, 5–7 mm. wide, apex acute or obtuse; seeds 1 or few in each pod, globose-reniform, 2–3 mm. diam.; testa dark brown or obscurely marbled, verruculose or smooth.

ssp. **leiosperma** (*Boiss.*) *Širjaev* in Beih. Bot. Centralbl., 49, 2: 590 (1932); Davis, Fl. Turkey, 3: 382 (1970); Osorio-Tafall et Seraphim, List Vasc. Plants Cyprus, 59 (1973).

O. leiosperma Boiss., Fl. Orient., 2: 57 (1872).

O. antiquorum L. var. *leiosperma* (Boiss.) Post, Fl. Pal., ed. 1 (1896), ed. 2, 1: 305 (1932).

[*O. antiquorum* (non L.) Kotschy in Unger et Kotschy, Die Insel Cypern, 373 (1865); Boiss., Fl. Orient., 2: 57 (1872) pro parte quoad spec. cypr.; Holmboe, Veg. Cypr., 103 (1914); Lindberg f., Iter Cypr., 20 (1946); Chapman, Cyprus Trees and Shrubs, 45 (1949); Rechinger f. in Arkiv för Bot., ser. 2, 1: 424 (1950).]

[*O. spinosa* L. ssp. *antiquorum* (non (L.) Briq.) Osorio-Tafall et Seraphim, List Vasc. Plants Cyprus, 59 (1973).]

Low, intricately branched subshrub, often with slender, flexuous branches and numerous spinose branchlets; leaves of mature shoots mostly 1-foliolate, small, leaflets narrowly oblong, less than 8 mm. long and 5 mm. wide, petiole very short, usually less than 6 mm. long, stipules 2–6 mm. long, 1–4 mm. wide, distinctly veined, obscurely denticulate, the apex broadly triangular; corolla 7–13 mm. long, usually equalling or shortly exceeding calyx, pink (f. *leiosperma*) or white (f. *albiflora* Lindberg f., Iter Cypr., 20). Seeds with a smooth testa, often marbled.

var. **leiosperma**

TYPE: " ad vias Syriae, in argillosis montis Amani prope Beilan (Ky. exs. 296), in Caelosyriâ et circa Damascum (Gaill.), in Persiâ australi ad radices montis Kuh Daëna (Ky. exs. 788) " (G, and *Kotschy* 788 K !).

Usually less than 40 cm. high; branches shortly glandular-pubescent or subglabrous, often pallid and markedly flexuous (zig-zag).

HAB.: In dry, stony fields, by roadsides, in waste ground or on dry open hillsides, occasionally on sandy seashores; sea-level to 4,300 ft. alt.; fl. May–Oct.

DISTR.: Divisions 2–4, 6–8, common in Divisions 2, 6, 7. Eastern Mediterranean region from Albania to Palestine, east to Iran.

2. Common; *Lindberg f.* s.n.; *Chapman* 117; *Casey* 904 ! *G. E. Atherton* 564 ! *D. P. Young* 7419 !
3. Moni near Limassol, 1905, *Holmboe* 641; Arkhimandrita, 1937, *Chapman* 300 !
4. Between Prastio and Kouklia, 1939, *Lindberg f.* s.n.
6. Common; *Syngrassides* 232 ! 987 ! *R. M. Nattrass* 535; *Lindberg f.* s.n. ! *P. H. Oswald* 17 !
7. Common; *Kotschy* 944; *Sintenis & Rigo* 649 ! *Chapman* 432; 662 ! *G. E. Atherton* 16 ! 346 !
8. Near Komi Kebir, 1912, *Haradjian* 324.

NOTES: Distinguishable from *O. spinosa* L. ssp. *antiquorum* (L.) Briq. solely by its smooth seeds, and in the absence of seeds really indeterminable at infraspecific level. All the fruiting material seen, except for one specimen, belongs to ssp. *leiosperma* and it may be assumed that this, if not the only subspecies present in Cyprus, is at least the predominant one; in the circumstances the author has taken the liberty of referring all non-fruiting material to this segregate. The one exceptional specimen is *G. E. Atherton* 131 from Kyrenia Range, " 9 miles W. end of Forest Road ", 2,000–2,500 ft. alt., July 13, 1955; the seeds of this, unfortunately very immature and imperfect, appear to be sparsely verruculose, but it may well be that they would be smooth in better material from the same individual plant. Ssp. *antiquorum* (L.) Briq., with distinctly verruculose seeds, should, however, be looked for in Cyprus.

var. **tomentosa** (*Boiss.*) *Širjaev* in Beih. Bot. Centralbl., 49, 2: 591 (1932).

O. leiosperma Boiss. var. *tomentosa* Boiss., Fl. Orient., 2: 57 (1872).

O. macracantha Clarke, Travels, ed. 2, 1: 354 (1812).

TYPE: Lebanon; " circa Berythum (Cadet !), in Libano circa Dimam (Bl. !), prope Sidonem (Bl. !) " (G).

Up to 80 cm. high, often rather rigidly erect; branches glandular-pubescent and ± densely clothed with much longer, multicellular, eglandular white hairs, not pallid or markedly flexuous.

HAB.: Roadsides, border of fields, sometimes in marshy ground; sea-level to 400 ft alt.; fl. April to ? Oct.

DISTR.: Divisions 2–4, rare. Turkey, Syria and Palestine, east to Iran.

2. Lefka, Karavostasi, 1937, *Syngrassides* 1545 !
3. Asomatos, 1962, *Meikle* 2919 !
4. Ormidhia, 1936, *Syngrassides* 1010 !

NOTES: In its extreme form very distinct, closely resembling *O. hircina* Jacq., but with the flowers solitary or very rarely in pairs. None of the Cyprus specimens has fruits, so that it is impossible to see whether the seeds have smooth or verruculose testas. Širjaev (*op. cit.*, 592) refers *Ussher* [" Ush "] 29 to var. *tomentosa*, but this specimen (at Kew) is typical var. *leiosperma*.

SUBSECT. 3. **Variegatae** *Širjaev* Annuals; leaves 1-foliolate, the floral leaves sometimes aphyllous; flowers solitary, corolla yellow, standard pilose. Seeds smooth or tuberculate.

9. O. variegata *L.*, Sp. Plant., ed. 1, 717 (1753); Boiss., Fl. Orient., 2: 62 (1872); Holmboe, Veg. Cypr., 103 (1914); Post, Fl. Pal. ed. 2, 1: 309 (1932); Širjaev in Beih. Bot. Centralbl., 49, 2: 613 (1932); Lindberg f., Iter Cypr., 20 (1946); Rechinger f. in Arkiv för Bot., ser. 2, 1: 424 (1950); Davis, Fl. Turkey, 3: 382 (1970); Zohary, Fl. Palaest., 2: 120, t. 175 (1972); Osorio-Tafall et Seraphim, List Vasc. Plants Cyprus, 59 (1973).

TYPE: " *in* Europae *australis maritimis* ".

Spreading or decumbent annual 4–12 (–15) cm. high, much branched from near base, branches ± densely pilose, the basal parts often buried in sand and aphyllous; stipules conspicuous, broadly oblong, 3–6 mm. long, 2–5 mm. wide, adnate to petiole for the greater part of their length, strongly veined, apex rather blunt or shortly acute, margins distinctly and sharply serrate, petiole very short, flattened, scarcely extending beyond adnate stipules, leaves 1-foliolate, leaflet narrowly obovate, 5–8 mm. long, 2–4 mm. wide, glaucous-green, thinly glandular-pubescent, conspicuously veined, with a rounded apex and sharply serrate margins; flowers in congested or lax racemose inflorescences near the tips of the branches; pedicels usually less than 3 mm. long, ± densely glandular-pubescent; calyx glandular-pubescent, 6–7 mm. long, teeth narrowly lanceolate, acute, 3–4 mm. long; corolla yellow, longer than calyx, standard obovate, thinly pilose externally, 8–10 mm. long, 6–8 mm. wide, with an apiculate apex and a short rather broad claw, wings broadly oblong, 8 mm. long, 3–4 mm. wide near the rounded apex, basal auricle blunt, conjunctival tooth small or wanting, keel 8–9 mm. long, 3 mm. wide, upcurved rather gradually to short bluntish beak; stamens 9 mm. long, staminal tube glabrous about 7 mm. long; anthers slightly unequal, about 0·4 mm. and 0·6 mm. long, 0·3 mm. wide; ovary oblong, about 2 mm. long, 1–1·2 mm. wide, thinly glandular-pubescent, style glabrous, 6–7 mm. long, stigma small, terminal, capitate. Pod oblong, 8–9 mm. long, 3 mm. wide, pale or dark brown, thinly pubescent, with a rather blunt apex; seeds reniform, about 1·5 mm. long, 1 mm. wide; testa bright brown, smooth.

HAB.: On sandy seashores and on sand-dunes; about sea-level; fl. March–May.

DISTR.: Divisions 3, 7, 8. Throughout the Mediterranean region.

3. Limassol, 1905, *Holmboe* 651; Cape Gata, 1941, *Davis* 3561 ! Akrotiri, 1963, *J. B. Suart* 102 !
7. Between Melanissiko and Ayios Epiktitos, 1880, *Sintenis & Rigo* 408A ! Ammos near Kyrenia, 1939, *Lindberg f.* s.n.; between Ayios Epiktitos and Ayios Amvrosios, 1941, *Davis* 3615 ! 6 miles E. of Kyrenia, 1951, *Casey* 1109 !
8. Near Cape Andreas, 1880, *Sintenis & Rigo* 408 ! Near Rizokarpaso, 1912, *Haradjian* 139 ! Galinoporni, 1962, *Meikle* 2497 !

SUBSECT. 4. **Diffusae** *Širjaev* Annuals; leaves 3-foliolate, floral leaves sometimes 1-foliolate or aphyllous; peduncles wanting; inflorescence usually elongating after anthesis; corolla pink or white, usually hairy externally. Seeds smooth or tuberculate.

10. O. diffusa *Ten.*, Fl. Nap. Prodr., XLI (1811–15), Fl. Nap. Syll. in Fl. Nap., 4: 100 (1830), Fl. Nap., 5: 98, t. 169, fig. 1—error fig. 2—(1835–36); Širjaev in Beih. Bot. Centralbl., 49, 2: 618 (1932); Rechinger f. in Arkiv för Bot., ser. 2, 1: 424 (1950); Osorio-Tafall et Seraphim, List Vasc. Plants Cyprus, 59 (1973).

O. serrata Forssk. var. *major* Lange in Vidensk. Meddelelser, 1865: 156 (1866); Boiss., Fl. Orient., 2: 63 (1872); Post, Fl. Pal., ed. 2, 1: 309 (1932).

O. serrata Forssk. ssp. *diffusa* Rouy et Fouc., Fl. France, 4: 268 (1897); Holmboe, Veg. Cypr., 103 (1914).

TYPE: Italy; " in aridis incultis prope Puteolos: sulla strada prima del monte delle brecce, Miseno; In Calabria, Piano della corona sopra Bagnara, e prima di Scrisi (Guss.) " (FI, ? NAP).

Erect or spreading annual (13–) 20–30 (–40) cm. high, often much branched, with glandular-pubescent branches; leaves rather distant, 3-foliolate; stipules ovate, 3–10 mm. long 2–5 mm. wide, with an acute apex and sharply serrate margins; petioles of lower leaves 1–2·5 cm. long, channelled above, glandular-pubescent; leaflets oblong-obovate, 1–2 cm. long, 0·5–1 cm. wide, shortly glandular, apex rounded or truncate, base broadly cuneate, margins sharply serrulate with 12–24 (–30) small teeth, terminal leaflet with a petiolule 2–7 mm. long; flowers crowded in dense terminal, spicate, oblong inflorescences, floral leaves usually 1-foliolate, often small and bract-like, pedicels very short, usually less than 3 mm. long, densely glandular-pubescent; calyx glandular, 7–8 mm. long, accrescent in fruit, teeth linear-lanceolate, acuminate, about 5 mm. long; corolla pink (with a whitish keel) distinctly longer than calyx, standard broadly ovate-orbicular, apiculate, 9–10 mm. long, 6–7 mm. wide, glandular externally, with a distinct, rather narrow claw, wings oblong, 8–9 mm. long, 2·5–3 mm. wide, apex rounded, base with a rather acute auricle and small, often flap-like conjunctival tooth, keel 9–10 mm. long, 3 mm. wide, upper (adaxial) margin almost straight, basal (abaxial) margin gently curved to a bluntish beak; stamens 10 mm. long, staminal tube glabrous 6–7 mm. long; anthers subequal, shortly oblong 0·3–0·4 mm. long, 0·2 mm. wide, filaments often distinctly dilated just below anther; ovary oblong, glandular-pubescent, about 2 mm. long and 0·8 mm. wide, style glabrous 6–7 mm. long, stigma terminal, capitate. Pod pale brown, ovate, compressed, about 9 mm. long, 5 mm. wide, apex acute; seeds orbicular-reniform, about 2 mm. diam.; testa dark brown, closely papillose-verruculose.

HAB.: Sandy seashores and sand-dunes; about sea-level; fl. March–June.

DISTR.: Divisions 4, 5, 8, rare. Throughout the Mediterranean region; Atlantic Islands.

4. Near Famagusta, 1912, *Haradjian* 123 ! Sand dunes near Ayios Memnon, 1948, *Mavromoustakis* s.n. !
5. Salamis, 1905, *Holmboe* 465.
8. Near Rizokarpaso, 1912, *Haradjian* 168.

11. O. serrata *Forssk.*, Fl. Aegypt-Arab., 130 (1775); Boiss., Fl. Orient., 2: 63 (1872)—excl. var. β; Sintenis in Oesterr. Bot. Zeitschr., 32: 364 (1882); Holmboe, Veg. Cypr., 103 (1914) pro parte; Post, Fl. Pal., ed. 2, 1: 309 (1932) pro parte; Širjaev in Beih. Bot. Centralbl., 49, 2: 624 (1932); Davis, Fl. Turkey, 3: 383 (1970); Zohary, Fl. Palaest., 2: 120, t. 177 (1972); Osorio-Tafall et Seraphim, List Vasc. Plants Cyprus, 59 (1973).

TYPE: Egypt; Alexandria, 1761, *Forsskål* (C).

Erect annual, 10–30 cm. high, sparingly or profusely branched, with spreading or erecto-patent, glandular-pubescent, purplish branches; leaves often more dense than those of *O. diffusa*, 3-foliolate; stipules narrowly ovate-oblong, 5–10 mm. long, 2–2·5 mm. wide, with an acuminate apex and entire or irregularly serrate margin; petioles of lower leaves 0·7–1·4 cm. long, channelled above, glandular-pubescent; leaflets narrowly oblong 6–18 mm. long, 2–7 mm. wide, shortly glandular, apex truncate or abruptly acute, margins sharply serrate with 10–20 conspicuous, spreading teeth, terminal leaflet with a petiolule 2–4 mm. long; flowers at first rather crowded, later in lax, slender, terminal, spicate inflorescences, floral leaves similar to those of *O. diffusa* but often rather better developed, occasionally 3-foliolate, together with the calyces densely viscid-glandular; pedicels very short or flowers subsessile; calyx 5–6 mm. long, somewhat accrescent in fruit, teeth linear-subulate, about 3 mm. long; corolla pink (with a whitish keel) distinctly longer than calyx, standard oblong-obovate, apiculate, 8–9 mm. long, 4–5 mm. wide, glandular externally, with a long rather narrow claw, wings narrowly oblong, 8 mm. long, 2 mm. wide, apex rounded or subacute, base with a conspicuous auricle, and short, blunt conjunctival tooth, keel about 8 mm. long, 2·5–3 mm. wide, upper (adaxial) margin almost straight, basal (abaxial) margin gently curved to a bluntish beak; stamens 8–9 mm. long, staminal tube glabrous about 6 mm. long; anthers small, ovoid-oblong, subequal, about 0·2 mm. long and 0·1 mm. wide, filaments distinctly dilated just below anther; ovary oblong, about 1·5 mm. long, 0·6 mm. wide, glandular-pubescent, style glabrous about 7 mm. long, stigma terminal, capitate. Pod oblong, 5–7 mm. long, 3–4 mm. wide, dull brown, much compressed with a blunt or rounded apex; seeds globose-reniform about 1·5 mm. diam.; testa pale brown, densely papillose-tuberculate. *Plate 19, figs. 11–12.*

HAB.: Sandy seashores and sand-dunes; about sea-level; fl. March–June.

DISTR.: Divisions 3–5, 7, 8. North Africa and Orient eastwards to Iran and Iraq, Turkey and Aegean Islands.

3. Limassol, 1905, *Holmboe* 652.
4. Coast near Pyla, 1934, *Syngrassides* 475 ! Famagusta, 1934, *Syngrassides* 1001 ! Sand dunes at Ayios Memnon, 1948, *Mavromoustakis* s.n. !
5. Salamis, 1949, *Casey* 554 ! and, 1962, *Meikle* 2566 !
7. St. Hilarion Castle, 1939, *Mavromoustakis* 14 !
8. Sandhills near Yialousa, 1880, *Sintenis & Rigo* 402 !

NOTES: Usually a plant of sandhills and sandy ground, and perhaps an introduction at St. Hilarion (*Mavromoustakis* 14), an unlikely habitat.

SUBSECT. 5. **Mitissimae** *Širjaev* Annuals; cauline leaves 3-foliolate, floral leaves bract-like, imbricate, whitish, coriaceous; flowers in congested terminal spicate inflorescences; corolla pink, standard glabrous. Seed tuberculate.

12. O. mitissima *L.*, Sp. Plant., ed. 1, 717 (1753); Boiss., Fl. Orient., 2: 64 (1872); Post, Fl. Pal., ed. 2, 1: 310 (1932); Širjaev in Beih. Bot. Centralbl., 49, 2: 638 (1932); A. K. Jackson in Kew Bull., 1937: 342 (1937); Rechinger f. in Arkiv för Bot., ser. 2, 1: 424 (1950); Davis, Fl. Turkey, 3: 384 (1970); Zohary, Fl. Palaest., 2: 122, t. 179 (1972); Osorio-Tafall et Seraphim, List Vasc. Plants Cyprus, 59 (1973).
TYPE: " *in* Barbados, Lusitania ".

Robust, erect or spreading annual, (15–) 20–60 cm. high; stems pale green, much branched, thinly glandular-hairy or glabrescent; cauline leaves 3-foliolate; petiole short, usually under 1 cm. long, channelled above, thinly glandular-pilose; stipules oblong, 5–10 mm. long, 2–5 mm. wide, glandular, apex acute, margins entire or shortly toothed above; leaflets obovate 1–2·5 cm. long, 0·5–1·5 cm. wide, shortly glandular, apex obtuse or rounded, margins finely and regularly serrulate; flowers crowded in terminal spikes; floral leaves much reduced, often bract-like, with conspicuous pale green or whitish stipules; pedicels wanting or very short; calyx 7–10 mm. long, tube

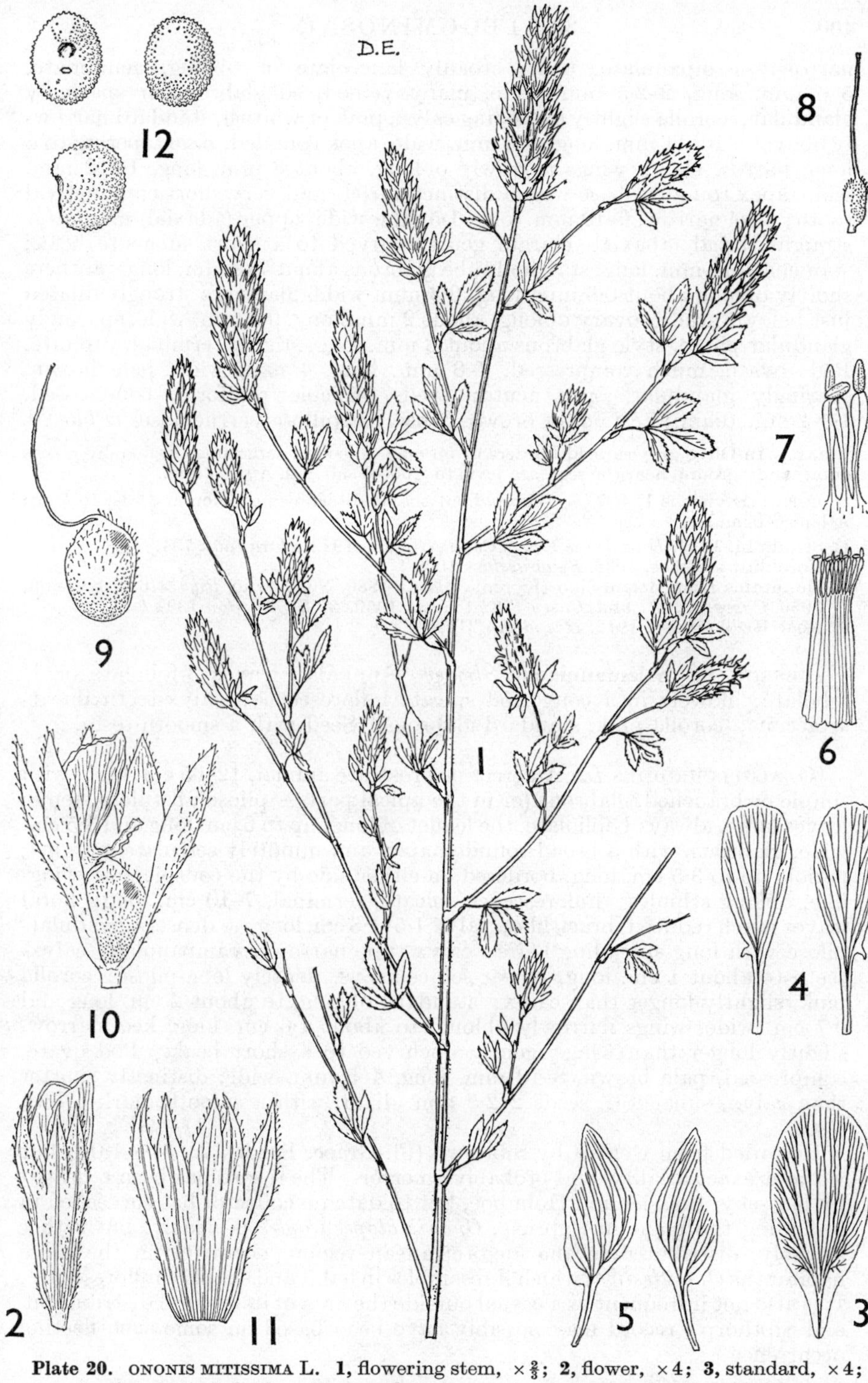

Plate 20. ONONIS MITISSIMA L. **1**, flowering stem, $\times\frac{2}{3}$; **2**, flower, $\times 4$; **3**, standard, $\times 4$; **4**, wing, $\times 4$; **5**, keel, $\times 4$; **6**, androecium, $\times 4$; **7**, stamen detail, $\times 12$; **8**, gynoecium, $\times 4$; **9**, pod, $\times 4$; **10**, pod within calyx, $\times 4$; **11**, calyx, $\times 4$; **12**, seed, $\times 6$. (**1–8, 10 & 12**, *Kotschy* 384; **9**, *Field & Lazar* 748.)

narrowly campanulate, teeth broadly lanceolate or oblong, acuminate, 5–6 mm. long, 2–2·5 mm. wide, many veined, subglabrous or sparingly glandular; corolla slightly exceeding calyx, pink or whitish, standard narrowly obovate, 10–11 mm. long, 4–5 mm. wide, apex rounded, base tapering to a long, narrow claw; wings narrowly oblong, about 8 mm. long, 1·5–2 mm. wide, apex rounded, base with a distinct auricle and very short conjunctival tooth; keel narrow, 9–10 mm. long, 1·5 mm. wide, upper (adaxial) margin ± straight, basal (abaxial) margin gently curved to a short subacute beak; stamens 9–10 mm. long, staminal tube glabrous about 7–8 mm. long; anthers shortly oblong about 0·5 mm. long, 0·3 mm. wide, filaments strongly dilated just below anther; ovary oblong, about 2 mm. long, 0·8 mm. wide, sparingly glandular-pilose, style glabrous about 8 mm. long, stigma terminal, capitate. Pods ovate, much compressed, 7–8 mm. long, 4 mm. wide, pale brown, sparingly glandular, apex acute; seeds orbicular-reniform, compressed, 1·5–2 mm. diam.; testa dark brown, densely papillose-verruculose. *Plate 20.*

HAB.: In Olive groves, field borders or on open hillsides, sometimes in marshy places or on sandy ground near the sea; sea-level to 2,000 ft. alt.; fl. April–June.

DISTR.: Divisions 1, 4, 7, 8. Throughout the Mediterranean region and east to Iraq; Atlantic Islands.

1. Stroumbi, 1913, *Haradjian* 778 ! Cape Arnauti, 1913, *Haradjian* 794.
4. Voroklini marshes, 1936, *Syngrassides* 1006 !
7. Mountains near Melanissiko (Kyrenia distr.) 1880, *Sintenis & Rigo* 646 ! Kyrenia, 1950, *Casey* 1026 ! and, *Casey* 1722 ! also, 1950, *G. E. Atherton* 1394 !
8. Near Rizokarpaso, 1912, *Haradjian* 214.

SUBSECT. 6. **Salzmannianae** *Širjaev* Annuals; leaves 3-foliolate or 1-foliolate; flowers in a congested spicate inflorescence; calyx-teeth linear-setaceous; corolla pink, standard glabrous. Seed with a smooth testa.

[O. ALOPECUROIDES *L.* An erect or prostrate annual, 12–65 cm. high, with simple or branched, glabrous (or in the apical part ± pilose), fistulose stems; leaves large, always 1-foliolate, the leaflet oblong, up to 6 cm. long and 3·5 cm. wide, glabrous, with a broad rounded apex and minutely serrulate margins; petiole up to 3·5 cm. long, bordered on either side by the conspicuous, wing-like, oblong stipules; inflorescence spicate, terminal, 7–10 cm. long; floral leaves much reduced, bract-like; calyx 1·5–1·8 cm. long, ± densely glandular-pilose with long spreading hairs; calyx-tube narrowly campanulate, calyx-teeth to about 1 cm. long, linear-setaceous, ± densely long-pilose; corolla pink, slightly longer than calyx; standard-oblong, to about 2 cm. long and 0·7 cm. wide, wings narrowly oblong, to about 1·6 cm. long, keel narrow, slightly longer than wings, gently upcurved to a short beak. Pod ovate, compressed, pale brown, 8–10 mm. long, 4–5 mm. wide, distinctly shorter than calyx, subacute; seeds 2–2·5 mm. diam. with a smooth, dark brown testa.

Recorded from Cyprus by Sibthorp (Fl. Graec. Prodr., 2: 57; 1813) but without exact locality, and probably an error. The record has been repeated by Kotschy, Boissier and Holmboe, but to date no collector has succeeded in re-finding the plant in Cyprus. *Ononis alopecuroides* is widely, but rather sparsely, distributed in the Mediterranean region, occurring in the same habitats as *O. mitissima*, which it resembles in habit and shape of inflorescence. It is also not infrequent as a casual outside the area of its natural distribution, and Sibthorp's record may possibly have been based on some such fleeting occurrence.]

7. TRIGONELLA *L.*

Sp. Plant., ed. 1, 776 (1753).
Gen. Plant., ed. 5, 338 (1754).
Širjaev in Publ. Fac. Sci. Univ. Masaryk, 1928–34.

Annual or, more rarely, perennial herbs, often strongly aromatic when dried; stems erect, diffuse or decumbent, glabrescent or ± pilose, rarely glandular; leaves pinnately 3-foliolate, leaflets obovate, obcuneate or rarely linear-lanceolate, denticulate or rarely lobed or entire, frequently glabrous above, ± pilose or sometimes glandular below, terminal leaflet petiolulate, lateral leaflets sessile or subsessile; stipules adnate to petiole; peduncle axillary, usually well developed; inflorescence capitate-racemose or subumbellate, or flowers few or solitary and apparently subsessile; bracts small, membranous or sometimes wanting; pedicels short, slender, often reflexed; calyx campanulate or subcylindrical, with 5 subequal or ± distinctly unequal, persistent teeth; corolla yellow, whitish, blue or violet; petals free from staminal tube, glabrous, caducous; standard obovate, suborbicular or oblong, rounded or emarginate at apex, usually longer than keel, wings oblong, usually shorter than keel, apex entire, rounded or sometimes bidentate, base auriculate and with or without a conjunctival tooth, keel obtuse, adaxial margins ± connate; vexillary stamen sometimes free, filaments not dilated at apex, anthers 2-thecous, uniform; ovary sessile or shortly stipitate, 1–many-ovulate, straight or almost straight, style glabrous, stigma small, terminal. Pod generally much exceeding calyx, linear, oblong, elliptic or ovate, straight or falcate, terete or flattened or rarely inflated, often reflexed, inarticulate or occasionally constricted between the seeds, dehiscing along the seed-bearing suture or rarely indehiscent or 2-valved, glabrous or pilose, rarely glandular; seeds estrophiolate; testa smooth or verruculose; cotyledons generally with a distinct basal articulation or node.

About 80 species mostly in Europe, the Mediterranean region, Atlantic Islands, and western Asia.

Inflorescence distinctly pedunculate, flowers in a short, often globose, raceme:
- Pods laterally flattened or strongly compressed:
  - Pods 1·3–2·3 cm. long, falcately upcurved, glabrous, obtuse or subacute — **1. T. balansae**
  - Pods 3–4 mm. long, deflexed, thinly pilose, with a slender, upcurved beak — **5. T. spicata**
- Pods terete or subterete, not strongly compressed or flattened:
  - Pods not constricted between the seeds, ± densely pubescent; leaflets pubescent — **2. T. sprunerana**
  - Pods constricted between the seeds, papery, thinly pubescent; leaflets glabrous above, thinly pubescent below - - - - - - **3. T. strangulata**

Inflorescence sessile or subsessile, not pedunculate, or flowers axillary, solitary or in pairs:
- Flowers in axillary clusters, minute, standard less than 5 mm. long:
  - Pods strongly falcate-upcurved, sometimes forming an almost complete circle, 40–60 mm. long - - - - - - - - - - **4. T. spinosa**
  - Pods straightish or slightly upcurved, 8–12 mm. long - - **6. T. monspeliaca**
- Flowers solitary or sometimes in pairs, axillary, not minute, standard generally exceeding 1 cm. in length:
  - Pod pubescent, 3–7 cm. long, narrowing abruptly to a long rigid beak — **7. T. cariensis**
  - Pod glabrous (or microscopically glandular) 7–12 cm. long, tapering gradually to an apical beak:
    - Valves or pod openly veined; plant 15–40 (–50) cm. high; petals yellowish-white — **8. T. foenum-graecum**
    - Valves of pod closely veined; plant usually much less than 15 cm. high; petals blue - - - - - - - - - - - - **9. T. berythea**

1. T. balansae *Boiss. et Reut.* in Boiss., Diagn., 2, 5: 79 (1856); Boiss., Fl. Orient., 2:

83 (1872); Širjaev in Publ. Fac. Sci. Univ. Masaryk, no. 110: 12 (1929); Davis, Fl. Turkey, 3: 460 (1970); Osorio-Tafall et Seraphim, List Vasc. Plants Cyprus, 60 (1973).

[*T. elatior* Sibth. et Sm., Fl. Graec. Prodr., 2: 108 (1813) pro parte non Fl. Graec., 8: 45, t. 762 (1838); Poech, Enum. Plant. Ins. Cypr., 38 (1842); Unger et Kotschy, Die Insel Cypern, 376 (1865); Boiss., Fl. Orient., 2: 83 (1872) pro parte quoad spec. cypr.]

[*T. corniculata* (non L.) Kotschy in Unger et Kotschy, Die Insel Cypern, 376 (1865); Boiss., Fl. Orient., 2: 83 (1872); Osorio-Tafall et Seraphim, List Vasc. Plants Cyprus, 60 (1973).]

TYPE: " in Graeciâ Boiss. pl. exs. 1842, in Asiâ minori propè *Smyrnam* Boiss. 1842, Balansa pl. exs. 1854 No. 177 sub *Tr. corniculatâ* " (G, K !).

An erect or diffuse annual 10–30 cm. high; stems glabrous, or thinly clothed with long hairs, usually much branched with elongate internodes; petioles 1–4 cm. long, channelled above; stipules ovate-acuminate, 3–9 mm. long, 1·5–3 mm. wide with sharply denticulate margins; leaflets subequal, oblong-obovate, 12–22 mm. long, 6–18 mm. wide, glabrous above, thinly pilose below, apex rounded or retuse, margins very shortly denticulate; peduncle 2–7 cm. long, usually much exceeding leaves, erect or erecto-patent; inflorescence a rather dense, subglobose raceme, 12–18 mm. diam.; pedicels slender, 1–3 mm. long, ± pilose; bracts narrowly deltoid-setaceous, membranous, a little shorter than pedicels, often thinly pilose; flowers distinctly reflexed; calyx narrowly campanulate, about 4 mm. long and 2 mm. wide, with linear-subulate teeth 1·5–2 mm. long, the 2 upper (adaxial) lobes distinctly longer than the 3 lower; petals bright yellow, standard oblong-obovate, 5–6 mm. long, 3·5–4 mm. wide, apex slightly retuse, base tapering to a short claw, wings oblong, about 5 mm. long, 1·5 mm. wide, apex truncate-rounded, base with a distinct auricle, an obscure conjunctival flap (not a tooth) and a long slender claw, keel about 6 mm. long, 2 mm. wide, curved to a short obtuse apex; stamens about 4·5 mm. long, all except the vexillary filament connate for about 3·5 mm.; anthers oblong about 0·4 mm. long and 0·2 mm. wide; ovary narrowly oblong about 2·5 mm. long, 0·5 mm. wide, style about 3 mm. long, stigma minute. Pod oblong, laterally flattened, falcately upcurved, about 1·3–2·3 cm. long, 0·3–0·4 cm. wide, glabrous, apex obtuse or subacute, valves distinctly veined; seeds oblong, compressed, 2–2·3 mm. long, 1–1·3 mm. wide; testa bright brown, verruculose.

HAB.: In cultivated fields and waste ground; sea-level to ? 1,000 ft. alt.; fl. March–June.

DISTR.: Division 6, rare. Also S. Greece, Crete, Aegean Islands and Turkey.

6. In stubble fields near Nicosia, 1859, *Kotschy* S.1020. Also recorded from Cyprus by *Sibthorp*.

NOTES: Not seen in recent years and possibly a casual occurrence. *T. balansae* is very closely allied to the western and central Mediterranean *T. corniculata* L., differing chiefly in its more compact, subglobose inflorescences and infructescences and blunter pods; it should probably be considered a subspecies of *T. corniculata*.

2. T. sprunerana *Boiss.*, Diagn., 1, 2: 17 (1843)—as *T. sprunneriana*, Fl. Orient., 2: 80 (1872); Sintenis in Oesterr. Bot. Zeitschr., 32: 398 (1882); Holmboe, Veg. Cypr., 104 (1914); Širjaev in Publ. Fac. Sci. Univ. Masaryk, no. 110: 24 (1929)—as *T. sprunneriana*; Davis, Fl. Turkey, 3: 461 (1970); Osorio-Tafall et Seraphim, List Vasc. Plants Cyprus, 60 (1973).

T. torulosa Griseb., Spic. Fl. Rum. et Bith., 1: 40 (1843).

T. sibthorpii Boiss., Diagn., 1, 9: 14 (1849), Fl. Orient., 2: 81 (1872); Holmboe, Veg. Cypr., 104 (1914); Širjaev in Publ. Fac. Sci. Univ. Masaryk, no. 110: 26 (1929); Post, Fl. Pal., ed. 2, 1: 316 (1932).

[*T. hamosa* (non L.) Sibth. et Sm., Fl. Graec. Prodr., 2: 108 (1813), Fl. Graec., 8: 47, t. 764 (1833); Unger et Kotschy, Die Insel Cypern, 376 (1865).]

TYPE: Greece; Attica, *Spruner* (K !).

Sprawling or prostrate annual 2–15 cm. high; stems much branched, especially from base, angular, ± densely pubescent; petioles 1–2 (–3·5) cm. long, channelled above, ± densely pubescent; stipules narrowly lanceolate-acuminate, 3–7 mm. long, 1–2·5 mm. wide, pubescent, with entire or subentire margins; leaflets subequal, obovate, 4–10 (–14) mm. long, 3–5 (–9) mm. wide, pubescent, apex rounded or truncate, margins minutely denticulate; peduncle 1–3 cm. long, equalling or exceeding leaves, erect or erecto-patent, ± pubescent, thickening and elongating in fruit; inflorescence a rather lax globose raceme, 12–17 mm. diam., rhachis pubescent or lanuginose, produced beyond the inflorescence as a short, rigid, persistent arista; pedicels 0·5–2 mm. long, pubescent; bracts narrowly linear-filiform, membranous, 0·5–1·5 mm. long; flowers patent and reflexed; calyx narrowly campanulate about 3 mm. long, 1·5 mm. wide, pubescent, teeth unequal, the 2 upper (adaxial) narrowly deltoid-subulate, about 1·5 mm. long, the 3 lower acutely deltoid about 0·8–1 mm. long; petals bright yellow, standard oblong-elliptic, 6–7 mm. long, about 3 mm. wide, apex rounded or retuse, base tapering to a distinct, rather narrow claw, wings 5–6 mm. long, about 1·5 mm. wide, apex rounded or truncate, base with a prominent auricle and long, slender claw, conjunctival tooth barely developed, keel 6–7 mm. long, 2·5 mm. wide, curved to a rounded apex; stamens about 6 mm. long, all except the vexillary filament connate for about 4 mm.; anthers oblong, about 0·4 mm. long, 0·2 mm. wide; ovary narrowly oblong, about 3–3·5 mm. long, densely pilose, tapering to a glabrous style 2·5–3·5 mm. long, stigma minute. Pod subterete, sharply reflexed with the apical half generally falcately upcurved, (1–) 1·3–2·5 cm. long, 0·2–0·25 cm. wide, ± angled, pubescent, obscurely veined, with a sharp, pointed beak; seeds oblong, 1·5–2 mm. long, about 1·2 mm. wide; testa pale brown, minutely granulose-verruculose.

HAB.: In coastal garigue; on rocky slopes by the sea, or on sand-dunes; sea-level to 100 ft. alt.; fl. Febr.–May.

DISTR.: Divisions 2, 3, 7, 8, common in Division 7. Also in Greece, Aegean Islands, Crete, Turkey, Syria, Palestine and eastwards to Iran.

2. Mile 42, Xeros–Polis road, 1957, *Merton* 2952 ! Mile 49, Xeros–Polis road, 1957, *Merton* 2956 ! Xeros–Polis road, near Kokkina Point, *D. R. Harris* in *Merton* 3103 !
3. Skarinou, in shingly river bed, 1941, *Davis* 2786 !
7. Common; *Syngrassides* 180 ! *Casey* 385 ! 1114 ! 1175 ! 1689 ! *G. E. Atherton* 1192 ! *P. H. Oswald* 122 !
8. Near Cape Andreas, 1880, *Sintenis & Rigo* 448 (see note) ! Ayios Philon near Rizokarpaso, 1941, *Davis* 2213 ! Near Cape Andreas, 1957, *Merton* 2914 !

NOTES: Indistinguishable from *T. sibthorpii*, which is rejected as a later synonym; the question of priority between the names *T. sprunerana* Boiss. (1843) and *T. torulosa* Griseb. (1843) has not been finally settled; Boissier's and Grisebach's publications both appeared in the same year, apparently in March (see Stafleu in Regnum Vegetabile, 52: 42, 184; 1967). The problem must be left to expert bibliographers.

Kew sheets of *Sintenis & Rigo* 448 are variously labelled " In mont. pr. Kythraeam 26/3 [1880] " and " In mont. pr. Cerignia. Majo. [1880] ". In fact it seems probable that all this material was collected near Cape Andreas on 26th April 1880; the mature condition of the specimens suggests that they were collected in late spring, and there is no evidence that *T. sprunerana* grows " on the mountains " anywhere in Cyprus.

3. T. strangulata *Boiss.*, Diagn., 1, 9: 17 (1849), Fl. Orient., 2: 79 (1872); Širjaev in Publ. Fac. Sci. Univ. Masaryk, no. 110: 35 (1929); Post, Fl. Pal., ed. 2, 1: 314 (1932); Davis, Fl. Turkey, 3: 466 (1970).

TYPE: Lebanon; " in collibus siccis Antilibani circa Rascheya " *Boissier* (G).

Erect or spreading annual 10–15 cm. high; stems usually much branched, somewhat flexuous, thinly crispate-pubescent; petioles slender, 1–2·5 cm. long, thinly pubescent or subglabrous; stipules deltoid-subulate, 1–2·5 mm. long, 0·5–1 mm. wide, pubescent or glabrescent; leaflets subequal, broadly obovate or suborbicular 7–12 mm. long, 5–10 mm. wide, glabrous above, thinly pubescent below, apex rounded or subtruncate, margins minutely denticulate; peduncle slender, distinctly angled, 1–2·5 cm. long, glabrescent or shortly crispate-pubescent, elongating slightly in fruit; inflorescence a lax subglobose raceme, rhachis pubescent, often produced beyond the inflorescence as a short arista; bracts minute, ovate-cuspidate, about 0·5 mm. long, membranous, caducous; pedicels very slender, about 1 mm. long; flowers patent or some reflexed; calyx campanulate, 2·5 mm. long, 1·5 mm. wide, thinly pubescent or subglabrous, teeth deltoid, the upper (adaxial) about 1 mm. long, the lower (abaxial) about 0·8 mm. long; petals yellow; standard obovate, about 6 mm. long, 3 mm. wide, apex rounded, base tapering to a short claw, wings narrowly oblong, about 6 mm. long, 1·2 mm. wide, apex rounded or subtruncate, base with a conspicuous auricle and slender claw, conjunctival tooth obsolete, keel about 6 mm. long, 1·6 mm. wide, upcurved to a blunt apex; stamens about 6 mm. long, filaments connate for about 4 mm.; anthers suborbicular, about 0·3 mm. diam.; ovary narrowly oblong, about 3·5 mm. long, 0·8 mm. wide, densely pubescent, narrowing rather suddenly to a glabrous style 2·5 mm. long, stigma minute. Pod reflexed, straight or ± upcurved, 1–1·8 cm. long, 0·3 cm. wide, terete, somewhat constricted between the seeds, apex narrowed to a short sharp beak, valves pale brown, papery, thinly pubescent, longitudinally veined; seeds oblong-ovoid, about 2·5 mm. long, 1·8 mm. wide; testa pale brown, closely covered with small, irregular dark brown warts.

HAB.: Dry rocky limestone hillside; 1,000–1,700 ft. alt.; fl. March–May.

DISTR.: Division 7, rare. Also Turkey, Syria, Lebanon and eastwards to Iraq.

7. Hillside S.W. of Kyrenia, 1949, *Casey* 647 ! Above Kyrenia Pass, S.E. aspect, 1953, *Casey* 1315 !

4. T. spinosa *L.*, Sp. Plant., ed. 1, 777 (1753); Boiss., Fl. Orient., 2: 75 (1872); Širjaev in Publ. Fac. Sci. Univ. Masaryk, no. 128: 5 (1930); Post, Fl. Pal., ed. 2, 1: 313 (1932); A. K. Jackson in Kew Bull., 1934: 271 (1934); Davis, Fl. Turkey, 3: 461 (1970); Zohary, Fl. Palaest., 2: 128, t. 188 (1972); Osorio-Tafall et Seraphim, List Vasc. Plants Cyprus, 60 (1973).

TYPE: " *in* Creta ".

Erect or sprawling annual, 5–20 cm. high; stems usually much branched, distinctly angled, subglabrous or thinly pilose; leaves rather distant; petioles very short, usually less than 7 mm. long, channelled above, thinly pilose; stipules ovate-acuminate, 3–4 mm. long, 2–3 mm. wide, with deeply serrate or laciniate margins; leaflets narrowly obovate-cuneate, 4–12 mm. long, 2–6 mm. wide, apex rounded, truncate or slighty retuse, margins of upper half of leaflet sharply denticulate, midrib ± pilose below; inflorescence sessile or subsessile, consisting of an axillary cluster of 2–6 flowers; rhachis produced into a short arista; pedicels recurved, 1–1·5 mm. long, bracts ovate-cuspidate, minute, less than 0·5 mm. long and about as wide; calyx narrowly campanulate, about 3·5 mm. long, 1–1·2 mm. wide, glabrous or subglabrous, teeth subequal, narrowly deltoid-subulate, about 1 mm. long; petals yellow, standard oblong, about 4 mm. long, 2 mm. wide, apex rounded or subacute, base tapering to a short claw, wings narrowly oblong, about 4 mm. long and 0·8 mm. wide, apex rounded, base with a prominent auricle and slender claw, conjunctival tooth reduced to a small swelling, keel about 4·5 mm. long, 1–1·3 mm. wide, curving upwards to a short blunt apex; stamens about 4·5 mm. long, vexillary stamen free, filaments of other stamens connate for about 3·5 mm.; anthers oblong, about 0·3 mm. long, 0·2 mm. wide; ovary narrowly oblong, about 4 mm. long, 0·5 mm. wide, glabrous, ± densely glandular, style 0·7 mm. long, glabrous, stigma minute. Pod linear, 40–60 mm. long, 1·5–2 mm. wide, strongly falcate-upcurved, sometimes forming an almost complete circle, glabrous, valves prominently transversely veined, apex blunt or subacute; seeds narrowly oblong, 2–2·5 mm. long, 1 mm. wide; testa pale brown, verruculose.

HAB.: In cultivated fields, gardens, roadsides etc., on open ground; about sea-level; fl. March–May.

DISTR.: Divisions 7, 8, rare. Crete, Rhodes, S. Turkey, Syria, Lebanon and Palestine; adventive elsewhere.

7. In nursery garden, Kyrenia, 1933, *Syngrassides* 182 ! 5 miles W. of Kyrenia, 1956, *Merton* 2670 !
8. Kantara Castle, 1880, *Sintenis & Rigo* 450.

5. T. spicata *Sibth. et Sm.*, Fl. Graec. Prodr., 2: 108 (1813), Fl. Graec., 8: 46, t. 763 (1833); Boiss., Fl. Orient., 2: 86 (1872); Holmboe, Veg. Cypr., 104 (1914); Širjaev in Publ. Fac. Sci. Univ. Masaryk, no. 128: 9 (1930); Post, Fl. Pal., ed. 2, 1: 318 (1932); Rechinger f. in Arkiv. för Bot., ser. 2, 1: 424 (1950); Davis, Fl. Turkey, 3: 477 (1970); Zohary, Fl. Palaest., 2: 134, t. 198 (1972); Osorio-Tafall et Seraphim, List Vasc. Plants Cyprus, 60 (1973).

TYPE: Aegean Islands; " in insulâ Seripho, ni fallar, legit Sibthorp " (OXF).

Erect, spreading or decumbent annual, 5–15 (–25) cm. high; stems usually much branched towards base, rarely simple, thinly adpressed-pilose, often reddish or purplish; leaves rather remote; petioles slender, 0·5–3 cm. long, slightly channelled above, thinly pilose; stipules setaceous, about 5 mm. long, 1–1·5 mm. wide at base, entire or subentire, thinly pilose; leaflets of the basal leaves broadly obcordate or suborbicular, about 8 mm. long, 7 mm. wide, glabrous above, thinly pilose along the midrib below, apex truncate, rounded or retuse, margins sharply denticulate, upper cauline leaflets oblong, 13–20 mm. long, 4–5 mm. wide, apex obtuse or subacute, margins often very conspicuously denticulate, and venation often prominent; flowers terminal and subterminal in dense shortly oblong or subglobose racemes; peduncles slender, erect, 1–3 cm. long, 4-angled, adpressed-pilose, usually exceeding subtending leaves; rhachis pilose, produced beyond raceme as a short arista; bracts filiform, membranous, about 1·3 mm. long; pedicels short, pilose, erect, about 1–2 mm. long; calyx nodding, narrowly campanulate, about 4 mm. long, 1·5 mm. wide, ± densely white-pilose, teeth subulate, very unequal, the 2 upper about 2·5 mm. long, the 2 median about 2 mm. long, the lowermost about 0·5 mm. long; petals yellow, standard narrowly obcordate, 5–6 mm. long, 2–2·5 mm. wide, apex retuse, base tapering gradually to a short claw, wings narrowly oblong, 5–5·5 mm. long, about 1 mm. wide, apex rounded, base with a conspicuous auricle, long slender claw, conjunctival tooth reduced to a slight swelling, keel about 5·5 mm. long, 1·2 mm. wide, upcurving gradually to a rounded or very blunt apex; stamens about 5 mm. long, the vexillary filament connate with the others for about 1 mm., the remainder connate for about 2 mm.; anthers oblong, about 0·4 mm. long, 0·2 mm. wide; ovary narrowly ovoid, about 1·5 mm. long, 0·7 mm. wide, densely silky-hairy, tapering to a glabrous style about 3·5 mm. long, stigma small, capitate. Pod strongly deflexed, ovoid, compressed, 3–4 mm. long, 2–2·5 mm. wide, valves thinly pilose, rather obscurely veined, apex produced into a long, slender, upcurved beak, about 4 mm. long (formed by the persistent style); seeds generally solitary, broadly ovoid, compressed, 1·8–2 mm. long, 1·5 mm. wide; testa dull olive-brown, densely verruculose.

HAB.: Dry hillside pastures; in garigue; occasionally in Olive groves or on sand-dunes; sea-level to 4,100 ft. alt.; fl. March–June.

DISTR.: Divisions 2, 3, 7, 8, very common in Division 7, common in Division 8. Also Bulgaria, Greece, Aegean Islands, Turkey, Syria, Lebanon, Palestine, and east to Iraq and Iran.

2. Kryos Potamos, 1938, *Kennedy* 1262 !
3. Limassol, 1930, *E. Wall* s.n.; Arminou, 1941, *Davis* 3442 ! 4 miles W. of Kandou, Limassol, 1963, *J. B. Suart* 16 !

7. Very common; *Sintenis & Rigo* 447 ! *Syngrassides* 1381 ! *Davis* 2919 ! *Casey* 558 ! 566 ! 618 ! 1176 ! 1297 ! *Kennedy* 1764 ! *Merton* 2084 !
8. Common; *Sintenis & Rigo*; *Haradjian* 168, 172; *Merton* 2926 ! *Meikle* 2480 ! 2997 !

6. T. monspeliaca *L.*, Sp. Plant., ed. 1, 777 (1753); Fl. Graec. Prodr. 2: 109 (1813), Fl. Graec., 8: 47, t. 765 (1833); Unger et Kotschy, Die Insel Cypern, 376 (1865); Boiss., Fl. Orient., 2: 76 (1872); Holmboe, Veg. Cypr., 104 (1914); Širjaev in Publ. Fac. Sci. Univ. Masaryk, no. 136: 20 (1931); Post, Fl. Pal., ed. 2, 1: 313 (1932); Lindberg f., Iter Cypr., 20 (1946); Davis, Fl. Turkey, 3: 476 (1970); Zohary, Fl. Palaest., 2: 133, t. 197 (1972); Osorio-Tafall et Seraphim, List Vasc. Plants Cyprus, 60 (1973).

TYPE: France; "Monspelii".

Prostrate or decumbent annual 1–12 cm. high, stems branched radiately from the base, subglabrous or thinly pilose; leaves rather distant; petioles slender, 1–2·5 cm. long, channelled above, subglabrous or thinly pilose; stipules ovate-caudate, 4–6 mm. long, 1–1·5 mm. wide, pubescent, margins deeply and sharply serrate-laciniate; leaflets grey-green, obovate or obcuneate, 2·5–8 mm. long, 2–7 mm. wide, pubescent on both sides, lateral veins prominent, apex truncate or rounded, margins sharply denticulate; flowers 3–8 in sessile or very shortly pedunculate axillary clusters; peduncle (when present) generally less than 3 mm. long, densely pubescent; rhachis pubescent, produced into a short persistent arista; bracts subulate, membranous, 1–2 mm. long; calyx narrowly campanulate, 1·5 mm. long, 0·6 mm. wide, pubescent, teeth subequal, subulate, about 0·8 mm. long; petals yellow, small, standard oblong-obovate with a very short claw, about 3–3·5 mm. long, 1·5 mm. wide, apex emarginate, wings oblong, 2·5–3 mm. long, 1 mm. wide, apex rounded or subacute, base with a conspicuous auricle and well-developed conjunctival tooth, keel very short, 2–2·5 mm. long, 0·8 mm. wide, upcurving to a very blunt apex; stamens 2 mm. long, vexillary stamen free, the others connate for 1·6 mm.; anthers broadly oblong, about 0·2 mm. long, 0·3 mm. wide; ovary oblong, 1·6 mm. long, 0·5 mm. wide, densely pubescent, tapering to a very short, glabrous, curved style about 0·4 mm. long, stigma terminal, relatively conspicuous. Pod deflexed, oblong, 8–12 mm. long, 1–1·5 mm. wide, strongly compressed, straightish or more usually slightly upcurved, apex blunt or subacute, tipped with the persistent upcurved style, valves thinly pubescent or glabrous, strongly transverse-veined; seeds oblong or oblong-rhomboid, compressed, 1–1·5 mm. long, 0·8–1 mm. wide; testa pale or bright brown, rugose-undulate (not tuberculate).

HAB.: Rocky hillsides; fallow or cultivated fields or on sandy seashores; sea-level to 3,750 ft. alt.; fl. March–June.

DISTR.: Divisions 1–5, 7, common in Division 7. Throughout the Mediterranean region, rare in Central Europe, and eastwards to Transcaucasia and Iran.

1. Karavopetres near Ayios Yeoryios (Akamas), 1962, *Meikle* 2065 !
2. Platres, 1938, *Kennedy* 1263 ! Kannaviou, in Ezuza R., 1939, *Lindberg f.* s.n.
3. Kalavasos, 1905, *Holmboe* 629.
4. Cape Pyla, 1880, *Sintenis & Rigo*.
5. Vatili, 1905, *Holmboe* 346; between Nisou and Stavrovouni, 1933, *Syngrassides* 26 ! Between Mia Milea and Mandres, 1933, *Syngrassides* 336 !
7. Common; *Sintenis & Rigo* 451 ! *Casey* 531 ! 1296 ! *Chapman* 87 ! 625 !

7. T. cariensis *Boiss.*, Diagn., 1, 2: 21 (1843), Fl. Orient., 2: 69 (1872); Sintenis in Oesterr. Bot. Zeitschr., 32: 398 (1882); Holmboe, Veg. Cypr., 104 (1914); Širjaev in Publ. Fac. Sci. Univ. Masaryk, no. 148: 27 (1932); Davis, Fl. Turkey, 3: 480 (1970); Osorio-Tafall et Seraphim, List Vasc. Plants Cyprus, 61 (1973).

T. foenum-graecum (non L.) Sibth. et Sm., Fl. Graec., 8: 48, t. 766 (1833).

TYPE: Turkey; "in *Cariâ* propè *Moglah*", *Aucher Éloy* 1160 (G, K !).

Sprawling or decumbent annual 5–20 (–25) cm. high; stems branched at base, usually unbranched above, thinly clothed with long, spreading hairs;

leaves rather remote; petioles 5–20 mm. long, flattened or narrowly winged, pilose; stipules ovate, acuminate, 5–12 mm. long, 2–3 mm. wide, irregularly serrate, thinly pilose; leaflets broadly obovate-cuneate, 5–14 mm. long, 4–12 mm. wide, glabrous (except along midrib below), glaucous, apex truncate, retuse or rounded, margins sharply denticulate; flowers erect, solitary or in pairs in the axils of the upper leaves; peduncle and bracts wanting, pedicels usually less than 4 mm. long, pilose; calyx cylindrical, about 12 mm. long, 3·5 mm. wide, thinly pilose, membranous with rather conspicuous nerves, teeth subulate, subequal, about 2·5 mm. long; petals yellowish-white stained bluish, standard oblong-obovate, about 25 mm. long, 10 mm. wide, the upper half sharply reflexed, the lower half tapering to a short claw, apex distinctly emarginate, wings oblong, 18–20 mm. long, 3–4 mm. wide, closely adherent to keel, apex rounded, base with a conspicuous hooked auricle and a very long slender claw, conjunctival tooth wanting, keel about 15 mm. long, 3·5 mm. wide, curving gently to a blunt, violet-blotched apex, claw very long and slender; stamens about 15 mm. long, vexillary stamen free for about 5 mm., the others free for about 2 mm.; anthers oblong, about 0·8 mm. long, 0·5 mm. wide; ovary narrowly oblong, about 10 mm. long, 1 mm. wide, very shortly pubescent, tapering to an erose-hairy beak about 5 mm. long, style glabrous, about 5 mm. long, stigma minute. Pod oblong, 3–7 cm. long, the seed-bearing portion 1·5–3 cm. long, 0·6–0·8 cm. wide, flattened, pubescent, distinctly reticulate-veined, narrowing rather suddenly above to form a straight rigid beak 2–4 cm. long, the adaxial margin ± straight, the abaxial much thickened and distinctly curved; seeds oblong, compressed, 3–4 mm. long, 2–3 mm. wide; testa bright brown, closely verruculose.

HAB.: Sandy seashore; about sea-level; fl. ? March–May.

DISTR.: Division 8, rare. Also Greece, Aegean Islands, Rhodes and Turkey.

8. North coast between Cape Andreas and Aphendrika, 1880, *Sintenis & Rigo.*

NOTES: Stated to be " very rare " by Sintenis & Rigo, and not seen since recorded by them. The plant is superficially very similar to *T. berythea*, but at once distinguished by its abruptly beaked fruits. The Sibthorp record for *T. foenum-graecum* L. " in Rhodo et Cypro insulis " may be an error for *T. cariensis*, but in view of the occurrence of true *T. foenum-graecum* on Cyprus, the identification must be considered dubious.

8. **T. foenum-graecum** *L.*, Sp. Plant., ed. 1, 777 (1753); Boiss., Fl. Orient., 2: 70 (1872); Širjaev in Publ. Fac. Sci. Univ. Masaryk, no. 148: 28 (1932); Post, Fl. Pal., ed. 2, 1: 312 (1932); Davis, Fl. Turkey, 3: 481 (1970); Zohary, Fl. Palaest., 2: 135, t. 199 (1972); Osorio-Tafall et Seraphim, List Vasc. Plants Cyprus, 61 (1973).
TYPE: France; " Monspelii ".

Erect or spreading annual, 15–40 (–50) cm. high; stems branched towards base, mostly unbranched above, glabrous or thinly pilose; leaves often rather remote; petioles generally short, less than 1 cm. long, but occasionally up to 4 cm. long, flattened, ± distinctly winged, subglabrous or thinly pilose; stipules narrowly ovate-acuminate, 5–7 mm. long, 1–2 mm. wide, entire or subentire, thinly pilose; leaflets deep bluish-green, those of the lower leaves broadly obovate and subentire, those of the upper leaves narrowly obovate-cuneate, acutely denticulate, 1–2·5 (–3·5) cm. long, 0·4–1 (–2) cm. wide, glabrous above, thinly pilose or subglabrous below, apex rounded or occasionally shallowly retuse; flowers erect, solitary, axillary; peduncle and bracts wanting; pedicels very short, usually less than 1 mm. long, pilose; calyx cylindrical-campanulate, membranous, 6–7 mm. long, 2 mm. wide, thinly pilose, teeth subulate, subequal, about 2–3 mm. long; petals yellowish-white, standard narrowly obovate 10–11 mm. long, about 5 mm. wide, upper half not reflexed, apex sharply emarginate, base tapering to a very short claw distinctly constricted about $\frac{1}{3}$ way up, wings oblong, about 10 mm. long, 2 mm. wide, apex blunt or rounded, base somewhat gibbous, with a very conspicuous auricle and slender claw, conjunctival tooth wanting, keel 5–6 mm. long, apical part rounded, almost orbicular, about 2 mm. diam., claw long and slender; stamens 5–6 mm. long, vexillary stamen (? always) free, other stamens connate for about 4–5 mm.; anthers small, oblong, about 0·4 mm. long, 0·3 mm. wide; ovary narrowly oblong, minutely glandular-papillose, about 2·5–3 mm. long, 0·8 mm. wide, style very short, about 1 mm. long, upcurved, stigma capitate, relatively conspicuous. Pod linear, straightish or curved downwards, 7–12 cm. long, 0·4–0·5 cm. wide, tapering gradually to a slender beak, dorsal margin distinctly thickened, valves glabrous, very thinly glandular, openly longitudinally veined; seeds oblong, 3–4·5 mm. long, 2·5–3 mm. wide, ± compressed; testa rich brown, rather sparsely verruculose.

HAB.: Cultivated fields; sea-level to 200 ft. alt.; fl. Febr.–April.

DISTR.: Division 8, rare. Throughout southern Europe and the Mediterranean region eastwards to Central Asia; adventive elsewhere; often cultivated and naturalized.

8. Ayios Philon, near Rizokarpaso, 1941, *Davis* 2237 ! Elisis near Galinoporni, 1962, *Meikle* 2491 !

NOTES: The status of *T. foenum-graecum* in Cyprus is obscure; Širjaev considers it indigenous only in Mesopotamia and Kurdistan, but it is a well-established, and presumably ancient, weed of cultivated land in other areas. The Sintenis and Rigo record for *T. cariensis* (see species 7 *supra*) is suspicious in view of the fact that *T. foenum-graecum* is now known to occur as a wild plant in the same area, but, as no voucher specimen has been located, it must be assumed that the determination is correct.

9. T. berythea *Boiss. et Bl.* in Boiss., Diagn., 2, 2: 10 (1856), Fl. Orient., 2: 70 (1872); Širjaev in Publ. Fac. Sci. Univ. Masaryk, no. 148: 38 (1932); Post, Fl. Pal., ed. 2, 1: 312 (1932); Zohary, Fl. Palaest., 2: 135, t. 199a (1972).

[*T. foenum-graecum* (non L.) Širjaev in Publ. Fac. Sci. Univ. Masaryk, no. 148: 37 (1932) pro parte quoad spec. cypr.]

TYPE: Lebanon; " ad promontorium *Raz Beyrout* propè *Beyrout* et in collibus ad radices *Libani* suprâ *Saida* Blanche " (G, K !).

Sprawling or decumbent annual, 4–10 (–15) cm. high; stems unbranched or branched towards the base, glabrescent or thinly clothed with long spreading hairs; leaves rather remote; petioles of lowermost leaves up to 5 cm long, flattened or very narrowly winged, glabrous or thinly pilose, those of upper leaves 1–2·5 cm. long, often distinctly pilose; stipules conspicuous, ovate, acute (not caudate-acuminate), 5–10 mm. long, 2–4 mm. wide, membranous, with reticulate greenish venation, subglabrous or thinly pilose; leaflets broadly obovate-cuneate, 6–20 mm. long, 5–18 mm. wide, glabrous above and glaucous with a distinct pale midrib, subglabrous or very thinly pilose below, apex rounded or truncate, subentire or with small, sharp, rather distant teeth; flowers solitary, axillary, sessile or with very short thick pedicel; peduncle and bracts wanting; calyx cylindrical, 6–8 mm. long, 2–2·5 mm. wide, membranous, pallid, thinly pilose, subulate, 2–3 mm. long, 0·8 mm. wide at base; petals pale steely blue, standard narrowly obovate, 12–14 mm. long, 5–6 mm. wide, folded longitudinally and adherent to wings and keel, apex emarginate, base tapering to a very short claw, with a marked constriction about $\frac{1}{3}$ way up; wings oblong, about 10 mm. long, 2·5 mm. wide, with a broad rounded apex, base gibbous with a conspicuous elongate auricle and a long very slender claw, conjunctival tooth wanting, keel about 7 mm. long, 2–2·5 mm. wide, curving to a short, blunt apex, and with a long slender claw; stamens about 7 mm. long, vexillary stamen connate for about 3 mm., or perhaps occasionally quite free, other stamens connate for about 5 mm.; anthers oblong, about 0·5 mm. long, 0·2 mm. wide; ovary narrowly oblong, minutely glandular-papillose, about 5 mm. long, 0·8 mm. wide, style very short, about 1 mm. long, upcurved, stigma capitate, relatively conspicuous. Pod linear, straightish or curved downwards, 7–12 cm. long, 0·4–0·5 cm.

wide, tapering gradually to a slender beak, dorsal margin distinctly thickened, valves glabrous or very thinly glandular, closely veined with longitudinal anastomosing veins; seeds oblong, 3·4·5 mm. long, 2·5–3 mm. wide, ± compressed; testa rich brown, closely and rather irregularly verruculose.

HAB.: On rock ledges or amongst garigue on limestone or, less often, igneous hills; 500–3,000 ft. alt.; fl. March–April.

DISTR.: Divisions 1, 6, 7. Also Syria, Lebanon and Palestine.

1. Smyies (Akamas), 1948, *Kennedy* 1605 ! Smyies, 1962, *Meikle* 2187 ! Neokhorio (Akamas), 1962, *Meikle* 2203 !
6. West of Orga, 1956, *Merton* 2582 !
7. Bellapais, 1880, *Sintenis & Rigo* 671 ! Ayios Dhimitrios, 1956, *G. E. Atherton* 1174 ! Forest Road West 4 miles, Kyrenia, 1956, *G. E. Atherton* 1280 !

8. MEDICAGO *L.*

Sp. Plant., ed. 1, 778 (1753).
Gen. Plant., ed. 5, 339 (1754).
Urban in Sitz. Verh. Bot. Ver. Brandenb., 15: 1–85 (1873).
C. C. Heyn, " The Annual Species of Medicago ", in Scripta Hierosolymitana, 12: 1–154 (1963).

Annual, biennial or perennial herbs or rarely subshrubs, not noticeably aromatic when dried; stems erect, diffuse or decumbent, glabrescent or ± pilose, rarely glandular; leaves pinnately 3-foliolate, leaflets obovate, obcuneate or oblong, denticulate or lobed, rarely entire, glabrous or ± pilose, terminal leaflet usually petiolulate, lateral leaflets sessile or subsessile; stipules adnate to petiole; peduncle axillary, usually well developed; inflorescence capitate-racemose or flowers few or solitary; bracts small, membranous; pedicels short, slender; calyx campanulate, with 5 subequal teeth; corolla yellow, rarely purple or violet; petals free from staminal tube, glabrous, caducous; standard obovate or oblong, rounded or retuse at apex, narrowed to a cuneate base, usually longer than keel; wings oblong usually longer than keel, apex generally rounded and entire, base auriculate with, rarely without, a distinct conjunctival tooth, keel obtuse; vexillary stamen free or partly connate, filaments not dilated at apex; anthers 2-thecous, uniform; ovary sessile or shortly stipitate, 1–many-ovulate, straightish or coiled, style glabrous, stigma terminal or oblique, capitate or subcapitate. Pod generally indehiscent or tardily dehiscent, exserted from persistent calyx, normally falcate-curved or coiled spirally, often spinose; seeds estrophiolate, testa smooth or tuberculate; cotyledons without a distinct basal articulation.

About 50 species, widely distributed in Europe and the Mediterranean region and extending eastwards to China. Often introduced and naturalized outside their natural area of distribution.

The generic limits of *Medicago*, *Trigonella* and *Melilotus* are by no means clearly defined (see Heyn, *op. cit.* pp. 19–22), nor are there any characters which in isolation can be used infallibly to distinguish the three genera. Apart from general appearance, the absence of an auriculation (or node) at the base of the cotyledon, the presence of a more or less well-developed floral tripping mechanism, together with the characteristic coiling of the pods, will usually suffice to distinguish *Medicago* from its allies.

Flowers 8–10 mm. long; pod almost straight, or falcate, or forming an open spiral; plants perennial; leaflets glabrous or thinly pilose:
- Pods almost straight or falcate; flowers yellow - - - - [M. FALCATA p. 411]
- Pods strongly falcate-circinnate, or coiled $1\frac{1}{2}$–$3\frac{1}{2}$ times:
  - Pod coiled $1\frac{1}{2}$–$3\frac{1}{2}$ times; flowers violet or purple - - - **1. M. sativa**
  - Pod falcate-circinnate, or coiled 1–$1\frac{1}{2}$ times; flowers yellow, violet, purple, greenish or whitish - - - - - - - - - **1a. M. × varia**

Flowers 2–8 mm. long; pod reniform or coiled into a close spiral; petals yellow:
Pods reniform, blackish when ripe; flowers in dense globose or oblong-cylindrical clusters - - - - - - - - - - - - **20. M. lupulina**
Pods spirally coiled:
Leaves and stems densely villose; inflorescences capitate-racemose, conspicuous; perennial plant of sandy seashores and sand-dunes - - - **2. M. marina**
Leaves and stems not densely villose; plants annual:
Pod without spines, flat, discoid, 12–16 mm. diam.; seeds with a tuberculate-rugulose testa - - - - - - - - - **3. M. orbicularis**
Pod not as above; seeds with a smooth testa:
Pods without spines, forming a depressed-globose rosette 10–17 mm. diam. **4. M. scutellata**
Pods not as above:
Pod without spines:
Stems and leaves glandular; pod depressed-globose or discoid:
Lateral surfaces of pod prominently reticulate-veined; standard about 4 mm. long - - - - - - - - **5. M. rugosa**
Lateral surface of pod smooth or obscurely veined; standard about 7 mm. long - - - - **6. M. blancheana** var. **bonarotiana**
Stems and leaves not glandular; pod cylindrical or barrel-shaped:
Peduncle long and slender, usually exceeding subtending leaf; inflorescence usually 3–8-flowered - - - - - **18. M. turbinata**
Peduncle rather short, not exceeding subtending leaf; inflorescence 1–3-flowered - - - - - **14. M. littoralis var. inermis**
Pod spinose (spines sometimes short and blunt):
Dorsal surface of pods, and spines, ± densely clothed with short multicellular hairs:
Spines short, thick, erect, pods 5–8 (–12) mm. diam. **16. M. rigidula**
Spines long, slender, spreading; pods 9–15 (–17) mm. diam. **19. M. intertexta var. ciliaris**
Dorsal surface of pods, and spines, glabrous or thinly clothed with simple hairs:
Fruiting peduncles long and slender, usually equalling or exceeding the subtending leaves:
Pods less then 5 mm. diam.; leaves and stems distinctly pilose; inflorescences capitate - - - - - - **8. M. coronata**
Pods more than 5 mm. diam.:
Spines spreading at a right angle to lateral surfaces of pod; leaflets narrowly obovate or elliptic, sometimes lobed or lacinate **7. M. rotata**
Spines erect; leaflets broadly obovate, obcuneate or obcordate, not lobed or laciniate:
Pods discoid or very shortly cylindrical, spines often long and reflexed; distal coils of pod often unarmed **11. M. disciformis**
Pods subglobose, ovoid or barrel-shaped, spines usually short and thick:
Inflorescence usually 1–2-flowered; standard about 4 mm. long; pod subglobose or broadly ovoid; small prostrate plant, stems usually less than 15 cm. long **17. M. constricta**
Inflorescence usually 3–8-flowered; standard about 6 mm. long; pod usually barrel-shaped; rather robust plant, stems 15–50 cm. long - **18. M. turbinata var. apiculata**
Fruiting peduncles generally shorter than subtending leaves:
Leaflets glabrous above, subglabrous or thinly pilose below:
Inflorescence 1–2-flowered; leaflets small, 2–6 mm. long **9. M. praecox**
Inflorescence 2–8 (–10)-flowered; leaflets large 5–20 (–30) mm. long:
Lateral surfaces of pods distinctly reticulate-veined; pod usually with erect spines; leaflets without blotches; stipules laciniate with slender filiform teeth; stems glabrous or subglabrous - - - - - - - - - **12. M. polymorpha**
Lateral surfaces of pods obscurely veined; pod with spreading spines; leaflets usually blotched; stipules deeply dentate or shortly laciniate with narrowly deltoid teeth; stems normally clothed with scattered multicellular hairs **13. M. arabica**
Leaflets ± pilose or pubescent on both surfaces:
Stipules entire or subentire; inflorescence usually 3–6-flowered; pods with slender spines - - - - - **10. M. minima**

Stipules dentate-laciniate; inflorescence 1–3-flowered; pods with rather short, thick spines:
Dorsal surface of pod glabrous with erect or suberect spines; dorsal suture not distinctly raised - **14. M. littoralis**
Dorsal surface of pod ± pilose with spreading spines; dorsal suture raised with a furrow on either side **15. M. truncatula**

SECT. 1. **Medicago** Perennial herbs. Pods falcate or forming an open spiral, often without spines; seeds 2 or more in each pod; testa smooth; radicle about half as long as cotyledons.

[M. FALCATA *L.*, Sp. Plant., ed. 1, 779 (1753); Boiss., Fl. Orient., 2: 93 (1872); Post, Fl. Pal., ed. 2, 1: 321 (1932); Osorio-Tafall et Seraphim, List Vasc. Plants Cyprus, 61 (1973).

M. sativa L. ssp. *falcata* (L.) Naeg. et Thell., in Viertelj. Naturf. Ges. Zürich, 50: 272 (1905); van Ooststroom et Reichgelt in Act. Bot. Neerl., 7: 97 (1958).
TYPE: " *in* Europae *pratis apricis, siccis* ".

Decumbent or sprawling perennial, 25–40 (–50) cm. high; stems glabrous or thinly pubescent, distinctly angular above; petioles slender, 4–10 mm. long, flattened above, stipules lanceolate-acuminate, 4–10 mm. long, 1–2 mm. wide, entire or with a few, irregular teeth, shortly adnate to petiole; leaflets narrowly oblong-oblanceolate 8–15 (–20) mm. long, 2–8 mm. wide, glabrous or glabrescent above, thinly pubescent below, apex rounded or truncate, the protruding midrib often forming a short cusp, margins minutely denticulate above, entire and tapering to the base; inflorescence a globose or capitate raceme; peduncle 1–2 cm. long, adpressed-pubescent, usually exceeding subtending leaves; pedicels very slender, erecto-patent, the lower 2–5 mm. long the upper usually shorter; bracts setaceous, membranous, 1–2 mm. long; calyx campanulate, about 4 mm. long, 2 mm. wide, thinly pubescent externally, teeth narrowly subulate, subequal, about 2 mm. long; petals yellow (rarely whitish), standard obovate, 8–10 mm. long, 5–6 mm. wide, apex shortly retuse, base cuneate; wings oblong, 6–7 mm. long, 2–3 mm. wide, apex rounded, base with a very conspicuous auricle, slender claw, and well-developed conjunctival tooth; keel oblong, about 7 mm. long, 1·5 mm. wide, with a rounded apex and long basal claw; stamens about 7 mm. long, the vexillary partly free but connate towards the base; anthers broadly oblong, about 0·4 mm. long, 0·3 mm. wide; ovary narrowly oblong, about 4 mm. long, 0·5 mm. wide, glabrescent or thinly pubescent; style glabrous, slightly upcurved, about 2 mm. long, stigma conspicuous, oblique. Pod oblong, straight or ± distinctly falcate, 6–8 mm. long, 2·5 mm. wide, apex obtuse or acute, valves fuscous, glabrous or thinly adpressed-pubescent, obscurely veined; seeds oblong 1·5–2 mm. long, 1·5 mm. wide; testa bright brown, smooth.

Not yet positively recorded from Cyprus, though the presence of the hybrid *M. falcata* x *sativa* (see below) suggests that it may occur, or may have occurred, in the Troödos area. *M. falcata* is widely distributed in Europe, and is said to occur eastwards through Central Asia to China, though most records require confirmation.]

1. M. sativa *L.*, Sp. Plant., ed. 1, 778 (1753); Boiss., Fl. Orient., 2: 94 (1872); Post, Fl. Pal., ed. 2, 1: 221 (1932); Rechinger f. in Arkiv för Bot., ser. 2, 1: 424 (1950); Davis, Fl. Turkey, 3: 488 (1970); Zohary, Fl. Palaest., 2: 140, t. 204 (1972); Osorio-Tafall et Seraphim, List Vasc. Plants Cyprus, 61 (1973).

M. sativa L. ssp. *sativa* (L.) Naeg. et Thell., in Viertelj. Naturf. Ges. Zürich, 50: 272 (1905); van Ooststroom et Reichgelt in Act. Bot. Neerl., 7: 97 (1958).
TYPE: " *in* Hispaniae, Galliae *apricis* ".

Closely resembling *M. falcata* L., but usually more robust and more erect to about 80 cm. high; leaflets to 30 mm. long and 12 mm. wide, narrowly oblong; peduncles commonly 2·5–4 cm. long; inflorescence usually an elongate oblong raceme, frequently 3–5 cm. long; calyx as in *M. falcata*, up to 6 mm. long; petals violet or purple, standard 12 mm. long, 4–5 mm. wide, keel and wings as in *M. falcata* but usually a little larger; ovary about 5 mm. long, 0·5 mm. wide, pubescent. Pod twisted spirally into 1½–3½ open coils; seeds as in *M. falcata* but often rather larger.

HAB.: Cultivated as a forage plant, and sometimes escaping to roadsides and waste ground; 500–? 3,000 ft. alt.; fl. May–July.

DISTR.: Divisions 1, 7. Widespread as a cultivated or naturalized plant in Europe and western Asia, probably indigenous in Central Asia and China.

1. Stroumbi near Paphos, 1913, *Haradjian* 724 !
7. Kythrea, cultivated, 1936, *Syngrassides* 1280 ! Lapithos, 1955, *Miss Mapple* 93 ! Kyrenia, 1955, *G. E. Atherton* 291 ! 356 !

NOTES: The above records can be referred to *M. sativa* with some certainty, but most of the Cyprus material so labelled is in fact the following hybrid, and it is questionable if *M. sativa* is found except as a cultivated plant or as an obvious escape from cultivation. The specimens cited by Holmboe (Veg. Cypr., 104) are also probably referrable to *M.* × *varia*.

1a. M. × varia *Martyn*, Fl. Rustica, 3: no. 87 (1792) (*M. falcata* L. × *M. sativa* L.).
M. sativa L. ssp. *varia* (Martyn) Naeg. et Thell. in Viertelj. Naturf. Ges. Zürich, 50: 273 (1905); van Ooststroom et Reichgelt in Act. Bot. Neerl., 7: 98 (1958).
[*M. sativa* (non L.) H. Stuart Thompson in Journ. Bot., 44: 205 (1906); ? Holmboe, Veg. Cypr., 104 (1914); Lindberg f., Iter Cypr., 19 (1946).]
TYPE: The illustration in Martyn, Flora Rustica, t. 87 !

Intermediate in character between *M. falcata* and *M. sativa*, with yellow, violet, purple, greenish or occasionally whitish flowers, or with the colour varying in young and aged inflorescences. Pods often sparingly developed, strongly falcate-circinnate, or forming 1–1½ complete coils, but more loosely coiled than in *M. sativa*.

HAB.: In cultivated or waste land; by roadsides; on dry hillsides amongst garigue or under Pines, sometimes by ditches and irrigation channels; 2,750–6,300 ft. alt.; fl. May–July.

DISTR.: Division 2, common in the Troödos area. Widely distributed in Europe and the Mediterranean region, ? Asia; often introduced.

2. Common; *Holmboe* 885; 1052; *Haradjian* 510 ! 902 ! 988 ! *Syngrassides* 744 ! *Kennedy* 550–556 ! *Davis* 1836 ! 3518 ! *Casey* 834 ! 886 ! *Merton* 2793–2795 !

NOTES: Possibly a spontaneous hybrid between introduced *M. sativa* and indigenous *M. falcata*, or possibly imported as a forage plant, like *M. sativa*. Whatever its origin, *M.* × *varia* is now thoroughly established in the Troödos area, though apparently less successful at altitudes below 3,000 ft. It would be interesting to know if genuine *M. falcata* can be found in the same area.

2. M. marina *L.*, Sp. Plant., ed. 1, 779 (1753); Sibth. et Sm., Fl. Graec. Prodr., 2: 112 (1813), Fl. Graec., 8: 51, t. 770 (1833); Unger et Kotschy, Die Insel Cypern, 375 (1865); Boiss., Fl. Orient., 2: 96 (1872); Holmboe, Veg. Cypr., 104 (1914); Post, Fl. Pal., ed. 2, 1: 321 (1932); Lindberg f., Iter Cypr., 19 (1946); Rechinger f. in Arkiv för Bot., ser. 2, 1: 424 (1950); Davis, Fl. Turkey, 3: 501 (1970); Zohary, Fl. Palaest., 2: 141, t. 205 (1972); Osorio-Tafall et Seraphim, List Vasc. Plants Cyprus, 61 (1973).
TYPE: " *in maris* Mediterranei *littoribus* ".

Prostrate perennial with a thick woody rootstock and long, tough, thong-like roots; stems densely white-villose, 10–30 cm. long, much branched from the base, with shorter lateral branchlets; petioles short, seldom exceeding 6 mm., densely villose; stipules ovate, acute or acuminate, about 4–5 mm. long, 1·5–2 mm. wide at base, densely villose; leaflets obovate-cuneate, 4–10 mm. long, 3–6 mm. wide, densely villose on both sides, apex rounded or slightly retuse, usually apiculate, margins entire or obscurely denticulate; inflorescences axillary in the upper half of the branches, capitate-racemose; peduncle 1–1·5 cm. long, densely villose; flowers sweetly scented; bracts subulate, dilated at base, membranous, about 1 mm. long; pedicels slender, generally less than 2 mm. long, sometimes almost wanting; calyx narrowly campanulate, 4–5 mm. long, 1–1·5 mm. wide, villose, teeth subulate about 1·5 mm. long; petals bright yellow, standard obovate, 7–8 mm. long, about 5 mm. wide, apex retuse, reflexed, margins often slightly undulate, tapering to a rather wide cuneate base, wings oblong, 5–6 mm. long, 1·5 mm. wide, apex rounded, base with a conspicuous auricle and slender claw, conjunctival tooth short, blunt, keel slightly shorter than wings, about 5 mm. long, 1·5 mm. wide, apex rounded or very blunt; stamens to about 5 mm. long, the vexillary free, about 4 mm. long, the remainder connate for 3–4 mm.; anthers oblong, 0·3 mm. long, 0·2 mm. wide; ovary narrowly oblong, about 3 mm. long, 0·7 mm. wide, thinly hairy on the abaxial surface, style thick, about 1·5 mm. long, glabrous, stigma capitate, terminal, rather large. Pod subglobose 6–8 mm. diam., with 3–5 rather loose, anticlockwise coils, dorsal suture hairy, with short distant spines, lateral surfaces subglabrous with loose anastomosing venation; seed oblong-reniform, compressed, about 3·5 mm. long, 2 mm. wide; testa smooth, brown.

HAB.: Sand-dunes and sandy seashores; about sea-level; fl. Febr.–April.

DISTR.: Divisions 1–8. Also coasts of western Europe south of the Loire estuary, and throughout the Mediterranean region.

1. Ktima, 1913, *Haradjian* 671 ! Amoudhi near Ayios Yeoryios (Akamas), 1962, *Meikle* 2077 !
2. Yialia, 1955, *Merton* 2106 !
3. Limassol, 1939, *Lindberg f.* s.n.; near Amathus, 1947, *Mavromoustakis* 22 ! Amathus 1963, *J. B. Suart* 6 !
4. Ayia Napa, 1905, *Holmboe* 8; near Famagusta, 1912, *Haradjian* 110 ! Famagusta, 1932, *Syngrassides* 373 !
5. Salamis, 1905, *Holmboe* 441; near Trikomo, 1939, *Lindberg f.* s.n. !
6. Cape Kormakiti, 1954, *Merton* 1880 !
7. 6 miles E. of Kyrenia, 1949, *Casey* 499 ! and, 1951, *Casey* 1133 ! Near Ayios Amvrosios, 19 miles E. of Kyrenia, 1949, *P. H. Oswald* 115 !
8. Near Yialousa, 1880, *Sintenis & Rigo* 479 ! Karpas, 1930, *C. B. Ussher* 107 ! Ayios Philon near Rizokarpaso, 1941, *Davis* 2265 ! Apostolos Andreas, 1957, *Merton* 3087 !

SECT. 2. **Orbiculares** *Urb.* Annual or perennial herbs. Pod coiled spirally, with more than 2 seeds in each complete coil; testa tuberculate; radicle as long as the cotyledons.

3. M. orbicularis (*L.*) *Bartal.*, Cat. Piant. Siena, 60 (1776); Boiss., Fl. Orient., 2: 97 (1872); Sintenis in Oesterr. Bot. Zeitschr., 31: 194, 288 (1881); Holmboe, Veg. Cypr., 104 (1914); Post, Fl. Pal., ed. 2, 1: 322 (1932); Heyn in Scripta Hierosolymitana, 12: 135 (1963); Davis, Fl. Turkey, 3: 486 (1970); Zohary, Fl. Palaest., 2: 139, t. 202 (1972); Osorio-Tafall et Seraphim, List Vasc. Plants Cyprus, 61 (1973).
M. polymorpha L. var. *orbicularis* L., Sp. Plant., ed. 1, 779 (1753).
TYPE: The illustration in Morison, Hist. Plant., 2: 152, t. 15, fig. 1 (1680).

Prostrate or straggling annual; stems usually much branched, glabrous, 10–40 (–50) cm. long; stipules laciniate, 2–6 mm. long, with narrow filiform, glabrous lobes; petioles slender, glabrous, 1–5 cm. long, flattened above; leaflets obovate or cuneate, 6–12 (–17) mm. long, 5–10 (–13) mm. wide, glabrous or thinly hairy below, apex truncate or emarginate, margins serrulate in the upper third, entire or subentire towards base; inflorescences scattered along the upper parts of the branches, 1–2 (–5)-flowered; peduncles slender, 8–15 mm. long, elongating in fruit, produced as a filiform cusp or awn beyond the inflorescence; bracts filiform, 1–1·5 mm. long, dilated at base, membranous, often tinged purplish; pedicels slender, patent, 2–3 mm. long, glabrous or subglabrous; calyx campanulate, 3·5 mm. long, 1·3 mm. wide, subglabrous or thinly adpressed-pilose, teeth subulate about 2 mm. long; petals yellow, standard broadly obovate, about 4·5 mm. long, 3·5 mm. wide, apex emarginate, base cuneate, wings broadly oblong, about 3 mm. long, 1·3 mm. wide, apex rounded, base with a distinct blunt auricle and slender claw, conjunctival tooth well developed, blunt; keel about 4 mm. long, 1·3 mm. wide, apex rounded or blunt; vexillary stamen free, about 3 mm. long, other stamens connate, about 4 mm. long; anthers broadly oblong, about 0·2 mm. long, 0·3 mm. wide; ovary narrowly oblong, about 3 mm. long, 0·5 mm. wide, minutely papillose, style thick, about 0·5 mm. long, stigma capitate, with a narrow, fimbriate marginal wing. Pod discoid, 12–16 mm. diam., greenish, purplish or dark brown when ripe, without spines, coiling anticlockwise (2½–) 3–4 (–7) times, coils flattened, especially towards margins, glabrous, the lateral surfaces conspicuously reticulate-veined, with a slightly thickened vein along the dorsal margin; seeds about 3–6 in each coil, oblong-deltoid, strongly compressed, about 3 mm. long 2·5 mm. wide; testa pale brown, distinctly rugulose-tuberculate.

HAB.: In cultivated and fallow fields, sometimes on rocky or stony slopes on limestone or igneous formations, usually on damp ground, but also occurring in dry sandy situations; sea-level to 3,750 ft. alt.; fl. Febr.–May.

DISTR.: Divisions 2–8. Throughout the Mediterranean region, and eastwards to Iran; adventive elsewhere.

2. Kambia, 1908, *Clement Reid* s.n. ! Platres, 1938, *Kennedy* 1267 ! Lefkara, 1941, *Davis* 2722 ! Platres, 1960, *N. Macdonald* 132 !
3. Alethriko, 1905, *Holmboe* 248.
4. Larnaca, 1880, *Sintenis & Rigo*; near Athienou, 1880, *Sintenis & Rigo* 478 ! Sotira, Ayios Antonios, 1950, *Chapman* 592 !
5. Salamis, 1960, *N. Macdonald* 114 !
6. Kormakiti, 1936, *Syngrassides* 1202 !
7. Kyrenia, 1933, *Syngrassides* 187 ! 1949, *Casey* 375 ! 1951, *Casey* 1168 ! and 1956, *G.E. Atherton* 1391 !
8. Platanisso, 1950, *Chapman* 205 ! Elisis near Galinoporni, 1962, *Meikle* 2496 !

SECT. 3. **Spirocarpos** *Ser.* Annual (rarely perennial) herbs. Pods forming a congested spiral usually of 2 or more coils, dorsal suture smooth or spinose; seeds never more than 2 in each coil; testa smooth; radicle about half as long as cotyledons.

4. **M. scutellata** (*L.*) *Mill.*, Gard. Dict., ed. 8, no. 2 (1768); Boiss., Fl. Orient., 2: 96 (1872); Post, Fl. Pal., ed. 2, 1: 321 (1932); A. K. Jackson in Kew Bull., 1934: 271 (1934); Heyn in Scripta Hierosolymitana, 12: 38 (1963).
M. polymorpha L. var. *scutellata* L., Sp. Plant., ed. 1, 779 (1753); Davis, Fl. Turkey, 3: 494 (1970); Zohary, Fl. Palaest., 2: 141, t. 206 (1972); Osorio-Tafall et Seraphim, List Vasc. Plants Cyprus, 61 (1973).
TYPE: The illustration in Morison, Hist. Plant., 2: 152, t. 15, fig. 3 (1680).

Suberect or sprawling annual 25–50 cm. high; stems usually much branched, angled or longitudinally rigid, ± densely glandular-pubescent; stipules ovate-deltoid, acuminate, 7–12 mm. long, 2·5–6 mm. wide, glandular-pubescent, margin shortly lobed or sharply 4–5-toothed; petiole 1–2·5 cm. long, channelled above, glandular; leaflets oblong-obovate, 10–25 mm. long, 5–15 mm. wide, glabrous above, glandular below, with intermixed, scattered, long, eglandular hairs, apex rounded or subacute, base narrowly cuneate, margins sharply serrulate in the upper half, entire below; inflorescence 1 (–3)-flowered; peduncle usually less than 1 cm. long, lengthening in fruit, produced into a slender awn beyond the terminal flower, glandular-pilose; bracts subulate-filiform, 1–1·5 mm. long, membranous, abruptly dilated at base; pedicels very short, about 1 mm. long, glandular; calyx campanulate, about 4 mm. long, 1·5 mm. wide, glandular-hairy, teeth linear-subulate about 2 mm. long; petals dull mustard-yellow, standard obovate, 4–5 mm. long, 3–4 mm. wide, apex emarginate, base broadly cuneate, wings 4 mm. long,

1 mm. wide, apex rounded, base with a prominent, blunt auricle, long, slender claw, and well-developed, blunt conjunctival tooth, keel about 4 mm. long, 1 mm. wide, apex blunt or rounded; vexillary stamen free, about 3 mm. long, other stamens connate, about 4 mm. long; anthers suborbicular, about 0·3 mm. diam.; ovary linear-oblong, slightly twisted, 3 mm. long, 0·5 mm. wide, subglabrous above, glandular-hairy below, style thick, upcurved, glabrous, about 1 mm. long, stigma capitate with narrow fimbriate wing. Pod greenish-brown or fuscous, depressed-globose, 1–1·7 cm. diam., coiling anticlockwise 5–7 (–8) times to form a rather loose rosette, lateral surfaces subglabrous or glabrous, distinctly reticulate-veined, dorsal suture without spines, thinly glandular-hairy, with a distinctly thickened peripheral vein; seeds reniform or semi-lunate, 5–7 mm. long, 2·5–4 mm. wide; testa smooth, light brown.

HAB.: In cultivated and fallow fields, and waste ground; sea-level to 2,000 ft. alt.; fl. Febr.–May.

DISTR.: Divisions 2, 4–7, common in Division 7. Throughout the Mediterranean region, adventive elsewhere.

2. Lefkara, 1941, *Davis* 2720 !
4. Larnaca, 1880, *Sintenis & Rigo* 821 !
5. Mile 2, Nicosia–Famagusta Road, 1950, *Chapman* 444 ! Mile 10, Nicosia–Famagusta Road, 1952, *Merton* 837 ! Salamis, 1960, *N. Macdonald* 113 !
6. Nicosia, 1957, *Merton* 3113 !
7. Common; *Syngrassides* 203 ! 451 ! *Casey* 325 ! 1156 ! *Meikle* 2376 !

5. **M. rugosa** *Desr.* in Lam., Encycl. Méth., 3: 632 (1972); Post, Fl. Pal., ed. 2, 1: 323 (1932); Heyn in Scripta Hierosolymitana, 12: 41 (1963); Davis, Fl. Turkey, 3: 494 (1970); Zohary, Fl. Palaest., 2: 142, t. 207 (1972).
M. elegans Jacq. ex Willd., Sp. Plant., 3: 1408 (1802); Boiss., Fl. Orient., 2: 98 (1872).
TYPE: France; " On le cultive au Jardin du Roi " (P).

Spreading, ascending or erect annual, 10–45 cm. high; stems branched, especially towards base, angular, clothed thinly or rather densely with glandular and eglandular hairs; stipules narrowly ovate-acuminate, slightly falcate, 4–8 mm. long, 1·5–3 mm. wide, with about 8 sharp, slender marginal teeth, upper surface glabrous, lower surface thinly glandular-pilose; petioles slender, 0·8–1·5 (–2·5) cm. long, glandular, slightly channelled above, leaflets oblong or obovate, 7–20 mm. long, 4–13 mm. wide, glabrous above, adpressed glandular-pilose below, apex rounded or subtruncate, margins sharply serrulate or sometimes laciniate above, entire towards base; inflorescences scattered along the upper half of the branches, (1–) 2–4 (–7)-flowered; peduncles slender, glandular, 0·8–1·5 (–2) cm. long, elongating in fruit, produced beyond the inflorescence as a slender awn; pedicels slender, glandular, 1–1·5 mm. long; bracts linear-subulate, about 1 mm. long, membranous; calyx campanulate, about 2·5 mm. long, 1·5 mm. wide, adpressed-pilose externally; teeth narrowly deltoid-subulate, about 1·5 mm. long; petals bright or dark yellow, standard broadly obovate, about 4 mm. long, 3 mm. wide, apex rounded or slightly retuse, base broadly cuneate, wings about 3 mm. long, 1 mm. wide, apex rounded, base with a prominent, broad auricle and a slender claw, conjunctival tooth short, conical, keel about 3 mm. long, apex rounded; vexillary stamen free, about 2 mm. long, other stamens connate, to about 3 mm. long; anthers broadly oblong, about 0·3 mm. long, 0·2 mm. wide; ovary oblong, about 1·5 mm. long, 0·5 mm. wide, abaxial margin glandular-pilose, style thick, slightly twisted, about 1 mm. long, stigma capitate with a papillose margin; young fruit protruding sideways from calyx. Pod pale brown orbicular-discoid, (5–) 7–9 (–10) mm. diam., coiling anticlockwise $2\frac{1}{2}$–$3\frac{1}{2}$ (–5) times, lateral surfaces glabrous, conspicuously reticulate-veined, the radial veins joining a thickened glandular vein near the unarmed dorsal suture; seed 1 in each coil, reniform or semi-lunate, 3 mm. long, 2·5 mm. wide; testa pale brown, smooth, shining.

HAB.: In cultivated and fallow fields, on grassy banks and slopes, and by roadsides and pathsides; sea-level to 2,200 ft. alt.; fl. Febr.–May.

DISTR.: Divisions 2, 4, 7. Throughout the Mediterranean region.

2. Lefkara, 1941, *Davis* 2728 !
4. Near Larnaca, 1880, *Sintenis & Rigo* 933 !
7. Near St. Hilarion, 1949, *Mercy Casey* in *Casey* 688 ! Kornos Vouno (Kyrenia range), 1954, *Casey* 1335 ! St. Hilarion Castle, 1963, *Townsend* 63/65 ! Tjiklos, 1974, *Osorio-Tafall et Seraphim* s.n. !

6. M. blancheana *Boiss.*, Diagn., 2, 5: 75 (1856), Fl. Orient., 2: 97 (1872); Post, Fl. Pal., ed. 2, 1: 322 (1932); Heyn in Scripta Hierosolymitana, 12: 44 (1963); Davis, Fl. Turkey, 3: 495 (1970); Zohary, Fl. Palaest., 2: 142, t. 208 (1972); Osorio-Tafall et Seraphim, List Vasc. Plants Cyprus, 61 (1973).

Erect or ascending annual (10–) 15–30 (–45) cm. high; stems strongly ribbed or angled, clothed with glandular and eglandular hairs, usually much branched, especially towards base; stipules lanceolate or narrowly ovate, 4–6 mm. long, 1–2·5 mm. wide, with several small sharp teeth near the base and a long acuminate, entire acumen, thinly glandular-pilose on both surfaces; petioles 0·5–2·5 cm. long, slender, channelled above, shortly glandular-pilose; leaflets obovate or elliptical 5–23 mm. long, 4–10 mm. wide, eglandular-pilose above, glandular-pilose below, the glands especially numerous along midrib and near margins, apex acute or subtruncate, often apiculate, margins obscurely serrulate towards apex (rarely laciniate) entire towards base; inflorescences scattered along the upper parts of the branches, 1–3 (–4)-flowered; peduncle long and slender, often 2–2·5 cm. long, usually longer than subtending petiole, ± densely glandular-pilose, generally deflexed or recurved, produced beyond inflorescence as a slender awn; pedicels about 1 mm. long, glandular-pilose; bracts subulate, about 1 mm. long, abruptly dilated towards base, ciliate; calyx campanulate, about 5 mm. long, 2·5 mm. wide, glandular-pilose externally, teeth subulate 2·5–3 mm. long; petals yellow or orange, conspicuous, standard obovate, about 7 mm. long, 5 mm. wide, apex emarginate, base broadly cuneate, wings about 5 mm. long, 2·5 mm. wide, apex rounded, base with a slender claw and a long, blunt auricle, conjunctival tooth well developed, keel about 6 mm. long, 1·5 mm. wide, apex blunt, base auriculate and with a long slender claw; vexillary stamen usually free, about 5 mm. long, other stamens connate, to about 6 mm. long; anthers oblong-orbicular, about 0·5 mm. long and almost as wide; ovary narrowly linear, about 3·5 mm. long, 0·4 mm. wide, glabrous or glandular below, style 2·5 mm. long, not very thick, usually upcurved, glabrous, stigma capitate, papillose; young pod coiled inside calyx. Pod fuscous, depressed-globose, 8–12 mm. diam., coiling anticlockwise 4–6 times, middle coil distinctly flattened, lateral surfaces finely and rather obscurely reticulate-veined, the radial veins most prominent at the spiny or unarmed dorsal suture; seeds 1–2 in each coil, reniform, 4–5 mm. long, 2·5–3 mm. wide; testa pale brown, smooth, shining.

var. bonarotiana (*Arcang.*) *Arcang.*, Fl. Ital., 61 (1882); Post, Fl. Pal., ed. 2, 1: 322 (1932) [as var. " *Bonartiana* "]; Heyn in Scripta Hierosolymitana, 12: 46 (1963); Davis, Fl. Turkey, 3: 495 (1970); Zohary, Fl. Palaest., 2: 143 (1972).
M. bonarotiana Arcang. in Nuov. Giorn. Bot. Ital., 8: 6 (1876).
M. blancheana Boiss. var. *inermis* Post, Fl. Pal., ed. 1, 227 (1896).

TYPE: Italy; " in agris et ad aggeres in praedio vulgo dicto *Podere del Fattojo*, nec non circa rurem illustris Bonnarotiae stirpis prope Septinianum, duo millia pasuum circiter Florentia ", *Arcangeli* (G).

Pod unarmed, at most rugose on account of the protruding radial veins.

HAB.: In cultivated and fallow fields; sea-level to 3,750 ft. alt.; fl. March–May.

DISTR.: Divisions 2, 7. Also Turkey, Syria, Lebanon and Palestine; adventive elsewhere.

2. Platres, 1938, *Kennedy* 1266 !
7. Kyrenia, 1949, *Casey* 460 ! Kythrea, 1950, *Chapman* 95 ! Kyrenia, 1951, *Casey* 1144 ! and, 1953, *Casey* 1292 !

NOTES: Typical var. *blancheana*, with numerous, short, blunt spreading spines along the dorsal suture, has not been found in Cyprus. Its general distribution is otherwise identical with that of var. *bonarotiana*.

7. **M. rotata** *Boiss.*, Diagn., ser. 1, 2: 24 (1843), Fl. Orient., 2: 97 (1872); Post, Fl. Pal., ed. 2, 1: 322 (1932); Heyn in Scripta Hierosolymitana, 12: 48 (1963); Davis, Fl. Turkey, 3: 495 (1970); Zohary, Fl. Palaest., 2: 143, t. 209 (1972); Osorio-Tafall et Seraphim, List Vasc. Plants Cyprus, 61 (1973).

TYPE: Palestine; *Aucher-Éloy* 937 (G, K !).

Erect, ascending or sprawling annual, 10–30 cm. high, ± densely clothed with long eglandular hairs (? or with a few glands occasionally intermixed); stems angular, much branched especially towards base; stipules lanceolate or narrowly ovate, 4–6 mm. long, 1–2·5 mm. wide, with a long acumen and several sharp teeth near the base as in *M. blancheana*; petioles 0·5–2 cm. long, slender, channelled above; leaflets narrowly obovate or elliptical, 5–20 mm. long, 4–10 mm. wide, thinly pilose on both surfaces, apex rounded or truncate, often apiculate, margins often irregularly lobed, obscurely serrulate towards apex, entire towards base; inflorescences 1–3 (–6)-flowered; peduncle long and slender, often more than 2 cm. long, usually longer than subtending petiole, often deflexed or recurved, produced beyond inflorescence as a slender awn; pedicels usually less than 1 mm. long, thinly pilose; bracts membranous, ciliate, about 0·5 mm. long, with a broad oblong base and a cuspidate apex; calyx campanulate, about 4 mm. long, 2·5 mm. wide, pilose externally, teeth narrowly deltoid, about 2 mm. long, 0·5 mm. wide at base; petals yellow or orange, conspicuous, standard ovate-elliptical, 6–7 mm. long, 3–4 mm. wide, apex emarginate, base broadly cuneate, wings about 4 mm. long, 1·5 mm. wide, apex blunt or rounded, base with a conspicuous auricle and long slender claw, conjunctival tooth well developed, keel about 5 mm. long, 1·2 mm. wide, apex blunt, base auriculate, with a long slender claw; vexillary stamen free or almost free, about 4 mm. long, other stamens connate, about 5 mm. long; anthers oblong, 0·3 mm. long, 0·2 mm. wide; ovary narrowly oblong, about 2·5 mm. long, 0·5 mm. wide, slightly twisted, papillose along both sutures, style glabrous, thickened, upcurved, about 2 mm. long, stigma capitate; young pod coiled inside calyx. Pod shortly cylindrical, flattened at both ends, 4–10 (–12) mm. high, 5–10 (–12) mm. diam., consisting of 3–5 (–7) anticlockwise coils, lateral surfaces glabrous, conspicuously and finely reticulate-veined, dorsal margin spinose with blunt, spreading spines 0·5–2 (–3) mm. long; seeds 1–2 in each coil, reniform, 3–5 mm. long; testa pale brown, smooth.

HAB.: In cultivated fields; sea-level to 500 ft. alt.; fl. March–May.

DISTR.: Divisions 5, 6. Also Turkey, Syria, Lebanon and Palestine; adventive elsewhere.

5. Mia Milea, 1950, *Chapman* 505 !
6. Near Nicosia, 1880, *Sintenis & Rigo* 473 !

NOTES: All the Cyprus specimens have the leaflets slightly lobed, and in this respect agree well with the Aucher-Éloy type material; var. *eliezeri* Eig, with numerous flowers and small, much-lobed leaflets has not been collected in our area. There is some gradation of characters between *M. blancheana* and *M. rotata*, possibly through hybridization (see Heyn, *loc. cit.*, 51); one specimen, *Chapman* 168 from Ayios Demetrianos, Kythrea (cited by Heyn under *M. rotata*) is almost certainly such a hybrid; it has the indumentum and leaves of *M. blancheana*, but young pods like those of *M. rotata*. *Medicago blancheana* is recorded from Ayios Demetrianos.

8. M. coronata (*L.*) *Bartal.*, Cat. Piant. Siena, 61 (1776); Unger et Kotschy, Die Insel Cypern, 375 (1865); Boiss., Fl. Orient., 2: 101 (1872); Holmboe, Veg. Cypr., 105 (1914); Post, Fl. Pal., ed. 2, 1: 326 (1932); Heyn in Scripta Hierosolymitana, 12: 51 (1963); Davis, Fl. Turkey, 3: 496 (1970); Zohary, Fl. Palaest., 2: 143, t. 210 (1972); Osorio-Tafall et Seraphim, List Vasc. Plants Cyprus, 61 (1973).

M. polymorpha L. var. *coronata* L., Sp. Plant., ed. 1, 780 (1753).

TYPE: The illustration of *Medica coronata cherleri parva* in J. Bauhin, Hist. Plant., 2: 386 (1651).

Erect or spreading annual 7–20 (–30) cm. high; stems much branched, especially towards base, obscurely angled or sulcate, pilose, with long spreading eglandular (or some glandular) hairs; stipules ovate, acuminate, 1–2·5 mm. long, 0·5–1·5 mm. wide, pilose, distinctly veined, subentire or with a few, irregular basal teeth; petioles 3–15 (–25) mm. long, slender, sulcate above; leaflets often broadly obovate, almost orbicular, sometimes cuneate, 2·5–7 (–9) mm. long, 2–6 (–7) mm. wide, pilose on both surfaces, apex truncate or emarginate, apiculate, margins sharply denticulate towards apex, entire towards base; inflorescences capitate, 3–12 (–16)-flowered, scattered along the upper parts of the branches; peduncle conspicuous, rather rigid, erecto-patent, 10–25 mm. long, adpressed white-pilose or glandular-pilose; bracts minute, long-cuspidate, with a strongly dilated base; pedicels generally less than 1·5 mm. long, pilose or glandular; calyx campanulate, about 1·5 mm. long, 1 mm. wide, glandular-pilose externally, teeth linear-subulate, about 0·5 mm. long; petals yellow, standard obovate, 4–5 mm. long, about 2·5 mm. wide, apex rounded or emarginate, base broadly cuneate, wings oblong, 2–3 mm. long, about 1 mm. wide, apex ± truncate or slightly lobed, base with a prominent auricle, slender claw, and well-developed conjunctival tooth, keel about 3 mm. long, 1 mm. wide, apex rounded, base shortly auriculate, with a rather short claw; vexillary stamen free, about 2 mm. long, other stamens connate, to about 3 mm. long; anthers suborbicular, about 0·3 mm. diam.; ovary linear-oblong, about 2 mm. long, 0·5 mm. wide, base distinctly stipitate, abaxial margin strongly papillose-glandular, style about 1·5 mm. long, upcurved, gibbous towards base, glabrous, stigma capitate, papillose; young pod protruding sideways from calyx. Pods pale brown, shortly cylindrical, 2–4 mm. long, about 3 mm. diam., coils ($1\frac{1}{2}$–) $2\frac{1}{2}$–$3\frac{1}{2}$, rather loose and open, anticlockwise, lateral surfaces glabrous, distinctly veined, the veins appearing concentric, dorsal margin spinose, the spines spreading, 0·5–1·5 mm. long; seeds 1–2 (–3) in each pod, oblong or semi-lunate, 1·5–2·5 mm. long, 0·5 mm. wide; testa smooth, bright or pale brown.

HAB.: Dry stony hillsides or waste ground; sea-level to 2,500 ft. alt.; fl. March–May.

DISTR.: Divisions 3–5, 7, 8, common in Divisions 5, 7. Widespread in the northern half of the Mediterranean region and eastwards to Iraq; adventive elsewhere.

3. Perapedhi, 1962, *Meikle* 2797 !
4. The Phaneromene, Larnaca, 1880, *Sintenis & Rigo*.
5. Common; *Holmboe* 341; *Sintenis & Rigo*; *Chapman* 36 ! 392 !
7. Common; *Kotschy* 446; *Davis* 2915 ! *Casey* 463 ! 1150 ! *Merton* 3221 !
8. Cape Andreas, 1880, *Sintenis & Rigo*; near Rizokarpaso, 1912, *Haradjian* 166 !

9. M. praecox *DC.*, Cat. Plant. Hort. Monsp., 123 (1813); Boiss., Fl. Orient., 2: 102 (1872); Heyn in Scripta Hierosolymitana, 12: 54 (1963); Davis, Fl. Turkey, 3: 497 (1970); Osorio-Tafall et Seraphim, List Vasc. Plants Cyprus, 61 (1973).

TYPE: France; " in locis sterilibus circa Forum-Julium (Fréjus) " *Balbis* (G?, TO).

Prostrate or decumbent annual (5–) 10–20 cm. high; stems much branched, especially at the base, angular, thinly pilose or glabrescent, often purplish; stipules 2–3 mm. long, 1–2 mm. wide, subglabrous above, sometimes thinly hairy below, margin sharply dentate-laciniate; petioles very slender 2–10 mm. long, glabrous or subglabrous, channelled above; leaflets small, cuneate, 2–6 mm. long, 2–4 mm. wide, glabrous above and distinctly veined, thinly pilose-strigose below, apex truncate or slightly emarginate, apiculate, margins minutely but sharply toothed above, entire towards base; inflorescences 1–2-flowered, scattered along the greater length of the branches; peduncle very short, usually less than 2 mm. long, glabrous or thinly pilose; bracts subulate-cuspidate, 0·5–1 mm. long, with a dilated (sometimes dentate) base; pedicels about 1 mm. long, pilose; calyx campanulate, 1·5–2 mm. long, about 1 mm. wide, thinly pilose externally, teeth narrowly deltoid, about 0·8 mm. long; petals small, yellow, standard suborbicular, about 2 mm. diam., wings about 1·6 mm. long, 0·8 mm. wide, apex rounded, base with an erect rounded auricle almost as large as the apical lobe, claw slender, conjunctival tooth well developed, keel 2 mm. long, 0·8 mm. wide, apex blunt, base scarcely auriculate; vexillary stamen about 1·5 mm. long (? always) sterile, other stamens connate, about 2 mm. long; anthers suborbicular, about 0·2 mm. diam.; ovary oblong, about 1 mm. long, 0·3 mm. wide, abaxial margin minutely papillose, style about 0·7 mm. long, relatively thick, glabrous, turned up in the middle at a right angle, stigma capitate, minutely papillose; young pod protruding sideways from calyx. Pod pale brown, shortly cylindrical, 3–4 mm. long, about 3 mm. diam., coils 2½–4 (–5), rather loose, anticlockwise, lateral surfaces thinly pilose or glabrous, strongly veined, dorsal margins spiny, the spines slender and spreading, 2–3 mm. long; seeds 1–2 in each coil, ellipsoid-reniform, about 2·5 mm. long, 1·3 mm. wide; testa smooth, pale brown.

HAB.: On grassy slopes and roadsides, sometimes in Pine forest; 500–4,500 ft. alt.; fl. March–May.

DISTR.: Divisions 1, 4, 5, rare. Widely distributed in the northern Mediterranean region from Spain to Lebanon, but usually rare; also in Crimea.

1. Limni Mine (?), 1928 or 1930, *Druce* 189 !
4. Sotira, Ayios Antonios, 1952, *Merton* 650 !
5. Malounda, 1950, *Chapman* 296 !

10. M. minima (*L.*) *Bartal.*, Cat. Piant. Siena, 61 (1776); Unger et Kotschy, Die Insel Cypern, 375 (1865); Boiss., Fl. Orient., 2: 103 (1872); Holmboe, Veg. Cypr., 105 (1914); Post, Fl. Pal., ed. 2, 1: 327 (1932); Heyn in Scripta Hierosolymitana, 12: 60 (1963); Davis, Fl. Turkey, 3: 497 (1970); Zohary, Fl. Palaest., 2: 145, t. 212 (1972); Osorio-Tafall et Seraphim, List Vasc. Plants Cyprus, 61 (1973).

M. polymorpha L. var. *minima* L., Sp. Plant., ed. 1, 780 (1753).

TYPE: The illustration of *Medica echinata minima* in J. Bauhin, Hist. Plant., 2: 386 (1651).

Decumbent, prostrate or ascending annual 5–25 (–30) cm. high; stems much branched near base, angular, ± densely pilose or glandular-pilose; stipules narrowly ovate, 3–6 mm. long, 1–2·5 mm. wide, acuminate, pilose on both sides, margins entire or obscurely denticulate; petioles 5–50 mm. long, slender, channelled above, ± densely pilose or glandular-pilose; leaflets obovate 3–10 mm. long, 2–7 mm. wide, pilose on both surfaces, apex rounded or obscurely emarginate, often apiculate, denticulate, basal half entire; inflorescences (2–) 3–6 (–8)-flowered, scattered along the greater length of the branches, peduncles usually very short, occasionally up to 12 mm. long, pilose, produced into a short pilose cusp; bracts membranous, cuspidate, about 0·5 mm. long, with a broad, oblong base; pedicels very short, often almost wanting, pilose; calyx narrowly campanulate, 2–3 mm. long, 1–2 mm. wide, villose externally, teeth subulate about 1–1·5 mm. long; petals bright yellow, standard obovate, about 4 mm. long, 2 mm. wide, apex emarginate, base broadly cuneate, wings oblong, about 3 mm. long, 1 mm. wide, apex rounded, base conspicuously auriculate, with a slender claw, conjunctival tooth very strongly developed, keel about 3·5 mm. long, 1 mm. wide, apex blunt; vexillary stamen free (? sterile) about 2·5 mm. long, other stamens connate, about 3 mm. long; anthers broadly oblong, about 0·3 mm. long and almost as wide; ovary narrowly oblong, about 2 mm. long, 0·5 mm. wide, ± densely adpressed-pilose; style thick, about 1 mm. long, glabrous, upcurved to an oblique capitate stigma; young pod protruding sideways from calyx. Pod dark brown, shortly cylindrical, 2–4 (–5) mm. high, 3–5 (–6) mm. diam., coils 3–5, anticlockwise, rather loose, lateral surfaces rather obscurely veined with a thickened dorsal margin, spines slender, erect or suberect, often hooked at the apex, up to 3 mm. long, but very variable in length, inserted some distance below the dorsal suture; seeds 1–2 in each coil, ellipsoid-reniform, 2–2·5 mm. long, 1–1·5 mm. wide; testa smooth, yellow or very pale brown.

HAB.: On rocky hillsides, in cultivated and fallow fields, by roadsides and on waste ground, occasionally in Pine forest; sea-level to 3,750 ft. alt.; fl. Febr.–May.

DISTR.: Divisions 1–5, 7, 8, common in Divisions 5, 7. Throughout the Mediterranean region and much of southern and central Europe; found as an adventive in may parts of the world.

1. Polis, 1928 or 1930, *Druce* 199 !
2. Pedhoulas, 1932, *Syngrassides* 49 ! Platres, 1938, *Kennedy* 1270 ! Yialia, 1955, *Merton* 2138 ! Kato Platres, 1958, *N. Macdonald* 67 !
3. Limassol, 1859, *Kotschy* S.984; Cape Gata near Ayios Nikolaos, 1862, *Kotschy* 607.
4. Larnaca, 1912, *Haradjian* 38 ! Famagusta, 1928 or 1930, *Druce* 198 !
5. Common; *Sintenis & Rigo*; *Holmboe* 326; *Syngrassides* 1168 ! *Merton* 630 ! 659 !
7. Common; *Sintenis & Rigo*; *Syngrassides* 185 ! *Davis* 2873 ! *Casey* 383 ! 464 ! 1146 ! *Chapman* 57 !
8. Cape Andreas, 1880, *Sintenis & Rigo*; Gastria, 1941, *Davis* 2420 ! Between Komi Kebir and Dhavlos, 1941, *Davis* 2441 !

11. M. disciformis *DC.*, Cat. Hort. Monsp., 124 (1813); Boiss., Fl. Orient., 2: 105 (1872); Sintenis in Oesterr. Bot. Zeitschr., 32: 123, 398 (1882); Rechinger f. in Arkiv för Bot., 2, 1: 424 (1950); Heyn in Scripta Hierosolymitana, 12: 68 (1963); Davis, Fl. Turkey, 3: 498 (1970); Osorio-Tafall et Seraphim, List Vasc. Plants Cyprus, 61 (1973).

Prostrate or ascending annual 10–20 cm. high; stems much branched especially at base, angular, ± densely pilose or glandular-pilose; stipules rather broadly ovate-acuminate, 4–6 mm. long, 2–3·5 mm. wide, pilose or glandular-pilose on both sides, margins usually sharply serrate; petioles slender, 5–20 (–40) mm. long, pilose, channelled above; leaflets obovate, 4–12 mm. long, 2–9 mm. wide, pilose on both surfaces, rounded or shortly retuse and apiculate at apex, base broadly or narrowly cuneate, margin sharply denticulate above, entire below; inflorescence lax, 1–3-flowered; peduncle 1–2·5 cm. long, erect or erecto-patent, pilose or glandular-pilose, produced as a distinct awn; bracts minute, cuspidate, less than 1 mm. long, with a broad oblong base; pedicels slender, 1–3 mm. long, pilose; calyx campanulate about 3 mm. long, 1 mm. wide, pilose externally, teeth subulate, about 1·5 mm. long; petals yellow, standard broadly obovate, 5–7 mm. long, 3–4·5 mm. wide, apex emarginate, base broadly cuneate, wings oblong, 3–4·5 mm. long, about 1·5 mm. wide, apex blunt, base with a very conspicuous elongate auricle and slender claw, conjunctival tooth well developed, keel 4–4·5 mm. long, apex blunt or rounded, auricle distinct, claw broad; stamens 3–4·5 mm. long, the vexillary free or ± united unilaterally with the others; anthers suborbicular, about 4 mm. diam.; ovary narrowly oblong, 2·5–3 mm. long, 0·4 mm. wide, pubescent or ± glandular, style thickened, glabrous, about 1 mm. long, curved upwards at a right angle to a distinct capitate stigma; young pod protruding sideways from calyx. Pod discoid or very shortly cylindrical, about 2–2·5 mm. long, 5–6 mm. diam., rich brown when ripe, coils 5–7, anticlockwise, closely and obscurely veined towards ventral margin, with a broad flat area towards dorsal margin, spines erect, or often reflexed, grooved, up to 4 mm. long, occasionally very short, the upper end of coil often unarmed; seeds 1–2 in each coil, oblong or slightly reniform, about 3·5 mm. long, 1·5 mm. wide; testa smooth, pale brown.

var. **disciformis**
TYPE: France; " prope *Castelnau* in agro Monspessulano " (? P, ? G).

Pods armed with well-developed spines 3–4 mm. long.

HAB.: On dry slopes and hillsides or in coastal garigue, occasionally in Pine forest; sea-level to 2,000 ft. alt.; fl. March–May.

DISTR.: Divisions 2, 4–8. Northern Mediterranean from Spain to Turkey.

2. Yialia, 1955, *Merton* 2131 !
4. Sotira, Ayios Antonios, 1952, *Merton* 652 !
5. Lefkoniko, 1880, *Sintenis & Rigo*.
6. Ayios Yeoryios near Liveras, 1962, *Meikle* 2408 !
7. Halevga, 1941, *Davis* 2913 ! Tjiklos cliff, Kyrenia, 1949, *Casey* 524 ! Kyrenia, 1951, *Casey* 1153 ! and, 1956, *G. E. Atherton* 1328 !
8. Cape Andreas, 1880, *Sintenis & Rigo* 472 (in part) !

var. **apiculata** *Urban* in Boiss., Fl. Orient., Suppl., 166 (1888); Holmboe, Veg. Cypr., 105 (1914); Heyn in Scripta Hierosolymitana, 12: 69 (1963).
TYPE: Cultivated from seeds of *Sintenis & Rigo* 472 (G).

Pods with very short triangular spines less than 1 mm. long.

HAB.: Dry hillsides and coastal garigue; sea-level to 1,600 ft. alt.; fl. March–May.

DISTR.: Divisions 7, 8, rare. Also Turkey.

7. Above Kyrenia Pass, 1953, *Casey* 1308 !
8. Cape Andreas, 1880, *Sintenis & Rigo* 472 (in part).

NOTES: Kew material of *Sintenis & Rigo* 472 includes only var. *disciformis* so it must be assumed either that this number consists of a mixture (as is frequent with these collectors) or that var. *apiculata* arose as a mutant in cultivation.

12. M. polymorpha *L.*, Sp. Plant., ed. 1, 779 (1753) emend. L. H. Shinners in Rhodora, 58: 5 (1956); Heyn in Scripta Hierosolymitana, 12: 71 (1963).

Prostrate or ascending annual, 10–40 cm. high; stems branching from the base, glabrous or very sparsely pilose; stipules ovate, acuminate, 4–7 mm. long, 2–3 mm. wide, margins laciniate with slender, filiform lobes; petioles 0·8–5 cm. long, glabrous, channelled above; leaflets obovate or obcordate, 5–20 mm. long, 4–16 mm. wide, glabrous above, sometimes thinly pilose below, apex retuse or emarginate, shortly apiculate, margins denticulate above, entire towards base; inflorescences (1–) 2–10-flowered, scattered along the greater length of the stems; peduncles glabrous, usually rather short, less than 1 cm. long, but sometimes to 2 cm. or more, without an awn, or with a very short one not exceeding the flowers; bracts minute, cuspidate, membranous, about 0·8 mm. long with a widely dilated base; pedicels very short, seldom esceeding 1 mm., glabrous or very thinly pilose; calyx campanulate, about 2·5 mm. long, 1·3 mm. wide, glabrous or thinly pilose, teeth deltoid, about 1 mm. long; petals yellow, standard obovate, about 3·5 mm. long, 2·5 mm. wide, apex emarginate, base very broadly cuneate, wings oblong, about 3 mm. long, 1 mm. wide, apex rounded, base with a small auricle and rather short wide claw, conjunctival tooth well developed, keel 2·5 mm. long, 1 mm. wide, apex blunt, base with a small auricle; vexillary stamen free, about 1·5 mm. long, other stamens connate, to about 2·5 mm. long; anthers suborbicular, about 0·3 mm. diam.; ovary oblong, falcately upcurved, about 2 mm. long, 0·5 mm. wide, thinly pilose, dorsal margins ± papillose, style short, thick, about 0·5 mm. long, sharply upcurved, stigma capitate; young pod protruding sideways from calyx. Pod dull greenish-brown, discoid or cylindrical, 2–12 mm. high, ($2\frac{1}{2}$–) 3–8 (–10) mm. diam., coils $1\frac{1}{2}$–6, anticlockwise, rather loose and usually not very much hardened, lateral surfaces conspicuously reticulate-veined, dorsal margins spinose, with spines up to 4 mm. long, or sometimes merely tuberculate or almost unarmed; seeds 1–2 in each coil, ellipsoid-oblong, about 3–4 mm. long, 2 mm. wide; testa smooth, pale brown.

var. **vulgaris** (*Benth.*) *Shinners* in Rhodora, 58: 310 (1956) emend. Heyn in Scripta Hierosolymitana, 12: 72, 75 (1963); Davis, Fl. Turkey, 3: 499 (1970); Zohary, Fl. Palaest., 2: 145, t. 213 (1972); Osorio-Tafall et Seraphim, List Vasc. Plants Cyprus, 61 (1973) (as ssp. *vulgaris*).

M. denticulata Willd., Sp. Plant., 3: 1414 (1802); Unger et Kotschy, Die Insel Cypern, 375 (1865).

M. denticulata Willd. var. *vulgaris* Benth., Cat. Pyr., 104 (1826).

M. denticulata Willd. var. *genuina* Boiss., Fl. Orient., 2: 102 (1872).

M. hispida Gaertn. var. *denticulata* (Willd.) Burnat, Fl. Alp.-Marit., 2: 106 (1896); Post, Fl. Pal., ed. 2, 1: 327 (1932); Lindberg f., Iter Cypr., 19 (1946).

M. hispida Gaertn. ssp. *denticulata* (Willd.) Holmboe, Veg. Cypr., 104 (1914).

[*M. lappacea* (non Desr.) Sintenis in Oesterr. Bot. Zeitschr., 31: 192, 194, 288 (1881).]

[*M. hispida* Gaertn. ssp. *lappacea* (non Desr.) Holmboe, Veg. Cypr., 104 (1914); Lindberg f., Iter Cypr., 19 (1946).]

TYPE: Based on *M. denticulata* Willd.; specimen in herb. Willdenow (B).

Inflorescence with (2–) 5–10 flowers. Pod discoid, (2½–) 3–5 (–7) mm. diam., coils 1½–3½ (–4), generally not much hardened even at maturity.

HAB.: In cultivated and fallow fields, in pastures or on grassy hillsides, on waste ground and by roadsides, occasionally on wet ground by streams; sea-level to 4,500 ft. alt., but usually lowland; fl. Febr.–Aug.

DISTR.: Divisions 1–5, 7, 8, common in Divisions 2, 4, 5, 7. Throughout the Mediterranean region and eastwards to the western Himalayas; introduced and naturalized almost everywhere in Europe, and in many other temperate countries.

1. Ktima, 1913, *Haradjian* 691 ! Yeroskipos, 1939, *Lindberg f.* s.n.; Ayios Neophytos Monastery, 1939, *Lindberg f.* s.n.; Dhrousha, 1957, *Merton* 3027 !
2. Common; *Kotschy* 843; *Syngrassides* 179 ! *Lindberg f.* s.n. ! *Merton* 2347 !
3. Cherkez, 1954, *Mavromoustakis* 14A !
4. Common; *Kotschy* 161; *Sintenis & Rigo* 471 ! 1016 ! *Holmboe* 253; *Chapman* 21 ! *Merton* 657 ! 2006 !
5. Common; *Sintenis & Rigo* 471B ! *Syngrassides* 72 ! *Chapman* 382 ! 445 ! *Merton* 629 !
7. Common; *Kotschy* 401; *Syngrassides* 179 ! *Davis* 3016A ! *Casey* 465 ! 1143 ! 1155 ! *Chapman* 96 ! *G.E. Atherton* 383 ! 836 ! 866 ! 950 ! 1120 ! 1155 !
8. Rizokarpaso, 1941, *Davis* 2388 ! Elisis near Galinoporni, 1962, *Meikle* 2494 !

NOTES: All the Cyprus material has been referred to var. *vulgaris*, and seems to fit here best, though many of the specimens have larger fruits than is usual in this variety, and tend to approach var. *polymorpha* (*M. lappacea* Desr.). No doubt it was on this account that Sintenis & Rigo and Holmboe recorded *lappacea* from Cyprus, but the extreme variant with many-coiled cylindrical fruits has not been found amongst the many sheets examined. As might be expected, there is considerable intergradation between var. *vulgaris* and var. *polymorpha* throughout their distributions.

13. M. arabica (*L.*) *Huds.*, Fl. Angl., ed. 1, 288 (1762); Druce in Rep. B.E.C., 9: 470 (1931); Post, Fl. Pal., ed. 2, 1: 327 (1932); Lindberg f., Iter Cypr., 19 (1946); Heyn in Scripta Hierosolymitana, 12: 81 (1963); Davis, Fl. Turkey, 3: 500 (1970); Osorio-Tafall et Seraphim, List Vasc. Plants Cyprus, 61 (1973).

M. polymorpha L. var. *arabica* L., Sp. Plant., ed. 1, 780 (1753).

M. maculata Willd., Sp. Plant., 3: 1412 (1802); Boiss., Fl. Orient., 2: 103 (1872).

TYPE: The illustration of *Medica cochleata minor polycarpos annua, capsula majore alba, folio cordato macula fusca* in Morison, Plant. Hist., 2: t. 13, fig. 12 (1680).

Prostrate or sprawling, rather robust annual; stems 20–50 (–60) cm. long, usually much branched, sharply angular, subglabrous or thinly pilose with long multicellular hairs (generally present in some abundance about the nodes); stipules ovate, acuminate, 5–12 mm. long, 3–5 mm. wide, margins deeply dentate or shortly laciniate, with sharp, narrowly deltoid teeth; petioles usually long and slender, 2–8 (–12) cm. long, thinly septate-hairy, flattened or slightly channelled above; leaflets broadly obovate, obcuneate or obcordate 0·7–2 (–3) cm. long and almost as wide, generally with a conspicuous dark central blotch, glabrous above, thinly pilose below, apex rounded, truncate or emarginate, margins shortly denticulate or entire; inflorescences (1–) 2–5 (–6)-flowered; peduncle rather short, very slender, 5–15 (–20) mm.

long, ± septate-pilose, produced as a short awn not exceeding the inflorescence; bracts ovate-acuminate about 1 mm. long, 0·5 mm. wide; pedicels generally not exceeding 1 mm., subglabrous or pilose; calyx campanulate, about 3 mm. long, 1·5 mm. wide, thinly hairy externally, teeth subulate-deltoid about 1·5 mm. long; petals yellow, standard broadly obovate, about 5 mm. long, 4 mm. wide, apex shortly emarginate, claw short and rather broad, wings 3 mm. long, 1–2 mm. wide, apex blunt, base with a conspicuous auricle and short slender claw, conjunctival tooth small, keel 4 mm. long, 1·2 mm. wide, apex blunt, base with a short auricle; vexillary stamen free, 2·5–3 mm. long, other stamens connate to 4 mm. long; anthers suborbicular, about 0·3 mm. diam.; ovary narrowly oblong, 2·5 mm. long, 0·5 mm. wide, base distinctly stipitate, dorsal (abaxial) margin papillose; style thick, about 1·5 mm. long, upcurved at a right angle to a conspicuous oblique stigma; young pod protruding sideways from calyx. Pod greenish brown, cylindrical or subglobose 5–7 mm. high and about the same in diameter, coils 3–7, anticlockwise, not hard or rigid, lateral surface pallid, closely but rather obscurely veined, dorsal margin spinose, spines spreading, grooved, up to 4 mm. long; seeds 1–2 in each coil, oblong-ellipsoid, 3–3·5 mm. long, 1·5–2 mm. wide; testa smooth, yellowish or light brown.

HAB.: Cultivated and waste ground, by roadsides, sometimes on shaded banks or by streamsides in Pine forest; sea-level to 4,400 ft. alt.; fl. March–July.

DISTR.: Divisions 2, 7, common in both Divisions. Europe and most of the Mediterranean region to eastern Asia; introduced elsewhere; often adventive.

2. Common; Lefka, Yialia, Prodhromos, etc.; *Syngrassides* 243 ! *Lindberg f.* s.n. ! *Merton* 2148 ! 2395 ! 2718 !

7. Common about Kyrenia; *Casey* 607 ! 625 ! 1177 ! *Merton* 740 !

14. M. littoralis *Rohde ex Lois.-Deslong.*, Not. Fl. France, 118 (1810); Boiss., Fl. Orient., 2: 98 (1872); Sintenis in Oesterr. Bot. Zeitschr., 32: 123, 398 (1882); Post, Fl. Pal., ed. 2, 1: 324 (1932); Rechinger f. in Arkiv för Bot., ser. 2, 1: 424 (1950); Heyn in Scripta Hierosolymitana, 12: 91 (1963); Davis, Fl. Turkey, 3: 502 (1970); Zohary, Fl. Palaest., 2: 147, t. 215 (1972); Osorio-Tafall et Seraphim, List Vasc. Plants Cyprus, 61 (1973).

Prostrate or ascending annual; stems 10–30 cm. long, much branched especially at the base, angular, ± pilose or glabrescent, often purplish; stipules narrowly ovate-acuminate, 3–5 mm. long, 1–2 mm. wide, margin sharply dentate-laciniate; petioles slender, 5–15 (–20) mm. long, pilose, channelled above; leaflets obovate or obcuneate, 3–7 (–15) mm. long, 2–5 (–12) mm. wide, adpressed-pilose on both surfaces, apex truncate, emarginate or rounded, sharply denticulate, base cuneate; inflorescences 1–3-flowered, scattered along the greater length of the stems; peduncles 5–15 mm. long, pilose, produced as a filiform awn 3–4 mm. long; bracts cuspidate, 1·5 mm. long, with a broad oblong base about 0·8 mm. wide; pedicels very short, less than 1 mm. long or almost wanting, ± densely pilose; calyx campanulate, 3–3·5 mm. long, 1 mm. wide, pilose, often purplish, teeth subulate, 1·5 mm. long; petals yellow or orange, standard obovate, 5–6 mm. long, 4–5 mm. wide, apex shortly emarginate, base cuneate, wings oblong, 4–5 mm. long, 1·5 mm. wide, apex rounded, base with a slender claw and prominent auricle, conjunctival tooth long, pointed, keel about 6 mm. long, 1·5 mm. wide, apex blunt, basal auricle fairly distinct; vexillary stamen about 4 mm. long, free above, connate below, other stamens about 5 mm. long; anthers suborbicular or broadly oblong, about 0·4 mm. long and almost as wide; ovary narrowly oblong, about 3·5 mm. long, 0·4 mm. wide, distinctly stipitate, basal margin ± papillose, glabrous, style about 2 mm. long, thickened, slightly upcurved, stigma capitate, oblique; young pod protruding sideways fron calyx. Pod shortly cylindrical, 3–10 mm. high, 3–7 mm. diam., glabrous, coils 2–6 (–8), usually clockwise, closely adpressed, lateral surfaces obscurely veined, dorsal surface flattish or with the dorsal suture a little raised or sunk, spinose, tuberculate or smooth; seeds 1–2 in each coil, reniform or semi-lunate, 2·5–3·5 mm. long, 1·5–2 mm. wide; testa smooth, pale brown or yellowish.

var. **littoralis**

TYPE: France; " dans le pays de Gênes, à Nice, et sur toutes les côtes de Provence ", *Rohde, Bertoloni, Suffren et Requien.*

Pods spinose, spines 1–4 mm. long, erect or occasionally loosely adpressed to pod.

HAB.: Sandy seashores or sand-dunes, or on dry rocky or stony pastures, usually maritime but sometimes inland; sea-level to 2,000 ft. alt.; fl. Febr.–May.

DISTR.: Divisions 2–8, common in Divisions 4, 7, 8. Throughout the Mediterranean region; Atlantic Islands; adventive elsewhere.

2. Yialia, 1955, *Merton* 2137 !
3. Akrotiri Forest, 1939, *Mavromoustakis* 117 ! Near Amathus, 1947, *Mavromoustakis* s.n. ! Amathus, 1960, *N. Macdonald* 20 !
4. Common; *Sintenis & Rigo* 1015 ! *Haradjian* 111 ! *Mavromoustakis* s.n. ! *Chapman* 2 ! *Merton* 2005 in part !
5. Near Lefkoniko, 1880, *Sintenis & Rigo* 482 ! By the bridge before Pyroi, 1936, *Syngrassides* 1049 ! Salamis, 1957, *Merton* 3212 !
6. Syrianokhori, 1935, *Syngrassides* 760 !
7. Common; *Davis* 2818B ! *Casey* 339 ! 1103 ! 1120 ! 1173 ! *Chapman* 58 ! 94A !
8. Common; *Davis* 2216 ! 2312 ! 2368 ! 2405 ! *Meikle* 2999 !

var. **inermis** *Moris*, Fl. Sard., 1: 439 (1837); Heyn in Scripta Hierosolymitana, 12: 94 (1963).

M. littoralis Rohde var. *subinermis* Boiss., Fl. Orient., 2: 99 (1872); Rechinger f. in Arkiv för Bot., ser. 2, 1: 424 (1950).

M. littoralis Rohde var. *tricycla* (DC.) Holmboe, Veg. Cypr., 104 (1914).

[*M. cylindracea* (non DC.) Kotschy in Unger et Kotschy, Die Insel Cypern, 375 (1865).]

TYPE: The illustration in Moris, Fl. Sard. Icon., t. 40, fig. B !

Pod unarmed or with very short tubercles, not spinose.

HAB.: Sandy ground by the sea, about sea-level; fl. Febr.–May.

DISTR.: Divisions 4, 8, rare. Throughout the Mediterranean region; Atlantic Islands; adventive elsewhere.

4. Near Larnaca, 1912, *Haradjian* 15 ! Larnaca Salt Lake, 1955, *Merton* 2005 in part !
8. Cape Andreas, 1880, *Sintenis & Rigo* 481 !

NOTES: The two varieties are barely worth recognition, being connected by a series of intermediates.

The close relationship between *Medicago littoralis* and *M. truncatula* is commented upon by Heyn (*op. cit.* pp. 100–103), and it is suggested that intermediates are the product of interbreeding. So close are the affinities between the two that it would perhaps be better to consider them subspecies of one aggregate species. It will be found that Cyprus specimens of *M. littoralis* have glabrous fruits, whereas in *M. truncatula* the fruits are almost invariably pilose, usually quite distinctly so. In view of the frequent confusion between the two species, no unexamined material has been cited above.

15. M. truncatula *Gaertn.*, Fruct., 2: 350 (1791) emend. Urban in Sitz. Verh. Bot. Ver. Brandenb., 67 (1873); Post, Fl. Pal., ed. 2, 1: 324 (1932); Heyn in Scripta Hierosolymitana, 12: 95 (1963); Davis, Fl. Turkey, 3: 502 (1970); Zohary, Fl. Palaest., 2: 148, t. 216 (1972); Osorio-Tafall et Seraphim, List Vasc. Plants Cyprus, 61 (1973).

M. tribuloides Desr. in Lam., Encycl. Méth., 3: 635 (1792); Boiss., Fl. Orient., 2: 99 (1872); Sintenis in Oesterr. Bot. Zeitschr., 32: 123, 398 (1882); Holmboe, Veg. Cypr., 104 (1914).

TYPE: The illustration in Gaertner, De Fruct., 2: t. 155 (1791).

Closely similar to *M. littoralis* Rohde in habit and appearance, but generally more robust; stems 15–30 (–40) cm. long; stipules as in *M. littoralis*, but often with the basal part deeply laciniate; leaflets usually obovate, 7–13 mm. long, 4–10 mm. wide, pilose on both surfaces, apex rounded or subacute, seldom emarginate, sharply denticulate; inflorescence 1–2 (–3)-flowered; peduncle slender, about as long as petiole, produced into a long filiform awn; bracts cuspidate, about 1·5 mm. long, exceeding the short pilose pedicels; calyx 4–5 mm. long, 1–1·5 mm. wide, pilose externally, teeth subulate 2·5–3 mm. long; petals yellow, standard 5–6 mm. long, 3–4 mm. wide, wings about 4 mm. long, 1–1·5 mm. wide, conjunctival tooth strongly developed, keel about 5 mm. long, 1–1·5 mm. wide, apex blunt; stamens about 5 mm. long, the vexillary stamen a little shorter, free almost to base; anthers about 0·4 mm. long and almost as wide; ovary 3–3·5 mm. long, 0·4–0·5 mm. wide, dorsal (abaxial) surface distinctly pilose; style thickened, upcurved, about 2 mm. long, stigma capitate, oblique; young pod protruding sideways from calyx. Pod cylindrical, 5–8 (–12) mm. high, 6–7 (–10) mm. diam., coils 4–5 (–8), strongly adpressed, usually clockwise, lateral surface glabrous, obscurely veined, dorsal surface spinose, the spines rather short and thick, and usually closely adpressed to pod, dorsal suture ± pilose with slender, simple hairs, usually distinctly raised with a well marked furrow on either side; seeds 1–2 in each coil, reniform, about 2·5 mm. long, 1·5 mm. wide; testa smooth, pale brown.

HAB.: Dry rocky pastures; margins of fields, usually inland; sea-level to 3,000 ft. alt.; fl. Febr.–May.

DISTR.: Divisions 2, 5, 7, 8, common in Division 7. Throughout the Mediterranean region; Atlantic Islands; introduced and naturalized elsewhere.

2. Khrysorroyiatissa, 1941, *Davis* 3360 ! Kakopetria, 1955, *Merton* 2345 !
5. Mia Milea, 1950, *Chapman* 490 ! and, 1952, *Merton* 628 !
7. Common; *Sintenis & Rigo* 480 ! 992 ! *Syngrassides* 905 ! 906 ! *Davis* 2818A ! 2877 ! *Chapman* 157 ! *Casey* 1141 ! 1147 ! *G. E. Atherton* 320 ! *Townsend* 63/68 !
8. Eleousa near Rizokarpaso, 1962, *Meikle* 2999B !

16. M. rigidula (*L.*) *All.*, Fl. Ped., 1: 316 (1758); Post, Fl. Pal., ed. 2, 1: 325 (1932); Heyn in Scripta Hierosolymitana, 12: 104 (1963); Davis, Fl. Turkey, 3: 503 (1970); Zohary, Fl. Palaest., 2: 149, t. 217 (1972); Osorio-Tafall et Seraphim, List Vasc. Plants Cyprus, 61 (1973).

M. polymorpha L. var. *rigidula* L., Sp. Plant., ed. 1, 780 (1753).

M. gerardi Waldst. et Kit. ex Willd., Sp. Plant., 3: 1415 (1802); Boiss., Fl. Orient., 2: 100 (1872); A. K. Jackson et Turrill in Kew Bull., 1939: 475 (1939) pro parte.

TYPE: The illustration of *Medica hirsuta echinis rigidioribus* in J. Bauhin, Hist. Plant., 2: 386 (1651).

Prostrate or ascending annual; stems 8–30 cm. long, much branched especially at the base, angular, pilose or glandular-pilose, often purplish; stipules narrowly ovate-acuminate, 4–5 mm. long, 1–2 mm. wide, pilose or glandular-pilose, margin coarsely dentate; petioles slender, short, usually less than 1 cm. long, pilose, channelled above; leaflets obcuneate, 5–9 mm. long, 3–6 mm. wide, pilose, apex truncate or slightly retuse, minutely denticulate, margins entire; inflorescences 1–2-flowered, scattered along the greater length of the branches; peduncles slender, 1–2 cm. long, pilose or glandular-pilose, produced into a filiform awn 3–4 mm. long; bracts minute, oblong, about 0·5 mm. long and almost as wide, apex acute but scarcely cuspidate; pedicels about 1 mm. long, pilose or glandular-pilose; calyx broadly campanulate, 4–5 mm. long, 1·5–2 mm. wide, thickly pilose externally, teeth deltoid-acuminate, 2–2·5 mm. long; petals yellow, standard broadly ovate or obovate, 5–6 mm. long, 4 mm. wide, apex shortly emarginate, base very broadly cuneate; wings oblong, about 4 mm. long, 1–1·3 mm. wide, apex rounded, base with a very slender filiform claw and conspicuous auricle, conjunctival tooth very well developed; keel 4–4·5 mm. long, apex blunt, base obscurely auriculate; stamens 3–4·5 mm. long, the vexillary ± free or united almost to half way; anthers suborbicular about 0·4 mm. diam.; ovary narrowly oblong, about 2·5 mm. long, 0·4 mm. wide, very shortly stipitate, glabrous or minutely glandular-papillose above, dorsal (abaxial) surface strongly barbellate, style not much thickened, about 1·5 mm. long, glabrous, slightly upcurved, stigma capitate, oblique; young pod concealed within calyx. Pod ovoid-cylindrical, 5–8 (–12) mm. high, 5–7 (–10) mm. diam., coils 4–7, anticlockwise, rather loose, lateral surfaces ± densely clothed with multicellular glandular and eglandular hairs, obscurely veined, dorsal surface convex or flattish, spinose, with thick, spreading, glandular-pilose spines 1–2 mm. long; seeds reniform, 1–2 in each coil, 3–4 mm. long, about 2 mm. wide; testa smooth, pale brown.

HAB.: Fallow fields, roadsides and waste ground; 1,000–4,500 ft. alt.; fl. April- June.

DISTR.: Division 2. Throughout the Mediterranean region, and eastwards to Iran.

2. Platres, 1938, *Kennedy* 1269 ! and, 1960, *N. Macdonald* 139 ! Prodhromos, 1955, *Merton* 2399 ! and, 1961, *D. P. Young* 7274 !

17. M. constricta *Dur.* in Act. Linn. Soc. Bordeaux, 29: XV (1873); Heyn in Scripta Hierosolymitana, 12: 108 (1963); Davis, Fl. Turkey, 3: 506 (1970); Zohary, Fl. Palaest., 2: 149, t. 218 (1972); Osorio-Tafall et Seraphim, List Vasc. Plants Cyprus, 61 (1973).

M. globosa (vix Presl) Urban in Sitz. Verh. Bot. Ver. Brandenb., 15: 71 (1873); Post, Fl. Pal., ed. 2, 1: 325 (1932).

TYPE: Rhodes; " dans les champs incultes, près *Trianda*, 1870 " *Bourgeau* (P, G).

Small prostrate annual; stems seldom much more than 15 cm. long, much branched especially towards base, angular, pilose, often purplish; stipules ovate, 3–4 mm. long, 1–2 mm. wide, thinly pilose, margins rather coarsely serrate; petioles very slender, 4–20 mm. long, thinly pilose, slightly channelled above; leaflets obcuneate, 2·5–4 mm. long, almost as wide, long-pilose on both surfaces, apex truncate or slightly emarginate, minutely denticulate, margins entire or subentire; inflorescence 1–2-flowered; peduncle slender, 1–2 cm. long, generally much exceeding the leaves, thinly pilose, produced as a short cusp; bracts minute, oblong-cuspidate, usually less than 1 mm. long and about as wide, often tinged purplish; pedicels 1·5–3 mm. long, filiform, often rather conspicuous; calyx campanulate, 2–2·5 mm. long, 1–1·3 mm. wide, pilose, teeth narrowly deltoid, about 1·5 mm. long; petals yellow; standard suborbicular, about 4 mm. diam., wings short, oblong, about 3 mm. long, 1·5–2 mm. wide, apex rounded, base with a prominent oblong auricle and short slender claw, conjunctival tooth well developed, keel about 4 mm. long, 1·3 mm. wide, apex rounded, base shortly auriculate; stamens subequal, 3·5–4 mm. long, the vexillary connate for about one third of its length; anthers suborbicular, about 0·3 mm. diam.; ovary narrowly oblong, 2–2·5 mm. long, 0·3 mm. wide, shortly stipitate, dorsal (abaxial) surface thinly pilose, style 1·5 mm. long, upcurved, a little thickened, stigma oblique, capitate; young pod concealed within calyx. Pod subglobose or ovoid, rounded at either end, 6–10 (–12) mm. high, 5–8 mm. diam., coils 6–8, closely adpressed, anticlockwise, at first glandular-pilose, later shortly puberulous or glabrescent, lateral surfaces rather distinctly veined with arching radial veins, dorsal surface flattish, spinose, with short, stout, erect spines, 1–2 mm. long, usually hooked at the apex; seeds 1–2 in each coil, reniform, 3–3·5 mm. long, 1·5–2 mm. wide; testa smooth, pale brown.

HAB.: Dry hillsides, sand-dunes, fallow fields, roadsides and waste ground; sea-level to 3,700 ft. alt.; fl. Febr.–May.

DISTR.: Divisions 2, 5–8, common in Division 7. Also Greece, Crete, Rhodes, Turkey, Syria, Lebanon and Palestine.

2. Platres, 1938, *Kennedy* 1268 ! 1271 !
5. Perakhorio, 1936, *Syngrassides* 1142 !
6. Mile 7, Nicosia–Troödos road, 1952, *Merton* 667 ! Mile 30, Nicosia–Troödos road, 1955, *Merton* 1961 in part !

7. Common; *Chapman* 94 ! *Casey* 1021 ! 1151 ! *G. E. Atherton* 941 ! 1091 !
8. Ronnas Bay near Rizokarpaso, 1941, *Davis* 2344 !

M. ACULEATA *Willd.* (*M. oliviformis* Guss.) which closely resembles the following species, *M. turbinata* (L.) All., differs chiefly in its 1–2-flowered inflorescence, smaller petals and larger, pilose or, rarely, subglabrous fruit; it has been recorded from Cyprus (Heyn, *op. cit.*, 114) evidently on the strength of a specimen collected by Sintenis and Rigo near Kythrea (no. 61 in the Naturhistorisches Museum, Vienna). No specimen of this number is represented in the Kew collections, but there are two sheets of *Sintenis and Rigo* 636 collected "in campis pr. Kythraeam", May 1880 and labelled "*Medicago olivaeformis* Guss." These specimens are without flowers, but have ripe fruits larger than is usual in *M. turbinata* (L.) All. and not unlike those of *M. aculeata* Willd. The fruits are, however, glabrous, and close examination shows that the inflorescences must have borne 4 or more flowers. Moreover the cusp which terminates the inflorescence is long and persistent, not caducous as is usual in *M. aculeata*. In the circumstances, it would appear that "*M. olivaeformis* Guss." is a misidentification and that *Sintenis and Rigo* 636 should be named *M. turbinata* (L.) All., a species abundantly represented in Cyprus. The Vienna specimen, cited by Heyn, is no doubt a similar mididentification, and the record for *M. aculeata* Willd. must be rejected. Since *M. aculeata* has been recorded from areas adjacent to Cyprus, it may yet be added to our flora.

18. M. turbinata (*L.*) *All.*, Fl. Ped., 1: 315 (1785); Heyn in Scripta Hierosolymitana, 12: 116 (1963); Davis, Fl. Turkey, 3: 507 (1970); Zohary, Fl. Palaest., 2: 150, t. 220 (1972); Osorio-Tafall et Seraphim, List Vasc. Plants Cyprus, 61 (1973).
M. polymorpha L. var. *turbinata* L., Sp. Plant., ed. 1, 780 (1753).
M. tuberculata Willd., Sp. Plant., 3: 1409 (1802); Boiss., Fl. Orient., 2: 99 1872); Sintenis in Oesterr. Bot. Zeitschr., 32: 262 (1882); Holmboe, Veg. Cypr., 104 (1914); Post, Fl. Pal., ed. 2, 1: 324 (1932); Lindberg f., Iter Cypr., 19 (1946).

A prostrate or ascending annual; stems 15–50 cm. long, usually much branched, especially from base, strongly angular, ± pilose with long simple hairs, often purplish or reddish; stipules rather large, ovate-acuminate, 5–10 mm. long, 2–4 mm. wide, thinly pilose, margins sharply laciniate, the basal laciniae long and slender, or rarely subentire; petioles rather slender 1–4 cm. long, thinly pilose, distinctly channelled above; leaflets obovate or broadly elliptic, occasionally obcuneate or lanceolate, 1–2·5 (–3·5) cm. long, 0·5–1·5 (–2·5) cm. wide, pilose above and below, apex rounded or acute, rarely truncate or emarginate, margins finely and sharply serrulate; inflorescence (1–) 3–8 (–12)-flowered, frequently capitate; peduncles 1·5–3·5 cm. long, usually longer than subtending petioles, thinly pilose, produced into a filiform, persistent cusp or awn; bracts cuspidate, about 1·5 mm. long, with an oblong base about 1 mm. wide; calyx campanulate, about 3·5 mm. long, 1·3 mm. wide, thinly pilose externally, teeth subulate, about 2 mm. long; petals golden- or orange-yellow, standard broadly obovate, about 6 mm. long, 5 mm. wide, apex rounded, often minutely apiculate, base broadly cuneate, wings 5 mm. long, 1·5 mm. wide, apex rounded, upcurved, base with a long narrow auricle, slender claw and well-developed conjunctival tooth, keel about 5·5 mm. long, 1·3 mm. wide, apex blunt, often truncate; stamens 6 mm. long, 0·8 mm. wide, vexillary stamen connate for about half its length; anthers suborbicular about 0·5 mm. diam.; ovary narrowly oblong, 4 mm. long, 0·5 mm. wide, distinctly stipitate, dorsal (abaxial) surface minutely papillose, style 2 mm. long, upcurved, slightly thickened, stigma oblique, small, capitate; young pod concealed within calyx. Pod ovoid, or barrel-shaped with flattened ends, 6–12 (–15) mm. high, 5–8 mm. diam., coils 5–7 (–9), closely adpressed, clockwise or anticlockwise, glabrous, lateral surfaces obscurely veined with a wide veinless margin, dorsal surface flattish, rugose, spinose or unarmed; seeds 1–2 in each coil, rather large, reniform, 5–6 mm. long, 3 mm. wide; testa smooth, pale brown.

var. **turbinata.**

TYPE: The illustration in J. Bauhin, Hist. Plant., 2: 385 (1651).

Inflorescence usually 5–8-flowered. Pods unarmed, almost smooth or at most very shortly rugose-tuberculate.

HAB.: Cultivated fields, roadsides and waste ground, sometimes in marshy places; sea-level to 3,800 ft. alt., usually lowland; fl. Jan.–April.

DISTR.: Divisions 1–4, 6–8, common in Divisions 2, 7. Widespread in the Mediterranean region and eastwards to Iraq.

1. Dhrousha, 1941, *Davis* 3216 !
2. Common; *Druce* 190 ! *Syngrassides* 252 ! *Lindberg f.* s.n.; *Davis* 2718 ! 3421 ! *Merton* 2108 ! *Meikle* 2244 !
3. Akhelia, 1939, *Lindberg f.* s.n.; Kolossi, 1971, *Townsend* 71/123 !
4. Famagusta, 1936, *Syngrassides* 1127 ! Mile 3, Larnaca–Famagusta road, 1950, *Chapman* 425 ! Near Famagusta, 1952, *Merton* 1367 !
6. Mile 7, Nicosia–Troödos road, 1952, *Merton* 668 !
7. Common; *Sintenis & Rigo* 636 ! *Syngrassides* 181 ! 1240 ! *Davis* 3016 ! *Casey* 461 ! 1145 ! *Merton* 2186 !
8. Between Rizokarpaso and Ronnas Bay, 1941, *Davis* 2379 ! Komi Kebir, 1941, *Davis* 2440 ! Platanisso, 1950, *Chapman* 239 ! Elisis near Galinoporni, 1962, *Meikle* 2495 !

var. **apiculata** (*Urb.*) *Heyn* in Scripta Hierosolymitana, 12: 118 (1963).

M. tuberculata Willd. var. *apiculata* Urb. in Sitz. Verh. Bot. Ver. Brandenb., 15: 72 (1873).

M. tuberculata Willd. var. *spinosa* H. Stuart Thompson in Journ. Bot., 44: 305 (1906) nomen nudum.

M. apiculata (non Willd.) Bast., Fl. Dep. Maine et Loire, 280 (1809).

TYPE: Probably an Urban specimen, now destroyed (see Heyn, *op. cit.*, 119).

Inflorescence 1–2 (–3)-flowered. Pods spinose, with short spines usually much thickened at the base.

HAB.: Dry hillsides; about 2,000 ft. alt.; fl. Jan.–April.

DISTR.: Division 7, rare. Widely but sparsely distributed in the northern Mediterranean region, and apparently more common in the east than in the west.

7. Kantara, 1880, *Sintenis & Rigo* 474 ! St. Hilarion, 1963, *Townsend* 63/64 !

19. M. intertexta (*L.*) *Mill.*, Gard. Dict., ed. 8, no. 4 (1768) emend. Heyn in Scripta Hierosolymitana, 12: 125 (1963); Davis, Fl. Turkey, 3: 509 (1970); Zohary, Fl. Palaest., 2: 152, t. 222 (1972); Osorio-Tafall et Seraphim, List Vasc. Plants Cyprus, 61 (1973).

A robust, ascending or decumbent annual, 30–50 cm. high; stems angular, glabrous or thinly clothed with simple, glandular or eglandular hairs, usually much branched; stipules large, 8–15 mm. long, 3–8 mm. wide, deeply dentate or laciniate, with 8 or more narrow, sharply pointed teeth; petioles 1–3 cm. long, flattened, sometimes narrowly winged; leaflets broadly or narrowly obovate, 6–20 mm. long, 5–15 mm. wide, glabrous on both sides or thinly hairy below, sometimes blotched red, apex acute, rounded or truncate, rarely retuse, margins minutely serrulate; inflorescences 1–4 (–10)-flowered, produced along the greater length of the branches; peduncle about as long as petiole, usually produced into a slender, filiform awn; bracts narrowly ovate, 1–2 mm. long, 0·5 mm. wide, with a long apical cusp; pedicels filiform about 2 mm. long; calyx campanulate, about 4 mm. long, 1·5–2 mm. wide, glabrous or thinly pilose, teeth subulate, about 2 mm. long; petals mustard-yellow; standard suborbicular, 5–6 mm. diam., apex retuse, base very shortly clawed, wings oblong, 4·5–6 mm. long, 2–2·5 mm. wide, apex blunt, base with an acute auricle and filiform claw, conjunctival tooth strongly developed, keel

5–6 mm. long, about 1·5–2 mm. wide, apex blunt or rounded, base with a long slender claw; stamens 5–6 mm. long, the vexillary free, the others connate almost to apex; anthers suborbicular about 0·4 mm. diam.; ovary narrowly oblong, slightly upcurved, about 4 mm. long, 0·5 mm. wide, glabrous or ± pubescent, style distinctly thickened, glabrous, 1–2 mm. long, stigma capitate; young pod protruding sideways from calyx. Pod spherical or broadly ovoid (rarely discoid), 9–15 (–17) mm. diam., with (3–) 6–10 anticlockwise coils, dorsal margin strongly spinose, with spreading, glabrous or pubescent spines 1–5 (–8) mm. long, lateral surfaces with a conspicuous network of anastomosing veins; seeds 1–2 in each coil, narrowly reniform, 4–5 mm. long, 2–2·5 mm. wide; testa smooth, dark blackish-brown.

var. ciliaris (*L.*) *Heyn* in Scripta Hierosolymitana, 12: 129 (1963).

M. polymorpha L. var. *ciliaris* L., Sp. Plant., ed. 1, 780 (1753).

M. ciliaris (L.) All., Fl. Ped., 1: 315 (1785); Boiss., Fl. Orient., 2: 104 (1872); Sintenis in Oesterr. Bot. Zeitschr., 31: 194 (1881); Post, Fl. Pal., ed. 2, 1: 328 (1932—as *M. ciliaris* (L.) Krock.); Rechinger f. in Arkiv för Bot., ser. 2, 1: 424 (1950).

TYPE: Based on " *Medicago leguminibus congestis globosis hispidis, stipulis ciliatis* " Sauvages, Meth. Fol., 186 (1751).

Leaflets broadly obovate, 15–20 mm. long, 8–15 mm. wide. Pods subglobose, 12–15 mm. diam., dorsal surface and spines rather densely clothed with multicellular hairs.

HAB.: Cultivated and waste ground, often in marshy situations; sea-level to 500 ft. alt.; fl. March–May.

DISTR.: Divisions 4, 5, 7. Widespread in the Mediterranean region; Atlantic Islands.

4. Near Famagusta, 1912, *Haradjian* 94; Kouklia, 1949, *Mercy Casey* in *Casey* 478 ! Mile 3, Larnaca–Famagusta road, 1950, *Chapman* 424 !

5. Near Kythrea, 1880, *Sintenis & Rigo* 475 !

7. Myrtou, 1932, *Syngrassides* 452 ! Kyrenia, 1951, *Casey* 1180 ! E. of Panagra, 1962, *Meikle* 2396 !

NOTES: Typical *M. intertexta* (L.) Mill., with larger, glabrous fruits, is absent from the eastern Mediterranean. The form with blotched leaves is occasionally cultivated in gardens as " Calvary Clover ", and is generally named *M. echinus* by seedsmen.

SECT. 4 **Lupularia** *Ser.* Annual, biennial or perennating herbs. Pods 1-seeded, reniform, distal portion coiled, without spines; seeds with a smooth testa; radicle about half as long as cotyledons.

20. M. lupulina *L.*, Sp. Plant., ed. 1, 779 (1753); Sibth. et Sm., Fl. Graec. Prodr., 2: 110 (1813); Unger et Kotschy, Die Insel Cypern, 375 (1865); Boiss., Fl. Orient., 2: 105 (1872); Holmboe, Veg. Cypr., 105 (1914); Post, Fl. Pal., ed. 2, 1: 329 (1932); Lindberg f., Iter Cypr., 19 (1946); Rechinger f. in Arkiv för Bot., ser. 2, 1: 424 (1950); Davis, Fl. Turkey, 3: 487 (1970); Zohary, Fl. Palaest., 2: 140, t. 203 (1972); Osorio-Tafall et Seraphim, List Vasc. Plants Cyprus, 61 (1973).

M. lupulina L. var. *willdenowiana* Koch, Syn. Fl. Germ. et Helv., 161 (1835); Lindberg f., Iter Cypr., 19 (1946)—as var. *willdenowii* Boenn.

TYPE: " *in* Europae *pratis* ".

Ascending or decumbent annual, biennial or short-lived perennial, 15–50 cm. high; stems much branched, angular, ± pubescent, occasionally glandular-pubescent; stipules ovate-acuminate, 3–12 mm. long, 2–4 mm. wide, subentire or minutely denticulate; petioles slender, 5–15 (–40) mm. long, channelled above, ± pubescent; leaflets broadly ovate- or obovate-elliptical, 5–20 mm. long, 4–15 mm. wide, adpressed-pubescent on both surfaces, apex shortly retuse, usually apiculate, margins minutely serrulate; inflorescences produced along the greater length of the stems, many-flowered, the small flowers forming dense globose or oblong-cylindrical clusters; peduncles usually much exceeding petioles, commonly 2–3 cm. long, slender, pubescent or glandular-pubescent; bracts minute, membranous, cuspidate, about 0·6 mm. long, less than 0·2 mm. wide at base; pedicels less than 1 mm. long, pubescent; calyx campanulate, pubescent, about 2 mm. long, 1 mm. wide, teeth subulate, about 1 mm. long; petals yellow, standard broadly obcordate about 2 mm. long and almost as wide, apex emarginate, base narrowed to a very short claw, wings oblong, about 1·5 mm. long, 0·6 mm. wide, apex blunt, base with a prominent auricle and slender claw, conjunctival tooth short, blunt, keel a little longer than wings, about 0·7 mm. wide, apex blunt, base with a rather broad claw; stamens about 1·5 mm. long, the vexillary free, the other connate almost to apex; anthers suborbicular, 0·2 mm. diam.; ovary stipitate, broadly ovoid, about 0·5 mm. long, 0·4 mm. wide, pubescent or ± glandular, style much thickened, upcurved, about 0·5 mm. long, stigma capitate; young pod protruding sideways from calyx. Pod blackish when ripe, reniform, 2–3 mm. wide, 1·8–2·3 mm. long, glabrous, pubescent or glandular (var. *willdenowiana* Koch), distinctly reticulate-veined, 1-seeded, the base concealed by the persistent calyx, the apex usually twisted clockwise; seeds reniform, about 1·8 mm. long, 1·5 mm. wide; testa smooth, pale or bright brown.

HAB.: By streamsides and irrigation ditches; less frequently on hillsides amongst garigue or in cultivated fields; 1,000–5,150 ft. alt.; fl. May–Sept.

DISTR.: Divisions 1, 2, 7, very common in Division 2. Widespread in Europe and Asia; introduced and naturalized in many temperate regions of the world.

1. Dhrousha, 1941, *Davis* 3237 ! Below Stroumbi, 1941, *Davis* 3377 !
2. Very common; *Haradjian* 424; 426; 976 ! *Syngrassides* 1093 ! *Kennedy* 1264 ! 1265 ! 1846 ! *Lindberg f.* s.n. ! *Casey* 864 ! 1201 ! *Merton* 2343 ! *G. E. Atherton* 520 !
7. Lapithos, 1939, *Lindberg f.* s.n. ! and, 1955, *Miss Mapple* 90 !

NOTES: Most of the material from Cyprus is to some degree glandular, and might be referred to var. *willdenowiana* Koch (*M. willdenowii* Boenn.); this variety, however, intergrades to such an extent with eglandular *M. lupulina* that it is scarcely deserving of taxonomic recognition.

9. FACTOROVSKYA *Eig*

Zionist Org. Inst. Agr. Nat. Hist., Bulletin 6: 11–19 (1927).
Post, Fl. Pal., ed. 2, 1: 319 (1932).

Prostrate annual; leaves pinnately 3-foliolate; inflorescence pedunculate, 1-flowered; flower small; petals yellow; stamens diadelphous, filaments not dilated at apex; ovary linear, 2-ovulate, slightly upcurved, stipitate, style falcate or geniculate; peduncle, pedicel and gynophore elongating after fertilization and burying the developing fruit underground. Ripe pod tomentose, consisting of 2 lenticular lobes joined by a very narrow neck, the distal lobe smaller and sometimes abortive; seed reniform-suborbicular, radicle parallel to dorsal suture of pod and directed towards base of pod, cotyledons obovate-oblong, tapering basally to a long petiole, micropyle situated below the funicle.

One species in Cyprus, Palestine, Syria, Iraq, S. Turkey, Egypt and Cyrenaica.

1. **F. aschersoniana** (*Urb.*) *Eig* in Zionist Org. Inst. Agr. Nat. Hist., Bulletin 6: 11–19 (1927); Post, Fl. Pal., ed. 2, 1: 319 (1932); Jones, Merton et al., Rep. Pasture Research, Survey and Development, 59 (1958); Davis, Fl. Turkey, 3: 482 (1970); Zohary, Fl. Palaest., 2: 136, t. 200 (1972); Osorio-Tafall et Seraphim, List Vasc. Plants Cyprus, 61 (1973).

Trigonella aschersoniana Urb. in Sitz. Verh. Bot. Ver. Brandenb., 23: 67 (1881).

TYPE: Egypt; " prope Mariut in collibus calcareis ", Febr. 1880, *Ascherson* (B).

Prostrate annual; stems much branched at base, forming loose or rather dense mats, 2–8 (–12) cm. long, greenish or purplish, glabrous; stipules ovate

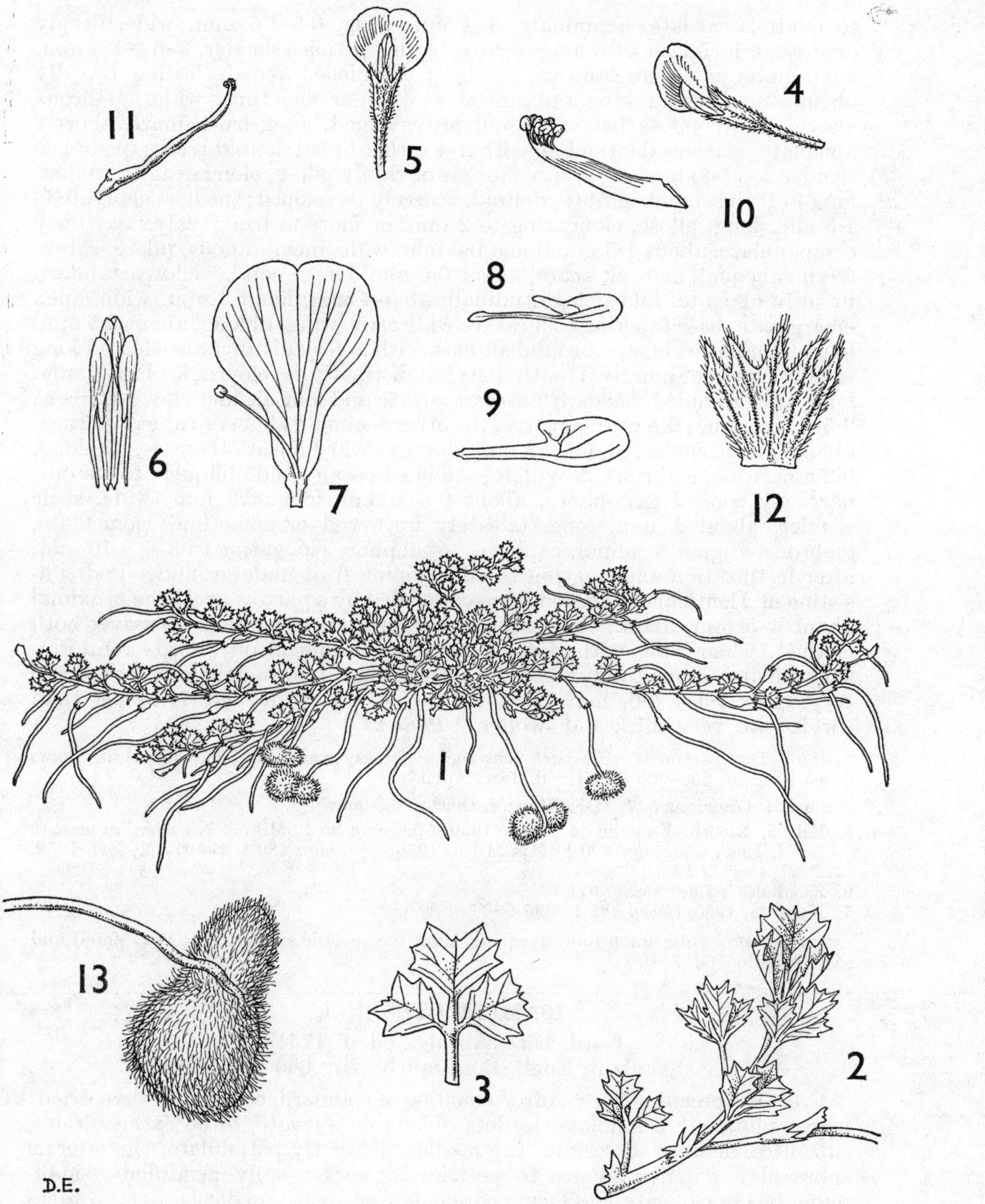

Plate 21. FACTOROVSKYA ASCHERSONIANA (Urb.) Eig **1,** habit, ×1; **2,** branch tip, ×4; **3,** leaf, ×4; **4–5,** flower, ×6; **6,** wings and keel, ×12; **7,** standard, ×12; **8,** wing, ×12; **9,** keel, ×12; **10,** androecium, ×12; **11,** gynoecium, ×12; **12,** calyx opened out, outer surface, ×12; **13,** fruit, ×3. (**1,** *Oswald* 120, Cyprus material; **2–13,** *Chapman* 35, Cyprus material.)

to ovate-lanceolate, acuminate, 1–4 mm. long, 0·5–1·5 mm. wide, deeply dentate or laciniate with long narrow teeth; petioles slender, 3–6 (–12) mm. long, glabrous or subglabrous, narrowly channelled above; leaflets broadly obcuneate, 2–4 mm. long and almost as wide, or sometimes wider, glabrous above, thinly pilose below, prominently veined, apex emarginate, shortly apiculate, margins denticulate with 2–4 rather broad deltoid teeth; peduncle slender 2–6 (–8) mm. long, subglabrous or thinly pilose, elongating to 20 mm. long in fruit; bract minute, deltoid, scarcely developed; pedicel short, 0·3–1·5 mm. long, pilose, elongating to 2 mm. or more in fruit; calyx narrowly campanulate, about 1·5 mm. long, 0·6 mm. wide, membranous, pilose, calyx-teeth subequal, deltoid, acute, about 0·5 mm. long; petals yellow, standard broadly obovate, folded lognitudinally, 3·5–4 mm. long, 3 mm. wide, apex emarginate, base tapering to a rather wide claw, wings oblong, about 2·5 mm. long, 0·6 mm. wide, apex rounded, base with a very distinct auricle and long slender claw, conjunctival tooth short but distinctly developed, keel 2·5–3 mm. long, apex rounded, base with a short auricle and long slender claw; stamens 2·5–3 mm. long, the vexillary free, the others connate almost to apex; anthers oblong-suborbicular, about 0·3 mm. diam.; ovary linear, about 1 mm. long, 0·2 mm. wide, glabrous, 2-ovulate, attached basally and obliquely to a stout, narrowly conical gynophore, about 0·8–1 mm. long, 0·3 mm. wide, style slender, about 1 mm. long, falcately upcurved or sometimes geniculate, glabrous, stigma terminal, capitate; gynophore elongating to 3–4 (–10) cm. after fertilization and burying the developing fruit underground. Pod consisting of 2 lenticular 1-seeded lobes connected by a narrow neck, the proximal about 4–5 mm. diam., the distal rather smaller, sometimes abortive, both clothed, when ripe, with dense tomentose indumentum; seeds reniform-suborbicular, much compressed, 2·5–4 mm. diam.; testa pale brown, smooth; cotyledons 3·5–4 (–6) mm. long, 1·5–2 (–4) mm. wide, radicle longer than cotyledons, very thick and swollen. *Plate 21.*

HAB.: Dry pastures; thin turf over rocks; grassy waste land and threshing floors (? sea-level to) 200–900 ft. alt.; fl. Jan.–April.

DISTR.: Divisions 5–7. Distribution that of the genus.

5. Mile 5, Nicosia–Famagusta Road, 1950, *Chapman* 35 ! Mile 3, Nicosia–Famagusta Road, 1950, *Chapman* 390 ! Mia Milea, 1950, *Chapman* 480 ! South of Xeri, 1959, *P. H. Oswald* 120 !
6. Kormakiti, 1950, *Casey* 971 !
7. Kyrenia, 1950, *Casey* 988 ! 995 !

NOTES: Probably much more frequent than the records suggest, but very small and easily overlooked.

10. MELILOTUS *Mill.*

Gard. Dict. Abridg., ed. 4 (1754).
O.E. Schulz in Engl., Bot. Jahrb., 29: 660–735 (1901).

Erect or spreading herbs, often smelling of coumarin especially when dried; leaves pinnately 3-foliolate, leaflets oblong or obovate, often rather fleshy, subentire, dentate or serrate, the median distinctly petiolulate, the laterals subsessile; stipules adnate to petiole; flowers usually pendulous, small, numerous in elongate, axillary, pedunculate racemes; pedicels short; bracts minute, subulate; calyx campanulate, toothed or shortly lobed; corolla yellow or white, rarely blotched blue, petals caducous, standard often folded longitudinally, wings oblong, obtuse, ± adherent to keel, but without a well-developed conjunctival tooth; stamens ± connate and monadelphous or diadelphous, filaments not dilated at apex; anthers uniform; ovary (1–) 2–8-ovulate, narrowed to a glabrous, upcurved style, stigma terminal. Pod small, containing 1–4 seeds, globose or ovoid, sometimes compressed, protruding from calyx, smooth or variously wrinkled or furrowed, tardily dehiscent. Seeds ovoid; testa smooth or tuberculate, endosperm scanty; cotyledons articulate at base.

About 20 species widely distributed in Europe, Asia and North Africa; found as an adventive almost throughout the world.

Petals white; racemes elongate often exceeding 5 cm. in length; pods reticulate-rugose, fuscescent when ripe - - - - - - - - - - **1. M. albus**
Petals yellow; racemes less than 5 cm. long; pods pale brown or olive-brown when ripe:
Pods irregularly reticulate-rugose; racemes long-pedunculate:
Pods with a distinct apical beak; flowers about 5 mm. long
M. NEAPOLITANUS p. 434
Pods rounded, without a beak; flowers about 2·5 mm. long - **2. M. indicus**
Pods with numerous close concentric ridges; racemes shortly pedunculate:
Pods 3–3·5 mm. long, obtusely oblong or subglobose; stipules laciniate
3. M. sulcatus
Pods 6–7 mm. long, acute; stipules shortly dentate - - **4. M. messanensis**

1. M. albus *Medik.* in Vorles Churpf. Phys.-ökon. Ges., 2: 382 (1787); Boiss., Fl. Orient., 2: 109 (1872); O. E. Schulz in Engl. Bot. Jahrb., 29: 694 (1901); Post, Fl. Pal., ed. 2, 1: 331 (1932); Davis, Fl. Turkey, 3: 451 (1970); Zohary, Fl. Palaest., 2: 154, t. 224 (1972); Osorio-Tafall et Seraphim, List Vasc. Plants Cyprus, 60 (1973).

Erect annual or biennial, 0·3–2 m. high; stems usually much branched, glabrous, angular; leaves rather distant; petioles 1–2 (–3) cm. long, somewhat flattened, channelled above, glabrous or very sparsely pilose; stipules filiform or narrowly subulate, 3–10 mm. long, entire, glabrous or subglabrous; leaflets oblong or obovate, 1–3 cm. long, 0·5–1·5 cm. wide, glabrous above, sometimes sparsely adpressed pubescent below, apex rounded, apiculate or subacute, margins rather irregularly serrate especially towards apex, sometimes almost entire; inflorescences usually elongate, often exceeding 6 cm. in length, dense or rather lax; peduncle well developed, 1·5–4 cm. long or occasionally longer, generally straight and erect; pedicels short, 0·5–1·5 mm. long, pendulous; calyx 1·5–2 mm. long, 1·2–1·5 mm. wide, glabrous, calyx-teeth deltoid, acute, subequal, 0·5–1 mm. long; petals white, standard obovate-emarginate, 3–4 (–5) mm. long, 2–4 mm. wide, tapering rather abruptly to a very short claw, wings oblong, a little shorter than standard, about 0·8–1·5 mm. wide, apex rounded, base with a distinct auricle and rather short claw; keel 2·5–3·5 (–4·5) mm. long, 1–1·8 mm. wide, apex blunt, base scarcely auriculate, claw rather short; stamens united for about half their length, 1·8–4 mm. long; anthers ovate, 0·3–0·5 mm. long, 0·2–0·4 mm. wide; ovary very shortly stipitate, narrowly ovoid, ± compressed, glabrous, 1–1·8 mm. long, 0·4–0·6 mm. wide, 2–4 (–5)-ovulate, style filiform, upcurved, 1·5–2·5 mm. long, stigma capitate. Pod ovoid, 3–4 mm. long, about 2 mm. wide, slightly compressed, irregularly reticulate-rugose, fuscescent when fully ripe, the persistent style-base forming a short apiculus; seeds 1–2, ovoid, slightly compressed, 2–2·5 mm. long, 1·5 mm. wide; testa light or dark brown, smooth.

var. **parviflorus** *Boiss.*, Fl. Orient., 2: 110 (1872); Post, Fl. Pal., ed. 2, 1: 331 (1932).
M. arborea Castagne ex Ser. in DC., Prodr., 2: 187 (1825).
M. alba Desr. ssp. [" proles "] *parviflora* (Boiss.) O. E. Schulz in Engl., Bot. Jahrb., 29: 696 (1901).

TYPE: Turkey; Constantinople, 1822, *Castagne* (G).

Flowers small, standard about 3 mm. long. Plant often tall and robust, said to reach 6 m. in height.

HAB.: In vineyards on limestone mountainside, 3,100 ft. alt.; fl. ? June–August.

DISTR.: Division 2, rare. Russia, Turkey, Syria, Palestine and eastwards to Central Asia. Sometimes cultivated.

2. Rare; Aphamis, 1937, *Kennedy* 557 !

NOTES: Possibly an introduction or casual. Typical *M. albus* Desr., with larger flowers, has not been recorded from Cyprus, though it very probably occurs, having a world-wide distribution as an adventive.

M. NEAPOLITANUS *Ten.*, Fl. Nap. Prodr., suppl. 1, LXII (1811–15); Boiss., Fl. Orient., 2: 107 (1872); O. E. Schulz in Engl. Bot. Jahrb., 29: 711 (1901); Holmboe, Veg. Cypr., 105 (1914); Osorio-Tafall et Seraphim, List Vasc. Plants Cyprus, 60 (1973).
Trifolium spicatum Sibth. et Sm., Fl. Graec. Prodr., 2: 93 (1813), Fl. Graec., 8: 32, t. 743 (1833).
Melilotus gracilis DC. in Lam. et DC., Fl. Franç., 5: 565 (1815); Unger et Kotschy, Die Insel Cypern, 377 (1865).

A slender erect annual, 20–30 cm. high; stems glabrescent, much branched; stipules entire, subulate, usually less than 5 mm. long; leaflets narrowly or broadly obovate, 0·8–2 cm. long, 0·3–1·2 cm. wide, glabrous above, glabrescent below, apex rounded or slightly emarginate, obscurely serrulate, base usually subentire or entire; inflorescences long-pedunculate, usually rather short and lax; pedicels very short; flowers yellow, about 5 mm. long, patent or slightly pendulous, standard about as long as wings and keel. Pod broadly ovoid or subglobose, 2–3 mm. diam., papery, glabrescent, strongly reticulate-rugose or foveate with a short sharp apical beak, pale brown when fully ripe, 1-seeded; seed broadly ovoid about 2 mm. long, 1·5 mm. wide; testa dull brown, minutely tuberculate.

Recorded from Cyprus by Sibthorp, but not seen since, and either a casual occurrence, or an error in localization. *M. neapolitanus* is widely distributed in the northern Mediterranean, from Spain to Rhodes (and eastwards through northern Turkey and the Caucasus to Iran); it may yet be refound in Cyprus.

2. M. indicus (*L.*) *All.*, Fl. Ped., 1: 308 (1785); O. E. Schulz in Engl., Bot. Jahrb., 29: 713 (1901); Holmboe, Veg. Cypr., 105 (1914); Post, Fl. Pal. ed. 2, 1: 330 (1932); Lindberg f., Iter Cypr., 19 (1946); Rechinger f. in Arkiv för Bot., ser. 2, 1: 424 1950); Davis, Fl. Turkey, 3: 448 (1970); Zohary, Fl. Palaest., 2: 156, t. 230 (1972); Osorio-Tafall et Seraphim, List Vasc. Plants Cyprus, 60 (1973).
Trifolium Melilotus indica L., Sp. Plant., ed. 1, 765 (1753).
Melilotus parviflora Desf., Fl. Atlant., 2: 192 (1800); Unger et Kotschy Die Insel Cypern, 376 (1865); Boiss., Fl. Orient., 2: 108 (1872).
TYPE: "*in* India, Africa".

Erect or sprawling annual or biennial (10–) 20–50 cm. high; stems glabrous or subglabrous, angled, much branched; stipules linear-subulate, 3–8 mm. long, entire or sometimes 1–2-toothed near the expanded base; petioles slender, 0·5–4 cm. long, channelled above, glabrous or subglabrous; leaflets obovate or oblanceolate 5–25 mm. long, 4–15 mm. wide, glabrous above, very thinly pilose below, apex truncate or shortly emarginate, often minutely apiculate, margins irregularly serrulate, especially towards apex; peduncles well developed, 1–3 cm. long, erect, glabrous or subglabrous; racemes many-flowered, often rather dense, 1–3 cm. long, elongating to 5 cm. or more in fruit; bracts subulate 0·3–0·5 mm. long; pedicels short, 0·5–0·8 mm. long, recurved; flowers spreading or sub-pendulous, very small; calyx 1–1·3 mm. long, glabrous or subglabrous, calyx-teeth deltoid, about 0·5 mm. long; petals yellow, standard narrowly obovate-emarginate, about 2·5 mm. long, 1·3 mm. wide, claw barely developed, wings about 2·2 mm. long, 0·7 mm. wide, apex rounded, base shortly auriculate with a rather long slender claw, keel about as long as wings, 0·8 mm. wide, apex blunt, base with a slender claw; stamens about 1·5 mm. long, the vexillary free, the others united for more than half their length; anthers suborbicular, about 0·2 mm. diam.; ovary ovoid-ellipsoid, compressed, about 1 mm. long, 0·5 mm. wide, glabrous, 2-ovulate, style filiform, upcurved, about 1 mm. long, stigma capitate. Pod broadly ovoid or subglobose, 2–2·5 mm. long, 2 mm. wide, pale brown when ripe, papery, irregularly reticulate-rugose, 1-seeded, apex rounded or with a very obscure apiculus; seed suborbicular, about 1·7 mm. diam.; testa dark brown, minutely tuberculate.

HAB.: Cultivated fields, gardens and waste ground, often by streams and irrigation channels, occasionally in river-beds or on sand dunes; sea-level to 4,500 ft. alt.; fl. March–June.

DISTR.: Divisions 1–8, common in Division 7. Throughout the Mediterranean region and eastwards to Central Asia and India; Atlantic Islands. Introduced and naturalized in many parts of the world.

1. Cape Arnauti, 1913, *Haradjian* 887.
2. Prodhromos, 1880, *Sintenis & Rigo* 833 ! and, 1939, *Lindberg f.* s.n.; Kannaviou, 1939, *Lindberg f.* s.n.
3. Yermasoyia, 1950, *F. J. F. Barrington* s.n. !
4. Ayios Memnon, 1948, *Mavromoustakis* s.n. !
5. Trikomo, 1939, *Lindberg f.* s.n.; Perivolia, 1939, *Lindberg f.* s.n. ! Salamis, 1962, *Meikle* 2563 !
6. Syrianokhori, 1938, *Syngrassides* 1815 ! Astromeritis, 1938, *R. M. Nattrass* 903; Nicosia, 1959, *P. H. Oswald* 136 !
7. Common; *Kotschy* 457; *Haradjian* 384; *Syngrassides* 208 ! 1361 ! *Casey* 626 ! 629 ! 630 !
8. Near Rizokarpaso, 1912, *Haradjian* 133.

3. M. sulcatus *Desf.*, Fl. Atlant., 2: 193 (1800); Unger et Kotschy, Die Insel Cypern, 377 (1865); Boiss., Fl. Orient., 2: 106 (1872); O. E. Schulz in Engl., Bot. Jahrb., 29: 721 (1901); Holmboe, Veg. Cypr., 105 (1914); Post, Fl. Pal., ed. 2, 1: 329 (1932); Lindberg f., Iter Cypr., 19 (1946); Davis, Fl. Turkey, 3: 449 (1970); Zohary, Fl. Palaest., 2: 155, tt. 226, 227 (1972); Osorio-Tafall et Seraphim, List Vasc. Plants Cyprus, 60 (1973).

TYPE: "Algeriâ in agris cultis" *Desfontaines* (P).

Erect or sprawling annual 10–30 cm. high; stems subglabrous or very sparsely pilose, angled, much branched; stipules ovate-acuminate, 5–8 mm. long, 1–3 mm. wide at base, laciniate, with narrow linear-filiform lobes; petioles slender, 8–25 mm. long, channelled above, thinly puberulous or subglabrous; leaflets obovate or oblanceolate, 10–25 mm. long, 5–13 mm. wide, glabrous above, thinly adpressed-pilose below, apex rounded, margins finely and regularly denticulate; peduncles short, 3–15 mm. long, erect, subglabrous or thinly pilose; racemes rather short, 1–2 cm. long at anthesis, dense or somewhat lax, flowers spreading or pendulous; rhachis produced into a short apical awn; bracts subulate-filiform, about 1 mm. long; pedicels 1–1·5 mm. long, recurved; calyx narrowly campanulate, 2 mm. long, 1 mm. wide, subglabrous, calyx-teeth narrowly deltoid, acute, about 1 mm. long; petals yellow, standard ovate, strongly folded longitudinally, about 3·5 mm. long, 2 mm. wide, apex subacute, often obscurely denticulate, base narrowed abruptly to a short distinct claw, wings oblong, about 3 mm. long, 0·6 mm. wide, apex rounded, base auriculate with a slender claw about 1·2 mm. long, keel about 3·5 mm. long, 0·8 mm. wide, apex blunt, base shortly auriculate and with a distinct claw; stamens about 3·4 mm. long, the vexillary free, the others connate for more than half their length; anthers broadly ovate, about 0·2 mm. long and almost as wide; ovary ovoid, compressed, subsessile, glabrous, about 1 mm. long, 0·5 mm. wide, 2-ovulate, style filiform, 2·5 mm. long, erect, upcurved at apex, stigma minute, capitate. Pod oblong or subglobose, 3–3·5 mm. long, about 3 mm. wide, 1–2-seeded, pale brown when ripe, concentrically ridged with conspicuous, close ridges, apex rounded, abaxial suture slightly keeled; seeds oblong-subglobose, not much compressed, 2·5 mm. long, 2 mm. wide; testa light olive-brown, minutely tuberculate.

HAB.: Cultivated fields and waste ground; roadsides; occasionally in open woodland, or in river beds; sea-level to 3,800 ft. alt.; fl. Febr.–June.

DISTR.: Divisions 1–5, 7, 8, common in Divisions 2–4, 7. Throughout the Mediterranean region; Atlantic Islands.

1. Paphos, 1928 or 1930, *Druce* 183 ! Ayios Neophytos Monastery, 1939, *Lindberg f.* s.n.; between Dhrousha and Kato Arodhes, 1941, *Davis* 3279 !
2. Common; *Lindberg f.* s.n.; *Kennedy* 1847 ! *Meikle* 2246 !
3. Common; *Kotschy* 562 ! *Holmboe* 239; *Lindberg f.* s.n.; *Davis* 2672 ! *Merton* 2747 ! *Mavromoustakis* 15 !
4. Common; *Haradjian* 69 ! *Lindberg f.* s.n.; *Chapman* 440 ! *Merton* 2001 ! 2631 !
5. Near Kythrea, 1880, *Sintenis & Rigo* 426 ! Mile 14, Nicosia–Limassol Road, 1950, *Chapman* 543 ! Salamis, 1962, *Meikle* 2572 ! and, 1971, *Townsend* 71/111 !
7. Common; *Davis* 2977 ! *Casey* 503 ! 631 ! *G. E. Atherton* 17 ! 937 ! 1081 !
8. Near Yialousa, 1941, *Davis* 2400 ! Komi Kebir, 1941, *Davis* 2439 ! Platanisso, 1950, *Chapman* 243 !

4. **M. messanensis** (*L.*) *All.*, Fl. Ped., 1: 309 (1785); Unger et Kotschy, Die Insel Cypern, 377 (1865); Boiss., Fl. Orient., 2: 107 (1872); O. E. Schulz in Engl., Bot. Jarhb., 29: 725 (1901); Holmboe, Veg. Cypr., 105 (1914); Davis, Fl. Turkey, 3: 449 (1970); Osorio-Tafall et Seraphim, List Vasc. Plants Cyprus, 60 (1973).
Trifolium messanense L., Mantissa Altera, 275 (1771); Sibth. et Sm., Fl. Graec., 8: 31, t. 741 (1833).
Trifolium Melilotus sicula Turra ex Vitm., Summa Plant., 4: 326 (1790).
Melilotus sicula (Turra ex Vitm.) B. D. Jackson in Index Kew. 2: 199 (1895); Post, Fl. Pal., ed. 2, 1: 330 (1932).
TYPE: " *in* Sicilia ".

Sprawling or decumbent annual 15–30 (–45) cm. high; stems glabrous, rather succulent, much branched especially near base; stipules ovate-acuminate, 5–8 mm. long, 2–5 mm. wide, glabrous or thinly pilose with shortly and irregularly dentate margins; petioles flattened, narrowly winged, 0·8–3 cm. long; leaflets obovate or oblanceolate, rather thick, 1–2 cm. long, 0·4–1·5 cm. wide, glabrous above, thinly adpressed-pilose below, apex rounded or shortly emarginate, base narrowly cuneate, margins sharply and somewhat irregularly serrulate; peduncle very short, usually less than 5 mm. long; raceme short, few-flowered, rhachis produced as a slender, filiform awn; bracts very short, subulate-deltoid, less than 0·5 mm. long; pedicels recurved, 0·8–1 mm. long; flowers spreading or pendulous; calyx campanulate, about 2 mm. long, 1·5 mm. wide, thinly pilose or subglabrous, calyx-teeth deltoid, acute, about 1 mm. long; petals yellow, standard narrowly oblong-obovate, about 4·5 mm. long, 1·5–2 mm. wide, apex shortly emarginate, base tapering to a distinct claw, wings oblong, about 4 mm. long, 0·8 mm. wide, apex rounded, base with a prominent auricle and slender claw, keel about 4·5 mm. long, 1·2 mm. wide, apex blunt, base with an inconspicuous auricle and slender claw; stamens diadelphous, or with the vexillary fused basally with the others, about 4 mm. long; anthers oblong, about 0·3 mm. long, 0·2 mm. wide; ovary narrowly ovoid, 1·5 mm. long, 0·8 mm. wide, glabrous, 2-ovulate, style 2·5–3 mm. long, stigma capitate. Pod ovoid, acute, 6–7 mm. long, 4–4·5 mm. wide, distinctly compressed, abaxially keeled, pale brown when ripe, closely marked with prominent concentric ridges; seeds 1–2 in each pod, oblong or subglobose, 2·5–4 mm. long, about 2·5 mm. wide; testa dark brown, minutely rugulose-tuberculate.

HAB.: Marshy or seasonally inundated ground, sometimes in brackish situations; sea-level to 650 ft. alt.; fl. March–April.

DISTR.: Divisions 3, 4, 6, 8, common in Division 4. Widespread in the Mediterranean region.

3. Near Mazotos, 1862, *Kotschy* 561 ! Asomatos, 1956, *Merton* 2748 !
4. Common; *Kotschy*; *Sintenis & Rigo*; *Holmboe* 398; *Druce* s.n. ! *Casey* 569 ! *Chapman* 441 !
6. Myrtou, 1932, *Syngrassides* 23 ! Myrtou and Panagra, 1934, *Syngrassides* 1356 ! Syrianokhori, 1935, *Syngrassides* 801 !
8. Near Ayios Andronikos, 1880, *Sintenis & Rigo* 425 ! Platanisso, 1950, *Chapman* 248 ! Galinoporni, 1962, *Meikle* 2493 !

11. TRIFOLIUM *L.*

Sp. Plant., ed. 1, 766 (1753) pro parte.
Gen. Plant., ed. 5, 337 (1754) pro parte.
M. Hossain in Notes Roy. Bot. Gard. Edinb., 23: 387–481 (1961).

Annual or perennial herbs; leaves palmately or rarely pinnately 3-foliolate (very rarely 5–7-foliolate); stipules mostly entire, adnate to the petiole; inflorescences axillary or pseudo-terminal, generally spicate, capitate or umbellate, flowers rarely solitary; bracts sometimes small or wanting, or sometimes well-developed and conspicuous; calyx persistent, calyx-teeth subequal or the lower longer than the upper; petals white, pink, purple or yellow, marcescent, persistent or deciduous, the claws ± united; wings without a conjunctival tooth; stamens diadelphous, the filaments adnate to the petals, often dilated at the apex; anthers uniform; ovary sessile or stipitate, style filiform, upcurved or sometimes sharply inflexed, stigma terminal, capitate, occasionally oblique. Pod generally small, often concealed by the persistent calyx and petals, 1–4 (rarely more)-seeded, walls mostly thin and membranous; seeds estrophiolate, testa usually smooth.

Approximately 300 species in Europe, Africa, Asia and North America, the greatest concentration of species being found in the Mediterranean region. Many Clovers have been introduced as fodder plants, and have subsequently become naturalized, in areas where the genus is not represented by indigenous species.

Calyx-tube becoming inflated, papery and conspicuous in fruit:
- Perennials with tough rootstocks:
  - Inflorescence with a distinct involucre of small bracts; standard 6–7·5 mm. long — **19. T. fragiferum**
  - Inflorescence without a distinct involucre; standard 9–11 mm. long — **20. T. physodes**
- Annuals:
  - Inflorescence 0·6–1·5 cm. diam.; corolla resupinate; seeds smooth:
    - Fruiting calyx narrowed to apex, subglabrous or thinly hairy, with conspicuous antenna-like calyx-teeth; peduncle equalling or exceeding subtending leaf (rarely shorter); standard 4·5–7 mm. long — **21. T. resupinatum**
    - Fruiting calyx subglobose, ± tomentose, calyx-teeth inconspicuous; peduncle shorter than subtending leaf; standard 3·5–4 mm. long — **22. T. tomentosum**
  - Inflorescence 2–3 cm. diam.; corolla not resupinate; seeds verruculose — **23. T. spumosum**

Calyx-tube not becoming inflated, papery or conspicuous in fruit:
- Perennials:
  - Inflorescences sessile, 2·5–3 cm. diam.; corolla rosy-purple; plants erect or sprawling, 30–70 (–100) cm. high - - - **1. T. pratense**
  - Inflorescences long-pedunculate, 1·4–2·5 cm. diam.; petals white or tinged pink; plants prostrate, rooting at the nodes - - - - **25. T. repens**
- Annuals:
  - Inflorescences sessile or subsessile:
    - Inflorescences densely clustered close to the ground; stem scarcely developed, usually subterranean; small tufted plants usually less then 3 cm. high — **28. T. suffocatum**
    - Inflorescences not clustered close to the ground; stems usually well developed, not subterranean:
      - Inflorescences terminal (pseudo-terminal) with an involucre of enlarged stipule-bracts:
        - Stipule-bracts suborbicular or broadly ovate; stipules with a short acumen; calyx-teeth about 4 mm. long - - - - **2. T. cherleri**
        - Stipule-bracts broadly ovate with a long cuspidate acumen; stipules elongate-acuminate; calyx-teeth 7–8 mm. long - - - **3. T. hirtum**
      - Inflorescences axillary, without an involucre of enlarged stipule-bracts:
        - Inflorescence ovoid, obovoid or oblong; plants pubescent:
          - Calyx-teeth lanceolate-acuminate, usually recurved in fruit; lateral veins of leaflets curving downwards at apex; claw of standard united with the claws of wings and keel - - - - **10. T. scabrum**

Calyx-teeth subulate-aristate, suberect in flower and fruit; lateral veins of leaflets running straight into margin of leaflet or very little curved; claw of standard free - - - - - - **15. T. striatum**

Inflorescence globose or subglobose; plant glabrous or subglabrous, usually very small; calyx-teeth ovate-acuminate, strongly recurved **27. T. glomeratum**

Inflorescences distinctly pedunculate:

Inner flowers of inflorescence sterile, without corollas, accrescent after anthesis; calyx-teeth becoming long and filiform, usually forming a tangled (medusoid) mass concealing the fruits:

Peduncle lengthening and recurving after anthesis and burying the inconspicuous infructescence - - - - - - **16. T. subterraneum**

Peduncle not noticeably accrescent; infructescence conspicuous, not buried:

Inflorescences terminal (pseudoterminal); fruiting heads often more than 2 cm. diam. - - - - - **17. T. globosum**

Inflorescences axillary; fruiting heads usually less than 1 cm. diam. **18. T. pilulare**

Inner flowers of inflorescence fertile, with corollas; calyx-teeth not tangled or medusoid in fruit:

Upper leaflets narrow, lanceolate, linear or narrowly oblong, at least twice as long as wide:

Uppermost leaves opposite:

Inflorescence very large, 2–3 cm. (or more) wide; calyx-teeth exceeding corolla - - - - - **12. T. dasyurum**

Inflorescence smaller, less than 2 cm. wide; calyx-teeth shorter than corolla:

Standard 8–10 mm. long; calyx-tube glabrous or thinly pilose, distinctly nerved - - - - - **8. T. echinatum**

Standard 5–6 mm. long; calyx-tube densely pilose, indistinctly nerved **9. T. leucanthum**

All the leaves alternate:

Corolla very small, about 2–3 mm. long, much shorter than the conspicuous, filiform calyx-teeth - - - - - **5. T. arvense**

Corolla 10–13 mm. long, usually equalling or exceeding calyx-teeth:

Corolla equalling or almost equalling the sharp-tipped calyx-teeth **6. T. angustifolium**

Corolla exceeding the blunt-tipped calyx-teeth - **7. T. pamphylicum**

All the leaflets orbicular, obcordate, shortly oblong or obcuneate, not narrow:

Uppermost leaves opposite; inflorescence terminal (pseudoterminal):

Corolla persistent, becoming dry and scarious; wings divaricate; seed with a minutely verruculose testa - - - **24. T. argutum**

Corolla deciduous or withering; wings not divaricate; seed with a smooth testa:

Calyx-teeth ovate or oblong, very unequal, strongly accrescent and spreading in fruit:

Corolla about 2–2·5 cm. long; stipules of uppermost (opposite) leaves connate for most of their length - - **13. T. clypeatum**

Corolla about 1–1·2 cm. long; stipules of uppermost (opposite) leaves connate only at base - - - **14. T. scutatum**

Calyx-teeth subulate or filiform, subequal, not markedly accrescent:

Calyx-tube 20-nerved; corolla usually shorter than calyx-teeth or only slightly exceeding it - - - **4. T. lappaceum**

Calyx-tube 10-nerved; corolla distinctly longer than calyx-teeth:

Standard 8–10 mm. long; calyx-tube glabrous or thinly pilose, distinctly nerved - - - **8. T. echinatum**

Standard 5–6 mm. long; calyx-tube densely pilose, indistinctly nerved - - - **9. T. leucanthum**

All the leaves alternate:

Inflorescence about 3 cm. diam., terminal (pseudoterminal); calyx-teeth 12–15 mm. long, conspicuous - - - - **11. T. stellatum**

Inflorescence much less than 3 cm. diam., lateral or axillary; calyx-teeth short, inconspicuous:

Flowers white or pale pink, calyx-tube 10-nerved **26. T. nigrescens ssp. petrisavii**

Flowers yellow (rarely mauve, purple or violet) soon becoming dry, brownish and scarious; calyx-tube 5-nerved:

Standard flattish or concave, not keeled or longitudinally folded over wings and keel:

Corolla 7–10 mm. long:

Petals violet or purple, becoming brownish
[T. SPECIOSUM p. 467]
Petals yellow, becoming brown - - **29. T. boissieri**
Corolla 3·5–5 mm. long:
Petals at first yellow, later brownish
30. T. campestre ssp. **campestre**
Petals at first violet or mauve, later brownish-yellow or brown
30. T. campestre ssp. **paphium**
Standard keeled or folded longitudinally over wings and keel; petals yellow - - - - - - - **31. T. dubium**

SECT. 1. **Trifolium** Annuals, biennials and perennials; inflorescence spicate or capitate; flowers sessile or subsessile, ebracteate; calyx-tube actinomorphic, usually pilose, throat mostly ± closed by a thickened annulus and/or a ring of hairs, calyx-teeth often rather rigid; petals generally with the claws united into a tube. Pod small, usually included within calyx-tube, 1–2-seeded.

1. T. pratense *L.*, Sp. Plant., ed. 1, 768 (1753); Boiss., Fl. Orient., 2: 115 (1872); Post, Fl. Pal., ed. 2, 1: 333 (1932); A. K. Jackson et Turrill in Kew Bull., 1938: 464 (1938); Rechinger f. in Arkiv för Bot., ser. 2, 1: 424 (1950); Hossain in Notes Roy. Bot. Gard. Edinb., 23: 409 (1961); Davis, Fl. Turkey, 3: 415 (1970); Osorio-Tafall et Seraphim, List Vasc. Plants Cyprus, 59 (1973).
TYPE: " *in* Europae *graminosis* ".

Erect or sprawling perennial, 30–70 (–100) cm. high; stems rather sparingly branched, chiefly at base, thinly pilose with long spreading hairs; stipules oblong-ovate, 1–2·5 cm. long, 0·3–0·5 cm. wide, subglabrous or thinly pilose, pallid, conspicuously veined with anastomosing green veins, abruptly narrowed to a slender filiform-subulate acumen; petioles of lower leaves usually very long, often exceeding 15 cm., becoming progressively shorter upwards, the uppermost leaves subsessile or very shortly petiolate; leaflets oblong-ovate, 1·5–5 cm. long, 1–3 cm. wide, entire, often white-blotched, thinly pilose or subglabrous above, a little more densely pilose below, apex obtuse or subacute, sometimes emarginate, base cuneate; inflorescence a dense pseudoterminal, subglobose, sessile spike, 2·5–3 cm. diam., subtended by 2 subopposite, reduced, bract-like leaves with very broad stipules; flowers sessile or subsessile; calyx-tube narrowly cylindrical-campanulate, about 3 mm. long, 1·4 mm. wide, pilose, obscurely 10-nerved, calyx-teeth filiform, rather rigid, thinly pilose, the lowermost about 7 mm. long, the others about 3·5 mm. long; petals rosy-purple, about 1·5 cm. long, united by their claws into a tube, limb of standard narrowly ovate-oblong, 6–7 mm. long, 3–3·5 mm. wide, glabrous, apex truncate, base shortly auriculate, claw about 8 mm. long, wings narrowly oblong, limb about 5 mm. long, 1·3 mm. wide, apex obtuse or subacute, base conspicuously auriculate, keel about 4 mm. long, 1·5 mm. wide, apex acute; stamens about 1·2 cm. long, all, except the vexillary, connate for about 1 cm., free portions of filaments distinctly dilated; anthers oblong, about 0·5 mm. long, 0·3 mm. wide; ovary oblong, 1·5 mm. long, 0·5 mm. wide, compressed, glabrous, 2-ovulate, style filiform, about 1 cm. long, stigma obliquely capitate. Pod completely concealed by persistent calyx and corolla; seeds ovoid, compressed, 1·7 mm. long, 1·2 mm. wide; testa smooth, rich brown.

HAB.: Damp shaded mountainsides; 5,000–6,000 ft. alt.; fl. June–Aug.

DISTR.: Division 2, rare. Widespread in Europe, western and central Asia, North Africa, Atlantic Islands; introduced elsewhere.

2. Rare; Troödos, 1912, *Haradjian* 474 ! Kryos Potamos, 1937, *Kennedy* 558 ! Troödos area, 1950, *F. M. Probyn* s.n. !

2. T. cherleri *L.*, Demonstr. Plant., 21 (1753); Boiss., Fl. Orient., 2: 119 (1872); Holmboe, Veg. Cypr., 105 (1914); Post, Fl. Pal., ed. 2, 1: 333 (1932); Lindberg f., Iter Cypr., 20 (1946); Hossain in Notes Roy. Bot. Gard. Edinb., 23: 411 (1961); Davis, Fl. Turkey, 3: 431 (1970); Zohary, Fl. Palaest., 2: 178, t. 252 (1972); Osorio-Tafall et Seraphim, List Vasc. Plants Cyprus, 60 (1973).

TYPE: Cultivated at Uppsala.

Decumbent annual; stems 4–15 cm. long, much branched at base, rather densely pilose with spreading hairs; stipules ovate-acuminate, 4–5 mm. long, 1·5–2·5 mm. wide, glabrous and whitish towards base, conspicuously veined, herbaceous and pilose towards apex, acumen short and rather broad; petioles 0·5–2 (–3) cm. long, slender, pilose, channelled above; leaflets obovate 2–10 mm. long, 1·5–6 mm. wide, pilose on both surfaces, apex emarginate, margins minutely denticulate, the veins running into the teeth, base narrowly cuneate; inflorescences pseudoterminal, globose, 1–1·7 cm. diam., sessile, subtended by a conspicuous involucre of whitish, prominently veined, broadly ovate or suborbicular, overlapping stipule-bracts, 5–7 mm. long and almost as wide; calyx narrowly campanulate, about 6–7 mm. long, 2–2·5 mm. wide, asymmetric, densely villose, calyx-tube c. 20-nerved, calyx-teeth setaceous, about 4 mm. long, villose, often purplish; petals whitish or very pale pink, becoming reddish with age, about 8–10 mm. long, united by their claws into a tube, limb of standard narrowly oblong, 5–6 mm. long, 1–1·5 mm. wide, longitudinally folded, glabrous, apex acute, sometimes mucronate, base scarcely auriculate, wings narrowly oblong, limb about 3 mm. long, apex obtuse, base prominently auriculate, keel about 2 mm. long, 1 mm. wide, apex acute, base scarcely auriculate; stamens 6–7 mm. long, all, except the vexillary, connate for about 5 mm., free portions of filaments slightly dilated; anthers oblong, about 4 mm. long, 3 mm. wide; ovary obovate, compressed, about 1·5 mm. long, 1 mm. wide, glabrous, 1–2-ovulate, style about 5 mm. long, glabrous, stigma terminal, capitate. Pod quite concealed by persistent calyx and corolla; seeds broadly ovoid or subglobose, 1·5–2 mm. long and almost as wide; testa smooth, pale brown. *Plate 22, figs. 10–11.*

HAB.: Pastures, fallow fields, roadsides, waste ground, and in garigue on dry, rocky hillsides; sea-level to 4,000 ft. alt.; fl. March–May.

DISTR.: Divisions 2, 4, 5, 7, 8, common in Divisions 2, 7. Widespread in the Mediterranean region and eastwards to W. Iran; Atlantic Islands.

2. Common; Lefka, Platres, Spilia etc.; *Kennedy* 1285 ! 1287 ! *Lindberg f.* s.n.; *Davis* 3105 ! *Meikle* 2811 !
4. Pergamos, 1955, *Merton* 1985 !
5. Peristerona, 1935, *Syngrassides* 775 !
7. Common; *Sintenis & Rigo* 413 ! *Davis* 2981 ! *Casey* 505 ! 609 ! 1709 ! *Chapman* 103 ! 177 ! 194 !
8. Cape Andreas, 1880, *Sintenis & Rigo*; Valia, 1905, *Holmboe* 483; Matakos near Galinoporni, 1962, *Meikle* 2442 !

3. T. hirtum *All.*, Auct. Fl. Ped., 20 (1789); Boiss., Fl. Orient., 2: 119 (1872); Post, Fl. Pal., ed. 2, 1: 333 (1932); A. K. Jackson et Turrill in Kew Bull., 1939: 475 (1939); Hossain in Notes Roy. Bot. Gard. Edinb., 23: 412 (1961); Davis, Fl. Turkey, 3: 430 (1970); Osorio-Tafall et Seraphim, List Vasc. Plants Cyprus, 59 (1973).

TYPE: Italy; " in *Monteferrato* secus agros ", *Allioni* (TO).

Closely resembles *T. cherleri* L., but usually more robust, with erect or decumbent, villose stems 10–20 (–35) cm. long; stipules 10–14 mm. long, 2–3 mm. wide, pallid, distinctly veined, pilose, the acumen elongate and slender, almost filiform; leaflets obovate 8–15 (–20) mm. long, 4–10 (–15) mm. wide, blotched or unblotched, pilose on both surfaces, apex usually rounded, base narrowly cuneate, margin minutely denticulate; inflorescences globose, 1·7–2 (–2·5) cm. diam., involucre less conspicuous than in *T. cherleri*, stipule-bracts broadly ovate, 5–11 mm. long, 5–8 mm. wide, pilose, with a long cuspidate acumen; calyx campanulate, 10–11 mm. long, 3 mm. wide, densely

Plate 22. TRIFOLIUM HIRTUM All. **1**, habit, $\times \frac{2}{3}$; **2**, inflorescence showing involucrate stipules, ×2; **3**, flower, ×5; **4**, calyx opened out, inner surface, ×4; **5**, standard, ×6; **6**, wing, ×6; **7**, keel, ×6; **8**, androecium, ×6; **9**, gynoecium, ×6. TRIFOLIUM CHERLERI L. **10**, inflorescence, showing involucrate stipules, ×2; **11**, standard, ×6. (**1–9**, *Davis* 19107, Turkish material; **10 & 11**, *Rawi* 28134.)

villose, c. 20-nerved, calyx-teeth 7–8 mm. long, setaceous; petals rose-pink or purplish, standard with a very narrowly oblong, longitudinally folded limb, 6–7 mm. long, about 1 mm. wide, apex acute, limb of wing oblong, 3 mm. long, 1 mm. wide, apex ovate, base distinctly auriculate, keel 3 mm. long, 0·8 mm. wide, apex acute or acuminate; stamens 7–7·5 mm. long, all, except the vexillary, connate for about 6 mm., free portions of filaments slightly dilated; anthers oblong, about 4 mm. long, 3 mm. wide; ovary oblong or obovate, 1–1·5 mm. long, 0·8–1 mm. wide, glabrous, 1–2-ovulate, style 6–7 mm. long, glabrous, stigma terminal, captitate. Pod quite concealed by persistent calyx and corolla; seeds broadly ovoid, 2–2·5 mm. long, 1·5–1·8 mm. wide; testa smooth, pale brown. *Plate 22, figs. 1–9.*

HAB.: Pastures, fallow fields and roadsides, or in garigue on dry rocky hillsides; 500–4,600 ft. alt.; fl. April–June.

DISTR.: Divisions 2, 3, common in Division 2. Widespread in the Mediterranean region and eastwards to Iraq.

2. Common; Platres, Prodhromos, Yialia etc.; *Kennedy* 1272 ! *Davis* 3187 ! *Casey* 722 ! *Merton* 2149 ! *D. P. Young* 7295 !
3. Michael Arkhangelos Monastery, west of Stavrovouni, 1957, *Merton* 3157 !

NOTES: All the Cyprus specimens of this species are dwarf and ± decumbent, superficially similar to *T. cherleri*, but readily distinguished by the elongate acumen of the stipules, and by the rosy-purple corollas.

4. T. lappaceum *L.*, Sp. Plant., ed. 1, 768 (1753); Sibth. et Sm., Fl. Graec. Prodr., 2: 96 (1813); Unger et Kotschy, Die Insel Cypern, 377 (1865); Boiss., Fl. Orient., 2: 119 (1872); Holmboe, Veg. Cypr., 105 (1914); Post, Fl. Pal., ed. 2, 1: 334 (1932); Rechinger f. in Arkiv för Bot., ser. 2, 1: 424 (1950); Hossain in Notes Roy. Bot. Gard. Edinb., 23: 413 (1961); Davis, Fl. Turkey, 3: 430 (1970); Zohary Fl. Palaest., 2: 177, t. 251 (1972); Osorio-Tafall et Seraphim, List Vasc. Plants Cyprus, 60 (1973).

T. lappaceum L. var. *brachyodon* Hausskn. in Mitt. Thur. Bot. Ver. N. F., 5: 73 (1893).

T. lappaceum L. var. *brevidens* Hal., Consp. Fl. Graec., 1: 382 (1900).

T. lappaceum L. ssp. *brevidens* (Hal.) Lindberg f., Iter Cypr., 20 (1946).

TYPE: France; " Monspelii ".

Erect or decumbent annual, 6–25 (–40) cm. high; stems usually much branched, especially towards the base, glabrous or thinly pilose especially towards apex; stipules pallid, papery, distinctly veined, 6–8 mm. long, 1–2 mm. wide, acumen linear-subulate, spreading, often thinly pilose; petioles slender, glabrous or thinly pilose, channelled above, 1–2 cm. long towards the base of the stem, shorter above, the uppermost leaves subsessile; leaflets obovate, 5–15 mm. long, 2–8 mm. wide, thinly pilose on both surfaces, apex rounded or very shortly emarginate, margins obscurely denticulate, base narrowly cuneate; inflorescences terminal, subglobose, 1–1·5 (–2) cm. diam., shortly pedunculate, without an involucre; calyx narrowly campanulate 5–7 mm. long, 2–2·5 mm. wide, tube glabrous or pilose, 20-nerved, with or rarely without a ring of hairs at the throat, calyx-teeth linear-setaceous, pilose, 3–4 mm. long at anthesis, elongating somewhat in fruit; petals whitish, shorter or longer than calyx; limb of standard oblong, about 3 mm. long, 2 mm. wide, apex rounded or truncate, base abruptly narrowed, limb of wings oblong, 2–2·5 mm. long, 1 mm. wide, apex blunt, base distinctly auriculate, keel oblong, about 2 mm. long, 0·8 mm. wide, apex rather blunt, base not auriculate; stamens 4·5–5 mm. long, the vexillary free, the others connate for about 4–4·5 mm., sometimes with the margins of the fused filaments joined near the apex and forming a short tube around the style; anthers suborbicular, about 0·2 mm. diam.; ovary ovoid, glabrous, about 1 mm. long, 0·4 mm. wide, slightly compressed, 2-ovulate, style glabrous 4–5 mm. long, stigma terminal, capitate. Pod quite concealed by persistent calyx; seeds subglobose or broadly ovoid, 1·3–1·8 mm. long, 1·3 mm. wide; testa mid-brown, glossy, smooth.

HAB.: In cultivated or fallow fields, on sandy ground by streams, or on dry hillsides under Pines; sea-level to 4,000 ft. alt.; fl. April–June.

DISTR.: Divisions 1, 2, 5, 7, 8. Widespread in the Mediterranean region, and sometimes as an adventive elsewhere.

1. Ktima, 1913, *Haradjian* 692; below Stroumbi, 1941, *Davis* 3373 !
2. Platres, 1938, *Kennedy* 1282 ! Kannaviou, 1939, *Lindberg f.* s.n. !
5. Mia Milea–Koutsovendis Road, 1974, *Meikle & Osorio-Tafall* in *Meikle* 4069 !
7. Kyrenia, 1880, *Sintenis & Rigo* 647 ! Larnaka tis Lapithou, 1941, *Davis* 2981A ! Kyrenia, 1955, *Merton* 2083 ! Kambyli, 1957, *Merton* 2982 !
8. Rizokarpaso, 1912, *Haradjian* 205.

NOTES: All the Cyprus specimens have rather short calyx-teeth, and could be referred to *T. lappaceum* L. var. *brachyodon* Hausskn., though, as remarked by Hossain (*loc. cit.*, 415), the value of the distinction is dubious, especially as the criteria for var. *brachyodon* do not always hold good. It is evident, however, that the variety has a definite geographical distribution in the eastern half of the Mediterranean region, and further research may show that it should not be merged with typical *lappaceum*.

5. T. arvense *L.*, Sp. Plant., ed. 1, 769 (1753); Sibth. et Sm., Fl. Graec. Prodr., 2: 99 (1813); Unger et Kotschy, Die Insel Cypern, 377 (1865); Boiss., Fl. Orient., 2: 120 (1872); Holmboe, Veg. Cypr., 106 (1914); Post, Fl. Pal., ed. 2, 1: 334 (1932); Lindberg f., Iter Cypr., 20 (1946); Hossain in Notes Roy. Bot. Gard. Edinb., 23: 415 (1961); Davis, Fl. Turkey, 3: 431 (1970); Zohary, Fl. Palaest., 2: 178, t. 253 (1972); Osorio-Tafall et Seraphim, List Vasc. Plants Cyprus, 60 (1973).
T. longisetum Boiss. et Bal. in Boiss., Diagn., 2, 6: 47 (1859).
T. arvense L. var. *longisetum* (Boiss. et Bal.) Boiss., Fl. Orient., 2: 120 (1872); Rechinger f. in Arkiv för Bot., ser. 2, 1: 424 (1950).
TYPE: " *in* Europa, America *septentrionali* ".

Erect or spreading annual, 5–30 cm. high; stems slender, usually much branched (escept in starved specimens), pubescent or villose; stipules ovate, 5–10 mm. long, pubescent, distinctly veined, with a slender filiform apex; petioles very slender, slightly channelled above, pubescent, up to 3·5 cm. long, becoming progressively shorter towards apex of stems, the uppermost leaves often subsessile; leaflets pubescent or villose on both surfaces, the lower broadly oblong, obovate or elliptical, 5–12 mm. long, 3–6 mm. wide, apex rounded or emarginate, upper leaflets narrowly linear-oblong, 7–20 mm. long, 1·5–4 mm. wide, apex subacute, cuspidate, often somewhat erose; peduncles erect, slender, pubescent, 1·5–4 (–6) cm. long; flowers densely crowded into oblong or ovoid spikes 1–3·5 (–4) cm. long, 1–1·5 cm. wide; involucre wanting; calyx narrowly campanulate, about 6 mm. long, 1 mm. wide, densely white-villose, tube 10-nerved, about 2 mm. long, not noticeably hairy at throat, calyx-teeth filiform, erect, or slightly spreading in fruit, subequal; petals whitish, becoming pink with age, much shorter than calyx-teeth, limb of standard oblong, 2·5 mm. long, 1·5 mm. wide, apex rounded or truncate, base scarcely auriculate, claw proportionately very short, limb of wings oblong, about 2 mm. long, 0·7 mm. wide, glabrous, apex rounded, base shortly auriculate, keel oblong, about 1·5 mm. long, 0·5 mm. wide, apex subacute; stamens about 3·5 mm. long, all, except the vexillary, connate for about 3 mm., the margins of the fused filaments joined near their apex to form a short tube around the style; anthers suborbicular about 0·2 mm. diam.; ovary ovoid, slightly compressed, 0·6 mm. long, 0·4 mm. wide, glabrous, 1 (–2)-ovulate, style glabrous, about 2·3 mm. long, upcurved at apex, stigma terminal, capitate. Pod completely concealed by persistent calyx; seed ovoid, 1·4 mm. long, 0·8 mm. wide; testa pale yellowish-brown, smooth.

HAB.: On dry mountainsides in garigue; in vineyards and on cultivated or waste ground; 500–4,500 ft. alt.; fl. April–July.

DISTR.: Divisions 2, 3, very common in Division 2. Most of Europe, Mediterranean region, Atlantic Islands and western Asia.

2. Very common, especially on pillow lavas; *Syngrassides* 1275 ! *Lindberg f.* s.n.; *Kennedy* 560 ! 561 ! 1275 ! 1276 ! *Davis* 1762 ! 3204 ! *Casey* 898 ! *Merton* 2129 ! *Meikle* 2753 !
3. Kalavasos, 1905, *Holmboe* 628; west of Stavrovouni Monastery, 1957, *Merton* 3155 !

6. T. angustifolium *L.*, Sp. Plant., ed. 1, 769 (1753); Boiss., Fl. Orient., 2: 122 (1872); Holmboe, Veg. Cypr., 105 (1914); Post, Fl. Pal., ed. 2, 1: 335 (1932); Hossain in Notes Roy. Bot. Gard. Edinb., 23: 418 (1961); Davis, Fl. Turkey, 3: 434 (1970); Zohary, Fl. Palaest., 2: 178, t. 254 (1972); Osorio-Tafall et Seraphim, List Vasc. Plants Cyprus, 60 (1973).

TYPE: " *in* G. Narbonensi, Italia ".

Erect annual, 8–20 (–30) cm. high; stems simple or sparingly branched at the base, adpressed-pilose; stipules narrowly oblong, connate and tubular towards base, 1–2·5 cm. long, 0·3–0·5 mm. wide, pallid, thinly pilose, conspicuously veined, apex produced into slender caudate appendages 5–10 mm. long; petioles 0·5–3·5 cm. long, slightly channelled, thinly pilose, the uppermost leaves often subsessile; leaflets narrowly oblong or linear, 15–50 mm. long, 1·5–5 mm. wide, thinly adpressed-pilose on both surfaces, the lower usually obtuse or rounded at apex, the upper acute or acuminate; inflorescence terminal, solitary, oblong-cylindrical, 1·5–6 cm. long, 1·5–2·5 cm. wide, peduncle rather short (usually less than 4 cm. long) densely adpressed-pilose; calyx narrowly campanulate, about 10 mm. long, 2 mm. wide, 10-nerved, densely pilose externally, glabrous internally with a distinct hairy annulus, calyx-teeth subulate, rather rigid, apex sharply pointed, abaxial teeth 5–6 mm. long, the others a little shorter; corolla purplish or pink, 10–12 mm. long, not much exceeding calyx, standard a little longer than wings, limb narrowly oblong, 6–7 mm. long, 3 mm. wide, longitudinally infolded, apex blunt or subacute, limb of wings oblong, about 3 mm. long, 1 mm. wide, apex blunt, base with a prominent auricle, claw very long and slender, keel about as long as wings, limb oblong, subacute, about 2·5 mm. long, 1 mm. wide; stamens 9–10 mm. long, connate almost to apex; anthers oblong-suborbicular, about 0·3–0·4 mm. long and almost as wide; ovary obovoid, 1–1·5 mm. long, 0·8 mm. wide, style glabrous, about 7 mm. long, stigma incurved. Pod concealed within persistent calyx; seed oblong, 2–2·5 mm. long, 1·5 mm. wide; testa smooth, yellowish.

HAB.: Fields and vineyards, roadsides and rocky hillsides; sea-level to 3,400 ft. alt; fl. April–May.

DISTR.: Divisions 1, 2, 3, 5, 7. Widespread in the Mediterranean region and east to Iran; Atlantic Islands.

1. Dhrousha, 1941, *Davis* 3254 !
2. Platres, 1938, *Kennedy* 1273 !
3. Near Souni, 1960, *N. Macdonald* 124 !
5. Near Vatili, 1905, *Holmboe* 334.
7. Kyrenia, 1949, *Casey* 604 ! and, 1955, *Miss Mapple* 65 ! Ayios Yeoryios, Kyrenia, 1955, *G. E. Atherton* 50 ! Kyrenia, 1956, *G. E. Atherton* 1344 !

7. T. pamphylicum *Boiss. et Heldr.* in Boiss., Diagn., 1, 9: 20 (1849), Fl. Orient., 2: 123 (1872); Rechinger f. in Arkiv för Bot., ser. 2, 1: 424 (1950); Hossain in Notes Roy. Bot. Gard. Edinb., 23: 421 (1961).

Erect or sprawling annual (3–) 8–15 (–25) cm. high; stems usually much branched at base, sometimes simple, rather densely clothed with erect or ± spreading hairs; stipules oblong, 5–10 mm. long, 3–6 mm. wide, pallid, pilose, distinctly veined, the acumen subulate or filiform, 2–6 mm. long; petiole flattened, slightly channelled, 8–15 mm. long; leaflets obovate, oblong or sub-linear, 5–20 (–30) mm. long, 2·5–6 (–7) mm. wide, unblotched, thinly pilose or subglabrous above, more densely adpressed-pilose below, apex rounded or acute, sometimes apiculate, base cuneate, margin entire; inflorescence terminal, usually solitary, oblong, 1·5–3 (–4) cm. long, 1–1·7 cm. wide at anthesis, peduncle densely adpressed-pilose, usually very short, but sometimes 2–4 cm. long; calyx narrowly campanulate or subcylindrical 5–11 mm. long, 2–3 mm. wide, 10-nerved, densely long-pilose externally, glabrous internally with a distinct ring of short hairs at the throat, teeth narrowly subulate, spreading in fruit, subequal, or the abaxial slightly longer than the others, 2·5–7 mm. long, densely pilose, apex shortly acute or often truncate; petals at first white or pale pink, becoming reddish with age, and often drying purplish, standard 10–13 mm. long, limb narrowly oblong, 2·5–3 mm. wide, apex rounded or subacute, often slightly erose, margins distinctly infolded especially towards base, limb of wing oblong, subacute, about 5 mm. long, 1·5–2 mm. wide, base distinctly auriculate, keel 6–10 mm. long, 1·5 mm. wide, apex acute or subobtuse, base indistinctly auriculate with a long slender claw; stamens to about 10 mm. long, filaments connate almost to apex, free portions slightly dilated at apex; anthers oblong, about 0·4 mm. long, 0·2 mm. wide; ovary oblong-ellipsoid, 0·7 mm. long, 0·4 mm. wide, glabrous, 2-ovulate, style 6–9 mm. long, glabrous, thickening towards apex, stigma oblique. Pod completely concealed within calyx; seeds oblong, compressed, 1–1·5 mm. long, 0·8–1 mm. wide; testa rich brown, smooth.

var. **pamphylicum.**

T. purpureum Loisel. ssp. *pamphylicum* (Boiss. et Heldr.) Holmboe, Veg. Cypr., 105 (1914); Osorio-Tafall et Seraphim, List Vasc. Plants Cyprus, 60 (1973).

T. purpureum Loisel. var. *pamphylicum* (Boiss. et Heldr.) Zohary in Davis, Fl. Turkey, 3: 436 (1970).

[*T. angustifolium* (non L.) Kotschy in Unger et Kotschy, Die Insel Cypern, 377 (1865); H. Stuart Thompson in Journ. Bot., 44: 306 (1906).]

[*T. desvauxii* (non Boiss. et Bl.) Sintenis in Oesterr. Bot. Zeitschr., 32: 290, 398 (1882); Boiss., Fl. Orient. Suppl., 167 (1888).]

[*T. desvauxii* Boiss. et Bl. var. *blancheanum* (non (Boiss.) Boiss.) Sintenis in Oesterr. Bot. Zeitschr., 32: 263 (1882) pro parte.]

TYPE: Turkey; " in arenosis Pamphyliae ad fluvium Melas infra Monargat ", April 1845, *Heldreich* (G, K !).

Leaflets linear-oblong, usually acute; stipules often with a long caudate apex; calyx-teeth 3–5 mm. long.

HAB.: Sand-dunes and sandy fields near the sea, less frequently inland on bare limestone or igneous slopes or in open garigue; sea-level to 2,000 ft. alt.; fl. March–May.

DISTR.: Divisions 1–8, common in Divisions 4, 6. Also Turkey and ? Syria.

1. Dhrousha, 1941, *Davis* 3217 !
2. Kambia, 1908, *Clement Reid* s.n. ! Near Philani, 1908, *Clement Reid* s.n. ! Galata (Kakopetria), 1941, *Davis* 3202 !
3. Limassol, 1930, *E. Wall*; Skarinou (Larnaca Distr.), 1941, *Davis* 2673 !
4. Common; *Kotschy* 303; *Haradjian* 8 ! *Druce* s.n. ! *Merton* 2000 !
5. Near Salamis, 1905, *Holmboe* 463; Mia Milea–Koutsovendis Road, 1974, *Meikle & Osorio-Tafall* in *Meikle* 4065 !
6. Common; *Syngrassides* 767 ! *Davis* 2589 ! *Chapman* 401 ! 576 ! *Meikle* 2389 !
7. Near Dhavlos, 1880, *Sintenis & Rigo* 419 ! Near Ayios Epiktitos, 1956, *Casey* 1705 !
8. Near Cape Andreas, 1880, *Sintenis & Rigo*; near Rizokarpaso, 1912, *Haradjian* 183.

var. **dolichodontium** *Hossain* in Notes Roy. Bot. Gard. Edinb., 23: 421 (1961).

TYPE: Cyprus; Platres, 1125 m. alt., May 28, 1938, *Kennedy* 1277 (K !).

Plants rather robust, usually suberect; leaflets acute, subacute or obtuse, often apiculate; calyx-teeth 5–7 mm. long; otherwise as in var. *pamphylicum*.

HAB.: Coastal sand-dunes and dry igneous mountainsides; sea-level to 4,000 ft. alt.; fl. March–May.

DISTR.: Divisions 2, 4. Seemingly endemic.

2. Platres, 1914, *G. T. Feilden* s.n. ! Platres, 1937, *Kennedy* 560A ! Platres, 1938, *Kennedy* 1277 ! Saïttas, 1950, *F. J. F. Barrington* 22 !
4. Ayios Memnon, 1948, *Mavromoustakis* s.n. !

NOTES: Scarcely more than a robust state of *T. pamphylicum* var. *pamphylicum* and probably not meriting distinct status.

var. **blancheanum** (*Boiss.*) *Meikle* **comb. nov.**

T. blancheanum Boiss., Diagn., 2, 2: 13 (1856); Zohary, Fl. Palaest., 2: 180 (1972).

T. desvauxii Boiss. et Bl. var. *blancheanum* (Boiss.) Boiss., Fl. Orient., 2: 124 (1872); Sintenis in Oesterr. Bot. Zeitschr., 32: 263 (1882) pro parte.

[*T. dichroanthum* (non Boiss.) Kotschy in Unger et Kotschy, Die Insel Cypern, 377 (1865); Holmboe, Veg. Cypr., 105 (1914).]

[*T. purpureum* Loisel. var. *roussaeanum* (non *T. roussaeanum* Boiss.) Holmboe, Veg. Cypr., 105 (1914).]

TYPE: Lebanon; "in arenosis ad vias circâ Beyrout, detexit Aprili 1850 clar. Blanche" (G).

Plants often sprawling or decumbent; leaflets oblong or obovate, apex rounded or obtuse; calyx-teeth subequal, usually not exceeding 4 mm. in length.

HAB.: By roadsides, on waste land, commonly on rocky ground near the sea; sea-level to 200 ft. alt.; fl. March–May.

DISTR.: Divisions 1, 3, 4, 6–8, very common in Division 7. ? Aegean Islands; Palestine.

1. Near Paphos, 1862, *Kotschy*.
3. Akrotiri, 1950, *F. J. F. Barrington* 28 !
4. Cape Greco, 1958, *N. Macdonald* 19 ! Famagusta, 1934, *Miss E. Haig* 28 !
6. Government House Gardens, 1935, *Syngrassides* 617 ! Nicosia, 1955, *E. W. Baker* 168 ! Liveras, 1974, *Meikle* 5013 !
7. Very common; *Sintenis & Rigo* 419 ! *Kotschy* 481; *Mavromoustakis* s.n. ! *Chapman* 273 ! 566 ! *Casey* 1172 ! 1715 ! *G. E. Atherton* 1255 ! 1301 !
8. Cape Andreas, 1948, *Mavromoustakis* s.n. ! Platanisso, 1950, *Chapman* 208 ! Eleousa, near Rizokarpaso, 1962, *Meikle* 2998 !

NOTES: Generally fairly easy to recognize on account of its short, proportionately broad leaflets; intermediate forms between this and var. *pamphylicum* do, however, occur, and it is doubtful if the two can be satisfactorily distinguished except as varieties of one species. A plant closely resembling var. *blancheanum*, and recorded by Hossain (*l.c.*, 420) as *T. purpureum* Loisel. var. *laxiusculum* (Boiss. et Bl.) Hossain, has been collected on the island of Samos by P. H. Davis (no. 1703 in herb. Kew !), otherwise the variety is apparently restricted to Cyprus and Palestine. It is commonly misidentified as *T. purpureum* Loisel. var. *desvauxii* (Boiss. et Bl.) Post, but the latter is at once distinguished by indumentum, leaflet-shape, calyx and corolla.

8. **T. echinatum** *M. Bieb.*, Fl. Taur.-Cauc., 2: 216 (1808); Holmboe, Veg. Cypr., 106 (1914); Post, Fl. Pal., ed. 2, 1: 338 (1932); Hossain in Notes Roy. Bot. Gard. Edinb., 23: 427 (1961); Zohary, Fl. Palaest., 2: 187, t. 268 (1972); Osorio-Tafall et Seraphim, List Vasc. Plants Cyprus, 60 (1973).

T. supinum Savi, Obs. Trif., 46, fig. 2 (1810); Kotschy, Die Insel Cypern, 378 (1865); Boiss., Fl. Orient., 2: 126 (1872).

[*T. leucanthum* (non M. Bieb.) Sintenis in Oesterr. Bot. Zeitschr., 32: 292 (1882).]

TYPE: Russia; "ad Caucasum" (LE).

Erect or sprawling annual (? or biennial) 5–25 cm. high; stems purplish, thinly pilose, usually much branched, stipules narrowly lanceolate, 10–12 mm. long, 1–2 mm. wide, subentire or shortly serrulate, pallid towards base with conspicuous purple veins, green above and thinly long-pilose; petiole rather short, usually less than 2 cm. long, narrowly canaliculate above, pilose, often purplish; uppermost leaves usually opposite, the lower alternate; leaflets obovate, 5–20 mm. long, 2·5–8 mm. wide, thinly adpressed-pilose on both surfaces, apex rounded or shortly emarginate, often mucronate, base cuneate, margins entire; inflorescence axillary, often apparently terminal, subglobose or very shortly oblong, about 1–2 cm. diam. at anthesis; peduncle conspicuous, 1–4 cm. long, ± densely adpressed-pilose; calyx narrowly campanulate 4–7 mm. long, 1·3–1·5 mm. wide, thinly pilose externally, glabrous internally with a ring of hairs at the throat, calyx-tube about 2 mm. long, distinctly 10-nerved, teeth narrowly subulate, suberect or spreading in fruit, thinly pilose, rather rigid, the three abaxial about 2–5 mm. long, distinctly longer than the two adaxial, apex sharply pointed; petals pink or purple, standard 8–10 mm. long, limb narrowly oblong, 1·5–2 mm. wide, apex rounded or truncate, sometimes erose; limb of wing narrowly oblong, subacute, 3·5–4 mm. long, 1 mm. wide, base strongly auriculate, with a swollen auricle; limb of keel 3 mm. long, 1·5 mm. wide, apex rather blunt, base indistinctly auriculate, with a long slender claw; stamens 5–7 mm. long, filaments connate nearly to apex, free portions slightly dilated at apex; anthers oblong, about 0·3 mm. long, 0·1 mm. wide; ovary suborbicular, compressed, glabrous, about 1 mm. long, 0·8 mm. wide, style 7–8 mm. long, glabrous, slightly thickened just above middle, stigma oblique. Pod completely concealed within calyx; seeds reniform, about 1·5 mm. long, 1·3 mm. wide; testa rich brown, smooth.

HAB.: Hillsides and roadside banks; sea-level to 4,500 ft. alt.; fl. April–July.

DISTR.: Divisions 2, 3, 7, 8. Italy, Yugoslavia, Greece, Turkey, Rhodes, Palestine, Iran and Iraq.

2. Between Prodhromos and Tris Elies, 1862, *Kotschy* 860A; Prodhromos, 1951, *Casey* 1207 !
3. Kolossi, 1971, *Townsend* 71/125 !
7. About Kyrenia, 1880, *Sintenis & Rigo* 414A ! Hill above Panagra, near Myrtou, 1954, *Casey* 1345 !
8. Between Ephtakomi and Leonarisso, 1880, *Sintenis & Rigo* 414 !

9. T. leucanthum *M. Bieb.*, Fl. Taur.-Cauc., 2: 214 (1808); Boiss., Fl. Orient., 2: 128 (1872); Post, Fl. Pal., ed. 2, 1: 339 (1932); A. K. Jackson et Turrill in Kew Bull., 1939: 475 (1939); Rechinger f. in Arkiv för Bot., ser. 2, 1: 425 (1950); Hossain in Notes Roy. Bot. Gard. Edinb., 23: 429 (1961); Davis, Fl. Turkey, 3: 441 (1970); Zohary, Fl. Palaest., 2: 186, t. 267 (1972); Osorio-Tafall et Seraphim, List Vasc. Plants Cyprus, 60 (1973).

TYPE: Crimea; " in Tauriae meridionalis collibus siccis " (LE).

Erect or sprawling annual 5–25 (–30) cm. high; stems usually much branched at base, occasionally simple, sparsely or rather densely villose; stipules narrowly oblong with long acuminate apices, about 15 mm. long, 2–3 mm. wide, shortly pubescent, the basal half distinctly veined; uppermost leaves opposite, the others alternate; petioles slender, 1·5–6 cm. long, pubescent, distinctly canaliculate above; leaflets narrowly oblong, 8–25 mm. long, 2–8 mm. wide, thinly adpressed-pilose on both surfaces, apex usually rounded or obtuse, margins entire or minutely serrulate towards apex; peduncles 3–8 cm. long, generally exceeding leaves, adpressed-pilose; inflorescences globose, 1–1·5 cm. diam.; calyx about 7 mm. long, pilose externally, glabrous internally with a ring of hairs at the throat, calyx-tube narrowly campanulate, about 2 mm. long, 1·5 mm. wide, densely pilose externally, obscurely 10-nerved, calyx-teeth subequal, subulate, rather rigid, about 3–5 mm. long, 0·8 mm. wide at base, suberect at anthesis, spreading in fruit; petals white, becoming pinkish with age, standard narrowly oblong, 5–6 mm. long, 1·5–2 mm. wide, apex obtuse or emarginate, base scarcely auriculate, claw short; limb of wings oblong, subacute, about 2 mm. long, 0·6 mm. wide, base prominently auriculate, claw about 2·5 mm. long, limb of keel oblong about 2 mm. long, 0·8 mm. wide, apex obtuse, claw about 2·5 mm. long; stamens about 4·5 mm. long, connate almost to apex; anthers oblong, about 0·3 mm. long, 0·2 mm. wide; ovary oblong-obovate, compressed, glabrous, about 0·8 mm. long, 0·6 mm. wide, style glabrous, about 4 mm. long, slender below, distinctly thickened above, stigma conspicuous, oblique. Pod completely concealed within calyx; seed oblong-suborbicular, 1·8–2 mm. long, 1·5 mm. wide; testa light brown, smooth.

HAB.: Rocky pastures and mountainsides, often in *Pinus* (or *Cedrus*) woodland; sea-level to 4,600 ft. alt.; fl. March–May.

DISTR.: Divisions 1–4, 7, 8, common in Division 2. Widespread in the Mediterranean region and eastwards to Iraq.

1. Fontana Amorosa, 1962, *Meikle* 2299 !
2. Common; *Kennedy* 1283 ! 1284 ! 1684 ! *Davis* 3150 ! 3483 ! *Casey* 725 ! *D. P. Young* 7250 !
3. Ayia Varvara (Stavrovouni), 1937, *Syngrassides* 1496 !
4. Pergamos, 1955, *Merton* 1980 !
7. Yaïla, 1941, *Davis* 2881 ! Kyrenia Range North, 1948, *Kennedy* 1657 ! Lakovounara Forest, 1950, *Chapman* 195 ! 364 !
8. Eleousa, near Rizokarpaso, 1962, *Meikle* 2545 !

10. T. scabrum *L.*, Sp. Plant., ed. 1, 770 (1753); Boiss., Fl. Orient., 2: 130 (1872); Holmboe, Veg. Cypr., 106 (1914); Rechinger f. in Arkiv för Bot., ser. 2, 1: 424 (1950); Hossain in Notes Roy. Bot. Gard. Edinb., 23: 430 (1961); Davis, Fl. Turkey, 3: 428 (1970); Zohary, Fl. Palaest., 2: 176, t. 250 (1972); Osorio-Tafall et Seraphim, List Vasc. Plants Cyprus, 59 (1973).

[*T. striatum* (non L.) Kotschy in Unger et Kotschy, Die Insel Cypern, 378 (1865).]

TYPE: " *in* Anglia, Gallia, Italia ".

Decumbent or prostrate annual; stems 4–20 cm. long, usually much branched, pilose, often stained purple, frequently ± zig-zag in fruiting specimens; stipules narrowly ovate, acuminate, 3–5 mm. long, 1–3·5 mm. wide at base, pubescent, acumen often recurved, basal part strongly nerved; petioles of basal leaves slender, 2–5 cm. long, pubescent, canaliculate, those of upper leaves progressively shorter and stouter, uppermost leaves subsessile with well-developed stipules; leaflets obovate, 3–8 (–14) mm. long, 2–5 (–8) mm. wide, thinly pilose on both surfaces, lateral veins curving downwards apically, apex of leaflet usually rounded, margin minutely denticulate or subentire; inflorescence subsessile, axillary, ovoid, 7–11 mm. long, 5–8 mm. wide, usually ebracteate; calyx 6–7 mm. long, densely pilose externally, glabrous internally with a pilose throat, calyx-tube cylindrical or narrowly campanulate, 2·5–3 mm. long, 1·5 mm. wide, distinctly 10-nerved, calyx-teeth lanceolate-acuminate, 3–4 mm. long, 1 mm. wide, the abaxial slightly larger than the others, all rather rigid, with a strong midrib, suberect or recurved in fruit; petals creamy-white becoming pinkish with age; standard 7–8 mm. long, glabrous, limb oblong, about 4 mm. long and 2 mm. wide, apex rounded or slightly emarginate, base scarcely auriculate, claw rather wide, limb of wings oblong, about 3 mm. long, 1·3 mm. wide, apex rounded, base prominently auriculate, claw slender about 4 mm. long, limb of keel oblong, about 2·6 mm. long, 1 mm. wide, apex obtuse, base shortly auriculate, claw about 4 mm. long; stamens 5–6 mm. long, filaments united almost to apex; anthers oblong, about 0·3 mm. long, 0·2 mm. wide; ovary ellipsoid, compressed, glabrous, about 1 mm. long, 0·5 mm. wide, style about 4 mm. long, glabrous, slender at base, distinctly thickened above, stigma conspicuous, oblique. Pod concealed within persistent calyx; seed oblong, about 1·5 mm. long, 0·9 mm. wide; testa yellowish or light brown, smooth.

HAB.: In stony or sandy pastures and fallows, sometimes in dried-up river beds or on rocky hillsides; sea-level to 3,750 ft. alt.; fl. March–April.

DISTR.: Divisions 1–8, common in Division 7. Western Europe and Mediterranean region eastwards to Iran.

1. Paphos, 1913, *Haradjian* 576; Cape Arnauti, 1913, *Haradjian* 808.
2. Platres, 1938, *Kennedy* 1278 !
3. Skarinou, 1941, *Davis* 2691 ! Petra tou Romiou, 1950, *F. J. F. Barrington* 19 !
4. Between Larnaca and the Phaneromene, 1862, *Kotschy* 77; Cape Greco, 1958, *N. Macdonald* 24 !
5. Mile 7, Nicosia–Famagusta Road, 1950, *Chapman* 371 !
6. Syrianokhori, 1935, *Syngrassides* 763 ! 764 !
7. Common; *Sintenis & Rigo*; *Davis* 2914 ! *Casey* 414 ! 608 ! 727 ! 1122 ! 1708 ! *Chapman* 278 !
8. Near Leonarisso, 1880, *Sintenis & Rigo*; near Cape Andreas, 1880, *Sintenis & Rigo*; Koma tou Yialou, 1950, *Chapman* 233 !

11. T. stellatum *L.*, Sp. Plant., ed. 1, 769 (1753); Sibth. et Sm., Fl. Graec. Prodr., 2: 99 (1813), Fl. Graec., 8: 36, t. 750 (1833); Unger et Kotschy, Die Insel Cypern, 378 (1865); Boiss., Fl. Orient., 2: 121 (1872); Holmboe, Veg. Cypr., 106 (1914); Post, Fl. Pal., ed. 2, 1: 335 (1932); Hossain in Notes Roy. Bot. Gard. Edinb., 23: 432 (1961); Davis, Fl. Turkey, 3: 424 (1970); Zohary, Fl. Palaest., 2: 176, t. 249 (1972); Osorio-Tafall et Seraphim, List Vasc. Plants Cyprus, 59 (1973).

Erect or sprawling annual (3–) 5–20 (–30) cm. high; stems usually much branched at base, sometimes simple, rather densely clothed with long, spreading silky hairs; stipules conspicuous, obovate, 5–15 mm. long, 3–10 mm. wide, villose, pallid except at margin, prominently veined, apex rounded or subacute often mucronate, margins entire or denticulate; petioles slender 2–8 cm. long thinly villose, canaliculate above; all the leaves alternate; leaflets obcuneate 4–15 mm. long, and almost as wide, thinly villose on both surfaces, apex truncate or emarginate, margins minutely denticulate; inflorescences broadly ovoid or subglobose about 3 cm. diam., terminal; peduncle usually well developed, to about 5 cm. long, ± densely clothed with spreading or adpressed silky hairs; calyx narrowly campanulate 12–15 mm. long, densely villose externally, glabrous or very shortly pubescent internally except at the densely hairy throat; calyx tube 3–5 mm. long, 2–3 mm. wide at throat, obscurely 10-veined, calyx-teeth subulate, subequal, 6–9 mm. long, 1–2 mm. wide at base, suberect at anthesis, spreading in fruit, petals at first creamy-white, later reddish-pink; limb of standard oblong, 7–8 mm. long, 2·5–3 mm. wide, glabrous, apex rounded or slightly emarginate, base scarcely auriculate, claw about 5 mm. long, limb of wings oblong, about 5 mm. long, 2 mm. wide, apex rounded, base prominently auriculate, claw 7–8 mm. long, limb of keel oblong about 3·5 mm. long, 1·5 mm. wide, apex subacute, base bluntly auriculate, claw about 8 mm. long; stamens about 10 mm. long, filaments united almost to apex; anthers oblong, about 0·5 mm. long, 0·3 mm. wide; ovary obovoid, 1 mm. long, 0·8 mm. wide, strongly compressed, glabrous, style about 10 mm. long, glabrous, slightly thickened above base, stigma conspicuous, oblique. Pod hidden within calyx; seed broadly oblong, about 2 mm. long, 1·5 mm. wide; testa light brown, smooth.

var. stellatum.

TYPE: " *in* Sicilia, Italia, G. Narbonensi ".

Peduncle with spreading hairs.

HAB.: In cultivated fields and on waste ground; in pastures or amongst shrubs on rocky hillsides, sea-level to 2,500 ft. alt.; fl. Febr.–April.

DISTR.: Divisions 1–8, common in Division 7. Western Europe and throughout the Mediterranean region eastwards to Iran; Atlantic Islands.

1. Smyies, 1962, *Meikle* 2188 ! Near Polis, 1962, *Meikle* 2266 !
2. Near Philani, 1862, *Kotschy*.
3. Stavrovouni, 1862, *Kotschy* 187; Cape Gata, 1862, *Kotschy* 609; Akrotiri, 1947, *Mavromoustakis* 7 !
4. Cape Greco, 1862, *Kotschy* 167; near Larnaca, 1881, *Barbey*; Ayia Napa, 1905, *Holmboe* 53A; Cape Greco, 1948, *N. Macdonald* 18 !
5. Mile 5, Nicosia–Famagusta Road, 1950, *Chapman* 32 !
6. Kokkini Trimithia, 1933, *Syngrassides* 333 !
7. Common; *Sintenis & Rigo*; *Syngrassides* 297 ! *Davis* 2853 ! *Casey* 407 ! 1687 ! *G. E. Atherton* 1134 ! 1259 !
8. Gastria, 1941, *Davis* 2428 !

var. adpressum *Turrill* in Kew Bull., 1924: 327 (1924); Hossain in Notes Roy. Bot. Gard. Edinb., 23: 433 (1961).

TYPE: Turkey; Gallipoli, Suvla and Anzac, May 16, 1924, *Kett* 116 (K !).

Peduncle with adpressed hairs.

HAB.: In fields on rocky soils; c. 1,500 ft. alt.; fl. March.

DISTR.: Division 2, rare. Also Turkey, Iraq and Iran.

2. Rare; Evrykhou, 1935, *Syngrassides* 831 !

NOTES: A second specimen of var. *adpressum* Turrill was collected, probably on the Troödos range, in 1952, by Mrs. F. M. Probyn (no. 70 in herb. Kew). It has not been possible to localize the specimen with certainty.

12. T. dasyurum *C. Presl*, Symb. Bot., 1: 53, t. 33 (1832); Davis, Fl. Turkey, 3: 437 (1970); Zohary, Fl. Palaest., 2: 181, t. 259 (1972); Osorio-Tafall et Seraphim, List Vasc. Plants Cyprus, 60 (1973).

T. formosum Urv. in Mém. Soc. Linn. Par., 1: 350 (1822); Boiss., Fl. Orient., 2: 124 (1872); Holmboe, Veg. Cypr., 106 (1914); Post, Fl. Pal., ed. 2, 1: 337 (1932); Hossain in Notes Roy. Bot. Gard. Edinb., 23: 434 (1961) non Savi (1810).

TYPE: " verosimiliter in Creta. *Kohaut* " (PR).

Erect annual (5–) 15–35 cm. high; stems simple or branched at the base, robust, ± densely villose with spreading hairs; stipules oblong with a long caudate acumen, 15–35 mm. long, 5–8 mm. wide, pallid, distinctly veined, shortly pubescent or thinly villose; uppermost leaves opposite, the others alternate; petioles 1–8 cm. long, becoming shorter towards apices of stems, pubescent or villose, deeply canaliculate above; leaflets lanceolate-oblong, often shortly petiolulate, 15–45 mm. long, 2–12 mm. wide, thinly pilose on both surfaces, apex rounded or acuminate, margins entire; inflorescences terminal, oblong, 2·5–4·5 cm. long, 2–3·5 cm. wide, peduncles usually well developed, to about 8 cm. long, densely villose, calyx about 1·8 cm. long, densely villose externally, glabrous internally except for a ring of hairs at the throat, calyx-tube about 5 mm. long, 2 mm. wide at throat, obscurely 10-veined, calyx-teeth subequal, very long and awn-like, up to 1·3 cm. in length, erect in flower, spreading in fruit; petals reddish-purple, standard 1·3–1·4 cm. long, limb glabrous, narrowly oblong, about 6 mm. long, 2 mm. wide, apex rounded, base scarcely auriculate, limb of wings oblong, about 4 mm. long, 1·3 mm. wide, apex acute, base distinctly auriculate, claw 8–9 mm. long, limb of keel oblong, about 3 mm. long, 1·2 mm. wide, apex blunt, base scarcely auriculate, claw 8–9 mm. long; stamens 1·3 mm. long, filaments united almost to apex; anthers shortly oblong, about 0·3 mm. long and almost as wide; ovary broadly ellipsoid, compressed, glabrous, about 1·5 mm. long and 1·2 mm. wide, style about 8 mm. long, slender, glabrous, stigma small, oblique. Pod concealed within calyx; seeds oblong, about 2·5 mm. long, 1·8 mm. wide; testa pale brown or yellowish, smooth.

HAB.: In cultivated fields, fallows, waste land and roadsides, occasionally on stony hillsides; sea-level to 2,400 ft. alt.; Fl. Febr.–April.

DISTR.: Divisions 2–7, common in Divisions 2, 7. Greece, Aegean Islands, Turkey, Syria, Palestine eastwards to Iran.

2. Common; *Kennedy* 1274 ! *Davis* 2711 ! 2725 ! 2756 ! *Chapman* 310 !
3. Akrotiri, 1927, *Rev. A. Huddle* 38 ! Stavrovouni, 1937, *Syngrassides* 1494 !
4. Near Larnaca, 1880, *Sintenis & Rigo* 817 ! Larnaca, 1905, *Holmboe* 113; Paralimni, Ayia Napa, 1936, *Syngrassides* 886 ! Pergamos, 1955, *Merton* 1977 !
5. Athalassa [" Thalassa "], 1927 or 1930, *Druce* 127 ! Mile 2, Nicosia–Famagusta Road, 1950, *Chapman* 443 !
6. Peristerona (Akaki), 1935, *Syngrassides* 779 !
7. Common; *Sintenis & Rigo* 503 ! *Davis* 2454 ! *Casey* 372 ! 501 ! 1706 ! *Chapman* 92 ! 176 ! *G. E. Atherton* 1133 ! 1185 ! *Meikle* 2421 !

13. T. clypeatum *L.*, Sp. Plant., ed. 1, 769 (1753); Sibth. et Sm., Fl. Graec. Prodr., 2: 99 (1813), Fl. Graec., 8: 37, t. 751 (1833); Unger et Kotschy, Die Insel Cypern, 378 (1865); Boiss., Fl. Orient., 2: 127 (1872); Holmboe, Veg. Cypr., 106 (1914); Post, Fl. Pal., ed. 2, 1: 341 (1932); Rechinger f. in Arkiv för Bot., ser. 2, 1: 424 (1950); Hossain in Notes Roy. Bot. Gard. Edinb., 23: 435 (1961); Davis, Fl. Turkey, 3: 442 (1970); Zohary, Fl. Palaest., 2: 188, t. 269 (1972); Osorio-Tafall et Seraphim, List Vasc. Plants Cyprus, 60 (1973); Elektra Megaw, Wild Flowers of Cyprus, 9, t. 8 (1973).

TYPE: " *in* Oriente ".

Robust sprawling or decumbent annual, 20–30 (–50) cm. high, stems much branched, thinly clothed with long spreading hairs; stipules broadly ovate, 10–20 mm. long, 8–15 mm. wide, the lower acute or acuminate, the upper subacute or rounded, all herbaceous, ciliate, with entire or irregularly denticulate margins, those of the uppermost pairs of leaves connate almost to apex; petioles up to 12 cm. long, thinly villose or subglabrous, slightly canaliculate above, uppermost leaves opposite, the rest alternate; leaflets broadly obcuneate, 1·4–3·5 cm. long, 0·8–3 cm. wide, thinly pilose on both surfaces, apex rounded or mucronate, margins subentire or minutely denticulate; inflorescence conspicuous, terminal, oblong-obovoid, 3–6 cm. long, 2–4·5 cm wide; peduncle usually well developed, to about 6 cm. long, thinly villose, often distinctly sulcate; calyx accrescent, at first about 1 cm. long, increasing to 1·6 cm. long in fruit, calyx-tube shortly cylindrical, about 4 mm. long at anthesis, thinly hairy externally, glabrous internally except at the shortly ciliate throat, distinctly 10-veined, increasing to 6 mm. long and 4 mm. wide in fruit and becoming conspicuously sulcate, calyx-teeth markedly unequal, erect at anthesis, spreading and veiny in fruit, the abaxial ovate-acuminate, about 6 mm. long, 3 mm. wide, thinly villose, increasing to 8–10 mm. long, 4–5 mm. wide in fruit, adaxial teeth shortly ovate-acuminate, about 2–2·5 mm. long, 1·5 mm. wide at anthesis, 3–6 mm. long and 2–4 mm. wide in fruit; petals at first creamy-white, turning deep pink with age, limb of standard oblong, about 12 mm. long, 7–8 mm. wide, apex emarginate, somewhat erose, base obscurely auriculate, claw 10–12 mm. long, limb of wings oblong, about 6 mm. long, 2·5 mm. wide, apex rounded, base strongly auriculate, claw about 13 mm. long, limb of keel oblong, about 5 mm. long, 2·5 mm. wide, apex subacute, base scarcely auriculate, claw about 13 mm. long; stamens about 18 mm. long, united almost to apex; anthers oblong, about 0·7 mm. long, 0·4 mm. wide; ovary oblong, compressed, glabrous, about 1·5 mm. long, 1 mm. wide, style about 13 mm. long, very slender below, thicker above, glabrous, stigma elongate, oblique. Pod concealed within accrescent calyx; seed oblong, 3–3·5 mm. long, about 2·5 mm. wide; testa smooth, pale brown.

HAB.: In cultivated and waste ground, by roadsides and streamsides, on rocky hillsides, usually in moist and slightly shaded sites; sea-level to 2,500 ft. alt.; fl. Febr.–May.

DISTR.: Divisions 1–4, 6–8, very common in Division 7. S. Greece, Aegean Islands, Turkey, Syria and Palestine.

1. Stroumbi, *Haradjian* 783 ! Near Ayios Yeoryios (Akamas), 1962, *Meikle* ! Near Polis, 1962, *Meikle* !
2. Near Evrykhou, 1859, *Kotschy* 478; near Xerarkaka, 1930, *C. Norman* 420; Lefka, 1932, *Syngrassides* 241 ! Stavros tis Psokas, 1959, *P. Oswald* 84 !
3. Yerasa, 1950, *F. J. F. Barrington* 36 !
4. Ayia Napa, 1905, *Holmboe* 69; near Famagusta, 1912, *Haradjian* 117; Famagusta, 1936, *Syngrassides* 880 !
6. Near Nicosia, 1930, *F. A. Rogers* 0670 ! Syrianokhori, 1935, *Syngrassides* 624 ! Near Dhiorios, 1962, *Meikle* !
7. Very common; *Kotschy*; *Sintenis & Rigo* 420 ! *Syngrassides* 298 ! *Kennedy* 1072 ! 1566 ! 1567 ! *Davis* 2885 ! *Casey* 409 ! 1688 ! *Chapman* 120 ! *G. E. Atherton* 949 ! 1312 ! *Meikle* 2365 !
8. Ayios Philon near Rizokarpaso, 1941, *Davis* 2205 ! Platanisso, 1950, *Chapman* 207 ! Near Akrades, 1962, *Meikle* !

14. T. scutatum *Boiss.*, Diagn., 1, 2: 27 (1843), Fl. Orient., 2: 129 (1872), Fl. Orient. Suppl., 168 (1883); Post, Fl. Pal., ed. 2, 1: 340 (1932); Hossain in Notes Roy. Bot. Gard. Edinb., 23: 436 (1961); Davis, Fl. Turkey, 3: 443 (1970); Zohary, Fl. Palaest., 2: 189, t. 270 (1972); Osorio-Tafall et Seraphim, List Vasc. Plants Cyprus, 60 (1973).

T. clypeatum L. ssp. *scutatum* (Boiss.) Holmboe, Veg. Cypr., 106 (1914).

TYPE: Turkey; " in montosis *Smyrnae* suprà *Bournabat*, unicum specimen floriferum fructiferumque Maio 1842 legi " *Boissier* (G).

Prostrate or sprawling annual; stems 10–30 cm. long, usually much branched at base, occasionally simple, thinly or rather densely villose, often tinged purplish; stipules narrowly oblong, 8–15 mm. long, 1·5–2·5 mm. wide, thinly villose with an elongate acute apex, base pallid, distinctly veined, those of the uppermost leaves shortly connate; uppermost leaves opposite, the rest alternate; petioles 0·5–12 cm. long, thinly villose, slightly channelled above; leaflets obcuneate, 6–18 mm. long, 4–15 mm. wide, thinly or rather densely villose on both surfaces, apex rounded or subtruncate, margins entire or minutely undulate; inflorescences terminal, subglobose, 1–1·5 cm. diam.; peduncle slender, 0·8–3 (–7) cm. long, ± densely subadpressed-villose; calyx about 5 mm. long, 1·5 mm. wide, thinly pilose externally, glabrous within except at the shortly hairy throat, calyx-tube narrowly cylindrical, about 2·5 mm. long, obscurely 10-nerved, calyx-teeth very unequal, strongly accrescent, erect at anthesis, spreading in fruit, the abaxial oblong-acute, about 2·5 mm. long, 1 mm. wide at anthesis, 4–5 mm. long, 2–2·5 mm. wide in fruit, adaxial about 0·6 mm. long, 0·4 mm. wide at anthesis, about 1·5 mm. long, 1 mm. wide in fruit; petals white or tinged pink, limb of standard about 5 mm. long, 2 mm. wide, folded longitudinally, apex blunt or subacute, base scarcely auriculate, claw about 5 mm. long, limb of wings about 3 mm. long, 1·2 mm. wide, apex blunt, base strongly auriculate, claw about 5 mm. long, limb of keel about 3 mm. long, 1·3 mm. wide, apex blunt, claw about 5 mm. long; stamens about 9 mm. long, united almost to apex; anthers oblong about 0·5 mm. long, 0·3 mm. wide; ovary ovoid, 0·7 mm. long, 0·5 mm. wide, glabrous, slightly compressed, style about 10 mm. long, glabrous, slightly swollen above, stigma rather short, oblique. Fruiting head ovoid, dense, about 1·5 cm. long, 1·2 cm. wide, covered with regularly imbricated enlarged calyx-teeth; seed oblong, about 1·4 mm. long, 1 mm. wide; testa smooth, yellowish or pale brown.

HAB.: Cultivated and fallow fields, roadsides and on stony hillsides, sea-level to 2,000 ft. alt.; fl. Jan.–April.

DISTR.: Divisions 4, 6–8, common in Division 7. Also Turkey and Palestine.

4. Pergamos, 1955, *Merton* 1984 !
6. Cape Kormakiti, 1956, *Merton* 2586 ! Ayios Yeoryios near Liveras, 1962, *Meikle* 2407 !
7. Common; *Syngrassides* 178 ! 296 ! *Casey* 144 ! 452 ! 1686 ! *Merton* 357 ! *G. E. Atherton* 835 ! 853B ! 1083 ! 1298 !
8. Near Cape Andreas, 1880, *Sintenis & Rigo* 415; Rizokarpaso, 1927 or 1930, *Druce* 137 !

15. T. striatum *L.*, Sp. Plant., ed. 1, 770 (1753); Boiss., Fl. Orient., 2: 130 (1872); A. K. Jackson et Turrill in Kew Bull., 1939: 475 (1939); Hossain in Notes Roy. Bot. Gard. Edinb., 23: 437 (1961); Davis, Fl. Turkey, 3: 426 (1970); Osorio-Tafall et Seraphim, List Vasc. Plants Cyprus, 59 (1973).

TYPE: "*in* Germania, Gallia, Hispania".

Prostrate or sprawling annual 5–25 (–30) cm. high; stems much branched, rarely simple, ± densely clothed with long spreading hairs; stipules oblong 4–6 mm. long, 2–3 mm. wide, membranous, villose, apex aristate; leaves usually all alternate, uppermost rarely opposite; petioles of lower leaves elongate, slender, to about 7 cm. long, slightly canaliculate above, thinly villose, those of the upper leaves becoming progressively shorter; leaflets narrowly obovate, 5–15 mm. long, 3–10 mm. wide, adpressed-villose on both surfaces, apex rounded or retuse, lateral nerves running straight into margin or very little curved at apex; inflorescences axillary and terminal, sessile or subsessile, shortly oblong, 8–12 mm. long, 7–10 mm. wide; calyx shortly cylindrical, about 4–5 mm. long, calyx-tube about 2·5 mm. long, 2 mm. wide, distinctly 10-nerved, villose externally, glabrous internally except at the shortly hairy throat, calyx-teeth subequal, subulate, about 2–3 mm. long, less than 0·5 mm. wide at base, narrowing above to a very slender aristate apex, suberect in flower and fruit; petals white or tinged pink, inconspicuous, limb of standard oblong, about 1·5 mm. long, 0·8 mm. wide, apex blunt or subacute, base scarcely auriculate, claw free (not united with staminal tube or with the claws of other petals) about 2 mm. long; limb of wing narrowly oblong, 1·5 mm. long, 0·5 mm. wide, apex obtuse or subacute, base minutely auriculate, claw about 2·5 mm. long, united with claw of keel, limb of keel oblong, 1·5 mm. long, 0·7 mm. wide, apex subacute, base indistinctly auriculate, claw about 2·5 mm. long; stamens 3·5 mm. long, filaments free for about 1 mm. from apex; anthers oblong, about 0·3 mm. long, 0·2 mm. wide; ovary obovoid, compressed, about 1·5 mm. long, 1·2 mm. wide, style sharply upcurved towards base, slender above, stigma small, oblique. Pod concealed within calyx; seed subglobose, about 1·8 mm. long and almost as wide; testa smooth, yellow or pale brown.

HAB.: In Cedar and Pine forest, on dry banks and in fallow fields; 500–4,400 ft. alt.; fl. March–May.

DISTR.: Division 2, rare. Widespread in Europe and the Mediterranean region eastwards to Iran; Atlantic Islands.

2. Platres, 1938, *Kennedy* 1280 ! Tripylos, 1941, *Davis* 3485 ! Yialia, 1955, *Merton* 2119 ! Prodhromos, 1955, *Merton* 2393 !

NOTES: All the Cyprus specimens have well-developed, slender calyx-teeth, and might be referred to *T. striatum* L. var. *spinescens* Lange (Pugillus, 363; 1865), though the calyx-teeth are not quite as long as in typical material of this variety. Such varieties are widely distributed in the Mediterranean region, and in some areas intergrade with typical *T. striatum*; such is the range of variation in the length of the calyx-teeth that it is hardly possible to employ the character, even at varietal level. *T. striatum* L. var. *macrodontum* Boiss. (Fl. Orient., 2: 130; 1872) is apparently indistinguishable from var. *spinescens* Lange.

SECT. 2. **Trichocephalum** *Koch* Annuals; inflorescence capitate; flowers sessile, ebracteate, reflexing after fertilization, the outer fertile, the inner sterile without corollas; calyx accrescent, the teeth filiform becoming ± medusoid in fruit; throat of calyx open; petals with the claws united into a tube. Pod small, usually included within calyx-tube.

16. T. subterraneum *L.*, Sp. Plant., ed. 1, 767 (1753); Boiss., Fl. Orient., 2: 133 (1872); Post, Fl. Pal., ed. 2, 1: 341 (1932); A. K. Jackson et Turrill in Kew Bull., 1938: 464 (1938); Hossain in Notes Roy. Bot. Gard. Edinb., 23: 439 (1961); Davis, Fl. Turkey, 3: 446 (1970); Zohary, Fl. Palaest., 2: 192, t. 275 (1972); Osorio-Tafall et Seraphim, List Vasc. Plants Cyprus, 60 (1973).

T. subterraneum L. var. *brachycladum* Gib. et Belli in Mem. Accad. Sci. Tor., ser. 2, 43: 183 (1893); Hossain in Notes Roy. Bot. Gard. Edinb., 23: 439 (1961).

TYPE: " *in* Gallia, Italia ".

Prostrate annual; stems usually much branched, sparsely or rather densely villose; stipules oblong-caudate, 4–14 mm. long, 2–4 mm. wide, membranous, subglabrous or thinly clothed with long hairs; leaves all alternate; petioles rather long and slender, up to 10 cm. long, subglabrous or thinly villose, slightly channelled above; leaflets broadly obovate-cuneate or obcordate, 3–17 mm. long, 3–22 mm. wide, subadpressed-hairy on both surfaces, apex usually emarginate, margins entire or obscurely denticulate; inflorescences axillary, with 1–5 fertile flowers; peduncle thinly or rather densely villose, at first 5–30 mm. long, lengthening and recurving after fertilization and burying the infructescence; calyx 4–5 mm. long, calyx-tube narrowly cylindrical, 1·5–2 mm. long, 1 mm. wide, subglabrous, often purplish, obscurely veined, calyx-teeth filiform, subequal, 2–3 mm. long, thinly villose; petals creamy-white, sometimes tinged pink, limb of standard oblong, 4–5 mm. long, 2–3 mm. wide, apex rounded or slightly emarginate, base obscurely auriculate, claw about 3 mm. long, limb of wings oblong, 3–3·5 mm. long, 1·3–1·5 mm. wide, apex rounded, base strongly auriculate, claw about 4 mm. long, limb of keel oblong about 2·5 mm. long, 1–1·3 mm. wide, apex blunt, base scarcely auriculate, claw about 4 mm. long; stamens 6–6·5 mm. long, filaments free for about 1–1·5 mm. from apex; ovary obovoid, 0·8–1 mm. long, 0·5–0·6 mm. wide, compressed, glabrous, style about 5 mm. long, glabrous, slighty thickened above, stigma small, oblique. Pod concealed by numerous reflexed sterile flowers which develop after fertilization, partly, but, not wholly, enclosed within enlarged fruiting calyx, minutely and obscurely veined; seed broadly oblong, 2·5–3 mm. long, 2–2·5 mm. wide; testa smooth, glossy, dark brown.

HAB.: On dry rocky hillsides and on grassy banks, generally confined to igneous formations or igneous intrusions in limestone formations; 1,700–3,150 ft. alt.; fl. March–May.

DISTR.: Divisions 2, 7. Western Europe from England southwards; widespread in the Mediterranean region and eastwards to Iran.

2. Kryos Potamos, 1937, *Kennedy* 564 ! Kanopetra Forest above Klirou, 1956, *Merton* 2492 ! Stavros tis Psokas, 1962, *Meikle* 2699 !

7. Kourtella Peak above Karmi village, 1949, *Casey* 357 ! and, 1953, *Casey* 1319 ! Kornos, 1954, *Merton* 1863 !

NOTES: Cyprus specimens show a great range of variation in size and general robustness; forms resembling var. *brachycladum* Gib. et Belli can be found in dry situations on poor soils, while typical *subterranean* will be found on better ground even in the same locality. It is difficult to believe that these dwarf compact plants and large trailing ones are more than states of a single species.

17. **T. globosum** *L.*, Sp. Plant., ed. 1, 767 (1753); Sibth. et Sm., Fl. Graec. Prodr., 2: 96 (1813), Fl. Graec., 8: 33 (1833); Unger et Kotschy, Die Insel Cypern, 378 (1865); Holmboe, Veg. Cypr., 106 (1914); Hossain in Notes Roy. Bot. Gard. Edinb., 23: 441 (1961); Davis, Fl. Turkey, 3: 445 (1970); Osorio-Tafall et Seraphim, List Vasc. Plants Cyprus, 60 (1973).

T. nidificum Griseb.. Spic. Fl. Rum. et Bith., 1: 32 (1843); Boiss., Fl. Orient., 2: 133 (1872).

TYPE: A cultivated specimen in herb. Hort. Cliff. (BM !)

Prostrate annual; stems 5–15 (–20) cm. long, much branched at base, thinly or densely villose; stipules broadly ovate-acuminate, 4–7 mm. long, 2–4 mm. wide, villose externally; leaves all alternate; petioles of basal leaves 3–10 mm. long, slender, thinly or densely villose, slightly canaliculate above, petioles of upper leaves progressively shorter, the uppermost leaves often subsessile; leaflets broadly obovate-cuneate 5–10 (–17) mm. long, 5–9 (–14) mm. wide, adpressed-villose on both surfaces, apex rounded or very slightly emarginate, margins minutely denticulate; inflorescences terminal (or pseudoterminal), globose, 1–1·5 cm. diam. at anthesis, 2 cm. or more diam. in fruit; peduncles usually distinct, 1–6 cm. long, thinly villose with spreading hairs; fertile flowers usually about 10; calyx about 7 mm. long, ± densely villose externally, calyx-tube narrowly cylindrical, about 3 mm. long, 1 mm. wide, obscurely nerved, calyx-teeth filiform, suberect or spreading at anthesis, subequal, about 4 mm. long, villose; petals white or pink, limb of standard oblong, about 4 mm. long, 2·5 mm. wide, apex emarginate, often shortly toothed, claw broad, about 3 mm. long, limb of wings oblong, 3–3·5 mm. long, 1·5 mm. wide, apex rounded, base strongly auriculate, claw about 4 mm. long, limb of keel oblong, 2·5–3 mm. long, 1·3 mm. wide, apex blunt or subacute, base scarcely auriculate, claw about 4 mm. long; stamens 6–7 mm. long, filaments free for about 2 mm. from apex; anthers oblong, about 0·5 mm. long, 0·3 mm. wide; ovary oblong, compressed, glabrous, about 1·5 mm. long, 0·6 mm. wide, style 6–7 mm. long, glabrous, tapering from a rather thick base, often irregularly bent, stigma small, oblique. Fruits concealed under a dense, globose head of villose sterile flowers; pod membranous, obscurely veined; seed oblong, 2–2·5 mm. long, about 2 mm. wide; testa smooth, dark brown.

HAB.: On rocky hillsides, or on sand-covered rocks by the seashore; sea-level to 2,500 ft. alt.; fl. March–May.

DISTR.: Divisions 1, 2, 8. Also Greece, Thasos, Thrace and Turkey.

1. Dhrousha, 1941, *Davis* 3213 !
2. Near Philani, 1908, *Clement Reid* s.n. ! Kambia, 1908, *Clement Reid* s.n. !
8. Shore W. of Cape Andreas, 1962, *Meikle* 2477 !

18. T. pilulare *Boiss.*, Diagn., 1, 2: 29 (1843); Fl. Orient., 2: 135 (1872); Post, Fl. Pal., ed. 2, 1: 342 (1932); A. K. Jackson in Kew Bull., 1937: 342 (1937); Rechinger f. in Arkiv för Bot., ser. 2, 1: 425 (1950); Hossain in Notes Roy. Bot. Gard. Edinb., 23: 443 (1961); Davis, Fl. Turkey, 3: 444 (1970); Zohary, Fl. Palaest., 2: 191, t. 273 (1972); Osorio-Tafall et Seraphim, List Vasc. Plants Cyprus, 60 (1973).

TYPE: Turkey; " in umbrosis graminosis montium *Lydiae* et *Cariae*, montes *Smyrnae*, *Mesogis*, *Tmolus* " May 1842, *Boissier* (G, K !).

Prostrate annual; stems (4–) 10–25 (–30) cm. long, much branched, angled, thinly or rather densely villose; stipules conspicuous, ovate, about 5 mm. long, 3 mm. wide, distinctly nerved with villose nerves, apex acute or shortly caudate; leaves all alternate; petioles slender, rather thinly adpressed-villose, those of the basal leaves to about 5 cm. long, those of the upper leaves much shorter, often less than 1 cm. long; leaflets broadly obovate-cuneate, 4–10 mm. long and about as wide, bright green often with a pale central spot, thinly hairy above and below, apex rounded or truncate; inflorescences axillary, globose, 8–10 mm. diam.; peduncle slender, adpressed-villose, about 8–15 mm. long; fertile flowers few, soon concealed by more numerous sterile flowers; calyx about 5 mm. long, calyx-tube cylindrical, about 2 mm. long, 1 mm. wide, densely villose externally, obscurely nerved, calyx-teeth filiform, subequal, villose, about 3 mm. long; petals whitish becoming pink with age, limb of standard oblong, 3–3·5 mm. long, 1·5 mm. wide, apex rounded or slightly emarginate, base obscurely auriculate, claw about 3 mm. long, limb of wings oblong, about 2 mm. long, 0·6 mm. wide, apex blunt, base prominently auriculate, claw about 4 mm. long, limb of keel oblong, 1·7 mm. long, 0·8 mm. wide, apex blunt, base scarcely auriculate, claw about 4 mm. long; stamens about 4·5 mm. long, filaments free for about 1 mm. from apex; anthers minute, about 0·3 mm. long, 0·2 mm. wide; ovary oblong, 0·8 mm. long, 0·5 mm. wide, compressed, glabrous, style about 3·5 mm. long; glabrous, tapering near apex, stigma minute, oblique. Fruits concealed under a dense globose head of villose sterile flowers; pod membranous, obscurely veined, seed shortly oblong or subglobose, compressed, about 2 mm. long, 1·7 mm. wide; testa smooth, almost black. *Plate 23.*

HAB.: On dry grassy or stony hillsides, sometimes in Pine forest, also by roadsides and margins of cultivated land, usually on igneous formations especially pillow lava; 500–4,600 ft. alt.; fl. March–June.

DISTR.: Divisions 1–3, 6, 7, common in Division 2. Aegean Islands, Turkey, Syria, Palestine and eastwards to Iraq.

1. Ayios Merkourios, 1962, *Meikle* 2285 !
2. Common; *Kennedy* 563 ! 1848 ! *Davis* 3071 ! *Mavromoustakis* s.n. ! *Casey* 723 ! *Chapman* 318 ! *Merton* 2161 !
3. Near Kouka, 1960, *N. Macdonald* 81 !
6. Koutraphas, 1936, *Syngrassides* 1015 ! South of Kato Mani, 1956, *Poore* in *Merton* 2760 !
7. Kornos, above Larnaka tis Lapithou, 1941, *Davis* 3040 ! Above Karmi, 1953, *Casey* 1320 !

SECT. 3. **Fragifera** *Koch* Annuals or perennials; inflorescence capitate; flowers sessile or shortly pedicellate, bracteate; calyx accrescent, reflexing

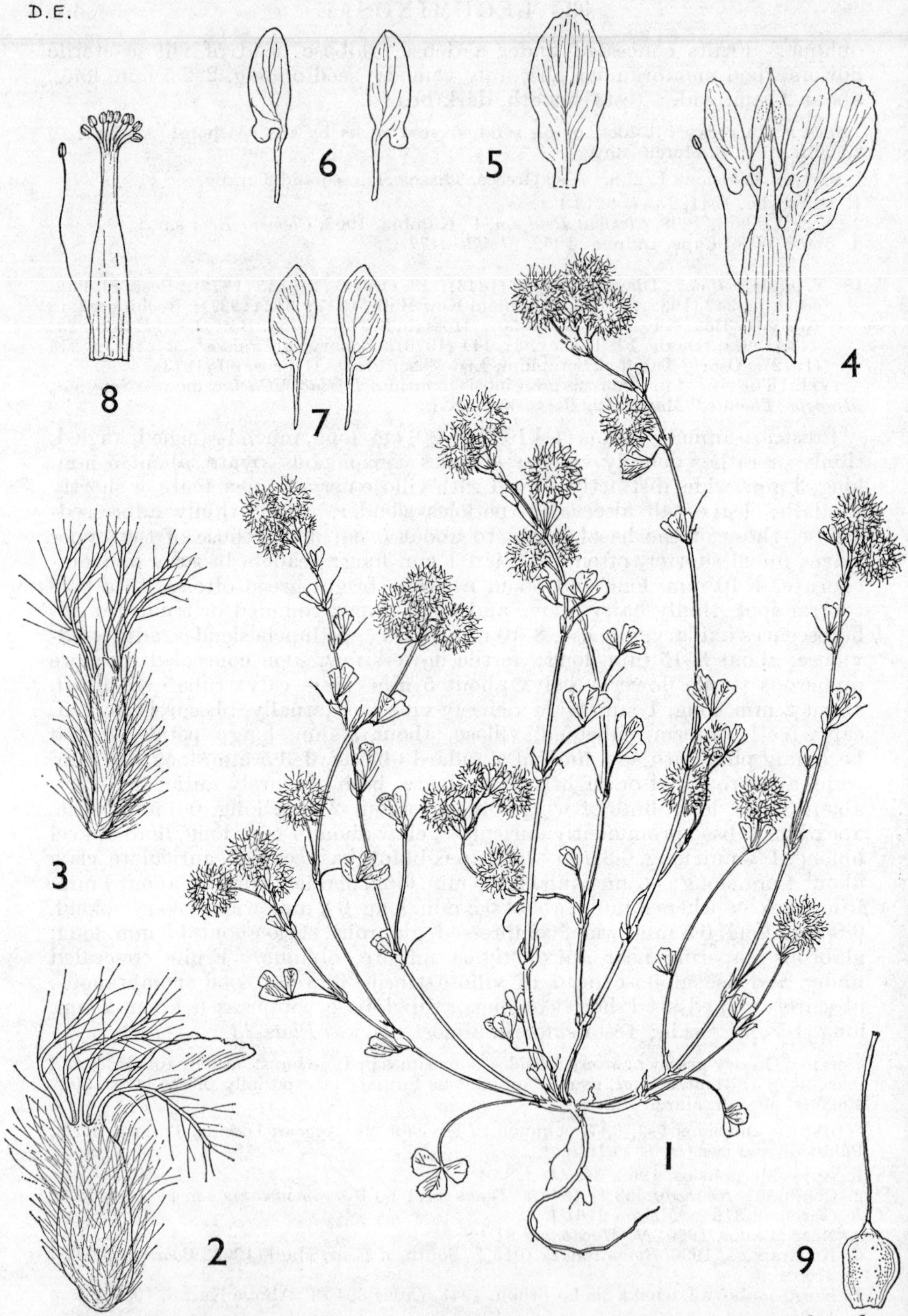

Plate 23. TRIFOLIUM PILULARE Boiss. **1,** habit, ×⅔; **2,** fertile flower, ×8; **3,** calyx of sterile flower, ×8; **4,** petals and vexillary stamen, ×8; **5, 6 & 7,** standard, wing and keel petals—free portion only, ×8; **8,** androecium, ×8; **9,** gynoecium, ×8. (*Bornmüller* 1141.)

and becoming inflated, scarious and reticulate in fruit, adaxial surface often hairy, becoming gibbous after fertilization, throat of calyx-tube without a thickened annulus; petals generally with united claws, sometimes resupinate. Pod ovoid, 1–2-seeded, included in the inflated calyx.

19. T. fragiferum *L.*, Sp. Plant., ed. 1, 772 (1753); Boiss., Fl. Orient., 2: 135 (1872); Holmboe, Veg. Cypr., 106 (1914); Post, Fl. Pal., ed. 2, 1: 343 (1932); Lindberg f., Iter Cypr., 20 (1946); Hossain in Notes Roy. Bot. Gard. Edinb., 23: 447 (1961); Davis, Fl. Turkey, 3: 409 (1970); Zohary, Fl. Palaest., 2: 169, t. 242 (1972); Osorio-Tafall et Seraphim, List Vasc. Plants Cyprus, 59 (1973).

TYPE: " *in* Suecia, Gallia, Anglia ".

Prostrate perennial with a tough rootstock; stems rooting at nodes, usually much branched, sometimes exceeding 30 cm. in length, glabrous or very sparsely hairy; stipules narrowly oblong, 10–25 (–30) mm. long, 2–6 mm. wide, glabrous, distinctly veined, the basal part pale and scarious, the apex long-acuminate, herbaceous, margins entire; leaves alternate; petioles up to 15 cm. long, often long and slender, glabrous or thinly hairy; leaflets oblong-obovate, shortly petiolulate, 6–25 mm. long, 4–18 mm. wide, glabrous above, subglabrous or thinly hairy below, apex rounded or shallowly emarginate, veins prominent, curved near the margins of the leaflets and excurrent as minute sharp teeth; inflorescences axillary, subglobose, 1–1·5 cm. diam. at anthesis, flowers numerous, crowded, very shortly pedicellate; peduncles 3—20 cm. long, usually exceeding petioles, glabrous or thinly hairy; involucre distinct, bracts narrowly oblong, about 2 mm. long, 0·5–0·8 mm. wide, acute, pallid with greenish veins, floral bracts much smaller, usually less than 1 mm. long and 0·5 mm. wide, pallid, cuspidate-acuminate; calyx (at anthesis) narrowly campanulate, about 3·5 mm. long, 1·5 mm. wide, calyx-tube about 2 mm. long, obscurely 10-veined, adaxial surface rather densely hairy, calyx-teeth erect, subequal, subulate, about 1·5 mm. long, the 2 adaxial teeth a little shorter and wider than the others; petals whitish or becoming pink with age, limb of standard bent upwards, 4–5 mm. long, 2 mm. wide, apex emarginate, base gradually narrowing to a broad claw about 2–2·5 mm. long, limb of wings narrowly oblong, about 3 mm. long, 0·8 mm. wide, apex blunt, base distinctly auriculate, claw 2·5–3 mm. long, limb of keel oblong, about 2 mm. long, 1 mm. wide, apex blunt, base scarcely auriculate, claw about 2·5 mm. long; stamens about 4·5 mm. long, filaments free for 1–1·5 mm. from apex; anthers broadly oblong, about 0·4 mm. long, 0·3 mm. wide; ovary narrowly oblong, about 1·5 mm. long, 0·6 mm. wide, glabrous, slightly compressed, style tapering from base to upcurved apex, glabrous, about 2·5–3 mm. long, stigma terminal, capitate. Fruiting calyx 5–7 mm. long, 4 mm. wide, reflexed, inflated, papery, distinctly reticulate, adaxial surface markedly gibbous; pod enclosed by calyx, brownish, membranous; seeds usually 2 in each pod, shortly oblong, about 1·5 mm. long, 1·2 mm. wide; testa olive-brown, flecked with small fuscous blotches.

HAB.: Margins of streams and water-channels, on marshy ground, occasionally by moist roadsides; 200–4,600 ft. alt.; fl. June–Oct.

DISTR.: Divisions 2, 5, 7, common in Division 2. Widespread in Europe and the Mediterranean region and eastwards to Afghanistan and Pakistan, also in N. Ethiopia.

2. Common; *Holmboe* 1020; *Lindberg f.* s.n.; *Casey* 827 ! 949 ! *G. E. Atherton* 536 ! *D. P. Young* 7294 !

5. Near Kythrea, 1939, *Lindberg f.* s.n. !

7. Lapithos, 1939, *Lindberg f.* s.n.; between Myrtou and Panagra, 1934, *Syngrassides* 613 !

20. T. physodes *Stev. ex M. Bieb.*, Fl. Taur.-Cauc., 2: 217 (1808); Boiss., Fl. Orient., 2: 136 (1872), Suppl., 168 (1888); Holmboe, Veg. Cypr., 106 (1914); Post, Fl. Pal., ed. 2, 1: 343 (1932); Hossain in Notes Roy. Bot. Gard. Edinb., 23: 449 (1961); Davis,

Fl. Turkey, 3: 409 (1970); Zohary, Fl. Palaest., 2: 169, t. 243 (1972); Osorio-Tafall et Seraphim, List Vasc. Plants Cyprus, 59 (1973).

T. ovatifolium Bory et Chaub., Nouv. Fl. Pélop., 51, t. XXVIII, fig. 1 (1838); Unger et Kotschy, Die Insel Cypern, 378 (1865).

TYPE: Caucasus; " in Iberiâ ", *Steven* (LE).

Prostrate or decumbent perennial with a tough rootstock; stems much branched at base, apparently not rooting at the nodes, up to 30 cm. long (or more) in robust specimens, subglabrous or thinly hairy; stipules narrowly oblong, 10–15 mm. long, 2–3 mm. wide, pallid and distinctly veined towards the base, apex greenish, acuminate-caudate; petioles up to 15 cm. long, slender, subglabrous or thinly hairy, narrowly channelled above; leaflets ovate-oblong, 5–25 mm. long, 4–16 mm. wide, glabrous or subglabrous above and below, with numerous close veins, apex rounded or rarely retuse, margins finely but rather irregularly serrulate; inflorescences axillary, globose, 1·2–2·5 cm. diam.; peduncle usually more than 3 cm. long, exceeding the leaves, subglabrous or thinly hairy; flowers shortly but distinctly pedicellate; bracts minute, deltoid, subulate or truncate, membranous, often gland-fringed, usually less than 1 mm. long; calyx (at anthesis) 4–5 mm. long, 2 mm. wide, calyx-tube cylindrical or narrowly campanulate, about 2 mm. long and about as wide, obscurely 10–13-veined, adaxial surface pubescent, distinctly gibbous, calyx-teeth unequal, the 2 adaxial about 3 mm. long, with a broadly deltoid base and subulate apex, the 3 abaxial about 2 mm. long, linear-subulate; petals white, becoming pink with age, standard free, limb of standard often strongly recurved, oblong, 6–7 mm. long, 2 mm. wide, apex rounded or slightly emarginate, base narrowing to a very broad claw about 3–4 mm. long, limb of wings narrowly oblong, about 6 mm. long, 1·3 mm. wide, apex blunt, base strongly auriculate, claw 4–4·5 mm. long, limb of keel oblong, 3·5 mm. long, 1·3 mm. wide, apex acute, almost rostrate, base abruptly narrowed to a claw 4–4·5 mm. long; stamens 6–7 mm. long, filaments free for about 2 mm. from apex; anthers oblong, about 0·5 mm. long, 0·4 mm. wide, apex shortly apiculate; ovary narrowly oblong, 2·5 mm. long, 0·6 mm. wide, compressed, glabrous or very thinly hairy at apex, style 4 mm. long, tapering from base to sharply upcurved apex, glabrous, stigma terminal, capitate. Fruiting calyx reflexed, inflated, papery, gibbous, conspicuously reticulate, about 8 mm. long, 6 mm. wide; pod concealed within calyx, 6 mm. long, 2 mm. wide, membranous, brownish, usually 2-seeded; seeds suborbicular, compressed, about 1·5 mm. diam.; testa smooth, dark brown. *Plate 24, figs. 1–10.*

HAB.: In garigue on igneous hillsides or in open Pine forest, sometimes by roadsides or by the margins of springs; 3,000–4,600 ft. alt.; fl. April–June.

DISTR.: Division 2, locally common Prodhromos–Platres–Papoutsa areas. Widely distributed in the northern and eastern Mediterranean region and eastwards to Iraq.

2. Common; *Kotschy* 704 ! *Sintenis & Rigo* 819 ! *Kennedy* 559 ! *Davis* 3107 ! 3153 ! *Mavromoustakis* 5A ! *Merton* 3136 ! *N. Macdonald* 95 ! *D. P. Young* 7294 !

21. T. resupinatum *L.*, Sp. Plant., ed. 1, 771 (1753); Boiss., Fl. Orient., 2: 137 (1872); Holmboe, Veg. Cypr., 106 (1914); Post, Fl. Pal., ed. 2, 344 (1932); Lindberg f., Iter Cypr., 20 (1946); Rechinger f. in Arkiv för Bot., ser. 2, 1: 425 (1950); Hossain in Notes Roy. Bot. Gard. Edinb., 23: 451 (1961); Davis, Fl. Turkey, 3: 412 (1970); Zohary, Fl. Palaest., 2: 170, t. 244 (1972); Osorio-Tafall et Seraphim, List Vasc. Plants Cyprus, 49 (1973).

An ascending, sprawling or prostrate, glabrous or subglabrous annual; stems 20–30 (–80) cm. long, usually much branched; stipules oblong, caudate, 10–17 mm. long, 2–4 mm. wide, basal part pallid and distinctly veined; petioles of lower leaves often more than 10 cm. long, slightly channelled above, those of the upper leaves progressively shorter; leaflets obovate,

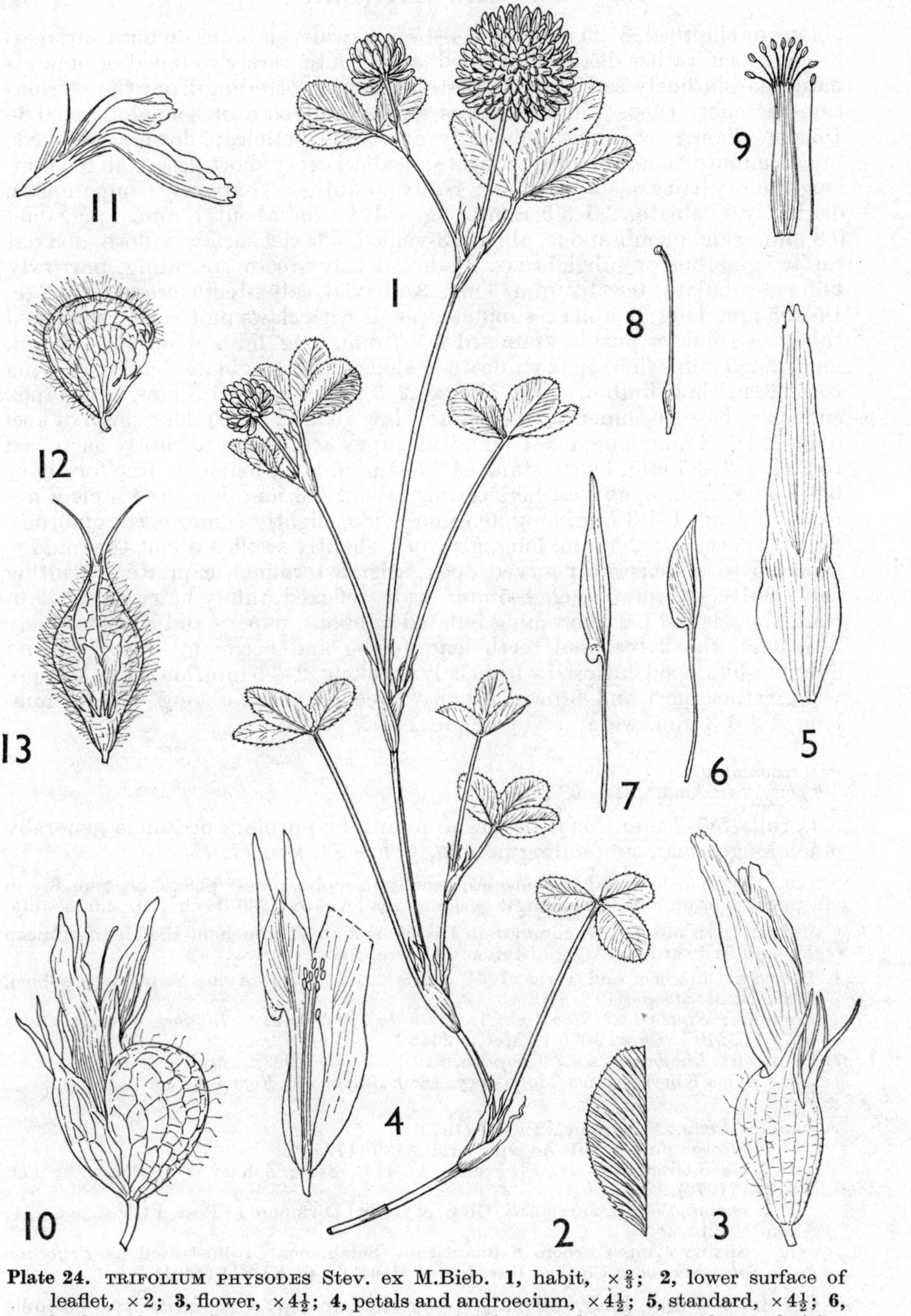

Plate 24. TRIFOLIUM PHYSODES Stev. ex M.Bieb. **1**, habit, $\times\frac{2}{3}$; **2**, lower surface of leaflet, $\times 2$; **3**, flower, $\times 4\frac{1}{2}$; **4**, petals and androecium, $\times 4\frac{1}{2}$; **5**, standard, $\times 4\frac{1}{2}$; **6**, keel, $\times 4\frac{1}{2}$; **7**, wing, $\times 4\frac{1}{2}$; **8**, gynoecium, $\times 4\frac{1}{2}$; **9**, androecium, $\times 4\frac{1}{2}$; **10**, fruiting calyx, $\times 4\frac{1}{2}$. TRIFOLIUM RESUPINATUM L. **11**, flower, $\times 4$; **13**, fruiting calyx, $\times 4$. TRIFOLIUM TOMENTOSUM L. **12**, fruiting calyx, $\times 4$. (**1–9**, *Alkas* 18690; **10**, *Rawi* 23471; **11**, *Helbaek* 862; **12**, *Robertson* S/1031; **13**, *Rogers* 0130.)

oblong or elliptical, 8–25 mm. long, 4–15 mm. wide, glabrous on both surfaces, bright green, rather distinctly veined, apex acute, rarely rounded or emarginate, margin finely serrulate, the lateral veins protruding from the serrulations as short cusps; inflorescences axillary, globose or semiglobose, 0·8–1·5 cm. diam.; peduncles generally exceeding petioles; flowers crowded; bracts minute, membranous, truncate; pedicels very short (less than 0·5 mm. long), thinly hairy or subglabrous; calyx (at anthesis) narrowly campanulate, distinctly bilabiate, 2·5–3·5 mm. long, calyx-tube about 1 mm. long, 0·6–0·8 mm. wide, membranous, about 13-veined, adaxial surface villose, abaxial surface glabrous or subglabrous, 2 adaxial calyx-teeth spreading, narrowly filiform-subulate, 0·8–1·2 mm. long, 3 abaxial calyx-teeth erect, subulate, 1·5–2·5 mm. long; corolla resupinate, petals with claws united into a twisted tube, rose-pink or purple, standard 4·5–7 mm. long, limb oblong, 3–4·5 mm. long, 2–2·5 mm. wide, apex truncate or shallowly emarginate, base narrowing to a broad claw, limb of wings oblong, 2–3 mm. long, 1–1·3 mm. wide, apex subacute, base prominently auriculate, claw about 2·5 mm. long, limb of keel oblong, 1·5–2 mm. long, 0·8–1 mm. wide, apex acute, base abruptly narrowed to a claw 2–2·5 mm. long; stamens 2·5–4 mm. long, filaments free for about 0·8–1 mm. from apex; anthers oblong, about 0·3 mm. long, 0·2 mm. wide; ovary oblong, 1–1·3 mm. long, 0·3 mm. wide, slightly compressed, glabrous, 2-ovulate, style 2–2·5 mm. long, glabrous, slightly swollen about the middle, tapering to a sharply upcurved apex, stigma terminal, capitate. Fruiting calyx 8–10 (–12) mm. long, 4–5 mm. wide, reflexed, thinly hairy or subglabrous, the adaxial part becoming inflated, gibbous, papery and conspicuously reticulate, the 2 terminal teeth lengthening and becoming upcurved and antenna-like; pod enclosed within calyx, oblong, 2–2·5 mm. long, 1·2–1·8 mm. wide, rather hard and brown, usually 1-seeded; seed oblong, 1·3–1·5 mm. long, 1·2–1·3 mm. wide; testa smooth, dark brown.

var. **resupinatum.**
TYPE: " *in* Anglia, Belgio ".

Corolla 5·5–7 mm. long, bright rose-pink or purple; peduncle generally much longer than subtending petiole. *Plate 24, figs. 11, 13.*

HAB.: By roadsides and streamsides, usually in moist, grassy places, occasionally in salt-marshes or on hillsides amongst garigue; sea-level to 4,600 ft. alt.; fl. March–July.

DISTR.: Divisions 1–3, 7, common in Divisions 2, 3. Throughout the Mediterranean region and eastwards to Afghanistan and Central Asia.

1. Between Dhrousha and Terra, 1941, *Davis* 3270 ! Near Ayios Yeoryios (Paphos), 1960, *N. Macdonald* 49 !
2. Common; *Sintenis & Rigo* 409A ! *Kennedy* 562 ! 1849 ! *Lindberg f.* s.n.; *Davis* 3175 ! 3270 ! *Casey* 1056 ! *Meikle* 2645 !
3. Common; *Lindberg f.* s.n.; *Chapman* 610 ! *Merton* 2063 ! *Meikle* 2802 !
7. Near Ayios Khrysostomos Monastery, 1880, *Sintenis & Rigo* 409 !

var. **minus** *Boiss.*, Fl. Orient., 2: 137 (1872).
T. bicorne Forssk., Fl. Aegypt.-Arab., 139 (1775).
T. clusii Gren. et Godr., Fl. France, 1: 414 (1849); Zohary, Fl. Palaest., 2: 171, t. 245 (1972).
T. resupinatum L. var. *clusii* (Gren. et Godr.) Dinsmore in Post, Fl. Pal., ed. 2, 1: 245 (1932).
TYPE: Spain; " circa urbem Salmanticam [Salamanca] ", illustrated as *Trifolium pratense Salmanticum* in Clusius, Rar. Plant. Hist., 6: CCXLVII (1601).

Corolla 4–5 mm. long, normally paler than in var. *resupinatum*; peduncle generally shorter than subtending petiole.

HAB.: Muddy ground by side of lake; c. 100 ft. alt.; fl. March–May.

DISTR.: Division 8, rare. Widespread in the Mediterranean region; Atlantic Islands; apparently not extending so far east as var. *resupinatum*.

8. Rare; Galatia, 1962, *Meikle* 2530 !

NOTES: The specimen cited is so nearly intermediate between *T. resupinatum* L. var. *resupinatum* and *T. tomentosum* L. that it would be reasonable to conclude that it, and other specimens of *T. resupinatum* var. *minus*, are hybrids between these two species, a conjecture supported by the fact that var. *minus* is commonly found in areas where both the putative parents are known to occur. It is, however, also interesting to note that var. *minus* is particularly common in certain areas (e.g. Atlantic Islands; Spain, Portugal, Egypt) while very rare or wanting in others (e.g. Turkey, Iraq, Iran, Afghanistan). This would suggest that the variety, if in fact a hybrid, is a fertile and relatively stable one.

In Cyprus *T. resupinatum* var. *minus* was seen in quantity with *T. tomentosum*, low-growing plants of the latter forming a distinct " understorey " below colonies of the relatively erect and robust var. *minus*. The two were quite distinct in flower and fruit, nor were any intermediates seen; the variety differed from typical *resupinatum* in flower colour, and in relative lengths of peduncle and petiole, again no intermediates were seen, nor is typical *resupinatum* recorded from the Karpas area. This strongly suggests that var. *minus* is an old member of the lakeside community at Galatia, and not a first generation hybrid.

Hossain's subdivision of *T. resupinatum* (Notes Roy. Bot. Gard. Edinb., 23: 451–452) into robust and slender varieties is not wholly satisfactory; examination of specimens cited by him under var. *minus* reveals a mixture of slender or starved *T. resupinatum* var. *resupinatum* as well as some specimens of true var. *minus* Boiss. Small, poor specimens of *resupinatum* are sometimes difficult to identify exactly, but var. *resupinatum* is almost always distinguishable by having the peduncles considerably longer than the petioles.

22. T. tomentosum *L.*, Sp. Plant., ed. 1, 771 (1753); Unger et Kotschy, Die Insel Cypern, 379 (1865); Boiss., Fl. Orient., 2: 138 (1872); Holmboe, Veg. Cypr., 106 (1914); Post, Fl. Pal., ed. 2, 1: 345 (1932); Lindberg f., Iter Cypr., 20 (1946); Hossain in Notes Roy. Bot. Gard. Edinb., 23: 453 (1961); Davis, Fl. Turkey, 3: 413 (1970); Zohary, Fl. Palaest., 2: 172, t. 246 (1972); Osorio-Tafall et Seraphim, List Vasc. Plants Cyprus, 59 (1973).

TYPE: " *in* G. Narbonensi, Hispania, Lusitania, *D. Sauvages. Loefling* ".

Prostrate or sprawling annual, closely resembling *T. resupinatum* in habit but generally less robust; stems glabrous or subglabrous, much branched, 5–25 cm. long; stipules narrowly oblong, caudate-acuminate, 10–15 mm. long, 2–3 mm. wide, basal part pallid and distinctly veined; petioles of lower leaves 2–5 cm. long, slender, glabrous or subglabrous, slightly channelled above, those of the upper leaves progressively shorter; leaflets obovate or sometimes almost elliptical, 5–17 mm. long, 3–9 mm. wide, glabrous above, occasionally thinly hairy below, rather distinctly veined, apex rounded or subacute, margin finely and sharply serrulate, the lateral veins protruding from the serrulations as short cusps; inflorescences axillary, globose or subglobose, 6–9 mm. diam.; peduncles generally shorter than petioles; flowers crowded; bracts minute, membranous, truncate or emarginate; pedicels very short (less than 0·5 mm. long) thinly hairy or subglabrous; calyx (at anthesis) narrowly campanulate, distinctly bilabiate, 2–2·5 mm. long, calyx-tube about 1–1·5 mm. long, 0·6–0·8 mm. wide, membranous, 12–13-veined, adaxial surface densely lanuginose, abaxial surface glabrous or subglabrous, 2 adaxial calyx-teeth erect or spreading, subulate, 0·7–1 mm. long, 3 abaxial calyx-teeth erect, subulate, 0·5–1 mm. long; corolla resupinate, petals with claws united into a twisted tube, pale or bright pink, standard 3·5–4 mm. long, limb oblong or narrowly obovate, about 2 mm. long, 1·5–2 mm. wide, apex emarginate, base narrowing to a broad claw, limb of wings oblong, 1·5 mm. long, 0·5–0·8 mm. wide, apex rounded, base distinctly auriculate, claw slender, about 2 mm. long, limb of keel oblong, 1–1·3 mm. long, 0·5–0·6 mm. wide, apex acute or obtuse, base abruptly narrowed to a claw 1·8–2 mm. long; stamens 2·5–3 mm. long, filaments free for about 0·5–0·8 mm. from apex; anthers oblong, about 0·3 mm. long and almost as wide; ovary oblong, 0·5–0·8 mm. long, 0·3–0·4 mm. wide, slightly compressed, glabrous, 2-ovulate, style 1·5–1·7 mm. long, glabrous, slightly swollen about the middle, tapering to a slightly upcurved apex, stigma terminal, capitate. Fruiting calyx 3–7 mm. long, 3–5 mm. wide, reflexed, lanuginose or subglabrous, the adaxial part strongly inflated, gibbous and papery, distinctly reticulate veined in subglabrous forms, the two terminal teeth hidden or rather short and upcurved; pod enclosed within calyx, oblong 1·5–2 mm. long, 1 mm. wide, membranous, 1-seeded; seed broadly oblong, 1·2–1·5 mm. long, 1–1·3 mm. wide; testa smooth, brown or yellowish. *Plate 24, fig. 12.*

HAB.: In cultivated and fallow fields, in waste ground and by roadsides and lake shores, occasionally on open grassy hillsides or in forest clearings; sea-level to 4,600 ft. alt.; fl. March–May.

DISTR.: Divisions 2–8, common in Divisions 2, 4, 7. Throughout the Mediterranean region and eastwards to Iran; Atlantic Islands.

2. Common; *Lindberg f.* s.n.; *G. E. Atherton* 601 ! *N. Macdonald* 35 ! 86 ! *D. P. Young* 7249 !
3. Limassol, 1859, *Kotschy* 1005; *Mazotos*, 1905, *Holmboe* 195; Akrotiri, 1927, *Rev. A. Huddle* 40 ! Yermasoyia, *F. J. F. Barrington* s.n. !
4. Common; *Kotschy* 27; *Sintenis & Rigo*; *Barbey*; *Holmboe* 35; *Chapman* 25 !
5. Perakhorio, 1936, *Syngrassides* 1138 !
6. Nicosia, 1902, *A. G. & M. E. Lascelles* s.n. ! Nicosia, 1927, *Rev. A. Huddle* 41 ! Syrianokhori, 1935, *Syngrassides* 761 ! Ayia Irini, 1941, *Davis* 2553 !
7. Common; *Sintenis & Rigo* 411 ! *Casey* 450 ! 1707 ! *G. E. Atherton* 1142 ! 1207 ! 1297 !
8. Near Cape Andreas, 1880, *Sintenis & Rigo* s.n. ! Galatia, 1962, *Meikle* 2529 !

NOTES: Many of the Cyprus specimens resemble *T. tomentosum* var. *curvisepalum* (V. Täckh.) Hossain in that the two adaxial calyx-teeth tend to be obvious and protruding even in ripe fruiting heads, and not concealed as in typical *tomentosum*. There is, however, considerable variation in this character, and it would be very difficult to discriminate satisfactorily between the two varieties. It is evident that var. *curvisepalum* is the prevalent variety in many parts of the Eastern Mediterranean and Middle East.

SECT. 4. **Mistyllus** (*Presl*) *Godr*. Annuals; inflorescence capitate; flowers sessile or subsessile, conspicuously bracteate; calyx accrescent, becoming ± inflated on all sides in fruit, longitudinally 20- or more-nerved, sometimes with transverse striations, throat not thickened, calyx-teeth ± awn-like, usually recurved; standard free from other petals. Pod ovoid or oblong, 1–4-seeded, included in the inflated calyx; seeds verruculose.

23. T. spumosum *L.*, Sp. Plant., ed. 1, 771 (1753); Sibth. et Sm., Fl. Graec. Prodr., 2: 100 (1813), Fl. Graec., 8: 38, t. 753 (1833); Unger et Kotschy, Die Insel Cypern, 379 (1865); Boiss., Fl. Orient., 2: 138 (1872); Holmboe, Veg. Cypr., 106 (1914); Post, Fl. Pal., ed. 2, 1: 345 (1932); Hossain in Notes Roy. Bot. Gard. Edinb., 23: 456 (1961); Davis, Fl. Turkey, 3: 407 (1970); Zohary, Fl. Palaest., 2: 167, t. 240 (1972); Osorio-Tafall et Seraphim, List Vasc. Plants Cyprus, 59 (1973).

TYPE: " *in* Gallia, Italia, Apulia ".

Robust, glabrous or subglabrous annual; stems prostrate or ascending, usually much branched, 8–25 (–40) cm. long; stipules conspicuous, ovate or oblong, membranous, 8–20 mm. long, 3–7 mm. wide, apex caudate-acuminate; uppermost leaves opposite, the rest alternate, petioles of lower leaves 6–14 cm. long, slightly channelled above, those of the upper leaves becoming progressively shorter; leaflets broadly obovate, 10–30 mm. long, 5–25 mm. wide, rather distinctly veined, apex rounded or occasionally somewhat emarginate, margins minutely serrulate, the lateral veins running into the serrulations; inflorescences terminal, globose, 2–3 cm. diam.; peduncle stout 1·5–4 cm. long; bracts ovate-aristate, 5–8 mm. long, 2·5–4 mm. wide, whitish, scarious; flowers crowded; calyx ovoid, inflated, 10–13 mm. long, 2–3·5 mm. wide, calyx-tube 8–9 mm. long, 20-nerved, pallid, membranous, with obscure transverse striations, calyx-teeth filiform-subulate, subequal, spreading, greenish, 2–4 mm. long; petals pink, limb of standard narrowly oblong, about

8 mm. long, 2·5 mm. wide, apex acute or acuminate, base tapering to a wide claw about 10 mm. long, limb of wings narrowly oblong, 4–5 mm. long, 1·5 mm. wide, apex acute, base strongly auriculate, claw about 10 mm. long, limb of keel oblong, 3·5 mm. long, 1·5 mm. wide, apex acute or shortly rostrate, base narrowed abruptly to a claw about 10 mm. long; stamens about 12 mm. long, filaments free for about 2 mm. from apex; anthers oblong, 0·5 mm. long, 0·3 mm. wide; ovary narrowly oblong, 3–4 mm. long, 0·5–0·7 mm. wide, shortly scabridulous, 4-ovulate, style 7–8 mm. long, scabridulous towards base, glabrous and sharply upcurved near apex, stigma somewhat elongate, oblique. Fruiting calyx inflated, whitish or pale brown, distinctly reticulate with well-marked transverse striations; pod membranous, about 5 mm. long, 2·5–3 mm. wide, compressed; seeds 3–4 in each pod, oblong or subglobose, 1·5–2 mm. long, 1–1·5 mm. wide; testa verruculose, bright brown.

HAB.: In pastures and Olive groves, by roadsides or edges of cultivated fields; sea-level to 800 ft. alt.; fl. March–April.

DISTR.: Divisions 1–7. Widespread in the Mediterranean region and eastwards to Iraq; Atlantic Islands.

1. Near Ayios Yeoryios (Akamas), *N. Macdonald* 50 !
2. Kambos [" Gambo "], 1927 or 1930, *Druce* 142 !
3. Stavrovouni, 1937, *Syngrassides* 1495 !
4. Larnaca aerodrome, 1950, *Chapman* 409 ! Pergamos, 1955, *Merton* 1976 ! Cape Kiti, 1955, *Merton* 2021 !
5. Trikomo, 1950, *Chapman* 217 ! Salamis, 1960, *N. Macdonald* 115 !
6. Ayios Yeoryios near Liveras, 1962, *Meikle* 2404 !
7. Near Kantara Monastery, 1880, *Sintenis & Rigo* ! Kyrenia, 1949, *Casey* 451 ! 1953, *Casey* 1293 ! and, 1956, *Casey* 1710 !

24. T. argutum *Banks et Sol.* in Russell, Nat. Hist. Aleppo, ed. 2, 2: 260 (1794); Hossain in Notes Roy. Bot. Gard. Edinb., 23: 458 (1961); Davis, Fl. Turkey, 3: 408 (1970); Zohary, Fl. Palaest., 2: 168, t. 241 (1972); Osorio-Tafall et Seraphim, List Vasc. Plants Cyprus, 59 (1973).

T. xerocephalum Fenzl, Pugill. Plant. Syr., 1: 5 (1842); Boiss., Fl. Orient., 2: 140 (1872).

TYPE: Syria; near Aleppo, *Russell* (BM).

Prostrate or ascending, glabrescent annual; stems much branched, 15–25 (–30) cm. long; stipules narrowly oblong, 6–13 mm. long, 1·5–3 mm. wide, distinctly nerved, apex elongate, subulate-caudate or almost filiform; uppermost leaves opposite, the rest alternate; petioles of the lower leaves 4–7 cm. long, shallowly channelled above, those of the upper leaves becoming progressively shorter; leaflets broadly obovate, 5–15 mm. long, 4·5–13 mm. wide, distinctly veined, veins down-curved towards margins of leaflets and excurrent as small sharp marginal serrulations, apex of leaflets usually rounded, occasionally slightly emarginate; inflorescence at first subglobose, later shortly oblong-ovoid, 1–2 cm. long, 1–1·5 cm. wide, flowers densely crowded; peduncles variable, 0·5–4 cm. or more long, sometimes almost wanting; bracts conspicuous, whitish, scarious, obovate, about 3 mm. long, 1·5–2 mm. wide, apex shortly cuspidate; calyx 4–5 mm. long, 1·5–2 mm. wide, calyx-tube shortly oblong, slightly inflated, about 2 mm. long, whitish, with about 45 close longitudinal nerves, calyx-teeth subequal, subulate, spreading, 1·5–2 mm. long; petals persistent becoming scarious, distinctly veined, the standard and wings bright pink or purplish, the keel paler, often whitish, limb of standard erect, ovate, 5–6 mm. long, 4 mm. wide, apex subacute or obtuse, margins distinctly recurved, base narrowed abruptly to a wide claw about 3 mm. long, limb of wings divaricately reflexed, oblong, 4 mm. long, 1·5–1·8 mm. wide, apex subacute or obtuse, base auriculate, abruptly narrowed to a slender claw about 4 mm. long, limb of keel erect, concave but not longitudinally plicate or boat-shaped, the two halves together broadly oblong, 3–4 mm. long, 2·5–3 mm. wide, apex obtuse or rounded, base abruptly narrowed or auriculate, claw about 4 mm. long; stamens about 7 mm. long, filaments free for about 2–2·5 mm. from apex; anthers oblong, about 0·5 mm. long, 0·4 mm. wide; ovary oblong, 1·2–1·3 mm. long, 0·6 mm. wide, sparsely scabridulous, 2-ovulate, style 5–6 mm. long, swollen and sparsely scabridulous towards base, glabrous, slender and upcurved towards apex, stigma small, capitate, terminal. Fruiting calyx scarcely at all accrescent; pod membranous, oblong, 2–3 mm. long, 1 mm. wide, 1-seeded; seed oblong, about 1·3 mm. long, 0·8 mm. wide; testa minutely verruculose, bright brown.

HAB.: On old stabilized sand-dunes; just above sea-level; fl. March–May.

DISTR.: Division 1, rare. Also in Turkey, Rhodes, Syria, Lebanon and Palestine.

1. Paphos, 1927 or 1930, *Druce* 131 ! Bay beyond Kissanerga and Ktima, 1955, *Merton* 2123 ! By the shore near the fort, Kato Paphos, 1974, *H. Morris* in *Meikle* 4033 !

SECT. 5. **Trifoliastrum** *Ser.* Annuals or perennials; inflorescence usually umbellate, with pedicellate flowers, rarely capitate, bracteate; calyx ± campanulate, scarcely accrescent, not becoming inflated, usually 10-nerved, throat glabrous, not thickened, calyx-teeth lanceolate or subulate, the 2 adaxial usually a little longer than the others; standard almost free from other petals; ovary 2–4 (or rarely 10)-ovulate.

25. **T. repens** *L.*, Sp. Plant., ed. 1, 767 (1753); Sibth. et Sm., Fl. Graec. Prodr., 2: 95 (1813); Unger et Kotschy, Die Insel Cypern, 378 (1865); Boiss., Fl. Orient., 2: 145 (1872); Holmboe, Veg. Cypr., 106 (1914); Post, Fl. Pal., ed. 2, 1: 348 (1932); Rechinger f. in Arkiv för Bot., ser. 2, 1: 425 (1950); Hossain in Notes Roy. Bot. Gard. Edinb., 23: 464 (1961); Davis, Fl. Turkey, 3: 392 (1970); Zohary, Fl. Palaest., 2: 162, t. 231 (1972); Osorio-Tafall et Seraphim, List Vasc. Plants Cyprus, 59 (1973).

TYPE: " *in* Europae pascuis ".

Prostrate, glabrescent perennial; stems 20–30 cm. (or more) long, much branched, rooting at the nodes; stipules connate below, narrowly oblong, about 10 mm. long, 1·5–2 mm. wide, the basal part membranous, distinctly veined, the apex subulate or almost filiform; leaves alternate; petioles slender, up to 20 cm. (or occasionally more) in length, slightly channelled above; leaflets obovate, oblong-elliptical or obcordate 4–20 (–40) mm. long, 3·5–12 (–20) mm. wide, distinctly veined, the veins excurrent into short marginal teeth, apex rounded, sometimes emarginate or subacute; flowers axillary; peduncles usually elongate, 4–30 (or sometimes more) cm. long; inflorescence globose, 1·4–2·5 cm. diam., flowers reflexed after fertilization; bracts membranous, lanceolate, 1·5–2 mm. long, 0·5 mm. wide, midrib excurrent as a short mucro; pedicels slender 1–4 (or more) cm. long; calyx narrowly campanulate, 4–5 mm. long, 1·5–2 mm. wide, calyx-tube 1·5–2 mm. long, membranous, 10-nerved, calyx-teeth subulate, the 2 adaxial about 2 mm. long, the others a little shorter; petals white or tinged pink with age, limb of standard oblong, 6–8 mm. long, about 4 mm. wide, apex obtuse, often denticulate, base narrowed to a wide claw about 2–3 mm. long, occasionally slightly auriculate, claw free almost to base, limb of wings oblong, about 4 mm. long, 2 mm. wide, apex rounded, base auriculate, claw about 3 mm. long, limb of keel oblong, about 2·5 mm. long, 1·5 mm. wide, apex rounded, base abruptly narrowed to a claw about 3 mm. long; stamens about 5 mm. long, filaments free for about 2 mm. from apex; anthers oblong, 0·5 mm. long, 0·3 mm. wide; ovary narrowly oblong, 2·5 mm. long, 0·4 mm. wide, glabrous, 4–5-ovulate, style 2–2·5 mm. long, glabrous, tapering from base to apex, stigma capitate, terminal. Pod oblong, about 4 mm. long, 1·8 mm.

wide, protruding from calyx tube, membranous, usually 2-seeded; seeds suborbicular-reniform, compressed, about 1·4 mm. diam.; testa smooth, yellowish-brown.

HAB.: Rocky mountainsides and streamsides at high altitudes; 5,000–6,000 ft. alt.; fl. June–July.

DISTR.: Division 2, rare. Widespread from Europe to Central Asia; introduced and naturalized in temperate regions elsewhere.

2. Khionistra, 1912, *Haradjian* 473; Kryos Potamos, 1938, *Kennedy* 1289 !

NOTES: Evidently very rare in Cyprus, and perhaps originally introduced as a fodder plant. Kotschy suggests that Sibthorp's record may be based on a misidentification of *T. nigrescens* Viv. ssp. *petrisavii* (Clem.) Holmboe.

26. T. nigrescens *Viv.*, Fragm. Fl. Ital., 1: 12, t. 13 (1808); Kotschy, Die Insel Cypern, 379 (1865); Lindberg f., Iter Cypr., 20 (1946).

Erect or sprawling, glabrous annual; stems 6–35 cm. long, usually much branched, angular, frequently tinged purple; stipules connate at base, narrowly oblong, 10–13 mm. long, 2–4 mm. wide, base membranous, distinctly veined, apex subulate-filiform; leaves alternate; petioles of lower leaves to 15 cm. long (sometimes more), shallowly channelled, those of the upper leaves becoming progressively shorter; leaflets broadly obovate or obcuneate, 3–20 mm. long, 3–17 mm. wide, distinctly veined, veins exserted from short marginal teeth, apex of leaflet usually rounded, sometimes truncate or slightly emarginate; inflorescences axillary, globose, 1–2 cm. diam.; flowers reflexed after fertilization; peduncles 1·5–5 cm. long; bracts narrowly lanceolate-acuminate, about 1 mm. long, 0·4 mm. wide; pedicels 1–3 mm. long; calyx narrowly campanulate, 4–4·5 mm. long, 1·5 mm. wide, calyx-tube pallid, membranous, 10-nerved, about 2 mm. long, calyx-teeth subulate, erect, green, the 2 adaxial about 2–2·5 mm. long, the rest rather shorter; petals white or pink-tinged, limb of standard oblong, about 5 mm. long, 3–4 mm. wide, folded longitudinally, apex blunt or emarginate, base tapering to a wide claw about 2 mm. long, limb of wings oblong, about 3 mm. long, 1·5 mm. wide, apex rounded, base conspicuously auriculate, claw about 2·5 mm. long, limb of keel oblong, about 2 mm. long, 1·3 mm. wide, apex rounded, base abruptly narrowed to a claw about 2·5 mm. long; stamens about 4 mm. long, filaments free for about 1 mm.; anthers suborbicular about 0·4 mm. diam.; ovary oblong, about 1·5 mm. long, 0·5–0·8 mm. wide, sparsely scabridulous, 2–4 (–6)-ovulate; style about 2·5 mm. long, ± uniformly thick, upcurved towards apex; stigma terminal, capitate. Pod oblong 2–4 mm. long, about 1–1·2 mm. wide, membranous, strongly or slightly constricted between the seeds; seeds oblong or suborbicular, 0·8–1·3 mm. long, 0·8 mm. wide; testa smooth, brown or olive-green.

ssp. **petrisavii** (*Clem.*) *Holmboe*, Veg. Cypr., 106 (1914); Hossain, Notes Roy. Bot. Gard. Edinb., 23: 467 (1961); Davis, Fl. Turkey, 3: 396 (1970); Zohary, Fl. Palaest., 2: 163, t. 232 (1972); Osorio-Tafall et Seraphim, List Vasc. Plants Cyprus, 59 (1973).

T. petrisavii Clem., Sert. Orient., 32, t. 7, fig. 2 (1855); Boiss., Fl. Orient., 2: 144 (1872); Post, Fl. Pal., ed. 2, 1: 348 (1932); Rechinger f. in Arkiv för Bot., ser. 2, 1: 425 (1950).

TYPE: Turkey; " Copiose provenit in collibus constantinopolitanis ultra *il Gran Campo* et tota aestate floret " (TO).

Ovules 2. Pod 1–2-seeded, strongly constricted between the seeds; seeds oblong, about 1·3 mm. long, 0·8 mm. wide. Otherwise as in ssp. *nigrescens*.

HAB.: Moist meadows and streamsides, damp slacks in sand-dunes, sometimes by roadsides, in olive groves, vineyards, or on damp hillsides; sea-level to 3,800 ft. alt.; fl. March–June.

DISTR.: Divisions 1, 2, 4, 6, 7, common in Division 7. Greece, Turkey, Aegean Islands, Syria, Lebanon and Palestine eastwards to Iraq.

1. Stroumbi, 1914, *Haradjian* 781; Dhrousha, 1941, *Davis* 3264 ! Near Neokhorio (Akamas), 1962, *Meikle* 2194 !
2. Pano Platres, 1930, *Wall* s.n.; Platres, 1938, *Kennedy* 1288 ! Mesopotamos Monastery, 1939, *Lindberg f.* s.n.
4. Cape Kiti, 1956, *Merton* 2020 !
6. Nicosia, 1930, *F. A. Rogers* 48 ! Syrianokhori, 1941, *Davis* 2618 !
7. Common; *Sintenis & Rigo* 410 ! 410A ! *Casey* 352 ! 1711 ! *G. E. Atherton* 1371 ! *Merton* 2986 !

NOTES: The species, in the aggregate sense, is widespread in the Mediterranean area, ssp. *nigrescens* in the western half and ssp. *petrisavii* in the eastern half; the latter alone seems to occur in Cyprus.

27. T. glomeratum *L.*, Sp. Plant., ed. 1, 770 (1753); Boiss., Fl. Orient., 2: 142 (1872); Post, Fl. Pal., ed. 2, 1: 347 (1932); A. K. Jackson et Turrill in Kew Bull., 1939: 475 (1939); Davis, Fl. Turkey, 3: 406 (1970); Osorio-Tafall et Seraphim, List Vasc. Plants Cyprus, 59 (1973).

TYPE: " *in* Anglia ".

Small prostrate annual; stems much branched, glabrous, often stained purple, 2–10 cm. long; stipules connate at base, oblong 4–10 mm. long, 1–2 mm. wide, membranous, obscurely veined, apex filiform; leaves alternate, glabrous or almost so; petioles to about 3 cm. long, usually short, distinctly channelled above; leaflets obovate, 3–12 mm. long, 2·5–7 mm. wide, venation distinct, veins down-curved, excurrent into short sharp marginal teeth, apex rounded or rarely subacute; inflorescences axillary, globose, 6–10 mm. diam., sessile, flowers densely crowded; stipules of floral leaves enlarged, broadly ovate, about 3 mm. wide, bract-like, membranous; bracts narrowly ovate or subulate, membranous, minute, about 0·6 mm. long; flowers sessile or subsessile; calyx campanulate or cylindrical, about 4 mm. long, 1·5 mm. wide, calyx-tube about 2·5 mm. long, whitish, glabrous, 10-nerved, calyx-teeth ovate-acuminate, about 1·5 mm. long, subequal, strongly recurved, green, 3-nerved with a narrow membranous margin, glabrous or thinly hairy; petals creamy white or pink-tinged, limb of standard oblong-ovate, about 2·5 mm. long, 2 mm. wide, infolded longitudinally, apex blunt, sometimes obscurely denticulate, base expanding into a very wide gibbous claw, about 2–2·5 mm. long and rather wider than the limb, limb of wings narrowly oblong, about 1·5 mm. long, 0·5 mm. wide, apex rounded, base shortly auriculate, claw 2–2·5 mm. long, limb of keel oblong, 1·4 mm. long, 0·6 mm. wide, apex blunt, base narrowed rather abruptly to claw about 2 mm. long; stamens about 3 mm. long, filaments free for about 0·5 mm. from apex; anthers minute, shortly oblong, about 0·2 mm. long and almost as wide; ovary oblong, about 1 mm. long, 0·4 mm. wide, glabrous, 2-ovulate, style about 2 mm. long, tapering from base to apex, stigma terminal, capitate. Pod shortly oblong, about 1·8 mm. long, 1·2 mm. wide, membranous, 1–2-seeded; seeds suborbicular or very shortly oblong, about 1 mm. diam.; testa minutely verruculose, bright yellow-brown.

HAB.: On stony slopes, by roadsides, on gravelly ground or in fallow fields; 2,500–4,500 ft. alt.; fl. April–June.

DISTR.: Division 2. S. England, S.W. Europe and widespread in the Mediterranean region eastwards to Iran; Atlantic Islands.

2. Platres, 1938, *Kennedy* 1279 ! 1281 ! Prodhromos, 1961, *D. P. Young* 7252 ! Stavros tis Psokas, 1962, *Meikle* 2762 !

28. T. suffocatum *L.*, Mantissa Altera, 276 (1771); Boiss., Fl. Orient., 2: 142 (1872); Post, Fl. Pal., ed. 2, 1: 347 (1932); Hossain in Notes Roy. Bot. Gard. Edinb., 23: 470 (1961); Davis, Fl. Turkey, 3: 406 (1970); Zohary, Fl. Palaest., 2: 167, t. 239 (1972); Osorio-Tafall et Seraphim, List Vasc. Plants Cyprus, 59 (1973).

TYPE: " *in* Siciliae *arenosis* ".

Very small tufted annual, usually less than 3 cm. high; stems often not developed, mostly subterranean, up to about 2 cm. long in exceptionally robust specimens, glabrous or thinly hairy; stipules narrowly oblong, 3–4 mm. long, 1 mm. wide, membranous, ± connate towards base, apex shortly filiform-caudate; leaves alternate; petioles 2–3 cm. long, glabrous, shallowly channelled or flattish above; leaflets obcordate or obcuneate 3–6 mm. long, 2–5 mm. wide, glabrous on both surfaces, veins rather obscure, produced into short marginal teeth, apex usually emarginate with a wide sinus, base cuneate; inflorescences ± globose, 4–5 mm. diam., densely clustered close to the ground at the base of the petioles; flowers sessile or subsessile; bracts minute, narrowly ovate, about 0·5 mm. long, 0·2 mm. wide, membranous; calyx cylindrical, subglabrous or thinly hairy, about 4 mm. long, 1 mm. wide, calyx-tube 2·5 mm. long, membranous, 10-nerved, calyx-teeth erect, lanceolate-acuminate, 1·5 mm. long, 0·5 mm. wide at base, 3-nerved with a narrow membranous margin; petals whitish, almost concealed within calyx; limb of standard oblong, acute or subacute, longitudinally infolded, about 2·5 mm. long, 1·3 mm. wide, base tapering to a rather slender claw about 1·5–1·7 mm. long, limb of wings oblong, about 1 mm. long, 0·4 mm. wide, apex rounded, base scarcely auriculate, narrowed abruptly to a slender claw about 2 mm. long, limb of keel oblong, a little more than 1 mm. long, about 0·4 mm. wide, apex rounded, base tapering to a claw about 2 mm. long; stamens about 2·5 mm. long, filaments free for about 0·4 mm. from apex; anthers oblong, minute, about 0·2 mm. long and 0·1 mm. wide; ovary oblong, glabrous, about 1 mm. long, 0·4 mm. wide, 2-ovulate, style 1·5 mm. long, tapering from base to apex, stigma capitate. Pod oblong, about 1·5 mm. long, 1 mm. wide, membranous, 2-seeded; seeds suborbicular or irregularly oblong, about 0·8–1 mm. diam.; testa minutely verruculose, bright yellow-brown.

HAB.: Pastures, open sandy ground and threshing floors; sea-level to 2,000 ft. alt.; fl. March–April.

DISTR.: Divisions 1, 2, 5, 8. Widespread in the Mediterranean region, extending north along the coasts of Europe to Britain and south to the Atlantic Islands.

1. Dhrousha, 1957, *Merton* 3026 !
2. Near Asinou Church, 1974, *Meikle* 4019 !
5. South of Xeri, 1959, *P. H. Oswald* 118 !
8. Platanisso, 1950, *Chapman* 258 !

NOTES: Probably widespread in the lowlands, but very easily overlooked on account of its small size and inconspicuous inflorescences.

SECT. 6. **Chronosemium** *Ser*. Annuals or perennials; inflorescence capitate, often ovoid; flowers shortly pedicellate, reflexed in fruit; bracts reduced to minute, red, glandular hairs; calyx ± campanulate, not accrescent, 5-nerved, throat glabrous, not thickened, calyx-teeth unequal, the 2 adaxial distinctly shorter than the 3 abaxial; petals yellow, mauve or pink, persistent and becoming brown and scarious in fruit, standard often free, with an orbicular or spathulate limb. Pod stipitate, usually exceeding the calyx, 1-seeded.

[T. SPECIOSUM *Willd.*, Sp. Plant., 2: 1382 (1798); Sibth. et Sm., Fl. Graec. Prodr., 2: 101 (1813), Fl. Graec., 8: 39 (1833); Boiss., Fl. Orient., 2: 151 (1872); Holmboe, Veg. Cypr., 107 (1914); Post, Fl. Pal., ed. 2, 1: 350 (1932); Davis, Fl. Turkey, 3: 398 (1970).

An erect or diffuse annual; stems usually much branched, 10–25 (–50) cm. long; leaves thinly clothed with long hairs, leaflets oblong, 10–25 mm. long, 5–9 mm. wide, apex truncate or emarginate, margins serrulate, terminal leaflet distinctly petiolulate; inflorescences axillary, globose or shortly ovoid, conspicuous, rather lax, 1·5–2 cm. diam.; peduncles 2–5 cm. long, much exceeding leaves, adpressed-pilose; pedicels longer than calyx-tube; flowers numerous, 7–10 mm. long, petals at first violet, soon becoming brown and scarious, standard broad, flattened, with a suborbicular or broad ovoid limb, 5–7 mm. wide, completely covering wings and keel in reflexed flower. Seed oblong-ellipsoid, about 1·5 mm. long; testa smooth, light brown.

Recorded from Cyprus by Sibthorp, but not seen since, and most probably an error in localization; *T. speciosum* is too conspicuous to be easily overlooked. It has a wide distribution from Italy eastwards to Iran, and is locally abundant in many parts of the eastern Mediterranean. In the circumstances it is just possible that the species may be refound in Cyprus, for which reason the above description is included in the Flora.]

29. T. boissieri *Guss. ex Soyer-Willemet et Godron* in Mém. Soc. Roy. Sci. Nancy, 1846: 220 (1847); Boiss., Fl. Orient., 2: 152 (1872), Suppl., 169 (1888); Holmboe, Veg. Cypr., 107 (1914); Post, Fl. Pal., ed. 2, 1: 350 (1932); Davis, Fl Turkey, 3: 400 (1970); I. Gruenberg-Fertig et W. Stearn in Israel Journ. Bot., 21: 5 (1972); Zohary, Fl. Palaest., 2: 165, t. 236 (1972); Osorio-Tafall et Seraphim, List Vasc. Plants Cyprus, 59 (1973).

TYPE: Greece; " in rupestribus calidis Argolidis et Atticae ", 1842, *Boissier* (? NAP, G).

Erect or diffuse annual; stems 8–30 cm. long, usually much branched, subglabrous or thinly clothed with long, spreading hairs; stipules narrowly oblong- or lanceolate-acuminate, very shortly connate at base, glabrous or with long-ciliate, obscurely toothed or subentire margins; leaves alternate; petioles slender, 8–15 mm. long, slightly channelled above; leaflets oblong-obovate, 8–20 mm. long, 3–12 mm. wide, subglabrous or thinly long-pilose, apex truncate or emarginate, margins minutely but distinctly serrulate, median leaflet conspicuously petiolulate; inflorescences axillary, globose, conspicuous, about 1·5 cm. diam., peduncles ± erect, 1·5–3 (–4) cm. long, generally exceeding the subtending leaves, thinly or rather densely clothed with long, spreading hairs; pedicel about 1·5 mm. long, glabrous or subglabrous; calyx about 3·5 mm. long, 2 mm. wide, glabrous or subglabrous, calyx-tube about 1 mm. long, calyx-teeth linear-subulate, thinly pilose, the 2 adaxial about 1 mm. long, the 3 abaxial about 2·5 mm. long; petals shortly connate at base, pale or bright yellow turning brown with age, standard ovate-oblong, about 8 mm. long, 6 mm. wide, somewhat keeled, apex ± rounded, margins shortly erose-ciliate, base narrowing gradually to a wide, short claw, about 2 mm. long, limb of wings oblong, minutely erose, about 4·5 mm. long, 2·5 mm. wide, apex rounded, base strongly auriculate, auricle narrowly oblong about 1·8 mm. long, about as long as the slender claw, with a conspicuous rounded sinus between claw and auricle, limb of keel oblong, about 4·5 mm. long, 2·3 mm. wide, apex acute, base obscurely auriculate, narrowed almost at a right angle to a claw about 2·5 mm. long; stamens 5·5 mm. long, filaments free for about 2·2 mm. from apex; anthers minute, oblong, about 0·4 mm. long, 0·3 mm. wide; ovary ovoid, compressed, glabrous, about 1–1·2 mm. long, 0·7 mm. wide, 2-ovulate, stipe about 2–2·5 mm. long, style 1·5 mm. long, sharply upcurved at apex, stigma terminal, capitate. Pod oblong, about 2·5 mm. long, 1·5 mm. wide; seed oblong, somewhat compressed, 1·5 mm. long, 1·3 mm. wide; testa smooth, yellowish.

HAB.: Rocky mountainsides, or in thin Pine forest, occasionally in olive groves, carob plantations or waste ground at low altitudes, generally (but not always) on calcareous formations; sea-level to 2,500 ft. alt.; fl. March–May.

DISTR.: Divisions 2, 7, common in Division 7. Also Greece, Turkey, Rhodes, Palestine and Iraq.

2. Stavros tis Psokas, 1962, *Meikle* 2746 !

7. Common; *Sintenis & Rigo* 418b ! 818b ! *Kennedy* 1290 ! *Davis* 2490 ! 2818 ! 2879 ! 2976 ! *Casey* 635 ! 1007 ! 1685 ! *G. E. Atherton* 1132 ! *Merton* 2027 ! *Casey & Meikle* in *Meikle* 2425 !

30. T. campestre *Schreb.* in Sturm, Deutschl. Fl., fasc. 16: 13 (1804); Post, Fl. Pal., ed. 2, 1: 350 (1932); Rechinger f. in Arkiv för Bot., ser. 2, 1: 425 (1950); Davis, Fl. Turkey, 3: 404 (1970); Zohary, Fl. Palaest., 2: 165, t. 235 (1972); Osorio-Tafall et Seraphim, List Vasc. Plants Cyprus, 59 (1973).

T. glaucescens Hausskn. in Mitt. Thür. Bot. Ver. N. F., 5: 79 (1893).

[*T. procumbens* (L. pro parte) Sibth. et Sm., Fl. Graec. Prodr., 2: 102 (1813); Unger et Kotschy, Die Insel Cypern, 379 (1865); Holmboe, Veg. Cypr., 107 (1914); Lindberg f., Iter Cypr., 20 (1946); Hossain in Notes Roy. Bot. Gard. Edinb., 23: 479 (1961).]

[*T. agrarium* (non L.) Boiss., Fl. Orient., 2: 153 (1872); H. Stuart Thompson in Journ. Bot., 44: 306 (1906); Holmboe, Veg. Cypr., 107 (1914).]

Prostrate, sprawling or ascending, occasionally erect annual; stems usually much branched, especially towards base, rarely simple, (3–) 5–20 (–30) cm. high, greenish or stained purple, rather thinly clothed with white adpressed hairs; leaves alternate; stipules oblong- or ovate-acuminate, 1·5–7 mm. long, 0·5–4 mm. wide, glabrous or sparsely pilose, especially along the entire or obscurely sinuate margins; petiole slender, rather short, usually less than 1 cm. long, subglabrous or thinly pilose, scarcely channelled above; leaflets variable, broadly or narrowly obovate or oblong, 3–18 mm. long, 1·5–10 mm. wide, glabrous or thinly pilose on both surfaces, sometimes glaucescent or tinged purple, apex rounded, truncate or slightly emarginate, base cuneate or rounded, terminal leaflet with a petiolule up to 5 mm. long; inflorescences axillary, globose or ovoid, 4–17 mm. long, 7–12 mm. wide; peduncle slender, to about 25 mm. long, or sometimes almost wanting, shorter than, equalling or occasionally longer than subtending leaf; flowers crowded, reflexed or spreading after anthesis; pedicels usually very short, less than 1 mm. long, or flowers subsessile; calyx membranous, 1–2 mm. long, 0·6–1 mm. wide, glabrous or with the calyx-teeth tipped with a few long hairs, calyx-tube about 0·5–1 mm. long, calyx-teeth unequal, the 2 adaxial deltoid or shortly linear, acute or obtuse, 0·3–0·5 mm. long, the 3 abaxial linear, obtuse, 0·5–1 mm. long; petals at first yellow or violet-mauve, becoming brown and scarious after anthesis, standard suborbicular, flattish or concave, rarely ± keeled, 3·5–5 mm. long, 2·5–4 mm. wide, apex rounded or shortly emarginate, margins distinctly erose-dentate especially towards base, claw very short and wide, generally less than 1·5 mm. long, limb of wings oblong, 2·5–2·8 mm. long, 1·3 mm. wide towards the expanded, rounded apex, margins minutely and obscurely erose, base strongly auriculate, auricle acute, about 0·5 mm. long, claw slender, about 1 mm. long, sinus between claw and auricle rounded or rather rectangular, limb of keel oblong, about 2 mm. long, 1 mm. wide, base slightly auriculate, narrowed abruptly to a claw about 0·8 mm. long, stamens 2–2·5 mm. long, filaments free for about 0·8–1 mm. from apex; anthers minute, suborbicular, about 0·2 mm. diam.; ovary stipitate, oblong, compressed, glabrous, about 0·5–0·8 mm. long, 2-ovulate, style 0·5–0·6 mm. long, sharply upcurved at apex, stigma terminal, capitate, stipe about 1 mm. long. Pod oblong-ellipsoid, compressed, membranous, about 1·3 mm. long, 1 mm. wide; seed oblong, about 1–1·2 mm. long, 0·6–0·8 mm. wide, testa smooth, yellowish-brown.

ssp. **campestre.**

TYPE: The illustration in Sturm, Deutschl. Fl., fasc. 16: 13 (1804).

Petals yellow at anthesis; stems up to 30 cm. long; leaves bright green or glaucescent; peduncle usually equalling or shorter than the subtending leaf; inflorescences becoming ovoid, 1–1·7 cm. long; abaxial calyx-teeth about 1 mm. long.

HAB.: Gardens, cultivated fields, roadsides, waste ground, pastures and grassy hillsides; sea-level to 4,000 ft. alt.; fl. Febr.–June.

DISTR.: Divisions 1–8, common in Divisions 2, 4, 7. Throughout Europe and the Mediterranean region eastwards to Iran; naturalized in many parts of the world.

1. Cape Arnauti, 1913, *Haradjian* 781.
2. Common; *Kotschy*; *Holmboe* 899; *Lindberg f.* s.n. ! *Davis* 3106 !
3. Kakomallis above Louvaras, 1962, *Meikle* 2082 !
4. Common; *Kotschy*; *Sintenis & Rigo*; *Holmboe* 47; 53b; 258; *Chapman* 1 ! 17 ! *Merton* 655 ! 1982 !
5. Laxia, 1932, *Syngrassides* 73 ! Pera Khorio, 1936, *Syngrassides* 1139 !
6. Nicosia, 1901, *A. G. & M. E. Lascelles* s.n. ! Ayia Irini, 1941, *Davis* 2546 !
7. Common; *Sintenis & Rigo* 418 ! *Syngrassides* 183 ! *Casey* 321 ! 1260 ! 1712 ! *Chapman* 59 ! *G. E. Atherton* 1069 ! 1092 !
8. Cape Andreas, 1880, *Sintenis & Rigo*; Rizokarpaso, 1912, *Haradjian* 206; Gastria, 1941, *Davis* 2429 !

NOTES: Varies greatly in habit, leaf and inflorescence; the specimens from Kakomallis (*Meikle* 2082) are uniformly unbranched and erect, with very small hairy leaflets; the flowers are, however, indistinguishable from those of typical *campestre*.

ssp. **paphium** *Meikle* in Hooker's Icon. Plant., 37: t. 3652 (1969).

TYPE: Cyprus; summit of Khorteri (Paphos Forest), above Stavros tis Psokas, 4,000 ft. alt., April 29, 1962, *Meikle* 2711 (K !).

Petals at first violet-mauve, later brownish-yellow; stems less than 12 cm. high, usually 5–8 cm. high, purplish; leaves stained purplish; peduncle 15–20 mm. long, much exceeding subtending leaf; inflorescences becoming globose, less than 1 cm. diam., abaxial calyx-teeth 0·5 mm. long.

HAB.: Dry, grassy mountain summit; 4,000 ft. alt.; fl. April–May.

DISTR.: Division 2, rare. Probably endemic.

2. Summit of Khorteri above Stavros tis Psokas, 1962, *Meikle* 2711 !

NOTES: Allied to *Trifolium procumbens* var. *erythranthum* Griseb. (Spec. Fl. Rum. Bith., 1: 36; 1843) and *T. lagrangei* Boiss., (which are scarcely inter-distinguishable) but smaller in all its parts, with narrow, cuneate, purple-tinged leaflets and conspicuously long-pedunculate inflorescences. The subspecies occurred uniformly, and in some abundance, on the summit of Khorteri, and may be found on other peaks in the Paphos Forest area.

31. T. dubium *Sibth.*, Fl. Oxon., 231 (1794); Post, Fl. Pal., ed. 2, 1: 351 (1932); Davis, Fl. Turkey, 3: 405 (1970); Osorio-Tafall et Seraphim, List Vasc. Plants Cyprus, 59 (1973).

[*T. procumbens* (L. pro parte) Boiss., Fl. Orient., 2: 154 (1872).]
[*T. filiforme* (non L.) Lindberg f., Iter Cypr., 20 (1946).]

TYPE: The illustration in Curtis, Flora Londinensis, t. 307 !

Prostrate or decumbent annual, much resembling *T. campestre* Schreb. (no. 30 *supra*); stems much branched, 3–30 (–45) cm. long, subglabrous or thinly subadpressed-pilose; stipules narrowly ovate-acuminate, 2–7 mm. long, 1–3 mm. wide, subglabrous, entire or obscurely denticulate; leaves alternate, shortly petiolate or subsessile, petiole slender, thinly pilose, generally less than 8 mm. long; leaflets narrowly obovate or obcuneate, 3–8 (–12) mm. long, 2–6 (–8) mm. wide, glabrous or very thinly pilose, apex rounded or truncate, sometimes emarginate, often distinctly denticulate, base cuneate, entire, median leaflet usually very shortly petiolulate; inflorescences axillary, subglobose, about 5–10 mm. diam.; flowers (4–) 5–12 (–20), reflexed and scarious after anthesis; peduncle slender, 2–20 (–35) mm. long, usually exceeding the subtending leaf, ± adpressed-pilose; pedicels very short or almost wanting; calyx narrowly campanulate, glabrous or subglabrous, 1·7–2·0 mm. long, about 1·0 mm. wide, calyx-tube membranous, 1 mm. long, calyx-teeth unequal, glabrous or tipped with hairs, 2 adaxial narrowly deltoid, blunt, about 0·5 mm. long, 3 abaxial linear, blunt, 0·7–1 mm. long; petals yellow, becoming brown with age, standard obovate, about 3·5–4 mm. long, 2–2·3 mm. wide, strongly concave, the sides infolded over the wings and keel, apex obtuse or slightly emarginate, base tapering to an indistinct keel less than 1 mm. long, limb of wings oblong, 2·5 mm. long, about 1 mm. wide towards the rounded, distinctly gibbous apex, base strongly auriculate, auricle slender, subulate, about 0·5 mm. long, claw slender about 1·2 mm. long, sinus between auricle and claw rounded, limb of keel oblong, 2 mm. long, 1 mm. wide, apex acute, base shortly auriculate, claw about 1 mm. long; stamens about 2·5 mm. long, filaments free from apex for about 0·8 mm; anthers minute, orbicular, about 0·2 mm. diam.; ovary oblong, stipitate, 0·8 mm. long, 0·5 mm. wide, glabrous, 2-ovulate, stipe about 1–1·2 mm. long, style 0·5 mm. long, sharply upcurved at apex, stigma terminal, capitate. Pod oblong, 1·8 mm. long, 1·3 mm. wide, membranous, 1-seeded; seed oblong, 1–1·3 mm. long, 0·8 mm. wide; testa smooth, light brown.

HAB.: Moist pastures and roadsides, or in temporarily flooded patches in Pine forest; 3,900–4,700 ft. alt.; fl. May–June.

DISTR.: Division 2, rare. Widespread in Europe, and eastwards to N. Iran; Atlantic Islands; rare in eastern Mediterranean region.

2. Platania, 1939, *Lindberg f.* s.n.; woods at Damaskinari near Prodhromos, 1956, *Merton* 2800 ! Khandria, 1962, *Meikle* 2837 !

NOTES: Possibly introduced with fodder, but now thoroughly established in the Troödos area.

TRIBE 4. **Loteae** *DC.* Herbs or rarely subshrubs; leaves usually exstipulate, or stipules very reduced, pinnate, or sometimes digitate through suppression of the rhachis; leaflets entire, the lowermost sometimes stipuliform; inflorescences axillary or terminal (pseudoterminal), usually capitate or umbellate, or flowers sometimes solitary or few; stamens 10, monadelphous or diadelphous, free parts of filaments alternately dilated, or sometimes all dilated, anthers uniform; ovary 2–many-ovulate. Pod inarticulate or rarely articulated, 2-valved or sometimes indehiscent.

12. PHYSANTHYLLIS *Boiss.*

Voy. Bot. Esp., 2: 162 (1839).
Anthyllis L., Sp. Plant., ed. 1, 719 (1753) pro parte.

Prostrate annual; leaves pinnate; flowers axillary in small clusters; calyx-tube membranous, becoming inflated after anthesis, calyx-teeth 5, subequal, standard tapering to a slender claw, wings adherent to keel; stamens diadelphous, free parts of filaments dilated. Pod membranous or papery, concealed within the persistent, inflated calyx, constricted in the middle and 2-locular; seeds with a verruculose or tuberculate testa.

One species widespread in the Mediterranean region.

1. **P. tetraphylla** (*L.*) *Boiss.*, Voy. Bot. Esp., 2: 162 (1839); Unger et Kotschy, Die Insel Cypern, 374 (1865); Boiss., Fl. Orient., 2: 159 (1872); Post, Fl. Pal., ed. 2, 1: 352 (1932); Zohary, Fl. Palaest., 2: 85, t. 123 (1972).

Anthyllis tetraphylla L., Sp. Plant., ed. 1, 719 (1932); Holmboe, Veg. Cypr., 107 (1914); Davis, Fl. Turkey, 3: 537 (1970); Osorio-Tafall et Seraphim, List Vasc. Plants Cyprus, 62 (1973).

TYPE: " *in* Italia, Sicilia ".

Prostrate annual; stems usually branched at base, rather robust, striate or ± angular, densely clothed with long, spreading hairs; stipules wanting, leaves glaucous, imparipinnate, leaflets 3–5, or occasionally reduced to 1 in the basal leaves, terminal leaflet very much larger than the others, obovate, 10–30 mm. long, 6–25 mm. wide, adpressed-pilose on both surfaces, apex rounded, obscurely apiculate, base tapering to a short flattened rhachis 8–15 mm. long, 2–6 mm. wide, lateral leaflets alternate or subopposite, unequal, the upper obovate or elliptic, 8–15 mm. long, 5–10 mm. wide, the lower often subulate and very small; inflorescences axillary, comprising 2–5 (–8) flowers in clusters; pedicels very short or wanting; calyx at first narrowly ovoid, 12–15 mm. long, 4–5 mm. wide, slightly inflated, membranous, densely adpressed-pilose, becoming more inflated after anthesis and up to 2 cm. long, 1 cm. wide in fruit, calyx-teeth subequal, subulate, about 4 mm. long, 0·8 mm. wide at base; standard pale sulphur-yellow, 1·7–2 cm. long, limb obcordate 6–8 mm. long, 5–6 mm. wide, pubescent externally, apex rounded or slightly emarginate, base tapering to a long claw, wings deeper yellow, about 1·5–1·8 mm. long, limb oblong, 4–5 mm. long, 1·5–2·5 mm. wide, apex rounded, base with a prominent crumpled auricle and conjunctival flap, keel sometimes tipped purple, narrowly oblong, about 5 mm. long, 1·5 mm. wide, upcurved to an acute apex, base narrowing to a slender claw about 12–15 mm. long; stamens 15–17 mm. long, the vexillary free, the others joined to within 2 mm. from apex, free part of filaments strongly dilated; anthers shortly oblong or suborbicular, to 0·4 mm. long, 0·3 mm. wide, connective conspicuous, reddish; ovary stipitate, linear or narrowly ellipsoid, 2·5–3 mm. long, 0·5–1 mm. wide, pubescent, 2-ovulate, stipe 2·5–3 mm. long, style about 10–12 mm. long, filiform, upcurved and slightly thickened towards apex, glabrous, stigma minute, terminal, capitate. Pod concealed within persistent calyx, oblong, 7–8 mm. long, 3–4 mm. wide, membranous, pubescent, with a marked median constriction and articulation, generally 2-seeded (or 1-seeded by abortion); seed suborbicular, compressed, about 2·5–3 mm. diam.; testa mid-brown, strongly verruculose.

HAB.: Sunny hillsides and dry, stony fields, occasionally in open forest glades; sea-level to 1,600 ft. alt.; fl. March–May.

DISTR.: Divisions 1, 3, 4, 6–8, common in Divisions 3, 7. Throughout the Mediterranean region.

1. Paphos, 1902, *A. G. & M. E. Lascelles* s.n. ! Pelathousa near Polis, 1962, *Meikle* 2273 !
3. Common; *Holmboe* 171; 615; *Syngrassides* 487 ! 1493 ! 1508 !
4. Near Ormidhia, 1880, *Sintenis & Rigo.*
6. Nicosia, 1894, *Post.*
7. Common; *Kotschy* 476a; *Sintenis & Rigo* 428 ! *Davis* 2436 ! *Casey* 986 ! *G. E. Atherton* 1139 !
8. Near Cape Andreas, 1880, *Sintenis & Rigo*; Koma tou Yialou, 1950, *Chapman* 242 !

13. HYMENOCARPOS *Savi*

Fl. Pisana, 2: 205 (1798) nom. cons.

Circinnus Medik., Vorles. Churpf. Phys. Oekon. Ges. 2: 384 (1787).

Annual herb; leaves imparipinnate, exstipulate (or with basal leaflets resembling stipules), leaflets entire; flowers in small bracteate clusters at the apex of a distinct peduncle, bracteoles wanting; calyx with 5 subequal teeth; petals yellow or orange, free from the staminal tube, shortly clawed, keel rostrate; stamens diadelphous, free parts of stamens dilated, anthers uniform; ovary sessile, 2-ovulate, style sharply upcurved, stigma terminal. Pod flattened, circinnate, indehiscent, 2-locular; seed with a smooth testa.

One species widely distributed in the Mediterranean region.

1. H. circinnatus (*L.*) *Savi*, Fl. Pisana, 2: 205 (1798); Boiss., Fl. Orient., 2: 159 (1872); Post, Fl. Pal., ed. 2, 1: 353 (1932); Davis, Fl. Turkey, 3: 533 (1970); Zohary, Fl. Palaest., 2: 86, t. 124 (1972); Osorio-Tafall et Seraphim, List Vasc. Plants Cyprus, 62 (1973).

Medicago circinnata L., Sp. Plant., ed. 1, 778 (1753); Sibth. et Sm., Fl. Graec. Prodr., 2: 110 (1813), Fl. Graec., 8: 50, t. 768 (1833); Unger et Kotschy, Die Insel Cypern, 375 (1865).

Circinnus circinnatus (L.) O. Ktze., Rev. Gen., 6: 171 (1891); Holmboe, Veg. Cypr., 107 (1914); Lindberg f., Iter Cypr., 19 (1946).

TYPE: " *in* Hispania, Italia ".

Prostrate or sprawling annual; stems 5–30 (–40) cm. long, much branched, somewhat angular, rather densely clothed with spreading hairs; leaves irregularly 3–7-foliolate, the lower often simple, spathulate-obovate, with a flattened petiole 1–2 cm. long, the upper usually pinnate, sessile or subsessile, the 2 basal leaflets simulating stipules, terminal leaflet spathulate or narrowly obovate-oblong, 1·5–4 (–6) cm. long, 0·3–2 (–3) cm. wide, thinly or rather densely adpressed-pilose on both surfaces, apex rounded or acute, often shortly apiculate, rhachis flattened, lateral leaflets usually much smaller than terminal, elliptic or oblong-acute, the lowermost occasionally minute and subulate; inflorescences axillary; peduncle 1–2 cm. long (lengthening to 4 cm. or more in fruit), slender, ± densely clothed with spreading hairs; bracts foliaceous, oblong or obovate-elliptic 3–8 mm. long, 1–4 mm. wide; flowers sessile or subsessile, 2–5 (occasionally more) in a cluster; calyx 4–5 mm. long, densely pilose externally, calyx-tube campanulate, short, 1–1·5 mm. long and almost as wide, calyx-teeth linear-subulate, subacute, 3–3·5 mm. long; petals yellow or orange, standard suborbicular or reniform, glabrous, 4–5 mm. long and almost as wide or a little wider, narrowed abruptly at the base to a claw 1–1·5 mm. long, limb of wing oblong, 4–5 mm. long, 2–2·5 mm. wide, apex rounded, base adherent to keel, auriculate, claw slender, 1–1·5 mm. long, limb of keel oblong-falcate, 4–4·5 mm. long, 1·5–2 mm. wide, apex upcurved, bluntly rostrate, adaxial margins connate for the greater part of their length; stamens 6–7 mm. long, all except vexillary united for about 3 mm., free parts of filaments strongly dilated just below apex; anthers oblong, 0·4 mm. long, 0·2–0·3 mm. wide, connective conspicuous, reddish; ovary reniform, 1·5–2 mm. long, 1 mm. wide, pubescent or subglabrous, 2-ovulate, abaxial margin often slightly crenate, style about 5 mm. long, glabrous, upcurved at almost a right angle just below middle, upper part slightly thickened, stigma small, terminal. Pod suborbicular or reniform, circinnate, flattened, 1–1·5 cm. wide, 0·8–1 cm. from apex to base, densely pubescent or subglabrous, obscurely reticulate-veined, becoming brown when fully ripe, dorsal margin entire or variously toothed or erose; seeds 1–2 in each pod, oblong or oblong-reniform, 3–3·5 mm. long, 2–2·5 mm. wide; testa smooth, rich dark brown. Old pod generally attached to rootstock, even of mature plants. *Plate 25.*

HAB.: In fields, on waste ground or dry hillsides, occasionally on sandy seashores; sea-level to 3,750 ft. alt.; fl. Febr.–June.

DISTR.: Divisions 1–8, common in Divisions 1, 2, 4, 7. Widespread and common in the Mediterranean region and eastwards to Iran.

1. Common; *Haradjian* 698 ! *Davis* 3356 ! *Paolides & Kennedy* s.n. ! *N. Macdonald* 25 ! *Meikle* !
2. Common; *Syngrassides* 815 ! *Kennedy* 565 ! 1291 ! *Lindberg f.* s.n.
3. Akrotiri, 1927, *Rev. A. Huddle* 42 ! Between Nisou and Stavrovouni, 1933, *Syngrassides* 17 ! Limassol, 1953, *Mavromoustakis* 5 !
4. Common; *Sintenis & Rigo*; *Haradjian* 10 ! *Chapman* 3 ! *Merton* 1986 ! *N. Macdonald* 108 !
5. Salamis, 1962, *Meikle* !
6. Dhiorios area, 1962, *Meikle* !
7. Common; *Sintenis & Rigo* 424 ! *Casey* 466 ! *Chapman* 84 ! *G. E. Atherton* 1095 ! 1138 ! 1241 ! 1392 !
8. Valia, 1905, *Holmboe* 484; Apostolos Andreas, 1941, *Davis* 2321 ! Koma tou Yialou, 1950, *Chapman* 203 ! Akrades, 1962, *Meikle* !

[CYTISOPSIS DORYCNIIFOLIA *Jaub. et Spach*, a woody subshrub 10–30 cm. high, with fascicled, oblanceolate, silvery-sericeous leaflets 1–2 cm. long, and solitary or geminate, golden-yellow flowers with cylindrical pinkish or brownish calyces about 1·5 cm. long, is recorded by F. A. Rogers (no. 0669) from " near Nicosia, Cyprus, April 2, 1930 " (specimen in Kew herbarium).

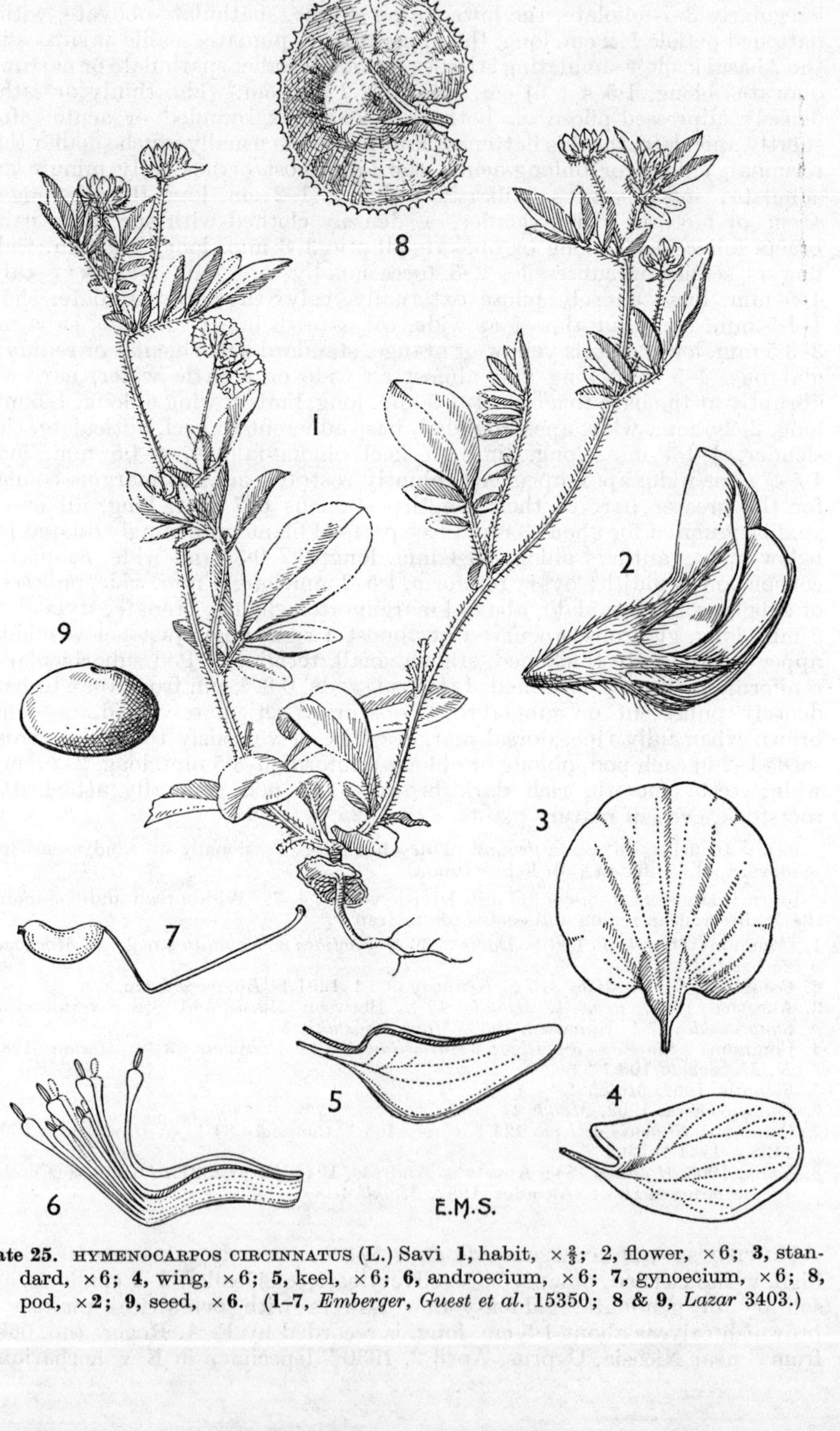

Plate 25. HYMENOCARPOS CIRCINNATUS (L.) Savi **1**, habit, $\times\frac{2}{3}$; **2**, flower, ×6; **3**, standard, ×6; **4**, wing, ×6; **5**, keel, ×6; **6**, androecium, ×6; **7**, gynoecium, ×6; **8**, pod, ×2; **9**, seed, ×6. (**1–7**, *Emberger, Guest et al.* 15350; **8 & 9**, *Lazar* 3403.)

The specimen is, however, part of a collecting from Alexandretta (Iskenderun), Turkey, and the record must be expunged. *Cytisus dorycniifolia* is locally common in parts of S. Turkey, Syria, Lebanon and Palestine, and may yet be found in Cyprus.]

14. DORYCNIUM *Mill.*

Gard. Dict. Abridg. ed. 4 (1754).

M. Rikli in Engl., Bot. Jahrb., 31: 314–404 (1901).

Herbs or subshrubs; leaves pinnate (4–) 5 (–9)-foliolate, leaflets entire, the basal sometimes stipuloid, true stipules wanting; inflorescence capitate or umbellate, axillary or pseudoterminal, often many-flowered; peduncle usually bearing 1–3 modified, bract-like leaves, true bracts minute, glandular, bracteoles wanting; calyx-tube campanulate or urceolate, calyx-teeth subequal or the abaxial somewhat longer than the adaxial; petals milky white tinged pink or purple, free from the staminal tube, standard narrowed to a short claw, wings sometimes connivent or coherent apically, keel blunt or subacute, not rostrate, often with a purple-blotched apex; stamens diadelphous, free parts of filaments alternately long and short, sometimes dilated apically, anthers uniform; ovary oblong, sessile, 2–many-ovulate, style upcurved, glabrous, stigma terminal. Pod ovoid, oblong or linear, terete or slightly compressed, 2-valved, continuous or septate within; seeds globose or compressed, estrophiolate.

About 12 species in the Mediterranean region, S. Europe and western Asia.

Standard about 15 mm. long; calyx 8–10 mm. long, densely clothed with long hairs; rhachis very short, leaflets crowded. Pod short, about 8–12 mm. long — D. HIRSUTUM p. 475

Standard 4–6 mm. long; calyx about 3 mm. long, shortly pilose:

- Rhachis well developed, lowermost leaflets broadly ovate or cordate, stipuloid. Pod linear-oblong, 10–12 mm. long - - - - - - - **1. D. rectum**
- Rhachis very short, leaflets crowded, the lowermost oblong or obovate, not stipuloid. Pod usually less than 10 mm. long, ovoid or shortly oblong - **2. D. graecum**

D. HIRSUTUM (*L.*) *Ser.* (*Lotus hirsutus* L.) An erect, subshrubby perennial, 20–50 cm. high; stems and leaves hirsute with long spreading hairs; leaflets crowded on a very short rhachis, the 3 upper narrowly obovate, obtuse or acute, 1–2·5 cm. long, 0·5–1·3 cm. wide, the 2 lowermost oblong or ovate, not stipuloid; inflorescences capitate or umbellate, 5–12 (rarely more)-flowered; calyx narrowly cylindrical-campanulate, 8–10 mm. long, 3–3·5 mm. wide, densely long-hirsute, calyx-teeth subequal, linear-subulate, 4–5 mm. long; standard oblong, about 15 mm. long, 5 mm. wide towards apex, narrowed basally to a short claw, wings oblong, 13–14 mm. long, keel purplish, 10–11 mm. long, subacute; filaments of the longer stamens distinctly thickened and slightly excavate at apex; stigma subglobose, rather large. Pod oblong 8–12 mm. long, valves dark shining brown; seeds subglobose or reniform, 2–2·5 mm. long, 1·5–2 mm. wide; testa shining brown, sometimes blotched dark blackish-brown.

Recorded from Cyprus by Sibthorp (Fl. Graec. Prodr., 2: 105; 1813, as *Lotus hirsutus* L.), but not seen since, and possibly an error in localization. Sibthorp's record is, however, somewhat authenticated by the existence of a genuine specimen of this plant, *Sintenis & Rigo* 650 in the Kew Herbarium, misnamed *Bonjeania recta* (L.) Reichb., and labelled " ad rivos in Kythraea Majo 1880 ". A second sheet of the same number, but labelled " ad rivulos pr. Kythraeam 7/6 [1880] " is correctly identified as *Bonjeania recta* (=*Dorycnium rectum*). Here again there is a strong possibility of error: *Dorycnium rectum* is known to grow at Kythrea, and was undoubtedly collected there by Sintenis and Rigo, but as these collectors apparently "bulked" dried specimens from different localities and subsequently distributed them under one number, it is not unlikely that the misnamed specimen of *Dorycnium hirsutum* has been wrongly localized, and may not have come from Cyprus at all. The specimen came to Kew as part of the Churchill Herbarium (bequeathed in 1906), and there is a further possibility that specimens and labels were interchanged when this herbarium was being mounted (at Kew). Whatever the facts, the occurrence of *Dorycnium hirsutum* in Cyprus remains dubious, and doubts can be removed only by the re-discovery of the plant on the island. *Dorycnium hirsutum* has a wide distribution in southern Europe and the Mediterranean region, and, from the phytogeographical viewpoint, there is no reason why it should not occur in Cyprus.

1. **D. rectum** (*L.*) *Ser.* in DC., Prodr., 2: 208 (1825); Boiss., Fl. Orient., 2: 161 (1872); Post, Fl. Pal., ed. 2, 1: 354 (1932); Lindberg f., Iter Cypr., 19 (1946); Rechinger f. in Arkiv för Bot., ser. 2, 1: 425 (1950); Davis, Fl. Turkey, 3: 513 (1970); Osorio-Tafall et Seraphim, List Vasc. Plants Cyprus, 62 (1973).
Lotus rectus L., Sp. Plant., ed. 1, 775 (1753)—as *L. recta*.
Bonjeania recta (L.) Reichb., Fl. Germ. Excurs., 507 (1832); Zohary, Fl. Palaest., 2: 96, t. 143 (1972).
TYPE: "*in* G. Narbonensi, Sicilia, Calabria".

Erect herbaceous perennial, up to 1·5 m. high; stems much-branched, striate or ± angular, glabrous or thinly pilose; leaves usually 5-foliolate, rhachis 5–10 mm. long, flattened or narrowly winged, 3 apical leaflets narrowly obovate, 15–30 mm. long, 7–15 mm. wide, glabrous or subglabrous above, often thinly pilose below, apex rounded, apiculate, base narrowly cuneate, 2 basal leaflets broadly ovate or cordate, 6–15 mm. long, 5–14 mm. wide, acute, very shortly petiolulate, stipuloid; inflorescences globose, dense, many-flowered, 13–17 mm. diam., arising from the axils of the upper leaves; peduncle 5–15 mm. long at anthesis, usually rather densely villose, elongating to 6 cm. or more in fruit, generally bearing, at or near its apex, 1 or more modified 1–3-foliolate leaves; bracts minute, reddish; pedicels 2–2·5 mm. long, villose; calyx 4–5 mm. long, 2 mm. wide, calyx-tube urceolate, constricted towards apex, 1·5–2 mm. long, ± villose, calyx-teeth subequal, subulate, villose, about 2 mm. long; petals milky-white, stained purple, standard oblong, 4 mm. long, 2 mm. wide at apex, glabrous, base gradually narrowing to a very short claw, limb of wing oblong, 3–3·5 mm. long, 1–1·5 mm. wide, longitudinally plicate, apex rounded, base shortly auriculate and with a small conjunctival flap, claw slender about 1·5 mm. long, limb of keel oblong, purple, about 3 mm. long, 1 mm. wide, apex slightly upcurved, obtuse or subacute, base scarcely auriculate, claw about 1·5 mm. long; stamens to about 4 mm. long, free parts of filaments alternately 1 and 2 mm. long, not noticeably dilated at apex; anthers oblong, about 0·4 mm. long, 0·2 mm. wide; ovary narrowly oblong, glabrous, about 2·5 mm. long, 0·5 mm. wide, about 10-ovulate, style glabrous, slightly upcurved, about 2 mm. long, stigma terminal, capitate. Pod linear-oblong, 10–12 mm. long, about 2 mm. wide, straight, subterete, crowned with the persistent style; valves shining dark brown, twisting spirally on dehiscence; seeds subglobose, compressed, about 1·2 mm. diam.; testa smooth, dark brown.

HAB.: Streamsides and wet ground by springs; sea-level to 2,500 ft. alt.; fl. May–July.

DISTR.: Divisions 5, 7. Widely distributed in the northern and eastern Mediterranean region and in western N. Africa.

5. Kythrea, 1880, *Sintenis & Rigo* 650 ! also, 1936, *Syngrassides* 1282 ! and, 1939, *Lindberg f.* s.n. !

7. Armenian Monastery, 1912, *Haradjian* 382 ! One mile E. of Kyrenia, 1949, *Casey* 801 ! 1956, *Casey* 1723 !

2. D. graecum (*L.*) *Ser.* in DC., Prodr., 2: 208 (1825); Davis, Fl. Turkey, 3: 514 (1970).
Lotus graecus L., Mantissa I, 104 (1767).
Dorycnium latifolium Willd., Sp. Plant., 3: 1397 (1802); Boiss., Fl. Orient., 2: 162 (1872).
D. kotschyi Boiss. et Reut. ex Boiss., Fl. Orient., 2: 161 (1872); Post, Fl. Pal., ed. 2, 1: 354 (1932).
TYPE: " *in* Oriente: *in* Graecia, Arabia ".

Erect perennial 20–60 cm. high with a woody rootstock; stems much branched at base, thinly villose with long spreading hairs; leaves (4–) 5–7 (–9)-foliolate, rhachis very short or almost wanting, leaflets crowded, apparently digitate, obovate, 8–30 mm. long, 4–15 mm. wide, thinly pilose above and below, apex rounded, apiculate, base cuneate, the lowermost leaflets similar to the upper but smaller, not stipuloid; inflorescences globose, dense, 1–1·6 cm. diam., many-flowered; flowers sweet-scented; peduncles 2–6 cm. long, usually longer than subtending leaves, bearing 1 or more bractlike leaves, true bracts minute, reddish; pedicels very short; calyx narrowly campanulate, about 4 mm. long, 1·5 mm. wide, ± densely villose, calyx-teeth subulate, about 2 mm. long; petals whitish, standard broadly obovate, about 5–6 mm. long, 3–4 mm. wide, apex rounded or subacute, often reflexed, base narrowing abruptly to a short claw, wings cohering at apex, limb oblong, about 3 mm. long, 1·5 mm. wide, somewhat gibbous, base shortly auriculate but without a distinct conjunctival flap, claw slender 1–1·5 mm. long, limb of keel purple, about 2 mm. long, 1·5 mm. wide, apex blunt, claw about 1·5 mm. long; stamens about 3 mm. long, free parts of filaments alternately about 0·5 and 0·8 mm. long; anthers oblong about 0·4 mm. long, 0·2 mm. wide; ovary narrowly oblong, glabrous or thinly pilose dorsally, about 2·5 mm. long, 0·6 mm. wide, style upcurved, glabrous, about 1·5 mm. long, stigma terminal, capitate. Pod oblong, about 4–5 mm. long, 2–3 mm. wide, apex blunt or acute, valves dark shining brown; seeds subglobose or oblong, 1·5–1·8 mm. long, 1·5 mm. wide; testa smooth, dark fuscous-brown.

HAB.: In woodland and maquis, occasionally by streams; sea-level to over 4,500 ft. alt.; fl. April–July.

DISTR.: " In insula Cypro ad rivulos ", Aug. 1831, *Aucher-Éloy* 1149 (fr.) ! Also Bulgaria, Greece, Aegean Islands, Turkey, Caucasus and ? Syria.

NOTES: This record seems to have been overlooked by Kotschy, Boissier, Holmboe and other writers on the flora of Cyprus, nor has the species been re-found on the island.

Aucher-Éloy visited Cyprus in August 1831, and traversed much of the southern and western half of the island. In a letter to his friend Coquebert de Montbret, dated Nov. 2, 1831, he refers to his visit, and comments " . . . mais déjà la saison était trop avancée, et je ne trouvai quelques plantes que sur les cimes " (Jaubert, Relations de Voyages en Orient, 2: 733; 1843). This suggests that *D. graecum* may have been found at some considerable altitude on the Troödos range, but the two specimens (in herb. Kew) are without precise locality.

Dorycnium graecum (incl. *D. kotschyi*) is known to grow on the Amanus range and about Iskenderun, so its occurrence in Cyprus is not surprising, though one would like to see fresh, and properly localized, material of the species. The two sheets of *Aucher-Éloy* 1149 at Kew are both in a very advanced state of maturity.

15. LOTUS *L.*

Sp. Plant., ed. 1, 773 (1753).
Gen. Plant., ed. 5, 338 (1754).
Brand in Engl., Bot. Jahrb., 25: 166–232 (1898).

Annual or perennial herbs or subshrubs; leaves (4–) 5-foliolate, without tendrils; stipules minute or (usually) wanting, the basal leaflets subsessile at the base of the rhachis and often stipuloid, upper leaflets crowded at the apex of the rhachis, so that the leaves appear to be 3-foliolate; inflorescences axillary, pedunculate, often umbellate, or flowers sometimes solitary or geminate; bracts generally foliaceous, bracteoles wanting; calyx-teeth subequal or the adaxial shorter than the abaxial, so that the calyx appears to be 2-lipped; petals commonly yellow or orange, occasionally red, pink or white, mostly conspicuous, free from the staminal sheath, keel usually rostrate; stamens diadelphous, the vexillary free, the others united into a sheath, free parts of filaments alternately long and short, the longer ones often distinctly dilated near the apex, anthers uniform; ovary sessile or shortly stipitate, ovules numerous, style upcurved, glabrous, sometimes gibbous or toothed below apex, stigma terminal or oblique. Pod oblong or linear, 2-valved, sometimes tardily dehiscent, unwinged, terete, subterete or laterally compressed or torulose, often septate within; seeds subglobose or compressed, estrophiolate.

About 60 species, found in most regions of the world (except tropical Asia), but most abundant in S. Europe, the Mediterranean region and western Asia.

Pod inflated, more than 5 mm. wide, adaxial suture deeply impressed, almost dividing the pod longitudinally - - - - - - - - - - **9. L. edulis**
Pod linear, slender, less than 5 mm. wide, adaxial suture not deeply impressed:
Pods laterally flattened, distinctly torulose; diffuse annual; inflorescence 1–4 (–5)-flowered - - - - - - - - - **8. L. ornithopodioides**
Pods terete or slightly compressed, not flattened, smooth or obscurely torulose:
Perennials with woody rootstocks; inflorescence umbellate, long-pedunculate, usually 2–8 (or more)-flowered; standard 8–15 mm. long:
Leaves adpressed-sericeous, silvery; beak of keel usually stained; plant of sandy seashores - - - - - - - - - - - **1. L. cytisoides**
Leaves green, subglabrous or thinly pilose, not silvery-sericeous:
Stems thinly villose with spreading hairs; calyx densely villose; pods 15–25 mm. long - - - - - - - - - **4. L. palustris**
Stems and calyces subglabrous or sparsely adpressed-pilose; pods 15–30 mm. long:
Calyx 2-lipped, calyx-teeth unequal, the 2 lateral shorter than the others; basal leaflets stipuloid, shorter and blunter than the apical leaflets **2. L. collinus**
Calyx not 2-lipped, calyx-teeth subequal; leaflets all similar, the basal scarcely stipuloid - - - - **3. L. corniculatus var. tenuifolius**
Annuals; inflorescences 1–2 (–4)-flowered; standard generally less than 8 mm. long:
Stems and calyces clothed with long spreading hairs; peduncles slender, elongate, up to 5 cm. long - - - - - - - - - **5. L. angustissimus**
Stems and calyces clothed with short, crispate subadpressed hairs; peduncles short, generally less than 2 cm. long:
Leaflets 10–15 mm. long, 5–10 mm. wide; peduncle generally shorter than subtending leaf; pod 2–4 mm. wide - - - - - **6. L. peregrinus**
Leaflets 2–10 (–14) mm. long, 1·5–5 mm. wide; peduncle usually equalling or exceeding subtending leaf; pod generally less than 2 mm. wide **7. L. halophilus**

SECT. 1. **Lotus** Calyx-teeth subequal or unequal, or calyx rarely somewhat 2-lipped; style simple, not appendiculate or toothed. Pod oblong or linear, straight or curved, subterete or compressed, without marginal wings, usually septate within, sometimes torulose or moniliform, adaxial suture not deeply impressed.

1. L. cytisoides *L.*, Sp. Plant., ed. 1, 776 (1753); Post, Fl. Pal., ed. 2, 1: 356 (1932); C. C. Heyn et I. Herrnstadt in Kew Bull., 21: 299–309 (1967); Davis, Fl. Turkey, 3: 521 (1970); Zohary, Fl. Palaest., 2: 89, t. 127 (1972); Osorio-Tafall et Seraphim, List Vasc. Plants Cyprus, 62 (1973).

L. creticus L. var. *cytisoides* (L.) Boiss., Fl. Orient., 2: 165 (1872).

L. creticus L. var. *genuinus* Boiss., Fl. Orient., 2: 165 (1872) pro parte quoad spec. cypr.

[*L. creticus* (non L.) Sibth. et Sm., Fl. Graec. Prodr., 2: 105 (1813), Fl. Graec.,

8: 42, t. 758 (1833); Brand in Engl., Bot. Jahrb., 25: 207 (1898); Holmboe, Veg. Cypr., 108 (1914); Post, Fl. Pal., ed. 2, 1: 355 (1932).]

TYPE: The illustration (t. 1031) in Barrellier, Plant. Gall. Hisp. Ital. (1714) !

Prostrate perennial; stems woody at base, much branched, 10–30 (or more) cm. long, often flexuous, obscurely striatulate, subglabrous or ± densely adpressed-sericeous; leaflets 5, subglabrous or ± densely adpressed-sericeous, the basal (stipuloid) pair ovate, 2–10 mm. long, 1·5–6 mm. wide, the others obovate or oblanceolate, 5–15 mm. long, 2–9 mm. wide, apex rounded or apiculate, sometimes acute; rhachis generally short and flattened in sericeous-leaved forms, slender and up to 8 mm. long in glabrous or subglabrous forms; inflorescences umbellate, 2–8-flowered; peduncle much exceeding leaves, 2–8 (or more) cm. long; bracts foliaceous, 3-lobed, lobes lanceolate or oblanceolate, 3–7 mm. long, 1·5–2·5 mm. wide; calyx subcylindrical or narrowly infundibuliform, 6–8 mm. long, 2·5–3 mm. wide, subglabrous or thinly adpressed-sericeous, calyx-teeth unequal, the abaxial (basal) linear, about 3–3·5 mm. long, 1·5 mm. wide, distinctly keeled, the 2 laterals flat, oblong-acute, about 1·5–2 mm. long, 1·5 mm. wide, the 2 adaxial (upper) narrowly deltoid, about 3 mm. long, 1·5 mm. wide; petals yellow, limb of standard about 6 mm. long, 7 mm. wide, glabrous, apex truncate or shallowly (rarely deeply) emarginate, claw about 6 mm. long, widely expanded at apex, tapering to a narrow base, margins strongly inflexed, limb of wings oblong, 8–9 mm. long, about 5 mm. wide, apex rounded, base often distinctly auriculate with an inconspicuous conjunctival swelling, claw slender, about 4 mm. long, limb of keel falcate, narrow, about 7 mm. long, 2 mm. wide, apex distinctly upcurved, bluntly rostrate, generally (? always) stained dark purple, base indistinctly auriculate, claw slender, about 4 mm. long; vexillary stamen free, the others connate for about 7 mm., free parts of filaments alternately about 4 mm. and 2 mm. long, the longer filaments strongly dilated and papillose at apex, the shorter ones slightly dilated, and obscurely papillose or smooth; anthers oblong, about 0·5 mm. long, 0·3–0·4 mm. wide; ovary linear, straight, 6–8 mm. long, 0·6 mm. wide, glabrous, slightly compressed, many-ovulate, the dorsal and ventral sutures slightly thickened, style upcurved, rather straight, slender, about 5 mm. long, the adaxial face channelled towards base, stigma small, capitate. Pod linear, subterete, usually straight, 20–35 mm. long, 2–3 mm. wide, apex obtuse or acute, often tipped with the persistent style-base, valves shining brown or fuscous at maturity, twisting spirally on dehiscence; seeds subglobose, slightly compressed, 1·5–2 mm. diam.; testa pale or rich brown, smooth, hilum minute.

HAB.: Sandy ground by the sea; near sea-level; fl. March–May.

DISTR.: Divisions 3, 8, rare. Throughout the Mediterranean region.

3. Akrotiri, 1959, *Mavromoustakis* 104 ! Cape Gata, 1962, *Meikle* 2915 !
8. Cape Andreas, 1880, *Sintenis & Rigo*.

NOTES: A very variable species, approaching, at one extreme, *Lotus creticus* L., and, at the other, *L. collinus* (Boiss.) Heldr. From the former it can be distinguished by its shorter, curved keel and more slender pods, from the latter by its shorter, blunter calyx-teeth and purple-tipped keel. To date only the sericeous-leaved variant (which resembles *L. creticus*) has been found in Cyprus, but it is quite likely that the subglabrous variant, which so closely resembles *L. collinus*, has been passed over in error for this very common plant and not collected.

2. **L. collinus** (*Boiss.*) *Heldr.*, Herb. Graec. Norm. no. 1320 (1897); Rechinger f. in Arkiv för Bot., ser. 2, 1: 425 (1950); Davis, Fl. Turkey, 3: 520 (1970); Zohary, Fl. Palaest., 2: 90, t. 128 (1972); Osorio-Tafall et Seraphim, List Vasc. Plants Cyprus, 62 (1973).

L. creticus L. var. *collinus* Boiss., Fl. Orient., 2: 165 (1872).

L. cytisoides L. ssp. *collinus* (Boiss.) Murb. in Act. Univ. Lund, 33: 68 (1897).

L. judaicus Boiss. ex Bornm. in Verh. Zool.-Bot. Ges. Wien, 48: 583 (1898); Post, Fl. Pal., ed. 2, 1: 356 (1932).

L. commutatus Guss. var. *collinus* (Boiss.) Brand in Engl., Bot. Jahrb. 25: 208 (1898).

L. creticus L. ssp. *collinus* (Boiss.) Holmboe, Veg. Cypr., 108 (1914).

TYPE: Greece; " in monte Malevo Laconiae prope Platanos (rare) " 1857, *Orphanides* 586 (G, K !).

Trailing or decumbent perennial with a thick, tough, woody rootstock; stems much branched, usually about 20 cm. long, subglabrous or very thinly adpressed-pilose, obscurely striatulate; leaflets 5, subglabrous or very thinly pilose, never sericeous, the basal (stipuloid) pair ovate or broadly elliptical, 3–8 mm. long, 2·5–6 mm. wide, the others obovate, 3–13 mm. long, 2·5–10 mm. wide, apex rounded, often shortly apiculate; rhachis short (usually less than 4 mm. long) but generally distinct, narrower than in sericeous forms of *L. cytisoides*, and slightly canaliculate above; inflorescence umbellate, 1–4 (–5)-flowered; peduncle generally very long and slender, often exceeding 4 cm. in length, much exceeding the subtending leaves; bracts foliaceous, 3-lobed, lobes obovate or oblanceolate, 3–7 mm. long, 2–3·5 mm. wide, calyx narrowly campanulate or infundibuliform, 7–10 mm. long, 2·5–3 mm. wide, somewhat 2-lipped, subglabrous or thinly adpressed-pilose, calyx-teeth narrowly deltoid, unequal, all ± keeled, the abaxial 3–6 mm. long, 1·5 mm. wide, the two lateral 2–3 mm. long, 1·5 mm. wide, the two adaxial 3–6 mm. long and about 2 mm. wide; petals golden-yellow, limb of standard suborbicular, 7–8 mm. long, and often a little wider, apex shortly emarginate, claw 4–7 mm. long, tapering from a wide apex, margins often strongly incurved, limb of wings broadly oblong, 7–9 mm. long, 4–5 mm. wide, apex rounded, base shortly auriculate with an indistinct conjunctival swelling, claw 2·5–3·5 mm. long, slender, limb of keel falcate, about 7 mm. long, 2·5–3 mm. wide, apex upcurved to a slender, concolorous (rarely purple-tinged) beak, base very shortly auriculate, claw 3–4 mm. long; vexillary stamen free, the others connate for 4–6 mm., free parts of filaments alternately 2 and 4 mm. long, apical part of longer filaments strongly dilated and papillose; anthers oblong, 0·5 mm. long, 0·3 mm. wide; ovary linear, somewhat compressed, glabrous, 4–6 mm. long, 0·6 mm. wide, dorsal and ventral sutures distinctly thickened, style upcurved at a right angle, slender, 4–4·5 mm. long, canaliculate adaxially at base, stigma small, terminal, capitate. Pod linear, subterete or somewhat compressed, usually straight, 20–30 mm. long, 2–3 mm. wide, apex blunt or acute, often tipped with the persistent style-base, valves shining brown or fuscous at maturity, with distinctly thickened sutures; seeds subglobose, about 1·5 mm. diam.; testa smooth, shining brown, hilum minute.

HAB.: On dry rocky slopes, on limestone or igneous formations, usually in garigue, sometimes under Pines, not infrequent on waste ground or by roadsides; near sea-level to 3,900 (or more?) ft. alt.; fl. Febr.–May.

DISTR.: 1–4, 7, 8, very common in Division 7. S. Turkey, Syria, Palestine, N. Africa, Spain, S. France, Greece.

1. Stroumbi, 1913, *Haradjian* 760 ! Near Inia (Paphos), 1957, *Merton* 3025 !
2. Platres, ? 1905, *Michaelides*; Platres, 1938, *Kennedy* 1293 !
3. Khoulou, 1905, *Holmboe* 757.
4. Cape Greco, 1960, *N. Macdonald* 106 !
7. Very common; *Sintenis & Rigo* 441 ! *Syngrassides* 13 ! 57 ! *Lindberg f.* s.n. ! *Mavromoustakis* 20 ! *Davis* 2857 ! *Chapman* 612 ! *G. E. Atherton* 165 ! 335 ! *Casey* 1716 !
8. Near Rizokarpaso, 1912, *Haradjian* 134 ! and, 1941, *Davis* 2363 ! Apostolos Andreas, 1948, *Mavromoustakis* s.n. ! Peristerges near Kilanemos, 1962, *Meikle* 2431 !

3. **L. corniculatus** *L.*, Sp. Plant., ed. 1, 775 (1753); Unger et Kotschy, Die Insel Cypern,

380 (1865); Boiss., Fl. Orient., 2: 165 (1872); Holmboe, Veg. Cypr., 108 pro parte; Post, Fl. Pal., ed. 2, 1: 356 (1932).

Trailing, decumbent or diffuse perennial, often closely resembling *L. collinus* (Boiss.) Heldr. in general appearance; leaves and stems glabrous or ± pilose, usually somewhat glaucous; leaflets obovate, ovate, elliptical, oblanceolate or sometimes almost linear, 3–10 (–20) mm. long, 2·5–8 (–10) mm. wide, rhachis generally well developed, slightly canaliculate above; inflorescences (1–) 2–6 (–8)-flowered; peduncle erect, rather stout, up to 8 cm. long, or sometimes longer; calyx campanulate or infundibuliform, about 5–6 mm. long, 2·5–3 mm. wide, glabrous or pilose, often tinged purple, calyx-teeth subequal, narrowly deltoid or subulate, 2–3 mm. long; petals golden yellow commonly becoming greenish when dried, often stained bright red, or the standard veined red, limb of standard suborbicular, 5–8 mm. diam., apex rounded or slightly emarginate, generally apiculate, claw 5–6 mm. long, tapering from a wide apex, margins usually strongly incurved, often shortly pubescent or papillose towards base, limb of wings oblong-obovate, 8–12 mm. long, 4–6 mm. wide, apex rounded, base strongly auriculate with a distinct conjunctival swelling, claw slender, about 2·5–3·5 mm. long, limb of keel falcate, 6–9 mm. long, 3–4 mm. wide, apex upcurved to a slender concolorous beak, adaxial margins distinctly gibbous but scarcely auriculate at base, claw 2·5–3 mm. long; vexillary stamen free, the others connate for 5–7 mm. with the sheath shortly pubescent at least towards base, free parts of filaments alternately 2–4 mm. and 4–6 mm. long, the longer ones strongly dilated and papillose towards apex; anthers oblong, 0·4–0·6 mm. long, 0·3–0·4 mm. wide; ovary linear, compressed, 6–7 mm. long, 0·6–0·8 mm. wide, style upcurved almost at a right angle, 3·5–7 mm. long, slender, adaxially canaliculate at base, stigma small, capitate. Pod linear, subterete, 15–30 mm. long, 2·5–3·5 mm. wide, usually straight, apex acute or obtuse, often tipped with the persistent style-base, valves shining brown; seeds oblong-suborbicular, about 1·5 mm. long, 1·2 mm. wide; testa smooth, dark brown, sometimes obscurely marbled, hilum minute.

var. **tenuifolius** *L.*, Sp. Plant., ed. 1, 776 (1753); Davis, Fl. Turkey, 3: 526 (1970).

L. tenuis Waldst. et Kit. ex Willd., Enum. Plant. Hort. Berol., 797 (1809); Rechinger f. in Arkiv för Bot., ser. 2, 1: 425 (1950); Zohary, Fl. Palaest., 2: 90, t. 128 (1972).

L. tenuifolius (L.) Reichb., Fl. Germ. Excurs., 506 (1832); Boiss., Fl. Orient., 2: 166 (1872); Lindberg f. Iter Cypr., 19 (1946).

L. corniculatus L. ssp. *tenuifolius* (L.) Hartm., Svensk och Norsk Excurs. Fl., 103 (1846); Holmboe, Veg. Cypr., 108 (1914); Osorio-Tafall et Seraphim, List Vasc. Plants Cyprus, 62 (1973).

[? *L. diffusus* Sibth. et Sm., Fl. Graec. Prodr., 2: 104 (1813) nec Fl. Graec., 8: 41, t. 757 (1833).]

[*L. corniculatus* L. var. *alpinus* (non Boiss.) Post in Mém. Herb. Boiss., 18: 92 (1900); H. Stuart Thompson in Journ. Bot., 44: 306 (1906).]

TYPE: Europe; " Burserus C. Bauhino hanc attulit ".

Plants diffuse, slender, stems up to 1 m. long, glabrous or very sparsely pilose; leaflets all (including basal) narrowly obovate, elliptic, oblanceolate or linear, up to 2 cm. long, generally less than 0·5 cm. wide, apex often acute, rhachis generally rather short; inflorescence 1–4-flowered, flowers often rather small; calyx-teeth narrowly subulate. Otherwise indistinguishable from var. *corniculatus*.

HAB.: In fresh and saline marshes, by irrigation channels and springs; sea-level to 4,500 ft. alt.; fl. March–Sept.

DISTR.: Divisions 1, 2, 4, 6–8, common in Divisions 2, 7. *L. corniculatus* aggr. widespread in Europe, Asia, N. and N.E. Africa, introduced elsewhere; var. *tenuifolius* throughout Europe and the Mediterranean region east to Afghanistan.

1. Cape Arnauti, 1913, *Haradjian* 802 !
2. Common; *Kotschy* 687; *Post*; *Kennedy* 566 ! 1292 ! *Lindberg f.* s.n.; *Casey* 876 ! *G. E. Atherton* 551 ! 584 ! 605 !
4. Near Larnaca, 1939, *Lindberg f.* s.n.; Kouklia (Mesaoria), 1952, *Merton* 1574 !
6. Nicosia, 1927, *Rev. A. Huddle* 43 ! Syrianokhori, 1935, *Syngrassides* 618 !
7. Common; Kotschy; *Sintenis & Rigo* 642 ! *Haradjian* 349; *Syngrassides* 101 ! *G. E. Atherton* 3 ! *Meikle* 2555 !
8. Ayios Andronikos, 1862, *Kotschy* 334.

NOTES: Typical *L. corniculatus* has not been satisfactorily recorded from Cyprus, all specimens so named appearing on examination to be either var. *tenuifolius* or *Lotus collinus* (Boiss.) Heldr. *L. corniculatus* var. *tenuifolius* is tolerably homogeneous and distinct in Cyprus, but merges with *L. corniculatus* var. *corniculatus* in other areas. It is commonly found near dwellings, and often has the appearance of a relatively recent introduction.

4. L. palustris *Willd.*, Sp. Plant., 3: 1394 (1802); Holmboe, Veg. Cypr., 108 (1914); Post, Fl. Pal., ed. 2, 1: 357 (1932); Rechinger f. in Arkiv för Bot., ser. 2, 1: 425 (1950); Davis, Fl. Turkey, 3: 524 (1970); Zohary, Fl. Palaest., 2: 91, tt. 130, 131 (1972); Osorio-Tafall et Seraphim, List Vasc. Plants Cyprus, 62 (1973).

L. lamprocarpus Boiss., Diagn., 1, 9: 33 (1849); Fl. Orient., 2: 166 (1872).

L. palustris Willd. var. *villosissimus* Lindberg f., Iter Cypr., 19 (1946).

TYPE: " *in* Creta " (B).

Suberect, diffuse or decumbent perennial with a woody rootstock; stems slender, up to 100 cm. long, subterete or obscurely striatulate, much branched, thinly villose with long spreading hairs; leaves 5-foliolate, thinly or rather densely villose on both surfaces, becoming livid on drying, rhachis up to 10 mm. long, but usually rather short, distinctly canaliculate above, basal leaflets stipuloid, asymmetrically ovate, 0·5–1 cm. long, 0·3–1 cm. wide, acute, upper leaflets oblong or narrowly obovate, up to 2 cm. long, 1 cm. wide, obtuse or subacute, shortly apiculate; inflorescence 2–4 (–5)-flowered; peduncle very slender, erect, thinly villose, up to 10 cm. long, much exceeding the subtending leaves; bracts foliaceous, lanceolate or narrowly obovate, to about 1·8 cm. long and 0·8 cm. wide; calyx narrowly campanulate or infundibuliform, 6–8 mm. long, 1·5–1·8 mm. wide, densely villose, calyx-teeth subequal, narrowly subulate or almost filiform, 3·5–5 mm. long, spreading or loosely adpressed to corolla; petals yellow, limb of standard obovate or suborbicular, 6–7 mm. long and almost as wide, apex rounded or obscurely emarginate, base tapering to a short claw about 2–3 mm. long, limb of wings broadly obovate-oblong, 6–7 mm. long, 3–4 mm. wide, apex rounded, base prominently but bluntly auriculate, with a distinct conjunctival swelling, limb of keel falcate, 6–7 mm. long, 3–3·5 mm. wide, apex upcurved to a rather bent, bluntish beak, base distinctly gibbous adaxially but scarcely auriculate, claw about 2 mm. long; vexillary stamen free, the others connate for about 4–5 mm., free parts of filaments alternately 2–3 mm. and 3–4 mm. long, apex of the longer filaments strongly dilated and papillose; anthers about 0·5 mm. long, 0·3 mm. wide; ovary linear, 4·5–5 mm. long, 0·6 mm. wide, glabrous, slightly compressed, style sharply upcurved, slender, 5–6 mm. long, stigma small, capitate. Pod linear, usually straight, 15–25 mm. long, 2–2·5 mm. wide, apex subacute, valves generally rich brown and shining; seeds suborbicular, about 1·3 mm. diam.; testa smooth, yellowish or pale brown, hilum minute.

HAB.: Wet ground by springs and streams; near sea-level to 4,500 ft. alt.; fl. May–Sept.

DISTR.: Divisions 1–3, 7, common in Division 2. Greece, Crete, Aegean Islands, Turkey, Syria, Palestine.

1. Cape Arnauti, 1913, *Haradjian* 799 !
2. Common; *Holmboe* 944; *Haradjian* 1008 ! *Kennedy* 567 ! 568 ! *Lindberg f.* s.n. ! *Casey* 862 ! 1208 ! *G. E. Atherton* 521 ! 550 !
3. Episkopi, 1905, *Holmboe* 698.
7. Myrtou, 1932, *Syngrassides* 198 ! Dhavlos, on cliffs below ramp, 1957, *H. Painter* 8 !

5. L. angustissimus *L.*, Sp. Plant., ed. 1, 774 (1753); Boiss., Fl. Orient., 2: 171 (1872); Post, Fl. Pal., ed. 2, 1: 359 (1932); A. K. Jackson et Turrill in Kew Bull., 1939: 475 (1939); Davis, Fl. Turkey, 3: 523 (1970); Zohary, Fl. Palaest., 2: 91, t. 132 (1972); Osorio-Tafall et Seraphim, List Vasc. Plants Cyprus, 62 (1973).

TYPE: France; "*in* G. [Gallia] Narbonensi ".

A slender, diffuse annual; stems 15–35 cm. long, subterete, obscurely striatulate, thinly villose, usually much branched especially towards base; leaves 5-foliolate, thinly villose; rhachis 3–6 mm. long, slender, canaliculate above; basal (stipuloid) leaflets narrowly ovate or lanceolate, 5–12 mm. long, 3–6 mm. wide, upper leaflets narrowly obovate or oblanceolate, to about 12 mm. long and 6 mm. wide, apex acute, generally very shortly mucronate; inflorescences 1–2 (–3)-flowered; peduncle very slender, erecto-patent, to about 5 cm. long; bracts foliaceous, to about 8 mm. long and 4 mm. wide; calyx narrowly campanulate, villose, about 6 mm. long, 2 mm. wide, calyx-teeth subequal, subulate, about 4 mm. long; petals yellow, the standard often veined crimson; limb of standard obovate, 4–4·5 mm. long, 3·5–4 mm. wide, apex rounded, shallowly emarginate, base tapering to a claw about 2–2·5 mm. long, limb of wings oblong, about 4 mm. long, 2 mm. wide, apex rounded, base prominently but bluntly auriculate, with a distinct conjunctival swelling, claw slender about 2 mm. long, limb of keel falcate, 4·5–5 mm. long, 2·5–3 mm. wide, apex upcurved at a right angle to a short rather blunt beak, adaxial margin distinctly gibbous at base, but scarcely auriculate, claw 2 mm. long; vexillary stamen free, others connate for about 4 mm., free parts of filaments alternately about 2 mm. and 1·5 mm. long, the longer filaments distinctly dilated and papillose at the apex; anthers oblong, about 0·3 mm. long, 0·2 mm. wide; ovary linear, about 4 mm. long, 0·4 mm. wide, distinctly compressed, style slender, upcurved at a right angle, about 3 mm. long, stigma minute, capitate. Pod linear, straight or slightly curved, compressed, 20–30 mm. long, 1·5 mm. wide, apex acute, tipped with the persistent style base, valves dark brown at maturity; seeds globose or shortly ovoid, 0·8–1 mm. long, 0·8 mm. wide; testa smooth, greenish or pale brown, hilum minute.

HAB.: Damp ground on mountainside; about 4,500 ft. alt.; fl. ? May–Sept.

DISTR.: Division 2, rare. Widespread in southern Europe, the Mediterranean region and eastwards to the Altai; Atlantic Islands.

2. Near Prodhromos, 1880, *Sintenis & Rigo* 815 !

6. L. peregrinus *L.*, Sp. Plant., ed. 1, 774 (1753); Unger et Kotschy, Die Insel Cypern, 380 (1865); Boiss., Fl. Orient., 2: 172 (1872); Holmboe, Veg. Cypr., 107 (1914); Post, Fl. Pal., ed. 2, 1: 359 (1932); Lindberg f. Iter Cypr., 19 (1946) pro parte; Davis, Fl. Turkey, 3: 522 (1970); Zohary, Fl. Palaest., 2: 92, tt. 133, 133a (1972); Osorio-Tafall et Seraphim, List Vasc. Plants Cyprus, 62 (1973).

[? *L. diffusus* (non Sm.) Sibth. et Sm., Fl. Graec., 8: t. 757 (1833) excl. descr. et synon.]

TYPE: "*in* Europa *australi* ".

Sprawling annual; stems usually much-branched, rarely simple in starved specimens, villose with whitish, crispate hairs; leaves 5-foliolate, crispate-villose, rhachis short, usually under 5 mm. long, distinctly flattened; basal (stipuloid) leaflets ovate, one often larger than the other, 3–10 mm. long, 2·5–8 mm. wide, upper leaflets obovate, 10–15 mm. long, 5–10 mm. wide, apex rounded or subacute, base cuneate; inflorescences 1–2 (–3)-flowered; peduncle short, generally less than 1 cm. long and shorter than subtending leaves; bracts foliaceous, narrowly obovate, to about 8 mm. long, 3 mm. wide; calyx cylindrical, about 5 mm. long, 2 mm. wide, thinly white-villous, calyx-teeth unequal, narrowly deltoid, margins incurved, the abaxial about 2·5 mm. long, the two lateral about 1·5 mm. long, the two adaxial about

2 mm. long; petals pale yellow, limb of standard suborbicular, about 3–4 mm. long and 3·5–4·5 mm. wide, apex distinctly emarginate, claw 2–3 mm. long, broadly dilated at apex, tapering to a rather broad base, margins somewhat incurved, limb of wings oblong, 4·5 mm. long, 2·5 mm. wide, apex rounded or subtruncate, base shortly and rather bluntly auriculate, claw slender, 2–3 mm. long, limb of keel falcate, about 4·5 mm. long, 1·5 mm. wide, gently upcurved to a subacute beak, adaxial margins distinctly gibbous near base, obscurely auriculate, claw slender, about 3 mm. long; vexillary stamen free, others connate for 4–5 mm., free parts of filaments unequal rarely subequal, alternately 1·5 and 2·5 mm. long, longer filaments dilated and papillose at apex; anthers oblong, about 0·5 mm. long, 0·3 mm. wide; ovary linear, somewhat compressed, about 5 mm. long, 0·5 mm. wide, style upcurved at a right angle, 2·5–3 mm. long, slender, stigma small, capitate. Pod linear, usually straight or slightly upcurved towards apex, 20–45 (–50) mm. long, 2–4 mm. wide, apex subacute, tipped with the persistent style-base, valves pale or mid-brown, obscurely torulose; seeds subglobose, somewhat compressed, 1·5–2 mm diam.; testa glossy, smooth, pale yellowish- or greenish-brown, hilum minute.

HAB.: Dry rocky slopes amongst garigue, on sand-dunes by the sea, occasionally by roadsides, in cultivated ground, or in shingly river-beds; sea-level to 2,500 ft. alt.; fl. Dec.–June.

DISTR.: Divisions 1–8, common in Divisions 4, 7. Greece, Aegean Islands, Turkey, Syria, Palestine, Egypt.

1. Karavopetres near Ayios Yeoryios (Akamas), 1962, *Meikle* 2059 !
2. Karavostasi-Lefka, 1935, *Syngrassides* 813 ! Valley above Pyrgos, 1957, *Merton* 2872 !
3. Near Amathus, 1947, *Mavromoustakis* s.n.
4. Common; *Kotschy* 129; *Holmboe* 87; *Haradjian* 34 ! *Syngrassides* 1110 ! *Lindberg f.* s.n.; *Merton* 2678 !
5. Salamis, Famagusta, 1959, *P. H. Oswald* 57 ! Salamis, 1962, *Meikle* 2588 !
6. Near Orga, 1956, *M. E. D. Poore* 2635 ! English School, Nicosia, 1959, *P. H. Oswald* 31 !
7. Very common; *Sintenis & Rigo* 446 ! 446A ! *Syngrassides* 51 ! 910 ! *Kennedy* 1294 ! *Lindberg f.* s.n.; *Casey* 388 ! 504 ! 726 ! *Chapman* 277 ! *G. E. Atherton* 915 ! 942 !
8. Koma tou Yialou, 1935, *Syngrassides* 622 ! and, 1950, *Chapman* 232 ! Ayios Philon near Rizokarpaso, 1941, *Davis* 2211 ! Apostolos Andreas, 1941, *Davis* 2325 !

NOTES: The illustration of *L. diffusus* in Fl. Graec. (8: t. 757) most probably represents this species, but the accompanying text and synonymy is confused, and includes references to several totally distinct entities.

7. **L. halophilus** *Boiss. et Spruner* in Boiss., Diagn., 1, 2: 37 (1843); Davis, Fl. Turkey, 3: 522 (1970); Zohary, Fl. Palaest., 2: 92, tt. 134, 134a (1972); Osorio-Tafall et Seraphim, List Vasc. Plants Cyprus, 62 (1973).

L. villosus Forssk., Fl. Aegypt. Arab., 71 (1775); Holmboe, Veg. Cypr., 108 (1914); Rechinger f. in Arkiv för Bot., ser. 2, 1: 425 (1950) non Burm. f. (1768) nom. illeg.

L. pusillus Viv., Fl. Lib. Spec., 47, t. 17, fig. 13 (1824); Boiss., Fl. Orient., 2: 173 (1872); Sintenis in Oesterr. Bot. Zeitschr., 32: 398 (1882); Post, Fl. Pal., ed. 2, 1: 359 (1932) non Medik. (1783) nom illeg.

L. perpusillus Sintenis in Oesterr. Bot. Zeitschr., 32: 364 (1882) nomen.

TYPE: Greece; " in arenis maritimis *Atticae* ad *Phalerum* ubi in consortio *Loti peregrini* crescit ", May, 1842, *Spruner* (G, K !).

Prostrate or decumbent annual; stems and leaves ± densely adpressed silvery-sericeous; stems much branched, 4–15 (–25) cm. long, subterete or obscurely sulcate; leaves 5-foliolate, rhachis usually very short, under 4 mm. long, sometimes almost wanting, basal (stipuloid) leaflets broadly ovate, 2–6 (–8) mm. long, 1·5–4 mm. wide, upper leaflets narrowly obovate or oblanceolate, 2–10 (–14) mm. long, 1·5–5 mm. wide, apex subacute, base cuneate; inflorescences 1–2 (–4)-flowered, flowers small; peduncle 6–15

(–30) mm. long, usually equalling or exceeding the subtending leaves; bracts foliaceous, to about 8 mm. long and 3 mm. wide; calyx narrowly campanulate or subcylindrical, 5–6 mm. long, 1·5–2 mm. wide, densely villose, calyx-teeth narrowly deltoid, unequal, but not markedly so, the abaxial tooth 2–2·5 mm. long, the two laterals about 1·5 mm. long, the two adaxial about 2 mm. long; petals yellow, limb of standard suborbicular, about 3 mm. diam., apex rounded or shortly emarginate, claw about 3 mm. long, strongly dilated at apex, tapering to base, limb of wings obovate-oblong, 4–5 mm. long, 2·5–3 mm wide, apex rounded, base with a prominent blunt auricle and slight conjunctival swelling, claw slender, 2–3 mm. long, limb of keel falcate, 3·5–4 mm. long, 1·5–2 mm. wide, gently upcurved to a rather slender, apiculate apex, adaxial margins gibbous but scarcely auriculate at base, claw 2·5–3·5 mm. long; vexillary stamen free, the others connate for 4–5 mm., free parts of filaments alternately 1 mm. and 1·5 mm. long, longer filaments very strongly dilated and papillose at apex; anthers oblong, 0·4 mm. long, 0·3 mm. wide; ovary linear, a little compressed, 4–5 mm. long, 0·4–0·5 mm. wide, style upcurved at a right angle, 2·5–3 mm. long, stigma capitate. Pod linear, straight or slightly upcurved, 20–30 mm. long, usually less than 2 mm. wide, apex acute, often tipped with the persistent style-base, valves dark brown or fuscous at maturity, generally somewhat torulose; seeds subglobose, a little compressed, about 1 mm. diam.; testa yellowish, smooth, hilum minute.

HAB.: Sandy seashores, rarely amongst garigue inland; sea-level to 550 ft. alt.; fl. Febr.–May.

DISTR.: Divisions 4–8, common in Division 8. Eastern Mediterranean region, from S. Italy and Sicily, eastwards to Iran; North Africa from Morocco to Egypt.

4. Larnaca, 1880, *Sintenis & Rigo*; Larnaca Salt Lake, 1939, *Lindberg f.*; near Dhekelia, 1956, *Merton* 2677 !
5. Salamis, 1905, *Holmboe* 439; Athalassa, 1950, *Chapman* 635 !
6. Syrianokhori, 1935, *Syngrassides* 786 !
7. Near Kyrenia, 1880, *Sintenis & Rigo.*
8. Common; *Sintenis & Rigo* 442 ! *Davis* 2214 ! 2406 !

8. **L. ornithopodioides** *L.*, Sp. Plant., ed. 1, 775 (1753); Boiss., Fl. Orient., 2: 173 (1872); Holmboe, Veg. Cypr., 107 (1914); Post, Fl. Pal., ed. 2, 1: 360 (1932); Rechinger f. in Arkiv för Bot., ser. 2, 1: 425 (1950); Davis, Fl. Turkey, 3: 523 (1970); Zohary, Fl. Palaest., 2: 93, t. 135 (1972); Osorio-Tafall et Seraphim, List Vasc. Plants Cyprus, 62 (1973).

[*Lotus diffusus* (non Sm.) Sibth. et Sm., Fl. Graec., 8: 41, excl. t. 757 (1833); Unger et Kotschy, Die Insel Cypern, 380 (1865).]

TYPE: " *in* Sicilia ".

Diffuse annual, 10–20 (–30) cm. high; stems usually much branched, especially towards base, slender, thinly villose or pubescent; leaves 5-foliolate, thinly villose; rhachis rather short, but generally conspicuous, less than 1 cm. long, distinctly flattened; basal (stipuloid) leaflets broadly ovate, often unequal, 3–9 mm. long, 3–8 mm. wide, upper leaflets obovate, 7–18 mm. long, 3–15 mm. wide, apex obtuse or subacute, base cuneate; inflorescences 1–4 (–5)-flowered; peduncle erecto-patent, 1–4 (–5) cm. long, somewhat exceeding the subtending leaves; bracts foliaceous, usually less than 1 cm. long and 0·5 cm. wide; calyx narrowly campanulate, 5–8 mm. long, 1·5–2·5 mm. wide, villose, calyx-teeth very unequal, the abaxial about 2·5–3·5 mm. long, subulate with strongly incurved margins, the two laterals oblong, blunt, about 1 mm. long, the two adaxial subulate about 2 mm. long; petals yellow, limb of standard obovate, 3–5 mm. long, 2·5–3·5 mm. wide, apex rounded or emarginate, claw 3–5 mm. long, tapering from a distinctly dilated apex, limb of wings oblong, 3·5–6 mm. long, 1·5–3·5 mm. wide, apex rounded, base shortly and bluntly auriculate, with a strong conjunctival swelling, claw slender 1·5–2·5 mm. long, limb of keel narrowly falcate, 3–4 mm. long, 1–1·5 mm. wide, gently upcurved to a subacute beak, adaxial margins distinctly gibbous towards base, very shortly auriculate, claw 2–3 mm. long; vexillary stamen free, minutely pubescent towards base of filament, other stamens connate into a sheath 3–5 mm. long, free parts of filaments alternately 1–1·5 and 2–2·5 mm. long, the longer ones distinctly dilated at apex; anthers oblong, 0·3–0·4 mm. long, 0·2–0·3 mm. wide; ovary linear, 3·5–5·5 mm. long, about 0·5 mm. wide, distinctly compressed, dorsal suture canaliculate, style 2·5–3·5 mm. long, upcurved at a right angle, stigma capitate. Pod linear, gently upcurved, 35–45 mm. long, about 3·5 mm. wide, flattened, conspicuously torulose with marked constrictions between the seeds, apex obtuse or shortly acute, valves pale or dark brown at maturity; seeds orbicular-lenticular, 1·5–1·8 mm. diam., faces with a distinct central umbo; testa shining olive- or dark-brown, smooth.

HAB.: Moist places by springs and streams on limestone mountainsides; c. 500–2,000 ft. alt.; fl. Febr.–May.

DISTR.: Divisions 1, 7, 8. Widespread in southern Europe and the Mediterranean region eastwards to Turkey; Atlantic Islands.

1. Ktima, 1913, *Haradjian* 683.
7. Kantara, 1880, *Sintenis & Rigo* 444 ! Larnaka tis Lapithou, 1936, *Syngrassides* 902 ! N. side of Sina Oros, 1967, *Merton* in ARI 788 !
8. Near Komi Kebir, 1912, *Haradjian* 305.

NOTES: Not normally a variable species; the specimens from Larnaka tis Lapithou (*Syngrassides* 902) are, however, exceptional in having small flowers (limb of standard about 3 mm. long) and the calyx heavily blotched violet. It would be interesting to know if this particular variant occurs elsewhere on the Northern Range, or in other parts of Cyprus.

SECT. 2. **Krokeria** (*Moench*) *Ser*. Calyx-teeth subequal, calyx campanulate; pod inflated, without marginal wings, almost 2-locular within, the adaxial suture very deeply impressed.

9. L. edulis *L*., Sp. Plant., ed. 1, 774 (1753); Sibth. et Sm., Fl. Graec. Prodr., 2: 103 (1813), Fl. Graec., 8: 41, t. 756 (1833); Unger et Kotschy, Die Insel Cypern, 380 (1865); Boiss., Fl. Orient., 2: 173 (1872); Holmboe, Veg. Cypr., 107 (1914); Post, Fl. Pal., ed. 2, 1: 360 (1932); Rechinger f. in Arkiv för Bot., ser. 2, 1: 425 (1950); Davis, Fl. Turkey, 3: 520 (1970); Zohary, Fl. Palaest., 2: 94, t. 140 (1972); Osorio-Tafall et Seraphim, List Vasc. Plants Cyprus, 62 (1973).
TYPE: " *in* Italia, Sicilia, Creta ".

Sprawling or prostrate annual; stems 3–30 cm. long, usually much branched, especially towards base, thinly pilose with spreading hairs; leaves 5-foliolate, rhachis 2–10 mm. long, distinctly flattened or very narrowly winged, basal (stipuloid) leaflets broadly ovate-suborbicular, 2–12 mm. long, 2–8 mm. wide, rounded or acute, upper leaflets obovate, 4–17 mm. long, 3–10 mm. wide, thinly pilose on both surfaces, apex blunt or subacute, base narrowly cuneate; inflorescence 1 (–2)-flowered, flowers rather large; peduncle up to 5 cm. long, usually much exceeding the subtending leaves, bracts foliaceous, to about 12 mm. long and 5 mm. wide; calyx 10–12 mm. long, about 8 mm. wide, rather densely pilose with spreading hairs, calyx-teeth subequal, lanceolate, 6–8 mm. long, 2 mm. wide; petals yellow often turning reddish with age, standard pencilled red, limb reniform or flabellate, about 8 mm. long, 10 mm. wide, apex very slightly emarginate, claw 7–8 mm. long, wide at apex, tapering to base, limb of wings narrowly obovate, about 11 mm. long, 5 mm. wide, apex rounded, base distinctly and subacutely auriculate with a slight conjunctival swelling, claw slender about 2 mm. long, limb of keel straightish, boat-shaped, about 10 mm. long, 4·5 mm. wide, apex shortly and subacutely rostrate, stained dark crimson, base bluntly auriculate, claw 2 mm. long; vexillary stamen free, others connate for about 6 mm., free parts of filaments alternately about 3 and 4 mm. long; anthers oblong, about 0·6 mm. long, 0·4 mm. wide; ovary shortly stipitate, linear-oblong, strongly compressed, about 4·5 mm. long, 1 mm. wide, glabrous, many-ovulate, style gently upcurved, about 7 mm. long, stigma small, capitate. Pod inflated, coriaceous, straightish or strongly falcate, 2–4 cm. long, 0·5–0·7 cm. wide, apex very shortly rostrate, ventral (adaxial) suture so strongly impressed that it almost divides the interior of the pod into 2 longitudinal loculi; seeds reniform, compressed, about 3 mm. wide, 2·5 mm. long; testa dark brown, closely verruculose; hilum minute.

HAB.: Fields, roadsides and amongst garigue on stony hillsides; sea-level to 2,000 ft. alt., usually lowland; fl. Febr.–May.

DISTR.: Divisions 1, 4, 6–8, common in Division 7. Throughout the Mediterranean region.

1. Ktima, 1913, *Haradjian* 661.
4. The Phaneromene, Larnaca, 1880, *Sintenis & Rigo*; Larnaca aerodrome, 1950, *Chapman* 20 ! Cape Greco, 1958, *N. Macdonald* 25 !
6. Nicosia, 1931, *F. A. Rogers* 0644 ! Kormakiti, 1936, *Syngrassides* 1199 ! Ayios Yeoryios church near Liveras, 1962, *Meikle* 2403 !
7. Common; *Sintenis & Rigo* 443 ! *Syngrassides* 1148 ! *Casey* 458 ! 512 ! 1121 ! *Chapman* 265 ! *G. E. Atherton* 1080 !
8. Cape Andreas, 1948, *Mavromoustakis* s.n. ! Platanisso, 1950, *Chapman* 240 !

16. TETRAGONOLOBUS *Scop.*

Fl. Carn., ed. 2, 2: 87 (1772) nom. cons.

Annual and perennial herbs, very similar to *Lotus* in habit and general appearance; leaves 5-foliolate, the 2 basal leaflets stipuloid and adnate to the rhachis along one margin, upper leaflets crowded at the apex of the rhachis, so that the leaves appear to be 3-foliolate; inflorescences axillary, flowers generally solitary or geminate, usually rather large and conspicuous; bracts foliaceous, bracteoles wanting; calyx-teeth subequal; petals yellow, orange, pink or red, keel rostrate; stamens as in *Lotus*; ovary many-ovulate, stipitate, style often gibbous near apex, stigma commonly oblique or even lateral. Pod narrowly oblong or linear, subterete, tardily dehiscent, distinctly 4-winged; seeds subglobose, estrophiolate.

Four (or five) species in Europe and the Mediterranean region.

1. T. purpureus *Moench*, Meth., 164 (1794); Unger et Kotschy, Die Insel Cypern, 380 (1865); Boiss., Fl. Orient., 2: 175 (1872); Post, Fl. Pal., ed. 2, 1: 361 (1932); Davis, Fl. Turkey, 3: 532 (1970); Osorio-Tafall et Seraphim, List Vasc. Plants Cyprus, 62 (1973).

Lotus tetragonolobus L., Sp. Plant., ed. 1, 773 (1753); Sibth. et Sm., Fl. Graec. Prodr., 2: 103 (1813), Fl. Graec., 8: 40, t. 755 (1833).

Tetragonolobus palaestinus (? Boiss. et Bl.) H. Stuart Thompson in Journ. Bot. 44: 306 (1906); Zohary, Fl. Palaest., 2: 95, t. 141 (1972).

Lotus tetragonolobus L. ssp. *palaestinus* Holmboe, Veg. Cypr., 107 (1914).

TYPE: " *in* Siciliae collibus ".

Sprawling or decumbent annual; stems usually much branched, 10–30 (or more) cm. long, sulcate, rather densely clothed with long, spreading hairs; rhachis 5–10 mm. long, pilose, canaliculate; basal (stipuloid) leaflets ovate, 5–20 mm. long, 4–10 mm. wide, apex acute or acuminate, base often pallid, adnate to the base of the rhachis; leaflets obovate, 8–30 mm. long, 6–20 mm. wide, adpressed-pilose on both surfaces, apex rounded or shortly acute, base cuneate; inflorescence 1 (–2)-flowered; peduncle 8–45 mm. long, shorter than or exceeding the subtending leaf, pilose with patent hairs; bracts foliaceous, to about 20 mm. long and 15 mm. wide; calyx narrowly campanulate, 8–15 mm. long, 5–7 mm. wide, pilose, often streaked dark violet externally, calyx-teeth linear or linear-lanceolate, 8–9 mm. long, 1–1·3 mm. wide at base;

petals intense crimson (occasionally yellowish), limb of standard broadly obovate or suborbicular, 10–13 mm. long and almost as wide, apex shortly emarginate, base rather abruptly narrowed to a claw about 8 mm. long, wings shortly connate along their apical margins, limb of wing oblong, 12–14 mm. long, 5–7 mm. wide, apex rounded, base with an elongate, blunt auricle and a distinct conjunctival flap, claw slender, about 3 mm. long, limb of keel straightish, about 6–7 mm. long, 3–4 mm. wide, apex rostrate, base scarcely auriculate, claw slender, about 6 mm. long; vexillary stamen free, the others connate into a sheath about 7 mm. long, free parts of filaments alternately about 4 mm. and 2 mm. long; anthers oblong, about 0·6 mm. long, 0·4 mm. wide; ovary linear, somewhat compressed, about 8 mm. long, 1 mm. wide, glabrous, shortly stipitate, many-ovulate, style gently upcurved, about 4 mm. long, rather swollen at the base, tapering towards apex, stigma lateral. Pod oblong, 25–60 mm. long, 4–8 mm. wide, blunt at apex and base, septate within, valves subterete, mid-brown with obscure, darker marbling, adaxial and abaxial sutures bordered by 4 well-developed, often undulate, wings; seeds subglobose, 3·5–4·5 mm. diam.; testa smooth, pale or dark brown; hilum very small.

HAB.: Fields, roadsides and waste ground, sometimes with other annuals amongst garigue; sea-level to 2,500 ft. alt., usually lowland; fl. Febr.–May.

DISTR.: Divisions 1, 3, 4, 6–8, very common in Division 7. Widespread in southern Europe and the Mediterranean region.

1. Between Ktima and Polis, 1950, *F. J. F. Barrington* 41 ! Ayios Yeoryios area (Akamas), 1962, *Meikle* ! Polis area, 1962, *Meikle* !
3. Kolossi near Limassol, 1862, *Kotschy*; Curium near Limassol, 1939, *Kennedy* 1342 ! 2 miles W. of Limassol, 1964, *J. B. Suart* 177 !
4. Larnaca, 1880, *Sintenis & Rigo*; the Phaneromene near Larnaca, 1905, *Holmboe* 138.
6. Nicosia, 1859, *Kotschy* 488; Dhiorios area, 1962, *Meikle* !
7. Very common; *Sintenis & Rigo*; *Syngrassides* 294 ! 1186 ! *Davis* 2876 ! *Casey* 335 ! 518 ! 1246 ! *G. E. Atherton* 913 ! *Meikle* 2375 !
8. Rizokarpaso, 1941, *Davis* 2366 ! Koma tou Yialou, 1950, *Chapman* 234 ! Akradhes, 1962, *Meikle* !

NOTES: Variable in general vigour and in flower-colour; the majority of specimens have rich, glowing crimson flowers (as in Fl. Graec. t. 755), but forms with copper-coloured or golden-yellow petals are occasionally recorded.

The claims of *T. palaestinus* Boiss. to recognition at specific level are very tenuous. It is true that many Palestine specimens have short peduncles, solitary flowers with narrow calyx-teeth and short petals, and narrowly winged pods, but such variants represent one extreme in a line of forms reaching its opposite extreme in Spain, where robust, large-flowered plants with broad-winged pods have been distinguished as *T. pseudo-purpureus* Uechtr. The type material of *Lotus tetragonolobus* L. falls somewhere between these two extremes, and the Cyprus plants, though smaller-flowered than typical *T. purpureus*, are on the whole distinctly larger-flowered than most of the Palestine specimens examined. At no point is it possible to single out, from any one area, specimens which would be deserving of recognition at specific or subspecific level, though one might grudgingly allow some of the smallest-flowered Palestine specimens varietal status.

TRIBE 5. **Galegeae** (*Bronn*) *Benth.* Herbs, shrubs, trees and climbers; leaves stipulate, generally imparipinnate, without tendrils; leaflets mostly entire; stamens 10, diadelphous, rarely monadelphous, filaments not dilated. Pod usually 2-valved, inarticulate, occasionally indehiscent and 1–2-seeded or membranous and inflated.

17. ASPALTHIUM *Medik.*

in Vorles. Churpf. Phys.-oekon. Ges., 2: 380 (1787).
Psoralea L., Sp. Plant., ed. 1, 762 (1753) pro parte.

Perennial herbs and subshrubs; leaves 3-foliolate, glandular-punctate; inflorescences axillary, capitate or spicate, pedunculate, bracteate, bracteoles wanting; calyx campanulate, calyx-teeth unequal; petals blue or violet, standard auriculate at base, keel upcurved, blunt; stamens monadelphous, filaments subequal, anthers uniform; ovary sessile, 1-ovulate, style dilated at the base, stigma terminal. Pod indehiscent, 1-seeded, with a long falciform beak; seed estrophiolate.

Two species in Europe, the Mediterranean region and western Asia.

1. **A. bituminosum** (*L.*) *Fourr.* in Ann. Soc. Linn. Lyon, ser. 2, 16: 365 (1868); O. Kuntze in Post et O. Ktze., Lex. Gen. Phan., 48 (1903) as *Asphaltium bituminosum*.

Psoralea bituminosa L., Sp. Plant., ed. 1, 763 (1753); Boiss., Fl. Orient., 2: 187 (1872); H. Stuart Thompson in Journ. Bot., 44: 306 (1906); Holmboe, Veg. Cypr., 108 (1914); Post, Fl. Pal., ed. 2, 1: 367 (1932); Lindberg f., Iter Cypr., 20 (1946); Davis, Fl. Turkey, 3: 264 (1970); Zohary, Fl. Palaest., 2: 50, t. 66 (1972); Osorio-Tafall et Seraphim, List Vasc. Plants Cyprus, 56 (1973).

TYPE: " *in* Siciliae, Italiae, Narbonae *collibus maritimis* ".

Erect perennial with a woody base; stems to about 1 m. high, branched chiefly at the base, conspicuously striate, adpressed-strigose; stipules subulate, 8–15 mm. long, 1–1·5 mm. wide at base, thinly adpressed-strigose externally; petioles of lower leaves often 10–12 cm. long, striate, channelled above, adpressed-strigose, those of the upper leaves progressively shorter; leaflets of lower leaves oblong or suborbicular, 3–5 cm. long, 1·5–4 cm. wide, entire, subglabrous or thinly strigose on both surfaces, apex blunt, rounded or slightly retuse, base usually rounded, terminal leaflet not much larger than the laterals, leaflets of upper leaves usually much narrower, oblong, lanceolate or even linear, 3–6 (or more) cm. long, 0·5–2 cm. wide, apex acute, cuspidate, base ± rounded; peduncles stout, erect, 8–30 cm. long, usually much longer than subtending leaves, striate, rather densely adpressed-strigose; inflorescence dense, 2–3·5 cm. diam., many-flowered; bracts ovate, long-acuminate, 5–15 mm. long, 3–5 mm. wide, adpressed-strigose externally, 2 or 3 often connate by their margins; flowers sessile or subsessile; calyx 12–18 mm. long, 3–4 mm. wide, clothed with adpressed, often fuscous, hairs externally, calyx-teeth long-acuminate, spinose-tipped, the 4 adaxial about 7–10 mm. long, the abaxial 10–13 mm. long; petals shorter than or exceeding the calyx-teeth; standard with an oblong limb, about 12 mm. long, 6 mm. wide, apex rounded or slightly emarginate, base distinctly auriculate, narrowing to a rather wide claw about 6 mm. long, wings with a narrowly oblong limb, about 10 mm. long, 3 mm. wide, apex rounded or subacute, base with a long, slender auricle, about 2–2·5 mm. long, claw slender about 6 mm. long, keel ± adherent to wings, limb oblong, 5–6 mm. long, 2 mm. wide, apex very blunt, blotched purple, base slightly auriculate, claw slender, about 5–6 mm. long; stamens about 12 mm. long, united almost to apex; anthers oblong, medifixed, about 5 mm. long, 3 mm. wide; ovary ovate, compressed, densely sericeous, 2–2·5 mm. long, 1·5 mm. wide, 1-ovulate, style 7–8 mm. long, strigose and dilated below, glabrous and filiform above, sharply upcurved and gibbous just below the capitate stigma. Pod flattened 18–23 mm. long, 3–4 mm. wide, villose below, with an acute or acuminate, falciform, adpressed-strigose beak; seed permanently invested with pericarp, oblong, compressed, about 6 mm. long, 3·5 mm. wide.

HAB.: Dry, rocky limestone hillsides; 1,300–2,500 ft. alt.; fl. April–June.

DISTR.: Division 7, common in St. Hilarion and Kyrenia areas, also at Halevga, Kornos, and Yaïla. S. Europe and Mediterranean region, Atlantic Islands.

7. Common; *Sintenis & Rigo* 690 ! *Holmboe* 817; *Syngrassides* 625 ! *Kennedy* 1071 ! *Lindberg f.* s.n. ! *Davis* s.n. ! *Casey* 728 ! 1581 ! *Chapman* 650.

NOTES: The Cyprus material of the species comprises robust, rather densely strigose plants with relatively large, many-flowered inflorescences; in some of the specimens the corolla is scarcely as long as the calyx, and the pods have a short rather straight, wide beak. Such characters are inconstant, nor does examination of the species throughout its range suggest that any useful purpose would be served by attempting to discriminate between the typical plant and varieties such as *Psoralea bituminosa* L. var. *plumosa* (Reichb.) Reichb. or var. *palaestina* (Gouan) Hal.

Robinia pseudoacacia *L.*, the Locust or False Acacia, a native of North America, is cultivated as an ornamental (G. Frangos in Cypr. Agric. Journ., 18: 85; 1923). It is a deciduous tree, up to about 25 m. in height, with roughly fissured bark and dark brown twigs often armed with stout stipular spines; the leaves are imparipinnate, up to 30 cm. in length, with 9–23 (or more) shortly petiolulate, glabrescent or sparsely pubescent, oblong or oblong-ovoid, entire, obtuse leaflets 2–5 cm. long, 1–3 cm. wide; the inflorescence is an erect or somewhat pendulous, rather lax, raceme bearing 20 or more white, sweet-scented flowers, each about 1·5–2 cm. long. The pods are linear-oblong, strongly flattened, about 7 cm. long and 1·3 cm. wide, blunt or rounded at the apex, with rather thin, brown valves, and with a conspicuous abaxial (or basal) keel or narrow wing; the seeds are reniform and compressed, about 4 mm. long, 3 mm. wide, with a smooth, brown or red-blotched testa. *Plate 26.*

Pink-flowered varieties are also in cultivation, as well as the smaller, shrubby *Robinia hispida* L., the Rose Acacia, a fragile, hispid plant with large, handsome rose-pink flowers.

Robinia spinosa, recorded from Limassol by Hume (in Walpole's Memoirs, 253; 1817), is clearly a misnomer. True *R. spinosa* L. (=*Caragana spinosa* (L.) DC.) is not found in Cyprus; perhaps *Calycotome villosa* or *Genista sphacelata* was intended.

Sesbania sesban (*L.*) *Merrill* (*Aeschynomene sesban* L.), widespread in the tropics and subtropics, is recorded, as a cultivated plant, from Kyrenia (*G. E. Atherton* 73 !). It is a small tree or shrub, rarely much more than 5 m. high, with erect, striate or angular, greenish branches and narrowly oblong, paripinnate leaves 10–20 (–30) cm. long, bearing 20–30 (or more) narrowly oblong, obtuse, subsessile, entire, glaucous leaflets, 13–20 mm. long, 3–5 mm. wide; the inflorescence is a very lax raceme, bearing 6–12 (sometimes more) distinctly pedicellate flowers, each about 2 cm. long, with yellow petals, the broad standard freckled with small brown spots. The dehiscent pod is narrowly linear, 20 cm. (or more) long, 0·4 cm. wide, often curved and slightly compressed with distinct constrictions between the seeds, the valves are yellowish or pale brown, with an acute or shortly rostrate apex, the seeds are oblong, 4–5 mm. long, about 2 mm. wide, with a smooth, brown testa.

A graceful plant, but not nearly so ornamental as the crimson-flowered *Sesbania grandiflora* (L.) Pers. from India and tropical Asia; this has flowers as much as 12 cm. long. It is cultivated in Arabia and should grow in lowland Cyprus.

18. ASTRAGALUS *L.*

Sp. Plant., ed. 1, 755 (1753).
Gen. Plant., ed. 5, 335 (1754).
Bunge in Mém. Acad. Sci. St. Pétersb. ser. 7, 11, no. 16 (1868) and ser. 7, 15, no. 1 (1869).
Eig, Syst. Stud. Astragali (1955).
Biserrula L., Gen. Plant., ed. 5, 336 (1754).

Annual, biennial and perennial herbs and subshrubs; leaves stipulate, imparipinnate or paripinnate, sometimes with the rhachis produced into a

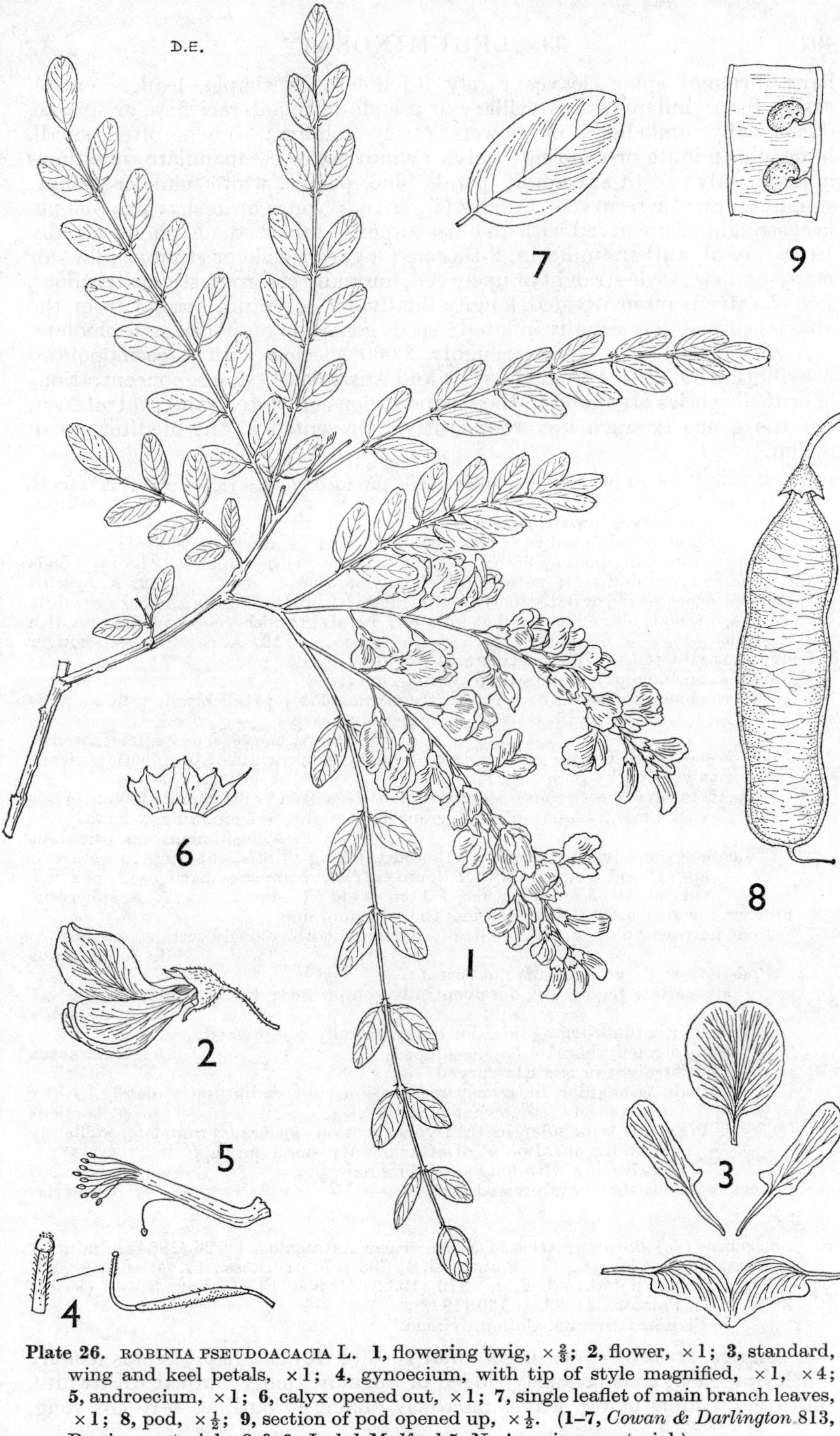

Plate 26. ROBINIA PSEUDOACACIA L. **1**, flowering twig, $\times \frac{2}{3}$; **2**, flower, $\times 1$; **3**, standard, wing and keel petals, $\times 1$; **4**, gynoecium, with tip of style magnified, $\times 1$, $\times 4$; **5**, androecium, $\times 1$; **6**, calyx opened out, $\times 1$; **7**, single leaflet of main branch leaves, $\times 1$; **8**, pod, $\times \frac{1}{2}$; **9**, section of pod opened up, $\times \frac{1}{2}$. (**1–7**, *Cowan & Darlington* 813, Persian material; **8 & 9**, *Isabel Mulford* 5, N. American material.)

hard terminal spine, leaves rarely 3-foliolate or simple, leaflets entire, exstipellate; inflorescences axillary or pseudo-terminal, racemose or spicate, occasionally umbellate, or flowers rarely solitary; bracts often small, bracteoles minute or wanting; calyx cylindrical or campanulate sometimes inflated, calyx-teeth subequal; petals blue, purple, white, pink or yellow, standard erect or recurved, usually larger than wings or keel, wings oblong, keel straight or upcurved with an obtuse apex; stamens diadelphous, usually 10 (rarely 5), anthers uniform, 2-thecous; ovary sessile or stipitate, few- or many-ovulate, style straight or upcurved, normally glabrous, stigma terminal. Pod 2-valved, often divided longitudinally by a septum arising from the abaxial suture, occasionally inflated; seeds generally reniform, estrophiolate.

A vast genus, comprising roughly 2,000 species, with a cosmopolitan distribution but absent from Malaysia and Australasia. Large concentrations of critical species are found in the steppe regions of western and central Asia, but the genus is much less abundantly represented in the Mediterranean region.

Plants spinose; leaves paripinnate, the rhachis produced into a rigid spine; flowers in small axillary clusters - - - - - - - - - **11. A. echinus**
Plants unarmed; leaves imparipinnate:
  Leaves all basal or sub-basal in tufts or rosettes, stems scarcely developed:
    Inflorescences long-pedunculate; leaflets silvery with adpressed hairs; pods narrowly cylindrical; petals pinkish or yellowish - - - **12. A. cyprius**
    Inflorescences sessile or indistinctly pedunculate; leaflets villose with long spreading hairs; pods oblong, somewhat inflated; petals greenish-yellow turning reddish with age - - - - - - - **10. A. caprinus** ssp. **laniger**
  Leaves not all basal or sub-basal; stems well developed:
    Flowers conspicuous; standard 15–30 mm. long:
      Inflorescences subsessile or very shortly pedunculate; petals bright yellow; pods large, inflated, with a corky or spongy pericarp **9. A. macrocarpus** ssp. **lefkarensis**
      Inflorescences distinctly pedunculate; petals creamy-white or purplish; pericarp not very corky or spongy:
        Leaflets silvery-sericeous; standard obovate with a broad rounded apex; pods very large, oblong, inflated, smooth, straight, 6–7 cm. long, 2·5 cm. wide **8. A. lusitanicus** ssp. **orientalis**
        Leaflets not silvery-sericeous; standard oblong-elliptic, tapering to a narrow apex; pod oblong-elliptic, dorsiventrally compressed, rugose, apex upcurved, 1·5–3 cm. long, 0·8–1·3 cm. wide - - - - **7. A. suberosus**
    Flowers inconspicuous; standard less than 15 mm. long:
      Pods narrowly oblong, dorsiventrally flattened with coarsely serrate margins **1. A. pelecinus**
      Pods not as above, margins not serrate:
        Pods cordate-triangular, dorsiventrally compressed; flowers minute **2. A. epiglottis**
        Pods not cordate-triangular, nor dorsiventrally compressed:
          Pods strongly hamate (sickle-shaped) - - - - - **5. A. hamosus**
          Pods straight or slightly curved:
            Pods triangular in transverse section, not radiating stellately, stiffly chartaceous, somewhat inflated - - - - - **6. A. boeticus**
            Pods not triangular in transverse section, generally radiating stellately from a central point of attachment; small annuals:
              Pods hirsute with long spreading hairs - - - **3. A. sinaicus**
              Pods shortly adpressed-strigose - - - - - - **4. A. asterias**

1. A. pelecinus (*L.*) *Barneby*, Atlas of N. American Astragalus, 1: 26 (1964) in adnot.
Biserrula pelecinus L., Sp. Plant., ed. 1, 762 (1753); Boiss., Fl. Orient., 2: 204 (1872); Post, Fl. Pal., ed. 2, 1: 372 (1932); Davis, Fl. Turkey, 3: 48 (1970); Zohary, Fl. Palaest., 2: 83, t. 120 (1972).
TYPE: " *in* Sicilia, Hispania, Galloprovincia ".

A slender, erect or sprawling annual 10–30 (–40) cm. high; stems usually much branched, conspicuously ridged or angular, thinly clothed with white, spreading, simple hairs; leaves narrowly oblong in outline, 4–10 cm. long,

0·8–1·8 cm. wide, imparipinnate with 7–15 pairs of leaflets; petiole very short, generally less than 5 mm. long; stipules ovate-lanceolate, 2–4 mm. long, 1–2·5 mm. wide, rather papery, entire, thinly white-pilose, adnate to the petiole; leaflets usually opposite, oblong or oblong-elliptic, 5–10 mm. long, 1·5–5 mm. wide, entire, thinly white-pilose, apex emarginate or truncate, base cuneate, tapering to a short but distinct petiolule; inflorescence an axillary, pedunculate, few-flowered raceme; peduncle up to 5 cm. long, generally shorter than the subtending leaf, angular, thinly pilose; flowers 2–8 (–11) in each raceme; bracts narrowly lanceolate, 1–1·5 mm. long, sparsely pilose with blackish hairs; pedicels very short, rarely exceeding 1 mm. at anthesis; calyx-tube campanulate, about 1–2 mm. long, clothed with blackish hairs, calyx-teeth subequal, linear-subulate, blackish-pilose, 1–2 mm. long; petals mauve or purplish, glabrous; standard oblong or obovate, 4–6 mm. long, 3–4 mm. wide, tapering to base, without appendages; wings oblong, about 3·5 mm. long, 1·5 mm. wide, apex rounded, base with a short auricle, claw about 1·5 mm. long, keel straightish, sharply upcurved to a rather blunt apex, limb about 3 mm. long, 2 mm. wide with a well-developed basal auricle, claw about 1·5 mm. long; stamens about 4 mm. long, the filaments united for about 3 mm.; anthers 5 (alternate free parts of filaments short and sterile), ovoid, about 0·4 mm. long, 0·3 mm. wide, yellow; ovary oblong-cylindric, straight or somewhat curved, glabrous or thinly hairy, about 4 mm. long, 0·7 mm. wide; style about 0·5 mm. long; stigma capitate. Pod narrowly oblong, straight or arcuate, 20–40 mm. long, 4–6 mm. wide, strongly compressed or flattened dorsiventrally with coarsely serrate margins, glabrous or adpressed-pubescent; seeds numerous, reniform, strongly compressed, about 2 mm. long, 1·5 mm. wide; testa light- or mid-brown, smooth or somewhat rugose.

HAB.: Sandy waste ground; near sea-level; fl. March–April.

DISTR.: Division 5, rare. Widespread in the Mediterranean region; Atlantic Islands; Ethiopia, and (ssp. *leiocarpa* (A. Rich.) Gillett) south to Tanzania.

5. Salamis, in the Forum, near some Stone Pines, April, 1967, *Mrs. I. Chatterton* s.n. !

NOTES: The reasons for uniting the monotypic genus *Biserrula* with *Astragalus* are given by Barneby (*loc. cit.*). It is possible that *A. pelecinus* may be a recent arrival in Cyprus, especially as it is unlikely that collectors would have overlooked such a distinctive plant.

2. **A. epiglottis** *L.*, Sp. Plant., ed. 1, 759 (1753); Sibth. et Sm., Fl. Graec. Prodr., 2: 88 (1813), Fl. Graec., 8: 23, t. 731 (1833); Boiss., Fl. Orient., 2: 223 (1872); Holmboe, Veg. Cypr., 108 (1914); Post, Fl. Pal., ed. 2, 1: 376 (1932); Davis, Fl. Turkey, 3: 60 (1970); Zohary, Fl. Palaest., 2: 58, t. 72 (1972); Osorio-Tafall et Seraphim, List Vasc. Plants Cyprus, 55 (1973).

TYPE: " *in* Hispania ". No authentic type material, but specimens in Hort. Cliff. herb. (no. 120) can be taken as representing Linnaeus's concept of this species.

Decumbent or prostrate annual; stems much branched at base, 2–15 cm. long, rather densely strigose with adpressed, medifixed hairs; stipules oblong, brownish, 2–4 mm. long, 1–2 mm. wide, strigose externally, with a slender caudate apex; petioles 5–10 mm. long, slightly channelled above, densely adpressed-strigose; leaves imparipinnate, 1–2·5 cm. long, 1–2 cm. wide, leaflets 3–7-jugate, narrowly oblong-elliptic, 4–10 mm. long, 1·5–3 mm. wide, densely silvery-strigose on both surfaces, apex blunt or acute; inflorescences axillary, crowded-spicate, subsessile or very shortly pedunculate; bracts minute, about 0·5 mm. long, pilose; pedicels very short and relatively stout; calyx campanulate, about 2 mm. long, 1·5 mm. wide, densely strigose, calyx-teeth linear, subacute, about 0·8 mm. long, fuscous-pilose; petals yellowish, standard obovate, 2–2·5 mm. long, 1·3 mm. wide, apex rounded, base tapering to a claw about 0·8 mm. long, limb of wing narrowly oblong, about 1·5 mm.

long, 0·5 mm. wide, apex rounded or obtuse, base strongly auriculate, claw slender about 0·5 mm. long, keel blunt, straightish, limb about 1·5 mm. long, 0·8 mm. wide, claw about 0·5 mm. long; stamens shorter than keel, and partly adnate to it; anthers apparently 5 (in the specimens examined), suborbicular, about 0·2 mm. diam.; ovary ovoid, compressed, abaxially sulcate, about 0·5 mm. long, 0·2 mm. wide, pubescent, 4-ovulate, style very short, upcurved, glabrous, stigma terminal. Pod cordate-triangular, dorsiventrally compressed, the two halves longitudinally deflexed, pale brown, 8–10 mm. long, 5 mm. wide; seeds reniform, flattened, about 2·5 mm. long, 1·8 mm. wide; testa pale brown, smooth or slightly rugose.

HAB.: Waste ground, roadsides and dry, open hillsides; sea-level to c. 1,000 ft. alt.; fl. March–April.

DISTR.: Divisions 3, 6. Widespread in the Mediterranean region.

3. Between Nisou and Stavrovouni, 1933, *Syngrassides* 28 ! Kakoradjia near Stavrovouni, 1938, *Syngrassides* 1809 !
6. Near Nicosia, 1880, *Sintenis & Rigo* 434 !

3. **A. sinaicus** *Boiss.*, Diagn., 1: 9: 57 (1849), Fl. Orient. Suppl., 174 (1888); Post, Fl. Pal., ed. 2, 1: 377 (1932); Eig, Syst. Stud. Astragali, 9 (1955); Davis, Fl. Turkey, 3: 61 (1970); Osorio-Tafall et Seraphim, List Vasc. Plants Cyprus, 55 (1973).

[*A. sesameus* (non L.) Sibth. et Sm., Fl. Graec. Prodr., 2: 87 (1813); Unger et Kotschy, Die Insel Cypern, 381 (1865).]

[*A. pseudostella* (non Del.) Boiss., Fl. Orient., 2: 225 (1872).]

[*A. tribuloides* (non Del.) Boiss., Fl. Orient. Suppl., 174 (1888); Holmboe, Veg. Cypr., 108 (1914).]

TYPE: " in deserto montis *Sinai* Aucher No. 1309 ex parte " (G).

Decumbent or prostrate annual; stems 3–10 (–15) cm. long, usually much branched at base, rather densely strigose; stipules narrowly deltoid or subulate, 1–3 mm. long, ± densely strigose; petioles 5–10 (–20) mm. long, slightly channelled above, strigose; leaves oblong, imparipinnate, 1·5–3 (–5) cm. long, 0·8–1·5 cm. wide, leaflets 5–8 (–12)-jugate, oblong-elliptic, 3–10 mm. long, 0·8–3 mm. wide, densely greyish-strigose, apex acute or obtuse; inflorescences 3–10-flowered, axillary, sessile or subsessile, crowded; bracts subulate, pilose, 2–2·5 mm. long; calyx cylindrical, about 5 mm. long, 1·5 mm. wide, densely pilose, calyx-teeth subequal, subulate, about 2·5 mm. long, fuscous-pilose towards apex; petals yellowish or purplish, standard oblong-obovate, about 5 mm. long, 2 mm. wide, apex rounded, base narrowing to a relatively wide claw, limb of wings narrowly oblong, about 2·5 mm. long, 0·8 mm. wide, apex blunt, base with a distinct falcate auricle, claw slender about 2 mm. long, limb of keel boat-shaped, about 2 mm. long, 1 mm. wide, apex blunt or subacute, base distinctly auriculate, claw slender, about 2 mm. long; stamens 4 mm. long, free from petals, all except the vexillary connate for about 3 mm. from base; anthers 10, suborbicular, 0·2–0·3 mm. diam.; ovary oblong, 2 mm. long, 0·6 mm. wide, densely pilose, many-ovulate, style straight, glabrous, about 1–1·5 mm. long, stigma terminal, capitate. Pods 4–6 (or more) radiating stellately, oblong, 10–12 mm. long, 3–4 mm. wide, apex acute, base slightly gibbous, valves hirsute with spreading hairs, abaxial suture deeply sulcate; seeds irregularly oblong, strongly compressed, about 2 mm. long, 1·5 mm. wide; testa bright brown, rather deeply pitted.

HAB.: Fields, waste ground, roadsides and dry rocky hillsides; c. 500 ft. alt.; fl. April–May.

DISTR.: Divisions 3, 4, 5. Greece, Aegean Islands, Cyrenaica, Sinai.

3. Between Nisou and Stavrovouni, 1933, *Syngrassides* 27 !
4. Near Larnaca, 1862, *Kotschy*.
5. Near Kythrea, 1880, *Sintenis & Rigo* 433b !

4. A. asterias *Stev. ex Ledeb.*, Fl. Ross., 1: 636 (1843); Davis, Fl. Turkey, 3: 62 (1970); Osorio-Tafall et Seraphim, List Vasc. Plants Cyprus, 55 (1973).

[*A. cruciatus* (non Link) Bunge in Mém. Acad. Sci. St. Pétersb., ser. 7, 15: 8 (1869); Boiss., Fl. Orient., 2: 225 (1872); Sintenis in Oesterr. Bot. Zeitschr., 32: 18 (1882); Boiss., Fl. Orient. Suppl., 174 (1888); Holmboe, Veg. Cypr., 108 (1914); Post, Fl. Pal., ed. 2, 1: 377 (1932).]

[? *A. stella* (non Gouan) Sibth. et Sm., Fl. Graec. Prodr., 2: 87 (1813); Unger et Kotschy, Die Insel Cypern, 381 (1865); Sintenis in Oesterr. Bot. Zeitschr., 31: 392 (1881).]

TYPE: U.S.S.R., Caucasus; " Iberia " *Steven* (LEN, K !).

Prostrate or decumbent annual, stems 3–20 (–30) cm. long, usually much branched at base, rather densely adpressed-strigose; stipules lanceolate-subulate, 3–4 mm. long, 1·5–2 mm. wide at base, adpressed-strigose; petioles 5–15 (–30) mm. long, strigose, slightly channelled above; leaves narrowly oblong, 2–6 cm. long, 0·5–1·5 cm. wide, imparipinnate, leaflets 6–12-jugate, narrowly oblong-elliptic, or oblong, 3–10 mm. long, 1·5–4 mm. wide, adpressed-strigose on both surfaces, apex acute or obtuse; inflorescences 5–8-flowered, axillary, loosely capitate; peduncle 1–5 cm. long, exceeding or shorter than the subtending leaf, densely adpressed-strigose; bracts subulate, strigose, 1–1·5 mm. long; calyx cylindrical, 4–7 mm. long, 1·5–2 mm. wide, fuscous-strigose, calyx-teeth subequal, linear-subulate, 1·5–2·5 mm. long, petals yellowish or purplish, standard narrowly oblong, 6–7 mm. long, 2–2·5 mm. wide, apex rounded, base tapering very gradually to a wide claw, not auriculate, limb of wings narrowly oblong, 2·5–3·5 mm. long, about 1 mm. wide, apex rounded, base distinctly auriculate, claw slender about 2–2·5 mm. long, limb of keel sharply boat-shaped, about 1·5–2 mm. long, 1–1·5 mm. wide, apex blunt, base narrowed abruptly, or slightly auriculate, claw slender, about 2 mm. long; stamens 4–5 mm. long, all except the vexillary connate for 3–4 mm.; anthers ovate, about 0·3 mm. long, 0·2 mm. wide; ovary oblong-cylindrical, 2–2·5 mm. long, 0·8–1 mm. wide, densely hirsute, many-ovulate, style straight or slightly upcurved, glabrous, 1–2 mm. long, stigma terminal, minute. Pods usually 5–6, radiating stellately, oblong, 10–15 mm. long, about 3 mm. wide, apex acute or subacute, base distinctly expanded, valves shortly adpressed-strigose with unequal hairs, abaxial suture very deep and narrow; seeds irregularly oblong, compressed, 2–2·5 mm. long, 1·5 mm. wide; testa brown or greenish, distinctly pitted.

HAB.: Fields, waste ground, roadsides and dry pastures; 500–2,000 ft. alt.; fl. March–April.

DISTR.: Divisions 2, ? 4, 5–7. Also Caucasus, Turkey, Syria, Palestine, Arabia and eastwards to Iran.

2. Evrykhou, 1935, *Syngrassides* 816 !
4. ? Larnaca, 1862, *Kotschy* 272.
5. Kythrea, 1880, *Sintenis & Rigo*.
6. Near Nicosia, 1880, *Sintenis & Rigo* 433 !
7. Kalyvakia, 1957, *M. E. D. Poore in Merton* 3093 !

NOTES: Boissier (Fl. Orient., 2: 225) identifies with this species the plant described and figured as *A. contortuplicatus* L. in Sibthorp and Smith, Fl. Graec., 8: 22, t. 729 (1833), and said to have been collected " ad campos in Asia minore, nec non in insula Cypro ". The plant figured in Fl. Graec. is not *A. contortuplicatus* L., nor, would it appear, *A. asterias* Stev. ex Ledeb., though it may possibly be a form of *A. pentaglottis* L. Nothing resembling it has since been found in Cyprus, and it must be concluded that the record is an error.

The records for *Astragalus stella* Gouan (Cyprus, 1787, *Sibthorp*; Larnaca 1862, *Kotschy* 272; Kythrea, 1880, *Sintenis & Rigo*) are also erroneous; this species is restricted to the western Mediterranean and does not occur in our area; the plants so named were most probably *A. asterias*, which *A. stella* superficially resembles.

Bunge's statement (*op. cit.*, 9) that Meyer erroneously confused *A. cruciatus* Link and *A. corrugatus* Bertol. is contradicted by Link's original description of *A. cruciatus* (Enum. Plant. Hort. Berol., 256; 1822) which agrees with the plant generally named *A. corrugatus*, and does not agree with *A. cruciatus* of Bunge and Boissier. It would in fact appear that the seeds sent to Meyer from Berlin were correctly identified, and that *A. asterias* Stev. ex Ledeb. must consequently be accepted as the next earliest legitimate name for the plant commonly named *A. cruciatus*.

5. **A. hamosus** *L.*, Sp. Plant., ed. 1, 758 (1753); Sibth. et Sm., Fl. Graec. Prodr., 2: 86 (1813); Unger et Kotschy, Die Insel Cypern, 381 (1865); Boiss., Fl. Orient., 2: 238 (1872); Holmboe, Veg. Cypr., 108 (1914); Post, Fl. Pal., ed. 2, 1: 381 (1932); Rechinger f. in Arkiv för Bot., ser. 2, 1: 425 (1950); Davis, Fl. Turkey, 3: 67 (1970); Zohary, Fl. Palaest., 2: 66, t. 87 (1972); Osorio-Tafall et Seraphim, List Vasc. Plants Cyprus, 55 (1973).

TYPE: " Messanae, Monspelii ".

Procumbent or sprawling annual or biennial with a tough, rather woody rootstock; stems 10–30 (–50) cm. high, thinly strigose with adpressed, unequally medifixed hairs, generally much branched at and above the base; stipules ovate-deltoid, 3–7 mm. long, 2·5–5 mm. wide, connate along part of the abaxial margins, ciliate, apex acuminate; petioles 1–5 cm. long, adpressed-strigose, slightly channelled above; leaves imparipinnate, narrowly oblong, 5–15 cm. long, 1·5–3 cm. wide, leaflets 5–12-jugate, oblong, 5–18 (–20) mm. long, 2–7 mm. wide, glabrous above, thinly adpressed-strigose below, apex truncate or emarginate, base cuneate; inflorescence a rather congested, 4–8 (–15)-flowered raceme; peduncle at first very short and densely strigose, lengthening to 5 cm. or more in fruit; bracts linear or linear-lanceolate, 2–4 mm. long, less than 0·5 mm. wide, membranous, thinly fuscous-strigose; calyx narrowly campanulate, 5–6 mm. long, 2 mm. wide, adpressed fuscous-strigose externally, calyx-teeth subequal, subulate, strigose, 2–2·5 mm. long; petals whitish, yellowish or very pale pink, standard obovate-oblong, about 6 mm. long, 3 mm. wide, apex emarginate with a short median apiculus, base tapering gradually to a wide claw about 2·5 mm. long, limb of wings narrowly oblong, about 3·5 mm. long, 1 mm. wide, apex rounded, base conspicuously auriculate, claw about 2–2·5 mm. long, limb of keel shortly boat-shaped, 3 mm. long, 1·5 mm. wide, apex rounded, base conspicuously auriculate, claw about 2 mm. long; stamens about 5 mm. long, all except the vexillary connate for 3–4 mm.; anthers oblong, about 0·4 mm. long, 0·3 mm. wide; ovary narrowly oblong, straight, about 4 mm. long, 0·8 mm. wide, style 1–1·5 mm. long, upcurved, glabrous, gradually thickening below, stigma terminal, capitate, thinly barbellate. Pod cylindrical, 30–40 mm. long, 3–3·5 mm. wide, strongly hamate with an upcurved, sharply pointed apex, slightly compressed laterally with a shallow abaxial suture, valves pale dull brown, thinly adpressed-pubescent; seeds rectangular or reniform, compressed, about 2 mm. long, 1 mm. wide; testa light brown, somewhat rugulose-pitted.

HAB.: In fields, gardens and waste ground; by roadsides and occasionally on rocky slopes near the sea; sea-level to 3,750 ft. alt.; fl. Febr.-April.

DISTR.: Divisions 1, 2, 4, 6–8, common in Division 7. Widespread in the Mediterranean region and eastwards to Iran; Atlantic Islands.

1. Ktima, 1913, *Haradjian* 685.
2. Platres, 1938, *Kennedy* 1297 ! Lefkara, 1941, *Davis* 2750 !
4. Near Larnaca, 1862, *Kotschy* 31; Larnaca, 1880, *Sintenis & Rigo*; Larnaca, 1881, *Barbey*.
6. Near Nicosia, 1880, *Sintenis & Rigo* 435 ! Nicosia, 1950, *Chapman* 522 ! Near Orga, 1956, *M. E. D. Poore* in *Merton* 2643 !
7. Common; *Sintenis & Rigo*; *Syngrassides* 184 ! *Davis* 2988 ! *Casey* 541 ! *Chapman* 272 !
8. Valia (Karpas), 1905, *Holmboe* 490; Cape Andreas, 1962, *Meikle* 2471 !

6. **A. boeticus** *L.*, Sp. Plant., ed. 1, 758 (1753); Sibth. et Sm., Fl. Graec. Prodr., 2: 87 (1813), Fl. Graec., 8: 22, t. 730 (1833); Unger et Kotschy, Die Insel Cypern, 382 (1865); Boiss., Fl. Orient., 2: 236 (1872); Holmboe, Veg. Cypr., 108 (1914); Post, Fl. Pal., ed. 2, 1: 381 (1932); Lindberg f., Iter Cypr., 19 (1946); Davis, Fl. Turkey,

3: 66 (1970); Zohary, Fl. Palaest., 2: 65, t. 85 (1972); Osorio-Tafall et Seraphim, List Vasc. Plants Cyprus, 55 (1973).

TYPE: " *in* Sicilia, Lusitania, Hispania ".

Erect or decumbent annual, 20–40 (–60) cm. high; stems robust, usually much branched from the base, frequently reddish, thinly clothed with basifixed or very unequally medifixed hairs; stipules broadly deltoid, 5–10 mm. long, 3–6 mm. wide, sparsely strigose externally, apex acuminate, abaxial margins free or connate; petioles 1–6 cm. long, strigose, flattened or slightly channelled above; leaves imparipinnate, narrowly oblong, 4–15 (–20) cm. long, 1–5 cm. wide; leaflets 8–12-jugate, oblong, 5–25 mm. long, 2·5–10 mm. wide, glabrous above, adpressed-strigose below, apex emarginate with a small median apiculus, base cuneate; inflorescences axillary, pedunculate, densely racemose; peduncle 1·5–7 cm. long, adpressed-strigose; bracts narrowly deltoid-ovate, 3–5 mm. long, 1–2·5 mm. wide, membranous, apex acuminate, margins ciliate, erose, sometimes sparsely glandular; 2 small, subulate bracteoles sometimes present at the base of the calyx; calyx narrowly cylindrical-campanulate, about 7 mm. long, 2–2·5 mm. wide, ± densely adpressed fuscous-strigose, calyx-teeth subequal, subulate, about 3 mm. long, 0·5 mm. wide at base; petals white or yellowish, standard oblong-obovate, about 9 mm. long, 4 mm. wide, apex shortly emarginate, apiculate, base tapering to a very wide claw, limb of wings narrowly oblong, about 4·5 mm. long, 1·5 mm. wide, apex rounded, base prominently auriculate with a rounded auricle, claw slender, about 2·5 mm. long, limb of keel shortly boat-shaped, about 3 mm. long, 2 mm. wide, apex blunt, base conspicuously auriculate, tapering to a broad claw about 2·5 mm. long; stamens 5 mm. long, all except the vexillary connate for about 4 mm.; anthers oblong, about 0·4 mm. long, 0·2 mm. wide; ovary oblong, about 4 mm. long, 1 mm. wide, adpressed-pubescent, many-ovulate; style about 2 mm. long, glabrous, upcurved, tapering from a wide base, stigma small, terminal. Pod indehiscent or tardily dehiscent, oblong, triangular in transverse section, about 2·5–3·5 cm. long, 0·7–0·8 cm. wide, stiffly chartaceous, somewhat inflated, thinly adpressed-strigose, often reddish with the style-base persisting as a hooked apical beak; seeds irregularly rectangular, 4–5 mm. long, 3–4 mm. wide, distinctly compressed; testa smooth, olive-green or brownish.

HAB.: In cultivated fields and gardens, on sand-dunes or on sandy river-margins; sea-level to 3,700 ft. alt.; fl. Febr.–April.

DISTR.: Divisions 2, 4–6, 8, common in Division 4. Widely distributed in the Mediterranean region and eastwards to Iran; Atlantic Islands; not infrequent elsewhere as an adventive.

2. Troödos, 1880, *Sintenis & Rigo* 1013 ! Platres, 1937, *Kennedy* 1296 !
4. Common; *Kotschy* 30; *Holmboe* 261; *Syngrassides* 1057 ! *Lindberg f.* s.n.; *Merton* 1998 !
5. Between Nicosia and Kythrea, 1880, *Sintenis & Rigo* 432 ! Salamis, 1962, *Meikle* 2571 !
6. Nicosia, 1934, *Syngrassides* 1386 ! Between Syrianokhori and Ayia Irini, 1941, *Davis* 2614 ! Nicosia, 1950, *Chapman* 523 !
8. Ayios Philon near Rizokarpaso, 1941, *Davis* 2231 !

7. A. suberosus *Banks et Sol.* in Russell, Nat. Hist. Aleppo, ed. 2, 2: 260 (1794); Eig in Syst. Stud. Astragali, 26 (1955); Davis, Fl. Turkey, 3: 68 (1970); Osorio-Tafall et Seraphim, List Vasc. Plants Cyprus, 55 (1973).

A. tuberculosus DC., Astragalogia, 33, t. 22 (1802); Boiss., Fl. Orient., 2: 239 (1872); Holmboe, Veg. Cypr., 108 (1914); Post, Fl. Pal., ed. 2, 1: 382 (1932).

A. tuberculosus DC. ssp. *mersinensis* Širj. in Feddes Repert., 53: 241 (1944).

Prostrate or decumbent biennial or perennial with a tough woody rootstock; stems 15–30 cm. long, much branched at base, thinly clothed with spreading white or black, slender hairs; stipules ovate-deltoid, herbaceous,

5–10 mm. long, 3–4 mm. wide, apex acute or acuminate, margins thinly glandular-ciliate, shortly connate abaxially; petioles usually short, often less than 1 cm. long, patently pilose, upper surface distinctly canaliculate; leaves imparipinnate, narrowly linear-oblong, 3–7 (–12) cm. long, 0·8–1·8 cm. wide; leaflets 10–18-jugate, or frequently subopposite or scattered, small, broadly oblong or suborbicular, 3–9 mm. long, 3–5 mm. wide, rather thick, glabrous above, thinly pilose below, apex rounded or emarginate, base rounded or cuneate; petiolule very short, often sparsely glandular; inflorescences axillary, conspicuously pedunculate; flowers 8–12 (or more) in a rather congested raceme; peduncle to about 10 cm. long, elongating in fruit, usually exceeding the subtending leaf, subglabrous or thinly clothed with spreading (often black) hairs; bracts ovate or lingulate, 3–5 mm. long, 1–2 mm. wide, subherbaceous, apex acuminate, shortly acute, sometimes rounded, margins erose, ciliate, often sparsely glandular; calyx cylindrical, 8–10 mm. long, 3–4 mm. wide, often reddish, ± densely clothed with black, spreading or suberect hairs, calyx-teeth linear-subulate, 2·5–3 mm. long, 1 mm. wide at base, apex subacute; petals creamy-white or purplish, standard oblong-elliptic, 17–20 mm. long, 8 mm. wide, tapering gradually to a blunt apex and basally to a wide claw 6–7 mm. long, limb of wings narrowly oblong, 9 mm. long, 2 mm. wide, apex rounded, base very conspicuously auriculate, claw slender about 6 mm. long, limb of keel boat-shaped, 7 mm. long, 3·5 mm. wide, apex rounded, slightly upcurved, base distinctly auriculate, claw slender about 6 mm. long; stamens about 12 mm. long, connate for about 9 mm.; anthers narrowly oblong, about 0·7 mm. long, 0·3 mm. wide; ovary narrowly oblong, 6 mm. long, 1·5 mm. wide, adpressed-pubescent or glabrescent, many-ovulate, style 7 mm. long, gently upcurved, thickening gradually towards base, stigma terminal, capitate. Pod oblong-elliptic, 1·5–3 cm. long, 0·8–1·3 cm. wide, dorsiventrally compressed, arcuate, apex terminating in a sharp recurved beak formed by the style base, valves rugose-tuberculate, pubescent or glabrous, pale brown; seeds irregularly rectangular, 2·5–3 mm. long and about as wide; testa mid-brown, somewhat striate-pitted.

var. suberosus.
TYPE: Syria; Aleppo, *Russell* (BM).

Pods pubescent.

HAB.: Cultivated and fallow fields; sea-level to 500 ft. alt.; fl. March–April.

DISTR.: Divisions 4, 6, rare. Also S. Turkey, Syria.

4. Ayia Napa, 1905, *Holmboe* 31.
6. West of Nicosia, 1905, *Holmboe* 272a; near the English School, Nicosia, 1963, *B. Trenbath* s.n. !

var. hartmannii (*Holmboe*) *Meikle* **comb. nov.**
A. tuberculosus DC. var. *hartmannii* Holmboe, Veg. Cypr., 109 (1914).
TYPE: Dry fields west of Nicosia, 1905, *Holmboe* 272b (O).

Pods glabrous, less rugose than in var. *suberosus*.

HAB.: Cultivated and fallow fields; c. 500 ft. alt.; fl. ? March–April.

DISTR.: Division 6, rare. ? Endemic.

6. Dry fields west of Nicosia, 1905, *Holmboe* 272b.

8. A. lusitanicus *Lam.*, Encycl. Méth., 1: 312 (1783); Boiss., Fl. Orient., 2: 267 (1872); Holmboe, Veg. Cypr., 109 (1914); Post, Fl. Pal., ed. 2, 1: 387 (1932); Lindberg f., Iter Cypr., 19 (1946).
Phaca boetica L., Sp. Plant., ed. 1, 755 (1753); Sibth. et Sm., Fl. Graec. Prodr., 2: 284 (1813), Fl. Graec., 8: 20, t. 727 (1833).
Erophaca boetica (L.) Boiss., Voy. Bot. Esp., 2: 177 (1839); Unger et Kotschy, Die Insel Cypern, 381 (1865).

Erect or sprawling perennial 30–100 cm. high, stems and leaves rather densely adpressed greyish-pubescent; stems robust usually branched below, unbranched or sparingly branched above, with distinct longitudinal ribs; stipules brownish, papery, conspicuous, connate abaxially, the lower broadly deltoid-ovate, 8–15 mm. long, 5–8 mm. wide, acute or acuminate, the upper deltoid-subulate, long-acuminate, 10–15 mm. long, 2–4 mm. wide at base; leaves imparipinnate, narrowly oblong, 10–20 cm. long, 4–7 cm. wide, petioles 2–3 cm. long, flattened or slightly sulcate above; leaflets 6–12-jugate, ovate-elliptic, 2·5–5 cm. long, 1–2·5 cm. wide, adpressed-sericeous below and glabrous or glabrescent above (ssp. *lusitanicus*) or sericeous on both surfaces (ssp. *orientalis*), dotted over, on the upper surface, with small brown, sessile glands, apex obtuse or subacute, mucronate, base generally rounded; petiolules very short, flanked by minute, fleshy, brown, glandular stipels; inflorescences axillary, erect, conspicuous, rather densely racemose, arising from the axils of the middle leaves; peduncles stout, 5–8 cm. long, shorter than the subtending leaves; racemes 6–9 cm. long, bracts membranous, subulate-filiform, to about 15 mm. long, 1·5 mm. wide at base, the basal margins fringed with reddish glands, bracteoles 2, apical on the pedicels, filiform, 3–8 mm. long, glandular at the base; pedicels 3–10 mm. long, densely fuscous-pubescent, usually recurved; calyx broadly campanulate with a rounded base, 10–13 mm. long, 5–7 mm. wide, thinly or densely adpressed fuscous-pubescent, calyx-teeth unequal, the adaxial pair about 3 mm. long, the others about 5 mm. long, all narrowly deltoid-acuminate; petals creamy- or greenish-white, standard obovate, about 2·5 cm. long, 1·3 cm. wide, apex rounded, margins strongly reflexed, base tapering gradually to a short, broad claw, limb of wings narrowly oblong, about 16 mm. long, 4 mm. wide, apex obtuse, base abruptly narrowed to a slender claw about 10 mm. long, but without any distinct auricle, limb of keel boat-shaped, about 17 mm. long, 9 mm. wide, apex rounded, somewhat upcurved, base distinctly auriculate, claw slender, about 10 mm. long; stamens unequal, the vexillary free, about 15 mm. long, the others connate for about 15–22 mm. from base, the free parts of the filaments 3–5 mm. long; anthers oblong, about 1 mm. long, 0·5 mm. wide; ovary oblong, compressed, many-ovulate, 5–7 mm. long, 1·5–2 mm. wide, densely adpressed-pilose, with a slender gynophore 1·5–2 mm. long, style glabrous 6–10 mm. long, tapering to apex, stigma minute, terminal. Pod conspicuous, pendulous, inflated, oblong, 6–7 cm. long, 2·5 cm. wide, apex acute, shortly rostrate, base narrowed to a short stipe, valves reddish-brown, coriaceous, ± densely clothed with short greyish hairs; seeds reniform, compressed, about 10 mm. long, 6–7 mm. wide; testa rich brown, sublustrous, minutely foveolate.

ssp. **orientalis** *Chater et Meikle* in Feddes Rep., 79: 48 (1968); Davis, Fl. Turkey, 3: 94 (1970); Osorio-Tafall et Seraphim, List Vasc. Plants Cyprus, 55 (1973).

TYPE: Greece; "in monte Malevo Laconiae prope Hagios Joannis, 900 m., 19-iv-1857", *Orphanides* 618 (K !).

Leaflets adpressed-sericeous on both surfaces; calyx rather densely fuscous-pilose; young pod densely greyish-felted.

HAB.: Dry mountain slopes, often under Pines, sometimes in garigue amongst *Cistus* etc.; almost confined to igneous formations, very rarely on chalk; 1,000–4,500 ft. alt.; fl. Febr.–May.

DISTR.: Divisions 2, 3, very common in Division 2. Ssp. *orientalis* in Greece, Aegean Islands, Turkey, Syria, Lebanon and Cyprus; ssp. *lusitanicus* in Spain, Portugal, Morocco, Algeria.

2. Very common; *Kotschy*; *Sintenis & Rigo* 713 ! *Holmboe* 804; *Syngrassides* 341 ! *Kennedy* 570 ! *Lindberg f.* s.n. ! *Davis* 3064 ! 3452 ! *Casey* 670 ! *Chapman* 328 ! *Meikle* 2263 ! etc.

3. Lefkara, 1941, *Davis* 2791 ! Below Stavrovouni, 1959, *C. E. H. Sparrow* 35 ! Trimiklini, 1963, *J. B. Suart* 108 !

NOTES: The two variants are usually easy to recognize, the western ssp. *lusitanicus* having the leaflets quite glabrous or soon glabrescent above, and the calyces thinly pubescent with pale hairs, whereas the leaflets of ssp. *orientalis* are almost equally pubescent on both surfaces, and the calyces are densely fuscous-pilose. A few specimens from Algeria are somewhat transitional, though clearly more closely allied to ssp. *lusitanicus*. It is strange that none of the authors dealing with the genus *Astragalus* seems to have noted this well-marked distinction.

9. A. macrocarpus *DC.*, Astragalogia, 143, t. 28 (1802); Boiss., Fl. Orient. 2: 273 (1872); Post, Fl. Pal., ed. 2, 1: 388 (1932).

Erect perennial 30–60 cm. high with a tough, fibrous tap-root; stem robust, longitudinally ridged, adpressed-pilose or densely lanuginose with long spreading hairs; basal leaves rather small, usually less than 10 cm. long and 2 cm. wide, with conspicuous oblong or ovate-acuminate, membranous, ciliate stipules 15–20 mm. long, 5–7 mm. wide at base; upper leaves much larger, 14–20 cm. long, 2–6 cm. wide, stipules subulate, 20–25 mm. long, 3–5 mm. wide at base, membranous, persistent, with lanuginose margins, petioles 3·5–7 cm. long, ± angular, lanuginose, leaflets 15–30, opposite or alternate, oblong, 7–27 mm. long, 4–12 mm. wide, glabrous or subglabrous above, thinly lanuginose below, apex emarginate or rounded, rarely subacute, base tapering to a short petiolule; inflorescences 3–9-flowered, congested, racemose, arising in the axils of the median leaves; bracts subulate-filiform, to about 20 mm. long, 1·5 mm. wide at base, membranous, thinly lanuginose; pedicels very short, usually less than 2 mm. long; calyx 14–20 mm. long, 5–6 mm. wide, thinly lanuginose, calyx-tube cylindrical or slightly urceolate, 8–12 mm. long, pallid, obscurely veined, calyx-teeth 5–10 mm. long, linear-filiform, irregularly recurved; petals bright yellow, standard narrowly oblong-obovate, glabrous, about 25–28 mm. long, 10–12 mm. wide, longitudinally folded, apex rounded, base tapering to an indistinct claw, wings narrowly oblong, 25–27 mm. long, 3–4 mm. wide, limb 18–20 mm. long, apex blunt or rounded, base with a conspicuous oblong auricle, 2–3 mm. long, claw very slender about 7 mm. long, keel 23–24 mm. long, slightly upcurved or almost straight, limb narrowly oblong, 14–15 mm. long, 4 mm. wide, apex rounded, base conspicuously auriculate, claw about 10 mm. long; stamens about 22 mm. long, filaments of all, except the vexillary, joined to within 4 mm. from apex; anthers oblong, about 1 mm. long, 0·5 mm. wide; ovary narrowly oblong, about 5 mm. long, 1·4 mm. wide, glabrous, many-ovulate, style 17–18 mm. long, glabrous, slender, stigma terminal, capitate-clavate. Pod large, oblong, inflated, 2·5–5 cm. long, 1·5–3 cm. wide, with a sharp straight beak 7–10 mm. long, pericarp corky or spongy, dark red, with rather obscure transverse venation, suture conspicuously bordered; seeds reniform, compressed, about 6 mm. long, 5 mm. wide; testa smooth, pale brown.

ssp. **lefkarensis** *Agerer-Kirchoff et Meikle* in Boissiera, 25: 120 (1976).

TYPE: Cyprus; above Pano Lefkara on steep chalk slope (abandoned vineyard), April 7, 1974, *Meikle* 5007 (K !).

Stems and leaves lanuginose with long, spreading, silvery hairs; calyx-teeth green; pod 2·5–3 cm. long, 1·5–2 cm. wide.

HAB.: In garigue on a steep chalky slope, with *Cistus*, *Lithospermum* and *Onobrychis*; 2,200 ft. alt.; fl. March, fr. May.

DISTR.: Division 2, very rare. Apparently endemic.

2. Chalk hill above Pano Lefkara, by path to Kato Dhrys, 1941, *Kennedy & Davis* in *Davis* 2710 ! Same locality, 1941, *Davis* 3551 ! and, 1974, *Meikle* 5007 !

NOTES: Very closely allied to *A. macrocarpus* DC. from Palestine and Syria; but differing in the long-haired, lanuginose indumentum of stems and leaves, the green (not yellowish) calyx-teeth, and in the relatively small number of rather large leaflets. The fruit is rather smaller and more distinctly veined transversely in ssp. *lefkarensis* than in ssp. *macrocarpus*, but the value of this character is difficult to judge in the absence of additional fruiting material of both species.

10. A. caprinus *L.*, Sp. Plant., ed. 2, 1071 (1763); DC., Astragalogia, 214 (1802).

Acaulescent or short-stemmed perennial with long, thick, tough roots; leaves imparipinnate, spreading to form a distinct rosette, narrowly oblong, (5–) 10–20 (–25) cm. long, 1–4 (–6) cm. wide; stipules oblong-acuminate, membranous, conspicuous, 10–30 mm. long, 3–8 mm. wide, glabrous or villose; petioles 2–9 (–13) cm. long, sulcate above, glabrous or villose with long white hairs; leaflets (10–) 13–16 (–20)-jugate, oblong or ovate, 4–30 mm. long, 3–15 mm. wide, glabrous or villose, apex rounded, retuse or subacute, base rounded; stipels wanting; inflorescences axillary, pedunculate or almost sessile and forming a radical cluster; peduncle up to 10 cm. long, glabrous or ± villose with spreading hairs; flowers 4–14 (or more) in dense, short, racemes; bracts narrowly lanceolate or almost filiform, 10–20 mm. long, 1–4 mm. wide at base, glabrous or villose; pedicels about 3 mm. long, glabrous or villose; bracteoles wanting; calyx cylindrical, 15–17 mm. long, 4–6 mm. wide, membranous, glabrous or thinly villose, calyx-teeth subequal, subulate, 6–7 mm. long, about 1 mm. wide at base; petals yellow or greenish turning reddish with age, standard narrowly obovate, 23–30 mm. long, 8–13 mm. wide, limb somewhat upcurved, apex emarginate, margins sharply reflexed, base tapering to a wide claw about 10 mm. long, limb of wing narrowly oblong, 11–15 mm. long, 3–4 mm. wide, apex rounded or subtruncate, base with a conspicuous oblong auricle, conjunctival flap distinctly developed, claw slender, 10–11 mm. long, limb of keel boat-shaped, 8–9 mm. long, 3–5 mm. wide, upcurved almost at a right angle, apex blunt, base strongly and deeply auriculate, claw slender 11–13 mm. long; stamens about 20 mm. long, axillary, free, the remainder united in 2 series, the free parts of the longer filaments about 3 mm. long, those of the shorter filaments about 2 mm. long; anthers oblong, about 0·8 mm. long, 0·3 mm. wide; ovary narrowly oblong, 4–6 mm. long, 1–1·5 mm. wide, adpressed-villose, many-ovulate, the base narrowed to a slender gynophore about 4–5 mm. long, style glabrous, about 12 mm. long, sharply upcurved towards apex, stigma capitate, somewhat oblique. Pod oblong, somewhat inflated, 1·5–3 cm. long, 1–1·7 cm. wide, ± stipitate, apex distinctly and sharply rostrate, valves cartilaginous, pale brown, glabrous or thinly villose externally, ± distinctly veined; seeds reniform, compressed, 4–6 mm. long, 3–4 mm. wide; testa glabrous or furfuraceous, rich brown, sometimes with scattered purple blotches.

ssp. **laniger** (*Desf.*) *Maire* in Jahandiez et Maire, Cat. Plant. Maroc., 2: 413 (1932)—as *A. caprinus* ssp. *lanigerus*.

A. laniger Desf., Fl. Atlant., 2: 181, t. 202 (1799); DC., Astragalogia, 172 (1802); Bunge in Mém. Acad. Sci. St. Pétersb., 15: 48 (1869); Davis, Fl. Turkey, 3: 108 (1970); Osorio-Tafall et Seraphim, List Vasc. Plants Cyprus, 56 (1973)—as *A. lanigerus*.

A. alexandrinus Boiss., Diagn., 1, 9: 75 (1849), Fl. Orient., 2: 284 (1872), Suppl. 179 (1888); Holmboe, Veg. Cypr., 109 (1914); Post, Fl. Pal., ed. 2, 1: 390 (1932); Lindberg f., Iter Cypr., 19 (1946); Rechinger f. in Arkiv för Bot., ser. 2, 1: 425 (1950); Zohary, Fl. Palaest., 2: 73, t. 101 (1972).

TYPE: Tunisia; " in arenis deserti prope Cafsam [Gafsa] ", *Desfontaines* (P).

Leaves, stipules, bracts and calyces generally villose with long spreading hairs; leaflets rather small, 4–18 mm. long, 3–10 mm. wide, apex usually rounded, truncate or retuse; inflorescences subsessile or shortly pedunculate, few (3–8)-flowered. Pods often shortly and indistinctly stipitate.

R

HAB.: Eroded hillsides, generally on calcareous or sandstone formations, sometimes in fields or in thin *Pinus* woodland; 500–3,600 ft. alt.; fl. March–May.

DISTR.: Divisions 2, 3, 7, 8, common in Division 7. Also Turkey, Syria, Iraq, N. Africa.

2. Between Pano Panayia and Khrysorroyiatissa Monastery, 1905, *Holmboe* 1062; Platres, 1937, *Kennedy* 571 ! Near Khrysorroyiatissa Monastery, 1941, *Davis* 3370 ! Maliades, between Mandria and Perapedhi, 1962, *Meikle* 2867 !
3. Alethriko, 1905, *Holmboe* 206.
7. Common; *Sintenis & Rigo* 431 ! *Lindberg f.* s.n.; *Davis* 2199 ! 2934 ! *Casey* 571 ! *Chapman* 667 !
8. Matakos, near Galinoporni, 1962, *Meikle* 2441 !

NOTES: None of the characters used by Boissier and Eig to distinguish *A. alexandrinus* from *A. laniger* can be considered wholly satisfactory, nor is it likely that a plant which ranges from Morocco to Cyrenaica should stop short at the borders of Egypt and be replaced by a distinct species around Alexandria. Some of the Cyprus specimens closely resemble typical *alexandrinus* in having small, subsessile pods; others approach *Astragalus bersabeënsis* Eig et Samuelsson in having larger, stipitate pods, but there can be no doubt that all the Cyprus material represents simply a range of variation within a single species. Examination of North African material shows a similar range of variation in *A. caprinus* and *A. laniger*, with such a broad overlap of characters as to justify Maire's opinion that the two can be distinguished only at infraspecific level, even if the extremes appear quite distinct.

[A. GLAUX *L.*, Sp. Plant., ed. 1, 759 (1753). Recorded, "in Cypri campestribus" by Sibthorp (Fl. Graec. Prodr., 2: 88; 1813), but undoubtedly an error. *A. glaux* is restricted to the western Mediterranean region, occurring in S. France, Spain, Morocco and Algeria; it is not known from any of the countries adjacent to Cyprus, and must be excluded from the flora of the island. The records for *A. glaux* in Poech (Enum. Plant. Ins. Cypr., 40; 1842) and in Unger et Kotschy (Die Insel Cypern, 381; 1865) are simply repetitions of Sibthorp's original reference.]

11. A. echinus *DC.*, Astragalogia, 197, t. 34 (1802); Boiss., Fl. Orient., 2: 338 (1872); Holmboe, Veg. Cypr., 109 (1914); Post, Fl. Pal., ed. 2, 1: 393 (1932); Lindberg f., Iter Cypr., 19 (1946); Chapman, Cyprus Trees and Shrubs, 45 (1949); Rechinger f. in Arkiv för Bot., ser. 2, 1: 425 (1950); Zohary, Fl. Palaest., 2: 76, t. 108 (1972); Osorio-Tafall et Seraphim, List Vasc. Plants Cyprus, 56 (1973).

[*A. aristatus* (non L'Hérit.) Sibth. et Sm., Fl. Graec. Prodr., 2: 90 (1813), Fl. Graec., 8: 25 (1833) pro parte excl. t. 735.]

Woody subshrub 12–60 cm. high, forming dense, much-branched, spiny cushions; lower parts of stems covered with the persistent remains of stipules and leaf-rhachises; stipules ovate-acuminate, 8–10 mm. long, 4–5 mm. wide, scarious, yellowish or pale brown, glabrous except for a fringe of white silky hairs near the base; leaves paripinnate, crowded, 1–3 (–5) cm. long, petiole up to 1·3 cm. long, sometimes very short or almost wanting, subglabrous or thinly villose; rhachis produced beyond leaflets as a hard, straight, yellowish spine; leaflets 2–5 (–7)-jugate, narrowly obovate-elliptical, 3–9 mm. long, 1–3 mm. wide, subglabrous or ± densely villose, apex with a pallid, cartilaginous cusp, base tapering to a short petiolule; inflorescence axillary, flowers 2–4 in a sessile cluster, partly concealed by the subtending stipules; bracts linear-oblong, about 5–6 mm. long, 1 mm. wide, densely villose externally, slightly keeled, apex acute or subacute; calyx ± campanulate, 6–8 mm. long, 3 mm. wide, densely villose externally, calyx-teeth erect, sharply pointed, about 4–5 mm. long, 0·5 mm. wide at base; petals milky-white or pale pink, standard glabrous, panduriform, 8–11 mm. long, 4–5 mm. wide near base of limb, claw wide, 2–4 mm. long, 1·5–2 mm. wide, limb of wings narrowly oblong, 3·5–4·5 mm. long, 0·8–1 mm. wide, apex rounded, base strongly auriculate, conjunctival flap well developed, claw slender, 4·5–6 mm. long, limb of keel oblong, straightish or slightly upcurved,

3·5–4·5 mm. long, 1·8–2 mm. wide, base auriculate, claw slender 5–6 mm. long; stamens 7·5–9 mm. long, all but the vexillary connate to within 1·5–2 mm. from apex; anthers oblong, about 0·7 mm. long, 0·3 mm. wide; ovary ovate, ± compressed, 2·5–3·5 mm. long, 1–2·5 mm. wide, densely villose, 6–7-ovulate, style 5–6 mm. long, glabrous or villose towards base, stigma minute, terminal. Pod small, ovoid, compressed, about 5 mm. long, 2·5–3 mm. wide, villose externally, usually 1-seeded; seed ovoid-oblong, about 3·5 mm. long, 2·5 mm. wide; testa dark brown.

var. **echinus.**
TYPE: " in Oriente. *Labillardière* " (G).

Petiole and rhachis 2–3 (–5) cm. long, stout; leaflets generally more than 5 mm. long, 2 mm. wide, rather thinly villose or occasionally subglabrous. Plants usually more than 10 cm. high, densely tufted.

HAB.: On dry mountainsides on igneous formations, 1,600–4,000 ft. alt.; fl. May–June.

DISTR.: Division 2, common at Evrykhou, Platres, also at Troödos and Galata. Also Syria, Lebanon, Palestine and Sinai.

2. Common; *Sintenis & Rigo* 814 ! *Holmboe* 848; *E. Chapman* 81 ! *Kennedy* 569 ! *Lindberg f.* s.n. ! *Meikle* 2861 !

var. **chionistrae** (*Lindberg f.*) *Meikle* **stat. nov.**
M. echinus DC. ssp. *chionistrae* Lindberg f., Iter Cypr., 19, fig. 15 (1946); Osorio-Tafall et Seraphim, List Vasc. Plants Cyprus, 56 (1973).
[*A. angustifolius* (non Lam.) Unger et Kotschy, Die Insel Cypern, 382 (1865); H. Stuart Thompson in Journ. Bot., 44: 306 (1906); Holmboe, Veg. Cypr., 109 (1914); Osorio-Tafall et Seraphim, List Vasc. Plants Cyprus, 56 (1973).]
TYPE: Cyprus; Troödos, on stony ground at the summit of Khionistra, June 16, 1939, *Lindberg f.* s.n. (K !).

Petiole and rhachis usually less than 2 cm. long, rather slender; leaflets usually less than 5 mm. long and 2 mm. wide, densely villose. Plants usually less than 10 cm. high, in very dense, low cushions.

HAB.: On rocky mountainsides and summits on igneous formations; 5,850–6,400 ft. alt.; fl. July–Sept.

DISTR.: Division 2, common, Prodhromos, Khionistra, Troödos area. Apparently endemic.

2. Common; *Kotschy* 781; *Post* s.n.; *Holmboe* 887; *Kennedy* 1298 ! 1299 ! *Lindberg f.* s.n. ! *Davis* 1819 ! *F. M. Probyn* 129 ! *Merton* 1922 !

NOTES: There is usually little difficulty in recognizing the var. *chionistrae*, on account of its dwarf habit and grey, densely woolly leaflets; the distinctions between this and typical var. *echinus* are, however, so slight, that it would seem more appropriate to distinguish them at varietal rather than subspecific level. It would be interesting to know if, as suggested by the records, there is in fact a gap in the vertical distribution of the two varieties, with var. *echinus* not occurring above 4,000 ft. alt., and var. *chionistrae* not descending below c. 5,500 ft. alt.

12. A. cyprius *Boiss.*, Fl. Orient., 2: 469 (1872); Holmboe, Veg. Cypr., 109 (1914); Lindberg f., Iter Cypr., 19 (1946); Rechinger f. in Arkiv för Bot., ser. 2, 1: 425 (1950); Osorio-Tafall et Seraphim, List Vasc. Plants Cyprus, 56 (1973).
[*A. incanus* (non L.) Sibth. et Sm., Fl. Graec. Prodr. 2: 89 (1813), Fl. Graec., 8: 23 (1833) pro parte quoad spec. cypr. excl. t. 732; Unger et Kotschy, Die Insel Cypern, 382 (1865); H. Stuart Thompson in Journ. Bot., 44: 306 (1906).]
[*A. spruneri* (non Boiss.) Post in Mém. Herb. Boiss., 18: 93 (1900); Holmboe, Veg. Cypr., 110 (1914); Osorio-Tafall et Seraphim, List Vasc. Plants Cyprus, 56 (1973).]
[*A. dyctiocarpus* (non Boiss.) Kotschy in Unger et Kotschy, Die Insel Cypern, 382 (1865).]
TYPE: Cyprus; " in argillosis insulae Cypri ad Buffavento alt. 1300′ ", 1862, *Kotschy* 255 (W).

Loosely tufted perennial with a tough woody rootstock; stems very short, usually less than 3 cm. long, densely covered within the decayed remains of leaves and stipules; leaves crowded, basal or sub-basal, narrowly oblong, imparipinnate, 2·5–15 cm. long, 0·5–2·5 cm. wide; stipules narrowly oblong-acuminate, 3–10 mm. long, 1·5–2·5 mm. wide, densely clothed with adpressed, medifixed silvery hairs; petiole 1–8 cm. long, adpressed-pilose, slightly sulcate above; leaflets oblong-obovate, 5–12-jugate, 2–20 mm. long, 1·5–10 mm. wide, silvery-grey on both surfaces with adpressed medifixed hairs, apex rounded or subacute, base rounded or cuneate; inflorescences axillary, pedunculate, racemose; peduncle 3–15 cm. long, often curved, usually exceeding the leaves; racemes lax or rather congested, 3–12 (–17)-flowered; bracts narrowly deltoid, 1–3 (–5) mm. long, 1–1·5 mm. wide, reddish or fuscescent, adpressed-pilose externally; pedicels very short or almost wanting; calyx cylindrical, 8–10 mm. long, 2·5–3 mm. wide, densely clothed with blackish adpressed hairs, calyx-teeth linear, subacute, 1·5–2·5 mm. long, 0·6 mm. wide, erect, subequal; petals dirty pinkish or yellowish, standard oblong-elliptic, 15–20 mm. long, 5–6 mm. wide, somewhat upcurved, limb folded longitudinally, with an elongate blunt apex and strongly reflexed margins, claw rather wide, 5–6 mm. long, limb of wings narrowly oblong, 6–8 mm. long, 1·5 mm. wide, apex rounded, base strongly auriculate, claw slender, 6–8 mm. long, limb of keel oblong, upcurved, 5–6 mm. long, 2 mm. wide, apex blunt, base auriculate with a small conjunctival pouch, claw 7–10 mm. long, stamens 12–15 mm. long, filaments of all except the vexillary united to within 4 mm. from apex, and forming a short ± glanduligerous basal cupule to the inner side of which the vexillary filament is attached; anthers oblong, 0·7 mm. long, 0·4 mm. wide; ovary narrowly oblong, 4–6 mm. long, 0·8–1 mm. wide, adpressed-pubescent, many-ovulate, style glabrous, 8–11 mm. long, upcurved at apex, stigma terminal, truncate, scarcely capitate. Pods erect or spreading, cylindrical, 20–30 mm. long, 4–5 mm. wide, straight or ± upcurved, scarcely compressed, valves hard, often reddish, thinly adpressed-strigose, apex tapering to a short sharp beak, base enclosed within the persistent (often split) calyx; seeds oblong-reniform, 3·5–4 mm. long, 2·5–3 mm. wide, strongly compressed; testa shining brown, conspicuously pitted.

HAB.: Dry hillsides on calcareous or sandstone formations; 200–3,000 ft. alt.; fl. Febr.–May.

DISTR.: Divisions 1–5, 7, 8, very common in Division 7, common in Divisions 3, 5, 8. Apparently endemic.

1. Between Ayios Neophytos Monastery and Stephani, 1939, *Lindberg f.* s.n.; Dhrousha, 1941, *Davis* 3291 !
2. Makheras Monastery, 1862, *Kotschy* 217; Lefkara, 1941, *Davis* 2761 ! Khrysorroyiatissa Monastery, 1941, *Davis* 3357 ! Statos, 1941, *Davis* 3397 !
3. Common; *Kotschy* 555; *Davis* 3087 ! 3146 ! *Mavromoustakis* s.n. ! *Meikle* 2798 !
4. Between Ormidhia and Cape Pyla, 1880, *Sintenis & Rigo*.
5. Common; *Sintenis & Rigo*; *Post*; *Syngrassides* 61 ! 923 ! 1169 ! *Lindberg f.* s.n. !
7. Very common; *Kotschy* 255; *Sintenis & Rigo*; *Lindberg f.* s.n. ! *Mavromoustakis* 19 ! *Davis* 2503 ! 2906 ! 2942 ! *Casey* 990 ! *Chapman* 56 ! *Kennedy* 1786 !
8. Common; *Sintenis & Rigo*; *Holmboe* 527; *Haradjian* 144; *Davis* 2299 ! 2418 ! *Meikle* 2443 !

NOTES: Very closely allied to *A. incanus* L. and *A. spruneri* Boiss., and likely to prove just one of many segregates which might best be ranked as subspecies under *A. incanus* aggr. It is distinguished from *A. spruneri* Boiss. by its long narrow pods with tapering apices, and from *A. incanus* L. by its usually rather lax and sparse-flowered racemes, small fuscescent bracts and black-hairy calyces. The Cyprus population of the species is, on the whole, very uniform.

[ASTRAGALUS DEPRESSUS *L.*, Sp. Plant., ed. 2, 1073 (1763) is recorded for Cyprus in Davis, Fl. Turkey, 3: 94 (1970) and in Osorio-Tafall et Seraphim, List Vasc. Plants Cyprus, 55 (1973). I have seen no localized record or specimen, and suspect both references are based on some misunderstanding.]

19. GLYCYRRHIZA *L.*

Sp. Plant., ed. 1, 741 (1753).
Gen. Plant., ed. 5, 330 (1754).

Perennial herbs, often woody at the base; roots long, tough, sometimes sweetly flavoured, leaves imparipinnate, generally many-foliolate; leaflets entire, usually gland-dotted; stipules narrow, membranous, caducous; stipels wanting; inflorescences axillary, racemose or spicate, often dense and many-flowered, pedunculate or sessile; bracts narrow, membranous, caducous; bracteoles wanting; calyx tubular or campanulate, calyx-teeth subequal or the 2 adaxial rather shorter than the others; petals purple, white or yellowish, standard erect, narrowly obovate or oblong, shortly clawed, wings narrowly oblong, acute or subacute, keel rather shorter than wings; apex acute or obtuse; stamens 10, filaments all joined into a sheath split adaxially, or the vexillary stamen free, anthers incompletely 2-thecous, the thecae confluent towards apex, alternately unequal; ovary sessile, 2–many-ovulate, style filiform or somewhat thickened, glabrous, upcurved, stigma terminal. Pod ovoid, oblong or linear, occasionally moniliform, smooth or glandular-muricate, indehiscent or dehiscent, continuous within; seeds reniform or subglobose, estrophiolate.

About 12 species in the Mediterranean area, temperate Asia, Australia and temperate North and South America; *G. glabra* often cultivated and subsequently naturalized.

1. **G. glabra** *L.*, Sp. Plant., ed. 1, 742 (1753); Boiss., Fl. Orient., 2: 202 (1872); Post Fl. Pal., ed. 2, 1: 371 (1932); Lindberg f., Iter Cypr., 19 (1946); Davis, Fl. Turkey, 3: 260 (1970); Zohary, Fl. Palaest., 2: 84, t. 121 (1972); Osorio-Tafall et Seraphim, List Vasc. Plants Cyprus, 56 (1973).
TYPE: "*in* Franconia, Gallia, Hispania, Italia".

Erect, subshrubby perennial to about 1 m. high; stems sparingly branched, glandular-puberulous, distinctly angled or sulcate; leaves 10–20 cm. long, 6–9 cm. wide, imparipinnate, leaflets 3–7-jugate; stipules linear-subulate, about 2 mm. long, 0·5 mm. wide, pubescent, caducous; petiole stout, ± triangular in transverse section, 1·5–3 cm. long, shortly glandular-pubescent; leaflets oblong-lanceolate, 2–4·5 cm. long, 0·8–2 cm. wide, glandular and glutinous above and below, apex acute or obtuse, margins entire, base rounded or cuneate; inflorescences axillary in the upper part of the stems; peduncle erecto-patent, 2–3 cm. long, shortly pubescent; raceme oblong, 5–7 cm. long, 1·5–2·5 cm. wide, many-flowered; bracts oblong-acuminate, 1·5–3 mm. long, 0·8–1 mm. wide, reddish, glandular, somewhat carinate; pedicels very short, pubescent; calyx cylindrical, about 5 mm. long, 2 mm. wide, sparsely glandular-puberulous, calyx-teeth lanceolate, subequal, about 2 mm. long, 1 mm. wide at base, acute or subacute; petals purplish or violet, standard ovate-oblong, about 10 mm. long, 5 mm. wide, folded longitudinally, apex obtuse or subacute, base narrowing rather abruptly to a short claw, generally less than 1·5 mm. long, limb of wings narrowly oblong, about 6 mm. long, 1·5 mm. wide, apex upcurved, acute, base shortly auriculate, claw short, about 2 mm. long, limb of keel oblong, about 5·5 mm. long, 1·8 mm. wide, apex shortly rostrate-acute, abruptly upcurved, base shortly auriculate, claw about 2·5 mm. long; stamens diadelphous, about 6 mm. long, all except the vexillary united for about 3 mm.; anthers unequal, alternately large and oblong and small and suborbicular, the larger anthers about 0·8 mm. long, 0·5 mm. wide, the smaller about 0·5 mm. diam.; ovary linear-oblong, compressed laterally, 3–4 mm. long, 0·8 mm. wide, glabrous, about 7-ovulate, style 3–4 mm. long, tapering from base to apex, glabrous, stigma small, terminal. Pod oblong, strongly compressed, 10–30 mm. long, 5–8 mm. wide, apex shortly rostrate, valves dark brown, glabrous, smooth (var. *glabra*) or glandular-hispid (var. *glandulifera* (W. et K.) Reg. et Herd.); seeds reniform, somewhat compressed, 3·5 mm. long, 2·8 mm. wide; testa dark brown, smooth. *Plate 27.*

HAB.: In sandy or marshy fields; near sea-level; fl. May–June.

DISTR.: Divisions 5, 6. Widely distributed in Europe, temperate Asia and North Africa; often an escape from cultivation.

5. Salamis, 1901, *A. G. & M. E. Lascelles* s.n. ! and, 1939, *Lindberg f.* s.n. ! and, 1971, *Townsend* 71/113 !
6. Syrianokhori, 1936, *Syngrassides* 975 !

NOTES: Not seen on undisturbed ground in Cyprus, and most probably a relict of cultivation. The Cyprus specimens all seem to belong to var. *glabra.*

TRIBE 6. **Hedysareae** *DC.* Herbs, shrubs or subshrubs; leaves generally imparipinnate, without tendrils, occasionally 1-foliolate or simple; stipules present, stipels wanting; inflorescences umbellate or racemose, axillary; stamens diadelphous or monadelphous, free part of all the filaments, or of alternate filaments dilated, anthers dimorphic or uniform. Pods articulated, usually breaking into 1-seeded joints, sometimes indehiscent; seeds generally estrophiolate.

20. SCORPIURUS *L.*

Sp. Plant., ed. 1, 744 (1753).
Gen. Plant., ed. 5, 332 (1754).

Decumbent or sprawling annuals; leaves simple, entire, lamina decurrent along the petiole; stipules adnate to petiole; inflorescences axillary, pedunculate, umbellate; flowers rather small, nodding; bracts minute, membranous, bracteoles wanting; calyx with the 2 adaxial teeth connate; petals yellow, with long claws, standard obovate or suborbicular, wings obliquely obovate or oblong, keel upcurved, rostrate-acuminate; stamens diadelphous, alternate filaments dilated towards apex, anthers uniform; ovary sessile, many-ovulate, style incurved, dilated about the middle, attenuate towards apex, stigma terminal, capitate. Pods elongate, subterete, circinnately twisted and conspicuously constricted between the seeds, indehiscent or tardily dehiscent, ± articulated or subcontinuous within, valves prominently ribbed longitudinally, the ribs frequently muricate or tuberculate; seeds ovoid-globose, estrophiolate, cotyledons elongate, twisted.

Two species in the Mediterranean region, western Asia, Atlantic Islands and N.E. Tropical Africa.

1. S. muricatus *L.*, Sp. Plant., ed. 1, 745 (1753) emend. Lam., Fl. Franç., ed. 1, 2: 582 (1778); Fiori et Paoletti, Fl. Anal. Ital., 2, 1: 89 (1900); Zohary, Fl. Palaest., 2: 98, t. 146 (1972); Osorio-Tafall et Seraphim, List Vasc. Plants Cyprus, 63 (1973).

Decumbent annual; stems angular, subglabrous or ± densely pilose with long, spreading or subadpressed hairs; stipules subulate-acuminate, 5–20 mm. long, 1–4 mm. wide at base, membranous with a greenish midrib, subglabrous or pilose, persistent, basal margin adnate to petiole; leaves obovate-spathulate, 2–12 cm. long, 0·5–3 cm. wide, subglabrous or thinly pilose on both surfaces, often with 3 rather conspicuous longitudinal nerves and rather obscure venation, apex acute or subacute, rarely rounded or emarginate, base tapering very gradually to a flattened or somewhat canaliculate petiole; inflorescence axillary, pedunculate, umbellate, (1–) 2–5-flowered; peduncle erect, rather stout, 3–15 (–25) cm. long, sulcate; bracts minute, membranous,

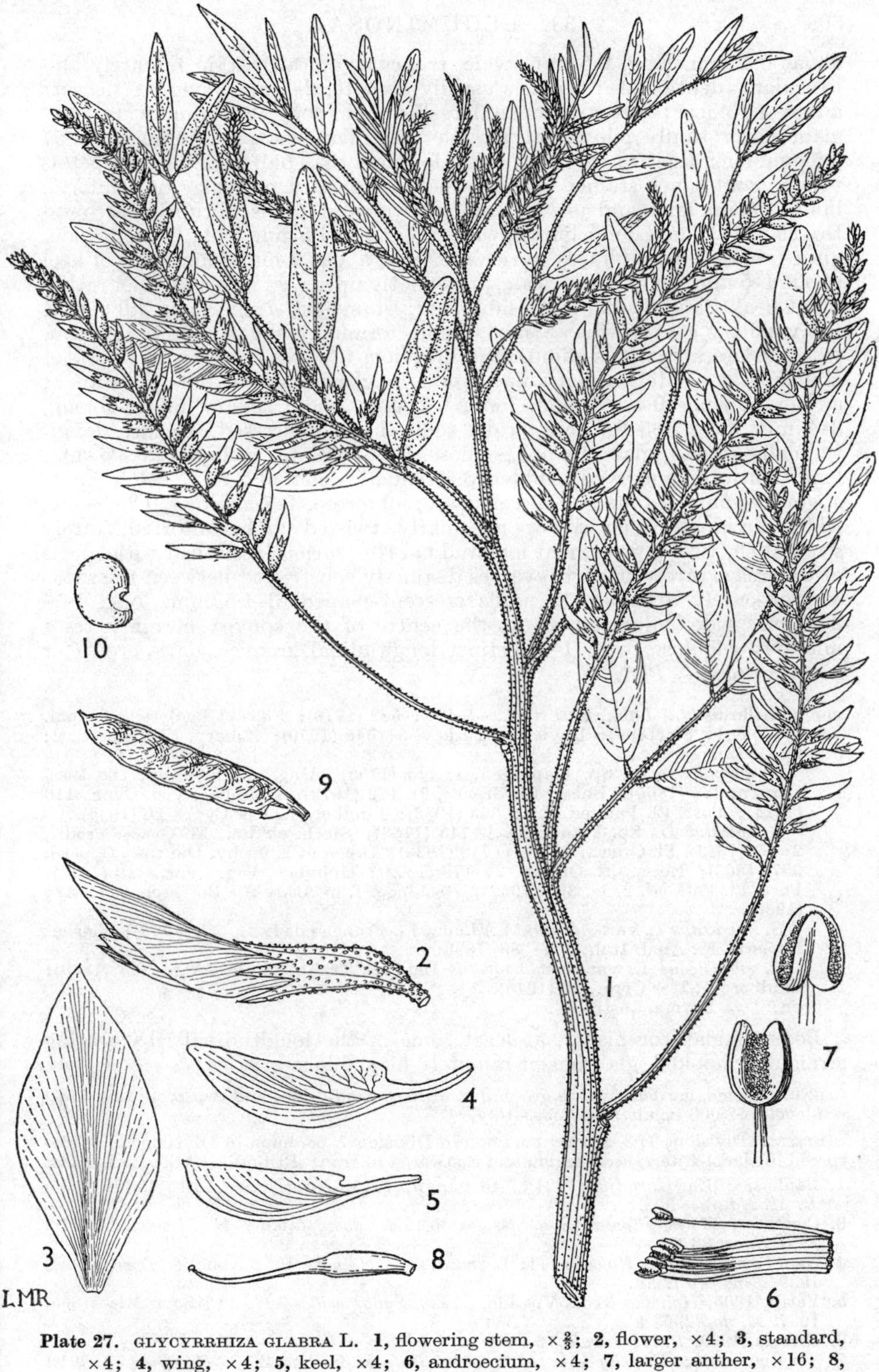

Plate 27. GLYCYRRHIZA GLABRA L. **1,** flowering stem, ×⅔; **2,** flower, ×4; **3,** standard, ×4; **4,** wing, ×4; **5,** keel, ×4; **6,** androecium, ×4; **7,** larger anther, ×16; **8,** gynoecium, ×4; **9,** pod, ×1; **10,** seed, ×4. (**1–8,** *Robertson* S/1031; **9–10,** *Field & Lazar* 962.)

about 0·4 mm. long, 0·5 mm. wide, truncate, apical margin minutely and irregularly denticulate; pedicels usually arcuate, 2–3 mm. long, ± densely adpressed-pilose; calyx campanulate 3–5 mm. long, 2·5 mm. wide, subglabrous or thinly pilose, calyx-teeth triangular, sharply acuminate, 1·5–2·5 mm. long, the two adaxial connate for more than half their length; petals yellow, sometimes stained red externally, standard orbicular or reniform, limb 4–5 mm. long, and as wide or rather wider, glabrous, abruptly narrowed to a claw 2–2·5 mm. long, limb of wings oblong, 4–4·5 mm. long, 1·5 mm. wide, apex rounded, base bluntly auriculate, claw about 2 mm. long, limb of keel about 4–5 mm. long, 2 mm. wide, apex gently upcurved into a distinct rather blunt beak, claw slender 2–2·5 mm. long; stamens 6–7 mm. long, all except the vexillary connate for about 3·5 mm., forming a slightly curved sheath, with the lateral margins minutely scabridulous towards the base, free parts of filaments alternately about 2 and 3 mm. long, strongly dilated at the apex; anthers oblong 0·4–0·5 mm. long, 0·2 mm. wide; ovary linear-oblong, 3–4 mm. long, 0·5–0·7 mm. wide, somewhat compressed, arcuate, longitudinally striate, glabrous or ± pilose, 5–10-ovulate, style about 3·5 mm. long, abruptly upcurved, suddenly attenuate about 0·5 mm. below apex, stigma terminal, capitate. Pod elongate, subterete, 3–5 cm. long, 0·2–0·4 cm. wide, circinnately curled or irregularly twisted and contorted, rarely straightish, with 9 prominent longitudinal ribs variously clothed with spines or tubercles, rarely smooth; valves distinctly constricted between the seeds and obscurely articulated; seeds crescent-shaped, 3–4·5 mm. long, 1·5–2·5 mm. wide, hilum small, at the centre of the convex margin; testa minutely rugulose, with 1–3 distinct longitudinal grooves, dark brown or reddish-brown. *Plate 28.*

var. **subvillosus** (*L.*) *Lam.*, Fl. Franç., ed. 1, 2: 582 (1778); Fiori et Paoletti, Fl. Anal. Ital., 2, 1: 89 (1899); Davis, Fl. Turkey, 3: 548 (1970); Zohary, Fl. Palaest., 2: 99 (1972).

S. subvillosus L., Sp. Plant., ed. 1, 745 (1753); Unger et Kotschy, Die Insel Cypern, 388 (1865); Boiss., Fl. Orient., 2: 179 (1872); Holmboe, Veg. Cypr., 110 (1914); Post, Fl. Pal., ed. 2, 1: 363 (1932); Lindberg f., Iter Cypr., 20 (1946).

S. sulcatus L., Sp. Plant., ed. 1, 745 (1753); Sibth. et Sm., Fl. Graec. Prodr., 2: 81 (1813), Fl. Graec., 8: 14, t. 719 (1833); Unger et Kotschy, Die Insel Cypern, 387 (1865); Boiss., Fl. Orient., 2: 179 (1872); Holmboe, Veg. Cypr., 110 (1914); Post, Fl. Pal., ed. 2, 1: 363 (1932); Rechinger f. in Arkiv för Bot., ser. 2, 1: 425 (1950).

S. muricatus L. var. *sulcatus* (L.) Lam., Fl. Franç., ed. 1, 2: 582 (1778); Fiori et Paoletti, Fl. Anal. Ital., 2, 1: 89 (1899).

S. subvillosus L. var. *breviaculeatus* Batt. et Trab., Fl. de l'Alg., 285 (1889); Lindberg f., Iter Cypr., 20 (1946).

TYPE: " *in* Europa australi ".

Pods echinate or hispid, at least some of the longitudinal ribs bearing straight or hooked, glabrous or minutely hispidulous bristles.

HAB.: Fields, gardens, waste ground, roadsides, sand-dunes and dry rocky hillsides; sea-level to 2,000 ft. alt.; fl. Febr.–July.

DISTR.: Divisions 1, 3–8, very common in Division 7, common in Division 3. Widespread in the Mediterranean region and eastwards to Iran; Ethiopia; Atlantic Islands.

1. Paphos, 1913, *Haradjian* 624 ! Stroumbi, 1913, *Haradjian* 715 ! Polis, 1935, *R. M. Nattrass* 328.
3. Common; *Kotschy* 550a; *Syngrassides* 36 ! *Lindberg f.* s.n.; *N. Macdonald* 121 ! *J. B. Suart* 38 !
4. Near Larnaca, 1862, *Kotschy* 144; Larnaca, 1905, *Holmboe* 105; Kouklia (Famagusta), 1939, *Lindberg f.* s.n.
5. Vatili, 1905, *Holmboe* 316. Vitsadha, 1936, *Syngrassides* 1228 ! Above Mia Milea, 1957, *Merton* 2963 !
6. Nicosia, 1959, *P. H. Oswald* 76 !
7. Very common; *Kotschy* 424; 476; 503; *Sintenis & Rigo* 423 ! *Casey* 644 ! 791 ! *Chapman* 93 ! 344 ! *G. E. Atherton* 52 ! 375 ! 990 !

Plate 28. SCORPIURUS MURICATUS L. **1**, habit, ×1; **2**, flower, ×6; **3**, standard, ×6; **4**, keel, ×6; **5**, wing, ×6; **6**, gynoecium, ×6; **7**, androecium, ×6; **8**, segment of fruit, ×5; **9**, seed, ×4. SCORPIURUS MURICATUS L. var. SUBVILLOSUS (L.) Lam. **10**, segment of fruit, ×5. (**1–9**, *Haines* W. 195; **10**, *Paranjpye* s.n.)

R 2

8. Near Apostolos Andreas Monastery, 1941, *Davis* 2329 !

NOTES: Typical *S. muricatus* L. (var. *muricatus*), with smooth or shortly tuberculate pods, has not been recorded from Cyprus.

21. ORNITHOPUS *L.*

Sp. Plant., ed. 1, 743 (1753).
Gen. Plant., ed. 5, 331 (1754).

Slender, prostrate or decumbent annuals; leaves imparipinnate, usually with numerous small, entire leaflets; stipules membranous, adnate to petiole, stipels wanting; inflorescences axillary, pedunculate, flowers in compact clusters often subtended by a well-developed floral leaf, bracts and bracteoles small; calyx cylindrical or campanulate, calyx-teeth subequal, the 2 adaxial connate for part of their length; petals small, yellow, white or pinkish, standard obovate or suborbicular, wings oblong, keel almost straight, short, obtuse; stamens diadelphous, alternate filaments dilated at apex, anthers uniform; ovary sessile, many-ovulate, style glabrous, upcurved, stigma capitate. Pod linear, compressed or subterete, straight or curved, often with a slender, uncinate apex, transversely articulate, the individual joints indehiscent; seeds oblong, ovate or subglobose, estrophiolate.

Five species in Europe, the Mediterranean region, western Asia, Atlantic Islands, sometimes elsewhere as introduced weeds.

1. O. compressus *L.*, Sp. Plant., ed. 1, 744 (1753); Boiss., Fl. Orient., 2: 178 (1872); Post, Fl. Pal., ed. 2, 1: 362 (1932); A. K. Jackson et Turrill in Kew Bull., 1938: 464 (1938); Rechinger f. in Arkiv för Bot., ser. 2, 1: 425 (1950); Davis, Fl. Turkey, 3: 544 (1970); Zohary, Fl. Palaest., 2: 100, t. 148 (1972); Osorio-Tafall et Seraphim, List Vasc. Plants Cyprus, 63 (1973).

TYPE: " *in* Italia, Sic[i]lia ".

Prostrate or decumbent, rarely suberect, pubescent annual; stems somewhat angular, 2·5–20 (–30) cm. long; leaves narrowly oblong, 2·5–7 cm. long, 0·5–1·5 cm. wide; petioles of basal leaves up to 2·5 cm. long, those of upper leaves very short, or leaf subsessile; stipules of basal leaves narrowly oblong, up to 7 mm. long, 1·5 mm. wide, pubescent-membranous, adnate along one margin to petiole, apex acute, free, recurved, stipules of cauline leaves minute or wanting; leaflets 6–15-jugate, oblong or suborbicular, sessile, entire, 2–8 mm. long, 1–3·5 mm. wide, pale green, pubescent on both surfaces, apex acute or obtuse, base rounded; inflorescences arising from the axils of the upper leaves, peduncle slender, 5–15 mm. long, shorter than subtending leaf, tipped by a small floral leaf bearing 5–11 acute leaflets; flowers 2–4 in small clusters; bracts minute, dark crimson, glandular, about 1 mm. wide, 0·5 mm. long, apex truncate or emarginate, ciliate, bracteoles wanting; calyx cylindrical, densely pubescent, 4–5 mm. long, 1 mm. wide, calyx-teeth subequal, lanceolate-acuminate, about 1–1·5 mm. long, 0·5 mm. wide, the 2 adaxial connate for a short distance from base; petals yellow, attached about half way up calyx-tube, standard obovate, about 5–6 mm. long, 2·5 mm. wide, apex truncate or slightly emarginate, base narrowed rather abruptly to a short broad claw about 1–1·5 mm. long, limb of wings oblong-obovate, about 3·5 mm. long, 1·5 mm. wide, apex rounded, base bluntly auriculate, claw about 1·5 mm. long, limb of keel about 1·5 mm. long, 1·2 mm. wide, upcurved at a right angle to a blunt apex, claw slender, about 2 mm. long; stamens attached at same level as petals, about 3·5 mm. long, filaments free for about 0·8 mm., all flattened, alternate ones rather more dilated towards apex; anthers ovoid, about 0·3 mm. long, 0·2 mm. wide; ovary linear, about 3 mm. long, 0·5 mm. wide, pubescent or glabrous, c. 7-ovulate, style glabrous, about 0·8 mm. long, upcurved at a right angle, tapering from base to apex, stigma terminal, capitate. Pod linear, compressed, downcurved, about 2·5–4 cm. long, 0·3 cm. wide, apex attenuate, uncinate, base tapering, sessile; valves dull brown, longitudinally rugose, subglabrous or thinly pubescent; seeds oblong, much-compressed, about 3 mm. long, 1·6 mm. wide; testa yellowish, smooth.

HAB.: In garigue on mountainsides, by roadsides, in open Pine forest or on grassy flats by streams; 350–3,750 ft. alt.; fl. Nov.–April.

DISTR.: Division 2, locally common at Platres and in the Pyrgos valley, also at Yialia. Mediterranean region; Atlantic Islands.

2. Locally common; *Kennedy* 571A ! B ! 1300 ! *Merton* 699 ! *P.H. Oswald* 97 ! 176 ! *N. Macdonald* 133 ! *Meikle* 2716 !

22. CORONILLA *L.*

Sp. Plant., ed. 1, 742 (1753).
Gen. Plant., ed. 5, 330 (1754).
Bonaveria Scop., Intr., 310 (1777).
Securigera DC. in Lam. et DC., Fl. Franç., 4: 609 (1805) nom. cons.
Uhrová in Beih. Bot. Centralbl., 53, B: 1–174 (1935).

Annual or perennial herbs or shrubs; leaves generally imparipinnate, rarely trifoliolate or simple, stipulate, leaflets entire, often glabrous and glaucous; stipules free or connate, lateral or leaf-opposed; inflorescences axillary, umbellate, pedunculate; bracts small, bracteoles wanting; calyx short, cup-shaped or slightly 2-lipped, calyx-teeth 5, subequal, often very small, the 2 adaxial ± connate; petals yellow, white, pink, purple or crimson, often long-clawed, keel upcurved, acuminate or rostrate; vexillary stamen generally free, the others connate, free parts of all or of alternate filaments slightly dilated at apex, anthers uniform or ± unequal; ovary sessile, ovules many, style upcurved, glabrous, stigma terminal, capitate. Pod terete, compressed or tetragonous, often curved, dehiscing transversely into 1-seeded joints or rarely indehiscent; seeds oblong, sometimes compressed, estrophiolate; hilum ventral.

About 20 species in Europe, the Mediterranean region, Atlantic Islands, western Asia, N.E. Tropical Africa.

Shrubby perennial; flowers large, the claws of the petals much longer than the calyx; pods inarticulate or very obscurely articulated - **1. C. emerus ssp. emeroides**
Herbaceous annuals; flowers rather small, the claws of the petals equalling or not much longer than the calyx:
 Pods distinctly articulated with a blunt or shortly acute apex, usually strongly arcuate or curled, not much compressed laterally:
  Leaves with 7–11 leaflets; stipules free, lateral; inflorescences 5–9 (or more)-flowered; petals yellow, pink or purplish - - - - **2. C. rostrata**
  Leaves with (1–) 3–5 (–9) leaflets; stipules connate, leaf-opposed; inflorescences 1–5-flowered; petals yellow:
   Upper cauline leaves subsessile with 1–3 leaflets - - - **4. C. scorpioides**
   Upper cauline leaves petiolate with 5 (–9) leaflets - - - - **5. C. repanda**
 Pods not (or very indistinctly) articulated, with a slender, curved rostrate apex, otherwise straight or very slightly curved, strongly compressed laterally; petals yellow - - - - - - - - - - - - - - **3. C. securidaca**

SECT. 1. **Emerus** *Desv.* Shrubs; stipules free, lateral inflorescences generally few-flowered, flowers yellow, rather large, claws of petals much exceeding calyx. Pods elongate, not articulated, or obscurely articulated.

1. C. emerus *L.*, Sp. Plant., ed. 1, 742 (1753); Uhrová in Beih. Bot. Centralbl., 53B: 35 (1935).

Erect, much-branched shrub, 0·5–1 (–2) m. high, older branches dark or pale brown, roughly fissured, young branches greenish, prominently angled;

leaves oblong, 3–7 cm. long, 1·5–5 cm. wide, imparipinnate, stipules oblong, 1–2·5 mm. long, 1–1·5 mm. wide, membranous, caducous, apex obtuse or subacute, pilose, usually dark crimson and glandular; petioles slender, 1–2·5 cm. long, subglabrous, slightly sulcate above; leaflets (3–) 5–7 (–9), oblong-obovate, subglabrous or glabrescent, green above, glaucous below, 5–30 mm. long, 4–18 mm. wide, apex rounded, truncate or slightly emarginate, sometimes apiculate; inflorescences axillary on short, condensed or elongate lateral shoots, 2–9-flowered; peduncle 3–9 cm. long, usually exceeding subtending leaf, elongating in fruit; bracts small, ovate, subacute, about 0·8–1 mm. long and almost as wide, membranous, often crimson-glandular, pilose at margins; pedicels 1–5 mm. long, thinly puberulous or glabrescent; calyx cup-shaped, gibbous adaxially, about 4–5 mm. long, 4 mm. wide, calyx-teeth obscure, broadly deltoid, less than 1 mm. long, shortly pubescent at apex, the 2 adaxial teeth connate for most of their length; petals yellow, attached with the stamens to a thickened concave disk adnate to the base of the calyx-tube, standard 1·5–2 cm. long, limb suborbicular, 8–12 mm. diam., apex truncate or sharply emarginate, claw about 10 mm. long, very slender and rigid, sulcate above and with a small marginal flap or tooth near the middle or towards the base, limb of wings obliquely oblong, 11–12 mm. long, 6–8 mm. wide, with a long conjunctival fold, apex obtuse or rounded, base shortly and bluntly auriculate, claw slender 8–10 mm. long, limb of keel 10–12 mm. long, 4–5 mm. wide, upcurved almost at a right angle to a slender, acute beak, base abruptly truncate-auriculate, conjunctival flap well developed, claw 8–10 mm. long; stamens 2–2·5 cm. long, all the filaments, except the vexillary, connate for 1·3–1·8 cm., free parts of filaments minutely papillose, alternately slightly shorter, the apices of the longer filaments a little more distinctly dilated than those of the shorter ones; anthers of longer filaments attached near the base, oblong, about 0·8 mm. long, 0·4 mm. wide, those of the shorter filaments attached at or near the middle, about 0·5 mm. long and almost as wide; ovary linear, compressed, glabrous, 13–17 mm. long, 0·5–0·8 mm. wide, 8–16-ovulate, style 6–7 mm. long, bent upwards at a right angle, stigma slightly oblique, capitate. Pod linear, ± arcuate, about 60–100 mm. long, 2·5 mm. wide, glabrous, ± compressed, somewhat constricted between the seeds, but not distinctly articulated, apex tapering to persistent style, valves pale brown, reticulately veined; seeds sausage-shaped, subterete, about 3–6 mm. long, 1–1·5 mm. wide with rounded extremities, hilum ± circular; testa dark brown, smooth.

ssp. **emeroides** (*Boiss. et Sprun.*) *Holmboe*, Veg. Cypr., 111 (1914); Uhrová in Beih. Bot. Centralbl., 53B: 47 (1935); Chapman, Cyprus Trees and Shrubs, 46 (1949); Davis, Fl. Turkey, 3: 538 (1970); Osorio-Tafall et Seraphim, List Vasc. Plants Cyprus, 63 (1973).

C. emeroides Boiss. et Sprun. in Boiss., Diagn., 1, 2: 100 (1843), Fl. Orient., 2: 179 (1872); Post, Fl. Pal., ed. 2, 1: 364 (1932).

TYPE: " in *Graecia* prope *Corinthum* (Boissier), in *Messenia* et *Achaia* Fl. Peloponn., insula *Chio* Aucher pl. exs. No. 1009 " (G).

Inflorescences (2–) 4–8 (–9)-flowered. Pods without any obvious articulations. Otherwise as in ssp. *emerus*.

HAB.: On shaded cliffs and in crevices of limestone rocks; 1,000–2,500 ft. alt.; fl. March–April.

DISTR.: Division 7, common in St. Hilarion area, also at Bellapais and Yaïla. Eastern Mediterranean region from S. Italy to W. Syria. Aggregate *C. emerus* is widespread in Europe and the Mediterranean region.

7. Locally common; *Sintenis & Rigo* 1014 ! *Holmboe* 815; *Syngrassides* 1537 ! *Davis* 1927 ! 2904 ! 3606 ! 3593 ! *Kennedy* 1662 ! *Casey* 533 ! *G. E. Atherton* 736 !

NOTES: Many of the Cyprus specimens have exceptionally large leaflets, but this characteristic may be no more than a response to a shaded habitat.

SECT. 2. **Coronilla** Annual or perennial herbs or subshrubs; stipules lateral or leaf-opposed, free or connate; inflorescences often many-flowered, flowers small or large, yellow, purple, pink or white, claws of petals not exceeding, or only a little longer than calyx. Pods elongated or rather short, generally with conspicuous articulations, very rarely without obvious articulations.

[C. MINIMA *L.*, recorded " in Peloponnesi, et insulae Cypri, locis campestribus " by Sibthorp (Fl. Graec. Prodr., 2: 78; 1813), but undoubtedly in error, since the species is restricted to western and central Europe and the western Mediterranean region. *C. minima* is a dwarf, much branched subshrub, usually less than 20 cm. high, with small, bluntly obovate, glaucous leaflets, less than 5 mm. long, stipules leaf-opposed, and 3–10-flowered umbels of small yellow flowers, rather resembling those of *C. rostrata* Boiss. et Sprun., with which species it may have been confused.]

[C. VARIA *L.*, also recorded " in Cypri insulae campestribus " by Sibthorp (Fl. Graec. Prodr., 2: 79; 1813) and most probably an error in localization, for this conspicuous plant has not since been seen on the island. *C. varia* is a diffuse herbaceous perennial, 30–50 cm. high, with narrowly oblong leaves bearing 11–25 rather large, bluntly oblong, apiculate leaflets; it has decorative (5–) 10–20-flowered, globose umbels, about 3 cm. diam., of pink and white or mauve and white flowers, and is not infrequently cultivated as an ornamental.

C. varia has a wide natural distribution in Europe and western Asia, from France eastwards to Iran, and occurs in S. Turkey and Syria, close to Cyprus. It is just possible that Sibthorp's record was based on the fleeting occurrence of the plant as a garden escape or accidental introduction. The references to *C. varia* in Poech (1842), Unger et Kotschy (1865), Holmboe (1914) and other authors are all based on Sibthorp's record.]

2. C. rostrata *Boiss. et Sprun.* in Boiss., Diagn., 1, 2: 100 (1843).

C. parviflora (non Moench) Willd., Sp. Plant., 3, 2: 1155 (1802); Unger et Kotschy, Die Insel Cypern, 388 (1865); Boiss., Fl. Orient., 2: 183 (1872); Holmboe, Veg. Cypr., 110 (1914); Post, Fl. Pal., ed. 2, 1: 365 (1932); Uhrová in Beih. Bot. Centralbl., 53B: 146 (1935); Davis, Fl. Turkey, 3: 542 (1970); Osorio-Tafall et Seraphim, List Vasc. Plants Cyprus, 63 (1973).

[*C. cretica* (non L.) Sintenis in Oesterr. Bot. Zeitschr., 32: 365, 398 (1882).]

TYPE: Greece; " ad littora Boeotiae non procul ab urbe Chalci Eubeae secus viam quae ad *Oropo* ducit " May, 1842, *Boissier* (G, K !); " in insula Rhodo " *Aucher* 1006 (G, K !).

Prostrate or decumbent annual 10–30 cm. high; stems much branched especially at base, subterete, glabrous or sparsely hispidulous with hyaline hairs, sometimes tinged purple; leaves oblong, 3–10 cm. long, 1–3·5 cm. wide; stipules lateral, oblong-lingulate, 1–2 mm. long, 0·8–1 mm. wide, apex rounded or subacute, often reddish-glandular, margins membranous; petioles 2–7 cm. long, glabrous or very sparsely hispidulous, sulcate above; leaflets 3–5-jugate, shortly petiolulate, oblong or narrowly obcuneate, 5–15 mm. long, 3–8 mm. wide, glabrous, somewhat glaucous, apex truncate or shortly retuse, apiculate; inflorescences arising from the upper nodes; peduncles erect, 2·5–10 cm. long, glabrous or thinly hispidulous near the base; bracts oblong, about 0·8 mm. long, 0·5 mm. wide, membranous, with reddish, often laciniate apices; flowers 5–9 or more attached to the thickened, pulvinate apex of peduncle, ± reflexed; pedicels 1–2 mm. long, glabrous; calyx about 1·5 mm. long and about as wide, glabrous or minutely papillose on the adaxial surface, or occasionally papillose all over, calyx-teeth minute, deltoid, the 2 adaxial a little larger than the others, up to about 0·5 mm. long; petals yellow, pink, mauve or purplish, standard oblong, 6–8 mm. long, 4–5 mm. wide, glabrous, apex emarginate, base auriculate, narrowing suddenly to a claw about 1·5–2 mm. long, limb of wings broadly oblong, about 5–6 mm. long, 2·5–3·5 mm. wide, apex rounded, base distinctly auriculate, conjunctival tooth small, claw 1–1·5 mm. long, limb of keel about 5 mm. long, 1·8 mm. wide, gradually curving upwards to a shortly rostrate, often dark-blotched, apex, base shortly auriculate, claw 1–1·5 mm. long; stamens about 6 mm. long, free parts of filaments alternately about 1·2 and 1·5 mm. long, all slightly dilated at apex; anthers suborbicular, about 0·3 mm. diam., those of the longer filaments basifixed, those of the shorter filaments versatile; ovary linear, compressed, glabrous, about 4 mm. long, 0·5 mm. wide, 4–7-ovulate, style about 2·5 mm. long, glabrous, upcurved and tapering from a rather swollen base, stigma capitate, terminal. Pod linear, arcuate, 30–50 mm. long, 2–2·5 mm. wide, slightly compressed, distinctly articulated, apex blunt, valves dull brown, glabrous or papillose; seeds oblong, compressed, about 5 mm. long, 2 mm. wide; testa rich brown, smooth.

HAB.: In pastures, waste ground, field borders, and on dry stony hillsides amongst garigue; sea-level to 1,000 ft. alt.; fl. Febr.–May.

DISTR.: Divisions 7, 8, common in Division 7. Also in Greece, Crete, Aegean Islands, Turkey, Syria, Lebanon.

7. Common; *Kotschy* 399 ! *Sintenis & Rigo* 422B ! *Syngrassides* 1029 ! 1533 ! *Davis* 2774 ! *Casey* 403 ! 513 ! 1126 ! 1152 ! *G. E. Atherton* 1011 ! 1115 ! 1191 ! *Poore* 6 ! 7 !

8. Aphendrika (Epiotissa), 1880, *Sintenis & Rigo* 422A ! Eleousa near Rizokarpaso, 1962, *Meikle* 2543 !

NOTES: The yellow-flowered form appears to be commoner in Cyprus than the pink or mauve-flowered variant described by de Candargy as *C. parviflora* Willd. var. *rubriflora*; both variants are however found in the neighbourhood of Kyrenia, occasionally (cf. *Poore* 6, 7) in the same locality. Flower colour apart it is impossible to distinguish between them, but further investigation of their distributions would be of interest.

3. C. securidaca *L.*, Sp. Plant., ed. 1, 743 (1753); Sibth. et Sm., Fl. Graec. Prodr., 2: 78 (1813), Fl. Graec., 8: 9, t. 712 (1833).

Bonaveria securidaca (L.) Scop., Intr., 310 (1777); Holmboe, Veg. Cypr., 107 (1914).

Securigera coronilla DC., in Lam. et DC., Fl. Franç., 4: 609 (1805); Boiss., Fl. Orient., 2: 176 (1872).

Securigera securidaca (L.) Deg. et Doerfl. in Denkschr. Akad. Wiss. Wien, 64: 718 (1897); Post, Fl. Pal., ed. 2, 1: 361 (1932); Rechinger f. in Arkiv för Bot., ser. 2, 1: 425 (1950); Davis, Fl. Turkey, 3: 537 (1970); Zohary, Fl. Palaest., 2: 98, t. 145 (1972); Osorio-Tafall et Seraphim, List Vasc. Plants Cyprus, 62 (1973).

TYPE: Spain; " *inter* Hispaniae *segetes* ".

Sprawling annual 30–40 (–50) cm. high; stems usually branched at base, striate, subglabrous or sparsely hispidulous, often tinged purple; leaves oblong, 3–15 cm. long, 1·8–4·5 cm. wide; stipules lateral, oblong-lingulate, 4–6 mm. long, 3–4 mm. wide, apex rounded or obtuse, sometimes reddish-glandular, margins narrowly membranous, often slightly erose; petioles 1–7 cm. long, striate, ± hispidulous; leaflets 5–9-jugate, shortly petiolulate, oblong, 8–20 mm. long, 3–12 mm. wide, glabrous, somewhat glaucous, apex truncate, shortly apiculate, base cuneate; inflorescences arising from the upper nodes; peduncles stout, erect, 4–20 cm. long, usually hispidulous near the base; bracts oblong, about 1·5 mm. long, 0·5 mm. wide, membranous, with reddish, often laciniate, apices; flowers 6–12 or more, attached to the thickened, pulvinate apex of peduncle, reflexed; pedicels usually less than 2 mm. long, glabrous; calyx 2–3·5 mm. long and almost as wide, adaxial surface of the short calyx-tube usually distinctly hispidulous, calyx-teeth subulate-deltoid 1·5–2·5 mm. long, 1–1·5 mm. wide at base; petals golden-

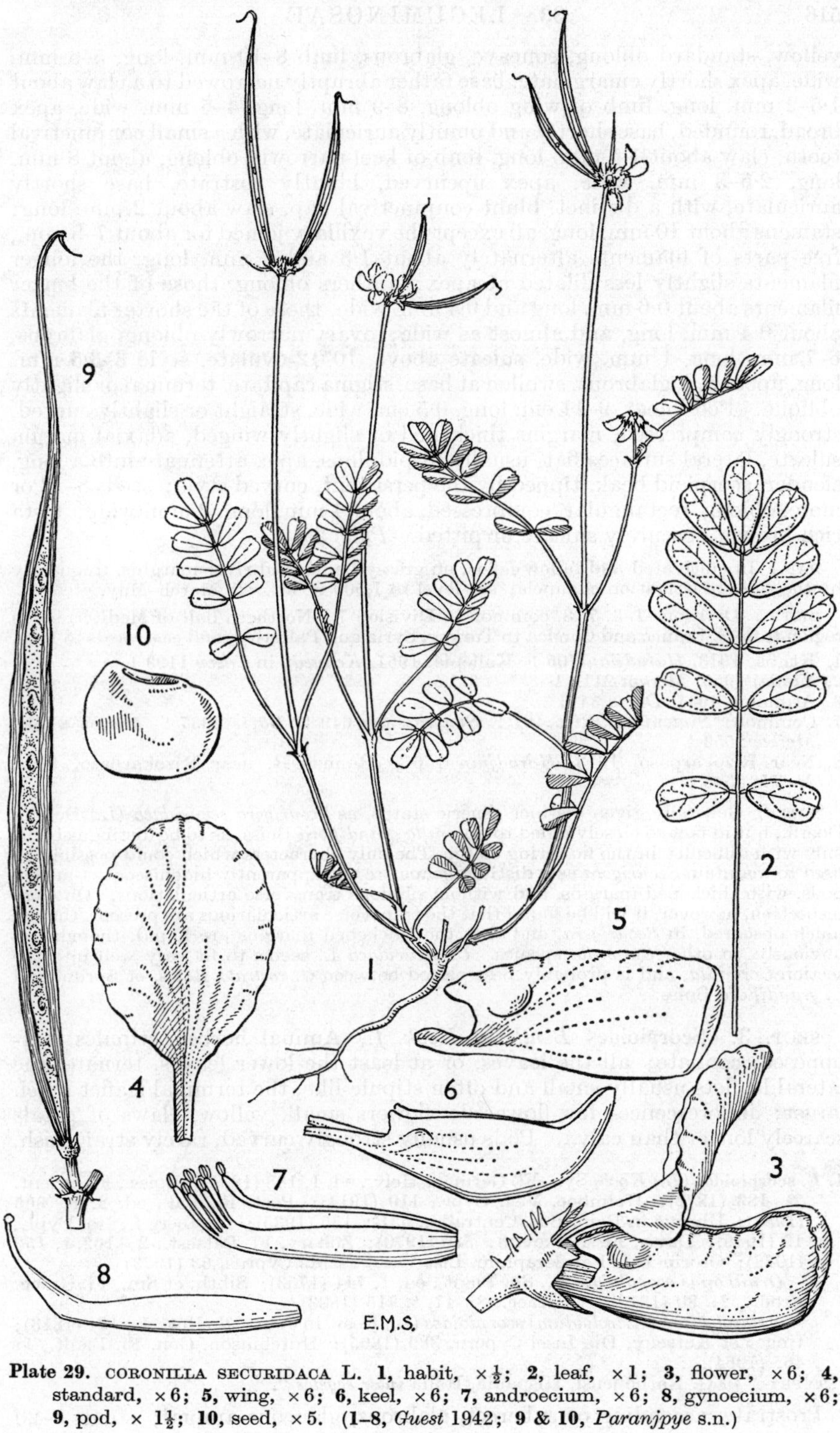

Plate 29. CORONILLA SECURIDACA L. **1**, habit, $\times\frac{1}{2}$; **2**, leaf, $\times 1$; **3**, flower, $\times 6$; **4**, standard, $\times 6$; **5**, wing, $\times 6$; **6**, keel, $\times 6$; **7**, androecium, $\times 6$; **8**, gynoecium, $\times 6$; **9**, pod, $\times 1\frac{1}{2}$; **10**, seed, $\times 5$. **(1–8,** *Guest* 1942; **9 & 10,** *Paranjpye* s.n.**)**

yellow, standard oblong, concave, glabrous, limb 8–10 mm. long, 5–6 mm. wide, apex shortly emarginate, base rather abruptly narrowed to a claw about 1·5–2 mm. long, limb of wing oblong, 8–9 mm. long, 4–5 mm. wide, apex broad, rounded, base shortly and bluntly auriculate, with a small conjunctival tooth, claw about 1·5 mm. long, limb of keel narrowly oblong, about 8 mm. long, 2·5–3 mm. wide, apex upcurved, bluntly rostrate, base shortly auriculate, with a distinct, blunt conjunctival flap, claw about 2 mm. long; stamens about 10 mm. long, all except the vexillary joined for about 7–8 mm., free parts of filaments alternately about 1·8 and 2 mm. long, the longer filaments slightly less dilated at apex; anthers oblong, those of the longer filaments about 0·6 mm. long and 0·4 mm. wide, those of the shorter filaments about 0·4 mm. long, and almost as wide; ovary narrowly oblong, glabrous, 6–7 mm. long, 1 mm. wide, sulcate above, 10–12-ovulate, style 3–3·5 mm. long, upcurved, glabrous, swollen at base, stigma capitate, terminal or slightly oblique. Pod linear, 9–11 cm. long, 0·5 cm. wide, straight or slightly curved, strongly compressed, margins thickened or slightly winged, adaxial margin sulcate, lateral surfaces flat, usually hispidulous, apex attenuate into a long, slender, scorpioid beak, tipped by the persistent, curved style; seeds 8–12 or more, oblong, rectangular, compressed, about 4 mm. long, 2 mm. wide; testa rich brown, obscurely sulcate or pitted. *Plate 29.*

HAB.: In cultivated and fallow fields, on grassy banks and river-margins, frequently by the sides of irrigation channels; sea-level to 1,500 ft. alt.; fl. March–May.

DISTR.: Divisions 1–3, 7, 8, common in Division 7. Northern half of Mediterranean region from S. France and Corsica to Turkey, Syria and Palestine, and eastwards to Iran.

1. Ktima, 1913, *Haradjian* 705 ! Kallepia, 1951, *Kennedy* in *Casey* 1193 !
2. Yialia, 1955, *Merton* 2112 !
3. Arminou, 1941, *Davis* 3411 !
7. Common; *Sintenis & Rigo* 401 ! *Syngrassides* 649 ! *Davis* 2937 ! *Merton* 2174 ! *Meikle* 2553 !
8. Near Rizokarpaso, 1913, *Haradjian* 156 ! Ronnas R. near Rizokarpaso, 1962, *Meikle* 2546 !

NOTES: Generally given distinct generic status, as *Securigera securidaca* (L.) Deg. et Doerfl., but in fact so closely allied to *Coronilla grandiflora* Boiss. as to be distinguishable only with difficulty in the flowering state. The only characters which could possibly be used to maintain *Securigera* as a distinct genus are the apparently indehiscent, 4-angled pods, with thickened margins, and without obvious transverse articulations. On close inspection, however, it will be found that the transverse articulations are present, though much obscured, in *Securigera*, and that the thickened margins are found, though less obviously, in other *Coronilla* species. *C. securidaca* L. seems to fit very well into the section *Coronilla*, and is probably best placed between *C. rostrata* Boiss. et Sprun. and *C. grandiflora* Boiss.

SECT. 3. **Scorpioides** *Benth. et Hook. f.* Annual herbs; stipules leaf-opposed, connate; all the leaves, or at least the lower leaves, ternate, the lateral leaflets usually small and often stipule-like, the terminal leaflet much larger; inflorescences few-flowered; flowers small, yellow, claws of petals scarcely longer than calyx. Pods usually strongly curved, rarely straightish.

4. C. scorpioides (*L.*) *Koch*, Syn. Fl. Germ. et Helv., ed. 1, 188 (1835); Boiss., Fl. Orient. 2: 183 (1872); Holmboe, Veg. Cypr., 110 (1914); Post, Fl. Pal., ed. 2, 1: 365 (1932); Uhrová in Beih. Bot. Centralbl., 53B: 155 (1935); Lindberg f., Iter Cypr., 19 (1946); Davis, Fl. Turkey, 3: 540 (1970); Zohary, Fl. Palaest., 2: 102, t. 152 (1972); Osorio-Tafall et Seraphim, List Vasc. Plants Cyprus, 63 (1973).

Ornithopus scorpioides L., Sp. Plant., ed. 1, 744 (1753); Sibth. et Sm., Fl. Graec. Prodr., 2: 80 (1813), Fl. Graec., 8: 11, t. 715 (1833).

Arthrolobium [*Artrolobium*] *scorpioides* (L.) Desv. in Journ. de Bot., 1: 121 (1813); Unger et Kotschy, Die Insel Cypern, 389 (1865); Hutchinson, Gen. Fl. Plant., 1: 465 (1964).

TYPE: " *in* G. Narbonensi, Hispania, Italia *inter segetes* ".

Prostrate, ascending or suberect, glabrous, glaucous annual; stems 5–20

(–30) cm. long, subterete or obscurely angled, usually much branched; leaves subsessile, all (1–) 3-foliolate; stipules leaf-opposed, connate, 2–4 mm. long, almost as wide, membranous, with a shortly bifid apex; terminal leaflet oblong, 0·8–4 cm. long, 0·3–2·8 cm. wide, apex rounded or subtruncate, base cordate or broadly cuneate, lateral leaflets much smaller than terminal, suborbicular or ear-shaped, 2–15 mm. diam.; inflorescences arising from the axils of the upper leaves, 2–4 (–5)-flowered; peduncle 0·5–3 cm. long, usually exceeding subtending leaf and lengthening in fruit; bracts connate, minute, deltoid or subulate, less than 0·5 mm. long, membranous; pedicels short, relatively stout, 1–2 mm. long; calyx cup-shaped, about 2 mm. long, 1·8 mm. wide, calyx-teeth obscure, very widely deltoid; petals yellow, the standard sometimes pencilled brownish, limb of standard ovate, 4–5 mm. long, 3–4 mm. wide, apex rounded or somewhat emarginate, base narrowed rather suddenly to a claw 1–1·5 mm. long; limb of wings strongly concave, about 4 mm. long, 2–2·5 mm. wide, apex rounded, base abruptly narrowed but scarcely auriculate, conjunctival tooth small but distinct, claw about 1·5 mm. long, limb of keel crescent-shaped, bluntly rostrate, about 3 mm. long, 1 mm. wide, with a distinct conjunctival flap, claw about 2 mm. long; stamens about 4·5 mm. long, all the filaments except the vexillary connate for about 3 mm., free parts of filaments subequal, slightly dilated at apex; anthers alternately basifixed and versatile, the basifixed about 0·3 mm. long and 0·1 mm. wide, the versatile slightly smaller and attached to slightly shorter, slightly less dilated filaments; ovary linear, compressed, slightly upcurved, 3–3·5 mm. long, 0·4 mm. wide, style 2 mm. long, tapering from a slightly swollen base, stigma small, terminal, capitate. Pod linear, usually much curved, 2–5 cm. long, 0·2–0·3 cm. wide, glabrous, prominently articulated, and often somewhat narrowed between the articulations, dorsal and ventral sutures well developed, apex acute or shortly rostrate; seeds sausage-shaped, slightly curved, about 4 mm. long, 1·2 mm. wide; testa dark brown, minutely verruculose. *Plate 30.*

HAB.: Cultivated and fallow fields, gardens and waste ground, occasionally on stony mountainsides or on fixed sand-dunes; sea-level to 3,750 ft. alt.; fl. Jan.–April.

DISTR.: Divisions 1–8, common in Divisions 3, 4, 7. Throughout the Mediterranean region and eastwards to Iran.

1. Dhrousha, 1941, *Davis* 3215 ! Near Polis, 1962, *Meikle* 2315 !
2. Platres, 1938, *Kennedy* 1302 ! East of Mandria, 1938, *Kennedy* 1303 !
3. Common; *Syngrassides* 15 ! *Lindberg f.* s.n.; *N. Macdonald* 90 ! *J. B. Suart* 37 !
4. Common; *Kotschy*; *Sintenis & Rigo*; *Holmboe* 12; *Merton* 1981 !
5. Mia Milea, 1933, *Syngrassides* 326 !
6. Nicosia, 1901, *A. G. & M. E. Lascelles* s.n. !
7. Common; *Casey* 561 ! *Chapman* 80 ! 266 ! *G. E. Atherton* 1008 ! 1131 !
8. Near Apostolos Andreas, 1941, *Davis* 2310 !

5. **C. repanda** (*Poir.*) *Guss.*, Fl. Sic. Syn., 2: 302 (1844); Post, Fl. Pal., ed. 2, 1: 365 (1932); Uhrová in Beih. Bot. Centralbl., 53B: 164 (1935); Zohary, Fl. Palaest., 2: 102, t. 151 (1972).
Ornithopus repandus Poir., Voy. en Barbarie, 2: 215 (1789).
TYPE: Algeria; " aux environs de la Calle ".

Very similar to *C. scorpioides* (L.) Koch in habit and general appearance. Lower leaves (1–) 3-foliolate, upper leaves shortly petiolate, 5 (–9)-foliolate; terminal leaflets of lower leaves rather small and narrowly oblong, 0·8–3·5 cm. long, 0·5–1·5 cm. wide, apex rounded or subacute, mucronulate, leaflets of upper leaves often very small, oblong or narrowly obovate, 5–20 mm. long, 3–8 mm. wide; inflorescence 1–3 (–4)-flowered; calyx 2–2·5 mm. long, 1·8–2 mm. wide, calyx-teeth deltoid, acute, very short, but more conspicuous than in *C. scorpioides*; petals yellow (often tinged red in dried specimens); limb of standard ovate, acute, about 5 mm. long, 4 mm. wide, claw about

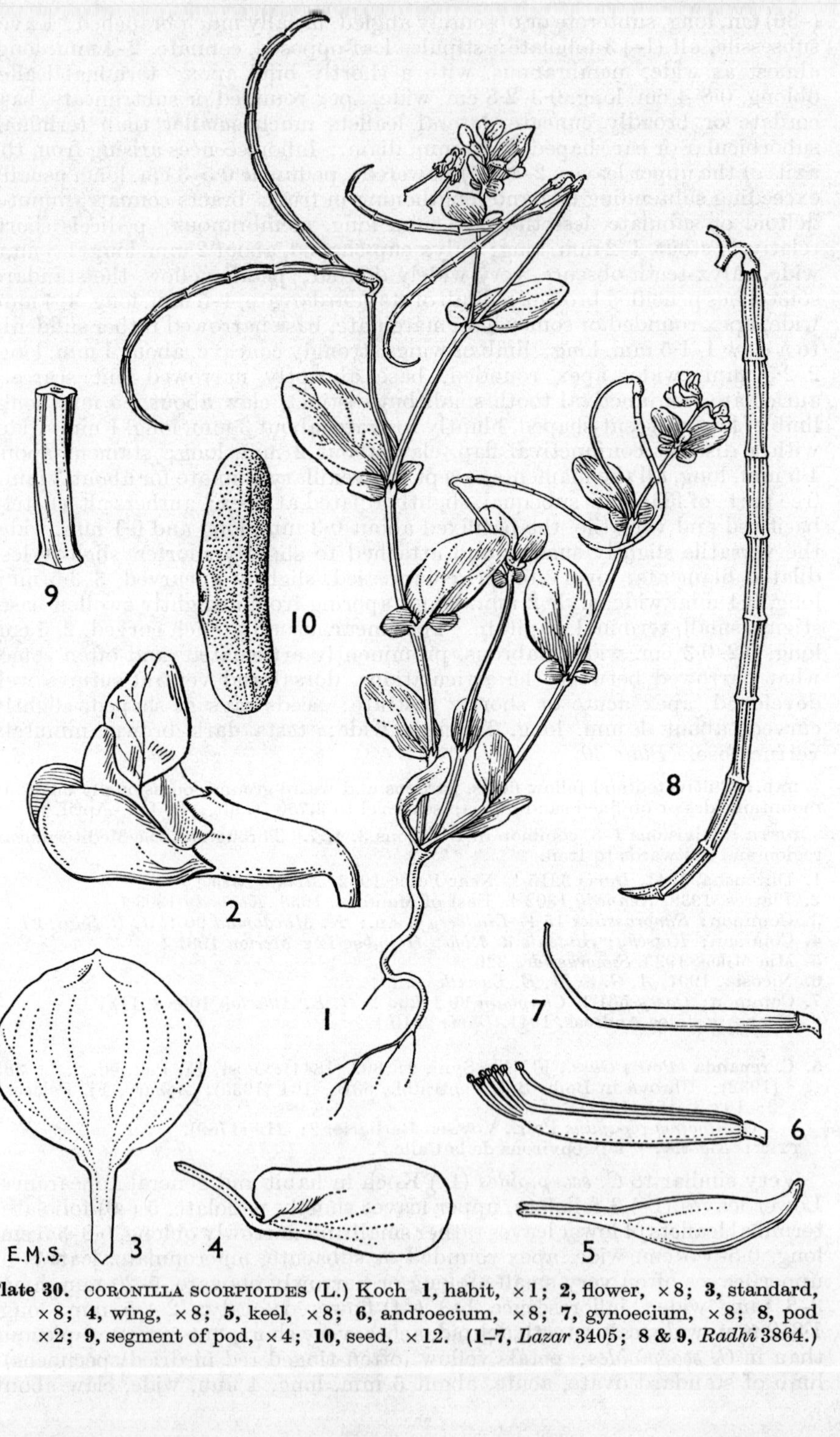

Plate 30. CORONILLA SCORPIOIDES (L.) Koch **1**, habit, ×1; **2**, flower, ×8; **3**, standard, ×8; **4**, wing, ×8; **5**, keel, ×8; **6**, androecium, ×8; **7**, gynoecium, ×8; **8**, pod ×2; **9**, segment of pod, ×4; **10**, seed, ×12. (**1–7**, *Lazar* 3405; **8 & 9**, *Radhi* 3864.)

2–2·5 mm. long, limb of wings obliquely oblong, concave, about 5 mm. long, 2·5 mm. wide, apex obtuse, base scarcely auriculate, conjunctival tooth very small or wanting, claw about 2 mm. long, limb of keel narrow, about 5–6 mm. long, 1–1·5 mm. wide, sharply upcurved to an obtuse beak, claw 2–2·5 mm. long; stamens all united, or the vexillary free, about 6 mm. long, connate for about 4 mm., free parts of filaments distinctly dilated at apex; anthers alternately basifixed and versatile, narrowly oblong, about 0·3 mm. long, 0·1 mm. wide, ovary linear, compressed, about 4·5 mm. long, 0·5 mm. wide, style slender, upcurved, about 3 mm. long, stigma terminal, capitate. Pod 30–50 mm. long, 1·5–2 mm. wide, strongly curled (generally curled downwards, while that of *C. scorpioides* is commonly curled upwards), conspicuously articulated and narrowed between the articulations, apex acute; seeds sausage-shaped, slightly curved, 2·5–3 mm. long, 0·4 mm. wide; testa dark brown, minutely verruculose.

HAB.: Sandy seashores; about sea-level; fl. Febr.–April.

DISTR.: Division 5, rare. Spain, Italy, Sicily, and N. Africa from Morocco to Libya, also in Palestine.

5. Salamis, 1936, *Syngrassides* 873 !

NOTES: The Cyprus specimens (like those from Palestine) belong to ssp. *repanda*, with the lower leaves 3-foliolate and the upper generally 5-foliolate; ssp. *dura* (Cav.) Maire, with the upper leaves 5–9-foliolate, and often very small, is apparently confined to Spain and western North Africa; the differences between the two subspecies are, however, blurred by the occurrence of intermediates.

23. HIPPOCREPIS *L.*

Sp. Plant., ed. 1, 744 (1753).
Gen. Plant., ed. 5, 331 (1754).

Annual or perennial herbs or subshrubs; leaves imparipinnate, stipules generally small, free, lateral, sometimes glanduligerous, stipels wanting, leaflets often numerous, entire; inflorescences axillary, umbellate, or flowers sometimes solitary, or few, in axillary clusters; bracts small, bracteoles wanting; calyx campanulate, with 5 acute teeth, the 2 adaxial teeth connate for part of their length; petals generally yellow or yellowish, long-clawed, standard suborbicular, keel upcurved, rostrate; stamens diadelphous, free parts of filaments in 2 series, dilated at the apex, anthers uniform, basifixed; ovary sessile, ovules generally numerous. Pod dehiscing into 1-seeded joints, generally much compressed, frequently curved, the adaxial margin deeply and regularly excavate at each seed; seeds strongly arcuate, estrophiolate, hilum ventral.

About 12 species in Europe, the Mediterranean region, Atlantic Islands and western Asia.

Flowers subsessile, solitary or in pairs, axillary:
- Stipules eglandular or minutely glanduligerous; pods glabrous or shortly papillose about the seed-bearing portion - - **1. H. unisiliquosa ssp. unisiliquosa**
- Stipules distinctly glanduligerous; pods barbellate, with long papillae about the seed-bearing portion - - - - - - - **1. H. unisiliquosa ssp. bisiliqua**

Flowers 1–6 in pedunculate umbels:
- Pods lobed along the convex (outer) margin, glabrous or shortly papillose about the seed-bearing portion - - - - - - - - - **2. H. multisiliquosa**
- Pods lobed along the concave (inner) margin, papillose-barbellate about the seed-bearing portion - - - - - - - - - - **3. H. ciliata**

1. H. unisiliquosa *L.*, Sp. Plant., ed. 1, 744 (1753); Sibth. et Sm., Fl. Graec. Prodr., 2: 80 (1813), Fl. Graec., 8: 12, t. 716 (1833); Unger et Kotschy, Die Insel Cypern, 388 (1865); Boiss., Fl. Orient., 2: 184 (1872); Holmboe, Veg. Cypr., 111 (1914); Bornm. in Notizbl. Bot. Gart. Berlin, 10: 421–439 (1928); Post, Fl. Pal., ed. 2, 1: 366 (1932); Rechinger f. in Arkiv för Bot., ser. 2, 1: 425 (1950); Davis, Fl. Turkey,

3: 545 (1970); Zohary, Fl. Palaest., 2: 103, t. 153 (1972); Osorio-Tafall et Seraphim, List Vasc. Plants Cyprus, 63 (1973).

Prostrate or suberect annual; stems usually much branched, especially towards base, sulcate, glabrous or thinly and shortly pilose; leaves narrowly oblong, 1–5 cm. long, 0·8–2 cm. wide, petiole 0·5–1·5 cm. long, sulcate, glabrous or thinly pilose; stipules oblong, or ovate-oblong, acute or acuminate, 2–4·5 mm. long, 1–1·5 mm. wide at base, membranous or herbaceous, the apical part usually ± ciliate and often irregularly denticulate, the basal part sometimes glanduligerous at the junction of stipule and petiole; leaflets 7–13, narrowly oblong, 3–12 mm. long, 1–4 mm. wide, glabrous or sparsely pilose, apex truncate or emarginate, base cuneate; flowers axillary, solitary or paired, subsessile or very shortly pedicellate, bracts minute, membranous, ciliate; calyx campanulate or almost infundibuliform, 2·5–3 mm. long, 1·8–2 mm. wide, subglabrous, calyx-teeth narrowly deltoid or subulate, acute, about 1–1·2 mm. long, 0·6–0·8 mm. wide at base, glabrous or thinly ciliate; petals yellow; limb of standard suborbicular, 3–5 mm. long and almost as wide, apex rounded or mucronulate, base subauriculate, narrowed abruptly to a claw about 2 mm. long, limb of wing oblong, 4–4·5 mm. long, 1·5–2 mm. wide, apex rounded, base abruptly narrowed or very obscurely auriculate, conjunctival flap wide and not very prominent, claw 1·8–2·5 mm. long, limb of keel about 3·5 mm. long, 1·5–2 mm. wide, upcurved to a rather slender pointed beak, base abruptly narrowed but not auriculate, with a wide conjunctival pleat, claw 2–2·5 mm. long; stamens 5–6·5 mm. long, filaments connate for 3·5–4·5 mm., free parts alternately about 1·5 and 2·5 mm. long, all dilated at apex; anthers oblong, about 0·3 mm. long, 0·1 mm. wide; ovary linear, compressed, about 3·5–4 mm. long, 0·4–0·6 mm. wide, style 2–2·5 mm. long, upcurved, tapering from a slightly swollen base; stigma minute, terminal. Pod flattened, straight or strongly curved, 25–45 (–55) mm. long, 3–5 mm. wide, the adaxial margin excavated into 3–9 circular sinuses, separated by ovoid-shaped, somewhat emarginate interstices, the seed-bearing base of the sinuses distinctly veined, glabrous or variously papillose or barbellate; seeds horseshoe-shaped or almost circular, 5–8 mm. long, about 1 mm. wide; testa brown, smooth.

ssp. **unisiliquosa.**

TYPE: "*in* Italia".

Stems usually glabrous or subglabrous; stipules eglandular or with a very minute gland, narrowly oblong-acuminate, often herbaceous; flowers mostly solitary; calyx-teeth deltoid, frequently rather short. Pod glabrous or with the seed-bearing part very shortly papillose, 4–5 mm. wide.

HAB.: Fields, waste ground and rocky hillsides, sometimes on wet ground by streams; sea-level to 2,500 ft. alt.; fl. March–May.

DISTR.: Divisions 1–4, 7, 8, common in Division 7. Widespread in the Mediterranean region and eastwards to Iran, but becoming rare east of Greece and Algeria.

1. Cape Arnauti, 1913, *Haradjian* 793.
2. Vouni Palace, 1974, *Meikle* 4021 !
3. Lefkara, 1941, *Davis* 2753 ! Episkopi, 1960, *N. Macdonald* 40 !
4. Cape Greco, 1862, *Kotschy* 164; Larnaca, 1928 or 1930, *Druce* 121 !
7. Common; *Syngrassides* 295 ! 901 ! *Davis* 2777A ! *Casey* 413 ! 1004 ! *G. E. Atherton* 1296 !
8. Ephtakomi, 1905, *Holmboe* 526; Elisis near Galinoporni, 1962, *Meikle* 2492 !

NOTES: Some doubt must attach to records based on literature, since most authors have failed to distinguish this subspecies from ssp. *bisiliqua* (Forssk.) Bornm., which largely replaces ssp. *unisiliquosa* in western Asia. Both subspecies are probably more or less evenly represented in Cyprus.

ssp. **bisiliqua** (*Forssk.*) *Bornm.* in Notizbl. Bot. Gart. Berlin, 10: 438 (1928); Davis, Fl. Turkey, 3: 546 (1970); Zohary, Fl. Palaest., 2: 104 (1972).

H. bisiliqua Forssk., Fl. Aegypt-Arab., LXXI (1775); Post, Fl. Pal., ed. 2, 1: 366 (1932).

H. biflora (non Spreng.) Boiss., Fl. Orient., 2: 185 (1872) pro parte; H. Stuart Thompson in Journ. Bot., 44: 306 (1906); Holmboe, Veg. Cypr., 111 (1914).

TYPE: Egypt; Alexandria, *Forsskål* (?).

Usually smaller and more slender than ssp. *unisiliquosa*; stems thinly pilose, stipules with a conspicuous dark gland (in dried specimens), ovate-oblong, acute or shortly acuminate, membranous; flowers often in pairs; calyx-teeth narrowly deltoid or subulate, about as long as calyx-tube. Pod with the seed-bearing part barbellate or long-papillose, 3–4 mm. wide.

HAB.: Fields, waste ground, hillsides, often in company with ssp. *unisiliquosa*; sea-level to 2,000 ft. alt.; fl. March–May.

DISTR.: Divisions 4, 6, 7, 8. Eastern Mediterranean region from Greece to Egypt, and eastwards to Iran.

4. Near Larnaca, 1880, *Sintenis & Rigo* 824 !
6. Nicosia, 1955, *E. W. Baker* 174 !
7. Pentadaktylos, 1902, *A. G. & M. E. Lascelles* s.n. ! Karmi, 1941, *Davis* 2777B ! Kyrenia, 1949, *Casey* 413 !
8. Koma tou Yialou, 1950, *Chapman* 231 !

NOTES: Almost replaces ssp. *unisiliquosa* in Iraq and Iran, and is generally the prevailing subspecies east of the coastal belt of the eastern Mediterranean. Most readily identified by its prominent stipular glands and barbellate pods, but intermediates sometimes occur in Sicily, Balkan Peninsula, Crete, Aegean Islands, Cyprus and Malta. *Sintenis and Rigo* 824 is such an intermediate, but is nearer to ssp. *bisiliqua* than to ssp. *unisiliquosa*. The characters given by Boissier and other authors to distinguish *H. unisiliquosa* and *H. biflora* are for the most part worthless, and have caused unnecessary confusion; normally the two subspecies are fairly easy to distinguish.

2. H. multisiliquosa *L.*, Sp. Plant., ed. 1, 744 (1753); Sibth. et Sm., Fl. Graec. Prodr., 2: 80 (1813), Fl. Graec., 8: 13, t. 717 (1833); Unger et Kotschy, Die Insel Cypern, 388 (1865); Boiss., Fl. Orient., 2: 185 (1872); Holmboe, Veg. Cypr., 111 (1914); Post, Fl. Pal., ed. 2, 1: 366 (1932); Davis, Fl. Turkey, 3: 547 (1970); Zohary, Fl. Palaest., 2: 104, tt. 154, 155 (1972); Osorio-Tafall et Seraphim, List Vasc. Plants Cyprus, 63 (1973).

TYPE: " *in* Angliae, G. Narbonensis, Hispaniae, Italiae *cretaceis* ".

Prostrate or diffuse, glabrous or subglabrous annual; stems (8–) 10–30 (–40) cm. long, distinctly striate or angular, usually much branched, especially towards base; leaves oblong, 1·5–6 cm. long, 1–2·5 cm. wide; stipules narrowly deltoid or subulate, acuminate, 1–5 mm. long, 0·5–2 mm. wide at base, with large conspicuous glands at point of junction with stem and petiole; petiole short, usually less than 1·5 cm. long, channelled above; leaflets 9–15, narrowly oblong or oblong-obovate 5–16 (–20) mm. long, 1·5–6 mm. wide, glabrous or subglabrous, apex truncate or emarginate, often shortly apiculate, base narrowly cuneate; inflorescence axillary, umbellate, 2–6-flowered, pedunculate, peduncle 5–45 (–50) mm. long; bracts small, membranous, subulate, about 0·5–1 mm. long, often irregularly denticulate, distinctly glandular at base; pedicels short, recurved; calyx narrowly campanulate or infundibuliform, about 3·5 mm. long, 2–2·5 mm. wide, glabrous, calyx-teeth subulate 1–1·5 mm. long; petals yellow, limb of standard suborbicular about 3·5 mm. long and almost as wide, apex rounded, apiculate, base abruptly narrowed or subauriculate, claw about 3 mm. long, with a small internal tooth near the base, limb of wings oblong, about 3·5 mm. long, 1·5 mm. wide, apex rounded, base auriculate, conjunctival flap wide and shallow, claw about 3 mm. long, limb of keel about 3·5 mm. long, 1·5–2 mm. wide, upcurving gently to a rostrate apex, base minutely auriculate, conjunctival pouch wide and shallow, claw about 3 mm. long; stamens 6–6·5 mm. long, filaments connate for about 4·5 mm., free parts of filaments alternately 1·5 and 2 mm. long, dilated at apex; anthers subequal, oblong, about 0·3 mm. long and 0·1 mm. wide; ovary linear, compressed, glabrous, 4–4·5 mm. long, 0·5 mm. wide, style sharply upcurved, about 2–2·5 mm. long, tapering from a slightly swollen base, stigma terminal, capitate. Pod flattened, often curled (downwards) almost into a circle, 20–35 mm. long, 3–4 mm. wide, the adaxial (convex) margin excavated into 5–9 circular sinuses, valves glabrous or more commonly papillose, especially about the seed-bearing base of the sinuses, seeds horseshoe-shaped, about 5 mm. long, 0·6 mm. wide; testa smooth, olive-brown.

HAB.: Cultivated fields and gardens, waste ground, hillsides or on damp sandy ground near the sea; sea-level to c. 1,000 ft. alt.; fl. Febr.–May.

DISTR.: Divisions 3–5, 7. S. Spain, Portugal, Balearics, Corsica, Sardinia, Malta, Sicily, Greece, Crete, Aegean Islands, Turkey, Syria, Lebanon, Palestine and North Africa from Morocco to Libya.

3. Near Limassol, 1862, *Kotschy*; Cape Gata, 1962, *Meikle* 2909 !
4. Near Larnaca, 1862, *Kotschy*; between Dhekelia and Ormidhia, 1880, *Sintenis & Rigo*; Famagusta, 1936, *Syngrassides* 874 !
5. Salamis, 1905, *Holmboe* 457.
7. Larnaka tis Lapithou, 1941, *Davis* 3024 !

3. H. ciliata *Willd.* in Ges. Naturf. Fr. Berlin Mag., 2: 173 (1808); Boiss., Fl. Orient., 2: 185 (1872); Sintenis in Oesterr. Bot. Zeitschr., 32: 398 (1882); Holmboe, Veg. Cypr., 111 (1914); Davis, Fl. Turkey, 3: 546 (1970); Osorio-Tafall et Seraphim, List Vasc. Plants Cyprus, 63 (1973).

TYPE: Spain; near Aranjuez (B).

Prostrate or suberect annual, 5–15 (–25) cm. high; stems slender, glabrous or subglabrous, obscurely sulcate, usually much branched towards the base; leaves oblong, 1·5–4 cm. long, 0·8–1·5 cm. wide, stipules narrowly deltoid or subulate, 1·5–2·5 mm. long, 0·8–1·3 mm. wide, base glanduligerous, margins usually ciliate, petioles slender, 3–20 mm. long, glabrous or subglabrous, sulcate above; leaflets 5–11, oblong or narrowly obovate, 2·5–10 mm. long, 1·5–5 mm. wide, glabrous above, subglabrous below, apex truncate or emarginate, rarely apiculate; inflorescences axillary, umbellate, 1–5-flowered, peduncle slender, 5–20 mm. long; bracts ovate, acute, minute, membranous, about 0·4 mm. long, 0·3 mm. wide, glanduligerous at the base; calyx campanulate, 2–2·5 mm. long and about as wide, calyx-teeth narrowly deltoid, 1–1·3 mm. long, 0·5 mm. wide at base, often thinly ciliate towards apex; petals yellow, limb of standard suborbicular, 2·5–3 mm. long and about as wide, apex rounded, base narrowed abruptly to a narrow claw 1·5–2 mm. long, very obscurely auriculate, limb of wings oblong, about 3 mm. long, 1·5 mm. wide, apex rounded, base obscurely auriculate, conjunctival flap wide and shallow, claw about 2 mm. long, limb of keel about 3 mm. long, 1·5 mm. wide, curved sharply upwards to a slender rostrate apex, conjunctival pouch distinct, claw about 2 mm. long; stamens 4–5 mm. long, all except the vexillary connate for 3–3·5 mm., free parts alternately about 0·8 and 1·3 mm. long, all distinctly dilated at apex; anthers oblong, about 0·2 mm. long, 0·1 mm. wide; ovary linear, compressed, 3–3·5 mm. long, about 0·5 mm. wide, style slender, about 1·5 mm. long, stigma terminal, capitate. Pod flattened, often curled (upwards) almost into a circle, 20–30 mm. long, 3–4 mm. wide, the adaxial (concave) margin excavated into 5–9 circular sinuses, valves papillose, or barbellate especially about the seed-bearing base of the sinuses; seeds horseshoe-shaped, about 4 mm. long, 0·5 mm. wide; testa smooth, pale brown.

HAB.: Rocky hillsides amongst garigue, sometimes in open *Pinus* woodland, or on moist sandy ground near the sea; sea-level to 1,000 ft. alt.; fl. March–May.

DISTR.: Divisions 1, 3–8. Widespread in the Mediterranean region and eastwards to Iraq.

1. Shore near Lara, 1974, *Meikle* 4004 !

3. Between Nisou and Stavrovouni, 1933, *Syngrassides* 25 ! Cape Gata, 1962, *Meikle* 2909A !
4. Near Larnaca, 1880, *Sintenis & Rigo* 823 !
5. Vatili, 1905, *Holmboe* 329, 338.
6. Nicosia, 1959, *P. H. Oswald* 81 !
7. Larnaka tis Lapithou, 1941, *Davis* 3024A ! Near Vokaris village, Kyrenia, 1950, *Casey* 1003 ! 10 m. E. of Kyrenia, 1951, *Casey* 1131 !
8. Near Cape Andreas, 1880, *Sintenis & Rigo*; Koma tou Yialou, 1950, *Chapman* 201 !

24. HEDYSARUM *L.*

Sp. Plant., ed. 1, 745 (1753).
Gen. Plant., ed. 5, 332 (1754).
Fedtschenko in Act. Hort. Petrop., 19: 185–342 (1902).

Perennial or annual herbs, subshrubs or rarely shrubs; leaves imparipinnate, leaflets entire, exstipellate, sometimes pellucid-punctate; stipules scarious, free or connate, lateral or leaf-opposed; inflorescence racemose, axillary, pedunculate; bracts scarious or setaceous, bracteoles present, setaceous; calyx-teeth subequal; petals purple, pink, whitish or yellowish, limb of standard obovate or obcordate, claw very short or almost wanting, limb of wings oblong, usually shorter than standard, keel usually longer than wings, apex obliquely truncate, less often subarcuate, obtuse; stamens diadelphous, anthers uniform; ovary subsessile, 4–5 (–many)-ovulate, style filiform, abruptly upcurved, stigma small, terminal. Pod flattened or strongly compressed, transversely articulated, joints rounded or ± quadrate, indehiscent, smooth or clothed with spines or bristles; seeds compressed, reniform, estrophiolate.

About 100 species in Europe, the Mediterranean region, Asia and North America.

Herbaceous annual; inflorescence 2–7-flowered; pods densely clothed with long, hooked bristles - - - - - - - - - - - - - **1. H. spinosissimum**
Subshrubby perennial; inflorescence many-flowered; pods reticulate-veined, without bristles - - - - - - - - - - - - - **2. H. cyprium**

1. **H. spinosissimum** *L.*, Sp. Plant., ed. 1, 750 (1753); Sibth. et Sm., Fl. Graec. Prodr., 2: 82 (1813), Fl. Graec., 8: 16, t. 721 (1833); Unger et Kotschy, Die Insel Cypern, 389 (1865); Boiss., Fl. Orient., 2: 513 (1872); Fedtschenko in Act. Hort. Petrop., 19: 226 (1902); Davis, Fl. Turkey, 3: 551 (1970); Zohary, Fl. Palaest, 2: 106, t. 157 (1972); Osorio-Tafall et Seraphim, List Vasc. Plants Cyprus, 63 (1973).
H. capitatum Desf. var. *pallens* Moris, Fl. Sard., 1: 548, t. 68B (1837).
H. pallens (Moris) Hal., Consp. Fl. Graec., 1: 453 (1901); Post, Fl. Pal., ed. 2, 1: 409 (1932).
H. spinosissimum L. ssp. *pallens* (Moris) Holmboe, Veg. Cypr., 111 (1914).
TYPE: "*in* Hispania".

Sprawling or suberect annual, (7–) 15–30 (–50) cm. high; stems much branched, sulcate, often purplish, thinly clothed with adpressed white hairs; leaves oblong, 2–7 cm. long, 1–3 cm. wide; stipules free, lateral, ovate-acuminate, 3–6 mm. long, 1·5–3 mm. wide, purplish, thinly adpressed-pilose externally; petioles up to 2·5 cm. long, those of the upper leaves usually very short, thinly adpressed-pilose, sulcate above; leaflets 9–17, narrowly oblong, 5–17 mm. long, 1·5–5 mm. wide, glabrous and dotted over with small reddish sessile glands above, thinly adpressed-pilose below, apex truncate or emarginate, usually minutely apiculate, base cuneate; inflorescences very shortly racemose or subcapitate, 2–7-flowered, arising from the axils of the upper leaves; peduncles 4–11 cm. long, generally exceeding subtending leaves, thinly adpressed-pilose; bracts ovate-caudate, 3–5 mm. long, 1–2 mm. wide, membranous, purple-striate, usually ciliate; pedicels very short or flowers subsessile; bracteoles filiform, ciliate, 2–4 mm. long; calyx narrowly campanulate, 4–5 mm. long, 2 mm. wide, thinly adpressed-pilose externally, calyx-teeth narrowly lanceolate, about 3–3·5 mm. long, 1 mm. wide at base, rather densely pilose internally; standard pink or pink-striped, wings and keel whitish; limb of standard narrowly obovate-oblong, 6–7 mm. long, 3–4 mm. wide, apex emarginate, apiculate, claw very short and wide, 1–2 mm. long; limb of wings oblong, 5–6 mm. long, 1–1·5 mm. wide, apex rounded, base distinctly auriculate, conjunctival flap obscure, claw 1–1·5 mm. long, limb of keel 6–7 mm. long, 2–2·5 mm. wide, margin rather abruptly upcurved to a broad, rounded apex, base abruptly narrowed but scarcely auriculate, claw 1–2 mm. long; stamens 7–10 mm. long, filaments, except the vexillary, connate for 5–8 mm., free parts alternately 1·5 and 2 mm. long, not dilated at apex; anthers all versatile, oblong, about 0·6 mm. long, 0·4 mm. wide; ovary linear, compressed, 3–4 mm. long, 0·5–0·6 mm. wide, glabrous, style 4–5 mm. long, stigma terminal, minute, capitate. Pod flattened, 20–30 mm. long, 7–10 mm. wide, articulated transversely into 2–3 (–4) orbicular, indehiscent joints, valves ± densely tomentellous and covered with long, rigid, purplish, uncinate bristles; seeds flattened, reniform, 3·5–4 mm. long, 2·5–3·5 mm. wide; testa smooth, rich shining brown.

HAB.: Amongst garigue on rocky or eroded hillsides, sometimes in open *Pinus* forest or on stabilized sand-dunes, occasionally in rocky or sandy fields; sea-level to 1,200 ft. alt.; fl. Febr.–April.

DISTR.: Divisions 1, 3–8, common in Divisions 3, 7. Widespread in the Mediterranean region.

1. Akamas, Ayia Nicola, 1941, *Davis* 3322 ! Near Polis, 1962, *Meikle* 2265 !
3. Common; *Kotschy* 556 ! *Holmboe* 231; 587; *N. Macdonald* 126 ! *J. B. Suart* 17 !
4. Near Larnaca, 1859, *Kotschy* 432; Ayia Napa, 1862, *Kotschy* 161; near Larnaca, 1881, *Barbey*; Kouklia (Famagusta), *Holmboe* 374.
5. Nisou, 1936, *Syngrassides* 1117 ! Athalassa, 1936, *Syngrassides* 1174 !
6. Nicosia, 1927, *Rev. A. Huddle* 45 ! Ayia Irini, 1941, *Davis* 2527 ! and, 1951, *Merton* 872 ! West of Dhiorios, 1952, *F. M. Probyn* 86 !
7. Common; *Sintenis & Rigo* 440 ! *Davis* 2921 ! *Casey* 537 ! *Chapman* 55 ! *G. E. Atherton* 1249 !
8. Near Apostolos Andreas, 1938, *Syngrassides* 1775 ! Apostolos Andreas, 1941, *Davis* 2319 !

[H. FLEXUOSUM *L.*, Sp. Plant., ed. 1, 750 (1753) is recorded, growing with *H. spinosissimum*, " in Cypri campestribus " by Sibthorp (Fl. Graec. Prodr., 2: 82; 1813), but is not represented by a specimen in the Sibthorp herbarium. The record must be rejected as erroneous. *H. flexuosum*, a robust annual, with 5–11-foliolate leaves, broadly oblong-obovate or suborbicular leaflets and long, dense racemes of conspicuous crimson flowers, is a native of S. Spain, Morocco, Algeria and Tunisia, and is not indigenous in any part of the eastern Mediterranean region. It is impossible to know exactly what Sibthorp had in mind, possibly *H. cyprium* Boiss. or an *Onobrychis* species. The later records for *H. flexuosum* in Poech, Enum. Plant. Ins. Cypr., 41 (1842) and in Unger et Kotschy, Die Insel Cypern, 389 (1865) are both based on Sibthorp's reputed discovery.]

2. H. cyprium *Boiss.*, Fl. Orient. Suppl., 189 (1888); Holmboe, Veg. Cypr., 111 (1914); Rechinger f. in Arkiv för Bot., ser. 2, 1: 425 (1950); Osorio-Tafall et Seraphim, List Vasc. Plants Cyprus, 63 (1973).

[*H. atomarium* (non Boiss.) Post in Mém. Herb. Boiss., no. 18: 93 (1900); H. Stuart Thompson in Journ. Bot., 44: 306 (1906); Holmboe, Veg. Cypr., 111 (1914).]

[*H. syriacum* (non Boiss.) Osorio-Tafall et Seraphim, List Vasc. Plants Cyprus, 63 (1973).]

[? *H. saxatile* (non L.) Sibth. et Sm., Fl. Graec. Prodr., 2: 83 (1813) pro parte quoad spec. cypr.; Sibth. in Walpole, Travels, 18 (1820).]

TYPE: Cyprus; " in montibus Cypri prope Kythraeam ", June 5, 1880, *Sintenis & Rigo* 637 (G, K !).

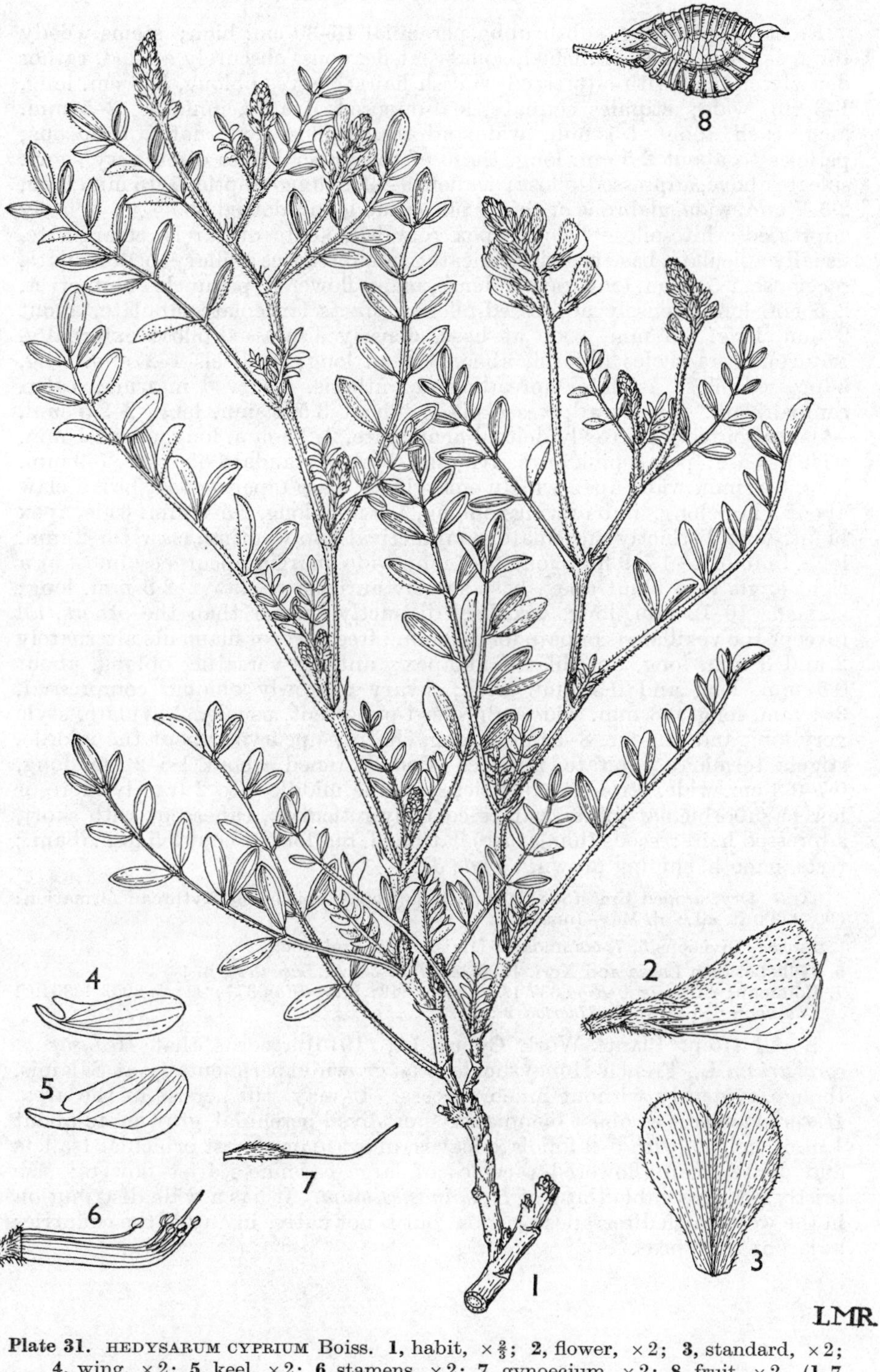

Plate 31. HEDYSARUM CYPRIUM Boiss. **1**, habit, $\times \frac{2}{3}$; **2**, flower, $\times 2$; **3**, standard, $\times 2$; **4**, wing, $\times 2$; **5**, keel, $\times 2$; **6**, stamens, $\times 2$; **7**, gynoecium, $\times 2$; **8**, fruit, $\times 2$. (**1–7** from *Chapman* 668; **8** from *Davis* 3640.)

Erect or spreading, subshrubby perennial 15–30 cm. high; stems woody towards base, much branched, somewhat flexuous, obscurely angled, rather densely clothed with adpressed whitish hairs; leaves oblong, 3–8 cm. long, 1–3 cm. wide; stipules connate, leaf-opposed, ovate-acuminate, 3–5 mm. long, each about 1·5 mm. wide, adpressed-pilose externally, caducous; petioles to about 2·5 cm. long, those of the upper leaves often very short, sulcate above, adpressed-pilose; leaflets 9–19, oblong-elliptic, 5–15 mm. long, 2·5–7 mm. wide, glabrous or thinly pilose and gland-dotted above, ± densely adpressed white-pilose below, apex rounded, acute or rarely emarginate, usually apiculate, base broadly cuneate; inflorescences axillary, pedunculate, racemose, 2·5–5 cm. long, rather dense, many-flowered, peduncle stout, erect, 2–5 cm. long, densely adpressed-pilose; bracts lanceolate-subulate, about 2 mm. long, 0·5 mm. wide at base, densely adpressed-pilose externally, caducous, bracteoles filiform, about 1 mm. long; pedicels 1–2 mm. long, adpressed-pilose, erect or spreading at anthesis, recurved in fruit; calyx campanulate, densely adpressed-pilose, about 3·5–4 mm. long, 2–2·5 mm. wide, calyx-teeth narrowly deltoid-acuminate, 2–2·5 mm. long, about 1 mm. wide at base; petals pink or mauve-pink, limb of standard obovate, 7–9 mm. long, 6–7 mm. wide, apex widely emarginate, base tapering to a broad claw about 2 mm. long, limb of wings oblong, 7–9 mm. long, 2·5–3 mm. wide, apex blunt, base distinctly auriculate, conjunctival flap wanting, claw 1·5–2 mm. long, limb of keel 7–9 mm. long, 5–6 mm. wide, margin upcurved almost at a right angle to a blunt apex, base shortly auriculate, claw 2–2·5 mm. long; stamens 10–12 mm. long, vexillary distinctly shorter than the others, all (except the vexillary) connate for 8–9 mm., free parts of filaments alternately 2 and 3 mm. long, not dilated at apex; anthers versatile, oblong, about 0·5 mm. long and 0·3 mm. wide; ovary narrowly oblong, compressed, 3–4 mm. long, 0·8 mm. wide, adpressed-pubescent, usually 2-ovulate, style very long and slender, 8–10 mm. long, sharply upcurved about the middle, stigma terminal, capitate, minute. Pod flattened, about 1·5–2 cm. long, 0·7–0·8 cm. wide, strongly constricted at the middle into 2 (rarely more or fewer) suborbicular joints, valves coarsely reticulate, canescent with short, adpressed hairs; seeds (immature) flattened, reniform, about 3·5 mm. diam.; testa smooth, shining brown. *Plate 31.*

HAB.: Dry, eroded limestone or sandstone slopes, chiefly on Kythrean formation; 600–2,000 ft. alt.; fl. May–June.

DISTR.: Divisions 5, 7, common in Division 7. Endemic.

5. Hills between Laxia and Xeri, 1900, *A. G. & M. E. Lascelles* s.n. !
7. Common; *Sintenis & Rigo* 637 ! *Holmboe* 843; *Haradjian* 377; *Davis* 2953 ! 3640 ! *Chapman* 668 ! *G. E. Atherton* 1 !

Bovill (Rep. Plant. Work Cypr., 14; 1915) records that *Hedysarum coronarium* L., French Honeysuckle, was grown experimentally at Salamis, though evidently without much success. It may still persist in the area. *H. coronarium* is a robust biennial or short-lived perennial, growing to about 1 m. in height, with 5–9-foliolate leaves, obovate or almost orbicular leaflets and dense, many-flowered racemes of large carmine-scarlet flowers; the bristly pods resemble those of *H. spinosissimum.* It has a wide distribution in the western Mediterranean region, but is not native in any of the countries adjacent to Cyprus.

25. ONOBRYCHIS *Mill.*

Gard. Dict. Abridg., ed. 4 (1754).
Širjaev in Publ. Fac. Sci. Univ. Masaryk, 56: 1–196 (1925); 76: 1–165 (1926); Bull. Soc. Bot. Bulg., 4: 1–24 (1931).

Annual or perennial herbs or subshrubs; stems generally unarmed, rarely spinose; stipules usually connate, leaf-opposed, membranous, ciliate or pilose, rarely free; leaves imparipinnate, leaflets usually numerous, entire, exstipellate, often densely covered with brown sessile glands on the upper surface of the lamina; inflorescences racemose, pedunculate, 2–many-flowered; bracts membranous, usually persistent, bracteoles 2, membranous, filiform, adnate to calyx-tube, calyx-teeth linear-subulate, the adaxial often shorter than the rest or rarely all subequal; petals pink, white, red, purple or yellowish, standard obovate, obcordate or elliptical, with a very short wide claw, keel blunt or obliquely truncate, longer or shorter than standard; stamens usually diadelphous, anthers versatile, uniform; ovary sessile or shortly stipitate, 1–2 (–3)-ovulate, style filiform, stigma small, terminal. Pod 1 (–2)-locular, stipitate or sessile, indehiscent, suborbicular or circinnately curled, generally cristate, valves pilose (rarely glabrous) reticulate-foveolate or spinose; seeds 1–2 (–3), reniform or oblong, estrophiolate.

About 130 species in Europe, the Mediterranean region, western and central Asia and Ethiopia.

Annuals; leaflets narrow, 1–4 mm. wide, without coloured veins; inflorescences few-flowered; pods rigidly spinose:
  Peduncles generally shorter than subtending leaves:
    Pods with a crest of slender spines; calyx-tube glabrous - **1. O. caput-galli**
    Pods with a crest of flattened, deltoid or truncate, denticulate lobes; calyx-tube ± pilose - - - - - - - - - - **3. O. crista-galli**
  Peduncles generally much longer than subtending leaves; pod with a crest of regular, deltoid lobes - - - - - - - - - - **2. O. aequidentata**
Perennials; leaflets broad, 3–20 mm. wide, with conspicuous bronze or purple veins; inflorescences many-flowered, rather congested; pods unarmed or feebly spinulose **4. O. venosa**

SUBGENUS 1. **Onobrychis** Petals glabrous, deciduous, or the standard very rarely with a few hairs; ovary 1–2 (–3)-ovulate, sessile. Pod sessile, seed-bearing suture straight.

1. O. caput-galli (*L.*) *Lam.*, Fl. Franç., ed. 1, 2: 651 (1778); Poech, Enum. Plant. Ins. Cypr., 42 (1842); Unger et Kotschy, Die Insel Cypern, 389 (1865); Boiss., Fl. Orient., 2: 529 (1872); Holmboe, Veg. Cypr., 111 (1914); Post, Fl. Pal., ed. 2, 1: 411 (1932); Rechinger f. in Arkiv för Bot., ser. 2, 1: 425 (1950); Davis, Fl. Turkey, 3: 565 (1970); Zohary, Fl. Palaest., 2: 110, t. 162 (1972); Osorio-Tafall et Seraphim, List Vasc. Plants Cyprus, 63 (1973).

Hedysarum caput-galli L., Sp. Plant., ed. 1, 751 (1753); Sibth. et Sm., Fl. Graec. Prodr., 2: 83 (1813), Fl. Graec., 8: 17, t. 723 (1833).

TYPE: France; " *in* Galloprovinciae *littoribus maris* ".

Erect or prostrate annual; stems 7–30 (–40) cm. long, angled, sulcate, glabrous or thinly pilose; leaves narrowly oblong, 2–10 cm. long, 1–3 cm. wide; stipules connate, leaf-opposed, membranous, each 5–9 mm. long, 1–2·5 mm. wide, aristate-acuminate with an exserted nerve; petioles of lower leaves up to 5 cm. long, those of upper leaves very short or almost wanting; leaflets 9–17, narrowly obovate, oblong, oblanceolate or elliptical, 4–15 mm. long, 1–3 mm. wide, glabrous or subglabrous and minutely gland-dotted above, thinly adpressed pilose below with rather distinct venation, apex acute or rounded, shortly cuspidate, base narrowing to a very short petiolule; inflorescence racemose, axillary, pedunculate, 2–8 (–12)-flowered;

peduncle up to 8 cm. long, generally shorter than the subtending leaf, glabrous or thinly pilose; flowers sessile or subsessile; bracts narrowly ovate, 1–2 mm. long, 1 mm. wide, aristate-acuminate, membranous, bracteoles minute, narrowly ovate-acuminate; calyx 6–7 mm. long, 1·5 mm. wide, calyx-tube glabrous, campanulate, calyx-teeth subulate, ciliate, about 5 mm. long, 0·8 mm. wide at base; petals magenta-pink or purple, standard oblong, about 5–6 mm. long, 3 mm. wide, apex truncate or slightly emarginate, base narrowed to a very short wide claw, limb of wings oblong, about 4 mm. long, 1·2 mm. wide, apex acute, base indistinctly auriculate, claw about 1 mm. long, limb of keel about 5 mm. long, 2 mm. wide, gently upcurved to a blunt truncate apex, claw about 2 mm. long; stamens diadelphous, about 6 mm. long, all but the vexillary connate for about 5 mm., free part of filaments ± pilose, not dilated; anthers oblong, about 0·5 mm. long, 0·3 mm. wide; ovary compressed, suborbicular, 1-ovulate, about 1 mm. diam., foveolate, style 4–5 mm. long, filiform, stigma terminal, capitate. Pod semi-orbicular, about 7–8 mm. diam., compressed, foveolate, rather densely adpressed-pubescent, with a crest of small sharp spines, the valves also generally armed with several series of similar, but smaller, spines; seeds reniform or subglobose, a little compressed, 2·5–3 mm. diam.; testa dark brown, smooth, shining.

HAB.: Amongst garigue on dry limestone or igneous hillsides; in cultivated fields; by pathsides or on waste ground, sometimes in open *Pinus* forest or on sand-dunes; sea-level to 3,300 ft. alt.; fl. Febr.–May.

DISTR.: Divisions 1–3, 5, 7, 8, common in Divisions 2, 7, 8. Widespread in the Mediterranean region and eastwards to Iraq.

1. Dhrousha, 1941, *Davis* 3251 !
2. Common; *Kennedy* 574 ! 575 ! 576 ! 1305 ! *D. R. Harris* 3105 !
3. Zagala above Trimiklini, 1941, *Davis* 3142 !
5. Pera Khorio, 1936, *Syngrassides* 1144 !
7. Common; *Sintenis & Rigo* 438 ! *Casey* 437 ! 656 ! *Chapman* 156 ! *Merton* 2076 ! *G. E. Atherton* 1219 ! 1360 !
8. Common; *Sintenis & Rigo* s.n.; *Haradjian* 145, 201; *Davis* 2407 ! *Meikle* 2429 ! 2524 !

NOTES: Most of the material examined is referable to *O. caput-galli* (L.) Lam. f. *brevispina* Hausskn. ex Širjaev (Publ. Fac. Sci. Univ. Masaryk, no. 56: 37; 1925); this has rather smaller, more weakly armed pods than typical *O. caput-galli*, and tends to be commoner in the eastern Mediterranean than the typical form. There are, however, many transitional forms connecting the two extremes, and it is questionable if the two can be maintained as distinct when a large range of specimens is examined.

2. O. aequidentata (*Sibth. et Sm.*) *Urv.* in Mém. Soc. Linn. Par., 1: 346 (1822); Unger et Kotschy, Die Insel Cypern, 390 (1865); Boiss., Fl. Orient., 2: 529 (1872); Holmboe, Veg. Cypr., 111 (1914); Post, Fl. Pal., ed. 2, 1: 411 (1932); Davis, Fl. Turkey, 3: 565 (1970); Osorio-Tafall et Seraphim, List Vasc. Plants Cyprus, 63 (1973).

Hedysarum aequidentatum Sibth. et Sm., Fl. Graec. Prodr., 2: 84 (1813), Fl. Graec., 8: 18, t. 725 (1833).

TYPE: Cyprus; " In insulâ Cypro ", *Sibthorp* (OXF).

Erect or spreading annual, 10–30 cm. high; stems branched at base, sometimes simple, adpressed-pubescent, obscurely sulcate; leaves oblong, 3–12 cm. long, 1–3 cm. wide; stipules connate, leaf-opposed, membranous, often stained purplish, each ovate-acuminate, 4–6 mm. long, 2–3 mm. wide; petioles of lower leaves up to 5 cm. long, those of upper leaves usually much shorter; leaflets 7–11, oblong or narrowly elliptical, 4–20 mm. long, 2–5 mm. wide, subglabrous and minutely gland-dotted above, thinly adpressed-pilose below, apex generally acute, apiculate, base tapering to a short petiolule; inflorescences axillary, racemose, flowers 2–5, often rather remote; peduncle stiff, straight, erect, 6–20 cm. long, generally much exceeding subtending leaves, distinctly sulcate, thinly adpressed-pilose; bracts ovate-acuminate or cuspidate, 2–5 mm. long, 1–2 mm. wide at base, membranous, often ciliate above with a clearly defined costa, bracteoles minute, filiform, about 1·5 mm. long, pilose; calyx about 10 mm. long, calyx-tube about 2 mm. wide, thinly pilose, calyx-teeth erect, subulate, about 8 mm. long, 1 mm. wide at base, pilose internally; petals bright magenta-purple; standard obovate, 9–11 mm. long, 5–6 mm. wide, apex shortly emarginate, base tapering to an indistinct claw, limb of wings narrowly oblong, about 8 mm. long, 2 mm. wide, apex subacute, base shortly but distinctly auriculate, claw about 1–1·5 mm. long, limb of keel 8–9 mm. long, 3·5–4 mm. wide, gently curving upwards to a rounded apex, claw about 1·5 mm. long; stamens about 10 mm. long, all (except the vexillary) connate for about 8·5 mm., free parts glabrous, not dilated; anthers oblong, versatile, about 0·5 mm. long, 0·3 mm. wide; ovary sessile, subglobose, 1–1·3 mm. diam., deeply foveolate, thinly pilose, 1-ovulate, style very slender, about 9 mm. long, stigma terminal, capitate. Pod suborbicular, strongly compressed, about 10 mm. diam., with a crest (like a cockscomb) of deltoid, entire acute spines each about 1–3 mm. long and 1–2·5 mm. wide at the base, valves adpressed-pubescent, conspicuously foveolate, unarmed or with 1 or 2 small spines; seeds large, reniform, compressed, about 4 mm. long and 5 mm. wide; testa smooth, rich brown.

HAB.: Amongst garigue on dry limestone or igneous hillsides or in open *Pinus* forest, sometimes by pathsides or in cultivated or fallow fields; 300–2,900 ft. alt.; fl. March–May.

DISTR.: Divisions 2, 3, 6–8, common in Division 7. Northern Mediterranean region from S. France to Turkey and eastwards to Iraq.

2. East of Mandria, 1938, *Kennedy* 1304 ! Khrysorroyiatissa Monastery, 1941, *Davis* 3358 !
3. Zagala above Trimiklini, 1941, *Davis* 3129 !
6. Karpasha, 1956, *Merton* 2567 !
7. Common; *Sintenis & Rigo* 439 ! 439A ! *C. H. Wyatt* 3 ! *Davis* 2871 !
8. Ayios Bakos near Yialousa, 1962, *Meikle* 2438 !

3. O. crista-galli (*Murr.*) *Lam.*, Fl. Franç., ed. 1, 2: 652 (1778); Unger et Kotschy, Die Insel Cypern, 389 (1865); Boiss., Fl. Orient., 2: 528 (1872); Holmboe, Veg. Cypr., 111 (1914); Post, Fl. Pal., ed. 2, 1: 410 (1932); Lindberg f., Iter Cypr., 19 (1946); Davis, Fl. Turkey, 3: 566 (1970); Zohary, Fl. Palaest., 2: 110, t. 163 (1972); Osorio-Tafall et Seraphim, List Vasc. Plants Cyprus, 63 (1973).

Hedysarum crista-galli Murr. in L., Syst. Veg., ed. 13, 563 (1774); Sibth. et Sm., Fl. Graec. Prodr., 2: 83 (1813), Fl. Graec., 8: 18, t. 724 (1833).

Onobrychis squarrosa Viv., Plant. Aegypt. Dec. 4: 14 (1830); Širjaev in Bull. Soc. Bot. Bulg., 4: 8 (1931).

O. gaertneriana Boiss., Diagn., 1, 9: 108 (1842), Fl. Orient., 2: 528 (1872); Freyn in Bull. Herb. Boiss., 5: 607 (1897); H. Stuart Thompson in Journ. Bot., 44: 306 (1906); Holmboe, Veg. Cypr., 111 (1914).

TYPE: Not designated (? herb. Murr.).

Prostrate or sprawling annual; stems usually much branched at base, 8–30 (–40) cm. long, angular-sulcate, thinly or rather densely clothed with long, spreading hairs; leaves narrowly oblong, 2–13 cm. long, 0·8–2·5 cm. wide; stipules connate, leaf-opposed, membranous, each ovate-oblong, acuminate, 3–7 mm. long, 1·5–2·5 mm. wide, with a distinct costa; petioles 1–4 cm. long, sulcate above, glabrous or thinly pilose; leaflets 9–21, narrowly oblong or lanceolate, 4–15 mm. long, 1–4 mm. wide, glabrous and gland-dotted above, thinly pilose below, apex emarginate, rounded or acute, shortly apiculate, base shortly petiolulate; inflorescence axillary, racemose, 2–5-flowered; peduncle erect, 1–4 (–6) cm. long, usually much shorter than subtending leaf, sulcate, ± pilose; bracts ovate, acute, membranous, ciliate, 2–3·5 mm. long, 1–1·5 mm. wide at base; bracteoles minute, filiform, about 1–2 mm. long, thinly pilose; calyx 6–8 mm. long, calyx-tube about 1·5–2 mm. wide, thinly pilose, calyx-teeth linear-setaceous, 4–5 mm. long, 0·5 mm. wide at base, pilose within; petals pale or dark pink, standard narrowly obovate,

6–7 mm. long, 3–4 mm. wide, apex emarginate, base tapering to an indistinct claw, limb of wings oblong, about 4·5 mm. long, 1·5 mm. wide, apex obtuse or rounded, base distinctly auriculate, claw about 1·5 mm. long, limb of keel about 4 mm. long, 2–2·5 mm. wide, upcurving sharply to a rounded or obtuse apex, base obscurely auriculate, claw about 2–2·5 mm. long; stamens 5–6 mm. long, all (except the vexillary) connate for about 4–5 mm., free parts glabrous, not dilated; anthers shortly oblong, about 0·4 mm. long, 0·3 mm. wide; ovary oblong, compressed, about 1·5 mm. long, 1 mm. wide, rugose, shortly pubescent, 2 (–3)-ovulate, style about 4 mm. long, rather thicker than in the previous species (*O. caput-galli*, *O. aequidentata*), stigma terminal, capitate, distinctly barbellate. Pod oblong or ± hemispherical, 12–15 mm. long, 8–12 mm. wide, strongly compressed, with a conspicuous crest of rather irregular, oblong or deltoid, acute or truncate, spinose-margined lobes 2–5 mm. long, 1–2·5 mm. wide; valves adpressed-pubescent, conspicuously foveolate, spinose; seeds usually 2, oblong, somewhat compressed, about 3–3·5 mm. long, 3 mm. wide; testa smooth, rich brown.

HAB.: Amongst garigue on dry rocky, calcareous or non-calcareous hillsides, in cultivated and fallow fields, by roadsides, on waste ground, occasionally in *Pinus* forest or on sand-dunes; sea-level to 2,900 ft. alt.; fl. Febr.–May.

DISTR.: Divisions 1, 3–8, very common in Division 7, common in Divisions 3, 4. Aegean Islands, Turkey, Syria, Lebanon, Palestine, Egypt and N. Africa, eastwards to Iran, occasionally as an introduction elsewhere.

1. Dhrousha, 1941, *Davis* 3253 !
3. Common; *Kotschy* 570 ! *Mavromoustakis* 7 ! 121 ! *Lindberg f.* s.n.; *J. B. Suart* 68 !
4. Common; *Kotschy* 415A; *Holmboe* 375; *Merton* 1999 ! *N. Macdonald* 21 ! 107 !
5. Mia Milea, 1933, *Syngrassides* 328 ! Pera Khorio, 1936, *Syngrassides* 1143 ! Mia Milea, 1950, *Chapman* 504 !
6. Nicosia, 1905, *Holmboe* 273, 289; Syrianokhori, 1935, *Syngrassides* 787 ! Ayios Yeoryios, Orga, 1956, *M. E. D. Poore* in *Merton* 2639 !
7. Very common; *Sintenis & Rigo* 438 ! *Syngrassides* 1181 ! *Mavromoustakis* 23 ! *Davis* 2830 ! *Casey* 477 ! 538 ! 1162 ! *Chapman* 52 ! 97 ! 158 ! *Merton* 641 ! 2076A ! 2077 ! *G. E. Atherton* 1244 !
8. Near Cape Andreas, 1880, *Sintenis & Rigo*; Gastria, 1941, *Davis* 2423 !

NOTES: I cannot find any valid distinction between *O. crista-galli* and *O. squarrosa* (*O. gaertneriana*), and have united all the Cyprus material under the earlier name.

It is interesting to note that the two (or sometimes three) seeds found in each pod of *O. crista-galli* do not always germinate simultaneously, but often in successive years. The withered remnants of the previous year's growth will frequently be found at the base of growing plants.

SUBGENUS 2. **Sisyrosema** (*Bunge ex Boiss.*) *Širjaev* Petals ± persistent, the standard generally pilose; ovary 1–2-ovulate, almost always curved and stipitate. Pod stipitate, seed-bearing suture curved.

4. **O. venosa** (*Desf.*) *Desv.*, Journ. de Bot., ser. 2, 3: 81 (1814); Unger et Kotschy, Die Insel Cypern, 390 (1865); Boiss., Fl. Orient., 2: 553 (1872); Durand ap. Bornm. in Magyar Bot. Lapok, 1: 133 (1902); Holmboe, Veg. Cypr., 112 (1914); Lindberg f., Iter Cypr., 19 (1946); Osorio-Tafall et Seraphim, List Vasc. Plants Cyprus, 63 (1973).

Hedysarum venosum Desf., Fl. Atlant., 2: 179, t. 201 (1800); Sibth. et Sm., Fl. Graec. Prodr., 2: 82 (1813), Fl. Graec., 8: 16, t. 722 (1833).

TYPE: Cyprus; probably collected by Labillardière in 1787 (P); but see note.

Tufted perennial with a long, tough taproot; stem very short or wanting, densely clothed with the decayed remains of stipules and petiole-bases; leaves oblong, 4–15 cm. long, 1–6 cm. wide; stipules free, lateral, ovate-acuminate, 5–10 mm. long, 2–5 mm. wide, glabrous internally, densely lanuginose externally; petioles 1·5–7 cm. long, thinly lanuginose, sulcate above; leaflets 5–9 (–11), ovate, ovate-oblong or suborbicular, 7–35 mm. long, 3–20 mm. wide, glabrous and minutely glandular-punctate above, with conspicuous purple or bronze venation, adpressed-sericeous below, apex obtuse or acute often minutely apiculate, base sessile or very shortly petiolulate; inflorescences axillary, rather densely racemose, many-flowered; peduncle 10–17 cm. long, erect, thinly lanuginose; bracts lanceolate-acuminate, 3–4 mm. long, 0·5–1 mm. wide, lanuginose externally; bracteoles minute, filiform; calyx about 5 mm. long, 3–4 mm. wide, softly pilose externally, calyx-teeth erect, subequal, ovate-acuminate, 3 mm. long, 1 mm. wide at base; petals pale creamy yellow pencilled purple or red, standard suborbicular, about 10 mm. diam., pubescent externally, apex emarginate, base tapering to a short, broad claw about 2 mm. long, limb of wings very small, usually less than 2 mm. long and about as wide, apex very blunt or rounded, base auriculate, claw broad about 1 mm. long, limb of keel about 8 mm. long, 5 mm. wide, upcurved almost at a right angle to a rounded apex, glabrous or very sparsely pubescent, base bluntly auriculate, claw 3 mm. long; stamens 10–11 mm. long, all the filaments except the vexillary connate for about 8 mm., free parts glabrous, not dilated; anthers oblong, about 0·8 mm. long, 0·5 mm. wide; ovary suborbicular, strongly compressed, shortly stipitate, about 1·5 mm. long and almost as wide, rather densely pilose, 1-ovulate, style filiform, glabrous, about 11 mm. long, stigma minute, terminal. Pod flattened, circinnately upcurved and almost orbicular, 1–1·5 cm. diam., ± tomentose, papery, the centre and radiately veined margins smooth or minutely spinulose; seed reniform, compressed, 4–5 mm. long, 3 mm. wide; testa smooth, dark brown. *Frontispiece.*

HAB.: Dry rocky slopes amongst garigue; sometimes on sandy ground near the sea; usually on calcareous soils; sea-level to 3,100 ft. alt.; fl. Febr.–May.

DISTR.: 1, 3–8, very common in Division 7, common in Divisions 3–5. Endemic.

1. Dhrousha, 1941, *Davis* 3290 ! Statos (Paphos), 1941, *Davis* 3399 !
3. Common; *Kotschy* 550 ! *Rev. A. Huddle* 46 ! *Davis* 3147 ! *Mavromoustakis* 4/57 !
4. Common; *Kotschy* 432 ! *Sintenis & Rigo* 436A ! *Holmboe* 81; *Lindberg f.* s.n.
5. Common; *Sintenis & Rigo*; *Syngrassides* 81, 1368 ! *Kennedy* 1344 ! 1693 ! 1694 ! 1695 !
6. Between Nicosia and Ayios Khrysostomos Monastery, 1889, *Pichler* ! Near Nicosia, 1930, *F. A. Rogers* 0659 ! Ayia Irini, 1936, *Syngrassides* 1039 ! and, 1941, *Davis* 2565 !
7. Very common; *Sintenis & Rigo* 436 ! *Holmboe* 540; *Kennedy* 1344 ! *Lindberg f.* s.n.; *Mavromoustakis* 17 ! *Davis* 2939 ! *Casey* 583 ! 712 ! *Chapman* 54 ! *G. E. Atherton* 1240 ! 1359 ! *Meikle* 2361 !
8. Karpas, 1930, *C. B. Ussher* 116 ! Gastria, 1941, *Davis* 2421 !

NOTES: Desfontaines (1800) erroneously believed that the type material of his *Hedysarum venosum* came from North Africa, " in montibus Sbibae apud Tunetanos ". It is now generally accepted that *O. venosa* is confined to Cyprus, and that, by some error in labelling, the Labillardière specimen, collected in Cyprus in 1787 (and represented by a duplicate at Kew) was confused with Tunisian collections examined by Desfontaines. Details concerning this error and consequent confusion will be found in the Bornmüller paper cited above (Magyar Bot. Lapok, 1: 129–133; 1902).

ONOBRYCHIS SAXATILIS (*L.*) *Lam.*, recorded from Cyprus (as *Hedysarum saxatile* L.) by Sibthorp (Fl. Graec. Prodr. 2: 183; 1813) and said to have been found on " the mountains to the left [E.] of Lapasis [Bellapais] " on April 21, 1787, is undoubtedly a misidentification. The plant found was almost certainly *Hedysarum cyprium* Boiss., which grows about Bellapais, and which was not recognized as a distinct species until a hundred years after Sibthorp's visit to Cyprus. True *O. saxatilis* (L.) Lam. is restricted to S.W. Europe and the western Mediterranean region.

26. ALHAGI *Adans.*

Fam. Plant., 2: 328, 514 (1763).

Perennial herbs or subshrubs; stems much branched, ultimate divisions spinose; leaves simple, entire, often small and caducous; stipules minute; flowers in lax, rather few-flowered axillary racemes, rhachis rigid, spine-tipped; bracts very small, bracteoles present, calyx-tube campanulate, calyx-teeth subequal, short or almost obsolete; petals pink, purple or crimson, standard obovate, cucullate with inflexed margins and a very short claw; keel upcurved to an obtuse apex, about as long as wings and standard; stamens diadelphous, anthers uniform; ovary subsessile, many-ovulate, style filiform, upcurved, glabrous, stigma small, terminal. Pod usually elongate, subterete or slightly compressed, indehiscent, frequently constricted between the seeds, but not transversely articulated, walls rather thick and corky; seeds reniform, estrophiolate; embryo enveloped in a thick gelatinous, translucent membrane.

Ovary sericeous; inflorescence 1–3-flowered; pod subterete or irregularly constricted between the seeds; calyx-teeth small, but generally distinct **1. A. graecorum**
Ovary glabrous; inflorescence 3–8-flowered; pod moniliform, strongly constricted between the seeds; calyx-teeth almost obsolete - - - **2. A. maurorum**

1. A. graecorum *Boiss.*, Diagn., 1, 9: 114 (1849), Fl. Orient., 2: 559 (1872).

A. maurorum DC. var. *karduchorum* Boiss., Fl. Orient., 2: 559 (1872).

A. mannifera Desv. in Journ. de Bot., ser. 2, 1: 120, t. 4, fig. 4 (1813) nom. nud.; Keller et Shaparenko in Sovetsk. Bot., 1933: 15 (1933); Davis, Fl. Turkey, 3: 597 (1970); Osorio-Tafall et Seraphim, List Vasc. Plants Cyprus, 63 (1973).

[*Hedysarum alhagi* (non L.) Sibth. et Sm., Fl. Graec. Prodr., 2: 82 (1813), Fl. Graec., 8: 15, t. 720 (1833); Clarke, Travels, ed. 2, 2: 720 (1815).]

[*Alhagi maurorum* (non Medik.) DC., Prodr., 2: 352 (1825); Boiss., Fl. Orient., 2: 558 (1872); Post, Fl. Pal., ed. 2, 1: 415 (1932); Rechinger f. in Arkiv för Bot., ser. 2, 1: 426 (1950).]

TYPE: "in littoribus *Atticae* ad *Sunium* promontorium et *Phalerum* (Spruner! Heldreich !), insulâ *Syra* (Boiss. Heldr.) *Samo* (Sibth.)" (G, K !).

Erect perennial 30–60 cm. high; stems woody at base, with pale brown, fissured bark, much branched above, glaucous, sulcate, thinly pubescent or glabrescent, ultimate divisions forming rigid divaricate spines, 2–4 cm. long; leaves oblanceolate or obovate, 5–20 mm. long, 2–10 mm. wide, rather succulent, thinly pubescent and gland-dotted above, more densely pubescent below, apex blunt or rounded, often apiculate; stipules subulate, about 2 mm. long, brown, pilose, caducous; petioles generally less than 3 mm. long; inflorescences 1–3-flowered, rhachis terminating in a slender, sharp, pale brown spine 1–1·5 cm. long; bracts minute, deltoid, fuscous, persistent, about 0·8 mm. long, 0·4 mm. wide at base; pedicels short, usually less than 2 mm. long, pubescent; bracteoles oblong, about 0·5 mm. long, 0·2 mm. wide, brownish, pilose; calyx campanulate, about 4 mm. long, 2 mm. wide, adpressed-pubescent externally, calyx-teeth minute, about 0·5 mm. long, deltoid or semicircular with a pale, ciliate margin; petals crimson, the standard often buff-yellow externally, limb of standard oblong-obovate about 12 mm. long, 7 mm. wide, apex cucullate, base narrowed abruptly to a claw about 1·5 mm. long, limb of wings narrowly oblong, about 8 mm. long, 2 mm. wide, apex rounded, base bluntly auriculate, claw about 3 mm. long, limb of keel about 7 mm. long, 3 mm. wide, apex rounded, base shortly auriculate, claw about 4 mm. long; stamens 11–12 mm. long, all the filaments (except the vexillary) united for 8–10 mm., free parts not dilated at apex; anthers oblong, about 0·6 mm. long, 0·5 mm. wide, versatile; ovary linear, compressed, sericeous, about 7 mm. long, 0·7 mm. wide, style glabrous, strongly upcurved, about 6 mm. long, stigma minute, terminal. Pod subterete or

Plate 32. ALHAGI GRAECORUM Boiss. **1,** flowering branch , ×1; **2,** flower, ×6; **3,** calyx opened out, inner surface, ×5; **4,** standard, ×5; **5,** standard (lateral view), ×5; **6,** wing, ×5; **7,** keel, ×5; **8,** androecium, ×5; **9,** gynoecium, ×5; **10,** pod, ×2. ALHAGI MAURORUM Medik. **11,** pod, ×2. (**1–10,** *Guest* 16144; **11,** *Furse* 8922, Persian material.)

slightly compressed, 20–40 mm. long, 2–4 mm. wide, irregularly constricted between the seeds, valves pale brown, glabrescent; seeds reniform, compressed, 3 mm. long, 2 mm. wide; testa smooth, dark olive-brown often marbled blackish. *Plate 32, figs. 1–10.*

HAB : Waste and cultivated ground; sand-dunes; just above sea-level; fl. June–August.

DISTR.: Division 4, common around Larnaca, also at Famagusta. Throughout the eastern Mediterranean region from Greece to Tripolitania, and eastwards to Iran.

4. Locally common; *Sintenis & Rigo* 673 ! 673A ! *Post* s.n. ! *Haradjian* 76 ! *Mavromoustakis* 36 !

NOTES: The Cyprus material is heterogeneous, and includes stout-spined plants (viz. *Mavromoustakis* 36) resembling *A. graecorum* from Egypt and Palestine, as well as more slender-spined specimens which are more like typical *A. graecorum* Boiss. or *A. maurorum* var. *karduchorum* Boiss. No clear distinction can be drawn between these variants, and I have decided to unite them all under *A. graecorum* pending detailed field studies of the aggregate. One suspects that the taxonomy of *Alhagi* may be complicated by hybridization, and by the fact that at least some of the forms found in Cyprus may be relatively recent introductions from other regions.

2. A. maurorum *Medik.* in Vorles. Churpf. Phys.-Oekon. Ges., 2: 397 (1787).
Hedysarum alhagi L., Sp. Plant., ed. 1, 745 (1753).
H. pseud-alhagi M. Bieb., Fl. Taur.-Cauc., 2: 174 (1808).
Alhagi camelorum Fisch., Cat. Jard. Gorenk., ed. 2, 72 (1812); Boiss., Fl. Orient., 2: 559 (1872); Post, Fl. Pal., ed. 2, 1: 415 (1932).
A pseudalhagi Desv. in Journ. de Bot., ser. 2, 1: 120 (1813) nom. nud.; Keller et Shaparenko in Sovetsk. Bot., 1933: 162 (1933); Davis, Fl. Turkey, 3: 596 (1970).

Superficially very similar to *A. graecorum*, but generally a more slender plant, with erecto-patent or suberect branches; inflorescences with 3–8 (sometimes more) flowers in a lax raceme; calyx 3 mm. long, 2·5 mm. wide, glabrous or sparsely adpressed-pubescent, calyx-teeth inconspicuous or almost obsolete, broadly and shortly deltoid, sparsely ciliate; petals as in *A. graecorum*, standard 8–12 mm. long, 4–8 mm. wide, wings 7–12 mm. long, 2–3 mm. wide, distinctly auriculate at base, keel 8–12 mm. long, 3–5 mm. wide, with a rounded apex and obscurely auriculate base; stamens 8–10 mm. long; ovary glabrous, compressed, 7–8 mm. long, 0·6–0·9 mm. wide, style sharply upcurved, 4–5 mm. long, glabrous, stigma terminal, minute. Pod conspicuously moniliform, 20–30 mm. long, 2·5–3·5 mm. wide, usually 3–6-seeded, occasionally 1-seeded by abortion, valves pale brown; seeds reniform, about 3 mm. long, 2·5 mm. wide; testa smooth, pale brown marbled blackish.

var. **maurorum.**

TYPE: Almost certainly from S.E. Russia, north of the Caspian (LINN !). See Notes.

Calyx glabrous, with the teeth almost obsolete, often turning blackish on drying; inflorescences with 5–8 (or more), rather small flowers; branches erect or suberect, slender. *Plate 32, fig. 11.*

HAB.: Dry fields and waste ground; near sea-level; fl. June–August.

DISTR.: Division 7, rare. Rhodes, Turkey, Iraq, Iran, Afghanistan, S.E. Russia and Central Asia.

7. Kyrenia, 1948, *Casey* 45 ! and, 1949, *Casey* 865 !

NOTES: The Linnean type mentioned is without flowers and fruits, but resembles in general habit the plant more commonly named *A. camelorum* Fisch. or *A. pseud-alhagi* Desv. An inscription (ⴿ) on the sheet almost certainly means that the specimen came from the region around Saratov on the lower reaches of the Volga, and that it was sent to Linnaeus during the period 1744–50 with other specimens collected in this area by Heinzelmann or Traugott Gerber. *A. maurorum* Medik. (*A. camelorum*, *A. pseud-alhagi*) is known to occur in this part of the U.S.S.R., and there can be little doubt that the epithet *maurorum*, however inappropriate, should be applied to the plant described above, rather than to *A. graecorum* Boiss., since the latter does not occur in the region. Cyprus would seem to be at the western limit of the range of *A. maurorum*, but there are reasons to suspect that the species may be an introduction, and not indigenous in the island.

var. **turcorum** (*Boiss.*) *Meikle* **comb. nov.**

A. turcorum Boiss., Diagn., 1, 9: 113 (1849).

A. camelorum Fisch. var. *turcorum* (Boiss.) Boiss., Fl. Orient., 2: 559 (1872); Post, Fl. Pal., ed. 2, 1: 415 (1932).

[*A. maurorum* (non Medik.) Gaudry, Recherches Sci., 187 (1855); Unger et Kotschy, Die Insel Cypern, 390 (1865); Holmboe, Veg. Cypr., 112 (1914); Lindberg f., Iter Cypr., 19 (1946).]

TYPE: " In *Asiâ minori* ad *Koniah* (Heldreich), *Syriâ* ad *Beyrout* (Boiss.), *Mesopotamiâ* Kotschy No. 406 " (G, K !).

Calyx thinly adpressed-pubescent, the teeth a little more distinct than in var. *maurorum*, not turning black on drying; inflorescence with 3–5 rather large flowers; branches more spreading than in var. *maurorum* and generally more robust, the ultimate (spinose) divisions often almost patent.

HAB.: Cultivated and waste ground, occasionally on sand-dunes; near sea-level; fl. June–August.

DISTR.: Divisions 4, 5, 8. Turkey, Syria, Lebanon, Palestine, ? Iran.

4. Paralimni, 1905, *Holmboe* 1164.
5. Between Trikomo and Syngrasis, 1853, *Gaudry*; sand dunes between Salamis and Trikomo, 1905, *Holmboe* 469; between Prastio and Lefkoniko, 1935, *Syngrassides* 840 ! Trikomo, 1939, *Lindberg f.* s.n. !
8. Boghaz (Famagusta), 1937, *Syngrassides* 1648 !

NOTES: Almost intermediate between *A. graecorum* and *A. maurorum* var. *maurorum*, and perhaps of hybrid origin. It is also very closely akin to *A. persarum* Boiss. et Buhse, but this normally has a glabrous calyx.

TRIBE 7. **Vicieae** (*Bronn*) *DC.* Annual or perennial herbs, often climbing; leaves frequently paripinnate, the rhachis terminating in a bristle or tendril, less commonly imparipinnate or rarely reduced to a phyllode; stipules often foliaceous, ear-like or semi-sagittate, stipels wanting; flowers axillary, solitary or in racemes; stamens diadelphous, or the vexillary filament more or less united with the others, free parts of filaments filiform or dilated towards apex, anthers uniform, versatile; ovary 2–many-ovulate, style glabrous or hairy. Pod 2-valved; seed with the funicle expanded above the hilum; cotyledons large, fleshy, often hypogeal.

27. CICER *L.*

Sp. Plant., ed. 1, 738 (1753).
Gen. Plant., ed. 5, 327 (1754).

Annual or perennial herbs, commonly with glandular-pubescent leaves and stems; leaves paripinnate, imparipinnate or trifoliolate, the rhachis sometimes terminating in a spine or short tendril; leaflets dentate or laciniate; stipules lateral, oblique, often dentate or laciniate; flowers solitary or in sparse axillary racemes; bracts small; bracteoles wanting; calyx-tube oblique or dorsally gibbous, calyx-teeth subequal or the 2 adaxial shorter and connivent; petals white, pink or bluish, standard with a broad claw, wings free from keel, obliquely obovate, keel broad, upcurved, blunt or acute; stamens diadelphous, free parts of filaments dilated towards apex; ovary sessile, ovules 2 or more, style filiform, upcurved or inflexed, not bearded, stigma terminal. Pod ovoid or oblong, turgid; seeds subglobose or irregularly obovoid, hilum small, funicle not dilated.

About 20 species in S.E. Europe, western and central Asia; *C. arietinum* L. widely distributed as an escape from cultivation.

1. C. arietinum *L.*, Sp. Plant., ed. 1, 738 (1753); Unger et Kotschy, Die Insel Cypern, 382 (1865); Boiss., Fl. Orient., 2: 560 (1872); Post, Fl. Pal., ed. 2, 1: 416 (1932); Lindberg f., Iter Cypr., 19 (1946); Davis, Fl. Turkey, 3: 272 (1970); Osorio-Tafall et Seraphim, List Vasc. Plants Cyprus, 56 (1973).

TYPE: " *inter* Hispaniae, Italiae *segetes* ".

Erect or sprawling annual, 15–40 cm. high, stems and leaves rather densely glandular-pubescent; leaves oblong, 4–8 cm. long, 1–4 cm. wide, imparipinnate, 13–17-foliolate; petiole short, usually less than 1 cm. long, canaliculate above; stipules ovate-acute, 2–12 mm. long, 1·5–6 mm. wide, foliaceous, with coarsely serrate-dentate margins; leaflets oblong-obovate, 5–20 mm. long, 2·5–10 mm. wide, apex rounded or acute, base cuneate, margins sharply serrate; flowers solitary, axillary; peduncle 8–15 mm. long; bracts 2–3, unequal, 0·5–3 mm. long, 0·2–0·5 mm. wide, linear, glandular-pubescent; pedicels 10–15 mm. long, deflexed after anthesis; calyx campanulate with a dorsal hump, 6–10 mm. long, 2–4 mm. wide, glandular-pubescent, calyx-teeth subequal, linear-subulate, 4–6 mm. long, 1–1·5 mm. wide at base; petals white, pink or mauve, glabrous, standard obovate, 9–10 mm. long, 7 mm. wide, apex rounded, base tapering to a wide claw about 2–2·5 mm. long, limb of wings oblong, about 6 mm. long, 2·5 mm. wide, apex rounded, base with a prominent curved auricle and a claw 1·5 mm. long, limb of keel about 5 mm. long, 3 mm. wide, sharply upcurved to a subacute apex, base with a distinct auricle and a slender claw about 2 mm. long; stamens diadelphous, the vexillary free, about 7 mm. long, all the filaments (except the vexillary) connate for about 5 mm., free parts distinctly dilated towards apex; anthers oblong, about 0·6 mm. long, 0·4 mm. wide, versatile; ovary oblong or suborbicular, compressed, densely glandular-pilose, about 3 mm. long, 2·5 mm. wide, 2-ovulate, style sharply upcurved at base, glabrous and filiform above, about 4 mm. long, stigma small, terminal. Pod oblong-ellipsoid, inflated, 2–3 cm. long, 1–1·5 cm. wide, apex acute or subacute, valves pale brown, shortly glandular-pubescent; seeds subglobose with a prominent radicle and a shallow median furrow, about 8 mm. diam.; testa pale brown, minutely and sparsely rugulose.

HAB.: Cultivated and waste ground; sea-level to 3,600 ft. alt.; fl. March–July.

DISTR.: Divisions 1–3, rare. Natural distribution obscure, widely cultivated in the Mediterranean region and eastwards to India, and found almost throughout the world as an escape from cultivation or as a casual.

1. Polis, 1936, *Lady Loch* 7 !
2. Prodhromos, 1939, *Lindberg f.* s.n.
3. Near Episkopi and Vrecha, 1862, *Kotschy.*

NOTES: Chick Pea, Garbanzo, Gram. Grown since antiquity for its edible seeds, and as a forage plant. Probably originated in Iran, Iraq or Afghanistan, but its natural distribution is now impossible to map. It is considered the most important pulse crop in India (Wealth of India, 2: 154; 1950) with 18 to 19 million acres devoted to its cultivation.

C. INCISUM (*Willd.*) *K. Malý*, a native of Greece, Crete, Turkey and Palestine, is recorded in Osorio-Tafall et Seraphim, List Vasc. Plants Cyprus, 56 (1973). This would appear to be the only report of its occurrence on the island.

28. VICIA *L.*

Sp. Plant., ed. 1, 734 (1753).
Gen. Plant., ed. 5, 327 (1754).

Annual or perennial herbs, often climbing by means of tendrils; leaves commonly paripinnate, the rhachis terminating in a branched or simple tendril or bristle, or rarely all or some of the leaves imparipinnate, without tendrils; leaflets entire or dentate, exstipellate; stipules generally semi-sagittate, often toothed; inflorescences racemose, axillary, or flowers sometimes solitary or in clusters in the leaf-axils; bracts generally small and caducous, bracteoles wanting; calyx often oblique and asymmetrical, the calyx-teeth subequal or the adaxial pair shorter than the others and ± connate; petals blue, violet, yellow or whitish, standard obovate or oblong, emarginate, tapering basally to a broad claw, wings obliquely obovate or oblong, adherent to the keel, or sometimes free, keel usually shorter than wings; stamens diadelphous, or the vexillary stamen more or less united with the others, staminal sheath usually oblique at apex, free parts of filaments not dilated towards apex, anthers uniform; ovary subsessile or shortly stipitate, ovules 2–many, style subterete or dorsiventrally compressed, seldom flattened at apex, pilose, or with a tuft of hairs at the apex, rarely quite glabrous, stigma terminal. Pod compressed, 2-valved, continuous within; seeds globose or rarely compressed, funicle dilated into a thin aril.

About 120 species widely distributed in temperate regions of the northern hemisphere, and in South America.

Inflorescence sessile or subsessile in the leaf axil, or very shortly pedunculate, the peduncle (or pedicel) generally shorter than the calyx and corolla:
- Petals yellow or yellowish:
  - Standard pilose on the adaxial (outer) surface; pod pilose or pubescent:
    - Flowers 2–3 (–5) in a congested raceme or cluster; standard about 2 cm. long; calyx-teeth subequal - - - - - - - - **1. V. pannonica**
    - Flowers solitary; standard about 3 cm. long; adaxial calyx-teeth distinctly shorter than the abaxial - - - - - - - - **2. V. hybrida**
  - Standard glabrous or shortly papillose on the adaxial (outer) surface:
    - Pods adpressed-pilose; flowers solitary or sometimes paired [V. SERICOCARPA p. 540]
    - Pods glabrous; flowers 1–5 - - - - - - - - - **3. V. assyriaca**
- Petals purple, crimson, bluish or whitish:
  - Plants erect or suberect, not sprawling or climbing; stems robust; leaflets broadly oblong, ovate, obovate or elliptical:
    - Pods 5–6 cm. long, papillose-ciliate along the sutures; tendrils usually well developed - - - - - - - - - - **8. V. narbonensis**
    - Pods exceeding 6 cm. in length, not papillose-ciliate along the sutures; tendrils usually reduced to a short awn - - - - - - [V. FABA p. 548]
  - Plants sprawling or climbing; stems slender, weak; leaflets narrowly oblong, obovate, elliptical or linear:
    - Standard more than 8 mm. long; seeds with a smooth, not papillose, testa:
      - Calyx-teeth narrowly subulate, 3–8 mm. long; stipules often rather large and toothed or laciniate:
        - Leaves (2–) 4–10-foliolate; leaflets usually with an emarginate apex; flowers subsessile or very shortly pedicellate - - - - - **4. V. sativa**
        - Leaves 2–4-foliolate; leaflets tapering to apex and base; flowers distinctly pedicellate, or 2 (–3) in short pedunculate racemes - **7. V. bithynica**
      - Calyx-teeth less than 3 mm. long, the 2 adaxial deltoid; stipules small, semi-sagittate:
        - Limb of standard deeply emarginate at apex; leaflets conspicuously truncate or emarginate at apex; flowers solitary - - - **6. V. peregrina**
        - Limb of standard rounded or shortly emarginate at apex; leaflets obtuse, acute or sometimes obscurely emarginate at apex; flowers sometimes in short, few-flowered racemes:
          - Leaves generally 6–8-foliolate; leaflets oblong-elliptic, 4–12 mm. long, 1·5–3 mm. wide; limb of standard usually somewhat recurved **11. V. cretica**
          - Leaves generally 8–14-foliolate; leaflets narrowly oblong, 5–20 mm. long, 1–6 mm. wide; limb of standard usually straight, not recurved **12. V. monantha**
    - Standard less than 8 mm. long; seeds with a papillose testa **5. V. lathyroides**

Inflorescence conspicuously pedunculate, peduncle exceeding the calyx and corolla; flowers in racemes:
- Pods moniliform, strongly constricted between the seeds; plants erect or spreading **18. V. ervilia**
- Pods not moniliform, not constricted between the seeds; plants sprawling or climbing:

Stipules dissimilar, one of each pair small and subulate, the other much larger, divided palmately into numerous filiform lobes - [V. ARTICULATA p. 554]
Stipules more or less uniform, not as above:
Leaflets of middle and upper leaves large, 3–6 cm. long, often exceeding 1 cm. in width:
Leaves 2–4-foliolate; racemes 1–3-flowered; calyx-teeth long, subulate **7. V. bithynica**
Leaves 5–10-foliolate; racemes many-flowered; calyx-teeth short, deltoid [V. DUMETORUM p. 548]
Leaflets of the middle and upper leaves usually less than 3 cm. long and 1 cm. wide:
Flowers small, less than 8 mm. long; petals commonly pale blue or whitish:
Pod 18–25 mm. long, 6–8 mm. wide, apex shortly rostrate **14. V. palaestina**
Pod 6–14 mm. long, 2·5–3·5 mm. wide, blunt or shortly acute at apex, not rostrate:
Pod (1–) 2-seeded; leaflets with a truncate or emarginate apex; flowers very small and pale - - - - - - **17. V. hirsuta**
Pod 3–8-seeded; leaflets acute or obtuse at apex, not truncate or emarginate:
Calyx-teeth narrowly deltoid, 0·8–1·5 mm. long; pod glabrous **19. V. laxiflora**
Calyx-teeth narrowly subulate, 2–2·5 mm. long; pod usually thinly pubescent - - - - - - - - **20. V. pubescens**
Flowers medium or large, more than 8 mm. long, variously coloured:
Racemes 1–5-flowered:
Leaflets shortly oblong, obovate or suborbicular, almost as broad as long; petals mauve-blue and white or mauve-blue and yellow:
Pod straight; petals mauve-blue and white - - **15. V. cypria**
Pod crescent-shaped (lunate); petals mauve-blue and yellow **16. V. lunata**
Leaflets narrowly oblong or oblong-elliptic, more than twice as long as wide:
Leaves generally 6–8-foliolate; leaflets oblong-elliptic, 4–12 mm. long, 1–3 mm. wide, limb of standard generally recurved **11. V. cretica**
Leaves generally 8–14-foliolate; leaflets narrowly oblong, 5–20 mm. long, 1–6 mm. wide, limb of standard not recurved **12. V. monantha**
Racemes (2–) 5–many-flowered:
Pods glabrous:
Perennial with tough woody rootstock; leaves linear or almost subulate; inflorescence (8–) 10–30-flowered, calyx scarcely gibbous at base **9. V. tenuifolia**
Annual; leaves narrowly oblong, apex rounded or shortly acute; inflorescence 2–10-flowered, very lax; calyx distinctly gibbous at base - - - - - - - - - - **10. V. cassia**
Pods pubescent; inflorescence 3–12 (–16)-flowered; calyx distinctly gibbous at base; plant usually perennial **13. V. villosa ssp. eriocarpa**

SECT. 1. **Vicia** Flowers usually 1–2 (rarely more), sessile or shortly pedicellate or in short congested racemes in the axils of the upper leaves; style bearded abaxially near apex; plants usually climbing or sprawling; leaves with few to many leaflets.

[V. SEPIUM *L.*, a common European species, with climbing or trailing stems, congested axillary racemes of 2–6 bluish or purplish flowers, and dull brown, glabrous pods, has been recorded " in insulae Cypri nemorosis " by Sibthorp (Fl. Graec. Prodr., 2: 73; 1813), but certainly in error, for the species is found nowhere on the island, nor in adjacent regions of the eastern Mediterranean.]

1. V. pannonica *Crantz*, Stirp. Austr., 5: 393 (1769); Boiss., Fl. Orient., 2: 569 (1872); Davis, Fl. Turkey, 3: 314 (1970); Osorio-Tafall et Seraphim, List Vasc. Plants Cyprus, 57 (1973).

TYPE: Austria; towards the borders of Hungary, *Grecsowski*; " ad aggerem Belveder ", *Crantz* (B, BP).

Climbing annual 30–60 cm. high; stems glabrescent or thinly pilose, conspicuously striate, usually much branched at base; leaves paripinnate, 3–9 cm. long, 1·5–4 cm. wide; petiole short and rather stout, usually under 1·5 cm. long; stipules oblong or ovate, acute, about 2–4 mm. long, 1–2 mm. wide, 2–3-nerved with a median, dark, pilose, glandular patch; rhachis produced into a branched tendril; leaflets 10–20, oblong, 6–20 mm. long, 2–5 mm. wide, thinly adpressed-villose on both surfaces or subglabrous above, apex truncate, emarginate, rounded or subacute, usually apiculate; flowers 2–3 (–5) pendulous, in a congested raceme or cluster; peduncle very short, pedicels 1–2 mm. long; bracts obsolete; calyx cylindrical, about 8–9 mm. long, 4 mm. wide, thinly villose externally, base slightly gibbous, calyx-teeth subequal, subulate, 2–2·5 mm. long, 0·5 mm. wide at base; petals yellowish or stained purplish, limb of standard suborbicular, about 10 mm. diam., villose on the adaxial (outer) surface, claw about 12 mm. long, 8–9 mm. wide, limb of wings oblong, about 10 mm. long, 5 mm. wide, apex rounded, base strongly but bluntly auriculate, claw about 10 mm. long, limb of keel suborbicular or very shortly oblong, 4–5 mm. long, 4 mm. wide, base auriculate, claw about 10 mm. long; stamens all ± united, about 11 mm. long, free parts of filaments about 1 mm. long; anthers oblong, about 0·5 mm. long, 0·3 mm. wide; ovary linear, compressed, about 12 mm. long, 1 mm. wide, adpressed-villose, c. 7-ovulate, base stipitate, apex abruptly narrowed to a perpendicular style about 2 mm. long, stigma terminal, concealed by the conspicuous stylar beard. Pod oblong, compressed, 2·5–3 cm. long, 0·8–1 cm. wide, apex abruptly narrowed, shortly rostrate, valves pale brown, thinly villose; seeds subglobose, 3–5 mm. diam.; testa smooth, dark brown.

HAB.: In cultivated fields; near sea-level; fl. April–June.

DISTR.: Division 7, rare. Central and southern Europe and Turkey; widely distributed as an adventive, and probably a casual occurrence in Cyprus.

7. In fields near Kyrenia, 1880, *Sintenis & Rigo* 652 !

NOTES: Easily confused with *V. hybrida* L. and perhaps overlooked on this account, but distinguished by 2–3 (or more) flowers, and not a solitary flower, in each axil.

2. V. hybrida *L.*, Sp. Plant., ed. 1, 737 (1753); Boiss., Fl. Orient., 2: 570 (1872); Holmboe, Veg. Cypr., 113 (1914); Post, Fl. Pal., ed. 2, 1: 417 (1932); Davis, Fl. Turkey, 3: 314 (1970); Zohary, Fl. Palaest., 2: 204, t. 293 (1972); Osorio-Tafall et Seraphim, List Vasc. Plants Cyprus, 57 (1973).

[*V. sericocarpa* Fenzl var. *microphylla* (non Boiss.) H. Stuart Thompson in Journ. Bot., 44: 307 (1906).]

TYPE: France; " Monspelii, Massiliae ".

Scrambling or climbing annual, 15–50 cm. high; stems glabrescent, sharply angular, usually much branched especially at base; leaves paripinnate, oblong, 2–5 cm. long, 1–2 cm. wide, the upper with long branched tendrils, the basal without tendrils, or with the rhachis produced into a short bristle; petiole very short, generally less than 1 cm. long; stipules semi-sagittate, 1–4 mm. long, 0·5–1·5 mm. wide, glabrescent, often with a small, dark, median gland; leaflets 6–14, oblong or obovate, 3–12 mm. long, 2–8 mm. wide, thinly pilose on both surfaces, apex truncate, emarginate or rounded, generally apiculate; flowers solitary in the axils of the upper leaves, pendulous; pedicel villose, usually less than 5 mm. long; bracts obsolete; calyx thinly pilose, cylindrical, about 10 mm. long, 4 mm. wide, base distinctly gibbous, calyx-teeth filiform-subulate, the 2 adaxial about 2·5 mm. long, the 3 abaxial about 3 mm. long, all frequently with dark blackish apices; petals sulphur-yellow, pencilled or stained dull brown, or rarely purplish; limb of standard obovate, pilose adaxially (externally), 15–19 mm. long, 13–15 mm. wide, apex shortly emarginate, margins reflexed, claw 13–15 mm. long, 6–8 mm. wide, limb of wings narrowly obovate 13–14 mm. long, 5–6 mm. wide; apex rounded, base strongly falcate-auriculate, claw about 12–13 mm. long, limb of keel shortly oblong, about 6 mm. long, 4·5 mm. wide, apex rounded, adaxial margin often with a small tooth, base distinctly auriculate, claw slender, about 15 mm. long; stamens about 20 mm. long, all united, free parts of filaments about 2–3 mm. long; anthers oblong, about 0·8 mm. long, 0·4 mm. wide, alternately basifixed and ± versatile; ovary linear-elliptic, strongly compressed, densely adpressed-pilose, about 18 mm. long, 2–3 mm. wide, base conspicuously long-stipitate, apex narrowed to base of style, ovules 7–8, style 3–4 mm. long, bearded near apex, stigma small, capitate. Pod oblong, strongly compressed, 20–30 mm. long, 9–14 mm. wide, apex acute, shortly rostrate, base narrowly cuneate, valves pale brown, thinly pilose; seeds subglobose, 5–6 mm. diam.; testa smooth, dark brown, sometimes obscurely marbled.

HAB.: In fields and on waste ground, by roadsides, amongst garigue on hillsides, by streams or forest-margins, occasionally on sand dunes, often in shaded, moist situations; sea-level to 2,500 ft. alt.; fl. Febr.–April.

DISTR.: Divisions 1–4, 6–8, common in Divisions 2, 7. Throughout the Mediterranean region and eastwards to Iraq and Iran.

1. Ayios Yeoryios (Akamas), 1962, *Meikle* ! Near Polis, 1962, *Meikle* !
2. Common; *Syngrassides* 102 ! 354 ! *Kennedy* 588 ! *Davis* 3357 ! *Merton* 91 !
3. Near Amathus, 1947, *Mavromoustakis* s.n. ! Episkopi, 1960, *N. Macdonald* 17 !
4. Near Pyrga, 1880, *Sintenis & Rigo* 940 ! Ayia Napa, 1905, *Holmboe* 74.
6. Near Nicosia, 1930, *Rev. F. A. Rogers* 0671 !
7. Common; *Sintenis & Rigo* ! *Syngrassides* 1366 ! *Casey* 416 ! 585 ! *Chapman* 119 ! *G. E. Atherton* 887 ! 1137 !
8. Ayios Philon near Rizokarpaso, 1941, *Davis* 2234 ! Ayios Theodhoros River near Yialousa, 1962, *Meikle* 2458 !

[V. SERICOCARPA *Fenzl*, Pugillus no. 4 (1842); Unger et Kotschy, Die Insel Cypern, 386 (1865); Boiss., Fl. Orient., 2: 571 (1872); Holmboe, Veg. Cypr., 113 (1914); Post, Fl. Pal., ed. 2, 1: 418 (1932); Osorio-Tafall et Seraphim, List Vasc. Plants Cyprus, 57 (1973).

Closely resembling *V. hybrida* L. in habit, foliage and flower, but with a glabrous or minutely papillose (or scabridulous) standard; the flowers are occasionally paired, especially in robust specimens. The pods are very similar to those of *V. hybrida*, but normally more densely adpressed-sericeous.

Recorded by Kotschy (*loc. cit.*) from Cape Greco (*Kotschy* 152), near Larnaca (*Kotschy* 295), Ayios Khrysostomos Monastery (*Kotschy* 415) and Bellapais (*Kotschy* 500a) and noted from Ayios Khrysostomos Monastery and Cape Andreas by Sintenis and Rigo (Oesterr. Bot. Zeitschr., 32: 54, 398; 1882). It is significant that neither of these authors mentions *V. hybrida*, despite its general abundance throughout the lowlands of Cyprus, and one may reasonably conclude that their records for *V. sericocarpa* are simply misidentifications of *V. hybrida*. It is rather more difficult to dispose of Holmboe's record for *V. sericocarpa* from Ayia Napa (*Holmboe* 74), and the specimen awaits re-examination. The specimen of *V. sericocarpa* var. *microphylla* cited by H. Stuart Thompson (Journ. Bot., 44: 307; 1906) turns out to be ordinary *V. hybrida*, and of all the material of *V. sericocarpa* examined, only one sheet from Cyprus is correctly named. This is *Rogers* 0651 (or K. 71) said to have been collected " Nr. Nicosia " on March 2, 1930. Since it is evident, however, that some of Rogers' specimens (cf. *Cytisopsis dorycniifolia*) are mislabelled, and come, not from Cyprus but from the neighbourhood of Alexandretta (Iskenderun), this sheet cannot be accepted as satisfactory proof of the occurrence of *V. sericocarpa* in Cyprus.]

3. V. assyriaca *Boiss.*, Diagn., 2, 9: 123 (1849), Fl. Orient., 2: 572 (1872); Davis, Fl. Turkey, 3: 304 (1970).

V. noëana Reut. ex Boiss., Fl. Orient., 2: 572 (1872); Post, Fl. Pal., ed. 2, 1: 418 (1932).

TYPE: Turkey; " in *Assyriâ* circa *Diarbekir* et in monte *Karadja Dagh Kotschy* 213 " (G, K !).

Scrambling annual; stems 20–40 cm. long, much-branched, angular, subglabrous or thinly adpressed-pubescent; leaves paripinnate, 3–5 cm. long, 1–3·5 cm. wide, the basal 6–10-foliolate, the upper 10–16-foliolate, tendril slender, branched or unbranched; stipules small, ovate-acuminate, 1–2 mm. long, 1 mm. wide, occasionally semi-sagittate, subglabrous; petioles very short, those of the lower leaves usually less than 1 cm. long, the upper leaves sessile or subsessile; leaflets narrow, linear-oblong, 5–20 mm. long, 1–3 mm. wide, thinly adpressed-villose below, glabrous or subglabrous above, apex truncate or emarginate, distinctly apiculate, base very shortly petiolulate or subsessile; flowers solitary or 2–5 in short axillary racemes; peduncle and rhachis generally less than 1 cm. long, adpressed-villose, pedicels very short; calyx narrowly campanulate or shortly cylindrical, 5–6 mm. long, 3–3·5 mm. wide, calyx-tube tinged purple or violet, adpressed-pubescent, oblique at apex, gibbous at base, calyx-teeth deltoid-subulate, 1–2 mm. long, the abaxial rather longer than the others; petals pale yellow, tinged brown with age, limb of standard glabrous, suborbicular, 7–10 mm. long and almost as wide, apex rounded or slightly emarginate, margins usually reflexed, claw broad, about 10 mm. long, 7 mm. wide at apex, limb of wings oblong, 9–11 mm. long, 4–5 mm. wide, apex rounded, base distinctly auriculate, with a short blunt auricle, claw slender 8–9 mm. long, limb of keel squarish, about 4–5 mm. long and almost as wide, apex rounded, adaxial margin with a short, violet-tinged tooth, base bluntly auriculate, claw slender, 8–9 mm. long; stamens 10–12 mm. long, free parts of filaments 2–3 mm. long; anthers oblong, about 0·8 mm. long, 0·4 mm. wide; ovary linear, compressed, 6–8 mm. long, 0·8–1 mm. wide, minutely verruculose, with a slender basal stipe about 5–6 mm. long, style twisted upwards at a right angle, 2–3 mm. long, densely bearded above, stigma capitate. Pod oblong, strongly compressed, 25–30 mm. long, 10–12 mm. wide, apex acute or shortly rostrate, valves glabrous, distinctly, but finely veined; seeds suborbicular, compressed, about 5 mm. diam.; testa smooth, dark brown.

HAB.: In cultivated field amongst *Vicia sativa*, 200 ft. alt.; probably introduced; fl. March–April.

DISTR.: Division 7, rare. Also Turkey, Syria, Palestine and Iraq.

7. Between Kyrenia and Bellapais, 1952, *Casey* 1245 ! and *Casey* 1265 !

NOTES: It is difficult to find any adequate reason for separating *V. assyriaca* and *V. noëana*; the characters given by Boissier in *Fl. Orientalis* do not hold if a range of specimens is examined. For this reason the two supposed species are united; *V. assyriaca* is scarcely more than a casual in our area.

4. V. sativa *L.*, Sp. Plant., ed. 1, 736 (1753); Boiss., Fl. Orient., 2: 574 (1872); Holmboe, Veg. Cypr., 113 (1914); Davis, Fl. Turkey, 3: 318 (1970); Zohary, Fl. Palaest., 2: 205, t. 295 (1972).

Diffuse or sprawling annual; stems much-branched, angular, subglabrous or sparingly clothed with short, spreading hairs; leaves oblong, paripinnate, 1·5–8 cm. long, 1–5 cm. wide, the upper with a well-developed, branched tendril, the lower bristle-tipped or sometimes without any tendril; stipules semisagittate, 3–8 mm. long, 2–6 mm. wide, with a conspicuous dark, glandular median blotch, margins acutely dentate-laciniate; petiole very short, generally less than 10 mm. long; leaflets (2–) 4–10, very variable in shape, sometimes broadly obovate or obcordate, sometimes narrowly oblong or linear, 3–30 mm. long, 2–15 mm. wide, thinly pilose on both surfaces or subglabrous, apex acute, rounded or emarginate, generally cuspidate or apiculate, base shortly but distinctly petiolulate; flowers solitary or sometimes paired, subsessile or very shortly pedicellate; calyx cylindrical, 8–15 mm. long, 2–5 mm. wide, subglabrous or thinly pilose, calyx-teeth subulate, subequal, 3–8 mm. long, 1–1·5 mm. wide at base, 3-nerved, often with a small glandular blotch, thinly pilose or ciliate; petals crimson, violet or bluish, the standard often paler than wings and keel; limb of standard glabrous, obovate, ovate or oblong, emarginate, 7–15 mm. long, 8–17 mm. wide, tapering abruptly to a claw 7–10 mm. long, 3·5–5 mm. wide, limb of wings oblong or obovate, 7–10 mm. long, 3–8 mm. wide, apex rounded, base produced into a spur-like auricle 2–3 mm. long, claw slender 6–8 mm. long, limb of keel shortly oblong, 4–5 mm. long, 3–4 mm. wide, abaxial margin rounded, apex blunt or apiculate, base obscurely auriculate, claw 8–9 mm. long; stamens 8–13 mm. long, free parts of filaments about 1·5 mm. long; anthers oblong, about 0·5 mm. long, 0·3 mm. wide; ovary linear, compressed, 8–12 mm. long, 1·5–2 mm. wide, subglabrous, minutely puberulous or distinctly pubescent, tapering to apex and base, ovules 7–10, style sharply upcurved, about 1·5 mm. long, glabrous below, strongly bearded at apex, stigma capitate. Pod narrowly oblong, strongly compressed 30–50 mm. long, 5–8 mm. wide, apex tapering to a distinct beak, valves subglabrous, dark or light brown; seeds subglobose or slightly compressed, 2·5–5 mm. diam.; testa dark brown, sometimes distinctly marbled or pruinose, occasionally pale.

var. sativa.

V. sativa L. var. *obovata* Ser. ex DC., Prodr., 2: 361 (1825); Unger et Kotschy, Die Insel Cypern, 385 (1865).

[*V. cordata* (? Wulfen) Sintenis in Oesterr. Bot. Zeitschr., 31: 392 (1881).]

TYPE: " *inter* Europae *segetes* ".

Plants robust with large leaves and leaflets; flowers rather large, the standard 20–25 mm. long, and 17 mm. wide, often conspicuously paler than wings and keel. Pods about 40–50 mm. long, 6–8 mm. wide; seeds 4–5 mm. diam., often pruinose.

HAB.: Cultivated and waste ground, or roadsides, an escape from cultivation; sea-level to 2,000 ft. alt.; fl. Jan.–April.

DISTR.: Divisions 2–4, 7, common in Division 7. Widespread as an escape from cultivation throughout most temperate, and many tropical regions of the world.

2. Khrysorroyiatissa Monastery, 1862, *Kotschy* 684b; Lefka, 1932, *Syngrassides* 242 ! Lefkara, 1941, *Davis & Kennedy* 2715 !
3. Kalavasos, 1905, *Holmboe* 634.
4. Near Larnaca, 1880, *Sintenis & Rigo*; near Larnaca, 1862, *Kotschy*; Larnaca, 1881, *Barbey*; Mile 3 Larnaca–Famagusta road, 1950, *Chapman* 426 !
7. Common; *Kotschy* 499a; *Syngrassides* 177 ! *Davis* 2460 ! 2780 ! *Casey* 322 ! 1001 ! *Chapman* 100 ! *G. E. Atherton* 848 ! 935 !

NOTES: The origin and natural distribution of *Vicia sativa* var. *sativa* is obscure; it is widely grown in lowland areas of Cyprus, and the specimens cited above are most probably strays from, or relics of, cultivation.

var. angustifolia *L.*, Fl. Suec., ed. 2, 255 (1755) in obs.

V. angustifolia (L.) L., Amoen. Acad., 4: 105 (1759); Boiss., Fl. Orient., 2: 574 (1872); Holmboe, Veg. Cypr., 113 (1914); Post, Fl. Pal., ed. 2, 1: 419 (1932); Lindberg f., Iter Cypr., 20 (1946); Rechinger f. in Arkiv för Bot., ser. 2, 1: 426 (1950).

V. sativa L. var. *nigra* L., Sp. Plant., ed. 2, 1037 (1763).

V. sativa L. ssp. *nigra* (L.) Ehrh. in Hannover. Mag., 1780 (15): 229 (1780); Davis, Fl. Turkey, 3: 318 (1970); Osorio-Tafall et Seraphim, List Vasc. Plants Cyprus, 57 (1973).

V. sativa L. var. *segetalis* (Thuill.) Ser. ex DC., Prodr., 2: 361 (1825); Unger et Kotschy, Die Insel Cypern, 385 (1865).

V. sativa L. ssp. *angustifolia* (non Gaudin) Aschers. et Graebn., Syn. Mitt.-Europ. Fl., 6 (2): 971 (1909); Zohary, Fl. Palaest., 2: 206 (1972).

TYPE: Not designated, but presumably the name was originally given to a Swedish specimen.

More slender and often more glabrous than var. *sativa*; leaves and leaflets rather small, the leaflets of the upper leaves often very narrow, but not always so; flowers smallish, the standard about 14–18 mm. long and 7–8 mm. wide, frequently bright crimson-purple or violet and not much paler than the wings or keel. Pods 30–40 mm. long, 5–6 mm. wide; seeds 2·5–4 mm. diam.; testa dark brown, often marbled.

HAB.: Waste ground, roadsides, amongst garigue on stony hillsides, on grassy slopes under Pines, occasionally on sand-dunes; sea-level to 5,000 ft. alt.; fl. Jan.–April.

DISTR.: Divisions 2–8, common in Divisions 2, 7. Widespread in Europe and the Mediterranean region and eastwards to Iran and Afghanistan; adventive elsewhere.

2. Common; *Kotschy* 799; *Lindberg f.* s.n.; *Davis* 3121 ! *Meikle* 2672 !
3. Zagala above Trimiklini, 1941, *Davis* 3140 ! Temple of Apollo near Episkopi, 1950, *Chapman* 565 !
4. Near Larnaca, 1880, *Sintenis & Rigo*.
5. Nicosia–Famagusta road, 1950, *Chapman* 399 ! Athalassa, 1957, *Merton* 2905 !
6. Beuyuk Kaimakli, 1934, *Syngrassides* 1350 ! Syrianokhori, 1935, *Syngrassides* 620 !
7. Common; *Sintenis & Rigo* 468 ! *Syngrassides* 616 ! *Davis* 2772 ! 2844 ! *G. E. Atherton* 1018 !
8. Cape Andreas, 1938, *Syngrassides* 1764 ! Ayios Philon near Rizokarpaso, 1941, *Davis* 2217 ! Apostolos Andreas, 1941, *Davis* 2317 ! Komi Kebir, 1941, *Davis* 2446 !

NOTES: Intergrades with var. *sativa*, but generally recognizable by its slender habit and bright flowers. Unlike var. *sativa*, it is undoubtedly indigenous in our area.

var. **amphicarpa** *Boiss.*, Voy. Bot. Esp., 2: 193 (1839).

V. subterranea Gérard ex Dorthes in Obs. Phys. Hist. Nat., 35: 131 (1789).

V. angustifolia L. var. *amphicarpa* (Boiss.) Alef. in Bot. Zeit., 20: 362 (1862); Boiss., Fl. Orient., 2: 575 (1872).

V. sativa L. ssp. *amphicarpa* (Boiss.) Aschers. et Graebn., Syn. Mitt.-Europ. Fl., 6(2): 974 (1909); Davis, Fl. Turkey, 3: 321 (1970); Zohary, Fl. Palaest., 2: 207 (1972).

[*V. amphicarpa* (non (L.) Lam.) Dinsmore in Post, Fl. Pal., ed. 2, 1: 419 (1932).]

TYPE: Spain; " in cultis prope Malaga ubi legit amic. Haenseler " (G).

Plant usually rather small and slender, often producing sterile autumnal growths with ecirrhose leaves and 1–2 pairs of very narrow, linear leaflets; spring growths resembling those of var. *angustifolia* L., but flowers of two kinds: those of the aerial stems petaloid and rather large, the standard 20 (–25) mm. long and 10 mm. wide; those of the leafless, basal subterranean stems apetalous and with a modified membranous calyx. Pods and seeds of petaloid flowers like those of var. *angustifolia*; pods of subterranean flowers ovoid or ellipsoid, compressed, pallid, about 10–15 mm. long and 6 mm. wide, seeds 1–2, rather large.

HAB.: On stony or grassy hillsides, often in Pine forest, 300–2,000 ft. alt.; fl. Jan.–April.

DISTR.: Divisions 5, 7, rare. Throughout the Mediterranean region, and eastwards to Iran.

5. Ayia Varvara near Nisou, 1936, *Syngrassides* 1137 !
7. St. Hilarion, 1941, *Davis* 2495 ! Side of Yaïla Peak 2 m. E. of Halevga, 1974, *Meikle* 4036 !

NOTES: A puzzling plant, which Boissier and other authors have considered to be little more than an ecological form occurring in different variants of *V. sativa*. While it is true that considerable variation is to be found in herbarium material labelled *amphicarpa*, it does appear that the geocarpic habit is very often associated with other distinctive morphological characters, e.g. the sterile, autumnal growths, and the relatively large flowers. It is just possible that in addition to geocarpic forms of *V. sativa* aggr. there is a distinct and fairly uniform geocarpic Vetch which may deserve recognition at specific (or possibly subspecific) level. Characteristic specimens of this distinctive *amphicarpa* (which is, incidentally, almost certainly the true *V. subterranea* Gérard ex Dorthes) are not uncommon on the slopes of the Northern Range, generally growing in a deep litter of fallen Pine needles.

5. V. lathyroides *L.*, Sp. Plant., ed. 1, 736 (1753); Sibth. et Sm., Fl. Graec. Prodr., 2: 71 (1813); Unger et Kotschy, Die Insel Cypern, 386 (1865); Boiss., Fl. Orient., 2: 574 (1872); Holmboe, Veg. Cypr., 113 (1914); Post, Fl. Pal., ed. 2, 1: 420 (1932); A. K. Jackson et Turrill in Kew Bull., 1938: 465 (1938); Davis, Fl. Turkey, 3: 317 (1970); Osorio-Tafall et Seraphim, List Vasc. Plants Cyprus, 57 (1973).

TYPE: " *in* Scothia, Lusatia ".

A small diffuse annual; stems usually much branched, distinctly angled, 2–10 cm. long, pubescent with short spreading hairs; leaves paripinnate, 1–2-foliolate, 6–15 mm. long, 5–20 mm. wide, without tendrils, or with the rhachis produced into a short, slender bristle; stipules semi-sagittate, 1–2·5 mm. long, 0·4–1 mm. wide, entire or subentire, rather densely pubescent; petiole usually distinctly developed, 1–6 mm. long; leaflets obcordate or narrowly linear-oblong, 2–10 mm. long, 1·5–3·5 mm. wide, ± pilose on both surfaces, apex emarginate or acute, generally apiculate, base obscurely petiolulate; flowers solitary, axillary, inconspicuous; pedicel very short, rather densely pilose; calyx cylindrical, about 4 mm. long, 2 mm. wide, pilose externally, calyx-teeth subulate, subequal, about 1·5 mm. long; petals bluish or mauve, limb of standard broadly obovate, 4 mm. long, 3·5 mm. wide, apex emarginate, often shortly apiculate, base tapering to a claw about 3 mm. long, 2 mm. wide, limb of wing narrowly oblong, about 4 mm. long, 1·5 mm. wide, apex rounded, papillose, base produced into a long narrow auricle, claw about 2·5 mm. long, limb of keel suborbicular, about 1·5 mm. long, 1·3 mm. wide, apex rounded, base distinctly auriculate, claw about 2 mm. long; stamens about 3 mm. long, free parts of filaments about 0·5 mm. long; anthers suborbicular, about 0·2 mm. diam.; ovary linear, compressed, about 3 mm. long, 0·8 mm. wide, glabrous or ± puberulous, 5–8-ovulate, style sharply upcurved, short, scarcely more than 0·5 mm. long, bearded near apex, stigma small, capitate. Pod linear, compressed, 15–25 mm. long, 3–4 mm. wide, apex acuminate-rostrate, valves dark brown, glabrous or subglabrous; seeds subglobose, or ± cube-shaped, about 1·5 mm. diam.; testa dark brown, distinctly and closely verruculose.

HAB.: Amongst garigue on dry mountainsides, on igneous formations; 2,000–3,500 ft. alt.; fl. March–April.

DISTR.: Divisions 2, [4], locally common in Division 2. Widespread in Europe and the Mediterranean region, and eastwards to Iraq.

2. Locally common; Platres, Makheras Mon. etc.; *Kennedy* 585 ! 587 ! *Casey* 285 ! *N. Macdonald* 33 ! 85 !

[4. Near Ormidhia, 1880, *Sintenis & Rigo*.]

NOTES: Some doubt must attach to Sibthorp's record " in Cypri arvis ", especially as the flowers are said to have been whitish, with a striped standard, a very unusual variation for this species, but characteristic in *V. peregrina* L. The Sintenis & Rigo record, from near Ormidhia, is also dubious, and it is much more likely that *V. sativa* var. *angustifolia* was the plant seen. *V. lathyroides* flowers early, and is easily overlooked; when in fruit (even in young fruit) it can at once be distinguished from small specimens of *V. sativa* var. *angustifolia* by its verruculose seeds.

6. V. peregrina *L.*, Sp. Plant., ed. 1, 737 (1753); Boiss., Fl. Orient., 2: 576 (1872); Holmboe, Veg. Cypr., 113 (1914); Post, Fl. Pal., ed. 2, 1: 420 (1932); Davis, Fl. Turkey, 3: 308 (1970); Zohary, Fl. Palaest., 2: 202, t. 290 (1972); Osorio-Tafall et Seraphim, List Vasc. Plants Cyprus, 57 (1973).

? *V. carnea* Kotschy in Unger et Kotschy, Die Insel Cypern, 386 (1865).

[*V. michauxii* (non Spreng.) Boiss., Fl. Orient., 2: 577 (1862) pro parte quoad plant.

cypr.; Osorio-Tafall et Seraphim, List Vasc. Plants Cyprus, 57 (1973) pro parte quoad plant. cypr.]
TYPE: " *in* Gallia D. Sauvages ".

Suberect or sprawling annual; stems (6–) 10–25 (–35) cm. long, much branched, slender, angular, subglabrous or very sparingly pilose, leaves paripinnate, 1·5–4 (–6) cm. long, 1–2·5 (–3) cm. wide, the basal often 2-foliolate, the upper 4–10 (–12)-foliolate; tendril long and slender, branched or unbranched; stipules small, semi-sagittate, 1–4 mm. long, usually under 1·5 mm. wide, deeply and acutely lobed, subglabrous or distinctly pilose; petioles very short, those of the lower leaves up to 1 cm. long, the upper leaves often sessile or subsessile; leaflets usually very narrow, 2–30 mm. long, 0·5–3·5 mm. wide, glabrous or thinly subadpressed-pilose, apex truncate or emarginate, usually apiculate, base very shortly petiolulate or subsessile; flowers solitary in the axils of the upper leaves; pedicel slender, 4–10 mm. long, adpressed-pilose, often recurved after anthesis; calyx campanulate, about 7 mm. long, 3·5 mm. wide, thinly pilose externally, base distinctly gibbous, calyx-teeth unequal, the 2 adaxial deltoid, about 1·5 mm. long, slightly recurved, the 3 abaxial subulate, about 2 mm. long, straight or almost straight; petals purple, mauve, lavender or whitish, the standard usually darker in colour than the wings or keel and often pencilled purple; limb of standard obovate, 9–11 mm. long, about 8 mm. wide, glabrous, apex deeply emarginate, base tapering to a claw about 5 mm. long and almost as wide, limb of wings oblong, 7–8 mm. long, 4 mm. wide, apex rounded, base shortly and bluntly auriculate, claw slender, 5–6 mm. long, limb of keel suborbicular, about 3 mm. diam., blotched purple adaxially, base distinctly but bluntly auriculate, claw 4–5 mm. long; stamens about 7 mm. long, free parts of filaments 1–1·5 mm. long; anthers oblong, 0·5 mm. long, 0·3 mm. wide; ovary ± linear, 5–6 mm. long, 1–1·5 mm. wide, compressed, adpressed-pubescent, adaxial margin distinctly gibbous, ovules about 7–8, style 2 mm. long, bearded above, stigma capitate, slightly oblique. Pod strongly compressed, oblong-elliptic, 20–35 mm. long, 5–8 mm. wide, tapering at both ends, valves pale brown, sometimes blotched purple, adpressed-pubescent; seeds subglobose or oblong, 2–3 mm. long, 2 mm. wide; testa smooth, dark chocolate-brown.

HAB.: Waste and cultivated ground, roadsides, pastures, and amongst garigue on rocky hillsides, usually on calcareous formations; sea-level to 2,700 ft. alt.; fl. Febr.–June.

DISTR.: Divisions 1–8, common in Divisions 3, 7. Southern Europe and the Mediterranean region eastwards to Central Asia; Atlantic Islands.

1. Ayios Neophytos Monastery, 1939, *Lindberg f.* s.n. ! Dhrousha, 1941, *Davis* 3214 ! Baths of Aphrodite, 1962, *Meikle* 2221 !
2. Khrysorroyiatissa Monastery, 1862, *Kotschy* 681; and, 1941, *Davis* 3358 !
3. Common; *Davis* 2758 ! 2785 ! 3144 ! 3090 ! *Mavromoustakis* s.n. ! *N. Macdonald* 87 !
4. Larnaca, 1881, *Barbey*; the Phaneromene near Larnaca, 1905, *Holmboe* 135.
5. Near Pyroi, 1936, *Syngrassides* 1149 ! Athalassa, 1957, *Merton* 2904 !
6. Nicosia, 1937, *Syngrassides* 1409 ! Nicosia airport, 1954, *Merton* 1841 ! Nicosia, 1957, *Merton* 3084 !
7. Common; *Sintenis & Rigo* 469b ! *Syngrassides* 1149 ! *Davis* 2496 ! 2828 ! *Casey* 364 ! 434 ! 516 ! *Chapman* 99 ! *G. E. Atherton* 1030 ! *Merton* 2522 !
8. Near Eleousa, 1880, *Sintenis & Rigo* 469 ! Koma tou Yialou, 1935, *Syngrassides* 623 ! Ayios Philon, 1941, *Davis* 2235 ! Komi Kebir, 1941, *Davis* 2442 !

NOTES: Boissier (Fl. Orient., 2: 577) seems to have been mistaken in considering *V. carnea* Kotschy synonymous with the yellow-flowered, large-seeded, *V. michauxii* Spreng. The type of Kotschy's species is from Khrysorroyiatissa, where *V. peregrina* is known to occur, and is no doubt just a pallid-flowered variant of this species.

7. **V. bithynica** (*L.*) *L.*, Syst. Nat., ed. 10, 1166 (1759); Boiss., Fl. Orient., 2: 587 (1872); Druce in Rep. B.E.C., 9: 470 (1931); Rechinger f. in Arkiv för Bot., ser. 2, 1: 426

(1950); Davis, Fl. Turkey, 3: 321 (1970); Zohary, Fl. Palaest., 2: 207, t. 296 (1972); Osorio-Tafall et Seraphim, List Vasc. Plants Cyprus, 57 (1973).

Lathyrus bithynicus L., Sp. Plant., ed. 1, 731 (1753).

TYPE: " *in* Italiae, Bavariae *arvis* ".

Sprawling or scrambling annual 25–60 cm. high; stems subglabrous, sharply angled, generally much branched especially towards base; leaves 2–4-foliolate; tendril slender, usually branched; stipules large and conspicuous, ovate-oblong, 8–15 mm. long, 3–8 mm. wide, glabrous or very thinly pilose, often with a glandular spot, apex acuminate, margins sharply laciniate-dentate; petiole 1–2 cm. long; leaflets subglabrous or thinly pilose on the lower surface, those of the basal leaves ovate-oblong, 1·5–3 cm. long, 0·5–1·3 cm. wide, apex obtuse or subacute, those of the upper leaves narrowly elliptical or lanceolate, 3–6 cm. long, 0·3–1 cm. wide; flowers solitary or in pairs, arising from the axils of the upper leaves; peduncle 1–3 cm. long, or sometimes almost wanting, pedicels slender, pilose, 2–5 mm. long; calyx campanulate, 9–11 mm. long, 3–4 mm. wide, thinly pilose, calyx-teeth subequal, subulate, 6–8 mm. long, 1·5 mm. wide at base; petals purplish (wings and keel often whitish), limb of standard glabrous, ovate-oblong, 10–12 mm. long, 8–10 mm. wide, apex blunt, shortly emarginate, margins reflexed, claw broad, 6–8 mm. long; limb of wings oblong, 9–10 mm. long, 3–5 mm. wide, apex rounded, base prominently auriculate, claw slender about 7 mm. long, limb of keel suborbicular, 3–4 mm. diam., base shortly but sharply auriculate, claw slender about 5–6 mm. long; stamens 9–10 mm. long, free parts of filaments about 1·5–2 mm. long; anthers oblong, about 0·5 mm. long, 0·3 mm. wide; ovary linear, compressed, 8–9 mm. long, 1·5 mm. wide, glabrescent with ciliate sutures, tapering towards apex and base, c. 6-ovulate, style about 1·5 mm. long, upcurved at a right angle, glabrous below, densely bearded at apex, stigma small, capitate. Pod oblong, flattened, 3–4·5 cm. long, 1–1·2 cm. wide, apex sharply rostrate, base abruptly narrowed, valves light or dark brown, obscurely venulose, thinly pilose or subglabrous; seeds subglobose, 3–5 mm. diam.; testa dark brown, smooth or obscurely pitted.

HAB.: By streams and rivers and on damp ground; near sea-level; fl. March–April.

DISTR.: Divisions 1, 3, 4, 6. Widespread in western and southern Europe; Aegean Islands; Turkey and eastwards to Afghanistan.

1. Ktima, 1913, *Haradjian* 702; Paphos, 1928 or 1930, *Druce*; Konia R. near Ktima, 1951, *Kennedy* in *Casey* 1195 !
3. Stavrovouni, 1928 or 1930, *Druce*.
4. Larnaca, 1928 or 1930, *Druce*.
6. Syrianokhori, 1935, *Syngrassides* 619 !

NOTES: The Stavrovouni record is a little doubtful; elsewhere in Cyprus *V. bithynica* is confined to the coastal strip, not ascending far above sea-level.

SECT. 2. **Faba** (*Adans.*) *Aschers. et Graebn.* Flowers usually few, in subsessile or shortly pedicellate axillary clusters, rather large; style bearded abaxially near apex; plants erect or suberect; leaves 2–6-foliolate with large leaflets.

8. **V. narbonensis** *L.*, Sp. Plant., ed. 1, 737 (1753); Boiss., Fl. Orient., 2: 577 (1872); Holmboe, Veg. Cypr., 113 (1914); Post, Fl. Pal., ed. 2, 1: 421 (1932); Davis, Fl. Turkey, 3: 321 (1970); Zohary, Fl. Palaest., 2: 208, t. 297 (1972); Osorio-Tafall et Seraphim, List Vasc. Plants Cyprus, 57 (1973).

Stout erect or suberect annual; stems (10–) 20–50 (–60) cm. high, rather sparingly branched at the base or unbranched, sharply angled, subglabrous or ± densely ciliate along the angles; basal leaves pinnately 2-foliolate, upper leaves paripinnately 4–6-foliolate, oblong, 1–7 cm. long, 2·5–8 (–10) cm. wide;

tendrils usually restricted to the uppermost leaves, slender, unbranched or 2–3-fid near apex; stipules oblong or broadly semi-sagittate, 5–18 mm. long, 3–9 mm. wide, obtuse or acute, often with a small dark glandular spot near apex, margin sometimes sharply toothed towards base, generally ciliate with bulbous-based hairs; petiole stout, but generally less than 2 cm. long; leaflets oblong, ovate, obovate or broadly elliptical, 1–5 cm. long, 0·6–3·5 cm. wide, glabrous or very thinly pilose on the upper and lower surfaces, the upper surface often sparsely glandular with small sessile glands, apex acute or rounded, often shortly apiculate, base subsessile or very shortly petiolulate, margins entire or bluntly toothed, ciliate with bulbous-based hairs; flowers solitary or in pairs in the axils of the upper leaves; pedicels very short, thinly adpressed-pilose; calyx cylindrical, 9–14 mm. long, 2·5–3 mm. wide, glabrous or very thinly pilose, base scarcely gibbous, calyx-teeth deltoid or narrowly ovate, acuminate, the adaxial about 2·5 mm. long and 1·5 mm. wide at base, the abaxial about 4·5 mm. long, 2·5 mm. wide at base, all with thinly ciliate margins; petals whitish, veined purple or purplish all over, limb of standard glabrous, obovate, 12–14 mm. long, 9–12 mm. wide, apex emarginate, base tapering to a broad claw 8–10 mm. long, limb of wings oblong, 8–9 mm. long, 4–5 mm. wide, often purple-blotched at the broad rounded apex, base distinctly but bluntly auriculate, claw slender about 10 mm. long, limb of keel oblong, 5–7 mm. long, 3–3·5 mm. wide, apex rounded, adaxial margin with a very short tooth, base distinctly but bluntly auriculate, claw 10–12 mm. long; stamens 15–17 mm. long, free parts of filaments not dilated, about 1·5–2 mm. long; anthers oblong, about 0·6 mm. long, 0·4 mm. wide; ovary linear, 12–15 mm. long, 1–1·5 mm. wide, strongly compressed, adaxial and abaxial margins densely pilose, surfaces thinly pilose, ovules 6–8, style bent upright at a right angle, 2·5–3 mm. long, densely bearded at apex, stigma small, somewhat oblique. Pod oblong, flattened, 5–6 cm. long, 1–1·3 cm. wide, apex acute-rostrate, valves glabrous or thinly pilose, margins conspicuously papillose-ciliate; seed suborbicular, about 5 mm. diam.; testa smooth, dark brown.

var. **narbonensis**.

TYPE: " *in* Gallia, Anglia ".

Leaflets entire, often rather large; plants usually robust.

HAB.: In cultivated fields or on waste ground, sometimes in garigue or in Pine forest; near sea-level to 3,000 ft. alt.; fl. Febr.–May.

DISTR.: Divisions 2, 4–7. Widely distributed in southern Europe and the Mediterranean region eastwards to Iran.

2. Lefkara, 1941, *Davis* 2717 ! Khrysorroyiatissa, 1941, *Davis* 3365 ! Stavros tis Psokas, 1962, *Meikle* 2751 !
4. Larnaca, 1880, *Sintenis & Rigo*; at the Phaneromene near Larnaca, 1905, *Holmboe* 131a; Paralimni, Ayia Napa, 1936, *Syngrassides* 1155 ! 1318 !
5. Lefkoniko, 1935, *Nattrass*.
6. Nicosia, 1937, *Syngrassides* 1408 ! Near Kykko (Mesaoria), 1952, *F. M. Probyn* 15 !
7. Pentadaktylos, 1880, *Sintenis & Rigo*; above Halevga, 1963, *J. B. Suart* 73 !

var. **serratifolia** (*Jacq.*) *Ser.* in DC., Prodr., 2: 365 (1825); Post, Fl. Pal., ed. 2, 1: 421 (1932); Davis, Fl. Turkey, 3: 322 (1970); Zohary, Fl. Palaest., 2: 209 (1972).

V. serratifolia Jacq., Fl. Austr., 5: 30, t. 8 (1778); Boiss., Fl. Orient., 2: 578 (1872).

V. narbonensis L. ssp. *serratifolia* (Jacq.) Nym., Consp. Fl. Europ., 209 (1878); Holmboe, Veg. Cypr., 113 (1914); Osorio-Tafall et Seraphim, List Vasc. Plants Cyprus, 57 (1973).

? *V. johannis* Tamamsch. sec. Schöfer in Kulturpflanze, 21: 261 (1973) quoad plant. cypr.

? *V. narbonensis* L. var. *salmonea* (Mout.) Schöfer in Kulturpflanze, 21: 253 (1973) quoad plant. cypr.

[*V. narbonensis* (non L.) Sibth. et Sm., Fl. Graec. Prodr., 2: 73 (1813); Unger et Kotschy, Die Insel Cypern, 386 (1865).]

TYPE: Hungary; " circa Sopronium [Sopron] Hungariae urbem " *Mygind* (W).

Leaflets, or some of them, distinctly toothed; usually a more slender plant than var. *narbonensis*, but many intermediates occur.

HAB.: Cultivated fields and waste ground; sea-level to 500 ft. alt.; fl. March–May.

DISTR.: Divisions 3–7. Southern Europe, Mediterranean region, and eastwards to Iran.

3. Akrotiri, 1927, *Rev. A. Huddle* 49 !
4. At the Phaneromene near Larnaca, 1905, *Holmboe* 131b.
5. Laxia, 1932, *Syngrassides* 80 !
6. Near Nicosia, 1880, *Sintenis & Rigo* 470 ! Nicosia, 1937, *Syngrassides* 1403 !
7. Kyrenia, 1949, *Casey* 345 ! and, 1958, *G. E. Atherton* 907 ! Near Ayios Epiktitos, 1959, *P. H. Oswald* 69 !

NOTES: Two of the specimens cited above (*Casey* 345 and *Atherton* 907) approach typical *V. narbonensis*, and have only very slightly toothed leaflets. Such plants are not infrequent in Cyprus. Holmboe records typical specimens of both varieties growing together near Larnaca.

V. FABA *L.* The cultivated Broad bean, widely grown in Cyprus, is not infrequently found on waste ground as an escape from cultivation. It is nowhere known as a truly wild plant, and its history and relationships remain conjectural, though in general habit and appearance it has an unmistakable affinity with *V. narbonensis.* The very large pods and seeds, and the thick, smooth (not ciliate) valves of the pods, provide the most obvious distinctions. *V. faba* sometimes has tendrils like those of *V. narbonensis*, but in most cultivated specimens the tendrils are either wanting or else very much reduced.

SECT. 3. **Cracca** *S. F. Gray* Inflorescence usually a pedunculate, many-flowered raceme; style bearded or pubescent all round near apex; leaves often with many leaflets.

[V. DUMETORUM *L.*, a widespread European species, is reported " in Cypern in Vorbergen über Omodos [Omodhos] 1859 " by Kotschy (Die Insel Cypern, 383). Holmboe (Veg. Cypr., 113) notes that no specimens of *V. dumetorum* could be found in Kotschy's collections of Cyprus plants at Vienna, and one must conclude that the record is somewhat dubious. In view, however, of Kotschy's general accuracy and reliability, a brief description of *V. dumetorum* is appended, in the hope that someone may succeed in re-discovering it in Cyprus.

A robust perennial climber often more than 1 m. high; stems glabrous, angular; leaves 5–10-foliolate, terminating in a long branched tendril; stipules large, semi-cordate, with sharply serrate margins; leaflets commonly alternate, ovate, oblong or elliptical, acute or obtuse, 1·3–4·5 cm. long, 0·8–3 cm. wide, glabrous or thinly puberulous; inflorescence a long-pedunculate, lax, many-flowered raceme; calyx glabrous, with rather short teeth; petals blue or purple, the standard about 1·5 cm. long. Pods glabrous, pale brown, oblong, 3–5 cm. long, 0·9–1·2 cm. wide, strongly compressed; seeds subglobose, 3–5 mm. diam.; testa dark brown or blackish, smooth; hilum almost half the circumference of the seed.]

[V. ONOBRYCHOIDES *L.*, a species locally common in many parts of southern Europe and western North Africa, is recorded from Cyprus by Sibthorp (Fl. Graec. Prodr., 2: 70), though Smith has added a note to the effect that specimens are not to be found in Sibthorp's herbarium. Most probably

Sibthorp confused *V. onobrychoides* with the annual *V. cassia* Boiss., which is often abundant on the Troödos massif, or possibly with the perennial *V. tenuifolia* Roth from the same area. *V. onobrychoides*, a perennial, is usually distinguishable by its long, stout racemes of 10 or more, large (17–24 mm. long), blue or purple flowers, by its laciniate-dentate stipules, and obtuse or truncate leaflets. No authentic specimens have been collected in Cyprus, the records cited by Poech, Kotschy and Stuart Thompson being based on Sibthorp's identification.]

9. V. tenuifolia *Roth*, Tent. Fl. Germ., 1: 309 (1788); Boiss., Fl. Orient., 2: 586 (1872); Post, Fl. Pal., ed. 2, 1: 423 (1932); Zohary, Fl. Palaest., 2: 196, t. 278 (1972).

V. cracca L. ssp. *tenuifolia* (Roth) Gaudin, Fl. Helv., 4: 507 (1829); Davis, Fl. Turkey, 3: 283 (1970); Osorio-Tafall et Seraphim, List Vasc. Plants Cyprus, 56 (1973).

Sprawling or scrambling perennial 30–100 cm. high; rootstock tough, woody; stems much branched at base, strongly ribbed, thinly subadpressed-pilose, rather flexuous; leaves oblong, 3–10 (or more) cm. long, 1·5–8 cm. wide, 8–30-foliolate; tendril branched, usually rather short and slender; stipules linear or semisagittate, 4–10 mm. long, with or without a spreading basal lobe 2–5 mm. long; petiole very short or wanting; leaflets linear or almost subulate, 8–35 mm. long, 0·5–6 (–8) mm. wide, subglabrous and rather greyish-green above, thinly subadpressed pilose below, apex acute or acuminate often shortly cuspidate, base shortly petiolulate with a hairy petiolule; inflorescences arising from the axils of the upper leaves, laxly or densely racemose, (8–) 10–30-flowered; peduncle 4–10 cm. long, often rather stout and erect, thinly adpressed-pilose; rhachis as long as, or sometimes exceeding the peduncle, thinly adpressed-pilose and sparsely glandular with small sessile glands; flowers somewhat deflexed; pedicel short, 1–2 mm. long; calyx shortly cylindrical, 3–4 mm. long, 2–2·5 mm. wide, thinly adpressed-pilose, base scarcely gibbous, calyx-teeth unequal, the abaxial narrowly deltoid-acuminate, 1·5–2·5 mm. long, the adaxial deltoid 1–1·5 mm. long, all ciliate with long eglandular and short glandular hairs; petals lilac, rich violet-blue or purple, limb of standard glabrous, obovate, 8–12 mm. long, 7–10 mm. wide, apex emarginate, claw 4–6 mm. long, about 3–4 mm. wide, and distinctly gibbous near the middle, limb of wings closely adherent to keel, oblong, 7–11 mm. long, 3–4 mm. wide, apex rounded, base sharply auriculate, claw slender 5–7 mm. long, limb of keel oblong, 4–4·5 mm. long, 3–4 mm. wide, curved to a rounded apex, base distinctly but bluntly auriculate, claw 4–5 mm. long; stamens 8–10 mm. long, free parts of filaments about 2–2·5 mm. long; anthers oblong, 0·5 mm. long, 0·3–0·4 mm. wide; ovary narrowly oblong, 6–8 mm. long, 0·8–1 mm. wide, strongly compressed, tapering to a stipitate base, glabrous or minutely glandular along the sutures, about 6-ovulate; style curved upwards at a right angle, 2–2·5 mm. long, pubescent all round near apex, stigma capitate. Pod oblong-elliptic, flattened, 20–35 mm. long, 6–9 mm. wide, gradually tapering to base, apex more abruptly acute, valves pale brown when ripe; seeds suborbicular, compressed, 3–5 mm. diam.; testa dark fuscous-brown, smooth; hilum about $\frac{1}{4}$–$\frac{1}{5}$ circumference of seed.

var. laxiflora *Griseb.*, Spic. Fl. Rum. et Bith., 1: 82 (1843).

V. elegans Guss., Fl. Sic. Prodr., 2: 438 (1828); Unger et Kotschy, Die Insel Cypern, 384 (1865).

V. tenuifolia Roth var. *stenophylla* Boiss., Fl. Orient., 2: 586 (1872); Lindberg f., Iter Cypr., 20 (1946).

V. cracca L. ssp. *stenophylla* Vel., Fl. Bulg., 163 (1891); Davis, Fl. Turkey, 3: 284 (1970); Osorio-Tafall et Seraphim, List Vasc. Plants Cyprus, 561 (1973).

V. cracca L. ssp. *elegans* (Guss.) Holmboe, Veg. Cypr., 113 (1914).

V. tenuifolia Roth var. *elegans* (Guss.) Dinsm. in Post, Fl. Pal., ed. 2, 1: 423 (1932).

[*V. cracca* (non L.) Sibth. et Sm., Fl. Graec. Prodr., 2: 70 (1813); Poech, Enum. Plant. Ins. Cypr., 40 (1842); Unger et Kotschy, Die Insel Cypern, 384 (1865); H. Stuart Thompson in Journ. Bot., 44: 307 (1906).]

TYPE: Greece; " in Thraciae et Macedoniae litoralis fruticetis alt. 0′–2,400′ frequens " *Grisebach* (GOET, K !).

Inflorescence lax, (8–) 10–20-flowered; flowers often rather large (15–18 cm. long); leaflets generally, but not always, narrowly linear-subulate with upcurved margins.

HAB.: On rocky hillsides amongst scrub or under Pines; by roadsides or on waste ground, often abundant in neglected vineyards; generally (? always) on igneous formations; 2,000–5,500 ft. alt.; fl. May–June.

DISTR.: Division 2, common around Prodhromos, Troödos, Platres etc., also at Kykko, Makheras Monastery and Zoöpiyi. S. Italy, Sicily, Balkan Peninsula, Turkey, Syria, Palestine, and eastwards to Afghanistan.

2. Common; *Sibthorp*; *Kotschy* 915 ! *Sintenis & Rigo* 813 ! *Holmboe* 915; *Syngrassides* 1088 ! 1602 ! *Kennedy* 580 ! 581 ! 1308 ! 1758 ! 1762 ! *Casey* 702 ! 742 ! *Meikle* 2944 !

NOTES: The Cyprus population of this species is tolerably uniform, with lax inflorescences and narrow leaflets. In other areas (particularly Crimea, Caucasus, N. Turkey, Iraq and Iran) var. *laxiflora* intergrades with typical *tenuifolia*, and it is doubtful if the two can be ranked as more than varieties of a single polymorphic species.

10. V. cassia *Boiss.*, Diagn., 1, 9: 119 (1849); Unger et Kotschy, Die Insel Cypern, 384 (1865); Boiss., Fl. Orient., 2: 587 (1872); Holmboe, Veg. Cypr., 114 (1914); Post, Fl. Pal., ed. 2, 1: 423 (1932); Davis, Fl. Turkey, 3: 293 (1970); Osorio-Tafall et Seraphim, List Vasc. Plants Cyprus, 56 (1973).

V. pinardii Boiss., Diagn., 1, 9: 120 (1849).

TYPE: Turkey; " in sylvaticis regionis inferioris montis Cassii [Akra Dağ] in Syriâ boreali ", June 1846, *Boissier* (G).

A slender trailing or scrambling annual; stems 30–100 cm. long, sharply angular, subglabrous or very thinly pilose, often dull bronze-green; leaves oblong, 3–7 cm. long, 1–3 cm. wide, 7–13-foliolate, tendril slender, branched; stipules semi-sagittate, 2–5 mm. long, about 1 mm. wide, subglabrous; petiole generally distinct, 1–1·5 cm. long; leaflets narrowly oblong, 5–15 mm. long, 1·5–3·5 mm. wide, subglabrous, apex rounded or sometimes acute, mucronate, base rounded or abruptly narrowed; inflorescence a lax 2–10-flowered axillary raceme; peduncle rather stout and erect, 2–5 cm. long, flowers distant along the rhachis, nodding; pedicels slender, about 2–4 mm. long, thinly pilose; calyx campanulate or shortly cylindrical, gibbous at base, 3–4 mm. long, 3 mm. wide, thinly adpressed-pilose, calyx-teeth unequal, the adaxial broadly deltoid, about 1 mm. long, the abaxial about 2 mm. long, subulate; standard rich purple, pencilled violet, wings and keel whitish or pale blue; limb of standard glabrous, rather narrowly oblong-obovate, 11–13 mm. long, 7–8 mm. wide, apex shortly emarginate, margins strongly reflexed, claw broad, about 10 mm. long, 7–8 mm. wide, limb of wings oblong, 11–12 mm. long, 4–5 mm. wide, apex rounded, base strongly but rather bluntly auriculate, claw slender about 9 mm. long, limb of keel oblong, about 6 mm. long, 4·5 mm. wide, apex upcurved, short, very blunt or rounded, base bluntly and shortly auriculate, claw 8–9 mm. long; stamens about 15 mm. long, free parts of filaments about 2–2·5 mm. long; anthers oblong, 0·5 mm. long, 0·3 mm. wide; ovary linear, about 14 mm. long, 1 mm. wide, tapering to apex and to an elongate, stipitate base, usually about 5-ovulate; style upcurved at a right-angle, about 3 mm. long, pubescent all round near apex, stigma capitate. Pod oblong-linear, about 20–30 mm. long, 6 mm. wide, somewhat compressed, tapering to a long apical beak and a slender stipitate base, valves pale grey-brown when ripe; seeds oblong or suborbicular,

3·5–4 mm. long and almost as wide; testa dark brown, smooth, hilum very short.

HAB.: On dry igneous mountainsides, in vineyards, waste and cultivated ground, in *Pinus* woodland, or amongst garigue; 1,500–5,000 ft. alt.; fl. March–May.

DISTR.: Divisions 2, 3, very common on the Troödos massif. Turkey, ? Syria.

2. Very common; Prodhromos, Platres, Platania, Yerasa, Makheras etc.; *Kotschy* 722 ! *Kennedy* 578 ! 579 ! 1306 ! 1760 ! 1761 ! *Syngrassides* 1547 ! *Davis* 3120 ! *Casey* 703 ! 717 ! *Merton* 387 ! 2351 ! 3190 ! *Meikle* 2702 !
3. Yerasa, 1950, *F. J. F. Barrington* 38 ! Trimiklini, 1963, *J. B. Suart* 107 !

11. V. cretica *Boiss. et Heldr.* in Boiss., Diagn., 1, 9: 118 (1849), Fl. Orient., 2: 588 (1872); Davis, Fl. Turkey, 3: 294 (1970).

TYPE: Crete; "inter dumos *Poterii spinosi* in Eparchia *Sitia* ad radices montis *Taroutsi* Heldr. pl. exs." (G).

Slender prostrate or scandent biennial or perennial; stems 10–20 (–40) cm. long, much branched especially towards base, angular, thinly subadpressed-pilose; leaves oblong, 2–3 cm. long, 1·0–2·5 cm. wide, (4–) 6–8 (–10)-foliolate, tendril slender, branched or unbranched; stipules semi-sagittate, about 2 mm. long, 1 mm. wide, subglabrous; petiole very short, usually less than 4 mm. long; leaflets narrowly oblong-elliptic, 4–12 mm. long, 1·5–3 mm. wide, glabrous above, thinly adpressed-pilose below, apex and base obtuse or rather shortly acute, the apex often shortly mucronate; inflorescence a rather lax 1–5-flowered axillary raceme; peduncle usually short, less than 2 cm. long, or almost wanting (especially in Cyprus specimens), flowers spreading or nodding; pedicels hairy, 1–2 mm. long; calyx campanulate, slightly gibbous at base, 3–4·5 mm. long, 2–2·5 mm. wide, thinly adpressed-pilose, calyx-teeth unequal, the adaxial very shortly deltoid, less than 0·5 mm. long, the abaxial subulate, about 2 mm. long; standard purplish, sometimes distinctly striped, limb oblong-obovate, 6–8 mm. long, 5–6 mm. wide, apex shortly emarginate, base scarcely narrowed, claw very broad, 5–6 mm. long, about 5 mm. wide, limb of wings oblong, about 5 mm. long, 2–3 mm. wide, apex very blunt, base distinctly but bluntly auriculate, claw slender 5–6 mm. long, limb of keel semicircular, about 4 mm. long, 2·5–3 mm. wide, curved upwards to a blunt apex, with distinct lateral swellings (or flaps) just below apex, base bluntly auriculate, claw 4–5 mm. long; stamens 7–9 mm. long, free parts of filaments 1–1·5 mm. long; anthers oblong, about 0·6 mm. long, 0·3 mm. wide; ovary narrowly oblong, compressed, glabrous, about 5–6 mm. long, 0·8–1 mm. wide, tapering to an elongate, stipitate base, and to a slender apex, usually 2–4-ovulate; style upcurved at a right angle, about 1·5 mm. long, villose all over towards apex, stigma small, capitate. Pod oblong or oblong-elliptic, 20–30 mm. long, about 7 mm. wide, strongly compressed, tapering to apex and base, valves pale brown, glabrous, obscurely veined; seeds suborbicular, 2·5–3 mm. diam., slightly compressed; testa smooth, dark brown.

HAB.: On rocky hillsides amongst *Sarcopoterium*; near sea-level to c. 1,000 ft. alt.; fl. March–May.

DISTR.: Divisions 1, 3, 7, rare. Also Greece, Crete, Aegean Islands and Turkey.

1. Toxeftera, near Ayios Yeoryios (Akamas), 1974, *Meikle* 4002 !
3. Stavrovouni, 1974, *Meikle* 4057 !
7. On hillside near Pentadaktylos, 1880, *Sintenis & Rigo* 645 ! Kazan, Kyrenia Pass, 1949, *Casey* 429 !

NOTES: The almost invariable association of this species with *Sarcopoterium spinosum* (L.) Spach invites further investigation.

12. V. monantha *Retz.*, Obs. Bot., 3: 39 (1783); Post, Fl. Pal., ed. 2, 1: 423 (1932); B. L. Burtt et P. Lewis in Kew Bull., 4: 497–515 (1950); Davis, Fl. Turkey, 3: 296 (1970); Zohary, Fl. Palaest., 2: 199, t. 284 (1972); Osorio-Tafall et Seraphim, List Vasc. Plants Cyprus, 57 (1973).

V. calcarata Desf., Fl. Atlant., 2: 166 (1799); Boiss., Fl. Orient., 2: 590 (1872).
TYPE: Unlocalized; " Demum sub nomine *Viciae syriacae* communicata " *Retzius* (S).

Trailing or climbing annual; stems much branched, especially at base (10–) 15–40 (–50) cm. long, thinly subadpressed-villose, sharply angular; leaves oblong, 2–4 (–5) cm. long, 1–3·5 cm. wide, (6–) 8–14 (–18)-foliolate; tendrils branched or unbranched; stipules semi-sagittate, 2–4 mm. long, 0·5–1 mm. wide, glabrous or thinly villose; petiole very short, generally less than 5 mm. long; leaflets oblong, 5–20 mm. long, 1–6 mm. wide, glabrous above, thinly villose below, apex rounded or obscurely emarginate, base cuneate, shortly petiolulate; inflorescence a lax 1–3 (–5)-flowered axillary raceme; peduncle erect, 7–30 (–60) mm. long, pedicels usually less than 2 mm. long, thinly villose; flowers spreading or slightly nodding, purple or pinkish; calyx campanulate 3–5 (–6·5) mm. long, 2–4 mm. wide, thinly villose, calyx-teeth deltoid, subequal, 1–2 mm. long, 0·8–1 mm. wide at base; limb of standard oblong, 6–10 mm. long, 4·5–8 mm. wide, slightly emarginate at apex, scarcely narrowed at base, claw very broad, 5–8 mm. long, 4·5–7 mm. wide, limb of wings oblong, 6–11 mm. long, 1·5–4 mm. wide, apex rounded, base strongly but bluntly auriculate, claw slender 4–7 mm. long, limb of keel oblong, 3·5–5 mm. long, 1·5–3 mm. wide, curved upwards to a blunt apex, base shortly and bluntly auriculate, claw 4·5–6 mm. long; stamens 8–14 mm. long, free part of filaments 1–1·5 mm. long; anthers small, broadly oblong or suborbicular about 0·3 mm. long and almost as wide; ovary narrowly oblong, 7–13·5 mm. long, 1–3 mm. wide, strongly compressed, glabrous, tapering to apex and base, 4–7-ovulate, style upcurved almost at a right angle, 1·5–2·5 mm. long, villose all round near apex, stigma capitate. Pod oblong, strongly compressed, 2·3–5 cm. long, 0·8–1·2 cm. wide, apex with a short, sharp upcurved or down-curved beak, valves pale brown when ripe, finely reticulate-veined; seeds shortly oblong or suborbicular, slightly compressed, 2·5–5 mm. long and almost as wide; testa smooth, minutely rugulose or pitted, dark or light brown, sometimes distinctly marbled.

HAB.: Waste ground, possibly a casual; c. 500 ft. alt.; fl. March–May.

DISTR.: Divisions 6, 8, rare. Widespread in the Mediterranean region and eastwards to Afghanistan and Pakistan.

6. Nicosia, 1937, *Syngrassides* 1410 !
8. Ayios Andronikos, 1880, *Sintenis & Rigo* 463 !

NOTES: All the Cyprus material appears to belong to typical ssp. *monantha*, with small flowers and pods, and few-flowered inflorescences. The more robust ssp. *triflora* (Ten.) Burtt et Lewis is not infrequent in Egypt, and may well occur as a casual in our area. The differences between the two subspecies are described in detail by Burtt and Lewis in Kew Bull., 4: 497–515 (1950).

13. V. villosa *Roth*, **Tent.** Fl. Germ., 2: 182 (1793); Boiss., Fl. Orient., 2: 591 (1872); Post, Fl. Pal., ed. 2, 1: 424 (1932); Davis, Fl. Turkey, 3: 290 (1970); Zohary, Fl. Palaest., 2: 197, t. 279 (1972).

Sprawling or climbing annual, biennial or perennial 15–100 (or more) cm. high; stems sharply angular, glabrous or ± densely villose with spreading hairs; leaves oblong, 1–12 cm long, 0·5–7 cm. wide, 6–16-foliolate, terminating in a branched (or rarely simple) tendril; petiole very short, usually less than 1 cm. long; stipules semi-sagittate, 1–12 mm. long, 0·5–6 mm. wide; leaflets linear, lanceolate, oblong, elliptic or suborbicular, 3–30 mm. long, 1·5–9 mm. wide, subglabrous or thinly villose, apex obtuse, acute or sharply mucronate, base shortly petiolulate or subsessile; inflorescences axillary, forming dense, many-flowered or lax, few-flowered racemes 1·3–7 (or more) cm. long; peduncle spreading, 3–10 cm. long, usually shorter than, but sometimes exceeding the subtending leaf; pedicels short, slender, 1–1·5 mm. long; flowers crimson-violet, purple or pale mauve, occasionally whitish or yellowish tinged mauve; calyx shortly cylindrical, distinctly gibbous at base, glabrous or thinly hairy, 5–10 mm. long, 2–2·5 mm. wide, calyx-teeth unequal, the 2 adaxial narrowly deltoid or subulate, usually less than 1 mm. long, the 3 abaxial subulate or filiform, 2–5 mm. long; limb of standard much smaller than claw, oblong, 4–7 mm. long, 4–6 mm. wide, recurved, apex shortly emarginate, claw 8–13 mm. long, 5–8 mm. wide, often slightly dilated about the middle or tapering to a rather broad base; limb of wings narrowly oblong, sometimes widening towards the blunt or rounded apex, 9–13 mm. long, 2–5 mm. wide, closely adherent to keel with a distinct conjunctival tooth, base auriculate, with a long, oblong, blunt auricle, claw slender 8–10 mm. long, limb of keel oblong, slightly upcurved, 4–4 mm. long, 2–3 mm. wide, apex blunt or subacute, blotched dark purple, base shortly but distinctly auriculate, claw rather broad, 7–8 mm. long; stamens 10–14 mm. long, free part of filaments 2–2·5 mm. long; anthers oblong, 0·4–0·5 mm. long, 0·3 mm. wide; ovary linear, glabrous or pubescent, 8–11 mm. long, 0·6–0·8 mm. wide, attenuate to an elongate-stipitate base, 4–6-ovulate, style sharply upcurved, about 2 mm. long, bearded for some distance below apex, stigma obliquely capitate. Pod oblong, flattened, 20–30 mm. long, 8–10 mm. wide, apex acute or shortly rostrate, base ± stipitate, narrowing rather abruptly; valves pale brown, with a distinct marginal rim, obscurely veined, glabrous or ± pubescent; seeds subglobose or somewhat compressed, 4–4·5 mm. diam.; testa dark blackish-brown, sometimes obscurely marbled, smooth; hilum small, about 1·5 mm. long.

ssp. **eriocarpa** (*Hausskn.*) *P. W. Ball* in Feddes Repert., 79: 45 (1968); Davis, Fl. Turkey, 3: 290 (1970); Osorio-Tafall et Seraphim, List Vasc. Plants Cyprus, 56 (1973).

? *V. salaminia* Heldr. et Sart. ssp. *macrophyllaria* Cand. in Bull. Soc. Bot. Fr., 45: 187 (1898).

V. varia Host var. *eriocarpa* Hausskn. in Mitt. Thür. Bot. Ver., n.s., 5: 87 (1893).

V. eriocarpa (Hausskn.) Hal., Consp. Fl. Graec., 1: 489 (1900); Rechinger f. in Arkiv för Bot., ser. 2, 1: 426 (1950); B. L. Burtt in Kew Bull., 9: 70 (1954).

[*V. microphylla* (non Urv.) Kotschy in Unger et Kotschy, Die Insel Cypern, 384 (1865); Boiss., Fl. Orient., 2: 592 (1872) pro parte; Holmboe, Veg. Cypr., 114 (1914); Rechinger f. in Arkiv för Bot., ser. 2, 1: 426 (1950).]

[*V. dasycarpa* (non Ten.) Lindberg f., Iter Cypr., 20 (1946).]

[*V. villosa* Roth ssp. *microphylla* (non (Urv.) P. W. Ball) Osorio-Tafall et Seraphim, List Vasc. Plants Cyprus, 56 (1973).]

TYPE: Greece; " in dumosis ad radices Hymetti; in collibus ad Akro- et Neo-Korinthum, supra Naupliam; in P.D. [Pindus Dolopicus] reg. infer. pr. mon. Korona ", *Haussknecht*.

Pods shortly pubescent, distinctly stipitate at base; stems sparingly adpressed-pubescent, leaves 8–14 (–16)-foliolate, leaflets oblong, 8–15 (–20) mm. long, 2–4 mm. wide; inflorescence rather lax, 3–12 (–16)-flowered, flowers crimson-purple or rich violet. Plant (? always) perennial.

HAB.: In garigue or by roadsides, on calcareous or igneous formations; sea-level to 3,500 ft. alt.; fl. March–June.

DISTR.: Divisions 1–3, 5–7, common in Division 2. Greece, Aegean Islands, Turkey, Syria, Lebanon and eastwards to Iraq.

1. Lyso, 1913, *Haradjian* 838, 844; Philousa near Lyso, 1962, *Meikle* 2766 !
2. Common; *Druce* 162 ! *Syngrassides* 811 ! *Kennedy* 577 ! *Lindberg f.* s.n.; *Merton* 2953 !
3. Curium (near Episkopi), 1862, *Kotschy* 622 !
5. Between Athalassa and Yeri, 1936, *Syngrassides* 1176 !
6. Near Liveras, 1974, *Meikle* 5012 !
7. Near Kyrenia, 1880, *Sintenis & Rigo* 651 ! Kyrenia, 1956, *G. E. Atherton* 1348 ! Temblos, 1956, *G. E. Atherton* 1398 !

NOTES: Aggregate *Vicia villosa* Roth comprises six (or possibly more) tolerably distinct segregates, variously ranked by authors as distinct species, subspecies or varieties, and differing chiefly in indumentum of stems, leaves and pods, shape of leaflets, number of flowers in each inflorescence, and in the colour of the petals. The relationships between all these segregates are so close that it seems best to regard them as geographical (and ecological) races of one polymorphic species, and to give them subspecific rank. In Cyprus, *V. villosa* ssp. *eriocarpa* with narrow leaflets and pubescent pods is alone satisfactorily recorded. Its nearest ally, *V. villosa* ssp. *microphylla*, has glabrous or pubescent pods, but differs in being a more slender plant, with smaller, elliptic or suborbicular leaflets, and in having laxer-, fewer-flowered inflorescences of rather larger flowers. Although recorded from Cyprus by Kotschy and other authors, it seems probable that all the specimens named *microphylla* should be referred to *V. villosa* ssp. *eriocarpa*.

14. V. palaestina *Boiss.*, Diagn., 1, 9: 116 (1849), Fl. Orient., 2: 592 (1872), Fl. Orient. Suppl., 194 (1888); Holmboe, Veg. Cypr., 114 (1914); Post, Fl. Pal., ed. 2, 1: 424 (1932); Davis, Fl. Turkey, 3: 294 (1970); Zohary, Fl. Palaest., 2: 197, t. 280 (1972); Osorio-Tafall et Seraphim, List Vasc. Plants Cyprus, 56 (1973).

TYPE: Palestine; "in rupestribus umbrosis Palaestinae circa *Kuriet el Enab*, *Hierosolymam*, montem *Ithaburem* etc." *Boissier* (G, K !).

Slender sprawling or scandent perennial 30–60 (–100) cm. high; stems sharply angular, usually much branched, subglabrous or sparsely pubescent, often thinly glandular with sessile glands; leaves oblong, 2·5–7 cm. long, 1–4 cm. wide, 9–20-foliolate, terminating in a slender branched tendril; stipules semi-sagittate, 2–5 mm. long, 1–3 mm. wide; petiole very short or almost wanting; leaflets narrowly oblong, 5–20 (–30) mm. long, 1–4 (–7) mm. wide, glabrous or sparsely pilose on both surfaces, ± densely covered with sessile glands on the upper surface, apex acute or acuminate, margins often distinctly but narrowly involute, base very shortly petiolulate; flowers 3–7 (occasionally more) in lax axillary racemes; peduncle slender, spreading or suberect, 1–4 cm. long, generally much shorter than subtending leaf; pedicels short, but distinct, usually about 1·5–2 mm. long; calyx campanulate, about 3 mm. long, 1·5 mm. wide, thinly hairy or subglabrous, calyx-teeth unequal, deltoid, the 2 adaxial less than 0·5 mm. long, the 3 abaxial about 1 mm. long; petals usually pale milky blue; limb of standard about 3·5 mm. long, 3·5 mm. wide, apex truncate and distinctly emarginate, base merging into a very wide claw about 3–3·5 mm. long, limb of wings oblong, about 4 mm. long, 2 mm. wide, apex rounded, base distinctly but bluntly auriculate, claw about 2 mm. long, limb of keel semicircular, about 2–2·5 mm. long and 2 mm. wide, apex very blunt, base obscurely auriculate, claw about 2 mm. long; stamens about 4·5 mm. long, free part of filaments about 1–1·5 mm. long; anthers suborbicular about 0·3 mm. diam.; ovary ellipsoid, flattened, glabrous, about 3 mm. long 1–1·5 mm. wide, 2–3-ovulate, base tapering to a distinct stipe about 1 mm. long, style upcurved about 1·5 mm. long, bearded towards apex, stigma capitate. Pod oblong, flattened, 1·8–2·5 cm. long, 0·6–0·8 mm. wide, apex shortly rostrate, base shortly stipitate, valves pale brown when mature, glabrous, often thinly glandular; seeds circular, strongly flattened, about 3–3·5 mm. diam.; testa dark brown, smooth, hilum about 2 mm. long.

HAB.: Amongst garigue on screes and rocky calcareous slopes; 250–2,200 ft. alt.; fl. Febr.–April.

DISTR.: Divisions 3, 4, 6–8, common in Division 7. Turkey, Syria, Lebanon and Palestine, eastwards to Iraq.

3. Kalavasos, 1905, *Holmboe* 633.
4. Ayia Napa, 1905, *Holmboe* 70.
6. W. of Orga, 1956, *Merton* 2587 !
7. Common; *Sintenis & Rigo* 465 ! 643 ! *Syngrassides* 909 ! 1019 ! *Davis* 2494 ! 2856 ! 3020 ! *Casey* 517 ! *Chapman* 125 ! *Meikle* 2372 !
8. Ayios Andronikos, 1880, *Sintenis & Rigo* 465A; Koma tou Yialou, 1935, *Syngrassides* 621 ! Rizokarpaso, Ronnas Valley, 1957, *Merton* 2917 ! Yioti, 1962, *Meikle* 2521 !

[V. ARTICULATA *Hornem.* (*V. monanthos* (L.) Desf. non Retz.) has been recorded by Sibthorp " in insulae Cypri, nec non in Peloponnesi, arvis " (Fl. Graec. Prodr., 2: 67 (1813)—under *Lathyrus monanthos* (L.) Willd.). It is readily distinguished from all but its nearest allies by the stipules, one of each pair being small and subulate, the other very much larger, and divided palmately into numerous slender filiform lobes, sometimes almost 1 cm. long. In general habit *V. articulata* resembles *V. palaestina*, but the racemes are normally 1-flowered, the individual flowers distinctly larger than in *palaestina* and the leaflets usually distinctly notched at the apex. It has not been refound in Cyprus, and all other records are based upon Sibthorp's original one. Since the species is rare in the eastern Mediterranean (though recorded from Greece and Turkey), Sibthorp's statement must be regarded with some scepticism, and the record omitted pending the re-discovery of the species. It should be noted that *Vicia articulata* Hornem. (*V. monanthos* (L.) Desf.) is quite distinct from *V. monantha* Retz. (*V. calcarata* Desf.), which latter is known to occur, though rarely, in Cyprus.]

15. V. cypria *Kotschy* in Unger et Kotschy, Die Insel Cypern, 384 (1865); Boiss., Fl. Orient., 2: 593 (1872); Holmboe, Veg. Cypr., 114 (1914); Davis, Fl. Turkey, 3: 295 (1970); Zohary, Fl. Palaest., 2: 199, t. 283 (1972); Osorio-Tafall et Seraphim, List Vasc. Plants Cyprus, 57 (1973); Elektra Megaw, Wild Flowers of Cyprus, 9, t. 9 (1973).

TYPE: Cyprus; " sehr häufig auf Felsboden des Cape Greco und des Pentadactylos ", March 30, 1862 and April 13, 1862, *Kotschy* 151 (K !) and 373.

A sprawling annual; stems usually much branched especially towards base, generally less than 30 cm. long, angular, glabrous or almost so; leaves oblong, 2–4 (–7) cm. long, 0·8–2 (–4) cm. wide, 8–12 (–16)-foliolate, terminating in a slender, branched or sometimes unbranched tendril; petiole distinct, 5–10 (–15) mm. long; stipules broadly ovate, or rarely semi-sagittate, 3–5 mm. long, 2–4 mm. wide, both (or occasionally one stipule of each pair) deeply cut into numerous slender, filiform lobes; leaflets shortly oblong, obovate or suborbicular, 3–10 (–18) mm. long, 2–8 (–15) mm. wide, glabrous, glandular on upper surface, apex truncate or shallowly emarginate, apiculate, base bluntly cuneate, subsessile; inflorescence a lax, axillary, 1–2-flowered raceme, peduncle erect, 15–25 mm. long, glabrous, generally terminating in a slender, awn-like appendage, pedicels 1·5–5 mm. long, often thinly hairy, especially towards base; calyx campanulate, 4–5 mm. long, 2–2·5 mm. wide, calyx-teeth subequal, subulate, 1·5–2·5 mm. long; petals milky white, limb of standard with a large median, mauve-blue blotch, obcordate, about 12 mm. long, 10 mm. wide, apex emarginate, base tapering to a wide claw, 4–5 mm. long, limb of wings oblong-obovate, 10–11 mm. long, 4–5 mm. wide, base distinctly but bluntly auriculate, conjunctival tooth strongly developed, claw slender, 5–6 mm. long, limb of keel 5–7 mm. long and about as wide, upcurved almost at a right-angle to a short, bluntly rostrate apex, base obscurely auriculate, claw 5–7 mm. long; stamens 12–15 mm. long, the vexillary usually free almost to base, the other filaments free for about 3–4 mm.; anthers oblong, 0·5 mm. long, 0·3 mm. wide; ovary narrowly oblong, 10–12 mm. long, 1·5–2 mm. wide, glabrous, distinctly stipitate at base, 4–5-ovulate, style slender, erect, about 5 mm. long, glabrous except at apex, stigma capitate. Pod narrowly oblong, flattened, about 25 mm. long, 5–6 mm. wide, apex sharply, but shortly rostrate, valves pale brown, rather thin and papery; seeds oblong, compressed, 4–5 mm. long, 3–4 mm. wide; testa dark brown, minutely rugulose-verruculose; hilum about 1 mm. long. *Plate 33.*

HAB.: On rocky calcareous slopes, and in Pine forest, usually in cool, moist situations; 500–2,800 ft. alt.; fl. March–April.

DISTR.: Divisions [4], 7, very common on the Northern Range. Also S. Turkey (Amanus) and Palestine.

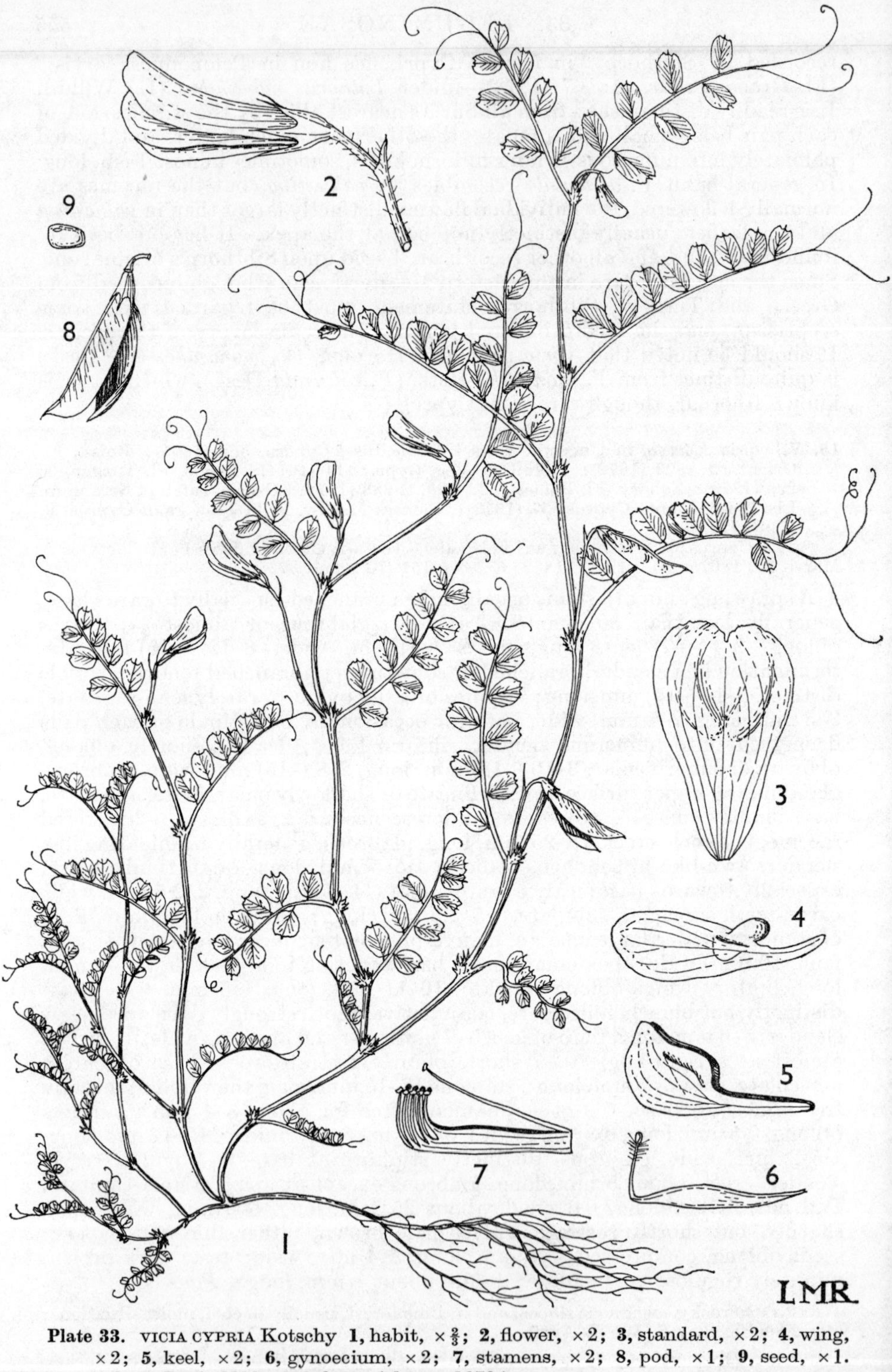

Plate 33. VICIA CYPRIA Kotschy **1,** habit, $\times\frac{2}{3}$; **2,** flower, $\times 2$; **3,** standard, $\times 2$; **4,** wing, $\times 2$; **5,** keel, $\times 2$; **6,** gynoecium, $\times 2$; **7,** stamens, $\times 2$; **8,** pod, $\times 1$; **9,** seed, $\times 1$. (**1** from *Davis* 2811; **2–9** from *Atherton* 1279.)

[4. Cape Greco, 1862, *Kotschy*.]
7. Very common; *Kotschy* 151 ! *Sintenis & Rigo* 464 ! *Syngrassides* 1150 ! 1442 ! *Mavromoustakis* s.n. ! *Davis* 2811 ! 2880 ! 2962 ! *Casey* 361 ! 637 ! 1136 ! 1166 ! *Merton* 1855 ! *G. E. Atherton* 1279 !

NOTES: Kotschy's record for *V. cypria* from Cape Greco is probably an error; no one else has succeeded in finding this beautiful species in Cyprus outside the neighbourhood of the Northern Range. Such, however, is the general reliability of Kotschy's statements, that the record is repeated in the hope that the plant may yet be re-discovered in this outlying station.

16. V. lunata (*Boiss. et Bal.*) *Boiss.*, Fl. Orient., 2: 594 (1872); A. K. Jackson in Kew Bull., 1937: 342 (1937); Davis, Fl. Turkey, 3: 302 (1970); Osorio-Tafall et Seraphim, List Vasc. Plants Cyprus, 57 (1973).

Ervum lunatum Boiss. et Bal. in Boiss., Diagn., 2, 5: 93 (1856).

V. lunata (Boiss. et Bal.) Boiss. var. *grandiflora* U. Plitm. in Davis, Fl. Turkey, 3: 598 (1970).

TYPE: Turkey; " in pinetis ad septentrionem spectantibus montis Nifé dagh ad orientem *Smyrnae* ", May 28, 1854, *Balansa* 210 (G, K !).

Superficially similar to the preceding species, *V. cypria* Kotschy, but differing in colour of flowers, number of flowers per inflorescence and shape of pods.

A slender trailing or sprawling annual; stems usually branched near base, 5–20 cm. long, distinctly angled, glabrous; leaves oblong, 1–4 cm. long, 0·7–1·2 cm. wide, 6–10-foliolate, the lower often ecirrhose, the upper with a slender branched or unbranched tendril; stipules rather unequal, ovate, 2–5 mm. long, 0·8–2 mm. wide, margin laciniate with narrow, subulate lobes; petiole distinct, 0·5–1·3 cm. long; leaflets obovate, or shortly oblong, often small, 1·5–8 mm. long, 1·5–5 mm. wide, glabrous or very thinly puberulous on midrib below, apex rounded or truncate, usually apiculate, base broadly cuneate, shortly petiolulate; flowers 2–5 in rather lax axillary racemes; peduncle up to 5 cm. long, often equalling or exceeding the subtending leaf; pedicels slender, to about 5 mm. long; calyx shortly cylindrical 3–5 mm. long, 1·8–3 mm. wide, glabrous, often stained violet-bronze, calyx-teeth subequal, subulate, 1·5–2·5 mm. long; limb of standard lavender-blue pencilled dark mauve, obovate, 9–12 mm. long, 8–10 mm. wide, apex emarginate, base tapering to a broad claw 4–5 mm. long; limb of wings bright yellow, oblong-obovate, 8–10 mm. long, 3–5 mm. wide, apex rounded, base shortly but sharply auriculate with a well-developed conjunctival tooth, claw about 3–6 mm. long, slender, limb of keel yellow, 6–7 mm. long and about as wide, curved upwards at a right-angle to a short blunt beak, base bluntly auriculate, claw 5–6 mm. long; stamens 14–17 mm. long, free part of filaments 4–5 mm. long; anthers oblong, 5–6 mm. long, about 3 mm. wide; ovary linear, flattened, 10–11 mm. long, 1–1·5 mm. wide, tapering to a long, stipitate base, usually 4–5-ovulate, style upcurved at a right-angle, slender, hairy near apex, stigma capitate. Pod crescent-shaped, flattened, 10–20 mm. long, 8–10 mm. wide, apex blunt, base shortly stipitate, valves pale brown, papery, obscurely reticulate; seeds oblong, compressed, about 4 mm. long, 3 mm. wide; testa dark brown, minutely rugulose-verruculose, obscurely marbled; hilum minute, barely 0·5 mm. long.

HAB.: On rocky, igneous mountainsides, in garigue or under Pines; 2,000–4,000 ft. alt.; fl. March–May.

DISTR.: Division 2, common on the Troödos massif. Turkey, ? Syria.

2. Common; Makheras, Kalopanayiotis, Stavros tis Psokas, Asinou, etc.; *Syngrassides* 949 ! *Kennedy* 582 ! 583 ! *Casey* 286 ! *Mrs. Grove* in *Casey* 1338 ! 1339 ! *Meikle* 2262 ! 2703 !

NOTES: Probably overlooked by Sibthorp and other collectors in Cyprus on account of its earliness in flowering; by the end of April it has almost everywhere gone to seed.

A most attractive species, and like *V. cypria*, worth consideration as a spring-flowering garden annual.

SECT. 4. **Ervum** (*L.*) *S. F. Gray* Flowers usually few, in distinctly pedunculate racemes, generally small; style glabrous or pubescent all round towards apex; plants climbing, sprawling or suberect; leaves usually with numerous leaflets. Pods small, with a shortly acute or blunt apex.

17. V. hirsuta (*L.*) *S. F. Gray*, Nat. Arrang. Br. Plants, 2: 614 (1821); Boiss., Fl. Orient., 2: 595 (1872); Druce in Rep. B.E.C., 9: 470 (1931); Rechinger f. in Arkiv för Bot., ser. 2, 1: 426 (1950); Davis, Fl. Turkey, 3: 301 (1970); Osorio-Tafall et Seraphim, List Vasc. Plants Cyprus, 57 (1973).
Ervum hirsutum L., Sp. Plant., ed. 1, 738 (1753).
TYPE: " *in* Europae *agris* ".

Trailing or climbing annual; stems slender, much-branched, 30–50 (–70) cm. long, glabrous or thinly pubescent, distinctly angular; leaves oblong, 3–6 cm. long, 1–3 cm. wide, 6–16-foliolate, terminating in a slender, branched tendril or the lowermost sometimes ecirrhose; petiole short, usually under 1 cm. long; stipules semi-sagittate, 2–4 mm. long, 1·5–2·5 mm. wide, acuminate, with a slender, subulate basal lobe; leaflets narrowly oblong or linear, 4–15 mm. long, 1·5–3 mm. wide, apex truncate or emarginate, distinctly apiculate, base sessile or shortly petiolulate; inflorescence 2–7-flowered, the flowers very small, usually rather crowded at the apex of a slender peduncle 1·5–4 cm. long; pedicels very short (less than 2 mm. long), pubescent, down-curved; calyx campanulate, about 2 mm. long, 1 mm. wide, pubescent, calyx-teeth subulate, subequal about 1·5 mm. long; petals milky white or pale blue, only slightly longer than calyx, limb of standard obovate, about 2·5 mm. long, 2 mm. wide, apex emarginate, base gradually narrowed to a very short wide claw about 0·5 mm. long, limb of wings oblong, about 1·8 mm. long, 1 mm. wide, apex rounded, base scarcely auriculate, claw slender about 1 mm. long, limb of keel about 1·2 mm. long and almost as wide, sharply upcurved to a blunt apex, base exauriculate, claw slender about 1 mm. long; stamens all united, about 2·5 mm. long, free part of filaments about 0·5 mm. long; anthers minute, suborbicular, about 0·2 mm. diam.; ovary oblong, about 2·5 mm. long, 1 mm. wide, somewhat compressed, hirsute, 2-ovulate, style about 0·5 mm. long, sharply upcurved, glabrous, stigma capitate. Pod oblong, 6–9 mm. long, 3·5 mm. wide, somewhat compressed, (1–) 2-seeded, apex blunt or shortly acute, not rostrate, base not stipitate, valves dark fuscous at maturity, thinly hirsute; seeds subglobose, about 2 mm. diam.; testa dark brown, smooth; hilum about 2 mm. long.

HAB.: Roadsides and waste ground, possibly an adventive; 2,000–3,500 ft. alt.; fl. April–May.

DISTR.: Division 2. Widespread in Europe, N. Africa, Asia and eastern Tropical Africa, introduced elsewhere.

2. Stavros tis Psokas, 1930, *Druce* s.n.; Pano Platres, 1930, *E. Wall*; Platania, 1955, *Merton* 2198 !

18. V. ervilia (*L.*) *Willd.*, Sp. Plant., 3: 1103 (1802); Boiss., Fl. Orient., 2: 595 (1872) Holmboe, Veg. Cypr., 114 (1914); Post, Fl. Pal., ed. 2, 1: 425 (1932); Lindberg f. Iter Cypr., 20 (1946); Davis, Fl. Turkey, 3: 298 (1970); Zohary, Fl. Palaest., 2: 200, t. 285 (1972); Osorio-Tafall et Seraphim, List Vasc. Plants Cyprus, 57 (1973).
Ervum ervilia L., Sp. Plant., ed. 1, 738 (1753); Unger et Kotschy, Die Insel Cypern, 383 (1865); H. Stuart Thompson in Journ. Bot., 44: 307 (1906).
TYPE: " *in* Gallia, Italia ".

Erect or spreading annual, 15–40 cm. high; stems usually much branched, subglabrous or thinly pubescent, angular; leaves narrowly oblong, 3–12 cm. long, 1–3 cm. wide, 10–30 (–40)-foliolate, ecirrhose or with a very short terminal awn; stipules semi-sagittate or irregularly laciniate, 2–7 mm. long, 1–3 mm. wide; petioles generally less than 1 cm. long; leaflets oblong, 5–15 mm. long, 1–4 mm. wide, subglabrous or glabrous, apex truncate or shortly emarginate, apiculate, base very shortly petiolulate or subsessile; inflorescences axillary from the upper leaves, forming lax, 1–3 (–4)-flowered racemes; flowers small, inconspicuous; peduncle erect, 1·5–2·5 cm. long, pubescent, much shorter than subtending leaf; pedicel curved downwards, pubescent, usually less than 2 mm. long; calyx shortly cylindrical, about 6 mm. long, 2 mm. wide, thinly pubescent externally, calyx-teeth setaceous, about 3–4 mm. long, subequal; petals whitish or very pale mauve or blue; limb of standard obovate about 8 mm. long, 5 mm. wide, apex slightly emarginate, mucronate, base tapering to a barely perceptible claw, limb of wings oblong, about 6 mm. long, 2·5 mm. wide, apex rounded, base with a conspicuous oblong, blunt auricle, conjunctival tooth well developed, claw about 3 mm. long, limb of keel about 3·5 mm. long, 2 mm. wide, curved to a short, dark-blotched, subacute beak, base obscurely auriculate, claw about 3 mm. long; stamens about 7 mm. long, free part of filaments about 2 mm. long; anthers oblong, about 0·5 mm. long, 0·3 mm. wide; ovary linear, compressed, about 6 mm. long, 0·8 mm. wide, minutely papillose, scarcely stipitate at base, 3 (–4)-ovulate, style sharply upcurved, about 3 mm. long, glabrous below, thinly hairy near apex, stigma capitate. Pod subterete, 15–20 mm. long, 5–6 mm. wide, 1–3-seeded, conspicuously constricted between the seeds, valves pale brown, rather papery; seeds bluntly angled or subglobose, about 4–5 mm. diam.; testa dull brown, marbled fuscous, smooth; hilum very small, about 1 mm. long.

HAB.: Roadsides, waste ground, or as a weed in cultivated fields, often a relict of cultivation; sea-level to 3,100 ft. alt.; fl. Febr.–June.

DISTR.: Divisions 2, 3, 5, 7, 8, common in Division 5. Mediterranean region and eastwards to Afghanistan.

2. Saïttas, 1932, *Syngrassides* 350 ! Kryos Potamos, 1937, *Kennedy* 589 !
3. Trimiklini, 1941, *Davis* 3127 !
5. Common; *Kotschy* 545; *Sintenis & Rigo* 452 ! *Syngrassides* 259 ! 477 ! *Lindberg f.* s.n. ! *Chapman* 506 !
7. Kyrenia, 1949, *Casey* 606 ! Temblos, 1936, *G. E. Atherton* 1165 !
8. Ephtakomi, 1905, *Holmboe* 516; Komi Kebir, 1941, *Davis* 2447 ! Ronnas Valley, Rizokarpaso, 1957, *Merton* 2918 !

19. V. laxiflora *Brot.*, Phytogr. Lusit., ed. 3, 1: 125, t. 52 (1816); Davis, Fl. Turkey, 3: 300 (1970); Osorio-Tafall et Seraphim, List Vasc. Plants Cyprus, 57 (1973).

V. gracilis (non Banks et Sol.) Loisel., Fl. Gall., 460, t. 12 (1807); Boiss., Fl. Orient., 2: 596 (1872); Holmboe, Veg. Cypr., 114 (1914); Rechinger f. in Arkiv för Bot., ser. 2, 1: 426 (1950).

Ervum gracile (Loisel.) DC., Cat. Hort. Monsp., 109 (1813); Sintenis in Oesterr. Bot. Zeitschr., 31: 392 (1881); 32: 54, 192, 194, 398 (1882) pro parte.

Vicia tenuissima Schinz et Thell. in Viertelj. Naturf. Ges. Zürich, 58: 70 (1913); Post, Fl. Pal., ed. 2, 1: 425 (1932)—non *Ervum tenuissimum* M. Bieb., Tabl. Casp., 185, app. no. 55 (1798).

TYPE: Portugal; " in solo calcareo vallium, colliumque prope Conimbricam, etiam in solo raro, circa *Setubal* et alibi in Beira, et Extremadura ".

Slender trailing or scandent annual; stems 15–40 cm. long, usually much branched, sharply angled, subglabrous or thinly pilose; leaves oblong, 1–3 cm. long, about as wide, 4–10-foliolate, with a slender, simple or branched tendril; petiole usually short, generally less than 5 mm. long especially in the upper leaves; stipules semi-sagittate, 2·5–6 mm. long, 1–2 mm. wide; leaflets narrowly linear-oblong, or those of the lowermost leaves sometimes broader or almost orbicular, 3–17 mm. long, 1·5–3 mm. wide, subglabrous or thinly clothed with rather long hairs, apex acute or acuminate, base tapering to a very short petiolule; inflorescence a lax, 1–5-flowered axillary raceme;

flowers rather small; peduncle erect, 3·5–6·5 cm. long, usually exceeding the subtending leaf and terminating in a distinct awn 1–3 mm. long; pedicels spreading or curved downwards, 2–4 mm. long, adpressed-pilose; calyx campanulate, 3–3·5 mm. long, 1–1·5 mm. wide, thinly adpressed-pilose, calyx-teeth narrowly deltoid, unequal, the 2 adaxial about 0·8 mm. long, the 3 abaxial about 1·5 mm. long; petals pale blue; limb of standard obovate, about 7 mm. long, 5 mm. wide, apex rounded, base tapering to a very short, indistinct claw, limb of wings oblong, about 5 mm. long, 1·5 mm. wide, apex rounded, base distinctly auriculate, with a pointed auricle, claw slender about 3 mm. long, conjunctival flap fairly well developed, limb of keel 3–3·5 mm. long about 2 mm. wide, curving upwards to a blunt apex, base indistinctly auriculate, claw about 3 mm. long; stamens about 7 mm. long, free part of filaments about 1 mm. long; anthers oblong, about 0·5 mm. long, 0·3 mm. wide; ovary linear, compressed, about 6 mm. long, 1 mm. wide, narrowed at base to a slender stipe about 2 mm. long, glabrous, 4–5-ovulate, style sharply upcurved, about 1·5 mm. long, glabrous except at apex, stigma capitate. Pod narrowly oblong, flattened, about 12–14 mm. long, 3 mm. wide, apex blunt or subacute, base very shortly stipitate, valves pale brown, glabrous, obscurely venulose; seeds 4–5, strongly compressed, suborbicular, about 1·5 mm. diam.; testa dark brown, smooth; hilum minute, about 0·4 mm. long.

HAB.: On rocky hillsides, in Pine forest, in garigue or by roadsides, sea-level to 2,000 ft. alt.; fl. Febr.–May.

DISTR.: Divisions 2–4, 7, 8. Widespread in western and southern Europe, the Mediterranean region, western and southern Turkey, Syria, Lebanon.

2. Ayios Ioannis, 1905, *Holmboe*.
3. Limassol, 1913, *Haradjian* 603 ! Cherkez, 1954, *Mavromoustakis* 3 !
4. Paralimni, 1928 or 1930, *Druce* 156 ! Mile 3, Larnaca–Famagusta road, 1950, *Chapman* 427 !
7. Near Kythrea, 1880, *Sintenis & Rigo* 454 ! Near Kantara, 1880, *Sintenis & Rigo* 454A ! Below Buffavento Castle, 1955, *Merton* 2028 !
8. Akradhes, 1962, *Meikle* 2514 !

20. V. pubescens (*DC.*) *Link*, Handbuch, 2: 190 (1831); Boiss., Fl. Orient., 2: 596 (1872); Holmboe, Veg. Cypr., 114 (1914); Post, Fl. Pal., ed. 2, 1: 425 (1932); Davis, Fl. Turkey, 3: 299 (1970); Zohary, Fl. Palaest., 2: 210, t. 299 (1972); Osorio-Tafall et Seraphim, List Vasc. Plants Cyprus, 57 (1973).

Ervum pubescens DC., Cat. Hort. Monsp., 109 (1813); Unger et Kotschy, Die Insel Cypern, 383 (1865).

[*Ervum gracile* (non (Loisel.) DC.) Sintenis in Oesterr. Bot. Zeitschr., 31: 392 (1881); 32: 54, 192, 194, 398 (1882) pro parte.]

TYPE: France; " in sepibus circa Olbias in Galloprovincia australiori " (G).

Superficially similar to *V. laxiflora* Brot. (*supra*), but easily distinguished by its long calyx-teeth and pubescent pods. Slender climbing or sprawling annual, 9–45 cm. high; stems usually much branched, angular, pubescent with rather long, scattered hairs; leaves oblong, 1–4·5 cm. long, 1–4 cm. wide, 4–10-foliolate, the lowermost often ecirrhose, the upper terminating in a slender branched tendril; stipules narrow, semi-sagittate, 1–5 mm. long, 0·5–1 mm. wide; petiole usually less than 1 cm. long, or leaves subsessile; leaflets ovate, oblong or narrowly elliptical, 3–20 mm. long, 1–6 mm. wide, thinly pilose or subglabrous above and below, apex blunt or shortly acute, obscurely mucronate; inflorescence a 1–5-flowered axillary raceme; flowers small, pale blue; peduncle 1–5 cm. long, equalling or sometimes exceeding the subtending leaf; pedicels very slender, usually less than 3 mm. long; calyx narrowly campanulate, 3–4 mm. long, 1·3–1·8 mm. wide, pilose externally, calyx-teeth subequal, narrowly subulate, 2–2·5 mm. long; limb of standard 3–3·5 mm. long and about as wide, apex shallowly emarginate, base tapering to a wide keel about 2 mm. long, limb of wings oblong, about 4 mm. long, 1·5 mm. wide, apex rounded, base auriculate with a narrow, acute auricle, conjunctival tooth reduced to a swelling, claw slender, about 2 mm. long, limb of keel oblong, 2–3 mm. long, 1·5–2 mm. wide, curved upwards to a blunt apex, base scarcely auriculate, claw slender, about 2 mm. long; stamens 5–6 mm. long, the vexillary usually free to the base, others with the free part of the filaments about 1 mm. long; anthers ovate, about 0·4 mm. long, 0·3 mm. wide; ovary linear, compressed, about 4 mm. long, 1 mm. wide, thinly pubescent, 4–6-ovulate, base narrowed to a stipe about 1 mm. long, style sharply upcurved, slender, about 1·5 mm. long, strongly pilose near apex, stigma capitate. Pod oblong, strongly compressed, 9–14 mm. long, 2·5–3·5 mm. wide, apex blunt or subacute, base shortly stipitate, valves pale brown, with a distinctly thickened margin, usually thinly pubescent, occasionally almost glabrous; seeds 3–6 (–8), suborbicular, somewhat compressed, about 1·5 mm. diam.; testa smooth, mid-brown often with darker marbling; hilum minute, about 0·2 mm. long.

HAB.: Clearings in Pine forest, or in open garigue, on calcareous or igneous formations; 100–2,500 ft. alt.; fl. March–May.

DISTR.: Divisions 1, 2, 7, 8, common in Division 7. S. Europe and northern half of Mediterranean region eastwards to Crimea, Turkey, Syria, Lebanon and Palestine.

1. Kako Skala, near Baths of Aphrodite, 1957, *Merton* 3047 !
2. Ayios Ioannis, 1905, *Holmboe* 559.
7. Common; *Kotschy* 399; *Sintenis & Rigo* 454b ! *Davis* 2882 ! *Casey* 386 ! *Merton* 713 ! 2142 ! 2955 !
8. Akradhes, 1962, *Meikle* 2513 !

NOTES: Frequently grows intermixed with *V. laxiflora* Brot., but seems to remain permanently distinct, though easily confused with the latter unless carefully examined.

29. LENS *Mill.*

Gard. Dict. Abridg., ed. 4 (1754) nom. cons.
Ervum L., Sp. Plant., ed. 1, 738 (1753) pro parte.

Slender annual herbs, stems generally suberect or sprawling; leaves paripinnate, the rhachis terminating in a short awn or longer, simple or (rarely) branched tendril; leaflets entire or emarginate, exstipellate; stipules linear, lanceolate, ovate or sometimes semi-sagittate, entire or dentate; inflorescences axillary, pedunculate, flowers solitary or few, in short racemes; bracts small and caducous, bracteoles wanting; calyx campanulate, not distinctly asymmetrical or gibbous, calyx-tube very short, calyx-teeth elongate, subulate, subequal; petals whitish, blue or pale violet, standard obovate or suborbicular with a short, broad claw, wings adherent to keel with a well-developed conjunctival gibbosity, obliquely obovate or oblong, keel shorter than wings, somewhat inflated and longitudinally plicate, apex acute or shortly rostrate; vexillary stamen free, the others connate forming a sheath with an oblique apex, anthers uniform; ovary subsessile, 2-ovulate, style somewhat flattened towards apex, with the adaxial surface bearded, the abaxial glabrous. Pod strongly compressed, 1–2-seeded, continuous within, the valves often rather thin and chartaceous; seeds distinctly compressed, funicle dilated into a thin aril.

Five species in southern Europe, the Mediterranean region and western Asia; the Lentil (*L. culinaris* Medik.) is widely distributed as a cultivated plant, or as a relict of cultivation.

Pod 10–14 (–17) mm. long, 8–10 mm. wide, seeds 5–6 mm. diam.; plant 20–35 cm. high; leaves 8–16-foliolate - - - - - - - - - - **1. L. culinaris**
Pod 5–9 mm. long, 3–5 mm. wide, seeds 2·5–3 mm. diam.; plant 7–25 (–30) cm. high, slender; leaves 4–10-foliolate:

Stipules semi-sagittate or dentate:
  Peduncle terminating in a well-developed awn; pod glabrous - **2. L. nigricans**
  Peduncle without a terminal awn, or with a very short one; pod pubescent all over or (rarely) ciliate - - - - - - - - - - - **4. L. ervoides**
Stipules ovate or lanceolate, entire; pod glabrous - - - - - **3. L. orientalis**

1. L. culinaris *Medik.* in Vorles. Churpf. Phys.-Oekon. Ges., 2: 361 (1787).
Ervum lens L., Sp. Plant., ed. 1, 738 (1753); Unger et Kotschy, Die Insel Cypern, 383 (1865); Boiss., Fl. Orient., 2: 598 (1872); Davis, Fl. Turkey, 3: 327 (1970); Zohary, Fl. Palaest., 2: 210, t. 299 (1972); Osorio-Tafall et Seraphim, List Vasc. Plants Cyprus, 57 (1973); D. Zohary in Chromosomes Today, 4: 317–319 (1973).
Lens esculenta Moench, Meth., 131 (1794); Post, Fl. Pal., ed. 2, 1: 426 (1932).
Vicia lens (L.) Coss. et Germ., Fl. Par., ed. 1, 143 (1845); Holmboe, Veg. Cypr., 114 (1914).
TYPE: France; " *inter* Galliae *segetes* ".

Erect or suberect annual 20–35 cm. high; stems usually much branched, angular, sparingly clothed with crispate hairs and small sessile glands; leaves oblong, 2–4 cm. long, 1–3 cm. wide, paripinnate, 8–16-foliolate, rhachis produced into a simple, or occasionally branched, tendril; stipules lanceolate, acuminate, 5–7 mm. long, 1·5–3 mm. wide, thinly villose, margin entire or subentire; petiole very short or almost wanting; leaflets oblong, 5–15 mm. long, 2–5 mm. wide, thinly subadpressed-villose on both surfaces, apex rounded or subacute; inflorescences arising from the axils of the upper leaves (1–) 2–3 (–4)-flowered; peduncle 1·5–3·5 cm. long usually rather shorter than subtending leaf, rhachis terminating in a distinct awn 2–5 mm. long, pedicels arcuate, 2–3 mm. long; calyx campanulate, thinly villose, 5–7 mm. long, 1·5–2 mm. wide, calyx-teeth subequal, filiform-subulate, 5–7 mm. long, 0·5–0·7 mm. wide at base; petals milky-white or pencilled with pale blue veins, limb of standard squarish or suborbicular, 5–6 mm. long and about as wide, apex deeply emarginate, base abruptly narrowed to a rectangular claw about 2 mm. long and 1·5 mm. wide, limb of wings narrowly obovate, 4–4·5 mm. long, 2·5 mm. wide, apex rounded, base shortly and sharply auriculate, with a distinct conjunctival gibbosity, claw slender, about 2 mm. long, limb of keel oblong, about 4 mm. long, 2 mm. wide, rather inflated and longitudinally plicate, apex subacute, base obscurely auriculate, claw about 2 mm. long; stamens about 6 mm. long, free parts of the connate filaments about 1·5 mm. long; anthers oblong, about 0·4 mm. long and 0·3 mm. wide; ovary strongly compressed, elliptical, glabrous, about 2·5 mm. long, 1 mm. wide, 2-ovulate, the base tapering to a distinct stipe 1–1·5 mm. long, the apex narrowed to a slender, sharply upcurved style about 2 mm. long, stigma capitate, slightly oblique. Pod flattened, oblong-rhomboid, 10–14 (–17) mm. long, 8–10 mm. wide, apex acute or very shortly rostrate, valves glabrous, pale brown; seeds usually 1, compressed-orbicular, 5–6 mm. diam.; testa smooth, reddish-brown. *Plate 34, figs. 10–11.*

HAB.: In cultivated fields and waste ground; sea-level to 1,000 ft. alt.; fl. March–May.

DISTR.: Divisions 5–7. All more or less obviously escapes or relicts of cultivation. Widely distributed as a cultivated plant in Europe, Asia and North Africa; natural distribution uncertain, apparently a cultivated derivative of *L. orientalis* (Boiss.) Hand.-Mazz.

5. Vatili, 1905, *Holmboe* 314.
6. Ayia Irini, 1941, *Davis* 2571 ! Near Dhiorios, 1962, *Meikle* 2346 !
7. Kyrenia, 1956, *G. E. Atherton* 1214 !

2. L. nigricans (*M. Bieb.*) *Godr.*, Fl. Lorraine, ed. 1, 1: 173 (1843); Post, Fl. Pal., ed. 2, 1: 426 (1932); Davis, Fl. Turkey, 3: 326 (1970); Osorio-Tafall et Seraphim, List Vasc. Plants Cyprus, 57 (1973).
Ervum nigricans M. Bieb., Fl. Taur.-Cauc., 2: 164 (1808); Boiss., Fl. Orient., 2: 598 (1872).

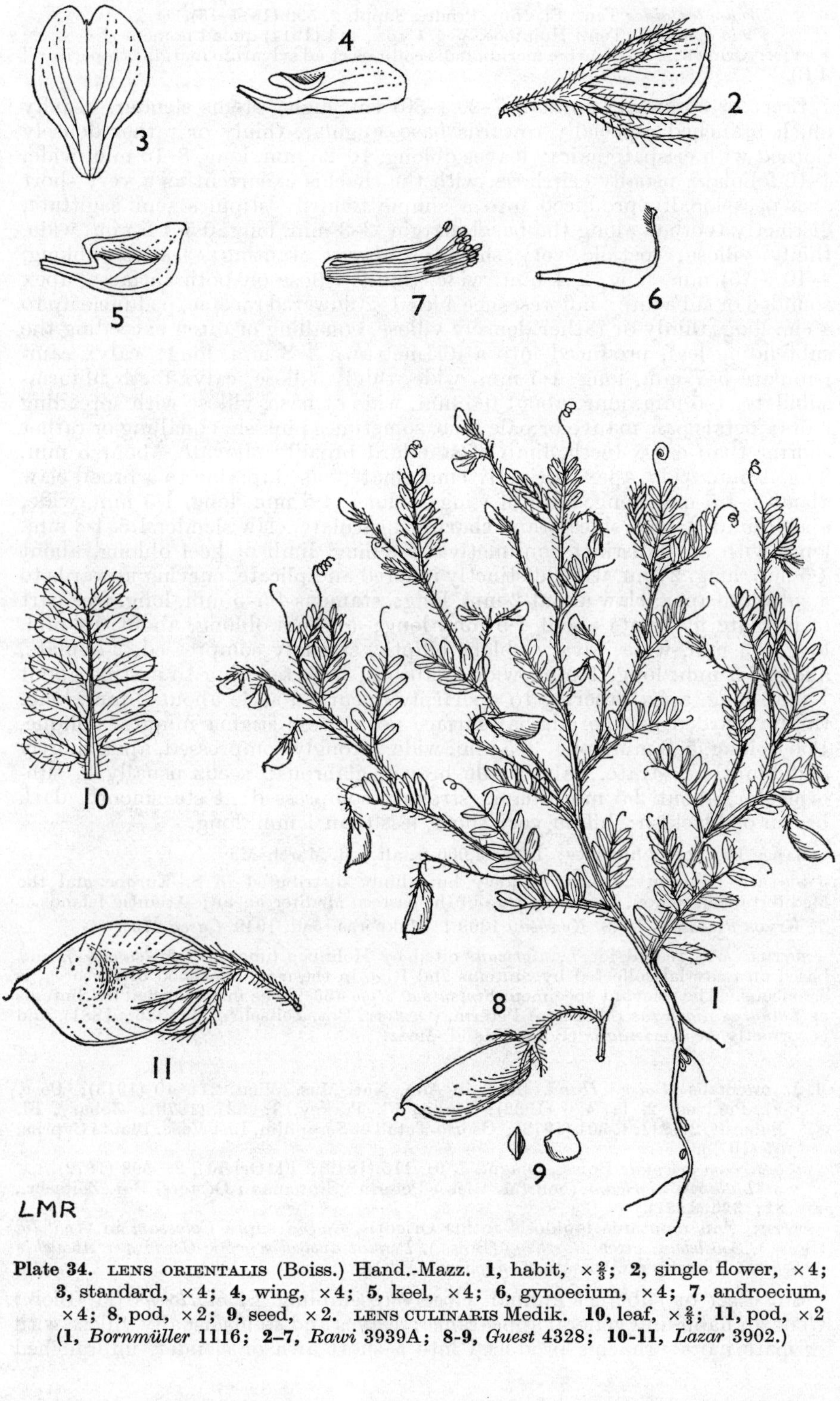

Plate 34. LENS ORIENTALIS (Boiss.) Hand.-Mazz. **1**, habit, $\times\frac{2}{3}$; **2**, single flower, ×4; **3**, standard, ×4; **4**, wing, ×4; **5**, keel, ×4; **6**, gynoecium, ×4; **7**, androecium, ×4; **8**, pod, ×2; **9**, seed, ×2. LENS CULINARIS Medik. **10**, leaf, $\times\frac{2}{3}$; **11**, pod, ×2 (**1**, *Bornmüller* 1116; **2–7**, *Rawi* 3939A; **8-9**, *Guest* 4328; **10-11**, *Lazar* 3902.)

Ervum lentoides Ten., Fl. Nap. Prodr., Suppl. 2, 596 (1811–15).
Vicia lentoides (Ten.) Holmboe, Veg. Cypr., 114 (1914) quoad nomen.

TYPE: Crimea; " in Tauriae meridionalis collibus et ad sylvarum margines reperitur " (LE).

Erect or sprawling annual, 7–20 (–30) cm. high; stems slender, usually much branched especially towards base, angular, thinly or rather densely clothed with crispate hairs; leaves oblong, 10–25 mm. long, 8–15 mm. wide, 4–10-foliolate, usually ecirrhose, with the rhachis excurrent as a very short awn occasionally produced into a simple tendril; stipules semi-sagittate, distinctly toothed along the basal margin, 2–3 mm. long, 0·5–1·5 mm. wide, thinly villose; petiole very short or almost wanting; leaflets oblong, 4–10 (–15) mm. long, 2–4 mm. wide, thinly villose on both surfaces, apex rounded or subacute; inflorescence a lax 1–2-flowered raceme, peduncle up to 4 cm. long, thinly or rather densely villose, equalling or often exceeding the subtending leaf, produced into a distinct awn 4–8 mm. long; calyx campanulate 5–7 mm. long, 1·5 mm. wide, thinly villose, calyx-teeth filiform-subulate, 4–5 mm. long, about 0·5 mm. wide at base, villose with spreading hairs; petals pale mauve or pale blue, sometimes pinkish, equalling or rather shorter than calyx-teeth, limb of standard broadly obovate, about 5 mm. long, 4 mm. wide, apex distinctly emarginate, base tapering to a broad claw about 1–1·5 mm. long, limb of wings oblong, 4·5 mm. long, 1·5 mm. wide, apex rounded, base shortly and sharply auriculate, claw slender 1·5–1·8 mm. long, with an indistinct conjunctival swelling, limb of keel oblong, about 4·5 mm. long, 2 mm. wide, distinctly inflated and plicate, curving upwards to a subacute apex, claw about 2 mm. long; stamens 4·5–5 mm. long, free part of connate filaments about 1·5 mm. long; anthers oblong, about 0·4 mm. long, 0·3 mm. wide; ovary oblong-elliptic, strongly compressed, 2-ovulate, about 2·5 mm. long, 1 mm. wide, glabrous, base tapering to a stipe about 1 mm. long, apex tapering to a sharply upcurved style about 2 mm. long, thinly bearded on the adaxial surface near apex, stigma minute, oblique. Pod oblong, 5–8 mm. long, 3–5 mm. wide, strongly compressed, apex shortly and bluntly rostrate, valves pale brown, glabrous; seeds usually 2, suborbicular, about 2·5 mm. diam., strongly compressed; testa smooth, dark brown or blackish; hilum very short, less than 1 mm. long.

HAB.: Dry rocky hillsides; 1,000–3,000 ft. alt.; fl. March–May.

DISTR.: Division 2, rare. Widely but thinly distributed in S. Europe and the Mediterranean region, becoming rare in the eastern Mediterranean; Atlantic Islands.

2. Kryos Potamos, 1938, *Kennedy* 1309 ! Makheras road, 1949, *Casey* 279 !

NOTES: The record for *L. nigricans* cited by Holmboe (under *Vicia lentoides*), and based on material collected by Sintenis and Rigo in the neighbourhood of Kythrea, is erroneous. The relevant specimen (*Sintenis & Rigo* 455 !) was misidentified by Sintenis as *Lathyrus nigricans* (M. Bieb.) Peterm. (Oesterr. Bot. Zeitschr., 31: 326; 1881), and is correctly *Lens orientalis* (Boiss.) Hand.-Mazz.

3. L. orientalis (*Boiss.*) *Hand.-Mazz.* in Ann. Nat. Mus. Wien, 27: 40 (1913); Post, Fl. Pal., ed. 2, 1: 426 (1932); Davis, Fl. Turkey, 3: 327 (1970); Zohary, Fl. Palaest., 2: 211, t. 301 (1972); Osorio-Tafall et Seraphim, List Vasc. Plants Cyprus, 57 (1973).

Ervum orientale Boiss., Diagn., 1, 9: 115 (1849); Fl. Orient., 2: 598 (1872).
[*Lathyrus nigricans* (non (M. Bieb.) Peterm.) Sintenis in Oesterr. Bot. Zeitschr., 31: 326 (1881).]

TYPE: " in montanis lapidosis totius Orientis, *Cadmi* supra *Colossam* in *Anatolia* (Boiss.), *Antilebani* circa *Rascheya* (Boiss.), *Persiae australis* prope *Gere* inter *Abouchir* et *Schiraz* Kotschy no. 59 sub. *E. nigricanti* " (G).

An erect or suberect annual, closely resembling *L. nigricans* (M. Bieb.) Godr. in habit and general appearance; leaves and stems usually villose with crispate hairs; rhachis produced into a short awn or slender, unbranched tendril; stipules ovate or lanceolate, acuminate, 2–3 mm. long, 1–2 mm. wide, basal margin entire, not toothed or auriculate; leaflets oblong sometimes with a slightly emarginate apex; inflorescence a lax 1–2 (rarely 3)-flowered raceme; peduncle filiform, 1–2·5 cm. long, equalling or exceeding the subtending leaf and produced into a slender terminal awn; pedicels about 2 mm. long; calyx campanulate, 3·5–4·5 mm. long, 1·5–2 mm. wide, calyx-teeth subulate, 3–4 mm. long, villose with subadpressed hairs; petals pale milky-blue, or sometimes bright blue, exceeding calyx-teeth, limb of standard suborbicular, about 5 mm. diam., apex distinctly emarginate, claw very short, rarely exceeding 1 mm. in length, and about 1·5 mm. wide, limb of wings oblong, about 4 mm. long, 2 mm. wide, apex rounded, base with a sharp auricle and prominent conjunctival gibbosity, claw slender, about 2 mm. long, limb of keel boat-shaped, somewhat inflated, about 3 mm. long, 2 mm. wide, curved upwards to a subacute apex, claw about 2 mm. long; stamens about 5 mm. long, free parts of connate filaments about 2·5–3 mm. long; anthers minute, suborbicular, about 0·2–0·3 mm. diam.; ovary 2-ovulate, narrowly elliptical, strongly compressed, about 2·5 mm. long, 1 mm. wide, glabrous, with a basal stipe about 1 mm. long, apex tapering to a sharply upcurved style about 2 mm. long, stigma obliquely capitate. Pod oblong, flattened, 1–2-seeded, about 9 mm. long, 4·5 mm. wide, apex very shortly rostrate, valves glabrous, pale brown; seeds orbicular, strongly compressed, about 2·5–3 mm. diam.; testa smooth, reddish-brown with fuscous blotches. *Plate 34, figs. 1–9.*

HAB.: On dry open hillsides, or in dry Pine forest; 500–2,000 ft. alt.; fl. March–April.

DISTR.: Divisions 6, 7, rare. Greece, Turkey, Palestine and eastwards to Iran.

6. Near the English School, Nicosia, 1959, *P. H. Oswald* 80 ! and, 1974, *Meikle* 5010 !
7. On mountains above Kythrea, 1880, *Sintenis & Rigo* 455 ! Slopes of Yaïla Peak, 1974, *Meikle* 4037 !

4. **L. ervoides** (*Brign.*) *Grande* in Bol. Ort. Bot. Nap., 5: 58 (1918); Davis, Fl. Turkey, 3: 326 (1970); Zohary, Fl. Palaest., 2: 210, t. 300 (1972); Osorio-Tafall et Seraphim, List Vasc. Plants Cyprus, 57 (1973).

Cicer ervoides Brign., Fasc. Rar. Plant. Forojul., 27 (1810).

Ervum lenticula Schreb. ex Hoppe in Sturm, Deutschl. Fl., part Ia, 8, Heft 32 (1812); Boiss., Fl. Orient., 2: 599 (1872).

Lens lenticula (Schreb. ex Hoppe) Alef. in Bonplandia, 9: 129 (1861); Post, Fl. Pal., ed. 2, 1: 426 (1932).

Vicia lenticula (Schreb. ex Hoppe) Janka in Termész Fuzetek, 9: 138 (1885); Holmboe, Veg. Cypr., 114 (1914).

TYPE: Italy; " inter difractos lapides ad arcem dirutam Montisfalconis, vulgo *la Rocca* ", *Berini* (BASSA).

Suberect, sprawling or subscandent annual 15–25 cm. high; stems very slender, angular, much branched, subglabrous or very thinly villose; leaves oblong, 1–2 cm. long, 1–3 cm. wide, 4–6-foliolate; stipules semi-sagittate, 1–2 mm. long, 0·5–1 mm. wide, subglabrous or thinly villose; petiole very short or almost wanting, rhachis produced into a short awn or slender, unbranched tendril; leaflets narrowly oblong or linear, 4–14 (–23) mm. long, 1–4 mm. wide, subglabrous or thinly villose, often distinctly veined, apex acute or rounded, usually shortly apiculate, rarely a little emarginate; inflorescence 1 (–2)-flowered; peduncle filiform, 1–4 cm. long, usually exceeding the subtending leaf, apex generally without an awn, or with a very short awn, pedicel adpressed-villose, 1–2 mm. long; calyx campanulate, 3–4 mm. long, 1·5 mm. wide, subglabrous or thinly adpressed-villose, calyx-teeth filiform-subulate, 2·5–3 mm. long, slightly shorter than the corolla; petals pale or bright lavender-blue, limb of standard suborbicular, about 3 mm. diam., apex very shortly emarginate, claw very short, about 1 mm. long and about as wide, limb of wings oblong, 2·5–3 mm. long, 1·5–2 mm.

wide, apex rounded, base sharply auriculate with a distinct conjunctival gibbosity, claw slender about 1·5 mm. long, limb of keel boat-shaped, about 2 mm. long, 1·5 mm. wide, somewhat inflated and longitudinally plicate, curved upwards to a subacute apex; base abruptly narrowed to a claw about 1·5 mm. long; stamens about 4 mm. long, free parts of connate filaments about 1·5 mm. long; anthers suborbicular about 0·3 mm. diam.; ovary 2-ovulate, narrowly oblong or elliptic, strongly compressed, pubescent, about 2 mm. long, 0·8 mm. wide, base tapering to a stipe about 0·8 mm. long, apex tapering to a sharply upcurved style about 1·5 mm. long; stigma oblique, capitate. Pod oblong, flattened, 7–9 mm. long, about 4 mm. wide, strongly compressed, apex very shortly rostrate or acute, valves pale brown, usually adpressed-pubescent, rarely subglabrous with a ciliate margin; seeds 1–2, suborbicular, compressed, about 2·5 mm. diam.; testa smooth, reddish brown with darker blotches, hilum about 1 mm. long.

HAB.: Rocky mountainsides, on igneous or calcareous formations, often under *Pinus brutia*; 500–3,750 ft. alt.; fl. March–May.

DISTR.: Divisions 2, 7, common in Division 2. Widespread in the Mediterranean region; Caucasus area; Ethiopia, Uganda.

2. Common; Saïttas, 1928 or 1930, *Druce* 160 ! Tripylos, 1941, *Davis* 3486 ! Yialia, 1955, *Merton* 2141 ! Platres, 1955, *Kennedy* 1845 ! Platres, 1960, *N. Macdonald* 179 ! Livadhi (Paphos Forest), 1962, *Meikle* 2673 !

7. Near Kantara, 1880, *Sintenis & Rigo* 457 ! At Ayios Ioannis near Ardhana, 1905, *Holmboe* 558; S.W. of Kyrenia, 1949, *Casey* 662 !

30. LATHYRUS *L.*

Sp. Plant., ed. 1, 729 (1753).
Gen. Plant., ed. 5, 326 (1754).

Annual or perennial herbs; stems climbing or diffuse, commonly ± winged; leaves generally paripinnate, terminating in a well-developed tendril, or short awn; leaflets usually rather few and distinctly parallel-veined, sometimes wanting or reduced to a tendril, or petiole and rhachis dilated and leaf-like; stipules usually large and semi-sagittate, persistent, rarely entire; inflorescence axillary, racemose, or flowers solitary; bracts very small, caducous, bracteoles wanting; calyx-tube frequently oblique and somewhat gibbous, calyx-teeth subequal, or the adaxial shorter than the abaxial; limb of standard usually broad with a short claw, wings oblong or falcately obovate, free or adherent to the blunt or shortly acute keel; vexillary stamen free or connate with the others, free parts of filaments filiform or dilated above, sheath usually truncate at apex, anthers uniform; ovary subsessile or stipitate, few- or many-ovulate; style inflexed, often flattened apically and commonly bearded along the adaxial surface, rarely glabrous, stigma terminal. Pod continuous within, dehiscing into 2 valves; seeds globose or angular, rarely compressed; testa smooth, rugulose or papillose.

About 130 species widely distributed in the northern hemisphere; also in Africa and South America.

Leaves wholly wanting, stipules much enlarged, looking like simple, opposite leaves; corolla yellow - - - - - - - - - - - **2. L. aphaca**
Leaves not wholly wanting, all or at least the upper with leaflets:
  Lower leaves without leaflets, consisting of much-expanded petioles and stipules, looking like simple alternate leaves; upper leaves usually with 2 leaflets; corolla yellow or yellowish - - - - - - - - - - **1. L. ochrus**
  All the leaves with leaflets; stipules not leaf-like:
    Pod (even when immature) with 2 distinct wings along the dorsal suture:
      Corolla blue or pinkish; pods glabrous; tendrils often branched **7. L. sativus**
      Corolla red or reddish; pods ciliate; tendrils generally simple **8. L. blepharicarpos**
    Pod without wings along the dorsal suture:

Leaves mostly 4-foliolate; leaflets generally truncate or emarginate; corolla very small, whitish, veined pale blue - - - - - - **11. L. saxatilis**
Leaves mostly 2-foliolate; corolla conspicuous, yellow, pink or red:
Pod narrowly linear (2·5–) 3·5–7 cm. long, 0·4–0·6 cm. wide, valves with numerous prominent longitudinal veins; corolla red **10. L. sphaericus**
Pod oblong, valves without prominent longitudinal veins:
Corolla small, usually less than 1 cm. long, bright red; leaflets narrowly linear-lanceolate, grass-like; petiole often very short; peduncles and pedicels slender, capillary; seeds bluntly papillose-tuberculate **9. L. setifolius**
Corolla rather large, more than 1 cm. long; leaflets not grass-like, petiole usually well developed, peduncles not capillary:
Stipules very narrow, the lobes subulate or filiform; racemes 1–3-flowered; seeds with a tuberculate or rugose testa:
Calyx eglandular externally; petals yellow or yellowish veined red; testa closely and strongly tuberculate - - - **3. L. annuus**
Calyx dotted over externally with sessile glands; corolla pink; testa with scattered blunt tubercles, the intervening spaces minutely scabridulous - - - - - - - **4. L. cassius**
Stipules generally rather broad, semi-sagittate or semi-hastate, usually 2–5 (–8) mm. wide; racemes always 1-flowered; seeds with a smooth (rarely very obscurely tuberculate) testa:
Standard generally less than 15 mm. long (orange or reddish or salmon-pink); limb of wings 10–11 mm. long, 4–5 mm. wide; calyx-teeth usually adpressed to petals at anthesis, deltoid or lanceolate **5. L. cicera**
Standard generally more than 15 mm. long (yellow or pale orange); limb of wings c. 13 mm. long, 4–5 mm. wide; calyx-teeth often patent or reflexed an anthesis, deltoid-subulate - **6. L. gorgonei**

SECT. 1. **Clymenum** (*Adans.*) *DC.* Standard with two basal gibbosities; style not twisted; lower leaves with expanded, foliaceous petioles, usually without leaflets.

1. **L. ochrus** (*L.*) *DC.* in Lam. et DC., Fl. Franç., 4: 578 (1805); Unger et Kotschy, Die Insel Cypern, 387 (1865); Boiss., Fl. Orient., 2: 601 (1872); Holmboe, Veg. Cypr., 114 (1914); Post, Fl. Pal., ed. 2, 1: 428 (1932); Davis, Fl. Turkey, 3: 365 (1970); Zohary, Fl. Palaest., 2: 221, t. 314 (1972); Osorio-Tafall et Seraphim, List Vasc. Plants Cyprus, 58 (1973).

Pisum ochrus L., Sp. Plant., ed. 1, 727 (1753); Sibth. et Sm., Fl. Graec. Prodr., 2: 62 (1813), Fl. Graec., 7: 80, t. 689 (1830).

TYPE: " *inter segetes* Cretae, Italiae ".

Sprawling or subscandent annual; stems 15–40 cm. long, unbranched or sparingly branched near the base, glabrous, glaucescent; stipules merged into leaves; lower leaves without leaflets, the lowermost narrowly oblong, acute, often less than 10 mm. long and 5 mm. wide, ecirrhose, the middle oblong, 5–8 cm. long, 1–3 cm. wide, glabrous, glaucescent, with long-decurrent bases and an emarginate apex usually terminating in 3 simple tendrils; upper leaves similar to the middle but terminating in a single branched tendril and generally bearing 1–4 (–5) ovate or oblong-elliptic acute leaflets, 1·3–4 cm. long and 0·8–2 cm. wide; flowers solitary (or rarely 2), axillary, pedicel 1–3 cm. long, transversely articulate about the middle, glabrous, usually curved at apex, elongating considerably in fruit; calyx campanulate, glabrous, calyx-tube about 4–5 mm. long and almost as wide, 2 adaxial calyx-teeth narrowly deltoid about 4 mm. long, 3 abaxial teeth about 5 mm. long, narrowly subulate; petals whitish or pale yellow, limb of standard suborbicular, about 10 mm. long, and 10–12 mm. wide, apex emarginate, base with 2 prominent gibbosities, claw about 5 mm. long and 5 mm. wide, limb of wings obovate, about 12 mm. long, 5 mm. wide, free or slightly adherent to keel, apex rounded, base scarcely auriculate, claw slender about 4 mm. long, limb of keel boat-shaped, about 8 mm. long, 4 mm. wide, curved upwards to an apiculate apex, claw about 3 mm. long; stamens all connate into a sheath, about 12 mm. long, free parts of filaments about 5 mm. long, filiform; anthers oblong, about 0·7 mm. long, 0·5 mm. wide; ovary oblong, compressed, glabrous, about 10 mm. long, 3 mm. wide, 6-ovulate; style sharply upcurved, about 5 mm. long, spathulate towards apex and bearded adaxially. Pod narrowly oblong, strongly compressed, 4–5 cm. long, 1–1·5 cm. wide, apex acute, often crowned with the persistent style, adaxial sutures narrowly winged; seeds subglobose, about 5–7 mm. diam.; testa smooth, dark or reddish-brown, hilum about 2–3 mm. long.

HAB.: Cultivated or waste ground; sea-level to 2,500 ft. alt.; fl. March–May. An escape or relict of cultivation.

DISTR.: Divisions 3–5, 7, 8. Widespread in S. Europe and the Mediterranean region; often as an escape from cultivation.

3. Mazotos, 1905, *Holmboe* 165; Perapedhi, 1938, *Kennedy* 1318 !
4. The Phaneromene near Larnaca, 1905, *Holmboe* 151.
5. Near Herakli and Politiko, 1862, *Kotschy* 227; near Kythrea, 1880, *Sintenis & Rigo* 466 ! Kythrea, 1933, *Nattrass* 401; Nisou, 1934, *Nattrass* 435.
7. Temblos, 1956, *G. E. Atherton* 1177 !
8. Yialousa, 1880, *Sintenis & Rigo.*

SECT. 2. **Aphaca** (*Adans.*) *Reichb.* Standard with basal gibbosities; style not twisted; leaves reduced to a tendril; stipules foliaceous.

2. L. aphaca *L.*, Sp. Plant., ed. 1, 729 (1753); Boiss., Fl. Orient., 2: 602 (1872); Holmboe, Veg. Cypr., 114 (1914); Post, Fl. Pal., ed. 2, 1: 428 (1932); Davis, Fl. Turkey, 3: 367 (1970); Zohary, Fl. Palaest., 2: 220, t. 313 (1972); Osorio-Tafall et Seraphim, List Vasc. Plants Cyprus, 58 (1973).

TYPE: " *in* Italia, Gallia, Anglia *inter segetes* " (LINN !).

Trailing or subscandent annual; stems 7–30 cm. long, usually much branched especially at the base, angular (but not winged), glabrous, glaucescent; leaves reduced to a simple, filiform tendril; stipules large, foliaceous, ovate-hastate, 5–30 mm. long, 5–20 mm. wide, glabrous, glaucous, apex acute often apiculate, base with spreading deltoid lobes; flowers solitary or sometimes (very rarely in Cyprus) paired; pedicel erecto-patent, filiform, glabrous, 10–50 mm. long, articulated about 5 mm. from apex; calyx campanulate, 5–10 mm. long, 2–3 mm. wide, glabrous, calyx-teeth subequal, linear-lanceolate, acuminate, 3–5 mm. long, 1–1·5 mm. wide at base, longitudinally nervose with a narrow scarious margin; petals pale or bright yellow; limb of standard suborbicular, 5–9 mm. diam., emarginate, often streaked with violet, smooth externally, densely and minutely papillose internally, with 2 distinct gibbous folds just above the claw; claw 3–4 mm. long and about as wide; limb of wings broadly spathulate, 5–7 mm. long, 3–4 mm. wide, papillose internally, apex rounded, base with a long narrow auricle and distinct conjunctival tooth, claw slender 2–3 mm. long, limb of keel boat-shaped, about 5 mm. long, 2·5–3 mm. wide, apex distinctly apiculate, base shortly auriculate, claw about 3 mm. long; stamens 7–8 mm. long, all joined into a sheath, free part of filaments filiform, 2–3 mm. long; anthers oblong, 0·4 mm. long, 0·3 mm. wide; ovary narrowly oblong, about 5 mm. long, 1·2 mm. wide, tapering to apex and base, minutely glandular, 5–8-ovulate; style slightly upcurved, 2·5–3 mm. long, strongly flattened towards apex and bearded adaxially. Pod narrowly oblong, 25–30 mm. long, about 7 mm. wide, straight or slightly curved, strongly compressed, apex acute or shortly rostrate, valves mid-brown at maturity, obscurely veined; seeds broadly oblong, 3·5–4 mm. long, 3–3·5 mm. wide, distinctly compressed; testa rich crimson-brown, often glossy, smooth; hilum short, 1–1·5 mm. long.

HAB.: Cultivated and fallow fields, roadsides, waste ground, not infrequent on rocky hillsides or in Pine forests on calcareous and igneous formations; sea-level to 3,200 ft. alt.; fl. Jan.–May.

DISTR.: Divisions 2–8, very common in Division 7. Widespread in Europe and the Mediterranean region, and eastwards to India; adventive elsewhere.

2. Between Troödhitissa Monastery and Trikoukkia, 1957, *Merton* 3177 ! N. of Stavros tis Psokas, 1959, *P. H. Oswald* 92 ! Saïttas, 1960, *N. Macdonald* 5 ! Stavros tis Psokas, 1962, *Meikle* 2750 !
3. Akrotiri, 1927, *Rev. A. Huddle* 52 ! Yermasoyia R., 1952, *Mavromoustakis* 29 !
4. Near Larnaca, 1880, *Sintenis & Rigo* 822 ! Near Larnaca, 1880, *C. & W. Barbey* s.n.; the Phaneromene near Larnaca, 1905, *Holmboe* 143.
5. Athalassa, 1957, *Merton* 2901 !
6. Skouriotissa, 1932, *Syngrassides* 103 ! Ayia Irini, 1941, *Davis* 2568 ! Near Strovolos, Nicosia, 1959, *P. H. Oswald* 32 !
7. Very common; *Sintenis & Rigo*; *Syngrassides* 186 ! *Casey* 387 ! 1243 ! *Chapman* 118 ! *G. E. Atherton* 928 ! 983 ! 1150 ! 1169 !
8. Ayios Philon near Rizokarpaso, 1941, *Davis* 2269 !

NOTES: Two distinct colour-forms occur in Cyprus, one with pale sulphur-yellow or almost whitish petals, the other with clear golden-yellow petals. The pale form tends to predominate in lowland areas, the darker yellow form in the mountains, but there is no clear demarcation, ecological or geographical, in the distribution of the two variants, nor does more minute examination of their morphology reveal any additional reasons for considering them distinct. Some of the Cyprus specimens (especially those coming from Pine forest in mountain areas) approach *L. pseudoaphaca* Boiss. in having small flowers and slender stems. It is difficult to discover on what characters *L. pseudoaphaca* is held to be specifically distinct from typical *L. aphaca*.

SECT. 3. **Cicercula** (*Medik.*) *Gren. et Godr.* Standard without distinct basal gibbosities; style twisted; leaves with leaflets; annual herbs.

3. L. annuus *L.*, Demonstr. Plant., 24 (1753); Sibth. et Sm., Fl. Graec. Prodr., 2: 67 (1813); Unger et Kotschy, Die Insel Cypern, 387 (1865); Boiss., Fl. Orient., 2: 603 (1872); Holmboe, Veg. Cypr., 115 (1914); Post, Fl. Pal., ed. 2, 1, 429 (1932); Davis, Fl. Turkey, 3: 356 (1970); Osorio-Tafall et Seraphim, List Vasc. Plants Cyprus, 58 (1973).

L. hierosolymitanus Boiss., Diagn., 1, 9: 127 (1859), Fl. Orient., 2: 604 (1872); Freyn in Bull. Herb. Boiss., 5: 608 (1897).

L. annuus L. var. *hierosolymitanus* (Boiss.) Post, Fl. Pal., ed. 1, 292 (1896).

TYPE: " *in* Hispania, Monspelii " (LINN !).

Climbing or sprawling annual, 20–60 cm. high; stems much branched, distinctly winged, often rather flexuose, glabrous; leaves normally 2-foliolate, the lowermost ecirrhose, the upper with a well developed, branched tendril; petioles winged, 5–30 (or more) cm. long; stipules semi-sagittate, those of the upper leaves 10–30 mm. long, with subulate or almost filiform lobes; leaflets linear or narrowly lanceolate, 1·5–15 cm. long, 0·2–2 cm. wide, glabrous, bright green, distinctly parallel-veined, tapering to a slender acumen; inflorescence a lax 1–3-flowered raceme; peduncle 1·5–10 cm. long, rigid, erecto-patent, glabrous or very sparsely glandular; pedicels about 5 mm. long, straight or arcuate, generally glabrous; calyx campanulate, 5–8 mm. long, 2–3 mm. wide, glabrous and distinctly veined externally, sometimes sparsely clothed with brown club-shaped glands internally, calyx-teeth subequal, narrowly deltoid-acuminate, 3–4 mm. long, 1·5 mm. wide at base, margins very narrowly scarious; petals yellow, orange or salmon-coloured, the keel often greenish; limb of standard flabellate, 8–10 mm. long, 12–14 mm. wide, glabrous, apex shallowly emarginate, base with 2 indistinct gibbosities, tapering to a wide claw about 2·5–3 mm. long, limb of wings oblong, about 8 mm. long, 3 mm. wide, apex rounded, base shortly auriculate, conjunctival tooth scarcely developed, claw slender, about 3 mm. long, limb of keel about 6 mm. long, 4 mm. wide, rather sharply upcurved to a blunt or rounded apex, base scarcely auriculate, claw slender about 5 mm. long; stamens about 12 mm. long, all united into a sheath, the free parts of the filaments about 4 mm. long, filiform; anthers oblong, about 0·6 mm. long, 0·4 mm. wide; ovary narrowly oblong, about 7–8 mm. long, 1·3 mm. wide, flattened, densely glandular, 6–8-ovulate, narrowing rather suddenly to a twisted style about 3 mm. long, base scarcely stipitate. Pod oblong, flattened, 5–7 cm. long, 0·7–1·4 cm. wide, apex acute or rostrate, dorsal and ventral sutures thickened but not winged, valves pale brown at maturity, thinly glandular, reticulately veined; seeds subglobose or cubical, about 4–5 mm. across; testa closely and very strongly tuberculate or rugose, dark brown, hilum small, broad, about 1·5 mm. long. *Plate 35, figs. 13–21.*

HAB.: Cultivated fields and waste ground; frequently in moist places by lakes or the margins of irrigation channels; sea-level to 2,000 ft. alt.; fl. Febr.–May.

DISTR.: Divisions 1, 3–8, common in Divisions 4, 7. Portugal and the Mediterranean region east to Iraq.

1. Paphos area, 1951, *Kennedy* 1196 !
3. Near Episkopi, 1862, *Kotschy* 299 ! Akrotiri, 1927, *Rev. A. Huddle* 51 ! Lefkara, 1941, *Davis* 2730 !
4. Common; *Sintenis & Rigo* 934 ! *Deschamps* 160, 161, 162; *Syngrassides* 1041 !
5. Athalassa, 1957, *Merton* 2903 !
6. Near Nicosia, 1859, *Kotschy* S1024; Nicosia, 1902, *A. G. & M. E. Lascelles* s.n. ! and, 1937, *Syngrassides* 1407 ! Ayia Irini, 1941, *Davis* 2563 !
7. Common; *Sintenis & Rigo* 459 ! *Davis* 2920 ! 3042 ! *Casey* 1002 ! 1238 ! 1684 ! *Meikle* 2554 !
8. Galatia, 1962, *Meikle* 2532 !

NOTES: Three of the specimens cited above (*Casey* 1002, 1238, 1684) resemble *L. hierosolymitanus* Boiss. in having small, narrow pods (4–4·5 cm. long, 0·7 cm. wide) and seeds with a bluntly rugose, not acutely tuberculate, testa; they differ, however, from Palestine material of this supposed species in having 1–2-flowered racemes of relatively large flowers. None of the additional characters given by authors for distinguishing between *L. annuus* and *L. hierosolymitanus* can be considered satisfactory, and until some other valid distinctions, apart from pod and seed, can be found, it seems best to consider both as variants of a single species, with the Cyprus specimens as connecting links between the two extremes.

4. **L. cassius** *Boiss.*, Diagn., 1, 9: 128 (1849), Fl. Orient., 2: 604 (1872), Suppl., 195 (1888); Post, Fl. Pal., ed. 2, 1: 429 (1932); Davis, Fl. Turkey, 3: 356 (1970); Zohary, Fl. Palaest., 2: 214 (1972); Osorio-Tafall et Seraphim, List Vasc. Plants Cyprus, 58 (1973).
L. annuus L. ssp. *cassius* (Boiss.) Holmboe, Veg. Cypr., 115 (1914).

TYPE: Turkey; "in dumosis jugi Cassii inter *Cassab* et *Suadieh*. Legi jun. 1846" *Boissier* (G, K !).

A sprawling or scandent annual, very similar to *L. annuus* L. in general appearance and habit; stems glabrous, distinctly winged, to 60 cm. or more in length; leaves with a well-developed, branched tendril, 2-foliolate; petiole 1–3 cm. long, winged; stipules very narrowly semi-sagittate as in *L. annuus*; leaflets linear, lanceolate or narrowly elliptical, 1–7 cm. long, 0·2–1·2 cm. wide, distinctly parallel-veined, glabrous, apex often shortly cuspidate; inflorescence a 1–2-flowered raceme, peduncle straight, rather stout, up to 8 cm. long, often exceeding the subtending petiole; bracts small, narrowly subulate; pedicel and calyx ± densely covered with sessile, brownish glands; calyx campanulate, 6–7 mm. long, about 2·5 mm. wide, calyx-teeth ovate or rather broadly deltoid, subequal, 2·5–3 mm. long, 1–1·5 mm. wide at base, acute or shortly acuminate; petals pink, limb of standard suborbicular, 8–9 mm. long and about as wide, apex rounded or very slightly emarginate, sometimes apiculate, basal gibbosities not developed, claw broad, 4–5 mm. long, 3–4 mm. wide, limb of wings oblong, 7–9 mm. long, 2·5–3·5 mm. wide, apex rounded, base with a long narrow auricle, conjunctival flap slightly developed, claw 2–3 mm. long, limb of keel about 6 mm. long, 6 mm. wide, upcurved at a right angle to a very blunt or rounded apex, base distinctly auriculate, claw 2–3 mm. long; stamens 10–11 mm. long, vexillary stamen free, the others connate for about 7 mm.; anthers oblong, about 7 mm. long, 3·5 mm. wide; ovary strongly compressed, narrowly oblong, about 6 mm.

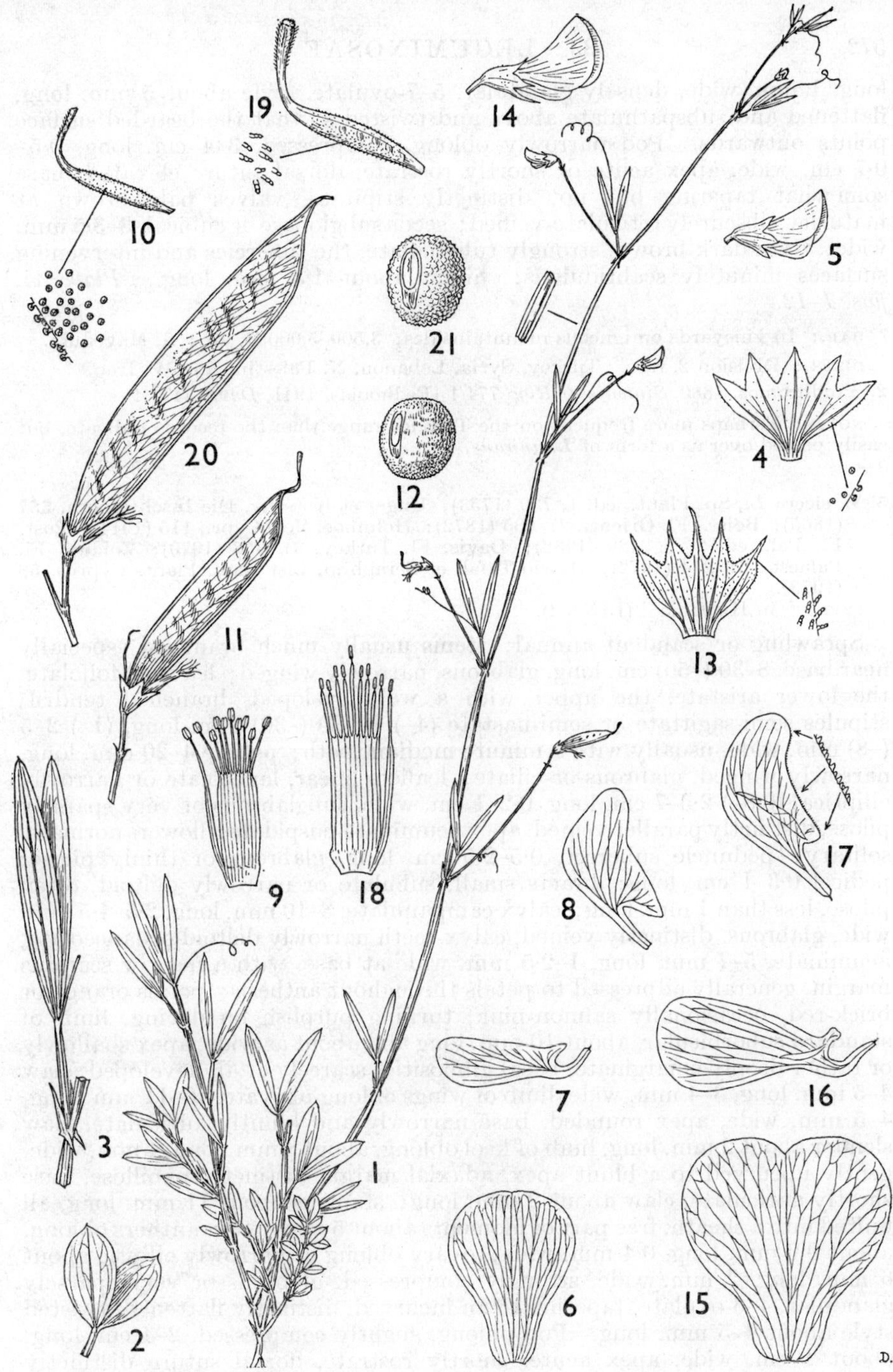

Plate 35. LATHYRUS CASSIUS Boiss. **1**, habit, $\times\frac{2}{3}$; **2–3**, basal and stem leaves, $\times 1$; **4**, calyx opened out, outer surface, $\times 3$; **5**, flower, $\times 1$; **6**, standard, $\times 2$; **7**, wing, $\times 2$; **8**, keel, $\times 2$; **9**, androecium, $\times 2$; **10**, gynoecium, $\times 2$; **11**, pod, $\times 1$; **12**, seed, $\times 3$. LATHYRUS ANNUUS L. **13**, calyx opened out, inner surface, $\times 3$; **14**, flower, $\times 1$; **15**, standard, $\times 2$; **16**, wing, $\times 2$; **17**, keel, $\times 2$; **18**, androecium, $\times 2$; **19**, gynoecium, $\times 2$; **20**, pod, $\times 1$; **21**, seed, $\times 3$. (**1–3 & 11**, *Bornmüller* 534, Lebanese material; **4–10**, *Davis* 4726, Palestinian material; **12**, *Haines* W. 525; **13–19**, *Davis* 2920, Cyprus material; **20–21**, *Lindsay* 824, Persian material.)

long, 1 mm. wide, densely glandular, 5–7-ovulate, style about 5 mm. long, flattened and subspathulate above and twisted so that the bearded surface points outwards. Pod narrowly oblong, compressed, 3–4 cm. long, 0·5–0·7 cm. wide, apex acute or shortly rostrate, dorsal suture elevated, base somewhat tapering but not distinctly stipitate, valves pale brown at maturity, obscurely reticulate-veined; seeds subglobose or cubical, 3–3·5 mm. wide; testa dark brown, strongly tuberculate, the tubercles and intervening surfaces minutely scabridulous; hilum about 1·5 mm. long. *Plate 35, figs. 1–12.*

HAB.: In vineyards on igneous mountainsides; 3,500–5,000 ft. alt.; fl. May–July.

DISTR.: Division 2, rare. Turkey, Syria, Lebanon, N. Palestine and W. Iraq.

2. Prodhromos, 1880, *Sintenis & Rigo* 774 ! Pedhoulas, 1941, *Davis* 3471 !

NOTES: Perhaps more frequent on the Troödos range than the records indicate, but easily passed over as a form of *L. annuus*.

5. **L. cicera** *L.*, Sp. Plant., ed. 1, 730 (1753); Unger et Kotschy, Die Insel Cypern, 387 (1865); Boiss., Fl. Orient., 2: 605 (1872); Holmboe, Veg. Cypr., 115 (1914); Post, Fl. Pal., ed. 2, 1: 429 (1932); Davis, Fl. Turkey, 3: 358 (1970); Zohary, Fl. Palaest., 2: 214 (1972); Osorio-Tafall et Seraphim, List Vasc. Plants Cyprus, 58 (1973).

TYPE: " in Hispania " (LINN !).

Sprawling or scandent annual; stems usually much branched especially near base, 8–30 (–50) cm. long, glabrous, narrowly winged; leaves 2-foliolate, the lower aristate, the upper with a well-developed, branched tendril; stipules semi-sagittate or semi-hastate (4–) 10–20 (–30) mm. long, (1–) 2–5 (–8) mm. wide, usually with a minute median tooth; petiole 4–20 mm. long, narrowly winged, glabrous or ciliate; leaflets linear, lanceolate or narrowly elliptical (0·8–) 2·5–7 cm. long, 0·2–1 cm. wide, subglabrous or very sparsely pilose, distinctly parallel-veined, apex acuminate-cuspidate; flowers normally solitary; peduncle suberect, 0·5–2·5 cm. long, glabrous or thinly pilose; pedicel 0·3–1 cm. long; bracts small, subulate or narrowly deltoid, often pilose, less than 1 mm. long; calyx campanulate, 8–10 mm. long, 2·5–4·5 mm. wide, glabrous, distinctly veined, calyx-teeth narrowly deltoid or lanceolate, acuminate, 5–7 mm. long, 1–2·5 mm. wide at base, with a narrow scarious margin, generally adpressed to petals throughout anthesis; petals orange or brick-red, occasionally salmon-pink, turning purplish on drying, limb of standard suborbicular, about 10 mm. long and about as wide, apex shallowly or rather deeply emarginate, basal gibbosities scarcely at all developed, claw 4–5 mm. long, 3–4 mm. wide, limb of wings oblong-obovate, 10–11 mm. long, 4–5 mm. wide, apex rounded, base narrowly and bluntly auriculate, claw slender about 2 mm. long, limb of keel oblong, about 8 mm. long, 4 mm. wide, gently upcurved to a blunt apex, adaxial margin distinctly papillose, base shortly auriculate, claw about 2 mm. long; stamens about 11 mm. long, all united into a sheath, free part of filaments about 5 mm. long; anthers oblong, about 0·6 mm. long, 0·4 mm. wide; ovary oblong or narrowly elliptic about 6 mm. long, 2 mm. wide, strongly compressed, glabrous or very sparsely glandular, 4–6-ovulate, tapering to an incurved, distinctly flattened, twisted style about 4–5 mm. long. Pod oblong, slightly compressed, 2–4 cm. long, about 1 cm. wide, apex acute, shortly rostrate, dorsal suture distinctly canaliculate, valves pale brown, often blotched purplish at maturity, obscurely reticulate-veined; seeds squarish in outline, about 5 mm. wide, ± wedge-shaped in cross-section; testa bright brown, blotched pale brown, smooth or minutely rugulose; hilum about 1·5 mm. long.

HAB.: Cultivated or fallow fields, waste ground, grassy hillsides; sea-level to 3,750 ft. alt.; fl. Febr.–May.

DISTR.: Divisions 2, 4–7, common in Divisions 2, 7. Widespread in southern Europe and the Mediterranean region eastwards to Iran.

2. Common; *Syngrassides* 104 ! *Kennedy* 592 ! 1314 ! 1315 ! *Davis* 2751 !
4. Near Larnaca, 1862, *Kotschy* 154.
5. Near Lefkoniko, 1880, *Sintenis & Rigo* 462 ! Near Pyroi, 1936, *Syngrassides* 1043 !
6. Near Nicosia, 1930, *F. A. Rogers* 0673 ! Beuyuk Kaimakli, 1934, *Syngrassides* 1349 !
7. Common; *Syngrassides* 1439 ! *Davis* 2461 ! *Casey* 993 ! 997 ! 999 ! 1680 ! *Chapman* 479 !

6. L. gorgonei *Parl.* in Giorn. Sci. Sic., 62: 3 (1838); Post, Fl. Pal., ed. 2, 1: 430 (1932); R. M. Nattrass, First List Cypr. Fungi, 39 (1937); Davis, Fl. Turkey, 3: 357 (1970); Zohary, Fl. Palaest., 2: 214, t. 304 (1972); Osorio-Tafall et Seraphim, List Vasc. Plants Cyprus, 58 (1973).

L. amoenus Fenzl in Russegg., Reisen, 1: 890, t. 2 (1842); Boiss., Fl. Orient., 2: 605 (1872); Sintenis in Oesterr. Bot. Zeitschr., 32: 54 (1882); J. Freyn in Bull. Herb. Boiss., 5: 608 (1897); Holmboe, Veg. Cypr., 115 (1914).

TYPE: Sicily; " Panormi alla Pizzuta " (FI).

Indistinguishable from *L. cicera* L. in habit and vegetative characters; flowers solitary, peduncle up to 3 cm. long, straight, suberect, pedicel usually about 5 mm. long, curved at apex, bracts filiform or narrowly subulate; calyx campanulate, 8–10 mm. long, 2–3 mm. wide, glabrous, distinctly veined, calyx-teeth deltoid-subulate, 4–6 mm. long, about 1·5 mm. wide at base, with a narrow scarious margin, generally patent or reflexed at anthesis; petals larger than in *L. cicera*, pale orange, or the wings and keel yellow and the standard orange, limb of standard suborbicular, about 15 mm. long and almost as wide, apex slightly emarginate, base tapering to a claw about 4 mm. long, 3 mm. wide, basal gibbosities not developed, limb of wings oblong or narrowly obovate, about 13 mm. long, 4–5 mm. wide, apex rounded, base distinctly auriculate with a long narrow auricle, claw about 2 mm. long, limb of keel broadly boat-shaped, about 12 mm. long, 7–8 mm. wide, upcurved to a blunt apex, adaxial margin minutely papillose, base very shortly auriculate, claw about 2·5 mm. long; stamens about 17 mm. long, united to form a sheath, free part of filaments about 7 mm. long; anthers oblong about 1 mm. long, 0·5 mm. wide; ovary narrowly linear, strongly compressed, ± densely brown-glandular, about 9 mm. long, 1·2 mm. wide, 5–8-ovulate, apex tapering to a narrowly spathulate, twisted style about 8 mm. long. Pod linear-oblong, 3–5 cm. long, 0·6–0·8 cm. wide, distinctly compressed, apex shortly rostrate, dorsal suture canaliculate, valves pale brown when mature, distinctly reticulate-veined; seeds almost cubical, about 3·5–4 mm. wide; testa dark brown, smooth; hilum about 1·5 mm. long, rather broad.

HAB.: Cultivated and fallow fields; waste ground, rarely in grassy pastures or garigue; sea-level to 4,500 ft. alt.; fl. Febr.–May.

DISTR.: Divisions 2–8, common in Division 7. Sicily, Sardinia, Turkey, Syria, Palestine, Egypt, Cyrenaica and eastwards to Iraq and Iran.

2. Perapedhi, 1938, *Kennedy* 1313 ! Platres, 1938, *Kennedy* 1311 ! Prodhromos, 1955, *G. E. Atherton* 598 !
3. Akrotiri, 1927, *Rev. A. Huddle* 50A ! Limassol, 1938, *Kennedy* 1312 ! Yermasoyia River, 1952, *Mavromoustakis* 28 !
4. Near Larnaca, 1880, *Sintenis & Rigo* 935 ! Larnaca, 1893, *Deschamps* s.n.; the Phaneromene near Larnaca, 1905, *Holmboe* 136; 1 m. S.E. of Goshi, 1974, *Meikle* 4094 !
5. Kythrea, 1932, *Syngrassides* 263 ! Lakatamia, 1932, *Syngrassides* 365 ! Vatili, 1936, *Nattrass* 761.
6. Nicosia airport, 1954, *Merton* 1830B !
7. Common; *Syngrassides* 176 ! *Casey* 323 ! 559 ! 996 ! 998 ! 1683 ! *Chapman* 98 ! *G. E. Atherton* 984 ! 1114 ! 1121 !
8. Komi Kebir, 1941, *Davis* 2444 ! Ovgoros (Famagusta), 1941, *Davis* 2481 !

7. L. sativus *L.*, Sp. Plant., ed. 1, 730 (1753); Boiss., Fl. Orient., 2: 606 (1872); Holmboe, Veg. Cypr., 115 (1914); Post, Fl. Pal., ed. 2, 1: 430 (1932); Rechinger f.

in Arkiv för Bot., ser. 2, 1: 426 (1950); Davis, Fl. Turkey, 3: 359 (1970); Osorio-Tafall et Seraphim, List Vasc. Plants Cyprus, 58 (1973).

TYPE: " *in* Hispania, Gallia ".

Sprawling or scandent annual, indistinguishable from *L. cicera* L. and *L. gorgonei* Parl. in habit and vegetative characters; flowers solitary; peduncles rather long, frequently exceeding 4 cm. in length, pedicels arcuate, 5–8 mm. long; calyx campanulate, 8–10 mm. long, 3–4 mm. wide, glabrous, rather indistinctly veined, calyx-teeth lanceolate, acuminate, about 5 mm. long, 1–1·5 mm. wide at base; petals bluish, the standard frequently tinged pink, the keel usually bright blue; limb of standard suborbicular or flabelliform, 12–15 mm. long, 15–20 mm. wide, apex emarginate, base without any distinct gibbosities, claw very short and broad, usually less than 2 mm. long and 4–6 mm. wide, limb of wings obovate, about 12 mm. long and 6 mm. wide, apex rounded, base with a prominent, acute auricle, claw short, about 2 mm. long, limb of keel deltoid, about 6 mm. long, 7 mm. wide, upcurved at a right angle to a subacute apex, adaxial margin minutely papillose, base distinctly auriculate, claw 3 mm. long; stamens about 12 mm. long, the vexillary generally free, the others connate for about 7 mm., free parts of filaments filiform; anthers oblong, about 1 mm. long, 0·4 mm. wide; ovary oblong-elliptic, strongly compressed, glabrous or sparsely and minutely glandular, about 7 mm. long, 2–2·5 mm. wide, 3–5-ovulate, apex tapering to an upcurved, narrowly spathulate style about 6 mm. long. Pod oblong, flattened, 2–5 cm. long, 1·5–2 cm. wide, apex shortly rostrate, dorsal sutures distinctly winged, valves pale brown when mature, distinctly reticulate-veined. Seeds squarish in outline, about 5–8 mm. wide, wedge-shaped in cross-section; testa pale creamy-brown, smooth; hilum about 1·5 mm. long.

HAB.: Cultivated fields, waste ground and roadsides; an escape from cultivation; sea-level to 800 ft. alt.; fl. Febr.–May.

DISTR.: Divisions 3–5, 7, 8. Widespread in central and southern Europe and in the Mediterranean region eastwards to India; temperate and tropical Africa and Atlantic Islands.

3. Limassol, 1913, *Haradjian* 604.
4. Sotira, between Larnaca and Famagusta, 1905, *Holmboe* 406; Larnaca, 1950, *Mrs. Scott-Moncrieff* 1287 !
5. Near Lefkoniko, 1938, *Syngrassides* 1799 ! Kythrea, 1950, *Chapman* 345 !
7. Ayios Yeoryios (Kyrenia), 1956, *G. E. Atherton* 1254 !
8. Komi Kebir, 1941, *Davis* 2445 ! Ovgoros (Famagusta), 1941, *Davis* 2480 ! Peristerges near Kilanemos, 1962, *Meikle* 2433 !

8. **L. blepharicarpos** *Boiss.*, Diagn., 1, 9: 126 (1849), Fl. Orient., 2: 607 (1872); Post, Fl. Pal., ed. 2, 1: 431 (1932); Holmboe, Veg. Cypr., 115 (1914); Davis, Fl. Turkey, 3: 360 (1970); Zohary, Fl. Palaest., 2: 217, t. 307 (1972); Osorio-Tafall et Seraphim, List Vasc. Plants Cyprus, 58 (1973).

Decumbent or sprawling annual; stems 5–30 (–50) cm. long, usually much-branched, strongly angled or narrowly winged, thinly pilose with patent hairs or occasionally subglabrous; leaves 2-foliolate, the lower ones usually ecirrhose, the upper with a slender, poorly developed, simple (or very rarely branched) tendril; stipules semi-sagittate or semi-hastate, 5–15 mm. long, 2–5 mm. wide, thinly pilose or subglabrous; petioles generally less than 1 cm. long, narrowly winged; leaflets elliptical, lanceolate or almost linear, 5–35 (–45) mm. long, 2–10 mm. wide, subglabrous above, thinly pilose (or rarely subglabrous) especially along the veins below, apex shortly cuspidate; flowers solitary; peduncles suberect, 1–3 cm. long, thinly pilose (rarely glabrous); bracts subulate, usually about 1 mm. long, caducous; pedicel arcuate 3–5 mm. long; calyx campanulate, 5–7 mm. long, 2·5–4 mm. wide, calyx-tube generally glabrous, calyx-teeth lanceolate, acuminate, about 3·5 mm. long, 1·5 mm. wide at base, margins generally ciliate with long hairs;

petals reddish-pink or terra-cotta red, limb of standard suborbicular or flabelliform, 12–14 mm. long, 10–12 mm. wide, apex shortly emarginate, base tapering to a short claw about 3 mm. long and 3 mm. wide, limb of wings obovate, about 11 mm. long, 4–5 mm. wide, apex rounded, base with a prominent oblong auricle, claw slender about 1·5 mm. long, limb of keel broadly boat-shaped, 6–7 mm. long, 3–4 mm. wide, upcurved to a blunt apex, base very shortly auriculate, adaxial margin minutely papillose, claw about 2 mm. long; stamens 8–10 mm. long, vexillary stamen free, others connate for about 5–6 mm., free part of filaments filiform; anthers oblong, about 1 mm. long, 0·4 mm. wide; ovary narrowly oblong, strongly compressed, about 5 mm. long, 1·5 mm. wide, sutures densely ciliate with long hairs, ovules 4, style sharply upcurved, 4–5 mm. long, distinctly spathulate with membranous, winged margins. Pod oblong, 2–2·5 cm. long, about 1 cm. wide, ciliate, dorsal sutures strongly winged, apex abruptly and shortly rostrate, valves pale brown at maturity, distinctly reticulate-veined; seeds 1–4, suborbicular, rather compressed, about 4·5 mm. diam.; testa reddish-brown, marbled black, smooth or obscurely rugulose-tuberculate; hilum small, about 1 mm. long.

var. **cyprius** *Meikle* in App. III, p. 806.

[*L. amphicarpos* (non L.) Sibth. et Sm., Fl. Graec. Prodr., 2: 65 (1813), Fl. Graec., 7: 62, t. 693 (1830); Unger et Kotschy, Die Insel Cypern, 387 (1865).]

TYPE: Cyprus; Akradhes, April 8, 1962, *Meikle* 2511 (K !).

Leaflets broadly lanceolate or elliptical, 5–20 mm. long, 2–10 mm. wide; otherwise indistinguishable from typical *L. blepharicarpos*.

HAB.: Stony hillsides, in pastures or amongst garigue, occasionally in open Pine forest, or on waste or cultivated land, or by roadsides; sea-level to c. 2,000 ft. alt.; fl. Febr.–April.

DISTR.: Divisions 3, 4, 7, 8, common in Divisions 3, 7. Turkey, Rhodes, Syria, Lebanon and Palestine, the variety perhaps restricted to Cyprus.

3. Common; *Mavromoustakis* 26 ! 33 ! *N. Macdonald* 89 ! *J. B. Suart* 154 !
4. Ayia Napa, 1905, *Holmboe* 73; Larnaca, 1905, *Holmboe* 89.
7. Common; *Sintenis & Rigo* 466 ! *Syngrassides* 1182 ! *Davis* 2510 ! *Casey* 356 ! 1139 ! 1681 ! *Chapman* 270 ! *G. E. Atherton* 926 !
8. Ayios Philon near Rizokarpaso, 1941, *Davis* 2264 ! Akradhes, 1962, *Meikle* 2511 !

NOTES: The Cyprus plants are distinguished from mainland specimens solely by their uniformly short and relatively broad leaflets, which give the plants a rather neat facies. The distinction, though of trivial importance, is sufficiently marked to warrant some recognition.

L. CYPRIUS *Rechinger f.* (in Svensk Bot. Tidskr., 43: 37; 1949) collected at Ktima, May 20, 1913, by Haradjian (no. 708, K, isotype !) is indistinguishable from the cultivated Sweet Pea, *Lathyrus odoratus* L. The original material may have been collected in a garden, or as a garden escape on waste ground. *L. odoratus*, a native of Sicily and S. Italy, has no claim to be considered an indigenous, or even a naturalized, member of the Cyprus flora.

SECT. 4. **Lathyrus** Standard without basal gibbosities; style twisted; leaves with leaflets; perennial herbs.

L. LATIFOLIUS *L.* (Sp. Plant., ed. 1, 733; 1753), a robust climbing perennial, with 2-foliolate leaves and long-pedunculate, 5–15-flowered racemes of pink or purple flowers, has been collected " in a plantation at the Summer Camp ", 5,000 ft. alt. on Troödos (July 25, 1938, *Kennedy* 1320 !). *L. latifolius*, the Everlasting Pea, is a native of central and southern Europe, commonly grown as an ornamental, and without doubt an introduction in Cyprus. It tends to persist when once established, and may yet survive at high altitudes on the Troödos range.

SECT. 5. **Orobastrum** *Boiss.* Standard with or without basal gibbosities; style not twisted, adaxially canaliculate towards base, leaves with leaflets; annual or perennial herbs.

9. L. setifolius *L.*, Sp. Plant., ed. 1, 731 (1753); Boiss., Fl. Orient., 2: 612 (1872); C. et W. Barbey, Herborisations au Levant, 99 (1880); Post, Fl. Pal., ed. 2, 1: 432 (1932); Davis, Fl. Turkey, 3: 352 (1970); Zohary, Fl. Palaest., 2: 219, t. 310 (1972).

TYPE: France; " Monspelii " (LINN !).

Climbing or trailing annual; stems slender 30–50 cm. (or more) long, usually much branched, especially towards the base, glabrous, narrowly winged; leaves 2-foliolate, the upper with a slender, branched tendril; petiole very short, 3–15 mm. long, glabrous, narrowly winged; stipules narrowly semi-sagittate, 3–10 mm. long, 1–2 mm. wide, glabrous, generally with a short median tooth; leaflets narrowly linear-lanceolate or almost filiform, 15–90 mm. long, 0·5–5 mm. wide, glabrous, distinctly parallel-veined, apex acuminate or shortly cuspidate; inflorescences 1-flowered; peduncle erect, filiform, 10–30 (–40) mm. long, without a terminal awn; pedicel 2–8 (–10) mm. long, thickening and arcuate in fruit; calyx campanulate, about 4 mm. long, 2·5 mm. wide, glabrous, calyx-teeth narrowly deltoid, about 2·5 mm. long, 1 mm. wide at base; petals brick-red, limb of standard broadly obovate, about 10 mm. long, 9 mm. wide, apex slightly emarginate, base tapering to a very short, broad claw, limb of wings narrowly obovate, about 7 mm. long, 3 mm. wide, apex rounded, base with a short sharp auricle, claw about 2 mm. long, limb of keel boat-shaped, about 6 mm. long, 3 mm. wide, apex rather blunt, base shortly auriculate, claw about 2 mm. long; stamens about 9 mm. long, connate for about 5 mm.; anthers oblong, about 0·6 mm. long, 0·4 mm. wide; ovary narrowly oblong, about 5 mm. long, 1·5 mm. wide, strongly compressed, minutely adpressed-pubescent, 2–4-ovulate, apex tapering to a sharply upcurved style about 4 mm. long, base shortly stipitate, stigma narrowly spathulate, not conspicuously flattened. Pod oblong, 2–3 cm. long, about 1 cm. wide, strongly compressed, apex abruptly rostrate, base distinctly stipitate, valves very pale brown, subglabrous or very thinly pubescent, especially along the dorsal suture; seeds subglobose, about 5 mm. diam.; testa bluntly papillose-tuberculate, fuscous; hilum about 1·2 mm. long.

HAB.: On waste ground or amongst garigue; about sea-level; fl. Febr.–April.

DISTR.: Division 4, rare. Widespread in the Mediterranean region, but more common in the western than the eastern half.

4. Larnaca, 1880, *C. & W. Barbey* 365.

NOTES: Since C. & W. Barbey record fruiting specimens, there can be little doubt about the accuracy of the identification, though it is strange that *L. setifolius* should have been overlooked by every other collector. Holmboe, though acquainted with Barbey's " Herborisations au Levant ", makes no mention of *L. setifolius*.

10. L. sphaericus *Retz.*, Obs. Bot., 3: 39 (1783); Boiss., Fl. Orient., 2: 613 (1872); Holmboe, Veg. Cypr., 115 (1914); Rechinger f. in Arkiv för Bot., ser. 2, 1: 426 (1950); Davis, Fl. Turkey, 3: 352 (1970); Osorio-Tafall et Seraphim, List Vasc. Plants Cyprus, 57 (1973).

L. sphaericus var. *pilosulus* Murb. in Lunds Univ. Årsskr., n.s. 2, 18: 76 (1922); Lindberg f., Iter Cypr., 19 (1946).

TYPE: No specimen or locality cited, perhaps cultivated at Lund (LD).

A suberect or spreading annual (6–) 10–30 cm. high; stems usually much branched towards base, glabrous or sparsely pilose, acutely angled or narrowly winged; leaves 2-foliolate; petioles very short, usually under 1 cm. long, narrowly winged, terminating in a short awn or (upper leaves) in a slender, unbranched tendril; stipules narrowly semi-sagittate, 3–13 mm. long, usually less than 1 mm. wide, the lobes subulate or almost filiform; leaflets linear, 10–80 mm. long, 1–3 mm. wide, glabrous or very sparsely pilose, with indistinct, parallel veins; flowers solitary, axillary; peduncle short, usually less than 10 mm. long, generally terminating in a rather long awn, pedicel slightly arcuate, about 5 mm. long, lengthening and thickening in fruit; calyx campanulate, 5–7 mm. long, 2·5–3 mm. wide, glabrous or subglabrous, calyx-teeth subulate, 3–4 mm. long, less than 1 mm. wide at base, with a very narrow scarious margin; petals bright orange-red, limb of standard obovate-oblong, 5–9 mm. long, 4–8 mm. wide, apex slightly emarginate, base narrowing and merging with a very broad claw 2–3 mm. long and about as wide, limb of wings oblong, 5–8 mm. long, 2·5–4·5 mm. wide, apex rounded, base with a short deltoid auricle, claw 3–4 mm. long, limb of keel boat-shaped, 4–5 mm. long, 2·5–3 mm. wide, apex rather abruptly upcurved to an apiculate apex, base distinctly auriculate, claw 3–4 mm. long; stamens 6–8 mm. long, free part of filaments 1·5–2 mm. long; anthers shortly oblong, to about 0·5 mm. in length and almost as wide; ovary linear, strongly compressed, 4–6 mm. long, 0·8–1 mm. wide, minutely papillose-pubescent, 8–12-ovulate, style 2–3 mm. long, apex slightly dilated. Pod linear (2·5–) 3·5–7 cm. long, 0·4–0·6 cm. wide, somewhat compressed, straight, apex distinctly rostrate, sutures thickened, valves pale brown (or tinged purplish) at maturity, with numerous, prominent parallel veins; seeds globose, 2·5–3 mm. diam.; testa dark brown, often pruinose, smooth; hilum about 1 mm. long.

HAB.: In Pine forest, or occasionally on open stony mountainsides; near sea-level to 4,000 ft. alt.; fl. Febr.–May.

DISTR.: Divisions 1, 2, 7, 8, common in Division 2. Widespread in southern Europe, but becoming rare in the eastern Mediterranean; eastwards to Afghanistan; N. Africa and Tropical E. Africa south to Kenya and Uganda; Atlantic Islands.

1. Paphos, 1928 or 1930, *Druce* 145 ! Ayios Neophytos, 1939, *Lindberg f.* s.n.
2. Common; *Kennedy* 590 ! 591 ! 1316 ! 1317 ! *Mavromoustakis* 50 ! *Merton* 3178 !
7. Ayios Ioannis (? Ayios Iakonos) between Trikomo and Akanthou, 1905, *Holmboe* 557.
8. Akradhes, 1962, *Meikle* 2512 !

NOTES: The very frequent occurrence of this species on the litter in Pine forests is noteworthy, also its apparently disjunct vertical distribution at high and low altitudes.

11. L. saxatilis (*Vent.*) *Vis.*, Fl. Dalm., 3: 330 (1852); Boiss., Fl. Orient., 2: 614 (1872), Suppl., 195 (1888); Sintenis in Oesterr. Bot. Zeitschr., 31: 393 (1881), 32: 192 (1882); Holmboe, Veg. Cypr., 115 (1914); Davis, Fl. Turkey, 3: 351 (1970); Osorio-Tafall et Seraphim, List Vasc. Plants Cyprus, 57 (1973).

Orobus saxatilis Vent., Jard. Cels., t. 94 (1803).

TYPE: The illustration in Ventenat, Jard. Cels., t. 94, based on material collected by Gerard " sur les collines arides et pierreuses du departement du Var " !

Prostrate or sprawling annual (5–) 9–20 (–35) cm. high; stems much branched at base, sharply angled, rather densely clothed with spreading hairs; leaves 2–4 (–6)-foliolate, ecirrhose, or with the tendril reduced to a short awn; stipules semi-sagittate, often minute, 1–4 mm. long, about 1 mm. wide with narrowly deltoid lobes; petiole generally very short, less than 5 mm. long, channelled above, but scarcely winged, thinly pilose; leaflets very variable, linear, oblong or narrowly obovate, rarely filiform, 3–30 mm. long, 0·5–5 mm. wide, thinly pilose on both sides, apex usually truncate or emarginate with the midrib produced into a short apiculus, base cuneate or tapering to a very short petiolule; flowers solitary, axillary, peduncle about 2–5 mm. long, lengthening considerably in fruit, sometimes terminating in a short (occasionally lobed) awn, pedicel equalling or a little shorter than peduncle; flowers small, calyx 3–3·5 mm. long, 2·5 mm. wide, glabrous or subglabrous externally, usually thinly pilose within, calyx-teeth deltoid, subequal, about 1–1·3 mm. long, 0·8–1 mm. wide at base; petals milky-white or yellowish, often pencilled pale blue; limb of standard broadly obovate, 3·5–4·5 mm. long and almost as wide, apex emarginate, base with two distinct gibbosities, tapering to a broad claw 2–2·5 mm. long and about as wide, limb of wings oblong, 4–4·5 mm. long, 2–2·5 mm. wide, apex rounded, base with a distinct blunt auricle, claw slender, about 2–2·5 mm. long, limb of keel broad, boat-shaped, 3·5–4 mm. long, 2–2·5 mm. wide, curving upwards to a subacute apex, base bluntly auriculate; stamens 6–7 mm. long, free part of filaments about 2 mm. long; anthers shortly oblong, about 0·4 mm. long, 0·3 mm. wide; ovary oblong, 5–6 mm. long, 1·5–2 mm. wide, strongly compressed, glabrous, c. 6-ovulate, base tapering to a short stipe, apex tapering rather abruptly to a sharply upcurved style 1·5–2 mm. long, apex of style bearded within, flattened but scarcely dilated. Pod oblong, 1·8–2·5 cm. long, about 0·6–0·7 cm. wide, strongly compressed, apex shortly rostrate, dorsal suture distinctly thickened, valves pale brown at maturity, chartaceous; seeds subglobose, somewhat compressed, about 2·5–2·8 mm. diam.; testa smooth, reddish-brown, usually pruinose; hilum about 1·2 mm. long.

HAB.: Rocky hillsides, on screes, or in thin Pine forest, on calcareous or igneous formations; 500–2,500 ft. alt.; fl. Febr.–April.

DISTR.: Divisions 3, 7. Widespread in the Mediterranean region, but becoming rare in the east, and apparently reaching the eastern limit of its distribution in Cyprus.

3. Stavrovouni, 1974, *Meikle* 4063 !

7. Pentadaktylos, 1880, *Sintenis & Rigo* 456 ! Kantara, 1880, *Sintenis & Rigo*; Yaïla, 1941, *Davis* 2839 ! Kyrenia, below Tjiklos cliff, 1949, *Casey* 430 !

31. PISUM *L.*

Sp. Plant., ed. 1, 727 (1753).
Gen. Plant., ed. 5, 324 (1754).

Sprawling or climbing annuals (or rarely perennials); stems terete or angled, not winged; leaves paripinnate, terminating in a well-developed, branched tendril (reduced to a short awn in *P. formosum* (Stev.) Alef.); stipules generally conspicuous, often foliaceous; leaflets usually few, entire or toothed; inflorescence a long-pedunculate, axillary, few-flowered raceme, or flowers solitary; bracts minute, caducous, bracteoles wanting; calyx campanulate, the tube sometimes slightly gibbous adaxially, calyx-teeth subequal or the 2 adaxial rather broader than the others; petals white, pink, purple or tawny-yellow, standard often broader than long, with a short broad claw, wings falcate-oblong, adherent to the relatively short, obtuse or subacute keel; stamens all united or the vexillary free, filament-sheath truncate at apex, free parts of filaments slightly dilated towards apex, anthers uniform; ovary subsessile or shortly stipitate, ovules numerous, style sharply upcurved, the winged margins strongly reflexed, connate above and fused with the apical portion of the style, upper part of style bearded adaxially, stigma subterminal, rather small. Pod compressed, generally glabrous; seeds angled or subglobose, the funicle dilated into a thin aril covering the oblong hilum.

Two (or possibly three) species in S. Europe, the Mediterranean region, western and central Asia and E. tropical Africa. *P. sativum* L. (including its numerous subspecies) cultivated and widely distributed as an adventive.

Seeds with a smooth testa:
Petals white - - - - - - - - - **P. sativum var. SATIVUM**
Petals pink or purple - - - - - - - **P. sativum var. ARVENSE**

Seeds with a papillose-tuberculate testa:
Standard 2 cm. or more wide; pods usually more than 5 cm. long; leaflets (2–) 4–6 (–8), often subentire:
Peduncle 2–4 times as long as stipules; plant usually robust **P. sativum ssp. elatius var. elatius**
Peduncle $\frac{1}{4}$–1 (–1$\frac{1}{2}$) times as long as stipules; inflorescence 1-flowered; plant slender, less than 80 cm. high **P. sativum ssp. elatius var. brevipedunculatum**
Standard less than 2 cm. wide; pods less than 5 cm. long; leaflets 2–4, commonly toothed - - - - - - - - **P. sativum ssp. elatius var. pumilio**

1. P. sativum *L.*, Sp. Plant., ed. 1, 727 (1753); Sibth. et Sm., Fl. Graec. Prodr., 2: 62 (1813); Unger et Kotschy, Die Insel Cypern, 383 (1865); Boiss., Fl. Orient., 2: 622 (1872); Holmboe, Veg. Cypr., 115 (1914); Post, Fl. Pal., ed. 2, 1: 434 (1932); Davis, Fl. Turkey, 3: 370 (1970); Osorio-Tafall et Seraphim, List Vasc. Plants Cyprus, 58 (1973), D. Zohary in Chromosomes Today, 4: 315–317 (1973).

Climbing or sprawling annual, 15–150 (–200) cm. high; stems glabrous, commonly glaucous, striate but not winged, simple or branched, especially towards base; leaves 2–6 (–8)-foliolate, up to 20 cm. (sometimes more) in length and 10 cm. in width, and terminating in a well-developed, branched tendril; stipules ovate or oblong, blunt or subacute, 1–6 cm. long, 0·5–5 cm. wide, glabrous, glaucous, the margin distinctly toothed near the base, entire or sometimes toothed towards apex; petiole 1–6 cm. long, flattened or slightly canaliculate above but not winged; leaflets oblong, elliptic or obovate, 1–5 cm. long, 0·5–3·5 cm. wide, glabrous, often glaucous, apex usually apiculate, margins entire or toothed; inflorescence a lax 1–2 (–3)-flowered raceme; peduncle rather stout, 1–15 (–20) cm. long, suberect, rhachis generally excurrent into an arista up to 1 cm. long, bracts deltoid or subulate, glabrous or thinly pilose, caducous, pedicels 0·5–1·5 cm. long; calyx campanulate, 7–10 (–15) mm. long, 4–7 mm. wide, glabrous or subglabrous externally, sometimes sparsely red-glandular within, calyx-teeth narrowly deltoid, acute, 4–7 mm. long, the 2 adaxial 2–4 mm. wide at base, the 3 abaxial distinctly narrower; petals white (var. *sativum*), lavender, pink or dark purple, the wings often with a dark purple limb (ssp. *elatius*; var. *arvense* (L.) Poir.); limb of standard broadly obcordate 1–1·5 cm. long, 1·2–2·5 cm. wide, apex emarginate, shortly apiculate, base with 2 distinct median gibbosities, claw 4–5 mm. long, 5–8 mm. wide, limb of wings suborbicular, 0·5–1·2 cm. long and about as wide, apex rounded, base with a distinct, blunt, gibbous auricle, claw 3–5 mm. long, limb of keel oblong, 8–10 mm. long, 4–6 mm. wide, upcurved almost at a right angle to a short acute apex, lower (connate) margin distinctly cristate, base of upper margins shortly auriculate, claw 3–4 mm. long; stamens 1–1·8 cm. long, free part of filaments 5–8 mm. long, slightly dilated towards apex; anthers oblong, 0·8–1 mm. long, 0·2–0·4 mm. wide; ovary narrowly oblong, flattened, glabrous, 8–11 mm. long, 1·5–3 mm. wide, c. 8-ovulate, base very shortly stipitate; style upcurved at a right angle, 6–8 mm. long. Pod linear-oblong, 3–8 cm. long, 0·8–1·8 cm. wide, compressed or subterete, apex shortly rostrate, valves pale brown and distinctly veined at maturity; seeds subglobose or angular, 0·5–1 cm. diam.; testa pale brown, dark brown, reddish or fuscous, sometimes greenish, smooth or minutely and closely papillose-tuberculate; hilum 1–2·5 mm. long.

var. SATIVUM.
TYPE: " *in* Europae *agris* ".

Petals white; seeds with a smooth, pale brown or greenish testa, usually rather large.

HAB.: In gardens and cultivated ground, or an escape from cultivation; fl. Febr.–May.

DISTR.: Recorded, "in arvis Argolicis, Messeniacis et Cypriis" by Sibthorp, but neither indigenous nor established as an alien in Cyprus.

ssp. **elatius** (*M. Bieb.*) *Aschers. et Graebn.*, Syn. Mitt.-Europ. Fl., 6: 2: 1064 (1910).

Standard pink or mauve, wings darker, purple or crimson; seeds with a closely and minutely papillose-tuberculate testa, dark brown or fuscous, usually rather small.

var. **elatius** (*M. Bieb.*) *Alefeld*, Landw. Fl., 44 (1866); Davis, Fl. Turkey, 3: 371 (1970).
P. elatius M. Bieb., Fl. Taur.-Cauc., 2: 151 (1808); Boiss., Fl. Orient., 2: 623 (1872); Post, Fl. Pal., ed. 2, 1: 434 (1932).
? *P. arvense* L. var. *variegatum* Guss., Fl. Sic. Syn., 2 (1): 279 (1844).
TYPE: Caucasus; "in Iberia, *D. Steven*" (LEN).

Plants robust, frequently 80 cm. or more in height; leaves often 4–6-foliolate, leaflets commonly subentire, oblong or elliptical; peduncles often robust and elongate; flowers large, the standard more than 2 cm. wide. Pods frequently more than 5 cm. long and proportionally rather narrow.

HAB.: Vineyards and waste ground; 3,000–4,000 ft. alt.; fl. April–May.

DISTR.: Division 2, rare. Widespread in S. Europe and the Mediterranean region, also in Turkey, Caucasia and Iran.

2. Agros, 1941, *Davis* 3061 !

NOTES: Very rare in our area; perhaps introduced as a food plant, and now surviving as an escape from cultivation.

var. **brevipedunculatum** *Davis et Meikle* in Notes Roy. Bot. Gard. Edinb., 29: 320 (1969); Davis, Fl. Turkey, 3: 371 (1970).
TYPE: Turkey; Ankara, Dikmen, June 8, 1933, *W. Kotte* 1113 (K !).

Plant usually less than 80 cm. high, rather slender; leaflets in (1–) 2 pairs, entire or subentire; inflorescences generally 1-flowered, the peduncle shorter than or not much longer than the stipules; standard 2–2·2 cm. wide. Pods usually 6–7 cm. long, 1·3–1·5 cm. wide.

HAB.: Vineyards and waste ground; 3,500–4,000 ft. alt.; fl. April–May.

DISTR.: Division 2. Also Turkey.

2. Platres, 1938, *Kennedy* 1319A ! Phini, 1938, *Kennedy* 1319 !

var. **pumilio** *Meikle* in Notes Roy. Bot. Gard. Edinb., 29: 320 (1969); Davis, Fl. Turkey, 3: 371 (1970).
P. humile Boiss. et Noë in Boiss., Diagn., 2, 2: 45 (1856), Fl. Orient., 2: 623 (1872); J. Freyn in Bull. Herb. Boiss., 5: 609 (1897); Post in Mém. Herb. Boiss., 18: 93 (1900); Post, Fl. Pal., ed. 2, 1: 435 (1932) non *P. humile* Mill., Gard. Dict., ed. 8 no. 2 (1768).
P. arvense L. ssp. *humile* (Boiss. et Noë) Holmboe, Veg. Cypr., 115 (1914).
P. sativum L. ssp. *syriacum* Berger in Hedrick et al., The Vegetables of New York, 1, 1: 17 (1928); H. S. Gentry in F.A.O. Plant Genetic Resources Newsletter, no. 25: 3–13 (1971).
P. pumilio (Meikle) Greuter in Ann. Mus. Goulandris, 1: 42 (1973).
[*P. arvense* (non L.) H. Stuart Thompson in Journ. Bot., 44: 307 (1906); Holmboe, Veg. Cypr., 115 (1914).]
[*P. elatius* (non M. Bieb.) H. Stuart Thompson in Journ. Bot., 44: 307 (1906).]
[*P. arvense* L. ssp. *elatius* (non M. Bieb.) Holmboe, Veg. Cypr., 115 (1914).]
[*P. fulvum* (vix Sibth. et Sm.) Sintenis in Oesterr. Bot. Zeitschr., 31: 194 (1881), 32: 54 (1882); Holmboe, Veg. Cypr., 115 (1914).]
TYPE: Turkey; "ad rupes Armeniae meridionalis prope Bakker Maaden. Maio 1852 cl. Noë." (G).

Plants rather slender and bushy, usually less than 80 cm. high; leaves 2–4-foliolate, leaflets commonly toothed, oblong or obovate; peduncles generally rather short; flowers smallish, the standard less than 2 cm. wide, the wings not so noticeably darker in colour as in var. *elatius*. Pods usually less than 4 cm. long and proportionately rather broader than those of var. *elatius*.

HAB.: Cereal fields, roadsides, gardens and waste ground; sea-level to 1,000 ft. alt.; fl. Febr.–May.

DISTR.: Divisions 3–8, common in Divisions 4, 7. Turkey, Caucasus area, Iran, Iraq, Syria, Palestine and Egypt.

3. Akrotiri, 1927, *Rev. A. Huddle* 55 !
4. Common; *Sintenis & Rigo*; *Deschamps* 163a, 163b, 163c; *Holmboe* 140; *Syngrassides* 1276 !
5. Xeri [Iseri], Nicosia, 1932, *Syngrassides* 75 ! Geunyeli near Nicosia, 1952, *Kennedy* 1763 !
6. Nicosia, 1894, *Post*; near Kykko (Mesaoria) 1952, *F. M. Probyn* 18 !
7. Common; *Sintenis & Rigo* 991 ! *Syngrassides* 1025 ! *Davis* 2459 ! *Casey* 487 ! *Chapman* 117 !
8. Ovgoros, Famagusta distr., 1941, *Davis* 2475 !

NOTES: Possibly the wild ancestor of *P. sativum* ssp. *sativum* and ssp. *elatius* var. *elatius*, and probably indigenous in our area, but so often a denizen of disturbed or cultivated ground that the limits of its natural distribution are obscure. Generally distinguishable from var. *elatius* by its low diffuse habit and smallish flowers and pods, but so closely allied that the two can scarcely be considered more than varieties of a single aggregate.

[P. FULVUM *Sibth. et Sm.*, Fl. Graec. Prodr., 2: 262 (1813) and Fl. Graec., 7: 79, t. 688 (1830), has been recorded from Cyprus by Sintenis (Oesterr. Bot. Zeitschr., 31: 194; 1881 and 32: 54; 1882). These records are correctly referable to *P. sativum* L. var. *pumilio* Meikle (*P. humile* auct.), to which variety a specimen at Kew labelled "*Pisum fulvum* fl. gr." by Sintenis and Rigo (S. & R. 453 !) also belongs.

True *P. fulvum* Sibth. et Sm. is at once distinguished from all other Peas by its yellow, tawny or copper-coloured petals. In other respects it much resembles *P. sativum* var. *pumilio*, though normally very slender, with 2-foliolate leaves and very distinctly toothed leaflets. Its claims to independent specific status are somewhat questionable. *P. fulvum* is fairly widely distributed in S. Turkey, Syria and Palestine, and may quite probably occur in Cyprus; satisfactory proof of its occurrence is, however, required before it can be admitted to the Flora.]

TRIBE 8. **Phaseoleae** (*Bronn*) *DC*. Annual or perennial herbs or climbers, occasionally woody; leaves pinnately 3-foliolate (or rarely unifoliolate or 5–7-foliolate), leaflets entire or lobed; stipules and stipels generally present; inflorescence usually racemose or paniculate, bracts mostly small and caducous, bracteoles sometimes present; limb of keel rostrate or erostrate, sometimes spirally twisted; stamens normally diadelphous, occasionally united into a sheath split adaxially, anthers mostly uniform; style bearded along the adaxial surface or around the stigma; ovules few to many. Pod 2-valved, open or septate within, not jointed transversely; seed strophiolate or estrophiolate, radicle inflexed.

PHASEOLUS VULGARIS *L.*, Sp. Plant., ed. 1, 723 (1753), the Kidney or Haricot bean, an erect or subscandent annual with 3-foliolate leaves, rhomboid-ovate, acuminate, subglabrous leaflets 5–10 (–15) cm. long, 3–8 cm. wide, 1- to few-flowered racemes of white or pink-purple flowers with spirally twisted keels, and linear-compressed, often slightly falcate, pods 9–15 cm. or more in length and 1–1·5 cm. wide, is commonly cultivated in Cyprus. The reniform or oblong, slightly compressed beans are about 15 mm. long and 8 mm. wide, with a smooth, milky white, brown, blackish or variously blotched testa. The dwarf var. *nanus* (L.) Aschers. seems to be more common in cultivation than the more robust, subscandent var. *vulgaris*. Both are probably natives of Central America.

PHASEOLUS COCCINEUS *L.*, Sp. Plant., ed. 1, 724 (1753), the Runner bean, another Central American species, is grown as a summer vegetable, the whole pod being cooked and eaten while the valves are green and succulent. It is a vigorous climber with many-flowered racemes of scarlet flowers and clusters of linear, pendulous pods frequently more than 30 cm. long. The flowers are of considerable decorative value.

PHASEOLUS LUNATUS *L.*, Sp. Plant., ed. 1, 724 (1753), the Lima bean, also from Central and South America, has the racemose inflorescences of *P. coccineus* but the flowers are small and usually white (or purple) as in *P. vulgaris*. It differs from both in having broad, relatively short, subfalcate pods, 8–10 (or more) cm. long, 1·5–2·5 cm. wide. Like *P. vulgaris* it is grown primarily for its seeds, which are exceptionally large in the Butter bean (var. *macrocarpus* Benth.).

PHASEOLUS AUREUS *Roxb.* and P. MUNGO *L.*, two closely related Asiatic species, are also reported from Cyprus. The first in the " Haricot Vert ", " Green Gram ", or " Mung "; the second the " Black Gram " or " Urd ". Both resemble *P. vulgaris* in habit and general appearance, but are at once distinguished by their peltate (not basally attached) stipules, and narrowly linear, somewhat torulose, more or less strigose pods, generally less than 8 cm. long and 0·5 cm. wide. The seeds are small and oblong, about 2·5–4·5 mm. long and 2·5–3 mm. wide, with a dark, blackish testa. Both are so distinct from the South and Central American representatives of the genus *Phaseolus* as to warrant their transfer to the genus *Vigna*, under the names *V. radiata* (L.) Wilczek (*Phaseolus aureus* Roxb.) and *V. mungo* (L.) Hepper. *V. radiata* is generally distinguishable from *V. mungo* by its shortly strigose pods and smaller seeds (usually less than 3·5 mm. long), but it is questionable if the two are more than distinct cultivars of one common ancestral species.

DOLICHOS LABLAB *L.*, Sp. Plant., ed. 1, 725 (1753)—*Dolichos purpureus* L., *Lablab purpureus* (L.) Sweet, *Lablab niger* Medik., *Lablab vulgaris* Savi—the Hyacinth, Bonavista or Egyptian bean—is recorded (under the name *Lablab vulgaris*) as a cultivated plant by Kotschy (Die Insel Cypern, 390). It is a robust, scandent or subscandent perennial (or rarely annual), resembling *Phaseolus* in general aspect, with pinnately 3-foliolate, long-petiolate leaves, and ovate-deltoid leaflets 2·5–15 cm. long, 1·5–14 cm. wide. The axillary, many-flowered, racemose inflorescences are borne on stout peduncles 2–40 cm. long, and are furnished with deciduous bracts and persistent, elliptic or suborbicular bracteoles. The flowers are conspicuous, with purple (or rarely whitish) petals, the standard oblate-orbicular, auriculate and with 2 distinct callosities at the base, and the keel is sharply upcurved. The pods are 2–5-seeded, oblong or oblong-falcate, 3·5–6 cm. long, 1–2 cm. wide, flattened, with a sharp beak formed by the persistent style. The seeds are strongly compressed and oblong, 5–9 (–15) mm. long, 4–7 mm. wide, with a dark (rarely pale) testa and a long, conspicuous, whitish strophiole.

Dolichos lablab, probably a native of tropical Africa, is now widely distributed throughout the warmer parts of the world. The young pods can be eaten, or the ripe beans, and the plant is also distinctly ornamental.

TRIBE 9. **Dalbergieae** *DC*. Trees, shrubs or woody climbers; leaves pinnate rarely trifoliate or unifoliolate; stipels often wanting; inflorescence cymose, paniculate or fasciculate-racemose; stamens all connate basally, or cleft into 2 phalanges, or the vexillary free; anthers generally uniform. Pod indehiscent, exceeding the calyx, membranous, coriaceous or woody, sometimes drupaceous or samaroid.

DALBERGIA SISSOO *Roxb.*, Fl. Ind., ed. Carey, 3: 223 (1832); G. Frangos in Cyprus Agric. Journ., 18: 86 (1923). A rather slender tree 3–10 m. high; branches pale brown, glabrous or shortly pubescent; leaves pinnate, (1–) 3–5-foliolate; stipules lanceolate-acuminate, about 7 mm. long, pubescent externally, caducous; stipels wanting; petiole about 2 cm. long, at first pubescent, later subglabrous; leaflets usually alternate, rather remote, suborbicular, 2–5 cm. diam., subglabrous or thinly pubescent below when mature, margins entire, apex cuspidate-acuminate, venation rather obscure; inflorescence a much-branched, rather condensed axillary panicle 4–6 cm. long; peduncle and pedicels pubescent; calyx tubular-campanulate, about 4·5 mm. long, with 5 unequal, rather blunt lobes, corolla about 9 mm. long, creamy-white or pale yellow. Pod narrowly oblong-elliptic, flattened, 5–7 cm. long, 1–1·5 cm. wide, 1–2-seeded, apex blunt or acute, base distinctly stipitate, valves pale brown, thinly coriaceous or chartaceous, finely reticulate-veined; seeds reniform, flattened, about 7 mm. long, 4 mm. wide; testa dark brown, sparsely scabridulous. *Plate 36*.

A native of India, widely cultivated in the tropics and subtropics and reported from Larnaca by Frangos.

TIPUANA TIPU (*Benth.*) *O. Ktze.*, Rev. Gen. Plant., 3, 3: 72 (1898).

Machaerium tipu Benth. in Hook., Kew Journ., 5: 267 (1853); G. Frangos in Cyprus Agric. Journ., 18: 87 (1923).

A broad-crowned tree, 7–35 m. high; branches striate, shortly pubescent when young; leaves alternate or subopposite, pinnate, 14–30-foliolate, 20–35 cm. long, 6–12 cm. wide; stipules minute, caducous, stipels wanting; petiole 1–3 cm. long; leaflets narrowly oblong, entire, 2·5–6 cm. long, 1–2 cm. wide, glabrous above, thinly adpressed-pubescent below, often with a purple median blotch, apex rounded or shortly emarginate, venation rather prominent below; inflorescence a lax axillary raceme, or sometimes 2 racemes in each axil, 8–20 cm. long, bearing 10–20 (or more) flowers; peduncle, rhachis and pedicels densely pubescent; pedicels 1–2·5 cm. long, articulate and minutely bracteolate near apex; calyx broadly campanulate, about 7 mm. long, 9 mm. wide, pubescent externally, shortly and rather irregularly lobed; petals yellow, 1·5–2·3 cm. long, the standard very broad and rounded. Pod samaroid, like the fruit of an *Acer*, oblong-elliptic, about 8 cm. long, 3 cm. wide, the basal, seed-bearing part woody and swollen, expanding above into a broad, pale brown, chartaceous wing; seed usually solitary, relatively small, ovoid, about 6 mm. long, 4 mm. wide, strongly compressed; testa dark brown, minutely rugulose.

A native of N. Argentina, widely planted in the tropics and subtropics. Recorded from Larnaca by Frangos, and from Kolossi Castle grounds, 1941 (*Davis* 3537 !) and 1947 (*T. A. Couppis*, comm. *R. R. Waterer* !).

TRIBE 10. **Sophoreae** (*Bronn*) *DC.* Mostly trees and shrubs (rarely subherbaceous); leaves pinnately 5–many-foliolate (rarely 1- or digitately 3-foliolate); leaflets entire; stipels generally wanting; inflorescence racemose or paniculate, usually terminal; bracteoles mostly small; corolla papilionaceous or sometimes almost actinomorphic with the adaxial petal outermost; stamens 10, filaments free or connate only at base, anthers uniform; ovary

Plate 36. DALBERGIA SISSOO Roxb. **1**, flowering branch, ×⅔; **2**, single flower, ×3; **3**, calyx opened out, inner surface, ×3; **4**, standard, ×3; **5**, wing, ×3; **6**, keel, ×3; **7**, androecium, ×3; **8**, gynoecium, ×3; **9**, pods, ×⅔. (**1–8**, *Hb. Hooker* s.n., Indian material; **9**, *Rogers* 277.)

1–several-ovulate; style glabrous, stigma terminal or lateral. Pod indehiscent or 2-valved, sometimes winged.

Sophora japonica *L.*, Mantissa 1, 67 (1767), the Japanese Pagoda tree, is said to be cultivated at Salamis and Nicosia (*Bovill*, *Nattrass* etc.) and is probably to be found elsewhere in Cyprus. It is a handsome tree, 18–30 m. high, with a rounded crown and rough greyish bark. The bright green leaves are pinnate, 9–15-foliolate, with shortly petiolulate, ovate-oblong leaflets 2·5–5 cm. long and 1·3–2·5 cm. wide. The inflorescence is a much-branched terminal panicle bearing numerous creamy-white flowers about 12 mm. long. The calyx is cupular, with short, blunt lobes, and the linear pods are 3–5 cm. long, 1–1·5 cm. wide, fleshy and indehiscent, with the valves constricted between the relatively large brown seeds.

Despite its name, the Japanese Pagoda tree, though long cultivated in Japan, is strictly a native of China. It was first introduced into European gardens towards the middle of the 18th century, and is now extensively planted as an ornamental.

subfamily 2. **Caesalpinioideae** Mostly trees or shrubs, or woody climbers; leaves pinnate or rarely bipinnate; flowers zygomorphic, but rarely as strongly so as in *Papilionoideae*, petals imbricate, the upper (adaxial) inside the laterals; stamens 10 (rarely 5) occasionally more numerous, anthers usually without an apical gland; seeds generally without an areole, radicle normally straight.

tribe 1. **Caesalpinieae** *Endl.* Leaves bipinnate (rarely simply pinnate); calyx-lobes free as far as the disk; petals generally 5, rather unequal; anthers versatile; ovary 2–many-ovulate, rarely 1-ovulate; gynophore not adnate to calyx-tube.

Caesalpinia gilliesii (*Hook.*) *Dietr.*, Syn. Plant. 2: 1495 (1840), (*Poinciana gilliesii* Hook., Bot. Misc., 1: 129, t. 34; 1829; Unger et Kotschy, Die Insel Cypern, 391; 1865), Bird of Paradise Flower. A lax, much-branched shrub or small tree, native of the Argentine, but now widely cultivated in warm-temperate and tropical regions. The bipinnate leaves are oblong, 15–25 cm. long, 7–10 cm. wide, 10–15-jugate, the pinnae with 8–12 pairs of small, narrowly oblong, obtuse, glaucescent leaflets, each 7–9 mm. long, 1·5–2·5 mm. wide. The inflorescence is a conspicuous, bracteate, terminal, densely glandular raceme bearing 8–15 (or more) showy, long-pedicellate flowers with 5 yellow, narrowly oblong or oblanceolate petals, 2·5–3 cm. long, 0·7–1 cm. wide, and 10, far-protruding, crimson filaments, 6–8 cm. (or more) in length. The mature pods are oblong, 5–9 cm. long, about 2 cm. wide, flattened, with pale brown, scabrid-glandular valves; they contain 5 or more ovate, flattened seeds, about 10 mm. long, 8 mm. wide, with a smooth, dark brown testa, and a very small hilum.

Recorded from near Larnaca by Kotschy, and not infrequent in gardens throughout the lowlands of Cyprus.

Caesalpinia coriaria (*Jacq.*) *Willd.*, Sp. Plant., 2: 532 (1800), (*Poinciana coriaria* Jacq., Select. Stirp. Amer. Hist., 123, t. 175, fig. 36; 1763), a small tree, generally less than 10 m. high, with leaves resembling those of *C. gilliesii*, but with terminal panicles of inconspicuous whitish or yellow, sweet-scented flowers, is said to have been planted at Salamis (Bovill, Rep. Plant. Work Cyprus, 14; 1915), but may not have survived. It is the " Divi-Divi " of

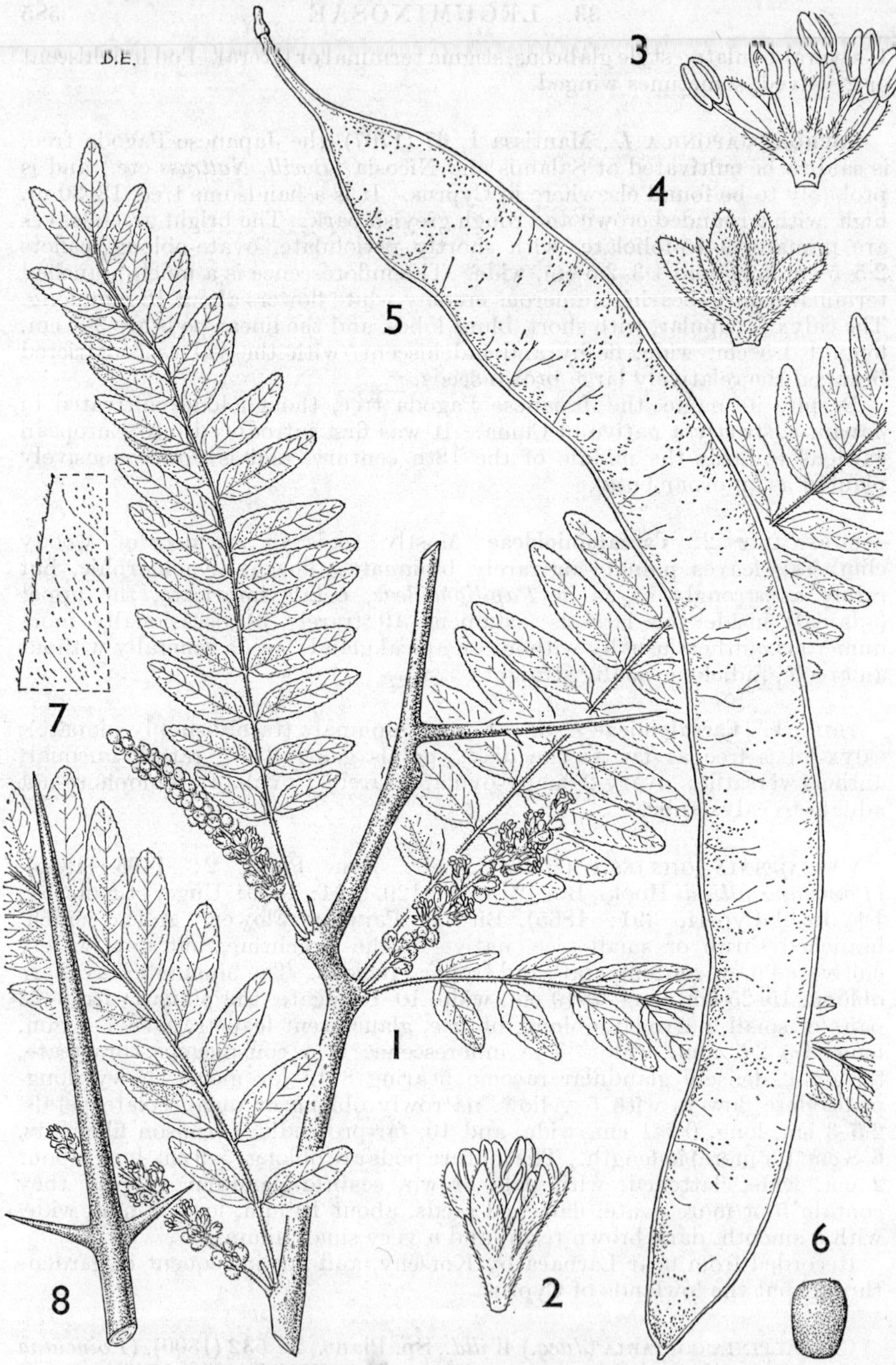

Plate 37. GLEDITSIA TRIACANTHOS L. **1,** flowering shoot, ×⅔; **2,** flower, ×4; **3,** flower opened out, ×4; **4,** flowers opened out (outer surface), ×4; **5,** pod, ×1; **6,** seed, ×1; **7,** enlarged margin of leaflet, ×5; **8,** spine, ×1. (**1, 7 & 8** *Gillett & Rawi* 7686; **2–4,** *Wettern* s.n., English cultivated material; **5,** *Nicholson* s.n., English cultivated material; **6,** *Herb. Mus. Florent.* s.n., Italian cultivated material.)

Central America and the West Indies, and has thick, glossy, blackish-brown pods which are valued in tanning.

GLEDITSIA TRIACANTHOS *L.*, Sp. Plant., ed. 1, 1056 (1753), the Honey Locust, a native of eastern North America, cultivated in Europe and Asia, is reported (Bovill, Rep. Plant. Work Cyprus, 14; 1915) to have been planted at Salamis, and may occur, as an introduction, in other parts of Cyprus. It is a handsome tree, not infrequently more than 30 m. high, the trunk and branches often armed with viciously sharp, rigid, dark brown, branched or simple spines commonly more than 30 cm. long; leaves bipinnate on robust growths or simply pinnate on less robust growths, leaflets 20–60 or more, lanceolate or narrowly oblong, obtuse, subglabrous or sparsely pubescent, 8–30 mm. long, 2·5–12 mm. wide; flowers small, greenish, polygamous, numerous in dense, cylindrical, spicate racemes from the apices of short lateral spurs; calyx densely adpressed-pilose; petals small, pilose, scarcely distinguishable from calyx-lobes; stamens usually 6. Pod narrowly oblong, 15–30 cm. long, 3–5 cm. wide, flattened, valves dark glossy brown; seeds numerous, bluntly elliptical, about 12 mm. long, 8 mm. wide, flattened; testa dark reddish-brown, smooth. *Plate 37.*

The pods contain a sugary pulp, and are said to be useful as a foodstuff for cattle. The spineless f. *inermis* (L.) Zabel is probably a better garden tree than the typical form, though the cruel spines of the latter are not without decorative value.

PARKINSONIA ACULEATA *L.*, Sp. Plant., ed. 1, 375 (1753). Jerusalem Thorn. A slender tree, 4–7 m. high; leaves bipinnate (often apparently simply pinnate) with a very short, spine-tipped rhachis, pinnae narrowly linear, 25–30 cm. long, the rhachilla (pinna-rhachis) flattened, 2·5–3 mm. wide, bordered along either margin by numerous, much-reduced, oblanceolate leaflets, seldom exceeding 5 mm. in length and usually much shorter; inflorescence a lax axillary raceme, 10–30 cm. long, bearing 10–30 (or more) long-pedicellate flowers; sepals glabrous, ovate, reflexed at anthesis; petals ovate-elliptical, 12–15 mm. long, 6–8 mm. wide, showy, yellow, the claw and the lower part of the lamina distinctly bearded; stamens 10, filaments bearded towards base; ovary many-ovulate. Pod linear, torulose or moniliform, 50–140 mm. long, about 8 mm. wide, valves pale- or mid-brown, distinctly striate-sulcate; seeds oblong, about 10 mm. long, 4–5 mm. wide, testa dark brown, smooth; hilum subapical, very small. *Plate 38.*

Generally believed to be a native of South America, but now widely cultivated throughout the tropics and subtropics, and better able to withstand long periods of drought than the majority of ornamental trees and shrubs.

Parkinsonia torreyana Wats., nowadays generally referred to *Cercidium floridum* Benth. in A. Gray, Plant. Wright., 1: 58 (1852) is said to have been planted at Salamis (Bovill, Rep. Plant. Work Cyprus 8, 14; 1915), and is known to be cultivated in Egypt and in other parts of the eastern Mediterranean. It is a native of Arizona, California and Mexico, and is a shrub, or small tree, 2–5 m. high, with pale green, shortly spinose branches; the leaves are small, bipinnate, generally 1-jugate, the pinnae with a slender rhachilla, bearing 4–6 blunt, oblong, glaucescent leaflets 5–10 mm. long, 2·5–4 mm. wide; the inflorescences are produced abundantly along the upper parts of the young branches, and are shortly racemose or corymbose, with 4–8 yellow flowers, not unlike those of *Parkinsonia aculeata* L., but much smaller. The pods are strongly compressed, linear or narrowly oblong, sometimes constricted between the seeds, about 60–80 mm. long and 12 mm. wide, with bright golden-brown, obscurely striate valves; the seeds are

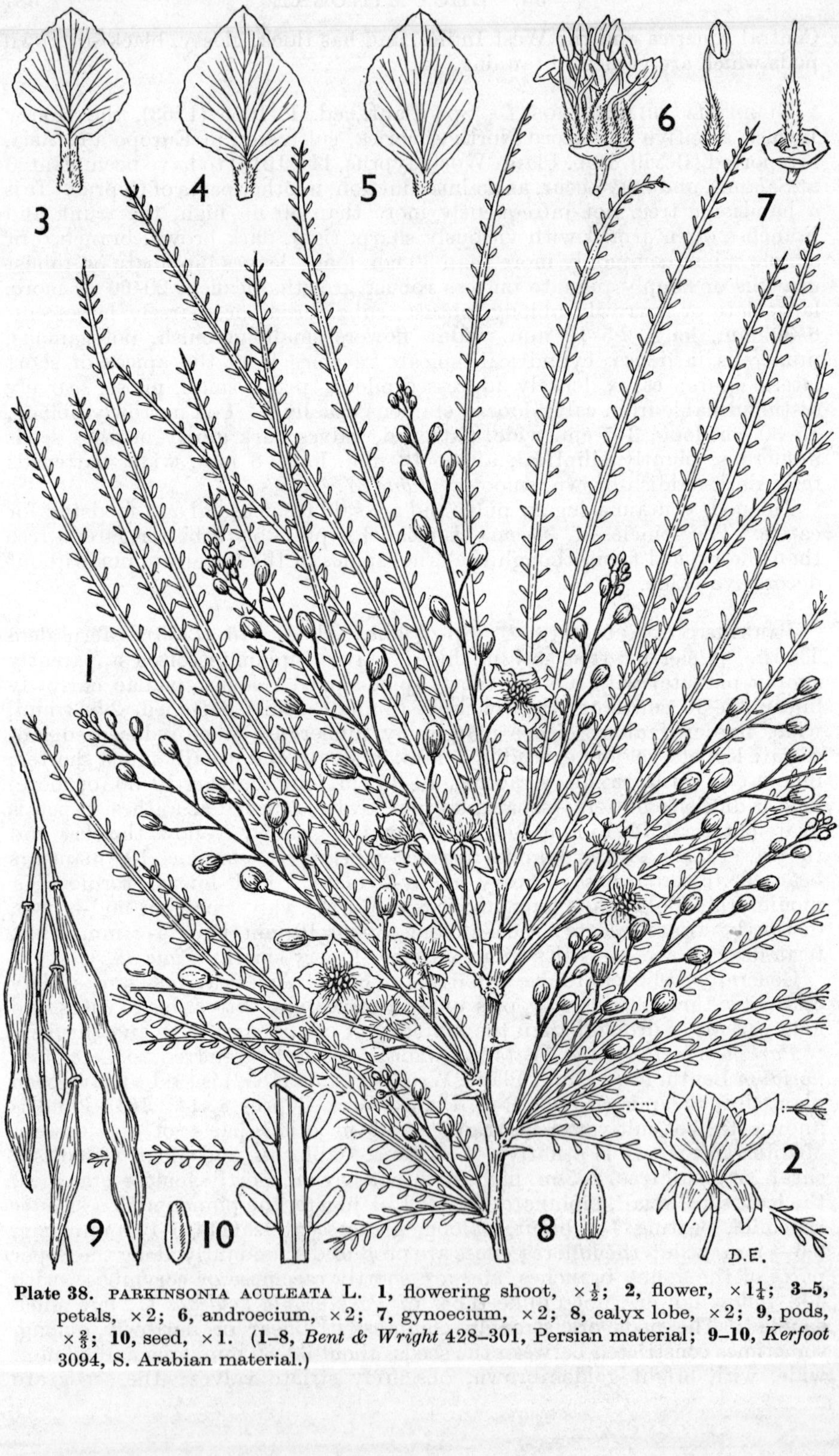

Plate 38. PARKINSONIA ACULEATA L. **1**, flowering shoot, $\times\frac{1}{2}$; **2**, flower, $\times 1\frac{1}{4}$; **3–5**, petals, $\times 2$; **6**, androecium, $\times 2$; **7**, gynoecium, $\times 2$; **8**, calyx lobe, $\times 2$; **9**, pods, $\times\frac{2}{3}$; **10**, seed, $\times 1$. (**1–8**, *Bent & Wright* 428–301, Persian material; **9–10**, *Kerfoot* 3094, S. Arabian material.)

oblong, compressed, about 12 mm. long and 7 mm. wide, with a dark brown, smooth testa.

TRIBE 2. **Cassieae** *Bronn* Leaves imparipinnate or paripinnate, rarely reduced to phyllodes; calyx-lobes free as far as the disk; petals 5 or fewer; anthers opening by pores, or if longitudinally dehiscent then usually basifixed and not versatile; ovary 2–many-ovulate, rarely 1-ovulate; gynophore not adnate to calyx-tube. Seeds with endosperm.

CASSIA ARTEMISIOIDES *Gaudich. ex DC.*, Prodr., 2: 495 (1825). An erect or diffuse shrub or small tree 1–2 (–4) m. high; twigs closely ashy-felted; leaves shortly petiolate, tomentellous, leaflets 3–8-jugate, narrowly linear, 15–40 mm. long, 1–2·5 mm. wide, subterete, rather fleshy; flowers yellow, in short, sub-corymbose, 4–8-flowered, axillary racemes; calyx-lobes blunt, deeply divided, about 4 mm. long; petals obovate-orbicular, shortly clawed, about 8–10 mm. long; anthers linear, about 3 mm. long, dark reddish-brown; ovary usually falciform. Pod narrowly oblong, flattened, about 5–8 cm. long, 0·8–1 cm. wide, apex blunt, apiculate, valves shining brown, coriaceous, septate within; seeds 10 or more, obovate, flattened, 6–7 mm. long, 4–5 mm. wide, with a distinct marginal rim; testa dark shining brown or blackish.

A native of eastern Australia, occasionally planted elsewhere and recorded from gardens at Kyrenia (Aug. 5, 1955, *G. E. Atherton* 296 !) and Nicosia (*G. Joscht* 6523 !).

Several *Cassia* species are popular ornamentals in tropical and subtropical gardens, and it is likely that some of the more showy ones are found in Cyprus. The Indian Pudding-pipe Tree, *C. fistula* L., grows to 10 m. or more, and has very large ovate-elliptical leaflets, long pendulous racemes of very showy, pale yellow flowers, and cylindrical candle-like pods. *C. didymobotrya* Fres., from Tropical Africa, is an erect shrub with many-foliolate leaves, narrow oblong-elliptic leaflets, and erect dense racemes of yellow flowers. The Senna-pods of commerce are mostly from the Alexandrian Senna, *C. senna* L., a native of Egypt, Arabia and N.E. Tropical Africa.

32. CERATONIA *L.*

Sp. Plant., ed. 1, 1026 (1753).
Gen. Plant., ed. 5, 450 (1754).

Large evergreen shrub or tree; leaves paripinnate; stipules minute or wanting; flowers small, apetalous, polygamous or dioecious or occasionally hermaphrodite, forming short, solitary or fasciculate, spike-like racemes on the year-old branches; bracts and bracteoles minute, deciduous; calyx-lobes 5, imbricate, shortly connate at base; stamens 5, opposite the calyx-lobes, with long filaments (dolichostemonous) or short filaments (brachystemonous) inserted below a large pulviniform disk; ovary shortly stipitate, ovules numerous. Pod strongly compressed, oblong, indehiscent, thickened at the sutures; seeds transverse, compressed, separated by areas of pulpy, sugary mesocarp.

One species only, native in the eastern Mediterranean, but cultivated throughout the Mediterranean region, and in other warm-temperate parts of the world. The pods provide a sweet and nutritious animal-food.

1. C. siliqua *L.*, Sp. Plant., ed. 1, 1026 (1753); Unger et Kotschy, Die Insel Cypern, 391 (1865); Boiss., Fl. Orient., 2: 632 (1872); P. Gennadius, The Carob-tree (1902); Holmboe, Veg. Cypr., 101 (1914); Post, Fl. Pal., ed. 2, 1: 440 (1932); Chapman, Cyprus Trees and Shrubs, 44 (1949); Davis, Fl. Turkey, 3: 7 (1970); Zohary,

Fl. Palaest., 2: 32, t. 45 (1972); Osorio-Tafall et Seraphim, List Vasc. Plants Cyprus, 54 (1973).

TYPE: " in Apulia, Sicilia, Creta, Cypro, Syria, Palaestina ".

Evergreen tree, to 10 m. high, with a broad, hemispherical crown sometimes 14 m. wide; bark rough, dark brown; branches sturdy; leaves oblong, usually pinnate (rarely bipinnate), 12–30 cm. long, 3–15 cm. wide, with 4–8 (–12) oblong or obovate leaflets; petiole and rhachis 7–15 cm. long, thinly tomentellous, often tinged red, distinctly sulcate above; petiolule usually less than 5 mm. long; leaflets 2–8 (or more) cm. long, 1–4·5 (or more) cm. wide, coriaceous, dark shining green above, dull and rather ashy-grey below, with slightly raised nervation, apex blunt or rounded, occasionally emarginate, base rounded or shortly cuneate; inflorescences cylindrical, rather rigid 3·5–15 (–25) cm. long, 0·8–2·5 cm. wide, bearing 20–60 (or more) flowers; rhachis and calyx reddish, pedicels very short and stout, generally less than 3 mm. long, minutely puberulous; calyx-lobes ovate or deltoid, 1·5–2·5 mm. long, about 1·5 mm. wide, puberulous, apex acute or obtuse, calyx-tube almost obsolete; stamens of dolichostemonous male flowers with glabrous filaments 6–7 mm. long; anthers narrowly oblong 2–2·5 mm. long, 1·5 mm. wide, disk flattish, crenulate, about 4 mm. diam., ovary rudimentary; brachystemonous male flowers similar but filaments less than 2 mm. long and anthers a little smaller, female flowers with minute, ovoid, subsessile staminodes scarcely 0·5 mm. long and 0·4 mm. wide, partly concealed by the disk, ovary 5–6 mm. long, 1·5 mm. wide, arcuate, minutely puberulous, c. 12-ovulate, with a slightly narrower stipe about 2 mm. long, stigma about 2·5–3 mm. diam., sessile or subsessile, convex with deflexed margins (like the pileus of an agaric); hermaphrodite flowers with well-developed stamens and pistils. Pod oblong, 10–30 (or more) cm. long, 1·5–3·5 (or more) cm. wide, compressed, straight or curved, dark shining brown, the base abruptly narrowed to a short stipe, apex blunt or subacute; seeds very hard, ovate-oblong, compressed, 8–10 mm. long, 7–8 mm. wide; testa smooth, dark brown; hilum minute; mesocarp glutinous, with a sweet, rather foetid smell.

HAB.: On dry hillsides in garigue, and in coastal and sub-maritime maquis; often a relic of cultivation or an escape; sea-level to 2,000 ft. alt. (exceptionally to 2,800 ft.), but generally lowland; fl. July–Nov.

DISTR.: Divisions 1–3, 6, 7.

1. Ayios Yeoryios (Akamas), 1962, *Meikle* ! Near Polis, 1962, *Meikle* !
2. Vavatsinia, 1905, *Holmboe* 1166; Stavros tis Psokas valley, *Chapman* 317.
3. Between Mazotos and Limassol, 1862, *Kotschy*; Limassol, *Michaelides*.
6. Dhiorios, 1962, *Meikle* !
7. About Kyrenia, 1862, *Kotschy*; and, 1955, *G. E. Atherton* 700 ! 711 ! 712 !

NOTES: Widely cultivated in lowland areas, except in the Mesaoria, and particularly successful in maritime and sub-maritime areas. The " Carob Survey " (Append. V of the Cyprus Government's " Report on Pasture Research, Survey and Development in Cyprus, 86–88; 1958) gives detailed consideration to the probable limits of economic production of Carobs in the island. Most of the records cited above relate to " wild " populations, but the exact distribution of the spontaneous (and almost certainly indigenous) Carob is still very imperfectly known. The plant is seldom collected, and most specimens are taken from cultivated trees.

In spite of numerous reports and memoranda, it is difficult to believe that all the facts concerning the floral morphology and fertilization of the Carob are as yet finally established. From literature (cited below), and from personal examination, it is evident that at least five types of inflorescence occur:

1. Male inflorescences, the flowers with long filaments and abortive pistils (Dolichostemonous male flowers).
2. Male inflorescences, the flowers with short filaments and abortive pistils (Brachystemonous male flowers).
3. Hermaphrodite inflorescences, the flowers with fully-developed stamens and pistils.
4. Female inflorescences, the flowers with abortive staminodes and fully-developed pistils.

5. Polygamous inflorescences, some of the flowers male, some female, and some hermaphrodite.

Specimens of (1) and (4) are evidently common in Cyprus, but no material with hermaphrodite inflorescences can be found in the very limited range of specimens available.

Since it is reported that isolated female trees can produce pods and viable seeds in the absence of any male trees, one must conclude that the segregation of the sexes (as in Willows and other dioecious plants) is seldom wholly complete, and that a sufficient number of male or hermaphrodite inflorescences are normally produced on an otherwise female tree to effect pollination and fertilization of the female flowers. This would also explain Gennadius' claim that the yield of pods is increased when a male tree is planted in a colony of females. In fact " wild " (indigenous or naturalized) male trees are probably so widespread in Cyprus, that, at least in the majority of areas where Carobs are grown, sufficient pollen is likely to be available to effect pollination of female, cultivated trees. But in the absence of reliable information, all this is surmise.

Apart from Gennadius' excellent paper (quoted above) detailed, and sometimes conflicting, accounts of the pollination and floral morphology of the Carob are to be found in the following publications:

Heikel, E. in Bull. Soc. Bot. Fr., 39: 354–359 (1892).
McLean Thompson, J. in Proc. Linn. Soc., 156: 46–69 (1944).
Albo, G. in Nuov. Giorn. Bot. Ital., n.s., 58: 60–72 (1951).
Francini, E. in Nuov. Giorn. Bot. Ital., n.s., 58: 423–440 (1951).

TRIBE 3. **Cerceae** *Bronn* Leaves simple, entire or 2-lobed, sometimes 2-foliolate; calyx-lobes united into a spathe or tube above the disk, or valvately partite, sometimes imbricate; petals 5; anthers versatile; gynophore free or adnate to calyx-tube; ovules 2–many, seeds with endosperm.

BAUHINIA VARIEGATA *L.*, Sp. Plant., ed. 1, 375 (1753), a native of India, is frequently planted in gardens. It is a tall shrub or small tree 3–5 m. high, with large, simple, suborbicular, bluntly 2-lobed leaves 6–15 cm. diam. The showy purple or mauve-pink flowers are borne in lateral clusters on the leafless branches (March–May). Each flower has a spathaceous, shortly pubescent calyx 2–2·5 cm. long; 5 narrowly obovate, long-clawed, slightly unequal petals 3–6·5 cm. long, 1·5–2·5 cm. wide; fertile stamens 5, with long filaments, ovary long-stipitate, thinly pilose. Pod linear, flattened, pendulous, 15–30 cm. long, 2–2·5 cm. wide, valves dark brown, rather woody.

B. PURPUREA *L.*, Sp. Plant., ed. 1, 375 (1753) is similar, but with more deeply lobed leaves, 3 fertile stamens and with the calyx irregularly split into 5 narrow lobes. In B. ACUMINATA *L.*, Sp. Plant., ed. 1, 375 (1753) the leaves are even more deeply divided, with narrow, subacute lobes; the petals are white and there are 10 fertile stamens. B. TOMENTOSA *L.*, Sp. Plant., ed. 1, 375 (1753), also with 10 fertile stamens, has smaller leaves, and pale sulphur-yellow flowers, the largest (uppermost) petal with a conspicuous maroon, basal blotch. B. GALPINII *N.E. Br.* in Hook. Icon. Plant., 20: t. 1994 (1891) is one of the most decorative species, with relatively small leaves and scarlet or crimson 3-stamened flowers. It is a native of South Africa.

CERCIS SILIQUASTRUM *L.*, Sp. Plant., ed. 1, 374 (1753), the Judas Tree or Redbud, a native of S.E. Europe and western Asia, is rather surprisingly absent from Cyprus as an indigenous plant, though fairly frequent in cultivation. It is a tall, cauliflorous, shrub or smallish tree, 6–10 m. high. Leaves suborbicular-reniform with a deep basal sinus, 6–13 cm. diam., glabrous or subglabrous; stipules small, deltoid, caducous; petiole stout, up to 6 cm. long; flowers in dense lateral clusters along mature twigs and branches, appearing (March–May) slightly in advance of the leaves; pedicels slender 1·5–2 cm. long; calyx cupular, about 6 mm. long, with 5 short, rounded, tomentose, imbricate lobes; petals 5, rosy-purple, papilionaceous, but the 2 abaxial petals free and not connate to form a keel, the adaxial ("standard") internal in bud; stamens 10, free, with long slender filaments, anthers uniform, versatile; ovary stipitate, free from calyx-tube, many-ovulate, style long, upcurved, slender, stigma capitate. Pod flattened, narrowly oblong, acute, 6–11 cm. long, 1·5–2 cm. wide, glabrous or shortly and densely pubescent (var. *hebecarpa* Bornm.) often reddish-purple, dorsal suture narrowly but distinctly winged; seeds 4–10 (or more), strongly compressed, oblong, about 6 mm. long, 4 mm. wide; testa dark fuscous, smooth; hilum small, circular. *Plate 39.*

In gardens at Kyrenia, Nicosia, Dhiorios, Makheras Monastery etc. (*Syngrassides* 1636 ! *G. E. Atherton* 183 !).

SUBFAMILY 3. **Mimosoideae** Mostly trees, shrubs and woody climbers, rarely herbs; leaves generally bipinnate; flowers generally in spikes or heads, or forming shortly pedicellate racemes, actinomorphic, usually 5-merous; petals valvate in bud; stamens as many as or twice as many as the petals, or often numerous, free or adnate to the corolla, the filaments sometimes united into a tube, generally exserted; anthers small, versatile, sometimes with an apical gland. Seeds commonly with a distinct areole (an area on each face of the seed bounded by a fine line or fissure); radicle straight.

The tribes into which *Mimosoideae* is subdivided by Bentham, Hutchinson and other authors cannot be regarded as satisfactory, and have been ignored in the present account. Only one member of the subfamily, *Prosopis farcta* (Banks et Sol.) Macbride, is indigenous in our area.

33. PROSOPIS *L.*

Mantissa 1, 10 (1767).

Spinose or unarmed trees or shrubs; leaves bipinnate, occasionally very reduced or wanting; rhachis glandular above at the insertion of the pinnae; leaflets often numerous, opposite or alternate; flowers in spikes, racemes or heads, hermaphrodite; calyx gamosepalous, with 5 lobes; petals 5, free or connate towards base; stamens 10, anthers with an apical gland (sometimes sessile and inconspicuous). Pod straight, curved or coiled, woody or coriaceous, subterete or ± compressed, internally septate between the seeds; seeds unwinged, with endosperm.

About 37 species, chiefly in America, with three outlying representatives in S. Europe, Mediterranean region, Africa and western Asia.

Sprawling [native] shrub, less than 1 m. high, with far-creeping underground rhizomes; pod thick, globose or oblong, dark brown, 1·5–4 cm. long - - **1. P. farcta**
Erect [introduced] shrub or tree to 15 m. high; pods not as above:
Pods linear, compressed, often torulose, not twisted - [P. JULIFLORA p. 594]
Pods spirally twisted into 6–12 (or more) coils - - [P. PUBESCENS p. 596]

1. P. farcta (*Banks et Sol.*) *Macbride* in Contr. Gray Herb., n.s. 59: 17 (1919); Davis, Fl. Turkey, 3: 9 (1970); Zohary, Fl. Palaest., 2: 30, t. 43 (1972).
Mimosa farcta Banks et Sol. in Russell, Nat. Hist. Aleppo, ed. 2, 2: 266 (1794).
Mimosa stephaniana M. Bieb., Beschreib. Casp., 205 (1800).
Lagonychium stephanianum (M. Bieb.) M. Bieb., Fl. Taur.-Cauc., 3: 288 (1819); Gaudry, Recherches Sci., 188 (1855); Kotschy in Oesterr. Bot. Zeitschr., 12: 277 (1862).
Prosopis stephaniana (M. Bieb.) Kunth ex Spreng., Syst. Veg., 2: 326 (1825); Unger et Kotschy, Die Insel Cypern, 392 (1865); Boiss., Fl. Orient., 2: 634 (1872); Holmboe, Veg. Cypr., 101 (1914); Post, Fl. Pal., ed. 2, 1: 442 (1932); Lindberg f., Iter Cypr., 20 (1946); Chapman, Cyprus Trees and Shrubs, 44 (1949).
TYPE: Syria; Aleppo, *Russell* (BM).

Sprawling or suberect, much-branched, spiny shrub 20–50 cm. high, with tough, far-creeping underground stems; branches slender, subterete, some-

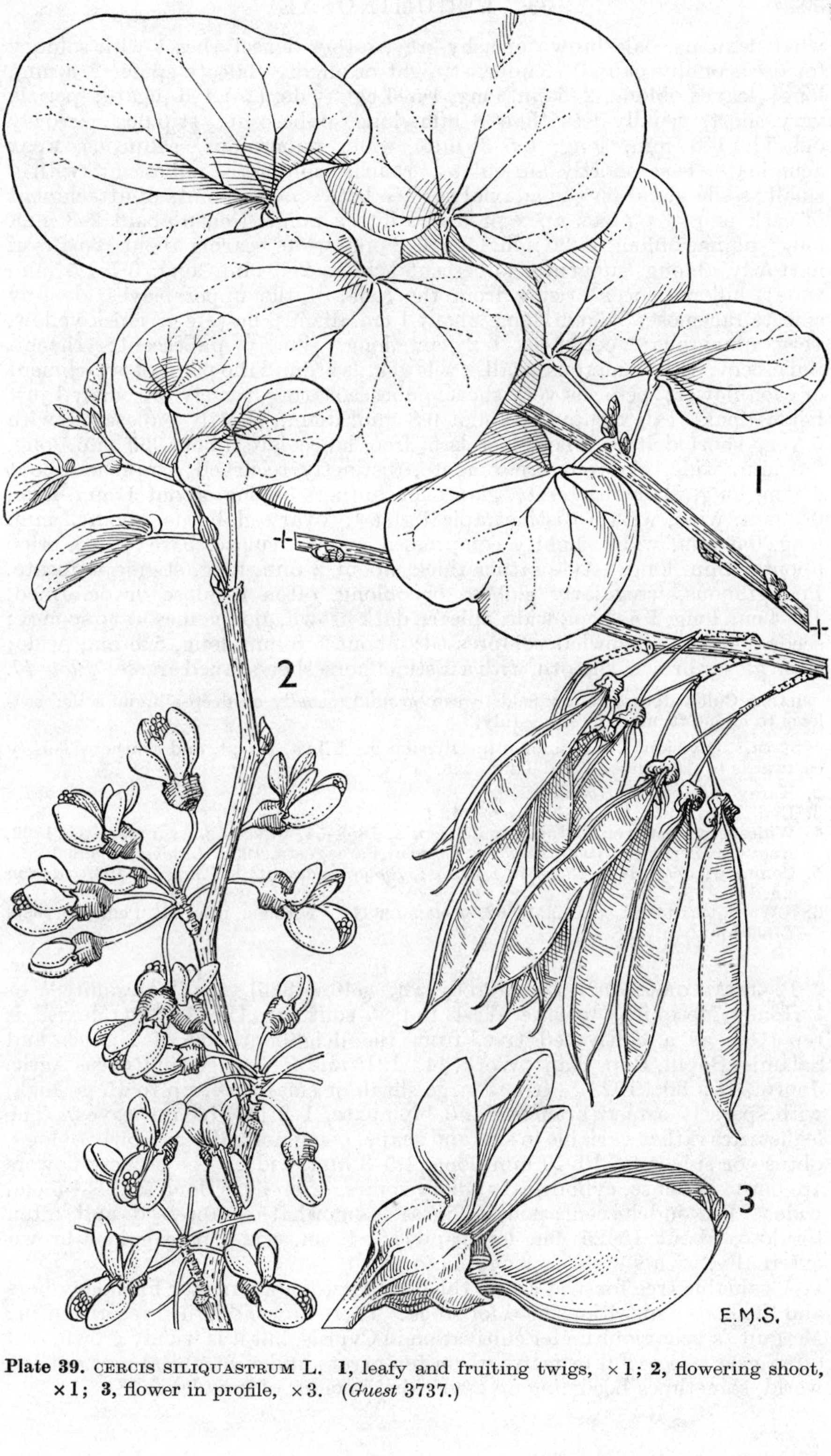

Plate 39. CERCIS SILIQUASTRUM L. **1**, leafy and fruiting twigs, ×1; **2**, flowering shoot, ×1; **3**, flower in profile, ×3. (*Guest* 3737.)

what flexuous, pale brown or ashy-grey, rather densely beset with solitary (or occasionally paired) slender, straight or slightly falcate spines 2–8 mm. long; leaves oblong, 2–5 cm. long, 1·5–5 cm. wide, (2–) 4–6-jugate; petiole very short, usually less than 8 mm. long, pubescent; stipules narrowly deltoid, 1–8 mm. long, 0·5–3 mm. wide, puberulous, caducous, apex acuminate, base shortly auriculate; rhachis subterete, pubescent, with a small sessile gland on the adaxial surface between the points of attachment of each pair of pinnae, apex produced into a pointed cusp about 2–3 mm. long; pinnae oblong, 8–20 mm. long, 3–8 mm. wide, bearing about 12 pairs of narrowly oblong, subacute, pubescent leaflets 2–6 mm. long, 0·7–1·5 mm. wide; inflorescences arising from the axils of the upper leaves, densely spicate-racemose, 3–5 cm. long, about 1 cm. diam.; flowers greenish-yellow, very numerous; peduncle 1–2 cm. long, shortly pubescent; rhachis pubescent, with numerous small sessile glands around the point of attachment of each flower; pedicels very short, pubescent, transversely articulated just below apex; calyx cupular, about 0·8 mm. diam., shortly pubescent, with 5 very short deltoid lobes; petals 5, free, lanceolate, about 3·5 mm. long, 0·8 mm. wide, glabrous, apex acute, distinctly recurved; stamens about 4 mm. long, filaments terete, glabrous; anthers oblong, about 1 mm. long, 0·7 mm. wide, with a distinct apical gland; ovary glabrous, about 2 mm. long, 0·5 mm. wide, slightly compressed, many-ovulate, base with a stipe about 1 mm. long; style rather thick, about 2 mm. long, stigma truncate. Pod gibbous, irregularly globose or oblong, often torulose or deformed, 1·5–4 cm. long, 1·5–2 cm. wide, epicarp dark brown, glossy, mesocarp spongy; seeds obovate, somewhat compressed; about 7–8 mm. long, 5–6 mm. wide; teata glossy brown, smooth, with a distinct horseshoe-shaped areole. *Plate 40.*

HAB.: Cultivated or fallow fields, waste ground, usually on deep alluvial soils; sea-level to c. 500 ft. alt.; fl. June–July.

DISTR.: Divisions 2–6, common in Division 5. Libya, Egypt, and southern Turkey eastwards to Afghanistan and Turkestan.

2. Karavostasi, 1905, *Holmboe* 810.
3. Evdhimou, 1959, *C. E. H. Sparrow* 42 !
4. Widespread between Larnaca and Nicosia, 1853–54, *Gaudry* s.n.; near Kiti, 1862, *Kotschy*; Perivolia, 1939, *Lindberg f.*; near Famagusta, 1939, *Lindberg f.* s.n. !
5. Common; *Sintenis & Rigo* 676 ! 676A ! *Syngrassides* 402 ! *Chapman* 102. *Merton* s.n. !
6. S.W. outskirts of Nicosia, 1862, *Kotschy* 983 ! Nicosia, near R. Pedieos, 1939, *Lindberg f.* s.n.

P. JULIFLORA (*Sw.*) *DC.*, Prodr., 2: 446 (1825), the “Mesquite” or “Honey Mesquite” of the West Indies, southern U.S.A. and Mexico is reported, as a cultivated tree, from the neighbourhood of Nicosia and Salamis (Bovill, Rep. Plant. Work, 14; 1915 and G. Frangos in Cyprus Agric. Journ., 18: 86; 1923). It is a large shrub or small tree, up to 15 m. high, with sparsely armed branches and bipinnate, 1–2 (–3)-jugate leaves. The leaflets are rather variable in size and shape, being normally narrowly oblong, obtuse or subacute, 10–40 mm. long, 1·5–3 mm. wide. The creamy flowers are borne in dense, cylindrical axillary spikes, often 6 cm. long and 1–1·5 cm. wide. The indehiscent pod is linear, somewhat compressed and often torulose, about 15–25 cm. long and 1–1·4 cm. wide, pale golden-brown externally, with a sweet, spongy mesocarp.

A valuable tree for arid areas, the flowers providing nectar for honey-bees and the pods a nutritious food for stock. There are no detailed records of the Mesquite's behaviour under cultivation in Cyprus, but it is widely grown, and often very successful in many warm-temperate and subtropical areas of the world, sometimes becoming an aggressive invader of grasslands.

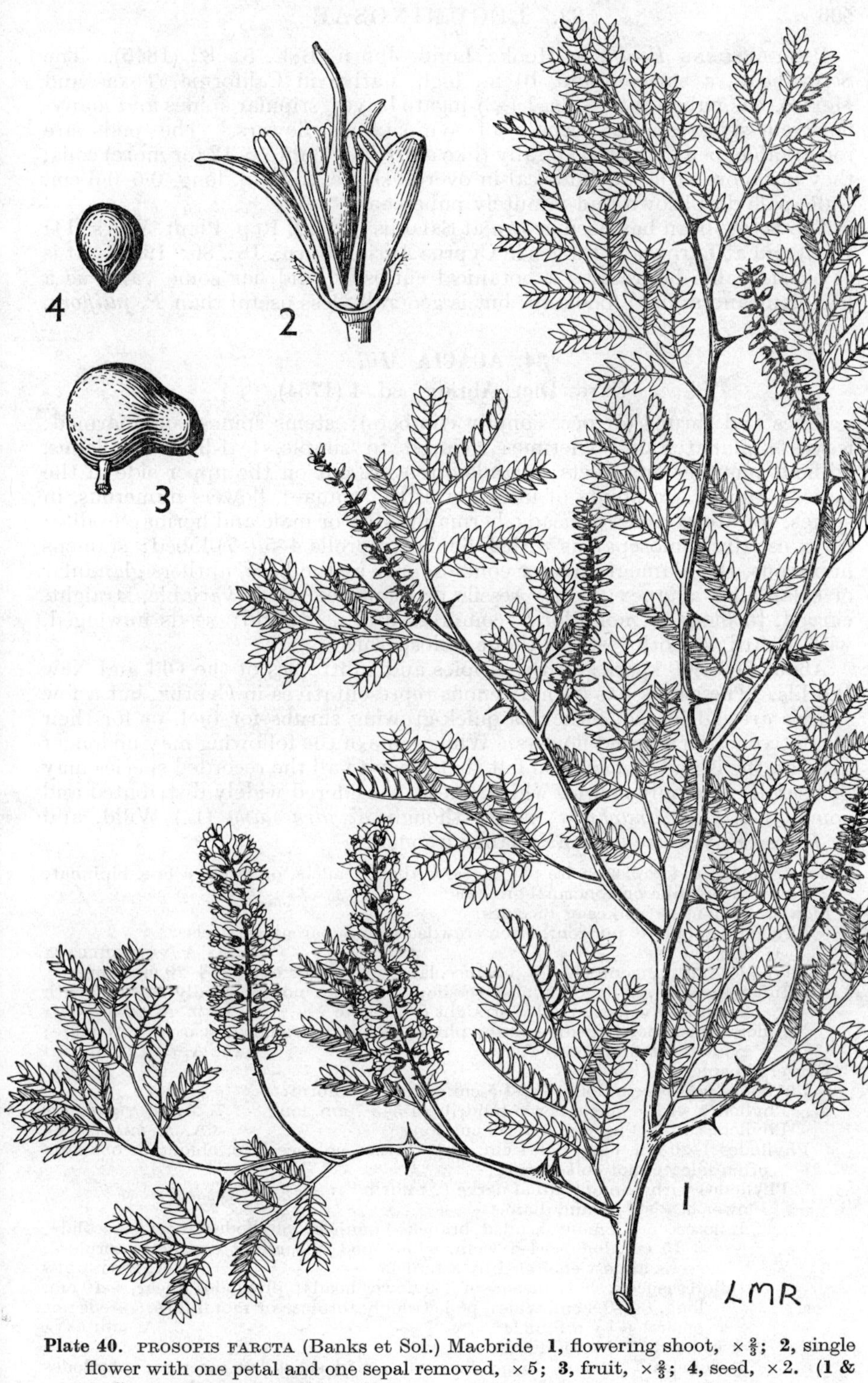

Plate 40. PROSOPIS FARCTA (Banks et Sol.) Macbride **1**, flowering shoot, $\times \frac{2}{3}$; **2**, single flower with one petal and one sepal removed, $\times 5$; **3**, fruit, $\times \frac{2}{3}$; **4**, seed, $\times 2$. (**1 & 2**, *Guest* 4401; **3 & 4**, *Guest* 490.)

P. PUBESCENS *Benth.* in Hook., Lond. Journ. Bot., 5: 82 (1845). The Screw-bean, a shrub about 10 m. high, native in California, Texas and Mexico, has pubescent stems, 1 (–2)-jugate leaves, stipular spines and many-flowered spicate inflorescences of sympetalous flowers. The pods are remarkable, being twisted spirally (like a *Medicago*) into 6–12 (or more) coils; they are more or less cylindrical in overall shape, 2–4 cm. long, 0·5–0·6 cm. wide, pale dull brown and minutely pubescent.

The Screw-bean has been grown at Salamis (Bovill, Rep. Plant. Work. 14; 1915) and at Larnaca (Frangos in Cyprus Agric. Journ., 18: 86; 1923). It is of considerable interest as a botanical curiosity, and has some value as a bee-plant and animal foodstuff, but is generally less useful than *P. juliflora*.

34. ACACIA *Mill.*

Gard. Dict. Abridg., ed. 4 (1754).

Trees and shrubs (or occasionally climbers); stems spinose or unarmed; leaves bipinnate, or sometimes reduced to simple, leaf-like, phyllodes, without pinnules or leaflets; glands often present on the upper side of the petiole, and at the points of insertion of the pinnae; flowers numerous, in spikes, spicate racemes or heads, hermaphrodite or male and hermaphrodite; calyx usually gamosepalous with 4–5 lobes; corolla 4–5 (–7)-lobed; stamens numerous, the filaments free or connate only at the base, anthers glandular or eglandular at apex; ovary sessile or stipitate. Pods variable, straight, curved, torulose or moniliform, compressed or ± turgid; seeds unwinged, with a hard, smooth testa, without endosperm.

About 700–800 species in the tropics and subtropics of the Old and New Worlds. The genus has no indigenous representatives in Cyprus, but a few species are cultivated, either as quick-growing shrubs for fuel, or for their attractive, often scented flowers. While some of the following may no longer exist on the island, it has been felt that a key to all the recorded species may be useful. The only species which can be considered widely distributed and common is *A. cyanophylla* Lindl., though *A. farnesiana* (L.) Willd. and *A. karroo* Hayne are evidently fairly frequent.

Leaves reduced to simple, entire phyllodes, without leaflets, or rarely a few, bipinnate leaves in juvenile or abnormal growths:
- Flowers in elongate spikes or racemes:
  - Phyllodes subulate, pungent, often crowded into apparent verticels - 1. A. VERTICILLATA
  - Phyllodes not pungent; linear, oblanceolate, obovate or oblong, 4–20 cm. long:
    - Inflorescences sessile or almost sessile; phyllodes not markedly falcate, with reticulate venation; pod straight or arcuate - - 9. A. LONGIFOLIA
    - Infloresceness shortly pedunculate; phyllodes falcate, venation scarcely reticulate; pods twisted and curled - - - - - - 10. A. CUNNINGHAMII
- Flowers in spherical heads:
  - Phyllodes 0·5–3·7 cm. long, 0·4–1·5 cm. wide, cultriform:
    - Phyllodes with a single nerve (midrib), 1·5–3·7 cm. long - 7. A. CULTRIFORMIS
    - Phyllodes with 2 nerves, 0·5–1·3 cm. long - - - - 6A. A. PRAVISSIMA
  - Phyllodes 4–20 cm. long, 0·2–4 cm. wide, linear, oblanceolate, oblong or obovate, often falcate, not cultriform:
    - Phyllodes with 1 longitudinal nerve (" midrib "):
      - Flower heads 3–5 mm. diam.:
        - Inflorescence a many-headed, branched panicle; phyllodes linear-lanceolate, 6–15 cm. long, 0·4–1·3 cm. wide; pod linear, not distinctly torulose; seeds almost encircled by a funicle - - - - 2. A. RETINODES
        - Inflorescence a short raceme of 2–6 flower-heads; phyllodes linear, 4–10 cm. long, 0·2–0·8 cm. wide; pod strongly torulose or moniliform; seeds not encircled by a funicle - - - - - - 5. A. LIGULATA
      - Flower-heads 7–9 mm. diam.:
        - Peduncles 7–20 mm. long; flower-heads in short lax racemes; phyllodes straight or slightly falcate - - - - - - 3. A. CYANOPHYLLA

Peduncles usually less than 5 mm. long; flower-heads in narrow elongate racemes; phyllodes strongly falcate - - - - 4. A. PYCNANTHA

Phyllodes with 3–5 (–7) longitudinal nerves:

Phyllodes 0·5–1·5 cm. wide; inflorescence of 1–4 flower-heads; pods 1–1·2 cm. wide (true leaves rarely, if ever, present) - - - - 7. A. CYCLOPS

Phyllodes 0·6–3·5 cm. wide; inflorescence of 1–6 flower-heads; pods 0·4–0·7 cm. wide (true bipinnate leaves occasionally produced on " reversion " shoots) 8. A. MELANOXYLON

Leaves all bipinnate, not reduced to simple phyllodes:

Flower-heads spherical:

Involucel concealed under flowers at apex of peduncle; pods thick, indehiscent:

Leaflets 2–7 mm. long, 0·7–1·7 mm. wide, with rather obscure venation; peduncles pubescent; pods glabrous, dark brown - - - 11. A. FARNESIANA

Leaflets 9–13 mm. long, 2–2·5 mm. wide, with prominent venation; peduncles glabrous; pods densely and closely grey-tomentose 12. A. ERIOLOBA

Involucel situated towards base or about the middle of peduncle, conspicuous; pods strongly compressed and dehiscent:

Stipular spines often 3–15 cm. long; flower-heads in lateral and terminal cylindrical or conical panicles, each head about 10 mm. diam.; pod not strongly torulose - - - - - - - - - 13. A. KARROO

Stipular spines to about 2 cm. long; flower-heads all lateral, about 8 mm. diam.; pods strongly torulose - - - - - - 14. A. CONSTRICTA

Flower-heads elongate, spiciform, not spherical:

Twigs armed with scattered prickles; stipules not spinose; flower-spikes 1·5–4 cm. long; pods flattened, light brown, often torulose or moniliform 15. A. GREGGII

Twigs without scattered prickles; stipules spinose; flower-spikes 3·5–14 cm. long; pods thick, curled but not torulose, orange when ripe - - 16. A. ALBIDA

1. A. VERTICILLATA (*L'Hérit.*) *Willd.*, Sp. Plant., 4: 1049 (1806); Bovill, Rep. Plant. Work Cyprus, 15 (1915).
Mimosa verticillata L'Hérit., Sert. Angl., fasc. 1: 30 (1789).

Erect shrub about 3–6 m. high; twigs sulcate, pubescent, densely clothed with pungent, subulate phyllodes, 10–15 mm. long, 0·8–3 mm. wide, in false whorls or irregularly scattered; inflorescences solitary, spicate, cylindrical, shortly pedunculate, about 15 mm. long, 4–5 mm. wide. Pods linear, straight or curved, strongly compressed, 20–70 mm. long, about 4 mm. wide, apex blunt or subacute, valves subglabrous or sparsley pubescent, dark brown; seeds oblong, 4 mm. long, 2 mm. wide; testa dark brown; funicle irregularly twisted, not circinnate.

A native of south Australia and Tasmania, frequently cultivated, and reported by Bovill to have been planted at Salamis.

2. A. RETINODES *Schlechtend.* in Linnaea, 20: 664 (1847); Frangos in Cyprus Agric. Journ., 18: 86 (1923); Davis, Fl. Turkey, 3: 9 (1970).

Erect shrub about 5 m. high; twigs glabrous, distinctly striate; phyllodes narrowly linear-lanceolate, straight or slightly falcate, glabrous, acute, 6–15 cm. long, 0·4–1·3 cm. wide, with 1 longitudinal nerve and numerous, obscure, lateral veins; inflorescence a branched panicle of spherical heads, each 3–5 mm. diam.; peduncles short, 3–6 mm. long. Pods linear, 60–85 mm. long, 4–7 mm. wide, strongly compressed, valves greyish, obscurely reticulate, glabrous; seeds oblong, about 5 mm. long, 3–4 mm. wide; testa dark fuscous-brown; funicle almost encircling seed.

A native of southern Australia, often cultivated, and recorded from the grounds of the Department of Agriculture, Nicosia. It much resembles *A. cyanophylla* Lindl., with which it is sometimes confused, but is distinguished from the latter by its narrower phyllodes, smaller flower-heads and shorter peduncles.

3. A. CYANOPHYLLA *Lindl.*, Bot. Reg., 25: append., 45 (1839); Hutchins, Rep. Cyprus Forestry, 73–75 (1909); Bovill, Rep. Plant. Work Cyprus, 13 (1915); Unwin,

Forests of Cyprus, 12 (1927); Druce in Rep. B.E.C., 9: 470 (1931); Waterer, Island Cyprus, 153 (1947); Davis, Fl. Turkey, 3: 9 (1970); Osorio-Tafall et Seraphim, List Vasc. Plants Cyprus, 54 (1973).

[? *A. saligna* (non (Labill.) Wendl.) Bovill, Rep. Plant. Work Cyprus, 14 (1915).]

Diffuse shrub 3–5 m. high; twigs striate, glabrous; phyllodes oblanceolate or almost linear, straight or slightly falcate, glaucous, 8–20 cm. long, 0·4–3 (–4) cm. wide, with 1 longitudinal nerve, venation rather obscure, apex blunt or acuminate; inflorescence a short rather lax axillary raceme, peduncles 7–20 mm. long, flower heads 3–9, spherical, each 7–9 mm. diam. Pod linear, strongly compressed, straight or curved, distinctly torulose, 6–12 cm. long, 0·5 cm. wide, valves dark brown with conspicuously thickened, pale brown sutures; seeds oblong, about 5 mm. long, 3 mm. wide; testa dark glossy brown; funicle short, curved, much thickened towards point of attachment to seed.

A native of south-west Australia, extensively planted elsewhere, and valuable as a sand-binder on maritime dunes, and as a source of fire-wood. Recorded from Ayios Memnon (Bovill), Salamis (Bovill), Syrianokhori (*Davis* 2599 !) Limassol (*Barrington* 29 !) Karokoumi (*G. E. Atherton* 1233 !), Yialia (*Meikle* !), but widespread and common as an introduced plant in most lowland areas of Cyprus.

4. A. PYCNANTHA *Benth.* in Hook., Lond. Journ. Bot., 1: 351 (1842); G. Frangos in Cyprus Agric. Journ., 18: 86 (1923); Osorio-Tafall et Seraphim, List Vasc. Plants Cyprus, 54 (1973).

Similar to *A. cyanophylla* Lindl. in general habit and appearance, but phyllodes more strongly falcate, the two longitudinal divisions of the lamina (on either side of the median nerve) often markedly unequal; peduncles very short, seldom exceeding 5 mm.; flower heads commonly 10–12 (–20) in narrow, spiciform racemes, spherical, each about 7–9 mm. diam. Pod linear, torulose, strongly compressed, 6–10 cm. long, 0·7–0·8 cm. wide; seeds as in *A. cyanophylla*, the funicle apparently somewhat longer and more tapering.

A native of south Australia, rather less widely introduced than *A. cyanophylla*, but recorded (by Frangos) from Government House grounds, Nicosia. It is a very ornamental species, said to furnish one of the richest of tanning barks, and to be useful as a sand-binder.

5. A. LIGULATA *A. Cunn. ex Benth.* in Hook., Lond. Journ. Bot., 1: 362 (1842); Osorio-Tafall et Seraphim, List Vasc. Plants Cyprus, 54 (1973).

A spreading shrub 1–3·5 m. high; twigs glabrous or glabrescent, distinctly angled when young; phyllodes linear, 4–10 cm. long, 0·2–0·8 mm. wide, glabrous, glaucous, with one longitudinal nerve, venation obscure, apex generally blunt with a short oblique mucro; inflorescence a very short axillary raceme, bearing 2–6 spherical flower-heads, each about 5 mm. diam.; peduncles about 5–8 mm. long. Pod linear, compressed, glabrous, often strongly torulose and breaking transversely into short 1-seeded joints, valves dull brown with rather thickened sutures; seeds oblong, about 4–6 mm. long, 2·5–4 mm. wide; testa smooth, dark brown; funicle twisted, yellow or red, forming a compact aril at the base of the seed.

Widespread in Australia; rarely cultivated but reported from the grounds of the Department of Agriculture, Nicosia (*Syngrassides* 1307 !); also evidently naturalized on " dry hillside " at " Sayers Beach ", ½ mile west of Kyrenia (*G. E. Atherton* 241 !); probably an escape from some garden in the vicinity. The Cyprus material is not quite typical, but too incomplete and imperfect for critical examination.

6. A. cultriformis *A. Cunn. ex G. Don*, Gen. Syst., 2: 406 (1832); Bovill, Rep. Plant. Work Cyprus, 15 (1915); G. Frangos, Cyprus Agric. Journ., 18: 85 (1923).

An erect or spreading shrub, 2–3 m. high; twigs glabrous; phyllodes cultriform (shaped like a short, broad scimitar), numerous, often glaucous, glabrous, 1-nerved with a prominent gland about midway along the adaxial margin, 1·5–2·5 (–3·7) cm. long, 0·4–1·5 cm. wide, apex obtuse or acute, mucronate, venation prominent, reticulate; inflorescences racemose, axillary, numerous along the upper parts of the stems; flower heads (5–) 10–12 (–18), spherical, 3–4 mm. diam.; peduncles short, usually less than 4 mm. long. Pods linear, compressed, torulose, 4–8 cm. long, 0·6–0·7 cm. wide, valves fuscous, with thickened, paler sutures; seeds oblong, about 5 mm. long, 3 mm. wide; testa dark shining brown; funicle about 5 mm. long, adpressed to seed, tapering to tip from point of attachment to seed.

Recorded from Salamis plantation and Nicosia, and probably grown elsewhere in Cyprus. A native of New South Wales, distinct on account of its numerous, asymmetric, rather small phyllodes.

6A. A. pravissima *F. Muell.* Fragm. Phytogr. Austral., 1: 5 (1858); Osorio-Tafall et Seraphim, List Vasc. Plants Cyprus, 54 (1973).

A spreading shrub 1–2 m. high or occasionally a small tree up to 6 m. high; twigs subglabrous or sparsely pubescent; phyllodes broadly cultriform, numerous, green, glabrous, 2-nerved, with a prominent gland about midway along the adaxial margin, 0·5–1·3 cm. long, 0·4–1·3 cm. wide, apex obtuse or acute, mucronate, venation usually obscure; inflorescences racemose, axillary, numerous along the upper parts of the stems, flower heads (5–) 10–15, spherical, 2–4 mm. diam.; peduncles short, usually less than 4 mm. long. Pods linear, compressed, rather torulose, 4–8 cm. long, 0·6–0·8 cm. wide, valves brownish or fuscous, with slightly thickened, paler, sutures; seeds oblong, about 5 mm. long, 3 mm. wide; testa blackish-brown; funicle about 3–4 mm. long, adpressed to seed, tapering to tip from point of attachment to seed.

Seen at Stavros tis Psokas and elsewhere on the island in 1974, but evidently a recent introduction. A native of Victoria and New South Wales, closely related to *A. cultriformis* but distinguished by its smaller, rather weakly 2-nerved phyllodes.

7. A. cyclops *A. Cunn. ex G. Don*, Gen. Syst., 2: 404 (1832) Bovill, Rep. Plant. Work Cyprus, 14 (1915); G. Frangos in Cyprus Agric. Journ., 18: 86 (1923); Osorio-Tafall et Seraphim, List Vasc. Plants Cyprus, 54 (1973).

A low shrub about 1–2 m. high; twigs glabrous, at first angular; phyllodes lanceolate or oblanceolate, straight or somewhat falcate, 5–10 cm. long, 0·5–1·5 cm. wide, glabrous, with 3–5 (–7) longitudinal nerves, venation distinct, reticulate, apex acute, usually twisted obliquely; flower heads solitary or forming a lax 2–4-flowered axillary raceme, spherical, 5–6 mm. diam.; peduncles 5–8 mm. long. Pods linear or narrowly oblong, strongly compressed, twisted circinnately, blunt or shortly acute, 5–10 cm. long, 1–1·2 cm. wide, valves dark brown, fissured transversely at maturity; seeds oblong, about 5 mm. long, 3·5 mm. wide; testa shining brown; funicle orange or scarlet, more than 2 cm. long, bent back on itself and almost encircling the seed.

A native of south-west Australia, occasionally cultivated elsewhere, and reported from the Department of Agriculture grounds, Nicosia, and from the plantations at Salamis. It may well occur in additional localities.

The conspicuous, waxy funicle, and the multinerved phyllodes serve to distinguish it from *A. cyanophylla* and its allies.

8. A. MELANOXYLON *R. Br.* in Aiton f., Hort. Kew., ed. 2, 5: 462 (1813); G. Frangos in Cyprus Agric. Journ., 18: 86 (1923).

A tall shrub or tree 4–20 m. high; twigs glabrous or pubescent, angular; leaves all reduced to phyllodes, or occasionally some of them bipinnately 1–3-jugate, with numerous small leaflets; phyllodes oblanceolate or narrowly obovate, 8–10 cm. long, 0·6–3·5 cm. wide, glabrous, longitudinally 3–5 (–7)-nerved, venation conspicuous, reticulate, apex acute or obtuse; true leaves with a narrowly winged rhachis, pinnae 4–9 cm. long, bearing about 40 narrowly oblong, obtuse leaflets about 8 mm. long and 2·5 mm. wide; flower heads 1–6 in short, rather congested axillary racemes, spherical, 5–8 mm. diam.; peduncles usually less than 4 mm. long. Pods linear, compressed, much twisted and curled, 10 cm. or more long, 0·4–0·7 cm. wide, apex blunt, valves dark brown with paler, thickened sutures; seeds oblong, about 4 mm. long, 2·5 mm. wide; testa dark shining brown; funicle more than 1 cm. long, bent back on itself and almost completely encircling the seed.

Widely distributed in eastern Australia and in Tasmania, and long in cultivation elsewhere. Stated by Frangos to have been planted in the grounds of the Agricultural School, Nicosia, also between Asinou Church and Ayios Theodhoros (*Meikle* 5014 !), and perhaps grown elsewhere in Cyprus. The two kinds of foliage are an interesting peculiarity of the tree.

9. A. LONGIFOLIA (*Andr.*) *Willd.*, Sp. Plant., 4: 1052 (1806); G. Frangos in Cyprus Agric. Journ., 18: 86 (1923); Davis, Fl. Turkey, 3: 9 (1970).
Mimosa longifolia Andr., Bot. Reposit., 3: t. 207 (1802).

Erect, spreading or pendulous shrub or small tree, 2–8 m. high; twigs glabrous or glabrescent, angular; phyllodes very variable, linear, oblanceolate, narrowly oblong or narrowly obovate, 4–20 cm. long, 0·2–4 cm. wide, glabrous, obtuse or acute, with 2–3 (or more) longitudinal nerves and obscure or prominent reticulate venation; inflorescences axillary, solitary or in clusters of 2–5 (or more), spicate, cylindrical, 2–5 cm. long, 0·4–0·5 cm. wide, lax or dense, consisting of numerous, small, crowded flower heads; peduncles wanting or very short. Pods linear, somewhat torulose, straight or arcuate, 5–9 cm. long, 0·5 cm. wide, valves dark brown, glabrous or minutely adpressed-strigillose, longitudinally rugose; seeds oblong, not much compressed, about 6 mm. long, 2·5 mm. wide; testa dark, shining blackish-brown; funicle twisted, compact, forming a whitish, waxy aril about base of seed.

Widespread in eastern and southern Australia, also in Tasmania, and not infrequently introduced elsewhere. Recorded from a garden in Nicosia by Frangos.

ACACIA SOPHORAE (*Labill.*) *R. Br.* in Aiton f., Hort. Kew. ed. 2, 5: 462 (1813)—*Mimosa sophorae* Labill., Nov. Holl. Plant., 2: 87, t. 237 (1806)—is recorded by Frangos (Cyprus Agric. Journ., 18: 86 (1923) from Nicosia. It is frequently listed as a variety of *A. longifolia* (var. *sophorae* (Labill.) F. Muell.), and differs from typical *longifolia* principally in having shorter, broader, obovate phyllodes with rather prominent, reticulate venation.

10. A. CUNNINGHAMII *Hook.* in Hook. Icon. Plant., 2: t. 165 (1837); G. Frangos in Cyprus Agric. Journ., 18: 86 (1923).

An erect shrub or small tree, to about 8 m. high; twigs strongly angled, glabrous or thinly pubescent; phyllodes oblanceolate, falcate, 7–20 cm. long, 0·7–4 cm. wide, apex acute or acuminate, longitudinal nerves 3 or more, venation distinct, mostly parallel to nerves, not reticulate; inflorescences axillary, solitary, or in clusters of 2 or more, spicate, cylindrical, up to 10 cm.

long, about 0·5 cm. wide, flowers generally congested; peduncles short, stout, about 5–7 mm. long. Pods linear, twisted and curled, somewhat torulose, 6–12 cm. long (sometimes longer), 0·3–0·5 mm. wide, subterete, valves dark brown, rugose; seeds oblong, about 5 mm. long, 2 mm. wide, not much compressed; testa dark shining blackish-brown; funicle twisted, compact, forming a whitish waxy aril about base of seed.

A common species in eastern and southern Australia, recorded by Frangos from nursery gardens at Nicosia, as *A. cunninghamiana*.

11. A. FARNESIANA (*L.*) *Willd.*, Sp. Plant., 4: 1083 (1806); Unger et Kotschy, Die Insel Cypern, 391 (1865); Osorio-Tafall et Seraphim, List Vasc. Plants Cyprus, 54 (1973).
Mimosa farnesiana L., Sp. Plant., ed. 1, 521 (1753).

Erect or spreading shrub, 2–5 m. high; twigs rather flexuous, dark grey-brown, longitudinally striate, somewhat furfuraceous and rough; stipules pungent, spinose, straight, 5–15 mm. long; leaves oblong, bipinnate, 3–7 cm. long, 1·5–3 cm. wide, pinnae 2–7-jugate; leaflets in 10–21 pairs, narrowly oblong, 2–7 mm. long, 0·7–1·7 mm. wide, subglabrous, nervation and venation rather indistinct; rhachis and rhachilla thinly pubescent; petiole slender, 0·5–1·5 cm. long. Inflorescences axillary, solitary or in small clusters; peduncle slender, pubescent, 2–4 cm. long with an involucel at its apex; flowers golden yellow, fragrant, in spherical heads 9–13 mm. diam. Pod thick, indehiscent, narrowly oblong, straight or curved, scarcely compressed, 4–7 cm. long, 0·9–1·5 (–2) cm. wide, valves glabrous, dark blackish brown, minutely striate longitudinally; seeds ellipsoid-oblong, 7–8 mm. long, 5–6 mm. wide; testa smooth, dark brown; funicle inconspicuous.

Generally considered a native of tropical America, but widely cultivated in the tropics and subtropics, the flowers being used in perfumery. Not infrequently planted in Cyprus gardens, and recorded from Larnaca (*Kotschy*), Syngrasis (*Syngrassides* 1005 !) Nicosia (*Syngrassides* 1311 !) and Kyrenia (*G. E. Atherton* 464 !). The thick, indehiscent pods are distinctive.

12. A. ERIOLOBA *E. Mey.*, Comment. Plant. Afr. Austr., 1: 171 (1836).
[*A. giraffae* (non Willd.) Bovill, Rep. Plant. Work Cyprus, 14 (1915) et auctt. mult.]

Similar in general aspect to *A. farnesiana* (L.) Willd., but altogether more robust, sometimes up to 15 m. high; stipules spinose, straight, pallid, 2–6 (or more) cm. long, often swollen and hollow at the base and forming conspicuous " ant-galls " (a feature of several African species); leaves bipinnate, pinnae 1–4-jugate, leaflets narrowly oblong, blunt, glaucous, glabrous, 9–13 mm. long, 2–2·5 mm. wide, venation rather prominent; inflorescences solitary or in small axillary clusters; peduncle glabrous, up to 6 cm. long with an involucel at its tip; flowers golden-yellow in spherical heads 10–13 mm. diam. Pod thick, oblong, compressed, indehiscent, usually falcate, 6–12 cm. long, 3–5 cm. wide, valves densely and closely grey-tomentose; seeds elliptical or suborbicular, 10–12 mm. long, up to 12 mm. wide; testa dark shining brown, funicle inconspicuous.

A common species in many parts of South Africa, not much cultivated, but recorded by Bovill from Salamis plantation, where it was probably grown experimentally. There are no recent records from Cyprus.

13. A. KARROO *Hayne*, Arzneykunde Gewächse, 10: t. 33 (1827).
A. armata (non R. Br.) G. Frangos in Cyprus Agric. Journ., 18: 85–86 (1923).

A shrub or small tree 2–7 m. high; twigs pale grey-brown or dark brown, longitudinally striatulate, often distinctly lenticellate; stipules spinose, often very large and wicked, whitish, straight, 3–15 cm. long; leaves bipinnate, pinnae 2–6-jugate, petiole short; rhachis subglabrous or glabrous, strongly sulcate above, with conspicuous peltate glands at the points of insertion of the pinnae; leaflets in 5–20 pairs, oblong, blunt, glabrous or subglabrous, 5–10 mm. long, 1·5–5 mm. wide; nervation and venation obscure; inflorescences terminal and lateral, the lateral in small axillary clusters, the terminal in elongate (15–20 cm. or more long) cylindrical or conical, many-flowered panicles; peduncle pubescent or subglabrous, glandular, about 2 cm. long, with a conspicuous involucel about 6–7 mm. from base; flowers golden-yellow, fragrant, in spherical heads about 1 cm. diam. Pod linear, compressed, dehiscent, strongly arcuate, and usually somewhat torulose, 6–15 cm. long, about 0·7 cm. wide, valves dark or mid-brown, glabrous, thinly glandular, obscurely striate longitudinally; seeds oblong, somewhat compressed, 6–8 mm. long, 4 mm. wide; testa smooth, olive-green; funicle slender, rather inconspicuous.

A native of South Africa, planted in Nicosia (*Syngrassides* 1309 !) Kyrenia (*G. E. Atherton* 25 !) and Larnaca (*Frangos*), and probably grown elsewhere in Cyprus.

The fragrant flowers are attractive, and the spiny bushes can be clipped to make an impenetrable hedge.

[ACACIA ARMATA *R. Br.* from Australia is quite distinct, with small, simple phyllodes, and without true leaves.]

14. A. CONSTRICTA *Benth.* in A. Gray, Plant. Wright., 1: 66 (1852); Bovill, Rep. Plant. Work Cyprus, 14 (1915).

A shrub or small tree 3–14 m. high; twigs dark brown, glandular, striatulate; stipules spinose, straight, whitish, to about 2 cm. long, sometimes wanting or very short; leaves bipinnate, 1·5–5 cm. long, 1–3 cm. wide, pinnae 4–6-jugate; petiole 5–10 mm. long; rhachis glabrous or pubescent, sulcate above and glandular at the points of insertion of the pinnae; leaflets in 8–10 pairs, small, oblong, blunt, about 1·5–2·5 mm. long, 0·8 mm. wide, glabrous or pubescent, nervation indistinct; inflorescences solitary or in clusters of 2–4, axillary; peduncle slender, 1·5–2·5 cm. long, glabrous or pubescent, glandular; involucel about half way along peduncle; flowers yellow, in spherical heads about 8 mm. diam. Pods linear, strongly torulose, straight or curved, somewhat compressed, 5–12 cm. long, 0·4–0·6 cm. wide, valves bright brown, longitudinally striate; seeds elliptical or ovate, compressed, 5–8 mm. long, 3–3·5 mm. wide; testa mottled dark and paler grey-brown, smooth; funicle minute.

A native of Mexico, said to have been planted at Salamis.

15. A. GREGGII *A. Gray*, Plant. Wright., 1: 65 (1852); Bovill, Rep. Plant. Work Cyprus, 14 (1915).

A spreading shrub 1·5–5 m. high; twigs greyish, distinctly striate, pubescent, with scattered, arcuate, broad-based prickles; stipules small, not spinose; leaves bipinnate, 1·5–2 cm. long, 1·5–3 cm. wide, pinnae 1–3-jugate; petiole 5–8 mm. long; rhachis sulcate above, distinctly glandular at points of insertion of pinnae, pubescent with spreading hairs; leaflets oblong, blunt, pubescent, 2–6 mm. long, 1–3·5 mm. wide, nervation obscure; inflorescences solitary or in axillary clusters of 2–4 heads; peduncle 1·5–2 cm. long, pubescent with spreading hairs, sparsely glandular, longitudinally striate, accrescent in fruit; flowers in cylindrical spikes 1·5–4 cm. long, 0·8 cm. wide. Pods flattened, oblong, usually twisted, often torulose or moniliform, 6–15 cm. long, 1·5–2·5 cm. wide, valves light brown, glabrous or thinly pubescent, reticulate-veined; seeds lenticular, 7–9 mm. diam.; testa dark brown or fuscous, smooth; funicle very slender, inconspicuous.

A native of southernmost U.S.A. and Mexico, said by Bovill to have been planted at Salamis.

16. A. ALBIDA *Del.*, Fl. Égypt. Illustr., 142, t. 52, fig. 3 (1813); G. Frangos in Cyprus Agric. Journ., 18: 85 (1923); Zohary, Fl. Palaest., 2: 27, t. 38 (1972); Osorio-Tafall et Seraphim, List Vasc. Plants Cyprus, 54 (1973).

Tree 6–30 m. high with a spreading crown and rough, dark grey-brown bark; twigs pallid, greyish or greenish-white, glabrous or pubescent; stipules spinose, often very short and conical, rarely up to 2 cm. long; leaves bipinnate, 5–7 cm. long, 3–6 cm. wide, pinnae 3–10-jugate; petiole short, usually under 1 cm. long; rhachis glabrous or pubescent, with a conspicuous gland at the point of attachment of the pinnae; leaflets in 6–23 pairs, oblong, blunt or shortly apiculate, glabrous or pubescent, 3–6 (–12) mm. long, 0·7–2·5 (–4) mm. wide, venation obscure; inflorescence solitary, axillary; peduncles glabrous or hairy, 1·3–4 cm. long; flowers creamy-white, in elongate spikes 3·5–14 cm. long, 0·8–1 cm. wide. Pods oblong, thick, indehiscent, glabrous or rarely puberulous, bright orange when ripe, 6–25 cm. long, 2–4 cm. wide, often falcate or circinnately curled; seeds oblong-orbicular, 9–11 mm. long, 6–8 mm. wide; testa dark brown, shining, funicle inconspicuous.

Widespread in tropical and subtropical Africa, also considered native in Egypt and Palestine. Its status in Cyprus is uncertain: several records from the Khrysokhou Valley (Skoulli, *Frangos*; Polis–Ktima road, *Syngrassides* 843 ! *Chapman* 145) suggest that it is established, or even possibly spontaneous in this area; but it is certainly an introduction at Famagusta (Ayia Ekaterina, *Frangos*) and may well be an introduced tree throughout our area. The pods provide a useful stock-food.

ALBIZIA *Durazz.*

in Magazz. Tosc., 3 (4) (vol. 12): 10, 13, illustr. (1772).

Trees, shrubs or rarely climbers; twigs generally unarmed; leaves bipinnate, pinnae with one to many pairs of leaflets; inflorescences in spherical heads or in spikes or spiciform racemes, axillary, often in clusters near the ends of the branches; flowers usually hermaphrodite, or the central flowers in each head sometimes larger than the others and male; calyx gamosepalous with (4–) 5 (–7) lobes; corolla gamopetalous, infundibuliform or campanulate, usually with 5 lobes; stamens numerous, the filaments connate in their lower part and forming a slender tube. Pods oblong, straight, flattened, usually dehiscent, continuous within; seeds usually flattened.

About 100–150 species widely distributed in the tropics and subtropics; not represented by any indigenous species in Cyprus, but some of the more ornamental species are widespread in cultivation.

Flower-heads spherical:
Filaments rose-pink above, whitish below; leaflets in 15–30 pairs along each pinna, each 6–20 mm. long, 2–5 mm. wide; pods 7–8 cm. long, 2·5–3·5 cm. wide
1. A. JULIBRISSIN
Filaments greenish-white or yellowish; leaflets in 3–11 pairs along each pinna, each 15–48 mm. long, 8–30 mm. wide; pods 15–33 cm. long, 3–6 cm. wide
2. A. LEBBECK
Flower-heads spiciform-racemose, cylindrical; filaments creamy-white; leaflets small, in 20–40 pairs along each pinna - - - - - - - 3. A. LOPHANTHA

1. A. JULIBRISSIN *Durazz.* in Magazz. Tosc., 3 (4): 11 (1772); Bovill, Rep. Plant. Work Cyprus, 15 (1915)—as *A. julibriscium*; G. Frangos in Cyprus Agric. Journ., 18: 86 (1923)—as *A. ghulimbrischium*; Davis, Fl. Turkey, 3: 10 (1970).

Acacia julibrissin Willd., Sp. Plant., 4: 1065 (1806); Unger et Kotschy, Die Insel Cypern, 391 (1865).

A broad-crowned tree, 10–15 m. high; branches glabrous, sometimes rather conspicuously lenticellate; leaves 15–30 cm. long, 10–20 cm. wide, pinnae 6–12-jugate; petiole 4–10 cm. long, adaxial surface slightly channelled and thinly pilose usually with a conspicuous gland near its base; leaflets in 15–30 pairs along each pinna, oblong-cultriform, 6–20 mm. long, 2–5 mm. wide, upper surface glabrous, margin, costa, and underside above costa thinly pilose, apex acute, or obtuse and mucronate; inflorescence a terminal panicle, comprising 10–50 (or more) spherical flower-heads; peduncles sulcate, thinly pilose, 4–6 cm. long, red-glandular at the slightly swollen apex; flowers sessile or subsessile, calyx and corolla infundibuliform, pale greenish-white, thinly pubescent, the corolla 7–8 mm. long; filaments 25–30 mm. long, connate for a short distance above base, silky, rose-pink towards apex; whitish near base; anthers minute. Pod narrowly oblong, conspicuously stipitate, 13–16 cm. long, 2–3 cm. wide, flattened, with conspicuously thickened sutures, valves pale brown, prominently veined, papery; seeds (5–) 9–15, flattened, elliptical, 7–8 mm. long, 2·5–3·5 mm. wide; testa smooth, olive-brown.

Apparently native about the southern coasts of the Caspian in S. Russia and N. Iran, but perhaps an ancient introduction, now fully naturalized: *A. julibrissin* is certainly indigenous in China, and possibly also in Japan. It is extensively grown in warm-temperate and subtropical countries, and is said by Kotschy to be a popular ornamental tree in Cyprus gardens. Specimens have been collected in Nicosia (*A. G. & M. E. Lascelles* ! *Frangos*) and in Salamis plantation (*Bovill*).

2. A. LEBBECK (*L.*) *Benth.* in Hook., Lond. Journ. Bot., 3: 87 (1844); G. Frangos in Cyprus Agric. Journ., 18: 86 (1923).
Mimosa lebbeck L., Sp. Plant., ed. 1, 516 (1753).

A spreading tree 3–15 m. high; branches subglabrous or pubescent; leaves 10–40 cm. long, 8–28 cm. wide, pinnae (1–) 2–4 (–5)-jugate; petiole 3–8 cm. long, glabrous or pubescent, generally with a large sessile gland near its base; leaflets in 3–11 pairs along each pinna, oblong or the terminal obovate, 1·5–4·8 cm. long, 0·8–3 cm. wide, glabrous or thinly pubescent above and below, apex rounded; inflorescences axillary towards the tips of the young shoots, solitary or in pairs; peduncles striate, subglabrous or pubescent, 5–10 cm. long; flower heads spherical, 3–7 cm. diam.; flowers numerous, fragrant; pedicels 1·5–5 (–7) mm. long, puberulous, bracteoles minute, caducous; calyx tubular-infundibuliform, 2·5–5 mm. long, pale green, shortly 5-lobed; corolla whitish, 5–9 mm. long, 5-lobed; filaments 15–30 mm. long, connate for a short distance above the base, silky, greenish-white or yellowish, whitish below; anthers minute. Pod oblong, flattened, 15–33 cm. long, 3–6 cm. wide, not or very shortly stipitate with conspicuously thickened sutures, valves pale brown, rather glossy, distinctly veined, papery or subcoriaceous, prominently umbonate about the seeds; seeds 4–12, sub-orbicular or oblong, strongly compressed, 6·5–10 mm. long, 6·5–7 mm. wide; testa smooth, dull brown. *Plate 41.*

A native of tropical Asia, now cultivated and widely naturalized in the tropics and subtropics, and a useful shade tree. Recorded by Frangos from gardens at Larnaca and Kolossi.

3. A. LOPHANTHA (*Willd.*) *Benth.* in Hook., Lond. Journ. Bot., 3: 86 (1844); G. Frangos in Cyprus Agric. Journ., 18: 86 (1923).
Acacia lophantha Willd., Sp. Plant., 4: 1070 (1806).

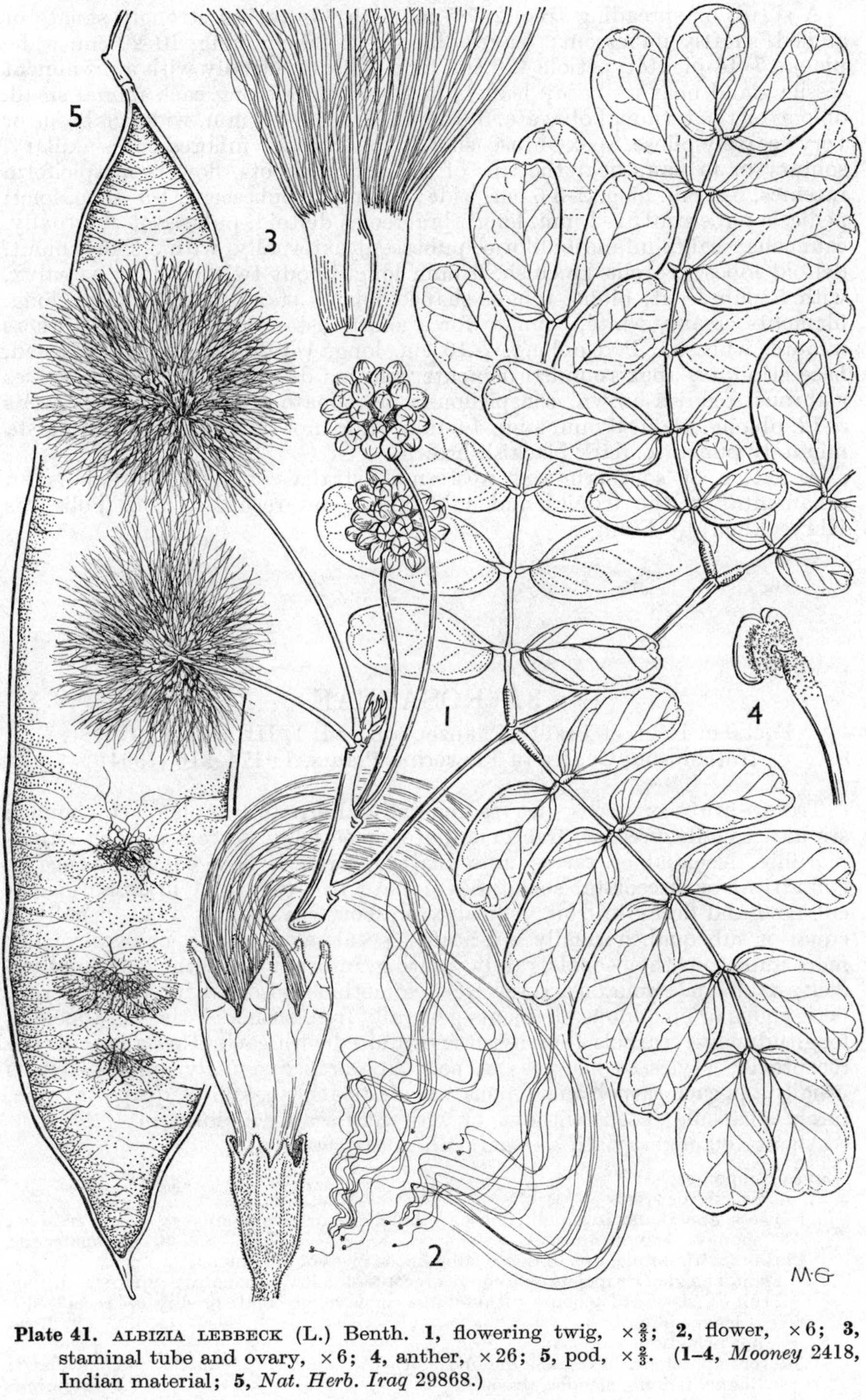

Plate 41. ALBIZIA LEBBECK (L.) Benth. **1**, flowering twig, $\times \frac{2}{3}$; **2**, flower, $\times 6$; **3**, staminal tube and ovary, $\times 6$; **4**, anther, $\times 26$; **5**, pod, $\times \frac{2}{3}$. (**1–4**, *Mooney* 2418, Indian material; **5**, *Nat. Herb. Iraq* 29868.)

A shrub or spreading tree, 2–7·5 m. high; branches strongly striate or angled, shortly pubescent; leaves oblong, 12–25 cm. long, 10–20 cm. wide, pinnae 7–14-jugate; petiole 4–6 cm. long, sulcate, usually with a prominent sessile gland near its base; leaflets in 20–40 pairs along each pinna, small, oblong or the terminal obovate, 5–15 mm. long, 2–4·5 mm. wide, glabrous or very sparsely pilose, apex obtuse, shortly mucronate; inflorescences axillary, solitary or in pairs near the tips of the young shoots, flowers in spiciform racemes, 5–7 cm. long, 2–3·5 cm. wide; peduncle pubescent, 1·5–4 cm. long; pedicels pubescent, 1–2 mm. long; bracteoles deltoid, pubescent externally, caducous; calyx infundibuliform, pubescent externally, with 5 short, blunt, deltoid lobes; corolla greenish, 6 mm. long, about twice as long as calyx, with 5 pubescent, blunt, tongue-shaped lobes; stamens 1·3–1·8 cm. long, filaments creamy-white, connate for a short distance above base; anthers minute, whitish. Pod oblong, 7–10 cm. long, 1·5–2·5 cm. wide, flattened, base stipitate, apex rounded or acute, sutures distinctly thickened, valves glabrous, reddish-brown, conspicuously umbonate about the seeds; seeds 5–12, oblong, about 8 mm. long, 4·5 mm. wide, not much compressed; testa minutely punctate, dark blackish-brown.

A native of southern and western Australia, often cultivated as an ornamental in the tropics and subtropics, and recorded from Polis [tis Khrysokhou] by Frangos.

34. ROSACEAE

Focke in Engl. et Prantl, Pflanzenfam., ed. 1, III, 3: 1–61 (1888); Hutchinson, Genera of Flowering Plants, 1: 174–216 (1964).

Trees, shrubs or herbs; leaves generally alternate, simple or compound; stipules usually present, free or adnate to petiole; flowers mostly actinomorphic, bisexual or rarely unisexual; receptacle flat, convex, concave, cylindrical or urceolate, sometimes adnate to ovary and in some genera enlarged and fleshy in fruit; epicalyx present or wanting; sepals usually 5, equal or subequal, normally imbricate; petals as many as, or occasionally more numerous than sepals, rarely wanting, inserted below the margin of the disk; stamens numerous, rarely few, sometimes inserted in several series, perigynous or epigynous, filaments generally free, anthers 2-thecous, opening longitudinally; carpels 1–many, free or united with each other and with the receptacle; styles as many as carpels, free or occasionally united; ovules usually 2 in each carpel, anatropous. Fruit an achene, follicle, drupe or pome, rarely a capsule; seeds without, or with very scanty endosperm.

About 100 genera with a cosmopolitan distribution.

Leaves compound:
- Stems prickly or spinose:
  - Flowers apetalous, in small, dense, subglobose or ovoid spikes; fruits reddish, spongy; leaves pinnate - - - - - - - - **8. Sarcopoterium**
  - Flowers with petals, not in dense subglobose or ovoid spikes:
    - Fruit a head of drupelets on a convex receptacle; leaves usually palmate **2. Rubus**
    - Fruit consisting of achenes enclosed in a concave, urceolate or ellipsoid receptacle; leaves pinnate - - - - - - - - - - - **9. Rosa**
- Stems not prickly or spinose:
  - Flowers apetalous, in dense subglobose or ovoid spikes, greenish or purple-tinged; filaments long, slender, drooping - - - - - - **7. Poterium**
  - Flowers with petals, not as above:

Flowers in an elongate, spike-like raceme; fruit campanulate, crowned with short hooked spines - - - - - - - **6. Agrimonia**
Flowers and fruits not as above:
Trees with pinnate leaves; flowers numerous in flat corymbs; fruits fleshy **12. Sorbus**
Herbs; flowers not in corymbs; fruits not fleshy:
Lower part of style persistent and accrescent in fruit; terminal leaflet of pinnate leaf usually much larger than the laterals - **3. Orthurus**
Lower part of style not persistent or accrescent in fruit; leaves palmate (in Cyprus species), leaflets not markedly unequal - - **4. Potentilla**
Leaves simple:
Trees and shrubs:
Flowers in cymes or corymbs:
Leaves entire, ashy-tomentose below - - - - **14. Cotoneaster**
Leaves toothed or lobed:
Carpel-wall cartilaginous; twigs unarmed; leaves serrate, ashy-lanuginose below - - - - - - - **12. Sorbus**
Carpel wall hard and bony; twigs usually spinose; leaves irregularly lobed, not lanuginose below - - - - - **13. Crataegus**
Flowers solitary or in fascicles or umbels:
Carpels united with the receptacle, at least in the lower half; fruit a pome:
Flowers solitary; leaves entire, downy [CYDONIA p. 633]
Flowers in fascicles or umbels; leaves toothed or serrulate - - **11. Pyrus**
Carpels free from the receptacle; fruit a drupe - - - **1. Prunus**
Herbs:
Flowers minute, in dense glomerules or cymes; stipules large, toothed or lobed; fruit small, not woody or spinose - - - - - **5. Aphanes**
Flower solitary, axillary; stipules minute or obsolete; fruit relatively large, woody, spinose - - - - - - - - **10. Neurada**

SUBFAMILY 1. **Prunoideae** Trees and shrubs; leaves simple; stipules present, sometimes caducous; receptacle flat, concave or tubular; carpel 1 (rarely more), free from receptacle; ovules 2 in each carpel. Fruit a 1-seeded drupe.

1. PRUNUS *L.*

Sp. Plant., ed. 1, 473 (1753).
Gen. Plant., ed. 5, 213 (1754).

Deciduous or evergreen trees or shrubs; buds covered with imbricate scales; leaves alternate, serrate or nearly entire, conduplicate or convolute in vernation, stipulate; flowers bisexual, solitary, or in fascicles or racemes; receptacle concave, cup-shaped, tubular or urceolate; sepals 5; petals 5, white, pink or crimson; stamens numerous, free; style elongate, stigma capitate or emarginate; ovules 2. Fruit a drupe, the epicarp and mesocarp sometimes splitting at maturity (*Amygdalus*).

About 200 species widely distributed in the temperate regions of the northern hemisphere, with some outlying representatives in the Andean regions of South America.

Many botanists subdivide the genus into smaller generic units—*Amygdalus*, *Cerasus*, *Laurocerasus*, *Padus* etc.—in the present account these are rated as subgenera.

Flowers sessile, subsessile or very shortly pedicellate; fruit usually velutinous or pubescent:
Leaves broadly ovate or subcordate; petiole long; fruit with a smooth or slightly rugose stone - - - - - - - - - [P. ARMENIACA p. 608]
Leaves oblanceolate or narrowly oblong; petiole short; fruit with a deeply rugose or pitted stone:
Leaves tapering at the base; petiole usually less than 1·5 cm. long; petals bright pink; stone deeply rugose - - - - - - - [P. PERSICA p. 609]
Leaves rounded or abruptly cuneate at base; petiole often more than 1·5 cm. long; petals pale pink or whitish; fruit with a pitted stone - [P. DULCIS p. 609]

Flowers distinctly pedicellate; fruit glabrous:
Leaves acute or obtuse with crenate-serrate or almost entire margins; fruit 3–8 (or more) cm. long, with a relatively short stalk; stone narrowly ovate or elliptic, strongly compressed - - - - - - - - **1. P. domestica**
Leaves cuspidate-acuminate, with strongly, unequally serrate margins; fruit less than 2·5 cm. long, pendulous on a relatively long stalk; stone subglobose, slightly compressed - - - - - - - - - - **2. P. avium**

SUBGENUS 1. **Prunus** Flowers solitary or fasciculate, usually pedicellate, appearing before or with the leaves; buds solitary in the leaf-axils, terminal bud wanting; leaves convolute (or rarely conduplicate) in bud. Fruit glabrous or pubescent (*P. armeniaca* and allies) grooved down one side; stone compressed, often longer than broad, smooth or slightly rugose or pitted.

1. P. domestica *L.*, Sp. Plant., ed. 1, 475 (1753); Unger et Kotschy, Die Insel Cypern, 372 (1865); Boiss., Fl. Orient., 2: 652 (1872); Post, Fl. Pal., ed. 2, 1: 452 (1932); Davis, Fl. Turkey, 4: 10 (1972); Osorio-Tafall et Seraphim, List Vasc. Plants Cyprus, 53 (1973).

TYPE: " *in* Europae *australioris locis elevatis* ".

A deciduous shrub or small tree, usually less than 10 m. high, often with sucker shoots; twigs pubescent when young, sometimes glabrescent later, becoming greyish or brownish at maturity, usually unarmed; leaves elliptic, oblong or obovate, up to 10 cm. long and 6 cm. wide, pubescent below at least on midrib and nerves, apex acute or obtuse, margins crenate-serrate or almost entire; petiole 5–20 mm. long; stipules linear-filiform, to about 4 mm. long, with sparsely glandular margins, caducous; flowers appearing with the leaves, or sometimes a little in advance of them, usually 1–3 together; pedicels 5–20 mm. long, pubescent or glabrous; receptacle cup-shaped, 2·5–3 mm. long, pubescent or glabrous; sepals reflexed in flower, oblong, about as long as the receptacle, apex obtuse or rounded, margins minutely glandular-ciliate; petals white, broadly obovate, 7–12 mm. long, 5–8 mm. wide, apex rounded, base very shortly clawed; stamens 5–7 mm. long, filaments slender; anthers suborbicular less than 1 mm. diam.; ovary glabrous, style about 6 mm. long, stigma capitate. Fruit globose, ovoid or oblong, 3–8 cm. long (or more in some cultivated varieties), epicarp purple, violet, red or yellowish-green, mesocarp greenish or yellowish, often sweet and juicy, stone narrowly ovate or elliptic, distinctly compressed, 1·5–4 cm. long, 1–2 cm. wide, slightly rugose or pitted.

HAB.: Cultivated in gardens; locally naturalized on rocky slopes or by streamsides, 2,500–5,600 ft. alt.; fl. March-May.

DISTR.: Division 2. Widespread as a cultivated tree throughout the temperate world, not infrequently naturalized in Europe, Asia and elsewhere.

2. Stavros tis Psokas, by Ayios Theodhoros R., 1962, *Meikle* 2695 ! Troödos, 1962, *Meikle* 2877 ! Common at intervals along the Ayia valley, 1974, *Meikle* !

NOTES: Generally regarded as of hybrid origin, with the European Sloe, *Prunus spinosa* L., as one parent and the western Asiatic Cherry Plum, *P. cerasifera* Ehrh., as the other. Several cultivars are recorded from Cyprus: Duhamel (Traité Arbr. Fruit., 2: 74, 82; 1768) lists two " Prunes de Chypre ", one with large blue fruits, the other with small blue fruits, like the European Damson, and probably identical with *Prunus cerasia* Blanche ex Post, Fl. Pal., ed. 1, 304 (1896). Kotschy (Die Insel Cypern, 372; 1865) mentions blue and yellow plums in Cyprus gardens, referring the former to *P. domestica* L. var. *juliana* DC. (Prodr., 2: 534; 1825) and the latter to var. *aubertiana* DC. (l.c., 534). There are no satisfactory records of *P. cerasifera* or *P. spinosa* as native or naturalized Cyprian plants.

[P. ARMENIACA *L.*, Sp. Plant., ed. 1, 474 (1753), the Apricot, a native of the Caucasus region, is widely cultivated in Cyprus, and recorded (sometimes as *Armeniaca vulgaris* Lam.) by Kotschy (Die Insel Cypern, 371; 1865) and others. Gennadius (Rep. Agric. Cyprus, part 2: 16; 1896) praises the quality of the early-maturing " Kaïssia " or " Tsaoulia ". The Apricot is a deciduous shrub or small tree, seldom more than 8 m. high, with glossy, reddish-brown twigs and broadly ovate or subcordate, glabrescent leaves; the white or pinkish flowers are solitary or occasionally paired, and appear with, or a little in advance of, the leaves; they are normally subsessile or very shortly pedicellate. The fruit is subglobose, often more than 5 cm. diam., pubescent, pale orange-yellow, frequently with a reddish cheek; the stone is ovate or elliptic, strongly compressed, slightly rugulose or almost smooth.]

SUBGENUS 2. **Amygdalus** (*L*). *Focke* Flowers sessile or shortly pedicellate, solitary or in 1–3-flowered fascicles, appearing before the leaves; buds often 3 in each axil (the 2 laterals flower-buds); leaves conduplicate in bud. Fruit pubescent or tomentose (rarely glabrous) sometimes dehiscent; stone rugose or almost smooth.

[P. PERSICA (*L.*) *Batsch*, Beytr. Entw. Pragm. Gesch. Natur-Reiche, 1: 30 (1801), the Peach, and its glabrous-fruited variant (var. *nectarina* (Ait.) Maxim.), the Nectarine, are both cultivated in Cyprus. *Prunus persica* (L.) Batsch (*Amygdalus persica* L., Sp. Pl., ed. 1, 472; 1753) is a spreading shrub or small tree, to about 8 m. high, with oblanceolate or narrowly oblong, glabrous or subglabrous, acuminate, glandular-serrate leaves, 5–15 cm. long, 1–3 (–4) cm. wide; the petiole is short, usually less than 1·5 cm. long; the flowers are solitary or sometimes in pairs, with a glabrous, campanulate receptacle, and short, ovate-oblong, bluntish sepals; the obovate petals are 10–13 mm. long and 7–8 mm. wide, rounded or shortly emarginate at the apex, and normally bright pink. The fruit is subglobose, 5–7 (or more) cm. diam., usually shortly tomentose or velutinous (except in var. *nectarina* (Ait.) Maxim.) greenish or yellowish with a red cheek, with a fleshy, juicy, greenish or yellowish mesocarp; the stone is broadly compressed-ellipsoid, about 1·5–2 cm. long and 0·8–1·3 cm. wide and deeply rugose.

The Peach, a native of China, has been cultivated in the Mediterranean region for at least 2,000 years; although so long established, and so widely grown in gardens, it is only rarely found as an escape from cultivation.]

[P. DULCIS (*Mill.*) *D. A. Webb* in Feddes Repert., 74: 24 (1967) (*Amygdalus communis* L., Sp. Plant., ed. 1, 473, 1753; *Prunus amygdalus* Batsch, Beytr. Entw. Pragm. Gesch. Natur-Reiche, 1: 30, 1801), the Almond, a native of western and central Asia, is very widely cultivated in Cyprus, and is sometimes found as an escape from cultivation. It much resembles *P. persica* in general appearance, but is usually more robust and tree-like, sometimes (especially in wild specimens) with spine-tipped twigs; the leaves are lanceolate or oblong, 3·5–6 (–10) cm. long, 1–2·5 cm. wide, glabrous, crenate-serrate, generally narrowing rather abruptly at the base, with a slender petiole 0·5–2 cm. long. The flowers are rather large and decorative, often crowded along the leafless twigs; the receptacle is campanulate, the sepals bluntly ovate-oblong, and the petals white or pale pink, obovate, 15–20 mm. long, 10–13 mm. wide, with a rounded or shortly emarginate apex. The fruit is ovoid-ellipsoid, 3–4 (–6) cm. long, 2–2·5 cm. wide, greyish-felted, the pericarp splitting at maturity; the stone is 2·5–3 cm. long, 1·5–2 cm. wide, slightly keeled and finely pitted.]

SUBGENUS 3. **Cerasus** *Pers.* Flowers solitary or in clusters or very short racemes, usually pedicellate and with conspicuous bracts, generally appearing with the leaves; buds solitary (occasionally in threes) in each axil;

leaves conduplicate in bud, style furrowed; stigma emarginate. Fruit usually glabrous; stone smooth, furrowed or pitted, often subglobose.

2. P. avium *L.*, Fl. Suec., ed. 2, 165 (1755); Post, Fl. Pal., ed. 2, 1: 452, vix *P. cerasus* L. var. ἰ *avium* L., Sp. Plant., ed. 1, 474 (1753).

Cerasus avium Moench, Meth., 672 (1794); Unger et Kotshy, Die Insel Cypern, 372 (1865); Boiss., Fl. Orient., 2: 649 (1872); H. Stuart Thompson in Journ. Bot., 44: 307 (1906); Davis, Fl. Turkey, 4: 18 (1972).

? *Prunus cerasus* Osorio-Tafall et Seraphim, List. Vasc. Plants Cyprus, 53 (1973).

TYPE: Unlocalized.

A deciduous tree or large bush, 5–24 m. high, with smooth, shining, rich brown bark, often peeling off in narrow papery strips; branches spreading or ascending; leaves oblong-elliptic or obovate-elliptic, 4–10 (–15) cm. long, 3–8 (–10) cm. wide, glabrous and dull green above, paler below and thinly villous, especially in the axils of the midrib and the prominent ascending lateral nerves, apex distinctly cuspidate-acuminate, base broadly cuneate or almost rounded, margins unequally, and often rather conspicuously serrate; petiole 3–6 cm. long, slender, glabrous or thinly villous, deeply channelled above, with 2 (sometimes 1) conspicuous glands near the apex; flowers in clusters of 2–4 (or sometimes more), springing from lateral and subterminal buds at the same time as the leaves; bracts oblong, blunt, 7–8 mm. long, 4–5 mm. wide, densely villose on the adaxial surface, with conspicuously glandular margins; pedicels slender, often deflexed, 1·5–5 (or more) cm. long, glabrous; receptacle shortly cylindrical, or slightly constricted below apex, about 5 mm. long, 3·5 mm. wide, glabrous, often reddish; sepals strongly reflexed, oblong, about 4–5 mm. long, 2·5 mm. wide, glabrous; petals pure white, spreading, obovate-cuneate, about 12 mm. long, 8 mm. wide, apex shortly emarginate; stamens numerous, unequal, filaments 3–10 mm. long, glabrous; anthers suborbicular, about 0·7 mm. diam.; ovary ovoid, glabrous, about 3 mm. long, 2 mm. wide, style about 10 mm. long, stigma capitate, slightly more than 1 mm. diam. Fruit globose, scarlet or crimson (or occasionally yellow) when ripe, about 10–15 mm. diam. in wild plants, larger in cultivated varieties, generally sweet when fully mature, stone subglobose somewhat compressed, about 9 mm. long, 8 mm. wide in wild specimens, with 3–4 shallow furrows down one margin, otherwise smooth; seed ovoid, a little compressed, about 7 mm. long, 5 mm. wide; testa brown, longitudinally veined, minutely papillose-verruculose.

HAB.: By springs and streamsides in mountainous areas, escapes from cultivation; 3,000–5,600 ft. alt.; fl. May.

DISTR.; Division 2. Widespread in Europe.

2. Troödos; Kryos Potamos below Government Cottage, 1957, *Merton* 3167 ! Kannoures Springs, 1962, *Meikle* 2878 !

NOTES: Extensively cultivated about Moutoullas and Pedhoulas, and not indigenous anywhere on the island.

SUBFAMILY 2. **Rosoideae** Herbs or shrubs. Stipules present, often conspicuous and persistent. Receptacle convex, concave, cylindrical or urceolate. Carpels usually numerous, free, sometimes enclosed in the receptacle; ovules 1–2 in each carpel. Fruit a group of achenes or a group of drupelets.

TRIBE 1. **Rubeae** *Dum.* Shrubs or subshrubs; stems often armed with prickles; calyx persistent, epicalyx wanting; receptacle convex, globose or conical; ovary with 2 pendulous ovules (1 aborting). Fruit an aggregate of drupelets.

2. RUBUS *L.*

Sp. Plant., ed. 1, 492 (1753).
Gen. Plant., ed. 5, 218 (1754).

Erect, sprawling, climbing or prostrate herbs and shrubs; stems often biennial or short-lived, frequently armed with sharp prickles or acicles; leaves palmate (in our area) or less commonly pinnate; flowers usually in branching axillary and terminal cymes, sometimes solitary or few; epicalyx wanting; sepals 5, imbricate; petals 5 (sometimes more), white or pink; stamens numerous, often with coloured filaments; receptacle convex or conical; carpels numerous, spirally arranged; ovules 2. Fruit compound, an aggregate of small 1-seeded drupelets.

A large and taxonomically very intricate genus, with more than 2,000 microspecies of the *R. fruticosus* L. aggregate recorded from Europe alone. Fortunately the bewildering proliferation of variants is not evident in Cyprus, nor in adjacent regions of the Mediterranean.

Most of the *Rubi* have edible, refreshing fruits, and the Raspberry (*R. idaeus* L.), with red or yellow fruits, is very highly esteemed in many temperate countries; it requires a cool, moist, situation, and is not likely to succeed in Cyprus, except possibly at very high altitudes on the Troödos range.

1. **R. sanctus** *Schreb.*, Icones et Descr. Plant., 1: 15, t. 8 (1766); Unger et Kotschy, Die Insel Cypern, 370 (1865); Post, Fl. Pal., ed. 2, 1: 462 (1932); Davis, Fl. Turkey, 4: 33 (1972); Osorio-Tafall et Seraphim, List Vasc. Plants Cyprus, 51 (1973).

R. ulmifolius Schott var. *anatolicus* Focke ex Heldreich, Herb. Graec. Norm. no. 844a (1884); Lindberg f., Iter Cypr., 18 (1947).

R. ulmifolius Schott ssp. *anatolicus* Focke in Abh. Nat. Ver. Brem., 9: 335 (1887); Holmboe, Veg. Cypr., 99 (1914).

R. anatolicus (Focke) Hausskn. in Mitt. Thür. Bot. Ver., 5: 90 (1893); Chapman, Cyprus Trees and Shrubs, 41 (1949).

[*R. candicans* (non Weihe) Kotschy in Unger et Kotschy, Die Insel Cypern, 370 (1865).]

[*R. discolor* (non Weihe et Nees) Boiss., Fl. Orient., 2: 695 (1872); Post in Mém. Herb. Boiss., no. 18: 93 (1900).]

[*R. ulmifolius* (? Schott) R. M. Nattrass, First List Cypr. Fungi, 13, 32 (1937); Chapman, Cyprus Trees and Shrubs, 41 (1949).]

TYPE: " *in* Creta, Palaestina Pocock[e] " (M !).

A trailing, climbing or sprawling, much-branched shrub; stems 1–2 m. long, distinctly angled, often purplish, canescent with adpressed or somewhat spreading, simple and compound hairs, and armed with numerous, sharp, straight or hooked, reddish or yellowish prickles; leaves or sterile stems (turions) palmately 5-foliolate, those of fertile stems 1–3-foliolate; stipules linear-filiform, 0·5–1·5 cm. long, thinly villose; petioles 1–4 (or more) cm. long, villose, flattened or slightly sulcate above, armed with small hooked prickles; lateral leaflets subsessile, terminal with a prickly petiolule 1 cm. or more in length; leaflets dull green and sparsely pilose or subglabrous above, densely and closely grey-felted below with prominent nervation; terminal leaflet ovate, obovate or suborbicular, 1·5–5 (or more) cm. long, 1–4 (or more) cm. wide, apex subacute, acute or shortly cuspidate, base broadly cuneate or rounded, margins sharply and often irregularly serrate, lateral leaflets similar, but often smaller and blunter; inflorescence a lax, pyramidal, or a dense, cylindrical-thyrsoid, branched cyme; rhachis and branches cinereous with spreading or subadpressed compound and simple hairs, usually armed with small hooked prickles; bracts linear-lanceolate, 8–10 mm. long, 2–3 mm. wide, villose, often deeply laciniate; pedicels 2–8 mm. long, thickened above, densely cinereous-felted, with scattered prickles; sepals ovate, 3–5 mm. long, 2–3 mm. wide, densely cinereous-felted on both surfaces, strongly reflexed after anthesis; petals pink or sometimes white, oblong, obovate or suborbicular, 8–12 mm. long, 4–8 mm. wide, apex rounded or slightly emarginate, base narrowing to a very short claw; stamens 2·5–4 mm. long, filaments usually pink, glabrous, 2·5–4 mm. long; anthers oblong, about 0·5 mm. long, 0·4 mm. wide, reddish, pilose; receptacle bluntly conical, pilose; carpels oblong-reniform, compressed, glabrous, about 1 mm. long, 0·8 mm. wide; style about 3 mm. long, slightly swollen at apex, stigma truncate. Drupelets shining black, rather small and dry, about 4 mm. long, 3 mm. wide, stone about 3 mm. long, 2 mm. wide, rather coarsely reticulate-foveolate. *Plate 42.*

HAB.: Streamsides, thickets, roadsides and on old walls, usually in moist places or near springs or streams; sea-level to 5,000 ft. alt.; fl. June to Oct.

DISTR. Divisions 1, 2, 3, 6, 7. Common in divisions 2 and 7. Greece, Aegean Islands, Turkey, Syria, Lebanon, Palestine, Egypt and eastwards to Afghanistan.

1. Skoulli, near Polis, 1939, *Lindberg f.* s.n.; Ayios Yeoryios (Akamas) 1962, *Meikle* ! Polis, 1962, *Meikle* !
2. Common in Troödos area (Troödos, Platres etc.) *Kotschy*; *Michaelides*; *C. B. Ussher* 27 ! *Lindberg f.* s.n.; *Chapman* 106; *N. Chiotellis* 452 ! *G. E. Atherton* 561 ! Between Evrykhou and Soli, *Kotschy* 917; below Makheras Monastery, *Kotschy.*
3. Mazotos, 1905, *Holmboe* 198; Episkopi, 1905, *Holmboe* 705.
6. Syrianokhori, 1936, *Syngrassides* 1245 !
7. Common in Kyrenia area; *Lindberg f.* s.n. ! *G. E. Atherton* 65 ! 160 ! *P. H. Oswald* 154 ! Lapithos, *Syngrassides* 216 ! Kythrea; *Sintenis & Rigo*; *Meikle* !

NOTES: It is questionable if the western and central European *R. ulmifolius* Schott can be maintained as a distinct species from *R. sanctus.*

TRIBE 2. **Potentilleae** *Spreng.* Herbs, shrubs or subshrubs; stems not armed; calyx persistent, epicalyx present; receptacle flat, convex or concave; carpels often numerous; ovule 1, erect or ascending; fruit an aggregate of achenes.

3. ORTHURUS *Juzepczuk*

in Fl. U.R.S.S., 10: 262, 616 (1941).

Geum. L. sect. *Orthurus* Boiss., Fl. Orient., 2: 698 (1872).

Perennial herbs; leaves imparipinnate with an enlarged terminal leaflet; inflorescence cymose; flowers bisexual, 5-merous; calyx infundibuliform, persistent; epicalyx present; stamens numerous; carpels rather few (usually less than 15) mostly borne at the apex of a distinct, accrescent gynophore; apex of each carpel produced into a straight (not hooked) beak, clothed subapically with retrorse bristles; style caducous. Achenes pilose, divergent at maturity.

Two species, one (*O. heterocarpus* (Boiss.) Juzepczuk) with a scattered distribution in Europe, N. Africa and Asia, the other (*O. kokanicus* (Regel et Schmalh.) Juzepczuk) restricted to Iran and Central Asia.

1. **O. heterocarpus** (*Boiss.*) *Juzepczuk* in Fl. U.R.S.S., 10: 262, 616 (1941); Davis, Fl. Turkey, 4: 73 (1972); Osorio-Tafall et Seraphim, List Vasc. Plants Cyprus, 52 (1973).

Geum heterocarpum Boiss. in Biblioth. Univ. Genève, 13: 408 (1838), Elenchus Plant. Nov., 40 (1838), Voy. Bot. Esp., 201, t. 58 (1840), Fl. Orient., 2: 698 (1872); Post, Fl. Pal., ed. 2, 1: 463 (1932); A. K. Jackson et Turrill in Kew Bull., 1938: 465 (1938); Lindberg f., Iter Cypr., 18 (1946); Rechinger f. in Arkiv för Bot., ser. 2, 1: 423 (1950).

TYPE: Spain; " in altis umbrosis *Sierra Tejeda* et *Nevada*, in dumetis densis *Berberidis Creticae* praecipué crescens. Alt. 4000′–6000′." *Boissier* (G).

Erect, tufted perennial 30–60 cm. high with a tough rhizome clothed with the dark brown decayed leaf-bases; flowering stems usually several, unbranched below inflorescence, thinly villose; radical leaves long-petiolate,

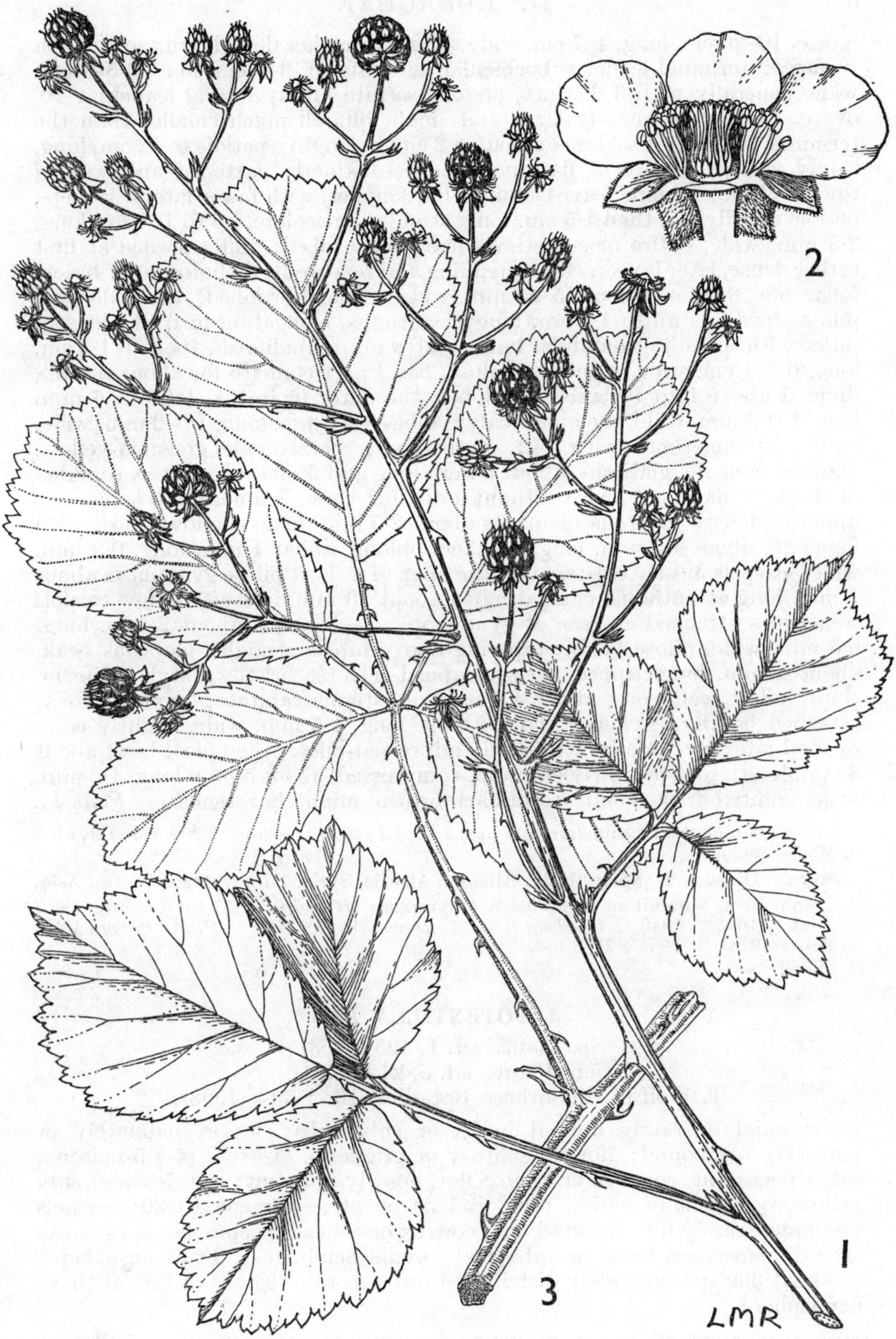

Plate 42. RUBUS SANCTUS Schreb. **1**, fruiting panicle, $\times \frac{2}{3}$ (*Guest* 12997); **2**, flower, $\times 2$ (*Guest* 2430); **3**, section of stem (current year's growth), with stem-leaf, $\times \frac{2}{3}$ (*Rogers* 0270.)

lyrate, 10–20 cm. long, 4–7 cm. wide, thinly or rather densely villose on both surfaces; terminal leaflet suborbicular or reniform, 3–7 cm. long, 3·5–8 cm. wide, generally with 3 distinct, crenate-serrate lobes; lateral leaflets 4–10, ovate, serrate, alternately large and small, but all much smaller than the terminal, the largest seldom exceeding 2 cm. in length; petiole to 12 cm. long, thinly villose, sulcate or flattened above, distinctly flattened and winged towards base; cauline leaves similar, but smaller, with fewer lateral leaflets, petiole usually less than 1·5 cm. long, stipules lanceolate, up to 12 mm. long, 2·5 mm. wide, entire or sometimes irregularly lobed; inflorescence at first rather dense, later lax, erect or spreading and repeatedly dichotomous; bracts foliaceous, 8–25 mm. long, 5–20 mm. wide, toothed or lobed, rather densely pilose; pedicels up to 1·5 cm. long at anthesis, elongating in fruit, densely pilose with spreading glandular hairs; calyx infundibuliform, about 1–1·5 cm. long, 0·8–1 cm. wide, glandular-pilose, basal part connate for about 5 mm., divided above into 10 lanceolate lobes, the outer (epicalyx) lobes 5–6 mm. long, 1–1·5 mm. wide, the inner (calyx) lobes 8–10 mm. long, 1·5–2 mm. wide, entire or sometimes with 1–2 short lobes; petals pale greenish-yellow, obovate-cuneate, glabrous, about 6 mm. long and 5 mm. wide, apex rounded or shortly emarginate, base without a distinct claw; stamens in 3 series, the uppermost with glabrous filaments about 2·5 mm. long, the lowermost with filaments about 4·5 mm. long; anthers oblong, about 1 mm. long, 0·8 mm. wide; carpels mostly clustered on the apex of a short pilose gynophore about 1 mm. long at anthesis, elongating to about 10 mm. in fruit; a few carpels sometimes attached at base of gynophore; each carpel about 2 mm. long, 0·5 mm. wide, pilose below, tapering above into a straight, glabrous beak, about 2 mm. long, tipped with subapical retrorse bristles; style fiiliform, thinly pilose, caducous, about 4 mm. long, stigma capitate; ovule 1, erect, attached basally. Achenes about 6 mm. long, 1·5 mm. wide, slightly compressed laterally, pale brown, thinly adpressed-pilose, the apical beak about 8 mm. long; seed narrowly ellipsoid, compressed, about 5 mm. long, 1·2 mm. wide, pointed at each end; testa dark brown, minutely reticulate. *Plate 43.*

HAB.: On rocky mountain summits in shade of *Juniperus* scrub; 6,200–6,400 ft. alt.; fl. May–June.

DISTR.: Division 2. Spain, Italy, Albania, Algeria, Syria, Turkey, Iran, Central Asia.

2. Common on summit of Khionistra (Olympus); *Haradjian* 477 ! *Kennedy* 595 ! 1044 ! 1045 ! 1046 ! *Lindberg f.* s.n. ! *Davis* 3164 ! 3499 ! *F. M. Probyn* 130 ! *Merton* 3181 ! *Meikle* 2778 !

4. POTENTILLA *L.*

Sp. Plant., ed. 1, 495 (1753).
Gen. Plant., ed. 5, 219 (1754).
T. Wolf in Bibliotheca Botanica, 16: 1–714 (1908).

Perennial or rarely annual herbs or subshrubs; leaves palmately or pinnately compound; flowers solitary or in cymes, bisexual (4–) 5-merous; calyx persistent, concave or almost flat, epicalyx present; petals commonly yellow, occasionally white, pink, red or purple; stamens 10–30; carpels numerous, rarely few, inserted on a convex or conical receptacle; style often lateral, sometimes basal or subapical; ovule pendulous. Fruit an achene.

About 300 species widely distributed in temperate regions of the northern hemisphere.

Stems erect or ascending; flowers numerous in branched cymes - **1. P. recta**
Stems prostrate, rooting at the nodes; flowers solitary - - - **2. P. reptans**

1. P. recta *L.*, Sp. Plant., ed. 1, 497 (1753); Tutin et al., Fl. Europ., 2: 42 (1968).

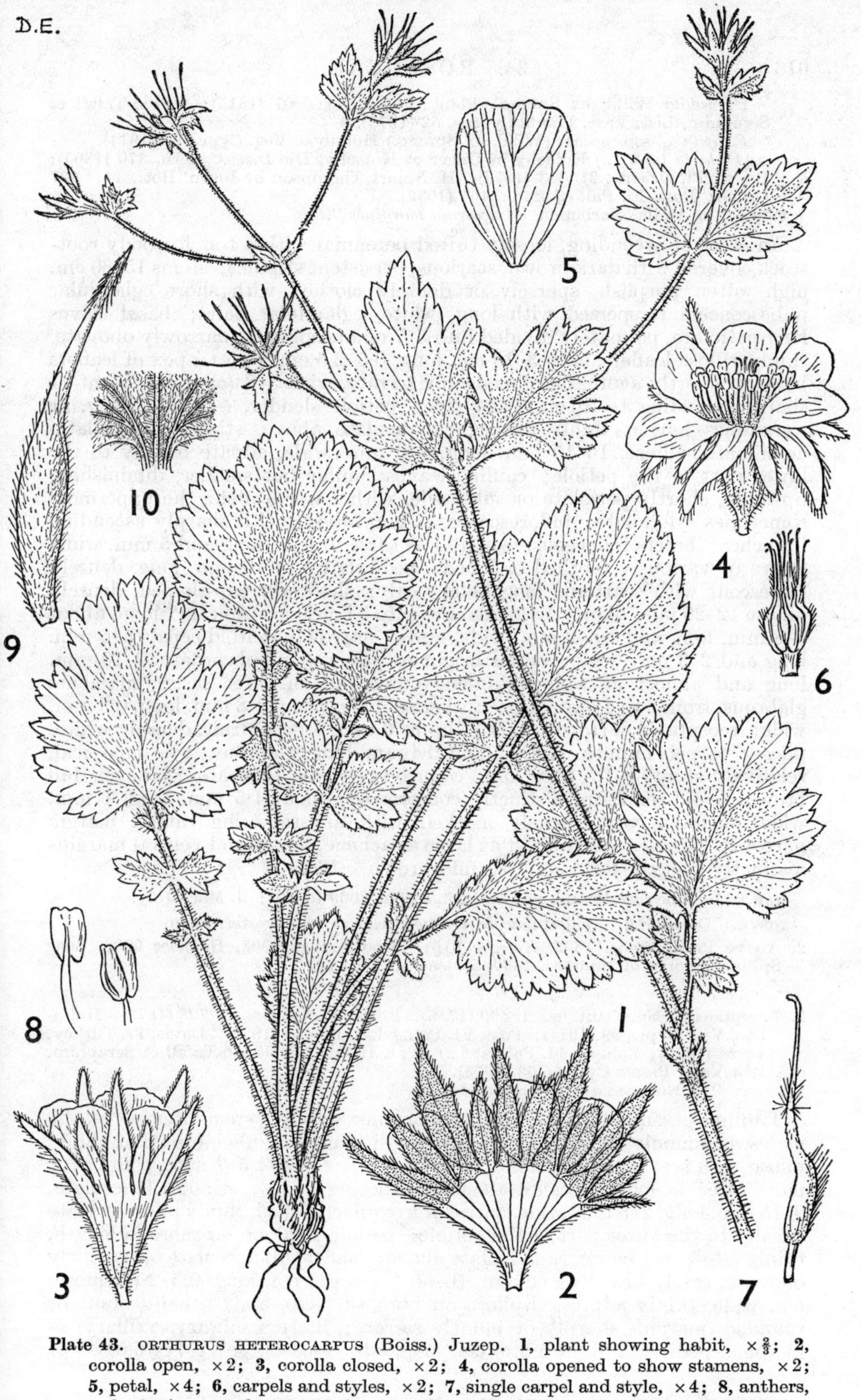

Plate 43. ORTHURUS HETEROCARPUS (Boiss.) Juzep. **1**, plant showing habit, $\times\frac{2}{3}$; **2**, corolla open, ×2; **3**, corolla closed, ×2; **4**, corolla opened to show stamens, ×2; **5**, petal, ×4; **6**, carpels and styles, ×2; **7**, single carpel and style, ×4; **8**, anthers, ×6; **9**, ripe carpel, ×4; **10**, lower surface of leaf base, ×3. (*Bornmüller* 1054.)

P. pedata Willd. ex Spreng., Plant. Pugill. Sec., 67 (1815); Osorio-Tafall et Seraphim, List. Vasc. Plants Cyprus, 52 (1973).

P. hirta L. ssp. *pedata* (Willd. ex Spreng.) Holmboe, Veg. Cypr., 99 (1914).

[*P. hirta* (non L.) Kotschy in Unger et Kotschy, Die Insel Cypern, 370 (1865); Boiss., Fl. Orient., 2: 743 (1872); H. Stuart Thompson in Journ. Bot., 44: 307 (1906); Post, Fl. Pal. ed. 2, 1: 464 (1932).]

TYPE: " *in* Italia, Narbonna, *ad agrorum margines* ".

An erect or ascending, loosely tufted perennial with a tough woody rootstock covered with dark brown, scarious, persistent stipules; stems 15–60 cm. high, often purplish, sparsely or densely clothed with short eglandular pubescence interspersed with long, white, eglandular hairs; basal leaves long-petiolate, palmately divided into 5–7 oblanceolate or narrowly obovate, thinly villose leaflets, each 3–10 cm. long, 0·5–2·5 cm. wide; apex of leaflets blunt or shortly acute, margins deeply serrate, with 3–8 (or more) blunt or sharp, ascending teeth on either side; petiole slender, 4–10 cm. long, or sometimes longer, thinly villose, canaliculate above, stipules lanceolate-acuminate, villose, 10–15 mm. long, 3–6 mm. wide, adnate basally to the lower part of the petiole; cauline leaves similar but smaller, diminishing upwards, shortly petiolate or subsessile, with fewer leaflets, the uppermost sometimes 1-foliolate; inflorescence ± corymbose, with sharply ascending branches; bracts foliaceous, usually less than 10 mm. long and 5 mm. wide, entire or variously toothed or lobed; pedicels up to 10 mm. long, densely pubescent with scattered long hairs; calyx shallowly cup-shaped, densely villose 12–20 mm. diam., divided above into 5 lanceolate epicalyx lobes about 5–8 mm. long and 1·5–2 mm. wide, and 5, narrowly deltoid sepals 5–6 mm. long and 2–3 mm. wide; petals bright or pale yellow, obcordate, 7–10 mm. long and almost as wide; stamens numerous, in three series, filaments glabrous, from 2 to 4 mm. long; anthers oblong, about 1·8 mm. long, 0·8 mm. wide; receptacle narrowly conical, up to 6 mm. long in fruit, densely white-pilose; carpels minute, ovoid, about 0·4 mm. long, glabrous, style subapical, relatively stout, 1–1·5 mm. long, caducous, stigma truncate. Achenes dull brown, ovoid-reniform, distinctly compressed, about 1·5 mm. long, 1 mm. wide, surfaces conspicuously marked with anastomosing ridges, margin narrowly winged; seed almost as large as achene, dorsal and ventral margins distinctly keeled; testa smooth, mid-brown. *Plate 44.*

HAB.: In Pine forests and vineyards, c. 4,000–5,000 ft. alt.; fl. Mar.–July.

DISTR.: Division 2. Widespread in southern Europe and western Asia.

2. Above Prodhromos, 1862, *Kotschy* 816; Prodhromos, 1905, *Holmboe* 920; near Spilia on slopes of Madhari, 1880, *Sintenis & Rigo* 722 !

2. P. reptans *L.*, Sp. Plant., ed. 1, 499 (1753); Boiss., Fl. Orient,. 2: 723 (1872); Holmboe, Veg. Cypr., 99 (1914); Post, Fl. Pal., ed. 2, 1: 464 (1932); Davis, Fl. Turkey, 4: 64 (1972); Zohary, Fl. Palaest., 2: 12, t. 19 (1972); Osorio-Tafall et Seraphim, List Vasc. Plants Cyprus, 52 (1973).

TYPE: " *in* Europae *apricis argillosis* ".

Trailing perennial; roots thick and rather woody; stems rooting at the nodes, commonly 30 cm. or more in length, angular, subglabrous, or thinly pilose with scattered long hairs; leaves palmately (3–) 5–7-nate, solitary or more often in loose fascicles about the rooting nodes; stipules lanceolate, 8–15 mm. long, 2–5 mm. wide, entire or irregularly lobed, thinly pilose, adnate basally to the lower part of the petiole; petiole 1–20 cm. or more in length, thinly adpressed-pilose, canaliculate above; leaflets oblanceolate or narrowly obovate, rarely broadly obovate, 10–50 (or more) cm. long, 0·5–2 (or more) cm. wide, thinly adpressed-pilose on both surfaces, apex usually blunt or rounded, margins sharply or bluntly serrate; flowers solitary, axillary, 5-merous; pedicels 2–8 (or more) cm. long, adpressed-pilose; calyx flattish or

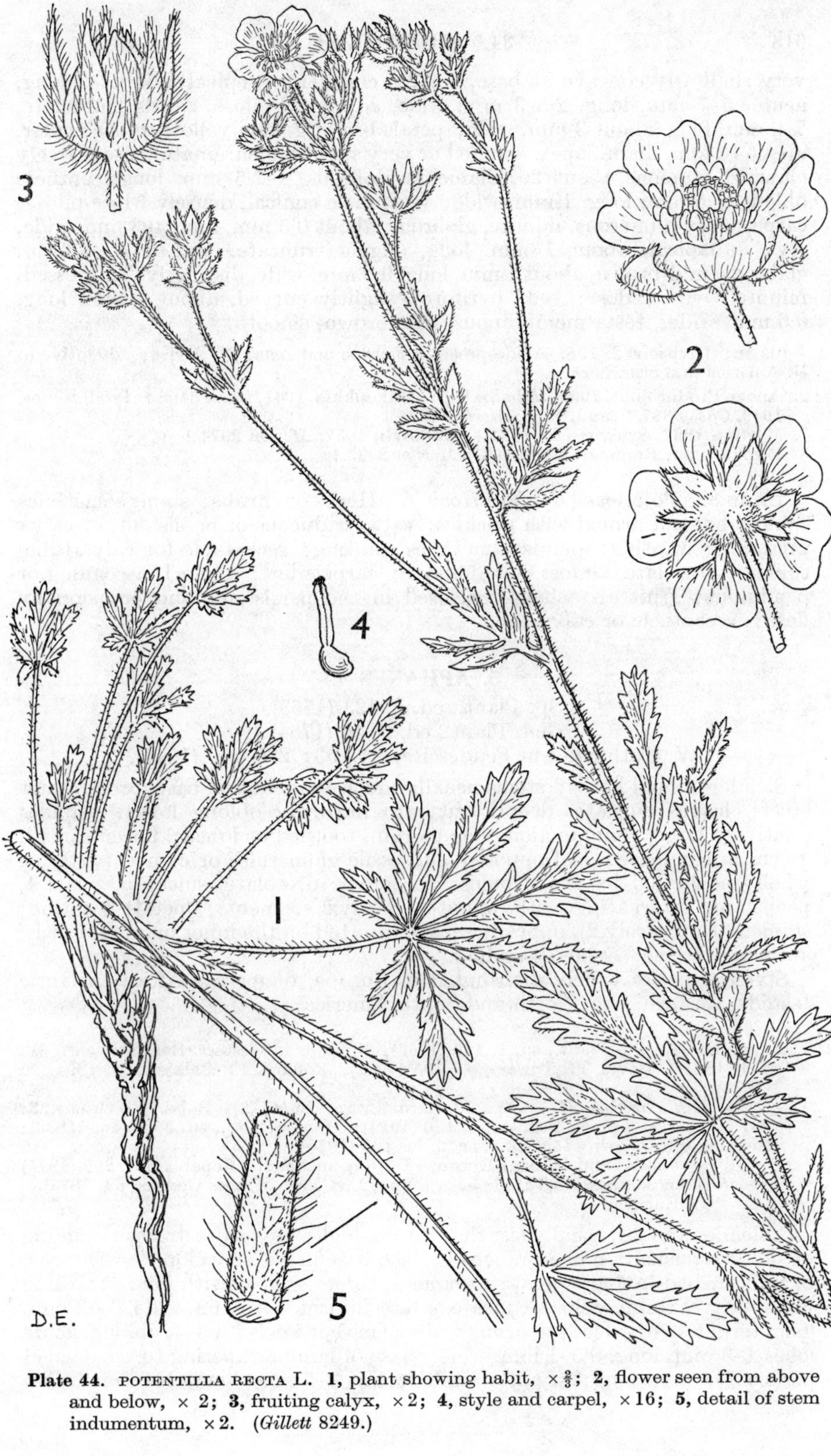

Plate 44. POTENTILLA RECTA L. **1,** plant showing habit, $\times \frac{2}{3}$; **2,** flower seen from above and below, $\times 2$; **3,** fruiting calyx, $\times 2$; **4,** style and carpel, $\times 16$; **5,** detail of stem indumentum, $\times 2$. (*Gillett* 8249.)

very shallowly concave at base, 1·5–2·5 cm. diam.; epicalyx lobes oblong, acute, 5–7 mm. long, 2·5–3 mm. wide, adpressed-pilose, sepals lanceolate, 7–9 mm. long, about 3 mm. wide; petals bright golden-yellow, suborbicular, about 10 mm. diam., apex rounded or very slightly emarginate, base scarcely clawed; stamens about 20, filaments glabrous, 1–2·5 mm. long; anthers oblong, 1·5 mm. long, 1 mm. wide; receptacle conical, densely white-pilose; carpels very numerous, minute, glabrous, about 0·5 mm. long, 0·3 mm. wide, style subapical, about 1 mm. long, stigma truncate. Achenes reniform, glabrous, mid-brown, about 1 mm. long, 0·8 mm. wide, distinctly compressed, minutely verruculose; seed pyriform, slightly curved, about 1 mm. long, 0·6 mm. wide; testa membranous, pale brown, smooth.

DISTR.: Divisions 2, 7, 8. Widespread in Europe and Asia; N. Africa; adventive in N. America and elsewhere.

2. Above Prodhromos, 1905, *Holmboe* 893; Trikoukkia, 1941, *Davis* 3465 ! Prodhromos, 1949, *Casey* 887 ! and, 1955, *Merton* 2398 !
7. Myrtou, 1937, *Syngrassides* 1534 ! Kambyli, 1957, *Merton* 2974 !
8. Rizokarpaso, Ronnas Valley, 1957, *Merton* 2932 !

TRIBE 3. **Poterieae** (*Reichb.*) *Hook. f.* Herbs or shrubs; stems sometimes spinose but not armed with prickles; calyx caducous or persistent, epicalyx present or wanting; petals sometimes wanting; receptacle (or calyx-tube) tubular, urceolate, almost closed above; carpels few; ovule 1, ascending or pendulous; fruit an achene, enclosed in the persistent, and occasionally fleshy, receptacle or calyx-tube.

5. APHANES *L.*

Sp. Plant., ed. 1, 123 (1753).
Gen. Plant., ed. 5, 59 (1754).
W. Rothmaler in Feddes Repert., 53: 265–270 (1944).

Slender annual herbs; stems usually much branched at base, prostrate or erect; leaves cuneate, deeply cut into narrowly oblong lobes; stipules relatively large, connate along one margin, toothed or lobed; flowers small, green, in condensed, leaf-opposed, subsessile glomerules or cymes, partly or wholly concealed by the stipules; receptacle urceolate, sulcate; sepals 4, persistent, alternating with minute epicalyx segments; petals wanting; stamens 1 (or rarely 2), opposite sepals, inserted on the inner margin of disk; anthers extrorse; ovule ascending.

Sixteen species widely distributed in Europe, temperate Africa, Atlantic Islands, western Asia, North and South America.

1. **A. arvensis** *L.*, Sp. Plant., ed. 1, 123 (1753); Sintenis in Oesterr. Bot. Zeitschr., 31: 257 (1881); Davis, Fl. Turkey, 4: 105 (1972); Zohary, Fl. Palaest., 2: 13, t. 20 (1972).

Alchemilla arvensis (L.) Scop., Fl. Carniol., 1: 115 (1772); Boiss., Fl. Orient., 2: 731 (1872); Holmboe, Veg. Cypr., 100 (1914); Post, Fl. Pal., ed. 2, 1: 466 (1932); Rechinger f. in Arkiv för Bot., ser. 2, 1: 423 (1950).

[*A. floribunda* (non (Murb.) Rothm.) Rothm. in Feddes Repert., 53: 266 (1944) quoad pl. cypr.; Osorio-Tafall et Seraphim, List Vasc. Plants Cyprus, 52 (1973).

TYPE: " *In* Europae *arvis* ".

A slender annual, usually less than 10 cm. high; stems prostrate, ascending or suberect, usually much-branched at the base in well-developed specimens, or unbranched in depauperate specimens, terete, villose with long spreading hairs; leaves thinly villose, cuneate or flabelliform, 4–15 mm. wide, 3–12 mm. long, ternately divided, the primary divisions cut into (2–) 3–4, oblong, acute lobes 1–3 mm. long, 0·5–1 mm. wide; base of lamina tapering to a flattened petiole 1–4 mm. long; stipules semicircular, 3–5 mm. long, 2–4 mm. wide, connate along their basal margins and adnate to the petiole-base, forming a shallow pouch, outer surface subglabrous or thinly villose, free margins ciliate, with 4–6 ovate-deltoid teeth 1–2 mm. long; flowers in glomerules of 6–10 or more, somewhat exserted from the stipular pouch; pedicels up to 1 mm. long, thinly villose; receptacle urceolate, villose, about 1 mm. long, 0·7 mm. wide; sepals 4, ovate-deltoid, acute, about 0·6–1 mm. long, ciliate with long hairs; epicalyx segments minute, deltoid, villose; stamen 1, filament about 0·2 mm. long; anther suborbicular, about 0·2 mm. diam.; style slender, about 1 mm. long, attached near base of ovary, stigma capitate. Fruit ovoid, distinctly compressed, about 1·2 mm. long, 1 mm. wide, glabrous, enclosed within the persistent and slightly accrescent calyx and receptacle.

HAB.: Roadsides, grassy clearings in forest, gravelly screes and summits; 1,800–4,600 ft. alt.; fl. April–June.

DISTR.: Divisions, 2, 3, 4, 7. Common in Division 2. Widespread in Europe, Turkey, Palestine and N. Iran.

2. Common, though mainly in west. *Holmboe* s.n.; *Davis* 3488 ! *D. P. Young* 7248 ! *Meikle* 2681 ! 2710 ! 2741 ! 2755 ! 2819 ! 2837 !
3. Khoulou, 1905, *Holmboe* 745.
4. Near Ormidhia, 1880, *Sintenis & Rigo*.
7. Above Bellapais, 1880, *Sintenis & Rigo* 520 !

NOTES: Rothmaler (Feddes Repert., 53: 266; 1944) lists Cyprus as one of the areas where *Aphanes floribunda* (Murb.) Rothm. is to be found. Unfortunately neither locality nor specimen is cited, though one supposes that the record must be based upon specimens collected by Sintenis & Rigo or Holmboe. None of the material seen in Cyprus, and none of the specimens collected, agrees with *A. floribunda*, which is readily recognizable by its subsessile upper leaves and large stipules, usually quite concealing the flowers. It seems probable that Rothmaler misidentified a depauperate specimen of *A. arvensis*, and until convincing evidence for the occurrence of *A. floribunda* is forthcoming, the record is rejected.

6. AGRIMONIA *L.*

Sp. Plant., ed. 1, 448 (1753).
Gen. Plant., ed. 5, 220 (1754).

Perennial herbs, generally glandular and fragrant; stems usually erect; leaves pinnate with alternating large and small leaflets; stipules adnate to the base of the petiole; inflorescence a spicate raceme; bracts and bracteoles present; flowers 5-merous, hermaphrodite; epicalyx wanting; petals generally yellow; stamens (5–) 10–20; receptacle turbinate, campanulate or hemispherical, often sulcate and covered or tipped with small spines which become rigid in fruit; carpels 2; styles terminal; stigma capitate; ovule 1, pendulous. Fruit indehiscent, 1 (–2)-seeded.

About 20 species, chiefly North American, but with approximately 5 representatives in Europe, Africa and Asia.

1. **A. eupatoria** *L.*, Sp. Plant., ed. 1, 448 (1753); Boiss., Fl. Orient., 2: 727 (1872); Holmboe, Veg. Cypr., 100 (1914); Post, Fl. Pal., ed. 2, 1: 465 (1932); Davis, Fl. Turkey, 4: 74 (1972); Zohary, Fl. Palaest., 2: 14, t. 21 (1972); Osorio-Tafall et Seraphim, List Vasc. Plants Cyprus, 52 (1973).
TYPE: " *in* Europae *pratis apricis argillaceis* ".

Erect perennial, 20–120 cm. high; rootstock woody, rhizomatous; stems rather densely clothed with long, spreading hairs, somewhat sulcate; leaves pinnate, 7–15 (–25) cm. long, 5–12 (–15) cm. wide, with 5–13 large, oblong, elliptic or broadly lanceolate, deeply and sharply serrate leaflets 1–7 cm. or more long, 0·5–3 cm. or more wide, alternating with pairs or groups of much smaller, ovate, entire or lobed leaflets; upper surface of lamina thinly adpressed-hairy, lower surface often rather densely hairy, with numerous, scattered, minute, shining sessile glands; petiole short, usually less than

2 cm. long, villose; stipules reniform, foliaceous, sharply serrate-lobed, 1·5–2·5 cm. or more long, 0·8–1·5 cm. or more wide; inflorescence a lax, spicate terminal raceme, 9–50 cm. long or longer, bearing 50 or more erect flowers; pedicels less than 1 cm. long, often almost wanting; bracts deeply 3-lobed, with linear-lanceolate, alternate lobes, up to 1 cm. long at the base of the inflorescence, diminishing upwards, bracteoles connate in pairs towards apex of pedicel, often deeply 3-lobed, with acuminate, pectinate-ciliate lobes, 2–3 mm. long, 0·3–0·7 mm. wide; receptacle campanulate, 2–3 mm. long, 2 mm. wide, longitudinally sulcate, adpressed-pilose, tipped with several rows of stiff, spreading spines; sepals oblong-ovate, blunt or subacute, about 3 mm. long, 2 mm. wide, distinctly 3-nerved, spreading in flower, persistent and erect in fruit; petals yellow or occasionally white, narrowly oblong or obovate, 3–5 (–7) mm. long, 2–3·5 mm. wide, apex rounded, base not unguiculate; stamens about 13–15, filaments glabrous, about 2–3 mm. long; anthers suborbicular, about 0·7 mm. diam.; disk strongly convex, about 1·2 mm. diam.; styles apical about 2·5 mm. long, stigmas small, capitate. Fruit strongly deflexed, campanulate, 5–7 mm. long, and about as wide across the top, the basal part hairy and distinctly sulcate, the apex crowned with short, rigid, erect or spreading, hooked spines; seeds 1 or 2, oblong, about 3 mm. long, 2·5 mm. wide; testa mid-brown, minutely rugulose. *Plate 45.*

HAB.: Dry mountainsides; 4,000–5,000 ft. alt.; fl. May–June.

DISTR.: Division 2. Throughout Europe, also in North Africa and western Asia.

2. About Prodhromos, 1880, *Sintenis & Rigo* 758 ! Prodhromos, 1905, *Holmboe* 908; Papoutsa, 1905, *Holmboe.*

7. POTERIUM *L.*

Sp. Plant., ed. 1, 994 (1753).
Gen. Plant., ed. 5, 430 (1754).

Perennial herbs with erect or ascending stems; leaves imparipinnate, the radical long-petiolate; stipules inconspicuous, adnate to petiole; flowers anemophilous, monoecious or polygamous, in dense ovoid or subglobose, bracteate spikes; epicalyx wanting; sepals 4, caducous, greenish or purplish, imbricate; petals wanting; stamens numerous, inserted within the disk, filaments long, slender, drooping; carpels (1–) 2 (–3), styles terminal, stigmas plumose; ovule solitary, pendulous. Fruit comprising 1–3 achenes enclosed in a persistent, accrescent and indurated receptacle.

About 25 species in Europe, North Africa and western Asia.

Fruits distinctly 4-angled, the surfaces reticulate; achenes normally 1–2 in each fruit **1. P. sanguisorba**

Fruits not angled, or obscurely angled, surfaces strongly verrucose-tuberculate; achenes normally 2–3 in each fruit - - - - - - - - - **2. P. verrucosum**

1. P. sanguisorba *L.*, Sp. Plant., ed. 1, 994 (1753); Boiss., Fl. Orient., 2: 733 (1872).
Sanguisorba minor Scop., Fl. Carniol., ed. 2, 1: 110 (1772); G. Nordborg in Opera Bot., 16: 98 (1967); Davis, Fl. Turkey, 4: 78 (1972).

Erect perennial, 20–45 cm. or more high; rootstock tough, woody; stems angular, glabrous or thinly villose above, rather more densely villose towards base; radical leaves 6–15 cm. long, 1–4 cm. wide, with 3–12 pairs of opposite, subopposite or irregularly spaced leaflets; leaflets shortly oblong or suborbicular, obtuse, rounded or truncate, rarely subacute, glabrous or thinly pilose, sharply serrate, green or glaucescent, 0·5–2 cm. long, 0·5–1·8 cm. wide; petiole up to 15 cm. long, thinly clothed with crisped, septate hairs, distinctly sulcate above; stipules membranous, adnate for the greater part of

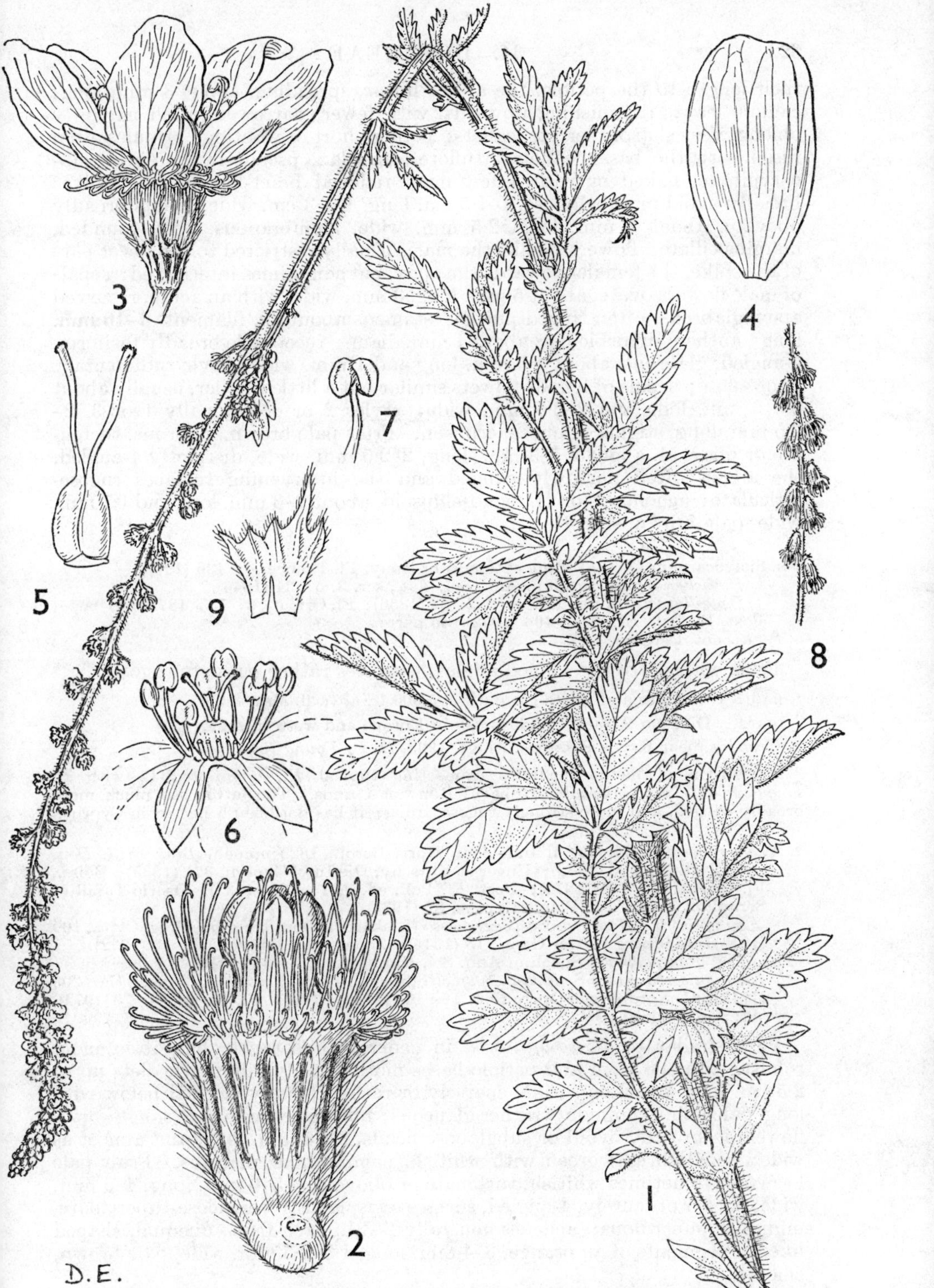

Plate 45. AGRIMONIA EUPATORIA L. **1**, plant showing habit, $\times\frac{2}{3}$; **2**, fruit, $\times 9$; **3**, flower, $\times 9$; **4**, petal, $\times 12$; **5**, ovary and styles, $\times 12$; **6**, stamens, ovary and styles, $\times 12$; **7**, anther, $\times 14$; **8**, fruits on stem, $\times\frac{2}{3}$; **9**, opened calyx, $\times 12$. (*Rawi* 22592.)

their length to the petiole, the apices linear, spreading, about 3 mm. long; cauline leaves diminishing upwards, with fewer, narrower (often narrowly oblong) very sharply serrate leaflets; petiole short or wanting; stipules often fused with the basal leaflets. Inflorescence lax, pseudo-racemose or sub-corymbose, naked or with a few much-reduced bract-like leaves; spikes broadly ovoid or subglobose, 1–1·5 cm. long, 1–1·3 cm. wide, bracts broadly obovate, about 3 mm. long, 2·5 mm. wide, membranous, apex rounded, margins ciliate; flowers sessile, the males usually restricted to the lower part of the spike, the females towards the apex, but sometimes intermixed; sepals of male flowers ovate, about 5 mm. long, 3 mm. wide, with an acute, recurved apex, glabrous, often tinged purple; stamens about 20, filaments 7–10 mm. long; anthers suborbicular, about 1 mm. diam.; receptacle broadly fusiform, 4-angled, glabrous, about 1·5 mm. long and 1 mm. wide; style rudimentary or obsolete; sepals of female flowers similar but a little smaller, usually about 2·5–3 mm. long and 2–2·5 mm. wide; styles 2 or occasionally 1 or 3, 2–2·5 mm. long, stigma plumose, crimson. Fruit pale brown, glabrous, turbinate or obovoid, about 2·5–3 mm. long, 2–2·5 mm. wide, distinctly 4-angled, the angles often narrowly winged and the intervening surfaces rugose-reticulate; achenes 2 (or 1), ovoid-ellipsoid, about 2–3 mm. long and 1·7 mm. wide, pale brown, smooth.

ssp. **dictyocarpum** (*Spach*) *Rouy et Camus* in Rouy, Fl. France, 6: 436 (1900).
P. dictyocarpum Spach in Ann. Sci. Nat., ser. 3, 5: 34 (1846).
P. gaillardotii Boiss., Diagn., 2, 2: 52 (1856); Fl. Orient., 2: 733 (1872) pro parte; Post, Fl. Pal., ed. 2, 1: 466 (1932) pro parte.
TYPE: not designated.

Fruit with narrowly winged angles; surfaces rather strongly rugose.

HAB.: Roadsides and waste ground; c. 5,000 ft. alt.; fl. May–June.

DISTR.: Division 2, rare. Widespread in Europe and western Asia.

2. Troödos, near Kryos Potamos bridge, 1963, *D. P. Young* 7843 !

NOTES: One of the intermediates connecting typical *P. sanguisorba* L. with *P. sanguisorba* L. ssp. *muricatum* (Spach) Rouy et Camus. The latter has much more broadly winged, and more strongly rugose, fruits; it has not been collected in Cyprus.

2. P. verrucosum [*Ehrenb.* in Ind. Sem. Hort. Berol., 1829 nomen] *Link ex G. Don*, Gen. Syst., 2: 595 (1832); Unger et Kotschy, Die Insel Cypern, 371 (1865); Boiss., Fl. Orient., 2: 734 (1872); Post, Fl. Pal., ed. 2, 1: 467 (1932); Osorio-Tafall et Seraphim, List Vasc. Plants Cyprus, 52 (1973).
Sanguisorba minor Scop. ssp. *verrucosa* (Link ex G. Don) Holmboe, Veg. Cypr., 100 (1914); Lindberg f., Iter Cypr., 18 (1946); Zohary, Fl. Palaest., 2: 14 (1972).
Poterium magnolii Spach in Ann. Sci. Nat., ser., 3, 5: 38 (1846).
Sanguisorba minor Scop. ssp. *magnolii* (Spach.) Briq., Prodr. Fl. Corse, 2 (1): 209 (1913); G. Nordborg in Opera Bot., 16 : 106 (1967); Davis, Fl. Turkey, 4: 79 (1972).
TYPE: Unlocalized (BM).

Very similar to *P. sanguisorba* in general appearance, but often more robust, with the stem and petiole bases more densely villose; leaflets up to 2·5 cm. long and 2 cm. wide, sparsely or rather densely clothed below with long, pale hairs, commonly rather glaucous; inflorescence as in *P. sanguisorba*, flowers in broadly ovoid or subglobose heads, 8–15 mm. long, and almost as wide; sepals bright green with whitish, membranous margins. Fruit pale brown or sometimes whitish, turbinate or obovoid, 4·5–6 mm. long, 4–5 mm. wide, not, or obscurely, 4-angled, surface conspicously verrucose-tuberculate, minutely puberulous; achenes generally 2–3 in each fruit, unequal, shaped like the segments of an orange, 3–4 mm. long, 1·5–2·5 mm. wide, pale brown, smooth.

HAB.: On hillsides in garigue; on grassy waste ground, field-margins and sides of irrigation channels; sea-level to 3,000 ft. alt.; fl. March–May.

DISTR.: Divisions 1–3, 5–7. Common in Division 7. Greece, Crete and eastwards to Iraq.

1. Ktima, 1894, *Post* s.n.; Polis, 1962, *Meikle* !
2. Prodhromos, 1939, *Lindberg f.* s.n.; Platania, 1939, *Lindberg f.* s.n.; Kykko Monastery, 1939, *Lindberg f.* s.n.; Khrysorroyiatissa Monastery, 1941, *Davis* 3437 !
3. Akhelia, 1939, *Lindberg f.* s.n.; Mile 22, Nicosia–Limassol road, 1950, *Chapman* 550 ! Limassol, 1953, *Mavromoustakis* 6 !
5. Salamis, 1905, *Holmboe* 466; near Salamis 1948, *Mavromoustakis* s.n. !
6. Near Panteleimon Monastery, Myrtou, 1862, *Kotschy* 940; Dhiorios, 1952, *F. M. Probyn* 82 !
7. Common; *Sintenis & Rigo* 45 ! 45A ! *Syngrassides* 264 ! *Lindberg f.* s.n.; *Mavromoustakis* 21 ! *G. E. Atherton* 1310 ! *Merton* 3090 ! *Meikle* 2557 !

POTERIUM SANGUISORBA *L.* (" Salamis Plantation ", Bovill, Rep. Plant. Work Cypr., 14; 1915) and P. OBTUSUM (*Maxim.*) *Franch. et Savat.* (" Kyrenia " C. E. Gresham in New Flora & Sylva, 5: 271; 1933) are both undoubted misidentifications. The first is almost certainly *P. verrucosum* Link ex G. Don; the second is wholly mysterious.

8. SARCOPOTERIUM *Spach*

Ann. Sci. Nat., ser. 3, 5: 43 (1846).
Hutchinson, Gen. Fl. Plant., 1: 207 (1964).

Low much-branched shrub; lateral branches repeatedly dichotomous, spine-tipped; leaves alternate, imparipinnate, leaflets subcoriaceous; stipules adnate to base of petiole; inflorescences leaf-opposed, spicate; flowers sessile, unisexual, those towards the base of the inflorescence usually male, the uppermost female; bracts 3, membranous; sepals 4, greenish or stained reddish, reflexed at anthesis; petals wanting; stamens numerous, exserted; carpels 2–4; stigma plumose, often reddish. Fruit globose, spongy, red, commonly crowned with the persistent calyx.

One species widely distributed in the eastern Mediterranean region, and westwards to Sardinia.

1. **S. spinosum** (*L*). *Spach* in Ann. Sci. Nat., ser. 3, 5: 43 (1846); Davis, Fl. Turkey, 4: 76 (1972); Zohary, Fl. Palaest., 2: 15, t. 23 (1972); Osorio-Tafall et Seraphim, List Vasc. Plants Cyprus, 52 (1973).

Poterium spinosum L., Sp. Plant., ed. 1, 994 (1753); Unger et Kotschy, Die Insel Cypern, 371 (1865); Boiss., Fl. Orient., 2: 734 (1872); Post, Fl. Pal., ed. 2, 1: 467 (1932); Chapman, Cyprus Trees and Shrubs, 42 (1949).

Sanguisorba spinosa (L.) Bertol., Fl. Ital., 2: 185 (1836); Holmboe, Veg. Cypr., 100 (1914); Lindberg f., Iter Cypr., 19 (1946).

TYPE: " *in* Creta, Libano ".

A low, much-branded shrub or subshrub, seldom more than 50 cm. high; branches repeatedly dichotomous terminating in slender, sharp spines; bark pale silvery-brown, peeling in irregular strips; leaves 10–60 mm. long, 1·5–15 mm. wide, imparipinnate, with 3–8 pairs of lateral leaflets; petiole short, generally less than 10 mm. long; stipules acuminate, somewhat falcate, 3–7 mm. long, 1–4 mm. wide, pubescent, those of exceptionally robust growths sometimes lobed and foliaceous; leaflets of two kinds: those of autumnal and winter growth small, ovate or suborbicular, usually less than 3 mm. long and 2 mm. wide, margins bluntly crenulate and strongly reflexed, upper surface of lamina densely and shortly glandular, with impressed nervation, lower surface tomentose; spring and summer leaflets up to 15 mm. long and 10 mm. wide, margins very slightly reflexed, often sharply serrate-laciniate, upper surface of lamina thinly glandular with scattered long hairs, lower surface thinly tomentose or adpressed-pilose, nervation not distinctly impressed; inflorescence spicate, 2–6 cm. long; peduncle 1–2 cm. long, shortly pubescent; flowers 8–12 or more, often crowded, sessile or subsessile; bracts 3, ovate-deltoid, membranous, pilose, the exterior about 4 mm. long, 2 mm. wide, the 2 interior about 3 mm. long, 2 mm. wide; sepals 4, ovate, glabrous except at the apex, concave, longitudinally 3-nerved, about 3·5 mm. long and 3 mm. wide, greenish or purple-tinged; stamens (in male flowers) numerous, filaments very slender, about 2–2·5 mm. long; anthers reniform, about 0·6 mm. long, 1 mm. wide; receptacle (in female flower) globose, fleshy, about 2 mm. diam., minutely puberulous; carpels (2–) 3–4, styles fiilform, about 0·5 mm. long; stigmas plumose. Fruit spongy, bright red, globose, 5–7 mm. diam.; achenes usually 3, shaped like the segments of an orange, about 3·5 mm. long, 1–1·5 mm. wide, smooth, pale brown.

HAB.: In batha and garigue; waste ground; edges of cultivated fields, or in the more open parts of forests; sea-level to 4,000 ft. alt.; fl. Febr.–April.

DISTR.: Divisions 1–8. Common in Divisions 5, 7, and 8. Distribution that of the genus.

1. Ayios Yeoryios (Akamas), 1962, *Meikle* ! Polis, 1962, *Meikle* !
2. Platres, 1937, *Kennedy* 534 ! 535 !
3. Stavrovouni, 1787, *Sibthorp*; Limassol, 1801, *Hume*.
4. Larnaca, 1862, *Kotschy* and 1880, *Sintenis & Rigo*; Ayia Napa, 1905, *Holmboe* 24.
5. Common; *Sintenis & Rigo* 46 ! *Rev. A. Huddle* 56 ! *Syngrassides* 1009 ! *Chapman* 379 !
6. About Dhiorios, 1962, *Meikle* !
7. Common about Kyrenia; *Kotschy*; *C. E. Gresham*; *Mrs. F. Tracey*; *Kennedy* 536 ! *C. H. Wyatt* 24 ! *Syngrassides* 270 ! *Lindberg f.* s.n.; *Miss Mapple* 56 ! *G. E. Atherton* 96 ! 489 ! 896 ! 1106 !
8. Kantara, 1880, *Sintenis & Rigo*; near Yialousa, 1880, *Sintenis & Rigo* 46a ! Koronia, ?1939–40, *Chapman* 7; Akrades, 1962, *Meikle* !

NOTES: Clarke (Travels in European and Asiatic Turkey, 2: 722; 1816) is one of the very few observers to comment on the very striking differences in the leaves produced by this plant at different seasons.

TRIBE 4. **Roseae** Shrubs; stems usually armed with prickles; calyx caducous or persistent, epicalyx wanting; receptacle (hypanthium or calyx-tube) urceolate, ellipsoid or subglobose, almost closed at apex; carpels numerous, free; ovule solitary. Fruit consisting of achenes enclosed within the enlarged, fleshy, often brightly coloured receptacle.

9. ROSA *L.*

Sp. Plant., ed. 1, 491 (1753).
Gen. Plant., ed. 5, 217 (1754).
Boulenger in Bull. Jard. Bot. Brux., 13: 165–266 (1935), 14: 114–221 (1936).

Erect, scrambling or trailing shrubs; stems armed with prickles, or rarely unarmed; leaves deciduous or occasionally evergreen, normally imparipinnate; flowers terminal, solitary or in corymbs, panicles or umbellate clusters, actinomorphic, (4–) 5-merous, hermaphrodite, often showy and fragrant; stamens numerous, inserted around a distinct (often conical) disk; styles free or occasionally united into a column; ovary 1-locular, containing a single pendulous ovule. Fruit consisting of numerous bony achenes enclosed within an accrescent, coloured, fleshy receptacle.

More than 200 species chiefly in temperate and subtropical regions of the northern hemisphere. The taxonomy of the genus is so complex, and the limits between species so obscure, that an accurate assessment of the number of species in the genus is virtually impossible.

Pedicels glandular-setose; leaflets sweet-scented, densely glandular on the undersurface; sepals conspicuously glandular externally; petals white - **1. R. chionistrae**
Pedicels glabrous or very sparsely setose; leaflets inodorous, not glandular on the under-

surface; sepals not conspicuously glandular externally; petals usually pink, sometimes white:

Leaflets glabrous or almost glabrous on both sides; styles distinctly pilose **2. R. canina var. canina**

Leaflets pubescent on the undersurface, at least along midrib and nerves; styles glabrous or thinly pilose - - - - - **2. R. canina var. dumetorum**

R. GALLICA *L.*, Sp. Plant., ed. 1, 492 (1753), the French Rose or Provins Rose, an erect rhizomatous shrub 0·5–1 m. high; branches armed with numerous weak prickles and acicles; leaves 3–7-foliolate, with rather large, blunt, somewhat rugose leaflets, is recorded by Druce (Rep. B.E.C., 9: 470; 1931) as " probably introduced " at Omodhos and Vasa. No specimens have been seen, and the record must be regarded as dubious, though the " double ", crimson-pink Apothecary's Rose (*Rosa gallica* var. *officinalis* Thory), a popular garden plant since remote antiquity, formerly used in medicine, is almost certainly cultivated in Cyprus. The wild, " single " *R. gallica*, widespread in central and southern Europe, does not seem to grow spontaneously in Cyprus.

R. DAMASCENA *Mill.*, Gard., Dict., ed. 8, no. 15 (1768), the Damask Rose, an erect or sprawling shrub, branches armed with mixed slender prickles and acicles; leaves 5–7-foliolate, with large ovate-oblong, glaucescent, subglabrous or shortly pubescent leaflets, and lax terminal panicles of sweetly scented pink flowers, with narrow glandular-setose receptacles and long-caudate sepals, is widely cultivated in Cyprus, and employed in the manufacture of rose-water and attar of roses. It has been collected at Kato Lefkara (1937, *Syngrassides* 1511 !) and as a relic of cultivation in *Quercus alnifolia* woodland near Troödos (1959, *P. H. Oswald* 166 !).

R. damascena, a very ancient garden rose, is nowadays considered to be of hybrid origin with *R. gallica* L. and *R. moschata* Herrm. or *R. phoenicia* Boiss. as its presumed parents.

R. NOISETTIANA *Thory* in Redouté, Les Roses, 2: 77 (1821), a tall, lax or semi-scandent shrub; branches uniformly and rather sparingly armed with rather stout prickles; leaves usually 7-foliolate, with subglabrous, glossy, ovate-elliptical, acuminate leaflets, and lax, many-flowered terminal clusters of rather small, pink or whitish flowers, with a narrow, thinly pubescent receptacle, narrow, almost entire sepals and long, free or slightly connate, hairy styles, has been recorded from Makheras Monastery garden (1937, *Syngrassides* 1626). It originated early in the 19th century, and is considered a hybrid between *R. chinensis* Jacq. and *R. moschata* Herrm.

1. R. chionistrae *Lindberg f.* in Soc. Sci. Fenn. Årsbok, 20B: fig. 13 (1942); Act. Soc. Sci. Fenn., n.s.B., 11: 18 (1946); Osorio-Tafall et Seraphim, List Vasc. Plants Cyprus, 52 (1973).

TYPE: Cyprus; " in glareosis in cacumine Chionistra, 1950 m.", June 17, 1939, *Lindberg f.* s.n. (H, K !).

An erect or arching shrub, 1·5–3 m. high, branches strongly armed with uniform, curved, rather stout prickles; leaves sweetly fragrant, 5–7-foliolate; petiole 1–3 cm. long, glandular, channelled above, sometimes bearing a few slender prickles; stipules narrowly oblong, 0·7–1·5 cm. long, 0·1–0·3 cm. wide, adnate to the petiole for the greater part of their length, glabrous above, thinly glandular below, margins fringed with conspicuous glands, apices acute, erect or slightly curved outwards; rhachis slender, rather densely glandular; leaflets ovate or broadly elliptical, 1–3 cm. long, 0·6–2·3 cm. wide, glabrous or very thinly hairy along the midrib above, conspicuously glandular below, margins rather coarsely glandular-biserrate, apex acute or sometimes obtuse, base broadly cuneate; flowers solitary or sometimes in clusters of 2 or 3; bracts conspicuous, ovate-acuminate, about 1·5 cm. long, 0·6 cm. wide, thinly glandular or subglabrous with glandular margins; pedicels short, seldom exceeding 1 cm. in length, glandular-setose with a few scattered hairs; receptacle narrowly ovoid or ellipsoid at anthesis, about 1 cm. long, 0·4–0·5 cm. wide, glabrous or thinly glandular-setose at the extreme base; sepals foliaceous with 4–6, narrowly lanceolate lateral lobes, about 1·5 cm. long, 0·4 cm. wide at base, conspicuously glandular externally, spreading or reflexed after anthesis, deciduous in fruit; petals white, broadly obcordate, 2–2·3 cm. long, about 2 cm. wide, apex shallowly emarginate; stamens numerous, filaments glabrous, 2·5–3 mm. long; anthers oblong, about 2 mm. long, 1·5 mm. wide; disk prominent, conical, glabrous, with an aperture about 0·7 mm. diam.; styles free, pilose, forming a low dome about 3·5 mm. diam., stigmas capitate. Fruit broadly ovoid or subglobose, about 1·5 cm. long and 1·2–1·5 cm. wide, scarlet; achenes bluntly angular, pale brown, about 5 mm. long, 3 mm. wide. *Plate 46.*

HAB.: Hedges, roadsides, screes and clearings in *Pinus nigra* woodland; 3,800–5,400 ft. alt.; fl. May–June.

DISTR.: Division 2, common. Apparently endemic.

2. Common in Troödos area; also found at Madhari; *Sintenis & Rigo* 777 ! *Chapman* 88 ! *Lindberg f.* s.n. ! *F. M. Probyn* s.n. ! *Merton* 510 ! 1929 ! 2388 ! *G. E. Atherton* 505 ! *D. P. Young* 7264 !

NOTES: Closely allied to *R. iberica* Stev., and perhaps better regarded as a subspecies of this rose, but differing in its coarsely serrate leaflets which are uniformly eglandular and subglabrous above. The calyx-lobes are longer and more conspicuously foliaceous than in typical *R. iberica*. It resembles even more closely *Rosa arabica* Crep., from Sinai, which Boulenger (loc. cit., 13: 225) considers synonymous with *R. iberica*, but again differs in the toothing and indumentum of the leaflets. To judge from the specimens seen *R. chionistrae* is consistently uniform in all its characters.

2. R. canina *L.*, Sp. Plant., ed. 1, 491 (1753); Boiss., Fl. Orient., 2: 685 (1872), Suppl., 213 (1888); Post, Fl. Pal., ed. 2, 1: 459 (1932); Lindberg f., Iter Cypr., 18 (1946); Davis, Fl. Turkey, 4: 124 (1972); Zohary, Fl. Palaest., 2: 16, t. 24 (1972); Osorio-Tafall et Seraphim. List Vasc. Plants Cyprus, 52 (1973).

Erect or sprawling shrub 1–4 m. or more in height; prickles curved or hooked, generally uniform, without acicles or bristles; leaves 5–12 cm. long, 5–7-foliolate, leaflets ovate or elliptical, 1·5–5 cm. long, 1–3 cm. wide, upper surface usually glabrous, often shining, lower surface glabrous, pubescent or sometimes a little glandular on midrib and nerves, apex acute, margins simply or doubly serrate, occasionally glandular; petiole and rhachis glabrous, pubescent or sometimes sparsely glandular; stipules often rather large, especially in the upper regions of the stem, bordered with glands and with divergent, acuminate lobes; flowers solitary or few in corymbose clusters; pedicels usually more than 1 cm. long, exceeding the bracts, smooth or sometimes glandular-hispid; receptacle narrowly ellipsoid, ovoid or subglobose, about 8 mm. long, smooth and glabrous or rarely sparsely glandular-hispid; sepals glabrous or sparsely glandular-pubescent, 1–2 cm. long, lanceolate-caudate, often with numerous, narrow lateral lobes, reflexed after flowering and generally falling before the fruit ripens. Petals white or pink, cuneate-obcordate, 1·5–3 cm. long and as wide or wider, apex broadly and shallowly emarginate, disk flattish, convex or conical, conspicuous, with a narrow aperture; styles free, glabrous or pilose, forming a prominent dome. Fruit globose, ovoid, ellipsoid or sub-urceolate, 1·5–2·5 cm. long, 1–2 cm. wide, scarlet, smooth or very sparsely glandular-hispid.

var. canina

R. canina L. var. *vulgaris* Koch, Syn. Fl. Germ. et Helv., 226 (1835); Boiss., Fl. Orient., 2: 685 (1872).

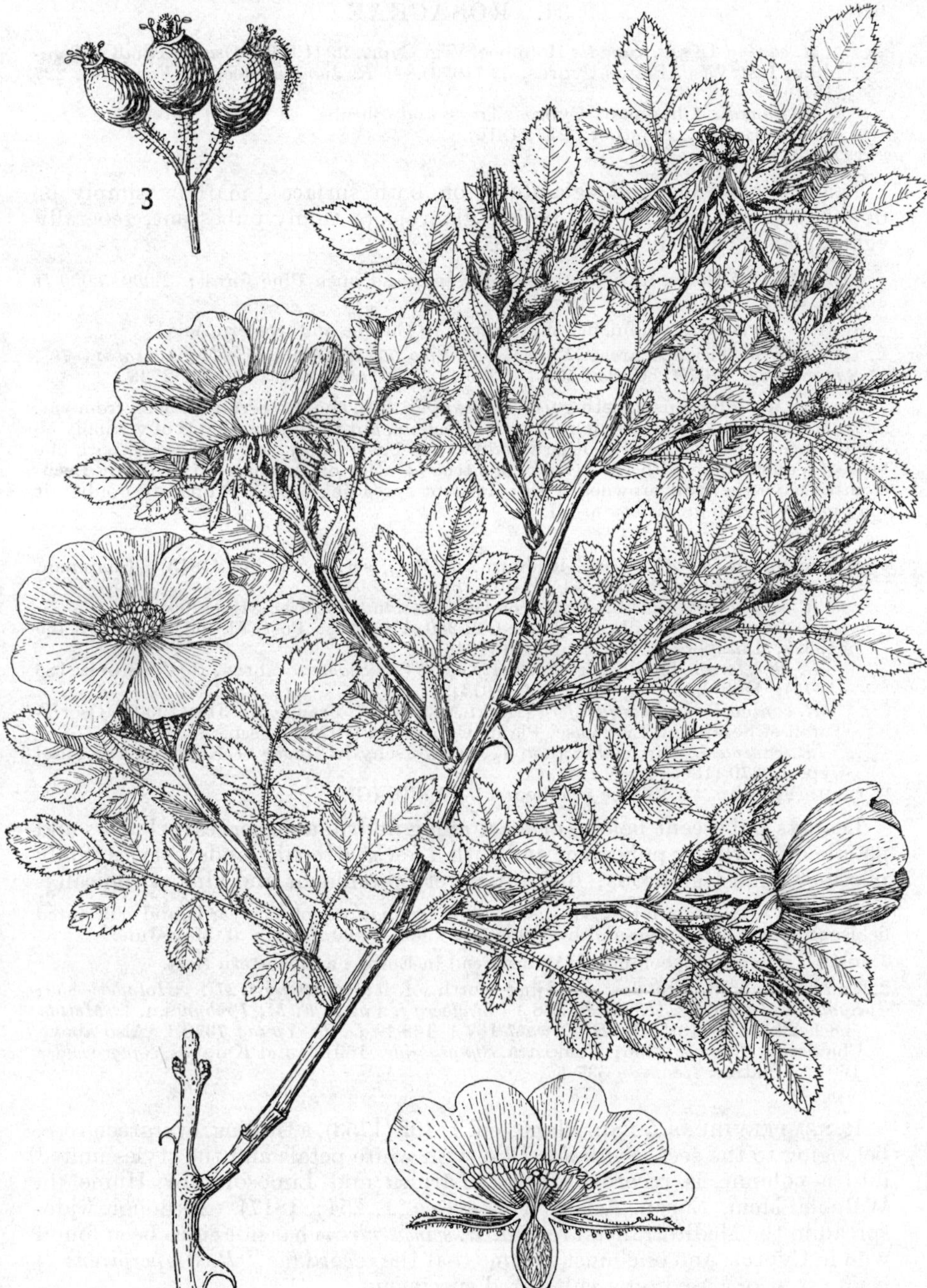

Plat 46. ROSA CHIONISTRAE Lindberg f. **1,** flowering branch, $\times\frac{2}{3}$; **2,** flower, $\times 1$; **3,** fruit, $\times\frac{2}{3}$. (**1, 2** from *Chapman* 88; **3** from *Atherton* 505.)

R. canina L. ssp. *dumalis* Holmboe, Veg. Cypr., 99 (1914); Osorio-Tafall et Seraphim, List Vasc. Plants Cyprus, 52 (1973).—? *R. dumalis* Bechst., Forstbot., 227 (1810).

R. dumalis Chapman, Cyprus Trees and Shrubs, 41 (1949) pro parte—? *R. dumalis* Bechst., Forstbot., 227 (1810).

TYPE: " *in* Europa ".

Leaflets glabrous and eglandular on both surfaces, margins simply or doubly serrate; rhachis and petiole glabrous or thinly pubescent, generally eglandular; styles distinctly pilose.

HAB.: By roadsides, in hedgerows, vineyards and open Pine forest; 2,000–5,000 ft alt.; fl. April–June.

DISTR.: Division 2, common. Widespread in Europe and western Asia.

2. Common in Troödos area; *Sintenis & Rigo* 778 ! *Holmboe* 874; *Chapman* 76 ! *Syngrassides* 1707 ! *Merton* 509 ! 2386 ! 3189 ! Chakistra, *Meikle* 2733 !

NOTES: Many botanists distinguish var. *canina*, with simply serrate leaflets, from var. *dumalis* (Bechst.) Dum. with doubly serrate leaflets, but the distinction is of virtually no significance in Cyprus, where otherwise identical bushes may be found, some with one type of leaflet, some with the other. The true identity of *Rosa dumalis*, said by Bechstein to have the fruit crowned with persistent spreading sepals, is open to doubt; it may not be a *canina* variety at all.

var. **dumetorum** (*Thuill.*) *Desv.* in Journ. de Bot., ser. 2, 2: 115 (1813); Chapman, Cyprus Trees and Shrubs, 41 (1949).

R. corymbifera Borkh., Vers. Forstbot. Beschr. Holzart., 319 (1790).

R. dumetorum Thuill., Fl. Pal., ed. 2, 250 (1798–99); Post, Fl. Pal., ed. 2, 1: 459 (1932); Lindberg f., Iter Cypr., 18 (1946).

R. dumetorum Thuill. var. *thuillieri* Christ f. *trichoneura* Christ, Ros. Schweiz, 185 (1873); Holmboe, Veg. Cypr., 99 (1914).

R. canina L. var. *corymbifera* (Borkh.) Rouy, Fl. France, 6: 314 (1900); Osorio-Tafall et Seraphim, List Vasc. Plants Cyprus, 52 (1973) [as ssp.].

[*R. canina* L. var. *glabra* (non Koch) Kotschy in Unger et Kotschy, Die Insel Cypern, 370 (1865)].

TYPE: France; " à Vilaine et ailleurs ", *Thuiller* (G).

Leaflets pubescent below, at least along midrib and nerves; rhachis and petiole often thinly pubescent and with a few scattered glands; styles thinly pilose or almost glabrous; disk often very prominent and almost conical.

HAB.: In open Pine forest; in hedgerows, by margins of vineyards and cultivated fields, on banks and screes and by roadsides, 2,000–6,400 ft. alt.; fl. May–June.

DISTR.: Division 2, common. Widespread in Europe and western Asia.

2. Common about Troödos and just north of it; *Kotschy* 727 ! *Holmboe* 859; *Syngrassides* 1097 ! *Kennedy* 538 ! *Lindberg f.* s.n. ! *F. M. Probyn* s.n. ! *Merton* 508 ! 2387 ! 2440 ! *P. H. Oswald* 167 ! 168 ! *D. P. Young* 7370 ! Also above Chakistra, *Holmboe* 1036; Chakistra, *Syngrassides* 1601 ! and Kambos, *Syngrassides* 1601 ! Platres, *Kennedy* 537 !

R. SEMPERVIRENS *L.*, Sp. Plant., ed. 1, 492 (1753), a trailing, evergreen rose, belonging to the sect. *Synstylae* DC., with white petals and the styles united into a column, is recorded from " Larnica and Limosol " by Hume (in Walpole, Mem. Europ. Asiatic Turkey, ed. 1, 254; 1817). Although widespread in the Mediterranean region, *R. sempervirens* has not since been found wild in Cyprus, and one must assume that the record (as " *R. sempervivens* ") is an error, or based on a cultivated specimen.

SUBFAMILY 3. **Neuradoideae** Herb; stipules minute; receptacle discoid, echinate, becoming woody in fruit, persistent as a collar around rootstock of young and mature plants; stamens 10; carpels 10, immersed radially in the thickened receptacle; ovules 1 in each carpel. Fruit woody, 10-valved. Comprises the single genus *Neurada*.

10. NEURADA *L.*

Sp. Plant., ed. 1, 441 (1753).
Gen. Plant., ed. 5, 119 (1754).

Annual herb; stems prostrate, unarmed; leaves alternate, lobed; stipules minute or obsolete; flowers axillary, solitary, pedicellate; receptacle discoid, flattened, echinate, persistent and accrescent in fruit; sepals 5, deltoid with 5 alternating epicalyx-lobes; petals 5, small, greenish; stamens 10; styles 10, subulate, more or less connivent to form a central cone, persistent and spinescent in fruit; stigmas obliquely capitate; carpels 10, immersed radially and horizontally in the thickened receptacle; ovules solitary, apparently horizontal, but actually pendulous in each carpel. Fruit woody, 10-valved; seeds curved.

One species in western Asia, Pakistan, North Africa, Eritrea and Mali.

1. **N. procumbens** *L.*, Sp. Plant., ed. 1, 441 (1753); Boiss., Fl. Orient., 2: 735 (1872), Suppl., 236 (1888), see note below; Holmboe, Veg. Cypr., 100 (1914); Osorio-Tafall et Seraphim, List Vasc. Plants Cyprus, 54 (1973).

N. procumbens L. var. *orbicularis* Del. [Fl. Égypt., t. 64, fig. 2, unpublished] ex C &. W. Barbey, Herborisations au Levant, 175–176 (1882).

TYPE: " *in* Aegypto, Arabia ".

A prostrate, densely tomentose annual, rarely much more than 14 cm. high, with numerous branches radiating from the rootstock; leaves 0·8–2·5 (–3) cm. long, 0·5–2 cm. wide, ovate, obtuse, with irregularly and bluntly lobed margins; petiole 0·5–1·5 cm. long; stipules minute, generally hidden by the dense whitish tomentum that covers stems, leaves and flowers; pedicel about 0·3–1 cm. long; receptacle about 0·5–1 cm. diam. at anthesis; sepals acute, about 2·5 mm. long and 1–1·5 mm. wide, about as long as, and a little wider than the alternating epicalyx-lobes; petals oblong-ovate, blunt, about 2·5–3 mm. long, 1·5–2 mm. wide, distinctly veined; stamens about 2 mm. long, with slender, glabrous filaments and broadly ovate anthers about 0·5 mm. long and almost as wide; styles rather longer than stamens, thinly tomentose, accrescemt amd indurated in fruit; stigmas resembling the anthers in general outline. Fruit 1–1·8 cm. diam., generally discoid, but sometimes pentagonal, tough and woody, the upper, convex surface densely echinate, the flattish lower surface mostly smooth; carpels dehiscing inwardly about the central pit; the fruit does not break up on the germination of the seeds, but remains attached as a collar around the rootstocks of young and mature plants; seed curved, about 2·5 mm. long, 1·5 mm. wide, with the distal end tapering and horn-like; testa rich brown, rather glossy. *Plate 47.*

HAB.: Sand-dunes at sea-level; fl. March–June.

DISTR.: Division 3. Distribution that of the genus.

3. On sand-dunes by the shore between Limassol and Amathus; 1905, *Holmboe* 644; also, 1941, *Davis* 3592 ! 1947, *Mavromoustakis* s.n. !

NOTES: Holmboe has already drawn attention to the ambiguity of the " Sarona " record in Boiss., Fl. Orient. Suppl., 236, and points out in a footnote that *Sintenis & Rigo* 920 is from Palestine and not from Cyprus.

All the Cyprus material examined appears to have discoid fruits, but it is doubtful if the distinction between plants with discoid, and those with pentagonal fruits, is worth maintaining, even if extreme forms are conspicuously distinct.

SUBFAMILY 4. **Pomoideae** Trees and shrubs; leaves simple or compound; stipules present, often small; receptacle (or calyx-tube) concave; carpels 1–5, united at least in the lower half with the receptacle; ovules usually 2 (occasionally fewer or more) in each carpel. Fruit a pome, often crowned with the persistent calyx.

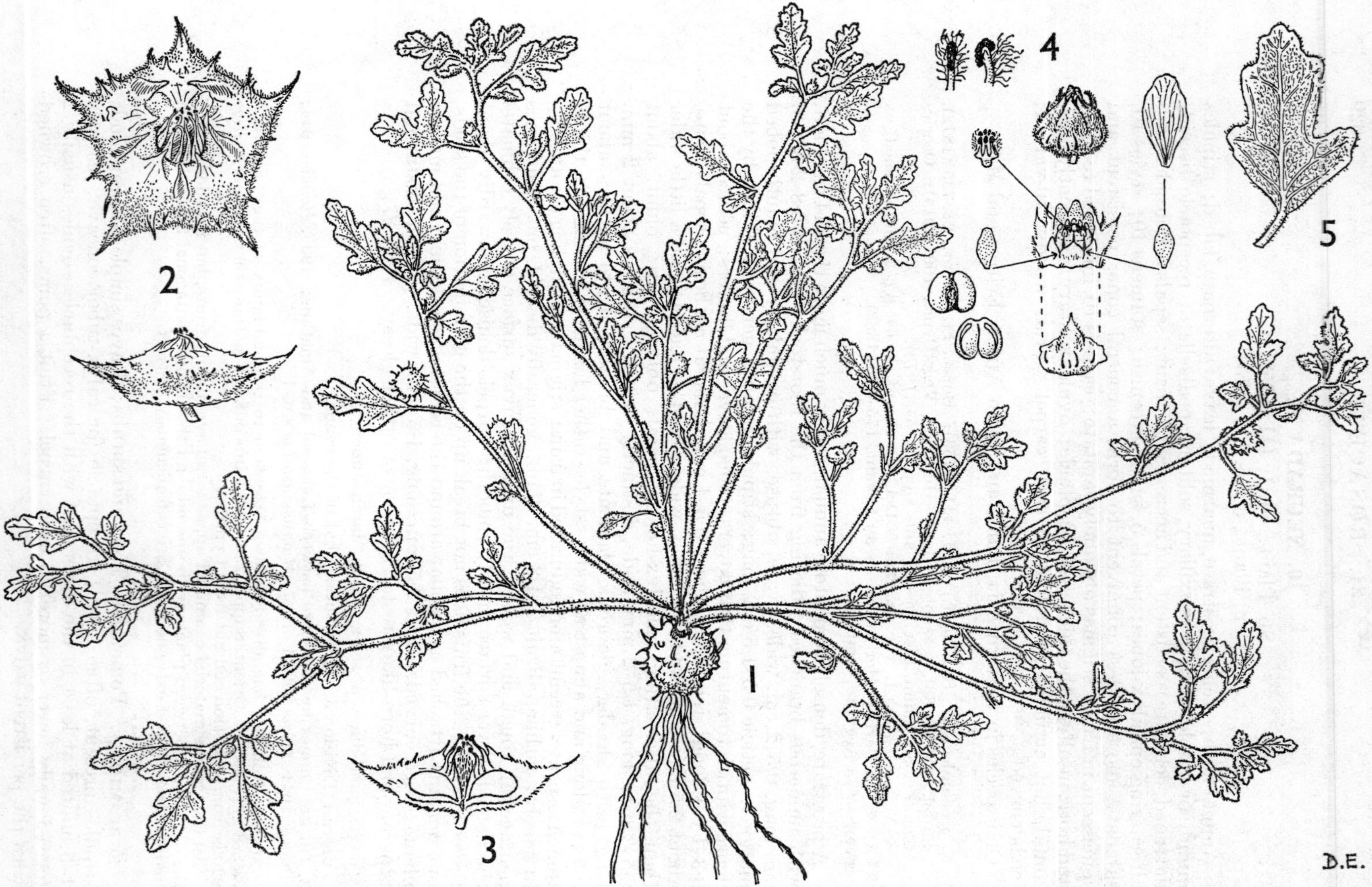

Plate 47. NEURADA PROCUMBENS L. **1**, plant showing habit, $\times\frac{2}{3}$; **2**, fruit viewed laterally and from above, $\times 2$; **3**, longitudinal section through fruit; **4**, dissection of flower, $\times 8$; **5**, leaf, $\times 2$. (*Guest & Rawi* 14244.)

11. PYRUS *L.*

Sp. Plant., ed. 1, 479 (1753).
Gen. Plant., ed. 5, 214 (1754).

Deciduous trees and shrubs; branches sometimes armed with thorns; leaves simple, entire or serrate; flowers in umbel-like clusters; petals 5, shortly clawed; stamens numerous; styles free or united at the base; receptacle turbinate or globose, accrescent; carpels 2–5, united with and enclosed within the receptacle, the walls becoming cartilaginous at maturity; ovules 2 in each loculus. Fruit brownish or greenish, sometimes yellowish or reddish when fully ripe, flesh juicy or woody, with or without grit cells; seeds generally with a dark, cartilaginous testa.

1. **P. syriaca** *Boiss.*, Diagn., 1, 10: 1 (1849); Unger et Kotschy, Die Insel Cypern, 368 (1865); Boiss., Fl. Orient., 2: 655 (1872); Holmboe, Veg. Cypr., 100 (1914)—as *Pirus syriaca*; Post, Fl. Pal., ed. 2, 1: 454 (1932); Chapman, Cyprus Trees and Shrubs, 42 (1949); Davis, Fl. Turkey, 4: 164 (1972); Zohary, Fl. Palaest., 2: 17, t. 26 (1972); Osorio-Tafall et Seraphim, List Vasc. Plants Cyprus, 53 (1973).

TYPE: Turkey–Syria; " in sylvaticis regionis inferioris montis *Cassii* supra *Cassab*, in Antilibano inter *Zebdani* et *Zachle*. (*Boiss.*) " (G).

An erect or spreading tree, 5–8 m. high; trunk fuscous with roughly fissured bark; branches greyish-brown, often armed with short, straight spines, especially in juvenile specimens; leaves narrowly ovate or ovate-lanceolate, 3·5–6 cm. long, 1·2–3 cm. wide, at first thinly floccose, soon glabrescent, dark glossy green above, paler below, apex acute or acuminate, base rounded or cuneate, margins finely crenate or serrulate; petiole slender, channelled above, up to 4 cm. long; stipules narrowly deltoid, 1–2 mm. long, membranous, thinly lanuginose, caducous; inflorescence a corymbose or umbel-like cluster of 10–15 or more flowers; pedicels 4–5 cm. long, thinly clothed with loose arachnoid tomentum; receptacles turbinate or subglobose, about 3 mm. long and about as wide, at first rather densely tomentose, later glabrescent; sepals ovate-acute, about 4 mm. long, 2·5–3 mm. wide, tomentose, often shed before fruit is fully ripe; petals white, obovate, or sometimes almost orbicular, about 1 cm. long, 0·8 cm. wide, apex rounded or retuse, base narrowing abruptly to a short claw; stamens unequal, some almost 5 mm. long, others distinctly shorter, filaments slender, glabrous; anthers dark crimson-purple, oblong or suborbicular, about 1 mm. diam.; styles 3–4 (–5), free, about 5 mm. long, stigmas subcapitate. Fruit turbinate or subglobose 2–3 cm. long, about 2 cm. wide, yellowish when ripe, often crowned with the persistent calyx, flesh sweetish but very gritty; seeds narrowly ovate, compressed, about 6 mm. long, 3 mm. wide; testa dark brown. *Plate 48.*

HAB.: Hillsides, river banks and fields; sometimes planted or used as a stock for cultivated pears; 500–5,000 ft. alt.; fl. March.

DISTR.: Divisions 1, 2 and 6. Also Turkey, Syria, Lebanon, Caucasus and Iran.

1. Between Kouklia and Paphos, and on the way from Ktima to Khrysokhou, 1862, *Kotschy* 668 ! Toxeftera near Ayios Yeoryios (Akamas), 1962, *Meikle* 2003 !

2. Dry slopes at Prodhromos, 1905, *Holmboe* 940; Platania, 1949, *E. Chapman* 119; Ayia Valley, 1906, *Merton* 2832 !

6. Myrtou, 1932, *Syngrassides* 459 ! Road from Myrtou to Panagra Gorge, 1956, *Merton* 2511 !

P. COMMUNIS *L.*, Sp. Plant. ed. 1, 479 (1753); Unger et Kotschy, Die Insel Cypern, 368 (1865), the Common Pear, differing from *P. syriaca* Boiss. in its broadly ovate, shortly acute leaves and larger fruits, is cultivated on a limited scale, especially in the more elevated valleys. Gennadius (Report

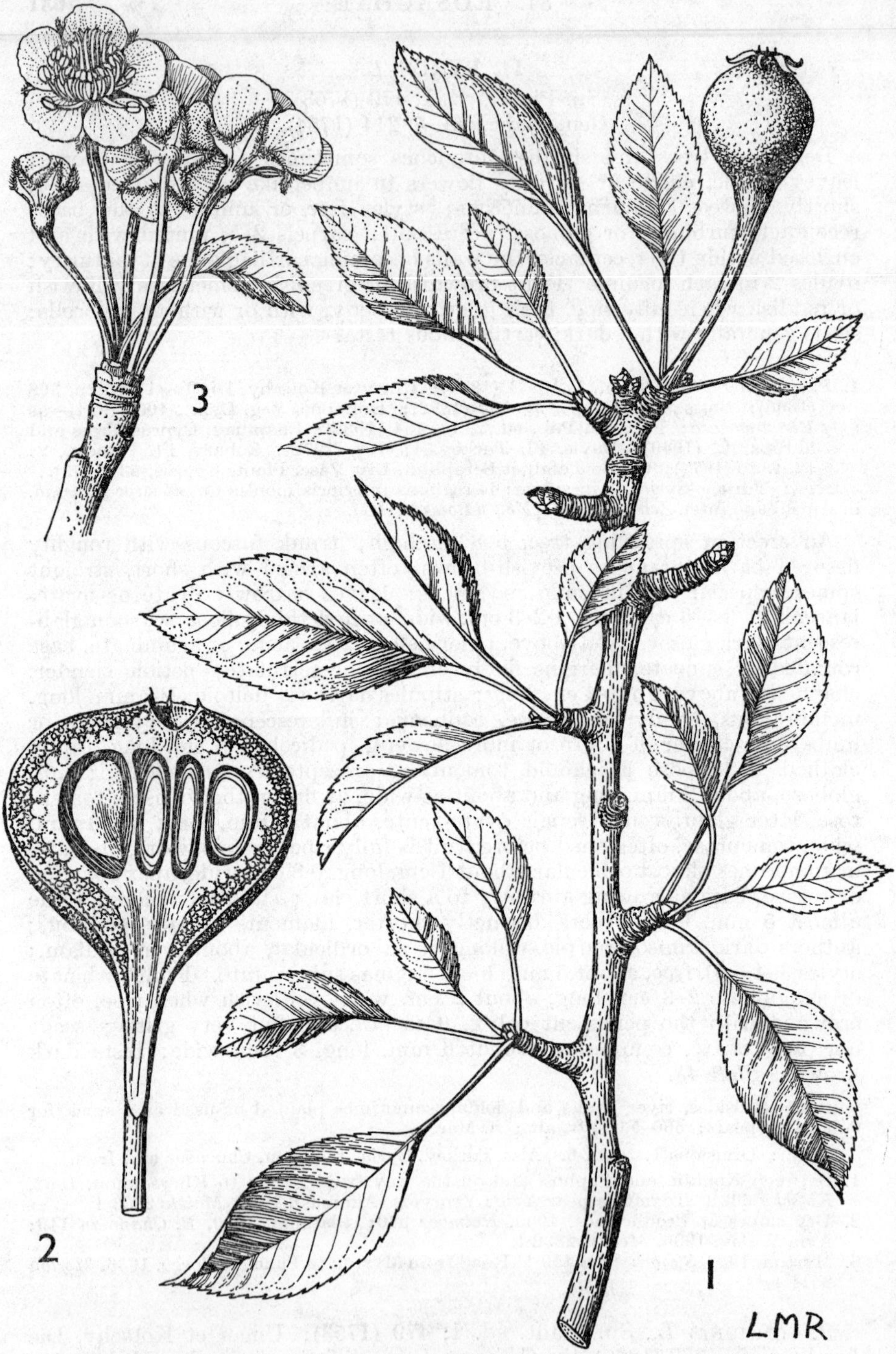

Plate 48. PYRUS SYRIACA Boiss. **1**, twig with fruit, ×1 (*Guest* 2749 & 13125); **2**, longitudinal section of fruit, ×2 (*Guest* 2749); **3**, inflorescence, ×1 (*Guest* 585.)

on the Agriculture of Cyprus, part 2: 16; 1896) mentions a large-fruited variety growing sparsely in the villages near the Marathasa river; his reference to Pliny (Nat. Hist., 15: 18) is however obscure, for no mention is made of a Cyprian pear in the passage cited.

P. MALUS *L.*, Sp. Plant., ed. 1, 479 (1753); Unger et Kotschy, Die Insel Cypern, 369 (1865). *Malus sylvestris* Mill., Gard. Dict., ed. 8, no. 1 (1768), the Apple, with pink or pink-tinged petals, yellow anthers, styles united near the base, and a depressed globose or subglobose fruit without grit cells, is extensively cultivated about Prodhromos and elsewhere at high altitudes. Self-sown plants are occasionally found, but the tree cannot be regarded as naturalized in Cyprus.

CYDONIA OBLONGA *Mill.*, Gard. Dict., ed. 8, no. 1 (1768). *C. vulgaris* Pers., Syn. Plant., 2: corrigenda (1807); Unger et Kotschy, Die Insel Cypern, 368 (1865), the Quince, with entire, greyish-downy leaves, solitary flowers with pink or white petals, and large, yellowish, subglobose or turbinate, aromatic, many-seeded fruits, is occasionally cultivated in Cyprus. It is noted from Agros (July 1933) and from Myrtou (Dec. 1933) by Nattrass (First List of Cyprus Fungi, 8, 41, 42, 46; 1937), but does not occur as a wild plant. *C. oblonga* is generally considered indigenous in Turkey, Iran and the Caucasus area.

12. SORBUS *L.*

Sp. Plant., ed. 1, 477 (1753).
Gen. Plant., ed. 5, 213 (1754).

Unarmed deciduous trees or shrubs; leaves simple or pinnately compound, entire or variously toothed and lobed; flowers in many-branched corymbose clusters, individually small; petals white or occasionally pink; stamens 15–25; styles free or connate at the base; carpels 2–5, inferior or semi-inferior, united at least to the middle, cartilaginous in fruit; ovules 2 (one usually abortive). Fruit a fleshy, red, yellow or brownish, often conspicuously lenticellate pome.

About 100 species, including many critical apomicts; the genus is widely distributed in temperate regions of the northern hemisphere.

Leaves compound, imparipinnate; fruit yellowish or brownish when ripe, 2–3 cm. diam. **1. S. domestica**
Leaves simple, broadly obovate or suborbicular, ashy-white below; fruit red or orange, 1·3–1·5 cm. diam. - - - - - - - - **2. S. aria ssp. cretica**

1. S. domestica *L.*, Sp. Plant., ed. 1, 477 (1753); Boiss., Fl. Orient., 2: 675 (1872); Rehder, Bibl. Cult. Trees and Shrubs, 255 (1949); Davis, Fl. Turkey, 4: 148 (1972).
TYPE: "*in* Europae *calidioribus*".

A round-headed tree, 10–20 m. high; trunk with dark, rough, scaling bark; buds brown, glabrescent; leaves imparipinnate, 10–15 (or more) cm. long, 7–10 cm. wide, with 13–18 (or more) narrowly oblong leaflets, 3–6 cm. long, 0·7–2·5 cm. wide, glabrous and bright green above, ± floccose-tomentose below at least when young, apex of leaflet usually sharply biserrate, basal half often entire; rhachis ± densely tomentose; petiole 3–5 cm. long, channelled above, thinly tomentose or lanuginose; stipules subulate, about 5 mm. long, membranous, brown, caducous; flowers numerous in flat branched corymbs; receptacle obconical, lanuginose, about 2 mm. long, 3 mm. wide at apex; sepals 5, acute, deltoid, about 1·5 mm. long, 1·5 mm. wide at base; petals white, broadly obovate or suborbicular, concave, 4–5 mm. long, 4 mm. wide, apex rounded, base shortly clawed; stamens numerous, styles about

4–5 mm. long, flattened; anthers oblong, about 1 mm. long, 0·8 mm. wide; styles 5, about as long as the stamens, densely woolly below, stigmas capitate. Fruit globose or turbinate, yellowish or brownish and about 2–3 cm. diam. when ripe; seeds D-shaped, strongly compressed, about 9 mm. long, 5 mm. wide; testa dark brown, smooth or very obscurely rugulose.

HAB.: In Hazel (*Corylus*) groves, almost certainly planted; c. 3,700 ft. alt.; fl. April–May.

DISTR.: Division 2, rare. Central and south Europe to Turkey and the Caucasus, N. Africa.

2. Rare; " *Sorbus terminalis* (Service Tree) one tree is to be found near Lagoudhera and another near Alona " 1923, *G. Frangos.* Also Lagoudhera, 1957, *Merton* 2992 (a single tree) !

NOTES: Formerly cultivated for its small, rather mealy, subacid fruits, but now seldom seen. The trees noted above are no doubt relics of cultivation. The Wild Service Tree (*Sorbus torminalis* L.) recorded in error by Frangos, with sharply lobed, but not compound leaves, is not known from Cyprus; the record unquestionably refers to the tree described above, the true Service Tree or Sorb.

2. S. aria (*L.*) *Crantz*, Stirp. Austr., 2: 46 (1763); Boiss,. Fl. Orient., 2: 658 (1872); Post, Fl. Pal., ed. 2, 1: 455 (1932); Chapman, Cyprus Trees and Shrubs, 42 (1949).
Crataegus aria L., Sp. Plant., ed. 1, 475 (1753).

A tall shrub or small tree generally less than 20 m. high; twigs dark brown, at first thinly floccose-lanuginose soon becoming glabrous, often distinctly lenticellate; buds ovoid, acute, thinly lanuginose; leaves simple, ovate, obovate, oblong or sub-elliptic, occasionally suborbicular or obcuneate, 4–11 cm. long, 3–8 cm. wide, dark green (or rarely subcinereous) above, pale ashy-white and lanuginose below; lateral nerves in (6–) 7–14 (–15) pairs, apex acute, obtuse or rounded, base rounded or cuneate, margins sharply biserrate, sometimes shallowly lobed; petiole 0·8–2·5 cm. long, often rather short, stout, thinly lanuginose, channelled above; stipules subulate, 4–5 mm. long, membranous, brownish, caducous; inflorescence densely corymbose, many-flowered; peduncles and pedicels lanuginose; bracts and bracteoles linear-subulate, brownish, membranous, caducous; receptacle ovoid or infundibuliform, 4–5 mm. long, 3–4 mm. wide; sepals deltoid, lanuginose, 2–3·5 mm. long, 2–2·5 mm. wide at base; petals white, obovate, concave, glabrous, 6–7 mm. long, 3–4 mm. wide, claw short, stamens numerous, filaments narrowly subulate, glabrous, about 5 mm. long; anthers yellow, oblong, about 1 mm. long, 0·8 mm. wide; styles 2, about 4 mm. long, densely lanuginose below, subglabrous above, stigmas capitate. Fruit subglobose, red (or orange) about 1·3–1·5 cm. diam., obscurely or conspicuously lenticellate, crowned with the persistent, connivent sepals; seed narrowly ellipsoid, 7–8 mm. long, 3–4 mm. wide; testa rich brown, glabrous, finely striatulate.

ssp. **cretica** (*Lindl.*) *Holmboe*, Veg. Cypr., 100 (1914); Warr et Gresham in Journ. R. H. S., 59: 71 (1934).
Pyrus aria Ehrh. var. *cretica* Lindl. in Trans. Hort. Soc. London, 7: 236 (1830).
Crataegus graeca Spach, Hist. Nat. Veg., 2: 102 (1834) pro parte quoad descr. excl. syn.
Sorbus graeca (Spach) Kotschy [in Oesterr. Bot. Zeitschr., 12: 279 (1862)] in Unger et Kotschy, Die Insel Cypern, 369 (1865) pro parte nec *Pyrus aria* Sibth. et Sm., Fl. Graec., t. 479; Hedlund in Svensk Bot. Handl., n.s., 35, 1: 75 (1901).
S. aria (L.) Crantz var. *graeca* Boiss., Fl. Orient., 2: 658 (1872) pro parte; Post, Fl. Pal., ed. 2, 1: 455 (1932) pro parte.
S. cretica (Lindl.) Fritsch et Rechinger in Kerner, Fl. Exsicc. Austro-Hung., no. 2448 (1896).
S. umbellata (Desf.) Fritsch var. *cretica* (Lindl.) Schneid., Illustr. Handb. Laubholzkunde, 1: 690 (1906); Lindberg f., Iter Cypr., 19 (1946); Davis, Fl. Turkey, 4: 154 (1972).
[*S. aria* (non (L.) Crantz) Chapman, Cyprus Trees and Shrubs, 42 (1949).]

[*S. umbellata* (non (Desf.) Fritsch) Osorio-Tafall et Seraphim, List Vasc. Plants Cyprus, 53 (1973).]

TYPE: Cultivated; "received from Messrs. Loddiges under the name of *Pyrus graeca*".

Leaves broadly and shortly obovate or almost suborbicular, lateral nerves in (6–) 7–10 pairs, leaf-margins sharply biserrate but scarcely lobed. Fruits red, subglobose, not or very sparsely lenticellate.

HAB.: Rocky mountainsides, (2,000–) 5,000–6,400 ft. alt.; fl. May–July.

DISTR.: Division 2, common. Italy, Sicily, S.E. Europe, Caucasus, Turkey, Iran, Iraq, ? Palestine.

2. Frequent about the summit of Khionistra, 5,000–6,400 ft. alt., less common at lower altitudes; restricted to this area. *Kotschy* 766 ! 779 ! *Sintenis & Rigo* 756 ! *G. Feilden* s.n. ! *Syngrassides* 727 ! *Kennedy* 1574 ! 1575 ! *Lindberg f.* s.n. ! *Davis* 1950 ! 3501 ! *Mavromoustakis* s.n. ! *Merton* 1923 ! *G. E. Atherton* 510 ! *D. P. Young* 7324 !

13. CRATAEGUS *L.*

Sp. Plant., ed. 1, 475 (1753).
Gen. Plant., ed. 5, 213 (1754).

Deciduous, often spinose, trees and shrubs; leaves serrate, pinnatifid or variously lobed; stipules commonly conspicuous, serrate-laciniate, sometimes almost foliaceous; flowers in corymbs or rarely solitary; petals 5, white or pinkish, generally concave and broadly obovate or suborbicular with a short claw; stamens 5–25; styles as many as carpels; carpels 1–5, free at apex, united along the inner margin, or at least basally, carpel-wall hard and bony at maturity forming a pyrene; ovules 2, the upper abortive. Fruit fleshy, globose or ovoid, usually crowned with the persistent triangular sepals; flesh mealy.

About 100 species in north temperate regions of the Old World, and possibly more than 1,000 highly critical species in North America. The species will often hybridize, and careful collections should be made of populations in which puzzling intermediates are found.

Styles usually 2–3 (–4); fruit large, 1·5–1·8 cm. diam., generally yellowish flushed red; young shoots densely pubescent; petiole usually less than 1 cm. long, often very short - - - - - - - - - - - - - **1. C. azarolus**
Style usually 1; fruit 5–7 mm. diam., dark red; young shoots glabrous or thinly hairy; petiole slender, 1–2 cm. long - - - - - - - - **2. C. monogyna**

1. C. azarolus *L.*, Sp. Plant., ed. 1, 477 (1753); Unger et Kotschy, Die Insel Cypern, 369 (1865)—as *C. aarzolus*; Boiss., Fl. Orient., 2: 662 (1872); Holmboe, Veg. Cypr., 101 (1914); Post, Fl. Pal., ed. 2, 1: 457 (1932); Lindberg f., Iter Cypr., 18 (1946); Chapman, Cyprus Trees and Shrubs, 42 (1949); Zohary, Fl. Palaest., 2: 19, t. 28 (1972).

C. aronia (L.) Bosc ex DC., Prodr., 2: 629 (1825); Unger et Kotschy, Die Insel Cypern, 370 (1865); Davis, Fl. Turkey, 4: 139 (1972); Zohary, Fl. Palaest., 2: 19, t. 28 (1972); Osorio-Tafall et Seraphim, List Vasc. Plants Cyprus, 53 (1973).

C. azarolus L. var. *β aronia* L., Sp. Plant., ed. 1, 477 (1753).

[*C. orientalis* (non Pall. ex M. Bieb.) Kotschy in Oesterr. Bot. Zeitschr., 12: 278 (1862).]

TYPE: "Florentiae, Monspelii" (not in herb. Linn.).

A small, round-headed tree to about 6 m. high, with a rugged, fissured trunk; branches usually armed with short, straight spines, growths of current year densely clothed with short, crisped, ashy-white hairs; leaves obovate or inversely cuneate, 2·5–3·5 (–5) cm. long, 1·5–3 (–5) cm. wide, generally with 3–5 (–7) narrowly oblong, ascending lobes, irregularly and bluntly lobed or toothed near the apex; lamina glabrous or subglabrous, or occasionally thinly pubescent above, pubescent with crisped hairs, or rarely almost glabrous below; petiole rather short and stout, usually not much more than 1 cm. long; stipules foliaceous, about 1–1·5 cm. long, margins deeply and rather sharply toothed or lobed; inflorescence a compact, terminal, 10–20-flowered corymb; flowers heavy-scented; pedicels usually less than 2 cm. long, relatively stout, ± densely pubescent; bracts linear, membranous, caducous, usually less than 4 mm. long; receptacle broadly infundibuliform, pubescent, about 4 mm. long, 3 mm. wide at apex; sepals triangular, acute, about 3 mm. long and 3 mm. wide at base, densely pubescent externally; petals white, orbicular, 7–8 mm. diam., claw very short or almost wanting; filaments glabrous, 2·5–3 mm. long; anthers shortly oblong, about 0·6 mm. long, 0·5 mm. wide; styles (1–) 2–3 (–4), a little longer than the stamens, stigmas capitate. Fruit globose, often solitary or few together, yellowish or tinged reddish when ripe, 1·5–1·8 cm. diam., crowned with the persistent reflexed sepals, subglabrous or thinly pubescent, especially around the base of the calyx; pyrene about 0·8 cm. long, 0·5–0·7 cm. wide, dorsal surface furrowed, ventral surface keeled or flat.

HAB.: Rocky mountainsides; also by field-margins and roadsides at lower altitudes, where it is often planted; sea-level to 3,500 ft. alt.; fl. March–April.

DISTR.: Divisions 1, 2, 4–7. Rare in Divisions 1 and 6, common in Division 2. Widespread in the Mediterranean region and eastwards to Iraq and Iran.

1. Polis area, 1962, *Meikle* !
2. Common; Prodhromos, Platres, Platania, Pharmakas, Livadhi etc.; *Kotschy* 730 ! *Sintenis & Rigo* 759 ! *Holmboe* 1132; *Kennedy* 540 ! *Chapman* 62; *Meikle* 2261 !
4. Kato Vlakhos near Larnaca, 1862, *Kotschy* 89, 328; Kouklia (Mesaoria), 1905, *Holmboe* 377.
5. Athalassa, 1937, *Syngrassides* 1515 ! 1939, *Lindberg f.* s.n.; Syngrasi, 1937, *Syngrassides* 1643 ! Salamis area, 1962, *Meikle* !
6. Dhiorios area, 1962, *Meikle* !
7. Pano Dhikomo, 1880, *Sintenis & Rigo* 44 ! St. Hilarion, 1939, *Lindberg f.* s.n.; Kyrenia Pass, 1939, *Lindberg f.* s.n. ! Kyrenia, 1956, *G. E. Atherton* 1082 !

NOTES: Holmboe (Veg. Cypr., 101) has already commented on the problem of identifying this species and *C. aronia* (L.) Bosc ex DC. (*C. azarolus β aronia* L.). In the absence of satisfactory type material of either species it is difficult to come to any definite conclusions, but it would not be unreasonable to conclude that all the specimens cited above (with leaves pubescent below and with large, yellowish or red-flushed fruits) pertain to *C. aronia* rather than to *C. azarolus sensu stricto*. The identity of the latter is more problematic: Linnaeus presumably intended to contrast his var. *aronia* with a plant which had the leaves glabrous below, and which grew about Montpellier and Florence. Such plants are to be found in Cyprus (and about Montpellier), and are here tentatively assigned to *C. × sinaica* Boiss. (? *C. ruscinonensis* Grenier et Blanc); there is some reason for regarding them as hybrids between *C. azarolus* (or *C. aronia*) and *C. monogyna* Jacq., but, in Cyprus at least, the evidence is not conclusive. It is also by no means certain, nor can it be ascertained, that Linnaeus intended to apply the name *azarolus* to the presumed hybrid; some authors maintain that he had the cultivated Azarole in mind; if this so, then his variety *aronia* is superfluous, since it is indistinguishable from the cultivated Azarole.

2. C. monogyna *Jacq.*, Fl. Austr., 3: 50 (1775); Unger et Kotschy, Die Insel Cypern, 369 (1865); Boiss., Fl. Orient., 2: 664 (1872); Holmboe, Veg. Cypr., 101 (1914); Post, Fl. Pal., ed. 2, 1: 457 (1932); Lindberg f., Iter Cypr., 18 (1946); Chapman, Cyprus Trees and Shrubs, 43 (1949); Davis, Fl. Turkey, 4: 145 (1972); Zohary, Fl. Palaest., 2: 20, t. 30 (1972); Osorio-Tafall et Seraphim, List Vasc. Plants Cyprus, 53 (1973).

[*C. calycina* (non Peterm.) Diapulis in Feddes Repert., 34: 65 (1933).]

TYPE: Austria; " passim et copiose ad margines viarum, in fruticetis, in sylvis Danubialibus, in montibus, atque alibi " *Jacquin* (W).

A spreading shrub or small tree 2–7 (–10) m. high; trunk often deeply fluted, bark rugged; twigs glabrous or thinly pubescent when young, greyish or dark reddish-brown, generally well armed with spines 1–2 cm. or more long, occasionally unarmed; leaves variable, ovate, oblong or obovate, 2–5 cm. long, 1·3–5 cm. wide, glabrous or subglabrous above and below at maturity, generally rather deeply divided into 3–5 (–7) acute, sharply dentate lobes, occasionally subentire or irregularly serrate-laciniate, apex usually acute, base broadly cuneate; petiole rather slender, 1–2 cm. long, canaliculate above; stipules ear-shaped, 5–8 mm. long, 3–8 mm. wide, dentate, sub-foliaceous; inflorescence lax, 3–6 cm. wide, 7–11 (or more)-flowered; flowers heavily scented; pedicels 6–15 mm. long, glabrous or thinly hairy; receptacle narrowly infundibuliform, 2–3 mm. long, 1·5–2 mm. wide, glabrous or thinly hairy; sepals broadly triangular, acute, about 1·5 mm. long, 2 mm. wide at base, reflexed in flower and persistent in fruit; petals white (or sometimes pinkish), orbicular, 3–5 mm. diam., very shortly clawed; stamens 15–20, 3–4 mm. long; anthers shortly oblong, crimson, about 1·2 mm. long, 1 mm. wide; style 1 (rarely 2), usually a little shorter than stamens; stigma capitate. Fruit red, ovoid, 8–10 mm. long, 5–7 mm. wide, glabrous; pyrene ovoid, slightly furrowed, about 6 mm. long, 4 mm. wide.

HAB.: On mountainsides, frequently by streams; by roadsides and field borders; 1,000–5,000 ft. alt.; fl. April–May.

DISTR.: Divisions 2 and 3. Common in Division 2, rare in Division 3. Widespread in Europe and western Asia.

2. Common about Prodhromos, Platres, Platania, Panayia, Perapedhi etc.; *Kotschy* 720 ! *Holmboe* 926; *Kennedy* 541 ! 542 ! *Chapman* 77; *Merton* 3194 ! *Meikle* 2805 !
3. Khoulou, 1905, *Holmboe* 759.

NOTES: Since it is by no means easy to distinguish between *C. monogyna* and the following, *C.* × *sinaica* it is possible that some of the specimens cited above, but not examined, should be referred to the latter rather than to the Hawthorn described above.

C. × **sinaica** *Boiss.*, Diagn., 2, 2: 48 (1856) pro spec.

? *C.* × *ruscinonensis* Grenier et Blanc in Billotia, 1: 71 (1864) pro spec.; Loret et Barrandon, Fl. de Montpellier, ed. 2, 609 (1886).

C. azarolus L. × *monogyna* Jacq. sec. Planch. in Compt. Rend. Acad. Sci. Par., 74: 613 (1872).

TYPE: Sinai; " in praeruptis montium *Sinaiticorum* Bové ! Schimper ! " (G. K !).

Leaves resembling those of *C. azarolus* L. but generally glabrous or subglabrous below; young shoots and inflorescences glabrous or thinly hairy; flowers larger than those of *C. monogyna*, in compact or rather lax corymbs. Fruits red, spherical or ovoid, containing 1 or 2 carpels.

HAB.: Amongst scrub on mountainsides, or by the edges of fields; 3,000–5,400 ft. alt.; fl. May–June.

DISTR.: Division 2. Sinai, Syria, Turkey, and probably elsewhere wherever the two parent species occur together.

2. Amiandos, 1934, *Syngrassides* 518 ! Kyperounda, 1937, *Syngrassides* 1708 ! Platania 1939, *Lindberg f.* s.n. ! Near Prodhromos, on road to Troödos, 1961, *D. P. Young* 7263 !

NOTES: So clearly intermediate between *C. azarolus* and *C. monogyna* in every respect that there is reason to suppose that *C.* × *sinaica* is a hybrid. A thorough field study of the hybrid and the two presumed parents would be of considerable interest.

14. COTONEASTER *Medik.*

Geschichte der Botanik, 85 (1793).

Erect or decumbent, deciduous or evergreen shrubs or small trees; leaves alternate, often somewhat coriaceous, generally entire; stipules subulate, caducous; flowers in lateral and terminal cymes, or occasionally solitary, usually rather small; sepals 5, generally short, persistent in fruit; petals 5, white or pink, erect or spreading; stamens numerous; styles 2–5 (as many as loculi of ovary); ovary with 2 erect ovules in each loculus. Fruit fleshy, globose or turbinate, containing 2–5 bony pyrenes.

About 70 species widely distributed in Europe and Asia, particularly abundant in China and the Himalayan region.

1. **C. racemiflorus** (*Desf.*) *C. Koch*, Dendrol., 1: 170 (1860); Schneid., Illustr. Handb. Laubholzkunde, 1: 754 (1906); Rehder, Bibl. Cult. Trees and Shrubs, 236 (1949); Osorio-Tafall et Seraphim, List Vasc. Plants Cyprus, 53 (1973).

Shrub 1–2 (–3) m. high; twigs dark purplish-brown, covered with thin whitish tomentum when young; leaves orbicular, ovate, obovate or elliptical, 1·5–4 cm. long, 1–3·5 cm. wide, apex acute, obtuse or rounded, generally mucronate, base rounded or shortly cuneate, margin entire, upper surface dark green, glabrous or thinly pilose, lower surface white- or grey-felted or rarely subglabrous; petiole short, usually less than 8 mm. long, pilose, channelled above; stipules subulate, 1–2 mm. long, brown, caducous; flowers (3–) 4–8 (or sometimes more) in congested cymes terminating short lateral branches; pedicels shortly tomentose 2–5 (–8) mm. long; receptacle campanulate, tomentose, about 3 mm. long, 3 mm. wide across apex; sepals triangular, about 1 mm. long, 1 mm. wide at base, tomentose externally; petals white, orbicular or suborbicular, spreading, 2–3 mm. diam., claw very short; stamens about 1·5 mm. long, glabrous; anthers suborbicular, about 0·3 mm. diam.; styles usually 2, glabrous, about 1·5 mm. long, stigmas capitate. Fruit red, turbinate or subglobose, 6–8 mm. long, 4–6 mm. wide, thinly tomentose even when ripe; pyrenes 2, obovate, about 5 mm. long and 3 mm. wide, light brown, lustrous, a little wrinkled.

var. **nummularius** (*Fisch. et Mey.*) *Dippel*, Handb. der Laubholzkunde, 3: 415 (1893); Rehder, Bibl. Cult. Trees and Shrubs, 236 (1949).

C. nummularia Fisch. et Mey., Ind. Sem. Hort. Petrop., 2: 34 (1835); Unger et Kotschy, Die Insel Cypern, 369 (1865); Boiss., Fl. Orient., 2: 666 (1872); Holmboe, Cypr., 101 (1914); Post, Fl. Pal., ed. 2, 1: 458 (1932); Lindberg f., Iter Cypr., 18 (1946); Chapman, Cyprus Trees and Shrubs, 43 (1949); Davis, Fl Turkey, 4: 131 (1972).

TYPE: Caucasus and Transcaucasia; " in montibus Talüsh, in Iberia et in montibus Samamisicis " (L, K !).

Leaves broadly and bluntly ovate or suborbicular, generally less than 1·8 cm. diam., apex shortly mucronate or occasionally emarginate, lamina generally ashy-tomentose below.

HAB.: Rocky mountainsides on igneous formations; 4,000–6,400 ft. alt.; fl. May–July.

DISTR.: Division 2, common. Widespread from Sicily and Greece eastwards to Afghanistan; N. Africa, perhaps also eastwards to N. India and China.

2. Common about Khionistra; also at Madhari; *Kotschy* 779; *Sintenis & Rigo* 757 ! *A. G. and M. E. Lascelles* s.n. ! *Haradjian* 986 ! *Feilden* s.n. ! *Kennedy* 543 ! 544 ! *C. Wyatt* 30 ! *Syngrassides* 1681 ! *Lindberg f.* s.n. ! *Davis* 1936 ! 3526 ! *Chapman* 93; *Mavromoustakis* s.n. ! *F. M. Probyn* s.n. ! *G. E. Atherton* 570 ! *D. P. Young* 7332 !

35. SAXIFRAGACEAE

Engler in Engl. et Prantl, Pflanzenfam., ed. 2, 18a: 74–226 (1930).

Perennial or, less frequently, annual herbs; leaves generally alternate, exstipulate; flowers actinomorphic in cymes or less often solitary or in racemes, usually hermaphrodite; sepals normally 4–5, imbricate or valvate; petals alternating with sepals, sometimes wanting, perigynous or epigynous; stamens generally twice as many as petals, rarely equal in number or fewer, inserted with the petals, filaments free, anthers 2-thecous; ovary superior to almost inferior, free or adnate to the campanulate or cylindrical receptacle; carpels 2, united below, divergent above; styles free, as many as the carpels;

ovules numerous on axile placentas. Fruit a capsule; seeds small, numerous, usually with abundant endosperm.

About 30 genera, mostly in temperate regions of the northern hemisphere. Only one genus in our area.

1. SAXIFRAGA *L.*

Sp. Plant., ed. 1, 398 (1753).
Gen. Plant., ed. 5, 189 (1754).
Engler et Irmscher in Engl., Pflanzenr., 67 (IV. 117): 1–709 (1919).

Perennial or (in our area) annual herbs, rarely suffruticose; leaves generally alternate, entire or variously toothed or lobed; flowers in cymes or racemes, sometimes solitary, usually pedicellate; sepals 5 or 4, normally persistent, imbricate; petals 5 or 4, or occasionally wanting, imbricate; stamens 10 or sometimes 8, anthers dehiscing laterally; ovary superior or almost inferior, of 2 (rarely more) partly united carpels, free or often adnate to the receptacle; styles free, usually divergent especially after anthesis. Seeds with copious endosperm.

About 300 species, widely distributed in the northern hemisphere.

Ovary superior; leaves oblate or reniform with broadly triangular lobes; stems flaccid, sprawling - - - - - - - - - - - - **1. S. hederacea**
Ovary inferior or almost so; leaves obovate or oblanceolate, entire or deeply and narrowly lobed; stems erect - - - - - - - **2. S. tridactylites**

1. S. hederacea *L.*, Sp. Plant., ed. 1, 405 (1753); Sibth. et Sm., Fl. Graec. Prodr., 278 (1806), Fl. Graec., 4: 72, t. 379 (1823); Unger et Kotschy, Die Insel Cypern, 314 (1865); Boiss., Fl. Orient., 2: 812 (1872); Holmboe, Veg. Cypr., 96 (1914); Post, Fl. Pal., ed. 2, 1: 497 (1932); Davis, Fl. Turkey, 4: 258 (1972); Zohary, Fl. Palaest., 2: 9, t. 15 (1972); Osorio-Tafall et Seraphim, List Vasc. Plants Cyprus, 51 (1973).

TYPE: Crete; the type of *Saxifraga Cretica, annua, minima, Hederaceo folio* of Tournefort, Cor. Inst. Rei Herb., 18 (1719) (P).

Erect or sprawling annual; stems slender, flaccid, up to 18 cm. long, subglabrous or sparsely hyaline-pilose; basal leaves alternate, oblate or reniform, 3–15 mm. long, 3–20 mm. wide, glabrous or subglabrous with 3–5 broadly triangular, blunt or acute marginal lobes; petioles slender, subglabrous, up to 2·5 cm. long; upper leaves similar, but becoming progressively smaller and more shortly petiolate, the uppermost often opposite and cuneate at the base; inflorescence a lax cyme; flowers small; pedicels fiilform up to 3 cm. long, minutely glandular; sepals 5, persistent, ovate-oblong, acute or obtuse, 1·5–2 mm. long, 0·8 mm. wide, thinly glandular externally; petals white, sparsely glandular, oblong, obtuse, 3·5 mm. long, 1·4 mm. wide, with a short basal claw; filaments 2–2·5 mm. long; anthers reniform, about 0·4 mm. wide; ovary superior, ovoid, about 2 mm. long and almost as wide, glabrous; styles 2, about 1 mm. long; stigmas capitate. Fruit ovoid, 3·5–4 mm. long, 2·5–3 mm. wide, crowned with the persistent styles and stigmas; seeds ovoid, about 0·3 mm. long, 0·25 mm. wide; testa dark brown, densely and bluntly papillose.

HAB.: Shady limestone rocks and cliff-faces; 300–2,500 ft. alt.; fl. March–May.

DISTR.: Divisions 4 and 7, common in Division 7. Also Sicily, Yugoslavia, Greece, Crete, Aegean Islands, Turkey, Syria, Palestine and Cyrenaica.

4. Cape Greco, 1859, *Kotschy*.
7. Common along the Northern Range (Panayia tou Kantara, St. Hilarion, Pentadaktylos, Trypa Vouno); *Kotschy* 364 ! *Sintenis & Rigo* 32 ! *Druce* 204 ! *Casey* 642 ! 1330 ! *Davis* 2516 !

2. S. tridactylites *L.*, Sp. Plant., ed. 1, 404 (1753); Boiss., Fl. Orient., 2: 808 (1872); Holmboe, Veg. Cypr., 96 (1914); Post, Fl. Pal., ed. 2, 1: 496 (1932); Davis, Fl.

Turkey, 4: 250 (1972); Osorio-Tafall et Seraphim, List Vasc. Plants Cyprus, 51 (1973).

TYPE: " *in* Europae *arenosis* ".

Slender annual; stems erect, often reddish, up to 11 cm. high, thinly glandular; basal leaves alternate, forming a loose winter rosette, obovate or oblanceolate, sparsely glandular, 8–15 mm. long, 3–10 mm. wide, occasionally entire but often deeply divided into 3 (–5) lanceolate or narrowly ovate, acute or obtuse lobes; petioles generally less than 10 mm. long; upper leaves sessile or subsessile, often entire and oblanceolate, the uppermost frequently opposite; inflorescence a lax cyme, flowers very small, pedicels filiform, generally upcurved, to almost 2 cm. long, minutely glandular; sepals 5, persistent, tongue-shaped, thinly glandular externally, about 0·8 mm. long, 0·6 mm. wide, apex rounded; petals white, oblanceolate, about 2·5 mm. long, 0·8 mm. wide, apex rounded or slightly emarginate, base tapering to a short claw; filaments 0·5 mm. long; anthers ovoid, 0·3 mm. long, 0·2 mm. wide; ovary inferior, broadly ovoid or subglobose, about 2·5 mm. long and almost as wide, adnate to the thinly glandular receptacle; styles strongly divergent, about 0·3 mm. long, stigmas relatively large, ovoid-capitate. Fruit ovoid or subglobose, about 3 mm. long, 2·5 mm. wide; seeds ovoid with a marked longitudinal ridge down one side, about 0·3 mm. long, 0·2 mm. wide; testa dark brown, bluntly papillose.

HAB: On limestone or igneous rocks, in open or shaded situations; 700–4,000 ft alt; fl. March–April.

DISTR.: Divisions 1, 2, 7, 8, common in Division 7. Widespread in Europe and the Mediterranean region.

1. Panormi near Peyia (Akamas), 1962, *Meikle* 2129 !
2. Kionia, 1941, *Kennedy & Davis* 2743 !
7. Common; Kantara, St. Hilarion, Buffavento, Klepini, Ayios Khrysostomos Monastery; *Sintenis & Rigo* 33 ! *Davis* 2517 ! 2808 ! *Casey* 1325 !
8. Kantara Castle, 1905, *Holmboe* 550.

36. CRASSULACEAE

A. Berger in Engl. et Prantl, Pflanzenfam., ed. 2, 18a: 352–483 (1930).

Herbs or subshrubs, often fleshy or succulent; leaves alternate or opposite, exstipulate, generally simple, glabrous or pilose; flowers usually in cymes, actinomorphic, mostly hermaphrodite, commonly 5-merous, less often 3-, 4- or up to 32-merous; sepals free or united towards the base, persistent in fruit; petals as many as sepals, free or variously united; stamens as many as, or twice as many as petals, perigynous or almost hypogynous, free or adnate to the petals, anthers 2-thecous, introrse, dehiscing longitudinally; hypogynous scales (or nectaries) present at the base of the carpels, the latter superior, as many as the petals, free or united at the base, 1-locular; styles subulate or filiform, stigmas capitate or truncate, often oblique; ovules usually numerous, attached in 2 (sometimes more) rows to the adaxial suture of the carpel, rarely few or solitary. Fruit a membranous or coriaceous follicle, often surrounded by the persistent perianth; seeds mostly very small, generally with fleshy endosperm.

About 30 genera distributed throughout the world; the generic limits are difficult to define, and estimates of the number of genera vary from author to author. Many of the *Crassulaceae* (e.g. *Sedum*, *Sempervivum*, *Aeonium*,

Kalanchoë, *Crassula* and *Echeveria*) are decorative, and easily grown in arid or semi-arid climates.

Leaves opposite, connate at base - - - - - - - - **1. Crassula**
Leaves not connate:
  Corolla tubular with short lobes; basal leaves suborbicular, often peltate **2. Umbilicus**
  Corolla deeply lobed or divided to base, lobes spreading:
    Tall suffruticose perennials; branches long, erect, naked, terminating in a flat rosette of leaves - - - - - - - AEONIUM p. 654
    Low perennials and annuals, not or scarcely suffruticose:
      Basal leaves in rosettes:
        Plants perennial, generally tufted, stems often bare and subligneous at the base - - - - - - - - - - **3. Rosularia**
        Plants annual or monocarpic, not tufted; stems generally unbranched **4. Sedum**
      Basal leaves not in rosettes:
        Calyx free; fruit a dehiscent many-seeded follicle - - **4. Sedum**
        Calyx adnate to rhachis; fruit indehiscent, 1-seeded - - **5. Telmissa**

1. CRASSULA *L.*

Sp. Plant., ed. 1, 282 (1753).
Gen. Plant., ed. 5, 136 (1754).

Tillaea L., Sp. Plant., ed. 1, 128 (1753); Gen. Plant., ed. 5, 62 (1754).
Bulliarda DC. in Bull. Sci. Soc. Philom., 3, no. 49: 1 (1801).

Perennial or (in our area) small annual herbs or subshrubs; stems and leaves usually rather fleshy; leaves opposite, simple, often sessile or connate, entire, glabrous or pilose; flowers usually small in cymes or heads, 3–5 (–9)-merous; petals white, pink or reddish, free or connate at the base; carpels 3–5, free; ovules few to many in each carpel; styles short, stigmas minute.

About 300 species widely distributed, but especially numerous in South Africa. The Cyprus representatives both belong to the group frequently classified as a distinct genus, *Tillaea* L.; they are characteristically small, not very fleshy, annuals, with minute, inconspicuous flowers.

Leaves and sepals acuminate, shortly awned; flowers in dense cymose clusters **1. C. alata**
Leaves and sepals obtuse; flowers solitary, axillary - - - **2. C. vaillantii**

1. C. alata (*Viv.*) *Berger* in Engl. et Prantl, Pflanzenfam., ed. 2, 18a: 389 (1930); Zohary, Fl. Palaest., 2: 2, t. 2 (1972).

Tillaea alata Viv., Plant. Aegypt. Dec., IV: 16 (1830); Post, Fl. Pal., ed. 2, 1: 488 (1932).
T. trichopoda Fenzl ex Boiss., Fl. Orient., 2: 767 (1872).
[*T. muscosa* (non L.) Holmboe, Veg. Cypr., 96 (1914).]
[*Crassula tillaea* (non Lest.-Garl.) Osorio-Tafall et Seraphim, List Vasc. Plants Cyprus, 50 (1973).]
TYPE: Egypt; " in paludosis prope Kahirum " (? GE).

Slender, erect or prostrate annual; stems usually less than 10 cm. long, leafy, often reddish, fleshy (becoming distinctly winged in dried specimens), glabrous; leaves connate at the base, subulate, about 4–7 mm. long, 1–2 mm. wide, glabrous, entire, apex shortly aristate; flowers in rather dense, cymose, bracteate clusters along almost the whole length of the stems; pedicels slender, 1–2·5 mm. long at anthesis, lengthening to 6 mm. or more in fruit; sepals 3–4, free to base, subulate-aristate, 1·5–2 mm. long; petals as many as sepals, membranous, narrowly ovate-acuminate, about 0·8 mm. long, 0·3 mm. wide; stamens 3–4, filaments about 0·5 mm. long; anthers minute, suborbicular; carpels 3–4, narrowly ovoid, about 0·6 mm. long, tapering to a short style and minute, truncate or subcapitate stigma; ovules 2 in each loculus, one or more of the carpels usually aborting. Fruit a membranous follicle containing 1–2 proportionately large, oblong seeds, 0·3–0·4 mm. long, 0·2 mm. wide; testa dark brown, rather regularly longitudinally rugulose.

HAB.: Moist sandy fields, or sandy depressions in *Pinus* woodland; sea-level to 900 ft. alt.; fl. Febr.–April.

DISTR.: Divisions 1, 4 and 6, rare. Aegean Islands, Syria, Palestine, Egypt, Libya and eastwards to Iraq and Iran.

1. Karavopetres (Akamas), 1974, *Meikle* 4007 ! Fabrika Rocks near Kato Paphos, 1974, *Meikle* 4011 !
4. Larnaca, 1905, *Holmboe* 83; Ayia Napa, 1905, *Holmboe*; Ayios Antonios, Sotira, 1952, *Merton* 654 !
6. Dhiorios, 1962, *Meikle* 2329 !

2. C. vaillantii (*Willd.*) *Roth*, Enum. Plant. Phaen., 1: 992 (1827); Berger in Engl. et Prantl, Pflanzenfam., ed. 2, 18a: 388 (1930); Osorio-Tafall et Seraphim, List Vasc. Plants Cyprus, 50 (1973).

Tillaea vaillantii Willd., Sp. Plant., 1: 720 (1798); Holmboe, Veg. Cypr., 96 (1914).

Bulliarda vaillantii DC., Plant. Succ. Hist., 74 (1801).

TYPE: France; " *in* Galliae *humidis* " (B). Figured in Vaillant, Bot. Paris., 182, t. 10, fig. 2 (1727) from specimens collected in the forest of Fontainebleau.

Small glabrous annual; stems erect or prostrate, simple or branched, often rooting at the nodes, to about 6 (–8) cm. long, rather fleshy, often tinged red; leaves connate at the base, narrowly oblong or oblanceolate, 2–3 mm. long, less than 1 mm. wide, rather fleshy and somewhat concave above with an obtuse apex; flowers solitary, axillary, ebracteate; pedicel 2–5 mm. long; sepals 4, connate for about half their length, the free lobes about 0·5 mm. long, broadly and bluntly triangular; petals 4, pink, ovate-oblong, obtuse, about 1·5 mm. long, 0·8 mm. wide; stamens 4, filaments about 1 mm. long; anthers minute, suborbicular; carpels 4, broadly ovoid, about 0·5 mm. long, rather abruptly narrowed to a short style and a minute stigma; ovules numerous in each loculus. Fruit a membranous follicle containing 9 or more small, oblong, slightly curved seeds, 0·3 mm. long, 0·1 mm. wide; testa dark brown, finely longitudinally rugulose.

HAB.: Moist muddy places by the side of rock pools and puddles; sea-level to about 500 ft. alt.; fl. Febr.–April.

DISTR.: Divisions 1 and 6, rare. S. Europe and the western Mediterranean region eastwards to Libya, Malta and Cyprus.

1. Near crossroads between Neokhorio and Baths of Aphrodite, 1962, *Meikle* 2235 !
6. Nicosia, near the High Commissioner's residence, 1905, *Holmboe* 312.

2. UMBILICUS *DC.*

in Bull. Sci. Soc. Philom., 3: 1 (1801); Prodr., 3: 400 (1828).

Perennial, glabrous, succulent herbs; rootstock tuberous or rhizomatous; basal leaves petiolate, lamina usually rather large, suborbicular, often peltate; flowers numerous in terminal bracteate racemes or panicles, 5-merous; sepals free or almost free, shorter than the sympetalous, cylindrical or campanulate corolla; corolla-lobes ± erect; stamens 10 (or rarely 5), epipetalous, usually in 2 ranks; filaments short; carpels free; style very short. Fruit a slender, many-seeded follicle.

About 15 species, chiefly in Europe, W. Asia, N. Africa and the Atlantic Islands.

Pedicels 2–10 mm. long; flowers pendulous; rhachis generally exceeding vegetative portion of stem; upper cauline leaves flabellate or suborbicular **1. U. rupestris**

Pedicels generally less than 2 mm. long, often almost wanting; flowers usually horizontal or slightly declinate; rhachis generally shorter than vegetative portion of stem; upper cauline leaves subulate - - - - - - - **2. U. horizontalis**

1. **U. rupestris** (*Salisb.*) *Dandy* in Riddelsd., Hedley et Price, Fl. Gloucester., 611 (1948); Davis, Fl. Turkey, 4: 212 (1972); Osorio-Tafall et Seraphim, List Vasc. Plants Cyprus, 50 (1973).
Cotyledon rupestris Salisb., Prodr. Hort. Chapel Allerton, 307 (1796).
Umbilicus pendulinus DC. in Lam. et DC., Fl. Franç., 4: 383 (1805), Plant. Succ. Hist., 162 (1829); Unger et Kotschy, Die Insel Cypern, 312 (1865); Boiss., Fl. Orient., 2: 769 (1872); Holmboe, Veg. Cypr., 94 (1914); Lindberg f., Iter Cypr., 18 (1946); Rechinger f. in Arkiv för Bot., ser. 2, 1: 423 (1950).
Cotyledon tuberosa (L.) Hal., Consp. Fl. Graec., 1: 577 (1901); Post, Fl. Pal., ed. 2, 1: 489 (1932).
TYPE: Not precisely indicated.

Erect, tuberous perennial; tuber subglobose or ovoid, 1–3 cm. diam., usually covered with a dense web of fine roots; stem 10–50 cm. high, glabrous, terete; basal leaves suborbicular, peltate, succulent, 3–10 cm. diam., usually concave and shallowly funnel-shaped, margins coarsely crenate; petioles up to 15 cm. long or sometimes longer; cauline leaves becoming progressively smaller and more shortly petiolate, the upper broadly cuneate-flabellate or suborbicular, not peltate; all the leaves shrivelling and dropping off at anthesis; flowers very numerous in elongate cylindrical racemes or sparingly branched panicles, somewhat pendulous, the rhachis usually exceeding in length the vegetative portion of the stem; bracts subulate, oblanceolate or obovate, 5–12 mm. long, 2–7 mm. wide, entire or obscurely serrate; pedicels 2–10 mm. long; calyx divided almost to base into 5 narrowly triangular lobes about 1·5 mm. long, 1 mm. wide at base; corolla greenish-yellow often finely streaked and spotted red, 5–8 mm. long, 2·5–5 mm. diam., divided at the apex into 5 overlapping, broadly ovate-acuminate erect lobes about 2–2·5 mm. long and almost as wide at base; stamens in 2 series; filaments about 0·8 mm. long, attached towards apex of corolla-tube; anthers suborbicular, about 0·8 mm. diam.; carpels 5, about 4 mm. long, 1 mm. wide; nectary-scales narrowly oblong, about 2 mm. long, 0·6 mm. wide, truncate at apex; style rather short; stigma obliquely capitate. Fruit a narrowly boat-shaped, membranous follicle, about 5·5 mm. long, 1·5 mm. wide; seeds numerous, narrowly ellipsoid, about 0·4 mm. long, 0·1 mm. wide; testa dark brown, closely and rather regularly erose-striate.

HAB.: Rocks and walls; sea-level to 5,600 ft. alt.; fl. March–June.

DISTR.: Divisions 1–3, 7, 8, common in Divisions 2 and 7. Southern and western Europe, Turkey, Syria, Palestine; western N. Africa, Azores, Madeira.

1. Tsadha (Paphos), 1905, *Holmboe* 732; Dhrousha, 1941, *Davis* 3273 !
2. Common; *Haradjian* 940 ! *Syngrassides* 1558 ! *Kennedy* 522 ! 523A ! 1060–1065 ! *Davis* 3173 ! etc.
3. Stavrovouni, 1880, *Sintenis & Rigo*.
7. Common; *Syngrassides* 684 ! 1446 ! 1447 ! *Kennedy* 1066 ! *Atherton* 744 ! 1209 ! 1352 ! *J. B. Suart* 72 !
8. Kantara, 1880, *Sintenis & Rigo* 211 ! Aphendrika [" Epiotissa "], 1880, *Sintenis & Rigo*.

NOTES: Some of the specimens (e.g. *Kennedy* 1063, *Haradjian* 940) from high altitudes on the Troödos range depart from typical *U. rupestris* in having unusually long pedicels and short corollas; there is, however, no one character which can be used to distinguish them at specific or infraspecific level, and they appear to intergrade with the type in other areas of the island.

2. **U. horizontalis** (*Guss.*) *DC.*., Prodr., 3: 400 (1828); Boiss., Fl. Orient., 2: 770 (1872); Lindberg f., Iter Cypr., 18 (1946); Davis, Fl. Turkey, 4: 213 (1972); Osorio-Tafall et Seraphim, List Vasc. Plants Cyprus, 50 (1973).
Cotyledon horizontalis Guss. in Ten., Fl. Nap. Prodr. App. 5, 13 (1826); Post, Fl. Pal., ed. 2, 1: 489 (1932).
[? *Umbilicus intermedius* (vix Boiss.) Rechinger f. in Arkiv för Bot., ser. 2, 1: 423 (1950).]
TYPE: Sicily; without precise locality, *Gussone* (NAP).

Very similar to *U. rupestris* in habit and foliage, but leaves diminishing abruptly so that the upper two-thirds of the stem is clothed only with subulate or narrowly lanceolate-accuminate, scale-like bracts generally less than 2 cm. long and 0·5 cm. wide at base; flowers numerous, subsessile or very shortly pedicellate, in narrow, cylindrical, congested, often unbranched, spikes or racemes, usually horizontal or slightly declinate, the rhachis normally shorter than the vegetative portion of the stem; upper cauline leaves subulate, about 5 mm. long, 1 mm. wide, generally entire; pedicels usually less than 2 mm. long, sometimes almost wanting; calyx divided to a little way above base into 5 rather broadly triangular, acuminate lobes 1–1·5 mm. long, 0·8–1 mm. wide at base; corolla greenish-yellow, often finely streaked and spotted red, 4·5–5·5 mm. long, 2–2·5 mm. diam., divided at the apex into 5 overlapping ovate-acuminate, erect lobes about 1·5 mm. long and almost as wide at base; stamens in 2 series; filaments about 0·3 mm. long, attached towards apex of corolla-tube; anthers suborbicuar, about. 0·5 mm. diam.; nectary-scales narrowly oblong, about 1·5 mm. lonlg, 0·4 mm wide, truncate at apex; carpels 5, about 4 mm. long, 1 mm. wide; style very short; stigma obliquely capitate. Fruit a narrowly boat-shaped, membranous, dark reddish, rugulose follicle, about 4·5 mm. long, 1·5 mm. wide; seeds numerous, narrowly ellipsoid, about 0·4 mm. long, 0·1 mm. wide; testa dark brown, closely and rather irregularly rugulose-striate.

HAB.: Crevices in limestone rocks and on old walls; sea-level to 1,200 ft. alt.; fl. April–May.

DISTR.: Divisions 4, 7, rare. Italy and Sicily eastwards to Turkey.

4. Near Famagusta, 1913, *Haradjian* 126; Sotira, Ayios Antonios, 1950, *Chapman* 591 !
7. St. Hilarion, 1939, *Lindberg f.* s.n.; ¾ m. E. of St. Hilarion, 1951, *Casey* 1189 !

3. ROSULARIA (*DC.*) *Stapf*

Bot. Mag., 149: t. 8985 (1923).

Umbilicus DC. sect. *Rosularia* DC., Prodr., 3: 399 (1828).

Perennial succulent herbs with a thick, often woody, rootstock; basal leaves sessile, forming a rosette, usually oblanceolate, obovate or spathulate; flowers usually in lax panicles, 5 (–6)-merous; sepals lobed almost to base, shorter than the gamopetalous, deeply lobed corolla; stamens twice as many as corolla-lobes; carpels free, erect; styles short. Fruit a small, many-seeded follicle.

About 16 species in the Aegean area, western and Central Asia.

Calyx-lobes broadly ovate, connivent, forming a distinctly 5-angled calyx; leaves usually entire and glandular-pubescent all over - - - - - - **1. R. cypria**
Calyx-lobes rather narrowly ovate, not connivent; leaves usually glabrous with papillose-denticulate margins - - - - - - - - - - **2. R. pallidiflora**

1. R. cypria (*Holmboe*) *Meikle* **comb. nov.**

Umbilicus cyprius Holmboe, Veg. Cypr., 94 (1914); Lindberg f., Iter Cypr., 18 (1946); R. Seligman in Bull. Alp. Gard. Soc., 19: 55 (1951), 20: 234 (1952); Osorio-Tafall et Seraphim, List Vasc. Plants Cyprus, 50 (1973).

[*Umbilicus pestalozzae* (non Boiss.) Kotschy in Oesterr. Bot. Zeitschr., 12: 275 (1862); Post in Mém. Herb. Boiss., 18: 93 (1905); H. Stuart Thompson in Journ. Bot., 44: 308 (1906).]

[*Umbilicus globulariifolius* (non Fenzl) Kotschy in Unger et Kotschy, Die Insel Cypern, 311 (1865) pro parte; H. Stuart Thompson in Journ. Bot., 44: 308 (1906) pro parte; Holmboe, Veg. Cypr., 94 (1914) pro parte.]

TYPE: Cyprus; St. Hilarion, *Holmboe* 824a (O).

Loosely tufted perennial, often with an elongate, leafless basal trunk or rootstock; basal leaves crowded into rather untidy rosettes, oblanceolate or narrowly spathulate, grey-green, about 3–4 cm. long and 0·7–1 cm. wide, densely, but shortly glandular-pubescent all over, aromatic, apex blunt or subacute, margins entire; flowering stems to about 20 cm. long, spreading or suberect, densely glandular-pubescent, the lower part rather densely clothed with overlapping oblanceolate or spathulate leaves; flowers numerous in narrow, rather condensed panicles 7–12 cm. long; bracts oblanceolate, foliaceous, to about 2 cm. long, 0·5 cm. wide; pedicels 3–5 mm. long, glandular-pubescent, calyx deeply 5-lobed, rather angular; lobes broadly ovate, acute, about 4 mm. long, 3 mm. wide, densely glandular, connivent; corolla-lobes 5, white, glandular, narrowly lanceolate-acuminate, distinctly recurved, 8–10 mm. long, about 1·5 mm. wide, free for about two-thirds of their total length; stamens indistinctly biseriate, attached towards base of corolla-tube, filaments about 3 mm. long; anthers minute, yellow, about 0·2 mm. diam.; nectary-scales narrowly oblong, about 0·8 mm. long, 0·4 mm. wide, apex distinctly emarginate; carpels narrowly boat-shaped, about 4 mm. long, 1 mm. wide, glabrous or very thinly glandular, style very short, stigma obliquely capitate. Fruit pale brown, papery; seeds ovoid-ellipsoid, about 0·4 mm. long, 0·3 mm. wide; testa mid-brown with numerous, regular, longitudinal ribs. *Plate 49, figs. 1–2.*

HAB.: Limestone rocks, cliffs and walls, usually in shady, north-facing aspects; 1,000–3,000 ft. alt.; fl. April–July.

DISTR.: Division 7, locally common. Endemic.

7. Locally common (St. Hilarion, Kyrenia, Buffavento, Yaïla, Lapithos, Pentadaktylos); *Kotschy* 488; *Holmboe* 824a; *Sintenis & Rigo* 598 (partly) ! *Lindberg f.* s.n. ! *Kennedy* 521 ! 1055–1057 ! 1488 ! 1577 ! *Davis* 1739 ! 2969 ! 3645 ! 3647 !

NOTES: Most early authors confuse this and the following species under the names *U. globulariifolius* or *U. pestalozzae*, from which both are quite distinct. Kotschy's record for *U. pestalozzae* from Cape Greco (Oesterr. Bot. Zeitschr., 12: 275; 1862) is a mystery. The record is not repeated or referred to in *Die Insel Cypern.*

2. R. pallidiflora (*Holmboe*) *Meikle* **comb. nov.**

Umbilicus pallidiflorus Holmboe, Veg. Cypr., 95 (1914); R. Seligman in Bull. Alp. Gard. Soc. 19: 55 (1951), 20: 234 (1952); Osorio-Tafall et Seraphim, List Vasc. Plants Cyprus, 50 (1973).

[*Umbilicus globulariifolius* [" *globulariaefolius* "] (non Fenzl) Kotschy in Unger et Kotschy, Die Insel Cypern, 311 (1865) pro parte; H. Stuart Thompson in Journ. Bot., 44: 308 (1906) pro parte; Holmboe, Veg. Cypr., 94 (1914) pro parte.]

[*Umbilicus pestalozzae* (non Boiss.) Post in Mém. Herb. Boiss., 18: 93 (1900); H. Stuart Thompson in Journ. Bot., 44: 308 (1906).]

TYPE: Cyprus; St. Hilarion, *Holmboe* 824b (O).

Tufted perennial, rootstock usually less branched than in *R. cypria*, commonly thick and distinctly swollen towards the apex; basal leaves numerous, crowded into a rather compact rosette, broadly oblanceolate or spathulate, grey-green, 1·5–3·5 cm. long, 0·4–1·2 cm. wide, generally glabrous, apex blunt, apiculate or subacute, often somewhat incurved, margins irregularly papillose-denticulate; flowering stems (3–) 5–11 (–16) cm. long, usually erect, densely glandular-pubescent, rather thinly clothed with slightly overlapping oblanceolate, obtuse, caducous leaves; flowers 8–35 or sometimes more in short, rather loose panicles 2–5 (–7) cm. long; bracts oblanceolate, small, rarely more than 4 mm. long, 2·5 mm. wide; pedicels 3–8 mm. long, glandular-pubescent, calyx deeply 5-lobed, not angular, lobes rather narrowly ovate-acuminate, 2·5–4 mm. long, 1·5–2 mm. wide, densely glandular, corolla-lobes 5, white, veined pink or red, glandular, narrowly lanceolate-acuminate, often distinctly recurved towards apex, 5–7 mm. long, 1·5–2 mm. wide, free for about one half their total length; stamens indistinctly biseriate, attached towards base of corolla-tube, filaments about 2·5–3·5 mm. long; anthers minute, yellow, about 0·2 mm. diam.; nectary-scales narrowly oblong, 0·8–1·2 mm. long, 0·4 mm. wide, apex distinctly emarginate; carpels boat-shaped, about 5–6 mm. long, 1·5–2 mm. wide, glabrous; style very short,

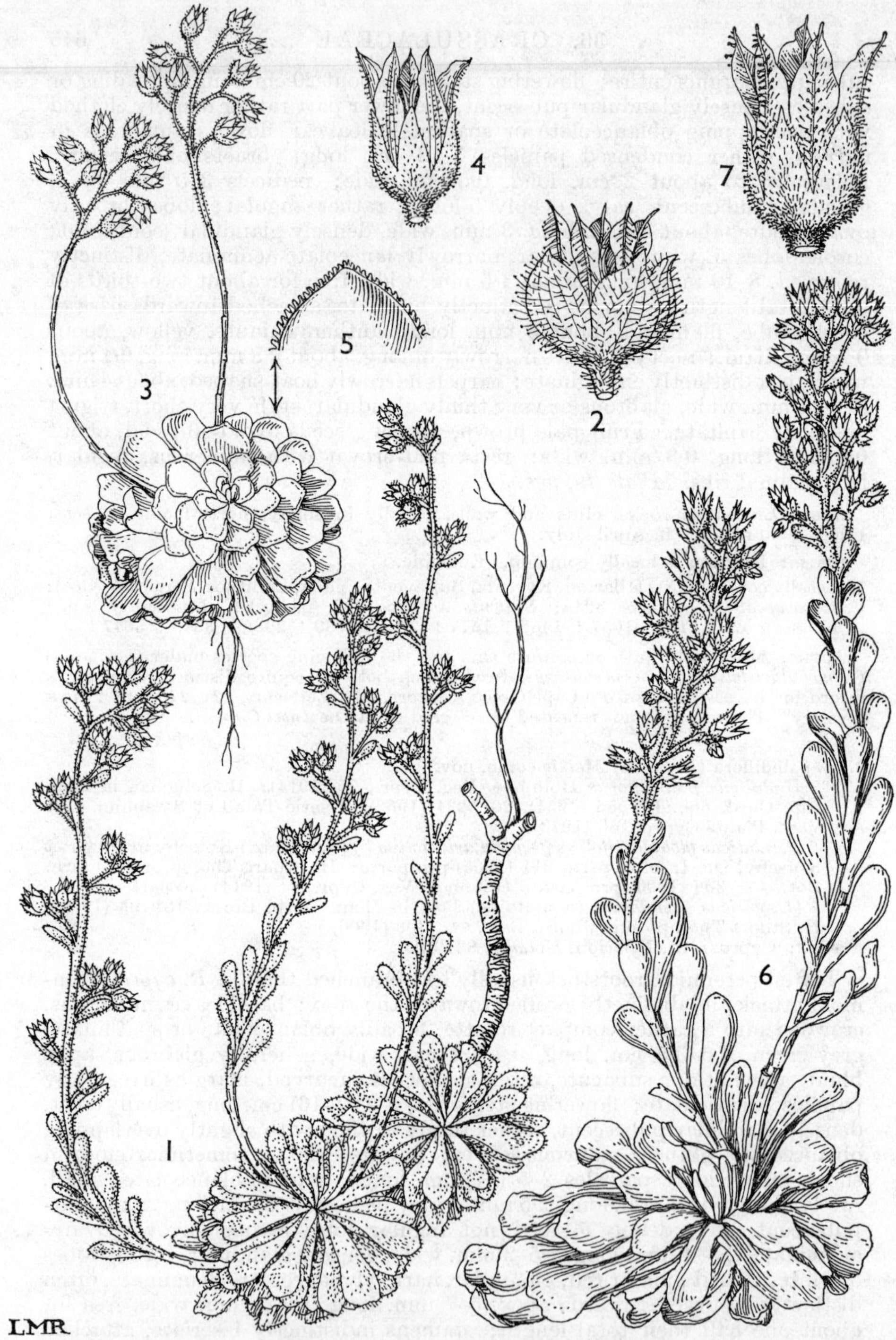

Plate 49. ROSULARIA CYPRIA (Holmboe) Meikle (figs. **1, 2**) **1,** habit, ×$\frac{2}{3}$; **2,** flower, ×3. R. PALLIDIFLORA (Holmboe) Meikle (figs. **3–5**); **3,** habit, ×$\frac{2}{3}$; **4,** flower, ×3; **5,** leaf margin, ×3. R. ? CYPRIA ×PALLIDIFLORA or R. PALLIDIFLORA form (figs. **6, 7**); **6,** habit, ×$\frac{2}{3}$; **7,** flower, ×3. (**1, 2** from *Kennedy* 1488; **3, 5** from *Kennedy* 1487; **4** from *Davis* 3644; **6, 7** from *Kennedy* 1576.)

stigma obliquely capitate. Fruit pale brown, papery; seeds narrowly ellipsoid, about 0·7 mm. long, 0·3 mm. wide; testa dark brown with numerous, regular, longitudinal ribs. *Plate 49, figs. 3–5.*

HAB.: Limestone rocks, cliffs and walls, frequently in sunny, south-facing aspects; 1,000–3,000 ft. alt.; fl. April–July.

DISTR.: Division 7, locally common. Endemic.

7. Locally common (St. Hilarion, Yaïla, Pentadaktylos, Bellapais); *Sintenis & Rigo* 598 (partly) ! *Holmboe* 824b; *Kennedy* 1053 ! 1054 ! 1487 ! 1576 ! *Davis* 2970 ! 3644 ! 3647 ! *Townsend* 63/71 !

NOTES: Generally quite distinct from *R. cypria*; two specimens (*Kennedy* 1576 and *Townsend* 63/71, both from St. Hilarion Castle) are somewhat intermediate and may conceivably be hybrids. The Townsend specimen resembles *R. cypria* in every respect save in inflorescence and calyx, which agree better with *R. pallidiflora*. The Kennedy specimen (*Plate 49, figs. 6–7*) has the inflorescence of *R. cypria*, but is otherwise *R. pallidiflora*.

4. SEDUM *L.*

Sp. Plant., ed. 1, 430 (1753).
Gen. Plant., ed. 5, 197 (1754).
R. Ll. Praeger in Journ. R. H. S. 46: 1–314 (1921).
J. A. Huber in Ber. Naturw. Ver. Landshut (Bayern), 1911–1928: 9–118 (1928).
H. Fröderström in Medd. Göteb. Bot. Trädg., 5–7, 10 (1930–36).

Annual or perennial herbs, or rarely subshrubs; leaves succulent, alternate, opposite or verticillate, entire or variously toothed or lobed; flowers generally in terminal corymbose or paniculate cymes, sometimes axillary, unisexual or hermaphrodite, 5-merous or rarely 4- or 6–9-merous; petals white, pink, yellow, blue, or rarely greenish or brownish, free or slightly connate at base; stamens usually twice as many as petals; carpels as many as petals. Fruit consisting of erect or spreading many-seeded follicles.

About 500 species throughout the northern hemisphere, also south of the Equator but only in tropical Africa and tropical America.

Basal leaves forming compact rosettes; plants monocarpic; flowers very numerous in cylindrical or narrowly pyramidal panicles:
  Basal leaves glabrous, rather broadly spathulate:
    Basal leaves with minutely papillose margins, 4–10 cm. long, 2–4 cm. wide - **2. S. lampusae**
    Basal leaves with entire margins, 3–6 cm. long, 1–2 cm. wide - **4. S. cyprium**
  Basal leaves thinly glandular-pubescent, linear-spathulate - **3. S. microstachyum**
Basal leaves not forming rosettes; flowers in lax cymes or in corymbs:
  Robust perennial 15–60 cm. high with sterile shoots; flowers pale yellowish in dense corymbs - - - - - - - - - - **1. S. sediforme**
  Slender annuals without sterile shoots, always less than 15 cm. high:
    Flowers white or pinkish; seeds longitudinally ribbed or costate:
      Follicles erect or spreading, rugulose-papillose; upper parts of stem and rhachis generally glandular:
        Follicles erect or suberect, about 1–1·5 mm. wide at base; inflorescence usually lax with spreading or recurved branches - - **5. S. porphyreum**
        Follicles spreading, about 1·8 mm. wide at base; inflorescence usually compact with erecto-patent or slightly recurved branches - - **6. S. rubens**
      Follicles stellate-patent, smooth with a conspicuous lateral nerve; stem and rhachis glabrous - - - - - - - - **7. S. caespitosum**
    Flowers yellow or yellowish; seeds not costate - - - - - **8. S. litoreum**

SECT. 1. **Sedum** Perennials with sterile shoots; flowers hermaphrodite; follicles many-seeded.

1. S. sediforme (*Jacq.*) *Pau* in Act. Mem. Prim. Congr. Nat. Esp. Zaragoza, 246 (1909); Post, Fl. Pal., ed. 2, 1: 492 (1932); Davis, Fl. Turkey, 4: 230 (1972); Osorio-Tafall et Seraphim, List Vasc. Plants Cyprus, 50 (1973).

Sempervivum sediforme Jacq., Hort. Vindob., 1: 35, t. 81 (1770); Hume in Walpole, Mem. Europ. Asiatic Turkey, ed. 1, 253 (1817).

Sedum altissimum Poir. in Lam., Encycl. Méth., 4: 634 (1796); Unger et Kotschy, Die Insel Cypern, 313 (1865); Boiss., Fl. Orient., 2: 785 (1872); Holmboe, Veg. Cypr., 63 (1914); Lindberg f., Iter Cypr., 18 (1946).

TYPE: Origin unknown; the illustration in Jacquin, *Hortus botanicus Vindobonensis*, t. 81, cited above !

A robust perennial; stems erect or ascending, 15–60 cm. high, often rather woody at the base, generally much branched; leaves succulent, glaucous, oblong-lanceolate or narrowly elliptical, 1·5–3 cm. long, 0·3–0·5 cm. wide, flattened above, glabrous, apex acuminate or apiculate, base shortly and bluntly spurred, margins entire; flowering stems usually unbranched, thinly leafy almost to apex; inflorescence terminal, corymbose, dense, 4–8 cm. diam., with 6 or more recurved branches; bracts small, narrowly ovate, foliaceous; pedicels usually less than 2 mm. long, glabrous; calyx divided almost to base into 5–6 ovate, acute, glabrous lobes 1·5–2 mm. long, 1–1·3 mm. wide; petals 5–6, pale yellow or whitish, narrowly oblong, obtuse, 5–6 mm. long, 2–2·5 mm. wide, concave, with a distinct midrib, spreading or suberect, connate only at the very base; stamens 10–12, attached at the base of the corolla, about 5 mm. long; filaments subulate, glabrous or thinly papillose towards base; anthers narrowly oblong, about 1·5 mm. long, 0·6 mm. wide; carpels 5–6, erect, shortly papillose, often spotted reddish, about 2·5–3 mm. long, 0·8–1 mm. wide; styles slender, about 1·5 mm. long; stigma minute; ovules numerous in each carpel. Follicles erect, about 5 mm. long, 2·5 mm. wide; seeds very narrowly ellipsoid, about 0·8 mm. long, 0·3 mm. wide; testa brown, longitudinally ribbed, the ribs connected by numerous, horizontal, parallel striae.

HAB.: Cliffs and rocky ground on limestone or igneous formations; sea-level to 3,000 ft. alt.; fl. April–July.

DISTR.: Divisions 1–3, 6–8. Widespread in the Mediterranean region.

1. Paphos, 1901, *A. G. & M. E. Lascelles* s.n. !
2. Between Prodhromos and Tris Elies, 1862, *Kotschy* 816a; Galata, 1880, *Sintenis & Rigo* 718 ! Kato Platres, 1937, *Kennedy* 1067 ! Galata, 1939, *Lindberg f.* s.n. ! Lefkara, 1940, *Davis* 1913 !
3. Limassol, 1801, *Hume* s.n.
6. Spilia tou Drakou near Kormakiti, 1949, *Casey* 868 !
7. Antiphonitis [Church], 1905, *Holmboe* 841; Kalyvakia, 1938, *Kennedy* 1068 ! Kyrenia Range—9 miles W. end of Forest Rd., 1955, *G. E. Atherton* 135 !
8. Koma tou Yialou, 1936, *Syngrassides* 1243 !

SECT. 2. **Cepaea** *Koch* Annual or monocarpic; stems often unbranched; leaves flat, the basal forming a rosette; inflorescence usually paniculate and many-flowered; flowers hermaphrodite; follicles many-seeded.

2. S. lampusae (*Kotschy*) *Boiss.*, Fl. Orient., 2: 787 (1872), Suppl., 246 (1888); Holmboe, Veg. Cypr., 93 (1914); A. K. Jackson et Turrill in Kew Bull., 1939: 477 (1939); Lindberg f., Iter Cypr., 19 (1946); Osorio-Tafall et Seraphim, List Vasc. Plants Cyprus, 50 (1973); Meikle in Bot. Mag., 180: n.s. t. 660 (1974).

Umbilicus lampusae Kotschy in Unger et Kotschy, Die Insel Cypern, 312 (1865).

[*Sempervivum globiferum* (non Jacq.) Sibth. et Sm. sec. Kotschy in Unger et Kotschy, Die Insel Cypern, 312 (1865).]

TYPE: Cyprus; Lapithos (formerly Lampusa), April 17, 1862, *Kotschy* 471 (W).

Erect monocarpic herb (15–) 20–50 cm. high; stem generally unbranched; leaves sessile, becoming dry and shrivelled at or soon after anthesis, the basal flat, spathulate, acute or obtuse, glaucous, glabrous or thinly pubescent with minutely papillose margins, 4–10 cm. long, 2–4 cm. wide, crowded into a neat rosette; cauline leaves becoming smaller and semi-terete upwards; inflorescence a narrow, unbranched, cylindrical panicle, often 30 cm. long, or sometimes pyramidal with several lateral branches towards the base; rhachis densely and shortly glandular; bracts small, lanceolate or subulate, glandular; flowers very numerous, crowded, greenish or brownish-green; pedicels slender, glandular, up to 5 mm. long; calyx divided almost to the base into 5 ovate, acuminate, glandular lobes, about 1·5 mm. long, 1 mm. wide; petals 5, narrowly ovate-acuminate, membranous, about 4 mm. long, 2 mm. wide; filaments glabrous, filiform, about 2·5 mm. long; anthers yellow or reddish, reniform, about 0·4 mm. long and wide; carpels usually 5, erect, narrowly ovoid, glandular, about 4 mm. long, 1·2 mm. wide, tapering upwards to a slender style about 1 mm. long; stigma minute, punctiform. Follicles erect, scarcely accrescent; seeds numerous, bluntly oblong-ellipsoid, about 0·7 mm. long, 0·4 mm. wide; testa mid-brown, with numerous, close, longitudinal ribs.

HAB.: Walls, rock-crevices and dry stony slopes on calcareous formations; 600–2,500 ft. alt.; fl. June–August.

DISTR.: Division 7, common along the Northern Range. Endemic.

7. Locally common; Syngrassides 64 ! *Kennedy* 520 ! 1047–1050 ! 1578a ! 1636 ! 1637a ! *Lindberg f.* s.n. ! *Atherton* 250 ! 743 !

NOTES: Apparently confined to the Northern Range; earlier records from the Troödos area are errors for *S. cyprium* or *S. microstachyum*.

3. S. microstachyum (*Kotschy*) *Boiss.*, Fl. Orient., 2: 787 (1872), Suppl., 246 (1888); Holmboe, Veg. Cypr., 93 (1914); A. K. Jackson et Turrill in Kew Bull. 1939: 478 (1939); Lindberg f., Iter Cypr., 18 (1946); Osorio-Tafall et Seraphim, List Vasc. Plants Cyprus, 50 (1973).

Umbilicus microstachyus Kotschy in Unger et Kotschy, Die Insel Cypern, 311 (1865).

Sedum lampusae (Kotschy) Boiss. var. *microstachyum* (Kotschy) Fröderström in Medd. Göteb. Bot. Trädg., 7: 17 (1932)—as var. *michrostachyum*.

TYPE: Cyprus; north side of Troödos peak, May 20, 1862, *Kotschy* 786 (W).

Erect monocarpic herb (9–) 25–40 cm. high; stem generally unbranched; leaves sessile, shrivelling soon after anthesis, the basal flat, linear-spathulate, subacute or obtuse, green in shade, reddish in sun, thinly glandular-pubescent with entire, narrowly scarious margins, 3–6 cm. long, 0·5–1·2 cm. wide, crowded into a loose rosette; cauline leaves numerous, flat, much shorter than the basal but proportionately rather wide, obtuse, densely glandular-pubescent, not diminishing very noticeably upwards, and continuing as bracts along the lower part of the inflorescence; inflorescence a narrow, generally unbranched, cylindrical panicle, generally about 6–13 cm. long, but sometimes up to 25 cm.; rhachis densely glandular; flowers very numerous, crowded, reddish or greenish; pedicels slender, glandular, 4–5 mm. long; calyx divided almost to the base into 5 narrowly ovate-deltoid, glandular lobes about 3 mm. long, 1·3 mm. wide; petals 5, narrowly ovate-acuminate, membranous, about 5 mm. long, 1·5 mm. wide; filaments glabrous, filiform, about 2·5 mm. long; anthers yellow or reddish, reniform, about 0·7 mm. long, 0·5 mm. wide; carpels usually 5, erect, narrowly ovoid, thinly glandular, about 2·5 mm. long, 1·5 mm. wide, tapering to a slender style about 1 mm. long; stigma minute, punctiform. Follicles erect, about 5 mm. long, 1·5 mm. wide; seeds numerous, bluntly oblong-ellipsoid, about 0·5 mm. long, 0·3 mm. wide; testa mid-brown with numerous, close, longitudinal ribs.

HAB.: Crevices of igneous rocks; 5,000–6,300 ft. alt.; fl. June–Sept.

DISTR.: Division 2, locally common. Endemic.

2. Locally common on Khionistra; *Sintenis & Rigo* 720 ! *Haradjian* 488 ! *Kennedy* 526 ! 527 ! 1051 ! 1248 !–1251 ! *Davis* 1946 ! 3550 ! *Mavromoustakis* 48 ! *D. P. Young* 7802 ! etc.

4. S. cyprium *A. K. Jackson et Turrill* in Kew Bull. 1939: 476 (1939); Lindberg f. Iter Cypr., 18 (1946); Rechinger f. in Arkiv för Bot., ser. 2, 1: 423 (1950); Osorio-Tafall et Seraphim, List Vasc. Plants Cyprus, 50 (1973).

TYPE: Cyprus; Kryos Potamos, in rock high above the stream; 860–980 m., June 16, 1937, *Kennedy* 524 (K !).

Erect monocarpic herb, similar in general appearance to *S. lampusae* and *S. microstachyum*; generally 9–15 cm. high, but occasionally attaining 30–35 cm.; leaves sessile, shrivelling at or soon after anthesis, the basal flat, rather broadly spathulate, obtuse or rounded, green in shade, reddish in sun, glabrous with entire narrowly scarious margins, 3–6 cm. long, 1–2 cm. wide, crowded into a rather loose rosette; cauline leaves numerous, flat, shorter than the basal but proportionately rather wide, obtuse, thinly glandular-pubescent, continuing as bracts along the lower part of the inflorescence, inflorescence generally unbranched and forming a cylindrical or narrowly conical panicle 6–15 (–20) cm. long; rhachis densely glandular; flowers very numerous, but less densely crowded than in *S. microstachyum*, reddish or greenish; pedicels slender, glandular, 3–4 mm. long; calyx divided almost to the base into 5 ovate-acuminate, glandular lobes about 1·5 mm. long, 0·8 mm. wide; petals 5, narrowly ovate-acuminate, keeled, about 2·5 mm. long, 1 mm. wide; filaments about 1·7 mm. long, minutely papillose towards the base; anthers suborbicular, greenish or yellowish, about 0·4 mm. diam.; carpels usually 5, erect, narrowly ovoid, glabrous or very thinly glandular, about 1·8 mm. long, 0·8 mm. wide, tapering to a slender style about 1 mm. long; stigma minute, punctiform. Follicles erect, about 2·5–3 mm. long, 1 mm. wide; seeds numerous, bluntly oblong-ellipsoid, about 0·4 mm. long, 0·2 mm. wide; testa mid-brown, with numerous, close, longitudinal ribs.

HAB.: Crevices of igneous (or rarely calcareous) rocks; 500–5,350 ft. alt.; fl. June–Sept.

DISTR.: Divisions 1 and 2, common in Division 2. Endemic.

1. Akamas, Smyies forest station, 1941, *Davis* 3320 ! Akamas, Ayia Nikola above Neokhorio, 1941, *Davis* 3321 !
2. Locally common; *Sintenis & Rigo* 893 ! 895 ! *Haradjian* 955 ! *Kennedy* 524 ! 525 ! 1252 ! 1551–1553 ! 1634 ! *Lindberg f.* s.n. ! *Davis* 1763 ! 1831 ! 1874 ! etc.

SECT. 3. **Epeteium** *Boiss.* Annual or monocarpic, without sterile shoots; leaves small, not forming rosettes; inflorescences not paniculate, rather few-flowered; flowers hermaphrodite. Follicles many-seeded.

5. S. porphyreum *Kotschy* in Unger et Kotschy, Die Insel Cypern, 313 (1865); Boiss., Fl. Orient., 2: 790 (1872); Rechinger f. in Arkiv för Bot., ser. 2, 1: 423 (1950); Osorio-Tafall et Seraphim, List Vasc. Plants Cyprus, 51 (1973).

S. porphyreum Kotschy var. *parviglandulosum* Lindberg f., Iter Cypr., 18 (1946).

TYPE: Cyprus; "Bei Cerinia in den Steinbrüchen an Kanten der Felsen", April 17, 1862, *Kotschy* 475 (lectotype W !).

Small erect annual, generally less than 6 cm. high; stem unbranched or sparingly branched, usually stained reddish-purple, glabrous; leaves oblong-lingulate or narrowly obovate, sessile, flat, 3–12 mm. long, 2–6 mm. wide, glabrous, usually stained reddish-purple, apex rounded or subacute, base produced into a very short blunt spur; inflorescence a lax or rather dense terminal cyme, 1·5–6 cm. diam., with 2–8 (rarely more) spreading or recurved branches becoming ± erect in fruit; rhachis generally thinly glandular (var. *parviglandulosum* Lindberg f.) occasionally quite glabrous; bracts foliaceous; flowers mostly secund, shortly pedicellate, pedicels stout, 1–1·5 mm. long, frequently glandular; calyx divided almost to base into 5 ovate, acute lobes 1–1·5 mm. long, 0·8–1 mm. wide; petals 5, lanceolate-acuminate, 3–4·5 mm. long, about 1 mm. wide, whitish with a purplish or reddish keel; stamens usually 7–8 (?–10), filaments glabrous, about 3·5 mm. long; anthers dark purple, suborbicular, about 0·3 mm. diam.; carpels erect, narrowly oblong, glabrous or thinly glandular, about 2 mm. long, 0·8 mm. wide, tapering to a slender style about 1 mm. long; stigma minutely capitate. Follicles erect or suberect, about 3·5 mm. long, 1–1·5 mm. wide, dark brown, generally densely rugulose-papillose, many-seeded; seeds broadly ovoid, about 0·6 mm. long, 0·3 mm. wide; testa dark brown with numerous, close, longitudinal ridges.

HAB.: Rock crevices, screes and dry rocky ground; sea-level to 1,800 ft. alt.; fl. March–May.

DISTR.: Divisions 1, 3–5, 7, 8, locally common in Division 7. Endemic.

1. Stroumbi, 1913, *Haradjian* 788 ! Dhrousha, 1941, *Davis* 3242 ! Akamas, Ayia Nikola above Neokhorio, 1941, *Davis* 3335 !
3. 4 miles N. of Kandou, 1963, *J. B. Suart* 55 ! 94 !
4. Sotira, 1956, *Merton* 733 ! Ayia Napa, 1970, *Hansen* s.n. !
5. Vatili, 1905, *Holmboe* 324; Milea-Piyi, 1936, *Syngrassides* 935 !
7. Locally common at Kyrenia; *Kotschy* 475; *Sintenis & Rigo* 663 ! *Syngrassides* 1 ! *Atherton* 1337 ! Kharcha [Hartcha], 1950, *Chapman* 336 !
8. Near Rizokarpaso, 1912, *Haradjian* 412; Ayios Theodhoros, 1939, *Lindberg f.* s.n. ! Cape Andreas, 1948, *Mavromoustakis* s.n. !

NOTES: Generally glandular, but *Haradjian* 788 is quite eglandular and the type (Kotschy 475) nearly so; I doubt if var. *parviglandulosum* Lindberg f. can be maintained even as a variety.

Kotschy 819, purporting to come from Prodhromos, seems to be correctly identified as *S. porphyreum*, though the specimens are very immature and poor. Since *S. porphyreum* is absent from high altitudes on the Troödos range, one must suspect an error in localization. *Kotschy* 817a, from between Prodhromos and Tris Elies, is also likely to be an error, though I have not seen the specimen.

6. S. rubens *L.*, Sp. Plant., ed. 1, 432 (1753); Unger et Kotschy, Die Insel Cypern, 312 (1865); Boiss., Fl. Orient., 2: 794 (1872), Suppl., 247 (1888); Holmboe, Veg. Cypr., 94 (1914); Post, Fl. Pal., ed. 2, 1: 494 (1932); Davis, Fl. Turkey, 4: 243 (1972); Zohary, Fl. Palaest., 2: 7, t. 10 (1972); Osorio-Tafall et Seraphim, List Vasc. Plants Cyprus, 51 (1973).

Crassula rubens (L.) L., Syst. Nat., ed. 10, 969 (1759); Sintenis in Oesterr. Bot. Zeitschr., 31: 393 (1881), 32: 397 (1882).

TYPE: " *in* Gallia, Italia ".

Small erect annual seldom exceeding 5 cm. in our area, often taller elsewhere (up to 14 cm.); stems simple or branched, subglabrous or thinly glandular below, often rather densely hispidulous-glandular above; leaves narrowly oblong, flattish or semi-terete, often tinged red, 8–10 mm. long, 2·5–4·5 mm. wide, apex blunt or rounded, base very shortly and bluntly spurred, margins entire; inflorescence a terminal corymbose cyme, 2–7 cm. diam., with 2–8 erecto-patent or slightly recurved branches, becoming erect in fruit; rhachis rather densely glandular (or very rarely subglabrous); bracts foliaceous, linear-oblong or lanceolate, blunt, about 0·5–1 cm. long, 0·2–0·3 cm. wide; flowers mostly secund, very shortly and stoutly pedicellate; calyx divided almost to base into 5 ovate, acute, glandular lobes 1·5 mm. long, 1 mm. wide; petals 5, narrowly lanceolate-acuminate, about 5 mm. long, 1 mm. wide, keeled, whitish or pinkish with a purplish or reddish keel; stamens 5 (or rarely 10), filaments glabrous, about 3·5 mm. long; anthers purple, suborbicular, about 0·2–0·3 mm. diam.; carpels rather spreading, 3–3·5 mm. long, 1·2–1·5 mm. wide, thinly glandular-papillose; style about 0·8–1 mm. long; stigma minutely capitate. Follicles spreading, about 3·5 mm. long, 1·8 mm. wide at base, usually purplish and strongly glandular-papillose, many-seeded; seeds ovoid-oblong, about 0·7 mm. long, 0·4 mm. wide; testa dark brown with numerous, close, longitudinal ridges.

HAB.: On rocks, walls and stony slopes, apparently confined to igneous formations; 2,000–4,500 ft. alt.; fl. April–June.

DISTR.: Division 2, locally common. Widespread in southern Europe; W. Asia; N. Africa, Atlantic Islands.

2. Prodhromos, 1880, *Sintenis & Rigo* 32 ! Kalopanayiotis Bridge, 1937, *Syngrassides* 1559 ! Platres, 1938, *Kennedy* 1058 ! 1059 ! Kyperounda, 1940, *Kennedy* 1583 ! 1584 ! 1585B ! Tripylos, 1941, *Davis* 3527 !

NOTES: In the past confused with *S. caespitosum* (Cav.) DC. and *S. porphyreum* Kotschy. It would appear to grow only at fairly high altitudes on the Troödos range, and Kotschy's record from Ayios Khrysostomos, as also those of Sintenis & Rigo from Cape Pyla, Cape Andreas and Pentadaktylos, are almost certainly erroneous. The Cyprus plant tends to be rather smaller than the general run of the species.

7. **S. caespitosum** (*Cav.*) *DC.*, Prodr., 3: 405 (1828); Boiss., Fl. Orient. Suppl., 247 (1888); Holmboe, Veg. Cypr., 94 (1914); Davis, Fl. Turkey, 4: 240 (1972); Zohary, Fl. Palaest., 2: 6, t. 8 (1972); Osorio-Tafall et Seraphim, List Vasc. Plants Cyprus, 51 (1973).

Tillaea rubra L., Sp. Plant., ed. 1, 129 (1753).

Crassula cespitosa Cav., Icones, 1: 50, t. 69, fig. 2 (1791).

Sedum rubrum (L.) Thell. in Feddes Repert., 10: 290 (1912); Post, Fl. Pal., ed. 2, 1: 494 (1932); Lindberg f., Iter Cypr., 18 (1946) non Royle ex Edgeworth (1846).

TYPE: Spain; in the neighbourhood of Madrid, *Cavanilles* (MA).

Small erect annual, seldom exceeding 4 cm. in height; stems often branched, glabrous; leaves loosely imbricate, ovate or broadly oblong, 4–5 mm. long, 3–4 mm. wide, glabrous, entire; apex obtuse or rounded, base shortly and bluntly spurred; inflorescence generally a terminal 2-branched cyme, with erect or suberect branches 2–3 cm. long; rhachis glabrous; bracts foliaceous, similar to the cauline leaves, but smaller and often acute or subacute; flowers attached on alternate sides of the rhachis, not secund, subsessile or very shortly pedicellate; calyx divided to almost half its length into (4–) 5 broadly deltoid lobes about 1 mm. long and 1 mm. wide at base; petals (4–) 5, narrowly lanceolate-acuminate, about 3 mm. long, 0·8 mm. wide, keeled, whitish or pinkish with a purplish or reddish keel; stamens usually 5, filaments glabrous, about 1·5 mm. long; anthers yellow, suborbicular, about 0·2–0·3 mm. diam.; carpels at first erect, about 2 mm. long, 0·8 mm. wide, glabrous, tapering above to a style about 0·3 mm. long; stigma minutely capitate. Follicles stellately patent, about 3·5–4 mm. long, 0·8–1 mm. wide at base, usually purplish, glabrous with a conspicuous lateral nerve, many-seeded; seeds ovoid-ellipsoid, about 0·6 mm. long, 0·4 mm. wide; testa dark brown with about 12 prominent longitudinal ridges.

HAB.: Dry hillsides and rocky ground on calcareous or igneous formations; sea-level to 4,000 ft. alt.; fl. Febr.–April.

DISTR.: Divisions 1–8. Southern Europe, western Asia.

1. Fabrika Rocks near Kato Paphos, 1974, *Meikle* 4010 !
2. Livadhia of Kyperounda, 1940, *Kennedy* 1581B ! Kyperounda, 1940, *Kennedy* 1582 !
3. Near Kouka, Perapedhi, 1960, *Macdonald* 80 !
4. Larnaca, 1905, *Holmboe*.
5. Milea–Piyi, 1936, *Syngrassides* 934 ! Mile 5, Nicosia–Famagusta Rd., 1950, *Chapman* 41 !
6. Kokkini Trimithia, 1935, *Syngrassides* 785 !
7. Pentadaktylos, 1880, *Sintenis & Rigo* 1002 ! Kyrenia, 1949, *Casey* 291 ! Kyrenia, 1951, *Casey* 1104 ! 1149 !
8. Ayios Theodhoros, 1939, *Lindberg f.* s.n.

8. **S. litoreum** *Guss.*, Plant. Rar., 185, t. 37, fig. 2 (1826); Kotschy in Unger et Kotschy, Die Insel Cypern, 313 (1865); Boiss., Fl Orient., 2: 793 (1872); Holmboe, Veg. Cypr., 93 (1914); Post, Fl. Pal., ed. 2, 1: 494 (1932); Lindberg f., Iter Cypr., 18 (1946); Davis, Fl. Turkey, 4: 239 (1972); Zohary, Fl. Palaest., 2: 7, t. 12 (1972); Osorio-Tafall et Seraphim, List Vasc. Plants Cyprus, 51 (1973).

[*S. pallidum* (non M. Bieb.) Boiss. ex Holmboe, Veg. Cypr., 94 (1914).]

[*S. palaestinum* (non Boiss.) H. Stuart Thompson in Journ. Bot., 44: 308 (1906); Holmboe, Veg. Cypr., 94 (1914).]

TYPE: Italy; " in ruderatis maritimis Calabriae orientalis; *alle Castella tra Catanzaro e Capo Rizzuto* " *Gussone* (NAP).

Small erect annual, seldom exceeding 10 cm. in height; stems often branched, glabrous, commonly tinged pink; leaves oblanceolate or narrowly spathulate, 3–20 mm. long, 2–5 mm. wide, the upper sessile or subsessile and shortly spurred at base, the lower often tapering to a slender petiole-like base, glabrous, distinctly flattened, apex obtuse, rounded or subacute, margins entire; inflorescence generally a terminal 2–3-branched cyme, with suberect or spreading branches 1·5–3 cm. long; rhachis glabrous; bracts foliaceous, similar to the upper cauline leaves but smaller; flowers attached on alternate sides of the rhachis, sessile or subsessile; calyx divided more than half way into 5 rather narrow deltoid lobes 1–2 mm. long, 0·8–1·3 mm. wide at base; petals 5, oblong-acuminate or cuspidate, 3–4 mm. long, about 1·3 mm. wide, keeled, pale or bright yellow, with a distinctly coloured keel; stamens usually 5, filaments glabrous, about 2·5 mm. long; anthers yellow, suborbicular, about 0·5 mm. diam.; carpels erect or somewhat spreading, about 3 mm. long, 1·5 mm. wide, glabrous, narrowed rather abruptly above to a style about 0·4 mm. long; stigma minutely capitate. Follicles erecto-patent, about 4·5 mm. long, 1·5 mm. wide, pale brown or whitish, persistent and remaining attached, and bleached, to the dead stems for a year or more, many-seeded; seeds ovoid-ellipsoid, somewhat compressed, about 0·8 mm. long, 0·4 mm. wide; testa mid-brown, closely rugulose-papillose, but not costate.

HAB.: Limestone and sandstone rocks and cliffs, usually near the sea; sea-level to 2,000 ft. alt.; fl. April–May.

DISTR.: Divisions 1, 4, 7, 8, common in Division 7. Mediterranean region eastwards from S. France; Turkey; Palestine.

1. Neokhorio–Smyies, Akamas, 1948, *Kennedy* 1590 ! 1591 ! Skala, above Peyia, 1962, *Meikle* 2037 ! 2038 !
4. Cape Greco, 1905, *Holmboe* 415.
7. Common; *Kotschy* 474 ! *A. G. & M. E. Lascelles* s.n. ! *Syngrassides* 2 ! 16 ! *Kennedy* 1579 ! 1580 ! 1581A ! 1585A ! *Davis* 2964 ! 2998 ! 3050 ! etc.
8. Cape Andreas, 1880, *Sintenis & Rigo* 31 ! Kantara Castle, 1940, *Kennedy* 1578 !

5. TELMISSA *Fenzl*

Pugill. Plant. Syr. no. 50 (1842).

Sedum L. sect. *Telmissa* (Fenzl) Schönland in Engler et Prantl, Pflanzenfam., ed. 1, III, 2a: 31 (1891).

Small annual plants, similar in general aspect to *Sedum* L. sect. *Epeteium* Boiss.; basal leaves not forming a rosette; flowers minute, extra-axillary, calyx partly adnate to rhachis; calyx-lobes, petals and stamens usually 4; carpels usually 3. Fruit indehiscent, 1-seeded.

One species only, confined to the eastern Mediterranean.

1. **T. microcarpa** (*Sm.*) *Boiss.*, Fl. Orient., 2: 795 (1872), Suppl., 247 (1888); Holmboe, Veg. Cypr., 96 (1914); Post, Fl. Pal., ed. 2, 1: 495 (1932); Davis, Fl. Turkey, 4: 244 (1972).

Crassula microcarpa Sm., Fl. Graec. Prodr., 1: 217 (1806); Poech, Enum. Plant. Ins. Cypr., 29 (1842); Kotschy in Unger et Kotschy, Die Insel Cypern, 311 (1865).

Telmissa sedoides Fenzl, Pugill. Plant. Syr., no. 50 (1842); Kotschy in Unger et Kotschy, Die Insel Cypern, 311 (1865); Sintenis in Oesterr. Bot. Zeitschr., 31: 393 (1881).

Sedum microcarpum (Sm.) Schönland in Engler et Prantl, Pflanzenfam. III, 2a: 31 (1891); Fröderström in Medd. Göteb. Bot. Trädg., 7: 95 (1931); Zohary, Fl. Palaest., 2: 8, t. 14 (1972); Osorio-Tafall et Seraphim, List Vasc. Plants Cyprus, 51 (1973).

TYPE: Cyprus; " In insulae Cypri rupibus ", *Sibthorp* (OXF).

Erect, sprawling or prostrate annual; stems glabrous, reddish, usually branched, about 2–8 cm. long; leaves succulent, semi-terete, sessile, spreading, narrowly oblong, glabrous, 10–12 mm. long, 1·5–2 mm. wide, apex obtuse, base very shortly and bluntly spurred, margins entire; inflorescence a lax 2–3-branched terminal cyme, branches erect or slightly recurved, 2–4 cm. long; rhachis glabrous, bracts foliaceous becoming progressively smaller upwards; flowers minute, remote, mostly secund, sessile; calyx unilaterally adnate to rhachis, lobes 4 (–5), reduced to minute apiculi; petals 4 (–5), white, oblong, 1–1·5 mm. long, 0·5 mm. wide, apex obtuse or subacute; stamens 4 (–5), filaments glabrous, about 0·7 mm. long; anthers purple, suborbicular, about 0·3 mm. diam.; carpels 3 (–4), the adaxial generally undeveloped, oblong-prismatic, glabrous, about 0·6 mm. long, 0·4 mm. wide at base, apex narrowing rather abruptly to a minute but distinct style; stigma capitate; ovule 1, pendulous. Fruits usually 3, indehiscent, erecto-patent, about 1·5 mm. long, 0·5 mm. wide, apiculate, glabrous, rugulose-papillose; seed 1, broadly or narrowly ovoid, tapering to a slender, pointed apex, almost 1·5 mm. long, 0·3–0·6mm. wide; testa membranous, pale brown, minutely and irregularly rugulose-reticulate.

HAB.: On rocks and stony grounds, usually on calcareous or sandstone formations; sea-level to 2,000 ft. alt.; fl. March–April.

DISTR.: Divisions, 1, 4–7. S.E. Turkey, Syria, Palestine.

1. Ayios Yeoryios (Akamas), 1962, *Meikle* ! Polis, 1962, *Meikle* !
4. Ayia Napa, 1862, *Kotschy* 101 !
5. Mile 5, Nicosia–Famagusta Rd., 1950, *Chapman* 45 ! Athalassa, 1962, *Meikle* 2989 !
6. Kokkini Trimithia, 1935, *Syngrassides* 784 !
7. Pentadaktylos, 1880, *Sintenis & Rigo* 30 ! and 1950, *Casey* 982 !

NOTES: Fröderström (*loc. cit. supra*, p. 95) described *T. microcarpa* as a subaquatic plant; in Cyprus it is generally found, sometimes with *Sedum* species, on dry, rocky ground.

AEONIUM ARBOREUM (*L.*) *Webb et Berth.* (*Sempervivum arboreum* L.), an erect, suffruticose perennial, up to 1 m. high, with long naked branches terminating in flattish rosettes of shining green leaves, and with large terminal panicles of yellow (9–) 10 (–11)-petalled flowers, has been recorded by Sibthorp (Walpole, Travels, 25; 1820) from Ktima [" Istinia "] " upon rubbish ground on the outskirts of the town "—May 11, 1787. The plant has been extensively cultivated in the Mediterranean region since Classical times, but is now considered a native of western North Africa (see Praeger, An Account of the Sempervivum Group, 159; 1932). The references in Unger & Kotschy, Holmboe etc. all trace back to Sibthorp's record, but the plant is still found in Cyprus, either in cultivation or as an escape from cultivation.

37. HAMAMELIDACEAE

Harms in Engl. et Prantl, Pflanzenfam., ed. 2, 18a: 303–345 (1930).
Hutchinson, Gen. Fl. Plant., 2: 93–103 (1967).
Incl. *Altingiaceae* Lindl., Veg. Kingd., 253 (1846) (subfam. *Liquidambaroideae* (Dum.) Harms).

Trees or shrubs; leaves generally alternate, deciduous or evergreen, simple, pinnately or palmately nerved, stipulate; inflorescences often capitate; flowers frequently small, bisexual or unisexual, generally actinomorphic; calyx adnate to ovary, lobes, when present, valvate or imbricate; petals 4 or more, or occasionally wanting, perigynous or epigynous, usually imbricate or valvate; stamens 4-many, generally perigynous and 1-seriate, sometimes in capitate clusters; filaments free; anthers 2-locular, opening by longitudinal slits or by valves, connective often produced; ovary consisting of 2 carpels, often free at the apex, 1- or many-ovulate; styles free, often recurved. Fruit a beaked, 2-locular, woody capsule, opening from the summit; seeds 1 to several in each loculus, sometimes hard and shining, endosperm thin, fleshy, embryo straight, often large with broad flat cotyledons.

25 genera widely distributed in temperate and tropical regions, absent from South America and the West Indies.

1. LIQUIDAMBAR *L.*

Sp. Plant., ed. 1, 999 (1753).
Gen. Plant. ed. 5, 434 (1754).

Balsamiferous trees; leaves alternate, deciduous, long-petiolate, palmatilobed, glandular-serrate, stipules deciduous; flowers generally unisexual, in globose heads, the male capitula in racemes, the female solitary; male flowers without calyx or corolla, stamens crowded; anthers basifixed, opening by longitudinal slits; female flowers connate, calyx-limb reduced to minute scales, petals wanting; ovary semi-inferior, 2-locular; styles and stigmas elongate, subulate, often recurved, persistent; ovules numerous. Fruit a woody, globular head comprising numerous beaked capsules; seeds compressed, shortly winged, with a crustaceous testa, many of them abortive.

Four (or perhaps fewer) species in North America and Asia.

1. **L. styraciflua** *L.*, Sp. Plant., ed. 1, 999 (1753); J. E. Smith in Trans. Linn. Soc., 12: 1–5 (1817); Sibthorp in Walpole, Travels, 17 (1820); Boiss., Fl. Orient., 2: 819 (1872) in sched.; Holmboe, Veg. Cypr., 96 (1914); Lindberg f., Iter Cypr., 18 (1946); Osorio-Tafall et Seraphim, List Vasc. Plants Cyprus, 51 (1973).

L. orientalis (? Mill.) Kotschy in Oesterr. Bot. Zeitschr., 12: 276 (1862); Chapman, Cyprus Trees and Shrubs, 39 (1949).

L. imberbe (? Aiton) Kotschy in Unger et Kotschy, Die Insel Cypern, 220 (1865).

TYPE: " *in* Virginia, Mexico ".

Tree to 12 m. high (in our area), exuding an aromatic gum when cut; bark grey, deeply fissured, twigs and shoots glabrous or glabrescent; buds large, ovoid, pointed, shining dark brown; leaves 4–10 cm. long, 6–15 cm. wide, deeply (3–) 5-lobed, subcordate at base, lobes broadly or narrowly deltoid, acute or acuminate, occasionally lobulate, margins bluntly glandular-serrulate, lamina glabrous above, subglabrous below with distinct tufts of hairs in the axils of the primary and sometimes of the secondary nerves; petioles slender, subterete, glabrous, up to 10 cm. long; stipules oblong-acuminate, about 8 mm. long, 5 mm. wide, thinly villose or subglabrous; male inflorescences narrowly pyramidal, racemose, composed of 10 or more shortly pedunculate, globose heads 5–8 mm. diam.; main axis and peduncles usually villose; stamens densely packed; filaments reddish, about 1 mm. long; anthers reddish, suborbicular, about 0·5 mm. diam.; female inflorescences solitary forming heads about 1 cm. diam.; peduncle 4–10 cm. long, at first villose, soon glabrescent; carpels glabrous, almost completely immersed in the connate calyces, about 1·5 mm. long, 1 mm. wide; ovules numerous, axile, mostly abortive; styles 2, thick, linear, about 5 mm. long, with the reddish stigma decurrent along the greater part of the inner surface, surrounded at the base by small, truncate, papillose scales and suborbicular staminodes. Fruit immersed in the accrescent woody calyx, surmounted by the persistent style-bases; seeds lanceolate, flattened, about 8 mm. long, 2 mm. wide, the distal end with a distinct wing; testa light brown, slightly rugulose. *Plate 50.*

HAB.: In cultivation only; near sea-level to 1,500 ft. alt.; fl. March–April.

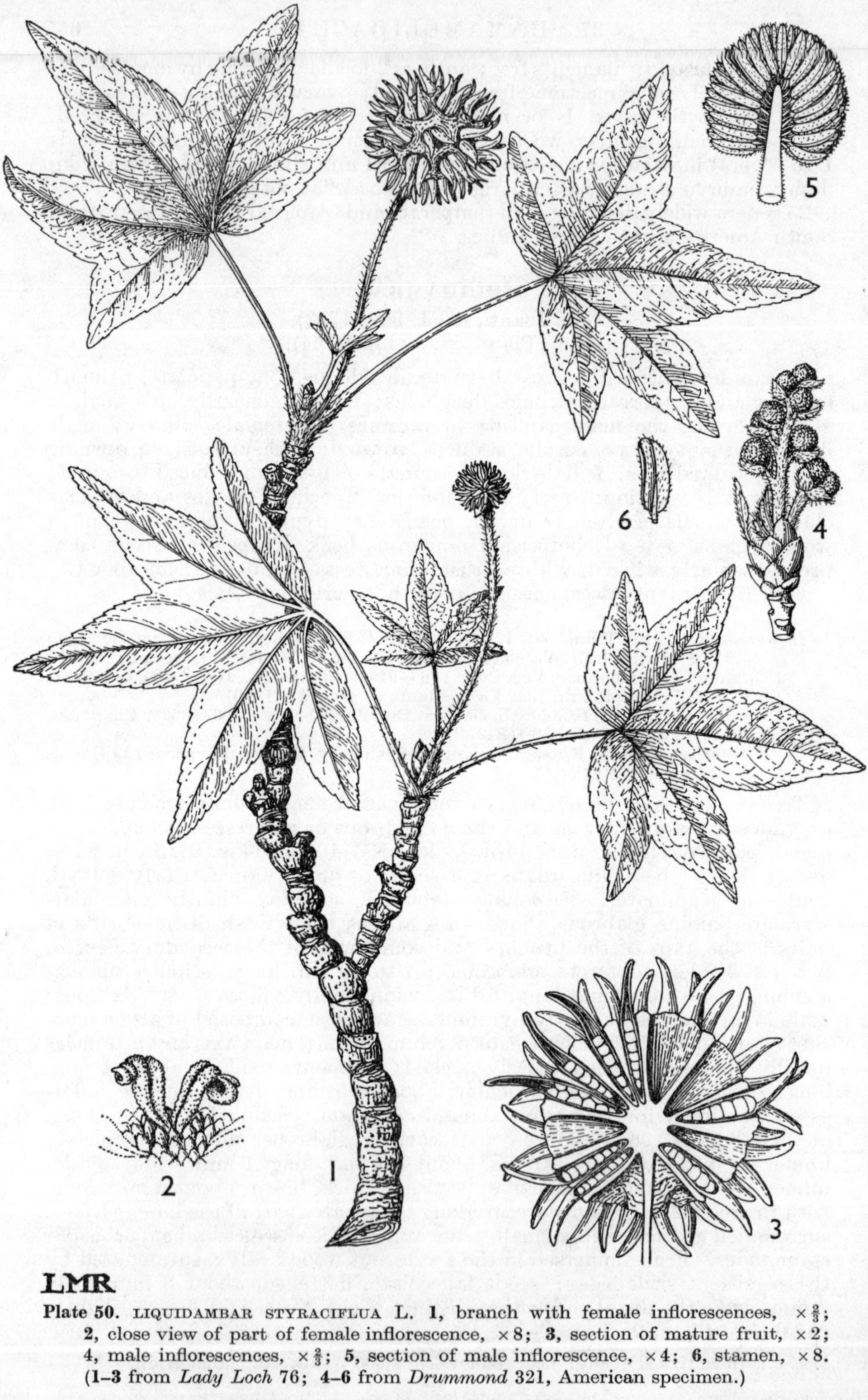

Plate 50. LIQUIDAMBAR STYRACIFLUA L. **1**, branch with female inflorescences, $\times\frac{2}{3}$; **2**, close view of part of female inflorescence, $\times 8$; **3**, section of mature fruit, $\times 2$; **4**, male inflorescences, $\times\frac{2}{3}$; **5**, section of male inflorescence, $\times 4$; **6**, stamen, $\times 8$. (**1–3** from *Lady Loch* 76; **4–6** from *Drummond* 321, American specimen.)

DISTR.: Divisions 1, 3, 7. Cultivated. Eastern and central United States south to Mexico, with a close ally in Rhodes and S.E. Turkey.

1. Stephani near Ayios Neophytos Monastery, 1862, *Kotschy* ! 1930, *C. B. Ussher* 92 ! 1939, *Lady Loch* 76 ! 1939, *Lindberg f.* s.n.; Tala (same locality), 1974, *Meikle* 4012A !
3. Kolossi Castle, *Chapman* s.n.
7. Antiphonitis Monastery, 1738, *Pococke*; 1787, *Sibthorp*; 1862, *Kotschy* 532 ! 1905, *Holmboe* 840; 1949, *Chapman* 427.

NOTES: There is little to add to the remarks made by Holmboe on this species (Veg. Cypr., 97–99). The Cyprus material comes rather closer to the American *L. styraciflua* L. than to the Asiatic *L. orientalis* Mill., but the distinctions between these two are so trifling that one cannot regard them as more than geographical races of a single species. Z. I. Makarova (in Bot. Zhurn. S.S.R., 42: 1182–1195; 1957) concludes that *L. styraciflua* and *L. orientalis* began to diverge from a common parent (*L. europaea*) some time about the beginning of the Pliocene. Their inter-relationship is obviously very close indeed, and, as Holmboe points out, the two cannot always be discriminated. The fruit-size is quite useless as a distinction, and the leaf-shape, though apparently important, becomes less useful when one is shown the range of variability of American *L. styraciflua* in this respect (see T. Holm in Rhodora, 32: 95–100).

It is curious that such a famous tree should not have been noted by visitors to Cyprus before 1738, and it would be interesting to know if research into church records could throw any light on the earlier history of the plant. The Ayios Neophytos legend, though significant, is of little scientific value without more exact historic records.

38. CALLITRICHACEAE

F. Hegelmaier, Mon. Gatt. Callitriche, 1–64 (1864).
D. H. Schotsman, Les Callitriches (Fl. de France 1), 1–152 (1967).

Small aquatic or limnicolous herbs; stems slender, often flaccid, leaves opposite often connate, decussate, entire or emarginate, exstipulate, frequently forming somewhat irregular, apical, floating rosettes; flowers unisexual, inconspicuous, generally solitary, axillary, naked or with two small, membranous, falcate bracteoles; male flowers with a single stamen; filament usually elongate, anther reniform, 2-thecous; female flower a sessile or pedunculate ovary, divided by false septa into 4 loculi; styles 2, erect or reflexing, sometimes caducous, stigmatiferous along most of their length; ovules solitary, pendulous, anatropous with a ventral raphe. Fruit suborbicular, flattened, distinctly 4-lobed; lobes sometimes keeled or winged, breaking into 4, 1-seeded mericarps at maturity; endosperm fleshy; embryo small, straight or slightly curved.

One genus only.

1. CALLITRICHE *L.*

Sp. Plant., ed. 1, 969 (1753).
Gen. Plant., ed. 5, 5 (1754).

Description that of the family.

About 25 species, some very critical, found almost throughout the world, excepting South Africa.

1. C. brutia *Petagna*, Instit. Bot., 2: 10 (1787); Gussone, Plant. Rar., 2, t. 1 (1826); D. H. Schotsman, Les Callitriches (Fl. de France 1), 84 (1967).

C. pedunculata DC. in Lam. et DC., Fl. Franç., 4: 415 (1805); Boiss., Fl. Orient., 2: 756 (1872); Litardière in Briquet, Prodr. Fl. Corse, 2 (2): 105 (1936).

C. hamulata Kuetz. var. *pedunculata* (DC.) Bab., Man. Brit. Bot., ed. 5, 294 (1862).

C. hamulata Kuetz. ssp. *pedunculata* (DC.) Syme, English Bot. ed. 3, 8: 121, t. 1274 (1868).

C. intermedia Hoffm. var. *pedunculata* (DC.) Druce, Fl. Berks, 223 (1897).

C. intermedia Hoffm. ssp. *pedunculata* (DC.) Clapham in Clapham, Tutin et Warburg, Fl. Brit. Isles, ed. 1, 622 (1952); Davis, Fl. Turkey, 4: 200 (1972).

TYPE: Italy; " inventa a Julio Candida in lacubus vulgo di Scrisà Calabriae ulterioris " (FI).

A very slender submerged or limnicolous annual; stems flaccid or rather rigid, 3–15 (or more) cm. long, glabrous except for a few scattered, lepidote, 8–15-celled microscopic hairs; leaves of two kinds: the submerged linear or strap-shaped, 4–10 mm. long, 0·5 mm. wide, glabrous, truncate or more or less emarginate, the aerial obovate-spathulate, 2·5–3·5 mm. long, 1·5–2 mm. wide, apex subentire or somewhat indented, base tapering to an apparent petiole about 0·5 mm. long, uppermost leaves sometimes forming an irregular, loose rosette; flowers naked or with poorly developed, caducous bracteoles; male flower very inconspicuous with a filament usually less than 0·5 mm. long; anther about 0·2 mm. diam., yellow; female flower at first shortly pedunculate, the peduncle lengthening to 5–15 (–20) mm. in fruit; ovary suborbicular, about 0·5–0·7 mm. diam.; styles reflexed, to 1·5 mm. long, caducous. Fruit pale or rather dark brown, 1–1·5 mm. diam., the lobes distinctly, though narrowly, winged; pericarp closely investing seed, rather spongy.

HAB.: In shallow, seasonal pools on sandstone ledges; c. 2,100 ft. alt.; fl. March–April.

DISTR.: Division 1, rare. Widely distributed in the Mediterranean area and western Europe, also recorded from the Caucasus.

1. Inia (Akamas), 1957, *Merton* 3039 !

NOTES: Questionably distinct from the plant generally named *C. hamulata* Kuetz. or *C. intermedia* Hoffm. by European botanists. If these are combined, then *C. brutia*, having priority in date of publication, must be regarded as the correct name for the aggregate. Gussone's illustration of *C. brutia* (see reference above) is very untypical, but Dr. D. H. Schotsman has seen the type material and is satisfied that it agrees with the plant subsequently named *C. pedunculata* by de Candolle.

39. MYRTACEAE

Niedenzu in Engl. et Prantl, Pflanzenfam., ed. 1, III, 7: 57–105 (1893).

Trees, shrubs and subshrubs; leaves simple, usually entire and opposite, glandular-punctate; stipules wanting or very small and caducous; flowers solitary or in racemes, or rarely in cymes, axillary or subterminal, mostly actinomorphic, bisexual or polygamous by abortions; hypanthium (or calyx-tube) generally adnate to the ovary; sepals 3 or more, imbricate, valvate or irregularly split; petals usually 4–5, sometimes wanting, inserted on the margin of the disk, often strongly imbricate, sometimes connivent (or connate) and falling off in a mass; stamens generally numerous, inserted on the disk; filaments free or connate at the base, sometimes in bundles opposite the petals, variously inflexed or folded in the bud, sometimes almost straight; anthers 2-locular, splitting lengthwise or opening by pores, connective often gland-tipped; ovary inferior, 1–many-locular, placentation generally axile, ovules numerous, rarely few or solitary; style generally simple, elongate; stigma often small. Fruit dehiscent or indehiscent, sometimes fleshy; seeds without endosperm, or with very little endosperm; embryo straight or variously curved.

Throughout the tropics and subtropics, becoming rarer in temperate regions; particularly well represented in Australasia.

Sepals and petals distinct, free; leaves opposite - - - - - 1. **Myrtus**
Sepals and petals united, forming a circumscissile operculum over the stamens; mature leaves usually alternate - - - - - - - - - 2. **Eucalyptus**

1. MYRTUS *L.*

Sp. Plant., ed. 1, 471 (1753).
Gen. Plant., ed. 5, 212 (1754).
Niedenzu in Engl. et Prantl, Pflanzenfam., ed. 1, III, 7: 66–69 (1893).

Shrubs or trees; leaves opposite, simple, entire, evergreen, gland-dotted, often aromatic; flowers solitary or occasionally in racemes, hypanthium turbinate or subglobose, enveloping the ovary; sepals 5 (rarely 4), somewhat imbricate in bud, soon becoming open; petals (4–) 5 or more, spreading; stamens numerous in several series, free, inflexed in bud; anthers dehiscing longitudinally; ovary 2–3-locular, ovules usually numerous; style simple, stigma usually small. Fruit a berry, generally crowned with the persistent calyx; seeds numerous, reniform; embryo curved.

About 100 species, mostly in tropical regions, especially in Australasia.

1. **M. communis** *L.*, Sp. Plant., ed. 1, 470 (1753); Hume in Walpole, Mem. Europ. Asiatic Turkey, ed. 1, 254 (1817); Unger et Kotschy, Die Insel Cypern, 368 (1865); Boiss., Fl. Orient., 2: 736 (1872); Holmboe, Veg. Cypr., 134 (1914); Post, Fl. Pal., ed. 2, 1: 468 (1932); Lindberg f., Iter Cypr., 24 (1946); Chapman, Cyprus Trees and Shrubs, 58 (1949); Davis, Fl. Turkey, 4: 172 (1972); Zohary, Fl. Palaest., 2: 371, t. 542 (1972); Osorio-Tafall et Seraphim, List Vasc. Plants Cyprus, 75 (1973).
TYPE: " *in* Europa *australi*, Asia, Africa ".

Dense evergreen shrub, up to 3 m. high; bark greyish, rather rough or leprose, sometimes exposing cinnamon-brown patches; twigs often distinctly 4-angled, or even somewhat winged, subglabrous or ± densely pubescent; leaves ovate or lanceolate, acute, acuminate, obtuse or even rounded at apex, rounded or cuneate at base, quite entire, 1·5–4 (–6) cm. long, 0·4–1·5 (–3) cm. wide, dark shining green above, rather paler below, conspicuously gland-dotted, aromatic, sometimes pubescent when young, soon glabrous or subglabrous, margins narrowly recurved; petiole very short, rarely exceeding 1·5 mm., slightly channelled above; flowers normally solitary, axillary; pedicels slender, 1–2·5 cm. long, glabrous or somewhat pubescent; bracteoles 2, linear-lanceolate, about 1·5 mm. long, 0·2–0·5 mm. wide, herbaceous, caducous; hypanthium turbinate, about 4 mm. long, 3 mm. wide at apex, usually glabrous or subglabrous; sepals broadly deltoid, about 1·5 mm. long, 2 mm. wide at base, spreading, glabrous or subglabrous with ciliate margins, petals white, obovate, 7–10 mm. long, 5–7 mm. wide, spreading, apex rounded, margins ciliate, stamens conspicuous; filaments 6–10 mm. long, glabrous, inserted on a pubescent disk; anthers ovoid, yellow, about 0·7 mm. long, 0·5 mm. wide, connective not noticeably produced; style filiform, about 12–14 mm. long; stigma truncate. Fruit a subglobose, black, pruinose (or rarely white), dryish berry, 8–10 mm. diam., often crowned with the persistent calyx; seeds reniform, often deeply excavate ventrally, 2·5–3 mm. long, 1·5–2 mm. wide; testa shining pale brown or whitish, minutely rugulose.

HAB.: Rocky slopes, banks and roadsides, frequently by streams, springs or rivers; sea-level to about 5,000 ft. alt.; fl. May–August.

DISTR.: Divisions 1–8, locally common in Division 2, common in Division 7. Throughout the Mediterranean region eastwards to Afghanistan and N.W. India; Azores and Madeira.

1. Near Paphos, 1862, *Kotschy*; Ayios Yeoryios, 1962, *Meikle* ! Polis, 1962 *Meikle* !
2. Locally common (Platres, Platania, Makheras Monastery, Stavros tis Psokas, Khionistra; *Kotschy*; *Holmboe* 1041 ! *Haradjian* 489 ! *Ussher* 16 ! 17 ! *Chapman* 112 ! *Kennedy* 657 !
3. Cape Gata, 1862, *Kotschy*; Limassol, 1801, *Hume*; Paramali, 1959, *Sparrow* 43 !
4. Larnaca, 1801, *Hume*.
5. Lefkoniko, 1880, *Sintenis & Rigo*.
6. Near Orga, 1957, *Merton* 2841 !
7. Common; *Kotschy*; *Sintenis & Rigo*; *Syngrassides* 190 ! 1667 ! *Lindberg f.* s.n. ! *Mapple* 27 ! *Atherton* 92 ! 215 ! 376 ! 710 ! *Merton* in ARI 40 !
8. Cape Andreas, 1880, *Sintenis & Rigo*; Akradhes, 1962, *Meikle* ! Leonarisso, 1966, *Economides* 4 !

NOTE: The whitish-fruited *M. communis* L., var. *leucocarpa* DC. is found occasionally in cultivation (*Syngrassides* 1667 !) and in the wild (*Merton* 2841 !); apart from its fruits, it differs from the general run of Cyprus *M. communis* in having relatively broad, blunt leaves. It would be interesting to know if this distinction always holds good and if the two Myrtles can be identified in the absence of fruits.

2. EUCALYPTUS *L'Hérit.*

Sert. Angl., 18 (1789), t. 20 (1790).

E. F. Chapman, The Cyprus Eucalypts, 56 pp. (1954).

Evergreen trees; leaves simple, entire, aromatic, often glaucous, generally dimorphic, opposite and often amplexicaul in juvenile growths, usually alternate and petiolate in mature growths; flowers solitary or in clusters or umbels; peduncles axillary, bracts membranous, caducous; hypanthium (or calyx-tube) turbinate or campanulate, becoming woody in fruit; perianth segments connate forming a conical or hemispherical, circumscissile operculum which covers the stamens and falls off at anthesis; stamens very numerous, free, with slender filaments; anthers versatile, dehiscing longitudinally; ovary 3–5-locular, many-ovulate; style slender, stigma small. Fruit a woody or leathery capsule, dehiscing loculicidally at the apex; seeds numerous, small, usually angular, with a membranous testa.

About 400 species in Australia and Malesia; frequently planted, and sometimes naturalized in other tropical and warm-temperate regions.

More than 70 species have been recorded as cultivated in Cyprus, and additional species are constantly being introduced. Only a very small fraction of the total number can be considered an established part of the flora; for descriptions of the more uncommon species reference should be made to E. F. Chapman's, " The Cyprus Eucalypts " (Cyprus Government Printing Office, 1954) or to G. Moggi " Guida al riconoscimento degli Eucalitti coltivati in Italia " (Pubblicazioni di Centro di Sperimentazione Agricola e Forestale, 7 : 149–220; 1964). Two other publications likely to be of help to anyone making a detailed study of the genus in Cyprus are: Stan Kelly, " Eucalypts " (Nelson, 1969) and L. D. Pryor & L. A. S. Johnson, " A Classification of the Eucalypts " (Australian National University Press, 1971). " Eucalypts for Planting " (F. A. O. Forestry and Forest Products Studies No. 11, pp. 403; 1955) furnishes much valuable information on the species and their cultivation, as well as an extensive bibliography.

Flowers red, pink or orange; operculum rostrate; hypanthium with a strongly corrugate base - - - - - - - - - - - - - - 8. E. TORQUATA

Flowers white or yellowish:

  Operculum dome-shaped, with the base distinctly wider than the apex of hypanthium 1. E. GOMPHOCEPHALA

  Operculum conical or horn-shaped, base not wider than apex of hypanthium:

    Operculum elongate-conical or horn-shaped, conspicuously longer than hypanthium:

      Operculum 2·5–3·5 cm. long; flowers 7–20, sessile in dense umbels; valves of fruit usually 3, connivent, much exserted, forming a slender beak 2. E. CORNUTA

Operculum less than 2 cm. long; flowers 2–12, subsessile or pedicellate; valves of fruit usually 4–5:
  Inflorescence 6–12-flowered; flowers subsessile; peduncle usually flattened or strap-shaped - - - - - - - 2. E. CORNUTA HYBRID
  Inflorescence 2–7-flowered; flowers pedicellate:
    Peduncle flattened or strap-shaped often recurved; valves of fruit shortly exserted - - - - - - - - - 3. E. OCCIDENTALIS
    Peduncle terete or angled, not flattened; valves of fruit distinctly exserted 4. E. TERETICORNIS
Operculum shortly conical or almost hemispherical, sometimes shortly beaked, not conspicuously longer than hypanthium:
  Valves of fruit distinctly exserted:
    Inflorescence 6–15 (–20)-flowered; pedicels slender, up to 10 mm. long; operculum often beaked - - - - - - 5. E. CAMALDULENSIS
    Inflorescence 3–8 (–10)-flowered; pedicels rather thick, usually less than 5 mm. long; operculum rarely beaked - - - - - 6. E. RUDIS
  Valves of fruit minute, sunken, inconspicuous; inflorescence 2–7-flowered 7. E. MELLIODORA

1. E. GOMPHOCEPHALA *DC.*, Prodr., 3: 220 (1828); Bovill, Rep. Plant. Work Cyprus, 15 (1915); Chapman, Cyprus Eucalypts, 31 (1954); Stan Kelly, Eucalypts, 21, pl. 47 (1969); Osorio-Tafall et Seraphim, List Vasc. Plants Cyprus, 75 (1973).

Tree 18–20 (–40) m. high with a well-developed crown; bark finely fibrous, soft, grey or brown; mature leaves lanceolate, acuminate, 10–18 cm. long, 2–4 cm. wide, coriaceous, dark green; petioles 2–3 cm. long; flowers 6–8 in axillary umbels, sessile or almost so; peduncle stout, strap-shaped, 2–3 cm. long; hypanthium campanulate or somewhat funnel-shaped, 10–12 mm. long, 8–9 mm. diam. at apex; operculum dome-shaped or very bluntly conical, reddish or orange, 8–10 mm. long, distinctly wider than the apex of the hypanthium. Fruit campanulate, slightly angled, about 15 mm. long, 12 mm. diam. at apex, with a distinct apical rim and 3–4, very slightly raised, valves.

Tuart or Black Eucalyptus. Said to be one of the most widely cultivated species in Cyprus, tolerating arid sites and a degree of salinity. Bovill records that it was planted at Salamis prior to 1915. It is a native of Western Australia, and is valued as a shade or street tree, also producing heavy, strong timber.

2. E. CORNUTA *Labill.*, Relat. Voy., 1: 403, t. 20 (1800); Chapman, Cyprus Eucalypts, 28 (1954); Stan Kelly, Eucalypts, 20, pl. 43 (1969).

Tree 9–18 m. high; bark rough, furrowed, dark brown or grey, branches usually smooth; mature leaves lanceolate, acute or shortly acuminate with a shortly mucronate, oblique apex, 7–10 cm. long, 1·5–2·5 cm. wide, coriaceous, dark green or glaucescent; petioles 1–2 cm. long; flowers 7–20, sessile or subsessile in dense axillary clusters; peduncles subterete 1–3 cm. long; hypanthium campanulate, 5–7 mm. long, about 5 mm. diam. at apex; operculum horn-shaped, terete, orange-buff, 2·5–3·5 cm. long; filaments up to 4 cm. long, lemon-yellow. Fruit broadly campanulate, 7–8 mm. long, 5–7 mm. wide at apex; valves usually 3, prominently exserted, connivent, forming a beak up to 10 mm. long.

Yate. A native of south-western Australia, valued for its exceptionally strong timber. Evidently very rare in Cyprus, where it is largely replaced by a hybrid or series of hybrids, between *E. cornuta*, *E. occidentalis* Endl. and possibly *E. platypus* Hook. These hybrids are variable, as might be expected, being large shrubs or trees up to 20 m. high with smooth or rough, pale or pinkish bark; the leaves are mostly broader and thicker than those of *E. cornuta*, broadly lanceolate or ovate, 6–10 cm. long, 2–3 cm. wide, subacute or obtuse, with apiculate or sometimes emarginate apices; the petioles are stout and erect; the flowers are subsessile in 6–12-flowered axillary umbels, with an hypanthium similar to that of *E. cornuta*, but with the horn-shaped operculum usually less than 1·5 cm. long; the peduncle is generally very stout and strap-shaped, up to 3 cm. long and 0·8 cm. wide, but is said to be very rarely almost terete. The fruit resembles that of *E. cornuta*, with 4–5 valves which are usually shortly exserted, but occasionally long and beaked.

The presumed hybrid is found in some quantity around Larnaca Salt Lake, and in other parts of Cyprus, and grows very satisfactorily on saline ground or in areas exposed to salt spray.

3. E. OCCIDENTALIS *Endl.*, Enum. Plant. Nov. Holl., 49 (1837); Chapman, Cyprus Eucalypts, 37 (1954); Stan Kelly, Eucalypts, 23, pl. 55 (1969); Osorio-Tafall et Seraphim, List Vasc. Plants Cyprus, 75 (1973).

Tree 10–25 m. high, often with a flat-topped, spreading crown; bark rough, fissured and grey on the main trunk, smooth and whitish or yellowish on the branches; mature leaves thick, coriaceous, variable in shape, ovate or lanceolate, acute, acuminate or obtuse, 6–12 cm. long, 1·5–3 (–8) cm. wide, straightish or falcate; petiole 1–2 cm. long; inflorescence a 2–7-flowered axillary umbel; peduncle stout or slender, usually strongly compressed or strap-shaped, 1–3 cm. long, often recurved in fruit; flowers distinctly pedicellate, with stout, compressed pedicels 1·5–5 mm. long; hypanthium campanulate, terete, 7–8 mm. long, about 7 mm. diam. at apex; operculum shortly horn-shaped or elongate-conical, 1–1·5 cm. long, greenish, often distinctly constricted just above the base; stamens pale yellow. Fruit campanulate, 9–15 mm. long, 6–10 mm. diam. at apex, rim often rather distinct; valves 4–5 inserted below the rim with the tips shortly exserted or (before dehiscence) forming a short beak.

Flat-topped Yate, a native of Western Australia, said to grow well on poor, salty soils and to have possible value as a fuel or shade tree. Apparently less common in Cyprus than the hybrid *E. cornuta* × *occidentalis* (see under 2. *E. cornuta*, above).

4. E. TERETICORNIS *Sm.*, Spec. Bot. New Holland., 41 (1975); Bovill, Rep. Plant. Work Cyprus, 8 (1915); Unwin in Cypr. Journ., 19: 7 (1924); Chapman, Prelim. Rep. Eucalypts Cypr., 36 (1951); Stan Kelly, Eucalypts, 32, pl. 95 (1969).
E. umbellata (non (Gaertn.) Domin) Chapman, Cyprus Eucalypts, 43 (1954).

A tall tree, up to 17 m. high with a spreading, rather open crown; bark decorticating in large irregular flakes leaving whitish or greyish-blue patches; branches usually smooth and whitish; mature leaves coriaceous, green or glaucescent, narrowly lanceolate-acuminate, 6–14 (–22) cm. long, 1–2·5 (–4) cm. wide, usually distinctly falcate; petiole 1–3 cm. long; inflorescence a 3–7-flowered axillary umbel; peduncle 1–1·5 cm. long, subterete or somewhat angled; pedicels distinct, 2–6 mm. long; hypanthium cup-shaped, 2–3 mm. long, 3–4 mm. diam. at apex, subterete; operculum conical or very shortly horn-shaped, 6–12 mm. long, reddish at first, becoming whitish or pale yellow at maturity; stamens whitish. Fruit hemispherical, about 6 mm. diam., a little constricted at apex; valves 4, broadly triangular, about 1·5–2 mm. long, distinctly exserted.

Forest Red Gum. Widely distributed along the east coast of Australia and as far north as Papua. It has been planted here and there in different parts of Cyprus, but is not considered as useful as the widely cultivated *E. camaldulensis*.

5. E. CAMALDULENSIS *Dehnhardt*, Cat. Plant. Hort. Camald., ed. 2, 20 (1832); Chapman, Cyprus Eucalypts, 27 (1954); Stan Kelly, Eucalypts, 34, pl. 102 (1969); Davis, Fl. Turkey, 4: 173 (1972); Osorio-Tafall et Seraphim, List Vasc. Plants Cyprus, 75 (1973).
E. rostrata Schlecht. in Linnaea, 20: 655 (1847); R. R. Waterer, Isl. Cyprus, 153 (1947).

Tall tree 15–24 (–30) m. high with a rather dense crown and pendulous branches; bark decorticating irregularly leaving whitish, yellowish or greyish patches; mature leaves narrowly lanceolate, 6–15 (–20) cm. long, 1–2·5 cm. wide, coriaceous, glaucous, usually falcate and elongate-acuminate; petiole 1·5–2·5 (–3) cm. long; inflorescence a (3–) 5–10 (–15)-flowered axillary or occasionally inter-axillary umbel; peduncle usually 1–2·5 cm. long, subterete or slightly angled; pedicels usually distinct, slender, up to 10 mm. long; hypanthium hemispherical 2–3 mm. long, 3–4 mm. diam. at apex, terete; operculum shortly conical or with a short, acute beak, 5–7 mm. long; stamens whitish. Fruit hemispherical with a distinct rim, about 5 mm. long, 7 mm. diam. at apex, coriaceous; valves 4 (–5) triangular, about 1–1·5 mm. long, distinctly exserted.

River Red Gum. Widely distributed in Australia, and said to be the most commonly cultivated eucalypt in the world; it has been extensively planted in Cyprus, where some specimens have reached majestic proportions. It is generally considered one of the most valuable of introduced timber trees.

6. E. RUDIS *Endl.*, Enum. Plant. Nov. Holl., 49 (1837); Chapman, Cyprus Eucalypts, 39 (1954); Stan Kelly, Eucalypts, 34, pl. 103 (1969).

Tree to 18 m. high; bark usually rough, fissured, dark grey; branches smooth, grey; mature leaves narrowly lanceolate-acuminate, 8–15 cm. long, 1·5–2·5 cm. wide, dull glaucous-green, often distinctly falcate; petioles 1·5–2 cm. long; inflorescence a 3–8 (–10)-flowered axillary umbel; peduncle 0·8–1·5 cm. long, rather slender, terete; pedicels short, rather thick, usually less than 5 mm. long; hypanthium broadly funnel-shaped, about 4 mm. long, 6 mm. diam. at apex; operculum shortly and bluntly conical, about 5–7 mm. long, rarely with a very short bluntish beak; stamens whitish. Fruit hemispherical, 6–7 mm. long, 8–9 mm. diam. at apex with a very conspicuous flattened rim; valves (3–) 4 (–5), triangular, 2–3 mm. long, often distinctly incurved.

Flooded Gum, Moitch. A native of Western Australia, similar in many respects to *E. camaldulensis* but with rough bark and fewer-flowered umbels of more shortly pedicellate flowers with bluntish opercula. It is evidently not so common in Cyprus, nor so useful, as *E. camaldulensis*.

7. E. MELLIODORA *A. Cunn. ex Schauer* in Walpers, Repert. Bot. Syst., 2: 924 (1843); Chapman, Cyprus Eucalypts, 36 (1954); Stan Kelly, Eucalypts, 65, pl. 215 (1969).

A tall tree to 20 m. high with a spreading crown and somewhat pendulous branches; bark variable, smooth, streaked with grey, or hard, blackish and scaly or yellow-brown and rather fibrous; branches usually smooth, white, cream or grey; mature leaves narrowly oblong-lanceolate, 6–12 cm. long, 1–2·5 cm. wide, dark green, apex acuminate, acute or obtuse; petiole 1–1·5 cm. long; inflorescence a 2–7-flowered axillary umbel; peduncle slender, subterete or somewhat angled, usually less than 10 mm. long; pedicels short, but distinct, slender, less than 5 mm. long; hypanthium cup-shaped or funnel-shaped, about 4 mm. long, 5 mm. diam. at apex, terete; operculum shortly and bluntly conical, or sometimes very shortly beaked, usually less than 4 mm. long; stamens whitish. Fruit small, cup-shaped or shortly turbinate, 4–7 mm. long, about 4–6 mm. diam. at apex, rim inconspicuous; valves triangular, minute, sunken.

Yellow Box. A native of Victoria and New South Wales, useful both as a source of timber and honey. It is said to be quite successful in the neighbourhood of Nicosia, but is evidently not commonly planted in Cyprus.

8. E. TORQUATA *Luehmann* in Victorian Naturalist, 13: 147 (1897); Stan Kelly, Eucalypts, 29, pl. 80 (1969).

Tree 8–11 cm. high with a spreading crown; bark rough, dark brown; branches smooth, dull brownish and reddish; mature leaves lanceolate-acuminate, slightly oblique at base, 8–14 cm. long, 1·3–3 cm. wide, glaucous, venation obscure; petiole 1·5–2·5 cm. long; inflorescence a pendulous, 5–9-flowered axillary umbel; peduncle 15–25 mm. long, slender, terete or somewhat flattened; pedicels slender, 10–20 mm. long, subterete or slightly angular; hypanthium reddish, cylindrical or often distinctly urceolate, 8–10 mm. long, 7–8 mm. diam. at apex, the base swollen and conspicuously corrugate, upper portion indistinctly ridged; operculum depressed-hemispherical, rostrate, the beak 7–9 mm. long, base strongly corrugate, wider than the apex of the hypanthium; filaments 8–12 mm. long, bright red or orange-red; anthers yellowish, oblong, about 1·2 mm. long, 0·6 mm. wide. Fruit urceolate-cylindrical, 13–15 mm. long, 10–12 mm. diam.; rim inconspicuous; valves minute, sunken.

A native of S.W. Australia becoming popular as a cultivated ornamental. It is not uncommon in parks and gardens about Nicosia and Larnaca, but appears to be a recent introduction, not being listed by Chapman (1954) or Moggi (1964).

[PSIDIUM GUAJAVA *L.* (Sp. Plant., ed. 1, 470; 1753), the Guava, a native of South America, is said to have been in cultivation at Morphou (G. Frangos in Cyprus Agric. Journ., 18 (3): 87; 1928). It is a bush or small tree, up to 10 m. high with strongly 4-angled, pubescent-tomentose twigs; leaves opposite, oblong, ovate or elliptic, acute or obtuse, shortly petiolate, 6–15 cm. long, 3·5–7 cm. wide, coriaceous, dark green above with strongly impressed nervation, paler below and pubescent; flowers usually solitary, axillary (rarely in few-flowered cymes); sepals thick, closely valvate in bud, often splitting irregularly at anthesis; petals 4–5, white, obovate, patent, about 1–1·5 cm. long, 0·8–1 cm. wide; stamens numerous, ovary (2–) 4–5 (–7)-locular; style long; stigma capitate. Fruit a large, fleshy, ovoid, oblong or subglobose berry, up to 10 cm. long, yellowish externally with whitish or reddish flesh; seeds usually numerous, reniform, with a hard testa.

Widely cultivated for its sweet, musky fruits, and naturalized (or even a troublesome weed) in some tropical countries. It would not appear to be commonly grown in Cyprus.]

40. LYTHRACEAE

E. Koehne in Engl. et Prantl, Pflanzenfam., ed. I, III, 7: 1–16 (1892), and in Engl., Pflanzenr., 17 (IV, 216): 1–326 (1903).

Trees, shrubs, or (in our area) annual or perennial herbs; leaves commonly opposite, decussate, rarely verticillate or alternate, simple, entire, stipulate or exstipulate; inflorescences axillary or rarely extra-axillary, forming dichasia, spikes or panicles, or flowers sometimes solitary; flowers actinomorphic or zygomorphic, generally hermaphrodite, (3–) 4–6 (–16)-merous; hypanthium cup-shaped or cylindrical; sepals usually valvate, often with an epicalyx; petals free, crumpled in bud, inserted near the top of the hypanthium, occasionally wanting; stamens usually 4 or 8, sometimes fewer or very numerous, inserted on the hypanthium; filaments sometimes very unequal; anthers 2-thecous, dehiscing longitudinally; ovary superior, free from hypanthium, sometimes stipitate, usually 2–6-locular; ovules usually numerous, axile, often not reaching to top of ovary; style simple; stigma generally capitate. Fruit usually a capsule, dehiscing transversely or by valves, or irregularly, rarely indehiscent; seeds generally numerous, without endosperm; cotyledons usually flat, rarely convolute.

About 25 genera chiefly in warm temperate and tropical regions.

1. LYTHRUM *L.*

Sp. Plant., ed. 1, 446 (1753).
Gen. Plant., ed. 5, 205 (1754).

J. Borja Carbonell in Anal. Inst. Bot. Cavanilles, 23: 145–162 (1968).

Annual, biennial or perennial herbs; leaves alternate, opposite or whorled; flowers sometimes heterostylous, solitary or in axillary cymes or terminal spikes; hypanthium usually tubular, costate; calyx usually 6-merous (sometimes 4–5-merous), epicalyx generally well developed; petals usually as many as sepals, often pink or purple and conspicuous, sometimes wanting; stamens (1–) 4–12, usually inserted at more than one level and unequal; ovary sessile or subsessile, 2-locular; style filiform; stigma capitate or subcapitate. Fruit generally a 2-valved, septicidal capsule, rarely dehiscing irregularly or by 4 apical teeth; seeds usually small, numerous.

About 25 species, chiefly in temperate regions of the northern hemisphere.

Petals about 6 mm. long, conspicuous; stamens 12, some of them usually exserted; leaves shortly and bluntly auriculate at base - - - **1. L. junceum**
Petals usually less than 3 mm. long, inconspicuous; stamens (1–) 2–6 (–8), included; leaves not auriculate at base:
Lobes of epicalyx narrowly deltoid, much longer than the sepals; leaves linear or narrowly oblong - - - - - - - - - **2. L. hyssopifolia**
Lobes of epicalyx minute, bluntly deltoid, rather shorter than the sepals; leaves narrowly ovate, obovate or oblong - - - - - **3. L. tribracteatum**

1. L. junceum *Banks et Sol.* in Russell, Nat. Hist. Aleppo, ed. 2, 2: 253 (1794); Davis, Fl. Turkey, 4: 176 (1972); Zohary, Fl. Palaest., 2: 370, t. 540 (1972); Osorio-Tafall et Seraphim, List Vasc. Plants Cyprus, 74 (1973).
L. graefferi Ten., Fl. Nap. Prodr., LXVIII (1811); Unger et Kotschy, Die Insel Cypern, 367 (1865); Boiss., Fl. Orient., 2: 739 (1872); Post, Fl. Pal., ed. 2, 1: 470 (1932).
[*L. flexuosum* (non Lag.) E. Koehne in Engl., Pflanzenr., 17 (IV, 216): 68 (1892); Holmboe, Veg. Cypr., 133 (1914); Lindberg f., Iter Cypr., 24 (1946).]
TYPE: Syria; near Aleppo. *Russell* (BM).

Decumbent or ascending, glabrous perennial, stems usually much branched, especially near the base, the branches 7–30 (–40) cm. long, usually lax and straggling, subterete or distinctly angled, often reddish-purple; leaves generally opposite towards base of plant, alternate above, subsessile or very shortly petiolate, narrowly oblong, (0·5–) 1–2 (–5) cm. long, (0·2–) 0·4–0·6 (–1) cm. wide, glaucous-green, apex rounded, obtuse, or subacute, base shortly and bluntly auriculate; flowers solitary, subsessile in the axils of the upper leaves, trimorphically heterostylous; hypanthium narrowly tubular, about 7 mm. long, 2 mm. wide at apex, tapering to base, often tinged reddish; bracteoles 2, narrowly subulate, membranous, 1–1·5 mm. long; epicalyx lobes 6, deltoid, acute, 0·7 mm. long, 0·5 mm. wide at base, as long as or longer than the sepals, thickish, opaque, often reddish; sepals 6, broadly deltoid, membranous; petals 6, inserted at the apex of the hypanthium, narrowly obovate, about 6 mm. long, 2 mm. wide below the rounded apex, base tapering to a very short claw; stamens 12, included or shortly exserted, attached at different levels towards the middle of the hypanthium; filaments very unequal; anthers yellow, broadly oblong, about 0·5 mm. long, 0·4 mm. wide;

ovary cylindrical, 2–2·5 mm. long, 0·5 mm. diam., style filiform, varying in length; stigma capitate, papillose. Capsule cylindrical, dark brown, about 5 mm. long, 1·5 mm. diam., included in persistent hypanthium; seeds broadly ovoid, about 0·8 mm. long, 0·6 mm. wide, gibbous or strongly convex on the dorsal surface, flattish or slightly convex on the ventral; testa shining mid-brown, minutely rugulose.

HAB.: Ditches, marshes, damp meadows, pools, and by streams and springs; sea-level to 5,500 ft. alt.; fl. March–September.

DISTR.: Divisions 1–8, common in Divisions 2 and 7, locally common in Division 3. Widespread in Mediterranean region; Atlantic Islands.

1. Ayios Neophytos, 1939, *Lindberg f.* s.n.
2. Common (Platres, Prodhromos, Troödos, Kannaviou, Moniatis); *Kotschy* 847; *Feilden* s.n. ! *Kennedy* 655 ! 656A ! *Mavromoustakis* 61 ! *Probyn* s.n. ! *Atherton* 559 !
3. Locally common (around Zakaki, Asomatos, Akrotiri, Episkopi, Kannaviou R. & Ezuza R.); *Holmboe* 393; *Mavromoustakis* 1 ! 58 ! 63 ! 80 ! *Lindberg f.* s.n.; *Meikle* 2922 ! *Merton* in ARI 697 !
4. Ayia Napa, 1862, *Kotschy* 108; Famagusta, 1931, *Tracey* 69a ! Near Akhyritou, 1948, *Mavromoustakis* s.n. !
5. Kythrea, 1880, *Sintenis & Rigo* 588 ! and, 1941, *Davis* 3641 !
6. Myrtou Valley, 1932, *Syngrassides* 100 ! Krini marshes, 1955, *Soteriadou* in *Merton* 2266 !
7. Common (Kyrenia, Thermia, Kambyli, Lapithos, Boghazi, Paleo Mylos); *Kotschy* 943; *Huddle* 58 ! *Lindberg f.* s.n. ! *Casey* 178 ! 785 ! *Atherton* 42 ! 486 ! 1340 ! *Merton* 2981 ! etc.
8. Kantara, 1880, *Sintenis & Rigo*; Gastria, 1905, *Holmboe* 474.

2. L. hyssopifolia *L.*, Sp. Plant., ed. 1, 447 (1753); Boiss., Fl. Orient., 2: 739 (1872); Holmboe, Veg. Cypr., 134 (1914); Post, Fl. Pal., ed. 2, 1: 470 (1932); Lindberg f., Iter Cypr., 24 (1946); Davis, Fl. Turkey, 4: 177 (1972); Zohary, Fl. Palaest., 2: 369, t. 538 (1972); Osorio-Tafall et Seraphim, List Vasc. Plants Cyprus, 74 (1973).

TYPE: "*in* Germaniae, Helvetiae, Angliae, Galliae *inundatis*".

Glabrous annual; stems erect or decumbent, 6–20 (–30) cm. long, terete or somewhat angled, often tinged reddish; leaves linear or narrowly oblong, 5–15 mm. long, 1–2·5 mm. wide, sessile or subsessile, glaucous green, sometimes papillose, apex obtuse or subacute, base not auriculate; flowers solitary, subsessile in the axils of the upper leaves, homostylous; hypanthium tubular or narrowly funnel-shaped, about 4·5 mm. long, 2 mm. wide at apex; bracteoles 2, subulate, 1–2·5 mm. long, membranous or herbaceous; epicalyx lobes 4–6, narrowly deltoid, about 0·7 mm. long, 0·4 mm. wide at base, thickish, opaque; sepals 4–6, very broadly deltoid, much shorter than epicalyx lobes, about 0·4 mm. long, 0·7 mm. wide at base, membranous; petals 4–6, magenta-pink, oblanceolate, 2·5–3 mm. long, 0·8 mm. wide, attached at apex of hypanthium; stamens (1–) 2–6 (–8), included; filaments about 1·5 mm. long; anthers shortly oblong, about 0·3–0·5 mm. long, 0·2–0·4 mm. wide; ovary cylindrical, 1·5–2 mm. long, 0·5 mm. diam.; style about 0·8 mm. long; stigma capitate, papillose. Fruit cylindrical, 4–4·5 mm. long, 0·7 mm. diam., about as long as the hypanthium, dark brown; seeds ovoid, about 0·8 mm. long, 0·6 mm. wide, almost equally convex on both surfaces, shortly winged at the distal end; testa mid-brown, closely, minutely, but distinctly, rugulose.

HAB.: Moist places by springs and streams; brackish pools and muddy hollows; sides of irrigation channels; sea-level to 5,200 ft. alt.; fl. May–July.

DISTR.: Divisions 1–4, 6. Europe (except the north) to Central Asia; North Africa, Atlantic Islands, also Australasia, parts of tropical Africa, South Africa, North and South America, but possibly adventive in these regions.

1. Akamas, 1905, *Holmboe* 781.
2. Troödos, 1894, *Post* s.n.; Evrykhou, 1880, *Sintenis & Rigo* 690 ! Mandria, 1937, *Kennedy* 656 ! Platania, 1939, *Lindberg f.* s.n.; Prodhromos, towards Khionistra, 1961, *D. P. Young* 7453 !
3. Akrotiri Forest, 1939, *Mavromoustakis* 55 !

4. Larnaca, 1801, *Hume*; Larnaca aerodrome, 1950, *Chapman* 697 ! and 1952, *Merton* 825 ! Mouth of Pedieos R., 1955, *Merton* 2262 !
6. Peristerona, 1936, *Syngrassides* 986 !

NOTES: Extraordinarily variable in the number of the floral parts; some Cyprus specimens have 6 sepals, 6 petals and 6 stamens; some 4 sepals, 4 petals and 2 (or 3) stamens; one specimen has 5 sepals, 5 petals and 8 stamens, another 4 sepals, 4 petals and 1 stamen. All the Cyprus material, with one exception, clearly belongs to typical *L. hyssopifolia* L. The exception (*D. P. Young* 7453), from Prodhromos, has numerous, overlapping, very narrow, somewhat papillose leaves, and small flowers (with 4 sepals, 4 petals and 2 stamens). It may be the critical *L. thymifolia* L., but unfortunately the specimens are stunted and depauperate, and unsatisfactory for accurate determination.

L. thymifolia L. (perhaps better ranked as *L. hyssopifolia* L. ssp. *thymifolia* (L.) Batt.) is locally frequent in S. France, Spain and North Africa, but is generally rare in the eastern Mediterranean region. Its presence at high altitudes on the Troödos Range would be of considerable interest, and it should be looked for in damp hollows in open parts of the Pine forests. While often markedly distinct by reason of its slender, erect habit, narrow (almost filiform), papillose leaves, and very small flowers and fruits, *L. thymifolia* sometimes intergrades with *L. hyssopifolia*; nor can it be distinguished from the latter by any one consistently reliable character.

3. L. tribracteatum *Salzm. ex Sprengel*, Syst. Veg., 4 (2): 190 (1827); E. Koehne in Engl., Pflanzenr., 17 (IV, 216): 64 (1892); Lindberg f., Iter Cypr., 24 (1946); Davis, Fl. Turkey, 4: 177 (1972); Zohary, Fl. Palaest., 2: 369, t. 537 (1972); Osorio-Tafall et Seraphim, List Vasc. Plants Cyprus, 74 (1973).

L. bibracteatum Boiss., Fl. Orient., 2: 740 (1872) pro parte quoad nomen; Post, Fl. Pal., ed. 2, 1: 471 (1932) pro parte quoad nomen.

TYPE: France; Montpellier.

Erect or spreading, glabrous annual, 3–18 cm. high, often forming dense colonies; stems distinctly 4-angled or even narrowly winged, commonly bright red or purple, with 2–6 pairs of opposite, patent branches; leaves opposite towards base of plant, alternate above, narrowly ovate, obovate or oblong, obtuse, sessile or subsessile, bright green, somewhat fleshy, 5–15 mm. long, 2–5 mm. wide; flowers solitary in the axils of the upper leaves, sessile or very shortly stalked, homostylous; hypanthium narrowly funnel-shaped or cylindrical, 5–6 mm. long, 1–1·5 mm. wide at apex, tapering to base, often reddish; bracteoles 2, subulate, membranous, usually very small (less than 1 mm. long) but said to be sometimes as long as hypanthium; epicalyx lobes 6, minute, bluntly deltoid, less than 0·5 mm. long, rather shorter than the sepals, herbaceous; sepals very broadly deltoid, less than 0·5 mm. long, membranous, usually inflexed in fruit; petals rosy-purple, oblong or somewhat obovate, 2·5–3 mm. long, about 1 mm. wide, apex rounded, base cuneate, scarcely clawed, attached at apex of hypanthium; stamens 5–6, inserted at different levels about the middle of hypanthium; filaments about 1·5 mm. long; anthers included, suborbicular, about 0·3 mm. diam.; ovary cylindrical, about 2 mm. long, 0·5 mm. diam.; style about 2 mm. long; stigma capitate, papillose. Fruit cylindrical or slightly swollen at base, about 5 mm. long, 1·3 mm. wide at base, about as long as the persistent hypanthium; seeds broadly ovoid, compressed, about 0·6 mm. long, 0·5 mm. wide, sometimes inconspicuously winged at the distal end; testa pale brown, minutely rugulose.

HAB.: Depressions and muddy or sandy pans in salt-marshes; about sea-level; fl. April–June.

DISTR.: Divisions 4 and 5, locally abundant. Also S. France, Spain, Italy, Aegean Islands, N. Africa.

4. Near Famagusta, 1939, *Lindberg f.* s.n.; Larnaca airport, 1952, *Merton* 824 ! 856 !
5. Near Salamis, 1962, *Meikle* 2567 !

NOTES: The name *L. tribracteatum* Salzm. ex Spreng. has been attached to two apparently distinct plants: (a) a plant with glabrous, smooth stems, leaves and hypanthia, and with minute, membranous bracteoles, barely visible without the aid of a lens. (b) a plant with scabrid-papillose stems, leaves and hypanthia, and with herbaceous bracteoles often half as long as, or sometimes as long as, the hypanthium. All the Cyprus material examined belongs to (a) which is unquestionably *L. salzmannii* Jord. (Obs. Plant. Nouv., 5: 44–47; 1847—*L. tribracteatum* Salzm. ex Spreng. var. *salzmannii* (Jord.) E. Koehne in Engl., Pflanzenr., 17 (IV, 216): 64; 1903) and which is here interpreted as typical *L. tribracteatum.* The second plant (b) is *L. tribracteatum* Salzm. ex Spreng. var. *candollei* E. Koehne (in Engl., Pflanzenr., 17 (IV, 216): 64; 1903), at least according to the description, though the citation of *L. thymifolia* var. *major* DC. (Prodr., 3: 81; 1828) as a synonym may mean that var. *candollei* is nomenclaturally identical with typical *L. tribracteatum* (var. *tribracteatum*). So consistent and obvious are the distinctions between (a) and (b) that it is difficult to regard them as other than distinct species, though the equally obvious similarities cannot be overlooked, together with the fact that both are not infrequently found in the same localities, sometimes apparently (to judge from the number of " mixed collectings ") intermixed in the same population. The problem calls for additional study in the field, and possibly for experimental work. It should be noted that *L. hyssopifolia* L. (including *L. thymifolia* L.) tends to vary in precisely the same way.

LAWSONIA INERMIS *L.* (Sp. Plant., ed. 1, 349; 1753), Henna, widely used in the East as a dye for the palms of the hands, fingers-nails and hair, is said to have been cultivated in ancient times as a commercial crop (see Pliny, Nat. Hist. XII, LI, 109), and is indeed supposed by some to have given its name (Greek, *Κύπρος*, Latin, *Cypros*) to the island, a suggestion now generally discounted.

Lawsonia inermis is a spinose or unarmed shrub or small tree, growing to about 6 m. in height, with opposite, obovate, oblong or elliptic, acute, entire, subsessile or shortly petiolate leaves, and large terminal panicles of numerous, small, creamy-white, 4-merous, scented flowers. The fruit is brownish or orange, spherical, indehiscent, about the size of a pea or slightly smaller, containing numerous angular seeds.

A native of tropical Africa, naturalized in other tropical countries, but nowhere wild or naturalized in Cyprus, where it would appear to have become rare, even in cultivation, by the nineteenth century. Kotschy (Die Insel Cypern, 367) records it from gardens in Nicosia, but adds that it is uncommon, while Sir Samuel Baker (Cyprus as I saw it in 1879, 187; 1879) says: " It is not a production of Cyprus, as has been erroneously stated by some authors; I made particular inquiries in all portions of the island, and of all classes, upon the subject. The henna, or shenna, is only to be met with in some few gardens, where it is cultivated as an ornamental shrub. . . ."

It may be worth noting that the *Κύπρος* or *Cypros* of Dioscorides and other early botanists would appear to be Privet, *Ligustrum vulgare* L., rather than Henna, and that the two shrubs are not infrequently confused in old literature; they are not dissimilar in foliage and flower.

40a. PUNICACEAE

Niedenzu in Engl. et Prantl, Pflanzenfam., ed. 1, III, 7: 22–25 (1892).

Shrubs and small trees; branches terete, sometimes spinose; leaves opposite or subopposite, simple, entire, exstipulate; flowers solitary or in small clusters, terminal, actinomorphic, large, showy; hypanthium broadly urceolate or funnel-shaped; sepals 5–8, valvate, deltoid, coriaceous, persistent; petals 5–8, free, imbricate in bud, obovate, shortly clawed, crimson or scarlet, inserted at apex of hypanthium; stamens very numerous, attached in several series below calyx, filaments free; anthers versatile; ovary adnate to hypanthium, multilocular, the loculi superposed; ovules numerous, placentation axile in the lower part of the ovary, parietal above; style simple, stigma capitate. Fruit large, globose or subglobose, crowned by the persistent erect sepals, indehiscent or irregularly dehiscent, exocarp coriaceous; seeds numerous, surrounded by translucent (often reddish), juicy pulp, without endosperm; cotyledons spirally convolute.

Includes only one genus (*Punica* L.) and two species, *P. granatum* L., the Pomegranate (west and central Asia) and *P. protopunica* Balf. f. (Socotra).

PUNICA GRANATUM *L.* (Sp. Plant., ed. 1, 472; 1753), the Pomegranate, a much-branched, sometimes spinose shrub or small tree, up to about 7 m. high, with oblanceolate, oblong or narrowly obovate leaves about 3–6 cm. long, and conspicuous scarlet flowers, is commonly cultivated in Cyprus gardens, being valued both as an ornamental, and for its large, juicy fruits.

According to legend (Eriphos in Athenaeus, Deipnos., II: 84c) Aphrodite first planted the tree on the island, a statement not altogether devoid of scientific interest, since it indicates that *Punica granatum* has been in cultivation in Cyprus, as elsewhere in the eastern Mediterranean region, from the earliest times, and that, as with the Phoenicians and Jews, it had acquired some significance in religious ritual. Though naturalized here and there in the Mediterranean region, its native home is to be sought further east, possibly in N.E. Turkey, and certainly in Iran and Afghanistan.

The only other member of the genus (and family) is *P. protopunica* Balf. f., endemic in Socotra, off the coast of southern Arabia. This has flowers and fruits closely resembling *P. granatum*, but is generally arborescent, with very much broader, almost orbicular, leathery leaves.

41. ONAGRACEAE

Raimann in Engl. et Prantl, Pflanzenfam., ed. I, III, 7: 199–223 (1893).

Annual or perennial herbs, or rarely shrubs or trees; leaves opposite or alternate, simple, generally exstipulate; flowers solitary, axillary or in spikes, racemes or panicles, hermaphrodite, actinomorphic or somewhat zygomorphic, usually 4-merous; hypanthium adnate to ovary, often elongate and cylindrical; sepals valvate; petals free, contorted or imbricate, rarely wanting; stamens as many as, or twice as many as sepals, in one or two whorls; anthers 2-thecous; ovary inferior or semi-inferior, 1–, 2–, 4– or 5– (–6)-locular; ovules 1–many; placentation axile; style simple; stigma capitate or ± deeply lobed. Fruit a loculicidal capsule, less often a berry or a nut; seeds generally without endosperm, or with very scanty endosperm, sometimes comose; embryo usually straight or almost straight.

About 20 genera with a cosmopolitan distribution.

1. EPILOBIUM *L.*

Sp. Plant., ed. 1, 347 (1753).
Gen. Plant., ed. 5, 164 (1754).
Haussknecht, Mon. Gatt. Epilobium, 316 pp. (1884).

Perennial herbs; leaves usually opposite towards the base of the stem, sometimes forming distinct overwintering rosettes, alternate above, or occasionally all alternate, rarely verticillate, exstipulate, the upper reduced to small bracts; flowers in simple or branched racemes, or solitary and axillary, actinomorphic or somewhat zygomorphic, 4-merous, hypanthial tube short or wanting; sepals free; petals white, pink or purple, often emarginate; stamens 8 in 2 whorls; ovary 4-locular; style generally distinct; stigma clavate or capitate, or divided into 4 lobes. Fruit a long, narrow capsule dehiscing into 4 valves; seeds numerous, small, with a chalazal coma or tuft of hairs.

About 200 species with a world-wide distribution, but chiefly in temperate regions; hybrids are often abundant, and tend to obscure the rather subtle distinctions between the species.

Lower leaves alternate; flowers numerous in terminal bracteate racemes, somewhat zygomorphic; petals entire or obscurely emarginate - - **1. E. angustifolium**
Lower leaves opposite; flowers solitary, axillary, actinomorphic; petals bilobed at apex:
Stigma 4-lobed; stems hairy or crispate-pubescent; fruits 30–80 mm. long:
Petals 10–20 mm. long; leaves sessile, amplexicaul - - **2. E. hirsutum**
Petals 4–8 mm. long; leaves not amplexicaul:
Stems clothed with spreading hairs; lower leaves sessile or very shortly petiolate **3. E. parviflorum**
Stems clothed with crispate hairs; lower leaves distinctly petiolate **4. E. lanceolatum**
Stigma clavate, without lobes; lower part of stem glabrous; fruits 70–110 mm. long **5. E. tetragonum**

SECT. 1. **Chamaenerion** *Tausch* Leaves all alternate; flowers somewhat zygomorphic, often showy; petals entire or very slightly emarginate. Robust suckering perennials, without winter rosettes.

1. E. angustifolium *L.*, Sp. Plant., ed. 1, 347 (1753); Hartmann in Mitt. Deutsch. Dendr. Ges., 14: 171 (1905); Holmboe, Veg. Cypr., 134 (1914); Post, Fl. Pal., ed. 2, 1: 472 (1932); Davis, Fl. Turkey, 4: 185 (1972); Osorio-Tafall et Seraphim, List Vasc. Plants Cyprus, 75 (1973).

Chamaenerion angustifolium (L.) Scop., Fl. Carniol., ed. 2, 1: 269 (1772); Lindberg f., Iter Cypr., 24 (1946).

Epilobium spicatum Lam., Fl. Franç., 3: 482 (1788); Boiss. Fl. Orient., 2: 745 (1872).

Chamaenerion spicatum (Lam.) S. F. Gray, Nat. Arrang. Br. Plants, 2: 559 (1821) nom. illeg.

Chamerion angustifolium (L.) Holub in Folia Geobot. et Phytotax., 7: 86 (1972).

TYPE: " *in* Europa *boreali* ".

Robust suckering perennial; stems erect, to 2·5 m. high, terete or subterete, crispate-pubescent, often reddish; leaves generally all alternate, sessile or subsessile, linear-lanceolate, 6–16 (–20) cm. long, 1–3·5 cm. wide, entire or subentire, bright or dark green and glabrous above, slightly greyish and sometimes thinly pubescent below, apex acuminate, base cuneate; flowers numerous in a lax, elongate terminal raceme, 15–35 cm. long at anthesis, elongating in fruit; bracts narrowly subulate or filiform, the lower often 1 cm. long, decreasing in length and breadth upwards; pedicels 5–15 mm. long, adpressed-pubescent; hypanthium cylindrical, about 10 mm. long, 1·5 mm. wide, densely greyish-pubescent; sepals narrowly oblong, 10–12 mm. long, 2–3 mm. wide, acute, pubescent externally, generally reddish-purple; petals vivid pink-purple, obovate, shortly clawed, the lower 10–15 mm. long, 8–10 mm. wide, the upper a little smaller and narrower, apex rounded or obscurely emarginate; filaments somewhat unequal, 6–9 mm. long, glabrous, dilated towards base, at first spreading or erecto-patent, becoming deflexed and flaccid after the anthers have dehisced; anthers narrowly oblong, 2·5–3 mm. long, 1 mm. wide, reddish-purple; style about 8 mm. long, hairy at the base, at first sharply deflexed with the immature stigma-lobes connivent and clavate, at maturity suberect with cruciform stigma-lobes 2·5 mm. long. Fruit (25–) 40–80 (–120) mm. long, 2·5 mm. wide, obscurely tetragonous, densely or sparsely adpressed crispate-pubescent; seeds narrowly oblong-ellipsoid, about 1·3 mm. long, 0·4 mm. wide, comal tuft about 1·5 cm. long; testa pale brown, rather glossy, smooth or slightly rugulose.

HAB.: Damp places and streamsides at 4,000–6,000 ft. alt.; probably introduced; fl. June–Sept.

DISTR.: Division 2, locally common. Widespread in Europe, Asia and N. America, but a recent introduction in many areas.

2. Locally common on Khionistra, 4,000–6,000 ft. alt. and apparently spreading; also recorded from Papoutsa; *Hartmann*; *Holmboe* 865; *Haradjian* 470 ! 990 ! *C. B. Ussher* 110 ! *Syngrassides* 753 ! *Kennedy* 660 ! 661 ! *Lindberg f.* ! *Davis* 1807 ! *Casey* 843 ! *Merton* 2790 ! *D. P. Young* 7445 !

NOTES: It is conceivable that *E. angustifolium* may be indigenous at high altitudes on the Troödos massif; the absence, however, of any records prior to 1905, and the fact that Hartmann (1905) clearly viewed it as an introduction, strongly suggest that it is a recent arrival on the island, now actively extending its distribution here as elsewhere.

SECT. 2. **Epilobium** Lower leaves opposite, upper alternate; flowers actinomorphic; petals commonly emarginate or 2-lobed. Tufted or stoloniferous herbs, often forming over-wintering rosettes.

2. **E. hirsutum** *L.*, Sp. Plant., ed. 1, 347 (1753); Boiss., Fl. Orient., 2: 746 (1872); Post in Mém. Herb. Boiss., 18: 93 (1900); H. Stuart Thompson in Journ. Bot., 44: 308 (1906); Holmboe, Veg. Cypr., 134 (1914); Post, Fl. Pal., ed. 2, 1: 473 (1932); Davis, Fl. Turkey, 4: 187 (1972); Zohary, Fl. Palaest., 2: 374, t. 545 (1972); Osorio-Tafall et Seraphim, List Vasc. Plants Cyprus, 75 (1973).

TYPE: " *in* Europae *humidiusculis* ".

Erect, much-branched rhizomatous perennial, up to 2 m. high or occasionally even more; stems terete or subterete, usually rather densely clothed with long whitish hairs and short glandular hairs, but occasionally almost glabrous; leaves lanceolate, amplexicaul, 5–12 (–15) cm. long, 0·8–3 (–4·5) cm. wide, sessile, opposite except at the tips of the shoots, whitish-villose or rarely subglabrous, apex acute, margins rather sharply serrulate, with alternate larger and smaller teeth; flowers solitary, axillary at the tips of the shoots, subtended by reduced leaves; pedicel erect, glandular-villose, to about 8 mm. long; hypanthium 2·5–4·5 cm. long, 0·2–0·3 cm. wide, rather densely glandular-villose; sepals lanceolate-acuminate, 8–10 mm. long, 2–3 mm. wide, villose or occasionally subglabrous, united for a short distance above hypanthium; petals rich pink-purple, broadly obcordate, deeply 2-lobed, suberect, overlapping, 10–20 mm. long, 12–15 mm. wide, hirsute at base, very shortly clawed; stamens very unequal, the longer with filaments 5–8 mm. long, the shorter with filaments 1·5–2·5 mm. long; anthers yellow, oblong, about 2·5 mm. long, 1 mm. wide; style glabrous, erect, 8–12 mm. long, stigmas deeply 4-lobed, lobes 2–3 mm. long, sometimes tinged pink or red. Fruit erect, linear, 60–80 mm. long, 2·5–3 mm. wide, usually rather densely villose; seeds oblong-ellipsoid, about 1 mm. long, 0·4 mm. wide; testa dark brown, densely papillose with glistening papillae; comal tuft 8–10 mm. long.

HAB.: Wet places and borders of irrigation-channels; sea-level to over 5,000 ft. alt.; fl. June–Oct.

DISTR.: Divisions 2, 5, 7, rare. Widespread in Europe, Asia and Africa; Atlantic Islands; introduced and often naturalized elsewhere.

2. Troödos, 1894, *Post* s.n.; Kato Mylos, between Agros and Limassol, 1937, *Syngrassides* 1700 !

5. Kythrea, 1934, *Syngrassides* 525 !

7. Akhiropitos, near Karavas, 1936, *Syngrassides* 1299 !

3. E. parviflorum *Schreber*, Spic. Fl. Lips., 155 (1771); Stokes in Withering, Bot. Arrangement, ed. 1, 1: 225 (1776); Poech, Enum. Plant. Ins. Cypr., 38 (1842); Unger et Kotschy, Die Insel Cypern, 367 (1865); Boiss., Fl. Orient., 2: 746 (1872); Holmboe, Veg. Cypr., 134 (1914); Post, Fl. Pal., ed. 2, 1: 473 (1932); Lindberg f., Iter Cypr., 24 (1946); Davis, Fl. Turkey, 4: 187 (1972); Zohary, Fl. Palaest., 2: 374, t. 546 (1972); Osorio-Tafall et Seraphim, List Vasc. Plants Cyprus, 75 (1973).
Chamaenerion parviflorum Schreber, Spic. Fl. Lips., 146 (1771).
TYPE: Germany; " in humidis ad *Schoenfeld, Altnauendorf* ", *Schreber* (? M).

Erect perennial 20–100 cm. high with overwintering rosettes; stems terete or subterete, sometimes branched at the base, but frequently unbranched except at the apex, commonly white-villose, but sometimes sparingly so; leaves mostly opposite, sessile, oblong, ovate or lanceolate, 1·5–10 cm. long, 0·5–2·5 cm. wide, sparingly or rather densely crispate-villose, apex acute or obtuse, margins rather remotely and irregularly serrulate; flowers solitary, axillary, subtended by much-reduced, alternate leaves; pedicel erecto-patent, about 5 mm. long; hypanthium slender 25–35 mm. long, about 1 mm. wide, tetragonous, often rather densely villose; sepals oblong, acute, about 5 mm. long, 1·5 mm. wide, thinly villose externally, connate for a short distance above the base; petals pale pink, obcordate, rather deeply and bluntly 2-lobed, about 6–8 mm. long, 3–4 mm. wide, narrowing to a very short claw; stamens very unequal, the longer filaments about 2 mm. long, the shorter about 0·5 mm. long; anthers yellow, shortly oblong, about 1·3 mm. long, 1 mm. wide; style 3–3·5 mm. long; stigmas deeply 4-lobed, lobes about 1·3 mm. long. Fruit erect, 30–80 mm. long, 1·5 mm. wide, tetragonous, often thinly villose; seeds oblong, 0·8–1 mm. long, 0·4 mm. wide; testa dark brown, minutely and densely papillose, comal tuft about 1 cm. long.

HAB.: Marshy ground, and wet places by streams and irrigation channels; 1,800–4,500 ft. alt.; fl. June–Oct.

DISTR.: Division 2, uncommon; widespread in Europe and Asia; N. Africa and Atlantic Islands.

2. Evrykhou, Prodhromos, Platania, Pharmakos, Kato Mylos, Kambos, Trikoukkia, Cedar valley (Tripylos); *Kotschy* (1840) 7; *Sintenis & Rigo* 704 ! *Holmboe* 935 ! 1128 ! *Syngrassides* 1086 ! 1701 ! *Kennedy* 659 ! *Lindberg f.* ! *Casey* 870 ! *G. E. Atherton* 542 ! *Merton* in ARI 67 ! *Economides* in ARI. 1203 ! 1208 !

4. E. lanceolatum *Seb. et Mauri*, Fl. Rom. Prodr., 138 (1818); Boiss., Fl. Orient., 2: 747 (1872); Holmboe, Veg. Cypr., 134 (1914); Post, Fl. Pal., ed. 2, 1: 473 (1932); Lindberg f., Iter Cypr., 24 (1946); Davis, Fl. Turkey, 4: 188 (1972); Osorio-Tafall et Seraphim, List Vasc. Plants Cyprus, 75 (1973).
E. montanum L. var. *lanceolatum* (Seb. et Mauri) Koch, Syn. Fl. Germ. et Helv., ed. 1, 1: 266 (1837); Unger et Kotschy, Die Insel Cypern, 366 (1865).
TYPE: Italy; " sepes circa Roman frequens ", *Sebastiani et Mauri*, figured in Fl. Rom. Prodr, t. 1, fig. 2 !

Erect perennial, usually less than 50 cm. high, with short stolons forming loose over-wintering rosettes; stems rather slender, subterete, sparsely crispate-pubescent, often reddish, occasionally simple, but usually with numerous very slender lateral branches; leaves mostly opposite, lanceolate, 1–8 (–10) cm. long, 0·3–2·5 (–3) cm. wide, subglabrous or thinly crispate-pubescent especially on midrib and nerves, dull green above, often reddish below, apex often rather blunt, margins remotely and irregularly serrulate (or sometimes subentire), base entire, cuneate; petiole generally distinct, especially in the lower leaves, 0·5–1 cm. long, pubescent, somewhat channelled above; flowers solitary, axillary, each subtended by a small, much-reduced leaf or bract, at first nodding, later erect; pedicel 0·8–3·5 cm. long, merging above with the base of the hypanthium, canescent, with crispate hairs intermixed with small, shining glands; hypanthium linear, often somewhat arcuate, 20–35 mm. long, 1–1·5 mm. wide at anthesis, crispate-canescent and shortly glandular; sepals oblong-lanceolate, 2·5–4 mm. long, 0·5–0·8 mm.

wide, pubescent externally, often reddish, connate for a short distance above base; petals at first whitish, later turning pale flesh- or salmon-pink, obcordate, 4–5 mm. long, 2–3 mm. wide, spreading, apex distinctly bilobed, base narrowing to a short glabrous or somewhat pubescent claw; stamens very unequal, the longer with filaments up to 2 mm. long, the shorter with filaments about 0·8 mm. long; anthers yellow, oblong, about 0·8 mm. long, 0·5 mm. wide; style relatively thick, about 2·5 mm. long; stigma divided into 4, short, blunt, suberect lobes about 1·3 mm. long. Fruit a slender, straight or slightly arcuate, tetragonous capsule 50–70 mm. long, about 1·5 mm. wide; seeds narrowly oblong-ellipsoid, about 1·2 mm. long, 0·3 mm. wide; testa dark or light brown, densely glistening-papillose, comal tuft about 1 cm. long.

HAB.: Streamsides, moist screes and damp ground in Pine forests at high altitudes; 3,500–5,500 ft. alt.; fl. May–July.

DISTR.: Division 2, locally abundant on the Troödos range; widespread in Europe and eastwards to N. Iran.

2. Troödos, Platania, Kykko Monastery, Kannoures Springs, Madhari, Prodhromos, etc.; *Kotschy* 830; *Sintenis & Rigo* 702 ! 704 p.p. ! *Haradjian* 447 ! 989 ! *Syngrassides* 1104 ! 1582 ! *Kennedy* 658 ! *Casey* 845 ! *Merton* 2419 ! *D. P. Young* 7321 ! 7434 !

4a. E. lanceolatum *Seb. et Mauri* × **E. parviflorum** *Schreb.*

E. × *aschersonianum* Hausskn., Mon. Gatt. Epilobium, 178 (1884).

Intermediate between the parents, with thinly hairy stems, shortly petiolate, subglabrous or thinly hairy, rather sharply serrulate leaves and sterile fruits.

Specimens of the hybrid have been identified by Dr. P. H. Raven amongst material of *E. parviflorum* Schreber collected by *Merton* (no. 2826) by the edge of a small swamp at Livadhi, on the S. side of Tripylos Peak, Sept. 27, 1956. Although *E. lanceolatum* has not been recorded from this locality, its presence is to be expected; the collector notes: " flower colour variable, pink to magenta; is this a hybrid swarm? " The hybrid, most easily recognized by its sterility, is likely to be found elsewhere on the Troödos range.

5. E. tetragonum *L.* Sp. Plant., ed. 1, 348 (1753); Boiss., Fl. Orient., 2: 748 (1872); Post, Fl. Pal., ed. 2, 1: 473 (1932); Davis, Fl. Turkey, 4: 189 (1972).

E. adnatum Griseb. in Bot. Zeit., 10: 854 (1852); Hausskn., Mon. Gatt. Epilobium, 97 (1884).

TYPE: " *in* Europa ".

Erect perennial 20–90 cm. high, with short stolons forming loose overwintering rosettes; stems often rigid and unbranched below, usually much-branched in the floriferous part, glabrous below, sparingly crispate-pubescent above, striatulate, the leaf bases continued downwards as distinct ridges; leaves mostly opposite except towards apex of plant, lanceolate or linear-oblong, 2·5–10 cm. long, 0·3–2 cm. wide, glabrous, acute or acuminate, margins rather coarsely and irregularly serrulate, sometimes shortly ciliate, tapering gradually to a decurrent base; flowers solitary in the axils of reduced leaves; pedicel 1–1·5 cm. long, thinly crispate-pubescent; hypanthium narrow, straight or slightly arcuate, 40–80 mm. long at anthesis, 1–1·5 mm. wide, tetragonous, thinly crispate-pubescent; sepals lanceolate, acuminate, 6–7 mm. long, 1 mm. wide, crispate-pubescent externally, connate for a short distance above the base; petals purplish-pink, spreading, oblong-obovate, 2-lobed, about 7 mm. long, 4 mm. wide, tapering to a very short basal claw; stamens very unequal, the longer with filaments 3·5–4 mm. long, the shorter with filaments 2 mm. long; anthers yellow, oblong, about 1·5–

1·8 mm. long, 0·7 mm. wide; style stout, 2·5–3 mm. long; stigma elongate-clavate, not lobed, 3·5–4 mm. long. Fruit 70–110 mm. long, not much more than 1·5 mm. wide; seeds narrowly obovoid, compressed, about 1 mm. long, 0·4 mm. wide; testa dark brown with close longitudinal rows of prominent papillae.

HAB.: Moist ground at high altitudes; c. 5,000 ft. alt.; fl. May–July.

DIST.: Division 2, rare. Europe, Mediterranean region, west and central Asia.

2. Prodhromos, " in faucibus ", June 30, 1880, *Sintenis & Rigo* 704 (as *E. parviflorum*) !

NOTES: The exact localization of this single specimen of *E. tetragonum* is rather dubious. Sintenis and Rigo labelled their 704 " Epilobium parviflorum L.", but the sheet at Kew comprises two plants, one *E. lanceolatum* Seb. et Mauri, the other *E. tetragonum* L. There are also in the Kew collections two sheets of *Sintenis & Rigo* 702, both consisting of *E. lanceolatum* specimens, but both, rather surprisingly, labelled *E. tetragonum* L. var. *lamyi* F. W. Schultz. These are said to come from " sylvis montis Troodos ". It may be that this label was originally intended to go with the specimens subsequently distributed as no. 704, or, equally likely, *Sintenis & Rigo* may have misidentified *E. lanceolatum* (which they do not record from Cyprus) as *E. tetragonum* var. *lamyi*. Whatever the truth, it is apparent that *E. tetragonum* occurs at high altitudes on the Troödos range. Its rediscovery and exact localization would be welcome.

[OENOTHERA MICRANTHA *Hornem.*, a native of California, is recorded by Hume at " Larnica and Limosol " in Walpole, Mem. Europ. Asiatic Turkey, ed. I, 254 (1817) under the illegitimate name *Oenothera hirta* Link. There is no description of the plant, nor, from the manner in which it is listed, is it altogether clear what Hume intends. As there are no subsequent records for this species, it should be expunged from the Cyprus flora.]

42. DATISCACEAE

Airy Shaw in Willis, Dict. Fl. Plants & Ferns, ed. 7, 333 (1966).

Erect, robust perennial herbs, superficially resembling *Cannabis*; leaves alternate, exstipulate, imparipinnate or pinnatisect; flowers numerous, rather small, in crowded fascicles on long leafy branches, unisexual or occasionally bisexual, actinomorphic; male flowers with 3–9 unequal sepals, petals usually absent, stamens 8–25, anthers conspicuous; female flowers apetalous, with an inferior, 1-locular ovary, stigmas 3–5, free, bipartite, placentation parietal, ovules numerous, anatropous. Fruit an oblong-ovoid, erect or pendulous capsule, dehiscing at the apex between the styles; seeds small, endosperm wanting or very scanty, embryo straight.

One genus only, represented in Crete, S.W. Asia, Himalayas and California.

1. DATISCA *L.*

Sp. Plant., ed. 1, 1037 (1753).
Gen. Plant., ed. 5, 460 (1754).

Description that of the family.

Two species, *D. cannabina* L. and *D. glomerata* (Presl) Baill., the former in Crete, S.W. Asia and Himalayas, the latter in California.

1. **D. cannabina** *L.*, Sp. Plant., ed. 1, 1037 (1753); Boiss., Fl. Orient., 2: 763 (1872); Holmboe, Veg. Cypr., 132 (1914); Post, Fl. Pal., ed. 2, 1: 484 (1932); Davis, Fl.

Turkey, 4: 422 (1972); Osorio-Tafall et Seraphim, List Vasc. Plants Cyprus, 73 (1973).
TYPE: " *in* Creta ".

Erect, glabrous perennial 1–2 m. high; stems longitudinally sulcate, virgate; leaves imparipinnate, 20–30 cm. long, with 3–11 lanceolate-acuminate, sharply serrate leaflets 8–13 cm. long, 0·6–3 cm. wide; leaf-nervation prominent on the lower surface of the lamina; petiole 3–8 cm. long; inflorescence much-branched, forming a large, pyramidal, leafy panicle, consisting of numerous fascicles of flowers subtended by leaves or simple, linear or linear-lanceolate, entire or serrate, subsessile or shortly petiolate, foliaceous bracts; male flowers shortly pedicellate; sepals lanceolate, sub-acute, very unequal, 0·5–2 mm. long, 0·3–0·5 mm. wide; anthers subsessile, oblong, yellow, about 2·5–3 mm. long, 0·8 mm. wide; female flowers at first very shortly pedicellate, pedicel lengthening and deflexing in fruit; hypan-thium 3–5 mm. long, 1·5–3 mm. wide, longitudinally sulcate; stigmas caudate, whitish, 5–9 mm. long, divided almost to the base; sepals minute, less than 0·5 mm. long, deltoid or subulate; placentas conspicuously thick-ened, swelling to gibbosities at the apex and closing, at anthesis, the otherwise slightly open ovary. Capsules at first pendulous, about 5–10 mm. long, 2·5–3·5 mm. wide, brownish, chartaceous, slightly sulcate longitudinally; seeds very numerous, oblong, about 1 mm. long, 0·4 mm. wide, with a small cup-shaped basal appendage; testa mid-brown, conspicuously and regularly reticulate.

HAB.: Shaded ground by streams; 2,000–4,000 ft. alt.; fl. June–Sept.

DISTR.: Division 2, rare. Also Crete, S.W. Asia, Himalayas.

2. Kykko, 1905, *Holmboe* 1012; Makheras, 1905, *Holmboe* 1122; near Platanistasa, N.W. of Palekhori, Jan. 1955, *Mrs. Scott-Moncrieff* in *Casey* 1673 !

43. CUCURBITACEAE
by C. Jeffrey.

C. Jeffrey in Willis, Dict. Fl. Plants & Ferns, ed. 7, 306–308 (1966).

Mostly herbs, climbing or trailing, generally with tendrils; leaves alter-nate, usually simple, palmately veined; tendrils lateral to the leaf-base, simple or branched; flowers commonly yellowish, actinomorphic, unisexual, monoecious or dioecious, the male usually racemose, the female often solitary; hypanthium usually campanulate or obconic, often short, especially in female flowers; sepals usually 5, often small; petals usually 5, ± united at least at the base, sometimes free; stamens variable in number and form, commonly 3, two double 2-thecous, one single 1-thecous, inserted on the hypanthium, free or variously united; anther-thecae often flexuous or con-torted; staminodes often present in female flowers; ovary inferior, syn-carpous, unilocular; ovules usually numerous, anatropous, most often horizontal; placentation usually parietal; styles 1, less often 3; stigmas usually 3, bilobed. Fruit a berry or firm-walled berry (*pepo*), or a fleshy capsule, sometimes dry, variously dehiscent or indehiscent; seeds large, without endosperm, sometimes winged.

About 100 genera, mainly in the tropical and subtropical regions, including some important economic plants—see Whitaker, T. W., & Davis, G. N.,

Cucurbits (Leonard Hill, London, 1962). The genera given in parentheses in the following key occur in Cyprus only as cultivated plants.

Plant prostrate, without tendrils; fruit hispid, pendulous at the apex of an erect peduncle, when ripe expelling the seeds forcibly through the point of detachment **2. Ecballium**

Plant climbing or trailing, with tendrils; fruit not as above:

Leaf-lamina with a pair of lateral glands at the base adjacent to the point of insertion of the petiole; petals white - - - - - [LAGENARIA p. 679]

Leaf-lamina without such glands; petals some shade of yellow:

Petals united for more than half their length; anthers united into an oblong head [CUCURBITA p. 679]

Petals free or united only near the base; anthers free, although sometimes coherent:

Fruits dry and fibrous, opening by an apical operculum; petals free, more than 25 mm. long; sepals ovate or triangular, 8–12 mm. long [LUFFA p. 679]

Fruits fleshy, indehiscent; petals united near the base, rarely exceeding 20 mm.; sepals subulate or dentiform, not exceeding 5 mm.:

Male and female flowers both usually in racemes or corymbs; fruits small, 8–10 mm. in diameter, soft-walled, red when ripe - - **3. Bryonia**

Male flowers solitary or 2–8 in small clusters; female flowers solitary; fruits much larger, firm-walled, greenish, yellowish or brownish when ripe:

Anther-connectives not produced beyond the thecae; female flowers without a ring-shaped disc surrounding the base of the style; seeds 2 mm. or more thick; leaves usually deeply, ± pinnately, divided **1. Citrullus**

Anther-connectives produced beyond the thecae; female flowers with a ring-shaped disc surrounding the base of the style; seeds up to 1·5 mm. thick; leaves palmately, often rather shallowly lobed **4. Cucumis**

1. CITRULLUS *Eckl. et Zeyh.*

Enum. Plant. Afr. Austr. Extratrop., 279 (1836) nom. cons.
Colocynthis Mill., Gard. Dict. Abridg., ed. 4 (1754) nom. rejic.

Annual or perennial, trailing or climbing herbs, usually with tendrils; leaves usually deeply, ± pinnately, lobed; tendrils simple or 2–4-branched; flowers monoecious, solitary; hypanthium short, obconic in male flowers, very shortly cylindrical in female flowers; petals yellow, united below the middle; stamens free; connectives broad, not produced apically; anther-thecae flexuous; ovary hairy, subglobose. Fruit large, ellipsoid or subglobose, greenish or yellowish, often mottled, firm-walled, fleshy, indehiscent, smooth; seeds numerous, ovate in outline, compressed, smooth or slightly rough.

3 species, 2 in the Kalahari region of south-western Africa, the third Saharo-Sindian.

1. C. colocynthis (*L.*) *Schrad.* in Linnaea 12: 414 (1838); Unger et Kotschy, Die Insel Cypern, 341 (1865); Gennadius, Rep. Agric. Cypr., 2: 19 (1896); H. Stuart Thompson in Journ. Bot., 44: 307 (1906); Rechinger f. in Arkiv för Bot. ser. 2, 1: 432 (1950); Davis, Fl. Turkey, 4: 202 (1972); Osorio-Tafall et Seraphim, List Vasc. Plants Cyprus, 100 (1973).

Cucumis colocynthis L., Sp. Plant., ed. 1, 1011 (1753).
Colocynthis vulgaris Schrad., Ind. Sem. Hort. Goett., 2 (1833).

TYPE: described from plants cultivated at Uppsala (LINN 1152/1 !).

Prostrate trailing perennial herb; rootstock deep, stout, rather woody; stems annual, angled, shortly retrorsely hairy, becoming scabrid; leaves petiolate, elongate-ovate in outline, scabrid-hairy beneath, scabrid-punctate above, 2·5–11 cm. long, 1·5–6·5 cm. wide, deeply 3–5-lobed, lobes ± pinnately lobulate, the central largest; petiole roughly retrorsely hairy, 1–6 cm. long; tendrils simple or less often bifid; male flowers on peduncles 1–2 cm. long; hypanthium green, broadly obconic, 1·5–2·5 mm. long; sepals lanceolate, 2–5 mm. long; petals pale yellow, 6–10 mm. long, 4–6 mm. wide; female flowers on peduncles 1–4·5 cm. long; ovary 7–9 mm. long; hypanthium very short; perianth as in male flowers. Fruit subglobose, smooth, green or yellow, often longitudinally variegated, becoming dry, yellow, hollow and brittle when old, 5–12 cm. diam.; flesh exceedingly bitter; seeds smooth, 6·5–10 mm. long, 3·5–5 mm. wide, 2–2·5 mm. thick.

HAB.: Dry sandy and rocky places, including foreshores and coastal dunes; sea-level to 500 ft. alt.; fl. May–July.

DISTR.: Divisions 1, 5, 7, locally common. Semi-desert and desert areas of North Africa, southern Europe and Asia, from the Canary Is. eastwards to India.

1. Ktima, 1913, *Haradjian* 672 !
5. Locally common, Athalassa, Dhali, Trikomo etc.; *Sintenis & Rigo* 810 ! *Syngrassides* 405 ! *Lindberg f.* ! *Oswald* 506/64 !
7. Locally common near Kyrenia; *Davis* 3621 ! *Casey* 141 ! 821 ! *Merton* 2290 ! *G. E. Atherton* 115 !

C. LANATUS (*Thunb.*) *Mats. et Nakai* (*C. vulgaris* Eckl. et Zeyh.; *C. edulis* Spach), which differs from the above in its annual habit, soft indumentum and larger, non-bitter fruits, is an important crop-plant grown for its edible fruits and seeds; it has been recorded from Varosha (near Famagusta) by Kotschy (Die Insel Cypern, 341: 1865).

2. ECBALLIUM *A. Rich.*

in Bory, Dict. Class. Hist. Nat., 6: 19 (1825) nom. cons.
Elaterium Mill., Gard. Dict. Abridg., ed. 4 (1754), non Jacq. (1760) nom. rejic.

Prostrate perennial herbs, without tendrils; leaves simple, cordate; flowers monoecious, the male racemose, the female solitary; hypanthium short, campanulate; petals yellow, united near the base; stamens free; connectives broad; anther-thecae flexuous; ovary minutely tuberculate, hispid, oblong. Fruit oblong-ellipsoid, greenish, glaucous, muricate, hispid, fleshy, at maturity coming away from the peduncle and forcibly expelling the seeds and pulp through the aperture so formed; seeds numerous, ovoid, obscurely margined.

One species only, in the Mediterranean region.

1. **E. elaterium** (*L.*) *A. Rich.* in Bory, Dict. Class. Hist. Nat., 6: 19 (1825); Boiss., Fl. Orient., 2: 760 (1872); Halácsy, Consp. Fl. Graec., 1: 549 (1901); Holmboe, Veg. Cypr., 176 (1914); Mathews, Lilies of the Field, 46, cum ic. (1968); Davis, Fl. Turkey, 4: 203 (1972); Osorio-Tafall et Seraphim, List Vasc. Plants Cyprus, 100 (1973).

Momordica elaterium L., Sp. Plant., ed. 1, 1010 (1753); Unger et Kotschy, Die Insel Cypern, 341 (1865); H. Stuart Thompson in Journ. Bot., 44: 307 (1906).

TYPE: described from cultivated plants (LINN 1150/10 !).

Prostrate perennial herb with thick fleshy root; stems short, stout, prostrate or ascending, hispid, pale green, up to 2·5 m. long; leaves petiolate, ovate-triangular, cordate, rather fleshy, stiff, deep green and scabrid above, pale green and densely hispid-tomentose beneath, sinuate and undulate at the margins, rather concave above, 4–14·5 cm. long, 3·5–17 cm. wide, shortly 5-lobed with the basal lobes ± erect; petiole stout, hispid, 5–15 cm. long; male racemes 3·5–37 cm. long, few to many flowered; peduncles erect, 2–10 cm. long; pedicels 1–3 cm. long; sepals linear-lanceolate, villous, 5–6 mm. long; petals yellow, 8–16 mm. long, 4–7 mm. wide; female flowers coaxillary with male; peduncles 1–8 cm. long; perianth rather smaller than in male flowers, with petals 6–12 mm. long and 4–7 mm. wide. Fruit pendulous at the apex of ± erect peduncles, 3–5 cm. long, 1·5–2·5 cm. diam.; pulp exceedingly bitter; seeds tumid, about 4 mm. long.

HAB.: Open grassy and rocky places, roadsides, banks and walls; sea-level to 2,900 ft. alt.; fl. Jan.–July.

DISTR.: Divisions 2, 4, 5, 7, common. Azores, Mediterranean region, S. Russia, Crimea, Caucasus.

2. Makheras Mon., 1927, *Syngrassides* 1635 !
4. Kouklia Reservoir, 1957, *Merton* in ARI 442 !
5. Kythrea, 1880, *Sintenis & Rigo* 5 ! Trikomo, 1956, *Merton* 2467 !
7. Common about Kyrenia; *Casey* 56 ! *Mapple* 113 ! *G. E. Atherton* 182 ! 893 !

3. BRYONIA *L.*

Sp. Plant., ed. 1, 1012 (1753).
Gen. Plant., ed. 5, 442 (1754).
C. Jeffrey in Kew Bull., 23: 441–461 (1969).

Perennial climbing herbs with annual stems and large subterranean tuberous rootstocks; tendrils present, simple; leaves simple, palmately lobed; flowers monoecious or dioecious, usually in axillary racemes or corymbs; hypanthium campanulate; petals yellow, united near the base; stamens free; connectives broad; anther-thecae flexuous; ovary smooth, globose. Fruit small, baccate, globose, usually red or black; seeds many, ovoid, little compressed.

About 11 species in Europe, the Mediterranian region, and south-west Asia, from Canary Is. eastwards to N.W. India.

1. **B. cretica** *L.*, Sp. Plant., ed. 1, 1013 (1753); Unger et Kotschy, Die Insel Cypern, 341 (1865); Boiss., Fl. Orient., 2: 760 (1872); Halácsy, Consp. Fl. Graec., 1: 550 (1901); Holmboe, Veg. Cypr., 176 (1914); Post, Fl. Pal., ed. 2, 1: 481 (1932); Tutin et al., Fl. Europ., 2: 297 (1968) quoad ssp. *creticam* tantum; Davis, Fl. Turkey, 4: 204 (1972); Osorio-Tafall et Seraphim, List Vasc. Plants Cyprus, 100 (1973).

[*B. dioica* (non Jacq.) Kotschy in Unger et Kotschy, Die Insel Cypern, 340 (1868); H. Stuart Thompson in Journ. Bot., 44: 307 (1906); Holmboe, Veg. Cypr., 176 (1914); Osorio-Tafall et Seraphim, List Vasc. Plants Cyprus, 100 (1973).]

[*B. multiflora* (non Boiss. et Heldr.) Sintenis in Oesterr. Bot. Zeitschr., 32: 365 (1882); Holmboe, Veg. Cypr., 176 (1914); Rechinger f. in Arkiv för Bot., ser. 2, 1: 432 (1950); Osorio-Tafall et Seraphim, List Vasc. Plants Cyprus, 100 (1973).]

TYPE: described from cultivated plants originating from Crete (BM, lectotype !).

Climbing perennial herb; rootstock a large subterranean tuber; stem annual, glabrous or coarsely spiculate, rarely glaucous; leaves petiolate, ovate-cordate or broadly so in outline, mid-green, herbaceous, scabrid, becoming scabrid-punctate above, glabrous to scabrid-spiculate beneath, 3–9 cm. long, 3·5–9 cm. wide, shallowly to deeply 5–7-lobed, the lobes ovate to lanceolate, obtusely to acutely sinuate or ± lobulate at the margins, the central largest; petiole glabrous or sparsely spiculate, sometimes glaucous, 0·5–5 cm. long; flowers dioecious; male racemes usually many-flowered, 4·5–13·5 cm. long, glabrous or almost so except for a few glands on the pedicels; sepals triangular-dentiform, 1–2 mm. long; petals yellow, 3–6 mm. long, 2·2–4 mm. wide; female flowers few to many in lax to congested corymbs or racemes, rather smaller than the male; stigmas coarsely hairy. Fruits red, 8–9 mm. in diameter; seeds 4–6 mm. long, 2–4 mm. wide, 1–2·5 mm. thick.

HAB.: Forests, scrub, hedges; sea-level to 1,000 ft. alt.; fl. Feb.–April.

DISTR.: Divisions 1, 3, 4, 7, 8, local. Eastern Mediterranean region, from Greece and Cyrenaica to Palestine.

1. Baths of Aphrodite, 1962, *Meikle* 2219 !
3. Near Khirokitia, 1903, *A. G. & M. E. Lascelles* s.n. !
4. Cape Greco, 1862, *Kotschy* 116 ! Near Famagusta 1946, *Mavromoustakis* 41 !
7. Lapithos, 1955, *Merton* 2242 !
8. Rizokarpaso, Eleousa Mon., 1880, *Sintenis & Rigo* 6 ! Near Komi Kebir, 1912, *Haradjian* 327 ! Ayios Theodhoros, 1934, *Syngrassides* 436 ! Beyond Rizokarpaso, 1957, *Merton* 2874 !

4. CUCUMIS *L.*

Sp. Plant., ed. 1, 1011 (1753).
Gen. Plant., ed. 5, 442 (1754).

Annual or perennial climbing or trailing herbs, with tendrils; leaves usually palmately, sometimes deeply lobed; tendrils simple; flowers usually monoecious, axillary, the male solitary or in small fascicles, the female solitary; hypanthium campanulate; petals yellow, united near the base; stamens free; connectives narrow, produced apically beyond the anther-thecae; anther-thecae flexuous; disc elevated, subglobose in male flowers, annular around base of style in female flowers; ovary subglobose to cylindrical, smooth and hairy or covered with apically setulose, soft spines or papillae. Fruit large, globose to cylindrical, greenish or yellowish, often striped, firm-walled, fleshy, indehiscent, smooth or covered in fleshy, apically setiferous spines or tubercles; seeds numerous, elliptic in outline, compressed, smooth, pallid.

About 25 species, mostly natives of tropical and southern Africa.

1. C. melo *L.*, Sp. Plant., ed. 1, 1011 (1753); Unger et Kotschy, Die Insel Cypern, 342 (1865); Post, Fl. Pal., ed. 2, 1: 480 (1932); Davis, Fl. Turkey, 4: 207 (1972).

C. dudaim L., Sp. Plant., ed. 1, 1011 (1753); Unger et Kotschy, Die Insel Cypern, 342 (1865) [sphalm. "*dunaim*"].

C. callosus (Rottl.) Cogn. in Engl., Pflanrenr., 88 (IV. 275, 2): 129 (1924); Post, Fl. Pal., ed. 2, 1: 479 (1932).

TYPE: described from cultivated plants, (BM, lectotype !).

Climbing or trailing annual herb; stems ± spreading-hairy, hispid to lanate; leaves petiolate, suborbicular, reniform or ovate in outline, cordate, hispid or scabrid-hairy, especially on veins beneath, 2·5–16 cm. long, 3–20 cm. wide, very shallowly to deeply palmately 3–7-lobed, the lobes usually broad, shallow and rounded; petiole spreading-setulose, 3–10 cm. long; male flowers usually 2–4 in small fascicles; pedicels slender, 2–15 mm. long; hypanthium pale green, 2·5–6 mm. long; sepals ovate-lanceolate to filiform, 1–6 mm. long; petals yellow, 4–22 mm. long; female flowers solitary; pedicels stouter, 2–50 mm. long; ovary ellipsoid to cylindrical, densely hairy, 5–60 mm. long; hypanthium and perianth as in male flowers. Fruit in wild variants ellipsoid, yellow, bitter, 2·5–4 cm. long, 1·8–2·8 cm. wide, in cultivated variants larger, not bitter, and very variable in size, shape and colour; seeds elliptic in outline, 3·7–8 mm. long, 1·8–4 mm. wide, 0·9–1·5 mm. thick.

HAB.: Cultivated fields, as a weed; fl. July.

DISTR.: Division 1, rare, probably of recent introduction. Throughout the tropics and subtropics of the Old World.

1. E. of Yeroskipos, 1967, *Economides* in ARI 1000 !

NOTES: *C. melo* L., the Melon, as represented by its cultivated variants is also an important crop plant, grown for its edible fruits.

C. sativus L., the Cucumber, which differs from the above in its triangular, acute leaf-lobes, ovary with apically setiferous papillae, and fruit with distinct though sometimes scattered and obscure apically setiferous tubercles, is also an important crop plant grown for its edible fruits.

Besides the plants mentioned above, five species of *Cucurbita* are crop plants cultivated for their fruits. *C. pepo* L., the Vegetable Marrow, is by far the most important, *C. maxima* Lam. less so. *C. moschata* (Lam.) Poiret, *C. mixta* Pangalo and *C. ficifolia* Bouche are grown only occasionally. For a key and descriptions distinguishing these species, see Whitaker & Davis, *op. cit.* pp. 44–50. *Lagenaria siceraria* (Molina) Standley (*L. vulgaris* Ser., *L. leucantha* Rusby), Calabash Gourd, a vigorous softly-hairy annual with white flowers and green, hard-shelled fruits, and *Luffa cylindrica* (L.) M. J. Roem. (*L. aegyptiaca* Mill.), the Loofah, a vigorous annual with yellow flowers and dry, internally fibrous fruits, are occasionally cultivated.

43a. CACTACEAE

D. R. Hunt in Hutchinson, Gen. Fl. Plant., 2: 427 (1967).

Fleshy perennials; stems polymorphic, terete, globular or compressed, smooth, sulcate or mammillate, often spinose and generally leafless; spines and flowers arising from cushion-like, axillary areoles; flowers usually solitary, sessile, hermaphrodite, actinomorphic; hypanthium generally exceeding the ovary and bearing the perianth segments and stamens; perianth segments usually numerous, in sepaloid and petaloid series; stamens numerous; anthers 2-thecous, dehiscing longitudinally; ovary almost always inferior with 3–many parietal placentas; ovules numerous; style single; stigmas 3–many, entire or rarely lobed, variously shaped. Fruits dry or juicy, indehiscent or dehiscent; seeds numerous, often arillate or strophiolate; endosperm present or absent; embryo curved or occasionally straight; cotyledons generally much reduced, rarely foliaceous.

A very distinctive family of about 84 genera, chiefly American, but with a few outlying representatives in Africa, Madagascar and Mascarene Islands, and Ceylon.

1. Opuntia *Mill.*

Gard. Dict. Abridg., ed. 4 (1754).

Shrubs or trees with jointed, cylindrical, subglobose or flattened branches; areoles glochidiate, raised on more or less prominent tubercles, usually 1–many-spinose; leaves caducous, terete or subulate, small; flowers solitary, sessile, lateral or subterminal, rarely terminal; hypanthium exceeding ovary, variously clothed with scales, areoles, glochids and spines; perianth segments numerous, spreading or erect, inserted on the rim of the hypanthium, sometimes separated from the stamens by a ring of hairs; stamens inserted in throat of hypanthium, often sensitive; ovary inferior, ovules with circinnate funicles. Fruit fleshy or dry, with areoles and glochids, sometimes spinose; seeds encased in a hard white aril; cotyledons foliaceous, fleshy.

A large genus, represented by indigenous species from southern Canada south to the Strait of Magellan, and in the West Indies and Galapagos islands; several species are extensively naturalized outside the Americas, and of these two have been recorded from Cyprus.

O. ficus-indica (*L.*) *Mill.*, Gard. Dict., ed. 8, no. 2 (1768); Unger et Kotschy, Die Insel Cypern, 342 (1865); Holmboe, Veg. Cypr., 132 (1914); Post., Fl. Pal., ed. 2, 1: 485 (1932); Davis, Fl. Turkey, 4: 209 (1972); Osorio-Tafall et Seraphim, List Vasc. Plants Cyprus, 74 (1973).

Cactus ficus-indica L., Sp. Plant., ed. 1, 468 (1753).

A robust shrub, 2–5 m. high, with spreading, jointed branches; joints flattened, oblong-obovate or oblong-elliptic, (10–) 20–40 (–50) cm. long, (7–) 10–20 cm. wide, dull glaucous; areoles and glochids present, pale or yellowish; spines usually absent, or, when present, pale and slender, up to 1·5 cm. long; flowers yellow or orange, 4–8 cm. diam.; perianth segments spreading or suberect. Fruit ovoid, truncate at apex, about 6–10 cm. long, fleshy, red, edible when the epicarp with its areoles and glochids has been removed.

Probably of Central American origin, now cultivated and naturalized in many parts of the Mediterranean region and elsewhere. Kotschy records it from Larnaca, and Michaelides from Limassol, but, as Holmboe notes, it is frequent in Cyprus near towns and villages, though perhaps sometimes confused with the following species.

O. VULGARIS *Mill.*, Gard. Dict. ed. 8, no. 1 (1768); Post., Fl. Pal., ed. 2, 1: 485 (1932); Nattrass, First List Cyprus Fungi, 4 (1937).

Erect or sprawling shrub up to 4 m. high; joints flattened, narrowly obovate 10–30 cm. long, 5–12 cm. wide, bright green, areoles prominent, densely woolly; glochids brownish; spines 1–2 per areole (or more on the trunk), stout, 2–5 (–10) cm. long, greyish with a brown tip; flowers golden yellow, 5–7 cm. diam.; perianth segments spreading, the outer striped reddish externally. Fruit obovoid, 5–7·5 cm. long, fleshy, dark purplish, edible; epicarp glochidiate, without spines.

A native of South America, introduced into many of the warmer areas of the Old World, and recorded by R. M. Nattrass (nos. 169 and 930) from Dhekelia and Peristerona. Nattrass, who was concerned primarily with the identification of fungi, does not cite authorities for the names of the host plants, and may possibly have intended *O. vulgaris* Tenore, an illegitimate synonym of *O. ficus-indica*.

44. AÏZOACEAE

Pax et Hoffmann in Engl. et Prantl, Pflanzenfam., ed. 2, 16c: 179–233 (1934).

Annual or perennial, commonly succulent, herbs and subshrubs; leaves opposite or alternate, stipulate or exstipulate; flowers solitary or in cymes, axillary or terminal, actinomorphic, hermaphrodite; hypanthium sometimes developed, free or adnate to the ovary; sepals 5–8, free or united below, imbricate or valvate, sometimes fleshy or coloured; petals small or wanting, or replaced by petaloid staminodes in 1 or more series, inserted at the base of the calyx or on the rim of the hypanthium; stamens 3–5 or numerous, often fasciculate; anthers dehiscing longitudinally; ovary superior, inferior or semi-inferior, 2–many-locular; ovules solitary to numerous; placentation axile, basal or parietal. Fruit a capsule, or indehiscent and fleshy, sometimes surrounded by the persistent hypanthium or calyx; seeds usually small; endosperm sparse or copious, farinaceous; embryo curved.

About 130 genera, widely distributed, but with an extensive representation of genera and species in South Africa. The more showy species of the old aggregate *Mesembryanthemum* (including *Carpobrotus*, *Dorotheanthus*, etc.) are popular garden plants in the Mediterranean area, and are sometimes found as escapes from cultivation.

Leaves exstipulate; stamens indefinite:
  Stems and leaves succulent, covered with glistening papillae:
    Calyx-tube (or hypanthium) adnate to ovary; petal-like staminodes numerous — **1. Mesembryanthemum**
    Calyx-tube (or hypanthium) free from ovary; petal-like staminodes absent — **2. Aïzoön**
  Stems and leaves not succulent, clothed with white, stellate hairs - - **3. Glinus**
Leaves with membranous, scarious stipules; stamens 5, opposite the sepals — **4. Telephium**

TRIBE 1. **Mesembryanthemeae** *Reichb.* Leaves exstipulate, succulent;

hypanthium adnate to ovary; tepals (or sepals) unequal; staminodes present, usually numerous and petaloid.

1. MESEMBRYANTHEMUM *L.*

Sp. Plant., ed. 1, 480 (1753) pro parte.
Gen. Plant., ed. 5, 215 (1754) pro parte.
H. Herre, The Genera of the Mesembryanthemaceae, 316 pp. (1971).

Annual or biennal, succulent herbs; stems and leaves covered with glistening papillae; leaves exstipulate, at first opposite, later alternate, flat or terete, sessile or shortly petiolate; flowers solitary or in cymes, sessile or shortly pedicellate, bracteate or ebracteate; sepals 4–5, unequal, fleshy, often foliaceous; petaloid staminodes numerous, linear, united at the base and forming a short tube; stamens numerous; filaments long; ovary semi-inferior, placentas axile; stigmas 4–5, erect, filiform. Capsule (4–) 5-locular, valves keeled; seeds numerous, small.

Leaves flat, obovate-rhomboid, 2·5–5 cm. wide - - - - **1. M. crystallinum**
Leaves subterete, linear, 0·2–0·3 cm. wide - - - - - **2. M. nodiflorum**

1. M. crystallinum *L.*, Sp. Plant., ed. 1, 480 (1753); Unger et Kotschy, Die Insel Cypern, 343 (1865); Boiss., Fl. Orient., 2: 764 (1872); Holmboe, Veg. Cypr., 67 (1914); Post, Fl. Pal., ed. 2, 1: 486 (1932); Osorio-Tafall et Seraphim, List Vasc. Plants Cyprus, 34 (1973).
TYPE: " *in* Africa ".

Sprawling annual, covered with large, translucent, glistening papillae; stems terete or 4-angled, 5–20 (–30) cm. long; lower leaves opposite, flat, obovate-rhomboid, 4–8 cm. long, 2·5–5 cm. wide, margins entire, strongly undulate, base tapering to a petiole 1–4·5 cm. long; upper leaves alternate, usually less than 5 cm. long and 2 cm. wide, sessile or subsessile; inflorescence much branched, flowers solitary in the axils of foliaceous bracts or reduced leaves, or terminal; pedicel short, stout, seldom more than 3 mm. long; hypanthium broadly cup-shaped, 8–10 mm. long and almost as wide; sepals 5, very unequal, oblong or semicircular, acute, obtuse or rounded, 2–4 mm. long, 2–6 mm. wide, green, fleshy, papillose externally; stamens numerous, united basally into a tube about 4 mm. high; petaloid staminodes numerous in several series, whitish, linear, about 4 mm. long, 0·3 mm. wide; free part of filaments 1·5–2 mm. long; anthers yellow, oblong, about 0·6 mm. long, 0·5 mm. wide; styles 5, filiform, about 3 mm. long; ovary 5-locular, ovules numerous in each loculus. Fruit a pentagonal, flattened capsule, about 12–14 mm. diam., closely invested with the persistent hypanthium and sepals; valves strongly keeled; seeds subglobose, 0·8–1 mm. diam.; testa smooth, unevenly blotched or stained dark brown or fuscous; embryo with a prominent curved radicle; cotyledons narrowly oblong; endosperm copious.

HAB.: Loose, open sandy or gravelly ground; near sea-level; fl. May–August.

DISTR.: Division 4, very rare and perhaps a casual; widespread in the Mediterranean region; Atlantic Islands.

4. Near Larnaca (" auf dem gelockerten Boden der Ruinen bei Larnaca "), 1862, *Kotschy* s.n.

NOTES: No recent records, though the plant is likely to be found in Cyprus. Sintenis & Rigo's record from Cape Andreas (Oesterr. Bot. Zeitschr., 32: 398; 1882) is a misidentification of *M. nodiflorum* L.

2. M. nodiflorum *L.*, Sp. Plant., ed. 1, 480 (1753); Unger et Kotschy, Die Insel Cypern, 343 (1865); Boiss., Fl. Orient., 2: 764 (1872); Holmboe, Veg. Cypr., 67 (1914); Post, Fl. Pal., ed. 2, 1: 486 (1932); Lindberg f., Iter Cypr., 13 (1946); Rechinger f. in Arkiv för Bot., ser. 2, 1: 420 (1950); Davis, Fl. Turkey, 2: 345 (1967); Osorio-Tafall et Seraphim, List Vasc. Plants Cyprus, 34 (1973).
TYPE: " *in* Aegypto, Neapoli ".

Sprawling annual, covered with translucent papillae as in *M. crystallinum*, but less noticeably glistening; stems 4–20 (–30) cm. long, terete; lower leaves opposite, sessile, subterete, linear, 1·5–4 cm. long, 0·2–0·3 cm. wide, apex blunt or subacute, margins entire; upper leaves alternate, otherwise very similar to the lower; inflorescence much branched, flowers solitary in the axils of foliaceous bracts or reduced leaves, or terminal, sessile or subsessile; hypanthium cup-shaped, about 5 mm. long and about as wide; sepals 5, very unequal, 3–5 mm. long, 1·5–3·5 mm. wide, narrowly oblong and herbaceous, or broad, sometimes semicircular, with hyaline membranous wings and a herbaceous dorsal cusp; stamens numerous, united basally into a tube about 3·5 mm. high; petaloid staminodes numerous in several series, white, linear, about 1·5–2 mm. long, 0·3–0·5 mm. wide; free part of filaments 1·5–2·5 mm. long; anthers yellow, oblong, 1–1·3 mm. long, 0·4 mm. wide; styles 5, filiform, about 1·5 mm. long, stigmatiferous for at least two-thirds of their length; ovary 5-locular, ovules numerous in each loculus. Fruit a pentagonal, flattened capsule, about 10 mm. wide (the margins becoming erect, and the whole capsule turbinate when dry), closely invested with the persistent hypanthium and sepals; valves strongly keeled; seeds reniform, 0·8–1 mm. long, 0·7–0·8 mm. wide; testa rich brown, with distinct, bluntly verruculose dorsal ridges; embryo curved; endosperm copious.

HAB.: Sandy fields and seashores; rock-crevices, gravel-banks and margins of salt lakes, always near the sea; sea-level; fl. April–July.

DISTR.: Divisions 1, 3, 4, 7, 8, locally common; widespread in the Mediterranean region and eastwards to Iraq and the Persian Gulf.

1. Paphos, 1901, *A. G. & M. E. Lascelles* s.n. ! Ktima, 1913, *Haradjian* 642.
3. Limassol, 1905, *Holmboe* 668.
4. Larnaca, 1862, *Kotschy* 86A ! also *Holmboe* 102; *Syngrassides* 979 ! *Chapman* 705 ! *Merton* in ARI 368 ! Famagusta, 1934, *Syngrassides* 508 ! Pyla-Dhekelia, 1936, *Syngrassides* 1158 ! Cape Greco, 1962, *Meikle* 2606 !
7. Kyrenia Castle, 1955, *G. E. Atherton* 196 ! also Kyrenia, 1956, *G. E. Atherton* 1356 !
8. Cape Andreas, 1880, *Sintenis & Rigo* 512 !

TRIBE 2 **Aïzoeae** Leaves exstipulate, succulent; hypanthium free from ovary, tepals ± equal; staminodes wanting.

2. AÏZOÖN *L.*

Sp. Plant., ed. 1, 488 (1753).
Gen. Plant., ed. 5, 216 (1754).

Fleshy herbs and subshrubs; leaves exstipulate, usually alternate, sessile or petiolate; flowers axillary, solitary or in leafy cymes; hypanthium turbinate or hemispherical, free from the ovary except towards base; calyx lobes 4–5, valvate or imbricate, often white or coloured within, at least towards apex; petals and petaloid staminodes wanting; stamens usually up to 20, sometimes more numerous, inserted in fascicles near the base of the hypanthium, between the calyx-lobes; ovary 4–5-locular, with 2–many ovules in each loculus; styles free, 4–5. Fruit a loculicidal (rarely septicidal) capsule; valves often distinctly keeled; seeds small, numerous, smooth or verruculose; embryo curved.

About 20 species in the Mediterranean region, S.W. Asia, Australia and South Africa.

1. **A. hispanicum** *L.*, Sp. Plant., ed. 1, 488 (1753); Unger et Kotschy, Die Insel Cypern, 343 (1865); Boiss., Fl. Orient., 2: 765 (1872); Holmboe, Veg. Cypr., 67 (1914); Post, Fl. Pal., ed. 2, 1: 487 (1932); Osorio-Tafall et Seraphim, List Vasc. Plants Cyprus, 34 (1973).

TYPE: " *in* Hispania ".

Spreading or suberect annual; leaves and stems somewhat succulent, covered with glistening papillae; stems much divided at the base, repeatedly dichotomous above, up to 20 cm. or more in length, terete; lowermost leaves usually opposite, the remainder alternate, narrowly oblanceolate, 1–4 cm. long, 0·3–0·7 cm. wide, apex obtuse or rounded, margins quite entire, sometimes revolute; flowers sessile or subsessile in leafy dichasia; hypanthium obconical, longitudinally sulcate, about 7–8 mm. long, 10 mm. wide, densely papillose externally; calyx-lobes 5, erect, narrowly deltoid-subulate, acute or subacute, ± equal, about 9–10 mm. long, 3 mm. wide at base, green externally, white internally, especially towards apex; stamens 15, inserted near base of hypanthium; filaments united in fascicles of 3 for a short distance above base, about 6·5 mm. long, flattened, thinly papillose; anthers yellow, oblong, about 1 mm. long, 0·8 mm. wide, minutely papillose; ovary obconical, 6–7 mm. long, about 8 mm. wide, expanding above into 5, triangular, acute, divaricate wings, densely papillose; styles 5, about 2 mm. long, with strongly revolute apices. Capsule chartaceous, about 10 mm. long, 12 mm. wide; valves prominently keeled; seeds numerous, subglobose, about 0·9 mm. diam.; testa dark brown, conspicuously and regularly sulcate on the lateral surfaces, verruculose dorsally, embryo strongly curved; endosperm copious, farinaceous.

HAB.: Denuded clay or marl slopes, or sandy ground; sea-level to c. 200 ft. alt.; fl. March–May.

DISTR.: Divisions 4, 8, rare; also S. Spain, N. Africa, Palestine, Arabia, Iran and Atlantic Islands.

4. Frequent on bare sandy ground East of Larnaca, 1862, *Kotschy* s.n.
8. Hillsides near Galinoporni, 1880, *Sintenis & Rigo* 513 ! Elisis near Galinoporni, on denuded marly slope, 1962, *Meikle* 2488 !

TRIBE 3. **Mollugineae** (*Fenzl*) *Benth. et Hook. f.* Leaves opposite or in whorls, exstipulate, not succulent; sepals 5; petals absent, sometimes replaced by petaloid staminodes; stamens 3–12–many.

3. GLINUS *Loefl. ex L.*

Sp. Plant., ed. 1, 463 (1753).
Gen. Plant., ed. 5, 208 (1754).
Loefling, Iter Hisp., 145–146 (1758).

Annual (or rarely perennial) herbs, leaves apparently opposite or whorled, exstipulate, simple, entire, often woolly or setose; flowers in axillary clusters; sepals 5, equal; petals wanting or replaced by ligulate, filiform or fimbriate staminodes; stamens 3–30, inserted at base of calyx, sometimes fasciculate; styles 3–5, straight; ovary 3–5-locular, ovules numerous. Fruit a loculicidal 3–5-locular capsule; seeds reniform, strophiolate.

About 12 species, chiefly in the warmer regions of the Old World.

1. **G. lotoides** *L.*, Sp. Plant., ed. 1, 463 (1753); Unger et Kotschy, Die Insel Cypern, 344 (1865); Boiss., Fl. Orient., 1: 755 (1867); Holmboe, Veg. Cypr., 67 (1914); Davis, Fl. Turkey, 2: 346 (1967); Osorio-Tafall et Seraphim, List Vasc. Plants Cyprus, 34 (1973).

TYPE: " *in* Sicilia, Hispania *ad fossas* ".

Much-branched, sprawling annual; stems terete, 20–40 cm. long, softly white, stellate-villose; leaves strictly alternate, but appearing to be in whorls of three or four, unequal, suborbicular or broadly and bluntly spathulate, 0·5–4 cm. long, 0·4–2·5 cm. wide, dull green, sparsely or rather densely stellate-villose on both surfaces; petiole usually distinct, slender, up to 1·5 cm. long; flowers in axillary clusters of (2–) 3–6; pedicels very unequal, sometimes almost absent, or sometimes up to 1·5 cm. long, filiform, thinly villose; sepals 5, narrowly ovate-acute, about 8 mm. long, 3 mm. wide, stellate-hairy externally, glabrous internally, somewhat keeled; petals 0; petaloid staminodes indefinite, greenish-white, strap-shaped, about 5 mm. long, 0·5 mm. wide, apex often deeply divided into 2 filiform lobes; stamens (9–) 15 (–30), filaments slender, about 4 mm. long; anthers yellow, oblong, about 1 mm. long, 0·8 mm. wide; ovary narrowly ovoid, about 6 mm. long, 3·5 mm. wide, incompletely 5-septate, longitudinally 5-sulcate, glabrous; styles about 0·5 mm. long, united; stigmas free, clavate, about 0·6 mm. long; placentation axile; ovaries very numerous. Capsule membranous; seeds numerous, reniform, about 0·5 mm. long, 0·4 mm. wide; testa rich brown, regularly papillose-verruculose; strophiole conspicuous, white, membranous, spreading, not enveloping the seed.

HAB.: Ditches, inundated places and deep, rich alluvial ground; sea-level to 200 ft. alt.; fl. (March–) July–Nov.

DISTR.: Division 5, very rare; throughout tropical and warm-temperate regions of the Old World; introduced into America.

5. With *Coronopus squamosus* on rich ground near Vatili [" Ivatli "], 1862, *Kotschy* s.n. Not recorded since.

TRIBE 4. **Telephieae** (*Bartl.*) *DC.* Leaves alternate (occasionally sub-opposite), stipulate, not fleshy; sepals, petals and stamens 5; styles 3, connate below.

4. TELEPHIUM *L.*

Sp. Plant., ed. 1, 271 (1753).
Gen. Plant., ed. 5, 131 (1754).
F. N. Williams in Journ. Bot., 44: 289–304 (1906).

Perennial herbs with a woody rootstock; stems usually branched at base, prostrate or sprawling; leaves alternate (or occasionally sub-opposite) often secund, entire, obscurely nerved, with small scarious stipules; flowers in dense terminal cymes; sepals 5; petals 5, white, free, entire, inserted on a small disk at base of calyx; stamens 5, united at base, hypogynous, opposite the sepals; ovary flask-shaped, 1-locular or very incompletely 3–4-locular, ovules numerous, attached to a basal placenta; styles 3, stigmas recurved. Fruit a flask-shaped, trigonous, loculicidal capsule surrounded by the persistent calyx; seeds reniform or subglobose, estrophiolate; embryo peripheral, curved; endosperm copious, farinaceous.

About 5 species in the Mediterranean region and western Asia.

1. **T. imperati** *L.*, Sp. Plant., ed. 1, 271 (1753).

Sprawling, glabrous perennial with a thick, tough rootstock; stems numerous, sometimes branched at base, generally unbranched above, (5–) 10–30 (–45) cm. long, slender, subterete or obscurely angular; leaves sessile, glaucous, obovate or oblong, rarely subelliptic, 5–25 mm. long, 3–10 mm. wide, apex rounded, obtuse, or occasionally acute, base narrowly cuneate; stipules ear-shaped, laciniate, scarious, about 1·5 mm. long, 1 mm. wide, persistent; flowers numerous, in dense terminal, corymbose or capitate cymes; sepals erect, imbricate, the 3 outer oblong, boat-shaped, 3·5–6 mm. long, 1–2 mm. wide, subacute, glabrous, with a narrow membranous margin, 2 inner narrower, 0·7–1 mm. wide, less concave, with a broader membranous margin; petals white, oblong, rounded or obtuse at apex, 4–5 mm. long, about 2 mm. wide, inserted on a small disk, claw scarcely developed; filaments glabrous, subulate, about 2·5 mm. long, united at base, free above;

anthers versatile, introrse, yellow, narrowly oblong, about 1·5 mm. long, 0·5 mm. wide; ovary flask-shaped, somewhat trigonous, glabrous, about 2·5 mm. long, 2 mm. wide; styles 3, about 0·8 mm. long, united at base, upper stigmatiferous part 3-lobed, with spreading or recurved lobes. Fruit partly exserted from persistent calyx, sharply trigonous, shining brown, shortly beaked, about 4–6 mm. long, 3–4 mm. wide; seeds compressed-subglobose, about 1·2–1·5 mm. diam.; testa black or dark brown, closely and regularly verruculose.

ssp. **orientale** (*Boiss.*) *Nyman*, Consp. Fl. Europ., 254 (1879); Holmboe, Veg. Cypr., 67 (1914); Lindberg f., Iter Cypr., 15 (1946); Davis, Fl. Turkey, 2: 97 (1967); Osorio-Tafall et Seraphim, List Vasc. Plants Cyprus, 39 (1973).

T. orientale Boiss., Diagn., 1, 10: 11 (1849).

T. imperati L. var. *orientale* (Boiss.) Boiss., Fl. Orient., 1: 754 (1867); F. N. Williams in Journ. Bot., 44: 295 (1906); H. Stuart Thompson in Journ. Bot., 44: 341 (1906); Post, Fl. Pal., ed. 2, 1: 219 (1932).

TYPE: Specimens from Crete, Turkey, Lebanon and Iran cited by Boissier, (G).

Smaller than the type in all its parts: leaves usually less than 10 mm. long, frequently rather narrow and acute; inflorescence often capitate or subglobose and less than 2·5 cm. diam.

HAB.: Rocks, screes and stony slopes at high altitudes; 4,000–6,400 ft. alt.; fl. June–August.

DISTR.: Division 2, local. Greece, Crete, Turkey, Palestine, Egypt eastwards to Afghanistan.

2. Khionistra 5,500–6,400 ft. alt., *Sintenis & Rigo* 719 ! *Holmboe* 972, 1085; *Kennedy* 413 ! 414 ! *Lindberg f.*; *Davis* 1800 ! etc. Madhari, 1940, *Davis* 1878 ! Above Spilia, 1962, *Meikle* 2818 !

NOTES: The affinities of *Telephium* in the *Centrospermae* are obscure, and I have simply followed Bentham and Hooker in allowing it to remain in the *Aïzoaceae*. Here, at least, it can be put without unduly disrupting the circumscription of the family. Its affinities with the *Illecebraceae*, particularly with *Corrigiola*, cannot fail to impress, were it not for the capsular, many-seeded fruit; while its alternate leaves, superficially very like those of *Andrachne*, make it anomalous in the *Caryophyllaceae* sensu stricto.

45. UMBELLIFERAE

O. Drude in Engl. et Prantl, Pflanzenfam., ed. 1, III, 8: 63–271 (1897).

H. Wolff in Engl., Pflanzenr., 43 (IV. 228) (1910); 6 (IV. 228) (1913); 90 (IV. 228) (1927).

V. H. Heywood [Ed.], The Biology and Chemistry of the Umbelliferae, in Journ. Linn. Soc. Bot., 64, suppl. 1, X+438 (1971).

Annual or perennial herbs, rarely shrubs or trees; stems commonly furrowed, often fistulose; leaves alternate, frequently much divided; petioles generally sheathing at the base; stipules mostly absent; flowers small, numerous, generally in compound or simple umbels, sometimes in spikes or heads or occasionally cymose, hermaphrodite or rarely unisexual; calyx adnate to ovary; sepals much reduced or absent; petals epigynous, inflexed in bud, valvate or somewhat imbricate, entire or lobed at the apex, equal, or the outer conspicuously larger than the inner (*radiant*), often caducous; stamens 5, alternate with petals; filaments inflexed in bud; anthers 2-thecous, dehiscing longitudinally; ovary inferior, usually 2-locular, the carpels generally attached to a central axis or *carpophore*; ovules solitary in each loculus, pendulous; styles generally 2, often divergent, and mostly arising from a thickened base or *stylopodium*. Fruit dry, often laterally or dorsally compressed, sometimes globose or terete, generally separating from the carpophore into 2 mericarps at maturity, each mericarp frequently with 5 primary and 4 secondary, smooth, pilose, spinose or winged ribs or ridges separated by furrows or *valleculae*; resin-canals (or *vittae*) normally present between or below the ridges on the outer surfaces of the mericarps and along their adjacent or *commissural* faces; seeds generally with abundant endosperm and a small embryo.

About 300 genera distributed throughout the world, but generally most abundant in temperate regions. The family includes many useful culinary herbs (Carrots, Parsnips, Celery, Parsley, Fennel, Dill, etc.) as well as a few very poisonous ones (Hemlock, Dropwort, etc.). It forms a very homogeneous group, normally at once recognizable by its umbellate inflorescences of small whitish or yellowish flowers.

Ripe fruits are usually essential for easy and accurate identification (unripe fruits are often misleading) and basal leaves should be collected whenever possible since they are frequently very different from the cauline.

The description of Umbellifer leaves, especially where these are elaborately and deeply compound, presents considerable difficulties; in this treatment *leaf* refers to the whole lamina, including all its divisions; *segments* are the divisions of deeply compound leaves; *ultimate segments* are the final divisions; *lobes* retain their usual meaning, and are used for the relatively shallow divisions of simple leaves or leaf-segments.

- Leaves simple with entire margins - - - - - - **21. Bupleurum**
- Leaves toothed, lobed, dissected or compound:
  - Flowers in dense heads or spikes, not in compound umbels:
    - Bracts and leaves spinose; sepals not pinnatisect - - - - **1. Eryngium**
    - Bracts and leaves not spinose; sepals pinnatisect - - - - **2. Lagoecia**
  - Flowers in compound umbels or spreading panicles:
    - Ultimate segments of leaves filiform, subulate or narrowly linear:
      - Flowers yellow or greenish-yellow:
        - Involucral bracts persistent, conspicuous:
          - Ovary verruculose-crustaceous or papillose - - - - **20. Cachrys**
          - Ovary smooth - - - - - - - - - - - **33. Ferulago**
        - Involucral bracts absent or inconspicuous and caducous:
          - Flowers polygamous, those of the central umbels hermaphrodite, fertile, subtended by long-stalked lateral umbels of male flowers; tall, robust plants with conspicuous, broad, inflated sheaths **32. Ferula**
          - Flowers not as above; plants slender with narrow leaf-sheaths:
            - Umbels mostly lateral, leaf-opposed:
              - Fruits not much compressed; leaf-sheaths conspicuous, up to 20 cm. long; plants perennial or biennial - - - - **29. Foeniculum**
              - Fruits strongly compressed or flattened dorsally; leaf-sheaths inconspicuous, much less than 20 cm. long; plants annual **30. Anethum**
            - Umbels mostly terminal - - - - - - - **23. Ridolfia**
      - Flowers white or whitish:
        - Petals of outermost flowers of umbel conspicuously enlarged and radiant:
          - Umbellules crowded, the central conspicuous, sterile, violet-purple; fruits winged - - - - - - - - - - - **12. Artedia**
          - Umbellules not crowded, the central not conspicuous or sterile; fruits globose, not winged - - - - - - - - - **13. Coriandrum**
        - Petals of outermost flowers of umbel not conspicuously enlarged or radiant:
          - Fruits globose or biglobose:
            - Fruits globose, mericarps not separating readily - - **13. Coriandrum**
            - Fruits biglobose, mericarps separating readily - - - **14. Bifora**
          - Fruits not globose or biglobose:
            - Involucral bracts few, inconspicuous; fruits narrowly oblong, 4–5 mm. long - - - - - - - - - - - **26. Bunium**
            - Involucral bracts numerous, conspicuous; fruits shortly oblong, about 2 mm. long - - - - - - - - - - - **24. Ammi**

Ultimate segments of leaves broader, not filiform, subulate or narrowly linear:
Flowers yellow or yellowish:
Involucral bracts large, conspicuous, lobed or pinnatisect - **11. Daucus**
Involucral bracts small, entire, or absent:
Leaves and lower parts of stems scabrid or hispid - - **34. Opopanax**
Leaves and lower parts of stems smooth, glabrous or subglabrous:
Ultimate segments of leaves linear or oblanceolate, more than 3 times as long as wide:
Ovary verruculose-crustaceous or papillose - - **20. Cachrys**
Ovary smooth:
Stems erect; styles long - - **33. Ferulago**
Stems sprawling or decumbent; styles short - - **28. Crithmum**
Ultimate segments of leaves ovate, obovate, oblong or elliptic, not 3 times as long as wide:
Bracts and bracteoles absent or obsolete; mericarps large, blackish, more than 2 mm. wide - - - - - - **16. Smyrnium**
Bracts and bracteoles present; mericarps small, brownish, less than 2 mm. wide:
Basal leaves oblong in outline, simply pinnate; umbels subsessile or shortly stalked - - - - - - - **22. Apium**
Basal leaves deltoid in outline, 3-pinnate; umbels long-stalked
PETROSELINUM (p. 738)
Flowers white, pink or purplish:
Basal and lower cauline leaves with few (less than 12) broad divisions or segments, or sometimes undivided:
Segments entire, glaucous, glabrous, subcoriaceous - **39. Glaucosciadium**
Segments toothed or crenate:
Leaves ternately divided, broadly deltoid in outline; segments large, subglabrous, shining green; fruit dark brown, conspicuously scabrid
19. Lecokia
Leaves pinnately divided, or undivided, not broadly deltoid in outline:
Plants glabrous or almost glabrous:
Petals of outermost flowers of umbel much enlarged and radiant, upper cauline leaves divided into numerous narrow segments; fruits globose - - - - - - - - **13. Coriandrum**
Petals of outermost flowers of umbel not much enlarged or radiant:
Plant creeping or decumbent, rooting at the nodes; leaves all rather uniform - - - - - - - - - **22. Apium**
Plant erect or spreading, not rooting at the nodes; basal and cauline leaves dissimilar - - - - - - - **27. Pimpinella**
Plants pubescent, hairy or hispid:
Involucre present; upper cauline leaves not divided into narrow segments:
Involucral bracts numerous, conspicuous, filiform, deflexed after anthesis - - - - - - - - **38. Ainsworthia**
Involucral bracts few, inconspicuous:
Peduncle and rays hispid with spreading bristles; fruit spinose-hispid without a thickened or winged margin; flowers usually reddish or pinkish - - - **8. Turgenia**
Peduncle and rays not, or very obscurely hispid; fruit with a conspicuous thickened or winged margin; flowers white or whitish:
Fruit with a winged margin; rays very unequal, becoming rigidly erect in fruit - - - **35. Synelcosciadium**
Fruit with a thickened, pithy margin; rays not as above
37. Tordylium
Involucre absent; upper cauline leaves divided into narrow segments
27. Pimpinella
Basal and lower cauline leaves divided into numerous (more than 12) small segments:
Involucral bracts lobed or pinnatisect:
Fruit with a long slender beak, even when young - - - **5. Scandix**
Fruit not beaked:
Stems and fruits glabrous, smooth - - - - - **24. Ammi**
Stems pubescent or hispid; fruits hispid or spinose:
Secondary ridges of fruit with a single row of spines; mericarps not strongly compressed dorsally; styles usually long and slender
11. Daucus

Secondary ridges of fruit with a double row of spines; mericarps almost flattened dorsally; styles short, thick - **10. Pseudorlaya**
Involucral bracts entire, not lobed or pinnatisect, or absent:
Fruit with a long slender beak, even when young - - - **5. Scandix**
Fruit not beaked, or with a short, inconspicuous beak:
Petals of outermost flowers of umbel conspicuously enlarged and radiant **37. Tordylium**
Petals of outermost flowers of umbel not conspicuously enlarged or radiant:
Stems pubescent, pilose or bristly:
Inflorescences much branched, forming a spreading, domed panicle; outer flowers of each umbellule sterile with accrescent, thickening pedicels - - - - - - **3. Echinophora**
Inflorescences and flowers not as above:
Fruits spinose or hispid, even when young:
Involucral bracts conspicuous, with membranous, ciliate margins - - - - - - - **9. Orlaya**
Involucral bracts absent or inconspicuous:
Stems distinctly ribbed or angled; leaves broadly deltoid in outline, very regularly and minutely dissected; fruit strongly ribbed with rows of hooked spines along the ribs - - - - - - - - **7. Caucalis**
Stems terete or obscurely ribbed; leaves not as above; fruits densely hispid-spinose - - - **6. Torilis**
Fruits shortly pubescent, not spinose or hispid **36. Zosima**
Stems glabrous or almost glabrous:
Fruits distinctly spinose, hispid, papillose or rugose, even when young:
Stems distinctly ribbed or angled:
Fruits transversely rugose; peduncles short - **31. Krubera**
Fruits spinose; peduncles usually long - **7. Caucalis**
Stems terete or very obscurely ribbed:
Fruit shortly beaked - - - - - - **4. Anthriscus**
Fruit not beaked - - - - - - - **6. Torilis**
Fruits smooth or ribbed, but not spinose, hispid, papillose or transversely rugose:
Leaves mostly basal, the cauline much smaller, or reduced to sheaths:
Involucre wanting or reduced to a single, subulate bract; fruit about 2 mm. wide - - - - **15. Scaligeria**
Involucre of 3–6 bracts; fruit about 5 mm. wide **17. Physospermum**
Leaves not all basal, the cauline well developed:
Fruits globose or biglobose:
Fruits globose - - - - - **13. Coriandrum**
Fruits biglobose, didymous - - - - **14. Bifora**
Fruits not globose or biglobose:
Stems more than 1 m. high, robust, fistulose, often purple-spotted; fruits broadly ovoid or suborbicular; plant with an unpleasant smell when crushed **18. Conium**
Stems usually less than 1 m. high, not fistulose or spotted; fruits oblong or oblong-ellipsoid; plant pleasantly aromatic or scentless when crushed - **25. Carum**

SUBFAM. 1. SANICULOIDEAE Flowers generally in simple umbels or dense heads. Fruit often spinose or scaly.

1. ERYNGIUM *L.*

Sp. Plant., ed. 1, 232 (1753).
Gen. Plant., ed. 5, 108 (1754).

Glabrous annual, biennial or perennial herbs, often with thick, fleshy roots; leaves usually pinnatisect, sometimes entire or almost so, segments generally rigid, spinose, often conspicuously veined; inflorescence usually branched; flowers crowded into compact hemispherical or cylindrical capitula, subtended by entire or lobed, spinose bracteoles and with a basal involucre of patent, spinose bracts; sepals rigid, usually longer than the erect, emarginate petals. Fruit ovoid or subglobose, generally scaly; carpophore absent.

Bracteoles entire; inflorescences narrow, thyrsoid; petioles with long basal sheaths **2. E. glomeratum**

Bracteoles 3-lobed at apex, with spinose lobes; petioles with short basal sheaths; inflorescences spreading:

Involucral bracts subulate, entire or remotely spinose, less than 3 mm. wide; capitula numerous in lax, slender-branched inflorescences - - - - **1. E. creticum**

Involucral bracts ovate, 0·5–2 cm. wide, coarsely spinose; capitula rather few in rigid, thick-branched inflorescences - - - - - - **3. E. maritimum**

1. E. creticum *Lam.*, Encycl. Méth., 4: 754 (1798); Unger et Kotschy, Die Insel Cypern, 298 (1865); Boiss., Fl. Orient., 2: 827 (1872); H. Wolff in Engl., Pflanzenr., 61 (IV. 228): 131 (1913); Holmboe, Veg. Cypr., 134 (1914); Post, Fl. Pal., ed. 2, 1: 507 (1932); Lindberg f., Iter Cypr., 25 (1946); Davis, Fl. Turkey, 4: 298 (1972); Zohary, Fl. Palaest., 2: 386, t. 558 (1972); Osorio-Tafall et Seraphim, List Vasc. Plants Cyprus, 75 (1973).

E. cyaneum Sibth. et Sm., Fl. Graec. Prodr., 1: 175 (1806), Fl. Graec., 3: 53, t. 258 (1819); Clarke, Travels, 2 (3): 719 (1816).

[? *E. pusillum* (non L.) Hume in Walpole, Mem. Europ. Asiatic Turkey, ed. 1, 254 (1817) nomen.]

TYPE: Crete, *Tournefort* (P).

Erect, rigid biennial; stems subterete or bluntly angled, 30–60 cm. high, usually unbranched towards base, much branched above, forming lax cylindrical or paniculate inflorescences, generally tinged metallic-blue at maturity; radical leaves withering long before anthesis, simple, oblong, 3·5–5 cm. long, 2·5–3 cm. wide, apex obtuse or rounded, margins irregularly crenate-serrate, base subcordate; cauline leaves pinnatisect, 2·5–7 cm. long, 2–4 cm. wide, segments narrowly oblong, spinose, becoming progressively narrower, and with longer spines, in the upper parts of the stem; petioles up to 10 cm. long in the basal leaves, upper leaves very shortly petiolate or subsessile, sheaths short, inconspicuous, strongly spinose in the upper cauline leaves; inflorescences usually lax and repeatedly branched; branches metallic blue; capitula subglobose, small, 0·5–1 cm. diam. at anthesis; involucral bracts 4–6, subulate, rigid, patent, 1–5 cm. long, less than 0·3 cm. wide, much exceeding the capitulum, entire or remotely spinose, especially towards base, tinged blue; bracteoles about 5 mm. long, equalling or slightly exceeding the flowers, the lower half linear-oblong, the upper regularly 3-lobed, with rigid, spreading, spinose lobes; sepals erect, oblong, obtuse, mucronate, about 1 mm. long, 0·8 mm. wide; petals narrowly oblong, about 3 mm. long, 0·7 mm. wide, distinctly costate, folded transversely about the middle, the upper half fused to the lower by a ventral wing; stamens inflexed; filaments about 2·5 mm. long; anthers narrowly oblong, about 1 mm. long, 0·5 mm. wide, pale yellow; ovary turbinate, about 2 mm. long, 1·2 mm. wide, scaly; styles suberect, about 3 mm. long; stigmas truncate or slightly capitate. Fruit oblong-prismatic, about 2 mm. long, 1 mm. wide, roughly scaly, crowned with the erect, persistent sepals.

HAB.: Cultivated fields and pastures, waste land, hillsides, roadsides etc., occasionally in fissures of limestone cliffs; sea-level to 2,000 ft. alt., usually lowland; fl. May–Nov.

DISTR.: Divisions 2–5, 7. Eastern Mediterranean region from Yugoslavia to Egypt and eastwards to Iraq.

2. Pano Panayia, 1905, *Holmboe* 1049.
3. Limassol, 1859, *Kotschy*; Lefkara, 1940, *Davis* 1915 ! Between Limassol and Amathus, 1941, *Davis* 3556 !
4. Near Athienou, 1862, *Kotschy* 973 !
5. Near Pyroi, 1934, *Syngrassides* 469 ! Athalassa, 1939, *Lindberg f.* s.n. ! Nicosia, 1950, *Chapman* 623 !

7. Above Kythrea, 1880, *Sintenis & Rigo* 836 ! Myrtou, 1932, *Syngrassides* 214 ! Kyrenia, 1936, *Kennedy* 664 ! Larnaka tis Lapithou, 1941, *Davis* 2088 ! Kazaphani, 1955, *G. E. Atherton* 113 !

NOTES: Probably much commoner than the records suggest; *E. creticum* flowers late, and tends to be missed by collectors.

2. E. glomeratum *Lam.*, Encycl. Méth., 4: 755 (1798); Boiss., Fl. Orient., 2: 823 (1872); H. Wolff in Engl., Pflanzenr., 61 (IV. 228): 146 (1913); Post, Fl. Pal., ed. 2, 1: 505 (1932); Lindberg f., Iter Cypr., 25 (1946); Davis, Fl. Turkey, 4: 301 (1972); Zohary, Fl. Palaest., 2: 385, t. 556 (1972); Osorio-Tafall et Seraphim, List Vasc. Plants Cyprus, 75 (1973).

TYPE: Crete, *Tournefort* (P).

Perennial with a thick woody rootstock; stems erect, rigid, 50–100 cm. high, terete, somewhat sulcate, unbranched below, shortly and fastigiately branched towards apex, forming a narrow cylindrical or thyrsoid inflorescence; leaves numerous along the whole length of the stem, spreading, the basal broadly ovate, up to 12 cm. long, 10 cm. wide, deeply bipinnatisect with narrow, linear, spine-pointed, spinulose-serrate segments; petiole short, usually less than 4 cm. long, broadly sheathing; upper leaves similar, but more finely divided into narrowly subulate, pungent segments; petioles broadly sheathing, ribbed, encircling and covering the internodes; capitula small, hemispherical or ovoid, about 1·3 cm. long, 1 cm. wide; bracts usually 5, rather unequal, spreading or suberect, subulate-pungent, entire or subentire, 1–5 cm. long, about 0·2 cm. wide at base; bracteoles narrowly lanceolate, entire, spine-tipped, about 5–6 mm. long, 1·5 mm. wide, rather concave, slightly incurved; sepals erect, deltoid, pungent, about 1·5 mm. long, 1 mm. wide at base; petals oblong, about 2·3 mm. long, 0·8 mm. wide, folded transversely as in *E. creticum*, the apparent apex deeply and narrowly emarginate; stamens erect or incurved; filaments about 3·5 mm. long; anthers narrowly oblong, about 1 mm. long, 0·4 mm. wide, pale yellow; styles erect, 2·5–3 mm. long; stigmas truncate; ovary about 2·5 mm. long, 2 mm. wide, densely acute-scaly, scarcely altering its size and shape in fruit.

HAB.: Hillsides and dry fields; sea-level to c. 800 ft. alt.; fl. June–Oct.

DISTR.: Divisions 1, 2, 7, apparently rare; also Crete, Aegean Islands, Turkey, Syria, Palestine and Iraq.

1. Yiolou, between Polis and Stroumbi, 1938, *Syngrassides* 1842 !
2. Simboulas tis Dillirgas, between Polis and Pomos, 1939, *Lindberg f.* s.n. !
7. Kyrenia, 1948, *Casey* 62 !

NOTES: Perhaps, as with *E. creticum*, commoner than the records suggest, but overlooked by collectors because of its late flowering.

[E. CAMPESTRE *L.* (Sp. Plant., ed. 1, 233; 1753) has entire bracteoles and entire or remotely spinulose bracts as in *E. glomeratum* Lam., but the capitula are in spreading, corymbose or domed cymes, not forming narrow, thyrsoid, inflorescences, the leaf-sheaths are short and inconspicuous, and the leaves are much more coarsely segmented. It differs from *E. creticum* Lam. in its entire bracteoles and greenish-yellow or glaucous, not metallic-blue inflorescences.

Kotschy (Die Insel Cypern, 298 (1865)) records *E. campestre* L., " nicht selten am Wege von Kuklia nach Hierokipos am 5 Mai ". The record is repeated by Stuart Thompson and Holmboe, but no other collector has refound the plant about Kouklia or Yeroskipos, nor anywhere else in Cyprus. Kotschy cites no numbered specimen, nor, on 5th May, was it likely that the plant would have been sufficiently mature to be worth collecting. Most probably it was a misidentification of juvenile material of *E. creticum* or *E. glomeratum*, though *E. campestre* is widely distributed in the eastern Mediterranean region, and may yet be re-discovered in Cyprus.]

3. **E. maritimum** *L.*, Sp. Plant., ed. 1, 233 (1753); Unger et Kotschy, Die Insel Cypern, 298 (1865); Boiss., Fl. Orient., 2: 829 (1872); H. Wolff in Engl., Pflanzenr., 61 (IV. 228): 121 (1913); Holmboe, Veg. Cypr., 134 (1914); Post, Fl. Pal., ed. 2, 1: 507 (1932); Lindberg f., Iter Cypr., 25 (1946); Davis, Fl. Turkey, 4: 294 (1972); Zohary, Fl. Palaest., 2: 386, t. 559 (1972); Osorio-Tafall et Seraphim, List Vasc. Plants Cyprus, 75 (1973).

TYPE: " *ad* Europae *littora arenosa maritima* ".

Loosely tufted glaucous perennial with thick, fleshy rhizomes; stems erect or sprawling, stout, rigid, obscurely sulcate, 13–50 cm. high, usually unbranched below, sparingly branched above, forming a spreading, blue-tinged inflorescence; leaves thick, glaucous, the basal suborbicular in outline, 8–12 cm. diam., divided into 3 broadly cuneate, spinose-lobed, distinctly reticulate-veined segments; petiole thick, 7–15 cm. long, with a short basal sheath; stem-leaves similar, but progressively smaller upwards, the lower distinctly petiolate, with inconspicuous sheaths, the upper sessile and amplexicaul; capitula rather few, broadly ovoid or oblong, about 1·5–2 cm. long, and almost as wide; bracts 5–6, narrowly or broadly ovate, 1·5–4 cm. long, 0·5–2 cm. wide, glaucous or tinged metallic-blue, coarsely spinose-lobed; bracteoles 8–10 mm. long, about as long as the flowers, linear below, deeply 3-lobed at apex, with spreading, spinose lobes; sepals spreading, ovate, about 4 mm. long, 1·5 mm. wide, spine-tipped and occasionally with a few spines along the margins; petals oblong, about 2·5 mm. long, 1 mm. wide, folded transversely as in *E. creticum* with the apparent apex deeply emarginate; stamens erect; filaments about 5 mm. long; anthers oblong, about 2 mm. long, 0·8 mm. wide, pale yellow; styles erect, about 2 mm. long; stigmas truncate; ovary about 2 mm. long, 1·5 mm. wide, densely clothed with narrow subulate scales. Fruit broadly ovoid, about 8 mm. long, 5 mm. wide, spongy, light brown, thinly subulate-scaly.

HAB.: Sandy seashores and sand-dunes; about sea-level; fl. June–August.

DISTR.: Divisions 1, 4, 5, 8, uncommon. Also coasts of western Europe, Mediterranean and Black Sea.

1. Paphos, 1787, *Sibthorp* ! Near Ayios Yeoryios (Akamas), 1962, *Meikle* ! Near Akhelia, 1967, *Economides* in ARI 1017 !
4. Ayia Napa, 1905, *Holmboe* 7; and 1948, *Mavromoustakis* s.n. !
5. Salamis, 1939, *Lindberg f.* s.n. ! and, 1962, *Meikle* ! Trikomo, 1939, *Lindberg f.* s.n.
8. Near Galinoporni, 1962, *Meikle* !

2. LAGOECIA *L.*

Sp. Plant., ed. 1, 203 (1753).
Gen. Plant., ed. 5, 95 (1754).

Slender annual; leaves narrow, simply pinnate; umbels dense, globose, many-flowered, at first nodding, later erect; bracts and bracteoles finely pectinate; pedicels short; umbellule 1-flowered; sepals 5, conspicuously filiform-pectinate; petals white, minute, caducous, with slender aristate lobes on either side of the inflexed apex; ovary 1-locular, 1-ovulate; style 1, short. Fruit of one mericarp, ovoid, blackish, clothed with minute hairs and crowned with the persistent calyx; vittae obscure, 2 commissural, 3 dorsal.

One species only, widely distributed in the Mediterranean region, but especially common from Greece to Palestine, and eastwards to Iraq and Iran.

1. **L. cuminoides** *L.*, Sp. Plant., ed. 1, 203 (1753); Clarke, Travels 2 (3): 720 (1816); Unger et Kotschy, Die Insel Cypern, 307 (1865); Boiss., Fl. Orient., 2: 833 (1872); H. Wolff in Engl., Pflanzenr., 61 (IV. 228): 271 (1913); Holmboe, Veg. Cypr., 135 (1914); Post, Fl. Pal., ed. 2, 1: 509 (1932); Davis, Fl. Turkey, 4: 304 (1972); Zohary, Fl. Palaest., 2: 387, t. 560 (1972); Osorio-Tafall et Seraphim, List Vasc. Plants Cyprus, 76 (1973).

TYPE: " *in* Creta, Lemno, Lysia, Galatia ".

Slender annual; stems erect, glabrous, 3–45 cm. high, simple or with numerous erect or spreading branches, shallowly sulcate; leaves rather few and distant, narrowly oblong or lanceolate, 1–16 cm. long, 0·3–1·5 cm. wide, with 7–20 or more alternate or subopposite pairs of broadly ovate or cuneate, glabrous, deeply aristate-dentate segments, 2–10 mm. long, 2–8 mm. wide, margins of segments distinctly thickened and scabridulous; basal sheaths narrow, papery, whitish, about 5–10 mm. long; petioles scarcely developed; umbels lateral and terminal, globose, about 1–1·8 cm. diam.; peduncles very slender, 0·5–2 cm. long; involucral bracts 6–8, ovate, 5–10 mm. long, 4–7 mm. wide, pectinate-incised, with numerous long-aristate teeth; pedicels filiform, 4–5 mm. long, with a whorl of 4 (–5) finely pectinate, ovate bracteoles 4–5 mm long, 2–3 mm. wide, about half way along their length; sepals 5, ovate, erect, about 2·5 mm. long, 1 mm. wide, pectinate with long, filiform teeth; petals 5, white, obovate, about 0·5 mm. long, 0·3 mm. wide, produced apically into 2 slender, scabridulous awns, about 1–1·5 mm. long; stamens 5; filaments about 0·8 mm. long; anthers pale yellow, suborbicular, about 0·3 mm. diam.; ovary narrowly ovoid, about 1·3 mm. long, 0·6 mm. wide; stylopodium prominently domed; style 0·3 mm. long, stigma capitate. Fruit surrounded by the persistent bracteoles and crowned with the calyx, narrowly ovoid, blackish, about 2 mm. long and 1 mm. wide, one side strongly convex and clothed with minute, patent, pin-headed hairs, the other (site of the abortive carpel) flat or concave and very sparsely pilose.

HAB.: Cultivated fields, pastures and waste ground, roadsides, and on grassy or rocky hillsides; sea-level to 3,400 ft. alt.; fl. March–May.

DISTR.: Divisions 1–8, common. World distribution that of the genus.

1. Paphos, 1862, *Kotschy*; Dhrousha, 1941, *Davis* 3234 !
2. Near Prodhromos, 1862, *Kotschy*; near Lefka, 1927, *Rev. A. Huddle* 60 ! Aphamis, 1939, *Kennedy* 1374 !
3. Kophinou, 1905, *Holmboe* 580; Limassol, 1913, *Haradjian* 583 ! Episkopi, 1960, *N. Macdonald* 83 ! and, 1963, *J. B. Suart* 34 !
4. Cape Greco, *Kotschy*; near Famagusta, 1947, *Mavromoustakis* 44 !
5. Common; *Holmboe* 309, 339; *Chapman* 385 ! *Merton* 2961 !
6. Near Kokkini Trimithia, 1933, *Syngrassides* 321 ! Arkhangelos Mon., 1936, *Syngrassides* 950 ! Vizakia, 1936, *Syngrassides* 1227 !
7. Common; *Kotschy*; *Sintenis & Rigo* 327 ! *Syngrassides* 27 ! *Kennedy* 1073 ! *Davis* 1073 !
8. Cape Andreas, 1948, *Mavromoustakis* s.n. !

SUBFAM. 2. APIOIDEAE Flowers generally in compound umbels. Fruit not scaly.

3. ECHINOPHORA *L.*

Sp. Plant., ed. 1, 239 (1753).
Gen. Plant., ed. 5, 111 (1754).

Erect, often rather rigid, biennials or perennials with solid stems; leaves pinnately or ternately compound, the ultimate segments various, sometimes filiform or spinescent; petiole shorter than the rhachis; inflorescence a terminal umbel or sometimes irregularly branched and not obviously umbellate; bracts and bracteoles present, sometimes spinose; outer pedicels often thickening and concrescent in fruit; central flower of umbellule sessile, fertile, hermaphrodite, the other pedicellate, male, or central umbellule occasionally wholly sterile; sepals present in outer flowers, often unequal, frequently absent in inner flowers; petals white or yellow, unequal, sometimes radiant. Fruit ovoid or pyriform, subterete, sometimes fused with the accrescent outer pedicels, one of the carpels commonly abortive; primary ridges usually distinct, secondary ridges absent; vittae 0; seed sometimes excavate on one surface.

About 10 species, ranging from the Mediterranean region eastwards to Iran.

1. E. tenuifolia *L.* Sp. Plant., ed. 1, 239 (1753); Tutin in Fl. Europ., 2: 324 (1968).

Erect, rigid perennial, 20–50 cm. high, with a long, tough taproot; stems pubescent or shortly canescent-tomentose, sulcate or almost smooth, branched or unbranched at the base, usually much branched above; basal leaves broadly ovate-deltoid in outline, 15–30 cm. long, 10–23 cm. wide, 2–3-pinnate with opposite pinnae, ultimate segments linear-lanceolate or ovate-cuneate, 2–5 mm. long, 1–4 mm. wide, entire or sharply incised-serrate, rather thick and rigid, subglabrous to thinly tomentellous; petiole 2–7 mm. long, gradually expanding into a wide, herbaceous sheath; inflorescence much branched, forming a spreading, domed panicle; umbels small, seldom more than 2·5 cm. diam., 3–6-rayed, rays spreading, unequal, about 3–10 mm. long, subglabrous to thinly tomentose; bracts subulate, 3–6 mm. long, 1 mm. wide at base, persistent, becoming rigid and spinescent, usually spreading or slightly reflexed; bracteoles similar, but seldom more than 3 mm. long; pedicels of lateral flowers usually less than 3 mm. long, subglabrous or thinly tomentose; outermost flowers quite sterile, with a conspicuous, yellow, pulviniform stylopodium, and 2–3- subulate, accrescent sepals about 1·5 mm. long at anthesis; inner flowers sterile, male, sepals minute, slightly accrescent; petals 5, yellow, membranous, pubescent, oblong, about 0·6 mm. long, with strongly inflexed apices; stamens 5, inflexed; filaments glabrous, about 1·5 mm. long; anthers yellow, shortly oblong, about 0·4 mm. long, 0·3 mm. wide; stylopodium conspicuous, yellow; central flower sessile, hermaphrodite; sepals 5, almost obsolete, a little accrescent in fruit; petals 5, membranous; stamens 5; stylopodium inconspicuous; ovary about 1·5 mm. long, 1·3 mm. wide, subglabrous or pubescent; styles 2, erect, glabrous or pubescent, about 1·3 mm. long, stigmas capitate, ovules 2. Fruit enclosed in a cage formed by the erect, incrassate, partly fused bracteoles and pedicels, narrowly flask-shaped, about 6 mm. long, 2·5 mm. wide, glabrous or pubescent, crowned with the persistent styles; seed 1, fusiform, terete, about 5 mm. long, 2 mm. wide, testa pale brown.

ssp. **sibthorpiana** (*Guss.*) *Holmboe*, Veg. Cypr., 135 (1914); Tutin in Feddes Repert., 74: 31 (1967) and in Fl. Europ., 2: 324 (1968); Davis, Fl. Turkey, 4: 309 (1972); Osorio-Tafall et Seraphim, List Vasc. Plants Cyprus, 76 (1973).

E. sibthorpiana Guss., Suppl. Fl. Sic. Prodr., 69 (1832); Boiss., Fl. Orient., 2: 949 (1872); Post in Mém. Herb. Boiss., 18: 93 (1900); H. Stuart Thompson in Journ. Bot., 44: 308 (1906).

[*E. tenuifolia* (non L.) Sibth. et Sm., Fl. Graec. Prodr., 1: 179 (1806), Fl. Graec., 3: 59, t. 266 (1821); Post, Fl. Pal., ed. 2, 1: 541 (1932); Lindberg f., Iter. Cypr., 25 (1946).]

TYPE: "in Peloponneso copiosè; etiam ad viam inter Smyrnam et Bursam", *Sibthorp* (OXF).

Stem densely pubescent or shortly tomentose, subterete or obscurely sulcate; ultimate segments of leaves ovate-cuneate, incised-serrate; fruit rather small, usually less than 6 mm. long; fruiting sepals minute, only slightly accrescent, the surrounding fused bracteoles and accrescent pedicels densely pubescent.

HAB.: Cultivated and fallow fields, vineyards and roadsides, occasionally on rocky or sandy shores; sea-level to 3,000 ft. alt.; fl. June–August.

DISTR.: Divisions 1, 2, 5, 7, not common. Also Sicily and south-eastern Europe, Turkey, Palestine and eastwards through Iraq and Iran to Afghanistan and Central Asia.

1. Yeroskipos, 1939, *Lindberg f.*, s.n. !

2. Pomos, 1934, *Syngrassides* 971 ! Khryssoroyiatissa Mon., 1941, *Davis* 3415 !
5. Dhali, 1880, *Sintenis & Rigo* 839 ! Nicosia, 1894, *Post*; Athalassa, 1935, *Syngrassides* 957 ! Outskirts of Nicosia, on Kyrenia road, 1966, *Merton* in ARI 44 !
7. 3 m. E. of Kyrenia, 1948, *Casey* 44 ! Glykyotissa Island, 1955, *G. E. Atherton* 273 !

NOTES: Perhaps Gussone was correct in regarding *E. sibthorpiana* as specifically distinct from *E. tenuifolia*, which, typically, is a subglabrous plant, with narrow, linear-lanceolate ultimate segments, and a large, subglabrous fruit, crowned with distinct accrescent sepals.

4. ANTHRISCUS *Pers.*

Syn. Plant., 1: 320 (1805) nom. cons.

Annuals or perennials; stems commonly fistulose; leaves 2–4-pinnate; umbels compound; bracts present or absent; bracteoles present, generally unequal; petals white, or sometimes pink, occasionally radiant; sepals minute or wanting; stylopodium small, conical or depressed. Fruit usually narrow, laterally compressed, oblong, ovoid, or narrowly flask-shaped, tapering to a long or short apical beak; mericarps terete or subterete, glabrous or bristly-scabrid, primary ridges inconspicuous, confined to apex; vittae inconspicuous, 4 dorsal, 2 commissural.

About 20 species, widely distributed in Europe, Asia and Africa; some of the species are aromatic, and Chervil (*A. cerefolium* (L.) Hoffm.) is sometimes cultivated as a pot-herb.

1. **A. caucalis** *M. Bieb.*, Fl. Taur.-Cauc., 1: 230 (1808); Davis, Fl. Turkey, 4: 324 (1972); Osorio-Tafall et Seraphim, List Vasc. Plants Cyprus, 76 (1973).
Scandix anthriscus L., Sp. Plant., ed. 1, 257 (1753); Sibth. et Sm., Fl. Graec. Prodr., 1: 198 (1806).
Anthriscus vulgaris Pers., Syn. Plant., 1: 320 (1805); Poech, Enum. Plant. Ins. Cypr., 29 (1842); Unger et Kotschy, Die Insel Cypern, 307 (1865); Boiss., Fl. Orient., 2: 912 (1872); Post, Fl. Pal., ed. 2, 1: 530 (1932)—non Bernh. (1800).
A. scandix Beck, Fl. Hernst., 211 (1884); Holmboe, Veg. Cypr., 135 (1914).
TYPE: " *in* Europae *aggeribus terrenis* ".

Erect or spreading annual 30–80 (–100) cm. high; stems usually much branched, fistulose, striate or shallowly sulcate, glabrous or subglabrous, sometimes tinged purple; leaves ovate-deltoid, 3-pinnate, 6–15 cm. long, 3·5–12 cm. wide, ultimate segments ovate-cuneate, small, usually less than 5 mm. long, 4 mm. wide, subglabrous to rather densely hispidulous, deeply lobed, with acute or cuspidate lobes; petioles of lower leaves often more than 10 cm. long, expanding into a membranous-margined basal sheath, those of upper leaves progressively shorter, or reduced to the sheathing base; umbels lateral and terminal, lax, 2–4 cm. diam., rather unequally 3–6-rayed; peduncle slender, 0·5–4 cm. long, usually shorter than the subtending leaf; bracts 0 (–2), usually absent; rays spreading, filiform, 0·8–2·5 cm. long; bracteoles 3–5, subequal, narrowly ovate-acuminate, about 3 mm. long, 1 mm. wide, ciliate or thinly hispid; umbellules 3–6-flowered; pedicels 1–2 mm. long at anthesis, lengthening somewhat in fruit, tipped with a collar of white, bristly hairs; flowers small, white; petals obovate, truncate or obscurely emarginate with inflexed apices, about 1–1·2 mm. long, 0·8 mm. wide; stamens 5; filaments glabrous, spreading, about 0·5 mm. long; anthers suborbicular, about 0·4 mm. diam., fuscous; ovary oblong, about 1·5 mm. long, 1·2 mm. wide, densely hispid (or rarely smooth); stylopodium very small, conical; styles very short, less than 0·5 mm. long; stigmas truncate. Fruit narrowly oblong, about 4 mm. long, 1·4 mm. wide, densely hispid with hooked, swollen-based bristles, rarely smooth, tapering at the apex into a very short, furrowed beak, becoming blackish when ripe.

HAB.: Waste ground and fallow fields; 2,000–5,400 ft. alt.; fl. May–June.

DISTR.: Division 2, rare. Widespread in Europe, also in N.W. Africa, Turkey, Syria.
2. Near Prodhromos and Galata, 1862, *Kotschy*; near Prodhromos, 1880, *Sintenis & Rigo* 771 !

5. SCANDIX *L.*

Sp. Plant., ed. 1, 256 (1753).
Gen. Plant., ed. 5, 124 (1754).

Slender annuals; leaves 1–3-pinnate, ultimate segments narrow, linear; umbels leaf-opposed, compound, usually with few rays; bracts 0–1, inconspicuous; bracteoles usually conspicuous, entire or lobed; flowers white, generally hermaphrodite; sepals absent; petals sometimes radiant, emarginate or subentire; stylopodium usually distinct, often with an undulate margin. Fruit linear or subulate, subterete or laterally compressed, with a long beak; primary ridges usually distinct, secondary absent; vittae very slender and obscure, 5 dorsal, 0–2 commissural.

Bracteoles deeply pinnatisect, foliaceous - - - - - - - - **1. S. stellata**
Bracteoles entire or bifid at apex, not foliaceous:
Beak of fruit strongly compressed dorsally; bracteoles lanceolate, apex usually 2-lobed or deeply emarginate - - - - - - - **2. S. pecten-veneris**
Beak of fruit subterete or slightly compressed laterally; bracteoles ovate-elliptic, entire or shortly bifid or emarginate at apex:
Umbels 1–2 (–3)-rayed; flowers all fertile and hermaphrodite, not markedly radiant; largest petal about 1–1·5 mm. long; styles usually less than 0·6 mm. long **3. S. australis**
Umbels 3 (–5)-rayed, rays often subdivided into secondary umbels; flowers partly male and sterile, markedly radiant; largest petal often 5 mm. long; styles frequently 1·5 mm. long - - - - - - - - **4. S. grandiflora**

1. S. stellata *Banks et Sol.* in Russell, Nat. Hist. Aleppo, ed. 2, 2: 249 (1794); Davis, Fl. Turkey, 4: 326 (1972); Zohary, Fl. Palaest., 2: 392, t. 567 (1972).
S. pinnatifida Vent., Hort. Cels., t. 14 (1800); Boiss., Fl. Orient., 2: 916 (1872); Post, Fl. Pal., ed. 2, 1: 531 (1932).
TYPE: Syria; from the environs of Aleppo (BM).

Erect or spreading annual, 7–20 (–50) cm. high; stems usually much branched, sometimes simple, subglabrous or ± densely white-pilose, often tinged purple, obscurely or rather distinctly sulcate; leaves deltoid in outline, 1·5–6 (–10) cm. long, 1–4 cm. wide, 3-pinnate, ultimate segments linear-filiform, 2–10 mm. long, 0·5–1 mm. wide, thinly and shortly pilose, canaliculate; petioles of basal leaves slender, 3–6 cm. long, expanding below into narrow, shortly pilose sheaths with narrow membranous margins; upper leaves shortly petiolate or subsessile; umbels rather irregularly 1–3-rayed; bracts absent; rays generally very short, less than 0·8 cm. long, shortly pilose, often purplish, thickening slightly in fruit; umbellules 2–8-flowered; bracteoles foliaceous, deeply pinnatisect with linear, ciliate lobes, 4–5 mm. long, persistent; pedicels very short, stout, usually less than 3 mm. long; flowers minute; petals whitish or tinged purple, obovate, about 0·5 mm. long, apex incurved, shortly acute; filaments glabrous, about 0·5 mm. long; anthers suborbicular, about 0·3 mm. diam.; ovary narrowly oblong, about 2 mm. long, 1 mm. wide, produced into a slender beak about 3 mm. long; stylopodium minute, truncate; styles erect, about 0·2 mm. long, often reddish; stigmas truncate or slightly capitate. Fruit subulate, 2–3 cm. long, often blackish or dark purple when ripe, basal seed-bearing portion about 6 mm. long, 1·5–2 mm. wide, subglabrous or sparsely scabridulous, beak slender, about 1·5–2·5 cm. long, shortly and rather densely scabrid-pilose along two sides, not much compressed.

HAB.: Rocky hillsides; 3,150 ft. alt.; fl. March–April.

DISTR.: Division 2, very rare. Widespread in the Mediterranean region and eastwards to Central Asia.

2. Phini, 1939, *Kennedy* 1386 !

2. S. pecten-veneris *L.*, Sp. Plant., ed. 1, 256 (1753); Sibth. et Sm., Fl. Graec. Prodr., 1: 197 (1806); Unger et Kotschy, Die Insel Cypern, 306 (1865); Boiss., Fl. Orient., 2: 914 (1872); Holmboe, Veg. Cypr., 135 (1914); Post, Fl. Pal., ed. 2, 1: 531 (1932); Zohary, Fl. Palaest., 2: 390, t. 564 (1972); Osorio-Tafall et Seraphim, List Vasc. Plants Cyprus, 76 (1973).

TYPE: " *inter* Germaniae & Europae *australioris segetes* ".

Erect or spreading annual, 4–30 (–50) cm. high; stems usually much branched, subglabrous or ± densely hispid-villose, shallowly sulcate; leaves narrowly deltoid 1–10 cm. long, 0·5–8 cm. wide, 3-pinnate, the ultimate segments linear or lanceolate, acute or obtuse, 2–5 mm. long, 0·5–1·5 mm. wide, lamina flattish, with narrowly recurved scabridulous margins; petiole of basal leaves slender, often exceeding 10 cm., canaliculate, expanding at the base into a villose-margined sheath 3–20 mm. long, upper leaves shortly petiolate or subsessile; umbels 1–2 (–3)-rayed, bracts usually solitary or absent, narrowly linear-lanceolate, 3–8 mm. long, 1 mm. wide, with villose margins; rays subequal 0·8–3 cm. long, accrescent and usually distinctly incrassate in fruit; bracteoles generally 5, lanceolate, 2–7 mm. long, 1–3 mm. wide, commonly reflexed, apex 2-lobed or deeply emarginate, less often entire and acuminate, margins ciliate; flowers 3–8 (–10) in each umbellule; pedicels very short, usually less than 3 mm. long, lengthening a little and thickening greatly in fruit; petals white, very unequal, somewhat radiant, the largest obovate, obtuse or rounded, 2–3 mm. long, 1–1·5 mm. wide, the others similar, but usually less than 1 mm. long and about 0·5 mm. wide; filaments about 1 mm. long; anthers yellow, suborbicular, about 0·3 mm. diam.; ovary linear-subulate, 3–4 mm. long, about 0·5–1 mm. wide, scabridulous along 2 lateral margins; stylopodium flattened with an undulate margin, usually tinged purple; styles about 1 mm. long, erect, stigmas truncate. Fruit linear-subulate, 3–5 cm. long, 0·2–0·4 mm. wide, seed-bearing part 0·8–1·5 mm. long, distinctly ridged, usually scabridulous, beak compressed dorsally, usually scabridulous along the margins.

HAB.: Cultivated and fallow fields, roadsides, occasionally on old walls or stony, bare hillsides; sea-level to 3,800 ft. alt.; fl. Jan.–April.

DISTR.: Divisions 1–8, common. Widespread in Europe, N. Africa and S.W. Asia to the borders of India.

1. Paphos, 1862, *Kotschy*; near Ayios Yeoryios (Akamas), 1962, *Meikle* 2086 !
2. Common; *Kennedy* 1385 ! *Chapman* 306 ! 331 ! *Merton* 3010 ! *Meikle* 2255 !
3. Common; *Mavromoustakis* 15A ! 105 ! *Merton* 2059 !
4. Larnaca, 1862, *Kotschy* 24a; 1905, *Holmboe* 148; Cape Greco, 1958, *N. Macdonald* 10 !
5. Common; *Sintenis & Rigo* 328 ! *Syngrassides* 57 ! 363 ! *Merton* 2560 ! 3100 !
6. Nicosia, 1959, *Oswald* 33 ! Between Kormakiti and Liveras, 1970, *Hecker* 40 !
7. Common; *Syngrassides* 1323 ! *Davis* 2509 ! *Casey* 405 ! *Chapman* 129 ! etc.
8. Valia, 1905, *Holmboe* 505 ! Rizokarpaso, 1941, *Davis* 2359 ! Leonarisso, 1968, *Economides* in ARI 1111 !

3. S. australis *L.*, Sp. Plant., ed. 1, 257 (1753); Sibth. et Sm., Fl. Graec. Prodr., 1: 197 (1806), Fl. Graec., 3: 78, t. 285 (1819); Unger et Kotschy, Die Insel Cypern, 306 (1865); Boiss., Fl. Orient., 2: 917 (1872); Holmboe, Veg. Cypr., 135 (1914); Post, Fl. Pal., ed. 2, 1: 532 (1932); Davis, Fl. Turkey, 4: 329 (1972); Osorio-Tafall et Seraphim, List Vasc. Plants Cyprus, 76 (1973).

[*S. macrorrhyncha* (non C. A. Mey.) Sintenis in Oesterr. Bot. Zeitschr., 31: 393 (1881); Holmboe, Veg. Cypr., 135 (1914).]

[*S. pecten-veneris* L. ssp. *macrorrhyncha* (non (C. A. Mey.) Rouy et Camus) Osorio-Tafall et Seraphim, List Vasc. Plants Cyprus, 76 (1973).]

TYPE: " *in* G. Narbonensi, Italia, Creta ".

Slender, erect or spreading annual (2·5–) 4–20 (–40) cm. high; stems usually much branched, subglabrous or hairy especially towards the base, obscurely sulcate; leaves oblong-deltoid, 1–6 cm. long, 0·8–2·5 cm. wide, 3-pinnate, ultimate segments linear or narrowly oblanceolate, acute or obtuse, 3–5 mm. long, 0·5–0·8 mm. wide, lamina flattish with narrowly recurved scabridulous margins; petiole of basal leaves slender, 2–5 (or more) cm. long, canaliculate, expanding basally into a thinly villose or villose-margined sheath 8–15 mm. long; upper leaves shortly petiolate or subsessile; umbels 1–2 (–3)-rayed, ebracteate; rays slender, spreading, subequal, 1–6 cm. long, not much accrescent in fruit; bracteoles 3–5, ovate-elliptic, 3–5 mm. long, 0·8–1·5 mm. wide, erect or spreading, apex entire or shortly bifid, margins narrowly membranous, ciliate; flowers 3–10 (–15) in each umbellule, all hermaphrodite and fertile; pedicels very short, generally less than 2 mm. long, scarcely lengthening in fruit; petals white, inconspicuous, very unequal, somewhat radiant, the largest obovate, obtuse or shortly emarginate about 1–1·5 mm. long, 0·8 mm. wide, the others similar, but less than 0·8 mm. long and 0·5 mm. wide; filaments about 0·5 mm. long; anthers suborbicular, about 0·2 mm. diam., yellowish or olive-green; ovary linear-subulate, 2–3 mm. long, about 0·8 mm. wide, scabridulous along 2 margins; stylopodium flattened with an undulate-crenate margin, usually tinged purple; styles erect, usually less than 0·6 mm. long, purplish; stigmas truncate. Fruit linear-subulate, 20–30 mm. long, 1–1·5 mm. wide, very slender, straight or slightly curved; seed-bearing portion 0·8–1 cm. long, distinctly ridged; beak subterete or a little compressed laterally, usually scabridulous along the margins.

HAB.: Hillsides and Pine forests, often in rather shaded sites; sea-level to c. 1,000 ft. alt.; fl. March–May.

DISTR.: Divisions 1, 7, 8, apparently rare. Widespread in the Mediterranean region and eastwards to the Caucasus, but chiefly in the western Mediterranean region.

1. Near Cape Arnauti, 1957, *Merton* 3044 ! Near Fontana Amorosa, 1962, *Meikle* 2297 !
7. Pentadaktylos, 1880, *Sintenis & Rigo* 329 !
8. Koma tou Yialou, 1950, *Chapman* 214 !

4. S. grandiflora *L.*, Sp. Plant., ed. 1, 257 (1753); Boiss., Fl. Orient., 2: 917 (1872); Post, Fl. Pal., ed. 2, 1: 532 (1932).

S. australis L. ssp. *grandiflora* (L.) Thell. in Hegi, Illustr. Fl. Mittel-Europa, ed. 1, 5 (2): 1035 (1926); Cannon in Fl. Europ., 2: 327 (1968); Davis, Fl. Turkey, 4: 329 (1972).

TYPE: " *in* Oriente ".

Very similar to *S. australis* in size, habit and foliage, but the umbels commonly 3 (–5)-rayed, often with one of the rays subdivided into a secondary umbel; rays subequal, very slender and filiform, commonly patent or reflexed and sometimes clothed with conspicuous spreading hairs; flowers partly fertile and hermaphrodite, but generally with some (occasionally all) male and sterile in each umbellule; bracteoles larger than in *S. australis*, with a broader membranous-ciliate margin, frequently rather conspicuously bifid at the apex and strongly reflexed; petals very markedly unequal, the larger conspicuous and radiant, often 5 mm. long and 4 mm. wide, the remainder very small and inconspicuous; styles frequently 1·5 mm. long and somewhat divergent. Fruit as in *S. australis* but usually more slender, and generally somewhat curved, occasionally clothed with conspicuous spreading hairs (var. *lasiactina* Boiss.).

HAB.: Hillside on limestone; c. 2,000 ft. alt.; fl. March–May.

DISTR.: Division 7, very rare. Balkan peninsula and eastwards to Caucasus, W. Iran and Palestine; sometimes adventive elsewhere.

7. Sina Oros near Kantara, 1970, *A. Hansen* 404 !

6. TORILIS *Adanson*

Fam. Plant., 2: 99, 612 (1763).

Annual or biennial herbs; leaves 1–3-pinnate, ultimate segments entire or more commonly toothed; umbels terminal and/or leaf-opposed, generally ebracteate; umbellules bracteolate; sepals inconspicuous; petals white, pink or purple, with inflexed apices, sometimes radiant. Fruit shortly linear, narrowly oblong or cylindrical, laterally compressed; primary ridges ciliate or adpressed-hispid; secondary ridges densely spinose, the spines usually concealing the fruit-surface; mericarps homomorphic or sometimes heteromorphic.

About 12 species, chiefly in Europe and western Asia, occasionally naturalized elsewhere.

Umbels sessile or subsessile, or peduncle usually less than 2 cm. long **1. T. nodosa**
Umbels distinctly pedunculate; peduncle usually more than 2 cm. long:
Umbels 4–12 (or more)-rayed; petals white or whitish:
Slender plant less than 50 cm. high, ultimate leaf-segments narrowly oblong or linear less than 1 mm. wide, petals very small, inconspicuous **6. T. tenella**
Robust plant 50–100 cm. high; ultimate leaf-segments oblong or ovate, toothed or lobed, more than 1 mm. wide; petals conspicuous, often distinctly radiant **2. T. arvensis ssp. neglecta**
Umbels 2–3 (–4)-rayed; petals small, not radiant, usually pink or purple or tinged with these colours:
Ultimate segments of basal and lower cauline leaves toothed or lobed, more than 1 mm. wide; umbels leaf-opposed and terminal; petals pink or purple:
Upper and lower cauline leaves similar, 2–3-pinnate; ultimate segments acute but not long-acuminate or caudate - - - - - - **3. T. purpurea**
Upper and lower cauline leaves dissimilar, the upper 3-foliolate or simple, long-acuminate, the uppermost frequently entire - - **4. T. heterophylla**
Ultimate segments of basal and lower cauline leaves entire, 0·5–1 mm. wide; umbels all leaf-opposed; petals whitish or tinged pink - - **5. T. leptophylla**

1. T. nodosa (*L.*) *Gaertn.*, Fruct., 1: 82 (1788); Unger et Kotschy, Die Insel Cypern, 306 (1865); Boiss., Fl. Orient., 2: 1083 (1872); Holmboe, Veg. Cypr., 135 (1914); Post, Fl. Pal., ed. 2, 1: 570 (1932); Rechinger f. in Arkiv för Bot., ser. 2, 1: 428 (1950); Zohary, Fl. Palaest., 2: 397, t. 573 (1972); Osorio-Tafall et Seraphim, List Vasc. Plants Cyprus, 80 (1973).

Tordylium nodosum L., Sp. Plant., ed. 1, 240 (1753).

Caucalis nodosa (L.) Scop., Fl. Carniol., ed. 2, 192 (1772); Sibth. et Sm., Fl. Graec. Prodr., 1: 183 (1806); Druce in Rep. B.E.C., 9: 470 (1931).

Spreading or decumbent annual, 20–50 cm. high; stems usually much branched, thinly retrorse-hispid, obscurely sulcate, with long internodes; leaves oblong-deltoid in outline, 5–15 cm. long, 3–8 cm. wide, bipinnate, the ultimate segments deeply pinnatisect with narrow, linear, acute, subglabrous or thinly hispid lobes, 2–10 mm. long, 1–2·5 mm. wide; petioles of the lower leaves up to 10 cm. long, canaliculate above, gradually expanding at the base into a narrow membranous-margined sheath; umbels leaf-opposed, small, generally less than 2 cm. diam., crowded, subsessile or shortly pedunculate, obsoletely 2–3-rayed; bracts usually absent; rays generally inconspicuous, occasionally up to 1 cm. long, sulcate, adpressed-hispid; flowers subsessile, 5–10 (or more) in each umbellule; bracteoles subulate, 2–5 mm. long, 0·5–1 mm. wide, thinly hispidulous; sepals obsolete or reduced to minute subulate-deltoid lobes; petals white, obovate, 0·5–0·8 mm. long, 0·3–0·6 mm. wide, hispidulous externally, apex strongly incurved; stamens 5, about 0·6 mm. long; anthers yellow, suborbicular, about 0·2 mm. diam.; stylopodium convex; styles very short, usually less than 0·5 mm. long; stigmas capitate, often subsessile; ovary 2–2·5 mm. long, 1·5–2 mm. wide, heterocarpous, with some of the mericarps glochidiate-hispid, and others verruculose and thinly white-hispid, or homocarpous, with all the mericarps glochidiate-hispid. Fruits small, oblong, about 2–3 mm. long, 2–2·5 mm. wide, glochidiate-hispid or verruculose.

f. **nodosa.**
TYPE: " *in* Gallia, Italia *ad vias* ".

Umbels subsessile; fruits heterocarpous, the outer either with both mericarps glochidiate-hispid, or with one mericarp hispid, the other verruculose, the inner (or central) fruits mostly with both mericarps verruculose; styles generally obsolete or very short.

HAB.: Fields and cultivated ground; sea-level to c. 1,000 ft. alt.; fl. April–Sept.

DISTR.: Divisions 2, 3, 7, apparently rare. Widespread in western and central Europe, S.W. Asia and eastwards to Central Asia, introduced elsewhere.

2. Argaka, 1967, *Economides* in ARI 243 !
3. Akrotiri, 1967, *Merton* in ARI 698 !
7. Kythrea, 1880, *Sintenis & Rigo* 346 ! Kyrenia, 1942, *Casey* 1267 !

f. **homoeocarpa** *Thellung* in Hegi, Illustr. Fl. Mittel-Europa, 5 (2): 1059 (1926).
Caucalis fallax Boiss. et Bl. var. *brevipes* Boiss., Fl. Orient., 2: 1086 (1872); Holmboe, Veg. Cypr., 135 (1914).
Torilis nodosa (L.) Gaertn. var. *bracteosa* Murbeck in Contrib. Fl. Nord-Ouest Afr., 1: 84 (1897) excl. syn.
TYPE: Not indicated; to be selected from amongst the material cited by Murbeck (*loc. cit.*).

Umbels distinctly pedunculate (peduncles sometimes exceeding 2 cm. in length); fruits homocarpous, densely clothed with long, often yellowish, scabrid-glochidiate bristles; styles short, but usually distinct.

HAB.: Gardens, waste ground, sometimes in open forest or maquis; sea-level to c. 500 ft. alt.; fl. March–May.

DISTR.: Divisions 3, 5, 7, apparently rare. Eastern Mediterranean region from Sicily to Palestine; North Africa, Atlantic Islands.

3. Between Mazotos and Moni, 1862, *Kotschy* 568.
5. Athalassa, 1936, *Syngrassides* 1114 !
7. Near Dhavlos, 1880, *Sintenis & Rigo* 346b ! Kyrenia, 1952, *Casey* 1252 ! 1956, *G. E. Atherton* 1327 ! Near Ayios Amvrosios, 1974, *Meikle* 4031 !

NOTES: Further investigation may show that *T. nodosa* f. *homoeocarpa* deserves a higher rank. It is readily recognized by the fruits and appears to have a definite distribution in the Mediterranean region. The specimen from Dhavlos (*Sintenis & Rigo* 346b) is cited by Murbeck under *T. nodosa* (L.) Gaertn. var. *bracteosa* (Bianca) Murb. (*T. nodosa* (L.) Gaertn. ssp. *bracteosa* (Bianca) Nym.), but it is questionable if *Torilis bracteosa* Bianca (in Att. Accad. Gisenia Sc. Nat., Catania, ser. 2, 3: 31; 1846), the basionym of Murbeck's var. *bracteosa*, is the same as the plant described above. It is true that *T. nodosa* f. *homoeocarpa* is found in Sicily (e.g. *Todaro* 887 from Palermo in herb. Kew) but the sole sheet I have seen labelled *T. bracteosa*, namely *E. & A. Huet du Pavillon* 100, collected at Arola, Sicily, on 19th April, 1856, appears to be no more than typical heterocarpous *T. nodosa* with shortly pedunculate umbels.

2. **T. arvensis** (*Huds.*) *Link*, Enum. Hort. Berol., 1: 265 (1821); Holmboe, Veg. Cypr., 135 (1914); Rechinger f. in Arkiv för Bot., ser. 2, 1: 428 (1950); Davis, Fl. Turkey, 4: 520 (1972); Zohary, Fl. Palaest., 2: 395, t. 571 (1972); Osorio-Tafall et Seraphim, List Vasc. Plants Cyprus, 80 (1973).
Caucalis arvensis Huds., Fl. Angl., 99 (1762).

Erect or spreading, annual or biennial herb, up to 100 cm. high; stems usually much branched, branches sometimes short and spreading, forming a low domed bush (ssp. *arvensis*; *T. helvetica* Gmel., *T. arvensis* (Huds.) Link ssp. *divaricata* (Moench) Thell.), sometimes long and erect, obscurely sulcate, subglabrous or thinly retrorse-hispid; leaves deltoid, 2–3-pinnate, 5–30 cm. long, 3–15 cm. wide, thinly adpressed-strigose, ultimate segments narrowly oblong or broadly ovate, acute or acuminate, sharply toothed or lobed; petiole rather short, seldom exceeding 2 cm., expanding rather suddenly into a short, membranous-margined basal sheath; umbels terminal and leaf-opposed, generally ebracteate, peduncle (2–) 4–10 (–15) cm. long, angular, sulcate, thinly retrorse-hispid; rays 2–12, unequal, 0·5–6 cm. long; bracteoles (3–) 5–8, linear-subulate, 2–5 (–10) mm. long, about 0·5 mm. wide, adpressed-strigose; flowers shortly pedicellate or subsessile, usually about 5–12 in each umbellule; petals obovate, white, small, subequal or distinctly unequal and radiant, 0·5–2 mm. long, 0·4–1 mm. wide; stamens usually 5; filaments filiform, about 0·5–1 mm. long; anthers suborbicular, about 0·3 mm. diam., whitish or pale yellow; ovary broadly ellipsoid, usually less than 2 mm. long and 1·4 mm. wide, densely covered with spreading, often blackish, glochidiate-scabrid bristles; styles short or up to 1·5 mm. long and patent or recurved; stigmas capitate. Mericarps narrowly ellipsoid, 4–5 mm. long, 1–2 mm. wide, sometimes densely hispid all over, sometimes partly hispid and partly scabridulous, or sometimes one of the mericarps hispid, the other scabridulous.

ssp. **neglecta** (*Sprengel*) *Thellung* in Hegi, Illustr. Fl. Mittel-Europa, 5 (2): 1055 (1926); Davis, Fl. Turkey, 4: 520 (1972).

T. radiata Moench, Meth., 103 (1794).

T. neglecta Sprengel in Roem. et Schult., Syst. Veg. 6: 484 (1820); Boiss., Fl. Orient., 2: 1083 (1872); Post, Fl. Pal., ed. 2, 1: 570 (1932).

T. arvensis (Huds.) Link var. *heterocarpa* (Batt.) Maire, Cat. Plant. Maroc., 2: 525 (1932); Lindberg f., Iter Cypr., 25 (1946).

T. infesta Hoffm. var. *heterocarpa* Batt., Fl. Alg., 375 (1889).

TYPE: The plate in Jacquin, Fl. Austr., 1: 28, t. 46 (1773) !

Stems tall, commonly 50–100 cm. high; branches erect; ultimate segments of leaves rather broadly ovate, acute but not acuminate or caudate; umbels long-pedunculate, 6–12 (or more)-rayed; flowers with unequal petals, distinctly radiant; styles often 1·5 mm. long, conspicuous immediately after anthesis. Fruits densely clothed with long greyish bristles, or heterocarpous, one mericarp with long bristles, the other more shortly hispid-scabridulous.

HAB.: Roadsides and cultivated ground; sea-level to 5,500 ft. alt.; fl. June–Sept.

DISTR.: Divisions 1, 2, 3, rare. Widespread in central and southern Europe, N. Africa and western Asia to Iraq and Iran.

1. Ktima, 1913, *Haradjian* 648.
2. Near Evrykhou, 1880, *Sintenis & Rigo* 842 ! Pano Panayia, 1913, *Haradjian* 896; Kambos, 1939, *Lindberg f.* s.n. ! Milikouri, 1939, *Lindberg f.* s.n. ! Prodhromos, 1952, *Casey* 1278 ! and, 1955, *G. E. Atherton* 629 !
3. Episkopi, 1905, *Holmboe* 699; Kolossi, 1939, *Mavromoustakis* 72 !

NOTES: Uniform in our area and distinct from the low, bushy, few-rayed ssp. *arvensis*, but it intergrades with the latter in many parts of central and southern Europe.

3. T. purpurea (*Ten.*) *Guss.*, Fl. Sic. Prodr., 1: 325 (1827); Lindberg f., Iter Cypr., 25 (1946).

Caucalis purpurea Ten. Corso Bot. Lez., ed. 2, 4 (1): 209 (1822); Fl. Nap. Prodr., Append. 4, 12 (1823); Fl. Nap., 3: 279, t. 131 (1824–29).

T. homophylla Stapf et Wettst. in Denkschr. Akad. Wiss. Wien, 51: 372 (1886); Post, Fl. Pal., ed. 2, 1: 570 (1932).

T. triradiata Post (non Boiss. et Heldr.) Fl. Pal., ed. 1, 375 (1896).

T. arvensis (Huds.) Link ssp. *purpurea* (Ten.) Hayek, Prodr. Fl. Pen. Balc., 1: 1057 (1927); Davis, Fl. Turkey, 4: 321 (1972). Osorio-Tafall et Seraphim, List Vasc. Plants Cyprus, 80 (1973).

T. arvensis (Huds.) Link ssp. *elongata* (Hoffm. et Link) Cannon in Feddes Repert., 79: 62 (1968); Osorio-Tafall et Seraphim, List Vasc. Plants Cyprus, 80 (1973).

TYPE: " in arvis prope Neapolim " *Tenore* (NAP).

Slender erect annual, 25–50 (–80) cm. high; stems usually unbranched at base, flexuous and much-branched, with divaricate branches above, usually glabrous, and shallowly sulcate below, thinly hispidulous with retrorse bristles towards apex; leaves as in *T. arvensis* (Huds.) Link ssp. *neglecta*

(Sprengel) Thellung, frequently rather small and narrowly lobed; umbels terminal and leaf-opposed, peduncle (2–) 4–10 (–15) cm. long, as in *T. arvensis* ssp. *neglecta*; rays 2–3 (–4) unequal, 0·5–3 cm. long; bracteoles 3–5, linear-subulate, 2–5 (–8) mm. long, about 0·5 mm. wide, adpressed-strigose; flowers subsessile, usually about 5–8 in each umbellule; petals obovate, pink or purple, very small, unequal, the largest about 1·3 mm. long, 1·2 mm. wide, deeply emarginate, the smallest about 0·6 mm. long and almost as wide; stamens 5; filaments filiform, about 0·4–0·5 mm. long; anthers suborbicular, about 0·2 mm. diam., whitish or pinkish; ovary broadly ovoid-ellipsoid, about 1·8 mm. long and 1·4 mm. wide, densely covered with suberect or spreading, blackish, glochidiate-scabrid bristles; styles less than 1 mm. long, inconspicuous, erect; stigmas truncate. Mericarps narrowly ellipsoid, about 5 mm. long, 1–1·5 mm. wide, densely glochidiate-bristly all over, except along the commissural faces.

HAB.: Dry rocky hillsides, or by roadsides, on igneous or calcareous formations, usually in slightly shaded places; sea-level to 5,000 ft. alt.; fl. April–July.

DISTR.: Divisions 1, 2, 7, locally frequent. Italy, Sicily, Yugoslavia, Albania, Greece, Crete, Turkey, Syria, Palestine, Caucasus and Iran.

1. Ktima, 1913, *Haradjian* 648 ! Baths of Aphrodite, 1957, *Merton* 3058 !
2. Locally frequent, about Prodhromos, Pedhoulas, Platania, Lefkara, etc.; *Haradjian* 896 ! *Lindberg f.* ! *Davis* 1823 ! 1909 ! *Casey* 903 ! *Merton* 2332 ! *D. P. Young* 7403 !
7. Locally frequent about St. Hilarion, Lapithos, Kantara, Akanthou, etc.; *Syngrassides* 218 ! *Lindberg f.* ! *Davis* 3625 ! *Merton* 765 ! *Miss Mapple* 88 ! *Merton* in ARI 776 !

NOTES: Though currently often regarded as a subspecies of *T. arvensis* (Huds.) Link, *T. purpurea* is uniformly distinct in Cyprus, and readily identified by its slender, erect habit, few-rayed umbels and purple or pink petals.

4. T. heterophylla *Guss.*, Fl. Sic. Prodr., 1: 326 (1827); Boiss., Fl. Orient., 2: 1082 (1872); Post, Fl. Pal., ed. 2, 1: 570 (1932).

[*T. triradiata* (non Post) Lindberg f., Iter Cypr., 25 (1946).]

TYPE: Sicily; " In dumetis glareosis montosis; *Monti di Mistretta, e di Cannata, Nocelle de Polizzi* " *Gussone* (? NAP).

A slender erect annual 20–50 cm. high, very similar to *T. purpurea* (Ten.) Guss. in habit and flower; lower stem leaves 2–3-pinnate, barely distinguishable from those of *T. arvensis* (Huds.) Link and *T. purpurea*; upper leaves trifoliolate or simple, linear, long-acuminate, 1–8 cm. long, 0·2–1 cm. wide, mostly with the margins sharply serrate-lobed, but the uppermost leaves often wholly entire, or entire in the lower half of the lamina; umbels 2–3-rayed as in *T. purpurea* with similar umbellules and very small pink or purple flowers. Fruit generally heterocarpous, one of the mericarps densely clothed with long spreading scabrid-glochidiate bristles as in *T. purpurea*, the other (? abortive) scabrid-hispid with very short bristles.

HAB.: Shaded places on rocky mountainsides; 1,000–5,000 ft. alt.; fl. March–June.

DISTR.: Divisions 2 and 7. Widespread in the northern Mediterranean region from Spain and France to the Aegean, also in Crimea, Turkey, Syria and Palestine.

2. Prodhromos, 1880, *Sintenis & Rigo* 920 ! Stavros tis Psokas, 1937, *Syngrassides* 1611 ! and, 1962, *Meikle* 2758 ! Platres, 1938, *Kennedy* 1396 ! 1397 ! Kryos Potamos, 2,900 ft. alt., 1938, *Kennedy* 1398 !
7. Pentadaktylos, 1880, *Sintenis & Rigo* 351 ! Bellapais, 1939, *Lindberg f.* s.n. ! Lapithos, 1974, *G. Joscht* 7152 !

5. T. leptophylla (*L.*) *Reichb.*, Icones Fl. Germ., 21: 83, t. 2010 (1867); Davis, Fl. Turkey, 4: 522 (1972); Osorio-Tafall et Seraphim, List Vasc. Plants Cyprus, 80 (1973).

Caucalis leptophylla L., Sp. Plant., ed. 1, 242 (1753); Unger et Kotschy, Die Insel Cypern, 306 (1865); Boiss., Fl. Orient., 2: 1084 (1872); Holmboe, Veg. Cypr., 135 (1914); Post, Fl. Pal., ed. 2, 1: 51 (1932).

Torilis erythrotricha Reichb., Icones Fl. Germ., 21: 84 (1867).
Caucalis erythrotricha (Reichb.) Boiss. et Hausskn. in Boiss., Fl. Orient., 2: 1085 (1872); Post, Fl. Pal., ed. 2, 1: 571 (1932).
Caucalis leptophylla L. var. *erythrotricha* (Reichb.) Post, Fl. Pal., ed. 1, 375 (1896); Druce in B.E.C. Rep., 9: 470 (1931).
TYPE: " *in* Gallia, Italia ".

Slender annual; stems erect or spreading, 5–30 cm. high, usually much branched, subglabrous or thinly retrorse-hispid, shallowly sulcate; leaves rather distant, the basal narrowly deltoid-oblong, 2–6 cm. long, 0·6–3 cm. wide, 2–3-pinnate, ultimate segments small, narrow, oblong, acute, 1–3 mm. long, 0·5–1 mm. wide, thinly adpressed-hispid; petiole slender, 3–5 cm. long, canaliculate above, thinly hispid; upper leaves similar to basal but smaller and more deltoid in outline, with short, but generally distinct, narrowly sheathing petioles; umbels leaf-opposed, generally 2-rayed, sometimes 3-rayed in robust specimens; peduncle divaricate, 1–4 (–5) cm. long, thinly adpressed-hispid or almost glabrous; rays usually less than 1 cm. long, occasionally up to 2·5 cm., rather thick and rigid; bracts wanting; bracteoles lanceolate, concave, acute, about 3 mm. long, 1 mm. wide, midrib adpressed-hispid, margins membranous; flowers 5–8 in each umbellule; pedicels unequal, up to 6 mm. long, adpressed-hispid; petals small, white, pink or purplish, unequal, broadly obovate-orbicular, deeply emarginate, the largest about 2 mm. long and almost as wide, the smallest about 1 mm. long; stamens 5, pinkish or whitish; filaments about 0·4 mm. long; anthers suborbicular, about 0·4 mm. diam.; ovary oblong, about 2 mm. long, 0·8 mm. wide, densely bristly; styles very short; stigmas truncate. Mericarps linear-oblong, about 6 mm. long, less than 1 mm. wide, prominently ribbed, the secondary ridges with numerous, long, spreading, pale or purple-tinged, glochidiate-scabrid bristles, the primary ridges with sparse, subadpressed hairs.

HAB.: Cultivated fields and waste ground; hillsides in garigue or in dry stony pastures; sea-level to 3,800 ft. alt.; fl. Febr.–May.

DISTR.: Divisions 1–8, locally common. Widespread in S. Europe and the Mediterranean region, and eastwards to Pakistan and Central Asia.

1. Near Inia, 1957, *Merton* 3023 ! Near Ayios Yeoryios (Akamas), 1960, *N. Macdonald* 51 !
2. Near Kalopanayiotis, 1937, *Syngrassides* 1562 ! Kryos Potamos, 3,000 ft. alt., 1937, *Kennedy* 678 ! Palekhori, 1957, *Merton* 3001 ! Platres, 1960, *N. Macdonald* 138 !
3. Kophinou, 1905, *Holmboe* 581; Stavrovouni, 1928 or 1930, *Druce* s.n.; Zagala above Trimiklini, 1941, *Davis* 3143 ! Lefkara, 1941, *Davis* 2754 ! Limassol, 1957, *Mavromoustakis* 9 ! Moni, 1964, *J. B. Suart* 216 !
4. Cape Greco, 1862, *Kotschy* 128, 407; Larnaca, 1905, *Holmboe* 85a.
5. Mile 5, Nicosia–Famagusta road, 1950, *Chapman* 38 ! Mia Milea, 1951, *Merton* 30a ! Klirou above Dheftera, 1952, *F. M. Probyn* 68 ! and, 1967, *Merton* in ARI 273 !
6. Peristerona, 1935, *Syngrassides* 781 ! and, 1936, *Syngrassides* 1212 ! Kormakiti, 1938, *Syngrassides* 1220 ! Near Dhiorios, 1962, *Meikle* 2355 !
7. Common; *Sintenis & Rigo* 348 ! 985 ! *Syngrassides* 274 ! *Casey* 426 ! 495 ! *Chapman* 326 ! *G. E. Atherton* 1204 ! 1232 !
8. Gastria, 1941, *Davis* 2422 !

NOTES: *T. leptophylla* (L.) Reichb. var. *erythrotricha* (Reichb.) Post, with purplish or reddish flowers and fruits, occurs here and there throughout the island, often intermixed and intergrading with typical pallid *T. leptophylla*. It is scarcely deserving of taxonomic recognition at any rank.

6. **T. tenella** (*Del.*) *Reichb.*, Icones Fl. Germ., 21: 84 (1867); Davis Fl. Turkey, 4: 523 (1972); Osorio-Tafall et Seraphim, List Vasc. Plants Cyprus, 80 (1973).
Caucalis tenella Del., Fl. Égypt., 58, t. 21, fig. 3 (1813); Unger et Kotschy, Die Insel Cypern, 306 (1865); Boiss., Fl. Orient., 2: 1084 (1872); Holmboe, Veg. Cypr., 135 (1914); Post, Fl. Pal., ed. 2, 1: 571 (1932).
TYPE: Egypt; " croît à Alexandrie . . . près des catacombes " (P).

Slender annual, stems erect, (6–) 20–30 (–50) cm. high, commonly unbranched in the lower half, subglabrous or sparsely adpressed-hispid, obscurely sulcate; leaves rather remote, narrowly deltoid or oblong-deltoid, 2–3-pinnate, 2–8 cm. long, 0·8–4 cm. wide, ultimate segments very small, narrowly oblong or linear, 2–4 mm. long, less than 1 mm. wide, acute, thinly adpressed-hispid; petioles distinct, slender, those of the basal leaves up to 6 cm. long, canaliculate, narrowly sheathed at the base, those of the upper leaves shorter and often sheathing throughout their length; umbels leaf-opposed, 4–8 (–10)-rayed, generally ebracteate; peduncle slender, erect, 2–8 (–14) cm. long, sulcate, thinly adpressed-hispid; rays unequal, 0·8–4 cm. long; bracteoles 3–5, filiform, up to 8 mm. long, about 0·5 mm. wide, thinly hispid with narrow membranous margins; flowers usually 6–10 in each umbellule; pedicels 1–2 mm. long at anthesis, lengthening to 7 mm. or more in fruit; petals 5, small, whitish, broadly obovate, subequal, about 0·7 mm. long, 0·6 mm. wide, apex slightly emarginate; stamens minute; filaments about 0·3 mm. long; anthers yellow, suborbicular, about 0·2 mm. diam.; ovary narrowly oblong, about 2·5 mm. long, 0·8 mm. wide, densely hispid, with long spreading bristles; styles about 0·3 mm. long; stigmas capitate. Mericarps narrowly oblong, about 4 mm. long, 1 mm. wide, secondary ridges furnished with a single row of rather distant, spreading, conspicuous, yellowish or greenish scabrid bristles 2–3 mm. long, primary ridges very sparsely adpressed-hispid.

HAB.: Cultivated fields and fallows, occasionally in open grassland or garigue; sea-level to 2,000 ft. alt.; fl. Febr.–May.

DISTR.: Divisions 1–8, locally frequent. Eastern Mediterranean region, from Greece to Egypt, and eastwards to Iran.

1. Above Peyia, 1962, *Meikle* 2032 !
2. 3 m. from Lefka, 1927, *Rev. A. Huddle* 63 !
3. Lefkara, 1967, *Merton* in ARI 639 !
4. Larnaca, 1862, *Kotschy* 83.
5. Frequent about Nicosia and Kythrea; *Sintenis & Rigo* 349 ! *Holmboe* 264; *Syngrassides* 1320 ! *Merton* 1306 ! and in ARI 204 !
6. Kokkini Trimithia, 1957, *Merton* 2960 !
7. Frequent; *Casey* 1183 ! *G. E. Atherton* 1151 ! *N. Macdonald* 70 ! *Merton* in ARI 333 !
8. Cape Andreas, 1880, *Sintenis & Rigo*.

7. CAUCALIS *L.*

Sp. Plant., ed. 1, 240 (1753).
Gen. Plant., ed. 5, 112 (1754).

Annual herb; leaves 2–3-pinnate, finely dissected; umbels compound, leaf-opposed and terminal; involucre usually small; involucel present; flowers white, pink or purple; petals obovate with incurved apical lobes, the outer sometimes larger than the inner, but not conspicuously radiant; calyx teeth distinct, persistent. Fruit oblong-ellipsoid, somewhat compressed laterally; primary ridges thinly hispid; secondary ridges conspicuously spinose, with smooth, usually uncinate, spines; carpophore generally terete, undivided; stylopodium conical; styles short; endosperm deeply sulcate on the commissural face, with revolute margins.

One species widely distributed in Europe, the Mediterranean region and eastwards to Iran and the Caucasus. The tropical African species included in the genus are probably deserving of independent generic status.

1. C. platycarpos *L.*, Sp. Plant., ed. 1, 241 (1753); Davis, Fl. Turkey, 4: 526 (1972); Osorio-Tafall et Seraphim, List Vasc. Plants Cyprus, 80 (1973).

C. daucoides L., Syst. Nat., ed. 12, 2: 205 (1767); Unger et Kotschy, Die Insel Cypern, 305 (1865); Boiss., Fl. Orient., 2: 1084 (1872); Holmboe, Veg. Cypr., 135 (1914); Post, Fl. Pal., ed. 2, 1: 571 (1932)—non L., Sp. Plant., ed. 1, 241 (1753).

TYPE: " *in* Italia, Monspelii ".

Erect annual; stems 10–40 (–50) cm. high, usually much branched, subglabrous or thinly hispid, especially at the nodes, distinctly ribbed; leaves broadly deltoid in outline, 1·5–7 cm. long and about as wide, 2–3-pinnate, the ultimate segments narrowly linear-oblong, 2–4 mm. long, 1–1·5 mm. wide, acute, subglabrous, margins usually recurved, giving the lamina a characteristic crispate appearance; basal leaves with slender canaliculate petioles up to 4 cm. long, upper leaves subsessile with a broad hispid-ciliate sheathing base; umbels leaf-opposed and terminal, unequally or subequally 2–5-rayed; peduncle up to 10 cm. long, stout, ribbed, rather divaricate; bracts 1–3, sometimes absent, unequal, deciduous, linear-subulate, 2–4 mm. long, 0·5–1 mm. wide, acute, margins hispid-ciliate, narrowly membranous; rays 0·8–3 mm. long at anthesis; bracteoles similar to bracts but generally rather persistent; flowers 6–12 in each umbellule, about half of them functionally male and sterile, without even an abortive ovary; pedicels short, to about 3 mm. long at anthesis, minutely scabridulous; sepals narrowly deltoid, 1–2·5 mm. long, 0·4–0·8 mm. wide in fertile flowers, rather smaller in sterile ones; petals white, pink or purple, broadly obovate, deeply emarginate at apex, the largest about 2·5 mm. long and about as wide, the smallest about 1 mm. long and 0·7 mm. wide; filaments short, about 1·5 mm. long; anthers yellow, broadly oblong, about 1 mm. long, 0·8 mm. wide; styles very short and stout, stigmas subcapitate; ovary oblong, about 2 mm. long, 1·5 mm. wide, densely covered with bristles. Mericarps 8–9 mm. long, about 2 mm. wide; crowned with the persistent calyx; primary ridges sparsely and shortly hispid; secondary ridges rigidly uncinate-spinose, spines about 3 mm. long, flattened and confluent at the base.

HAB.: A weed of cultivated ground, or of open stony places; 4,000–5,000 ft. alt.; fl. May–Sept.

DISTR.: Division 2, one record only. Distribution that of the genus.

2. Prodhromos, 1862, *Kotschy* 807.

NOTES: Voucher specimen not examined, but in view of the fact that Kotschy distinguishes the deceptively similar *Orlaya daucoides*, with which *C. platycarpos* is repeatedly confused, there is little doubt about the accuracy of the record. The plant should be searched for in gardens, vineyards and waste ground at high altitudes in the Troödos area.

8. TURGENIA *Hoffm.*

Gen. Umb., ed. 1, 59 (1814).

Caucalis L. subgenus *Turgenia* (Hoffm.) Benth. in Benth. et Hook. f., Gen. Plant., 1: 929 (1865); Drude in Engl. et Prantl, Pflanzenfam., ed. 1, III. 8: 157 (1898).

Scabrid or bristly annuals; leaves simply pinnate, with broad, toothed segments; umbels compound, terminal and leaf-opposed; bracts and bracteoles with wide, scarious or membranous margins; calyx generally distinct and persistent; petals crimson, pink or whitish, unequal, broadly obovate with an incurved, emarginate apex; outer petals usually much larger than the inner, sometimes conspicuously radiant; disk inconspicuous; stylopodium conical; styles rather short and stout. Mericarps narrowly oblong, somewhat compressed laterally, papillose; primary ridges obscure with a single row of glochidiate spines; secondary ridges prominent with 2 rows of conspicuous glochidiate-scabrid spines; carpophore terete or shortly bifid; endosperm deeply sulcate along the commissural face, with revolute margins.

Two species in Europe and the Mediterranean region east to Central Asia.

1. T. latifolia (*L.*) *Hoffm.*, Gen. Umb., ed. 1, 59 (1814); Poech, Enum. Plant. Ins. Cypr.,

28 (1842); Unger et Kotschy, Die Insel Cypern, 306 (1865); Boiss., Fl. Orient., 2: 1087 (1872); Holmboe, Veg. Cypr., 135 (1914); Post, Fl. Pal., ed. 2, 1: 572 (1932); Davis, Fl. Turkey, 4: 527 (1972); Zohary, Fl. Palaest., 2: 398, t. 575 (1972); Osorio-Tafall et Seraphim, List Vasc. Plants Cyprus, 80 (1973).

Caucalis latifolia L., Syst. Nat., ed. 12, 2: 205 (1767); Sibth. et Sm., Fl. Graec. Prodr., 1: 182 (1806).

TYPE: " *in* Agro Cantabrigiensi Angliae, *in* Agro Narbonensi *circa Monspelium* & *Nemausium* ".

Erect annual 12–40 (–60) cm. high; stems usually much branched, prominently ribbed, shortly setose, with longer scattered bristles; leaves oblong in outline, 4–15 cm. long, 3–10 cm. wide, divided into 3–6 pairs of opposite, linear-oblong, coarsely serrate-lobed, acute, hispid segments 0 5–5 (–7) cm. long, 0·3–2 cm. wide; upper leaves subsessile with a sheathing membranous-margined base; lower leaves with sheathing base and canaliculate petioles up to 10 cm. long; umbels subequally or unequally 2–5-rayed; peduncle hispid, sulcate, to 13 cm. long (or sometimes longer); involucre of 2–5, narrowly oblong, membranous-margined, persistent bracts about 6 mm. long and 2 mm. wide; rays hispid, spreading, 1–6 (–10) cm. long; bracteoles persistent, similar to bracts, but usually with broader membranous margins, and often with a hispid midrib; flowers 8–12 in each umbellule, about half of them functionally male and sterile; fertile flowers sessile or subsessile at anthesis; ovary about 3 mm. long, 1·8–2 mm. wide, densely spinose; calyx of 5, small, unequal, deltoid sepals, the largest about 0·5 mm. long; petals crimson, pink or whitish, broadly obovate, deeply 2-lobed, the outer about 2·5 mm. long and 5 mm. wide, a little radiant, the inner about 1 mm. long, 1·5 mm. wide; stamens about 1 mm. long; anthers yellow, suborbicular, about 0·8 mm. long and almost as wide; styles thick, less than 1 mm. long; stigmas truncate; sterile flowers shortly pedicellate, without an ovary; petals and sepals smaller than those of the fertile flowers. Fruits shortly pedicellate; mericarps narrowly oblong, about 8–12 mm. long, 2–3 mm. wide; primary and secondary ribs variously spinose, sometimes with a single row, or sometimes with a double row of stout scabrid-glochidiate spines, about 3 mm. long.

HAB.: Cultivated fields and vineyards, occasionally in garigue on hillsides; 500–3,500 ft. alt.; fl. May–June.

DISTR.: Divisions 1, 2, 5, rare. Distribution that of the genus.

1. Near Tsadha, 1905, *Holmboe* 730.
2. Near Khryssoroyiatissa Mon., 1862, *Kotschy*; near Spilia, 1880, *Sintenis & Rigo* 837 ! Kryos Potamos, 3,100 ft. alt., 1937, *Kennedy* 680 !
5. Dhikomo area, 1950, *Chapman* 684 !

NOTES: First recorded (1787) by Sibthorp, but without exact indication of locality. Kotschy also records it from " Slewra ", a name which cannot be traced either in his own, or any other, map. Perhaps Skylloura (" Scluri " of Kotschy) was intended.

9. ORLAYA *Hoffm.*

Gen. Umb., ed. 1, 58 (1814).

Subglabrous or pilose annuals; leaves 2–3-pinnate, with the ultimate segments small and narrow; umbels compound, bracts and bracteoles conspicuous, persistent, usually with membranous margins; calyx small; petals white, the outer often much enlarged and radiant, all broadly obovate with a deeply emarginate apex. Mericarps strongly compressed dorsally; primary ridges shortly hispid; secondary ridges conspicuously spinose, the bases of the spines flattened laterally and confluent, the spine-tips uncinate; stylopodium conical; styles long, slender, divergent; stigmas truncate or subcapitate; carpophore terete; endosperm flattish along the commissural face.

Three species in Europe, the Mediterranean region and Asia east to Iran.

1. **O. daucoides** (*L.*) *Greuter* in Boissiera, 13: 92 (1967); Davis, Fl. Turkey, 4: 530 (1972); Zohary, Fl. Palaest., 2: 399, t. 576 (1972); Osorio-Tafall et Seraphim, List Vasc. Plants Cyprus, 80 (1973).
Caucalis daucoides L., Sp. Plant., ed. 1, 241 (1753)—non L., Syst. Nat., ed. 12, 2: 205 (1767).
Orlaya platycarpos Koch in Nov. Act. Nat. Cur., 12, 1: 79 (1824); Unger et Kotschy, Die Insel Cypern, 304 (1865); Boiss., Fl. Orient., 2: 1071 (1872); Holmboe, Veg. Cypr., 136 (1914); Post, Fl. Pal., ed. 2, 1: 563 (1932); Lindberg f., Iter Cypr., 25 (1946)—non *Caucalis platycarpos* L., Sp. Plant., ed. 1, 241 (1753) nec Sp. Plant., ed. 2, 1: 347 (1762).
O. kochii Heywood in Agron. Lusit., 22: 13 (1960).
TYPE: France; " Monspelii *D*: *Sauvages* ".

Erect or spreading annual; stems 10–40 (–50) cm. high, usually much branched, shallowly sulcate, thinly hispid-pilose; leaves oblong or narrowly deltoid in outline, 2–8 cm. long, 1·5–5 cm. wide, 2–3-pinnate, finely dissected, thinly hispid, the ultimate segments narrowly oblong-lanceolate, acute, about 0·5–5 mm. long, 0·3–2 mm. wide; lower leaves with slender petioles up to 10 cm. long, upper leaves with a short, sheathing, narrowly membranous-margined, ciliate base; umbels terminal and leaf-opposed, 2–3 (–4)-rayed; peduncle 1–8 (–12) cm. long, subglabrous, conspicuously ribbed; bracts 2–3, lanceolate, 10–14 mm. long, 1·5–3 mm. wide, acute or acuminate, with broad, long-ciliate, membranous margins; rays 5–20 mm. long at anthesis, lengthening somewhat in fruit; bracteoles 3–5, very unequal, the larger resembling the bracts, though proportionately somewhat broader, the smaller barely 5 mm. long, narrowly linear-subulate, with very narrow membranous margins; flowers numerous in each umbellule, 2–3 fertile, the remainder functionally male and sterile; pedicels slender, about 2 mm. long; ovary of fertile flowers about 2 mm. long and 1·3 mm. wide, densely spinose; calyx with 5 minute, deltoid-subulate lobes, barely 0·3 mm. long; petals white, unequal, the outer radiant ones about 3 mm. long and about as wide, deeply 2-lobed, the inner less than 1·5 mm. long and 1–1·5 mm. wide; filaments 1·5–2 mm. long; anthers yellow, broadly ovoid, about 0·5 mm. long and almost as wide; styles straight, erect, about 0·8 mm. long at anthesis, lengthening in fruit; stigmas small, usually truncate, sometimes a little swollen and subcapitate; sterile flowers with equal or unequal (sometimes radiant) petals, calyx very small, with minute, deltoid teeth. Fruit compressed-ellipsoid, about 12–14 mm. long, 7–8 mm. wide; mericarps almost flat, the primary ridges shortly hispid, with numerous bristles; the secondary ridges spinose, with hooked spines about 2·5 mm. long.

HAB.: In garigue on dry, rocky hillsides; in vineyards and fallow fields and by roadsides; sea-level to 5,000 ft. alt., generally between 1,000 and 3,000 ft. alt.; fl. March–May.

DISTR.: Divisions 2, 3, 7, locally frequent. S. Europe, Mediterranean region and eastwards to Iran.

2. Frequent; Prodhromos, Platania, Platres, Pedhoulas, Kakopetria, Lagoudhera, Stavros tis Psokas, etc.; *Kotschy*; *Lindberg f.*; *Kennedy* 679 ! *Syngrassides* 1614 ! *Davis* 3470 ! *Merton* 2415A ! *Meikle* 2848 !
3. Kalavasos, 1905, *Holmboe* 620; Skarinou, 1941, *Davis* 2671 ! Zakaki, 1955, *Merton* 2067 !
7. Above Bellapais, 1880, *Sintenis & Rigo* 593 ! St. Hilarion, 1949, *Casey* 394 !

10. PSEUDORLAYA *Murb.*

Acta Univ. Lund, 33, 12: 86 (1897).

Dwarf maritime annuals; stems and leaves usually rather densely hispid; leaves 2–3-pinnate as in *Orlaya*; umbels compound, leaf-opposed and terminal; rays very unequal; involucre and involucel of simple or pinnatisect, persistent, herbaceous bracts and bracteoles; flowers small; calyx minute; petals often tinged pink or purple, scarcely radiant; stylopodium hemispherical or conical; styles short, thick; stigmas truncate or subcapitate. Fruit strongly compressed dorsally; primary ridges hispid; secondary ridges spinose with 2 rows of straight (not hooked) glochidiate spines which, on the lateral ribs, are conspicuously flattened and confluent at the base; carpophore entire, terete; endosperm as in *Orlaya*.

One (or possibly two) species widely distributed in the Mediterranean region.

1. P. pumila (*L.*) *Grande* in Nuov. Giorn. Bot. Ital., n.s., 32: 86 (1925); Davis, Fl. Turkey, 4: 536 (1972); Zohary, Fl. Palaest., 2: 400, t. 577 (1972); Osorio-Tafall et Seraphim, List Vasc. Plants Cyprus, 81 (1973).
Caucalis pumila L., Syst. Nat., ed. 10, 955 (1759).
C. maritima Gouan, Hort. Reg. Monsp., 135 (1762).
Orlaya maritima (Gouan) Koch in Nov. Act. Nat. Cur., 12, 1: 79 (1824); Unger et Kotschy, Die Insel Cypern, 304 (1865); Boiss., Fl. Orient., 2: 1071 (1872); Holmboe, Veg. Cypr., 136 (1914).
Orlaya pumila (L.) Hal., Consp. Fl. Graec., 1: 622 (1901); Post, Fl. Pal., ed. 2, 1: 563 (1932).

TYPE: Probably collected in Spain by Alströmer.

Dwarf, greyish-hispid annual, usually 5–15 cm. high, but occasionally up to 30 cm.; stems normally much branched at the base, spreading, distinctly ribbed; leaves oblong-deltoid, 1·5–6 cm. long, 1–3·5 cm. wide, 2–3-pinnate, ultimate segments linear-oblong, 2–5 mm. long, 0·8–2 mm. wide, usually rather crowded, obtuse or shortly acute; petioles usually well developed, up to 6 cm. long, rather stout, flattish or canaliculate above, base shortly sheathing, herbaceous or with narrow membranous-ciliate margins; the petioles of the lower leaves are commonly buried in sand, and are then very elongate and blanched; umbels leaf-opposed and terminal, with 3–7 rays; peduncles 1–8 cm. long at anthesis, up to 16 cm. long in fruit, stout, distinctly sulcate and hispid; bracts variable, sometimes linear-subulate and entire, sometimes lobed or pinnatisect, herbaceous, hispid, 5–15 mm. long, 1–10 mm. wide; rays very unequal, usually less than 1 cm. long at anthesis, up to 4 cm. long in fruit; bracteoles 3–5, similar to the bracts, but smaller and usually entire or very shortly lobed, spreading or somewhat reflexed in fruit; flowers very small, shortly pedicellate or subsessile, crowded, about 8–12 in each umbellule, rather more than half of them functionally male and sterile; fertile flowers with a broadly oblong, densely hispid ovary about 2 mm. long, 1·8 mm. wide; calyx divided into 5, unequal, deltoid-acuminate sepals about 0·5–1 mm. long, 0·3–0·5 mm. wide; petals pink or purplish, subequal, broadly obovate, about 1·5 mm. long and almost as wide, apex widely emarginate; filaments slender, glabrous, about 0·8 mm. long; anthers broadly ovate or suborbicular, coloured like the petals, about 0·7 mm. long and almost as wide; stylopodium hemispherical or conical; styles thick, erect, about 0·4 mm. long, stigmas truncate; sterile flowers more distinctly pedicellate than the fertile, with a very small calyx and a conspicuous, yellowish, hemispherical stylopodium. Fruits ellipsoid, strongly compressed dorsally, about 8–10 mm. long, 4–5 mm. wide; primary ridges hispid; secondary ridges with conspicuous, rigid, straight, glochidiate spines, often tinged red at the base.

HAB.: Sand-dunes and sandy seashores; about sea-level; fl. Febr.–April.

DISTR.: Divisions 1–3, 5–8, widely distributed around the coast. General distribution that of the genus.

1. Near Paphos, 1862, *Kotschy* 664a.
2. Yialia, 1955, *Merton* 2101 !
3. Near Amathus, 1862, *Kotschy* 574.

5. Salamis, 1905, *Holmboe* 446, and, 1957, *Merton* 3211 !
6. Syrianokhori, 1935, *Syngrassides* 768 ! Ayia Irini, 1941, *Davis* 2545 ! also, 1953, *Kennedy* 1802 ! and, 1962, *Meikle* 2382 !
7. Near Dhavlos, 1880, *Sintenis & Rigo* 345 ! Kyrenia, 1951, *Casey* 1132 !
8. Eleousa near Rizokarpaso, 1880, *Sintenis & Rigo*; Cape Andreas, 1880, *Sintenis & Rigo*; Ayios Philon, 1941, *Davis* 2207 !

11. DAUCUS *L.*

Sp. Plant., ed. 1, 242 (1753).
Gen. Plant., ed. 5, 113 (1754).

Annual or biennial herbs; stems often hispid; leaves 2–3-pinnate, the ultimate segments usually small and numerous; umbels compound; involucre often conspicuous, of lobed or pinnatisect, or, rarely, entire bracts; involucel of simple or variously lobed bracteoles; flowers white, pinkish or yellowish, those at the centre of the umbel or of the umbellules frequently much modified, sterile and dark blackish-crimson; outer petals sometimes radiant; disk small, inconspicuous; stylopodium convex or shortly conical; styles often long and divaricate. Fruit subterete or dorsally compressed, ovoid-oblong or ellipsoid; primary ridges hispidulous, secondary ridges prominent, usually with a single row of conspicuous spines; carpophore terete, entire; endosperm flattish along its commissural face; vittae solitary, under the secondary ridges.

About 30 species, mostly in Europe and the Mediterranean region.

Umbels sessile or subsessile, without obvious stalks - - - - **7. D. durieua**
Umbels distinctly stalked:
Stems decumbent or sprawling, 2–20 (–35) cm. high; leaves rather fleshy with blunt segments; umbels with few, conspicuously unequal rays; bracts and bracteoles entire or shortly 3-fid - - - - - - - - - **5. D. glaber**
Stems erect or ascending; leaves and umbels not as above:
Stems usually less than 10 cm. high, shorter than the long (6–16 cm.) naked peduncles; umbels with few (3–5) rays; bracts pinnatisect, much exceeding umbellules at anthesis; petals small, subequal - - **6. D. involucratus**
Stems usually more than 10 cm. high, longer than the peduncles; umbels 6–many-rayed; petals unequal, often radiant:
Leaf-segments and bracts narrowly linear or filiform; flowers and fruits becoming yellow when dried; central fruit of each umbellule subsessile and tuberculate **2. D. aureus**
Leaf-segments and bracts broader, flowers rarely yellowish when dried, fruits reddish-purple or pale brown, central fruit of each umbellule not as above:
Spines of mature fruit not or very narrowly confluent-winged at base; involucre often conspicuous:
Plants robust, up to 200 cm. high; rays usually very numerous, becoming erect and crowded in fruit; fruit-spines short, generally less than 2 mm. long, obscurely glochidiate - - - - - **1. D. carota**
Plants slender, usually 10–30 cm. high, rays 6–12, not becoming very erect or crowded in fruit; fruit spines 2–3 mm. long, distinctly glochidiate **4. D. guttatus**
Spines of mature fruit distinctly confluent-winged at the base; plants generally 15–30 (–50) cm. high, usually much branched; involucre not conspicuous **3. D. broteri**

1. D. carota *L.*, Sp. Plant., ed. 1, 242 (1753); Davis, Fl. Turkey, 4: 531 (1972); Zohary, Fl. Palaest., 2: 452, t. 655 (1972); Osorio-Tafall et Seraphim, List Vasc. Plants Cyprus, 81 (1973).

Polymorphic biennial; often with a thickened, frequently reddish, aromatic taproot; stems erect or diffuse, 10–200 cm. high, usually branched, retrorsely hispid or sometimes almost glabrous, distinctly sulcate; leaves oblong or oblong-deltoid in outline, 1·5–30 cm. long, 1–20 cm. wide, 2–3-pinnate, the ultimate segments linear, linear-lanceolate, oblong or broadly obovate-cuneate, entire or toothed, hispid or ± glabrous; lower leaves with long, canaliculate petioles, upper leaves subsessile, with a conspicuous, membranous-margined, sheathing base; umbels leaf-opposed and terminal, convex or flattish, few (c. 8) to many rayed, becoming congested and concave in fruit; peduncle usually well developed, up to 40 cm. long, sulcate, hispid or subglabrous; involucre usually conspicuous and persistent, sometimes exceeding the flowers; bracts usually pinnatisect with narrow linear-subulate, oblong or oblanceolate lobes; rays slender, unequal, up to 5 cm. long at anthesis; bracteoles simple or lobed, linear or lanceolate, usually with membranous-ciliate margins; flowers white or tinged pink, those at the centre of the umbel (or occasionally at the centre of each umbellule) sterile, rather fleshy, and dark blackish-crimson; sepals minute, deltoid-acuminate; outermost petals in each umbellule often distinctly radiant; filaments about 1 mm. long, glabrous; anthers broadly ovoid or suborbicular, usually yellow, about 0·5 mm. long, 0·4 mm. wide; stylopodium convex; styles divergent, about 0·8 mm. long, rather slender; stigmas subcapitate. Fruit oblong, 3–5 mm. long, 1·5–2·5 mm. wide; mericarps somewhat compressed dorsally; primary ridges indistinct, thinly hispidulous; secondary ridges very prominent, ± winged, bearing a single row of straight, spreading, pallid spines 0·8–1·5 cm. long; tips of spines usually rather obscurely glochidiate, or with the glochidia reduced to a single, minute barb; bases of spines not or scarcely confluent; carpophore terete, entire or shortly 2-cleft at apex; endosperm flattish along its commissural face.

ssp. **carota.**

D. carota Gaudry, Recherches Sci., 198 (1853–54); Unger et Kotschy, Die Insel Cypern, 305 (1865); Boiss., Fl. Orient., 2: 1076 (1872); Holmboe, Veg. Cypr., 142 (1914); Post, Fl. Pal., ed. 2, 1: 566 (1932); Lindberg f., Iter Cypr., 25 (1946).

TYPE: " *in* Europae *campis exaridis* ".

Stems rather slender, usually less than 100 cm. high, often distinctly flexuous, retrorse-hispid or subglabrous; ultimate segments of leaflets usually small, narrow, less than 5 mm. long and 3 mm. wide, those of the upper leaves often very narrow and elongated.

HAB.: Sandy and stony pastures, seashores, cultivated and waste ground; sea-level to 4,500 ft. alt.; fl. April–Sept.

DISTR.: Divisions 1, 2, 4, 5, 7, not common; widespread in Europe, Mediterranean region and western Asia, and often as an escape from cultivation elsewhere.

1. Khlorakas, 1967, *Economides* in ARI 968 !
2. Prodhromos, 1853, *Gaudry*, also, from the same area, *Lindberg f.* s.n.; *Merton* 2400 ! *G. E. Atherton* 544 ! Mandria, 1938, *Kennedy* 1395 !
4. Larnaca aerodrome, 1950, *Chapman* 694 ! and, 1952, *Merton* 845 ! Cape Greco, 1958, *N. Macdonald* 31 !
5. Alambra, 1932, *Syngrassides* 387 !
7. Kyrenia, 1955, *Miss Mapple* 111 ! and *G. E. Atherton* 795 !

NOTES: Probably as escape from cultivation in some of these localities, but unquestionably native along the coast of Cyprus.

ssp. **maximus** (*Desf.*) *Ball* in Journ. Linn. Soc. Bot., 16: 476 (1878); Holmboe, Veg. Cypr., 142 (1914); Lindberg f., Iter Cypr., 25 (1946); Osorio-Tafall et Seraphim, List Vasc. Plants Cyprus, 81 (1973).

D. maximus Desf., Fl. Atlant., 1: 241 (1798); Unger et Kotschy, Die Insel Cypern, 305 (1865); Boiss., Fl. Orient., 2: 1076 (1872); Post, Fl. Pal., ed. 2, 1: 566 (1932).

TYPE: " Algeria ad limites agrorum " (P).

Stems robust, erect, usually more than 100 cm. high, not distinctly flexuous, glabrous or thinly retrorse-hispid; ultimate segments of leaves coarsely cuneate-ovate or cuneate-rhomboid, 8–20 mm. long, 5–13 mm. wide, shortly acute or obtuse, sharply serrate, those of the upper leaves elongate, but coarser and more obviously toothed than in ssp. *carota*.

HAB.: Cultivated fields, roadsides and streamsides, sea-level to 4,500 ft. alt.; fl. April–Sept.

DISTR.: 1, 2, 5, 6, 7, rather rare. Mediterranean region.

1. Cape Arnauti, 1913, *Haradjian* 810 ! Limni Mine, 1938, *Syngrassides* 1840 ! Skoulli, 1939, *Lindberg f.* s.n.
2. Pano Panayia, 1905, *Holmboe* 1050; Platres, 1929, *C. B. Ussher* 12 ! Prodhromos, 1949, *Casey* 900 !
5. Strongylos, 1967, *Merton* in ARI 468 !
6. Morphou, 1955, *N. Histellis* 471 !
7. Panteleimon Mon. near Paleomylos, 1862, *Kotschy* 960; Lapithos, 1880, *Sintenis & Rigo* 609 ! Kyrenia, 1939, *Lindberg f.* ! also, 1955, *Miss Mapple* 97 ! and *G. E. Atherton* 97 !

NOTES: Generally, but not always, distinguished from ssp. *carota* by its robustness, and by the coarsely segmented leaves.

2. D. aureus *Desf.*, Fl. Atlant., 1: 242, t. 61 (1798); Boiss., Fl. Orient., 2: 1076 (1872); Post, Fl. Pal., ed. 2, 1: 566 (1932); Lindberg f., Iter Cypr., 25 (1946); Zohary, Fl. Palaest., 2: 451, t. 654 (1972); Osorio-Tafall et Seraphim, List Vasc. Plants Cyprus, 80 (1973).

TYPE: Algeria; " inter segetes circa Mascar " (P).

Erect biennial; stems 60–150 cm. high, frequently branched only in the upper half, subglabrous, or less often rather sparsely retrorse-hispid, shallowly sulcate; leaves oblong or oblong-deltoid in outline, 4–12 cm. long, 1·5–8 cm. wide, 3-pinnate, the ultimate segments narrowly linear-acuminate or filiform (towards the top of the stem) 2–25 mm. long, 0·5–2 mm. wide, subglabrous or very thinly hispid; lower leaves with canaliculate petioles up to 6 cm. long; upper leaves subsessile, with a narrow, membranous-margined basal sheath about 1·5–2 cm. long; umbels convex or flattish at anthesis, terminal and leaf-opposed, many-rayed; peduncles erect, up to 20 cm. long, distinctly sulcate; involucre conspicuous, persistent, reflexed in fruit, of 5 or more pinnatisect bracts 2–3 cm. or more long, divided into slender, filiform lobes; rays 1–6 cm. long at anthesis, rather unequal, accrescent in fruit, and becoming suberect and crowded; bracteoles similar to bracts, but much smaller, rarely exceeding 1 cm. in length, and sometimes narrowly membranous-ciliate along the margins; pedicels very slender, about 2 mm. long at anthesis, lengthening to 5 mm. or more in fruit; flowers whitish, becoming yellow when dried, generally fertile, without dark, modified, sterile flowers at the centre of the umbel; calyx minute; petals unequal, the outer larger, but scarcely radiant, to about 1·5 mm. long, 1·3 mm. wide, deeply bilobed, the inner similar but usually less than 1 mm. long; filaments about 1·3 mm. long; anthers oblong, about 0·8 mm. long, 0·4 mm. wide, yellow or orange; stylopodium convex; styles 0·8–1·3 mm. long; stigmas subcapitate. Fruit oblong, 4–5 mm. long, 1·5–2 mm. wide, not much compressed dorsally, primary ridges with very short, scattered spines; secondary ridges conspicuously yellow-spinose, the spines in a single row, patent, straight, scabridulous, about 2·5–3 mm. long, apex glochidiate with several barbs, bases strongly flattened and confluent; endosperm distinctly sulcate along the commissural face; central fruit of each umbellule subsessile, with all the ribs papillose-tuberculate, without spines; carpophore slender, terete, entire.

HAB.: Cultivated ground; sea-level to 100 ft. alt.; fl. May–July.

DISTR.: Divisions 4, 5, 7, rare. Spain, Italy, Sicily, N. Africa, Palestine.

4. Kouklia, 1952, *Merton* 849 !
5. Trikomo, 1939, *Lindberg f.* s.n. !
7. Kyrenia, 1949, *Casey* 764 !

3. D. broteri *Ten.*, Fl. Nap., 4, syll. app. 3: IV, t. 222 (1830); Boiss., Fl. Orient., 2: 1073 (1872); Post in Mém. Herb. Boiss., 18: 94 (1900); Holmboe, Veg. Cypr., 142

(1914); Post, Fl. Pal., ed. 2, 1: 564 (1932); Davis, Fl. Turkey, 4: 533 (1972); Osorio-Tafall et Seraphim, List Vasc. Plants Cyprus, 80 (1973).

TYPE: Italy; " in collibus argillosis Aprutii: *Turri Valignano presso Chieti* " (NAP).

Erect or spreading annual; stems 15–30 (–50) cm. high, flexuous, slender, usually much branched, subglabrous or thinly retrorse-hispid, obscurely sulcate; leaves rather few and remote, oblong-deltoid in outline, 2–3-pinnate, 1·5–7 cm. long, 1–3 cm. wide, ultimate segments lanceolate-acute, generally about 5 mm. long and 1–3 mm. wide, entire or toothed; lower leaves with slender, canaliculate petioles up to 5 mm. long, upper leaves subsessile, all with inconspicuous, narrowly membranous-margined, subglabrous or thinly hispid basal sheaths 5–10 mm. long; umbels leaf-opposed and terminal, flattish, 6–12-rayed; peduncles erect, slender, up to 14 cm. long at anthesis, lengthening in fruit; involucre conspicuous, persistent, generally shorter than the rays; bracts pinnatisect, with rather few linear segments 5–15 mm. long; rays slender, rather unequal, 2–15 mm. long at anthesis, lengthening and becoming suberect in fruit; bracteoles heteromorphous, the 3 outer linear-subulate, to about 5 mm. long, 1–1·5 mm. wide, herbaceous, hispid with a very narrow membranous margin, the 2–3 inner lanceolate-cuspidate, somewhat keeled, about 3–4 mm. long, 1–1·5 mm. wide, membranous-ciliate except for the median nerve; flowers white (or yellowish when dried) some of them functionally male and sterile, or occasionally wholly sterile and much modified with thickened blackish-crimson petals; pedicels very short at anthesis, lengthening to about 5 mm. in fruit; calyx minute; petals very small, unequal, obovate, 2-lobed, the outermost in each umbellule about 1–1·5 mm. long and almost as wide, slightly radiant, the inner less than 1 mm. long; filaments about 1 mm. long; anthers yellow or orange, oblong, about 0·4 mm. long, 0·3 mm. wide; stylopodium convex; styles about 1·4 mm. long when fully developed, spreading or recurved; stigmas subcapitate. Fruit narrowly oblong-ellipsoid, about 3–5 mm. long, 1·5 mm. wide; mericarps somewhat compressed dorsally; primary ridges shortly setose with two rows of short, spreading bristles; secondary ridges conspicuously spinose, spines in a single row 4–5 mm. long, flattened and confluent at the base, and forming a distinct wing, apices glochidiate with several barbs, sometimes tinged reddish-purple, especially when immature; carpophore slender, terete, entire; endosperm flattish along the commissural face.

HAB.: Cultivated and waste ground, vineyards, roadsides; 100–4,500 ft. alt.; fl. May–Sept.

DISTR.: Divisions ? 1, 2, 4, apparently rare. Italy, ? Balkan peninsula, Aegean islands, Turkey, Syria, Palestine, Egypt.

1. Stroumbi, 1913, *Haradjian* 726 ! An immature and untypical specimen, but probably *D. broteri*.
2. Galata, 1880, *Sintenis & Rigo* 923 (as *D. muricatus*) ! Platres–Perapedhi, 1935, *Syngrassides* 756 ! and, same area, 1941, *Davis* 1843 ! Prodhromos, 1949, *Casey* 926 ! and, 1955, *G. E. Atherton* 591 ! 595 !
4. Kouklia, 1939, *Lindberg f.* s.n.; near Akhyritou, 1948, *Mavromoustakis* s.n. !

NOTES: The first published record, unfortunately without precise indication of locality, is to be found in Post's list (*loc. cit. supra*); *D. broteri* has seldom been collected since, perhaps because it flowers late in the season when most field work is drawing to a close. The species is often confused with the following, *D. guttatus* Sibth. et Sm., but the two are distinguishable by their fruits, and usually, but not always, by the longer, filiform or setaceous bracts of *D. guttatus*.

It is worth nothing that, with one exception, all the Cyprus material of *D. broteri* agrees with the typical plant in lacking the dark crimson sterile flowers so characteristic of *D. guttatus*.

In the Flora of Turkey, 4: 535 (1972), following Boissier, *D. bicolor* Sibth. et Sm. is treated as a variety of *D. broteri*. If this were correct, then nomenclaturally, *D. broteri* (1830) would have to be regarded as a later synonym of *D. bicolor* (1806); it would seem, however, that Flora Europaea, 2: 373 (1968) is more correct in referring *D. bicolor* to *D. guttatus*. The distinctions between *D. broteri* and *D. guttatus*, though evident enough in some areas, are not always satisfactory.

4. **D. guttatus** *Sibth. et Sm.*, Fl. Graec. Prodr., 1: 184 (1806), Fl. Graec., 3: 63, t. 269 (1819); Davis, Fl. Turkey, 4: 533 (1972); Zohary, Fl. Palaest., 2: 451, t. 652 (1972); Osorio-Tafall et Seraphim, List Vasc. Plants Cyprus, 80 (1973).

D. setulosus Guss. in DC., Prodr., 4: 211 (1830); Boiss., Fl. Orient., 2: 1075 (1872); Post, Fl. Pal., ed. 2, 1: 565 (1932).

D. setulosus Guss. var. *brachylaenus* Boiss., Fl. Orient., 2: 1075 (1872).

D. guttatus Sibth. et Sm. var. *brachylaenus* (Boiss.) Hayek, Prodr. Fl. Pen. Balc., 1: 1051 (1927); Lindberg f., Iter Cypr., 25 (1946).

TYPE: " In insulis Graeciae frequens; etiam in Asia minore " (OXF).

Erect or spreading annual; stems 10–30 (–50) cm. high, rather flexuous, slender, usually much branched, subglabrous or thinly retrorse-hispid, obscurely sulcate; leaves often rather few, narrowly oblong-deltoid in outline, 3-pinnate, 1·5–8 cm. long, 0·8–4 cm. wide, ultimate segments lanceolate or linear-lanceolate, acute, 2–8 mm. long, 1–2 mm. wide, generally entire; lower leaves with slender, canaliculate petioles up to 6 cm. long, upper leaves subsessile, all with inconspicuous, narrowly membranous margined, subglabrous or thinly hispid basal sheaths 5–20 mm. long; umbels leaf-opposed and terminal, flattish, 6–12-rayed; peduncles erect, slender, up to 14 cm. long at anthesis, lengthening in fruit; involucre generally very conspicuous, persistent, often (but not always) longer than the rays; bracts pinnatisect, with few to many filiform, setaceous or narrowly linear, hispidulous segments 8–25 (–30) mm. long; rays slender, unequal, 5–20 mm. long at anthesis, lengthening and becoming suberect in fruit; bracteoles heteromorphous, 3–4 outer setaceous, to about 15 mm. long, 1–1·5 mm. wide, herbaceous, hispidulous, with a very narrow membranous margin, the 2–3 inner linear-acuminate, somewhat keeled, about 5 mm. long, 1–1·5 mm. wide, membranous-ciliate except for the median nerve; flowers white, some of them functionately male and sterile, and those at the centre of the umbel and umbellules usually much modified, wholly sterile and dark blackish-crimson; pedicels slender, hispidulous, up to 8 mm. long at anthesis, lengthening in fruit; calyx and petals as in *D. broteri*; stylopodium usually conical; styles long, slender, up to 2 mm., divergent; stigmas subcapitate. Fruit ellipsoid, somewhat compressed dorsally, about 2–3 mm. long, 1·5–2 mm. wide; primary ridges of mericarp thinly pilose-hispidulous, with 2 irregular ranks of very slender, short bristles, secondary ridges spinose with a single row of slender, glochidiate spines 2–3 mm. long, bases of spines flattened and confluent, forming a very narrow wing, but not conspicuously winged as in *D. broteri*; carpophore slender, terete, entire; endosperm flattish along commissural face.

HAB.: Dry sandy or stony ground, maritime cliffs and sand-dunes; about sea-level; fl. May–Sept.

DISTR.: Division 4, [7], very rare. S. Italy, Balkan peninsula, Crete, Aegean islands, Turkey, ? Iran.

4. Near Larnaca, on sandy shore, 1939, *Lindberg f.* s.n. !
[7. Dhavlos, on sandy cliffs, 1957, *H. Painter* 7.]

NOTES: The one specimen seen corresponds with typical *D. guttatus* Sibth. et Sm. var. *guttatus* (*D. guttatus* var. *brachylaenus* (Boiss.) Hayek) in having relatively short involucral bracts, distinctly shorter than the rays; the more common variety (*D. setulosus* Guss.) with long bracts (reminding one of *Nigella damascena*) has not yet been collected in Cyprus.

The Dhavlos record is likely, but, in the absence of the specimen, cannot be regarded as wholly trustworthy.

5. **D. glaber** (*Forssk.*) *Thell.* in Mém. Soc. Nat. Cherbourg, 38: 407 (1912).

Caucalis glabra Forssk., Fl. Aegypt.-Arab., 206 (1775).

Daucus littoralis Sibth. et Sm., Fl. Graec. Prod., 1: 185 (1806), Fl. Graec., 3: 65, t. 271 (1819); Boiss., Fl. Orient., 2: 1074 (1872); Holmboe, Veg. Cypr., 142 (1914); Post, Fl. Pal., ed. 2, 1: 565 (1932); Lindberg f., Iter Cypr., 25 (1946); Osorio-Tafall et Seraphim, List Vasc. Plants Cyprus, 80 (1973).

D. littoralis Sibth. et Sm. var. *genuinus* Boiss., Fl. Orient., 2: 1074 (1872).

D. littoralis Sibth. et Sm. var. *forskahlei* Boiss., Fl. Orient., 2: 1074 (1872).

D. muricatus L. var. *littoralis* (Sibth. et Sm.) H. Stuart Thompson in Journ. Bot., 44: 308 (1906).

[*D. muricatus* (non L.) Kotschy in Unger et Kotschy, Die Insel Cypern, 305 (1865).]

[*D. maritimus* (non Lam.) Kotschy in Unger et Kotschy, Die Insel Cypern, 305 (1865); Sintenis in Oesterr. Bot. Zeitschr., 32: 398 (1882).]

TYPE: Egypt; Alexandria, *Forsskål*.

Decumbent or sprawling annual or biennial; stems 2–20 (–35) cm. high, generally much branched, especially at the base, glabrous (and often bright yellowish where buried in sand), subglabrous, or ± densely retrorse-hispid, rather fleshy, obscurely sulcate; leaves narrowly oblong in outline, usually 2-pinnate, 1–6 cm. long, 0·4–3·5 cm. wide, subglabrous or scabridulous, rather glaucous and fleshy, ultimate segments obovate-cuneate, to about 5 mm. long and 3 mm. wide, rather bluntly lobed or toothed; petioles of basal leaves up to 5 cm. long; upper leaves shortly petiolate or subsessile; leaf-sheaths narrow, to about 12 mm. long, often densely scabridulous with very narrow, membranous-ciliate margins; umbels leaf-opposed and terminal, flattish, 5–7 (–12)-rayed; peduncle proportionately very long, up to 15 cm. or more, rather stout, subglabrous or thinly scabridulous; involucre small, consisting of 3–6 simple, oblanceolate-acute or shortly 3-fid, fleshy, spreading bracts 2–10 mm. long; rays very unequal; some 3 cm. long at anthesis, others so short as to be barely visible, so that at least one or two of the umbellules usually appear to be sessile, while others are thrust out far from the centre of the umbel; bracteoles usually simple, lanceolate, the larger 5–6 mm. long, herbaceous, scabridulous, the smaller 2–4 mm. long, somewhat carinate, membranous-ciliate except for a scabridulous midrib; pedicels very short, generally less than 2 mm. long; flowers white or tinged pink, partly perfect and partly male and sterile, but wholly sterile flowers with modified dark crimson petals are apparently absent; ovary oblong, about 1 mm. long and 0·8 mm. wide, whitish hispid; calyx minute with deltoid, acute teeth; petals unequal, the outer broadly obovate and slightly radiant, about 1·5–2 mm. long, 2–2·5 mm. wide, deeply bilobed, the others much smaller, seldom exceeding 1 mm. in length; filaments about 2 mm. long; anthers oblong, yellowish or reddish, about 0·4 mm. long and almost as wide; stylopodium conical; fully developed styles up to 2 mm. long, often rather thick; stigmas distinctly capitate. Fruit oblong, about 7 mm. long, 2 mm. wide; mericarps slightly compressed dorsally; primary ridges rather prominent, rather densely and irregularly hispid with short whitish bristles; secondary ridges conspicuously spinose, with a single row of rather stout, straight, glochidiate spines 0·5–1·5 mm. long, bases of spines strongly compressed and confluent, forming a narrow, or rather wide wing; carpophore slender, terete, entire; endosperm flat along its commissural face.

HAB.: Sandy seashores and sand-dunes; near sea-level; fl. April–June.

DISTR.: Divisions 1, 2, 7, 8, rather rare. Also Palestine and Egypt.

1. Ktima, 1862, *Kotschy* 664 ! Khrysokhou, 1862, *Kotschy* 672 !
2. Yialia, 1955, *Merton* 2101b !
7. Shore near Dhavlos, 1880, *Sintenis & Rigo* 344 !
8. Near Rizokarpaso, 1912, *Haradjian* 195 !

NOTES: Often growing with *Pseudorlaya pumilis*, and superficially very much like it, but distinguishable by its more numerous rays, white or whitish petals, and very different fruits.

The plant was first found in Cyprus (1787) by Sibthorp, and subsequently described as

Daucus littoralis in Fl. Graec. Prodr. and Fl. Graec.; unfortunately neither publication gives any detailed information as to source of the type material.

6. D. involucratus *Sibth. et Sm.*, Fl. Graec. Prodr., 184 (1806), Fl. Graec., 3: 65, t. 271 (1819); Poech, Enum. Plant. Ins. Cypr., 28 (1842); Unger et Kotschy, Die Insel Cypern, 305 (1865); Boiss., Fl. Orient., 2: 1075 (1872), Suppl., 272 (1888); Holmboe, Veg. Cypr., 142 (1914); Davis, Fl. Turkey, 4: 534 (1972); Osorio-Tafall et Seraphim, List Vasc. Plants Cyprus, 81 (1973).

Durieua graeca Boiss. in Ann. Sci. Nat., ser. 3, 2: 50 (1844).

TYPE: " In maritimis insulae Cypri ", *Sibthorp* (OXF).

Slender annual; stems (excluding peduncles) usually less than 10 cm. high, unbranched or sparingly branched, subglabrous or thinly clothed with white hairs or bristles, subterete or obscurely sulcate; leaves rather few, narrowly oblong in outline, 1·5–5 (–7) cm. long, 0·7–1·5 (–3) cm. wide, 2–3-pinnate, ultimate segments linear-lanceolate, 2–5 mm. long, 1–1·5 mm. wide, acute, subglabrous or thinly hispidulous; basal leaves with slender, canaliculate petioles 1–8 (–10) cm. long; upper leaves more shortly petiolate or subsessile; leaf-sheaths narrow, inconspicuous, frequently ciliate-hispid with long white hairs; umbels terminal and leaf-opposed, usually 3–5-rayed; peduncles erect, very long and slender, 6–16 cm. long, subglabrous or very sparsely hispidulous, obscurely sulcate; bracts conspicuous, pinnatisect or bi-pinnatisect with linear-subulate, acute lobes, often 2 cm. long and considerably exceeding the rays at anthesis; rays short, slender, usually less than 8 mm. long at anthesis, at first spreading, becoming suberect in fruit; bracteoles setaceous, usually entire, 4–8 mm. long, about 1 mm. wide, herbaceous or membranous, hispidulous or ciliate; pedicels short, usually less than 3 mm. long at anthesis; flowers 8–12 in each umbellule, some of them functionally male and sterile, but none wholly sterile or with modified blackish-crimson petals; calyx minute; petals white, tipped pink or red, not very unequal nor radiant, 0·5–0·8 mm. long, 0·8–1 mm. wide, deeply 2-lobed; filaments 0·5 mm. long; anthers yellow, 0·3 mm. long, 0·2 mm. wide; stylopodium convex; styles very short, seldom more than 0·3 mm. long, stigmas subcapitate. Fruits narrowly oblong-ellipsoid, about 3·5 mm. long, 1–1·5 mm. wide; mericarps somewhat compressed dorsally, at first purplish, becoming pallid when ripe; primary ridges inconspicuous with 2 rows of rather regular, short, spreading bristles; secondary ridges conspicuously spinose, with a single row of slender, spreading, glochidiate spines about 2 mm. long, bases of spines flattened, confluent, forming a narrow wing; carpophore slender, terete, entire; endosperm flattish along the commissural face.

HAB.: In garigue or in Pine forest on dry, rocky hillsides; sometimes in dry pastures near the sea; sea-level to 4,000 ft. alt.; fl. April–July.

DISTR.: Divisions 1–5, 7, 8, locally common. Greece, Crete, Aegean Islands, Turkey.

1. About the ruins of Paphos, 1862, *Kotschy* 657.
2. Prodhromos, 1862, *Kotschy* 907; Phini, 1880, *Sintenis & Rigo* 335 ! and, same locality, 1938, *Kennedy* 1391 ! Platres, 1938, *Kennedy* 1392 ! Mandria, 1938, *Kennedy* 1393 ! Perapedhi, 1938, *Kennedy* 1394 ! and, 1960, *N. Macdonald* 149 ! Yialia, 1955, *Merton* 2151 ! Khionistra, 1957, *Merton* 3142 !
3. Between Yerasa and Kalokhorio, 1962, *Meikle* 2897 !
4. Athna Forest, 1952, *Merton* 820 !
5. Near Alambra, 1967, *Merton* in ARI 609 !
7. Above Kythrea, 1880, *Sintenis & Rigo* 335a ! N.W. of Bellapais, 1952, *Casey* 1262 ! Above Kyrenia, 1953, *Casey* 1313 ! Coast 2 m. W. of Kyrenia, 1956, *Poore* in *Merton* 2695 ! Kambyli, 1957, *Merton* 3150 !
8. Cape Andreas, 1880, *Sintenis & Rigo* 335 !

7. D. durieua *Lange* in Willk. et Lange, Prodr. Fl. Hisp., 3: 23 (1874).

D. subsessilis Boiss., Fl. Orient. Suppl., 272 (1888); Holmboe, Veg. Cypr., 142

(1914); Post, Fl. Pal., ed. 2, 1. 565 (1932); Zohary, Fl. Palaest., 2: 451, t. 653 (1972); Osorio-Tafall et Seraphim, List Vasc. Plants Cyprus, 80 (1973).

TYPE: " dans l'Herbier d'Isnard, croît vraisemblablement en Espagne " (P-JU).

Erect, unbranched or sparingly branched annual 6–16 cm. high; stems shortly hispidulous, subterete or obscurely sulcate, internodes slightly swelling upwards to each node; leaves sparse and rather remote, oblong in outline, 1–2 cm. long, 0·8–1·5 cm. wide, 2–3-pinnate, rather densely hispidulous, the ultimate segments linear-lanceolate, acute, very small, 1–2 (–3) mm. long, 0·5–1 mm. wide; petioles very slender, canaliculate, hispidulous, those of the basal leaves up to 3 cm. long, those of the upper leaves shorter, but generally distinct; leaf-sheaths narrow, hispidulous, with narrowly membranous-ciliate margins; umbels leaf-opposed, sessile, 3–5-rayed; involucral bracts resembling the leaves but smaller, exceeding the umbellules; rays very unequal, sometimes almost wanting, sometimes conspicuous and incrassate, up to 2 cm. long becoming erect in fruit; bracteoles similar to bracts but smaller; flowers few (about 5) in each umbellule; pedicels very unequal, incrassate, up to 5 mm. long; calyx minute; petals small, suborbicular, 0·5 mm. diam., not radiant, tinged red, sparsely hispidulous externally; stamens 0·5 mm. long; anthers reddish, suborbicular, about 0·3 mm. diam.; stylopodium conical; styles very short, less than 0·5 mm. long; stigmas subcapitate. Fruits oblong, about 4 mm. long, 1·5 mm. wide; mericarps compressed dorsally, at first purplish, later pallid, primary ridges with numerous, irregularly spreading, short, whitish bristles; secondary ridges conspicuously spinose with a single row of rather stout, spreading, hispidulous-papillose, glochidiate spines about 2 mm. long, bases of spines flattened, confluent, forming a narrow wing; carpophore slender, terete, entire; endosperm flattish along its commissural face.

HAB.: Dry, stony hillsides, 500–1,000 ft. alt.; fl. March–April.

DISTR.: Division 7, very rare. Also Spain, Portugal, N. Africa, Syria and Palestine.

7. Between Kythrea and Ayios Khrysostomos Mon., 1880, *Sintenis & Rigo* 347 ! Lakovounara, near Koutsovendis, 1974, *Osorio-Tafall & Meikle* in *Meikle* 4064 !

12. ARTEDIA *L.*

Sp. Plant., ed. 1, 242 (1753).
Gen. Plant., ed. 5, 112 (1754).

Annual herb; leaves finely 3-pinnate with very slender setaceous or filiform segments; umbels leaf-opposed and terminal, compound, many-rayed; involucre conspicuous, 1–2-pinnate with long filiform segments; involucel resembling involucre, but smaller; umbellules many-flowered, flowers partly hermaphrodite and fertile, partly male, those at the centre of the umbel wholly sterile and transformed into a penicillate tuft of blackish-violet hairs; calyx obsolete; petals white, conspicuously heteromorphous, those of the outermost flowers of the outer umbellules very much enlarged and radiant, those of the inner flowers subequal, small and incurved-emarginate; stylopodium small, conical; styles rather long, erect or somewhat divergent; stigmas subcapitate. Fruit ovate-elliptic or suborbicular; mericarps strongly compressed dorsally, the commissural face longitudinally ridged and rugose with numerous transverse wrinkles; primary ridges and the 2 dorsal secondary ridges slender, smooth, filiform, the 2 lateral secondary ridges broadly winged, the papery wings divided into 5–6 obovate-spathulate lobes; carpophore very slender, filiform; endosperm flattish along its commissural face; vittae solitary beneath the secondary ridges, very obscure.

One species in western Asia.

1. A. squamata *L.*, Sp. Plant., ed. 1, 242 (1753); DC. Prodr., 4 (1): 209 (1830); Unger et Kotschy, Die Insel Cypern, 304 (1865); Boiss., Fl. Orient., 2: 1070 (1872); Holmboe, Veg. Cypr., 142 (1914); Post, Fl. Pal., ed. 2, 1: 562 (1932); Davis, Fl. Turkey, 4: 536 (1972); Zohary, Fl. Palaest., 2: 448, t. 649 (1972); Osorio-Tafall et Seraphim, List Vasc. Plants Cyprus, 81 (1973).

TYPE: " *in* Libano ".

Erect glabrous annual; stems 20–50 cm. high, usually much branched, sulcate; leaves narrowly oblong towards the base of the stem, becoming broader and more deltoid in outline upwards, 1·5–15 cm. long, 1–4 cm. wide, pinnately divided into numerous slender, filiform segments; lower leaves distinctly petiolate, with petioles 7–8 cm. long, upper leaves sessile or subsessile; leaf-sheaths conspicuous, with broad scarious margins, oblong, truncate-auriculate at apex, 1–3 cm. long, 0·8–1 cm. wide, auricles adnate to the basal pinnae of the upper cauline leaves; umbels very showy, 3–7 cm. diam., flat-topped; peduncles 8–20 (–30) cm. long; involucral bracts spreading at anthesis, strongly reflexed in fruit, exceeding the outermost umbellules, dissected into filiform segments (like those of *Nigella damascena*), with scarious basal appendages like the leaf-sheaths but smaller; radiant petals narrowly obovate, 8–12 mm. long, 3–6 mm. wide, non-radiant petals subequal, broadly obovate-emarginate, about 0·7–1 mm. long, 0·8–1·2 mm. wide; filaments 1·5–2 mm. long; anthers suborbicular, yellow, about 0·4 mm. long and almost as wide; styles about 0·5 mm. long at anthesis, lengthening a little in fruit. Mericarps (including wings) 10–12 mm. long, 8–12 mm. wide, central seed-bearing part 6–8 mm. long, 4–6 mm. wide with regularly undulate margins, wings 10–12-lobed, lobes subtruncate at apex, thickened and often stained reddish at base.

HAB.: Cultivated fields, roadsides, waste ground or dry stony hillsides in degraded garigue; at low altitudes; fl. April–July.

DISTR.: Division 3, very rare or possibly extinct. Turkey, Syria, Palestine east to Iraq and Iran.

3. Kolossi, 1862, *Kotschy* 618a.

NOTES: De Candolle (Prodr. *loc. cit. supra*) records *Artedia squamata* from " Cypro (Sibth.) ", but for no obvious reason, since Sibthorp and Smith make no reference to Cyprus in any of their published records of the plant. Authority for its inclusion in the Cyprus flora rests solely upon Kotschy's record, and, in view of the fact that no collector has since found this beautiful and very obvious flower, one must suspect that it was at most of fleeting occurrence on the island—a surprising situation in view of its abundance on adjacent shores.

13. CORIANDRUM *L.*

Sp. Plant., ed. 1, 256 (1753).
Gen. Plant., ed. 5, 124 (1754).

Erect, glabrous, odoriferous annuals; leaves 1–3-pinnate or pinnatisect; umbels compound, leaf-opposed and terminal; involucre often absent; involucel of a few, entire, subulate or filiform bracteoles; calyx well developed, with unequal, persistent lobes; petals white or tinged pink, unequal, obovate, the outermost much enlarged and radiant; stylopodium conical; styles rather long, divergent; stigmas subcapitate. Fruit globose, mericarps hemispherical; primary ridges obscure, smooth, flexuous; secondary ridges more prominent, straight; carpophore very slender, bipartite; endosperm hemispherical, concave along its commissural face; vittae solitary below the secondary ridges, inconspicuous.

Two species, both probably natives of western Asia; *C. sativum* L., Coriander, is cultivated for use as a condiment almost throughout the world.

1. C. sativum *L.*, Sp. Plant., ed. 1, 256 (1753); Sibth. et Sm., Fl. Graec. Prodr., 1: 196

(1806), Fl. Graec., 3: 76, t. 283 (1819); Unger et Kotschy, Die Insel Cypern, 309 (1865); Boiss., Fl. Orient., 2: 921 (1872); Holmboe, Veg. Cypr., 136 (1914); Post, Fl. Pal., ed. 2, 1: 533 (1932); Lindberg f., Iter Cypr., 25 (1946); Davis, Fl. Turkey, 4: 330 (1972); Zohary, Fl. Palaest., 2: 401, t. 579 (1972); Osorio-Tafall et Seraphim, List Vasc. Plants Cyprus, 76 (1973).

TYPE: " *in* Italiae *agris* ".

Erect, rather slender herb 15–30 (–50) cm. high; stems much branched, distinctly sulcate; leaves heteromorphous, the basal simply pinnate or pinnatisect, oblong in outline, to 10 cm. long and 6 cm. wide, with 2–3 pairs of broadly cuneate-obovate, toothed and lobed segments, and a rather larger, more deeply lobed terminal segment; upper leaves becoming progressively more deeply dissected, the uppermost 3-pinnate, deltoid in outline, 1–3 cm. long and almost as wide, with linear, blunt or subacute segments 2–10 mm. long, 0·3–1·5 mm. wide; basal leaves with slender petioles up to 10 cm. long, upper leaves sessile or subsessile; leaf-sheaths narrow, 1–2 cm. long with rather wide, membranous margins; umbels 3–7-rayed; peduncle 1–5 cm. long; involucre usually absent, sometimes represented by a single, linear-subulate bract about 4 mm. long; rays 5–15 mm. long at anthesis; bracteoles 3–5, subulate or filiform, unequal, 2–5 mm. long; pedicels slender 1–3 mm. long at anthesis, lengthening in fruit; calyx-lobes ovate-deltoid, unequal, the largest about 2 mm. long, the smallest scarcely 0·5 mm.; petals often tinged pink, the outermost conspicuously enlarged and radiant, 2-lobed almost to the base, the lobes up to 6 mm. long and 3 mm. wide; inner petals broadly obovate, rounded or incurved-emarginate at the apex, about 1 mm. long, 0·8 mm. wide; filaments about 1·5 mm. long; anthers yellow, broadly oblong, about 0·7 mm. long, 0·6 mm. wide; styles filiform, about 1·5–2 mm. long. Fruit globose, light brown, about 3–3·5 mm. diam., crowned with the persistent calyx-lobes, unpleasantly foetid when green, sweetly aromatic when ripe and dry; mericarps not separating at maturity except under pressure.

HAB.: Roadsides, waste and cultivated ground; sea-level to 4,000 ft. alt., probably always an escape from cultivation.

DISTR.: Divisions 1–6. Widespread as an escape from cultivation in Europe and Asia, also, less commonly, in Africa and the Americas.

1. Paphos, 1862, *Kotschy*; Ayios Neophytos Monastery, 1939, *Lindberg f.*
2. Platres, 1938, *Kennedy* 1380 !
3. Kalavasos, 1905, *Holmboe.*
4. Larnaca, 1880, *Sintenis & Rigo* 1000 !
5. Kythrea, 1939, *Lindberg f.*
6. Koutraphas, 1936, *Syngrassides* 1016 !

14. BIFORA *Hoffm.*

Gen. Umb., ed. 2, 191 (1816) nom. cons.

Slender, erect, glabrous annuals; leaves 2–3-pinnate or pinnatisect, ultimate segments usually narrow, sometimes filiform; umbels compound, leaf-opposed and terminal; involucre usually absent; involucel of a few, entire, subulate or filiform bracteoles; calyx with distinct or obsolete lobes; petals white, equal or unequal, the outermost sometimes radiant; disk generally inconspicuous; stylopodium shortly and bluntly conical or convex; styles recurved or spreading; stigmas subcapitate. Fruit didymous, mericarps with narrow, excavate commissural faces, subglobose, rugulose, usually without distinct ridges; carpophore bipartite; endosperm almost hemispherical with a deeply concave commissural face.

Three (or four) species in Europe, Asia, N. Africa and N. America.

1. **B. testiculata** (*L.*) *DC.*, Bot. Gall., 218 (1828), Prodr., 4: 249 (1830); Unger et Kotschy, Die Insel Cypern, 309 (1865); Boiss., Fl. Orient., 2: 291 (1872); Holmboe,

Veg. Cypr., 136 (1914); Post, Fl. Pal., ed. 2, 1: 534 (1932); Davis, Fl. Turkey, 4: 332 (1972); Zohary, Fl. Palaest., 402, t. 580 (1972); Osorio-Tafall et Seraphim, List Vasc. Plants Cyprus, 76 (1973).

Coriandrum testiculatum L., Sp. Plant., ed. 1, 256 (1753).

Biforis testiculata Spreng. ex Schultes in Roemer et Schultes, Syst. Veg., 6: XXXVIII & 448 (1820); Roth, Enum. Pl. Phaen., 1: 888 (1827).

TYPE: " *in* Europae *australis agris* ".

Stems angular or prominently ridged, (7–) 15–30 (–50) cm. high, usually much branched; leaves oblong-deltoid in outline, 2–3-pinnate, ultimate segments narrowly oblong or linear, 6–10 mm. long, 0·5–2 mm. wide, acute, flat or with narrowly revolute margins; lower leaves with slender petioles 4–6 or more cm. long; upper leaves sessile or subsessile; leaf-sheaths narrow, 0·5–1·5 mm. long, glabrous with wide membranous margins; umbels leaf opposed and terminal, irregularly (1–) 2–3 (–5)-rayed; peduncle slender, 1–2 cm. long at anthesis, lengthening to 6 cm. or more in fruit; involucre usually absent, sometimes represented by a single subulate or filiform bract about 8 mm. long; rays at first 3–4 mm. long, lengthening a little in fruit; bracteoles filiform, usually less than 5 mm. long, sometimes wanting; pedicels 2–3 mm. long; flowers few (c. 2–6), mostly fertile, sometimes one or two functionally male at the centre of each umbellule; calyx obsolete; petals white, subequal, minute, fugacious, about 1 mm. long, 0·8 mm. wide, apex incurved-emarginate; filaments about 1·2 mm. long; anthers oblong, purple, about 0·4 mm. long, 0·3 mm. wide; stylopodium shortly conical; styles very short, recurved, subcapitate; mericarps 3·5–4 mm. diam., greenish or pale brown, much constricted along the commissure with a prominent, somewhat enlarged stylopodium, separating and falling from the bipartite carpophore as soon as ripe.

HAB.: Cultivated fields and fallow land; sea-level to 3,400 ft. alt.; fl. Febr.–May.

DISTR.: Divisions 2–7. S. Europe, Mediterranean region and eastwards to Caucasus and Iran.

2. Platres, 1937, *Kennedy* 1378 ! Perapedhi, 1938, *Kennedy* 1376 ! Mandria, 1938, *Kennedy* 1377 ! Pomos Point, 1962, *Meikle* 2256 !
3. Alethriko, 1905, *Holmboe* 241; Lefkara, 1967, *Merton* in ARI 637 !
4. Between Larnaca and Cape Pyla, 1880, *Sintenis & Rigo*.
5. Near Pyroi, 1934, *Syngrassides* 465 !
6. Kokkini Tremithia, 1859, *Kotschy*; Myrtou, 1936, *Syngrassides* 1201 ! Ayia Irini, 1941, *Davis* 2540 !
7. Near Ardhana, 1880, *Sintenis & Rigo* 336 ! Kyrenia, 1932, *Syngrassides* 273 ! and, 1949, *Casey* 404 !

15. SCALIGERIA *DC.*

Mém. Fam. Ombellifères, 70, t. 1, B (1829) nom. cons.
Engstrand in Bot. Notiser, 123: 505–511 (1970).

Erect perennial or biennial herbs with thickened tuberous or rhizomatous roots; leaves various, simple, 3-fid or 1–3-pinnate or pinnatisect; umbels terminal, compound, often forming branched inflorescences; bracts and bracteoles present or absent, inconspicuous, linear; flowers mostly hermaphrodite; calyx obsolete; petals white, not radiant; disk obscure or crenate-undulate and conspicuous; stylopodium flattish, convex or conical; styles sometimes very long and strongly reflexed after anthesis; stigmas capitate. Fruit ovoid or subglobose, smooth, glabrous, sometimes distinctly didymous, mericarps rounded dorsally; ridges slender, often indistinct; endocarp flat or concave along its commissural face; vittae conspicuous or slender and interrupted.

About 14 species in the Mediterranean area and western Asia.

1. **S. cretica** (*Mill.*) *Boiss.*, Diagn., 1, 10: 52 (1849); Unger et Kotschy, Die Insel

Cypern, 309 (1865); Boiss., Fl. Orient., 2: 875 (1872), Suppl., 255 (1888); Holmboe, Veg. Cypr., 136 (1914); Post, Fl. Pal., ed. 2, 1: 521 (1932); Lindberg f., Iter Cypr., 25 (1946); Rechinger f. in Arkiv för Bot., ser. 2, 1: 428 (1950); Tutin in Fl. Europ., 2: 328 (1968).

Bunium creticum Mill., Gard. Dict., ed. 8, no. 2 (1768); Urv. in Mém. Soc. Linn. Par., 1: 287 (1822).

B. napiforme Willd. ex Spreng., Spec. Umb., 95 (1818).

Scaligeria napiformis (Willd. ex Spreng.) Grande in Bol. Orto Bot. Nap., 4: 188 (1914); Greuter in Boissiera, 13: 94 (1967); Engstrand in Bot. Notiser 123: 506 (1970); Davis, Fl. Turkey, 4: 334 (1972); Zohary, Fl. Palaest., 2: 428, t. 618 (1972); Osorio-Tafall et Seraphim, List Vasc. Plants Cyprus, 76 (1973).

TYPE: Crete; *Tournefort* (P).

Erect perennial; root tuberous, fusiform or occasionally globose, to 3 cm. diam.; stems bright green, glabrous, very obscurely sulcate, 30–70 (–100) cm. high, generally unbranched in the lower part, much branched above; leaves glabrous, deltoid in outline, 6–15 cm. long, 4–10 cm. wide, bipinnate; ultimate segments of the basal and lower cauline leaves ovate or rhomboid, 0·4–2·5 cm. long, 0·3–1·5 cm. wide, acute with sharply serrate or irregularly lobulate margins; upper cauline leaves smaller with much narrower, elongate, often linear and subentire segments, the uppermost reduced to acuminate sheaths; petioles of the basal leaves up to 15 cm. long, slender, canaliculate above; leaf-sheaths narrow, 3–4 cm. long, with very narrow membranous margins and small membranous apical auricles; inflorescence repeatedly bifurcate; peduncles 4–7 cm. long; involucre wanting or occasionally reduced to a single subulate bract less than 1 cm. long; umbels lax, 5–12-rayed; rays very slender, spreading, 2–5 cm. long; bracteoles 1–3 (–5), sometimes absent, inconspicuous, subulate, 3–4 mm. long, 1–2 mm. wide, with narrow membranous margins; pedicels up to 8 mm. long at anthesis; flowers (5–) 10–15 (–20) in each umbellule; calyx obsolete; petals subequal, broadly oblong-obovate, about 1·3 mm. long and almost as wide, apex shallowly incurved-emarginate; filaments about 1·5 mm. long; anthers oblong, about 0·8 mm. long, 0·5 mm. wide, yellow; stylopodium conspicuous, conical; styles about 0·8 mm. long and erect at anthesis, lengthening to 1·2 mm. or more in fruit and becoming sharply reflexed; stigmas capitate. Fruit broadly ovoid-subglobose, about 2 mm. long and about as wide or a little wider, a little compressed laterally, distinctly didymous, base distinctly cordate; mericarps blackish or dark brown when ripe, rugulose, glabrous; ridges very obscure except near the apex of the mericarp; carpophore bipartite; endosperm concave along its commissural face.

HAB.: In Pine forest, or amongst shrubs on dry, rocky, igneous or calcareous hillsides; near sea-level to 3,000 ft. alt.; fl. April–June.

DISTR.: Divisions 1–3, 7, 8. Yugoslavia, Greece, Crete, Aegean region, W. Turkey, Lebanon, Palestine and Libya.

1. Stroumbi, 1913, *Haradjian* 750; near Lyso, 1913, *Haradjian* 852 ! Akamas, 1948, *Kennedy* 1643 ! Ayios Nikolaos, Akamas, 1957, *Merton* 3046 ! 3054 !
2. Roudhkias valley between Pano Panayia and Kykko, 1941, *Davis* 3386 ! Ayii Anargyri near Phini, 1962, *Meikle* 2872 !
3. Khoulou, 1905, *Holmboe* 754; Yerasa, 1941, *Davis* 3075 ! Lefkara, 1967, *Merton* in ARI 648 !
7. Frequent; Paleomylos near Panteleimon Mon., Akanthou, Pentadaktylos, Halevga, etc.; *Kotschy* 935; *Sintenis & Rigo* 326 ! *Haradjian* 390 ! *Davis* 1728 ! *Casey* 1046 !
8. Near Rizokarpaso, 1912, *Haradjian* 178, 179; Ronnas bay, 1941, *Davis* 2358 ! Eleousa Forest, 1957, *Merton* 2881 !

NOTES: It has been argued (Greuter, 1967, *loc. cit.*) that the correct name for this species is *Scaligeria napiformis* (Willd. ex Spreng.) Grande (1914), on the grounds that the older combination *S. cretica* (Urv.) Boiss. (1849) is inadmissible under Art. 72 of the International Code, being based upon *Bunium creticum* Urv. (1822), an illegitimate later homonym of *B. creticum* Mill. (1768). This argument is tenable so long as one regards *B. creticum* Mill. and *B. creticum* Urv. as having different types. In my opinion, however, all the names concerned, viz. *Bunium creticum* Mill., *B. napiforme* Willd. ex Spreng., *B. creticum* Urv., *Scaligeria tournefortii* Boiss., *S. cretica* Boiss. and *S. napiformis* (Willd. ex Spreng.) Grande are based upon *Bulbocastanum Creticum, radice napiformi* of Tournefort, and share the same type. In the circumstances, I regard the name *Bunium napiforme* Willd. ex Spreng. as illegitimate and superfluous under Art. 63 of the Code, since it includes the type of *B. creticum* Mill. Boissier's *Scaligeria tournefortii* is likewise inadmissible since the epithet *creticum* was available at the date (1844) of publication. To conclude, one may either regard the combination *Scaligeria cretica* as technically inadmissible, because it would *appear* to have been based on the illegitimate *Bunium creticum* Urv., or, as I prefer, allow that Boissier *may* have been aware of *Bunium creticum* Mill. by giving the plant the name and author citation adopted here. This seems the most sensible solution to the problem; the alternative is to regard *Scaligeria cretica* as a new name, to be attributed solely to Boissier, and to date from 1849.

16. SMYRNIUM *L.*

Sp. Plant., ed. 1, 262 (1753).
Gen. Plant., ed. 5, 127 (1754).

Erect, glabrous, biennial or perennial herbs with thick taproots; stems often robust and rather succulent; leaves coarsely ternate-pinnate, or the upper sometimes undivided; inflorescences branched; umbels compound, without (or with very reduced) bracts and bracteoles; flowers hermaphrodite or polygamous; calyx obsolete; petals yellow, shortly inflexed-emarginate; stylopodium conical or flattened. Fruit large, often distinctly didymous; mericarps ovoid, laterally compressed, blackish or dark brown and rather coriaceous when ripe, with 3 (or rarely 5) distinct dorsal ridges; endosperm excavate with revolute margins along its commissural face; vittae numerous in the valleculae and on the commissural faces.

About 8 species in Europe, the Mediterranean region and western Asia.

Upper stem leaves not connate or amplexicaul, usually ternate; mericarps with prominent dorsal ridges - - - - - - - - **1. S. olusatrum**
Upper stem leaves connate or amplexicaul, usually undivided and conspicuously different from the basal leaves; mericarps obscurely ridged:
Upper stem leaves alternate, amplexicaul with deeply cordate bases **2. S. rotundifolium**
Upper stem leaves opposite, connate along their basal margins - **3. S. connatum**

1. S. olusatrum *L.*, Sp. Plant., ed. 1, 262 (1753); Sibth. et Sm., Fl. Graec. Prodr., 1: 203 (1806); Unger et Kotschy, Die Insel Cypern, 308 (1865); Boiss., Fl. Orient., 2: 927 (1872); Holmboe, Veg. Cypr., 138 (1914); Post, Fl. Pal., ed. 2, 1: 536 (1932); Lindberg f., Iter Cypr., 25 (1946); Davis, Fl. Turkey, 4: 337 (1972); Zohary, Fl. Palaest., 2: 404, t. 582 (1972); Osorio-Tafall et Seraphim, List Vasc. Plants Cyprus, 76 (1973).
TYPE: " *in* Scotia, Wallia, Gallia ".

Robust erect perennial 0·5–1·7 m. high; stems glabrous, rather fleshy, fistulose, conspicuously ribbed, usually much branched; basal leaves broadly deltoid in outline, 25–50 cm. long and about as wide, ternately pinnate or sub-bipinnate, segments broadly oblong-obovate, 3–8 cm. long, 1·5–5 cm. wide, glossy dark green, apex acute or obtuse, base cuneate, margins serrate, or with serrate lobes, petioles 15–20 cm. long, canaliculate, fleshy, expanding at the base into a long narrow sheath; upper leaves alternate or the uppermost opposite, ternate with undivided or deeply 3-fid segments; petiole very short or wanting; basal sheath conspicuous, papyraceous, prominently nerved, glabrous except at the broad, auriculate, lanuginose apex; inflorescence branched; umbels hemispherical or strongly convex, (5–) 10–20 (–30)-rayed; peduncles rather short and stout, strongly ribbed, usually less than 10 cm. long at anthesis, lengthening considerably in fruit; involucre absent, or reduced to a few, short filiform bracts; rays 2–4 cm. long, sparingly lanuginose at the base; bracteoles 2–4, minute, deltoid, membranous, about

0·5 mm. long; flowers numerous, crowded, predominantly hermaphrodite and fertile in each umbellule; pedicels 2–4 mm. long at anthesis, minutely scabridulous; petals subequal, oblong-obovate, 1·5–2 mm. long, 1–1·5 mm. wide; filaments 2–2·5 mm. long; anthers suborbicular, about 0·5 mm. diam., greenish; stylopodium at first flattish, later strongly convex or conical; styles at first erect, less than 1 mm. long, lengthening to 2 mm. in fruit; stigmas truncate or slightly swollen. Fruit broadly ovate or suborbicular, about 1 cm. diam., strongly didymous, black and coriaceous when ripe; mericarps separating at maturity, more or less prismatic with a very narrow commissure; dorsal ridges 3, very prominent; valleculae rugulose; carpophore bipartite; endosperm very deeply excavate with revolute commissural margins.

HAB.: Moist shady places under trees or by irrigation channels; near sea-level to 2,000 ft. alt.; fl. Febr.–April.

DISTR.: Divisions 2–8, local. S. Europe (W. Europe, introduced), Mediterranean region, Atlantic Islands.

2. Makheras Mon., 1862, *Kotschy*; Lefka, 1932, *Syngrassides* 250 ! Perapedhi, 1932, *Nattrass* 179.
3. Mazotos, 1862, *Kotschy* 546 !
4. Famagusta, 1905, *Holmboe* 432 ! Kouklia, 1967, *Merton* in ARI 450 !
5. Kythrea, 1880, *Sintenis & Rigo*; Dhali, 1937, *Syngrassides* 1518 !
6. Ayia Irini, 1941, *Davis* 2592 !
7. St. Hilarion, 1901, *A. G. & M. E. Lascelles*, also *Lindberg f.*; *Davis* 3599 ! *Casey* 1191 ! Templos, 1956, *G. E. Atherton* 1112 ! Paleosophos, 1968, *Economides* in ARI 1065 !
8. Panaia tou Kantara, 1880, *Sintenis & Rigo*; Komi Kebir, 1936, *Syngrassides* 1063 !

2. S. rotundifolium *Mill.*, Gard. Dict., ed. 8, no. 2 (1768); Boiss., Fl. Orient., 2: 925 (1872); Davis, Fl. Turkey, 4: 338 (1972).

S. perfoliatum (non L.) Sibth. et Sm., Fl. Graec. Prodr., 1: 202 (1806), Fl. Graec., 3: 82 quoad plant. Cypr.; Unger et Kotschy, Die Insel Cypern, 308 (1865); Boiss., Fl. Orient., 2: 925 (1872) quoad plant. Cypr.; Holmboe, Veg. Cypr., 137 (1914); Osorio-Tafall et Seraphim, List Vasc. Plants Cyprus, 76 (1973).

TYPE: Cultivated in Chelsea Physic Garden, origin Sicily or Crete (BM).

Erect biennial 30–100 cm. high with a short, thick, fusiform taproot; stems generally unbranched, glabrous, sulcate; basal leaves deltoid in outline, ternate-bipinnate, 5–20 cm. long and about as wide, ultimate segments small, obovate-rhomboid, 0·5–2 cm. long, 0·4–1·5 cm. wide, bright shining green, apex acute or obtuse, base cuneate, margins irregularly toothed or shortly lobed; petioles up to 10 cm. long, slender, canaliculate, basal sheath conspicuous, papyraceous, broadly oblong-auriculate, 1–2·5 cm. long, prominently nerved, glabrous or with a small lanuginose tuft at the apex; lower stem leaves similar to basal, but with fewer, larger segments; upper stem leaves sessile, amplexicaul, bluntly orbicular-cordate or with 3 shallow, rounded lobes, 2–8 cm. long, 1·5–7 cm. wide, margins entire or very faintly crenulate; inflorescence branched; umbels flattish or convex, 7–12-rayed; peduncles short, seldom exceeding 2 cm. at anthesis; bracts and bracteoles absent; rays 0·8–2·5 cm. long at anthesis, lengthening a little in fruit; flowers numerous and crowded in each umbellule, mostly fertile and hermaphrodite; petals yellow, obovate, about 1·5 mm. long, 1 mm. wide; filaments about 2 mm. long; anthers oblong, about 0·8 mm. long, 0·5 mm. wide; stylopodium at first flattish, becoming conical in fruit; styles erect at anthesis, about 0·4 mm. long, lengthening to 1·5 mm. in fruit and becoming sharply reflexed; stigmas truncate or slightly swollen. Fruit about 2 mm. long, 3·5 mm. wide, conspicuously didymous; mericarps broadly ovoid or subglobose, blackish-brown and rugulose when ripe, dorsal ridges obsolete or obscure except at apex of mericarp; endosperm strongly excavate along its commissural face.

HAB.: Moist shaded ground in forests and thickets; 3,000–4,500 ft. alt.; fl. May–June.

DISTR.: Division 2, rare. Corsica, Sardinia, S. Italy, Sicily, Greece, Crete, Aegean Islands, W. Turkey.

2. Near Troödhitissa Mon., 1787, *Sibthorp*; Prodhromos, 1862, *Kotschy* 731 ! Askas, in Hazel groves, 1957, *Merton* 3121 ! By Yironas R. below Madhari, 1962, *Meikle* 2833 !

NOTES: Much confused with *S. perfoliatum* L., a plant with a more northern distribution, differing in its angular, strongly ribbed or winged stems and ovate-amplexicaul, usually acute and distinctly crenate-dentate stem-leaves.

3. S. connatum *Boiss. et Kotschy* in Unger et Kotschy, Die Insel Cypern, 309 (1865); Boiss. Fl. Orient., 2: 926 (1872); Holmboe, Veg. Cypr., 137 (1914); Post, Fl. Pal., ed. 2, 1: 536 (1932); Davis, Fl. Turkey, 4: 338 (1972); Zohary, Fl. Palaest., 2: 404, t. 583 (1972); Osorio-Tafall et Seraphim, List Vasc. Plants Cyprus, 76 (1973).
S. cognatum Kotschy in Oesterr. Bot. Zeitschr., 12: 276 (1862) nomen.

TYPE: Turkey; "in monte Lokmanni "Dschebbel Nur" lectae", 1859, *Kotschy* 19 (K !).

Robust erect biennial or perennial, 0·8–1·5 m. high; taproot thick, elongate-fusiform or branched; stems usually unbranched in the lower part, glabrous, strongly ribbed or angular, fistulose; basal leaves deltoid in outline, 14–20 cm. long, 8–16 cm. wide, ternate-bipinnate, shining dark green; ultimate segments ovate-oblong or obovate, 1–3·5 cm. long, 0·8–2·5 cm. wide, apex acute or obtuse, base truncate, cuneate or rounded, margins crenate-dentate or serrate, sometimes irregularly lobed; petioles up to 15 cm. long, glabrous, canaliculate; basal sheaths conspicuous, papery or membranous; lower stem leaves alternate, similar to basal, but with fewer, larger (up to 5 cm. long, 4·5 cm. wide) segments; petiole short; basal sheaths often very large, auriculate, up to 10 cm. long, 8 cm. wide, yellowish-green, distinctly veined; upper stem leaves opposite, obtuse, connate along their basal margins, forming a shallow, elliptic-suborbicular, prominently veined, vivid yellow-green cup around the stem, 4–15 cm. long, 2–10 cm. wide, margins subentire or obscurely crenulate-dentate; inflorescence much branched; umbels flattish or convex, 10–20-rayed; peduncles short, usually less than 3 cm. long at anthesis; bracts and bracteoles absent; rays spreading, angular, 1–3 cm. long at anthesis, flowers numerous, crowded, mostly fertile and hermaphrodite; pedicels 1–3 mm. long; petals yellow, obovate, about 1 mm. long, 0·8 mm. wide; filaments 2–3 mm. long; anthers oblong, greenish, about 0·8 mm. long, 0·5 mm. wide; stylopodium flattish at first, becoming convex or shortly conical in fruit; styles suberect, about 0·8 mm. long at anthesis, lengthening to 1·5 mm. and becoming strongly reflexed in fruit; stigmas clavate or subcapitate. Fruit 3 mm. long, 4·5 mm. wide, conspicuously didymous, blackish when ripe; mericarps subglobose, rugulose, indistinctly ridged; endosperm very strongly excavate.

HAB.: Rocky mountainsides on igneous and calcareous formations, 1,200–5,700 ft. alt.; fl. April–May.

DISTR.: Divisions 2, 7, not common. Turkey, Syria, Palestine.

2. Madhari (Adelphi), 1880, *Sintenis & Rigo* 835 ! Kykko, 1905, *Holmboe* 1005 ! and, 1937, *Syngrassides* 1578 ! Papoutsa, 1941, *Davis* 3066 ! Stavros tis Psokas, 1951, *Merton* 1188 ! Khionistra, 1956, *Kennedy* 1878 ! Lazania, 1967, *Merton* in ARI 266 !
7. Buffavento, 1862, *Kotschy* 344 ! also, same locality, 1880, *Sintenis & Rigo* 370 ! Kyrenia, 1949, *Casey* 659 !

17. PHYSOSPERMUM *Cusson ex Jussieu*

Mém. Soc. Roy. Méd. Par., 5: 279 (1787).

Perennial herbs with thick woody rootstocks; basal leaves long-petiolate, ternate-pinnate; stem leaves absent or very reduced and bract-like;

inflorescences branched; bracts and bracteoles present, small, entire; umbels compound; sepals small but distinct; petals white, subequal, not radiant; stylopodium ultimately conical. Fruits didymous, blackish, shining and rather fleshy when ripe; mericarps subglobose or broadly ovoid, laterally somewhat compressed with a very narrow commissure, dorsal ridges obscure, filiform; carpophore bipartite; valleculae 1-vittate, commissure 2-vittate; endosperm conduplicate.

Two species in Europe, Mediterranean region and W. Asia.

1. **P. cornubiense** (*L.*) *DC.*, Prodr., 4: 246 (1830); Post, Fl. Pal., ed. 2, 1: 535 (1932); Davis, Fl. Turkey, 4: 378 (1972); Osorio-Tafall et Seraphim, List Vasc. Plants Cyprus, 77 (1973).

Ligusticum cornubiense L., Cent. II Plant., 13 (1756).
Coriandrum aquilegifolium All., Misc. Taur., 5: 81 (1774).
Danaä aquilegifolia (All.) All., Fl. Ped., 2: 34 (1785).
Physospermum aquilegifolium (All.) Koch in Nov. Act. Nat. Cur., 12 (1): 134 (1824); Unger et Kotschy, Die Insel Cypern, 308 (1865); Boiss., Fl. Orient., 2: 923 (1872); H. Stuart Thompson in Journ. Bot., 44: 308 (1906); Holmboe, Veg. Cypr., 138 (1914).

TYPE: British Isles; "*in* Cornubia".

Erect perennial 0·8–1·5 m. high; stems usually unbranched except near apex, glabrous, shallowly sulcate; basal leaves broadly deltoid in outline, 15–20 cm. long, as wide or rather wider, ternate-pinnate, the ultimate segments deeply and acutely pinnatisect, broadly ovate, 1·5–6 cm. long, 1–5 cm. wide, glabrous above and below, often rather distinctly reticulate-veined with narrowly revolute, scabridulous margins; petioles 17–25 cm. or more long, glabrous, canaliculate above; stem leaves usually absent or reduced to short lanceolate-subulate sheaths 1–3 cm. long, occasionally developed and resembling the basal leaves, but smaller; inflorescences lax, branched; peduncles 6–10 cm. or more long; umbels 10–15-rayed; bracts 3–6, narrowly oblong-lanceolate, 5–6 mm. long, 1·5–2 mm. wide; rays unequal, 1–6 cm. long at anthesis, slender, spreading or suberect; bracteoles 2–5, oblong-linear, 2–3 mm. long, about 1 mm. wide; flowers 10–20 in each umbellule, about half of them hermaphrodite and fertile, the other half functionally male and sterile; pedicels filiform, 4–8 mm. long; sepals bluntly deltoid, about 0·5 mm. long and as wide or rather wider; petals oblong-obovate, about 1·5 mm. long, 1 mm. wide, apex strongly incurved-emarginate; filaments about 1·5 mm. long; anthers oblong, yellow, about 0·8 mm. long, 0·6 mm. wide; stylopodium at first convex, later becoming shortly conical; styles up to 2 mm. long at anthesis, at first erect, becoming strongly reflexed in fruit; stigmas capitate. Fruits conspicuously didymous, about 4 mm. long, 5 mm. wide; mericarps separating readily at maturity.

HAB.: In woodland or on shaded hillsides under trees or bushes, about 1,000 ft. alt.; fl. June–July.

DISTR.: Division 7, very rare. Widespread in S. Europe and eastwards to the Caucasus and N. Iran.

7. Near Paleomylos, not far from Panteleimon Mon., 1862, *Kotschy* 935.

NOTES: Only one record, and the specimen has not been examined, though, as the species was known to Kotschy, one may reasonably assume that his identification is correct. *Physospermum cornubiense* is found in areas of Turkey and Syria adjacent to Cyprus.

18. CONIUM *L.*

Sp. Plant., ed. 1, 243 (1753).
Gen. Plant., ed. 5, 114 (1754).

Erect, glabrous biennial herbs; stems fistulose, often spotted purple, usually much branched; leaves large, 2–3-pinnate; ultimate segments small;

inflorescences branched; umbels compound; bracts and bracteoles small; calyx obsolete; petals white, subequal, not radiant, apex usually incurved-emarginate; stylopodium flattish or convex; styles rather short, becoming recurved, not very persistent. Fruit broadly ovoid or suborbicular, laterally compressed; carpophore bipartite; mericarps separating readily at maturity, conspicuously 5-ribbed with a narrow commissure; endosperm conduplicate along its commissural face; vittae obsolete.

Two species, one in Europe and Asia (and introduced elsewhere), the other in South Africa.

1. C. maculatum *L.*, Sp. Plant., ed. 1, 243 (1753); Unger et Kotschy, Die Insel Cypern, 308 (1865); Boiss., Fl. Orient., 2: 922 (1872); Holmboe, Veg. Cypr., 138 (1914); Post, Fl. Pal., ed. 2, 1: 534 (1932); Lindberg f., Iter Cypr., 25 (1946); Davis, Fl. Turkey, 4: 380 (1972); Zohary, Fl. Palaest., 2: 405, t. 585 (1972); Osorio-Tafall et Seraphim, List Vasc. Plants Cyprus, 77 (1973).

TYPE: " *in* Europae *cultis, agris, ruderatis* ".

A robust, erect, much-branched, glabrous biennial 1–2·5 m. high with a strong, unpleasant mouse-like odour when crushed; lower parts of stems usually spotted or blotched purple, shallowly sulcate; lower cauline leaves broadly deltoid in outline, finely 3-pinnate, often 30–50 cm. long and as wide or somewhat wider; ultimate segments cuneate-oblanceolate or narrowly obovate, 0·5–1·5 cm. long, 0·2–0·7 cm. wide, acute, sharply and rather deeply serrate or lobed with inconspicuous venation; petiole 13–20 cm. long, expanding at the base into a narrow, close-fitting sheath; upper leaves similar but much smaller, sessile or subsessile; inflorescence large, much branched; peduncles rather short and slender generally less than 8 cm. long; umbels 10–20-rayed, flattish or convex; bracts narrowly ovate-acuminate, entire, 2–3 mm. long, 1–1·5 mm. wide, margins whitish, membranous; rays unequal, spreading or suberect, 1–2·5 cm. long at anthesis; flowers about 20 in each umbellule, crowded, almost all hermaphrodite and fertile, bracteoles resembling bracts but a little smaller; pedicels usually less than 3 mm. long at anthesis; petals broadly obovate, emarginate, about 1–1·5 mm. long and as wide or a little wider; filaments about 1·5 mm. long; anthers ovoid-oblong, yellow, about 0·8 mm. long, 0·6 mm. wide; styles about 0·5 mm. long; stigmas subcapitate. Mericarps greyish-brown when ripe, with 5 conspicuous pallid ribs, about 3·5 mm. long, 2·5 mm. wide, commissural face narrow, dorsal surface broad, rounded.

HAB.: Roadsides, waste ground, gardens etc., sometimes on rocky hillsides in the open or under Pines; 1,000–4,500 ft. alt.; fl. May–June.

DISTR.: Division 2, locally abundant. Widespread in Europe and western Asia.

2. Near Khrysorroyiatissa Mon., 1862, *Kotschy* 695; Evrykhou, 1880, *Sintenis & Rigo* 841 ! Kannaviou, 1905, *Holmboe* 763; Kykko Mon., 1913, *Haradjian* 959 ! and, 1937, *Syngrassides* 1604 !

NOTES: Common according to Holmboe, but, if so, rarely collected, and then only from one area of the island. It may be that familiarity with the plant elsewhere has caused collectors to ignore it in Cyprus.

19. LECOKIA *DC.*

Mém. Fam. Ombellifères, 67 (1829).

Perennial herb with a thick, fleshy rhizome; leaves large, coarsely ternate-pinnate or ternate-bipinnate with broad oblong-elliptic, serrate segments, the basal long-petiolate, the cauline subsessile, frequently opposite or sub-opposite; inflorescences branched; umbels compound; bracts and bracteoles small, linear, inconspicuous or sometimes absent; flowers hermaphrodite or functionally male and sterile; calyx obsolete, or with a few, very reduced, blunt lobes; petals white, subequal, not radiant, incurved-emarginate at the apex; stylopodium conical; styles long; stigmas subcapitate. Fruit large, oblong, densely covered with hooked bristles; mericarps separating readily at maturity, somewhat compressed laterally, with a narrow commissure, dorsal surface rounded, with deep, narrow valleculae and 5 broad, rounded ridges; endosperm deeply excavate; vittae conspicuous, numerous.

One species ranging from Crete eastwards to the Caucasus and Iran.

1. L. cretica (*Lam.*) *DC.*, Mém. Fam. Ombellifères, 67, t. II, L (1829); Poech, Enum. Plant. Ins. Cypr., 29 (1842); Kotschy in Oesterr. Bot. Zeitschr., 12: 276 (1862); Unger et Kotschy, Die Insel Cypern, 307 (1865); Boiss., Fl. Orient., 2: 931 (1872); Holmboe, Veg. Cypr., 138 (1914); Post, Fl. Pal., ed. 2, 1: 537 (1932); Lindberg f., Iter Cypr., 25 (1946); Davis, Fl. Turkey, 4: 381 (1972); Zohary, Fl. Palaest., 2: 406, t. 586 (1972); Osorio-Tafall et Seraphim, List Vasc. Plants Cyprus, 77 (1973).

Cachrys cretica Lam., Encycl. Méth., 1: 259 (1783).

Scandix latifolia Sibth. et Sm., Fl. Graec. Prodr., 1: 197 (1806), Fl. Graec., 3: 77, t. 284 (1819).

TYPE: Crete; *Tournefort, in herb. Jussieu* (P).

Erect, glabrous herb 40–100 cm. high; stems robust, often streaked with purple, conspicuously ridged; basal leaves broadly deltoid in outline, 15–20 cm. long, 20–30 cm. wide; ultimate segments oblong-elliptic, 3–8 cm. long, 1–4 cm. wide, subglabrous, shining green, apex acute, base cuneate, margins distinctly and somewhat irregularly serrate; petiole 8–14 cm. long; basal sheath 5–8 cm. long, 1·5–2 cm. wide, prominently nerved, auriculate with conspicuous membranous or papery auricles; stem leaves smaller, sessile or subsessile, often opposite or in irregular whorls with conspicuous, auriculate basal sheaths; inflorescence branches rather rigidly erect; peduncles sulcate, up to 18 cm. long at anthesis, scabrid-papillose at apex; umbels convex or flattish, often rather congested, 6–15-rayed; bracts few or sometimes absent, linear-subulate, about 5–10 mm. long, 0·5–1 mm. wide; rays unequal, 1–4 cm. long at anthesis, lengthening greatly (to 9 cm. or more) in fruit, usually scabrid-papillose at the apex; bracteoles similar to bracts but rather smaller; pedicels 2–8 mm. long, lengthening to 1–3 cm. in fruit; flowers of the central (or main) umbel mostly hermaphrodite and fertile, those of the lateral (or subordinate) umbels usually male-functional and sterile; petals obovate-emarginate, about 2 mm. long, 1–1·3 mm. wide; filaments 3·5 mm. long; anthers oblong, about 0·7 mm. long, 0·5 mm. wide, yellow; styles 2·5–3 mm. long, erect or slightly recurved, persistent. Ripe fruit dark brown, conspicuously scabrid, about 1·3–1·5 cm. long and up to 1 cm. wide, crowned with the conical stylopodium and persistent styles.

HAB.: Rocky ground and hillsides, commonly in shade and by streams or rivulets; 600–6,000 ft. alt.; fl. Jan.–April.

DISTR.: Divisions 1, 2, 7, 8, locally common. Distribution that of the genus.

1. Ayios Minas near Smyies, 1962, *Meikle* 2168 !
2. Frequent; Prodhromos, Khionistra, Kionia, Tripylos, etc.; *Kotschy*; *Kennedy* 666 ! 667 ! *Lindberg f.* ! *Davis* 2697 ! 3462 ! etc.
7. Frequent about Kyrenia, Bellapais, Kazaphani, Vasilia etc.; *Kotschy* 732 ! *Syngrassides* 1449 ! *Davis* 2092 ! *Casey* 1134 ! *G. E. Atherton* 854 ! etc.
8. Near Kantara, 1880, *Sintenis & Rigo* 341 ! and, 1941, *Davis* 2435 !

20. CACHRYS *L.*

Sp. Plant., ed. 1, 246 (1753).
Gen. Plant., ed. 5, 117 (1754).
Hippomarathrum Hoffmsgg. et Link, Fl. Port., 2: 411 (1820) non Gaertn., Mey. et Scherb. (1799).

Robust perennial herbs; stems much branched; leaves 3–4-pinnate, ultimate segments filiform or subulate; inflorescence much branched; bracts and bracteoles present, usually simple, linear or subulate; umbels compound; calyx with distinct lobes; petals greenish-yellow, incurved-emarginate; disk conspicuous, crenellate-crispate; stylopodium convex or shortly conical; styles usually long; stigmas capitate. Fruit usually rather large, oblong or subglobose, distinctly didymous; mericarps somewhat compressed laterally, densely verruculose-crustaceous with broad, rounded, conspicuous or obscure dorsal ridges; carpophore bipartite; endosperm deeply excavate along its commissural face; vittae numerous, slender.

Seven or eight species in S. Europe, Mediterranean region and western Asia.

Leaves finely dissected, ultimate segments very numerous 2–50 mm. long; fruit suborbicular, about 5 mm. diam. - - - - - - - **1. C. scabra**

Leaves coarsely dissected, ultimate segments few, thick, rigid, 15–80 mm. long; fruit broadly oblong, about 10 mm. long, 9 mm. wide - - - - **2. C. crassiloba**

1. C. scabra (*Fenzl*) *Meikle* **comb. nov.**

Ferula scabra Fenzl in Flora, 26: 461 (1843).

Hippomarathrum scabrum (Fenzl) Boiss., Fl. Orient., 2: 933 (1872); Davis, Fl. Turkey, 4: 392 (1972).

[*Cachrys sicula* (non L.) Sibth. et Sm., Fl. Graec. Prodr., 1: 190 (1806), Fl. Graec., 3: 71 (1819) pro parte quoad spec. Cypr.]

[*Cachrys pterochlaena* (non DC.) Poech, Enum. Plant. Ins. Cypr., 29 (1842); Kotschy in Unger et Kotschy, Die Insel Cypern, 307 (1865).]

[*Hippomarathrum cristatum* (non *Cachrys cristata* DC.) Boiss., Fl. Orient., 2: 932 (1872) pro parte quoad spec. Cypr.]

[*H. boissieri* Reuter et Hausskn. ex Boiss., Fl. Orient., 2: 933 (1872) pro parte excl. spec. Joppensi; Boiss., Fl. Orient. Suppl., 260 (1888) pro parte quoad spec. Cypr.; Holmboe, Veg. Cypr., 138 (1914); Lindberg f., Iter Cypr., 25 (1946).]

TYPE: Turkey; " in Mesopotamia inter Diarbekir et Mardin—*Kotschy* " (K !).

Stout erect perennial to 1·5 m. high; stems pleasantly aromatic when crushed, usually much branched above, sulcate, glabrous or ± densely scabridulous-pubescent, sometimes dotted over with small, round papillae-clusters; basal leaves large, forming a loose rosette, withering at or before anthesis, broadly oblong-deltoid in outline, 3–4-pinnate, up to 50 cm. long and 30 cm. wide, the primary and secondary divisions sulcate-canaliculate; ultimate segments variable, linear-filiform or subulate, 2–50 mm. long, 0·2–2 mm. wide, acute, flattish or subterete, glabrous or scabridulous or with scabridulous margins; petiole 15–20 cm. long, 1–2·5 cm. wide, flattened, sheathing at the base; cauline leaves much smaller, sessile, deltoid, with fewer, often longer and coarser ultimate segments; inflorescence repeatedly branched with rather rigid, erecto-patent branches often arising in irregular whorls; umbels flattish or convex, 2–12-rayed; peduncles unequal 1–10 (–15) cm. long; bracts linear or subulate, erect or somewhat recurved, 5–25 mm. long, 0·5–1·5 mm. wide; rays unequal, 1–7 cm. long; bracteoles similar to bracts but usually smaller; flowers mostly hermaphrodite and fertile; calyx lobes distinct, broadly deltoid, acute or obtuse, about 0·5 mm. long and almost as wide at the base, commonly papillose-pubescent; petals obovate, greenish-yellow, 1–1·5 mm. long, 0·8–1 mm. wide, strongly incurved, often pubescent externally; filaments 1·8–2 mm. long; anthers yellow, broadly oblong, about 0·8 mm. long and almost as wide; styles 1·5–2 mm. long, at first erect, later reflexed; stigmas capitate. Fruit suborbicular, about 5 mm. diam., pale brown when ripe; dorsal surface of mericarps rounded, with obscure or obsolete ridges, densely verruculose or verruculose-scabridulous.

HAB.: Sandy banks and field borders, or shingle banks, generally near the sea; sea-level to 100 ft. alt.; fl. April–June.

DISTR.: Divisions 4, 5, rare. S.E. Turkey, Syria, Palestine and Iraq.

4. Between Vatili and Troulli, 1862, *Kotschy* 540 ! Near Larnaca, 1880, *Sintenis & Rigo* 982 ! By the coast near Pyla, 1934, *Syngrassides* 476 ! and, 1936, *Syngrassides* 1007 !

Larnaca, 1939, *Lindberg f.* ! and, 1952, *Merton* 833 ! Kouklia, 1951, *Merton* 439 ! 5. Salamis, 1962, *Meikle* 2632 !

NOTES: Sibthorp's record of *Cachrys sicula* L. (" in petrosis Boeotiae et insulae Cypri ") may, I think, be dismissed as a misidentification of the above species; likewise Boissier's record for *Hippomarathrum cristatum*, partly based on Sibthorp's record and on the Kotschy specimen cited above. Neither species has since been found in Cyprus. The records for *Hippomarathrum boissieri* Reuter et Hausskn. ex Boiss. depend upon the typification of this name. Boissier included two elements in his description of *H. boissieri*, one (from Syria, Lebanon and Anti-Lebanon) apparently conspecific with *C. scabra*, the other (from Joppa) distinct. I prefer to regard the Joppa plant (coll. *Sieber*) as the type of *H. boissieri*; it does not appear to occur in Cyprus.

C. scabra is, I think, very variable in its indumentum and leaves, being sometimes glabrous or subglabrous, sometimes rather densely scabridulous-pubescent; the ultimate segments of the leaves may be long and almost filiform, or short, rigid and rather broadly subulate. The number of umbel-rays is also variable and should not be used to distinguish *C. scabra* from *C. microcarpa* or *C. cristata*.

The specimen from Salamis (*Meikle* 2632) is remarkable for its exceptionally well-developed bracts and bracteoles; it is, unfortunately, without fruits, and, together with *Syngrassides* 1007, is tentatively referred to *C. scabra*.

On the mainland of western Asia, *C. scabra* would appear to grow at high altitudes, generally between 4,000 and 6,000 ft. On Cyprus it is strictly lowland, and predominantly maritime. Further investigation may show that it is a relatively recent introduction, and perhaps not strictly indigenous, in our area.

2. C. crassiloba (*Boiss.*) *Meikle* **comb. nov.**

Hippomarathrum crassilobum Boiss., Diagn., 1, 10: 56 (1849); Holmboe, Veg. Cypr., 138 (1914); Davis, Fl. Turkey, 4: 392 (1972).

? *Peucedanum veneris* Kotschy in Unger et Kotschy, Die Insel Cypern, 302 (1865).

TYPE: Turkey; " in campis *Pamphyliae* prope ruinas urbis Perge ", 1845, *Heldreich* (K !).

Erect perennial, up to 1·5 m. high; rootstock thick, with a dense fibrous collar; stems stout, glabrous, sulcate, grey-green, not much branched except in the upper half, usually somewhat flexuous; basal leaves broadly oblong-deltoid, 10–20 cm. long and about as wide, 3–4-pinnate, ultimate segments rather few and remote, rigid, succulent, glabrous, about 1·5–8 cm. long, 0·3–0·8 cm. wide, flat or slightly concave above, apex acute, pungent; petioles 5–12 cm. long, deeply canaliculate with involute margins; basal sheath 6–8 cm. long, 1–2 cm. wide; cauline leaves commonly biternate, smaller than the basal and more shortly petiolate, the uppermost frequently simple, linear or subulate and subsessile; inflorescence much branched, the branches generally opposite or in whorls of 3 or more, rigid, erecto-patent; umbels convex, 4–8-rayed; peduncles rigid, sulcate, 2–6 cm. long; bracts subulate, 3–10 mm. long, 1–2 mm. wide, thick, at first spreading, often reflexed in fruit; rays subequal, 2·5–5 cm. long at anthesis; bracteoles similar to bracts but smaller; flowers mostly hermaphrodite and fertile; pedicels 2–8 mm. long; calyx-lobes broadly deltoid, about 1 mm. long, 1·5 mm. wide at base, minutely pubescent externally; petals bright yellow, oblong, strongly incurved, puberulous externally, about 2 mm. long, 1·3 mm. wide; disk very conspicuous with undulate-crenulate margins; filaments about 2 mm. long; anthers broadly oblong, yellow, about 0·8 mm. long and almost as wide; stylopodium at first flattish, later shortly conical; styles short and thick, less than 1 mm. long at anthesis, at first erect, becoming reflexed in fruit; stigmas subcapitate. Fruit large, broadly oblong, up to 1 cm. long, 0·9 cm. wide, distinctly didymous, pale brown when ripe, conspicuously scabrid with short, flattened cristate prominences; carpophore bipartite; mericarps rounded dorsally, ridges obscure or obsolete, commissural face rather broad, up to 4·5 mm. wide, endosperm deeply excavate along the commissural face, with revolute margins.

HAB.: In garigue or in vineyards on calcareous or igneous mountainsides; 2,500–4,000 ft. alt.; fl. July–August.

DISTR.: Division [? 1], 2. Also in S. Turkey.

[? 1. On the plain between Kouklia and Yeroskipos, 1862, *Kotschy* 632.]
2. Between Mandria and Perapedhi, 1937, *Kennedy* 668 ! and, 1940, *Davis* 1855 ! Valley of the Kryos Potamos, 3,400 ft. alt., 1937, *Kennedy* 669 ! Aphamis, 1937, *Kennedy* 670 ! Between Platres and Aphamis, 1941, *Davis* 3136 !

21. BUPLEURUM *L.*

Sp. Plant., ed. 1, 236 (1753).
Gen. Plant., ed. 5, 110 (1754).
H. Wolff in Engler, Pflanzenr., 43 (IV. 228): 36–173 (1910).

Annual or perennial (or occasionally biennial) herbs, or rarely subshrubs; leaves alternate, simple, entire, the lower generally petiolate, the upper often sessile or sometimes perfoliate; umbels compound, with few to many rays; involucre and involucel present, or involucre occasionally wanting; flowers generally yellow or greenish; calyx-lobes obsolete or inconspicuous; petals incurved, often cucullate, smooth or papillose, entire or variously toothed; stylopodium flattish, convex or rarely shortly conical, the margin usually entire. Fruit generally ovoid, oblong or ellipsoid, rarely subglobose, somewhat compressed laterally, more or less didymous; mericarps separating at maturity from the bipartite or undivided carpophore, or very rarely remaining united; dorsal surface of mericarps rounded, often with 5 prominent primary ridges; commissural face rather broad; endosperm flattish or a little excavate along the commissural face; vittae generally conspicuous in the valleculae, variable in number and form.

About 70 species in temperate and subtropical regions of the northern hemisphere, chiefly in Europe and western Asia; one species (*B. mundtii* Cham. et Schlechtendal) in South Africa.

Upper leaves broad, perfoliate; bracteoles broadly ovate:
  Flowers 6–16 in each umbellule; styles minute, much shorter than stylopodium, leaves generally with a tapering acumen - - - - - - **1. B. lancifolium**
  Flowers 15–25 in each umbellule; styles distinct, equalling or exceeding stylopodium; leaves acute but without a tapering acumen - - - **2. B. subovatum**
Upper leaves not perfoliate, linear, lanceolate or subulate:
  Bracteoles lanceolate or ovate, concave, conspicuous, membranous or translucent between the nerves, becoming erect and enveloping the fruits:
    Bracteoles with conspicuous ascending lateral veins between the nerves **4. B. odontites**
    Bracteoles without lateral veins, or with the longitudinal nerves at most very shortly and irregularly branched:
      Umbels and umbellules subsessile or very shortly stalked, stalks generally less than 0·5 cm. - - - - - - - - - **3. B. nodiflorum**
      Umbels and umbellules distinctly stalked; stalks up to 3–4 cm. **5. B. gracile**
  Bracteoles linear-subulate, entirely herbaceous and opaque, not enveloping the fruits:
    Umbels subsessile or very shortly stalked, stalk generally less than 1·5 mm.; fruits papillose or clothed with hooked bristles:
      Fruits covered with short, pallid papillae:
        Bracts 4; bracteoles conspicuous, up to 7 mm. long **6. B. semicompositum**
        Bracts 3; bracteoles inconspicuous, less than 3 mm. long - **7. B. orientale**
      Fruits covered with longitudinal rows of pale hooked bristles - **8. B. sintenisii**
    Umbels distinctly stalked; stalks up to 4 cm. long; fruits smooth, without papillae or bristles:
      Bracteoles usually 5; flowers usually 6 in each umbellule - **9. B. gerardii**
      Bracteoles usually 2–3; flowers usually 2–3 in each umbellule **10. B. trichopodum**

1. **B. lancifolium** *Hornem.*, Hort. Hafn., 1: 267 (1813); Davis, Fl. Turkey, 4: 399 (1972); Zohary, Fl. Palaest., 2: 411, t. 592 (1972); Osorio-Tafall et Seraphim, List Vasc. Plants Cyprus, 78 (1973).

B. heterophyllum Link, Enum. Hort. Berol., 1: 262 (1821); Boiss., Fl. Orient. Suppl., 251 (1888).

B. protractum Hoffmsgg. et Link var. *heterophyllum* (Link) Boiss., Fl. Orient., 2: 836 (1872).

B. subovatum Link ex Spreng. var. *heterophyllum* (Link) H. Wolff in Engl., Pflanzenr., 43 (IV. 228): 48 (1910); Post, Fl. Pal., ed. 2, 1: 510 (1932).

TYPE: Cultivated in the Botanic Garden, Copenhagen, from seeds originally taken from an Egyptian plant (C).

Erect or spreading, glabrous, glaucous annual, 5–30 (–40) cm. high; stems usually branched, terete, obscurely sulcate; basal leaves usually withered before anthesis, linear-lanceolate, sessile, long-acuminate, 3–8 cm. long, 0·3–1 cm. wide; upper leaves narrowly ovate or linear-lanceolate, perfoliate, apiculate, generally with a long, gradually tapering acumen, 1·5–10 cm. long, 0·3–2·5 cm. wide; umbels arising from the uppermost leaves, (1–) 2–3 (–4)-rayed; peduncles 1–3 cm. long; bracts absent; rays rather unequal, generally less than 1 cm. long; bracteoles usually 5, occasionally 6–9, greenish, somewhat unequal, broadly ovate, 2·5–10 (–13) mm. long, 2–7 mm. wide, apex sharply acuminate-cuspidate, bases connate for 1–2 mm.; flowers yellow or orange-yellow, mostly hermaphrodite and fertile, 6–16 in each umbellule; pedicels very short, rarely much exceeding 1 mm. at anthesis; petals broadly oblong, about 0·4 mm. long and as wide or slightly wider, strongly incurved; filaments very short, generally less than 0·3 mm. long at anthesis; anthers yellow, suborbicular, about 0·3 mm. diam.; stylopodium flattish, about 0·5 mm. diam. at anthesis, increasing to about 1 mm. in fruit; styles minute and barely visible at anthesis, lengthening to 0·2–0·3 mm. in fruit, much shorter than the radius of the stylopodium; stigmas subcapitate. Fruits oblong, 2–2·5 mm. long, 1·8 mm. wide, somewhat pruinose; mericarps rounded dorsally, not much compressed, conspicuously verruculose-tuberculate with 5 prominent, sharp ridges; endosperm flattish along the commissural face.

HAB.: Cultivated and fallow fields; sea-level to 1,000 ft. alt.; fl. March–May.

DISTR.: Divisions 4–7. S. Spain, N. Africa and Aegean area eastwards to Iran and Transcaucasia.

4. Athna forest, 1952, *Merton* 799 !
5. Near Ayios Seryios, 1962, *Meikle* 2582 !
6. Ayia Irini, 1941, *Davis* 2579 !
7. Kythrea, 1880, *Sintenis & Rigo* 531 (see note) ! Pentadaktylos, 1902, *A. G. & M. E. Lascelles* s.n. ! Kythrea, fields above Kephalovryso, 1967, *Merton* in ARI 334 !

NOTES: In the past much confused with *B. subovatum* Link ex Spreng. (*B. protractum* Hoffmsgg. et Link). Two duplicate sheets of the Sintenis & Rigo collections at Kew (nos. 333 and 531) consist of mixed gatherings of *B. lancifolium* and *B. subovatum*. One sheet (no. 333) is labelled " in agris pr. Palaeo Khori ", the other " Kythraea ", but since these collectors made a practice of amalgamating specimens of what they thought were the same species from different localities, it is now impossible to be quite certain whether the two species were growing intermixed, or were subsequently mixed.

As *B. lancifolium* has since been collected near Kythrea, I have assumed that the Sintenis & Rigo material of this species came from this locality. The " Palaeo Khori " mentioned by Sintenis & Rigo is not the well-known Palekhori in Pitsilia, but a small village situated between Rizokarpaso and Apostolos Andreas Monastery.

2. B. subovatum *Link ex Spreng.*, Sp. Umb., 19 (1818); H. Wolff in Engl., Pflanzenr., 43 (IV. 288): 46 (1910); Post, Fl. Pal., ed. 2, 1: 509 (1932).

B. protractum Hoffmsgg. et Link, Fl. Port., 2: 387 (1820); Unger et Kotschy, Die Insel Cypern, 300 (1865); Boiss., Fl. Orient., 2: 836 (1872); Holmboe, Veg. Cypr., 138 (1914).

B. intermedium (Lois. ex DC.) Steudel, Nomen. Bot., ed. 1, 128 (1821); Davis, Fl. Turkey, 4: 399 (1972); Osorio-Tafall et Seraphim, List Vasc. Plants Cyprus, 78 (1973).

B. rotundifolium L. var. *intermedium* Lois. ex DC. in Lam. et DC., Fl. Franç., 6: 514 (1815); Poir. in Lam., Encycl. Méth. Suppl., 5: 585 (1817).

B. rotundifolium (non L.) Sibth. et Sm., Fl. Graec. Prodr., 1: 176 (1806) quoad pl. Cypr.; Poech, Enum. Plant. Ins. Cypr., 27 (1842); Unger et Kotschy, Die Insel Cypern, 300 (1865); H. Stuart Thompson in Journ. Bot., 44: 308 (1906).

TYPE: " in Lusitania, Aegypto et Lenam Sibiriae " (B).

Erect, glaucous annual very similar to *B. lancifolium* Hornem. (*supra*) in size, habit and general appearance, but with the basal leaves oblong, lanceolate, oblanceolate or elliptic, 3–8 cm. long, 1–2·5 cm. wide, acute, but not long-acuminate; upper leaves broadly ovate, 2–8 cm. long, 1·5–3 cm. wide, acute, apiculate, but without a gradually tapering acumen, base conspicuously perfoliate; umbels 3–5-rayed; rays rather unequal, 1–2·5 cm. long; bracteoles usually 5, subequal, broadly ovate, 4–15 mm. long, 3–10 mm. wide, bright yellow-green, overlapping, apex abruptly apiculate, bases very shortly connate; flowers yellow or orange, mostly hermaphrodite and fertile, 15–25 in each umbellule; petals very broadly oblong, 0·5–0·7 mm. long, up to 0·8 mm. wide, strongly incurved; filaments very short; anthers yellow, suborbicular, 0·4–0·5 mm. diam.; stylopodium flattish, about 1–1·5 mm. diam. at anthesis; styles lengthening to 0·9 mm. in fruit, almost as long as the radius of the stylopodium, or sometimes longer. Fruits oblong, 3·5–4·5 (–5) mm. long, 3–3·5 mm. wide, occasionally somewhat pruinose, otherwise as in *B. lancifolium.*

HAB.: Cultivated and fallow fields; sea-level to 3,000 ft. alt.; fl. Febr.–May.

DISTR.: Divisions 1–8, frequent. Southern, western and central Europe; N.W. Africa, Turkey, Syria, Palestine; often as a casual in other areas of the world.

1. Between Ktima and Khrysokhou, 1862, *Kotschy* 621; Stroumbi, 1941, *Davis* 3355 !
2. Phiti, 1905, *Holmboe* 768; Perapedhi, 1937, *Kennedy* 671 !
3. Evdhimou, 1950, *Chapman* 609 ! Souni, 1960, *N. Macdonald* 120 ! Moni, 1964, *J. B. Suart* 213 !
4. Perivolia, 1938, *Syngrassides* 1824 ! and, 1955, *Merton* 2009 !
5. Salamis, 1905, *Holmboe* 467; Strongylos, 1905, *Holmboe* 359; Kythrea, 1927, *Rev. A. Huddle* 62 ! Dhali, 1936, *Syngrassides* 1136 ! Apostolos Varnavas Mon., 1948, *Mavromoustakis* s.n. !
6. Karpasha, 1959, *P. H. Oswald* 106 !
7. Kyrenia, 1936, *Mrs. Dray* 10 ! and, 1956, *G. E. Atherton* 1244 ! Panagra, 1962, *Meikle* 2366 !
8. Gastria, 1936, *Syngrassides* 1278 !

3. B. nodiflorum *Sibth. et Sm.*, Fl. Graec. Prodr., 1: 177 (1806), Fl. Graec., 3: 55, t. 260 (1819); Unger et Kotschy, Die Insel Cypern, 300 (1865); Boiss., Fl. Orient., 2: 840 (1872); Holmboe, Veg. Cypr., 139 (1914); Post, Fl. Pal., ed. 2, 1: 510 (1932); Davis, Fl. Turkey, 4: 400 (1972); Zohary, Fl. Palaest., 2: 413, t. 594 (1972); Osorio-Tafall et Seraphim, List Vasc. Plants Cyprus, 78 (1973).

TYPE: Cyprus; " In arvis insulae Cypri "; t. 260 in Fl. Graeca !

Erect or spreading, glabrous annual 5–25 (–35) cm. high; stems rigid, with rather irregularly dichotomous, patent, subterete or more or less angular, distinctly ridged branches; lower leaves linear-lanceolate, long acuminate, 5–12 cm. long, 0·2–0·5 cm. wide with prominent parallel nervation; upper leaves very similar, but usually shorter and narrower, amplexicaul but not perfoliate at the base; umbels terminating the main axis and the branches, subsessile or very shortly pedunculate, 4–7-rayed; bracts 3–4, shortly connate, lanceolate-acuminate, 5–18 mm. long, 2–4 mm. wide, prominently 3-nerved with narrow membranous margins; rays unequal, short, seldom exceeding 5 mm.; bracteoles 4–5, subequal, lanceolate-acuminate, concave at the base, very shortly connate, 7–8 mm. long, 2·5–3 mm. wide, prominently 3-nerved, membranous between the nerves, spreading at anthesis, erect in fruit; flowers 6–12 in each umbellule, mostly fertile; pedicels about 0·5 mm. long; petals pallid, oblong, distinctly costate with a narrow truncate-emarginate, sharply incurved apex, about 0·6–0·8 mm. long, 0·5 mm. wide; filaments very short; anthers yellow, suborbicular, about 0·4 mm. diam.; stylopodium flattish, about 0·5 mm. diam.; styles minute at anthesis, lengthening to 0·5 mm. in fruit; stigmas subtruncate. Fruit narrowly oblong, 1·5–2 mm. long, 1–1·2 mm. wide, dark brown when ripe; mericarps separating readily at maturity, rounded dorsally, with 5, pale, filiform ridges, and 5 dark, slightly rugulose longitudinal stripes between the ridges.

HAB.: Cultivated and fallow fields; sea-level to 1,000 ft. alt.; fl. April–June.

DISTR.: Divisions 1 and 3, rare. Also Turkey, Syria, Palestine, Egypt.

1. Below Stroumbi, 1941, *Davis* 3378 !
3. Limassol, 1905, *Holmboe* 653; Parekklisha, 1950, *F. J. F. Barrington* 47 ! Near Mamonia, 1960, *N. Macdonald* 155 !

4. B. odontites *L.*, Sp. Plant., ed. 1, 237 (1753); Sibth. et Sm., Fl. Graec. Prodr., 1: 177 (1806); Unger et Kotschy, Die Insel Cypern, 299 (1865); Boiss., Fl. Orient., 2: 839 (1872); Davis, Fl. Turkey, 4: 400 (1972); Zohary, Fl. Palaest., 2: 412, t. 593 (1972); Osorio-Tafall et Seraphim, List Vasc. Plants Cyprus, 78 (1973).

B. fontanesii Guss., Fl. Sic. Prodr., 313 (1827); Holmboe, Veg. Cypr., 138 (1914); Post, Fl. Pal., ed. 2, 1: 510 (1932).

TYPE: " *in alpibus* Vallesiae " (LINN)—an error in localization according to Snogerup (in Davis, Fl. Turkey, 4: 402; 1972).

An erect or spreading, glabrous annual 8–50 cm. high; stems somewhat flexuous, much branched, distinctly sulcate; lower leaves linear-lanceolate, 4–10 (–17) cm. long, 0·3–1 cm. wide, prominently parallel-nerved, upper leaves similar, usually smaller, amplexicaul but not perfoliate at the base; umbels terminating the branches, 4–7-rayed; peduncles short, stout, up to 1·5 cm. long; bracts 4–5, almost free, lanceolate-acuminate, 1–4·5 cm. long, 0·2–0·5 cm. wide, distinctly 5-nerved with numerous ascending lateral veins, margins very narrowly membranous, together with the dorsal surface of the nerves, minutely scabridulous; rays very unequal, 0·5–35 mm. long; bracteoles usually 5, very shortly connate, narrowly lanceolate-acuminate, 5–20 mm. long, 1·5–4 mm. wide, 3-nerved, membranous between the nerves with conspicuous, ascending venation, erect in fruit; flowers 8–13 in each umbellule, mostly fertile; pedicels slender, hair-like, unequal, up to 8 mm. long; petals pallid, rather narrowly oblong, about 0·8 mm. long, 0·3 mm. wide, costate, strongly incurved with a narrower, truncate-emarginate apex; filaments very short; anthers yellow, broadly oblong or suborbicular, about 0·4 mm. diam.; stylopodium flattish, about 0·5 mm. diam.; styles at first minute, lengthening to about 0·2 mm. in fruit; stigmas subcapitate. Fruit dark brown when ripe, oblong, about 1·5 mm. long, 1 mm. wide; mericarps separating readily at maturity, dorsal surface rounded, smooth, rather glossy, without any obvious ridges.

HAB.: Open ground, or as a weed in cultivated fields; probably at low altitudes; fl. April–July.

DISTR.: Only once recorded for Cyprus, without precise locality, by Sibthorp. Widespread in southern Europe and the Mediterranean region and frequent as an introduced weed or casual in many parts of the world.

NOTES: Sibthorp's record is almost certainly correct, especially as the plant is stated to have been growing with the then newly discovered *B. nodiflorum*, which it somewhat resembles. Improved agricultural methods may well have contributed to a diminution, or perhaps to the disappearance, of this attractive weed.

5. B. gracile *Urv.* in Mém. Soc. Linn. Par., 1: 286 (1822); Snogerup in Bot. Notiser 115: 363 (1962); Davis, Fl. Turkey, 4: 403 (1972); Osorio-Tafall et Seraphim, List Vasc. Plants Cyprus, 78 (1973).

B. glumaceum Sibth. et Sm., Fl. Graec. Prodr., 1: 177 (1806); Unger et Kotschy, Die Insel Cypern, 299 (1865); Boiss., Fl. Orient., 2: 337 (1872); H. Wolff in Engl., Pflanzenr., 43 (IV. 228): 61 (1910); Holmboe, Veg. Cypr., 138 (1914) omnes pro parte quoad pl. Cypr.

TYPE: Astipalaia; " In sterilibus Astypalaeae vulgaris ", May, 1819, *Dumont d'Urville* (P, K !).

Erect, glabrous annual 10–30 (–40) cm. high; stems slender, usually much branched, especially in the upper half, often flexuous, obscurely sulcate or angled; basal leaves linear-oblanceolate or sometimes narrowly elliptic with an elongate filiform base, 20–100 mm. long, 1–5 mm. wide, acute, acuminate or apiculate, margins entire, basal sheath with a narrow membranous margin; cauline leaves similar but often shorter and narrower, generally rather sparse and distant; inflorescence repeatedly branched; peduncles erecto-patent, 0·5–4·5 cm. long, slender, distinctly ribbed or angled; umbels (2–) 3–4 (–5)-rayed; bracts 2–3 (–4), narrowly ovate-acuminate, 4–8 mm. long, 0·5–2·5 mm. wide, concave, 3-nerved with narrow membranous margins; rays very unequal, up to 30 mm. long, suberect, angular; bracteoles 5, ovate-acuminate or aristate, 4–8 mm. long, 2–4 mm. wide, concave, yellowish, membranous except for the 3 prominent, simply or shortly branched, longitudinal nerves, becoming erect and enclosing the fruit; flowers 4–8 in each umbellule, mostly fertile; pedicels very short, usually less than 1 mm. long; petals yellowish, oblong, distinctly costate, 0·5–0·6 mm. long, 0·4–0·6 mm. wide, with a narrower, sharply incurved, emarginate apical appendage; filaments very short; anthers yellow, suborbicular, 0·2–0·3 mm. diam.; stylopodium flattish, barely 0·4 mm. diam.; styles barely visible in flower or fruit; stigmas truncate. Fruit oblong, 1·5 mm. long, about 1 mm. wide; mericarps dark brown, smooth, dorsal surface rounded, obscurely ridged.

HAB.: In garigue or under Pines on dry rocky calcareous or non-calcareous hillsides; sea-level to 2,000 ft. alt.; fl. April–June.

DISTR.: Divisions 1, 3. Crete, Greece, Aegean area, Turkey, Lebanon.

1. Ktima, 1913, *Haradjian* 665 ! Dhrousha, 1941, *Davis* 3235 ! Akamas, 1941, *Davis* 3312 ! 3332 ! Kakoskala (Akamas), 1954, *Casey* 1344 ! Cape Arnauti, 1957, *Merton* 3042 !

3. Between Pissouri and Kouklia, 1862, *Kotschy* 623 ! Between Limassol and Akrotiri, 1905, *Holmboe* 667; between Pendalia and Amargeti, 1941, *Davis* 3394 ! Kandou, 1950, *F. J. F. Barrington* 53 ! Episkopi, 1954, *Merton* 1887 ! Trimiklini, 1954, *Mavromoustakis* 19 ! Souni, 1960, *N. Macdonald* 151 ! Pissouri, 1962, *Meikle* 2649 !

6. **B. semicompositum** *L.*, Demonstr. Plant., 7 (1753); Sibth. et Sm., Fl. Graec. Prodr., 1: 178 (1806), Fl. Graec., 3: 55, t. 261 (1819); Unger et Kotschy, Die Insel Cypern, 299 (1865); Holmboe, Veg. Cypr., 139 (1914); Davis, Fl. Turkey, 4: 405 (1972); Zohary, Fl. Palaest., 2: 415, t. 599 (1972); Osorio-Tafall et Seraphim, List Vasc. Plants Cyprus, 78 (1973).

B. glaucum Robill. et Cast. in Lam. et DC., Fl. Franç., 6: 515 (1815); Boiss., Fl. Orient., 2: 842 (1872); Post, Fl. Pal., ed. 2, 1: 511 (1932).

B. semicompositum L. var. *glaucum* (Robill. et Cast.) H. Wolff in Engl., Pflanzenr., 43 (IV. 228): 106 (1910).

TYPE: Spain; " in Hispania " *Loefling* (LINN).

Erect or spreading, glabrous, glaucous annual, (3–) 8–30 (–40) cm. high; stems slender, obscurely ridged or angled, repeatedly branched either from the base or in the upper half, sometimes forming loose tufts; basal leaves linear-subulate or narrowly oblanceolate, 1·5–6 cm. long, 0·1–0·6 cm. wide, generally with 3 distinct parallel nerves, apex acute or acuminate, base attenuate, margins entire; basal sheath inconspicuous with a very narrow membranous margin; upper leaves similar, but smaller and narrower, frequently subulate, rather few and remote; umbels terminal and lateral, often rather crowded, 3–4 (–6)-rayed, the lateral sessile, subsessile or very shortly pedunculate, the terminal with slender, angular peduncles up to 1·5 cm. long; bracts 4, linear-subulate, herbaceous, unequal, 5–12 mm. long, 0·5–1·5 mm. wide, concave, 3-nerved; rays very unequal, 1–15 (–22) mm. long, angular, scabridulous; bracteoles usually 5, similar to bracts but rarely more than 7 mm. long, 1 mm. wide, suberect or spreading in fruit, margins and nerves often minutely scabridulous; flowers 4–7 in each umbellule, mostly fertile; pedicels unequal, very short, rarely exceeding 2 mm.; petals oblong, brownish, strongly incurved, very small, fugacious, about 0·3–0·4 mm. long and almost as wide, smooth or indistinctly papillose, with an obscure midrib; filaments very short; anthers yellow, suborbicular, about 0·2 mm. diam.; stylopodium flattish, about 0·4 mm. diam.; styles about 0·2 mm. long, spreading; stigmas subcapitate. Fruit subglobose, laterally compressed, about 1 mm. diam., distinctly didymous; mericarps rounded dorsally, dark blackish-brown with numerous small white papillae.

HAB.: Sandy banks and pastures, or on the drier parts of salt-marshes, generally on slightly saline soils; sea-level to 500 ft. alt.; fl. April–May.

DISTR.: Divisions 1, 3, 4–6, 8. Widespread in the Mediterranean region and east to Iran.

1. Arodhes [? " Arora "], between Ktima and Khrysokhou, 1787, *Sibthorp*.
3. Akrotiri, 1905, *Holmboe* 684 !
4. Larnaca, 1862–1938, *Kotschy* 317; *Post* 401 ! *Syngrassides* 1836 ! Ayios Memnon, 1948, *Mavromoustakis* s.n. ! Mile 3, Larnaca–Famagusta road, 1950, *Chapman* 435 ! Near mouth of Pedieos River, 1955, *Merton* 2258 !
5. Salamis, 1962, *Meikle* 2587 ! 2592 !
6. Mile 3, Nicosia–Myrtou road, 1950, *Chapman* 580 ! Nicosia, 1962, *Meikle* 2946 !
8. Aphendrika, 1880, *Sintenis & Rigo* 332 ! Cape Andreas, 1962, *Meikle* 2461 !

NOTES: Very variable in size and habit. *Meikle* 2587 and 2592, both from the same salt-marsh near Salamis, represent extremes, one (2587) 22 cm. high, very lax, with bracteoles up to 9 mm. long, the other (2592) dwarf, about 5 cm. high, compact and rather congested, the bracteoles barely 4 mm. long; such variations appear to be in direct response to the amount of moisture available; the lush plant is sometimes distinguished as var. *pseudodontites* (Rouy et Camus) H. Wolff.

7. **B. orientale** *Snogerup* in Bot. Notiser, 124: 365, fig. 5 (1971); Davis, Fl. Turkey, 4: 407 (1972); Osorio-Tafall et Seraphim, List Vasc. Plants Cyprus, 78 (1973).
[*B. tenuissimum* L. ssp. *gracile* (nec *Odontites gracilis* M. Bieb.) H. Wolff in Engl., Pflanzenr., 43 (IV. 228): 104, 201 (1910) pro parte quoad spec. Cypr.]
[*B. gracile* (non (M. Bieb.) DC.) Lindberg f., Iter Cypr., 25 (1946).]
TYPE: Cyprus; " Salzsteppe bei Larnaka, 12-9-1880. *Sintenis* 916 p.p." (LD).

Erect, glabrous, rather glaucous annual 15–50 cm. high; stems slender, much branched, with spreading or suberect, distinctly ridged or angular branches; basal leaves narrowly linear-oblanceolate, 4–7 cm. long, 0·2–0·4 cm. wide, straight or slightly falcate, with 3 prominent parallel nerves, apex acuminate, base attenuate, margins entire; basal sheath flattish with a broad membranous margin; cauline leaves similar, but shorter and narrower; umbels lateral and terminal, sessile or subsessile, forming lax, spiciform inflorescences, 1–4-rayed; bracts generally 3, subulate-lanceolate, 1·5–3 mm. long, about 1 mm. wide, rigid, smooth; rays unequal, generally very short (less than 4 mm.) occasionally up to 15 mm., slender, angular; bracteoles 3–5, very similar to bracts but usually a little smaller; flowers 1–6 in each umbellule, rather crowded, mostly fertile; pedicels very short (less than 1 mm. long) or almost absent; petals brownish, oblong, about 0·5 mm. long, 0·3 mm. wide, apex strongly incurved, slightly emarginate; filaments very short; anthers yellow, suborbicular, about 0·2 mm. diam.; stylopodium flattish, about 0·4 mm. diam.; styles very short, spreading; stigmas subcapitate. Immature fruit oblong, about 1·6 mm. long, 1 mm. wide, dark brown, densely covered with short, pallid papillae; mericarps rounded dorsally, apparently without any obvious ridges.

HAB.: Sandy fields and salt-marshes, usually near the coast; about sea-level; fl. July–Sept.

DISTR.: Divisions 3, 4, locally abundant. Also S. Turkey, Syria, Palestine.

3. Asomatos marshes, 1939 & 1940, *Mavromoustakis* 3 ! 60 ! 76 !
4. Larnaca salt lake, 1880, *Sintenis & Rigo* 916, 980 ! Near Famagusta, 1939, *Lindberg f.* s.n. ! Between Prastio and Kouklia, 1939, *Lindberg f.* s.n.; near mouth of Pedieos River, 1955, *Merton* 2259 !

8. **B. sintenisii** *Aschers. et Urban ex Huter* in Oesterr. Bot. Zeitschr., 55: 360 (1905); H. Wolff in Engl., Pflanzenr., 43 (IV. 228): 109 (1910); Holmboe, Veg. Cypr., 139 (1914); Lindberg f., Iter Cypr., 25 (1946); Osorio-Tafall et Seraphim, List Vasc. Plants Cyprus, 78 (1973).

Glochidopleurum sintenisii (Aschers. et Urban ex Huter) Kozo-Polj. in Journ. Russe Bot., 1913: 6, t. 5 (1913).

B. rigoi (sphalm. *ripoi*) Huter ex David, Etude Anat. Genre *Bupleurum*, 85 (1901) nomen non rite publ.

TYPE: Cyprus; " in collibus inter Nicosia et Kythraea 25 Majo 1880 " *Sintenis & Rigo* 594 (K !).

Diminutive, glabrous, glaucous annual (0·8–) 1–3 (–5) cm. high; stems erect or with a few spreading branches, distinctly angled; leaves linear or very narrowly linear-oblanceolate, 10–20 mm. long, 0·5–1·3 mm. wide, rather rigid, acute, margins entire or sometimes scabridulous, base scarcely attenuate, basal sheath obscure with a very narrow pallid margin; umbels lateral and terminal, 2–5-rayed; peduncle very short, angular, usually less than 2 mm. long; bracts 2–3, narrowly oblanceolate, rigid, 4–7 mm. long, 0·5–1 mm. wide, acute, often with scabridulous margins; rays unequal, short, generally less than 5 mm. long, angular; bracteoles 3–5, oblanceolate, concave, somewhat keeled, 2–4 mm. long, 0·5–1 mm. wide, margins and keel minutely scabridulous; flowers 4–10 in each umbellule, rather crowded, mostly fertile; pedicels to 1·7 mm. long, scabridulous; petals yellowish-brown, oblong, cucullate, about 0·5 mm. long, 0·4 mm. wide, incurved apex short, truncate or somewhat emarginate; filaments very short; anthers orange, suborbicular, about 0·4 mm. diam.; stylopodium flattish, about 0·6 mm. diam.; styles minute, suberect; stigmas truncate. Fruits (immature) oblong, about 2 mm. long, 1·2 mm. wide; mericarps dark brown, rounded dorsally, primary and secondary ridges covered with longitudinal rows of short, pale, hooked bristles with wide, ± confluent, laterally flattened bases.

HAB.: Dry, bare pastures and rocky hillsides; 350–700 ft. alt.; fl. May–July.

DISTR.: Divisions 4, 5, rare. Endemic.

4. Pergamos, 1955, *Merton* 2216 ! Troulli, Ayios Yeoryios, 1962, *Meikle* 2960 !
5. Between Nicosia and Kythrea, 1880, *Sintenis & Rigo* 594 ! Athalassa, 1939, *Lindberg f.* s.n. ! Lymbia, near Arkhangelos Mon., 1956, *Poore* 10 !

9. **B. gerardii** *All.*, Auct. Syn. Meth. Stirp. Hort. Reg. Taur., 81 (1774); Boiss., Fl. Orient., 2: 845 (1872); H. Wolff in Engl., Pflanzenr., 43 (IV. 228): 88 (1910); Holmboe, Veg. Cypr., 139 (1914); Post, Fl. Pal., ed. 2, 1: 512 (1932); Davis, Fl. Turkey, 4: 414 (1972); Zohary, Fl. Palaest., 2: 414, t. 596 (1972); Osorio-Tafall et Seraphim, List Vasc. Plants Cyprus, 78 (1973).

TYPE: France; " in gallopr. australis sterilibus, campestribus ", figured in Gérard, *Flora Gallo-Provincialis*, Fig. 9 (1761).

Erect, glabrous, glaucous annual (6–) 9–16 (–25) cm. high; stems slender, usually unbranched, conspicuously ribbed or almost angular; lower leaves linear, 10–50 mm. long, 1–4 mm. wide, acute, prominently 3-nerved, amplexicaul but scarcely sheathing at the base; upper leaves a little shorter, generally with a long-attenuate, acuminate apex; inflorescence usually branched; umbels terminal and leaf-opposed (3–) 5–6-rayed; peduncles slender, erect, 0·8–3 cm. long; bracts usually 3–5, subulate, unequal, 3–13 mm. long, 0·5–1 mm. wide, margins minutely scabridulous, rays unequal, very slender, 3–25 mm. long, distinctly angled; bracteoles usually 5, similar to bracts, but rarely more than 6 mm. long; flowers usually 6 in each umbellule, mostly fertile; pedicels 0·5–2 mm. long; petals yellowish, broadly oblong, about 0·5 mm. long, 0·4 mm. wide, sometimes a little papillose dorsally, with a short, sharply incurved, emarginate apex; filaments very short; anthers yellow, suborbicular, about 0·2 mm. diam.; stylopodium flattish, about 0·5 mm. diam.; styles minute, reflexed; stigmas truncate.

Fruit oblong, about 2 mm. long, 1·3 mm. wide; carpophore bipartite; mericarps dark brown, smooth, with 5 prominent longitudinal ribs.

HAB.: In dry, open Pine forest; in garigue on rocky mountainsides, or sometimes in fallow fields; 3,000–4,600 ft. alt.; fl. May–June.

DISTR.: Division 2. Widespread in southern and south-central Europe eastwards to Turkey, Syria and W. Iran.

2. Troödos, 1880, *Sintenis & Rigo* 838 ! Below Prodhromos, 1905, *Holmboe* 924; Kykko Mon., 1913, *Haradjian* 943; Platres, 1938, *Kennedy* 1379 ! Above Khrysorroyiatissa Mon., 1941, *Davis* 3430 ! Trikoukkia, 1941, *Davis* 3465 ! Ayia Irini, above Platres, 1960, *N. Macdonald* 159 ! Prodhromos, 1961, *D. P. Young* 7347 !

NOTES: A variable species; the measurements given above are taken from Cyprus specimens, which tend to be smaller in all their parts than typical *B. gerardii* from western Europe.

10. B. trichopodum *Boiss. et Spruner* in Ann. Sci. Nat., ser. 3, 1: 145, Fig. 9 (1844); Boiss., Fl. Orient., 2: 846 (1872); Post, Fl. Pal., ed. 2, 1: 513 (1932); Davis, Fl. Turkey, 4: 415 (1972); Osorio-Tafall et Seraphim, List Vasc. Plants Cyprus, 78 (1973).

B. trichopodum Boiss. et Spruner var. *depauperatum* Boiss., Fl. Orient., 2: 846 (1872) et Suppl., 252 (1888); Holmboe, Veg. Cypr., 139 (1914).

TYPE: Boissier & Spruner cite 5 syntypes from Greece and Turkey (G).

Very slender, erect, glabrous annual, 7–30 (–40) cm. high; stems usually unbranched in the lower half, subterete or obscurely sulcate; lower leaves linear or linear-oblanceolate, 10–60 (–80) mm. long, 1·5–6 mm. wide, 1–3-nerved, margins entire, apex generally attenuate-acuminate, base amplexicaul but scarcely sheathing; upper leaves similar, but smaller, sometimes subulate with a broad amplexicaul base; inflorescence much branched, with almost filiform branches; umbels terminal and leaf-opposed, very lax, 2–7-rayed; peduncle filiform, 0·8–4 cm. long; bracts 1–3, sometimes absent, subulate, unequal, 3–12 mm. long, usually less than 1 mm. wide, with a very slender, awn-like apex; rays unequal, filiform, spreading, 6–30 mm. long; bracteoles usually 2–3, very narrowly subulate or awn-like, to about 5 mm. long, less than 0·5 mm. wide, equalling or exceeding the flowers, margin entire; flowers usually 2–3 in each umbellule; pedicels 1–2 mm. long at anthesis; petals yellowish or tinged red, caducous, about 5 mm. long and about as wide, with a sharply incurved, emarginate apex; filaments very short; anthers yellow, about 0·2 mm. long, 0·15 mm. wide; stylopodium flattish, about 0·4 mm. diam.; styles about 0·2 mm. long, strongly reflexed; stigmas subcapitate. Fruit oblong, 2·5–3 mm. long, 1·5 mm. wide, carpophore bipartite; mericarps dark brown, faintly rugulose with 5 prominent ribs.

HAB.: On rocky slopes in garigue, usually in slightly shaded places; generally on limestone or chalk, but occasionally on serpentine; 2,000–3,000 ft. alt.; fl. April–May.

DISTR.: Divisions 3, 7, rare. Greece and Aegean area eastwards to Turkey.

3. Zagala above Trimiklini, 1941, *Davis* 3133 ! Kakomallis above Louvaras, 1962, *Meikle* 2898 !

7. Pentadaktylos, 1880, *Sintenis & Rigo* 590 ! St. Hilarion, 1941, *Davis* 3619 ! also, same area, 1948, *Kennedy* 1664 ! and *Casey* 1028 !

22. APIUM *L.*

Sp. Plant., ed. 1, 264 (1753).
Gen. Plant., ed. 5, 128 (1754).
H. Wolff in Engl., Pflanzenr., 90 (IV. 228): 26 (1927).
Helosciadium Koch in Nov. Act. Nat. Cur., 12: 125 (1824).

Glabrous annual, biennial or perennial herbs; leaves pinnate or ternate-pinnate; umbels compound, often leaf-opposed; bracts present or absent, bracteoles usually present; flowers mostly hermaphrodite and fertile; sepals minute or obsolete; petals subequal, white or greenish, acute or sometimes with an inflexed, emarginate apex; stylopodium flattish or shortly conical, with an entire or crenulate margin. Fruit ovoid or subglobose, laterally compressed and somewhat didymous, glabrous or very occasionally setulose; carpophore entire or bifid; mericarp with 5 equal, or occasionally somewhat unequal ridges; valleculae univittate; endosperm flattish along the commissural face.

About 30 species, some aquatic or amphibious, found almost throughout the world.

Stems erect; umbels lateral and terminal; leaflets usually 3-lobed and distinctly stalked **1. A. graveolens**
Stems creeping and rooting at the basal nodes; umbels lateral, leaf-opposed; leaflets (except the terminal) not lobed, sessile or subsessile - - **2. A. nodiflorum**

1. A. graveolens *L.*, Sp. Plant., ed. 1, 264 (1753); Boiss., Fl. Orient., 2: 856 (1872); Post, Fl. Pal., ed. 2, 1: 515 (1932); A. K. Jackson in Kew Bull., 1937: 342 (1937); R. M. Nattrass, First List of Cyprus Fungi, 31, 44 (1937); Lindberg f., Iter Cypr., 25 (1946); Rechinger f. in Arkiv för Bot., ser. 2, 1: 428 (1950); Davis, Fl. Turkey, 4: 422 (1972); Zohary, Fl. Palaest., 2: 416, t. 600 (1972); Osorio-Tafall et Seraphim, List Vasc. Plants Cyprus, 78 (1973).

TYPE: " in Europae *humectis, praesertim maritimis* ".

Erect biennial with a strong, aromatic smell when crushed; stems 30–60 cm. high, glabrous, strongly sulcate, usually much branched especially in the upper half; basal leaves simply pinnate, generally with 5 leaflets, oblong in outline, 4–25 cm. long, 1·5–18 cm. wide; petiole slender, canaliculate above, sometimes exceeding 25 cm. in length, dilated and broadly sheathing at the base; leaflets broadly cuneate-ovate, 0·5–6 cm. long, 1–7 cm. wide, usually 3-lobed, with the lobes coarsely and rather irregularly mucronate-dentate; petiolules usually distinct, 0·3–3 cm. long; upper stem leaves similar, but with fewer, smaller leaflets, and more shortly petiolate, uppermost leaves often with 1 deeply 3-lobed leaflet; umbels lateral and terminal, often terminating short axillary shoots, 3–12-rayed; peduncle short or wanting, usually less than 1 cm. long; rays unequal, spreading, 1–2·5 cm. long, lengthening a little in fruit; bracts and bracteoles absent; flowers 10–15 (–20) in each umbellule, mostly fertile; pedicels 1–2 mm. long; petals greenish, broadly ovate, 0·4–0·5 mm. long and about as wide, apex somewhat incurved; filaments 0·7 mm. long; anthers suborbicular, about 0·3 mm. diam., yellow; stylopodium flattish 0·5 mm. diam., styles at first inconspicuous, lengthening to about 0·5 mm. in fruit and becoming reflexed or arcuate-spreading; stigmas subcapitate. Fruit subglobose, about 1·5 mm. diam., not very distinctly didymous; mericarps dark brown when ripe, with 5 very prominent, pallid ridges.

HAB.: Ditches, stream-beds and marshy ground, usually under saline conditions, or near the sea; sea-level to 2,400 ft. alt.; fl. May–June.

DISTR.: Divisions 1, 2, 6, 7, common in Division 6. Widespread in Europe, the Mediterranean region and S. Africa; commonly cultivated and often found as a garden escape in other parts of the world. The asterisked Cyprus records probably refer to cultivated plants.

1. Ktima, 1913, *Haradjian* 644 ! Paphos,* 1931, *Nattrass* 42.
2. Kykko Mon. *Lindberg f.* s.n.; Kakopetria, 1955, *Merton* 2315 !
6. Common; Morphou,* 1932, *Nattrass* 174; Nicosia,* 1935, *Nattrass* 553; Dheftera,* 1935, *Nattrass* 591; Myrtou, 1936, *Syngrassides* 1204 ! Syrianokhori marshes, 1952, *Merton* 808 ! Yerolakkos, 1967, *Economides* 1021 !
7. Lapithos, 1939, *Lindberg f.* s.n. !

2. A. nodiflorum (*L.*) *Lag.*, Amen. Nat., 1: 101 (1821); H. Wolff in Engl., Pflanzenr., 90 (IV. 228): 37 (1927); Post, Fl. Pal., ed. 2, 1: 515 (1932); Davis, Fl. Turkey, 4:

422 (1972); Zohary, Fl. Palaest., 2: 416, t. 601 (1972); Osorio-Tafall et Seraphim, List Vasc. Plants Cyprus, 78 (1973).

Sium nodiflorum L., Sp. Plant., ed. 1, 251 (1753).

Helosciadium nodiflorum (L.) Koch in Nov. Act. Nat. Cur. 12: 126 (1824); Boiss., Fl. Orient., 2: 856 (1872); Holmboe, Veg. Cypr., 139 (1914); Lindberg f., Iter Cypr., 25 (1946).

TYPE: " *in* Europa *ad ripas fluviorum* ".

Creeping perennial, rooting at the lower nodes; flowering stems erect, very variable in height and general robustness, sometimes very slender and less than 10 cm. long, sometimes thick and exceeding 1 m., glabrous, sulcate, but less markedly so than in *A. graveolens*; leaves pinnate, oblong in outline, 2–20 cm. (or more) in length, 0·8–8 cm. or more in width; leaflets opposite, usually in 3–5 pairs, ovate or ovate-elliptic, sessile or subsessile, glossy green, 0·4–7 cm. long, 0·3–3 cm. wide, terminal leaflet often 3-lobed, the lateral acute or subacute with shortly serrate margins and a rounded or cuneate base; petiole 0·5–15 cm. long, canaliculate above, expanding at the base and forming a conspicuous, membranous-margined sheath; umbels leaf-opposed, 3–8 (–12)-rayed; peduncle usually very short, normally less than 2 cm. long, sometimes almost wanting, bracts often absent, sometimes 1–2, ovate, acute, 2–3 mm. long, about 2 mm. wide; rays very unequal, 0·3–2 cm. long; bracteoles 5–7, narrowly ovate, 3–8 mm. long, 1–3 mm. wide, with membranous margins; flowers about 12 in each umbellule, mostly hermaphrodite and fertile; pedicels 1–1·5 mm. long; petals white, suborbicular or broadly oblong, about 0·5–0·7 mm. diam.; filaments 0·4 mm. long; anthers ovoid, about 0·3 mm. long, 0·2 mm. wide, purplish; stylopodium about 0·3 mm. diam., styles at first minute, lengthening to 0·5 mm. in fruit, and often becoming reflexed; stigmas capitate. Fruit oblong-ellipsoid, about 2–2·5 mm. long, 1–1·5 mm. wide, rather distinctly didymous; carpophore shortly bifid at apex; mericarps dark brown when ripe, with 5 very prominent, pallid ridges.

HAB.: Wet ground, in ditches or by streams and water-channels; sea-level to 3,700 ft. alt.; fl. April–June.

DISTR.: Divisions 2, 5–7. Europe, Mediterranean region and Atlantic Islands eastwards to Afghanistan and Central Asia; naturalized in North and South America.

2. Evrykhou, 1905, *Holmboe* 849; Kakopetria, 1936, *Syngrassides* 1100 ! and, 1949, *Casey* 829 ! Platania, 1939, *Lindberg f.* s.n.; Platres, *Michaelides*.
5. Kythrea, 1939, *Lindberg f.* s.n.
6. Myrtou, 1932, *Syngrassides* 94 !
7. Kythrea, 1880, *Sintenis & Rigo* 834 ! Lapithos, 1939, *Lindberg f.* s.n. ! Lambousa, 1955, *G. E. Atherton* 343 ! Kambyli, 1957, *Merton* 2977 !

PETROSELINUM CRISPUM (*Mill.*) *A. W. Hill*, Parsley, a popular culinary herb, valued for its aromatic, pleasantly flavoured foliage, has been collected by a roadside at Prodhromos (1955, *Atherton* 622 !) and no doubt occurs elsewhere in Cyprus as an escape from gardens. It is an erect, glabrous biennial, with deltoid-ovate, 3-pinnate, long-stalked leaves, and sharply dentate or crisped leaf-segments. The greenish flowers are borne in long-stalked, terminal, bracteate, 7–20-rayed umbels, and the fruits are ovate, somewhat tapering towards the apex, with a conical stylopodium and persistent reflexed styles. The mericarps are dark brown and distinctly ridged as in *Apium graveolens*, but the carpophore is bipartite almost to the base. The natural home of Parsley is uncertain, though most probably in S.E. Europe or western Asia; unlike Celery (*Apium graveolens* supra) it prefers to grow in dry, stony situations, often in the crevices of cliffs or on old walls.

23. RIDOLFIA *Moris*

Enum. Sem. Hort. Reg. Taurin., 43 (1841).
Fl. Sardoa, 2: 212 (1842).

Erect biennial herb; leaves 4–5-pinnate, the ultimate segments capillary or filiform; umbels terminal or rarely lateral, pedunculate; bracts and bracteoles absent; flowers numerous in each umbellule, mostly hermaphrodite and fertile; sepals obsolete; petals golden-yellow, strongly incurved, truncate-emarginate at apex; stylopodium conspicuous, flattish, margins at first entire, later crenulate; styles ultimately as long as the radius of the stylopodium; stigmas capitate. Fruit oblong, laterally compressed, not distinctly didymous; carpophore deeply bipartite; mericarps obscurely ridged, smooth; dorsal vittae solitary, slender, commissural vittae 2; endosperm flattish along its commissural face.

One species, widely distributed in the Mediterranean region.

1. R. segetum (*Guss.*) *Moris*, Enum. Sem. Hort. Reg. Taurin., 43 (1841), Fl. Sardoa, 2: 212, t. 75 (1842); Boiss., Fl. Orient., 2: 858 (1872); Sintenis in Oesterr. Bot. Zeitschr., 32: 293, 316 (1882); Holmboe, Veg. Cypr., 139 (1914); Post, Fl. Pal., ed. 2, 1: 516 (1932); Lindberg f., Iter Cypr., 25 (1946); Rechinger f. in Arkiv för Bot., ser. 2, 1: 428 (1950); Davis, Fl. Turkey, 4: 424 (1972); Zohary, Fl. Palaest., 2: 419, t. 605 (1972); Osorio-Tafall et Seraphim, List Vasc. Plants Cyprus, 78 (1973).

Meum segetum Guss., Fl. Sic. Prodr., 1: 346 (1827).

TYPE: Sicily; " Inter segetes, et in arvis argillosis " *Gussone* (NAP).

Glaucous biennial 30–80 (–100) cm. high; stems erect, finely and obscurely sulcate, often unbranched except in the region of the inflorescence; leaves unpleasant-smelling when crushed, deltoid-ovate in outline, 5–25 cm. long, 4–25 cm. wide, dissected pinnately into numerous very slender filiform segments 3–30 mm. long, 0·5–1 mm. wide, with shortly subulate-acute apices; petioles often exceeding 15 cm., longitudinally striate, canaliculate above, expanding towards base into a narrow, membranous-auriculate sheath 1–4 cm. long; inflorescence much branched, peduncles up to 20 cm. long; umbels (15–) 25–50 (–60)-rayed, rays slender, subequal, curving upwards from a patent base, 2–8 cm. long; flowers 30 or more in each umbellule; pedicels 1·5–10 mm. long; petals yellow, oblong, about 0·8 mm. long, 0·5 mm. wide, apex strongly incurved, emarginate-truncate; filaments about 0·5 mm. long; anthers suborbicular, about 0·4 mm. diam., yellow; stylopodium conspicuous, about 0·4–0·5 mm. diam., styles at first inconspicuous, lengthening to 0·4 mm. soon after anthesis. Fruit rather narrowly oblong, about 2 mm. long, 1 mm. wide; carpophore very slender and setaceous, bipartite; mericarps dark olive-brown, smooth, rather lustrous, dorsal ridges very obscure, filiform, lateral ridges much thicker, pallid.

HAB.: Cultivated fields; sea-level to 2,500 ft. alt.; fl. April–June.

DISTR.: Divisions 1–3, 5–8, common in Division 7. Widespread in S. Europe and the Mediterranean region.

1. Polis, 1913, *Haradjian* 800 !
2. Below Pano Panayia, 1941, *Davis* 3349 ! Khrysorroyiatissa, 1941, *Davis* 3434 ! Kakopetria, 1955, *Merton* 2316 !
3. Akhelia, 1939, *Lindberg f.* s.n.; Kellaki, 1967, *Merton* 663 !
5. Trikomo, 1939, *Lindberg f.* s.n. ! Salamis, 1962, *Meikle* 2580 !
6. Myrtou, 1932, *Syngrassides* 993 ! Peristerona, 1967, *Merton* 804 ! Karpasha, 1959, *Oswald* 137 !
7. Common; Kyrenia, 1880, *Sintenis & Rigo* 352 ! and, 1949, *Casey* 812 ! Ayios Yeoryios, 1880, *Sintenis & Rigo* 352 ! Dhikomo, 1936, *Syngrassides* 993 ! Aghirda, 1955, *Merton* 2240 !
8. Near Yialousa, 1880, *Sintenis & Rigo* s.n.; near Rizokarpaso, 1912, *Haradjian* 180.

24. AMMI *L.*

Sp. Plant., ed. 1, 243 (1753).
Gen. Plant., ed. 5, 113 (1754).
H. Wolff in Engl., Pflanzenr., 90 (IV. 228): 115 (1927).

Annual or biennial, glabrous herbs; leaves pinnate or pinnatisect; umbels terminal, compound, generally many-rayed; bracts numerous, conspicuous, generally pinnate or pinnatisect; bracteoles numerous, usually entire. Flowers hermaphrodite, mostly fertile; sepals minute or obsolete; petals whitish or yellowish, obcordate with a bilobed or emarginate apex and a short basal claw; stylopodium flattish or shortly conical with an entire or crenulate margin; styles short. Fruit ovoid-oblong, somewhat compressed laterally; mericarps convex dorsally with blunt (sometimes obscure) ridges; carpophore entire or bipartite; valleculae 1-vittate, commissure 2-vittate; endosperm flattish or slightly concave along its commissural face.

Six species in Europe, Asia, Africa and the Atlantic Islands, sometimes occurring as adventives elsewhere.

Rays of the umbel arising from a conspicuously thickened disk or pulvinus; ultimate segments of upper leaves linear or filiform, entire - - - **1. A. visnaga**
Rays not arising from a conspicuous disk or pulvinus, ultimate segments of upper leaves linear-lanceolate or narrowly oblong, serrate - - - - **2. A. majus**

1. A. visnaga (*L.*) *Lam.*, Fl. Franç., ed. 1, 3: 462 (1778); Boiss., Fl. Orient., 2: 892 (1872); Post, Fl. Pal., ed. 2, 1: 526 (1932); A. K. Jackson in Kew Bull., 1937: 342 (1937); Lindberg f., Iter Cypr., 25 (1946); Rechinger f. in Arkiv för Bot., ser. 2, 1: 428 (1950); Davis, Fl. Turkey, 4: 426 (1972); Zohary, Fl. Palaest., 2: 418, t. 604 (1972); Osorio-Tafall et Seraphim, List Vasc. Plants Cyprus, 78 (1973).
Daucus visnaga L., Sp. Plant., ed. 1, 242 (1753).
TYPE: " *in* Europa *australi* ".

Robust erect biennial; stems up to 1 m. high, often more than 12 mm. diam., conspicuously but shallowly sulcate, commonly unbranched except in the upper part; basal leaves often withered at anthesis, deltoid-ovate in outline, 15–25 cm. long, 10–20 cm. wide, 4-pinnate, ultimate segments linear, entire, sharply acute, about 10 mm. long, 1·5 mm. wide; petiole thick, canaliculate above, often exceeding 5 cm., expanding rather suddenly into a short, conspicuously nerved basal sheath; upper leaves sessile or subsessile, broadly deltoid, ultimate segments narrowly linear or almost filiform, often exceeding 1 cm. in length, usually less than 1 mm. wide; inflorescence usually branched, with stout, erect branches; umbels terminal, usually more than 50-rayed; peduncle stout, conspicuously sulcate, often exceeding 15 cm. in length and terminating in a distinct disk or pulvinus, often more than 1·5 cm. diam.; involucral bracts numerous, often more than 2·5 cm. long, 2–3-pinnate, with long filiform segments, base conspicuously flattened and attached to the rim of the pulvinus; rays subequal, spreading, 3–5 cm. long, forming a flattish or shallowly convex umbel; bracteoles filiform, entire, often less than 0·5 mm. wide, usually spreading; flowers very numerous in each umbellule; pedicels unequal, slender, up to 7 mm. long; calyx obsolete; petals white, about 1·8 mm. long and about as wide, deeply and unequally bilobed, apex not much incurved; filaments 1·5 mm. long; anthers orbicular, about 0·5 mm. diam., pinkish; stylopodium flattish, about 0·8 mm. diam. at anthesis; styles at first erect, about 0·5 mm. long, lengthening to about 0·8 mm., and becoming reflexed in fruit; stigmas subcapitate. Rays of fruiting umbel becoming erect and connivent; fruit oblong, about 2 mm. long, 1 mm. wide, somewhat compressed laterally; carpophore entire, rigid; mericarps dark brown, convex dorsally with 5 fairly prominent blunt ridges; endosperm flattish along its commissural face.

HAB.: Cultivated and fallow fields; sea-level to c. 800 ft. alt.; fl. May–August.

DISTR.: Divisions 1, 3, 4, 6. Mediterranean region eastwards to Iran; Atlantic Islands; Ethiopia; an introduced weed in many other parts of the world.

1. Cape Arnauti near Polis, 1913, *Haradjian* 801 ! Between Polis and Khrysokhou, 1934, *Syngrassides* 967 ! Yeroskipos, 1939, *Lindberg f.* s.n. ! Sarama, 1941, *Davis* 3348 !
3. Kholetria, 1941, *Davis* 3462 !
4. Larnaca, 1880, *Sintenis & Rigo* 900 ! Livadhia, 1935, *Syngrassides* 836 ! Between Prastio and Kouklia, 1939, *Lindberg f.* s.n.
6. Dheftera, 1880, *Sintenis & Rigo* 900 !

2. A. majus *L.*, Sp. Plant., ed. 1, 243 (1753); Unger et Kotschy, Die Insel Cypern, 298 (1865); Boiss., Fl. Orient., 2: 891 (1872); Holmboe, Veg. Cypr. 139 (1914); Post, Fl. Pal., ed. 2, 1: 525 (1932); Lindberg f., Iter Cypr., 25 (1946); Davis, Fl. Turkey, 4: 427 (1972); Zohary, Fl. Palaest., 2: 417, t. 603 (1972); Osorio-Tafall et Seraphim, List Vasc. Plants Cyprus, 78 (1973).

TYPE: " *in* Europa *australi* ".

Erect annual or biennial (8–) 20–100 cm. high; stems subterete below, distinctly sulcate above, generally branched only in the upper half; basal leaves usually withered at anthesis, often simply pinnate, with 3–5 oblong-lanceolate or obovate, serrate leaflets 1·3–6 cm. long, 0·3–2 cm. wide; petioles up to 20 cm. long, canaliculate above; basal sheath short, inconspicuous, narrowly membranous-margined; upper leaves 2–3-pinnate, deltoid-ovate in outline, 2–15 cm. long, 0·8–13 cm. wide, ultimate segments linear-lanceolate, lanceolate or narrowly oblong, sharply serrate; petioles becoming progressively shorter upwards, basal sheaths inconspicuous; inflorescence with spreading branches; umbels terminal, 8–50 (or more)-rayed; peduncle sulcate, 2–16 cm. long; bracts pinnate or bipinnate, up to 5 cm. long, divided into slender, filiform or narrowly linear segments, not conspicuous or much flattened at the base; rays subequal, up to 6 cm. long, spreading in flower and fruit, often sparsely papillose at the base; bracteoles linear or filiform, 2–6 mm. long, 0·5–1 mm. wide, entire, apex shortly aristate; flowers very numerous in each umbellule; pedicels slender, unequal, up to 9 mm. long; petals white, very unequally 2-lobed, about 1·5 mm. long, 1·3 mm. wide, apex not much incurved; filaments slender, 1·5 mm. long; anthers broadly oblong, about 0·4 mm. long, whitish or tinged pink; stylopodium flattish or somewhat convex, about 0·5 mm. diam.; styles at first erect, about 0·4 mm. long, lengthening to about 0·6 mm. and becoming reflexed in fruit; stigmas capitate. Fruits oblong, somewhat compressed laterally, about 1·8 mm. long, 1·2 mm. wide; carpophore slender, entire or deeply bipartite; mericarps dark brown when ripe, strongly convex dorsally with 5 slender, but distinct, pallid ridges; valleculae sometimes with a median row of pale papillae; endosperm flattish along its commissural face.

HAB.: Cultivated and fallow fields; sea-level to c. 3,000 ft. alt.; fl. April–June.

DISTR.: Divisions 1–8. S. Europe and Mediterranean region eastwards to Iran; Atlantic Islands; common as an adventive in many other areas.

1. Cape Arnauti, 1913, *Haradjian* 817 ! Polis, 1938, *Syngrassides* 1845 ! Between Terra and Dhrousha, 1941, *Davis* 3271 ! and 3272 !
2. Pano Panayia, 1905, *Holmboe* 1053; near Mesopotamos Mon., 1939, *Lindberg f.* s.n.; Kambos, 1939, *Lindberg f.* s.n.
3. Limassol, 1947, *Mavromoustakis* 14 ! Cape Gata, 1962, *Meikle* 2910 !
4. Kiti near Larnaca, 1862, *Kotschy* s.n.
5. Athalassa, 1939, *Lindberg f.* s.n. ! Salamis, 1962, *Meikle* 2635 !
6. Near Nicosia, 1862, *Kotschy* 968 ! also, 1937, *Syngrassides* 1618 ! and 1619 !
7. Ayios Khrysostomos Mon., 1880, *Sintenis & Rigo* s.n.; Kythrea, 1880, *Sintenis & Rigo* 338 !
8. Ephtakomi, 1880, *Sintenis & Rigo* 338 ! Leonarisso, 1880, *Sintenis & Rigo* s.n.; Apostolos Andreas Mon., 1880, *Sintenis & Rigo* s.n.

25. CARUM *L.*

Sp. Plant., ed. 1, 263 (1753).
Gen. Plant., ed. 5, 127 (1754).
incl. *Hellenocarum* H. Wolff in Engl., Pflanzenr., 90 (IV. 228): 167 (1927).

Glabrous biennial or perennial herbs, with a fusiform rootstock; leaves 2–4-pinnate or pinnatisect; bracts absent or much reduced; bracteoles present; umbels compound, terminal and lateral, usually pedunculate; flowers mostly hermaphrodite; sepals very small or obsolete; petals whitish or tinged pink or yellow, obovate, emarginate, incurved at the apex, the outer sometimes slightly radiant; stylopodium flattish or shortly conical; styles often rather long and reflexed. Fruit oblong-ellipsoid, compressed laterally; carpophore usually bifid or bipartite; mericarps convex dorsally, with 5 rather prominent ridges; valleculae usually 1-vittate; commissure 2-vittate; endosperm flattish along its commissural face.

About 30 species, chiefly in temperate Europe and Asia.

Bracteoles present; rays 15–40, spreading, subequal - - - **1. C. multiflorum**
Bracteoles absent; rays 5–10, becoming suberect, distinctly unequal C. CARVI p. 743

1. C. multiflorum (*Sibth. et Sm.*) *Boiss.*, Fl. Orient., 2: 882 (1872); Holmboe, Veg. Cypr., 139 (1914); Davis, Fl. Turkey, 4: 349 (1972); Osorio-Tafall et Seraphim, List Vasc. Plants Cyprus, 78 (1973).

Athamanta multiflora Sibth. et Sm., Fl. Graec. Prodr., 1: 188 (1806), Fl. Graec., 3: 69, t. 276 (1819).

Ligusticum cyprium Spreng., Sp. Umb., 125 (1818); Unger et Kotschy, Die Insel Cypern, 300 (1865).

Hellenocarum multiflorum H. Wolff in Engl., Pflanzenr., 90 (IV. 228): 168 (1927).

TYPE: " in insulâ Cypro ", *Sibthorp* (but see note).

Robust erect biennial or perennial, up to 1 m. high with a thick, broadly fusiform root; stems shallowly sulcate, usually much branched above; basal leaves broadly deltoid-ovate, 2–3-pinnate, up to 25 cm. long and about as wide, ultimate segments small, cuneate, about 1 cm. long, 0·3–0·5 cm. wide, bluntly lobed, glabrous; petiole up to 10 cm. long, canaliculate above; basal sheath narrow, not very conspicuous; upper stem leaves similar, but smaller and usually subsessile or very shortly petiolate; umbels mostly terminal, 15–40-rayed; peduncle stout, up to 15 cm. long; bracts usually numerous, reflexed, lanceolate or linear, 5–10 mm. long, 2–3 mm. wide; rays subequal, 1·5–4 cm. long; bracteoles similar to bracts, but rarely more than 5 mm. long and 2 mm. wide; flowers numerous in each umbellule; pedicels short, rarely exceeding 3 mm.; calyx obsolete; petals whitish, oblong-obovate, about 1 mm. long, 0·8 mm. wide, apex emarginate, incurved; filaments about 1·2 mm. long; anthers whitish or yellowish, oblong, about 0·4 mm. long, 0·3 mm. wide; stylopodium shortly but distinctly conical, about 1 mm. diam.; styles slender, reflexed or spreading, about 1·5–2·5 mm. long; stigmas capitate. Fruit narrowly oblong-ellipsoid, about 2–2·5 mm. long, 1–1·5 mm. wide, laterally compressed, distinctly didymous; mericarps dark brown when ripe, with 5 conspicuous, filiform or slightly winged ridges.

HAB.: Usually on rocky calcareous slopes and in rock fissures; up to 7,000 ft. alt. in Greece; fl. May–August.

DISTR.: Cyprus, without precise locality, 1787, *Sibthorp*. Also S.E. Italy, Crete, Balkan peninsula and Turkey.

NOTES: The occurrence of *C. multiflorum* in Cyprus must be regarded as questionable, the more so as many of the records in the *Flora Graeca* are now known to be erroneous. The plant has not been seen on the island by any recent collector, and its nearest stations are in Rhodes and western Turkey (about Izmir). It is included here solely because there is still a remote possibility that it may be refound, possibly in crevices on the Northern Range, as it is normally a plant of limestone formations.

C. CARVI *L.*, Sp. Plant., ed. 1, 263 (1753), Caraway, frequently cultivated for its aromatic fruits, probably occurs as an escape from gardens. It is an erect biennial, up to 1 m. high, with a fusiform root, narrowly oblong, 3-pinnatisect basal leaves, with numerous, narrowly cuneate, lobulate, ultimate segments; the stems are very sparsely leafy, and the terminal umbels have 5–10, very unequal, ultimately suberect rays; the bracts are often wanting, and, when present, are much reduced, narrowly linear or filiform; the bracteoles are normally absent. Caraway has white or pinkish flowers, and the fruiting pedicels, like the rays, are unequal and suberect. The fruits are oblong, 2·5–4 mm. long, 2–3 mm. wide, distinctly didymous, with an entire or shortly bifid carpophore, and with dark brown, characteristically aromatic, distinctly ridged mericarps.

Carum carvi L. appears to be native over a wide area of northern and central Europe.

26. BUNIUM *L.*

Sp. Plant., ed. 1, 243 (1753).
Gen. Plant., ed. 5, 114 (1754).
H. Wolff in Engl., Pflanzenr., 90 (IV. 228): 186 (1927).

Glabrous perennial herbs with globose, tuberous roots, and with the subterranean part of the stem slender and flexuous; leaves 2–3-pinnate or pinnatisect, often rather sparse and mostly basal or apparently basal, ultimate segments usually numerous, small, narrow; umbels compound, terminal; bracts usually wanting or much reduced; bracteoles present; flowers hermaphrodite or polygamous; sepals obsolete; petals white or pink, obcordate, emarginate with an incurved apex; stylopodium flattish or shortly conical; styles usually rather long and spreading. Fruits ovoid, ellipsoid, oblong or narrowly prismatic-oblong, glabrous, usually quite smooth; carpophore generally bifid; mericarps with slender, usually conspicuous, ridges, valleculae 1 (or sometimes 3)-vittate; endosperm flattish along its commissural face; seedling usually with 1 cotyledon, the other abortive.

About 30 species in Europe, the Mediterranean region and eastwards to Central Asia.

1. B. ferulaceum *Sibth. et Sm.*, Fl. Graec. Prodr., 1: 186 (1806); Holmboe, Veg. Cypr., 139 (1914); Post, Fl. Pal., ed. 2, 1: 524 (1932); Davis, Fl. Turkey, 4: 346 (1972); Zohary, Fl. Palaest., 2: 421, t. 608 (1972); Osorio-Tafall et Seraphim, List Vasc. Plants Cyprus, 78 (1973).

B. ferulifolium [*ferulaefolium*] Desf. in Ann. Mus. Hist. Nat. Par., 11: 275, t. 30 (1808); Unger et Kotschy, Die Insel Cypern, 299 (1865); Sintenis in Oesterr. Bot. Zeitschr., 32: 400 (1882).

Carum ferulifolium [*ferulaefolium*] (Desf.) Boiss., Diagn., 1, 10: 22 (1849), Fl. Orient., 2: 886 (1872).

TYPE: " In insulae Cypri arvis ", 1787, *Sibthorp* (OXF).

Spreading or erect perennial 15–30 (–50) cm. high; stems usually much branched, rather slender, distinctly sulcate, arising from a deeply buried, globose tuber, 2·5–5 or more cm. diam.; basal leaves broadly deltoid in outline, 5–15 cm. long, 2–12 cm. wide, ternate-bipinnate, the 3 primary segments divided into numerous linear, subacute or obtuse segments 5–25 mm. long, 1–1·5 mm. wide; petioles mostly subterranean, often exceeding 10 cm., very slender and flexuous, basal sheaths inconspicuous, forming a scaly mass at the top of the tuber; upper stem leaves very few, much reduced, sessile or subsessile, consisting of a few elongate-linear segments, basal sheath inconspicuous, about 1–1·5 cm. long, with a narrow scarious margin; umbels lax, usually 7–15-rayed; peduncle frequently

10–15 cm. long; bracts generally reduced to 1–0, small, subulate, reflexed appendages; rays spreading, unequal or subequal at anthesis, 2·5–4 cm. long; bracteoles 3–5, subulate, herbaceous, usually less than 5 mm. long, 1 mm. wide; flowers rather numerous, mostly hermaphrodite and fertile; pedicels unequal, spreading, 2–8 mm. long, becoming thick and rigid in fruit; petals white, obovate, about 1·2 mm. long, 1 mm. wide, deeply incurved-emarginate at apex; filaments about 1 mm. long; anthers purplish, oblong, about 0·5 mm. long, 0·4 mm. wide; stylopodium about 0·7 mm. diam., flattish at first, later somewhat convex; styles spreading, about 0·7 mm. long; stigmas capitate. Fruits narrowly oblong, 4–5 mm. long, 1·5 mm. wide, not much compressed laterally; mericarps dark blackish-brown when ripe, with distinct, but not very prominent ridges.

HAB.: Cultivated fields, fallow land and roadsides, occasionally in open Pine forest; sea-level to 3,000 ft. alt.; fl. March–May.

DISTR.: Divisions 1–4, 6–8, locally common in Division 7 around Kyrenia. Yugoslavia, Greece, Crete, Bulgaria, Aegean Islands, Turkey, Syria, Palestine.

1. Arodhes [" Arora "], 1862, *Kotschy* 666; between Dhrousha and Kato Arodhes, 1941, *Davis* 3278 ! Kakoscaliou Island, Akamas, 1948, *Kennedy* 1644 !
2. Perapedhi, 1938, *Kennedy* 1382 ! Mandria, 1938, *Kennedy* 1383 ! Kryos Potamos, 1938, *Kennedy* 1384 ! Near Klirou, 1967, *Merton* 283 !
3. Limassol, 1938, *Kennedy* 1381 ! Near Souni, 1960, *N. Macdonald* 1301 !
4. Near Larnaca, 1862, *Kotschy* 214; between Avgorou and Phrenaros, 1905, *Holmboe* s.n.; Dherinia, 1970, *A. Hansen* 294 !
6. Myrtou, 1956, *Merton* 2626 ! also, 1962, *Meikle* 2423 ! Orga, 1956, *Poore* 2658 !
7. Common; Kyrenia, Northern range, 1932, *Syngrassides* 685 ! Kyrenia, 1932, *Nattrass* 209, also, 1949, *Casey* 507 ! also, 1956, *G. E. Atherton* 1268 ! Larnaka tis Lapithou, 1936, *Syngrassides* 920 ! also, 1941, *Davis* 2983 ! Karmi, 1941, *Davis* 2779 ! Akhna forest, 1952, *Merton* 727 ! Temblos, 1956, *G. E. Atherton* 1396 ! Ayios Epiktitos, 1967, *Merton* 567 !
8. Komi Kebir, 1880, *Sintenis & Rigo* 330 ! Rizokarpaso, 1957, *Merton* 2937 !

27. PIMPINELLA *L.*

Sp. Plant., ed. 1, 263 (1753).
Gen. Plant., ed. 5, 128 (1754).
H. Wolff in Engl., Pflanzenr., 90 (IV. 228): 219 (1927).

Perennial, or, less commonly, annual or biennial herbs; leaves often heteromorphous, the lower undivided or 1–2-pinnate with broad segments, the upper narrowly dissected; umbels compound, terminal and/or leaf-opposed; bracts generally absent; bracteoles usually present; flowers mostly hermaphrodite and fertile; sepals small or obsolete; petals white, yellow, pink or reddish, variously shaped, glabrous or dorsally pilose; the outer sometimes somewhat radiant, apex entire or lobed, flat or incurved; stylopodium convex or conical; styles short or long, generally patent or reflexed. Fruit polymorphous, ovoid or subglobose, ovoid-oblong or oblong, frequently attenuate towards the apex, obscurely or sometimes distinctly didymous, somewhat compressed laterally, but usually with a broad commissure, glabrous, hairy or variously papillose or scabridulous; carpophore shortly bifid or bipartite; mericarps subterete or 5-angled with distinct, but usually slender dorsal ridges; valleculae generally 3-vittate, sometimes 2- or 4-vittate; endosperm flattish or slightly concave along its commissural face.

About 150 species, widely distributed in the Old World.

Perennial with a thick woody rootstock; leaf-segments prominently reticulate-nerved below - - - - - - - - - - **4. P. cypria**
Annuals or biennials, without a thick woody rootstock:
Basal leaves usually present at anthesis, commonly undivided, suborbicular or flabelliform; plants annual, 5–70 cm. high; umbels 4–20-rayed, usually erect immediately before anthesis; fruits puberulous:

Basal leaves coarsely and irregularly toothed; fruit about 3 mm. long; umbels 10–20-rayed; petals not or obscurely radiant - - - - **1. P. anisum**

Basal leaves crenate or bluntly crenate-dentate; fruit about 1·8 mm. long; umbels 4–12-rayed; outer petals distinctly radiant - - - - - **2. P. cretica**

Basal leaves usually withered at anthesis; plants biennial, 50–100 cm. high; umbels 12–60-rayed, usually drooping immediately before anthesis; fruits hispid with spreading bristles - - - - - - - - - - - **3. P. peregrina**

1. P. anisum *L.*, Sp. Plant., ed. 1, 264 (1753); Unger et Kotschy, Die Insel Cypern, 299 (1865); Boiss., Fl. Orient., 2: 866 (1872); Holmboe, Veg. Cypr., 139 (1914); Post, Fl. Pal., ed. 2, 1: 518 (1932); Davis, Fl. Turkey, 4: 355 (1972); Zohary, Fl. Palaest., 2: 425, t. 615 (1972); Osorio-Tafall et Seraphim, List Vasc. Plants Cyprus, 77 (1973).

TYPE: "*in* Aegypto".

Robust, erect annual, 20–70 cm. high; stems striate or obscurely sulcate, usually much branched in the upper half, unbranched or sparsely branched below; radical leaves generally undivided, broadly obovate-cuneate or flabelliform, 1·5–3·5 cm. long, 1·5–3 cm. wide, coarsely and irregularly toothed or sub-lacinate, subglabrous or very sparingly pubescent, sometimes ternate with obovate-cuneate, toothed segments; petioles well developed, sometimes 15 cm. long, flattish or canaliculate above; basal sheath narrow, inconspicuous; upper stem leaves simply pinnate, with 3–5, obovate-cuneate, laciniate-dentate segments, similar to the basal leaves, but more coarsely toothed and with shorter petioles; uppermost leaves subsessile, usually bipinnatisect, with narrow, oblanceolate-acuminate segments 1·5–4 cm. long, 0·1–0·4 cm. wide; umbels terminal, 10–20-rayed; peduncle usually well developed, 5–10 cm. long, puberulous; bracts and bracteoles generally absent, or, when present, usually solitary and much reduced, subulate or filiform, less than 5 mm. long, minutely puberulous; rays slender, spreading or suberect, subequal, 0·8–4 cm. long; flowers mostly hermaphrodite and fertile, numerous in each umbellule; pedicels 3–8 mm. long, generally subequal; sepals obsolete; petals oblong-obovate, 1–1·5 mm. long, 0·8–1 mm. wide, white or pinkish, pilose dorsally, apex shortly bifid, not much incurved; filaments 1·5–2 mm. long; anthers oblong, yellow, about 0·5 mm. long, 0·4 mm. wide; stylopodium convex or very shortly conical, about 0·6 mm. diam., with a crenulate margin; styles slender, 1·3–2·5 mm. long, at length reflexed; stigmas subcapitate. Fruit pyriform, with an attenuate apex, about 3 mm. long, 2 mm. wide, distinctly puberulous; commissure entire or shortly bifid; mericarps brown when ripe, usually separating readily, convex dorsally with 5 distinct but slender, pallid ridges; valleculae 4–8-vittate, the vittae almost coalescing into an annulus; endosperm somewhat concave along its commissural face.

HAB.: Cultivated fields, gardens and waste ground, an escape from cultivation; sea-level to ? 1,000 ft. alt.; fl. April–August.

DISTR.: Division 1, not native. Widespread in Europe, the Mediterranean region and western Asia, but perhaps always as an escape; also occurring as an adventive in many other regions of the world.

1. Near Yeroskipos, 1862, *Kotschy* 637 !

2. P. cretica *Poir.* in Lam., Encycl. Méth. Suppl., 1: 684 (1810); Boiss., Fl. Orient., 2: 866 (1872); Sintenis in Oesterr. Bot. Zeitschr., 32: 398 (1882); Post, Fl. Pal., ed. 2, 1: 518 (1932); A. K. Jackson in Kew Bull., 1937: 343 (1937); Rechinger f. in Arkiv för Bot., ser. 2, 1: 428 (1950); Davis, Fl. Turkey, 4: 356 (1972); Zohary, Fl. Palaest., 2: 426, t. 617, 617a (1972); Osorio-Tafall et Seraphim, List Vasc. Plants Cyprus, 77 (1973).

TYPE: Crete; probably first collected by Tournefort (P).

Slender erect or spreading annual (5–) 10–40 cm. high; stems branched or unbranched near base, usually branched above, somewhat flexuous, subterete or obscurely sulcate, subglabrous or minutely puberulous, especially in the lower parts, often tinged purple; basal leaves undivided, suborbicular, 5–25 mm. diam. with a cordate base and crenate or crenate-dentate margins, glabrous or very obscurely puberulous; petiole very slender, up to 5 cm. long; basal sheath inconspicuous with a narrow membranous-ciliate margin; upper leaves either similar to basal, or more often ternate, with broadly ovate or rhomboid segments up to 2·5 cm. long, 1·5 cm. wide, base cuneate, margins irregularly dentate or serrate, commonly 3-lobed; uppermost leaves usually sessile or subsessile, ternate, with linear or filiform, entire or serrate margins, 5–35 mm. long, 0·5–5 mm. wide, basal sheath narrow, closely adpressed to stem; lamina sometimes wanting or replaced by a short arista at the apex of the stems; umbels leaf-opposed, 4–12-rayed; peduncle slender, erecto-patent, up to 5 cm. long; bracts and bracteoles generally absent; rays spreading, subequal, very slender, to about 3 cm. long; flowers about 12 in each umbellule, mostly hermaphrodite and fertile; pedicels up to 4 mm. long; sepals obsolete; petals white or tinged pinkish, glabrous, the exterior distinctly radiant, obovate, very unequally 2-lobed, flattish, up to 1·8 mm. long, 1·3 mm. wide, apex of lobes rounded; filaments up to 2 mm. long; anthers oblong, pallid or purplish, about 0·4 mm. long, 0·3 mm. wide; stylopodium minute, shortly conical; styles very slender, about 1·5 mm. long, at first erect, later reflexed, generally falling off before the fruit ripens; stigmas capitate. Fruit ovoid, somewhat attenuate towards apex, about 1·8 mm. long, 1·2 mm. wide, distinctly adpressed-puberulous, not much compressed laterally; carpophore deeply bipartite; mericarps falling apart readily when ripe, dark purplish-brown, convex dorsally, with 5 indistinct or sometimes almost obsolete ridges; endosperm flattish along its commissural face.

HAB.: Rough, rocky pastures, sometimes in the open parts of Juniper forest; sea-level to 4,300 ft. alt.; fl. April–June.

DISTR.: Divisions 1–5, 7, 8. Common in Division 7 and locally common near Cape Andreas in Division 8. Also Greece, Crete, Aegean Islands, Turkey, Syria, Palestine and Arabia.

1. Lyso, 1913, *Haradjian* 858.
2. Kykko Mon., 1913, *Haradjian* 937 !
3. Limassol, 1913, *Haradjian* 569; Nicosia–Limassol Rd., Mile 24, 1962, *Meikle* 2958 !
4. Near Akhyritou, 1948, *Mavromoustakis* s.n. !
5. Mia Milia, 1950, *Chapman* 680 ! also 1954, *Merton* 1900 !
7. Common; Armenian Mon., 1912, *Haradjian* 369 ! Kyrenia, 1938, *C. H. Wyatt* 19 ! Yaïla Peak, 1941, *Davis* 3637 ! Near St. Hilarion, 1950, *Casey* 1051 ! Near Akanthou, 1970, *A. Hansen* 355 !
8. Locally common; Cape Andreas, 1880, *Sintenis & Rigo* 337 ! Near Rizokarpaso, 1912, *Haradjian* 191 ! Cape Andreas, *Mavromoustakis* s.n. ! Apostolos Andreas, 1962, *Meikle* 2458 ! Damalaes, Apostolos Andreas Forest, 1962, *Meikle* 2482 !

3. P. peregrina *L.*, Sp. Plant., ed. 1, 264 (1753); Boiss., Fl. Orient., 2: 867 (1872); Holmboe, Veg. Cypr., 139 (1914); Post, Fl. Pal., ed. 2, 1: 519 (1932); Lindberg f., Iter Cypr., 25 (1946); Rechinger f. in Arkiv för Bot., ser. 2, 1: 428 (1950); Davis, Fl. Turkey, 4: 356 (1972); Zohary, Fl. Palaest., 2: 424, t. 612 (1972); Osorio-Tafall et Seraphim, List Vasc. Plants Cyprus, 77 (1973).

TYPE: " in Italiae *pascuis sterilibus* ".

Erect biennial 50–100 cm. high with a fusiform taproot; stems subglabrous (var. *peregrina*) or conspicuously pilose or hispid (var. *hispida* (Lois.) Thell.) generally unbranched in the lower half and rather sparingly branched above, striate or inconspicuously sulcate; branches suberect; basal leaves usually withered at anthesis; undivided or ternate, with suborbicular, cordate, pubescent or puberulous, irregularly toothed segments, 3–5 cm. diam.; petioles 10–15 cm. long, subglabrous or hispid, canaliculate above; basal sheath inconspicuous; upper leaves pinnate, narrowly oblong in outline, up to 15 cm. long, 9 cm. wide, with 5–7 bluntly ovate, irregularly toothed segments 2–5 cm. long, 1·5–4 cm. wide, the terminal segment often deeply 3-lobed; uppermost leaves sessile or subsessile, bipinnately dissected into numerous narrowly ovate-cuneate or linear, entire or serrate segments 0·5–3 cm. long, 0·1–0·6 cm. wide; umbels leaf-opposed and terminal, drooping before anthesis, 12–60-rayed; peduncle 3–30 cm. long; bracts and bracteoles absent, or much reduced and filiform, generally somewhat hispid, rays spreading, slender, 3–7 cm. long; flowers numerous in each umbellule, mostly hermaphrodite and fertile; pedicels very slender, up to 1 cm. long; sepals obsolete; petals white, not, or scarcely, radiant, broadly obovate or suborbicular, unequally lobed, thinly pilose dorsally, about 0·8 mm. long and almost as wide; filaments about 1 mm. long; anthers yellow, suborbicular, about 0·4 mm. diam.; stylopodium conical, about 0·5 mm. diam.; styles slender, about 1·3 mm. long, at first erect, spreading or somewhat recurved in fruit; stigmas capitate. Fruit broadly ovoid, 1·3–1·5 mm. long, 1·2 mm. wide, not much compressed or attenuate, rather densely hispid with spreading bristles; carpophore slender, bipartite; mericarps brown when ripe, convex dorsally, ridges very obscure.

HAB.: Fallow fields, shady banks and roadsides; 200–4,000 ft. alt.; fl. May–July.

DISTR.: Divisions 2, 7. Widely distributed in the northern and eastern Mediterranean region and eastwards to Central Asia.

2. Galata, 1880, *Sintenis & Rigo* 611B ! Kykko Mon., 1905, *Holmboe* 1015; Panayia, 1913, *Haradjian* 912 ! Arminou, 1941, *Davis* 3420 ! Khrysorroyiatissa, 1941, *Davis* 3429 !

7. Near Lapithos, 1880, *Sintenis & Rigo* 611A ! Lapithos, 1939, *Lindberg f.* s.n. ! Larnaka tis Lapithou, 1941, *Davis* 3021 !

4. **P. cypria** *Boiss.*, Fl. Orient. Suppl., 253 (1888); Holmboe, Veg. Cypr., 139 (1914); Lindberg f., Iter Cypr., 25 (1946); Rechinger f. in Arkiv för Bot., ser. 2, 1: 428 (1950); Osorio-Tafall et Seraphim, List Vasc. Plants Cyprus, 77 (1973).

P. tragium Vill. var. *cypria* [" *cyprica* "] (Boiss.) H. Wolff in Engl., Pflanzenr., 90 (IV. 228): 251 (1927).

[*P. tragium* Vill. var. *pseudotragium* (non (DC.) Boiss.) Post in Mém. Herb. Boiss., 18: 93 (1900).]

[*P. tragium* (non Vill.) H. Stuart Thompson in Journ. Bot., 44: 308 (1906).]

TYPE: Cyprus; " ad rupes supra Bellapais ", 1880, *Sintenis & Rigo* 592 (G, K !).

Erect perennial 10–70 cm. high with a stout, woody rootstock clothed in the upper part with the remains of withered and indurated leaf-sheaths; stems conspicuously sulcate, rather densely (or rarely sparingly) clothed with soft spreading hairs, not much branched except in the region of the inflorescence; basal leaves oblong in outline, simply pinnate with 5 (less commonly 3) segments, or rarely undivided and flabelliform, up to 12 cm. long and 7 cm. wide; segments broadly ovate-cuneate, up to 4 cm. long, 3 cm. wide, thinly puberulous above, more densely pubescent below with prominent reticulate nervation, apex obtuse, margins bluntly toothed and irregularly lobed, the terminal segment often deeply 3-lobed; petioles up to 12 cm. long, densely pubescent, canaliculate above, basal sheaths conspicuous, overlapping, thick, distinctly nerved; stem leaves very sparse, similar to basal or sometimes divided into narrow, linear or oblanceolate, pinnatisect or laciniate segments, shortly petiolate or subsessile; umbels terminal, in lax, spreading panicles, 6–14-rayed; peduncles 2–9 cm. long; bracts and bracteoles wanting; rays shortly hispidulous, spreading, subequal, up to 3 cm. long at anthesis; flowers mostly hermaphrodite and fertile, numerous and rather dense in each umbellule; pedicels hispidulous, up to 5 mm. long at anthesis; sepals obsolete; petals dirty white, broadly oblong-obovate, unequally 2-lobulate, flattish, about 1 mm. long, 0·9 mm. wide, conspicuously pilose dorsally; filaments up to 1·5 mm. long, glabrous, inflexed; anthers yellow, oblong, about 0·5 mm. long, 0·4 mm. wide; stylopodium convex or shortly conical, about 0·6 mm. diam. with a crenulate margin; styles filiform, spreading, about 2 mm. long; stigmas capitate. Fruit narrowly ovoid-ellipsoid, about 4 mm. long, 2·5 mm. wide, laterally compressed, conspicuously white-pilose; carpophore deeply bipartite; mericarps dark brown when ripe, convex dorsally, with 5 distinct, pallid, filiform ridges, attenuate towards apex; endosperm flattish along its commissural face.

HAB.: Fissures of shaded, north-facing limestone cliffs; 1,100–2,500 ft. alt.; fl. April–May.

DISTR.: Divisions 7, 8, very common in Division 7. Endemic.

7. Bellapais, 1880, *Sintenis & Rigo* 592 ! St. Hilarion, 1894, *Post*; also, 1938, *Kennedy* 1069 ! also, 1939, *Lindberg f.* s.n. ! also, 1941, *Davis* 3598 ! Near the Armenian Monastery, 1912, *Haradjian* 389; Halevga, 1940, *Davis* 1930 ! Pentadaktylos, 1940, *Davis* 1933; Kornos (Kyrenia range), 1941, *Davis* 3037 ! Yaïla, 1941, *Davis* 3611 ! Cliff S.E. of St. Hilarion, 1949, *Casey* 612 !

8. Near Komi Kebir, 1912, *Haradjian* 329.

28. CRITHMUM *L.*

Sp. Plant., ed. 1, 246 (1753).
Gen. Plant., ed. 5, 116 (1754).

Glabrous perennial with a woody rootstock; leaves glaucous, thick, fleshy, aromatic when crushed, 1–2-pinnate with narrow segments; umbels compound, bracts and bracteoles present; flowers mostly hermaphrodite and fertile; sepals obsolete; petals greenish-yellow, not radiant; stylopodium prominent, conical; styles very short; stigmas subcapitate. Fruit oblong, compressed dorsally with a broad commissure; pericarp spongy, separating readily from seed; carpophore bipartite to base; mericarps pale brown when ripe, dorsal ridges prominent; vittae numerous, contiguous; endosperm concave or flattish along its commissural face.

One species on the coasts of western Europe and the Mediterranean.

1. **C. maritimum** *L.*, Sp. Plant., ed. 1, 246 (1753); Kotschy in Oesterr. Bot. Zeitschr., 12: 277 (1862); Unger et Kotschy, Die Insel Cypern, 301 (1865); Boiss., Fl. Orient., 2: 977 (1872); Holmboe, Veg. Cypr., 140 (1914); Post, Fl. Pal., ed. 2, 1: 545 (1932); Davis, Fl. Turkey, 4: 367 (1972); Zohary, Fl. Palaest., 2: 429, t. 621 (1972); Osorio-Tafall et Seraphim, List Vasc. Plants Cyprus, 77 (1973).
TYPE: " *ad oceani* Europaei *litora* ".

Decumbent or sprawling, much branched perennial, forming loose tufts or carpets; stems up to 50 cm. long, striatulate or indistinctly sulcate; basal and upper stem leaves similar, deltoid in outline, (5–) 10–15 (–18) cm. long, (4–) 6–10 (–15) cm. wide, usually 2-pinnate, the ultimate segments narrowly oblanceolate, fleshy, about 1·5–6 cm. long, 0·2–1 cm. wide, apex acute, margins entire, base gradually attenuate; petiole up to 7 cm. long, flattish or canaliculate above; basal sheaths conspicuous, 8–10 mm. long, persistent, auriculate at apex; umbels terminal, 10–40-rayed; peduncle stout, to about 8 cm. long; bracts 8 or more, narrowly oblong, fleshy, about 5–8 mm. long, 1·5–2·5 mm. wide, acute or subacute, persistent, becoming reflexed in fruit; rays subequal, spreading, 1·5–3·5 cm. long at anthesis; bracteoles 6–8, ovate or oblong, acute, about 2–4 mm. long, 1–2·5 mm. wide; flowers 15–20 in each umbellule; pedicels rather short and thick, to about 4 mm. long at anthesis; petals suborbicular, about 0·8 mm. diam., distinctly incurved with a shortly emarginate apex; filaments about 1·5 mm. long; anthers yellow, oblong, about 0·5 mm long, 0·4 mm. diam.; stylopodium about 0·6 mm. diam.; styles at first barely visible, lengthening to 0·3 mm. as the fruit develops. Fruit 8–9 mm. long, 4–6 mm. wide; mericarps separating fairly readily at maturity.

HAB.: Rocky (or rarely sandy) seashores; about sea-level; fl. July–November.

DISTR.: Divisions 1, 4, 6, 7. General distribution that of the genus.

1. Neighbourhood of Paphos, 1787, *Sibthorp*, and, 1862, *Kotschy* s.n.
4. Cape Greco, 1958, *N. Macdonald* 1 ! and 59 !
6. Coast north of Myrtou, 1862, *Kotschy* s.n.
7. Kyrenia, 1936, *Kennedy* 672 ! also, 1951, *Casey* 1212 ! also, 1955, *Merton* 225 ! 702 ! and 752 !

SESELI ANNUUM L., Sp. Plant., ed. 1, 260 (1753) (*S. coloratum* Ehrh.) recorded from Cyprus by Sibthorp (Fl. Graec. Prodr., 1: 199; 1806, " in insulae Cypri montibus ") is certainly an error. The plant has a wide distribution in northern and central Europe, but does not occur in the eastern Mediterranean region, and has not since been seen in Cyprus. Poech's record (Enum. Plant. Ins. Cypr., 28: 1842) for *S. coloratum,* also cited by Kotschy (Die Insel Cypern, 300; 1865) is based upon Sibthorp's error.

29. FOENICULUM *Mill.*

Gard. Dict. Abridg., ed. 4 (1754).

Erect, glabrous, aromatic biennials and perennials; leaves 3–5-pinnate, ultimate segments slender, filiform; basal sheaths usually broad and conspicuous; umbels compound, lateral, leaf-opposed; bracts and bracteoles absent; flowers mostly hermaphrodite and fertile; sepals obsolete; petals yellow, incurved, lobulate or emarginate; stylopodium conspicuous, convex or conical; styles short; stigmas capitate. Fruit clearly didymous, oblong or ovoid, not much compressed laterally; carpophore bipartite; mericarps with a broad commissure and 5 prominent ridges; valleculae 1-vittate, commissure 2-vittate; endosperm flattish along its commissural face.

One species widespread in Europe, Asia and Africa, and cultivated as a culinary herb throughout temperate and warm temperate regions of the world.

1. F. vulgare *Mill.*, Gard. Dict., ed. 8, no. 1 (1768); Holmboe, Veg. Cypr., 140 (1914); Lindberg f., Iter Cypr., 25 (1946); Davis, Fl. Turkey, 4: 376 (1972); Zohary, Fl. Palaest., 2: 432, t. 626 (1972); Osorio-Tafall et Seraphim, List Vasc. Plants Cyprus, 77 (1973).

Anethum foeniculum L., Sp. Plant., ed. 1, 263 (1753).

Foeniculum officinale All., Fl. Ped., 2: 25 (1785); Boiss., Fl. Orient., 2: 975 (1872).

Tall, erect perennial or biennial up to 2 (–2·5) m. high; stems glaucous or glaucescent, finely sulcate or striatulate, usually branched, especially in the upper part, smelling pleasantly of aniseed when crushed; branches erect; lower leaves broadly deltoid in outline, up to 30 cm. long and almost as wide, divided into numerous filiform segments 5–40 (or more) mm. long, 0·3–0·7 mm. wide with sharply subulate-acute apices; petiole usually short, sometimes wanting, seldom more than 5 cm. long, canaliculate above, basal sheath conspicuous, ribbed, up to 20 cm. long and 3·5 cm. wide; apex auriculate, margins narrowly membranous; upper leaves few and much reduced, sometimes consisting solely of a basal sheath with a short, aristate terminal appendage; umbels usually numerous, 3–30-rayed; rays subequal or unequal, spreading, 1–7 cm. long; flowers 15 or more in each umbellule; pedicels up to 1 cm. long at anthesis, petals oblong, 1 mm. long, 0·8 mm. wide, distinctly incurved with a shortly emarginate or lobulate apex; stylopodium at first convex, later shortly conical; filaments 1·5 mm. long; anthers yellow, suborbicular, up to 0·6 mm. diam.; styles barely visible at anthesis, lengthening to 0·4 mm. in fruit, spreading. Fruit oblong, 4–6 mm. long, 2·5–3 mm. wide; mericarps separating readily at maturity.

ssp. **vulgare**.

F. capillaceum Gilib., Fl. Lithuan., 2: 40 (1781); Rechinger f. in Arkiv för Bot., ser. 2, 1: 428 (1950).

F. vulgare Mill. ssp. *capillaceum* (Gilib.) Holmboe, Veg. Cypr., 140 (1914).

TYPE: Probably cultivated at Chelsea Physic Garden.

Plant robust, commonly exceeding 1·5 m., leafy; leaves plumose with fine, flaccid segments often 2 cm. or more in length; umbels 15–30-rayed; fruit sweet to the taste.

HAB.: Fields and gardens, perhaps always an escape from cultivation; sea-level to 5,600 ft. alt.; fl. April–Sept.

DISTR.: Divisions 2, 3, 6, apparently uncommon. General distribution that of the genus.

2. Troödos, 1939, *Lindberg f.* s.n.; Kykko Mon., 1939, *Lindberg f.* s.n.
3. Moni, 1905, *Holmboe* 640.
6. Myrtou, 1955, *G. E. Atherton* 406 !

ssp. **piperitum** (*Ucria*) *Coutinho*, Fl. Port., ed. I, 450 (1913); Holmboe, Veg. Cypr., 140 (1914); Osorio-Tafall et Seraphim, List Vasc. Plants Cyprus, 77 (1973).

Anethum piperitum Ucria in Nuov. Racc. Opusc. Aut. Sic., 6: 250 (1793).

Foeniculum piperitum (Ucria) Presl, Fl. Sic., 1: XXVI (1826); Unger et Kotschy, Die Insel Cypern, 300 (1865); Boiss., Fl. Orient., 2: 975 (1872); Post, Fl. Pal., ed. 2, 1: 545 (1932); Lindberg f., Iter Cypr., 25 (1946).

TYPE: Sicily; precise locality not indicated.

Plant rather slender and scoparioid, generally less than 1·5 m.; leaves sparse with rigid, rather fleshy segments usually less than 1·5 cm. long; umbels 3–15-rayed; fruit acrid to the taste.

HAB.: Sandy shores and sand dunes, dry stony fields and hillsides, dry river-beds; sea-level to 4,000 ft. alt.; fl. June–Oct.

DISTR.: Divisions 1, 2, 4, 5, 7, 8. Widespread in the Mediterranean region.

1. Between Kouklia and Yeroskipos, 1862, *Kotschy*; Polis, 1936, *Lady Loch* 6 ! Near Kissonerga, 1967, *Economides* 989 !
2. Pano Panayia, 1905, *Holmboe* 1065; Pharmakas, 1905, *Holmboe* 1136; Platres, 1937, *Kennedy* 673 !
4. Near Larnaca, 1880, *Sintenis & Rigo* 981 ! Famagusta, 1939, *Lindberg f.* s.n. !
5. Pedieos river near Govt. House gardens, 1935, *Syngrassides* 662 !
7. Kyrenia, 1948, *Casey* 55 ! Kyrenia range, 1955, *Miss Mapple* 108 ! also, 1955, *G. E. Atherton* 127 ! and, 357 ! Lapithos, 1955, *G. E. Atherton* 691 !
8. Valia, 1953, *Merton* 1057 !

NOTES: While typical specimens of *F. vulgare* ssp. *vulgare* and *F. vulgare* ssp. *piperitum* are easy to distinguish, one is always left with a residue of specimens intermediate in character between the two subspecies. The subspecific epithets are, however, generally useful in Cyprus for labelling cultivated (or escaped) Fennel from its indigenous counterpart.

If *Foeniculum divaricatum* Griseb. (*F. piperitum* var. *pluriradiatum* Boiss.) is regarded as part of the *piperitum* complex rather than a variant of *F. vulgare* ssp. *vulgare*, then it should be noted that, at subspecific rank, the epithet *divaricatum* has priority over *piperitum*, as it has been employed at this rank since 1879 in the combination *F. piperitum* (Ucria) Presl ssp. *divaricatum* (Griseb.) Nym., Consp. Fl. Europ., 292.

30. ANETHUM *L.*

Sp. Plant., ed. 1, 263 (1753).
Gen. Plant., ed. 5, 127 (1754).

Erect, glabrous, aromatic annual; leaves 3–5-pinnate, ultimate segments slender, filiform; basal sheaths shorter than in *Foeniculum* and less conspicuous; umbels compound, lateral, leaf-opposed; bracts and bracteoles absent; flowers mostly hermaphrodite and fertile; sepals obsolete; petals yellow, incurved, lobulate or emarginate; stylopodium convex or shortly conical with a crenulate base; styles short, reflexed in fruit; stigmas subcapitate. Fruit oblong-elliptic, dorsally flattened with a broad commissure; carpophore bipartite; mericarps separating readily at maturity; primary ridges conspicuous, the lateral distinctly winged; valleculae 1-vittate; endosperm flattish along its commissural face.

One species widely distributed in Europe, Asia and Africa, but usually as a cultivated plant or as an escape from cultivation.

1. **A. graveolens** *L.*, Sp. Plant., ed. 1, 263 (1753); Unger et Kotschy, Die Insel Cypern, 303 (1865); Boiss., Fl. Orient., 2: 1026 (1872); Holmboe, Veg. Cypr., 140 (1914); Post, Fl. Pal., ed. 2, 1: 554 (1932); Davis, Fl. Turkey, 4: 377 (1972); Zohary, Fl. Palaest., 2: 433, t. 627 (1972); Osorio-Tafall et Seraphim, List Vasc. Plants Cyprus, 77 (1973).

TYPE: " *inter* Lusitaniae & Hispaniae *segetes* ".

Stems erect, commonly unbranched in the lower half, 30–70 (–100) cm. high, subterete or inconspicuously sulcate; leaves deltoid in outline, 3–15 cm. long, 2·5–13 cm. wide, the lower usually with rather short ultimate segments, generally less than 1 cm. long, the upper with elongate-filiform segments, often 2–3 cm. long and less than 1 mm. wide, apex of segments sharply subulate-acute; sheaths up to 10 cm. long, auriculate at apex with narrowly membranous margins; inflorescence usually much branched; peduncles erect, 6–12 cm. long; umbels (6–) 18–40-rayed; rays subequal, spreading, 1–4 (–8) cm. long at anthesis; flowers 15–30 in each umbellule, pedicels to about 8 mm. long at anthesis; petals yellow, oblong, about 0·6 mm. long, 0·4 mm. wide, strongly incurved, apex widely emarginate; filaments about 1 mm. long; anthers suborbicular, about 0·5 mm. diam.; stylopodium about 0·6 mm. diam.; styles at first barely visible, lengthening to about 0·2 mm. in fruit and strongly reflexed. Fruit 3–6 mm. long, 2–3 mm. wide; mericarps dark brown when ripe, with conspicuous pallid dorsal ridges and marginal wings.

HAB.: Gardens and waste ground; near sea-level; fl. April–July.

DISTR.: Division 3, an escape from cultivation. General distribution that of the genus.

3. Near Limassol, 1862, *Kotschy*.

NOTES: Very similar to *Foeniculum* in general appearance, apart from the distinct fruits. It can generally be distinguished by its more slender, annual habit, shorter leaf-sheaths and many-rayed umbels.

31. KRUBERA *Hoffm.*

Gen. Umb., 103 (1814).

Erect or sprawling annual; stems much branched, conspicuously sulcate; leaves 3–4-pinnate, ultimate segments small, narrowly oblong or linear; basal sheaths inconspicuous; umbels lateral, leaf-opposed, rarely terminal, 2–5-rayed; peduncle short; bracts and bracteoles present, small, entire, deciduous; rays short, subequal, thickening in fruit; flowers few in each umbellule, subsessile, hermaphrodite and fertile; sepals minute but distinct; petals white, incurved, shallowly emarginate; stylopodium shortly conical; styles short, rather thick, becoming reflexed in fruit; stigmas truncate or subcapitate. Fruit oblong-ellipsoid, strongly compressed dorsally with a wide commissure; carpophore bipartite; mericarps separating tardily, dark brown when ripe; dorsal ridges very prominent, rugulose, the marginal strongly thickened; vittae apparently absent; endosperm flattish along its commissural face.

One species with a wide distribution in the Mediterranean region and Atlantic Islands, often united with *Capnophyllum* Gaertn. (type: *C. africanum* (L.) Gaertn.) but differing considerably in fruit and floral characters.

1. **K. peregrina** (*L.*) *Hoffm.*, Gen. Umb., ed. 1, 104 (1814); Boiss., Fl. Orient., 2: 1027 (1872).

Tordylium peregrinum L., Mantissa I, 55 (1767).

Krubera leptophylla Hoffm., Gen. Umb., ed. 2, 104, t. 3 (1816); Unger et Kotschy, Die Insel Cypern, 301 (1865); Sintenis in Oesterr. Bot. Zeitschr., 32: 292 (1882).

Capnophyllum peregrinum (L.) Lange in Willk. et Lange, Prodr. Fl. Hisp., 3: 33 (1874); Holmboe, Veg. Cypr., 140 (1914); Post, Fl. Pal., ed. 2, 1: 555 (1932); Davis, Fl. Turkey, 4: 438 (1972); Zohary, Fl. Palaest., 2: 434, t. 628 (1972); Osorio-Tafall et Seraphim, List Vasc. Plants Cyprus, 79 (1973).

TYPE: " *in* Oriente. D. Arduini ".

Annual; stems 20–40 cm. high, glabrous or sometimes very sparsely scabridulous, usually much branched from near base, conspicuously sulcate; leaves deltoid in outline, 3–15 cm. long, 2–12 cm. wide, finely dissected, glabrous, ultimate segments linear-oblong, about 2–3 mm. long, 1–1·5 mm. wide, apex acute, mucronate; basal leaves with canaliculate petioles up to 10 cm. long, upper leaves sessile or subsessile; basal sheaths small, rarely more than 1·5 cm. long, with fairly wide membranous margins; umbels mostly lateral; peduncle short, stout, usually less than 1·5 cm. long, distinctly sulcate, sometimes almost wanting; bracts 1–3, narrowly deltoid or subulate, about 2–2·5 mm. long, 0·5 mm. wide with narrow membranous margins; rays 5–12 mm. long at anthesis, lengthening and becoming thick and rigid in fruit; bracteoles 3–4, ovate, acute, about 1–1·5 mm. long, 0·5–1 mm. wide; pedicels very short, becoming very thick in fruit, and forming a distinct pulvinus at the apex of the ray; sepals narrowly deltoid, acute, less than 0·3 mm. long; petals suborbicular, about 1·2 mm. diam., apex emarginate with an incurved lobe; filaments about 1 mm. long at anthesis; anthers broadly ovoid or suborbicular, yellow, about 0·5 mm. diam.; styles at first erect and minute, lengthening to about 0·5 mm. in fruit and becoming reflexed. Fruit 6 mm. long, 4·5–5 mm. wide, strongly compressed dorsally; mericarps transversely rugose, with 3 conspicuous dorsal ridges and 2 thick, conspicuously rugose lateral ridges.

HAB.: Cultivated fields; sea-level to 500 ft. alt.; fl. March–May.

DISTR.: Divisions 4, 5, 8, rare. General distribution that of genus.

4. Near Larnaca, 1877, *J. Ball* s.n. !
5. Near Vatili, 1862, *Kotschy* 541 !
8. Between Ephtakomi and Leonarisso, 1880, *Sintenis & Rigo* 339 !

32. FERULA *L.*

Sp. Plant., ed. 1, 246 (1753).
Gen. Plant., ed. 5, 117 (1754).
E. Korovin, Gen. Fer. Mon. Illustr., 83 pp., 48 pls. (1947).

Robust perennial herbs; rootstock thick with a fibrous collar; leaves usually large, 2–6-pinnate, ultimate segments slender, filiform or broader and toothed; basal sheaths often large and conspicuous; inflorescence corymbose-paniculate; umbels terminal and lateral, the terminal frequently short-pedunculate, with hermaphrodite, fertile flowers, the lateral often long-pedunculate, functionally male and sterile; bracts absent; bracteoles absent or small and caducous; sepals minute or obsolete; petals yellow, entire or subentire with an incurved acumen; stylopodium flattish or shortly conical, arising from a broad, crenulate disk. Fruits suborbicular, obovate or oblong, strongly compressed dorsally; carpophore bipartite; mericarps laterally winged, with 3 slender dorsal ridges; valleculae 1–several-vittate, commissural vittae 2 or more; endosperm flattish or slightly concave along its commissural face.

About 130 species in the Mediterranean region, western and central Asia.

Fruit 12–15 mm. long, 9–12 mm. wide, marginal wing 1·5–2 mm. wide; plant very robust, 2–5 m. high - - - - - - - - - - - **1. F. communis**

Fruit about 9 mm. long, 5 mm. wide, marginal wing very narrow, less than 0·5 mm. wide; plant less than 2 m. high - - - - - - - - - **2. F. cypria**

1. F. communis *L.*, Sp. Plant., ed. 1, 246 (1753); Sibth. et Sm., Fl. Graec. Prodr., 1: 190 (1806); Poech, Enum. Plant. Ins. Cypr., 28 (1842); Boiss., Fl. Orient., 2: 991 (1872), Suppl., 264 (1888); Holmboe, Veg. Cypr., 140 (1914); Post, Fl. Pal., ed. 2, 1: 547 (1932); E. Korovin, Gen. Fer. Mon. Illustr., 48 (1947); Davis, Fl. Turkey, 4: 444 (1972); Zohary, Fl. Palaest., 2: 437, t. 633 (1972); Osorio-Tafall et Seraphim, List Vasc. Plants Cyprus, 79 (1973).

F. nodiflora L., Sp. Plant., ed. 1, 247 (1753); Sibth. et Sm., Fl. Graec. Prodr., 1: 190 (1806); Fl. Graec., 3: 70, t. 279 (1819).

F. graeca Sibth. in Walpole, Mem. Europ. Asiat. Turkey, ed. 1, 284 (1817) nomen.

F. communis L. var. *anatriches* Kotschy in Unger et Kotschy, Die Insel Cypern, 302 (1865); Holmboe, Veg. Cypr., 140 (1914).

F. anatriches (Kotschy) Sintenis in Oesterr. Bot. Zeitschr., 31: 192 (1881).

TYPE: " *in* Europa *australi* ".

Robust, erect, glabrous perennial 2–5 m. high; stems subterete or obscurely sulcate, sometimes 10 cm. or more diam. near base, unbranched or sometimes branched; basal leaves broadly deltoid, 4–6-pinnate, lamina often more than 30 cm. long and as wide or wider; ultimate segments filiform, 15–50 mm. long, 0·8–1·5 mm. wide, green on both sides (ssp. *communis*) or glaucous below (ssp. *glauca* (L.) Rouy et Camus), apex subulate-acute; stem leaves becoming progressively smaller upwards, the uppermost with the lamina reduced to a few filiform segments; basal leaves with a canaliculate petiole often more than 15 cm. long; upper leaves sessile or subsessile with very large, conspicuous, inflated basal sheaths, commonly more than 15 cm. long and 5 cm. wide; inflorescence corymbose-paniculate, repeatedly branched and frequently bearing 40 or more umbels; terminal umbel of each branch with a short, stout peduncle often less than 5 cm. long, 12–25-rayed, rays subequal, spreading, 3–7 cm. long at anthesis, lengthening to 15 cm. or more in fruit; flowers 12–20 in each umbellule, mostly hermaphrodite and fertile; pedicels slender, 3–8 mm. long at anthesis; lateral umbels smaller, peduncles 5–15 cm. or more long; rays 3–5 cm. long, flowers functionally male and sterile; petals yellow, ovate, about 2 mm. long, 1·5 mm. wide, with an incurved acute or acuminate apex; filaments 2·5–3 mm. long; anthers yellow, oblong, about 0·8 mm. long, 0·6 mm. wide; stylopodium of fertile flower shortly conical, 1·5–2 mm. diam., expanding basally to form a distinct crenulate disk; style slender, about 1·5 mm. long at anthesis, strongly reflexed in fruit; stigma capitate. Fruit suborbicular or oblong, 12–15 mm. long, 8–12 mm. wide, strongly compressed or flattened; mericarps separating readily at maturity, light brown, with 3 slender dorsal ridges and winged margins; valleculae 1–3-vittate (commonly 1-vittate in Cyprus specimens), commissural vittae 4–6; endosperm flattish along its commissural face.

HAB.: Waste ground, roadsides, by irrigation ditches, etc.; usually in deep, rich soils, but sometimes on sandy or rocky ground; sea-level to 1,500 ft. alt.; fl. March–May.

DISTR.: Divisions 4–8; common in Divisions 4 and 7. Widespread in the Mediterranean region.

4. Common near Larnaca, 1862, *Kotschy* 322 ! also, 1880, *Sintenis & Rigo* s.n.; Phaneromene, 1880, *Sintenis & Rigo* s.n.; nr. Livadhia, 1880, *Sintenis & Rigo* s.n. Kouklia, 1967, *Merton* 451 !
5. Athienou, 1880, *Sintenis & Rigo* s.n.; Salamis, 1905, *Holmboe*, also, 1931, *Nattrass*, and 1962, *Meikle* !
6. Myrtou, 1932, *Syngrassides* 201 !
7. Common; Panteleimon Mon. near Myrtou, 1862, *Kotschy* s.n.; Akanthou, 1880, *Sintenis & Rigo* 591 ! St. Hilarion, 1903, *A. G. & M. E. Lascelles* ! also, 1941, *Davis* 3600 ! Kyrenia, 1936, *Nattrass* 827; also, 1956, *G. E. Atherton* 1119 ! Near Panagra, 1962, *Meikle* !
8. Near Kantara, 1880, *Sintenis & Rigo* s.n.; Komi Kebir, 1880, *Sintenis & Rigo* s.n.; Rizokarpaso, 1880, *Sintenis & Rigo* s.n.

NOTES: Kotschy (Die Insel Cypern, 302) distinguishes var. *anatriches* chiefly on account of its 1-vittate valleculae, and, from an examination of a very limited range of fruiting specimens, it would appear that Cyprus plants differ from most populations of *F. communis* in this respect. On one sheet however (Myrtou, *Syngrassides* 201), the valleculae are partly (and incompletely) 2-vittate and one suspects that a reasonable sample of Cyprus specimens would show that the distinction is not altogether satisfactory. Since most herbarium sheets cannot be identified with respect to this character, I have, for the present, preferred to disregard what may be a valid distinction.

2. F. cypria *Post* in Mém. Herb. Boiss., 18: 94 (1900); H. Stuart Thompson in Journ. Bot., 44: 308 (1906); Holmboe, Veg. Cypr., 140 (1914); E. Korovin, Gen. Fer. Mon. Illustr., 47 (1947).

Ferulago cypria (Post) Lindberg f., Iter Cypr., 25 (1946) quoad nomen non H. Wolff in Fedde Repert., 20: 67 (1924).

TYPE: Cyprus; " ad Castello Sancti Hilarionis Cypri; fructiferat Julio [*Post*] No. 905 (BE, BM !).

Stems robust, erect, to 1 m. or more in height, terete or obscurely sulcate, becoming lustrous light brown in fruiting specimens; basal leaves not seen; stem leaves (according to Post) biternate with linear segments about 30 mm. long, 2 mm. wide; basal sheaths conspicuous, inflated, coriaceous; inflorescence repeatedly branched; terminal umbel subsessile, 10–15-rayed, rays 28–40 mm. long; pedicels rather shorter than fruit; mericarps oblong, dark brown when ripe, about 9 mm. long, 5 mm. wide, strongly compressed dorsally; 3 dorsal ridges filiform, rather prominent; marginal wing very narrow; valleculae 1-vittate; commissure 2-vittate; endosperm flattish or slightly concave along its commissural face.

HAB.: Dry, rocky calcareous hillsides; 1,000–2,000 ft. alt.; fl. May–June.

DISTR.: Division 7, very rare. ? S.E. Turkey, Syria, Lebanon, Palestine (see notes).

7. St. Hilarion, 1898, *Post* 905 !

NOTES: Isotype material at BM comprises a portion of the stem and several fruiting umbels; there are no leaves, but the bases of the large, inflated leaf-sheaths remain. The fruit is indistinguishable from that of *Ferula amani* Zohary et Davis (Kew Bull. 1947: 90) which (Fl. Turkey, 4: 449; 1972) is now united with *Ferula meifolia* (Fenzl) Boiss. and *F. cassii* Zohary et Davis under *F. elaeochytris* Korovin (Gen. Fer. Mon. Illustr., 61; 1947). If all these should prove conspecific with *F. cypria*, then the latter name will have priority, but in the absence of complete material of *F. cypria* any conclusion would be premature. The brief description given above is largely taken from that given by Post in his original account of the species.

33. FERULAGO *Koch*

Nov. Act. Nat. Cur., 12 (1): 97 (1824).

Erect perennial; rootstock with a fibrous collar; leaves 3–6-pinnate, ultimate segments generally small, lanceolate or filiform-setaceous; basal sheaths usually inconspicuous; inflorescence branched, bracts and bracteoles generally well developed and persistent; umbels compound, terminal and lateral, the terminal often larger than the lateral with hermaphrodite, fertile flowers, the lateral sometimes functionally male and sterile; sepals minute; petals yellow, strongly incurved; stylopodium flattish or shortly conical, arising from a broad, crenate-margined disk; styles often long and reflexed in fruit; stigmas capitate or truncate. Fruit suborbicular, ellipsoid or oblong, dorsally compressed with a wide commissure; mericarps with prominent filiform or winged dorsal ridges and a winged margin; valleculae and commissure generally with numerous vittae; endosperm somewhat concave along its commissural face.

About 50 species in southern Europe, Mediterranean region and eastwards to Central Asia.

Stems distinctly angular, often rather flexuous; ultimate segments of leaves linear or narrowly oblong, smooth - - - - - - 1. F. cypria

Stems terete, straight, rigid; ultimate segments of leaves narrowly oblong-lanceolate, scabrid - - - - - - - - - - - - - **2. F. syriaca**

1. F. cypria *H. Wolff* in Feddes Repert., 20: 67 (1924); Lindberg f., Iter Cypr., 25 (1946) quoad specimen; Osorio-Tafall et Seraphim, List Vasc. Plants Cyprus, 79 (1973).

TYPE: Cyprus; " in montibus inter Cerignia et Pano Dikomo ", June 1, 1880, *Sintenis & Rigo* 610 (K !).

Erect perennial to about 1 m. high; rootstock thick with a dense fibrous collar; stems angular, shallowly sulcate, often rather flexuous; leaves rather crowded near base of stem, broadly deltoid in outline, 25–45 cm. long, 20–30 cm. wide, 4-pinnate with 6–10, lax, remote pinnae; ultimate segments linear or narrowly oblong, 3–8 mm. long, 0·5–1 mm. wide, glabrous, smooth, rather fleshy, flattish or canaliculate, apex acute or apiculate; petiole short, usually less than 5 cm. long; basal sheath about 6 cm. long, 1·5 cm. wide, distinctly ribbed, minutely scabridulous-papillose, obscurely auriculate at apex; inflorescence pyramidal, repeatedly branched, branches lax, divaricate, the lower up to 50 cm. long, becoming shorter upwards, subtended by much modified leaves, or by small tongue-like bracts less than 1 cm. long and 0·6 cm. wide; umbels lax, 4–7-rayed, the terminal and lateral umbels not always clearly distinguishable; peduncles angular, 1·5–5 cm. long; bracts narrowly ovate, 3–4 mm. long, 1·5–2 mm. wide, thickish, opaque; rays rather unequal, 2·5–4 cm. long; bracteoles 3–5, ovate-lanceolate, 2–5 mm. long, 1–1·5 mm. wide; flowers about 5–7 in each umbellule; pedicels less than 3 mm. long at anthesis, lengthening to 12 mm. or more in fruit; ovary glabrous, deeply sulcate, about 1 mm. long and almost as wide; sepals obsolete; petals yellow, strongly incurved, about 1 mm. long and as wide or slightly wider, apex truncate or shallowly emarginate; filaments curved, about 1·2 mm. long; anthers yellow, suborbicular, about 0·9 mm. diam.; stylopodium flattish or slightly convex, about 1·8 mm. diam.; styles at first erect, about 0·8 mm. long, ? lengthening and becoming reflexed in fruit; stigmas truncate. Fruit (imperfect) oblong, strongly compressed, about 10 mm. long, 8 mm. wide, 3 dorsal ridges very prominent, almost winged, lateral wings 2–2·5 mm. wide; vittae evidently obscure. *Plate 51.*

HAB.: Dry, rocky, limestone slopes; 600–2,000 ft. alt.; fl. May–June.

DISTR.: Division 7, rare. Endemic.

7. Between Kyrenia and Pano Dhikomo, 1880, *Sintenis & Rigo* 610 ! St. Hilarion, 1939, *Lindberg f.* s.n. ! Lefkoniko Pass, 1957, *Merton* 3222 ! and, 1967, *Merton* in ARI 805 !

NOTES: Allied to *F. syriaca* Boiss., but differing markedly in its angular stems, also its foliage and inflorescences. No good, well-developed fruits have been seen; those examined having been removed from the remains of old, fungus-infected infructescences.

2. F. syriaca *Boiss.*, Diagn. 1, 10: 38 (1849); Fl. Orient., 2: 1002 (1872); Holmboe, Veg. Cypr., 140 (1914); Post, Fl. Pal., ed. 2, 1: 550 (1932); Lindberg f., Iter Cypr., 25 (1946); Davis, Fl. Turkey, 4: 465 (1972); Zohary, Fl. Palaest., 2: 440, t. 638 (1972); Osorio-Tafall et Seraphim, List Vasc. Plants Cyprus, 79 (1973).

[*Thapsia foetida* (non L.) Sibth. et Sm., Fl. Graec. Prodr., 1: 201 (1806).]

TYPE: " in fissures rupium montium pone *Antischiam* sitorum, prope *Suadie* ad ostium *Orontis* et in Antilibano inter *Damascum* et *Berythum* " *Boissier* (G, K !).

Erect, glabrous, glaucous perennial 90–200 cm. high; rootstock thick with a coarsely fibrous collar; stem usually unbranched in the lower part, rigid, terete, obscurely sulcate; leaves rather crowded near base of stem, sparse above, broadly deltoid in outline, 15–30 cm. long, 8–25 cm. wide, 4-pinnate with 6–10 lax, remote pinnae; ultimate segments narrowly oblong-lanceolate, 5–8 mm. long, 1–2 mm. wide, glabrous, distinctly scabrid along margins, midrib and apical parts of rhachis, rather fleshy, flattish with slightly thickened margins, apex obtuse or subacute, apiculate; petiole 5–10 cm. long, gradually expanding into a narrow, glabrous, ribbed basal sheath;

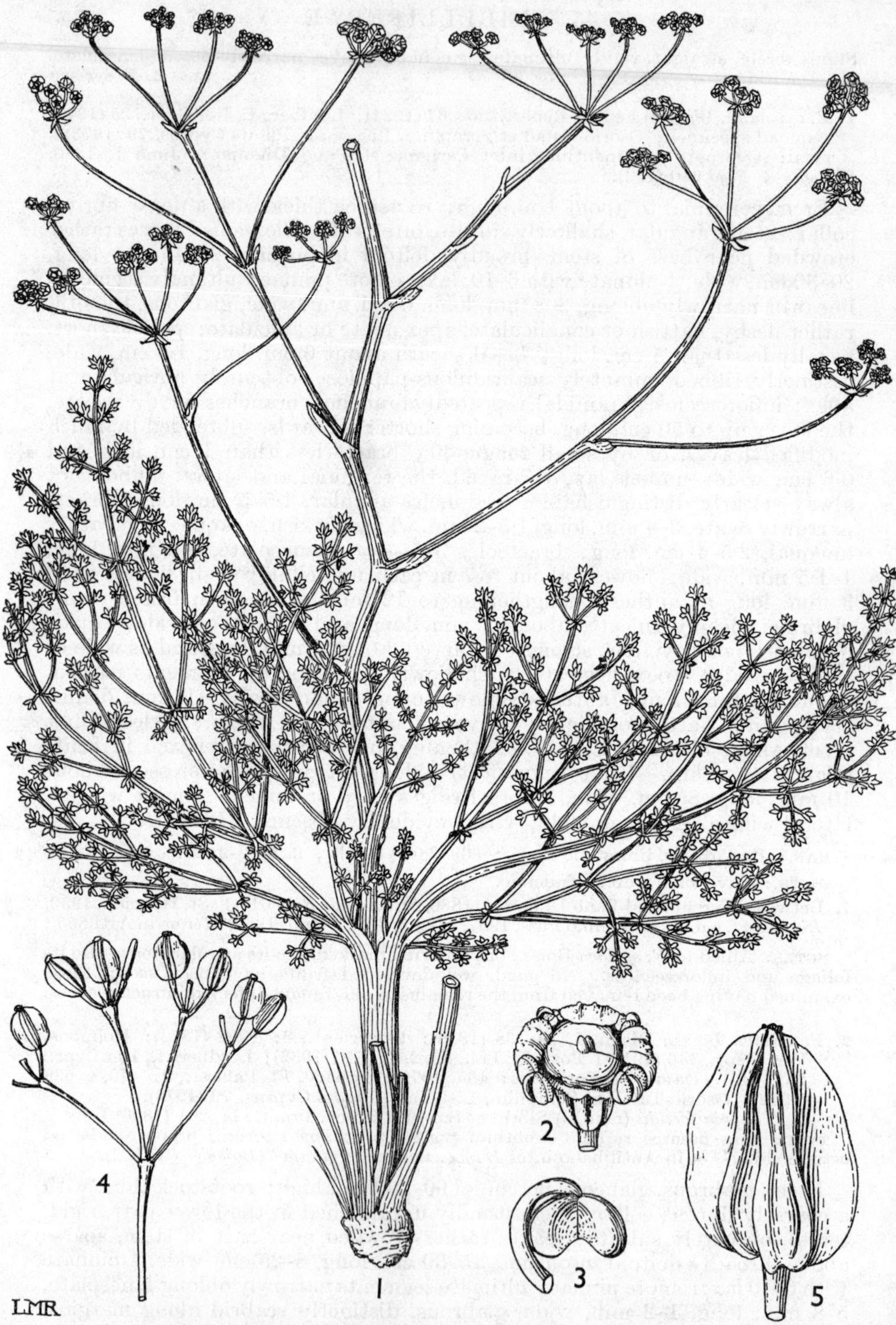

Plate 51. FERULAGO CYPRIA H. Wolff **1**, habit, ×⅔; **2**, flower with 2 petals and 1 stamen removed, ×8; **3**, stamen, ×16; **4**, infructescence, ×⅔; **5**, mature fruit, ×4. (**1–3** from *Lindberg f.* s.n.; **4, 5** from *Merton* 3222.)

inflorescence broadly corymbose-pyramidal, repeatedly branched, branches lax, divaricate, the lower up to 30 cm. long, becoming shorter upwards, subtended by narrow tongue-like bracts up to 2 cm. long and 0·6 cm. wide; umbels lax (5–) 6–8 (–12)-rayed, the terminal and lateral umbels not always distinguishable; peduncles terete, 3–8 cm. long; bracts oblong or ovate 5–6 mm. long, 1–2 mm. wide, thickish, opaque, spreading; rays rather unequal, 1–6 cm. long; bracteoles 3–5, oblong-lanceolate, 2–6 mm. long, 1–2 mm. wide; flowers about 12–14 in each umbellule, rather crowded; pedicels 2–7 mm. long at anthesis, lengthening to 12 mm. or more in fruit; ovary glabrous, deeply sulcate, 1–2 mm. long and almost as wide; sepals obsolete; petals yellow, glabrous, strongly incurved, about 1 mm. long and as wide or slightly wider, apex truncate or obscurely emarginate; filaments curved, about 1 mm. long; anthers yellow, broadly oblong, about 0·8 mm. long, 0·7 mm. wide; stylopodium flattish or slightly convex, about 1·8 mm. diam.; styles at first erect and about 1 mm. long, lengthening to 2 mm. or more in fruit and becoming irregularly reflexed; stigmas subcapitate. Fruit oblong-elliptic, strongly compressed, about 9 mm. long, 5 mm. wide, 3 dorsal ridges somewhat elevated, thick, blunt; lateral wings about 1 mm. wide; vittae 10 or more.

HAB.: Dry, rocky limestone slopes, 500–1,000 ft. alt.; fl. May–June.

DISTR.: Divisions 1, 7, rare. Also S. Turkey and Palestine.

1. Akamas, on the slopes of a mountain (? Kakoskala) near Fontana Amorosa, May 13, 1787, *Sibthorp*; between Khrysokhou and Fontana Amorosa, 1905, *Holmboe* 786 !
7. Ammos, by a streamside, July 7, 1939, *Lindberg f.*

34. OPOPANAX *Koch*

Nov. Act. Nat. Cur. 12 (1): 96 (1824).

Robust erect perennials; rootstock thick, with a fibrous collar; leaves pinnate or 2-pinnate, ultimate segments large, broad, ovate or oblong, crenate or dentate, commonly hispid-pilose; petioles and rhachis often conspicuously hispid; inflorescence much branched; umbels compound, terminal and lateral, the flowers of the terminal umbel hermaphrodite and fertile, those of the lateral functionally male and sterile, or sometimes all fertile; bracts and bracteoles few, small, narrow, often caducous; sepals obsolete; petals yellow, strongly incurved, broad, entire or emarginate, sometimes carinate; stylopodium flattish or slightly convex, expanding at the base into a broad, crenulate disk; styles long, reflexed in fruit; stigmas capitate. Fruit almost flattened dorsally, oblong or obovate, carpophore bipartite; mericarps with 3 slender dorsal ridges and conspicuous, often whitish, marginal wings; valleculae 1–3-vittate; endosperm flattish along its commissural face.

Three species in southern Europe, Mediterranean region and western Asia.

1. O. hispidus (*Friv.*) *Griseb.*, Spic. Fl. Rum. et Bith., 1: 378 (1843); Holmboe, Veg. Cypr., 140 (1914); Post, Fl. Pal., ed. 2, 1: 560 (1932); Lindberg f., Iter Cypr., 25 (1946); Davis, Fl. Turkey, 4: 472 (1972); Osorio-Tafall et Seraphim, List Vasc. Plants Cyprus, 79 (1973).

Pastinaca opopanax L., Sp. Plant., ed. 1, 262 (1753); Sibth. et Sm., Fl. Graec. Prodr., 1: 202 (1806), Fl. Graec., 3: 81, t. 288 (1819).

Ferula hispida Friv. in Flora, 18: 333 (1835).

Opopanax orientale Boiss. in Ann. Sci. Nat., ser. 3, 1: 330 (1844), Fl. Orient., 2: 1059 (1872); Unger et Kotschy, Die Insel Cypern, 301 (1865).

TYPE: Bulgaria; " prope Philippopolim [Plovdiv]", *Frivaldszky* (BP).

Stout erect perennial, up to 1·8 m. high; rootstock crowned with numerous narrow, ribbon-like fibres; stem subterete, obscurely sulcate, rather densely retrorse hispid towards base of plant, thinly hispid or glabrous above, usually not much branched except in region of the inflorescence; leaves 2-pinnate, broadly deltoid in outline, up to 50 cm. long and almost as wide, segments oblong or narrowly ovate, up to 10 cm. long, 2·5 cm. wide, thinly hispidulous with furcate bristles on both sides, apex acute or obtuse, base usually decurrent along rhachis, margins finely crenulate-dentate; petiole 8–20 cm. long, expanding basally to form a fairly conspicuous sheath 2–5 cm. long; inflorescence broadly paniculate with numerous spreading, opposite, verticillate or alternate branches, subtended by narrow, linear, modified leaves or bracts 1–1·5 cm. long; umbels 10–15 (–20)-rayed; peduncles 1–15 cm. long; bracts 3 or more, or sometimes absent, linear or narrowly oblong, acute, 3–10 mm. long, 0·5–2·5 mm. wide, caducous or persistent; rays spreading, subequal, 2–7 cm. long; bracteoles 2–5, sometimes absent, linear-subulate, less than 5 mm. long and 1 mm. wide; flowers about 10 in each umbellule, those of the terminal umbels usually hermaphrodite and fertile, often subtended by sterile umbels, or with some of the umbellules transformed into sterile secondary umbels; petals bright yellow, broad, about 1·2 mm. long, 1 mm. wide, strongly incurved with a truncate or somewhat emarginate apex; filaments about 1 mm. long; anthers yellow, suborbicular, about 0·8 mm. diam.; stylopodium about 1 mm. diam., somewhat convex; styles at first erect, about 0·5 mm. long, lengthening to 1·5 mm. in fruit and strongly reflexed. Fruit dorsally flattened, oblong-obovate, 10–12 mm. long, 8–9 mm. wide; mericarps glabrous or thinly hispidulous, with 3 slender, filiform dorsal ridges, marginal wing 1–1·5 mm. wide, whitish, conspicuous; valleculae conspicuously 1–2-vittate, commissure 6–8-vittate; endosperm flat along its commissural face.

HAB.: Cultivated fields, vineyards, roadsides or in garigue on dry calcareous or igneous hillsides; 300–4,000 ft. alt.; fl. May–June.

DISTR.: Divisions 1, 2, 4, 5, 7, 8. Common in Divisions 2 and 7. Northern Mediterranean region and west Asia from Italy to W. Iran.

1. Tsadha, 1905, *Holmboe* 739; Paphos, 1933, *Nattrass* 328; near Stroumbi, 1938, *Syngrassides* 1850 ! Between Terra and Dhrousha, 1941, *Davis* 3268 !
2. Common; between Chakistra and Kambos, 1937, *Syngrassides* 1612 ! Platres, 1938, *Kennedy* 1388 ! and 1387 ! Arminou, 1941, *Davis* 3409 ! Khrysorroyiatissa, 1941, *Davis* 3413 ! Mountain above Spilia, 1962, *Meikle* 2823A !
4. Between Athienou and Troulli, 1862, *Kotschy* 537, 573.
5. Peristerona, 1967, *Merton* 803 !
7. Common; near Kythrea, 1880, *Sintenis & Rigo* 597 ! Larnaka tis Lapithou, 1936, *Syngrassides* 1225 ! St. Hilarion, 1938, *Kennedy* 1074 ! and, 1939, *Lindberg f.* s.n., and, 1941, *Davis* 3596 ! Kythrea, 1941, *Davis* 3643 !
8. Komi Kebir, 1912, *Haradjian* 274 !

PEUCEDANUM VENERIS *Kotschy* (in Unger et Kotschy, Die Insel Cypern, 302; 1865), recorded as not uncommon between Kouklia and Yeroskipos (5th May, 1862, *Kotschy* 632) cannot be satisfactorily identified. Only the basal part of the plant and the radical leaves were seen by Kotschy, and the type material cannot now be traced. Holmboe suggests *Cachrys crassiloba* Boiss. (see p. 728) as a possible identification. This seems likely enough from the description, were it not for the fact that *C. crassiloba* is generally found at altitudes above 2,000 ft., whereas Kotschy makes it clear that *Peucedanum veneris* grew on the plains.

PEUCEDANUM OSTRUTHIUM (*L.*) *Koch*, recorded by Sibthorp (as *Imperatoria ostruthium* L.) from waste ground near Troödhitissa Monastery, 30th April, 1787 (in Walpole, Travels, 22; 1820) and included, but without exact locality in Fl. Graec. Prodr., 199 (1806). The record is not repeated in Flora Graeca, and is, as Holmboe comments, almost certainly an error, for the plant does not grow in the eastern Mediterranean, and has not since been collected in Cyprus.

35. SYNELCOSCIADIUM *Boiss.*

Ann. Sci. Nat., ser. 3, 1: 345 (1844).

Erect hispid annual; stems usually much branched, sulcate or somewhat angular towards base, subterete above; leaves pinnate with broad, oblong, ovate or obovate coarsely toothed segments; umbels unequally 3–7-rayed, the rays becoming erect in fruit; bracts and bracteoles present; flowers polygamous, some hermaphrodite and fertile, some evidently male and sterile and some female; sepals distinct, subulate, caducous; petals white, hispidulous externally, deeply bilobed, the outer conspicuously enlarged and radiant; stylopodium shortly conical, disk obsolete; styles rather long, erect, persistent; stigmas subcapitate. Fruit suborbicular, flattened dorsally, strigulose; carpophore very slender, bipartite; mericarps separating readily at maturity; dorsal ridges obscure, very slender; valleculae 1-vittate; commissure 2-vittate; endosperm flattish along its commissural face.

One species with a restricted distribution in Cyprus, Syria and Palestine.

1. S. carmeli (*Labill.*) *Boiss.* in Ann. Sci. Nat., ser. 3, 1: 346 (1844); Boiss., Fl. Orient., 2: 1050 (1872); Post, Fl. Pal., ed. 2, 1: 559 (1932); Rechinger f. in Arkiv för Bot., ser. 2, 1: 428 (1950); Zohary, Fl. Palaest., 2: 444, t. 645 (1972); Osorio-Tafall et Seraphim, List Vasc. Plants Cyprus, 79 (1973).

Heracleum carmeli Labill., Icones Plant. Syr. Rar., Dec. 5: 3, t. 1 (1812).

TYPE: Palestine; " in monte Carmelo Syriae " *Labillardière* (G).

Erect, lank annual 60–150 cm. high; stems conspicuously hispid with long retrorse hairs towards base of plant, subglabrous or with shorter adpressed bristles in the upper part, obscurely angular or rather shallowly sulcate; leaves sparse, remote, oblong in outline, 8–16 cm. long, 4–8 cm. wide, imparipinnate with 5–9 hispidulous leaflets 1–5 cm. long, 0·5–2·5 cm. wide, leaf margins irregularly and coarsely serrate-dentate; petioles 3–4 cm. long, hispid, canaliculate above, expanding towards base into a narrow, inconspicuous sheath 3–4 cm. long; inflorescence lax, much branched; umbels terminal; peduncles 6–12 cm. long, thinly clothed with adpressed retrorse bristles and becoming somewhat incrassate at apex in fruit; bracts 4–6, inconspicuous, subulate, 2–5 mm. long, less than 1 mm. wide, hispidulous; rays very unequal, 0·5–2·5 cm. long at anthesis, lengthening to 4 cm. and becoming rigidly erect in fruit; bracteoles similar to bracts but smaller and narrower; pedicels densely hispidulous, up to 2 mm. long; sepals erect, subulate, about 0·5 mm. long; petals white, deeply 2-lobed, the outer radiant, up to 4 mm. long and almost as wide, the inner 1·5–2·5 mm. long; filaments to 1·5 mm. long; anthers yellow, suborbicular, about 0·7 mm. diam.; styles erect, hispidulous, about 1·5 mm. long; stigmas subcapitate. Fruit suborbicular, about 7 mm. diam., dorsally flattened or strongly compressed; mericarps pale brown when ripe, thinly adpressed-strigulose; marginal wing about 1 mm. wide, slightly thickened; vittae conspicuous.

HAB.: Waste ground, rocky fields and hillsides; sea-level to 500 ft. alt.; fl. May–June.

DISTR.: Division 1, very rare. General distribution that of genus.

1. Ktima, 1913, *Haradjian* 663.

36. ZOSIMA *Hoffm.*

Gen. Umb., ed. 1, 145 (1814).

Perennial herbs with thick taproots crowned with a fibrous or scaly collar; leaves 1–4-pinnate, ultimate segments narrow and entire or broad and toothed; umbels compound, terminal; bracts and bracteoles present, sometimes numerous, but not very large or conspicuous; flowers mostly hermaphrodite and fertile; sepals obsolete; petals whitish, with an incurved emarginate apex, the outer a little larger than the inner but not conspicuously radiant; stylopodium small, flattish, expanding basally into a conspicuous flattish or slightly convex, undulate or crenulate disk; styles long, slender, reflexed in fruit; stigmas capitate. Fruit suborbicular, elliptic, oblong or obovate, flattened dorsally, apex sometimes shortly emarginate; carpophore slender, bipartite; mericarps separating readily at maturity, dorsal ridges slender, filiform, marginal wing broad, conspicuous, somewhat thickened; valleculae 1-vittate, with broad conspicuous vittae, commissure 2-vittate, or vittae fused into an unbroken zone; endosperm flattish along its commissural face.

Ten species in western and central Asia.

1. Z. absinthiifolia (*Vent.*) *Link*, Enum. Hort. Berol., 1: 274 (1821); Unger et Kotschy, Die Insel Cypern, 303 (1865); Boiss., Fl. Orient., 2: 1037 (1872); Holmboe, Veg. Cypr., 140 (1914); Post, Fl. Pal., ed. 2, 1: 558 (1932); Lindberg f., Iter Cypr., 25 (1946); Davis, Fl. Turkey, 4: 503 (1972); Zohary, Fl. Palaest., 2: 445, t. 646 (1972); Osorio-Tafall et Seraphim, List Vasc. Plants Cyprus, 79 (1973).

Heracleum absinthifolium Vent., Choix des Plantes, t. 7 (1803); Sibth. et Sm., Fl. Graec., 3: 74, t. 281 (1819).

Heracleum tomentosum Sibth. et Sm., Fl. Graec. Prodr., 1: 192 (1806).

TYPE: Cultivated at Cels from seeds collected by Bruguière and Olivier, " sur la route de Bagdad à Kermanchah " (G).

Robust erect perennial, 0·6–1 m. high; rootstock thick, crowned with a conspicuous fibrous collar; stems branched or unbranched, deeply and conspicuously sulcate, canescent with long, greyish hairs, exuding a heavy, unpleasant odour when crushed; leaves canescent, 4-pinnate, deltoid in outline, up to 30 cm. long and 25 cm. wide, ultimate segments narrowly oblong, 4–6 mm. long, 0·8–1 mm. wide, acute or subacute; petioles up to 20 cm. long, canaliculate above, expanding basally into a short, inconspicuous sheath; inflorescence rather sparingly branched; peduncles stout, erect, up to 20 cm. long, deeply sulcate; umbels 10–30-rayed; bracts numerous, subulate, tomentose, up to 10 mm. long and 1·5 mm. wide at base; rays spreading, subequal, 2·5–8 cm. long, thinly canescent; bracteoles similar to bracts, but often broader and more lanceolate, rather densely tomentose with membranous margins; flowers very numerous and crowded in each umbellule; pedicels slender, up to 8 mm. long; ovary tomentellous; petals milky-white, the outermost broadly obovate, 1·8–2 mm. long, 1·8 mm. wide, apex emarginate and incurved, the inner similar but smaller, usually not exceeding 1·5 mm.; filaments 2 mm. long; anthers yellow, suborbicular, about 0·8 mm. diam.; stylopodium and disk conspicuous, about 1·5 mm. diam.; styles at first erect, about 1·5 mm. long, lengthening a little and becoming reflexed in fruit; stigmas obscurely capitate. Fruit flattened, oblong-elliptic, about 1 cm. long, 0·8 cm. wide, thinly pubescent; mericarps pale brown when ripe, dorsal ridges very slender and obscure, margins somewhat thickened, forming pallid wings 2–2·5 mm. wide; vittae united into a uniform unbroken zone around the seed.

HAB.: Dry, rocky hillsides on calcareous or non-calcareous formations; 300–3,150 ft. alt.; fl. Febr.–April.

DISTR.: Divisions 2–4, 6, 7. Common in Division 7. South-west and central Asia from Turkey to Pakistan.

2. Kryos Potamos, 1937, *Kennedy* 677 ! Lefkara, 1941, *Davis* 2763 ! and 3552 ! Agros, 1941, *Davis* 3063 !

3. Mazotos, 1862, *Kotschy* 549, and, 1905, *Holmboe* 184.

4. Akhna Forest, 1967, *Merton* 430 !

6. Panagra, 1962, *Meikle* 2398 !
7. Very common; Kythrea, 1862, *Kotschy*, and, 1880, *Sintenis & Rigo* 342 ! Buffavento, 1862, *Kotschy* 441; St. Hilarion, 1933, *Syngrassides* 679 ! and, 1939, *Lindberg f.* s.n. ! and, 1941, *Davis* 2498 ! Kyrenia, 1956, *G. E. Atherton* 1034 !

37. TORDYLIUM *L.*

Sp. Plant., ed. 1, 239 (1753).
Gen. Plant., ed. 5, 111 (1754).
Runemark in Bot. Notiser, 121: 233–258 (1968).

Pubescent or villose annuals or biennials; stems usually branched; leaves undivided or pinnate with broad, variously toothed or lobed segments; umbels compound, terminal or leaf-opposed; bracts and bracteoles usually present, often conspicuous, rarely obsolete or caducous; flowers hermaphrodite and unisexual in the same umbellule; sepals usually present, often unequal; petals white, yellowish or purplish, the outermost often very conspicuously radiant, equally or unequally 2-lobed with an incurved apex; stylopodium and disk variable; styles short or long, deciduous or persistent. Fruit ovate-elliptic or suborbicular, sometimes dimorphic, the peripheral in each umbellule strongly flattened or compressed dorsally and 2-carpellate, the inner deeply concave or hemispherical and 1-carpellate; flattened mericarps with obscure dorsal ridges and a conspicuous, thickened, often beaded or moniliform margin; carpophore slender, bipartite; valleculae 1 (or rarely 3)-vittate; commissure 2–many-vittate; endosperm usually flattish along its commissural face.

About 18 species in Europe, the Mediterranean region and western Asia.

Leaf segments of upper cauline leaves crenate, toothed or shortly lobed, not pinnatisect; fruits uniform, all flattened and disk-like:
  Bracteoles very long and conspicuous, up to 20 mm. long, greatly exceeding the umbellules; mericarps with thickened, moniliform wings - - **3. T. syriacum**
  Bracteoles short, less than 10 mm. long, inconspicuous, shorter than, or not much exceeding the umbellules:
    Umbels conspicuously radiant; segments of uppermost leaves short, obcuneate or oblanceolate; mericarps with thickened, moniliform wings - **1. T. apulum**
    Umbels not conspicuously radiant; segments of uppermost leaves long, narrowly oblong-caudate; mericarps with smooth thickened wings - **2. T. maximum**
Leaf segments of upper cauline leaves deeply pinnatisect; fruits dimorphic, the outer flattened, disk-like, the inner hemispherical or cup-shaped; umbels conspicuously radiant - - - - - - - - - - - **4. T. aegyptiacum**

1. T. apulum *L.*, Sp. Plant., ed. 1, 239 (1753); Boiss., Fl. Orient., 2: 1034 (1872); Post, Fl. Pal., ed. 2, 1: 556 (1932); Davis, Fl. Turkey, 4: 505 (1972); Osorio-Tafall et Seraphim, List Vasc. Plants Cyprus, 79 (1973).
TYPE: Italy; " *in* Italiae, Apuliae *incultis* ".

Pubescent or villose annual, 10–60 cm. high; stems usually much branched, erect or spreading, distinctly sulcate or somewhat angular; leaves pinnate, the basal and lower cauline with (3–) 5–9 (–11) broadly ovate or suborbicular segments 1–2·5 cm. long, 0·8–1·5 cm. wide, apex of segments obtuse or rounded, base truncate or broadly cuneate, margins crenate and/or bluntly lobed, segments of upper leaves oblanceolate or obcuneate, usually less than 0·8 cm. wide, less pilose than those of the lower leaves, with sharply acuminate marginal lobes; umbels (2–) 5–8 (–10)-rayed, mostly terminal; peduncle 3–12 cm. long; bracts 4–6, linear-subulate, 2–10 mm. long, usually less than 1 mm. wide, thinly hispidulous, spreading at anthesis, persistent and often somewhat reflexed in fruit; rays subequal, 1–4 cm. long, spreading or suberect in fruit; bracteoles similar to bracts but rather smaller; pedicels slender, hispidulous, usually less than 4 mm. long at anthesis; sepals small, obscure, deltoid, usually less than 0·4 mm. long; petals white, the outermost conspicuously radiant, obovate, deeply and equally 2-lobed, 5–9 mm. long,

4–7 mm. wide, inner broadly obovate or oblong, 1·5–2 mm. long and almost as wide, apex blunt or emarginate, strongly incurved; filaments 0·8–1 mm. long; anthers suborbicular, about 0·4 mm. diam., greenish; disk about 1 mm. diam. with obscurely crenulate margins; stylopodium obsolete; styles very short at anthesis, usually less than 0·8 mm. long and reflexed in fruit; stigmas subcapitate. Fruits uniform, broadly elliptic or suborbicular, flattened dorsally, 6–10 mm. long and almost as wide; mericarps rather pale brown when ripe, apex shortly emarginate, dorsal surface thinly hispidulous-papillose, margins pallid, moniliform, much thickened; dorsal ridges 9–13, very slender and obscure; valleculae 1–3-vittate, commissure 9–10-vittate; endosperm flattish along its commissural face.

HAB.: On rocky slopes and stony seashores; sea-level to 50 ft. alt.; fl. March–April.

DISTR.: Divisions 6, 7, rare. S. Europe, Mediterranean region.

6. Near Orga, 1956, *M. E. D. Poore* 2636 ! Ayios Yeoryios near Liveras, 1962, *Meikle* 2405 !

7. Kharcha, 1950, *Chapman* 276 !

2. T. maximum *L.*, Sp. Plant., ed. 1, 240 (1753); Boiss., Fl. Orient., 2: 1031 (1872); Holmboe, Veg. Cypr., 141 (1914); Davis, Fl. Turkey, 4: 506 (1972); Osorio-Tafall et Seraphim, List Vasc. Plants Cyprus, 79 (1973).

TYPE: Italy; " *in* Italiae *ruderatis sepibus* ".

Erect annual or biennial 30–100 (–120) cm. high; stems conspicuously sulcate or somewhat angular, hispid with retrorse bristles, rather sparingly branched chiefly in the upper half, with erect branches; basal and lower cauline leaves oblong in outline, 6–15 cm. long, 4–10 cm. wide, pinnate with 5–9 ovate, strigose segments 1·5–8 cm. long, 1–5 cm. wide, apex of segments acute or obtuse, base cuneate, margins bluntly dentate and/or lobed; petioles hispid, up to 10 cm. long, canaliculate above, expanding basally into narrow, rather inconspicuous, thinly herbaceous or membranous sheaths; upper cauline leaves subsessile or very shortly petiolate, with narrowly oblong-caudate, deeply lobulate-toothed, acuminate segments up to 10 cm. long and 2 cm. wide; umbels 5–15-rayed, mostly lateral and leaf-opposed; peduncles stout, 5–25 cm. long; bracts 5–7, linear-subulate, 8–12 mm. long, less than 1 mm. wide, hispidulous, spreading or reflexed; rays unequal, ascending, 0·5–2·5 cm. at anthesis, lengthening to 3 cm. or more in fruit; bracteoles similar to bracts, 5–8 mm. long; flowers numerous, rather crowded, mostly hermaphrodite and fertile; pedicels very short at anthesis, lengthening to 2–4 mm. in fruit; sepals narrowly deltoid, about 1 mm. long, 0·5 mm. wide at base; petals white or purplish, adpressed-hispid externally, the outermost 2–4 mm. long, 1·5–3 mm. wide, obovate, unequally 2-lobed, somewhat radiant but usually not conspicuously so, inner petals usually less than 2 mm. long, and about as wide; filaments about 1 mm. long; anthers yellow, oblong, about 0·5 mm. long, 0·4 mm. wide; stylopodium and disk inconspicuous; styles erect, deciduous, about 0·5 mm. long, hispidulous; stigma subcapitate. Fruit suborbicular, about 6 mm. diam., strongly compressed or flattened; carpophore very slender, bipartite; mericarps pale brown when ripe, adpressed hispid; dorsal ridges obsolete, margin with a thickened, smooth, pallid wing about 1 mm. wide; valleculae 1-vittate for their whole length, commissure 2-vittate; endosperm flattish along its commissural face.

HAB.: On waste ground; c. 1,500 ft. alt.; fl. May–July.

DISTR.: Division 1, very rare, and adventive. Europe and N. Turkey to Caucasus and N. Iran.

1. On heaps of rubbish near the monastery of Ayios Neophytos, 1905, *Holmboe* 741.

NOTES: Far outside the natural range of this species, and certainly an alien in Cyprus. It has not been recorded by any other collector.

3. **T. syriacum** *L.*, Sp. Plant., ed. 1, 239 (1753); Unger et Kotschy, Die Insel Cypern, 303 (1865); Boiss., Fl. Orient., 2: 1030 (1872), Suppl., 268 (1888); Holmboe, Veg. Cypr., 141 (1914); Post, Fl. Pal., ed. 2, 1: 556 (1932); Rechinger f. in Arkiv för Bot., ser. 2, 1: 428 (1950); Davis, Fl. Turkey, 4: 510 (1972); Zohary, Fl. Palaest., 2: 442, t. 642 (1972); Osorio-Tafall et Seraphim, List Vasc. Plants Cyprus, 79 (1973).

TYPE: " *in* Syria ".

Erect or spreading, hispidulous annual (5–) 10–25 (–40) cm. high; stems conspicuously sulcate or angular, usually branched near the base, sometimes unbranched; basal leaves (commonly withered at anthesis) simple, 3-lobed or sometimes 3-foliolate, segments suborbicular or oblong, rounded, 1–4 cm. long, 1–3 cm. wide, margins irregularly crenate; petiole slender, canaliculate above, up to 10 cm. long, expanding basally into a narrow, inconspicuous sheath; upper leaves pinnate, usually with 5 segments, or rarely 3-lobed or simple, segments broadly obovate, 1–4·5 cm. long, 0·8–3·5 cm. wide, apex obtuse or rounded, base cuneate, margins crenate, dentate or irregularly crenate-lobed; petioles generally well developed; basal sheaths narrow, strigose with long-ciliate margins. Umbels leaf-opposed, 5–10-rayed; peduncle divaricate, 4–12 cm. long, rather stout, sulcate; bracts 2–5 (or more), very unequal, linear-subulate, 5–20 mm. long (or sometimes longer) usually less than 1·5 mm. wide at base, herbaceous, persistent, usually reflexed in fruit; rays spreading, very unequal, 2–18 mm. long at anthesis, lengthening to 30 mm. or more in fruit; bracteoles generally 5, very unequal, the 2 inner often minute and filiform, the 3 outer linear-subulate or narrowly oblanceolate, up to 20 mm. long, 2 mm. wide, scabridulous, conspicuous and persistent in fruit; flowers 8–12 in each umbellule, about half of them hermaphrodite and fertile, pedicels very short and slender, rarely exceeding 3 mm.; sepals minute, narrowly deltoid; petals white, unequal, the outermost obovate, subequally 2-lobed, up to 2·5 mm. long, 2 mm. wide, glabrous, somewhat radiant, the inner unequally incurved-emarginate, about 1 mm. long and almost as wide; filaments 1·5 mm. long; anthers violet, oblong, about 0·5 mm. long, 0·4 mm. wide; stylopodium and disk very small, convex or shortly conical; styles 1·5–2 mm. long at anthesis, spreading, rather persistent; stigma subcapitate. Fruit suborbicular or very broadly oblong-elliptic, 9–11 mm. long, 8–10 mm. wide, strongly compressed dorsally; carpophore very slender, bipartite; mericarps pale brown when ripe, closely papillose with a few scattered bristles, dorsal ridges obscure, margin with a very conspicuous, pallid, thickened, somewhat moniliform wing about 1·5 mm. wide; valleculae completely or incompletely 1-vittate; commissure 2-vittate; endosperm flattish along its commissural face.

HAB.: Cultivated and fallow fields, occasionally in maquis or on dry stony hillsides; sea-level to 1,500 ft. alt.; fl. Febr.–April.

DISTR.: Divisions 4, 5, 7, 8, fairly common. Also Turkey, Syria, Lebanon.

4. Between Larnaca and Livadhia, 1862, *Kotschy* 256a; Ayia Napa, 1952, *Merton* 707 ! Malounda (Famagusta), 1963, *J. B. Suart* 90 !

5. Kythrea, 1880, *Sintenis & Rigo* 325 ! and, 1956, *G. E. Atherton* 1065 ! and, 1963, *Merton* 340 ! Syngrasis reservoir, 1936, *Syngrassides* 926 !

7. Pentadaktylos, 1880, *Sintenis & Rigo* s.n.; near Ayios Khrysostomos Mon., 1880, *Sintenis & Rigo* s.n.; Akanthou, 1938, *Lady Loch* 35 ! Halevga Forest, 1950, *Chapman* 651 !

8. Near Rizokarpaso, 1912, *Haradjian* 194; Ayios Philon near Rizokarpaso, 1941, *Davis* 2222 ! Ronnas Bay near Rizokarpaso, 1941, *Davis* 2354 ! Peristerges near Kilamenos, 1962, *Meikle* 2430 !

4. **T. aegyptiacum** (*L.*) *Poir.* in Lam., Encycl. Méth., 7: 711 (1806); Boiss., Fl. Orient., 2: 1030 (1872); Holmboe, Veg. Cypr., 140 (1914); Post, Fl. Pal., ed. 2, 1: 556 (1932); Davis, Fl. Turkey, 4: 511 (1972); Zohary, Fl. Palaest., 2: 441, t. 640, 641 (1972); Osorio-Tafall et Seraphim, List Vasc. Plants Cyprus, 79 (1973).

Hasselquistia aegyptiaca L., Cent. I Plant., 9 (1755); Unger et Kotschy, Die Insel Cypern, 303 (1865).

TYPE: " in Arabia " *Hasselquist* (LINN).

Erect annual 7–40 (–60) cm. high; stems thinly villose, deeply sulcate or angular, branched or unbranched, or frequently branched only in the upper part; basal and lower cauline leaves usually withered at anthesis, oblong in outline, 3–8 cm. long, 2–4 cm. wide, pinnate with 2–3 deeply and bluntly pinnatisect, pubescent segments 1–2 cm. long, 0·5–1 cm. wide; petiole slender, 3–8 cm. long, thinly villose, canaliculate above, expanding basally into a narrow, inconspicuous sheath; upper leaves similar, but often larger, up to 16 cm. long and 9 cm. wide, segments more sharply pinnatisect, petiole often very short or wanting, basal sheath 2–3 cm. long, 1 cm. wide; umbels lateral, leaf-opposed and terminal, 5–15-rayed; peduncles rather slender, 2·5–7 cm. long, sulcate, thinly hispidulous; bracts 2–7, sometimes obsolete, very unequal, filiform-subulate, up to 10 mm. long, less than 0·5 mm. wide at base, subglabrous or scabrid-ciliate; rays unequal, 0·5–5 cm. long, spreading; bracteoles 5, very unequal, 3 outer filiform-subulate, up to 5 mm. long, 2 inner very small, less than 1 mm. long, narrowly deltoid; flowers about 15 in each umbellule, most of the central ones male-functional and sterile, the central umbellule completely sterile and forming a dense violet-black capitulum; pedicels slender, up to 5 mm. long; sepals of fertile flowers distinct, deltoid, about 1·2 mm. long, 1 mm. wide at base, those of sterile flowers smaller; petals white, glabrous, unequal, the outermost obovate, about 4–5 mm. long and almost as wide, equally or unequally 2-lobed, conspicuously radiant, inner about 1–1·5 mm. long and almost as wide, unequally 2-lobed; filaments 1·5–2 mm. long; anthers oblong, greenish, 0·6 mm. long, 0·5 mm. wide; stylopodium obsolete, disk flattish about 1 mm. diam.; styles slender, 1 mm. long, reflexed in fruit; stigma capitate. Fruits of two kinds, the outermost generally orbicular or broadly elliptic-orbicular, flattened, 2-carpellate, about 10 mm. long, 8–9 mm. wide, mericarps pale brown when ripe, papillose, thinly hispidulous, dorsal ridges very obscure, margin with a distinct, somewhat moniliform, pallid wing about 0·8 mm. wide; valleculae 1-vittate, vittae complete or incomplete, commissure 2-vittate, endosperm flattish along its commissural face; inner fruits 1-carpellate, cup-shaped or hemispherical, about 5 mm. diam., with incurved, winged margins, pale brown when ripe, closely papillose, 4-vittate.

HAB.: Cultivated and fallow fields; sea-level to 1,000 ft. alt.; fl. Febr.–April.

DISTR.: Divisions 1, 3–8. Common in Divisions 4 and 5. Very common in the Mesaoria. Also Turkey, Syria, Palestine, Iraq, Egypt.

1. Ktima, 1862, *Kotschy* 666a; Ayios Yeoryios (Akamas), 1962, *Meikle* ! Polis, 1962, *Meikle* !
3. Kolossi, 1862, *Kotschy* s.n; Limassol, 1862, *Kotschy* s.n.
4. Common; Larnaca, 1862, *Kotschy* 47; 1880, *Sintenis & Rigo* s.n.; 1881, *C. & W. Barbey*, and, 1905, *Holmboe* 130; Livadhia, 1880, *Sintenis & Rigo*; Ayia Napa, 1936, *Syngrassides* 922 !
5. Common; Kythrea, 1880, *Sintenis & Rigo* 334 ! and, 1950, *Chapman* 340 ! Strongylos, 1905, *Holmboe* 360; before Pyroi from Nicosia, 1934, *Syngrassides* 466 ! Syngrasis, 1936, *Syngrassides* 925 ! Mia Milea, 1950, *Chapman* 502 ! and, 1957, *Merton* 2969 !
6. Ayia Irini, 1941, *Davis* 2541 ! Near Kondemenos, 1967, *Merton* 107 !
7. Karakoumi, 1968, *Economides* 1173 !
8. Near Komi Kebir, 1880, *Sintenis & Rigo* s.n.; Gastria, 1941, *Davis* 2413 ! Leonarisso, 1969, *Economides* 1069 !

38. AINSWORTHIA *Boiss.*

Ann. Sci. Nat., ser. 3, 1: 343 (1844).
Alava in Notes Roy. Bot. Gard. Edinb., 31: 109–118 (1971).

Scabridulous or hispidulous annuals, very similar to *Tordylium* in general aspect; basal leaves simple, cordate, long-petiolate; cauline leaves either resembling basal or pinnate with 3–5 broad, blunt segments; umbels terminal and leaf-opposed; rays usually numerous, unequal, sometimes becoming erect in fruit; bracts numerous, conspicuous, filiform, deflexed in fruit; bracteoles 5 or more, filiform; flowers mostly hermaphrodite and fertile; sepals obsolete; petals white, glabrous, 2-lobed, the outermost conspicuously enlarged and radiant; stylopodium small, shortly conical; styles slender, reflexed in fruit, rather persistent; stigmas capitate. Fruits uniform or dimorphic, either strongly compressed dorsally and disk-like, or cup-shaped, rather small; carpophore slender, bipartite; mericarps with conspicuous, thickened margins; valleculae 1-vittate, commissure 2-vittate; endosperm flattish along its commissural face.

Two species in the eastern Mediterranean region and further eastwards to Iraq.

1. A. trachycarpa *Boiss.*, Diagn., 1, 10: 43 (1849); Fl. Orient., 2: 1035 (1872); J. Freyn in Bull. Herb. Boiss., 5: 621 (1897); Post, Fl. Pal., ed. 2, 1: 557 (1932); Davis, Fl. Turkey, 4: 512 (1972); Zohary, Fl. Palaest., 2: 443, t. 643 (1972); Osorio-Tafall et Seraphim, List Vasc. Plants Cyprus, 79 (1973).

A. cordata Boiss. in Ann. Sci. Nat., ser. 3, 1: 343 (1844); Unger et Kotschy, Die Insel Cypern, 304 (1865); Boiss., Fl. Orient., 2: 1034 (1872); Post, Fl. Pal., ed. 2, 1: 557 (1932) quoad descr. nec *A. cordata* (Jacq.) Boiss. (≡*Hasselquistia cordata* Jacq., Hort. Vindob., 2: 91; 1771).

Tordylium cordatum Holmboe, Veg. Cypr., 141 (1914) nec *T. cordatum* (Jacq.) Poir.

T. cordatum (Jacq.) Poir. ssp. *trachycarpum* (Boiss.) Holmboe, Veg. Cypr., 141 (1914).

TYPE: Palestine; " in collibus et cultis Judaeae, prope Hierosolymam, S. Saba et in descensu ad *mare Mortuum* Aprili 1846 " *Boissier* (G).

Erect annual 20–70 (–100) cm. high; stems slender, deeply sulcate or angled, thinly or rather densely hispid-villose with long spreading or retrorse hairs, usually much branched; basal leaves cordate, 1·5–7 (–10) cm. long and about as wide, thinly hispidulous, apex acute or obtuse, margins crenate or indistinctly lobed; petiole 5–15 cm. long, canaliculate above, expanding basally into a narrow, inconspicuous sheath; cauline leaves usually pinnate (occasionally simple) with 3 or 5 obovate, ovate or cordate segments with crenate or irregularly lobed margins; petiole usually short or obsolete; umbels 10–25 (–30)-rayed; peduncles erect 3–20 cm. or more long, scabridulous, deeply sulcate; bracts 12–18, filiform, scabridulous, 8–15 mm. long, deflexed in fruit; rays unequal, up to 3·5 cm. long, spreading at anthesis, becoming erect in fruit, closely scabrid-papillose on the adaxial surface, thinly hispidulous on the abaxial; bracteoles 5–7, filiform, scabridulous, unequal, the 3 outer 7–18 mm. long, the inner much shorter; flowers 15–20 in each umbellule, mostly fertile; pedicels very slender, up to 6 mm. long; sepals obsolete; petals white, 2-lobed, glabrous, the outermost up to 6 mm. long and 5 mm. wide, distinctly radiant, the inner about 0·8 mm. long and about as wide, rather strongly incurved; filaments about 1·3 mm. long; anthers yellow, suborbicular, about 0·4 mm. diam.; styles 0·6 mm. long, persistent, recurved in fruit; stigmas capitate. Fruits uniform, suborbicular or broadly oblong-elliptic about 3–4 mm. long, 3 mm. wide, strongly compressed or flattened dorsally; mericarps dull brown when ripe, sparsely and minutely scabridulous on the dorsal surface with a conspicuous pale, smooth, thickened wing about 0·7 mm. wide; valleculae 1-vittate; commissure 2-vittate, endosperm flattish along its commissural face.

HAB.: Cultivated and fallow fields, pastures, roadsides, waste ground and grassy slopes; sea-level to 2,000 ft. alt.; fl. Febr.–May.

DISTR.: Divisions 3–8, common in Divisions 5 and 7. Also Turkey, Syria, Palestine and Iraq.

3. Kalavasos, 1905, *Holmboe* 612; Sotira, Ayios Antonios, 1950, *Chapman* 596 !
4. Paralimni, *N. MacDonald* 50 !
5. Common; Strongylos, 1905, *Holmboe* 36 ! Ayia Katerina, 1948, *Mavromoustakis* s.n. ! Mile 3, Nicosia–Famagusta Rd., 1950, *Chapman* 381 ! 4 miles E. of Nicosia, 1967, *Merton* 479 !
6. Myrtou, 1932, *Syngrassides* 753 ! Ayia Irini, 1941, *Davis* 2506 ! and, 1962, *Meikle* 2394 ! Syrianokhori marshes, 1952, *Merton* 776 !
7. Common. Lapithos, 1862, *Kotschy*, and, 1939, *C. B. Ussher* 109 ! Between Akanthou and Phlamoudhi, 1880, *Sintenis & Rigo* 324 ! Ayios Epiktitos, 1936, *Syngrassides* 1253 ! and, 1967, *Merton* 584 ! Kyrenia, 1938, *C. H. Wyatt* 29 ! 1951, *Casey* 1184 ! 1955, *Merton* 2085 ! 1955, *Miss Mapple* 55 ! 1956, *G. E. Atherton* 1009 ! 1262 ! 1206 !
8. Valia near Ayios Theodhoros, 1905, *Holmboe* 496; Cape Andreas, 1948, *Mavromoustakis* s.n. !

39. GLAUCOSCIADIUM *Burtt et Davis*
in Kew Bull. 1949: 225–230 (1949).

Robust, erect, glabrous, glaucous perennial; rootstock thick, tough, rhizomatous with a scale-covered apex; stems solid, much branched, subterete or obscurely sulcate; leaves mostly basal, long-petiolate, 1–2-pinnate with large, entire, cordate, ovate, elliptic or suborbicular segments; leaf-sheaths conspicuous, persistent; umbels compound, terminal, 7–15-rayed; peduncles rather slender; bracts and bracteoles obsolete or much reduced and caducous; rays unequal, spreading; flowers 10–30 or more in each umbellule, rather crowded, the outermost hermaphrodite and fertile, the inner functionally male and sterile; pedicels short; sepals obsolete; petals creamy-white, subequal, oblong, emarginate, strongly incurved, not radiant; stylopodium shortly conical, expanding below into a crenulate disk; styles slender, persistent, reflexed in fruit; stigmas subcapitate. Fruit oblong, flattened dorsally; carpophore bipartite; mericarps separating readily at maturity, dull brown or purplish with a pale marginal wing; dorsal ridges slender, not very prominent; valleculae completely or incompletely 1–4-vittate, commissural vittae absent; endosperm flattish along its commissural face.

One species with a limited distribution in southern and central Turkey and Cyprus.

1. **G. cordifolium** (*Boiss.*) *Burtt et Davis* in Kew Bull., 1949: 229 (1949); Davis, Fl. Turkey, 4: 514 (1972); Osorio-Tafall et Seraphim, List Vasc. Plants Cyprus, 79 (1973).

Siler cordifolium Boiss., Fl. Orient. Suppl., 263 (1888); Post in Mém. Herb. Boiss., 18: 94 (1900); H. Stuart Thompson in Journ. Bot., 44: 308 (1906); Holmboe, Veg. Cypr., 141 (1914); A. K. Jackson et Turrill in Kew Bull., 1938: 465 (1938).

Laser (?) *cordifolium* (Boiss.) Thellung in Le Monde des Plantes, no. 38–153: 4 (1925).

TYPE: Turkey; " in regione inferiore montis Cadmi Cariae supra Denisleh et Mesogis prope Dervent (Boiss., exempl. sterilia !), in montibus prope Ermenek Ciliciae Tracheae (Pironin, floriferum !) " (G, LD, P).

Description as for genus. Acrid-smelling perennial; stems 40–120 (–150) cm. high; basal leaves deltoid in outline, up to 30 cm. long and 25 cm. wide, pinnately divided into 3–11 broad, sessile or petiolulate segments, the lowermost pinnae often ternately subdivided; segments glaucous, subcoriaceous (often dark red when beginning to develop), acute or obtuse, commonly apiculate, up to 8 cm. long and 6 cm. wide; petiole stout, up to 20 cm. long, expanding basally into a narrow, but conspicuous, glabrous sheath; cauline leaves few, remote, with 3 small, ovate-elliptic segments, or often reduced to a leafless, lingulate basal sheath; peduncles generally 5–6 cm. long at anthesis, up to 20 cm. long in fruit; bracts usually absent, or few, narrowly deltoid, about 2 mm. long, 1 mm. wide, soon deciduous; rays 1–3·5 cm. long; petals about 0·8 mm. long, 0·6 mm. wide; filaments 1–1·5 mm. long; anthers yellow, orbicular, about 0·4 mm. diam.; styles about 0·6 mm. long at anthesis, lengthening a little in fruit. Mericarps 10–12 mm. long, 5–6·5 mm. wide; valleculae completely or incompletely 1–4-vittate, vittae slender, rather obscure; commissural vittae absent.

HAB.: In garigue on limestone or igneous mountainsides; 3,000–4,000 ft. alt.; fl. August–Sept.

DISTR.: Division 2; locally common; general distribution that of the genus.

2. Locally common; Perapedhi, 1894, *Post*, and, 1940, *Davis* 1835 ! Between Pano Panayia and Khrysorroyiatissa Mon., 1905, *Holmboe* 1061; Aphamis, 1937, *Kennedy* 674 ! 675 ! 675a ! 1939, *Kennedy* 1390A ! 1939, *Davis* 1979 ! 1940, *Davis* 1979K ! Platres, 1937, *Kennedy* 676 ! 1938, *Kennedy* 1389 ! 1399 ! 1940, *Davis* 1848 ! Khrysorroyiatissa, 1941, *Davis* 3368 ! Between Kalousa and Kakopetria, 1952, *Probyn* 139 !

46. ARALIACEAE

Hutchinson, Genera of Flowering Plants, 2: 52–81 (1967).

Trees, shrubs and climbers, sometimes with aerial roots, very rarely herbs; indumentum generally of stellate hairs, sometimes prickly; leaves alternate, rarely opposite, simple or often palmately or pinnately lobed or compound; stipules usually present, adnate to petiole or variously connate, rarely free; inflorescences umbellate, capitate, spicate or racemose, or flowers rarely solitary; bracts commonly resembling stipules, rarely foliaceous, often caducous; bracteoles present or absent; pedicels sometimes articulated at the apex and expanded to form a small calyculus; flowers small, hermaphrodite, polygamous or unisexual, actinomorphic; hypanthium adnate to ovary; sepals present or absent, when present usually as many as petals; petals mostly 5, sometimes fewer or numerous, valvate or somewhat imbricate, free or united into a calyptra, deciduous or persistent, usually with a broad base; stamens as many as petals and alternating with them, rarely more numerous; filaments free, often short, thick and incurved; anthers dorsifixed, ovoid or oblong, dehiscing longitudinally; disk epigynous, flattish, pulviniform or conical, often confluent with the thickened style-bases and marked with radiating ridges; ovary 1–many-locular; styles as many as ovary-loculi; stigmas terminal or decurrent on the inner face of the styles; ovules usually solitary in each loculus, pendulous, anatropous. Fruit a berry or a drupe, with a fleshy (rarely a membranous) pericarp; endocarp divided into 1–many, bony, crustaceous, cartilaginous or rarely membranous pyrenes; endosperm usually copious, fleshy or cartilaginous, occasionally ruminate; embryo small; cotyledons ovate or oblong.

About 84 genera and 920 species widely distributed in the tropics, becoming less frequent in temperate regions.

1. HEDERA *L.*

Sp. Plant., ed. 1, 202 (1753).
Gen. Plant., ed. 5, 94 (1754).
F. Tobler, Die Gattung Hedera, iv + 151 pp. (1912).

Climbing evergreen shrubs, often with aerial roots; leaves alternate, undivided or palmately lobed with entire margins; stipules absent; inflorescence a panicle of umbels; bracts minute and caducous or absent; pedicels not or obscurely articulated; flowers polygamous or hermaphrodite; sepals minute, deltoid; petals 5, free, valvate, deltoid with a broad base, greenish-yellow, usually recurved at anthesis; stamens 5, alternating with petals; disk pulviniform or shortly conical, nectariferous; ovary 5-locular; styles united; stigmas sessile, apical. Fruit subglobose, berry-like; pyrenes ovoid, bluntly trigonous, membranous or chartaceous; endosperm ruminate.

Species 6 (or probably fewer), widely distributed in Europe, Asia and N. Africa.

1. **H. helix** *L.*, Sp. Plant., ed. 1, 202 (1753); Poech, Enum. Plant. Ins. Cypr., 29 (1842); Unger et Kotschy, Die Insel Cypern, 310 (1865); Boiss., Fl. Orient., 2: 1090 (1872); Holmboe, Veg. Cypr., 134 (1914); Post, Fl. Pal., ed. 2, 1: 574 (1932); Lindberg f., Iter Cypr., 24 (1946); Chapman, Cyprus Trees and Shrubs, 59 (1949); Davis, Fl. Turkey, 4: 538 (1972); Zohary, Fl. Palaest., 2: 378, t. 551 (1972); Osorio-Tafall et Seraphim, List Vasc. Plants Cyprus, 75 (1973).

TYPE: " *in* Europae *arboribus putrescentibus, inque sepibus* ".

Woody climber, attaching itself to tree trunks and rocks by aerial rootlets; stems with rough, sulcate and fissured, grey-brown bark, young shoots subglabrous or thinly stellate-pubescent; leaves of sterile growths cordate or deltoid, 2–10 (–15) cm. long, 1·3–7 (–13) cm. wide, lustrous dark green above with paler venation, uniformly paler green below, bluntly and often obscurely 3-lobed in our area (often deeply 5–7- or more lobed elsewhere); petioles 1–8 (or more) cm. long, subterete, glabrous or sparsely stellate-pubescent; leaves of flowering growths generally ovate, elliptic or broadly lanceolate, base rounded or tapering, margins without lobes; inflorescences terminal, branched, the branches ± densely stellate-pubescent; bracts minute, broadly deltoid, caducous, stellate-pubescent; umbels 6–25-flowered; pedicels slender, pubescent, 5–13 mm. long; hypanthium shallowly obconical, about 2 mm. long, 3 mm. diam. at apex, stellate-pubescent; sepals minute, inconspicuous; petals greenish-yellow, rather fleshy, about 3 mm. long, 2 mm. wide at base, strongly recurved at anthesis, stellate-pubescent externally; filaments glabrous, flattened, about 2·5 mm. long; anthers yellow, oblong, about 1·3 mm. long, 0·8 mm. wide; disk conspicuous, slightly conical, about 2·5 mm. diam.; united styles about 1 mm. long. Fruit globose, 7–9 mm. diam., dull blackish or yellow (ssp. *poetarum* (Bertol.) Nym.) when ripe, apex distinctly flattened, usually bearing the persistent styles; pyrenes ovoid-prismatic or trigonous, about 5 mm. long, 3 mm. wide; endocarp pale, papery; seed brownish, conspicuously ruminate.

HAB.: Wooded valleys, often close to streams, generally at altitudes above 1,500 ft. but descending to 300 ft. in the Karpas peninsula; fl. Aug.–Oct.; fr. Nov.–March.

DISTR.: Divisions 2, 7, 8. Widespread in temperate Europe and Asia.

2. Near Troödhitissa Mon., 1840, *Kotschy*; near Philani, 1862, *Kotschy*; Makheras Mon., 1905, *Holmboe* 1117; above Lefka, 1927, *Rev. A. Huddle* 64 ! Agros, 1930, *Druce* s.n. ! Kakopetria, 1930, *C. B. Ussher* 123 ! Above Kakopetria, 1935, *Syngrassides* 866 ! Kryos Potamos, 3,100 ft. alt., 1937, *Kennedy* 662 ! Prodhromos, 1939, *Lindberg f.* s.n. ! Near Kykko Mon., 1939, *Lindberg f.* s.n. ! Near Stavros tis Psokas, 1939, *Lindberg f.* s.n. ! Near Ayia sawmill, 1939, *Lindberg f.* s.n.; between Pano Panayia and Kykko, 1941, *Davis* 3392 ! Prodhromos, 1949, *Casey* 873 ! 1955, *G. E. Atherton* 539 !

7. Lapithos, 1905, *Holmboe* 834 !
8. Ronnas valley, Rizokarpaso, 1957, *Merton* 2938 !

NOTES: In the absence of relevant data, it is impossible to ascertain whether the Cyprus plant is always *Hedera helix* L. ssp. *helix* (with blackish ripe fruits) or sometimes *H. helix* L. ssp. *poetarum* (Bertol.) Nym. (with yellowish fruits). The latter is recorded from Greece and Turkey, and may well occur in our area. A specimen with blackish fruits has been collected near Polystipos, 1974 (*Meikle* 4095 !).

The leaves of the Cyprus plant are noticeably less lobed than is usual in western European material of the species; this is also characteristic of ssp. *poetarum*.

47. CAPRIFOLIACEAE

Shrubs and climbers, rarely herbs or small trees; wood often soft and with broad pith; leaves opposite, simple, entire or lobed; stipules inconspicuous or absent; inflorescences usually cymose; flowers hermaphrodite, actinomorphic or zygomorphic; calyx-tube (or hypanthium) adnate to ovary; calyx-lobes usually 5, equal or unequal; corolla gamopetalous, epigynous, rotate, campanulate, infundibuliform or narrowly tubular, sometimes distinctly 2-lipped, base occasionally gibbous; corolla-lobes imbricate; stamens inserted on the corolla-tube, equal or unequal, dehiscing longitudinally; ovary 2–5-locular, ovules 1–several in each loculus; style often long and slender; stigma generally capitate. Fruit a berry or rarely a capsule; seeds with a crustaceous or leathery testa; endosperm copious; embryo often very small, usually straight.

About 13 genera, represented chiefly in temperate regions of the northern hemisphere.

1. LONICERA *L.*

Sp. Plant., ed. 1, 173 (1753).
Gen. Plant., ed. 5, 80 (1754).
Rehder in Rep. Missouri Bot. Gard., 14: 27–232 (1903).

Erect or twining, rarely creeping, shrubs; stems solid or fistulose; leaves opposite, sessile or petiolate, sometimes connate at the base, entire or pinnately lobed; interpetiolar stipules sometimes present; flowers in simple 2–3-flowered axillary cymes, or in terminal heads or panicles, subtended by bracts and bracteoles, generally zygomorphic; hypanthium ovoid or subglobose with 5 short lobes; corolla white, yellow, pink, purple or red, often distinctly bilabiate; anthers commonly exserted; ovary 2–3 (–5)-locular; ovules 3–8 in each loculus, pendulous; style usually elongate; stigma capitate. Fruit a berry or syncarp; seeds ovoid, compressed; cotyledons ovate.

About 150 species, mostly in temperate and subtropical regions of the northern hemisphere.

1. L. etrusca *Santi*, Viagg. Montam., 1: 113, t. 1 (1795); Unger et Kotschy, Die Insel Cypern, 262 (1865); Boiss., Fl. Orient., 3: 5 (1875); Holmboe, Veg. Cypr., 173 (1914); Post, Fl. Pal., ed. 2, 1: 577 (1932); Chapman, Cyprus Trees and Shrubs, 77 (1949); Davis, Fl. Turkey, 4: 548 (1972); Osorio-Tafall et Seraphim, List Vasc. Plants Cyprus, 98 (1973).

L. etrusca Santi var. *roeseri* Heldr. ex Boiss., Diagn., 2, 2: 107 (1856), Fl. Orient., 3: 5 (1875); Holmboe, Veg. Cypr., 173 (1914); Post, Fl. Pal., ed. 2, 1: 577 (1932); Lindberg f., Iter Cypr., 33 (1946); Chapman, Cyprus Trees and Shrubs, 77 (1949).

L. etrusca Santi ssp. *roeseri* (Heldr. ex Boiss.) Osorio-Tafall et Seraphim, List Vasc. Plants Cyprus, 98 (1973) nom. invalid.

[*L. periclymenum* (non L.) Sibth. et Sm., Fl. Graec. Prodr., 148 (1806); Poech, Enum. Plant. Ins. Cypr., 23 (1842); Unger et Kotschy, Die Insel Cypern, 263 (1865); Boiss., Fl. Orient., 3: 5 (1875) quoad plant. Cypr.]

[*L. implexa* (non Ait.) Gaudry, Recherches Sci., 197 (1855).]

TYPE: Italy; Castel del piano near Ciaccine, *Santi* (? PI).

Woody climber up to 3 m. high with twining stems; old growths covered with lightly fissured grey-brown bark; growths of current season commonly purplish, somewhat pruinose, glabrous or rarely thinly pilose, smooth, terete; leaves obovate, entire, obtuse or rounded at apex, often minutely apiculate, glaucous-green, the lower shortly petiolate, with a petiole less than 3 mm. long, glabrous (var. *roeseri* Heldr. ex Boiss., var. *glabra* Lowe) or thinly pubescent-pilose (var. *etrusca*, var. *hispidula* Boiss.); upper leaves larger than lower, 4–8 cm. long, 2–4 cm. wide, generally glabrous, the uppermost connate, with the fused bases often distinctly reflexed; inflorescence a lax, branched, terminal panicle; flowers sweet-scented, sessile in 4–12-flowered, pedunculate heads; peduncles erect or spreading, up to 4 cm. long, usually glabrous; bracteoles broadly oblong or suborbicular, blunt, up to 2·5 mm. long, shorter than the hypanthia, generally glabrous and purplish; hypanthium broadly ovoid or subglobose, 2·5–3 mm. long, 2–2·5 mm. wide; calyx-lobes minute, blunt, usually less than 0·4 mm. long; corolla creamy-yellow, narrowly tubular, zygomorphic, 2-lipped, tube 2·5–4 cm. long, not much expanded at apex, lobes 1–1·5 cm. long, the lower lip usually simple, strap-shaped, about 2 mm. wide, the upper 3–4 (–5)-toothed, with short, broad, blunt teeth; stamens conspicuously exserted with glabrous filaments up to 2 cm. long; anthers linear, curved, up to 4 mm. long, less than 0·5 mm. wide; style 4–5 cm. long, stigma capitate. Fruit a subglobose scarlet berry, about 1 cm. diam., containing several broadly oblong-ovate, flattened, plano-convex seeds about 5 mm. long, 3·5–4 mm. wide; testa pale brown, obscurely pitted.

HAB.: Climbing through shrubs and trees, sometimes amongst rocks on calcareous and igneous formations; 1,000–5,000 ft. alt.; fl. May–July.

DISTR.: Divisions 1, 2, 7, locally abundant. Widespread in the Mediterranean region.

1. Tsadha, 1905, *Holmboe* 735; Ayios Neophytos Mon., 1905, *Holmboe* 742; Lyso, 1913, *Haradjian* 868 !
2. Common; Troödos, Prodhromos, Galata, Anadhiou, Platania, Ayia, Stavros tis Psokas, Kakopetria, Makheras Mon., Kykko Mon., Platres, Trikoukkia, etc. *Kotschy* 920 ! *Sintenis & Rigo* 707 ! *Chapman* 99 ! 154 ! 196 ! *Syngrassides* 1085 ! 1621 ! *Kennedy* 905 ! 906 ! *Davis* 3458 ! etc.
7. Above Bellapais, 1880, *Sintenis & Rigo* 515 ! Above Kyrenia, 1955, *Miss Mapple* 46 !

NOTES: Cyprus material consists mostly of glabrous-leaved specimens, but since the lower parts of the stems of *L. etrusca* are seldom collected, this is to be expected. In view of the fact that several collectors note the occurrence of glabrous and pubescent plants in the same population, the distinction is scarcely worth maintaining.

L. JAPONICA *Thunb.*, with ovate, shortly petiolate leaves and axillary 2-flowered clusters of white, yellow or reddish, sweet-scented flowers subtended by relatively large (1–1·5 cm. long) narrowly ovate-elliptic, foliaceous bracts, has been found near an irrigation canal at Prodhromos (Sept. 6, 1955, *G. E. Atherton* 543 !). A native of China and Japan, widely cultivated in Europe and America, and no doubt an escape from cultivation in Cyprus. It has become a serious pest in California and elsewhere. *L. japonica* has small black fruits.

48. SAMBUCACEAE

Trees, shrubs and perennials; stems with much pith; leaves opposite, imparipinnate; stipules present or absent, or reduced to a fringe of glandular hairs; inflorescence a terminal, many-flowered, corymbose or paniculate cyme; flowers small, actinomorphic, hermaphrodite; hypanthium ovoid or turbinate; calyx-lobes 3–5, small; corolla rotate or campanulate with a very short tube, lobes 3–5, imbricate or valvate; stamens usually 5, inserted towards base of corolla; filaments subulate, rather short; anthers oblong, extrorse; ovary inferior, 3–5 locular; style short, thick; stigmas 3–5, capitate; ovules solitary, pendulous in each loculus. Fruit a berry-like drupe containing 3–5 pyrenes; seeds oblong, compressed, with a membranous testa and fleshy endosperm; embryo straight, cotyledons ovoid.

One genus with a world-wide distribution, though best represented in temperate and subtropical regions.

1. SAMBUCUS *L.*

Sp. Plant., ed. 1, 269 (1753).
Gen. Plant., ed. 5, 130 (1754).
F. von Schwerin in Mitt. Deutsch. Dendr. Ges., 1920: 194–231 (1920).

Characters those of the family.

About 20 species; distribution that of the family.

Herb; stipules well developed, foliaceous, generally more than 1 cm. long **1. S. ebulus**
Tree or large shrub; stipules absent or reduced to a fringe of hairs - **2. S. nigra**

1. **S. ebulus** *L.*, Sp. Plant., ed. 1, 269 (1753); Boiss., Fl. Orient., 3: 2 (1875); Sintenis in Oesterr. Bot. Zeitschr., 31: 326 (1881); Post, Fl. Pal., ed. 2, 1: 576 (1932); Lindberg f., Iter Cypr., 33 (1946); Davis, Fl. Turkey, 4: 542 (1972); Osorio-Tafall et Seraphim, List Vasc. Plants Cyprus, 98 (1973).

TYPE: " *in* Europa ".

Erect, robust, malodorous perennial herb, 0·5–2 m. high, with a creeping rhizome; stems branched or unbranched, glabrous or thinly glandular-puberulous, conspicuously sulcate; leaflets in 3–6 pairs, the lowermost sometimes subdivided, oblong-lanceolate, 6–15 cm. long, 1–4 cm. wide, thinly glandular-puberulous, base rounded-cuneate, unequal-sided, apex acute or acuminate, margins sharply and closely serrate or biserrate; petiole stout, 4–6 cm. long, canaliculate above; stipules ovate, often large and foliaceous, 0·6–4 cm. long, 0·5–2 cm. wide; inflorescence corymbose, 8–20 cm. diam., with 3 or more primary branches; flowers white, crowded; hypanthium turbinate, glabrous, 1·5–1·8 mm. long, 1·2–1·3 mm. wide; calyx-lobes 5, broadly deltoid, about 0·8 mm. long and almost as wide; corolla campanulate, divided almost to the base into 5 narrowly oblong, acute, rather thick, reflexing lobes about 4 mm. long, 1·5 mm. wide; filaments thickish, nodulose, about as long as corolla-lobes; anthers oblong, violet-purple, about 1·2 mm. long, 0·8 mm. wide; stigmas usually 3, more or less connate, subsessile. Fruit globose, shining purple-black, about 7 mm. diam.; pyrenes oblong-prismatic, about 3–3·5 mm. long, 1·8 mm. wide, distinctly rugulose.

HAB.: Roadsides and by borders of streams and water-channels; 200–500 ft. alt.; fl. June–August.

DISTR.: Division 5, introduced. Widespread in Europe and western Asia.

5. Kythrea, 1880–1970, *Sintenis & Rigo*; *Druce* ! *Nattrass*; *Syngrassides* 839 ! *Lindberg f.*; *A. Hansen* 666 ! etc.

2. S. nigra *L.*, Sp. Plant., ed. 1, 269 (1753); Unger et Kotschy, Die Insel Cypern, 263 (1865); Boiss., Fl. Orient., 3: 2 (1875); Post, Fl. Pal., ed. 2, 1: 576 (1932); Davis, Fl. Turkey, 4: 542 (1972); Osorio-Tafall et Seraphim, List Vasc. Plants Cyprus, 98 (1973).

TYPE: " *in* Germania ".

Large shrub or small tree up to 10 m. high; bark light grey-brown, coarsely fissured; stems glabrous, prominently lenticellate, inconspicuously sulcate; leaflets in 1–3 pairs, the lowermost occasionally subdivided, oblong-elliptic, 4–10 (–17) cm. long, 1·5–5 (–10) cm. wide, glabrous or thinly strigose along the midrib and nerves, base cuneate, generally unequal-sided, apex acuminate-cuspidate, margins closely and regularly serrate; petiole stout, 2·5–6 (–10) cm. long, canaliculate above; stipules reduced to a line of inconspicuous hairs or wanting; inflorescence corymbose, 9–15 (–25) cm. diam. with 5 or more primary branches; flowers white, crowded; hypanthium turbinate, glabrous, about 0·8 mm. long and almost as wide; calyx-lobes 5, deltoid, about 0·5 mm. long and almost as wide; corolla rotate, divided almost to the base into 5 shortly rounded-oblong or suborbicular lobes about 2 mm. long, 1·8 mm. wide; filaments slender, about 2 mm. long; anthers oblong, yellow, about 1 mm. long, 0·8 mm. wide; styles very short, thick, connivent; stigmas usually 3, capitate, more or less connate. Fruit globose, shining purple-black, 5–7 mm. diam.; pyrenes oblong, obscurely prismatic, about 3–3·5 mm. long, 1·8 mm. wide, distinctly rugulose.

HAB.: Cultivated and waste ground, roadsides, almost certainly introduced; sea-level to 4,000 ft. alt.; fl. April–July.

DISTR.: Divisions 2–6, ? planted. Widespread in Europe and western Asia.

2. Kykko Mon., 1914, *Feilden* s.n.!
3. Stavrovouni, 1,000 ft. alt.; 1937, *Syngrassides* 1482 !
4. Larnaca, 1862, *Kotschy*.
5. Kythrea, 1931, *Nattrass* 134.
6. Nicosia, 1931, *Nattrass* 52.

49. RUBIACEAE

Verdcourt in Bull. Jard. Bot. Brux., 28: 209–281 (1958).
Bremekamp in Act. Bot. Neerl., 15: 1–33 (1966).

Trees, shrubs and herbs; leaves simple, opposite, decussate or whorled (or sometimes apparently whorled where there are foliaceous stipules), almost always entire; stipules usually conspicuous and persistent, interpetiolar or intrapetiolar, free or variously connate, sometimes leaf-like (notably in the tribe *Rubieae*); inflorescence generally cymose; flowers usually hermaphrodite, actinomorphic; hypanthium adnate to ovary; calyx-lobes (3–) 4–5 (–8), sometimes very minute or obsolete, mostly equal or subequal and open in aestivation; corolla gamopetalous, variable in shape, commonly infundibuliform or rotate, lobes usually (3–) 4–5 (–11), valvate, imbricate or contorted; stamens as many as corolla-lobes and alternate with them, inserted on or at the apex of the corolla-tube; anthers mostly 2-locular and dehiscing longitudinally, introrse; ovary inferior, commonly 2-locular (sometimes 1 or 3 or more -locular); ovules 1–many in each loculus; placentation axile, apical, basal or rarely parietal; style often elongate and slender; stigmas united or 2–many-lobed. Fruit a capsule, berry or drupe; seed generally with abundant endosperm; embryo straight or curved.

About 500 genera, with a world-wide distribution, but most abundantly represented in tropical regions. The family includes many large, broad-leaved evergreen shrubs, and the Cyprus representatives (all, except *Putoria*, belonging to the tribe *Rubieae*) are uncharacteristic of the group as a whole.

Leaves opposite:
Dwarf malodorous shrub with leathery leaves and clusters of pink flowers - **1. Putoria**
Annual herbs; flowers whitish - - - - - - - - **4. Galium**
Leaves (or apparent leaves) in whorls of 3–6 or more:
Corolla-tube conspicuous, longer than lobes:
Flowers in narrow, caudate, conspicuously bracteate spikes - - **7. Crucianella**
Flowers in cymes, verticels or capitula, not in narrow caudate spikes:
Calyx minute or absent - - - - - - - - **6. Asperula**
Calyx well developed, persistent and accrescent in fruit - - **8. Sherardia**
Corolla-tube inconspicuous, shorter than lobes:
Fruit concealed by the recurved, strongly accrescent, spinulose peduncle and pedicels - - - - - - - - - - **2. Valantia**
Fruit not concealed by an accrescent peduncle and pedicels:
Leaves persistent, leathery; corolla usually 5-lobed; fruit rather fleshy, black and shining - - - - - - - - - **3. Rubia**
Leaves not persistent and leathery; corolla usually 4-lobed; fruit not fleshy or shining black:
Peduncles and pedicels deflexed after flowering; lateral flowers in each cyme sterile and functionally male, central flower hermaphrodite **5. Cruciata**
Peduncles and pedicels not deflexed after flowering; flowers hermaphrodite:
Pedicels bracteolate - - - - - - - - - **6. Asperula**
Pedicels ebracteolate - - - - - - - - - **4. Galium**

1. PUTORIA *Pers.*

Syn. Plant., 1: 524 (1805).

Dwarf, much-branched, malodorous shrubs; leaves opposite, small, rather coriaceous; stipules interpetiolar, small; flowers solitary or in clusters at the tips of the branches; corolla with a long, narrowly infundibuliform tube and 4 recurved, valvate lobes; stamens inserted at apex of corolla-tube; anthers linear, versatile, exserted; ovary 2-locular, ovules solitary in each loculus; style slender, exserted, stigmas 2. Fruit a drupe, with 2 cartilaginous pyrenes; cotyledons linear, flat.

Two, or possibly three, species in the Mediterranean area.

1. P. calabrica (*L. f.*) *DC.*, Prodr., 4: 577 (1830); Boiss., Fl. Orient., 3: 12 (1875); Holmboe, Veg. Cypr., 170 (1914); Post, Fl. Pal., ed. 2, 1: 580 (1932); Lindberg f., Iter Cypr., 32 (1946); Chapman, Cyprus Trees and Shrubs, 76 (1949); Osorio-Tafall et Seraphim, List Vasc. Plants Cyprus, 97 (1973).
Asperula calabrica L. f., Suppl. Plant., 120 (1781).
Ernodea montana Sibth. et Sm., Fl. Graec. Prodr., 1: 98 (1806), Fl. Graec., 2: 33, t. 143 (1813).
TYPE: Italy; " in Calabria, La Tourette " (LINN).

Dwarf, sprawling, much-branched shrub, malodorous when crushed, generally less than 15 cm. high, the primary branches covered with flaking, pale brown bark; stems bluntly 4-angled, thinly puberulous; leaves narrowly oblong-elliptic, 1–2 cm. long, 0·2–0·5 cm. wide, obtuse or subacute, coriaceous, with narrowly recurved, minutely papillose margins; stipules deltoid, about 1·5 mm. long, 1 mm. wide at base, chartaceous with erose-papillose margins; petioles indistinct, less than 0·8 mm. long; inflorescence a terminal cluster of (2–) 4–12 (–15) flowers; peduncles generally less than 5 mm. long, densely hispidulous-papillose, bearing one or more narrowly deltoid, papillose bracts 0·5–1·5 mm. long; pedicels very short or almost obsolete, hispidulous-papillose, subtended by minute subulate bracteoles; hypanthium narrowly tubular-infundibuliform, 4–5 mm. long, 1–2 mm. wide, sparsely covered with raphides; calyx-lobes usually 4, subequal, bluntly deltoid, about 0·5 mm. long and about as wide; corolla pink, the tube slender, narrowly infundibuliform, 7–10 mm. long, lobes usually 4, narrowly oblong or strap-shaped, recurved, about 3 mm. long, 0·8 mm. wide, glabrous or minutely papillose internally near the base; stamens exserted, filaments rather thick, about 2 mm. long; anthers linear-oblong, about 1·5 mm. long, 0·5 mm. wide; style 1·8–2 cm. long, divided apically into 2 short, acute, erect or spreading stigmatic lobes about 0·8 mm. long. Fruit red or blackish when ripe, oblong or ellipsoid, about 7 mm. long, 2–3·5 mm. wide, crowned with the persistent calyx; pyrenes narrowly oblong, about 6 mm. long, 1·5 mm. wide, rounded at either end, convex dorsally, flat ventrally, pale brown; seed narrowly oblong, about 5 mm. long, 1 mm. wide; testa brown, membranous; endosperm copious; embryo minute, about 0·5 mm. long, straight.

HAB.: In garigue or on dry stony slopes, generally on igneous formations; 1,500–5,000 ft. alt.; fl. May–August.

DISTR.: Divisions 1–2, locally abundant; widespread in the Mediterranean region.

1. Fissures of rocks near Ayios Neophytos Mon., 1939, *Lindberg f.* s.n.
2. Locally common on the Troödos range: Prodhromos, Kakopetria, Platres, Platania, Palekhori, Kalokhorio, Perapedhi, Kambos, etc. *Sintenis & Rigo* 901! *Haradjian* 514! *C. B. Ussher* 46! *Chapman* 91! *Kennedy* 868–872! *Syngrassides* 1705! *Davis* 1826! *Probyn* 121! *G. E. Atherton* 567! *P. H. Oswald* 143! *D. P. Young* 7396! *Meikle* 2945! *Economides* in ARI 1237! etc.

NOTES: Holmboe (Veg. Cypr., 16) remarks that this is probably the " lizare " recorded from Cyprus by Reinhard in 1768, and said to be used by the Cypriots " in great parts of the island . . . for dyeing purposes ". Is it also the " root called Fuy, which is a sort of madder ", recorded from Cyprus by Pococke in 1738?

2. VALANTIA *L.*

Sp. Plant., ed. 1, 1051 (1753).
Gen. Plant., ed. 5, 471 (1754).
Vaillantia G. F. Hoffmann, Fl. Germ. Compend., 2: 274 (1804).

Small annual herbs; stems tetragonous; leaves (or apparent leaves) in whorls of 4, lanceolate, oblong or obovate; flowers minute, whitish or yellowish, in groups of 3 in the axils of the uppermost leaves, the lateral unisexual and male, the median hermaphrodite; hypanthium obsolete in male flowers, subglobose in hermaphrodite flowers; calyx-lobes obsolete; corolla rotate, 3-lobed in male flowers, 4-lobed in hermaphrodite; lobes glabrous, valvate; stamens 3–4, alternate with the corolla-lobes; filaments short; anthers minute, suborbicular or reniform; ovary 2-locular, or with 1 fertile loculus; style-branches 2, stigmas capitate. Fruit small, indehiscent, dry, concealed by the accrescent, recurved, spinulose peduncle and pedicels; mericarps 1–2; seeds subglobose; endosperm horny; embryo curved, with ovate or cordate, foliaceous cotyledons.

Four species in the Mediterranean region and western Asia.

Fruiting peduncle with a conspicuous, snout-like appendage; mericarps glabrous **1. V. muralis**

Fruiting peduncle without a dorsal appendage; mericarps papillose-hispidulous **2. V. hispida**

1. **V. muralis** *L.*, Sp. Plant., ed. 1, 1051 (1753); Unger et Kotschy, Die Insel Cypern, 259 (1865); Boiss., Fl. Orient., 3: 82 (1875); Holmboe, Veg. Cypr., 172 (1914); Post, Fl. Pal., ed. 2, 1: 601 (1932); Osorio-Tafall et Seraphim, List Vasc. Plants Cyprus, 98 (1973).

TYPE: " *in* G. Narbonensis, Italiae *muris* ".

Small, erect or spreading annual 2–15 (–20) cm. high; stems branched or unbranched, distinctly tetragonous, glabrous in the lower half, subglabrous or sometimes densely hispidulous above; leaves oblong or obovate, 2–10 mm. long, 1–5 mm. wide, obtuse or subacute, glabrous, or the upper sometimes thinly adpressed-hispidulous; petiole very short or obsolete; flowers subsessile in the axils of the upper leaves, minute and inconspicuous; hypanthium of hermaphrodite flower about 0·4 mm. diam.; corolla-lobes of hermaphrodite and male flowers greenish-yellow, ovate, about 0·4 mm. long; filaments shorter than corolla-lobes; anthers suborbicular or broadly reniform, about 0·1 mm. wide; style divided almost to base. Fruit concealed within the pale brown, recurved, accrescent, thinly spinulose peduncle and pedicels, the covering with a conspicuous, cylindrical, fimbriate, dorsal, snout-like appendage; mericarp reniform, dark brown, about 1 mm. long, 0·8 mm. wide, glabrous except for a band of rhaphides around the hilum.

HAB.: Rocky ground, frequently near the sea; sea-level to 1,000 ft. alt.; fl. March–April.

DISTR.: Divisions 4, 7, rare. Widespread in the Mediterranean region, but commoner in the north and west.

4. Cape Greco, 1862, *Kotschy* 165.
7. Ayios Khrysostomos Mon., 1862, *Kotschy* 448; coast at Kyrenia, 1932, *Syngrassides* 15 ! and, 1949, *Casey* 338 !

NOTES: Apparently much less frequent than *V. hispida* L., but perhaps overlooked on account of its close superficial resemblance to the latter.

2. V. hispida *L.*, Syst. Nat., ed. 10, 1307 (1759); Unger et Kotschy, Die Insel Cypern, 259 (1865); Boiss., Fl. Orient., 3: 82 (1875); Post in Mém. Herb. Boiss., 18: 94 (1900); Holmboe, Veg. Cypr., 172 (1914); Post, Fl. Pal., ed. 2, 1: 602 (1932); Lindberg f., Iter Cypr., 33 (1946); Osorio-Tafall et Seraphim, List Vasc. Plants Cyprus, 98 (1973).
[*V. muralis* (non L.) Sintenis in Oesterr. Bot. Zeitschr., 31: 192 (1881), 32: 193, 397 (1882).]

TYPE: " *in* Europa *australi* ".

A small annual, superficially very similar to *V. muralis*, but usually rather more robust (up to 20 cm. high in luxuriant specimens); stems commonly hispidulous throughout their length, often densely so in the upper half; leaves oblong or obovate, 2–14 mm. long, 1–5 mm. wide, obtuse or subacute, glabrous or thinly and shortly hispidulous, occasionally fleshy and shining green in maritime forms; petiole very short or obsolete; flowers subsessile in the axils of the upper leaves, closely resembling those of *V. muralis*. Fruit concealed within the recurved accrescent peduncle and pedicels, the covering densely or sparsely spinulose-hispidulous, lacking the conspicuous snout-like dorsal appendage characteristic of *V. muralis*; mericarp reniform, dark brown, about 1·2 mm. long, 1 mm. wide, thinly papillose-hispidulous with a conspicuous band of rhaphides around the hilum.

HAB.: Rocky hillsides, rocky or sandy seashores, occasionally as a weed in sandy cultivated or waste ground; sea-level to 5,450 ft. alt.; fl. March–April.

DISTR.: Divisions 1–8, common. Widely distributed in the Mediterranean region and particularly common in the eastern half.

1. Near Ayios Neophytos Mon., 1939, *Lindberg f.* s.n.; Amoudhi near Ayios Yeoryios, 1962, *Meikle* 2079 !
2. Lefka, 1932, *Syngrassides* 681 ! Khionistra, 1938, *Kennedy* 1429 ! Phini, 1938, *Kennedy* 1430 !
3. Akrotiri, 1905, *Holmboe* 682; Limassol, 1939, *Lindberg f.* s.n.; near Kouklia, 1960, *N. Macdonald* 22 !
4. Cape Greco, 1862, *Kotschy*; near Larnaca, 1880, *Sintenis & Rigo* 188 ! Ayia Napa, 1905, *Holmboe* 20; Psevdas, 1908, *Clement Reid* s.n. ! Larnaca Salt Lake, 1967, *Merton* in ARI 410 !
5. Mile 5, Nicosia–Famagusta road, 1950, *Chapman* 39 !
6. Cape Kormakiti, 1967, *Merton* in ARI 127 !
7. Pentadaktylos, 1880, *Sintenis & Rigo* 188A ! St. Hilarion, 1939, *Lindberg f.* s.n.; Kyrenia, 1949, *Casey* 318 ! 514 ! Kharcha, 1950, *Chapman* 261 ! Kazaphani, 1956,

G. E. Atherton 1158 ! Ayios Yeoryios, 1955, *G. E. Atherton* 56 ! Ayios Epiktitos, 1967, *Merton* in ARI 580 ! etc.

8. Ayios Philon near Rizokarpaso, 1941, *Davis* 2258 ! Koma tou Yialou, 1950, *Chapman* 255 !

NOTES: Rather variable in robustness, leaf-shape and fruit indumentum. Three specimens (*Kennedy* 1430 ! *Merton* in ARI 127 ! and 410 !) have exceptionally hirsute-hispid fruiting peduncles and pedicels, and in this respect show some approach to *Valantia columella* (Ehrenb. ex Boiss.) Meikle **comb. nov.** (*Galium columella* Ehrenb. ex Boiss., Fl. Orient., 3: 81; 1875; *Valantia lanata* Del. ex Coss.) from Egypt and Cyrenaica; in this latter the indumentum is almost lanuginose.

3. RUBIA *L.*

Sp. Plant., ed. 1, 109 (1753).
Gen. Plant., ed. 5, 47 (1754).

Perennial herbs and subshrubs; stems tetragonous, sometimes scrambling, often scabrid with recurved prickles; leaves (or apparent leaves) in whorls of 4–6, or very occasionally opposite and stipulate, sessile or petiolate, generally with scabrid margins; flowers small, in axillary and terminal cymes, usually hermaphrodite; hypanthium ovoid or globose; calyx-lobes obsolete; corolla yellowish or greenish, rotate or infundibuliform, with (4–) 5 valvate lobes; stamens 5, inserted in the corolla-tube; filaments short; styles 2, free or partly connate; stigmas capitate. Fruit fleshy, indehiscent, 2-locular or usually 1-locular by abortion; seeds solitary in each loculus; embryo curved with foliaceous cotyledons.

About 60 species with a wide distribution in Europe, temperate Asia, Africa and America. Madder (*Rubia tinctorum* L.) formerly much valued for its red dye, is commonly found as a persistent escape or relict of cultivation.

Leaves tapering upwards from a wide base - - - - 3. **R. laurae**
Leaves tapering at both ends:
  Inflorescences lax, terminal, many-flowered, exceeding the leaves; leaves 3–8 cm. long, 0·5–3 cm. wide - - - - - - - - - 1. **R. tinctorum**
  Inflorescences axillary, few-flowered, usually shorter than, or equalling the leaves; leaves 0·8–5 cm. long, 0·2–1·5 cm. wide - - - - - 2. **R. tenuifolia**

1. **R. tinctorum** *L.*, Sp. Plant., ed. 1, 109 (1753); Unger et Kotschy, Die Insel Cypern, 261 (1865); Boiss., Fl. Orient., 3: 17 (1875); Holmboe, Veg. Cypr., 172 (1914); Post, Fl. Pal., ed. 2, 1: 581 (1932); Lindberg f., Iter Cypr., 33 (1946); Osorio-Tafall et Seraphim, List Vasc. Plants Cyprus, 98 (1973).

TYPE: " Monspelii & *ad* Danubium ".

Sprawling or scrambling perennial herb up to 1 m. high; rhizomes tough, stringy, stained red; stems sharply tetragonous, pallid, scabrid with small recurved prickles along the angles; leaves (2–) 4 (–6)-whorled, lanceolate or elliptic, 3–8 cm. long, 0·5–3 cm. wide, rich green, subcoriaceous, apex acute or acuminate, base cuneate, tapering to a short, indistinct petiole, margins and underside of midrib scabrid with recurved prickles; inflorescences lax, terminal, much-branched, the branches subtended by lanceolate, oblanceolate or lanceolate-subulate, opposite, bracts; flowers numerous; pedicels slender, generally less than 5 mm. long, glabrous; hypanthium subglobose, about 0·8 mm. diam.; corolla pale greenish-yellow, infundibuliform, about 2 mm. long, deeply divided into 5 lanceolate or narrowly ovate, acuminate lobes; filaments very short; anthers linear-oblong, 0·6–0·8 mm. long; 0·2 mm. wide, not exserted; style branches 0·6–0·8 mm. long; stigmas capitate. Fruit black, fleshy, globose, about 5–6 mm. diam., covered with numerous pallid rhaphides.

HAB.: Hedgerows, waste ground and margins of cultivated fields; sea-level to 3,000 ft. alt.; fl. May–July.

DISTR.: Divisions 1–6, a relict of cultivation; widely distributed in central and southern Europe, but probably indigenous only in western Asia.

1. Paphos, 1901, *A. G. & M. E. Lascelles* s.n. ! and, 1913, *Haradjian* 638 ! Yeroskipos, 1939, *Lindberg f.* s.n. !
2. Near Kyperounda, 1880, *Sintenis & Rigo* 802 !
3. Kalavasos, 1905, *Holmboe* 635; Kouklia, 1939, *Lindberg f.* s.n.
4. Near Paralimni, 1862, *Kotschy* s.n.
5. Salamis, 1939, *Lindberg f.* s.n., and, 1962, *Meikle* 2583 !
6. Near Morphou, 1862, *Kotschy* s.n.; Syrianokhori, 1936, *Syngrassides* 983 !

NOTES: Formerly much cultivated as a dye plant in Cyprus, and almost certainly nowhere truly wild on the island.

2. R. tenuifolia *Urv.*, Enum. Plant., 17 (1822); Post, Fl. Pal., ed. 2, 1: 582 (1932); Osorio-Tafall et Seraphim, List Vasc. Plants Cyprus, 98 (1973).

R. olivieri A. Rich. in Mém. Soc. Hist. Nat. Par., 5: 132 (1834); Boiss., Fl. Orient., 3: 17 (1875); Holmboe, Veg. Cypr., 172 (1914); Chapman, Cyprus Trees and Shrubs, 77 (1949) [as " *R. oliviera* "].

R. doniettii Griseb., Spic. Fl. Rum. et Bith., 2: 156 (1844).

R. brachypoda Boiss., Diagn. 1, 10: 57 (1849); Unger et Kotschy, Die Insel Cypern, 261 (1865); Sintenis in Oesterr. Bot. Zeitschr., 31: 393 (1881); 32: 193, 293 (1882).

R. olivieri A. Rich. var. *elliptica* Boiss., Fl. Orient., 3: 18 (1875).

R. olivieri A. Rich. var. *stenophylla* Boiss., Fl. Orient., 3: 18 (1875); Post in Mém. Herb. Boiss., 18: 94 (1900).

R. olivieri A. Rich. ssp. *brachypoda* (Boiss.) Holmboe, Veg. Cypr., 172 (1914); Lindberg f., Iter Cypr., 32 (1946); Chapman, Cyprus Trees and Shrubs, 77 (1949).

R. olivieri A. Rich. ssp. *doniettii* (Griseb.) Holmboe, Veg. Cypr., 173 (1914); Chapman, Cyprus Trees and Shrubs, 77 (1949).

[*R. lucida* (non L.) Sibth. et Sm., Fl. Graec. Prodr., 1: 97 (1806); Fl. Graec., 2: 32, t. 142 (1813); Poech, Enum. Plant. Ins. Cypr., 23 (1842); Kotschy in Oesterr. Bot. Zeitschr., 12: 279 (1862).]

TYPE: Cos; " in summis montibus insulae Coi ", 1819, *d'Urville* (P).

Sprawling or scrambling, subshrubby perennial; stems up to 1 m. high, woody with rough, fissured, greyish bark at the base, 4–6-angled, subglabrous or shortly puberulous, the angles scabrid with short recurved prickles; leaves 4–6-whorled, varying from broadly ovate, obovate or elliptic to narrowly oblong or lanceolate; sessile or subsessile, 0·8–5 cm. long, 0·2–1·5 cm. wide, glabrous, coriaceous, apex shortly and sharply mucronate, base cuneate, margins narrowly recurved, scabrid with short recurved prickles; inflorescences axillary, relatively few-flowered, shorter than the subtending leaves; peduncles and pedicels slender, subglabrous or shortly pubescent; bracts and bracteoles subulate, generally less than 5 mm. long and 2 mm. wide; hypanthium subglobose, about 0·8 mm. diam.; corolla yellowish-green sometimes stained purplish, rotate or broadly infundibuliform, 3–5 mm. diam., deeply divided into 5 narrowly ovate, acuminate or shortly caudate lobes; stamens minute; filaments about 0·8 mm. long; anthers reniform, about 0·2 mm. wide; style branches about 0·8 mm. long; stigmas capitate. Fruit globose, glossy black, fleshy, about 5 mm. diam., without any obvious rhaphides. *Plate 52.*

HAB.: Rocky banks and hillsides, usually scrambling through shrubs, sometimes on bare rocks or cliffs; sea-level to 5,000 ft. alt.; fl. March–June.

DISTR.: Divisions 1–3, 7, 8, locally common. Greece, Crete, Aegean islands, Turkey, Syria, Palestine, Iraq.

1. Rigopoulos, N.E. of Neokhorio, 1957, *Merton* 3018 ! Kritou Terra, 1957, *Merton* 3029 ! Above Peyia, 1962, *Meikle* 2034 ! Erimidhes, near Ayios Yeoryios, 1962, *Meikle* 2138 !
2. Locally common: Galata, Prodhromos, Kykko, Platania, Kakopetria, Khionistra, Platres, Perapedhi, Stavros tis Psokas, Varisha, Pyrgos, Ayii Vavatsinias, etc.; *Kotschy* 922; *Holmboe* 957; *Haradjian* 924 ! *Chapman* 63 ! *Syngrassides* 1263 ! *Kennedy* 873–878 ! *Davis* 1770 ! *Casey* 913 ! *G. E. Atherton* 547 ! *Merton* 2843 ! 2858 ! *D. P. Young* 7255 ! *J. B. Suart* 49A ! *Merton* in ARI 719 !

Plate 52. RUBIA TENUIFOLIA Urv. **1**, habit, $\times\frac{2}{3}$; **2**, fruiting branch, $\times\frac{2}{3}$; **3**, flower, $\times 6$. (*Kennedy* 876, 877.)

3. Mazotos, 1905, *Holmboe* 204; near Stavrovouni Mon., 1937, *Syngrassides* 1478! Zagala above Trimiklini, 1941, *Davis* 3139! Yerasa, 1941, *Davis* 3081!
7. Locally common: Kyrenia, Pentadaktylos, Myrtou, Larnaka tis Lapithou, St. Hilarion, Orga, Kornos, Lefkoniko Pass, Lapithos, etc.; *Kotschy* 371; *Syngrassides* 212! 908! *Chapman* 218! *Lindberg f.*! *Mavromoustakis* 30! *Davis* 2106! 3044! 3624! *Casey* 562! *G. E. Atherton* 118! 955! 1311! *Merton* 2897! *Merton* in ARI 529!
8. Near Yialousa, 1880, *Sintenis & Rigo* 189! Near Rizokarpaso, 1912, *Haradjian* 255! Apostolos Andreas Mon., 1933, *Chapman* 16! Koma tou Yialou, 1950, *Chapman* 252! Ayios Theodhoros, 1962, *Meikle* 2450!

NOTES: A polymorphic plant: most of the Cyprus material matches the relatively narrow-leaved var. *elliptica* Boiss. (*R. brachypoda* Boiss.) which prevails in Turkey, Syria and Palestine, but some specimens from Division 7 (e.g. *Casey* 562! *Merton* in ARI 529!) have the short, broad, obovate leaves of *R. tenuifolia* from Greece and the Aegean islands. Plants from Divisions 1 and 2 (and some from Division 8) are frequently very slender, with 4 (–5) (not 5–6) narrow, oblong or lanceolate leaves; some could be referred to var. *stenophylla* Boiss., but as with var. *elliptica* it is questionable if any satisfactory demarcation can be drawn between these variants. For this reason, the species has been regarded as an aggregate in the present treatment; it may be that detailed study would show that the Northern Range material and that from other parts of the island (especially from Divisions 1 and 2) can be subdivided infraspecifically.

3. R. laurae (*Holmboe*) *Airy Shaw* in Kew Bull., 1937: 343 (1937); Lindberg f., Iter Cypr., 32 (1946); Rechinger f. in Arkiv för Bot., ser. 2, 1: 431 (1950); Osorio-Tafall et Seraphim, List Vasc. Plants Cyprus, 98 (1973).

Galium laurae Holmboe, Veg. Cypr., 171 (1914).

TYPE: Cyprus; Mari, east of Limassol, 1905, *Holmboe* 639 (O).

Trailing or scrambling perennial with a tough woody rootstock; stems (5–) 10–40 (–50) cm. long, distinctly angled, glabrous or minutely papillose-asperulous, branched or unbranched, sometimes woody towards the base with greyish, fissured bark; leaves (2–) 4-whorled, sessile, narrowly ovate-acuminate from a wide base, 0·8–3 cm. long, 0·2–0·8 cm. wide, somewhat deflexed, glabrous, glaucous, coriaceous, obscurely nerved, margins irregularly and remotely serrulate, commonly revolute in dried specimens, apex shortly mucronate; inflorescences terminal on short lateral branches, forming lax 10–15-flowered cymes; bracts lanceolate or narrowly ovate, 2–4 mm. long, 1–2 mm. wide, coriaceous; pedicels slender, glabrous, 2–6 mm. long; hypanthium cup-shaped, glabrous, about 0·8 mm. long and almost as wide; corolla yellowish-brown or stained purplish, divided almost to the base into 5 ovate-acuminate lobes about 2 mm. long, 1·4 mm. wide; stamens minute; filaments about 0·4 mm. long; anthers oblong, about 0·6 mm. long, 0·4 mm. wide; style branches divaricate, about 0·7 mm. long; stigmas capitate. Fruit globose, glossy black, fleshy, about 5 mm. diam., without any obvious rhaphides.

HAB.: Rocky slopes on calcareous or non-calcareous formations, in shade under Pines, or in open garigue; sea-level to 4,000 ft. alt.; fl. May–August.

DISTR.: Divisions 1–3, 7, locally frequent. Endemic.

1. Lyso, 1913, *Haradjian* 825!
2. Platres, 1937, *Kennedy* 879! 880! Aphamis, 1937, *Kennedy* 881! Stavros tis Psokas, 1937, *Syngrassides* 1609! Mesopotamos, 1939, *Lindberg f.* s.n.! Perapedhi, 1940, *Davis* 1845! Arminou, 1941, *Davis* 2423!
3. Mari, 1905, *Holmboe* 639; Kakomallis above Louvaras, 1962, *Meikle* 2943!
7. Pentadaktylos, 1880, *Sintenis & Rigo* 526! and same locality, 1939, *Lindberg f.* s.n.; Armenian Mon., 1912, *Haradjian* 366! and, 1939, *Lindberg f.* s.n.; St. Hilarion, 1934, *Syngrassides* 671! and 1939, *Lindberg f.* s.n.; 1940, *Davis* 2039! Kyrenia, 1955, *G. E. Atherton* 39b! 54! 74! 284! Halevga, 1940, *Davis* 1747! Yaïla Peak, 1941, *Davis* 3609!

4. GALIUM *L.*

Sp. Plant., ed. 1, 105 (1753).
Gen. Plant., ed. 5, 46 (1754).

Annual or perennial herbs or subshrubs; stems generally 4-angled, sometimes terete, often retrorsely aculeolate; leaves opposite or in whorls; ultimate branches of inflorescence frequently without bracts and always ebracteolate; pedicels generally longer than hypanthium or fruit; flowers usually hermaphrodite, 3–4-merous; calyx minute or absent; corolla rotate or a little cup-shaped, rarely tubular or infundibuliform; stigma capitate. Fruit usually dry, sometimes covered with hooked hairs.

About 400 species, distributed throughout the world.

Perennials:
- Leaves canescent-villose; flowers reddish, purplish or pink - - **1. G. canum**
- Leaves not canescent-villose; flowers white or whitish **2. G. humifusum var. lasiocarpum**

Annuals:
- Stems sprawling or scrambling with adhesive retrorse bristles; fruits 3–6 mm. diam.:
  - Fruits bristly or scabrid-papillose; leaf-margins retrorsely aculeolate:
    - Fruits covered with hooked bristles; pedicels divaricate in fruit **5. G. aparine**
    - Fruits scabrid-papillose; pedicels strongly recurved in fruit **7. G. tricornutum**
  - Fruits smooth or bluntly verruculose; leaf-margins with forward-pointing bristles:
    - Fruit smooth, 5–6 mm. diam. - - - - - - - **6. G. pisiferum**
    - Fruit conspicuously verruculose, 4–5 mm. diam. - - **8. G. verrucosum**
- Stems erect or diffuse, not markedly adhesive; fruits less than 3 mm. diam.:
  - Peduncles and pedicels very short, inflorescence as a whole subspicate, with distant verticels of condensed flower-clusters:
    - Fruits subglobose - - - - - - - - - - **13. G. recurvum**
    - Fruits oblong, distinctly longer than wide - - - - **14. G. murale**
  - Peduncles and pedicels well developed; inflorescences lax, not subspicate:
    - Leaves opposite or 3–4-whorled, ovate or oblong-ovate, obtuse **4. G. peplidifolium**
    - Leaves (4–) 5–10-whorled, setaceous, linear, lanceolate or almost subulate, acute, mucronate or shortly awned:
      - Inflorescence arising from the upper nodes of stems and branches, forming well-marked, terminal, corymbose cymes:
        - Bracts exceeding or equalling partial inflorescences - - **3. G. setaceum**
        - Bracts shorter than partial inflorescences - - - - **9. G. floribundum**
      - Inflorescences terminal and lateral arising from all but the lowermost nodes and forming ill-defined oblong, pyramidal or obcuneate inflorescence-structures:
        - Ultimate branches, peduncles and pedicels conspicuously slender and capillary:
          - Pedicels up to 20 mm. long, very elongate and slender **12. G. tenuissimum**
          - Pedicels usually less than 2·5 mm. long - - - - **11. G. divaricatum**
        - Ultimate branches, peduncles and pedicels not conspicuously slender or capillary - - - - - - - - - - **10. G. parisiense**

G. RUBIOIDES *L.*, Sp. Plant., ed. 1, 105 (1753) from central and eastern Europe, extending to the Caucasus, is erroneously recorded by Hume " at Larnica and Limosol " (in Walpole, Mem. Europ. Asiat. Turkey, ed. 1, 254; 1817). In the absence of a specimen it is impossible to re-identify Hume's plant, but it may possibly have been *Rubia tinctorum* L., which bears some superficial resemblance to *G. rubioides*.

1. G. canum *Req.* in DC., Prodr., 4: 602 (1830); Unger et Kotschy, Die Insel Cypern, 259 (1865); Boiss., Fl. Orient., 4: 65 (1875); Holmboe, Veg. Cypr., 171 (1914); Post, Fl. Pal., ed. 2, 1: 593 (1932); Chapman, Cyprus Trees and Shrubs, 76 (1949); Osorio-Tafall et Seraphim, List Vasc. Plants Cyprus, 97 (1973).

G. musciforme Boiss., Diagn., 1, 3: 44 (1843).

G. canum Req. var. *musciforme* (Boiss.) Boiss., Fl. Orient., 4: 66 (1875); Post in Mém. Herb. Boiss., no. 18: 94 (1900); Holmboe, Veg. Cypr., 171 (1914); Post, Fl. Pal., ed. 2, 1: 594 (1932); Lindberg f., Iter Cypr., 32 (1946); Chapman, Cyprus Trees and Shrubs, 77 (1949); Osorio-Tafall et Seraphim, List Vasc. Plants Cyprus, 97 (1973).

G. canum Req. ssp. *hilarionis* Lindberg f., Iter Cypr., 32 (1946).

TYPE: " in Syria legerunt Donati et Labillardière " (G).

Loosely tufted perennial with a woody rootstock and pendulous or trailing branches; stems 10–30 cm. long, densely white-villose, usually much branched and distinctly tetragonous; leaves sessile or subsessile, (2–) 4–6-whorled, ovate, 2–12 mm. long, 1–6 mm. wide, often apparently narrow-oblong or linear because of the strongly revolute margins, thinly villose above and below, nervation obscure, apex obtuse or subacute, margins entire; flowers very numerous in much-branched terminal and lateral panicles forming cylindrical or narrowly pyramidal, nebulous inflorescences, 3–15 cm. long; peduncles filiform, thinly villose or subglabrous up to 10 mm. long, generally terminating in 1–2 lanceolate foliaceous bracts 2–4 mm. long; pedicels very slender, divaricate, about 2–5 mm. long; hypanthium subglobose, about 0·4 mm. diam., densely white-villose; corolla reddish-purple, mauve or pinkish, rotate, divided almost to the base into 4 ovate lobes 0·8 mm. long, 0·6 mm. wide, lobes thinly villose externally, minutely verruculose internally with a conspicuous cuspidate-aristate apex; filaments about 0·1 mm. long; anthers purplish, oblong, about 0·2 mm. long, 0·1 mm. wide; styles about 0·2 mm. long, united to above the middle; stigmas capitate. Fruit subglobose, laterally somewhat compressed, distinctly didymous, about 1·5 mm. diam., blackish, thinly hirsute or subglabrous.

HAB.: Crevices of limestone cliffs, rocks and walls; sea-level to 2,500 ft. alt.; fl. May–Sept.

DISTR.: Divisions 4, 7, 8, locally abundant. Also Aegean Islands, Turkey, Syria and Palestine.

4. Cape Greco, 1862, *Kotschy* 160, 363, also, 1905, *Holmboe* 413.
7. Locally abundant about Kyrenia and the Northern Range; Pentadaktylos, Dhikomo, Ayios Epiktitos, St. Hilarion, Halevga, Buffavento, Yaïla, Lapithos, Ayios Khrysostomos Mon., etc.; *Kotschy* 359, 408, 465; *Sintenis & Rigo* 525 ! *Haradjian* 370 ! *Syngrassides* 450 ! *Kennedy* 888 ! 889 ! *Chapman* 297 ! *Lindberg f.* ! *Davis* 1743 ! 3610 ! *Casey* 784 ! 793 ! etc.
8. Ephtakomi, 1905, *Holmboe* 524; Komi Kebir, 1912, *Haradjian* 301 !

2. G. humifusum *M. Bieb.*, Fl. Taur.-Cauc., 1: 104 (1808).

Asperula humifusa (M. Bieb.) Besser, Cat. Hort. Crem., Suppl. 3: 4 (1811), Cat. Hort. Crem., 17 (1816); Boiss., Fl. Orient., 3: 44 (1875); Post, Fl. Pal., ed. 2, 1: 588 (1932).

Sprawling or ascending perennial 20–100 cm. high; stems usually much branched, especially towards base, sharply angular or tetragonous, pilose with long spreading or subadpressed whitish hairs or sometimes glabrous and shining; leaves 4–6 (–9)-whorled, often deflexed, variable, linear or oblanceolate, 5–30 mm. long, 1–6 mm. wide, thinly scabrid-pilose or glabrous, sublustrous, rather coriaceous, becoming blackish when dried, apex acute or subacute, shortly spinulose-apiculate, margins recurved or revolute, scabrid with small forward-pointing teeth; inflorescence a narrow, elongate, interrupted panicle; bracts foliaceous, resembling the leaves but smaller, sometimes rather densely crowded, glabrous or hispid; flowers solitary or in few (2–4)-flowered cymes; pedicels 1–1·5 mm. long, glabrous or hispidulous; hypanthium shortly oblong, about 0·5 mm. long, 0·4 cm. wide, glabrous or hispidulous; calyx obsolete; corolla white, infundibuliform, glabrous or pilose externally, about 2 mm. long, 1·5 mm. wide at apex, divided more than half its length into 4 spreading or suberect, oblong-ovate, concave, subacute lobes; stamens inserted just below base of lobes; filaments very short;

anthers oblong, about 0·5 mm. long, 0·3 mm. wide, styles slender, about 0·8 mm. long, united for about half their length; stigmas capitate. Fruit globose or subglobose, about 1·8 mm. diam., usually consisting of 2 mericarps; pericarp smooth, glabrous or pilose, usually streaked with small, pale rhaphides.

var. **lasiocarpum** (*Boiss.*) *Meikle* **comb. nov.**

Asperula pycnantha Boiss., Diagn., 1, 10: 64 (1849).

A. humifusa (M. Bieb.) Besser var. *pycnantha* (Boiss.) Boiss., Fl. Orient., 3: 45 (1875); Post, Fl. Pal., ed. 2, 1: 588 (1932); Rechinger f. in Arkiv för Bot., ser. 2, 1: 431 (1950).

A. pycnantha Boiss. var. *lasiocarpa* Boiss., Diagn. 1, 10: 64 (1849); A. K. Jackson et Turrill in Kew Bull., 1939: 478 (1939).

TYPE: Syria; near Damascus, *Aucher-Éloy* 712 (G, K !).

Altogether more robust than *G. humifusum* var. *humifusum*; stems sprawling or scrambling; cauline leaves large, frequently 20–30 mm. long, 3–6 mm. wide; inflorescence-branches generally remote; flowers usually in congested, leafy clusters.

HAB.: Edges of cultivated fields, waste ground, gravelly or grassy banks; fl. June–Aug.

DISTR.: Divisions 4, 8, rare. The species has a wide distribution in S.E. Europe and Western Asia, from Hungary eastwards to the Urals and Afghanistan; the variety seems to be restricted to E. Turkey, Syria, Cyprus, ? Iraq, ? Iran, ? Palestine.

4. Near Larnaca, 1913, *Haradjian* 1012 !

8. Ayios Andronikos, 1937, *Syngrassides* 1656 !

3. **G. setaceum** *Lam.*, Encycl. Méth., 2: 584 (1788); Unger et Kotschy, Die Insel Cypern, 260 (1865); Boiss., Fl. Orient., 4: 77 (1875); Holmboe, Veg. Cypr., 172 (1914); Post, Fl. Pal., ed. 2, 1: 599 (1932); Osorio-Tafall et Seraphim, List Vasc. Plants Cyprus, 98 (1973).

Slender, erect, much-branched annual (4–) 10–25 (–30) cm. high; stems bluntly tetragonous, glabrous or subglabrous, often stained bronze-purple; leaves (4–) 6–8 (–10)-whorled, narrowly linear or setaceous, sessile, patent, 3–15 mm. long, usually less than 1 mm. wide, glabrous or thinly hispid above, apex obtuse or subacute, margin narrowly recurved, minutely serrulate or almost smooth; inflorescence lax, nebulous, spreading, consisting of many 4–6-flowered lateral and terminal cymes; peduncles filiform, 1–1·5 cm. long; bracts 1–2, similar to the leaves but generally narrower; pedicels very slender and hair-like, 2–5 mm. long, often overtopped by the filiform bracts; hypanthium subglobose, about 0·8 mm. diam., densely clothed with white hooked hairs or occasionally glabrous or subglabrous; corolla purplish or whitish, minute, divided almost to the base into 4 erect, glabrous, oblong-ovate lobes about 0·8 mm. long, 0·5 mm. wide, apex of lobes acute or shortly cuspidate, distinctly incurved; stamens minute; filaments about 0·1 mm. long; anthers oblong, about 0·2 mm. long, 0·5 mm. wide; styles about 0·3 mm. long, united for almost half their length; stigmas capitate. Fruit subglobose about 1·3 mm. diam., blackish, distinctly didymous, clothed with hooked hairs or glabrous and speckled with rhaphides.

var. **setaceum.**

TYPE: Spain; without locality. Collected by Vahl (P).

Fruits setose with numerous hooked hairs.

HAB.: In rock crevices, dry banks or hillsides, sometimes in dried-up river beds; sea-level to 3,300 ft. alt.; fl. March–May.

DISTR.: Divisions 2–8, widespread and locally abundant. Widely distributed in the Mediterranean region eastwards to Afghanistan.

2. Vrecha, 1862, *Kotschy* 603; Kryos Potamos, 3,300 ft. alt., 1937, *Kennedy* 899 ! Khionistra, 1957, *Merton* 3141 !

3. Skarinou, 1941, *Davis* 2676 ! Near Stavrovouni, 1957, *Merton* 3156 !
4. Athna Forest, 1952, *Merton* 816 !
5. Vatili, 1905, *Holmboe* 330; Athalassa, 1936, *Syngrassides* 1236 ! Perakhorio, 1955, *Merton* 2042 ! Mile 10, Nicosia–Limassol road, 1967, *Merton* in ARI 595 !
6. Mile 3, Nicosia–Myrtou road, 1950, *Chapman* 575; near Liveras, 1956, *Merton* 2619 ! Dhiorios, 1962, *Meikle* 2322A !
7. Ayios Khrysostomos Mon., 1862, *Kotschy* 449; Kantara, 1880, *Sintenis & Rigo* 184 ! Larnaca tis Lapithou, 1936, *Syngrassides* 913 ! and, 1941, *Davis* 2997 ! Bellapais, 1938, *Kennedy* in K. 2270 ! Yaïla, 1941, *Davis* 3612 ! Kyrenia, 1949, *Casey* 426 ! St. Hilarion, 1949, *Casey* 693 !
8. Ephtakomi, 1905, *Holmboe* 525; near Cape Andreas, 1880, *Sintenis & Rigo*.

var. **urvillei** (*Req.*) *Hal.* in Oesterr. Bot. Zeitschr., 45: 338 (1895)—as var. *urvillaei*.

G. urvillei Req. in DC., Prodr., 4: 607 (1830).

G. setaceum Lam. f. *leiocarpum* Bornm. in Verh. zool.-bot. Ges. Wien, 48: 596 (1898); Post, Fl. Pal., ed. 2, 1: 599 (1932).

TYPE: Aegean Islands; " in aridis insulae Archipelagi Astipalaeae " (G).

Fruit glabrous, or subglabrous with a few short papillae or bristles, speckled with rhaphides, otherwise as in var. *setaceum*.

HAB.: Schistose hillside; about 100 ft. alt.; fl. Febr.–March.

DISTR.: Division 8, rare. Greece, Crete, Aegean area, Turkey, Syria, Palestine and perhaps eastwards to Afghanistan.

8. Gastria, 1941, *Davis* 2417 !

NOTES: Populations of *G. setaceum* Lam. from Spain and the western Mediterranean appear to be predominantly, if not invariably, hispid-fruited. From southern Greece eastwards the number of glabrous-fruited specimens increases, until in Iran and Afghanistan, the typical variety with hispid fruits becomes exceptional. Much of this eastern glabrous-fruited material differs markedly from typical *setaceum* in its diffuse, flexuous habit, very slender stems and leaves, and few-flowered cymes with exceptionally long, capillary pedicels. The basal stem-leaves are often broadly ovate or spathulate, and very different from the upper leaves. Such plants are commonly named *G. decaisnei* Boiss., and have a much better claim to recognition than the glabrous-fruited plants from Greece, the Aegean and Cyprus, which, fruits apart, are quite indistinguishable from typical *setaceum*.

Bornmüller, and Ehrendorfer (in Biol. Meddel. Kong. Danske Vid. Selsk., 10: 144; 1958), both question the claims of *G. decaisnei* to species rank; the latter uniting it with *G. setaceum*, as a subspecies of the latter. While it is true that the two taxa are interlinked by numerous specimens, there are distinguishable correlations between fruit and vegetative characters, which would appear to have been partly obscured by the occurrence of the glabrous-fruited variety of *G. setaceum* mentioned above. I can find no very obvious or satisfactory distinction in the shape of the corolla-lobe apex, a character cited by both Boissier and Ehrendorfer to separate *setaceum* from *decaisnei*.

4. **G. peplidifolium** *Boiss.*, Diagn., 1, 3: 46 (1843); Unger et Kotschy, Die Insel Cypern, 261 (1865); Boiss., Fl. Orient., 3: 71 (1875); Holmboe, Veg. Cypr., 172 (1914); Post, Fl. Pal., ed. 2, 1: 596 (1932); Lindberg f., Iter Cypr., 32 (1946); Osorio-Tafall et Seraphim, List Vasc. Plants Cyprus, 98 (1973).

G. peplidifolium Boiss. f. *pygmaea* Kotschy in Unger et Kotschy, Die Insel Cypern, 261 (1865) nomen.

TYPE: Amongst the several specimens from different parts of Turkey cited by Boissier (G).

Slender, dwarf, much-branched annual; stems spreading, 5–10 (–18) cm. long, sharply tetragonous, glabrous or subglabrous, often stained purple; leaves ovate or oblong-ovate, opposite or 3–4-whorled, subsessile or very shortly petiolate, 2–12 mm. long, 1·5–5 mm. wide, glabrous or minutely strigulose above and below, sometimes stained purple, apex obtuse, margins minutely serrulate, base cuneate; inflorescences mostly terminal, forming a lax spreading cyme; peduncles 2–10 mm. long; bracts foliaceous, resembling leaves, but smaller and narrower; pedicels very slender, 2–8 mm. long, glabrous or very sparingly strigulose; hypanthium subglobose, less than 1 mm. diam., glabrous or subglabrous; corolla whitish, deeply divided into 4 ovate, suberect, glabrous, acute or subacute lobes about 0·8 mm. long,

0·5 mm. wide; stamens minute; filaments about 0·1 mm. long; anthers shortly oblong, about 0·1 mm. long and a little less in width; styles about 0·2 mm. long, united for a short way above the base; stigmas capitate. Fruit conspicuously didymous, about 2 mm. wide, 1·5 mm. high, glabrous or subglabrous, rather fleshy, often sparsely covered with pallid rhaphides.

HAB.: Rocky ground, usually in moist shade near springs or streams at high altitudes; 4,000–6,400 ft. alt.; fl. April–June.

DISTR.: Division 2, locally frequent. Also E. Aegean Islands, Turkey, Syria, Palestine.

2. Locally frequent at high altitudes on the Troödos Range: Khionistra, Tripylos, Papoutsa; *Kotschy* 715 ! *Sintenis & Rigo* 903 ! *Holmboe* 872; *Kennedy* 890 ! 891 ! 892 ! *Lindberg f.* s.n. ! *Davis* 1802 ! 3119 ! 3169 ! 3209 ! 3480 ! 3516 ! *N. Macdonald* 184 !

5. **G. aparine** *L.*, Sp. Plant., ed. 1, 108 (1753); Boiss., Fl. Orient., 3: 68 (1875); Holmboe, Veg. Cypr., 172 (1914); Post, Fl. Pal., ed. 2, 1: 595 (1932); Lindberg f., Iter Cypr., 32 (1946); Osorio-Tafall et Seraphim, List Vasc. Plants Cyprus, 98 (1973).

G. aparine L. var. *macrocarpum* Boiss. et Kotschy in Unger et Kotschy, Die Insel Cypern, 260 (1865) nomen.

[*G. vaillantii* (non DC.) Kotschy in Unger et Kotschy, Die Insel Cypern, 260 (1865).]

[*G. spurium* L. var. *vaillantii* (non (DC.) Gren. et Godr.) Holmboe, Veg. Cypr., 172 (1914).]

[*G. spurium* L. var. *tenerum* (non (Schleicher ex Gaud.) Gren. et Godr.) Holmboe, Veg. Cypr., 172 (1914).]

[*G. spurium* L. ssp. *vaillantii* (non (DC.) Gaud.) Lindberg f., Iter Cypr., 32 (1946).]

[*G. spurium* L. ssp. *tenerum* (non *G. tenerum* Schleicher ex Gaud.) Osorio-Tafall et Seraphim, List Vasc. Plants Cyprus, 98 (1973).

TYPE: " *in* Europae *cultis & ruderatis* ".

Trailing, sprawling or scrambling annual; stems up to 1 m. long in our area (often longer in Europe) branched or unbranched, sharply tetragonous and rough with retrorse bristles, often distinctly swollen at the nodes; leaves (4–) 6–9 (–10)-whorled, oblanceolate or narrowly obovate, 10–50 mm. long, 1·5–12 mm. wide, thinly adpressed-strigose above, subglabrous except along midrib below, apex rounded or obtuse, shortly cuspidate, margins bristly, narrowing to a slender tapering base; inflorescences lateral from the middle and upper nodes, few (1–4)-flowered, divaricate; peduncles up to 2·5 cm. long; bracts 1–3, resembling leaves but usually less than 10 mm. long, 2·5 mm. wide; pedicels usually straight, to about 12 mm. long; hypanthium subglobose, about 1 mm. diam., densely setose with hooked bristles; corolla minute, whitish, divided almost to the base into 4 spreading ovate-cuspidate lobes about 1·5 mm. long, 0·8 mm. wide; filaments about 0·5 mm. long; anthers minute, oblong, about 0·2 mm. long, 0·1 mm. wide; styles about 0·4 mm. long, divided almost to base; stigmas capitate. Fruit conspicuously didymous, with 2 subglobose mericarps (2·5–) 3–4 mm. diam., rather densely clothed with stiff, spreading, hooked bristles.

HAB.: Waste ground, cultivated fields and gardens, bare stony hillsides or gravelly seashores; sea-level to 6,400 ft. alt.; fl. March–July.

DISTR.: Division 1–4, 7, 8. Widespread in Europe and western Asia.

1. Peyia, 1962, *Meikle* 2049 ! Paphos, 1974, *Meikle* 5015 !
2. Khionistra, 1905, *Holmboe* 964, and, 1912, *Haradjian* 519 ! 1937, *Kennedy* 896 ! 897 ! 1939, *Lindberg f.* s.n. !
3. Akrotiri, 1964, *J. B. Suart* 188 ! Polemidhia, 1964, *J. B. Suart* 194 !
4. Larnaca, 1862, *Kotschy* 36; Cape Greco, 1862, *Kotschy*; Perivolia, 1905, *Holmboe* 157a; Mile 3, Larnaca–Famagusta road, 1950, *Chapman* 423 !
7. Kantara, 1880, *Sintenis & Rigo*; Dhavlos, 1880, *Sintenis & Rigo*; Kyrenia, 1947, *Casey* 367 ! 1952, *Casey* 1234 ! Halevga, 1950, *Chapman* 130 !
8. Cape Andreas, 1880, *Sintenis & Rigo*.

NOTES: The occurrence of *G. spurium* L. in Cyprus must be regarded as doubtful.

All the material so named proves, on examination, to be *G. aparine* L. or *G. tricornutum* Dandy. Much of the Cyprus material of *G. aparine* consists of small, unbranched specimens, very dissimilar from the robust scrambler of western European hedgerows. Such depauperate forms are, however, by no means uncommon in Europe, and, in other respects, especially in leaf-shape and fruit, the Cyprus plant is referable to *G. aparine*.

6. G. pisiferum *Boiss.*, Diagn., 1, 10: 67 (1849), Fl. Orient., 3: 69 (1875); Unger et Kotschy, Die Insel Cypern, 261 (1865); Holmboe, Veg. Cypr., 172 (1914); Post, Fl. Pal., ed. 2, 1: 596 (1932); Osorio-Tafall et Seraphim, List Vasc. Plants Cyprus, 98 (1973).

TYPE: Palestine; " legi in monte Carmelo et circa Rascheya ad radices Antilibani ", 1846, *Boissier* (G, K !).

Sprawling or scrambling annual, very similar to *G. aparine* L. in habit and general appearance; stems up to 1 m. long, glabrous, sharply tetragonous with retrorse-spinulose margins, a little thickened at the nodes; leaves (3–) 5–6-whorled, oblanceolate, 10–50 mm. long, 2·5–8 mm. wide, subglabrous or thinly adpressed-setose above, glabrous below except along midrib, apex obtuse or shortly cuspidate, margins aculeolate with patent- or forward-pointing bristles, base gradually attenuate-cuneate; inflorescences lateral and terminal from the upper nodes, lax, much-branched, and usually forming many-flowered, open cymes; peduncles 1–2·5 (–6) cm. long, usually exceeding the leaves; bracts usually 2, resembling the leaves, 3–20 mm. long, 1–8 mm. wide; pedicels 1·5–10 mm. long, hispid, spreading or reflexed in fruit; hypanthium subglobose, glabrous, about 1 mm. diam.; corolla white, divided almost to the base into 4 spreading oblong-ovate lobes about 1·5 mm. long, 1 mm. wide, apex of lobes shortly incurved-cuspidate; filaments about 0·8 mm. long; anthers oblong, about 0·3 mm. long, 0·1 mm. wide; styles about 0·8 mm. long, divided almost to the base; stigmas capitate. Fruit globose, glabrous, smooth, about 5–6 mm. diam., resembling a small pea, one of the mericarps normally abortive; seed separating readily from the papery pericarp, globose, about 5 mm. diam., dark brown when ripe; testa conspicuously foveolate.

HAB.: In crevices of limestone rocks; near sea-level to 400 ft. alt.; fl. March–April.

DISTR.: Division 4, rare. Also Syria and Palestine.

4. Cape Greco, 1862, *Kotschy* 139 ! Sotira, Ayios Antonios, 1950, *Chapman* 711 !

7. G. tricornutum *Dandy* in Watsonia, 4: 47 (1957); Osorio-Tafall et Seraphim, List Vasc. Plants Cyprus, 97 (1973).

G. tricorne Stokes in With., Bot. Arrangement, ed. 2, 1: 153 (1781) pro parte excl. basionym.; Unger et Kotschy, Die Insel Cypern, 260 (1865); Boiss., Fl. Orient., 3: 67 (1875); Holmboe, Veg. Cypr., 171 (1914); Post, Fl. Pal., ed. 2, 1: 595 (1932); Lindberg f., Iter Cypr., 32 (1946); Rechinger f. in Arkiv för Bot., ser. 2, 1: 431 (1950) nomen illeg.

TYPE: Great Britain; Isle of Wight, in fields near Carisbrooke, 1806, *D. Turner & W. Borrer* (BM).

Sprawling or scrambling annual; stems usually much branched especially at or towards the base, sharply tetragonous, glabrous except at the retrorsely spinulose angles; leaves 5–6 (–8)-whorled, linear or narrowly oblanceolate, 10–25 mm. long, 1–3 mm. wide, glabrous except along the midrib below, apex acuminate or cuspidate, margins strongly retrorse-spinulose; inflorescences lateral, usually 3-flowered; peduncle up to 2·5 cm. long, generally shorter than the subtending leaves, rather densely spinulose; bracts 1–2, foliaceous, subulate to rather broadly lanceolate, 3–8 mm. long, 0·5–1·5 mm. wide; pedicels up to 5 mm. long, minutely spinulose, distinctly recurved in fruit; hypanthium subglobose, about 1 mm. diam.; corolla white, divided almost to the base into 4 ovate lobes about 0·8 mm. long, 0·5 mm. wide; filaments minute, about 0·1 mm. long; anthers oblong, yellow, about 0·2 mm. long, 0·1 mm. wide; styles very short; stigmas capitate; lateral flowers in each inflorescence sometimes male or functionally male. Fruit didymous, or with only one mericarp through abortion; mericarps subglobose, about 3 mm. diam., greenish when ripe, densely scabrid-papillose; pericarp rather closely adherent to seed; testa brown, regularly reticulate-foveolate.

HAB.: Cultivated or fallow fields or waste ground; sea-level to 5,600 ft. alt.; fl. Febr.–June.

DISTR.: Divisions 2–5, 7, 8. Widespread in Europe and the Mediterranean region eastwards to Afghanistan and Baluchistan.

2. Makheras Mon., 1862, *Kotschy* 234 ! Prodhromos, 1939, *Lindberg f.* s.n. ! Troödos, 1939, *Lindberg f.* s.n., and, 1963, *D. P. Young* 7827 ! Gourri, 1950, *Chapman* 332 !
3. Two miles W. of Limassol, 1964, *J. B. Suart* 175 ! Limassol, 1930, *E. Wall* s.n.
4. Larnaca, 1880, *Sintenis & Rigo*; Cape Pyla, 1880, *Sintenis & Rigo*; Aradhippou, 1934, *Syngrassides* 463 ! Near Famagusta, 1912, *Haradjian* 78.
5. Pera, 1908, *Clement Reid* s.n. ! Lakatamia, 1939, *Syngrassides* 361 ! Mora, 1956, *Merton* 2535 ! Salamis, 1962, *Meikle* 2577 ! Strongylos, 1967, *Merton* in ARI 439 !
7. Near Kythrea, 1880, *Sintenis & Rigo* 185b, and, 1950, *Chapman* 101 ! Kantara, 1880, *Sintenis & Rigo*; Dhavlos, 1880, *Sintenis & Rigo*; Kyrenia, 1949, *Casey* 637A ! Ayios Epiktitos, 1951, *Casey* 1111 ! and, 1968, *Economides* 1051 !
8. Near Cape Andreas, 1880, *Sintenis & Rigo*; Ayios Philon near Rizokarpaso, 1941, *Davis* 2232 !

8. G. verrucosum *Huds.* in Phil. Trans. Roy. Soc., B, 56: 251 (1767); Sibth et Sm., Fl. Graec. Prodr., 1: 93 (1806); Fl. Graec., 2: 24, t. 133 (1813).

Valantia aparine L., Sp. Plant., ed. 1, 1051 (1753) nec *Galium aparine* L., Sp. Plant., ed. 1, 108 (1753).

Galium valantia Weber in Wigg., Primit. Fl. Holsat., 12 (1780); Dandy in Watsonia, 4: 48 (1957); Osorio-Tafall et Seraphim, List Vasc. Plants Cyprus, 97 (1973).

G. saccharatum All., Fl. Ped., 1: 9 (1785); Unger et Kotschy, Die Insel Cypern, 260 (1865); Boiss., Fl. Orient., 3: 67 (1875); Holmboe, Veg. Cypr., 171 (1914); Post, Fl. Pal., ed. 2, 1: 594 (1932).

TYPE: Cultivated in Chelsea Physic Garden (BM !).

Spreading or sprawling annual; stems generally less than 15 cm. long, usually branched at base, sharply tetragonous, glabrous or with the angles sparsely and rather remotely retrorse-aculeolate; leaves generally 5–6-whorled, oblanceolate, 4–15 mm. long, 1·5–5 mm. wide, apex acute or shortly cuspidate, margins rough with forward-pointing bristles; inflorescences lateral, usually 3-flowered, the lateral flowers mostly sterile and functionally male; peduncles up to 10 mm. long, generally deflexed after anthesis, angular, glabrous; bracts wanting; pedicels short, slender, usually less than 3 mm. long; hypanthium subglobose, about 1 mm. diam., verrucose; corolla whitish, deeply divided into 4 ovate, acute, spreading lobes about 1·5 mm. long, 0·8 mm. wide; filaments about 0·8 mm. long; anthers shortly oblong or suborbicular, about 0·3 mm. diam.; styles very short; stigma punctiform. Fruit consisting of 2 globose mericarps, 4–5 mm. diam., or commonly of one mericarp through abortion; pericarp rather pale brown, coarsely and conspicuously verrucose-papillose.

HAB.: Cultivated and waste ground, roadsides, stony hillsides, shingly beaches or crevices of rocks or walls; sea-level to 5,000 ft. alt.; fl. Febr.–April.

DISTR.: Divisions 1–4, 7. Widespread in southern Europe and the Mediterranean region.

1. Fabrika Rocks near Kato Paphos, 1974, *Meikle* 4009 !
2. Prodhromos, 1862, *Kotschy* 849.
3. Amathus, 1964, *J. B. Suart* 157 !
4. Cape Greco, 1862, *Kotschy* 139A.
7. Pentadaktylos, 1880, *Sintenis & Rigo* 183 ! and, 1950, *Casey* 976 ! Kyrenia, 1932, *Syngrassides* 687 ! and, 1956, *G. E. Atherton* 389 ! 1959, *Casey* 1130 ! Klepini, 1960, *N. Macdonald* 66 !

9. G. floribundum *Sibth. et Sm.*, Fl. Graec. Prodr., 1: 94 (1806), Fl. Graec., 2: 25, t. 134 (1813); Poech, Enum. Plant. Ins. Cypr., 22 (1842); Unger et Kotschy, Die Insel

Cypern, 261 (1865); Boiss., Fl. Orient., 3: 74 (1875); Holmboe, Veg. Cypr., 172 (1914); Osorio-Tafall et Seraphim, List Vasc. Plants Cyprus, 98 (1973).
? *G. cyprium* Sibth. in Walpole, Travels, 25 (1820) nomen.
TYPE: " In insulâ Cypro ", 1787, *Sibthorp* (OXF).

Erect or spreading annual 3–30 cm. high; stems repeatedly branched, rather obscurely tetragonous, thinly pilose with spreading hairs or glabrous, often rather distinctly swollen at the nodes; leaves usually 6–8 (–10)-whorled, linear, 5–15 mm. long, 1–1·5 mm. wide, glabrous or thinly pilose, apex acute or mucronate, margins strongly revolute; inflorescences terminal, much branched, forming nebulous, rounded, many-flowered cymes; peduncles very slender, filiform, glabrous or spreading-pilose, up to 1 cm. long; bracts subulate, usually less than 5 mm. long, solitary or sometimes paired; pedicels capillary, 3–4 mm. long; hypanthium globose, 0·5–0·8 mm. diam., rather densely pilose, or occasionally glabrous; corolla reddish-brown, glabrous or pilose externally, deeply divided into 4 broadly ovate lobes about 0·8 mm. long, 0·5 mm. wide, apex of lobes acute or terminating in a filiform awn sometimes 5 mm. long; filaments about 0·3 mm. long; anthers oblong, reddish, about 0·2 mm. long, 0·1 mm. wide; styles very short; stigmas punctiform. Fruit subglobose, about 1 mm. diam., inconspicuously didymous; pericarp minutely papillose, glabrous or clothed with numerous incurved hairs.

HAB.: Fallow fields, gravelly ground, dry rocky slopes or dried-up stream beds; sea-level to 4,500 ft. alt.; fl. May–July.

DISTR.: ? Division 1, very rare. Aegean Islands, Turkey, Syria, Iran.

1. Near Paphos, 1787, *Sibthorp* (*teste* Kotschy and Holmboe—see Notes).

NOTES: I have not been able to discover Kotschy's reason for stating that the type locality for *G. floribundum* is near Paphos. Sibthorp (in Walpole's Travels, 25; 1820) records " the Galium Cyprium on the diamond rocks " near Ktima [" Iftinia "], and it may be that this is the plant subsequently named *G. floribundum*, though I have, as yet, found no positive grounds for such an identification. In view of the fact that *G. floribundum* has not since been collected in Cyprus, one tends to suspect, though perhaps unjustifiably, that the type material was incorrectly localized.

10. G. parisiense *L.*, Sp. Plant., ed. 1, 108 (1753); Boiss., Fl. Orient., 3: 72 (1875); Lindberg f., Iter Cypr., 32 (1946); Osorio-Tafall et Seraphim, List Vasc. Plants Cyprus, 97 (1973).
TYPE: " *in* Anglia, Gallia ".

Slender, sprawling, intricately branched annual; stems 5–35 cm. long, sharply tetragonous, retrorsely spinulose along the angles, otherwise glabrous or subglabrous; leaves 5–7-whorled, lanceolate, 3–7 (–10) mm. long, 1–2·5 mm. wide, glabrous or sparsely strigose, apex acute or mucronate, margins flattish or somewhat recurved, with short, forward-pointing bristles; inflorescences terminal and lateral, arising from a little way above the base of the plant, the lateral branches commonly subequal, so that the whole inflorescence is elongate and narrowly oblong or narrowly pyramidal in outline; peduncles usually 3-flowered, divaricate, 2–5 mm. long; bracts solitary or in pairs, foliaceous, linear-lanceolate, usually less than 2 mm. long and 1 mm. wide; pedicels slender, 0·5–2·5 mm. long, widely divaricate; hypanthium subglobose, 0·5 mm. diam., pilose (var. *parisiense*) or glabrous (var. *leiocarpum* Tausch); corolla greenish-white or tinged red, divided almost to the base into 4 erect or somewhat incurved, broadly deltoid-ovate lobes about 0·5 mm. long and almost as wide, lobes glabrous or thinly adpressed-strigose externally, apex acute or subacute; filaments and anthers minute; styles very short; stigmas punctiform. Fruits subglobose about 1 mm. diam., not conspicuously didymous; pericarp minutely papillose, glabrous or clothed with short pale, hooked bristles.

HAB.: Dry stony slopes, walls, roadsides; about 4,000–5,000 ft. alt.; fl. June–July.

DISTR.: Division 2, rare. Widespread in central, western and southern Europe, the Mediterranean region; also Caucasia, Turkey and Iran; Atlantic Islands.

2. Prodhromos, 1880, *Sintenis & Rigo* 905 ! and, 1961, *D. P. Young* 7306 ! Platania, 1939, *Lindberg f.* s.n.

NOTES: All the Cyprus material has hispid fruits.

11. G. divaricatum *Pourret ex Lam.*, Encycl. Méth., 2: 580 (1788); Boiss., Fl. Orient., 3: 72 (1875); Post, Fl. Pal., ed. 2, 1: 597 (1932).

TYPE: France; " M. Desfontaines l'a rencontrée dans le Berry; & M. l'Abbé Pourret nous en communiqué des individus qu'il a trouvés dans le Languedoc " (P).

Erect or sprawling annual 5–12 (–30) cm. high; stems slender, usually much branched near the base, sharply tetragonous with retrorsely aculeolate angles; leaves 5–7-whorled, linear-lanceolate or almost subulate, 2–10 mm. long, 0·5–1·5 mm. wide, glabrous or sparsely adpressed-strigose, apex acute, mucronate or shortly awned, margins often strongly recurved or almost revolute, with forward-pointing bristles; inflorescence lax, nebulous, with very slender, capillary peduncles and pedicels, the lateral branches rather unequal, with the upper often much longer than the lower, making the inflorescence as a whole broadly obconical in outline and widely diffuse; bracts subulate, foliaceous, usually less than 5 mm. long; peduncles (1–) 2–3-flowered, 0·5–2·5 (–5) mm. long; pedicels very slender, usually less than 2·5 mm. long, spreading or somewhat deflexed after anthesis; hypanthium subglobose, about 0·5 mm. diam., glabrous; corolla whitish or yellowish, sometimes tinged red, divided almost to the base into 4 erect, deltoid-ovate, subacute, glabrous lobes about 0·4 mm. long and almost as wide; filaments and anthers minute; styles very short, stigmas punctiform. Fruit subglobose, less than 1 mm. diam., distinctly didymous with the mericarps separating readily at maturity; pericarp glabrous, minutely granulate-papillose.

HAB.: Dry banks on limestone under Pines; c. 900 ft. alt.; fl. March–June.

DISTR.: Division 6, rare. Southern Europe, Mediterranean region.

6. Near Dhiorios, 1962, *Meikle* 2322 !

12. G. tenuissimum *M. Bieb.*, Fl. Taur.-Cauc., 1: 104 (1808); Boiss., Fl. Orient., 3: 73 (1875), Fl. Orient. Suppl., 283 (1888); Holmboe, Veg. Cypr., 172 (1914); Post, Fl. Pal., ed. 2, 1: 597 (1932); Osorio-Tafall et Seraphim, List Vasc. Plants Cyprus, 98 (1973).

TYPE: " in Tauriae et Caucasi etiam Iberiae incultis et rupestribus " (LE).

Slender, erect or sprawling annual (10–) 15–30 (–50) cm. high; stems generally unbranched except in the region of the inflorescence, glabrous, sharply tetragonous with the angles sparingly retrorse-aculeolate; leaves (4–) 6–8-whorled, linear-oblanceolate, 5–10 mm. long, 1–2 mm. wide, glabrous, apex acute or shortly awned, margins somewhat recurved, with forward-pointing bristles; inflorescence lax, nebulous, with very slender peduncles and pedicels, the lateral branches spreading, ultimately forming a broadly oblong, pyramidal or obconical, much-divided cyme; bracts subulate, foliaceous, usually less than 5 mm. long; peduncles 2–3-flowered, 4–15 (–20) mm. long; pedicels very slender, 0·5–20 mm. long, widely divaricate after anthesis; hypanthium subglobose, glabrous, about 0·5 mm. diam.; corolla greenish, divided almost to the base into 4 erect, ovate, acute, glabrous lobes, about 0·4 mm. long, 0·3 mm. wide; filaments and anthers minute; styles very short; stigmas punctiform or subcapitate. Fruit subglobose, about 1·2 mm. diam., not very distinctly didymous; mericarps apparently not separating until fully mature; pericarp glabrous, minutely and bluntly granulate-papillose, often streaked with conspicuous, pallid rhaphides.

HAB.: Stony mountainsides, vineyards and similar open ground at high altitudes; about 4,000 ft. alt.; fl. May–June.

DISTR.: Division 2, rare. South-eastern Europe, Crimea, Caucasia, Turkey, Syria, Lebanon eastwards to Central Asia.

2. Madhari (Mt. Adelphi), 1880, *Sintenis & Rigo* 904 ! Prodhromos, 1905, *Holmboe* 921; Platres, 1960, *N. Macdonald* 178 !

13. G. recurvum *Req.* in DC., Prodr., 4: 609 (1830); Boiss., Fl. Orient., 3: 68 (1875), Fl. Orient. Suppl., 283 (1888); Holmboe, Veg. Cypr., 171 (1914); Lindberg f., Iter Cypr., 32 (1946); Osorio-Tafall et Seraphim, List Vasc. Plants Cyprus, 97 (1973).

G. pauciflorum Kotschy in Unger et Kotschy, Die Insel Cypern, 260 (1865) non Bunge (1833).

G. recurvum Req. var. *pauciflorum* Boiss., Fl. Orient., 3: 68 (1875).

G. recurvum Req. var. *glabratum* Holmboe, Veg. Cypr., 171 (1914); Lindberg f., Iter Cypr., 32 (1946).

TYPE: Aegean Islands; Thira, " ad montem ignivomum ins. novae Camini [Nea Kaimeni] legit D'Urville " (P, G).

Erect or spreading annual 3–20 (–30) cm. high; stems usually much branched at the base, sparingly branched above, slender, tetragonous, subglabrous or thinly hispid with sparsely retrorse-aculeolate angles; leaves 4–6-whorled, oblanceolate or narrowly obovate, 2–7 mm. long, 1–3 mm. wide, subglabrous or hispidulous, often purplish, apex acute, or rounded with a short mucro, margins slightly recurved, with forward-pointing bristles; flowers in short, condensed lateral cymes forming a narrow, interrupted subspicate inflorescence; peduncles very short, usually less than 3 mm. long, (1–) 2–3 (–5)-flowered; bracts absent; pedicels 1–3 mm. long, hispidulous or glabrous, spreading or usually somewhat recurved after anthesis; hypanthium subglobose, less than 1 mm. diam., glabrous or hispid; corolla greenish-white, divided into 4 erect, ovate, glabrous or hispidulous lobes, about 0·5 mm. long and almost as wide; filaments and anthers minute; styles very short, subcapitate. Fruit subglobose, slightly more than 1 mm. diam., distinctly didymous; mericarps separating readily at maturity, pericarp hispid with pale hooked hairs or glabrous and minutely granulate-papillose (var. *glabratum* Holmboe).

HAB.: Screes and stony slopes, bare ground in Pine forest or by roadsides at high altitudes; 4,500–6,200 ft. alt.; fl. May–June.

DISTR.: Division 2, (? 7); Aegean Islands, W. Turkey.

2. Khionistra, 1880, *Sintenis & Rigo* 906a ! 906b (var. *glabratum* Holmboe) ! and, 1937, *Kennedy* 895 ! 1939, *Lindberg f.* s.n. (vars. *recurvum* and *glabratum* Holmboe) ! 1941, *Davis* 3523 (var. *glabratum* Holmboe) ! *Davis* 1816 (var. *glabratum* Holmboe) ! Prodhromos, 1905, *Holmboe* 962, and, 1961, *D. P. Young* 7260 (var. *glabratum* Holmboe) ! 7271 (var. *glabratum* Holmboe) ! and 7304 !

[7. " Bei Lapethus [Lapithos] n. 487 Auch sonst in Kalkbergen ", 1862, *Kotschy* 487.]

NOTES: There is reason to question the accuracy of Kotschy's localization. *G. recurvum* has not since been collected on or near the Northern Range, and appears to be confined to the serpentines of the Troödos Range, at altitudes above 4,000 ft. It is true that a very few mountain plants (e.g. *Anemone blanda*) occur on both ranges, but an error is more probable in this instance.

14. G. murale *(L.) All.*, Fl. Ped., 1: 8 (1785); Unger et Kotschy, Die Insel Cypern, 260 (1865); Boiss., Fl. Orient., 3: 78 (1875); Holmboe, Veg. Cypr., 172 (1914); Post, Fl. Pal., ed. 2, 1: 600 (1932); Lindberg f., Iter Cypr., 32 (1946); Osorio-Tafall et Seraphim, List Vasc. Plants Cypr., 98 (1973).

Sherardia muralis L., Sp. Plant., ed. 1, 103 (1753).

Galium murale (L.) All. var. *troödi* Lindberg f., Iter Cypr., 32 (1946).

TYPE: " *in Italia* ".

Sprawling, diffuse annual (3–) 5–15 (–30) cm. high; stems much branched, especially towards base, slender, tetragonous, glabrous or hispid with retrorse bristles; leaves 3–5-whorled, or paired towards apices of stems, oblanceolate or lanceolate, 3–8 mm. long, 1–2 mm. wide, subglabrous or hispidulous, apex acute or obtuse and mucronate, margins recurved, with forward-pointing bristles; flowers in short 1–3-flowered lateral cymes, forming a long, interrupted, subspicate inflorescence; peduncles usually less than 2 mm. long, hispid or subglabrous; pedicels 1–1·5 (–2) mm. long, hispid or subglabrous, conspicuously recurved or deflexed after anthesis; hypanthium oblong, about 1 mm. long, 0·5–0·8 mm. wide, glabrous or hispidulous; corolla greenish, minute, divided almost to the base into 4 erect, acute, ovate, glabrous or hispidulous lobes about 0·2 mm. long and almost as wide; filaments and anthers minute; styles very short; stigmas subcapitate. Fruit oblong, about 1·5 mm. long, 0·8 mm. wide; mericarps separating readily at maturity, often unequal, pericarp hispid with pale hooked bristles or subglabrous except at the apex (var. *troödi* Lindberg f.), minutely granulate-papillose.

HAB.: Dry banks and fallows, vineyards, stony hillsides and walls; sea-level to 5,000 ft. alt.; fl. March–June.

DISTR.: Divisions 1, 2, 4–8. Widespread in the Mediterranean region; Atlantic Islands.

1. Dhrousha, 1941, *Davis* 3282 !
2. Prodhromos, 1862, *Kotschy* 715, and, 1905, *Holmboe* 923, also, 1961, *D. P. Young* 7346 ! Palekhori, 1957, *Merton* 3012 !
4. Near Ormidhia, 1880, *Sintenis & Rigo*; Famagusta, 1971, *Townsend* 71/106 !
5. Vatili, 1905, *Holmboe* 343.
6. Dhiorios, 1962, *Meikle* 2321 !
7. Near Kythrea, 1880, *Sintenis & Rigo* 186 ! Kyrenia, 1949, *Casey* 433 ! Lakovounara, 1950, *Chapman* 520 ! Kharcha, 1950, *Chapman* 284 ! Ayios Epiktitos, 1967, *Merton* in ARI 310 ! Kantara–Mersinnikki Road, 1967, *Merton* in ARI 754 !
8. Cape Andreas, 1880, *Sintenis & Rigo*; Valia, 1905, *Holmboe* 480 !

5. CRUCIATA *Mill.*

Gard. Dict. Abridg., ed. 4 (1754).
Ehrendorfer in Notes Roy. Bot. Gard. Edinb., 22: 393 (1958).

Annual or perennial herbs superficially resembling *Galium*; leaves 4-whorled, with 1–3 primary nerves; inflorescence narrow, verticillate, consisting of superimposed whorls of short-stalked axillary cymes; central flower in each cyme hermaphrodite, lateral flowers (when present) sterile and functionally male; peduncles and pedicels deflexed after anthesis but not markedly accrescent or coalescent; calyx absent; corolla yellowish, rotate, 4-lobed; stigmas capitate. Fruit small, dry, glabrous or pilose, of 1–2 mericarps.

Six species in Europe, the Mediterranean region and western Asia.

Upper cauline and floral leaves cordate, accrescent, strongly deflexed; inflorescence-whorls closely superimposed - - - - - - - - - - 1. **C. articulata**
Upper cauline leaves and floral leaves oblong-elliptic, not noticeably accrescent or strongly deflexed; inflorescence-whorls usually remote - 2. **C. pedemontana**

1. **C. articulata** (*L.*) *Ehrend.* in Notes Roy. Bot. Gard. Edinb., 22: 396 (1958).
Valantia articulata L., Sp. Plant., ed. 1, 1052 (1753).
Galium articulatum (L.) Roem. et Schultes, Syst. Veg., 3: 250 (1818); Boiss., Fl. Orient., 3: 81 (1875); Post, Fl. Pal., ed. 2, 1: 601 (1932).
TYPE: " *in* Aegypto, Syria ".

Erect annual, 3–15 cm. high; stems usually branched at or near the base, unbranched above, subglabrous or thinly hispidulous, sharply tetragonous; basal and lower cauline leaves obovate, 6–12 mm. long, 3–6 mm. wide, glabrous, base tapering, sub-petiolate, apex obtuse, margins entire; upper cauline leaves (in the region of the inflorescences) strongly deflexed, cordate, sessile, at first 5–15 mm. long, 3–8 mm. wide, strongly accrescent and up to 16 mm. long, and almost as wide in fruit, 1-nerved, glabrous or with minutely scabridulous margins, entire, acute or subacute; verticels of flowers closely superimposed; cymes usually 5–7-flowered; peduncles 1–3 mm. long, scabridulous; pedicels usually less than 3 mm. long, slender, scabridulous; corolla lobes greenish, ovate, sub-acute, glabrous, spreading, about 0·8 mm. long, 0·5 mm. wide; filaments about 0·2 mm. long; anthers minute, yellow, reniform or suborbicular; styles about 0·5 mm. long, divided almost to base; stigmas capitate. Fruit concealed under the persistent, deflexed, papyraceous floral leaves, generally comprising only one fully developed, reniform, somewhat compressed mericarp about 2 mm. long, 1·5 mm. wide; pericarp glabrous, dull olive-brown, covered with conspicuous pale rhaphides.

HAB.: Cultivated fields, ? 500–1,000 ft. alt.; fl. April–May.

DISTR.: Division 5 (or 7), very rare. Transcaucasia, Turkey, Syria, Palestine, Egypt, Iran.

5 (or 7). Near Kythrea, 1880, *Sintenis & Rigo* 530 !

NOTES: This record, the only one for Cyprus, was not published by Sintenis, and was, on this account, overlooked by subsequent authors. The plant was collected by Sintenis " in campis pr. Kythraeam " on 23rd May, 1880; it was fully mature and beginning to ripen its seeds at this date. *Cruciata articulata* should be looked for about the margins of cornfields, or in neglected vineyards and gardens; the subspicate inflorescences and overlapping deflexed floral leaves make it conspicuous and unmistakable.

2. C. pedemontana (*Bell.*) *Ehrend.* in Notes Roy. Bot. Gard. Edinb., 22: 396 (1958).
Valantia pedemontana Bell., Osserv. Bot., 61 (1788).
Galium pedemontanum (Bell.) All., Auct. Fl. Ped., 2 (1789); Boiss., Fl. Orient., 3: 80 (1875); Post, Fl. Pal., ed. 2, 1: 601 (1932); Lindberg f., Iter Cypr., 32 (1946); Osorio-Tafall et Seraphim, List Vasc. Plants Cyprus, 98 (1973).
TYPE: Italy; recorded from several localities by Bellardi (TO, K !).

Slender, diffuse annual, 6–30 cm. high; stems unbranched or branched near the base, sharply tetragonous, glabrous or thinly hirsute, the angles usually retrorse-aculeolate; leaves 4-whorled, the lower whorls usually remote; lamina oblong-elliptic, 5–8 mm. long, 2–5 mm. wide, thinly strigose-hirsute, apex obtuse or subacute, margins sometimes distinctly recurved; flowers solitary or sometimes in short 2–3-flowered cymes; peduncles and pedicels very slender, thinly villose, usually less than 3 mm. long, much shorter than the leaves, strongly deflexed after anthesis; corolla-lobes ovate, about 0·4 mm. long, 0·3 mm. wide, erect, glabrous, yellowish; stamens minute; anthers suborbicular, yellow; styles about 0·2 mm. long, divided almost to the base; stigmas capitate. Fruit of 1 or 2 mericarps, often concealed under the persistent, loosely deflexed floral leaves; mericarps subglobose, about 1 mm. diam., dark brown, glabrous, often streaked with pale rhaphides.

HAB.: Rocky slopes and moist, shaded rocks; 5,000–6,000 ft. alt.; fl. June–July.

DISTR.: Division 2, very rare. Widespread in south and south-central Europe, also Transcaucasia, Turkey, Palestine.

2. Near Kannoures Springs, 1939, *Lindberg f.* s.n.

6. ASPERULA *L.*

Sp. Plant., ed. 1, 103 (1753).
Gen. Plant., ed. 5, 45 (1754).

Annual or perennial herbs or subshrubs, stems usually tetragonous but not retrorsely aculeolate; leaves sessile or subsessile, opposite or 4–6 (–10)-whorled; flowers in branched panicles or sometimes in capitula, 3–4 (–5)-

merous; bracts (and frequently bracteoles) present; corolla cup-shaped, funnel-shaped or tubular, rarely rotate; styles bifid; stigmas capitate. Fruit oblong, globose or biglobose, of 2 mericarps or one by abortion, dry or somewhat fleshy, glabrous or pilose, but without hooked bristles.

About 200 species, mostly in Europe, the Mediterranean region and western Asia, but extending to central Asia and the Himalayas. The Australian and New Zealand species are perhaps generically distinct.

Corolla cup-shaped or almost rotate, divided almost to base into 4 (–5) lobes; a woody subshrub 30–60 cm. high - - - - - - - - - **1. A. cypria**
Corolla narrowly tubular with short lobes:
 Flowers whitish or yellowish (tinged brown or pink) in narrow, elongate terminal spikes; perennial - - - - - - - - - - **2. A. stricta**
 Flowers pale or bright blue in hemispherical glomerules; annual **3. A. arvensis**

1. A. cypria *Ehrend.* in Plant Systematics and Evolution, no. 126: 406 (1977).
Galium suberosum Sibth. et Sm., Fl. Graec. Prodr., 1: 91 (1806), Fl. Graec., 2: 21, t. 128 (1813); Unger et Kotschy, Die Insel Cypern, 259 (1865); Boiss., Fl. Orient., 3: 61 (1875), Fl. Orient. Suppl., 282 (1888); Holmboe, Veg. Cypr., 171 (1914); Lindberg f., Iter Cypr., 32 (1946); Chapman, Cyprus Trees & Shrubs, 76 (1949); Osorio-Tafall et Seraphim, List Vasc. Plants Cyprus, 97 (1973)—non *Asperula suberosa* Sibth. et Sm.

TYPE: "In Cretae montibus", *Sibthorp* (OXF)—an erroneous localization; the species has not been found in Crete.

Erect or spreading, woody subshrub 30–60 cm. high; stems much-branched, erect, rather crowded, the older parts covered with rough, pale brown, corky bark; young stems sharply tetragonous, glabrous or occasionally minutely scabridulous; leaves sparse, often much reduced, remote, opposite or 4–10-whorled, linear or narrowly oblanceolate, 2–20 mm. long, 1–2·5 mm. wide, coriaceous, often reddish in exposed situations, glabrous or minutely scabridulous, acute or obtuse, margins recurved or revolute; inflorescence terminal, much branched, the flowers subsessile in distinct whorls or clusters; bracts tongue-shaped, concave, usually glabrous, about 2–3 mm. long, 1–2 mm. wide, obtuse or subacute; bracteoles similar, but often proportionately broader and more concave, with membranous, ciliate margins; hypanthium turbinate, glabrous, about 1 mm. long, 1·3 mm. wide at apex; calyx obsolete; corolla cup-shaped or subrotate, whitish or tinged pink or red, divided almost to the base into 4 (–5) erect, or slightly spreading, triangular, glabrous or minutely papillose lobes about 1 mm. long and 1 mm. wide at base, apex of lobes commonly apiculate and incurved; filaments about 0·5 mm. long; anthers oblong, reddish or brownish, about 0·5 mm. long, 0·4 mm. wide; styles very short, stigmas capitate. Fruit of 1 or 2 globose mericarps about 1–1·5 mm. diam.; pericarp rather fleshy, often stained dark reddish or brownish, minutely and regularly papillose.

HAB.: Dry, rocky limestone or igneous hillsides, sometimes under Pines, 500–4,000 ft. alt.; fl. May–June.

DISTR.: Divisions 1–5, 7, 8, locally common. Endemic.

1. Stroumbi near Paphos, 1913, *Haradjian* 757 ! Lyso, 1913, *Haradjian* 843 !
2. Yialia, 1905, *Holmboe* 794; Kykko, 1913, *Haradjian* 935 ! Platres, 1937, *Kennedy* 882 ! 883 ! 884 ! 885 ! 886 ! Kakopetria, 1939, *Lindberg f.* s.n. ! Ayia valley, *Chapman* 166; Nikos, 1955, *Merton* 2390 !
3. Kophinou, 1905, *Holmboe* 570; Lefkara, 1940, *Davis* 1908 ! Episkopi Forest, 1950, *Barrington* 58 ! Near Souni, 1960, *N. Macdonald* 188 !
4. Between Phourni and Athienou, 1862, *Kotschy* 966 !
5. Athalassa, 1936, *Syngrassides* 1233 ! 1235 ! Also, 1939, *Lindberg f.* s.n. !
7. Rather common: above Kythrea, between Lefkoniko and Akanthou, Kyrenia, Armenian Monastery, Kantara, etc.; *A. G. & M. E. Lascelles* 8 ! *Sintenis & Rigo* 524 ! *Syngrassides* 230 ! *Kennedy* 887 ! *Davis* 1732 ! *Casey* 786 ! *G. E. Atherton* 256 ! etc.
8. Komi Kebir, 1912, *Haradjian* 325 !

2. A. stricta *Boiss.*, Diagn., 1, 3: 33 (1843), Fl. Orient., 3: 35 (1875); Post in Mém. Herb. Boiss., 18: 94 (1900); H. Stuart Thompson in Journ. Bot., 44: 309 (1906); Holmboe, Veg. Cypr., 170 (1914); Post, Fl. Pal., ed. 2, 1: 586 (1932); Osorio-Tafall et Seraphim, List Vasc. Plants Cyprus, 97 (1973).

TYPE: Based on a number of syntypes from Greece and Turkey (G).

Erect or spreading perennial with a tough, woody rootstock; stems 20–50 cm. high, much branched at the base, usually unbranched above, obscurely tetragonous, smooth or scabridulous, pruinose; leaves 4–6-whorled, linear or linear-subulate, 5–20 mm. long, 0·5–2 mm. wide, patent or somewhat deflexed, smooth or scabridulous, pruinose, apex shortly spinulose, margins strongly revolute; inflorescences terminal, lax, paniculate; flowers sessile or subsessile in elongate, terminal, bracteate spikes; bracts ovate-concave or cymbiform, very shortly connate at base, sharply acuminate at apex, smooth or scabridulous with ciliate-lacerate margins; bracteoles similar, but smaller, often with a rostrate-acuminate apex; hypanthium oblong, about 1 mm. long, 0·7 mm. wide, glabrous or minutely papillose; calyx obsolete; corolla whitish or yellowish, sometimes tinged brown or pink, narrowly tubular, 5–7 mm. long, glabrous or shortly pubescent externally, lobes generally 5, suberect, ovate, about 1–1·2 mm. long, 0·8 mm. wide, obtuse with a sharply inflexed caudate apical appendage about 0·8 mm. long; stamens attached a little below base of lobes; filaments very short; anthers linear-oblong, about 1·2–1·5 mm. long, 0·4 mm. wide, fuscous or brownish; styles about 3·5 mm. long, divided almost to base; stigmas large, globose-capitate. Fruit globose, about 2 mm. diam., usually of 2 mericarps; pericarp somewhat fleshy, densely papillose.

HAB.: Dry rocky limestone slopes; about 3,000 ft. alt.; fl. June–August.

DISTR.: Division 7, rare. Widespread in the eastern Mediterranean region, from Greece to Palestine.

7. Buffavento, July 1894, *Post* s.n.

3. A. arvensis *L.*, Sp. Plant., ed. 1, 103 (1753); Unger et Kotschy, Die Insel Cypern, 262 (1865); Boiss., Fl. Orient., 3: 30 (1875); Holmboe, Veg. Cypr., 170 (1914); Post, Fl. Pal., ed. 2, 1: 586 (1932); Lindberg f., Iter Cypr., 32 (1946); Osorio-Tafall et Seraphim, List Vasc. Plants Cyprus, 97 (1973).

TYPE: " *in* Gallia, Flandria, Germania, Anglia ".

Slender erect annual; stems 10–30 cm. high, branched or unbranched, sharply tetragonous, glabrous or minutely scabridulous; cotyledons large (up to 1 cm. long) broadly oblong-ovate, commonly persisting until anthesis; lower cauline leaves usually 4–5-whorled, upper 6–8-whorled, narrowly oblanceolate or linear, 6–23 mm. long, 1·5–6 mm. wide, glabrous or sparsely scabridulous especially along the midrib, apex obtuse or rounded, margins flat or narrowly revolute; flowers subsessile, crowded in terminal, hemispherical glomerules 1·3–3 cm. diam.; bracts forming a distinct involucre, foliaceous, narrowly oblong or linear, 5–18 mm. long, 1–2·5 mm. wide, apex subacute, margins scabridulous above, densely ciliate with long whitish or reddish bristles towards base; bracteoles similar but smaller and narrower, with a distinctly tapering base; hypanthium suborbicular, about 1·2 mm. diam., glabrous, sparingly dotted over with small rhaphides; calyx obsolete; corolla bright or pale blue, narrowly tubular, about 5–7 mm. long, divided apically into 4 lingulate, spreading lobes about 1·4 mm. long, 1 mm. wide; stamens inserted about 2 mm. below mouth of corolla-tube; filaments very short; anthers brownish, narrowly linear, about 1·2 mm. long, less than 0·5 mm. wide; style about 4 mm. long, bifid for about 1 mm.; stigmas club-shaped. Fruit of 1 or 2 depressed-globose mericarps 2·8–3·5 mm. diam., ventral surface deeply umbilicate, pericarp dark brown with numerous pale rhaphides.

HAB.: Cultivated fields and vineyards; 200–4,500 ft. alt.; fl. March–May.

DISTR.: Divisions 2–5. Widespread as a weed in Europe, the Mediterranean region and western Asia eastwards to Iran.

2. Prodhromos, 1862, *Kotschy* 746 ! and, 1905, *Holmboe* 910, and, 1949, *Casey* 739 ! Platres, 1937, *Kennedy* 900 ! 901 ! and, 1960, *N. Macdonald* 117 ! Troödos, 1939, *Lindberg f.* s.n.; Kalokhorio (Makheras), 1950, *Chapman* 303 !
3. Yerasa, 1948, *Mavromoustakis* s.n. ! Above Kellaki, 1956, *Merton* 2557 !
4. The Phaneromene near Larnaca, 1905, *Holmboe* 125.
5. About Kythrea, 1880, *Sintenis & Rigo* 527 ! Mora, 1956, *Merton* 2533 ! 2534 ! (see note).

NOTES: A specimen from Mora (*Merton* 2534 !) resembles *A. setosa* Jaub. et Spach in having small, pale flowers. It was, however, found with typical *A. arvensis*, and, on close examination, is evidently no more than a minor variant of this species. True *A. setosa*, an eastern species, ranging from Turkey eastwards to central Asia and Chitral, is not reliably recorded from Cyprus; the inclusion of the species in Osorio-Tafall & Seraphim, List Vasc. Plants Cyprus, 97 (1973) is possibly based upon the erroneous Mora record.

7. CRUCIANELLA *L.*

Sp. Plant., ed. 1, 108 (1753).
Gen. Plant., ed. 5, 47 (1754).

Annuals or perennials, sometimes woody; leaves (and leaf-like stipules) in whorls of 4–6 or more; inflorescences terminal and lateral, narrow, spicate; flowers usually sessile, sometimes shortly pedicellate, hermaphrodite, subtended by a conspicuous, often membranous-margined bract and 2 bracteoles; calyx obsolete; corolla narrowly infundibuliform or tubular, 4–5-lobed, the lobes generally with an incurved apical awn; style unequally bifid; stigmas capitate. Fruit dry, generally comprising 2 narrowly ovoid or oblong, 1-seeded, glabrous mericarps.

About 30 species in S. Europe, the Mediterranean region, western and central Asia.

Corolla-tube equalling or only a little longer than subtending bract:
  Pairs of bracts connate for almost 2 mm. above base - - - **1. C. latifolia**
  Pairs of bracts free or almost free to base:
    Bracts smooth or minutely scabridulous only at apex, broadly ovate, cuspidate-acute - - - - - - - - - - - - - **3. C. imbricata**
    Bracts conspicuously scabridulous-ciliate, lanceolate or narrowly ovate-acuminate **4. C. aegyptiaca**
Corolla-tube much longer than subtending bract - - - - **2. C. macrostachya**

1. C. latifolia *L.*, Sp. Plant., ed. 1, 109 (1753); Boiss., Fl. Orient., 3: 20 (1875); Holmboe, Veg. Cypr., 170 (1914); Post, Fl. Pal., ed. 2, 1: 583 (1932); Rechinger f. in Arkiv för Bot., ser. 2, 1: 431 (1950); Osorio-Tafall et Seraphim, List Vasc. Plants Cyprus, 97 (1973).

TYPE: "*in* Creta & Monspelii".

Erect annual 10–40 cm. high; stems usually branched at or near base, unbranched or sparingly branched above, occasionally simple, markedly tetragonous, thinly retrorse-hispidulous or subglabrous; leaves 4–6-whorled, sessile or subsessile, usually narrowly lanceolate, occasionally broadly lanceolate or ovate-elliptic, 10–30 mm. long, 2–8 mm. wide, glabrous or sparsely hispidulous, apex usually acute or acuminate, margins recurved or revolute, aculeolate with forward-pointing teeth; inflorescences terminal, solitary or geminate; peduncle up to 10 cm. long, erect, sharply tetragonous; spike narrowly cylindrical or caudate, 4–15 cm. long, 0·2–0·3 cm. wide; bracts rigidly erect, narrowly ovate, concave, about 5 mm. long, 2·5–3 mm. wide, acute, basal margins connate for about 2 mm., midrib conspicuous, much thickened, margins broadly membranous, scabridulous-ciliate; bracteoles linear, concave, keeled, free to base, about 4 mm. long, 0·8 mm. wide, keel herbaceous, scabridulous, margins membranous, scabridulous-ciliate; flower sessile; hypanthium oblong, glabrous, about 1·5 mm. long, 0·7 mm. wide; corolla creamy-white or pinkish, narrowly tubular, about 4 mm. long, 0·5 mm. wide, tube glabrous, lobes 4–5, erect, ovate-cucullate, about 0·7 mm. long, 0·5 mm. wide, papillose externally with an inflexed apical awn about 0·7 mm. long; stamens almost included; filaments very short; anthers brownish, narrowly oblong, about 1·3 mm. long, 0·4 mm. wide; styles very unequal, one about 3 mm. long, the other about 1·8 mm. long, united for more than 1 mm.; stigmas oblong-capitate. Fruit of 2 free, narrowly oblong mericarps about 3 mm. long, 1 mm. wide, enclosed within the persistent, hardened bract, ventral surface of mericarps longitudinally sulcate; pericarp dark brown, minutely rugulose-reticulate, streaked with a few pale rhaphides.

HAB.: Rocky hillsides in Pine forest or garigue, sometimes by roadsides or on waste ground; sea-level to 2,500 ft. alt.; fl. April–June.

DISTR.: Divisions 1, 3, 5, 7, 8, locally common. Widespread in the Mediterranean region, especially in the eastern half.

1. Akamas, 1941, *Davis* 3301 !
3. Akrotiri, 1963, *J. B. Suart* 32 !
5. Mia Milea, 1967, *Economides* in ARI 937 !
7. Locally common; Kyrenia, St. Hilarion, Halevga, Bellapais, etc.; *Sintenis & Rigo* 528 ! *Syngrassides* 676 ! *Davis* 3602 ! 3620 ! *Chapman* 649 ! *Casey* 1257 ! *Merton* 2221 ! etc.
8. Valia, 1905, *Holmboe* 508 ! Rizokarpaso, 1912, *Haradjian* 170 ! Kantara, 1948, *Mavromoustakis* s.n. ! Koma tou Yialou, 1950, *Chapman* 236 !

2. C. macrostachya *Boiss.*, Diagn., 1, 3: 27 (1843), Fl. Orient., 3: 20 (1875); Holmboe, Veg. Cypr., 171 (1914) *pro parte*; Post, Fl. Pal., ed. 2, 1: 583 (1932); Lindberg f., Iter Cypr., 32 (1946); Rechinger f. in Arkiv för Bot., ser. 2, 1: 431 (1950); Osorio-Tafall et Seraphim, List Vasc. Plants Cyprus, 97 (1973).

[*C. angustifolia* (non L.) Kotschy in Unger et Kotschy, Die Insel Cypern, 262 (1865).]

TYPE: " in Libano Aucher No. 679 " (G, K !).

Erect or spreading annual 10–50 (–70) cm. high; stems generally much branched, rarely simple, glabrous or thinly retrorse-hispidulous, obscurely tetragonous, with long internodes; leaves (3–) 4–6-whorled, lanceolate or narrowly elliptic; 8–30 mm. long, 2–5 mm. wide, thinly scabridulous above, apex sharply acuminate, margins recurved or revolute, sparsely scabridulous with forward-pointing teeth; inflorescences lateral and terminal, or terminal only in depauperate specimens, narrowly caudate-spicate, 5–18 cm. long, 0·3–0·4 cm. wide, peduncle to 8 cm. long, or sometimes almost wanting in lateral inflorescences; rhachis scabridulous or hispid, rarely subglabrous; bracts imbricate, rigidly erect, narrowly oblong-ovate, concave, about 6 mm. long, 1·5–2 mm. wide, acute, connate only at the very base, midrib conspicuous, prominent, minutely scabridulous towards apex, margins broadly membranous, scabridulous-ciliate; bracteoles very similar to bracts, but rather shorter and narrower, minutely papillose externally; flowers sessile; hypanthium oblong, about 1 mm. long, 0·7 mm. wide, covered with pallid rhaphides; corolla creamy or yellowish, narrowly tubular, about 9 mm. long, 1 mm. wide, lobes 4–5, suberect, ovate, cucullate, about 0·8 mm. long, 0·5 mm. wide, papillose externally, apical awn 2–2·5 mm. long; stamens inserted just below base of lobes, almost wholly included; filaments very short; anthers linear, brownish, about 1·5 mm. long, 0·2 mm. wide; styles unequal, the longer 4–5 mm. long, the shorter about 3 mm. long, connate for about 2 mm. from base; stigmas globose-capitate. Fruit spindle-shaped, about 3 mm. long, 1·2 mm. diam. towards apex, comprising only 1 mericarp in all the specimens seen, ventral surface deeply sulcate; pericarp dark brown, minutely and regularly papillose, streaked with pale rhaphides.

HAB.: Dry, stony hillsides, sometimes on waste ground, by roadsides or in Olive groves; sea-level to 2,000 ft. alt.; fl. May–July.

DISTR.: Divisions 1, 3, 7. Also Crete, Aegean Islands, Turkey, Syria, Palestine.

1. Paphos, 1787, *Sibthorp* teste Kotschy; Stroumbi, 1913, *Haradjian* 746 ! Limni mine, 1938, *Syngrassides* 1841 !
3. Limassol aerodrome, 1950, *Chapman* 559 !
7. Locally common about Kyrenia, *Sintenis & Rigo* 529 ! *Kennedy* 904 ! *Lindberg f.* s.n. ! *Casey* 1052 ! 1258 ! 1274 ! etc.

NOTES: Holmboe's record from Prodhromos, as well as the Kotschy record of *C. latifolia* from the same locality, are almost certainly both errors for *C. imbricata* Boiss., which is locally frequent in the Prodhromos area, and which is not mentioned either by Kotschy or Holmboe.

3. **C. imbricata** *Boiss.*, Diagn., 1, 10: 59 (1849), Fl. Orient., 3: 22 (1875); Post, Fl. Pal., ed. 2, 1: 584 (1932); Rechinger f. in Arkiv för Bot., ser. 2, 1: 431 (1950); Osorio-Tafall et Seraphim, List Vasc. Plants Cyprus, 97 (1973).
[*C. latifolia* (non L.) Kotschy in Unger et Kotschy, Die Insel Cypern, 262 (1865).]
[*C. macrostachya* (non Boiss.) Holmboe, Veg. Cypr., 171 (1914) pro parte.]
TYPE: " *in Syriâ, loci proprii non memini* ", *Boissier* (G).

A small erect or suberect annual, 6–15 (–20) cm. high; stems often much branched at or near the base, usually unbranched above, simple in depauperate specimens, sharply tetragonous, subglabrous or minutely scabridulous, with long internodes; leaves commonly 4–whorled, sometimes opposite or 3- or 5-whorled, linear or lanceolate, 5–15 mm. long, 0·8–3·5 mm. wide, subglabrous or thinly scabridulous above, apex sharply acuminate, margins recurved or revolute, scabridulous with forward-pointing teeth; inflorescences terminal, mostly solitary, shortly caudate-spicate, (1–) 2–6 (–8) cm. long, 0·3–0·4 cm. wide; peduncle 1–8 cm. long, rarely wanting; rhachis glabrous or subglabrous; bracts imbricate, at first rather loosely erect, becoming more rigid and adpressed in fruit, rather broadly ovate-oblong, concave, about 5 mm. long, 2·5–3 mm. wide, sharply cuspidate-acute, connate only at the very base, midrib prominent, smooth or minutely scabridulous towards apex, margins very broadly membranous, scabridulous-ciliate; bracteoles narrowly lanceolate-acuminate, about 5 mm. long, 1 mm. wide, connate along one margin for a short way above base, keel and margins minutely scabridulous; flowers sessile; hypanthium oblong, about 1·2 mm. long, 1 mm. wide, covered with pallid rhaphides; corolla creamy or yellowish, narrowly tubular, about 3·5–4 mm. long, 0·5 mm. wide; lobes usually 4, erect, ovate, cucullate, about 0·4 mm. long, 0·3 mm. wide, papillose externally, apical awn about 0·4 mm. long; stamens inserted just below base of lobes, almost wholly included; filaments very short; anthers linear, brownish, about 0·6 mm. long, 0·2 mm. wide; styles unequal, the longer about 2 mm. long, the shorter about 1·5 mm. long, connate for about 1 mm. above base; stigmas broadly clavate or subglobose. Fruit oblong or somewhat spindle-shaped, about 2 mm. long, 0·9 mm. wide, usually comprising 2 free mericarps; ventral surface of mericarp deeply sulcate; pericarp olive-brown or dark brown, minutely papillose-verruculose, streaked with pale rhaphides.

HAB.: Dry rocky slopes generally on igneous formations, frequently under *Pinus nigra* ssp. *pallasiana*; 2,000–5,500 ft. alt.; fl. May–July.

DISTR.: Divisions 1, 2 only. Also Crete, Turkey, Syria.

1. Lyso, 1913, *Haradjian* 857.
2. Troödos, 1912, *Haradjian* 429 ! Moniatis R., 1937, *Kennedy* 902 ! Platres, 1937, *Kennedy* 903 ! Prodhromos, 1951, *Casey* 1209 ! and, 1961, *D. P. Young* 7276 ! Below Madhari on ridge W. of Sarandi, 1962, *Meikle* 2825 !

4. **C. aegyptiaca** *L.*, Mantissa I, 38 (1767); Sibthorp in Walpole, Travels, 25 (1820); A. K. Jackson in Kew Bull., 1937: 343 (1937).
C. herbacea Forssk., Fl. Aegypt.-Arab., 30 (1775); Boiss., Fl. Orient., 3: 22

(1875); Post, Fl. Pal., ed. 2, 1: 584 (1932); Osorio-Tafall et Seraphim, List Vasc. Plants Cyprus, 97 (1973).
TYPE: " *in* Aegypto Forskåhl " (cultivated at Uppsala).

Erect or spreading annual, (5–) 10–30 (–40) cm. high; stems usually much branched, even in depauperate specimens, tetragonous, thinly hispidulous with retrorse bristles, internodes often short; leaves 4–6-whorled, linear-lanceolate or subulate, 8–20 mm. long, 1–2 mm. wide, thinly hispidulous, apex sharply acuminate, margins strongly recurved or revolute; inflorescences numerous, terminal and lateral, caudate-spicate, 2–10 cm. long, 0·5–0·7 cm. wide; peduncle usually very short or wanting, occasionally up to 4 cm. long, sharply tetragonous, scabridulous; rhachis hispidulous or scabridulous; bracts rather loosely imbricate, even in fruit, lanceolate or narrowly ovate, 6–8 mm. long, about 2 mm. wide, acuminate, connate only at the very base, midrib prominent, scabridulous, margins broadly and conspicuously membranous, distinctly scabridulous-ciliate; bracteoles boat-shaped, strongly keeled, about 5 mm. long, 2 mm. wide, keel and membranous margins conspicuously scabridulous-ciliate, flowers sessile; hypanthium broadly oblong or somewhat obovate, about 1 mm. long and almost as wide; corolla creamy-white or yellowish, narrowly tubular, about 5 mm. long, 1 mm. wide; lobes usually 5, erect, ovate, cucullate, about 0·5 mm. long, 0·4 mm. wide, papillose externally, apical awn about 0·4 mm. long; stamens inserted just below base of lobes, included; filaments very short; anthers brownish, linear, about 0·5 mm. long, 0·2 mm. wide; styles unequal, the longer about 4 mm. long, the shorter about 2·5 mm. long, connate for about 1 mm. above base; stigmas subglobose. Fruit (unripe) oblong or slightly spindle-shaped, about 2 mm. long, 1 mm. wide, comprising 2 free mericarps; ventral surface of mericarp deeply sulcate; pericarp olive-brown, minutely papillose-verruculose.

HAB.: Maritime sands or marshy ground near the sea; sea-level to 100 ft. alt.; fl. April–July.

DISTR.: Divisions 1, 4, 5, rare. Also Libya, Egypt, Palestine.

1. Near Paphos, May 11, 1787, *Sibthorp*.
4. Near Perivolia (Cape Kiti) towards the lighthouse, 1936, *Syngrassides* 1217 !
5. East of Trikomo, on maritime sands, 1970, *A. Hansen* 643 !

8. SHERARDIA *L.*

Sp. Plant., ed. 1, 102 (1753).
Gen. Plant., ed. 5 (1754).

Annual herb; leaves sessile, 4–6-whorled; flowers in terminal, involucrate heads; calyx distinct, 4–6-toothed, persistent and conspicuous in fruit; corolla narrowly tubular with 4 apical lobes; styles bifid; stigmas capitate. Fruit dry, comprising 2, 1-seeded mericarps.

One species widely distributed in Europe, the Mediterranean region and western Asia.

1. **S. arvensis** *L.*, Sp. Plant., ed. 1, 102 (1753); Unger et Kotschy, Die Insel Cypern, 262 (1865); Boiss., Fl. Orient., 3: 19 (1875); Holmboe, Veg. Cypr., 170 (1914); Post, Fl. Pal., ed. 2, 1: 583 (1932); Lindberg f., Iter Cypr., 33 (1946); Osorio-Tafall et Seraphim, List Vasc. Plants Cyprus, 97 (1973).
TYPE: " *in arvis* Scaniae, Germaniae, Angliae ".

Procumbent or diffuse annual; stems much branched, sharply tetragonous, glabrous, smooth or sparsely scabridulous along the angles, 4–20 cm. long; leaves 4–6-whorled, the lowermost broadly obovate or suborbicular, 3–5 mm. long, 3–4 mm. wide, obtuse or cuspidate, sparsely scabridulous with forward-pointing bristles, becoming larger and narrower upwards, the upper lanceolate or narrowly elliptic, up to 15 mm. long and 5 mm. wide, apex acuminate, margins narrowly cartilaginous, sometimes sparingly scabridulous, usually not much recurved; inflorescences terminal and lateral, capitulate, 8–20 mm. diam.; peduncles 3–20 mm. long, glabrous, sharply tetragonous; bracts 6–8 (or more), lanceolate, 5–15 mm. long, 2–4 mm. wide, rather rigid, spreading, smooth or minutely and sparsely scabridulous, connate for 2–3 mm. above base and forming a distinct involucre, apex sharply acuminate, margins cartilaginous but scarcely recurved, base often pallid; flowers 6–12 or more, crowded; hypanthium turbinate, glabrous, about 1·5 mm. long, 1 mm. wide; calyx conspicuous, with 6 erect, subulate, hispidulous lobes, 1–1·5 mm. long and less than 0·5 mm. wide at base; corolla bright lavender-blue, narrowly tubular, about 5–6 mm. long; corolla-lobes spreading, ovate, subacute, about 2 mm. long, 0·8 mm. wide; stamens inserted just below base of corolla-lobes; filaments distinct, about 0·5 mm. long; anthers exserted, oblong, blackish, about 0·5 mm. long, 0·3 mm. wide; styles about 3·5 mm. long, a little unequal, united for about 2 mm.; stigmas broadly clavate or subcapitate. Fruit about 3 mm. long, 2·5 mm. wide, comprising 2, 1-seeded mericarps free except at apex and base and crowned with the persistent, somewhat accrescent, spreading, papery calyx-lobes; seeds oblong, about 2 mm. long, 1·3 mm. wide, deeply sulcate longitudinally along ventral surface; testa olive-brown, minutely foveolate-reticulate.

HAB.: Cultivated and fallow fields, occasionally on dry open hillsides or waste ground; sea-level to 5,000 ft. alt.; fl. Febr.–May.

DISTR.: All Divisions. Almost throughout Europe and western Asia; introduced and naturalized in many parts of the world.

1. Ayios Yeoryios (Akamas), 1962, *Meikle* ! Polis, 1962, *Meikle* !
2. Platres, 1938, *Kennedy* 1428 ! Prodhromos, 1939, *Lindberg f.* s.n. ! Mesopotamos Mon., 1939, *Lindberg f.* s.n. !
3. Limassol, 1964, *J. B. Suart* 176 !
4. Larnaca, 1880, *Sintenis & Rigo*; Ormidhia, 1880, *Sintenis & Rigo* !
5. Vatili, 1905, *Holmboe* 325 ! Salamis, 1962, *Meikle* !
6. Strovolos, 1901, *A. G. & M. E. Lascelles* s.n. ! Dhiorios, 1962, *Meikle* !
7. Ayios Khrysostomos Mon., 1862, *Kotschy* 983 ! Kyrenia, 1932, *Syngrassides* 688 ! also *Casey* 317 ! *G. E. Atherton* 883 ! 1017 ! Lakovounara, 1950, *Chapman* 519 !
8. Komi Kebir, 1936, *Syngrassides* 1337 ! Ayios Philon, 1941, *Davis* 2280 ! Gastria, 1941, *Davis* 2425 ! Koma tou Yialou, 1950, *Chapman* 227 ! 228 ! Akradhes, 1962, *Meikle* !

50. THELIGONACEAE

RUBIACEAE tribe THELIGONEAE *Wunderlich*
R. Wunderlich in Oesterr. Bot. Zeitschr., 119: 329–394 (1971).

Annual and perennial anemophilous herbs with rather succulent stems and leaves, and unisexual flowers; leaves simple, entire, the lower opposite, the upper usually alternate through the suppression of one leaf of each pair; interpetiolar stipules present; flowers monoecious, sessile, in small few-flowered unisexual or bisexual axillary cymes; perianth of male flowers closed in bud, valvately 2–5-partite and revolute at anthesis; stamens variable in number, (2–) 7–12 (–30), inserted at base of perianth; filaments long, slender, free; anthers linear, 2-thecous, dehiscing longitudinally, erect in bud, pendulous at anthesis; pistillodes absent; perianth of female flowers membranous, tubular, at first enclosing the ovary, but becoming lateral as the ovary develops, shortly toothed at apex; staminodes absent; ovary monocarpellary, containing 1 basal, campylotropous ovule; style gynobasic, exserted from mouth of perianth-tube. Fruit a subglobose, nut-like drupe, speckled with pale rhaphides; endosperm fleshy, copious; embryo strongly curved.

One genus only, represented in the Mediterranean region, China and Japan.

Wunderlich (*op. cit. supra*) lists a number of reasons for regarding *Theligonaceae* as a tribe of *Rubiaceae*. Other authors have detected affinities with *Urticaceae*, *Haloragidaceae* and the *Centrospermae*.

1. THELIGONUM *L.*

Sp. Plant., ed. 1, 993 (1753).
Gen. Plant., ed. 5, 430 (1754).
Ulbrich in Engl. et Prantl, Pflanzenfam., ed. 2, 16c: 368–378 (1934).

Description as for family.

1. **T. cynocrambe** *L.*, Sp. Plant., ed. 1, 993 (1753); Unger et Kotschy, Die Insel Cypern, 219 (1865); Holmboe, Veg. Cypr., 66 (1914); Lindberg f., Iter Cypr., 13 (1946); Rechinger f. in Arkiv för Bot., ser. 2, 1: 420 (1950); Osorio-Tafall et Seraphim, List Vasc. Plants Cyprus, 33 (1973).

Cynocrambe prostrata Gaertn., Fruct., 1: 362, t. 75 (1788); Boiss., Fl. Orient., 4: 897 (1879); Post, Fl. Pal., ed. 2, 2: 426 (1933).

TYPE: "*in* Italia, Monspelii".

Prostrate, diffuse or rarely erect annual 4–30 cm. high; stems usually much branched from near the base, occasionally simple, glabrous or very thinly strigillose; leaves rather fleshy, the lower opposite, the upper generally alternate, ovate, 1·5–5 cm. long, 0·8–3 cm. wide, glabrous or subglabrous with obscure nervation, apex acute or subacute, margins scabridulous with forward-pointing teeth, base rounded, broadly cuneate or subtruncate; petioles 1–5 cm. long, distinctly flattened dorsiventrally; stipules conspicuous, membranous; inflorescences in the axils of all but the lowermost leaves, very inconspicuous; perianth of male flowers membranous, divided into 2–3, broadly strap-shaped revolute lobes 2–3 mm. long, 1·5–2 mm. wide; filaments thread-like, very slender and flaccid, papillose, about 2 mm. long; anthers yellow, narrowly linear, about 2·5 mm. long, 0·2 mm. wide; perianth of female flowers membranous, obliquely tubular, about 1 mm. long, irregularly ruptured at the apex; ovary subglobose or broadly obovoid, about 1 mm. long, 0·7 mm. wide; style and linear stigma about 1·6 mm. long, exserted from apex of perianth-tube. Fruit subglobose, about 2 mm. diam., olive-brown, speckled with pale rhaphides, and with a waxy, orange-tinged basal caruncle; seed subglobose, comprising 2 closely adpressed hemispherical lobes; testa orange-brown, minutely foveolate.

HAB.: Walls, rock-fissures, occasionally on open stony or sandy ground; sea-level to 4,300 ft. alt.; fl. Febr.–June.

DISTR.: Divisions 1, 2, 4–8, common. Widespread in the Mediterranean region and eastwards to Iraq; Atlantic Islands.

1. Stroumbi, 1913, *Haradjian* 770 ! Lyso, 1913, *Haradjian* 819 ! Ayios Yeoryios (Akamas), 1962, *Meikle* ! Polis, 1962, *Meikle* !
2. Xerokolymbos, 4,300 ft. alt., 1937, *Kennedy* 323 ! Kryos Potamos, 4,100 ft. alt., 1938, *Kennedy* 1156 ! Vouni, 1957, *Merton* 2864 ! Near Pyrgos, 1959, *P. H. Oswald* 178 !
4. Cape Greco, 1862, *Kotschy* 136 ! Larnaca, 1880, *Sintenis & Rigo*.
5. Salamis, 1962, *Meikle* !
6. Dhiorios, 1962, *Meikle* 2327 (see Notes) !
7. Pentadaktylos, 1862, *Kotschy* 204a; above Kythrea, 1880, *Sintenis & Rigo* 49 ! Near Kantara, 1880, *Sintenis & Rigo*; St. Hilarion, 1939, *Lindberg f.* s.n.; Kyrenia, 1949, *Casey* 238 !

8. Cape Andreas, 1880, *Sintenis & Rigo*; Komi Kebir, 1912, *Haradjian* 330 ! and, 1936, *Syngrassides* 1061 ! Akradhes, 1962, *Meikle* !

NOTES: The specimen from Dhiorios (*Meikle* 2327), growing on flat open pasture under *Pistacia lentiscus* bushes, has simple, erect stems and very small leaves. It agrees exactly with *Theligonum cynocrambe* L. var. *minor* Ulbrich (in Notizbl. Bot. Gart. Berlin-Dahlem, 11: 899; 1933), but may be nothing more than a depauperate condition of the typical plant.

APPENDIX I

A SHORT LIST OF LITERATURE CITED IN FLORA OF CYPRUS, VOL. 1

BOOKS

Aschers. et Graebn., *Syn. Mitt.-Europ. Fl.*—P. F. A. Ascherson et K. O. R. P. P. Graebner, *Synopsis der Mittel-europaeischen Flora.* Leipzig. Ed. 1, 12 vols., 1896–1939; ed. 2, 1912–1920; (continued from vol. 12(2), 1930 by P. Graebner and P. Graebner f.).

Benth. et Hook. f., *Gen. Plant.*—G. Bentham et J. D. Hooker, *Genera Plantarum, . . . etc.* London. 3 vols., 1862–1883.

Boiss., *Diagn.*—P. E. Boissier, *Diagnoses Plantarum Orientalium Novarum . . . etc.* Leipzig et Paris. 3 vols., 1842–1859. Published in 3 *volumes*, comprising 2 *series*; series 1 includes 13 *fascicles*; series 2 includes 6 *fascicles*. Cited as " Diagn., [series] 1, [fasc.] 3: 46 (1843) ", etc.

Boiss., *Fl. Orient.*—P. E. Boissier, *Flora Orientalis, . . . etc.* Bâsle, Geneva and Leiden. 5 vols., 1867–1884. *Suppl.* 1888.

Bovill, *Report Plant. Work Cyprus*—A. K. Bovill, *A Report on Plantation Work in Cyprus from 1879–1914.* Nicosia. 1915.

Chapman, *Cyprus Trees and Shrubs*—E. F. Chapman, *Cyprus Trees and Shrubs.* Nicosia. 1949.

Davis, *Fl. Turkey*—P. H. Davis, *Flora of Turkey . . . etc.* Edinburgh. 5 vols. to date. 1965 +.

DC., *Prodr.*—A. P. de Candolle, *Prodromus Systematis Naturalis Regni Vegetabilis . . . etc.* 17 vols., 1824–1873.

DC., *Regn. Veg. Syst. Nat.*—A. P. de Candolle, *Regni Vegetabilis Systema Naturale, . . . etc.* Paris. 2 vols., 1817–1821.

Desf., *Fl. Atlant.*—R. L. Desfontaines, *Flora Atlantica . . . etc.* Paris. 2 vols., 1798–1799.

Engl., *Pflanzenr.*—H. G. A. Engler, *Das Pflanzenreich. Regni Vegetabilis Conspectus . . . etc.* Leipzig. 1900 +.

Engl. et Prantl, *Pflanzenfam.*—H. G. A. Engler et K. A. E. Prantl, *Die natürlichen Pflanzenfamilien . . . etc.* Ed. 1, Leipzig. 1887–1915 incl. *Nachträgen.* Ed. 2, Leipzig et Berlin. 1925 +.

Fisch. et Mey., *Ind. Sem. Hort. Petrop.*—F. E. L. von Fischer et C. A. von Meyer, *Index Seminum . . . Hortus Botanicus Imperialis Petropolitanus . . . etc.* St. Petersburg. 1835–1873.

Forssk., *Fl. Aegypt.-Arab.*—P. Forsskål, *Flora Aegyptiaco-Arabica, . . . etc.* Copenhagen. 1775.

Gaudry, *Recherches Sci.*—J. Gaudry, *Recherches Scientifiques en Orient . . . etc.* Paris. 1855.

Hal., *Consp. Fl. Graec.*—E. von Halácsy, *Conspectus Florae Graecae.* Leipzig. 3 vols., 1900–1904. 2 *supplements*, 1908–1912.

Holmboe, *Veg. Cypr.*—J. Holmboe, *Studies on the Vegetation of Cyprus . . . etc.* Bergen. 1914. [Also in *Bergens Mus. Skr. n.s.*, 1(2), 1914.]

L., *Gen. Plant.*—C. von Linnaeus, *Genera Plantarum . . . etc.* Stockholm. Ed. 5, 1754.

L., *Sp. Plant.*—C. von Linnaeus, *Species Plantarum . . . etc.* Stockholm. Ed. 1, 1753; ed. 2, 1762–1763. (Ed. 4, see Willdenow.)

Lam., *Encycl. Méth.*—J. B. A. P. Monet de Lamarck, *Encyclopédie méthodique . . . etc. Botanique.* Paris. 8 vols., 1783–1808. (Vols. 5–8 and supplements continued by Poiret; for supplements see under Poiret.)

Lindberg f., *Iter Cypr.*—H. Lindberg, *Iter Cyprium . . . etc.* [in *Acta Soc. Sci. Fenn.*, n.s.B. 2]. Helsinki. 1946.

M. Bieb., *Fl. Taur.-Cauc.*—F. A. Marschall von Bieberstein, *Flora Taurico-Caucasica . . . etc.* Kharkov. 3 vols., 1808–1819.

Mill., *Gard. Dict.*—P. Miller, *The Gardener's Dictionary.* London. Ed. 8, 1768.

Mill., *Gard. Dict. Abridg.*—P. Miller, *The Gardener's Dictionary . . . Abridged.* London. Ed. 4, 1754.

Moench, *Meth.*—C. Moench, *Methodus Plantas Horti Botanici . . . etc.* Marburg. 1794.

Nattrass, *First List Cyprus Fungi*—R. M. Nattrass, *A First List of Cyprus Fungi.* Nicosia. 1937.

Nym., *Consp. Fl. Europ.*—C. F. Nyman, *Conspectus Florae Europeae . . . etc.* Örebro. 1878–1882. *Supplements*, 1883–1890.

Osorio-Tafall et Seraphim, *List Vasc. Plants Cyprus*—B. F. Osorio-Tafall et G. M. Seraphim, *List of the Vascular Plants of Cyprus*. Nicosia. 1973.

Pers., *Syn. Plant.*—C. H. Persoon, *Synopsis Plantarum . . . etc.* Paris and Tübingen. 2 vols., 1805–1807.

Poech, *Enum. Plant. Ins. Cypr.*—J. Poech, *Enumeratio Plantarum hucusque cognitarum Insulae Cypri*. Vienna. 1842.

Poiret, *Encycl. Méth. Suppl.*—J. L. M. Poiret, *Encyclopédie méthodique. Botanique. Supplements* 1–5 (vols. 9–13 of Lam., *Encycl. Méth.*). Paris. 1822.

Post, *Fl. Pal.*—G. E. Post, *Flora of Syria, Palestine and Sinai . . . etc.* Ed. 1, 1896; ed. 2 by J. E. Dinsmore, 2 vols., 1932–1933.

Rechinger f., *Fl. Aegaea*—K. H. Rechinger f., *Flora Aegaea . . . etc.* [in *Denkschr. Akad. Wiss. Math.-Nat. Kl.* (*Wien*), 105(1)]. Vienna. 1943.

Sibth. et Sm., *Fl. Graec.*—J. Sibthorp et J. E. Smith, *Flora Graeca . . . etc.* London. 10 vols., 1806–1840.

Sibth. et Sm., *Fl. Graec. Prodr.*—J. Sibthorp et J. E. Smith, *Florae Graecae Prodromus . . . etc.* London. 2 vols., 1806–1816.

Ten., *Fl. Nap.*—M. Tenore, *Flora Napolitana . . . etc.* Napoli. 1811–1838. (This includes *Fl. Nap. Prodr.* et *Fl. Nap. Syll.*)

Unger et Kotschy, *Die Insel Cypern*—F. J. A. N. Unger et T. Kotschy, *Die Insel Cypern*. Vienna. 1865.

Urv., *Enum. Plant.*—J. S. C. Dumont d'Urville, *Enumeratio Plantarum . . . etc.* Paris. 1822.

Walpole, *Mem. Europ. Asiat. Turkey*—R. Walpole, *Memoirs Relating to European and Asiatic Turkey . . . etc.* London. Ed. 1, 1817; ed. 2, 1818.

Walpole, *Travels*—R. Walpole, *Travels in Various Countries of the East, . . . etc.* London. 1820.

Willd., *Sp. Plant.*—C. L. Willdenow, *Species Plantarum . . . etc.* Berlin. Vols. 1–5(1), 1797–1810.

Zohary, *Fl. Palaest.*—M. Zohary, *Flora Palaestina*. Jerusalem. 2 vols. to date. 1966 +.

JOURNALS

Ann. Sci. Nat.—Annales des Sciences Naturelles. Paris. 1824 + (including variant titles).

Arkiv för Bot.—Arkiv för Botanik. Uppsala. 1903 +.

Bull. Alp. Gard. Soc.—[Quarterly] Bulletin of the Alpine Garden Society of Great Britain. London, etc. 1930 +.

Bull. Herb. Boiss.—Bulletin de l'Herbier Boissier. Geneva. 1893–1908.

Bull. Soc. Bot. Fr.—Bulletin de la Société Botanique de France. Paris. 1854 +.

Cypr. [**Agric.**] **Journ.**—Cyprus Agricultural Journal . . . etc. Nicosia. 1904–1940.

Feddes Repert.—Repertorium Specierum Novarum Regni Vegetabilis. Berlin. 1905 + (as *Fedde Repertorium* before 1943).

Flora—Flora oder allgemeine botanische Zeitung. Regensburg. 1818 +.

Journ. Bot.—The Journal of Botany, British and Foreign. London. 1863–1942.

Journ. Linn. Soc. Bot.—The Journal of the Linnean Society. Botany. London. 1856 +.

Kew Bull.—Bulletin of Miscellaneous Information. Royal Gardens, Kew. London. 1887–1941; continued as *Kew Bulletin* 1946 +.

Mém. Herb. Boiss.—Mémoires de l'Herbier Boissier. Geneva. 1900 +.

Notes Roy. Bot. Gard. Edinb.—Notes from the Royal Botanic Garden, Edinburgh. Edinburgh and Glasgow. 1900 +.

Oesterr. Bot. Zeitschr.—Oesterreichische botanische Zeitschrift . . . etc. Vienna. 1858 +.

Pflanzenfam.—see Books, Engl. et Prantl, Pflanzenfam.

Pflanzenr.—see Books, Engl., Pflanzenr.

Rep. B.E.C.—Report of the Botanical Society and Exchange Club of the British Isles. London, etc. 1867–1947 (including variant titles).

Trans. Linn. Soc.—Transactions of the Linnean Society of London. London. 1791 + (including variant titles).

(A full list of literature references will be given in Vol. 2.)

APPENDIX II

COLLECTORS CITED IN VOL. I OF THE FLORA

	Collected
Antoniades	1939
Atherton, Miss G. E.	1955–1956
Aucher-Éloy, Pierre Martin Rémi (1793–1838)	1831
Baker, E. W.	1955
Baker, Sir Samuel White (1821–1893)	1879
Ball, John (1818–1889)	1877
Barbey, Caroline (1847–1918)	1880
Barbey, William (1842–1914)	1880
Barclay, Sir Colville (b. 1913)	1968
Barrington, F. J. F.	1950
Bauer, Ferdinand Lucas (1760–1826)	1787
Biddulph, Sir Robert (1835–1918)	1882
Bovill, Alfred Karslake	1882–1920
Casey, Edwards C.	1947–1958
Casey, Mercy M.	1947–1958
Chapman, Mrs. E. F.	1933–1951
Chapman, G. W.	1933–1951
Chatterton, Mrs. I.	1967
Chicken, Eric	1970, 1973
Chiotellis, N.	1955
Clarke, Edward Daniel (1769–1822)	1801
Davis, Peter Hadland (b. 1918)	1940–1941
Deschamps, E.	1892–1894
Dickson, Mrs. H. R. P.	1951
Dray, Mrs. M. E.	1936
Druce, George Claridge (1850–1932)	1928, 1930
Drummond, Alexander	1745, 1750
Duke-Woolley, H.	1948
Economides, C. V.	1966–1970
Feilden, G. T.	1914
Foggie, A.	1933
Frangos, G.	1923
Gaudry, Jean Albert (1827–1908)	1853, 1854
Gennadius, Panagiotes G.	1895–1901
Genneou, A.	1969–1970
Giuseppi, Paul Leon (d. 1947)	1932
Godman, Miss C. Edith	1937
Graves, Mrs.	1954
Gresham, Major C. E.	1933
Grove, Mrs.	1954
Guichard, Kenneth M.	1971
Hadjichristodoulou	1970
Haig, Miss Elizabeth	1934
Hanbury, Daniel (1825–1875)	1860
Hansen, Alfred (b. 1925)	1970
Haradjian, Manoog (d. ? 1914)	1912–1913

Harris, D. R.	1907
Hartmann, Ernst	1904–1905
Hasselquist, Fredrik (1722–1752)	1751
Haussknecht, Heinrich Carl (1838–1903)	1866
Hawkins, John (? 1758–1841)	1787
Hecker, Mrs. I. M.	1968
Holmboe, Jens (1880–1943)	1905
Hooker, Joseph Dalton (1817–1911)	1860
Huddle, Rev. A.	1927
Hughes, Mrs. Jean	1955
Hume, Sir Abraham (1749–1838)	1801
Huovila, Olli	1965–1966
Ingram, Collingwood (b. 1880)	1948
Joscht, Mrs. Gerda	1971–1974
Kennedy, Mrs. Elizabeth Warren (d. 1964)	1936–1955
Kotschy, Karl Georg Theodor (1813–1866)	1840, 1859, 1862
Labillardière, Jacques-Julien Houtton de (1755–1834)	1787
Lascelles, A. G.	1900–1902
Lascelles, Miss M. E.	1900–1902
Leontiades, L.	1964
Lindberg, Harald (1871–1963)	1939
Loch, Margaret, Lady	1936–1952
Macdonald, Sq. Ldr. and Mrs. N.	1958
Mapple, Miss Beatrice M.	1955
Mariti, Giovanni	1760
Matthews, Miss Ann	1965–1974
Mavromoustakis, G. A.	1939–1948, 1957
Meikle, Robert Desmond (b. 1923)	1962, 1974
Merton, Lionel Francis Herbert (1919–1974)	1951–1957, 1966
Michaelides, Michael G.	1905
Morris, Hubert Meredydd	1930–1951, 1974
Mouillefert, Pierre (1846–1903)	1893
Nattrass, Roland Marshall (b. 1895)	1935
Nicolaides, George	1961
Norman, Cecil (1872–1947)	1930
Osorio-Tafall, Bibiano F. (b. 1903)	1967–1974
Oswald, Philip	1959, 1960
Painter, H.	1957
Paolides	1954
Papadakis	1964
Pichler, Thomas (1828–1903)	1889
Pierides, Luke	1905
Pococke, Rev. Richard (1704–1765)	1738
Polunin, Oleg Vladimirovitsch	1961
Poore, Martin Edward Duncan (b. 1925)	1956
Post, George Edward (1838–1909)	1894
Price, William Robert (1886–1975)	1972
Probyn, Mrs. F. M.	1947–1952
Reid, Clement (1853–1916)	1908
Rigo, Gregorio (1841–1922)	1880
Rogers, Rev. Frederick Arundel (1876–1944)	1930
Roger-Smith, H.	1936–1937
Samson, Miss E. A.	1904
Scott-Moncrieff, Mrs. Gundred	1950, 1954, 1955
Seligman, Richard Joseph Simon	1954

Seraphim, George M.	1960–1974
Sestini, Domenico (1750–1832)	1782
Sibthorp, John (1758–1796)	1787
Sintenis, Paul Ernst Emil (1847–1907)	1880
Soteriadou, Mrs. A. C.	1951–1974
Sparrow, Brigadier C. H. E.	1959
Stagg, Mrs.	1937
Suart, J. B.	1963, 1964
Syngrassides, Anastasios	1932–1939
Townsend, Clifford Charles (b. 1926)	1971
Tracey, J. A.	1931
Trenbath, Brian Rahr	1963
Unger, Franz Joseph Andreas Nicolaus (1800–1870)	1862
Ussher, C. B.	1929–1930
Wall, Erik (1871–1959)	1930
Wyatt, C. H.	1937, 1938
Young, Donald Peter (1917–1972)	1961, 1963

Collectors' numbers are usually cited immediately after their names, except where the number is an institutional one, as with the Agricultural Research Institute (near Nicosia); here the specimens are cited as, for example, " *Economides* in ARI 960 ". The letters " K ", " G ", etc. following specimen citations are herbarium abbreviations listed in *Index Herbariorum* Part I, ed. 6, compiled by P. K. Holmgren and W. Keuken, and published in *Regnum Vegetabile* vol. 92: 303–354 (1974).

APPENDIX III

DESCRIPTIONS OF NEW SUBSPECIES, VARIETIES, ETC., AND A SUMMARY OF NEW NAMES AND COMBINATIONS, ETC. INCLUDED IN VOL. 1 OF THE FLORA OF CYPRUS

DESCRIPTIONS:

Page 160 **Malcolmia nana** (*DC.*) *Boiss.* var. **glabra** *Meikle* **var. nov.** a varietate typica caulibus, foliis et fructibus glabris vel subglabris differt. HOLOTYPUS: Cyprus; Yialia, on the seashore; March 15, 1962, *Meikle* 2195 (K !).

Page 195 **Helianthemum salicifolium** (*L.*) *Mill.* var. **glabrum** *Meikle* **var. nov.** a varietate typica caulibus, foliis, etc. omnino glabris, seminibus ut videtur minoribus differt. HOLOTYPUS: Cyprus; Kokkini Trimithia, Apr. 7, 1933, *Syngrassides* 331 (K !).

Page 384 **Genista sphacelata** *Decne.* var. **crudelis** *Meikle* **var. nov.** a varietate typica habitu nano, congesto, differt; planta tegetes amplas vel pulvinos hemisphericos efformans, ramulis rigide spinosis, intricate divisis, 5–17 cm. longis, obtecta; floribus plerumque eisdem var. *sphacelatae* minoribus, vulgo minus quam 1 cm. longis, etiam distinguitur. HOLOTYPUS: Cyprus; above Amiandos, 5,000 ft. alt., June 21, 1935, *A. Foggie* 361 (K !).

Page 575 **Lathyrus blepharicarpus** *Boiss.* var. **cyprius** *Meikle* **var. nov.** a varietate typica foliolis late lanceolatis vel ellipticis, 5–20 mm. longis, 2–10 mm. latis differt. HOLOTYPUS: Cyprus; Akradhes, April 8, 1962, *Meikle* 2511 (K !).

NEW NAMES, COMBINATIONS, ETC.:

Page 22 **Cedrus libani** *A. Rich.* ssp. **brevifolia** (*Hook. f.*) *Meikle* **comb. nov.** BASIONYM: *Cedrus libani* A. Rich. var. *brevifolia* Hook. f. in Journ. Linn. Soc. Bot., **17**: 518 (1880).

Page 215 **Dianthus strictus** *Banks et Sol.* var. **troodi** (*Post*) *S. S. Hooper* **comb. nov.** BASIONYM: *D. multipunctatus* var. *troodi* Post in Mém. Herb. Boiss., **18**: 91 (1900).

Page 220 **Petrorhagia kennedyae** (*A. K. Jackson et Turrill*) *Meikle* **comb. nov.** BASIONYM: *Tunica kennedyae* A. K. Jackson et Turrill in Kew Bull., 1938: 462 (1938).

Page 286 **Scleranthus annuus** *L.* ssp. **delortii** (*Gren.*) *Meikle* **comb. nov.** BASIONYM: *S. delortii* Gren. in F. W. Schultz (ed.), Arch. de Flore, 206 (1852).

Page 323 **Linum** *L.* sect. **Halolinum** (*Planch.*) *Meikle* **stat. nov.** BASIONYM: *Linum* L. series *Halolinum* Planch. in Hook., Lond. Journ. Bot., 6: 598 (1847).

Page 358 **Ziziphus zizyphus** (*L.*) *Meikle* **comb. nov.** BASIONYM: *Rhamnus zizyphus* L., Sp. Plant., ed. 1, 194 (1753).

Page 392 **Ononis** *L.* sect. **Natrix** *DC.* emend. *Griseb.* subsect. **Pubescentes** (*Širjaev*) *Meikle* **stat. nov.** BASIONYM: *Ononis* L. sect. *Natrix* DC. emend. Griseb. series *Pubescentes* Širjaev in Beih. Bot. Centralbl., 49, 2: 518 (1932).

Page 446 **Trifolium pamphylicum** *Boiss. et Heldr.* var. **blancheanum** (*Boiss.*) *Meikle* **comb. nov.** BASIONYM: *T. blancheanum* Boiss., Diagn., 2, 2: 13 (1856).

Page 498 **Astragalus suberosus** *Banks et Sol.* var. **hartmannii** (*Holmboe*) *Meikle* **comb. nov.** BASIONYM: *A. tuberculosus* DC. var. *hartmannii* Holmboe, Veg. Cypr., 109 (1914).

Page 503 **Astragalus echinus** *DC.* var. **chionistrae** (*Lindberg f.*) *Meikle* **stat. nov.** BASIONYM: *A. echinus* DC. ssp. *chionistrae* Lindberg f. in Act. Soc. Sci. Fenn. n.s.B., 2(7): 19 (1946).

Page 535 **Alhagi maurorum** *Medik.* var. **turcorum** (*Boiss.*) *Meikle* **comb. nov.** BASIONYM: *A. turcorum* Boiss., Diagn. 1, 9: 113 (1849).

Page 644 **Rosularia cypria** (*Holmboe*) *Meikle* **comb. nov.** BASIONYM: *Umbilicus cyprius* Holmboe, Veg. Cypr., 94 (1914).

Page 645 **Rosularia pallidiflora** (*Holmboe*) *Meikle* **comb. nov.** BASIONYM: *Umbilicus pallidiflorus* Holmboe, Veg. Cypr., 95 (1914).

Page 727 **Cachrys scabra** (*Fenzl*) *Meikle* **comb. nov.** BASIONYM: *Ferula scabra* Fenzl in Flora, 26: 461 (1843).

Page 728 **Cachrys crassiloba** (*Boiss.*) *Meikle* **comb. nov.** BASIONYM: *Hippomarathrum crassilobum* Boiss., Diagn., 1, 10: 56 (1849).

Page 776 **Valantia columella** (*Ehrenb. ex Boiss.*) *Meikle* **comb. nov.** BASIONYM: *Galium columella* Ehrenb. ex Boiss., Fl. Orient., 3: 81 (1875).

Page 782 **Galium humifusum** *M. Bieb.* var. **lasiocarpum** (*Boiss.*) *Meikle* **comb. nov.** BASIONYM: *Asperula pycnantha* Boiss. var. *lasiocarpa* Boiss., Diagn., 1, 10: 64 (1849).

INDEX

Accepted names are in roman type; synonyms and misidentifications in italic. Page numbers in bold face are references to detailed treatment of the taxa concerned; subsidiary references are in roman type.